Conversion Factors

Mass	$1\ \text{kg} = 1000\ \text{g}$
	$1\ \text{g} = 10^{-3}\ \text{kg}$
	$1\ \text{u} = 1.66 \times 10^{-27}\ \text{kg} \leftrightarrow 931.5\ \text{MeV}/c^2$
	$1\ \text{slug} = 14.6\ \text{kg}$
Length	$1\ \text{m} = 100\ \text{cm} = 1000\ \text{mm}$
	$1\ \text{foot} = 1\ \text{ft} = 0.305\ \text{m}$
	$1\ \text{inch} = 1\ \text{in.} = 2.54\ \text{cm}$
	$1\ \text{mile} = 1\ \text{mi} = 5280\ \text{ft} = 1609\ \text{m}$
	$1\ \text{light-year (ly)} = 9.46 \times 10^{15}\ \text{m}$
Time	$1\ \text{min} = 60\ \text{s}$
	$1\ \text{h} = 60\ \text{min} = 3600\ \text{s}$
	$1\ \text{day} = 24\ \text{h} = 8.64 \times 10^4\ \text{s}$
	$1\ \text{yr} = 365.24\ \text{days} = 3.16 \times 10^7\ \text{s}$
Force	$1\ \text{N} = 0.225\ \text{lb}$
	$1\ \text{dyne} = 10^{-5}\ \text{N}$
	$1\ \text{lb} = 4.45\ \text{N} = 16\ \text{oz}$
Energy	$1\ \text{J} = 10^7\ \text{erg} = 0.239\ \text{cal}$
	$1\ \text{cal} = 4.186\ \text{J}$
	$1\ \text{kW} \cdot \text{h} = 3.60 \times 10^6\ \text{J}$
	$1\ \text{eV} = 1.60 \times 10^{-19}\ \text{J}$
Power	$1\ \text{W} = 1\ \text{J/s}$
	$1\ \text{hp} = 745.7\ \text{W}$
Pressure	$1\ \text{Pa} = 1\ \text{N/m}^2$
	$1\ \text{atmosphere} = 1\ \text{atm} = 1.01 \times 10^5\ \text{Pa} = 14.7\ \text{lb/in.}^2$
	$1\ \text{millimeter of Hg} = 1\ \text{mm Hg} = 133\ \text{Pa}$
	$1\ \text{bar} = 10^5\ \text{Pa}$
	$1\ \text{lb/in.}^2 = 6.89 \times 10^3\ \text{Pa}$
Speed/velocity	$1\ \text{m/s} = 3.60\ \text{km/h}$
	$1\ \text{mi/h} = 0.447\ \text{m/s}$
	$100\ \text{mi/h} \approx 45\ \text{m/s}$
Temperature	$T(\text{K}) = T(^\circ\text{C}) + 273.15$
	$T(^\circ\text{F}) = \frac{9}{5}T(^\circ\text{C}) + 32$
	$T(^\circ\text{C}) = \frac{5}{9}[T(^\circ\text{F}) - 32]$
Angle	$180^\circ = \pi\ \text{rad}$

Useful Physical Data

Acceleration due to gravity near the Earth's surface	$g = 9.8 \text{ m/s}^2$
Atmospheric pressure	$1.01 \times 10^5 \text{ Pa}$
Density of water (4°C)	$1000 \text{ kg/m}^3 = 1 \text{ g/cm}^3$
Density of air (20°C, 1 atm)	1.20 kg/m^3
Absolute zero of temperature	$-273.15°\text{C}$
Speed of sound in air (20°C, 1 atm)	343 m/s
Radius of the Earth	$6.37 \times 10^3 \text{ km} = 6.37 \times 10^6 \text{ m}$
Mass of the Earth	$5.97 \times 10^{24} \text{ kg}$
Radius of the Moon	1740 km
Mass of the Moon	$7.35 \times 10^{22} \text{ kg}$
Average Earth–Moon distance	$3.84 \times 10^5 \text{ km}$
Radius of the Sun	$6.95 \times 10^5 \text{ km}$
Mass of the Sun	$1.99 \times 10^{30} \text{ kg}$
Average Earth–Sun distance	$1.50 \times 10^8 \text{ km}$

College Physics
REASONING AND RELATIONSHIPS

Nicholas J. Giordano

AUDIENCE RESPONSE TECHNOLOGY
Conceptual Learning / Applications
JoinIn™ Student Response System

ONLINE HOMEWORK MANAGEMENT
Practice / Visualization
Enhanced WebAssign

FOUNDATION: TEXT
Theory / Practice / Application

ENHANCED WebAssign

College Physics: Reasoning and Relationships, motivates student learning by building a solid foundation for true understanding by recognizing the relationships between physics and the real world, utilizing a consistent problem solving approach and promoting conceptual understanding through reasoning exercises.

With over one million users and a half a billion submissions, **WebAssign** is the leading homework and grading solution for math and science. Created by physics teachers for physics teachers, **WebAssign** was designed to be easy to use and works on all computers using any browser, with click-and-assign simplicity, an easy-to-navigate grade book, and tools to analyze student answers and return appropriate feedback.

Access to an enhanced version of **WebAssign** can be packaged with Giordano's *College Physics: Reasoning and Relationships*. Using pedagogy and content from the book, **Enhanced WebAssign** will help your students see physics in action. Whether you're an experienced veteran or new to online homework, **Enhanced WebAssign** is the perfect way to meet your teaching objectives and your students' learning needs.

PencilPad® ▶

Enhanced WebAssign's newest feature, PencilPad®, is a simple drawing tool for students to use to answer questions. Students can draw physical situations, sketch trends of plots, show steps to solutions, or even create small art pieces!

It's simple. There is a pencil and an eraser and one or more lined pages on which to draw or write. WebAssign stores your students' work and allows you to grade their submissions just like essays, if you choose.

End-of-chapter problems ▶

Selected end-of-chapter problems, algorithmically driven where appropriate and containing feedback when an incorrect answer is entered.

Concept Checks ▶

Available for assignment in multiple-choice format.

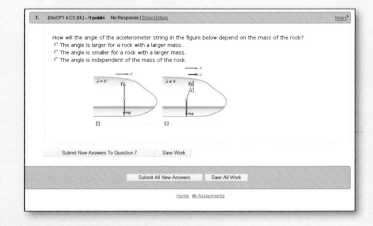

ENHANCED WebAssign

Reasoning and Relationship Problems ▶

Students are at risk of missing which crucial quantities must be estimated. A coached solution can help students learn how to attack these problems and arrive at a sensible answer.

Reasoning Problem 5.30

Consider a Ferris wheel in which the chairs hang down from the main wheel via a cable. The cable is 2.2 m long, and the radius of the wheel is 13 m. When a chair is in the orientation shown in the figure below, the cable attached to the chair makes an angle of $\theta = 23°$ with the vertical. Find the speed of the chair.

Part 1 of 4 - Recognize the Principles

Which of the following are critical to solving this problem?

- ☑ equations of motion
- ☐ terminal velocity
- ☑ Newton's second law
- ☐ centripetal acceleration

Remember that the car is undergoing uniform circular motion.

[Submit] [Skip]

Part 2 of 4 - Sketch the Problem

Which of the following free body diagrams best shows the forces acting upon the Ferris wheel cart?

- ○ A
- ◉ B
- ○ C
- ○ D

Check your angle!

[Submit] [Skip]

Part 3 of 4 - Identify the Relationships

Along the x axis, how should Newton's 2nd law read?

- ○ $\Sigma F_{center} = T \cos \theta = \dfrac{mv^2}{r}$
- ○ $\Sigma F_{center} = T \cos \theta = 0$
- ○ $\Sigma F_{center} = T \sin \theta = \dfrac{mv^2}{r}$
- ○ $\Sigma F_{center} = T \sin \theta + \dfrac{mv^2}{r} = 0$

Along the y axis, how should Newton's 2nd law read?

- ○ $\Sigma F_{y\ axis} = T \cos \theta - mg = 0$
- ○ $\Sigma F_{y\ axis} = T \cos \theta + mg = 0$
- ○ $\Sigma F_{y\ axis} = T \sin \theta = mg$
- ○ $\Sigma F_{y\ axis} = T \sin \theta = 0$

Part 4 of 4 - Solve

Estimates

Within an order of magnitude, approximate the radius of the wheel - chair system.

[15] m Good. This value is correct to an order of magnitude and is adequate to carry through the remainder of the problem.

Calculations (use your estimates)

Calculate the speed of the car.

[22] m/s You will have to simultaneously solve the two Newton's 2nd law equations we derived earlier to obtain the velocity.

[Submit] [Skip]

WebAssign at a glance...

Program and Technical Features

- Simple, friendly user interface

- Course creation in two easy steps

- Assignment creation in under five minutes

- Instructor gradebook with weighted categories and scores

- Automatic grading of online homework assignments

- Intuitive Question Editor that allows questions to be edited or added without learning a programming language

- Proper display of mathematical expressions

- Timed, secure, password- and IP-restricted exercises

- Student forums, bulletin boards, messaging

- Built-in calendar and communication tools

- Compatible with all recent operating systems (Windows®, Apple® Macintosh, Linux®, Sun™ Solaris™)

- Works with most web browsers (e.g., Mozilla Firefox®, Internet Explorer®, Safari™)

- No proprietary plug-ins required

COLLEGE PHYSICS

Reasoning and Relationships

COLLEGE PHYSICS

Reasoning and Relationships

Nicholas J. Giordano

PURDUE UNIVERSITY

BROOKS/COLE
CENGAGE Learning

Australia • Brazil • Japan • Korea • Mexico • Singapore • Spain • United Kingdom • United States

SCI
G431
2010
t9

College Physics: Reasoning and Relationships, First Edition

Nicholas J. Giordano

Editor-in-Chief: Michelle Julet

Publisher: Mary Finch

Managing Editor: Peggy Williams

Senior Development Editor: Susan Dust Pashos

Development Project Manager: Terri Mynatt

Development Editor: Ed Dodd

Development Editor Art Program: Alyssa White

Associate Development Editor: Brandi Kirksey

Editorial Assistant: Stephanie Beeck

Media Editors: Sam Subity/ Rebecca Berardy Schwartz

Marketing Manager: Nicole Mollica

Marketing Coordinator: Kevin Carroll

Marketing Communications Manager: Belinda Krohmer

Project Manager, Editorial Production: Teresa L. Trego

Creative Director: Rob Hugel

Art Director: John Walker

Print Buyer: Rebecca Cross

Permissions Editor: Mollika Basu

Production Service: Lachina Publishing Services

Text Designer: Brian Salisbury

Art Editor: Steve McEntee

Photo Researcher: Dena Digilio Betz

Copy Editor: Kathleen Lafferty

Illustrator: Dartmouth Publishing Inc., 2064 Design/Greg Gambino

Cover Designer: William Stanton

Cover Image: Ted Kinsman/Photo Researchers, Inc.

Compositor: Lachina Publishing Services

For product information and technology assistance, contact us at **Cengage Learning Customer & Sales Support, 1-800-354-9706.**

For permission to use material from this text or product, submit all requests online at **www.cengage.com/permissions.** Further permissions questions can be e-mailed to **permissionrequest@cengage.com.**

Library of Congress Control Number: 2008931521

ISBN-13: 978-0-534-42471-8
ISBN-10: 0-534-42471-6

Volume 1
ISBN-13: 978-0-534-46243-7
ISBN-10: 0-534-46243-X

Volume 2
ISBN-13: 978-0-534-46244-4
ISBN-10: 0-534-46244-8

Brooks/Cole
10 Davis Drive
Belmont, CA 94002-3098
USA

Cengage Learning is a leading provider of customized learning solutions with office locations around the globe, including Singapore, the United Kingdom, Australia, Mexico, Brazil, and Japan. Locate your local office at: **international.cengage.com/region**

Cengage Learning products are represented in Canada by Nelson Education, Ltd.

For your course and learning solutions, visit **www.cengage.com**

Purchase any of our products at your local college store or at our preferred online store **www.ichapters.com**

Printed in Canada
1 2 3 4 5 6 7 12 11 10 09

Brief Table of Contents

Table of Contents

© comstock Images/Jupiterimages

© Joseph Van Os/Getty Images

© Radius Images/Jupiterimages

© Andy Moore/photolibrary/Jupiterimages

Chapter 20

Magnetic Fields and Forces   644

© Cengage Learning/Charles D. Winters

Chapter 21

Magnetic Induction   687

Chapter 22

Alternating-Current Circuits and Machines   723

© Charles Gupton/Stone/Getty Images

NASA/JPL-Caltech/University of Arizona/STScl

Chapter 27

Relativity 917

Chapter 28

Quantum Theory 954

Chapter 29

Atomic Theory 986

© Photo Researchers/Alamy

Preface

College Physics: Reasoning and Relationships is designed for the many students who take a college physics course. The majority of these students are not physics majors (and don't want to be) and their college physics course is the only physics class they will ever take. The topics covered in a typical college physics course have changed little in recent years, and even decades. Indeed, except for many of the applications, much of the physics covered here was well established more than a century ago. Although the basic material covered may not be changing much, the way it is taught should not necessarily stay the same.

GOALS OF THIS BOOK

Reasoning and Relationships

Students often view physics as merely a collection of loosely related equations. We who teach physics work hard to overcome this perception and help students understand how our subject is part of a broader science context. But what does "understanding" in this context really mean?

Many physics textbooks assume understanding will result if a solid problem-solving methodology is introduced early and followed strictly. Students in this model can be viewed as successful if they deal with a representative collection of quantitative problems. However, physics education research has shown that students can succeed in such narrow problem-solving tasks and at the same time have fundamentally flawed notions of the basic principles of physics. For these students, physics *is* simply a collection of equations and facts without a firm connection to the way the world works. Although students do need a solid problem-solving framework, I believe such a framework is only one component to learning physics. For real learning to occur, students must know how to reason and must see the relationships between the ideas of physics and their direct experiences. Until the reasoning is sound and the relationships are clear, fundamental learning will remain illusive.

The central theme of this book is to weave reasoning and relationships into the way we teach introductory physics. Three important results of this approach are the following:

1. Establishing the relationship between forces and motion.
2. A systematic approach to problem solving.
3. Reasoning and relationship problems.

1. Establishing the relationship between forces and motion.

All of Chapter 2 is devoted to Newton's laws of motion and what they tell us about the way force and motion are related. This is the central thread of mechanics. Armed with an understanding of the proper relationship between kinematics and forces, students can then reason about a variety of problems in mechanics such as "nonideal" cases in which the acceleration is not constant, as found for projectiles with air drag.

From Chapter 2, page 45

2.5 | WHY DID IT TAKE NEWTON TO DISCOVER NEWTON'S LAWS?

Newton's second law tells us that the acceleration of an object is given by $\vec{a} = (\Sigma \vec{F})/m$, where $\Sigma \vec{F}$ is the total force acting on the object. In the simplest situations, there may only be one or two forces acting on an object, and $\Sigma \vec{F}$ is then the sum of these few forces. In some cases, however, there may be a very large number of forces acting on an object. Multiple forces can make things appear to be very complicated, which is perhaps why the correct laws of motion—Newton's laws—were not discovered sooner.

Forces on a Swimming Bacterium ⊗

Figure 2.32 shows a photo of the single-celled bacterium *Escherichia coli*, usually referred to as *E. coli*. An individual *E. coli* propels itself by moving thin strands of protein that extend away from its body (rather like a tail) called flagella. Most *E. coli* possess several flagella as in the photo in Figure 2.32A, but to understand their function, we first consider the forces associated with a single flagellum as sketched in Figure 2.32B. A flagellum is fairly rigid, and because it has a spiral shape, one can think of it as a small propeller. An *E. coli* bacterium moves about by rotating this propeller, thereby exerting a force $\vec{F}_w$ on the nearby water. According to Newton's third law, the water exerts a force $\vec{F}_E$ of equal magnitude and opposite direction on the *E. coli*, as sketched in Figure 2.32B. One might be tempted to apply Newton's second law with the force $\vec{F}_E$ and conclude that the *E. coli* will move with an acceleration that is proportional to this force. However, this is incorrect because we have not included the forces from the water on the body of the *E. coli*. These forces are also indicated in Figure 2.32B; to properly describe the total force from the water, we must draw in many force vectors, pushing the *E. coli* in virtually all directions. At the molecular level, we can understand these forces as follows. We know that water is composed of molecules that are in constant motion, and these water molecules bombard the *E. coli* from all sides. Each time a water molecule collides with the *E. coli*, the molecule exerts a force on the bacterium, much like the collision of the baseball and bat in Figure 2.30. As we saw in that case, the two colliding objects both experience a recoil force, another example of action–reaction forces. So, in the present case, the *E. coli* and the water molecule exert forces on each other. An individual *E. coli* is not very large, but a water molecule is much smaller than the bacterium, and the force from one such collision will have only a small effect on the

Figure 2.32 Ⓐ *E. coli* use the action–reaction principle to propel themselves. An individual *E. coli*

2. A systematic approach to problem solving.

Every worked example follows a five-step format. The first step is to "*Recognize the principle*" that is key to the problem. This step helps students see the "big picture" the problem illustrates. The other steps in the problem-solving process are "*Sketch the problem*," "*Identify the relationships*," "*Solve*," and "*What does it mean?*" The last step emphasizes the key principles once more and often describes how the problem relates to the real world. Explicit problem-solving strategies are also given for major classes of quantitative problems, such as conservation of mechanical energy.

From Chapter 21, page 698

Figure 21.12 Example 21.3.

EXAMPLE 21.3 Using Lenz's Law to Find the Direction of an Induced Current

Consider the sliding metal bar in Figure 21.12. The magnetic field is constant and is directed into the plane of the drawing. If the bar is sliding to the right, use Lenz's law to find the direction of the induced current.

RECOGNIZE THE PRINCIPLE

According to Lenz's law, the induced current produces a field that opposes the change in flux through the circuit loop.

SKETCH THE PROBLEM

Following our "Applying Lenz's Law" problem-solving strategy (step 2), the sketch in Figure 21.12 shows a dashed, rectangular path. We are interested in the current induced in this circuit, so we must consider the flux through this rectangle.

IDENTIFY THE RELATIONSHIPS

Step 3: The magnetic field is directed into the plane of this drawing, so the flux through the rectangular surface is directed into the plane. The area of this chosen surface is wL, and the magnetic flux through the surface is $\Phi_B = BwL$. Because the bar is sliding to the right, Φ_B is increasing and is downward.

SOLVE

The induced emf produces an induced current that opposes the downward increase in the flux, so the induced magnetic field must be directed *upward*. Applying right-hand rule number 1 (Chapter 20), this field direction is produced by a *counterclockwise* induced current.

What does it mean?
The flux through a given area may be "upward" or "downward," and its magnitude may be increasing or decreasing with time. The induced emf always *opposes any changes* in the flux.

Applying Lenz's Law: Finding the Direction of the Induced emf

1. **RECOGNIZE THE PRINCIPLE.** The induced emf always opposes changes in flux through the Lenz's law loop or path.

2. **SKETCH THE PROBLEM,** showing a closed path that runs along the perimeter of a surface crossed by magnetic field lines.

3. **IDENTIFY** if the magnetic flux through the surface is increasing or decreasing with time.

4. **SOLVE.** Treat the perimeter of the surface as a wire loop; suppose there is a current in this loop and determine the direction of the resulting magnetic field. Find the current direction for which this induced magnetic field opposes the change in magnetic flux. This current direction gives the sign (i.e., the "direction") of the induced emf.

5. Always *consider what your answer means* and check that it makes sense.

From Chapter 21, page 698

Dealing with Reasoning and Relationships Problems

1. **RECOGNIZE THE PRINCIPLE.** Determine the key physics ideas that are central to the problem and that connect the quantity you want to calculate with the quantities you know. In the examples found in this section, this physics involves motion with constant acceleration.

2. **SKETCH THE PROBLEM.** Make a drawing that shows all the given information and everything else that you know about the problem. For problems in mechanics, your drawing should include all the forces, velocities, and so forth.

3. **IDENTIFY THE RELATIONSHIPS.** Identify the important physics relationships; for problems concerning motion with constant acceleration, they are the relationships between position, velocity, and acceleration in Table 3.1. For many reasoning and relationships problems, values for some of the essential unknown quantities may not be given. You must then use common sense to make reasonable

estimates for these quantities. Don't worry or spend time trying to obtain precise values of every quantity (such as the amount that the child's knees flex in Fig. 3.23). Accuracy to within a factor of 3 or even 10 is usually fine because the goal is to calculate the quantity of interest to within an order of magnitude (a factor of 10). Don't hesitate to use the Internet, the library, and (especially) your own intuition and experiences.

4. **SOLVE.** Since an exact mathematical solution is not required, cast the problem into one that is easy to solve mathematically. In the examples in this section, we were able to use the results for motion with constant acceleration.

5. Always *consider what your answer means* and check that it makes sense.

As is often the case, practice is a very useful teacher.

3. Reasoning and relationship problems.

Many real-world applications require an estimation of certain key parameters. For example, the approximate force on a car bumper during a collision can be found by making a few simplifying assumptions about the collision and the way the bumper deforms, and estimating the mass of the car. Physicists find these "back-of-the-envelope" calculations very useful for gaining an intuitive understanding of a situation. The ability to deal with such problems requires a good understanding of the key relationships in the problem and how fundamental principles can be applied. Most textbooks completely ignore such problems, but I believe students can, with a careful amount of coaching and practice, learn to master the skills needed to be successful with these problems. This kind of creative problem solving is a valuable skill for students in all areas.

From Chapter 3, page 78

EXAMPLE 3.6 Ⓡ Cars and Bumpers and Walls

Consider a car of mass 1000 kg colliding with a rigid concrete wall at a speed of 2.5 m/s (about 5 mi/h). This impact is a fairly low-speed collision, and the bumpers on a modern car should be able to handle it without much damage to the car. Estimate the force exerted by the wall on the car's bumper.

RECOGNIZE THE PRINCIPLE

The motion we want to analyze starts when the car's bumper first touches the wall and ends when the car is stopped. To treat the problem approximately, we *assume* the force on the bumper is constant during the collision period, so the acceleration is also constant. We can then use our expressions from Table 3.1 to analyze the motion. Our strategy is to first find the car's acceleration and then use it to calculate the associated force exerted by the wall on the car from Newton's second law.

SKETCH THE PROBLEM

Figure 3.24 shows a sketch of the car along with the force exerted by the wall on the car. There are also two vertical forces—the force of gravity on the car and the normal force exerted by the road on the car—but we have not shown them because we are concerned here with the car's horizontal (x) motion, which we can treat using $\sum F = ma$ for the components of force and acceleration along x.

IDENTIFY THE RELATIONSHIPS

To find the car's acceleration, we need to estimate either the stopping time or the distance Δx traveled during this time. Let's take the latter approach. We are given the initial velocity ($v_0 = 2.5$ m/s) and the final velocity ($v = 0$). Both of these quantities are in Equation 3.4:

$$v^2 = v_0^2 + 2a(x - x_0) \tag{1}$$

At start of collision

A

After collision, bumper has compressed a distance Δx

B

Figure 3.24 Example 3.6.
A When a car collides with a wall, the wall exerts a force F on the bumper. This force provides the acceleration that stops the car.
B During the collision and before the car comes to rest, the bumper deforms by an amount Δx. The car travels this distance while it comes to a complete stop.

From Chapter 3, page 78

Changing the Way Students View Physics

The *relationships* between physics and other areas of science are rapidly becoming stronger and are transforming the way all fields of science are understood and practiced. Examples of this transformation abound, particularly in the life sciences. Many students of college physics are engaged in majors relating to the life sciences, and the manner in which they need and will use physics differs from only a few years ago. For their benefit, and for the benefit of students in virtually all technical and even nontechnical disciplines, textbooks must place a greater emphasis on how to apply the *reasoning* of physics to real-world examples. Such examples come quite naturally from the life sciences, but many everyday objects are filled with good applications of fundamental physics principles as well. For instance, my discussion of molecular motors in the context of work and energy in Chapter 6 is unique, as are discussions of photosynthesis as a thermodynamic process in Chapter 16 and electricity in the atmosphere (lightning) in Chapter 19. Students must be made to see that physics is relevant to their daily lives and to the things they find interesting.

Figure 6.34 Some molecular motors move by "walking" along long strands of a protein called actin. These motors are the subject of much current research. We can use work–energy principles to understand their behavior.

From Chapter 6, page 195

6.8 | ⊗ WORK, ENERGY, AND MOLECULAR MOTORS

In Section 6.7, we discussed how our ideas about work, energy, and power can be used to understand the behavior of motors and similar devices. The same ideas apply to all types of motors, including the molecular motors that transport materials within and between cells. Several different types of molecular motors have been discovered, one example of which is sketched in Figure 6.34. This motor is based on a molecule called myosin that moves along long filaments composed of actin molecules.

Myosin is believed to move by "walking" along a filament in steps, much like a person walks on two legs. This process takes place in muscles, where the filaments form bundles. A myosin molecule is also attached to its own muscle filament, so as it walks, the myosin drags one muscle filament relative to another. The operation of your muscles is produced by these molecular motors.

Calculating the Force Exerted by a Molecular Motor

The precise biochemical reactions involved in the myosin walking motion are not completely understood. However, we do know that each step has a length of approximately 5 nm (5×10^{-9} m) and that the energy for this motor comes from a chemical

Although much can be gained by bringing many new and current examples into the text, traditional physics examples such as inclined planes, pulleys, and resistors in series or parallel can still be useful pedagogical tools. A good example, however, must do more than just illustrate a particular principle of physics; students should also see clearly how the example can be expanded and generalized to other (and, I hope, interesting) situations. The block-and-tackle example in Chapter 3 is one such case, illustrating pulleys and tension forces in a traditional way but going on to describe how this device can amplify forces. This theme of force amplification is revisited in future chapters in discussions of torque and levers, work, hydraulics, and conservation of energy, and it is also applied to the mechanical function of the human ear. Returning to key themes throughout the text gives students a deeper understanding of fundamental physics principles and their relationship to real-world applications.

Encouraging student curiosity. Many important and fundamental ideas about the world are ignored in most textbooks. By devoting some time to these ideas, this book helps students see that physics can be extremely exciting and interesting. Such issues include the following. (1) Why is the inertial mass equal to the gravitational mass? (This question is mentioned in Chapter 3, revisited in the discussion of gravitation, and mentioned again in Chapter 27 on relativity.) (2) What is "action-at-a-distance," and how does it really work? (The concept of a *field* is mentioned in several places, including but not limited to the sections on gravitation and Coulomb's law.) (3) How do we know the structure of Earth's core? (4) What does color vision tell us about the nature of light? These issues and others like them are essentially unmentioned in current texts, yet they get to the heart of physics and can stimulate student curiosity.

Starting where students are and going farther than you imagined possible. Many students come to their college physics course with a common set of pre-Newtonian misconceptions about physics. I believe the best way to help students overcome these misconceptions is to address them directly and help students see where and how their pre-Newtonian ideas fail. For this reason, *College Physics: Reasoning and Relationships* devotes Chapter 2 to the fundamental relationships between force and motion as Newton's laws of motion are introduced. The key ideas are then reinforced in Chapter 3 with careful discussions of several applications of Newton's laws in one dimension. This approach allows us to get to the more interesting material faster, and, in my experience, the students are more prepared for it then.

Building on prior knowledge. A good way to learn is to build from what is already known and understood. (Learning scientists call this "scaffolding.") This book therefore revisits and builds on selected examples with a layered development, deepening and extending the analysis as new physical principles are introduced. In typical cases, a topic is revisited two or three times, both within a chapter and across several chapters. One example is the theme of amplifying forces, which begins in Chapter 3. This theme reappears in a number of additional topics, including the mechanics of the ear and the concepts of work and energy. Layered or scaffolded development of concepts, examples, and problem topics helps students see relationships between various physical principles.

From Chapter 3, page 74

Using Pulleys to Redirect a Force

Cables and ropes are an efficient way to transmit force from one place to another, but they have an important limitation: they can only "pull," and this force must be directed along the direction in which the cable lies. In many situations, we need to change the direction of a force, which can be accomplished by using an extremely useful mechanical device called a pulley. A simple pulley is shown in Figure 3.21A; it is just a wheel free to spin on an axle through its center, and it is arranged so that a rope or cable runs along its edge without slipping. For simplicity, we assume both the rope and the pulley are massless. Typically, a person pulls on one end of the rope so as to lift an object connected to the other end. The pulley simply changes the direction of the force associated with the tension in the rope as illustrated in Figure 3.21B, which shows the rope "straightened out" (i.e., with the pulley removed). In either case—with or without the pulley in place—the person exerts a force F on one end of the rope, and this force is equal to the tension. The tension is the same everywhere along this massless rope, so the other end of the rope exerts a force of magnitude T on the object. A comparison of the two arrangements in Figure 3.21—one with the pulley and one without—suggests the tension in the rope is the same in the two cases, and in both cases the rope transmits a force of magnitude $T = F$ from the person to the object.

Figure 3.21 Ⓐ A simple pulley. If the person exerts a force F on a massless string, there is a tension T in the string, and this tension force can be used to lift an object. Ⓑ The string "straightened out." The pulley simply redirects the force.

Work, Energy, and Amplifying Forces

In Chapter 3, we encountered a device called the block and tackle and showed how it can amplify forces. Figure 6.8 shows a block and tackle that amplifies forces by a factor of two. We saw in Chapter 3 that if a person applies a force F to the rope, the tension in the rope is $T = F$. Because the pulley is suspended by two portions of the rope, the upward force on the pulley is $2T$ and the pulley exerts a total force $2T$ on the object to which it is connected, that is, the crate in Figure 6.8. Let's now consider how this process of force amplification affects the work done by the person. Suppose the person lifts his end of the rope through a distance L. That will raise the pulley by half that amount, that is, a distance of $L/2$. You can see why by noticing that when the pulley moves upward through a distance $L/2$, the sections of the rope on both sides become shorter by this amount, so the end of the rope held by the person must move a distance L.

When the pulley moves upward by a distance $L/2$, the crate is displaced by the same amount. The work done *by the pulley on the crate* is equal to the total force of the pulley on the crate ($2T$) multiplied by the displacement of the crate, which is $L/2$:

$$W_{\text{on crate}} = 2T(L/2) = TL$$

At the same time, the person does work on the end of the rope since he exerts a force $F = T$ and the displacement of the end of the rope is L. The work done *by the person on the rope* is equal to the force that he exerts on the rope (T) multiplied by the displacement of the rope, which is L, so

$$W_{\text{on rope}} = FL = TL$$

From Chapter 6, page 173 Thus, the work done on the rope is precisely equal to the work done on the crate.

Tension = $2T$

Crate

Figure 6.8 The person applies a force T to the rope. The block and tackle amplifies this force, and the total force applied to the crate by the rope is $2T$. However, when he moves the end of the rope a distance L, the crate moves a distance of only $L/2$.

Going the extra step: reasoning and relationship problems. Many interesting real-world physics problems cannot be solved exactly with the mathematics appropriate for a college physics course, but they can often be handled in an approximate way using the simple methods (based on algebra and trigonometry) developed in such a course. Professional physicists are familiar with these back-of-the-envelope calculations. For instance, we may want to know the approximate force on a skydiver's knees when she hits the ground. The precise value of the force depends on the details of the landing, but we are often interested in only an approximate (usually order-of-magnitude) answer. We call such problems *reasoning and relationship* problems because solving them requires us to identify the key physics relationships and quantities needed for the problem and that we also estimate values of some important quantities (such as the mass of the skydiver and how she flexes her knees) based on experience and common sense.

Reasoning and relationship problems provide physical insight and a chance to practice critical thinking (reasoning), and they can help students see more clearly the fundamental principles associated with a problem. A truly unique feature of this book is the inclusion of these problems in both the worked examples and the end-of-chapter problems. The ability to deal with this class of problems is an extremely useful skill for all students, in all fields.

Problem solving: a key component to understanding. Although reasoning and relationship problems are used to help students develop a broad understanding of physics, this book also contains a strong component of traditional quantitative problem solving. Quantitative problems are a component of virtually all college physics courses, and students can benefit by developing a systematic approach to such problems. *College Physics: Reasoning and Relationships* therefore places extra emphasis on step-by-step approaches students can use in a wide range of situations. This approach can be seen in the worked examples, which use a five-step solution process: (1) recognize the physics principles central to the problem, (2) draw a sketch showing the problem and all the given information, (3) identify the relationships between the known and unknown quantities, (4) solve for the desired quantity, and (5) ask what the answer means and if it makes sense. Explicit problem-solving strategies are also given for major classes of quantitative problems, such as applying the conservation of mechanical energy.

From Chapter 17, page 535

PROBLEM SOLVING | Calculating Forces with Coulomb's Law

1. **RECOGNIZE THE PRINCIPLE.** The electric force on a charged particle can be found using Coulomb's law together with the principle of superposition.

2. **SKETCH THE PROBLEM.** Construct a drawing (including a coordinate system) and show the location and charge for each object in the problem. Your drawing should also show the directions of all the electric forces—$\vec{F}_1$, $\vec{F}_2$, and so forth—on the particle(s) of interest.

3. **IDENTIFY THE RELATIONSHIPS.** Use Coulomb's law (Eq. 17.3 or 17.5) to find the magnitude of the forces $\vec{F}_1$, $\vec{F}_2$, ... acting on the particle(s)

4. **SOLVE.** The total force on a particle is the sum (the superposition) of all the individual forces from steps 2 and 3. Add these forces *as vectors* to get the total force. When adding these vectors, it is usually simplest to work in terms of the components of $\vec{F}_1$, $\vec{F}_2$, ... along the coordinate axes.

5. Always *consider what your answer means* and check that it makes sense.

PROBLEM SOLVING | Plan of Attack for Problems in Statics

1. **RECOGNIZE THE PRINCIPLE.** For an object to be in static equilibrium, the sum of all the forces on the object must be zero. This principle leads to Equation 4.2, which can be applied to calculate any unknown forces in the problem.

2. **SKETCH THE PROBLEM.** It is usually a good idea to show the given information in a picture, which should include a coordinate system. Figures 4.1 through 4.3 and the following examples provide guidance and advice on choosing coordinate axes.

3. **IDENTIFY THE RELATIONSHIPS.**
   • Find all the forces acting on the object that is (or should be) in equilibrium and construct a free-body diagram showing all the forces on the object.
   • Express all the forces on the object in terms of their components along x and y.
   • Apply the conditions[3] $\sum F_x$ and $\sum F_y = 0$.

4. **SOLVE.** Solve the equations resulting from step 3 for the unknown quantities. The number of equations must equal the number of unknown quantities.

5. Always *consider what your answer means* and check that it makes sense.

From Chapter 4, page 95

ORGANIZATION AND CONTENT

Translational motion. The organization of topics in this book follows largely traditional lines with one exception. Forces and Newton's laws of motion are introduced in Chapter 2 along with basic defining relationships from kinematics. In almost all other texts, kinematic equations are covered first in the absence of Newton's laws, which obscures the cause of motion. This text presents the central thread of all mechanics from the beginning, allowing students to see and appreciate the motivations for many kinematic relationships. Students can then address and overcome common misconceptions early in the course. They are also able to deal sooner with interesting and realistic problems that do not involve a constant acceleration. When forces and Newton's laws are introduced early, we can discuss issues such as air drag and terminal velocity at an earlier stage, which avoids giving students the impression that physics problems are limited to the mathematics of ideal cases. Chapters 2 and 3 are limited to one-dimensional problems for simplicity before moving on to two dimensions in Chapters 4 and 5. Major conservation principles (of energy and momentum) are introduced in Chapters 6 and 7 before moving on to rotational motion.

Rotational motion. In the same way that force is connected to acceleration in Chapters 2 and 3 while introducing the variables of translational motion, torque is connected to angular acceleration while the variables of rotational motion are introduced in Chapter 8. Parallel development of topics in translational and rotational motion is direct and deliberate. The central thread in Chapters 8 and 9 is once again Newton's laws of motion, this time in rotational form.

Fluids. Chapter 10 discusses fluids, including the principles of Pascal, Archimedes, and Bernoulli.

Waves. Chapters 11 through 13 cover harmonic motion, waves, and sound. Waves—moving disturbances that transport energy without transporting matter—provide a link to later topics in electromagnetism, light, and quantum physics.

Thermal physics. Chapters 14 through 16 on thermal physics have conservation of energy as their central thread. Thermodynamics is about the transfer of energy between systems of particles and tells how changes in the energy of a system can affect the system's properties.

Electricity and magnetism. Chapters 17 through 23 keep conservation of energy as an important thread, with additional development of concepts introduced earlier such as that of a field (action-at-a-distance). The topic of magnetism brings in the new concept of a velocity-dependent force. The importance of Maxwell's theory of electromagnetism is emphasized without undue mathematical details.

Light and optics. In Chapters 24 through 26, students can compare and contrast properties of light that depend on its wave nature with properties that require a particle (or ray) approach. Students can apply the principles of optics to model how the human eye works, including the mechanism of color vision.

Twentieth-century physics. Students are introduced to the modern concepts of relativity, quantum, atomic and nuclear physics in Chapters 27 through 31. Quantum physics reveals that matter, like light and other electromagnetic radiation, has both particle and wave properties.

Worked Examples

Worked examples are problems that are solved quantitatively within a chapter's main text. They are designed to teach sound problem-solving skills, and each involves a principle or result that has just been introduced. The worked examples also have other attributes:

- Extra emphasis is placed on step-by-step approaches that students can use in a wide range of situations. All worked examples use a five-step solution process: (1) *recognize* the physics principles central to the problem, (2) draw a *sketch* showing the problem and all the given information, (3) *identify the relationships* between the known and unknown quantities, (4) *solve* for the desired quantity, and (5) *ask what the answer means* and if it makes sense. Answers are boxed for clarity, and all examples emphasize the key final step of asking what the answer means, whether it makes sense, or what can be learned from it.

- Some worked examples are designated with the symbol ⑬ as *reasoning and relationship problems,* designed for back-of-the-envelope solutions. A reasoning and relationship problem requires an approximate mathematical solution, a rough estimate of one or two key unknown quantities, or both. These examples and corresponding homework problems begin in Chapter 3, where Section 3.6 introduces and explains the notions of estimating and reasoning. Reasoning and relationship problems and examples are distributed throughout later chapters.

- Worked examples of special interest to life science students are designated with the symbol ⓧ.

Problem-Solving Strategies

The five-step problem-solving method is adapted to suit broad classes of problems students will encounter, such as when applying Newton's second law, using the principles of conservation of momentum and energy, or finding the current in two branches of a DC circuit. For these classes of problems, a problem-solving strategy is highlighted within the chapter for special study emphasis.

Concept Checks

From Chapter 6, page 187

At various points in the chapter are conceptual questions, called "Concept Checks." These questions are designed to make the student reflect on a fundamental issue. They may involve interpreting the content of a graph or drawing a new graph to predict a relationship between quantities. Many Concept Check questions are in multiple-choice format to facilitate their use in audience response systems. Answers to Concept Checks are given at the end of the book. Full explanations of each answer are given in the *Instructor's Solutions Manual.*

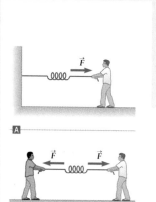

Figure 6.26 Concept Check 6.5.

CONCEPT CHECK 6.5 | Spring Forces and Newton's Third Law

Figure 6.26 shows two identical springs. In both cases, a person exerts a force of magnitude F on the right end of the spring. The left end of the spring in Figure 6.26A is attached to a wall, while the left end of the spring in Figure 6.26B is held by another person, who exerts a force of magnitude F to the left. Which statement is true?

(1) The spring in Figure 6.26A is stretched half as much as the spring in Figure 6.26B.
(2) The spring in Figure 6.26A is stretched twice as much as the spring in Figure 6.26B.
(3) The two springs are stretched the same amount.

Insights

Each chapter contains several special marginal comments called "Insights" that add greater depth to a key idea or reinforce an important message. For instance, Insight 3.3 emphasizes the distinction between weight and mass, and Insight 16.1 explains why diesel engines are inherently more efficient than conventional gasoline internal combustion engines.

Diagrams with Additional Explanatory Labeling

Every college physics textbook contains line art with labeling. This book adds another layer of labeling that explains the phenomenon being illustrated, much as an instructor would explain a process or relationship in class. This additional labeling is set off in a different style.

From Chapter 3, page 80

Figure 21.8 Different ways to produce an induced emf.

Figure 3.26 The skydiver's motion is initially like free fall; compare with Figure 3.15 A. Eventually, however, air drag becomes as large as the force of gravity and the skydiver reaches her terminal velocity v_{term}.

From Chapter 21, page 696

Insight 16.1
EFFICIENCY OF A DIESEL ENGINE
A diesel engine is similar to a gasoline internal combustion engine (Example 16.8). One difference is that a gasoline engine ignites the fuel mixture with a spark from a spark plug, whereas a diesel engine ignites the fuel mixture purely "by compression" (without a spark). The compression of the fuel mixture is therefore much greater in a diesel engine, which leads to a higher temperature in the hot reservoir. According to Equation 16.21 and Example 16.8, this higher temperature gives a higher theoretical limit on the efficiency of a diesel engine.

From Chapter 16, page 512

Chapter Summaries

To make the text more usable as a study tool, chapter summaries are presented in a modified "study card" format. Concepts are classified into two major groups:

Key Concepts and Principles

Applications

Each concept is described in its own panel, often with an explanatory diagram. This format helps students organize information for review and further study.

Relation between the electric field and the electric potential
Suppose the potential changes by an amount ΔV over a distance Δx. The component of the electric field along this direction is then

$$E = -\frac{\Delta V}{\Delta x} \qquad \text{(18.18) (page 573)}$$

The electric field thus has units of volts per meter (V/m).

Capacitors
Two parallel metal plates form a *capacitor*. The *capacitance* C of this structure determines how easily charge can be stored on the plates. The charge on a capacitor is related to the magnitude of the *potential difference* between the plates by

$$\Delta V = \frac{Q}{C} \qquad \text{(18.30) (page 581)}$$

This relation holds for any type of capacitor. For a parallel-plate capacitor, the capacitance is

$$C = \frac{\varepsilon_0 A}{d} \quad \text{(parallel-plate capacitor)} \qquad \text{(18.31) (page 581)}$$

From Chapter 18, page 594

End-of-Chapter Questions

Approximately 20 questions at the end of each chapter ask students to reflect on and strengthen their understanding of conceptual issues. These questions are suitable for use in recitation sessions or other group work. Answers to questions designated SSM are provided in the *Student Companion & Problem-Solving Guide*, and all questions are answered in the *Instructor's Solutions Manual*.

13. Two workers are carrying a long, heavy steel beam (Fig. Q8.13). Which one is exerting a larger force on the object? How can you tell?

From Chapter 8, page 273　　　　**Figure Q8.13**

End-of-Chapter Problems

Homework problems are designed to match the examples that are worked throughout the chapter. Most of these problems are grouped according to the matching chapter section. A final list of "Additional Problems" contains problems that bring together ideas from across the chapter or from multiple chapters. Unmarked problems are straightforward, and intermediate and challenging problems are indicated. Problems of special interest to life science students ⊗, reasoning and relationship problems ⓡ, and problems whose solutions appear in the *Student Companion & Problem-Solving Guide* SSM are so indicated. Answers to odd-numbered problems appear at the end of the book.

6. ⊠ A bar magnet is thrust into a current loop as sketched in Figure P21.6. Before the magnet reaches the center of the loop, what is the direction of the induced current as seen by the observer on the right, clockwise or counterclockwise?

From Chapter 21, page 718

Observer

Figure P21.6 Problems 6, 7, and 8.

ANCILLARIES

Using Technology to Enhance Learning

Enhanced WebAssign is the perfect solution to your homework management needs. Designed by physicists for physicists, this system is a reliable and user-friendly teaching companion. Enhanced WebAssign is available for *College Physics: Reasoning and Relationships,* giving you the freedom to assign

- Selected end-of-chapter problems, algorithmically driven where appropriate and containing an example of the student solution.

- Reasoning and relationship problems. Students are at risk of missing which crucial quantities must be estimated. A coached solution can help students learn how to attack these problems and arrive at a sensible answer.
- Concept Checks from the chapter, available in multiple-choice format for assignment.
- Algorithmically generated versions of the worked examples from the text. These can be assigned to students to help them prepare for homework problems.

Please visit **www.webassign.net/brookscole** to view an interactive demonstration of Enhanced WebAssign.

PowerLecture™ CD-ROM is an easy-to-use multimedia tool allowing instructors to assemble art with notes to create fluid lectures quickly. The CD-ROM includes prepared PowerPoint® lectures and digital art from the text as well as editable electronic files of the *Instructor's Solutions Manual* and the *Test Bank*. The CD also includes the ExamView® Computerized Test Bank, giving you the ability to build tests featuring an unlimited number of new questions or any of the existing questions from the preloaded *Test Bank*. Finally, the CD includes audience response system content specific to the textbook. Contact your local sales representative to find out about our audience response software and hardware.

Additional Instructor Resources

Instructor's Solutions Manual by Michael Meyer (Michigan Technological University), David Sokoloff (University of Oregon) and Raymond Hall (California State University, Fresno). This two-volume publication provides full explanations of Concept Check answers, answers to end-of-chapter questions, and complete solutions to end-of-chapter problems using the five-step problem-solving methodology developed in the text.

Test Bank by Ed Oberhofer (University of North Carolina at Charlotte and Lake-Sumter Community College) is available on the *PowerLecture*™ CD-ROM as editable electronic files or via the ExamView® test software. The file contains questions in multiple-choice format for all chapters of the text. Instructors may print and duplicate pages for distribution to students.

Student Resources

Student Companion & Problem-Solving Guide by Richard Grant (Roanoke College) will prove to be an essential study resource. For each chapter, it contains a summary of problem-solving techniques (following the text's methodology), a list of frequently-asked questions students often have when attempting homework assignments, selected solutions to end-of-chapter problems, solved Capstone Problems representing typical exam questions, and a set of MCAT review questions with explanation of strategies behind the answers.

Physics Laboratory Manual, third edition by David Loyd (Angelo State University) supplements the learning of basic physical principles while introducing laboratory procedures and equipment. Each chapter includes a prelaboratory assignment, objectives, an equipment list, the theory behind the experiment, experimental procedures, graphing exercises and questions. A laboratory report form is included with each experiment so that the student can record data, calculations, and experimental results. Students are encouraged to apply statistical analysis to their data. A complete *Instructor's Manual* is also available to facilitate use of this lab manual.

ACKNOWLEDGMENTS

Creating a new textbook is an enormous job requiring the assistance of many people. To all these people, I extend my sincere thanks.

Contributors

Raymond Hall of California State University, Fresno and Richard Grant of Roanoke College contributed many interesting and creative end-of-chapter questions and problems. Ray Hall's ideas also led to the design of the book's cover.

Accuracy Reviewers

David Bannon, *Oregon State University*
Ken Bolland, *The Ohio State University*
Stephane Coutu, *The Pennsylvania State University*
Stephen D. Druger, *University of Massachusetts—Lowell*
A. J. Haija, *Indiana University of Pennsylvania*
John Hopkins, *The Pennsylvania State University*
David Lind, *Florida State University*
Edwin Lo
Dan Mazilu, *Virginia Tech*
Tom Oder, *Youngstown State University*
Brad Orr, *University of Michigan*
Chun Fu Su, *Mississippi State University*

Manuscript Reviewers

A special thanks is due to Amy Pope of Clemson University for her thoughtful reading of the entire manuscript.

Jeffrey Adams, *Montana State University*
Anthony Aguirre, *University of California, Santa Cruz*
David Balogh, *Fresno City College*
David Bannon, *Oregon State University*
Phil Baringer, *University of Kansas*
Natalie Batalha, *San Jose State University*
Mark Blachly, *Arsenal Technical High School*
Gary Blanpied, *University of South Carolina*
Ken Bolland, *The Ohio State University*
Scott Bonham, *Western Kentucky University*
Marc Caffee, *Purdue University*
Lee Chow, *University of Central Florida*
Song Chung, *William Patterson University*
Alice Churukian, *Concordia College*
Thomas Colbert, *Augusta State University*
David Cole, *Northern Arizona University*
Sergio Conetti, *University of Virginia*
Gary Copeland, *Old Dominion University*
Doug Copely, *Sacramento City College*

Robert Corey, *South Dakota School of Mines & Technology*
Andrew Cornelius, *University of Nevada, Las Vegas*
Carl Covatto, *Arizona State University*
Nimbus Couzin, *Indiana University, Southeast*
Thomas Cravens, *University of Kansas*
Sridhara Dasu, *University of Wisconsin, Madison*
Timir Datta, *University of South Carolina*
Susan DiFranzo, *Hudson Valley Community College*
David Donnelly, *Texas State University*
Sandra Doty, *The Ohio State University*
Steve Ellis, *University of Kentucky*
Len Finegold, *Drexel University*
Carl Fredrickson, *University of Central Arkansas*
Joe Gallant, *Kent State University, Warren Campus*
Kent Gee, *Brigham Young University*
Bernard Gerstman, *Florida International University*
James Goff, *Pima Community College*
Richard Grant, *Roanoke College*

William Gregg, *Louisiana State University*

James Guinn, *Georgia Perimeter College, Clarkston*

Richard Heinz, *Indiana University, Bloomington*

John Hopkins, *Pennsylvania State University*

Karim Hossain, *Edinboro University of Pennsylvania*

Linda Jones, *College of Charleston*

Alex Kamenev, *University of Minnesota*

Daniel Kennefick, *University of Arkansas*

Aslam Khalil, *Portland State University*

Jeremy King, *Clemson University*

Randy Kobes, *University of Winnipeg*

Raman Kolluri, *Camden County College*

Ilkka Koskelo, *San Francisco State University*

Fred Kuttner, *University of California, Santa Cruz*

Richard Ledet, *University of Louisiana, Lafayette*

Alexander Lisyansky, *Queens College, City University of New York*

Carl Lundstedt, *University of Nebraska, Lincoln*

Donald Luttermoser, *East Tennessee State University*

Steven Matsik, *Georgia State University*

Sylvio May, *North Dakota State University*

Bill Mayes, *University of Houston*

Arthur McGum, *Western Michigan University*

Roger McNeil, *Louisiana State University*

Rahul Mehta, *University of Central Arkansas*

Charles Meitzler, *Sam Houston State University*

Michael Meyer, *Michigan Technological University*

Vesna Milosevic-Zdjelar, *University of Winnipeg*

John Milsom, *University of Arizona*

Wouter Montfrooij, *University of Missouri*

Ted Morishige, *University of Central Oklahoma*

Halina Opyrchal, *New Jersey Institute of Technology*

Michelle Ouellette, *California Polytechnic State University, San Louis Obispo*

Kenneth Park, *Baylor University*

Galen Pickett, *California State University, Long Beach*

Dinko Pocanic, *University of Virginia*

Amy Pope, *Clemson University*

Michael Pravica, *University of Nevada, Las Vegas*

Laura Pyrak-Nolte, *Purdue University*

Mark Riley, *Florida State University*

Mahdi Sanati, *Texas Tech University*

Cheryl Schaefer, *Missouri State University*

Alicia Serfaty de Markus, *Miami Dade College, Kendall Campus*

Marc Sher, *College of William & Mary*

Douglas Sherman, *San Jose State University*

Marllin Simon, *Auburn University*

Chandralekha Singh, *University of Pittsburgh*

David Sokoloff, *University of Oregon*

Noel Stanton, *Kansas State University*

Donna Stokes, *University of Houston*

Carey Stronach, *Virginia State University*

Chun Fu Su, *Mississippi State University*

Daniel Suson, *Texas A & M University, Kingsville*

Doug Tussey, *Pennsylvania State University*

John Allen Underwood, *Austin Community College, Rio Grande Campus*

James Wetzel, *Indiana University-Purdue University, Fort Wayne*

Lisa Will, *Arizona State University*

Gerald T. Woods, *University of South Florida*

Guoliang Yang, *Drexel University*

David Young, *Louisiana State University*

Michael Yurko, *Indiana University-Purdue University, Indianapolis*

Hao Zeng, *State University of New York at Buffalo*

Nouredine Zettili, *Jacksonville State University*

Focus Group Participants

Edward Adelson, *The Ohio State University*

Mark Boley, *Western Illinois University*

Abdelkrim Boukahil, *University of Wisconsin, Whitewater*

Larry Browning, *South Dakota State University*

Thomas Colbert, *Augusta State University*

Susan DiFranzo, *Hudson Valley Community College*

Hector Dimas, *Mercer County Community College*

David Donnelly, *Texas State University*

Taner Edis, *Truman State University*

Kevin Fairchild, *La Costa Canyon High School*

Joseph Finck, *Central Michigan University*

David Groh, *Gannon University*

Kathleen Harper, *The Ohio State University*

John Hill, *Iowa State University*

Karim Hossain, *Edinboro University of Pennsylvania*

Debora Katz, *United States Naval Academy*

Larry Kirkpatrick, *Montana State University*

Terence Kite, *Pepperdine University*

Lois Krause, *Clemson University*

Mani Manivannen, *Missouri State University*

Michael Meyer, *Michigan Technological University*

John Milsom, *University of Arizona*

M. Sultan Parvez, *Louisiana State University, Alexandria*

Amy Pope, *Clemson University*

Michael Pravica, *University of Nevada, Las Vegas*

Joseph Priest, *Miami University*

Shafiqur Rahman, *Allegheny College*

Kelly Roos, *Bradley University*
David Sokoloff, *University of Oregon*
Jian Q. Wang, *Binghamton University*
Lisa Will, *San Diego City College*

David Young, *Louisiana State University*
Michael Ziegler, *The Ohio State University*
Nouredine Zettili, *Jacksonville State University*

Class Testers

Abdelkrim Boukahil, *University of Wisconsin, Whitewater*
David Groh, *Gannon University*
Tom Moffet, *Purdue University*
Nouredine Zettili, *Jacksonville State University*

Extra Credit

Special thanks go to Lachina Publishing Services, especially Katherine Wilson, Jeanne Lewandowski, and Aaron Kantor, for keeping up with innumerable changes during the production stages of this book. Kathleen M. Lafferty diligently edited the copy. Steve McEntee expertly guided the art program, and Sarah Palmer coordinated the efforts of the art studio at Dartmouth Publishing Services. Greg Gambino created the "Stick Dude" art for worked examples and end-of-chapter problems that proved so popular with reviewers and class testers. Dena Digilio Betz went the extra mile to find photos from unusual sources. Ed Dodd skillfully guided the arrangement of pages during the hectic final phase of production.

I would like to thank all the team at Cengage Learning, including Michelle Julet, Mary Finch, Peggy Williams, John Walker, Teresa L. Trego, Sam Subity, Terri Mynatt, Alyssa White, Brandi Kirksey, Stefanie Beeck, and Rebecca Berardy Schwartz, for giving me the opportunity to undertake this project. I also want to give special thanks to Susan Pashos for her unwavering and tireless support, and to my wife for all her patience.

No textbook is perfect for every student or for every instructor. It is my hope that both students and instructors will find some useful, stimulating, and even exciting material in this book and that you will all enjoy physics as much as I do.

Nicholas J. Giordano

About the Author

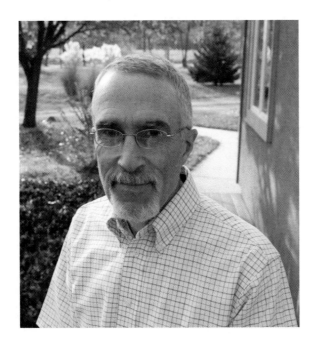

Nicholas J. Giordano obtained his B.S. at Purdue University and his Ph.D. at Yale University. He has been on the faculty at Purdue since 1979, served as an Assistant Dean of Science from 2000 to 2003, and in 2004 was named the Hubert James Distinguished Professor of Physics. His research interests include the properties of nanoscale metal structures, nanofluidics, science education, and biophysics, along with musical acoustics and the physics of the piano. Dr. Giordano earned a Computational Science Education Award from the Department of Energy in 1997, and he was named Indiana Professor of the Year by the Carnegie Foundation for the Advancement of Teaching and the Council for the Advancement and Support of Education in 2004. His hobbies include distance running and restoring antique pianos, and he is an avid baseball fan.

COLLEGE PHYSICS

Reasoning and Relationships

Introduction

The laws of physics describe the behavior of things that are very large, such as the planet Saturn and its rings (which are about 300 million meters in diameter) and things that are very small, such as a red blood cell (which is about five millionths of a meter in diameter), and everything in between. (Top: Courtesy NASA/JPL-Caltech; Bottom: © Dr. David M. Phillips/Visuals Unlimited)

Figure 1.1 Physics is the science of matter and energy, and the interactions between them. At the center of this galaxy lies a massive black hole, containing a very large amount of matter and energy.

Figure 1.2 Isaac Newton (1642−1727) developed the laws of mechanics we study in the first part of this book. Newton was also a great mathematician and invented much of calculus.

1.1 | THE PURPOSE OF PHYSICS

This book is about the field of science known as physics. Let's therefore begin by considering what the word *physics* means. One popular definition is

> *physics*: **the science of matter and energy, and the interactions between them.**

Matter and energy are fundamental to all areas of science; thus, physics is truly a foundational subject. The principles of physics form the basis for understanding chemistry, biology, and essentially all other areas of science. These principles enable us to understand phenomena ranging from the very small (atoms, molecules, and cells) to the very large (planets and galaxies; Fig. 1.1). Indeed, the word *physics* has its origin in the Greek word for "nature." For this reason, an alternative and much broader definition is

> *physics*: **the study of the natural or material world and phenomena; natural philosophy.**

An important part of this definition is the term *natural philosophy*. In a sense, physics is the oldest of the sciences, and at one time *all* scientists were physicists. Many famous thinkers, including Aristotle, Galileo, and Isaac Newton (Fig. 1.2), were such natural philosophers/physicists. This slightly historical perspective is important because there will be times when an understanding of the historical origin and context of a physics principle can be very helpful.

So, if physics is the science of matter and energy, how does one actually study and learn physics? Our primary objective is to learn how to predict and understand the way matter and energy behave; that is, we want to predict and understand how the universe works. Physics is organized around a collection of **physical laws**. In the first part of this book, we learn about Newton's laws, which are concerned with the motion of mechanical objects such as cars, baseballs, and planets. Later we'll encounter physical laws associated with a variety of other phenomena including heat, electricity, magnetism, and light. Our job is to learn about these physical laws (sometimes called **laws of nature**) and how to use them to predict the workings of the universe: how objects move, how electricity flows, how light travels, and more. These physical laws are usually expressed mathematically, so much of our work will involve mathematics. However, good physics is more than just good mathematics; an appreciation of the basic concepts and how they fit together is essential.

In addition to *predicting how* the world works, we would also like to *understand why* it works the way it does. Making predictions requires us to apply the physical laws to a particular situation and work out the associated mathematics to arrive at specific predictions. Understanding the world is more difficult because it involves understanding where the physical laws "come from." This "come from" question is a very difficult one since physical law comes into being in the following way. Initially, someone formulates a hypothesis that describes all that is known about a particular phenomenon. For example, Newton probably formulated such an initial hypothesis for the laws of motion. One must then show that this hypothesis correctly describes *all* known phenomena. For Newton, it meant that his proposed laws of motion had to account correctly for the motion of apples, rocks, arrows, and all other terrestrial objects. In addition, a successful hypothesis is often able to explain things that were previously not understood. In Newton's case, his hypothesis was able to explain celestial motion (the motion of the planets and the Moon, as sketched in Fig. 1.3), a problem that was unsolved before Newton's time. Only after a hypothesis passes such tests does it qualify as a law or principle of physics.

This process through which a hypothesis becomes a law of physics means that there is no way to prove that such a physical law is correct. It is always possible that a future experiment or observation will reveal a flaw or limitation in a particular

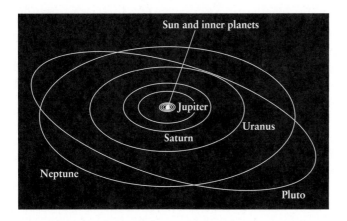

Figure 1.3 Newton showed that the motion of the planets around our Sun can be explained by the same laws of motion that describe the movement of terrestrial objects, such as baseballs and automobiles. This discovery unified human understanding of the motion of terrestrial and celestial objects.

"law." This step is part of the scientific process because the discovery of such flaws leads to the discovery of new laws and new insights into nature.

Although this process of constructing and testing hypotheses can lead us to the laws of physics, it does not necessarily give us an understanding of why these laws are correct. That is, why has nature chosen a particular set of physical laws instead of a different set? Insight into this question can often be gained by examining the form of a physical law and the predictions it leads to. We'll do that as we proceed through this book and in this way get important glimpses into how nature works. The important point is that the problem of *predicting how* the world works is different from *understanding why* it works the way it does. Our goal is to do both.

1.2 | PROBLEM SOLVING IN PHYSICS: REASONING AND RELATIONSHIPS

According to our definition in Section 1.1, physics involves predicting how matter and energy behave in different situations. Making such predictions usually requires that we apply a general physical law to a particular case, a process known as *problem solving*. Problem solving is an essential part of physics, and we'll spend a great deal of time learning how to do it in many different situations. In some ways, learning the art of problem solving is like learning how to play the piano or hit a golf ball: it takes practice. Just as in playing a musical instrument or learning a sport, however, certain practices are the keys to good problem solving. Following these practices consistently will help lead to a thorough understanding of the underlying concepts.

Physical laws are usually expressed in mathematical form, so problem solving often involves some mathematical calculations. Many problems we encounter are *quantitative problems*, which give quantitative information about a situation and require a precise mathematical calculation using that information. An example is to calculate the time it takes an apple to fall from a tree to the ground below, given the initial height of the apple. A problem may also involve the application of a key concept in a nonmathematical way. We use such *concept checks* to test your general understanding of a particular physical law or concept and how it is applied. We will also encounter *reasoning and relationship problems*, which require you to identify important information that might be "missing" from an initial description of the problem. For example, you might be asked to calculate the forces acting on two cars when they collide, given only the speed of the cars just before the collision. From an understanding of the relationship between force and motion, you would need to recognize that additional information is needed to solve this problem. (In this problem, that additional information is the mass of the cars and the way the car bumpers deform in the collision.) It is your job to use common sense and experience to estimate realistic values of these "missing" quantities for typical cars and then use these values to compute an answer. Because such estimated values vary from case to case (e.g., not all cars have the same mass), an approximate mathematical solution and an approximate numerical answer are usually sufficient for such reasoning and relationship problems.

Many different types of problems are found in the "real world," so an ability to deal with these three different types of problems is essential to gaining a thorough understanding of physics and for success in any science- or technology-related field.

PROBLEM SOLVING | Problem-Solving Strategies

We'll encounter problems involving different laws of physics, including Newton's laws of mechanics, the laws of electricity and magnetism, and quantum theory. Although these problems involve many different situations, they can all be attacked using the same basic problem-solving strategy.

1. **RECOGNIZE THE** key physics **PRINCIPLES** that are central to the problem. For example, one problem might involve the principle of conservation of energy, whereas another might require Newton's action–reaction principle. The ability to recognize the central principles requires a conceptual understanding of the laws of physics, how they are applied, and how they are interrelated. Such knowledge and skill are obtained from experience, practice, and careful study.

2. **SKETCH THE PROBLEM.** A diagram showing all the given information, the directions of any forces, and so forth is valuable for organizing your thoughts. A good diagram will usually contain a coordinate system to be used in measuring the position of an object and other important quantities.

3. **IDENTIFY THE** important **RELATIONSHIPS** among the known and unknown quantities. For example, Newton's second law gives a relationship between force and motion, and is thus the key to analyzing the motion of an object. This step in the problem-solving process may involve several parts (substeps), depending on the nature of the problem. For example, problems involving collisions may involve steps that aren't needed or necessary for a problem in magnetism. When dealing with a reasoning and relationship problem, one of these substeps will involve identifying the "missing" information or quantities and then estimating their values.

4. **SOLVE** for the unknown quantities using the relationships in step 3.

5. **WHAT DOES IT MEAN?** Does your answer make sense? Take a moment to think about your answer and reflect on the general lessons to be learned from the problem.

1.3 | DEALING WITH NUMBERS

Scientific Notation

During the course of problem solving, we will often encounter numbers that are very large or very small, and it is useful to have an efficient and precise way to express such numbers. *Scientific notation* was invented as a convenient way to abbreviate extremely large or extremely small numbers. We can understand how scientific notation works by using it to express some of the numbers found in Table 1.1, which lists some important lengths and distances. As you probably know, lengths and distances can be measured in units of *meters*. We'll say more about meters and other units of measure in the next section. For now, we can rely on your intuitive notion of length and that 1 meter (abbreviated "m") is approximately the distance from the floor to the doorknob for a typical door. The height of a typical adult male is about 1.8 m, and the diameter of a compact disc (CD) is about 0.12 m (Fig. 1.4).

The numerical values in Table 1.1 span a tremendous range. The largest is the distance from Earth to the Sun, which is 150,000,000,000 m. This number is so large and contains so many zeros to the left of the decimal point[1] that in Table 1.1

---

[1]In numbers like this one, the decimal point is usually not written explicitly. If we were to include the decimal point, however, this number would be written as 150,000,000,000. m.

Table 1.1 Some Common Lengths and Distances

QUANTITY	LENGTH OR DISTANCE IN METERS (m)	QUANTITY	LENGTH OR DISTANCE IN METERS (m)
Diameter of a proton	1×10^{-15}	Height of the author	1.80
Diameter of a red blood cell*	8×10^{-6}	Height of the Empire State Building	443.2
Diameter of a human hair (removed from the author)	5.5×10^{-5}	Distance from New York City to Chicago	1,268,000
Thickness of a sheet of paper	6.4×10^{-5}	Circumference of the Earth	4.00×10^{7}
Diameter of a compact disc	0.12	Distance from Earth to the Sun	1.5×10^{11}

*Red blood cells are not spherical, but are shaped more like flat plates. This is the approximate diameter of the plate.

we have written it in scientific notation. When written in this way, the distance from Earth to the Sun is 1.5×10^{11} m. The exponent has the value 11 because the decimal point in the number 150,000,000,000 is 11 places to the right of its location in the number 1.5. Likewise, the distance from New York to Chicago is 1,268,000 m, which is 1.268×10^{6} m. Here the decimal point in the number 1,268,000 is six places to the right of its location in the number 1.268. Scientific notation is also a useful way to write very small numbers. For example, the diameter of a hair taken from the author's head is 0.000055 m, which is written as 5.5×10^{-5} m in scientific notation (Table 1.1). Here the exponent has the value −5 because the decimal point in 0.000055 is five places to the left of its location in the number 5.5.

These rules for expressing a number in scientific notation can be summarized as follows. Move the decimal point in the original number to obtain a new number between 1 and 10. Count the number of places the decimal point has been moved; this number will become the exponent of 10 in scientific notation. If you started with a number greater than 10 (such as 150,000,000,000), the exponent of 10 is positive (1.5×10^{11}). If you started with a number less than 1 (such as 0.000055), the exponent is negative (5.5×10^{-5}).

Writing numbers in scientific notation

Figure 1.4 A compact disc has a diameter of approximately 0.12 m.

EXAMPLE 1.1	Population of Earth

The number of people living on Earth in 2007 is estimated to have been approximately 6,600,000,000. Express this number in scientific notation.

RECOGNIZE THE PRINCIPLE

To write this number using scientific notation, we must compare the location of the decimal point in the number 6,600,000,000 with its location in the number 6.6.

SKETCH THE PROBLEM

We must move the decimal point nine places to get a number between 1 and 10:

$$6,600,000,000$$
$$\underbrace{\qquad\qquad\qquad}_{9 \text{ places}}$$

IDENTIFY THE RELATIONSHIPS AND SOLVE

We started with a number greater than 10. Using our rules for expressing a number in scientific notion, the population of Earth in 2007 was

$$6,600,000,000 = \boxed{6.6 \times 10^{9}}$$

What have we learned?
The exponent in this answer is 9 because the decimal point in the number 6,600,000,000 is nine places to the right of its location in the number 6.6.

Significant Figures

When performing a mathematical calculation, it is important to pay attention to accuracy. Suppose you are asked to measure the width of a standard piece of paper such as a page in this book. The answer is approximately 0.216 meter, as you can confirm for yourself using a ruler. However, you should realize that this value is not exact. Different pieces of paper will be slightly different in size, and your measurement will also have some uncertainty (some experimental error). There is an uncertainty associated with all such measured values, including all the values in Table 1.1, and these uncertainties affect the way we write a numerical value. For our piece of paper, we have written the width using three *significant figures*. The term *significant* refers to the number of digits that are meaningful with regard to the accuracy of the value. In this particular case, writing the value as 0.216 m implies that the true value lies between 0.210 m and 0.220 m, and that it is likely to be close to 0.216 m. We should not be surprised to find a value of 0.215 m or 0.217 m, though.

When writing a number in scientific notation, the number of digits that are written depends on the number of significant figures. For our piece of paper, we would write 2.16×10^{-1} m and would say that all three digits are significant. There can be some ambiguity here when dealing with digits that are zero. For example, the length of a particular race in the sport of track and field is 100 meters (the distance between the starting line and the finish). In terms of significant figures, you might think that this number has only one significant figure, which would imply that the length of the race falls within the rather large range of 0 to 200 m. In this example, however, the distance between the start and finish lines will be measured quite precisely, so we should suspect the value to have at least three significant figures, with the true value lying between 99 and 101 meters. If this accuracy is in fact correct, the value "100 meters" actually contains three significant figures; the zeros here are significant because they are indicative of the accuracy. In such cases, we must determine the number of significant figures from the context of the problem or from other information. This ambiguity does not arise with scientific notation; the length of our race would be written as 1.00×10^2 m, which indicates explicitly that the zeros are significant.

Significant figures are also important in calculations. As an example, Table 1.1 lists the thickness of a sheet of paper as 6.4×10^{-5} m. This value—which has two significant figures—was measured by the author for a piece of paper of the kind used in this book. Suppose a book were to contain 976 such sheets. We could then compute the thickness of the book (without the covers) as

$$\text{thickness of book} = \text{number of sheets of paper} \times \text{thickness of one sheet}$$
$$= 976 \times 0.000064 \text{ m} = 0.062464 \text{ m} \qquad (1.1)$$

The final result in Equation 1.1 is written with five significant figures, as you might read from a calculator used to do this multiplication. Because the number of significant figures indicates the expected accuracy of a value, writing this answer with five significant figures implies that we know the thickness of the book to very high accuracy. However, since the thickness of one page is known with an accuracy of only two significant figures, the final value of this calculation—the thickness of the entire book—will actually have an accuracy of only two significant figures. So, we should round the result in Equation 1.1 to two significant figures:

$$0.062464 \text{ m rounded to two significant figures} = 0.062 \text{ m} \qquad (1.2)$$

Rules for significant figures in calculations involving *multiplication and division*.

1. Use the full accuracy of all known quantities when doing the computation. In Equation 1.1, we used the number of sheets as given to three

---

Insight 1.1
SIGNIFICANT FIGURES
The number of significant figures in the value 0.062464 is *five*. For a number smaller than 1, the zeros immediately to the right of the decimal point are *not* significant figures. Hence, the number 0.000064 has just two significant figures.

Determining the number of significant figures in a calculation involving multiplication or division significant figures and the thickness per sheet, which is known to two significant figures.

2. At the *end* of the calculation, round the answer to the number of significant figures present in the *least* accurate starting quantity. In Equation 1.1, the least accurate starting value was the thickness of a single sheet, which was known to two significant figures, so we rounded our final answer to two significant figures in Equation 1.2.

Notice that the rounding in Equation 1.2 took place at the *end* of the calculation. Some calculations involve a sequence of several separate computations, and rounding can sometimes cause trouble if it is applied at an intermediate stage of the computation. For example, suppose we want to use the result from the calculation in Equation 1.2 to find the height of a stack of 12 such books. We could multiply the answer from Equation 1.2 by 12 to get the height of the stack. We thus have

$$\text{height of stack} = 12 \times (0.062 \text{ m}) = 0.744 \text{ m}$$

$$= 0.74 \text{ m (rounding to two significant figures)} \qquad (1.3)$$

Here we have rounded our final answer to two significant figures because the starting value—the thickness of one book—is known to two significant figures. On the other hand, if we use the unrounded value of the thickness of one book from Equation 1.1, we get

$$\text{height} = 12 \times (0.062464 \text{ m}) = 0.749568 \text{ m}$$

$$= 0.75 \text{ m (rounding to two significant figures)} \qquad (1.4)$$

In the last step in Equation 1.4, we again rounded to two significant figures. Comparing the results in Equations 1.3 and 1.4, we see that the final answers differ by 0.01 m. This difference is due to **roundoff error**, which can happen when we round an answer too soon in the course of a multistep calculation. Such errors can be avoided by carrying along an extra significant figure through intermediate steps in a computation and then performing the final rounding at the very end. In this example, we should keep three significant figures for the answer from Equation 1.1 (carrying an extra digit); using this three-significant-figure value in Equation 1.3 would then have given us the correct answer of 0.75 m.

The procedures we have just described for dealing with significant figures apply to calculations that involve multiplication and division. Computations involving addition or subtraction require a different approach to determine the final accuracy.

Rule for significant figures in calculations involving *addition or subtraction*.

The location of the least significant digit in the answer is determined by the location of the least significant digit in the starting quantity that is known with the *least accuracy*.

Determining the number of significant figures in a calculation involving addition or subtraction

Consider the addition of the numbers 4.52 and 1.2. The least accurate of these numbers is 1.2, so the least significant digit here and in the final answer is one place to the right of the decimal point. Pictorially, we have

| Least significant digit in the number 1.2, and in the final sum, is one place to the right of the decimal point. | 4.52 +1.2 ——— 5.7 | The value of this digit is unknown, so the answer is not known to this accuracy. |

Hence, in this example, the final answer has two significant figures.

An example involving subtraction is shown below, where we consider the calculation $4.52 - 1.2$. The starting numbers contain three and two significant figures, respectively, and the final answer contains two. Pictorially,

| Least significant digit in the number 1.2, and in the final answer, is one place to the right of the decimal point. | $\begin{array}{r} 4.52 \\ -1.2 \\ \hline 3.3 \\ \uparrow \end{array}$ | The value of this digit is unknown, so the answer is not known to this accuracy. |

As another example, consider the subtraction of two numbers whose difference is very small, such as $4.52 - 4.1$. Now we have

| Least significant digit in the final answer. | $\begin{array}{r} 4.52 \\ -4.1 \\ \hline 0.4 \\ \uparrow \end{array}$ | The value of this digit is unknown, so the answer is not known to this accuracy. |

In this case, the number of significant figures in the final answer is *smaller* than the number of significant figures in either starting number.

Some numbers that appear in a calculation are known exactly. For example, the number of seconds in 1 minute is exactly 60. Although at first glance it might appear that this value contains two significant figures, it is an *exact* value, so it really should be thought of as 60.000000... in a calculation. The number of significant figures in a calculation with such numbers is then determined by the accuracy (i.e., the number of significant figures) with which other quantities in the calculation are known.

In most of the calculations and problems in this book, we'll round answers to two significant figures.

EXAMPLE 1.2 | ⊗ Volume of a Blood Cell

Blood contains several different types of cells: red blood cells and white blood cells (named according to their color), and colorless cells called lymphocytes.

(a) Consider a lymphocyte that is approximately spherical in shape with a radius of $r_1 = 5.0 \times 10^{-6}$ m. What is the volume of this cell?

(b) A second lymphocyte has a radius of $r_2 = 5.1 \times 10^{-6}$ m. What is the *difference* in the volumes of the two cells?

Be sure to express your answers with the correct numbers of significant figures.

RECOGNIZE THE PRINCIPLE

The volume of a sphere of radius r is $V = \frac{4}{3}\pi r^3$. We must evaluate V for both values of the radius, r_1 and r_2. Since r_1 and r_2 are both given with two significant figures, the results for V must also have two significant figures.

SKETCH THE PROBLEM

The radii r_1 and r_2 are almost the same.

IDENTIFY THE RELATIONSHIPS AND SOLVE

(a) Using the value of r_1 given above, the volume of the first cell V_1 is[2]

$$V_1 = \frac{4}{3}\pi r_1^3 = \frac{4}{3}\pi (5.0 \times 10^{-6} \text{ m})^3 = 5.236 \times 10^{-16} \text{ m}^3$$

---

[2]Notice that this answer is given in units of m³ = cubic meters. We'll say more about this and other such units below.

Insight 1.2

A MISUSE OF SIGNIFICANT FIGURES
It is often quoted that normal body temperature (for a human) is 98.6°F. This value is measured with the Fahrenheit temperature scale, in "Fahrenheit units." This temperature was measured originally in experiments that employed the Celsius temperature scale. When measured in "Celsius units," it was found that healthy people exhibit temperatures in the range of 36°C to 38°C, with 37°C being typical. So, the value for body temperature in Celsius units is 37°C and thus has *two* significant figures. However, when this value was converted to the Fahrenheit scale it was specified as 98.6°F, with *three* significant figures. To be consistent with significant figures (and with the accuracy of the original experimental value and the variation from person to person), this value should actually be rounded to two significant figures, which would give 99°F. Using the value of 98.6°F with the *incorrect* number of significant figures implies that a difference of 0.1°F is significant, but that is not the case at all!

The value of r_1 is given to two significant figures, so we must round our final answer to two significant figures:

$$V_1 = \boxed{5.2 \times 10^{-16} \text{ m}^3}$$

(b) A similar calculation gives the volume of the second cell:

$$V_2 = \tfrac{4}{3}\pi r_2^3 = \tfrac{4}{3}\pi (5.1 \times 10^{-6} \text{ m})^3 = 5.6 \times 10^{-16} \text{ m}^3$$

where we have again rounded to two significant figures. The difference in the volumes of the two cells is

$$\Delta V = V_2 - V_1 = (5.6 \times 10^{-16} - 5.2 \times 10^{-16}) \text{ m}^3$$

$$\Delta V = 0.4 \times 10^{-16} \text{ m}^3 = \boxed{4 \times 10^{-17} \text{ m}^3}$$

What have we learned?
The given values of r_1 and r_2 contained two significant figures, so the results for V_1 and V_2 also contained two significant figures. However, according to the rule for significant figures in subtraction, the value for ΔV has only one significant figure.

CONCEPT CHECK 1.1 | Significant Figures
Give the number of significant figures for all the lengths and distances in Table 1.1.

1.4 | PHYSICAL QUANTITIES AND UNITS OF MEASURE

To conduct experiments and test physical theories, we must be able to *measure* various physical quantities. Typical quantities of importance in physics include the *distance* traveled by a baseball, the *time* it takes an apple to fall from a tree, and the *mass* of the Earth. These three types of quantities—distance (or equivalently, length), time, and mass—play central roles in physics. To be able to measure a particular physical quantity, we must have a *unit of measure* for that quantity. Consider, for example, the quantity length. Table 1.1 lists a variety of lengths and distances, with values given in *meters*, a unit of measure widely used in scientific work. The meter was initially defined in terms of the circumference of the Earth and was subsequently redefined and maintained using a carefully constructed metal bar composed of platinum and iridium (Fig. 1.5). In 1960, the meter was redefined again, in terms of the wavelength of light emitted by krypton atoms. Other units, such as inches and feet, can be used to measure length, and scientists involved in metrology (the science of measurement) have determined how the many different units of length, including meters, inches, and feet, are related.

Figure 1.5 Prior to 1960, the length of the standard meter was maintained using a platinum–iridium bar like the ones shown here from the National Institute of Standards and Technology.

It is often necessary to convert the value of a particular length from one unit to another. For example, the width of a page of this book is approximately 0.216 m. We can express this width in inches using the appropriate *conversion factor*, which relates the two units of interest. For instance, it has been established that a length of 1 inch (abbreviated as "in.") is equal to 2.54×10^{-2} m. The conversion between inches and meters can be expressed in equation form as 1 in. = 2.54×10^{-2} m or in fractional form as

$$\frac{1 \text{ in.}}{2.54 \times 10^{-2} \text{ m}}$$

If a page is 0.216 m wide, its width in inches is

$$0.216 \text{ m} = 0.216 \text{ m} \times \overbrace{\frac{1 \text{ in.}}{2.54 \times 10^{-2} \text{ m}}}^{\text{conversion factor}} = 8.50 \text{ in.} \qquad (1.5)$$

In terms of only the units, the conversion in Equation 1.5 has the schematic form

$$\text{meters} \times \frac{\text{inches}}{\text{meters}} = \cancel{\text{meters}} \times \frac{\text{inches}}{\cancel{\text{meters}}} = \text{inches} \qquad (1.6)$$

Converting from one unit to another

so the "unwanted" units—in this case meters—cancel. The conversion of units always has this form, with the unwanted units canceling to leave the desired unit.

This example of units conversion involves units of length, but the same approach applies to other types of units (e.g., converting from hours to seconds). A set of commonly used conversion factors is given inside the front cover of this book. These conversion factors are either exact numbers or are known with very high precision, so they can generally be treated as exact numbers when determining the number of significant figures in a calculation.

EXAMPLE 1.3 | Converting Units

Find the number of inches in 1 mile.

RECOGNIZE THE PRINCIPLE

Inches and miles are two units used to measure length. They are not used often in scientific work, but are often encountered in everyday activities. To convert from miles to inches, we use our knowledge that 1 mile is equal to exactly 5280 feet and that there are exactly 12 inches in 1 foot.

SKETCH THE PROBLEM

No sketch is needed.

IDENTIFY THE RELATIONSHIPS AND SOLVE

The conversion from miles to inches looks like

$$1 \text{ mile} \times \frac{5280 \text{ feet}}{1 \text{ mile}} \times \frac{12 \text{ inches}}{1 \text{ foot}} = \boxed{63{,}360 \text{ inches}} \qquad (1)$$

In terms of the units, this conversion has the form

$$1 \text{ mile} = \cancel{\text{mile}} \times \frac{\cancel{\text{feet}}}{\cancel{\text{mile}}} \times \frac{\text{inches}}{\cancel{\text{foot}}} = \text{inches}$$

What have we learned?
The result in the Equation (1) answer is *exact* because the number of feet in 1 mile and the number of inches in 1 foot are both exact values, determined by the definitions of these units and not by measurement.

Units of Time and Mass

A particularly important physical quantity is *time*. As you probably know, time is measured in units of seconds, minutes, hours, and so forth. Table 1.2 lists the values of a number of important time intervals. Since many of these intervals are very small or very large, scientific notation is again a convenient way to express these values. In most scientific work, time is measured in units of seconds (denoted by "s"). The value of the second is based on the frequency of light emitted by cesium atoms. Figure 1.6 shows a cesium atomic clock from the National Institute of Standards and Technology.

A third important physical quantity is *mass*. Intuitively, mass is related to the amount of material contained in an object; a slightly better definition will be given when we discuss Newton's laws of motion in Chapter 2. Mass can be measured in units called kilograms and in several other units, including grams and slugs. Table 1.3 lists the masses of several common objects in units of kilograms (abbreviated

© 2005 Geoffrey Wheeler Photography

Figure 1.6 Some of the most accurate clocks are based on the properties of cesium atoms and the light they emit. These clocks gain or lose less than 10^{-8} s each day.

Table 1.2 Some Common Time Intervals	
QUANTITY	TIME IN SECONDS (s)
Time for light to travel 1 meter	3.34×10^{-9}
Time between heartbeats (approximate)	1
Time for light to travel from the Sun to Earth	500
1 day	86,400
1 month (30 days)	2,592,000
Human life span (approximate)	2×10^9
Age of the universe	5×10^{17}

Table 1.3 Some Common Masses	
OBJECT	MASS IN KILOGRAMS (kg)
Electron	9.1×10^{-31}
Proton	1.7×10^{-27}
Red blood cell	1×10^{-13}
Mosquito	1×10^{-5}
Typical person	70
Automobile	1200
Earth	6.0×10^{24}
Sun	2.0×10^{30}

"kg"). By international agreement, the current standard of mass is a piece of a specially configured metal alloy composed of platinum and iridium stored in Paris under extremely well controlled conditions (Fig. 1.7).

The SI System of Units

To be useful, units of measure must be standardized. Everyone should agree on the length of 1 meter, the length of 1 second, and the amount of mass in 1 kilogram. Without such an agreement, there would be no way to conduct science or commerce because we could not accurately compare quantities measured by different individuals. This standardization has been established by international agreements. One such agreement established the *Système International d'Unités* set of units, usually referred to as the SI system of units. It employs meters, seconds, and kilograms as the primary units of length, time, and mass. We generally use the SI system of units throughout this book and usually abbreviate meters as m, seconds as s, and kilograms as kg.

Two other systems of units are often encountered. The CGS system uses the units of centimeters (cm) for length, grams (g) for mass, and seconds (s) for time. Another common system, the U.S. customary system, measures length in feet, mass in slugs, and time in seconds. These systems are summarized in Table 1.4.

In this book, we will almost always work with SI units. However, since we often encounter other units (such as miles per hour) in everyday life we will also consider how to deal with such units in many of our problems.

Powers of 10 and Prefixes

We have already seen how to use scientific notation to work with numbers that are very large or very small. Another way to deal with such numbers is to add a *prefix* to the unit of measure. Table 1.5 lists the common prefixes; when attached to a unit, a prefix multiplies the unit by the amount indicated in the table. These prefixes can be applied to any unit of measure. For example, the prefix *milli* is equivalent to a factor of 10^{-3}, so 1 millimeter is 10^{-3} m = 0.001 m. Another common prefix is *kilo*, which is equivalent to a factor of $10^3 = 1000$. One kilometer is thus 1000 m,

Figure 1.7 The standard kilogram is a cylinder of metal composed of an alloy of platinum and iridium, and is stored (for very safe keeping) in France. This photo shows a copy that is kept in a vacuum inside two separate glass jars. (It is treated very carefully!)

Table 1.4 Units of Length, Mass, and Time			
DIMENSION	SI	CGS	U.S. CUSTOMARY UNITS
length	meter (m)	centimeter (cm)	foot (ft)
mass	kilogram (kg)	gram (g)	slug
time	second (s)	second (s)	second (s)

Table 1.5 Prefixes That Can Be Attached to Units of Measure

Prefix	Power of 10	Symbol
atto	10^{-18}	a
femto	10^{-15}	f
pico	10^{-12}	p
nano	10^{-9}	n
micro	10^{-6}	μ
milli	10^{-3}	m
centi	10^{-2}	c
deci	10^{-1}	d
deka	10^{1}	da
hecto	10^{2}	h
kilo	10^{3}	k
mega	10^{6}	M
giga	10^{9}	G
tera	10^{12}	T
peta	10^{15}	P
exa	10^{18}	E

whereas 1 kilosecond is 1000 s. Indeed, you can see that the SI unit of mass contains this prefix since 1 kilogram = 1000 grams.

EXAMPLE 1.4 Units Conversion and Scientific Notation

How many seconds are in 1 year? Express your answer in scientific notation.

RECOGNIZE THE PRINCIPLE

To compute the number of seconds in 1 year, we use our knowledge that there are 365 days in a year (except for leap years), 24 hours in 1 day, 60 minutes in 1 hour, and 60 seconds in 1 minute.

SKETCH THE PROBLEM

No sketch is needed.

IDENTIFY THE RELATIONSHIPS AND SOLVE

The conversion from years to seconds is

$$1 \text{ year} = 1 \text{ year} \times \frac{365 \text{ days}}{1 \text{ year}} \times \frac{24 \text{ hours}}{1 \text{ day}} \times \frac{60 \text{ minutes}}{1 \text{ hour}} \times \frac{60 \text{ seconds}}{1 \text{ minute}}$$

$$1 \text{ year} = 31{,}536{,}000 \text{ s}$$

Expressed in scientific notation,

$$\boxed{1 \text{ year} = 3.1536 \times 10^7 \text{ s}}$$

What have we learned?
Note that this value is exact (i.e., there are exactly 60 seconds in 1 minute, etc.).

CONCEPT CHECK 1.2 | Using Prefixes and Powers of 10

Write the values of all the lengths in Table 1.1 using a prefix from Table 1.5.

1.5 | DIMENSIONS AND UNITS

We have discussed the physical quantities length, time, and mass, and their units in some detail, but they are only three of the many different types of physical quantities we encounter in physics. In many cases, the units for these other quantities are *derived* from the units of length, time, and mass. In the SI system, *all* the units needed for the study of mechanics can be derived from the **primary units** meters, seconds, and kilograms. For example, the volume of an object is measured in units of cubic meters, which is written as m³. Hence, the unit of volume is a derived unit since it can be expressed in terms of the primary unit meters. Likewise, the density of an object is equal to its mass divided by its volume, so density is measured in units of

$$\text{density} = \frac{\text{mass}}{\text{volume}} \Rightarrow \frac{\text{kg}}{\text{m}^3} \tag{1.7}$$

The units of density are thus derived from the SI units kilogram and meter.

This distinction between derived units and the primary units of meters, seconds, and kilograms also carries over to our fundamental view of physical quantities. For example, what is length? We all have a clear intuitive notion of length, but it is very difficult to give a formal definition of the term *length* that does not rely on some other physical quantity or idea.[3] Likewise, it is difficult to give a satisfactory formal

---

[3]Defining length is like trying to give a definition of every word in a dictionary. The first definitions have to use words that are not yet defined!

definition of the quantity we call time. The best we can do is take a set of primary physical quantities as "given" and then use them to build our theories of physics. For our studies of mechanics, we need only three primary physical quantities—length, time, and mass—and we will build all other necessary quantities—such as force, velocity, energy, pressure, and work—from these primary quantities.

In later chapters, we will learn about four more primary units; they are associated with electricity and magnetism, and heat. There are thus seven primary units in the SI system. All other units can be derived from the seven primary units.

Dimensions and Dimensional Analysis

In our calculation of the volume of a blood cell in Example 1.2, we expressed the answer in units of cubic meters. Hence, we calculated both a numerical value and the units of the answer. This pattern is followed for most calculations because most answers are meaningless unless a unit of measure is specified. Calculating and then checking the units of an answer is an important part of the problem-solving process. This checking process, called *dimensional analysis*, can be carried out in a very general way using the primary physical quantities. For example, the distance from New York City to Chicago has the *dimensions* of length and can be measured in units of meters. If we denote these primary dimensional quantities as **L** (length), **T** (time), and **M** (mass), we can express the dimensions of any combination of these quantities in terms of **L**, **T**, and **M**. Suppose the answer for a particular problem can be written as a length divided by a time. The SI units of the answer would then be the units for length (meters) divided by the units for time (seconds), or m/s. We would express the dimensions of this answer as **L/T**. The dimensions of an answer are independent of the particular units used to measure length, time, and so forth.

| **EXAMPLE 1.5** | Using Dimensional Analysis to Check an Answer |

Suppose we have performed a calculation of the position of an object (x) as a function of time (t) and arrived at the result

$$x = vt \tag{1}$$

Use dimensional analysis to find the dimensions of the quantity v.

RECOGNIZE THE PRINCIPLE

The dimensions must be the same on both sides of the equal sign in Equation (1), so the dimensions of x must equal the dimensions of the product vt. If we rearrange the equation to solve for v, we get $v = x/t$.

Hence, the dimensions of v must equal the dimensions of x divided by t.

SKETCH THE PROBLEM

No sketch is necessary.

IDENTIFY THE RELATIONSHIP AND SOLVE

The dimensions of position x (i.e., distance) is **L**, and the dimensions of t is **T**. The dimensions of v must therefore be

$$\text{dimensions of } v = \frac{\text{L}}{\text{T}}$$

What have we learned?
In the SI system, v could be measured in units of meters per second, or m/s. It is always a good idea to check the dimensions of an answer; this check can sometimes reveal errors in a calculation.

The problem solving we encounter in this book involves three basic types of mathematics: algebra, trigonometry, and vectors. The next few sections are intended as quick refreshers on these topics. A mathematical review is also given in Appendix B.

When solving a problem, we often need to deal with one or two equations containing one or two unknown quantities. Our job is usually to solve for the unknown quantities. Suppose we are given a single equation containing two variables a and T and are asked to find a in terms of T. The equation might be

$$5a = T - 10$$

The goal is to isolate a on one side of the equal sign (usually the left). In this example, we can divide both sides by the constant factor 5 to get the solution

$$a = \frac{T}{5} - 2$$

The numerical value of T might be already known so that the value of a can now be found, or a result expressed in terms of T may be sufficient, depending on the problem.

A slightly more complicated case is with two equations,

$$5a = T - 10 \tag{1.8}$$

$$7a = T + 24 \tag{1.9}$$

A set of equations like this one might arise in a problem in which there are two unknown quantities, a and T. The goal now is to solve for both unknowns. According to the rules of algebra, to solve for two unknown quantities we must have (at least) two equations, so Equations 1.8 and 1.9 should contain all the mathematical information we need to solve for these unknowns. To work out a solution, we can first rearrange Equation 1.8 to obtain a in terms of T:

$$a = \frac{T}{5} - 2 \tag{1.10}$$

Substituting into Equation 1.9 gives

$$7a = T + 24$$

$$7\left(\frac{T}{5} - 2\right) = T + 24$$

$$\frac{7T}{5} - 14 = T + 24 \tag{1.11}$$

We now have a single equation containing only the unknown T. We can find T by moving all the terms involving T to one side to get

$$\frac{7T}{5} - T = 24 + 14$$

$$\frac{7T - 5T}{5} = 38$$

$$\frac{2T}{5} = 38$$

$$T = \frac{5 \times 38}{2} = 95 \tag{1.12}$$

This value of T can then be inserted back into Equation 1.10 to find the value of a:

$$a = \frac{T}{5} - 2 = \frac{95}{5} - 2 = 17$$

Checking the Units of an Answer

Algebra can be used to solve an equation or a set of equations to find the values of variables such as a and T in the examples above. In physics, these variables represent physical quantities, such as mass, or force, or velocity. Physical quantities possess units, and the algebra we use to compute a and T will also give the value of the associated unit. For example, suppose we have computed the density of an object (represented by the variable ρ) and arrived at the answer

$$\rho = \frac{M}{V} \qquad (1.13)$$

Here, M and V are both physical quantities, so they both have units, and their units combine to determine the units of ρ. If we focus only on the units in Equation 1.16, we have the relation

$$\text{units of } \rho = \frac{\text{units of } M}{\text{units of } V}$$

Suppose M is the mass of an object, so its units are kilograms, and V is a volume with units of cubic meters. The units of ρ are then

$$\text{units of } \rho = \frac{\text{units of } M}{\text{units of } V} = \frac{\text{kg}}{\text{m}^3} \qquad (1.14)$$

This answer is indeed the correct unit for density (compare with Equation 1.7), so this result is not a surprise.

In Example 1.5, we saw how dimensional analysis can be used to check an answer, and exactly the same approach can be used when calculating and checking units. For example, in our calculation of the density, we might have made an error and obtained the result

$$\rho = \frac{M^2}{V} \quad (\textbf{wrong!!}) \qquad (1.15)$$

The units in this case would work out as

$$\text{units of } \rho = \frac{\text{units of } M^2}{\text{units of } V} = \frac{\text{kg}^2}{\text{m}^3} \quad (\textbf{wrong!!})$$

Since we know that this answer is not the correct unit for density, this tells us immediately that there must be an error in the calculation that led to Equation 1.15. It is *always* useful to check the units (and dimensions) of your answers.

1.7 | TRIGONOMETRY

We often need to deal with angles and the location of an object within a coordinate system. This is most easily done with trigonometry, right triangles, and the Pythagorean theorem. Figure 1.8 shows a right triangle with sides x, y, and r, where r is the hypotenuse. According to the Pythagorean theorem, we know that

$$x^2 + y^2 = r^2 \qquad (1.16)$$

The trigonometric functions sine, cosine, and tangent are defined as

$$\sin \theta = y/r \qquad (1.17)$$
$$\cos \theta = x/r \qquad (1.18)$$
$$\tan \theta = y/x \qquad (1.19)$$

The Pythagorean theorem (Equation 1.16) implies that

$$\sin^2 \theta + \cos^2 \theta = 1 \qquad (1.20)$$

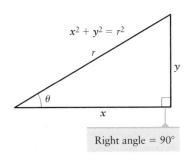

Right angle = 90°

Figure 1.8 The Pythagorean theorem is a relation between the lengths of the sides of a right triangle.

for any value of the angle θ. Equation 1.20 also leads to a number of other trigonometric relations (called **identities**), some of which are listed inside the back cover of this book and in Appendix B.

Equations 1.17 through 1.19 tell us how to calculate the sine, cosine, and tangent of an angle from the lengths of the sides of the associated right triangle. We can also use this approach to find the value of the angle. For example, if we know the values of y and r, we can solve for the angle:

$$\sin \theta = y/r$$

$$\theta = \sin^{-1}(y/r) \tag{1.21}$$

The function $\sin^{-1}$ is the **inverse sine** function, also known as the **arcsine** (= arcsin $= \sin^{-1}$). In words, Equation 1.21 says that θ equals the angle whose sine is y/r. Scientific calculators all contain this function (as a single "button"). If the values of y and r are known, you only take the ratio y/r and then compute the arcsine of this value to find θ in Equation 1.21. Likewise, there are also functions for the inverse cosine ($\cos^{-1}$ = arccos) and inverse tangent ($\tan^{-1}$ = arctan). From Equations 1.18 and 1.19, we have

$$\theta = \cos^{-1}(x/r) \tag{1.22}$$

and

$$\theta = \tan^{-1}(y/x) \tag{1.23}$$

We often need to deal with triangles like the one in Figure 1.9, where we have a right triangle with interior angles α and β. The sum of a triangle's three interior angles is always 180°. For the triangle in Figure 1.9, one of the angles is 90° (because it is a right triangle), so

$$\alpha + \beta = 90° \tag{1.24}$$

Two angles whose sum is 90° are called complementary angles. From the definitions of the sine and cosine functions in Equations 1.17 and 1.18, we have the relations

$$\sin \alpha = \cos \beta$$

$$\cos \alpha = \sin \beta$$

for any pair of complementary angles α and β.

Measuring Angles

The value of an angle is often measured in units of **degrees**; indeed, we have already used this unit when discussing trigonometry. Another common unit for measuring angles is the **radian** (abbreviated "rad"). The conversion between radians and degrees is accomplished just like any other units conversion problem, using that 360 degrees (= 360°) is equivalent to 2π rad. That is,

$$360° = 2\pi \text{ rad}$$

or

$$1° = \frac{2\pi}{360} \text{ rad} \tag{1.25}$$

At this point, you might ask why we need two different units—degrees and radians—to measure angles. To explain the mathematical origin of the radian unit, Figure 1.10 shows a circle of radius r, and we have also drawn a "radius line" that makes an angle θ with the x axis. The arc length s is measured *along the circle* from the x axis to the point where the radius line meets the circle. The angle θ when measured in radians is equal to the ratio of the arc length to the radius:

$$\theta = \frac{s}{r} \quad (\theta \text{ measured in radians}) \tag{1.26}$$

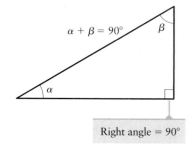

Figure 1.9 The sum of the three interior angles of a triangle is 180°. Hence, $\alpha + \beta = 90°$.

An angle of approximately 57° corresponds to 1 radian.

The angle θ here is approximately 1 radian.

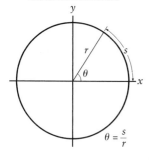

Figure 1.10 The angle θ equals the ratio of the length s measured along the circular arc, divided by the radius r of the circle. This ratio gives the value of θ in units of radians.

This connection between the angle and the length measured along a circular arc is very useful in work on circular motion (in which a particle moves along a circular path) and rotational motion (in which an object spins about an axis). The simple relation between s, r, and θ in Equation 1.26 only holds when s and r are measured in the same units (e.g., both measured in meters) and the angle is measured in radians. Most calculators can be set to work with angles in either degrees or radians, but you should always be clear about which way your calculator is set! Converting between degrees and radians is discussed in Example 1.6.

EXAMPLE 1.6 | The 3-4-5 Right Triangle and Designing a Roof

The roof of a house has the form of a right triangle with sides of length 3.0 m, 4.0 m, and 5.0 m (Fig. 1.11). Find the interior angles of the triangle. Express your answers in degrees and radians.

RECOGNIZE THE PRINCIPLE

Given the lengths of the sides of a right triangle, we can find the values of the interior angles using the $\sin^{-1}$, $\cos^{-1}$, and $\tan^{-1}$ functions, Equations 1.21 through 1.23. The interior angles of a right triangle are also complementary (Eq. 1.24).

SKETCH THE PROBLEM

Figure 1.11 shows the problem.

Figure 1.11 Example 1.6.

IDENTIFY THE RELATIONSHIPS AND SOLVE

The right triangle in Figure 1.11 has sides $x = 4.0$ m, $y = 3.0$ m, and $r = 5.0$ m. We can compute the value of θ using the inverse sine function ($\sin^{-1}$) in Equation 1.21:

$$\theta = \sin^{-1}(y/r) = \sin^{-1}(3.0/5.0) = \boxed{37°} \tag{1}$$

where we have rounded to two significant figures. To express this value in radians, we convert units:

$$\theta = 37° \times \frac{2\pi \text{ rad}}{360°} = \boxed{0.64 \text{ rad}}$$

The other interior angle is complementary to θ (see Eq. 1.24) and has a value

$$90 - \theta = 53° = 53° \times \frac{2\pi \text{ rad}}{360°} = \boxed{0.93 \text{ rad}}$$

What have we learned?

In computing the $\sin^{-1}$ function in Equation (1), you must be sure that you know how your calculator is set and whether it is giving an answer in degrees or radians.

Insight 1.4
INVERSE TRIGONOMETRIC FUNCTIONS
The inverse sine function used in Example 1.6 can be understood as follows. If $\theta = \sin^{-1}(w)$, then w is equal to the sine of the angle θ; that is, $w = \sin\theta$. When computing an inverse trigonometric function, be sure to take note of the units; that is, you should know if your calculator is set for degrees or for radians.

1.8 | VECTORS

In this book, we deal with two different types of mathematical quantities. One type includes quantities such as time and mass, which are described by a simple number (with a unit, of course) and are called *scalars*. Quantities described by a simple *magnitude*, such as the mass of a car or the number of seconds in 1 year, are scalars. Some physical quantities, however, cannot be described mathematically in this way. For example, the velocity with which the wind is blowing is a *vector*. A vector quantity has both a magnitude and a *direction*. In diagrams and figures, it is often useful to represent a vector quantity by an arrow. The arrow's length indicates the magnitude of the vector (e.g., how fast the wind is blowing), while the arrow's direction indicates the direction of the vector relative to a chosen coordinate system (e.g.,

Vectors have a magnitude and a direction.

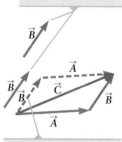

Figure 1.12 The vector $\vec{B}$ can be drawn in different locations, and it is still the same vector. Drawing the tail of $\vec{B}$ so that it coincides with the tip of $\vec{A}$ is a convenient way to add vectors. In this case, $\vec{C} = \vec{A} + \vec{B}$.

the direction in which the wind is blowing). Many quantities in physics are vectors, and we often need to add, subtract, and perform multiplication with vectors.

Adding Vectors

Representing vectors as arrows leads to a convenient way to think about the addition of two vectors. Figure 1.12 shows two vectors $\vec{A}$ and $\vec{B}$; here we use a notation in which vector variables are indicated by placing arrows over a boldface symbol. A vector has a magnitude and a direction, so both the lengths and the directions of the arrows in Figure 1.12 are important. However, when we draw a vector in this manner, the location of the arrow is somewhat arbitrary. We can place the arrow in different locations, and as long as it has the same length and the same direction, it is still the same vector. In Figure 1.12, we have drawn $\vec{B}$ several times; in one case, the "tail" of $\vec{B}$ is located at the "tip" of $\vec{A}$. The sum of $\vec{A}$ and $\vec{B}$ is then the vector $\vec{C}$, which connects the tail of $\vec{A}$ with the tip of $\vec{B}$. Mathematically,

$$\vec{C} = \vec{A} + \vec{B} \tag{1.27}$$

A useful way to understand Equation 1.27 is to think of these vectors as displacements or movements. The vector $\vec{A}$ then represents movement from the tail of $\vec{A}$ to the tip of $\vec{A}$, while the vector $\vec{B}$ represents movement from the tail of $\vec{B}$ to the tip of $\vec{B}$. When these two movements are added together, we get the total displacement represented by $\vec{C}$.

Multiplying a Vector by a Scalar and Subtracting Vectors

Before dealing with vector subtraction, it is useful to first consider the multiplication of a vector by a scalar. Scalars are simply numbers like 2 or 5.7. Multiplication of a vector by a scalar changes the vector's length. Figure 1.13 shows a vector $\vec{A}$ multiplied by a scalar K:

$$\vec{B} = K\vec{A} \tag{1.28}$$

If the scalar K is greater than 1 ($K > 1$), the vector $\vec{B}$ is longer than $\vec{A}$, while if K is positive and less than 1, $\vec{B}$ is shorter than $\vec{A}$. If the scalar K is negative, then $\vec{A}$ and $\vec{B}$ are in *opposite directions*.

We can now see how to do vector subtraction. Subtracting the vector $\vec{B}$ from the vector $\vec{A}$ is equivalent to adding the vectors $\vec{A}$ and $-\vec{B}$ (see Fig. 1.14):

$$\vec{A} - \vec{B} = \vec{A} + (-\vec{B}) \tag{1.29}$$

Vectors and Components

Representing vectors as arrows and performing addition, subtraction, and multiplication graphically as in Figures 1.12 through 1.14 is a helpful way to gain an intuitive understanding of vector arithmetic. However when performing calculations involving vector quantities it is often useful to deal with vectors in their component form. Figure 1.15 shows a vector $\vec{A}$ drawn with its tail at the origin of an $x-y$ coordinate system. This vector has **components** along the x and y directions, as shown. The components A_x and A_y are the projections of $\vec{A}$ along x and y.

Working with vector components often involves trigonometry. The vector $\vec{A}$ in Figure 1.15 can be described mathematically in two different ways. The first involves the angle θ this vector makes with the x axis along with the length of $\vec{A}$,

Figure 1.13 When a vector (such as $\vec{A}$) is multiplied by a scalar, the result is a vector that is parallel to the original vector but with a different length.

$\vec{B} = 0.5 \times \vec{A}$

$\vec{A}$

$\vec{B} = 2 \times \vec{A}$

Some examples of $\vec{B} = K\vec{A}$ with different values of K.

$\vec{B} = (-1) \times \vec{A} = -\vec{A}$

which we can denote either by A or by $|\vec{A}|$. The second is to give its components along x and y. According to Figure 1.15, the components of $\vec{A}$ are

$$A_x = A\cos\theta = |\vec{A}|\cos\theta \tag{1.30}$$

$$A_y = A\sin\theta = |\vec{A}|\sin\theta \tag{1.31}$$

Notice again that $A\ (=|\vec{A}|)$ is the length of the vector $\vec{A}$. With Equations 1.30 and 1.31, we can calculate the components if we know the magnitude (i.e., length) and direction (θ) of a vector. We can also work backward from the components to find the magnitude and direction. Applying some trigonometry to Figure 1.15 gives

$$A = |\vec{A}| = \sqrt{A_x^2 + A_y^2} \tag{1.32}$$

$$\tan\theta = A_y/A_x$$

or

$$\theta = \tan^{-1}(A_y/A_x) \tag{1.33}$$

Hence, if we know the components A_x and A_y, we can find A (the length of the vector) and θ (the direction).

Let's again consider the addition of two vectors, but now work in terms of the components. Figure 1.16 shows two vectors $\vec{A}$ and $\vec{B}$ added graphically, and we have also indicated the values of each vector's components. To add two vectors, we simply add their components. That is, to compute the vector sum $\vec{C} = \vec{A} + \vec{B}$ we add components so that

$$C_x = A_x + B_x \quad \text{and} \quad C_y = A_y + B_y \tag{1.34}$$

Multiplication of a vector by a scalar K can also be done using components. If

$$\vec{B} = K\vec{A}$$

we then have, in terms of components (Fig. 1.17),

$$B_x = KA_x \quad \text{and} \quad B_y = KA_y \tag{1.35}$$

The corresponding result for subtraction is

$$\vec{C} = \vec{A} - \vec{B}$$

$$C_x = A_x - B_x \quad \text{and} \quad C_y = A_y - B_y \tag{1.36}$$

In our discussion of vector components, we have so far dealt with only the x and y components. However, these results can be generalized to include a third coordinate direction, z, for cases in which we must deal with three-dimensional space. For example, when adding two vectors, Equation 1.34 becomes

$$C_x = A_x + B_x \quad C_y = A_y + B_y \quad C_z = A_z + B_z \tag{1.37}$$

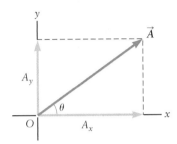

Figure 1.14 Vector subtraction. Calculating $\vec{A} - \vec{B}$ is equivalent to computing the sum of $\vec{A}$ and $-\vec{B}$.

Figure 1.15 The components of a vector are the projections of the vector onto the coordinate axes. Here we show a vector that lies in the $x-y$ plane, so it has components along the x and y axes. Some vectors have components along the z axis (not shown here).

Figure 1.16 To compute the sum of two vectors, we add their components. Here we add the x components of $\vec{A}$ and $\vec{B}$ to get the x component of $\vec{C}$. We follow the same procedure to find the y component of $\vec{C}$.

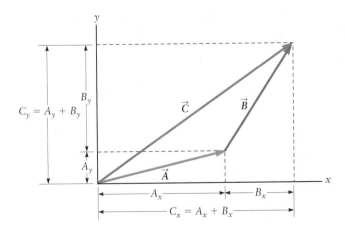

Figure 1.17 To multiply a vector $\vec{A}$ by a scalar K, we multiply each component of the vector by K.

EXAMPLE 1.7 | Hiking and Vectors

A hiker is walking through the woods. He starts from his campsite, which is at the origin in Figure 1.18. He initially travels along the x axis in Figure 1.18A, and his initial displacement of 1200 m takes him to the tip of vector $\vec{A}$; hence, $\vec{A}$ has a length of 1200 m and is parallel to the x direction. He then turns and moves along the path described by vector $\vec{B}$ in Figure 1.18A. This vector has a length of 1500 m and makes an angle $\theta_B = 30°$ with the x axis. When the hiker stops, how far is he from the campsite?

RECOGNIZE THE PRINCIPLE

To find the final location of the hiker, we must add the vectors $\vec{A}$ and $\vec{B}$. We can do so by adding the components of these vectors: $\vec{C} = \vec{A} + \vec{B}$ (Eq. 1.34).

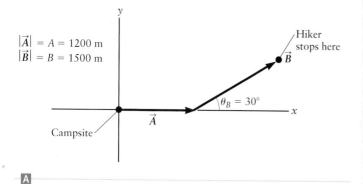

$|\vec{A}| = A = 1200$ m
$|\vec{B}| = B = 1500$ m

A

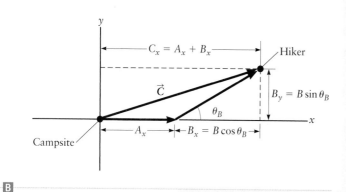

B

Figure 1.18 Example 1.7.

SKETCH THE PROBLEM

Figure 1.18 illustrates the problem. Figure 1.18B also shows the components of the various vectors, which will be needed in the solution.

IDENTIFY THE RELATIONSHIPS

According to Figure 1.18B, we can find the distance of the hiker to the campsite (i.e., the origin) if we know the components of the vectors $\vec{A}$ and $\vec{B}$ along the x and y axes. The vector $\vec{A}$ is parallel to x, so (see Fig. 1.18B)

$$A_x = A = 1200 \text{ m} \quad A_y = 0$$

where we again denote the length of the vector $\vec{A}$ by A. For $\vec{B}$, we use the trigonometry sketched in Figure 1.18B to get

$$B_x = B \cos \theta_B$$

$$B_y = B \sin \theta_B$$

The hiker's final location is described by the vector $\vec{C}$, which is equal to the sum of the vectors $\vec{A}$ and $\vec{B}$:

$$\vec{C} = \vec{A} + \vec{B}$$

In terms of components,

$$C_x = A_x + B_x \quad C_y = A_y + B_y$$

SOLVE

Inserting our values for the components of $\vec{A}$ and $\vec{B}$ gives

$$C_x = A_x + B_x = A_x + B \cos \theta_B = 1200 \text{ m} + (1500 \text{ m})\cos(30°) = 2500 \text{ m}$$

and

$$C_y = A_y + B_y = 0 + B \sin \theta_B = (1500 \text{ m})\sin(30°) = 750 \text{ m}$$

The distance from the hiker's final location to the origin is equal to the length of the vector $\vec{C}$. From the Pythagorean theorem (Eq. 1.32), we get

$$C = \sqrt{C_x^2 + C_y^2} = \sqrt{(2500 \text{ m})^2 + (750 \text{ m})^2} = \boxed{2600 \text{ m}}$$

What have we learned?
We'll work with vectors and their components in many problems. At this time, it might be valuable to review your trigonometry (see Appendix B).

What is physics?

Physics is the science of matter and energy, and the interactions between them. The laws of physics enable us to predict and understand the way the universe works.

Problem solving

Problem solving is an essential part of physics. Problems may be quantitative with a precise answer, or they may be conceptual. For some problems, you may need to use commonsense reasoning to estimate the values of important quantities.

Scientific notation and significant figures

Scientific notation is used to express very large and very small numbers. We can also use prefixes to write very large or very small numbers, such as 1 micrometer = 1 μm = 1×10^{-6} meters and 1 Megasecond = 1 Ms = 1×10^{6} seconds.

The accuracy of a quantity is reflected in the number of significant figures used to express its value. The result of a calculation should be expressed with an appropriate number of *significant figures*.

Units of measure

The *primary units* of mechanics involve length, time, and mass. Most scientific work employs the SI system of units, in which length is measured in *meters*, time is measured in *seconds*, and mass is measured in *kilograms*. The values of the standard meter, the standard second, and the standard kilogram are established by international agreement and are a foundation for all scientific work.

Physical quantities and dimensions

We cannot give a "definition" of the concepts of length, time, and mass, but must instead treat them as "givens." The definitions of all other physical quantities encountered in mechanics can be derived from these three primary quantities. A total of seven primary physical quantities are needed to describe all of physics; the other four are connected with electricity and magnetism, and heat.

The *dimensions* of all quantities in mechanics can be expressed in terms of length $\mathbf{L}$, mass $\mathbf{M}$, and time $\mathbf{T}$. Dimensional analysis involves checking that the dimensions of an answer correctly match the dimensions of the quantity being calculated.

The mathematics of physics

Several types of mathematics are used in this book: algebra, trigonometry, and vectors. Algebra is essential for solving systems of equations. When dealing with motion or other problems in physics, we often need to express position or movement in terms of a *coordinate system* (usually the x–y–z set of coordinate axes). Trigonometry and vectors are extremely useful in such calculations. A *vector* quantity has both a *magnitude* and a *direction*. Vectors can be added graphically or in terms of their *components*. Appendix B contains a quick review of algebra, trigonometry, and vectors.

1. Suppose a friend told you that the density of a sphere is $\frac{4}{3}\pi r^4$, where r is the radius of the sphere. Compute the dimensions of this expression and show that it cannot be correct.

2. Discuss the difficulties of giving a definition of the concept of "time" without using any other scientific terms.

3. SSM Which of the following are units of volume?

cubic meters   acres/m²
mm × mi²   hours × mm³/s
kiloseconds × ft²   mm × cm × mm²/ft
kg² × cm

4. What physical property, process, or situation is described by the following combinations of primary units? (For example, the combination of units kilograms per square meter can be used to measure the mass per unit area, sometimes called the areal density, of a thin sheet of metal.)
m/s   m/s²
m³/s   m²/s
kg/m

5. The standard meter used to be a metal bar composed of platinum and iridium. The current standard is based on the measured speed of light and an optical device called an interferometer. What advantages does the new standard have over the old?

6. SSM Which of the following quantities have the properties of a vector and which have the properties of a scalar?
mass   density
velocity   temperature
displacement (change in position)

7. Discuss the advantages and disadvantages of the SI system compared with the U.S. customary system of units.

8. **Astronomical distances.** Spiral galaxies like our own Milky Way measure approximately 2×10^5 light-years in diameter, where 1 light-year (ly) equals the distance traveled by light in 1 year. Our nearest-neighbor galaxy is the great spiral galaxy Andromeda, which has been determined to be approximately 2.5 million ly away. Construct a scale model to show the sizes of the Milky Way and Andromeda galaxies using two pie tins each 20 cm in diameter. To keep the model to scale, how far apart would these pie tins need to be separated?

9. A football field in the United States is 120 yards long (including the end zones). With how many significant figures do you think this value should be written?

1.3 DEALING WITH NUMBERS

1. Ⓡ The thickness of a typical piece of paper is 6×10^{-5} m. Suppose a large stack of papers is assembled, a stack as tall as the Empire State Building. How many pieces of paper are in the stack?

2. You take a long walk on a sandy beach. When you return home, you dump the sand out of one of your shoes and find that the pile has a volume of 1.5 cm³. If the volume of one grain of sand is approximately 0.1 mm³, estimate the number of grains of sand that were in your shoe.

3. Ⓡ You decide to walk from Chicago to New York. Estimate the number of steps you will take during this journey.

4. ⊗ A red blood cell has a diameter of approximately 8×10^{-6} m. A large number of red blood cells are laid side by side in a line of length 1.0 m. How many cells are in the line?

5. Determine the number of significant figures for all the values listed in Tables 1.2 and 1.3 that are not exact values.

6. A bucket contains several rocks with a combined mass (including the bucket) of 4.55 kg. Another rock of mass 0.224 kg is added to the bucket. What is the total mass of the bucket plus rocks? Be sure to give the correct number of significant figures in your answer.

7. Points A, B, C, and D in Figure P1.7 lie along a line. The distance from point A to point B is 3.45 km, and the distance from B to C is 5.4 km. What is the distance from A to C? Be sure to give the correct number of significant figures in your answer.

Figure P1.7 Problems 7 and 8.

8. The distance from point B to point D in Figure P1.7 is 3.15 km while the distance from A to B is 3.45 km. What is the distance from A to D? Be sure to give the correct number of significant figures in your answer.

9. The mass of an object is $m = 23$ kg, and its volume is $V = 0.005$ m³. What is its density, m/V? Be sure to give the correct number of significant figures in your answer.

10. The magnitude of the momentum of an object is the product of its mass m and speed v. If $m = 6.5$ kg and $v = 1.54$ m/s, what is the magnitude of the momentum? Be sure to give the correct number of significant figures in your answer.

11. The speed of light is $c = 299{,}792{,}458$ m/s. (a) Write this value in scientific notation. (b) What is the value of c to four significant figures?

12. SSM For each of the following formulas and associated values, find and express the solution with the appropriate number of significant figures and appropriate units. (a) The area A of a rectangle of length $L = 2.34$ m and width $w = 1.874$ m. (b) The area A of a circle of radius $r = 0.0034$ m. (c) The volume V of a cylinder with height $h = 1.94 \times 10^{-2}$ m and radius $r = 1.878 \times 10^{-4}$ m. (d) The perimeter P of a rectangle of length $L = 207.1$ m and width $w = 28.07$ m.

13. Ⓡ A hydrogen atom has a diameter of approximately 10^{-10} m. If this atom were expanded to the size of an apple, how big would a real apple be if it were expanded by the same amount? Can you think of an object whose size is comparable to this gigantic apple?

14. ⊗ Ⓡ (a) Estimate the number of times your heart beats in 1 year. Express your answer in scientific notation. (b) Some people have (jokingly) suggested that each person has only a certain number of heartbeats before he or she dies. The average life expectancy for a person born in the United States in 2004 is 77.9 years. If the average person's heart rate was reduced by

12%, how long would he or she live? Be sure to give the correct number of significant figures in your answer.

15. Write the following values using scientific notation. (a) The radius of the Earth, 6,370,000 m. (b) The distance from Earth to the Moon, 384,000 km. (c) The mass of the Earth, 5,970,000,000,000,000,000,000,000 kg.

1.4 PHYSICAL QUANTITIES AND UNITS OF MEASURE

1.5 DIMENSIONS AND UNITS

16. Express the distance from Chicago to New York in (a) miles, (b) inches, (c) kilometers, and (d) micrometers.

17. Ⓡ Calculate the volume of a typical 1000-page textbook. Express your answer in cubic meters (m^3) and in cubic centimeters (cm^3). *Hint*: You will need to use a ruler (or some other means) to find the dimensions of the book.

18. Find the number of millimeters in 1 mile.

19. Ⓡ What is the volume of a basketball? *Hint*: You will have to estimate, measure, or find a basketball's diameter.

20. SSM A U.S. football field is 120 yards long (including the end zones). How long is the field in (a) meters, (b) millimeters, (c) feet, and (d) inches?

21. Table 1.2 lists the age of the universe in seconds. Express this age in years.

22. Express 5.0 m in units of (a) centimeters, (b) feet, (c) inches, and (d) miles.

23. Express 75 ft in units of (a) meters and (b) millimeters.

24. An object has a mass of 273 g. What is its mass in kilograms?

25. A card table has a top surface area of exactly 1 square meter (1 m^2). What is its area measured in (a) square centimeters and (b) square millimeters?

26. A bottle has a volume of 1.2 L. (a) What is its volume in cubic centimeters? (b) In cubic meters?

27. Michael Jordan is approximately 6.5 ft tall. What is his height in millimeters?

28. The distance between two towns is 4.5 km. What is the distance in miles?

29. A cardboard box can hold exactly 1 cubic meter of volume (1 m^3). Use the conversion factors 1 m = 100 cm = 1000 mm to find this volume expressed in units of (a) cubic centimeters and (b) cubic millimeters.

30. How many hours are in 1 million seconds?

31. **Sports cars.** A Corvette C4 (circa 1971) came equipped with a huge 454 V8 engine, where the volume of the combustion chambers summed to 454 cubic inches ($in.^3$). In modern engines, this volume is measured in SI units. For example, a 2005 Corvette Z06 touts a 7.0-liter V8 engine. Which engine has the larger combustion volume? Is it surprising that the 2005 Corvette engine is 19% more powerful than the C4's 454 V8?

32. Consider the equation $mgh = \frac{1}{2}mv^2$, where m has units of mass (kilograms), g has units of length/time^2 (m/s^2), h has units of length (meters), and v has units of length/time (m/s). (a) Is this equation dimensionally correct or incorrect? (b) Is the equation $v^2/h = g$ dimensionally correct or incorrect? (c) What are the units of the combination g/v^2?

33. The variable m has dimensions of mass, h and y have dimensions of length, and t_1 and t_2 have dimensions of time. What are the dimensions of the following quantities? (a) my/t_2, (b) hy/t_1^2, (c) $y^3/(ht_1)$

1.6 ALGEBRA AND SIMULTANEOUS EQUATIONS

34. Consider the equation $5x - 17 = 10x - 27$. Solve for x.

35. The speed v of a bullet satisfies the relation $vt = 95$ m, where $t = 0.25$ s. Find v.

36. SSM ⭐ Consider the equations

$$5x + 2y = 13 \qquad -3x + 7y = 25$$

(a) Put each equation into the standard format for a straight line ($y = mx + b$) and plot both lines. Use your graph to find the point (x, y) where these lines intersect. (b) Use algebra to find the values of x and y that simultaneously satisfy these two equations. (c) How do your answers to parts (a) and (b) compare?

37. ⭐ A parent has exactly $25 to spend and needs to supply birthday treats to his child's third-grade class of 32 students. Cupcakes come in boxes of four for $3.80, and doughnuts come six to a bag for $3.00. If each student is to get one such treat (donut or cupcake) and the parent spends exactly $25, determine the number of boxes of cupcakes (x) and bags of doughnuts (y) that need to be purchased for the party. *Hint*: Set up two simultaneous equations like those in Problem 36.

38. ⭐ Two well-known organic chemicals are benzene, C_6H_6, and propane, C_3H_8. Benzene has a molecular weight of 78.11 g/mol and propane 44.096 g/mol, where the molecular weight is a useful measure in chemistry and is defined as the mass of 1 mole (Avogadro's number) of molecules or atoms. Determine the molecular weight (in g/mol) of pure hydrogen and pure carbon. *Hint*: The subscripts in the molecular formula are the number of atoms of that type per molecule. Set up two simultaneous equations to solve for the molecular weights.

1.7 TRIGONOMETRY

39. SSM ⭐ Consider the motion of the hour hand of a clock. What angle does the hour hand make with respect to the vertical 12 o'clock position when it is (a) 3 o'clock, (b) 6:00, (c) 6:30, (d) 9:00, and (e) 11:10? Express each angle in both degrees and radians, and measure angles going clockwise from the vertical axis.

40. A right triangle (Fig. 1.8) has sides of length $x = 4.5$ m and $y = 3.7$ m. (a) Find the value of r (the length of hypotenuse). (b) Find the values of the interior angles. Express your answers in degrees and radians. (c) Find the sine, cosine, and tangent of the smaller interior angle. (d) Find the sine, cosine, and tangent of the larger interior angle.

41. A right triangle has an interior angle of 25°. What are the values of the other interior angles?

42. ⭐ Find values of x, y, and r in Figure 1.8 that will give a right triangle in which one of the triangle's interior angles is 0.70 radian.

43. An angle has a value of 73°. Find its value in radians.

44. An angle has a value of 1.25 radians. Find its value in degrees.

45. A ladder of length 2.5 m is leaned against a wall (Fig. P1.45). If the top of the ladder is 1.7 m above the floor, what is the angle the ladder makes with the wall? What angle does it make with the floor?

$L = 2.5$ m

Figure P1.45

46. Steep grade. A mountain road makes an angle $\theta = 8.5°$ with the horizontal direction (Fig. P1.46). If the road has a total length of 3.5 km, how much does it climb? That is, find h.

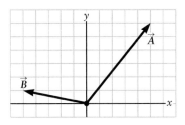

Figure P1.46

1.8 VECTORS

47. Consider the vectors $\vec{A}$ and $\vec{B}$ in Figure P1.47. (a) Draw a picture that graphically constructs the sum of these two vectors $\vec{A} + \vec{B}$ (i.e., similar to what was done in Fig. 1.12). Also construct $\vec{A} - \vec{B}$, $\vec{B} - \vec{A}$, $\vec{A} - 5\vec{B}$, and $-1.5\vec{A}$. (b) Suppose $\vec{A}$ has a length of 2.3 m and $\vec{B}$ has a length of 1.5 m. Use your graph to estimate the lengths of all the vectors in part (a).

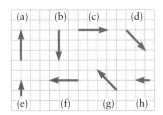

Figure P1.47

48. A vector has components $A_x = 54$ m and $A_y = 23$ m. Find the length of the vector $\vec{A}$ and the angle it makes with the x axis.

49. A vector of length of 2.5 m makes an angle of 38° with the y axis. Find the components of the vector.

50. Which of the vectors in Figure P1.50 have approximately the same magnitude?

Figure P1.50 Problems 50 and 51.

51. Which of the vectors in Figure P1.50 have approximately the same direction?

52. Which vector in Figure P1.52 might equal $\vec{A} + \vec{B}$?

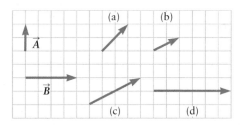

Figure P1.52 Problems 52 and 56.

53. ☆ A vector in the $x{-}y$ plane has a length of 4.5 m and a y component of -2.7 m. What angle does this vector make with the x axis? Why does this problem have two answers?

54. ☆ Consider a map oriented so that the x axis runs east–west (with east being the "positive" direction) and y runs north–south (with north "positive"). A person drives 15 km to the west, then turns and drives 45 km to the south. Find the x and y components of the total movement of the person.

55. ☆ For Problem 54, find the magnitude and direction of the total movement of the driver. Express the direction as an angle measured clockwise from the positive x axis.

56. Consider the vector drawn in part (c) of Figure P1.52 and denote that vector by $\vec{C}$. Construct a drawing that shows the vector $\vec{C}$ along with the vectors $4\vec{C}$, $-\vec{C}$, and $-3\vec{C}$.

57. A vector $\vec{A}$ has a magnitude of 15 (in some unspecified units) and makes an angle of 25° with the x axis, and a vector $\vec{B}$ has a length of 25 and makes an angle of 70° with the x axis. Compute the magnitudes and directions of the vectors:
(a) $\vec{C} = \vec{A} + \vec{B}$ (b) $\vec{C} = \vec{A} - \vec{B}$
(c) $\vec{C} = \vec{A} + 4\vec{B}$ (d) $\vec{C} = -\vec{A} - 7\vec{B}$.

58. Find the components of the vector $\vec{C}$ in parts (a) through (d) of Problem 57.

59. A boater travels from some initial point P_1 to a final point P_2 along the dashed line shown in Figure P1.59. Find θ and the total distance traveled.

60. ☆ A car starts from the origin and is driven 1.2 km south, then 3.1 km in a direction 53° north of east. Relative to the origin, what is the car's final location? Express your answer in terms of an angle and a distance.

61. SSM ☆ A mountaineer uses a global positioning system receiver to measure his displacement from base camp to the top of Mount McKinley. The coordinates of base camp are $x = 0$, $y = 0$, and $z = 4300$ m (here z denotes altitude), while those of Mount McKinley are $x = 1600$ m, $y = 4200$ m, and $z = 6200$ m. What is the displacement in going from base camp to the top?

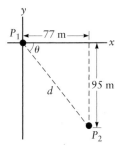

Figure P1.59

---

62. ⊗ ⓡ Estimate the number of times a student's heart will beat during the 4 years she is a college student.

63. SSM ☆ ⓡ **The solar system in perspective.** The diameters of the Earth and the Sun are approximately 1.3×10^7 m and 1.4×10^9 m respectively, and the average Sun–Earth distance is 1.5×10^{11} m. Consider a scale model of the solar system, where the Earth is represented by a peppercorn 3.7 mm in diameter.

(a) Calculate the diameter of the Sun in this model. Name an object of this size that could be used in the model. (b) To keep the model in scale, how far would the peppercorn Earth be from the center of the model Sun? (c) Pluto's orbit can take it as far as 5.9×10^{12} m away from the Sun. A grain of sand could represent Pluto in our model. How far away from the model Sun would this grain of sand be?

64. The unit of area called the acre comes from an old system of measurement, in which 1 acre is the area of a rectangle whose length is 220 yards and whose width is 22 yards. Determine the number of acres in 1 square mile. How many square feet are in 1 acre? In 1 square mile?

65. ✪ You wish to replace the flooring in your kitchen with square linoleum tiles that measure 12 inches on a side. If your kitchen has a floor area of 10.7 square yards, how many such tiles will you need to buy?

66. ⓡ Consider the following three vectors, described here by their length and angle with respect to the x axis: $\vec{A}$: 5.0 cm, 30°; $\vec{B}$: 7.5 cm, 0°; and $\vec{C}$: 10.0 cm, 350°. (a) Using the tip-to-tail graphical method for vector addition, show that the order in which these vectors are added does not change the final displacement, $\vec{D}$. That is, show that the following vector equation holds: $\vec{D} = (\vec{A} + \vec{B}) + \vec{C} = \vec{A} + (\vec{B} + \vec{C})$. (b) Use your graph to estimate the length of $\vec{D}$ and the angle it makes with the positive x axis.

67. ✪ A river flows with a current of 5.0 mi/h directly south. A small boat has a maximum speed in still water (i.e., on a quiet lake) of 10 mi/h. The pilot points the boat due east and sets the engine at full throttle to move the boat at maximum speed. In what direction does the boat actually move? *Hint*: The velocities of the river and the boat in still water can both be described as vectors, each having a magnitude and direction. The total velocity of the boat when traveling in the river is the sum of these two vectors.

68. ✪ ⓧ ⓡ Approximately how many cells are there in the human body? *Hint*: The volume of an average cell is about 1×10^{-16} m³.

69. ✪ ⓧ ⓡ Approximately how fast does human hair grow? Express your answer in m/s.

70. [SSM] ✪ ⓧ ⓡ At room temperature, 1.0 g of water has a volume of 1.0 cm³. (a) What is the approximate volume of one water molecule? (b) The human body is composed mainly of water. Assuming for simplicity your body is only water (and nothing else), find the approximate number of water molecules in your body.

71. A person stretches a rope of length 55 m between the top of a building to the base of a tree 25 m away (Fig. P1.71). How tall is the building?

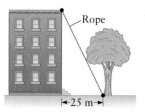

Figure P1.71

72. The thickness of a $1 bill is 0.11 mm. If you have a stack of $1 bills 350 m tall, how much money do you have?

73. ✪ The radius of Jupiter is 11 times as large as the radius of the Earth. What is the ratio of the volume of Jupiter to the volume of the Earth?

74. ✪ **If Saturn could fit in a bathtub, would it float?** Use the planetary data in Appendix A to calculate the density (mass/volume) of the Earth, Mars, Jupiter, and Saturn. What are the differences? Can you explain them?

Chapter 2

Motion, Forces, and Newton's Laws

The laws of mechanics describe the motion of objects such as this snowboarder. (© comstock Images/Jupiterimages)

We begin our study of physics with the field known as ***mechanics***. This area of physics is concerned with the motion of objects such as rocks, balloons, cars, water, planets, and snowboarders. To understand mechanics, we must be able to answer two questions. First, what *causes* motion? Second, given a particular situation, *how* will an object move?

The laws of physics that deal with the motion of terrestrial objects were developed over the course of many centuries, culminating with the work of Isaac Newton and the formulation of what are known as Newton's laws of motion. Newton's laws are a cornerstone of physics and are the basis for nearly everything we do in the first part of this book. Even so, it is useful to begin with a brief discussion of some of the ideas that dominated physics before Newton's time. Those ideas originated many centuries ago with the Greek philosopher–scientist Aristotle.[1] While Aristotle was a leading scientist of his era and was certainly a very deep thinker, we now know that many of his ideas about how objects move were not

---

[1] Aristotle, who lived during the fourth century BC, was also well-known as a teacher. One of his most famous students was the man we now know as Alexander the Great.

correct. Why, then, should we now be concerned with such historical (and incorrect) notions about motion? We examine Aristotle's ideas because many students begin this course with some of the *same* incorrect notions. A good way to reach a thorough understanding of the physics of motion is to consider the origins of Aristotle's ideas and identify where they go wrong.

2.1 | ARISTOTLE'S MECHANICS

Aristotle began by identifying two general types of motion: (1) celestial motion, the motion of things like the planets, the Moon, and the stars; and (2) terrestrial motion, the motion of "everyday" objects such as rocks and arrows. He believed that celestial motion is fundamentally different from the motion of rocks and arrows. We all know that while rocks and other terrestrial objects can move in a variety of ways, they usually come to rest eventually. On the other hand, it appears that celestial objects never come to a stop. According to Aristotle, terrestrial objects move only when something acts on them directly. A rock moves because some other object, such as a person's hand, acts on it to make it move. In contrast, there does not seem (at least to Aristotle) to be anything acting on the Moon to cause its motion. Aristotle thus raised a key point: the motions of celestial and terrestrial objects look very different. A theory of mechanics must explain why.

Let's first focus on the motion of terrestrial objects. Aristotle asserted that the "natural" state for a terrestrial object is a state of rest and that this is why terrestrial objects move only when acted on by another object. In modern terminology, Aristotle claimed that an object only moves when acted on by a *force*. A force is simply a push or a pull, and Aristotle believed that a push or a pull on one object is always produced by a second object. Furthermore, he claimed that a force could only exist if the two objects are in direct contact.

Hence, Aristotle believed that (1) *motion is caused by forces*, and (2) *forces are produced by contact with other objects*. These ideas *seem* quite reasonable when we consider something like a refrigerator being pushed along a level floor as in Figure 2.1. Our everyday experience is that the refrigerator only moves while someone is pushing on it. If the person stops pushing, the refrigerator quickly comes to a stop. The person pushing on the refrigerator exerts a force directly on the refrigerator, and the person is in contact with the refrigerator. So far, so good.

A key aspect of mechanics is the notion of force. We have already noted that a force is simply a push or a pull on an object. A force has both a magnitude and a direction, so *force is a vector quantity*, often denoted by $\vec{F}$. The magnitude of a force is the strength of the push or the pull, while the direction of this vector is the direction of the push or the pull.

We next need to consider how to describe motion. One quantity we normally associate with motion is *velocity*. Velocity is also a vector quantity and is usually represented by the symbol $\vec{v}$. A careful mathematical definition of velocity is given in the next section. For our discussion of Aristotle's ideas, it is enough to know that the magnitude of $\vec{v}$ is the distance that an object travels (as might be measured in meters) per second. The direction of the vector $\vec{v}$ is the direction of motion (Fig. 2.2).

Aristotle proposed that force and velocity are directly connected. In the mathematical language of today he would have written

$$\vec{v} = \frac{\vec{F}}{R} \quad \text{(INCORRECT!)} \tag{2.1}$$

In words, this relationship says that the velocity $\vec{v}$ of an object is proportional to the force $\vec{F}$ that acts on it, with the constant of proportionality being related to a quantity R, which is the resistance to motion. We'll refer to this relationship as "Aristotle's law of motion." We also caution that this relationship, Equation 2.1,

Figure 2.1 This person is exerting a force, denoted by $\vec{F}$, on the refrigerator.

Force is a *vector* quantity. It has a magnitude and a direction.

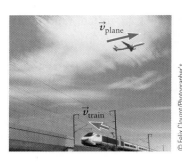

Figure 2.2 The velocities of an airplane and a train are vectors, having both magnitude and direction.

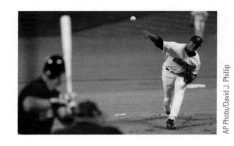

Figure 2.3 After it leaves the pitcher's hand, the only forces acting on the baseball are a force from gravity and a force from the air through which it moves. This second force is called air drag.

is *incorrect*. It is *not* a correct law of physics. It does, however, seem to explain the motion of the refrigerator in Figure 2.1. If no force is exerted on the refrigerator, then $\vec{F} = 0$. According to Equation 2.1, the velocity $\vec{v}$ is then also zero and hence the refrigerator does not move. When a person pushes on the refrigerator, $\vec{F}$ is nonzero and Aristotle's theory (Eq. 2.1) predicts that $\vec{v}$ is therefore also nonzero. Since $\vec{v}$ and $\vec{F}$ are both vectors, Equation 2.1 asserts that the velocity and the force are always in the same direction.

The Failures of Aristotle's Ideas about Mechanics

Aristotle's law of motion—Equation 2.1—*seems* to explain the motion of the refrigerator in Figure 2.1, but it does *not* work as well in other situations. Consider the motion of a thrown baseball (Fig. 2.3). We know that after a baseball leaves the thrower's hand it continues to move until it hits the ground, is struck by a bat, or is caught by someone. This very familiar motion is *not* consistent with Equation 2.1. Recall Aristotle's assertion that forces are always caused by direct contact with another object. While the ball is traveling through the air, there does not appear to be another object in contact with it, so according to this reasoning the force on the baseball is zero after it leaves the thrower. If the force is zero, Equation 2.1 says that the ball's velocity should also be zero. However, we know from experience that a thrown baseball continues to move! One way to attempt an escape from this dilemma is to recognize that the ball also experiences a force due to gravity. Unfortunately (at least for Aristotle), the gravitational force does not fit into Aristotle's ideas about forces because there is nothing in "direct" contact with the baseball to produce this force. Even if we overlook this difficulty, we still have to consider that the gravitational force on the ball is directed downward (toward the surface of the Earth) and hence the gravitational force and the velocity are not in the same direction for the ball in Figure 2.3. Figures 2.3 and 2.4 both illustrate situations that contradict Equation 2.1, which predicts that force and velocity are always parallel. It is particularly puzzling (at least for Aristotle) that objects such as baseballs and arrows can be traveling *upward* when the forces acting on them are directed *downward*.

More problems with Aristotle's law of motion can be seen when we consider falling objects. If someone simply drops a baseball, it will fall to the ground (or whatever is below). There is nothing in "direct contact" with the ball to cause it to fall. What, then, makes it move from the person's hand? Because the case of falling objects is quite common, Aristotle gave it special attention. He proposed that in this case there is a special force that does not require contact with anything. This force is the ***weight*** of the object, which Aristotle believed was equivalent to what we now call mass.[2] Aristotle also thought that the resistance factor R in Equation 2.1 is a

Figure 2.4 An archer shoots an arrow straight upward. While the arrow is traveling on the upward part of its trajectory, the forces acting on it—from gravity and from air drag—are both directed downward. Hence, these forces are not parallel to the velocity of the arrow while it is traveling upward.

---

[2]We'll see in the next chapter that mass and weight are *not* the same. Recall that mass is a fundamental physical quantity, as explained in Chapter 1. Weight is defined in Chapter 3.

property of the substance through which the object moves. For an object dropped near the Earth's surface, this substance is air. This reasoning leads one to expect that if Equation 2.1 is correct, a heavy object will fall faster than a light one when both are dropped in the same medium. Thanks to experiments performed by Galileo (Fig. 2.5), we know that this is *not* the case. Galileo showed that light objects fall at the same rate as heavy objects.

Despite all these difficulties, Aristotle's ideas have a certain appeal. They are based on the notion that an object's velocity is directly proportional to the force acting on the object, which at first glance seems plausible. However, we have seen a number of very simple situations in which this notion fails badly. In the next few sections, we discuss the connection between force and motion in more detail and arrive at the laws of motion discovered by Isaac Newton. We'll then see how Newton's laws overcome the difficulties that frustrated Aristotle.

Although we must reject Aristotle's theory of motion, it is still very instructive to understand where his theory goes wrong and why. Some of the everyday ideas that *you* have brought to this course may be similar to those of Aristotle. Understanding precisely when and why those ideas fail will help you understand and appreciate the correct way to describe motion.

CONCEPT CHECK 2.1 | Force and Motion

Do the following examples of motion appear to be *consistent* or *inconsistent* with Aristotle's law of motion?
  (a) A hockey puck sliding along an icy horizontal surface
  (b) A car coasting along a level road
  (c) A piano that is pushed across a room

2.2 | WHAT IS MOTION?

We have used the term *motion* quite a bit. It is now time to consider how motion is described and measured in a precise, mathematical sense. The concept of motion is more complicated than you might first guess, and we'll need several quantities— *position*, *velocity* (which we have already encountered), and *acceleration*—to describe it fully. Our next job is to give careful definitions for these quantities.

Let's first consider motion along a straight line, called *one-dimensional motion*. A hockey puck sliding on a horizontal icy surface is a good example of this type of motion, and the top portion of Figure 2.6A shows what would happen if we took a multiple-exposure photo of such a hockey puck. These exposures are captured at evenly spaced time intervals, and we can use them to construct the graph of the puck's position as a function of time shown in Figure 2.6B. Here we measure position as the distance from the origin on the x axis to the center of the hockey puck. For one-dimensional motion, the value of this distance, which we denote by x, completely specifies the position of the object. Figure 2.6A is sometimes called a *motion diagram*, a multiple-exposure photograph or similar sketch that shows the location of an object at regularly spaced instants in time. Notice in Figure 2.6B that the x axis is now vertical as we plot the position (x) as a function of time (t). In such a position–time plot, it is conventional to plot time along the horizontal axis.

Velocity and Speed

The distance between adjacent dots on the x axis in Figure 2.6A shows how far the puck has moved during each time interval. We have already mentioned that velocity $\vec{v}$ is a vector quantity. The magnitude of $\vec{v}$ is called the *speed*, the distance traveled per unit of time, while the direction of $\vec{v}$ gives the direction of the motion (in this example, $\vec{v}$ is directed to the right). Speed is a *scalar* quantity; it does not have a direction. In SI units, position is measured in meters and time is measured in seconds, so velocity and speed are both measured in meters per second, or simply m/s.

Figure 2.5 Galileo is reported to have used the Leaning Tower of Pisa in Italy in his studies of falling objects. Although the story may not be strictly accurate, Galileo certainly did conduct experiments showing that light and heavy objects fall at the same rate.

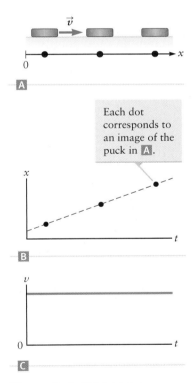

Figure 2.6 Ⓐ Multiple images of a hockey puck traveling across an icy surface. Ⓑ Plot of the position x of the puck as a function of time. The dots correspond to the images of the puck in Ⓐ. Ⓒ Velocity v of the puck as a function of time.

Speed and velocity are related quantities, but they are *not* the same. Speed tells how fast an object is moving; here it is the distance traveled per second, which is always a positive quantity (or perhaps zero). The velocity contains this information and in addition tells the *direction* of motion. For the hockey puck in Figure 2.6A, the direction may be positive (motion to the right, toward larger or more positive values of x) or negative (motion to the left, toward smaller or more negative values of x). If we recall that velocity is a *vector*, we should recognize that for one-dimensional motion the direction of the velocity vector must lie parallel to the x axis. Thus, in this particular example and in other cases involving one-dimensional motion, we need only deal with the *component* of the velocity parallel to x. This component can be either positive, negative, or zero.

In Chapter 1, we saw that vector quantities are usually written with arrows overhead, so the velocity vector is generally written as $\vec{v}$. For one-dimensional situations, the velocity vectors have only one component, and we can simplify the notation and refer to this component as simply v, without an arrow. The *sign* of v (either positive or negative) then gives the direction of the velocity. Thus, for motion in one dimension,

$$\text{speed} = |v| \quad \text{(one dimension)} \tag{2.2}$$

In words, this expression says that speed is equal to the magnitude of the velocity. For two- or three-dimensional motion,

$$\text{speed} = |\vec{v}| \quad \text{(two or three dimensions)} \tag{2.3}$$

where here the vertical bars again indicate that the speed is equal to the magnitude of the velocity vector. In either case (Eq. 2.2 or 2.3), speed is always a positive quantity or zero, but never negative.

The hockey puck in Figure 2.6A is sliding at a constant speed, so its velocity also has a constant value as shown in the velocity–time (v–t) graph in Figure 2.6C. In this case, the velocity is positive, which means that the direction of motion is toward increasing values of x (i.e., to the right).

How Is an Object's Velocity Related to Its Position?

Velocity is the change in position per unit time. Since time is measured in seconds, it is natural to think about the position at 1-s time intervals as suggested by the dots in Figure 2.6B. Alternatively, we could consider a particular time interval that begins at time t_1 and ends at time t_2 so that the size of the time interval is $\Delta t = t_2 - t_1$. The change in position during a particular time interval is called the **displacement**. For one-dimensional motion, displacement is denoted by $\Delta x = x_2 - x_1$, where x_1 is the **initial position**, the position at the beginning of the time interval (t_1), and x_2 is the **final position**, the position at the end of the interval (t_2). The **average velocity** during this time interval is

$$v_{\text{ave}} = \frac{x_{\text{final}} - x_{\text{initial}}}{t_{\text{final}} - t_{\text{initial}}} = \frac{x_2 - x_1}{t_2 - t_1}$$

$$v_{\text{ave}} = \frac{\Delta x}{\Delta t} \tag{2.4}$$

When an object moves with a constant speed, the average velocity is constant throughout the motion and the position–time graph has a constant slope as in Figure 2.6B. For more general cases, the average velocity is the slope of the line segment that connects the positions at the beginning and end of the time interval. This is illustrated in the hypothetical x–t graph in Figure 2.7.

Another example of motion along a line is the case of a rocket-powered car traveling on a flat road. Let's assume the car is initially at rest; here, "initially" means that the car is not moving when our clock reads zero. At $t = 0$, the driver turns on the rocket engine and the car begins to move in a straight-line path, along a horizontal axis that we denote as x. Figure 2.8A is a motion diagram showing the

Insight 2.1
SPEED AND VELOCITY
Speed is equal to the *magnitude* of the velocity. For motion in two or three dimensions, velocity is a vector. In these cases, speed is the *magnitude* of the velocity vector $\vec{v}$.

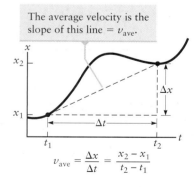

Figure 2.7 Hypothetical plot of an object's position as a function of time. The average velocity during the time interval from t_1 to t_2 is the slope of the line connecting the two corresponding points on the x–t curve.

position of the car at evenly spaced instants in time. The x axis is along the road, and the position of the car at a particular instant is the distance from the origin to the center of the car. The corresponding position–time graph for the car is shown in Figure 2.8B, where we have again used dots to mark the car's position at evenly spaced time intervals. In this case, the dots are not equally spaced along the x axis. Instead, their spacing increases as the car travels. This means that the car moves a greater distance during each successive (and equal) time interval and hence the speed of the car increases with time. In this example, the car moves toward increasing values of x, so the velocity is again positive and v increases smoothly with time as shown in Figure 2.8C. The precise shape of the v–t curve will depend on the way the engine fires; we'll learn how to deal with that problem in Chapter 3.

Average Velocity and Instantaneous Velocity

Figure 2.9 shows the position as a function of time for a hypothetical object moving in a straight line. We have again used dots to mark the position at the beginning and end of a particular time interval that starts at $t = 1.0$ s and ends at $t = 2.0$ s. According to Equation 2.4, the average velocity during this time interval is just the displacement during the interval divided by the length of the interval. Figure 2.9A shows that this average velocity is the *average slope* of the position–time curve (i.e., the slope of the line connecting the start to the end of the entire interval).

With this approach, though, we lose all details about what happens in the middle of the interval. In Figure 2.9A, the slope of the x–t curve varies considerably as we move through the interval from $t = 1.0$ s to $t = 2.0$ s. If we want to get a more accurate description of the object's motion at a particular instant within this time interval, say at $t = 1.5$ s, it is better to use a smaller interval. How small an interval should we use? In Figure 2.9B, we consider slopes over a succession of smaller time intervals. Intuitively, we expect that using a smaller interval will give a better measure of the motion at a particular point in time. From Figure 2.9B, we see that as we take ever smaller time intervals we are actually calculating the slope of the position–time curve at the point of interest (here at $t = 1.5$ s). This slope of the position–time curve is called the **instantaneous velocity**. For one-dimensional motion, the instantaneous velocity v is the *slope of the position–time (x–t) curve* and is given by[3]

$$v = \lim_{\Delta t \to 0} \frac{\Delta x}{\Delta t}$$

(2.5)

Definition of instantaneous velocity

For the example shown in Figure 2.9, v is not constant. Rather, it varies with time over the interval from $t = 1.0$ s to $t = 2.0$ s.

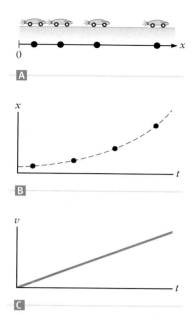

Figure 2.8 **A** "Multiexposure" sketch of a rocket-propelled car traveling along a horizontal road. **B** Position as a function of time for the rocket-powered car. **C** Velocity of the car as a function of time.

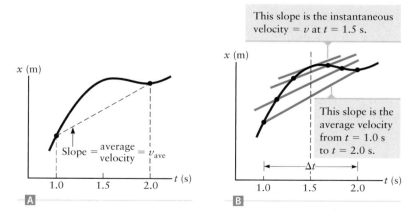

Figure 2.9 **A** The average velocity during a particular time interval is the slope of the line connecting the start of the interval to the end of the interval. **B** The instantaneous velocity at a particular time is the slope of the x–t curve at that time. The instantaneous velocity in the middle of a time interval is not necessarily equal to the average velocity during the interval. The instantaneous velocity is defined as the limit of the slope over a time interval Δt as $\Delta t \to 0$.

[3]Here the term "lim" (limit) means to take the ratio $\Delta x/\Delta t$ as the quantity Δt approaches zero.

The difference between the average and instantaneous values can be understood in analogy with a car's speedometer. The speedometer reading gives your instantaneous speed, the magnitude of your instantaneous velocity at a particular moment in time. If you are taking a long drive, your average speed will generally be different because the average value will include periods at which you are stopped in traffic, passing other cars, and so forth. In many cases, such as in discussions with a police officer, the instantaneous value will be of greatest interest.

The instantaneous velocity gives a mathematically precise measure of how the position is changing at a particular moment, making it much more useful than the average velocity. For this reason, from now on in this book we refer to the instantaneous velocity as simply the "velocity," and we denote it by v as in Equation 2.5.

CONCEPT CHECK 2.2 | Estimating the Instantaneous Velocity

Is there an instant in time in Figure 2.9 at which the instantaneous velocity is zero? If so, what is the approximate value of t at which $v = 0$?

EXAMPLE 2.1 | Average Velocity of a Bicycle

Consider the multiple images in Figure 2.10, showing a bicyclist moving along a level road. Find the average velocity of the bicyclist during the interval from $t = 2.0$ s to $t = 3.0$ s.

Figure 2.10 Example 2.1.

RECOGNIZE THE PRINCIPLE

The average velocity during a particular time interval is the slope of the position–time graph during that interval. This means that $v_{ave} = \Delta x / \Delta t$ (Eq. 2.4). The time interval is given, so we need to deduce the displacement Δx from Figure 2.10.

SKETCH THE PROBLEM

Our first step is to draw a picture that contains all the relevant information; this is essential for organizing our thoughts and seeing connections. The images in Figure 2.10 form the heart of the picture, but because we want to extract some quantitative information, we have added the x axis. We have chosen the origin of this axis to be at the bicyclist's position at $t = 1.0$ s. Using this coordinate axis, we can read off the value of the bicycle's position at the times of interest.

IDENTIFY THE RELATIONSHIPS

The average velocity is the bicyclist's displacement during a particular time interval divided by the length of the interval (Eq. 2.4). We have therefore added arrows to our picture that mark the positions at the start ($t = 2.0$ s) and end ($t = 3.0$ s) of the interval of interest. We have

$$v_{ave} = \frac{\Delta x}{\Delta t} = \frac{x_{final} - x_{initial}}{t_{final} - t_{initial}} \tag{1}$$

SOLVE

Inserting values from Figure 2.10 into Equation (1) we get

$$v_{ave} = \frac{x(t = 3.0\text{ s}) - x(t = 2.0\text{ s})}{t_{final} - t_{initial}} = \frac{(12\text{ m} - 5\text{ m})}{(3.0\text{ s} - 2.0\text{ s})} = \boxed{7\text{ m/s}}$$

What have we learned?
Each bicycle in Figure 2.10 corresponds to a point on the x–t graph, with coordinates given by reading off the values of x and t. Once we had the values of x and t, we then found the average velocity through the relation $v_{ave} = \Delta x / \Delta t$.

In Example 2.1, we saw how to use observations of the position as a function of time to compute an object's average velocity. As you might expect, it is also very useful to be able to deduce the instantaneous velocity from the position–time behavior. We can do so by estimating the slope of the position–time curve at different values of t and using these estimates to make a qualitative plot of the velocity–time relation. This graphical approach is illustrated in Example 2.2.

| EXAMPLE 2.2 | Computing Velocity Using a Graphical Method |

A hypothetical object moves according to the x–t graph shown in Figure 2.11A. This object is initially (when t is near t_1) moving to the right, in the "positive" x direction. The object reverses direction near t_2 and t_3, and it is again moving to the right at the end when t is near t_4. **(a)** Sketch the qualitative behavior of the velocity of the object as a function of time using a graphical approach. **(b)** Estimate the average velocity during the interval between $t_1 = 1.0$ s and $t_2 = 2.5$ s.

RECOGNIZE THE PRINCIPLE

For part (a), we want to find the velocity—which means the *instantaneous* velocity—so we need to estimate the slope of the x–t curve as a function of time. For part (b), we use the fact that the average velocity over the interval $t = 1.0$ s to $t = 2.5$ s is the slope of the x–t curve during this interval.

SKETCH THE PROBLEM

Figure 2.11B shows the x–t graph again, this time with lines drawn tangent to the x–t curve at various instants. The slopes of these tangent lines are the velocities at particular times $t_1, t_2, \ldots$ as indicated in Figure 2.11A.

IDENTIFY THE RELATIONSHIPS AND SOLVE

(a) At $t = t_1$, the x–t slope is large and positive, so our result for v in Figure 2.12A is large and positive at $t = t_1$. At $t = t_2$, the x–t slope is approximately zero and hence v is near zero. At $t = t_3$, the object is moving toward smaller values of position x, so the slope of the x–t curve and hence also the velocity are negative. Finally, at $t = t_4$, the object is again moving to the right as x is increasing with time, so v is again positive. After estimating the x–t slope at these places, we can construct the smooth v–t curve shown in Figure 2.12A. This figure shows the *qualitative* behavior of the object's velocity as a function of time.

(b) To estimate the average velocity between t_1 and t_2, we refer to Figure 2.12B, which shows a line segment that connects these two points on the position–time graph. The slope of this segment is the average velocity:

$$v_{ave} = \frac{\Delta x}{\Delta t} = \frac{x_2 - x_1}{t_2 - t_1}$$

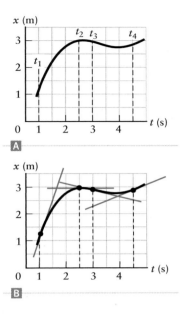

Figure 2.11 Example 2.2.
A Hypothetical position–time graph. The slopes of the tangent lines in B are equal to the velocity at various instants in time.

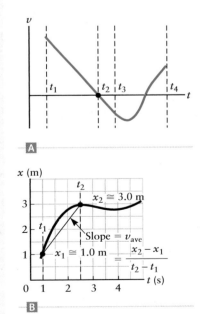

Figure 2.12 Example 2.2.
A Qualitative plot of the velocities obtained from the slopes in Figure 2.11B. B Calculation of the average velocity during the interval between $t_1 = 1.0$ s and $t_2 = 2.5$ s.

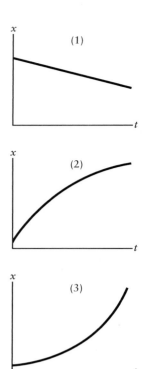

Figure 2.13 Concept Check 2.3.

Reading the values of x_1, x_2, t_1, and t_2 from that graph, we find

$$v_{\text{ave}} = \frac{x_2 - x_1}{t_2 - t_1} = \frac{(3.0 \text{ m}) - (1.0 \text{ m})}{(2.5 \text{ s}) - (1.0 \text{ s})} = \boxed{1.3 \text{ m/s}}$$

What have we learned?
The (instantaneous) velocity is the slope of the position–time graph. To find the qualitative behavior of the velocity, we found approximate values by drawing lines tangent to the x–t curve at several places and estimating their slopes. The value of v at a particular value of t is always equal to the slope of the x–t curve at that time.

CONCEPT CHECK 2.3 | The Relation between Velocity and Position

For which of the position–time graphs in Figure 2.13 are the following statements true?
   (a) The velocity increases with time.
   (b) The velocity decreases with time.
   (c) The velocity is constant (does not change with time).

Acceleration

We have seen that two quantities—position and velocity—are important for describing the instantaneous state of motion of an object, but are they sufficient? Are any other quantities needed to describe the physics of motion? One additional quantity, *acceleration*, will play a central role in our theory of motion. Acceleration is related to how the velocity changes with time. Consider again our rocket-powered car from Figure 2.8. In Figure 2.14, we resketch the v–t plot that we derived in Figure 2.8C and notice again that the car's velocity increases as time proceeds. Acceleration is defined as the rate at which the velocity is changing. If the velocity changes by an amount Δv over the time interval Δt, the *average acceleration* during this interval is

Definition of average acceleration

$$a_{\text{ave}} = \frac{\Delta v}{\Delta t} \tag{2.6}$$

As with the velocity, we are usually concerned with the acceleration at a particular instant in time, which leads us to consider the acceleration in the limit of very small time intervals. We thus define the *instantaneous acceleration* as

Definition of instantaneous acceleration

$$a = \lim_{\Delta t \to 0} \frac{\Delta v}{\Delta t} \tag{2.7}$$

The instantaneous acceleration a equals the *slope* of the v–t curve at a particular instant in time. The SI unit of acceleration is m/s^2 (meters per second squared).

We have now introduced several quantities associated with motion, including position, displacement, velocity, and acceleration. These quantities are connected in a mathematical sense through Equations 2.5 and 2.7. We also showed important graphical relationships: velocity is the slope of the position–time graph, while acceleration is the slope of the velocity–time graph. You might now be wondering if we'll continue this progression and consider the slope of the acceleration and so forth. The answer is that acceleration is as far as we need to go; x, v, and a are all we need in our formulation of a complete theory of motion. Newton's laws of motion (Section 2.4) will show us why.

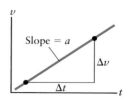

Figure 2.14 Acceleration is the slope of the velocity–time curve. For the case shown here, in which v varies linearly with time, the average acceleration is equal to the instantaneous acceleration.

EXAMPLE 2.3 | ⊗ Acceleration of a Sprinter

Consider a sprinter (Fig. 2.15A) running a 100-m dash. Figure 2.15B shows the velocity–time graph for the sprinter. Use a graphical approach to calculate the corresponding acceleration–time graph. What is the approximate value of the sprinter's maximum acceleration, and when does it occur?

Figure 2.15 Example 2.3. The velocity–time graph of this sprinter **A** is shown in **B**.

RECOGNIZE THE PRINCIPLE

Acceleration is the slope of the velocity–time graph, so we must estimate this slope at enough different values of t to be able to make a qualitative plot of the sprinter's acceleration as a function of time.

SKETCH THE PROBLEM

In Figure 2.15B, we have drawn in several lines tangent to the v–t curve at various times. The slopes of these tangent lines give the acceleration.

IDENTIFY THE RELATIONSHIPS AND SOLVE

We have measured approximate values of the slopes of the tangent lines in Figure 2.15B, and the results are plotted in Figure 2.16. The resulting acceleration–time graph is only qualitative (approximate). More accurate results would be possible had we started with a more detailed graph of the velocity. (Such examples are waiting for you in the end-of-chapter problems.)

What have we learned?
Acceleration is the slope of the velocity–time graph, so given the v–t graph we can derive the qualitative behavior of the acceleration. The qualitative shape of the acceleration–time curve in Figure 2.16 is consistent with expectations. The largest value of the acceleration (about 10 m/s^2) occurs at the start of the race when the sprinter is "getting up to speed"; this is where the v–t slope is largest. At the end of the race as the runner crosses the finish line, she slows down and eventually comes to a stop with $v = 0$ at the far right in Figure 2.15B. During that time, her velocity is decreasing with time so her acceleration is negative.

Figure 2.16 Example 2.3. The corresponding acceleration–time graph is given qualitatively.

The Relation between Velocity and Acceleration

An interesting feature of the graphs in Example 2.3 is that the maximum velocity and the maximum acceleration do *not* occur at the same time. It is tempting to think that if the "motion" is large, both v and a will be large, but this notion is *incorrect*. Acceleration is the slope—the rate of change—of the velocity with respect to time. The time at which the rate of change of velocity is greatest may not be the time at which the velocity itself is greatest.

CONCEPT CHECK 2.4 | Analyzing a Position–Time Graph

The position–time curve of a hypothetical object is shown in Figure 2.17. Which of the following scenarios might be described by this x–t graph? (Notice that we have not shown numerical values for x or t. No hints here!)

Figure 2.17 Concept Check 2.4. What type of motion is described by this position–time graph?

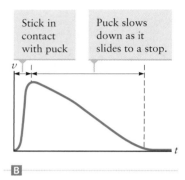

Stick in contact with puck

Puck slows down as it slides to a stop.

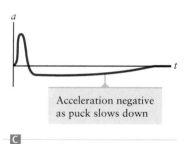

Acceleration negative as puck slows down

Figure 2.18 Example 2.4. Motion of a hockey puck as described by graphs of its position, velocity, and acceleration versus time.

(a) A car starting from rest when a traffic light turns green.
(b) A car coming slowing to a stop when a traffic light turns red.
(c) A bowling ball rolling toward the pins.
(d) A runner slowing from top speed to a stop at the end of a race.

EXAMPLE 2.4 | Sliding to a Stop

Consider a hockey puck that starts from rest. At $t = 0$, the puck is struck by a hockey stick, which starts the puck into motion with a high velocity. The puck then slides a very long distance before coming to rest. Draw qualitative plots of the puck's position, velocity, and acceleration as functions of time.

RECOGNIZE THE PRINCIPLES

We first use our experience with hockey pucks and other sliding objects to deduce the qualitative shape of the velocity–time (v–t) curve. The puck starts from rest, so initially $v = 0$. The velocity then increases quickly to a high value when the stick is in contact with the puck. After the puck leaves the stick, the velocity gradually decreases as the puck slides to a stop. From the behavior of the v–t curve, we can get the position and acceleration as functions of time.

SKETCH THE PROBLEM

In Figure 2.18, imagine drawing part B first (qualitative behavior of velocity). The puck reaches a high velocity very quickly and then falls to zero at long times.

IDENTIFY THE RELATIONSHIPS

As explained above, the velocity as a function of time is plotted in Figure 2.18B. To find the acceleration, we must find the slope of this v–t curve as a function of time. To get the position, we must find an x–t curve whose slope gives this velocity–time graph.

SOLVE

The acceleration–time graph in Figure 2.18C was obtained from estimates of the slope of the curve in part (b) at different times. Notice that the acceleration is *positive* and large when the stick is in contact with the puck. The acceleration is *negative* as the puck slides to a stop since the velocity is then decreasing with time and the v–t slope is negative. To deduce the x–t result, we must work backward from Figure 2.18B so as to make the slope of the position–time graph correspond to the v–t curve. We can do so by noting that the x–t slope is greatest when v is greatest and that this slope is small when v is small. Once v reaches zero, the position no longer changes with time.

What does it mean?
These graphs of the position and velocity can also be appreciated using the "time-lapsed" sketches in Figure 2.19. These sketches show the location of the puck at different evenly spaced instants in time, beginning at $t_0 = 0$ when it is struck by the hockey stick and then at $t_1, t_2, \ldots$ until the puck comes to rest at t_4. These times t_1, t_2, t_3, and t_4 are equally spaced along the time axis (Fig. 2.18A), so the time intervals $\Delta t = t_1 - t_0 = t_2 - t_1, \ldots$ are all equal. However, the displacements during these intervals are *not* equal. For example, the displacement during the first interval (from t_0 to t_1) is much larger than the displacement during the second interval. In words, the puck moves farther—because it has a higher velocity—during the first interval. Likewise, at the end of the time period, the puck's velocity is smaller, so the distance traveled between t_3 and t_4 is much smaller than the distance traveled between t_0 and t_1.

Figure 2.19 Example 2.4. Time-lapse sketches of the hockey puck's motion.

| EXAMPLE 2.5 | Stopping in a Hurry

A driver is in a hurry, and her car is traveling along a straight road at a velocity of $v_0 = 20$ m/s (about 40 mi/h) when she spots a problem in the road ahead and applies the brakes. The car then slows to a stop according to the velocity–time curve in Figure 2.20. Find the average acceleration during the interval from $t = 0.0$ s to $t = 4.0$ s and also estimate the instantaneous acceleration at $t = 2.0$ s.

RECOGNIZE THE PRINCIPLE

The average acceleration is the average slope of the velocity–time curve *during the time interval* of interest. The instantaneous acceleration is the slope of the v–t curve *at the point* of interest.

SKETCH THE PROBLEM

Figure 2.20 describes the problem and contains all the information we need to solve it.

IDENTIFY THE RELATIONSHIPS

From the definition of the average acceleration in Equation 2.6, we have

$$a_{ave} = \frac{\Delta v}{\Delta t} \qquad (1)$$

We can read the values of Δv and Δt from Figure 2.20.

SOLVE

Applying Equation (1) over the interval from $t = 0.0$ s to $t = 4.0$ s gives

$$a_{ave} = \frac{\Delta v}{\Delta t} = \frac{v(t = 4.0 \text{ s}) - v(t = 0.0 \text{ s})}{\Delta t}$$

Inserting the values from Figure 2.20 of $v(4.0$ s$) = 0$ and $v(0.0$ s$) = 20$ m/s along with $\Delta t = 4.0$ s, we find

$$a_{ave} = \frac{(0 - 20) \text{ m/s}}{4.0 \text{ s}} = \boxed{-5.0 \text{ m/s}^2}$$

This result is the average acceleration, so it is the average slope of the v–t curve during the entire interval shown in Figure 2.20. Since this curve is approximately a straight

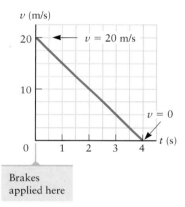

Figure 2.20 Example 2.5.

line, the average slope is approximately equal to the "instantaneous" slope at all times in the interval. The instantaneous slope of the v–t curve is the instantaneous acceleration a, so we also get

$$a \approx \boxed{-5.0 \ \text{m/s}^2}$$

What does it mean?

The average and instantaneous acceleration in this example are both negative because the velocity *decreases* during the time interval of interest. The velocity is positive (Fig. 2.20) during the entire interval, so the velocity and acceleration are in opposite directions; here the velocity is positive (along the x direction) while the acceleration is negative (along the $-x$ direction).

2.3 | THE PRINCIPLE OF INERTIA

We spent much of Section 2.1 discussing Aristotle's ideas about motion, and you should now be convinced that his theory of motion has some serious flaws. Nevertheless, it is worthwhile to consider how Aristotle would have explained the motion of the rocket-powered car in Figure 2.8. He probably would have claimed that when the rocket engine is turned on, the car moves by virtue of the force exerted on it by the engine, with a velocity $v = F/R$ as predicted from Equation 2.1. If the rocket engine is then turned off or runs out of fuel, however, the force would vanish ($F = 0$) and, according to the same argument, the car would stop immediately. Your intuition should tell you that this prediction is not correct; the car would instead continue along, for at least a short period, after the engine is turned off. That is, the car will coast for a while before coming to rest.

How would Aristotle, or you, explain this dilemma? The difficulties with Aristotle's ideas about motion all come down to his belief that force and velocity are directly linked. This linkage is expressed in Equation 2.1, which implies that if there is a nonzero velocity, it must be caused by a force. That is, if an object is found to have a nonzero velocity, then according to Aristotle there must be a nonzero force acting on the object at that time. However, our example with a rocket-powered car shows that force and velocity are *not* linked in this way because it is possible for the car to have a nonzero velocity even when the force is zero.[4] The correct connection between force and motion is instead based on a direct linkage between force and *acceleration*. This connection to acceleration is at the heart of Newton's laws of motion.

Before we can really appreciate Newton, however, we must first consider the work of Galileo. Although Galileo did not arrive at the correct laws of motion, his experiments on the motion of terrestrial objects showed that an object can move even if there is *zero* total force acting on it. This result led to the discovery of the *principle of inertia*, which is a cornerstone of Newton's laws.

Galileo's Experiments on Motion

Aristotle was not able to explain how an object could move when there did not appear to be any force acting on it. A good example is the somewhat idealized case of a hockey puck sliding on a very smooth, horizontal, icy surface. Our intuition tells us that the puck will slide a very long way before coming to rest; in fact, if we could somehow make the surface "perfectly" icy so that there is absolutely no friction whatsoever, our intuition suggests that the puck would slide forever. You might

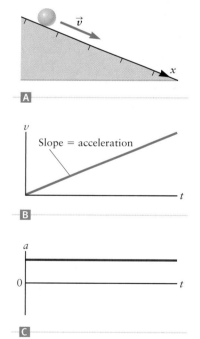

Figure 2.21 Ⓐ Ball rolling down a simple incline. Ⓑ Galileo found that the velocity of the ball increases linearly with time. The slope of this line is the acceleration in Ⓒ.

---

[4]Here, we are concerned with motion along a horizontal direction (x). The car's velocity along the horizontal direction is not necessarily zero, while the force along this direction is zero if the car's engine is turned off.

object to this argument on the grounds that it is impossible to make a surface completely frictionless. Even so, we can still think about what would happen in such a situation. In fact, thinking about such idealized cases is a very useful practice in physics because these cases can often help us see to the heart of a problem.

Galileo did not carry out experiments with a hockey puck on an icy surface. Instead, his experiments involved a ball rolling on an incline. If the ball is very hard, like a billiard ball, and the surface of the incline is also very hard and smooth, the effect of friction on the ball's motion is very small. Hence, this situation is actually quite similar to that of our hockey puck.

Galileo experimented with the motion of a ball on an incline as sketched in Figure 2.21. He found that when the ball is released from rest, its velocity varies with time as shown in Figure 2.21B. This motion is another example of one-dimensional motion, with the direction of motion denoted by an x axis lying along the incline as shown. The velocity of the ball increases as the ball rolls down the incline, and Galileo found that the magnitude of the velocity increases *linearly* with time. Since acceleration is the slope of the v–t curve, the acceleration is *constant* and *positive*.

Galileo then repeated this experiment, but this time with the ball rolling *up* an incline with the same tilt angle (Fig. 2.22A). When he gave the ball some initial velocity, he found that it rolled up the incline with the v–t curve shown in Figure 2.22B. Again he observed that the velocity varies *linearly* with time, but now the slope is *negative*. In fact, the slope in this case is equal in magnitude to the slope found for the down-tilted incline. Hence, the acceleration was opposite in sign but equal in magnitude to that found when the ball rolled down the incline. Galileo performed this experiment with many inclines and with many different tilt angles, and he always observed that the velocity varied linearly with time, with the slope of the v–t graph being determined by the tilt angle. Moreover, the acceleration (the slope of the v–t relation) when a ball rolled up a particular incline was always equal in magnitude, but opposite in sign, when compared with the acceleration when the ball rolled down the same incline. He then reasoned that if the tilt of the incline were precisely zero (a perfectly level surface), the slope of the v–t line and hence the acceleration would be zero. Galileo therefore asserted that on a level surface the ball would roll with a *constant velocity*.

This result may seem obvious to you. For example, it means that a ball placed at rest on a horizontal surface will remain in balance and hence be motionless, with zero velocity. Such a ball has a constant velocity because $v = 0$ is a constant! It was, however, the genius of Galileo to realize that his experiment implied another result, which is sketched in Figure 2.23. Here, a ball is rolling along a different sort of ramp. The initial portion of the ramp is sloped like the incline in Figure 2.21, but the final portion is perfectly horizontal and extremely long, with only part of it shown here. Galileo realized that for this type of ramp the ball would have a positive velocity when it completes the first portion of the ramp. Then, on the second (horizontal) part of the ramp, the v–t relation would be a straight line with a slope of zero. In other words, the ball's velocity on the final portion of this ramp would have the constant value produced during the initial part of the ramp, and the ball would *maintain* this velocity. Hence, for this idealized case that ends with a perfectly horizontal ramp, Galileo proposed that the ball would roll *forever*.

Galileo's experiment demonstrates the **principle of inertia**. According to this principle, an object will *maintain its state of motion—its velocity—unless it is acted on by a force*. On a horizontal surface, there is no force in the direction of motion, so the velocity is constant and the ball rolls forever. With his discovery of the principle of inertia, Galileo broke Aristotle's link between velocity and force. Galileo showed that one can have motion (a nonzero velocity) *without* a force. This discovery does not answer the question of how force is linked to motion, however. The answer to that question was provided by Newton. (It is interesting that Newton was born in 1643, the year after Galileo died. In a very direct sense, Newton thus built on the work of Galileo.)

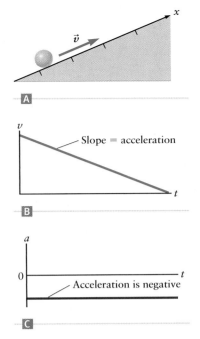

Figure 2.22 Ⓐ Ball rolling up an incline with the same angle as in Figure 2.21. Ⓑ Galileo found that now the velocity of the ball decreases linearly with time. Ⓒ The slope of this line—the acceleration—has the same magnitude, but the opposite sign, as the acceleration when the ball rolls down the incline (Fig. 2.21Ⓒ).

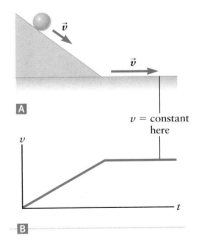

Figure 2.23 When a ball rolls along this two-part incline, its velocity increases linearly with time while on the initial (sloped) portion, and v is then constant on the final (flat) portion.

2.4 | NEWTON'S LAWS OF MOTION

Newton's laws of motion are three separate statements about how things move.

Newton's First Law

Newton's first law is a careful statement about the principle of inertia that we encountered in the previous section in connection with Galileo's experiments.

> **Newton's first law of motion: If the total force acting on an object is zero, the object will maintain its velocity forever.**

According to Newton's first law, if there is no *total* force acting an object—that is, if the total force acting on it is *zero*—the object will move with a constant velocity. In other words, such an object will move with a constant speed along a particular direction and will continue this motion—with the same speed in the same direction—forever, or as long as the total force acting on it is zero.[5] This is another way of stating the principle of inertia, and you can see that it is essentially what Galileo found in his experiments with balls moving along inclined or flat surfaces. The total force acting on one of Galileo's balls rolling on a horizontal surface is zero, so in this ideal case the ball will roll forever.

Inertia and Mass

Now that we have introduced the principle of inertia, you should be curious about the term *inertia* and precisely what it means. The inertia of an object is a measure of its *resistance to changes in motion*. This resistance to change depends on the object's *mass*. It is not possible to give a "first principles" definition of the term *mass*, but your intuitive notion is very useful. The mass of an object is a measure of the amount of matter it contains. Objects that contain a large amount of matter have a larger mass and a greater inertia than objects containing a small amount of matter. The SI unit of mass is the kilogram (kg). Mass is an intrinsic property of an object. For example, the mass m of an object is independent of its location; m is the same on the Earth's surface as on the Moon and in distant space. In addition, the mass does not depend on an object's velocity or acceleration. Although an object's mass does not appear explicitly in Newton's first law, it has an essential role in Newton's second law.

Newton's first law may seem surprising to you. After all, terrestrial objects always come to rest eventually. The difference here, and a key to appreciating Newton's first law, is that an object will move with a constant velocity only in the ideal case that the total force is precisely zero. For terrestrial objects, it is very difficult to find such ideal cases because it is difficult to completely eliminate all forces, especially frictional forces. Hence, terrestrial objects come to rest because of the forces that act on them.

Two other hypothetical cases may help make Newton's first law fit with your intuition. Case 1: Imagine a frozen lake with an extremely smooth surface. A hockey puck sliding on such a perfectly flat and icy surface experiences a very tiny frictional force and will therefore slide for a very long distance before coming to rest. If this frictional force could be made to vanish, the puck would slide forever (if the lake were large enough). Of course, actual icy surfaces are not perfect; there will always be a small amount of friction, and this small amount of friction would make the puck eventually come to rest. Case 2: Imagine a spaceship that is coasting (i.e., with its engines turned off) someplace in the universe very far from any stars or planets. Such a spaceship would experience only a very small gravitational force (from the nearest stars and planets). If the nearest stars and planets were very far

---

[5]This result may be surprising because it may *appear* to contradict your intuition. This result only holds if the total force on an object is *precisely zero*, however, and that can be hard to achieve in practice. See below.

away, this force would be negligible and the spaceship would move with a constant velocity, in accord with Newton's first law.

Newton's Second Law

Newton's second law of motion: **In many situations, several different forces are acting on an object simultaneously. The total force on the object is the sum of these individual forces, $\vec{F}_{\text{total}} = \sum \vec{F}$. The acceleration of an object with mass m is then given by**

$$\vec{a} = \frac{\sum \vec{F}}{m} \qquad (2.8)$$

Newton's second law tells us how an object will move when acted on by a force or by a collection of forces. *This law is our link between force and motion.* The acceleration of an object is directly proportional to the total force that acts on it. Newton's second law, Equation 2.8, is often written in the equivalent form

$$\sum \vec{F} = m\vec{a}$$

Keep in mind that the term $\sum \vec{F}$ in Newton's second law is the *total* force on the object, from *all* sources. As you might imagine, and as we'll see in some examples, in most cases there are several forces acting on an object. So, we have to add them all up, and that resulting vector sum is the force $\sum \vec{F}$ in Newton's second law (see Fig. 2.24). You should also recall that vectors must be added according to the vector arithmetic procedures described in Chapter 1. The direction of the acceleration $\vec{a}$ is then parallel to the direction of the total force $\sum \vec{F}$.

In the SI system of units, force is measured in units called newtons (N). We can use Newton's second law to express this unit in terms of the primary SI units. Mass is measured in units of kilograms, while acceleration has the units meters per second squared. In terms of only the units, Equation 2.8 can be rearranged to read

$$\text{force} = \text{mass} \times \text{acceleration}$$

$$\text{newtons} = \text{kg} \times \frac{\text{m}}{\text{s}^2}$$

The value of the newton as a unit of force is therefore

$$1\,\text{N} = 1\,\text{kg} \cdot \text{m/s}^2 \qquad (2.9)$$

We'll spend the next dozen chapters or so exploring applications of Newton's second law. Many applications start by determining the total force acting on an object from all sources. The object's acceleration can then be calculated using Equation 2.8. Acceleration is the change in velocity per unit time, so it is possible to use the acceleration to deduce the velocity. In a similar manner, since velocity is the change in position per unit time, one can use the velocity to find the object's position as a function of time. In this way, an object's acceleration, velocity, and position can all be found.

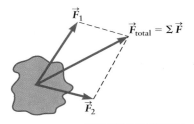

The acceleration $\vec{a}$ is parallel to $\sum \vec{F}$.

Figure 2.24 When several forces act on an object, the vector sum of these forces $\sum \vec{F}$ determines the acceleration according to Newton's second law.

| EXAMPLE 2.6 | Using Newton's Second Law |

A single force of magnitude 6.0 N acts on a stone of mass 1.1 kg. Find the acceleration of the stone.

RECOGNIZE THE PRINCIPLE

The force on the stone and its acceleration are related through Newton's second law,

$$\vec{a} = \frac{\sum \vec{F}}{m} \qquad (1)$$

Here, $\sum \vec{F}$ is the total force on the stone, which has a magnitude of 6.0 N.

SKETCH THE PROBLEM

We begin by drawing a picture, showing all the forces acting on the stone (Fig. 2.25). Since there is only a single force in this example, it is also the total force.

IDENTIFY THE RELATIONSHIPS AND SOLVE

Using Newton's second law, Equation (1), the magnitudes of the acceleration and force are related by

$$a = \frac{\sum F}{m} = \frac{6.0\ \text{N}}{1.1\ \text{kg}} = \frac{6.0\ \text{kg} \cdot \text{m/s}^2}{1.1\ \text{kg}} = \boxed{5.5\ \text{m/s}^2} \tag{2}$$

Figure 2.25 Example 2.6.

The direction of the stone's acceleration is parallel to that of the total force, so the direction of $\vec{F}$ in Figure 2.25 gives the direction of $\vec{a}$.

What does it mean?
Notice how the units in Equation (2) combine, as the factors of kilograms on the top and bottom cancel to give an answer with units of meters per second squared, as expected for acceleration. You should always check the units of an answer in this way. An incorrect result for the units usually indicates an error in the calculation.

EXAMPLE 2.7 | Motion of a Falling Object

A ball is dropped from a bridge onto the ground below. The height of the ball above the ground as a function of time is shown in Figure 2.26. Use a graphical approach to find, as functions of time, the qualitative behavior of (a) the velocity of the ball, (b) the acceleration of the ball, and (c) the total force on the ball.

RECOGNIZE THE PRINCIPLE

Velocity is the slope of the position–time curve, so we can find v from the slope of the y–t curve in Figure 2.26. Notice that here we use y to measure the position, taking the place of x in previous examples. Acceleration is the slope of the velocity–time curve, so we can find the behavior of a once we have the behavior of the velocity. Once we have the acceleration, the total force on the ball can then be found through Newton's second law, $\sum \vec{F} = m\vec{a}$.

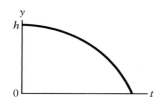

Figure 2.26 Example 2.7. Position (y) above the ground as a function of time for a ball that falls from a bridge.

SKETCH THE PROBLEM

Figure 2.26 shows how the position of the ball varies with time, and Figure 2.27 shows a picture of the ball as it falls from the bridge. The motion of the ball is one-dimensional, falling directly downward from the bridge to the ground, and its position can be measured by its height y above the ground. This notation follows the common practice of using y to represent position along a vertical direction. Figure 2.27 also shows the y coordinate axis, with its origin at ground level.

IDENTIFY THE RELATIONSHIPS AND SOLVE

(a) Velocity is the change in position per unit time, so the value of v at any particular time is the slope of the y–t graph at that instant (since in this example y is our position variable). We can obtain this slope graphically by following the approach in Example 2.2 and drawing tangent lines to the y–t curve at various times in Figure 2.28A. Using estimates for the slopes of these lines leads to the qualitative velocity–time graph shown alongside the y–t curve. Note that the velocity is always negative since y decreases monotonically with time. Also, the magnitude of the velocity, which is the speed, becomes larger and larger as the ball falls. (The plot here shows v as a function of time up until just before the ball reaches the ground.)

Figure 2.27 Example 2.7. A ball is dropped from a bridge, starting from rest ($v = 0$) at a height h above the ground. The motion of this ball is described by the y–t graph in Figure 2.26.

A	**B** **C**

(b) Acceleration is the slope of the *v–t* curve; that slope is constant in Figure 2.28A, leading to the qualitative acceleration–time graph in Figure 2.28B. The acceleration is negative because the value of v is decreasing (the value of v becomes more negative with time). The magnitude of a is approximately constant during this period.

(c) To compute the force on the ball, we use Newton's second law. According to Equation 2.8, the total force is proportional to the acceleration. Even if we do not know the sources of all the forces on the ball, we can still compute the total force from Newton's second law. We can rearrange Newton's second law (Eq. 2.8) as $\sum \vec{F} = m\vec{a}$ and thus arrive at the qualitative force–time graph in Figure 2.28C.

What have we learned?
Given the behavior of the position as a function of time, we can deduce (by estimating slopes) the qualitative behavior of both the velocity and the acceleration as functions of time. The behavior of the force can then be found by using Newton's second law. For this falling ball, the total force is negative (Fig. 2.28C), meaning that the force is directed downward, along the −y direction. This force is just the gravitational force acting on the ball.

Figure 2.28 Example 2.7.
A Position as a function of time for the falling ball from Figure 2.26. The slopes of the tangent lines give the ball's velocity at three instants in time. The velocity of the ball as a function of time is obtained from plotting these slopes. **B** The ball's acceleration is constant. **C** The force on the ball is proportional to the ball's acceleration.

Newton's Second Law and the Directions of $\vec{v}$ and $\vec{a}$

Newton's second law can be written as $\sum \vec{F} = m\vec{a}$, and we have already mentioned that this relation means that the acceleration of an object $\vec{a}$ is always parallel to the total force $\sum \vec{F}$ acting on the object. We have also seen examples in which the acceleration and velocity are in different directions, which means that *the velocity and total force need not be in the same direction*. For example, when an object such as the arrow in Figure 2.29 is fired upward, its velocity is "upward," along the +y direction in the figure. The total force on this object is due almost entirely to gravity (we'll discuss this more in later chapters) and is downward, along −y. Hence, in this case, $\vec{v}$ and $\sum \vec{F}$ are in *opposite* directions. A key point is that although the force and acceleration are *always* parallel, the direction of the velocity can be different.

Newton's Third Law

The term $\sum \vec{F}$ appearing in Newton's second law is the total force acting on the object whose motion we are studying or calculating. This force must come from somewhere; in fact, it is *always* produced by other objects. For example, when a baseball is struck by a bat, the force on the ball is due to the action of the bat. Newton's third law is a statement about what happens to the "other" object, the bat.

> *Newton's third law of motion:* **When one object exerts a force on a second object, the second object exerts a force of the same magnitude and opposite direction on the first object.**

Newton's third law is often called the *action–reaction principle.*
The forces exerted between a baseball bat and a ball (Fig. 2.30) provide an excellent illustration of Newton's third law. When the bat is in contact with the baseball,

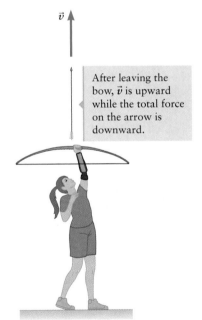

After leaving the bow, $\vec{v}$ is upward while the total force on the arrow is downward.

Figure 2.29 As it travels upward, the velocity of this arrow is upward, but the total force on the arrow is directed downward, so $\vec{v}$ and $\sum \vec{F}$ are in opposite directions.

A

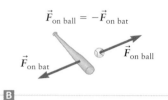

$\vec{F}_{\text{on ball}} = -\vec{F}_{\text{on bat}}$

$\vec{F}_{\text{on bat}}$ $\vec{F}_{\text{on ball}}$

B

Figure 2.30 A When a bat strikes a baseball, the bat exerts a large force on the ball. B According to Newton's third law, the ball exerts a reaction force—a force of equal magnitude and opposite direction—on the bat.

the ball experiences a force that leads to its acceleration, as can be calculated from Newton's second law. At the same time, the bat experiences a force acting on it that comes from the ball. You may have already encountered this force; it is most noticeable when you hit the ball a little away from the "sweet spot" of the bat. The point of Newton's third law is that forces *always* come in such pairs. The force exerted by the bat on the ball and the force exerted by the ball acting back on the bat are known as an action–reaction pair of forces. According to Newton's third law, these two forces are always *equal in magnitude* and *opposite in direction*, and they must act on *different* objects.

Which Law Do We Use?

Before we go any further, it is worthwhile to "look ahead" to how we will actually use Newton's laws to calculate the motion of various objects. Newton's second law tells us how to calculate a particular object's acceleration. For example, if we want to study the motion of a baseball, we need to calculate the total force on the ball and then insert this $\sum \vec{F}$ into Newton's second law (Eq. 2.8) to find the ball's acceleration. This total force will often have contributions from several sources; for a baseball, there may be a force from the impact with a bat along with other forces. The main points are that the total force $\sum \vec{F}$ in Equation 2.8 is the total force on *just the ball* and that Newton's second law enables us to calculate the acceleration of *just the ball*. On the other hand, Newton's third law tells us that forces always come in pairs. Hence, if we were somehow able to calculate or measure the force exerted by a baseball bat on a ball, Newton's third law tells us that there must be a corresponding reaction force that acts *on the bat*. It is extremely useful to know about this reaction force since it will contribute to the total force *on the bat*.

Newton's three laws of motion are the foundation for nearly everything that we do in the first part of this book. In this section, we have given some background for each of the three laws so that you can appreciate what they tell us about the nature of motion. We'll show in subsequent chapters that Newton's laws contain many other ideas and concepts, such as energy and momentum, that are essential to our thinking about the physical world.

EXAMPLE 2.8 | Action–Reaction

A person pushes a refrigerator across the floor of a room. The person exerts a force $\vec{F}_1$ on the refrigerator. From Newton's third law, we know that $\vec{F}_1$ is part of an action–reaction pair of forces. What is the reaction force to $\vec{F}_1$?

RECOGNIZE THE PRINCIPLE

The force on the refrigerator $\vec{F}_1$ is caused by the action of the person *on the refrigerator*. According to Newton's third law, the reaction force must be equal in magnitude to $\vec{F}_1$ but in the opposite direction. The reaction force must also act on a different object (it cannot act on the refrigerator).

SKETCH THE PROBLEM

Figure 2.31 shows the problem.

The forces in an action–reaction pair act on different objects.

$\vec{F}_2$ $\vec{F}_1$

Figure 2.31 Example 2.8. Newton's third law, the action–reaction principle, applies to people and refrigerators, too.

IDENTIFY THE RELATIONSHIPS AND SOLVE

The reaction force $\vec{F}_2$ is the action of the refrigerator back *on the person* as shown in Figure 2.31. According to Newton's third law, the forces in an action–reaction pair have equal magnitudes and are in opposite directions, so $\vec{F}_2 = -\vec{F}_1$.

What does it mean?
Forces *always* come in action–reaction pairs. The two forces in an action–reaction pair always act on different objects.

Which of the following is *not* an action–reaction pair of forces? (More than one answer may be correct.)

    (a) The force exerted by a pitcher on a baseball and the force exerted by the ball when it hits the bat

    (b) When you lean against a wall, the force exerted by your hands on the wall and the force exerted by the wall on your hands

    (c) In Figure 2.30, the force exerted by the ball on the bat and the force exerted by the bat on the player's hands

2.5 | WHY DID IT TAKE NEWTON TO DISCOVER NEWTON'S LAWS?

Newton's second law tells us that the acceleration of an object is given by $\vec{a} = (\sum \vec{F})/m$, where $\sum \vec{F}$ is the total force acting on the object. In the simplest situations, there may only be one or two forces acting on an object, and $\sum \vec{F}$ is then the sum of these few forces. In some cases, however, there may be a very large number of forces acting on an object. Multiple forces can make things appear to be very complicated, which is perhaps why the correct laws of motion—Newton's laws—were not discovered sooner.

Forces on a Swimming Bacterium ⊗

Figure 2.32 shows a photo of the single-celled bacterium *Escherichia coli*, usually referred to as *E. coli*. An individual *E. coli* propels itself by moving thin strands of protein that extend away from its body (rather like a tail) called flagella. Most *E. coli* possess several flagella as in the photo in Figure 2.32A, but to understand their function, we first consider the forces associated with a single flagellum as sketched in Figure 2.32B. A flagellum is fairly rigid, and because it has a spiral shape, one can think of it as a small propeller. An *E. coli* bacterium moves about by rotating this propeller, thereby exerting a force $\vec{F}_w$ on the nearby water. According to Newton's third law, the water exerts a force $\vec{F}_E$ of equal magnitude and opposite direction on the *E. coli*, as sketched in Figure 2.32B. One might be tempted to apply Newton's second law with the force $\vec{F}_E$ and conclude that the *E. coli* will move with an acceleration that is proportional to this force. However, this is incorrect because we have not included the forces from the water on the body of the *E. coli*. These forces are also indicated in Figure 2.32B; to properly describe the total force from the water, we must draw in many force vectors, pushing the *E. coli* in virtually all directions. At the molecular level, we can understand these forces as follows. We know that water is composed of molecules that are in constant motion, and these water molecules bombard the *E. coli* from all sides. Each time a water molecule collides with the *E. coli*, the molecule exerts a force on the bacterium, much like the collision of the baseball and bat in Figure 2.30. As we saw in that case, the two colliding objects both experience a recoil force, another example of action–reaction forces. So, in the present case, the *E. coli* and the water molecule exert forces on each other. An individual *E. coli* is not very large, but a water molecule is much smaller than the bacterium, and the force from one such collision will have only a small effect on the *E. coli*. However, because there are many water molecules and many such collisions the sum of the forces from all the water molecule collisions is quite substantial.

    Suppose the flagellum is rotating so that the bacterium is moving toward the left in Figure 2.32B. There will then be more collisions with the water molecules ahead of it (on the left) than with those behind it, resulting in a larger collision force on the bacterium from the left than from the right. This extra force on the front edge of the bacterium is the same type of force that you feel when you try to move your body while swimming through water. This resistive force, due to collisions with the water molecules, brings the *E. coli* to a stop when the flagellum stops rotating.

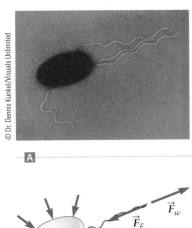

© Dr. Dennis Kunkel/Visuals Unlimited

A

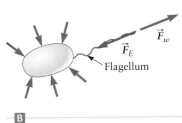

B

Figure 2.32 Ⓐ *E. coli* use the action–reaction principle to propel themselves. An individual *E. coli* bacterium is a few micrometers in diameter (1 μm = 0.000001 m). Ⓑ The flagellum exerts a force $\vec{F}_w$ on the water, and the water exerts a force $\vec{F}_E = -\vec{F}_w$ on the *E. coli*. There are also forces from the water molecules (indicated by the other red arrows) that act on the cell body.

This example with *E. coli* illustrates two important points. First, the force in Newton's second law is the *total* force on the object. It is essential that we account for *all* the forces on an object when applying Newton's second law. Second, the real world can be very complicated. Accounting for and calculating all the collision forces on an *E. coli* bacterium is a very complex problem. In fact, we'll learn about other ways to deal with the motion of an *E. coli* and similar problems when we consider the process of *diffusion* in Chapter 15.

2.6 | THINKING ABOUT THE LAWS OF NATURE

In Chapter 1, we described in general terms how ideas and theories can eventually lead to the discovery of a "law" of physics. Let's now discuss this process in a little more detail and consider how Newton's laws came to be.

Discovery of a New Law of Physics

During his lifetime and for many years after, Aristotle's ideas were, in a sense, laws of physics. As we have seen, however, these ideas were not able to explain some important observations, which ultimately led to Galileo's work on the principle of inertia and to Newton. Although we'll never know for sure how Newton arrived at his discoveries, it seems likely that he developed his ideas by first postulating or hypothesizing his laws of motion. He then tested his postulates by comparing their predictions with the behavior observed in the real world, as in the experiments of Galileo and others. When Newton found discrepancies, he would rework his theories until eventually they were able to describe correctly the motion of everything that had been studied up to that time. Newton also found that his theories were able to explain the motion of the Moon and the planets. This finding was a tremendous result because prior to Newton there was really no satisfactory theory or explanation of celestial motion. After thus showing that his theory could successfully predict the motion of a wide variety of objects under a wide range of conditions, Newton proposed that they are true "laws" of nature. Other scientists then used Newton's ideas to predict motion in situations that had not yet been studied and tested these predictions with new experiments. When Newton's theories passed these tests, they were accepted as "the" laws of physics, replacing what had come before.

After Newton, What Next?

After Newton's laws were accepted as the "true" laws of motion, what then? Newton's laws are limited to describing how things move. Although this is a broad area of physics, many problems concerning matter and energy are not covered by Newton's laws. Such problems include the behavior of light and electricity, and we require additional theories and laws to describe these phenomena. Also, while we may accept Newton's three laws of motion as the laws of physics that describe mechanics, it is important to continue testing these laws in new situations. In the vast majority of cases, Newton's laws have passed these tests, but around the beginning of the 20th century it was found that Newton's laws do not correctly predict the behavior of electrons and protons, and of atoms and molecules. This monumental discovery eventually led to the development of new laws of physics known as quantum mechanics. What, then, happened to Newton's laws? They were not discarded; instead, physicists realized that Newton's ideas work extremely well for describing motion in what is now known as the *classical* regime. This regime includes terrestrial-scale objects, such as rocks and cars, and it even extends to *E. coli* and to planets. Newton's laws break down in the *quantum* regime of electrons, protons, and atoms, however, and in that regime a different law of nature—quantum mechanics—must be used.

If Newton's laws break down when applied in certain regimes or to certain types of objects, are we still justified in referring to them as laws of nature? One could take the point of view that a law of nature must always hold true and apply to all situations, with no exceptions. Unfortunately, *no* current law of physics passes this test! All the presently known laws of physics are known to fail or to be inadequate in some regime or another. A more widely accepted viewpoint is that a law of physics must correctly describe all behavior in a particular regime of nature. Newton's laws are restricted to the regime of classical physics, which includes terrestrial objects and extends to things like planets, the Sun, and individual cells. It is believed that Newton's laws provide an accurate description of all motion in this regime.

Although the process of testing and retesting the laws of physics may lead to the discovery of new laws, it is usually found that the law being tested works fine. Such testing, however, can still lead to a better understanding of physics and can reveal important new insights. For example, Newton's three laws do not mention anything about the notion of *energy*. Nevertheless, we'll see that the concept of energy and many other useful ideas are contained within Newton's laws. Applying Newton's laws to new situations can help us discover such unanticipated results.

SUMMARY | Chapter 2

KEY CONCEPTS AND PRINCIPLES

Motion

A complete description of an object's motion involves its *position*, *velocity*, and *acceleration*. For motion in one dimension (along a line), position can be specified by a single quantity, *x*. The *instantaneous velocity* is related to changes in position by

$$v = \lim_{\Delta t \to 0} \frac{\Delta x}{\Delta t} \qquad \text{(2.5) (page 31)}$$

The instantaneous velocity is usually referred to as simply the "velocity."

The *instantaneous acceleration* is given by

$$a = \lim_{\Delta t \to 0} \frac{\Delta v}{\Delta t} \qquad \text{(2.7) (page 34)}$$

and is usually referred to simply as the "acceleration."

Position, velocity, and acceleration are vector quantities. For motion in two or three dimensions, we must consider how the directions of these quantities relate to our chosen coordinate axes. This topic is explored further in Chapter 4.

Newton's laws of motion

The connection between *force* and motion is at the heart of *mechanics*. A force is a push or a pull. Force is a vector. Given the forces acting on an object, we can calculate how it will move using Newton's three laws of motion.

- *Newton's first law*: If the total force acting on an object is zero, the object will move with a constant velocity.

(Continued)

- **Newton's second law**:

$$\vec{a} = \frac{\sum \vec{F}}{m} \qquad\qquad \text{(2.8) (page 41)}$$

Forces thus cause acceleration.

- **Newton's third law**: For every action (force), there is a reaction (force) of equal magnitude and opposite direction.

All forces come in **action–reaction pairs**. The two forces in an action–reaction pair act on different objects.

The relationships between position, velocity, and acceleration

The **instantaneous velocity** at a particular time t is the slope of the position–time curve at that time (Fig. 2.33).

The **average velocity** during a particular time interval is equal to the slope of the line connecting the start and end of that interval on the position–time curve (Fig. 2.33).

The **instantaneous acceleration** equals the slope of a plot of the instantaneous velocity v as a function of time (Fig. 2.34).

The **average acceleration** during a time interval is equal to the slope of the line connecting the start and end of that interval on the velocity–time curve (Fig. 2.34).

Figure 2.33

Figure 2.34

QUESTIONS

1. Make a hypothetical sketch of a velocity–time graph in which the velocity is always positive, but the acceleration is always negative. Give a physical example of such motion.

2. In Example 2.7, we considered the motion of a falling ball while it was in the air. We ended the example and the plot in Figure 2.26 at the moment just *before* the ball hit the ground. Figure Q2.2 shows the position–time graph of the ball, including a short time after it hits the ground. Use this graph to

Figure Q2.2

deduce the acceleration when the ball first makes contact with the ground. Explain why a is positive immediately after the ball hits the ground, while it is coming to a stop. How can a positive acceleration make the velocity become zero?

3. When a car collides with a wall, a force on the car causes it to stop. Identify an action–reaction pair of forces involving the car.

4. You push on a refrigerator, as in Figure 2.1, but the refrigerator does not move. Hence, even though you are applying a nonzero force, the acceleration is still zero. Explain why this does not contradict Newton's second law.

5. A car is traveling on an icy road that is extremely slippery. The driver finds that she is not able to stop or turn the car. Explain this situation in terms of the principle of inertia (Newton's first law).

6. A car is initially at rest on an icy road that is extremely slippery. The driver finds that he is unable to get the car to drive away because the wheels simply spin when he tries to accelerate. Explain this situation in terms of the principle of inertia (Newton's first law).

7. Give an example of motion for which the average velocity is zero, but the speed is never zero.

8. Give an example of motion for which the average velocity is equal to the instantaneous velocity. *Hint*: Sketch the velocity–time graph.

9. [SSM] **Abracadabra!** A magician pulls a tablecloth off of a set table with one swift, graceful motion. Amazingly, the fine china, glassware, and silverware are practically undisturbed. Although amazing, this feat is not an illusion. Describe the behavior of the plates and glasses in terms of the principle of inertia (Newton's first law).

10. Give three examples of motion in which (a) the velocity is positive and the acceleration is positive, (b) the velocity is negative and the acceleration is negative, and (c) the velocity is positive and the acceleration is negative.

11. A car starts from the origin at $t = 0$. At some later time, is it possible for the car's velocity to be positive but its displacement from the origin to be equal to zero? Explain and give an example.

12. An object moves in such a way that its instantaneous acceleration always equals its average acceleration. (a) Draw the acceleration–time graph that describes this motion. (b) Draw the corresponding velocity–time graph. (c) Does the instantaneous velocity equal the average velocity? Explain why or why not.

13. Consider again Example 2.7. Draw plots of the position, velocity, and acceleration as functions of time, starting from when the ball is released and ending *after* the ball hits the ground. Indicate the time at which the ball hits the ground.

14. Make a qualitative sketch of the position y as a function of time for a yo-yo. Also make sketches of the velocity and acceleration as functions of time. Is the total force on the yo-yo zero or nonzero? Explain how you can tell from your graphs.

15. [SSM] Figure Q2.15 shows a motion diagram for a rocket-powered car. The photos are taken at 1.0-s intervals. Make qualitative plots of the position, velocity, acceleration, and force on the car as functions of time.

16. A person stands on level ground and throws a baseball straight upward, into the air. (a) Does the person exert a force on the ball while it is in his hand? After it leaves his hand? (b) Does the

Figure Q2.15

ball ever exert a force on the person? If so, what is the direction of this force? (c) If your answer to part (b) is yes, why does the person not accelerate?

17. Consider the motion of the Moon as it orbits Earth. (a) Is the Moon's acceleration zero or nonzero? Explain. (b) If the Moon has a nonzero acceleration, what force is responsible?

18. Consider the motion of a marble as it falls to the bottom of a jar of honey. Experiments show (see also Chapter 3) that the marble moves with a constant velocity. Applying Newton's first law, does that mean that no forces are acting on the marble?

19. According to Newton's first law (the principle of inertia), if there is no force exerted on an object, the object will move with constant velocity. Consider the following examples of motion. Is the velocity of the object constant? What are the forces acting on each object? If the velocity is constant, explain why Newton's first law applies.
(a) A hockey puck sliding on ice
(b) A mug of root beer sliding down a bar
(c) A car skidding on a flat, desert highway

20. Three blocks rest on a table as shown in Figure Q2.20. Identify three action–reaction pairs of forces.

Figure Q2.20

21. Two football players start running at opposite ends of a football field (opposite goal lines), run toward each other, and then collide at the center of the field. They start from rest and are running at top speed when they collide. (a) Draw a graph showing the position as a function of time for both players. (b) Draw a graph showing their velocities as functions of time.

22. A ball rolls up a hill, just barely reaches the top, and then rolls down the other side. Draw a qualitative sketch of how the speed of the ball varies with time.

PROBLEMS

[SSM] = solution in Student Companion & Problem-Solving Guide
⭐ = intermediate   ✪ = challenging

Ⓧ = life science application
Ⓡ = reasoning and relationships problem

2.2 WHAT IS MOTION?

1. In SI units, velocity is measured in units of meters per second (m/s). Which of the following combinations of units can also be used to measure velocity?
(a) cm/s
(b) cm/s^2
(c) m^3/(mm^2 s^2)
(d) km/s
(e) miles per hour
(f) km/hour
(g) miles/cm

2. A typical airplane can fly at a speed of 400 miles per hour. What is its speed in meters per second?

3. In SI units, acceleration is measured in units of meters per second squared (m/s^2). Which of the following combinations of units can also be used to measure acceleration?
(a) cm/s
(b) cm/s^2
(c) m^3/(mm^2 · s^2)
(d) km/s^2
(e) miles per hour
(f) miles/(minutes)^2

4. A falling baseball has an acceleration of magnitude 9.8 m/s². What is its acceleration in feet per second squared?

5. An elite runner can run 100 m in 10 s. What is his average speed?

6. ✴ ✖ Consider the motion of a sprinter running a 100-m dash. When it is run outdoors, this race is run along a straight-line portion of a track, so this is an example of motion in one dimension. Draw qualitative plots of the position, velocity, and acceleration as functions of time for the sprinter. Start your plots just before the race starts and end them when the runner comes to a complete stop.

7. ✇ A jogger maintains a speed of 3.0 m/s for 200 m until he encounters a stoplight, and he abruptly stops and waits 30 s for the light to change. He then resumes his exercise and maintains a speed of 3.5 m/s for the remaining 50 m to his home. (a) What was his average velocity for this entire time interval? (b) What were his maximum and minimum velocities, and how do they compare with this average?

8. ✴ A hockey puck that is sliding on an icy surface will eventually come to rest. The (horizontal) force that makes it stop is due to friction between the puck and the ice. Draw qualitative plots of the position, velocity, and acceleration as functions of time for the puck. Pay special attention to the sign of the acceleration.

9. ✇ Consider a marble falling through a very thick fluid, such as molasses. Draw qualitative plots of the position, velocity, and acceleration as functions of time for the marble, starting from the moment it is released in the molasses.

10. ✴ A person riding on a skateboard is initially coasting on level ground. He then uses his feet to push on the ground so that he speeds up for a few seconds, and then he coasts again. Draw qualitative plots of his position, velocity, and acceleration as functions of time.

11. ✇ **In the drop zone.** Consider a skydiver who jumps from an airplane. In a typical case, the skydiver will wait for a (short) time before opening her parachute. Draw qualitative plots of the position, velocity, and acceleration as functions of time for the skydiver, starting from the time she jumps from the plane, assuming (somewhat unrealistically) that she falls straight down. Here, the position, y, is her vertical height above the ground. Be sure to indicate the time at which she opens the parachute. *Hint*: Consider how the velocity will vary with time after the chute is open for a long time.

12. ✴ Consider a skier who coasts up to the top of a hill and then continues down the other side. Draw a qualitative plot of what the skier's speed might look like.

13. SSM ✴ Figure P2.13 shows three motion diagrams, where each dot represents the location of an object between equal time intervals. Assume left-to-right motion. For each motion diagram, sketch the appropriate position–time, velocity–time, and acceleration–time graphs.

Figure P2.13

14. ✴ A bicycle is moving initially with a constant velocity along a level road. The bicyclist then decides to slow down, so she applies her brakes over a period of several seconds. Thereafter, she again travels with a constant velocity. Draw qualitative sketches of her position, velocity, and acceleration as functions of time.

15. ✴ Figure P2.15 shows several hypothetical position–time graphs. For each graph, sketch qualitatively the corresponding velocity–time graph.

Figure P2.15 Problems 15 and 16.

16. ✴ The position–time graphs in Figure P2.15 each describe the possible motion of a particular object. Give at least one example of what the object and motion might be in each case.

17. ✴ Figure P2.17 shows several hypothetical velocity–time graphs. For each case, sketch qualitatively the corresponding acceleration–time graph.

Figure P2.17 Problems 17, 18, and 19.

18. ✇ Figure P2.17 shows several hypothetical velocity–time graphs. For each case, sketch qualitatively the corresponding position–time graph.

19. ✴ Give examples of objects whose motion might be described by the graphs in Figure P2.17.

20. ✴ Figure P2.20 shows several hypothetical acceleration–time graphs. For each case, sketch qualitatively the corresponding velocity–time graph.

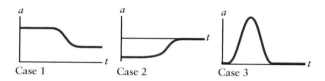

Figure P2.20 Problems 20 and 21.

21. ✴ Give examples of objects whose motion is described by the plots in Figure P2.20.

22. ✴ Match each of the following examples of motion to one of the position–time graphs in Figure P2.22.
   (a) A person at the beginning of a race, starting from rest
   (b) A runner near the end of a race, just after crossing the finish line

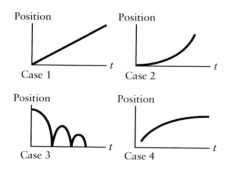

Figure P2.22

(c) A ball dropped from a window, hitting the ground below and then bouncing several times

(d) A bowling ball as it rolls down a lane, just after it leaves the bowler's hand

23. ★ Match each of the examples of motion in Problem 22 to one of the velocity–time graphs shown in Figure P2.23.

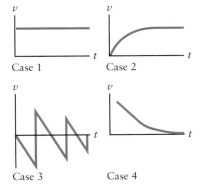

Figure P2.23

24. ★ Consider the position–time graph shown in Figure P2.24. Make a careful graphical estimate of the velocity as a function of time by measuring the slopes of tangent lines. What is an approximate value of the maximum velocity of the object?

Figure P2.24 Problems 24 and 25.

25. For the object described by Figure P2.24, estimate the average velocity (a) over the interval from $t = 2.0$ s to $t = 4.0$ s and (b) over the interval from $t = 1.0$ s to $t = 5.0$ s.

26. Repeat Problem 24 using the position–time graph in Figure P2.26.

Figure P2.26

27. Figure P2.27 shows the velocity–time curve of a falling brick. Make a careful estimate of the slope to find the acceleration of the brick at $t = 3.0$ s.

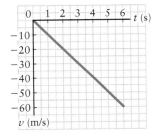

Figure P2.27

28. ★ Using a graphical approach (i.e., by estimating the slope at various points), find the qualitative behavior of the acceleration as a function of time for the object described by the velocity–time graph in Figure P2.28.

Figure P2.28 Problems 28 and 30.

29. For the object described by the velocity–time graph in Figure P2.29, estimate the average acceleration over the interval from $t = 0$ s to $t = 50$ s and over the interval from $t = 100$ s to $t = 200$ s.

Figure P2.29 Problems 29 and 30.

30. ✪ Draw a position–time graph for an object whose velocity as a function of time is described by (a) Figure P2.28 and (b) Figure P2.29.

31. ★ Draw a graph showing the position as a function of time for an object whose acceleration is (a) constant and positive, (b) constant and negative, and (c) positive and increasing with time.

32. A car travels along a straight, level road. The car begins a distance $x = 25$ m from the origin at $t = 0.0$ s. At $t = 5.0$ s, the car is at $x = 100$ m; at $t = 8.0$ s, it is at $x = 300$ m. Find the average velocity of the car during the interval from $t = 0.0$ s to $t = 5.0$ s and during the interval from $t = 5.0$ s to $t = 8.0$ s.

33. SSM ★ A squirrel falls from a very tall tree. Initially (at $t = 0$), the squirrel is at the top of the tree, a distance $y = 50$ m above the ground. At $t = 1.0$ s, the squirrel is at $y = 45$ m, and at $t = 2.0$ s, it is at $y = 30$ m. Estimate the average velocity of the squirrel during the intervals from $t = 0.0$ s to $t = 1.0$ s and from $t = 1.0$ s to $t = 2.0$ s. Use these results to estimate the average acceleration of the squirrel during this time. (No squirrels were harmed during the writing of this problem.)

34. The space shuttle takes off from Cape Canaveral in Florida and orbits Earth 18 times before landing back at Cape Canaveral 24 hours and 15 minutes later. During this time, it moves in a circular orbit with radius 6.7×10^6 m. (a) What is the average speed of the space shuttle during its journey? (b) What is the average velocity?

35. ★ Figure P2.35 shows the velocity as a function of time for an object. (a) What is the average acceleration during the interval

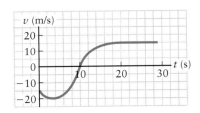

Figure P2.35

from $t = 0$ to $t = 20$ s? (b) Estimate the instantaneous acceleration at $t = 5$ s, (c) at $t = 10$ s, and (d) at $t = 20$ s.

36. ⭐ Figure P2.36 shows the acceleration as a function of time for an object. (a) If the object starts from rest at $t = 0$, what is the velocity of the object as a function of time? (b) If the object instead has a velocity of +40 m/s at $t = 0$, how does your result for part (a) change?

Figure P2.36

37. ✪ A rabbit runs in a straight line with a velocity of +1.5 m/s for a period of time, rests for 10 s, and then runs again along the same line at +0.60 m/s for an unknown amount of time. The rabbit travels a total distance of 1200 m, and its average speed is 0.80 m/s. (a) What is the total time the rabbit spends running at 1.5 m/s? (b) How long does it spend running at 0.60 m/s?

2.3 THE PRINCIPLE OF INERTIA

38. SSM ⭐ Figure 2.22 shows one of Galileo's experiments in which a ball rolls up an incline. A ball that is initially rolling up the incline will roll up to some maximum height and then roll back down the incline. Draw qualitative plots of the position and velocity as functions of time for the ball. Take $x = 0$ at the bottom of the ramp.

39. ✪ Consider the inclined plane (i.e., ramp) in Figure 2.23A, and assume it is covered with a layer of ice so that it is extremely slippery (frictionless). The horizontal surface of the bottom is not covered with ice, however. A box is released from the same height on the ramp for a number of trials and slides down the ramp. For each trial, the box encounters a different horizontal surface at the bottom of the smooth ramp. Draw qualitative plots to compare the velocity as a function of time for each case: (a) a grass field, (b) an asphalt street, (c) a hardwood floor, (d) an ice rink, and (e) an ideal, frictionless surface.

40. ✪ ✖ **An unsafe way to transport your penguin.** A flatbed truck hauls a block of ice on top of which stands a penguin (see Fig. P2.40). Assume the block of ice is frozen solid to the bed of the truck, but the top surface of the ice is extremely slippery. The truck skids to a stop from an initial velocity of 20 m/s. Describe what happens to the penguin in terms of Newton's first law. How does it differ from what happens to the driver of the truck, who is wearing a seat belt? Draw and compare the velocity–time graphs of the penguin and truck driver.

Figure P2.40

41. ⭐ A flatbed railcar is moving at a slow but constant velocity. A man stands in the railcar, facing sideways (perpendicular) to the motion of the railcar. The man holds a baseball at arm's length and drops it onto the railcar bed. Where does the principle of inertia predict that the ball will land, (a) directly below the drop point on the railcar, (b) slightly in front of the drop point on

the railcar, or (c) slightly behind the drop point on the railcar? Explain.

2.4 NEWTON'S LAWS OF MOTION

42. The person shown in Figure 2.1 is pushing on a refrigerator that sits on a level floor. Assume the floor is very slippery so that the only horizontal force on the refrigerator is due to the person. If the force exerted by the person has a magnitude of 120 N and the refrigerator has a mass of 180 kg, what is the magnitude of the refrigerator's acceleration?

43. ⭐ **Action and reaction.** Use Newton's third law along with his second law to find the acceleration of the man in Problem 42 as he pushes the refrigerator. Assume the man has a mass of 60 kg and the floor is very slippery so that there is no frictional force between the floor and the man.

44. An object is found to move with an acceleration of magnitude 12 m/s² when it is subjected to a force of magnitude 200 N. Find the mass of the object.

45. In SI units, force is measured in newtons, with $1 \text{ N} = 1 \text{ kg} \cdot \text{m/s}^2$. Which of the following combinations of units can also be used to measure force?
(a) $\text{g} \cdot \text{m/s}$
(b) $\text{g} \cdot \text{cm}^2/\text{s}$
(c) $\text{kg} \cdot \text{m}^4/(\text{s}^2 \cdot \text{cm}^3)$
(d) $\text{g} \cdot \text{cm/s}^2$
(e) $\text{kg} \cdot \text{miles}/(\text{minutes})^2$

46. In the U.S. customary system of units, mass is measured in units called slugs. Suppose an object has a mass of 15 kg. Use the conversion factors inside the front cover of this book to express the mass of the object in slugs.

47. In the U.S. customary system of units, force is measured in units of pounds (abbreviated lb). Suppose the force on an object is 150 lb. Using the conversion factors inside the front cover of this book, express this force in units of newtons.

48. A force is found to be 240 g · cm/s². Convert this value into units of newtons.

49. SSM ⭐ A cannon is fired horizontally from a platform (Fig. P2.49). The platform rests on a flat, icy, frictionless surface. Just after the shell is fired and while it is moving through the barrel of the gun, the shell (mass 3.2 kg) has an acceleration of +2500 m/s². At the same time, the cannon has an acceleration of −0.76 m/s². What is the mass of the cannon?

Cannon shell

Ice

Figure P2.49

50. According to Newton's third law, for every force there is always a reaction force of equal magnitude and opposite direction. In each of the examples below, identify an action–reaction pair of forces.
(a) A tennis racket hits a tennis ball, exerting a force on the ball.
(b) Two ice skaters are initially at rest and in contact. One of the skaters then pushes on the other skater's back, exerting a force on that skater.
(c) A car is moving at high speed and runs into a tree, exerting a force on the tree.
(d) Two cars are moving in opposite directions and collide head-on.
(e) A person leans on a wall, exerting a force on the wall.
(f) A hammer hits a nail, exerting a force on the nail.
(g) A mass hangs by a string tied to a ceiling, with the string exerting a force on the mass.
(h) A bird sits on a telephone pole, exerting a force on the pole.

51. ⊠ Tom has two ways he could drive home from work. He could take Highway 99 for 45 miles with a speed limit of 65 mi/h, or he could take Interstate 5, which would take him a bit out of the way at 57 miles, but with a speed limit of 75 mi/h. (a) Assuming Tom misses the rush hour and drives at an average speed equal to the maximum speed limit, which route gets him home the fastest? How much time does he save? (b) If Tom breaks the law and travels at 75 mi/h on Highway 99, how much time would he save (if he is not stopped to get a speeding ticket) compared to driving at the legal speed limit on Highway 99?

52. SSM ⊠ **Throwing heat.** In professional baseball, pitchers can throw a fastball at a speed of 90 mi/h. (a) Given that the regulation distance from the pitcher's mound to home plate is 60.5 ft, how long does it take the ball to reach home plate after the ball leaves the pitcher's hand? (b) It takes (on average) 0.20 s for the batter to get the tip of his bat over home plate. How much time does that give him to react? (c) The average fast pitch in professional women's softball is about 60 mi/h, where the regulation distance from the pitcher's mound to home plate is 40.0 ft. How do the travel time of a pitched ball and reaction time of a batter in softball compare with those in baseball? For simplicity, assume the pitcher releases the ball just above the center of the pitcher's mound. Keep three significant figures in your calculation.

53. ⊠ Figure P2.53 shows the position as a function of time for an object. (a) What is the average velocity during the period from $t = 0.0$ s to $t = 10.0$ s? (b) What is the average velocity between $t = 0.0$ s and $t = 5.0$ s? (c) Between $t = 5.0$ s and $t = 10.0$ s? (d) Explain why and how your answers to parts (a), (b), and (c) are related.

Figure P2.53

54. SSM Ⓡ On your vacation, you fly from Atlanta to San Francisco (a total distance of 3400 km) in 4.0 h. (a) Draw a qualitative sketch of how the speed of your airplane varies with time. (b) What is the average speed during your trip? (c) Estimate the top speed during your trip. *Hint:* You reach your top speed about 10 minutes after taking off. (d) What is your average acceleration during the first 10 minutes of your trip? (e) What is the average acceleration during the central hour of your trip?

55. ✪ Figure P2.55 shows the position as a function of time for an apple that falls from a very tall tree. (a) At what time does the apple hit the ground? (b) Use a graphical approach to estimate and plot the acceleration as a function of time. (c) Make a sketch of the velocity of the apple as a function of its height above the ground.

56. ⊠ Ⓧ A cheetah runs a distance of 100 m at a speed of 25 m/s and then runs another 100 m in the same direction at a speed of 35 m/s. What is the cheetah's average speed?

57. **Astronaut frequent-flyer miles.** The average speed of the space shuttle while in orbit is about 8900 m/s. How far does the space shuttle travel during a mission that lasts 7.5 days?

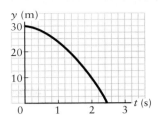

Figure P2.55

58. ✪ A cat is being chased by a dog. Both are running in a straight line at constant speeds. The cat has a head start of 3.5 m. The dog is running with a speed of 8.5 m/s and catches the cat after 6.5 s. How fast did the cat run?

59. ✪ Ⓧ **Predator and prey.** The author's cat enjoys chasing chipmunks in the front yard. In this game, the cat sits at one edge of a yard that is $w = 30$ m across (Fig. P2.59), watching as chipmunks move toward the center. The cat can run faster than a chipmunk, and when a chipmunk moves more than a certain distance L from the far end of the yard, the cat knows that it can catch the chipmunk before the chipmunk disappears into the nearby woods. If the cat's top speed is 7.5 m/s and a chipmunk's top speed is 4.5 m/s, find L. Assume the chipmunk and cat move along the same straight-line path.

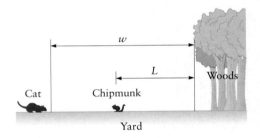

Figure P2.59

60. ✪ A thief is trying to escape from a parking garage after completing a robbery, and the thief's car is speeding ($v = 12$ m/s) toward the door of the parking garage (Fig. P2.60). When the thief is $L = 30$ m from the door, a police officer flips a switch to close the garage door. The door starts at a height of 2.0 m and moves downward at 0.20 m/s. If the thief's car is 1.4 m tall, will the thief escape?

Figure P2.60

Chapter 3

Forces and Motion in One Dimension

These railroad tracks carry trains across the desert in southern California. The tracks are very straight and very long, so the motion of the trains they carry is indeed one-dimensional motion. These tracks can also be viewed as a coordinate axis complete with "tick marks"! (© Richard T. Norwitz/National Geographic/Getty Images)

In Chapter 2, we presented Newton's laws of motion and explained what they mean from a qualitative point of view. We also introduced a number of quantities, including displacement, velocity, and acceleration, that are essential for describing motion. Our next job is to apply Newton's laws to calculate how things move in various situations. In this chapter, we consider motion in one dimension, that is, motion along a straight line. You might wonder why we devote an entire chapter to such an idealized case; after all, the world is not one dimensional, and in most situations, objects move along fully three-dimensional trajectories. Even so, it is very useful to start with one-dimensional examples. The mathematics is simpler in this case, and the basic ideas and approaches we develop for dealing with motion in one dimension can be applied quite directly to motion in higher dimensions.

3.1 | MOTION OF A SPACECRAFT IN INTERSTELLAR SPACE

Figure 3.1 shows a spacecraft traveling somewhere in distant space. This hypothetical spacecraft is traveling along a straight-line path from one galaxy to another. Since the spacecraft is moving along a line, this is an example of one-dimensional motion. To describe this motion, we must first choose a coordinate system with which to measure the displacement, velocity, and acceleration of the spacecraft. The spacecraft's motion is along the line that connects the two galaxies, so we choose that line to be our coordinate axis. We have drawn this axis in Figure 3.1 and labeled it as the x axis. We have also placed the origin at the starting galaxy and taken the "positive" direction along the x axis as headed toward the destination.

If our spacecraft is very far from any planets or stars, the force of gravity on the spacecraft will be very small. To make this example as simple as possible, let's assume this gravitational force is zero. We now want to calculate how the spacecraft will move in two different cases.

In the first case, the engine is turned off so that the spacecraft is "coasting." In more scientific terms, we can say that because the engine is off and the gravitational force is zero, there are no forces acting on the spacecraft. Newton's first law tells us that when the total force acting on an object is zero, the object has a constant velocity. Hence, our spacecraft moves with a constant velocity; that is, it moves in a straight-line path with a constant speed. Examples of such constant-velocity motion are given in Figures 3.2 and 3.3, which show graphs of how the velocity and position might look as functions of time. We learned in Chapter 2 that displacement and velocity are vectors, having both magnitude and direction. For the cases involving one-dimensional motion that we consider here and throughout this chapter, these vectors all lie along the chosen coordinate axis. In general, we could write these quantities using our usual vector notation with vector arrows, but with one-dimensional motion, we can simplify the notation and drop the vector arrows because we specify direction by the sign of each quantity (+ or −). This notation corresponds to specifying the *components* of the displacement and velocity along the coordinate axis.

Figure 3.2 shows the behavior of the velocity and position of our spacecraft when the velocity is zero, while Figure 3.3 shows the behavior for a constant nonzero velocity. We know that v is the slope of the position–time curve; in both Figures 3.2 and 3.3, the velocity is constant, so the x–t relations are both linear. Although Newton's first law tells us that the velocity is constant, it does not tell us the value of v. This value is determined by forces applied to the spacecraft at earlier times. In this problem, a force might have been applied to the spacecraft before the engine was turned off, giving a nonzero acceleration prior to $t = 0$ and causing the velocity in Figure 3.3 to be nonzero.

Behavior of the Velocity and Position for Constant Nonzero Acceleration

Let's next consider what happens when the spacecraft's engine is turned on. There is now a force exerted on the spacecraft, so we need to use Newton's second law to determine the resulting motion. Recall that according to Newton's second law the acceleration is given by $a = \sum F/m$. For our case of one-dimensional motion, both the total force $\sum F$ and the acceleration a are parallel to the x axis in Figure 3.1 (so these are the components of the force and acceleration along x). In this example, there is only a single force (due to the engine); in the simplest case, the engine will be designed to produce a constant force on the spacecraft, so the force term $\sum F$ in Newton's second law is a constant. For simplicity, we also assume the mass of the spacecraft m is constant.[1] We can then use Newton's second law to calculate the acceleration and find that a is also a constant. Since acceleration is the slope of the

Figure 3.1 This hypothetical spacecraft is following a one-dimensional trajectory along the x axis.

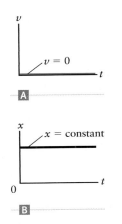

Figure 3.2 If the spacecraft's velocity is zero, its position x is a constant. This is one example of motion with a constant velocity.

Figure 3.3 Motion with a constant and nonzero velocity. Since v is equal to the slope of the position–time curve, the graph of x as a function of t is a straight line.

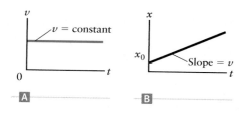

A **B**

velocity–time curve, the v–t relation must be a *straight line*. This relationship is shown in Figure 3.4A; we have

$$v = v_0 + at \qquad (3.1)$$

In Equation 3.1, v_0 is the "initial" velocity of the spacecraft, that is, the velocity at $t = 0$; it is useful to think of v_0 as the velocity at the moment the stopwatch used to describe this problem is started. The value of v_0 depends on what happened prior to $t = 0$; for example, one might imagine that the engine was turned on for a while to launch the spacecraft and then turned off before $t = 0$. In any event, the value of v_0 is determined by the prior history of the spacecraft. In most situations (and problems!), we must rely on someone else (such as the person who writes the problem or the pilot of the spacecraft) to provide the value of v_0.

To give a complete description of the spacecraft's motion, we must also derive its position as a function of time. We already have the velocity as a function of time from Equation 3.1, and we know that v is the slope of the x–t curve. There are several ways to derive the x–t relationship; one approach is shown in Figure 3.4B. In Chapter 2, we discussed the difference between the instantaneous velocity and the average velocity. The instantaneous velocity v is the slope of the x–t curve at a particular point on the curve, whereas the average velocity over a particular time interval is found by taking the slope of the line that spans the interval. One such time interval is shown in Figure 3.4B. The slope of the line connecting $t = 0$ with some general time t is

$$v_{\text{ave}} = \frac{x - x_0}{t}$$

where x_0 is the value of x at $t = 0$ (the "initial" position). We can also calculate the average velocity from Equation 3.1; since v varies linearly with time, the average velocity during a particular time interval is the average of the instantaneous velocities at the start and end of the interval. Hence,

$$v_{\text{ave}} = \frac{v(t = 0) + v(t)}{2}$$

We can now insert our result for $v(t)$ from Equation 3.1:

$$v_{\text{ave}} = \frac{v(t = 0) + v(t)}{2} = \frac{v_0 + v_0 + at}{2}$$

$$v_{\text{ave}} = v_0 + \tfrac{1}{2}at \qquad (3.2)$$

Equating this expression to our previous result $v_{\text{ave}} = (x - x_0)/t$ and doing a little rearranging leads to

$$v_{\text{ave}} = \frac{x - x_0}{t} = v_0 + \tfrac{1}{2}at$$

$$x - x_0 = v_0 t + \tfrac{1}{2}at^2$$

$$x = x_0 + v_0 t + \tfrac{1}{2}at^2 \qquad (3.3)$$

Behavior of the velocity for motion with a constant acceleration

Slope = a

at

v_0

v

v_0

A

The slope of this line is the average velocity.

$$x = x_0 + v_0 t + \frac{1}{2}at^2$$

Slope = v

x_0

Slope = v_0

B

Figure 3.4 **A** The slope of the velocity–time graph is equal to the acceleration, so when the acceleration a is constant, the v–t plot is a straight line. **B** The corresponding graph of position versus time. The slope of the x–t plot at a certain point on the curve is equal to the instantaneous velocity v.

Behavior of the position for motion with a constant acceleration

---

[1]Most rocket engines work by expelling gas, so the total mass of the spacecraft, including the engine and fuel, will generally decrease with time. Here, however, we ignore that effect.

This result tells us how the position of our spacecraft varies with time and is sketched in Figure 3.4B.

Relations for Motion with Constant Acceleration

It is useful to recap how we used Newton's second law to analyze motion when the total force acting on an object is constant. The first step was to determine the total force on the object. For our spacecraft problem, this force was simply a given; for example, the engine designer might have told us the value. We then used Newton's second law $a = \sum F/m$ to calculate the acceleration a. Because the total force and m were both constants, a is also constant. Acceleration is the slope of the v–t relation; hence, the velocity is a linear function of t. The final step was to deduce the x–t relation; that analysis was sketched in Figure 3.4B and led to Equation 3.3. Keep in mind that these results for the position and velocity of an object with constant acceleration (Eqs. 3.3 and 3.1) are not "new" laws of physics; both are a direct result of Newton's second law.

It is rare to find situations in real life in which the acceleration is *precisely* constant, so Equations 3.1 and 3.3 will usually not give an exact description of the motion of an object. However, in many interesting examples the acceleration is very nearly constant, and these examples are well described by the constant-acceleration relations in Equations 3.1 and 3.3. The constant-acceleration case is also quite instructive because the mathematical forms of x and v as functions of time are especially simple, making many quantitative calculations possible.

For future reference, Table 3.1 collects these results, and also lists one more very useful relationship. Equations 3.1 and 3.3 give the velocity and position as functions of time. We can eliminate t from these two relations to get a single equation that relates x and v without direct reference to t. We first use Equation 3.1 to express t in terms of v and a:

$$v = v_0 + at$$

$$t = \frac{v - v_0}{a}$$

We next insert this into Equation 3.3:

$$x = x_0 + v_0 t + \frac{1}{2} at^2 = x_0 + v_0\left(\frac{v - v_0}{a}\right) + \frac{1}{2} a\left(\frac{v - v_0}{a}\right)^2$$

Collecting and combining terms gives

$$x = x_0 + \frac{v_0 v}{a} - \frac{v_0^2}{a} + \frac{1}{2}\frac{v^2}{a} - \frac{v_0 v}{a} + \frac{1}{2}\frac{v_0^2}{a} = x_0 + \frac{1}{2a}(v^2 - v_0^2)$$

We can now rearrange to reach our desired result:

$$v^2 = v_0^2 + 2a(x - x_0) \tag{3.4}$$

which is also listed in Table 3.1. This relation is interesting because it does not contain time (t); it only involves position, velocity, and acceleration. We'll encounter many cases where this result is very handy.

Table 3.1 Equations for Motion with Constant Acceleration

EQUATION NUMBER	MATHEMATICAL RELATION	VARIABLES	
3.3	$x = x_0 + v_0 t + \frac{1}{2}at^2$	position, acceleration, and time	Relations between x, v, a, and t for motion with a constant acceleration
3.1	$v = v_0 + at$	velocity, acceleration, and time	
3.4	$v^2 = v_0^2 + 2a(x - x_0)$	position, velocity, and acceleration	

Note: In cases in which the position variable in a problem is y, the same equations apply with y inserted in place of x.

Figure 3.5 Example 3.1. Sketch of a spacecraft moving in response to a constant force F along the $+x$ direction.

Insight 3.2
WORKING WITH UNITS
In our calculations in Example 3.1, we canceled and combined units as reviewed in Chapter 1. When dealing with the units of force, it is simplest to express the newton in terms of the fundamental units of meters, kilograms, and seconds before canceling and combining units. (Recall that $1\,\text{N} = 1\,\text{kg} \cdot \text{m/s}^2$.) You should always include the units in your calculations and check that they correctly match the quantity being calculated.

EXAMPLE 3.1 | Accelerating into Space

Let's see how the basic relations between x, v, and t in Table 3.1 can be used to calculate the motion of our spacecraft in a particular case. We assume the spacecraft has a mass $m = 200$ kg and is initially at rest. The engine is turned on at $t = 0$, and from this time forward there is a constant force $F = 2000$ N on the spacecraft. **(a)** What is the velocity of the spacecraft at $t = 40$ s? **(b)** How far does the spacecraft travel during this time? **(c)** What is the velocity of the spacecraft when it reaches a distance 1000 m from where it started?

RECOGNIZE THE PRINCIPLE

The only force on the spacecraft is the force F from the engine, so this force is equal to $\sum F$ in Newton's second law. The force is constant; hence, the acceleration is also constant and we can find its value using Newton's second law. We can then use our relations for motion with constant acceleration from Table 3.1.

SKETCH THE PROBLEM

Figure 3.5 shows the spacecraft moving along the x direction, with a force F from the engine. The force and therefore also the acceleration are both along the $+x$ direction. We have also noted that the spacecraft is initially at rest (hence, $v_0 = 0$) and have chosen the origin for the x axis to be the spacecraft's location when $t = 0$, so $x_0 = 0$.

IDENTIFY THE RELATIONSHIPS

Using Newton's second law along with the given values of the mass and the component of the force along x gives

$$a = \frac{\sum F}{m} = \frac{2000 \text{ N}}{200 \text{ kg}}$$

The unit of force is newtons (N), and we saw in Chapter 2 that $1\,\text{N} = 1\,\text{kg} \cdot \text{m/s}^2$. We thus have

$$a = 10\,\frac{\text{kg} \cdot \text{m/s}^2}{\text{kg}}$$

Canceling the unit of kilograms that appears in both the numerator and denominator we get

$$a = 10 \text{ m/s}^2$$

We can now use our relations for motion with a constant acceleration from Table 3.1. For part (a) of the problem, we are given t and asked to find the velocity. Equation 3.1 involves v, t, and other variables whose values we already know, namely a and v_0, so we can use it to solve for v. Likewise, for parts (b) and (c) of this problem, we use the relations in Table 3.1 that contain the quantity we want to find, along with the quantities whose values we know.

SOLVE

(a) We compute the velocity at $t = 40$ s using Equation 3.1:

$$v = v_0 + at = 0 + (10 \text{ m/s}^2)(40 \text{ s}) = \boxed{400 \text{ m/s}}$$

(b) We are given the value of t (= 40 s) and asked to find x. We use Equation 3.3 because it contains x, t, and other quantities we know, namely a and v_0. So,

$$x = x_0 + v_0 t + \tfrac{1}{2}at^2 = 0 + 0 + \tfrac{1}{2}(10 \text{ m/s}^2)(40 \text{ s})^2 = \boxed{8000 \text{ m}}$$

(c) To deal with this part of the problem, we note that we are given the value of the position and asked to find the velocity; there is no mention of time. So, we can use Equation 3.4 in Table 3.1 to get

$$v^2 = v_0^2 + 2a(x - x_0) = 0 + 2(10 \text{ m/s}^2)(1000 \text{ m}) = 2.0 \times 10^4 \text{ m}^2/\text{s}^2$$

$$v = \boxed{140 \text{ m/s}}$$

Problem solving with the relations for motion with constant acceleration involves one or more of Equations 3.1, 3.3, and 3.4. Choosing the appropriate equation depends on what quantities are known and what you wish to calculate. It is usually best to choose the equation that contains only one unknown quantity.

3.2 | NORMAL FORCES AND WEIGHT

In this section, we consider some examples in which the force of gravity plays an important role. Let's first think about the case sketched in Figure 3.6A, which shows a person standing on the floor of a room. The person is standing still, so her velocity and acceleration are both zero. Since the acceleration is zero and, according to Newton's second law, $\sum \vec{F} = m\vec{a}$, you should strongly suspect that the total force on the person is zero. This statement is true, but it is not the whole story.

In this case, there are actually two opposing forces acting on the person. One is the gravitational force exerted by the Earth; this force is called the **weight** of the person and is denoted in Figure 3.6 by the vector $\vec{F}_{grav}$. When an object of mass m is located near the surface of the Earth, the gravitational force on the object is directed downward (toward the Earth) and has a magnitude

$$|\vec{F}_{grav}| = mg \qquad (3.5)$$

Weight is the force of gravity exerted by the Earth on an object.

We consider g and the expression for weight in Equation 3.5 in detail in Chapter 5. There we'll see how this result for the force of gravity near the Earth's surface is a consequence of Newton's law of universal gravitation. For now, we note that (1) the value of g is approximately the same for all locations near the surface of the Earth, with $g \approx 9.8$ m/s^2; (2) because weight is due to the gravitational attraction of the Earth (or whatever planet, etc., on which the object is located), the weight of an object will be different if it is taken to another planet or to the Moon; (3) the value of g is independent of the mass of the object, so the weight of an object (Eq. 3.5) is proportional to its mass; and (4) for reasons we'll see shortly, g is commonly referred to as the **acceleration due to gravity**.

The weight $\vec{F}_{grav}$ of an object is a force and can thus be measured in units of newtons. It is often convenient to use the coordinate system shown in Figure 3.6, where we follow common convention and label the vertical axis as y (reserving the x axis for the horizontal direction). For this choice of coordinates, the gravitational force lies along the vertical direction with a component along y given by

$$F_{grav} = -mg \qquad (3.6)$$

The negative sign here indicates that the force is in the "negative" y direction, or *down*, toward the center of the Earth.

Besides the gravitational force, the other force acting on the person in Figure 3.6 is exerted by the floor on the bottoms of her feet. This force is called a **normal force** because it acts in a direction perpendicular ("normal") to the plane of contact with the floor. Normal forces are very common in nature, occurring whenever the surfaces of two objects come into contact. In this example, the normal force is directed upward, as indicated by the upward arrow labeled $\vec{N}$. The person in Figure 3.6 is at rest with an acceleration $a = 0$. Using Newton's second law for components of the force and acceleration along y, we can then write

$$\sum F = -mg + N = ma = 0$$

Hence, in this case $N = mg$. In words, the normal force is equal in magnitude and opposite in direction to the person's weight.

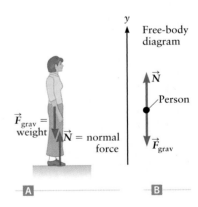

Figure 3.6 A Person standing on a level floor. B Free-body diagram for the person, showing all the forces acting on her: the normal force $\vec{N}$ and the force of gravity $\vec{F}_{grav}$. In a free-body diagram, the object is often denoted by a "dot."

There are two forces acting on the person in Figure 3.6A: the force of gravity $\vec{F}_{grav}$ and $\vec{N}$, the normal force from the floor. These forces are equal in magnitude and opposite in direction. Are they an action–reaction pair? Explain why or why not. If not, identify the reaction force to $\vec{N}$.

Free-Body Diagrams

We have already mentioned the importance of drawing a picture as part of the problem-solving process. This picture should include coordinate axes, along with other information such as the values of forces and the initial velocity. In preparation for an analysis using Newton's second law, you should also construct a simplified diagram showing all the forces acting on each object involved in the problem. In Figure 3.6, these forces are the normal force and the force of gravity acting on the person. The diagram that shows them, Figure 3.6B, is called a *free-body diagram*. The recommended procedures for constructing a free-body diagram are as follows.

Always draw a free-body diagram.

PROBLEM SOLVING | Constructing Free-Body Diagrams and Applying Newton's Laws

1. **RECOGNIZE** the objects of interest. List all the forces acting on each one.

2. **SKETCH THE PROBLEM.**
   - Start with a drawing that shows all the objects of interest in the problem along with all the forces acting on each one, as in Figure 3.6A.
   - Make a separate sketch showing the forces acting on each object, which is the free-body diagram for that object. For clarity, you can represent each object by a simple dot.
   - Forces in a free-body diagram should be represented by arrows, with the direction of the arrow showing the direction of the force. In each free-body diagram, show only the forces acting on that particular object.

3. **IDENTIFY THE RELATIONSHIPS.** A force in a free-body diagram may be known (i.e., its direction and magnitude may be given), or it may be unknown. Represent unknown quantities by a symbol, such as $\vec{N}$ in Figure 3.6. Typically, equations derived from Newton's second law are used to solve for these unknowns.

4. **SOLVE.** The information contained in a free-body diagram can usually be used directly in writing Newton's second law for an object. A few algebraic steps will then lead to values for the unknown quantities.

5. Always *consider what your answer means* and check that it makes sense.

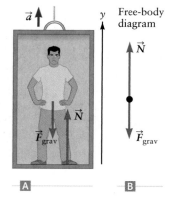

Figure 3.7 �**A** When a person stands in an elevator, there are only two forces acting on him: $\vec{N}$ and $\vec{F}_{grav}$. � **B** Free-body diagram for the person.

Acceleration and Apparent Weight

We showed that the normal force acting on the person in Figure 3.6 is equal in magnitude to her weight (mg). This result is often found for objects at rest on a level surface. However, in many situations the normal force on an object is not equal to its weight. Such a situation is sketched in Figure 3.7A, which shows a person standing in a moving elevator. There are again two forces acting on the person (Fig. 3.7B), the force of gravity (the person's weight) and the normal force exerted by the floor of the elevator on the bottoms of his feet. However, this situation is not quite the same as in our previous example because the elevator in Figure 3.7A has an acceleration $\vec{a}$, which is also the acceleration of the person inside.

Applying Newton's second law to the motion of the person gives

$$m\vec{a} = \sum \vec{F} = \vec{N} + \vec{F}_{grav}$$

where m is the person's mass. Because all these vectors are along y, it is customary to write this expression in terms of the y components of the acceleration and force vectors:

$$ma = +N - mg \tag{3.7}$$

The signs here indicate the directions of the normal force and the force of gravity. As usual, these directions are defined by our choice of the coordinate axis (y) in Figure 3.7, which shows that the "positive" direction is upward. This choice for the positive direction also applies to the acceleration. Solving Equation 3.7 for the normal force, we find

$$N = mg + ma$$

Hence, if the acceleration is positive (directed upward in Fig. 3.7), the normal force is greater than the weight mg, while if the acceleration is negative (downward), the normal force is less than mg. You have probably experienced this result yourself. When you are in an elevator that is just starting to move to a higher floor, the acceleration is positive (directed upward) and you "feel" heavier. This feeling comes from your feet because you perceive that the force exerted by the floor of the elevator on the bottoms of your feet is greater than your true weight mg. Likewise, if you are in an elevator that is just beginning to move to a lower floor, your acceleration is negative (downward), so N is smaller than mg and you feel lighter.

The normal force acting on the bottoms of a person's feet as in Figure 3.7 is called the *apparent weight*. Loosely speaking, the only way that you are sensitive to the value of your weight is through the normal force that supports you. Consider (hypothetically) what would happen if the elevator in Figure 3.7 were to malfunction and fall freely down the elevator shaft. This kind of arrangement is similar to that used by NASA to train astronauts; NASA uses a special airplane (with a padded interior; Fig. 3.8) that flies in a long dive. During this dive, the plane and the astronauts fall together, so the normal force on the astronauts is zero. Their apparent weight is then zero, and the astronauts feel "weightless." During such a dive, their apparent weight is zero even though their true weight is not zero (and the force of gravity is also *not* zero).

All Forces Come from Interactions

Aristotle believed that a force must come from the direct action of one object on another. (See Chapter 2.) Newton's third law implies a similar worldview: every force on an object arises from an interaction with a second object. This notion that forces come from interactions leads to the concept of an action–reaction pair of forces. Newton also understood that two objects could exert forces on each other even when they are not in direct contact. The gravitational force is a good example. The Earth exerts a gravitational force on objects such as the person in the elevator in Figure 3.7 or the young people in a free-fall dive in Figure 3.8, even though these people are not in contact with the planet. The exertion of forces between objects that are not in direct contact is known as *action-at-a-distance*, an extremely important idea in physics. You probably know that the gravitational force from the Sun is responsible for the Earth's orbital motion. The space between the Sun and Earth is largely a vacuum, and it is certainly not obvious how this force can make itself felt through such an empty region. Action-at-a-distance is a fundamental aspect of nature that bothered many scientists after it was proposed by Newton, and it took the genius of Einstein to explain how this works for the gravitational force. The notion of action-at-a-distance also applies to other forces, including electric and magnetic forces, as we'll discuss later.

Another important idea is contained in the notion of "direct contact" between the floor and the person's feet in Figure 3.6, where it certainly seems like the floor and the person are in contact. But what does this mean at a microscopic scale? We know that the floor and the person's feet are composed of atoms. Can two atoms actually "touch"? Contact forces are a result of electric forces between atoms that are in very close proximity. This atom–atom interaction cannot be described by Newton's laws, but must be treated using quantum mechanics (Chapters 28 and 29). The normal forces in Figures 3.6 and 3.7 are caused by these electric interaction forces.

Figure 3.8 These young people are experiencing apparent "weightlessness" in a special NASA plane. Notice the padding.

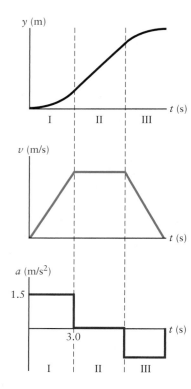

Figure 3.9 Example 3.2. Position, velocity, and acceleration versus time for an elevator as it travels from the first to the fifth floor of a building. In period I, the elevator is accelerating from rest. In period II, the acceleration is zero and the elevator moves with a constant velocity. During period III, the elevator slows to a stop.

EXAMPLE 3.2 Traveling Up

An elevator begins from rest at the first floor of a building. It starts a journey to the fifth floor by moving with an acceleration of 1.5 m/s² for 3.0 s; thereafter, it moves with a constant velocity until it gets near to its stopping point. Find the apparent weight of a person of mass m who is traveling in the elevator at different times during the motion.

RECOGNIZE THE PRINCIPLE

The person's apparent weight will depend on the acceleration. During its journey to the fifth floor, the elevator must go through three stages of motion: (I) a period of positive acceleration with ($a = 1.5$ m/s²), (II) a period of constant velocity (with $a = 0$), and (III) a period of negative acceleration while it slows to a stop.

SKETCH THE PROBLEM

Figure 3.7 shows a person in an elevator and defines our coordinate system; the motion is again along y (the vertical direction). Figure 3.7B is our free-body diagram.

IDENTIFY THE RELATIONSHIPS

Figure 3.9 contains qualitative plots of the position, velocity, and acceleration of the elevator versus time during the three phases of the motion. In each phase, we apply the relations for motion with constant acceleration from Table 3.1.

The apparent weight of a person in the elevator depends on the acceleration, so it is different in periods I, II, and III in Figure 3.9. Using Newton's second law and the components of the forces along y, as sketched in the free-body diagram in Figure 3.7B gives

$$ma = \sum F = +N + F_{grav} = +N - mg \qquad (1)$$

The negative sign in front of the last term indicates (again) that the force of gravity is directed downward.

SOLVE

Rearranging Equation (1) gives the person's apparent weight as $N = ma + mg$. The person's true weight is mg, while his apparent weight during period I is

$$N = m(a + g) = m(1.5 \text{ m/s}^2 + 9.8 \text{ m/s}^2) = m(11.3 \text{ m/s}^2)$$

Comparing N to the true weight, we find

$$\frac{N}{mg} = \frac{m(11.3 \text{ m/s}^2)}{mg} = \frac{11.3 \text{ m/s}^2}{g} = \frac{11.3 \text{ m/s}^2}{9.8 \text{ m/s}^2} \approx 1.2$$

or

$$N = \boxed{1.2(mg)}$$

Hence, during the acceleration period (I) the person's apparent weight N is approximately 20% larger than his true weight.

During period II, the elevator's acceleration is zero, so $N = mg$ and the person's apparent weight is equal to his true weight. During the negative acceleration phase (III), a is negative. The normal force during this period is again $N = ma + mg$, and since a is now negative, the normal force and thus also the person's apparent weight are smaller than mg. The precise value of N depends on the value of a while the elevator is coming to a stop (during period III in Fig. 3.9).

What does it mean?
The value of the apparent weight depends on an object's acceleration. The apparent weight can be greater than, equal to, or less than the true weight.

What Is "Mass"?

The force of gravity has many interesting features that we'll explore in more detail in Chapter 5. In this section, we have only considered the force of gravity on terrestrial objects, and we have seen that this force has a very simple form

$$F_{grav} = -mg \tag{3.8}$$

where (as usual) we take the "positive" direction to be upward. However, this simplicity tends to hide a very mysterious fact. According to Equation 3.8, the force of gravity depends on the mass of the object. You will recall that we originally encountered mass in connection with the concept of inertia and Newton's laws of motion. According to Newton's second law, the mass of an object determines how it will move in response to forces. Here is the puzzle: if mass is the property of an object that determines how it moves (as calculated using Newton's second law), why does the *same* quantity determine the magnitude of the force of gravity? Even Newton was not able to answer this profound question, which we'll discuss more in Chapters 5 and 27. The quantity called "mass" that enters into Newton's second law is called the **inertial mass** of an object, while the "mass" that enters into the gravitational force is referred to as the object's **gravitational mass**. The connection between Newton's laws of motion and gravitation thus amounts to the statement that the inertial mass is precisely equal to the gravitational mass. It is not immediately obvious that this should be true, but it does suggest a fundamental connection between gravitation and Newton's laws of motion. We'll see why when we study relativity in Chapter 27.

3.3 | ADDING FRICTION TO THE MIX

One of Aristotle's central beliefs was that terrestrial objects all tend toward a natural state of rest. Indeed, that is what we commonly observe in the world around us because objects generally do spend most of their time at rest relative to us as observers. However, Newton's explanation of this behavior was quite different from the one provided by Aristotle. According to Newton, when an object comes to a state of rest, it does so because a force acts on it, and this force is often due to a phenomenon called friction.

Kinetic Friction

Figure 3.10 shows a hockey puck sliding on an icy surface. While this surface is very slippery, there is a small amount of friction, and we now consider how the force of friction eventually causes the puck to stop.

Figure 3.10B shows a free-body diagram, containing all the forces acting on our hockey puck.[2] The puck's motion is along the horizontal direction x, and we assume the puck is sliding toward the right, in the $+x$ direction. There is only one force along the horizontal, the force of friction. Because the puck's velocity is along $+x$ (to the right in Fig. 3.10), the force of friction will oppose this motion and hence be directed along $-x$ (to the left). There are also two forces in the vertical direction, y: the weight of the puck (the force of gravity) and the normal force exerted by the icy surface acting on the puck. This example is our first one involving multiple forces acting in more than one direction, and we have not yet discussed how this affects the use of Newton's laws. For now, we appeal to your intuition and argue that the two forces directed along y, the normal force and the weight of the puck, will cancel each other. This is very similar to the cancellation we found in connection with Figure 3.6. We thus have $N - mg = 0$, or $N = mg$ for the hockey puck.

A

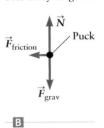

Free-body diagram

B

Figure 3.10 Ⓐ Three forces act on this hockey puck: the force of gravity (its weight), the normal force from the ice, and the force of friction. Ⓑ Free-body diagram for the puck.

---

[2]Actually, we have omitted the force from air drag (sometimes called air resistance), but that will usually be very small for a hockey puck.

If the motion of our puck is along x, why did we mention forces along y, especially when these forces cancel each other? The reason is that the normal force is intimately connected with the force of friction. Experiments show that for a sliding object, the magnitudes of these two forces are related by

Force of kinetic friction

$$F_{\text{friction}} = \mu_K N \qquad (3.9)$$

The quantity μ_K is called the *coefficient of kinetic friction*. The frictional force in Equation 3.9 describes cases in which two surfaces are in motion (slipping) with respect to each other. As a "coefficient," μ_K is a pure number without dimensions or units, and its value depends on the surface properties. For a hockey puck on an icy surface, μ_K is relatively small and the frictional force is very small; the value of μ_K depends on the smoothness of both the ice and the hockey puck, and might typically be $\mu_K = 0.05$. The coefficient of kinetic friction for two rough surfaces would be much larger, with $\mu_K = 0.3$ being a common value. Table 3.2 lists coefficients of kinetic and static friction (discussed below) for some common materials. These measured values of μ_K show that the frictional force depends on the nature of the surfaces that are in contact.

Note that the magnitude of the frictional force in Equation 3.9 depends on the normal force. If the normal force is increased—say, by piling a large additional mass on top of our hockey puck—the frictional force is larger than if the extra mass is not present. We must emphasize that Equation 3.9 is not a "law" of nature; it is merely an approximate relation that has been found to work well in a wide variety of cases. To derive a fundamental law or theory of friction would require us to consider in detail the atomic interactions that occur when two surfaces are in contact. This problem is very complicated and is currently a topic of much research.

Analyzing Motion in the Presence of Friction

We are now ready to analyze the motion of the hockey puck in Figure 3.10. Suppose we know the value of μ_K and that the puck is given some initial velocity v_0. How long will the puck slide before coming to rest? Once we have drawn the free-body diagram in Figure 3.10B, we can apply Newton's second law to calculate the puck's acceleration. In this problem, the motion in which we are interested is along x, so we need to focus on the forces along the x direction. Writing Newton's second law for motion along x, we get

$$ma = \sum F_x = F_{\text{friction}} = -\mu_K N$$

All these terms involve components of acceleration and force along the horizontal (x) direction. The minus sign in the last term indicates that the frictional force is directed along $-x$ because friction acts to oppose the motion of the puck. The velocity in Figure 3.10 is to the right, so the frictional force acts to the left. We already know that in this case the normal force is equal to mg, and we have

$$ma = -\mu_K N = -\mu_K mg$$
$$a = -\mu_K g$$

Table 3.2 Typical Values for the Coefficients of Kinetic Friction and Static Friction for Some Common Materials

SURFACES	μ_K	μ_S	SURFACES	μ_K	μ_S
rubber on dry concrete	0.8	0.9	ski on snow	0.05	0.10
rubber on wet concrete	0.5	0.6	glass on glass	0.4	0.9
rubber on glass	0.7	0.9	Teflon on Teflon	0.04	0.04
hockey puck on ice	0.05	0.10	metal on metal	0.5	0.6
wood on wood	0.3	0.5	bone on bone (with joint fluid)	0.015	0.02

Notes: The values for metal on metal are only approximate. The coefficients in all cases depend on the smoothness of the surfaces.

Since μ_K and g are both constants, the puck's acceleration is constant, so we can apply our general relations for motion with constant acceleration in Table 3.1. However, we must decide which one of these relations to use. In this problem, we are asked to find when the puck stops, and we'll know that the puck has come to a stop when $v = 0$. We therefore need a relation that involves v (the component of the velocity along x) and t. Examining Table 3.1, we see that Equation 3.1 involves these quantities. Using this relation, we have

$$v = v_0 + at = v_0 - \mu_K g t = 0$$

and solving for t gives

$$t = \frac{v_0}{\mu_K g} \qquad (3.10)$$

It is always a good idea to work out the units of an answer, even if no numbers are given. The coefficient of friction is dimensionless, so the units of the term on the right in Equation 3.10 are $(m/s)/(m/s^2) = s$, as expected because the result is a time. Notice also that we did not need to know the mass of the puck; the value of m canceled when calculating the acceleration.

Another question that we might have asked is, "How far will the puck slide before coming to a stop?" We could find the answer in two different ways: (1) by using the value of t found in Equation 3.10 and inserting it into Equation 3.3 in Table 3.1 or (2) by using Equation 3.4. Let's use the second approach. When the puck stops, its velocity is $v = 0$, and inserting this into Equation 3.4 gives

$$v^2 = v_0^2 + 2a(x - x_0) = 0$$

We can then solve for the distance traveled:

$$2a(x - x_0) = -v_0^2$$

$$(x - x_0) = -\frac{v_0^2}{2a} = \frac{v_0^2}{2\mu_K g}$$

Static Friction

The case of the sliding hockey puck in Figure 3.10 involves surfaces that are in motion (slipping) relative to each other, and we mentioned that the frictional force in such cases is referred to as kinetic friction. However, in many situations the relevant surfaces are not slipping. Such cases involve what is known as *static friction*. An example involving static friction is given in Figure 3.11, which shows a refrigerator at rest on a level floor. This situation is similar to the hockey puck in Figure 3.10, but with an additional force exerted by a person pushing on the refrigerator. We know intuitively that when the force exerted by the person is small, the refrigerator

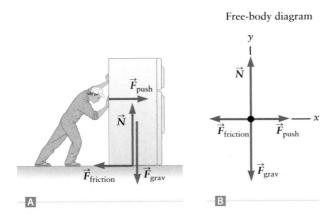

Free-body diagram

Figure 3.11 ◻A Four forces act on this refrigerator. The force of static friction acts along the horizontal and opposes the movement of the refrigerator relative to the floor. ◻B Free-body diagram for the refrigerator. Here we assume the force F_{push} is small enough that the refrigerator does not tip.

will not move. Since the acceleration is then zero, applying Newton's second law along the horizontal direction leads to

$$ma = \sum F_x = F_{push} + F_{friction} = 0$$

In this case, the force exerted by the person and the force of friction cancel. The force of static friction is sufficiently strong that no relative motion (no slipping) occurs. In terms of magnitudes, we have

$$|F_{push}| = |F_{friction}|$$

However, the value of F_{push} can vary and the refrigerator will still remain at rest. That is, we can push a lot or a little or not at all, without the refrigerator starting into motion. In all these cases, the frictional force exactly cancels F_{push}. The only way this can happen is if the magnitude of the frictional force varies depending on the value of F_{push}, as described by

Force of static friction

$$|F_{friction}| \leq \mu_S N \qquad (3.11)$$

where μ_S is the ***coefficient of static friction***.

The term *static* again means that the two surfaces (the floor and the bottom of the refrigerator) are not moving relative to each other. According to Equation 3.11, the magnitude of the force of static friction can take any value up to a maximum of $\mu_S N$. If F_{push} in Figure 3.11 is small, the force of static friction is small and will precisely cancel F_{push} so that the total horizontal force is zero. If F_{push} is increased, the force of static friction increases and again precisely cancels F_{push}. However, the magnitude of the static friction force has an upper limit of $\mu_S N$. The magnitude of the force of static friction cannot be greater than this upper limit.

Comparing Kinetic Friction and Static Friction

To describe friction fully, we must consider both the kinetic and static cases, with two different coefficients of friction. In the kinetic case, the two surfaces are slipping relative to each other and the frictional force has the simple value $F_{friction} = \mu_K N$ (Eq. 3.9). The static case is a bit more complicated; here, the frictional force varies so as to cancel the other force(s) on the object, up to a maximum value of $\mu_S N$ (Eq. 3.11). For a given combination of surfaces, it usually happens that $\mu_S > \mu_K$ (see Table 3.2), so the maximum force of static friction is greater than the force of kinetic friction. This confirms an experiment you have probably done yourself. The force required to start something into motion—that is, the minimum value of F_{push} required to just barely move the refrigerator in Figure 3.11—is larger than the force required to keep it moving subsequently at a constant velocity.

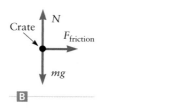

Free-body diagram

Figure 3.12 Example 3.3. Ⓐ The force of friction on the crate is directed to the right, in the direction of motion of the truck and the crate. The frictional force acts to prevent the crate from slipping relative to the truck. Ⓑ Free-body diagram for the crate.

EXAMPLE 3.3 | Sliding Away

A wooden crate sits on the floor of a flatbed truck that is initially at rest, and the coefficient of static friction between the crate and the floor of the truck is $\mu_S = 0.25$. The driver of the truck wants to accelerate to a velocity of 20 m/s within 4.0 s. If she does so, will the crate slide off the back of the truck?

RECOGNIZE THE PRINCIPLE

Figure 3.12A shows that the only horizontal force on the crate is the force of static friction. The static frictional force in Figure 3.12 is not opposing any other force. Instead, static friction is opposing the tendency of the crate to slide along the floor of the truck. That is, static friction opposes any *relative motion* of the two surfaces. If there is no relative motion and the crate does not slip, it will be moving along with the truck. Hence, in this example the force of friction is what causes the crate to accelerate! Since there is an upper limit to the magnitude of the static frictional force (Eq. 3.11), there is an upper limit to this horizontal force on the crate. In other words, there is an upper limit to the acceleration that static friction can provide to the crate. If this acceleration limit is smaller than the truck's actual acceleration (as determined by the driver), the crate will not be able to keep up with the truck and the crate will slide off the back.

SKETCH THE PROBLEM

Figure 3.12B shows a free-body diagram for the crate. The vertical forces on the crate are its weight (directed downward) and the normal force from the floor of the truck (directed upward). As the truck accelerates, there is also a horizontal force on the crate caused by static friction from the floor of the truck.

IDENTIFY THE RELATIONSHIPS

Using Newton's second law, the force of static friction leads to a maximum acceleration of the crate in the x direction of

$$a_{\text{crate, max}} = \frac{F_{\text{friction, max}}}{m} = \frac{\mu_s N}{m} \tag{1}$$

Calculating the normal force between the crate and floor of the truck is done in the same way as with the hockey puck in Figure 3.10. Hence, the magnitude of the normal force acting on the crate is $N = mg$, where m is the mass of the crate. Substituting into Equation (1) leads to

$$a_{\text{crate, max}} = \frac{\mu_s N}{m} = \frac{\mu_s mg}{m} = \mu_s g \tag{2}$$

To calculate the acceleration of the truck as determined by the driver, we note that it is undergoing motion with constant acceleration and choose Equation 3.1 from Table 3.1 because this relation involves velocity, acceleration, and time. The truck starts from rest, so $v_0 = 0$. Inserting this into Equation 3.1 gives

$$v = v_0 + a_{\text{truck}}t = a_{\text{truck}}t$$

$$a_{\text{truck}} = \frac{v}{t} \tag{3}$$

SOLVE

We now need to calculate the value of the truck's acceleration in Equation (3) and compare it to the maximum possible acceleration of the crate in Equation (2). Solving for a_{truck}, we find

$$a_{\text{truck}} = \frac{v}{t} = \frac{20 \text{ m/s}}{4.0 \text{ s}} = 5.0 \text{ m/s}^2$$

Evaluating Equation (2) for the maximum possible acceleration of the crate gives

$$a_{\text{crate, max}} = \mu_s g = (0.25)(9.8 \text{ m/s}^2) = 2.5 \text{ m/s}^2$$

The acceleration of the truck is thus greater than the maximum possible acceleration of the crate, so the crate will *not* be able to keep up with the truck and it will slide off the back.

What does it mean?

Normally, friction either prevents an object from moving or brings it to a stop. However, in this case friction is responsible for the motion of the crate because friction always opposes the relative motion of the surfaces in contact.

CONCEPT CHECK 3.2 | Trucks and Crates

The driver of the truck in Example 3.3 is in a hurry and wants to drive at the highest speed possible to her destination without losing the crate. Is there a maximum speed at which she can drive without having the crate slide off the truck?

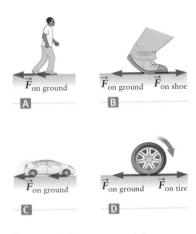

Figure 3.13 Frictional forces are essential for walking and rolling.

The Role of Friction in Walking and Rolling

It is interesting that the force of static friction on the crate in Figure 3.12 actually causes the crate's motion. Let's now consider the role of friction in two other cases. Figure 3.13A shows a person walking on level ground. As the person "pushes" off during each step, the bottoms of his shoes exert a force $\vec{F}_{\text{on ground}}$ on the ground. If the shoes do not slip, this force is due to static friction since the shoes do not move relative to the ground. According to Newton's third law, there is a reaction force $\vec{F}_{\text{on shoe}}$ on the person's shoes (Fig. 3.13B), and this force "propels" a person as he walks or runs. If the surface were extremely slippery and there were no frictional force, then $\vec{F}_{\text{on ground}}$ and $\vec{F}_{\text{on shoe}}$ would both be zero and the person would slip when he attempted to move. Hence, the force of friction makes walking and running possible.

A similar set of forces is found when a wheel rolls along level ground. If a car's tire does not slip, there is a force of friction between the tire and the ground leading to a force $\vec{F}_{\text{on ground}}$ on the ground (Fig. 3.13C). There is a reaction force on the tire $\vec{F}_{\text{on tire}}$ (Fig. 3.13D), and it is this force that propels the car forward. Friction thus also plays a key role in rolling motion.

3.4 | FREE FALL

In Chapter 2, we discussed different ways to characterize the motion of an object. We saw that Aristotle was primarily concerned with an object's velocity, whereas Newton recognized the fundamental connection between force and acceleration that is captured by his second law. The relationship and differences between velocity and acceleration are absolutely crucial for understanding motion, as we now illustrate for a type of motion known as *free fall*. An example of free fall is the motion of a ball that is thrown upward, reaches some maximum height, and then falls back down to Earth. After the ball leaves the thrower's hand, it follows the familiar trajectory sketched in Figure 3.14. Let's analyze this trajectory and consider how the acceleration, velocity, and position of the ball all vary with time.

We begin with the acceleration. Before the ball leaves the thrower's hand, there are forces from gravity and from her hand acting on the ball, and these forces lead to an upward acceleration as she gives the ball an upward velocity. In Figure 3.14, we take $t = 0$ to be the instant just *after* the ball leaves her hand. From that instant until the ball hits the ground, the only forces acting on the ball are the force of gravity and a force due to air drag. In many cases, the force from air drag is very small, so we'll ignore it here.

To analyze the motion during free fall, we choose a coordinate system that measures the position as the height y of the ball above the ground, with the origin chosen to be at ground level. Using Newton's second law, we have

$$a = \frac{F_{\text{grav}}}{m} = \frac{-mg}{m} = -g \tag{3.12}$$

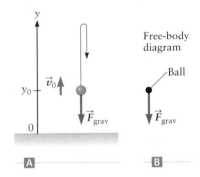

Figure 3.14 A ball is thrown upward beginning at an initial height y_0 with initial velocity $\vec{v}_0$. After leaving the thrower's hand, the ball is in free fall and the only force acting on the ball is the force of gravity. Even though this force is directed downward, the ball still moves up initially due to the initial velocity imparted by the thrower.

where the minus sign comes from the fact that the force of gravity is directed downward, in the $-y$ direction. We can again use our relations for motion with constant acceleration (Eqs. 3.1 and 3.3) and find the velocity and position of the ball as functions of time:

$$v = v_0 + at = v_0 - gt \tag{3.13}$$

$$y = y_0 + v_0 t + \tfrac{1}{2}at^2 = y_0 + v_0 t - \tfrac{1}{2}gt^2 \tag{3.14}$$

Sketches of how v and y vary with time are shown in Figure 3.15. The velocity varies linearly with time with a slope of $-g$ and an intercept equal to v_0, the velocity of the ball at $t = 0$ (just after it leaves the thrower's hand). The position varies quadratically with t, with a positive intercept y_0 that is the height of the ball when it leaves the thrower's hand.

The sketches of velocity and position in Figure 3.15 are simply plots of Equations 3.13 and 3.14, with some important features worth emphasizing. Initially, the ball's velocity is *positive*, whereas its acceleration is *negative*. That is, the velocity vector is directed *upward*, and the acceleration vector is directed *downward*. In fact, the ball's acceleration is downward during the *entire* time (Fig. 3.15 and Eq. 3.12). According to Newton, the force on the ball and therefore its acceleration are both negative; they are both directed along $-y$ throughout the motion. Since the acceleration is the slope of the v–t curve, the v–t relation must have a negative and constant slope throughout as shown in Figure 3.15A. While this slope is always negative, the velocity itself can have a value that is positive, negative, or zero. The velocity is positive while the ball is on the upward (initial) part of its trajectory and negative while it is on the downward (final) part. The velocity is zero, shown by the v–t curve crossing through zero in Figure 3.15A, at the instant the ball reaches its highest point along y (the highest point on its trajectory).

To understand the behavior of the displacement, recall again that v is the slope of the y–t curve, so while the ball is on the way up and v is positive, the y–t curve has a positive slope. The ball reaches its highest point when the velocity is zero, and the y–t slope at that instant is zero. While the ball is on the way down, the velocity is negative, so the slope of the y–t curve is negative.

An extremely important result from this analysis is that the acceleration and velocity can be in *different* directions; that is, they can have different signs. Depending on the situation, they can have either the same signs (both positive or both negative) or one can be positive while the other is negative.

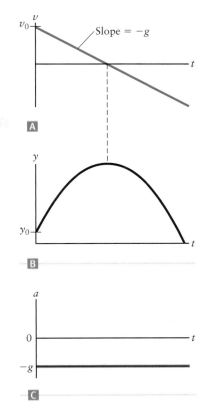

Figure 3.15 **A** For a ball in free fall, the acceleration is constant, so the velocity varies linearly with time. The slope of the v–t relation is $-g$. **B** Height of the ball y versus time. The ball reaches its maximum height when $v = 0$. **C** The acceleration of the ball is constant and nonzero ($a = -g$) during the entire time, even at the highest point on the trajectory.

CONCEPT CHECK 3.3 | The Direction of Velocity and Acceleration

For which of the following situations is the velocity parallel to the acceleration, and for which are $\vec{v}$ and $\vec{a}$ in opposite directions?
  (a) The hockey puck sliding to a stop in Figure 3.10.
  (b) The crate in the flat bed truck in Figure 3.12.
  (c) An apple falling from a tree.

EXAMPLE 3.4 | Maximum Height of a Projectile

Consider the motion of the ball in Figure 3.14. A typical person can throw a ball with an initial speed of at least 20 m/s (about 40 mi/h). Calculate **(a)** the time it takes for such a ball to reach its maximum height, **(b)** the maximum height reached by the ball, **(c)** the time the ball spends in the air, and **(d)** the ball's velocity just prior to hitting the ground. For simplicity, assume the initial position of the ball is at ground level.

RECOGNIZE THE PRINCIPLE

Since the acceleration is constant for free fall, we can use the relations from Table 3.1 or, equivalently, Equations 3.13 and 3.14. For each part of the problem, we select the relation that contains the quantity we want to find along with quantities that we know. For part (a), we know that the velocity is zero when the ball reaches the maximum height, so we select Equation 3.13. We attack the other parts of the problem in a similar way.

SKETCH THE PROBLEM

Figure 3.14 contains all the information we need, including a free-body diagram. This free-body diagram applies at *all* points on the trajectory. While the ball is traveling through the air, the only force exerted on it is the force of gravity.[3]

---

[3]We again assume the force exerted by air drag is very small and can be neglected.

IDENTIFY THE RELATIONSHIPS

The ball starts at ground level, so $y_0 = 0$. The initial velocity of the ball is $v_0 = 20$ m/s.

(a) At the top of the trajectory, the velocity is $v = 0$. From Equation 3.13, we get

$$v_{top} = v_0 - gt_{top} = 0$$

$$t_{top} = \frac{v_0}{g}$$

(b) Once we know the time needed to get to the top of the trajectory, we can find the ball's position at the top of its path using Equation 3.14:

$$y_{max} = y_0 + v_0 t_{top} - \tfrac{1}{2} g t_{top}^2$$

SOLVE

(a) Inserting the given value of v_0 into our result for t_{top} yields

$$t_{top} = \frac{20 \text{ m/s}}{9.8 \text{ m/s}^2} = \boxed{2.0 \text{ s}}$$

(b) Substituting the value we just found for t_{top} into the equation for y_{max} gives

$$y_{max} = 0 + (20 \text{ m/s})(2.0 \text{ s}) - \tfrac{1}{2}(9.8 \text{ m/s}^2)(2.0 \text{ s})^2 = \boxed{20 \text{ m}}$$

IDENTIFY THE RELATIONSHIPS

(c) To calculate the time needed for the ball to reach the ground, we note that ground level corresponds to $y = 0$. Again using Equation 3.14, we have

$$y = y_0 + v_0 t - \tfrac{1}{2} g t^2 = 0 \qquad (1)$$

We set $t = t_{ground}$ and take $y_0 = 0$:

$$0 + v_0 t_{ground} - \tfrac{1}{2} g t_{ground}^2 = 0$$

$$t_{ground} = \frac{2v_0}{g}$$

(d) To find the velocity of the ball just before it lands, we use Equation 3.13 because it contains v and t. We want the velocity at time $t = t_{ground}$.

SOLVE

(c) Inserting the given value for v_0 into our result for t_{ground} yields

$$t_{ground} = \frac{2(20 \text{ m/s})}{9.8 \text{ m/s}^2} = \boxed{4.1 \text{ s}}$$

(d) Substituting the value we just found for t_{ground} into Equation 3.13 for the velocity gives

$$v = v_0 - gt_{ground} = 20 \text{ m/s} - (9.8 \text{ m/s}^2)(4.1 \text{ s}) = \boxed{-20 \text{ m/s}}$$

What does it mean?

In this example, we assumed the initial position of the ball was $y_0 = 0$. A nonzero value would be more realistic, but would complicate (slightly) the mathematics in part (c) since solving Equation (1) with a nonzero value of y_0 would require us to solve a quadratic equation. (See Problem 53 at the end of this chapter for a similar problem that requires use of the quadratic formula.)

Motion of a Dropped Ball

Example 3.4 gave us an interesting result. The ball's speed just before it hit the ground is equal to the ball's initial speed (although the ball's *velocity* is positive at the start and negative at the end). This result is no accident; it is true because the position y is a symmetric function with respect to the time at which the ball reaches its highest point (see Fig. 3.15B). The time spent on the way up is equal to the time spent falling back to the ground. We can see why by comparing the results for t_{top} and t_{ground} in Example 3.4, where we found

$$t_{\text{top}} = \frac{v_0}{g} \quad \text{and} \quad t_{\text{ground}} = \frac{2v_0}{g}$$

Hence, t_{ground} is precisely[4] twice t_{top}. This symmetry is an important property of motion with constant acceleration, although it is not found if the acceleration varies with time. For example, when air drag is important, the acceleration is not constant and the position–time relation is no longer symmetric with respect to t_{top}.

EXAMPLE 3.5 | Symmetry of Free Fall

Consider an apple that drops from a tall tree. To make this problem more interesting, let's assume the apple falls from a height that is the same as the maximum height reached in the previous example. Calculate (a) how long it takes the apple to reach the ground and (b) its velocity the instant before it hits the ground.

RECOGNIZE THE PRINCIPLE

Since the apple is in free fall, the only force on it is due to gravity and the acceleration is constant. We can thus (yet again) use the expressions for motion with a constant acceleration in Table 3.1.

SKETCH THE PROBLEM

Figure 3.16 shows the problem. The apple starts at $y_0 = 20$ m (the maximum height reached by the ball in Example 3.4), with an initial velocity of zero. It undergoes free fall until it reaches the ground.

IDENTIFY THE RELATIONSHIPS

(a) The apple reaches the ground when $y = 0$, so to calculate when the apple reaches this point we choose the relation that contains both position and time (Eq. 3.14, or its equivalent Eq. 3.3). Applying Equation 3.14, we have

$$y = y_0 + v_0 t - \tfrac{1}{2}gt^2 \qquad (1)$$

The initial velocity is $v_0 = 0$ because the apple starts from rest.

SOLVE

(a) Substituting $v_0 = 0$ and $y = 0$ into Equation (1) and solving for t gives

$$y = y_0 - \tfrac{1}{2}gt_{\text{ground}}^2 = 0$$

$$t_{\text{ground}} = \sqrt{\frac{2y_0}{g}} = \sqrt{\frac{2(20 \text{ m})}{9.8 \text{ m/s}^2}} = \boxed{2.0 \text{ s}}$$

IDENTIFY THE RELATIONSHIPS

(b) We now need to evaluate the velocity at the value of t found in part (a). For the apple's velocity the instant before it hits the ground, we have (Eq. 3.13)

$$v = v_0 - gt_{\text{ground}} = -gt_{\text{ground}}$$

where again $v_0 = 0$.

Figure 3.16 Example 3.5. The acceleration of an apple during free fall is the same as for the thrown ball in Figure 3.14, but the initial position ($y = y_0$) and initial velocity ($v = 0$) are different.

---

[4]The numerical value we found for t_{ground} in Example 3.4 was 4.1 s, whereas for t_{top} we got 2.0 s. However, those values were obtained after rounding to two significant figures as we worked through the example. We can see from the results before numerical evaluation that $t_{\text{ground}} = 2t_{\text{top}}$.

(b) Inserting the value of t from part (a) gives

$$v = -(9.8 \text{ m/s}^2)(2.0 \text{ s}) = \boxed{-20 \text{ m/s}}$$

What have we learned?
This result for the velocity is the same as we found for the thrown ball in Example 3.4, part (d). That is no accident: the motion *on the way down* is precisely the same in the two cases.

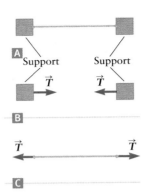

CONCEPT CHECK 3.4 | Free Fall

Consider an acorn as it falls from the top of a tall tree. Take the y axis to be in the vertical direction, with "up" being the $+y$ direction. Which of the following statements is true?

(a) The velocity is negative, the acceleration is negative, and the speed is increasing.
(b) The velocity is negative, the acceleration is positive, and the speed is decreasing.
(c) The velocity is negative and the speed is negative.

Figure 3.17 **A** When a rope is stretched between two supports, it **B** exerts a force of magnitude T on both supports, the one on the left and the one on the right. **C** According to Newton's third law, each support exerts a force of magnitude T on the rope.

3.5 | CABLES, STRINGS, AND PULLEYS: TRANSMITTING FORCES FROM HERE TO THERE

Strings (and also cables and ropes) exert a force on the objects they are connected to due to the strings' **tension**. To understand these tension forces, we begin by considering an elevator supported by a cable. If you were in the elevator design business, you might be given the job of choosing the correct cable to use for a particular elevator. To choose a cable that is strong enough for the job, you need to understand the forces acting on the cable. When a cable (or rope or string, etc.) is pulled tight, it exerts a force on whatever it is connected to. This force is illustrated Figure 3.17, which shows a rope stretched between two supports. The ends of the rope both exert a force of magnitude T on the supports to which they are connected (Fig. 3.17B). At the same time, Newton's third law (the action–reaction principle) tells us that the supports both exert a reaction force of magnitude T on the ends of the rope (Fig. 3.17C). In words, T is equal to the tension in the rope.

The reasoning in Figure 3.17 also applies to the elevator in Figure 3.18; the cable exerts an upward force T on the elevator compartment and a separate (downward) force of magnitude T on whatever it is connected to at the top of the elevator shaft.

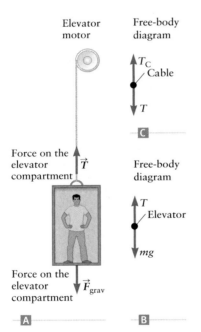

Figure 3.18 **A** When an elevator is lifted by a cable, there are two forces on the elevator compartment, the gravitational force and the tension from the cable. This tension force is part of an action–reaction pair; the other force in the pair acts downward on the cable. **B** Free-body diagram for the elevator. **C** Free-body diagram for the cable.

Tension Forces

Let's focus on the motion of the elevator compartment in Figure 3.18. From the free-body diagram in Figure 3.18B, we see that there are two forces acting on the compartment, the force of gravity and a force from the cable that has a magnitude T (recall that T is the tension in the cable). If the elevator has an acceleration a along the vertical direction and m is the mass of the compartment plus passengers, we can apply Newton's second law for motion along the vertical direction to write

$$ma = \sum F = F_{\text{grav}} + T = -mg + T$$

The tension in the cable is thus

$$T = mg + ma$$

This answer is quite reasonable; if the acceleration is zero, the force exerted by the cable must only support the weight of the elevator compartment and $T = mg$. If a is positive, a larger tension is required.

We next consider the forces acting on the cable. We have just seen that the cable exerts an upward force T on the elevator. According to Newton's third law—the action–reaction principle—the elevator must exert a force of equal magnitude and opposite direction on the cable. This is the origin of the downward force T acting on the cable in the free-body diagram in Figure 3.18C. There is a second force T_C acting on the cable, which comes from its connection to the top of the elevator shaft (to a motor, etc., located there). Usually the mass of a cable is very small when compared with the mass of other parts of the system, so for simplicity we will assume the cable is massless. If the mass of the cable is zero, the force of gravity on the cable will also be zero. Applying Newton's second law to our elevator cable in Figure 3.18C and assuming its mass is zero, we find

$$m_{cable} a = (0)a = +T_C - T = 0$$
$$T_C = T$$

Hence, for this cable, and for all such massless cables, the forces acting on the two ends are equal and both equal T, the tension in the cable. (Put another way, for a massless cable the tension is the same at all points along the cable.) The strength of a cable is usually specified by the maximum tension the cable can withstand without breaking. Notice that tension has units of force.

Let's now take on the more complicated situation in Figure 3.19, which shows a hanging spider holding onto its dinner (a bug). Here we have two "cables," both composed of spider silk, and we need to deal with the tensions in these two strands of silk. Figure 3.19 also shows free-body diagrams for the spider, the bug, and the two pieces of silk. The tension in the lower piece of silk is T_2. This piece is attached to both the spider and the bug, so it exerts forces of magnitude T_2 on both. There is a force of magnitude T_2 directed downward on the spider, and there is also a corresponding reaction force T_2 directed upward on the silk. The silk exerts a separate force of magnitude T_2 directed upward on the bug, and there is a reaction force T_2 acting downward on the silk. There are similar forces due to the tension T_1 in the upper piece of silk, which are also shown in the figure. We must include all these forces when we apply Newton's second law to this situation.

The spider and bug have masses m_{spider} and m_{bug} and a common acceleration a. Let's now find the tension in each piece of silk, assuming the silk is massless. We begin by applying Newton's second law to the bug in Figure 3.19. The free-body diagram for the bug shows that the forces in this case are just the force of gravity and the tension T_2, so from Newton's second law for motion along y we have

$$m_{bug} a = \sum F = F_{grav,\ bug} + T_2 = -m_{bug} g + T_2$$
$$T_2 = m_{bug}(g + a)$$

We next write Newton's second law for the spider:

$$m_{spider} a = \sum F = F_{grav,\ spider} + T_1 - T_2 = -m_{spider} g + T_1 - T_2$$
$$T_1 = m_{spider}(g + a) + T_2$$

Inserting our result for T_2 leads to

$$T_1 = m_{spider}(g + a) + m_{bug}(g + a) = (m_{spider} + m_{bug})(g + a)$$

This result for T_1 could also have been obtained by realizing that because the spider and the bug are connected and move together, we can think them as just a single object of mass $(m_{spider} + m_{bug})$. Applying Newton's second law to this composite object leads to the same result.

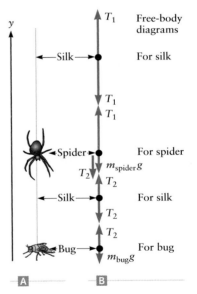

Figure 3.19 Ⓐ A spider hanging and holding onto a bug using a strand of silk. Ⓑ Free-body diagrams showing the forces acting on the spider, the bug, and the two pieces of silk employed by the spider. The upward force T_1 on the upper piece of silk comes from whatever that strand is attached to at its upper end. There are several action–reaction force pairs in this problem.

Figure 3.20 ◆ A box attached to a cable. This cable has a non-zero mass, so there is a force on the cable due to gravity. ◆ Free-body diagrams for the cable and for the hanging box.

Some Cables Are Not Massless

So far, we have assumed cables and ropes are massless, and we justified this assumption with the claim that the mass of a realistic cable or rope is usually very small compared with that of other parts of the system. Let's now see how to deal with cases when a cable's mass is not zero.

Consider the cable of mass m_{cable} in Figure 3.20, holding up a box of mass m. There are now three forces acting on the cable: one tension force from the support at the top (T_1), a second tension force from the box tied to the bottom end (T_2), and a third force due to the weight of the cable. Writing Newton's second law for the cable for motion along the y direction gives

$$m_{cable}a = \sum F = +T_1 - T_2 - m_{cable}g$$

If the cable is not accelerating, then $a = 0$ and

$$T_1 = T_2 + m_{cable}g \tag{3.15}$$

so

$$T_1 > T_2$$

Hence, we now have two different values for the tension in the cable: T_1 is the tension at the top end of the cable, and T_2 is the tension at the bottom. According to Equation 3.15, the tension at the top is larger than the tension at the bottom, which makes sense because the tension at the top must support the weight of the cable. We can also see what is required for a cable to be approximated as "massless." From Equation 3.15, the difference in the tensions at the top and bottom is equal to the weight of the cable. A cable can thus be approximated as massless if its weight is small compared to either tension.

Using Pulleys to Redirect a Force

Cables and ropes are an efficient way to transmit force from one place to another, but they have an important limitation: they can only "pull," and this force must be directed along the direction in which the cable lies. In many situations, we need to change the direction of a force, which can be accomplished by using an extremely useful mechanical device called a pulley. A simple pulley is shown in Figure 3.21A; it is just a wheel free to spin on an axle through its center, and it is arranged so that a rope or cable runs along its edge without slipping. For simplicity, we assume both the rope and the pulley are massless. Typically, a person pulls on one end of the rope so as to lift an object connected to the other end. The pulley simply changes the direction of the force associated with the tension in the rope as illustrated in Figure 3.21B, which shows the rope "straightened out" (i.e., with the pulley removed). In either case—with or without the pulley in place—the person exerts a force F on one end of the rope, and this force is equal to the tension. The tension is the same everywhere along this massless rope, so the other end of the rope exerts a force of magnitude T on the object. A comparison of the two arrangements in Figure 3.21—one with the pulley and one without—suggests the tension in the rope is the same in the two cases, and in both cases the rope transmits a force of magnitude $T = F$ from the person to the object.

Figure 3.21 ◆ A simple pulley. If the person exerts a force F on a massless string, there is a tension T in the string, and this tension force can be used to lift an object. ◆ The string "straightened out." The pulley simply redirects the force.

Amplifying Forces

A pulley can do much more than simply redirect a force. Figure 3.22 shows a pulley configured as a block and tackle, a device used to lift heavy objects. To analyze this case, we have to think carefully about the force between the string and the surface of the pulley. We have already mentioned that the rope does not slip along this surface. There is a frictional force between the rope and the surface of the pulley wherever the two are in contact, which is all along the bottom half of the pulley in Figure 3.22. Since the rope does not slip relative to the pulley, the rope exerts a force on the pulley as if the rope and pulley were actually attached to each other. First consider the part of the rope held by the person: it exerts an upward force on the pulley, equal in magnitude to the tension T (Fig. 3.22B). The same thing happens with the section of the rope on the left that is tied to the ceiling; this part of the rope also exerts a force on the pulley, and this force is again equal to the tension. Hence, there is a force $2T$ acting upward on the pulley. The interesting point is that the person exerts a force of only T on the rope, but the block and tackle converts it into a force of magnitude $2T$ *on the pulley*. This force on the pulley can then be used to lift an object attached to the pulley, such as the wooden crate that hangs from the pulley's axle in Figure 3.22A. Hence, a block and tackle can *amplify forces*. This particular arrangement enables the person to lift twice as much weight as would be possible if he used the simpler pulley arrangement in Figure 3.21.

The block and tackle in Figure 3.22 amplifies forces by a factor of two, and more complex block-and-tackle devices can amplify an applied force by even greater factors. This result might disturb you a little because it appears to violate an intuitive notion that "you can't get something for nothing." Although this statement is not a formal law of physics, you might still wonder what Newton would have thought. A key feature of the block and tackle in Figure 3.22 is that while it amplifies the force by a factor of two, the displacement of the object is *reduced* by a factor of two. If the person pulls the end of the rope upward through a distance L in Figure 3.22A, the body of the pulley and hence also the mass attached at the bottom move a distance of only $L/2$. In fact, this reduction of the displacement is a feature of all block and tackles. While it is possible to amplify a force by a factor that can be much larger than unity, the corresponding displacement is always reduced by the same factor. You do *not* get something for nothing.

A

Figure 3.22 **A** A block-and-tackle arrangement. The upward force on the pulley, from both sides of the rope, is $2T$. This force can be used to lift an object attached to the pulley. **B** Close-up of rope in contact with the pulley.

3.6 | REASONING AND RELATIONSHIPS: FINDING THE MISSING PIECE

You may have the impression that physics always involves a lot of precise mathematics. However, it is sometimes very useful to solve a problem approximately, perhaps because some important quantities (such as the initial velocity of an object) are not known precisely or because an approximate answer is sufficient. Moreover, the reasoning involved in an approximate analysis can often give clear insight into a problem, without the "distraction" of a lot of mathematics. In this section, we practice such *reasoning and relationships problems*.

Dealing with these problems involves two challenges that have not been present in the problems we have faced so far. First, we may need to identify important information that is "missing" from an initial description of the problem. For example, you might be asked to calculate the forces acting on two cars when they collide, given only the speed of the cars just before the collision. From an understanding of the relationship between force and motion, you would have to *recognize that additional information is needed* to solve this problem. In this problem, that information is the mass of the cars and the way the car bumpers deform in the collision. (We'll actually consider this problem later in this section.) It would then be your job to use common sense to *make reasonable estimates* of these "missing" quantities for typical cars and use your values to compute an answer. Because such estimated values vary from case to case (not all cars have the same mass or the same type of bumper), an approximate mathematical solution and an approximate numerical answer is usually sufficient. Let's now consider a few such reasoning and relationships problems.

Jumping off a Ladder

We first consider the case of a child who jumps off a ladder or playground structure for fun. We want to find the approximate force between the child and the ground

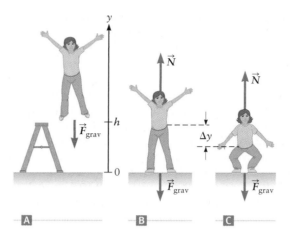

Figure 3.23 ◻A When a child jumps from a ladder, she undergoes free fall until she hits the ground. ◻B When she reaches the ground, she experiences a normal force N. ◻C The upper part of her body then travels a distance Δy while coming to a stop.

when she hits the ground and what strategy she should follow to minimize this force so as to make the landing as "soft" as possible. This type of problem would be of interest to someone who designs playgrounds or to a doctor who deals with children's injuries. The same sort of problem arises in various sports and activities, such as basketball and skydiving. For example, a skydiver might want to know how much force her legs will have to withstand on landing.

As usual, we begin by making a drawing for the problem, Figure 3.23. The initial part of the motion is just free fall as the child falls from the ladder to the ground. When she strikes the ground, the ground exerts a force (the normal force N) on the child, which causes her to stop. Let's first deal with the initial motion, the free-fall phase. For simplicity, we assume the child leaves the ladder with an initial speed of zero ($v_0 = 0$). While she is falling, her motion is described by our expressions for constant acceleration in Table 3.1, and we use Equation 3.4 with an acceleration $a = -g$ to calculate her velocity just before she hits the ground. We choose the origin ($y = 0$) to be at ground level as in Figure 3.23A and the child's initial position to be $y_0 = h$, where h is the height of the ladder. We can thus calculate v_{land}, the speed of the child just before she hits the ground at $y = 0$:

$$v_{land}^2 = v_0^2 + 2a(y - y_0) = 0 + 2(-g)(0 - h) = 2gh$$

$$v_{land} = \sqrt{2gh}$$

(3.16)

This result is her *speed* at the end of the free-fall period; her velocity will, of course, be directed downward.

When the child reaches the ground, she begins the second phase of her motion. There are now two forces acting on her (Fig. 3.23B): gravity and the normal force from the ground N, a quantity that we want to calculate. The precise value of N and how it varies with time depends on how the child lands. To get an approximate value, we *assume* N is constant during this period. This is only an approximation to the true behavior, but it should lead to a good approximate value for the average normal force. Our goal is obtain a value for N that is accurate to within an order of magnitude (i.e., to within a factor of 10).

Identifying the "Missing" Information and Making an Estimate

We now need to consider in a little more detail the child's motion just after hitting the ground. During this time, her legs will flex at the knees, and the upper portion of her body travels a distance Δy while coming to a stop. We know intuitively that if she lands with stiff knees (very little bending), the normal force will be much higher and will act over a shorter time than if her knees flex a lot or if she tumbles down. To make an estimate of her stopping distance, we note that the maximum flex possible is roughly the distance from her feet to her knees, which is about $\Delta y \approx 0.3$ m for an average child. We can use Equation 3.4 again to describe her motion while she comes to a stop. Since the distance traveled during the deceleration phase is $y - y_0 = -\Delta y$ (see Figs. 3.23B and C), we have

$$v^2 = v_0^2 + 2a(y - y_0) = v_0^2 + 2a(-\Delta y) \tag{3.17}$$

Here v is the speed at the end of the deceleration phase, which is zero because the child comes to a stop. In Equation 3.17, v_0 is the speed at the *start* of the deceleration phase and is thus equal to her speed at the *end* of the free-fall phase, v_{land} from Equation 3.16. With these values, we can use Equation 3.17 to calculate the acceleration during this part of the motion. Once we have the acceleration, we can use Newton's second law to find the normal force N exerted by the ground.

Rearranging Equation 3.17 to solve for the acceleration, we find

$$v^2 = 0 = v_0^2 + 2a(y - y_0) = v_{land}^2 + 2a(-\Delta y)$$

$$a = \frac{v_{land}^2}{2\,\Delta y} \tag{3.18}$$

This result is the acceleration of the child just after she hits the ground and before she comes to a complete stop. This acceleration is caused by the forces acting on the child during the deceleration period. Using Newton's second law during this period gives

$$ma = \Sigma F = F_{grav} + N = -mg + N$$

$$N = mg + ma$$

Although it seems natural to take m to be the mass of the child, it may actually be more accurate to let m be the mass of only her upper body because that is what is decelerating. However, because we are aiming for an approximate answer we'll let m be her total mass. Inserting the result for the acceleration from Equation 3.18, we can solve for the normal force acting on the child:

$$N = mg + ma = mg + \frac{mv_{land}^2}{2\,\Delta y}$$

According to Equation 3.16, $v_{land}^2 = 2gh$. Therefore,

$$N = mg + \frac{mgh}{\Delta y} = mg\left(1 + \frac{h}{\Delta y}\right) \tag{3.19}$$

For a typical "experiment" using a ladder of height $h = 2$ m and a child with $\Delta y = 0.3$ m, Equation 3.19 predicts

$$N = mg\left(1 + \frac{h}{\Delta y}\right) = mg\left(1 + \frac{2}{0.3}\right) \approx mg \times 8$$

The normal force is thus about eight times the weight (mg) of the child.

We emphasize that this is only an approximate solution. In our calculation, we assumed the normal force is constant (and hence that the stopping acceleration is constant), and in practice that may not be exactly the case. Another key to the analysis was our estimate of the stopping distance Δy; our estimated value of Δy was certainly not exact, but should be accurate to better than a factor of 10. Recall that our estimate was $\Delta y = 0.3$ m; a value of 0.03 m (about 1 inch) seems much too small, whereas a value of 3 m is unrealistically large. This level of accuracy for Δy should lead to similar accuracy in the final answer. Here and in most reasoning and relationships problems, our goal is to get the correct order of magnitude for the answer.

Besides giving a numerical estimate of the magnitude of the impact force, Equation 3.19 also provides insight on how to minimize that force. We see that a larger stopping distance Δy leads to a smaller value of N. This is why playgrounds and playground equipment are made of soft materials and why the steering column of a car is designed to collapse on impact.

The example in Figure 3.23 is intended to give some insight into how to deal with reasoning and relationships problems. A general strategy is presented on page 78.

Dealing with Reasoning and Relationships Problems

1. **RECOGNIZE THE PRINCIPLE.** Determine the key physics ideas that are central to the problem and that connect the quantity you want to calculate with the quantities you know. In the examples found in this section, this physics involves motion with constant acceleration.

2. **SKETCH THE PROBLEM.** Make a drawing that shows all the given information and everything else that you know about the problem. For problems in mechanics, your drawing should include all the forces, velocities, and so forth.

3. **IDENTIFY THE RELATIONSHIPS.** Identify the important physics relationships; for problems concerning motion with constant acceleration, they are the relationships between position, velocity, and acceleration in Table 3.1. For many reasoning and relationships problems, values for some of the essential unknown quantities may not be given. You must then use common sense to make reasonable

estimates for these quantities. Don't worry or spend time trying to obtain precise values of every quantity (such as the amount that the child's knees flex in Fig. 3.23). Accuracy to within a factor of 3 or even 10 is usually fine because the goal is to calculate the quantity of interest to within an order of magnitude (a factor of 10). Don't hesitate to use the Internet, the library, and (especially) your own intuition and experiences.

4. **SOLVE.** Since an exact mathematical solution is not required, cast the problem into one that is easy to solve mathematically. In the examples in this section, we were able to use the results for motion with constant acceleration.

5. Always *consider what your answer means* and check that it makes sense.

As is often the case, practice is a very useful teacher.

EXAMPLE 3.6 ® Cars and Bumpers and Walls

Consider a car of mass 1000 kg colliding with a rigid concrete wall at a speed of 2.5 m/s (about 5 mi/h). This impact is a fairly low-speed collision, and the bumpers on a modern car should be able to handle it without much damage to the car. Estimate the force exerted by the wall on the car's bumper.

RECOGNIZE THE PRINCIPLE

The motion we want to analyze starts when the car's bumper first touches the wall and ends when the car is stopped. To treat the problem approximately, we *assume* the force on the bumper is constant during the collision period, so the acceleration is also constant. We can then use our expressions from Table 3.1 to analyze the motion. Our strategy is to first find the car's acceleration and then use it to calculate the associated force exerted by the wall on the car from Newton's second law.

SKETCH THE PROBLEM

Figure 3.24 shows a sketch of the car along with the force exerted by the wall on the car. There are also two vertical forces—the force of gravity on the car and the normal force exerted by the road on the car—but we have not shown them because we are concerned here with the car's horizontal (x) motion, which we can treat using $\sum F = ma$ for the components of force and acceleration along x.

IDENTIFY THE RELATIONSHIPS

To find the car's acceleration, we need to estimate either the stopping time or the distance Δx traveled during this time. Let's take the latter approach. We are given the initial velocity ($v_0 = 2.5$ m/s) and the final velocity ($v = 0$). Both of these quantities are in Equation 3.4:

$$v^2 = v_0^2 + 2a(x - x_0) \tag{1}$$

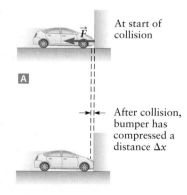

At start of collision

$\vec{F}$

A

After collision, bumper has compressed a distance Δx

B

Figure 3.24 Example 3.6.
A When a car collides with a wall, the wall exerts a force F on the bumper. This force provides the acceleration that stops the car.
B During the collision and before the car comes to rest, the bumper deforms by an amount Δx. The car travels this distance while it comes to a complete stop.

We can solve this to find the acceleration, provided we can make an estimate of $\Delta x = x - x_0$. Intuition tells us that for this low-speed collision, the bumper will deform only a little. We'll guess that a deformation of $\Delta x \approx 0.1$ m is a reasonable estimate and expect it to be correct to within an order of magnitude. We do not suggest that you try this experiment at home, but you have likely witnessed it! A value $\Delta x = 0.01$ m = 1 cm seems too small, whereas $\Delta x = 1$ m is much too large; our estimate of 0.1 m is somewhere in the middle.

SOLVE

The final velocity is $v = 0$ because the car comes to rest at the end of the collision. Inserting this into Equation (1) leads to

$$v^2 = v_0^2 + 2a(\Delta x) = 0$$

$$a = -\frac{v_0^2}{2\,\Delta x}$$

Notice that the acceleration is negative because it is directed opposite to the initial velocity. Using $v_0 = 2.5$ m/s and our estimate $\Delta x = 0.1$ m, the magnitude of the acceleration is

$$|a| = \frac{v_0^2}{2\,\Delta x} = \frac{(2.5 \text{ m/s})^2}{2(0.1 \text{ m})} = 30 \text{ m/s}^2$$

Applying Newton's second law for the mass $m = 1000$ kg, this acceleration requires a force of magnitude

$$|F| = m|a| = \boxed{3 \times 10^4 \text{ N}}$$

What have we learned?

A key to this problem was our estimate of the stopping distance Δx. Using some intuition and common sense, we estimated Δx with an accuracy of about an order of magnitude; this level of accuracy is usually sufficient for a reasoning and relationships problem.

CONCEPT CHECK 3.5 | Force on a Car Bumper

How large is the force exerted by the wall on the car bumper in Example 3.7 compared with the weight of the car?

3.7 | PARACHUTES, AIR DRAG, AND TERMINAL VELOCITY

We have spent considerable time discussing problems in which the acceleration is constant. Such cases are very useful because the mathematics of how the position and velocity vary with time is fairly simple and also because these cases are a good approximation to many real-life examples. In this section, we consider some interesting and important cases in which the acceleration is *not* constant. Included are cases in which an object is moving through a fluid (a liquid or gas); such an object is subject to a resistive drag force arising from contact with the fluid molecules.

In our discussions of the motion of a ball through the air, we claimed that the force of air drag can often be neglected, with the excuse that it is generally small compared with the force of gravity for objects moving slowly through our atmosphere. However, as the speed of an object increases, the magnitude of the air drag force increases rapidly. You have probably felt the force of air drag by holding your hand outside the window of a moving car. This force is barely noticeable at low speeds, but can be large at highway speeds. You probably also noticed that the drag

Insight 3.4

DRAG FORCES DEPEND ON THE SHAPE OF THE OBJECT

According to Equation 3.20, the force due to air drag is given by $F_{\text{drag}} = \frac{1}{2}\rho A v^2$. The drag force thus depends on the size of the object through its frontal area A. This expression for the drag force is only approximate. In a more accurate treatment, the drag force also depends on the aerodynamics of the object. For example, a smoothly shaped object like a race car will have a smaller drag force than a truck with a boxy shape having the same area and speed. This difference can be accounted for through the improved relation for F_{drag}:

$$F_{\text{drag}} = \frac{1}{2}C_D\rho A v^2$$

where C_D is called the "drag coefficient." The value of C_D depends on the aerodynamic shape. For a "normal" shape, such as a boxy truck, $C_D \approx 1$, but for more streamlined shapes, the drag coefficient can be much smaller than 1.

force depends on how you hold your hand; that is, the force depends on the exposed surface area of the moving object. For these reasons, the drag force is generally most significant for objects moving at high speeds, or with large surface areas perpendicular to the direction of motion, or both.

To get a rough idea of the importance of air drag, let's consider once again the motion of a ball. To be specific, consider a baseball hit directly upward with an initial speed of 45 m/s, which is typical of what a professional player can do (45 m/s is approximately 100 mi/h). If we ignore air drag, we can use results from Section 3.4 for the motion of a thrown ball (Fig. 3.14 and Eq. 3.14) and predict that in this case the ball will spend approximately 9.2 s in the air. For a real ball, the correct answer is about half as long. (We encourage you to observe this situation for yourself at a future baseball game.) The error is due to neglect of the force from air drag.

A high-accuracy calculation of the force due to air drag is extremely complicated, and there is no simple general "law" or formula that can be used. For common cases such as a baseball moving with speed v through the air, a good estimate of the magnitude of the drag force is

Force due to air drag

$$F_{\text{drag}} = \tfrac{1}{2}\rho A v^2 \qquad (3.20)$$

where ρ is the density of the air and A is the exposed area of the moving object; that is, A is the cross-sectional area perpendicular to the direction of motion. The direction of the drag force is always opposite to the direction of the velocity. We caution that Equation 3.20 is only approximate, but it is generally useful for speeds between about 1 m/s and 100 m/s.

The form of Equation 3.20 makes good intuitive sense. The drag force is proportional to the density of the fluid through which the object is moving. Here this fluid is air, but if we were considering a fluid such as water, the density and hence the drag force would increase. The drag force also increases as the exposed cross-sectional area A of the object is made larger (larger parachutes are better than small ones) and increases rapidly with the speed of the object.

Skydiving and Air Drag

Although the relation for the air drag force in Equation 3.20 has a fairly simple dependence on v, it is not possible to give simple mathematical expressions for position and velocity in the presence of air drag. Even so, we can calculate several interesting quantities.

Consider a skydiver who uses a parachute (Fig. 3.25). The free-body diagram shows two forces acting on the skydiver, the force of gravity (her weight) and the force of air drag. Using Newton's second law for motion along y, we find

$$ma = \sum F = F_{\text{grav}} + F_{\text{drag}}$$

Inserting the expression from Equation 3.20 for F_{drag} gives

$$ma = -mg + \tfrac{1}{2}\rho A v^2 \qquad (3.21)$$

There is no simple way to solve this to get the velocity as a function of time, but we can understand the general behavior as sketched in Figure 3.26. We assume the skydiver opens her parachute immediately after exiting the plane at $t = 0$, so the drag force is acting on her throughout the jump. When she leaves the plane, her velocity in the vertical direction is initially very small, so we have $v \approx 0$ in Equation 3.21; this leads to

$$ma = -mg + \tfrac{1}{2}\rho A v^2 \approx -mg$$

$$a \approx -g$$

Hence, the initial motion is approximately the same as if no air drag were present. We can then use our standard result for motion with a constant acceleration (Eq. 3.1), with $v_0 = 0$ and $a = -g$, to get $v \approx -gt$. The skydiver's speed thus increases

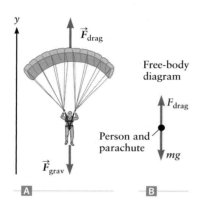

Figure 3.25 A There are two forces acting on this skydiver: the force of gravity and an upward force due to air drag. B Free-body diagram for the skydiver.

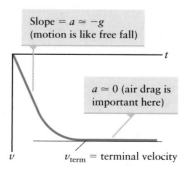

Figure 3.26 The skydiver's motion is initially like free fall; compare with Figure 3.15 A. Eventually, however, air drag becomes as large as the force of gravity and the skydiver reaches her terminal velocity v_{term}.

with time. However, as this speed increases she will eventually reach a time when the drag force is not small, and we must then include the effect of air drag. In fact, the speed will eventually become large enough that F_{drag} is nearly equal in magnitude, and opposite in direction, to the force of gravity so that the total force on the skydiver is very close to zero. From Newton's second law, we know that her acceleration is then also approximately zero and her velocity is constant as sketched in Figure 3.26. This constant velocity is known as the **terminal velocity**, and the skydiver maintains this velocity for the remainder of her jump. (The word *terminal* refers to the end of the skydiver's jump and not to the end of something else.) We can calculate v_{term} from Equation 3.21. When the skydiver reaches her terminal velocity, the total force is zero and she is moving at a constant speed equal to v_{term}, and

$$\sum F = 0 = F_{grav} + F_{drag} = -mg + \tfrac{1}{2}\rho A v_{term}^2$$

Solving for the magnitude of v_{term} we find

$$v_{term} = \sqrt{\frac{2mg}{\rho A}} \qquad (3.22)$$

For a skydiver of mass 60 kg with a parachute of area 100 m², and using the density of air ($\rho = 1.3$ kg/m³), we find a terminal velocity $v_{term} = 3.0$ m/s. This is approximately 7 mi/h, which should be an acceptably gentle landing.

The message here is that air drag *cannot* always be neglected. The drag force increases rapidly with velocity. Air drag is generally small for slowly moving objects, but it is very important in some cases.

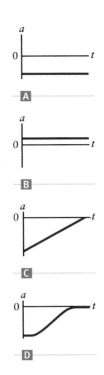

Figure 3.27 Concept Check 3.6.

CONCEPT CHECK 3.6 | Acceleration of a Skydiver

Figure 3.26 shows a skydiver's velocity as a function of time. Which of the plots in Figure 3.27 shows the qualitative behavior of the skydiver's acceleration as a function of time?

   (a) Plot (A)   (b) Plot (B)   (c) Plot (C)   (d) Plot (D)

EXAMPLE 3.7 | Ⓡ Air Drag on a Car

Estimate the drag force on a car moving at 50 mi/h and at 150 mi/h and compare it to the weight of a typical car with $m = 1000$ kg. The density of air is $\rho = 1.3$ kg/m³. *Hint*: Treat this as a reasoning and relationships problem and consider what essential information is missing from the description of the problem.

RECOGNIZE THE PRINCIPLE

The drag force depends on the car's speed v, the density of the air ρ, and frontal area A (the cross-sectional area perpendicular to the direction of motion, Fig. 3.28). We are given v and ρ, but not A. Hence, A is the "missing" information, so we must estimate a typical car's frontal area.

SKETCH THE PROBLEM

The problem is shown in Figure 3.28.

IDENTIFY THE RELATIONSHIPS

The frontal area A depends on the car's size and shape; we estimate A to be approximately 3 m² for a typical car. We arrive at this estimate since when viewed from the front, a typical car is about 2 m wide and 1.5 m tall.

SOLVE

The magnitude of the drag force is given by Equation 3.20,

$$F_{drag} = \tfrac{1}{2}\rho A v^2$$

Figure 3.28 Example 3.7. The drag force on a car depends on the frontal area A of a cross section perpendicular to the car's motion. The drag force always opposes the velocity.

To evaluate F_{drag}, it is convenient to express the speed in units of meters per second. For 50 mi/h, we find

$$50\,\frac{mi}{h} = 50\,\frac{mi}{h} \times \frac{1600\text{ m}}{1\text{ mi}} \times \frac{1\text{ h}}{3600\text{ s}} = 22\text{ m/s}$$

Inserting this value into our expression for F_{drag} along with the values of ρ and A given above leads to

$$F_{drag} = \tfrac{1}{2}\rho A v^2 = \tfrac{1}{2}(1.3\text{ kg/m}^3)(3\text{ m}^2)(22\text{ m/s})^2 = \boxed{1000\text{ N}}$$

The mass of the car is 1000 kg, so its weight is $mg = 9800$ N, which is about 10 times larger than this drag force. Repeating the calculation for a speed of 150 mi/h ($= 67$ m/s), we find $F_{drag} = \boxed{10,000\text{ N}}$. This is about a factor of 10 larger than the result at 50 mi/h; the difference can be traced to the strong dependence of the drag force on the velocity. At 150 mi/h, the air drag force on a car is approximately equal to the car's weight.

What does it mean?

These estimates for the drag force ignore the aerodynamics of the car because our simple expression for the drag force (Eq. 3.20) does not account for the effect of an object's aerodynamic shape. (See Insight 3.4.) For this reason, Equation 3.20 provides only an approximate value of F_{drag}. Even so, our results show that the air drag force on a car is quite substantial at high speeds. That is why a car's fuel efficiency decreases significantly at speeds greater than about 50 mi/h.

3.8 | ⊗ LIFE AS A BACTERIUM

In this section, we consider the motion of a small object in a liquid. We have in mind an *E. coli* bacterium as discussed in Chapter 2. Bacteria do not move as fast as baseballs or skydivers, and the expression for the drag force in Equation 3.20 is not very accurate for a swimming *E. coli*. For this case—and for other situations involving small things moving slowly (less than about 1 m/s or so)—the drag force is described much more accurately by an expression worked out in the 19th century by George Stokes (a famous physicist *and* mathematician). He applied Newton's laws to this problem and showed that the drag force on a spherical object moving slowly through a fluid is given by

$$\vec{F}_{drag} = -Cr\vec{v} \tag{3.23}$$

where C is a constant that depends on the properties of the fluid and r is the radius of the object. For water, $C \approx 0.02$ N·s/m^2. The negative sign in Equation 3.23 indicates that the drag force is always directed opposite to the velocity. Strictly speaking, Equation 3.23 applies only to spherical objects. Of course, most bacteria are not spherical (!), but we can usually approximate their shape at least roughly in this way.

Let's now consider how *E. coli* move about in a fluid. Recall from Section 2.5 that an *E. coli* possesses one or more flagella that it rotates to propel itself (Fig. 3.29). Writing Newton's second law for motion along the horizontal direction in Figure 3.29, we have

$$ma = \Sigma F = F_{flagellum} + F_{drag}$$

$$ma = F_{flagellum} - Cr v \tag{3.24}$$

The drag force on a bacterium is very large compared to its weight, so these microbes cannot move very rapidly. In fact, their acceleration is always so small that the two terms on the right-hand side of Equation 3.24 are both much larger in magnitude

Drag force in a liquid

Free-body diagram

Figure 3.29 The two dominant forces on an *E. coli* are from its flagellum (propeller) and from the drag force. These forces are in opposite directions.

than *ma* in any realistic situation. In other words, the forces on the bacterium are both large, but they are in opposite directions so that they almost exactly cancel. To a very good approximation, we can therefore simply set the right-hand side equal to zero and solve for the velocity, with the result

$$v = \frac{F_{\text{flagellum}}}{Cr} \tag{3.25}$$

Hence, in this case the velocity is proportional to the force exerted by the bacterium. When the flagellum stops turning and the force stops, the bacterium stops moving immediately; it does not coast!

It is common for thrown balls and sliding hockey pucks to coast since the frictional and drag forces they encounter are often small. However, for a bacterium in water the drag force is substantial and leads to very different behavior. In either case, though, the object's motion is described correctly and accurately by Newton's laws. While it may not be a simple matter to calculate all the forces exactly or to solve Newton's second law to determine the velocity and position as functions of time, Newton's laws are still the correct way to think about and describe the motion of an object.

SUMMARY | Chapter 3

KEY CONCEPTS AND PRINCIPLES

Forces and motion
A force is a push or a pull exerted by one object on another. Force is a *vector* quantity.

All calculations of motion start with Newton's second law. For one-dimensional motion, the forces and acceleration are directed along a line, and Newton's second law reads

$$\sum F = ma$$

A *free-body diagram* is a very useful first step in the application of Newton's second law.

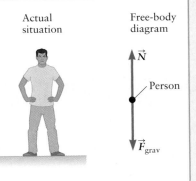

Actual situation   Free-body diagram

$\vec{N}$

Person

$\vec{F}_{\text{grav}}$

APPLICATIONS

Motion with constant acceleration
A common and extremely important case is motion with constant acceleration. When *a* is constant, the position, velocity, and acceleration are related by

$$x = x_0 + v_0 t + \tfrac{1}{2}at^2 \qquad \text{(3.3) (page 56)}$$

$$v = v_0 + at \qquad \text{(3.1) (page 56)}$$

$$v^2 = v_0^2 + 2a(x - x_0) \qquad \text{(3.4) (page 57)}$$

(Continued)

Common types of forces

Gravity
Near the surface of the Earth, the gravitational force on an object of mass m is directed along the vertical direction and is given by

$$F_{grav} = -mg \qquad \text{(3.8) (page 63)}$$

The minus sign here indicates that F_{grav} is downward. This force is also called the **weight** of the object. When an object is moving freely under the action of gravity, the motion is called **free fall** and the acceleration is $a = -g$.

Normal force
When two surfaces are in contact, there is a normal force between them. This force is perpendicular to the contact surface.

Friction
When two surfaces are in contact, a frictional force opposes motion of the surfaces relative to each other. If the surfaces are slipping relative to each other, the opposing force is **kinetic friction** given by

$$F_{friction} = \mu_K N \qquad \text{(3.9) (page 64)}$$

where N is the normal force. If the surfaces are not slipping, the resistive force is **static friction** given by

$$|F_{friction}| \leq \mu_S N \qquad \text{(3.11) (page 66)}$$

Drag forces
When an object moves through a fluid such as air or water, there is a drag force arising from contact between the fluid molecules and the moving object. We consider drag forces in two regimes:

- For motion through the air, the magnitude of the drag force is described approximately by

$$F_{drag} = \tfrac{1}{2}\rho A v^2 \qquad \text{(3.20) (page 80)}$$

  This regime is appropriate for skydivers and cars.
- For motion of cells and bacteria through water, the drag force is given by

$$\vec{F}_{drag} = -Cr\vec{v} \qquad \text{(3.23) (page 82)}$$

QUESTIONS

SSM = answer in Student Companion & Problem-Solving Guide   X = life science application

1. Identify an action–reaction pair of forces in each of the following situations.
   (a) A person pushing on a wall
   (b) A book resting on a table
   (c) A hockey puck sliding across an icy surface
   (d) A car accelerating from rest (*Hint*: Consider the tires.)
   (e) An object undergoing free fall in a vacuum
   (f) A basketball player jumping to dunk a basketball
   (g) A person throwing a baseball

2. Give an example of motion in which the acceleration and the velocity are in opposite directions.

3. Give an example of motion in one dimension for an object that starts at the origin at $t = 0$. At some time later, the displacement from the origin is zero but the velocity is not zero. Draw the corresponding position–time and velocity–time graphs.

4. Two objects are released simultaneously from the same height. The objects are both spherical with the same radius, but their masses differ by a factor of two. (a) Ignoring air drag, which object strikes the ground first? Or do they strike at the same time? Explain. (b) Including air drag, which object strikes the ground first? Or do they strike at the same time? Explain. (c) Find two household objects that might be used to demonstrate part (b).

5. Two mountain climbers are suspended by a rope as shown in Figure Q3.5. Which rope is most in danger of breaking? That is, which rope has the greatest tension?

Figure Q3.5

6. You are riding in a car that starts from rest, accelerates for a short distance, and then moves with constant velocity. Explain why you feel a force from the back of your seat only while the car is accelerating and not while you are moving with a constant velocity.

7. Two balls are thrown from a tall bridge. One is thrown upward with an initial velocity $+v_0$, while the other has an initial velocity $-v_0$. Which one has the greater speed just before it hits the ground?

8. A race-car driver wants to get his car started very quickly and "burns rubber" as he leaves the starting line. Explain where the force on the car comes from. Is it due to static friction or kinetic friction? Between what surfaces?

9. SSM A tire rolls without slipping on a horizontal road. Explain the role friction plays in this motion. Is it static friction or kinetic friction?

10. Normally, the coefficient of kinetic friction between two surfaces is smaller than the coefficient of static friction for the same two surfaces. Assuming that is true for the case of friction between a car tire and a road, explain why the car will accelerate faster if the driver does not "burn rubber."

11. The lower piece of silk in Figure 3.19 is acted on by two forces, $+T_2$ at the upper end and $-T_2$ at the lower end. These two forces are equal in magnitude and opposite in direction. Are they an action–reaction force pair? Explain why or why not.

12. Devise a block-and-tackle arrangement that amplifies the applied force by a factor of three.

13. Imagine that you are a passenger on the International Space Station. While on the station, you are "weightless" and so is everything else, including wrenches and furniture. Even though everything appears weightless, explain why mass still matters. Why does it still take longer for a given force to move a massive object a certain distance than to move an object with a very small mass the same distance?

14. Imagine a skydiver who waits a long time before opening her parachute. When she finally opens it, she is traveling faster than her terminal velocity with the parachute open. Explain the direction and magnitude of the force she feels from the parachute when she finally opens it. Why does she feel she is being pulled upward?

15. ✖ Explain why the air bags in a car reduce the forces on a passenger in the event of an accident.

16. For the books in Figure Q3.16, there is a force F from the table supporting the book on the bottom. On what object does the reaction force to F act?
(a) The table
(c) The Earth
(b) The book on top
(d) The floor

Figure Q3.16 Question 16 and Problems 30 and 31.

17. ✖ The bacterium in Figure 3.29 experiences a force due to drag from the surrounding fluid. Does the reaction force to F_{drag} act on (a) the flagellum, (b) the body of the bacterium, or (c) the fluid?

18. SSM Two balls of the same diameter are dropped simultaneously from a very tall bridge. One ball is solid lead, and the other is hollow plastic and has a much smaller mass than the solid lead ball. Use a free-body diagram to explain why the solid lead ball reaches the ground first. *Hint*: Include the air drag force in your analysis.

19. Consider a string with one end tied to a tall ceiling, while the other end hangs freely. Explain why the tension at the bottom of the string is smaller than the tension at the top.

20. In our discussion of the block and tackle in Figure 3.22, we claimed that when the right end of the string is lifted through a distance L, the body of the pulley is lifted a distance $L/2$. Give a geometric argument to prove this claim.

21. Explain and discuss the difference between mass and weight.

PROBLEMS

SSM = solution in Student Companion & Problem-Solving Guide
★ = intermediate   ✪ = challenging

✖ = life science application
® = reasoning and relationships problem

3.1 MOTION OF A SPACECRAFT IN INTERSTELLAR SPACE

1. A spacecraft that is initially at rest turns on its engine at $t = 0$. If its mass is $m = 3000$ kg and the force from the engine is 45 N, what is the acceleration of the spacecraft?

2. A hockey puck moves on an icy surface that is frictionless, with a constant speed of 30 m/s. How long does it take the puck to travel the length of the hockey rink (60 m)?

3. An ice skater moves without friction on a frozen pond. While traveling at 8.0 m/s, she finds that it takes 17 s to travel the length of the pond. How long is the pond?

4. An object moves with a constant acceleration of 4.0 m/s². If it starts with an initial speed of 30 m/s, how long does it take to reach a velocity of 250 m/s?

5. A car has a velocity of 10 m/s at $t = 7.0$ s. It then accelerates uniformly and reaches a velocity of 42 m/s at $t = 12.0$ s. What is its acceleration during this period?

6. ★ A constant force of 400 N acts on a spacecraft of mass 8000 kg that has an initial velocity of 30 m/s. How far has the spacecraft traveled when it reaches a velocity of 5000 m/s?

7. ★ A rocket-powered sled of mass 3500 kg travels on a level snow-covered surface with an acceleration of +3.5 m/s²

(Fig. P3.7). What are the magnitude and direction of the force on the sled?

Figure P3.7

8. ★ Your car has a dead battery. It is initially at rest, and you push it along a level road with a force of 120 N and find that it reaches a velocity of 2.0 m/s in 50 s. What is the mass of the car? Ignore friction.

9. ★ You are a newly graduated astronaut preparing for your first trip into space. Plans call for your spacecraft to reach a velocity of 500 m/s after 2.4 minutes. If your mass is 75 kg, what force will be exerted on your body? Assume the acceleration is constant.

10. ★ ✖ **Pulling g's.** Suppose again you are the astronaut in Problem 9. When most people are subjected to an acceleration greater than about $5 \times g$, they will usually become unconscious ("black out"). Will you be in danger of blacking out?

11. [SSM] ✸ An object with an initial velocity of 12 m/s accelerates uniformly for 25 s. (a) If the final velocity is 45 m/s, what is the acceleration? (b) How far does the object travel during this time?

12. ✸ A hockey puck has an initial velocity of 50 m/s and a final velocity of 35 m/s. (a) If it travels 35 m during this time, what is the acceleration? (b) If the mass of the puck is 0.11 kg, what is the horizontal force on the puck?

13. ✪ An elevator is moving at 1.2 m/s as it approaches its destination floor from below. When the elevator is a distance h from its destination, it accelerates with $a = -0.50$ m/s², where the negative sign indicates a downward vertical direction. ("Up" is positive.) Find h.

14. ✸ An airplane must reach a speed of 200 mi/h to take off. If the runway is 500 m long, what is the minimum value of the acceleration that will allow the airplane to take off successfully?

15. ✪ A drag racer is able to complete the 0.25-mi course in 6.1 s. (a) If her acceleration is constant, what is a? (b) What is her speed when she is halfway to the finish line?

16. ✖ Consider a sprinter who starts at rest, accelerates to a maximum speed v_{max}, and then slows to a stop after crossing the finish line. Draw qualitative plots of the acceleration, velocity, and position as functions of time. Indicate the location of the finish line on the x–t plot.

17. ✸ Draw a qualitative plot of the total force acting on the ball in Figure 3.14 as a function of time. Begin your plot while the ball is still in the thrower's hand and end it after the ball comes to rest back on the ground.

18. ✪ A barge on a still lake is moving toward a bridge at 10.0 m/s. When the bridge is 40.0 m away, the pilot slows the boat with a constant acceleration of -0.73 m/s². (a) Use Equation 3.3 to find the time it takes the barge to reach the bridge. Note that you will obtain two answers! Do both calculated times correspond to possible physical situations? (b) Find the final velocity for each time using Equation 3.1. (c) For each set of answers for t and v, sketch plots of velocity versus time and position versus time, and use them to describe the two situations.

19. ✸ To measure the height of a tree, you throw a rock directly upward, with a speed just fast enough that the rock brushes against the uppermost leaves and then falls back to the ground. If the rock is in the air for 3.6 s, how tall is the tree?

20. ✸ Your car is initially traveling at a speed of 25 m/s. As you approach an intersection, you spot a dog in the road 30 m ahead and immediately apply your brakes. (a) If you stop the instant before you reach the dog, what was the magnitude of your acceleration? (b) If your velocity is positive while you are slowing to a stop, is your acceleration positive or negative?

21. ✸ A car is traveling at 25 m/s when the driver spots a large pothole in the road a distance 30 m ahead. She immediately applies her brakes. If her acceleration is -27 m/s², does she manage to stop before reaching the pothole?

22. ✪ A bullet is fired upward with a speed v_0 from the edge of a cliff of height h (Fig. P3.22). (a) What is the speed of the bullet when it passes by the cliff on its way down? (b) What is the speed of the bullet just before it strikes the ground? (c) If the bullet is instead fired downward with the same initial speed v_0, what is its speed just before it strikes the ground? Express your answers in terms of v_0, h, and g. Ignore air drag. Assume the bullet is fired straight up in (a) and (b) and straight down in (c).

23. [SSM] ✸ A ball is thrown upward with a speed of 35 m/s from the

Figure P3.22
Problems 22 and 23.

edge of a cliff of height $h = 15$ m (Fig. P3.22). (a) What is the speed of the ball when it passes by the cliff on its way down to the ground? (b) What is the speed of the ball when it hits the ground? Ignore air drag. Assume the ball is thrown straight up.

24. ✪ A rock is dropped from a tree of height 25 m into a lake (depth 5.0 m) below. After entering the water, the rock then floats gently down through the water at a speed of 1.5 m/s to the bottom of the lake. What is the total elapsed time?

25. ✸ You are a passenger ($m = 110$ kg) in an airplane that is accelerating on the runway, beginning to take off. The force between your back and your seat is 400 N. Starting from rest, how far does the plane travel as it accelerates to a takeoff speed of 130 m/s?

26. ✸ A motivated mule can accelerate an empty cart of mass $m = 180$ kg from rest to 5.0 m/s in 10 s. If the cart is loaded with 540 kg of wood, how long will it take the mule to get the cart to 5.0 m/s? Assume a constant acceleration and that the mule exerts the same force as when the cart is empty.

3.2 NORMAL FORCES AND WEIGHT

27. A person has a weight of 500 N. What is her mass?

28. A book of mass 3.0 kg sits at rest on a horizontal table. What is the normal force exerted by the table on the book?

29. **Under siege.** Twenty soldiers hold a long beam of heavy wood against a fortified castle door and attempt to push it open. If each soldier is able to supply a forward force on the beam of 80 N, what normal force does the castle door exert on the beam if the door does not open?

30. ✸ Two books of mass $m_1 = 8.0$ kg and $m_2 = 5.5$ kg are stacked on a table as shown in Figure Q3.16. Find the normal force acting between the table and the bottom book.

31. ✸ The table in Figure Q3.16 is now sitting in an elevator, with $m_1 = 9.5$ kg and $m_2 = 2.5$ kg. The normal force between the floor and the bottom book is 70 N. Find the magnitude and direction of the elevator's acceleration.

32. A grand piano with three legs has a mass of 350 kg and is at rest on a level floor. (a) Draw a free-body diagram for the piano. Show the force of the floor on each leg as a separate force in your diagram. (b) What is the total force of the piano on the floor?

33. ✪ Your friend's car is broken and you volunteer to push it to the nearest repair shop which is 2.0 km away. You carefully move your car so that the bumpers of the two cars are in contact and then slowly accelerate to a speed of 2.5 m/s over the course of 1 min. (a) If the mass of your friend's car is 1200 kg, what is the normal force between the two bumpers? (b) If you then maintain a speed of 2.5 m/s, how long does it take you to reach the repair shop?

34. [SSM] ✸ A tall strongman of mass $m = 95$ kg stands upon a scale while at the same time pushing on the ceiling in a small room. Draw a free-body diagram of the strongman (Fig. P3.34) and indicate all normal forces acting on him. If the scale reads 1100 N (about 240 lb), what is the magnitude of the normal force that the ceiling exerts on the strongman?

Figure P3.34

35. ✪ ✖ A bodybuilder configures a leg-press apparatus (Fig. P3.35) to a resistance of 230 lb (approximately 1000 N). She pushes the weight to her full extension and comes to rest. (a) What is the normal force on each of her feet? (b) Assume she is able to move the push-plate of the leg-press machine at an acceleration of 0.50 m/s² for the first half of the displacement. What is the normal force on each foot during this period of acceleration? Assume the resistive force is due to a weight of 230 lb hanging from the end of the cable.

Figure P3.35

36. ✪ Three blocks rest on a frictionless, horizontal table (Fig. P3.36), with $m_1 = 10$ kg and $m_3 = 15$ kg. A horizontal force $F = 110$ N is applied to block 1, and the acceleration of all three blocks is found to be 3.3 m/s². (a) Find m_2. (b) What is the normal force between blocks 2 and 3?

Figure P3.36

3.3 ADDING FRICTION TO THE MIX

37. ⊡ A car is moving with a velocity of 20 m/s when the brakes are applied and the wheels lock (stop spinning). The car then slides to a stop in 40 m. Find the coefficient of kinetic friction between the tires and the road.

38. A hockey puck slides with an initial speed of 50 m/s on a large frozen lake. If the coefficient of kinetic friction between the puck and the ice is 0.030, what is the speed of the puck after 10 s?

39. ✪ Your moving company runs out of rope and hand trucks, so you are forced to push two crates along the floor as shown in Figure P3.39. The crates are moving at constant velocity, their masses are $m_1 = 45$ kg and $m_2 = 22$ kg, and the coefficients of kinetic friction between both crates and the floor are 0.35. Find the normal force between the two crates.

Figure P3.39
Problems 39 and 94.

40. You are given the job of moving a refrigerator of mass 100 kg across a horizontal floor. The coefficient of static friction between the refrigerator and the floor is 0.25. What is the minimum force that is required to just set the refrigerator into motion?

41. SSM ⊡ The coefficient of kinetic friction between a refrigerator (mass 100 kg) and the floor is 0.20, and the coefficient of static friction is 0.25. If you apply the minimum force needed to get the refrigerator to move, what will the acceleration then be?

42. ⊡ After struggling to move the refrigerator in the Problem 41, you finally apply enough force to get it moving. What is the minimum force required to keep it moving with a constant velocity? Assume $\mu_K = 0.20$.

43. ✪ You are trying to slide a refrigerator across a horizontal floor. The mass of the refrigerator is 200 kg, and you need to exert a force of 350 N to make it just begin to move. (a) What is the coefficient of static friction between the floor and the refrigerator? (b) After it starts moving, the refrigerator reaches a speed of 2.0 m/s after 5.0 s. What is the coefficient of kinetic friction between the refrigerator and the floor?

44. ✪ A driver makes an emergency stop and inadvertently locks up the brakes of the car, which skids to a stop on dry concrete. Consider the effect of rain on this scenario. Using the values in Table 3.2, how much farther would the car skid (expressed in percentage of the dry-weather skid) if the concrete were instead wet?

45. ⊡ Jeff Gordon (a race-car driver) discovers that he can accelerate at 4.0 m/s² without spinning his tires, but if he tries to accelerate more rapidly, he always "burns rubber." (a) Find the coefficient of friction between his tires and the road. Assume the force from the engine is applied to only the two rear tires. (b) Have you calculated the coefficient of static friction or kinetic friction?

46. ✪ **Antilock brakes.** A car travels at 65 mi/h when the brakes are suddenly applied. Consider how the tires of a moving car come in contact with the road. When the car goes into a skid (with wheels locked up), the rubber of the tire is moving with respect to the road; otherwise, when the tires roll, normally the point where the tire contacts the road is stationary. Compare the distance required to bring the car to a full stop when (a) the car is skidding and (b) when the wheels are not locked up. Use the coefficients of kinetic and static friction from Table 3.2 and assume the tires are rubber and the road is dry concrete. (c) How much farther does the car go if the wheels lock into a skidding stop? Give your answer as a distance in meters and as a percent of the nonskid stopping distance. (d) Can antilock brakes make a big difference in emergency stops? Explain.

47. ✪ A hockey puck slides along a rough, icy surface. It has an initial velocity of 35 m/s and slides to a stop after traveling a distance of 95 m. Find the coefficient of kinetic friction between the puck and the ice.

3.4 FREE FALL

48. A rock is dropped from a very tall tower. If it takes 4.5 s for the rock to reach the ground, what is the height of the tower?

49. A baseball is hit directly upward with an initial speed of 45 m/s. Find the velocity of the ball when it is at a height of 40 m. Is there one correct answer for v or two? Explain why.

50. A squirrel is resting in a tall tree when it slips from a branch that is 50 m above the ground. It is a very agile squirrel and manages to land safely on another branch after only 0.50 s. What is the height of the branch it lands on?

51. ⊡ **Basketball on the Moon.** If LeBron James can jump 1.5 m high on Earth, how high could he jump on the Moon (assume an indoor court), where $g = 1.6$ m/s²?

52. ⊡ An apple falls from a branch near the top of a tall tree. If the branch is 12 m above the ground, what is the apple's speed just before it hits the ground?

53. SSM ⊡ A ball is thrown directly upward with an initial velocity of 15 m/s. If the ball starts at an initial height of 3.5 m, how long is the ball in the air? Ignore air drag.

54. ✪ Two children are playing on a 150-m-tall bridge. One child drops a rock (initial velocity zero) at $t = 0$. The other waits 1.0 s and then throws a rock downward with an initial speed v_0. If the two rocks hit the ground at the same time, what is v_0?

55. ⊡ A rock is dropped from a tall bridge into the water below. If the rock begins with a speed of zero and has a speed of 12 m/s just before it hits the water, what is the height of the bridge?

56. ✪ A roofing tile falls from rest off the roof of a building. An observer from across the street notices that it takes 0.43 s for the tile to pass between two windowsills that are 2.5 m apart. How far is the sill of the upper window from the roof of the building?

57. ✪ You are standing at the top of a deep, vertical cave and want to determine the depth of the cave. Unfortunately, all you have is a rock and a stopwatch. You drop the rock into the cave and measure the time that passes until you hear the rock hitting the floor of the cave far below. If the elapsed time is 8.0 s, how deep is the cave? *Hints*: (1) Sound travels at a constant speed of 340 m/s. (2) Consider two separate time periods. During the first period, the rock undergoes free fall and lands at the bottom of the cave. During the second period, sound travels at a constant velocity back up the cave.

58. ⭐ Your friend is an environmentalist who is living in a tree for the summer. You are helping provide her with food, and you do so by throwing small packages up to her tree house. If her tree house is 30 m above the ground, what is the minimum (initial) speed you must use when throwing packages up to her?

59. ✪ You are standing across the street from a tall building when the top of the building (h = 80 m) is hit by lightning and a brick is knocked loose. You see the lightning strike and immediately see that the brick will fall to hit a person standing at the base of the building. You then run toward the person and push him out of the path of the brick, the instant before the brick reaches him. If you are initially 25 m from the building, how fast do you have to run?

3.5 CABLES, STRINGS, AND PULLEYS: TRANSMITTING FORCES FROM HERE TO THERE

60. A cable attached to a block of mass 12 kg pulls the block along a horizontal floor at a constant velocity. If the tension in the cable is 5.0 N, what is the coefficient of kinetic friction between the block and the floor?

61. ⭐ A crate of mass 55 kg is attached to one end of a string, and the other end of the string runs over a pulley and is held by a person as in Figure 3.21. If the person pulls with a force of 85 N, what is the crate's acceleration?

62. ⭐ A car of mass 1200 kg is being moved by a large crane in preparation for lowering it onto a junk pile. If the car is accelerating downward at 0.20 m/s², what is the tension in the cable?

63. SSM ⭐ You work for a moving company and are given the job of pulling two large boxes of mass m_1 = 120 kg and m_2 = 290 kg using ropes as shown in Figure P3.63. You pull very hard, and the boxes are accelerating with a = 0.22 m/s². What is the tension in each rope? Assume there is no friction between the boxes and the floor.

Figure P3.63 Problems 63 and 90.

64. ✪ ✪ **In traction.** When a large bone such as the femur is broken, the two pieces are often pulled out of alignment by the complicated combination of tension and compression forces that arise from the muscles and tendons in the leg (see the X-ray image in Figure P3.64A). To realign the bones and allow proper healing, these forces must be compensated for. A method called traction is often employed. If a total tension force of 400 N is applied to the leg as depicted in Figure P3.64B to realign the parts of the femur, how much mass m must be attached to the bottom pulley?

Figure P3.64

65. A mass with M = 102 kg is attached to the bottom of a block-and-tackle pulley system as depicted in Figure P3.65. How much tension force is needed to keep the mass at its current position?

Figure P3.65

3.6 REASONING AND RELATIONSHIPS: FINDING THE MISSING PIECE

66. ⭐ Ⓡ What is the approximate speed at which the force of air drag on a car is equal to the weight of the car? *Hint:* Start by estimating the mass and the frontal area of the car.

67. Ⓡ A bullet of mass 10 g leaves the barrel of a rifle at 300 m/s. Assuming the force on the bullet is constant while it is in the barrel, find the magnitude of this force. *Hint:* Start by estimating the length of the barrel.

68. ✪ Ⓡ When an airplane takes off, it accelerates on the runway starting from rest. What is the magnitude of this runway acceleration for a passenger jet? *Hint:* You will need to find (or estimate) one or more quantities such as the takeoff velocity, the distance traveled on the runway, and the time spent on the runway.

69. ✪ ✪ Calculate the terminal velocity for a pollen grain falling through the air using the drag force Equation 3.20. Assume the pollen grain has a diameter of 1 μm and a density of 0.2 g/cm³. If this grain is released from the top of a tree (height 10 m), estimate the time it will take to fall to the ground. *Hint:* The pollen grain will reach its terminal velocity very quickly and will have this velocity for essentially the entire motion. Your answer will explain why pollen stays in the air for a very long time.

70. SSM ✪ ✪ Ⓡ **Hang time.** LeBron James decides to jump high enough to dunk a basketball. What is the approximate force between the floor and his feet while he is jumping from the floor? *Hint:* Start by estimating the distance he moves while in contact with the floor and the height he must jump for his hands to reach just above the rim.

3.7 PARACHUTES, AIR DRAG, AND TERMINAL VELOCITY

71. ⭐ ✪ Ⓡ Consider a skydiver who lands on the ground with a speed of 3 m/s. What is the approximate force on the skydiver's legs when she lands?

72. Calculate the terminal velocity for a baseball. A baseball's diameter is approximately d = 0.070 m, and its mass is m = 0.14 kg. Express your answer in meters per second and miles per hour.

73. ✪ Ⓡ A skydiver opens her parachute immediately after jumping from an airplane. Approximately how long does it take to reach her terminal velocity? *Hint:* Use half the maximum acceleration as an estimate of the average acceleration during this time period.

74. Calculate the force of air drag on a hockey puck moving at 30 m/s. A hockey puck is approximately 3.0 cm tall and 8.0 cm in diameter.

75. Ⓡ The following items are dropped from an airplane. Rank them in order from lowest terminal velocity to highest and justify your ranking.
(a) Bowling ball
(b) Beach ball
(c) Spear or javelin (pointing downward)
(d) Watermelon
(e) Cantaloupe
(f) Apple

76. ⭐ A Styrofoam ball of radius 28 cm falls with a terminal velocity of 5.0 m/s. What is the mass of the ball?

77. SSM ⭐ You are a secret agent and find that you have been pushed out of an airplane without a parachute. Fortunately, you are wearing a large overcoat (as secret agents often do). Thinking quickly, you are able to spread out and hold the overcoat so that you increase your overall area by a factor of four. If your terminal velocity would be 43 m/s without the coat, what is your new terminal velocity with the coat? Could you survive impact with the ground? How about over a lake?

78. Calculate the drag force on a bullet that is 5.0 mm in diameter and moving at a speed of 600 m/s. If the mass of the bullet is 10 g, compare this force to the weight of the bullet and to your own weight. Assume the drag force on the bullet is given by Equation 3.20.

79. Ⓡ Hail forms high in the atmosphere and can be accelerated to a high speed before it reaches the ground. Estimate the terminal velocity of a spherical hailstone that has a diameter of 2.0 cm.

Hint: The mass of a piece of hail that has a volume of 1 cm³ is about 1 g.

3.8 LIFE AS A BACTERIUM

80. ✪ ⓧ Ⓡ Calculate the terminal velocity for a bacterium in water. Use the expression for the drag force in Equation 3.23. The values of other quantities you need are given in Example 3.10. How long does it take a bacterium to fall from the top of a lake of depth 5.0 m to the bottom?

81. SSM ✪ ⓧ Ⓡ The force exerted on a bacterium by its flagellum is 4×10^{-13} N. Find the velocity of the bacterium in water. Assume a size $r = 1 \ \mu m$.

ADDITIONAL PROBLEMS

82. ✪ The kick experienced when firing a rifle can be explained by Newton's third law. A .22-caliber rifle has a mass of $M = 5.2$ kg, and a bullet with a mass $m = 3.0$ g leaves the barrel of the gun at a velocity of 320 m/s. (a) If the bullet starts from rest and leaves the gun barrel after $t = 0.010$ s, what was the acceleration of the bullet? (b) What was the force on the bullet? (c) What was the magnitude of the force exerted on the gun? (d) What acceleration did the gun experience? (e) Compare the ratio of M and m to the ratio of the acceleration of each object.

83. ✪ **What is your reaction time?** The following simple method can be employed to determine reaction time. A partner holds a meter stick by pinching it at the top and letting it hang vertically. To measure your reaction time, place your thumb and forefinger just below the base of the meter stick, ready to pinch it when it falls. Without signaling, your partner releases the meter stick; it accelerates due to gravity at a rate of 9.8 m/s², and you grab it as fast as possible. (a) If your thumb pinches the meter stick at the 45-cm mark, what was your reaction time? Using a similar calculation, one can calibrate a "Grab-it Gauge" such as that shown in Figure P3.83. (b) Calculate the distance to draw each line from the bottom starting point for reactions times of 0.14, 0.16, 0.18, 0.20, and 0.22 s.

Figure P3.83

84. ✪ **Deer in the headlights.** There are two important time intervals to consider when coming to an emergency stop while driving. The first is the driver's reaction time to get a foot on the brake pedal, and the second is the time it takes to decelerate the car to rest. Consider a car moving at 30 m/s (about 65 mi/h) when the driver sees a deer in the road ahead and applies the brakes. (a) If the driver's reaction time is 1.1 s, how far the does the car travel before the brakes are applied? (b) If the deer is 100 m away when the driver sees it, what acceleration is needed to stop the car without hitting the deer? (c) If the concrete streets are wet, will the car be able to stop without hitting the deer? (d) If the car cannot stop in time, how fast will it be going

when it strikes the deer? If the car can stop in time, how far away from the deer will it come to rest?

85. ✪ A boy pushes a 3.1-kg book against a vertical wall with a horizontal force of 40 N. What is the minimum coefficient of friction that will keep the book in place without sliding?

86. ✪ An impish young lad stands on a bridge 10 m above a lake and drops a water balloon on a boat of unsuspecting tourists. Although the boat is traveling at a speed of 7.5 m/s, the boy manages to land the balloon right on the deck of the boat. How far away from the base of the bridge was the boat when the boy released the balloon? Assume he just lets the balloon go without throwing it (i.e., he simply drops it).

87. ✪ Ⓡ Two mischievous children drop water balloons from a bridge as depicted in Figure P3.87. If each water balloon is approximately 30 cm in diameter, what is the time interval between when the first balloon was let go and the second balloon was dropped? Assume both balloons were let go exactly above the bridge railing. Take measurements directly from the figure and scale appropriately.

Figure P3.87

88. ✪ A subway train is designed with a maximum acceleration of ± 0.20 m/s², which allows for both passenger safety and comfort. (a) If subway stations are 1.2 km apart, what is the maximum velocity that can be obtained between stations? (b) How long does it take to travel between two stations? (c) The train stops for a total of 45 s at each station. What is the overall average velocity of the train from station to station?

89. ✪ A spring scale indicates that a helium balloon tied to it produces a tension of 0.20 N in the string. The string is then cut, and the balloon rises until it comes to rest on the ceiling. (a) Draw a free-body diagram of the balloon on the ceiling. (b) What is the normal force exerted by the ceiling on the balloon?

90. (a) Draw free-body diagrams for the two blocks and for the person in Figure P3.63. (b) For each horizontal force in part (a), identify the corresponding reaction force.

91. ✪ A car is outfitted with a flat piece of plywood mounted vertically on its front bumper. As seen in Figure P3.91, a block of wood is simply placed in front of the car just as the car begins to accelerate. (a) If the coefficient of static friction between the block and the plywood is $\mu_S = 0.90$, what acceleration is needed to keep the block from falling? (b) Safety concerns limit the maximum speed of the car to 50 m/s (about 110 mi/h). How long can the car keep the block from falling this way? (c) If the mass of the wooden block is doubled, how does the answer to part (a) change?

Figure P3.91

92. ✪ Two spheres, one of wood and one of steel, have the same mass. (a) How many times greater is the terminal velocity of the steel sphere of diameter 10 cm, than that of the wood sphere of diameter 40 cm? (b) Another two spheres, again one of wood and one of steel, are this time exactly the same size. Find the ratio of the terminal velocity of the steel sphere of mass 40 kg to that of the wood sphere of mass 10 kg.

93. ✪ A block of mass $M_1 = 3.0$ kg rests on top of a second block of mass $M_2 = 5.0$ kg, and the second block sits on a surface that is so slippery that the friction can be assumed to be zero (see Fig. P3.93). (a) If the coefficient of static friction between the blocks is $\mu_S = 0.21$, how much force can be applied to the top block without the blocks slipping apart? (b) How much force can be applied to the bottom block for the same result?

Figure P3.93

94. ☆ Draw free-body diagrams for the two crates in Figure P3.39. For each force, identify the corresponding reaction force.

95. ☆ A bullet of mass $m = 10$ g and velocity 300 m/s is shot into a block of wood that is firmly attached to the ground. The bullet comes to rest in the wood at a depth of 8.1 cm. (a) What was the acceleration of the bullet? (b) How long did it take for the bullet to come to rest once it entered the wood? (c) What was the frictional force exerted by the wood on the bullet?

96. ✪ Ⓡ Consider a small sailboat with a triangular sail of height 10 m and width at the base of 5.0 m. (a) Assuming a wind speed of 15 mi/h relative to the boat, estimate the force exerted by the wind on the sail. (b) If the sailboat is moving with a constant speed, what is the drag force due to the water? (c) Suppose the speed of the wind relative to the boat is doubled to 30 mi/h. By what factor does the speed of the boat *relative to the water* increase? *Hint*: Assume as in Stokes's expression (Eq. 3.23) that the drag force due to the water is proportional to the speed of the boat relative to the water.

97. SSM ☆ **High dive.** The cliff-divers of Acapulco are famous for diving from steep cliffs that overlook the ocean into places where the water is very shallow. (a) Suppose a cliff-diver jumps from a cliff that is 25 m above the water. What is the speed of the diver just before he enters the water? (b) If the water is 4.0 m deep, what is the acceleration of the diver after he enters the water? Assume this acceleration is constant and it begins at the moment his hands enter the water.

98. ☆ Ⓡ The pedestrian walkway on the Golden Gate Bridge is about 75 m above the water below. This bridge is (unfortunately) a popular spot for some unhappy people, who attempt to jump off. Ignore air drag and calculate (a) the time it takes for an object to fall from the bridge to the water and (b) the speed of the object just before it hits the water. Express your answer to (b) in meters per second and miles per hour. (c) Use Equation 3.22 to calculate the terminal velocity of a person. Will air drag be important for a person who jumps off the Golden Gate Bridge?

99. ✪ ✖ Ⓡ The surfaces where bones meet are lubricated by a fluidlike substance, which makes the coefficient of friction between bones very small (see Table 3.2). What is the approximate lateral (i.e., horizontal) force required to make the bones in a typical knee joint slide across each other? For simplicity, assume the surfaces in the knee are flat and horizontal, and consider an adult of average mass.

Forces and Motion in Two and Three Dimensions

These Adelie penguins diving from an iceberg demonstrate motion in two dimensions. In this chapter, we learn how to describe and predict their motion. (Joseph Van Os, Getty Images)

In Chapter 3, we considered how to apply Newton's laws to deal with motion in one dimension (along a line). We now build on our understanding of motion in one dimension and deal with forces and motion in two and three dimensions. Newton's laws are again our foundation, and we continue to describe motion in terms of displacement, velocity, and acceleration. Now, though, we must allow for the vector nature of force, displacement, velocity, and acceleration. Recall that vector quantities have both a magnitude and a direction. For one-dimensional motion, we chose a positive direction at the start of a problem and used an algebraic sign (+ or −) to denote the directions of the displacement, velocity, and acceleration. For motion in two or three dimensions, we must express the directions of these vectors according to our chosen coordinate axes. Once that is done, we can use the same principles and problem-solving techniques we developed in Chapter 3.

4.1 | STATICS

We begin with Newton's second law in vector form,

Newton's second law

$$\sum \vec{F} = m\vec{a} \qquad (4.1)$$

where the left-hand side of this relation, $\sum \vec{F}$, is the total force acting on the object. When applying Newton's second law, we generally start by determining all the individual forces acting on the object. We can then construct a free-body diagram, showing the forces in graphical form. These individual forces must be added *as vectors* to determine the total force $\sum \vec{F}$ acting on the object. Once the total force is known, Equation 4.1 can be used to compute the object's acceleration, which then leads to its velocity and displacement.

Conditions for Translational Equilibrium

For simplicity, let's first consider problems in which the velocity and acceleration are both zero. This area of mechanics is known as *statics*, and in such situations we say that an object is in *translational equilibrium*. If these cases seem uninteresting to you, imagine being on a bridge or in a building that is not in translational equilibrium, but is accelerating instead! The problem of when an object is or is not in translational equilibrium is thus very important for many engineering applications. To simplify our terminology, we often drop the word *translational* and refer to such objects as being in "static equilibrium" or just "equilibrium."

From Equation 4.1, we see that if the acceleration is zero, the total force on the object $\sum \vec{F}$ must also be zero. Since there may be several separate forces $\vec{F}_1, \vec{F}_2, \vec{F}_3, \ldots$ acting on an object, the condition $\sum \vec{F} = 0$ means that the sum of all these forces must obey

$$\sum \vec{F} = \vec{F}_1 + \vec{F}_2 + \vec{F}_3 + \cdots = 0 \qquad (4.2)$$

which is known as the *condition for translational equilibrium*.[1]

Let's apply the condition for translational equilibrium to a situation we encountered in Chapter 3. Figure 4.1 shows a person attempting to push a refrigerator across a level floor.[2] The person is applying only a small force, so the refrigerator is at rest with zero acceleration. There are four forces acting on the refrigerator. Two of these forces are in the vertical direction (y): the force of gravity (i.e., the refrigerator's weight) and the normal force acting between the floor and the refrigerator. Two other forces are acting in the horizontal direction (x): the force exerted by the person and the force of static friction.

To apply Equation 4.2 to this case, we first draw a free-body diagram (Fig. 4.1B). This diagram also shows our coordinate system, with the usual horizontal and vertical axes denoted as x and y. Our next step is to express the forces in terms of their components along x and y. In this example, that work is already done for us since each of the four forces is directed along either x or y. We can then proceed to apply Equation 4.2. In this case, Equation 4.2 actually contains two relations, one for the x components of the forces and another for the y components:

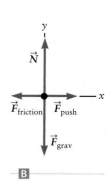

$$\sum F_x = 0 \quad \text{and} \quad \sum F_y = 0 \qquad (4.3)$$

In words, this equation says that for the refrigerator to be in equilibrium, the sum of all the forces along x must be zero and the sum of all forces along y must be zero. Applying this concept to Figure 4.1, we have $N - mg = 0$ for the y components and $F_{\text{push}} - F_{\text{friction}} = 0$ for the x components. These results are identical to what we

Figure 4.1 ◻ A Four forces are acting on this refrigerator. Two of them, the normal force $\vec{N}$ exerted by the floor and the force of gravity $\vec{F}_{\text{grav}}$, lie along the vertical direction (y). The other two forces, the force $\vec{F}_{\text{push}}$ exerted by the person and the force of friction $\vec{F}_{\text{friction}}$, lie along the horizontal direction (x). ◻ B Free-body diagram for the refrigerator.

---

[1]When we discuss objects in translational equilibrium, we will also assume that their velocity is zero. If the object is free to rotate, there is a separate condition for rotational equilibrium that we describe in Chapter 8.

[2]For simplicity, we assume the refrigerator does not tip.

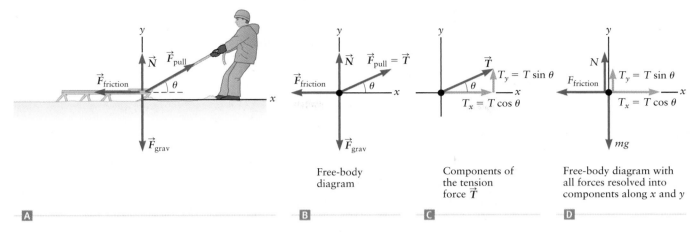

Free-body
diagram

Components of
the tension
force $\vec{T}$

Free-body diagram with
all forces resolved into
components along x and y

A B C D

derived in Section 3.3. This analysis could then be the first step in finding the minimum value of F_{push} needed to just break the equilibrium condition and accelerate the refrigerator.

Figure 4.2 shows a case in which the forces do not all align with the x or y directions. Here a sled is stuck in the snow, and a child is attempting to extract it by pulling with a rope that makes an angle θ with respect to the x axis. If the child pulls with a force $\vec{F}_{pull}$, the force on the sled due to the tension in the rope is $\vec{T} = \vec{F}_{pull}$. A full free-body diagram showing all the forces exerted on the sled is given in Figure 4.2B, and we see that several of the forces are parallel to our chosen coordinate axes. The normal force is parallel to y, the force of gravity is along $-y$, and the force of friction is parallel to the x axis. However, the force exerted by the rope on the sled has components along both x and y. These components are shown in Figure 4.2C, where we also show how the vector $\vec{T}$ and its components along x and y form a right triangle. From this triangle, we find that the component of $\vec{T}$ along x is $T_x = T \cos \theta$ whereas the component along y is $T_y = T \sin \theta$.

In Figure 4.2D, we have redrawn the free-body diagram with $\vec{T}$ replaced by its components, so all the forces in the free-body diagram are now parallel to the coordinate axes. We next apply the conditions for equilibrium in two dimensions (Eq. 4.3). For the forces along x,

$$\sum F_x = T_x - F_{friction} = T \cos \theta - F_{friction} = 0$$

whereas for the forces along y,

$$\sum F_y = N - mg + T_y = N - mg + T \sin \theta = 0 \qquad (4.4)$$

In Example 4.1, we show how this line of analysis can be used.

Figure 4.2 A The force exerted by the rope on the sled has components along the horizontal and the vertical directions. B Free-body diagram for the sled. C Resolving the tension force into x and y components. D Modified free-body diagram that can be used to write equations for the force components in both the x and y directions.

EXAMPLE 4.1 | Trying to Move a Sled

Suppose the sled in Figure 4.2 has a mass of 12 kg and the child is exerting a force of magnitude $F_{pull} = T = 200$ N. Find the angle θ at which the normal force N becomes zero. What is happening physically when $N = 0$?

RECOGNIZE THE PRINCIPLE

We need to apply the conditions for static equilibrium to the sled. Given the mass of the sled and the magnitude of F_{pull}, these conditions cannot be satisfied if the angle θ is too large.

SKETCH THE PROBLEM

Figure 4.2 describes the problem; it shows all the forces on the sled and their components along x and y.

IDENTIFY THE RELATIONSHIPS

From Equation 4.4, for the forces along the vertical (y),

$$\sum F_y = +N - mg + T\sin\theta = N - mg + F_{pull}\sin\theta = 0$$

We want to find the value of the angle at which $N = 0$. Inserting this condition leads to

$$F_{pull}\sin\theta = mg$$

$$\sin\theta = \frac{mg}{F_{pull}}$$

SOLVE

We can find the value of the angle θ using the inverse sine function ($\sin^{-1}$):

$$\theta = \sin^{-1}\left(\frac{mg}{F_{pull}}\right)$$

Inserting the given values of m and F_{pull} gives us

$$\theta = \sin^{-1}\left(\frac{mg}{F_{pull}}\right) = \sin^{-1}\left[\frac{(12\ kg)(9.8\ m/s^2)}{200\ N}\right]$$

$$\theta = \sin^{-1}(0.59) = \boxed{36°}$$

What does it mean?

The normal force is due to contact between the sled and the ground. When the child pulls at an angle of 36°, the normal force goes to zero, indicating that contact between the sled and the ground is being lost.

CONCEPT CHECK 4.1 | Pulling on the Sled

What will happen in Example 4.1 if the angle θ is made even larger than 36°?
(a) The sled will not move.
(b) The sled will be lifted off the ground.
(c) The rope will break.

A Tightrope Walker in Equilibrium

Figure 4.3 shows an interesting equilibrium situation in which several forces are involved. The photo in part (a) shows a person standing at the middle of a tightrope. Suppose the walker has a mass $m = 60$ kg and the tension in the rope is $T = 800$ N. What angle θ does the rope make with the x axis?

We construct, as usual, a free-body diagram (Fig. 4.3B). Since the tightrope walker and the rope are both at rest (at least the tightrope walker hopes so), we can apply our conditions for translational equilibrium to the small piece of rope on which the tightrope walker stands. Imagine the rope is composed of two separate pieces, left and right, tied together with a small knot, and the tightrope walker is standing on the knot. These two pieces of rope have tensions T_{right} and T_{left}. The walker is located at the midpoint of the rope, so you should suspect that these tensions are equal. (In the end, we'll set both of them equal to T.) However, when analyzing the equilibrium conditions it is useful to distinguish between the tensions on both sides of the knot.

The free-body diagram in Figure 4.3B shows the forces acting on the knot. There are tension forces from the left and right sections of the rope as well as a downward force $\vec{F}_{grav}$ whose magnitude is equal to the weight of the tightrope walker (mg). Our next step is to pick a coordinate system; we choose the usual x–y coordinate axes along the horizontal and vertical directions, respectively. We then express all the forces in terms of their components along x and y as shown in Figure 4.3B.

A

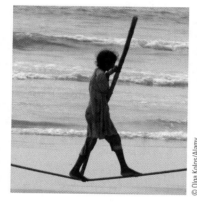

Free-body diagram

Tightrope walker

B

Figure 4.3 ⒶＢoth sections of the rope exert a tension force on the portion at the center where the tightrope walker is standing. Ｂ Free-body diagram for a tightrope walker located at the rope's center point.

With these components, we can apply the conditions for translational equilibrium in Equation 4.3 to get

$$\sum F_x = +T_{\text{right}} \cos\theta - T_{\text{left}} \cos\theta = 0 \tag{4.5}$$

$$\sum F_y = +T_{\text{right}} \sin\theta + T_{\text{left}} \sin\theta - mg = 0 \tag{4.6}$$

From Equation 4.5,

$$T_{\text{right}} \cos\theta = T_{\text{left}} \cos\theta$$

and hence

$$T_{\text{right}} = T_{\text{left}}$$

The tensions in the two sides of the rope are thus equal (as we had already expected), and we can write $T_{\text{right}} = T_{\text{left}} = T$, where T is "the" tension in the rope. From Equation 4.6, we see that there are two vertical components of the tension forces, one from the left side of the rope and one from the right. These forces act to support the tightrope walker. Inserting $T_{\text{right}} = T_{\text{left}} = T$ into Equation 4.6 gives

$$+T_{\text{right}} \sin\theta + T_{\text{left}} \sin\theta = 2T \sin\theta = mg$$

We can now solve for θ

$$\sin\theta = \frac{mg}{2T}$$

$$\theta = \sin^{-1}\left(\frac{mg}{2T}\right)$$

Inserting the given values of m and T leads to

$$\theta = \sin^{-1}\left(\frac{mg}{2T}\right) = \sin^{-1}\left[\frac{(60 \text{ kg})(9.8 \text{ m/s}^2)}{2(800 \text{ N})}\right] = 22°$$

PROBLEM SOLVING | Plan of Attack for Problems in Statics

1. **RECOGNIZE THE PRINCIPLE.** For an object to be in static equilibrium, the sum of all the forces on the object must be zero. This principle leads to Equation 4.2, which can be applied to calculate any unknown forces in the problem.

2. **SKETCH THE PROBLEM.** It is usually a good idea to show the given information in a picture, which should include a coordinate system. Figures 4.1 through 4.3 and the following examples provide guidance and advice on choosing coordinate axes.

3. **IDENTIFY THE RELATIONSHIPS.**
   - Find all the forces acting on the object that is (or should be) in equilibrium and construct a free-body diagram showing all the forces on the object.
   - Express all the forces on the object in terms of their components along x and y.
   - Apply the conditions[3] $\sum F_x$ and $\sum F_y = 0$.

4. **SOLVE.** Solve the equations resulting from step 3 for the unknown quantities. The number of equations must equal the number of unknown quantities.

5. Always *consider what your answer means* and check that it makes sense.

---

[3]Forces along the z direction, if any, should also be included in your free-body diagram and the additional condition $\sum F_z = 0$ must be applied.

Free-body diagram

Figure 4.4 Example 4.2. **A** Pulling on a car. **B** Free-body diagram for the "imaginary knot" at the center of the cable. These drawings show top views of the predicament. (Note the angle θ is not drawn to scale.)

EXAMPLE 4.2 | Stuck in the Mud

Your car is stuck in the mud and you ask a friend to help you pull it free using a cable. You tie one end of the cable to your car and then pull on the other end with a force of 1000 N. Unfortunately, the car does not move. Your friend then suggests you tie the other end of the cable to a tree as shown in Figure 4.4. Although you are skeptical that your friend's idea will help, you try it anyway and find that when a force $\vec{F}$ with the same magnitude (1000 N) is applied to the middle of the cable in the direction shown in Figure 4.4, you are able to pull the car free. Why does this work? Assume an angle $\theta = 10°$ as indicated in Figure 4.4.

RECOGNIZE THE PRINCIPLE

Our objective is to calculate the tension in the cable given the force $\vec{F}$ exerted by you and your friend as shown in Figure 4.4. The value that we find for the tension T will *not* necessarily be the same as the magnitude of $\vec{F}$. We imagine the car is just on the verge of moving, so we can apply our conditions for equilibrium to this situation. We need to think carefully in selecting the object to which we should apply these conditions. We could apply them to the car, but that would not help calculate T. However, if we consider the forces acting on the cable at point O (the point at which you and your friend exert your force) we can then find the tension in terms of the applied force $\vec{F}$. This approach is similar to the problem of the tightrope walker in Figure 4.3, where we considered the forces applied to an imaginary knot in the rope.

SKETCH THE PROBLEM

We follow the steps listed in the Plan of Attack for Problems in Statics and apply the conditions for static equilibrium to the cable at point O. Figure 4.4A shows the layout of the problem, with one end of the cable tied to the car and the other end to a tree, and part (b) shows a free-body diagram with the forces acting on the knot at point O.

IDENTIFY THE RELATIONSHIPS

Three forces act on the cable at point O: the force $\vec{F}$ and the tension forces from each side of the cable. Choosing the x–y coordinate system shown in the figure, we calculate the components of these three forces along x and y, and then apply Equation 4.3. Along y,

$$\sum F_y = +T_{\text{right}} \sin \theta + T_{\text{left}} \sin \theta - F = 0 \tag{1}$$

Since we have a single continuous cable and the angles on the two sides are equal, the tensions in the left and right portions of the cable are the same, $T_{\text{right}} = T_{\text{left}} = T$. Inserting this result into Equation (1) leads to

$$2T \sin \theta - F = 0 \tag{2}$$

SOLVE

Rearranging Equation (2) to solve for T, we find

$$T = \frac{F}{2 \sin \theta}$$

Inserting the value of θ (Fig. 4.4) gives

$$T = \frac{F}{2 \sin \theta} = \frac{F}{2 \sin(10°)} \approx \boxed{2.9 \times F}$$

What does it mean?
The tension in the cable is thus *larger* than F! By tying the cable to a tree, T is larger than it would have been had you simply pulled on one end of the cable with the same force F. Hence, your friend was right. This arrangement essentially *amplifies*

the force applied to the car. Recall that we encountered another example of amplifying forces in Chapter 3 when we discussed the block-and-tackle device. We'll say more about amplifying forces when we discuss work and energy in Chapter 6.

Static Equilibrium and Frictional Forces

In all our examples so far, we have considered objects on a level (horizontal) floor or road. Real life contains hills, so let's consider the situation in Figure 4.5 in which a car is parked on a steep hill. Imagine it is winter, when the surface of the hill might be slippery, and we want to determine if we can park the car safely or if it will instead slide down the hill. That is, we want to know if the car will be in equilibrium. Following our general strategy for problems in statics (in the Plan of Attack for Problems in Statics), Figure 4.5B shows the free-body diagram for the car. This diagram contains three forces: the force from gravity $\vec{F}_{\text{grav}}$, the normal force exerted by the surface $\vec{N}$, and the force of friction $\vec{F}_{\text{friction}}$, which is directed up the hill. Our next step is to choose a coordinate system. It might seem natural to choose coordinate axes that are along the horizontal and the vertical. Although this choice is a reasonable one, let's instead choose axes that are parallel and perpendicular to the plane defined by the hill as shown in Figure 4.5. With this choice of coordinate axes, the normal force is purely along y, whereas the frictional force is along x; we only have to worry about finding the components of the gravitational force.[4] Another reason for choosing x to be parallel to the hill is that were the car to move (as might happen in future problems), its velocity and acceleration would be purely along x. The trigonometry needed to find the components of the gravitational force $\vec{F}_{\text{grav}}$ along x and y is shown in Figure 4.5B.

Using the forces and components from Figure 4.5B, we apply the conditions for translational equilibrium. Along x,

$$\sum F_x = F_{\text{grav}} \sin \theta - F_{\text{friction}} = 0$$

where we have taken the $+x$ direction to be downward along the hill. The magnitude of the force of gravity is $F_{\text{grav}} = mg$; inserting this we get

$$mg \sin \theta - F_{\text{friction}} = 0 \tag{4.7}$$

Along y,

$$\sum F_y = N - F_{\text{grav}} \cos \theta = N - mg \cos \theta = 0$$

which leads to

$$N = mg \cos \theta \tag{4.8}$$

Unlike many of our previous problems, the normal force here is *not* equal to mg. We see from Equation 4.8 that the value of N depends on the angle of the hill; N will be much smaller than the weight if θ is large.

If we know the coefficient of friction for the car's tires with the hill and also the angle of the hill, we can determine if the car is in danger, that is, whether it will slip. From Equation 4.7, we find that the minimum frictional force required to keep the car from slipping is

$$F_{\text{friction}}(\text{required}) = mg \sin \theta \tag{4.9}$$

This result is static friction, and we know from Chapter 3 that the force of static friction can be as large as $\mu_S N$ but no larger. If the required force of friction in Equation 4.9 is greater than $\mu_S N$, the car will slip. On the other hand, if the required frictional

---

[4]Using coordinate axes oriented along the vertical and horizontal directions would lead to precisely the *same* answer as found with the axes chosen in Figure 4.5, although the algebra corresponding to Equations 4.7 and 4.8 would be a little different.

A

Free-body diagram

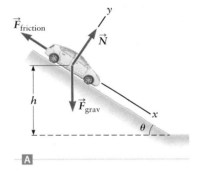

B

Figure 4.5 Three forces are acting on the car. For the car to be in translational equilibrium, the three forces must add to zero. We choose the x and y axes to be parallel and perpendicular (respectively) to the plane defined by the hill. We then express all the forces in terms of their components along these two coordinate directions. Notice that the magnitude of the force of gravity is mg.

force in Equation 4.9 is equal to $\mu_S N$, the car will just barely be in equilibrium. In this case,

$$\mu_S N = \mu_S mg \cos \theta = F_{\text{friction}}(\text{required}) = mg \sin \theta$$

where we have used the result for N from Equation 4.8. Solving for θ gives

$$\mu_S mg \cos \theta = mg \sin \theta$$

$$\frac{\sin \theta}{\cos \theta} = \frac{\mu_S mg}{mg}$$

$$\tan \theta = \mu_S \qquad (4.10)$$

Let's estimate the value of θ for a car just barely in equilibrium on a snowy street. The coefficient of static friction for rubber tires with snow varies with the way the snow is compacted, but is typically in the neighborhood of $\mu_S \approx 0.2$. Inserting this value into Equation 4.10 leads to $\theta \approx 10°$. This angle corresponds to a steep but realistic street, so our result suggests that a car parked on such a street might not always be in equilibrium under snowy conditions.

A

Free-body diagram

B

Figure 4.6 Example 4.3. **A** Forces acting on a flag suspended at a top corner by a cable in a stiff wind. **B** Free-body diagram for the flag.

| **EXAMPLE 4.3** | Supporting a Flag |

A flag of mass $m = 3.5$ kg is suspended by a single cable attached at the top corner as shown in Figure 4.6A. Suppose a stiff breeze exerts a horizontal force of 75 N on the flag. Find the angle θ at which the cable hangs from the flagpole. For simplicity, we assume that the flag stays flat as sketched in Figure 4.6A.

RECOGNIZE THE PRINCIPLE

We want the flag to be in static equilibrium. The unknowns in this problem are the magnitude of the tension T and the angle θ. Since we have two unknowns, we must use the conditions for equilibrium in two dimensions (Eq. 4.3) to obtain two equations to solve the problem.

SKETCH THE PROBLEM

Figure 4.6 shows the forces acting on the flag: the tension force from the cable $\vec{T}$, the force from the wind $\vec{F}_{\text{wind}}$, and the force of gravity on the flag $\vec{F}_{\text{grav}}$. The picture shows our choice of coordinate axes with x and y along the horizontal and vertical, respectively. The free-body diagram shows the components of all the forces along x and y.

IDENTIFY THE RELATIONSHIPS

Writing the condition for equilibrium for the force components along x gives us

$$\sum F_x = F_{\text{wind}} - T \sin \theta = 0$$

$$T \sin \theta = F_{\text{wind}} \qquad (1)$$

Along y,

$$\sum F_y = T \cos \theta - F_{\text{grav}} = 0$$

$$T \cos \theta = F_{\text{grav}} = mg \qquad (2)$$

SOLVE

We can solve for θ by taking the ratio of Equations (1) and (2):

$$\frac{T \sin \theta}{T \cos \theta} = \frac{F_{\text{wind}}}{mg}$$

$$\tan \theta = \frac{F_{\text{wind}}}{mg}$$

Inserting the given values of the force exerted by the wind and of m leads to

$$\theta = \tan^{-1}\left(\frac{F_{\text{wind}}}{mg}\right) = \tan^{-1}\left[\frac{75 \text{ N}}{(3.5 \text{ kg})(9.8 \text{ m/s}^2)}\right] = \boxed{65°}$$

What have we learned?
Our work on this problem began with a general picture together with a free-body diagram. The same approach will be useful when we apply Newton's second law to calculate the motion of objects, beginning with the motion of projectiles in Section 4.2.

4.2 | PROJECTILE MOTION

In Section 4.1, we considered various objects in static equilibrium in two and three dimensions. In those cases, the velocity and acceleration were both zero and we only had to deal with the forces acting on the object. We're now ready to consider objects that are in motion and learn how to calculate their displacement, velocity, and acceleration. In this section, we use *projectile motion* to illustrate the basic ideas. Projectile motion is the motion of a baseball, arrow, rock, or similar object that is thrown (i.e., *projected*), typically through the air. For the simplest type of projectile motion, the only force acting on the projectile is the force of gravity, and we ignore (at least for now) the force from air drag.

It is convenient to use the coordinate system sketched in Figure 4.7, with x along the horizontal and y along the vertical direction. The force of gravity has components

$$F_{\text{grav}, x} = 0 \qquad F_{\text{grav}, y} = -mg \qquad \text{(4.11)}$$

Writing Newton's second law (Eq. 4.1) in component form, we have

$$\sum F_x = ma_x \qquad \sum F_y = ma_y \qquad \text{(4.12)}$$

For the case of simple projectile motion, the only force is due to gravity, so Equation 4.11 gives the total force. Inserting this into Newton's second law (Eq. 4.12) gives

$$\sum F_x = 0 = ma_x$$
$$a_x = 0 \qquad \text{(4.13)}$$

and

$$\sum F_y = -mg = ma_y$$
$$a_y = -g \qquad \text{(4.14)}$$

Hence, the acceleration is constant along both x and y. In fact, the component of the acceleration along x is zero.

Now comes a crucial point: the motions along x and y are *independent* of each other. According to Equation 4.12, the acceleration along x is determined by the force along x, while the acceleration along y is determined by the force along y. For the gravitational force, the components of the force along x and y are independent of each other (Eq. 4.11), so the accelerations along x and y are also independent. According to Equation 4.13, the motion of a projectile along x is completely equivalent to one-dimensional motion along the x axis with an acceleration $a_x = 0$. Hence, the results found in Chapter 3 for the behavior of the position and velocity for motion in one dimension can be applied directly to give the x component of the position and the x component of the velocity for a projectile. Likewise, according to Equation 4.14, the motion of our projectile along the y direction is equivalent to one-dimensional motion along y with an acceleration $a_y = -g$. Hence, we can again use our results for one-dimensional motion to calculate the y component of the position and the y component of the velocity of a projectile. In fact, the motion along y is the same as simple free fall, which we studied in Chapter 3. As a result, *projectile motion*

Free-body diagram

Figure 4.7 **A** The dashed line shows the trajectory of a thrown or batted baseball. **B** Free-body diagram for the ball. Throughout the entire course of the motion, the force on the ball $\vec{F}_{\text{grav}}$ is directed downward with a magnitude equal to mg.

Figure 4.8 This car's motion is an example of projectile motion. The car's velocity vector is always tangent to the trajectory.

is simply two cases of motion with constant acceleration: one along x and one along y. The relationships among displacement, velocity, acceleration, and time given in Table 3.1 for constant acceleration thus apply directly to projectile motion.

Rolling Off a Cliff

An example of projectile motion is shown in Figure 4.8, which shows a car rolling off a cliff. Suppose the car leaves the cliff with a velocity of 10 m/s directed along the horizontal. If the cliff has a height $h = 20$ m, where and when will the car land? To attack this problem, we begin, as usual, with a picture that establishes our coordinate system and contains the starting information. In Figure 4.8, we choose x and y to be horizontal and vertical, with the origin at the bottom of the cliff. We then apply our relations for motion with constant acceleration from Table 3.1. Here we are given information on where the car starts and are asked to find where and when it lands, so we choose the relation that involves position and time. Writing this equation twice, once for the horizontal motion (x) and once for the vertical motion (y), we have

$$x = x_0 + v_{0x}t + \tfrac{1}{2}a_xt^2 = 0 + v_{0x}t + 0 = v_{0x}t \qquad (4.15)$$

$$y = y_0 + v_{0y}t + \tfrac{1}{2}a_yt^2 = h + 0 - \tfrac{1}{2}gt^2 = h - \tfrac{1}{2}gt^2 \qquad (4.16)$$

where we have used the fact that the car is initially at $x_0 = 0$ and that $y_0 = h$ (the height of the cliff). The initial velocity components are $v_{0x} = 10$ m/s and $v_{0y} = 0$ since the car is moving horizontally at the moment it leaves the cliff. Notice again that the acceleration along x is zero (so $a_x = 0$ and v_x is constant) and that $a_y = -g$. Equations 4.15 and 4.16 give the position of the car as a function of time. This position is a vector quantity and has components x and y that vary with time. We can calculate when the car lands by computing the value of t when the car reaches $y = 0$ because this point is ground level (the bottom of the cliff). From Equation 4.16, we have

$$y = h - \tfrac{1}{2}gt^2 = 0$$

$$t = \sqrt{\frac{2h}{g}} = \sqrt{\frac{2(20 \text{ m})}{9.8 \text{ m/s}^2}} = 2.0 \text{ s} \qquad (4.17)$$

To find where the car lands, we use this value of t in Equation 4.15:

$$x = v_{0x}t = (10 \text{ m/s})(2.0 \text{ s}) = 20 \text{ m}$$

EXAMPLE 4.4 | Driving Off a Cliff

For the car in Figure 4.8, find the x and y components of the velocity and also the speed of the car just before it hits the ground.

RECOGNIZE THE PRINCIPLE

We use the relations for motion with constant acceleration from Table 3.1. We are interested in the velocity, and we have to deal with two equations, one for the horizontal component of the velocity v_x, and another for the vertical component v_y. So,

$$v_x = v_{0x} + a_xt = v_{0x} \qquad (1)$$

$$v_y = v_{0y} + a_yt = v_{0y} - gt \qquad (2)$$

Here we have also inserted values for the components of the acceleration, $a_x = 0$ and $a_y = -g$, for our projectile.

SKETCH THE PROBLEM

Figure 4.8 shows the trajectory of the car and also the velocity and its components along x and y just before the car hits the ground. Notice that the direction of $\vec{v}$ is *parallel* (i.e., tangent) to the trajectory curve.

IDENTIFY THE RELATIONSHIPS

To find the velocity when the car reaches the ground, we evaluate v_x and v_y in Equations (1) and (2) at the value of t at which the car hits the ground as found in Equation 4.17. We also use the given values of the initial velocity, $v_{0x} = 10$ m/s and $v_{0y} = 0$.

SOLVE

We find

$$v_x = v_{0x} = \boxed{10 \text{ m/s}}$$

and

$$v_y = v_{0y} - gt = 0 - (9.8 \text{ m/s}^2)(2.0 \text{ s}) = \boxed{-20 \text{ m/s}}$$

These are the components of the velocity. The speed v of the car is the magnitude of the velocity vector:

$$v = \sqrt{v_x^2 + v_y^2} = \sqrt{(10 \text{ m/s})^2 + (-20 \text{ m/s})^2} = \boxed{22 \text{ m/s}}$$

What does it mean?
Notice that we have calculated the velocity and speed the instant *before* the car strikes the ground. When it contacts the ground, there will be additional forces on the car—a normal force exerted by the ground and also a frictional force—and the car's behavior will no longer be described by simple projectile motion.

Independence of the Vertical and Horizontal Motion of Projectiles

The independence of a projectile's vertical and horizontal motions leads to some interesting results. In the example with the car in Figure 4.8, this independence means that the time it takes the car to reach the ground (Eq. 4.17) is independent of the motion along x. This time is the same if the car is initially traveling very fast (large v_{0x}) or very slow, or even if the car is just nudged from the cliff ($v_{0x} \approx 0$).

This result is demonstrated in the time-lapse photos shown in Figure 4.9 showing the motion of two falling balls that are released from the same height at the same time. The ball on the left is simply dropped, while the ball on the right is given an initial velocity along the horizontal direction x. The ball on the left falls directly downward, whereas the ball on the right follows a parabolic path characteristic of projectile motion. While the balls have very different velocities along the horizontal direction, at each "flashpoint" in the photo the two balls are always at the same height, which shows that their displacements and velocities along y are the same. This result confirms that the motion along y does *not* depend on the velocities along the horizontal direction, in accordance with Equations 4.13 through 4.16.

This independence of the motion along x and y is also central to the following hypothetical (and, we hope, amusing) situation. A monkey has escaped from a zoo and is hiding in a tree. The zookeeper wants to capture the monkey without hurting him, so the zookeeper plans to shoot a tranquilizer dart at the monkey (Fig. 4.10). This very clever monkey is watching closely as the rifle is aimed at him. When the monkey observes a flash of light from the rifle (indicating that it has been fired), the monkey releases his grip on the tree and begins to fall. The monkey does this so that he will fall some distance vertically while the dart is traveling on its way to him. The monkey believes his falling will cause the dart to pass over his head and thereby miss him, but will the dart really miss the monkey?

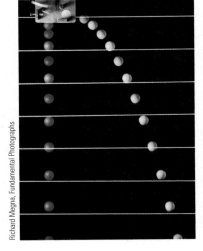

Figure 4.9 Time-lapse photos of two falling balls. The ball on the left falls straight downward, whereas the ball on the right has a nonzero component of velocity along the horizontal direction. The two balls strike the ground at the same time.

Gravity
"off"

A

Real case

$\Delta y = -\frac{1}{2}gt^2$

$\vec{v}$

B

Figure 4.10 A If gravity were "turned off," this dart would follow a straight-line path to the monkey. B When gravity is turned on (which is usually the case!), the dart and the monkey are both accelerated downward and they fall the same distance Δy. The dart thus hits the monkey, but at a location below the monkey's initial position.

Figure 4.10 shows the monkey and his predicament. Part A of the figure shows what would happen if we could (hypothetically) turn off gravity for the duration of the problem. That is, we assume $g = 0$. With this assumption, the acceleration of both the dart and the monkey are zero along both x and y, and both the dart and the monkey move with constant velocities. The dart thus moves in a straight line that takes it from the rifle to the monkey (as sketched), while the monkey does not move at all since his initial velocity is zero. Hence, with gravity turned off, the dart would hit the monkey, as you should have expected.

Now consider the case with gravity turned on. The trajectories of the dart and the monkey in this case are shown in Figure 4.10B. They have the same acceleration: $a_x = 0$, $a_y = -g$. The vertical motions of the dart and the monkey are both described by

$$y = y_0 + v_{0y}t - \tfrac{1}{2}gt^2$$

In words, we can write this result for y as

$$y = \underbrace{y_0 + v_{0y}t}_{\substack{\text{displacement} \\ \text{without gravity}}} \underbrace{- \tfrac{1}{2}gt^2}_{\substack{\text{displacement} \\ \text{due to gravity}}} \tag{4.18}$$

Equation 4.18 emphasizes that the effect of gravity on y is the *same* for the bullet and the monkey. The trajectories of both fall a distance $\Delta y = -\tfrac{1}{2}gt^2$ below the trajectories in the absence of gravity in Figure 4.10A. Hence, gravity causes the dart and the monkey to fall the same distance, so their trajectories still intersect. The dart *will* hit the monkey. The point of this story is that in projectile motion, the motion along the vertical (y) is *independent* of the horizontal motion. The dart and the monkey fall the same amount due to the acceleration caused by gravity, even though their horizontal velocities are very different.

Projectile Motion and Target Practice

Serious sharpshooters understand the independence of vertical and horizontal motion. Figure 4.11 shows a bullet fired horizontally at a target a distance L away. From our previous arguments, we know that the bullet will not travel on a straight-line path to the target. The acceleration caused by the gravitational force makes the bullet "fall" on its way to the target. According to Equation 4.18, the bullet falls a distance $\Delta y = -\tfrac{1}{2}gt^2$ below the straight-line trajectory it would have followed were gravity (hypothetically) turned off.

It is interesting to work out the value of Δy for a typical case. On a shooting range, the distance to the target might be 100 m or more, so let's take $L = 100$ m. The speed of the bullet when it leaves the rifle is typically in the range of 400 m/s to 800 m/s; we'll assume a value in the middle of this range and take $v_{0x} = 600$ m/s (about 1200 mi/h) for our calculation. To compute Δy, we must first calculate the time it takes the bullet to reach the target. Assuming simple projectile motion (so there is no force due to air drag), we can use Equation 4.15 to find

$$x = v_{0x}t = L$$

which leads to

$$t = \frac{L}{v_{0x}} = \frac{100 \text{ m}}{600 \text{ m/s}} = 0.17 \text{ s}$$

Figure 4.11 This bullet's initial velocity is horizontal. Gravity, however, causes the bullet to "fall" a distance Δy while it travels to the target.

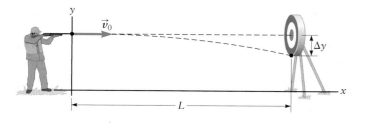

For Δy, we then get

$$\Delta y = -\tfrac{1}{2}gt^2 = -\tfrac{1}{2}(9.8 \text{ m/s}^2)(0.17 \text{ s})^2 = -0.14 \text{ m}$$

The distance the bullet falls on the way to the target is thus quite significant for a serious sharpshooter. In fact, you may now wonder how a sharpshooter can ever hit the bull's-eye. If he aims directly at the center of the target, the bullet will in this example pass 0.14 m $=$ 14 cm below the bull's-eye. The resolution of this "paradox" is that a sharpshooter does not actually aim directly at the target. The sighting telescope on a rifle is calibrated so that when the rifle is "sighted" on the bull's-eye, the barrel is in reality aimed a distance Δy *above* the target. Since the value of Δy depends on the distance to the target and the speed of the bullet, the sighting telescope must be adjusted for the particular target distance and bullet in use. Hence, the rifle "knows" its physics (or, rather, the rifle maker and sharpshooter do).

Motion of a Baseball: Calculating the Trajectory and the Velocity

Let's now consider the motion of a batted baseball. Our projectile—the baseball—starts from some initial position with a specified initial velocity, and we want to calculate its trajectory. The problem is sketched in Figure 4.12. The ball has an initial position with $x_0 = 0$ and $y_0 = h$, where h is the height of the ball when it leaves the bat. The ball is hit with an initial speed v_0 at an angle θ with respect to the horizontal, so the initial components of its velocity are $v_{0x} = v_0 \cos \theta$ and $v_{0y} = v_0 \sin \theta$. Our relations for motion with constant acceleration then have the form

$$x = x_0 + v_{0x}t + \tfrac{1}{2}a_x t^2 = 0 + v_{0x}t + 0 = v_0(\cos \theta)t \tag{4.19}$$

and

$$y = y_0 + v_{0y}t + \tfrac{1}{2}a_y t^2 = h + v_{0y}t - \tfrac{1}{2}gt^2 = h + v_0(\sin \theta)t - \tfrac{1}{2}gt^2 \tag{4.20}$$

To completely describe the motion, we also need the velocity:

$$v_x = v_{0x} + a_x t = v_0 \cos \theta \tag{4.21}$$

$$v_y = v_{0y} + a_y t = v_0 \sin \theta - gt \tag{4.22}$$

It is useful to plot these results in several different ways. Figures 4.13 and 4.14 show the components of the ball's position and velocity as functions of time. The results for the x direction are similar to what we would find for the bullet in Figures 4.10 and 4.11. Since there is no acceleration along the horizontal direction, v_x is constant and x varies linearly with t. For the motion along y, we have an acceleration of $-g$, hence y varies quadratically with t. This behavior for x and y leads to a parabolic trajectory; the path followed by the baseball is a parabola in the x–y plane as sketched in Figure 4.12. This path is symmetric in the following sense: the trajectory followed on the way toward the highest point is a mirror image of the trajectory on the way down. Moreover, the time the ball spends traveling from some initial height (such as h) to the top is the same as the time it spends traveling back down to that same height.

For the velocity, we find that v_x is simply a constant (as already mentioned) since the acceleration along x is zero, while v_y varies linearly with t, with slope $-g$. It is instructive to show the velocity components at different points along the trajectory as plotted in Figure 4.15. This is just another way of displaying the results contained in Equations 4.21 and 4.22 and in Figure 4.14. The velocity also displays a symmetry like that found for the trajectory. For example, the value of v_y at some point on the way up (i.e., at some particular height) is equal in magnitude but opposite in sign compared with v_y when the ball is at the same height on the way down. In other words, if a baseball begins its motion with speed v_0 at an initial height $y = h$, it will have the same speed when it is at this height on the way back down.

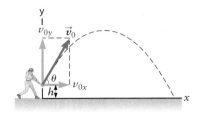

Figure 4.12 Trajectory of a batted baseball.

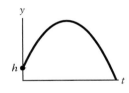

Figure 4.13 Plots of the x and y coordinates of a baseball as it travels along its trajectory.

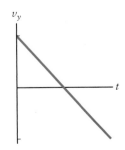

Figure 4.14 Plots of the velocity components v_x and v_y as a baseball travels along its trajectory.

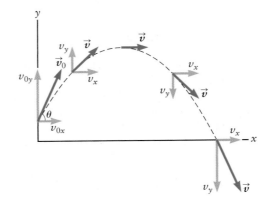

Figure 4.15 Components of the velocity at various points along the trajectory of a baseball. Compare with Figure 4.14.

CONCEPT CHECK 4.2 | Trajectory of a Baseball

As the baseball in Figure 4.12 travels through the air, at what point(s) on its trajectory is

    (a) the acceleration zero?
    (b) the vertical component of the velocity zero?
    (c) the force on the ball zero?
    (d) the speed zero?

Motion of a Baseball: Analyzing the Results

Let's now consider a few quantitative examples involving the baseball's trajectory in Figures 4.12 through 4.15. To be somewhat realistic, we assume the ball has an initial speed of 45 m/s (about 100 mi/h) with $\theta = 30°$ and it starts at an initial height of $h = 1.0$ m.

Top of the trajectory (maximum height). Suppose we want to find both the height and the speed of the ball when it reaches the uppermost point on its trajectory. We can find the time at which it reaches this point by knowing that $v_y = 0$ at that instant. Inserting this into Equation 4.22 gives

$$v_y = 0 = v_0 \sin \theta - gt_{\text{top}}$$

$$t_{\text{top}} = \frac{v_0 \sin \theta}{g} \tag{4.23}$$

Using Equation 4.20, the ball's height at this moment is

$$y_{\text{top}} = h + v_0(\sin \theta)t_{\text{top}} - \tfrac{1}{2}gt_{\text{top}}^2 \tag{4.24}$$

Inserting our initial conditions into these results, we get

$$t_{\text{top}} = \frac{v_0 \sin \theta}{g} = \frac{(45 \text{ m/s})\sin(30°)}{9.8 \text{ m/s}^2} = 2.3 \text{ s}$$

and

$$y_{\text{top}} = h + v_0(\sin \theta)t_{\text{top}} - \tfrac{1}{2}gt_{\text{top}}^2$$

$$y_{\text{top}} = (1.0 \text{ m}) + (45 \text{ m/s})\sin(30°)(2.3 \text{ s}) - \tfrac{1}{2}(9.8 \text{ m/s}^2)(2.3 \text{ s})^2 = 27 \text{ m}$$

The ball's speed at this moment will be

$$v_{\text{top}} = \sqrt{v_x^2 + v_y^2} = \sqrt{v_x^2 + 0} = v_x = v_0 \cos \theta \tag{4.25}$$

since $v_y = 0$ at the top. Indeed, we could have obtained this result for the speed directly from Figure 4.14. Evaluating Equation 4.25, we find

$$v_{\text{top}} = v_0 \cos \theta = (45 \text{ m/s})\cos(30°) = 39 \text{ m/s}$$

Landing (end of trajectory). To find the velocity of the baseball just before it hits the ground, we can use the fact that ground level is at $y = 0$. Using Equation 4.20, we get

$$y_{\text{lands}} = 0 = h + (v_0 \sin \theta)t_{\text{lands}} - \tfrac{1}{2}gt^2_{\text{lands}} \tag{4.26}$$

which gives us a quadratic equation to solve for t_{lands}. The solution[5] is $t_{\text{lands}} = 4.64$ s. (Here we are keeping three significant figures to minimize rounding errors in the next calculation.) To find the speed at this moment, we first need to find the component of the velocity along y, which we can obtain from Equation 4.22:

$$v_{\text{lands}, y} = v_0 \sin \theta - gt_{\text{lands}}$$

$$v_{\text{lands}, y} = (45 \text{ m/s})\sin(30°) - (9.8 \text{ m/s}^2)(4.64 \text{ s}) = -23 \text{ m/s}$$

We can now compute the speed using $v_{\text{lands}} = \sqrt{v^2_{\text{lands}, x} + v^2_{\text{lands}, y}}$. The component v_x is constant (see Eq. 4.21 and Fig. 4.14), so $v_{\text{lands},x} = v_x = v_0 \cos \theta = 39$ m/s as calculated above. Inserting these results we find

$$v_{\text{lands}} = \sqrt{v^2_{\text{lands}, x} + v^2_{\text{lands}, y}} = \sqrt{(39 \text{ m/s})^2 + (-23 \text{ m/s})^2} = 45 \text{ m/s}$$

We can use these results to check for the symmetry of the trajectory mentioned earlier. For a symmetric trajectory, t_{lands} should be twice the time it takes to reach the top of the trajectory found from Equation 4.23, and the speed just before the ball reaches the ground should be equal to the initial speed. To be strictly correct, we should notice that in this example the trajectory of the ball does not end at its initial height because it starts a small distance above the ground at $y_0 = h = 1.0$ m but lands at $y = 0$. Here, however, h is sufficiently small that to within the two significant figures used in most of our numerical evaluations, t_{lands} is equal to twice the time it takes to reach the peak of the trajectory (t_{top}) and the speed at landing is equal to the initial speed. In Example 4.5, we show that the time of flight is exactly symmetric for objects projected from ground level.

Range of a projectile. In our example with the baseball, we assumed the ball was hit with a particular value of the initial angle θ. It is interesting to consider how the trajectory varies as a function of θ. If you were a baseball coach, you might want to instruct your players on the value of θ they should use to get the ball to travel as far as possible. (This would be applied physics at its best.) This horizontal distance traveled by the ball is called the **range**. To solve this problem, we must calculate the distance traveled by a projectile as a function of angle θ and then find the value of θ that maximizes this distance. For simplicity, let's assume the baseball starts from ground level so that $y_0 = h = 0$. The time the baseball spends in the air can be found from Equation 4.26:

$$y_{\text{lands}} = 0 = h + (v_0 \sin \theta)t_{\text{lands}} - \tfrac{1}{2}gt^2_{\text{lands}} = (v_0 \sin \theta)t_{\text{lands}} - \tfrac{1}{2}gt^2_{\text{lands}}$$

$$t_{\text{lands}} = \frac{2v_0 \sin \theta}{g} \tag{4.27}$$

The ball then lands at the location

$$x_{\text{lands}} = v_{0x}t_{\text{lands}} = v_0(\cos \theta)\frac{2v_0 \sin \theta}{g} = \frac{2v_0^2 \sin \theta \cos \theta}{g}$$

and x_{lands} is also equal to the range. Using the trigonometric identity $\sin \theta \cos \theta = \tfrac{1}{2}\sin(2\theta)$ from Appendix B, we find

$$x_{\text{lands}} = \text{range} = \frac{v_0^2 \sin(2\theta)}{g} \tag{4.28}$$

---

[5]Question 9 at the end of this chapter asks you to use the quadratic formula to solve for t_{lands}.

for a trajectory starting from $x_0 = y_0 = 0$. For a given value of the initial speed v_0, the largest value of the range will be found when $\sin(2\theta) = 1$ since that is the largest possible value of the sine function. So, $2\theta = 90°$, or $\theta = 45°$ gives the maximum range. This result applies to all projectiles that start and end at the same height, provided that the only force acting on the projectile is the force of gravity. When the force from air drag is significant, the range is no longer given by Equation 4.28.

EXAMPLE 4.5 | Symmetry of Projectile Motion

Consider a projectile that starts and ends at ground level. Show that the time spent traveling to the point of maximum height y_{top} is equal to the time spent moving from y_{top} back to the ground.

RECOGNIZE THE PRINCIPLE

We want to prove this result in the general case, so we return to Equations 4.23 and 4.27. We need to compare t_{top} and t_{lands} and show that t_{lands} is exactly twice t_{top}.

SKETCH THE PROBLEM

Figures 4.13 and 4.14 describe the problem. For this example, the initial height is $y_0 = 0$ because the projectile starts at ground level.

IDENTIFY THE RELATIONSHIPS

From Equation 4.23, we have

$$t_{top} = \frac{v_0 \sin \theta}{g} \qquad (1)$$

Applying Equation 4.27 to find when the projectile lands gives

$$t_{lands} = \frac{2v_0 \sin \theta}{g} \qquad (2)$$

SOLVE

By comparing Equations (1) and (2), we see that $t_{lands} = 2t_{top}$. Therefore, the time spent traveling down is *precisely* equal to the time spent traveling to the top.

What does it mean?
This result confirms our earlier claim that the trajectory for a projectile that begins and ends at the same height is symmetric.

EXAMPLE 4.6 | Dropping a Payload

An airplane is carrying relief supplies to a person stranded on an island. The island is too small to land on, so the pilot decides to drop the package of supplies as she flies horizontally over the island. (a) Find the time the package spends in the air; that is, how long does it take the package to travel from the airplane to the island? (b) If the airplane is flying horizontally at an altitude h at speed v_{plane}, where should the pilot release the package?

RECOGNIZE THE PRINCIPLE

For this projectile to hit the target, the pilot knows she must release the package *before* she is directly over the island. We first compute how long it takes the package to fall a distance h and land on the island. This answer will give us the time the package spends in the air. We can then calculate how far along x the package travels during this time, and that will tell us how far in advance of the island the pilot should release the package.

SKETCH THE PROBLEM

The trajectory of the package is sketched in Figure 4.16, which shows snapshots of the airplane and the package at several instants after the package is released. Just after its release (Fig. 4.16A), the package has a velocity along the horizontal equal to the speed of the airplane. Since the package is a projectile, the horizontal (x) component of its velocity is constant while the package falls, so the velocity of the package along the x direction is always the same as the airplane's velocity. The package is therefore always directly under the airplane as it falls to the island as shown in Figure 4.16. This example again shows the independence of the horizontal and vertical motions for a projectile.

IDENTIFY THE RELATIONSHIPS AND SOLVE

(a) The time it takes the package to reach the ground can be calculated from the displacement along y:

$$y = y_0 + v_{0y}t + \tfrac{1}{2}a_y t^2$$

The initial height of the package is $y_0 = h$, and at the instant the package is released, $v_{0y} = 0$. The package's acceleration is $a_y = -g$. We want to find when the package reaches $y = 0$, so we have

$$y = h - \tfrac{1}{2}g t_{\text{lands}}^2 = 0$$

Solving for the time gives

$$t_{\text{lands}} = \sqrt{\frac{2h}{g}}$$

(b) During the time the package is in the air, the airplane and the package both travel a horizontal distance

$$x = x_0 + v_{0x}t = v_{\text{plane}}\, t_{\text{lands}} = \boxed{v_{\text{plane}}\sqrt{\frac{2h}{g}}}$$

What does it mean?
The pilot should release the package when she is this distance (horizontally) from the island.

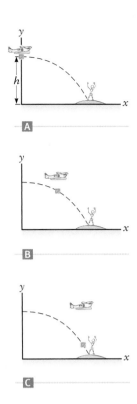

Figure 4.16 Example 4.6. As viewed by the recipient on the island, the package dropped by the airplane is a projectile that has an initial velocity equal to the velocity of the airplane. The package then follows the parabolic trajectory as sketched.

4.3 | A FIRST LOOK AT REFERENCE FRAMES AND RELATIVE VELOCITY

In Example 4.6, we considered the motion of the airplane and the falling package from the point of view of an observer on the ground. According to such an observer, the airplane is moving along the $+x$ direction with a velocity $\vec{v}_{\text{plane}}$ and the package moves along the parabolic path shown in Figure 4.16. Another way to view the problem is from the point of view of the pilot. According to her, the package simply falls straight down to Earth below (Fig. 4.17), while, according to the pilot, the island is moving along $-x$; that is, the island moves (!) relative to the airplane. Although the trajectories viewed by these two observers are different, the time the package takes to fall (t_{lands}) is the *same* from either point of view.

This problem is an example of using two different reference frames. A *reference frame* is an observer's choice of coordinate system, including an origin, for making measurements. In this example, one reference frame (and any observer in it) is at rest relative to the ground and the other reference frame (containing the airplane and its pilot) is moving with a constant velocity relative to the ground. Newton's laws give a

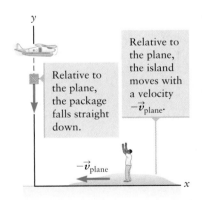

Figure 4.17 As viewed by the pilot of the airplane, the package falls directly downward because the package is always directly beneath the airplane. When viewed in the pilot's reference frame, the island is moving with a velocity $-\vec{v}_0 = -\vec{v}_{\text{plane}}$.

Figure 4.18 **A** Two cars moving along a straight road (i.e., in one dimension). **B** The velocity of car 2 relative to car 1 is $\vec{v}_2 - \vec{v}_1$.

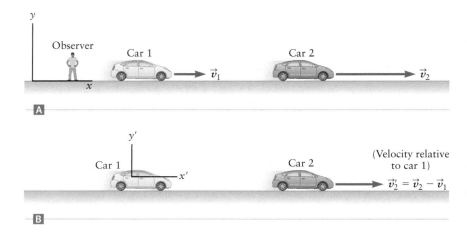

correct description of the motion in any reference frame that moves with a constant velocity. We discuss what happens in other types of reference frames in Section 4.5.

Relative Velocity

When we discuss or calculate the velocity of an object such as the package in Figures 4.16 and 4.17, we are always considering the velocity *relative* to a particular coordinate system or observer; that is, we consider the velocity in a particular reference frame. So, given $\vec{v}$ in one reference frame, how do we find the velocity in a different reference frame? Let's first consider this problem for motion in one dimension.

Figure 4.18A shows two cars (labeled car 1 and car 2) traveling at constant velocities along the x direction. According to an observer (and reference frame) at rest on the sidewalk, the cars have velocities $\vec{v}_1$ and $\vec{v}_2$. Now consider a reference frame at rest with respect to car 1; this reference frame is defined by the coordinate axes x' and y' in Figure 4.18B that "travel along" with car 1. What is the velocity of car 2 in this new reference frame with axes x' and y'? That is, what is the velocity of car 2 as seen by the driver of car 1? The velocity of car 2 *relative to car 1* is just $\vec{v}_2' = \vec{v}_2 - \vec{v}_1$. (You have probably already done this experiment while driving on a highway.) In other words,

velocity of car 2 relative to observer = velocity of car 2 relative to car 1

+ velocity of car 1 relative to observer

This is the general way velocities in different reference frames—that is, the velocities seen by different observers—are related.

The same ideas concerning relative velocities apply for motion in two dimensions. Figure 4.19 shows two people playing in a river: one is floating along with the current, while the other is swimming from one riverbank to the other. Relative to an observer on the shore, the water in this river moves at a velocity $\vec{v}_{current}$ along the x direction, and this current affects the motion of the two people in the river. The person floating along with the current (Fig. 4.19A) moves with a velocity equal to the velocity of the water. In terms of the velocity components in Figure 4.19A,

$$v_{floater,\,x} = v_{current}$$
$$v_{floater,\,y} = 0$$

The person in Figure 4.19B is swimming so as to travel across the river. Let's assume this swimmer is moving relative to the water with a velocity of magnitude v_0 in the y direction. Since the swimmer is carried along with the current, she will also have a nonzero velocity along x, and this velocity component will be equal to $v_{current}$. In words,

velocity of swimmer relative = velocity of swimmer relative to water
to observer on shore   + velocity of water relative to observer

A

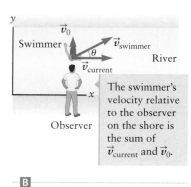

B

Figure 4.19 **A** Relative to an observer on the shore, a person floating along in a river has a velocity equal to the velocity of the river's current $\vec{v}_{current}$. **B** As viewed by the observer on the shore, this swimmer's velocity is equal to the sum of the swimmer's velocity relative to the water plus the velocity of the river current $\vec{v}_{current}$.

In terms of components, we thus have

$$v_{\text{swimmer}, x} = v_{\text{current}} \qquad (4.29)$$

$$v_{\text{swimmer}, y} = v_0$$

According to an observer on the shore in Figure 4.19B, the swimmer's velocity is the sum of the two components in Equation 4.29, so the total speed of the swimmer is

$$v_{\text{swimmer}} = \sqrt{v_{\text{swimmer}, x}^2 + v_{\text{swimmer}, y}^2} = \sqrt{v_{\text{current}}^2 + v_0^2} \qquad (4.30)$$

Moreover, *relative to this observer*, the swimmer does not travel directly across the river, but instead moves at an angle θ. From Figure 4.19B, this angle is given through the relation

$$\tan \theta = \frac{v_{\text{swimmer}, y}}{v_{\text{swimmer}, x}} \qquad (4.31)$$

The swimmer's velocity is thus different for an observer on the shore (Eqs. 4.30 and 4.31) than for the observer who floats along with the water current in Figure 4.19A.

EXAMPLE 4.7 | Swimming Across a River

Suppose the river's current in Figure 4.20 has a speed $v_{\text{current}} = 2.0$ m/s. To travel directly across the river *as viewed by an observer on the shore*, how should the swimmer swim?

RECOGNIZE THE PRINCIPLE

Let the components of the swimmer's velocity *relative to the water* be v_x and v_y (see Fig. 4.20). The velocity of the swimmer relative to an observer on the shore is equal to the velocity of the swimmer relative to the water *plus* the velocity of the water relative to the observer. These velocity vectors are shown in Figure 4.20, which also shows their vector sum. Relative to the observer on the shore, we thus have

$$v_{\text{swimmer}, x} = v_{\text{current}} + v_x \qquad (1)$$

$$v_{\text{swimmer}, y} = v_y$$

To swim directly across the river, we must have $v_{\text{swimmer}, x} = 0$.

SKETCH THE PROBLEM

Figure 4.20 shows the problem, including the components of the swimmer's velocity.

IDENTIFY THE RELATIONSHIPS AND SOLVE

Inserting $v_{\text{swimmer}, x} = 0$ into Equation (1), we find

$$v_{\text{current}} + v_x = 0$$

$$v_x = -v_{\text{current}}$$

Using the given value of v_{current}, we get

$$v_x = \boxed{-2.0 \text{ m/s}} \qquad (2)$$

What does it mean?
The minus sign in Equation (2) means that the swimmer must swim against the current. Notice that the value of v_y is not important; as long as it is not zero, the swimmer will be able to swim across the river.

Figure 4.20 Example 4.7.

Free-body diagram

Figure 4.21 Hypothetical spacecraft moving through deep space. The force from the engine $\vec{F}_E$ causes an acceleration along the y direction.

4.4 | FURTHER APPLICATIONS OF NEWTON'S LAWS

In Sections 4.2 and 4.3, we dealt with motion that results from the force of gravity near Earth's surface. There are many other forces in the universe, and in this section we consider some of these forces and the resulting motions. We start by revisiting the motion of a spacecraft as it travels between different solar systems. We assume the spacecraft is far from any planets or stars, so gravitational forces are extremely small. If the spacecraft's engine is initially turned off, the total force on the spacecraft is zero, and from Newton's second law we know that the acceleration is also zero. Hence, the spacecraft will move with a constant velocity; that is, it will move in a straight line with a constant speed. Suppose the pilot suddenly realizes that he is not traveling in the correct direction and wants to fire his engine to turn the spacecraft so that it is aimed at the desired destination. If the force from the engine is $\vec{F}_E$, how should the pilot fire the engine to change the direction of the spacecraft by a particular angle θ?

In the most general case, our job would be to help the pilot decide how long to fire the engine and in what direction to fire it. For simplicity, let's assume the engine is fired in a direction perpendicular to the initial velocity (see Fig. 4.21) and this force has a constant magnitude. We need to calculate the length of time t_E the engine should be turned on. Our first step is to establish a coordinate system as sketched in Figure 4.21. We pick the x axis to be parallel to the initial velocity of the spacecraft; the force from the engine is perpendicular to this axis; hence, $\vec{F}_E$ is along y. Writing Newton's second law for a spacecraft of mass m gives

$$\sum F_x = ma_x = 0$$
$$a_x = 0$$

and

$$\sum F_y = ma_y = F_E$$
$$a_y = \frac{F_E}{m}$$

Since the force from the engine is constant, we have a case of motion with constant acceleration (assuming the mass m of the spacecraft is constant). If the engine is turned on at $t = 0$, the components of the velocity are given by

$$v_x = v_{0x} = v_0$$
$$v_y = v_{0y} + a_y t = 0 + \frac{F_E}{m}t = \frac{F_E}{m}t \qquad (4.32)$$

where we have used the fact that there is no acceleration along x and the initial velocity along y is zero (see Fig. 4.21). The angle through which the spacecraft's velocity vector has turned is equal to the angle that the final velocity makes with the x direction. This angle is given by

$$\tan \theta = \frac{v_y}{v_x} = \frac{F_E t / m}{v_0} \qquad (4.33)$$

The longer the engine is turned on, the larger the turn angle will be. Rearranging Equation 4.33, we find that if the pilot wants to obtain a particular turn angle θ, he should fire the engine for a time t_E, where

$$t_E = \frac{mv_0 \tan \theta}{F_E}$$

If the engine is then turned off, the spacecraft will resume its straight-line, constant-velocity motion in this new direction.

We have now worked through enough applications of Newton's second law that we can formulate a strategy for attacking these problems.

PROBLEM SOLVING | with Newton's Second Law

1. **RECOGNIZE THE PRINCIPLE.** To use Newton's second law to find the acceleration, velocity, and position of an object, we need to consider all the forces acting on an object and compute the total force. For an object in static equilibrium, the total force must be zero; now, for an object that is not in equilibrium, the total force is equal to $m\vec{a}$.

2. **SKETCH THE PROBLEM.** Your picture should define a coordinate system and contain all the forces in the problem. It is usually a good idea to also show all the given information.

3. **IDENTIFY THE RELATIONSHIPS.**
   - Find all the forces acting on the object of interest (the object whose motion you wish to describe) and construct a free-body diagram.

- Express all the forces in terms of their components along x and y.
- Apply Newton's second law (Eq. 4.1) in component form:

$$\sum F_x = ma_x \quad \text{and} \quad \sum F_y = ma_y$$

- If the acceleration is constant along x or y (or both), you can apply the kinematic equations from Table 3.1.

4. **SOLVE.** Solve the equations resulting from step 3 for the unknown quantities in terms of the known quantities. The number of equations must equal the number of unknown quantities.

5. Always *consider what your answer means* and check that it makes sense.

Traveling Down a Hill

Another typical problem involving Newton's second law is sketched in Figure 4.22, which shows a sled traveling down a snowy hillside. If the sled starts from rest at the top of the hill, how long does it take to reach the bottom? For simplicity, we ignore friction between the sled and the snow, and we assume the hill makes a constant angle θ with the horizontal. There are then only two forces acting on the sled, the force of gravity and the normal force exerted by the hill's surface.

Following our problem-solving strategy, we next add a coordinate system to our diagram. We know that the sled's motion is along the hill, so we choose x to be directed parallel to the hill, with x increasing downward. The y axis is then perpendicular to the hill surface. We next draw a free-body diagram (Fig. 4.22B), which shows the components of the forces along x and y. The normal force is already along y, but the force of gravity on the sled has components along both x and y. Using some trigonometry, we can get the components of the forces along x and y (Fig. 4.22B). We now apply Newton's second law for motion along x:

$$\sum F_x = mg \sin \theta = ma_x \tag{4.34}$$

Applying Newton's second law along y, we have

$$\sum F_y = N - mg \cos \theta = ma_y \tag{4.35}$$

Since the sled is moving purely along the x direction, the acceleration along y must be zero. Hence, $a_y = 0$, and we could use Equation 4.35 to calculate the magnitude of the normal force, N. However, in this particular problem we are interested in the motion along the hillside (along x), so all we need is the acceleration a_x. We can get this acceleration by solving Equation 4.34:

$$a_x = g \sin \theta \tag{4.36}$$

This acceleration is constant, and we can use our relation for motion with constant acceleration:

$$x = x_0 + v_{0x}t + \tfrac{1}{2}a_x t^2$$

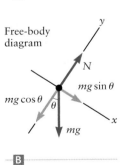

Figure 4.22 ◿ If this hill is frictionless, only two forces are acting on the sled: a normal force $\vec{N}$ exerted by the hill and the force of gravity $\vec{F}_{\text{grav}}$. ◾ Free-body diagram for the sled.

The sled starts at $x_0 = 0$ and begins from rest, so $v_{0x} = 0$. Inserting the acceleration from Equation 4.36 gives

$$x = \tfrac{1}{2}a_x t^2 = \tfrac{1}{2}g(\sin\theta)t^2 \qquad (4.37)$$

If the hill has a vertical height h, the distance to the bottom of the hill as measured along the slope is $h/\sin\theta$ (see Fig. 4.23A). Inserting this into Equation 4.37 we find

$$x = \tfrac{1}{2}g(\sin\theta)t^2 = \frac{h}{\sin\theta}$$

when the sled is at the bottom of the hill. The time it takes to reach the bottom is then

$$t_{\text{bottom}} = \frac{\sqrt{2h/g}}{\sin\theta} \qquad (4.38)$$

EXAMPLE 4.8 | Sledding Down the Hill

For the sled in Figure 4.22, find the velocity when it reaches the bottom of the hill.

RECOGNIZE THE PRINCIPLE

We wish to find the components of the sled's velocity along the x and y directions as defined by the coordinate axes in Figure 4.22. Here, the x direction is parallel to the hill and y is perpendicular. From Newton's second law, we can find the acceleration along x (Eq. 4.36) and then apply the relations for motion with a constant acceleration (Table 3.1) to get the velocity.

SKETCH THE PROBLEM

Figure 4.22 describes the problem.

IDENTIFY THE RELATIONSHIPS

The sled is moving along the hill (along x), so the component of the velocity along y (i.e., perpendicular to the hill) is zero; that is,

$$\boxed{v_y = 0}$$

We already calculated the acceleration along x in Equation 4.36, where we found $a_x = g\sin\theta$. The acceleration is constant, so the velocity along x is again given by one of our relations for motion with constant acceleration,

$$v_x = v_{0x} + a_x t$$

Inserting the result for a_x from Equation 4.36 gives

$$v_x = v_{0x} + a_x t = g(\sin\theta)t \qquad (1)$$

SOLVE

The sled reaches the bottom at t_{bottom}, which we found in Equation 4.38. Inserting t_{bottom} into Equation (1) leads to

$$v_{x,\text{bottom}} = g(\sin\theta)t_{\text{bottom}} = g\sin\theta\,\frac{\sqrt{2h/g}}{\sin\theta} = \boxed{\sqrt{2gh}}$$

What does it mean?

While the speed at the bottom depends on the height h of the hill, it is *independent* of the angle θ of the hill. We show in Chapter 6 that this is a general property of the gravitational force.

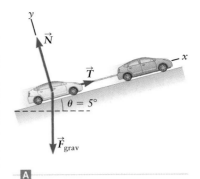

EXAMPLE 4.9 | Towing a Car

Your car breaks down, and you ask a friend to help you move it to the nearest repair shop. You need to move your car up a hill, so your friend is going to use his car to pull your vehicle using a rope as shown in Figure 4.23A. Your friend has a rather flimsy rope, though, and you are worried that it will break and lead to more trouble, so you decide to calculate ahead of time if this attempt will be successful. If the angle of the hill is $\theta = 5.0°$ and you pull your car ($m = 1200$ kg) with a constant velocity, what is the tension in the rope? Assume all friction can be neglected.

RECOGNIZE THE PRINCIPLE

We can use Newton's second law to calculate the motion of the cars (Fig. 4.23A). Since the cars are moving with constant velocity, the acceleration along the hill (the x direction) is $a_x = 0$; hence, the sum of all the forces along the hill must be zero. From this we can solve for the tension T in the rope and compare T with the maximum tension the rope can withstand.

SKETCH THE PROBLEM

Figure 4.23B shows a free-body diagram for the car being towed. The forces acting on it are the tension from the rope, the force of gravity, and the normal force exerted by the hill's surface. The figure also shows the components of these forces along the hill and perpendicular to the hill. The $+x$ direction is along the hill's surface and up the incline, while y is perpendicular to x.

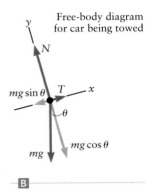

IDENTIFY THE RELATIONSHIPS

Writing Newton's second law for motion along the x direction (along the hill), we have

$$\sum F_x = -mg \sin \theta + T = ma_x$$

Since the velocity is constant, a_x is zero, which leads to

$$-mg \sin \theta + T = 0$$
$$T = mg \sin \theta$$

Figure 4.23 Example 4.9.
A Three forces are acting on this car: a normal force $\vec{N}$ exerted by the road, the force of gravity $\vec{F}_{grav}$, and a force due to the cable $\vec{T}$.
B Free-body diagram for the car.

SOLVE

Inserting the given values of m and θ gives

$$T = (1200 \text{ kg})(9.8 \text{ m/s}^2)\sin(5.0°) = \boxed{1000 \text{ N}}$$

What does it mean?
This tow rope only needs to withstand 1000 N (about 250 lb) to pull the car at a 5.0° angle without breaking, which is much less than the weight of the car.

Adding the Frictional Force

In the previous examples involving motion along a slope, we have assumed friction is negligible. Now let's see how to include it. Returning to our problem of a sled sliding down a snowy hill (Fig. 4.22), we add friction between the sled and the hill. The revised free-body diagram is shown in Figure 4.24, which contains a frictional force of magnitude $F_{friction} = \mu_K N$ directed up the incline (along $-x$). This force is kinetic friction because the sled is slipping relative to the hill. To calculate the frictional force, we must first find the normal force N, which we can calculate from Equation 4.35:

$$\sum F_y = N - mg \cos \theta = ma_y$$

Since the acceleration along y (perpendicular to the hill) is zero,

$$\sum F_y = N - mg \cos \theta = 0$$
$$N = mg \cos \theta \qquad (4.39)$$

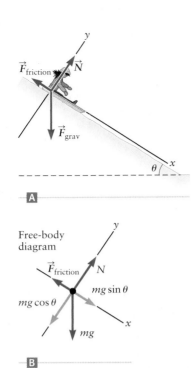

Figure 4.24 For this sled, there is a frictional force $\vec{F}_{\text{friction}}$ from the sled slipping on the snow. Compare with Figure 4.22.

We next apply Newton's second law for motion along the hill:

$$\sum F_x = mg \sin \theta - F_{\text{friction}} = mg \sin \theta - \mu_K N = ma_x$$

The acceleration down the incline thus depends on the normal force. Using the result for the normal force from Equation 4.39 leads to

$$a_x = g \sin \theta - \frac{\mu_K N}{m} = g \sin \theta - \frac{\mu_K mg \cos \theta}{m} = g(\sin \theta - \mu_K \cos \theta) \quad \textbf{(4.40)}$$

Let's now estimate quantitatively the effect of friction. Suppose we have a hill with $\theta = 10°$, which is a typical value for a reasonably steep (but safe) sledding hill. Without friction, $\mu_K = 0$ in Equation 4.40 and we get $a_x = g \sin \theta = 1.7$ m/s². A typical value for the coefficient of kinetic friction between a wooden ski and snow is $\mu_K \approx 0.05$, and using this value in Equation 4.40 we find $a_x \approx 1.2$ m/s². Hence, friction reduces the acceleration (and therefore also the speed at any moment) by about 30%.

Pulleys and Cables

In Chapter 3, we found that pulleys could be used to change the direction of a force and also to amplify forces. Let's now consider the associated motion in more detail. Figure 4.25 shows a common situation of two crates connected by a rope that passes over a pulley. For simplicity, we assume the rope is very light and the pulley is frictionless and massless. We want to find how the crates move; that is, we want to calculate their acceleration.

As usual, we need to apply Newton's second law. The forces on each crate are shown in Figure 4.25. Recall that for an ideal (massless) string, the tension is the same throughout the string, so the magnitudes of the tension forces acting on the two crates are the same. We next need to establish a coordinate system, and for this problem the choice requires some thought. Because the two crates are connected by the string, they must move together; otherwise, the string would break. Therefore, the magnitudes of their displacements, velocities, and accelerations must be the same. With that in mind, we choose the x coordinate direction to be *parallel to the string* as sketched in Figure 4.25B. This coordinate axis *follows the string* as it loops around the pulley. It may be useful to think of this coordinate "axis" as a sort of "railroad track" along which the two crates move. If we take the "positive" direction for this track to be upward for m_1, it will then be downward for m_2 as shown in the figure. Although the "positive" directions are thus opposite for the two crates, the motion is still essentially one dimensional, so we can use our familiar results for problems involving one-dimensional motion. This choice of a coordinate axis whose direction is different for the two crates will make our mathematical solution of Newton's second law a bit simpler because now the acceleration of the two crates will be the same from the start.

To find the acceleration, we first write Newton's second law for m_1. Since the forces are all along the string direction, the acceleration will also be along this direction, which we have labeled as the x direction. We denote the total force on m_1 as $\sum F_1$, so Newton's second law for this crate reads

$$\sum F_1 = m_1 a = +T - m_1 g \quad \textbf{(4.41)}$$

where we have written the acceleration as simply a (because the acceleration is the same for m_1 and m_2). The forces on the right side of Equation 4.41 are $+T$ (since the tension force is directed along the $+x$ direction for m_1) and $-m_1 g$ (since the force of gravity is along $-x$). For m_2,

$$\sum F_2 = m_2 a = -T + m_2 g \quad \textbf{(4.42)}$$

Here the tension term is $-T$ because the tension force on m_2 is directed along $-x$ for this crate. The sign is reversed with respect to Equation 4.41 because the x direction is reversed for m_2. We now have two equations and two unknowns—the acceleration and the tension—so we can solve for both a and T. To solve for the acceleration, we add Equations 4.41 and 4.42, which eliminates T and gives

$$m_1a + m_2a = +T - m_1g - T + m_2g$$

$$(m_1 + m_2)a = -m_1g + m_2g$$

$$a = \frac{(m_2 - m_1)g}{m_1 + m_2} \qquad (4.43)$$

To solve for the tension, we can insert this result for a back into Equation 4.41 to find

$$T = m_1a + m_1g = m_1(a + g)$$

$$T = m_1\left[\frac{(m_2 - m_1)g}{m_1 + m_2} + g\right] = m_1g\left[\frac{m_2 - m_1 + m_2 + m_1}{m_1 + m_2}\right]$$

$$T = \frac{2m_1m_2g}{m_1 + m_2}$$

If you are uneasy about our choice of the x directions in this problem, look again at Equation 4.43. There we see that if $m_2 > m_1$, the acceleration is positive and the block with the larger mass (m_2) moves downward as expected. On the other hand, if $m_2 < m_1$, Equation 4.43 tells us that the acceleration is negative, which simply means that m_2 now accelerates upward, again as we should expect. The *physical answer* does not depend on how we choose the "positive" direction.

EXAMPLE 4.10 | More Crates and a Pulley

A crate sits on a frictionless table and is connected to a second crate by a string that passes over a pulley as shown in Figure 4.26. If the pulley is frictionless and massless and the string is also massless, find the acceleration of the crates and the tension in the string.

RECOGNIZE THE PRINCIPLE

We follow our problem-solving strategy for applying Newton's laws. To find the acceleration of the crates, we must consider all the forces acting on them. Our sketch (Fig. 4.26) shows all these forces. We write Newton's second law for each crate and solve for the unknown quantities.

SKETCH THE PROBLEM

Figure 4.26 shows the forces acting on both crates and indicates our choice of coordinate system. As in Figure 4.25, we take the x direction to follow the string around the pulley. That is, we choose the $+x$ direction to be horizontal and to the right for the crate on the table, and down along the string for the crate that is suspended in midair.

IDENTIFY THE RELATIONSHIPS

Using the forces shown in Figure 4.26, we write Newton's second law for the crate on the table. We find

$$\sum F_x = m_1a = +T \qquad (1)$$

For the crate that hangs from the string, Newton's second law gives us

$$\sum F_x = m_2a = m_2g - T \qquad (2)$$

SOLVE

To find the acceleration, we eliminate T by adding Equations (1) and (2):

$$m_1a + m_2a = +T + m_2g - T = m_2g$$

$$a = \boxed{\frac{m_2g}{m_1 + m_2}}$$

A

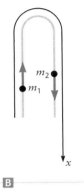

B

Figure 4.25 **A** Forces on two crates connected by a rope-and-pulley system. **B** The coordinate axis (the x axis) follows the string as it bends around the pulley.

Figure 4.26 Example 4.10. Another arrangement of crates and a pulley. We assume there is no frictional force between the table and crate 1. Notice that we have chosen the x direction to follow the string.

Solving for T then gives, from Equation (1),

$$m_1 a = +T$$

$$T = m_1 a = \boxed{\frac{m_1 m_2 g}{m_1 + m_2}}$$

What have we learned?

Our approach to this problem follows our usual pattern. We start with a picture and then construct free-body diagrams for the objects of interest (the two crates). We then write Newton's second law for each crate and solve for the unknowns (a and T).

EXAMPLE 4.11 Crates with Friction

Suppose there is a frictional force between the crate m_1 and the table in Example 4.10 (Fig. 4.26). Write the new Newton's second law equation for this crate's acceleration along the horizontal direction.

RECOGNIZE THE PRINCIPLE

We follow the same approach as taken in Example 4.10. Now our sketch (Fig. 4.27) contains an additional force due to friction on crate 1. This frictional force has a magnitude of $\mu_K N_1$, where N_1 is the normal force between crate 1 and the table.

SKETCH THE PROBLEM

Figure 4.27 shows the new situation along with free-body diagrams for both crates. This frictional force acts in the $-x$ direction since it opposes the motion of crate 1.

Figure 4.27 Example 4.11. We add friction to the system in Figure 4.26, so there is now a frictional force between crate 1 and the table.

IDENTIFY THE RELATIONSHIPS AND SOLVE

When we add the frictional force, Newton's second law for the crate on the table becomes

$$\sum F_x = m_1 a = +T - \mu_K N_1$$

What does it mean?

In this problem we have an additional unknown, the normal force N_1. This normal force can be found by considering the forces on m_1 along the y direction, leading to $N_1 = m_1 g$.

CONCEPT CHECK 4.3 | Tension in a String

Consider the two pulley arrangements in Figure 4.28. In both cases, a crate of mass m_1 is attached to one end of a string. In Figure 4.28A, the other end of the string is tied to a crate of mass m_2, so there is a force from gravity of magnitude $F_{2, \text{grav}} = m_2 g$ acting on this crate. In Figure 4.28B, crate m_2 is replaced by a person who pulls on the string, and we suppose he pulls with a force *equal* to $F_{2, \text{grav}} = m_2 g$. Will the acceleration of m_1 be the same in these two cases?

4.5 | ✖ DETECTING ACCELERATION: REFERENCE FRAMES AND THE WORKINGS OF THE EAR

Figure 4.28 Concept Check 4.3. Do these two systems have the same acceleration?

Suppose you are traveling in an airplane initially flying at a constant horizontal velocity and the pilot decides to accelerate slightly. You immediately sense this acceleration. How did you do it? The human ear contains structures that act as sensitive acceleration sensors, and we can understand how they work using Newton's second law.

Before we consider the workings of an actual ear, let's first consider a simple mechanical device you could build to detect and measure an airplane's acceleration. Our device, shown in Figure 4.29, consists of a rock tied to one end of a string, with the other end of the string fastened to the airplane's ceiling. We can see how this device would work as an accelerometer by comparing its behavior when the airplane moves at constant velocity as in Figure 4.29A with the behavior found when the airplane accelerates (Fig. 4.29B).

When the acceleration is zero, the string hangs vertically, which is consistent with a Newton's law analysis. Two forces are acting on the rock—the force of gravity and the tension force from the string—and in Figure 4.29A these forces are both along the vertical (y). The acceleration along y is zero, so these forces must add to zero with the result $T = mg$. There are no forces along the horizontal; the acceleration along x is zero, and the rock moves with a constant velocity.

When the airplane is accelerating along the horizontal (x), the rock will have the same acceleration as the airplane and the string hangs at an angle θ as sketched in Figure 4.29B. The value of θ depends on the airplane's acceleration. Writing Newton's laws for motion along x and y leads to

$$\sum F_x = ma_x = T \sin \theta$$
$$\sum F_y = ma_y = T \cos \theta - mg = 0 \tag{4.44}$$

where we have used the fact that the acceleration along y is again zero. We can use this relation for a_y to find the tension in the string,

$$T \cos \theta = mg$$
$$T = \frac{mg}{\cos \theta}$$

and then get the acceleration along x:

$$a_x = \frac{T \sin \theta}{m} = \frac{mg}{\cos \theta} \frac{\sin \theta}{m} = g \tan \theta \tag{4.45}$$

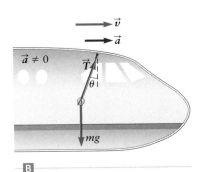

Figure 4.29 A rock hanging by a string acts as an accelerometer, a device that measures acceleration. A When the system (here an airplane cabin) moves with a constant velocity, the string hangs vertically. B When the system is accelerating, the string hangs at a nonzero angle θ.

Whenever the airplane has a nonzero acceleration, the angle θ is nonzero and the string hangs at an angle from the vertical. By measuring this angle, we can use Equation 4.45 to find the value of the acceleration a_x.

CONCEPT CHECK 4.4 | Effect of Mass on an Accelerometer

How will the angle of the accelerometer string in Figure 4.29 depend on the *mass* of the rock?
 (a) The angle is larger for a rock with a larger mass.
 (b) The angle is independent of the mass of the rock.
 (c) The angle is smaller for a rock with a larger mass.

Figure 4.30 A Hair cells within the utricle in a human ear. These cells act like the accelerometer in Figure 4.29. B When your head is accelerated, the hair cells are deflected from the vertical, and this deflection produces a signal that is sent to your brain.

Kinocilium Stereocilia

Otoliths
Gelatinous layer
Hair cells
Supporting cells
Sensory nerve fibers

A

$\vec{a}$

B

The Accelerometer in Your Ear

Your ears contain a structure that acts very much like the simple accelerometer in Figure 4.29. A region within the ear called the utricle (Fig. 4.30) contains hair cells that project up into a gelatinous layer filled with a very thick fluid. This gelatinous layer also contains some small masses called otoliths (also known as "ear stones"). When your head—that is, the utricle—accelerates along the horizontal, the gelatinous layer lags behind a small amount in the same way the string in Figure 4.29B hangs behind when the airplane accelerates. This displacement of the gelatinous layer causes the hair cells to deflect (again, just like the string in Fig. 4.29B), and the hair cells send signals to the brain when they are bent in this way. Hence, the brain is "told" that the ear is being accelerated. Although the ear is certainly more complex than a rock on a string, the accelerometers in Figures 4.29 and 4.30 are remarkably similar.

Inertial Reference Frames

Let's analyze the accelerometer in Figure 4.29 once more, this time from a slightly different point of view. In Equation 4.44, we applied Newton's second law from the point of view of an observer outside the airplane, a person who might have viewed the rock and string through a window as the airplane flew past. Now imagine how things look to an observer sitting inside the airplane. We first consider the situation when the acceleration is zero and the string hangs vertically as in Figure 4.29A. An observer inside the airplane would see two forces acting vertically on the rock: gravity and the tension in the string. This observer would say that the acceleration along y is zero and hence that $T = mg$. For the motion along the horizontal, the observer inside the airplane would say that there is no force along x and there is also no acceleration along x, so Newton's second law works just fine. Hence, when the airplane moves with a constant velocity, an observer inside the airplane would be in complete agreement with the observer outside the airplane. Both would find that Newton's second law describes things perfectly.

The situation is different when the airplane has a nonzero acceleration relative to an observer outside the airplane. Free-body diagrams for this case are shown in Figure 4.31. Our two observers—one outside the airplane and one inside—would draw the same free-body diagrams. However, they would disagree when they use these free-body diagrams to apply Newton's second law:

$$\sum F_x = T \sin \theta = ma_x \tag{4.46}$$

The outside observer would say that the acceleration a_x is nonzero and that Newton's second law works fine since θ is nonzero (Fig. 4.31A). The observer inside the airplane would also see that the string hangs at an angle and hence would conclude that there is a force $T \sin \theta$ along the x direction. According to this observer, though, there is *no acceleration along x*. That is, *as viewed by an observer riding inside the airplane*, the rock is simply at rest; there is no velocity or acceleration along the x direction *relative to this observer.* The observer inside the airplane would therefore say that there is a force along x but no acceleration along x, so the left-hand side of Equation 4.46 is not zero, whereas the right-hand side is zero. Our inside observer would thus be tempted to conclude that Newton's second law does not work inside airplanes!

The correct interpretation of this situation is that the fault lies not with Newton, but with the observer inside the airplane who is viewing things in an *accelerating reference frame.* Newton's second law can only be applied in nonaccelerated reference frames, also called **inertial reference frames.** An inertial reference frame can be moving, but it must be moving with a constant velocity relative to other inertial frames. Suppose an observer in one inertial reference frame (frame 1) conducts a test of Newton's second law by measuring the forces on an object and the acceleration of the object. He would then (in agreement with Newton's second law) find that

$$\vec{F}_{\text{measured}} = m\vec{a}_{\text{observed}} \tag{4.47}$$

A different observer in frame 2 might conduct the same test, measuring the force and acceleration from the point of view of a different inertial reference frame. We assume observer 2 is moving at a constant velocity relative to frame 1, so his reference frame (frame 2) is an inertial frame. Observer 2 would measure the same horizontal force as observer 1 because he will observe the tension in the string in Figure 4.31 to be the same. Observer 2 will also measure the *same* acceleration as observer 1; since acceleration is the rate of change of velocity, adding the constant velocity of frame 2 does not change the acceleration. Hence, Equation 4.47 would still be satisfied. However, if an observer in a *noninertial* reference frame (frame 3) were to conduct the same test he would measure the same force but a *different* acceleration. As in the case of the airplane in Figure 4.31, the acceleration in frame 3 would differ from that found in frame 1 by an amount equal to the acceleration of the noninertial frame. Hence, to the observer in frame 3, it would appear that Newton's laws fail. Again, Newton's laws can only be used in inertial reference frames.

The problem of inertial and noninertial reference frames has a deep connection to Newton's laws and to other theories of motion. We'll encounter reference frames again when we discuss gravitation and relativity in Chapter 27.

| **CONCEPT CHECK 4.5** | Newton's Laws and Noninertial Reference Frames

Give an example showing that Newton's first law—the principle of inertia—does not hold in a noninertial reference frame.

4.6 | PROJECTILE MOTION REVISITED: THE EFFECT OF AIR DRAG

In Section 4.2, we discussed "ideal" projectile motion in which the only force acting on a projectile is the force of gravity. Real projectiles generally also experience a force due to air drag, and in some cases this drag force can have a very significant effect. Let's consider the drag force in a few cases and estimate its effects.

We begin with the motion of a batted baseball and calculate how far it would travel in the absence of air drag. We worked through this problem in Section 4.2, where we saw that the range of the ball is (Eq. 4.28)

$$x_{\text{range}} = \text{range} = \frac{v_0^2 \sin(2\theta)}{g}$$

We also showed that for a given value of the initial speed v_0, the maximum range occurs when $\theta = 45°$. A skilled baseball player can set a ball into motion with an initial speed of $v_0 = 45$ m/s (about 100 mi/h). Inserting this value into our relation for x_{range} gives a range of 210 m, or about 680 ft. This distance is far longer than a batted

Observer outside airplane: Inertial frame.

A

Observer inside airplane: Noninertial frame.

B

Figure 4.31 **A** An observer in an inertial reference frame (i.e., an observer at rest, outside the airplane) finds that there is a horizontal component to the total force (because the string hangs at an angle) and also measures that the rock is accelerating, in agreement with Newton's second law. **B** An observer inside the airplane accelerates along with the rock. According to this noninertial observer, the rock's acceleration is zero since the rock is at rest relative to this observer. However, this observer still finds that there is a horizontal component to the total force on the rock. According to this observer, there is a horizontal force on the rock but no horizontal acceleration, so this observer claims that Newton's second law does not work!

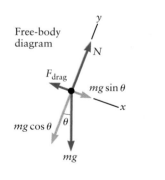

Insight 4.2
APPLYING NEWTON'S SECOND LAW IN THREE DIMENSIONS

In our applications of Newton's second law, we have dealt with many examples, such as projectiles, that move in two dimensions. For such cases, Newton's second law can be written as

$$\sum F_x = ma_x \quad \text{and} \quad \sum F_y = ma_y$$

and we can deal with forces and acceleration along x and y. To deal with a problem that involves forces or motion in three dimensions, we must also use

$$\sum F_z = ma_z$$

The basic approach to such problems is the same as in two dimensions.

Free-body diagram

Figure 4.32 ◢ When a bicycle coasts down a long hill, the bicycle eventually reaches its terminal velocity. The forces acting on the bicycle are the normal force $\vec{N}$ exerted by the hill's surface, the force of gravity $\vec{F}_{grav}$, and the drag force $\vec{F}_{drag}$. ◢ Free-body diagram with forces resolved along x and y.

---

baseball has ever traveled. (For a typical baseball field, the distance from home plate to the centerfield fence is about 120 m, and only the longest home runs travel this far.) Our calculation must have omitted something important: the force of air drag.

In Chapter 3, we saw that for an object like a baseball, a bullet, a cannon shell, or an airplane, the magnitude of the air drag force is given approximately by

$$F_{drag} = \tfrac{1}{2}\rho A v^2 \tag{4.48}$$

where ρ is the density of air (1.3 kg/m³), A is the cross-sectional area of the object, and v is its speed. (See Insight 3.4 for more on the drag force.) For a baseball[6] moving at 45 m/s, Equation 4.48 gives $F_{drag} = 5.4$ N. The baseball's weight is approximately $mg = 1.4$ N, so the drag force here is much *greater* than the force of gravity on the ball. We certainly cannot ignore the drag force in this case.

Although we have a mathematical expression for the drag force in Equation 4.48, it is difficult to work out a simple expression for the trajectory of a projectile (such as a baseball) when this force is included. However, such trajectories can be readily calculated with computer simulation methods, so we can describe how drag affects the shape of the trajectory and the range. We have already seen that for an ideal projectile (i.e., one without air drag), the maximum range occurs when the initial velocity makes an angle of 45° with the horizontal. Drag can substantially alter this result, and in either direction. For a baseball, the maximum range is found with a much smaller angle. The "best" angle depends on the initial speed of the ball and is typically around 35°. On the other hand, for a very powerful artillery gun, the artillery shell can travel very high into the atmosphere where the density of the air is much less than at ground level, which greatly reduces the drag force on the shell. To achieve maximum range, these guns are usually aimed well above 45°, and the artillery shell spends a large amount of its time at high altitudes where the air density is lower than near sea level.

Effect of Air Drag on a Bicycle

Air drag is very important for vehicles such as cars and bicycles. Indeed, most of us have felt the force from air drag when riding a bicycle even at rather low speeds. Consider the bicyclist in Figure 4.32 riding downhill on a frictionless bicycle so that the only forces along the incline are due to gravity and air drag. If the slope of the hill is $\theta = 5.0°$, how fast will the bicycle coast?

Figure 4.32 shows the forces acting on the bicycle. We take the x direction to be along the incline. Writing Newton's second law for motion along x, we get

$$\sum F_x = F_{grav,\,x} + F_{drag}$$

The component of the gravitational force along x is

$$F_{grav,\,x} = mg \sin \theta \tag{4.49}$$

Using Equation 4.48, the drag force is

$$F_{drag} = -\tfrac{1}{2}\rho A v_x^2 \tag{4.50}$$

where the negative sign indicates that F_{drag} is directed up the incline (along $-x$). When the bicyclist is coasting at her terminal velocity, v is constant, so the total force along x must be zero. Using Equations 4.49 and 4.50, we thus find

$$\sum F_x = mg \sin \theta - \tfrac{1}{2}\rho A v_x^2 = 0$$

$$v_x = \sqrt{\frac{2mg \sin \theta}{\rho A}}$$

---

[6]The frontal area of the baseball is $A = \pi r^2$, where the radius of a baseball is $r = 3.6$ cm.

To get a feeling for the value of this terminal (coasting) velocity, we consider a bicycle plus rider of mass 100 kg, with an approximate frontal area of 1 m². For a hill with slope angle $\theta = 5.0°$, a moderately steep hill, we have a speed of

$$v_x = \sqrt{\frac{2mg \sin \theta}{\rho A}} = \sqrt{\frac{2(100 \text{ kg})(9.8 \text{ m/s}^2)\sin(5.0°)}{(1.3 \text{ kg/m}^3)(1 \text{ m}^2)}} = 10 \text{ m/s}$$

(Here we keep just one significant figure, since the area is only estimated with this accuracy.) This is about 25 mi/h. How does this answer compare with your experience?

SUMMARY | Chapter 4

KEY CONCEPTS AND PRINCIPLES

Translational equilibrium
When the total force on an object is zero, Newton's second law tells us that the acceleration is also zero. In other words, all of the vector components of $\vec{F}_{total} = \sum \vec{F}$ and $\vec{a}$ are zero. Such an object is in *translational equilibrium*.

Analyzing motion in two and three dimensions
We use Newton's second law to calculate the acceleration of an object:

$$\sum \vec{F} = m\vec{a} \qquad \text{(4.1) (page 92)}$$

The acceleration can then be used to find the velocity and displacement.

Inertial and noninertial reference frames
A *reference frame* is an observer's choice of coordinate system, including an origin, for making measurements. An *inertial reference frame* is one that moves with a constant velocity, whereas a *noninertial* reference frame is one that is accelerating. Newton's laws are only obeyed in inertial reference frames. They are not obeyed in noninertial reference frames.

APPLICATIONS

Projectile motion
In many cases, the force of air drag is small. The force on a projectile near Earth's surface, such as a baseball, is then just the force of gravity, and the acceleration has a constant magnitude g directed downward. The displacement and velocity for such a projectile are described by the relations for motion with constant acceleration.

For projectile motion, the vertical and horizontal motions are independent. Two objects can have very different horizontal velocities, but they still fall at the same rate.

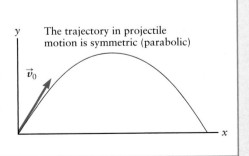

(Continued)

The trajectory for simple projectile motion starting from the origin is symmetric in two ways:

- It is symmetric in time. The time spent traveling to the point of maximum height is equal to the time spent falling back down to the initial height.
- It is symmetric in space. The trajectory has a parabolic shape.

How the body detects acceleration

The ear contains a sensitive acceleration detector that enables the brain to know when the head is accelerating.

QUESTIONS

SSM = answer in Student Companion & Problem-Solving Guide   Ⓧ = life science application

1. At what angle should a ball be thrown so that it has the maximum range? Ignore air drag.

2. Consider the range of a projectile as calculated in Equation 4.28 (i.e., ignoring the effect of air drag). We have already shown that the maximum range occurs at a launch angle of $\theta = 45°$. Consider the range of a projectile launched at a smaller angle α. Show that there is a corresponding launch angle β greater than 45° that yields the same value of the range and find a general relation between α and β.

3. An object has an initial velocity $\vec{v}_i$ and a final velocity $\vec{v}_f$ as sketched in Figure Q4.3. Sketch the approximate direction of the acceleration $\vec{a}$ of the object during this time interval.

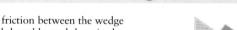

Figure Q4.3

4. During a particular time interval, an object has an acceleration $\vec{a}$ in the direction shown in Figure Q4.4. If $\vec{v}_i$ is the initial velocity of the object, draw a vector $\vec{v}_f$ that might be the velocity of the object at the end of the interval.

Figure Q4.4

5. The arrangement in Example 4.2 makes it possible to amplify a force. (Compare this example to the block and tackle in Fig. 3.22.) Amplifying forces in this way comes at a price. To appreciate this price, imagine that the force on the car in Figure 4.4 is just barely large enough to move the car. If the car begins to move, point O on the cable also begins to move. Calculate the ratio of the distance moved by the car to the distance moved by point O and compare it to the force amplification factor found in Example 4.2. Assume the car moves in the direction of the tension force; that is, it moves along the direction defined by the cable.

6. Consider a wedge that rests on a horizontal table as shown in Figure Q4.6 and assume there is no friction between the wedge and the table. A force is now applied vertically as shown in the figure. You might expect this force will cause the wedge to move to the left. Explain why the wedge does not move, even though there is no friction between it and the table.

Frictionless table

Figure Q4.6

7. SSM Consider again the wedge in Question 6, but now assume a block is placed onto it as shown in Figure Q4.7. There is again no friction between the wedge and the table, and there is also no friction between the block and the wedge. Explain why in this case the wedge *will* be accelerated to the left. How does this situation differ from that in Question 6? *Hint*: Compare the free-body diagrams for the wedge in the two cases.

Frictionless table

Figure Q4.7

8. The rock-on-a-string accelerometer in Figure 4.29 can be used to measure acceleration along a horizontal direction. Discuss how you could use it to also measure the vertical component of the acceleration.

9. **Time of flight.** Equation 4.26 gives a relation for the time at which a batted baseball lands. Use the quadratic formula to find the solutions for t_{lands}. (See Appendix B if you don't remember the quadratic formula.) Since you are solving a quadratic equation, there are *two* solutions for t_{lands}. Find numerical values for both solutions and give a physical interpretation of the results. *Hint*: Consider how the problem would be changed if the ball was initially directed "backward" along the trajectory in Figure 4.12. Assume $h = 1.0$ m, $v_0 = 45$ m/s, and $\theta = 30°$.

10. Explain why the string in Figure 4.29 hangs vertically if the velocity is constant. Consider both an airplane moving horizontally and one moving with a nonzero velocity component along y.

11. Ⓧ Explain why you cannot run unless there is a frictional force between your feet and the ground.

12. **Peel out!** The author has a small pickup truck. He finds that it is much easier to "burn rubber" (i.e., spin the back wheels so that they slip relative to the road surface) when the truck is empty than when it is carrying a heavy load. Explain why.

13. Give an example in which the magnitude of the instantaneous velocity is always larger than the average velocity.

14. Give an example in which the acceleration is perpendicular to the velocity.

15. Give an example of motion in which the instantaneous velocity is zero but the acceleration is not zero.

16. A rock is thrown up in the air in such a way that its velocity is zero at the top of its trajectory. Where does the rock land?

17. Give three examples in which the force of friction on an object has the same direction as the velocity of the object.

18. Ball 1 is thrown with an initial speed of v_0 at an angle of θ relative to the horizontal (x) axis. Ball 2 is thrown at the same angle, but with an initial speed of $2v_0$. (a) If ball 1 is in the air for a time t_1, how long is ball 2 in the air? (b) If ball 1 reaches a maximum height h_1, what is the maximum height reached by ball 2 in terms of h_1?

19. Ball 1 rolls off the edge of a table and falls to the floor below, while ball 2 is dropped from the same height (Fig. Q4.19). Which ball reaches the ground first? Explain why your answer does, or does not, depend on the velocity of ball 1 just before it leaves the table.

Figure Q4.19

20. [SSM] Two balls are thrown into the air with the same initial speed, directed at the same initial angle with respect to the horizontal. Ball 1 has a mass five times larger than the mass of ball 2, and the force of air drag is negligible.
(a) Which ball has the larger acceleration as it moves through the air?
(b) Which ball lands first?
(c) Which ball reaches the greatest height?
(d) For which ball is the force of gravity larger at the top of the trajectory?

4.1 STATICS

1. Two ropes are attached to a skater as sketched in Figure P4.1 and exert forces on her as shown. Find the magnitude and direction of the total force exerted by the ropes on the skater.

Figure P4.1

2. The three forces shown in Figure P4.2 act on a particle. If the particle is in translational equilibrium, find F_3 (the magnitude of force 3) and the angle θ_3.

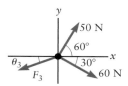

Figure P4.2

3. [SSM] ★ Several forces act on a particle as shown in Figure P4.3. If the particle is in translational equilibrium, what are the values of F_3 (the magnitude of force 3) and θ_3 (the angle that force 3 makes with the x axis)?

4. A man is lazily floating on an air mattress in a swimming pool (Fig. P4.4). (a) Draw a free-body diagram for the man and for the mattress. (b) Identify the reaction forces for all the forces in your free-body diagrams in part (a). (c) If the mass of the man is 110 kg and the mass of the mattress is 2.5 kg, what is the upward force of the water on the mattress? What is the force that the man exerts on the mattress? *Note:* Keep three significant figures in your answer.

Figure P4.3

Figure P4.4

5. A person leans against a wall (Fig. P4.5). Draw a free-body diagram for the person.

Figure P4.5

6. The sled in Figure 4.2 is stuck in the snow. A child pulls on the rope and finds that the sled just barely begins to move when he pulls with a force of 25 N, with the rope at an angle of 30° with respect to the horizontal. (a) Sketch the sled and all the forces acting on it. Also choose a coordinate system. (b) Determine the components of all the forces on the sled along the coordinate axes. (c) Write the conditions for static equilibrium along your two coordinate directions. (d) If the sled has a mass of 12 kg, what is the coefficient of friction between the sled and the snow? (e) Is this the coefficient of static friction or the coefficient of kinetic friction?

7. ★ A system of cables is used to support a crate of mass $m = 45$ kg as shown in Figure P4.7. Find the tensions in all three cables.

8. **Balancing act.** The tightrope walker in Figure P4.8 gets tired and decides to stop for a rest. During this rest period, she is in translational equilibrium. She stops at middle of the rope and finds that both sides of the rope make an angle of $\theta = 15°$ with the horizontal. (a) Sketch the walker and the rope. Show all the forces acting on the point of the rope on which the walker stands (call this point P). Also include a coordinate system. (b) Determine the components of all the forces on point P along the coordinate axes. (c) Write the conditions for static equilibrium along your two coordinate directions. (d) If the mass of the tightrope walker is 60 kg, what is the tension in the rope?

Figure P4.7

Figure P4.8

9. A flag of mass 2.5 kg is supported by a single rope as shown in Figure P4.9. A strong horizontal wind exerts a force of 12 N on

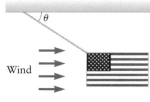

Figure P4.9

the flag. Find the tension in the rope and the angle θ the rope makes with the horizontal.

10. ⭐ A car of mass 1400 kg is parked on a very slippery hillside (Fig. P4.10). To keep it from sliding down the hill, the owner attaches a cable. (a) Sketch all the forces on the car. Include coordinate axes in your sketch. (b) Determine the components of all the forces on the car along the coordinate axes. (c) Write the conditions for static equilibrium along your two coordinate directions. (d) If there is no frictional force between the road and the tires, what is the tension in the cable?

Figure P4.10

11. ✳ A crate is placed on an inclined board as shown in Figure P4.11. One end of the board is hinged so that the angle θ is adjustable. If the coefficient of static friction between the crate and the board is $\mu_S = 0.30$, what is the value of θ at which the crate just begins to slip?

Figure P4.11

12. ✳ Two blocks of mass $m_1 = 45$ kg and $m_2 = 12$ kg are connected by a massless string that passes over a pulley as shown in Figure P4.12. The coefficient of static friction between m_1 and the table is $\mu_S = 0.45$. (a) Will this system be in static equilibrium? Assume the pulley is frictionless. (b) Find the tension in the string.

Figure P4.12 Problems 12, 13, and 44.

13. ⭐ For the system in Problem 12 and Figure P4.12, how large can m_2 be made without the system starting into motion?

14. ⭐ **Stemming a chimney.** A rock climber of mass 60 kg wants to make her way up the crack between two rocks as shown in Figure P4.14. The coefficient of friction between her shoes and the rock surface is $\mu_S = 0.90$. What is the smallest normal force she can apply to both surfaces without slipping? Assume the rock walls are vertical. *Hint*: Why is the normal force between the climber and the rock on the left equal to the normal force between her and the rock on the right?

© Greg Epperson/Jupiterimages

Figure P4.14

4.2 PROJECTILE MOTION

For all the problems in this section, ignore the force due to air drag on the projectile.

15. A rock is thrown horizontally with a speed of 20 m/s from a vertical cliff of height 25 m. (a) How long does it take the rock to reach the horizontal ground below? (b) How far will it land from the base of the cliff? (c) What is the velocity (magnitude and direction) of the rock just before it hits the ground?

16. A hockey puck is given an initial velocity such that $v_x = 12$ m/s and $v_y = 18$ m/s, where the x–y plane is horizontal. (a) What is the initial speed of the puck? (b) What angle does the initial velocity make with the x axis? (c) What angle does the initial velocity make with the y axis?

17. ⭐ A quarterback is asked to throw a football to a receiver who is 35 m away. What is the minimum speed the football must have when it leaves the quarterback's hand? Assume the ball is caught at the same height as it is thrown.

18. ⭐ Which of the graphs in Figure P4.18 might be a plot of the vertical component of the velocity of a projectile that is thrown from the top of a building?

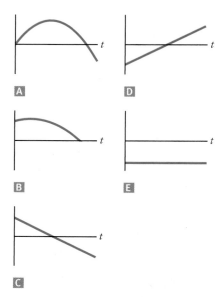

Figure P4.18

19. A soccer ball is kicked with an initial speed of 30 m/s at an angle of 25° with respect to the horizontal. Find (a) the maximum height reached by the ball and (b) the speed of the ball when it is at the highest point on its trajectory. (c) Where does the ball land? That is, what is the range of the ball? Assume level ground.

20. Consider a rock thrown off a bridge of height 75 m at an angle $\theta = 25°$ with respect to the horizontal as shown in Figure P4.20. The initial speed of the rock is 15 m/s. Find the following quantities: (a) the maximum height reached by the rock, (b) the time it takes the rock to reach its maximum height, (c) the place where the rock lands, (d) the time at which the rock lands, and (e) the velocity of the rock (magnitude and direction) just before it lands.

Figure P4.20

21. ✪ A football player wants to kick a ball through the uprights as shown in Figure P4.21. The ball is kicked from a distance of 30 m (he's playing metric football) with a velocity of magnitude 25 m/s at an angle of 30°. Will the ball make it over the crossbar (height 3.1 m)?

Figure P4.21

22. [SSM] ✪ An airplane is flying horizontally with a constant velocity of 200 m/s at an altitude of 5000 m when it releases a package. (a) How long does it take the package to reach the ground? (b) What is the distance between the airplane and the package when the package hits the ground? (c) How far ahead of the target along the x direction should the airplane be when it releases the package?

23. ✪ A horizontal rifle is fired at a target 50 m away. The bullet falls vertically a distance of 12 cm on the way to the target. (a) What is the speed of the bullet just after it leaves the rifle? (b) What is the speed of the bullet just before it hits the target?

24. ✪ A bullet of mass 0.024 kg is fired horizontally with a speed of 75 m/s from a tall bridge. If the bullet is in the air for 2.3 s, how far from the base of the bridge does it land?

25. ✪ Consider the game of baseball. A pitcher throws a ball to the catcher at a speed of 100 mi/h (45 m/s). If the velocity of the ball is horizontal when it leaves the pitcher's hand, how far (vertically) will it fall on the way to the catcher? The distance from the pitcher to the catcher is 60.5 ft. Express your answer in meters and in feet.

26. ✪ A batted baseball is hit with a speed of 45 m/s starting from an initial height of 1 m. Find how high the ball travels in two cases: (a) a ball hit directly upward and (b) a ball hit at an angle of 70° with respect to the horizontal. Also find how long the ball stays in the air in each case.

27. ✪ A juvenile delinquent wants to break a window near the top of a tall building using a rock. The window is 30 m above ground level, and the base of the building is 20 m from his hiding spot behind a bush. Find the minimum speed the rock must have when it leaves his hand. Also find the corresponding launch angle of the rock. *Hint*: Graph how the initial velocity changes as a function of launch angle to find the minimum value needed and the corresponding launch angle.

28. ✪ Ⓡ You are a serious basketball player and want to use physics to improve your free throw shooting. Do an approximate calculation of the minimum speed the ball must have in order to travel from your hand to the basket in a successful free throw. You will have to estimate or find several quantities, including the distance from your hand to the basket, the height of the ball when it leaves your hand, and the height of the basket. Graph how the initial velocity changes as a function of launch angle to find the minimum initial speed of the ball and particular launch angle.

29. ✪ Ⓡ Consider the problem of kicking a soccer ball past a goalkeeper into the goal (Fig. P4.29). You are 25 m away from

Figure P4.29

the goal and kick the ball at an angle of 30° with respect to the horizontal, and the ball just passes over the goalkeeper's hands. Find the initial speed of the ball. Also, calculate how long it takes the ball to get from you to the goal. *Hint*: You will have to estimate the height of the goalkeeper.

30. A ball is thrown straight up and rises to a maximum height of 24 m. At what height is the speed of the ball equal to half its initial value? Assume the ball starts at a height of 2.0 m above the ground.

31. ✪ Two rocks are thrown off a cliff. One rock (1) is thrown horizontally with a speed of 20 m/s. The other rock (2) is thrown at an angle θ relative to the horizontal with a speed of 30 m/s. While the two rocks are in the air, rock 2 is always directly above 1. Find θ.

32. ✪ A bullet is fired from a rifle with a speed v_0 at an angle θ with respect to the horizontal axis (Fig. P4.32) from a cliff that is a height h above the ground below. (a) Calculate the speed of the bullet when it strikes the ground. Express your answer in terms of v_0, h, g, and θ. (b) Explain why your result is independent of the value of θ.

Figure P4.32

33. A high jumper can run horizontally with a top speed of 10.0 m/s. (a) If he can "convert" this velocity to the vertical direction when he leaves the ground, what is the theoretical limit on the height of his jump? (b) How does your result to part (a) compare with the current high-jump record of approximately 2.5 m? Can you explain the difference?

34. ✪ A baseball is thrown with an initial velocity of magnitude v_0 at an angle of 60° with respect to the horizontal (x) direction. At the same time, a second ball is thrown with the same initial speed at an angle θ with respect to x. If the two balls land at the same spot, what is θ?

35. [SSM] ✪ A golf ball is hit with an initial velocity of magnitude 60 m/s at an angle of 65° with respect to the horizontal (x) direction. At the same time, a second golf ball is hit with an initial speed v_0 at an angle 35° with respect to x. If the two balls land at the same time, what is v_0?

4.3 A FIRST LOOK AT REFERENCE FRAMES AND RELATIVE VELOCITY

36. Consider a swimmer who wants to swim directly across a river as in Example 4.7. If the speed of the current is 0.30 m/s and the swimmer's speed relative to the water is 0.60 m/s, how long will it take her to cross a river that is 15 m wide?

37. Suppose the swimmer in Figure 4.20 has a swimming speed relative to the water of 0.45 m/s and the current's speed is 2.5 m/s. If it takes the swimmer 200 s to cross the river, how wide is the river?

38. An airplane has a velocity relative to the air of 200 m/s in a westerly direction. If the wind has a speed relative to the ground of 60 m/s directed to the north, what is the speed of the airplane relative to the ground?

39. Consider again the airplane in Problem 38, but now suppose the wind is directed in an easterly direction. How long does it take the airplane to travel a distance between two cities that are 300 km apart?

40. [SSM] ✪ An airplane flies from Boston to San Francisco (a distance of 5000 km) in the morning and then immediately returns to Boston. The airplane's speed relative to the air is 250 m/s. The wind is blowing at 50 m/s from west to east, so it is "in the face" of the airplane on the way to San Francisco and it is a tailwind on the way back. (a) What is the average speed of the airplane relative to the ground on the way to San Francisco? (b) What is the average speed relative to the ground on the way back to Boston? (c) What is the average speed for the entire trip? (d) Why is the average of the average speeds for the two legs of the trip not equal to the average speed for the entire trip?

41. ⭐ **Round trip.** An airplane flies from Chicago to New Orleans (a distance of 1500 km) in the morning and then immediately returns to Chicago. The airplane's speed relative to the air is 250 m/s. The wind is blowing at 50 m/s from west to east, so it is perpendicular to the line that connects the two cities. (a) What is the average speed of the airplane relative to the ground on the way to New Orleans? (b) What is the average speed relative to the ground on the way back to Chicago? (c) What is the average speed for the entire trip?

42. A 35-m-wide river flows in a straight line (the x direction) with a speed of 0.25 m/s. A boat is rowed such that it travels directly across the river (along y). If the boat takes exactly 4 minutes to cross the river, what is the speed of the boat relative to the water?

4.4 FURTHER APPLICATIONS OF NEWTON'S LAWS

43. Two children pull a sled of mass 15 kg along a frictionless surface as shown in Figure P4.43. (a) Find the magnitude and direction of the sled's acceleration. (b) How long does it take the sled to reach a speed of 10 m/s?

Figure P4.43

44. Consider the crates in Problem 12 (Fig. P4.12). Assume there is no friction between m_1 and the table. (a) Sketch all the forces on the crates. Include a coordinate system. (b) Determine the components of the forces along the coordinate axes. (c) Write Newton's second law for motion along both coordinate directions. (d) Solve the equations from part (c) to find the acceleration of the crates and the tension in the string.

45. ✪ When a car is traveling at 25 m/s on level ground, the stopping distance is found to be 22 m. This distance is measured by pushing hard on the brakes so that the wheels skid on the pavement. The same car is driving at the same speed down an incline that makes an angle of 8.0° with the horizontal. What is the stopping distance now, as measured along the incline?

46. A block slides down an inclined plane that makes an angle of 40° with the horizontal. There is friction between the block and the plane. (a) Sketch all the forces on the block. Include a coordinate system. (b) Determine the components of the forces along the coordinate axes. (c) Write Newton's second law for motion along both coordinate directions. (d) If the acceleration of the block is 2.4 m/s², what is the coefficient of kinetic friction between the block and the plane?

47. ⭐ A sled is pulled with a horizontal force of 20 N along a level trail, and the acceleration is found to be 0.40 m/s². An extra mass $m = 4.5$ kg is placed on the sled. If the same force is just barely able to keep the sled moving, what is the coefficient of kinetic friction between the sled and the trail?

48. Two crates are connected by a massless rope that passes over a pulley as shown in Figure 4.25. If the crates have mass 35 kg and 85 kg, what is their acceleration? If the system begins at rest, with the more massive crate a distance 12 m above the floor, how long does it take the more massive crate to reach the floor? Assume the pulley is massless and frictionless.

49. ⭐ A hockey puck is sliding down an inclined plane with angle $\theta = 15°$ as shown in Figure P4.49. If the puck is moving with a

Figure P4.49

constant speed, what is the coefficient of kinetic friction between the puck and the plane?

50. ✪ ® Consider a baseball player who wants to catch the fly ball in Problem 26(b). Estimate the force exerted by the ball on the player's glove as the ball is coming to rest. *Hint*: You will have to estimate the acceleration of the ball, and to do so it is useful to estimate the distance the ball travels while coming to rest in the glove. The mass of a baseball is 0.14 kg.

51. Two forces are acting on an object. One force has a magnitude of 45 N and is directed along the $+x$ axis. The other force has a magnitude of 30 N and is directed along the $+y$ axis. What is the direction of the object's acceleration? Express your answer by giving the angle between $\vec{a}$ and the x axis.

52. ⭐ A car is traveling down a hill that makes an angle of 20° with the horizontal. The driver applies her brakes, and the wheels lock so that the car begins to skid. The coefficient of kinetic friction between the tires and the road is $\mu_K = 0.45$. (a) Find the acceleration of the car. (b) How long does the car take to skid to a stop if its initial speed is 30 mi/h (14 m/s)?

53. SSM ⭐ A skier is traveling at a speed of 40 m/s when she reaches the base of a frictionless ski hill. This hill makes an angle of 10° with the horizontal. She then coasts up the hill as far as possible. What height (measured vertically above the base of the hill) does she reach?

54. ✪ Consider again the skier in Problem 53, but now assume there is friction with $\mu_K = 0.10$ between the skis and the snow. What height does she reach now?

55. ✪ ® An airplane of mass 25,000 kg (approximately the size of a Boeing 737) is coming in for a landing at a speed of 70 m/s. Estimate the normal force on the landing gear when the airplane lands. *Hint*: You will have to estimate the angle that the landing velocity makes with the horizontal and also the compression (vertical displacement) of the landing gear shock absorbers when the plane contacts the ground.

4.5 DETECTING ACCELERATION: REFERENCE FRAMES AND THE WORKINGS OF THE EAR

56. Consider the rock-on-a-string accelerometer in Figure 4.29. What is the accelerometer angle if the airplane has a horizontal acceleration of 1.5 m/s²?

57. ✪ Consider the accelerometer design shown in Figure P4.57, in which a small sphere sits in a hemispherical dish. Suppose the dish has an acceleration of a_x along the horizontal and a small amount of friction causes the sphere to reach a stable position relative to the dish as shown in the figure. Find the position of the sphere (i.e., find the angle θ) as a function of a_x.

Figure P4.57

58. SSM ✪ ® Consider a commercial passenger jet as it accelerates on the runway during takeoff. This jet has a rock-on-a-string accelerometer installed (as in Fig. 4.29). Estimate the angle of the string during takeoff.

4.6 PROJECTILE MOTION REVISITED: THE EFFECT OF AIR DRAG

59. SSM ⭐ ® Estimate the terminal velocity for a golf ball.

60. ⭐ Make qualitative sketches of the trajectory of a batted baseball ($v_0 = 50$ m/s) with and without air drag.

61. ⭐ ® A skier travels down a steep, frictionless slope of angle 20° with the horizontal. Assuming she has reached her terminal velocity, estimate her speed.

62. ✪ Ⓡ Consider the motion of a bicycle with air drag included. We saw how to deal with the motion on a hill in connection with Figure 4.32. Now assume the bicycle is coasting on level ground and is being pushed along by a tailwind of 10 mi/h (4.5 m/s). If the bicycle starts from rest at $t = 0$, what is the acceleration of the bicycle at that moment? Assume that there are no frictional forces (e.g., from the bicycle's tires or bearings, or any other source) opposing the motion of the bicycle.

63. ✪ Ⓡ What is the approximate magnitude of the drag force on an airplane (such as a commercial jet) traveling at 250 m/s?

ADDITIONAL PROBLEMS

64. SSM ✪ Consider the system of blocks in Figure P4.64, with $m_2 = 5.0$ kg and $\theta = 35°$. If the coefficient of static friction between block 1 and the inclined plane is $\mu_S = 0.25$, what is the largest mass m_1 for which the blocks will remain at rest?

Figure P4.64

65. ✪ A water skier (Fig. P4.65) of mass $m = 65$ kg is pulled behind a boat at a constant speed of 25 mi/h. If the tension in the horizontal rope is 1000 N (approximately 225 lbs), what are the magnitude and direction of the force the water exerts on the skier's ski?

Figure P4.65

66. ✪ An abstract sculpture is constructed by suspending a sphere of polished rock along the vertical with four cables connected to a square wooden frame 1.2 m on a side as shown in Figure P4.66. If the stone has mass $m = 8.0$ kg and the bottom two cables are under a tension of 25 N each, what is the tension in each top cable?

Figure P4.66

67. ✪ An airplane pulls a 25-kg banner on a cable as depicted in Figure P4.67. When the airplane has a constant cruising velocity, the cable makes an angle of $\theta = 20°$. (a) What is the drag force exerted on the banner by the air? (b) The weather changes abruptly and the pilot finds herself in a 20-minute rainstorm. After the storm, she finds that the cable now makes an angle of $\theta_{water} = 30°$. Assuming the velocity and drag force are the same after the storm, find the mass of the water absorbed by the banner's fabric during the storm.

Figure P4.67

68. ✪ A piece of wood with mass $m = 2.4$ kg is held in a vise sandwiched between two wooden jaws as shown in Figure P4.68. A blow from a hammer drives a nail that exerts a force of 450 N on the wood. If the coefficient of static friction between the wood surfaces is 0.67, what minimum normal force must each jaw of the vise exert on the wood block to hold the block in place?

Figure P4.68

69. ✪ **Stomp rocket.** A "stomp rocket" is a toy projectile launcher illustrated in Figure P4.69i, which uses a blast of air to propel a dart made of plastic. The air blast is produced by jumping on a plastic bladder as shown in Figure P4.69ii. It is found that when a 190-lb father jumps on the bladder, the rocket dart will fly on average a horizontal distance of 500 ft when launched at an angle of 45°. (a) If the rocket dart is instead pointed straight up and the same father jumps on the launcher, what would be the maximum height obtained? (b) What is the initial velocity of the rocket dart as it leaves the launch tube? Would you recommend the use of safety goggles? Ignore air resistance.

Figure P4.69

70. ✪ An archer lines up an arrow on the horizontal exactly dead center on a target as drawn in Figure P4.70. She releases the arrow, and it strikes the target a distance $h = 7.6$ cm below the center. If the target is 10 m from the archer, (a) how long was the arrow in the air before striking the target? (b) What was the initial velocity of the arrow as it left the bow? Report your answers in meters per second, feet per second, and miles per hour. Neglect air resistance.

Figure P4.70

71. ✪ Consider once again the swimmer in Example 4.7. Assume she can swim at a velocity of 0.30 m/s and the river is 15 m wide. She needs to get across the river as quickly as possible. (a) What direction should her velocity vector make relative to the water to get her across in the shortest time? (b) What is her overall velocity as measured from the shore? (c) What is the direction of her total velocity as seen from the shore? (d) How long does it take her to cross the river? (e) How far downstream does she end up?

72. ✪ A bartender slides a mug of root beer with mass $m = 2.6$ kg down a bar top of length $L = 2.0$ m to an inattentive patron who lets it fall a height $h = 1.1$ m to the floor. The bar top (Fig. P4.72) is smooth, but it still has a coefficient of kinetic friction of $\mu_K = 0.080$. (a) If the bartender gave the mug an initial velocity of 2.5 m/s, at what distance D from the bottom of the bar will the mug hit the floor? (b) What is the mug's velocity (magnitude and direction) as it impacts the floor? (c) Draw velocity–time graphs for both the x and y directions for the mug, from the time the bartender lets go of the mug to the time it hits the floor.

Figure P4.72

73. ✪ Two workers must slide a crate designed to be pushed and pulled at the same time as shown in Figure P4.73. Joe can always exert twice as much force as Paul, and Paul can exert 160 N of force. The crate has a mass $m = 45$ kg, the angle $\theta = 25°$, and the coefficient of kinetic friction between the floor and crate is 0.56. If we want to move the crate as quickly as possible, is it better to have Paul push and Joe pull, or vice versa? Does it matter? Calculate the acceleration for both cases to find out.

Figure P4.73

74. ✪ ⊗ **Inner ear.** A student constructs a model of the utricle of the ear by attaching wooden balls ($m = 0.080$ kg) by strings to the bottom of a fish tank and then submerging them in water so that they float with the strings in a vertical direction. She finds that the wooden balls float with a buoyant force that exerts a tension of 0.20 N on each string as shown in Figure P4.74i. She then has her mother take her on a drive. She notes that as her mother applies the gas pedal, the strings make an angle $\theta = 16°$ with respect to the vertical as shown in Figure P4.74ii. (a) What was the acceleration of the car? (b) If the car started from rest, how long would it take to get to freeway speed of 65 mi/h? During the acceleration, the water also makes an angle with the horizontal. Are the wooden balls even needed? Explain. Ignore any forces due to the pressure in the water.

Figure P4.74

75. ✪ A vintage sports car accelerates down a slope of $\theta = 17°$ as depicted in Figure P4.75. The driver notices that the strings of the fuzzy dice hanging from his rearview mirror are perpendicular to the roof of the car. What is the car's acceleration?

Figure P4.75

76. ✦ A spacecraft is traveling through space parallel to the x direction with an initial velocity of magnitude $v_0 = 3500$ m/s. At $t = 0$, the pilot turns on the engines, producing an acceleration of magnitude $a = 150$ m/s^2 along the y direction. Find the speed and the angle the velocity makes with the x direction at (a) $t = 15$ s and (b) 45 s.

77. A baseball is hit with an initial speed of 45 m/s at an angle of 35° relative to the x (horizontal) axis. (a) Find the speed of the ball at $t = 3.2$ s. (b) What angle does $\vec{v}$ make with the x axis at this moment? (c) While the ball is in the air, at what value of t is the speed the smallest?

78. A cannon shell is fired from a battleship with an initial speed of 700 m/s. (a) What is the maximum range of the shell? (b) For the firing angle that gives the maximum range, what is the maximum height reached by the shell? Ignore air drag.

79. SSM ✪ Two blocks of mass m_1 and m_2 are sliding down an inclined plane (Fig. P4.79). (a) If the plane is frictionless, what is the normal force between the two masses? (b) If the coefficient of kinetic friction between m_1 and the plane is $\mu_1 = 0.25$ and between m_2 and the plane is $\mu_2 = 0.25$, what is the normal force between the two masses? (c) If $\mu_1 = 0.15$ and $\mu_2 = 0.25$, what is the normal force?

Figure P4.79

80. **Daredevil stunt.** Once upon a time, a stuntman named Evel Knievel boasted that he would jump across the Grand Canyon on a motorcycle. (This stunt was never attempted.) The width of the Grand Canyon varies from place to place, but suppose Mr. Knievel attempted to jump across at a point where the width is about 5 mi and the two rims are at the same altitude. What minimum initial speed would be required to make it safely across the canyon? Express your answer in meters per second and miles per hour. Ignore air drag.

81. ⓡ Although Evel Knievel never succeeded in jumping over the Grand Canyon (see Problem 80), he was famous for jumping (with the help of his motorcycle) over, among other things, 14 large trucks. This jump covered an approximate distance of 135 ft. What was Mr. Knievel's minimum initial velocity for this jump? Ignore air drag.

82. ✦ A person travels on a ski lift (Fig. P4.82). (a) If the support strut on the ski lift makes an angle $\theta = 15°$, what is the horizontal acceleration of the person? (b) If the person plus the lift chair have a combined mass $m = 120$ kg, what is the tension force along the support strut? Assume the ski lift is moving horizontally. For simplicity, when considering forces involving the support strut, you can treat the strut in the same way as you would treat a massless cable.

Figure P4.82

83. ⭐ A block slides up a frictionless, inclined plane with $\theta = 25°$ (Fig. P4.83). (a) If the block reaches a maximum height $h = 4.5$ m, what was the initial speed of the block? (b) Now consider a second inclined plane with the same tilt angle and for which the coefficient of kinetic friction between the block and the plane is $\mu_K = 0.15$. What initial speed is now required for the block to reach the same maximum height?

Figure P4.83

84. A golf ball is hit by a golfer and travels down the fairway. Sketch the ball's trajectory and show qualitatively the direction and magnitude of the velocity at various points.

85. ⭐ A car travels along a level road with a speed of $v = 25$ m/s (about 50 mi/h). The coefficient of kinetic friction between the tires and the pavement is $\mu_K = 0.55$. (a) If the driver applies the brakes and the tires "lock up" so that they skid along the road, how far does the car travel before it comes to a stop? (b) If the same car is traveling downhill along a road that makes an angle $\theta = 12°$, how much does the car's stopping distance increase? (c) What is the stopping distance on an uphill road with $\theta = 12°$?

86. ✪ A person riding a bicycle on level ground at a speed of 10 m/s throws a baseball forward at a speed of 15 m/s relative to the bicycle at an angle of 35° relative to the horizontal (x) direction. (a) If the ball is released from a height of 1.5 m, how far does the ball travel horizontally, as measured from the spot where it is released? (b) How far apart are the bicycle and the ball when the ball lands? Ignore air drag.

87. ✪ The pilot of an airplane with an open cockpit throws a ball horizontally with a speed of 50 m/s relative to the airplane, and the airplane travels horizontally at a speed of 150 m/s relative to the ground. If the ball's speed relative to the ground is 175 m/s, what is the angle between the ball's velocity and the velocity vector of the airplane?

88. ⭐ A canoe travels across a 40-m-wide river that has a current of speed 5.0 m/s. The canoe has a speed of 0.80 m/s relative to the water (Fig. P4.88). The person paddles the canoe so as to cross

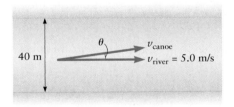

Figure P4.88

the river as fast as possible. (a) What is the canoe's velocity relative to an observer on shore? Give the magnitude and direction of the velocity vector. (b) How far downstream does the canoe travel as it crosses from one side of the river to the other?

89. Two crates of mass $m_1 = 35$ kg and $m_2 = 15$ kg are stacked on the back of a truck (Fig. P4.89). The frictional forces are strong enough that the crates do not slide off the truck. Assume the truck is accelerating with $a = 1.7$ m/s². Draw free-body diagrams for both crates and find the values of all forces in your diagrams.

Figure P4.89

90. ✪ A projectile is fired uphill as sketched in Figure P4.90. If $v_0 = 150$ m/s and $\theta = 30°$, what is L?

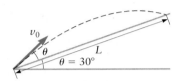

Figure P4.90

91. ✪ Ⓡ When airplanes land or take off, they always travel along a runway in the direction that is "into" the wind because the "lift force" on an airplane wing depends on the speed of the airplane relative to the air (v_{rel}). We'll see in Chapter 10 that the lift force is proportional to v_{rel}^2. Consider a case in which a runway is parallel to the x axis and the wind velocity is $v_{wind} = +20$ mi/h (directed along $+x$). (a) In what direction should the airplane travel to take off (along $+x$ or $-x$)? (b) Estimate the speed of a typical airplane (e.g., a Boeing 737) relative to the runway as it takes off. (c) Find the ratio of the lift force for an airplane that travels into the wind to the lift force when it travels in the opposite direction. (d) Is this factor large enough to make a significant difference?

92. ✪ Ⓡ Consider again the game of baseball. A pitcher throws a pitch to the catcher with a speed of 100 mi/h, releasing the ball from approximately shoulder height. At what angle with respect to the horizontal (x) should he throw the ball so that it crosses the plate waist high for the batter? The distance from the pitcher to the catcher is 60.5 feet. Ignore air drag. *Hint:* Assume the angle is small so that the horizontal component of the velocity is approximately equal to the speed.

Chapter 5

Circular Motion and Gravitation

The motion of this merry-go-round is an example of circular motion. Other examples explored in this chapter are the orbits of planets and moons and the rotation of the Earth about its axis. (© H. Mark Weidman Photography/Workbook Stock/Jupiterimages)

In the past few chapters, we studied the connection between forces and motion, and we learned how to apply Newton's laws in a variety of situations. In many cases, those situations involved motion in one dimension. While we have also studied two- and three-dimensional motion, most of those cases could be treated as a combination of one-dimensional problems. For example, in our work on projectile motion, we were able to treat the horizontal and vertical motions as essentially separate one-dimensional problems. In this chapter, we consider a different type of motion, called **circular motion**, in which the acceleration is far from constant and which *cannot* be reduced to a one-dimensional problem. Circular motion is found in many situations, such as a car traveling around a turn, a roller coaster traveling near the top or bottom of its track, a centrifuge, and the Earth orbiting the Sun. Our studies of orbital motion and the forces that make it possible will also lead us to explore the gravitational force in more detail. We'll then encounter Kepler's laws of orbital motion and see how they describe the motion of planets, moons, and satellites. These studies of gravitation will give us new insights into the quantity *g*, the acceleration due to gravity near the Earth's surface, and we'll explore the connection between *g* and the motion of the Moon as it orbits the Earth.

5.1 | UNIFORM CIRCULAR MOTION

Figure 5.1 shows a top-down view of a person running around a circular track. For simplicity, let's assume our runner is moving with a constant speed so that the magnitude of her velocity is constant. The direction of her velocity $\vec{v}$, however, changes as she moves around the track since the velocity vector is always directed tangent to the circle. Because the direction of $\vec{v}$ changes as she runs, her velocity is not constant, even though her speed is constant. Circular motion at constant speed is known as **uniform circular motion.**

The runner in Figure 5.1 is moving at a constant speed v, so the time it takes her to travel once around the track (to run one complete lap) is equal to the circumference of the track divided by v. If the track has a radius r, the circumference is $2\pi r$ and the time to complete one lap is

$$T = \frac{2\pi r}{v} \tag{5.1}$$

Period for an object undergoing uniform circular motion

This time T is called the **period** of the motion.

| **EXAMPLE 5.1** | **A Bug on a Compact Disc** |

A bug is sitting on the edge of a compact disc of radius $r = 6.0$ cm (Fig. 5.2). The bug undergoes uniform circular motion as the CD spins. **(a)** If the bug traverses this circle exactly six times in precisely 1 s, what is the period of the motion? **(b)** What is the bug's speed?

RECOGNIZE THE PRINCIPLE

The period T is the time it takes to travel once around the circle, so we can find T from the given information. We can then use the relation between period and v in Equation 5.1 to find the bug's speed.

SKETCH THE PROBLEM

Figure 5.2 shows the problem. The bug moves in a circle of radius r.

IDENTIFY THE RELATIONSHIPS AND SOLVE

(a) Since the bug completes six trips each 1 s, we have

$$T = \frac{1.0 \text{ s}}{6 \text{ trips around circle}} = \frac{0.17 \text{ s}}{1 \text{ trip}}$$

The period is the time to complete one trip around the circle, so the bug's period is

$$T = \boxed{0.17 \text{ s}}$$

(b) The bug is undergoing uniform circular motion, so the speed v is constant and is equal to the total distance traveled divided by the time. A distance d equal to six times the circumference is traveled in 1 s, so $d = 6 \times (2\pi r)$ while $t = 1.0$ s. Hence,

$$v = \frac{d}{t} = \frac{6(2\pi)(0.060 \text{ m})}{1.0 \text{ s}} = \boxed{2.3 \text{ m/s}}$$

What have we learned?
The period of circular motion is equal to the time it takes to move around the circle once. Hence, T does not depend on the bug's location on the CD; the period is the same for a bug at the edge of the CD or near the center. The speed of the bug, however, does depend on its location (the value of r). The speed is largest at the edge (a large value of r) and decreases as the bug moves toward the center.

Top view

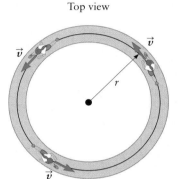

Figure 5.1 This runner is moving with a constant speed around a circular track. While her speed is constant, her velocity is not constant. The direction of $\vec{v}$ changes as she moves around the track.

Figure 5.2 Example 5.1. The bug sitting on this CD is undergoing uniform circular motion.

Expanded view

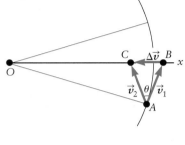

Figure 5.3 ◩ The velocity $\vec{v}$ of an object undergoing uniform circular motion is shown at two different locations along the circular path. The distance traveled in going from point P_1 (time t_1) to point P_2 (time t_2) is s. ◪ Expanded view of the velocity vectors at P_1 and P_2. The difference $\Delta\vec{v}$ is directed toward the center of the circle.

Centripetal Acceleration

While the speed is constant for a particle that undergoes uniform circular motion, the velocity $\vec{v}$ is not constant. Recall from Chapter 2 that the acceleration $\vec{a}$ of an object is proportional to the change in $\vec{v}$ over the course of a time interval Δt:

$$\vec{a} = \lim_{\Delta t \to 0} \frac{\Delta\vec{v}}{\Delta t} \tag{5.2}$$

For an object undergoing uniform circular motion, the velocity is changing with time, so $\Delta\vec{v}$ is definitely not zero. Uniform circular motion therefore involves a *nonzero acceleration*.

To understand and calculate this acceleration, let's consider how the velocity of the runner in Figure 5.1 changes as she moves around the track. Figure 5.3A shows her velocity vectors at two nearby locations P_1 and P_2, and Figure 5.3B shows the corresponding change in velocity $\Delta\vec{v}$ over a short time interval Δt. We can use the geometry in Figure 5.3B to estimate the average acceleration $\vec{a}_{ave} = \Delta\vec{v}/\Delta t$ during this interval. If Δt is small, the instantaneous acceleration (Eq. 5.2) is approximately equal to $\vec{a}_{ave}$. We can therefore write

$$\vec{a} \approx \vec{a}_{ave} = \frac{\Delta\vec{v}}{\Delta t} \tag{5.3}$$

We now want to use Figure 5.3 to find both the direction and magnitude of $\vec{a}$. Consider first the direction: Figure 5.3A shows the velocity vector at two closely spaced values of time for our runner. During the time interval $\Delta t = t_2 - t_1$, she travels a distance $s = v\,\Delta t$ along the circumference of the circle. (Recall that the speed v is the magnitude of both $\vec{v}_1$ and $\vec{v}_2$.) The corresponding velocity vectors $\vec{v}_1$ and $\vec{v}_2$ are also shown; both are tangent to the circle and hence both are perpendicular to the radius lines drawn at P_1 and P_2. To compute $\Delta\vec{v} = \vec{v}_2 - \vec{v}_1$, it is useful to redraw the velocity vectors so that they have a common "tail point" as shown in Figure 5.3B. When drawn in this way, the velocity vectors $\vec{v}_1$ and $\vec{v}_2$ form two sides of the triangle labeled ABC. The third side of this triangle is $\Delta\vec{v} = \vec{v}_2 - \vec{v}_1$ and $\Delta\vec{v}$ is directed toward the center of the circle. This result is true at all points along our runner's circular path. Since the acceleration vector is parallel to $\Delta\vec{v}$ (see Eq. 5.3), the acceleration of an object undergoing uniform circular motion is always directed *toward the center of the circle*.

To compute the magnitude of this acceleration, we again make use of triangle ABC in Figure 5.3B. Two sides of this triangle are formed by $\vec{v}_1$ and $\vec{v}_2$, so these sides are both of length v. The other side of this triangle has a length Δv. The triangle ABC has the same interior angles as the triangle defined by O, P_1, and P_2 in Figure 5.3A, so these triangles are similar. Two sides of triangle OP_1P_2 are along the radius of the circle and are thus of length r, while the other side (between points P_1 and P_2) has a length s, and when Δt is small, this arc approaches a straight line. The distance traveled by the runner from P_1 to P_2 is related to the speed by $s = v\,\Delta t$. The triangles ABC and OP_1P_2 are similar, so the ratios of their corresponding sides are equal; hence,

$$\frac{\Delta v}{v} = \frac{s}{r} = \frac{v(\Delta t)}{r}$$

$$\Delta v = \frac{v^2(\Delta t)}{r} \tag{5.4}$$

According to Equation 5.3, the magnitude of the acceleration is equal to the ratio $\Delta v/\Delta t$, so we can rearrange the result in Equation 5.4 to get

$$a_c = \frac{\Delta v}{\Delta t} = \frac{v^2(\Delta t)/r}{\Delta t}$$

Magnitude of the centripetal acceleration

$$a_c = \frac{v^2}{r} \tag{5.5}$$

This acceleration, called the *centripetal acceleration*,[1] is usually denoted by the symbol a_c. Although this derivation started with the average acceleration in Equation 5.3, the result becomes exact for a very small time interval $\Delta t \to 0$; hence, Equation 5.5 gives the instantaneous centripetal acceleration.

Circular Motion and Forces

The result for the acceleration in Equation 5.5 applies to any object undergoing circular motion. Such an object has an acceleration of magnitude $a_c = v^2/r$, and this acceleration is directed toward the center of the circle. From Newton's second law, we know that accelerations are caused by forces, so for an object undergoing uniform circular motion, we have

$$\sum \vec{F} = m\vec{a}$$

In terms of magnitudes,

$$\sum F = ma_c = \frac{mv^2}{r} \tag{5.6}$$

Force required for uniform circular motion

which is the force required to make an object of mass m travel with speed v in a circle of radius r. This force must be directed toward the center of the circle.

As an example, consider a rock tied to a string and suppose a person is twirling the string so that the rock moves in a circle. For simplicity, let's assume this experiment is being done by an astronaut in distant space (Fig. 5.4), so the only force on the rock comes from the string (i.e., all gravitational forces are negligible). This rock is undergoing circular motion; hence, its acceleration is given by Equation 5.5. According to Newton's second law and Equation 5.6, this acceleration is caused by a force of magnitude mv^2/r. This force comes from the tension T in the string with

$$\sum F = T = \frac{mv^2}{r}$$

For the rock to travel in a circle, the tension must have this value. (Note again that T here is the tension in the string, and not the period of the motion.)

When an object undergoes circular motion, there must be a force of magnitude mv^2/r acting on it. For the rock in Figure 5.4A, this force is the tension in the string, but in other situations, the force might be due to gravity or friction or some other source. Without such a force, the object *cannot* undergo circular motion.

What happens if the string in Figure 5.4 suddenly breaks? After the string breaks (at the lower right in Fig. 5.4B), the force on the rock is zero. According to Newton's first law, the rock will then move away in a straight-line path—that is, with a constant velocity. The rock does not move radially outward, nor does it "remember" its circular trajectory. The only way the rock can move in a circle is if there is a force that makes it do so, and that force is provided by the string.

> **CONCEPT CHECK 5.1 | Velocity and Acceleration in Circular Motion**
>
> Consider the rock in Figure 5.4 as it undergoes uniform circular motion (before the string breaks). Which of the following statements is correct? (More than one statement may be correct.)
> (a) The direction of $\vec{a}$ changes as the rock moves around the circle.
> (b) The direction of $\vec{v}$ changes as the rock moves around the circle.
> (c) $\vec{a}$ and $\vec{v}$ are always perpendicular.
> (d) $\vec{a}$ and $\vec{v}$ are always parallel.

| EXAMPLE 5.2 | Centripetal Acceleration of a Compact Disc

Consider again the bug in Example 5.1. Suppose this bug has a mass $m = 5.0$ g and sits on the edge of a compact disc of radius 6.0 cm that is spinning such that the bug

A

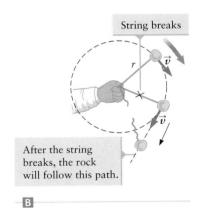

String breaks

After the string breaks, the rock will follow this path.

B

Figure 5.4 Ⓐ An astronaut in distant space twirls a rock on a string. The only force on the rock is due to the tension in the string (all gravitational forces are negligible). Ⓑ If $T = mv^2/r$, the rock will undergo uniform circular motion. If the string breaks, the rock will move along a straight line (obeying Newton's first law).

---

[1]The adjective "centripetal" means "center seeking."

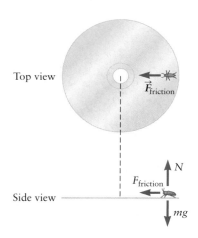

Top view

Side view

Figure 5.5 Example 5.2. There are three forces acting on the bug on this CD. The force responsible for the centripetal acceleration is provided by the force of static friction, and this frictional force is directed toward the center of the circle.

travels around its circular path three times per second. Find (a) the centripetal acceleration of the bug and (b) the total force on the bug. Also identify the *origin* of the force that enables the bug to move in a circle.

RECOGNIZE THE PRINCIPLE

Because the bug's acceleration along the vertical direction is zero, the normal force N and the force of gravity mg must cancel; hence, $N = mg$. The bug is undergoing circular motion, so we know a third force must provide the force required to produce the centripetal acceleration a_c. This third force keeps the bug from slipping relative to the CD and is just the frictional force. To make the bug undergo uniform circular motion, this force must be directed toward the center of the circle.

SKETCH THE PROBLEM

The forces acting on our bug are shown at the bottom of Figure 5.5. The force of gravity (the bug's weight) is directed downward (vertically), and there is a normal force N directed upward.

IDENTIFY THE RELATIONSHIPS

To find the centripetal acceleration, we need to determine the speed of the bug, which we can find from its period and the radius of its circular path. We can then compute the magnitude of a_c and the force needed to produce it.

SOLVE

(a) The bug traverses the circle three times in 1 s, so it travels a distance equal to three times the circumference each second. Its speed is thus

$$v = \frac{3 \times (2\pi r)}{t} = \frac{3(2\pi)(6.0 \text{ cm})}{1.0 \text{ s}} = 110 \text{ cm/s} = 1.1 \text{ m/s}$$

From Equation 5.5, the centripetal acceleration is

$$a_c = \frac{v^2}{r} = \frac{(1.1 \text{ m/s})^2}{0.060 \text{ m}} = \boxed{20 \text{ m/s}^2}$$

(b) The required force is (Eq. 5.6)

$$F = ma_c = (5.0 \text{ g})(20 \text{ m/s}^2) = (5.0 \times 10^{-3} \text{ kg})(20 \text{ m/s}^2)$$

$$F = 0.10(\text{kg} \cdot \text{m/s}^2) = \boxed{0.10 \text{ N}}$$

This force is produced by friction between the bug's feet and the surface of the CD.

What does it mean?
Even though the bug is moving, the force in part (b) is static friction because the bug is not slipping relative to the CD's surface.

Now that we have analyzed some examples of circular motion, it is time to outline a general way to approach such problems.

PROBLEM SOLVING | Analyzing Circular Motion

1. **RECOGNIZE THE PRINCIPLE.** An object can move in a circle only if there is a force $F = mv^2/r$ directed toward the center of the circle.

2. **SKETCH THE PROBLEM.** Make a drawing that shows the path followed by the object of interest. This drawing should identify the circular part of the

path, the radius of this circle, and the center of the circle.

3. IDENTIFY THE RELATIONSHIPS.

- Find all the forces acting on the object; as in our applications of Newton's laws in Chapters 3 and 4, a free-body diagram is often very useful.
- Using your drawing and free-body diagram, find the components of the forces that are directed *toward the center* of the circle and find the components perpendicular to this direction.
- Apply Newton's second law $\sum F = ma$ for motion toward the center of the circle and (if necessary)

in the perpendicular direction. The total force directed toward the center of the circle is (Eq. 5.6)

$$\sum F_{\text{center}} = ma_c = \frac{mv^2}{r}$$

4. SOLVE FOR THE QUANTITIES OF INTEREST. For example, the centripetal acceleration is (Eq. 5.5)

$$a_c = \frac{v^2}{r}$$

5. Always *consider what your answer means* and check that it makes sense.

When discussing the centripetal acceleration and the associated forces, it is often convenient to use a coordinate system in which the positive direction is toward the center of the circle. The "positive" direction changes as the object moves around the circle (as for the bug in Fig. 5.5). An advantage of this approach is that the centripetal acceleration is always positive, which can eliminate some extra minus signs in your equations.

Centripetal Acceleration of a Turning Car: What Are the Forces?

The difficulties faced by the bug in Example 5.2 are similar to those encountered by the driver of a car. While cars do not usually travel in perfect circles, a car making a turn or rounding a bend in a road is traveling in an approximate circle. Over a short distance, any curved path can be approximated by a circle, and the radius of this circle is called the ***radius of curvature***. Let's analyze the motion of a car that travels along a short section of such a circle as sketched in Figure 5.6. As it travels around a bend in this road, the car travels in a circular path, so there must be a force on the car directed toward the center of the circle (point C in Figure 5.6A). Continuing with our problem-solving strategy for circular motion (step 2) the sketch in Figure 5.6A shows the circular path followed by the car as the dashed curve. The center of this circle is at point C, and its radius is r. For step 3, the sketch in Figure 5.6B shows all the forces on the car along with a free-body diagram. There are two forces along the vertical (y) direction, the force of gravity and the normal force exerted by the road on the car, while the only horizontal force is the force of friction F_{friction} from the road on the car. This force is static friction because we assume the tires are not slipping relative to the road. Applying Newton's second law along y, we get

$$\sum F_y = +N - mg = ma_y$$

Since the acceleration along y is zero, this result leads to $N = mg$. The only force directed toward the center of the circle is F_{friction}, and inserting it into Equation 5.6 gives

$$F_{\text{friction}} = \frac{mv^2}{r}$$

Given the values of m, v, and r, we can now calculate the force required to make the turn successfully, without leaving the road.

Notice that the required force increases as the car's speed increases. There is an upper limit on the force of static friction; hence, there is a maximum speed at which the car can make the turn without slipping. The maximum frictional force is $F_{\text{friction, max}} = \mu_S N = \mu_S mg$, so the condition for this maximum speed is

$$\frac{mv^2}{r} = F_{\text{friction, max}} = \mu_S mg$$

Top view

A

End view

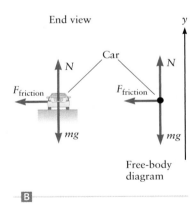

Free-body diagram

B

Figure 5.6 This car moves in an approximately circular path as it travels along a curve in the road. Friction between the car's tires and the road provides the force that causes the centripetal acceleration. **A** Top view. **B** End view showing forces on car and free-body diagram.

Solving for the maximum speed gives

$$\frac{mv^2}{r} = \mu_s mg$$

$$v = \sqrt{\mu_s gr} \tag{5.7}$$

Let's put in some realistic numbers to see how the result in Equation 5.7 compares with our everyday experience. For a typical intersection in a residential neighborhood, the effective radius of a turn (the radius of curvature) might be $r = 10$ m, and the coefficient of friction between rubber and a road surface is typically $\mu_s = 0.6$. Inserting these values into Equation 5.7 gives

$$v = \sqrt{\mu_s gr} = \sqrt{0.6(9.8 \text{ m/s}^2)(10 \text{ m})} = 8 \text{ m/s}$$

which is approximately 15 mi/h. Hence, our physics does indeed give a good description of "real-life" driving.

CONCEPT CHECK 5.2 | High-Speed Driving

Consider two vehicles, a compact car and a large truck, that are both traveling around the turn in the road in Figure 5.6 and assume the coefficient of friction between the tires and the road is the same for both. Which vehicle can drive through the turn at the largest speed without slipping, (a) the car, (b) the truck, or (c) it would be a tie because the nonslipping speed is the same for both vehicles?

A Car on a Banked Turn: Analyzing the Forces

The maximum speed for making a turn successfully can be increased by banking the turn as shown in Figure 5.7. Let's analyze this situation, but for simplicity, we assume there is no friction between the tires and the road (we'll add friction to this problem in Example 5.3). We want to calculate the speed at which a car can successfully travel through such a turn without slipping off the road.

We again follow our problem-solving strategy for circular motion. The car travels in a circle, so there must be a force $F = ma_c$ directed toward the center of its circular path. For step 2, the car is shown in Figure 5.7A. The overall circular path is the same as for the car in Figure 5.6, and we again take r to be the radius of curvature. For step 3, the forces on the car are shown Figure 5.7, which also gives a free-body diagram. We have assumed there is no friction, so the only forces on the car are due to gravity and the normal force exerted by the road on the car. We determine the components of the force along the vertical and along the direction toward the center of the circle (horizontal). Due to the banking, there is now a component of the normal force directed toward the center of the circle. To find the maximum safe speed, we apply Newton's second law for the centripetal acceleration and the associated force (Eq. 5.6). From the trigonometry in Figure 5.7B, the component of the force directed toward the center of the circle is $N \sin \theta$, and this must be equal to the mass of the car times the centripetal acceleration. For a car of mass m traveling at a speed v, we have

$$F = ma_c = \frac{mv^2}{r} = N \sin \theta$$

which can be solved for the speed of the car:

$$v = \sqrt{\frac{Nr \sin \theta}{m}} \tag{5.8}$$

At this speed, the car will just be able to negotiate the turn without sliding up or down the banked road. To complete our calculation of the speed v, we must know the value of the normal force, which we can determine by considering the forces

If there is no friction between the tires and the road, these forces are the only ones acting on the car.

Free-body diagram

Figure 5.7 ◻ On a very slippery—frictionless—incline, the only forces on this car are the force of gravity and the normal force from the incline. ◻ Free-body diagram with forces resolved into horizontal and vertical components. The horizontal component of the normal force is $N \sin \theta$. This is the only horizontal force on the car, and it provides the centripetal acceleration that enables the car to move in a circular path as it travels around a banked curve.

along the vertical direction in Figure 5.7B. Since the acceleration along the vertical direction (y) is zero, the total force along y must be zero:

$$\sum F_y = N \cos \theta - mg = 0$$

Hence,

$$N = \frac{mg}{\cos \theta}$$

Inserting this result for N into Equation 5.8, we find

$$v = \sqrt{\frac{Nr \sin \theta}{m}} = \sqrt{\frac{mg}{\cos \theta} \frac{r \sin \theta}{m}} = \sqrt{gr \tan \theta}$$

When $\theta = 0$, this road is flat, and the result for the maximum speed is then zero because $\tan(0) = 0$. Therefore, you cannot turn on a very icy (frictionless) road without slipping, as you probably already knew.

| EXAMPLE 5.3 | Traveling through a Banked Turn with Friction

Consider again the problem of a car making a turn on a banked road, but now let's add friction to the situation. If the coefficient of static friction between the tires and the road is μ_S, what is the maximum speed at which the car can safely negotiate a turn of radius r with a banking angle θ?

RECOGNIZE THE PRINCIPLE

Because the car is traveling in a circle, the total force on the car must be $\sum F = mv^2/r$ directed toward the center of the circle. We must now include the force of friction when computing $\sum F$.

SKETCH THE PROBLEM

Following our problem-solving strategy for circular motion (steps 2 and 3), Figure 5.8A shows our car along with all the forces—the force of gravity, the normal force from the road, and the force of static friction—acting on it. Since we want to calculate the maximum safe speed, the force of friction will be directed *down* the incline so as to oppose any slipping of the car that would otherwise be directed up the plane.

IDENTIFY THE RELATIONSHIPS

The components of the forces are shown in parts (b) and (c) of Figure 5.8. We have two unknown quantities, the speed of the car and the normal force. We can get two relations by applying Newton's second law along the vertical and horizontal directions, just as we did for the case without friction. We can then solve for N and v.

SOLVE

The car is not slipping up or down the incline, so the acceleration along y is zero. Hence, the total force along y must be zero, and from Figure 5.8, we have

$$\sum F_y = 0 = +N \cos \theta - F_{\text{friction}} \sin \theta - mg$$

If the car is on the verge of slipping, the frictional force will have a magnitude of $F_{\text{friction}} = \mu_S N$. Hence,

$$N \cos \theta - \mu_S N \sin \theta - mg = 0$$

and we can rearrange to find the normal force

$$N = \frac{mg}{\cos \theta - \mu_S \sin \theta}$$

Figure 5.8 Example 5.3. ⒜ When friction is added to the road surface in Figure 5.7, there is an additional force (static friction) parallel to the incline. The horizontal and vertical components of ⒝ the frictional force and ⒞ the normal force are shown. The total horizontal force on the car is the sum of the horizontal components of the frictional force and the normal force.

The total force along the horizontal direction provides the centripetal acceleration; in the coordinate system in Figure 5.8, we find

$$\sum F_x = -N \sin \theta - F_{\text{friction}} \cos \theta = -N \sin \theta - \mu_S N \cos \theta$$

The minus signs indicate that these forces are along the $-x$ direction. However, the center of the circle is also in this direction so we have

$$N \sin \theta + \mu_S N \cos \theta = ma_c = \frac{mv^2}{r}$$

Solving for v then gives

$$v = \sqrt{\frac{Nr(\sin \theta + \mu_S \cos \theta)}{m}}$$

Inserting our result for the normal force leads to

$$v = \sqrt{\frac{mg}{\cos \theta - \mu_S \sin \theta} \frac{r(\sin \theta + \mu_S \cos \theta)}{m}} = \boxed{\sqrt{gr \frac{\mu_S \cos \theta + \sin \theta}{\cos \theta - \mu_S \sin \theta}}}$$

What does it mean?

It is interesting to put some realistic numbers into this result. For an automobile racetrack, a banking of 20° is common, and the tracks are quite large; the radius of the turn could be 200 m. Inserting these values along with a coefficient of friction of $\mu_S = 0.6$, we find $v = 50$ m/s, which is approximately 110 mi/h; this is a typical speed in auto racing, so our model seems reasonable. To obtain an even higher speed would require a larger coefficient of friction, a larger banking angle, or both.

5.2 | EXAMPLES OF CIRCULAR MOTION

In our derivation of the centripetal acceleration (Eq. 5.5) and in all our examples so far, we have assumed the speed of the object undergoing circular motion is constant. For this special case, the centripetal acceleration is the total acceleration. In many situations, however, the speed is not constant, resulting in *nonuniform* circular motion. In such cases, the total acceleration has two components as indicated in Figure 5.9. One component is directed along the circumference, that is, tangent to the circle. The magnitude of the tangential acceleration is nonzero only when v is changing with time. The other component of the acceleration is directed toward the center of the circle and is just the centripetal acceleration we derived in Equation 5.5. Even when v is not constant, our results from Section 5.1 for the centripetal acceleration and the associated force are still valid. Whenever an object is moving in a circle, there must be a component of the acceleration equal to a_c directed toward the center of the circle, so there must also be an associated force equal to mv^2/r. Hence, we can use the results from Section 5.1 to analyze a great many situations involving circular motion; we are *not* limited to situations in which the speed is constant. In this section, we consider a variety of such problems.

Twirling a Rock on a String: What Is the Tension in the String?

A simple experiment in circular motion you have probably performed yourself—a rock tied to string is twirled in a vertical circle—is sketched in Figure 5.10A. The force of gravity acting on the rock makes it very difficult to keep the speed constant. Intuitively, we know that the speed will tend to be highest when the rock is at the bottom of the circle, and lowest when the rock is at the top. Because the speed varies with time, the tangential acceleration is nonzero. Even so, the centripetal accelera-

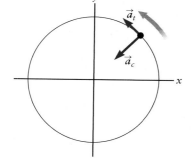

Figure 5.9 In general, an object traveling in a circle might not move with a constant speed. If the speed is not constant, there is a nonzero tangential acceleration $\vec{a}_t$ directed tangent to the circle. There is still a centripetal acceleration $\vec{a}_c$ directed toward the center of the circle, with a magnitude $a_c = mv^2/r$.

tion, the component of the acceleration directed toward the center of the circle, is still equal to $a_c = v^2/r$ provided that it is computed using the speed v at the point of interest.

Let's analyze the situation in Figure 5.10 in a little more detail. We assume we know the speed v of the rock and now consider how to calculate the tension T in the string. (Notice that T here is *not* the period of the motion!) The answer for T will depend on the rock's location, so let's first deal with the case when the rock is at the bottom of the circle. The forces on the rock for this case are shown in Figure 5.10B, and we see that the two forces, gravity and tension, are in opposite directions. These two forces together must add up to produce the centripetal acceleration

$$\sum F_{\text{center}} = +T_{\text{bottom}} - mg = \frac{mv^2}{r}$$

Solving for the tension gives

$$T_{\text{bottom}} = m\left(\frac{v^2}{r} + g\right) \tag{5.9}$$

Hence, the tension when the rock is at the bottom of the circle is large enough to overcome the gravitational force (mg) and also provide the required centripetal acceleration (mv^2/r).

When the rock is at the top of the circle as shown in Figure 5.10C, the gravitational force is parallel to the tension. The total force directed toward the center of the circle is now (recall that we take the positive direction to be toward the center of the circle)

$$\sum F_{\text{center}} = T_{\text{top}} + mg$$

The centripetal acceleration is also directed toward the center, so

$$\sum F_{\text{center}} = T_{\text{top}} + mg = \frac{mv^2}{r}$$

which leads to

$$T_{\text{top}} = m\left(\frac{v^2}{r} - g\right) \tag{5.10}$$

Comparing the results for T_{bottom} and T_{top} we see that if the speed v is the same at the top and bottom, the tension is smaller at the top. From Equation 5.10, we can also see that there is a minimum value of the speed that will keep the string taut. The tension in a string cannot be negative because a string cannot "push" on an object. As v is made smaller and smaller, however, the tension T_{top} in Equation 5.10 will eventually become zero. That will happen when

$$T_{\text{top}} = 0 = m\left(\frac{v^2}{r} - g\right)$$

$$\frac{v^2}{r} = g$$

$$v = \sqrt{gr} \tag{5.11}$$

If the speed is made smaller than this value, the string will become slack and circular motion is no longer possible.

Circular Motion and Amusement Park Activities: Maximum Speed of a Roller Coaster

The analysis of the rock's motion in Figure 5.10 can be applied to various amusement park activities. For example, when a roller coaster is near the minimum or maximum points on its track, its path is approximately circular. At these points, there must be an acceleration of magnitude v^2/r directed toward the center of the

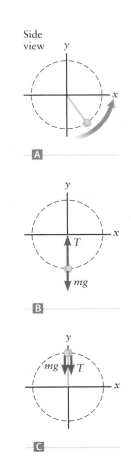

Side view

Figure 5.10 A This rock is moving in a vertical circle. There are two forces acting on it: the force of gravity and the tension from the string. B Forces on the rock when it is at the bottom of the circle. C Forces on the rock when it is at the top of its circular path.

circle, where r is the track's radius of curvature. Let's apply our problem-solving strategy for circular motion to the roller coaster in Figure 5.11. This roller coaster is at the highest point on the track, and we want to calculate the maximum speed it can have without leaving the track. (This calculation would presumably be of interest to the riders.) Figure 5.11 shows the circular path followed by the roller coaster as well as the forces in the problem, gravity and the normal force exerted by the track on the roller coaster. The gravitational force is toward the center of the circle while the normal force is directed upward, parallel to the $+y$ direction, and is thus away from the center of the circle. We next apply Equation 5.6. The total force directed toward the center of the circle is

$$\sum F_{\text{center}} = mg - N$$

which is related to the centripetal acceleration by

$$\sum F_{\text{center}} = mg - N = ma_c = \frac{mv^2}{r} \tag{5.12}$$

To find the maximum safe speed, we first rearrange Equation 5.12 to get the normal force:

$$N = m\left(g - \frac{v^2}{r}\right)$$

From this expression, we can see that as the speed v increases, the normal force N decreases. Eventually, we will reach a value of v at which N is zero:

$$N = m\left(g - \frac{v^2}{r}\right) = 0$$

which occurs when the speed satisfies

$$g - \frac{v^2}{r} = 0$$

$$v = \sqrt{gr} \tag{5.13}$$

If the speed is increased beyond this value, the value of N at the top of the track would have to be *negative*. Since a normal force can only push (but not pull), this scenario is impossible. Such a negative solution for N tells us that the roller coaster will leave the track if its speed exceeds $\sqrt{gr}$. Notice also that the normal force is the apparent weight of the roller coaster, and it will be your apparent weight if you are a passenger. Hence, when the speed is $v = \sqrt{gr}$, you will feel "weightless."

You might notice that our result (Eq. 5.13) for the speed at which the roller coaster leaves the track is the same as that found for the rock-on-a-string example in Figure 5.10C. There we saw (Eq. 5.11) that the string will go slack when the speed at the top is less than $v = \sqrt{gr}$. It is no accident that these two results are the same; in both cases, we are dealing with circular motion (so the acceleration is v^2/r for both), and in both cases, we have special situations in which the only force is gravity (because the tension in the string is zero and the normal force from the track is zero). In that sense, these problems are the "same."

Figure 5.11 This roller coaster is traveling in an approximately circular path at the top of the track. The forces acting on the roller coaster are also shown.

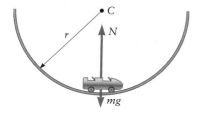

Figure 5.12 Example 5.4. Forces on a roller coaster at the bottom of its track.

EXAMPLE 5.4 | Apparent Weight on a Roller Coaster

Consider the roller coaster in Figure 5.12, which is at the lowest point on its track, where the radius of curvature is $r = 20$ m. If the apparent weight of a passenger on the roller coaster is 3.0 times her true weight, what is the speed of the roller coaster?

RECOGNIZE THE PRINCIPLE

The apparent weight is the normal force on the object (or person) undergoing circular motion. So, we need to find the value of v at which $N = 3.0 \times mg$.

SKETCH THE PROBLEM

We follow our problem-solving strategy for analyzing circular motion. Figure 5.12 shows the circle followed by the roller coaster (step 2); also shown (step 3) are the forces on the roller coaster, gravity and the normal force. Since the roller coaster is at the low point of the track, the normal force is directed toward the center of the circular arc defined by the track, and gravity is downward, away from the center.

IDENTIFY THE RELATIONSHIPS

We apply Newton's second law to relate the normal force to the speed of the roller coaster. Appling our relation for circular motion (Eq. 5.6), we have

$$\sum F_{center} = +N - mg = ma_c = \frac{mv^2}{r}$$

The apparent weight is three times the true weight, so $N = 3.0 \times mg$ and

$$N - mg = 3.0(mg) - mg = 2.0(mg) = \frac{mv^2}{r}$$

SOLVE

Solving for v, we find

$$v = \sqrt{2.0(gr)}$$

For a track of radius of $r = 20$ m, this result gives

$$v = \sqrt{2.0(gr)} = \sqrt{2.0(9.8 \text{ m/s}^2)(20 \text{ m})} = \boxed{20 \text{ m/s}}$$

What does it mean?
Our result for v is about 45 mi/h, which is enough for a nice thrill. Notice the similarity of this problem to the rock-on-a-string example in Figure 5.10B. The normal force on the roller coaster plays the role of tension in the string.

"Artificial Gravity" and a Rotating Space Station

A practical use of circular motion is sketched in Figure 5.13, which shows a hypothetical space station that uses circular motion to produce "artificial gravity" for its inhabitants. The station spins about a central axis so that the edge of the station, where the passengers spend most of their time, undergoes circular motion. The passengers in Figure 5.13 all experience a centripetal acceleration directed toward the center of the station (i.e., toward the center of the circle). This acceleration is produced by the normal force from the station floor on each passenger; this normal force provides the force required to make the passengers undergo circular motion. Hence, for each passenger, $N = mv^2/r$, where v is the speed of the station edge and r is the radius of the station.

The passengers would have a slightly different interpretation of the situation. If they do not look out the windows, they would not realize that they are undergoing circular motion. They would still be aware of the normal force acting on the bottoms of their shoes, but to them this force would be like the gravitational force they feel when on the surface of Earth. In fact, if $N = mv^2/r = mg$, this force of "artificial gravity" would be the same as the real gravitational force on Earth and the passengers would feel right at "home"!

| EXAMPLE 5.5 | Designing a Space Station |

Consider a rotating space station similar to the one in Figure 5.13. If the radius of the station is $r = 40$ m, how many times per minute must the station rotate to produce a force due to "artificial gravity" equal to 30% of that found on Earth?

A

B

Figure 5.13 A circular space station that rotates about its central axis (C) generates "artificial gravity" for its passengers.

RECOGNIZE THE PROBLEM

Figure 5.13A shows the circular path that the passengers follow. The only force is the normal force N, and it is directed toward the center of the circle. We can therefore apply Equation 5.6 to get

$$\sum F = N = \frac{mv^2}{r} \tag{1}$$

We want the normal force to be 30% of the gravitational force on Earth, which means $N = 0.30 \times mg$. We can use this expression together with Equation (1) to find the required speed v.

SKETCH THE PROBLEM

Figure 5.13 describes the problem and shows the forces on a passenger.

IDENTIFY THE RELATIONSHIPS AND SOLVE

Inserting the desired value of N into Equation (1) gives

$$N = 0.30(mg) = \frac{mv^2}{r}$$

and solving for v we find

$$v = \sqrt{0.30(gr)} \tag{2}$$

The time T for one revolution is equal to the distance traveled divided by the speed. The distance traveled is just the circumference of the station, so

$$T = \frac{2\pi r}{v} = \frac{2\pi r}{\sqrt{0.30(gr)}} = 2\pi\sqrt{\frac{r}{0.30g}} = 2\pi\sqrt{\frac{40 \text{ m}}{0.30(9.8 \text{ m/s}^2)}} = 23 \text{ s}$$

The number of revolutions per minute (rpm) is then

$$\frac{(60 \text{ s/min})}{T} = \frac{(60 \text{ s/min})}{(23 \text{ s})} = \boxed{2.6 \text{ rpm}}$$

What does it mean?

The speed in Equation (2) is independent of the person's mass. Hence, all passengers (regardless of their mass) experience an "artificial gravity" that is 30% of their weight on Earth.

ⓧ Physics of a Centrifuge

While a space station with "artificial gravity" has not yet been built (except in movies), a similar device called a ***centrifuge*** is used in many laboratories and factories. A centrifuge can be used to remove or separate particles or molecules that are in suspension in a liquid. A simple centrifuge design is shown in Figure 5.14. A test tube containing the liquid of interest is rotated so as to undergo circular motion; when compared to the space station in Figure 5.13, the test tube plays the part of a spoke on the station.

Suppose the liquid in Figure 5.14 contains small particles and we want to remove the particles for individual analysis. The liquid might be blood, and the "particles" could be a particular type of cell carried in the bloodstream. How long will it take to separate the cells from the blood? That is, how long will it take a cell to move from an initial location near the center of the centrifuge tube to the outer end of the tube?

Consider a cell that is already located at the end of the test tube so that it is in contact with the bottom of the tube. The circular motion of this cell is just like the

Figure 5.14 A centrifuge rotates at an extremely high rate about its axis at C, producing a large centripetal acceleration for the contents of the centrifuge. Here, the contents are a liquid that might contain cells in suspension. The result is that the cells move to the outer end of the tube.

motion of a passenger in the space station in Figure 5.13; hence, there is a normal force $N = mv^2/r$ exerted by the end of the tube on the cell. The cell thus experiences a force of "artificial gravity" $F_{AG} = mv^2/r$ much like the artificial gravity felt by a passenger on the space station. What's more, this "effective force" due to artificial gravity acts on the cell even when it is not at the end of the test tube. The effective force F_{AG} will thus cause a cell to move toward the bottom of the test tube. When viewed by an observer moving along with the test tube, it seems as if the cell is simply falling "vertically" to the bottom of the tube in response to the force of artificial gravity (Fig. 5.15A).

We have called F_{AG} an "effective force" because things appear different when viewed by an observer who is not rotating with the centrifuge. From the point of view of a stationary observer, the motion of the cell appears as shown in Figure 5.15B. A cell that is initially located away from the end of the test tube moves in an expanding arclike trajectory since there is nothing to provide the force required to make it move in a circle (until it reaches the end of the tube). The cell would move in a straight line (compare to the rock in Fig. 5.4), if not for the walls of the test tube and the drag force from the liquid.

Inertial and Noninertial Reference Frames Applied to a Centrifuge

While the motion of the cell in a centrifuge may be interpreted differently by different observers (as in Fig. 5.15), all observers will agree that the cell moves outward along the test tube, eventually coming to rest at the end of the tube. We have seen, however, that one observer (the cell) will deduce the presence of a force F_{AG}, whereas the other (stationary) observer will not. Which interpretation is correct? Is there actually a force F_{AG} or not? We mentioned in Chapter 4 that Newton's laws of motion should be used only by observers in *inertial reference frames*. You will recall that an inertial reference frame is one that moves with a constant velocity. In this example, the stationary observer is in an inertial reference frame, while Figure 5.15A corresponds to a noninertial reference frame. Only the stationary observer can use Newton's second law, and that observer's conclusion is correct: there is *no force* pushing the cell outward along the test tube. For this reason, F_{AG} as observed by the cell—that is, by the noninertial observer—is called a ***fictitious force***. Observers in rotating, and hence noninertial, reference frames often describe motion in terms of such fictitious forces.

We can thus explain the motion of a cell in the centrifuge in two different ways: (1) from the point of view of the rotating observer (the cell), according to which there is a fictitious force F_{AG} pushing the cell to the outer end of the test tube; or (2) from the point of view of a stationary (inertial) observer, who interprets the motion of the cell in terms of our usual picture of circular motion. Although the interpretations are different, the two observers agree on the cell's actual motion.

Why should we ever want to use a noninertial reference frame? Why not stick to inertial frames, where we know that Newton's laws can be used? The answer is that sometimes motion appears simpler when viewed in a noninertial frame. For example, with our cell in a centrifuge, the path seen by a noninertial observer (Fig. 5.15A) is very simple. The cell just undergoes a sort of free fall to the bottom of the test tube, and it moves *as if* there were a force F_{AG} directed along the "vertical" (i.e., along the tube). This motion is certainly simpler than the curved path seen by an inertial observer as sketched in Figure 5.15B. For this reason, it is sometimes useful to analyze motion from a noninertial reference frame. However, in such an analysis one must always account for the fictitious forces, such as F_{AG}, because these forces are necessary to make the inertial and noninertial results agree.

We can now calculate how long it will take for the cell to be removed from the solution. Let's do this analysis from the point of view of the cell as it falls to the bottom of the test tube due to the force of artificial gravity, $F_{AG} = mv^2/r$. We must recognize that another force on the cell opposes its fall: the drag force we

Figure 5.15 **A** A cell in the centrifuge moves as if there is an acceleration due to "artificial gravity" equal to the centripetal acceleration. **B** When viewed by an observer who is not spinning with the centrifuge, the cell moves in a path that spirals outward as indicated by the dashed line.

encountered in Chapter 3. There we saw that for a spherical object moving slowly through a liquid, there is a drag force described by Stokes's law (Eq. 3.23),

$$\vec{F}_{drag} = -Cr_{cell}\vec{v}_{cell} \tag{5.14}$$

where r_{cell} is the cell's radius and v_{cell} is the cell's speed along the test tube. Notice that v_{cell} is *not* the same as the speed v of the end of the test tube (v is the speed of the outer rim of the spinning centrifuge). Here, C is a constant that depends on the viscosity (the resistance to flow) of the fluid. For blood, C has a value of approximately 0.075 kg/(m · s). The drag force becomes larger as the speed v_{cell} of the cell increases, and the situation is very similar to the problem of a falling skydiver we encountered in Chapter 3. There we found that the skydiver reaches a terminal velocity at which the drag force balances the force of gravity. For the cell in a centrifuge, the drag force in Equation 5.14 will balance F_{AG} and the cell will move at a constant speed along the test tube. This balance condition is

$$F_{AG} + F_{drag} = F_{AG} - Cr_{cell}v_{cell} = 0$$

Since $F_{AG} = mv^2/r$, we have

$$\frac{mv^2}{r} = Cr_{cell}v_{cell}$$

Solving for the speed along the test tube—that is, the "terminal velocity" of the cell—gives

$$v_{cell} = \frac{mv^2}{Cr_{cell}r} \tag{5.15}$$

It is important to distinguish the tangential speed of the centrifuge (v) from the settling speed of the cell (v_{cell}), and the radius of the cell (r_{cell}) from the radius of the centrifuge (r).

CONCEPT CHECK 5.3 | Inertial and Noninertial Reference Frames

We have seen that an observer in an inertial reference frame can use Newton's laws, while observers in a noninertial frame must include fictitious forces. Identify the following reference frames as either inertial or noninertial and explain your answers.
  (a) A car traveling on a level road at a constant velocity
  (b) A car traveling up a steep mountain road at a constant velocity
  (c) A child sitting on a rotating merry-go-round
  (d) An apple falling from a tree

EXAMPLE 5.6 | ⊗ Operation of a Centrifuge

Calculate the settling speed v_{cell} of a cell of radius 10 μm and mass 10^{-5} mg ($= 10^{-11}$ kg) in blood using Equation 5.15. Assume the centrifuge has a radius of 10 cm and a rotation rate of 3000 revolutions per minute. How long must the centrifuge run to make the cell settle out?

RECOGNIZE THE PRINCIPLE

Many quantities involved in this problem, such as the radius and mass of a cell, are never known precisely and vary from cell to cell. Moreover, the Stokes's relation for the drag force, Equation 5.14, applies exactly only for a spherical object, and cells are usually not precisely spherical! Even so, we can use Equation 5.15 to get an approximate value for the speed at which a cell moves. The only unknown quantity in Equation 5.15 is v, the speed of the centrifuge. We can find v from the period of the centrifuge, along with the relation between period and v in Equation 5.1.

• C

Figure 5.16 Example 5.6. A cell in a centrifuge moves as if there is a force of artificial gravity acting "downward" on the cell.

SKETCH THE PROBLEM

Figure 5.16 shows the problem from the point of view of the cell. According to this noninertial observer, the force of artificial gravity ($\vec{F}_{AG}$) causes an acceleration a_c "downward" along the centrifuge tube.

IDENTIFY THE RELATIONSHIPS AND SOLVE

Our centrifuge makes 3000 revolutions in 1 min, so the time to make 1 revolution is $T = 1 \text{ min}/3000 = 0.020$ s, and the circumferential speed is

$$v = \frac{2\pi r}{T} = \frac{2\pi(0.10 \text{ m})}{0.020 \text{ s}} = 31 \text{ m/s}$$

Inserting this in Equation 5.15 gives

$$v_{cell} = \frac{mv_c^2}{Cr_{cell}r} = \frac{(1 \times 10^{-11} \text{ kg})(31 \text{ m/s})^2}{[0.075 \text{ kg/(m·s)}](1.0 \times 10^{-5} \text{ m})(0.10 \text{ m})} = \boxed{0.1 \text{ m/s}}$$

Since the centrifuge tube has a length equal to the radius of the centrifuge (see Fig. 5.14), the cell must travel a distance of at most $L = 10$ cm $= 0.10$ m. Using the value found for v_{cell}, the time required for the cell to move this distance is

$$t = \frac{L}{v_{cell}} = \frac{0.10 \text{ m}}{0.1 \text{ m/s}} = \boxed{1 \text{ s}}$$

What does it mean?

A centrifuge with a similar speed is used to separate blood cells from plasma. Allowing for variations in cell size, our analysis shows that it only takes a few seconds to completely separate the cells from the plasma.

5.3 | NEWTON'S LAW OF GRAVITATION

The orbital motions of planets and moons are circular (to a good approximation) in many cases, and the force responsible for this circular motion is gravity. We have already encountered the force of gravity as it acts on objects near the Earth's surface. Now we need to consider the gravitational force in more general situations.

Newton's three laws of motion are the foundation of our study of mechanics. However, Newton discovered another law of nature, the law of gravitation, that is perhaps just as important as his laws of motion. Newton's law of gravitation plays a key role in physics for two reasons. First, it allows us to calculate and understand the motions of a wide variety of objects, including baseballs, apples, and planets. Second, Newton's application of his law of gravitation to the motion of planets and moons was the first time that physics was applied (successfully) to describe the motion of the solar system. His work showed for the first time that the laws of physics apply to *all* objects and had a profound effect on how people viewed the universe.

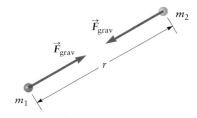

Figure 5.17 The gravitational force between two point masses m_1 and m_2 that are separated by a distance r is given by Equation 5.16. Notice that here the distance r is not the radius of a circle.

Newton's law of gravitation: **There is a gravitational attraction between *any* two objects. If the objects are point masses m_1 and m_2, separated by a distance r (Fig. 5.17), the magnitude of the gravitational force is**

$$F_{grav} = \frac{Gm_1m_2}{r^2} \tag{5.16}$$

Newton's law of gravitation

where $G = 6.67 \times 10^{-11}$ N·m²/kg² is a constant of nature known as the universal gravitational constant. The gravitational force is *always attractive*. Every mass attracts every other mass.

The gravitational force law, Equation 5.16, is "symmetric." That is, the magnitude of the gravitational force exerted by mass 1 on mass 2 is equal to the magnitude

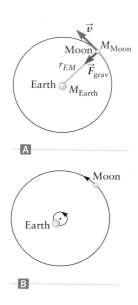

of the gravitational force exerted by mass 2 on mass 1. Since the forces are both attractive, this result is precisely what we would expect from Newton's third law; the two gravitational forces are an *action–reaction pair* because they are equal in magnitude and opposite in direction and they act on different members of the pair of objects.

Gravitation and the Orbital Motion of the Moon

The gravitational force is responsible for the motion of planets, moons, asteroids, and comets and for the motion of many terrestrial objects such as falling baseballs and other projectiles. For example, the Moon follows an approximately circular path as it orbits Earth, as sketched in Figure 5.18. This circular motion involves a centripetal acceleration, and that acceleration requires a force, which is provided by gravity. As a check, we can use Equation 5.6 to calculate the force required to make the Moon move in its circular orbit and then compare it with the gravitational force calculated from Newton's law of gravitation, Equation 5.16. The force required to make the Moon move in a circle is (from Eq. 5.6)

$$F = \frac{M_{Moon}v^2}{r_{EM}}$$

where M_{Moon} is the mass of the Moon and r_{EM} is the distance from Earth to the Moon (the radius of the Moon's orbit). To find the speed v of the Moon, we notice that it completes one orbit and travels a distance of $2\pi r_{EM}$ in approximately $T = 27.3$ days;[2] hence,

$$v = \frac{2\pi r_{EM}}{27.3 \text{ days}} = \frac{2\pi r_{EM}}{T}$$

The required force is thus

$$F = \frac{M_{Moon}v^2}{r_{EM}} = \frac{M_{Moon}(2\pi r_{EM}/T)^2}{r_{EM}} = \frac{4\pi^2 M_{Moon}r_{EM}}{T^2} \qquad (5.17)$$

Inserting the known values (see Table 5.1) $M_{Moon} = 7.35 \times 10^{22}$ kg and $r_{EM} = 3.85 \times 10^8$ m along with $T = 27.3$ days $= 2.36 \times 10^6$ s, we find

$$F = 2.0 \times 10^{20} \text{ N} \qquad (5.18)$$

This is the magnitude of the force needed to make the Moon follow its observed circular orbit. To confirm that this force is actually provided by gravity, let's calculate the gravitational force exerted by Earth on the Moon from Equation 5.16 and show that it is indeed equal to the result for F in Equation 5.18. The mass of the Earth is (from Table 5.1) $M_{Earth} = 5.98 \times 10^{24}$ kg, and inserting the other values given above into Equation 5.16 gives

$$F_{grav} = \frac{GM_{Earth}M_{Moon}}{r_{EM}^2} = 2.0 \times 10^{20} \text{ N} \qquad (5.19)$$

which does agree with the expected value of F in Equation 5.18.

When he was developing his theories of motion and gravitation, Newton almost certainly carried out these same calculations and obtained the results in Equations 5.18 and 5.19. That the force of gravity on the Moon is precisely equal to the force required to make the Moon move in its circular orbit must have been very strong evidence for Newton that his theories were indeed correct.

Applying Newton's Law of Gravitation: Calculating the Value of *g*

In Chapter 3, we introduced the "constant" *g* and learned that the gravitational force on an object near the Earth's surface has the magnitude $F_{grav} = mg$. Let's now

Figure 5.18 ◻ The Moon's orbit around the Earth is approximately circular. The centripetal acceleration is provided by Earth's gravitational force. To a fairly good approximation, the Earth in this picture can be assumed to be fixed in space. ◻ In a more accurate description, the Earth also moves in a circular "orbit" due to the gravitational force of the Moon on the Earth. This sketch is not to scale; the center of this orbit is inside the Earth.

Insight 5.2

ORBITAL MOTION OF THE MOON AND EARTH

In this analysis of the Moon's orbital motion, we have assumed the Moon orbits around a "fixed" Earth. That is, we have assumed the Earth does not move at all as the Moon moves in a circle of radius r_{EM}. Although that is a good approximation, it is only an approximation because it ignores two aspects of Earth's motion. (1) The Earth orbits the Sun. That orbital motion is much slower than the Moon's orbital speed, so it still makes sense (it is mathematically extremely accurate) in Equation 5.17 to treat the Earth as fixed. (2) The Earth also "orbits" the Moon. That is, as the Moon moves in its circular orbit, the Earth moves in a corresponding circular orbit as shown in Figure 5.18B. Earth's circular orbit is much smaller than that of the Moon. In fact, the center of the Earth's circle is inside the Earth(!) and is located at a spot called the *center of mass* that we'll discuss in Chapter 7.

[2]In this example and others involving planetary and satellite motion, we will need to carry three significant figures through the calculation (one more than usual in this book) to avoid problems due to rounding errors.

Table 5.1

Table 5.1 Solar System Data: Properties of Several Objects in the Solar System, Including the Planets and Two of the Largest Dwarf Planets (Pluto and Eris)

NAME	MEAN ORBITAL RADIUS ($\times 10^{11}$ m)	ORBITAL PERIOD (years)	RADIUS ($\times 10^{6}$ m)	MASS ($\times 10^{24}$ kg)	ORBITAL ECCENTRICITY
Mercury	0.579	0.240	2.44	0.330	0.21
Venus	1.08	0.615	6.05	4.87	0.007
Earth	1.50	1.00	6.37	5.98	0.017
Mars	2.28	1.88	3.39	0.644	0.093
Jupiter	7.78	11.9	71.5	1900	0.048
Saturn	14.3	29.4	60.3	568	0.054
Uranus	28.7	83.8	25.6	86.6	0.047
Neptune	45.0	164	24.8	102	0.009
Pluto	59.1	248	1.14	0.125	0.25
Eris	100	560	2.4	—	0.44
Earth's Moon	3.85×10^{8} m	27.3 days	1.74	0.0735	0.055
Sun			6.96×10^{8} m	1.99×10^{30} kg	

see how this result is contained in the universal law of gravitation. Strictly speaking, Equation 5.16 applies only to the case of two "point" masses, two objects that are very small compared to the distance between them. This assumption works for our calculation of the gravitational force between Earth and the Moon, but we certainly cannot consider the Earth to be a point mass in relation to an object on its surface. To appreciate the problem, consider the force exerted by Earth's gravity on a person standing on the surface as sketched in Figure 5.19A. We would like to use Equation 5.16 to calculate this force, but what value should we use for the distance r? Some parts of the Earth are very close to the person (just beneath his feet), whereas other parts are quite far away (as much as twice the radius of the Earth). The answer is that when dealing with a spherical object with a constant density, one can calculate the gravitational force that it exerts on another object *as if* all the sphere's mass were located at its center.

The Earth's shape is very close to spherical, so the gravitational force exerted by the Earth on the person in Figure 5.19 can be calculated as if all the Earth's mass is at its center as sketched in Figure 5.19B. We can therefore use Equation 5.16 to calculate this force, with the separation r equal to the distance from the person to the center of the Earth; this separation is just the Earth's radius, r_E. We thus have

$$F_{\text{grav}} = \frac{GM_{\text{Earth}}M_{\text{person}}}{r_E^2} \qquad (5.20)$$

This force is just the weight of the person, which until now we have denoted by $M_{\text{person}}g$, where g is the "acceleration due to the Earth's gravity." Hence, we can use Equation 5.20 to calculate the value of g. Examining Equation 5.20, we see that it has the form $M_{\text{person}}g$ provided that g is given by

$$g = \frac{GM_{\text{Earth}}}{r_E^2} \qquad (5.21)$$

The value of g is a function of only the Earth's mass and radius, and the universal gravitational constant G. The value of g is thus the same for all terrestrial objects

Figure 5.19 Assuming a spherical Earth (which is a good approximation), the gravitational force exerted by the Earth on an object (i.e., a person) on the surface as shown in ▣ is equal to the force one would find if all the mass of the Earth were located at its center as shown in ▣.

near the Earth's surface. Inserting the known values of M_{Earth} and r_E (Table 5.1) along with G, we find

$$g = \frac{GM_{Earth}}{r_E^2} = \frac{(6.67 \times 10^{-11}\,\text{N}\cdot\text{m}^2/\text{kg}^2)(5.98 \times 10^{24}\,\text{kg})}{(6.37 \times 10^6\,\text{m})^2}$$

$$g = 9.8\,\text{m/s}^2$$

g and Newton's law of gravitation

which is the value we have already been using for g.

This calculation shows us where the value of g "comes from," and it also shows that g is not really a constant. The Earth happens to be approximately spherical, so all objects on the surface of the planet (i.e., all terrestrial objects) are approximately the same distance from the center and hence all have approximately the same value of g. Equation 5.21, however, shows that g—and hence the weight of an object—changes if the object is moved farther from the Earth (e.g., by climbing a mountain or traveling in a balloon). It also shows that the weight of an object will be different on another planet or on the Moon than it is on Earth.

Insight 5.3
THE GRAVITATIONAL FORCE FROM THE EARTH
According to Figure 5.19, the gravitational force of a uniform spherical object can be calculated by assuming all the mass is at a point at the center of the object. This result also holds when the density varies with depth, as long as the density depends only on the distance from the center. Hence, the result applies to a more realistic model of the Earth, consisting of a core with a different density than the crust.

EXAMPLE 5.7 | What Would You Weigh on the Moon?

What is your weight on the Moon? Compare it with your weight on Earth.

RECOGNIZE THE PRINCIPLE

Your weight on the Moon is just the Moon's gravitational force when you are located on its surface. Hence, we need to evaluate Equation 5.21 using the mass of the Moon and the radius of the Moon in place of those quantities for the Earth (these data are listed in Table 5.1).

SKETCH THE PROBLEM

This problem is described by Figure 5.19, but with the Earth replaced by the Moon.

IDENTIFY THE RELATIONSHIPS AND SOLVE

The author has a mass of approximately $M_{author} = 70$ kg; inserting this value into Equation 5.20 gives

$$F_{grav} = \frac{GM_{moon}M_{author}}{r_M^2} = \frac{(6.67 \times 10^{-11}\,\text{N}\cdot\text{m}^2/\text{kg}^2)(7.35 \times 10^{22}\,\text{kg})(70\,\text{kg})}{(1.74 \times 10^6\,\text{m})^2}$$

$$F_{grav} = 110\,\text{N}$$

The author's weight on Earth is $M_{author}g = (70\,\text{kg})g = 690$ N, so his weight on the Moon is approximately one-sixth of his weight on Earth.

What does it mean?
The ratio of the weight on the Moon to the weight on Earth is *independent* of the mass of the object because both are proportional to the mass of the object. Hence, all objects weigh less on the Moon by a factor of approximately one-sixth.

EXAMPLE 5.8 | The Force of Gravity in a Very Tall Building

One of the tallest buildings in the world is the Empire State Building in New York City, with the top being a distance $h = 440$ m above the bottom. Your weight at the top of this building is slightly less than when you are at ground level. Find the ratio of your weight at the top of the building to your weight at ground level.

RECOGNIZE THE PRINCIPLE

At the top of the Empire State Building, your distance from the center of the Earth is $r_E + h$, where r_E is Earth's radius. Hence, the force of gravity (your weight) is slightly smaller than when you are on the ground. Note that because h is much smaller than r_E, we must carry extra significant figures in this calculation.

SKETCH THE PROBLEM

We want to compare the force of gravity at two locations as sketched in Figure 5.20. The force of gravity depends on distance from the center of the Earth.

IDENTIFY THE RELATIONSHIPS

At ground level, you are a distance r_E from the center of the Earth, so your weight is

$$F_{\text{grav}}(\text{ground level}) = \frac{GM_{\text{Earth}}M_{\text{person}}}{r_E^2}$$

At the top of the building you are a distance $r_E + h$ from the center of the Earth so

$$F_{\text{grav}}(\text{top}) = \frac{GM_{\text{Earth}}M_{\text{person}}}{(r_E + h)^2}$$

The ratio of these two forces is

$$\frac{F_{\text{grav}}(\text{top})}{F_{\text{grav}}(\text{ground level})} = \frac{GM_{\text{Earth}}M_{\text{person}}/(r_E + h)^2}{GM_{\text{Earth}}M_{\text{person}}/r_E^2} = \left(\frac{r_E}{r_E + h}\right)^2$$

SOLVE

Inserting values for Earth's radius and the height of the Empire State Building, we find

$$\frac{F_{\text{grav}}(\text{top})}{F_{\text{grav}}(\text{ground level})} = \left(\frac{r_E}{r_E + h}\right)^2 = \left[\frac{6.37 \times 10^6 \text{ m}}{(6.37 \times 10^6 + 440) \text{ m}}\right]^2 = \boxed{0.99986}$$

What does it mean?
This result is independent of the mass of the person, so all persons (and all objects) weigh less at the top of the Empire State Building by approximately 0.014%. Sensitive instruments called gravimeters can measure changes as small as 1 part in 10^9 in the gravitational force, so this difference can be easily measured.

CONCEPT CHECK 5.4 | Gravity on Another Planet

You travel to another planet in our solar system and find that your weight is twice as large as it is on Earth. If the radius of this planet is twice the Earth's radius, is the mass of the planet (a) two times the Earth's mass, (b) four times the Earth's mass, (c) half of the Earth's mass, or (d) eight times the Earth's mass?

Measuring *G*: The Cavendish Experiment

The first precision measurement of the force of gravity between two terrestrial objects was carried out in a famous experiment by Henry Cavendish. He wanted to measure the gravitational force using two large lead spheres. This experiment is very difficult for the following reason. Suppose the spheres each have a diameter of 1 m, as was approximately the case in Cavendish's experiment. The mass of a single lead sphere of this size is approximately 5900 kg. We have already seen that the gravitational force exerted by a spherical mass can be calculated as if all the mass is at the center. With that in mind, let's calculate the gravitational force between two point particles of mass 5900 kg, separated by a distance $r = 1.0$ m. Using Equation 5.16 with $m_1 = m_2 = 5900$ kg gives

$$F_{\text{grav}} = \frac{Gm_1m_2}{r^2} = \frac{(6.67 \times 10^{-11} \text{ N} \cdot \text{m}^2/\text{kg}^2)(5900 \text{ kg})(5900 \text{ kg})}{(1.0 \text{ m})^2}$$

$$F_{\text{grav}} = 2.3 \times 10^{-3} \text{ N} \tag{5.22}$$

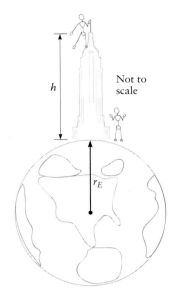

Figure 5.20 Example 5.8. A person at the top of the Empire State Building is a distance $r_E + h$ from the center of the Earth, where h is the height of the building.

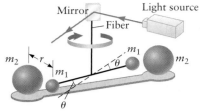

Figure 5.21 Cavendish used an apparatus like this one to measure the force of gravity between terrestrial objects.

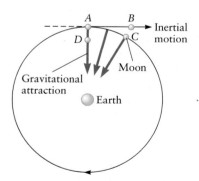

Figure 5.22 The Moon "falls" toward Earth as it travels in its orbit. Here, the Moon's acceleration due to gravity causes it to "fall" from B to C and thereby travel in a circular orbit. If the Moon's initial velocity were zero, it would simply fall from A to D, in the same way that Newton's apple fell from its tree.

This force is a rather small, only a little larger than the weight of a mosquito! That's why a high-precision measurement of the gravitational force between two terrestrial objects is a very challenging experiment. Indeed, that's why Cavendish chose the spheres in his experiment to be as large as possible and why he arranged to place them very close together: these choices make the gravitational force as large as possible.

Cavendish carried out his experiment more than 200 years ago, but the same basic design is still used today to study gravitational forces. The Cavendish apparatus (Fig. 5.21) uses a "dumbbell" with two spheres each of mass m_1 at the ends, suspended at the middle by a very thin wire. Another pair of spheres, each of mass m_2, is then brought very close to the dumbbell masses, and the force of gravity causes the dumbbell rod to rotate. The rotation angle θ in Figure 5.21 depends on the force, so by measuring θ, Cavendish was able to determine the gravitational force. Since the values of m_1 and m_2 can also be measured, along with the separation, Cavendish was able to deduce the value of G from Equation 5.16.

The quantity G is our first encounter with a *constant of nature*. Laws of physics, such as Newton's law of gravitation, often contain such constants, and the only way to know their values is through experimentation. You might think that we can determine the value of G from a measurement of the Moon's orbital motion (essentially using Eqs. 5.17 and 5.18), but that is difficult because the masses of the Moon and Earth are hard to measure. Cavendish's method for measuring G is still the basis for most experimental studies of gravitation.

Newton's Apple

Newton's discovery of the law of gravitation had a profound impact on our view of the universe. To appreciate this impact, it is useful to recall the allegory of Newton's apple. The story has several versions, but the general idea is that Newton was sitting under an apple tree, thinking about the motion of the Moon, when an apple fell from the tree and struck him on the head. According to the story, this jolt led to the "discovery" of gravity. The point of this story is not simply that falling apples undergo the motion we have called "free fall." Rather, it is that the acceleration of an apple falling from a tree and the acceleration of the Moon as it moves in a circular orbit around the Earth are *basically the same* (Fig. 5.22). They are caused by the same force (gravity), which has the same direction (toward the center of the Earth), and if the Moon and the apple were at the same distance from Earth, their accelerations would be equal. This connection is not obvious because for circular motion the acceleration is perpendicular to the velocity, whereas the acceleration of the apple is parallel to $\vec{v}$. It took the genius of Newton to make this connection.

Prior to Newton, it was widely believed that the motion of celestial objects, such as the Moon, is fundamentally different from the motion of terrestrial objects, such as apples. Newton showed that the motion of falling apples and the motion of the Moon are caused by the same force and governed by the same laws of motion. So, one can use Newton's laws to understand the motion of the solar system and beyond, a regime that had previously been "off-limits" to such scientific study.

5.4 | PLANETARY MOTION AND KEPLER'S LAWS

Studies of the motion of the Moon, the planets, and other celestial bodies provided valuable information for Newton in his work on gravitation. Of the many important astronomers prior to the time of Newton, one of the most famous is Johannes Kepler (1571–1630). During and prior to his lifetime, a number of astronomers recorded the positions of the Moon, planets, comets, and so forth as functions of time and showed that bodies in the solar system move in an extremely precise fashion. Kepler studied these results very carefully and found that the motion of the Moon and planets could be described by what are now known as *Kepler's laws* of planetary motion. Kepler's laws are not "laws of nature" in the sense of Newton's laws of motion or Newton's law of gravitation. Rather, Kepler's laws are mathematical rules that describe motion in the solar system. These rules were inferred by Kepler from the available astronomical observations, but he was not able to give a scientific explanation or derivation. It remained for Newton to show how his three fundamental laws of motion together with his law of gravitation explain Kepler's laws.

Kepler's First Law

Kepler's first law is a statement about the shapes of orbits. Aristotle and many other early scientists believed that orbits are circular. The main reason for this belief was that circles are the most "perfect" shape for such a curve, and nature must be perfect. By Kepler's time, however, it was well established from many observations of the planets[3] that while planetary orbits are approximately circular, they are definitely not perfect circles. Kepler showed that all the planetary orbits are *elliptical*. Furthermore, for the motion of planets around the Sun, the Sun is not at the center of the ellipse; instead, it is at one of the foci of the ellipse as illustrated in Figure 5.23. This discovery must have been quite a shock because the off-center placement of the Sun would seem to violate nature's tendency for symmetry.

For the planets known to Kepler, as well for the Moon, the orbits deviate only a small amount from perfect circles (by much less than the orbit in Fig. 5.23). For example, the distance between Earth and the Sun varies by only ±3% during the course of an orbit. Nevertheless, the deviation from a perfect circle is certainly real, and it had been accurately measured by Kepler's time.

Kepler's first law is the statement that planetary orbits are elliptical. This statement also applies to the Moon because the Moon is really just a "planet" belonging to Earth, and Kepler's first law also applies to artificial satellites that have been launched into orbit around Earth. Because a circle is a special case of an ellipse, Kepler's first law also allows circles, but real orbits generally deviate at least a small amount from a perfect circular shape. Soon after he developed his law of gravitation, Newton was able to use it to derive Kepler's first law. In this way, Newton provided an explanation for the shapes of planetary orbits.[4]

> *Kepler's first law of planetary motion:* **Planets move in elliptical orbits.**

We have already mentioned that the orbits of the inner planets are not far from circular. Many bodies in the solar system have highly elliptical orbits, however. Comets are a good example of elliptical orbits, and perhaps the best known case is Halley's comet. A scale drawing of the orbit of Halley's comet is shown in Figure 5.24. This figure includes the orbit of the dwarf planet Pluto, which is also noticeably elliptical. In fact, Pluto's orbit is such that it actually spends a substantial amount of time inside Neptune's orbit.

CONCEPT CHECK 5.5 | Acceleration of a Comet

Sketch the direction of the acceleration vector for Halley's comet at various points in its orbit in Figure 5.24.

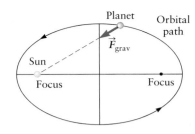

Figure 5.23 According to Kepler's first law, the planetary orbits are elliptical. An ellipse has two foci, which are offset from the center of the ellipse. The Sun is located at one of the foci.

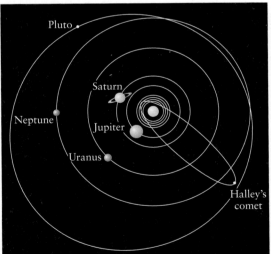

Figure 5.24 Scale drawing of the planetary orbits in our solar system. The orbits of most planets are nearly circular. The orbit of the dwarf planet Pluto deviates substantially from a circle. The orbit of Halley's comet, which is highly elliptical, is also shown.

---

[3]At that time, only the six innermost planets in our solar system had been discovered: Mercury, Venus, Earth, Mars, Jupiter, and Saturn.

[4]Newton also showed that other nonclosed trajectories are possible. For example, objects that pass near the Sun once and never return move along either a parabolic or a hyperbolic path. These trajectories, including the elliptical orbits of the planets, are *conic sections*. Kepler's first law is sometimes worded to reference these nonclosed trajectories. The only closed trajectories—that is, the only true orbits—are elliptical, however.

Kepler's Second Law

Kepler's second law concerns the speed of a planet as it moves around its orbit. (This law applies to a planet in our solar system moving about our Sun as well as to a planet in another solar system moving around its sun.) For a perfectly circular orbit, this speed is constant. However, for an elliptical orbit the speed is smallest when the planet is farthest from its sun, whereas v is largest when the planet is nearest the sun (Fig. 5.25). To understand Kepler's second law, it is useful to consider a line drawn from the sun to the planet. This line moves along with the planet and sweeps out area as the planet moves. According to Kepler's second law, this line sweeps out area at a constant rate. Let A_1 be the area swept out in time Δt when the planet is near the sun and A_2 be the amount of area swept out in the same amount of time Δt when the planet is somewhere else in its orbit. Kepler's second law then states that $A_1 = A_2$. For this statement to be true, the planet must (as we have already noted) speed up when it is nearest the sun. We'll discuss Kepler's second law again in Chapter 9, where we'll see that it is closely connected with the *angular momentum* of the planet.

Kepler's second law of planetary motion: **A line connecting a planet to its sun sweeps out equal areas in equal times as the planet moves around its orbit.**

Kepler's Third Law

Kepler's third law relates the timing of an orbit to the size of the orbit. It is simplest to derive Kepler's third law for the special case of a perfectly circular orbit, although it also applies to elliptical orbits. For a circular orbit, the speed of the planet is constant. If the orbit has a radius r, the time it takes to complete one orbit equals the distance traveled (the circumference) divided by the speed:

$$T = \frac{2\pi r}{v} \tag{5.23}$$

Since the gravitational force of the sun on the planet is equal to ma_c, where a_c is the centripetal acceleration, we get (compare with Eqs. 5.16 and 5.17)

$$F_{grav} = \frac{GM_{sun}M_{planet}}{r^2} = ma_c = \frac{M_{planet}v^2}{r} \tag{5.24}$$

From Equation 5.23, we have $v = 2\pi r/T$, and substituting for v gives

$$\frac{M_{planet}v^2}{r} = \frac{M_{planet}(2\pi r/T)^2}{r} = \frac{4\pi^2 M_{planet}r}{T^2} \tag{5.25}$$

We can now combine Equations 5.24 and 5.25 to get a relation between the orbital time (the period T) and the orbital radius r. We find

$$\frac{GM_{sun}M_{planet}}{r^2} = \frac{4\pi^2 M_{planet}r}{T^2}$$

and solving for T gives

$$T^2 = \left(\frac{4\pi^2}{GM_{sun}}\right)r^3 \tag{5.26}$$

which is Kepler's third law. In words, Equation 5.26 states that the square of the orbital period is proportional to the cube of the orbital radius. An important feature of this result is that it does not depend on the mass of the planet. It applies to the orbital motion of all bodies, including planets, comets, and spacecraft. It also applies to moons. Jupiter and Saturn, for example, possess many moons, and

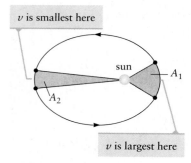

v is smallest here

sun — A_1

A_2

v is largest here

Figure 5.25 Kepler's second law is called the "equal areas" law. A line that extends from the sun to a planet sweeps out equal areas during equal time intervals as a planet moves around its orbit. If the time required for the planet to sweep out area A_1 is equal to the time associated with area A_2, the areas will be equal.

the orbital periods and radii of these moons must obey Equation 5.26, with M_{sun} replaced by the mass of the appropriate planet.

> *Kepler's third law:* **The square of the period of an orbit is proportional to the cube of the orbital radius.**

Kepler's third law of planetary motion

EXAMPLE 5.9 | Jupiter's Orbit

The Earth completes one orbit about the Sun in 1 year and has an orbital radius of 1.50×10^{11} m (see Table 5.1). If the orbital radius of Jupiter is 7.78×10^{11} m, what is the period of Jupiter's orbit?

RECOGNIZE THE PRINCIPLE

We could solve this problem by simply evaluating Equation 5.26 using the given radius of Jupiter's orbit. Here, though, we take a different approach by applying Equation 5.26 first to Earth and next to Jupiter, and then computing the ratio of their orbital periods. Notice that the orbits of the Earth and Jupiter are both very close to circular.

SKETCH THE PROBLEM

No sketch is needed.

IDENTIFY THE RELATIONSHIPS

Applying Equation 5.26 to Jupiter, we have

$$T_{Jupiter}^2 = \left(\frac{4\pi^2}{GM_{Sun}}\right) r_{Jupiter}^3$$

with a similar result for Earth. Taking the ratio gives

$$\frac{T_{Jupiter}^2}{T_{Earth}^2} = \frac{\left(\frac{4\pi^2}{GM_{Sun}}\right) r_{Jupiter}^3}{\left(\frac{4\pi^2}{GM_{Sun}}\right) r_{Earth}^3} = \frac{r_{Jupiter}^3}{r_{Earth}^3} \qquad (1)$$

SOLVE

Inserting the orbital radius of Jupiter and of the Earth from Table 5.1 gives

$$\frac{T_{Jupiter}^2}{T_{Earth}^2} = \left(\frac{7.78 \times 10^{11} \text{ m}}{1.50 \times 10^{11} \text{ m}}\right)^3 = 140$$

$$T_{Jupiter}^2 = 140 \times T_{Earth}^2$$

Taking the square root of each side, we get

$$T_{Jupiter} = \boxed{12 \times T_{Earth}}$$

Since $T_{Earth} = 1$ year, it takes Jupiter approximately $\boxed{12 \text{ years}}$ to complete one orbit.

What have we learned?
When doing a calculation, it is sometimes mathematically simpler to use the ratio of two similar quantities. In this example, the factors of G and M_{Sun} canceled when the ratio was taken in Equation (1), simplifying the form of the final answer.

Satellite Orbits around Earth

Many satellites, including most that carry astronauts, are in what is often called a "low Earth" orbit. You may wonder why any orbit would be called "low," but the reason for this expression becomes clear when we calculate the radius of a typical orbit. We can do so using Kepler's third law (Eq. 5.26) if we know the period of the

Figure 5.26 Scale drawing of a "low Earth" orbit. The orbit is shown in red. On this scale, it is barely distinguishable from the surface of the Earth.

orbit. The period for a satellite in low Earth orbit, such as the International Space Station, is approximately 90 min, so $T = (90 \text{ min})(60 \text{ s/min}) = 5400$ s. Kepler's third law, with the mass of the Earth as that of the central body, gives

$$T^2 = \left(\frac{4\pi^2}{GM_{\text{Earth}}}\right)r^3$$

Solving for the radius of the orbit, we find

$$r = \left(\frac{GM_{\text{Earth}}T^2}{4\pi^2}\right)^{1/3} \tag{5.27}$$

Inserting values for the mass of the Earth and T, we get

$$r = \left(\frac{GM_{\text{Earth}}T^2}{4\pi^2}\right)^{1/3} = \left[\frac{(6.67 \times 10^{-11} \text{ N} \cdot \text{m}^2/\text{kg}^2)(5.98 \times 10^{24} \text{ kg})(5400 \text{ s})^2}{4\pi^2}\right]^{1/3}$$

$$r = 6.66 \times 10^6 \text{ m} \tag{5.28}$$

The radius of the Earth is $r_E = 6.37 \times 10^6$ m, so these satellites orbit at a height above Earth's surface that is only 5% of the Earth's radius, which is why it is called a "low" orbit. This orbit is illustrated in the scale drawing in Figure 5.26. Even though it may seem low when compared to the Earth's radius, the tallest mountain on Earth (Mount Everest) is only 8840 m or approximately 0.1% of r_E above sea level, so the satellites are in no danger of colliding with any mountains!

EXAMPLE 5.10 Geosynchronous Orbits

Earth-orbiting satellites used for transmitting telephone or television signals travel in *geosynchronous* orbits. These orbits have a period of 1 day, so these satellites move in synchrony with the Earth's rotation and are thus always at the same position in the sky relative to a person on the ground. Sending signals to and from these satellites is thus greatly simplified because an antenna such as the author's satellite TV dish can be aligned only once and then needs no further adjustment. Calculate the orbital radius of a satellite in a geosynchronous orbit.

RECOGNIZE THE PRINCIPLE

We have already done a very similar problem in our discussion of a satellite in low Earth orbit. We can thus apply Equation 5.27 to the geosynchronous case using a period $T = 1$ day.

SKETCH THE PROBLEM

No sketch is needed.

IDENTIFY THE RELATIONSHIPS AND SOLVE

The period of a geosynchronous orbit is 1 day. Hence, the period $T = 1$ day = (1 day)(24 h/day)(3600 s/h) = 86,000 s. Inserting this result into Equation 5.27 gives

$$r = \left(\frac{GM_{\text{Earth}}T^2}{4\pi^2}\right)^{1/3}$$

$$= \left[\frac{(6.67 \times 10^{-11} \text{ N} \cdot \text{m}^2/\text{kg}^2)(5.98 \times 10^{24} \text{ kg})(8.6 \times 10^4 \text{ s})^2}{4\pi^2}\right]^{1/3}$$

$$r = \boxed{4.2 \times 10^7 \text{ m}}$$

What have we learned?
The radius of the Earth is 6.37×10^6 m, so the radius of a geosynchronous orbit is about seven times larger than the radius of the Earth. It is much larger than the low Earth orbit in Figure 5.26, which is why much more fuel is required to launch a satellite into geosynchronous orbit than into low Earth orbit.

Kepler's Laws, Putting a Satellite into Orbit, and the Origin of the Solar System

Kepler's three laws of planetary motion are statements about the shape, timing, and size of orbits, and about the orbital speed. They apply to all types of gravitationally produced orbital motion, including the motion of planets and comets about the Sun and the motion of moons and satellites about a planet. Indeed, one can also think of a freely falling apple as another example of gravitationally produced motion. It is fascinating that such a variety of motion can be produced by a single force. These different types of motion result from the different ways in which these objects are initially set into motion. The dropped apple is released from rest, so its initial velocity is zero. For the Moon to follow a nearly circular orbit, it must have been given the proper "initial" velocity at some point in the distant past. The same is true of planets and comets. That is, for planets such as Earth and Jupiter to be moving in orbits about the Sun that are now approximately circular, it was necessary that the planets be set into motion with the proper velocity. This problem brings us to some very interesting questions concerning the origin and evolution of the solar system, topics of much current research. It is now believed that the solar system was originally a rotating mass of gas and that this gas gradually condensed to form the planets. The rotational motion of the original gas cloud then led to the approximately circular orbits we now observe.

The problem of "establishing" or setting up an orbit is also encountered when a satellite is launched from Earth. NASA usually launches satellites into equatorial orbits. These orbits are approximately parallel to Earth's equator as sketched in Figure 5.27. We have always considered or computed orbital speeds as measured with respect to a stationary observer. The Earth's surface is not stationary, however; rather, it moves with a substantial speed v_R due to the Earth's rotation. When a satellite is launched toward the east in an equatorial orbit, it starts with a speed v_R and the rocket engines then add to this speed as the satellite is put into orbit. If the satellite were launched into a different orbit—say, into an equatorial orbit to the west—it would not be able to take advantage of v_R and the launch would require more fuel (and therefore more money).

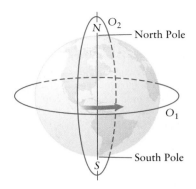

Figure 5.27 It is possible to place a satellite into an equatorial orbit (O_1) or a polar orbit (O_2). A polar orbit takes a satellite over the Earth's North and South Poles.

CONCEPT CHECK 5.6 | Apparent Weight and Earth's Rotational Motion

A person standing at the Earth's equator moves in a circle due to the rotational motion of the Earth. How does this circular motion affect the person's apparent weight?
  (a) The circular motion causes the apparent weight to be *larger* than it would be if the Earth were not rotating.
  (b) The circular motion causes the apparent weight to be *smaller* than it would be if the Earth were not rotating.
  (c) There is no effect on the apparent weight.

5.5 | MOONS AND TIDES

The Origin of Tides

As you have probably noticed, the level of Earth's oceans fluctuates up and down every day; these fluctuations are called tides (Fig. 5.28). The ocean level is high, a so-called high tide, whenever the Moon is overhead. Tides are due to the Moon's gravitational force on the oceans. Although the Moon exerts a gravitational force on both the ocean water and the solid Earth, the gravitational force decreases with distance, so the resulting acceleration is slightly higher for the water nearest the Moon (Fig. 5.29A). Thus, the water nearest the Moon "falls toward" the Moon slightly faster than does the solid Earth. The result is a "bulge" in the ocean, which is just a high tide.

Figure 5.28 An example of tides. The ocean level can vary substantially between periods of low and high tide. There are generally two high tides every 24 h (although in some regions, factors such as the shape of the ocean basin lead to only one high tide and one low tide each day).

Figure 5.29A explains why there is a high tide when the Moon is overhead. However, there are actually *two* high tides every day in most parts of the world, one when the Moon is overhead and another 12 h later when it is on the opposite side of the Earth. The origin of this second high tide is explained in Figure 5.29B. Because the gravitational force of the Moon decreases with increasing distance, the acceleration of the solid Earth is larger than the acceleration of the ocean on the far side of the Earth. As a result, the solid Earth "falls toward" the Moon faster than the ocean water on the far side and the ocean now bulges away from the Moon, producing another high tide.

The Sun also affects the tides, although its effect is smaller than the Moon's. When the Sun and the Moon are aligned and are on the same side of the Earth, their gravitational forces add, giving a higher high tide than produced by the Moon alone.

5.6 | DEEP NOTIONS CONTAINED IN NEWTON'S LAW OF GRAVITATION

The Inverse Square Law

We have seen how Newton's law of gravitation, together with an understanding of the dynamics of circular motion, enables us to explain and predict the motion of a variety of objects, ranging from roller coasters to moons and planets. The law of gravitation also provides deep insights into the workings of nature.

A key feature of the gravitational force is the manner in which it varies with distance. According to Equation 5.16, the gravitational force between two objects falls off as the square of the distance between them:

$$F_{\text{grav}} = \frac{Gm_1m_2}{r^2} \propto \frac{1}{r^2} \qquad (5.29)$$

Mathematically, Equation 5.29 is called an "inverse square" law. A number of other forces in nature, such as the force between two electric charges, fall off as the square of the distance and are thus also described by inverse square laws. Why do many natural forces follow this pattern? One way to explain the origin of inverse square behavior is sketched in Figure 5.30. We imagine that an object—for example, the Sun—possesses gravitational "field lines" that emanate out radially from it. We also imagine that the number of these lines is proportional to the mass of the object. When these lines intersect another object—for example, Earth—there is a force on that object directed parallel to the lines and hence toward the original object; for example, there is a force on the Earth directed toward the Sun.

This picture explains the inverse square dependence in Equation 5.29. Because the force lines emanate in three-dimensional space, the number of lines over a given area—that is, the number of lines that intercept the Earth in Figure 5.30—falls off with distance as $1/r^2$. This result is very appealing: it implies that gravity follows an inverse square law because we live in a three-dimensional space. It also means that we should expect other forces described by a field line picture to have the same inverse square dependence, which does seem to be the way nature works.

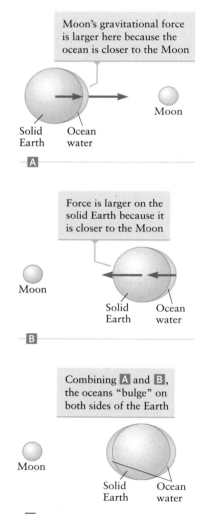

Moon's gravitational force is larger here because the ocean is closer to the Moon

Solid Earth   Ocean water   Moon

A

Force is larger on the solid Earth because it is closer to the Moon

Moon   Solid Earth   Ocean water

B

Combining **A** and **B**, the oceans "bulge" on both sides of the Earth

Moon   Solid Earth   Ocean water

C

Figure 5.29 The Moon's gravitational force on the oceans is responsible for the tides. (Not to scale.)

Do such field lines really exist? So far, no one has devised a way to "see" the lines; the best we can do is to observe the resulting force that the lines are presumed to cause. The field line picture explains the general form of Newton's law of gravitation, so there is strong evidence in favor of this model. However, while the field line picture is very attractive, it leads to other questions. The field lines that emanate from the Sun in Figure 5.30 and eventually reach Earth must travel through the nearly perfect vacuum of space. Other observations indicate that the gravitational force is felt even when a perfect vacuum separates two objects. How can "something," like a gravitational field line, exist in a vacuum? And what exactly is a field line anyway? Does it have a mass? How fast can it move? If the Sun were to suddenly move, how fast would the corresponding change in the gravitational field be felt on Earth?

The gravitational force is an example of what is called **action at a distance**. Newton's theory of gravitation tells us that action at a distance does indeed occur, but it does not tell us *how* it occurs. The field line picture was invented to answer this "how" question, but it does not provide a complete answer. For some forces, such as the electric force, there is a theoretical explanation of the nature of the associated field lines. Einstein's theories of relativity answer many of these questions in the case of the gravitational field, as we'll describe later in this book (Chapter 27).

Gravitation and Mass

A very important feature of the gravitational force law is that F_{grav} is linearly proportional to the mass m of each object. This quantity m is sometimes referred to as the **gravitational mass** of an object. We first encountered the concept of mass in Newton's second law of motion ($\sum \vec{F} = m\vec{a}$), and the m in Newton's second law is often called the **inertial mass** of the object. As far as physicists can tell, the gravitational mass of an object is *precisely equal* to the inertial mass. That $m_{grav} = m_{inertial}$ suggests a deep connection between gravitation and inertia (and hence motion), which is indeed the case. Newton did not understand why there should be a connection, and it was not explained until the work of Einstein. The theories of relativity developed by Einstein explain why the inertial and gravitational masses are the same and also explain the existence of many additional phenomena.

One such phenomenon concerns the effect of a gravitational field on the motion of light. Consider how we would apply Newton's second law to the motion of a "particle" of mass m_p that experiences a gravitational force caused by another object, such as the Sun. We would write Newton's second law as

$$m_p a_p = \frac{GM_{Sun}m_p}{r^2}$$

where a_p is the acceleration of the particle and r is the distance from it to the Sun. Strictly speaking, the mass m_p on the left side is the inertial mass of the particle, whereas the mass m_p on the right side is the gravitational mass. Because they are equal, however, we can cancel these factors to get

$$a_p = \frac{GM_{Sun}}{r^2} \qquad (5.30)$$

Hence, the acceleration is independent of the mass of the particle. We have seen this result many times before (e.g., in free fall), but there is a new point to make. Given that the acceleration is independent of the mass, does this result also apply to a "particle" that has no mass? As we'll see in Chapter 28, light can be described in terms of particles called **photons** that have no mass. Even though a photon's mass is zero, Equation 5.30 suggests that it is still accelerated by gravity. That is indeed the case, although a correct theoretical description of this acceleration requires Einstein's general theory of relativity. We'll say more about this acceleration and about the equivalence of gravitational and inertial mass in Chapter 27.

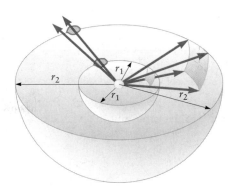

Figure 5.30 Lines of gravitational force. According to the force line picture, these lines emanate from all objects. The gravitational force exerted by one object on another is proportional to the number of lines intersecting the second object. This number falls as $1/r^2$ and thus accounts for the inverse square force law in Equation 5.29. Here, the object shown in blue is located at two different distances from the central object (yellow). The number of lines intercepted by the blue mass is smaller when the separation increases because the field lines emanate radially and expand into a larger area (tan squares).

Circular motion and centripetal acceleration
An object moving in a circle of radius r at a constant speed has an acceleration

$$a_c = \frac{v^2}{r}$$ (5.5) (page 132)

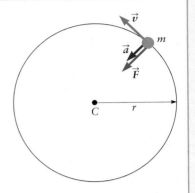

directed toward the center of the circle. The quantity a_c is called the **centripetal acceleration.** According to Newton's second law, this acceleration must be caused by a total force of magnitude $\sum F = ma_c$, so

$$\sum F = \frac{mv^2}{r}$$ (5.6) (page 133)

This force is directed toward the center of the circle.

Newton's law of gravitation
There is a gravitational force

$$F_{\text{grav}} = \frac{Gm_1 m_2}{r^2}$$ (5.16) (page 145)

between any two objects. This force is always attractive.

Newton's law of gravitation is an example of an **inverse square law**. This $1/r^2$ dependence suggests a force line model of gravity and tells us something about the geometry of the universe. Gravitation is also an example of "action at a distance." Other forces, including electric forces, exhibit this property.

There are many examples of circular motion, including roller coasters, cars traveling on a curved road, and centrifuges. In all cases, the total force on an object undergoing uniform circular motion must equal $F = mv^2/r$ and be directed toward the center of the circle.

The motion of the Moon as it orbits Earth and of the planets as they orbit the Sun are examples of circular motion. The force in these cases is due to gravitation. Near the Earth's surface, the magnitude of the gravitational force on an object of mass m is $F_{\text{grav}} = mg$ and is directed toward the center of the Earth.

Kepler's laws
Kepler deduced three laws of planetary motion: (1) planetary orbits are elliptical, (2) a planetary orbit sweeps out equal areas in equal times, and (3) the square of the orbital period is proportional to the cube of the average orbital radius. These laws apply to planets orbiting a sun and also to satellites and moons orbiting a planet.

Kepler's laws, and hence the motions of the planets, moons, comets, and other objects in the solar system, are all explained by Newton's law of gravitation, together with Newton's laws of motion. The circular motion of the Moon and the free fall of an apple look different, but they are due to the same force.

SSM = answer in Student Companion & Problem-Solving Guide   ⊗ = life science application

1. Give an example of motion in which the magnitude of the instantaneous velocity is always larger than the average velocity.

2. SSM In Example 5.3, we considered a car traveling on a banked turn with friction. Draw free-body diagrams for the car when the speed is low and when the speed is high, and explain why they are different. *Hint*: Consider the direction of the frictional force in the two cases.

3. In a reference listing found on the Internet, it is stated that $g = 9.80665$ m/s^2. Discuss why it is not correct to think that the "exact" value of g can be given with this accuracy. Indeed, is there an "exact" value of g?

4. Consider the Cavendish experiment in Figure 5.21. When he designed this experiment, Cavendish had to decide how large to make the spheres. If they are made larger, they will have a larger mass, which, according to Equation 5.16, will make the gravitational attraction larger and the force therefore easier to measure. If the spheres are made larger, however, the distance between their centers will necessarily increase, which makes the force smaller. Assuming the spheres in Figure 5.21 all have the same radii, suppose the value of r is increased by a factor of 2. Will that change increase or decrease the force, and by what factor? Assume that the spheres all have the same constant density.

5. Explain why a geosynchronous satellite cannot remain directly overhead for an observer in Boston. *Hint*: Consider first an observer at the North Pole.

6. Kepler's second law, the statement that a planet sweeps out equal areas in equal times, can be derived by a geometrical argument. To see how one might construct such a geometrical proof, consider the simpler case of a planet moving with constant velocity as shown in Figure Q5.6. The points A, B, C, . . . are spaced at equal time intervals. Show that this planet obeys Kepler's second law; that is, show that it sweeps out equal areas in equal times. *Hint*: Calculate the areas of triangles OAB, OBC, and so on.

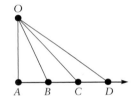

Figure Q5.6

7. It is sometimes claimed by astrologers (but not by astronomers!) that because the Moon dramatically affects the seas of the world, as evidenced by the tides, the Moon must also affect individual people because more than 60% of an average adult's mass is water. Does that claim make sense?

8. What force makes it possible for a car to move along a curved road? A straight flat road?

9. Explain why the high tides occur somewhat after the Moon passes overhead. *Hint*: Consider the inertia of water and the friction associated with its motion.

10. Explain why telecommunications satellites are in very "high" orbits (orbits that keep them far from Earth), but spy satellites are in very low orbits.

11. A double star system is a combination of two nearby stars that move in circular orbits around each other. Describe these orbits when the stars have equal mass. How does the orbital radius compare with the distance between the stars? That is, where is the center of the orbit? Assume the stars both move in circular orbits.

12. When a planet orbits around a star, the star also moves in an "orbit." Since it is much more massive than the planet, the star's orbital radius r_{star} is much smaller than that of the planet r_{planet}. Work out how the ratio r_{star}/r_{planet} depends on the ratio of the masses. Discuss how this effect could be used to detect the presence of planets in distant solar systems. For simplicity, assume circular orbits.

13. NASA uses a specially equipped airplane (called the "Vomit Comet") to provide a simulated zero-gravity environment for training and experiments. This airplane flies in a long, parabolic path. Explain how a passenger can feel "weightless" near the top of the parabola.

14. You are a prospector looking for gold by taking high-precision measurements of the acceleration due to gravity, g, at different points on the Earth's surface. In one region, you find that g is slightly higher than its average value. Are you standing over what might be a deposit of gold or over an underground lake? Explain.

15. How does your weight on a ship in the middle of the ocean compare with your weight when you are standing on solid ground? Explain why they are not the same.

16. The Sun exerts an overall force on Earth many times greater than that of the Moon. Then how can it be that the ocean tides are primarily due to the influence of the Moon and, to a much lesser extent, the Sun?

17. SSM The difference in the gravitational force is only about 10% less on an object that is in a low Earth orbit than it is for the same object on the ground. Why is it that an astronaut in orbit experiences weightlessness?

18. An astronaut on the peak of a mountain on the Moon fires a rifle along the horizontal direction. Is it possible, given a sufficient initial speed for the bullet, that the bullet might hit her in the back? Explain how it could happen.

19. **Pluto's mass.** In 1978, it was discovered that Pluto had a moon of its own. The moon was given the name Charon (now known to be one of three; see Fig. Q5.19). After the discovery of this moon, the hitherto unknown mass of Pluto was calculated to a precision of less than 1%. How did the discovery of Charon allow the mass to be determined?

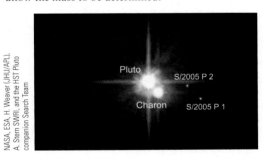

Figure Q5.19 Pluto and its three moons imaged by the Hubble Space Telescope.

20. **A coffee centrifuge?** In one popular demonstration, a full cup of hot coffee is placed on a platform suspended by strings to the lecturer's hand as seen in Figure Q5.20. With some practice, the platform, coffee and all, can be made to rotate in vertical circle. How does the coffee stay in the cup? If the rotation of the coffee and cup is sustained for some time, what would happen to any grinds that happen to be mixed in the coffee?

Figure Q5.20

5.1 UNIFORM CIRCULAR MOTION

1. A bicycle wheel of radius 0.30 m is spinning at a rate of 60 revolutions per minute. (a) What is the centripetal acceleration of a point on the edge of the wheel? (b) What is the period of the wheel's motion?

2. For the bicycle wheel in Problem 1, what is the centripetal acceleration of a point that is 0.10 m from the edge? Explain why this value is different from the answer to part (a) of Problem 1.

3. ★ Ⓡ The Earth has a radius of 6.4×10^6 m and completes one revolution about its axis in 24 h. (a) Find the speed of a point at the equator. (b) Find the speed of New York City.

4. ★ Ⓡ In the game of baseball, a pitcher throws a curve ball with as much spin as possible. This spin makes the ball "curve" on its way to the batter. In a typical case, the ball spins at about 30 revolutions per second. What is the maximum centripetal acceleration of a point on the edge of the baseball?

5. A jogger is running around a circular track of circumference 400 m. If the jogger has a speed of 12 km/h, what is the centripetal acceleration of the jogger?

6. Consider the motion of the hand of a mechanical clock. If the minute hand of the clock has a length of 6.0 cm, what is the centripetal acceleration of a point at the end of the hand?

7. If the circular track in Figure 5.1 has a radius of 100 m and the runner has a speed of 5.0 m/s, what is the period of the motion?

8. What is the acceleration of the Moon as it moves in its circular orbit around the Earth? *Hint:* You will find some useful data in Table 5.1.

9. ★ Consider points on the Earth's surface as sketched in Figure P5.9. Because of the Earth's rotation, these points undergo uniform circular motion. Compute the centripetal acceleration of (a) a point at the equator, and (b) at a latitude of 30°.

Figure P5.9

10. In the days before compact discs (ancient history!), music was recorded in scratches in the surface of vinyl-coated disks called records. In a typical record player, the record rotated with a period of 1.8 s. Find the centripetal acceleration of a point on the edge of the record. Assume a radius of 15 cm.

11. SSM ★ A compact disc spins at 2.5 revolutions per second. An ant is walking on the CD and finds that it just begins to slide off the CD when it reaches a point 3.0 cm from the CD's center. (a) What is the coefficient of friction between the ant and the CD? (b) Is this the coefficient of static friction or kinetic friction?

12. ✖ When a fighter pilot makes a very quick turn, he experiences a centripetal acceleration. When this acceleration is greater than about $8 \times g$, the pilot will usually lose consciousness ("black out"). Consider a pilot flying at a speed of 900 m/s who wants to make a very sharp turn. What is the minimum radius of curvature he can take without blacking out?

13. ★ ✖ Ⓡ At a practice for a recent automobile race, officials found that the drivers were nearly "blacking out," which led to cancellation of the race. The cars were traveling at about 240 mi/h, and the track was approximately 1.5 mi long. Find the centripetal acceleration during a turn and compare it with the physiological limit of approximately $8 \times g$ discussed in Problem 12. Assume the track was circular.

14. ★ Ⓡ The Daytona 500 stock car race is held on a track that is approximately 2.5 mi long, and the turns are banked at an angle of 31°. It is currently possible for cars to travel through the turns at a speed of about 180 mi/h. Assuming these cars are on the verge of slipping into the outer wall of the racetrack (because they are racing!), find the coefficient of static friction between the tires and the track.

15. ✪ Consider again the problem of a car traveling along a banked turn. Sometimes roads have a "reversed" banking angle. That is, the road is tilted "away" from the center of curvature of the road. If the coefficient of static friction between the tires and the road is $\mu_S = 0.50$, the radius of curvature is 15 m, and the banking angle is 10°, what is the maximum speed at which a car can safely navigate such a turn?

5.2 EXAMPLES OF CIRCULAR MOTION

16. Consider the motion of a rock tied to a string of length 0.50 m. The string is spun so that the rock travels in a vertical circle as shown in Figure P5.16. The mass of the rock is 1.5 kg, and it is twirling at constant speed with a period of 0.33 s.

Figure P5.16
Problems 16 and 17.

(a) Draw free-body diagrams for the rock when it is at the top and when it is at the bottom of the circle. Your diagrams should include the tension in the string, but the value of T is not yet known.

(b) What is the total force on the rock directed toward the center of the circle?

(c) Find the tension in the string when the rock is at the top and when it is at the bottom of the circle.

17. Consider the motion of the rock in Figure P5.16. What is the minimum speed the rock can have without the string becoming "slack"?

18. A stone of mass 0.30 kg is tied to a string of length 0.75 m and is swung in a horizontal circle with speed v. The string has a breaking-point force of 50 N. What is the largest value v can have without the string breaking? Ignore any effects due to gravity.

19. ★ The track near the top of your favorite roller coaster has a circular shape with a diameter of 20 m. When you are at the top, you feel as if your weight is only one-third your true weight. What is the speed of the roller coaster?

20. A roller coaster track is designed so that the car travels upside down on a certain portion of the track as shown in Figure P5.20. What is the minimum speed the roller coaster can have without falling from the track? Assume the track has a radius of curvature of 30 m.

Figure P5.20

21. SSM ★ A car of mass 1000 kg is traveling over the top of a hill as shown in Figure P5.21. (a) If the hill has a radius of curvature of 40 m and the car is traveling at 15 m/s, what is the normal force between the hill and the car at the top of the hill? (b) If the driver increases her speed sufficiently, the car will leave the ground at the top of the hill. What is the speed required to make that happen?

Figure P5.21

22. On a popular amusement park ride, the rider sits in a chair suspended by a cable as shown in Figure P5.22. The top end of the cable is tied to a rotating frame that spins, hence moving the chair in a horizontal circle with $r = 10$ m. The ride makes one complete revolution every 10 s.

Figure P5.22

(a) Draw pictures showing the path followed by the chair. Give both a side view and a top view.

(b) In your diagrams in part (a), indicate all the forces on the chair. Also draw a free-body diagram for the chair.

(c) Find the components of the forces in the vertical direction and in the direction toward the center of the chair's circular path. Express your answers in terms of the tension in the cable, the angle θ, and the mass of the chair m.

(d) Apply Newton's second law in both the vertical and radial directions. What is the acceleration of the chair along y?

(e) Find the angle θ the cable makes with the vertical.

23. ⭐ Consider a roller coaster as it travels near the bottom of its track as sketched in Figure P5.23. At this point, the normal force on the roller coaster is three times its weight. If the speed of the roller coaster is 20 m/s, what is the radius of curvature of the track?

Figure P5.23

24. ⭐ A coin is sitting on a record as sketched in Figure P5.24. It is found that the coin slips off the record when the rotation rate is 0.30 rev/s. What is the coefficient of static friction between the coin and the record?

Figure P5.24

25. A rock is tied to a string and spun in a circle of radius 1.5 m as shown in Figure P5.25. The speed of the rock is 10 m/s.

Figure P5.25

(a) Draw a picture giving both a top view and a side view of the motion of the rock.

(b) What are all the forces acting on the rock? Add them to your pictures in part (a). Then draw a free-body diagram for the rock.

(c) What is the total force on the rock directed toward the center of its circular path? Express your answer in terms of the (unknown) tension in the string T.

(d) Apply Newton's second law along both the vertical and the horizontal direction and find the angle θ the string makes with the horizontal.

26. Spin out! An interesting amusement park activity involves a cylindrical room that spins about a vertical axis (Fig. P5.26). Participants in the "ride" are in contact with the wall of the room, and the circular motion of the room results in a normal force from the wall on the riders. When the room spins sufficiently fast, the floor is retracted and the frictional force from the wall keeps the people "stuck" to the wall. Assume the room has a radius of 2.0 m and the coefficient of static friction between the people and the wall is $\mu_S = 0.50$.

Figure P5.26

(a) Draw pictures showing the motion of a "rider." Give both a side view and a top view.

(b) What are all the forces acting on a rider? Add them to your pictures in part (a). Then draw a free-body diagram for a rider.

(c) What are the components of all the forces directed toward the center of the circle (the radial direction)?

(d) Apply Newton's second law along both the vertical and the radial directions. Find the minimum rotation rate for which the riders do not slip down the wall.

27. ⭐ Consider the circular space station in Figure 5.13. Suppose the station has a radius of 15 m and is designed to have an acceleration due to "artificial gravity" of $g/2$. Find the speed of the rim of the space station.

28. A rock of mass $m = 2.5$ kg is tied to the end of a string of length $L = 1.2$ m. The other end of the string is fastened to a ceiling, and the rock is set into motion so that it travels in a horizontal circle of radius $r = 0.70$ m as sketched in Figure P5.28.

Figure P5.28

(a) Draw a picture showing the motion of the rock. Give both a top view and a side view.

(b) What are all the forces on the rock? Add them to your pictures in part (a). Then draw a free-body diagram for the rock.

(c) Find the vertical and horizontal components of all the forces on the rock.

(d) Apply Newton's second law in both the vertical (y) and the horizontal directions. What is the acceleration along y? Find the tension in the string.

29. ⭐ A car of mass 1700 kg is traveling without slipping on a flat, curved road with a radius of curvature of 35 m. If the car's speed is 12 m/s, what is the frictional force between the road and the tires?

30. ✪ Ⓡ Consider a Ferris wheel in which the chairs hang down from the main wheel via a cable. The cable is 2.0 m long, and the radius of the wheel is 12 m (see Fig. P5.30). When a chair is in the orientation shown in Figure P5.30 (the "3 o'clock" position), the cable attached to the chair makes an angle of $\theta = 20°$ with the vertical. Find the speed of the chair.

Figure P5.30

31. ⭐ A rock of mass $m = 1.5$ kg is tied to a string of length $L = 2.0$ m and is twirled in a vertical circle as shown in Figure 5.10. The speed v of the rock is constant; that is, it is the same at the top and the bottom of the circle. If the tension in the string is zero when the rock is at its highest point (so that the string just barely goes slack), what is the tension when the rock is at the bottom?

32. ⭐ ✖ Ⓡ We saw in Example 5.6 how a centrifuge can be used to separate cells from a liquid. To increase the rate at which objects can be separated from solution, it is useful to make the centrifuge's speed as large as possible. If you want to design a centrifuge of diameter 50 cm to have a force of 10^6 times the force of Earth's gravity, what is the speed of the outer edge of the centrifuge? Such a device is called an *ultra*centrifuge.

33. ✪ ✖ Ⓡ A centrifuge can be used to separate DNA molecules from solution. Estimate how long it will take the centrifuge in Example 5.6 to separate a DNA molecule from water. Assume the centrifuge tube is 2.0 cm long. For this case, the drag coefficient in Stokes's law (Eq. 5.14) is $C = 0.020$ N · s/m^2.

34. SSM ⭐ ✖ NASA has built centrifuges to enable astronauts to train in conditions in which the acceleration is very large. The device in Figure P5.34 shows one of these "human centrifuges." If the device has a radius of 8.0 m and attains accelerations as large as $5.0 \times g$, what is the rotation rate?

Figure P5.34

5.3 NEWTON'S LAW OF GRAVITATION

35. Two small objects of mass 20 kg and 30 kg are a distance 1.5 m apart. What is the gravitational force of one of these objects on the other?

36. If the masses of the objects in Problem 35 are both increased by a factor of $\sqrt{5}$, by what factor does the gravitational force change? Do not use a calculator to solve this problem!

37. ⭐ Three lead balls of mass $m_1 = 15$ kg, $m_2 = 25$ kg, and $m_3 = 9.0$ kg are arranged as shown in Figure P5.37. Find the total gravitational force exerted by balls 1 and 2 on ball 3. Be sure to give the magnitude and the direction of this force.

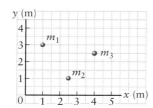

Figure P5.37

38. ⭐ **Travel and lose pounds!** Your apparent weight is the force you feel on the bottoms of your feet when you are standing. Due to the Earth's rotation, your apparent weight is slightly more when you are at the South Pole than when you are at the equator. What is the ratio of your apparent weights at these two locations? Carry three significant figures in your calculation.

39. Find the gravitational force of the Sun on the Earth.

40. Calculate the acceleration due to gravity on the surface of Mars.

41. ⭐ Ⓡ Estimate the gravitational force between two bowling balls that are nearly touching.

42. ✪ Ⓡ Suppose the bowling balls in Problem 41 are increased in size (radius) by a factor of two, but their density does not change. By what factor does the gravitational force change? *Hint*: When the radius is changed, both the mass and the separation will change.

43. SSM ⭐ When a spacecraft travels from Earth to the Moon, the gravitational force from Earth initially opposes this journey. Eventually, the spacecraft reaches a point where the Moon's gravitational attraction overcomes the Earth's gravity. How far from Earth must the spacecraft be for the gravitational forces from the Moon and Earth to just cancel?

44. Find the ratio of your weight on Earth to your weight on the surface of the Sun.

45. ⭐ Some communications and television towers are much taller than any buildings. These towers have been used to study how the Earth's gravitational force varies with distance from the center of the Earth. Calculate the ratio of the acceleration due to Earth's gravity at the top of a tower that is 600 m tall to the value of g at the Earth's surface. *Hint*: Keep four significant figures in your calculation.

46. Suppose the density of the Earth was somehow reduced from its actual value to 1000 kg/m³ (the density of water). Find the value

of g, the acceleration due to gravity, on this new planet. Assume the radius does not change.

47. You are an astronaut ($m = 95$ kg) and travel to a planet that is the same radius and mass size as Earth, but it has a rotational period of only 2 h. What is your apparent weight at the equator of this planet?

48. ⭐ In Section 5.4, we showed that the radius of a geosynchronous orbit about the Earth is 4.2×10^7 m, compared with the radius of the Earth, which is 6.4×10^6 m. By what factor is the force of gravity smaller when you are in geosynchronous orbit than when you are on the Earth's surface?

5.4 PLANETARY MOTION AND KEPLER'S LAWS

49. ⭐ Saturn makes one complete orbit of the Sun every 29.4 years. Calculate the radius of the orbit of Saturn. *Hint*: It is a very good approximation to assume this orbit is circular.

50. ⭐ The region of the solar system between Mars and Jupiter contains many asteroids that orbit the Sun. Consider an asteroid in a circular orbit of radius 5.0×10^{11} m. Find the period of the orbit.

51. SSM ⭐ A newly discovered planet is found to have a circular orbit, with a radius equal to 27 times the radius of Earth's orbit. How long does this planet take to complete one orbit around the Sun?

52. ✪ In recent years, a number of nearby stars have been found to possess planets. Suppose the orbital radius of such a planet is found to be 4.0×10^{11} m, with a period of 1100 days. Find the mass of the star.

53. ✪ Mars has two moons, Phobos and Deimos. It is known that the larger moon, Phobos, has an orbital radius of 9.4×10^6 m and a mass of 1.1×10^{16} kg. Find its orbital period.

54. ⭐ In our derivation of Kepler's laws, we assumed the only force on a planet is due to the Sun. In a real solar system, however, the gravitational forces from the other planets can sometimes be important. Calculate the gravitational force of Jupiter on the Earth and compare it to the magnitude of the force from the Sun. Do the calculations for the cases when Jupiter is both closest to and farthest from Earth (Fig. P5.54).

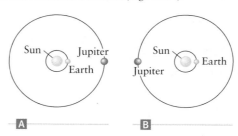

Figure P5.54

55. ⭐ What is the speed of a satellite in a geosynchronous orbit about Earth? Compare it with the speed of the Earth as it orbits the Sun.

5.5 MOONS AND TIDES

56. ⭐ **Syzygy.** We have seen that the normal tides are due to the gravitational force exerted by the Moon on Earth's oceans. When the Moon, Sun, and Earth are aligned as shown in Figure P5.56, the magnitude of the tide increases due to the gravitational force exerted by the Sun on the oceans (at the times of

Figure P5.56

the "new" Moon and the "full" Moon during the course of a month). Calculate the ratio of the gravitational force of the Sun to that of the Moon on the oceans. This "extra" force from the Sun does make a difference!

57. ☆ ⓡ In Figure 5.29, we saw that the tides on Earth are due to the variation of the Moon's gravitational force with distance. Find the approximate ratio of the Moon's gravitational force on two portions of the ocean, one nearest the Moon and one on the opposite side of the Earth.

58. ☆ Your weight is due to the gravitational attraction of the Earth. The Moon, though, also exerts a gravitational force on you, and when it is overhead, your weight decreases by a small amount. Calculate the effect of the Moon on your weight. Express your result as a percentage change for the cases of the Moon overhead and the Moon on the opposite side of the Earth.

59. SSM ☆ During an eclipse, the Sun, Earth, and Moon are arranged in a line as shown in Figure P5.59. There are two types of eclipse: (a) a lunar eclipse, when Earth is between the Sun and the Moon, and (b) a solar eclipse, when the Moon is between the Sun and Earth. Calculate the percentage change in your weight when going from one type of eclipse to the other.

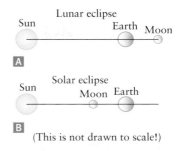

Figure P5.59

ADDITIONAL PROBLEMS

60. ✪ In the film *Mission to Mars* (released in 2000), the spacecraft (see Fig. P5.60) features a rotating section to provide artificial gravity for the long voyage. A physicist viewing a scene from the interior of the spacecraft notices that the diameter of the rotating portion of ship is about five times the height of an astronaut walking in that section (or about 10 m). Later, in a scene showing the spacecraft from the exterior, she notices that the living quarters of the ship rotate with a period of about 30 s. Did the movie get the physics right? Compare the centripetal acceleration of a 1.7-m-tall astronaut at his feet to that at his head. Compare these accelerations to g.

Figure P5.60 Not to scale.

61. ☆ Will your apparent weight at the top of Mount Everest (altitude = 8850 m = 29,035 ft) be more or less than at sea level at the same latitude (27.98° N)? What is the ratio of your apparent weight at these two locations? For simplicity, consider only the effect of altitude and ignore the spinning of the Earth.

62. ✪ A man stands 6.0 ft tall at sea level on the North Pole as shown in Figure P5.62. (a) What is the difference in the value of g (the gravitational acceleration) between his head and his feet? (b) The man is now put in a space suit and transported to a location one Earth radius away from a black hole of mass equal to 20 times the mass of our Sun. If the man's feet are pointing in the direction of the black hole, what is the difference in the gravitational acceleration between his head and his feet? Would this difference in acceleration be harmful, just noticeable, or somewhere in between?

63. ✪ ⓡ Proponents of astrology claim that the positions of the planets at the time of a baby's birth will affect the life of that person in important ways. Some assert that this effect is due to gravity. Examine this claim with Newton's law of gravity. Calculate the maximum force exerted on the baby by the planet Mars and compare that with the force of gravity exerted by the mass of the doctor's head as she delivers the baby. Assume Mars is at its closest approach to Earth at the time. You will need to estimate the mass of a newborn as well as the mass and distance between the doctor's head and the baby.

64. SSM ✪ ⓡ An ancient and deadly weapon, a *sling* consists of two braided cords, each about half an arm's length long, attached to a leather pocket. The pocket is loaded with a projectile made of lead, carved rock, or clay and made to swing in a vertical circle as shown in Figure P5.64. The projectile is released by letting go of one end of the cord. (a) If a Roman soldier can swing the sling at a rate of 7.5 rotations per second, what is the maximum range of his 100-g projectile? (Ignore air drag.) (b) What is the maximum tension in each cord during the rotation?

Figure P5.64 The proper use of a sling.

65. ✪ A popular circus act features daredevil motorcycle riders encased in the "Globe of Death" (Fig. P5.65), a spherical metal cage of diameter 16 ft. (a) A rider of mass 65 kg on a 125-cc (95-kg) motorcycle keeps his bike horizontal as he rides

Figure P5.65 The "Globe of Death."

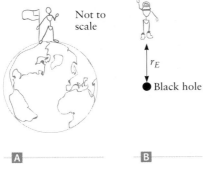

Figure P5.62

around the "equator" of the globe. What coefficient of friction is needed between his tire and the cage to keep him in place? (b) How many loops will the rider make per second? (c) The same rider performs vertical loops in the globe. What force does the cage need to withstand at the top and the bottom of the rider's loop? Assume a speed of 20 mi/h for both tricks.

66. ✪ Ⓡ **Asteroid satellite.** While on its way to Jupiter in 1993, the *Galileo* spacecraft made a flyby of asteroid Ida. Images captured (Fig. P5.66) of Ida showed that the asteroid has a tiny moon of its own, since given the name Dactyl. Measurements found Ida to be about 56 × 24 × 21 km (35 × 15 × 13 mi) in size, and Dactyl's orbital period and radius are approximately 27 h and 95 km, respectively. From these data, determine Ida's approximate mass and density.

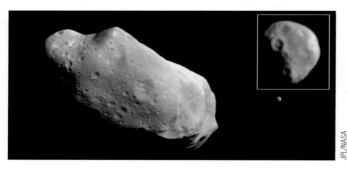

Figure P5.66 Ida and Dactyl from the *Galileo* spacecraft.

67. ✪ Ⓡ **The death spiral.** An Olympic pair figure-skating routine features an element called the death spiral shown in Figure P5.67. In this routine, the male skater swings his female partner in a circle. If the rotation rate is three-fourths a rotation per second, estimate the tension in the arms that the skaters' grip must withstand when performing this element. *Hint*: Approximate the female skater as a point mass located at her waist. It may be useful to take measurements directly from Figure P5.67 and scale appropriately.

Figure P5.67

68. ✪ Consider a hypothetical extrasolar world, planet Tungsten, that has twice the radius of the Earth and twice its density. (a) What is the acceleration due to gravity on the surface of planet Tungsten? (b) An interstellar astronaut lands on the equator of this planet and finds that his apparent weight matches his weight on Earth. What is the period of rotation of planet Tungsten?

69. ✪ The movie *2001: A Space Odyssey* (released in 1968) features a massive rotating space station of radius 100 m, similar to the one in Figure 5.13B. (a) What period of rotation is needed to provide an artificial gravity of g at the rim? (b) At what speed is the rim moving? (c) What is your apparent weight if you run along the rim at 4.2 m/s opposite the rotation direction? (d) What is your apparent weight if you instead run in the direction of rota-

tion? (e) In which direction would you run (either with or against the rotation) to get the best workout? Would it matter?

70. ✪ An astronaut stands on the surface of Vesta, which, with an average radius of 270 m, makes it the third largest object in the asteroid belt. The astronaut picks up a rock and drops it from a height of 1.5 m. He times the fall and finds that the rock strikes the ground after 3.2 s. (a) Determine the acceleration due to gravity at the surface of Vesta. (b) Find the mass of Vesta. (c) If the astronaut can jump to a height of 82 cm while wearing his space suit on Earth, how high could he jump on Vesta? Assume any rotation of Vesta can be ignored.

71. 🔲 The International Space Station orbits at an average height of 350 km above sea level. (a) Determine the acceleration due to gravity at that height and find the orbital velocity and the period of the space station. (b) The Hubble Space Telescope orbits at 600 km. What is the telescope's orbital velocity and period?

72. ✪ Ⓡ **Oil exploration.** When searching for gold, measurements of g can be used to find regions within the Earth where the density is larger than that of normal soil. Such measurements can also be used to find regions in which the density of the Earth is smaller than normal soil; these regions might contain a valuable fluid (oil). Consider a deposit of oil that is 300 m in diameter and just below the surface of the Earth. For simplicity, assume the deposit is spherical. Estimate the change in the acceleration due to gravity on the surface above this deposit. Assume the density of the oil is 1500 kg/m³ and the density of normal soil and rock is 2000 kg/m³. *Note*: Companies that search for valuable minerals actually use this method.

73. SSM ✪ A rock of mass m is tied to a string of length L and swung in a horizontal circle of radius r. The string can withstand a maximum tension T_{max} before it breaks. (a) What is the maximum speed v_{max} the rock can have without the string breaking? (b) The speed of the rock is now increased to $3v_{max}$. The original single string is then replaced by N pieces that are all identical to the original string. What is the minimum value of N required so that the strings do not break? Ignore the force of gravity on the rock.

74. 🔲 On which of the planets in our solar system would you weigh the most?

75. ✪ Experiments have shown that riders in a car begin to feel uncomfortable while traveling around a turn if their acceleration is greater than about 0.40 × g. Use this fact to calculate the minimum radius of curvature for turns at (a) 10 m/s (appropriate for driving in town) and (b) 30 m/s (highway driving).

76. ✪ In Insight 5.2, we discussed how, because of the force of gravity from the Moon, the Earth moves in an orbit around a point that lies between it and the Moon. (a) Find the radius of the Earth's orbit. (b) Where does this point lie relative to the Earth's surface? That is, does it lie inside or outside the Earth itself?

77. ✪ Ⓡ **Gravitational tractor.** In some science fiction stories, a "tractor beam" is used to pull an object from one point in space to another. That may not be just science fiction. It has been proposed that a "gravitational tractor" could be used to "tow" an asteroid. In theory, this tractor could be used to deflect asteroids that would otherwise collide with the Earth. A typical asteroid that could significantly damage life on Earth might have a radius 100 m and density 2000 kg/m³. The "tractor" would be just a very massive spacecraft; in current designs, it would have a mass of about 2 × 10⁴ kg. (a) What is the maximum gravitational force the tractor could exert on the asteroid? (b) What would be the acceleration of the asteroid? (c) If the tractor stayed near the asteroid for 1 year, what would be the deflection of the asteroid? (d) Consider an asteroid initially headed straight for Earth with a speed of 3 × 10⁴ m/s. If the tractor first comes close to the asteroid when it is 6 × 10¹¹ m from Earth (near the orbit of Jupiter), will the asteroid hit or miss the Earth? For simplicity, ignore the gravitational force of the Earth on the asteroid. *Note*: NASA is seriously considering the development of such a gravitational tractor.

Work and Energy

When this archer releases his bow string, energy stored in the bow is transferred to the now-moving arrow. In this chapter, we explore the connection between mechanical energy and motion. (© Bob Torrez/Stone/Getty Images)

So far in this book, our discussions of motion have been based on very direct applications of Newton's laws. For example, to predict the motion of an object we used Newton's second law ($\sum \vec{F} = m\vec{a}$) to calculate the acceleration; from there, we worked out the velocity and position, and how they vary with time. Such direct applications of Newton's laws can take us a long way, but there is a lot more to mechanics than simply forces and acceleration. In this and the next few chapters, we will explore some very important concepts and principles that are, in a sense, "hidden" just beneath the surface of Newton's second law. This chapter is based on the concepts of **work** and **energy**, and we'll see how Newton's second law leads to definitions of these quantities. We will also consider the concept of energy as applied to an individual particle and to systems of particles or objects. We'll find that the total energy of an isolated system is constant with time. This result is connected to the general principle of **conservation of energy**. This principle has an important role in many fields, including engineering, physiology, and physics.

6.1 | FORCE, DISPLACEMENT, AND WORK

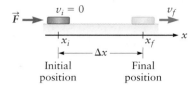

Figure 6.1 When a force $\vec{F}$ acts on this hockey puck, the puck accelerates. We also say that this force does work on the puck.

According to Newton's second law ($\sum \vec{F} = m\vec{a}$), the acceleration of a particle is proportional to the total applied force. A large force thus gives a large acceleration, while a small force gives a small acceleration. If we are limited so that we can only apply a small force, we can still produce a large effect (i.e., a large velocity or displacement) by simply applying the force for a long time, but how long a time is required? The notion of "work" gives a way to answer this and other similar questions.

Consider a hockey puck of mass m placed on a horizontal surface as sketched in Figure 6.1. Let's assume this surface is extremely icy and slippery, so the frictional force on the puck is zero. The puck is initially at rest, and we want to consider how a constant force applied horizontally will set the puck into motion. To be specific, suppose the magnitude F of the force is given and we wish to calculate how long it takes the puck to reach a particular value of the velocity v_{final}. We have already encountered this problem in Chapter 3, and we know the basic solution.

This problem involves motion in one dimension: the acceleration, velocity, and displacement are all along x, and we can write $F = ma$, where F is the total horizontal force on the puck. The acceleration is thus $a = F/m$. The acceleration is constant, so we can apply our results for motion with constant acceleration from Chapter 3 to deduce the velocity and displacement as functions of time. The puck starts at rest, so its initial velocity is $v_i = 0$. (Here we change from our usual notation and denote the initial velocity as v_i instead of v_0. We do so to emphasize that v_i is the *initial* velocity.) After a time t, the velocity of the puck is

$$v = v_i + at = at = \frac{F}{m} t \tag{6.1}$$

If t_f is the time it takes to reach a certain final velocity v_f,

$$v_f = \frac{F}{m} t_f$$

Solving for t_f gives

$$t_f = \frac{mv_f}{F} \tag{6.2}$$

At this time, the position of the puck is x_f, so the puck has traveled a distance

$$\Delta x = x_f - x_i = v_i t + \tfrac{1}{2} at_f^2$$

Using $v_i = 0$ along with $a = F/m$ and the result for t_f from Equation 6.2, we can solve for the displacement Δx

$$\Delta x = \frac{1}{2} at_f^2 = \frac{1}{2} \frac{F}{m} t_f^2 = \frac{F}{2m} \left(\frac{mv_f}{F} \right)^2$$

$$\Delta x = \frac{mv_f^2}{2F} \tag{6.3}$$

Now suppose we repeat the experiment in Figure 6.1, but with a force that is only half as large; that is, we use a force of magnitude $F/2$. We can then see from Equation 6.2 that it will take twice as long for the puck to reach the same final velocity because t_f will increase by a factor of two. According to Equation 6.3, Δx will increase by a factor of two, so the puck will travel twice as far during this time. We can repeat this calculation for other values of the force, using Equations 6.2 and 6.3 to find how the time and distance traveled vary if we change F by different factors. Interestingly, the product of the force and the distance traveled is *constant*. Combining Equations 6.2 and 6.3, the product $F \Delta x$ is always given by

$$F \Delta x = F \left(\frac{mv_f^2}{2F} \right) = \frac{mv_f^2}{2}$$

independent of the magnitude of the force.

In words, this result means that if we want to accelerate an object to a particular velocity, we can exert a large force over a short distance or a small force over a large distance. As long as the product of force and displacement is the same, the object will reach the same final velocity. The product $F \Delta x$ is called **work**. In mathematical terms, the work done by the applied force F on the hockey puck in Figure 6.1 is

$$W = F \Delta x \qquad (6.4)$$

In SI units, we have

$$\text{work} = \text{force} \times \text{displacement} = \text{newtons} \cdot \text{meters} = \text{N} \cdot \text{m}$$

The unit N · m is also called a joule (abbreviated "J").

The definition of work given in Equation 6.4 only applies for one-dimensional motion in which a constant force F is applied along the direction of motion. In two or three dimensions, we must consider that force $\vec{F}$ and displacement $\Delta \vec{r}$ are both vectors. In this case (Fig. 6.2), the work done on the particle is

$$W = F(\Delta r)\cos \theta \qquad (6.5)$$

where θ is the angle between the force and the particle's displacement. The factor F in Equation 6.5 denotes the magnitude of the force, and Δr is the magnitude of the displacement. Notice that although force and displacement are both vectors, work is a scalar. While W can be positive or negative, it does not have a direction.

W Depends on the Direction of the Force Relative to the Displacement

Figure 6.2 shows that the factor $F \cos \theta$ in Equation 6.5 is equal to the component of the force along the direction of the displacement. Hence, the work W equals the component of the force *along* the displacement multiplied by the magnitude of the displacement. In Figure 6.2, this component of the force is parallel to $\Delta \vec{r}$, and W is positive. A similar case in one dimension is shown in Figure 6.3A, in which the angle between the applied force and the displacement is zero (which was also the case for the hockey puck in Fig. 6.1). The $\cos \theta$ factor in Equation 6.5 is then equal to unity—because $\cos(0) = 1$—and this expression is equivalent to the relation for work in one dimension (Eq. 6.4).

Figure 6.3B shows a situation in which the force on an object and the object's displacement are in opposite directions; that is, $\vec{F}$ and $\Delta \vec{r}$ are *anti*parallel. This object might be a hockey puck sliding across a rough surface (a surface with friction), so the force due to friction is directed opposite to the displacement. We then have $\theta = 180°$, giving $\cos \theta = -1$ in Equation 6.5, and the work done by friction on the puck is *negative*. Another important case is shown in Figure 6.3C, in which the force is perpendicular to the displacement. This situation often arises in circular motion because, as we learned in Chapter 5, the force in that case is perpendicular to the direction of motion. When $\vec{F}$ and $\Delta \vec{r}$ are perpendicular, $\theta = 90°$ and $\cos \theta = 0$ in Equation 6.5; in this case, the work done on the object is $W = 0$.

The relationship between force, displacement, and work are central to this chapter and can be summarized as follows.

Key concepts: The relationship between force, displacement, and work

- Work is done *by* a force $\vec{F}$ acting *on* an object.
- The work W depends on the force acting on the object and on the object's displacement, according to Equation 6.5.
- The value of W depends on the direction of $\vec{F}$ relative to the object's displacement.
- W may be positive, negative, or zero, depending on the angle θ between the force and the displacement. W is a scalar; it is *not* a vector.
- If the displacement is zero (the object does not move), then $W = 0$, even though the force may be very large.

Figure 6.2 The force $\vec{F}$ acts on this particle while the particle moves through a displacement $\Delta \vec{r} = \vec{r}_f - \vec{r}_i$. The work done by the force is given by $W = F(\Delta r)\cos \theta$.

Definition of work

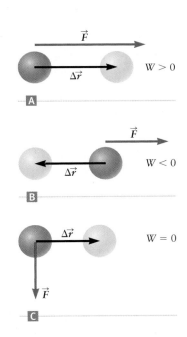

Figure 6.3 ◢ The work done on an object is positive when the applied force and object's displacement are parallel. ◢ W is negative if the force and displacement are antiparallel. ◢ When the force is perpendicular to the displacement, as in uniform circular motion, W is zero.

How Physics Uses the Term *Work*

The term *work* is used in everyday conversation. It is crucial to see how the everyday definition of this term differs from the "physics definition" in Equations 6.4 and 6.5. One important difference is that W (the "physics" work) can be negative. An example with $W < 0$ is sketched in Figure 6.3B, where the value of W is negative because the force and displacement are in opposite directions. As a result, the final speed of the object is less than its initial speed. In general, we can say that if $W > 0$, an object will "speed up"; if $W < 0$, it will "slow down."

Although we have not yet defined the concept of energy, we'll soon see how the motion of an object is related to its energy. Furthermore, your intuition should suggest that this energy can change when a force acts on the object as in Figure 6.3. We will find that situations in which the force *increases* the energy of the object correspond to a positive value of W. Conversely, it is possible for a force to *reduce* the energy of an object (as with friction and the hockey puck in Fig. 6.3B); in such cases, the work done on the object is negative.

> **CONCEPT CHECK 6.1** | Force, Displacement, and Work
>
> Figure 6.3 shows hypothetical cases in which the force is (a) parallel, (b) antiparallel, and (c) perpendicular to the displacement. Identify which case applies to the following situations.
>   (1) An apple falling from a tree
>   (2) A satellite in geosynchronous orbit around the Earth
>   (3) A car skidding to a stop on a horizontal road

EXAMPLE 6.1 | Towing a Car

A tow truck pulls a car of mass 1200 kg to a repair shop several miles away. As part of this journey, the tow truck and car must travel on a highway, and when entering the highway they accelerate uniformly as they merge into traffic. They have an initial speed of 10 m/s when they enter the merge ramp and a final speed of 30 m/s when they leave the merge lane and enter the highway. The length of the merge lane is 20 m. Find the work done on the car.

RECOGNIZE THE PRINCIPLE

To calculate W, we need to know the displacement of the car, the force on the car, and the angle between the force and the displacement. The displacement is equal to the length of the merge lane, whereas the force is directed along the lane; hence, $\theta = 0$ in Equation 6.5. We can obtain the force on the car from the acceleration together with Newton's second law.

SKETCH THE PROBLEM

Figure 6.4 shows the problem. When the car and tow truck are in the merge lane, they have a constant acceleration along the direction of the road. This is a one-dimensional problem because the displacement and the force are both directed along the merge lane, which we can take as the x direction.

Figure 6.4 Example 6.1. The work done in towing this car along a straight road is equal to $F\,\Delta x$.

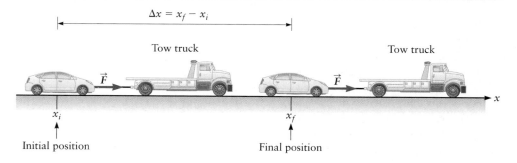

IDENTIFY THE RELATIONSHIPS

To calculate the acceleration, we use the relations for motion with constant acceleration from Chapter 3. We denote the initial velocity as v_i and the initial position as x_i. From Equation 3.4, we have

$$v_f^2 = v_i^2 + 2a(x_f - x_i)$$

The initial speed is given as $v_i = 10$ m/s and the final speed as $v_f = 30$ m/s, and the displacement during this time is $(x_f - x_i) = 20$ m. Solving for the acceleration, we get

$$a = \frac{v_f^2 - v_i^2}{2(x_f - x_i)} = \frac{(30 \text{ m/s})^2 - (10 \text{ m/s})^2}{2(20 \text{ m})} = 20 \text{ m/s}^2$$

We now use Newton's second law to find the force needed to produce this acceleration:

$$\Sigma F = ma = (1200 \text{ kg})(20 \text{ m/s}^2) = 2.4 \times 10^4 \text{ N}$$

Here the total force is equal to the total horizontal force, which we can simply denote by F.

SOLVE

The work done on the car is the product of F times the displacement:

$$W = F\,\Delta x = F(x_f - x_i) = (2.4 \times 10^4 \text{ N})(20 \text{ m})$$

$$W = 4.8 \times 10^5 \text{ N} \cdot \text{m} = \boxed{4.8 \times 10^5 \text{ J}}$$

Notice that the units of work are joules (J).

What does it mean?
This result is the work done by the tow truck on the car. Notice that W is positive and that the car's final speed is greater than its initial speed.

What Does the Work?

In Example 6.1, we calculated the total work done on the car. For simplicity, we assumed the only horizontal force acting on the car was the force F of the tow truck. According to our calculation, this force does a work $W = 4.8 \times 10^5$ J on the car, and we could say that the tow truck does this amount of work W *on the car*. We might also ask about the work done *by the car on the tow truck*. The work done on the tow truck is also given by Equation 6.5, where F is now the force exerted by the car on the truck. According to Newton's third law (the action–reaction principle), the force on the tow truck is equal in magnitude and opposite in direction to the force on the car. The force on the truck is thus opposite to its displacement, so the work done on the truck is *negative*. In general, when an agent applies a force to an object and does an amount of work W *on that object*, the object will do an amount of work equal to $-W$ "back" *on the agent*.

We can use the same ideas to discuss the work done in cases in which several forces, from different agents, act on an object. It is often useful to talk about the work done by each separate agent, that is, the work done separately by each force. These separate values of the work can again be calculated using Equation 6.5, with F being the force exerted by the particular agent of interest.

Graphical Analysis and Work Done by a Variable Force

In our discussions of work, we have so far assumed the force is constant. There are, of course, many real-world situations in which the force is not constant, and we must also consider how to deal with these cases. With that in mind, it is useful to plot the force as a function of the displacement. When F is constant (Fig. 6.5A), this

$W = F_0 \Delta x$

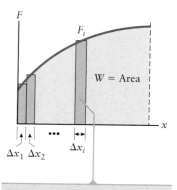

$W = \text{Area}$

$\Delta x_1 \ \Delta x_2 \quad \cdots \quad \Delta x_i$

The work done during each small step Δx_i is equal to $F \cdot \Delta x_i$, which is just the area of a shaded box.

The total work done is the total area under the curve.

Figure 6.5 Work is equal to the area under a graph of force versus displacement. **A** The force is constant and equals the area of the shaded rectangle. **B** The force is not constant, but the work done by this force is still equal to the area under the curve of F versus x.

graph is simply a horizontal line. For simplicity, let's assume we have a case of one-dimensional motion, so the force and displacement are both along the x axis. The work done is then equal to the force times the displacement, and W is equal to the area under this force–displacement plot. Now consider a case in which the force is not constant as sketched in Figure 6.5B. In this case, we can (mathematically) think of the motion as a sequence of displacements Δx_1, Δx_2, Δx_3, and so forth, with the force approximately constant during each of these displacements. For each displacement, we can calculate the work using Equation 6.4 and then sum the results to get the total work done. We can see from Figure 6.5B that this process is equivalent to calculating the area under the entire force–displacement curve. Hence, when dealing with a nonconstant force, the work W is equal to the area under the force–displacement graph. In practice (see Problems 13 and 14 at the end of this chapter), this area can be estimated by dividing the area under the force–displacement graph into a series of rectangles as in Figure 6.5B and then computing the areas of these rectangles.

6.2 | KINETIC ENERGY AND THE WORK–ENERGY THEOREM

We have seen that for the simplest case of one-dimensional motion, with the force directed parallel to the displacement, the work done by a force on an object is equal to the magnitude of the force times the object's displacement: $W = F \Delta x$. Using Newton's second law ($F = ma$ for this one-dimensional case), we can write this expression as

$$W = F \Delta x = ma \Delta x \qquad (6.6)$$

For simplicity, let's assume the force is constant so that the acceleration is also constant. We can then use one of our relations for motion with constant acceleration (Eq. 3.4) to write

$$v^2 = v_i^2 + 2a(x - x_i) \qquad (6.7)$$

Notice again that we are using the subscript i to indicate the *initial* velocity and position. The displacement is just $\Delta x = (x - x_i)$; inserting this into Equation 6.7 and rearranging, we find

$$a \Delta x = a(x - x_i) = \frac{v^2 - v_i^2}{2} \qquad (6.8)$$

We now want to calculate the work done on the object as it moves from the initial position x_i to a final position x_f. Combining Equations 6.8 and 6.6 leads to

$$W = ma(x_f - x_i) = m\left(\frac{v_f^2 - v_i^2}{2}\right)$$

$$W = \tfrac{1}{2}mv_f^2 - \tfrac{1}{2}mv_i^2 \qquad (6.9)$$

The expression $\tfrac{1}{2}mv^2$ that appears on the right-hand side of Equation 6.9 is an *extremely* important quantity. When an object of mass m has a speed v, it has an energy $\tfrac{1}{2}mv^2$ due to its motion. We call this **kinetic energy** (**KE**):

Kinetic energy

$$KE = \tfrac{1}{2}mv^2 \qquad (6.10)$$

"Energy" is a new concept for us in mechanics, and we can use Equation 6.9 to gain some understanding of it. Equation 6.9 tells us that the kinetic energy of an object can be changed by doing work on the object. If we write the change in kinetic energy as ΔKE, Equation 6.9 can be written

Work–energy theorem

$$W = \tfrac{1}{2}mv_f^2 - \tfrac{1}{2}mv_i^2 = KE_f - KE_i$$

$$W = \Delta KE \qquad (6.11)$$

The work done on an object is thus equal to the *change* in its kinetic energy. This relation tells us how work, and hence also force and displacement, are connected to the kinetic energy of an object. Equation 6.11 is called the **work–energy theorem**. According to this result, the units of work and energy are the same, so energy is measured in joules. Another commonly used unit of energy is the calorie,[1] with 1 cal equal to 4.186 J.

EXAMPLE 6.2 | Kinetic Energy of a Falling Object

A rock of mass m is dropped from the top of a tall building of height h. Find **(a)** the kinetic energy and **(b)** the speed of the rock just before it reaches the ground.

RECOGNIZE THE PRINCIPLE

The only force on the rock is the gravitational force, and as shown in the coordinate system of Figure 6.6, this force and the rock's displacement are both along the $-y$ direction. We have $F_{grav} = -mg$ where the negative sign indicates that the direction of this force is downward. This is a case of one-dimensional motion with the force parallel to the displacement, so we can apply Equation 6.4 to find the work done on the rock. We can then use the work–energy theorem to find the final kinetic energy and speed of the rock.

SKETCH THE PROBLEM

Figure 6.6 shows the problem. We have also drawn in a coordinate system along with a free-body diagram for the rock.

IDENTIFY THE RELATIONSHIPS

The initial position of the rock is at $y_i = h$, and the final location is at ground level, $y_f = 0$, so the displacement of the rock is

$$\Delta y = y_f - y_i = 0 - h = -h$$

The work done on the rock is equal to the force times the displacement (Eq. 6.4):

$$W = F_{grav}\,\Delta y = (-mg)(-h) = mgh$$

Using the work–energy theorem (Eq. 6.11), we can write

$$W = \Delta KE = KE_f - KE_i$$

SOLVE

(a) The rock is initially at rest ($v_i = 0$), so its initial kinetic energy is $KE_i = \tfrac{1}{2}mv_i^2 = 0$, which leads to

$$W = mgh = \Delta KE = KE_f - KE_i = KE_f - 0$$

$$KE_f = \boxed{mgh} \tag{1}$$

Notice that because the force and displacement are parallel (both downward), W is positive.

(b) The final speed of the rock v_f is related to its final kinetic energy by (Eq. 6.10):

$$KE_f = \tfrac{1}{2}mv_f^2$$

Inserting the result for KE_f from Equation (1) and solving for v_f gives

$$W = mgh = \tfrac{1}{2}mv_f^2$$

$$v_f^2 = 2gh$$

$$v_f = \boxed{\sqrt{2gh}} \tag{2}$$

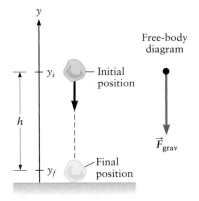

Figure 6.6 Example 6.2. When a rock falls vertically, the gravitational force is parallel to the rock's displacement. The work done by gravity on the rock is thus positive, with $W = mgh$.

---

[1]The calorie originated in studies of heat and temperature, as we'll describe in Chapter 14.

These results could also have been obtained using the methods for dealing with forces and acceleration that we developed in Chapter 3. Because the acceleration is constant, the displacement and velocity are related by (Eq. 3.4):

$$v_f^2 = v_i^2 + 2a(y_f - y_i) \tag{3}$$

The acceleration of the rock is $a = -g$, and (as we have already noted) $v_i = 0$, $y_i = h$, and $y_f = 0$. Inserting into Equation (3) gives

$$v_f = \sqrt{2gh}$$

This result is precisely the same as we found using the work–energy approach (Eq. 2). For this particular problem, both approaches are straightforward to carry out, but we'll soon encounter cases that are much easier to solve with the work–energy method.

CONCEPT CHECK 6.2 | Dependence of the Kinetic Energy on Mass and Speed

Consider a rock of mass m traveling at speed v. Suppose the mass is doubled and the speed is cut in half. Which of the following statements is true?
 (a) The kinetic energy is the same.
 (b) The kinetic energy increases.
 (c) The kinetic energy decreases.

EXAMPLE 6.3 | Speed of a Skier

Consider a person of mass m skiing down an extremely icy, frictionless ski slope. This is a specially constructed ski slope having the shape of an inclined plane of angle α as shown in Figure 6.7. If the skier starts from an initial height h, what is the work done by gravity on the skier while she skis to the bottom of the slope? Use this result to find the speed of the skier when she reaches the bottom, assuming she starts from rest.

RECOGNIZE THE PRINCIPLE

From Equation 6.5 and Figure 6.2, the work done by gravity on the skier equals the component of the gravitational force along the slope (parallel to the displacement) multiplied by the displacement. The force makes an angle θ with the displacement (see Fig. 6.7). This angle is *not* the same as the angle of the incline α.

SKETCH THE PROBLEM

Figure 6.7 shows the problem. Because the motion of the skier is along the incline, we take the coordinate axis to be parallel to the slope, with the positive x direction downward, parallel to the displacement of the skier.

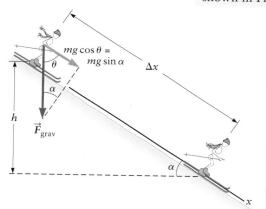

Figure 6.7 Example 6.3. The gravitational force is not parallel to this skier's displacement. The work done by gravity on the skier is equal to the component of the gravitational force along the slope multiplied by the skier's displacement.

IDENTIFY THE RELATIONSHIPS

The work done by the gravitational force is (Eq. 6.5)

$$W = F_{grav}\, \Delta x \cos \theta \tag{1}$$

The angle θ the force makes with the direction of the skier's displacement is related to the angle of the slope by $\theta = 90° - \alpha$ because they are two of the interior angles of a right triangle. The displacement of the skier Δx is the distance traveled along the slope. From the trigonometry in Figure 6.7, we have

$$\frac{h}{\Delta x} = \sin \alpha$$

$$\Delta x = \frac{h}{\sin \alpha}$$

SOLVE

Inserting this result for Δx into Equation (1) and using $\theta = 90° - \alpha$, we find

$$W = F_{\text{grav}} \, \Delta x \cos \theta = mg\left(\frac{h}{\sin \alpha}\right)\cos(90° - \alpha)$$

This result can be simplified if we use the trigonometric identity $\cos(90° - \alpha) = \sin \alpha$:

$$W = mg \frac{h}{\sin \alpha} \cos(90° - \alpha) = mg \frac{h}{\sin \alpha} \sin \alpha = \boxed{mgh} \qquad (2)$$

To find the final speed of the skier, we use the work–energy theorem and the fact that her initial speed and hence also her initial kinetic energy are zero:

$$W = KE_f - KE_i = \tfrac{1}{2}mv_f^2$$

Using our result for W ($= mgh$) and solving for v_f, we find

$$W = mgh = \tfrac{1}{2}mv_f^2$$

$$v_f = \boxed{\sqrt{2gh}}$$

What does it mean?
Our result for the work W done by gravity in Equation (2) has *precisely* the same functional form as the work done on the rock in Example 6.2. The work done by the gravitational force on a rock that falls vertically through a height h (Example 6.2) is precisely equal to the work done by gravity on a skier who moves along a slope through the same vertical height h. In the next section, we'll see that this result is no accident and that it has some profound consequences.

Work, Energy, and Amplifying Forces

In Chapter 3, we encountered a device called the block and tackle and showed how it can amplify forces. Figure 6.8 shows a block and tackle that amplifies forces by a factor of two. We saw in Chapter 3 that if a person applies a force F to the rope, the tension in the rope is $T = F$. Because the pulley is suspended by two portions of the rope, the upward force on the pulley is $2T$ and the pulley exerts a total force $2T$ on the object to which it is connected, that is, the crate in Figure 6.8. Let's now consider how this process of force amplification affects the work done by the person. Suppose the person lifts his end of the rope through a distance L. That will raise the pulley by half that amount, that is, a distance of $L/2$. You can see why by noticing that when the pulley moves upward through a distance $L/2$, the sections of the rope on both sides become shorter by this amount, so the end of the rope held by the person must move a distance L.

When the pulley moves upward by a distance $L/2$, the crate is displaced by the same amount. The work done *by the pulley on the crate* is equal to the total force of the pulley on the crate ($2T$) multiplied by the displacement of the crate, which is $L/2$:

$$W_{\text{on crate}} = 2T(L/2) = TL$$

At the same time, the person does work on the end of the rope since he exerts a force $F = T$ and the displacement of the end of the rope is L. The work done *by the person on the rope* is equal to the force that he exerts on the rope (T) multiplied by the displacement of the rope, which is L, so

$$W_{\text{on rope}} = FL = TL$$

Thus, the work done on the rope is precisely equal to the work done on the crate.

Figure 6.8 The person applies a force T to the rope. The block and tackle amplifies this force, and the total force applied to the crate by the rope is $2T$. However, when he moves the end of the rope a distance L, the crate moves a distance of only $L/2$.

There are several important messages from this example.

1. The work done by the person is effectively "transferred" to the crate.
2. Forces can be amplified, but work cannot be increased in this way. The force exerted by the person is increased by the block and tackle such that the force on the crate is amplified by a factor of two. However, the work is *not* amplified.
3. We'll see that this result—that work cannot be amplified—is a consequence of the principle of conservation of energy.

These points are all in accord with our observation in Chapter 3 that you "cannot get something for nothing." There, we also observed that pulleys can amplify a force, but they always do so by decreasing the associated displacement. Now we can go a bit further. According to the work–energy theorem, $W = \Delta KE$, which suggests we can "convert" or "trade" work to kinetic energy and vice versa. That work cannot be amplified suggests that this trade does not increase the amount of energy that is available or possible. These qualitative ideas are leading us to a very important fundamental principle of physics, the principle of conservation of energy. We'll spend the rest of this chapter exploring and applying that principle.

6.3 | POTENTIAL ENERGY

In Examples 6.2 and 6.3, we calculated the work done by the gravitational force in two different situations. In one case (Example 6.2), described by the path in Figure 6.9A, a rock of mass m fell vertically from a height h to ground level ($y = 0$), and we found that the work done by gravity on the rock is $W = mgh$. In the second case (Example 6.3), a skier of mass m skied down a slope, starting from an initial height h above the bottom of the slope, and we found that the work done by gravity is given by $W = mgh$. So, the work done by gravity was precisely the same, even though the rock and the skier followed very different paths. In fact, when an object of mass m follows *any* path that moves through a vertical distance h, the work done by the gravitational force is *always* equal to $W = mgh$. This amazing fact is connected with the notion of *potential energy*.

Figure 6.9 shows several different paths that might be followed by an object as it moves between the same initial and final positions. This figure is designed to illustrate that the work done by the gravitational force as an object moves from a particular initial location (*i*) to a particular final location (*f*) is completely *independent of the path* the object takes to move from *i* to *f*. This statement is true for all conceivable paths (only a few of which are shown in Fig. 6.9) and can be understood as follows. An object near the Earth's surface has a potential energy (*PE*) that depends only on the object's height h. The *PE* thus depends only on where the object is located and does not depend on how the object got there. In Figure 6.9, this potential energy is a result of the gravitational force between the Earth and the object. Strictly speaking, both the force and the potential energy are properties of the "system" that is composed of the Earth and the object. Although we will often speak of the gravitational potential energy of an object, it should be understood that this *PE* is a property of the object and Earth together. A general feature of potential energy is that it is always a property of the *system* of particles or objects involved in the underlying force.

The potential energy associated with a particular force is related to the work done by that force on an object as the object moves from one position to another. If this work is W, the change in the potential energy ΔPE is defined as

$$\Delta PE = PE_f - PE_i = -W \tag{6.12}$$

Since W is a scalar, potential energy is also a scalar. For the case of the gravitational force, we can thus find the change in the potential energy by calculating the work done in moving from an initial height y_i to a final height y_f. The work done is independent of the path, so we might as well use the simple path sketched in Figure

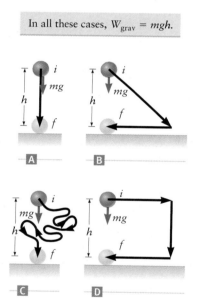

In all these cases, $W_{grav} = mgh$.

Figure 6.9 The work done by gravity is independent of the path taken. If the initial and final points are the same, the work done is the same. In addition, the work done by gravity on objects near the Earth's surface, W_{grav}, depends only on the change in height h. In all these examples, the work done by gravity is the same because the change in height is the same.

6.10 for the calculation of W. This path moves from our initial point to the final point in two steps. In the first step, we start at y_i and move downward (vertically) to the height y_f, while the second step moves horizontally to the final point. The work done by gravity during the second step is zero because the gravitational force and the object's displacement are perpendicular. We are thus left only with the work done during the first step, which is

$$W = F_{grav} \, \Delta y$$

The force is $F_{grav} = -mg$ (since the gravitational force is downward, along the $-y$ direction), and the displacement is $\Delta y = y_f - y_i$, so

$$W = F_{grav} \, \Delta y = (-mg)(y_f - y_i) \tag{6.13}$$

Using our relation for potential energy in Equation 6.12, the change in the potential energy is

$$\Delta PE = PE_f - PE_i = -W$$

$$\Delta PE = mg(y_f - y_i) \tag{6.14}$$

Another way to write this result is to say that the initial potential energy is $PE_i = mgy_i$ and the final potential energy is $PE_f = mgy_f$. Or, we could simply say that the gravitational potential energy of the object when it is at a height y is

$$PE_{grav} = mgy \tag{6.15}$$

Our derivation of Equation 6.15 used the fact that $F_{grav} = -mg$, so this result for the gravitational potential energy applies only for objects near the surface of the Earth. In Section 6.4, we'll consider the gravitational potential energy of planets, moons, and other objects in the solar system.

Potential Energy Is Stored Energy

The result for the gravitational potential energy in Equation 6.15 tells us that the potential energy increases as the object is moved higher and decreases when the object is moved lower, which agrees with our intuition. If an object is moved to a greater height, it is capable of producing a greater kinetic energy if it were to fall back to a lower height, where its potential energy is smaller. This example also shows that potential energy is *stored energy*. By "allowing" an object to fall, the energy stored as gravitational potential energy can be converted to kinetic energy. We can increase the gravitational *PE* of an object by increasing its height, that is, by increasing y in Equation 6.15. This process is sketched in Figure 6.11. As the object moves higher, the work done *by gravity* is negative because the gravitational force is directed downward while the displacement of the object is upward. Now, the object moves higher because there is another force acting on it. We call this an "external" force F_{ext}, and it might, for instance, be produced by your hand as you lift the object. This external force is directed upward, so the work done by F_{ext} on the object is positive. In fact, the work done by this external force, W_{ext}, is equal in magnitude but opposite in sign compared to the work done by gravity. Hence, we have stored an amount of energy W_{ext} as gravitational potential energy. Moreover, we can recover this energy by letting the object fall back down to its initial height, thus gaining kinetic energy.

Potential Energy and Conservative Forces

So far, our discussion of potential energy has been concerned mainly with gravitational potential energy, which is given by the potential energy function $PE = mgy$ (Eq. 6.15). Potential energy can also be associated with other forces, besides gravity. Such forces are described by different potential energy functions, and, like gravity, they can be used to store energy as potential energy. All such forces associated with

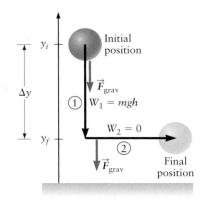

Figure 6.10 The work done by gravity along part 1 of this path is $W_1 = mgh$, whereas the work along part 2 is $W_2 = 0$ because the gravitational force and object's displacement are perpendicular along this path.

Gravitational potential energy for an object near the Earth's surface

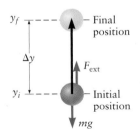

Figure 6.11 An external agent (perhaps a person's hand) lifts an object by exerting a force F_{ext} on the object. This agent does an amount of work $W_{ext} = F_{ext} \, \Delta y$ on the object. This energy is stored as potential energy of the object.

potential energy functions are called ***conservative forces***. Other types of forces do not have potential energy functions and are called ***nonconservative forces***. We'll spend the rest of this section learning how to apply and understand the gravitational potential energy in Equation 6.15. We then explore how these ideas apply to other types of conservative and nonconservative forces in subsequent sections.

We have introduced several new and important concepts connected with potential energy, and it is worth summarizing them now.

Key Facts about Potential Energy and Conservative Forces

- Potential energy is a result of the force(s) that act on an object. Since forces always come from the interaction of two objects, *PE* is a property of the objects (the "system") involved in the force.
- Potential energy is energy that an object or system has by virtue of its *position*.
- Potential energy is *stored energy*; it can be converted to kinetic energy.
- Potential energy is a scalar; its value can be positive or negative, but it does not have a direction.
- Forces that are associated with a potential energy (and hence have a potential energy function) are called *conservative forces*.
- Some forces, such as friction and air drag, are *nonconservative forces*. These forces do not have potential energy functions and cannot be used to store energy.

Potential Energy, the Work–Energy Theorem, and Conservation of Energy

Let's now reexamine the work–energy theorem and consider how potential energy fits into our ideas about work and energy. We first wrote the work–energy theorem as (compare with Eq. 6.11)

$$W = \Delta KE = KE_f - KE_i \tag{6.16}$$

Here, W is the work done by *all* the forces acting on the object of interest. In many cases, some of or all these forces are associated with a potential energy. If so, the potential energy associated with a particular force is related to work done by that force, with $\Delta PE = -W$ (Eq. 6.12). If *all* the work is done by a single type of conservative force, such as gravity, in Equation 6.16 we can replace the work done by that force with the negative of the change in the potential energy associated with the force. Then,

$$W = -\Delta PE = -(PE_f - PE_i) = KE_f - KE_i$$

which can be rearranged to read

Conservation of mechanical energy

$$KE_i + PE_i = KE_f + PE_f \tag{6.17}$$

This version of the work–energy theorem applies to situations in which all the forces are conservative forces, and it contains potential energy and kinetic energy together. In words, Equation 6.17 says that the sum of the potential and kinetic energies of an object is *conserved*. That is, the sum of the potential and kinetic energies when the object is at its initial location (*i*) is equal to the sum of the potential and kinetic energies at the final location (*f*). This sum of the potential and kinetic energies is called the ***total mechanical energy***.

Equation 6.17 is thus a statement of the ***conservation of mechanical energy***. In our derivation, we assumed only one type of force was acting on the object. This result also applies when many forces are involved as long as they are all conservative forces; that is, as long as all the forces possess a potential energy function. Frictional forces do not satisfy this requirement, and we'll learn how to deal with them in Section 6.5. Nevertheless, in many situations all the forces in a problem are connected with potential energy. In these cases, the principle of conservation of mechanical energy in Equation 6.17 is a very powerful tool for understanding, analyzing, and predicting motion.

Conservation of Mechanical Energy and the Speed of a Snowboarder

Let's apply the conservation of energy relation in Equation 6.17 to analyze the motion of a snowboarder who is sliding down an extremely icy (frictionless) hill (Fig. 6.12). If she starts from a height $h = 20$ m above the bottom of the hill with an initial speed $v_i = 15$ m/s, how fast will she be traveling at the bottom of the hill?

The only forces acting on the snowboarder are the force of gravity and the normal force from the hill. (Recall that the hill is very slippery and the frictional force is zero.) The normal force is perpendicular to the displacement of the snowboarder, so it does no work on her. The only remaining force is gravity, which is a conservative force, so the mechanical energy of the snowboarder is conserved. As the snowboarder moves down the hill, her potential energy is converted to kinetic energy, resulting in a larger velocity at the bottom of the hill. Applying Equation 6.17, we have

$$KE_i + PE_i = KE_f + PE_f$$

$$\tfrac{1}{2}mv_i^2 + mgy_i = \tfrac{1}{2}mv_f^2 + mgy_f \qquad (6.18)$$

where v_f is her velocity at the bottom of the hill and y_f is her height at the bottom. If we place the origin of the y axis at the bottom of the hill, then $y_i = h$ and $y_f = 0$. Inserting these values into Equation 6.18 and solving for v_f, we find

$$\tfrac{1}{2}mv_i^2 + mgh = \tfrac{1}{2}mv_f^2 + mg(0)$$

$$\tfrac{1}{2}v_i^2 + gh = \tfrac{1}{2}v_f^2$$

$$v_f = \sqrt{v_i^2 + 2gh} = \sqrt{(15 \text{ m/s})^2 + 2(9.8 \text{ m/s}^2)(20 \text{ m})} = 25 \text{ m/s}$$

Notice that v_f does *not* depend on the angle of the hill; it only depends on the height of the hill. That is because the gravitational potential energy depends only on the snowboarder's height, and changes in this potential energy are independent of the path she follows. So, the final speed will be the same for the two different hills in Figure 6.12, provided they have the same height.

Charting the Energy

A useful way to illustrate the conservation of energy is with bar charts. The bar charts in Figure 6.12C show the kinetic and potential energies of the snowboarder at the start and the end of the slope. These charts show how the initial potential energy of the snowboarder is converted to kinetic energy. Notice also how the sum of the kinetic and potential energy is same at the start and end, which is just another way to represent the conservation of energy.

Why Is the Principle of Conservation of Energy Useful?

Using conservation of energy principles made the calculation with the snowboarder in Figure 6.12 very straightforward. For the hill with the simple shape in Figure 6.12A, we could have done this problem in Chapter 4 by applying Newton's second law ($\sum \vec{F} = m\vec{a}$). Because this hill is just a ramp, the component of the gravitational force along the hill is constant; the acceleration is therefore also constant, so we could have used our relations for motion with constant acceleration to find the final velocity. We did not really need to use conservation of energy principles as expressed by Equation 6.17 to solve this problem. Unfortunately(!), most snowboard hills do not have such a simple shape. The hill in Figure 6.12B has a much more interesting and realistic shape. The slope of this hill is not constant, as it undulates several times as the snowboarder travels to the bottom. Because the hill's angle is not constant, the acceleration is definitely not constant and the methods we employed in Chapter 4 will not work here. However, this complicated hill can easily be handled with the conservation of energy approach. To find the final velocity of the snowboarder, we only need to know the height of the hill; the shape of the hill

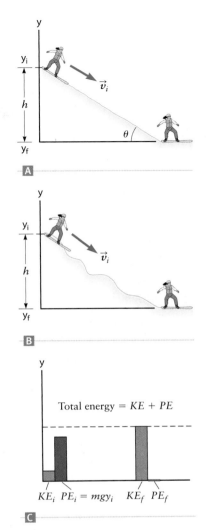

Figure 6.12 **A** and **B** These two hills have the same height h, so even though they have very different shapes, the work done by gravity in the two cases is the same: $W[\text{hill } \mathbf{A}] = W[\text{hill } \mathbf{B}]$. The change in the potential energy is therefore the same in the two cases ($\Delta PE_a = \Delta PE_b$). **C** These bar charts show the initial and final kinetic and potential energies of the snowboarder.

does not have any effect on the final velocity, and our calculation thus applies to any hill, regardless of its shape. This example illustrates the power of the conservation of energy approach; this approach will allow us to tackle many problems that cannot be handled using the methods of Chapter 4.

EXAMPLE 6.4 | Roller-Coaster Physics

Consider the roller coaster in Figure 6.13. The car starts at position *A* with an initial height $h_A = 20$ m and speed $v_A = 15$ m/s. Find **(a)** the speed of the roller coaster when it reaches the top of the track at position *B*, where $h_B = 25$ m, and **(b)** the speed of the roller coaster when it reaches the bottom of the track, where $h_C = 0$. **(c)** Make a qualitative sketch of the roller-coaster car's kinetic energy and potential energy as a function of position along the track. Ignore the rotational motion of the wheels and assume all frictional drag on the roller coaster is negligible.

RECOGNIZE THE PRINCIPLE

We follow the steps outlined in the problem-solving strategy on conservation of energy. Step 1: The object of interest is the roller-coaster car; its mechanical energy is conserved because the only force that does work on the car is gravity, which is a conservative force.

SKETCH THE PROBLEM

Step 2: Figure 6.13 shows the initial position of the car along with the other positions of interest (points *B* and *C*). We take the origin of the *y* axis to be at the bottom of the track.

IDENTIFY THE RELATIONSHIPS

Step 3: We next apply the conservation of energy relation (Eqs. 6.17 and 6.18) and then (step 4) solve for the speed at *B* and *C*.

Figure 6.13 Example 6.4.

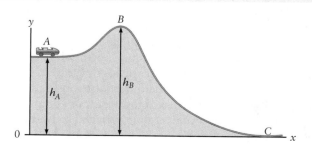

SOLVE

(a) Applying the conservation of energy relation (Eq. 6.17) at locations A and B gives

$$KE_A + PE_A = KE_B + PE_B$$

$$\tfrac{1}{2}mv_A^2 + mgh_A = \tfrac{1}{2}mv_B^2 + mgh_B$$

We now see that the mass of the roller coaster cancels. Solving for v_B, we find

$$\tfrac{1}{2}v_A^2 + gh_A = \tfrac{1}{2}v_B^2 + gh_B$$

$$v_B^2 = v_A^2 + 2gh_A - 2gh_B = v_A^2 + 2g(h_A - h_B)$$

$$v_B = \sqrt{v_A^2 + 2g(h_A - h_B)}$$

Inserting the given values of v_A, h_A, and h_B leads to

$$v_B = \sqrt{(15 \text{ m/s})^2 + 2(9.8 \text{ m/s}^2)(20 \text{ m} - 25 \text{ m})} = \boxed{11 \text{ m/s}}$$

(b) We again use Equation 6.17, but now we apply it at locations A and C, which leads to

$$v_C = \sqrt{v_A^2 + 2g(h_A - h_C)} = \sqrt{(15 \text{ m/s})^2 + 2(9.8 \text{ m/s}^2)(20 \text{ m} - 0)} = \boxed{25 \text{ m/s}}$$

(c) Figure 6.14 shows the qualitative behavior of the kinetic and potential energies as functions of x, the position along the track. The kinetic energy is largest at the lowest point on the track (C), while the potential energy is largest when the roller-coaster car is at the highest point (B). The sum $KE + PE$ is the total mechanical energy of the car and is constant throughout.

What does it mean?
The value of the final velocity v_C is independent of the shape of the roller-coaster track (compare with the snowboarder in Fig. 6.12) and depends only on the initial and final heights. However, this result assumes the roller coaster will actually make it past the top of the track. If the height at location B were made much larger, the roller coaster would not make it to the top of the track, and hence not be able to travel to location C.

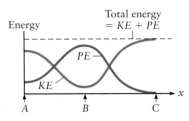

Figure 6.14 Behavior of the kinetic and potential energies of the roller-coaster car in Example 6.4.

CONCEPT CHECK 6.3 | Roller Coasters and Conservation of Energy

Consider the roller coaster in Example 6.4. How would you use conservation of energy ideas to calculate the value of h_B for which the roller coaster would barely make it over the highest point of the track?

EXAMPLE 6.5 | Driving through a Loop-the-Loop Track

The author's daughter likes to relax by driving fast. One particularly relaxing case involves loop-the-loop tracks like the one shown in Figure 6.15. Although few roads for automobiles are constructed with this shape, it is a common design for roller coasters. The middle of the track has a circular shape with a radius of $r = 15$ m, and the car enters and leaves along the horizontal portions at ground level. If a driver wants to coast all the way through without losing contact with the track when she reaches the top, what is the smallest velocity she can have when she enters the track at point A? As was the case with the roller coaster in Figure 6.13, ignore the rotational motion of the car's wheels and assume all friction is negligible.

RECOGNIZE THE PRINCIPLE

The only force that does work on the car is gravity (because the normal force exerted by the track is always perpendicular to the car's path), so the car's mechanical energy is conserved. We therefore follow the problem-solving strategy for problems in which energy is conserved as applied (step 1) to the car. Step 2: Figure 6.15 shows the car at

Free-body diagram for roller coaster on top of track

Figure 6.15 Example 6.5. A car enters this loop-the-loop track from the left at point A and coasts all the way through. The middle portion of the track has a circular shape, with radius r.

the initial point of its trip (point A). You should suspect that the most dangerous part of the trip will be when the car is upside down at the top of the track (point B), so we take this point as the final position. We can then follow steps 3 and 4 of the problem-solving strategy (see the solution below) to find the speed of the car at the top of the track v_B. However, just knowing v_B will not tell us if the car will stay on the track. For the car to stay on the track at point B, the normal force N exerted by the track on the car must be greater than zero. If the car has a very high speed at the top of the track, then N will be large. As this speed is reduced, N decreases. Eventually, if v_B is too small, the normal force is zero, signaling that the car has just lost contact. We can use the condition that N = 0 to find the minimum velocity at the top and then use conservation of energy principles to find the corresponding velocity at the bottom of the track at point A.

SKETCH THE PROBLEM

Figure 6.15 shows the car at its initial and final positions, along with a free-body diagram.

IDENTIFY THE RELATIONSHIPS

There are two forces acting on the car at point B, the normal force exerted by the track and the gravitational force, and both are directed downward. For an object undergoing uniform circular motion, the total force must be directed toward the center of the circle and have a magnitude mv^2/r. We thus have

$$\frac{mv_B^2}{r} = N + mg$$

Since the car is just barely staying on the track, the normal force will (as just discussed) be very small, so we can take N = 0. Solving for v_B then gives

$$v_B = \sqrt{gr}$$

This is the speed at the top that will just keep the car on the track. We now use the conservation of energy condition to work back and find the corresponding speed at the bottom of the track. Using conservation of mechanical energy, we have

$$PE_A + KE_A = PE_B + KE_B$$

$$mgy_A + \tfrac{1}{2}mv_A^2 = mgy_B + \tfrac{1}{2}mv_B^2 \tag{1}$$

If we take y = 0 at the bottom of the track, then $y_A = 0$ and $y_B = 2r$ (see Fig. 6.15).

SOLVE

Inserting these values of y_A and y_B and our result for v_B into Equation (1) gives

$$\tfrac{1}{2}mv_A^2 = mgy_B + \tfrac{1}{2}mv_B^2 = mg(2r) + \tfrac{1}{2}m(gr) = \tfrac{5}{2}mgr$$

and solving for v_A gives

$$v_A^2 = 5gr$$

$$v_A = \sqrt{5gr}$$

We then insert the value of the radius of the track to get

$$v_A = \sqrt{5gr} = \sqrt{5(9.8 \text{ m/s}^2)(15 \text{ m})} = \boxed{27 \text{ m/s}}$$

What does it mean?

This speed is approximately 60 mi/h, which is a typical speed for a roller coaster. This also suggests that this stunt should be possible for an ordinary automobile, but the author's daughter has (supposedly) not yet attempted it.

Projectile Motion and Conservation of Energy

It is instructive to apply conservation of energy principles to projectile motion and consider how the kinetic and potential energies vary as an object moves along its trajectory. Figure 6.16A shows an example we considered several times in Chapter 3: an object, such as a baseball, is thrown upward and then returns to the Earth. When the position y of the ball is plotted as a function of time t, we find the familiar parabolic relation between y and t sketched in Figure 6.16B. The behavior of the ball's energy as a function of time is shown in Figure 6.16C. Since the gravitational potential energy is $PE = mgy$, the potential energy also varies parabolically with time. Figure 6.16C shows the kinetic energy and the total mechanical energy, which is just the sum $PE + KE$. The total mechanical energy is *conserved*, so the sum $KE + PE$ is constant and does not change during the course of the motion. The kinetic energy thus varies as shown, in an "inverted" parabolic manner. Other problems of this type are explored in the end-of-chapter problems.

Only Changes in Potential Energy Matter

In our discussion of the gravitational potential energy, we were led to the result that $PE_{grav} = mgh$, where h is the height of the object (Eq. 6.15). However, in our calculations involving potential energy we have always ended up computing changes in the potential energy as an object moves from an initial position to a final position. Recall the conservation of energy condition (Eq. 6.17):

$$KE_i + PE_i = KE_f + PE_f$$

When this relation is used to compute either the final kinetic energy or the change in the kinetic energy, the answer depends only on the potential energy difference $\Delta PE = PE_f - PE_i$. Indeed, our initial definition of potential energy in Equation 6.12 was in terms of the work done by the associated force in moving an object from one place to another, and this original definition only referred to *changes* in potential energy.

The idea that only changes in potential energy are physically relevant can also be appreciated if we consider the variation of the potential energy as a ball is released from a height h above the ground and falls to ground level (Fig. 6.17). If we use a coordinate axis with the origin at ground level (the coordinate axis on the left in Fig. 6.17), we have $y_i = h$ and $y_f = 0$, so the change in potential energy is

$$\Delta PE = PE_f - PE_i = mgy_f - mgy_i = mg(0 - h) = -mgh \qquad (6.19)$$

On the other hand, we could just as well have chosen the origin of our coordinate system to be at the initial position of the ball (the axis shown on the right in Fig. 6.17). In this case, we would have $y_i = 0$ and $y_f = -h$, and the change in potential energy would be

$$\Delta PE = PE_f - PE_i = mgy_f - mgy_i = mg(-h - 0) = -mgh \qquad (6.20)$$

The change in the potential energy in Equations 6.19 and 6.20 is precisely the same, and we can use either coordinate system to calculate the speed of the ball just before it reaches the ground or, in fact, in any other calculation. Choosing a

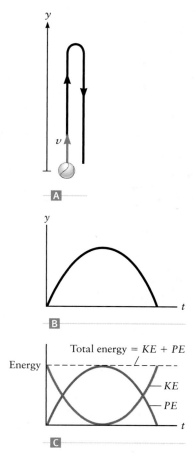

Figure 6.16 **A** When a ball is thrown upward (as a simple projectile), the height of the ball varies parabolically with time as sketched in **B**, the position versus time graph. **C** The KE, PE, and total mechanical energy vary with time as shown. Since the total mechanical energy is conserved, the sum $KE + PE$ is constant.

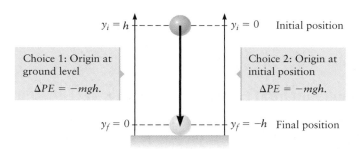

Figure 6.17 Only changes in the potential energy matter. Here, we consider the change in the gravitational potential energy using two different choices for the origin on the vertical axis. On the left, we take the origin at ground level, while on the right, the origin is at the initial position of the object. The change in the potential energy is the same; it does not depend on the choice of the origin.

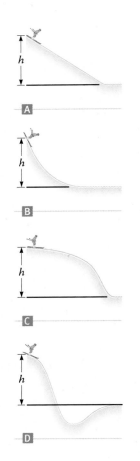

Figure 6.18 Concept Check 6.4.

different origin for our coordinate system is very similar to adding or subtracting a constant from the potential energy at all locations. The final results obtained from the application of the conservation of energy condition, Equation 6.17, depends only on changes in the potential energy, never on your choice of the "zero level" of the potential energy.

CONCEPT CHECK 6.4 | Skiing Down a Hill

Four expert skiers travel down the ski trails shown in Figure 6.18. All have the same initial height (at the start of the trail) and the same final height (at the bottom, on the far right in each sketch), and there is no friction on any of the trails. Which of the following statements is true? Assume air drag can be neglected.
  (1) The skier in Figure 6.18B probably arrives at the far right first.
  (2) The skiers all arrive at the bottom with the same speed.
  (3) The skier in Figure 6.18C probably arrives at the far right last.

6.4 | MORE POTENTIAL ENERGY FUNCTIONS

In the last section, we discussed the potential energy associated with the gravitational force near the Earth's surface and used this case to illustrate some the key ideas connected with potential energy. Let's now consider the potential energy functions associated with some other forces.

Gravitational Potential Energy in the Solar System

For objects near the surface of the Earth, the force of gravity exerted by the Earth is $F_{grav} = -mg$, and in the last section we showed that the potential energy associated with this force is $PE_{grav} = mgy$ (Eq. 6.15). In Chapter 5, we introduced Newton's law of gravitation, which describes the more general case of the gravitational force that exists between two objects separated by *any* distance. According to Newton's law of gravitation, the gravitational force between two objects of mass m_1 and m_2 separated by a distance r is (Eq. 5.16)

$$F_{grav} = -\frac{Gm_1m_2}{r^2} \tag{6.21}$$

The negative sign here indicates that the force is attractive. In a typical example, the mass m_1 might be the Earth, and m_2 could be a spacecraft that is either leaving the Earth or approaching the Earth as sketched in Figure 6.19. If this spacecraft is initially a distance r_i from the Earth and moves closer to a final distance r_f, the gravitational force does work on the spacecraft. According to the connection between work and potential energy (Eq. 6.12), there is a change in the gravitational potential energy of the spacecraft. While we will not give a detailed derivation here, the gravitational potential energy of two objects separated by a distance r is

$$PE_{grav} = -\frac{Gm_1m_2}{r} \tag{6.22}$$

The negative sign here means that the potential energy is lowered as the objects are brought closer together. The change in potential energy is then

$$\Delta PE_{grav} = -\frac{Gm_1m_2}{r_f} - \left(-\frac{Gm_1m_2}{r_i} \right) = -\frac{Gm_1m_2}{r_f} + \frac{Gm_1m_2}{r_i}$$

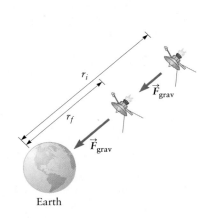

Figure 6.19 The potential energy associated with the gravitational force exerted by the Earth on this spacecraft changes as the separation r changes. This potential energy is given by Equation 6.22.

Gravitational Potential Energy: Launching a Satellite into Space

The result for the gravitational potential energy in Equation 6.22 can be used to analyze the motion of planets, moons, and satellites, including a calculation of the *escape velocity* of a satellite. Suppose you want to launch a projectile of mass m_p into distant space. NASA currently does so by attaching a rocket engine to the projectile. However, several scientists have considered the possibility of simply "firing" the projectile from a very large cannon.[2] What is the minimum initial velocity required so that a projectile will completely escape from the Earth? To calculate this velocity, we ignore the drag force due to the atmosphere, so the only force acting on the projectile after it leaves the cannon is the gravitational force. This is a conservative force, so we can apply the principle of conservation of energy to the projectile's motion. The instant after it leaves the cannon, the projectile is just above the Earth's surface with a speed v_i at a distance r_i from the Earth's center (Fig. 6.20). In this case, v_i is the unknown quantity we wish to calculate. The projectile eventually travels very far from the Earth; if it just barely escapes, the final speed is $v_f \approx 0$ and the final distance is very large, so $r_f \approx \infty$.

We now apply the conservation of energy condition with the potential energy given by Equation 6.22. Let M_{Earth} be the mass of the Earth and m_p the mass of the projectile. We then have

$$KE_i + PE_i = KE_f + PE_f$$

$$\frac{1}{2}m_p v_i^2 - \frac{GM_{\text{Earth}}m_p}{r_i} = \frac{1}{2}m_p v_f^2 - \frac{GM_{\text{Earth}}m_p}{r_f} \qquad (6.23)$$

The projectile begins on the surface of the Earth, so it is initially a distance $r_i = r_E$ from the center of the Earth, where r_E is the radius of the Earth. Since $r_f \approx \infty$, the final potential energy is

$$PE_f = -\frac{GM_{\text{Earth}}m_p}{r_f} \approx 0$$

Inserting these results along with the final speed $v_f \approx 0$ into the conservation of energy relation Equation 6.23, we find

$$\frac{1}{2}m_p v_i^2 - \frac{GM_{\text{Earth}}m_p}{r_E} = \frac{1}{2}m_p v_f^2 - \frac{GM_{\text{Earth}}m_p}{r_f} = 0$$

$$\frac{1}{2}m_p v_i^2 = \frac{GM_{\text{Earth}}m_p}{r_E}$$

$$v_i = \sqrt{\frac{2GM_{\text{Earth}}}{r_E}} \qquad (6.24)$$

which is called the escape velocity. This is the minimum velocity an object at the Earth's surface must have (ignoring air drag) to completely escape from the gravitational attraction of the Earth. Inserting values for the mass and radius of the Earth gives

$$v_i = \sqrt{\frac{2GM_{\text{Earth}}}{r_E}} = \sqrt{\frac{2(6.67 \times 10^{-11}\,\text{N}\cdot\text{m}^2/\text{kg}^2)(6.0 \times 10^{24}\,\text{kg})}{6.4 \times 10^6\,\text{m}}} = 1.1 \times 10^4\,\text{m/s}$$

which is approximately 24,000 mi/h.

The escape velocity is different for objects that are fired from different planets or from the Moon because it depends on the mass and radius of the planet or the Moon. Of course, to launch a real satellite, NASA should not rely on our simple calculation because the effect of air drag on such a projectile as it moves through the atmosphere cannot be ignored.

Elastic Forces and Potential Energy: Springs

One important type of potential energy is associated with springs and other elastic objects. You are probably familiar with a simple spring such as a tight coil of wire as sketched in Figure 6.21. In its "relaxed" state with no force applied to its end, the

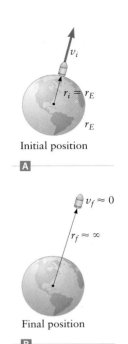

Figure 6.20 Calculating the escape velocity. **A** Initial position and speed of the projectile. **B** Final position and speed.

Escape velocity

---

[2]This possibility was also discussed in a story by Jules Verne.

spring rests as shown in Figure 6.21A. We might now pull on the spring with a force F_{pull}, causing the spring to stretch as shown in Figure 6.21B. When it is stretched, the spring itself exerts a force F_{spring} that opposes the stretching. Likewise, if you were to push on the spring with a force F_{push} so as to compress it (Fig. 6.21C), you would discover that the spring exerts a force back on you in opposition. The force exerted by the spring when it is either stretched or compressed has the form

Hooke's law

$$F_{spring} = -kx \qquad (6.25)$$

where x is the amount the end of the spring is displaced from its equilibrium position, which is assumed to be at $x = 0$. The quantity k is called the **spring constant**, and it has units of newtons per meter. The value of k depends on the properties of the spring (the number of coils, the stiffness of the wire, etc.) and is thus different for different springs. The negative sign in Equation 6.25 indicates that the force exerted by the spring always *opposes* the displacement of the end of the spring. If the end of the spring is pulled toward $+x$, the force exerted by the spring is directed along $-x$, whereas if the spring is pushed toward $-x$, the spring exerts a force along $+x$.

Equation 6.25, called **Hooke's law**, provides an accurate description of spring forces in a wide variety of situations. Notice that Hooke's law is not a "law" of physics in the sense of Newton's laws. Instead, Hooke's law is an empirical relation that experiments show to work well for most springs. Springs like the simple coils of wire sketched in parts (a) through (c) of Figure 6.21 are rather idealized; real springs can be far more complex. Hooke's law, however, applies to a variety of elastic objects. For example, a diving board is a type of spring: it exerts a force on a diver given by Equation 6.25, with x equal to the distance the diving board is deflected (Fig. 6.21D). A tennis racket also acts like a spring: the force exerted by the racket on a tennis ball is again given by an expression like Equation 6.25, with x now equal to the deflection of the strings (Fig. 6.21E).

Potential Energy Stored in a Spring

According to Hooke's law, the force exerted by a spring is proportional to x, the amount the string is stretched or compressed. This fact leads to some general results with regards to potential energy. To calculate the potential energy associated with the force exerted by a spring, we return to the definition of potential energy (Eq. 6.12),

$$\Delta PE = -W$$

with W now the work done by the Hooke's law force (Eq. 6.25). When you pull or push on a spring to stretch it or compress it as in part (B) or (C) of Figure 6.21, W is the work done *by the spring* on your hand. Because F_{spring} is not a constant force, we will calculate the work done using the graphical method explained in Section 6.1.

In Figure 6.5, we showed that the work done by a variable force is equal to the area under the force–displacement curve. This curve for a spring is shown in Figure 6.22. We assume the spring is initially in its relaxed state (not stretched or compressed) so that $x_i = 0$. If the spring is then stretched an amount $x = x_s$, the work done by the spring is the area under the curve $F_{spring} = -kx$. This area is just a triangle of height $F_s = kx_s$, and the area of this triangle is $\frac{1}{2}F_s x_s$ (see Fig. 6.22). Inserting the expression for F_{spring} from Hooke's law, the area of the triangle is $\frac{1}{2}F_s x_s = \frac{1}{2}kx_s^2$. We must now be careful to get the correct sign for the work done by the spring. In this example, the spring is being stretched, so according to Figure 6.21B the force exerted by the spring is directed to the left, whereas the displacement of the end of the spring is to the right. Since the force and the displacement are in opposite directions, the work done by the spring is negative, and

$$W = -\tfrac{1}{2}kx_s^2$$

Figure 6.21 The force exerted by a spring is always opposite to the displacement of the end of the spring. **A** When a spring is unstretched and uncompressed, the force exerted by the spring is zero. **B** This spring is stretched by pulling it to the right, so the force exerted by the spring is to the left. **C** The spring is compressed, and the force exerted by the spring is to the right. **D** and **E** Other objects, such as a diving board and a tennis racket, also obey Hooke's law (Eq. 6.25), with the force proportional to the displacement x.

Here, W is the work done by the spring as it is *stretched* an amount x_s starting from its relaxed state. From Figure 6.22, we see that the work done in *compressing* the spring is equal to the area under a similar triangular region, the triangle in the upper left in Figure 6.22. The work done by the spring when it is compressed is again negative because (as sketched in Fig. 6.21C) the force exerted by the spring is directed opposite to the displacement of the end of the spring. Hence, if a spring is stretched or compressed an amount x, the work done by the spring is

$$W_{spring} = -\tfrac{1}{2}kx^2$$

Using the definition of potential energy, we then get

$$PE_{spring} = \tfrac{1}{2}kx^2 \qquad (6.26)$$

Spring Forces and Potential Energy: A Recap

Our results for the elastic (spring) force in Equation 6.25 and potential energy in Equation 6.26 are collected in Figure 6.23, which shows how F_{spring} and PE_{spring} vary as a spring is stretched and compressed. The force exerted by the spring always opposes the displacement, so F_{spring} can be either positive or negative, depending on whether the spring is stretched or compressed. The elastic potential energy is smallest ($PE_{spring} = 0$) when a spring is in its relaxed state ($x = 0$), and PE_{spring} always increases as a spring is either stretched or compressed.

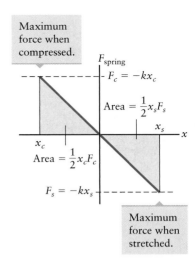

Figure 6.22 The work done by a variable force is equal to the area under the force–displacement graph. For a spring, this area is a triangle. When the spring is stretched, $x = x_s$ and W is the area of the triangle at the lower right. When the spring is compressed, $x = x_c$ and W is the area of the triangle at the upper left.

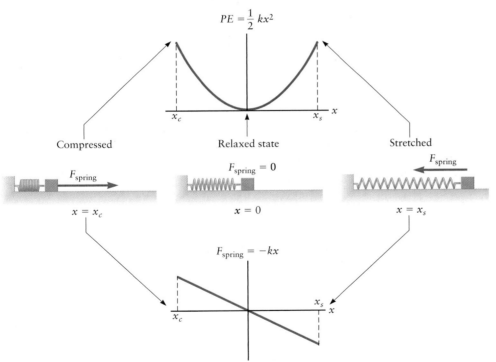

Figure 6.23 Variation of the potential energy and force as a spring is compressed and stretched.

EXAMPLE 6.6 | Force Exerted by a Bungee Cord

A bungee cord is a kind of spring, and the force exerted by the cord on a person attached to it is given by $F_{spring} = -kh$, where h is the amount the cord is stretched (Hooke's law, Eq. 6.25). Consider a person of mass $m = 75$ kg attached to a bungee cord with spring constant $k = 50$ N/m. When the jump is completed, the person is at rest, suspended by the cord (Fig. 6.24B). How much is the cord stretched at that time?

Figure 6.24 Example 6.6.
A The bungee cord in its relaxed state (neither stretched or compressed). **B** The bungee jumper is hanging at rest, suspended by the cord.

RECOGNIZE THE PRINCIPLE

At the end of the jump and after the person has stopped oscillating up and down, the person is at rest, so his acceleration is zero. Hence, according to Newton's second law, the total force must be zero.

SKETCH THE PROBLEM

Figure 6.24B shows the forces acting on the person when the cord is stretched: the force of gravity and the spring force from the bungee cord.

IDENTIFY THE RELATIONSHIPS

Applying Newton's second law for motion in the vertical direction for the person in Figure 6.24B, we have

$$\sum F = kh + F_{grav} = kh - mg = ma = 0 \qquad (1)$$

Notice that in this case the spring force is $+kh$, where h is the amount the bungee cord is stretched; the positive sign is here because this force is directed upward.

SOLVE

Solving Equation (1) for h, we get

$$kh = mg$$

$$h = \frac{mg}{k} = \frac{(75 \text{ kg})(9.8 \text{ m/s}^2)}{50 \text{ N/m}} = \boxed{15 \text{ m}}$$

What does it mean?
This result should be of interest to the bungee cord company. For safety reasons, it would want to be sure that their customer is suspended well above the ground.

EXAMPLE 6.7 | ℝ **Spring Constant of an Archer's Bow**

Consider a professional archer engaged in target practice. An archer's bow behaves like a spring and is described by Hooke's law. The force exerted by the bow on the arrow is proportional to the displacement of the bow string as shown in Figure 6.25. Suppose the archer is able to shoot an arrow of mass $m = 0.050$ kg with a velocity of 100 m/s just as it leaves the bow. What is the spring constant of her bow?

Figure 6.25 Example 6.7.
A A bow acts like a spring, and its potential energy is given by $PE = \frac{1}{2}kx^2$, where x is the distance the bowstring is displaced. **B** Bar charts showing the initial and final kinetic and potential energies of the arrow and bow.

RECOGNIZE THE PRINCIPLE

We want to apply the conservation of energy condition to the combination of the bow plus arrow. The initial state has the bow fully stretched (the bow string pulled back a distance x_i) and the arrow at rest. The final state has the bow in its relaxed state ($x_f = 0$) and the arrow leaving the bow with a velocity $v_f = 100$ m/s.

SKETCH THE PROBLEM

The problem is shown in Figure 6.25A.

IDENTIFY THE RELATIONSHIPS

Writing the conservation of energy condition with the elastic potential energy of the bow, we have

$$KE_i + PE_i = KE_f + PE_f$$

$$\tfrac{1}{2}mv_i^2 + \tfrac{1}{2}kx_i^2 = \tfrac{1}{2}mv_f^2 + \tfrac{1}{2}kx_f^2 \tag{1}$$

SOLVE

Inserting our values for the initial arrow velocity ($v_i = 0$) and final bow stretch ($x_f = 0$) into Equation (1) leaves us with an equation involving k and the initial bow stretch x_i. Solving for the spring constant k, we find

$$0 + \tfrac{1}{2}kx_i^2 = \tfrac{1}{2}mv_f^2 + 0$$

$$k = \frac{mv_f^2}{x_i^2} \tag{2}$$

We are given the final speed of the arrow v_f. To find the spring constant k, we need to also know the amount the bow stretches x_i, but this information was not given. Relying on intuition and experience, we estimate $x_i \approx 0.1$ m. A value of $x = 0.01$ m $= 1$ cm seems much too small, whereas $x = 1$ m is too large. Our estimate is in the middle. Evaluating Equation (2) for k then gives

$$k = \frac{mv_f^2}{x_i^2} = \frac{(0.050 \text{ kg})(100 \text{ m/s})^2}{(0.1 \text{ m})^2} = \boxed{5 \times 10^4 \text{ N/m}}$$

What did we learn?

Notice how we followed the general steps in our strategy for dealing with conservation of energy problems. We began with a picture and identified the initial and final states of the system (bow plus arrow). We then wrote the conservation of energy relation, Equation (1), and solved for the quantity of interest (k).

Figure 6.25B shows bar charts for the initial and final kinetic and potential energies. The initial potential energy is converted into the final potential energy, and the total energy $KE + PE$ is conserved.

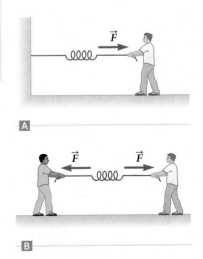

CONCEPT CHECK 6.5 | Spring Forces and Newton's Third Law

Figure 6.26 shows two identical springs. In both cases, a person exerts a force of magnitude F on the right end of the spring. The left end of the spring in Figure 6.26A is attached to a wall, while the left end of the spring in Figure 6.26B is held by another person, who exerts a force of magnitude F to the left. Which statement is true?

(1) The spring in Figure 6.26A is stretched half as much as the spring in Figure 6.26B.

(2) The spring in Figure 6.26A is stretched twice as much as the spring in Figure 6.26B.

(3) The two springs are stretched the same amount.

Figure 6.26 Concept Check 6.5.

Total Potential Energy with Multiple Forces

The conservation of energy relation tells us that

$$KE_i + PE_i = KE_f + PE_f$$

In all our problems so far, there has only been one type of force and hence only one type of potential energy function to deal with. Often, however, several different forces—such as a gravitational force and a spring force—may be acting together on an object. In these cases, we must be sure to account for the total potential energy from all forces. For example, if both gravity near the Earth's surface and a spring force are acting on an object, the total potential energy is

$$PE_{\text{total}} = PE_{\text{grav}} + PE_{\text{spring}} = mgh + \tfrac{1}{2}kx^2$$

This type of situation is illustrated in the next example.

EXAMPLE 6.8 | Potential Energy from a Spring Plus Gravity

Consider again the bungee jumper from Example 6.6. He is initially standing on the bungee platform with the cord unstretched (Fig. 6.27A). He then jumps from the platform and travels downward (Fig. 6.27B). How much is the cord stretched when the jumper reaches his lowest point? The mass of the jumper is $m = 75$ kg, and the spring constant of the bungee cord is $k = 50$ N/m.

RECOGNIZE THE PRINCIPLE

We again apply the principle of conservation of energy, but now we have to deal with the potential energy associated with both gravity and the bungee cord. Initially, this cord is in its relaxed state, so all the potential energy is from gravity. In the final configuration, the cord is stretched, so we have to include the contributions to the potential energy from both this "spring" and from gravity.

SKETCH THE PROBLEM

Figure 6.27 describes the problem. This sketch shows both the initial and final states of the system.

IDENTIFY THE RELATIONSHIPS

According to the conservation of energy condition,

$$KE_i + PE_{\text{spring},\,i} + PE_{\text{grav},\,i} = KE_f + PE_{\text{spring},\,f} + PE_{\text{grav},\,f}$$

where we have written the gravitational and cord (i.e., "spring") potential energies as separate terms. The jumper begins at rest ($v_i = 0$) and also ends at rest ($v_f = 0$), so the initial and final kinetic energies are both zero. The cord is initially relaxed, so the initial potential energy of the cord is zero. We take the origin ($y = 0$) at the initial position of the jumper, so the initial gravitational potential energy is also zero. Using the coordinate system in Figure 6.27, the final position of the jumper is at $y = y_f$. Inserting these values into the conservation of energy relation leads to

$$PE_{\text{spring},\,i} + PE_{\text{grav},\,i} = PE_{\text{spring},\,f} + PE_{\text{grav},\,f}$$

$$0 + 0 = \tfrac{1}{2}ky_f^2 + mgy_f$$

SOLVE

Solving for y_f, we get

$$\tfrac{1}{2}ky_f^2 = -mgy_f$$

$$y_f = -\frac{2mg}{k} = -\frac{2(75 \text{ kg})(9.8 \text{ m/s}^2)}{50 \text{ N/m}} = \boxed{-29 \text{ m}}$$

Figure 6.27 Example 6.8.
A This bungee jumper is initially at rest with the cord unstretched.
B At his lowest point, the bungee jumper's position is a distance y_f below the starting point.

The value of y_f is negative because the final position of the jumper is below the starting point at $y = 0$. Comparing this result with Example 6.6, we see that the jumper in Figure 6.27B is at a lower point than the person in Figure 6.24B. What are the velocity and acceleration of the jumper when he reaches the lowest point in Figure 6.27B? Is this lowest point a point of static equilibrium?

ⓧ Elastic Forces and the "Feeling" of Holding a Heavy Object

According to the definition of mechanical work (Eq. 6.4), W is zero if the displacement is zero, even though the force on the object might be very large. This definition is quite different from our "everyday" concept of work. Consider what happens when you hold an object in your hand (Fig. 6.28A). If the object is at rest, the displacement is zero, and, according to the physics definition of W, the work done by your hand on the object is zero. Hence, $W = 0$ if the object has a very small mass or even a very large mass. However, you also know intuitively that supporting a very massive object will eventually make your hand "tired," and in everyday language we would say that holding something requires an expenditure of "energy" by your hand. How can we reconcile this notion with the definition of mechanical work? The answer is indicated in Figure 6.28B. Your hand is flexible and is "compressed" a small amount, just like a spring, leading to an elastic force on the object. If your skin and muscles deform an amount x (Fig. 6.28B), the force exerted by your hand on the object is, according to Hooke's law, proportional to x.

If the "physics" work done on the object is zero, however, why do your muscles get tired? One reason is that your muscles consist of many parallel fibers, and when they apply a force, these fibers are constantly slipping relative to one another. Muscles also contain small motorlike molecules that pull and move each fiber relative to its neighbors to counteract this slipping (we say more about these motors in Section 6.8). Because the molecular motors exert a force on the muscle fibers as they move along a fiber, the motors do a nonzero amount of "physics" work, and the associated energy comes from chemical reactions in the muscle. You sense this expenditure of chemical energy in your muscles, and that is why you must exert "energy" to support an object.

When holding a massive object, your skin and muscles are compressed like a spring.

Figure 6.28 When you support an object with your hand, your muscles deform much like a spring, producing an elastic force on the object.

6.5 | CONSERVATIVE VERSUS NONCONSERVATIVE FORCES AND CONSERVATION OF ENERGY

In our derivation of the potential energy associated with gravity near the surface of the Earth (Eq. 6.15), we mentioned that W is independent of the path taken by the object. This path independence is an essential characteristic of the work done by a conservative force. The corresponding change in the potential energy must depend only on the object's initial and final positions. Changes in the potential energy, and hence in the value of W, are independent of the path taken. Many forces satisfy this requirement of path independence, including gravitation and spring (Hooke's law) forces. As already mentioned, such forces are called conservative forces.

Although potential energy is an extremely useful concept, some forces—called *nonconservative forces*—cannot be associated with a potential energy. Unlike conservative forces, the work done by a nonconservative force *does* depend on the path taken. In this section, we'll deal with nonconservative forces.

The work done by a conservative force is independent of the path.

The work done by a nonconservative force does depend on the path taken.

The Work Done by Friction Depends on the Path

Friction is a good example of a nonconservative force. We can see that friction is nonconservative by considering the work done by it as an object moves along the path sketched in Figure 6.29, where a crate is pushed across a level table from an initial location x_i to a final location x_f. This can be done by pushing the crate

Figure 6.29 The work done by friction is larger for path *B* than it is for path *A* because path *B* is longer. Since the work done by friction depends on the path taken, friction is a nonconservative force.

directly from x_i to x_f (path *A* in Fig. 6.29) or by taking an indirect path (path *B*). For path *A*, we can calculate the work done by friction in the usual way. Since we are dealing with kinetic friction, $F_{\text{friction}} = -\mu_K N$, where N is the normal force between the table and the crate, leading to

$$W_A = F_{\text{friction}} \Delta x = -\mu_K N(x_f - x_i)$$

Here, W_A is negative because friction opposes the motion. We can calculate the work done in moving along path *B* in the same way. If the total length of path *B* is Δx_B, the work done is

$$W_B = -\mu_K N \Delta x_B$$

The length of path *B* is much longer than that of path *A* (see Fig. 6.29), so W_A and W_B are certainly not equal. Hence, the work done by friction depends on the path taken by the object. From Equation 6.12, it is therefore impossible to have a single value for the change in a corresponding potential energy, so we conclude that such a potential energy cannot exist. It is thus impossible to have a potential energy associated with the frictional force. Other nonconservative forces have the same property: the work done by a nonconservative force depends on the path taken by the object, and for this reason it is not possible to have a corresponding potential energy function.

The Work–Energy Theorem Revisited: Including Nonconservative Forces

If we cannot define potential energy for a nonconservative force, how do we think about conservation of energy in the presence of nonconservative forces? To answer this question, it is simplest to return to our original version of the work–energy theorem (Eq. 6.11),

$$W_{\text{total}} = \Delta KE = KE_f - KE_i \qquad (6.27)$$

which tells us that the change in kinetic energy is equal to the *total* work done on the object. This total work can be due to several different forces. Let's split it into two terms, one for the work done by all the conservative forces and another for the work done by the nonconservative forces:

$$W_{\text{total}} = W_{\text{con}} + W_{\text{noncon}}$$

According to the definition of potential energy (Eq. 6.12), the work done by the conservative forces is related to the change in the potential energy:

$$-W_{\text{con}} = \Delta PE = PE_f - PE_i$$

Collecting these results and inserting them into Equation 6.27 leads to

$$KE_i + PE_i + W_{\text{noncon}} = KE_f + PE_f \qquad (6.28)$$

Work–energy theorem including nonconservative forces

which looks just like our previous conservation of energy result (Eq. 6.17) but with an additional term on the left for the work done by nonconservative forces. Equation 6.28 is our general conservation of energy/work–energy theorem with nonconservative forces included. The sum of the kinetic and potential energies is the total *mechanical energy* of an object. According to Equation 6.28, the final mechanical energy ($KE_f + PE_f$) is equal to the initial mechanical energy ($KE_i + PE_i$) plus the work done by any nonconservative forces that act on the object.

EXAMPLE 6.9	Work Done by Friction on a Skier

A skier of mass $m = 60$ kg starts from rest at the top of a ski hill of height $h = 50$ m (Fig. 6.30). When she reaches the bottom, she has a speed $v_f = 25$ m/s. How much work is done by friction on the skier as she travels down the hill?

RECOGNIZE THE PRINCIPLE

Since the ski hill does not have a simple shape (it is not a simple inclined plane), we cannot calculate the work done by friction using the definition of work in Equation 6.4 ($W_{friction} = F_{friction} \, \Delta x$). Friction is the only nonconservative force in the problem, however, so $W_{friction} = W_{noncon}$, and we can find the work done by friction using the conservation of energy relation

$$KE_i + PE_i + W_{friction} = KE_f + PE_f \qquad (1)$$

SKETCH THE PROBLEM

Figure 6.30 shows the initial and final states of the skier.

IDENTIFY THE RELATIONSHIPS

We first rearrange Equation (1):

$$W_{friction} = KE_f + PE_f - KE_i + PE_i \qquad (2)$$

The potential energy of the skier is due to gravity; choosing the origin of the vertical axis at the bottom of the hill, we get $PE_i = mgh$ and $PE_f = 0$. The skier starts from rest, so the initial kinetic energy is zero and the final kinetic energy is $\frac{1}{2}mv_f^2$.

SOLVE

Inserting all this information into Equation (2) leads to

$$W_{friction} = \tfrac{1}{2}mv_f^2 - mgh = \tfrac{1}{2}(60 \text{ kg})(25 \text{ m/s})^2 - (60 \text{ kg})(9.8 \text{ m/s}^2)(50 \text{ m})$$

$$W_{friction} = \boxed{-1.1 \times 10^4 \text{ J}}$$

What does it mean?

The work done by friction is negative because the frictional force opposes the skier's motion along the hill. Friction thus causes the final speed of the skier to be less than it would have been without friction.

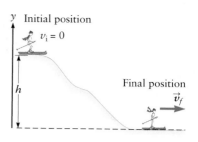

Figure 6.30 Example 6.9.

Conservation of Energy of a System

We have mentioned the concept of "conservation of energy" several times, in several different contexts. We now consider the principle of conservation of energy from a very general point of view. Suppose a system of particles or objects exert forces on one another as they move about. These forces may be conservative (perhaps gravity and elastic forces) or nonconservative (friction or air drag). While there are many forces involving particles within this system, let's assume there are no forces on any of these particles from outside the system so that the total energy of the system will be conserved. As the particles move around, we would expect some potential energy to be converted to kinetic energy (or vice versa). It is also possible for mechanical energy (i.e., elastic potential energy or kinetic energy) to be converted to another form of energy such as heat energy, electrical energy, or chemical energy that we describe later in this book. Nevertheless, the total energy of the entire system will be conserved.

Of course, some agent from outside this system may exert a force on one of the particles within the system. If this outside agent does an amount of work W_{ext} on the particle, the total energy of the system will change by an amount W_{ext}. This change in the system's energy might be positive or negative, depending on the sign of W_{ext}. In this way, energy can be exchanged between an external agent and the system. If a certain amount of energy is added to the system, the same amount of energy must be removed from the agent, so we again arrive at the general principle of *conservation of total energy.*

6.6 | THE NATURE OF NONCONSERVATIVE FORCES: WHAT IS FRICTION ANYWAY?

We have seen that when all the forces acting on an object are conservative forces, the total mechanical energy is conserved:

$$KE_i + PE_i = KE_f + PE_f$$

Intuitively, the idea that energy is conserved should not be surprising. If energy were not conserved, it would be possible to do some very incredible things—such as generate limitless amounts of energy and construct mechanical perpetual motion machines—and physicists have found no evidence that such things are possible.

We have also seen that when a nonconservative force such as friction is present, the principle of conservation of energy can be expressed as (Eq. 6.28)

$$KE_i + PE_i + W_{noncon} = KE_f + PE_f$$

or

$$\underbrace{(KE_f + PE_f)}_{\substack{\text{final} \\ \text{mechanical energy}}} - \underbrace{(KE_i + PE_i)}_{\substack{\text{initial} \\ \text{mechanical energy}}} = W_{noncon} \qquad (6.29)$$

In most situations, the work done by friction is negative because the frictional force is usually directed opposite to the displacement. So, Equation 6.29 tells us that in such cases the final mechanical energy is less than the initial mechanical energy. A simple illustration of this situation is given in Figure 6.31, which shows a block sliding along a rough surface. The block has some initial kinetic energy, so the initial mechanical energy is not zero. Friction causes the block to come to a stop, so the final kinetic energy and the final mechanical energy are both zero. Hence, the mechanical energy of the block is not conserved; by the end of this process, the mechanical energy of the block has changed by an amount W_{noncon}. Let's now consider precisely where the energy associated with W_{noncon} goes.

For the sliding block in Figure 6.31, the energy associated with W_{noncon} goes into heating up the surface of the block and the surface of the floor. This increase in the temperature of the two surfaces can be understood in terms of the motion of the atoms at the surfaces. All substances are composed of atoms in constant motion. The chemical bonds that hold the atoms together are analogous to springs, which are stretched and compressed as the atoms move and vibrate. When a substance is heated and its temperature is increased, these vibrations become larger in size and the atoms move faster. Since they move faster, the atoms have an increased kinetic energy, and there is also an increase in the potential energy of the chemical bonds/springs. Hence, the initial mechanical energy of the block in Figure 6.31 does not disappear, but shows up as an increase in the energy of the atoms at the surfaces. Even in the presence of friction, the *total* energy is still conserved.

This conclusion leads to another, and deeper, question. If the energy W_{noncon} has simply gone into the mechanical energy (kinetic and potential) of the atoms, is it possible to somehow return all this energy back to the block? That is, can we convert this added mechanical energy of the heated atoms back into the mechanical energy of the block? We'll see in Chapter 16 that it is possible to transfer *some* of the energy from the heated atoms into kinetic energy of the block. However, it is *not* possible to transfer *all* this energy back to the block.

This energy that flows in connection with a temperature difference is called *heat energy*, and it plays a central role in the branch of physics called *thermodynamics*, which we will explore in Chapters 14 through 16. Although mechanical energy is easily converted completely into heat as with the sliding block in Figure 6.31, it is not nearly as simple to convert heat energy back into mechanical energy. After all, if the block in Figure 6.31 were initially at rest and we then heated it, we would not expect the block to begin sliding along the floor!

Friction causes these surfaces to heat up, and their temperature increases.

Figure 6.31 When friction does work on a sliding block, the block's mechanical energy is converted into heat energy, resulting in a temperature increase at the surfaces of the block and the floor.

These observations also suggest why friction is called a "nonconservative" force. Conservative forces are associated with potential energy, and we have seen that potential energy is stored energy that can be "retrieved" and then converted freely into other forms of energy. The heat energy produced by friction cannot be retrieved in this manner, and it cannot all be transferred back to other forms of energy.

6.7 | POWER

In Section 6.2, we found that a block and tackle is able to amplify the force that can be applied to an object. However, this amplification comes at the cost of a reduction in the object's displacement. For the particular block and tackle in Figure 6.8, the force amplification factor is two and the displacement is reduced by the same factor. The work done on the object is therefore the *same* as the work done by the person as he pulls on the rope. With our understanding of the connection between work and energy, we see that this is just another way of stating that mechanical energy is conserved. We can view the person and the crate in Figure 6.8 as two "systems"; the block and tackle "transfers" energy from the person to the crate, and according to the principle of conservation of energy, the energy expended by the person—that is, the work done by the person—must equal the energy gained by the crate.

More complicated block-and-tackle arrangements can give much greater force amplification factors, so in principle a person could use such a device to lift an extremely massive object. While the force amplification factor can be very large, the displacement is always "deamplified" by the same factor. Hence, it may take a very long time to lift a massive object through an appreciable distance. Time enters into work–energy ideas through the concept of *power*. Consider the person in Figure 6.32 who is using a rope to lift a crate very slowly. Work is done on the crate as it moves from the floor to a height h, which changes the potential energy of the crate by an amount mgh. This energy comes from the person as he pulls on the other end of the rope and does an amount of work $W = mgh$ on the rope. If this work is expended during a time t, the average power exerted by the person is defined as

$$P_{ave} = \frac{W}{t} \qquad (6.30)$$

The units of power are joules per second; this unit has been given the name watt (W) after James Watt (1736–1819), a developer of the steam engine. In terms of the units we have

$$1 \text{ W} = 1 \text{ J/s}$$

According to Equation 6.30, power is the *rate* at which work is done. Since work is a way to transfer energy from one system to another, power is also equal to the energy output per unit time of a device. For example, the specifications for a motor or engine often include the maximum power the engine is able to produce. For historical reasons, the power of an automobile is usually given in terms of an odd unit called "horsepower." Presumably, 1 horsepower (hp) is equal to the power that a typical ("standard"?) horse is able to expend. You can convert between watts and horsepower using the relation

$$1 \text{ hp} = 745.7 \text{ W} = 745.7 \text{ J/s}$$

In addition to mechanical energy, there are other types of energy, including chemical energy and electrical energy, and they must be added to our conservation of energy relation for situations involving chemical reactions or electrical forces. The concept of power also applies to chemical and electrical processes and devices. Indeed, there is a power output or input associated with many of the electrical devices in your house. For example, a typical lightbulb is rated at 100 W = 100 J/s. This rating means that the lightbulb consumes 100 J of electrical energy for every second it is turned on. Likewise, devices such as computers and CD players also

$\vec{T}$

m

Crate

h

Figure 6.32 This person is expending a certain amount of power as he lifts the crate.

Table 6.1 Typical Values for the Power Output and Consumption of Some Common Appliances and Devices

DEVICE	POWER OUTPUT OR CONSUMPTION
Lightbulb	100 W
Personal computer	300 W
Automobile engine (small car)	100 hp = 7.5×10^4 W
Automobile engine (race car)	700 hp = 5.2×10^5 W
Portable CD player	20 W
Elite bicycle racer	400 W

have a "power rating." Table 6.1 lists the power output and consumption of a number of common devices. The distinction between power "consumption" and power "output" is important. For example, a lightbulb consumes a certain amount of electrical energy (for which you pay the utility company), and it outputs a certain amount of energy in the form of visible light along with a certain amount of energy as heat.

EXAMPLE 6.10 Comparing Mechanical and Electrical Energy

Consider a lightbulb rated for a power consumption of P_{bulb} = 100 W. This lightbulb consumes a certain amount of electrical energy in 1 second of use. Suppose we want to generate the same amount of energy by dropping the lightbulb from a height h to ground level. If the mass of the lightbulb is m = 0.050 kg, what is h?

RECOGNIZE THE PRINCIPLE

We apply conservation of energy principles to the lightbulb. The change in potential energy of the lightbulb is $\Delta PE = -mgh$. This energy must equal (in magnitude) the energy used by the lightbulb in 1 s.

SKETCH THE PROBLEM

Figure 6.33 describes the problem.

IDENTIFY THE RELATIONSHIPS

Using the definition of power (Eq. 6.30), the energy consumed by the lightbulb in time t is

$$W = P_{bulb}t$$

Setting this equal to the change in potential energy when the lightbulb falls gives

$$mgh = P_{bulb}t$$

SOLVE

Solving for h, we find

$$h = \frac{P_{bulb}t}{mg} = \frac{(100 \text{ W})(1.0 \text{ s})}{(0.050 \text{ kg})(9.8 \text{ m/s}^2)} = \boxed{200 \text{ m}}$$

What does it mean?
This result is a surprisingly large height: what we normally consider to be a rather small amount of electrical power consumption (by the lightbulb) corresponds to a substantial amount of mechanical energy. Imagine what h would be if the lightbulb were turned on for 1 day!

Figure 6.33 Example 6.10. A falling lightbulb.

Power and Velocity

Consider a "device" such as a motor or a person that is moving an object from one place to another. This device is expending an amount of power given by Equation 6.30. For one-dimensional motion with the force parallel to the displacement Δx, we can write this relation as

$$P_{ave} = \frac{W}{t} = \frac{F \, \Delta x}{t}$$

Since the average velocity is equal to $v_{ave} = \Delta x / t$, we have

$$P_{ave} = Fv_{ave}$$

This relationship between power and velocity also holds for the instantaneous velocity v and the instantaneous power P:

$$P = Fv \tag{6.31}$$

If the device, such as a motor, is capable of expending a certain amount of power, Equation 6.31 shows that there is a trade-off between force and velocity. That is, with a fixed value of P, the motor can exert a large force while the object moves at a small velocity or it can exert a small force while moving at a large velocity. Such trade-offs involving force and velocity are quite common.

Power, Force, and Efficiency

According to the relation between power, force, and velocity (Eq. 6.31), there is a trade-off between force and velocity. This trade-off limits the *maximum* possible force consistent with Newton's laws and conservation of mechanical energy. *Real* motors and other devices usually do not attain this theoretical limit, however. For example, an automobile engine might stop running ("stall") at a very low velocity, or there may be a maximum force a bicyclist can exert on the pedals, or there might be some frictional forces in the problem. To account for this behavior, we define the efficiency ε by

$$F = \varepsilon \frac{P}{v} \tag{6.32}$$

The efficiency cannot be larger than unity. We'll see an example involving efficiency in the next section, when we discuss the operation of molecular motors.

6.8 | ✪ WORK, ENERGY, AND MOLECULAR MOTORS

In Section 6.7, we discussed how our ideas about work, energy, and power can be used to understand the behavior of motors and similar devices. The same ideas apply to all types of motors, including the molecular motors that transport materials within and between cells. Several different types of molecular motors have been discovered, one example of which is sketched in Figure 6.34. This motor is based on a molecule called myosin that moves along long filaments composed of actin molecules.

Myosin is believed to move by "walking" along a filament in steps, much like a person walks on two legs. This process takes place in muscles, where the filaments form bundles. A myosin molecule is also attached to its own muscle filament, so as it walks, the myosin drags one muscle filament relative to another. The operation of your muscles is produced by these molecular motors.

Calculating the Force Exerted by a Molecular Motor

The precise biochemical reactions involved in the myosin walking motion are not completely understood. However, we do know that each step has a length of approximately 5 nm (5×10^{-9} m) and that the energy for this motor comes from a chemical

Figure 6.34 Some molecular motors move by "walking" along long strands of a protein called actin. These motors are the subject of much current research. We can use work–energy principles to understand their behavior.

reaction (hydrolysis) in which the molecule ATP reacts to form ADP. (ATP is short-hand for adenosine triphosphate, and ADP is adenosine diphosphate.) The energy released when a single ATP molecule breaks down is $E_{ATP} \approx 5 \times 10^{-20}$ J. With this information, we can use work–energy ideas to analyze the behavior of this motor. In particular, let's calculate the maximum force this motor can achieve.

The energy E_{ATP} is stored as chemical potential energy. One ATP molecule is consumed during each step, so the molecular motor uses an amount of potential energy E_{ATP} during each step. During the course of a single step, the potential energy of the motor decreases by E_{ATP} while it does an amount of work W_{motor}. Energy is conserved, so

$$W_{motor} = E_{ATP}$$

The work done by the motor is also equal to the motor force times the length of a single step Δx:

$$W_{motor} = F_{motor} \Delta x = E_{ATP} \qquad (6.33)$$

$$F_{motor} = \frac{E_{ATP}}{\Delta x}$$

Inserting the values of E_{ATP} and Δx given above, we find

$$F_{motor} = \frac{E_{ATP}}{\Delta x} = \frac{5 \times 10^{-20} \text{ J}}{5 \times 10^{-9} \text{ m}} = 1 \times 10^{-11} \text{ N}$$

for the force generated by the motor. This force is 10 piconewtons, which may seem like a very small force. However, it is actually very substantial compared with other "molecular-scale" forces such as the weight of a typical molecule.

Our calculation of F_{motor} assumes all the ATP potential energy is converted to work by the motor, that is, that the efficiency of the motor is $\varepsilon = 1$ (a "perfect" motor). Real motors will have a smaller efficiency as some of the chemical potential energy will go into vibrations and other molecular motions of the actin filament, the molecules in solution, and so forth. For this reason, our value for F_{motor} is an upper limit on the force that can be produced by a myosin motor. Evidence suggests that real molecular motors operate with a typical efficiency $\varepsilon > 0.5$, so our result for F_{motor} provides a fairly good description of real motors.

SUMMARY | Chapter 6

KEY CONCEPTS AND PRINCIPLES

Work

When a force acts on an object, the force does **work** on the object:

$$W = F(\Delta r)\cos \theta \qquad \textbf{(6.5)} \text{ (page 167)}$$

The work done by an applied force can be interpreted in a graphical manner as the area under the force–displacement curve.

Kinetic energy

Kinetic energy is energy of motion and is given by

$$KE = \tfrac{1}{2}mv^2 \qquad \textbf{(6.10)} \text{ (page 170)}$$

Work and kinetic energy are related through the **work–energy theorem**

$$W = \Delta KE \qquad \textbf{(6.11)} \text{ (page 170)}$$

where W is the work done on the object.

(Continued)

Potential energy

Potential energy is stored energy associated with a particular force or forces. The potential energy associated with a force is related to the work done by this force:

$$\Delta PE = PE_f - PE_i = -W \qquad \textbf{(6.12)} \text{ (page 174)}$$

The work done by a **conservative force** is independent of the path.

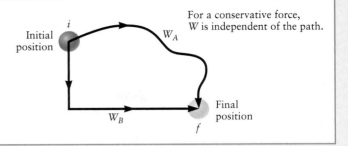

For a conservative force, W is independent of the path.

Conservation of energy

The principle of conservation of energy can be written as

$$KE_i + PE_i + W_{noncon} = KE_f + PE_f \qquad \textbf{(6.28)} \text{ (page 190)}$$

where W_{noncon} is the work done on the system by nonconservative forces.

APPLICATIONS

Hooke's law

The force exerted by a spring or similar elastic object is

$$F = -kx \qquad \textbf{(6.25)} \text{ (page 184)}$$

where k is the "spring constant" and x is the amount that the spring is stretched or compressed.

Potential energy functions

Gravity near the Earth's surface:

$$PE_{grav} = mgy \qquad \textbf{(6.15)} \text{ (page 175)}$$

Gravitational potential energy of two masses (general case):

$$PE_{grav} = -\frac{Gm_1m_2}{r} \qquad \textbf{(6.22)} \text{ (page 182)}$$

Elastic potential energy:

$$PE_{spring} = \tfrac{1}{2}kx^2 \qquad \textbf{(6.26)} \text{ (page 185)}$$

QUESTIONS

SSM = answer in Student Companion & Problem-Solving Guide   ⊗ = life science application

1. Consider an object undergoing uniform circular motion. We know from Chapter 5 that circular motion is caused by a force that is directed toward the center of the circle. What is the work done by this force during circular motion?

2. Explain why kinetic energy can never be negative, but the potential energy can be either positive or negative.

3. A heavy crate is pushed across a level floor at a constant velocity. What is the total work done on the crate by all forces? How does your answer change if the crate is being pushed up a hill?

4. Consider a satellite in an elliptical orbit around the Earth. What is the sign (positive or negative) of the power associated with the gravitational force of the Earth on the satellite? *Hint:* The answer depends on the position of the satellite along its orbit. Where in the orbit is the power zero?

5. Consider an apple that falls from a branch to the ground below. At what moment is the kinetic energy of the apple largest? At what moment is the gravitational potential energy largest?

6. SSM For the bungee jumper in Example 6.8, plot the following quantities as a function of time: (a) the gravitational potential energy of the jumper, (b) the kinetic energy of the jumper, (c) the potential energy associated with the bungee cord, and (d) the total mechanical energy.

7. The work done by friction is usually negative, as in the case of an object that slides to a stop. The work done by friction does not have to be negative, however. Give an example in which the work done by friction is positive.

8. Hooke's law (Eq. 6.25) is an empirical relationship, and as discussed in Section 6.4, this relationship has been found to be widely applicable. Discuss under what circumstances and extremes this relationship may not hold.

9. According to Equation 6.31, power is related to force and velocity by $P = Fv$. Is it possible for P to be negative? Give an example and explain what it means for P to be negative.

10. Give two examples in which the work done on an object by friction is negative.

11. Give two examples in which a nonzero force acts on an object, but the work done by that force is zero.

12. Explain how a mosquito can have a greater kinetic energy than a baseball.

13. A hockey puck slides along a level surface, eventually coming to rest. Is the energy of the hockey puck conserved? Is the total energy of the universe conserved? Discuss where the initial kinetic energy of the puck "goes."

14. A spring is initially compressed. It is then released, sending an object flying off the spring. Discuss this process in terms of conservation of energy. What and "where" is the initial energy, and where is the energy found at the end?

15. Tarzan starts at rest near the top of a tall tree. He then swings on a vine and reaches the base of the tree, where he jumps off the vine and comes to rest on the ground. Discuss and compare Tarzan's total energy at (a) the top of the tree, (b) just before he jumps off the vine at the bottom, and (c) when he is standing on the ground. Is Tarzan's total energy the same in all three places? Explain how total energy is conserved in going from (a) to (b) to (c).

16. During the course of each day, you move from place to place and do work on various objects. Where does the energy for these everyday activities come from? How does your body conserve energy (in the physics sense)? Explain.

17. SSM When a rubber ball is dropped onto a concrete floor, it bounces to a height that is slightly lower than its initial height. Compare the ball's (a) initial mechanical energy to (b) its energy just before hitting the ground, to (c) the total energy just after bouncing off the ground, and to (d) the total energy when it reaches it final height. Is the ball's mechanical energy conserved? Explain and discuss using the principle of conservation of total energy.

18. Construct bar charts showing the kinetic and potential energy of the roller coaster in Example 6.4 at points A, B, and C in Figure 6.13. Explain how these charts illustrate the principle of conservation of mechanical energy.

19. Construct bar charts showing the kinetic and potential energy for the bungee jumper in Example 6.8 at various points during his jump. Explain how these charts illustrate the principle of conservation of mechanical energy.

PROBLEMS

SSM = solution in Student Companion & Problem-Solving Guide
☆ = intermediate ✪ = challenging

Ⓧ = life science application
Ⓡ = reasoning and relationships problem

6.1 FORCE, DISPLACEMENT, AND WORK

1. A crate of mass 50 kg is pushed across a level floor by a person. If the person exerts a force of 25 N in the horizontal direction and moves the crate a distance of 10 m, what is the work done by the person?

2. A sphere of mass 35 kg is attached to one end of a rope as shown in Figure P6.2. It is found that the rope does an amount of work $W = 550$ J in pulling the sphere upward through a distance 1.5 m. Find the tension in the rope.

Figure P6.2
Problems 2 and 3.

3. ☆ For the sphere in Problem 2, find the work done by the force of gravity on the sphere.

4. SSM ☆ A hockey puck of mass 0.25 kg is sliding along a slippery frozen lake, with an initial speed of 60 m/s. The coefficient of friction between the ice and the puck is $\mu_K = 0.030$. Friction eventually causes the puck to slide to a stop. Find the work done by friction.

5. A crate of mass 24 kg is pushed up a frictionless ramp by a person as shown in Figure P6.5. Calculate the work done by the person in pushing the crate a distance of 20 m as measured along the ramp. Assume the crate moves at a constant velocity.

Figure P6.5
Problems 5, 6, and 7.

6. ☆ Consider again the crate in Problem 5, but now include friction. Assume the coefficient of kinetic friction between the crate and the ramp is $\mu_K = 0.25$. What is the work done by the person as he pushes the crate up the ramp?

7. ☆ Find the work done by gravity on the crate in Problem 6.

8. ☆ A block of mass $m = 5.0$ kg is pulled along a rough horizontal surface by a rope as sketched in Figure P6.8. The tension in the rope is 40 N, and the coefficient of kinetic friction between the block and the surface is $\mu_K = 0.25$. (a) If the block travels a distance of 4.5 m along the surface, what is the work done by the rope? (b) Find the work done by friction on the block.

Figure P6.8

9. A car ($m = 1200$ kg) is traveling at an initial speed of 15 m/s. It then slows to a stop over a distance of 80 m due to a force from the brakes. How much work is done on the car? Assume the acceleration is constant.

10. A snowboarder of mass 80 kg slides down the trail shown in Figure P6.10. The first part of the trail is a ramp that makes an angle of 20° with the horizontal, and the final portion of the trail is flat. Find the work done by gravity on the snowboarder as she travels from the beginning to the end of this trail.

Figure P6.10

11. Two railroad cars, each of mass 2.0×10^4 kg, are connected by a cable to each other, and the car in front is connected by

Figure P6.11

a cable to the engine as shown in Figure P6.11. The cars start from rest and accelerate to a speed of 1.5 m/s after 1 min. (a) Find the work done by cable 1 on the car in the back. (b) Find the work done by cable 1 on the car in front. (c) Find the work done by cable 2 on the car in front.

12. A person pushes a broom at an angle of 60° with respect to the floor (Fig. P6.12). If the person exerts a force of 30 N directed along the broom handle, what is the work done by the person on the broom as he pushes it a distance of 5.0 m?

Figure P6.12

13. A car is pushed along a long road that is straight, flat, and parallel to the x direction. (a) The horizontal force on the car varies with x as shown in Figure P6.13. What is the work done on the car by this force? (b) What is the work done by gravity on the car?

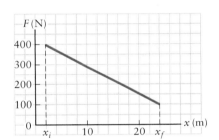

Figure P6.13

14. The force on an object varies with position as shown in Figure P6.14. Estimate the work done on the object as it moves from the origin to x = 5.0 m. Assume the motion is one dimensional.

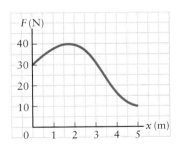

Figure P6.14

6.2 KINETIC ENERGY AND THE WORK–ENERGY THEOREM

15. A tomato of mass 0.22 kg is dropped from a tall bridge. If the tomato has a speed of 12 m/s just before it hits the ground, what is the kinetic energy of the tomato?

16. Two objects have the same kinetic energy. One has a speed that is 2.5 times greater than the speed of the other. What is the ratio of their masses?

17. A car of mass 1500 kg is initially traveling at a speed of 10 m/s. The driver then accelerates to a speed of 25 m/s over a distance of 200 m. Calculate the change in the kinetic energy of the car.

18. ⭐ A rock of mass 0.050 kg is thrown upward with an initial speed of 25 m/s. Find the work done by gravity on the rock from the time it leaves the thrower's hand until it reaches the highest point on its trajectory. *Hint*: You do not have to calculate the rock's maximum height.

19. A horizontal force of 15 N pulls a block of mass 3.9 kg across a level floor. The coefficient of kinetic friction between the block and the floor is $\mu_K = 0.25$. If the block begins with a speed of 8.0 m/s and is pulled for a distance of 12 m, what is final speed of the block?

20. You are pushing a refrigerator across the floor of your kitchen. You exert a horizontal force of 300 N for 7.0 s, during which time the refrigerator moves a distance of 3.0 m at a constant velocity. (a) What is the total work (by all forces) done on the refrigerator? (b) What is the work done by friction?

21. An archer is able to fire an arrow (mass 0.020 kg) at a speed of 250 m/s. If a baseball (mass 0.14 kg) is given the same kinetic energy, what is its speed?

22. A baseball pitcher can throw a baseball at a speed of 50 m/s. If the mass of the ball is 0.14 kg and the pitcher has it in his hand over a distance of 2.0 m, what is the average force exerted by the pitcher on the ball?

23. A softball pitcher can exert a force of 100 N on a softball. If the mass of the ball is 0.19 kg and the pitcher has it in her hand over a distance of 1.5 m, what is the speed of the ball when it leaves her hand?

24. ⭐ Consider a skydiver of mass 70 kg who jumps from an airplane flying at an altitude of 1500 m. With her parachute open, her terminal velocity is 8.0 m/s. (a) What is the work done by gravity and (b) what is the average force of air drag during the course of her jump?

25. Consider a small car of mass 1200 kg and a large sport utility vehicle (SUV) of mass 4000 kg. The SUV is traveling at the speed limit ($v = 35$ m/s). The driver of the small car travels so as to have the same kinetic energy as the SUV. Find the speed of the small car.

26. A truck of mass 9000 kg is traveling along a level road at an initial speed of 30 m/s and then slows to a final speed of 12 m/s. Find the total work done on the truck.

27. ⭐ You use an elevator to travel from the first floor ($h = 0$) to the fourth floor of a building ($h = 12$ m). If you start from rest and you end up at rest on the fourth floor, what is the work done by gravity on you? Assume you have a mass of 70 kg.

28. SSM ⭐ A car of mass $m = 1500$ kg is pushed off a cliff of height $h = 24$ m (Fig. P6.28). If the car lands a distance of 10 m from the base of the cliff, what was the kinetic energy of the car the instant after it left the cliff?

Figure P6.28

29. ⭐ The force on an object varies with position as shown in Figure P6.29. The object begins at rest from the origin and has a speed of 5.0 m/s at x = 4.0 m. What is the mass of the object?

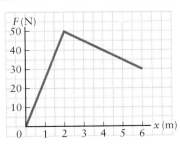

Figure P6.29 Problems 29 and 30.

30. Suppose the object in Figure P6.29 has a mass of 2.5 kg and a speed of 9.0 m/s at x = 0. What is its speed at x = 5.0 m?

6.3 POTENTIAL ENERGY

31. A skier of mass 110 kg travels down a frictionless ski trail. (a) If the top of the trail is a height 200 m above the bottom, what is the work done by gravity on the skier? (b) Find the velocity of the skier when he reaches the bottom of the ski trail. Assume he starts from rest.

32. Suppose the ski trail in Problem 31 is *not* frictionless. Find the work done by gravity on the skier in this case.

33. Consider a roller coaster that moves along the track shown in Figure P6.33. Assume all friction is negligible. (a) Is the mechanical energy of the roller coaster conserved? (b) Add a coordinate system to the sketch in Figure P6.33. (c) If the roller coaster starts from rest at location A, what is its total mechanical energy at point A? (d) If v_B is the speed at point B, what is the total mechanical energy at point B? (e) Find the speed of the roller coaster when it reaches locations B and C.

Figure P6.33 Problems 33 and 34.

34. Consider again the roller coaster in Figure P6.33, but now assume the roller coaster starts with a speed of 12 m/s at point A. Find the speed of the roller coaster when it reaches locations B and C.

35. ⭐ Mount McKinley (also called Denali) is the tallest mountain in North America, with a height of 6200 m above sea level. If a person of mass 120 kg walks from sea level to the top of Mount McKinley, how much work is done by gravity on the person?

36. ✪ The roller coaster in Figure P6.36 starts with a velocity of 15 m/s. One of the riders is a small girl of mass 30 kg. Find her apparent weight when the roller coaster is at locations B and C. At these two locations, the track is circular, with the radii of curvature given in the figure. The heights at points A, B, and C are $h_A = 25$ m, $h_B = 35$ m, and $h_C = 0$.

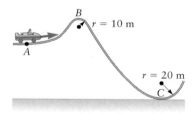

Figure P6.36

37. SSM ⭐ A roller coaster (Fig. P6.37) starts at the top of its track (point A) with a speed of 12 m/s. If it reaches point B traveling at 16 m/s, what is the vertical distance (h) between A and B? Assume friction is negligible and ignore the kinetic energy of the wheels.

Figure P6.37

38. ⭐ A skateboarder starts at point A on the track (Fig. P6.38) with a speed of 15 m/s. Will he reach point B? Assume friction

Figure P6.38

is negligible and ignore the kinetic energy of the skateboard's wheels.

39. ✪ **Frontside air.** A skateboarder is practicing on the "half-pipe" shown in Figure P6.39, using a special frictionless skateboard. (You can also ignore the kinetic energy of the skateboard's wheels.) (a) If she starts from rest at the top of the half-pipe, what is her speed at the bottom? (b) If the skateboarder has a mass $m = 55$ kg, what is her apparent weight at the bottom of the half-pipe? (c) What speed does she then have when she reaches the top edge on the other side of the half-pipe? (d) Now suppose she has a speed of 11 m/s at the bottom of the half-pipe. What is the highest point she can reach? *Hint:* This point may be above the edge of the half-pipe.

Figure P6.39 Problems 39 and 40.

40. Consider again the skateboarder in Figure P6.39. Assuming she starts at the top of the half-pipe, make qualitative sketches of her kinetic energy and potential energy as a function of height in the half-pipe.

41. ⭐ A hockey puck of mass 0.25 kg starts from rest and slides down the frictionless ramp shown in Figure P6.41. The bottom end of the ramp is horizontal. After it leaves the ramp, the puck travels as a projectile and lands a distance L from the ramp. (a) Is the puck's mechanical energy at the top of the ramp equal to its mechanical energy when it leaves the ramp and is in the air? Explain why or why not. (b) Add a coordinate system to the sketch in Figure P6.41. Where is a convenient place to put the origin of the vertical (y) axis? Do you have to put it there? (c) What is the initial mechanical energy of the puck? (d) What is the mechanical energy of the puck the instant it leaves the ramp? (e) What is the speed of the puck the instant it leaves the ramp? (f) How far from the base of the ramp does the puck land?

Figure P6.41

42. SSM ⭐ A rock of mass 12 kg is tied to a string of length 2.4 m, with the other end of the string fastened to the ceiling of a tall room (Fig. P6.42). While hanging vertically, the rock is given an initial horizontal velocity of 2.5 m/s. (a) Add a coordinate system to the sketch in Figure P6.42. Where is a convenient place to choose the origin of the vertical (y) axis? (b) What are the initial kinetic and potential energies of the rock? (c) If the rock swings to a height h above its initial point, what is its potential energy? What is its total mechanical energy at that point? (d) How high will the rock swing? Express your answer in terms of the angle θ that the string makes with the vertical when the rock is at its highest point. (e) Make qualitative sketches of the kinetic energy and potential energies as functions of height h.

Figure P6.42

43. ✪ Consider the rock and string in Figure P6.43. The string is fastened to a hinge that allows it to swing completely around in a vertical circle. If the rock starts at the lowest point on this circle and is given an initial speed v_i, what is the smallest value of v_i that will allow the rock to travel completely around the circle without the string becoming slack at the top? Assume the length of the string is 1.5 m.

Figure P6.43

44. 🅡 A rock climber ($m = 90$ kg) starts at the base of a cliff and climbs to the top ($h = 35$ m). He then walks along the plateau at the top for a distance of $L = 40$ m. Find the work done by gravity and the change in the gravitational potential energy of the rock climber.

45. Two crates of mass $m_1 = 40$ kg and $m_2 = 15$ kg are connected by a massless rope that passes over a massless, frictionless pulley as shown in Figure P6.45. The crates start from rest. (a) Add a coordinate system to Figure P6.45. (b) What are the initial kinetic and potential energies of each crate? (c) If the crate on the left (m_1) moves downward through a distance h, what is the change in the potential energy of m_1 and m_2? Express your answers in terms of h. (d) What is the speed of the crates after they have moved a distance $h = 2.5$ m?

Figure P6.45
Problems 45 and 46.

46. ✪ Consider again the crates in Problem 45, but now assume the rope has a mass of 2 kg. If the tops of the crates start at the same initial height and the rope has a length of 10 m, what is the final speed?

47. Sketch how the potential energy, kinetic energy, and total mechanical energy vary with time for an apple that drops from a tree.

48. Sketch how the potential energy, kinetic energy, and total mechanical energy vary with time for a cannon shell that is fired from a large cannon. Ignore air drag.

49. 🔲 A rock of mass 3.3 kg is tied to a string of length 1.2 m. The rock is held at rest as shown in Figure P6.49 so that the string is initially tight, and then it is released. (a) Find the speed of the rock when it reaches the lowest point of its trajectory. (b) What is the maximum tension in the string?

Figure P6.49

50. 🔲 A ball of mass 1.5 kg is tied to a string of length 6.0 m as shown in Figure P6.50. The ball is initially hanging vertically and is given an initial velocity of 5.0 m/s in the horizontal direction. The ball then follows a circular arc as determined by the string. What is the speed of the ball when the string makes an angle of 30° with the vertical?

Figure P6.50

51. A bullet is fired from a rifle with a speed v_0 at an angle θ with respect to the horizontal axis (Fig. P6.51) from a cliff that is a

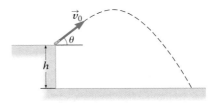

Figure P6.51

height h above the ground below. (a) Use conservation of energy principles to calculate the speed of the bullet when it strikes the ground. Express your answer in terms of v_0, h, g, and θ, and ignore air drag. (b) Explain why your result is independent of the angle θ.

6.4 MORE POTENTIAL ENERGY FUNCTIONS

52. Calculate the escape velocity for an object on the Moon.

53. An engineer at NASA decides to save money on fuel by launching satellites from a very tall mountain. (a) Calculate the escape velocity for a satellite launched from an altitude of 8000 m above sea level (a mountain comparable to Mount Everest). (b) If the amount of fuel required is proportional to the initial kinetic energy, by what fraction is the fuel load reduced?

54. SSM 🔲 Calculate the velocity needed for an object starting at the Earth's surface to just barely reach a satellite in a geosynchronous orbit. Ignore air drag.

55. Sketch how the potential energy, kinetic energy, and total mechanical energy vary with position for a projectile that is fired from the Earth's surface. Assume the projectile's initial velocity is equal to the escape velocity and ignore air drag.

56. ✪ Consider a spacecraft that is to be launched from the Earth to the Moon. Calculate the minimum velocity needed for the spacecraft to just make it to the Moon's surface. Ignore air drag from the Earth's atmosphere. *Hint:* The spacecraft will not have zero velocity when it reaches the Moon.

57. ✪ A space station is orbiting the Earth. It moves in a circular orbit with a radius equal to twice the Earth's radius. A supply satellite is designed to travel to the station and dock smoothly when it arrives. If the supply satellite is fired as a simple projectile from the Earth's surface, what is the mimimum initial speed required for it to reach the space station and have a speed equal to the station's speed when it arrives? Ignore both air drag and the rotational motion of the Earth.

58. A mass and spring are arranged on a horizontal, frictionless table as shown in Figure P6.58. The spring constant is $k = 500$ N/m, and the mass is 4.5 kg. The block is pushed against the spring so that the spring is compressed an amount 0.35 m, and then it is released. Find the velocity of the mass when it leaves the spring.

Figure P6.58

59. A block is dropped onto a spring with $k = 30$ N/m. The block has a speed of 3.3 m/s just before it strikes the spring. If the spring compresses an amount 0.12 m before bringing the block to rest, what is the mass of the block?

60. A block of mass 35 kg is sitting on a platform as shown in Figure P6.60. The platform sits on a spring with $k = 2000$ N/m. The mass is initially at rest. (a) Add a coordinate system to this sketch. Where is a convenient place to choose the origin of the vertical (y)

Figure P6.60

axis? (b) If the platform is depressed a distance 0.50 m and held there, what is the mechanical energy of the system? (c) The platform is then released, and the mass moves upward and eventually flies off the platform. Assume the spring returns to its relaxed state (unstretched and uncompressed) after the mass leaves the platform. What are the kinetic energy, the gravitational potential energy, and the elastic potential energy of the system when the mass reaches its highest point? (d) How high will the block go? Measure this distance from the height of the uncompressed platform.

61. An archer's bow can be treated as a spring with $k = 3000$ N/m. If the bow is pulled back a distance 0.12 m before releasing the arrow, what is the kinetic energy of the arrow when it leaves the bow?

62. For the arrow in Problem 61, what is the work done on the bow as the bow string is pulled back into position?

63. ☆ Ⓡ The shock absorbers on a car are fancy springs. Because one is attached to each wheel, your car is supported by four of these springs. Estimate the spring constant for a shock absorber.

64. ☆ Ⓡ A tennis racket can be treated as a spring. The displacement of the spring (what we commonly call x) is the displacement of the strings at the center of the racket. Estimate the spring constant of this spring. Assume a typical tennis ball has a mass of about 57 g and a speed of 50 m/s when it leaves the racket.

65. ☆ Ⓡ A diving board acts like a spring and obeys Hooke's law (Fig. 6.21D). Estimate the spring constant of a diving board.

66. ☆ Ⓡ A tennis ball is a flexible, elastic object. If a person of average size stands on a tennis ball, the ball will compress to about half its original (non-compressed) diameter. What is the approximate spring constant of a tennis ball?

6.5 CONSERVATIVE VERSUS NONCONSERVATIVE FORCES AND CONSERVATION OF ENERGY

67. SSM ☆ A tennis ball ($m = 57$ g) is projected vertically with an initial speed of 8.8 m/s. (a) If the ball rises to a maximum height of 3.7 m, how much kinetic energy was dissipated by the drag force of air resistance? (b) How much higher would the ball have gone in a vacuum?

68. ☆ A snowboarder of mass 80 kg travels down the slope of height 150 m shown in Figure P6.68. If she starts from rest at the top and has a velocity of 12 m/s when she reaches the bottom, what is the work done on her by friction? *Note*: Figure not drawn to scale.

Figure P6.68

69. ✪ **Big bounce.** A Super Ball is a toy ball made from the synthetic rubber polymer polybutadiene vulcanized with sulfur. Manufactured by Wham-o since 1965, these toys have a "super" elastic property such that they bounce to 90% of the height from which they are dropped. A Super Ball of mass 100 g is dropped from a height of 2.0 m. How much kinetic energy is dissipated by the collision with the ground? A lump of clay of equal mass is dropped from a similar height and sticks to the ground without a bounce. Where did the kinetic energy go in the case of the clay? What about in the case of the Super Ball?

70. Consider the skateboarder in Figure P6.70. If she has a mass of 55 kg, an initial velocity of 20 m/s, and a velocity of 12 m/s at the top of the ramp, what is the work done by friction on the skateboarder? Ignore the kinetic energy of the wheels of the skateboard.

Figure P6.70

71. A skier of mass 100 kg starts from rest at the top of the ski slope of height 100 m shown in Figure P6.71. If the total work done by friction is 3.0×10^4 J, what is the skier's speed when he reaches the bottom of the slope?

Figure P6.71

72. ☆ Ⓡ The main frictional force on a good bicycle is the force of air drag (Eq. 3.20). Consider a bicyclist coasting at a constant velocity on a road that is sloped at 5.0° with respect to the horizontal. Find her speed. How much work is done by air drag as she travels a distance of 5.0 km?

6.7 POWER

73. Every year, there is a foot race to the top of the Empire State Building. The vertical distance traveled (using the stairs!) is about 440 m. The current record for this race is about 9 min 30 s. If the record holder has a mass of 60 kg, what was his average power output during the race?

74. ✪ Ⓡ A bicyclist travels on the hill shown in Figure P6.74. She begins from rest at the top of the hill, coasts all the way to the bottom, and finds that she gets to a point approximately 60% of the way up the other side before the bicycle comes (instantaneously) to rest. The total mass of the bicyclist plus the bicycle is 100 kg. (a) Estimate the total work done on the bicycle by all frictional forces (including air drag). (b) Assuming the main frictional force is due to air drag, find the approximate magnitude of this force throughout the motion. *Hint*: Calculate the power due to air drag by assuming her speed during the entire time is a constant and approximately equal to v_{ave}.

Figure P6.74

75. In Example 6.10, we compared the gravitational potential energy of a 100-W lightbulb to the electrical energy consumed when it is turned on. Repeat that calculation, but now assume the lightbulb is turned on for one day. Find the height h through which the lightbulb would have to fall for the change in gravitational potential energy to equal the electrical energy consumed by the lightbulb. Compare this answer to the height of Mount Everest. Assume you can use the relation $PE_{grav} = mgh$.

76. The rate of energy use in a typical house is about 2.0 kW. If the kinetic energy of a car (mass 1400 kg) is equal to the total energy used by the house in 1 min, what is the car's speed?

77. An electric motor is rated to have a maximum power output of 0.75 hp. If this motor is being used to lift a crate of mass 200 kg, how fast (i.e., at what speed) can it lift the crate? *Hint:* Assume the only other force on the crate is the force of gravity.

78. SSM ✪ For a car moving with a speed v, the force of air drag on a car is proportional v^2. If the power output of the car's engine is doubled, by what factor does the speed of the car increase?

79. ✪ Ⓡ A typical car has a power rating of 150 hp. Estimate the car's maximum speed on level ground.

6.8 WORK, ENERGY, AND MOLECULAR MOTORS

80. Ⓧ Consider a molecular motor that has an efficiency of 50%. Find the maximum force the motor can produce. Assume a step size of 4 nm. Also assume the energy consumed for each step is 5×10^{-20} J (as given for an ATP-powered motor in Section 6.8).

81. Ⓧ Suppose a molecular motor consumes 3 ATP molecules in every step. If this motor has a step size of 8.0 nm, what is the maximum force the motor could exert?

82. ✪ Ⓧ Ⓡ Molecular motors are often studied by attaching them to small plastic spheres. These spheres are much larger than the motor and enable the motor's motion to be observed with a microscope. Consider a molecular motor to which a plastic sphere of radius 20 nm is attached. As the motor moves through water, a drag force given by Equation 3.23 acts on the sphere. If the molecular motor is able to produce a power output of 1×10^{-17} W, what is the speed of the motor as it pulls the sphere through water?

83. Ⓧ Consider a molecular motor that consumes the energy from 100 ATP molecules per second. What is the power output of this motor?

84. Ⓧ If the motor in Problem 83 takes a step of length 6.0 nm for each ATP molecule it consumes, what are the speed of the motor and the force it is able to produce?

85. SSM ✪ Ⓧ Ⓡ In Section 6.8, we calculated the force produced by a myosin molecular motor and found that it is approximately 10 piconewtons. Compare this force to the weight of a typical amino acid.

ADDITIONAL PROBLEMS

86. SSM ✪ Ⓧ **Conceptualizing units.** A small frozen burrito, like those bought at your local convenience store, has a mass of approximately 100 g, which gives it a weight of roughly 1 newton. (a) From what height would you need to drop the burrito to give it 1 J of kinetic energy when it hits the ground? How many burritos would you have to drop from that height to equal the kinetic energy of (b) the arrow in Example 6.7 and (c) the snowboarder ($m = 65$ kg) in Figure 6.12?

87. ✪ Ⓧ Ⓡ **Counting calories.** The chemical potential energy in foods is measured in units called Calories (notice the capital C). This "food Calorie" is equal to 1000 "physics" calories. For example, a typical apple contains approximately 75 Calories, which is actually 75 *kilo*calories. If all the energy in an apple were converted to gravitational potential energy, how high would you be lifted?

88. ✪ Ⓧ Ⓡ Repeat Problem 87, but replace the apple with a typical fast-food hamburger (approximately 400 C).

89. SSM ✪ Ⓧ Ⓡ A doughnut (Fig. P6.89) contains roughly 350 C (1.5×10^6 J) of potential energy locked in chemical bonds. Find the ratio of the potential energy in the doughnut to that in an equivalent volume of TNT (trinitrotoluene), which has a density of 1.65 g/cm^3 and releases 2.7×10^6 J per kg of explosive. Aren't you glad the doughnut does not release all its potential energy at once?

Figure P6.89

90. ✪ One gallon of gasoline contains 3.1×10^7 calories of potential energy that are released during combustion. If 1 gal of gasoline can provide the force that moves a car though a displacement of 25 mi, what is the average force produced by the gasoline?

91. ✪ Tarzan ($m = 74$ kg) commutes to work swinging from vine to vine. He leaves the platform of his tree house and swings on the end of a vine of length $L = 8.0$ m. (a) If the platform is 1.9 m above the lowest point in the swing, what is the tension in the vine at the lowest point in the swing? (b) Tarzan again takes to his morning commute, but this time a monkey of mass $m_M = 23$ kg hitches a ride by jumping onto Tarzan's back. If a vine can withstand a maximum tension of 1200 N, will it snap under the tension of the added passenger? If so, at what angle with respect to the vertical does the vine break?

92. ✪ A toy gun shoots spherical plastic projectiles by means of a spring. A typical projectile has a mass of $m = 25$ g. The spring used has a spring constant of $k = 15$ N/m, and when put under load, it is displaced 6.0 cm as shown in Figure P6.92. The barrel of the gun exerts a slight frictional force of magnitude $F_{friction} = 0.074$ N on the pellet as it moves down the barrel from its starting point a total length of $L = 15$ cm. If the toy gun is fired in a horizontal position, (a) at what position measured from the starting point does the pellet reach maximum velocity? *Hint:* Consider an equilibrium condition between the spring force and the friction force on the pellet before it leaves the spring. (b) What is the maximum speed achieved by the pellet as it is fired from the gun? (c) At what speed does the pellet leave the barrel?

Figure P6.92

93. ✪ A pole vaulter of mass 70 kg can run horizontally with a top speed of 10.0 m/s. The current record height for the pole vault is about 6.2 m. (a) Approximately how much energy must be stored in the pole just before the vaulter leaves the ground? Assume the vaulter's speed is zero just prior to taking off. (b) The energy in part (a) is the maximum energy stored in the pole, so we denote it by PE_{max}. How does PE_{max} compare to the maximum kinetic energy KE_{run} of the pole vaulter before he takes off? (c) Explain how PE_{max} can be greater than KE_{run}.

94. ✪ A spring is mounted at an angle of $\theta = 35°$ on a frictionless incline as illustrated in Figure P6.94. The spring is compressed to 15 cm where it is allowed to propel a mass of 5.5 kg up the incline. (a) If the spring constant is 550 N/m, how fast is the mass moving when it leaves the spring? (b) To what maximum distance from the starting point will the mass rise up the incline?

Figure P6.94
Problems 94 and 95.

95. ✪ Consider the system in Problem 94 and assume the coefficient of kinetic friction between the mass and the incline is 0.17. (a) At what position, measured from the starting point, does the maximum velocity occur? (b) What is the maximum velocity attained by the mass at this point? (c) What is now the maximum distance from the starting point that the mass will rise up the incline?

96. ✭ Ⓡ **Spring shoes.** Tae-Hyuk Yoon of South Korea is the inventor of a novel method of human locomotion: the Poweriser spring boot shown in Figure P6.96. Promotional materials claim that a 180-lb man can jump up to 6 ft high while wearing these devices. Estimate the spring constant of the leaf springs on the Poweriser. The maximum compression can be estimated from the figure.

Figure P6.96 Poweriser spring boots and their application.

97. ✪ A chairlift rises in elevation 2000 ft up a ski slope and has a total of 200 chairs evenly spaced 40 ft apart. On a busy weekend, the ski lift is at full capacity (two people per seat), and skiers come off the lift every 10 s. Assuming a skier weighs 160 lb on average, what is the power output required for the ski lift to operate? State your answer in both watts and horsepower.

98. ✭ A 70-kg skydiver reaches a terminal velocity of 50 m/s. At what rate is the drag force dissipating his kinetic energy? How does this rate compare with his rate of change of potential energy?

99. ✪ The world record for the highest-altitude skydive was made by Joseph Kittinger in 1960. Kittinger jumped from a high-altitude balloon (Fig.

Figure P6.99

P6.99) at a height of 102,800 ft (31,330 m) above sea level, and according to reports, attained speeds up to 624 mi/h allowed by the thin air of the stratosphere. Calculate his maximum speed (i.e., neglect air drag) for his 14,500 ft of free fall. How does this speed compare with the reported value?

100. ✪ The NEAR spacecraft flew by the asteroid Eros in 1997 (Fig. P6.100). The flyby allowed the craft to measure the asteroid's density at 2.7 g/cm³. Assume most asteroids have similar composition and that you personally can jump 1.0 m high on the Earth. (a) What is the radius of the largest spherical asteroid you could literally jump off of (in other words, that you could jump from with an initial velocity equal to the escape velocity of the asteroid)? (b) Eros has a mass of 6.7×10^{15} kg. Could you jump off Eros? If not, how far from the surface would you go before descending back to the asteroid?

Figure P6.100 Image of Eros asteroid from NEAR encounter.

101. ✪ ✘ **Aerobic workout.** After a 5-min workout on a Climb Max stair machine, the readout panel (Fig. P6.101) reports that the 75-kg user burned 19.7 Calories (see Problem 87) and climbed a total of 180 steps. (a) If each step is 15 cm in height, how much would the potential energy of the user have changed if she were actually climbing a stairway? (b) What was her rate of change in potential energy? (c) At what rate was her metabolism consuming energy? (d) What is her efficiency in converting the energy in the chemical bonds in her food into potential energy? (e) She really didn't change her vertical displacement on this machine, so where did that energy go?

Figure P6.101 Results from a workout.

102. ✪ The expression for the force due to air drag described in Chapter 3 (Eq. 3.20) ignores the aerodynamic shape of an object. Some objects, such as a race car, are shaped so as to minimize the drag force. This can be accounted for by adding a factor called the drag coefficient C_D. The drag force is then

$$F_{drag} = -\tfrac{1}{2} C_D \rho A v^2$$

A boxy car might have a drag coefficient $C_D \approx 1.0$, whereas a racecar might have $C_D \approx 0.25$. If all else is the same (the power produced by the engine, etc.) and there is no other source of friction or drag, what is the ratio of the race car's top speed to the boxy car's top speed?

Momentum, Impulse, and Collisions

Collisions are involved in many important situations, including the game of pool. (© David Leah/Stone/Getty Images)

In the past few chapters, we introduced Newton's laws of motion and used them to understand and predict the motion of various objects. Until now, we have treated all objects as "point particles," objects whose mass is located at a single point in space. The point particle assumption is very useful and is the correct way to deal with many situations. However, some important kinds of motion cannot be dealt with in this way. For example, objects such as baseballs and boomerangs can spin or rotate as they move along their trajectory. To deal with such motion, we must consider the ball or boomerang as an extended object, taking size and shape into account. In this and the next two chapters, we show how Newton's laws can be used to treat the motion of such extended objects. Conceptually, we will imagine that our baseball is composed of many very small "pieces." If the pieces are small enough, we can treat each as a point particle using the methods developed in previous chapters. The different pieces of an object exert forces on one another—when you push on one side of a baseball, the rest of the ball moves, too—so we must also deal with the forces that inevitably exist in such a *system* of interacting particles. We first encountered the notion of a system in Chapter 6, where we found that the total energy of a closed system (i.e., a system that is not acted on by any outside forces) is conserved. In this chapter, we introduce another property called **momentum** and discover that the total momentum of a closed system of particles is also conserved. This principle of **conservation of momentum** provides a very powerful tool in the analysis of motion in a wide variety of situations. It will lead us to the concept of **center of mass** and provide a general way to deal with collisions between particles.

7.1 | MOMENTUM

The variables velocity and acceleration are needed to describe the motion of a single particle. When dealing with a collection of particles, it is useful to define an additional quantity, **momentum**. The momentum of a single particle of mass m moving with a velocity $\vec{v}$ is defined as

The momentum of a particle depends on its mass and velocity.

$$\vec{p} = m\vec{v} \tag{7.1}$$

The momentum of a particle $\vec{p}$ is proportional to its velocity, so the momentum vector is along the same direction as the velocity. Notice also that $\vec{p}$ is proportional to the mass of the particle. We'll soon see how this relation for the momentum arises from Newton's second law.

EXAMPLE 7.1 | The Momentum of a Bullet

Consider a bullet that has a mass of 10 g and a speed of 600 m/s. (a) Find the momentum of the bullet. (b) Find the speed of a baseball ($m = 0.14$ kg) that has the same momentum as the bullet.

RECOGNIZE THE PRINCIPLE

The magnitude of the momentum can be found directly from Equation 7.1, given the mass and speed of the objects of interest. The goal of this example is to get a sense of the momentum of two "everyday" objects.

SKETCH THE PROBLEM

No figure is necessary.

IDENTIFY THE RELATIONSHIPS

The speed and mass of the bullet are both given, so we can apply Equation 7.1

$$\vec{p} = m\vec{v}$$

SOLVE

(a) In terms of magnitudes the momentum of the bullet is

$$p = (0.010 \text{ kg})(600 \text{ m/s}) = \boxed{6.0 \text{ kg} \cdot \text{m/s}}$$

The direction of $\vec{p}$ is parallel to the velocity of the bullet.

(b) If a baseball has the same momentum as this bullet, we have

$$\vec{p}_{\text{ball}} = \vec{p}_{\text{bullet}}$$

In terms of the magnitudes, we then get

$$m_{\text{ball}}v_{\text{ball}} = m_{\text{bullet}}v_{\text{bullet}}$$

Hence, the speed of the ball is

$$v_{ball} = \frac{m_{bullet} v_{bullet}}{m_{ball}}$$

Inserting the numerical values given above, we find

$$v_{ball} = \frac{m_{bullet} v_{bullet}}{m_{ball}} = \frac{(0.010 \text{ kg})(600 \text{ m/s})}{0.14 \text{ kg}} = \boxed{43 \text{ m/s}}$$

What have we learned?

This speed is approximately 100 mi/h and is about the speed that can be reached by a professional baseball pitcher. A key point is that the momentum of a particle is the *product* of its mass and velocity. A particular value of $\vec{p}$ can be obtained with a small mass moving at a high velocity or with a large mass moving with a smaller velocity. The momentum carried by a baseball thrown by a major-league pitcher is about the same as the momentum of a bullet!

Momentum of a System of Particles

Equation 7.1 tells us how to calculate the momentum of a single particle or object. To find the total momentum of a system of particles, we take the sum of the momenta of all the individual pieces of the system. For example, if we want to find the total momentum of a boomerang, we can imagine it as a collection of many small pieces as illustrated in Figure 7.1A. Each piece can be treated as a point particle, and the momentum of each piece is given by Equation 7.1. The total momentum of a collection of such pieces is then

$$\vec{p}_{total} = \sum_i \vec{p}_i = \sum_i m_i \vec{v}_i \qquad (7.2)$$

where the sums are over all the individual particles (pieces) in the system. Such a system of particles may be the pieces of a solid object, such as a boomerang, or pieces of a nonsolid object, such as the molecules in a gas (Fig. 7.1B). In either case, the total momentum of the system, as calculated by Equation 7.2, is a very useful physical quantity.

| EXAMPLE 7.2 | Momentum of Two Cars |

Consider two cars that are on course for a collision in one dimension (Fig. 7.2). If the small car has a mass of 1200 kg and a speed of 20 m/s and the large car has a mass of 2000 kg and a speed of 15 m/s in the opposite direction, what is the total momentum of the two cars?

RECOGNIZE THE PRINCIPLE

We can find the momentum of each car using Equation 7.1 and then add the results to get the total momentum. We must include the sign of the velocity (+ or −) of each car when calculating their momenta.

SKETCH THE PROBLEM

Figure 7.2 describes the problem. The cars are both moving along the x axis.

IDENTIFY THE RELATIONSHIPS

To calculate the total momentum, we can treat each car as a point particle. We then have (Eq. 7.2)

$$\vec{p}_{total} = \sum_i \vec{p}_i = \sum_i m_i \vec{v}_i = m_1 \vec{v}_1 + m_2 \vec{v}_2$$

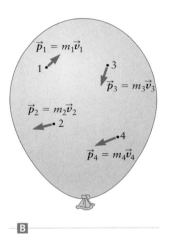

The total momentum is equal to the sum of the momenta of all the individual pieces.

Figure 7.1 **A** A solid object such as a boomerang can be thought of as a system composed of very small pieces, that is, a collection of point particles. **B** The gas in a balloon is a system of molecules. The total momentum is the sum of the individual momenta of all the molecules.

Figure 7.2 Example 7.2.

where m_1 and m_2 are the masses of the cars and $\vec{v}_1$ and $\vec{v}_2$ are the cars' velocities. In this example, the motion is one dimensional, so we can work in terms of the x components of $\vec{p}$ and $\vec{v}$. From Figure 7.2, we have $v_1 = +20$ m/s for the smaller car and $v_2 = -15$ m/s for the larger car. Here, the signs *do* matter.

SOLVE

The total momentum of our system of two cars is thus

$$p_{total} = m_1 v_1 + m_2 v_2$$

$$p_{total} = (1200 \text{ kg})(20 \text{ m/s}) + (2000 \text{ kg})(-15 \text{ m/s}) = \boxed{-6000 \text{ kg} \cdot \text{m/s}}$$

What does it mean?

Even though the smaller car has a higher speed, the magnitude of the momentum of the larger car is greater because of its greater mass. For this reason, the total momentum is in the $-x$ direction, that is, toward the left in Figure 7.2.

CONCEPT CHECK 7.1 | Momentum and Kinetic Energy

Consider a rock of mass m moving with a speed v. Suppose the mass of the rock is doubled, while the speed is cut in half. Which of the following statements is true?

(a) The momentum of the rock is the same, but its kinetic energy is smaller.
(b) The momentum and kinetic energy of the rock are both unchanged.
(c) The momentum of the rock is greater, and its kinetic energy is smaller.

7.2 | FORCE AND IMPULSE

According to Newton's first law, if there is no force acting on a particle, the velocity of the particle will be constant. Hence, for a simple particle with a mass m, the momentum $\vec{p} = m\vec{v}$ is also constant. However, we also know that if a net force is applied to the particle, its velocity will change in response to this applied force. So, let's consider a force $\vec{F}$ acting on a particle with a constant mass m and calculate how the momentum changes.

Using Newton's second law, we have $\sum \vec{F} = m\vec{a}$. For simplicity, we assume the force and hence also the acceleration are constant. (We'll soon generalize our results to cases in which the force is not constant.) For ease of notation, we denote the total force on the particle as simply $\vec{F}$. During the course of a very short time interval Δt, the acceleration is related to changes in the velocity according to $\vec{a} = \Delta\vec{v}/\Delta t$. Combining these two facts gives

$$\vec{F} = m\vec{a} = m\frac{\Delta\vec{v}}{\Delta t} = m\frac{\vec{v}_f - \vec{v}_i}{\Delta t} \tag{7.3}$$

Here, we imagine that the particle has a velocity $\vec{v}_i$ just before the force is applied and that after a time Δt the velocity is $\vec{v}_f$. From the definition of momentum in Equation 7.1, the initial momentum of the particle is $\vec{p}_i = m\vec{v}_i$ and the final momentum is $\vec{p}_f = m\vec{v}_f$. Inserting this all into Equation 7.3 and rearranging gives

$$\vec{F} \Delta t = m(\vec{v}_f - \vec{v}_i) = \vec{p}_f - \vec{p}_i = \Delta\vec{p} \tag{7.4}$$

where $\Delta\vec{p}$ is the change in the momentum. The product $\vec{F} \Delta t$ is called the *impulse*. According to Equation 7.4, when a force acts on a particle, the change of the momentum of the particle is equal to the impulse:

Impulse theorem

$$\text{impulse} = \vec{F} \Delta t = \Delta\vec{p} \tag{7.5}$$

Equation 7.5 is called the *impulse theorem*, and it follows directly from Newton's laws. Notice that impulse is a vector; its direction is parallel to the total force on the particle. Because impulse is the product of force and time, a particular value of the impulse can be obtained with a small force acting for a long time or with a large force acting for a short time.

Impulse Associated with a Variable Force

In our derivation of the relation between impulse and momentum in Equation 7.5, we assumed the force acts for a very short time Δt and that the force is constant during this time interval as in Figure 7.3A. In many cases involving impulse and momentum, however, the force is not constant. Such a case is illustrated in Figure 7.4, which shows a tennis ball being struck by a racket. The force exerted by the racket on the ball has the form sketched in Figure 7.3B: the magnitude of the force grows rapidly from zero to some maximum value and then decreases to zero as the ball leaves the racket. For the simpler case of a constant force, the impulse is simply $\vec{F} \Delta t$ (Eq. 7.5), which is the area under the corresponding force–time curve in Figure 7.3A. When the force is not constant, we can calculate the impulse by dividing the entire time period into many small intervals and using Equation 7.5 for each interval. The total impulse is then the sum of the impulses during each interval. The case of the interaction between a tennis racket and ball is illustrated in Figure 7.3B. The total impulse is equal to the area under the complete force–time curve

$$\text{impulse} = \text{area under the force–time curve} = \Delta \vec{p} \qquad (7.6)$$

In both cases—with a constant force as in Figure 7.3A or a force that varies with time as in Figure 7.3B—the impulse is equal to the change in the momentum of the object. Hence, the impulse theorem we initially derived in Equation 7.5 for the case of a constant force also applies for the impulse produced by a force that varies with time.

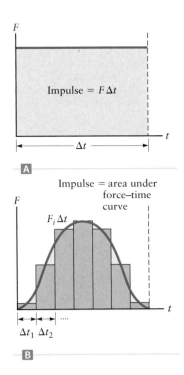

Figure 7.3 Impulse is equal to the area under the force–time graph when the force is constant, as in **A**, and when the force varies with time, as in **B**. When the force varies with time, the impulse during a time interval Δt_i is approximately equal to the area of a rectangular bar of height F_i.

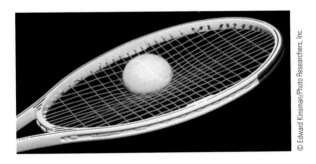

Figure 7.4 The force exerted by a tennis racket on a ball is usually not constant. In such cases, the impulse is still equal to the area under the force–time graph, as illustrated in Figure 7.3B.

EXAMPLE 7.3 | Impulse and a Home Run

Consider a baseball player who hits a baseball as sketched in Figure 7.5. The ball travels from the pitcher toward the batter with a speed of 45 m/s (about 100 mi/h). The batter then hits it directly back at the pitcher with a speed of 50 m/s. If the ball has a mass of 0.14 kg, what is the impulse imparted by the bat to the baseball?

RECOGNIZE THE PRINCIPLE

We are given the initial and final velocities of the baseball, so we can find the impulse using Equation 7.5. We must be careful to use the correct signs (+ and −) for the velocities.

SKETCH THE PROBLEM

Figure 7.5 shows the problem. The initial velocity is along −x because the ball is moving to the left, while the final velocity is positive (to the right along +x).

Figure 7.5 Example 7.3. **A** The ball has some initial momentum as it travels to the batter and **B** some final value of momentum after being hit by the bat. The impulse imparted to the baseball equals the change in its momentum. Here, $\Delta\vec{p}$ is equal to the final momentum minus the initial momentum.

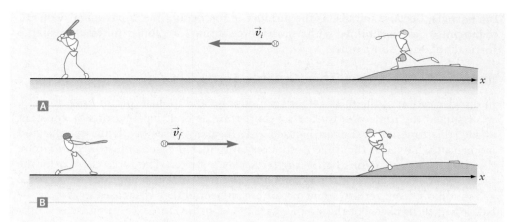

IDENTIFY THE RELATIONSHIPS

We begin with the impulse theorem:

$$\text{impulse} = \Delta\vec{p} = m\vec{v}_f - m\vec{v}_i \qquad (1)$$

This problem is one dimensional since the initial and final velocities are both along the line that extends from the batter to the pitcher, the x direction in Figure 7.5. The force exerted by the bat on the ball is also parallel to this direction. We can thus work simply in terms of the components of the velocity and momentum vectors along x. Using the coordinate system in Figure 7.5, we have $v_i = -45$ m/s and $v_f = +50$ m/s.

SOLVE

Inserting our values of m, v_i, and v_f into Equation (1), the impulse is

$$\Delta p = mv_f - mv_i = (0.14 \text{ kg})(50 \text{ m/s}) - (0.14 \text{ kg})(-45 \text{ m/s}) = \boxed{13 \text{ kg} \cdot \text{m/s}}$$

What does it mean?

The initial velocity and the initial momentum are both in the $-x$ direction. The bat then imparts a large positive impulse to the ball, so the final momentum of the ball is positive, directed toward the right in Figure 7.5B.

Figure 7.6 A baseball is in contact with a bat for approximately 1 ms (= 0.001 s). During this short time, the force exerted by the bat on the ball is very large.

© Chuck Savage/Corbis

Impulse and the Average Force

For cases such as the force exerted by a tennis racket on a ball in Figure 7.4, it is usually very difficult to calculate the precise form of the force–time curve, so it is also difficult to apply Equation 7.6 to find the impulse. However, analyzing the impulse in a graphical manner is very useful in a different way. Often, the interaction time—the length of time during which the force is nonzero—is very short. For example, when a baseball is hit by a bat, high-speed photography (Fig. 7.6) shows that the bat and ball are in contact for approximately $\Delta t = 0.001$ s. We can approximate the force–time curve by assuming F is constant during this short interaction time and write

$$\text{impulse} = \vec{F}_{\text{ave}} \Delta t = \Delta\vec{p} \qquad (7.7)$$

where $\vec{F}_{\text{ave}}$ is the *average* interaction force as indicated graphically in Figure 7.7. The value of F_{ave} is chosen so that the area under the rectangular region bounded by F_{ave} in Figure 7.7 is equal to the area under the true force–time curve. In many cases, this average force is all we are interested in. For example, if we assume the baseball in Example 7.3 is in contact with the bat for a time $\Delta t = 0.001$ s, the average force is related to the change in momentum by

$$\vec{F}_{\text{ave}} \Delta t = \Delta\vec{p}$$

We can then use the result for the impulse found in Example 7.3 to calculate the size of this force. In terms of the magnitudes, we get

$$F_{ave} = \frac{\Delta p}{\Delta t} = \frac{13 \text{ kg} \cdot \text{m/s}}{0.001 \text{ s}} \approx 1 \times 10^4 \text{ N} \qquad (7.8)$$

The force on the baseball is thus *extremely* large. The weight of the ball is $mg = (0.14 \text{ kg})g = 1.4$ N, so the force exerted by the bat in Equation 7.8 is about 10,000 times the ball's weight! Such "collision" forces are often very large because the corresponding interaction times are usually very short.

Figure 7.7 We can find the average force associated with an impulse by approximating the force–time relation by a constant (F_{ave}) during the time interval over which the force acts. The dashed horizontal line shows the (constant) average force over this time interval.

Minimizing Collision Forces

Now that we have introduced the concept of impulse, it is worthwhile to revisit a problem from Chapter 3. In Section 3.6, we considered a child jumping from a tall ladder to the ground below, and we analyzed the forces on the child's legs during the landing. We can also view this example as an impulse problem. The child has a certain momentum (her "initial" momentum) just before she hits the ground, and her final momentum is zero because she comes to rest. The impulse exerted by the ground on the child can be calculated in the same way we found the impulse on the batted baseball in Example 7.3. In Chapter 3, we discussed strategies for minimizing the force on the child, and we found that it is safest for her to allow her legs to bend and flex as much as possible. By doing so, she is making the interaction time Δt as long as possible. Our understanding of impulse and force (Eq. 7.7) tells us that increasing Δt will reduce the average force F_{ave}. The effect of Δt on the force is illustrated graphically in Figure 7.8. A particular value of the impulse (the area under the force–time curve) can be obtained with a large force acting for a short time or with a small force acting for a long time. There are many other examples of this relationship between force and collision time. For example, the air bags in your car are designed to take advantage of this trade-off. When your car stops suddenly, as in an accident, the air bags inflate and fill the region between you (the driver) and the steering wheel and front windshield. Your collision with an air bag then involves a much longer interaction time than if you were to collide with the steering column or windshield. This longer collision time leads to a smaller force, and a safer accident.

Figure 7.8 Impulse is equal to the area under the force–time curve. A particular value of the impulse can be produced by a small force acting for a long time or by a large force acting for a short time.

> **CONCEPT CHECK 7.2 | Impulse on a Bouncing Ball**
>
> Consider the collision of a bouncing ball with a floor as sketched in Figure 7.9. What is the direction of the impulse imparted by the floor to the ball?
>   (a) Horizontal and to the right in Figure 7.9, $\Delta \vec{p} = \rightarrow$
>   (b) Vertical and upward, $\Delta \vec{p} = \uparrow$
>   (c) At 45° from the vertical direction, $\Delta \vec{p} = \nearrow$

Figure 7.9 Concept Check 7.2.

7.3 | CONSERVATION OF MOMENTUM

The concept of a system played an important role in our discussion of energy and the principle of conservation of energy in Chapter 6. We can also apply our ideas about momentum and the impulse theorem to the motion of a system of particles. Consider the system comprised of just two colliding point particles sketched in Figure 7.10. For the moment, we ignore any rotational or spinning motion. With this assumption, let's use the notion of impulse to analyze how a collision affects the subsequent motion of the two particles.

When two objects collide, there is an associated collision force. For example, if the objects are two hockey pucks or two billiard balls, the collision force is a normal (contact) force. Let $\vec{F}_1$ be the force exerted by object 2 on object 1 and $\vec{F}_2$ be the force exerted by object 1 on object 2. These forces are an action–reaction pair,

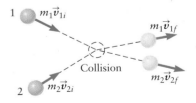

Figure 7.10 When two objects collide, the total momentum just after the collision is equal to the total momentum just before the collision. Momentum is conserved.

so according to Newton's third law, they must be equal in magnitude and opposite in direction. Hence,

$$\vec{F}_1 = -\vec{F}_2 \tag{7.9}$$

If this collision takes place over a time interval Δt, the impulse imparted to object 1 is $\vec{F}_1 \Delta t$. According to the impulse theorem, this impulse equals the change in momentum of object 1, so we have

$$\Delta \vec{p}_1 = \vec{p}_{1f} - \vec{p}_{1i} = \vec{F}_1 \Delta t \tag{7.10}$$

Here, $\vec{p}_{1i}$ is the (initial) momentum of object 1 just before the collision and $\vec{p}_{1f}$ is the (final) momentum just after the collision. Likewise, the impulse imparted to object 2 is $\vec{F}_2 \Delta t$, which equals the change in the momentum of object 2:

$$\Delta \vec{p}_2 = \vec{p}_{2f} - \vec{p}_{2i} = \vec{F}_2 \Delta t \tag{7.11}$$

Since $\vec{F}_1 = -\vec{F}_2$ and the interaction times Δt are the same in Equations 7.10 and 7.11, the impulse imparted to object 1 is equal in magnitude but opposite in sign to the impulse imparted to object 2. We thus have

$$\Delta \vec{p}_1 = \vec{F}_1 \Delta t = -\vec{F}_2 \Delta t = -\Delta \vec{p}_2$$

$$\Delta \vec{p}_1 = -\Delta \vec{p}_2 \tag{7.12}$$

We can also write Equation 7.12 as

Conservation of momentum in a collision

$$\Delta \vec{p}_1 + \Delta \vec{p}_2 = \Delta \vec{p}_{\text{total}} = 0 \tag{7.13}$$

Hence, the collision does *not* change the *total* momentum of the two particles. Whatever momentum is lost or gained by one particle in the collision is gained or lost by the other. The *total momentum of the system is conserved.*

Conservation of Momentum for a System of Many Particles

Our derivation of the conservation of momentum in Equation 7.13 was based on a simple system of just two particles undergoing a single collision, but the result is much more general. We can apply the same argument to a system containing a very large number of particles, in which the particles undergo many collisions with one another. Further, we assume this system is *closed*, meaning that there are no external forces exerted on any of the particles; that is, the only forces on a particle are exerted by other particles within the system. These internal forces are due to collisions between particles in the system; each collision can be viewed as in Figure 7.10, and the sum of the momenta of the two particles is the same before and after the collision (Eq. 7.13). The same argument applies to all the collisions between particles in the system, so we conclude that the total momentum of the entire system is conserved.

This result is very similar to what we found when we considered the total energy of a closed system in Chapter 6. Recall that the particles in a closed system can exchange energy among one another in various ways, but the particles do not gain or lose energy to objects outside the system. Energy that is gained or lost by one particle is thus lost or gained by other particles in the system, and the total energy of the system is conserved. The same notion applies to the total momentum. The particles in a closed system do not experience any forces from objects external to the system. The momentum gained or lost by one particle in a collision is lost or gained by other particles in the system, and the total momentum of the system is conserved.

This analysis has so far assumed the system is composed of a collection of particles that are traveling more or less independently, except for occasional collisions. It also applies to a solid object such as a boomerang, a football, or a planet. We can think of a solid object as being composed of many point particles; these individual particles are subject to strong forces from other particles in the system, and these forces hold the object together. These forces will always come in action−reaction pairs, so for any two interacting particles, we always have $\vec{F}_1 = -\vec{F}_2$, just as in Equation 7.9. The same reasoning then tells us that any momentum lost by one particle will be gained by the other. Hence, we again conclude that the total momentum of our system, in this case a solid object, is conserved, provided there are no external forces acting on the object.

Momentum Conservation and External Forces

We have just shown that for a closed system of particles, the interaction forces between particles do not change the total momentum of the system. These interactions cause some particles to lose momentum while others gain momentum, but the total momentum of the system stays the same. However, these arguments apply only to the internal forces, that is, the forces that act between particles within the system. There may also be forces acting from outside the system, called external forces. As an example, consider again the two-particle system in Figure 7.10 and assume these two particles are asteroids in outer space. If these two asteroids are near a star, the gravitational force from the star is an external force on this two-asteroid system. This external force causes the asteroids to accelerate toward the star and hence changes the total momentum of the two asteroids. Although collisions between the asteroids do not alter the momentum of this system, the external force from the star does change the total momentum of the system of two asteroids.

In practice, it is difficult to find situations in which the external forces on a system are exactly zero. For example, when dealing with any system of terrestrial objects, the Earth's gravitational force is always present as an external force. However, our analysis of the collision between a baseball and a bat (Eq. 7.8) showed that collision forces are often much larger than this gravitational force. For the baseball−bat collision, the collision force was about 10,000 times larger than the weight of the ball. Thus, while the total momentum of the bat−baseball system is not conserved precisely during this collision, it will change by only about 0.01% (1 part in 10,000). For this reason, the notion of momentum conservation is still a very useful way to analyze most collisions.

7.4 | COLLISIONS

In Section 7.3, we analyzed a system of two particles and showed that the total momentum of the particles before a collision is equal to the total momentum of the particles after the collision. In this section, we apply this principle of momentum conservation to analyze several different types of collisions. In general, a collision changes the particle velocities; the final velocities (as found just after the collision) are different from the initial velocities (from just before the collision). The kinetic energy of a particle is proportional to the square of the speed, so the kinetic energy of any particular particle is also changed by the collision. Collisions fall into two general types, depending on what happens to the total kinetic energy of the entire system (e.g., the two particles involved in the collision). In an *elastic collision*, the system's kinetic energy is conserved: the total kinetic energy of the two particles after the collision is equal to their total kinetic energy before the collision. In contrast, *inelastic collisions* are collisions in which some kinetic energy is lost. In both types of collisions, *momentum is conserved*.

The term *elastic* can help us understand how and why a collision affects the kinetic energy. For example, an extremely elastic ball (such as a Super Ball) is compressed in a collision, and this compression stores energy in the ball just as energy is stored in a compressed spring. In an ideal Super Ball, all this potential energy is turned back into kinetic energy when the ball uncompresses ("springs back") at the end of the collision. Collisions involving such objects will therefore tend to be elastic. On the other hand, consider a collision involving a ball composed of soft putty or clay. Such a ball does not "spring back" at the end of the collision, causing the kinetic energy after the collision to be smaller than the initial kinetic energy. The collision is thus inelastic.

We next analyze a few examples of these different types of collisions and show how conservation principles regarding momentum (for all collisions) and kinetic energy (for elastic collisions) can be used to predict the motions of two colliding particles. Our analyses will be guided by the following general approach.

PROBLEM SOLVING | Analyzing a Collision

1. **RECOGNIZE THE PRINCIPLE.** The momentum of a system is conserved only when the external forces are zero. For a collision between two particles, the system is just the two particles. The principle of momentum conservation can be applied when the collision force between the particles is much larger than the external forces.

2. **SKETCH THE PROBLEM.** Make a sketch of the system, showing the coordinate axes and (where possible) the initial and final velocities of the particles in the system.

3. **IDENTIFY THE RELATIONSHIPS.**
   • Express the conservation of momentum condition (Eq. 7.13) for the system.
   • Is kinetic energy conserved?

Yes → Elastic collision

No → Inelastic collision

Elastic collision

4. **SOLVE.**
   • Express the conservation of kinetic energy for both particles.
   • Use the equations describing conservation of momentum and kinetic energy to solve for unknown quantities, such as the final velocities.

Inelastic collision

4. **SOLVE** for unknown quantities using the equations derived from the conservation of momentum.

5. Always *consider what your answer means* and check that it makes sense.

Insight 7.1

ALWAYS BEGIN BY IDENTIFYING THE SYSTEM

When applying the principle of conservation of momentum, it is important to first identify the system. It is usually best to choose the system so that all the important forces act between different parts of the system and the external forces are zero or at least small. If the external forces are zero, the total momentum of the system will be conserved exactly.

Elastic Collisions in One Dimension

Let's apply the above strategy for dealing with collisions to an elastic collision in one dimension involving particles with masses m_1 and m_2 that are subject to no external forces. Step 1: Because the external forces are zero, the momentum of the system (the two colliding particles) will be conserved. Step 2 is the sketch in Figure 7.11, which shows the particles and their velocities before and after the collision. The particles are moving along the x coordinate axis before the collision with velocities v_{1i} and v_{2i}; since this problem is one dimensional, we only need to be concerned with the x components of the velocities. After the collision, the velocities are v_{1f} and v_{2f}. In a typical problem, the values of the initial or final velocities might be given. The total momentum of the two particles is conserved, so we have

$$\overbrace{m_1 v_{1i} + m_2 v_{2i}}^{\text{initial momentum}} = \overbrace{m_1 v_{1f} + m_2 v_{2f}}^{\text{final momentum}} \tag{7.14}$$

Elastic collision in one dimension

This is an elastic collision, so the total kinetic energy of the system is also conserved, which gives

$$\overbrace{\tfrac{1}{2} m_1 v_{1i}^2 + \tfrac{1}{2} m_2 v_{2i}^2}^{\text{initial kinetic energy}} = \overbrace{\tfrac{1}{2} m_1 v_{1f}^2 + \tfrac{1}{2} m_2 v_{2f}^2}^{\text{final kinetic energy}} \tag{7.15}$$

In a typical collision, the initial velocities are known, and we have two equations—Equations 7.14 and 7.15—to solve for the two unknowns, the final velocities v_{1f} and v_{2f}.

A Collision between Two Billiard Balls

Let's now use Equations 7.14 and 7.15 to analyze a one-dimensional elastic collision between two billiard balls. We assume one of the balls is initially at rest and the other has an initial speed v_0 as sketched in Figure 7.12A. Because our problem is one dimensional, the collision is "head-on" and the initial and final velocities are all along the same direction. Experienced players know that the spin of a billiard ball can have an extremely important effect on a collision. For simplicity, though, we'll ignore all effects of spin in this problem. In the game of billiards, all balls have the same mass m. Inserting $v_{1i} = v_0$ and $v_{2i} = 0$ into our relations for the momentum (Eq. 7.14) and the kinetic energy (Eq. 7.15), we find

$$mv_0 = mv_{1f} + mv_{2f} \tag{7.16}$$

$$\tfrac{1}{2}mv_0^2 = \tfrac{1}{2}mv_{1f}^2 + \tfrac{1}{2}mv_{2f}^2 \tag{7.17}$$

Solving for v_{2f} in Equation 7.16 gives

$$v_{2f} = v_0 - v_{1f} \tag{7.18}$$

and inserting this into Equation 7.17, we get

$$\tfrac{1}{2}mv_0^2 = \tfrac{1}{2}mv_{1f}^2 + \tfrac{1}{2}m(v_0 - v_{1f})^2$$

Canceling the factors of $\tfrac{1}{2}m$ and solving for v_{1f} leads to

$$v_0^2 = v_{1f}^2 + (v_0 - v_{1f})^2 = v_{1f}^2 + v_0^2 - 2v_0v_{1f} + v_{1f}^2$$

Subtracting v_0^2 from each side and collecting terms gives

$$0 = v_{1f}^2 - 2v_0v_{1f} + v_{1f}^2 = 2v_{1f}^2 - 2v_0v_{1f}$$

$$2v_{1f}^2 = 2v_0v_{1f} \tag{7.19}$$

This equation has two solutions. One solution is $v_{1f} = 0$, and using Equation 7.18 then gives $v_{2f} = v_0$. Physically, ball 1 comes to a complete stop after the collision and ball 2 moves away with a velocity of v_0 afterward (Fig. 7.12B). This solution is familiar to many pool players.

The second solution of Equation 7.19 is $v_{1f} = v_0$, and using Equation 7.18 leads to $v_{2f} = 0$. Hence, with this solution, ball 2 stays at rest and ball 1 simply passes through it. You can think of this as a collision between two transparent balls. While you may not have expected this solution for a collision problem, it certainly does conserve both kinetic energy and momentum. This solution would only apply if the collision force between the two balls is zero, and that does not occur in any real collision between two objects.

EXAMPLE 7.4 | **Elastic Collision between Balls of Different Mass**

Consider again an elastic collision between two balls with one initially at rest (Fig. 7.12A), but this time assume the balls have different masses. Repeat the analysis of this collision and find the final velocities of both balls.

RECOGNIZE THE PRINCIPLE

We can again apply conservation of momentum (Eq. 7.14) and conservation of kinetic energy (Eq. 7.15). Our goal is to understand how the relative masses of the particles affect the final velocities. We follow the steps outlined in the problem-solving box on analyzing a collision. Step 1: The external forces are zero, so the momentum of the colliding particles will be conserved.

Before collision

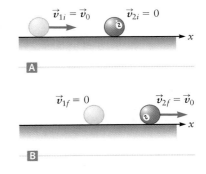

A

After collision

B

Figure 7.11 In an elastic collision, both momentum and kinetic energy are conserved. **A** Velocities just before a one-dimensional collision. **B** Velocities just after the collision.

Figure 7.12 Elastic collision in one dimension between two billiard balls. **A** One of the balls is initially at rest. **B** After the collision, the other ball is at rest.

Figure 7.13 Example 7.4.
Analyzing an elastic collision.

SKETCH THE PROBLEM

Step 2: Figure 7.13 shows the problem. We denote the masses as m_1 and m_2. One of them is initially at rest, so the velocities before the collision are $v_{1i} = v_0$ and $v_{2i} = 0$. The final velocities (denoted v_{1f} and v_{2f}) are unknowns that we wish to find. This information is all collected in Figure 7.13.

IDENTIFY THE RELATIONSHIPS

We continue with our problem-solving strategy. Step 3: The conservation of momentum condition gives

$$\overbrace{m_1 v_{1i} + m_2 v_{2i}}^{\text{initial momentum}} = m_1 v_0 = \overbrace{m_1 v_{1f} + m_2 v_{2f}}^{\text{final momentum}} \tag{1}$$

We are given that kinetic energy is conserved, so we have an elastic collision.

SOLVE

Step 4: Setting the initial and final kinetic energies equal, we find

$$\overbrace{\tfrac{1}{2}m_1 v_{1i}^2 + \tfrac{1}{2}m_2 v_{2i}^2}^{\text{initial } KE} = \tfrac{1}{2}m_1 v_0^2 = \overbrace{\tfrac{1}{2}m_1 v_{1f}^2 + \tfrac{1}{2}m_2 v_{2f}^2}^{\text{final } KE} \tag{2}$$

We proceed by using the momentum conservation relation (Eq. 1) to solve for the final velocity of mass 2:

$$v_{2f} = \frac{m_1(v_0 - v_{1f})}{m_2} \tag{3}$$

Next, we substitute this result into the conservation of kinetic energy expression, Equation (2):

$$\frac{1}{2}m_1 v_0^2 = \frac{1}{2}m_1 v_{1f}^2 + \frac{1}{2}m_2 \left[\frac{m_1(v_0 - v_{1f})}{m_2} \right]^2$$

We can now solve for the final velocity of mass 1 and then use Equation (3) to find the final velocity of mass 2. We'll leave the algebra for an end-of-chapter question (Question 9). We find two solutions. Solution 1 is

$$v_{1f} = v_0 \quad \text{and} \quad v_{2f} = 0$$

In this solution, there is no collision at all because mass 1 simply passes through 2; that is, the final velocity of mass 1 is equal to its initial velocity, while 2 is at rest before and after the collision. These results for the final velocities do conserve kinetic energy and momentum, so they certainly satisfy our conditions for conservation of total momentum and total kinetic energy. However, this particular solution is only possible if the collision force between the two balls is zero, which does not occur in a real collision. There is a second solution for the final velocities that does describe a real collision. It is

$$v_{1f} = \left(\frac{m_1 - m_2}{m_1 + m_2} \right) v_0 \quad \text{and} \quad v_{2f} = \left(\frac{2m_1}{m_1 + m_2} \right) v_0$$

What does it mean?

These results for the final velocities of the two balls correspond to an actual collision, and we explore some of their consequences in the end-of-chapter problems. We note one prediction here, however: if m_2 is very small compared with m_1, then

$$v_{1f} = \left(\frac{m_1 - m_2}{m_1 + m_2} \right) v_0 \approx \left(\frac{m_1 - 0}{m_1 + 0} \right) v_0 = v_0$$

Hence, the final velocity of a very massive particle (m_1) is approximately unchanged by the collision with a very light particle (m_2). When a car collides with a mosquito, the car's velocity barely changes.

The Power of Conservation Principles

In dealing with the elastic collisions in Figures 7.11 through 7.13, we started with the conditions for conservation of momentum and conservation of kinetic energy. For a one-dimensional collision, this process leads to two equations (Eqs. 7.14 and 7.15). In a typical problem, we might know the initial velocities of the two objects and be asked to find their final velocities. Hence, our two equations are all we need to solve for these two unknowns. This result is remarkable: it means that we can work out the complete outcome of the collision without knowing *anything* about the force that acts between the two objects during the collision. For an elastic collision in one dimension, the conditions for conservation of kinetic energy and momentum *completely* determine the results. The nature of the interaction force, the collision time, and so forth all have no effect on the outcome of the collision. This example illustrates that conservation laws are an extremely powerful way to get from general principles to specific results.

Inelastic Collisions in One Dimension

We have just analyzed some collisions in which kinetic energy was conserved. However, in many collisions the total kinetic energy after a collision is smaller than the kinetic energy just prior to the collision. These collisions are called *inelastic*. Although some portion of the kinetic energy of the two colliding objects is "lost" in such a collision, the *total* energy of the universe is still conserved. This "lost" kinetic energy goes into other forms of energy, such as potential energy and heat energy. Although kinetic energy is not conserved in an inelastic collision, the total momentum of a closed system (a system not subject to external forces) is still conserved. Hence, for a one-dimensional inelastic collision, we can still use the conservation of momentum relation

$$m_1 v_{1i} + m_2 v_{2i} = m_1 v_{1f} + m_2 v_{2f} \qquad (7.20)$$

However, this equation does not give enough information to allow us to solve for the final velocities of both objects. We have only one relation, but two unknown quantities, v_{1f} and v_{2f}. So, to analyze an inelastic collision fully, we need additional information.

Completely Inelastic Collisions

The simplest type of inelastic collision is one in which the two objects stick together after the collision so that the objects have the same final velocity. This is called a ***completely inelastic collision***. A good example of a completely inelastic collision is one in which two cars lock bumpers. Figure 7.14A shows two cars of mass m_1 and m_2 coasting on a straight, one-dimensional road with velocities v_{1i} and v_{2i}. The cars collide and lock bumpers, and they have the same velocity v_f after the collision (Fig. 7.14B). The cars are moving in a horizontal direction (x), and if they are coasting, there are no external forces on the cars in this direction. Hence, the total momentum along x is conserved and we can apply our problem-solving strategy for dealing with collisions. Writing the condition for conservation of momentum along x gives

$$m_1 v_{1i} + m_2 v_{2i} = m_1 v_f + m_2 v_f \qquad (7.21)$$

In this case, we have only one unknown, the final velocity v_f, so the conservation of momentum relation in Equation 7.21 is all we need. Solving for v_f, we have

$$m_1 v_{1i} + m_2 v_{2i} = (m_1 + m_2) v_f$$

$$v_f = \frac{m_1 v_{1i} + m_2 v_{2i}}{m_1 + m_2} \qquad (7.22)$$

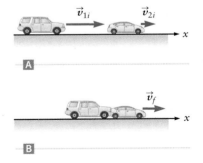

Figure 7.14 **A** Two cars on a collision course in one dimension. **B** In a completely inelastic collision, the two objects stick together after the collision.

Completely inelastic collision in one dimension

EXAMPLE 7.5 | A Completely Inelastic Collision between Two Cars

Two cars, an SUV of mass 2500 kg and a compact model of mass 1200 kg, are coasting along a straight road. Their initial velocities along this coordinate direction are 40 m/s and 10 m/s, respectively, so the SUV is overtaking the compact car. When they collide, the cars lock bumpers. Find the cars' velocity just after the collision.

RECOGNIZE THE PRINCIPLE

The cars "stick together" after the collision, so this is another example of a completely inelastic collision, and we can use our problem-solving strategy for collisions. We first (step 1) note that the two cars are the system. Since they are both coasting, there are no external forces along the horizontal direction (x), and the total momentum along x is conserved.

SKETCH THE PROBLEM

Step 2: The collision is described in Figure 7.14. It is a one-dimensional collision, so the given initial velocities are the velocity components along our coordinate direction (x), which we choose as parallel to the road.

IDENTIFY THE RELATIONSHIPS

Step 3: The cars have masses $m_1 = 2500$ kg (the SUV) and $m_2 = 1200$ kg (the compact car), with initial velocities $v_{1i} = 40$ m/s and $v_{2i} = 10$ m/s (Fig. 7.14A). The final velocity of the two cars together is unknown, and we denote it by v_f. Continuing with our problem-solving strategy, conservation of momentum leads to the results in Equations 7.21 and 7.22. This collision is inelastic, so the kinetic energy of the two cars is not conserved.

SOLVE

Inserting the given values of the masses and initial velocities into Equation 7.22 leads to

$$v_f = \frac{m_1 v_{1i} + m_2 v_{2i}}{m_1 + m_2} = \frac{(2500 \text{ kg})(40 \text{ m/s}) + (1200 \text{ kg})(10 \text{ m/s})}{2500 \text{ kg} + 1200 \text{ kg}} = \boxed{30 \text{ m/s}}$$

What have we learned?
The final velocity lies between the two initial velocities, but v_f is closer to the initial velocity of the larger car because it had a larger fraction of the initial momentum.

CONCEPT CHECK 7.3 | Hitting a Parked Car

A careless driver is coasting at 20 m/s (about 40 mi/h) along a straight, level road in a car of mass 1400 kg when she collides with a parked car. The parked car is in neutral, and its brakes are off. The two cars lock bumpers and move off together. If the final velocity is 10 m/s, what is the mass of the parked car?
  (a) 700 kg   (b) 1400 kg   (c) 2800 kg

Inelastic Collisions: What Happens to the Kinetic Energy?

We have already mentioned that kinetic energy is not conserved in an inelastic collision. Let's compare the initial and final kinetic energies for the specific case in Example 7.5. The kinetic energy before the collision is

$$KE_i = \tfrac{1}{2}m_1 v_{1i}^2 + \tfrac{1}{2}m_2 v_{2i}^2$$

Inserting the values of the masses and velocities from Example 7.5 leads to

$$KE_i = \tfrac{1}{2}m_1 v_{1i}^2 + \tfrac{1}{2}m_2 v_{2i}^2 = \tfrac{1}{2}(2500 \text{ kg})(40 \text{ m/s})^2 + \tfrac{1}{2}(1200 \text{ kg})(10 \text{ m/s})^2$$

$$KE_i = 2.1 \times 10^6 \text{ J}$$

The cars stick together after the collision, so they have the same final velocity v_f. The final kinetic energy is therefore

$$KE_f = \tfrac{1}{2}m_1 v_{1f}^2 + \tfrac{1}{2}m_2 v_{2f}^2 = \tfrac{1}{2}(m_1 + m_2)v_f^2$$

Inserting the value of the final velocity from Example 7.5 gives

$$KE_f = \tfrac{1}{2}(m_1 + m_2)v_f^2 = \tfrac{1}{2}(2500 \text{ kg} + 1200 \text{ kg})(30 \text{ m/s})^2 = 1.7 \times 10^6 \text{ J}$$

Comparing the values of KE_i and KE_f, we see that the final kinetic energy is about 20% smaller than the initial kinetic energy. In this example, this "missing" energy would be converted to several other forms, including heat, sound, and elastic energy associated with the compression of the bumpers.

Now consider a completely inelastic "head-on" collision involving two objects. For example, suppose two cars of equal mass are moving with equal speeds, but in opposite directions. In this case, $v_{1i} = -v_{2i}$, and we also have $m_1 = m_2 = m$. Inserting these values into the result for v_f for a completely inelastic collision (Eq. 7.22) gives

$$v_f = \frac{m_1 v_{1i} + m_2 v_{2i}}{m_1 + m_2} = \frac{m(v_{1i} - v_{1i})}{2m} = 0$$

Hence, the two objects come to a complete stop after the collision, and the final kinetic energy is zero. *All* the initial kinetic energy is converted into other forms of energy.

Collisions in Two Dimensions

Dealing with collisions in two dimensions involves the same basic ideas as in one dimension. Now, however, the final velocity of each object involves two unknowns: the two components of the velocity vector. Hence, for two colliding objects, there are usually four unknowns (two velocity components for each object). The momentum conservation relation then becomes

$$\vec{p}_{1i} + \vec{p}_{2i} = \vec{p}_{1f} + \vec{p}_{2f}$$

where $\vec{p}_{1i}$ is the initial momentum of object 1, $\vec{p}_{1f}$ is its final momentum, and so forth. It is useful to write this relation in terms of components:

$$p_{1ix} + p_{2ix} = p_{1fx} + p_{2fx} \quad \text{and} \quad p_{1iy} + p_{2iy} = p_{1fy} + p_{2fy} \tag{7.23}$$

To analyze a two-dimensional collision, we can again follow the basic approach outlined in our problem-solving strategy for analyzing a collision. The momentum conservation relation now includes both components of the momentum (Eq. 7.23). We must then include whatever additional information is available. This additional information might be that the collision is elastic and hence that total kinetic energy is conserved, or we might have some information regarding the final velocity of one of the objects.

A Collision in Two Dimensions: A Rocket, an Asteroid, and Saving the Earth

Our planet is constantly being bombarded with asteroids. These asteroids are usually very small and disintegrate completely as they enter the Earth's atmosphere. On occasion, however, a very large asteroid reaches the Earth's surface and causes a great deal of damage. It is believed that such a collision formed Meteor Crater, an impact crater 1.2 km in diameter located in eastern Arizona (Fig. 7.15). The collision between the Earth and this asteroid involved a very large amount of energy, and it has been proposed that a similar collision involving a much larger asteroid led to the extinction of the dinosaurs. For this reason, NASA is concerned about what might happen to life on the Earth when another large asteroid strikes the Earth at some future time. NASA scientists (and others) have therefore considered how to use rockets to deflect incoming asteroids before they reach the Earth. The collision between

Figure 7.15 This crater (known as Meteor Crater) was produced when an asteroid struck the Earth in eastern Arizona about 49,000 years ago. This crater is about 1.2 km across.

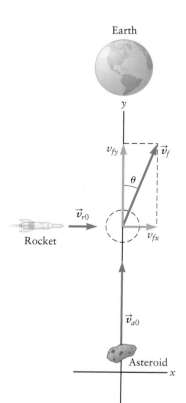

Figure 7.16 The collision between a rocket and an asteroid could be used to deflect an asteroid that is heading toward the Earth. If the rocket embeds (sticks) in the asteroid, the collision would be completely inelastic.

such a rocket and an asteroid is sketched in Figure 7.16. For simplicity, we assume the rocket hits the asteroid and then becomes embedded in it as was the case when the asteroid that produced Meteor Crater struck the Earth. Hence, this collision is completely inelastic. Given the masses of the rocket and asteroid and their initial velocities, we want to calculate the common final velocity of the two. From that result, we can determine if the asteroid will, or will not, subsequently strike the Earth.

We can analyze this collision using our problem-solving strategy for dealing with collisions. Step 1: The system consists of the two colliding "particles," the rocket and the asteroid. The external forces on this system are due to gravity from the Sun and the Earth, which will be much smaller than the collision force when the rocket hits the asteroid. Hence, to very good accuracy, the momentum of this system will be conserved. Step 2: We choose the coordinate system in Figure 7.16 with the initial velocity of the asteroid in the $+y$ direction, aimed directly toward the Earth. The asteroid's initial velocity is then $v_{ax} = 0$, $v_{ay} = v_{a0}$. The rocket approaches the asteroid with an initial velocity in the $+x$ direction, so the rocket's initial velocity is $v_{rx} = v_{r0}$, $v_{ry} = 0$. The rocket "sticks to" the asteroid after the collision, so the rocket and asteroid have the same final velocity, and we denote the components of this final velocity by v_{fx} and v_{fy}. Step 3: We can now write the momentum conservation relations for our system of the rocket plus the asteroid. For the momentum along the x direction, we have

$$\underbrace{m_a v_{ax} + m_r v_{rx}}_{\text{initial momentum along } x} = m_r v_{r0} = \underbrace{(m_a + m_r) v_{fx}}_{\text{final momentum along } x}$$

The corresponding result for the momentum along y is

$$\underbrace{m_a v_{ay} + m_r v_{ry}}_{\text{initial momentum along } y} = m_a v_{a0} = \underbrace{(m_a + m_r) v_{fy}}_{\text{final momentum along } y}$$

Step 4: Solving for the components of the final velocity, we find

$$v_{fx} = \frac{m_r v_{r0}}{m_a + m_r} \tag{7.24}$$

$$v_{fy} = \frac{m_a v_{a0}}{m_a + m_r} \tag{7.25}$$

We can now calculate the final speed using $v_f = \sqrt{v_{fx}^2 + v_{fy}^2}$ and express the final direction in terms of the angle $\theta = \tan^{-1}(v_{fx}/v_{fy})$ (see Fig. 7.16).

Now let's put in some realistic values to see if this asteroid deflection plan can actually work. A rocket of reasonable size has a mass of approximately $m_r = 2 \times 10^6$ kg, and a large asteroid has a mass of $m_a = 1 \times 10^{18}$ kg. If we assume the approach velocity of the rocket is similar to the velocity attained when NASA sent the Apollo spacecrafts to the Moon, then $v_{r0} = 1000$ m/s. The velocity of an approaching asteroid depends on its orbital motion about the Sun and how that orbit intersects the Earth's path, so this velocity will depend on the asteroid's trajectory as it approaches the Earth. To get an estimate, let's assume the asteroid's initial velocity is equal to the Earth's orbital velocity relative to the Sun, which is approximately 3×10^4 m/s. Inserting all these values into Equations 7.24 and 7.25 leads to

$$v_{fx} = \frac{m_r v_{r0}}{m_a + m_r} = \frac{(2 \times 10^6 \text{ kg})(1000 \text{ m/s})}{(1 \times 10^{18} + 2 \times 10^6) \text{ kg}} = 2 \times 10^{-9} \text{ m/s} \qquad (7.26)$$

and

$$v_{fy} = \frac{m_a v_{a0}}{m_a + m_r} = \frac{(1 \times 10^{18} \text{ kg})(3 \times 10^4 \text{ m/s})}{(1 \times 10^{18} + 2 \times 10^6) \text{ kg}} = 3 \times 10^4 \text{ m/s}$$

These values for the final velocity of the asteroid plus rocket show that asteroid deflection is a very tough problem. The value of v_{fx} is *extremely* small because the mass of the asteroid is much, much larger than the mass of the rocket. In addition, v_{fy} is essentially equal to the initial velocity of the asteroid, again because the rocket is much smaller than the asteroid. The real test of success (or failure) can be determined by calculating the angle through which the asteroid is deflected. That angle is

$$\theta = \tan^{-1}\left(\frac{v_{fx}}{v_{fy}}\right) = \tan^{-1}\left(\frac{2 \times 10^{-9}}{3 \times 10^4}\right)$$

$$\theta = 7 \times 10^{-14} \text{ rad} = 4 \times 10^{-12} \text{ degrees} \qquad (7.27)$$

This angle is very small, so the deflection will be quite small. It is not easy to deflect an asteroid.

CONCEPT CHECK 7.4 | Saving the Earth

The deflection angle in the rocket−asteroid collision in Figure 7.16 (and Eq. 7.27) is very small. Which of the following changes would increase the deflection angle? More than one choice may apply.
(a) Increase the mass of the rocket.
(b) Decrease the mass of the rocket.
(c) Increase the speed of the rocket.
(d) Decrease the speed of the rocket.
(e) Pick a slower asteroid.
(f) Options (a), (c), and (e) would all work.
(g) Options (b) and (d) would both work.

CONCEPT CHECK 7.5 | Meteor Crater

The collision between the Earth and the asteroid that caused Meteor Crater was an inelastic collision. Where did the kinetic energy go?
(a) Sound
(b) Heat energy
(c) Seismic waves in the Earth
(d) All the above
(e) None of the above

EXAMPLE 7.6 | ⓑ Forces in an Automobile Collision

Two cars approach an intersection as shown schematically in Figure 7.17. One of the cars ignores a stop sign, resulting in a right-angle collision with the cars locked together after the collision. What is the approximate average force between the two cars during the collision? Assume they each have a mass of $m = 1000$ kg and they both are traveling with an initial speed of 20 m/s (about 45 mi/h).

RECOGNIZE THE PRINCIPLE

This problem requires that we combine what we know about force and impulse together with the principle of conservation of momentum. The geometry is similar to the rocket−asteroid problem in Figure 7.16, and we can take the same approach to calculating the final velocity of the two cars $\vec{v}_f$. From $\vec{v}_f$, we can use the impulse theorem to get the impulse exerted by one car on the other. If we then estimate the collision time, we can get the average force exerted by one car on the other during the collision.

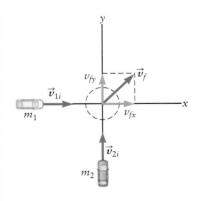

Figure 7.17 Example 7.6.

SKETCH THE PROBLEM

Figure 7.17 shows the problem. The initial velocities are along both x (car 1) and y (car 2), so we will have to account for momentum conservation in two dimensions.

IDENTIFY THE RELATIONSHIPS

Noticing the directions of $\vec{v}_{1i}$ and $\vec{v}_{2i}$ in Figure 7.17, we can write the components of the initial velocities of the two cars as $v_{1x} = 20$ m/s, $v_{1y} = 0$, and $v_{2x} = 0$, $v_{2y} = 20$ m/s. The collision is a completely inelastic right-angle collision, just as with the rocket and asteroid in Figure 7.16, so we can modify Equations 7.24 and 7.25 to get

$$v_{fx} = \frac{mv_{1x}}{m + m} = \frac{v_{1x}}{2}$$

and

$$v_{fy} = \frac{mv_{2y}}{m + m} = \frac{v_{2y}}{2}$$

Inserting the given values for the initial velocities, we find

$$v_{fx} = v_{fy} = 10 \text{ m/s} \tag{1}$$

We can now use the impulse theorem to relate the collision force exerted on one of the cars to the change in the momentum of that car.

SOLVE

Writing the impulse theorem (Eq. 7.5) for the x component of the momentum of car 2, we have

$$\text{impulse on car 2 along } x = \Delta p_{2x}$$

Expressing the impulse along x in terms of the average force along x (Eq. 7.7) gives

$$F_{\text{ave}, x} \, \Delta t = \Delta p_{2x}$$

The initial momentum of car 2 along x is zero, which leads to

$$F_{\text{ave}, x} \Delta t = \Delta p_{2x} = mv_{fx} - mv_{ix} = mv_{fx}$$

$$F_{\text{ave}, x} = \frac{mv_{fx}}{\Delta t}$$

For our final step, we need to know the collision time Δt. We estimate a value of approximately $\Delta t = 0.1$ s. As usual, we aim here for accuracy to within a factor of 10. A value of $\Delta t = 1$ s seems too long, whereas 0.01 s seems too short. We found v_{fx} in Equation (1). Using that result, we get

$$F_{\text{ave}, x} = \frac{mv_{fx}}{\Delta t} = \frac{(1000 \text{ kg})(10 \text{ m/s})}{(0.1 \text{ s})} = \boxed{1 \times 10^5 \text{ N}}$$

What does it mean?
This force is quite large, as we should expect; cars are massive objects. Notice also that this is just the x component of the collision force. We'll leave the calculation of the force along y to the end-of-chapter problems; from the symmetry of this problem in terms of x and y, however, you should suspect that the y component of the collision force equals $F_{\text{ave}, x}$.

CONCEPT CHECK 7.6 | How Big Is the Collision Force?

In Example 7.6, we estimated the collision force in an automobile accident. Which of the following statements is correct?
  (a) The magnitude of the collision force is much larger than a car's weight.
  (b) The magnitude of the collision force is much smaller than a car's weight.
  (c) The magnitude of the collision force is approximately equal to a car's weight.

7.5 | USING MOMENTUM CONSERVATION TO ANALYZE INELASTIC EVENTS

Momentum is the product of mass and velocity, and in all our examples so far, the mass has been fixed (that is, unchanging). When the momentum of an object or system with a constant mass is conserved, the velocity is constant. In some situations, though, the mass of an object or system can change with time, and it is interesting to consider how the principle of conservation of momentum applies in such cases.

Figure 7.18 shows an empty railroad car coasting at a speed v_0 along a level track. The track is straight and lies along the x direction. A load of gravel is then dropped into the car as it moves along. Given the masses of the empty car m_c and the load of gravel m_g, what is the speed of the railroad car after the gravel is dropped into it?

This problem can be analyzed in two ways: (1) we can view the railroad car as an object whose mass increases abruptly at some point during its motion, or (2) we can treat the problem as a "collision" involving the gravel and the railroad car. In either case, we can use the principle of momentum conservation.

Let's first consider this problem as involving an object (the railroad car) whose mass changes. The railroad car moves on a straight track, so its motion is one dimensional, along the x direction in Figure 7.18. If we assume there is no frictional force on the car (from bearings in the wheels, etc.), there are no external forces along x, and the component of the momentum along this direction is conserved. The initial momentum of the car along x is $m_c v_0$. Prior to landing in the car, the gravel has no momentum along x, so the total initial momentum of the car plus the gravel along x is also $m_c v_0$. After the gravel is added to the car, the car plus gravel has a velocity v_f, so the momentum conservation relation has the form

$$\underbrace{m_c v_0}_{\substack{\text{initial momentum} \\ \text{along } x}} = \underbrace{(m_c + m_g)v_f}_{\substack{\text{final momentum} \\ \text{along } x}} \qquad (7.28)$$

Solving for the final velocity gives

$$v_f = \frac{m_c v_0}{m_c + m_g}$$

You can see from this result that v_f is always smaller than v_0, so the railroad car will always slow down after the gravel is dropped into it.

We can also arrive at Equation 7.28 by treating this problem as a completely inelastic collision between the railroad car and the gravel. Prior to the collision, the gravel's momentum along x is zero, while that of the car is $m_c v_0$. The system of railroad car plus gravel thus has an initial momentum along x, which is $m_c v_0$. Applying momentum conservation along x then leads to Equation 7.28.

Our analysis so far has only considered momentum along x; we should also consider what happens to the momentum along y during this collision. Your first expectation may be that the momentum along y should also be conserved. Before the collision, the gravel has momentum along the $-y$ direction, whereas the car has no momentum along y. After the collision, the motion is solely along x, so the final momentum along y is zero. Hence, the momentum of the gravel plus the car along y is *not* conserved in this example. Why not? The answer can be seen by considering the forces on the gravel and the car during the collision, as the gravel lands in the car. When the gravel hits the car, a vertical force is exerted by the car on the gravel, bringing the gravel to a stop along the y direction in the car. A corresponding normal force exerted by the railroad track on the car prevents the car from falling through the track and moving along y. The force exerted by the track on our system (the car plus gravel) is an external force; it imparts an impulse to the system and hence produces a change in the system's momentum along y. For this

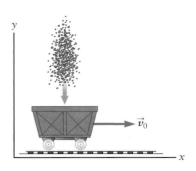

Figure 7.18 This system consists of the railroad car and the load of gravel dropped into it. The momentum of this system along the x direction is conserved.

reason, the momentum of this system along y is not conserved. The momentum of a system is conserved *only* if the forces from outside the system are negligible. In this example, there is an external force along the y direction, so the momentum of the system (the car plus the gravel) along this direction is not conserved. However, the momentum of the system along the x direction is still conserved since there are no external forces along x.

Applying the Principle of Conservation of Momentum to Inelastic Events

The principle of momentum conservation can be applied to inelastic events, such as the railroad car in Figure 7.18, using the same problem-solving strategy we have used for dealing with collisions. That strategy was introduced with collisions in mind, so it is useful to state a new version written specifically for inelastic events. The basic ideas are the same: whenever the external forces on a system are zero or negligible, the total momentum of the system is conserved.

PROBLEM SOLVING | Analyzing an Inelastic Event

1. **RECOGNIZE THE PRINCIPLE.** The momentum of a system in a particular direction is conserved only when the net external force in that direction is zero or negligible.

2. **SKETCH THE PROBLEM.** Make a sketch of the system, including a coordinate system, and use the given information to determine (where possible) the initial and final velocities of the system's components.

3. **IDENTIFY THE RELATIONSHIPS.**
   - Express the conservation of momentum condition for the direction(s) identified in step 1.
   - Use any given information to determine the increase or decrease of the kinetic energy.

4. **SOLVE** for the unknown quantities, such as the final velocities or the force exerted on one of the particles in the system.

5. Always *consider what your answer means* and check that it makes sense.

EXAMPLE 7.7 | Momentum on Ice

Two angry children are fighting while standing on a very slippery icy surface (Fig. 7.19). Their masses are $m_1 = 25$ kg and $m_2 = 40$ kg, and they are initially at rest. The larger child pushes the smaller child, who then slides away to the left with a speed of 1.5 m/s. What is the final velocity of the larger child?

RECOGNIZE THE PRINCIPLE

We follow the strategy outlined in the problem-solving box on inelastic events. Step 1: The two children are the "system." The ice is very slippery, so the external forces in the horizontal (x) direction are negligible and the total momentum along x is conserved.

SKETCH THE PROBLEM

Continuing with our problem-solving strategy (step 2), the problem is sketched in Figure 7.19, which shows the initial and final velocities. Initially, the children are at rest, so $v_{1i} = v_{2i} = 0$, and we denote their final velocities as v_{1f} and v_{2f}.

IDENTIFY THE RELATIONSHIPS

Step 3: The total initial momentum of this system along the horizontal (x) is zero, and the final momentum along x is $m_1 v_{1f} + m_2 v_{2f}$. The final momentum must be equal to the initial momentum, so

Figure 7.19 Example 7.7.

$$\overbrace{0}^{\text{initial momentum}} = \overbrace{m_1 v_{1f} + m_2 v_{2f}}^{\text{final momentum}}$$

SOLVE

Solving for the final velocity of child 2 gives

$$v_{2f} = -\frac{m_1 v_{1f}}{m_2} = -\frac{(25\text{ kg})(-1.5\text{ m/s})}{40\text{ kg}} = \boxed{0.94\text{ m/s}}$$

What does it mean?
The final velocity of child 1 is to the left, so his velocity is negative, which leads to a positive velocity (to the right) for child 2.

Inelastic Processes Are Similar to Collisions

Most inelastic processes are very similar to collisions. Consider again the two children in Example 7.7. In Figure 7.20A, the children are initially at rest, whereas after pushing on each other they are traveling in opposite directions (Fig. 7.20B). A closely related situation is sketched in parts C and D of Figure 7.20. Here, the children are initially moving toward each other; then they collide and stick together, which is just a completely inelastic collision. Comparing these two scenarios, we see that the inelastic event in which the children push each other apart is just like a completely inelastic collision in "reverse." In both scenarios, the total momentum is conserved.

> **CONCEPT CHECK 7.7** | Children and Forces in an Inelastic Event
>
> Consider again the two children in Example 7.7, who push on each other as they fight. If the push lasts for 0.50 s, what is the magnitude of the average force exerted by the larger child on the smaller one?
>   (a) 25 N   (c) 75 N
>   (b) 38 N   (d) Zero

Figure 7.20 **A** and **B**: An inelastic event in which two children on ice push on each other and move apart. **C** and **D**: An inelastic collision in which the two children come together and "collide."

Splitting Asteroids

Let's now take another look at the daunting task of deflecting an asteroid that threatens the Earth. In our previous analysis, we tried to deflect the asteroid by arranging for an inelastic collision with a rocket (Fig. 7.16) and found that it is difficult to deflect an asteroid very much with a rocket of reasonable size. We will therefore try a different approach that has been employed in several recent (science-fiction) movies. This approach uses a large explosive device (a bomb) to split the asteroid into two pieces as illustrated in Figure 7.21. It is another example of an inelastic event, and we can attack it using our problem-solving strategy for inelastic events. The two pieces of the asteroid are the system, and because the external forces are very small (we assume we are far from the Sun, so its gravitational force is small), the momentum of this system is conserved. For simplicity, we assume the two pieces are of equal mass m and choose the coordinate axis x to pass through the center of the asteroid (Fig. 7.21A). Before the explosion, the x component of the velocity is zero, so the momentum along x is also zero. Since there are no external forces on the system (the asteroid plus the explosive device), the final momentum along x is also zero. Hence, the two pieces will separate after the explosion with velocities that are equal in magnitude and opposite in direction. We therefore take $\pm v_f$ to be the x components of the velocity of the two pieces of the asteroid after the explosion (Fig. 7.21B).

To calculate v_f, we need to know something about the final kinetic energy. Suppose an energy E_{exp} is released in the explosion. If we are optimistic, we might hope

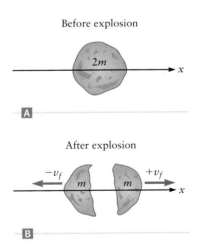

Figure 7.21 When this asteroid is split into two pieces by an explosion, the explosion releases energy, so the kinetic energy of the system—the pieces of the asteroid—is larger after the explosion. However, the forces from outside the system are zero, so the total momentum is still conserved.

that this energy will all go into the final kinetic energy.[1] Assuming that to be true, we have

$$\tfrac{1}{2}mv_f^2 + \tfrac{1}{2}mv_f^2 = mv_f^2 = E_{\text{exp}}$$

Here, there are two terms equal to $\tfrac{1}{2}mv_f^2$ because the asteroid is split into two pieces, each of mass m. We can now solve for the final velocity and find

$$v_f = \sqrt{\frac{E_{\text{exp}}}{m}} \tag{7.29}$$

To decide if this approach will succeed, we need to estimate the value of v_f with some realistic numbers. The mass of a typical large asteroid is 1×10^{18} kg, so if the asteroid is split into two equal-sized pieces, we would have $m = 5 \times 10^{17}$ kg. We must next consider the energy we can obtain from the bomb because it will be providing the final kinetic energy of the two asteroid pieces. The energy released by one of the atomic bombs that was dropped in World War II was approximately 1×10^{17} J. Let's assume that with a contemporary thermonuclear explosive we can increase that value by a factor of 10, so $E_{\text{exp}} = 1 \times 10^{18}$ J. Inserting these values into Equation 7.29 gives

$$v_f = \sqrt{\frac{E_{\text{exp}}}{m}} = \sqrt{\frac{1 \times 10^{18} \text{ J}}{5 \times 10^{17} \text{ kg}}} \approx 1 \text{ m/s}$$

which is *much* larger than the lateral velocity we found for a collision with a rocket (Eq. 7.26). Hence, splitting the asteroid with a bomb is the better approach for achieving a significant deflection. Even in this case, though, v_f is not very large, so the task is still not easy. Notice that the momentum of our asteroid system is conserved since the *external* forces on it are zero. Although the bomb's explosion certainly does exert forces on the asteroid, the bomb is part of the system so these are *internal* forces. The kinetic energy is certainly *not* conserved because the energy of the bomb goes into the final kinetic energy of the system, the two pieces of the asteroid.

7.6 | CENTER OF MASS

In the previous sections, we distinguished between *internal* and *external* forces, that is, between forces that act between particles within a system and forces that come from outside the system. This distinction is important for understanding the motion of a composite object, that is, an object composed of many point particles such as the boomerang in Figure 7.1A. Before we attempt to deal with an object as complicated as a boomerang, let's consider the simpler case of two point masses connected by a light rod (Fig. 7.22A). We can write Newton's second law for the two combined masses as

$$\sum \vec{F} = \vec{F}_1 + \vec{F}_2 = m_1\vec{a}_1 + m_2\vec{a}_2 \tag{7.30}$$

where $\vec{F}_1$ is the total force acting on m_1 and $\vec{F}_2$ is the total force acting on m_2. A portion of the force acting on m_1 will come from its interaction with m_2 and is an internal force. The rest of the force on m_1 will come from outside the system and is thus an external force. The same argument applies to the force on m_2. Let's therefore write the total force on the left-hand side of Equation 7.30 as a sum of the external and internal forces:

$$\underset{\substack{\text{total} \\ \text{force}}}{\sum \vec{F}} = \underset{\substack{\text{external} \\ \text{forces}}}{\sum \vec{F}_{\text{ext}}} + \underset{\substack{\text{internal} \\ \text{forces}}}{\sum \vec{F}_{\text{int}}}$$

The force exerted by m_1 on m_2 and the force exerted by m_2 on m_1 are *internal* forces.

x_{CM} is the "balance point" of the system.

Figure 7.22 **A** A system consisting of two particles connected by a massless rod. The forces exerted by each particle on the other (through the rod) are internal forces. **B** The center of mass of the two point particles is the "balance" point along a line that connects them.

---

[1] In most explosions, some of the energy also goes into heat.

The internal forces always come in action–reaction pairs; for the example in Figure 7.22A, the force exerted by m_1 on m_2 is equal in magnitude and opposite in sign to the force exerted by m_2 on m_1. Hence, when we consider the entire system, the internal forces will cancel so that

$$\sum \vec{F}_{\text{int}} = 0 \qquad (7.31)$$

While we have derived this result for a system of just two particles, Equation 7.31 is true for a system with any number of particles (including a boomerang). The internal forces always come in action–reaction pairs, so the sum of all the internal forces is always zero. Thus, Equation 7.30 becomes

$$\sum \vec{F}_{\text{ext}} = m_1 \vec{a}_1 + m_2 \vec{a}_2$$

If we denote the total mass of the system as $m_1 + m_2 = M_{\text{tot}}$, we can write this relation as

$$\sum \vec{F}_{\text{ext}} = m_1 \vec{a}_1 + m_2 \vec{a}_2 = M_{\text{tot}} \vec{a}_{CM} \qquad (7.32)$$

where $\vec{a}_{CM}$ is the acceleration of what is called the **center of mass** of the system. Equation 7.32 has precisely the same form as Newton's second law for a point particle. In words, it says that the total external force ($\sum \vec{F}_{\text{ext}}$) on a system of particles is equal to the mass of the system (M_{tot}) times a new quantity called the acceleration of the center of mass.

What Is the Center of Mass and How Is It Useful?

So what is the "center of mass"? According to Equation 7.32, the acceleration of the center of mass is related to the accelerations of all the pieces of the system. We can rearrange Equation 7.32 to find $\vec{a}_{CM}$ for a system of just two particles moving in one dimension. Solving for $\vec{a}_{CM}$ gives

$$\vec{a}_{CM} = \frac{m_1 \vec{a}_1 + m_2 \vec{a}_2}{M_{\text{tot}}}$$

This relation suggests that we can express the position of the center of mass along the x coordinate direction in terms of the positions of all the pieces of the system:

$$x_{CM} = \frac{m_1 x_1 + m_2 x_2}{M_{\text{tot}}} \qquad (7.33)$$

Here, x_1 and x_2 are the positions of the two particles measured along the x axis in Figure 7.22A, and x_{CM} is the position of the center of mass along x. If the two particles have equal masses m, according to Equation 7.33 the center of mass is at

$$x_{CM} = \frac{m_1 x_1 + m_2 x_2}{M_{\text{tot}}} = \frac{m x_1 + m x_2}{2m} = \frac{x_1 + x_2}{2}$$

The center of mass of this system is thus at the point midway between the two particles. On the other hand, if one particle is more massive than the other, the center of mass is closer to the larger one. Intuitively, you can think of the center of mass as the "balance point"; that is, the center of mass is the point on the connecting rod in Figure 7.22B where the system would be "in balance."

This result for the center of mass for two particles can be generalized to a system of many particles in two or three dimensions. The result is

$$x_{CM} = \frac{\sum_i m_i x_i}{\sum_i m_i} = \frac{\sum_i m_i x_i}{M_{\text{tot}}}$$

$$y_{CM} = \frac{\sum_i m_i y_i}{\sum_i m_i} = \frac{\sum_i m_i y_i}{M_{\text{tot}}}$$

(7.34)   Center of mass

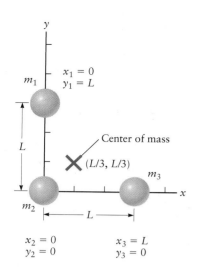

Figure 7.23 The center of mass of a collection of particles is given by Equation 7.34.

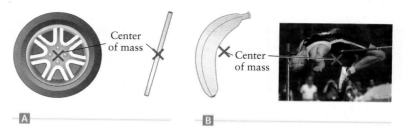

Figure 7.24 Approximate location of the center of mass of various objects. **A** For a symmetric object, the center of mass is located at the center of symmetry (i.e., the center of the object). **B** For an object with a complicated shape, the center of mass may not lie inside the object.

with a similar expression for z_{CM}. Here, the sums are over all the particles (i) in the system. The total mass of the system is just

$$M_{tot} = \sum_i m_i$$

When using Equation 7.34 to find the location of the center of mass, you must first choose a coordinate system (with an origin). The values of x_{CM} and y_{CM} then refer to the same coordinate system.

For a collection of several masses, computing the center of mass requires a sum involving the positions and masses of the objects. For example, if there are three particles of equal mass m arranged in plane as shown in Figure 7.23, the coordinates of the center of mass are given by

$$x_{CM} = \frac{\sum_i m_i x_i}{\sum_i m_i} = \frac{m_1 x_1 + m_2 x_2 + m_3 x_3}{m_1 + m_2 + m_3} = \frac{m(0) + m(0) + m(L)}{3m} = \frac{L}{3}$$

and

$$y_{CM} = \frac{\sum_i m_i y_i}{\sum_i m_i} = \frac{m_1 y_1 + m_2 y_2 + m_3 y_3}{m_1 + m_2 + m_3} = \frac{m(L) + m(0) + m(0)}{3m} = \frac{L}{3}$$

The calculation of the center of mass of a solid object, such as a baseball bat or a boomerang, requires a sum over all the pieces of the object; that is, a sum over all the individual "point particles" that make up the object is needed. Such a calculation can be complicated, especially if the mass is not distributed uniformly (as is usually the case for a baseball bat) or if the shape is complex (as for a boomerang); even so, the intuitive notion of "balance point" can be very helpful in locating the center of mass. Thus, for a symmetric object such as a simple rod or wheel, the center of mass is at the center of symmetry of the object. Notice also that the center of mass need not be located within an object. The center of mass of a banana may not be inside the banana (Fig. 7.24B).

EXAMPLE 7.8 | Center of Mass of a Bracket

You work for a carpenter and are building a bracket composed of two straight sections of wood as sketched in Figure 7.25. You want to show your boss that physics can be useful, so you decide to calculate the position of the center of mass of the bracket. (Knowing this position might help in supporting the bracket during a construction project.) If the two sections of wood have lengths $L_1 = 1.2$ m and $L_2 = 2.0$ m and masses $m_1 = 0.50$ kg and $m_2 = 1.0$ kg, what is the location of the center of mass of the entire bracket?

RECOGNIZE THE PRINCIPLE

We could find the center of mass of this system of particles by applying the definition in Equation 7.34 and summing over all the pieces that make up the bracket. However, a simpler approach uses the notion of the center of mass as a balance point. With that in mind, we first find the center of mass of each straight section of the bracket. We

Figure 7.25 Example 7.8. This bracket is composed of two separate pieces. Each piece is a simple board, one lies along x and one along y, and they overlap near the origin.

can treat these two pieces as particles of mass m_1 and m_2 located at their respective centers of mass and then compute the location of the overall center of mass of this two-particle system.

SKETCH THE PROBLEM

Figure 7.25 shows the problem.

IDENTIFY THE RELATIONSHIPS

For each piece of wood, we use the balance-point notion of center of mass to tell us that the center of mass of each piece will be at each respective center as indicated in Figure 7.25. Thus, the center of mass of piece 1 is at

$$x_{1,CM} = 0 \quad \text{and} \quad y_{1,CM} = \frac{L_1}{2} = 0.60 \text{ m}$$

and for piece 2, we have

$$x_{2,CM} = \frac{L_2}{2} = 1.0 \text{ m} \quad \text{and} \quad y_{2,CM} = 0$$

SOLVE

We can now treat this bracket as a system of two particles of mass m_1 and m_2 at these two locations and use the relations for the center of mass coordinates in Equation 7.34 to find the center of mass of the entire bracket:

$$x_{CM} = \frac{m_1 x_{1,CM} + m_2 x_{2,CM}}{m_1 + m_2} = \frac{(0.50 \text{ kg})(0) + (1.0 \text{ kg})(1.0 \text{ m})}{(0.50 \text{ kg}) + (1.0 \text{ kg})} = \boxed{0.67 \text{ m}}$$

$$y_{CM} = \frac{m_1 y_{1,CM} + m_2 y_{2,CM}}{m_1 + m_2} = \frac{(0.50 \text{ kg})(0.60 \text{ m}) + (1.0 \text{ kg})(0)}{(0.50 \text{ kg}) + (1.0 \text{ kg})} = \boxed{0.20 \text{ m}}$$

The center of mass is outside the bracket.

What have we learned?
When an object has a simple shape (e.g., a wheel, ball, or a straight piece of wood), the center of mass is at the center of symmetry of the object. For more complicated shapes, Equation 7.34 must be used to find the center of mass.

CONCEPT CHECK 7.8 | Finding the Center of Mass

A sledgehammer has a massive metal head attached to a much lighter wooden handle (Fig. 7.26). Which location is most likely to be the sledgehammer's center of mass?

Motion of the Center of Mass

Now that we have seen how to calculate the location of the center of mass, let's consider how it *moves*. Figure 7.27 shows two ice skaters; they are initially very close together, so they form a system with the center of mass initially located as shown in part A. We also assume the skaters begin at rest, so their center of mass is initially at rest. If the skaters then push very forcefully on each other, they will move apart as sketched in Figure 7.27B.

If the ice is very slippery (i.e., frictionless), the skaters will move with constant velocities after they separate since there are no forces on either skater along the horizontal (x) direction. Moreover, the *external* force on this system along x is zero the entire time (before, during, and after the skaters push each other), so the x component of the total momentum of the two skaters is conserved. The initial velocities of the two skaters are zero, so the initial momentum of the system (before the push) is zero and the final momentum must also be zero. The final momentum

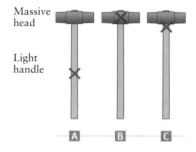

Massive head

Light handle

A   B   C

Figure 7.26 Concept Check 7.8.

Initial center
of mass

A

Final center
of mass

B

Figure 7.27 **A** These two skaters are initially at rest, and there are no external forces along the horizontal direction (*x*). **B** After they push on each other, the skaters move apart, but the center of mass does not move.

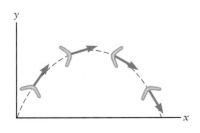

Figure 7.28 An object such as a boomerang may rotate or tumble when it is thrown, but the center of mass follows a parabolic trajectory characteristic of projectile motion.

of the system can be written in terms of the final velocities of the two skaters (see Fig. 7.27), which gives

$$p_{\text{final}} = m_1 v_{1f} + m_2 v_{2f} = 0 \qquad (7.35)$$

We have already noted that the skaters move with constant velocities, so if we assume they both start at the origin (they are initially very close together), their positions are then given by $x_1 = v_{1f}t$ and $x_2 = v_{2f}t$ and we have

$$m_1 v_{1f} + m_2 v_{2f} = m_1 \frac{x_1}{t} + m_2 \frac{x_2}{t}$$

Combining this with Equation 7.35, we find

$$m_1 \frac{x_1}{t} + m_2 \frac{x_2}{t} = 0$$

If we now cancel the factor of *t*, we obtain

$$m_1 x_1 + m_2 x_2 = 0 \qquad (7.36)$$

According to Equation 7.34, the center of mass of the two skaters is given by

$$x_{CM} = \frac{m_1 x_1 + m_2 x_2}{m_1 + m_2}$$

Notice that according to Equation 7.36, the numerator in this expression for x_{CM} is equal to zero. So, the center of mass of the two skaters is at $x_{CM} = 0$. Moreover, this result applies at all values of *t*. Hence, even though the skaters move after the push, their center of mass does not move. It stays in the same location, before and after the push.

We could have anticipated this result from our original derivation of the center of mass with Equation 7.32, where we found

$$\sum \vec{F}_{\text{ext}} = M_{\text{tot}} \vec{a}_{CM} \qquad (7.37)$$

which is simply Newton's second law written for the center of mass. In words, Equation 7.37 says that a system of particles moves as if all the mass (M_{tot}) is located at the center of mass and that the center of mass motion is caused *only by the external forces* acting on the system.

Translational Motion of a System

Consider the motion of a complicated object such as a boomerang (Fig. 7.28). A boomerang has an asymmetrical shape, which causes it to spin in a complicated manner when it is thrown. This complicated motion can be viewed as a combination of *translational motion* of the center of mass together with *rotational motion* as the boomerang spins about an axis that extends through its center of mass. Translational motion is often referred to as *linear motion*. Here the term *linear* does not mean motion along a straight line; rather, it means motion in which the center of mass moves along a path such as the parabolic trajectories we found in projectile motion. Translational motion is different from rotational motion, in which an object spins or rotates about a particular axis. According to Equation 7.37, the translational motion of any system of particles is described by Newton's second law as applied to an equivalent particle of mass M_{tot}. This equivalent particle is located at the center of mass, and for the purposes of the translational motion, we can treat the motion as if all the mass were located at the center of mass. We can then use everything we know about the motion of a point particle to understand the motion of more complicated objects such as boomerangs. For example, we know that if the force due to air drag is small, the center of mass of the boomerang will move in a simple parabolic trajectory as calculated in Chapter 4. Even though the boomerang may appear to spin in a very complicated manner as it moves through the air, the center of mass motion will be *precisely the same* as that of a simple point particle.

The concept of center of mass allows us to deal with the translational motion of any complicated object. According to Equation 7.37, the center of mass always moves as if all the mass were located at the center of mass of the object. The rotational motion of the object can then be treated separately; that is a problem we'll tackle in Chapters 8 and 9.

7.7 | A BOUNCING BALL AND MOMENTUM CONSERVATION

The impulse theorem and the principle of momentum conservation can be used to treat many different situations. In this section, we consider several more examples to illustrate the power of this approach. The first example involves one aspect of the game of pool. A pool player often wants to bounce the cue ball off a "rail" at the edge of the table as sketched in Figure 7.29. A good player must be able to predict how a ball will bounce off the rail. The ball approaches the rail at a certain angle θ_i (Fig. 7.29) and we want to calculate the outgoing angle θ_f. We can attack this problem using the principle of momentum conservation.

When the ball collides with the rail, the force exerted by the rail changes the momentum of the ball. For a typical ball and table, the cue ball has nearly the same speed after the collision as it had before hitting the rail, so for simplicity let's assume the initial and final speeds are equal. The final kinetic energy of the ball will therefore equal the initial kinetic energy, so we could also think of this collision as an elastic process. The force of the rail on the cue ball is just a normal force and is directed perpendicular to the rail. Using the coordinate system in Figure 7.29, this normal force is along the y direction and will therefore impart an impulse to the ball in the y direction. There is no force on the ball along the x direction, so the momentum of the ball along x is conserved,[2] and we can write

$$mv_{ix} = mv_{fx}$$

Hence, the initial and final components of the velocity along x are equal:

$$v_{ix} = v_{fx} \tag{7.38}$$

Since the initial and final kinetic energies of the ball are equal, we also have

$$\tfrac{1}{2}mv_i^2 = \tfrac{1}{2}mv_f^2$$

$$\tfrac{1}{2}m\sqrt{v_{ix}^2 + v_{iy}^2} = \tfrac{1}{2}m\sqrt{v_{fx}^2 + v_{fy}^2} \tag{7.39}$$

which leads to

$$\sqrt{v_{ix}^2 + v_{iy}^2} = \sqrt{v_{fx}^2 + v_{fy}^2}$$

$$v_{ix}^2 + v_{iy}^2 = v_{fx}^2 + v_{fy}^2 \tag{7.40}$$

From Equation 7.38, we have $v_{ix} = v_{fx}$, so the terms involving these quantities in Equation 7.40 cancel, leaving

$$v_{iy}^2 = v_{fy}^2$$

which gives

$$v_{fy} = \pm v_{iy}$$

We thus have two solutions for the final y component of the velocity, and we have to decide which one describes the reflection in Figure 7.29. The cue ball bounces from the rail,[3] so the final velocity is directed opposite to the initial velocity and

Top view of a cue ball bouncing from the edge of a pool table.

Figure 7.29 When this cue ball bounces from a wall, the force $\vec{F}$ of the wall on the ball is along y, so the momentum of the ball along x is conserved. Here, $\vec{F}$ is a normal force directed perpendicular to the wall.

---

[2]Things will be more complicated if the ball has sidespin; as is well known by experienced pool players, sidespin can lead to a force on the ball along x.

[3]The other solution corresponds a ball that passes through the rail, much like some of the solutions we saw in Section 7.4 in which two particles "pass through" each other instead of colliding.

the solution that describes the reflection is $v_{fy} = -v_{iy}$. The direction of the cue ball's final velocity (see Fig. 7.29) is then

$$\theta_f = \tan^{-1}\left|\frac{v_{fx}}{v_{fy}}\right| = \tan^{-1}\left|\frac{v_{ix}}{v_{iy}}\right| = \theta_i$$

The outgoing angle is thus equal to the incoming angle. This result, called a *specular reflection*, is no surprise to an experienced pool player This kind of reflection will come up again when we study the reflection of light from a mirror in Chapter 24.

7.8 | THE IMPORTANCE OF CONSERVATION PRINCIPLES IN PHYSICS

Conservation of momentum is the second example of a *conservation principle* we have encountered in this book; conservation of energy was a key part of Chapter 6. Conservation principles are important for several reasons. First, they allow us to analyze problems in a very general and powerful way. For example, we saw that for a one-dimensional elastic collision, the combination of conservation of momentum and conservation of kinetic energy completely determines the outcome of the collision. Analyzing the collision in this way shows that the outcome is a result of these general conservation principles rather than the details of the interaction forces.

Second, conservation principles are extremely general statements about the physical world. While we derived conservation of momentum starting from Newton's laws, this principle is much more general than that. In later chapters, we will study situations in which Newton's laws fail, such as the quantum regime of atoms and subatomic particles, and the astrophysical regime of massive stars and black holes. Although Newton's laws fail in those situations, momentum is still conserved. Indeed, the principle of conservation of momentum is believed to be an exact law of physics, valid in all regimes.

Third, careful tests of conservation principles can sometimes lead to new discoveries. The discovery of an elementary particle called the *neutrino* was based quite closely on ideas from this chapter. A subatomic particle called the *neutron* was discovered in the early 1930s and was soon recognized as an important component of the atomic nucleus. It was also found that an isolated neutron—that is, a neutron removed from a nucleus—is an unstable particle. When the decay of an isolated neutron was first studied, the details of the decay process were very puzzling. It was initially believed that a neutron decays to leave behind just a proton and an electron:

$$\text{neutron} \rightarrow \text{proton} + \text{electron} \qquad \text{(incomplete!)} \qquad (7.41)$$

We have labeled this reaction as "incomplete" for the following reason. This decay process, in which one particle (the neutron) decays to form two other particles (a proton and an electron), is very similar to the exploding asteroid in Figure 7.21. The decaying neutron releases a certain amount of energy E_{decay} (analogous to the bomb that splits the asteroid), and this energy goes into the kinetic energy of the reaction products. There are no external forces acting on the system, so we also know that momentum must be conserved. With conservation of momentum together with knowledge of the reaction energy, we can calculate the final velocities of the proton and electron, just as we calculated the final speeds of the asteroid's pieces in Figure 7.21B.

Let's now analyze the neutron decay reaction in a little more detail.[4] To conserve momentum, the proton and electron in Figure 7.30 must travel away in opposite directions, along a common axis we call x. The problem is thus one dimensional, and we only need to worry about the components of the proton and electron velocities along x. For simplicity, we assume the neutron was at rest prior to its decay, so the total momentum of the proton plus electron after the collision must be zero. If

Figure 7.30 Diagram of a hypothetical decay of a neutron into a proton and electron. **A** The neutron is initially at rest and then **B** decays into a proton and an electron. If there are only two outgoing particles, they must be emitted in opposite directions; otherwise, momentum cannot be conserved. This leads to a single value for the speed of the outgoing electron. Neutrons do not decay in this way.

---

[4]We will learn in Chapter 27 that our Newton's law expressions for momentum and kinetic energy begin to break down when applied to the motion of neutrons, protons, electrons, and neutrinos moving at high speeds. However, these expressions are still a useful approximation and will take us to the correct conclusions regarding the decay of the neutron.

the electron's velocity along x is v_e and the proton's velocity along x is v_p, conservation of momentum along x gives

$$m_e v_e + m_p v_p = 0 \tag{7.42}$$

The conservation of energy condition can be written as

$$\tfrac{1}{2} m_e v_e^2 + \tfrac{1}{2} m_p v_p^2 = E_{\text{decay}} \tag{7.43}$$

We thus have two conservation conditions (two equations) and two unknowns, the final velocities of the electron and the proton. Hence, these final velocities are fully determined: their precise values are just the solution of these two equations. If Equation 7.41 is really the correct decay reaction, all the electrons must have the same outgoing speed.

When the neutron decay reaction was studied very carefully, a very different result was observed as sketched in Figure 7.31. The speed of the outgoing electron was *not* always a single, precise value as expected from the analysis in Equations 7.42 and 7.43. Instead, a range of speeds was observed. This result was very puzzling, and some physicists suggested that either conservation of momentum or conservation of energy might not apply to subatomic particles. The correct explanation was not this drastic, however. It was eventually found that a neutron actually decays into *three* particles:

neutron → proton + electron + antineutrino   (correct!)

The third particle was an entirely new type of particle that is now called an *antineutrino*. Adding the contributions of the antineutrino to the final momentum and kinetic energy in Equations 7.42 and 7.43 leads to a range of possible values for the final velocity of the electron, as found in the experiments. In fact, this experiment, together with its analysis in terms of conservation laws, marked the discovery of the antineutrino and led to the observation of a closely related particle called the neutrino. The point of this story is that a very careful test of conservation principles led to an entirely unexpected, and extremely important, discovery.

Real neutron decay

Figure 7.31 Diagram of the actual decay of a neutron. There are three outgoing particles: a proton, an electron, and an antineutrino. The speed of the outgoing electrons is not the same for every electron.

SUMMARY | Chapter 7

KEY CONCEPTS AND PRINCIPLES

Momentum
The *momentum* of a particle is given by

$$\vec{p} = m\vec{v} \qquad \text{(7.1) (page 206)}$$

The total momentum of a system of particles is equal to the sum of the momenta of the individual particles.

Impulse theorem
When a constant force acts on an object, it imparts an *impulse* to the object that is equal to $\vec{F}\,\Delta t$. The impulse is equal to the area under the force–time curve. The impulse imparted to an object is equal to the change in the momentum of the object:

$$\text{impulse} = \vec{F}\,\Delta t = \Delta\vec{p} \qquad \text{(7.5) (page 208)}$$

Conservation of momentum
If there is no external force on a system of particles, the total momentum of the system is constant. In such a case, momentum is *conserved*.

(Continued)

APPLICATIONS

Collisions

The two objects involved in a collision can be thought of as a "system." During a collision, the forces external to the system are usually very small, and the total momentum of the system after the collision is equal to the total momentum of the system before the collision. The momentum is conserved.

- In an *elastic collision*, both momentum and kinetic energy are conserved.
- In an *inelastic collision*, the momentum is conserved, but kinetic energy is not conserved.
- In a *completely inelastic collision*, the two objects stick together after the collision. Momentum is still conserved.

Center of mass

Most real objects have a size and shape, and they can be thought of as being composed of many separate pieces. Such extended objects possess a *center of mass*, and the motion of this point is very simple. For any object, no matter how complicated, the center of mass moves in response to the total external force acting on the object. This motion follows Newton's laws as if all the mass were located at the center of mass. For a system of two particles, the x coordinate of the center of mass is given by

$$x_{CM} = \frac{m_1 x_1 + m_2 x_2}{m_1 + m_2} \qquad (7.33) \text{ (page 227)}$$

with a similar expression for y_{CM} and z_{CM}. For a system of many particles, the result is

$$x_{CM} = \frac{\sum_i m_i x_i}{\sum_i m_i} = \frac{\sum_i m_i x_i}{M_{tot}} \qquad (7.34) \text{ (page 227)}$$

where M_{tot} is the total mass of the system.

QUESTIONS

SSM = answer in Student Companion & Problem-Solving Guide   ⊗ = life science application

1. **SSM** A bomb that is initially at rest breaks into several pieces of approximately equal mass, two of which are shown with their velocity vectors in Figure Q7.1. Use conservation of momentum to determine if there might be other pieces of the bomb not shown in the figure. Assume there is only one missing piece and estimate the direction it is traveling after the explosion.

Figure Q7.1

2. ℝ What is the approximate location of the center of mass of the boomerang in Figure Q7.2?

3. Explain why the center of mass of an object (such as the boomerang in Fig. Q7.2) may not be "inside" the object. We discussed how the center of mass can be thought of as the "balance point" of an object. How might you use the idea of a balance point to locate the center of mass in cases in which the center of mass is outside the object?

Figure Q7.2
Questions 2 and 3.

4. If the magnitude of the momentum of an object is increased by a factor of three, by what factor is the kinetic energy changed? Assume the mass of the object is constant.

5. Two objects of different mass have the same kinetic energy. Which one has the larger momentum?

6. A baseball has a certain momentum p_{ball} and a certain kinetic energy KE_{ball}. One golf ball (ball 1) is given a momentum equal to p_{ball}, and a second golf ball (ball 2) is given a kinetic energy equal to KE_{ball}. Which golf ball has the higher speed?

7. Consider the momentum of an object that is undergoing projectile motion. Is its momentum conserved? If the answer is no, is any component of the momentum conserved? Explain.

8. In which case is the momentum change largest, (a) a pitched baseball caught by a catcher, (b) a baseball thrown by a pitcher, or (c) a baseball hit by a batter? Explain your answer.

9. Finish the analysis of a one-dimensional elastic collision in Example 7.4. Fill in the steps that lead to the results for the final velocities of the two balls after the collision.

10. **Deep impact?** Consider the collision between an asteroid and a rocket as discussed in connection with Figure 7.16. We have already calculated the deflection angle that results from the collision with the rocket (Eq. 7.27), but we need to also determine

Photodisc/Getty Images

if this deflection will actually save the Earth. If we assume the collision takes place near the Moon, will the asteroid miss the Earth? Ignore the effect of the Earth's gravitational force on the asteroid.

11. Finish the analysis of the force in the automobile collision in Example 7.6. Find the average collision force along y and the magnitude of the total collision force.

12. A cue ball bounces off the rail of a pool table as shown in Figure Q7.12. The momentum of the ball changes. What is the source of the impulse? That is, what is the source of the force?

13. Two crates are on an icy, frictionless, horizontal surface. One of the crates is then given a push and collides with the other. If the crates stick together after the collision, which of the following quantities is conserved in the collision, (a) horizontal momentum, (b) kinetic energy, (c) both, or (d) neither?

14. **SSM** A tennis ball and a ball of soft clay are dropped from the same height onto the floor below. The force from the floor produces an impulse on each ball. If the balls have the same mass, which impulse is larger? Explain.

Figure Q7.12
Question 12 and Problems 8 and 9.

15. Newton's second law states that the total force on an object equals the object's mass times its acceleration. Express Newton's second law in terms of the momentum of an object instead of its acceleration.

16. When you fire a gun, you experience a recoil force due to impulse produced by the bullet on the barrel of the gun. Is it possible to design a gun in which the recoil force is zero? If so, explain how you could do it. Why do you think guns are not made in this way?

17. Two masses m_1 and m_2 have the same momentum. If the masses are not equal, which statement(s) are correct?
(a) The velocities are the same.
(b) The velocities are in the same direction.
(c) The kinetic energies are the same.

18. ⊗ In event of a two-car, head-on collision, would you rather be riding in a heavy car (large mass) or a small car? Justify your answer in terms of the impulse and force on a car during a collision.

19. ⊗ Explain why a padded dashboard makes a car safer in the event of an accident.

20. As an apple falls from a tree, its speed increases, so the magnitude of its momentum also increases. Explain why this is consistent with the principle of conservation of momentum.

7.1 MOMENTUM

1. What is the magnitude of the momentum of a baseball ($m = 0.14$ kg) traveling at a speed of 100 mi/h? Express your answer in SI units.

2. Consider two cars, one a compact model ($m = 800$ kg) and the other an SUV ($m = 2500$ kg). The SUV is traveling at 10 m/s, while the compact car is traveling at an unknown speed v. If the two cars have the same momentum, what is v?

3. Consider two cars that are on course for a head-on collision. If they have masses of $m_1 = 1200$ kg and $m_2 = 1800$ kg and are both traveling at 30 m/s, what is the magnitude of the total momentum?

4. **SSM** Two particles of mass $m_1 = 1.2$ kg and $m_2 = 2.9$ kg are traveling as shown in Figure P7.4. What is the total momentum of this system? Be sure to give the magnitude and direction of the momentum.

Figure P7.4

7.2 FORCE AND IMPULSE

5. ⭐ **Slap shot!** A hockey player strikes a puck that is initially at rest. The force exerted by the stick on the puck is 1000 N, and the stick is in contact with the puck for 5.0 ms (0.0050 s). (a) Find the impulse imparted by the stick to the puck. (b) What is the speed of the puck ($m = 0.12$ kg) just after it leaves the hockey stick?

6. A constant force of magnitude 25 N acts on an object for 3.0 s. What is the magnitude of the impulse?

7. A rubber ball (mass 0.25 kg) is dropped from a height of 1.5 m onto the floor. Just after bouncing from the floor, the ball has a speed of 4.0 m/s. (a) What is the magnitude and direction of the impulse imparted by the floor to the ball? (b) If the average force of the floor on the ball is 18 N, how long is the ball in contact with the floor?

8. ⭐ For the cue ball in Figure Q7.12, make a qualitative sketch of how the force exerted by the rail on the ball varies with time. Be sure to plot both the x and y components of the force and indicate when the ball is in contact with the rail.

9. ⭐ Consider again the cue ball in Question 12 and Figure Q7.12. (a) What is the impulse imparted to the ball? Be sure to give the magnitude and direction. (b) What is the change in momentum of the ball? Be sure to give the magnitude and direction.

10. ⭐ A rubber ball is dropped and bounces vertically from a horizontal concrete floor. If the ball has a speed of 3.0 m/s just before striking the floor and a speed of 2.5 m/s just after bouncing, what is the average force of the floor on the ball? Assume the ball is in contact with the floor for 0.12 s and the mass of the ball is 0.15 kg.

11. **SSM** ⭐ A baseball player hits a baseball ($m = 0.14$ kg) as shown in Figure P7.11. The ball is initially traveling horizontally with speed of 40 m/s. The batter hits a fly ball as shown, with a speed $v_f = 55$ m/s. (a) What is the magnitude and direction of the impulse imparted to the ball? (b) If the ball ($m = 0.14$ kg) and bat are in contact for a time of 8.0 ms, what is the magnitude of the average force of the bat on the ball? Compare this answer to the weight of the ball. (c) What is the impulse imparted to the bat?

Figure P7.11
Problems 11 and 12.

12. The baseball in Problem 11 is replaced by a very flexible rubber ball of equal mass. (a) Does the contact time increase or decrease? (b) If the contact time changes by a factor of 50 but the initial and final velocities are the same, by what factor does the force of the bat on the ball change?

13. ⭐ An egg and a tomato of equal mass are dropped from the roof of a three-story building onto the sidewalk below. The sidewalk imparts an impulse to both as they splatter. (a) Is the impulse imparted to the egg greater than, less than, or equal to the impulse imparted to the tomato? (b) Is the average force exerted on the egg greater than, less than, or equal to the force on the tomato?

14. ⭐ Ⓡ Consider an archer who fires an arrow. The arrow has a mass of 0.030 kg and leaves the bow with a horizontal velocity of 80 m/s. (a) What is the impulse imparted to the arrow? Give both the magnitude and direction of the impulse. (b) What is the approximate average force of the bow string on the arrow?

15. ⭐ Ⓡ Consider a tennis player who hits a serve at 140 mi/h (63 m/s). Estimate the average force exerted by the racket on the ball ($m = 57$ g).

16. ⭐ Ⓡ Consider the problem of returning a serve in tennis. A serve is hit at a speed of 63 m/s and is hit back to the server with a speed of 40 m/s. (a) What is the impulse imparted to the ball? (b) What is the approximate average force on the ball?

17. Ⓡ Suppose the force on an object varies with time as shown in Figure P7.17. Estimate the impulse imparted to the object. Assume the motion is one dimensional.

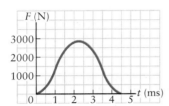

Figure P7.17

18. ⭐ ✖ Ⓡ **Uppercut!** A boxer hits an opponent on the chin and imparts an impulse of 500 N · s. Estimate the average force.

19. SSM ⭐ Ⓡ A golf ball is hit from the tee and travels a distance of 300 yards. Estimate the magnitude of the impulse imparted to the golf ball. Ignore air drag in your analysis.

7.3 AND 7.4 CONSERVATION OF MOMENTUM AND COLLISIONS

20. ⭐ Two objects of mass m and $3m$ undergo a completely inelastic collision in one dimension. If the two objects are at rest after the collision, what was the ratio of their speeds before the collision?

21. Two particles of mass $m_1 = 1.5$ kg and $m_2 = 3.5$ kg undergo a one-dimensional head-on collision as shown in Figure P7.21. Their initial velocities along x are $v_{1i} = 12$ m/s and $v_{2i} = -7.5$ m/s. The two particles stick together after the collision (a completely inelastic collision). (a) Find the velocity after the collision. (b) How much kinetic energy is lost in the collision?

Figure P7.21 Problems 21 and 23.

22. Two skaters are studying collisions on an ice-covered (frictionless) lake. Skater 1 ($m_1 = 85$ kg) is initially traveling with a speed of 5.0 m/s, and skater 2 ($m_2 = 120$ kg) is initially at rest. Skater 1 then "collides" with skater 2, and they lock arms and travel away together. (a) Identify a "system" whose momentum is conserved. (b) Draw a sketch of your system in part (a) before the collision. Include a coordinate system and identify the initial velocities of the different parts of the system. (c) Express the conservation condition for your system. (d) Does your system undergo an elastic collision or an inelastic collision? (e) Solve for the final velocity of the two skaters.

23. The particles in Figure P7.21 ($m_1 = 1.5$ kg and $m_2 = 3.5$ kg) undergo an elastic collision in one dimension. Their velocities before the collision are $v_{1i} = 12$ m/s and $v_{2i} = -7.5$ m/s. Find the velocities of the two particles after the collision.

24. Two hockey pucks approach each other as shown in Figure P7.24. Puck 1 has an initial speed of 20 m/s, and puck 2 has an initial speed of 15 m/s. They collide and stick together. (a) If the two pucks form a "system," is the momentum of this system along x or y conserved? (b) Find the components along x and y of the initial velocities of both particles. (c) Express the conservation of momentum condition for motion along x and y. (d) Is this collision elastic or inelastic? (e) Find the final velocity of the two pucks after the collision. (f) What fraction of the initial kinetic energy is lost in the collision?

Figure P7.24
Problems 24 and 26.

25. SSM ⭐ Consider again the collision between two hockey pucks in Figure P7.24, but now they do not stick together. Their speeds before the collision are $v_{1i} = 20$ m/s and $v_{2i} = 15$ m/s. It is found that after the collision one of the pucks is moving along x with a speed of 10 m/s. What is the final velocity of the other puck?

26. ✪ In Example 7.4, we considered an elastic collision in one dimension in which one of the objects is initially at rest. Notice that mass 2 is the one that is initially at rest. Use the results for v_{1f} and v_{2f} to answer the following problems. (a) Assume $m_1 = 10m_2$ so that the incoming object is much more massive than the one at rest. Find v_{1f}/v_{1i} and v_{2f}/v_{1i}. Explain why the final velocity of the smaller object is much larger (in magnitude) than the final velocity of the bigger object. (b) Assume $m_1 = m_2/10$. Find v_{1f}/v_{1i} and v_{2f}/v_{1i}. Explain why the final velocity of the smaller object is opposite its initial direction.

27. ⭐ Two billiard balls collide as shown in Figure P7.27. Ball 1 is initially traveling along x with a speed of 10 m/s, and ball 2 is at rest. After the collision, ball 1 moves away with a speed of 4.7 m/s at an angle $\theta = 60°$. (a) Find the speed of ball 2 after the collision. (b) What angle does the final velocity of ball 2 make with the x axis?

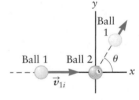

Figure P7.27

28. Consider an elastic collision in one dimension that involves objects of mass 2.5 kg and 4.5 kg. The larger mass is initially at rest, and the smaller one has an initial velocity of 12 m/s. Find the final velocities of the two objects after the collision.

29. ⭐ Two hockey players are traveling at velocities of $v_1 = 12$ m/s and $v_2 = -18$ m/s when they undergo a head-on collision. After the collision, they grab each other and slide away together with a velocity of -4.0 m/s. Hockey player 1 has a mass of 120 kg. What is the mass of the other player?

30. SSM ⭐ Two cars of equal mass are traveling as shown in Figure P7.30 just before undergoing a collision. Before the collision, one of the cars has a speed of 18 m/s along $+x$, and the other has a speed of 25 m/s along $+y$. The cars lock bumpers and then slide away together after the collision. What are the magnitude and direction of their final velocity?

Figure P7.30

31. Two cars collide head-on and lock bumpers on an icy (frictionless) road. One car has a mass of 800 kg and an initial speed of 12 m/s and is moving toward the north. The other car has a mass of 1200 kg. If the speed of the cars after the collision is 4.5 m/s and they are moving north, what is the initial velocity of the large car?

7.5 USING MOMENTUM CONSERVATION TO ANALYZE INELASTIC EVENTS

32. A railroad car ($m = 3000$ kg) is coasting along a level track with an initial speed of 25 m/s. A load of coal is then dropped into the car as sketched in Figure 7.18. (a) Treat the railroad car plus coal as a system. Is the momentum of this system along the horizontal (x) or vertical (y) conserved? (b) What are the velocities of the car and the coal before the collision along the direction found in part (a)? (c) Express the conservation of momentum condition for the system. (d) If the final speed of the car plus coal is 20 m/s, what is the mass of the coal?

33. A baseball pitcher ($m = 80$ kg) is initially standing at rest on an extremely slippery, icy surface. He then throws a baseball ($m = 0.14$ kg) with a horizontal velocity of 50 m/s. What is the recoil velocity of the pitcher?

34. ⭐ A military gun is mounted on a railroad car ($m = 1500$ kg) as shown in Figure P7.34. There is no frictional force on the car, the track is horizontal, and the car is initially at rest. The gun then fires a shell of mass 30 kg with a velocity of 300 m/s at an angle of 40° with respect to the horizontal. Find the recoil velocity of the car.

Figure P7.34

35. An open railroad car of mass 2500 kg is coasting with an initial speed of 14 m/s on a frictionless, horizontal track. It is raining, and water begins to accumulate in the car. After some time, it is found that the speed of the car is only 11 m/s. How much water (in kilograms) has accumulated in the car?

36. ⭐ The M79 grenade launcher was first used by the U.S. Army in 1961. It fires a grenade with an approximate mass $m \approx 7.0$ kg at a speed of 75 m/s. (a) If a soldier fires the M79 horizontally while standing on a very slippery surface, what is the recoil speed of the soldier? Take the mass of the soldier plus M79 to be 100 kg. (b) The barrel of the M79 is 36 cm long. Estimate the average recoil force on the soldier.

37. A skateboarder ($m = 85$ kg) takes a running jump onto a friend's skateboard that is initially at rest as sketched in Figure P7.37. The friend is standing on the skateboard (mass of friend plus skateboard = 110 kg). After landing on the skateboard, the velocity of the board plus the two skateboarders is 3.0 m/s. What was the horizontal component of the velocity of the jumping skateboarder just before he landed on the skateboard?

Figure P7.37

38. ✪ A thief is traveling with stolen goods on his bicycle when he spots a police officer approaching. The thief wants to throw the stolen goods away before the officer arrives and wants the goods to land as far away as possible from him. The thief is traveling along the +x direction and throws the goods with a certain speed v_0 relative to the bicycle and an angle $\theta = 45°$ relative to the x axis. Should he throw the goods (a) in the +x direction or (b) in the −x direction as sketched in Figure P7.38? Does it matter? Assume the bicycle is coasting throughout.

Figure P7.38

39. ✪ An airplane is sent on a rescue mission to help a stranded explorer near the South Pole. The plane cannot land, but instead drops a supply package of mass $m = 60$ kg while it is flying horizontally with a speed of $v_p = 120$ m/s. The supply package drops onto a waiting sled that is initially at rest (Fig. P7.39). After the package lands in the sled, the speed of the sled plus package is found to be 30 m/s. What is the mass of the sled?

Figure P7.39

40. SSM ⭐ A railroad car containing explosive material is initially traveling south at a speed of 5.0 m/s on level ground. The total mass of the car plus explosives is 3.0×10^4 kg. An accidental spark ignites the explosive, and the car breaks into two pieces, which then roll away along the same track. If one piece has a mass of 2.0×10^4 kg and a final speed of 2.5 m/s toward the south, what is the final speed of the other piece?

7.6 CENTER OF MASS

41. Find the center of mass of the two particles in Figure P7.41.

Figure P7.41

42. Find the center of mass of the four particles in Figure P7.42. Their masses are $m_1 = 12$ kg, $m_2 = 18$ kg, $m_3 = 7.9$ kg, and $m_4 = 23$ kg.

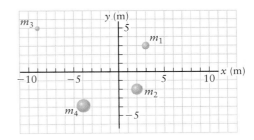

Figure P7.42

43. Two objects are moving along the x axis with velocities of 35 m/s (object 1) and −25 m/s (object 2). (a) If the center of mass has a velocity of +10 m/s, which object has the greater mass? (b) What is the ratio of their masses?

44. SSM ⭐ Consider the motion of the two ice skaters in Figure 7.27 and assume they have masses of 60 kg and 100 kg. If the larger skater has moved a distance of 12 m from his initial position, where is the smaller skater?

45. ⭐ ⓡ Estimate the location of the center of mass of the thin metal plate in Figure P7.45.

46. Consider the Sun and the Earth as a "system" of two "particles". The center of mass of this system lies a distance L from the center of the Sun. Find L and compare it to the radius of the Sun.

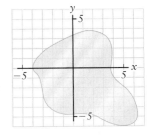

Figure P7.45

47. **Finding the barycenter.** How far is the center of mass of the Earth−Moon system from the center of the Earth? Is it below the surface of the Earth?

48. ⭐ A person of mass 90 kg standing on a horizontal, frictionless surface fires a bullet of mass 40 g in a horizontal direction. After the bullet has traveled a distance of 50 m, what distance has the person moved?

49. ⭐ Ⓡ Consider the motion of an apple and the Earth as the apple falls to the ground. If the apple begins at the top of a tree of height 15 m, how much does the Earth move?

7.7 A BOUNCING BALL AND MOMENTUM CONSERVATION

50. A bomb that is initially at rest at the origin explodes and breaks into two pieces of mass $m_1 = 1500$ kg and $m_2 = 2500$ kg as shown in Figure P7.50. If piece 1 has a velocity of $+35$ m/s directed along x, what is the velocity of piece 2 after the explosion?

Figure P7.50

51. SSM ⭐ A hand grenade is thrown with a speed of $v_0 = 30$ m/s, as sketched in Figure P7.51, just prior to exploding. It breaks into two pieces of equal mass after the explosion. One piece has a final velocity of 40 m/s along x. Find the velocity of the other piece after the explosion.

Figure P7.51

52. ✪ **The ballistic pendulum.** A bullet of mass $m = 30$ g is fired into a wooden block of mass $M = 5.0$ kg as shown in Figure P7.52. The block is attached to a string of length 1.5 m. The bullet is embedded in the block, causing the block to then swing as shown in the figure. If the block reaches a maximum height of $h = 0.30$ m, what was the initial speed of the bullet?

Figure P7.52

53. ✪ Ⓡ A rock of mass 30 g strikes a car windshield while the car is traveling on a highway. Estimate the force of the rock on the windshield. *Hint*: You will need to estimate the collision time (i.e., the time that the rock is in contact with the windshield). For comparison, the collision time for a baseball–bat collision is approximately 0.001 s, while the time for a tennis racket–ball collision is about 0.01 s.

54. ✪ Two particles of mass 2.3 kg and mass 4.3 kg that are free to move on a horizontal track are initially held at rest so that they compress a spring as shown in Figure P7.54. The spring has a spring constant $k = 400$ N/m and is compressed 0.12 m. Find the final velocities of the two particles.

Figure P7.54

55. ✪ Ⓡ A tennis ball of mass 0.12 kg is dropped vertically onto a hard concrete floor from a height of 1.5 m. The ball then bounces up to some maximum height h. If the ball bounces up to a height $h = 1.0$ m, what is the approximate average force of the floor on the ball?

56. Consider a reaction that involves the creation of two elementary particles as in Figure P7.56. One particle is a proton of mass m_p, and the other has a mass of $m_2 = m_p/1800$. If the proton leaves the reaction with a speed v_p, what is the final speed of the other particle? Express your answer in terms of v_p.

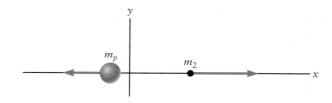

Figure P7.56

ADDITIONAL PROBLEMS

57. ✪ A bullet of mass 15 g is fired with an initial speed of 300 m/s into a wooden block that is initially at rest. The bullet becomes lodged in the block, and the two slide together on the floor for a distance of 1.5 m before coming to rest (Fig. P7.57). If the coefficient of friction between the block and the floor is 0.40, what is the mass of the block?

Figure P7.57

58. ⭐ A father ($m_F = 75$ kg) and his daughter ($m_D = 35$ kg) stand on a flat, frozen lake of negligible friction. They hold a 10-m-long rope stretched between them. The father and daughter then pull the rope to bring them together. If the father is initially standing at the origin, how far from the origin will they meet?

59. ✪ Twin brothers, each of mass 50 kg, sit in a symmetric canoe of mass 40 kg. Each boy sits 1.5 m away from the center of the canoe as they toss a basketball (mass = 0.60 kg) back and forth. (a) How much does the center of the canoe move as the ball moves from one brother to the other? (b) How much would the canoe shift on each throw if instead they used a medicine ball of mass 11 kg?

60. ✪ A block of mass 3.5 kg is initially at rest on a wedge of mass 20 kg, height 0.30 m, and an incline angle of $\theta = 35°$ as shown in Figure P7.60. There is no friction between the wedge and the floor. Starting at the top of the incline, the block is released and slides toward the bottom of the wedge. At the same time, the wedge "recoils" and slides some distance L to the right. Find L when the block has reached the bottom of the wedge. *Hint:* The center of mass of a right triangle of height h is a distance $h/3$ above the base of the triangle.

Figure P7.60

61. ✪ Consider three scenarios for a one-dimensional elastic collision with a stationary target. For each scenario, determine the final velocities of both the projectile mass, v_{1f}, and the target mass, v_{2f}, in terms the projectile's initial velocity, v_0. Assume the masses are sliding on a horizontal surface of negligible friction. Case (a): Both the projectile and target are bowling balls; $m_1 = m_2 = 7.25$ kg. Case (b): The target is one of the bowling balls, but the projectile is a Styrofoam ball of the same diameter; $m_1 = 140$ g, $m_2 = 7.25$ kg. Case (c): Now make the Styrofoam ball the target and the bowling ball the projectile; $m_2 = 140$ g, $m_1 = 7.25$ kg. Compare parts (b) and (c) to the following extremes: (d) when the target is the Earth and the projectile is a tennis ball, and (e) when the target is a frozen pea and the projectile is a truck's windshield.

62. ✪ **Fun with a Super Ball.** A Super Ball has a coefficient of restitution of $\alpha = 0.90$. This means that the y components of the ball's velocity just before and after hitting the floor are related by $v_{fy} = -\alpha v_{if}$. It is dropped from a height of 1.8 m such that it goes under a table of height 70 cm. The ball bounces back and forth from the underside of the table and the ground (Fig. P7.62). How many total times will the ball bounce off the bottom of the table?

Figure P7.62

63. ✪ ℝ Consider again the discussion in Section 7.2 concerning air bags and impulse for a car colliding with a tree. The car's velocity of 40 mi/h is reduced to zero very quickly, but the head of a passenger is still moving at 40 mi/h (about 18 m/s). Calculate the average force exerted on the passenger's head coming to rest for the case (a) when it is stopped by the dashboard in 5.0 ms and (b) when it is stopped by an air bag that compresses in 45 ms. (c) Air bags must deploy quickly. If the distance between the passenger's head and the dashboard is 60 cm and it takes the air bag 45 ms to safely bring the passenger to rest, how much time is there for the air bag to inflate?

64. ✪ ✪ **Crumple zones.** A life-saving development in automobile manufacture is the invention of crumple zones, areas of the body and frame of a car deliberately made to collapse in a collision (Fig. P7.64) such that the passenger compartment will not. Introduced in 1955 on the Heckflosse made by Mercedes, crumple zones became mandatory on all cars sold in the United States in 1967. For a head-on collision with a tree at 40 mi/h (about 18 m/s), find the ratio of the average force on a car with a stiff frame (stop time of 0.010 s) to that on a car with a crumple zone (stop time of 0.25 s).

Figure P7.64 Front crumple zone on a Volvo.

65. ✪ A projectile of mass 10 kg is launched at an angle of 55° and an initial speed of 87 m/s. Just as the object reaches the maximum height in its trajectory, a small explosion along the

Figure P7.65

horizontal blows off the projectile's back portion equal to one-fourth its original mass. The smaller piece lands exactly at the launch point as shown in Figure P7.65. How far from the launch point does the larger portion land?

66. ✪ A cart of mass $m_1 = 10$ kg slides down a frictionless ramp and is made to collide with a second cart of mass $m_2 = 20$ kg, which then heads into a vertical loop of radius 0.25 m as shown in Figure P7.66. (a) Determine the height h from which cart 1 would need to start to make sure that cart 2 completes the loop without leaving the track. Assume an elastic collision. (b) Find the height needed if instead the more massive cart is allowed to slide down the ramp into the smaller cart.

Figure P7.66 Problems 66 and 67.

67. ✪ Again consider the track and carts in Figure P7.66 with $m_1 = 10$ kg and $m_2 = 20$ kg, but this time the carts stick together after the collision. (a) Find the height h from which cart 1 would need to start to make sure that both carts complete the loop without leaving the track. (b) Find the height needed if instead the more massive cart is made to collide with the smaller cart.

68. ✪ A block of wood ($M_{block} = 5.2$ kg) rests on a horizontal surface for which the coefficient of kinetic friction is 0.28. A bullet ($m_{bullet} = 32$ g) is shot along the horizontal direction and embeds in the block, displacing the block 87 cm. What was the initial velocity of the bullet before impact?

69. ✪ A firefighter directs a horizontal stream of water toward a fire at an angle of 40° measured from the horizontal. The fire hydrant and hose deliver 95 gallons per minute (1 gal/min = 6.31×10^{-4} m³/s), and the nozzle projects the water with a speed of 7.5 m/s. How much horizontal force must the firefighter exert on the hose to keep it stationary?

70. ✪ ℝ A machine gun directs a flow of bullets at a steel target mounted on a cart with negligible friction and of mass 600 kg. In this case, the gun is an Uzi, which uses 9-mm (8.0-g) bullets and can fire at a rate of 10 rounds/s, and the bullets have an average muzzle velocity of 400 m/s. (a) Assuming each bullet's ricochet off the target can be approximated by an elastic collision, with what final velocity would the cart be moving if all 50 rounds in the Uzi's magazine were fired at the maximum rate at the target? (b) How fast would the cart be moving if the target were made of soft wood such that all 50 bullets became embedded? (c) The Uzi is now firmly attached to the cart while at rest. The trigger is then pulled, and all 50 rounds are discharged horizontally into the air. What is the final velocity of the cart?

71. SSM ✕ Rocket engines work by expelling gas at a high speed as illustrated in Figure P7.71. We wish to design an engine that expels an amount of gas $m_g = 50$ kg each second, and we want the engine to exert a force $F = 20,000$ N on a rocket. At what speed v_g should the gas be expelled? Assume the mass of the rocket is very large.

72. ✕ Comet Shoemaker–Levy struck Jupiter in 1994. Just before the collision, the comet broke into several pieces. The total mass of all the pieces was estimated to be 3.8×10^8 kg, and they all had a velocity relative to Jupiter of 60 km/s. How much energy was deposited on Jupiter?

Figure P7.71

Chapter 8

Rotational Motion

Wheels and other kinds of rotating objects have played major roles in science and technology for many years. (© Daniel Dempster Photography/Alamy)

We have learned how to apply Newton's laws to deal with the motion of objects in many different situations. Up to this point, however, we have assumed our objects are "point particles." A point particle is one in which all the mass is located at an infinitesimal point in space. Certain elementary particles, such as electrons and positrons, are believed to be true point particles, but terrestrial objects such as soccer balls, cars, and people are definitely not. For these objects, the mass is distributed over a certain region of space; that is, they have a certain size and shape. The point particle treatment is the correct way to describe the *translational* motion of all objects. Thus, everything we have learned so far about Newton's laws applies to point particles and to all other types of objects as well. However, the point particle approximation does not allow us to deal with the **rotational motion** of an object. Our job in this chapter and Chapter 9 is to learn how to use Newton's laws to describe and analyze rotational motion. To do so, we have to treat objects more realistically, by accounting for their size and shape.

8.1 | DESCRIBING ROTATIONAL MOTION

The first step in our studies of translational motion was to introduce the concepts of position, velocity, and acceleration. Because translational motion usually involves motion along a trajectory or line in space, it is commonly referred to as *linear motion*. Hence, the translational velocity and acceleration are often referred to as the linear velocity and linear acceleration. To deal with rotational motion, we now need to define the analogous rotational quantities: *angular position*, *angular velocity*, and *angular acceleration*. We encountered angular position in Chapter 5 when we studied circular motion. Circular motion is related to rotational motion: in both cases, an object or a point on an object moves in a circular path. A simple particle undergoing circular motion as in Chapter 5 thus closely resembles a point on a rotating wheel or cylinder.

With problems involving translational motion, one of our first steps is to set up coordinate axes in one, two, or three dimensions. For problems involving rotational motion, the corresponding step is to choose the *rotation axis*. Figure 8.1 shows two different objects undergoing rotational motion. The rod in Figure 8.1A is hinged at one end so that it can rotate about the *z* axis. The angle θ gives the angular position of the rod, that is, the angle the rod makes with the *x* axis. Figure 8.1B shows a compact disc spinning about an axis passing through its center, and we again take this rotation axis to be along the *z* direction. The angular position of the CD can be specified by the angle θ that a reference line on the CD makes with the *x* axis. For both the hinged rod and the CD, the angle θ tells us all we need to know about the object's angular position.

Figure 8.2 shows the motion of our hinged rod in a little more detail. As it rotates, the end of the rod sweeps out a circle of radius *r*, where *r* is equal to the length of the rod. Suppose the moving end of the rod travels a distance *s* along this circular path. At the same time, the rod sweeps out an angle θ as sketched in Figure 8.2. The distance *s* and the angle θ are related by

$$\theta = \frac{s}{r} \tag{8.1}$$

where the angle θ is measured in *radians* (rad). Of course, angles can also be measured in degrees, and we use both units throughout this book. An angle of 360° is equivalent to 2π radians (one complete circle). (The measurement of angles using units of degrees and radians was discussed in Chapter 1. You might want to review that material now.)

| CONCEPT CHECK 8.1 | Measuring Angles in Degrees and Radians

Figure 8.2 shows the angle θ between the rod and the *x* axis. Does this angle have an approximate value of

(a) 90°   (b) 60°   (c) 0°
(d) π (e) $\pi/2$ (f) $\pi/3$
(g) both values (a) and (e)   (h) both values (b) and (f)?

Angular Velocity and Acceleration

To describe the rotational motion of a hinged rod or a CD, we need a variable that describes how the angular position is changing with time. That variable is the *angular velocity*, denoted by ω. Consider again Figure 8.2 and imagine that the end of our hinged rod is moving with a linear speed *v* along its circular path. We can describe the motion of just the *end* of the rod by its linear velocity $\vec{v}$ and its linear acceleration $\vec{a}$. We can describe the rotational motion of the *entire rod* using the corresponding angular variables.

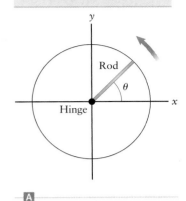

Top view of a rod rotating in the *x–y* plane. The rotation axis *z* is perpendicular to the plane of this figure.

A

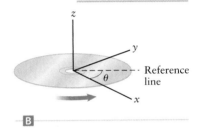

Perspective view of a CD in the *x–y* plane. The rotation axis is along *z*.

B

Figure 8.1 For both **A** a rotating rod and **B** a spinning compact disc, the angular position is described by the angle θ. In both of these examples, the rotation axis is the *z* coordinate axis.

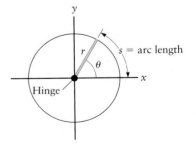

Figure 8.2 The arc length *s* measured along the circle is proportional to the angle $\theta = s/r$, measured in radians.

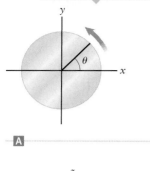

θ increases with time ⇒ ω > 0 (counter-clockwise motion).

A

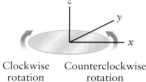

Clockwise rotation ω < 0   Counterclockwise rotation ω > 0

B

Figure 8.3 An increasing value of θ corresponds to counterclockwise motion. Hence, the angular velocity is positive if the object is rotating counterclockwise.

Definition of angular acceleration

© Cengage Learning/Charles D. Winters

Figure 8.4 When this ceiling fan is turned on, the blades undergo rotational motion.

The angular velocity ω is a measure of how rapidly the angle θ is changing with time. If we are interested in the rotational motion over a time interval Δt, we can define the average angular velocity during this interval as

$$\omega_{\text{ave}} = \frac{\Delta\theta}{\Delta t} \tag{8.2}$$

where $\Delta\theta$ is the change in the angle θ during this time interval. We can also define an instantaneous angular velocity given by

$$\omega = \lim_{\Delta t \to 0} \frac{\Delta\theta}{\Delta t} \tag{8.3}$$

In words, ω is the slope of the θ–t curve, which can be calculated by considering the change $\Delta\theta$ during a time interval Δt as this interval is allowed to shrink to zero ($\Delta t \to 0$). As with linear motion, the instantaneous angular velocity equals the average angular velocity when ω is constant. Angular velocity ω is usually measured in units of radians per second (rad/s), but can also be expressed in terms of revolutions per minute (rpm), where one revolution about the rotation axis corresponds to an angular change of 360° (2π rad).

Because ω is called the angular *velocity*, you should expect it to be a vector quantity, just as the linear velocity $\vec{v}$ is a vector. All vectors have both a magnitude and a direction. The magnitude of the angular velocity is given by Equation 8.3; how do we determine its direction? The direction of ω (and of any other quantity describing angular motion) is determined by the rotation axis. Figure 8.3 shows a rotating circular disc lying in the x–y plane, with a rotation axis is along $+z$. If the angle θ increases with time, then according to the definition in Equation 8.3 the angular velocity is positive. Hence, a counterclockwise rotation corresponds to a positive value of ω. A clockwise rotation would then give a negative value for ω.

Acceleration is a very important quantity when dealing with translational motion, and the same is true with rotational motion. We now define the **angular acceleration**, denoted by α. The average angular acceleration during a time interval Δt is related to changes in ω by

$$\alpha_{\text{ave}} = \frac{\Delta\omega}{\Delta t} \tag{8.4}$$

while the instantaneous angular acceleration is defined by

$$\alpha = \lim_{\Delta t \to 0} \frac{\Delta\omega}{\Delta t} \tag{8.5}$$

Thus, α is the slope of the ω–t curve and has units of rad/s². As with the linear motion, the instantaneous angular acceleration equals the average angular acceleration when α is constant.

A good example of rotational motion is a ceiling fan found in many homes and offices (Fig. 8.4). Suppose the fan is spinning (i.e., rotating) at a constant angular velocity of 100 rpm. Let's calculate the angular velocity in rad/s and also find the total rotation angle after exactly 5 minutes.

To convert from units of rpm to rad/s, we know that one complete revolution corresponds to 2π radians, so calculating ω gives (to two significant figures):

$$\omega = 100 \, \frac{\text{rev}}{\text{min}} \times \frac{2\pi \, \text{rad}}{1 \, \text{rev}} \times \frac{1 \, \text{min}}{60 \, \text{s}} = 10 \, \text{rad/s}$$

In this example, ω is constant, so the instantaneous angular velocity equals the average angular velocity. We can therefore rearrange Equation 8.3 to get the angular displacement $\Delta\theta$:

$$\Delta\theta = \omega \, \Delta t$$

Inserting our value for ω and using $\Delta t = 5 \, \text{min} = 300 \, \text{s}$, we have

$$\Delta\theta = \omega \, \Delta t = \left(\frac{10 \, \text{rad}}{\text{s}}\right)(300 \, \text{s}) = 3000 \, \text{rad}$$

Note that $\Delta\theta = 3000$ rad corresponds to (3000 rad) $\times$ (1 rev/2π rad) = 480 revolutions.

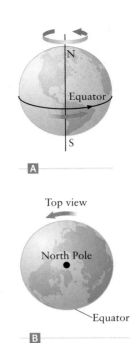

Figure 8.5 Example 8.1. The Earth undergoes rotational motion as it spins about its axis.

| EXAMPLE 8.1 | Angular Velocity of the Earth

The Earth rotates about the axis that connects the North and South Poles as sketched in Figure 8.5A. Find the Earth's angular velocity in rad/s.

RECOGNIZE THE PRINCIPLE

The angular velocity of the Earth is constant, so $\omega = \omega_{ave}$, which can be calculated from the angle $\Delta\theta$ through which the Earth rotates in a given amount of time Δt using Equation 8.2 or 8.3.

SKETCH THE PROBLEM

Figure 8.5B shows a top view of how the Earth's equator rotates (compare with Figs. 8.1 through 8.3).

IDENTIFY THE RELATIONSHIPS

The Earth completes one complete rotation ($\Delta\theta = 1$ rev = 2π rad) every day, so its angular velocity is simply 1.0 rev/day.

SOLVE

Expressing 1.0 rev/day in rad/s, we find

$$\omega = \frac{\Delta\theta}{\Delta t} = \frac{1.0 \text{ rev}}{1 \text{ day}} \times \frac{1 \text{ day}}{24 \text{ h}} \times \frac{1 \text{ h}}{60 \text{ min}} \times \frac{1 \text{ min}}{60 \text{ s}} \times \frac{2\pi \text{ rad}}{\text{rev}} = \boxed{7.3 \times 10^{-5} \text{ rad/s}}$$

What does it mean?

The Earth's angular velocity is constant, so the average angular velocity equals the instantaneous angular velocity. Also, a point on the Earth's equator moves in a circle with constant speed and hence undergoes uniform circular motion as studied in Chapter 5.

| EXAMPLE 8.2 | Angular Acceleration of a Compact Disc

A compact disc used to play music does not spin with a constant angular velocity. Instead, it spins most rapidly when music is being read from regions nearest the rotation axis (the inner "tracks" near the center of the disc) and slowest when music is played from regions near the edge of the disc (the outer "tracks"). If the angular velocity decreases uniformly from $\omega = 500$ rpm to 200 rpm as the CD is scanned from the innermost to the outermost track, what is the angular acceleration? Assume the CD contains 60 minutes of music.

RECOGNIZE THE PRINCIPLE

We are given that the angular velocity decreases uniformly, which means that the angular acceleration is constant; hence, $\alpha = \alpha_{ave}$. We can compute α from the change in ω during the 60 minutes it takes to play the CD.

SKETCH THE PROBLEM

Figure 8.6 shows the tracks on a CD. These tracks form one very long "spiral" that begins near the center of the disc and spirals out to the edge.

IDENTIFY THE RELATIONSHIPS

The angular velocity ω decreases from 500 rpm to 200 rpm as the CD player moves from the inside to the outside tracks. Because α is constant, we can write Equation 8.5

Figure 8.6 Example 8.2. The information on a compact disc is stored in a very long spiral "track."

with a time interval that is not infinitesimally small. The angular acceleration is then given by

$$\alpha = \frac{\Delta \omega}{\Delta t} = \frac{\omega_f - \omega_i}{\Delta t}$$

SOLVE

The initial angular velocity is $\omega_i = 500$ rpm and the final angular velocity is $\omega_f = 200$ rpm, which leads to

$$\alpha = \frac{\omega_f - \omega_i}{\Delta t} = \frac{(200 - 500) \text{ rpm}}{60 \text{ min}} = -5.0 \text{ rev/min}^2$$

We now need to convert the units and express the answer in rad/s²:

$$\alpha = \frac{-5.0 \text{ rev}}{\text{min}^2} \times \frac{2\pi \text{ rad}}{\text{rev}} \times \left(\frac{1 \text{ min}}{60 \text{ s}}\right)^2 = \boxed{-8.7 \times 10^{-3} \text{ rad/s}^2}$$

What does it mean?

The information on a CD is stored in a very "tight" spiral (Fig. 8.6), so at any particular point on the CD, the path containing the musical information is extremely close to being a perfect circle. During one revolution, the length of each of these approximately circular paths is just the circle's circumference, a distance smaller near the center of the CD than for regions near the edge. The amount of music stored in a circular track is proportional to the length of a track, so a circle near the center contains less music than a circle near the edge. To compensate for this difference, the CD player spins faster when reading music from near the center since the CD player must read (and also play) music at a constant rate.

Angular and Centripetal Acceleration Are Different

The angular acceleration α defined in Equation 8.5 is *different* from the centripetal acceleration associated with circular motion discussed in Chapter 5. This difference can be appreciated from Figure 8.2. Let's suppose this figure is showing a particle moving around the circle with a constant linear speed v. Such a particle is undergoing uniform circular motion. The particle's angular position θ increases at a constant rate; hence, its angular velocity is constant, and a constant value for ω means the angular acceleration $\alpha = 0$ (according to Eq. 8.5). At the same time, our particle is undergoing uniform circular motion, so from Chapter 5 we know that it also has centripetal acceleration $a_c = v^2/r$ (Eq. 5.5). A key point is that the centripetal acceleration refers to the linear motion (and the linear acceleration) of the particle, whereas the angular acceleration α is concerned with the associated angular motion. Hence, the centripetal acceleration a_c is *not* zero, even though the angular acceleration α does equal zero.

The Period of Rotational Motion

When an object such as a compact disc or the Earth is rotating with a particular value of ω, the angular velocity is the *same* for all points on the object. This is one property that makes ω such a useful quantity for describing rotational motion. While the angular velocity is the same for all points on an object, the linear velocity can be different for different points on the object. To illustrate, we consider again the motion of a compact disc as it rotates about its axis in Figure 8.7. If we focus on the motion of a general point P on the CD, we can relate the magnitude of the linear velocity v of this point to the angular velocity ω. If the angular velocity is constant, the CD moves through ω radians every second. Since one revolution corresponds to 2π radians, the CD moves through $\omega/(2\pi)$ complete revolutions every second. Equivalently, we can

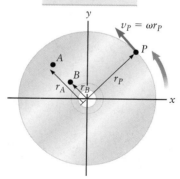

The rotation axis is perpendicular to the page.

Figure 8.7 When an object such as this compact disc undergoes rotational motion, the linear velocity of a point on the object depends on its distance from the rotation axis, but ω is the same for all points on the object.

say that each complete revolution takes $2\pi/\omega$ seconds. The time required to complete one revolution is called the **period** of the motion, denoted by T:

$$\text{period} = T = \frac{2\pi}{\omega} \tag{8.6}$$

Period and angular velocity

We encountered the notion of period in Chapter 5 in connection with circular motion. The period T is a very useful quantity whenever the motion "repeats itself" as time progresses, as is the case for circular and rotational motion, oscillatory motion, and wave motion. The relation between period and angular velocity in Equation 8.6 assumes ω is constant over the time it takes to complete one revolution (one period). This assumption is usually a good one.

The Connection between Linear and Rotational Motion

Point P on our CD in Figure 8.7 lies on a circle of radius r_P. As the CD rotates, this point completes one revolution of the circle during each period of the motion, so it travels a distance $2\pi r_P$ every T seconds. Its velocity thus has a magnitude of $2\pi r_P/T$. Using Equation 8.6, we have

$$v_P = \frac{2\pi r_P}{T} = \frac{2\pi r_P}{2\pi/\omega} = \omega r_P$$

We derived this result for point P, but it applies just as well to any other point on the CD. The *linear* velocity of a point that lies a distance r from the rotation axis is related to the *angular* velocity by

$$v = \omega r \tag{8.7}$$

Relation between linear speed and angular speed

Equation 8.7 gives a very general and useful relation between the linear velocity and the angular velocity. To be precise, this relation really only involves the linear and angular *speeds* because here we have not worried about the directions of these two types of velocities.

Now that we have Equation 8.7, we can compare the linear speeds of any two points, such as points A and B on the CD sketched in Figure 8.7. These points are located distances r_A and r_B, respectively, from the axis of the CD. Using Equation 8.7, we find $v_A = \omega r_A$ and $v_B = \omega r_B$. Point A is farther from the axis (so $r_A > r_B$), so the linear speed of point A will be *greater* than that of point B, even though they are both attached to the same CD! Physically, the reason for this difference is the different values of the radius. Point A is farther from the rotation axis, so it travels on a larger circle than point B. The motions of the two points have the same period, however, so point A must travel faster to complete its circle in the same amount of time (T). While the linear speeds at points A and B are thus *not* equal, the angular velocities at these two points are the *same*.

The relationship between the angular and translational speeds in Equation 8.7 can be extended to the accelerations. For linear motion, we showed in Chapter 2 that when the acceleration is constant, the instantaneous acceleration is related to changes in the velocity by (Eqs. 2.6 and 2.7):

$$a = \frac{\Delta v}{\Delta t}$$

If a is constant, the angular acceleration is also constant, so we can write (from Eq. 8.5)

$$\alpha = \frac{\Delta \omega}{\Delta t}$$

The ratio a/α is therefore

$$\frac{a}{\alpha} = \frac{\Delta v/\Delta t}{\Delta \omega/\Delta t} = \frac{\Delta v}{\Delta \omega}$$

From Equation 8.7, we can write $\omega = v/r$, which leads to $\Delta\omega = \Delta v/r$. We thus get

$$\frac{a}{\alpha} = \frac{\Delta v}{\Delta\omega} = \frac{\Delta v}{\Delta v/r} = r$$

The *angular* acceleration α and the *linear* acceleration a of a point a distance r from the rotation axis are therefore related by

$$a = \alpha r \qquad (8.8)$$

Relation between linear acceleration and angular acceleration

Figure 8.8 Concept Check 8.2.

CONCEPT CHECK 8.2 | Angular Velocity and Period

Consider two merry-go-rounds (Fig. 8.8) whose radii differ by a factor of two. If they have the same angular velocities, how do their periods compare?

(a) The larger merry-go-round has a shorter period.
(b) The larger merry-go-round has a longer period.
(c) The periods are the same.

EXAMPLE 8.3 | Ⓡ Speed of a Merry-Go-Round

Estimate the angular velocity of a typical merry-go-round and find the corresponding linear speed of a point on the outer edge of the merry-go-round.

RECOGNIZE THE PRINCIPLE

The angular velocity is related to the period of the motion (Eq. 8.6), and it is also related to the linear speed v of a point on a rotating object (the merry-go-round) through Equation 8.7.

SKETCH THE PROBLEM

Figure 8.9 shows a top view of a rotating merry-go-round. A point on the edge moves in a circle, with r equal to the radius of the merry-go-round.

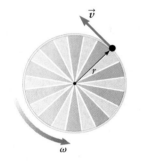

Figure 8.9 Example 8.3.

IDENTIFY THE RELATIONSHIPS

According to Equation 8.6, the angular velocity and period are related by

$$\omega = \frac{2\pi}{T}$$

To find ω we must therefore estimate the value of T for a typical merry-go-round. The linear speed is related to ω by

$$v = \omega r$$

Hence, to find v, we must know the radius r of the merry-go-round. The value of r is not given, so we must also estimate r for a typical case.

SOLVE

A typical merry-go-round (Fig. 8.8) might have a radius of 3 m and move at a rate such that it completes one revolution in about 5 s. Hence, the period of the motion is $T = 5$ s. Working to one significant figure (to match our estimates of r and T), these values lead to

$$\omega = \frac{2\pi}{T} = \frac{2\pi}{5 \text{ s}} = \boxed{1 \text{ rad/s}}$$

The linear speed of a point on the edge of the merry-go-round is then

$$v = \omega r = (1 \text{ rad/s})(3 \text{ m}) = \boxed{3 \text{ m/s}}$$

8.2 | TORQUE AND NEWTON'S LAWS FOR ROTATIONAL MOTION

Now that we have the quantities and relationships we need to describe rotational motion, our next job is to explore the connection between force and rotational motion. In particular, we want to know how forces give rise to angular acceleration; this approach is very similar to the one we took in Chapter 2 when we dealt with the linear (i.e., translational) motion of a particle.

Torque and Lever Arm

Let's analyze another experiment from the playground. Figure 8.10 shows a child sitting on one end of a seesaw, while his friend prepares to jump onto the opposite side. The seesaw is supported near the center by a hinge that allows it to rotate about its point of support P. Hence, both ends of the seesaw, and indeed all points along it, move on circular arcs as the seesaw rotates. When the child jumps on board, her weight gives a force on the seesaw, and when the seesaw is horizontal (as drawn here), this force is directed perpendicular to the seesaw. We know intuitively that this force makes the seesaw rotate and that it will cause an angular acceleration of the seesaw. In analogy with Newton's second law, we might expect that if we double the force, we will double the angular acceleration and so forth. To predict the angular acceleration, however, we must also consider *where* the force is applied to the seesaw.

From our playground experiences, we know that applying a force far from the support point P is more effective than applying the same force close to P. We therefore define a quantity called *torque*, which equals the product of the applied *force* and the *distance* it is applied from the support point. Since the seesaw is able to rotate about this point, point P is often called the *pivot point*. In the simplest cases, such as this seesaw, the force is directed *perpendicular* to a line connecting its point of application to the pivot point. We then define the *lever arm* as the distance between the pivot point and the place where the force acts. Later, we'll describe how the lever arm is defined in more general situations. The torque in this simple case is given by

$$\tau = Fr \qquad (8.9)$$

where r is the length of the lever arm. Torque is the fundamental "force-type" quantity with regards to rotational motion; it is analogous to force in translational motion. A key feature of torque is that it is the *product* of force and the distance to the pivot point. The units of τ are thus N · m (force times distance).

We can use the seesaw in Figure 8.10 to understand why torque depends on both the force and the lever arm. For a force of a given magnitude, it is easier to rotate the seesaw if r is large than if r is small. That is, a given force produces a larger angular acceleration when it is applied farther from the pivot point.

Relating Torque and Angular Acceleration

Our next job is to derive the relation between torque and the angular motion it produces. We go back to basics and apply Newton's second law to the sketch in Figure 8.11, which shows a hinged rod to which a ball of mass m is attached at one end. The other end of the rod at point P is held in place, but the hinge allows the rest of the rod

Figure 8.10 Torque is the product of the force and the lever arm. Here the force $\vec{F}$ produces a torque of magnitude Fr.

Figure 8.11 Only a perpendicular force can make an object rotate. The force $\vec{F}_{applied}$ accelerates this ball in the perpendicular direction.

to rotate about P. If a force $\vec{F}_{applied}$ is applied to the ball, we expect the rod and ball will rotate. Let's now apply Newton's second law to analyze this motion.

For simplicity, we assume the rod is extremely light (massless), so all the rotating mass is concentrated in the ball. Applying Newton's second law to the ball, we have

$$\sum \vec{F} = m\vec{a}$$

where $\sum \vec{F}$ is the total force on the ball. In this example, there are two contributions to this total force: one from our applied force $\vec{F}_{applied}$ and another force $\vec{F}_{rod}$ from the rod acting on the ball. We thus have

$$\sum \vec{F} = \vec{F}_{applied} + \vec{F}_{rod} = m\vec{a} \qquad (8.10)$$

Let's begin with the case in which $\vec{F}_{applied}$ is directed perpendicular to the rod as shown in Figure 8.11. The rod is rigid, so it cannot stretch or bend, and the ball will move in a circular arc along the direction of $\vec{F}_{applied}$. The force $\vec{F}_{rod}$ is along the direction of the rod, perpendicular to the ball's circular path. As a result, $\vec{F}_{rod}$ does not contribute to the ball's angular acceleration. *Only forces with a component perpendicular to the rod can contribute to the angular acceleration.* With that in mind, consider only the perpendicular components in Equation 8.10 of the force on the ball and its acceleration along this circular path. We call this acceleration $a_\perp$ because it denotes the acceleration in the direction perpendicular to the rod (see Fig. 8.11). Likewise, we denote the perpendicular component of the applied force as $F_\perp$. The components of Equation 8.10 along this perpendicular direction then read

$$F_\perp = ma_\perp$$

The component $a_\perp$ is a linear (translational) acceleration, and we saw in Section 8.1 that this linear acceleration is proportional to the angular acceleration of the ball as it moves along its circular path. Using that result (Eq. 8.8) here leads to

$$F_\perp = ma_\perp = m\alpha r$$

where r is the distance from the hinge to the ball. Multiplying both sides by r and do a little rearranging gives

$$F_\perp r = (m\alpha r)r = (mr^2)\alpha \qquad (8.11)$$

Comparing with Equation 8.9, you should recognize that the left-hand side of Equation 8.11 is the torque on the ball; it is the product of the force and the lever arm r. The right-hand side contains mass and (angular) acceleration, so Equation 8.11 looks a lot like Newton's second law. Indeed, it *is* Newton's second law for rotational motion, and it is usually written as

Newton's second law for rotational motion

$$\sum \tau = I\alpha \qquad (8.12)$$

where $\sum \tau$ is the total torque, that is, the sum of all the different torques acting on the object. Here we have introduced a new quantity I called the ***moment of inertia***. In rotational motion, I plays the role that mass does in translational motion. For the system considered here, consisting of a ball attached to a massless hinged rod, we can compare Equations 8.11 and 8.12 to find that the moment of inertia is

$$I = mr^2 \quad \text{(massless hinged rod/ball)} \qquad (8.13)$$

Newton's Second Law for Rotational Motion and the Analogy with Translational Motion

Even though the examples of the seesaw in Figure 8.10 and the mass on a hinged rod in Figure 8.11 involve simple cases, they enabled us to arrive at a very deep and fundamental result: Newton's second law for rotational motion, $\sum \tau = I\alpha$. This relation and the associated definitions of torque and moment of inertia have important implications and consequences. However, before exploring them, let's review and recap three major points.

First, *torque* plays the role of force in rotational motion. Torque depends on the magnitude of the force and on *where* the force is applied relative to the pivot point. Hence, the rotation axis's location greatly affects the torque produced by a given force, and, from Newton's second law (Eq. 8.12), it will affect the resulting angular acceleration. There may be multiple forces acting on an object; hence, there may be multiple lever arms and multiple torques, all involving a single pivot point.

Second, the *moment of inertia* enters into rotational motion in the same way that mass enters into translational motion. We found (Eq. 8.13) that the moment of inertia of a ball at the end of a massless rod of length r is $I = mr^2$. For an object composed of many pieces of mass m_i located at distances r_i from the pivot point, the total moment of inertia *of the object* is just the sum of the moments of inertia of the individual pieces (Fig. 8.12); hence,

$$I = \sum_i m_i r_i^2 \qquad (8.14)$$

The moment of inertia of an object thus depends on the object's mass and on how that mass is distributed relative to the rotation axis.

Equation 8.14 looks complicated, but it is usually not necessary to perform such a sum over the pieces of an object. We'll apply Equation 8.14 and consider the moment of inertia for various types of objects in Section 8.4.

Third, Newton's second law for translational motion leads to *Newton's second law for rotational motion*:

$$\sum \tau = I\alpha$$

We'll spend the rest of this chapter exploring how to apply this relation.

The similarities between translational and rotational motion are reinforced in Table 8.1, where we list some of the quantities used to describe each.

Torques and Lever Arms Revisited: A More General Definition

We have defined torque for cases in which the applied force is purely in the perpendicular direction (Figs. 8.10 and 8.11), but, in general, the force may be applied in a nonperpendicular direction. One such example is shown in Figure 8.13, again for our hinged rod with a ball attached at the end. Here the applied force $\vec{F}_{applied}$ acts at an angle ϕ with respect to the rod. Since the ball is attached to the rod, the ball is again restricted to move along the circular arc sketched in Figure 8.13. Only the forces directed along this arc (perpendicular to the rod) contribute to the torque. For this reason, we can ignore the force of the rod on the ball when calculating the torque; this force does not contribute to the angular acceleration. The perpendicular force is $F_\perp = F_{applied} \sin \phi$, and the torque in this case is

$$\tau = F_\perp r = (F_{applied} \sin \phi)r$$

$$\tau = F_{applied}\, r \sin \phi \qquad (8.15)$$

Here $F_\perp$ is the component of the force that is perpendicular to the rod and r is the distance between the pivot point and the point where the force acts. Equation 8.15 is our general definition of torque, and it applies even when the force is not directed

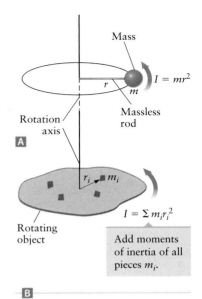

Figure 8.12 ◼ Moment of inertia of a single point mass rotating about an axis: $I = mr^2$. ◼ The moment of inertia of an extended object is just the sum of the moments of inertia of all the individual pieces of the object. The moment of inertia of each piece can be calculated as in part ◼.

Definition of torque

Figure 8.13 Only the perpendicular component of $\vec{F}_{applied}$ contributes to the torque. Any forces or force components parallel to the rod do not contribute to the torque.

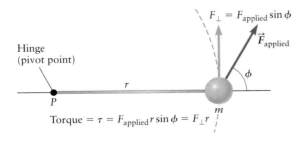

Table 8.1 Quantities Used to Describe Linear Motion and Rotational Motion

LINEAR MOTION	ROTATIONAL MOTION
Linear position = x	Angular position = θ
Linear velocity = v	Angular velocity = ω
Linear acceleration = a	Angular acceleration = α
Force = F	Torque = τ
Mass = m	Moment of inertia = I

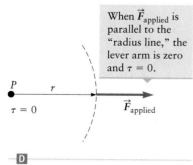

Figure 8.14 Torque is given by $\tau = F_{\text{applied}}(r \sin \phi) = F_\perp r = F_{\text{applied}} r_\perp$.

perpendicular to the lever arm. When the force is directed perpendicular to the lever arm, $\phi = 90°$; hence, $\sin \phi = 1$ and $F_\perp = F_{\text{applied}}$. When the force is parallel to the lever arm, $\phi = 0$ and the torque is zero. This conclusion makes sense because a force directed parallel to the lever arm ($\phi = 0$) cannot cause an object to rotate.

Two Ways to Think about Torque

The general relation for the torque in Equation 8.15 can be illustrated graphically in two different ways. We have already noted that the perpendicular component of the applied force is $F_\perp = F_{\text{applied}} \sin \phi$, so we can group the terms in Equation 8.15 as

$$\tau = \overbrace{(F_{\text{applied}} \sin \phi)}^{F_\perp} r = F_\perp r \tag{8.16}$$

which is illustrated in Figure 8.14A; in words, this says that torque equals the product of the perpendicular component of the force ($F_\perp$) and the distance from the pivot point to the point where the force acts on the object (r). This product of perpendicular force and distance is equivalent to our definition of torque in the simpler cases in Figures 8.10 and 8.11. Figure 8.14A also shows that you can find the angle ϕ by extending the "radius line" beyond the point where the force acts; this extension is shown as the dashed line in Figure 8.14A, and ϕ is the angle between this extension and the force vector.

On the other hand, we can group the terms in Equation 8.15 as

$$\tau = F_{\text{applied}} \overbrace{(r \sin \phi)}^{r_\perp} = F_{\text{applied}} r_\perp \tag{8.17}$$

which is illustrated in Figure 8.14B and C; in words, this equation says that torque equals the product of the applied force (F_{applied}) and the perpendicular distance $r_\perp$. This perpendicular distance is the distance from the line of action of the force (see Fig. 8.14B) to the rotation axis as measured on a line that is perpendicular to both. The perpendicular distance $r_\perp = r \sin \phi$ is the lever arm in the general case of a force applied at an angle. These two ways of writing the torque (Eqs. 8.16 and 8.17) are completely equivalent and simply correspond to the two different ways of thinking about the torque. The general expression for the lever arm is

$$r_\perp = r \sin \phi \tag{8.18}$$

When the force is purely perpendicular and $\phi = 90°$, this expression reduces to just r as in the cases we considered in Figures 8.10 and 8.11. On the other hand, if $\phi = 0$, the lever arm is zero, corresponding to a force applied parallel to the "radius line" (Fig. 8.14D). Such a parallel force cannot cause an object to rotate and thus cannot produce a torque.

Figure 8.15 Concept Check 8.3. Forces on a torque wrench.

CONCEPT CHECK 8.3 | Forces on a Torque Wrench

A torque wrench is a special type of wrench that indicates the magnitude of the torque it is applying to a bolt, helping the user avoid breaking a bolt or stripping a thread. Consider the torque wrench shown in Figure 8.15. It has a long handle, and we suppose one user applies a force $F_1 = 50$ N at a distance of 10 cm from the head of the wrench. If a second user applies a force F_2 at a point 25 cm from the head, what value of F_2 will produce the same torque? Assume forces F_1 and F_2 are both applied perpendicular to the wrench handle.

(a) 10 N   (b) 20 N   (c) 50 N   (d) 100 N   (e) 125 N

Center of Gravity, Center of Mass, and the Direction of Torque

As an example of torques and lever arms, consider the motion of the hands of a large outdoor clock such as one might find near the top of a clock tower as represented very schematically in Figure 8.16. Let's calculate the torque on one clock hand due to the force of gravity. For simplicity, we first assume (unrealistically) that all the mass m is located at the end of the hand, which makes the clock hand just like the ball attached to a massless rod in Figures 8.11 and 8.13. Using Equation 8.15, the magnitude of the torque is

$$\tau = F_{grav} r \sin \phi$$

Figure 8.16 shows the forces and angles when the clock hand is at two different positions.

When the hand is at "3 o'clock" (i.e., horizontal), the gravitational force is perpendicular to the hand and $\phi = 90°$. If L is the length of the clock hand, we have

$$\tau = F_{grav} r \sin \phi = mgL \sin(90°)$$
$$\tau = mgL \tag{8.19}$$

which is the *magnitude* of the torque. We must also pay attention to its *direction*. If we have a single rotation axis, as is the case here, the direction of the torque is specified by its sign. Recall our convention for the positive and negative directions for rotations; see Figure 8.3. Those conventions also give the positive and negative directions for τ. A *positive* torque is one that would produce a counterclockwise rotation, whereas a *negative* torque would produce a clockwise rotation. In this example, with the clock hand at 3 o'clock, the torque from gravity is negative since this torque would, if it acted alone, cause the clock hand to move downward and hence rotate clockwise. If the clock hand were instead at the 9 o'clock position (Fig. 8.16B), the torque would have the same magnitude as found in Equation 8.19 (because the angle ϕ is again a right angle), but τ would now be positive because it would cause a counterclockwise rotation of the hand.

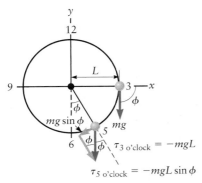

Clockwise direction = negative torque

A

Counterclockwise direction = positive torque

B

Figure 8.16 **A** When all mass of this clock hand is at the end, the torque due to gravity is $\tau = -mgL \sin \phi$. When the clock hand is at 3 o'clock, $\phi = 90°$, whereas when the hand is at 5 o'clock, this angle is smaller, so the magnitude of the torque is smaller. **B** When the clock hand is at 9 o'clock, the torque is positive because it would (if acting alone) make the hand rotate clockwise.

Figure 8.17 When the mass of the clock hand is distributed uniformly, the torque is calculated by letting the force of gravity act at the center of mass.

Having all the mass located at the end of the clock hand is rather unrealistic, even for a physics problem, so let's now make the more reasonable assumption that the mass is distributed uniformly along the clock hand. This situation is shown in Figure 8.17, with the clock hand located at 3 o'clock. To calculate the torque, we want to apply our relation for τ in Equation 8.15, but we immediately run into a problem. Torque involves the product of force and lever arm, and in this case, with the force in a purely perpendicular direction, the lever arm lies along the clock hand. The length of the lever arm depends on *where* the force acts. For the gravitational force acting on the clock hand, the force acts wherever there is mass, so it acts at all points along the hand. So we are left with two choices for how to calculate the torque. We could imagine the hand to be broken up (figuratively, of course) into many infinitesimally small pieces and then add up the torques on each piece to get the total torque, but a much more convenient approach is to use of the concept of **center of gravity**.

For the purposes of calculating the torque due to the gravitational force, one can assume all the force acts at a single location called the object's center of gravity. For our clock hand and for all other examples we encounter in this book, the center of gravity of an object is located at its center of mass,[1] defined in Chapter 7. For our uniform clock hand, the center of mass is at the center of the hand. The length of the lever arm is thus $L/2$ as indicated in Figure 8.17. The torque with the clock hand at 3 o'clock is therefore

$$\tau = -mgr \sin \phi = -mg\left(\frac{L}{2}\right)\sin(90°) = -\frac{mgL}{2}$$

because again the angle $\phi = 90°$. The negative signs here accounts for gravity's tendency to make the clock hand rotate clockwise, which, according to our convention, is the negative direction.

CONCEPT CHECK 8.4 | Determining the Sign of the Torque

Figure 8.18 shows several rotating objects and the forces applied to them. According to our convention for finding the sign of τ, which of these torques are positive and which are negative?

8.3 | ROTATIONAL EQUILIBRIUM

We showed in Chapter 4 that for an object to be in equilibrium with respect to its translational motion, the linear acceleration $\vec{a}$ must be zero. According to Newton's second law, $\Sigma \vec{F} = m\vec{a}$, so if an object is in translational equilibrium, the total force must be zero, $\Sigma \vec{F} = 0$. Also, since the acceleration is zero, an object in translational equilibrium will have a constant velocity $\vec{v}$. If that velocity is zero, we say that the object is in static equilibrium with respect to its translational motion.

We now want to extend these ideas to the notion of **rotational equilibrium**. An object that is in equilibrium with regard to both its translational and its rotational motion must have a linear acceleration of zero *and* an angular acceleration that is zero. The condition that the total force is zero, $\Sigma \vec{F} = 0$, is *not* sufficient to guarantee that both these requirements are satisfied. See, for instance, Figure 8.19, which shows two forces applied to an airplane propeller. The forces $\vec{F}_1$ and $\vec{F}_2$ are applied horizontally, and we suppose they are equal in magnitude but opposite in direction, so their sum is zero. Although the condition for translational equilibrium along x is thus satisfied, it is clear that this propeller will not be in rotational equilibrium. Even though the sum of these two forces is zero, they still produce a net torque on the propeller and lead to an angular acceleration in the clockwise direction.

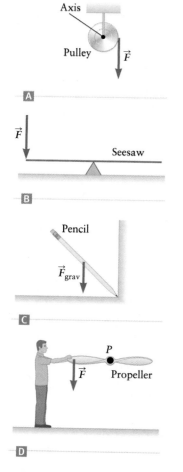

Figure 8.18 Concept Check 8.4.

[1]The center of gravity and the center of mass of an object are usually at the same point, but that is not the case when the force of gravity varies with position, as in planetary motion. We do not treat such cases in this text.

For an object to be in complete equilibrium, we also require the angular acceleration to be zero; from Newton's second law for rotational motion (Eq. 8.12), this requirement means that the total torque must be zero:

$$\sum \tau = \tau_1 + \tau_2 + \tau_3 + \cdots = 0 \qquad (8.20)$$

This is a necessary condition for rotational equilibrium. For the problems we consider in this book, all the torques in this equation refer to a particular rotation axis. The same ideas can be applied to the rotation about two or three different axes simultaneously, but we must leave that for a more advanced course.

To apply the condition for rotational equilibrium in Equation 8.20, we must calculate the total torque on an object from all the applied forces. The following careful procedure describes how to calculate the torque and apply the conditions for equilibrium in any situation.

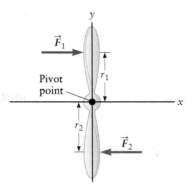

Figure 8.19 Although the horizontal forces on this propeller may add up to zero, they will still produce a nonzero torque. Hence, the propeller will not be in rotational equilibrium.

PROBLEM SOLVING | Applying the Conditions for Translational and Rotational Equilibrium

1. **RECOGNIZE THE PRINCIPLE.** An object is in translational static equilibrium if its linear acceleration and linear velocity are both zero. An object is in rotational static equilibrium if its angular acceleration and angular velocity are both zero.

2. **SKETCH THE PROBLEM.** Make a drawing showing the object of interest along with all the forces that act on it. This sketch should contain a set of coordinate axes.

3. **IDENTIFY THE RELATIONSHIPS.** Find the rotation axis and pivot point. They are necessary for calculating the torques and will depend on the problem. The torque calculations can be broken into several steps:
   • To calculate the torque from each force, first determine the lever arm, the perpendicular distance from the pivot point to the point at which the force acts.
   • Calculate the magnitude of the torque using $\tau = Fr \sin \phi$ (Eq. 8.15). Here, F is the magnitude of the force.
   • Determine the sign of τ. If this force *acting alone* would produce a *counterclockwise* rotation, the torque is *positive*. If the force would cause a *clockwise* rotation, the torque is *negative*.
   • Add the torques from each force to get the total torque. Be sure to include the proper sign for each individual torque.

4. **SOLVE** for the unknowns by applying the condition for rotational equilibrium $\sum \tau = 0$ and, if necessary, the condition for translational equilibrium $\sum \vec{F} = 0$.

5. Always *consider what your answer means* and check that it makes sense.

These steps follow our usual problem-solving strategy, but with several new considerations to keep in mind. First, your sketch should always *show where the forces act*. When we applied Newton's laws to the translational motion of a point particle (in Chapters 3 and 4), we began with a free-body diagram showing all the forces acting on the particle. When dealing with rotational motion, we also start with a force diagram, but we cannot simply show the object as a point. We must consider all the forces on the object and the *places* on the object where these forces act. This information is needed to determine the torque associated with each force.

Second, you should think carefully about where to put the pivot point. Choosing a pivot point is an essential part of dealing with torques and rotational motion, especially in equilibrium problems. In many cases, such as the seesaw in Figure 8.10, there is a "natural" choice for the pivot point, but sometimes there may be more than one plausible choice. Fortunately, for an object in rotational equilibrium, you may choose *any* spot to be the pivot point without affecting the final answer.

However, some choices will lead to a simpler torque equation. Since torque is equal to the product of force and lever arm, forces whose lever arms are zero will not contribute to the torque. These forces will then not appear in the torque equation. This is illustrated in Example 8.4.

Rotational Equilibrium of a Lever: Amplifying Forces

We return to our seesaw theme in Figure 8.20, which shows a lever being used to lift a very heavy rock of mass m. Let's use our method for dealing with objects in rotational equilibrium to calculate the minimum force needed to just begin lifting the rock and then find how it compares to the rock's weight.

Since the rock is just barely being lifted from the ground, both its acceleration and its angular acceleration are essentially zero and we can apply the conditions for equilibrium to the lever. Figure 8.20 serves as our drawing (step 2 in our problem-solving procedure). It shows all the forces on the lever and the places at which they act. For simplicity, we assume the lever is much lighter than the rock, so we can ignore the force of gravity on the lever. That still leaves three forces acting on the lever: the force from the rock on the lever $\vec{F}_{\text{rock}}$ (the weight of the rock), the force applied by the person on the other end of the lever $\vec{F}_{\text{person}}$, and the force of the support on the lever $\vec{F}_N$ (at point P). We next (step 3) identify the pivot point: we choose the pivot point to be the spot at which the lever is supported (point P in Fig. 8.20), with the rotation axis perpendicular to the plane of the picture. The lever arm associated with $\vec{F}_{\text{rock}}$ is L_{rock}, while the lever arm for the torque associated with $\vec{F}_{\text{person}}$ is L_{person}. The lever arm for the force from the support is zero because this force acts at the pivot point.

Using the results from Figure 8.20 for the lever arms and forces, we can write the condition for rotational equilibrium as

$$\Sigma\,\tau = \tau_{\text{rock}} + \tau_{\text{person}} + \tau_N = 0$$

$$+|F_{\text{rock}}|L_{\text{rock}} - |F_{\text{person}}|L_{\text{person}} + 0 = 0 \qquad (8.21)$$

The signs in front of these terms come from our convention for positive and negative torques. The force from the rock has a magnitude $|F_{\text{rock}}|$ and is directed downward, so it gives a counterclockwise (positive) torque (hence the positive sign in front of this term in Eq. 8.21). To balance this torque, the force exerted by the person on the lever's other end must be directed downward, and this force has a magnitude $|F_{\text{person}}|$. This force produces a clockwise (negative) torque (hence the negative sign in front of that term in Eq. 8.21). Since F_{rock} is due to the weight of the rock, $|F_{\text{rock}}| = mg$. Inserting into Equation 8.21 leads to

$$mgL_{\text{rock}} - |F_{\text{person}}|L_{\text{person}} = 0 \qquad (8.22)$$

We want to find the force F_{person} that is just sufficient to balance the rock on the lever, and we can find its value by solving Equation 8.22 (step 4):

$$|F_{\text{person}}| = mg\,\frac{L_{\text{rock}}}{L_{\text{person}}}$$

Figure 8.20 There are three forces acting on the lever (assuming it is very light so that its weight can be ignored). The normal force F_N exerted by the support acts at the pivot point P, so this force does not produce a torque about P.

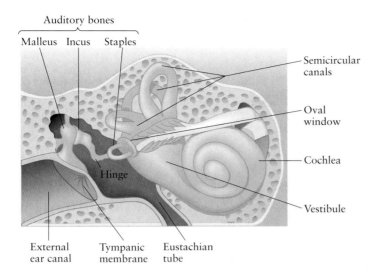

Figure 8.21 The incus bone in the ear acts as a lever as it transmits forces to the inner ear. Notice the location of the "hinge," which is the pivot point for the incus lever.

Auditory bones

Malleus Incus Staples

Semicircular canals

Oval window

Cochlea

Vestibule

Hinge

External ear canal

Tympanic membrane

Eustachian tube

This expression gives us the magnitude of F_{person}, but we also need to know its sign (i.e., direction). From Figure 8.20, we can see that the person exerts a downward force on the lever (as we also assumed in our calculation of the torque), so F_{person} must be negative. We get

$$F_{person} = -mg \frac{L_{rock}}{L_{person}} \tag{8.23}$$

What does this result mean? Equation 8.23 shows that the magnitude of the force exerted by the person (F_{person}) can be smaller than the weight of the rock. If the lever arm extending to the person is longer than the lever arm to the rock (i.e., if $L_{person} > L_{rock}$), this simple lever will *amplify* the force exerted by the person. Of course, this result is probably what you expected if you have ever used a screwdriver to pry something open.

⊗ Amplification of Forces in the Ear

Many mechanical devices use a simple lever to amplify forces. The same principle is at work in the human ear. The ear is a complicated device that uses physics in many interesting ways (including the accelerometer described in Chapter 4). As you may know, three tiny bones help carry sound through the middle ear. These bones are shown in Figure 8.21. One of them, the incus, is supported by a sort of hinge so that it is able to transmit forces from one end to the other, just like the simple lever in Figure 8.20. The analysis given above applies also to the forces at the ends of the incus in Figure 8.21. The lever arm from this hinge to the end in contact with the stapes bone is about a factor of three shorter than the end connected to the malleus bone. Hence, when applied to the ear, the ratio of the lever arms in Equation 8.23 is approximately a factor of three, and this lever amplifies the forces associated with a sound vibration by the same factor.

| EXAMPLE 8.4 | Painters and Physics |

A housepainter is using a board of length $L = 3.0$ m that sits on two supports as a scaffolding on one side of a house (Fig. 8.22). The painter often wants to reach very far to one side of the board so that he can paint as much area as possible without moving the scaffolding. **(a)** How far can the painter in Figure 8.22 walk to the right-hand side of the board before it tips? The mass of the painter is 80 kg, and the mass of the board is 30 kg. **(b)** What is the force of the right-hand support on the board when the board just begins to tip? Assume the painter is always standing straight up, so his center of mass is always above his feet.

RECOGNIZE THE PRINCIPLE

The board does not quite tip, so it is always in rotational equilibrium. The total torque must therefore equal zero. The board is also in translational equilibrium, so the total force on the board must also equal zero.

SKETCH THE PROBLEM

Figure 8.22 shows all the forces on the board: the forces exerted by both supports F_R and F_L, the force of gravity on the board F_B, and the force of gravity on the painter F_P. The figure also shows where each force acts, information that is needed to calculate the associated torques.

IDENTIFY THE RELATIONSHIPS

We choose the pivot point to be at *P* because that is the point around which the board would rotate if the painter moves too far to the right. When the painter is standing as far to the right as possible, the board will just barely begin to tip, which means that F_L, the force exerted by the left-hand support, is *zero*.

We next compute the torques using the information in Figure 8.22 and apply the condition for rotational equilibrium $\sum \tau = 0$. Using the lever arms from Figure 8.22 and $F_B = -m_B g$ (the weight of the board) and $F_P = -m_P g$ (the weight of the painter), we have

$$\tau = 0 = -F_L\left(\frac{3}{4}L\right) + F_R(0) + m_B g\left(\frac{L}{4}\right) - m_P g(x) \tag{1}$$

The signs of these torque terms are determined using the convention that a counter-clockwise rotation is in the positive direction. For example, the last term on the right in Equation (1) is the torque from the weight of the painter; this torque would produce a clockwise rotation and thus is negative.

The next step is to write the condition for translational equilibrium $\sum \vec{F} = 0$. The forces on the board are all along the *y* direction. The force condition is then

$$\sum F = F_L + F_R + F_B + F_P = 0 \tag{2}$$

Notice that some of these terms, such as $F_B = -m_B g$ and $F_P = -m_P g$, are themselves negative. Inserting $F_L = 0$ and our values of F_B and F_P into Equation (2) gives

$$F_R - m_B g - m_P g = 0 \tag{3}$$

SOLVE

(a) We can now solve Equation (1) to find *x*, the location of the painter. Inserting $F_L = 0$ (because the board is on the verge of tipping), we find

$$m_B g\left(\frac{L}{4}\right) - m_P g(x) = 0$$

$$x = \frac{m_B g(L/4)}{m_P g} = \frac{m_B L}{4 m_P}$$

Inserting the given values of the masses of the board and painter, along with L, gives

$$x = \frac{m_B L}{4m_P} = \frac{(30 \text{ kg})(3.0 \text{ m})}{4(80 \text{ kg})} = \boxed{0.28 \text{ m}}$$

The painter would thus be well-advised to remain at a distance somewhat less than 0.28 m from the pivot point.

(b) To find the force from the right-hand support, we rearrange Equation (2) and get

$$F_R = m_B g + m_P g = (m_B + m_P)g = (80 \text{ kg} + 30 \text{ kg})(9.8 \text{ m/s}^2)$$

$$F_R = \boxed{1100 \text{ N}}$$

What does it mean?
Since F_L is zero, the right-hand support must support the entire weight of the system. Hence, the force from the right-hand support must equal the weight of the board plus the weight of the painter, as found in the answer to part (b).

EXAMPLE 8.5 | Leaning on a Wall

We continue our theme of housepainters and physics from Example 8.4 and consider the stability of a ladder of mass m and length L leaning against a wall as shown in Figure 8.23. The painter wants to be sure it will not slip across the floor. He knows that the frictional force exerted by the floor on the bottom end of the ladder is needed to keep the ladder in equilibrium. He therefore performs a careful experiment and finds that the coefficient of static friction between the ladder and the floor is $\mu_S = 0.50$. He now wants to calculate the angle ϕ at which the ladder will just become unstable and slide across the floor. For simplicity, he assumes there is no frictional force between the ladder and the wall.

RECOGNIZE THE PRINCIPLE

The ladder is in equilibrium, so we can apply the general conditions for static equilibrium $\Sigma \tau = 0$ and $\Sigma \vec{F} = 0$. The angle ϕ plays a key role: if we take the bottom of the ladder (point P) as our pivot point, the torque on the ladder due to gravity increases as ϕ becomes smaller because the lever arm increases. At some critical value of the angle, the torque will be large enough that the ladder will start to move.

SKETCH THE PROBLEM

Figure 8.23 shows all the forces acting on the ladder along with their components along x and y. There are normal forces exerted by the floor (N_F) and the wall (N_W), the force of friction exerted by the floor (F_{friction}), and the ladder's weight (mg). The force from the floor acts on the end of the ladder at point P, while the force of gravity acts at the ladder's center of mass.

IDENTIFY THE RELATIONSHIPS

Let's choose the base of the ladder at P as the pivot point. By doing so, we make the torques due to the forces from the floor (N_F and F_{friction}) zero because these forces act at P and hence their lever arms are zero. Using the lever arms and angles from Figure 8.23, the torque condition is

$$\Sigma \tau = 0 = +N_W L \sin \phi - mg\left(\frac{L}{2}\right)\sin(90° - \phi)$$

$$0 = +N_W L \sin \phi - mg\left(\frac{L}{2}\right)\cos(\phi) \tag{1}$$

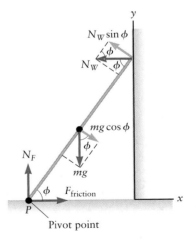

Figure 8.23 Example 8.5. Forces acting a leaning ladder. We assume there is no friction between the wall and the top end of the ladder.

where we have used the general definition of torque (Eq. 8.15) along with the trigonometric identity $\sin(90° − \phi) = \cos(\phi)$. The appearance of the cosine term in Equation (1) can be understood from the expression for torque in Equation 8.16. The torque due to the ladder's weight is the product of $L/2$ (the distance from the center of mass to the pivot point) and the component of the force of gravity *perpendicular* to the ladder. Figure 8.23 shows that this component involves $\cos \phi$.

Since the ladder is in static equilibrium, the sum of all forces along both x and y must be zero, which leads to

$$\sum F_x = 0 = −N_W + F_{\text{friction}} \quad (2)$$

$$\sum F_y = 0 = N_F − mg = 0 \quad (3)$$

SOLVE

From Equation (3), we can immediately solve for the normal force from the floor and get $N_F = mg$. We also know that the maximum frictional force is $F_{\text{friction}} = \mu_S N_F$, so we have

$$F_{\text{friction}} = \mu_S N_F = \mu_S mg$$

and, according to Equation (2), this result is also equal to N_W. Inserting into Equation (1) leads to

$$+N_W L \sin \phi − mg\left(\frac{L}{2}\right)\cos \phi = 0$$

$$\mu_S mg L \sin \phi = mg\left(\frac{L}{2}\right)\cos \phi$$

$$\tan \phi = \frac{1}{2\mu_S}$$

Using the value $\mu_S = 0.50$ measured by the painter gives

$$\tan \phi = \frac{1}{2\mu_S} = \frac{1}{2(0.50)} = 1.0 \quad (4)$$

$$\phi = \boxed{45°}$$

What does it mean?

For a smaller angle, the frictional force would not be able to keep the ladder in equilibrium and the ladder would slip. Most sensible painters would probably choose a steeper angle than 45°.

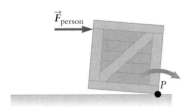

A

B

Figure 8.24 **A** Forces on a crate. As the person pushes harder and harder, the crate will eventually tip and **B** rotate about a pivot point P at the corner of the crate.

CONCEPT CHECK 8.5 | Adding Friction between the Ladder and the Wall

Suppose we add friction between the wall and the upper end of the ladder in Example 8.5. How would that change the result for ϕ?
  (a) The minimum safe value of ϕ would increase.
  (b) The minimum safe value of ϕ would decrease.
  (c) The minimum safe value of ϕ would stay the same.

Pushing on a Crate: When Will It Tip?

Another example of an object in static equilibrium is shown in Figure 8.24A. Here we show a cubical crate of length L sitting at rest on a level floor as a person pushes on one side of the crate. If the person applies a very large force of magnitude F_{person} horizontally at the top edge and the frictional force F_{friction} at the bottom edge is large, the crate will tip as sketched in Figure 8.24B. Let's now consider how to calculate the value of F_{person} at which the crate just begins to tip.

When the crate is on the verge of tipping, it is still in static equilibrium, so the total torque and the total force are both zero. There are four forces acting on the crate: (1) its weight (the force of gravity), which acts at the center of mass; (2) the force of static friction, which acts along the bottom surface of the crate; (3) the normal force exerted by the floor, which acts at the surface in contact with the floor; and (4) the force applied by the person, F_{person}. Any tipping will involve rotation about the point P located at the corner of the crate, so let's choose that as our pivot point. The corresponding lever arms are indicated in Figure 8.24A; the lever arm associated with the frictional force is zero, so it will not contribute to the torque. If the crate is on the verge of rotating about P, the normal force exerted by the floor will then be acting along a line passing through P and its lever arm is also zero. The total torque on the crate is then

$$\sum \tau = 0 = \overbrace{mg(L/2)}^{\text{gravity}} - \overbrace{F_{person}(L)}^{\text{person}} + \overbrace{N(0)}^{\substack{\text{normal}\\\text{force}}} + \overbrace{F_{friction}(0)}^{\text{friction}} = 0 \qquad (8.24)$$

where the corresponding lever arm factors are all given in parentheses. The lever arm for the gravitational force is $L/2$ because this force acts at the center of mass, which is in the middle of the crate, a distance $L/2$ from the side of the crate. The lever arm for the torque from F_{person} is L because the person is pushing at the top edge of the crate. The signs in Equation 8.24 can again be understood using our convention for positive and negative torque. For example, a positive value of the applied force (F_{person} directed to the right, along $+x$), gives a clockwise rotation and hence a *negative* torque. We can solve Equation 8.24 for F_{person} and find

$$mg(L/2) = F_{person}(L)$$

$$F_{person} = \frac{mg}{2}$$

which is half the weight of the crate. If F_{person} exceeds this value, the crate will tip.

This problem of a tipping crate is very similar to the problem of a car rolling over in a traffic accident. This subject is of great interest to automobile designers and is explored in the next example.

EXAMPLE 8.6 | ⓡ Cars and Static Equilibrium

Figure 8.25 shows the forces acting on a car in a collision. The force $\vec{F}_{collision}$ comes from another car, and for simplicity we assume this force is directed horizontally and acts at the center of the body of the car. Estimate the magnitude of $\vec{F}_{collision}$ for which the car in Figure 8.25 will just roll over. Express the answer as a ratio $|\vec{F}_{collision}|/mg$, where m is the mass of the car, and use the dimensions of a typical car in your calculation.

RECOGNIZE THE PRINCIPLE

We wish to calculate the minimum force $\vec{F}_{collision}$ that will just barely cause the car to tip. This problem is very similar to the tipping crate in Figure 8.24. When the car tips, it will rotate about point P in Figure 8.25, so we take that as the pivot point.

SKETCH THE PROBLEM

Figure 8.25 shows all the forces on the car, including the force of friction between the tires and road ($F_{friction}$) and the force of gravity ($F_{grav} = mg$). It also shows the points at which these forces act.

IDENTIFY THE RELATIONSHIPS

Although this problem is similar to the tipping crate in Figure 8.24, there are some important differences. Because the car sits on four wheels, we must account for the

Figure 8.25 Example 8.6. If a force $\vec{F}_{collision}$ is applied horizontally to a car, the car may tip, that is, rotate about the point P.

height of the wheels when finding the car's center of mass. Indeed, we are not given any information about the dimensions of the car's body (w and h in Fig. 8.25) or the wheels (d), so we'll also have to estimate their values. First, though, let's express the conditions for equilibrium in terms of w, h, and d. We will make estimates for their values below.

We treat the car as a simple "box" of width w and height h that is a distance d off the ground and ignore the mass in the wheels. As with the tipping crate, there are four forces on the car as indicated in Figure 8.25. The condition for rotational equilibrium is

$$\sum \tau = 0 = -F_{\text{collision}}\overbrace{\left(d + \frac{h}{2} \right)}^{\text{collision force}} + \overbrace{mg\left(\frac{w}{2} \right)}^{\text{gravity}} + \overbrace{N(0)}^{\substack{\text{normal} \\ \text{force}}} + \overbrace{F_{\text{friction}}(0)}^{\text{friction}} \quad (1)$$

where the lever arm factors are given in parentheses. The term involving $F_{\text{collision}}$ is the torque produced by the collision force. The lever arm in this case is the distance from the center of the car to the ground $d + h/2$ since this force acts at the middle of the car, at its center of mass. This force is directed to the right (along $+x$), and it would, if it acted alone, produce a clockwise rotation; hence, this torque is *negative*.

SOLVE

We can rearrange Equation (1) to get

$$F_{\text{collision}}\left(d + \frac{h}{2} \right) = mg\left(\frac{w}{2} \right)$$

Dividing $F_{\text{collision}}$ by the weight of the car leads to

$$\boxed{\frac{F_{\text{collision}}}{mg} = \frac{w}{2d + h}} \quad (1)$$

What does it mean?

If $F_{\text{collision}}/mg$ is small, a (relatively) small force is enough to cause a rollover accident, whereas a large ratio means that a larger impact force is needed to cause a rollover. Equation (1) shows that a large value of d—that is, a large ground clearance—will make a rollover more likely. For a typical compact car that sits relatively low to the ground, we estimate[2] $w = 170$ cm, $d = 20$ cm, and $h = 120$ cm, which gives the ratio

$$\frac{F_{\text{collision}}}{mg} = \frac{170 \text{ cm}}{2(20 \text{ cm}) + 120 \text{ cm}} = 1.1 \quad \text{(compact car)}$$

For an SUV, we estimate $w = 170$ cm, $d = 40$ cm, and $h = 140$ cm, which leads to

$$\frac{F_{\text{collision}}}{mg} = \frac{170 \text{ cm}}{2(40 \text{ cm}) + 140 \text{ cm}} = 0.77 \quad \text{(SUV)}$$

Hence, according to this measure, the SUV is more likely to roll over. Of course, real accidents are more complicated than this simple model. Current government rollover ratings are based in part on a similar calculation.

8.4 | MOMENT OF INERTIA

In Section 8.2, we introduced the rotational form of Newton's second law:

$$\sum \tau = I\alpha \quad (8.25)$$

---

[2]These estimates of w, d, and h were obtained from the specifications of actual vehicles.

In all our work on rotational equilibrium, $\alpha = 0$, so the right-hand side of Equation 8.25 was zero and we did not have to worry about the moment of inertia I. Now we are ready to tackle cases in which the angular acceleration is not zero, so let's consider the moment of inertia in more detail.

Equation 8.14 defines the moment of inertia for a general object composed of many pieces of mass m_i as

$$I = \sum_i m_i r_i^2 \qquad (8.26)$$

where each piece i is a distance r_i from the axis of rotation. The moment of inertia of an object depends on its mass *and* on how this mass is distributed with respect to the rotation axis. The value of I for a particular object thus depends on the choice of rotation axis, which can be illustrated using the simple "object" in Figure 8.26. Two point particles, both with mass m, are connected by a very light (massless) rod of length L. Figure 8.26A shows a rotation about the center of the rod. Applying Equation 8.26, the moment of inertia in this case is

$$I_a = \sum_i m_i r_i^2 = m_1 r_1^2 + m_2 r_2^2 \qquad (8.27)$$

Both particles have mass m; hence, $m_1 = m_2 = m$, and they are both a distance $L/2$ from the rotation axis, so $r_1 = r_2 = L/2$. Inserting this information into Equation 8.27 leads to

$$I_a = m(L/2)^2 + m(L/2)^2 = mL^2/2 \qquad (8.28)$$

This object can also rotate about an axis that passes through one end of the rod (Fig. 8.26B). The moment of inertia in this case is again given by Equation 8.27, but now $r_2 = 0$ (because the axis runs through mass 2) while $r_1 = L$. The moment of inertia is thus

$$I_b = m_1 r_1^2 + m_2 r_2^2 = m(L)^2 + m(0)^2 = mL^2 \qquad (8.29)$$

The moments of inertia I_a and I_b (Eqs. 8.28 and 8.29) are different, showing that the moment of inertia of an object does indeed depend on the choice of axis.

Using Equation 8.26 to find the moment of inertia of a complicated object can be difficult. Fortunately, though, the results for a variety of common shapes are known, and some are listed in Table 8.2. For example, the moment of inertia of a solid sphere of mass m and radius R is $I_{sphere} = 2mR^2/5$ for an axis that passes through the center of the sphere. This result applies to *all* solid spheres, so we can use it whenever a sphere is rotating about such an axis. Likewise, the moment of inertia of a disc or solid cylinder rotating about its central axis is $I_{disc} = mR^2/2$, where m is the total mass and R is the radius. The result is different from the moment of inertia of a hoop, which is $I_{hoop} = mR^2$ for an axis perpendicular to the plane of the hoop and passing through its center as indicated in Table 8.2. This case with a hoop is particularly simple because all the mass is a distance R from the rotation axis (Fig. 8.27). In this case, the sum in Equation 8.26 can be evaluated as

$$I_{hoop} = \sum_i m_i r_i^2 = \left(\sum_i m_i \right) R^2 \qquad (8.30)$$

Here all the mass of the hoop is at the same distance R from the rotation axis at the center, so we can pull this factor out of the summation. The sum $\sum_i m_i$ is the total mass m of the hoop, so Equation 8.30 becomes $I_{hoop} = mR^2$.

Table 8.2 also lists the moment of inertia for a solid uniform rod of length L and shows how the moment of inertia depends on the axis of rotation, just as we found for the simple object in Figure 8.26. If a solid rod rotates about a perpendicular axis that runs through the center of the rod, then $I_{rod\ center} = mL^2/12$, whereas if the rotation axis runs through one end of the rod, the moment of inertia is $I_{rod\ end} = mL^2/3$.

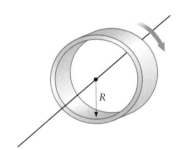

Figure 8.26 The moment of inertia depends on the choice of rotation axis.

Figure 8.27 For a simple hoop rotating about its geometric center, all the mass is a distance R from the rotation axis.

Table 8.2 | Moment of Inertia for Some Common Objects[a]

Object	Shape	Object	Shape
Hoop $I = mR^2$		Rod pivoted at center $I = \frac{1}{12}mL^2$	
Solid sphere $I = \frac{2}{5}mR^2$		Pulley/cylinder/disc $I = \frac{1}{2}mR^2$	
Spherical shell $I = \frac{2}{3}mR^2$		Wheel or hollow cylinder $I = \frac{1}{2}m(R_{max}^2 + R_{min}^2)$	
Rod pivoted at one end $I = \frac{1}{3}mL^2$		Solid square plate with axis perpendicular to plate $I = \frac{1}{6}mL^2$	

[a]In each case, m is the total mass of the object.

As an example, let's calculate the moment of inertia of a compact disc. A typical CD has a mass of about 9.0 g and a diameter of 12 cm. If we ignore the relatively small area of the hole in its center, a CD is just a disc of mass 0.0090 kg and radius 0.060 m. Inserting these values into the expression for a disc in Table 8.2 gives

$$I_{disc} = \frac{mR^2}{2} = \frac{(0.0090 \text{ kg})(0.060 \text{ m})^2}{2} \tag{8.31}$$

$$I_{disc} = 1.6 \times 10^{-5} \text{ kg} \cdot \text{m}^2 \tag{8.32}$$

This is the moment of inertia for rotations about an axis that runs through the center and is perpendicular to the plane of the disc. It is the axis appropriate for a CD player.

EXAMPLE 8.7 | Moment of Inertia of a Dirty CD

You are on a hike in the woods with your CD player when a rather large bug of mass 5.0 g manages to land on the edge of your favorite CD (mass 9.0 g and radius 6.0 cm). How does that affect the CD's moment of inertia? (Ignore the hole in the center of the CD.)

RECOGNIZE THE PRINCIPLE

To calculate the total moment of inertia of the bug plus the CD, we start with the general definition of the moment of inertia in Equation 8.26:

$$I_{total} = \sum_i m_i r_i^2$$

We now split this expression into two terms, one for the clean CD and one for the contribution of the bug:

$$I_{\text{total}} = \sum_i m_i r_i^2 = \left(\sum_i m_i r_i^2 \right)_{\text{CD}} + \left(\sum_i m_i r_i^2 \right)_{\text{bug}}$$

We calculated the first term (for the clean CD) in Equation 8.32. The bug (the second term) acts just as a point mass, as treated in connection with the object in Figure 8.11.

SKETCH THE PROBLEM
The CD, bug included, is shown in Figure 8.28.

IDENTIFY THE RELATIONSHIPS
A clean CD is simply a disc; its moment of inertia is $I_{\text{CD}} = m_{\text{CD}} R_{\text{CD}}^2 / 2$ and was calculated in Equations 8.31 and 8.32. For the contribution of the bug, we notice that all the bug's mass is located at the edge of the CD, so according to Equation 8.26 $I_{\text{bug}} = m_{\text{bug}} r_{\text{bug}}^2 = m_{\text{bug}} R_{\text{CD}}^2$. Combining these results gives

$$I_{\text{total}} = \frac{m_{\text{CD}} R_{\text{CD}}^2}{2} + m_{\text{bug}} R_{\text{CD}}^2$$

SOLVE
Inserting all the given values, we find

$$I_{\text{total}} = \frac{(0.0090 \text{ kg})(0.060 \text{ m})^2}{2} + (0.0050 \text{ kg})(0.060 \text{ m})^2 = \boxed{3.4 \times 10^{-5} \text{ kg} \cdot \text{m}^2}$$

What does it mean?
Comparing this result with the moment of inertia of a clean CD (Eq. 8.32), we see that the bug increases the moment of inertia by about a factor of two. This increase could well affect the way your CD player is able to spin a CD. It pays to keep your CDs clean.

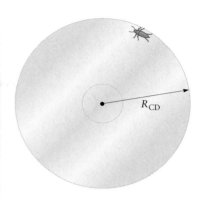

Figure 8.28 Example 8.7. The total moment of inertia of this "dirty" compact disc is the sum of the moment of inertia of the clean CD plus the moment of inertia of the bug.

8.5 | ROTATIONAL DYNAMICS

With our understanding of the moment of inertia from the last section, we are now ready to tackle some problems involving rotational dynamics. The basis for all our calculations will be

$$\sum \tau = I\alpha \qquad (8.33)$$

We can predict the rotational motion of an object by calculating both the total torque and the moment of inertia and using Equation 8.33 to find the angular acceleration α. We can then find the angular velocity and angular displacement of the object. Often, the angular acceleration is constant, as in cases in which the applied torque is constant. The behavior of ω and θ then have particularly simple forms. In fact, these mathematical forms are already quite familiar to us. You should recall that α is the slope of the ω–t curve (see Eq. 8.5). If the angular acceleration has the constant value α, then ω must vary linearly with time:

$$\omega = \omega_0 + \alpha t \qquad (8.34)$$

Equation 8.34 should remind you of the behavior of the linear velocity v as a function of time in situations in which the linear acceleration a is constant, as we studied in Chapter 3. Likewise, for this case of constant angular acceleration, the angular displacement has the familiar form

$$\theta = \theta_0 + \omega_0 t + \tfrac{1}{2}\alpha t^2 \qquad (8.35)$$

Rotational motion with constant angular acceleration

Table 8.3 Kinematic Relations for Constant Angular Acceleration α and Corresponding Relations for Linear Motion with Constant Acceleration a

ROTATIONAL MOTION	EQUATION NUMBER	TRANSLATIONAL MOTION
$\theta = \theta_0 + \omega_0 t + \frac{1}{2}\alpha t^2$	8.35	$x = x_0 + v_0 t + \frac{1}{2}at^2$
$\omega = \omega_0 + \alpha t$	8.34	$v = v_0 + at$
$\omega^2 = \omega_0^2 + 2\alpha(\theta - \theta_0)$	8.36	$v^2 = v_0^2 + 2a(x - x_0)$

We can also combine Equations 8.34 and 8.35 to get a relation that does not involve time; the result is

$$\omega^2 = \omega_0^2 + 2\alpha(\theta - \theta_0) \tag{8.36}$$

These relations, Equations 8.34 through 8.36, are very similar to the results for linear velocity and displacement we encountered for translational motion when the linear acceleration is constant. They are extremely useful, so we have collected them together in Table 8.3.

Angular Motion of a Compact Disc

Let's now use the relations in Table 8.3 to tackle a few problems involving rotational motion. Consider yet again the motion of a compact disc. Suppose the CD player is turned on at $t = 0$ and accelerates uniformly to reach its operational angular velocity of 500 rpm in 5.0 s. Let's calculate (a) the angular acceleration of the CD, (b) the total angular displacement of the CD during this period, and (c) the torque the CD player must exert on the CD.

To find the angular acceleration, we note that the initial angular velocity is zero ($\omega_0 = 0$) and that the final angular velocity is

$$\omega = 500 \text{ rpm} = 500 \frac{\text{rev}}{\text{min}} \times \frac{2\pi \text{ rad}}{1 \text{ rev}} \times \frac{1 \text{ min}}{60 \text{ s}} = 52 \text{ rad/s}$$

Inserting this result into Equation 8.34 and solving for α gives

$$\alpha = \frac{\omega - \omega_0}{t} = \frac{(52 \text{ rad/s}) - 0}{5.0 \text{ s}} = 10 \text{ rad/s}^2 \tag{8.37}$$

The angular displacement during this interval can be obtained from Equation 8.35

$$\theta = \theta_0 + \omega_0 t + \frac{1}{2}\alpha t^2 = 0 + 0 + \frac{1}{2}(10 \text{ rad/s}^2)(5.0 \text{ s})^2 = 130 \text{ rad}$$

The torque that must be supplied by the CD player can be found from Equation 8.33,

$$\sum \tau = I\alpha$$

We have already calculated the moment of inertia of a CD (Eq. 8.32). Using that result together with the angular acceleration from Equation 8.37, we find

$$\sum \tau = I\alpha = (1.6 \times 10^{-5} \text{ kg} \cdot \text{m}^2)(10 \text{ rad/s}^2) = 1.6 \times 10^{-4} \text{ N} \cdot \text{m}$$

EXAMPLE 8.8 ℝ Motion of a Frisbee

You and a friend are playing catch with a Frisbee (Fig. 8.29). Estimate the total angular displacement of the Frisbee as it travels from you to your friend. Assume the Frisbee's angular velocity is constant.

RECOGNIZE THE PRINCIPLE

We are given that the Frisbee moves with a constant angular velocity as it travels through the air; this makes sense, since we know from experience that when a Frisbee

© Digital Vision/Alamy

Figure 8.29 Example 8.8. A Frisbee spins about its axis with an angular velocity that is approximately constant during the course of a "throw." This rotational motion is independent of the Frisbee's translational motion.

reaches its destination, it is usually still spinning quite rapidly. If ω is constant, the angular acceleration $\alpha = 0$. We can then use the relations in Table 8.3 to find the total angular displacement.

SKETCH THE PROBLEM

The problem is shown in Figure 8.29. The top of the figure shows the axis of the Frisbee and the angular displacement θ.

IDENTIFY THE RELATIONSHIPS

We can find the total angular displacement from Equation 8.35, provided we can estimate the initial angular velocity ω_0 and the time the Frisbee spends in the air. (We already know that $\alpha = 0$.) Based on experience, we estimate that the Frisbee spends about 5 s in the air. It is a bit more difficult to estimate the Frisbee's angular velocity, but a reasonable guess is 10 rev/s. This value should be correct to within an order of magnitude; a value of 1 rev/s seems much too slow, and 100 rev/s would be quite fast. (The maximum rotation rate for most car engines is about 100 rev/s, and we don't expect a Frisbee to rotate that rapidly.)

SOLVE

Our estimate for the angular velocity is thus

$$\omega_0 \approx 10 \text{ rev/s } = 10 \, \frac{\text{rev}}{\text{s}} \times \frac{2\pi \text{ rad}}{\text{rev}} = 60 \text{ rad/s}$$

(Notice that our value for ω_0 contains only one significant figure because it is only an estimate.) Using Equation 8.35, in a time $t = 5$ s the Frisbee will undergo an angular displacement of

$$\Delta\theta = \theta - \theta_0 = \omega_0 t = (60 \text{ rad/s})(5 \text{ s}) = \boxed{300 \text{ rad}}$$

What does it mean?
One revolution corresponds to $2\pi \approx 6.3$ rad, so 300 rad is about 50 revolutions.

Pulling on a Pulley: Real Pulleys with Mass

In previous chapters, we discussed pulleys and explained how they can be used to amplify forces. However, our previous treatments of pulleys have assumed the pulley was extremely light (massless). Now we are ready to deal with the more realistic case of a pulley with mass.

Figure 8.30 shows a crate suspended by a very light (massless) rope. The rope is wound around a pulley, so as the crate is accelerated by the force of gravity, the pulley rotates. We now want to calculate the motion of the crate and pulley.

To calculate the linear acceleration of the crate and the angular acceleration of the pulley, we apply Newton's second law for linear motion ($\sum \vec{F} = m\vec{a}$) and Newton's second law for rotational motion ($\sum \tau = I\alpha$, Eq. 8.33). We take a systematic approach, similar to the method we used to deal with problems involving rotational equilibrium.

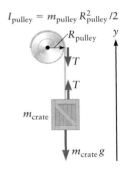

Figure 8.30 The torque on this pulley is due to the tension in the rope. The rotation axis is the axle of the pulley, so the torque equals the tension multiplied by the radius of the pulley.

3. **IDENTIFY THE RELATIONSHIPS.** Determine the rotation axis and pivot point for calculating the torques for objects that may rotate (these will depend on the problem). Then

- Find the total torque on the objects that are undergoing rotational motion. Calculate the torque using the method employed in our work on rotational statics. This torque will then be used in Newton's second law for rotational motion, $\sum \tau = I\alpha$.
- Calculate the sum of the forces on the objects that are undergoing linear motion for use in Newton's second law, $\sum \vec{F} = m\vec{a}$.

- Check for a relation between the linear acceleration and the angular acceleration. There will usually be a connection between a and α through a relation such as Equation 8.8.

4. **SOLVE** for the quantities of interest (which might be a, α, or both) using Newton's second law for linear motion ($\sum \vec{F} = m\vec{a}$) and rotational motion ($\sum \tau = I\alpha$).

5. Always *consider what your answer means* and check that it makes sense.

Motion of a Pulley and Crate: Example of Combined Translational and Rotational Motion

We demonstrate the above problem-solving approach by applying it to calculate the crate's translational motion and the pulley's rotational motion in Figure 8.30. This figure serves as our sketch; it shows all the forces on the crate (gravity and the force from the tension in the rope). Since the same rope is attached to the pulley, the torque on the pulley is also due to this tension. (We omit the force of gravity on the pulley because it is balanced by the force from the pulley's support, and neither of these forces produces a torque on the pulley.) The pulley rotates about its axle, so we pick that as the rotation axis.

To find the torque on the pulley, we first note that the tension in the rope will make the pulley rotate clockwise. According to our convention, the torque is negative. The force acts at the edge of the pulley, so the lever arm equals R_{pulley}. We thus have

$$\sum \tau = -TR_{\text{pulley}} = I_{\text{pulley}}\alpha \qquad (8.38)$$

The pulley is a disc; the axis of rotation is perpendicular to the plane of the disc and through the center, so the moment of inertia is $I_{\text{pulley}} = m_{\text{pulley}}R_{\text{pulley}}^2/2$ (see Table 8.2).

We next write Newton's second law for the translational motion of the crate. Taking the $+y$ direction as the "positive" direction for forces, we have

$$\sum F_y = +T - m_{\text{crate}}g = m_{\text{crate}}a \qquad (8.39)$$

Here, a is the acceleration of the crate along the y direction.

There are three unknowns—the tension in the rope T, the acceleration of the crate a, and the angular acceleration of the pulley α—but only two equations (Eqs. 8.38 and 8.39). We therefore need one more relation among these unknowns. We now recognize that the linear acceleration of the crate equals the linear acceleration of a point on the edge of the pulley. We have already discussed this situation (in connection with Eq. 8.8) and found that a and α are related by

$$a = \alpha R_{\text{pulley}} \qquad (8.40)$$

Equation 8.40 is actually a relation between the magnitudes of the linear and angular accelerations. In this problem, we must also be careful about the signs of a and α. Equation 8.40 treats these signs correctly for the crate and the pulley in Figure 8.30: a negative acceleration of the crate (i.e., a negative value of a) corresponds to a clockwise angular acceleration of the pulley and hence a negative value of α.

We can now solve these three relations (Eqs. 8.38 through 8.40) to find the accelerations of the crate and pulley. We first use Equation 8.40 to eliminate a from Equation 8.39:

$$T - m_{\text{crate}}g = m_{\text{crate}}a = m_{\text{crate}}\alpha R_{\text{pulley}} \qquad (8.41)$$

Equation 8.38 can be simplified by inserting the expression for the pulley's moment inertia,

$$-TR_{\text{pulley}} = I_{\text{pulley}}\alpha = \tfrac{1}{2}m_{\text{pulley}}R_{\text{pulley}}^2\alpha$$

and canceling one factor of R_{pulley}:

$$T = -\tfrac{1}{2}m_{\text{pulley}}R_{\text{pulley}}\alpha$$

Using this result, we can eliminate T from Equation 8.41:

$$\overbrace{-\tfrac{1}{2}m_{\text{pulley}}R_{\text{pulley}}\alpha}^{T} - m_{\text{crate}}g = m_{\text{crate}}\alpha R_{\text{pulley}}$$

Now we can solve for α:

$$\alpha(m_{\text{crate}}R_{\text{pulley}} + \tfrac{1}{2}m_{\text{pulley}}R_{\text{pulley}}) = -m_{\text{crate}}g$$

$$\alpha = \frac{-m_{\text{crate}}g}{m_{\text{crate}}R_{\text{pulley}} + \tfrac{1}{2}m_{\text{pulley}}R_{\text{pulley}}} = \frac{-g/R_{\text{pulley}}}{1 + [m_{\text{pulley}}/(2m_{\text{crate}})]}$$

This result for the angular acceleration of the pulley can be used in Equations 8.40 and 8.39 to find the linear acceleration of the crate and the tension, and thus complete the problem. Since α is constant, the angular velocity and displacement of the pulley is given by the relations for motion with a constant angular acceleration (Equations 8.34 through 8.36).

CONCEPT CHECK 8.6 | Angular Acceleration of a Rotating Rod

A rod of mass m and length L is hinged at one end, and a force F is applied at one end as shown in Figure 8.31 for rod 1. An identical rod is mounted on an axle so that it rotates about its center, and a force F is applied to its end as shown for rod 2 in Figure 8.31. If the forces are the same, which rod has the greater angular acceleration? What is the ratio of the angular accelerations?
  (a) The angular accelerations are the same, $\alpha_1/\alpha_2 = 1$.
  (b) Rod 1 has the larger angular acceleration, with $\alpha_1/\alpha_2 = 2$.
  (c) Rod 1 has the larger angular acceleration, with $\alpha_1/\alpha_2 = 4$.
  (d) Rod 2 has the larger angular acceleration, with $\alpha_2/\alpha_1 = 2$.
  (e) Rod 2 has the larger angular acceleration, with $\alpha_2/\alpha_1 = 4$.

Figure 8.31 Concept Check 8.6.

8.6 | COMBINED ROTATIONAL AND TRANSLATIONAL MOTION

The combination of rotational and translational motion arises in many common situations, including a rolling wheel and the motion of a baseball bat.

Rolling Motion

An extremely common situation that combines translation and rotation is **rolling motion**. Figure 8.32 shows a simple wheel as it rolls along the ground. We assume the wheel "as a whole" is moving with a constant linear speed v; that is, the center of wheel is moving horizontally with a constant velocity of magnitude v. However, a point on the edge of the wheel is certainly not moving with a constant velocity. In fact, if the wheel is rolling without slipping, the point of the wheel in contact with the ground is *at rest* during the instant it is in contact.

As a wheel rolls, it undergoes rotational motion about an axis that runs through its center (the axle), and this rotational motion is described by an angular velocity ω. To understand how this angular velocity is related to the linear velocity of the center of the wheel v, let's consider a series of "snapshots" of the wheel as it rolls as sketched in Figure 8.33. A succession of points along the wheel are labeled, and

Figure 8.32 A rolling object undergoes translational motion with a center of mass velocity v and simultaneous rotational motion with angular velocity ω.

Figure 8.33 Successive "snapshots" as a wheel rolls to the right. A wheel that rolls without slipping travels a linear distance of $2\pi R$ during each complete revolution of the wheel. Each complete revolution occurs in a time equal to the period, T. This leads to the relation $v = \omega R$ between the translational velocity of the center of the wheel and the angular velocity.

Figure 8.34 Concept Check 8.7. On some cars, the front and rear wheels have different radii. How are the angular velocities of the two different-sized wheels related?

corresponding labels where these points touch the ground as the wheel rolls are also shown. The wheel begins (Fig. 8.33A) with point 1 in contact with the ground, and when it completes one full revolution, this point is again touching the ground (Fig. 8.33F). Since the wheel is always in contact with the ground and does not slip, the distance traveled *as measured along the ground* must equal the *circumference* of the wheel, $2\pi R$. This distance is also equal to the distance traveled by the center of the wheel, so if the rotation period is T, the speed of the center is

$$v = \frac{2\pi R}{T} \tag{8.42}$$

At the same time, the wheel has completed one full revolution (2π rad), so its angular velocity is

$$\omega = \frac{2\pi}{T}$$

Combing this with Equation 8.42 gives

$$v = \omega R \quad \text{(rolling motion)} \tag{8.43}$$

This relation connects the linear velocity of the center of the wheel with its angular velocity and is true for any object that rolls without slipping. It is interesting that this relation (Eq. 8.43) has the same form as Equation 8.7, but even though these equations are related, the ideas are slightly different. The result in Equation 8.7 involves the angular velocity of a rotating object and the linear speed of a *point on the edge* of the object. The result for rolling motion in Equation 8.43 connects the angular velocity with the velocity of the *center* (i.e., the axis) of a rolling object. The arguments contained in Figure 8.33 can also be used to relate the acceleration of the center of a rolling object to its angular acceleration:

$$a = \alpha R \quad \text{(rolling motion)} \tag{8.44}$$

It is interesting that friction plays an essential role in rolling motion. Indeed, an object cannot roll without it! It may seem a bit strange to claim that an object cannot move *without* friction, but you have probably encountered this situation many times. Consider a car that is initially at rest on an icy, frictionless surface. If you try to drive away in the car, the wheels will spin freely because there is no frictional force to prevent them from slipping relative to the ice. On normal pavement (*with* friction), the force of friction between the road and the surface of the wheels enables the car to accelerate when the wheels turn. Usually, the wheels do not slip relative to the pavement, so this is static friction; if the driver tries to accelerate very quickly, however, the wheels may slip relative to the pavement bringing kinetic friction into play. Hence, for rolling motion to occur, friction is *required*.

CONCEPT CHECK 8.7 | Angular Velocity of Rolling Wheels

Consider a car whose rear wheels are much larger than the front wheels as shown in Figure 8.34. If the car is traveling up a hill, which wheels will have a larger angular velocity?
  (a) The front wheels will.
  (b) The rear wheels will.
  (c) The front and the rear wheels will have the same angular velocity.

EXAMPLE 8.9 | ℝ Angular Velocity of an Automobile Wheel

Consider a car traveling on a highway at a speed of 50 mi/h (about 23 m/s). Estimate the angular velocity of the car's wheels.

RECOGNIZE THE PRINCIPLE

A car's wheels are rolling without slipping (we hope!), so we can apply Equation 8.43 to relate the linear velocity v and the angular velocity ω.

SKETCH THE PROBLEM

This problem is illustrated in Figures 8.32 and 8.33.

IDENTIFY THE RELATIONSHIPS

The velocity of the center of a wheel equals the overall speed of the car, $v = 23$ m/s. The angular velocity of a wheel can then be obtained using the relation $v = \omega R$ (Eq. 8.43). This relation involves the radius R of the wheel, so we need to estimate the value of R. For a typical car, we estimate $R \approx 40$ cm $= 0.40$ m.

SOLVE

Using these values gives

$$\omega = \frac{v}{R} = \frac{23 \text{ m/s}}{0.4 \text{ m}} = \boxed{60 \text{ rad/s}}$$

What does it mean?

This angular velocity corresponds to

$$\omega = (60 \text{ rad/s}) \times (1 \text{ rev}/2\pi \text{ rad}) = 10 \text{ rev/s}$$

so the wheels undergo approximately 10 revolutions each second.

Sweet Spot of a Baseball Bat

The general motion of any object can be described as a combination of linear and rotational motion. The center of mass plays an important role in problems involving such combined motions. The linear motion of an object can be calculated from Newton's second law,

$$\sum \vec{F} = m\vec{a}$$

with $\vec{a}$ being the acceleration of the center of mass of the object. As far as the linear (i.e., translational) motion is concerned, the object moves as if it were a point particle with all its mass located at the center of mass. The rotational motion then follows from

$$\sum \tau = I\alpha$$

An interesting example of this principle involves a baseball bat and the problem of how best to hit a baseball or softball. You may know from experience that if you hit a ball off the very end of a bat, you feel significant and often uncomfortable vibrations at your hands. Likewise, if you hit the ball very near the handle of the bat—that is, near your hands—there are also strong vibrations. There is, however, a particular spot on the bat called the "sweet spot," and if the ball strikes the bat at that location, your hands feel essentially no vibrations. A ball hit from the sweet spot seems to rebound effortlessly from the bat and, all else being equal, will travel farther than a ball that hits the bat away from this location. Suppose a baseball-playing friend wants to use physics to calculate the location of the sweet spot on his favorite bat. How would you help him?

Figure 8.35 shows the problem. As the player swings, the bat undergoes rotational motion about the batter's hands at point P. In practice, it is not quite "pure" rotational motion because a batter will often move the bat handle forward a small amount while he swings. Since this point P usually moves very little compared with the other end of the bat, it is a good approximation to assume P is fixed while the bat simply rotates about P. When the ball strikes the bat, the ball exerts a force F_b on the bat due to the ball–bat collision, producing a torque on the bat. In Figure 8.35A, this force F_b is shown acting at the sweet spot, a distance d from the center of the bat.

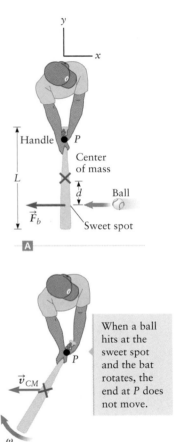

Figure 8.35 **A** When a baseball hits at the sweet spot of the bat, the velocity at the handle (point P) is zero. **B** The handle's velocity will be zero if the velocity v_{CM} of the center of mass and the angular velocity ω are related by $v_{CM} = \omega R = \omega(L/2)$, which will only be true if the linear acceleration and the angular acceleration are related by $a_{CM} = \alpha(L/2)$.

The special property of the sweet spot is that the force F_b does not cause the handle of the bat to move; your hands do not experience any "recoil" or other motion of the bat when a ball strikes at the sweet spot. If the batter holds the bat *very lightly*, the force exerted by his hands on the bat is approximately zero and the only significant horizontal force on the bat is the force F_b exerted by the ball. This force will cause the bat's center of the mass to move to the left in Figure 8.35B, and at the same time will make the bat rotate. If the bat's translational and rotational motions are related in just the right way (by Eqs. 8.43 and 8.44), the bat will undergo a sort of rolling motion with point P just like the point of a wheel in contact with the ground. This behavior is found only if the force is applied at the sweet spot. In this case, no force is needed by the player's hands to keep the end of the bat at P from moving.

To find the location of the sweet spot (the value of d in Fig. 8.35A), we need to calculate the linear acceleration of the center of mass a_{CM} and the angular acceleration α of the bat. Newton's second law for the bat's translational motion reads

$$\Sigma F = F_b = ma_{CM} \tag{8.45}$$

For simplicity, we consider only motion along the direction of F_b, so Equation 8.45 involves just the components of the force and acceleration along this direction. If we assume a bat with a uniform cross section (a simple rod, another physics simplification), the center of mass is at the center of the bat. The force exerted by the ball on the bat also produces a torque on the bat. To analyze the rotational motion, let's *not* use point P in Figure 8.35 as the pivot point. Instead we take the pivot point to be at the bat's center of mass so that the lever arm is d in Figure 8.35A. Newton's second law for the rotational motion is then

$$\Sigma \tau = F_b d = I\alpha \tag{8.46}$$

For point P on the bat handle in Figure 8.35 to stay at rest, the bat must rotate in the same way as the rolling wheel in Figure 8.33, which happens only if the linear speed of the center of mass "matches" the angular velocity according to $\omega = v_{CM}/R$. Here, R is the distance from the center of mass to point P, so $R = L/2$, where L is the length of the bat, and

$$\omega = \frac{v_{CM}}{R} = \frac{v_{CM}}{L/2}$$

If the velocities are related in this way, the angular and linear accelerations will also be related in the same way (see Eq. 8.44):

$$\alpha = \frac{a_{CM}}{L/2} \tag{8.47}$$

If we model the bat as a rod pivoted about its center, its moment of inertia from Table 8.2 is $I = mL^2/12$. Inserting our results for I and α into Equation 8.46 leads to

$$\Sigma \tau = F_b d = I\alpha = \frac{mL^2}{12} \frac{a_{CM}}{(L/2)} = \frac{mLa_{CM}}{6}$$

Combining this result with Equation 8.45 gives

$$F_b d = (ma_{CM})d = \frac{mLa_{CM}}{6}$$

which we can solve to find the location of the sweet spot:

$$d = \frac{L}{6}$$

Hence, the sweet spot for a uniform bat is located a distance $L/6$ from the center of the bat. In Question 3, we ask you to consider a more realistic shape for the bat and think about how this new assumption will affect the location of the sweet spot.

SUMMARY | Chapter 8

Angular displacement, velocity, and acceleration

Rotational motion is described by an **angular displacement** θ (measured in radians), an **angular velocity** ω (in rad/s), and an **angular acceleration** α (in rad/s^2). They are related to the linear motion of a point on the object by

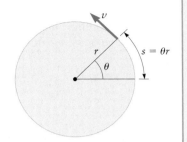

$$\theta = \frac{s}{r} \qquad \text{(8.1) (page 241)}$$

$$v = \omega r \qquad \text{(8.7) (page 245)}$$

$$a = \alpha r \qquad \text{(8.8) (page 246)}$$

Torque

Torque plays the role of force in cases of rotational motion. The torque produced by an applied force is

$$\tau = Fr \sin \phi \qquad \text{(8.15) (page 249)}$$

This relation for the torque can be written in two equivalent ways. One way is

$$\tau = F_\perp r \qquad \text{(8.16) (page 250)}$$

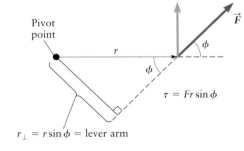

where $F_\perp$ is the component of the force perpendicular to the line defined by r, the distance from the rotation axis to the point on the object where the force acts. The torque can also be written as

$$\tau = Fr_\perp \qquad \text{(8.17) (page 250)}$$

where F is the applied force and

$$r_\perp = r \sin \phi \qquad \text{(8.18) (page 250)}$$

The distance $r_\perp$ is called the **lever arm**; it is the distance from the rotation axis to the line of action of the force, measured along a line that is perpendicular to both.

Newton's second law for rotational motion

$$\sum \tau = I\alpha \qquad \text{(8.12) (page 248)}$$

The **moment of inertia** I of an object depends on its mass and shape. For a collection of particles of mass m_i, the moment of inertia is

$$I = \sum_i m_i r_i^2 \qquad \text{(8.14) (page 249)}$$

where r_i is the distance from the rotation axis to m_i. The moment of inertia is proportional to the total mass and depends on its distribution relative to the rotation axis. Values of I for some common shapes are listed in Table 8.2.

Rotational equilibrium

An object is in translational equilibrium and rotational equilibrium if its linear acceleration and angular acceleration are both zero. For that to be true, both the total force and the total torque must be zero.

(Continued)

Rotational motion

Calculations of the rotational dynamics start with Newton's second law for rotational motion,

$$\sum \tau = I\alpha \qquad \textbf{(8.12)} \text{ (page 248)}$$

When the angular acceleration is constant, the solutions for the angular displacement and velocity as functions of time are (Table 8.3)

$$\theta = \theta_0 + \omega_0 t + \tfrac{1}{2}\alpha t^2 \qquad \textbf{(8.35)} \text{ (page 263)}$$

$$\omega = \omega_0 + \alpha t \qquad \textbf{(8.34)} \text{ (page 263)}$$

$$\omega^2 = \omega_0^2 + 2\alpha(\theta - \theta_0) \qquad \textbf{(8.36)} \text{ (page 264)}$$

These relations are very similar to the results we found for linear motion with constant linear acceleration in Chapter 3.

Rolling motion

Rolling motion is an example of combined translational and rotational motion. If an object rolls without slipping, the angular velocity and angular acceleration are related to the linear velocity and acceleration of the center of mass by

$$v = \omega R \qquad \textbf{(8.43)} \text{ (page 268)}$$

$$a = \alpha R \qquad \textbf{(8.44)} \text{ (page 268)}$$

Rolling motion

$$\omega = \frac{v}{R}$$
$$\alpha = \frac{v}{R}$$

QUESTIONS

⊗ = life science application

1. Consider a wheel that rolls without slipping such as the wheel in Figure 8.32. We have seen that slipping is prevented by the force of static friction between the wheel and the pavement. Can this force do work on the wheel?

2. A tennis racket possesses a sweet spot in close analogy to that of a baseball bat. One might expect that rackets are designed so that the sweet spot is in the center of the strings. Somewhat surprisingly, that is not the case. The approximate location of a racket's sweet spot can be found from the simple experiment sketched in Figure Q8.2. The racket is suspended by string tied to the handle and is then struck (gently) by a ball or a light blow. When this force is exerted at the sweet spot, the handle will not rebound. Perform this experiment yourself and show that the sweet spot is rather far from the center of the strings. Explain how you could add or subtract mass from different spots on the racket to move the sweet spot to this desired location.

Figure Q8.2

3. In Section 8.6, we found that for a uniform baseball bat, the sweet spot is a distance $L/6$ from the center of the bat. A real bat is not uniform, but is much thinner in the handle than in the barrel (the end of the bat farthest from point P in Fig. 8.35). Estimate how that affects the location of the sweet spot.

4. According to Table 8.2, the moment of inertia for a rod that rotates about an axis perpendicular to the rod and passing through one end is $I = ML^2/3$; if the axis passes through the center of the rod, then $I = ML^2/12$. Give a physical explanation for this difference in terms of the way the mass of the rod is distributed with respect to the axis in the two cases.

5. ⊗ The tree in Figure Q8.5 experiences a large torque due to the force of gravity, yet it is in static equilibrium. What additional torque balances the torque due to gravity?

Figure Q8.5
Question 5 and
Problem 23.

6. In our analysis of a rollover car accident in Example 8.6, we treated the car in a very simplified way. Develop a more realistic model of a car with special attention to the height of its center of mass and use it to compute the force required to tip over the car. (See Fig. 8.25.)

7. SSM Two golfers team up to win a golf tournament and are awarded a solid gold golf club. To divide their winnings, they balance the club on a finger and then cut the club into two pieces at the balance point with a hacksaw. Was the winning gold split evenly? Why or why not?

8. A mechanic's toolbox contains the two wrenches and two screwdrivers shown in Figure Q8.8. While working on her vintage car, the mechanic comes across a stubborn and rusted bolt. Which wrench would be the most useful in breaking the bolt free? Why is one better than the other? She also encounters

Wrench 1   Wrench 2

Screwdriver 1   Screwdriver 2

Figure Q8.8

a screw that is similarly stuck tight. Which screwdriver should she choose? Is one better than the other? Why?

9. A snap of the wrist will usually detach a paper towel from a roll. Why does that almost always work for a full roll, yet an almost empty roll often gives much more than a single sheet?

10. Two metersticks, one with a weight attached to its end, are held with one of their ends in the corner formed by the floor and a wall as depicted in Figure Q8.10. If the ends of the metersticks are let go at the same time and start with the same angle with respect to the floor, which one will hit the floor first? Why?

Figure Q8.10

11. **Three types of levers.** Levers can be classified as first, second, or third class, depending on how the load force ($\vec{F}_L$) and effort force ($\vec{F}_E$) are configured with respect to the pivot point or fulcrum as shown in Figure Q8.11. Determine the lever classification of the following items: (a) a crowbar, (b) a pair of pliers, (c) a clawed hammer pulling a nail, (d) a wheelbarrow (Fig. P8.35), (e) a seesaw (Fig. 8.10), (f) a fishing pole, (g) a light switch, (h) a gate with closing spring attached to hinge, and (i) biceps raising the forearm (Fig. P8.79).

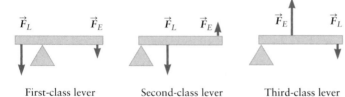

First-class lever   Second-class lever   Third-class lever

Figure Q8.11

12. In about 200 BC, Archimedes made the statement, "Give me a lever long enough and a place to stand, and I will move the

Earth." Assume he can exert a force equal to his weight on the Earth (about 900 N). Design a lever that would enable him to give the Earth an acceleration of 1 m/s^2.

13. Two workers are carrying a long, heavy steel beam (Fig. Q8.13). Which one is exerting a larger force on the object? How can you tell?

Figure Q8.13

14. ⊗ A tree has a large branch that grows horizontally out from the trunk. If the branch doubles in length and doubles in diameter, by what factor does the torque on the branch due to gravity increase?

15. **SSM** ⊗ A tree has two large branches that grow out horizontally from the trunk in opposite directions. One branch has a length L and diameter d, whereas the other has a length $L/2$. If the magnitude of the torque due to gravity is the same on the two branches, what is the diameter of the shorter branch?

16. A force of magnitude F is applied to a rod at various spots and at various angles (Fig. Q8.16). In which case, (1), (2), or (3), is the magnitude of the torque largest? Smallest?

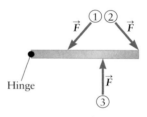

Figure Q8.16

17. ⊗ Bodybuilders do an exercise called "curls" in which they lift a weight by bending at the elbow. Explain why the torque on the elbow is largest when the arm is horizontal.

18. Explain why it is easier (it requires a smaller force) to open a large jar than a small one.

19. ⊗ Explain why placing your hands behind your head makes sit-ups more difficult than having your hands on your stomach.

PROBLEMS

8.1 DESCRIBING ROTATIONAL MOTION

1. An angle has a value of 85°. What is the angle in radians?

2. The shaft on an engine turns through an angle of 45 rad. (a) How many revolutions does the shaft make? (b) What is this angle in degrees?

3. An old-fashioned (vinyl) record rotates at a constant rate of 33 rpm. (a) What is its angular velocity in rad/s? (b) If the record has a radius of 18 cm, what is the linear speed of a point on its edge?

4. The angular speed of the shaft of a car's engine is 360 rad/s. How many revolutions does it complete in 1 hour?

5. Consider again the record in Problem 3. If it plays music for 20 min, how far does a point on the edge travel in meters?

6. The tachometer in the author's car has a maximum reading of 8000 rpm. (A tachometer is a gauge that shows the angular velocity of a car's engine.) Express this angular velocity in rad/s.

7. A car engine is initially rotating at 200 rad/s. It is then turned off and takes 3.0 s to come to a complete stop. What is the angular acceleration of the engine? Assume α is constant.

8. ★ What is the angular velocity of the Earth's center of mass as it orbits the Sun?

9. ⭐ What is the angular velocity of a person standing (a) at the equator? (b) At a latitude of 45°?

10. ⭐ Mercury spins about its axis with a period of approximately 58 days. What is the rotational angular velocity of Mercury?

11. ⭐ Mercury orbits the Sun with a period of approximately 88 days. What is the angular velocity of Mercury's orbital motion?

12. ⭐ An adult is riding on a carousel and finds that he is feeling a bit nauseated. He then decides to move from the outer edge of the carousel to a point a distance r_2 from the center so that his linear speed is reduced by a factor of 2.5. If the carousel has a constant angular velocity of 1.0 rad/s and a radius of 3.3 m, what is r_2?

13. SSM ⭐ What is the angular velocity and the period of the second hand of a clock? If the linear speed of the end of the hand is 5.0 mm/s, what is the clock's radius?

14. Construct a graph of the angular displacement of the minute hand of a clock as a function of time. The axes of your graph should include quantitative scales (i.e., numbers and units). What is the slope of the relation between θ and t?

15. Consider a race car that moves with a speed of 200 mi/h. If it travels on a circular race track of circumference 2.5 mi, what is the car's angular velocity?

8.2 TORQUE AND NEWTON'S LAWS FOR ROTATIONAL MOTION

16. A diver stands at rest at the end of a massless diving board as shown in Figure P8.16. (a) If the mass of the diver is 120 kg and the board is 4.0 m long, what is the torque due to gravity on the diving board with a pivot point at the fixed end of the board? (b) According to our convention for positive and negative torques, is the torque on the diving board positive or negative?

Figure P8.16 Problems 16 and 17.

17. ⭐ Consider again the diver in Figure P8.16. Assume the diving board now has a mass of 30 kg. Find the total torque due to gravity on the diving board.

18. SSM ⭐ Consider the clock in Figure 8.17. Calculate the magnitude and sign of the torque due to gravity on the hour hand of the clock at 4 o'clock. Assume the hand has a mass of 15 kg, a length of 1.5 m, and the mass is uniformly distributed.

19. ⭐ A rod of length 3.8 m is hinged at one end, and a force of magnitude $F = 10$ N is applied at the other (Fig. P8.19). (a) If the magnitude of the torque associated with this force is 18 N·m, what is the angle ϕ? (b) What is the sign of the torque?

Figure P8.19

20. ⭐ A person carries a long pole of mass 12 kg and length 4.5 m as shown in Figure P8.20. Find the magnitude of the torque on the pole due to gravity.

21. ⭐ If the length of the pole in Problem 20 is increased by a factor of three and its mass is increased by a factor of two, by what factor does the torque change?

Figure P8.20 Problems 20 and 21.

22. ✪ ✗ Ⓡ A difficult maneuver for a gymnast is the "iron cross," in which he holds himself as shown in Figure P8.22. Estimate the torque on the gymnast due to gravity. Choose a rotation axis that passes through one of his hands and assume each hand must support half of his weight.

Figure P8.22

23. SSM ⭐ ✗ Ⓡ A tree grows at an angle of 50° to the ground as shown in Figure Q8.5. If the tree is 25 m from its base to its top, and has a mass of 500 kg, what is the approximate magnitude of the torque on the tree due to the force of gravity? Take the base of the tree as the pivot point. (The answer is one reason trees need roots.)

24. ⭐ Ⓡ A baseball player holds a bat at the end of the handle. If the bat is held horizontally, what is the approximate torque due to the force of gravity on the bat, with a pivot point at the batter's hands?

8.3 ROTATIONAL EQUILIBRIUM

25. ✪ Suppose a crate of mass 7.5 kg is placed on the plank in Figure P8.25 at a distance 3.9 m from the left end. Find the forces exerted by the two supports on the plank.

Figure P8.25 Problems 25 and 26.

26. A uniform wooden plank of mass 12 kg rests on two supports as shown in Figure P8.25. The plank is at rest, so it is in translational equilibrium and rotational equilibrium. (a) Make a sketch showing all the forces on the plank. Be sure to show where the forces act. (b) Choose the left-hand support for the pivot point and calculate the torque from each of the forces in part (a). (c) Express the conditions for translational and rotational equilibrium. (d) Solve to find the forces exerted by the two supports on the plank.

27. ⭐ Consider the flagpole in Figure P8.27. If the flagpole has a mass of 20 kg and length 10 m and the angle the cable makes with the pole is $\phi = 25°$, what are the magnitude and direction of the force exerted by the hinge (at point P) on the flagpole? Assume the mass of the pole is distributed uniformly.

Figure P8.27

28. ⭐ The seesaw in Figure P8.28 is 4.5 m long. Its mass of 20 kg is uniformly distributed. The child on the left end has a mass of 14 kg and is a distance of 1.4 m from the pivot point while a

Figure P8.28

second child of mass 30 kg stands a distance L from the pivot point, keeping the seesaw at rest. (a) Is the seesaw in rotational equilibrium? (b) Make a sketch showing all the forces on the seesaw. Be sure to show where each force acts. (c) Express the conditions for translational and rotational equilibrium for the seesaw. (d) Solve to find L, the location of the child on the right.

29. ✪ Consider the ladder in Example 8.5 and assume there is friction between the vertical wall and the ladder, with $\mu_S = 0.30$. Find the angle ϕ at which the ladder just begins to slip.

30. ✪ A painter is standing on the ladder (mass 40 kg and length 2.5 m) in Figure P8.30. There is friction between the bottom of the ladder and the floor with $\mu_S = 0.30$, but there is no friction between the ladder and the wall. A painter of mass of 70 kg is standing a distance of 0.60 m from the top. (a) Make a sketch showing all the forces on the ladder. Be sure to show where each force acts. (b) Express the conditions for translational and rotational equilibrium for the ladder. (c) If $\theta = 60°$, use the results from part (b) and determine whether or not this ladder will slip.

Figure P8.30
Problems 30 and 31.

31. Consider again the ladder in Problem 30. What is the sign of the torque on the ladder due to the force from the wall?

32. ⭐ A solid cube of mass 40 kg and edge length 0.30 m rests on a horizontal floor as shown in Figure P8.32. A person then pushes on the upper edge of the cube with a horizontal force of magnitude F. At what value of F will the cube start to tip? Assume the frictional force from the floor is large enough to prevent the cube from sliding.

Figure P8.32
Problems 32, 33, and 34.

33. ⭐ Consider again the cube in Problem 32 (Fig. P8.32), but now assume the force is applied along a horizontal line that is 0.20 m above the floor. At what value of F will the crate begin to tip?

34. ✪ Repeat Problem 32, but assume the force F is applied at the corner of the cube and at an angle of 30° above the horizontal direction. At what value of F will the crate begin to tip?

35. ✪ Ⓡ Wheelbarrows are designed so that a person can move a massive object more easily than if he simply picked it up. If the person using the wheelbarrow in Figure P8.35 is able to apply a total vertical (upward) force of 200 N on the handle, what is the approximate mass of the heaviest object he could move with the wheelbarrow?

Figure P8.35 Problem 35 and Question 11.

36. ✪ Ⓡ One way to put a ladder in place on a wall is to "walk" the ladder to the wall. One end of the ladder is held fixed (perhaps at the base of the wall), and a person slides his hands along the ladder as he walks toward the fixed end as sketched in Figure P8.36. Anyone who has raised a ladder in this way knows that the amount of force the person must apply grows

Figure P8.36
Problems 36 and 37.

larger as the person moves toward the end that is fixed. Assuming the force from the person is perpendicular to the ladder, find the force the person must apply to hold the ladder at (a) an angle of 10° and (b) an angle of 20°.

37. ✪ Ⓡ Consider again the ladder in Figure P8.36. We saw in Problem 8.36 that the force supplied by the person increases with increasing angle ϕ as the ladder is initially lifted off the ground. When the ladder is sufficiently high (the angle ϕ is sufficiently large), though, it becomes easier to raise the ladder higher (i.e., the necessary applied force becomes smaller). Find the angle ϕ at which that happens.

38. SSM ⭐ A flagpole of length 12 m and mass 30 kg is hinged at one end, where it is connected to a wall as sketched in Figure P8.38. The pole is held up by a cable attached to the end. Find the tension in the cable.

Figure P8.38

39. ✪ A long, *nonuniform* board of length 8.0 m and mass $m = 12$ kg is suspended by two ropes as shown in Figure P8.39. If the tensions in the ropes are $mg/3$ (on the left) and $2mg/3$ (on the right), what is the location of the board's center of mass?

40. ✪ Ⓡ Consider again the problem of a tipping car in Example 8.6. This time, instead of applying a force F to the car, assume the car is traveling around a curve on a level road. Let the radius of curvature of the turn be r and assume the car's speed v is just barely fast enough to make the car's inside wheels lift off the ground. (a) Find v. Express your answer in terms of r and the variables w, d, and h from Figure 8.25. (b) Use realistic values to get a numerical estimate for v for $r = 80$ m. *Hint:* Use the car's center of mass as your pivot point.

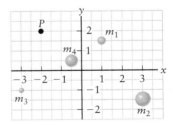

Figure P8.39

8.4 MOMENT OF INERTIA

41. Two particles of mass $m_1 = 15$ kg and $m_2 = 25$ kg are connected by a massless rod of length 1.2 m. Find the moment of inertia of this system for rotations about the following pivot points: (a) the center of the rod, (b) the end at m_1, (c) the end at m_2, and (d) the center of mass. Assume the rotation axis is perpendicular to the rod.

42. SSM ⭐ Four particles with masses $m_1 = 15$ kg, $m_2 = 25$ kg, $m_3 = 10$ kg, and $m_4 = 20$ kg sit on a very light (massless) metal sheet and are arranged as shown in Figure P8.42. Find the moment of inertia of this system with the pivot point (a) at the origin and (b) at point P. Assume the rotation axis is parallel to the z direction, perpendicular to the plane of the drawing.

Figure P8.42

43. ⭐ Consider a merry-go-round that has the form of a disc with radius 5.0 m and mass 100 kg. If three children, each of mass 20 kg, sit on the outer edge of the merry-go-round, what is the total moment of inertia?

44. Which has a larger moment of inertia, a disc of mass 20 kg or a hoop of mass 15 kg with the same radius?

45. ⭐ Ⓡ Estimate the moment of inertia of a bicycle wheel.

46. ★ If the mass of a wheel is increased by a factor of 2 and the radius is increased by a factor of 1.5, by what factor is the moment of inertia increased? Model the wheel as a solid disc.

47. ★ Consider two cylinders with the same density and the same length. If the ratio of their diameters is 1.5/1, what is the ratio of their moments of inertia?

48. ℝ In our examples with CDs (such as Example 8.7), we usually ignored the hole in the center. Calculate the moment of inertia of a CD, including the effect of the hole. For a CD of radius 6.0 cm, estimate the percentage change in the moment of inertia due to a hole of radius 7 mm.

49. ★ The moment of inertia for a square plate of mass M and length L that rotates about an axis perpendicular to the plane of the plate and passing through its center is $ML^2/6$ (Fig. P8.49A). What is the moment of inertia of the same plate when it is rotated about an axis that lies along one edge of the plate (Fig. P8.49B)?

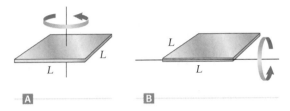

Figure P8.49

50. ✪ ℝ Estimate the moment of inertia of a baseball bat for the following cases.

(a) Assume the bat is a wooden rod of uniform diameter and take the pivot point to be (i) in the middle of the bat, (ii) at the end of the bat, and (iii) 5 cm from the end of the handle, where a batter who "chokes up" on the bat would place his hands. *Hint*: Consider the bat as two rods held together at the pivot point.

(b) Assume a realistic bat, which is narrower at the handle than at the opposite end, and consider the three locations from part (a).

8.5 ROTATIONAL DYNAMICS

51. Figure P8.51 shows the angular displacement of an object as a function of time. What is the approximate angular velocity of the object?

52. SSM ★ Figure P8.52 shows the angular displacement of an object as a function of time. (a) What is the approximate angular velocity of the object at $t = 0$? (b) What is the approximate angular velocity at $t = 0.10$ s? (c) Estimate the angular acceleration at $t = 0.050$ s.

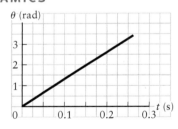

Figure P8.51

53. ★ A ceiling fan of radius 45 cm runs at 90 rev/min. How far does the tip of the fan blade travel in 1 hour?

Figure P8.52

54. ★ ℝ Construct a plot of the angular displacement of a CD as a function of time. Use information from this chapter to attach quantitative labels (numbers and units) to your graph. Take $t = 0$ to be the time

when the CD player is first turned on. End your graph when the CD player is turned off and the CD stops spinning.

55. A potter's wheel of radius 0.20 m is turned on at $t = 0$ and accelerates uniformly. After 60 s, the wheel has an angular velocity of 2.0 rad/s. Find the angular acceleration and the total angular displacement of the wheel during this time.

56. ★ Construct a graph of the angular velocity of a car wheel as a function of time. (a) Assume the wheel starts from rest and moves with a constant (center of mass) velocity. (b) Assume the car starts from rest and accelerates uniformly. (c) Assume the car comes to a stop uniformly at a traffic light.

57. Two crates of mass 5.0 kg and 9.0 kg are connected by a rope that runs over a pulley of mass 4.0 kg as shown in Figure P8.57. (a) Make a sketch showing all the forces on both crates and the pulley. (b) Express Newton's second law for the crates (translational motion) and for the pulley (rotational motion). The linear acceleration a of the crates, the angular acceleration α of the pulley, and the tensions in the right and left portions of the rope are unknowns. (c) What is the relation between a and α? (d) Find the acceleration of the crates. (e) Find the tensions in the right and left portions of the rope.

Figure P8.57

58. SSM ★ Two crates of mass $m_1 = 15$ kg and $m_2 = 25$ kg are connected by a cable that is strung over a pulley of mass $m_{pulley} = 20$ kg as shown in Figure P8.58. There is no friction between crate 1 and the table. (a) Make a sketch showing all the forces on both crates and the pulley. (b) Express Newton's second law for the crates (translational motion) and for the pulley (rotational motion). The linear acceleration a of the crates, the angular acceleration α of the pulley, and the tensions in the right and left portions of the rope are unknowns. (c) What is the relation between a and α? (d) Find the acceleration of the crates. (e) Find the tensions in the right and left portions of the rope.

Figure P8.58
Problems 58 and 59.

59. ✪ For the system in Figure P8.58, if the coefficient of friction between crate 1 and the table is $\mu_K = 0.15$, what is the acceleration?

60. ★ ℝ Consider the CD in Example 8.2. This CD player is turned on at $t = 0$ and very quickly starts to play music so that the CD has an angular velocity of 500 rpm just a few seconds after $t = 0$. The CD then plays for 1 hour, during which time it has a constant angular acceleration and a final angular velocity of 200 rpm. Estimate the total distance traveled by a point on the edge of the CD. Assume the CD has a radius of 6.0 cm.

61. ✪ Consider again the CD in Problem 60. As was explained in Example 8.2, the data on the CD are encoded in a long spiral "track," where the spacing between each turn of the track is 1.6 μm and the inner and outer radii of the program area are about 25 mm and 58 mm, respectively. Estimate the length of this spiral.

62. ★ You like to swim at a nearby lake. On one side of the lake is a cliff, and the top of the cliff is 6.5 m above the surface of the lake. You dive off the cliff doing somersaults so as to have an angular speed of 2.2 rev/s. How many revolutions do you make before you hit the water? Assume your initial center of mass velocity is horizontal and you begin from a standing position. Do you enter the water head first or feet first?

63. A Ferris wheel is moving at an initial angular velocity of 1.0 rev/30 s. If the operator then brings it to a stop in 3.0 min, what is the angular acceleration of the Ferris wheel? Express your answer in rad/s². Through how many revolutions will the Ferris wheel move while coming to a stop?

64. ✪ A board of length 1.5 m is attached to a floor with a hinge at one end as shown in Figure P8.64. The board is initially at rest and makes an angle of 40° with the floor. If the board is then released, what is its angular acceleration?

Figure P8.64

8.6 COMBINED ROTATIONAL AND TRANSLATIONAL MOTION

65. ⭐ A car starts from rest and then accelerates uniformly to a linear speed of 15 m/s in 40 s. If the tires have a radius of 25 cm, what are (a) the average linear acceleration of a tire, (b) the angular acceleration of a tire, (c) the angular displacement of a tire from $t = 0$ to 40 s, and (d) the total distance traveled by a point on the edge of a tire?

66. A bicycle wheel of radius 25 cm rolls a distance of 3.0 km. How many revolutions does the wheel make during this journey?

67. SSM ⭐ Consider a tennis ball that is hit by a player at the baseline with a horizontal velocity of 45 m/s (about 100 mi/h). The ball travels as a projectile to the player's opponent on the opposite baseline, 24 m away, and makes 25 complete revolutions during this time. What is the ball's angular velocity?

68. ✪ A cylinder rolls without slipping down an incline that makes a 30° angle with the horizontal. What is the acceleration of the cylinder?

69. ✪ A string is rolled around a cylinder ($m = 4.0$ kg) as shown in Figure P8.69. A person pulls on the string, causing the cylinder to roll without slipping along the table. If there is a tension of 30 N in the string, what is the acceleration of the cylinder?

Figure P8.69

70. ⭐ A truck accelerates from rest to a speed of 12 m/s in 5.0 s. If the tires have a radius of 40 cm, how many revolutions do the tires make during this time?

71. A car is traveling with a speed of 20 m/s. If the tires have an angular speed of 62 rad/s, what is the radius of the tires?

72. A car has a top speed of 70 m/s (about 150 mi/h) and has wheels of radius 30 cm. If the car starts from rest and reaches its operating speed after 9.0 s, what is the angular acceleration of the wheels? Assume the car accelerates uniformly.

73. ✪ The wheels on a special bicycle have different radii R_f and R_b (Fig. P8.73). The bicycle starts from rest at $t = 0$ and accelerates uniformly, reaching a linear speed of 10 m/s at $t = 9.0$ s. (a) What is the bicycle's linear acceleration? (b) At $t = 4.5$ s, the angular speed of the back wheel is 11 rad/s, whereas the angular speed of the front wheel is greater by a factor of three. Find R_b and R_f.

Figure P8.73

---

74. ✪ Two uniform density bricks are stacked as shown in Figure P8.74. What is the maximum overhang distance d that can be achieved? *Hint*: Consider how the torque on each brick's center of mass must be in equilibrium about the balance point (the edge of the ledge).

Figure P8.74
Problems 74 and 75.

75. ✪ Consider again the problem of overhanging bricks in Figure P8.74. Find the maximum overhang distance, d, for systems of (a) three bricks and (b) four bricks. (c) How many bricks are needed for the top brick to have an overhang of $d = 2L$?

76. SSM ⭐ Three gears of a mechanism are meshed as shown in Figure P8.76 and are rotating with constant angular velocities. The ratio of the diameters of gears 1 through 3 is 3.5/1.0/2.0. A torque of 20 N·m is applied to gear 3. If gear 2 has a radius of 10 cm, what is the torque on gears 1 and 2?

Figure P8.76

77. ✪ A chain drive (Fig. P8.77) transfers the rotational motion of a tractor transmission to a grain elevator. The drive shaft places a counterclockwise torque on the small sprocket of radius 2.2 cm so that it can overcome the resistance in the mechanism of the auger, which puts a clockwise torque of 200 N·m on the larger sprocket of radius 17 cm. If the gears turn at a constant rate, (a) what is the tension in the chain and (b) what is the torque applied to the small sprocket by the drive shaft?

78. ✪ Bicycles like the one shown in Figure P8.78 typically have 24 possible gear ratios (three gears on the front and eight on the back). Suppose the diameters of the largest and smallest sprockets are 17 cm and 9.0 cm on the front cluster and 11 cm and

© Joe Fox/Alamy

Figure P8.77

5.0 cm on the rear cluster. The crank measures 18 cm from the center of the pedal shaft axis to the center of the front sprocket, and the back wheel (including the tire) has a diameter of 68 cm. Assume the cyclist is traveling at constant speed. If a downward force of 320 N is applied to the pedal when the crank is in the horizontal position, what is the horizontal force the rear wheel

© Cengage Learning/Charles D. Winters

Figure P8.78

exerts on the road for (a) the lowest and (b) the highest gear combinations? Also find the torque on the rear axle in both cases. Assuming rolling without slipping, how far does the bike move forward for each full pedal stroke (each full 360° rotation) for (c) the lowest and (d) the highest gear ratios?

79. ✪ ✖ A bodybuilder is in training and performs curls with a 10-kg barbell to strengthen her biceps muscle. At the top of the stroke, her arm is configured as in Figure P8.79, with her forearm at 35° with respect to the horizontal. Approximate her forearm and hand as a uniform rod of mass 1.4 kg and length 32 cm (from pivot point at the elbow to center of barbell). The biceps muscle is attached 3.1 cm from the pivot point at the elbow joint and produces a force that is directed 85° away from the forearm. What force must the biceps muscle exert to keep the weight at its current position?

Figure P8.79 Problem 79 and Question 11.

80. ✪ ✖ ℝ **Runner's gait.** The amount of torque required by an external applied force (other than gravity) to move a leg forward is less than that required to move it backward if the leg is bent as shown in Figure P8.80. The lengths of the vectors $\vec{r}_1$, $\vec{r}_2$,

Figure P8.80

and $\vec{r}_3$ indicate the position of the center of mass of the upper leg, lower leg, and foot from the axis of rotation. The ratio of the mass of the upper leg/lower leg/foot is approximately 5/3/1. Find the ratio of the moment of inertia for the leg moving forward to that of the leg moving backward. Approximate the three parts of the leg as point masses.

81. ✪ A wheel of mass 11 kg is pulled up a step by a horizontal rope as depicted in Figure P8.81. If the height of the step is equal to one-third the radius of the wheel, $h = \frac{1}{3}R$, what minimum tension is needed on the rope to start the wheel moving up the step? *Hint:* The wheel will start to move just as the normal force on the very bottom of the wheel becomes zero.

Figure P8.81

82. ✪ The end of a pencil of mass m and length L rests in a corner as the pencil makes an angle θ with the horizontal (x) direction (Fig. P8.82). If the pencil is released, it will rotate about point O with an acceleration α_1. Suppose the length of the pencil is increased by a factor of two but its mass is kept the same; the pencil would then have an angular acceleration α_2 when released. Find the ratio α_1/α_2.

Figure P8.82

83. ☆ A uniform wooden plank of length L and mass m is supported by an axle that passes through its center. A small package of mass m_p is placed at one end (Fig. P8.83). Find the angular acceleration of the system. Express your answer in terms of L, m, and m_p.

Figure P8.83

Energy and Momentum of Rotational Motion

In Chapter 8, we introduced the notion of torque as the rotational analog of force. Torques give rise to rotational motion in much the same way that forces give rise to linear motion. In this chapter, we consider the kinetic energy associated with rotation and derive a work–energy theorem that relates torque and rotational kinetic energy. We then explore situations involving the conservation of mechanical energy, this time with the rotational kinetic energy included. We also introduce the idea of angular momentum, the rotational analog of linear momentum. In Chapter 7, we saw that the general principle of momentum conservation (for linear momentum) has a fundamental place in physics. Angular momentum has a similar role: the conservation of angular momentum is important in many situations, ranging from planetary orbits to the structure of atoms and the periodic table, and it also explains the mystery of how cats are almost always able to land on their feet.

This cat uses the conservation of angular momentum to rotate its body so as to land on its feet. (© Biosphoto/Labat J.-M. & Roquette F./Peter Arnold, Inc.)

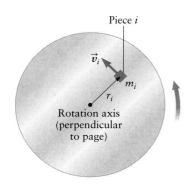

Figure 9.1 When an object rotates, each "piece" (labeled here as i) has a velocity $\vec{v}_i$ and hence a kinetic energy $\frac{1}{2}m_i v_i^2$. When the kinetic energies of all the pieces are added up, the total equals $\frac{1}{2}I\omega^2$.

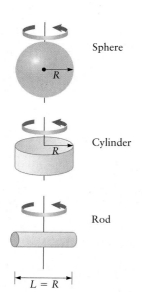

Figure 9.2 Concept Check 9.1.

Figure 9.3 When an object such as a wheel rolls without slipping, its total KE is the sum of the translational KE associated with the center of mass velocity ($\frac{1}{2}mv_{CM}^2$) and the KE of rotation about the center of mass ($\frac{1}{2}I\omega^2$). For rolling motion, $\omega = v_{CM}/R$.

9.1 | KINETIC ENERGY OF ROTATION

In Chapter 6, we defined kinetic energy, the energy associated with the motion of an object. For a simple point particle of mass m moving with a linear speed v, the kinetic energy is

$$KE = \tfrac{1}{2}mv^2 \tag{9.1}$$

Rotational motion is concerned with an extended object—that is, an object with some size and shape—and we now wish to generalize Equation 9.1 to include the contribution of rotation to the kinetic energy. We can view an extended object as being composed of many small pieces, and each piece can be thought of as a point particle with a kinetic energy given by Equation 9.1. The total kinetic energy can be found by adding up the kinetic energies of all the pieces. If each piece has a mass m_i and moves with a linear speed v_i, the total kinetic energy of all the pieces is

$$KE = \sum_i \tfrac{1}{2} m_i v_i^2 \tag{9.2}$$

This is the total kinetic energy of any moving object, including objects that are undergoing rotational motion. This result can be simplified if we assume the pieces in Equation 9.2 are part of a rigid object undergoing simple rotational motion. For an object rotating about a fixed axis with an angular velocity ω, we have, from Chapter 8 (Eq. 8.7), $v_i = \omega r_i$ where r_i is the distance of the ith piece from the rotation axis (Fig. 9.1). Because this object is rigid, all pieces have the same angular velocity, and we can rearrange Equation 9.2 to get

$$KE_{\text{rot}} = \sum_i \tfrac{1}{2} m_i v_i^2 = \sum_i \tfrac{1}{2} m_i (\omega r_i)^2 = \sum_i \tfrac{1}{2} m_i r_i^2 \omega^2$$

$$KE_{\text{rot}} = \tfrac{1}{2}\left(\sum_i m_i r_i^2 \right)\omega^2 \tag{9.3}$$

In Equation 9.3, we pulled the factor of ω^2 out of the summation since ω is the same for all pieces of the object. The term in parentheses in Equation 9.3 is the moment of inertia I (recall that $I = \sum_i m_i r_i^2$, Eq. 8.14). We thus have

$$KE_{\text{rot}} = \tfrac{1}{2}I\omega^2 \tag{9.4}$$

This kinetic energy is due to the rotational motion of an extended object. You should see, however, that KE_{rot} has as its origin the translational kinetic energies of all the individual pieces of the object, as given by Equation 9.1.

CONCEPT CHECK 9.1 | Rotational Kinetic Energy

A sphere of radius R, a cylinder of radius R, and a rod of length $L = R$ all have the same mass, and all three are rotating about an axis through their centers (Fig. 9.2). If the objects all have the same angular velocity, which one has the greatest rotational kinetic energy, (a) the sphere, (b) the cylinder, or (c) the rod?

The Total Kinetic Energy of an Object Is the Sum of the Rotational and Translational Kinetic Energies

The rotational kinetic energy of an object undergoing "pure" rotation about a fixed axis is $KE_{\text{rot}} = \frac{1}{2}I\omega^2$ (Eq. 9.4). Rolling objects such as the wheel in Figure 9.3 undergo both translational and rotational motions. In this case, the *total* kinetic energy has contributions from both the rotational motion and the overall translational motion. In practice, it is usually desirable to use an axis that passes through the center of mass. For an object of mass m whose center of mass moves with a linear speed v_{CM}, the total kinetic energy is then given by

$$\overbrace{KE_{total} = \tfrac{1}{2}mv_{CM}^2}^{\text{translational } KE} + \overbrace{\tfrac{1}{2}I\omega^2}^{\text{rotational } KE} \qquad (9.5)$$

Total kinetic energy of translation and rotation for a rotation axis through the center of mass

which is the total kinetic of *any* moving object; it applies to rolling wheels, spinning boomerangs, Frisbees, and baseballs. The first term on the right of the equal sign in Equation 9.5 is the kinetic energy of translation, whereas the second term is the rotational kinetic energy. In Equation 9.5, ω is the angular velocity about a rotation axis that passes through the center of mass.

EXAMPLE 9.1 | Energy of a Wind Farm

A promising and environmentally friendly method of generating electricity is a wind turbine (Fig. 9.4A). Wind turns the propellers, which in turn drive an electrical generator that produces electricity. A typical propeller unit has three blades (usually composed of a lightweight carbon fiber composite material) that are about 30 m long and rotate at 30 rpm. Calculate the rotational kinetic energy of this propeller unit. Approximate each blade as a uniform rod of mass 300 kg.

RECOGNIZE THE PRINCIPLE

The total kinetic energy of the propeller unit equals the sum of the rotational kinetic energies of the three blades. The kinetic energy of each blade can be found from Equation 9.4, using the moment of inertia for a uniform rod rotating about an axis that passes through one end, $I_{rod} = \tfrac{1}{3}mL^2$, where m is the mass of a blade and L is its length (Table 8.2).

SKETCH THE PROBLEM

The problem is sketched in Figure 9.4B. The three blades have the same angular velocity.

IDENTIFY THE RELATIONSHIPS

The total rotational kinetic energy of the propeller unit is

$$KE_{rot} = \tfrac{1}{2}I_{tot}\omega^2 \qquad (1)$$

where I_{tot} is the total moment of inertia of all three blades. We treat each blade as a uniform rod that rotates about an axis through one end. The total moment of inertia of three propeller blades is thus

$$I_{tot} = 3(mL^2/3) = mL^2 = (300 \text{ kg})(30 \text{ m})^2 = 2.7 \times 10^5 \text{ kg} \cdot \text{m}^2$$

where we have inserted the given values for m and L. The angular velocity of the wind turbine is

$$\omega = 30 \text{ rpm} = 30 \frac{\text{rev}}{\text{min}} \times \frac{2\pi \text{ rad}}{\text{rev}} \times \frac{1 \text{ min}}{60 \text{ s}} = 3.1 \text{ rad/s}$$

A

SOLVE

Inserting these values into our expression for the rotational kinetic energy (Eq. 9.4) gives

$$KE_{rot} = \tfrac{1}{2}I_{tot}\omega^2 = \tfrac{1}{2}(2.7 \times 10^5 \text{ kg} \cdot \text{m}^2)(3.1 \text{ rad/s})^2 = \boxed{1.3 \times 10^6 \text{ J}}$$

What does it mean?
Notice that the units here are joules. If this much energy can be extracted each second, the power generated by the wind turbine will be

$$P = \frac{KE_{rot}}{\Delta t} = \frac{1.3 \times 10^6 \text{ J}}{1 \text{ s}} = 1.3 \times 10^6 \text{ W} = 1.3 \text{ MW}$$

which is enough energy to power more than 1000 homes.

$I_1 = \tfrac{1}{3}mL^2$

B

Figure 9.4 Example 9.1.

Rolling Motion and the Distribution of Kinetic Energy

The rolling motion of an object such as a wheel combines translational and rotational motions. Let's now consider how the kinetic energy of a rolling wheel is distributed. For example, if you are designing the wheels for a skateboard and want the skateboard to go as fast as possible, you will want to minimize the fraction of kinetic energy that goes into the rotational motion of each wheel. Doing so will leave more energy for the translational motion of the wheel and hence the skateboard.

To analyze this problem, consider the wheel in Figure 9.3. The total kinetic energy has contributions from the translational kinetic energy ($\frac{1}{2}mv_{CM}^2$) and the rotational kinetic energy ($\frac{1}{2}I\omega^2$). If the wheel has the shape of a disk, the moment of inertia is $I = mR^2/2$, where m is the wheel's total mass and R is its radius (Table 8.2). Combining all these results, the total kinetic energy is

$$KE_{tot} = \frac{1}{2}mv_{CM}^2 + \frac{1}{2}I\omega^2 = \frac{1}{2}mv_{CM}^2 + \frac{1}{2}\left(\frac{mR^2}{2}\right)\omega^2 = \frac{1}{2}mv_{CM}^2 + \frac{1}{4}mR^2\omega^2$$

In our analysis of rolling motion in Chapter 8, we found that the wheel's angular velocity and center of mass velocity are related by $\omega = v_{CM}/R$ (Eq. 8.43). Inserting this information leads to

$$KE_{tot} = \frac{1}{2}mv_{CM}^2 + \frac{1}{4}mR^2\omega^2 = \frac{1}{2}mv_{CM}^2 + \frac{1}{4}mR^2\left(\frac{v_{CM}}{R}\right)^2$$

$$\underbrace{KE_{tot}}_{} = \underbrace{\tfrac{1}{2}mv_{CM}^2}_{\text{translational } KE} + \underbrace{\tfrac{1}{4}mv_{CM}^2}_{\text{rotational } KE} = \underbrace{\tfrac{3}{4}mv_{CM}^2}_{\text{total}} \tag{9.6}$$

The wheel's rotational kinetic energy is thus $\frac{1}{4}mv_{CM}^2$; it is half as large as the translational kinetic energy and one third as large as the total kinetic energy. These fractions are *independent* of the wheel's radius since all factors of R have canceled in deriving Equation 9.6. The only assumption we made is that the moment of inertia is given by $I = mR^2/2$, so our results are true for any object that has the shape of a wheel or disk. One might have expected that for a smaller wheel (a smaller radius R), less of the total kinetic energy would go into rotation, but that is not the case. For rolling motion, we have $\omega = v_{CM}/R$, so a smaller wheel will have a larger angular velocity and the *fraction* of the total kinetic energy associated with rotation is *independent* of R. The only way to reduce this fraction is to change the moment of inertia, that is, to change the wheel's shape so as to reduce its moment of inertia.

The calculation that led to Equation 9.6 was motivated by the goal of computing the total kinetic energy of a skateboard, including the wheels. Notice that the distribution of the kinetic energy in Equation 9.6 applies only to the wheels; the rest of a skateboard (the part you stand on) does not rotate, and the kinetic energy of that part of the skateboard comes just from translational motion ($KE = \frac{1}{2}mv_{CM}^2$).

Torque and Rotational Kinetic Energy: Rotational Version of the Work–Energy Theorem

In Chapter 6, we introduced the notion of **work** and showed that the work done on an object equals the change in the object's translational kinetic energy. Let's now derive a similar result involving torque and rotational kinetic energy.

Consider a one-dimensional situation in which a force of magnitude F pushes an object of mass m through a distance s as sketched in Figure 9.5A. Since the force here is parallel to the displacement, the work equals the product of the force and the displacement; hence, $W = Fs$. Applying the work–energy theorem (Eq. 6.11), W is equal to the change in the kinetic energy of the object:

$$W = Fs = \Delta KE = \tfrac{1}{2}mv_f^2 - \tfrac{1}{2}mv_i^2 \tag{9.7}$$

where v_i and v_f are the object's initial and final speeds. Now suppose this object is held by a very light (massless) rod so that the object rotates as sketched in Figure

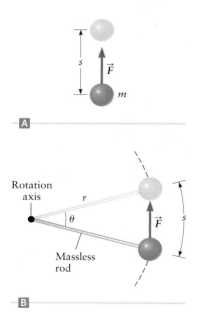

Figure 9.5 **A** When this mass moves a distance s while acted on by the force F, the work done is $W = Fs$. **B** If the mass is attached to a light (massless) rod so that the mass and the rod now rotate, the work done is also equal to $\tau\theta$.

9.5B, with a rotational radius of r. If the angle θ is very small, the force F produces a torque $\tau = Fr$ about this axis. If the object moves a linear distance s, there is a corresponding angular displacement $\theta = s/r$ according to the definitions of angles and radians (Eq. 1.26). If we insert a factor of r in the denominator and numerator of Equation 9.7, we get

$$W = Fs = \overbrace{(Fr)}^{\tau}\overbrace{\left(\frac{s}{r}\right)}^{\theta}$$

$$W = \Delta KE = \tau\theta \tag{9.8}$$

Because it is moving along a circle, the object is now undergoing rotational motion with an angular velocity related to its linear velocity by $v_i = \omega_i r$ and $v_f = \omega_f r$ for the initial and final velocities, respectively. Inserting into Equation 9.7 gives

$$\Delta KE = \tfrac{1}{2}mv_f^2 - \tfrac{1}{2}mv_i^2 = \tfrac{1}{2}m(\omega_f r)^2 - \tfrac{1}{2}m(\omega_i r)^2$$

$$\Delta KE = \tfrac{1}{2}(mr^2)\omega_f^2 - \tfrac{1}{2}(mr^2)\omega_i^2$$

The term mr^2 is just the moment of inertia of the object (from Eq. 8.13), so the change in kinetic energy is

$$\Delta KE = \tfrac{1}{2}I\omega_f^2 - \tfrac{1}{2}I\omega_i^2$$

which we should have expected from our earlier results for rotational kinetic energy (Eq. 9.4). Combining this expression with Equation 9.8 gives

$$W = \tau\theta = \Delta KE = \tfrac{1}{2}I\omega_f^2 - \tfrac{1}{2}I\omega_i^2 \tag{9.9}$$

Work–energy theorem for rotational motion

This result is the **work–energy theorem** for rotational motion. It says that the product of the torque τ and the angular displacement θ equals the change in the rotational kinetic energy.

EXAMPLE 9.2 | Physics on the Playground

Consider a merry-go-round that is initially at rest. A child is given the job of getting the merry-go-round up to an enjoyable speed by pushing on its edge as she runs along the outer circumference as sketched in Figure 9.6. If she pushes with a force of 30 N, how far will she have to run to get the merry-go-round to rotate at an angular velocity of $\omega_f = 0.20$ rev/s (which is an angular velocity of 1 rev every 5.0 s = 1.3 rad/s)? Assume the merry-go-round is a disk of mass $m = 100$ kg and a radius $R = 2.0$ m. Express your answer in terms of the child's angular displacement during her run.

RECOGNIZE THE PRINCIPLE

Since the force and radius of the merry-go-round are given, we can find the applied torque. The initial and final angular velocities are also given, so we can calculate the change in the rotational kinetic energy. The rotational work–energy theorem (Eq. 9.9) then relates the change in kinetic energy to the angular displacement, which is what we want to find.

SKETCH THE PROBLEM

Figure 9.6 shows the problem.

IDENTIFY THE RELATIONSHIPS

The torque exerted by the child is $\tau = FR$ since the associated lever arm is the radius R of the merry-go-round. (See Fig. 9.6.) If the child travels an angular displacement θ as she pushes the merry-go-round, the work done is $W = \tau\theta$, which also equals the change in the merry-go-round's kinetic energy. The initial kinetic energy is zero

TOP VIEW OF
MERRY-GO-ROUND

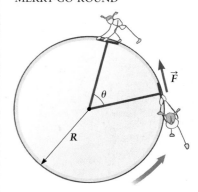

Figure 9.6 Example 9.2. This child pushes on the merry-go-round with a force F while it moves through an angular displacement θ.

(because it starts from rest), whereas the final kinetic energy is $\frac{1}{2}I\omega_f^2$. Inserting all these results into the work–energy theorem (Eq. 9.9) leads to

$$W = \tau\theta = FR\theta = \Delta KE = \tfrac{1}{2}I\omega_f^2$$

$$\theta = \frac{I\omega_f^2}{2FR}$$

SOLVE

The merry-go-round is a disk, so its moment of inertia is $I = mR^2/2$. (See Table 8.2.) Inserting this equation and the given values of F, R, and ω_f gives

$$\theta = \frac{I\omega_f^2}{2FR} = \frac{(mR^2/2)\omega_f^2}{2FR} = \frac{mR\omega_f^2}{4F} = \frac{(100 \text{ kg})(2.0 \text{ m})(1.3 \text{ rad/s})^2}{4(30 \text{ N})} = \boxed{2.8 \text{ rad}}$$

What does it mean?
This answer is less than half of a revolution (since 1 rev corresponds to $2\pi \approx$ 6.2 rad), so the child will not have to run very far.

9.2 | CONSERVATION OF ENERGY AND ROTATIONAL MOTION

In Chapter 6, we introduced the notion of *conservative forces*. We showed that if all the forces that do work on an object are conservative forces, the total mechanical energy of the object is conserved. In such cases, we have

Conservation of total mechanical energy

$$\overbrace{KE_i + PE_i}^{\text{initial mechanical energy}} = \overbrace{KE_f + PE_f}^{\text{final mechanical energy}} \qquad (9.10)$$

Here KE denotes the *total* kinetic energy (including translation and rotation) given by Equation 9.5. The form of the potential energy term PE depends on the forces that are involved.

As an example of how to apply the principle of conservation of mechanical energy to a rotating object, consider a solid ball (a sphere) that starts from rest and rolls down a hill as sketched in Figure 9.7. We want to calculate its velocity when it reaches the bottom. In this example, there are three forces acting on the ball: the normal force exerted by the hill, the force of friction between the ball and the hill, and the force of gravity. The normal force does not do any work on the ball because this force is always directed perpendicular to the ball's displacement. The force of friction is necessary for the ball to roll without slipping. However, friction does not do any work on the ball because the contact point where the ball meets the surface does not slip. Hence, friction does not play a role in the conservation of energy condition (Eq. 9.10) when an object rolls without slipping (see also Insight 9.1). The only force that does work on the ball in Figure 9.7 is the force of gravity.

We saw in Chapter 6 that the work done by gravity equals the change in the gravitational potential energy $PE_{\text{grav}} = mgh$. Hence, the conservation of mechanical energy condition for the ball becomes

$$KE_i + PE_i = KE_f + PE_f$$

$$\tfrac{1}{2}mv_i^2 + \tfrac{1}{2}I\omega_i^2 + mgh_i = \tfrac{1}{2}mv_f^2 + \tfrac{1}{2}I\omega_f^2 + mgh_f \qquad (9.11)$$

where the subscript i denotes the ball's initial state (at the top of the hill) and the subscript f denotes the ball's final state (at the bottom). The velocities v_i and v_f in Equation 9.11 are the center of mass velocities of the ball, which (until this point) we have denoted by v_{CM}. (To avoid the excessive use of subscripts, in the rest of this chapter we often drop the subscript "CM" in places where it should be obvious.)

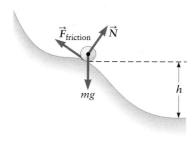

Figure 9.7 When an object rolls without slipping, its total mechanical energy is conserved.

The ball started from rest, so $v_i = \omega_i = 0$. The ball is also rolling without slipping, so $\omega_f = v_f/R$, where R is the ball's radius. Inserting this expression into Equation 9.11 and taking $h_f = 0$ and $h_i = h$ (the height of the hill), we find

$$\overbrace{\tfrac{1}{2}mv_i^2}^{0} + \overbrace{\tfrac{1}{2}I\omega_i^2}^{0} + \overbrace{mgh_i}^{mgh} = \overbrace{\tfrac{1}{2}mv_f^2 + \tfrac{1}{2}I\omega_f^2 + mgh_f}^{0}$$

$$mgh = \tfrac{1}{2}\,mv_f^2 + \tfrac{1}{2}\,I\omega_f^2 = \tfrac{1}{2}\,mv_f^2 + \tfrac{1}{2}I\left(\frac{v_f}{R}\right)^2 \qquad (9.12)$$

Inserting the moment of inertia of a solid sphere ($I = 2mR^2/5$) from Table 8.2 leads to

$$mgh = \frac{1}{2}\,mv_f^2 + \frac{1}{2}\,I\left(\frac{v_f}{R}\right)^2 = \frac{1}{2}\,mv_f^2 + \frac{1}{2}\,\frac{2mR^2}{5}\left(\frac{v_f}{R}\right)^2$$

$$mgh = \frac{1}{2}\,mv_f^2 + \frac{1}{5}\,mv_f^2 = \frac{7}{10}\,mv_f^2$$

Solving for the speed v_f at the bottom, we have

$$v_f = \sqrt{\frac{10gh}{7}} \qquad (9.13)$$

In getting this result, we have relied only on the fact that the mechanical energy of the ball is conserved. We did not need to know anything about the hill other than its height h. Alternatively, we might try to tackle this problem using the rotational form of Newton's second law ($\sum \tau = I\alpha$), but to calculate the forces and torque on the ball, we would need to take into account the shape of the hill. Thus, for a realistically shaped hill, the Newton's second law approach would be very complicated mathematically.

Our application of the principle of conservation of energy to the rolling ball in Figure 9.7 is similar to the way we used conservation of energy principles in Chapter 6. Our general strategy for dealing with these problems is as follows.

PROBLEM SOLVING	Applying the Principle of Conservation of Energy to Problems of Rotational Motion

1. **RECOGNIZE THE PRINCIPLE.** The mechanical energy of an object is conserved only if all the forces that do work on the object are conservative forces. Only then can we apply the conservation of mechanical energy condition in Equation 9.10.

2. **SKETCH THE PROBLEM.** Always use a sketch to collect your information concerning the initial and final states of the system.

3. **IDENTIFY THE RELATIONSHIPS.** Find the initial and final kinetic and potential energies of the object. They will usually involve the initial and final linear and angular velocities, and the initial and final

heights, as for the rolling ball in Figure 9.7. You will usually also need to find the moment of inertia.

4. **SOLVE** for the unknown quantities using the conservation of energy condition,

$$KE_i + PE_i = KE_f + PE_f$$

The kinetic energy terms must account for both the translational and the rotational kinetic energies (Eq. 9.5)

$$KE_{\text{total}} = \overbrace{\tfrac{1}{2}mv_{\text{CM}}^2}^{\text{translational } KE} + \overbrace{\tfrac{1}{2}I\omega^2}^{\text{rotational } KE}$$

5. Always *consider what your answer means* and check that it makes sense.

CONCEPT CHECK 9.2 | Rolling versus Sliding

It is interesting to compare the result for the rolling ball in Equation 9.13 with those we found in Chapter 6 for an object (such as a skier) that slides down a frictionless hill. Which of the following statements is correct?

(a) The ball has a higher linear speed because it does not slip.

(b) The sliding object has a higher speed because the hill is frictionless in that case.

(c) The ball and skier have the same speed at the bottom of the hill.

(d) The ball has a smaller linear speed because some of its kinetic energy goes into rotational motion.

EXAMPLE 9.3 | Motion of a Yo-Yo

Consider the yo-yo in Figure 9.8. The yo-yo starts from rest and then moves downward in the manner you have probably practiced at least a few times. Find the velocity of the yo-yo after it has moved a distance of 50 cm below its starting location. For simplicity, assume the yo-yo is a solid disk with the string wound around the outer edge.

RECOGNIZE THE PRINCIPLE

There are two forces acting on the yo-yo: the force exerted by the string and the force of gravity. The force exerted by the string plays a role very similar to that of friction for the rolling ball in Figure 9.7, and the string does no work on the yo-yo. That is, the yo-yo "rolls down" the string in the same way that the ball rolls without slipping down the hill. The only force that does work on the yo-yo is the force of gravity. Since gravity is a conservative force, we can apply the principle of conservation of energy.

SKETCH THE PROBLEM

Figure 9.8 shows the initial and final positions of the yo-yo.

IDENTIFY THE RELATIONSHIPS

The conservation of energy relation for the yo-yo is

$$KE_i + PE_i = KE_f + PE_f$$

The kinetic energy has two contributions, one from translational motion and another from rotational motion:

$$\tfrac{1}{2}mv_i^2 + \tfrac{1}{2}I\omega_i^2 + PE_i = \tfrac{1}{2}mv_f^2 + \tfrac{1}{2}I\omega_f^2 + PE_f$$

The potential energy is due to gravity; taking the starting height to be $h_i = 0$, the final height is $h_f = -h$ (= -0.50 m as given above). This yo-yo is a disk of mass m and radius R, so its moment of inertia is $I = mR^2/2$ (from Table 8.2). Inserting this expression into the conservation of energy relation leads to

$$\frac{1}{2}\,mv_i^2 + \frac{1}{2}\left(\frac{mR^2}{2}\right)\omega_i^2 = \frac{1}{2}\,mv_f^2 + \frac{1}{2}\left(\frac{mR^2}{2}\right)\omega_f^2 + mg(-h) \qquad (1)$$

The yo-yo starts from rest, so $v_i = \omega_i = 0$. The yo-yo is also (as already noted) "rolling down" the string, so the final linear and angular velocities are related by $\omega_f = v_f/R$.

SOLVE

Putting all this information into Equation (1), we find

$$0 = \frac{1}{2}\,mv_f^2 + \frac{1}{2}\left(\frac{mR^2}{2}\right)\left(\frac{v_f}{R}\right)^2 - mgh = \frac{1}{2}\,mv_f^2 + \frac{1}{4}\,mv_f^2 - mgh$$

$$0 = \frac{3}{4}\,mv_f^2 - mgh$$

$$\frac{3}{4}\,mv_f^2 = mgh$$

$$v_f = \sqrt{\frac{4gh}{3}}$$

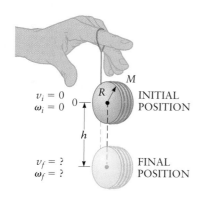

$v_i = 0$
$\omega_i = 0$ INITIAL POSITION

h

$v_f = ?$
$\omega_f = ?$ FINAL POSITION

Figure 9.8 Example 9.3. A yo-yo traveling down a string.

Inserting the given value of h, we find

$$v_f = \sqrt{\frac{4gh}{3}} = \sqrt{\frac{4(9.8 \text{ m/s}^2)(0.50 \text{ m})}{3}} = \boxed{2.6 \text{ m/s}}$$

What does it mean?
In this result for the velocity, notice that the mass and radius of the yo-yo have canceled. Hence, a large yo-yo and a small yo-yo should move at the same velocity; that is, they should "fall" at the same rate. This rate is similar to the speed of an object that undergoes free fall (Chapter 6): the speed depends on the distance the object falls, but not on its mass. Hence, Galileo could have used yo-yos in his experiments at the Leaning Tower of Pisa (Chapter 2).

CONCEPT CHECK 9.3 | Designing a Faster Yo-Yo

You want to design and market a faster yo-yo. That is, you want to make a yo-yo that will have a higher velocity than that found in Example 9.3. How could you do it?
 (a) Make the yo-yo radius larger.
 (b) Make the yo-yo radius smaller.
 (c) Add mass near the rotation axis of the yo-yo, so that the yo-yo is shaped more like a sphere.
 (d) Hollow out the center so that the yo-yo is shaped like a cylindrical shell.

9.3 | ANGULAR MOMENTUM

There are many parallels in the way we describe translational and rotational motions and in the way Newton's laws read for the two cases. In this section, we explore another similarity, connected with momentum. In Chapter 7, we defined the *linear momentum* for a point particle of mass m moving with a velocity $\vec{v}$ as

$$\vec{p} = m\vec{v} \tag{9.14}$$

We showed that the momentum of a system of particles is conserved provided there is no net external force on the system. It should not surprise you now to find that a rotating object has a property called angular momentum. In nearly all the rotational motion examples we have discussed so far, the angular velocity has been defined with respect to a *single rotation axis* that does not change direction during the motion. In such cases, the *angular momentum* is given by

$$L = I\omega \tag{9.15}$$

Angular momentum for rotation about a fixed axis

This equation looks very much like Equation 9.14: the moment of inertia I is analogous to mass, and angular velocity is analogous to linear velocity. One difference you may notice is that the linear momentum $\vec{p}$ is a vector, whereas L in Equation 9.15 is not. This difference is due to our simplified treatment of rotations. In Section 9.4, we'll generalize Equation 9.15 to a vector form with a true angular velocity vector when we discuss gyroscopes. In this section, we assume the rotation axis keeps a fixed direction so that we can use the scalar (nonvector) form in Equation 9.15. Even these "simple" cases contain a lot of interesting physics.

The definition of angular momentum in Equation 9.15 involves ω, and we have always associated an angular velocity with objects that rotate about a fixed axis. However, the notion of angular momentum also applies to more general cases. Figure 9.9 shows a particle moving freely through space. Even though this particle is not constrained to move in a rotational fashion, it can still have an angular momentum. For example, if we choose a particular pivot point P, the particle will at any given instant be moving tangent to the circular arc of radius r sketched in Figure

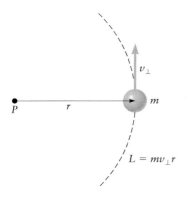

Figure 9.9 The angular momentum of a moving particle about a rotation axis that passes through point P is $L = mv_\perp r$, where $v_\perp$ is the component of the velocity perpendicular to the line that connects the particle to the pivot point P. Here the rotation axis is perpendicular to the plane of the drawing.

9.9. At this instant, the motion of the particle is the same as that of an object that rotates along this circular arc. We can therefore define an angular velocity $\omega = v_\perp/r$, where $v_\perp$ is the component of the particle's velocity perpendicular to the radius of the arc and r is the distance from the particle to the point P. The moment of inertia is given by $I = mr^2$ (Eq. 8.14). Inserting into our definition of angular momentum (Eq. 9.15), we have

$$L = I\omega = mr^2\left(\frac{v_\perp}{r}\right)$$

$$L = mrv_\perp \tag{9.16}$$

This result for L applies to any moving particle. We can always choose a hypothetical pivot point and rotation axis, even if the particle is not moving in a complete closed path about the axis. The angular momentum with respect to that axis is then proportional to $v_\perp$, the component of the velocity perpendicular to the line that connects the particle to the pivot point. The result for the angular momentum in Equation 9.16 will be important when we wish to calculate the total angular momentum of a system of particles and apply the principle of conservation of angular momentum.

Conservation of Angular Momentum and a Spinning Skater

A rotating object with an angular momentum $L = I\omega$ will maintain its angular momentum provided no external torques act on the object. We say that such an object's angular momentum is *conserved*. This notion of angular momentum conservation also applies to a system of objects. If there are no *external* torques on a system, the total angular momentum of the system will be conserved.

A familiar example in which angular momentum conservation plays a key role is the motion of the figure skater in Figure 9.10. Suppose the skater is spinning about a vertical axis with some initial angular velocity ω_i as in Figure 9.10A. If she then pulls her arms and legs in very close to the midline of her body, we know that her spin rate will increase; that is, her angular velocity increases so that her final angular velocity ω_f is much larger than her initial angular velocity ω_i. This behavior can be understood in terms of the skater's angular momentum.

We first notice that during the time she is spinning, no torques are acting on the skater. The only forces acting on her are gravity and the normal force from the ice, and they do not exert a torque of the skater. (We assume the ice is very slippery, so the frictional force is negligible.) The skater's angular momentum is therefore conserved:

$$L_i = L_f$$

Her initial angular momentum is $L_i = I_i\omega_i$, where the subscript i indicates the skater's initial state. When she pulls her arms and legs close to her midline, she is redistributing her mass so that it is closer to the axis of rotation, changing her moment of inertia to a new value I_f. Recall that the moment of inertia is equal to a sum over all the individual pieces of an object:

$$I = \sum mr^2$$

If an object changes shape so that some of the individual pieces are moved closer to the axis of rotation, the values of r for these pieces are reduced and the overall moment of inertia decreases. That is precisely what happens with the skater in Figure 9.10B. When she pulls her arms and legs in close to her rotation axis, these "pieces" have smaller values of r and her moment of inertia becomes smaller. Her total angular momentum is conserved; thus,

$$L_i = I_i\omega_i = L_f = I_f\omega_f \tag{9.17}$$

Since pulling her arms in close to her body reduces her moment of inertia, $I_f < I_i$ and hence $\omega_f > \omega_i$. She spins *faster*. The magnitude of this increase in her angular veloc-

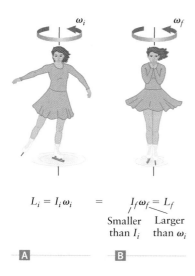

$L_i = I_i\omega_i$ $=$ $I_f\omega_f = L_f$

Smaller   Larger
than I_i than ω_i

A   **B**

Figure 9.10 There are no external torques acting on the skater, so her angular momentum is conserved. Her angular velocity increases when she pulls her arms and legs close to her body, but her angular momentum stays the same.

ity depends on how much she can reduce her moment of inertia; that is, it depends on how closely she can pull her arms and legs in line with the rest of her body.

Problem Solving with Angular Momentum

We saw in Chapter 7 that linear momentum and momentum conservation are extremely useful when analyzing collisions and other situations that involve a system of particles. The same is true of angular momentum. Typically, a system such as the skater in Figure 9.10 is moving in some manner with an initial angular momentum. There is then some sort of "change" in the system (such as the skater pulling in her arms and legs) that leaves the system in a final state of motion. If the external torque on the system is zero, the total angular momentum must be conserved. Hence, $L(\text{final}) = L(\text{initial})$, which we can use to find the final angular velocity or some other quantity. The general approach to such problems is as follows.

PROBLEM SOLVING | Applying the Principle of Conservation of Angular Momentum

1. **RECOGNIZE THE PRINCIPLE.** If the external torque on a system is zero, the angular momentum is conserved.

2. **SKETCH THE PROBLEM.** Always use a sketch to collect your information concerning the initial and final states of the system.

3. **IDENTIFY THE RELATIONSHIPS.**
   • The system of interest and its initial and final states depend on the problem. The system might be a single object (such as the skater in Figure 9.10) or a collection of objects.

   • Express the initial and final angular velocities and moments of inertia.
   • Apply any information concerning the initial and final mechanical energies. (Such information will only be available in some cases.)

4. **SOLVE** for the quantities of interest using the principle of conservation of angular momentum, $L_i = I_i\omega_i = L_f = I_f\omega_f$.

5. Always *consider what your answer means* and check that it makes sense.

This general approach is very similar to the way we tackled problems involving the conservation of linear momentum in Chapter 7. Let's now apply it to two examples.

EXAMPLE 9.4 | Angular Momentum on the Playground

Consider a merry-go-round of mass m_{mgr} and radius R_{mgr} that is rotating freely as sketched in Figure 9.11. A child is eager to enjoy a ride, so she steps onto the edge of the merry-go-round. After she gets on, she notices that the angular velocity of the merry-go-round has decreased slightly. If the child's mass is m_c, what is the ratio of the final angular velocity to the initial angular velocity?

RECOGNIZE THE PRINCIPLE

The merry-go-round and the child are a "system," and although there are external forces acting on them (such as gravity), for a well-oiled merry-go-round the torque on this system will be negligible. Hence, the angular momentum of the system is conserved.

SKETCH THE PROBLEM

Figure 9.11 shows the system just before the child steps onto the merry-go-round (the initial state) and just after (the final state).

Figure 9.11 Example 9.4. A child steps onto a rotating merry-go-round.

IDENTIFY THE RELATIONSHIPS

Let I_{mgr} be the moment of inertia of the merry-go-round and ω_i be its angular velocity before the child gets on. We also let I_c be the moment of inertia of the child, while her initial angular velocity is zero. The initial angular momentum is then

$$L_i = I_{mgr}\omega_i + I_c(0) = I_{mgr}\omega_i$$

We can approximate the merry-go-round as a large disk, with moment of inertia $I = mR^2/2$, so we have

$$I_{mgr} = \tfrac{1}{2}m_{mgr}R_{mgr}^2$$

The child has a mass m_c and is located a distance R_{mgr} from the axis of rotation, so her moment of inertia is

$$I_c = \sum mr^2 = m_c R_{mgr}^2$$

After the child steps onto the merry-go-round, the angular momentum of the system is

$$L_f = I_{mgr}\omega_f + I_c\omega_f$$

where ω_f is the (common) final angular velocity. (If the child is now on the merry-go-round, both the child and the merry-go-round have the same angular velocity.)

SOLVE

Angular momentum is conserved, so

$$L_i = L_f$$

which gives

$$I_{mgr}\omega_i = I_{mgr}\omega_f + I_c\omega_f \qquad (1)$$

Solving for the final angular velocity, we find

$$I_{mgr}\omega_i = (I_{mgr} + I_c)\omega_f$$

$$\omega_f = \frac{I_{mgr}\omega_i}{I_{mgr} + I_c}$$

The ratio of the initial and final angular velocities is then

$$\frac{\omega_f}{\omega_i} = \frac{I_{mgr}}{I_{mgr} + I_c} \qquad (2)$$

Inserting the results for the moments of inertia into Equation (2) gives

$$\frac{\omega_f}{\omega_i} = \frac{I_{mgr}}{I_{mgr} + I_c} = \frac{m_{mgr}R_{mgr}^2/2}{(m_{mgr}R_{mgr}^2/2) + m_c R_{mgr}^2} = \boxed{\frac{m_{mgr}}{m_{mgr} + 2m_c}}$$

What does it mean?

This example is similar to an inelastic collision between two objects as we encountered in Chapter 7. The two "objects"—the merry-go-round and the child—stick together and move with the same angular velocity after the "collision," just like the two objects in a completely inelastic collision stick together after colliding.

EXAMPLE 9.5 | Jumping off a Merry-Go-Round

The child plus merry-go-round system in Example 9.4 can be used to illustrate several other important aspects of angular momentum and its conservation. Suppose the child is initially standing on the edge of the merry-go-round and that both are initially at rest. The child then jumps off the merry-go-round, which begins to rotate. If she jumps in a direction tangent to the edge of the merry-go-round as in Figure 9.12A and has a speed of 3.0 m/s while she is in the air, what is the final angular velocity of the merry-go-round? Assume the mass of the child is $m_c = 30$ kg, while the merry-go-round has a mass $m_{mgr} = 100$ kg and its radius is $R_{mgr} = 2.0$ m. Also assume that the merry-go-round is frictionless.

RECOGNIZE THE PRINCIPLE

Just as in Example 9.4, the external torques on the merry-go-round and child in Figure 9.12 are negligible (since the merry-go-round is frictionless). Hence, the angular momentum of this system is conserved.

SKETCH THE PROBLEM

Figure 9.12 shows the initial and final states of the system. In Figure 9.12B, the child has just jumped off the merry-go-round and is still moving through the air (she has not landed on the ground yet).

IDENTIFY THE RELATIONSHIPS

The initial state of the system is just prior to when the child jumps; everything is at rest at that moment, so the initial angular momentum is zero. The final state is the instant just after the child has jumped and before she reaches the ground. Notice that *when the child is moving through the air as a projectile, she still has angular momentum* as described in Figure 9.9 and Equation 9.16, which we must account for when we compute the final angular momentum of the system.

If the final angular velocity of the merry-go-round is ω_{mgr}, its final angular momentum is $I_{mgr}\omega_{mgr}$, where the moment of inertia of the merry-go-round is as described in Example 9.4. The final angular momentum of the child can be found from Equation 9.16. The distance from the child to the rotation axis is R_{mgr}, and her velocity perpendicular to the radius of the merry-go-round is $v_\perp$ (= 3.0 m/s as given in the problem statement). The child's final angular momentum is thus

$$L_c = m_c R_{mgr} v_\perp$$

Combining all these results, the final angular momentum of the system is the sum of the child's angular momentum plus that of the merry-go-round:

$$L_f = L_{mgr} + L_c = I_{mgr}\omega_{mgr} + m_c R_{mgr} v_\perp$$

Inserting the moment of inertia of the merry-go-round ($I_{mgr} = \frac{1}{2}m_{mgr}R_{mgr}^2$) leads to

$$L_f = \tfrac{1}{2}m_{mgr}R_{mgr}^2\omega_{mgr} + m_c R_{mgr} v_\perp$$

SOLVE

The final angular momentum equals the initial angular momentum, which is zero, so

$$\tfrac{1}{2}m_{mgr}R_{mgr}^2\omega_{mgr} + m_c R_{mgr} v_\perp = 0$$

Solving for the final angular velocity of the merry-go-round leads to

$$\omega_{mgr} = -\frac{m_c R_{mgr} v_\perp}{m_{mgr}R_{mgr}^2/2} = -\frac{2m_c v_\perp}{m_{mgr}R_{mgr}}$$

The negative sign here indicates that the final angular velocities of the child and merry-go-round are in opposite directions. Inserting the given values of the various quantities, we find

$$\omega_{mgr} = -\frac{2m_c v_\perp}{m_{mgr}R_{mgr}} = -\frac{2(30 \text{ kg})(3.0 \text{ m/s})}{(100 \text{ kg})(2.0 \text{ m})} = \boxed{-0.90 \text{ rad/s}}$$

What does it mean?
An important point of this example is that an object moving in a straight line (here it is the child after she jumps from the merry-go-round) can still have angular momentum with respect to a particular rotation axis.

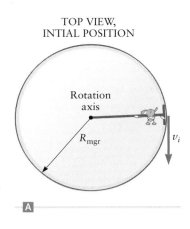

TOP VIEW, INITIAL POSITION

Rotation axis

R_{mgr}

v_i

A

TOP VIEW, FINAL POSITION

ω_f

Final angular momentum of child = $m_c R_{mgr} v_\perp$

$v_c = v_\perp$

B

Figure 9.12 Example 9.5. A child jumps off a stationary merry-go-round.

CONCEPT CHECK 9.4 | Where Does the Torque Come From?

In Example 9.5, we treated the merry-go-round plus child as a system (Fig. 9.12) and argued that the external torque on this system is zero so that its angular momentum is conserved. However, we can also consider the rotational motion of only the merry-go-round. The merry-go-round is initially at rest, so its initial angular momentum is zero, while at the end it is rotating, so its final angular momentum is nonzero. The angular momentum of the merry-go-round thus changes, which can only happen if there is a torque on the merry-go-round. What force provides this torque?

(a) The force of gravity on the merry-go-round
(b) The force of air drag on the merry-go-round
(c) The force produced by the child when she jumps off
(d) The force of friction at the axle of the merry-go-round

Angular Momentum and Kinetic Energy

Let's now return to the problem of the spinning ice skater in Figure 9.10 and consider how her rotational kinetic energy varies during the spin. Recall that our skater begins a spin with her arms and legs extended away from her body's rotation axis and then pulls them in close to her body, increasing her angular velocity. In terms of her angular momentum, we wrote (Eq. 9.17)

$$L_i = I_i \omega_i = L_f = I_f \omega_f$$

which can be rearranged to give

$$\omega_f = \frac{I_i}{I_f} \omega_i \qquad (9.18)$$

Since her final moment of inertia (with her arms and legs in) is smaller than her initial moment of inertia (arms and legs out), $\omega_f > \omega_i$ and she spins faster at the end.

Rotational kinetic energy is $KE = \frac{1}{2}I\omega^2$, so the skater's final rotational kinetic energy is

$$KE_f = \frac{1}{2}I_f\omega_f^2$$

Inserting ω_f from Equation 9.18 and using the skater's initial kinetic energy of $KE_i = \frac{1}{2}I_i\omega_i^2$ leads to

$$KE_f = \frac{1}{2} I_f\omega_f^2 = \frac{1}{2} I_f \left(\frac{I_i}{I_f} \omega_i\right)^2 = \frac{I_i}{I_f}\left(\frac{1}{2} I_i\omega_i^2\right)$$

$$KE_f = \frac{I_i}{I_f} KE_i \qquad (9.19)$$

Because the skater's initial moment of inertia is larger than her final moment of inertia ($I_i > I_f$), Equation 9.19 shows that her final kinetic energy is *greater* than her initial kinetic energy. Even though her angular momentum is conserved, her kinetic energy is *not conserved*.

This result may at first seem surprising, but it is actually quite similar to situations we encountered in connection with linear momentum and kinetic energy. In Chapter 7, we considered examples in which an asteroid was split into two pieces by an explosion inside the asteroid. That explosion resulted from forces acting *within* the system, that is, from forces between different pieces of the asteroid. Because there were no *outside* (external) forces on the asteroid, the total momentum was conserved. The explosion released energy, however, so the total kinetic energy of all the pieces was greater after the explosion. Just as with the spinning ice skater, the total linear momentum of the asteroid was conserved, but the kinetic energy was not. Where does the additional kinetic energy of the skater come from? That is, what force does work on the skater? This force comes from the skater herself as she pulls on her arms. Her arms undergo a displacement as they are pulled inward, and the work done (W) equals the product of this displacement and the component of the force parallel to the displacement. From the work–energy theorem, W equals the increase in the skater's kinetic energy.

9.4 | ANGULAR MOMENTUM AND KEPLER'S SECOND LAW OF PLANETARY MOTION

In Chapter 5, we learned about Kepler's three laws of planetary motion. According to Kepler's first law, planets follow elliptical orbits about the Sun, and his second law states that a planet moving about its orbit sweeps out equal areas in equal times. If we observe the motion of the planet during two separate time intervals of equal duration Δt, we can consider the areas A_1 and A_2 swept out during these intervals as sketched in Figure 9.13. During an interval Δt, the planet travels a distance $v \, \Delta t$. When the planet is far from the Sun (at point 1 in Fig. 9.13), its speed is smaller than when the planet is close to the Sun (at point 2). However, Kepler's second law states that the areas that are swept out are the same; the areas A_1 and A_2 in Figure 9.13 are equal. This equality holds for all points along the orbit; the area swept out by the planet in a time Δt is the same for all positions along the orbit. We'll now show that Kepler's second law is intimately connected with the angular momentum of the planet. Indeed, all orbiting planets, asteroids, and spacecraft must obey Kepler's second law as a direct result of the conservation of angular momentum.

Angular Momentum of an Orbiting Planet

The planet orbiting the Sun in Figure 9.13 has an angular momentum, just like the moving particle in Figure 9.9. Applying Equation 9.16, the planet's angular momentum is given by

$$L_{\text{planet}} = m_{\text{planet}} r v_{\perp} \qquad (9.20)$$

where r is the distance from the planet to the Sun and $v_{\perp}$ is the component of the velocity perpendicular to the line connecting the planet and Sun. As the planet moves around its orbit, both r and $v_{\perp}$ change but the planet's angular momentum stays constant for the reason shown in Figure 9.14. The planet is now our system, and the only force on this system is the force of gravity from the Sun. The gravitational force is always directed along the radial line that runs from the planet to the Sun. The axis for the planet's rotational motion passes through the Sun, so a force that acts along this radial line cannot exert a torque on the planet. The torque due

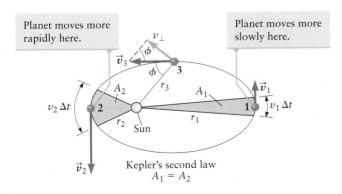

Figure 9.13 When a planet follows an elliptical orbit, the velocity $\vec{v}$ varies as the planet moves along the orbit, and the speed is largest when the planet is closest to the Sun. According to Kepler's second law, a planet sweeps out equal areas in equal time intervals as it moves about the Sun. Hence, $A_1 = A_2$.

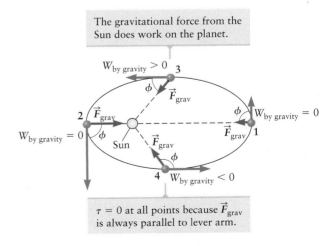

Figure 9.14 Since the force of gravity on the planet is directed along the orbital radius, this force cannot exert a torque on the planet, but it can do work on the planet.

to the gravitational force is thus zero, and the angular momentum of the planet is conserved.

Let's now show how conservation of the angular momentum of the planet leads to Kepler's second law. For simplicity, we consider the locations along the orbit where the planet is farthest and closest to the Sun, at points 1 and 2 in Figure 9.13. At these positions, the planet's velocity is perpendicular to the line between the Sun and the planet, so $v_\perp = v_1$ at point 1 and $v_\perp = v_2$ at point 2. Using Equation 9.20, we get

$$L_1 = m_{\text{planet}}r_1v_1 = m_{\text{planet}}r_2v_2 = L_2 \qquad (9.21)$$

$$r_1v_1 = r_2v_2 \qquad (9.22)$$

The regions A_1 and A_2 swept out by the planet in Figure 9.13 are approximately triangular,[1] so the area A_1 is given by

$$A_1 = \tfrac{1}{2}(\text{base})(\text{height}) = \tfrac{1}{2}(v_1\,\Delta t)\,(r_1)$$

Using Equation 9.22, we now have

$$A_1 = \tfrac{1}{2}(\Delta t)r_1v_1 = \tfrac{1}{2}(\Delta t)r_2v_2 = A_2$$

The areas A_1 and A_2 are thus equal; this relation is Kepler's second law.

We have just shown that Kepler's second law follows directly from the conservation of a planet's orbital angular momentum, but what role do energy considerations play in planetary motion? We have already noted that because the gravitational force from the Sun acts along the planet's orbital radius, this force cannot exert a torque on the planet. This result is true for all points along the orbit. Gravity, however, *can* do work on the planet. The work done during a time interval Δt is equal to

$$W_{\text{by gravity}} = Fd \cos \phi$$

where d is the displacement during this interval and the angle ϕ is indicated in Figure 9.14. At positions 1 and 2, this angle is 90°, so $\cos \phi = 0$ and the work done is zero, but at other places along the orbit, such as locations 3 and 4, the work done is not zero. The work done by gravity causes the planet to speed up (thus increasing its kinetic energy) as it approaches the Sun. As the planet moves farther from the Sun during the course of its orbit, the work done by gravity is negative (at point 4 in Fig. 9.14), causing the planet to slow down.

9.5 | THE VECTOR NATURE OF ROTATIONAL MOTION: GYROSCOPES

To this point, we have treated rotational quantities such as angular velocity as simple scalars because we always dealt with a single rotation axis whose orientation did not change during the course of the motion. However, to deal with more complex types of rotational motion we must recognize that many angular motion quantities—including angular velocity, angular acceleration, angular momentum, and torque—are all vectors. These quantities have both a magnitude and a direction, and their vector nature can make the rotational motion of a complex object, such as a boomerang, very complicated. We will therefore restrict our discussion to a few of the simplest examples in which the vector nature of angular motion plays a crucial role.

A rotating disk or wheel has an angular velocity whose direction is along the axle, as indicated in Figure 9.15. The **right-hand rule** provides a convenient way

Figure 9.15 Application of the right-hand rule to determine the direction of the angular velocity $\vec{\omega}$ and angular momentum $\vec{L}$ of a rotating disk or wheel.

[1]This approximation for the area becomes more and more accurate as the time interval Δt is made smaller and smaller.

to determine this direction. According to this rule, if you allow the fingers on your right hand to curl in the direction of motion of the edge of the wheel, your outstretched right thumb will then point in the direction of $\vec{\omega}$, the angular velocity vector. The direction of other angular quantities such as the angular momentum $\vec{L}$ is also given by the right-hand rule.[2] For the cases we consider in this book, the angular momentum is parallel to the angular velocity, with

$$\vec{L} = I\vec{\omega} \qquad (9.23)$$

When we say that angular momentum is conserved, we mean that both the *magnitude* and *direction* of $\vec{L}$ are conserved. That is the principle behind the operation of a *gyroscope*. There are many ways to build a gyroscope; one design (Fig. 9.16) employs a spinning wheel mounted on a special frame that allows the axle to rotate freely. Because of the way the wheel is mounted in this frame, the torque on the wheel is zero, even when the frame is moved or rotated. Hence, the wheel's angular momentum $\vec{L}$ maintains a constant magnitude and direction. Even if the gyroscope frame is moved or rotated, the wheel's orientation does not change because the angular momentum vector is conserved: $\vec{L}$ of the wheel maintains this constant direction. The orientation of a gyroscope thus provides a "direction finder," a tool extremely useful to pilots and navigators. All airplanes have gyroscopes so that pilots can be certain of their orientation and also the direction in which their plane is moving.[3]

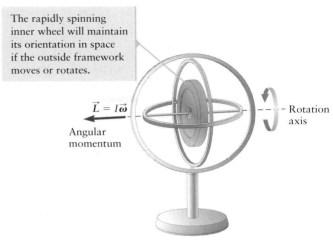

The rapidly spinning inner wheel will maintain its orientation in space if the outside framework moves or rotates.

$\vec{L} = I\vec{\omega}$

Angular momentum

Rotation axis

Figure 9.16 The rotating wheel in this gyroscope is mounted on a system of axles that are free to rotate independently, even when the base of the gyroscope is moved or rotated. Conservation of angular momentum keeps the direction of the inner wheel's axis fixed in space.

Earth as a Gyroscope

The Earth itself acts as a gyroscope, rotating once each day about an axis that runs between the geographic North and South Poles. This rotational motion is often referred to as "spin," and it produces the "spin angular momentum" of the Earth, which should be distinguished from the orbital angular momentum we discussed in connection with Kepler's second law (Fig. 9.13). The Earth moves in a nearly circular orbit around the Sun, and the Earth's rotational axis is not perpendicular to the orbital plane. Instead, the Earth's spin rotation axis is tilted approximately 23.5° away from the normal to this plane as shown in Figure 9.17. Applying the right-hand rule to the Earth's rotational motion in Figure 9.17 shows that the Earth's spin angular momentum is directed along its rotation axis and points in the direction of the North Pole. There are no external torques on the Earth; its spin angular momentum is therefore conserved, and the rotation axis remains tilted at a fixed angle with respect to the orbital plane. So, for part of each year, the northern half of the Earth tilts toward the Sun, and during the other part of the year, the southern half tilts toward the Sun. The part of the Earth that tilts toward the Sun receives slightly more energy from the sunlight, leading to seasons on Earth.

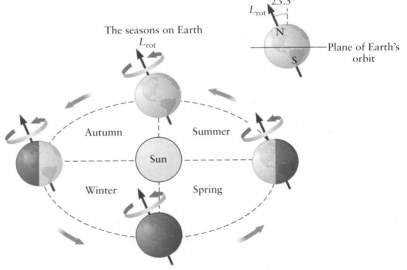

The seasons on Earth

L_{rot}

Autumn

Summer

Sun

Winter

Spring

23.5°

L_{rot}

N

S

Plane of Earth's orbit

Figure 9.17 The Earth's spin angular momentum $\vec{L}_{rot}$ lies along a direction parallel to its north–south axis. The vector $\vec{L}_{rot}$ is tilted at an angle of approximately 23.5° away from the normal to the Earth's orbital plane. The seasons indicated in the figure are those in the Northern hemisphere.

---

[2]We'll encounter another (very similar) right-hand rule when we study magnetism.

[3]There are many famous examples of airplane pilots who have become disoriented and flown for great distances in the wrong direction. One of the earliest flights across the Atlantic Ocean was made by a pilot who had intended to fly from New York to California.

Angular Momentum and the Stability of a Spinning Wheel

If you have a bicycle wheel that is not mounted on bicycle frame, you can perform an experiment that demonstrates gyroscopic motion. When the wheel is not rotating, it is extremely unstable. That is, a nonrotating wheel is very hard to "balance" on its edge and will quickly fall over if left alone. A rolling wheel, however, is much more stable and can travel for quite a long distance and quite a long time before it eventually tips over and falls to one side. The extra stability of a rolling wheel is caused by its angular momentum. According to the right-hand rule, the angular momentum $\vec{L}$ of the wheel is directed along the rotation axis (i.e., it would be horizontal). If the external torque on the rolling wheel were exactly zero, then $\vec{L}$ would remain in this direction forever and the wheel would never fall over. In reality, however, there is always some small external torque (due to slight deviations of the wheel from a perfect circular shape, from friction, and other sources), causing the wheel to fall eventually. Even so, its angular momentum makes a rolling wheel much more stable than a nonrotating wheel. This experiment is one you *should* try at home!

Precession

In some cases, the external torque on a rotating object can be quite substantial, leading to an effect called *precession*. A popular device that demonstrates this effect is shown in Figure 9.18A; it is a bicycle wheel mounted on an axle, with one end of the axle resting on a rotatable metal cup attached to a vertical pole. As you might expect, this "system" (composed of the wheel plus axle) can be extremely unstable; if the wheel is not rotating, the wheel and its axle will immediately fall over. On the other hand, if the wheel is spinning about the axle, the stability is greatly enhanced. The angular momentum of the spinning wheel prevents it from falling over, even though the force of gravity exerts a large external torque on the system.

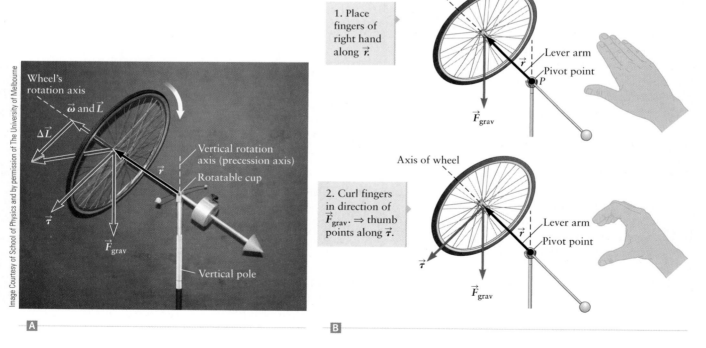

Figure 9.18 ◳ There is a torque on this bicycle wheel due to the force of gravity. The direction of the torque is given by applying the right-hand rule. This torque makes the axis of the wheel precess. Note that the change in angular momentum $\Delta \vec{L}$ is parallel to $\vec{\tau}$. ◳ Application of the right-hand rule to find the direction of the torque due to gravity.

We can calculate the torque on the wheel and axle system by assuming the entire gravitational force acts at the center of mass (Chapter 8). Most of the mass of the system in Figure 9.18A is in the wheel, so the system's center of mass is close to the center of the wheel as shown in the figure. Let's now compute the direction of the gravitational torque $\vec{\tau}$ with respect to the vertical rotation axis shown in Figure 9.18A. We use another version of the right-hand rule to compute $\vec{\tau}$, applying the procedure sketched in Figure 9.18B. (1) Start from the pivot point, which in this case is the point P on the rotatable cup at the base of the axle. With P as an origin, place the fingers of your right hand along the rotation axis of the wheel, that is, along the vector $\vec{r}$ in Figure 9.18B. Point your fingers toward the center of the wheel, which is the place where the force $\vec{F}_{\text{grav}}$ acts. (2) Curl your fingers in the direction of the force $\vec{F}_{\text{grav}}$. Your outstretched right thumb then points in the direction of $\vec{\tau}$ as indicated in Figure 9.18B.

We must next consider how this torque affects the angular momentum of this system (wheel plus axle). In Chapter 7, we showed that an applied force $\vec{F}$ produces a change in the linear momentum $\vec{p}$ of an object. This leads to the *impulse theorem*, involving force and impulse (Eq. 7.5):

$$\vec{F}\,\Delta t = \Delta \vec{p}$$

An analogous result is true for rotational motion: an applied torque produces a change in the angular momentum. We have

$$\vec{\tau}\,\Delta t = \Delta \vec{L} \qquad (9.24)$$

so if a constant torque $\vec{\tau}$ acts for a time Δt, there is a change $\Delta \vec{L}$ in the angular momentum. Applying this result to the wheel plus axle in Figure 9.18, we see that the torque due to the gravitational force is perpendicular to the rotation axis of the wheel (the axle), which is also the direction of the angular momentum vector $\vec{L}$. This torque causes the wheel's angular momentum to turn in the direction of $\vec{\tau}$ as indicated in the figure. However, as the rotation axis of the wheel turns, the direction of the torque also changes. In fact, $\vec{\tau}$ and hence also $\Delta \vec{L}$ from Equation 9.24 are *always* perpendicular to $\vec{L}$. As a result, the wheel plus axle rotate continuously about a vertical rotation axis (the pole in Fig. 9.18). This movement is called *precession* and is an example of the type of motion that can occur when the angular momentum and the applied torque are not parallel.

9.6 | CATS AND OTHER ROTATING OBJECTS

In this section, we discuss two interesting and well-known examples of rotational motion based on the principle of conservation of angular momentum.

Rotating Cats

A widely known piece of folklore is the claim that a falling cat always lands on its feet. Although we certainly do not want to encourage the unethical treatment of animals, you have probably observed that a falling cat is usually able to rotate its body so as to land feetfirst, even if it begins its fall with its feet pointing upward. This example of rotational motion is shown in Figure 9.19 and in the opening photo for this chapter. The cat in Figure 9.19A is not rotating when it begins its fall, so its initial angular momentum is zero. Since there are no external torques on the cat while it is falling, its angular momentum must stay constant and hence is always zero during the course of its fall. How can the cat manage to rotate even though it has no angular momentum?

Cats accomplish this feat by changing their shape while they are falling. To understand this process, consider separately the motion of the cat's front section (near its head) and its hindquarters (near its tail). In Figure 9.19B, the cat has pulled its front legs close to its body; just as for the figure skater in Figure 9.10, this reduces

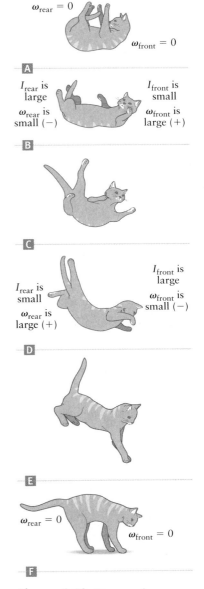

$\omega_{\text{rear}} = 0$

$\omega_{\text{front}} = 0$

A

I_{rear} is large

I_{front} is small

ω_{rear} is small $(-)$

ω_{front} is large $(+)$

B

C

I_{rear} is small

I_{front} is large

ω_{rear} is large $(+)$

ω_{front} is small $(-)$

D

E

$\omega_{\text{rear}} = 0$

$\omega_{\text{front}} = 0$

F

Figure 9.19 The angular momentum of this cat is conserved as it falls to the ground. The cat is still able to rotate its body so as to land safely, however.

the moment of inertia of the cat's front section (I_{front} is small). At the same time, the cat's tail and rear legs are extended away from its body, so the moment of inertia of the rear part of its body is large (I_{rear} is large). The cat then "swivels" at its middle as shown in Figure 9.19C, with the front half of the body rotating in one direction and the hindquarters in the *opposite* direction; the total angular momentum is still zero. Because the front and rear moments of inertia are different, the angular velocities of the front and rear parts of the cat's body are different. Since $I_{rear} > I_{front}$, we know from the example with the skater that $\omega_{rear} < \omega_{front}$. As a result, the rotation angle for the front half is much larger than for the hindquarters. Figure 9.19D shows the cat's upper body rotating by approximately 180°, whereas the lower body has rotated much less than this amount.

The cat then repeats the same process in Figure 9.19E. Here the cat pulls its hind legs and tail close to its body while extending its front legs. Then $I_{rear} < I_{front}$, and the cat can now swivel at the middle with $\omega_{rear} > \omega_{front}$. This swivel allows the cat to rotate its hindquarters through approximately 180° as in Figure 9.19E while rotating the front part of its body very little. The cat is then able to extend its legs and land safely in Figure 9.19F.

The point of this example is not that cats have a keen understanding of angular momentum, but rather that a system whose total angular momentum is zero can still rotate in some interesting ways. All we needed to explain this cat's motion was the concept that angular momentum is the product of the moment of inertia and the angular velocity.

EXAMPLE 9.6 | An Astronaut's Treadmill

An astronaut is working on a circular space station as sketched in Figure 9.20. The space station is initially at rest (i.e., not rotating). The astronaut then begins his daily exercise in which he runs laps around the station with a linear speed of $v_{astro} = 3.0$ m/s; that is the speed that would be measured by a stationary observer outside the station and is directed *tangentially*, along the edge of the station. Find the resulting angular velocity of the station. The mass of the astronaut is 80 kg, the mass of the station is 1.0×10^5 kg, and (for simplicity) assume the station has the shape of a simple disk, with a radius $R = 30$ m.

RECOGNIZE THE PRINCIPLE

The astronaut and the space station form a system. There are no external torques, so its angular momentum must be conserved. Just as with the cat in Figure 9.19, if part of the system starts to rotate in one direction (i.e., the astronaut runs in one direction), another part of the system must rotate in the opposite direction (the space station). We can use the principle of conservation of angular momentum to find how the angular velocities of the astronaut and the station are related.

SKETCH THE PROBLEM

Figure 9.20 shows the problem. The astronaut and station rotate in opposite directions.

IDENTIFY THE RELATIONSHIPS

The station and astronaut are both initially at rest relative to the observer in Figure 9.20, so the system's total angular momentum is zero throughout the astronaut's workout. Let $\omega_{station}$ be the angular velocity of the station when the astronaut is running. As he runs, the astronaut's motion is similar to that of the child on (or near) the edge of the merry-go-round in Figure 9.12 (compare also with Fig. 9.9). We can therefore find the astronaut's angular momentum L_{astro} from Equation 9.16, which gives

$$L_{astro} = m_{astro}Rv_{astro}$$

The total angular momentum of the astronaut plus the station is then

$$L_{total} = L_{astro} + L_{station} = m_{astro}Rv_{astro} + I_{station}\omega_{station} = 0 \qquad (1)$$

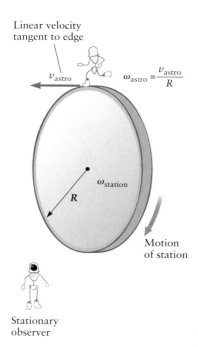

Linear velocity tangent to edge

v_{astro}

$\omega_{astro} = \dfrac{v_{astro}}{R}$

$\omega_{station}$

R

Motion of station

Stationary observer

Figure 9.20 Example 9.6.

We approximate the station as a disk, so its moment of inertia (from Table 8.2) is

$$I_{station} = I_{disk} = \tfrac{1}{2}m_{station}R^2$$

SOLVE

Inserting this result into Equation (1) leads to

$$0 = m_{astro}Rv_{astro} + \tfrac{1}{2}m_{station}R^2\omega_{station}$$

$$\omega_{station} = -\frac{m_{astro}Rv_{astro}}{m_{station}R^2/2} = -\frac{2m_{astro}v_{astro}}{m_{station}R}$$

Inserting the values of the various quantities gives

$$\omega_{station} = -\frac{2(80 \text{ kg})(3.0 \text{ m/s})}{(1.0 \times 10^5 \text{ kg})(30 \text{ m})} = \boxed{-1.6 \times 10^{-4} \text{ rad/s}}$$

What does it mean?
The angular velocity of the astronaut is

$$\omega_{astro} = \frac{v_{astro}}{R} = \frac{3.0 \text{ m/s}}{30 \text{ m}} = +0.10 \text{ rad/s}$$

The angular velocity of the station is much smaller because the station is much more massive than the astronaut and has a larger moment of inertia. Notice also that $\omega_{station}$ is negative because the station rotates in the opposite direction as the astronaut.

CONCEPT CHECK 9.5 | Rotational Motion When the Total Angular Momentum Is Zero

Explain what aspects of the motion in Example 9.6 are similar to the rotating cat in Figure 9.19.

Angular Momentum and Motorcycles

As a final example of angular momentum conservation, we consider the motion of a motorcycle stunt rider who jumps through the air as shown in Figure 9.21. Let's examine the combined translational and rotational motions of the system consisting of the motorcycle plus the rider. In both of the jumps shown in Figure 9.21, the system's center of mass moves as a projectile, following a parabolic trajectory as we established in Chapter 4. The rotational motions are different in the two cases, however. In Figure 9.21A, the motorcycle frame maintains a fixed angular orientation throughout the jump, and the motorcycle lands on its rear wheel. Throughout the jump, the angular velocity of the system (the motorcycle and rider) is zero, so the angular displacement of the motorcycle frame is also zero. This example shows conservation of angular momentum: the angular momentum of the entire system, and also of the motorcycle frame, is both zero and constant throughout the jump.

In the jump shown in Figure 9.21B, the motorcycle begins with zero angular velocity, but then *during the jump* the motorcycle frame rotates. This rotation allows the motorcycle to land simultaneously on both wheels, with the motorcycle frame "sloping down," leading to a gentler (and presumably safer) landing. How did the rider cause the angular velocity of the motorcycle frame to change from its value of zero at the start of the jump to a nonzero value during the jump?

The answer to this question is well known to stunt riders: during the jump, they adjust the throttle so as to change the angular velocity of the rear wheel as it rotates about its axle, thus changing the wheel's angular momentum. There are no external torques on the system, so the total angular momentum of the system must remain constant. The total angular momentum is equal to the sum of the angular momentum

Figure 9.21 ▣ In this jump, the orientation of the motorcycle frame changes very little during the course of the jump. ▣ Here the rider reduced the angular velocity of the rear wheel after leaving the ground (by adjusting the throttle). To conserve angular momentum about the axis that runs along the rear axle, the frame then began to rotate counterclockwise.

START   FINISH

A

ω_{frame}

ω_{wheel}

Rotation axis

If ω_{wheel} decreases, then ω_{frame} must increase.

START   FINISH

B

Figure 9.22 The total angular momentum of a motorcycle is conserved during a jump. The total angular momentum has contributions from the wheels and the frame.

of the wheels plus the angular momentum of the frame plus the angular momentum of the rider. In this case, we take the rotation axis to be the rear axle, which is perpendicular to the plane of the figure (see Fig. 9.22). If the angular momentum of the rear wheel is decreased (by adjusting the throttle), the remaining angular momentum must increase. This causes the frame to rotate clockwise about the rear axle as in Figures 9.21B and 9.22, and is another nice application of physics!

SUMMARY | Chapter 9

KEY CONCEPTS AND PRINCIPLES

Rotational kinetic energy
The *rotational kinetic energy* of an object is

$$KE_{rot} = \tfrac{1}{2}I\omega^2 \qquad \text{(9.4) (page 280)}$$

where I is the moment of inertia. This is the kinetic energy associated with just the rotational motion. The total kinetic energy of an object is the sum of this rotational kinetic energy plus the kinetic energy associated with the translational motion:

$KE_{rot} = \tfrac{1}{2}I\omega^2$

I

$$KE_{total} = \overbrace{\tfrac{1}{2}mv_{CM}^2}^{\text{translational } KE} + \overbrace{\tfrac{1}{2}I\omega^2}^{\text{rotational } KE} \qquad \text{(9.5) (page 281)}$$

If the rotation axis passes through the center of mass, then v_{CM} in Equation 9.5 is the center of mass velocity of the object.

(Continued)

Work–energy theorem for rotational motion

The work done by an applied torque is

$$W = \tau\theta \qquad \text{(9.8) (page 283)}$$

where θ is the angular displacement. The work–energy theorem applied to rotational motion is

$$W = \tau\theta = \Delta KE = \tfrac{1}{2}I\omega_f^2 - \tfrac{1}{2}I\omega_i^2 \qquad \text{(9.9) (page 283)}$$

Angular momentum

A rotating object has an **angular momentum** L. If the direction of the rotation axis is fixed, L is given by

$$L = I\omega \qquad \text{(9.15) (page 287)}$$

If the external torque on a system is zero, the total angular momentum of the system is conserved.

Vector nature of angular momentum

The direction of the angular velocity $\vec{\omega}$ is given by the **right-hand rule**, and the angular momentum vector $\vec{L}$ is

$$\vec{L} = I\vec{\omega} \qquad \text{(9.23) (page 295)}$$

APPLICATIONS

Conservation of energy

Conservation of energy principles can be applied to rotational motion. When all the forces that do work on an object are conservative forces, the total mechanical energy is conserved. When the potential energy is due to gravity, as for an object that rolls up or down an incline, conservation of mechanical energy leads to

$$KE_i + PE_i = KE_f + PE_f$$

$$\tfrac{1}{2}mv_i^2 + \tfrac{1}{2}I\omega_i^2 + mgh_i = \tfrac{1}{2}mv_f^2 + \tfrac{1}{2}I\omega_f^2 + mgh_f \quad \text{(9.11) (page 284)}$$

The kinetic energy given here is the total kinetic energy, including the rotational contribution.

Conservation of angular momentum

Angular momentum and its conservation play important roles in the motion of many objects, including figure skaters, cats, and motorcycles.

QUESTIONS

SSM = answer in Student Companion & Problem-Solving Guide $\bigotimes$ = life science application

1. Two balls (solid spheres) have the same radius and the same rotational kinetic energy. (a) If the ratio of their masses is 3:1, what is the ratio of their angular velocities? (b) What is the ratio of their angular momenta?

2. A pitcher throws a baseball of mass $m = 0.14$ kg with a linear speed v and an angular velocity ω. Which of the following statements is correct? Explain your answer. (a) The total kinetic energy of the baseball can be calculated knowing only v. (b) The total kinetic energy of the baseball can be calculated knowing only ω. (c) To calculate the total kinetic energy of the baseball, we need to know both v and ω.

3. When a spinning figure skater pulls her arms close to her body (Fig. 9.10), she spins faster. If she then pushes her arms away from her body, her angular velocity decreases. Her rotational kinetic energy also decreases. Explain why her kinetic energy decreases. What force or torque does the corresponding work?

4. [SSM] Consider the translational and rotational kinetic energies of a disc that rolls without slipping. Show that the ratio of these energies is independent of the size (the radius) of the wheel.

5. You are given two objects and a ramp. Explain how you could measure the ratio of the moments of inertia of the two objects. Assume the two objects are both round (so that both roll), have the same mass and radius, and are solid (no holes in the center, etc.). Consider only objects that meet these criteria and with the shapes in Table 8.2.

6. Devise a way to make a ball that rolls yet has no angular momentum. *Hint*: Consider having some moving parts inside the ball.

7. Tightrope walkers often carry a long pole (Fig. Q9.7). Explain why.

Figure Q9.7

© Associated Press/AP Images

8. Is it possible for the total force on a system to be nonzero but the torque to be zero? Give an example.

9. You are given the job of designing a new high-tech skateboard. For it to travel as fast as possible, you want to minimize the kinetic energy in its wheels. All else being equal (i.e., the mass and radius of the wheels), is it better to make the wheels in the shape of a disk, a sphere, or a conventional wheel with spokes?

10. Explain why the rim brakes on a bicycle wheel are located at the outer edge of the wheel instead of near the hub. *Hint*: Consider the work–energy theorem (Eq. 9.9) and how the magnitude of the torque produced by a given force varies when the force is applied at different distances from the axis of rotation.

11. The flywheel within a car engine is a form of gyroscope, and when rotating at high rpm, it carries significant angular momentum. (A car flywheel is just a disk attached to the engine's main rotating shaft.) In most cases, the engine of a race car is mounted such that the axis of rotation points in the forward direction and $\vec{L}$ is in the forward direction. Because angular momentum (magnitude and direction) is conserved, what happens to the car as it rounds the end of an oval track in a counterclockwise direction (turning left)? About what axis will the car tend to twist? If instead the engine were mounted such that the axis of rotation was parallel to the axles of the car, along which axis would the car then tend to twist? *Hint*: Consider the gyroscope and the rotation of precession. In auto racing, the cars typically go around the oval track in a counterclockwise direction as viewed from above. Would race cars be engineered differently if races were conducted in a clockwise direction?

12. Discuss how and if the moment inertia of the Earth associated with its orbital motion around the Sun changes as the Earth moves through its orbit. How are these changes connected to changes in the Earth's speed?

13. Many kilograms of asteroids land on the Earth each year. Discuss how they affect the length of a day.

14. Consider a child who is jumping on a trampoline as shown in Figure Q9.14. Explain how the child can have zero angular momentum, but still rotate to either the right or the left while she is in the air.

Figure Q9.14

© Inspirestock Inc./Alamy

15. The cat in Figure 9.19 undergoes free fall due to the force of gravity. Explain why we did not consider gravity when we discussed how the cat in Figure 9.19 managed to land on its feet.

16. The barrel of a rifle contains spiral grooves that impart some spin to a bullet as it leaves the barrel. Explain why this spin improves the rifle's accuracy.

17. [SSM] (a) Why do most small helicopters have a rotor on their tail (Fig. Q9.17)? (b) Many large helicopters have two large rotors. Why do these helicopters not need a tail rotor? Do you think the two large propellers rotate in the same direction? Explain.

Figure Q9.17

U.S. Navy/Mass Communication Specialist Seaman Jon Dasbach

18. ⊗ Why is it possible for a gymnast to perform more rotations during an airborne somersault when she is in a tuck position than when her body is straight (called the layout position)?

19. Some early spacecrafts used tape recorders to store data. These recorders have spools of magnetic tape that are rotated in spindles to move the tape past "heads" that record or read data. Explain why a tape recorder can affect a spacecraft's motion.

20. Give an example of an object, or collection of several objects, that has a nonzero rotational kinetic energy, but zero angular momentum.

PROBLEMS

[SSM] = solution in Student Companion & Problem-Solving Guide   ⊗ = life science application
★ = intermediate   ⊗ = challenging   ℝ = reasoning and relationships problem

9.1 KINETIC ENERGY OF ROTATION

1. A baseball (mass 0.14 kg, radius 3.7 cm) is spinning with an angular velocity of 60 rad/s. What is its rotational kinetic energy?

2. If the baseball in Problem 1 has a linear velocity of 45 m/s, what is the ratio of its rotational and translational kinetic energies?

3. Consider a quarter rolling down an incline. What fraction of its kinetic energy is associated with its rotational motion? Assume it rolls without slipping.

4. ★ ℝ Estimate the maximum rotational kinetic energy of a yo-yo. Use a rotation axis that passes through the yo-yo's center.

5. ★ A wheel of mass 0.50 kg and radius 45 cm is spinning with an angular velocity of 20 rad/s. You then push your hand against the edge of the wheel, exerting a force F on the wheel as shown in Figure P9.5. If the wheel comes to a stop after traveling 1/4 of a turn, what is F?

Figure P9.5

6. ✷ Ⓡ It has been proposed that large flywheels could be used to store energy. Consider a flywheel made of concrete (density 2300 kg/m³) in the shape of a solid disk, with a radius of 10 m and a thickness of 2.0 m. If its rotational kinetic energy is 100 MJ (1.0×10^8 J), what is the angular velocity of the flywheel?

7. ✷ Ⓡ Consider a wheel on a racing bicycle. If the wheel has a mass of 0.40 kg and an angular speed of 15 rad/s, what is the rotational kinetic energy of the wheel?

8. ✷ Ⓡ An automobile wheel has a mass of 18 kg and a diameter of 0.40 m. What is the total kinetic energy of one wheel when the car is traveling at 20 m/s?

9. ✷ Ⓡ Estimate the rotational kinetic energy of an airplane propeller.

10. ✪ Ⓡ Estimate the kinetic energy of the spinning top in Figure P9.10. Assume it is made of wood and is rotating with $\omega = 30$ rad/s.

11. ✷ Ⓡ Estimate the fraction of a bicycle's total kinetic energy associated with the rotational motion of the wheels.

12. SSM ✷ What is the rotational kinetic energy of the Earth as it spins about its axis?

$h = 3.0$ cm

Figure P9.10

13. ✪ What is the ratio of the rotational kinetic energy of the Earth to the rotational kinetic energy of the Moon as they spin about their axes?

9.2 CONSERVATION OF ENERGY AND ROTATIONAL MOTION

14. Two crates of mass $m_1 = 15$ kg and $m_2 = 9.0$ kg are connected by a rope that passes over a frictionless pulley of mass $m_p = 8.0$ kg and radius 0.20 m as shown in Figure P9.14. (a) What force(s) can do work on the masses and the pulley? Identify a system whose mechanical energy will be conserved. (b) The crates are released from rest. Crate 1 falls a distance of 2.0 m, at which time its speed is v_f. Make a sketch showing the initial and final states of the system. (c) What are the total initial and final kinetic energies of the system? Express your answers in terms of v_f. (d) What are the initial and final potential energies of the system? (e) What is the final speed v_f of the crates? (f) What is the final angular velocity of the pulley?

m_{pulley}

m_1

m_2

Figure P9.14 Problems 14, 16, and 17.

15. A marble (radius 1.0 cm and mass 8.0 g) rolls without slipping down a ramp of vertical height 20 cm. What is the speed of the marble when it reaches the bottom of the ramp?

16. ✷ For the system of two crates and a pulley in Figure P9.14, what fraction of the total kinetic energy resides in the pulley?

17. ✪ Consider again the system in Figure P9.14. Use the result from Problem 9.14 to find the acceleration of one of the crates. *Hint*: Is the acceleration constant?

18. Consider a hoop of mass 3.0 kg and radius 0.50 m that rolls without slipping down an incline. The hoop starts at rest from

a height $h = 2.5$ m above the bottom of the incline and then rolls to the bottom. (a) What forces act on the hoop? Which of these forces do work on the hoop? (b) Is the mechanical energy of the hoop conserved? (c) Make a sketch showing the hoop at the top and the bottom of the incline. (d) What is the total initial mechanical energy of the hoop? Take the zero of potential energy to be at the bottom of the incline. (e) If the hoop's speed at the bottom of the incline is v_f, what is its mechanical energy? Express your answer in terms of v_f. (f) Solve for v_f.

19. A marble rolls on the track shown in Figure P9.19, with $h_B = 25$ cm and $h_C = 15$ cm. If the marble has a speed of 2.0 m/s at point A, what is its speed at points B and C?

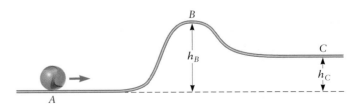

Figure P9.19

20. ✷ A yo-yo falls through a distance of 0.50 m. If it starts from rest, what is its speed?

21. ✪ A sphere rolls down the loop-the-loop track shown in Figure P9.21, starting from rest at a height h above the bottom. The ball travels around the inside of the circular portion of the track (radius $r = 5.0$ m). Notice that the ball is upside down when it reaches the top of the circular part of the track. It is found that the ball is just barely able to travel around the inside of the track without losing contact at the top. Find h.

Figure P9.21

22. A metal rod of length $L = 3.0$ m is free to rotate about a frictionless hinge at one end (Fig. P9.22). The rod is initially at rest and oriented horizontally and then released so that it swings to the bottom. (a) What are all the forces acting on the rod? Which force or forces do work on the rod as it swings? (b) Is the mechanical energy of the rod conserved? (c) Make a sketch showing the rod in its initial orientation and when it is at the bottom of its swing. (d) What is the initial mechanical energy of the rod? Choose the zero of potential energy so that the final PE is zero. (e) If the speed of the end of the rod when it is at its lowest point is v_f, what is the final mechanical energy of the rod? Express your answer in terms of v_f. (f) Find v_f.

L

Hinge (pivot point)

Figure P9.22 Problems 22, 23, and 25.

23. ✷ Consider the rod in Figure P9.22, but now assume it starts from rest in the "up" position (i.e., pointing vertically upward from the hinge). If the rod is then given a very small push, what is the linear speed of the end of the rod when it reaches the bottom of its circular path?

24. A ball rolls down the track shown in Figure P9.24, starting from rest at a height $h_1 = 20$ m as shown. The ball is traveling horizontally when it leaves the bottom of the track, which has a height $h_2 = 9.0$ m. Find where the ball hits the ground; that is, find L.

Figure P9.24

25. ✪ Repeat Problem 9.22(f) (that is, find v_f), but now assume the rod is initially oriented an angle of 30° above the horizontal.

26. ✪ A pencil (mass 10 g and length 15 cm) is initially balanced so that it is sitting vertically on a flat table as shown in Figure P9.26. If the pencil then falls, what is its angular velocity just before it strikes the tabletop? Assume the mass of the pencil is distributed uniformly and the end of the pencil in contact with the table does not slip.

Figure P9.26
Problems 26, 27, and 36.

27. ✪ Suppose the table in Figure P9.26 is extremely slippery so that there is no friction between the table and the end of the pencil. As a result, the pencil slips as it tips over. Let $\omega_{\text{no friction}}$ denote the angular velocity of the pencil just before it hits the table, and ω_{friction} denote the angular velocity found in Problem 26. What is the ratio $\omega_{\text{no friction}}/\omega_{\text{friction}}$?

28. ✪ A rod of mass 4.0 kg and length 1.5 m hangs from a hinge as shown in Figure P9.28. The end of the rod is then given a "kick" so that it is moving at a speed of 5 m/s. How high will the rod swing? Express your answer in terms of the angle the rod makes with the vertical.

Hinge

Steel rod

$v_i = 5.0$ m/s

Figure P9.28

29. A cylinder of mass 7.0 kg and radius 0.25 m rolls along a level floor. Its center of mass has a speed of 1.5 m/s. Find (a) the kinetic energy of translation and (b) the kinetic energy of rotation.

30. A bowling ball rolls up an incline as shown in Figure P9.30. If the ball has a speed of 12 m/s at the base of the incline, to what height h on the incline will the bottom edge of the ball reach?

$h = ?$

Figure P9.30

31. SSM ✪ A bucket filled with dirt of mass 20 kg is suspended by a rope that hangs over a pulley of mass 30 kg and radius 0.25 m (Fig. P9.31). Everything is initially at rest, but someone is careless and lets go of the pulley and the bucket then begins to move downward. What is the speed of the bucket when it has fallen a distance of 2.5 m?

32. ✪ ® A bicycle is moving at a speed of 15 m/s when the rider applies her brakes. The bicycle comes to a stop after moving forward another 20 m. Estimate the force of one of the bicycle brake pads on the rim of a wheel. Be sure to include that there are four brake pads (two for each wheel).

Figure P9.31

33. ✪ Two solid spheres, each of mass m, are rolling with a speed v. If the spheres have radii R_1 and R_2 with $R_1/R_2 = 3.0$, what is the ratio of their total kinetic energies?

9.3 ANGULAR MOMENTUM

34. ✪ ® Estimate the angular momentum of a CD that is playing your favorite music.

35. A bowling ball has a mass of 6.0 kg and a radius of 22 cm. If it is rolling down a lane with a speed of 9.0 m/s, what is the angular momentum of the ball?

36. ✪ Consider the pencil in Problem 26. What is the angular momentum of the pencil just before it strikes the table?

37. ✪ ® Estimate the angular momentum of a figure skater of mass 50 kg who is spinning at 300 rpm (Fig. 9.10). Assume her arms and legs are pulled in very close to her rotation axis.

38. A child ($m = 40$ kg) is playing on a merry-go-round ($m = 200$ kg, $R = 2.0$ m) that is initially at rest. The child then jumps off in a direction tangent to the edge of the merry-go-round as shown in Figure P9.38. The child has a speed of 5.0 m/s just before she lands on the ground. (a) Identify a system for which the total external torque is zero. (b) What is the initial angular momentum of your system? (c) Make a sketch showing your system just before and just after the child jumps off the merry-go-round. (d) If the final angular velocity of the merry-go-round is ω_f, what is the final angular momentum of the merry-go-round and of the child? Express your answers in terms of ω_f. (e) What is the final angular velocity of the merry-go-round?

$\vec{v}$

R

Figure P9.38

39. ✪ A puck (mass $m_1 = 0.50$ kg) slides on a frictionless table as shown in Figure P9.39. The puck is tied to a string than runs through a hole in the table and is attached to a mass $m_2 = 1.5$ kg. The mass m_2 is initially at a height of $h = 1.5$ m above the floor with the puck traveling in a circle of radius $r = 0.40$ m with a speed of 1.5 m/s. The force of gravity then causes mass m_2 to move downward a distance 0.15 m. (a) What is the new speed of the puck? (b) What is the change in the kinetic energy of the puck?

Figure P9.39

40. ✪ ® A figure skater begins a spin at an angular velocity of 200 rpm with her arms and legs out away from her body (Fig. 9.10A). She then pulls her arms and legs in close to her body, and her angular velocity increases. Estimate her final angular velocity.

41. A bug of mass 3.0 g is sitting at the edge of a CD of radius 8.0 cm. If the CD is spinning at 300 rpm, what is the angular momentum of the bug?

42. A child of mass 50 kg jumps onto the edge of a merry-go-round of mass 150 kg and radius 2.0 m that is initially at rest as sketched in Figure 9.11. While in the air (during her jump), the child's linear velocity in the direction tangent to the edge of the merry-go-round is 10 m/s. What is the angular velocity of the merry-go-round plus child after the child jumps onto the merry-go-round?

43. [SSM] [✪] [®] Consider a person who is sitting on a friction-less rotating stool as in Figure P9.43. The person initially has his arms outstretched and is rotating with an angular speed of 5.0 rad/s. He then pulls his arms close to his body. (a) Estimate his final angular speed. (b) Estimate the kinetic energy before and after the person pulls his arms into his body.

ω_i ω_f

Initial position   **Final position**

Figure P9.43 Problems 43 and 44.

44. [✪] [®] Repeat Problem 43, but now assume the person is holding a 5.0-kg mass in each hand.

45. [✪] [®] Estimate the angular momentum of an airplane propeller.

46. Two disks are located on an axle as shown in Figure P9.46. The lower disk is initially spinning at 50 rad/s, and the upper one is not spinning. The upper disk then falls onto the lower disk, and they stick together. (a) Is there a torque on the lower disk? On the upper disk? (b) Identify a system for which the external torque is zero so that the angular momentum is conserved. (c) What is the initial angular momentum of your system? (d) What is the final angular momentum of your system? Express your answer in terms of ω_f, the final angular velocity of the two disks. (e) Find ω_f.

$m_2 = 7.0$ kg
$R_2 = 1.0$ m

$m_1 = 20$ kg
$R_1 = 2.0$ m

$\omega_f = ?$

Initial position   **Final position**

Figure P9.46

47. A particle of mass 3.0 kg moves with a horizontal velocity of 20 m/s as shown in Figure P9.47. (a) What is the angular momentum of the particle about an axis that runs through point P and is directed out of the page? (b) What is L about an axis that runs through Q? (c) Explain why the answers to parts (a) and (b) are the same (or different).

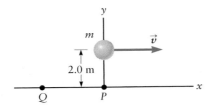

y

m $\vec{v}$

2.0 m

Q P x

Figure P9.47 Problems 47 and 53.

48. [SSM] [✪] A meteor with a volume of 1.0 km³ strikes the Earth at the equator as shown in Figure P9.48, and all the fragments stick to the surface. (a) What is the magnitude of the change in

the angular momentum of the Earth? (b) What is the change in the length of the day? *Hint*: Assume the average density of the asteroid is 1.0×10^4 kg/m³.

49. A CD (radius 6.0 cm) is spinning freely with an angular velocity of 400 rpm when a bug drops onto the CD a distance 4.5 cm from the center. If the CD slows to 300 rpm, what is the ratio of the bug's mass to the mass of the CD? (Ignore the effect of the hole in the center of the CD.)

N
Earth
S
Meteor

Figure P9.48

9.4 ANGULAR MOMENTUM AND KEPLER'S SECOND LAW OF PLANETARY MOTION

50. [SSM] Halley's comet moves about the Sun in a highly elliptical orbit. At its closest approach, it is a distance of 8.9×10^{10} m from the Sun and has a speed of 54 km/s. When it is farthest from the Sun, the two are separated by 5.3×10^{12} m. Find the comet's speed at that point in its orbit.

51. [®] Estimate the angular momentum associated with the Moon's orbital motion around the Earth.

52. What is the ratio of the angular momentum of the Moon due to its orbital motion around the Earth to the Earth's orbital angular momentum around the Sun?

9.5 THE VECTOR NATURE OF ROTATIONAL MOTION: GYROSCOPES

53. For the particle in Problem 47, what is the direction of the angular velocity of the particle?

54. Use the right-hand rule to find the direction of the orbital angular momentum of the Earth as it orbits the Sun as sketched in Figure 9.17.

55. Consider the rotational motion of the Earth as it spins about the axis that runs from the North Pole to the South Pole. Find the magnitude and direction of the angular momentum $\vec{L}$ associated with the spinning Earth. Assume the Earth is a sphere of uniform density.

56. [SSM] [✪] A ball of mass 3.0 kg is tied to a string of length 2.0 m. The other end of the string is fastened to a ceiling, and the ball is set into circular motion as shown in Figure P9.56. If $\theta = 30°$, what is the magnitude of the angular momentum of the ball with respect to the vertical axis that passes through the center of the circle (shown dashed in the figure)?

θ

Figure P9.56

57. Consider the rotational motion of the wheels of a car when the car is in motion. Use a drawing to show the direction of the angular momentum of each wheel.

58. Consider the yo-yo in Figure 9.8. What is the direction of the angular momentum vector of the yo-yo?

9.6 CATS AND OTHER ROTATING OBJECTS

59. [SSM] [✪] A child of mass 35 kg stands at the edge of a merry-go-round of mass 140 kg and radius 2.5 m, and both are initially at rest. The child then walks along the edge of the merry-go-round until she reaches a point opposite her starting point as measured on the ground (point B in Fig. P9.59). How far does the child walk as measured relative to the merry-go-round?

60. [✪] A boy of mass m_{boy} stands at the edge of a merry-go-round of radius $R = 3.5$ m and mass m_{mgr}, and both are initially at rest. The boy then walks along the edge of the merry-go-round.

Start

Finish

B •   •   • A

Figure P9.59

Figure P9.61

After walking a distance of 21 m relative to the merry-go-round, the boy finds that the merry-go-round has rotated through an angle of 20°. Find the ratio of m_{boy} to m_{mgr}.

61. ⚡ ✗ Ⓡ Consider the rotating diver shown in Figure P9.61. Suppose she has an angular velocity of 60 rpm = 6.3 rad/s when she has her arms and legs pulled close to her body as shown (called the tuck position). Estimate her angular velocity just before she enters the water, when her arms and legs are fully extended.

ADDITIONAL PROBLEMS

62. ⚡ The flywheel of an automobile engine is a solid disk with a mass of 8.7 kg and a radius of 22 cm and is rotating at 4000 rpm. If the entire rotational kinetic energy of the flywheel were converted to gravitational potential energy, to what height would the flywheel be lifted? Assume the flywheel is a uniform disk.

63. ⚡ **Good approximation!** Often when we apply equations to calculate moment of inertia, we are making a useful approximation, such as when we approximate a wagon wheel as a disk without a hole in the center. A manufactured item exists that is extraordinarily spherical, where the approximation of a uniform sphere is astonishingly close. Gravity Probe B is a satellite-borne experiment that uses ultraprecise gyroscopes to detect the warping of space–time due to the rotation of Earth's gravitational field as predicted by Einstein. The gyroscopes are spheres made out of fused quartz (density of 2.2 g/cm³), 38 mm in diameter, and are so spherical that if they were scaled to the size of the Earth, the difference between hills and valleys would be less than 4 m! The gyroscope operates at 10,000 rpm. (a) What is the kinetic energy of the gyroscope when it is turned on? (b) What is the angular momentum of the gyroscope?

64. SSM ⚡ **Better approximation!** There also exists an item in nature that is extraordinarily spherical, where the approximation of a uniform sphere is astonishingly close. A pulsar is a rotating neutron star, the remnant of a supernova of a star between six and eight solar masses. A typical pulsar rotates at a rate of 600 rpm, has an average mass of twice that of the Sun, and is only 20 km in diameter, or about as big across as Washington, D.C. (a) What is the rotational kinetic energy of an average pulsar? (b) What is the corresponding angular momentum?

65. ✪ A model of solar system formation is that the Sun condensed from a disk of gas and dust through mutual gravitational attraction (Fig. P9.65). Consider a uniform disk of gas and dust that starts with the mass of the Sun and has a radius equal to that of Pluto's mean orbital radius. The theory is that this disk had an initial tiny rotation ($T = 250,000$ years), and collapsed to the size of the Sun. (The mass of all planets and other objects in the solar system add up to less than 0.2% of the mass of the Sun.) (a) Assuming we can ignore the mass of the planets (all the mass in the disk collapses to the solid spherical Sun), what period of rotation for the Sun is predicted by conservation of angular momentum? (b) This model turns out to be faulty; the

Figure P9.65

Sun rotates about once in every 27 days. Repeat your conservation of angular momentum calculation, this time including the angular momentum of Jupiter and Saturn at their appropriate orbital radii. What is the new predicted period of rotation for the Sun?

66. ✪ The mechanical governor shown in Figure P9.66 was invented by James Watt (Chapter 6) to regulate and limit, through feedback, a steam engine's maximum speed. The device consists of two spherical masses connected via lightweight metal arms that can simultaneously rotate about a shaft and pivot outward from a hinge at the top, allowing, depending on

Figure P9.66

the rotation rate, the masses to rise. Two more rods connect the arms to a collar that can slide up and down the shaft, controlling a lever that can be connected to a switch or valve that regulates the flow of fuel to the engine. In this case, the governor is designed to cut off power when the shaft rotates at 150 rpm (or higher). (a) If the lever activates a switch when the mass arm makes an angle of $\phi = 75°$ with respect to the shaft, what length L should the connecting rod have to make the governor work as planned? Assume the connecting rods are made of very lightweight material. (b) If each mass is 250 g, approximately how much rotational kinetic energy does the governor have at the time of cutoff? (c) What is the angular momentum of the governor at this time? How would weights of larger mass affect the governor's response time to changes in rotational speed?

67. ✪ A large iron cylinder of mass 270 kg rests so that its circular face rolls along the edge of a table as shown in Figure P9.67. A smaller disk (radius 11 cm) is fastened to the circular face with its center axis aligned with the large cylinder, such that the small disk extends beyond the edge of the table. A cable is wrapped around the smaller disk, which has half the radius and one-fourth the mass of the cylinder. With the entire assembly at rest, a mass of 2.3 kg is attached to the free end of the cable and allowed to descend through a distance $h = 35$ cm. (a) Calculate the moment of inertia and the total mass of the entire assembly. (b) What is the total kinetic energy of the system after the attached mass descends through a height h? (c) If the entire assembly rolls without slipping, what is its velocity at this time? (d) How far has the system moved along the horizontal?

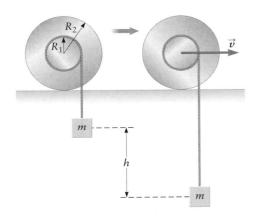

Figure P9.67

68. ✪ **Skee Ball amusement.** The popular arcade sport of Skee Ball involves bowling a wooden ball 3 in. (7.6 cm) in diameter down an alley 9 ft long, where the ball is then launched off a short ramp as shown in Figure P9.68. Highest points are awarded to those who can launch the ball such that it lands in the topmost cylindrical hole; landing in lower and larger holes is worth

Figure P9.68

fewer points. The launch ramp is 8.5 cm high and slopes at an angle of $\theta = 40°$. From the top of the launch ramp, the location of the score cups for the 30-point hole and 50-point hole are $(x_{30}, y_{30}) = (25$ cm, 6.0 cm$)$ and $(x_{50}, y_{50}) = (42$ cm, 12 cm$)$, respectively, as shown in Figure P9.68. (a) If the ball rolls without slipping, what initial horizontal velocity must the ball have to make it into the 30-point hole? (b) An expert makes a 50-point score with one ball. What were the initial linear and angular velocities of the ball? (c) Is it likely that the ball actually does not slip while it rolls?

69. ✪ ✪ **Extinction.** Sixty-seven million years ago, a large asteroid impacted Earth. It is widely accepted in the field of paleontology that this impact played a large role in a mass extinction that included all dinosaurs and 60% of the species living at the time. Such collisions could also alter fundamental aspects of the Earth's orbit and rotation. Consider an impact of a similar-sized asteroid of mass 8.9×10^{15} kg and a velocity with respect to the Earth's surface of 90,000 km/h. If the asteroid were to strike Earth at the equator as shown in Figure P9.69, (a) what is the change in the angular velocity of the Earth? (b) What is the maximum possible change in the length of a day?

Figure P9.69

70. ✪ Some passenger jets have one engine attached to each wing. The core of a jet engine is a rotating turbine that spins at a rate of 9,000 rpm. (a) Estimate the angular momentum of the airplane due to the rotation of the turbines in both engines. These turbines are cylindrical, are usually made of titanium alloys, and can be modeled as a cylinder 1.0 m in diameter and 4.0 m long. Assume the components of each engine are rotating in a counterclockwise direction as seen from the front of the aircraft so that a significant amount of angular momentum is pointing along the forward direction (in the positive direction along the roll axis). (b) If the pilot executes a left turn, about which axis (*roll, pitch,* or *yaw* as seen in Fig. P9.70) would the angular momentum of the engines cause the plane to rotate? (c) About which axis would a rotation occur if the plane flew upward in a vertical circle?

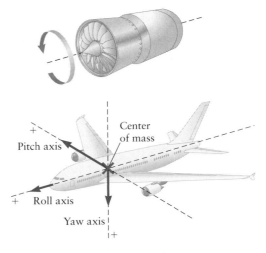

Figure P9.70

71. The hard drive in a laptop computer or some MP3 players contains one or more platters (spinning disks), each having an angular speed of typically 10,000 rpm, a radius of 4.0 cm, and a mass of about 50 g. (a) What is the kinetic energy of one platter? (b) What is the angular momentum of one platter? (c) Can you notice this angular momentum when you rotate your computer or MP3 player? Explain why or why not.

72. ⭐ A mass m_1 is connected to a pulley of radius R and mass m_2 as shown in Figure P9.72. The mass m_1 starts from rest and then falls through a distance h. (a) Use conservation of energy to calculate the linear speed of m_1. (b) What is the final angular velocity of the pulley? (c) What is the final angular momentum of the pulley? (d) What was the source of the torque that produced this angular momentum?

Figure P9.72

73. ⭐ A bullet of mass m_b = 25 g is fired with a speed of 250 m/s at a target that is a sheet of metal (mass m_t = 500 g). The target is square with sides of length L = 20 cm. The target is hinged along its top edge so that it can swing freely (Fig. P9.73). If the bullet hits in the center of the target and then becomes stuck, what is the angular velocity of the target plus bullet just after the bullet hits?

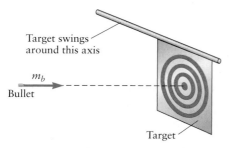

Figure P9.73

74. ⭐ A cylinder rolls down a ramp of height h = 1.5 m with θ = 30°, starting from rest at the top (Fig. P9.74). How long does it take to reach the bottom of the ramp?

Figure P9.74

75. A figure skater (Fig. 9.10) has an initial angular velocity of 3.8 rad/s. She then pulls her arms in, and her angular velocity increases to 9.5 rad/s. What is the ratio of her final kinetic energy to her initial kinetic energy?

76. ⭐ An object is rolling without slipping, with a translational kinetic energy of KE_{trans} and a rotational kinetic energy of KE_{rot}. If $KE_{\text{trans}}/KE_{\text{rot}}$ = 3/2, what might this object be, (a) a hoop, (b) a spherical shell, (c) a solid cylinder, or (d) a solid sphere?

77. A solid sphere and a solid cylinder, both of mass m and radius R, are rolling with a speed v. Find (a) the ratio of the angular momentum of the sphere to that of the cylinder and (b) the ratio of the total kinetic energy of the sphere to that of the cylinder.

78. ✪ **Electrical power produced by a wind turbine.** The force of the wind on the blades of a wind turbine (Fig. 9.4A) is proportional to v^2 where v is the speed of the wind (see Chapter 3). The angular velocity of the turbine is also proportional to the wind speed. (a) Show that the mechanical power exerted by the wind on a turbine varies as $P \propto v^\alpha$ and find the value of α. (b) At some promising sites for wind energy, the wind speed averages about 20 mi/h (about 9 m/s). If a wind turbine at this site produces 1.0 MW with a wind speed of 20 mi/h, what power would it produce when the wind speed is 25 mi/h? Your answer should show why it is important to place wind turbines at places where the wind speed is fastest.

79. ✪ Ⓡ The ice cap at the North Pole has an area of about 6 × 10^6 km^2 and an average thickness of about 25 m. If all this ice melts, what will be the approximate effect on the Earth's rotation rate?

Fluids

In this chapter, we use Newton's laws to describe and predict the motion of fluids, such as the water in this river or the air above it. (© Corbis/Jupiterimages)

In previous chapters, we learned how to apply Newton's laws to the motion of solid objects, first dealing with point particles and then with extended objects that may rotate. In this chapter, we turn our attention to the behavior of *fluids*. A fluid may be either a liquid or a gas. We say that a fluid *flows* from one place to another, and the shape of a fluid varies according to its container. To describe the mechanics of fluids, we need several new quantities, but our analysis is still based on Newton's laws of motion. Understanding the mechanics of fluids is certainly an important topic; most of the Earth's surface is covered with a liquid (water), and the atmosphere is a gas (air). Moreover, the motion of fluids (including the flow of blood in the body and the movement of water within plants) is essential to most living things.

10.1 | PRESSURE AND DENSITY

We want to apply Newton's second law to understand and predict the behavior of a fluid. Recall that Newton's second law is

$$\sum \vec{F} = m\vec{a}$$

When we apply this relation to describe the motion of a particle, m is the mass of the particle, $\vec{a}$ is its acceleration, and $\sum \vec{F}$ is the total force acting on the particle. To treat the motion of a fluid, what should we use for the mass m? If we want to describe the water in a river, should m be the mass of the entire river? The idea of using a single acceleration variable $\vec{a}$ to describe the entire river does not make sense either since different parts of the river might be moving in different directions.

We must therefore rewrite Newton's second law in a different form, a form designed with fluids in mind. Although everything we do in this chapter is based on Newton's laws, we need to define some new quantities to use in place of our familiar variables force and mass. The first of these new quantities is *pressure*. The pressure in a fluid is connected with the force that the fluid exerts on a particular surface. This concept is illustrated in Figure 10.1A, which shows the window of a hypothetical submarine. Outside the window is a fluid, seawater, that exerts a force on the window. This force is related to the pressure P in the seawater by

Definition of pressure

$$P = \frac{F}{A} \tag{10.1}$$

where F is the magnitude of the force exerted by the seawater on the window and A is the area of the window. This force is *perpendicular* to the window's surface. From the units on the right side of Equation 10.1, the units of pressure correspond to force divided by area, which in SI units is newtons/square meter (N/m^2). This unit is called the ***pascal*** (abbreviated Pa), in honor of French physicist and mathematician Blaise Pascal. In the U.S. customary system of units, pressure is measured in pounds per square inch ($lb/in.^2$). You may have encountered that unit when you last checked the pressure in an automobile or bicycle tire.

Although force is a vector quantity, pressure is not. The definition of pressure (Eq. 10.1) involves only the component of the force perpendicular to the surface in question, in this case the window. That is because a stationary fluid, such as the seawater outside the submarine in Figure 10.1, can only *push* on a surface. When an object is submerged in a static fluid, the fluid exerts a force on the object that is always directed perpendicular to the object's surface. As a result, there is never any doubt about the direction of the force in Equation 10.1. This point is illustrated in Figure 10.1B, which shows a small, sealed box filled with a liquid at pressure P. The fluid within the box produces an outward force on *all* the sides of the box. If the box is small, the pressure will be approximately constant throughout the box.[1] For any

Figure 10.1 🅐 The pressure P in the water around this submarine produces a force on the submarine's shell. This force is everywhere directed perpendicular to the surface of the submarine. The force on a submarine window is $F = PA$, where A is the window's area. 🅑 When a container is filled with a fluid at a pressure P, the fluid produces an outward force that is perpendicular to the walls of the container.

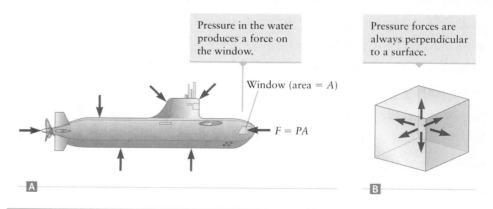

Pressure in the water produces a force on the window.

Window (area = A)

$F = PA$

Pressure forces are always perpendicular to a surface.

Ⓐ   Ⓑ

[1]The pressure in a fluid actually varies from place to place within the container, according to depth, as we'll see in Section 10.2. For a small box, however, this variation will be small and P will be nearly constant.

particular wall of the box (area A), the total force on the wall is equal to $F = PA$. The force associated with the pressure in a fluid is always normal (perpendicular) to the surface under consideration.

The pressure in the Earth's atmosphere near sea level has the approximate value

$$P_{\text{atm}} = 1.01 \times 10^5 \text{ Pa} \qquad (10.2)$$

Value of atmospheric pressure

This value of P_{atm} is approximate because the pressure in the atmosphere varies in response to changes in the weather and other factors (such as altitude).

Atmospheric Pressure

Recall that 1 Pa is equal to 1 N/m². Hence, atmospheric pressure (Eq. 10.2) corresponds to very large forces for areas you encounter every day. For example, a sheet of paper (like a page of this book) has an area of about $A = 500 \text{ cm}^2 = 0.05 \text{ m}^2$. Rearranging Equation 10.1, the force on one side of this sheet of paper is

$$F = PA = P_{\text{atm}}A = (1.01 \times 10^5 \text{ Pa})(0.05 \text{ m}^2)$$

$$F = 5 \times 10^3 \left(\frac{\text{N}}{\text{m}^2}\right)\text{m}^2 = 5 \times 10^3 \text{ N}$$

which is a very substantial force; it is equal to the combined weight of several large people! You might at first be skeptical of this result and ask why a sheet of paper does not fold or bend under such a large force. The answer is that the air exerts a force on *both sides* of the sheet of paper as sketched in Figure 10.2. The pressure in the air on both sides is equal to P_{atm}. Because these forces are both directed perpendicular to the paper's surface, they are in opposite directions and the total force on the paper is zero. This cancellation occurs in many situations, including the force due to air pressure on your body.

Vacuum and the Magdeburg Experiment

The balance of forces associated with air pressure can be altered by removing air from inside an enclosed volume with a device called a vacuum pump. This was first demonstrated 400 years ago by Otto von Guericke. He constructed two metal hemispheres as shown in Figure 10.3A; when fitted together, they form a Magdeburg sphere, so named in honor of the town where von Guericke lived. One hemisphere contained a small tube connected to a valve through which air was removed from inside the two hemispheres by a pump that von Guericke invented. The region inside where the pressure was zero, or very small, is known as a ***vacuum***.

To understand von Guericke's experiment, let's analyze the somewhat simpler case of two Magdeburg plates sketched in Figure 10.3B. If we consider the effect of atmospheric pressure on one of the plates, we see that the air outside produces an inward force equal to $P_{\text{atm}}A$, where A is the plate's area. Because there is no air on the inside, $P_{\text{inside}} \approx 0$ and there is no outward force from air pressure. There will be a similar inward force $P_{\text{atm}}A$ on the other plate, and the net result is that there is a

Figure 10.2 The atmosphere produces a downward force of magnitude $P_{\text{atm}}A$ on this sheet of paper, where A is the area of the sheet. The air underneath the sheet produces a force of the same magnitude directed upward.

There is no force on the inside surface because $P \approx 0$ inside.

Vacuum seal (rubber)

$P \approx 0$

$F = P_{\text{atm}}A$ $F = P_{\text{atm}}A$

The Granger Collection

A   **B**

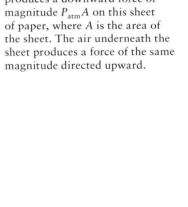

Figure 10.3 **A** Otto von Guericke demonstrated the force associated with atmospheric pressure using this apparatus. He made two tight-fitting hemispheres and then removed the air from the inside using a vacuum pump, which he invented. Von Guericke employed hemispheres about 1 m in diameter and used horses to (attempt to) pull them apart! **B** A similar von Guericke apparatus, using two flat plates in place of the hemispheres.

substantial force holding the plates together. The force required to pull the plates apart is calculated in Example 10.1.

EXAMPLE 10.1 | Force in the Magdeburg Demonstration

Suppose the Magdeburg plates in Figure 10.3B have an area of $A = 1.0$ m^2. What force is required to pull the plates apart?

RECOGNIZE THE PRINCIPLE

This problem involves an application of the relation between pressure, area, and force. The force due to air pressure on one plate is $F = P_{atm}A$ (Eq. 10.1).

SKETCH THE PROBLEM

Figure 10.4 shows the horizontal forces acting on one of the plates.

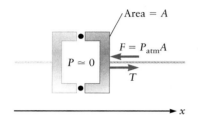

IDENTIFY THE RELATIONSHIPS

A cable with tension T is attached to the Magdeburg plate on the right in Figure 10.4. There are two forces: a force $P_{atm}A$ to the left due to atmospheric pressure and a force T to the right. (We can ignore the weight of the plate because it will be much smaller than the other two forces.) If the plates are just barely being pulled apart, we can take the acceleration of the plate on the right to be zero.

Figure 10.4 Example 10.1.

SOLVE

Applying Newton's second law gives

$$\Sigma F = +T - P_{atm}A = ma = 0$$

where these forces are along the horizontal. This leads to

$$T = P_{atm}A = (1.01 \times 10^5 \text{ Pa})(1.0 \text{ m}^2) = \boxed{1.0 \times 10^5 \text{ N}}$$

What does it mean?
This force is the combined weight of about 10 cars! Its no wonder the painting in Figure 10.3A shows a sizable team of horses pulling on the Magdeburg plates.

Figure 10.5 Concept Check 10.1.

CONCEPT CHECK 10.1 | Magdeburg Plates and Newton's Third Law

When we calculated the force required to pull apart the Magdeburg plates in Example 10.1, we actually calculated the force exerted by the atmosphere on *one* of the plates and found that $F = 1.0 \times 10^5$ N. Suppose the other plate is attached to a rigid wall as shown in Figure 10.5. Now what tension T is required in the cable on the right to pull apart the plates?
(a) 1.0×10^5 N   (b) 2.0×10^5 N   (c) 0

Gauge Pressure versus Absolute Pressure

The quantity P we defined in Equation 10.1 is often referred to as the **absolute** pressure. A common approach to measuring pressure leads to a related quantity called the **gauge pressure**. To understand the difference, consider the simple pressure-measuring device sketched in Figure 10.6. This device determines the pressure by measuring the force on a movable plate. One side of this plate is exposed to the fluid whose pressure is unknown, so the force on this side of the plate is equal to the product of the unknown pressure P_{abs} and the plate's area. Notice that P_{abs} is the *absolute* pressure in the fluid. It is often convenient to expose the other side of the plate to the atmosphere, that is, to atmospheric pressure. For example, a gauge used to measure the pressure in an automobile tire is constructed in this way. The

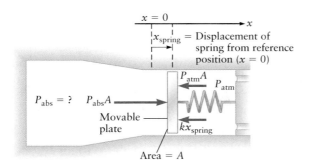

movable plate is also attached to a spring (with spring constant k), and this spring exerts a force $-kx_{spring}$ on the plate, according to Hooke's law (Chapter 6). Here x_{spring} is the displacement of the plate from a reference position where the spring is unstretched and uncompressed. The total force on the plate inside the gauge is due to the pressures on both sides plus the force exerted by the spring. If we take the $+x$ direction to be to the right in the figure, we have

$$\Sigma F = +P_{abs}A - P_{atm}A - kx_{spring}$$

If the gauge plate is in equilibrium, the total force must be zero and we get

$$P_{abs}A - P_{atm}A - kx_{spring} = 0$$

$$x_{spring} = (P_{abs} - P_{atm})\left(\frac{A}{k}\right)$$

The deflection of the gauge x_{spring} is thus proportional to the *difference* between the pressure one wants to measure (P_{abs}) and atmospheric pressure. This difference is called the gauge pressure. Hence, the absolute pressure is related to the gauge pressure by

$$P_{abs} = P_{gauge} + P_{atm}$$

Gauge pressure

Many common pressure gauges work this way and thus measure the gauge pressure. For the remainder of our work in this chapter, the term *pressure* means the absolute pressure unless we specifically note otherwise.

CONCEPT CHECK 10.2 | Pressure in an Automobile Tire

The pressure in an automobile tire is usually quoted (in the United States) in units of pounds per square inch (lb/in.²), and in a typical tire, the value the gauge pressure is $P_{gauge} = 25$ lb/in.². What is the value of the gauge pressure in pascals?
(a) 2.7×10^3 Pa   (b) 1.0×10^5 Pa   (c) 1.7×10^5 Pa   (d) 110 Pa

Density

Another quantity we need to describe fluids is the *density*, denoted by ρ. The density of a substance is equal to its mass per unit volume,

$$\rho = \frac{M}{V} \qquad (10.3)$$

where M is the mass of a sample of the substance and V is the sample volume. This definition of ρ assumes the density is constant throughout the volume V.

The density of a substance can vary according to the temperature and may also depend on the pressure. It is thus possible for the density of a substance to vary from place to place within the substance. However, for most liquids—including water and oil—the density is approximately independent of the pressure. Such fluids are called *incompressible*. The densities of some common substances are listed in Table 10.1.

Although most liquids are incompressible, that is not the case for gases. The density of a gas depends strongly on the pressure; that is, gases are *compressible*, and

Table 10.1 Densities of Some Common Solids, Liquids, and Gases[a]

SUBSTANCE	DENSITY (kg/m³)
Solids	
Ice (at 0°C)	917
Aluminum	2,700
Lead	11,300
Platinum	21,500
Gold	19,300
Tungsten	19,300
Steel	7,800
Concrete	2,000
Bone	1,500–2,000
Polystyrene	100
Glass	2,500
Wood (spruce)	400
Wood (balsa wood)	120
Wood (oak)	750
Liquids	
Water (0°C)	999.8
Water (4°C)	1,000.0
Water (20°C)	999.9
Seawater	1,025
Ethanol	790
Mercury	13,600
Oil	700–900
Gases	
Air (0°C)	1.29
Air (100°C)	0.84
Helium (0°C)	0.18
Hydrogen (0°C)	0.090
Carbon dioxide (25°C)	1.80

[a]Unless noted, all values are for room temperature and pressure.

The pressure in a fluid depends on the depth.

we'll show in Section 10.3 how that affects their behavior. The densities of some common gases at atmospheric pressure are listed in Table 10.1.

The SI unit of density is the kilogram per cubic meter, or kg/m³ (= mass/volume). Another way to express the density of a substance is in terms of its *specific gravity*, which is the ratio of the density of the substance to the density of water. We'll see that specific gravity is important for determining the buoyancy of an object, that is, whether or not it will float in water.

10.2 | FLUIDS AND THE EFFECT OF GRAVITY

Let's now consider how the pressure varies from place to place within a fluid. Figure 10.7 shows a liquid at rest in a container. Since this liquid is at rest, we can apply the conditions for static equilibrium that we learned in Chapter 4. In particular, consider the forces acting on a region of the fluid within the imaginary cubical "box" in Figure 10.7. The fluid is free to flow through the walls of this imaginary box, but the system is at rest, so the fluid inside the box will stay there and we can treat the fluid cube as a "simple" mass. For this cubical box, each wall (or face) has an edge length h and an area $A = h^2$.

We now apply our condition for translational equilibrium to the mass of fluid within the box. Because this mass is at rest, the total force on it must be zero. There are contributions to the total force from the pressure outside the box. These forces are all directed inward on the walls of the box as indicated in Figure 10.7. The forces along the horizontal directions are equal in magnitude and opposite in direction, so they cancel. However, the forces on the top and bottom of the box do not cancel. Although these forces are in opposite directions, they cannot be equal in magnitude because the pressure force at the bottom must support both the pressure force at the top and the weight of the fluid in the box. Hence, the pressure at the bottom must be greater than the pressure at the top. In general, the *pressure always increases as one goes deeper into a fluid*.

To calculate how the pressure depends on depth, note that the force on the bottom of the box is upward with $F_{bot} = P_{bot}A$, where P_{bot} is the pressure at the bottom face of the box. The force on the top of the box is $F_{top} = -P_{top}A$, where the negative sign indicates that this force is downward. There is also a gravitational force on the fluid mass due to its weight. This force is along the vertical with $F_{grav} = -mg$, where m is the fluid mass inside the box. The total force along y is then

$$\Sigma F_y = P_{bot}A - P_{top}A - mg = 0 \tag{10.4}$$

This total force is zero because the system is in translational equilibrium. From the definition of density (Eq. 10.3), the mass of the fluid in the box is $m = \rho V$, and the volume of the box is equal to the area A times the box height h, so $V = Ah = h^3$ and

$$m = \rho V = \rho Ah = \rho h^3$$

Inserting this into Equation 10.4 leads to

$$P_{bot}A - P_{top}A - mg = P_{bot}h^2 - P_{top}h^2 - \rho h^3 g = 0$$

$$P_{bot} - P_{top} = \rho gh$$

We can also write this result as

$$P = P_0 + \rho gh \tag{10.5}$$

Here P_0 is the pressure at a "reference point" in the fluid and P is the pressure at a depth h relative to the reference point as sketched in Figure 10.8. We have derived Equation 10.5 with the use of a cubical box, but this result is not limited to fluid "boxes." In words, the pressure increases as one goes to greater depths because a particular section of fluid must support the weight of the fluid above it.

We will often consider fluids having a surface open to the atmosphere, and it is usually convenient to take the reference point at this surface. In this case, $P_0 = P_{atm}$

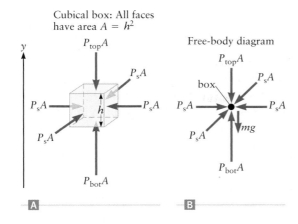

Cubical box: All faces have area $A = h^2$

Free-body diagram

box

A   **B**

as shown in Figure 10.8B. No matter where the reference point is chosen (Fig. 10.8A or B), a positive value of h in Equation 10.5 corresponds to a location that is *deeper* than the reference location.

Equation 10.5 is a general relation that tells how the pressure varies with depth within an *incompressible* fluid since our derivation assumed the density of the fluid is constant; that is, ρ does not vary with depth within the fluid. Equation 10.5 can therefore be applied only to liquids, not to gases. We'll see how to deal with compressible fluids (gases) after we consider a few examples involving liquids.

$P_A = P_0 + \rho g h$

$P_B = P_{atm} + \rho g h$

A   **B**

Figure 10.8 Illustration of the relation $P = P_0 + \rho g h$. In 🄰, the reference point is at a particular point inside the liquid, whereas in 🄱, the reference point is at the surface in contact with the atmosphere, so $P_0 = P_{atm}$. In both cases, h is the depth relative to the reference point.

EXAMPLE 10.2 | Pressure in the Marianas Trench

The Marianas Trench, a region at the bottom of the Pacific Ocean, is believed to be the deepest spot on the ocean floor. In 1960, a U.S. Navy submersible vehicle went to a spot in this trench that is 10,700 m below sea level. Suppose this submarine had a tiny window of area 1.0 cm^2 so that sailors could enjoy the view. Find the force on this window due to the water pressure.

RECOGNIZE THE PRINCIPLE

The force on the window depends on the pressure through $F = PA$. We thus need to find the pressure in the Marianas Trench, which we can do using the dependence of pressure on depth (Eq. 10.5).

SKETCH THE PROBLEM

Figure 10.8B describes the problem, with the submarine at point B.

IDENTIFY THE RELATIONSHIPS

We take the ocean surface as the reference point for calculating the pressure using Equation 10.5. Therefore, $P_0 = P_{atm}$, the depth is $h = +10{,}700$ m, and the density of seawater is given in Table 10.1.

SOLVE

From Equation 10.5 and using $P_0 = P_{atm}$, we find the pressure to be

$$P = P_{atm} + \rho g h = (1.01 \times 10^5 \text{ Pa}) + (1025 \text{ kg/m}^3)(9.8 \text{ m/s}^2)(10{,}700 \text{ m})$$

$$P = 1.1 \times 10^8 \text{ Pa}$$

which is about 1000 times atmospheric pressure. The force on the submarine's window is

$$F = PA = (1.1 \times 10^8 \text{ Pa})(1.0 \text{ cm}^2) = (1.1 \times 10^8 \text{ N/m}^2)(1.0 \times 10^{-4} \text{ m}^2)$$

$$F = \boxed{1.1 \times 10^4 \text{ N}}$$

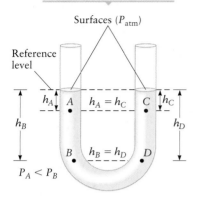

Surfaces (P_{atm})

Reference level

h_A A $h_A = h_C$ C h_C

h_B h_D

B $h_B = h_D$ D

$P_A < P_B$

Figure 10.9 This U-tube contains a single liquid. The pressure in the U-tube depends only on the depth below a reference level; here the reference level is at the liquid's surface.

CONCEPT CHECK 10.3 | Pressure at the Bottom of a Swimming Pool

The pressure increases as one goes deeper within a liquid, and in Example 10.2, we found that the pressure increase can be quite large in deep locations in the oceans. This pressure change can also be quite noticeable in a swimming pool. When you dive to the bottom of a pool, you can sense the pressure increase with your ears. For a pool with a depth of 3.0 m, what is the difference between the pressure at the bottom of the pool and atmospheric pressure?
(a) 2.9×10^4 Pa   (b) 3.0×10^5 Pa   (c) 3.0×10^3 Pa

Pressure in a U-tube

In the arrangement shown in Figure 10.9, a liquid is contained in a glass tube that has a "U" shape; this container is called, not surprisingly, a U-tube. U-tubes are very handy for measuring the pressure in a fluid and for other fluid experiments.

In a U-tube, different regions of the fluid can be widely separated in space. However, this separation does not affect our general expression for how the pressure varies with depth, Equation 10.5. According to that result, the pressure in the liquid in Figure 10.9 depends only on the depth h below a reference level. The pressures are the *same* at two spots on different sides of the U-tube that are at the *same* depth below the surface; hence, $P_A = P_C$ and $P_B = P_D$ in Figure 10.9. This result is true as long as the U-tube is filled with only one type of liquid. Example 10.3 shows how to apply the result in Equation 10.5 to a U-tube containing two different liquids.

EXAMPLE 10.3 | Two Liquids in a U-Tube

Consider the U-tube in Figure 10.10, with both ends open to the atmosphere. The tube is filled with two different liquids, of different densities. We also assume the two liquids are immiscible; that is, they do not mix. If one liquid (liquid 1) has a higher density than the other (liquid 2), the level of liquid 2 is higher than the level of liquid 1 by a distance d. Suppose liquid 1 is water ($\rho_1 = 1000$ kg/m^3); liquid 2 is a type of oil with $\rho_2 = 700$ kg/m^3 and is filled to a height $y_R = 0.10$ m above the oil-water boundary (Fig. 10.9). Find d.

RECOGNIZE THE PRINCIPLE

For the U-tube containing a single type of fluid in Figure 10.9, the pressure depends only on the depth, so at any particular depth below the surface, the pressure on the two sides of the tube is the same. This result is true only when the U-tube is filled with a single liquid so that the density is constant throughout, a key assumption in the derivation of Equation 10.5. In Figure 10.10, the density changes as we pass from one liquid to the other, so we *cannot* use Equation 10.5 to compare the pressures in the different liquids. We can, however, use Equation 10.5 to compare the pressures at different points within the same connected piece of liquid. Consider the pressure at point B in Figure 10.10. This point is inside liquid 1 and is just below the boundary that separates the two liquids. Because we consider point B to be at the same depth as point A (also located in liquid 1, but on the other side of the U-tube), the pressures at points A and B are the same.

We next consider point C; this point is in liquid 2, and just above the boundary between the liquids. Because points B and C are extremely close together, the pressures at these two locations must be the same. You can see how that is true by imagining a thin (massless) plastic sheet stretched along the boundary between liquids 1 and 2. Because the system is at rest, this sheet is in translational equilibrium; hence, the

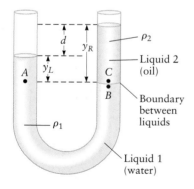

d y_R ρ_2

A y_L C   Liquid 2 (oil)

B   Boundary between liquids

ρ_1

Liquid 1 (water)

Figure 10.10 Example 10.3. When a U-tube contains more than one type of liquid, one must apply Equation 10.5 separately within each liquid region. Points B and C are on opposite sides of the boundary between the fluids. These points are extremely close together and are at the same height as point A.

total force on the sheet must be zero. That can only be true if the pressure is the same on the top and bottom surfaces of the sheet. We have thus concluded that the pressure is the same at points A, B, and C; that is, $P_A = P_B = P_C$. This condition will allow us to find the value of d.

SKETCH THE PROBLEM

Figure 10.10 describes the problem.

IDENTIFY THE RELATIONSHIPS

In liquid 1, the pressure at the surface is atmospheric pressure since this surface is in contact with the atmosphere; let's take this surface as our reference point in applying Equation 10.5. If y_L is the distance from the point A to the surface, we get

$$P_A = P_{atm} + \rho_1 g y_L$$

Likewise, in liquid 2,

$$P_C = P_{atm} + \rho_2 g y_R$$

where y_R is the distance from point C to the surface of liquid 2. Combining these results and using $P_A = P_B = P_C$ gives

$$P_A = P_{atm} + \rho_1 g y_L = P_C = P_{atm} + \rho_2 g y_R$$

$$\rho_1 g y_L = \rho_2 g y_R$$

$$y_L = y_R \frac{\rho_2}{\rho_1}$$

The difference in heights of the two surfaces is then

$$d = y_R - y_L = y_R\left(1 - \frac{\rho_2}{\rho_1}\right) = y_R \frac{\rho_1 - \rho_2}{\rho_1}$$

SOLVE

Inserting the values of the densities of water (ρ_1) and oil (ρ_2) given above and using the given value of y_R, we find

$$d = y_R \frac{\rho_1 - \rho_2}{\rho_1} = (0.10 \text{ m})\frac{(1000 - 700)(\text{kg/m}^3)}{1000 \text{ kg/m}^3} = \boxed{0.030 \text{ m}}$$

What have we learned?
When two fluids are in contact with each other and are at rest, the pressure must be the same on the two sides of the interface.

Barometers, Vacuums, and Measuring Pressure

U-tubes and similar types of containers can be used in many different ways. Example 10.3 shows that when a U-tube is filled with two different (and immiscible) liquids, the open surfaces of the two liquids will sit at different heights. This result can be used to measure the density of a liquid. If the density ρ_1 of liquid 1 in Figure 10.10 is known, a measurement of d as calculated in Example 10.3 then allows us to find the value of ρ_2.

A U-tube can also be used to measure pressure differences. For example, consider the situation in Figure 10.11, which shows a U-tube whose ends are connected to two different vessels. These vessels might be containers of gas at different pressures P_L and P_R. When partially filled with liquid, the U-tube can be used to measure the pressure difference $P_L - P_R$ by applying Equation 10.5. Taking our reference point to be the top surface of the liquid on the right side of the U-tube, we find

$$P_L = P_R + \rho g h$$

Figure 10.11 This U-tube can be used to measure the difference in pressure, $\Delta P = P_L - P_R$.

$P_{top} = 0$
(vacuum)

h

P_{atm}

Figure 10.12 Design of a simple barometer.

where h is the difference in the heights of the two surfaces of the liquid. Hence, the difference in pressures is

$$\Delta P = P_L - P_R = \rho g h \quad (10.6)$$

Figure 10.12 shows another way to measure pressure. Here we have a straight tube, partially filled with a liquid and closed at the top end, with the open end at the bottom inserted into a dish containing the same liquid. The pressure at the top of the tube is zero (this can be done with a vacuum pump), while the open surface of the liquid in the dish is exposed to the atmosphere. Applying Equation 10.5 and using the top surface of the liquid in the tube as our reference point, we have

$$P_{atm} = P_{top} + \rho g h$$

where h is the height of the liquid column in the tube. Since $P_{top} = 0$, we find

$$P_{atm} = \rho g h \quad (10.7)$$

Hence, if we measure h, we can use the device in Figure 10.12 as a **barometer** and measure the absolute pressure of the atmosphere.

Let's now design such a barometer and assume we are going to use water as the liquid. We want to calculate how tall the tube must be to measure the pressure in the atmosphere. We can rearrange Equation 10.7 to solve for the height of the water column h; inserting the values of atmospheric pressure and the density of water from Table 10.1 gives

$$h = \frac{P_{atm}}{\rho g} \quad (10.8)$$

$$h = \frac{1.01 \times 10^5 \text{ Pa}}{(1000 \text{ kg/m}^3)(9.8 \text{ m/s}^2)} = 10 \text{ m} \quad (10.9)$$

Hence, this barometer would require quite a long tube to measure atmospheric pressure; it would certainly not fit into a normal-sized room. That is why most barometers do not employ water as the working liquid. Instead, they use mercury, which has a much higher density than water, giving a smaller value of h in Figure 10.12 and Equation 10.8.

EXAMPLE 10.4 Designing a Mercury Barometer

Consider again the barometer in Figure 10.12, but now assume the liquid in the tube and dish is mercury. If the pressure to be measured is atmospheric pressure, what is h?

RECOGNIZE THE PRINCIPLE

We can use the results for the pressure in an incompressible fluid and the reasoning that led to Equation 10.8, but now we replace water with mercury, so we have a different value for the density.

SKETCH THE PROBLEM

Figure 10.12 describes the problem.

IDENTIFY THE RELATIONSHIPS AND SOLVE

Using the density of mercury (which is a liquid at room temperature) from Table 10.1 along with Equation 10.8 leads to

$$h = \frac{P_{atm}}{\rho g} = \frac{1.01 \times 10^5 \text{ Pa}}{(13,600 \text{ kg/m}^3)(9.8 \text{ m/s}^2)} = \boxed{0.76 \text{ m}}$$

Units for Measuring Pressure

Many different units are used for measuring pressure. For most scientific work, *P* is measured in pascals, and we have seen that 1 Pa = 1 N/m². In the United States, we often encounter the unit pounds per square inch (lb/in.²), especially when dealing with automobile or bicycle tires. The pressure unit "inches" is often heard on a TV weather report. This unit originates with a barometer like the one in Example 10.4, where the pressure in the atmosphere can be determined by giving the value of *h* (measured in inches) for a mercury-filled barometer. Likewise, the barometer height *h* can be measured in millimeters, which leads to the unit "millimeters of mercury," (sometimes referred to as simply mm). Another common unit is the atmosphere (atm), which is the pressure in the Earth's atmosphere at sea level under normal weather conditions. These units are summarized in Table 10.2, which also gives the value of normal atmospheric pressure in the different units.

Pumping a Liquid

Our discussion of barometers led us to Equation 10.9, showing that a water-filled barometer is not very practical. This result also has an important implication for pumping water from place to place. Figure 10.13 shows a well that brings water to the surface from deep underground. If we are going to use a pump to move the water, where is the best place to put the pump? Two possibilities are shown in Figure 10.13; in one case, the pump is located at the top of the well, and in the other, it is located underground at the bottom of the well. Placing it at the top of the well has several advantages; for example, it is simpler to fix the pump if it breaks down. There is a limitation with this location, however. If the pump is placed at the top, it must draw water up from the ground by establishing a very low pressure in the top of the pipe that carries water to the surface. This arrangement is just like the barometer in Figure 10.12, where a very low pressure (a vacuum) in the top of the tube and the atmospheric pressure at the dish act together to push water up the column. Suppose the water pump in Figure 10.13A generates a perfect vacuum at the top ($P_{top} = 0$). Since soil is porous to air, the pressure at the underground surface of the water is approximately atmospheric pressure. Following the approach we used in deriving Equation 10.8, we have, from Figure 10.13A,

$$P_{bot} = P_{top} + \rho g h$$
$$P_{atm} = 0 + \rho g h$$

which leads to

$$h = \frac{P_{atm}}{\rho g}$$

For water, then, *h* = 10 m as we found in Equation 10.9. This is the *maximum* height to which a vacuum pump can "draw" water. The design in Figure 10.13A can thus only be used if the well depth is less than 10 m.

Most water wells, however, are much deeper than 10 m. For example, the well at the author's home has a depth of approximately 50 m. These cases are handled by placing the pump at the bottom of the well as in Figure 10.13B. Now the pump is not a vacuum pump, but instead generates a large pressure, much greater than atmospheric pressure. We can again apply the relation for the pressure in an incompressible fluid (Eq. 10.5) and get

$$P_{bot} = P_{top} + \rho g h$$

Table 10.2 Units Used for Measuring Pressure

PRESSURE UNIT	VALUE OF ATMOSPHERIC PRESSURE
Pascal	1.01×10^5 Pa
Pounds per square inch	14.7 lb/in.²
Inches	29.9 in.
Millimeters of mercury	760 mm
Atmosphere	1 atm

A

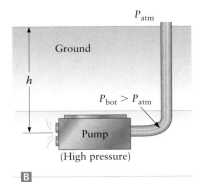

B

Figure 10.13 **A** A vacuum pump is used to pump water from a well (underground) to the surface. Since soil is porous, the pressure at the underground surface of the water is approximately P_{atm}. **B** Most wells use a pump placed at the bottom of the well. This pump produces a high pressure that pushes the water to the surface.

Pressure here must support the weight of the column $F = P_{atm} \cdot A$.

Column of atmosphere

Area A

Figure 10.14 For a fluid at rest in a gravitational field, the pressure at a particular point in the fluid must support the weight of the fluid above that point. For this reason, the pressure in a fluid always increases with depth within the fluid. This result applies to both liquids and gases.

where h is again the depth, and the pressure at the top, P_{top}, is now equal to atmospheric pressure. To satisfy this relation, the pressure at the bottom of the well pipe must be larger than P_{atm}. There is no theoretical limit on the magnitude of P_{bot}; a realistic pump can generate pressures of many times atmospheric pressure, so the design in Figure 10.13B can easily pump water from typical well depths. However, in the design of a water system there are other limiting factors, such as the strength of the pipe that carries the water to the surface. These considerations are important in other situations in which water must be moved or pumped to different vertical levels, such as in the water system of a tall building. (Many buildings are much taller than 10 m!)

Pressure in a Compressible Fluid

Equation 10.5 tells how the pressure varies with depth, but it applies only to an incompressible fluid, that is, a liquid such as water or mercury. Gases (such as the air in the atmosphere) are compressible fluids and follow a different relation between pressure and depth. For this problem, the important difference between a liquid and a gas is that the density of a gas depends greatly on the pressure, whereas the density of a liquid is approximately constant. The pressure in a compressible fluid still increases with depth as explained in Figure 10.14. The pressure at the Earth's surface is the usual atmospheric pressure P_{atm}. This pressure must support the weight of all the fluid (air) above it. So, if we consider a column of fluid of area A, the upward force supporting this column is $F = P_{atm}A$. For a liquid with a constant density, this analysis leads to Equation 10.5, which gives $P = P_{atm} + \rho g h$, where h is the *depth* in the fluid measured from the surface. In Figure 10.14, we denote the *height* above the surface as y, which is related to the depth by $y = -h$; we thus have for a liquid

$$P = P_0 - \rho g y \tag{10.10}$$

which is plotted as the straight line in Figure 10.15.

This result does *not* apply to a gas because as y increases and the pressure drops, the density becomes smaller. The weight of the fluid column is therefore smaller for a compressible fluid, and the pressure decreases more slowly than for an incompressible fluid. The pressure as a function of height y for a compressible fluid is also plotted in Figure 10.15. This result can be derived using Newton's laws and is

$$P = P_0 \exp\left(\frac{-y}{y_0}\right) \tag{10.11}$$

where P_0 is the pressure at a reference point, y is the height as measured from the reference point, and y_0 is a constant that depends on the gas. For air with a reference point at sea level, $P_0 = P_{atm}$ and $y_0 \approx 1.0 \times 10^4$ m. The height y_0 is the altitude at which the pressure falls to $e^{-1} = 0.37 = 37\%$ of its value at the surface. This result applies for low regions of the atmosphere, where the temperature is approximately constant. Dealing with the pressure variations over a wider altitude range is more complicated since the air temperature can vary substantially and the air density depends on temperature.

Figure 10.15 Variation of pressure in the Earth's atmosphere with height y above sea level.

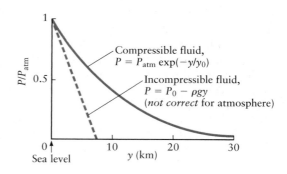

Compressible fluid, $P = P_{atm} \exp(-y/y_0)$

Incompressible fluid, $P = P_0 - \rho g y$ (*not correct* for atmosphere)

P/P_{atm}

Sea level

y (km)

Whether we are dealing with a liquid or a gas, the pressure in a fluid must vary with depth and P must increase as one goes deeper into the fluid. This variation is given by Equations 10.5 and 10.10 for a liquid (an incompressible fluid) or by Equation 10.11 for a gas (a compressible fluid). In both cases, the pressure difference between the top and bottom surfaces of an arbitrary volume of fluid (Fig. 10.7) must support the weight of the fluid within the volume, so the pressure increases as the depth increases.

EXAMPLE 10.5 | Pressure at the Top of Mount Everest

The peak of Mount Everest is 8850 m above sea level. Estimate the pressure in the atmosphere at this location.

RECOGNIZE THE PRINCIPLE

This example is a direct application of the result in Equation 10.11 for the pressure as a function of altitude in a compressible fluid (air).

SKETCH THE PROBLEM

Figure 10.15 shows the variation of pressure with altitude, y, in the atmosphere as described mathematically by Equation 10.11.

IDENTIFY THE RELATIONSHIP AND SOLVE

The altitude at the top of Mount Everest is $y = 8850$ m. Using this value in Equation 10.11 with $y_0 = 1.0 \times 10^4$ m as given above leads to

$$P = P_{atm} \exp\left(\frac{-y}{y_0}\right) = (1.01 \times 10^5 \text{ Pa})\exp\left(\frac{-8850 \text{ m}}{1.0 \times 10^4 \text{ m}}\right) = \boxed{4.2 \times 10^4 \text{ Pa}}$$

What does it mean?

This result is about 40% of atmospheric pressure. According to the kinetic theory of gases (Chapter 15), the density of a gas is proportional to the pressure, assuming its temperature is constant. The temperature at the top of Mount Everest is usually a bit below (!) the temperature at sea level, which reduces the pressure on Mount Everest by about another 15%. Hence, the air density at the top of Mount Everest is less than 40% of the density at sea level. That is why mountain climbers usually bring their own oxygen when scaling this mountain.

10.3 | HYDRAULICS AND PASCAL'S PRINCIPLE

For a liquid at rest (that is, in static equilibrium), the pressure varies with depth according to

$$P = P_0 + \rho g h \tag{10.12}$$

Now imagine that something is done to change the pressure at a particular point in the liquid; to be specific, suppose the pressure P_0 is changed by an amount ΔP. According to Equation 10.12, the pressure everywhere in the liquid must change by an equal amount. This result is known as *Pascal's principle*.

> *Pascal's principle:* If the pressure at one location in a closed container of fluid changes, this change is transmitted equally to *all* locations in the fluid.

Hydraulics: Pascal's principle

While Equation 10.12 (which is the same as Eq. 10.5) was derived for a liquid (an incompressible fluid), Pascal's principle is also true for a gas (a compressible fluid) as long as the gas is held in a closed container and is in static equilibrium. Pascal discovered this principle before the work of Newton. Looking back at our derivation of Equation 10.5, however, we see that this principle is a direct consequence of Newton's laws.

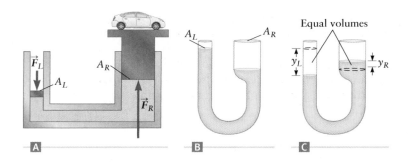

Figure 10.16 A hydraulic jack uses a liquid (usually oil) to transmit a pressure (and therefore a force) from one location to another. This device can amplify an applied force. Here the force on the right-hand piston is greater than the force applied to the piston on the left. **B** and **C** show how the liquid moves as the left-hand piston is depressed. The volume of liquid that leaves the left-hand tube must equal the volume that appears in the right-hand tube.

Designing a Hydraulic Lift: Amplifying Forces

Pascal's principle is the basis of **hydraulics**. A simple version of a hydraulic lift is shown in Figure 10.16A. It is shaped like a U-tube and is typically filled with a liquid such as oil, with pistons at both ends of the U-tube. These pistons are movable, but they fit tightly to keep the oil from leaking out. The key point is that the two ends of the tube have different areas. Here the area A_L on the left is much smaller than the area A_R on the right.

Now imagine that a very large object such as a car sits on the piston on the right. To keep the car and its piston from moving, or to lift the car very slowly off the ground, we apply a force F_L to the left as sketched in Figure 10.16A. What force F_L is required to keep the system in static equilibrium? If the car has a mass m, the car's weight produces a downward force on the right-hand piston of magnitude $F_R = mg$. From the general relation between force and pressure (Eq. 10.1), this force will lead to an increase in the pressure on the right, with

$$\Delta P = \frac{F_R}{A_R}$$

According to Pascal's principle, this pressure change is transmitted throughout the fluid, giving an increased force of magnitude $(\Delta P)A_L$ on the left-hand piston. To keep this piston from moving, there must be an external force of equal magnitude F_L applied down on the piston. Hence,

$$F_L = (\Delta P)A_L$$

or

$$F_L = F_R \frac{A_L}{A_R} \tag{10.13}$$

The piston on the right is much larger than the one on the left ($A_R > A_L$), so

$$F_L < F_R$$

The force on the left-hand piston is thus smaller than the force on the right-hand piston. A relatively small force applied on the left can therefore lift a very massive object (the car) on the right. In other words, the hydraulic device in Figure 10.16 *amplifies forces*. This principle is used by auto mechanics every day.

The hypothetical hydraulic system in Figure 10.16 illustrates the principles of hydraulics, but practical systems are usually made as sketched in Figure 10.17, where a pump is used to produce a high pressure in the liquid in a tube connected to an elevator. The tube leading from the pump plays the role of the left-hand tube in Figure 10.16 and is smaller than the elevator tube. Here the large object to be lifted is the elevator compartment.

Work–Energy Analysis of a Hydraulic System

Now let's analyze the hydraulic lift in terms of work and energy. We start with the configuration in Figure 10.16B and suppose the piston on the right moves upward

Figure 10.17 Schematic of a hydraulic system used for an elevator.

through a small distance y_R. The work done by the right-hand piston on the automobile is equal to the product of the force and the distance traveled:

$$W_R = F_R y_R \tag{10.14}$$

We can also consider the work done by the force F_L on the left-hand piston. That piston moves downward a distance y_L, so this work is

$$W_L = F_L y_L \tag{10.15}$$

Since we expect that the hydraulic lift will conserve mechanical energy, you should suspect that the work done *on* the left-hand piston should equal the work done *by* the right-hand piston; that is, $W_L = W_R$. Let's now show that this is indeed true.

When the two pistons move, there is a corresponding displacement of liquid (oil) from one side of the U-tube to the other as shown in parts B and C of Figure 10.16. Because the liquid is incompressible, the volume of oil that travels to the right must equal the volume that leaves from the left. These volumes are equal to the area of the tube times the height of the oil, so

$$A_L y_L = A_R y_R \tag{10.16}$$

and

$$y_L = y_R \frac{A_R}{A_L} \tag{10.17}$$

The piston with the *smaller* area (on the left) thus moves through a *larger* distance. Combining this result with our result for the forces from Equation 10.13 leads to

$$F_R y_R = \left(F_L \frac{A_R}{A_L} \right) \left(y_L \frac{A_L}{A_R} \right) = F_L y_L$$

The term on the far left, $F_R y_R$, equals W_R (Eq. 10.14), while the term on the far right, $F_L y_L$, is W_L (Eq. 10.15). We thus have

$$W_R = W_L$$

Hence, the work done *on the piston on the left* equals the work done *by the piston on the right*, and mechanical energy is indeed conserved. A hydraulic device can amplify forces (Eq. 10.13), but in doing so, it must "de-amplify" the displacement (Eq. 10.17).

EXAMPLE 10.6 | Designing a Hydraulic Lift

You are given the job of designing a hydraulic device for lifting large objects as in Figure 10.16. You want this device to be able to lift an SUV of mass $m_{SUV} = 2000$ kg, when operated with at force of just 750 N (the approximate weight of the author). The two pistons of the hydraulic lift are cylindrical, and suppose the piston lifting the SUV has a radius of 0.50 m. Find the radius of the other piston.

RECOGNIZE THE PRINCIPLE

From Pascal's principle, the pressure increases on the two sides of a hydraulic tube are equal. This leads to Equation 10.13, which (in words) states that the ratio of the forces equals the ratio of the piston areas. Given the ratio of the weight of the SUV and the weight of the author, we can thus find the ratio of the piston areas.

SKETCH THE PROBLEM

Figure 10.16 shows the problem. The piston lifting the SUV must be the one with the larger area.

To lift the SUV, we must amplify the applied force of 750 N by a factor of $m_{SUV}g/(750\text{ N}) = (2000\text{ kg})(9.8\text{ m/s}^2)/(750\text{ N}) = 26$, which is the required value of the ratio of the forces F_R/F_L in Figure 10.16A. Rearranging Equation 10.13 leads to

$$\frac{F_R}{F_L} = \frac{A_R}{A_L} = 26$$

The sides of the lift are cylindrical, so for each piston $A = \pi r^2$, and

$$\frac{A_R}{A_L} = \frac{\pi r_R^2}{\pi r_L^2} = \frac{r_R^2}{r_L^2} = 26$$

SOLVE

Solving for the radius of the left-hand piston gives

$$r_L = \frac{r_R}{\sqrt{26}} = \frac{0.50\text{ m}}{\sqrt{26}} = \boxed{0.098\text{ m}}$$

What does it mean?
This hydraulic system will amplify force by a factor of 26, and will de-amplify displacement by the same factor. The operator will have to displace 26 m of oil through the left-hand piston to lift the car through 1 m.

10.4 | BUOYANCY AND ARCHIMEDES'S PRINCIPLE

We established in Section 10.2 that the force of gravity causes the pressure in a fluid to vary with depth. This fact leads to *Archimedes's principle*, which applies to objects immersed in fluids.

Archimedes lived in the third century BC and is known for several contributions to science and mathematics. While his work on fluids was done long before the time of Newton, Archimedes made some very important discoveries involving forces in fluids that anticipated Newton's laws. Consider an object that is either fully immersed or is floating with a portion exposed above the surface of the fluid as sketched in Figure 10.18. In either case, the following statement applies.

> *Archimedes's principle:* **When an object is immersed in a fluid, the fluid exerts an upward force on the object that is equal to the weight of the fluid displaced by the object.**

This upward force on the object is called the *buoyant force*, and is due to the increase of pressure with depth in a fluid. The pressure at the bottom surface of an object is always greater than the pressure at the top, leading to a larger (upward) force on the bottom surface. The difference between these two forces is the buoyant force.

We can calculate the buoyant force (and thus prove Archimedes's principle) by applying our relation for the variation of the pressure with depth (Eq. 10.5) to the object in Figure 10.18C. Here, a solid box is completely submerged in a liquid of density ρ_L. For simplicity, we assume this box is rectangular with height h and area A. There are pressure forces on all faces of the box, the sides as well as the top and bottom. The horizontal forces on opposite sides of the box are equal in magnitude and opposite in direction and thus cancel, but the forces on the bottom and top of the box do not cancel. The pressure on the top is downward with magnitude $P_{top}A$, whereas the force on the bottom is upward and of magnitude $P_{bot}A$. The total force *exerted by the liquid* on the box is along the vertical direction, with

$$F = +P_{bot}A - P_{top}A$$

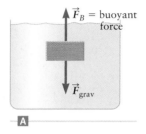

$\vec{F}_B$ = buoyant force

$\vec{F}_{grav}$

A

Only the *displaced* fluid volume (below the surface) contributes to the buoyant force.

$\vec{F}_B$

$\vec{F}_{grav}$

B

Area = A $F_{top} = P_{top} \cdot A$

h

$F_{bot} = P_{bot} \cdot A$

C

Figure 10.18 According to Archimedes's principle, an object that is immersed in a fluid will experience a buoyant force that is directed upward. That is true for **A** an object that is fully immersed in the fluid or **B** an object that floats at the surface and is partially exposed. In both cases, the buoyant force on the object equals the weight of the fluid displaced by the object. **C** The buoyant force is due to the increase in pressure with depth.

If we assume the fluid is a liquid of density ρ_L, according to Equation 10.5 the pressures at the bottom and top are related by

$$P_{bot} = P_{top} + \rho_L gh$$

where h is the height of the box. Combining these results gives

$$F = +P_{bot}A - P_{top}A = (P_{top} + \rho_L gh)A - P_{top}A = \rho_L hAg \qquad (10.18)$$

Hence, the force exerted by the liquid is upward; this is the buoyant force. Since the factor (hA) in Equation 10.18 equals the volume V of the box, this result is usually written as

$$F_B = \rho_L Vg \qquad (10.19)$$

The mass of the liquid displaced by the box is $\rho_L V$, so the term $\rho_L Vg$ in Equation 10.19 equals the weight of the displaced liquid. In words, Equation 10.19 says that the buoyant force F_B equals the weight of the *liquid* displaced by the box. This is another statement of Archimedes's principle.

Examples and Applications of Archimedes's Principle

Figure 10.18 shows Archimedes's principle at work in two different cases. In Figure 10.18A, a box of volume V is fully immersed in the fluid. The buoyant force equals the weight of the displaced fluid, and because the box is completely immersed, the volume of displaced fluid equals the volume of the box V. The weight of the displaced fluid is thus $F_B = \rho_L gV$, where ρ_L is the density of the fluid. If the weight of this box is exactly equal to F_B, the box will remain where it is, fully submerged in the liquid. If the box is heavier than F_B, it will sink to the bottom. It might also happen that the weight of the box is smaller than the weight of the liquid it displaces, which could happen with an empty box or with a boat. In such a case, the buoyant force for a fully submerged box would be *greater* than the weight of the box, and the box would rise to the surface and float (Fig. 10.18B). A portion of the box would then be exposed above the surface of the liquid so that the weight of the *displaced* liquid (the amount below the "waterline") equals the weight of the box. For that to occur, the average density of the box in Figure 10.18B must be less than the density of the liquid (just as the average density of a boat, including all volume below the waterline, must be less than the density of water).

As an application of Archimedes's principle, consider the situation in Figure 10.19. This box might be very heavy and submerged in a lake, and you might have the job of lifting it to the surface. You happen to know that the box is filled with lead and want to calculate how much force will be required to lift it. The minimum force required to lift the box is called the **apparent weight**.

Figure 10.19 shows the forces acting on the box. They are the force of gravity (the weight of the box), the force F_{lift} you are exerting as you lift the box, and the buoyant force F_B from the water. According to Archimedes's principle, F_B is the total force exerted by the fluid on the box. Since F_{lift} is the minimum force necessary to barely lift the box, we can take the acceleration of the box to be zero. Newton's second law for the box's motion along the vertical direction is then

$$\sum F = F_{grav} + F_B + F_{lift} = -mg + \rho_L Vg + F_{lift} = 0$$

where the buoyant force F_B is given by Archimedes's principle, Equation 10.19. The apparent weight of the box is equal to F_{lift}. Solving for F_{lift}, we find

$$F_{lift} = mg - \rho_L Vg \qquad (10.20)$$

The apparent weight is thus smaller than the true weight (mg) by an amount equal to the buoyant force. You have probably already done this type of experiment in a swimming pool. To estimate the magnitude of the buoyant force, assume the box in Figure 10.19 has a volume of 0.10 m^3 (about the size of a small suitcase). If it is filled with lead, its weight is

$$\text{true weight} = mg = \rho_{lead} Vg$$

Figure 10.19 The buoyant force causes the apparent weight (F_{lift}) to be smaller than the true weight (mg).

since the mass of the box is equal to the density of lead (ρ_{lead}) times the volume (V). Using the density of lead from Table 10.1, we find that a lead box of volume 0.10 m³ has a true weight of

$$\text{true weight} = \rho_{\text{lead}}Vg = (11{,}300 \text{ kg/m}^3)(0.10 \text{ m}^3)(9.8 \text{ m/s}^2) = 1.1 \times 10^4 \text{ N}$$

From Equation 10.20, the apparent weight of the lead-filled box equals

$$\text{apparent weight} = \rho_{\text{lead}}Vg - \rho_L Vg = (\rho_{\text{lead}} - \rho_L)Vg$$

where ρ_L is the density of the liquid. If we assume the liquid is water (which has a density of 1000 kg/m³), we find

$$\text{apparent weight} = (\rho_{\text{lead}} - \rho_L)Vg = (11{,}300 - 1000)(\text{kg/m}^3)(0.10 \text{ m}^3)(9.8 \text{ m/s}^2)$$

$$\text{apparent weight} = 1.0 \times 10^4 \text{ N}$$

Hence, in this case, the buoyant force makes the apparent weight about 10% smaller than the true weight.

EXAMPLE 10.7 | Floating in Water

A rectangular block of wood is floating at the surface of a container filled with water as shown in Figure 10.18B. You observe that 80% of the block is under water, whereas the rest is above the surface. Find the density of the wood.

RECOGNIZE THE PRINCIPLE

According to Archimedes's principle, the buoyant force equals the weight of the water displaced by the block. This upward buoyant force must equal the weight of the block in order for the block to remain in translational equilibrium.

SKETCH THE PROBLEM

Figure 10.20 shows the problem. The block of wood has a total height h. Since 80% of the block is submerged, the height of the portion under water is $d = 0.80h$.

IDENTIFY THE RELATIONSHIPS

The top and bottom of the block each have an area A, and the volume of the block that is under water is $V_{\text{under}} = dA$. The block thus displaces a volume of water equal to V_{under}, and the buoyant force equals the weight of this displaced water:

$$F_B = \rho_{\text{water}}V_{\text{under}}g = \rho_{\text{water}}dAg$$

The block is floating, so this buoyant force must equal (in magnitude) the gravitational force on the block (the block's true weight),

$$F_{\text{grav}} = m_{\text{wood}}g = \rho_{\text{wood}}V_{\text{total}}g$$

where $V_{\text{total}} = hA$ is the total volume of the block.

SOLVE

Setting F_{grav} equal to the buoyant force F_B leads to

$$F_B = \rho_{\text{water}}dAg = \rho_{\text{wood}}V_{\text{total}}g = \rho_{\text{wood}}hAg$$

$$\rho_{\text{water}}d = \rho_{\text{wood}}h$$

$$\rho_{\text{wood}} = \rho_{\text{water}}\frac{d}{h} \qquad (1)$$

Using $d = 0.80h$ together with the density of water from Table 10.1, we find

$$\rho_{\text{wood}} = \rho_{\text{water}}\frac{d}{h} = (1000 \text{ kg/m}^3)(0.80) = \boxed{800 \text{ kg/m}^3}$$

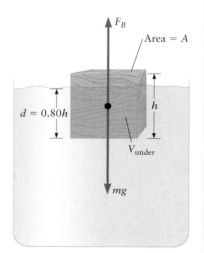

F_B

Area = A

$d = 0.80h$

h

V_{under}

mg

Figure 10.20 Example 10.7.

What does it mean?

We see from Equation (1) that if an object is floating in a liquid, it must have a density that is less than the density of the liquid (as we claimed in our discussion of Fig. 10.18B). Recall that the specific gravity of a substance equals the ratio of its density to the density of water. Hence, a substance will float in water if its specific gravity is less than 1.

Archimedes's Principle Holds for Objects of Any Shape, in Both Incompressible and Compressible Fluids

When we derived Archimedes's result for the buoyant force, Equation 10.19, we assumed the submerged (or floating) object has the shape of a simple rectangular block (as in Figs. 10.18 and 10.20). We also used our relation for the variation of the pressure with depth in a liquid ($P = P_0 + \rho g h$). However, Archimedes's principle applies to objects of *all* shapes that are immersed in gases as well as in liquids. Archimedes's principle thus applies to a very wide variety of situations.

EXAMPLE 10.8 | Operation of a Helium-Filled Balloon

NASA is experimenting with balloons for use on ultralong-duration (100-day) flights around the Earth. Consider a balloon that has the shape of a cylindrical chamber (Fig. 10.21) that is $h = 35$ m tall, has a radius $r = 30$ m, and is made of extremely light plastic. (The balloon in Figure 10.22 has a more realistic shape.) If our model cylindrical balloon is filled with helium gas, what is the mass of the maximum payload it could lift from the Earth's surface?

RECOGNIZE THE PRINCIPLE

The weight of the maximum payload equals the buoyant force on the balloon minus the weight of the balloon itself. We can find the buoyant force using Archimedes's principle from the volume of the balloon and the density of the air it displaces.

SKETCH THE PROBLEM

This problem is described by Figure 10.21. The balloon is fully immersed in the atmosphere.

IDENTIFY THE RELATIONSHIPS

The buoyant force on the balloon equals to the weight of the air the balloon displaces. If the balloon has a volume V, this weight is

$$\text{weight of air displaced} = \rho_{air} V g = F_B$$

The "skin" of the balloon is composed of a very light plastic, so we'll assume its mass is very small compared with both the mass of the payload m_P and the mass of the helium gas inside the balloon. This helium gas has a mass m_{He}, which is given by $m_{He} = \rho_{He} V$, where ρ_{He} is the density of helium gas. The buoyant force must support the weight of the helium and the weight of the payload, so

$$F_B = \rho_{air} V g = m_{He} g + m_P g = \rho_{He} V g + m_P g$$

$$m_P g = \rho_{air} V g - \rho_{He} V g$$

$$m_P = (\rho_{air} - \rho_{He}) V$$

The volume of a cylindrical balloon is equal to the area of the base times the height, $V = \pi r^2 h$.

Figure 10.21 Example 10.8.

Figure 10.22 Example 10.8. This large NASA balloon is filled with helium gas and used to lift scientific payloads to high altitudes.

SOLVE

Using the densities of air and helium from Table 10.1 along with the dimensions of the balloon, we find

$$m_P = (\rho_{air} - \rho_{He})V = (\rho_{air} - \rho_{He})(\pi r^2 h) = (1.29 - 0.18)(\text{kg/m}^3)\pi(30\text{ m})^2(35\text{ m})$$

$$m_P = \boxed{1.1 \times 10^5 \text{ kg}}$$

What does it mean?
This payload is quite large; most scientific payloads would have a much smaller mass. Why would NASA make the balloon so large? In this example, we have calculated the mass of the largest possible payload for the balloon at sea level. Air is a compressible fluid, so its density decreases as the balloon moves up in the atmosphere and the lower density at high altitudes decreases the lifting capability of the balloon. NASA designs the balloon to be able to lift the payload at the desired "cruising" altitude, so the balloon will need to have a much greater lift force at ground level.

CONCEPT CHECK 10.4 | Buoyant Force and a Floating Ship

A ship floats because of the buoyant force on it. A typical ship is composed of steel, and we see from Table 10.1 that the density of steel is much *greater* than the density of water. How can the buoyant force support a steel ship?

   (a) Real ships are made from special low-density metals.
   (b) A ship will float if its average density (including both the steel and the air inside) is less than the density of water.
   (c) A ship floats because it is watertight and does not leak.

10.5 | FLUIDS IN MOTION: CONTINUITY AND BERNOULLI'S EQUATION

So far in this chapter, we have only considered fluids that are at rest. We are now ready to tackle some problems in which a fluid is in motion. The general theory of the mechanics of a moving fluid is very complicated because the velocity $\vec{v}$ of a fluid usually varies from place to place within the fluid. This velocity is, of course, a vector, so one has to deal with three vector components that are each a function of position, leading quickly to some very involved mathematics.

To keep our discussion of fluid dynamics as simple as possible, we therefore make the following assumptions. (1) The fluid density is constant; that is, the fluid is incompressible. (2) The flow is steady; that is, the fluid velocity is independent of time (although it can still vary from point to point in space). (3) There is no friction, that is, no *viscosity*. (We'll discuss viscosity later in this chapter.) (4) We assume there are no "smoke rings" or other kinds of complex flow patterns such as turbulence. Fluids that satisfy these assumptions are called ***ideal fluids***. Although we thus consider only a few of the simplest situations, they will still illustrate some key aspects of fluid dynamics.

Figure 10.23 shows a fluid flowing through a pipe. Because of the simple geometry of this pipe, we need only consider the velocity of the fluid along the pipe, in the x direction in Figure 10.23. We denote the fluid's speed at the pipe inlet (on the left) as v_L and the speed at the pipe outlet (on the right) as v_R. We now want to consider how v_L and v_R are related.

The amount of fluid that flows through the pipe must be *conserved*; this means that the mass of fluid flowing in from the left during a particular time interval Δt must equal the mass flowing out at the right during this interval. In other words, the rate at which the fluid flows into the pipe must equal the rate at which the fluid flows out of the pipe. This is called the ***principle of continuity***.

Figure 10.23 According to the principle of continuity, the rate at which liquid flows into a pipe must equal the rate at which it flows out.

The basic idea behind the principle of continuity is that "what goes in must come out." This simple idea has important implications for the fluid's speed. For example, the pipe in Figure 10.23 has a larger diameter on the left (the inlet) than on the right (the outlet). According to the principle of continuity, the amount of fluid that leaves the pipe each second must equal the amount that flows in. Since the outlet is smaller than the inlet, the outlet speed (v_R) must be larger than the inlet speed. In other words, a small-diameter flow stream with a large speed can carry the same amount of fluid per unit time as a large-diameter flow stream with a small speed. That is why the speed of water coming from a garden hose increases when you put your thumb over part of the end of the hose (thus decreasing the outlet area).

Let's now give a more careful analysis of the principle of continuity. For an incompressible fluid, the *volume* of fluid flowing into a pipe must equal the volume of fluid that flows out (because mass is equal to density times volume, and the density is constant). This inflowing volume is shown in Figure 10.23; it equals the volume of fluid that flows into the pipe from the left during a time interval Δt, filling a cylinder of length $v_L \Delta t$. Since the inlet area is A_L, the volume of fluid flowing into the pipe is

$$V_{in} = (v_L \Delta t) A_L = v_L A_L \Delta t \qquad (10.21)$$

Notice that here the capital V is a *volume*; the lowercase v is a *speed*. Using the same argument, the fluid flowing out of the pipe is

$$V_{out} = (v_R \Delta t) A_R = v_R A_R \Delta t \qquad (10.22)$$

According to the principle of continuity, the amount that flows into the pipe must equal the amount that flows out, so V_{in} and V_{out} are equal, and

$$V_{in} = V_{out}$$
$$v_L A_L \Delta t = v_R A_R \Delta t$$

which leads to

$$v_L A_L = v_R A_R \qquad (10.23) \qquad \text{Principle of continuity}$$

This result is called the **equation of continuity**; it applies not only to pipes as in Figure 10.23, but to many situations involving fluid flow.

The product of fluid speed and area in Equation 10.23 is also called the **flow rate**, which we denote as Q:

$$\text{flow rate} = Q = vA \qquad (10.24)$$

Hence, the equation of continuity (Eq. 10.23) says that the incoming flow rate equals the outgoing flow rate. The SI units of flow rate are cubic meters per second (m^3/s), but liters per second are also widely used. In U.S. customary units, one might use gallons per minute (gal/min). The flow rate from a garden hose is typically 2 to 5 gal/min.

CONCEPT CHECK 10.5 | Flow Rate in the Aorta ⊗

The flow rate of blood through the aorta is typically 0.080 liter/s = 80 cm^3/s. If it takes 1 minute for all your blood to pass through the aorta, is the volume of blood in your body (a) 4.8 liters, (b) 10 liters, (c) 8.0 liters, or (d) 1.0 liter?

Bernoulli's Equation

Let's again consider an ideal fluid moving through a pipe as sketched in Figure 10.24 and examine the flow from the point of view of work and energy. The fluid in Figure 10.24 is flowing from left to right. We denote the pressure and speed of the fluid at the far left as P_1 and v_1, and we suppose the left end of the pipe is at a height h_1. Likewise, the pressure, fluid speed, and height at the far right are P_2, v_2, and h_2. Figure 10.24 shows two snapshots of the fluid as it moves through the pipe. The first snapshot (Fig. 10.24A) is at $t = 0$, and the second snapshot (Fig. 10.24B) is a short time Δt later. During this time, the "boundaries" of this portion of the fluid have moved a certain distance; the boundary on the left moved a distance Δx_1, and the boundary on the right moved a distance Δx_2.

Figure 10.24 Bernoulli's equation tells how the pressure, speed, and height at one location in an ideal fluid are related to P, v, and h at any other location. As the fluid flows through this pipe, the net effect is to transport the shaded volume from the left end in **A** to the right end in **B**. The change in the energy of the entire system is then just the difference between the energies of these two shaded volumes.

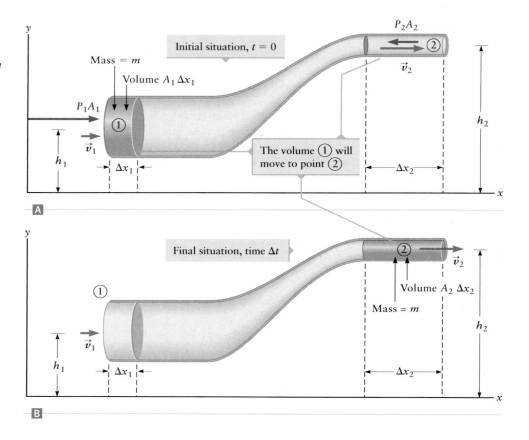

We now apply the work–energy theorem; our goal is to get a relationship between the pressure, speed, and height of the fluid on the two sides in Figure 10.24. There are external forces on the fluid associated with the pressure on the far left end and with the pressure on the far right end. The work done on the fluid by the pressure on the left equals the force on the boundary at the left multiplied by the distance this boundary travels. This force F_1 is related to the pressure on the left by $F_1 = P_1 A_1$, so the work done on the fluid is

$$W_1 = F_1 \, \Delta x_1 = P_1 A_1 \, \Delta x_1$$

To find the work done on the fluid by the pressure on the far right end, we note that because the force is a pressure force, it is directed to the left (in the $-x$ direction). The right-hand boundary of the fluid moves to the right, so the force and displacement are in opposite directions and thus the work done on the fluid is negative. We have

$$W_2 = -F_2 \, \Delta x_2 = -P_2 A_2 \, \Delta x_2$$

The total work done on the fluid is the sum of W_1 and W_2, and according to the work–energy theorem, this sum equals the change in the mechanical energy of the fluid:

$$W_{\text{total}} = W_1 + W_2 = P_1 A_1 \, \Delta x_1 - P_2 A_2 \, \Delta x_2 = \Delta KE + \Delta PE \qquad (10.25)$$

Figure 10.24 shows snapshots of the initial and final states of the system. The *only* difference between these two snapshots is that the volume of fluid shown as shaded has been moved from the left end of the pipe to the right end. So, the change in energy of the fluid in going from the snapshot in Figure 10.24A to that in Figure 10.24B is equal to the change in the energy of just this shaded region. If the shaded volume of fluid has a mass m, its initial mechanical energy (kinetic plus potential) is

$$KE_i + PE_i = \tfrac{1}{2}mv_1^2 + mgh_1$$

whereas the final mechanical energy is

$$KE_f + PE_f = \tfrac{1}{2}mv_2^2 + mgh_2$$

The change in the mechanical energy is therefore

$$\Delta KE + \Delta PE = \tfrac{1}{2}mv_2^2 + mgh_2 - (\tfrac{1}{2}mv_1^2 + mgh_1) \qquad (10.26)$$

Inserting this expression into Equation 10.25 leads to

$$W_{total} = P_1 A_1\,\Delta x_1 - P_2 A_2\,\Delta x_2 = \Delta KE + \Delta PE$$

$$P_1 A_1\,\Delta x_1 - P_2 A_2\,\Delta x_2 = \tfrac{1}{2}mv_2^2 + mgh_2 - (\tfrac{1}{2}mv_1^2 + mgh_1)$$

$$P_1 A_1\,\Delta x_1 + \tfrac{1}{2}mv_1^2 + mgh_1 = P_2 A_2\,\Delta x_2 + \tfrac{1}{2}mv_2^2 + mgh_2 \qquad (10.27)$$

The factor $A_1\,\Delta x_1$ is the volume of the shaded region on the left in Figure 10.24A, and by the principle of continuity, it must equal the volume of the shaded region on the right in Figure 10.24B. Because these volumes are equal, we'll call them both V. Dividing both sides of Equation 10.27 by V gives

$$P_1 + \frac{1}{2}\frac{m}{V}v_1^2 + \frac{m}{V}gh_1 = P_2 + \frac{1}{2}\frac{m}{V}v_2^2 + \frac{m}{V}gh_2$$

The ratio of the mass to the volume, m/V, is just the density, which is constant in an incompressible fluid. This leads to our final result

$$P_1 + \tfrac{1}{2}\rho v_1^2 + \rho gh_1 = P_2 + \tfrac{1}{2}\rho v_2^2 + \rho gh_2 \qquad (10.28) \qquad \text{Bernoulli's equation}$$

which is **Bernoulli's equation**. It tells how the pressure and speed vary from place to place within a *moving* fluid. You may notice that when $v = 0$, Bernoulli's equation gives our previous result (Eq. 10.5) for the variation of the pressure with depth in a stationary liquid.

Interpreting Bernoulli's Equation

From our derivation, you should see that Bernoulli's equation is a direct result of the conservation of energy. The terms in Equation 10.28 that have the form $\tfrac{1}{2}\rho v^2$ are the kinetic energies of a unit volume of the fluid, while the terms ρgh are gravitational potential energies per unit volume. How do we interpret the terms involving P? The pressure terms represent energy associated with pressure in the fluid; as you might expect, a higher value of P corresponds to a larger energy because a larger pressure means that the fluid can exert a larger force ($F = PA$). We can thus write Bernoulli's equation as

energy due to pressure at point 1	kinetic energy at point 1	potential energy at point 1		energy due to pressure at point 2	kinetic energy at point 2	potential energy at point 2
P_1	$+\ \tfrac{1}{2}\rho v_1^2$	$+\ \rho gh_1$	$=$	P_2	$+\ \tfrac{1}{2}\rho v_2^2$	$+\ \rho gh_2$

In words, Bernoulli's equation says that the total mechanical energy of the fluid is conserved as it travels from place to place (point 1 to point 2), but some of this energy can be "converted" from kinetic energy to potential energy and vice versa as the fluid moves.

Applications of Bernoulli's Equation

One of the most important applications of Bernoulli's equation involves the operation of an airplane wing. Figure 10.25 shows the pattern of airflow near a wing. Most airplane wings have an asymmetric shape, with a "bulge" on the top surface. This bulge causes the air speed to be slightly greater over the top of the wing compared with the air speed over the bottom of the wing. To see why the speed of the air on the top is greater, notice that the air flowing over the top must travel a greater distance to get around the wing (so it must travel faster) than the air that flows around the bottom. According to Bernoulli's equation (Eq. 10.28), an *increase* in the fluid speed leads to a *decrease* in the pressure. Hence, the pressure on the bottom of the wing is greater than the pressure on the top. The total force due to air pressure on the wing is

$$F_{total} = P_{bot}A - P_{top}A = (P_{bot} - P_{top})A \qquad (10.29)$$

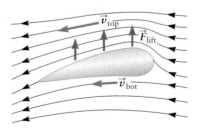

Figure 10.25 Schematic of the airflow pattern around an airplane wing. Because of the shape of the wing, the air speed is smaller under the bottom of the wing than over the top. According to Bernoulli's equation, this difference leads to a larger pressure on the bottom and hence to an upward force called "lift."

Since $P_{\text{bot}} > P_{\text{top}}$, this force is *upward*; it is the "lift" force on a wing that enables an airplane to fly.

Let's now consider the lift on an airplane wing in a little more detail. Suppose the area of the wing is 60 m², the air speed over the top surface of the wing is v_{top} = 250 m/s (about 500 mph), and the air speed under the bottom of the wing is v_{bot} = 200 m/s; these are typical values for a commercial jet. To calculate the lift using Equation 10.29, we need to find the difference in pressure between the bottom and top of the wing. Applying Bernoulli's equation to a point at the bottom of the wing (point 1 in Eq. 10.28) and a point at the top (point 2), we have

$$P_{\text{bot}} + \tfrac{1}{2}\rho v_{\text{bot}}^2 + \rho g h_{\text{bot}} = P_{\text{top}} + \tfrac{1}{2}\rho v_{\text{top}}^2 + \rho g h_{\text{top}}$$

which can be rearranged to give the pressure difference across the wing:

$$P_{\text{bot}} - P_{\text{top}} = \tfrac{1}{2}\rho v_{\text{top}}^2 - \tfrac{1}{2}\rho v_{\text{bot}}^2 + \rho g h_{\text{top}} - \rho g h_{\text{bot}} = \tfrac{1}{2}\rho(v_{top}^2 - v_{bot}^2) + \rho g(h_{top} - h_{bot})$$

Here ρ is the density of air, which we know from Table 10.1. (For simplicity, we use the value of ρ at sea level and 0°C; to be more accurate, we would need to know the precise altitude and temperature.) We also know the speeds; our last step is to deal with the difference in height from top to bottom. For a typical wing of a large commercial jet, this difference is about 1 m, so we take $h_{\text{top}} - h_{\text{bot}}$ = 1 m. Inserting all these values gives

$$P_{\text{bot}} - P_{\text{top}} = \tfrac{1}{2}\rho(v_{top}^2 - v_{bot}^2) + \rho g(h_{top} - h_{bot})$$

$$= \tfrac{1}{2}(1.29 \text{ kg/m}^3)[(250 \text{ m/s})^2 - (200 \text{ m/s})^2] + (1.29 \text{ kg/m}^3)(9.8 \text{ m/s}^2)(1 \text{ m})$$

$$P_{\text{bot}} - P_{\text{top}} = 1.5 \times 10^4 \text{ Pa} + 13 \text{ Pa} \qquad (10.30)$$

In the last step, we have written the pressure difference as a sum of two separate terms; the first is due to the difference in air velocities, and the second term is due to the difference in heights between the top and bottom of the wing. For this rapidly moving airplane wing, where the air velocities are large and the height difference is small, the kinetic energy terms ($\tfrac{1}{2}\rho v^2$) dominate Bernoulli's equation. Hence, the pressure difference in Equation 10.30 and thus also the lift are due almost entirely to the speed terms in Bernoulli's equation.

To complete our calculation of the lift on the wing, we insert the pressure difference from Equation 10.30 into our expression for the lift (Eq. 10.29). We find

$$F_{\text{lift}} = (P_{\text{bot}} - P_{\text{top}})A = (1.5 \times 10^4 \text{ Pa})(60 \text{ m}^2) = 9.0 \times 10^5 \text{ N}$$

The mass of a medium-sized commercial jet is about 5×10^4 kg (weight = $mg \approx 5 \times 10^5$ N), so this lift force is indeed sufficient for the jet to fly.

The problem of lift on a real airplane wing is complicated. The Bernoulli effect (Eq. 10.28) is responsible for much of the lift, but other effects are important and must be taken into account in the design of a real airplane. To measure the lift on prototype designs, many airplane makers employ wind tunnels (Fig. 10.26).

Insight 10.1

AIRPLANE WINGS IN AIR

In our derivation of Bernoulli's equation (Eq. 10.28), we assumed the fluid is incompressible. We then applied Bernoulli's equation to calculate the lift on an airplane wing (Eq. 10.29). The fluid in that case is air, which is compressible, but Bernoulli's equation can still be used to get an approximate value of the lift for the following reason. The pressure difference between the top and bottom of the wing (Eq. 10.30) is about 15% of atmospheric pressure and there is a corresponding 15% variation in the air density between the bottom and top. Hence, to within this accuracy, air under these conditions will have an approximately constant density. Our estimate for the lift force should thus be reliable to about this accuracy.

Figure 10.26 The aerodynamic forces on airplanes and other objects are often studied in wind tunnels. The Wright brothers used this wind tunnel to study the forces on wings they designed.

National Museum of USAF/U.S. Air Force Photo

EXAMPLE 10.9 | ® Speed of Water from a Faucet

Figure 10.27 shows the water system for a typical house. The source is a water tower located at some height above the houses it serves. In this particular water tower, the surface of the water in the tower is open to the atmosphere and is 30 m above the home shown in the figure. A faucet in the home is then opened. What is the speed of the water as it comes out of the faucet?

Figure 10.27 Example 10.9.

RECOGNIZE THE PRINCIPLE

We use Bernoulli's equation (Eq. 10.28) to find the speed of the water coming out of the faucet. We take point 1 to be at the surface of the water in the tower and point 2 to be at the faucet.

SKETCH THE PROBLEM

Figure 10.27 describes the problem.

IDENTIFY THE RELATIONSHIPS

Bernoulli's equation reads

$$P_1 + \tfrac{1}{2}\rho v_1^2 + \rho g h_1 = P_2 + \tfrac{1}{2}\rho v_2^2 + \rho g h_2$$

and we take locations 1 and 2 as shown in Figure 10.27. We wish to solve for the speed at the faucet v_2, so we must find values for the pressures at the tower and faucet (P_1 and P_2), the heights of the tower and faucet (h_1 and h_2), and the speed at the tower (v_1). The water tower is open to the atmosphere, so the pressure there is atmospheric pressure and $P_1 = P_{atm}$. Likewise, the outlet of the faucet is open to the atmosphere, so it is also at atmospheric pressure and $P_2 = P_{atm}$. We can take the origin of our y coordinate axis to be at the faucet, leading to $h_2 = 0$ and $h_1 = 30$ m. We must now deal with the v_1, the speed of the top surface of the water in the tower. The area of this top surface is much larger than the cross-sectional area of the faucet. As a result, this surface will fall very slowly as water flows from the faucet, and it is a good approximation to take $v_1 = 0$. This argument is equivalent to using the equation of continuity (Eq. 10.23) with the area of the water surface in the tower being much greater than the cross-sectional area of the faucet.

SOLVE

Putting all these quantities into Bernoulli's equation, we find

$$P_{atm} + 0 + \rho g h_1 = P_{atm} + \tfrac{1}{2}\rho v_2^2 + 0$$

$$\rho g h_1 = \tfrac{1}{2}\rho v_2^2$$

$$v_2 = \sqrt{2 g h_1} \qquad (1)$$

Inserting our value for the height of the water tower gives

$$v_2 = \sqrt{2 g h_1} = \sqrt{2(9.8 \text{ m/s}^2)(30 \text{ m})} = \boxed{24 \text{ m/s}}$$

What does it mean?
You should notice that our result for v_2 in Equation (1) is precisely the same as the speed of a rock (or other object) that falls freely from a height of h_1, as found in Chapters 3 and 6. That is no accident. The water that flows from the water tower to the faucet in Figure 10.27 is also undergoing free fall, and we should expect it to have the same final speed we would find for a rock or a baseball falling through the same distance.

Figure 10.28 shows another example of water undergoing free fall as it flows from a faucet. Notice that the water stream becomes narrower (a smaller cross-sectional area) the farther one gets from the faucet. This narrowing is a result of

Figure 10.28 The narrowing of this stream of water as it flows from a faucet can be understood using Bernoulli's equation and the principle of continuity.

the principle of continuity together with Bernoulli's equation. According to the equation of continuity (Eq. 10.23), we have

$$v_1 A_1 = v_2 A_2$$

for any two points in a flowing stream of liquid. The speed is larger at the bottom of the stream in Figure 10.28 than at the top, so the cross-sectional area must be smaller.

Figure 10.29 Concept Check 10.6.

CONCEPT CHECK 10.6 | Pressure in a Garden Hose

Water flows through two garden hoses, one large and one small, connected as shown in Figure 10.29. How are the pressures in the two hoses related?

(a) $P_1 > P_2$　　(b)　$P_1 = P_2$　　(c)　$P_1 < P_2$

10.6 | REAL FLUIDS: A MOLECULAR VIEW

In all our work on moving fluids in Section 10.5, we ignored frictional forces internal to the fluid and between the fluid and the walls of the container (i.e., a pipe). These *viscous* forces depend on the fluid velocity and are often negligible when this velocity is small. Bernoulli's equation therefore works well when the fluid viscosity is small (e.g., when the fluid is air), when the fluid is flowing slowly, or both. However, in many situations viscosity is important. To understand to origin of viscosity and several other effects in fluids, it is necessary to consider fluids at a molecular scale.

Viscosity and Poiseuille's Law

We have seen that friction can play a large role in the motion of many terrestrial objects. Frictional forces can also play an important role in fluid motion. Consider a fluid flowing through a tube as sketched in Figure 10.30. The fluid molecules near the wall of the tube experience strong forces from the molecules in the wall. In fact, these forces are so strong that in most cases the fluid molecules next to the wall are almost bound to it. As a result, molecules close to the wall move much more slowly than molecules that are far from the wall. It is useful to think of the fluid in the tube as a series of cylindrical layers, a few of which are sketched in Figure 10.30A. The fluid molecules in layers near the wall move relatively slowly due to their attraction to the wall, whereas those in layers far from the wall move with a higher velocity.

The motion of each fluid layer is strongly influenced by the adjacent layers. Whenever the layers move at different velocities, there is a force between them. This force is very similar to the force of kinetic friction between two solid surfaces that are sliding past each other (e.g., a hockey puck sliding across a horizontal surface). The fluid layer denoted as layer 2 in Figure 10.30A is moving faster than the adjacent layer 1. As a result, layer 2 exerts a force on layer 1 that acts to "drag" layer 1 along with layer 2. This force is due to a property of the fluid called *viscosity*. In a "thick" fluid such as honey, these forces are large and these fluids are said to have a large viscosity. "Thin" fluids such as water or methanol have a much smaller viscosity.

Figure 10.30 Real fluids have viscosity, leading to a frictional force between regions of the fluid that move with different velocities. **A** When a fluid flows through a tube or pipe, the fluid near the center has a higher velocity than the fluid near the walls, as plotted schematically in **B**.

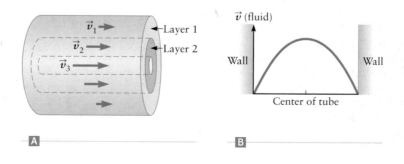

The viscous forces between adjacent fluid layers lead to a pattern of flow speeds shown in Figure 10.30B, where the fluid speed is plotted as a function of position within a pipe. Layers near the center of the tube have the highest speeds, and the flow speed approaches zero at the walls. The overall variation of speed with position is parabolic.

For a viscous fluid to move through a tube, the force due to the pressure difference across the ends of the tube must overcome the viscous force. If the pipe has a circular cross section with radius r and length L and the pressures at the ends are P_1 and P_2, a careful application of Newton's second law to the flow in Figure 10.31 shows that the *average* fluid speed in the tube is given by *Poiseuille's law*,

$$v_{ave} = \frac{P_1 - P_2}{8\eta L} r^2 \qquad (10.31)$$

Here η (the Greek letter eta) is the viscosity of the fluid. Values of the viscosity for some common fluids are given in Table 10.3. According to Poiseuille's law, increasing the pressure difference $(P_1 - P_2)$ increases v_{ave} and hence increases the flow rate. Most importantly, decreasing the tube radius r decreases the flow speed; in a smaller tube, a greater fraction of the fluid is close to the walls, so the viscous drag from the walls has a larger effect.

Poiseuille's law is useful for understanding fluid flow in various situations, including the flow of cooling water in an engine, the flow of crude oil through a long pipeline, and the flow of blood through the body. In many cases, one is interested in the flow rate Q (which we defined in Equation 10.24 in connection with the equation of continuity); here Q is the rate at which fluid passes into or out of a tube or pipe. We have already discussed both the volume flow rate in our discussion of the principle of continuity and the volume flow into and out of a pipe, V_{in} and V_{out}. In that discussion (in connection with Eqs. 10.21 and 10.22), we assumed the flow velocity was constant across the radius of the tube and found that the volume flow rate is $Q = vA$. For the case in Figure 10.30, in which the flow speed is not constant, we must use the average flow speed v_{ave}. Hence, we have

$$Q = v_{ave}A \qquad (10.32)$$

where A is the area of the tube or pipe.

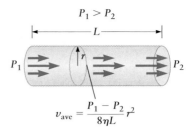

$$v_{ave} = \frac{P_1 - P_2}{8\eta L} r^2$$

Figure 10.31 Poiseuille's law tells how the average flow speed depends on the dimensions of the tube, the viscosity, and the pressure difference $P_1 - P_2$.

Table 10.3 Viscosities of Some Common Fluids

Fluid	Viscosity ($N \cdot s/m^2$)
Water (20°C)	1.0×10^{-3}
Blood (at body temperature, 37°C)	2.7×10^{-3}
Air (20°C)	1.8×10^{-5}
Honey (20°C)	1000
Methanol (30°C)	5.1×10^{-3}
Ethanol (30°C)	1.0×10^{-3}

EXAMPLE 10.10 | ✖ Blood Flow through an Artery

Consider the flow of blood through the large artery that extends from the heart to the lungs. The radius of this artery is typically 2.0 mm, and it is about 10 cm long. **(a)** If the pressure difference across the ends of the artery is 400 Pa (about 0.4% of atmospheric pressure), what is the average speed of the blood? **(b)** Now suppose this artery becomes somewhat narrower, as often happens with age. If the radius is reduced to 1.5 mm, what pressure difference is required to maintain the same average speed as in part (a)?

RECOGNIZE THE PRINCIPLE

This is a direct application of Poiseuille's law (Eq. 10.31), the relation between pressure difference and average speed for the flow of a viscous fluid (blood).

SKETCH THE PROBLEM

Figure 10.31 describes the problem.

IDENTIFY THE RELATIONSHIPS

According to Poiseuille's law, the average flow speed is related to the pressure across the artery $(P_1 - P_2)$ by

$$v_{ave} = \frac{P_1 - P_2}{8\eta L} r^2$$

where r is the radius of the artery and η is the viscosity.

SOLVE

(a) For blood (viscosity $\eta = 0.0027$ N · s/m² as listed in Table 10.3) flowing in an artery of radius $r = 2.0$ mm with a pressure difference $P_1 - P_2 = 400$ Pa, we find

$$v_{ave} = \frac{P_1 - P_2}{8\eta L} r^2 = \frac{(400\ \text{Pa})}{8(0.0027\ \text{N}\cdot\text{s/m}^2)(0.10\ \text{m})}(0.0020\ \text{m})^2 = \boxed{0.74\ \text{m/s}}$$

(b) To find the effect of changing the radius, we rearrange Poiseuille's law to solve for the pressure difference:

$$v_{ave} = \frac{P_1 - P_2}{8\eta L} r^2$$

$$P_1 - P_2 = \frac{8\eta L v_{ave}}{r^2}$$

Now we insert our value for the desired flow velocity—the value of v_{ave} found in part (a)—and the new radius:

$$P_1 - P_2 = \frac{8\eta L v_{ave}}{r^2} = \frac{8(0.0027\ \text{N}\cdot\text{s/m}^2)(0.10\ \text{m})(0.74\ \text{m/s})}{(0.0015\ \text{m})^2} = \boxed{710\ \text{Pa}}$$

What does it mean?
When the radius is decreased, the pressure must increase from 400 Pa to 710 Pa—quite a substantial change—to maintain the same flow speed. This example shows how what might seem to be a small change in an artery's size can lead to high blood pressure and the associated health risks.

Viscosity and Stokes's Law

Viscosity leads to a frictional force between adjacent "layers" of a moving fluid and also to a frictional force on the wall in Figure 10.30. The wall attracts molecules of the fluid, and the fluid molecules in turn exert a reaction force on the wall (Newton's third law at work again). This force on the wall is also the source of the drag force we considered in Chapter 3. According to Stokes's law (Eq. 3.23), the drag force on a spherical object of radius r that moves slowly through a fluid with a viscosity η is

$$\vec{F}_{drag} = -6\pi\eta r\vec{v} \tag{10.33}$$

where $\vec{v}$ is now the velocity of the object in the fluid. In Chapter 3, we wrote this relation as $\vec{F}_{drag} = -Cr\vec{v}$, where C is a constant; now we see that the drag coefficient C is proportional to the viscosity η. This is the *same* viscosity that enters into Poiseuille's law. The negative sign in Equation 10.33 indicates that the drag force is opposite to the object's velocity.

In Chapter 3, we applied Stokes's law to the motion of objects such as an *E. coli*. Stokes's law is also the key to understanding an important method for separating molecules according to their size. This method, based on a process called ***electrophoresis***, is shown schematically in Figure 10.32. Here we imagine a fluid containing a collection of molecules of different sizes; in a typical application, they are biological molecules such as DNA, RNA, or proteins. An electric force is then applied that pushes the molecules through the fluid such that the magnitude of the electric force is the same on each molecule.[2] If that were the only force, the molecules would be accelerated according to Newton's second law. However, each molecule also

The electric forces are the same, but the drag coefficients are different, so the terminal velocity is *larger* for the *smaller* molecule.

Figure 10.32 When an electric field is applied to a typical biomolecule in solution, the electric force F_E on the molecule is usually the same for all molecules because they usually have the same electric charge. The molecules also experience a drag force due to the viscosity of the fluid. This drag force depends on the size and shape of the molecules, so different molecules move with different terminal velocities. This analysis is very similar to our calculation of the terminal velocity of a skydiver in Chapter 3.

---

[2]The origin of this force is explained in Chapter 17, where we discuss how an electric field exerts a force on an ion.

experiences a Stokes drag force. The Stokes's law force in Equation 10.33 applies to spherical objects, so this functional relation does not apply precisely to a bio-molecule, but the basic physics of the viscous drag force certainly *does* apply to all molecules. The molecules in Figure 10.32 experience a drag force whose magnitude depends on the molecule's size and shape. For these biomolecules, the drag force is given by the relation

$$\vec{F}_{drag} = -B\vec{v} \qquad (10.34)$$

where B is a factor that depends on the molecule (especially its size). The value of B is a constant for any given kind of molecule—that is, for a particular type of DNA or RNA or for a particular virus—but it is different for different molecules.

If the electric force on a molecule is F_E, Newton's second law becomes

$$\Sigma F = F_E - Bv = ma$$

where we assume one-dimensional motion so that the forces and v are all along a single direction. This net force accelerates a molecule until it reaches a terminal velocity (as discussed in Chapter 3), at which time the molecule's velocity is constant and its acceleration is zero, leading to

$$\Sigma F = F_E - Bv_{term} = 0$$

$$v_{term} = \frac{F_E}{B}$$

Since B is different for different molecules, the terminal velocity v_{term} also varies. Hence, different molecules travel through the fluid with different velocities, providing a way to separate molecules by type. This difference is the basis for one method of DNA "fingerprinting."

Surface Tension

In discussing Figure 10.30, we described how the molecules in a fluid experience attractive forces from a solid container wall. It is also interesting to consider what happens at the free surface of a fluid. Consider a small amount of a liquid placed on the surface of a solid such as water sitting on a sheet of glass. Parts A and B of Figure 10.33 show two possible results; in part A, the water forms a spherical droplet, whereas in part B, the water spreads out as a flat layer. Which of these cases is found in real life? *Both* are possible, depending on the intermolecular forces. In most liquids, the liquid molecules bind more tightly to one another than to air. That is, the molecules in the liquid "prefer" to be in contact with one another, rather than in contact with the molecules in the air. For this reason, the liquid forms a shape giving the smallest possible surface area with the air, and that shape is a spherical drop.

Figure 10.33 Possible result for a small amount of liquid sitting on a solid surface. ☐ A spherical drop. ☐ A thin layer (film). ☐ and ☐ Surface tension allows certain creatures to "walk" on the surface of a liquid.

Table 10.4 Surface Tensions of Some Common Liquids at Room Temperature	
LIQUID	SURFACE TENSION (J/m²)
Water	7.3×10^{-2}
Ethanol	2.3×10^{-2}
Mercury	48×10^{-2}

This result can be expressed mathematically by saying that the fluid has a *surface tension* γ, where γ is a measure of the energy required to increase the surface area. The surface tensions of some typical fluids are given in Table 10.4. The SI unit of surface tension is joules per square meter (J/m^2).

Surface tension is responsible for effects such as the one shown in Figure 10.33C, where a bug is "walking" on water. As with the case of a spherical drop, the surface of the water takes on a shape that keeps the surface area as small as possible, given that it also must support the bug's weight. Here, that shape is a nearly flat surface, which supports the bug's feet rather than letting them penetrate deep under the original surface. The weight of the bug is very small, so the forces that bind the water molecules together at the surface are able to overcome the added force due to the bug's weight.

Capillary Pressure

The spherical shape of the liquid drop in Figure 10.33A is caused by the forces between the molecules of the liquid. The forces between the liquid's molecules and a wall can also affect the shape of a free liquid surface. We have just stated that the molecules of a liquid such as water are attracted more strongly to other water molecules than to air molecules. When a liquid is placed in a very narrow tube, however, the shape of the surface also depends on the forces between the molecules in the liquid and the walls of the tube. Figure 10.34 shows liquid in a small glass tube, called a capillary tube. The surface of the liquid in a capillary tube is *not* flat. Figure 10.34A shows the case of water in a glass capillary tube; here the water surface, which is called the meniscus, is concave upward. Figure 10.34A also shows a molecular-scale view. The water molecules are attracted to the glass, and this attraction pulls the surface of the water up a small amount at the walls of the tube, producing the concave shape of the meniscus.

The molecules in some liquids do not bind well to the surfaces of a capillary tube. The case of mercury is shown in Figure 10.34B; here the mercury atoms are attracted more strongly to one another than to the walls, leading to a convex meniscus, much like the shape of the water droplet in Figure 10.33B. Hence, the shape of the meniscus in a capillary tube depends on the molecular forces between the liquid and the walls of the tube.

The attraction of the water molecules in Figure 10.34A to the walls of a glass capillary tube leads to a force that "pulls" the water upward in the tube. This force, which is associated with the *capillary pressure*, causes the curvature of the water's surface (the meniscus), so the magnitude of this effect is proportional to the surface tension. In the simplest case, like the one shown in Figure 10.34A, the capillary pressure for a liquid in a tube of radius r is given by

$$P_{cap} = \frac{2\gamma}{r} \qquad (10.35)$$

where γ is again the surface tension (Table 10.4). This result applies to water in a glass tube, where the fluid molecules are strongly attracted to the walls of the tube. For other cases, such as mercury in a glass tube (Fig. 10.34B), the capillary pressure is given by a slightly more complicated relation than Equation 10.35 and is more than we want to discuss in this book.

How Plants Use Capillary Pressure ⊗

As an example of capillary pressure in action, let's consider how a tree moves water from its roots to its top branches. In a sense, this problem is similar to the one we considered in Figure 10.13, where we discussed how to pump water from an underground well to the surface. There we saw that with a pressure difference of P_{atm}, one can only pump water up to a maximum height h, where $h = (P_{atm}/\rho g) = 10$ m.

A tree is faced with a similar problem as it "pumps" water from its roots to its leaves. A tree also has a maximum pressure difference of P_{atm} to work with, hence one might suspect a tree to have the same limitation on the maximum pumping height.

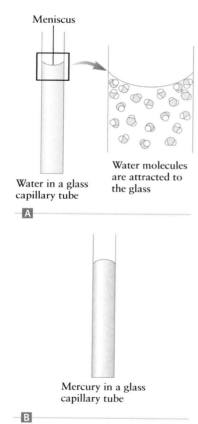

Meniscus

Water in a glass capillary tube

A

Water molecules are attracted to the glass

Mercury in a glass capillary tube

B

Figure 10.34 ◭ When a thin glass capillary tube contains water, the shape of the water surface (the meniscus) is concave upward. This shape is caused by the attraction of the water molecules to the surface of the tube. ◳ For mercury in a glass tube, the meniscus is convex.

So, with 10 m equal to about 30 ft, a tree cannot be taller than 30 ft and still move water from its roots to its top branches in this way. We know, though, that trees can be taller than 30 ft. How do they do it? The answer is that they use capillary pressure. The tubes that take water from tree's roots to its branches vary in size, but a radius of 1 μm (= 10^{-6} m) is quite common. According to Equation 10.35, and consulting Table 10.4 for the surface tension of water, we find a capillary pressure of

$$P_{\text{cap}} = \frac{2\gamma}{r} = \frac{2(7.3 \times 10^{-2} \text{ J/m}^2)}{1 \times 10^{-6} \text{ m}} = 1 \times 10^5 \text{ Pa} \qquad (10.36)$$

The case of a tree pumping water to its branches is then similar to the pump in Figure 10.13A, but now the capillary pressure adds to the atmospheric pressure, which enables the tree to pump water to a greater height. Recall that $P_{\text{atm}} = 1.01 \times 10^5$ Pa, so the capillary pressure in Equation 10.36 is just as great as atmospheric pressure. Narrower tubes can produce larger pressures. That is how trees do it.[3]

10.7 | TURBULENCE

In our discussion of moving fluids and Bernoulli's equation, we explained how the pattern of airflow around an airplane wing produces lift on the wing, and we drew a hypothetical flow pattern in Figure 10.25. The flow pattern shown there is particularly simple. In many situations, fluid flow patterns can be much more complex and very difficult to predict. Indeed, in the development of the airplane, designers such as the Wright brothers used experiments in wind tunnels (Fig. 10.26) to study lift forces on wings of different shapes. Similar studies can now be done with large-scale computer simulations, but wind tunnels are still widely used to explore the aerodynamic forces on cars, bicycles, airplanes, and many other objects.

Figure 10.35 shows an example of the types of complex flow patterns that occur near even simple objects. This photo shows the flow of air around a baseball. When the speed of the incoming air is greater than about 25 m/s (about 50 mi/h), *vortices* are created in the baseball's wake. These vortices are circulating patterns of air, much like small tornados.

When fluid velocities are large, the flow patterns can be very complex and fluctuate erratically with time, a behavior known as *turbulence*. Such complex flows play a role in a wide range of problems, including combustion, the weather (tornados and hurricanes), and the noise produced by rapidly moving cars and airplanes. Turbulence is the topic of much current research.

University of Notre Dame Archives

Figure 10.35 When a baseball travels at a typical speed, the airflow around the ball is turbulent. This baseball is moving from left to right.

---

[3]The precise details of this process in trees are still a matter of much research. Although our discussion is certainly a simplified version, capillary pressure is the basic key to the process.

SUMMARY | Chapter 10

KEY CONCEPTS AND PRINCIPLES

Pressure and density
The mechanics of fluids is described using the quantities *pressure* and *density*. Pressure is equal to the force per unit area,

$$P = \frac{F}{A} \qquad \text{(10.1) (page 310)}$$

(Continued)

Pressure can be associated with the force on the surface of a container, or it can involve a surface contained within the fluid. The pressure in the atmosphere at the Earth's surface is approximately

$$P_{atm} = 1.01 \times 10^5 \text{ Pa} \qquad \textbf{(10.2)} \text{ (page 311)}$$

The density of a substance is given by

$$\rho = \frac{M}{V} \qquad \textbf{(10.3)} \text{ (page 313)}$$

An *incompressible* fluid is one for which the density is independent of pressure; most liquids are incompressible. The density of a gas varies with pressure; hence, gases are *compressible* fluids.

Dependence of pressure on depth
The pressure in a liquid varies with depth as

$$P = P_0 + \rho g h \qquad \textbf{(10.5)} \text{ (page 314)}$$

The pressure increases as one moves deeper in a fluid because the pressure below must support the weight of the fluid above.

Real fluids
Fluids are composed of molecules, which helps explain the origin of *viscosity* (i.e., friction), the *surface tension*, and the *capillary pressure*. *Poiseuille's law* relates the average flow speed to the viscosity of the fluid and the pressure difference across the ends of a pipe:

$$v_{ave} = \frac{P_1 - P_2}{8\eta L} r^2 \qquad \textbf{(10.31)} \text{ (page 335)}$$

where $P_1 - P_2$ is the pressure difference and η is the viscosity. The corresponding flow rate is

$$Q = v_{ave} A \qquad \textbf{(10.32)} \text{ (page 335)}$$

The surface tension γ is the energy needed to increase the surface area of a fluid. The surface tension leads to an extra pressure when a fluid is confined to a narrow capillary. This capillary pressure is

$$P_{cap} = \frac{2\gamma}{r} \qquad \textbf{(10.35)} \text{ (page 338)}$$

APPLICATIONS

Pascal's principle
Pressure changes at one location in a closed container of fluid are transmitted equally to all locations in the fluid. This principle is the basis of *hydraulics*. Hydraulic devices, such as a lift, are able to amplify forces. However, they de-amplify the corresponding displacement so that the work done is conserved.

(Continued)

Archimedes's principle

When an object is immersed in a fluid, the fluid exerts an upward force on the object that is equal to the weight of the fluid that is displaced by the object. This upward force is called the **buoyant force**.

Archimedes's principle and buoyancy

Buoyant force = F_B = weight of fluid *displaced* by object

Principle of continuity

When a fluid flows through a pipe (or similar geometry), the **equation of continuity** tells how the speed v of a fluid is related to the cross-sectional area A of the pipe:

$$v_1 A_1 = v_2 A_2 \qquad \textbf{(10.23)} \text{ (page 329)}$$

where the subscripts 1 and 2 denote two different sections of the pipe.

Conservation of energy leads to **Bernoulli's equation**, which relates the speed, pressure, and height at two different locations in an ideal fluid

$$P_1 + \tfrac{1}{2}\rho v_1^2 + \rho g h_1 = P_2 + \tfrac{1}{2}\rho v_2^2 + \rho g h_2 \qquad \textbf{(10.28)} \text{ (page 331)}$$

According to Bernoulli's equation, when the flow speed increases, the pressure must decrease, and vice versa. This leads to the lift force on an airplane wing.

QUESTIONS

SSM = answer in Student Companion & Problem-Solving Guide

⊗ = life science application

1. Design a hydraulic lift that could be used to move a car. Assume the input piston the person would push on has an area of 0.10 m². If you want a force of 300 N applied by the person to lift a car, estimate the area of the piston supporting the car.

2. ⊗ Explain how Archimedes's principle can be used to measure a person's percentage body fat. *Hint*: Fat has a different density than other body tissue.

3. What are the units of specific gravity?

4. SSM Figure Q10.4 is a photograph of a graduated cylinder filled with four fluids. Starting with mercury on the bottom and going up, we have salt water, water, and vegetable oil. In addition, a solid object rests at the interface between each liquid. At the bottom is a steel ball bearing, next an egg, followed by a block of wood, and a table-tennis ball on top. (a) Rank these eight substances in terms of their densities from the highest to the lowest. (b) Which items have a specific gravity greater than 1? Less than 1?

Courtesy of Henry Leap and Jim Lehman

Figure Q10.4

5. Explain why an ice cube floats in water. For ordinary water (i.e., not salt water) at room temperature, what fraction of an ice cube is above the water?

6. Why does the "lava" in a Lavalamp (Fig. Q10.6) rise and then fall?

7. Consider a swimming pool and a tall graduated cylinder (like the one in Fig. Q10.4), both containing water and filled to the same height. (a) Is the pressure at the bottom of the pool greater than, less than, or equal to the pressure at the bottom of the graduated cylinder? (b) Is the total force exerted by the water on the bottom of the pool greater than, less than, or equal to the total force on the bottom of the graduated cylinder?

8. Why are dams thicker at the bottom than at the top?

9. Consider a glass tube that has the shape shown in Figure Q10.9. It is filled with a fluid and is closed at the end on the left. If P_A is the pressure at point A, P_B the pressure at point B, and so forth, rank the pressures P_A, P_B, ... from lowest to highest.

© David Harrison/Index Stock Imagery/ Jupiterimages

Figure Q10.6

Figure Q10.9

10. Two blocks with the same volume but different densities are in a lake as shown in Figure Q10.10. Block m_1 floats at the surface with a portion above the water level, and block m_2 floats underwater. (a) Which block has the greatest mass? (b) For which block is the buoyant force greatest?

Figure Q10.10

11. A cargo ship travels from the Mississippi River into the Gulf of Mexico. Will the ship sink or rise with respect to the waterline as it moves from the river to ocean water? Why?

12. Explain how the specific gravity of a substance is connected with the buoyant force when an object is immersed in water.

13. Explain how there can be dust on a moving fan blade.

14. ⊗ Explain why some people are able to float in ocean water but not in a freshwater lake (see Fig. Q10.14). (A freshwater lake contains rainwater.)

Figure Q10.14

15. When traveling in a commercial airplane, you are sometimes given snacks, such as peanuts, in a sealed bag. The bag is often "bulging" much more than when you purchase a bag of peanuts in a store. Explain why.

16. In Section 10.4, we proved Archimedes's principle, assuming the object is completely immersed in a liquid. Consider the case of an object that is floating so that some portion of the object is above the waterline. Starting with Equation 10.5, prove Archimedes's principle for this case.

17. SSM Figure Q10.17 shows a popular demonstration involving a moving fluid. Here an air "jet" is aimed upward and is used to levitate a small object such as a table-tennis ball. The ball is drawn *to the center of the jet*, where the velocity is greatest. This behavior is found when the air jet is directed vertically (Fig. Q10.17 left) and also when the jet is directed at an angle (Fig. Q10.17 right). Use Bernoulli's equation to explain this behavior.

18. An altimeter is a device used to measure altitude. Most altimeters are based on measurements of air pressure. Explain why they work and estimate how much the pressure changes for a

Figure Q10.17 A "jet" of air from a nozzle will "levitate" this table-tennis ball.

100-m change in altitude. Discuss how changes in the weather can affect an altimeter reading.

19. In preparation for a bike tour of several countries, you ship your bicycle by air freight to Europe. When you arrive, you find that the tires of your bicycle have both blown out. Explain why that happened.

20. Three containers are filled with water to the same height (Fig. Q10.20). For which container is the pressure at the bottom the greatest? Or, are the pressures the same? Explain.

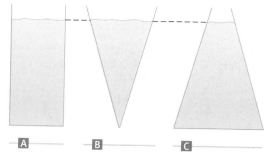

Figure Q10.20

21. The region near the South Pole in Antarctica has an altitude of about 3000 m, making the air pressure lower than at sea level. Explain why it is more difficult for an airplane to take off from the South Pole than from an airfield at sea level.

22. The author once had a small car with a "hatchback," a rear door that opens upward. One day, he forgot to latch it before driving on the highway, and he found that at high speeds the rear door lifted up as in Figure Q10.22B. Explain why.

$v = 0$ $v \neq 0$

Figure Q10.22

10.1 PRESSURE AND DENSITY

1. What is the average density of the Earth?

2. In a neutron star, all the mass has "collapsed" into a relatively small volume. (a) If a particular neutron star has a radius

of 1000 m and a mass of 2.0×10^{28} kg, what is its density? (b) Compare this value with the density of the Sun and with the density of steel.

3. ★ A piece of gold has a mass of 20 g. If this gold were formed into a very thin sheet of thickness 0.50 μm, find the area of the sheet. Assume that it has the same density as listed in Table 10.1.

4. Consider a highway composed of concrete. If the road is 2.0 km long, 10 m wide, and 0.50 m thick, what is its mass?

5. ✪ ⓡ Estimate the mass of water in a typical bathtub. Assume the tub contains just enough water that you could be completely submerged.

6. Consider a steel sphere of mass 20 kg. Find its radius.

7. ✪ ✪ ⓡ Knowing the density of water, estimate the volume of your body.

8. ✪ Consider an airplane window (area 0.30 m²). If the pressure inside the plane is atmospheric pressure and the pressure outside is 20% of P_{atm}, what is the force on the window? Ignore the velocity of the airplane (i.e., ignore Bernoulli's effect).

9. Consider a box with a lid 1.5 m wide and 0.70 m long. If the inside of the box is evacuated (i.e., its pressure is zero), how much force is required to open the lid?

10. ✪ The pressure in the atmosphere is not a constant, but fluctuates as the weather changes. If the pressure outside a window drops by 5% while the pressure inside does not change, what is the force on the window? Assume the area of the window is 1.5 m².

11. A block of mass 3.5 kg sits at rest on a thin plate that floats on the surface of a fluid as shown in Figure P10.11. If the area of the plate is 0.30 m², what is the pressure in the fluid just below the plate due to the presence of the mass?

m   Plate

Figure P10.11

12. SSM ✪ A suction cup works by virtue of a vacuum that is created within the cup. When the cup is pressed against a flat surface, most of the air is forced out, leaving a region of very low pressure. If a suction cup of area 1.0 cm² attached to a ceiling is able to support an object with a mass as large as 0.70 kg, what is the pressure inside the suction cup?

13. The four tires on the author's car are inflated to an absolute pressure of 2.5×10^5 Pa. If the car has a mass of 1800 kg, what is the contact area between the each tire and the ground?

14. ✪ ⓡ Estimate the pressure exerted by your feet when you stand upright on the floor. Compare it to the pressure when you are wearing "stiletto" heels (use the value at the heel).

15. ✪ ⓡ When a typical 1200-page textbook is sitting on a table, it exerts a force on the table. Estimate the associated pressure.

16. The recommended pressure in your car's tires is usually specified as a gauge pressure. If a tire has a gauge pressure of 15 lb/in.², what is the absolute pressure in pascals?

17. Express a pressure of 5.0×10^6 Pa in units of pounds per square inch.

18. The water in the pipes of a typical house is at a gauge pressure of 3.0 times atmospheric pressure. What is the absolute pressure in pascals and in units of pounds per square inch?

19. ✪ ⓡ Each tire of a car contacts the road over an area of approximately 150 cm². Estimate the mass of the car.

20. Add a new column to Table 10.1, listing the specific gravity of each substance.

10.2 FLUIDS AND THE EFFECT OF GRAVITY

21. Find the pressure at the bottom of a swimming pool that is 2.5 m deep.

22. Consider a rectangular swimming pool of depth 2.5 m with a width and length of 15 m. What is the total force on the bottom of the pool due to the water pressure?

23. ✪ Consider again the swimming pool in Problem 22. What is the force exerted by the water on one of the sidewalls of the pool?

24. A diver is swimming at a depth of 10 m. What is the pressure in the water around him?

25. ✪ ⓡ Estimate the force exerted by the water on the Hoover Dam.

26. ✪ ✪ ⓡ Consider a fish of average size swimming near the bottom of a lake of depth 20 m. (a) What is the water pressure next to the fish? (b) Estimate the total force exerted by the water on the outside surface of the fish.

27. SSM ✪ A U-tube contains two fluids with densities $\rho_1 = 1000$ kg/m³ and $\rho_2 = 600$ kg/m³ as sketched in Figure P10.27. What is the difference d in the heights of the top surfaces of the two fluids?

$d = ?$

y_L

$y_R = 15$ cm

Figure P10.27

28. Figure 10.12 shows a device that can be used to measure atmospheric pressure. If $h = 6.5$ m, what is the density of the liquid?

29. An engineer is given the job of designing the water system for a tall building. He decides to locate the pumps at ground level and feed the entire building from there. If the top floor of the building is 200 m above ground level, what must the pressure be at pump level?

30. ✪ A bucket is filled with a combination of water (density 1000 kg/m³) and oil (density 700 kg/m³). These fluids do not mix. (The oil will float on top of the water.) If the layer of oil is 20 cm tall, what is the pressure at the interface between the oil and the water?

31. ✪ ✪ ⓡ The blood pressure in a person's feet is greater that the pressure in his head. Estimate this pressure difference and compare it with normal blood pressure.

32. ✪ ✪ When a person is given intravenous fluid, the bag containing the fluid is typically held about 1.0 m above the person's body. If the pressure in this fluid is just barely able to push the fluid into the person, what is the pressure in the person's blood?

33. ✪ ⓡ The water pressure in the pipes in the basement of a tall house is 5.0×10^5 Pa. Estimate the water pressure on the second floor.

34. A swimming pool of depth 2.5 m is filled with ordinary (pure) water ($\rho = 1000$ kg/m³). (a) What is the pressure at the bottom of the pool? (b) When the pool is filled with very salty water, the pressure changes by 5.0×10^3 Pa. What is the difference between the density of the salt water and the density of pure water?

35. ✪ The water system in a town is designed by a careless engineer. He places the new water tower at the top of an extremely tall hill, overlooking the mayor's house. The mayor then finds that some of the pipes in his house leak, which never happened with the old system. Estimate the height of the water level in the tower relative to the mayor's house. *Hint:* The water pipes in a typical house leak when the pressure exceeds about five times atmospheric pressure.

10.3 HYDRAULICS AND PASCAL'S PRINCIPLE

36. ✪ A syringe has an area of 1.0 cm² in the barrel and then narrows down to an area of 0.10 mm² at the needle end. If a force of 5.0 N is applied to the syringe, what is the force produced at the tip of the needle?

37. Cars use a hydraulic system to transmit the force from the brake pedal to the actual brakes. If this system amplifies the force of your foot by a factor of 10, what is the ratio of the area of the hydraulic tube at the pedal to the area of the tube at a wheel?

38. SSM ✪ A block of mass 20 g sits at rest on a plate that is at the top of the fluid on one side of a U-tube as shown in Figure

P10.38. The U-tube contains two different fluids with densities $\rho_1 = 1000 \text{ kg/m}^3$ and $\rho_2 = 600 \text{ kg/m}^3$ and has a cross sectional area of $A = 5.0 \times 10^{-4} \text{ m}^2$. If the surfaces are offset by an amount h as shown, (a) which side of the U-tube contains the fluid with the greater density? (b) Find h. (Assume h is positive.)

$m = 20$ g

$h = ?$

20 cm

Figure P10.38

39. A car of mass 1400 kg rests on a hydraulic lift with a piston of radius 0.25 m that is connected to a second piston of radius 0.030 m. If the car is just barely lifted off the ground, what is the force at the second piston?

10.4 BUOYANCY AND ARCHIMEDES'S PRINCIPLE

40. A large suitcase full of gold sits at the bottom of a lake. A recovery diver finds that as he lifts the suitcase to the surface, he is exerting a force of 3.0×10^5 N. What is the volume of the suitcase?

41. A wooden log is found to float in a freshwater lake. If 35% of the log's volume is above the surface, what is the log's density?

42. ⭐ Under normal conditions, you are just barely able to lift a mass of 90 kg. Your friend drops a box of volume 2.5 m³ into a lake. If you are just able to lift it to the surface, what is the mass of the box?

43. A rectangular block of wood (density 750 kg/m³) of height 25 cm floats in a lake. What is the height of the wood above the water level?

44. ⭐ ✖ Ⓡ Estimate your apparent weight when you are under water.

45. An object has a weight of 20 N in air and an apparent weight of 5.0 N when submerged in water. What is its density? Will it float?

46. ✪ A submarine uses tanks that can be filled with either water or air so that the submarine can have "neutral buoyancy." That means that its average density is equal to the density of the water and it will remain at a desired level under water. The density of water changes a small amount with depth in the ocean due to the changes in pressure and temperature. How much additional mass must a submarine take on to sit at the ocean bottom at a depth of 500 m? At this depth, the density of water is about 0.25% larger than at atmospheric pressure. Assume the volume of the submarine is 3.0×10^4 m³ and it is initially floating at the surface, fully covered with water.

47. ✪ Consider a wooden block of density 700 kg/m³ and volume 2.0 m³. What force is required to hold it completely under water?

48. SSM ⭐ You want to build a raft that can hold you (80 kg) plus some supplies (40 kg) for a long trip. You decide to make the raft out of logs that are each 20 cm in diameter and 3.0 m long. If the density of each log is 600 kg/m³, how many logs do you need?

49. ⭐ You want to design a helium-filled balloon that will lift a total payload of 1500 kg. What volume of helium is needed to just barely lift this payload?

50. Two metal blocks are attached by a cable as shown in Figure P10.50 and are submerged at rest near the surface of a lake. If the volumes are $V_1 = 2.0$ m³ and $V_2 = 1.2$ m³, what is the total buoyant force on the two blocks?

V_1

V_2

Figure P10.50 Problems 50, 51, and 52.

51. ⭐ For the blocks in Figure P10.50, suppose instead of being near the surface of the lake they are just above the bottom of the lake. Find the buoyant force in this case.

52. ✪ For the blocks in Problem 50, assume they are barely submerged in water so that block 1 is just below the surface. If block 2 has a density twice that of block 1, what is the density of block 2?

53. A chest of volume 1.2 m³ filled with gold sits at the bottom of a lake of depth 12 m. How much force is required to lift the chest to the surface?

54. ⭐ A solid object is made by gluing together two pieces of equal mass. If one of the pieces is made from a material with density 500 kg/m³, while the other has a density 1400 kg/m³, will the object float?

55. ⭐ A piece of balsa wood is placed into a tub full of ethanol. What fraction of the balsa wood is above the surface of the ethanol?

56. ✪ A piece of aluminum is attached to a string and suspended in a pool of oil with density 750 kg/m³. If the apparent weight of the aluminum is 540 N, what is the volume of the aluminum?

10.5 FLUIDS IN MOTION: CONTINUITY AND BERNOULLI'S EQUATION

For Problems 57 through 67, assume an ideal fluid (i.e., a fluid with no viscosity).

57. Water is flowing through a pipe (area 4.0 cm²) that connects to a faucet adjusted to have an opening of 0.50 cm². If the water is flowing at a speed of 5.0 m/s in the pipe, what is its speed as it leaves the faucet?

58. ⭐ For the faucet in the Problem 57, how long does it take for water from the faucet to fill a bucket of volume 0.12 m³?

59. SSM ⭐ ✖ The blood flow rate through the aorta is typically 100 cm³/s, and a typical adult has about 5.0 liters of blood. (a) How long does it take for all your blood to pass through the aorta? (b) If your aorta has a diameter of 2.0 cm, what is the speed of blood as it flows through the aorta?

60. ⭐ Ⓡ It takes the author about 60 s to fill his truck with gasoline (48 liters, about 12 gal). (a) What is the flow rate in cubic meters per second? (b) Estimate the speed of the gasoline as it flows through the nozzle of the gasoline pump.

61. ✪ A tank of water sits at the edge of a table of height 1.2 m (Fig. P10.61). The tank springs a very small leak at its base, and water sprays out a distance of 1.5 m from the edge of the table. What is the water level h in the tank?

$h = ?$

Figure P10.61

62. Ⓡ A house sits in a valley and is 30 m lower than the water tower that serves it. What is the speed of water when it sprays out of a shower in the house?

63. ✖ Consider a large blood vessel that carries blood to the heart. If this blood vessel has a cross-sectional area of 1.5 cm² and the blood speed is 30 cm/s, what volume of blood is delivered in 1.0 s?

64. Consider an airplane wing of area 20 m². If the airflow speed over the top of the wing is 200 m/s, while the speed across the bottom is 150 m/s, what is the lift force on the wing at sea level at 0° C?

65. SSM ⭐ Ⓡ A Frisbee is observed to fly nearly horizontally, which implies that the lift force must be approximately equal to the weight of the Frisbee. If the air speed over the top of the Frisbee is 9.0 m/s, what is the flow speed across the bottom? Assume the Frisbee has a mass of 0.15 kg.

66. A fire extinguisher contains a high-pressure liquid so that it can spray the liquid out very quickly when needed. If the fluid leaves a fire extinguisher at a speed of 20 m/s, what is the pressure inside?

67. ★ The water in the author's house is provided by a well. Water is pumped from the well to a holding tank in the basement that is pressurized so that water can flow from the tank to the upper floors of the house. The pressure in the tank is approximately 3.0×10^5 Pa. If a bathroom on the second floor is 12 m above the basement, what is the speed at which water sprays out of the shower?

10.6 REAL FLUIDS: A MOLECULAR VIEW

68. ✖ A blood vessel is 20 cm long and has a radius of 2.0 mm. If blood is flowing through it with a volume flow rate of 2.0×10^{-7} m³/s, what is the difference in the pressures at the two ends of the blood vessel?

69. An air duct in a building has a radius of 10 cm and a length of 10 m. What pressure is required to push air through the duct at a flow rate of 1.0 m³/s?

70. ★ A glass sphere of radius 1.0 mm is dropped into a lake. What is the terminal velocity of the sphere?

71. ✪ Ⓡ Estimate the terminal velocity for a dust particle in air. Assume it has a radius of 1.0 μm and a density similar to that of balsa wood. Your answer will tell you why dust particles remain "suspended" in air for long periods.

72. SSM What is the capillary pressure for water in a vertical tube of diameter 0.10 mm?

73. ★ Repeat Problem 72, but now for a tube of radius 10 nm. Express your answer in pascals and as a ratio relative to atmospheric pressure.

74. ✪ A glass capillary tube with a radius of 0.050 mm is inserted into an open container filled with an unknown liquid, and the meniscus has an appearance similar to Figure 10.34A. It is found that the surface tension causes the liquid in the tube to rise a distance of 2.5 cm relative to the surface of the liquid outside the tube. If the density of the liquid is 1000 kg/m³, what is the surface tension?

75. ✖ The flow rate through a blood vessel is found to be 1.5×10^{-7} m³/s. Express this flow rate in units of cubic centimeters per second.

76. ★ ✖ Ⓡ Estimate the terminal velocity of a DNA molecule in water. Assume the only forces on the molecule are gravity and the Stokes drag force (Eq. 10.33).

77. ★ ✖ Blood flows through an artery of diameter d and length L, and the pressure difference between the ends of the artery is ΔP. If the diameter is reduced by a factor of two and the pressure difference is kept the same, what happens to the flow rate?

ADDITIONAL PROBLEMS

78. ★ A steel ball bearing will float in mercury as shown in Figure P10.78. What fraction of the ball bearing's volume is above the surface of the mercury?

79. ✪ What will happen if water is poured on top of the mercury and ball bearing of Problem 78 such that the ball bearing is completely submerged in the water? Will the ball bearing (a) sink farther into the mercury, (b) rise up from its former position, or (c) stay at the same position? Why? Calculate the fraction x of the ball bearing above the surface of the mercury under these new circumstances. *Hint*: If a fraction x of the volume is in the water, a fraction $(1 - x)$ of the volume will be in the mercury.

Figure P10.78 Problems 78 and 79.

© Cengage Learning/Charles D. Winters

80. ✪ A small, square plank of oak floats in a beaker half full of water. The piece of oak is 6.0 cm on a side and 3.0 cm thick and floats on its side as shown in Figure P10.80. (a) Find the location of the plank's center of mass with respect to the water's surface. (b) A very light vegetable oil is poured slowly into the beaker so that the oil floats on the water without mixing and until the oak plank is few centimeters below the surface of the oil. If the density of the oil is 700 kg/m³, what is the new position of the center of mass of the plank with respect to the water's surface?

Center of mass

Figure P10.80

81. ★ ✖ **Drawing (up) the longest straw.** Consider a contest where the challenge is to find the greatest height from which the contestant in Figure P10.81 is able to drink water through a straw. Find the theoretical limit on the vertical distance h between the open top surface of the liquid and the top of the straw. *Hint*: Assume (however unlikely) that the contestant can create a perfect vacuum at the top of the straw.

82. ✪ **Before fuel injection.** Some automobile engines (mainly older ones) use a carburetor to make the liquid fuel into vapor and mix it with air for combustion. The basic principle of carburetion is shown in Figure P10.82. A piston moves down in the cylinder, thereby drawing air from the outside through an air filter and into the carburetor. The filtered air enters from the left in Figure P10.82 and moves into the main intake, a tube of diameter 6.0 cm, with speed $v = 7.0$ m/s. The air must pass through a region of the intake that has a smaller diameter. Determine what diameter would be needed to cause a change in pressure such that fuel from the reservoir is pulled into the airflow. The surface of the fuel in the reservoir is $h = 42$ cm below the bottom of the intake, and the density of the fuel is 0.72 that of water.

Figure P10.81

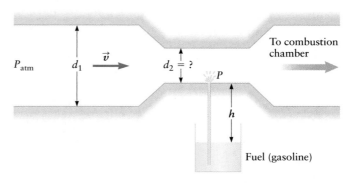

Figure P10.82

83. ✪ Consider a faucet with water running out in a smooth stream. How could one find the flow rate in cubic centimeters per second by simply taking measurements with a ruler? Following the diagram in Figure P10.83, find the rate of flow if $d_1 = 1.2$ cm, $d_2 = 0.70$ cm, and $h = 7.0$ cm.

Figure P10.83

84. ✪ 🅧 **Long snorkel.** Inhalation of a breath occurs when the muscles surrounding the human lungs move to increase the volume of the lungs, thereby reducing the air pressure there. The difference between the reduced pressure and outside atmospheric pressure causes a flow of air into the lungs. The maximum reduction in air pressure that chest muscles can produce in the lungs against the surrounding air pressure on the chest and body is about 3500 Pa. Consider the longest snorkel a person can operate. The minimum pressure difference needed to take in a breath is about 200 Pa. Find the depth h that a person could swim to and still breathe with this snorkel (see Fig. P10.84).

Figure P10.84

85. ⭐ 🅧 A scuba diver swims in a freshwater lake (see Fig. P10.85). (a) If the maximum pressure in her tanks is 10 atm, to what depth h could she swim and still breathe? (b) Could she go deeper in the ocean? Calculate your answers to three significant figures.

Figure P10.85

86. ✪ 🅧 **Snorkeling elephants.** An elephant can use its long trunk as a snorkel while swimming or walking across the bottom of rivers (see Fig. P10.86). The lining of an elephant's lung is reinforced with connective tissue not present in other mammals, allowing elephants to generate larger pressure differences when breathing. If an elephant's lungs are 2.0 m below the surface and the minimum pressure difference needed for it to inhale is 200 Pa, what is the maximum pressure difference an elephant's lungs can generate?

Figure P10.86 An elephant in the Chobe River, Botswana.

Compare your answer to the result for human lungs discussed in Problem 84.

87. ✪ 🅧 🅡 **Flail chest.** A dangerous condition resulting from severe blunt trauma to the upper torso occurs when a portion of the rib cage breaks free and is no longer held rigidly in place. This condition, known as flail chest, severely limits the movement of air and oxygen into the lungs. The diaphragm and chest muscles draw in air by increasing the volume of the lungs, creating a pressure in the lungs lower than that of atmospheric pressure. The resulting pressure difference, equal to approximately 400 Pa for a resting inhalation, draws air in through the trachea. In the condition of flail chest, when the muscles attempt to increase the volume, the broken ribs are pushed into the body as shown in Figure P10.87A, compromising the amount of volume increase. Instead of flowing in from the outside, air just moves from one lung to the other (Fig. P10.87B) during a respiration cycle. Consider a case of flail chest in which four ribs have been broken at both ends as shown in the figure. Estimate the maximum force on the area of injury due to inhalation. This force is the amount of traction necessary to apply to the patient's chest to restore normal breathing.

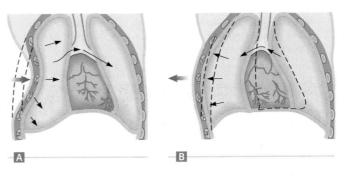

Figure P10.87 Flail chest during Ⓐ inhalation and Ⓑ exhalation.

88. ✪ 🅧 **Pulmonary surfactant.** Oxygen enters the blood in the hollow and roughly spherical alveoli of the lungs. The alveoli are small, approximately 170 μm in radius (see Fig. P10.88), yet some 300 million of them make up 86% of the volume of each lung. The inner lining of the alveoli is, of course, wet as all living tissue must be. Thus, there is essentially a bubble of liquid just inside the lining of an alveolus having a surface tension that acts to compress the bubble, effectively generating an inward pressure that can be described by Equation 10.35. In this case, r is the radius of the sphere and the P_{cap} is instead $P_{collapse}$, the inward pressure on the sphere. (a) If the liquid lining an alveolus is pure water, what is $P_{collapse}$? (b) What is the ratio of this pressure to the pressure difference generated when breathing? (c) A surfactant, a component of detergents, is a substance that greatly reduces the surface tension of a liquid. The alveoli have special cells that secrete a combination of phospholipids and proteins that give the fluid a surface tension of $\gamma = 1.8 \times 10^{-2}$ J/m^2.

Figure P10.88

Calculate $P_{collapse}$ for the new surface tension and compare it with the pressure difference needed for normal breathing described in Problem 84.

89. ✪ ⓧ **Sphygmomanometer.** Blood pressure is measured at the brachial artery in the upper arm (see Fig. P10.89). An inflatable strap is pumped with air until the artery is collapsed, stopping all flow. The pressure in the cuff is then slowly reduced while the operator listens with a stethoscope for a pulse downstream of the cuff. The pressure at which the pulse resumes is the *systolic* (maximum) pressure. The brachial artery has a radius of 0.50 cm, and the cuff is usually applied at a point 0.30 m away from the aorta at the heart. If the flow rate though the brachial artery is 1.5×10^{-5} m³/s, what is the pressure drop between the point of measurement and the aorta? Is a measurement from the brachial artery in the arm a good representation of the pressure in the aorta at the heart? (Assume the heart and arm are at the same height.)

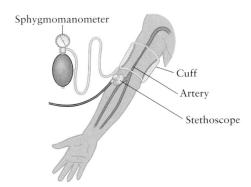

Sphygmomanometer

Cuff

Artery

Stethoscope

Figure P10.89

90. ★ ⓧ In Example 10.10, we considered how the pressure ΔP across the ends of a narrowed blood vessel must increase to keep the average flow speed constant. If we instead want the smaller blood vessel to have the same flow rate as the larger blood vessel, what value must ΔP have?

91. SSM ★ ⓧ The density of air decreases with height above the Earth, and when the density is too low, people get sick or even become unconscious. That is why the cabins of commercial airplanes are "pressurized" during flight. Cabin pressure, however, is lower than the pressure at sea level. Cabins typically have a pressure equal to the air pressure at an altitude of about 8000 ft above the Earth's surface. This reduced pressure is one reason

some people feel fatigued after a long flight. (a) Use Equation 10.11 to calculate this cabin pressure. (b) Estimate the total force on the walls of a Boeing 737 flying at 35,000 ft. Approximate the airplane as a cylinder of length 30 m and a diameter of 4 m. (c) What would the force on the walls be if the cabin pressure equals the pressure at the Earth's surface?

92. ★ You can make a rough estimate of the thickness of the Earth's atmosphere by assuming its density is constant with the value 1.3 kg/m³. How tall must the atmosphere be to give the observed pressure at the Earth's surface?

93. ★ A window in a house is rectangular, with height 40 cm and width 30 cm. The pressure inside the house is normal atmospheric pressure when a hurricane arrives and the wind speed outside is 75 m/s (about 165 mi/h). (a) Does the pressure difference between the inside and outside of the window tend to push the window into the house or out of the house? (b) What is the force on the window?

94. ★ A backyard waterfall used in landscaping (Fig. P10.94) has water cascading down at a rate of 18 liters/min. (a) If the waterfall is fed using a hose of diameter 2.5 cm, what is the speed of water in the hose? (b) If the hose is 15 m long, what is the pressure difference between the two ends of the hose?

Figure P10.94

95. ★ (a) Water flows out of a shower in a typical house at a rate of 20 liters/min. If this shower is fed through a pipe that is 2.0 cm in diameter and 15 m from the high-pressure water tank, what is the difference between the pressure in the tank and atmospheric pressure? Assume the tank is at the same height as the shower and assume an ideal fluid. (b) What pressure difference is needed for an energy-efficient shower head, which limits the flow rate to 4.0 liters/min?

Chapter 11

Harmonic Motion and Elasticity

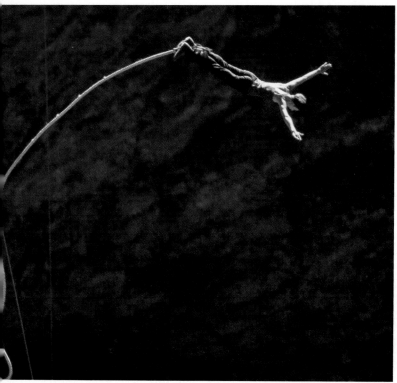

The physics of oscillations and elasticity are keys to the operation of this bungee cord (and the safety of the jumper!). (© Pure Stock/Getty Images)

In this chapter, we consider **harmonic motion**, a type of motion in which an object moves along a repeating path over and over and over again. The universe is filled with examples of harmonic motion, such as the pendulum motion of a child on a playground swing and the motion of the Earth as it orbits the Sun. While there are many specific examples of harmonic motion, certain features involving force and energy are common to all. We'll also find a close connection to many features of circular and rotational motion we have studied in previous chapters.

11.1 | GENERAL FEATURES OF HARMONIC MOTION

A familiar example of harmonic motion is sketched Figure 11.1A, which shows a person hanging from a spring. This spring is a little bit like the bungee cord in the photo at the start of this chapter, except that the spring in Fig. 11.1 is tightly coiled so that the force from the spring is the dominant force acting on the person (e.g., this force dominates over the force of gravity). Figure 11.1A shows several "snapshots" of the person as he "vibrates" up and down due to the force from the bungee spring.

We can describe the position of the person on the bungee spring by measuring his location y on a vertical axis, and Figure 11.1B shows a plot of y as a function of time. Notice how each point along the y–t plot matches the corresponding physical location of the person in Figure 11.1A. The jumper's position thus varies in a repeating fashion as he moves up and down. This is an example of *oscillatory motion*. With a perfect frictionless bungee spring and ignoring air drag, this repeating motion would continue forever.

To describe oscillatory motion, we use the notions of position, velocity, and acceleration, but we also need some additional quantities that relate to the repeating nature of the motion. One such variable is the *period T*, which is the *repeat time* of the motion. We have already encountered this quantity in our discussion of circular motion in Chapter 5. The period of our bungee jumper is the time it takes him to travel through one complete "cycle" of the motion, so it is the time interval between adjacent maxima of the position y when plotted as a function of time (Fig. 11.1B). The period T is also equal to the time interval between adjacent minima of y, or the interval between any two "equivalent" points on the y–t curve.

Another way to characterize harmonic motion is in terms of its *frequency*. The frequency f is the number of oscillations that occur in one unit of time; hence, f is the number of cycles completed in 1 second. Since the period is the number of seconds required to perform one cycle, frequency and period are related by

$$f = \frac{1}{T} \tag{11.1}$$

The units of frequency are cycles per second, and this unit has the name hertz (Hz).

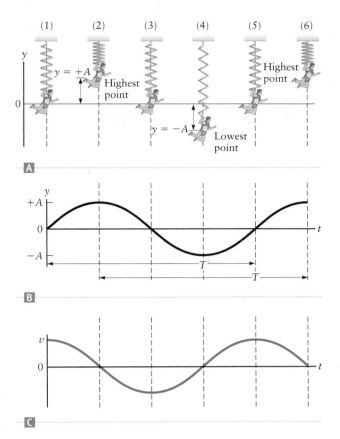

Figure 11.1 ◻ This bungee jumper undergoes simple harmonic motion as he oscillates up and down. ◻ Plot of the jumper's position y versus time. The period of the motion, T, is the repeat time, the time it takes to make one complete up-and-down oscillation. ◻ Plot of the jumper's velocity versus time.

EXAMPLE 11.1	Measuring the Period

Consider a hypothetical system for which the position y oscillates as sketched in Figure 11.2. Estimate the period and frequency of the motion.

RECOGNIZE THE PRINCIPLE

The period T is the time required for one *complete* oscillation, and we can estimate T from Figure 11.2. The frequency is then given by $f = 1/T$.

SKETCH THE PROBLEM

Figure 11.2 describes the problem.

IDENTIFY THE RELATIONSHIPS

We can estimate the period from the spacing along the time axis between two corresponding (i.e., "equivalent") points of the plot in Figure 11.2. For example, we could choose two adjacent maxima, such as points A and D, or we could choose two adjacent minima (points B and E) or any other pair of adjacent "repeated" points (such as C and F).

SOLVE

From Figure 11.2, we estimate that the time between points A and D is approximately

$$T \approx \boxed{1.7 \text{ s}}$$

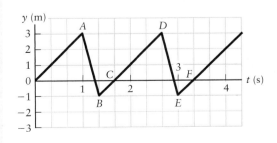

Figure 11.2 Example 11.1.

The frequency is then

$$f = \frac{1}{T} = \frac{1}{1.7\text{ s}} = \boxed{0.59\text{ Hz}}$$

What does it mean?
Notice that the units of f are hertz.

Simple Harmonic Motion

The key property of an oscillator is that its motion is periodic, that is, repeating. In many cases, this motion is described by a simple sinusoidal variation with time,

$$y = A\sin(2\pi f t) \tag{11.2}$$

For our bungee jumper (Fig. 11.1A), y is the location along the vertical axis, and Equation 11.2 describes how y varies with time in Figure 11.1B. Systems that oscillate in a sinusoidal manner as in Equation 11.2 are called **simple harmonic oscillators**, and they exhibit **simple harmonic motion**. You might think that the behavior described by Equation 11.2 is too simple to describe any real or interesting system, but we'll see that simple harmonic motion is, in fact, extremely common.

The frequency f tells us the rate at which the oscillator repeats its motion. The overall range of this motion depends on the quantity A in Equation 11.2, which is called the **amplitude** of the motion. The bungee jumper in Figure 11.1A moves back and forth between the maximum and minimum values $y = \pm A$ as indicated in the figure.

Figure 11.1C shows how the velocity of the bungee jumper varies with time. The velocity also "oscillates" between positive and negative values as the bungee jumper moves, and the frequency of these oscillations is the same as the frequency associated with the position y. However, the maximum values of the velocity do *not* occur when y has its maximum values. Instead, the largest values of v occur when $y = 0$. Likewise, the jumper's velocity is zero when he is at his maximum and minimum heights, at $y = \pm A$ (points 2, 4, and 6 in Fig. 11.1A).

CONCEPT CHECK 11.1 | Acceleration of an Object Undergoing
Simple Harmonic Motion

Consider the acceleration of an object that is undergoing simple harmonic motion. Is this acceleration zero or nonzero? If it is nonzero, there must be a nonzero force acting on the object. What forces act on the bungee jumper in Figure 11.1?

The Connection between Simple Harmonic Motion and Circular Motion

Consider the motion of a compact disc. A CD spinning at a constant rate is another example of periodic, repeating motion. There is a close relationship between uniform circular motion as followed by a point on the edge of the CD and simple harmonic motion.

A particle at the edge of the CD moves at a constant speed we call v_c as it moves around a circle of radius A (Fig. 11.3). We can specify the particle's position using a reference line drawn on the CD. If θ is the angle this reference line makes with the x axis, then θ increases with time according to

$$\theta = \omega t \tag{11.3}$$

where ω is the angular velocity of the corresponding rotational motion (Chapter 8). As the particle completes one full trip around the circle, θ varies from zero to 2π

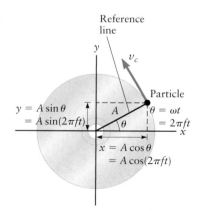

Figure 11.3 When a particle moves in a circle with constant speed v_c, both the x and y components of the particle's position undergo simple harmonic motion.

(measured in radians), corresponding to one full period of this harmonic motion. From Equation 11.3, we thus have after one full period ($t = T$):

$$\theta = 2\pi = \omega t = \omega T$$

Solving for T gives

$$T = \frac{2\pi}{\omega}$$

Comparing this result with our definition of frequency in Equation 11.1, we see that harmonic motion can be described in terms of either an angular velocity ω or an equivalent frequency f, with

$$f = \frac{\omega}{2\pi} \qquad (11.4)$$

The frequency f is the number of cycles a system completes in 1 second. The CD in Figure 11.3 moves through an angular displacement of 2π during each cycle, so Equation 11.4 says (in words) that the system completes an angular displacement of ω each second. For this reason, ω is also called the *angular frequency*.

Simple Harmonic Motion: Position as a Function of Time

The connection between circular motion and simple harmonic motion becomes even clearer when we describe the motion of the particle in Figure 11.3 in terms of its coordinates in the x–y plane. For the moment, let's follow the y coordinate. According to the trigonometry in Figure 11.3, the y component of the particle's position is $y = A \sin \theta$. Since $\theta = \omega t$ (Eq. 11.3), we have

$$y = A \sin \theta = A \sin(\omega t)$$

Using $\omega = 2\pi f$ (Eq. 11.4), we also get

$$y = A \sin(2\pi f t) \qquad (11.5)$$

which is precisely the same as the motion of a simple harmonic oscillator in Equation 11.2 with amplitude A.

There is nothing special about the y direction. We could just as easily make the connection between circular motion and simple harmonic motion using the x component of the particle's position. The x coordinate of the particle in Figure 11.3 is $x = A \cos \theta = A \cos(\omega t)$, which leads to

$$x = A \cos(2\pi f t) \qquad (11.6)$$

Although this equation looks a bit different from our original expression for simple harmonic motion (Eqs. 11.2 and 11.5), it describes the same basic motion. This can be seen from Figure 11.4, which shows both a sine function (Eqs. 11.2 and 11.5) and a cosine function (Eq. 11.6). Both functions describe the same basic simple harmonic motion with the same period $T = 1/f$. The only difference is a "shift" along the time axis. One expression (the sine function) begins at $y = 0$ when $t = 0$, while the other (the cosine function) starts at a maximum when $t = 0$. In terms of the actual motion of our bungee jumper in Figure 11.1, the sine function describes a jumper who starts at the "midpoint" ($y = 0$) at $t = 0$, whereas the cosine function describes a jumper who starts at the highest point ($y = +A$). The only difference is in the way in which the system is initially set into motion.

Simple Harmonic Motion: Velocity as a Function of Time

We can also use the connection with circular motion to derive the mathematical form of the velocity of a simple harmonic oscillator. Let's again follow the y coordinate of the particle (Fig. 11.5), so the velocity of our oscillator is v_y, the y component of the particle's velocity. Although the speed of the particle along the circular

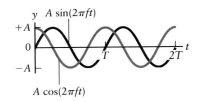

Both curves describe simple harmonic motion.

Figure 11.4 A sine-function time dependence (Eq. 11.5) and a cosine-function time dependence (Eq. 11.6) both describe simple harmonic motion. They differ only by a shift along the horizontal (time) axis.

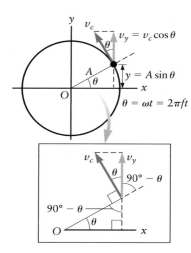

Figure 11.5 The speed v_c of the particle as it moves along the circle is constant, and the component of the velocity along y is $v_y = v_c \cos \theta$. Hence, v_y varies with time according to $v_y = v_c \cos(\omega t) = v_c \cos(2\pi f t)$.

Position and velocity of a simple harmonic oscillator

path in Figure 11.5 is constant, the y component v_y of the velocity is *not constant*. From Figure 11.5, we get

$$v_y = v_c \cos \theta = v_c \cos(\omega t)$$

$$v_y = v_c \cos(2\pi f t) \qquad (11.7)$$

The particle travels once around the circle (a distance of $2\pi A$) in a time equal to one period, so its speed is $v_c = 2\pi A/T$. Using Equation 11.1 relating the period and frequency gives

$$v_c = \frac{2\pi A}{T} = 2\pi f A \qquad (11.8)$$

Inserting this into Equation 11.7, we obtain

$$v_y = 2\pi f A \cos(2\pi f t)$$

Here v_y is the velocity of our simple harmonic oscillator, which we can simply call v.

We thus find that the position and velocity of a simple harmonic oscillator can be described by

$$y = A \sin(2\pi f t) \qquad v = 2\pi f A \cos(2\pi f t) \qquad (11.9)$$

(Compare Eq. 11.9 with the plots in Fig. 11.1.) The result in Equation 11.9 for the velocity applies when we use the y component of circular motion to describe simple harmonic motion. We could also have used the x component, which leads to (see Eq. 11.6)

$$x = A \cos(2\pi f t) \qquad v = -2\pi f A \sin(2\pi f t) \qquad (11.10)$$

The relations in Equation 11.9 and 11.10 both describe simple harmonic motion. The only difference is in how the oscillator is initially set into motion.

11.2 | EXAMPLES OF SIMPLE HARMONIC MOTION

Mass on a Spring

In Section 11.1, we described the simple harmonic motion of a person hanging vertically from a spring. An even simpler example of simple harmonic motion is sketched in Figure 11.6, which shows a block of mass m attached to a spring, with the opposite end of the spring attached to a wall. This is just like the person in Figure 11.1 except the block in Figure 11.6 moves on a horizontal, frictionless surface (unlike the person in Figure 11.1, who moves vertically).

Let's apply Newton's second law to the block in Figure 11.6. Along the vertical direction, the forces on the block are gravity and the normal force exerted by the table. These forces cancel and are not shown in Figure 11.6. There is also a horizontal force exerted by the spring. Recall from Chapter 6 that the force exerted by a spring is given by Hooke's law,

$$F_{\text{spring}} = -kx \qquad (11.11)$$

where x is the amount the spring is stretched or compressed away from its "natural" relaxed length. The "spring constant" k is a measure of the spring's strength and will be different for different springs. However, the general relation in Equation 11.11 is the same for all springs, provided only that they are not stretched or compressed an excessive amount.

The equilibrium position of the block in Figure 11.6 is at $x = 0$; when the block is at this position, the string is in its relaxed state and the force exerted by the spring on the block is zero. If the block is displaced to the right so that $x > 0$, the force exerted by the spring on the block, according to Hooke's law (Eq. 11.11), will be in the negative direction (i.e., to the left). Likewise, if the block is displaced to the left so that $x < 0$, the spring force will be to the right. This force is called a **restoring force** because it always opposes the displacement away from the equilibrium position. Whenever there is a restoring force described by Equation 11.11, the system will exhibit simple harmonic motion.

To calculate the frequency of the mass on a spring, we use Newton's second law, $F = ma$, where F is the total force in the horizontal direction (parallel to x) in Figure 11.6. The force is given by Hooke's law (Eq. 11.11), so we have

$$F = -kx = ma \qquad (11.12)$$

We have already used the connection with circular motion to find the behavior of the displacement. This connection is shown again in Figure 11.7, where we show a particle moving at constant speed v_c in a circle of radius A. The x component of the particle's position is $x = A \cos \theta = A \cos(2\pi f t)$ as in Equation 11.10. To apply Equation 11.12, we also need to find the particle's acceleration. It is moving in a circle, so it has a centripetal acceleration $a_c = v_c^2/A$ directed toward the center of the circle (because A is the radius of the circle; see Chapter 5). Here the simple harmonic motion is described by the particle's position along x, so we need to find the x component of the acceleration; from Figure 11.7, we get

$$a = -a_c \cos \theta = -\frac{v_c^2}{A} \cos \theta$$

Inserting this expression and our result for x into Equation 11.12 gives

$$-kx = -k(A \cos \theta) = ma = m\left(-\frac{v_c^2}{A} \cos \theta\right)$$

$$kA = \frac{mv_c^2}{A} \qquad (11.13)$$

From Equation 11.8, we have $v_c = 2\pi f A$; using this in Equation 11.13 leads to

$$kA = \frac{mv_c^2}{A} = \frac{m(2\pi f A)^2}{A} = 4\pi^2 mf^2 A$$

We can now solve for the frequency f of this simple harmonic oscillator:

$$f = \frac{1}{2\pi} \sqrt{\frac{k}{m}} \qquad (11.14)$$

The frequency thus increases as the spring constant k is increased (a "stiffer" spring), whereas it gets smaller as the mass m is increased. Notice also that f is *independent of the amplitude*.

As an application of Equation 11.14, consider the motion of a small block attached to a Slinky, a common child's toy. A Slinky is a rather loosely coiled spring (Fig. 11.8) with a spring constant of about $k = 1$ N/m. A small toy might have a mass of 300 g, so we take $m = 0.3$ kg. Inserting these values into Equation 11.14, we get

$$f = \frac{1}{2\pi} \sqrt{\frac{k}{m}} = \frac{1}{2\pi} \sqrt{\frac{1 \text{ N/m}}{0.3 \text{ kg}}}$$

The units here are

$$\sqrt{\frac{\text{N}}{\text{m} \cdot \text{kg}}} = \sqrt{\frac{\text{kg} \cdot \text{m/s}^2}{\text{m} \cdot \text{kg}}} = \sqrt{\frac{1}{\text{s}^2}} = \frac{1}{\text{s}}$$

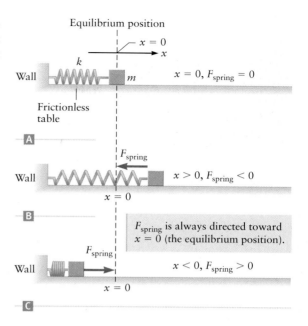

Figure 11.6 This mass on a spring is a simple harmonic oscillator. **A** When the block is at the equilibrium position ($x = 0$), the force exerted by the spring on the block is zero. **B** When the block is displaced to the right ($x > 0$), the force on the block is directed to the left. **C** When the block is displaced to the left ($x < 0$), the force is directed to the right. The force exerted by the spring on the block is thus always directed toward $x = 0$, the equilibrium point of the system.

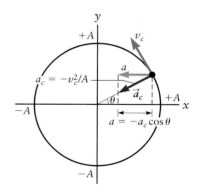

Figure 11.7 This particle undergoes uniform circular motion; its acceleration has a magnitude $a_c = v_c^2/A$ and is directed toward the center of the circle. The acceleration of the corresponding simple harmonic motion is the component of $\vec{a}_c$ along one of the axes; here we use the x direction.

Figure 11.8 A Slinky has a spring constant of about $k = 1$ N/m.

which is just the unit of frequency, hertz. Our result for the frequency is thus

$$f = \frac{1}{2\pi}\sqrt{\frac{1}{0.3}} \text{ Hz} = 0.3 \text{ Hz}$$

with a corresponding period of

$$T = \frac{1}{f} = \frac{1}{0.3 \text{ Hz}} = 3 \text{ s}$$

This period is rather long, which makes a Slinky an attractive toy for holding a child's attention.

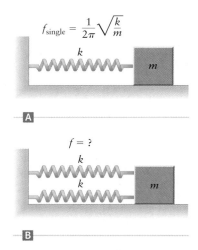

Figure 11.9 Example 11.2.

EXAMPLE 11.2 | Frequency of a Simple Harmonic Oscillator

The mass in Figure 11.9A is attached to a single spring with spring constant k, and according to Equation 11.14 it will undergo simple harmonic motion with a frequency

$$f_{\text{single}} = \frac{1}{2\pi}\sqrt{\frac{k}{m}}$$

If a second identical spring is then added as in Figure 11.9B, what is the new oscillation frequency?

RECOGNIZE THE PRINCIPLE

We need to consider how the restoring force on the mass depends on its displacement. If we can show that this restoring force has the same form as Hooke's law, we can then adapt Equation 11.14, which gives the frequency for the case of a single spring.

SKETCH THE PROBLEM

Figure 11.9 shows the problem.

IDENTIFY THE RELATIONSHIPS

We let the position of the mass be $x = 0$ at the location where the two springs are both in their relaxed state; that is the equilibrium point. If the block in Figure 11.9B is then displaced to one side by an amount x, *each* spring will exert a force on the block according to Hooke's law. Hence, the *total* horizontal force on the block is

$$F_{\text{both springs}} = \Sigma F = -2kx$$

This expression has precisely the same form as for the case of a single spring (compare with Eq. 11.11), so we can think of the two-spring combination as just a single "effective" spring with constant $k_{\text{new}} = 2k$.

SOLVE

For a single spring with spring constant k_{new}, we can calculate the frequency using Equation 11.14 and find

$$f_{\text{both springs}} = \frac{1}{2\pi}\sqrt{\frac{k_{\text{new}}}{m}} = \boxed{\frac{1}{2\pi}\sqrt{\frac{2k}{m}}}$$

What does it mean?
The frequency for the case of a single spring is

$$f_{\text{single}} = \frac{1}{2\pi}\sqrt{\frac{k}{m}}$$

so we can write our result for the two-spring combination as

$$f_{\text{both springs}} = \sqrt{2}\left(\frac{1}{2\pi}\sqrt{\frac{k}{m}}\right) = \boxed{\sqrt{2}f_{\text{single}}}$$

The frequency with two springs is thus $\sqrt{2}$ times higher than with a single spring.

Mass on a Vertical Spring: Bungee Jumping Revisited

The result for the frequency in Equation 11.14 was derived for a mass–spring system in which the mass moves along a frictionless, *horizontal* surface. The bungee jumper in Figure 11.1 is just like a mass on a spring, but in this case the mass hangs *vertically* and there are now two forces acting on the mass, the force from the spring and the force of gravity. You might now ask how this affects the oscillation frequency. Interestingly, the oscillation frequency for a mass on a vertical spring is precisely the *same* as the frequency for a mass on a horizontal spring. A mass hanging on a vertical spring is another example of simple harmonic motion, with an oscillation frequency given by Equation 11.14.

EXAMPLE 11.3 | ⓡ Analyzing a Bungee Oscillator

Consider the person attached to a bungee cord as shown in the photo at the start of this chapter. During the first part of the jump, the bungee cord is slack and the person undergoes (enjoys) free fall, but after a while, the person oscillates up and down in response to the forces of the bungee spring and gravity. Assume the mass of the cord is small compared with the mass of the person so that the bungee plus person can be modeled as a mass on a vertical, massless spring. For a person of average mass (e.g., your own mass), estimate (a) the frequency of the motion and (b) the spring constant of the bungee cord.

RECOGNIZE THE PRINCIPLE

The frequency of the motion is related to the period (the repeat time) by $f = 1/T$. To estimate the period, we appeal to your intuition (or experience) with bungee jumping. Once we have an approximate value for the frequency, we can use the relation for the frequency in Equation 11.14 along with the typical mass of a person to find k. This problem thus requires you to estimate two key quantities. The ability to make such estimates gives insight into how physics applies to real-world situations.

SKETCH THE PROBLEM

Figure 11.10 shows the person oscillating up and down and indicates the amplitude of the oscillation.

IDENTIFY THE RELATIONSHIPS

(a) The period of a bungee system depends on the length of the cord and the height of the jump. In a "typical" jump at a carnival, the person travels a distance $A = 10$ m below the equilibrium point (Fig. 11.10) and the period is about 5 s (based on the author's experience). These values are only approximate; other bungee jump experiments could be different.

(b) The mass of a typical person is about $m = 80$ kg.

SOLVE

(a) If the period of the oscillation is $T = 5$ s, the frequency is (Eq. 11.1)

$$f = \frac{1}{T} = \frac{1}{5\text{ s}} = \boxed{0.2\text{ Hz}}$$

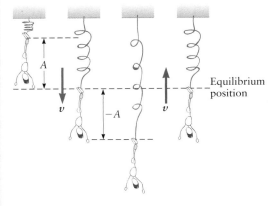

Figure 11.10 Example 11.3.

(b) A bungee oscillator is just a mass on a spring, and its frequency is (Eq. 11.14)

$$f = \frac{1}{2\pi} \sqrt{\frac{k}{m}}$$

Rearranging to solve for k, we find

$$\sqrt{\frac{k}{m}} = 2\pi f$$

$$k = 4\pi^2 f^2 m$$

Inserting our values for f and m gives

$$k = 4\pi^2 f^2 m = 4\pi^2 (0.2 \text{ Hz})^2 (80 \text{ kg}) = \boxed{100 \text{ N/m}}$$

Notice that we include only one significant figure in our answer because the period and frequency were only known to one significant figure.

What does it mean?
Other bungee jumps will have different heights, and their bungee cords will have different values of k. However, we can always use our experience (and common sense) to estimate quantities such as the period and frequency, and thus apply our results for simple harmonic motion.

The Simple Pendulum

A pendulum is another example of a simple harmonic oscillator. Pendulums can be constructed in several ways; a child on a playground swing is one example (Fig. 11.11). A pendulum can also be made by tying a rock of mass m to one end of a string, with the other end of the string fastened to a support as sketched in Figure 11.12. If the diameter of the rock is much smaller than the length L of the string and the string is massless, this system is a **simple pendulum**.

Figure 11.12A shows sketches of the rock (also called the pendulum "bob") as it moves along its path, a circular arc of radius L. It is convenient to measure the bob's displacement in terms of y, the displacement along this circular arc. To examine the pendulum motion in terms of Newton's second law, we have drawn the two forces acting on the bob: gravity and tension from the string. (We ignore the air drag force.) The total force is along (parallel to) the circular arc and is responsible for the simple harmonic motion. Figure 11.12A shows F_{parallel} at three spots along the trajectory. The bottom point on the trajectory ($y = 0$) is the equilibrium point of the pendulum, the point where $F_{\text{parallel}} = 0$. If the bob were placed at that location and given no initial velocity, it would remain there, in translational (and rotational) equilibrium (Chapters 4 and 8). As the pendulum oscillates, the bob moves back and forth past this equilibrium point, and the magnitude and direction of F_{parallel} vary with time. When the bob is on the right in Figure 11.12A, where $y > 0$, the force is directed to the left (in the negative direction). Likewise, if the bob is on the left, where $y < 0$, then F_{parallel} is to the right, the positive direction. Because this force always opposes the bob's displacement away from the equilibrium point, this is another example of a restoring force. (Compare with the restoring force on the mass in Fig. 11.6.)

For a simple pendulum, the restoring force is due to the component of the gravitational force directed along the bob's path. Figure 11.12B shows that for a pendulum of length L with a bob of mass m, this restoring force is

$$F_{\text{restore}} = F_{\text{parallel}} = -mg \sin \theta \qquad (11.15)$$

If the angle θ is small, the function $\sin \theta$ is approximately equal to θ when the angle is measured in radians. That is, $\sin \theta \approx \theta$. In addition, this angle is related to the

Figure 11.11 A child's swing is a good approximation to a simple pendulum.

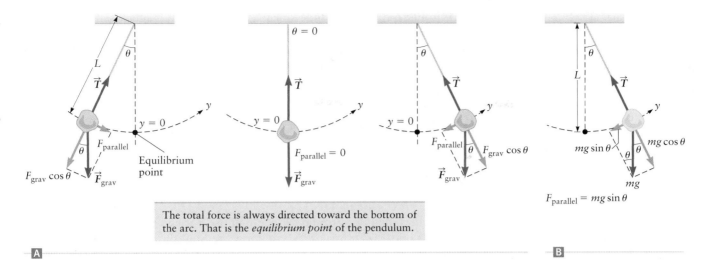

The total force is always directed toward the bottom of the arc. That is the *equilibrium point* of the pendulum.

Figure 11.12 ◭ This simple pendulum consists of a pendulum bob of mass m tied to the end of a very light string. There are two forces on the bob, from gravity and from the tension in the string. ◳ Their vector sum is directed tangent to the circular arc along which the bob swings. This total force is always directed back toward the bottom of the arc, the equilibrium point of the pendulum.

displacement by $\theta = y/L$ according to the general definition of angle as measured in radians. Inserting all this into Equation 11.15 leads to

$$F_{\text{parallel}} = -mg\theta = -\frac{mgy}{L} \tag{11.16}$$

We can now apply Newton's second law for motion along the circular arc ($F_{\text{parallel}} = ma$):

$$F_{\text{parallel}} = -\frac{mgy}{L} = ma \tag{11.17}$$

This relation has exactly the same form we found in Equation 11.12 when we applied Newton's second law to the mass on a spring. In both cases, the restoring force is proportional to the displacement from equilibrium. Comparing Equations 11.12 and 11.17, the only difference is that the spring constant k for the mass on a spring is replaced by the factor mg/L. We can therefore use the result for the frequency of a mass on a spring (Eq. 11.14) if we replace k by (mg/L), which leads to

$$f = \frac{1}{2\pi} \sqrt{\frac{k}{m}} = \frac{1}{2\pi} \sqrt{\frac{mg/L}{m}}$$

$$f = \frac{1}{2\pi} \sqrt{\frac{g}{L}} \tag{11.18}$$

Frequency of a simple pendulum

Hence, given the length of the pendulum, the frequency can be calculated. This frequency is *independent* of the mass of the bob, which may at first seem surprising; you might have expected that a heavier bob will move more slowly and have a lower frequency than a lighter bob. However, the restoring force is due to gravity and is thus proportional to the bob's mass (since its weight is equal to mg). The result is that the pendulum motion is independent of m for the same reason the velocity of a freely falling object is independent of the mass (as we learned in Chapter 3).

Another important result contained in Equation 11.18 is that the frequency of the pendulum is *independent of the amplitude* of the motion. That is, f does not depend on the amplitude A in Equation 11.9. Again, you might have expected the pendulum

Insight 11.1

HARMONIC MOTION AND THE SMALL ANGLE APPROXIMATION

For a simple pendulum, the restoring force is (Eq. 11.15) $F_{restore} = -mg \sin \theta$. When the pendulum angle θ is small, this force is given approximately by (Eq. 11.16) $F_{restore} = -mg \sin \theta \approx -mg\theta$, with $\sin \theta \approx \theta$ when θ is measured in radians. You can verify this approximation yourself by using a calculator to compare θ and $\sin \theta$, and you'll find that they differ by less than 10% when θ is smaller than about 45°. The approximation gets better and better as θ gets smaller.

frequency to depend on amplitude since if the amplitude is increased (by starting the pendulum farther from $y = 0$), the pendulum bob must travel over a greater distance. If, however, the amplitude A is increased, the bob's velocity at all points along its path is increased by the same amount because A enters in the results for y (displacement) and v (velocity) in the same way. Hence, *the period and frequency of a simple pendulum are independent of both the mass and the amplitude.* Because we assumed the angle θ is small when we derived Equation 11.18, this result holds as long as the amplitude is not too large. What angles are small enough? In typical cases, pendulum angles less than approximately 30° are small enough to make Equation 11.18 a very good approximation. In our work with pendulum oscillators, we'll always assume the angle is small so that Equation 11.18 applies.

CONCEPT CHECK 11.2 | Varying the Mass of a Playground Swing

Consider a child on a playground swing. Suppose a friend then jumps onto the swing so that both children are swinging together and the mass of the pendulum is increased by about a factor of two. How does that change the period of the swing?

(a) The period increases.
(b) The period decreases.
(c) The period stays the same.

EXAMPLE 11.4 The Frequency of a Playground Swing

Consider the pendulum swing on a nearby playground. If this swing has a length $L = 3.0$ m, what are the frequency and the period of the swing? Assume the pendulum angle is small.

RECOGNIZE THE PRINCIPLE

A pendulum swing is approximately a simple pendulum as in Figures 11.11 and 11.12, and its frequency is given by (Eq. 11.18)

$$f = \frac{1}{2\pi} \sqrt{\frac{g}{L}} \qquad (1)$$

SKETCH THE PROBLEM

Figure 11.12 describes the problem.

IDENTIFY THE RELATIONSHIPS

We are given the pendulum length, so we can use Equation (1) to find the frequency. The period is related to the frequency by $T = 1/f$ (Eq. 11.1).

SOLVE

Substituting the given value of L into Equation (1) gives

$$f = \frac{1}{2\pi} \sqrt{\frac{g}{L}} = \frac{1}{2\pi} \sqrt{\frac{9.8 \text{ m/s}^2}{3.0 \text{ m}}} = 0.29 \left(\frac{1}{s} \right) = \boxed{0.29 \text{ Hz}}$$

The period can be found from our general relation between frequency and period,

$$T = \frac{1}{f} = \frac{1}{0.29 \text{ Hz}} = \boxed{3.4 \text{ s}}$$

What does it mean?

These results are, as we have already noted, independent of the mass, so children of different size will swing at the same frequency.

The Human Arm as a Pendulum ⊗

Example 11.4 involves a simple calculation of the period of a pendulum. The same ideas can be used to understand the way your arm swings when you walk. If your arm swings freely, it acts as a sort of pendulum. Let's now estimate the period of the oscillation. The arm is not quite a simple pendulum because all the mass is not at the end. We can, however, get an approximate value of the period by assuming all the mass is at the center of mass, roughly halfway along your arm (Fig. 11.13). An adult's arm is about 60 cm long, so the center of mass is about $L = 30$ cm $= 0.30$ m from the shoulder. Taking 0.30 m as the length of a pendulum, the frequency is (Eq. 11.18)

$$f = \frac{1}{2\pi}\sqrt{\frac{g}{L}} = \frac{1}{2\pi}\sqrt{\frac{9.8 \text{ m/s}^2}{0.30 \text{ m}}} = 0.91 \text{ Hz}$$

which corresponds to a period

$$T = \frac{1}{f} = 1.1 \text{ s}$$

The next time you go for a walk, you should compare this result with the way you swing your arms.

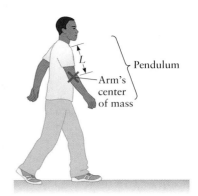

Figure 11.13 When you walk, each of your arms swings like a pendulum.

The Torsional Oscillator

Another system displaying simple harmonic motion is the **torsional oscillator**. Figure 11.14A shows one example, a circular disk suspended by a thin wire attached to the center of the disk, and Figure 11.14B shows another example, a rod attached to masses at each end, with the rod suspended by a wire attached to its center. The wire used in both of these examples is called a **torsion fiber**, and when it is twisted, it exerts a torque τ on whatever is connected to it, the disk in Figure 11.14A or the rod in Figure 11.14B. The torque is proportional to the twist angle θ

$$\tau = -\kappa\theta \tag{11.19}$$

Here, κ (the Greek letter kappa) is called the **torsion constant** and is a property of the particular torsion fiber. However, the mathematical form of the relation between the torque and the twist angle is the same for all fibers. The negative sign in Equation 11.19 indicates that it is a *restoring* torque. When the disk is rotated to positive angles, the torque tends to make the disk turn back toward the "negative" angular direction, whereas if the disk is rotated to negative angles, the torque acts toward the direction of positive angles. This is just like the restoring force for the mass on a spring in Figure 11.6 and for the pendulum in Figure 11.12. Since Equation 11.19 has the same mathematical form as Hooke's law for the spring (Eq. 11.11), the torsional pendulum is also a simple harmonic oscillator. The frequency of a torsional oscillator is given by

$$f = \frac{1}{2\pi}\sqrt{\frac{\kappa}{I}} \tag{11.20}$$

where I is the moment of inertia of the object attached to the fiber (the disk in Fig. 11.14A or the two masses and rod in Fig. 11.14B). This result is similar to what we found for a mass on a spring (compare with Eq. 11.14), but the frequency of the torsional oscillator depends on the moment of inertia of the disk I instead of on its mass.

Note that we have already encountered torsional oscillators in this book. The experiment of Cavendish (Chapter 5) used the basic arrangement in Figure 11.14B to study the gravitational force, and we'll return to this subject in Section 11.6.

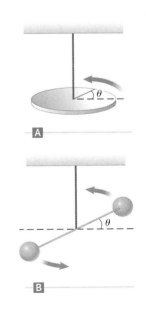

Figure 11.14 Two examples of torsional oscillators. These objects rotate about a vertical (y) axis, and the torque exerted by the fiber tends to restore the systems to their equilibrium positions.

Features Common to All Simple Harmonic Oscillators

All simple harmonic oscillators have a number of features in common. First, their motion exhibits a sinusoidal time dependence as in Equation 11.9 or, equivalently,

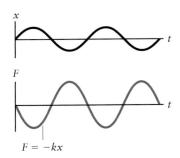

Figure 11.15 Plots of the position x and restoring force F as functions of time for a simple harmonic oscillator that starts at $x = 0$ at $t = 0$. Notice that $F = -kx$ at all times.

Properties of a restoring force

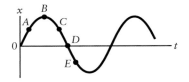

Figure 11.16 Concept Check 11.3.

Equation 11.10. The variables in these equations are called "position" (x or y) and "velocity" (v). For a simple pendulum and a mass on a spring, x and v are the ordinary position and velocity, whereas for the torsional oscillator, they are the angular position (the angle θ in Fig. 11.14) and the corresponding angular velocity. The position and velocity of *all* simple harmonic oscillators have these mathematical forms.

A second feature common to all simple harmonic oscillators is that they all involve a restoring force. Figure 11.15 shows the time dependence of both the position (which we denote by x) and the restoring force F. For the horizontal mass on a spring, F is the force exerted by the spring on the mass, whereas for a simple pendulum, F is the component of the gravitational force along the path of the pendulum bob. For the torsional oscillator, we would instead plot the torque exerted by the torsion fiber. Both the displacement and the restoring force or torque vary sinusoidally with time, and F always has a sign *opposite* that of the displacement. This is the same negative sign we encountered in Hooke's law (Eq. 11.11) and for the pendulum (Eq. 11.16). This is a key feature of a restoring force and is essential for producing harmonic motion. The magnitude of the restoring force is also proportional to the displacement (see again Eqs. 11.11 and 11.16), which leads us to the following general result: Whenever an object experiences a *restoring* force with a magnitude proportional to the displacement from the equilibrium position, the object will undergo simple harmonic motion.

The Frequency of a Simple Harmonic Oscillator Is Independent of the Amplitude

A very important feature of all simple harmonic oscillators is that the frequency is *independent* of the amplitude A. The frequency depends on quantities such as the length of the pendulum string (Fig. 11.12), the mass and spring constant in Figure 11.6, or the moment of inertia and torsion constant in Figure 11.14. Hence, this frequency does depend on the properties of the oscillator, but f does *not* depend on the amplitude of the motion. The amplitude is determined entirely by how the oscillator is initially set into motion.

> **CONCEPT CHECK 11.3 | Connection between Velocity and Force for a Simple Harmonic Oscillator**
>
> Figure 11.16 shows the displacement of a simple harmonic oscillator as a function of time. (a) At which point(s) is the velocity zero? (b) At which point(s) is the magnitude of the force a maximum?

11.3 | HARMONIC MOTION AND ENERGY

Let's next consider how the kinetic and potential energy of a simple harmonic oscillator vary with time. Figure 11.17 shows a mass on a horizontal spring as it oscillates back and forth. The mechanical energy is the sum of the kinetic energy ($\frac{1}{2}mv^2$) and potential energy ($\frac{1}{2}kx^2$, the spring potential energy). When the mass is at $x = 0$ (Fig. 11.17A and C), the potential energy is zero because the spring is relaxed, but the speed is a maximum; hence, the kinetic energy is large. When the displacement is largest, $x = \pm A$ (Fig. 11.17B and D), and the potential energy is large, the speed is zero and so the kinetic energy is zero. The energy thus "oscillates" back and forth between the kinetic and potential energies as shown in Figure 11.17E.

For an ideal oscillator, there is no friction, so the total mechanical energy is conserved. Hence, the sum of the kinetic and potential energy in Figure 11.17E is constant. This also means that the maximum kinetic energy (Fig. 11.17A) must equal the maximum potential energy (Fig. 11.17B). If the maximum speed is v_{max}, the maximum kinetic energy is $\frac{1}{2}mv_{max}^2$ and the maximum potential energy is $\frac{1}{2}kA^2$ (because the maximum displacement is $x_{max} = A$). We thus have

$$\tfrac{1}{2}mv_{max}^2 = KE_{max} = PE_{max} = \tfrac{1}{2}kA^2 \tag{11.21}$$

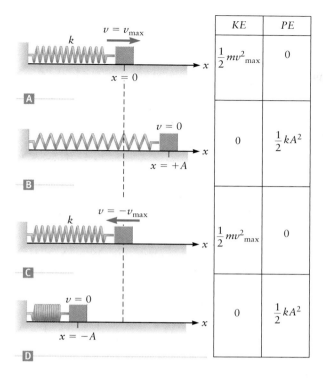

	KE	PE
A	$\frac{1}{2}mv^2_{max}$	0
B	0	$\frac{1}{2}kA^2$
C	$\frac{1}{2}mv^2_{max}$	0
D	0	$\frac{1}{2}kA^2$

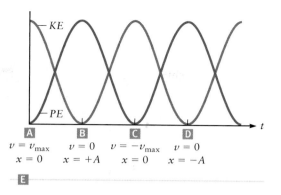

A — $v = v_{max}$, $x = 0$ B — $v = 0$, $x = +A$ C — $v = -v_{max}$, $x = 0$ D — $v = 0$, $x = -A$

E

Figure 11.17 The kinetic and potential energy of a mass on a spring oscillate with time. The sum $KE + PE$ is constant, as the mechanical energy of a simple harmonic oscillator is conserved.

which gives a useful relation between the maximum speed and displacement of the oscillator.

These results for the time dependence of the kinetic energy and potential energy have been derived for a mass on a horizontal spring, but they apply to all simple harmonic oscillators. The KE and PE of any simple harmonic oscillator vary with time in the manner sketched in Figure 11.17, and the total mechanical energy is conserved.

CONCEPT CHECK 11.4 | Energy of a Simple Harmonic Oscillator

Figure 11.18 shows the displacement of a simple harmonic oscillator as a function of time. The graph in Figure 11.18B shows either the kinetic or the potential energy. Which one?

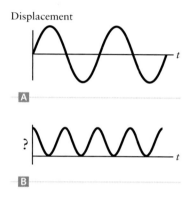

Figure 11.18 Concept Check 11.4.

EXAMPLE 11.5 | Maximum Speed of a Simple Harmonic Oscillator

Consider a simple harmonic oscillator consisting of a mass on a horizontal spring with $m = 3.0$ kg, $k = 500$ N/m, and $x = 0$ being the equilibrium position. The spring is stretched so that the mass is at $x_i = 0.10$ m and then released (Fig. 11.19A). What is the speed of the mass when it reaches $x = 0$?

RECOGNIZE THE PRINCIPLE

Just as the mass is released, its speed is zero, so its kinetic energy is zero. The total energy thus equals the potential energy at that moment (Fig. 11.19A). When the mass is at $x = 0$, the potential energy is zero because the spring is unstretched and uncompressed, so the total energy equals the kinetic energy at that moment (Fig. 11.19B). We thus have (compare with Eq. 11.21)

$$\tfrac{1}{2}mv^2 = KE_{max} = PE_{max} = \tfrac{1}{2}kx_i^2 \qquad (1)$$

We can use this relation to solve for v.

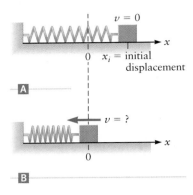

Figure 11.19 Example 11.5.

Figure 11.19 shows the system when the block is released and when it passes through $x = 0$.

IDENTIFY THE RELATIONSHIPS

We can rearrange Equation (1) to get

$$mv^2 = kx_i^2$$

$$v^2 = \frac{k}{m}x_i^2$$

$$v = \sqrt{\frac{k}{m}}x_i$$

SOLVE

The initial displacement is $x_i = 0.10$ m, and that is also the amplitude of the oscillation. Substituting with the given values of m and k leads to

$$v = \sqrt{\frac{k}{m}}x_i = \sqrt{\frac{500 \text{ N/m}}{3.0 \text{ kg}}}(0.10 \text{ m}) = \boxed{1.3 \text{ m/s}}$$

What does it mean?
The maximum potential energy occurs when the displacement is largest, whereas the maximum kinetic energy is found when the displacement is zero. This result applies to all simple harmonic oscillators, including pendulums and torsional oscillators.

11.4 | STRESS, STRAIN, AND HOOKE'S LAW

According to Hooke's law, the force exerted by a spring is directly proportional to the amount that the spring is stretched or compressed, with

$$F = -kx \tag{11.22}$$

Hooke's law applies not only to the "ideal" springs found in physics books, but to many real situations as well. Consider a solid bar of metal as sketched in Figure 11.20. If you "squeeze" the bar an amount x, it will "push back." The magnitude of this "push-back" force depends on how much you squeeze and is given by Hooke's law (Eq. 11.22). The value of the constant k depends on the size of the bar and the material it is made of. This push-back force also means that virtually all objects act as springs when squeezed, leading to many applications of our results for simple harmonic motion.

In Figure 11.20, a force of magnitude F is applied to each end of the metal bar, and the bar changes length by an amount ΔL. This force is a *compressive* force, making ΔL negative. If the force directions are reversed so that the bar is stretched (ΔL positive), we call F a *tensile* force. The value of ΔL depends on the material; a soft material compresses or stretches more than a stiff material. "Stiffness" is characterized by *Young's modulus*, Y. In Figure 11.20, the area of the end of the bar is $A = wh$, and L_0 is the bar's original length. For a wide variety of materials, the relation between the magnitude of the force F and the change in length ΔL is

$$\frac{F}{A} = Y\frac{\Delta L}{L_0} \tag{11.23}$$

In words, Equation 11.23 says that the change in length is directly proportional to the applied force. What's more, if the material is pulled so that it is elongated by an amount ΔL (i.e., if the directions of the forces in Fig. 11.20 are reversed), Equation

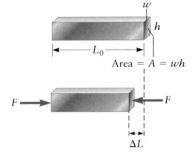

Figure 11.20 The forces F applied to the ends of this metal bar compress the bar by an amount ΔL. Tensile forces would stretch the bar instead.

Stress, strain, and Hooke's law

Table 11.1 Values of Young's Modulus, Shear Modulus, and Bulk Modulus for Some Common Materials

MATERIAL	YOUNG'S MODULUS, Y (PA)	SHEAR MODULUS, S (PA)	BULK MODULUS, B (PA)
Aluminum	7.0×10^{10}	2.4×10^{10}	7.1×10^{10}
Steel	2.0×10^{11}	8.1×10^{10}	1.4×10^{11}
Copper	1.3×10^{11}	4.8×10^{10}	1.4×10^{11}
Glass (Pyrex)	6.2×10^{10}	2.7×10^{9}	2.6×10^{10}
Human bone	9×10^{9} (compression) 2×10^{10} (when stretched)	8×10^{10}	
Diamond	1.2×10^{12}	4.8×10^{10}	6.2×10^{11}
Quartz	1.0×10^{11}	3.1×10^{10}	3.6×10^{10}
Teflon	4×10^{8}		6×10^{9}
Ice			9×10^{9}
Air (at atmospheric pressure)			1.0×10^{5}
Water			2.2×10^{9}
Mercury (liquid)			2.5×10^{10}

Note: These values generally vary slowly with temperature. The numbers given here are for room temperature.

11.23 still holds. Because the amount the bar is stretched or compressed is directly proportional to the magnitude of the applied force, this relation is the same as found for a simple spring that obeys Hooke's law (Eq. 11.22).

The ratio F/A that appears in Equation 11.23 is called the *stress*. By considering the force applied per unit of cross-sectional area, we can more easily compare the properties of bars of different size. For example, if we consider a second bar of the same material and length whose area is larger by a factor of two, a force twice as large is required to produce the same change in length ΔL. The stress F/A would be the same in the two cases, though. Likewise, the ratio $\Delta L/L_0$ is called the *strain*. This ratio is the fractional change in the length. The value of Young's modulus depends on the material; values for some common materials are listed in Table 11.1. Since the ratio $\Delta L/L$ on the right-hand side of Equation 11.23 is unitless, Y has units of force/area = N/m^2 = Pa.

Elastic versus Plastic Deformations

Young's modulus describes how a material stretches or compresses in response to an applied force. We call Y an *elastic constant* because, according to Equation 11.23, the material (such as the bar in Fig. 11.20) will return to its original length when the force F is removed. That is what is meant by "elastic" behavior. Such behavior is found for most materials provided the strain is not too large. If the strain exceeds a certain value, however, the simple linear relation between stress and strain in Equation 11.23 will no longer be found. The value of the strain at which that happens is called the *elastic limit*, and depends on the material. When the strain is pushed beyond this point, the material will no longer return to its original length even when F is completely removed. A typical plot of stress versus strain is shown in Figure 11.21. In intuitive terms, when the elastic limit is exceeded, the material will be "bent" or otherwise permanently deformed.

Insight 11.2
ELASTIC FORCES AND HOOKE'S LAW
Hooke's law for a spring (Eq. 11.22) contains a negative sign, whereas Equation 11.23 does not, but we have still called them both "Hooke's law." The difference is due to how F is defined in the two cases. For the spring described by Equation 11.22, F is the force exerted *by* the spring, whereas in the definition of the Young's modulus, Equation 11.23, F is the force *on* the bar. According to Newton's third law, the force exerted *by* the bar will equal $-F$, so Equation 11.23 is indeed equivalent to Hooke's law for a spring.

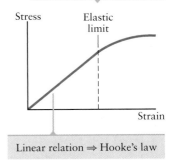

When the elastic limit is exceeded, the stress–strain curve is no longer linear.

Linear relation ⇒ Hooke's law

Figure 11.21 When a material is strained (i.e., compressed, stretched, or sheared), it will initially follow a linear stress–strain relation, or Hooke's law; the magnitude of the applied force is proportional to the amount that the material is strained. When the strain exceeds a certain value, called the elastic limit, this linear relation is no longer followed. Such large stresses usually cause the object to be permanently deformed.

EXAMPLE 11.6 | Ⓡ Stretching a Guitar String

When expressed in SI units, the numerical values of elastic constants are typically very large (see Table 11.1), making it difficult to get an intuitive feeling for how much an object will stretch or compress in response to typical forces. So, consider the following "everyday" problem. A guitar string is composed of steel and is under tension when mounted on the instrument. How much does the string stretch in response to this tension, assuming a tension of 200 N?

RECOGNIZE THE PRINCIPLE

When a force F is applied to the string, it becomes longer by an amount ΔL with (Eq. 11.23)

$$\frac{F}{A} = Y\frac{\Delta L}{L_0} \tag{1}$$

To use Equation (1) to find ΔL, we must have values for the length and area of the string (the value of Young's modulus is given in Table 11.1, and F is just the tension, which is given).

SKETCH THE PROBLEM

We can use Figure 11.20, with the bar replaced by a guitar string.

IDENTIFY THE RELATIONSHIPS

Based on experience, the length of a typical guitar string is about $L_0 = 0.6$ m and the radius is about $r = 0.5$ mm.

SOLVE

Rearranging Equation (1) to find ΔL, we get

$$\Delta L = \frac{FL_0}{YA} = \frac{FL_0}{Y\pi r^2}$$

where $A = \pi r^2$ is the cross-sectional area of the string and F is the tensile force, given as 200 N. Inserting the value of Young's modulus for steel from Table 11.1 along with the other values estimated above, we get

$$\Delta L = \frac{FL_0}{Y\pi r^2} = \frac{(200 \text{ N})(0.6 \text{ m})}{(2.0 \times 10^{11} \text{ N/m}^2)\pi(0.0005 \text{ m})^2} = 8 \times 10^{-4} \text{ m} = \boxed{0.8 \text{ mm}}$$

What does it mean?
Although 0.8 mm is not a large value, it is certainly noticeable to a guitarist turning the tuning peg of a guitar.

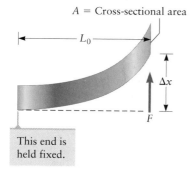

A = Cross-sectional area

This end is held fixed.

Figure 11.22 A shear force (related to the shear stress) acts to bend this bar.

The Shear Modulus

An applied force may deform a material in ways besides compressing or stretching, leading to other elastic constants besides Young's modulus. Consider a shearing force applied to a metal bar as shown in Figure 11.22; this force tends to make two parallel faces of the bar shift laterally by an amount Δx. The shear stress is defined as F/A, where A is the cross-sectional area of the end of the bar, and the shear strain is defined as $\Delta x/L_0$, where L_0 is the length of the bar (see Fig. 11.22). The shear stress and shear strain are related by

$$\frac{F}{A} = S\frac{\Delta x}{L_0} \tag{11.24}$$

where S is the *shear modulus*.

This relation is very similar to the expression that involves a compressive stress and Young's modulus (Eq. 11.23). In both cases, the deformation of the bar is directly proportional to the stress, so both relations are equivalent to Hooke's law. These relations between elastic stress and strain describe many materials, so many material objects will act as simple Hooke's law springs when acted on by a compressive or tensile force or by a shear force. We can therefore apply our results for simple springs to a variety of situations. Table 11.1 lists the shear modulus for several materials.

The Bulk Modulus

In another common situation, an object is subject to a force that tends to change the object's overall size. Consider an object immersed in fluid as sketched in Figure 11.23. The fluid might be a liquid or a gas. In either case, the pressure P in the fluid leads to a force on the surface of the object. This force is directed normal to the object's surface and causes the object's volume to change by an amount ΔV. If V_0 is the original volume of the object, the pressure and volume change are related by

$$P = -B \frac{\Delta V}{V_0} \tag{11.25}$$

where B is an elastic constant called the **bulk modulus**. Note that the pressure and volume change are directly proportional. The presence of the negative sign in Equation 11.25 means that a positive pressure will produce a negative value of ΔV, as we should expect.

In our discussions of the Young's and shear moduli, we have assumed the object in question is a solid, not a liquid or a gas. Compressive, tensile, and shear stresses (Figs. 11.20 and 11.22) can only be applied to an object that maintains its shape without the help of a container. In contrast, the concept of a bulk modulus applies to solids, liquids, and gases because all change volume in response to changes in pressure.

Elastic Properties and Simple Harmonic Motion

We learned in Section 11.2 that when an object is subject to a restoring force proportional to the displacement, the object can undergo simple harmonic motion. The same type of motion is found for the metal bars in Figures 11.20 and 11.22. If the end of the bar is displaced—that is, deformed an amount ΔL by either compressing or stretching the bar—it experiences a restoring force proportional to ΔL. If the displacement is caused by a compressive or tensile stress, the restoring force involves Young's modulus Y; for a shear displacement, the restoring force involves the shear modulus S. Some distortions produce both compression and shear, so the total restoring force may involve a combination of Y and S. Even in these mixed cases, the restoring force is still proportional to the displacement and thus is still described by Hooke's law. For this reason, a vibrating metal bar is closely analogous to a mass on a spring and generally behaves as a simple harmonic oscillator.

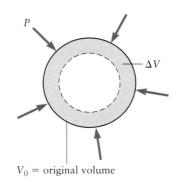

Figure 11.23 When an object is subject to a pressure (as when it is immersed in a fluid), the pressure force will compress the object, making the volume smaller by an amount ΔV.

| EXAMPLE 11.7 | ® Oscillation Frequency of a Diving Board |

When a diving board bends, the restoring force due to the board's elastic properties can be written as

$$F_{\text{board}} = -kx \tag{1}$$

where x is the displacement of the end of the board (Fig. 11.24) and k is the "spring" constant of the diving board. Estimate k for a typical diving board and use your result to find the oscillation frequency when a diver is standing at the end of the board.

Figure 11.24 Example 11.7.

RECOGNIZE THE PRINCIPLE

We can use Equation (1) to get k if we know how much a diving board will bend in response to a known force. Assuming the mass of the board is small, when a diver of mass m stands at the end of the board, the force on the board is about mg.

SKETCH THE PROBLEM

Figure 11.24 shows the diving board bent (displaced) an amount x as the diver stands at the end of the board.

IDENTIFY THE RELATIONSHIPS

A typical person has a mass of about $m = 80$ kg. When he or she stands at rest at the end of a diving board, the displacement of the board is typically $x = 10$ cm (Fig. 11.24). Of course, this value will vary from board to board, but $x = 1$ m seems too large and $x = 1$ cm is much smaller than common experience indicates. Applying Newton's second law to the diver in Figure 11.24 leads to

$$\sum F = F_{\text{board}} - mg = -kx - mg = 0$$

$$k = -\frac{mg}{x}$$

SOLVE

Using $m = 80$ kg and taking $x = -0.1$ m (a negative value because the board is displaced downward), we find

$$k = -\frac{(80 \text{ kg})(9.8 \text{ m/s}^2)}{(-0.1 \text{ m})} = \boxed{8000 \text{ N/m}}$$

If we assume the mass of the diving board is much smaller than the mass of the person (which is only a very rough approximation), the oscillation frequency is the same as that of a simple mass-on-a-spring system (Eq. 11.14),

$$f = \frac{1}{2\pi}\sqrt{\frac{k}{m}}$$

Inserting our values for k and m gives

$$f = \boxed{2 \text{ Hz}}$$

What does it mean?
The period for the diver's oscillation is thus about $T = 1/f = 0.5$ s. You should compare this value with the value you expected based on your intuition or the last time you visited a swimming pool.

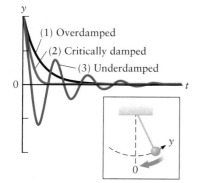

Figure 11.25 When a damped oscillator is given a nonzero displacement at $t = 0$ and then released, it can exhibit three different types of behavior: (1) overdamped, (2) critically damped, and (3) underdamped.

11.5 | DAMPING AND RESONANCE

So far, we have ignored the effect of friction on the motion of a harmonic oscillator. If there is no friction, an oscillator set into motion will oscillate forever. Of course, most oscillators eventually come to rest as their vibrations decay with time. The friction in an oscillating system is referred to as *damping*, and we now consider the motion of a *damped harmonic oscillator.*

Newton's second law can be used to derive the mathematical form for the displacement y of a damped oscillator as a function of time. However, the mathematics is more complicated than we can tackle here, so let's focus on the qualitative behavior. Consider a pendulum swing moving through air. Curve 3 in Figure 11.25 shows how the displacement varies with time when the damping is weak, that is, when there is only a small amount of friction. This plot applies to any weakly damped harmonic oscillator, so it describes the back-and-forth swinging of a pendulum or the motion of a mass on a horizontal surface (Fig. 11.6) when the surface is very slippery. The system still oscillates because the displacement alternates between positive and negative values, but the amplitude of the oscillation gradually decreases with time. Eventually, the amplitude goes to zero, but the system undergoes many oscillations before damping eventually brings it to rest. This type of motion is called an *underdamped* oscillation.

CONCEPT CHECK 11.5 | Sources of Damping in a Real Oscillator

Consider a real pendulum (Fig. 11.25) or a real mass-on-a-spring system (Fig. 11.6). What are some possible sources of damping in these systems?

When friction is very large, the oscillator is said to be *overdamped*. The resulting displacement as a function of time in this case is shown as curve 1 in Figure 11.25. This type of motion would be found if the pendulum mass were moving through a very thick fluid (such as honey) rather than air. When the bob of an overdamped pendulum is pulled to one side and then released, the bob swings *very* slowly back to the bottom ($y = 0$) as plotted in Figure 11.25 (curve 1).

The underdamped and overdamped regimes are distinguished in the following way. In the underdamped case, the displacement always passes through zero at least once, and usually many times, before the system comes to rest. Hence, for an underdamped pendulum, the bob always swings past the bottom at least once. In contrast, an overdamped system released from rest does not swing past the bottom, but moves to the equilibrium point without going past it. At the boundary between these two cases the system is *critically damped*. There, the displacement falls to zero as rapidly as possible *without* moving past the equilibrium position ($y = 0$). The displacement as a function of time for the critically damped case is illustrated in curve 2 in Figure 11.25.

These different categories of damping have important implications for applications. For example, the shock absorbers in a car are essentially springs that support the car's body as shown in Figure 11.26. Shock absorbers enable the tires to move up and down according to bumps in the road without directly passing these vibrations to the car's body (and the passengers). When the car hits a bump, the shock absorber springs are compressed, corresponding to the initial value of the displacement in Figure 11.25. To make the ride as comfortable as possible, the shock absorbers are designed to be critically damped, which allows the body of the car to return to its original height as quickly as possible. With underdamped (worn-out!) shock absorbers, a car oscillates up and down, whereas overdamped shock absorbers give a soft "spongy" ride with poor steering response and handling.

The Driven Oscillator

An oscillator such as a pendulum can be set into motion by pulling the pendulum bob to one side and then letting go or by imparting an initial impulse as in giving the bob an initial "kick." In many applications, however, an oscillator is subjected to a force at regular intervals, or even continuously. A familiar example—a child being pushed on a swing—is shown in Figure 11.27. This particular pendulum is "driven" by the force that the parent exerts on the swing. We have already seen (Eq. 11.18) that the pendulum–swing has an oscillation frequency of

$$f_{\text{nat}} = \frac{1}{2\pi} \sqrt{\frac{g}{L}}$$

which is often referred to as the *natural frequency* of the swing; it is not necessarily the same as the frequency associated with the driving force. Suppose the parent applies a small push to the swing at regular time intervals. If the period of this pushing is T_{drive}, we can define a *driving frequency* $f_{\text{drive}} = 1/T_{\text{drive}}$. The amplitude of the

Figure 11.26 The shock absorbers on a car act as springs with damping. The goal is usually to have a car respond to bumps in the road as a critically damped oscillator.

Figure 11.27 When the parent pushes this child, the swing acts as a driven oscillator.

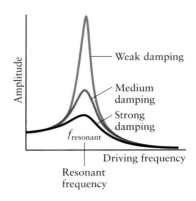

Figure 11.28 The amplitude of a driven oscillator depends on the damping. For weak damping, the oscillator will have a large amplitude when the driving frequency is near the resonant frequency, f_{resonant}. For a weakly damped oscillator, the resonant frequency is very close to the natural frequency.

oscillations of the pendulum–swing depend on the driving frequency as sketched in Figure 11.28.

If the frequency of the force pulses from the parent does not match the pendulum's natural frequency—that is, if f_{drive} is not close to f_{nat}—the pendulum amplitude is small, even if the parent applies a large force. The amplitude is largest when the driving frequency is close to the natural frequency of a system. That happens when the parent in Figure 11.27 times her pushing to match the natural swinging frequency of the pendulum. This phenomenon is called **resonance**. Whenever the frequency of the driving force matches the natural frequency of an oscillator, the amplitude of oscillation will be large. Figure 11.28 shows the behavior for cases of weak, medium, and strong damping. The resonance curve is narrowest and tallest when the damping is weakest, and it becomes lower and wider as the damping is increased. The peak of the resonance curve occurs at the **resonant frequency**; for weak damping, the resonant frequency is very close to the natural frequency f_{nat} of the oscillator.

A harmonic oscillator for which damping is important is the vibrating element in a loudspeaker, the so-called speaker cone. Different musical notes have different frequencies, and when a loudspeaker is used to play music, the speaker cone is subject to a force whose frequency is determined by the frequencies present in the music. According to Figure 11.28, if the speaker cone oscillator is weakly damped, it will respond strongly to music near its natural frequency and relatively weakly to frequencies that are far from f_{nat}. A weakly damped speaker will thus produce loud sounds at some frequencies and weak sounds at others, which is *not* how one would like a loudspeaker to work. Instead, a speaker should have a frequency response that is as constant (uniform) as possible. One way to accomplish that is to make the speaker cone heavily damped. The cone of a good loudspeaker, however, should also move fairly easily so that it produces sound efficiently, hence the damping cannot be too large. This trade-off is one of the challenges faced by the speaker designer. Similar design challenges are faced in other applications of harmonic oscillators, such as the shock absorbers for a car.

CONCEPT CHECK 11.6 | Damping of a Bridge

A structure such as a bridge or a building can oscillate in response to forces from the wind or an earthquake. Do you think these structures should be designed to be (a) underdamped, (b) critically damped, or (c) overdamped?

11.6 | DETECTING SMALL FORCES

Because the Hooke's law description of elastic forces leads directly to the restoring force necessary for producing simple harmonic motion, the phenomena of oscillatory motion and elasticity are closely connected. Oscillations and elasticity also come together in the problem of detecting small forces.

The Cavendish Experiment

Let's consider an experiment to measure the force of gravity between two terrestrial objects. In Chapter 5 we found that the force of gravity between two large (1 m in diameter) and closely spaced lead spheres is about $F_{\text{grav}} = 2 \times 10^{-3}$ N (Eq. 5.22). Even though those spheres were quite large, this is a small force, and we need to be clever if we want to measure it experimentally. We might try to measure this force by connecting m_1 to a spring as sketched in Figure 11.29. The gravitational force exerted by m_2 on m_1 will cause the spring to stretch an amount x, and by measuring x, we can determine the magnitude of the force. Using Hooke's law for the spring, we have (considering magnitudes only) $F_{\text{grav}} = kx$. If the spring is similar to a Slinky spring, the spring constant k is approximately $k = 1$ N/m. Using this value and $F_{grav} = 2 \times 10^{-3}$ N, we can calculate the expected value of x:

$$x = \frac{F_{grav}}{k} = \frac{2 \times 10^{-3}\ \text{N}}{1\ \text{N/m}} = 0.002\ \text{m} = 2\ \text{mm}$$

If we want to measure F_{grav} with an accuracy of 1%, we would need to measure x with a precision of 0.02 mm. This is about the diameter of a human hair, so this method is not a very promising approach for a high precision measurement of F_{grav}. We need a different strategy.

Figure 11.30 shows how Henry Cavendish tackled this problem in an experiment he conducted about 200 years ago. He mounted two metal spheres, each of mass m_1, on a light rod and suspended the rod by a thin wire. This setup is very similar to the torsional oscillator in Figure 11.14B. Cavendish then brought two other objects, each of mass m_2, near the original spheres as shown in Figure 11.30, and the force of gravity exerted by the objects m_2 on the spheres m_1 causes the torsion fiber to twist. We saw in Section 11.2 that when a torsion fiber is twisted by an angle θ, it exerts a torque $\tau = -\kappa\theta$, where κ is the torsion constant of the fiber. Torsion fibers can have very small diameters; they can be as thin as a human hair (about 20 μm in diameter), making them *extremely* easy to twist. In fact, a value of $\kappa = 1 \times 10^{-9}$ N·m is quite common for a wire of this diameter.

With such small values of the torsion constant, Cavendish was able to measure F_{grav} accurately. However, Cavendish also had to deal with the fact that the system in Figure 11.30 acts as a torsional *pendulum*. When the two masses m_2 are brought close to the spheres on the torsion fiber, the gravitational force will set this pendulum into oscillation. The oscillation frequency of a torsional pendulum is (Eq. 11.20)

$$f = \frac{1}{2\pi}\sqrt{\frac{\kappa}{I}} \qquad (11.26)$$

where I is the moment of inertia of the rotating part of the apparatus. Recall that the torsion constant κ is very small—that is what makes the Cavendish experiment so sensitive to small forces—but a very small value of κ in Equation 11.26 gives a very low frequency. In fact, frequencies of 0.0001 Hz are typical! That is *quite* a low frequency. The period of such an oscillator is $T = 1/f \approx 1 \times 10^4$ s, which is about 4 hours. Such a torsion pendulum is also a weakly damped system (because the friction in a torsion fiber is very small), so its oscillations will decay with time as in the plot for the underdamped case in Figure 11.25. An experimenter will thus have to wait many periods before the oscillations stop and the twist angle can be measured. These very slow and weakly damped oscillations make the Cavendish experiment quite challenging. Legend has it that Cavendish performed his experiments very late on Sunday evenings, when the interfering vibrations from traffic on neighboring roads were smallest.

The Atomic Force Microscope

We next consider a modern invention that uses a type of spring to measure very small forces. This invention is called an **atomic force microscope** (AFM) and is illustrated in Figure 11.31A. A small bar, typically a fraction of a millimeter in length (about the diameter of the period at the end of this sentence), is fashioned so that it has an even smaller sharp tip at one end. This bar, called a cantilever, bends in response to any forces that act on it. These forces usually act at the tip, so the cantilever bends up and down as shown in the figure. This microscope works by scanning the tip of the cantilever over the surface of a material that is to be studied. The tip is *very* sharp, with a width at the end of only a few atoms. Although the tip usually does not directly touch the surface atoms, it is attracted to them, and the size of this attractive force depends on the distance between the tip and the nearest

The gravitational force exerted by m_2 on m_1 will stretch the spring a distance x.

Figure 11.29 In theory, one could measure the gravitational force between two masses m_1 and m_2 by attaching m_1 to a spring and then observing how much the spring is stretched (x) when m_2 is brought nearby. This experimental design is not recommended because the displacement of the spring will be extremely small.

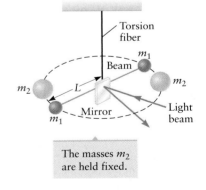

The masses m_2 are held fixed.

Figure 11.30 The Cavendish apparatus. The force of gravity exerted by the large masses m_2 on the small masses m_1 causes the torsion fiber to twist. The twist angle can be measured by attaching a small mirror and observing the reflection of a light beam.

Figure 11.31 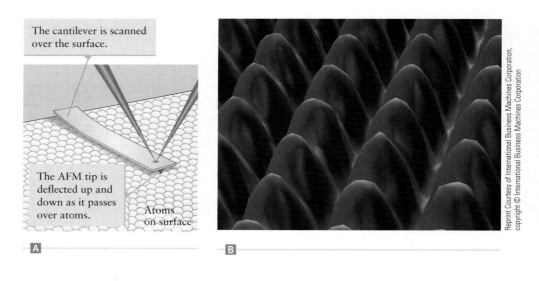 **A** Schematic design of an atomic force microscope (AFM). A very sharp tip is mounted on the end of a flexible bar called a "cantilever." The cantilever moves up and down as the atoms on the surface exert forces on the tip. The vertical deflection of the end of the cantilever is then used to construct an image of the surface profile. The red beams denote laser light reflected from the cantilever to measure its deflection. **B** An AFM image showing the atoms on a nickel surface. Each bump is an individual nickel atom.

The cantilever is scanned over the surface.

The AFM tip is deflected up and down as it passes over atoms.

Atoms on surface

surface atom. If an atom is "high" (i.e., sticks out and is close to the tip), the force on the tip is large, causing the cantilever to be deflected. By measuring this deflection as a function of the cantilever's position, one can make an image of the surface as shown in Figure 11.31B. An atomic force microscope is extremely sensitive to small forces. This sensitivity, along with the sharpness of the tip, makes it possible to resolve features as small as single atoms.

The cantilever behaves just like the diving board in Example 11.7 and can thus be treated as a simple spring obeying Hooke's law. A force of magnitude $F = kx$ is thus required to deflect the tip by a distance x. The value of k can be estimated from the shear modulus of whatever material is used to make the cantilever and from the dimensions of the cantilever (using Eq. 11.24). For a typical cantilever, $k = 0.1$ N/m. The force between the tip and the surface[1] is typically 1×10^{-9} N, so the deflection of the cantilever is

$$x = \frac{F}{k} = \frac{1 \times 10^{-9}\ \text{N}}{0.1\ \text{N/m}} = 1 \times 10^{-8}\ \text{m}$$

This very small deflection can be measured by reflecting a laser beam from the cantilever as shown in Figure 11.32. An AFM thus uses what is basically a simple spring to measure very small forces and displacements and thereby obtain images of the atoms on the surface of a solid.

Figure 11.32 The deflection of an atomic force microscope cantilever is often measured by using the deflection of a laser beam. The location of the laser spot on a distant screen depends on the angle θ of the cantilever tip from the horizontal. By measuring the location of the laser spot, you can thus measure how much the cantilever is bent when it approaches atoms on the surface being studied.

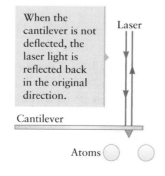

When the cantilever is not deflected, the laser light is reflected back in the original direction.

Laser

Cantilever

Atoms

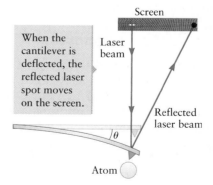

Screen

When the cantilever is deflected, the reflected laser spot moves on the screen.

Laser beam

Reflected laser beam

θ

Atom

[1]We'll discuss this force in Chapter 29.

SUMMARY | Chapter 11

Frequency and period

Any motion that repeats with time is **harmonic**. The repeat time is called the **period**. The **frequency** (measured in hertz) is equal to the number of cycles that are completed each second. Frequency and period are related by

$$f = \frac{1}{T}$$

(11.1) (page 349)

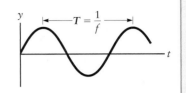

Simple harmonic motion

The position and velocity of a simple harmonic oscillator vary with time according to

$$y = A \sin(2\pi f t) \qquad v = 2\pi f A \cos(2\pi f t)$$

(11.9) (page 352)

The total mechanical energy of a simple harmonic oscillator is conserved. The kinetic energy and potential energy oscillate with time, but their sum is constant.

Stress, strain, and Hooke's law

Elastic objects stretch or compress in response to applied forces. For a compressive or tensile force, the deformation is

$$\frac{F}{A} = Y \frac{\Delta L}{L_0}$$

(11.23) (page 362)

where Y is an elastic constant called **Young's modulus**. The ratio F/A is called the **stress**, and $\Delta L/L_0$ is called the **strain**. Other types of deformations (such as shear stresses) are described by other elastic constants.

Damping and resonance

Most real harmonic oscillators are affected by friction, causing the oscillations to be damped. There are three classes of damped oscillators: **underdamped** (weak damping), **overdamped** (large damping), and **critically damped**. When a damped harmonic oscillator is subject to a time-dependent driving force, the amplitude will be largest when the driving frequency is close to the natural frequency of the oscillator. This is called **resonance**.

Mass connected to a spring

A mass connected to a spring with spring constant k oscillates with a frequency

$$f = \frac{1}{2\pi} \sqrt{\frac{k}{m}}$$

(11.14) (page 353)

(*Continued*)

QUESTIONS

SSM = answer in Student Companion & Problem-Solving Guide

 = life science application

1. You are given the job of designing a swing using two ropes attached to the branch of a tree as sketched in Figure Q11.1. Explain why it is important that both of the ropes be the same length. *Hint:* Consider how the period of a simple pendulum depends on the length of the string.

L_1 L_2
Tree
swing

Figure Q11.1 Why is it desirable to have $L_1 = L_2$?

2. A mass hangs from a vertical spring and is initially at rest. A person then pulls down on the mass, stretching the spring. Does the total mechanical energy of this system (the mass plus the spring) increase, decrease, or stay the same? Explain your answer.

3. Design an experiment to measure the torsion constant κ of a torsion fiber. The value of κ is needed in the analysis of the Cavendish experiment (Sec. 11.6).

4. SSM Consider a simple pendulum that is used as a clock. (a) What should the length of the pendulum be to make one oscillation (one "tick" of the clock) every second when it is at sea level? (b) Will this clock "speed up" or "slow down" when it is taken to the top of Mount Everest? (c) How long will it take this clock to make 60 ticks on Mount Everest?

5. If a spring with spring constant k_0 is cut in half, what is the spring constant of one of the pieces?

6. A car mounted on shock absorbers is like a mass on a spring. If you ignore damping, how will the frequency of the oscillations change if passengers (or a heavy load) are added to the car? Will the frequency increase, decrease, or stay the same?

7. A basketball player dribbles the ball as she moves along the court. What is a typical value for the frequency of this oscillatory motion?

8. When a group of marching soldiers reach a bridge, they often "break stride" and do not walk "in step" across the bridge. Explain why.

9. Two metal rods with the same length and cross-sectional area are both subjected to a compressive force F. If the length of rod 1 changes by more than the length of rod 2, which one has the larger Young's modulus?

10. Two solid rods are made of the same material and have the same cross-sectional areas, and their lengths differ by a factor of two. If a compressive force F is applied to both, what is ratio of their changes in length?
(a) The longer rod changes length by more, by a factor of two.
(b) The shorter rod changes length by more, by a factor of two.
(c) The lengths change by the same amount.

11. Consider the variation of the kinetic energy of a harmonic oscillator. The period of the oscillations of *KE* can be determined from the separation in time of adjacent maxima in a plot of *KE* as a function of time. Show that the frequency of the oscillations of *KE* is equal to twice the frequency of the displacement oscillations (Eq. 11.14). Explain why.

12. In Section 11.3, we discussed the total mechanical energy of a mass-on-a-spring oscillator. The result in Equation 11.21 shows that the total energy is proportional to the square of the amplitude. Derive the corresponding results for the total mechanical energy of a simple pendulum and for a torsional oscillator and show that they are also proportional to the square of the amplitude.

13. ⃞SSM Use energy considerations to derive the oscillation frequency for a mass-on-a-spring oscillator. *Hint*: The maximum potential energy stored in the spring must be equal to the maximum kinetic energy of the mass.

14. Derive the frequency of oscillation of a torsional oscillator, Equation 11.20. *Hint*: Consider how the restoring force depends on the twist angle and use Newton's second law for rotational motion ($\tau = I\alpha$).

15. Pendulum clocks use a pendulum to keep time (Fig. Q11.15). This type of clock can be adjusted by varying the length of the pendulum. How should the length be adjusted in the following cases?
(a) The clock is running slow.
(b) The clock is running fast.
(c) The clock is running fine at sea level, but needs to be moved to Denver.

Figure Q11.15

16. Many musicians use a metronome to help keep time when playing music. Some metronomes use a simple pendulum. If a metronome clicks 100 times per minute, what is the length of the pendulum?

17. In Chapter 3, we learned how the apparent weight of a person in an elevator depends on the acceleration of the elevator. Consider a simple pendulum that is placed in an elevator. Does the elevator's acceleration affect the pendulum's period? If the pendulum is accelerating downward with magnitude g, what is the period of the pendulum?

18. The windshield wipers on your car are an example of periodic motion. What is the approximate period of the motion?

19. Arrange these materials according to their Young's modulus from smallest to largest: (a) Jello, (b) steel, (c) wood, (d) diamond.

20. Discuss the difference between the Young's modulus of a material and the strength of a material. Does a large value of Y guarantee a strong material?

PROBLEMS

⃞SSM = solution in Student Companion & Problem-Solving Guide
★ = intermediate ◆ = challenging

ⓧ = life science application
Ⓡ = reasoning and relationships problem

11.1 GENERAL FEATURES OF HARMONIC MOTION

1. A simple harmonic oscillator takes 15 seconds to undergo 25 complete oscillations. (a) What is the period of the oscillator? (b) What is the frequency of the oscillator?

2. ⓧ A person's heart beats 70 times in 1 minute. What is the average frequency of this oscillation?

3. ★ Figure P11.3 shows the displacement of a simple harmonic oscillator as a function of time. (a) Find the period, frequency, and amplitude of the oscillator. (b) At what time(s) does the acceleration have its largest positive value? (c) When does the acceleration have its most negative value?

Figure P11.3 Problems 3, 5, and 15.

4. ★ Figure P11.4 shows a plot of the position as a function of time for a particle undergoing simple harmonic motion. Identify the following points on this graph: (a) Where is the speed largest? (b) Where is the velocity positive with its magnitude being largest? (c) Where is the

Figure P11.4 Problems 4 and 34.

magnitude of the acceleration largest? (d) Where is the force on the particle zero?

5. (a) For the simple harmonic oscillator in Figure P11.3, estimate the velocity at $t = 4.0$ s and at 7.0 s. (b) At what time(s) does the velocity have its largest positive value?

6. The displacement of a harmonic oscillator is given by $y = 3.4 \sin(25t)$, where the units of y are meters and t is measured in seconds. Find (a) the amplitude and (b) the frequency.

7. ★ The displacement of a harmonic oscillator is given by $y = 9.4 \sin(15t)$, where the units of y are meters and t is measured in seconds. What is its maximum velocity?

8. The displacement of a harmonic oscillator is given by $y = 7.1 \sin(48t)$, where the units of y are meters and t is measured in seconds. Give three values of t at which the oscillator has (a) its largest (and positive) displacement and (b) its largest (and positive) velocity.

9. Figure P11.9 shows the velocity as a function of time for an oscillator. Is it a simple harmonic oscillator? Explain why or why not. What is its frequency?

Figure P11.9

10. ⃞SSM ★ A simple harmonic oscillator has a frequency of 300 Hz and an amplitude of 0.10 m. What is the maximum velocity of the oscillator?

11. Figure P11.11 shows the velocity of a simple harmonic oscillator as a function of time. (a) Estimate the position when $v = 0$. (b) Estimate the frequency.

Figure P11.11

11.2 EXAMPLES OF SIMPLE HARMONIC MOTION

12. SSM ★ ® Estimate the spring constant for a trampoline. Assume a person is standing on the trampoline and oscillating up and down without leaving the trampoline. *Hint*: Begin by estimating the mass of the oscillator and the period of the motion.

13. Consider the mass on a spring in Figure 11.6. If the spring constant is $k = 30$ N/m and the mass is $m = 2.5$ kg, what is the period?

14. ★ A child plays on a bungee cord and oscillates with a certain frequency f. An adult with a mass that is five times greater than that of the child then uses the same bungee cord. What is the ratio of the frequency with the adult to the frequency with the child?

15. ✪ Suppose Figure P11.3 describes the displacement of a mass-on-a-spring harmonic oscillator. (a) If the mass is $m = 2.0$ kg, what is the spring constant? (b) Estimate the velocity at $t = 4.0$ s and at 5.0 s. (c) Estimate the acceleration at 1.5 s. (d) Estimate the maximum acceleration.

16. SSM ★ A mass $m = 4.5$ kg is attached to a vertical spring with $k = 200$ N/m and is set into motion. (a) What is the frequency of the oscillation? (b) If the amplitude of the oscillation is 3.5 cm, what is the maximum value of the velocity? (c) How long does it take the mass to move from $y = 1.5$ cm to $y = 2.5$ cm? (d) If the mass is oscillating with a maximum speed of 45 m/s, what is the amplitude? (e) If the spring constant is increased by a factor of two and the maximum kinetic energy of the mass is the same, by what factor does the amplitude change?

17. ✪ A mass $m = 2.4$ kg is attached to two springs, and the springs are fastened to two walls as shown in Figure P11.17. The springs both have $k = 400$ N/m and are both in their relaxed states (unstretched and uncompressed) when the mass is centered between the two walls. What is the frequency of this simple harmonic oscillator? (Consider only the horizontal motion and ignore the effect of gravity.)

Figure P11.17

18. The period of oscillation for a mass-on-a-spring system is 0.22 s. If $m = 3.5$ kg, what is the spring constant of the spring?

19. ★ A mass-on-a-spring system has $m = 50$ kg and $k = 200$ N/m. The mass is pulled a distance 0.25 m from its equilibrium position and then released. (a) What is the maximum acceleration of the mass? (b) What is its maximum velocity?

20. ★ Consider the mass-on-a-spring system in Figure P11.20. Three identical springs, with the same spring constant $k = 40$ N/m, are used to connect the mass ($m = 20$ kg) to a ceiling. What is the frequency of this simple harmonic oscillator?

Figure P11.20

21. ✪ ® Estimate the maximum acceleration of a bungee jumper. Is it larger or smaller than the acceleration due to gravity during free fall?

22. Consider a simple pendulum that consists of a rock of mass 3.5 kg tied to the end of a (massless) string of length 1.5 m. (a) What is the frequency of the pendulum? (b) What is the period of the pendulum?

23. A simple pendulum oscillates with a period of 3.5 s. What is its length?

24. ★ A simple pendulum has a length of 2.5 m and is pulled a distance $y = 0.25$ m to one side and then released. (a) What is the speed of the pendulum when it passes through the lowest point on its trajectory? (b) What is its acceleration in the direction along its circular arc when it passes through the lowest point on its trajectory?

25. ★ A simple pendulum has a period of 2.5 s on the Earth. An astronaut then takes it to the surface of the Moon. What period does the astronaut measure there?

26. ★ A simple pendulum uses a steel wire as the "string." The length of this wire is 1.5 m at room temperature. If the temperature is increased by 10°C, the length of the wire increases by 0.18 mm. What is the change in the period of the pendulum?

27. ★ The length of a simple pendulum is increased by a factor of three. By what factor does the frequency change?

28. SSM ★ Two playground swings are side by side. The children using the swings notice that one of the swings (swing 1) has a period of exactly 4.5 s. They also find that when the first swing has completed 10 oscillations, the other swing (swing 2) has completed 11 oscillations. (a) Is swing 2 longer or shorter than swing 1? (b) Find the length of swing 2.

29. A torsional oscillator has a moment of inertia of 4.5 kg · m² and a torsion constant of $\kappa = 0.15$ N · m. What is the frequency of the oscillator?

30. ✪ Consider the torsional oscillator in Figure P11.30. The rotating mass consists of two small spheres, each of mass $m = 2.0$ kg, that are attached to the ends of a massless rod of length $L = 1.5$ m. The frequency of this torsional oscillator is found to be 0.035 Hz. What is the value of the torsion constant κ for the torsion fiber?

Figure P11.30
Problems 30 and 31.

31. Consider a torsional oscillator like the one in Figure P11.30 and suppose it has a frequency of 0.045 Hz. If the length of the rod is increased by a factor of three, what is the new frequency of the oscillator?

32. ✪ Figure P11.32 shows the displacement of a mass on a spring as a function of time. (a) At what point is the acceleration positive and the velocity negative? (b) Are there any times during the oscillation when the acceleration and the velocity are both positive and nonzero?

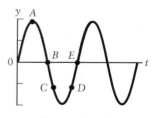

Figure P11.32

33. ✪ The cone of a loudspeaker oscillates with an amplitude of 2.0 mm. If the frequency is 2.5 kHz, what are the maximum velocity and the maximum acceleration of the cone? Assume the speaker cone moves as a simple harmonic oscillator.

11.3 HARMONIC MOTION AND ENERGY

34. ✪ Figure P11.4 shows the position as a function of time for a mass attached to a spring. At what points are (a) the magnitude of the momentum of the mass largest, (b) the kinetic energy largest, (c) the potential energy largest, and (d) the total energy largest?

35. ★ A mass-on-a-spring system has a spring constant $k = 400$ N/m and a mass $m = 0.50$ kg. (a) If it is given an initial displacement of 0.25 m and then released, what is the initial potential energy of the oscillator? (b) What is the maximum kinetic energy of the oscillator?

36. ✪ The position of a mass-on-a-spring oscillator is given by $y = A \sin(25t)$ where the value of t is in seconds, and $A = 0.35$ m. (a) What is the maximum kinetic energy of an oscillator of mass 1.7 kg? (b) Suppose the amplitude is increased so that the maximum kinetic energy is doubled. What is the new value of A?

37. SSM ✪ A particle attached to a spring with $k = 50$ N/m is undergoing simple harmonic motion, and its position is described by the equation $x = (5.7 \text{ m})\cos(7.5t)$, with t measured in seconds. (a) What is the mass of the particle? (b) What is the period of the motion? (c) What is the maximum speed of the particle? (d) What is the maximum potential energy? (e) What is the total energy?

38. Consider a simple harmonic oscillator whose position as a function of time is given by Figure P11.38. At what times in this plot is the kinetic energy a maximum? At what time(s) is the potential energy a maximum?

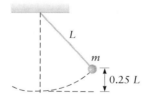

Figure P11.38

39. ✪ The position of a simple harmonic oscillator is given by $y = A \sin(2\pi f t)$, with $f = 400$ Hz. Find a value of t at which the potential energy is one quarter of its maximum value.

40. Consider a simple pendulum of length 1.2 m with $m = 2.0$ kg. If the amplitude of the oscillation is 0.15 m, what is the total mechanical energy of the oscillator?

41. Consider a simple pendulum with $L = 1.2$ m and $m = 2.0$ kg. Suppose the mass is initially at rest at the lowest point on its trajectory when it is given an impulse such that it then has velocity of 0.10 m/s. What is the total mechanical energy of the oscillator?

42. ✪ A simple pendulum has a length L and a mass m. At its highest point, the pendulum mass is $0.25L$ above its lowest point (Fig. P11.42). What is the speed of the mass when it is at its lowest point? Express your answer in terms of m, L, and g.

43. ✪ Consider a simple pendulum with $L = 5.2$ m and $m = 3.3$ kg. It is given an initial displacement $y = +0.15$ m (as measured along the circular arc along which it moves) and an initial velocity of $+0.20$ m/s. (a) What is the total mechanical energy of the oscillator? (b) What is the amplitude of the motion as measured along the circular arc followed by the pendulum mass? (c) What is the speed of the mass when it passes through the lowest point on its trajectory? (d) Suppose the initial velocity is instead equal to -0.20 m/s. How would the answers to parts (a), (b), and (c) change?

44. ✪ A mass-on-a-spring oscillator has $m = 7.0$ kg and $k = 300$ N/m and is oscillating with an amplitude of 12 cm. If the value of the mass is increased by a factor of two but the amplitude is kept fixed, is the maximum potential energy the same or different? Find the original value and the new value of the maximum potential energy.

45. ⭐ The bumper on a car is made of flexible plastic so that it will compress on impact and "cushion" the occupants during a collision. Suppose a car of mass $m = 1200$ kg is traveling at 2.5 m/s when it strikes a brick wall. It is found that the bumper compresses a maximum amount 5.0 cm. What is the spring constant of the bumper?

11.4 STRESS, STRAIN, AND HOOKE'S LAW

46. An aluminum bar has a length of 2.0 m and transverse dimensions of 3.5 mm and 6.9 mm. (a) What force is required to stretch the bar 1.2 mm? (b) What is the strain?

47. ⭐ Two solid rods having the same length are made of the same material with circular cross sections. Rod 1 has a radius r, and rod 2 has a radius $2r$. If a compressive force F is applied to both rods, their lengths are reduced by ΔL_1 and ΔL_2. Is the ratio $\Delta L_1/\Delta L_2$ (a) 4, (b) 2, (c) 1, (d) $\frac{1}{2}$, or (e) $\frac{1}{4}$?

48. Consider a steel rod of diameter 5.0 mm and length 3.5 m. If a compressive force of 5000 N is applied to each end, what is the change in the length of the rod?

49. ⭐ A thick copper wire of length 10 m is observed to stretch 0.10 cm. What is the strain? If the applied force is 50,000 N, what is the diameter of the wire?

50. A shear force of 3.0×10^5 N is applied to an aluminum bar of length $L = 20$ cm, width $w = 5.0$ cm, and height $h = 2.0$ cm as shown in Figure 11.22. What is the shear deformation Δx?

51. A solid glass sphere with an initial diameter $d = 25.00$ cm is subject to a pressure of $1000 \times P_{atm}$, where atmospheric pressure is $P_{atm} = 1.01 \times 10^5$ Pa. What is the new diameter of the sphere? The change in diameter is small, so keep at least four significant figures in your calculation.

52. A guitar string of diameter 0.5 mm and length 0.75 m is subject to a tension of 150 N. If the string stretches an amount 0.40 mm, what is Young's modulus of the string?

53. A steel cylinder of length 11 cm and radius $R = 3.5$ cm (Fig. P11.53A) is subjected to a compressive force of 5000 N. By what amount ΔL is the cylinder compressed?

Figure P11.53 Problems 53 and 54.

54. ⭐ The steel cylinder in Figure P11.53A is replaced by a steel tube (Fig. P11.53B). The outer radius of the tube is $R_1 = 3.5$ cm, and the inner radius is $R_2 = 1.5$ cm. If the compressive force is $F = 5000$ N, by what amount ΔL is the tube compressed?

55. ⚙ Young's modulus for bone under compression is about 9×10^9 Pa, and the breaking stress is about 2×10^8 Pa. (a) A leg bone of length 30 cm when unstressed is stressed to its breaking point. What is the change in length of the bone? (b) If the bone has a diameter of 3 cm, how much weight can it support just before it breaks?

56. ✪ Two bars, one composed of aluminum and one of steel, are placed end to end as shown in Figure P11.56. The bars are both initially of length $L = 25$ cm, and both have a square cross section with $h = 2.0$ cm. If a compressive force of $F = 10,000$ N is applied to each end, what is the change in length of each bar?

Figure P11.56

57. SSM ⭐ An aluminum sphere has a radius of 45 cm on the Earth. It is then taken to a distant planet where the atmospheric pressure P is much larger than on the Earth. If the sphere has a radius of 43 cm on that planet, what is P?

58. ✪ You want to use a steel cable to tow your car ($m = 2000$ kg). The cable is solid steel with a diameter of 1.0 cm and a length

of 10 m. If you are pulling the car such that its acceleration is 1.5 m/s², how much will the cable stretch?

59. ⭐ A steel string (diameter 1.0 mm) hanging vertically has a length of 0.25 m. A child of mass m grabs the lower end of the string and pulls her feet off the ground. She finds that the string stretches so that its new length is 0.18 mm longer than before. What is the mass of the child?

11.5 DAMPING AND RESONANCE

60. SSM A damped harmonic oscillator is displaced from equilibrium and then released. The oscillator displacement as a function of time is shown in Figure P11.60. Is this oscillator underdamped or overdamped?

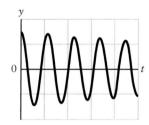

Figure P11.60

61. ✪ Ⓡ Figure P11.61 shows the displacement as a function of time for an underdamped harmonic oscillator. Estimate the fraction of the mechanical energy that is "lost" to friction during one cycle.

Figure P11.61

62. Figure P11.62 shows the amplitude as a function of frequency for a driven, damped oscillator. Estimate the resonant frequency.

Figure P11.62

11.6 DETECTING SMALL FORCES

63. ✪ Consider a Cavendish apparatus that employs a torsion fiber with $\kappa = 1.0 \times 10^{-8}$ N · m. Suppose the smallest twist angle that can be measured is $\theta = 2.0°$. If the distance from one of the masses m_1 to the torsion fiber is 10 cm (see Fig. 11.30), what is the smallest force F that can be measured? Assume a force F is applied to *both* of the masses m_2. Compare your result for F to the weight of a mosquito ($m \approx 10^{-3}$ g).

ADDITIONAL PROBLEMS

64. SSM ⭐ **Get new shocks!** A car with worn-out (underdamped) shock absorbers is observed driving over a dip in the road. The car bounces up and down a total of three times in a period of 5 seconds after hitting the dip. If an average automobile shock absorber spring has a spring constant of 6.0 kN/m, what is the approximate mass of the observed car? *Note*: The body of a car is supported on four shock absorbers.

65. ✪ A mass of 5.0 kg slides down a frictionless slope into a spring with spring constant $k = 4.9$ kN/m as depicted in Figure P11.65. (a) If the spring experiences a maximum compression of 20 cm, what is the height h of the initial release point? (b) What is the velocity of the mass when the spring has been compressed 15 cm? (c) If the mass sticks to the end of the spring, what is the period and amplitude of the oscillations that take place? (d) Determine the expressions for position as a function of time, $x(t)$, and velocity as a function of time, $v(t)$, for this oscillator. (e) What is the maximum velocity of the mass during any oscillation? (f) If the spring were to break while the mass is at the equilibrium position and moving to the right, to what maximum height would the mass rise up the slope? (g) Sketch or plot the position–time graph and velocity–time graph for the first second of the periodic motion.

Figure P11.65

66. ✪ A mass of 3.4 kg is released down a frictionless slope from a height of $h = 2.0$ m. When it reaches the bottom of the slope, it undergoes a completely inelastic collision with a mass of 1.1 kg attached to a spring of spring constant $k = 5.0$ kN/m as illustrated in Figure P11.66. The combined mass then undergoes periodic motion. Calculate (a) the maximum velocity, (b) the frequency, (c) the maximum amplitude, and (d) the period. (e) Determine expressions for position as a function of time and velocity as a function of time for the combined mass of this oscillating system.

Figure P11.66

67. ✪ A mass of 3.5 kg is attached to a spring with a spring constant of $k = 70$ N/m and then compressed by 49 cm from its equilibrium position (Fig. P11.67, top) and released. At exactly the same time, to ensure a collision at the equilibrium position of the spring (Fig. P11.67, bottom), an identical mass is released down a frictionless slope inclined at an angle of $\theta = 40°$ from the horizontal. The system is designed to produce periodic motion for both masses such that they always end up colliding in the same position. Assume there is no friction in the system and the collision is perfectly elastic. (a) What is the period of this oscillating system? (b) From what height h must the second mass be released such that it will collide with the mass on the

spring? *Hint:* Consider the period of the motion of the first mass and the acceleration of the second mass in determining the needed release point up the slope. (c) Sketch the first 2.0 s of the displacement–time graph for the horizontal (x) displacement of the first mass. On the same axes, draw the displacement–time graph for the vertical (y) displacement of the second mass.

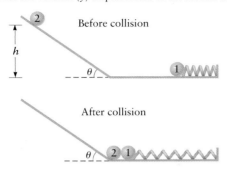

Figure P11.67

68. ✪ A pendulum of length $L = 65$ cm has a stiff frame and negligible mass, and it supports a mass of 3.4 kg on a small platform as shown in Figure P11.68, step 1. When the pendulum is raised and released (Fig. P11.68, step 2), the mass will slide off the platform and come into contact with a spring of constant $k = 820$ N/m at the bottom of the swing (Fig. P11.68, step 3). Next, the spring compresses and then sends the mass back onto the pendulum (Fig. P11.68, step 4), where, assuming negligible friction, the process repeats indefinitely. (a) What is the period for this repeating process? (b) If the mass on the pendulum is raised to a height of 20 cm, what is the maximum compression of the spring? (c) Graph the potential energy as a function of time for the first 2.0 s of operation. (d) What larger replacement mass would make the time spent on the spring equal to the time spent on the pendulum? (e) What is the maximum spring compression for this mass?

Figure P11.68

69. ✪ **Bobble-head.** A football bobble-head figure sits on your car dashboard. The 85-g head of the figure is attached to a spring with a spring constant of $k = 21$ N/m (Fig. P11.69). You are driving along a straight road when you notice a series of hoses lying across the road, perpendicular to your motion, that cannot be avoided, and you drive over them. If the hoses are spaced exactly 4.0 m apart, what speed should you avoid so your bobble-head figure does not sustain damage from its head vibrating too hard at maximum amplitude?

Figure P11.69

70. ✪ A soft synthetic rubber ball of radius 10 cm is submerged to 8.0 m beneath the surface of a lake. (a) If the bulk modulus of the rubber compound is $B_{\text{rubber}} = 4.0 \times 10^6$ Pa, what is the diameter of the ball at that depth? (b) A similar ball of equal radius, but made of thin rubber inflated with air, is also submerged to the same depth. If the bulk modulus of the air inside the ball under these conditions is $B_{\text{air}} = 1.4 \times 10^5$ Pa, what is the radius of this ball at that depth? (Ignore any effects of temperature.)

71. ✪ **Air spring.** The elasticity and bulk modulus of air can be measured using an apparatus consisting of a vertical glass tube and a ball bearing that just fits in the inner diameter of the tube. A 110-g ball bearing is dropped into the top of a 47-cm-high glass tube of diameter 3.0 cm. The ball is observed to oscillate as if the air column were a spring, 59 times in 10 s giving a period of 0.17 s. (a) Find the effective spring constant of the air column. (b) Show that this effective spring constant is related to the bulk modulus by $k = A^2B/V$, where A is the cross-sectional area of the tube and V is the volume of the tube. *Hint:* Start with Equation 11.25 and express the change in volume in terms of the displacement of the mass. (c) Determine the bulk modulus of air.

72. ★ Young's modulus of granite is about 5.0×10^{10} Pa. Assuming the elastic limit of granite corresponds to a compression of 10%, estimate how tall a mountain would have to be for the granite at the bottom to just reach its elastic limit.

Chapter 12

Waves

There are many types of waves, including waves on the surface of the ocean. (©Royalty-Free/Corbis/Jupiterimages)

In Chapter 11, we explored some of the important properties of simple harmonic motion. In this chapter, we build on that work as we study **wave motion**. A wave is *a moving disturbance that transports energy from one place to another without transporting matter*. This definition of a wave leaves two basic questions unanswered: (1) *what* is it that is "disturbed" by the wave, and (2) *how* is it disturbed? These questions will be answered as we proceed through this chapter. The motion associated with a wave disturbance often has a repeating form, so wave motion has much in common with the simple harmonic oscillators studied in Chapter 11.

In this chapter, we concentrate on aspects of wave motion that are common to all types of waves. The next chapter will then focus on a particularly common type of wave, *sound*. Chapters 23 through 26 are devoted to electromagnetic waves (which includes light), so many of the ideas that we discuss now will be revisited and developed further in those chapters.

Figure 12.1 **A** A long string. **B** When the end of a string is shaken once up and down (vertically), a wave pulse that travels horizontally along the string is generated. **C** Repeated shaking of the string generates a periodic wave.

12.1 | WHAT IS A WAVE?

Figure 12.1A shows a person holding one end of a flexible string, with the other end tied to a wall. When the person shakes his end of the string up and down, the string is displaced from its initial, undisturbed profile, creating a "disturbance" that moves horizontally along the string. The shape of the disturbance depends on how the person shakes the string. In Figure 12.1B, he gives the end of the string a single shake, producing a single *wave pulse*. (Sometimes we simply call it a "pulse.") In Figure 12.1C, the end of the string is shaken up and down in a simple harmonic manner, creating a *periodic wave* (described by a sine or cosine function). These disturbances are examples of *waves*. Since portions of the string are moving, these portions have a nonzero velocity; hence, kinetic energy is associated with the wave. There is also elastic potential energy in the string as it stretches while the wave pulse travels by. The wave carries this energy as it travels.

All waves carry energy from one location to another. In the example in Figure 12.1, a wave on the string gains energy from the person's hand as the force exerted by the hand does work on the string. The wave then carries this energy to the right as the wave moves along the string. Although a wave thus transmits energy from one place to another, it does *not* transport matter along with the energy. Pieces of the string do not move horizontally along with the wave disturbance. In particular, the left end of the string in Figure 12.1 does not move to the right along with the wave.

Figure 12.2 shows a wave pulse as it travels along a string. As with the wave pulse in Figure 12.1A, we suppose a person is holding the end of the string on the far left (out of the picture). We then imagine that the person gives one large up-and-down shake to the left end of the string, creating a single pulse that moves to the right as sketched in Figure 12.2. We say that this pulse *propagates* to the right. Parts A through C of Figure 12.2 show "snapshots" of the string's shape at three successive times. Several points along the string are labeled to show how they move upward and then back down to the original, undisturbed configuration as the wave pulse travels by.

Let's now focus on the motion of point D on the string as the pulse moves by. The displacement of point D is plotted as a function of time in Figure 12.2D. This displacement is along the y direction and is thus *perpendicular* to the direction of propagation. At early times, when $t < t_1$, point D is at rest as the wave pulse has not yet arrived. When the pulse arrives, point D on the string first moves upward (see its displacement in the snapshot at $t = t_2$, Fig. 12.2B) and then moves back down to its original, undisturbed position. Other points on the string move in a similar manner. While points on the string are set into motion when the wave travels by, these points do not travel horizontally along with the wave. Hence, we again see that this wave *transports energy without transporting matter*. This is a key feature of *all* waves.

Before the wave pulse in Figure 12.2 arrived or was generated, the string was in a "relaxed" state, lying parallel to the x (horizontal) axis. When the pulse arrives,

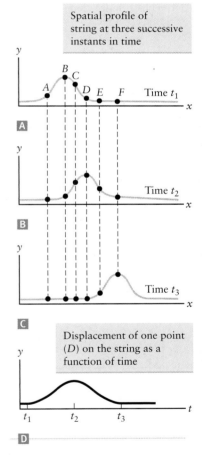

Figure 12.2 **A**, **B**, and **C**: Successive snapshots of a wave pulse traveling along a string with $t_3 > t_2 > t_1$. The wave pulse maintains its shape as it moves. This wave on a string is a transverse wave. Any particular point on the string (such as point D) moves in a direction perpendicular to the velocity of the wave. **D** Displacement of point D on the string as a function of time.

Figure 12.3 A Slinky is a loosely coiled spring. Here we consider a Slinky stretched and held horizontally. When one end of the Slinky is shaken back and forth in the horizontal direction, a longitudinal wave is generated. At some places, the coils of the Slinky are compressed an extra amount relative to an undisturbed Slinky, whereas at other spots, the Slinky coils are looser. These tight and loose places move along the Slinky, forming the wave.

it displaces the string from this state, so we can say that it disturbs the string. All waves are a "disturbance" of some type, and the "thing" that is disturbed by the wave is called its *medium*. In many cases, this medium is a material substance such as the string in Figure 12.1, and these waves are called *mechanical waves*. It is also possible for the wave's medium to be a field, such as the gravitational field or the electromagnetic field we'll learn about in later chapters. In either case, a wave must disturb something, and that something is the medium in which the wave travels. For a mechanical wave, the medium is a substance we can describe using Newton's laws of motion.

We used Figure 12.2 to compare and contrast the motion of the medium (the string in that case) with the motion of the wave. In Figure 12.2, pieces of the string move along the vertical direction (y) while the wave travels along a horizontal direction (x). Waves like this one, in which the motion of the medium is perpendicular to the direction of wave propagation, are called *transverse waves*.

If the motion of the medium is parallel to the wave's propagation direction, the waves are called *longitudinal*. For example, it is possible to generate longitudinal waves on a spring such as a Slinky as illustrated in Figure 12.3. Here the person holding the Slinky shakes the end back and forth along the x direction, generating a disturbance in which the loops of the Slinky alternate between being tighter (more compressed) and looser (more stretched) than normally found when the Slinky is at rest.

In nearly all cases, waves can be classified as either transverse or longitudinal, according to the motion of the medium.

EXAMPLE 12.1 | Determining the Velocity of a Wave

Figure 12.4 shows three snapshots of a wave pulse traveling along a string. This pulse was generated by shaking the left end of the string, generating a disturbance that propagated to the right. **(a)** Use Figure 12.4 to estimate the speed of the wave. **(b)** Estimate the speed of point A on the string at $t = 1.5$ s as the wave pulse passed by that point.

RECOGNIZE THE PRINCIPLE

(a) As was the case with the wave in Figure 12.2, the wave in Figure 12.4 is a single pulse. We can determine the velocity of this pulse by noting the position of its peak along the x direction as a function of time. **(b)** To estimate the velocity of a point *on the string*, we examine the motion of that point along the vertical (y) direction.

SKETCH THE PROBLEM

Figure 12.4 shows the wave pulse at three different times as well as the position (y) of point A.

IDENTIFY THE RELATIONSHIPS AND SOLVE

(a) From Figure 12.4, the peak of the wave pulse is at $x \approx 1.8$ m when $t = 1.0$ s and has moved to $x \approx 3.8$ m when $t = 3.0$ s. The average velocity during this time interval is thus along the $+x$ direction with

$$v = \frac{\Delta x}{\Delta t} = \frac{(3.8 - 1.8)\ \text{m}}{(3.0 - 1.0)\ \text{s}} = \boxed{1.0\ \text{m/s}}$$

which is the speed *of the wave*.

(b) Comparing the snapshots in Figure 12.4 at $t = 1.0$ s and 2.0 s, we see that point A on the string is near its undisturbed position with $y \approx 0.03$ m at $t = 1.0$ s and that this point is at the peak of the wave when $t = 2.0$ s, where we find $y \approx 0.20$ m. The velocity of point A at $t = 1.5$ s is thus approximately

$$v_A = \frac{\Delta y}{\Delta t} = \frac{(0.20 - 0.03)\ \text{m}}{(2.0 - 1.0)\ \text{s}} = \boxed{0.17\ \text{m/s}}$$

Figure 12.4 Example 12.1.

This wave is a transverse wave because the velocity of point *A* on the string is along *y*, which is perpendicular to the velocity of the wave. In general, the velocity of a point on the string will *not* have the same magnitude as the velocity of the wave. The velocity of a point on the string also will not be constant; here we have made an estimate of v_A during the time the wave is passing the point of interest. Before and after the wave pulse arrives, the velocity of points on the string will be zero.

12.2 | DESCRIBING WAVES

Many examples in the last section involved a wave pulse generated by applying a single "shake" to the end of a string (Figs. 12.2 and 12.4). Such wave pulses are often found in nature, but it is also common to find periodic waves. A periodic wave can be generated by shaking the end of the string up and down repeatedly, in a periodic fashion. In the case shown in Figure 12.5, we imagine that a person (not shown in the figure) is shaking the far left end of the string in such a way that this end is undergoing simple harmonic motion along the *y* (vertical) direction. Figure 12.5 then shows snapshots of the resulting periodic wave as it travels along a string. These successive snapshots show the "front edge" of the wave, making the overall motion of the wave clear. However, if we had waited until this front edge moves out of the picture to the right, all we would see is a repeating wave disturbance with a "sine-wave" form. If we examine just a single snapshot, we find places where the string displacement *y* is largest and positive, called wave *crests*. Likewise, at some places called wave *troughs*, *y* is equally large in magnitude but negative in sign.

Two wave crests in Figure 12.5 are marked with large dots to emphasize the motion of the wave. The wave's speed can be measured by following the position of a particular wave crest as a function of time, similar to what we did in Example 12.1.

We can classify waves into two general categories: periodic and nonperiodic. The wave pulses in Figure 12.2 and 12.4 are examples of nonperiodic waves. In these cases, the wave disturbance is limited to a small region of space. In contrast, a periodic wave as in Figure 12.5 extends over a very wide region of space, and the displacement of the medium varies in a repeating and often sinusoidal manner as a function of both position and time. For example, if we examine a single snapshot of a periodic wave—say, in Figure 12.5C—we see that the string displacement *y* varies in a periodic manner as a function of *position* along the string *x*. We can also monitor the string displacement at a particular point on the string and find that the displacement also varies in a periodic manner as a function of *time*. Hence, *a periodic wave involves repeating motion as a function of both space and time.*

The "Equation" of a Wave

The wave in Figure 12.5 was generated by shaking the left end of the string up and down. For simplicity, let's assume the displacement at the left end vibrates as a simple harmonic oscillator with $y_{\text{end}} = A \sin(2\pi ft)$, where *f* is the frequency of this harmonic motion. This periodic shaking generates a periodic wave that travels along the string as sketched in Figure 12.5. The string's displacement is given by the relation

$$y = A \sin\left(2\pi ft - \frac{2\pi x}{\lambda}\right) \qquad (12.1)$$

where λ (the Greek letter lambda) is called the *wavelength* as explained below.

Equation 12.1 is a mathematical description of a periodic wave. For a wave on a string, it tells how the transverse displacement *y* of a point on the string varies as a function of both time *t* and location *x* along the string. It applies to many different types of waves and describes how a wave travels (propagates) away from its source. For the wave on a string in Figure 12.5, the source is the person's hand at the left

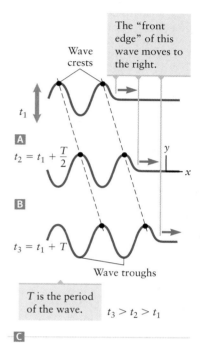

The "front edge" of this wave moves to the right.

Wave crests

t_1

A

$t_2 = t_1 + \dfrac{T}{2}$

B

$t_3 = t_1 + T$

Wave troughs

T is the period of the wave. $\quad t_3 > t_2 > t_1$

C

Figure 12.5 Snapshots of a periodic wave on a string. The places where the displacement is largest (and positive) are wave crests. The wave crests move along with the wave as time increases in going from **A** to **B** to **C**. We imagine that this wave was generated by a person (not shown here) who is shaking the left end of the string up and down.

Equation of a periodic wave

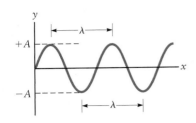

Figure 12.6 Snapshot of a periodic wave on a string. This is just a plot of Equation 12.1 showing the displacement associated with a wave as a function of x at a particular (and fixed) value of t. The wavelength λ is the "repeat distance." The wavelength can be measured from any point on the wave to the next "equivalent" point. The displacement oscillates between $+A$ and $-A$, where A is the amplitude.

end of the string. Equation 12.1 does not include what happens when the wave reaches the right end of the string. We'll learn about that in Sections 12.6 and 12.7, but first let's explore the meaning of Equation 12.1 in a little more detail.

Equation 12.1 has much in common with the displacement of a simple harmonic oscillator (Eq. 11.2). Like a simple harmonic oscillator, a periodic wave has a frequency f and an amplitude A. The frequency is related to the "repeat time" for the wave and can be understood using the snapshots in Figure 12.5. A point on the string that is at a crest in Figure 12.5A moves down to a trough in snapshot B and then returns to a crest in snapshot C, constituting one period T of the motion. The frequency of the wave is then given by $f = 1/T$. Since the sine function in Equation 12.1 repeats itself as the value of t increases, this behavior is all contained in Equation 12.1.

The meaning of the wavelength λ can be understood from Figure 12.6. Here again we show a snapshot of a wave on a string. This plot corresponds to Equation 12.1, showing the transverse displacement y as a function of position x along the string at a particular value of t. Figure 12.6 shows that y is a repeating function of position. That is, if we advance x by a distance equal to λ, the value of y is repeated. One can thus think of λ as the *repeat distance* of the wave. Since the sine function in Equation 12.1 repeats itself as x varies to either increasing or decreasing values, this behavior is contained in Equation 12.1. Periodic waves thus have both a repeat *time* and a repeat *distance*. A periodic wave is a combination of two simple harmonic motions: one as a function of time and the other as a function of space.

Figure 12.6 also illustrates the meaning of the amplitude A of a wave. The wave displacement y varies between $\pm A$, so the wave crests have a height $y = +A$ and, at the troughs, $y = -A$.

Speed of a Wave

The mathematical description of a wave in Equation 12.1 contains three important wave properties: the frequency, wavelength, and amplitude. Let's now consider how these are connected with the velocity of the wave. The snapshots in Figure 12.5 show that the wave moves a distance equal to $\Delta x = \lambda$ during one period of the motion ($\Delta t = T$), which also follows from the definitions of wavelength and period. The speed of the wave is therefore given by

$$v = \frac{\Delta x}{\Delta t} = \frac{\lambda}{T}$$

Because frequency is related to the period by $f = 1/T$, we have

$$v = \frac{\lambda}{1/f}$$

or

$$v = f\lambda \qquad (12.2)$$

The speed of a periodic wave is related to its frequency and wavelength.

Information regarding the *direction* of the wave is also contained in Equation 12.1. To see how, we focus on the motion of a wave crest. A crest occurs whenever the sine function in Equation 12.1 has its maximum value, and, because $\sin(\pi/2) = 1$, one of these maxima will occur when the argument of the sine function is equal to $\pi/2$. So,

$$2\pi f t - \frac{2\pi x}{\lambda} = \frac{\pi}{2} \qquad (12.3)$$

Hence, at $t = 0$, a crest will be found at

$$2\pi f(0) - \frac{2\pi x_{\text{crest}}}{\lambda} = \frac{\pi}{2}$$

$$x_{\text{crest}} = -\frac{\lambda}{4} \qquad (12.4)$$

After one period, $t = T$, and using Equation 12.3 we get

$$2\pi fT - \frac{2\pi x_{\text{crest}}}{\lambda} = \frac{\pi}{2}$$

$$x_{\text{crest}} = -\frac{\lambda}{4} + f\lambda T = -\frac{\lambda}{4} + vT$$

where we have also used the relation $v = f\lambda$. Comparing this expression with Equation 12.4, we see that x_{crest} has become larger, indicating that the wave has moved to the right. Hence, the velocity of the wave described by Equation 12.1 is positive.

A similar analysis shows how to write the equation of a wave that moves to the left. Such a wave has a negative velocity and is described by

$$y = A \sin\left(2\pi ft + \frac{2\pi x}{\lambda}\right) \tag{12.5}$$

Equations 12.1 and 12.5 describe periodic waves that move in opposite directions. Other similar functions can also be used to describe such traveling waves. For example,

$$y = A \cos\left(2\pi ft - \frac{2\pi x}{\lambda}\right) \tag{12.6}$$

describes a wave that moves in the $+x$ direction, whereas

$$y = A \cos\left(2\pi ft + \frac{2\pi x}{\lambda}\right) \tag{12.7}$$

is a wave that travels in the $-x$ direction. All four cases (Eqs. 12.1, 12.5, 12.6, and 12.7) describe waves with the same frequency, wavelength, speed, and amplitude. The only differences involve the direction of wave propagation (the sign of v) and the shape of the string at $t = 0$ (either a sine function or a cosine).

Interpreting the equation of a periodic wave

Many properties of a periodic wave can be read directly from the equation of the wave:

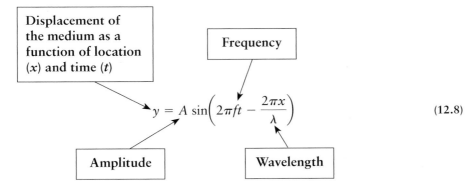

$$y = A \sin\left(2\pi ft - \frac{2\pi x}{\lambda}\right) \tag{12.8}$$

CONCEPT CHECK 12.1 | Waves on a String

A certain wave on a string is found to have a frequency of 500 Hz and a wavelength of 0.60 m. If a new wave of frequency 250 Hz is then established on this string and the wave speed does not change, is the new wavelength (a) 0.60 m, (b) 0.30 m, or (c) 1.2 m? *Hint*: You do not need a calculator!

EXAMPLE 12.2 Finding the Properties of a Wave

Figure 12.7 shows snapshots of a wave on a string at two very closely spaced times, separated by $\Delta t = 0.050$ s. Use the information in these snapshots to determine the

Snapshot at t_1

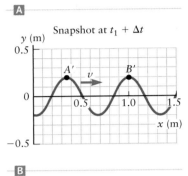

Snapshot at $t_1 + \Delta t$

Figure 12.7 Example 12.2. Snapshots of a wave on a string at time t_1 and at a short time later, $t_1 + \Delta t$.

wavelength, amplitude, frequency, and velocity of the wave. Assume the wave is moving to the right.

RECOGNIZE THE PRINCIPLE

The wavelength λ is the distance between wave crests, which we can read from Figure 12.7. We can also read the amplitude from the figure. By comparing the two snapshots, we can deduce the wave speed v, and we can then find the frequency from the relation $f\lambda = v$.

SKETCH THE PROBLEM

Figure 12.7 describes the problem. We have also labeled two wave crests A and B, which can be used to follow how the wave moves.

IDENTIFY THE RELATIONSHIPS AND SOLVE

From Figure 12.7, the distance between the wave crests at A and B is approximately

$$\lambda = \boxed{0.65 \text{ m}}$$

We can also estimate the amplitude from the height of a crest (or trough) and find

$$A = \boxed{0.20 \text{ m}}$$

To find the velocity, we must compare snapshots at different times. We are told that these snapshots were taken at very closely spaced times, so we can assume the wave has only had time to move a small amount. Wave crest A in the first snapshot (at t_1) then corresponds to crest A' in the second snapshot. From Figure 12.7, this crest has moved a distance of approximately 0.25 m in a time $t_2 - t_1 = \Delta t = 0.050$ s, so the wave velocity is

$$v = \frac{\Delta x}{\Delta t} = \frac{0.25 \text{ m}}{0.050 \text{ s}} = \boxed{5.0 \text{ m/s}}$$

The velocity is positive because the wave is moving along the $+x$ direction. We can find the frequency using Equation 12.2. We get

$$f = \frac{v}{\lambda} = \frac{5.0 \text{ m/s}}{0.65 \text{ m}} = \boxed{7.7 \text{ Hz}}$$

What does it mean?
All the properties of a wave can be read off from snapshots like those in Figure 12.7 if you know what to look for!

EXAMPLE 12.3 Analyzing the Equation of a Wave

Suppose a wave on a string is described by the equation

$$y = (0.040 \text{ m})\cos(25t - 30x) \tag{1}$$

where the units of t are seconds and x is given in meters. Find the frequency, wavelength, and velocity of this wave.

RECOGNIZE THE PRINCIPLE

We can read all the properties of the wave in Equation (1) by following the outline in Equation 12.8.

SKETCH THE PROBLEM

No figure is needed.

IDENTIFY THE RELATIONSHIPS AND SOLVE

We compare Equation (1) with Equation 12.8 and read off values for the corresponding quantities. The factor that multiplies t in Equation (1) equals $2\pi f$, so we have

$$2\pi f = 25 \text{ Hz}$$

The units here are hertz because that is the unit for frequency f in the SI system. Solving for f gives

$$f = \frac{25 \text{ Hz}}{2\pi} = \boxed{4.0 \text{ Hz}}$$

Likewise, the factor multiplying x in both Equation 12.8 and Equation (1) equals $2\pi/\lambda$. Setting $2\pi/\lambda = 30 \text{ m}^{-1}$ and solving for the wavelength, we find

$$\lambda = \frac{2\pi}{30 \text{ m}^{-1}} = \boxed{0.21 \text{ m}}$$

According to Equation 12.2 the speed of the wave is

$$v = f\lambda = (4.0 \text{ Hz})(0.21 \text{ m}) = \boxed{0.84 \text{ m/s}}$$

What does it mean?
Comparing the given equation again with Equation 12.1, we see that the velocity of this wave is positive; it moves in the $+x$ direction.

12.3 | EXAMPLES OF WAVES

We have discussed some general properties of waves and how they behave in time and space. Let's next consider some specific types of waves.

Waves on a String

Waves on a string are mechanical waves, and the medium that is disturbed is the string. For a transverse wave on a string, the velocity depends on the tension in the string and the string's mass per unit length, which is denoted by μ. This makes intuitive sense; we expect that a wave will travel faster in a lighter string (one with smaller μ) since a lighter string will have a larger acceleration (and velocity) for a given value of the tension. Likewise, the wave speed is increased when the tension is made larger (when the string is more "taut").

Let's denote the tension in the string by F_T, deviating from our usual symbol for tension so as not to confuse it with the period of the wave. The speed of a wave on a string can be calculated using Newton's laws of motion and is

$$v = \sqrt{\frac{F_T}{\mu}} \qquad (12.9)$$

The speed of this wave does *not* depend on its frequency. The frequency is determined by how rapidly the end of the string is shaken, but the wave speed in Equation 12.9 is completely independent of f. In fact, the speed of transverse waves on a string is the *same* for periodic and nonperiodic waves (wave pulses).

EXAMPLE 12.4 | Speed of a Wave on a Piano String

A typical string in the author's grand piano is composed of steel with a radius $r = 0.50$ mm and has a tension of about 650 N. What is the speed of a wave on such a string?

RECOGNIZE THE PRINCIPLE

We can find the wave speed from $v = \sqrt{F_T/\mu}$ (Eq. 12.9). The tension F_T is given, and we can find the mass per unit length μ from the known density of steel (Table 10.1).

SKETCH THE PROBLEM

No figure is needed.

IDENTIFY THE RELATIONSHIPS

A string is just a long cylinder; if it has a length L and radius r, the volume is $\pi r^2 L$. Density ρ is mass per unit volume, so the string's mass is

$$m = \pi r^2 L \rho$$

From Table 10.1, the density of steel is 7800 kg/m³. The mass per unit length equals m/L:

$$\mu = \frac{m}{L} = \pi r^2 \rho = \pi (5.0 \times 10^{-4}\ \text{m})^2 (7800\ \text{kg/m}^3) = 6.1 \times 10^{-3}\ \text{kg/m}$$

SOLVE

Inserting this value of μ into Equation 12.9 along with the given value of the tension, we find

$$v = \sqrt{\frac{F_T}{\mu}} = \sqrt{\frac{650\ \text{N}}{6.1 \times 10^{-3}\ \text{kg/m}}} = \boxed{330\ \text{m/s}}$$

What does it mean?

This speed is more than 700 miles per hour! The speed is high because of the high tension. (You should compare the tension of 650 N to your own weight.)

CONCEPT CHECK 12.2 | Waves on a Violin String

The tension in a violin string is 60 N, and the speed of a wave on this string is 300 m/s. If the tension is reduced to 30 N, will the new speed of a wave on this string be (a) 150 m/s, (b) 210 m/s, or (c) 600 m/s?

Sound

Sound is a mechanical wave that can travel through virtually any material substance, including gases, liquids, and solids. By far, the most familiar case is sound in a gas such as air. Sound waves can be excited in a gas by a loudspeaker as sketched in Figure 12.8. Here the surface of the speaker moves back-and-forth along the horizontal direction (x). As it moves, the speaker collides with nearby air molecules, and because the speaker is moving parallel to x, these collisions affect the x component of the velocity of the air molecules. The displacement of the air molecules associated with the sound wave is thus also along x. This displacement alternates between the positive and negative x directions due to the back-and-forth motion of the speaker. The displacement of the medium (the air molecules) is parallel to the wave's propagation direction, so this disturbance of the air molecules travels as a longitudinal wave. The speed of sound depends on the properties of the medium, such as the gas density. For air at room temperature, the speed of sound is approximately 343 m/s. This speed is independent of the frequency of the sound wave and applies to both periodic waves (steady tones) and nonperiodic waves (short "clicks" or "beeps").

The surface of this speaker moves back-and-forth horizontally.

Air density is higher than average.

Air density is lower than average.

Direction of wave
x

Figure 12.8 Sound is a longitudinal wave. Within this wave are regions where the density of air molecules is large, separated by regions where it is smaller. Compare with the compressed and loose regions of the Slinky in Figure 12.3.

Sound in a liquid or a solid also travels as longitudinal waves, with the displacements of the molecules in the liquid or solid being parallel to the propagation direction. The speed of sound is generally smallest for gases and highest for solids.

Wave Propagation in a Solid

We have just mentioned that sound (longitudinal waves) can propagate in a solid. It is also possible to have transverse waves in a solid. To understand how these two types of waves are related, let's first consider waves in a particular type of solid, a Slinky. Transverse waves on a Slinky are very similar to waves on a string, and their wave speed is given by Equation 12.9. It is also possible to excite longitudinal waves on a Slinky as we saw in Figure 12.3, and they travel at a different speed than transverse waves.

As with a Slinky, both longitudinal and transverse mechanical waves can travel through a solid such as steel or rock. The motions associated with each of these waves are sketched in Figure 12.9. The longitudinal waves involve motion of the atoms parallel to the direction of the wave (compare Fig. 12.9A with Fig. 12.8), and such waves are therefore also called "sound." In fact, when a sound wave from air impinges on a solid, a corresponding sound wave propagates into the solid. The velocity of sound waves in a solid depends on the solid's elastic properties, and sound travels faster in "stiffer" materials. For a thin bar of material, the speed of sound is given by

$$v = \sqrt{\frac{Y}{\rho}} \qquad (12.10)$$

where Y is Young's modulus (Chapter 11) and ρ is the density of the material.

Transverse waves can also travel through a solid, just as we have seen with a Slinky. Transverse waves involve motion of the atoms in directions perpendicular to the wave velocity; the velocity of these transverse waves is not given by Equation 12.9, but has a different (and slightly more complicated) dependence on the shear modulus and other elastic constants (Chapter 11). In general, the speed of these transverse waves is slower than the speed of longitudinal waves. The key points here are that *both* transverse *and* longitudinal waves can propagate in a solid, and that the longitudinal waves are called *sound* because they are just like the sound waves that propagate in a gas or a liquid.

While transverse waves can travel through a solid (Fig. 12.9B) they *cannot* travel through liquids or gases. The displacements associated with a transverse wave involve a shearing motion (a shear stress). When a solid is subject to a shear stress, it is bent as in Figure 12.9B and then "springs back." Liquids and gases *flow* (according to the definition of a fluid) and thus do not spring back, so there is no restoring force to produce the oscillations necessary for a transverse wave. Hence, only longitudinal waves (called sound) can propagate in a liquid or a gas. This fact has some important consequences that we'll explore in Section 12.9.

Figure 12.9 A solid, such as a metal bar, can carry ▲ longitudinal waves or ▣ transverse waves. In this sketch of a transverse wave, the amplitude is greatly exaggerated.

Only longitudinal (sound) waves can travel through a liquid or gas.

| EXAMPLE 12.5 | Speed of Sound in a Solid |

Calculate the speed of longitudinal waves (sound) in a steel bar.

RECOGNIZE THE PRINCIPLE

To find the speed of sound in steel, we apply Equation 12.10, $v = \sqrt{Y/\rho}$, where Y is Young's modulus.

SKETCH THE PROBLEM

No figure is necessary.

IDENTIFY THE RELATIONSHIPS

From Tables 10.1 and 11.1, the density of steel is $\rho = 7800$ kg/m^3, and Young's modulus is $Y = 2.0 \times 10^{11}$ Pa.

Visible Light and Other Electromagnetic Waves

Visible light is a type of *electromagnetic wave*. Other types include radio waves, X-rays, gamma rays, and microwaves. These are not mechanical waves, but are electrical and magnetic disturbances that can propagate even in a vacuum, requiring no mechanical medium. Because the electric and magnetic disturbances (the electric and magnetic fields) are always perpendicular to the propagation direction, electromagnetic waves are transverse.

Electromagnetic waves are classified according to their frequency, so visible light waves fall into a different range of frequencies than radio waves, X-rays, and so forth. The speed of an electromagnetic wave in a vacuum c is independent of the frequency of the wave and has the value $c = 3.00 \times 10^8$ m/s. We'll discuss the properties of electromagnetic waves in detail in Chapter 23.

Water Waves

All the waves discussed so far can be neatly classified as either transverse or longitudinal, according to the motion of the disturbance relative to the direction of propagation of the wave. A few types of waves, however, cannot be classified as easily. A wave can be generated on the surface of water by simply dropping a rock into a still body of water. After the rock strikes the water (Fig. 12.10A), waves propagate outward as shown in Figure 12.10B. The up-and-down motion of the water's surface is probably familiar to you. If you were to carefully watch a bug floating on the surface, though, you would find that the motion of the water's surface is actually *both* transverse *and* longitudinal as illustrated in Figure 12.11. A water surface wave is one the few types of waves that cannot be classified as purely transverse or longitudinal. Most waves fall into one of these two categories, making it a useful classification scheme.

CONCEPT CHECK 12.3 | What Determines the Properties of a Wave?

What determines the (1) frequency, (2) speed, (3) amplitude, and (4) wavelength of a wave, (a) the properties of the wave's medium or (b) the way the wave is generated (i.e., the wave's source)?

Insight 12.1
THE VELOCITY OF MOST WAVES IS INDEPENDENT OF FREQUENCY AND AMPLITUDE
The velocity of a wave depends on the properties of the medium through which it travels. This velocity varies widely, from the relatively low values found for a wave on a Slinky or string to the very high value of the speed of light. In most cases, the velocity is independent of both the frequency and amplitude. This velocity is also the same for periodic and nonperiodic waves (wave pulses). Although the wave velocity is usually independent of frequency, when we study light and optics we will see some important cases in which the speed does depend on *f*.

Figure 12.10 Ⓐ Dropping a rock into a pool of water generates a water wave. Ⓑ Waves generated by a disturbance on the surface of a pool of water. The wave crests and troughs move outward as expanding circles.

© Alex Buckingham/Stockbyte/Getty Images (both)

Spherical Waves

Consider a sound wave generated at the center of a large room by a small explosion. The resulting sound wave travels away from the source in a three-dimensional fashion as sketched in Figure 12.12A. The wave crests form concentric spheres centered at the source, so this is called a *spherical wave*. These crests are also called *wave fronts*. The direction of wave propagation is always perpendicular to the surface of a wave front. This direction is indicated by arrows in Figure 12.12, and these arrows are called *rays*.

Each wave crest carries a certain amount of energy as it travels away from the source. Since the spherical surface defined by the wave crest of a spherical wave expands as it moves away from the origin, the energy of a particular crest spreads over a larger and larger sphere as the wave crest moves outward. Several quantities provide a measure of the amount of energy carried by the wave, one of which is the *power*. The total power carried by a wave equals the energy emitted by the source (and carried away by the wave) per unit time. The units of power are thus the units of energy divided by time, or joules/second = watts, just as we found for mechanical power in Chapter 6. A closely related quantity is the *intensity* of the wave, which is defined as the power carried by the wave over a unit area of wave front. Hence, the units of intensity are the units of power/area = watts/m^2.

Let's now follow one spherical wave front as it travels away from the source in Figure 12.12A. The amount of energy carried by a particular wave front is determined when it is created at the source, so once the wave front is emitted, the total energy it contains is fixed. As the wave front then travels away from the origin and its spherical surface expands, this fixed amount of energy (and also power) is spread over a larger and larger surface. The power divided by the area thus becomes smaller, and the intensity of the wave falls as one moves away from the source. At a distance r from the source, the area of a spherical wave front is equal to $4\pi r^2$ (the surface area of a sphere of radius r), so we have

$$I = \frac{\text{power}}{\text{area}} = \frac{P}{4\pi r^2}$$

The intensity of a spherical wave thus falls with distance as

$$I \propto \frac{1}{r^2} \quad \text{(spherical wave)} \tag{12.11}$$

This dependence of the wave intensity on r is found whenever the wave front expands spherically in three-dimensional space, and it arises in many cases involving sound (as in this example) and light (such as the light waves that emanate from the Sun).

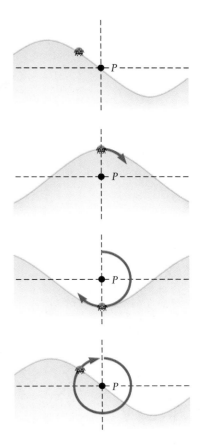

Figure 12.11 The motion associated with a wave on the surface of water has both transverse and longitudinal components. A bug on the surface of the water moves up and down—and also forward and backward—as the wave passes by, which can be seen by comparing the bug's position with that of point *P*.

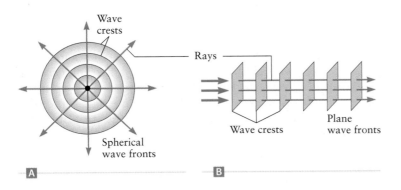

Figure 12.12 The geometry of a wave can be described by the shape of its wave fronts. **A** A spherical wave has spherical wave fronts that are concentric with the source of the wave. **B** The wave fronts of an ideal plane wave are parallel, flat, and infinite in extent. This idealization is not realized in practice, but is approximated by the light from a laser. We'll see how to generate such plane waves when we study optics in Chapter 24.

Plane Waves

Wave fronts are not always spherical. Another important case is a *plane wave*, in which the wave fronts are (flat) planes as sketched in Figure 12.12B. For a "perfect" plane wave, each wave crest and trough extend over an infinite plane in space, and these planes are perpendicular to the direction of propagation. Of course, such a perfect plane wave is an idealization; a real wave will not be spread over an infinite region of space. The light wave produced by a laser, however, is a very good approximation to an ideal plane wave. (We'll see other ways to generate approximate plane waves when we study geometrical optics in Chapter 24.) The wave fronts produced by a laser are very nearly flat and extend over the width of the laser beam. These wave fronts spread very little as they travel, so the wave intensity is approximately constant over long propagation distances. In contrast to the case of a spherical wave, where the intensity falls off as the square of the distance from the source (Eq. 12.11), the intensity of an *ideal* plane wave is constant, independent of distance from the source.

Intensity and Amplitude of a Wave

For all types of waves (and all types of wave fronts), the intensity of a wave is related to its amplitude A by

$$I \propto A^2$$

The reason I is proportional to A^2 (and not A or A^3) can be seen by considering waves on a Slinky. The Slinky is just a spring, and the potential energy associated with a spring equals $\frac{1}{2}kx^2$, where x is the amount the spring is stretched or compressed. For a wave on a Slinky, the magnitude of this stretching and compressing is equal to the displacement of the Slinky from its undisturbed state and is thus proportional to the amplitude A of the wave. Hence, the energy carried by a wave and its intensity are both proportional to A^2.

12.5 | SUPERPOSITION AND INTERFERENCE

All our discussions so far have involved the case of a single wave propagating by itself. We now want to consider what happens when two or more waves travel at the same time through the same medium. Or, in other words, what happens when two waves collide?

When two particles collide, the momentum of each particle changes (although the total momentum of the two particles is conserved; see Chapter 7). Hence, a collision has a big effect on the motion of the particles as illustrated in Figure 12.13A, but the behavior of two waves is quite different. Waves almost always propagate *independently* of one another. A wave such as a sound wave or light can travel through a particular region of space without affecting the motion of another wave traveling through the same region. This behavior is illustrated in Figure 12.13B, which shows sound waves crossing in space (as in a typical classroom). Persons A and B are having a conversation, and their sound waves travel independently of the sound waves of persons C and D. The same sort of behavior is found if persons A and C shine flashlights toward their partners. The light from the flashlight at A travels independently of the light from the flashlight at C. This independence is a result of the *principle of superposition*. This principle comes into play in several areas of physics; we'll encounter it again in our work on electric and magnetic fields. When applied to wave motion, this principle can be stated as follows.

Principle of superposition
When two (or more) waves are present, the displacement of the medium is equal to the sum of the displacements of the individual waves.

Collision of two particles

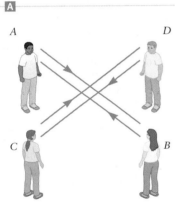

Waves (sound or light) pass through each other.

Figure 12.13 ◢ When two particles collide, their momenta change. ◢ If we replace the particles by waves (such as sound or light), the two waves will pass through each other. The sound waves involved in two conversations (between persons A and B and between persons C and D) do not affect each other; they propagate independently of each other.

Principle of superposition

For example, if we have two periodic waves with displacements y_1 and y_2 that are each described by a relation such as Equation 12.1, the total displacement of the medium is the sum $y_1 + y_2$. These two displacements each describe traveling waves, and the presence of one of the waves does not affect the frequency, amplitude, or velocity of the other. Hence, the two waves travel independently.

Interference

Figure 12.14 shows an example in which two wave pulses travel in opposite directions. These pulses have equal and positive amplitudes, and the successive snapshots of the string show how the two pulses pass through each other. By the time of the final snapshot, the pulses are again isolated. Notice that the shapes of the final pulses in snapshot E are the same as the initial pulse shapes in snapshot A. According to the superposition principle, the displacement of the medium is always equal to the sum of the displacements produced by the two individual pulses, so when the two pulses overlap completely as in Figure 12.14C, the maximum displacement of the string is twice the amplitude of a single pulse. This is an example of *constructive interference*. The effects of the two waves *on the medium* add together to produce a displacement that is larger than the displacements of either of the individual waves.

Figure 12.15 shows snapshots of a similar experiment with two wave pulses on a string, but this time one pulse has a negative amplitude; that is, the displacement of one string is downward and that of the other is upward. Here again the two waves pass through each other, and the pulse shapes at the end in snapshot E are the same as the initial pulse shapes in snapshot A. The displacement of the string is again (according to the superposition principle) equal to the sum of the displacements produced by the individual waves. However, since the waves have opposite amplitudes when the pulses overlap in snapshot C, the total displacement of the string is *zero*! This is an example of *destructive interference*. The total displacement is smaller than the individual displacements of the two waves.

It is interesting to consider the effect of interference on the energy carried by the two waves in Figure 12.15. If you were to examine the string at the moment of perfect destructive interference in Figure 12.15C, you might think that there are no waves present and hence no propagation of energy, but even though there is a moment when the displacement of the string is zero, its velocity is *not* zero. At the instant of snapshot C, all the energy of each wave is contained in the kinetic energy of the medium (the string). In this way, two waves can interfere destructively and still carry energy independently. The term *interference* is thus a bit misleading

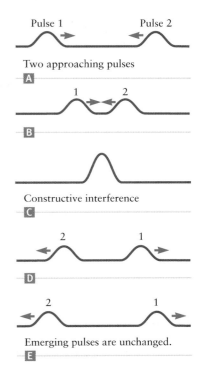

Figure 12.14 When two waves travel in the same region, the displacement of the medium is the sum of the displacements associated with the two waves. This process is called interference even though the waves pass through each other. Shown are successive snapshots (from top to bottom) of a string containing two pulses with equal (and positive) amplitudes traveling in opposite directions.

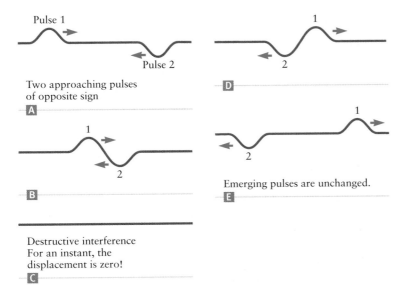

Figure 12.15 When the amplitudes of two interfering wave pulses are equal in magnitude but opposite in sign, the total displacement will be zero for an instant (snapshot C). This illustration is an example of destructive interference.

Figure 12.16 Ⓐ Interference of two overlapping water surface waves. Each source emits a circular pattern of wave crests and troughs. Ⓑ Photo of interfering water waves on the surface of still water.

Ⓐ   Ⓑ

because the presence of one wave does not "interfere" with the propagation of the other in any way.

Interference of Periodic Waves

Figure 12.16 shows an example of superposition with two periodic waves. Here two water waves are produced by dropping two rocks a short distance apart onto the surface of a still pool. Each rock produces a pattern of circular wave crests and troughs that travel away from the initial disturbance. Figure 12.16A shows the wave crests of these two waves separately. According to the principle of superposition, the total displacement of the water surface equals the sum of the displacements produced by the individual waves. Hence, there is constructive interference at all points where two wave crests in Figure 12.16A overlap, and the displacement of the surface is largest at these points. Likewise, there is destructive interference at points where a crest of one wave overlaps with a trough of the other, and at these points the displacement of the surface is small or zero. The resulting surface pattern is shown in Figure 12.16B. This *interference pattern* has an interesting geometry, with regions of constructive and destructive interference forming lines that extend away from the wave sources.

12.6 | REFLECTION

Consider a wave pulse traveling on a string as sketched in Figure 12.17. Initially, this pulse is traveling to the right with a positive amplitude. It then *reflects* from the right-hand end of the string, and in the final snapshots it is traveling to the left. Another example of reflection is given in Figure 12.18, which shows a light wave from a laser reflecting from a mirror. The incoming wave approaches from the lower left; reflection from the mirror causes the wave to travel away toward the upper left edge of the photo.

In both examples, the wave on a string in Figure 12.17 and the light wave in Figure 12.18, reflection changes the propagation direction of the wave. It is useful to show this change using rays to indicate the direction of energy flow (compare with Fig. 12.12). Rays for the light wave in Figure 12.18 are shown in more detail in Figure 12.19. The rays change direction when a wave reflects from the boundary of the medium.

The rays in Figure 12.19 make an incoming (initial) angle of θ_i and an outgoing (reflected) angle θ_r with a line drawn perpendicular to the surface. To understand how a wave changes direction when it is reflected, consider the component of a wave's velocity perpendicular to the surface, along with the component parallel to the surface. (This reflection is similar to the problem of a particle that collides elastically with a surface, discussed in Chapter 7.) The perpendicular component of the wave velocity reverses direction; that is, the component of the wave that was initially traveling to the right in Figure 12.19 travels to the left after being reflected. This is just like the case with the string in Figure 12.17, where the wave is reflected back

End of string is held fixed.

Wave pulse is inverted after reflection.

Figure 12.17 When a wave reaches a boundary (such as the end of a string), it can be reflected. When the end of the string is held fixed by tying it to a wall, the displacement of the reflected wave is reversed; here the displacement is changed from positive to negative.

in the opposite direction. On the other hand, the parallel component of the wave in Figure 12.19 is not affected by the reflection. The final result is then as sketched in Figure 12.19, with the angle of incidence equal to the angle of reflection

$$\theta_i = \theta_r \qquad (12.12)$$

Waves are usually reflected when they reach the boundaries of the medium, and this reflection process can sometimes (but not always) *invert* the wave's displacement. For example, in Figure 12.17, the wave reflected from the end of a string that is fastened to the wall (at the right) is inverted; the reflected wave pulse (at the bottom in Fig. 12.17) has a negative amplitude, whereas the incoming wave (at the top in Fig. 12.17) had a positive amplitude.

While waves are often inverted on reflection, that is not always the case. Figure 12.20 shows a wave on a string where the end of the string is not fastened rigidly to the wall, but is able to slide up and down. Now the reflected wave is not inverted. In this example and for other cases of wave reflection, the properties of the medium at the boundary determine if the reflected wave will be inverted or not.

Radar

The reflection of waves is important in many situations. One interesting application is *radar*,[1] a technique that uses radio waves to detect objects remotely and determine their distance (Fig. 12.21). A radio wave pulse is sent from a transmitting antenna and reflects from some distant object such as an airplane. If the reflecting surface of the airplane is large and flat, the wave will be reflected at a single angle as in Figure 12.19. For radar applications, however, the surfaces on the object are usually smaller than (or comparable to) the wavelength of the radio wave, and in this case, reflected waves travel outward in all directions. A portion of the reflected wave will thus arrive back at the original transmitter, where it is detected. Radar determines the distance to an object by measuring the time delay Δt between the original outgoing wave pulse and the returning one. If the speed of the wave is v, the total round-trip distance traveled by the wave is $L_{tot} = v\,\Delta t$, which is twice the distance from the transmitter to the object. So, by measuring Δt, a radar system uses wave reflection to measure the distance to a remote object. By using a rotating antenna, the direction of the object can also be determined. The amplitude of the reflected wave gives

Figure 12.18 A light wave can be reflected when it strikes a boundary such as a mirror.

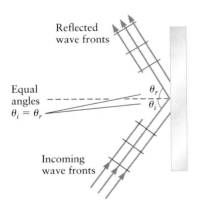

Figure 12.19 When a plane wave such as light, a water wave, or a sound wave reflects from a flat boundary, the angle of incidence θ_i is equal to the angle of reflection θ_r.

Figure 12.20 When the end of a string is attached to a ring that can slide up and down (e.g., along a pole), the reflected wave is not inverted.

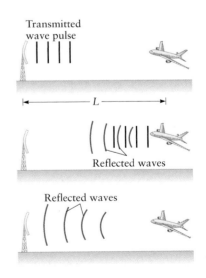

Figure 12.21 Radar uses reflected waves to measure the location of a distant object.

---

[1]An abbreviation for *ra*dio *d*etection *a*nd *r*anging. Similar techniques use sound waves in either water (sonar, short for *so*und *na*vigation *a*nd *r*anging) or air.

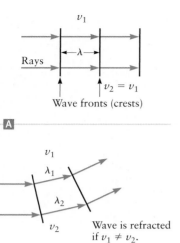

v_1

Rays $\longleftrightarrow \lambda \longrightarrow$

$v_2 = v_1$

Wave fronts (crests)

A

v_1

λ_1

λ_2

v_2 Wave is refracted if $v_1 \neq v_2$.

$v_2 > v_1$ $\lambda_2 > \lambda_1$

B

Figure 12.22 **A** When the wave speed is the same throughout the medium, the rays follow straight lines. **B** When the wave speed is different in different regions (here $v_2 > v_1$), the rays are "bent" and the waves are refracted.

Standing waves: nodes and antinodes

information about the object's size; a large object reflects more of the wave energy than a small object and thus gives a larger signal at the detecting antenna.

12.7 | REFRACTION

In our discussions of wave fronts (Section 12.4), we assumed the speed of the wave is the same at all places in the medium. This assumption is usually satisfied and means that the rays follow straight lines. Figure 12.22A shows such an example for a plane sound wave traveling through a solid. The wave speed is the same at the top (v_1) and bottom (v_2), and this plane wave propagates along a straight line. However, in some cases the wave speed may be different at different locations. A hypothetical case is sketched in Figure 12.22B, which shows several wave crests and rays. By definition, wave crests are always separated by a wavelength. The wave frequency is the same everywhere (its value is determined by the frequency of the source), and because $\lambda = v/f$, the wavelength and hence the distance between crests is greater where the speed is greatest. The ray direction and the propagation direction are always perpendicular to the wave fronts, so the rays no longer follow straight lines but follow bent paths instead (Fig. 12.22B) in the process called **refraction**. Refraction is the *change in direction of a wave due to a change in its velocity*. We'll see examples of refraction when we study seismic waves in the Earth (Section 12.9) and again when we study optics.

12.8 | STANDING WAVES

Consider waves traveling back-and-forth along a string of length L with both ends held fixed (i.e., tied to walls as sketched in Figure 12.23A). What will the string displacement look like as a function of both time t and position x along the string? For a very long string carrying a single periodic wave traveling to the right, the string displacement y is given by a relation such as Equation 12.1. Because the ends of the string in Figure 12.23A are fastened to rigid supports, however, the displacement must be zero at both ends. That is, the string displacement y must be zero at both $x = 0$ and $x = L$. These conditions can be satisfied by a periodic wave such as Equation 12.1 only for certain wavelengths. For example, suppose the wavelength is equal to precisely twice the length of the string so that $\lambda = 2L$. The wave will then "fit" onto the string as shown in Figure 12.23B, which shows several superimposed snapshots of the vibrating string. While the two ends are at rest with $y = 0$, the middle portion of the string oscillates up and down with a large amplitude.

The vibration in Figure 12.23B is called a **standing wave** because the outline of the wave appears to stand still. Such a wave pattern can be obtained by the interference of two waves, both with wavelength $\lambda = 2L$, traveling in *opposite* directions. These waves travel back-and-forth along the string and are reflected when they reach either end of the string. According to the principle of superposition, the total string displacement is the sum of the displacements of these two waves, and that sum is equal to the pattern shown in Figure 12.23B. Points where the string displacement is always zero are called **nodes** of the standing wave. The points where the displacement is largest are called **antinodes**. In Figure 12.23B, there is one antinode at the center of the string and nodes at both ends.

The wave with $\lambda = 2L$ in Figure 12.23B is not the only standing wave that "fits" onto the string in this way. Other possible standing waves are shown in parts C and D of Figure 12.23. For the wave in Figure 12.23C we have $\lambda = L$, whereas in Figure 12.23D, we find $\lambda = 2L/3$. We can generate many other standing waves by employing smaller and smaller wavelengths. The longest possible wavelength for a standing wave is the one shown in Figure 12.23B, and this longest possible wavelength corresponds to the *smallest* possible standing wave frequency. This is called the **fundamental frequency** of the string and is denoted by f_1. The frequency of a standing wave is related to its wavelength by $f = v/\lambda$ (Eq. 12.2). According to Figure 12.23B, the wavelength corresponding to the fundamental frequency equals $2L$, so we have

$$f_1 = \frac{v}{2L} \qquad (12.13)$$

The next longest wavelength for a standing wave is shown in Figure 12.23C. This vibration is called the **second harmonic**, and its frequency is $f_2 = v/L$. There is a pattern to the possible wavelengths, which is

$$\lambda_n = \frac{2L}{n} \quad \text{with } n = 1, 2, 3, \ldots \quad \text{(standing waves on a string with fixed ends)} \quad (12.14)$$

In Equation 12.14, $n = 1$ gives the wavelength of the fundamental, $n = 2$ gives the second harmonic, and so forth. The corresponding pattern for the frequencies is

$$f_n = \frac{v}{\lambda_n} \qquad (12.15)$$

which leads to $f_n = \dfrac{nv}{2L}$ and

$$f_n = nf_1 \quad n = 1, 2, 3, \ldots \quad \text{(standing waves on a string with fixed ends)} \quad (12.16)$$

The allowed standing wave frequencies are thus *integer multiples* of the fundamental frequency f_1.

CONCEPT CHECK 12.4 | Nodes and Antinodes

Consider a standing wave on a string with a frequency equal to the fourth harmonic (Eq. 12.16 with $n = 4$). How many nodes and antinodes does this standing wave possess?

Musical Tones

Standing waves on strings are closely connected to musical tones. Many musical instruments, including guitars, violins, and pianos, employ strings as a vibrating element. A guitar string is stretched under tension from the tuning peg at the top of the neck to a point where it is fastened to the bridge and top plate of the instrument as shown in Figure 12.24. The player uses his or her finger to press the string down against a metal strip called a "fret" so that a length L of the string extending from the bridge to the fret is free to vibrate. The end of the string at the fret is thus held rigidly and the end fastened to the bridge is also held fixed, so the section of the guitar string between the fret and the bridge is similar to the string in Figure 12.23.

A plucked guitar string vibrates according to all the possible standing wave patterns with frequencies $f_1, f_2, f_3, \ldots$ (Equation 12.16) all at the same time. The total string motion is, according to the superposition principle, a sum of the vibrations at these individual frequencies, although the vibrations at the lowest frequencies are usually largest. These string vibrations cause the body of the guitar to vibrate, producing the sound we hear, and the frequencies of this sound are a combination of the different standing wave frequencies of the guitar string. The sound of a single note is thus composed of many individual frequencies, which are related according to the harmonic sequence in Equation 12.16.

The **pitch** of a note is determined by its fundamental frequency. A guitar player can change the pitch by changing the length L of the vibrating section of the string, which can be done by holding the string against a different fret. According to our relation between the fundamental frequency and L, Equation 12.13, the pitch is inversely proportional to the length of the string. Two notes whose fundamental frequencies differ by a factor of two are said to be separated by an *octave*. According to Equation 12.13, reducing L by a factor of two increases the frequency by one octave. This relation between musical pitch, the notes of a musical scale, and the

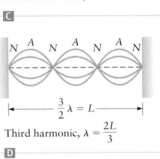

Figure 12.23 Standing waves on a string, such as on a guitar or a piano. **A** The string at rest. **B** through **D** Standing waves with different wavelengths. Here, N and A denote nodes and antinodes of the standing waves.

Tuning pegs

Strings

Frets

Bridge

©Comstock/Jupiterimages

Figure 12.24 A guitar uses standing waves on a string to generate particular frequencies.

length of a vibrating string was first discovered by Pythagoras, the mathematician famous for his work with right triangles.

EXAMPLE 12.6 Playing a Guitar

Guitarists often play several notes simultaneously by plucking or strumming two or more strings at the same time. Two notes whose fundamental frequencies are in the ratio 1.5/1 = 3/2 are known to sound pleasing when played together. This musical interval is known as a "perfect fifth." Suppose you wish to play these notes on two similar guitar strings, that is, strings that have the same value for the wave speed, v. What is the ratio of the vibrating lengths of the two strings?

RECOGNIZE THE PRINCIPLE

The fundamental frequency of a guitar string is $f_1 = v/2L$. We can apply this relation to both strings and then calculate the ratio of the lengths needed to get a frequency ratio of 3/2.

SKETCH THE PROBLEM

The problem is described by Figures 12.23 and 12.24.

IDENTIFY THE RELATIONSHIPS

Let L and L' be the vibrating lengths of the two strings. According to the relation between fundamental frequency and string length (Eq. 12.13) the fundamental frequencies are $f_1 = v/2L$ and $f_1' = v/2L'$. The ratio of the fundamental frequencies is thus

$$\frac{f_1'}{f_1} = \frac{v/2L'}{v/2L} = \frac{L}{L'}$$

SOLVE

Inserting the desired value for the ratio of the frequencies, we get

$$\frac{f_1'}{f_1} = \frac{3}{2} = \frac{L}{L'} = \boxed{1.5}$$

What does it mean?
This type of analysis can also be used to calculate how to position the frets on a guitar (see Question 8 at the end of this chapter). Changing the vibrating length of a guitar string by "one fret" changes the frequency so as to take you to the next note in the musical scale (e.g., from B to C or from C to C-sharp).

12.9 | SEISMIC WAVES AND THE STRUCTURE OF THE EARTH

Waves can be used to probe the structure of the Earth. Any large mechanical disturbance such as an earthquake or nuclear explosion acts as a source of *seismic waves*, another term for waves that propagate through the Earth. There are three main types of seismic waves as sketched in Figure 12.25A. One is called an S wave (S for "shear"). Like shear waves in a metal bar, S waves are transverse waves (Fig. 12.9B), with the displacement of the solid Earth perpendicular to the direction of propagation. Another type of seismic wave is called a P wave (P for "pressure"). P waves are longitudinal sound waves (see Fig. 12.9A and Section 12.3). Although S and P waves have different velocities, they both propagate into the body of the Earth. A third type of wave travels on the surface. These surface waves travel more slowly than S and P waves and are similar to water waves on the surface of a lake, which we described in Section 12.3.

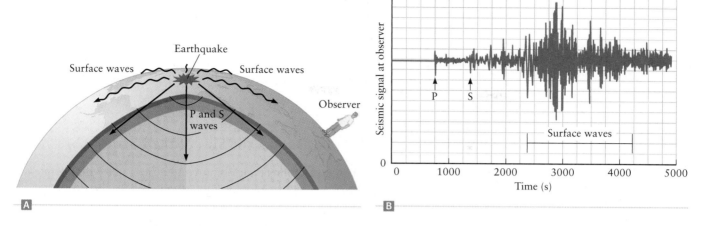

Figure 12.25 ◭ When an earthquake occurs, several different types of seismic waves are generated, including P waves and S waves (both of which travel through the Earth) and surface waves. ◳ These waves have different velocities, so they reach an observer at different times.

Seismic waves can be detected by an instrument called a *seismograph*, which measures the displacement of the Earth as a function of time. A typical seismograph signal is shown in Figure 12.25B. The P wave pulses arrive first because they travel fastest, followed by the S waves and then the surface waves. The seismograph can also measure the direction of the displacements associated with each of these wave pulses, which can be used to distinguish the different types of waves. By knowing the velocities of the three types of waves, the distance from the observer to the earthquake in Figure 12.25A can be determined.

This picture correctly describes things when the earthquake is not too far from the observer. When the earthquake and the observer are on opposite sides of the Earth, however, the observer's seismograph does not detect any S wave component. The reason for this is shown in Figure 12.26A. The center of the Earth contains a region called the *core*, and the area of the Earth where the S waves do not reach is in the "shadow" of the core. So, S waves do not propagate through the core, whereas P waves can. S waves are shear (transverse) waves, and we saw in Section 12.3 that they cannot travel in a liquid. The existence of the S wave shadow provided the first evidence that the Earth's core contains a liquid. It is now known that the core contains liquid metals, such as iron, at a very high temperature. Seismographic recordings provide the only way to measure the diameter of the core and study its properties.

Figure 12.26B shows in greater detail, in a more realistic sketch, how S and P waves propagate through the Earth. We again see the S wave shadow made by the core and also notice that the S and P waves do *not* travel in straight lines. Instead, they follow curved paths that bend toward the Earth's surface because of refraction (Section 12.7). The velocities of S and P waves depend on the composition and density of the Earth, and they change (increase) as one goes deeper into the Earth. Refraction causes the seismic rays to "bend" toward the surface of the Earth, in the direction of smaller wave velocity as shown in Figure 12.22B.

A similar refraction effect occurs for P waves when they pass into and out of the core because their velocity is different in the (liquid) core than it is in the surrounding solid Earth. This change in velocity produces a change in the direction of propagation of the P waves when they cross the boundary of the core (not shown in Fig. 12.26B). We'll encounter a similar phenomenon when we study the propagation of light waves in Chapter 24.

The amount of bending of the P and S waves depends on how the wave velocities vary from place to place within the Earth. Because it is impossible to directly measure the composition deep in the interior of the Earth, it is not possible to predict ahead of time precisely how the waves will travel. However, the data taken with

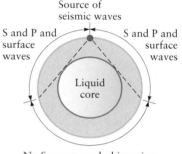

Simplified picture of S and P wave propagation

No S waves reach this region (P and surface waves only).

◭

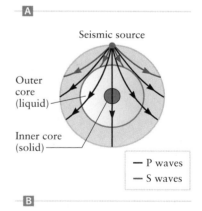

Figure 12.26 ◭ The core of the Earth contains a liquid, and transverse waves do not propagate through a liquid. Therefore, S waves do not travel through the core, and observers who are in the core's "shadow" do not record S waves from a source on the opposite side of the Earth. ◳ Due to refraction, P and S waves do not travel in straight lines.

seismographs at the Earth's surface can be used to infer the paths actually taken by S and P waves. These inferred paths give information on how the composition of the Earth varies with depth. In this way, it was discovered that the Earth is composed of several different solid layers, two of which are called the *crust* and the *mantle*, with different compositions. It has also been found that the core is not simply a spherical fluid region. Rather, the core has a solid central region (the inner core) surrounded by a fluid outer core (Fig. 12.26B). This surrounding fluid produces the S wave shadow described above.

| EXAMPLE 12.7 | Ⓡ Measuring the Size of the Earth's Core |

From extensive seismograph measurements, it is known that the angle θ that defines the edge of the S wave shadow in Figure 12.27 is approximately 39°. Use this angle to estimate the radius of the Earth's liquid core. For simplicity, ignore the refraction (bending) of the S and P waves and assume they travel in straight lines.

RECOGNIZE THE PRINCIPLE

We are assuming straight-line propagation, so we can apply the geometry in Figure 12.27. S waves that just graze the outer edge of the core follow a ray from A to B. The triangle ABC is a right triangle, and the side BC is the radius of the core, R_{core}, which is what we want to find.

SKETCH THE PROBLEM

Figure 12.27 shows the geometry of the problem.

IDENTIFY THE RELATIONSHIPS

The hypotenuse of triangle ABC is the radius of the Earth R_E, so we have

$$R_{core} = R_E \sin \theta$$

SOLVE

Inserting the known radius of the Earth ($R_E = 6.4 \times 10^6$ m, Table A.3), along with the given value of θ we find

$$R_{core} = R_E \sin \theta = (6.4 \times 10^6 \text{ m})\sin(39°) = \boxed{4.0 \times 10^6 \text{ m}}$$

What does it mean?

Our result for R_{core} is about 60% of the Earth's radius. The core thus makes up a relatively large fraction of the Earth's volume.

Figure 12.27 Example 12.7.

S wave shadow

SUMMARY | Chapter 12

KEY CONCEPTS AND PRINCIPLES

Waves

A *wave* is a disturbance that transports energy from one place to another without transporting matter. Waves travel through a medium; for mechanical waves, the medium is a material substance. For *transverse waves*, the displacement of the medium is perpendicular to the velocity of the wave. With *longitudinal waves*, the displacement is parallel to the wave velocity. Fluids (liquids and gases) cannot support transverse waves.

(Continued)

Frequency and period

A wave can consist of a "single" disturbance—a wave pulse—or the disturbance can have a repetitive form. A periodic wave has a characteristic repeat time called the **period** T. The **frequency** of a periodic wave is related to the period by $f = 1/T$.

Displacement of a wave

The displacement associated with a periodic wave is described by

$$y = A \sin\left(2\pi ft - \frac{2\pi x}{\lambda}\right) \qquad \textbf{(12.1)} \text{ (page 381)}$$

Here A is the amplitude of the wave, f is the frequency, and λ is the wavelength (the repeat distance).

The frequency, wavelength, and velocity of a wave are related by

$$v = f\lambda \qquad \textbf{(12.2)} \text{ (page 382)}$$

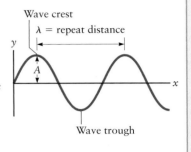

Superposition

According to the **principle of superposition**, when two waves are present simultaneously, the total displacement of the medium is the sum of the displacements of the individual waves. One result of this principle is that two waves can "pass through" each other.

Wave fronts and rays

A **wave front** is a surface that describes how the energy carried by a wave travels through space. **Rays** indicate the direction of energy flow and are perpendicular to the wave fronts.

Reflection and refraction

Waves can be reflected when they reach the boundary of the medium in which they travel. **Refraction** is the change in direction of a wave due to a change in velocity. The wave velocity can change at a boundary or within the medium.

APPLICATIONS

Examples of waves

There are many examples of waves. The velocity of a wave depends on the properties of the medium.

- **Waves on a string** are transverse waves and are employed in many musical instruments.
- **Sound** is a longitudinal wave that can exist in solids, liquids, and gases. (See also Chapter 13.)
- **Light** is an electromagnetic wave. Because they are not mechanical waves, light and other electromagnetic waves can travel through a vacuum as well as through a material medium such as air.

(Continued)

Interference and standing waves

When two waves overlap in space, they are said to *interfere*. One example of interference is the phenomenon of *standing waves*. A point along a standing wave where the displacement of the medium is always zero is called a *node*, and the points where the displacement is largest are *antinodes*. Standing waves are used to produce musical tones in stringed instruments such as guitars and pianos. The lowest possible frequency of the standing wave (corresponding to the longest possible wavelength) is called the *fundamental frequency*. For a string held rigidly at both ends, the standing waves form a series of *harmonics* with frequencies

$$f_n = nf_1 \quad \text{with } n = 1, 2, 3, \ldots \quad \text{(standing waves on a string with fixed ends)}$$

(12.16) (page 395)

where $f_1 = v/(2L)$ is the fundamental frequency.

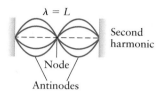

Standing waves on a string

$\lambda = 2L$ — Fundamental

$\lambda = L$ — Second harmonic

Node

Antinodes

Seismic waves

Seismic waves provide information about the Earth's inner structure.

QUESTIONS

1. Figure Q12.1 shows snapshots of several different waves on a string. Which wave has (a) the longest wavelength, (b) the highest frequency, and (c) the smallest amplitude?

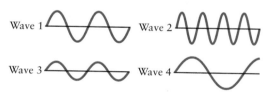

Wave 1   Wave 2
Wave 3   Wave 4

Figure Q12.1

2. Identify the medium for each of the following waves:
(a) Sound waves in a room
(b) Waves on the surface of a pond
(c) Radio waves
(d) Light waves
(e) Waves created when a guitar is plucked
(f) Seismic waves in the Earth

3. Identify the following waves as either longitudinal or transverse or both:
(a) Sound waves in a room
(b) Waves on a guitar string
(c) Seismic waves in the Earth
(d) Waves on the surface of a pond

4. Complete the following sentence: "In a longitudinal wave, the displacement of the medium is _____ to the propagation direction."

5. A key property of a wave is that it transports energy without transporting mass. Explain how this is possible, using work–energy ideas (i.e., $W = F\,\Delta x$). *Hint*: Consider the "leading edge" of a wave.

6. Devise or describe an experiment that uses standing waves to measure the velocity of a wave.

7. A bass guitar has longer strings than a regular guitar. Explain why these longer strings allow a bass guitarist to play "lower" notes.

8. [SSM] The neck of a guitar is designed with *frets* as shown in Figure Q12.8. A player can hold a string against one of the frets and thus shorten the vibrating length of the string L. In this way, a particular string can be used to play notes with different frequencies. The frequency of a note is determined by the fundamental standing wave mode (Fig. 12.23), and this frequency depends on L. In a musical scale, the frequencies of adjacent notes (i.e., between C and C-sharp) are in the ratio of $2^{1/12}/1 \approx 1.059$; this ratio produces the commonly used 12-tone musical scale. The artisans who make guitars position the frets according to the "rule of 18ths." According to this rule, each fret is positioned so that depressing adjacent frets shorten the vibrating length of the string by $L/18$. (See Fig. Q12.8.) Show that the rule of 18ths produces note frequencies that fit approximately on the desired musical scale.

Frets

© Tetraimages/Jupiterimages

Figure Q12.8

9. When a guitar string is plucked, in what ways are the vibrations in the string different from the vibrations produced in the air? In what ways are these two vibrations the same or similar?

10. Figure Q12.10 shows a child holding one end of a long Slinky. If the child wants to generate a longitudinal wave in the Slinky, in what direction should he shake the end of the Slinky, (a) in a horizontal direction or (b) in a vertical direction?

11. A student sends a transverse wave pulse down a stretched piece of string attached to a wall. A moment later, she sends a second wave pulse. Is there any way she could alter the conditions to make the second wave overtake the first wave (i.e., before it is reflected back)? What would happen if she pulls harder on the string?

12. A rope of linear density of about 1 kg per meter is hung from the ceiling. The lower free end is then oscillated producing waves that travel up the rope. Describe the velocity of the wave as it travels up the rope. Does it change? Why or why not?

13. **SSM** A woman wakes up just after 2 AM. Wondering what woke her, she goes to the window and looks around. Noting nothing out of the ordinary, she sits back down on her bed and all of a sudden a loud explosion rattles her window. The next day, she reads that an underground munitions dump exploded about 10 km from her home. She thinks that perhaps she woke up because she experienced a psychic precognition of the explosion. Can you think of a physical explanation for what might have awoken her the few moments before the sound from the blast hit her window? (See Problem 67.)

14. When a sound wave passes from air into a solid object, which of the following wave properties stays the same and which change? Explain.
   (a) The amplitude
   (b) The frequency
   (c) The wave speed
   (d) The wavelength

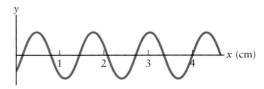

Figure Q12.10

15. Equation 12.1 describes a wave that is traveling to the right, and Equation 12.5 corresponds to a wave of the same frequency and wavelength that is moving to the left. Show that the sum of these two waves gives a standing wave. That is, show that this sum has nodes and antinodes as sketched in the standing waves in Figure 12.23.

16. Consider a water wave as sketched in Figure 12.10. This wave front is circular. How does the intensity of this wave vary as it moves away from its source? *Hint:* You should find that $I \propto r^p$. What is the value of p?

17. Which property of a wave is independent of its amplitude, (a) frequency, (b) wavelength, or (c) speed?

18. If the frequency of a wave is doubled, how does the wavelength change?

19. When you tune a guitar, what causes the frequency to change?

20. For guitars and violins, some strings are "simple" strings, and others contain a winding around a string core. Which type of string, simple or wound, is used for the low-frequency notes? Explain why.

21. **Seismic waves in the Earth.** Surface seismic waves generated by an earthquake arrive at a distant observer after the arrival of P waves and S waves (Fig. 12.25B), partly because surface waves travel more slowly than P and S waves. Even if all these waves had the same velocities, however, the surface waves would still arrive last. Explain why.

PROBLEMS

SSM = solution in Student Companion & Problem-Solving Guide   ⓧ = life science application
⭐ = intermediate   ✪ = challenging   ⓡ = reasoning and relationships problem

12.1 WHAT IS A WAVE?

1. **SSM** Figure P12.1 shows several snapshots of a wave pulse as it travels along a string. Estimate the speed of the wave pulse.

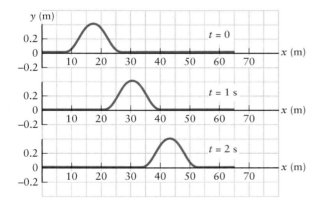

Figure P12.1 Problems 1, 2, and 3.

2. Is the wave on a string in Figure P12.1 transverse or longitudinal?

3. ⓡ For the wave in Figure P12.1, what is the approximate magnitude of the velocity *of a section of the string* as the wave pulse passes by? What is the direction of the string velocity relative to the velocity of the wave?

4. ✪ ⓡ Consider a clothesline stretched between two supports. If one end of the line is "wiggled," a wave will be produced. Estimate the speed of this wave on a clothesline.

5. A rough rule of thumb is that light from the Sun reaches the Earth in about 8 minutes. Use this estimate along with the known speed of light to get an approximate value for the distance from the Earth to the Sun.

12.2 DESCRIBING WAVES

6. ⭐ Figure P12.6 shows a snapshot of a portion of a periodic wave as it travels along a string. Estimate the wavelength of the wave. Is there enough information in this figure for you to also estimate the speed of the wave? Explain.

Figure P12.6

7. Figure P12.7 shows a snapshot of a wave on a string. If the frequency of this wave is 20 Hz, what is the approximate speed of the wave?

Figure P12.7

8. ★ A wave on a Slinky has a frequency of 2.3 Hz. It takes 3.5 s to travel the length of the Slinky (0.30 m). What is the wavelength?

9. Sound waves in water travel at a speed of approximately 1500 m/s. What is the wavelength of a sound wave that has a frequency of 440 Hz?

10. SSM ★ A wave on a string is described by the relation $y = A \sin(35t - 0.025x)$, where t is measured in seconds and x in meters, with $A = 0.15$ m. Find the frequency, wavelength, and speed of this wave.

11. ✪ A wave on a string is described by the relation $y = A \sin(72t - 0.045x)$, where t is measured in seconds and x in meters, with $A = 0.15$ m. What is the maximum speed of a point on the string?

12. ★ Using Equation 12.1, write the equation of a wave on string that has an amplitude of 0.30 m, a frequency of 350 Hz, and a velocity of 250 m/s and travels to the right.

13. ★ Using Equation 12.1, write the equation of a wave on string that has an amplitude of 0.50 m, a wavelength of 8.4 m, and a velocity of 330 m/s and travels to the left.

14. Consider a wave that has a frequency of 3.0 kHz and a wavelength of 0.10 m. Write an equation that might describe this wave.

15. ★ Figure P12.15 shows a snapshot of a wave on a string. This wave has a frequency of 200 Hz. Approximately how long does it take the wave crest located at point A in this snapshot to move to point B?

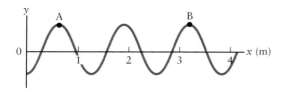

Figure P12.15

16. Draw a snapshot of a wave on a string that obeys the equation $y = (4.2 \text{ mm})\cos(3000t - 0.45x)$. Assume the units of length are millimeters and the units of time are seconds. Make your drawing at $t = 1.5$ s.

17. ★ A wave on a string is described by the equation $y = (4.2 \text{ mm})\cos(3000t - 0.45x)$ where t is measured in s and x is measured in mm. What are the frequency and wavelength of the wave?

18. ★ A wave on a string is described by the equation $y = (4.2 \text{ mm})\cos(3000t - 0.45x)$ where t is measured in s and x is measured in mm. Make a graph of the displacement of the string at $x = 0.35$ mm as a function of time. Show two complete periods of the wave.

19. Consider a wave that has a frequency of 3.0 kHz and a wavelength of 0.10 m. What is the speed of this wave?

20. ★ Two sound waves in air have frequencies that differ by a factor of five. What is the ratio of their wavelengths?

12.3 EXAMPLES OF WAVES

21. A wave on a string has a speed of 340 m/s. If the tension in the string is 150 N and its length is 2.0 m, what is the mass of the string?

22. Consider a piano string made of steel with a diameter of 1.0 mm. If the speed of a wave on this string is 280 m/s, what is the tension in the string? *Hint:* The density of steel is 7800 kg/m³.

23. Waves on a violin string travel at 900 m/s. If the string has a tension of 50 N and is 0.33 m long, what is its mass?

24. ★ A wave on a string has a speed of 200 m/s. If the diameter of the string is increased by a factor of two and the tension is held fixed, what is the speed of the wave on this new string?

25. SSM ★ Ⓡ Consider a violin string with a vibrating length of 30 cm. If the speed of a wave on this string is 250 m/s, what is the tension in the string? *Hint:* You will have to estimate the diameter of the string. Assume the density is the same as that of steel, 7800 kg/m³.

26. ★ The speed of a wave on a guitar string is $v = 120$ m/s. If the tension is increased from 70 N to 110 N, what is the new value of v?

27. The speed of sound in water is 1500 m/s. Find the wavelength of a sound wave whose frequency is 300 Hz.

28. ★ Spruce is a type of wood that is widely used in making musical instruments (such as violins and guitars) because its Young's modulus is high and its density is low. If $Y = 1.5 \times 10^{10}$ Pa and $\rho = 420$ kg/m³, what is the velocity of sound in spruce? How does that compare with the speed of sound in steel?

29. ★ Consider the speed of sound in a hypothetical metal. If the density is reduced by a factor of three, by what factor is the speed of sound changed?

30. The speed of longitudinal waves in a metal is 3000 m/s. If the density is 12,000 kg/m³, what is the Young's modulus of the metal?

31. ★ Radio waves are electromagnetic waves and travel at $c = 3.00 \times 10^8$ m/s in a vacuum. How long does it take a radio signal to travel from Mars to the Earth when the two planets are nearest each other? *Hint:* Assume their closest separation is equal to the difference in their mean orbital radii.

32. Suppose your favorite FM radio station broadcasts at a frequency of 101 MHz. What is the wavelength of this signal?

33. Your favorite television station transmits using an electromagnetic wave with a wavelength of approximately 1.5 m. What is the frequency?

34. ✪ Ⓡ Electromagnetic waves are used to transmit television signals from a satellite that is in orbit around the Earth to the author's home. Estimate how long such a signal takes to travel from the satellite to the ground. *Hint:* This satellite completes one orbit each day and is thus in a geosynchronous orbit (Chapter 5).

12.4 THE GEOMETRY OF A WAVE: WAVE FRONTS

35. SSM ★ A lightbulb emits a spherical wave. If the intensity of the emitted light is 1.0 W/m² at a distance of 2.5 m from the bulb, what is the intensity at a distance of 4.0 m?

36. ✪ For the spherical wave emitted by a small lightbulb, what is the ratio of the amplitudes of the wave at $r = 4.0$ m and $r = 2.5$ m from the lightbulb?

37. ✪ ✪ You are at a rock concert and find that when you stand 10 m from a speaker, the intensity is so high that it hurts your ears. You decide to move away from the stage so that your ears are more comfortable. If you want to reduce the intensity at your ears by a factor of three, how far from the speaker should you stand? Assume the sound emitted by a speaker at a rock concert has a spherical wave front. How would your answer change if the wave front has the shape of a hemisphere (Fig. P12.37)?

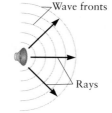

Figure P12.37

12.5 SUPERPOSITION AND INTERFERENCE

38. SSM ★ Two waves of equal frequency are traveling in opposite directions on a string. Constructive interference is found

at two spots on the string that are separated by a distance of 1.5 m. (a) What is the longest possible wavelength of the waves? (b) Give two other possible values of the wavelength.

39. Figure P12.39 shows an experiment in which the sound waves (of equal frequency) from two speakers interfere. If the interference at point *P* is constructive, what is a possible value of the frequency?

Figure P12.39

12.6 REFLECTION

40. Consider a radar system in an air traffic control center that is monitoring the distance to an approaching airplane. If the airplane is 4.0 km from the control center, what is the time delay between a transmitted wave pulse and the return of the reflected pulse?

41. Sonar is a system similar to radar except that sound is used in place of radio waves. Sonar systems are generally operated under water, where the speed of sound is approximately 1500 m/s. Suppose you are on a ship monitoring the distance to a submarine using sonar, and you find that the sonar reflections return in 24 s. How far away is the submarine?

42. SSM ★ A radar system can measure the time delay between the transmitted and reflected wave pulses from a distant airplane with an accuracy of 1.0 picosecond. What is the accuracy in the calculated distance to the plane?

12.7 REFRACTION

43. SSM Waves in a solid are found to refract as shown in Figure P12.43. Is the wave speed highest in region 1 or region 2? That is, is v_1 greater than or less than v_2?

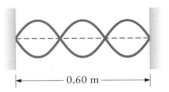

Figure P12.43

12.8 STANDING WAVES

44. What is the distance between two nodes of a standing wave?
(a) λ/4   (d) 2λ   (g) choices (c), (d), and (f)
(b) λ/2   (e) 3λ/2   (h) choices (b) and (e)
(c) λ   (f) 3λ

45. What is the distance between a node and an antinode of a standing wave?
(a) λ/4   (d) 2λ   (g) choices (c), (d), and (f)
(b) λ/2   (e) 3λ/2   (h) choices (b) and (e)
(c) λ   (f) 3λ

46. The antinode of a standing wave is 0.057 m away from the nearest node. What is the wavelength of the waves that combine to make the standing wave?

47. SSM ★ Consider a guitar string that has a tension of 90 N, a length of 65 cm, and a mass per unit length of 7.9×10^{-4} kg/m. What is the fundamental frequency of the string? That is, what is the frequency of the standing wave with the lowest possible frequency? What note does this string emit when played open (i.e., without being held against a fret)? *Hint:* You may want to consult a table that gives musical notes and their corresponding frequencies for standard tuning.

48. ✪ You are studying guitar strings and are able to measure a few of (but not all) the standing wave frequencies for a particular string. If you measure frequencies of 450 Hz, 750 Hz, 900 Hz, and 1200 Hz, what is the highest possible value for the fundamental frequency?

49. ★ A violin string is 35 cm long and has a fundamental standing wave frequency of 440 Hz. If the tension is 75 N, what is the mass of the string?

50. ★ A vibrating string has a fundamental frequency of 250 Hz. The tension is then increased by a factor of 1.9. What is the new value of the fundamental frequency?

51. ★ A violin string has an initial tension of 45 N. When playing the note A on the violin, the frequency is found to be 435 Hz. This note should actually have the frequency 440 Hz, so the player increases the tension to get the correct frequency. What is the new value of the tension? Give your answer to three significant figures.

52. ★ A steel piano string for the note middle C is about 0.60 m long and is under a tension of 600 N. If the fundamental frequency is 262 Hz, what is the string's diameter?

53. ★ The vibrating length of a guitar string is reduced by a factor of 2.5 when a player holds the string against a fret. What is the change in the frequency of the fundamental standing wave on the string?

54. ★ Two nearby nodes of a standing wave are separated by 3.5 m. If the frequency of the wave is 150 Hz, what is the speed of the wave?

55. ★ Consider a standing wave that has a frequency of 2000 Hz and a speed of 300 m/s. What is the distance between a node and the nearest antinode?

56. ★ Figure P12.56 shows a standing wave on a string. If the frequency of the wave is 440 Hz, what is the speed of a wave on this string?

Figure P12.56

57. ★ You want to set up a standing wave on a string that has a length of 3.5 m. You find that the lowest frequency that will work is 20 Hz. What is the speed of a wave on this string?

58. Two notes on a guitar that differ by an octave have fundamental frequencies that differ by a factor of exactly two. If these notes are played on strings that have the same diameter and tension, what is the ratio of the lengths of the strings?

59. ★ Consider two notes on a guitar whose fundamental frequencies are in the ratio 5/4. These notes differ by a musical interval called a "major third." If these notes are played on strings that are made of the same material, and have the same diameter and tension, what is the ratio of the lengths of the strings?

12.9 SEISMIC WAVES AND THE STRUCTURE OF THE EARTH

60. SSM You are operating a seismic laboratory when an earthquake occurs. Your seismographs indicate that the P wave generated by the earthquake arrives at precisely 1:55 PM, whereas the S wave arrives 5 minutes and 35 seconds later. If the velocity of the P wave is 6.0 km/s and the velocity of the S wave is 3.5 km/s, how far away is the epicenter of the earthquake?

61. ✪ An earthquake occurs at the South Pole. If the speed of the P waves is 6.0 km/s and the speed of the surface waves is 2.2 km/s, how long will it take these waves to reach New York City?

62. ⭐ A lift mechanic taps on an end of the main cable of a ski lift 2000 ft (610 m) long. He finds it takes 7.7 s for the pulse to reflect off the opposite end and return. If the steel cable is 3.8 cm in diameter, what is the tension in the cable?

63. ✪ Two cables of equal diameter and length, one made of aluminum and the other made of steel, are welded together and strung horizontally under tension. When the aluminum end is struck, a wave pulse propagates from the end at a velocity of 15 m/s. (a) The wave pulse crosses over the weld and onto the steel part of the cable. What is the velocity of the wave as it travels in the part made of steel? (b) If the total length of the cable is 10 m, how long does it take the wave pulse to reflect off the far end and return? *Hint*: Note that these are not sound waves, but are transverse waves like those on a string. Also, you may want to consult Table 10.1 for values of the densities of aluminum and steel.

64. ✪ A wave pulse moves along a strip of rubber with a velocity of 10 cm/s as depicted in Figure P12.64. Five equally spaced points are marked on the rubber strip labeled *A* though *E* as shown. (a) Rank the points, from highest to lowest, according to the average vertical speed of each point during the next 0.50 s of the wave pulse's travel. (b) Calculate the average velocity of point *C* over the same 0.50-s time interval.

Figure P12.64

65. ⭐ A guitar sting that should produce 110 Hz (the note A) instead produces 105 Hz. (a) What change in tension is needed to bring the string into tune? (b) Sometimes a guitar is tuned to "drop-D" tuning, where the low E string is tuned so that it plays the note D instead. What decrease in tension is needed to bring the string from E (82.4 Hz) to D (73.4 Hz)? Give your answer to three significant figures.

66. ✪ Two neighbors communicate with a homemade string phone (Fig. P12.66). The simple phone consists of two ends of a string (total mass of 18 g) attached to the bottom of two paper cups. The string is stretched between the two homes at a tension of 10 N over the distance of 35 m. (a) How long does it take communication to travel on the string from one home to the

Figure P12.66

other? (b) How long would it take to communicate by yelling out the window? (c) How long would it take to communicate over cell phones if the nearest cell tower is 5.5 km away? (Consider only the propagation time of the radio waves and assume that the call is already connected.)

67. ⭐ An explosion takes place in an underground mine 15 km away from a home located in a valley where the soil consists of a clay material with a Young's modulus of $Y = 2.9 \times 10^9$ Pa. The average density of the soil in that area is 3.1 g/cm³. How long after the P waves have shaken the home does the sound of the blast finally arrive? (See also Question 13.)

68. ✪ A standard guitar is strung with six strings of increasing diameter and all of length 65 cm. Each string is to have the

Table P12.68 Standard Tuning and Medium-Gauge Steel String Diameters for a Guitar

String	Note	Frequency (Hz)	Diameter (mm)
First	E	330	0.28
Second	B	247	0.36
Third	G	196	0.46
Fourth	D	147	0.71
Fifth	A	110	0.97
Sixth	E	82.4	1.24

fundamental frequency of each note (E B G D A E) for standard tuning. (a) If the strings are made of a steel alloy of density $\rho = 6.3$ g/cm³, what is the total force on the guitar neck from the combined tension in the strings? The string diameters and frequencies are listed in Table P12.68. (b) Rank the strings from highest to lowest according to the tension in each.

69. ✪ In moderate winds, high-tension power lines sometimes "sing" or hum as a result of transverse waves on the lines that winds can generate. The cables that carry high-voltage electricity are made of aluminum with a steel core added for strength. The cross-sectional areas of the steel and aluminum in the line are 39 mm² and 220 mm², respectively. The cable is 1.8 cm in diameter and is under a tension of 4.9×10^4 N. Typically, the towers are placed about 0.20 km apart. (See Figure P12.69.) (a) What is the fundamental frequency of such a power line? (The densities of steel and aluminum are given in Table 10.1.) (b) The wind can induce vibrations when flowing around a cylinder. The frequency of the vibrations induced this way is modeled well by the equation

$$f_{wind} = \frac{1}{5}\left(\frac{v_{wind}}{d}\right)$$

here v_{wind} is the speed of the wind and *d* is the diameter of the cylinder. The wind will induce vibration this way only if the frequency of the induced oscillation matches a natural frequency of the power line. It is found that the power line hums when the wind blows at a steady 6.7 m/s. What harmonic of the line is this induced vibration exciting? (c) If the line is humming at the 152nd harmonic, how fast is the wind blowing?

Figure P12.69

70. SSM ★ Figure P12.70A shows two wave pulses at time $t = 0$ s. The pulses travel at 5 cm/s in opposite directions. (a) Sketch the resulting waveform for the times $t = 1$ s, 2 s, and 3 s. (b) Figure 12.70B shows a different set of two pulses. Sketch the resulting waveform for these pulses at the times $t = 1$ s, 2 s, and 3 s.

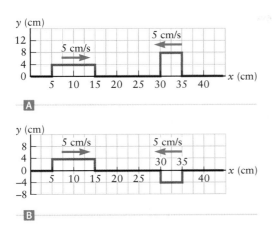

Figure P12.70

71. Wolfgang Mozart was not only an outstanding composer, but was also an excellent violinist. His violin teacher was his father, Leopold, and historical records show that Leopold advised his students to tune all four strings of the violin to the same tension. A typical violin string tension is about 45 N, and the four

Figure P12.71

strings of the violin (Fig. P12.71) are all 0.33 m long. Their lowest standing wave frequencies are 196 Hz (the note G), 294 Hz (D), 440 Hz (A), and 659 Hz (E). If the E string has a mass per unit length μ_E, what are the ratios μ_G/μ_E, μ_D/μ_E, and μ_A/μ_E? (Here, μ_G denotes the mass per unit length of the G string and so forth.)

72. ✪ Waves are produced in a medium from two sources. Each source can produce the sinusoidal traveling waves A, B, or C as shown in Figure P12.72i. If both sources are turned on at the same time, interference will occur and produce a combined wave pattern. (a) Using the principle of superposition, construct the shape of the wave that is formed by producing the waves A and B in the medium at the same time. *Hint:* Construct (sketch) the new wave A + B by considering small steps in the x direction and adding the displacements from the vertical as shown in Figure P12.72ii for each increment. (b) Construct the wave produced if waves A and C are applied simultaneously. (c) Find the wave produced by the combination of waves B and C. (d) If the grid lines denote intervals of 10 cm in x and 1.0 cm in y, what is the wavelength for waves A, B and C? (e) What are the wavelengths of the wave found in parts (a) through (c)? (f) When combining waves, is there always a final wavelength to measure?

Figure P12.72

Chapter 13

Sound

Sound is an important part of everyday life, especially on college weekends. (© M Stock/Alamy)

In Chapter 12, we discussed aspects of wave motion that are common to all types of waves. In this chapter, we focus on one particular type of wave, ***sound***. Sound is an especially important example of waves; it is central to hearing, speech, music (MP3 players), and many other daily activities. We'll consider how sound waves are generated, how they propagate, and how they are detected by the ear. We also describe some medical applications of sound.

13.1 | SOUND IS A LONGITUDINAL WAVE

Sound waves can travel through gases, liquids, and solids. Our everyday experience is mainly with sound that travels through air, so let's begin with a molecular-scale view of sound in a gas. Air is composed mainly of oxygen and nitrogen molecules that are in constant random motion, even in "still" air. A typical air molecule at room temperature has a speed of several hundred meters per second. We call this speed the "background velocity" and denote it by $\vec{v}_0$; the direction of this velocity is different for every molecule.

Figure 13.1 shows a tube filled with air, with a loudspeaker at one end of the tube. When the loudspeaker is turned off, the density of air molecules is the same throughout the tube (Fig. 13.1A). When the loudspeaker is turned on, a surface of the loudspeaker

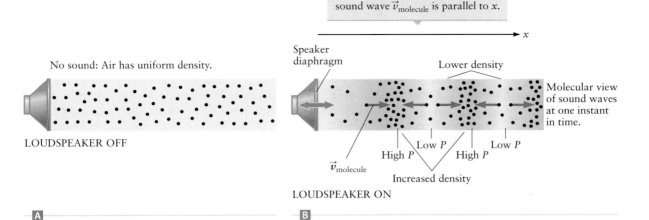

The molecular motion due to the sound wave $\vec{v}_{molecule}$ is parallel to x.

No sound: Air has uniform density.

LOUDSPEAKER OFF

A

Speaker diaphragm

Lower density

Molecular view of sound waves at one instant in time.

Low P Low P

High P High P

$\vec{v}_{molecule}$

Increased density

LOUDSPEAKER ON

B

Figure 13.1 Molecular-scale view of sound in a gas. **A** An air column with no sound present. The molecules are in motion, but their velocities are in random directions and the air molecules are distributed uniformly throughout the container. **B** A sound wave generated by a loudspeaker. Each molecule now has an additional contribution to its total velocity. This contribution is denoted by $\vec{v}_{molecule}$ and is parallel to x, the direction in which the sound wave propagates. The net effect is that the air density and pressure also vary along this direction; hence, sound can also be viewed as a pressure wave.

(called the diaphragm) moves back-and-forth along the x direction, generating a sound wave (Fig. 13.1B). As the diaphragm moves, it collides with air molecules, imparting a small extra velocity to the molecules, $\vec{v}_{molecule}$. Because the diaphragm moves back-and-forth along the x direction, $\vec{v}_{molecule}$ alternates along $\pm x$. At some instants, $\vec{v}_{molecule}$ at a particular location is directed to the right; at other instants, $\vec{v}_{molecule}$ at this same location is toward the left. The total velocity of a molecule is the sum of its background velocity $\vec{v}_0$ and the extra velocity associated with the sound wave $\vec{v}_{molecule}$. This extra velocity is *not* the velocity of the sound wave; $\vec{v}_{molecule}$ is the *velocity of the medium* (the air) associated with the sound wave.

The green arrows in Figure 13.1B show $\vec{v}_{molecule}$ at several points along the wave. This snapshot only applies at a particular instant in time. Since $\vec{v}_{molecule}$ is parallel to x, the "extra" motion of the molecules—that is, the displacement of the medium—is parallel to the direction of propagation of the wave. For this reason, sound is a *longitudinal wave*.

Because the displacement alternates between the $+x$ and $-x$ directions as we move along the sound wave (along the x axis in Fig. 13.1B), the density of molecules is increased in some regions and decreased in others. This density variation is another way to view a sound wave. Oscillations in the air density cause oscillations in the air pressure, so we can also view sound as a *pressure wave*.

A plot of the pressure associated with a sound wave at a particular instant in time is shown in Figure 13.2. The regions of high density and high pressure (the wave crests in Fig. 13.2B) are called regions of **condensation**. Places of low density and low pressure (the wave troughs) are called regions of **rarefaction**. Adjacent regions of high pressure are separated by a distance equal to the wavelength λ, and adjacent regions of low pressure are separated by the same distance. The speed of the sound wave is related to its frequency and wavelength by

$$v_{sound} = f\lambda \tag{13.1}$$

It is crucial to distinguish the velocity of the wave (v_{sound} in Eq. 13.1) from the velocity of particles within the medium $\vec{v}_{molecule}$. They are *not* the same.

Density and pressure vary along the tube, shown here at one instant in time.

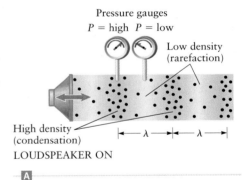

Pressure gauges
P = high P = low

Low density (rarefaction)

High density (condensation)

λ λ

LOUDSPEAKER ON

A

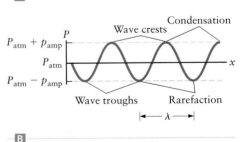

Condensation

Wave crests

P

$P_{atm} + p_{amp}$

P_{atm}

$P_{atm} - p_{amp}$

Wave troughs   Rarefaction

λ

B

Figure 13.2 **A** The pressure in a gas is largest in the regions with the highest density. **B** For a sound wave in air, the total pressure is the sum of atmospheric pressure P_{atm} and the oscillating pressure associated with the sound wave.

Table 13.1 Speed of Sound in Some Common Materials

Material	Speed of Sound (m/s)
Air	343
Helium gas	1000
Hydrogen gas	1330
Oxygen gas	330
Carbon dioxide gas	269
Water	1480
Seawater	1520
Ethyl alcohol	1160
Aluminum	5100
Steel	5790
Lead	1960
Concrete	3100
Glass	5600

Note: All values for 20°C and a pressure of 1 atm.

Figures 13.1 and 13.2 describe a sound wave in air, and the picture is similar for sound traveling through a solid or a liquid. The sound wave causes a displacement of the molecules in the solid or liquid parallel to the propagation direction, so sound is a longitudinal wave in solids and liquids as well as in gases.

In a solid, there can also be transverse waves that propagate through the solid, in which the molecular motion is perpendicular to the direction of wave propagation (such as some of the seismic waves described in Chapter 12). Such transverse waves are *not* possible in a gas or a liquid, however. The force associated with a sound wave is due to the pressure in the medium and this pressure force is always in the direction of propagation, so the motion of the medium must also be in that direction. Transverse waves are possible in solids because molecules in a solid experience transverse forces due to the bonds between neighboring atoms. The molecules in a gas or liquid are not bound in this way, so there can be *no transverse waves in a gas or liquid.*

The Speed of Sound

The speed of sound v_{sound} depends on the medium (Table 13.1). In general, the speed of sound is highest in solids and lowest in gases. The speed of sound in a solid was discussed in Chapter 12 and is given by (Eq. 12.10)

$$v_{sound} = \sqrt{\frac{Y}{\rho}} \quad \text{(speed of sound in a solid)} \tag{13.2}$$

where Y is Young's modulus and ρ is the density. Young's modulus is an elastic constant that measures the stiffness of a material. Hence, Equation 13.2 can also be written as

$$v_{sound} = \sqrt{\frac{stiffness}{density}} \tag{13.3}$$

Equation 13.3 explains qualitatively how the speed of sound varies when we compare solids, liquids, and gases. According to Hooke's law, a stiffer material (larger Young's modulus) generates a larger restoring force for a given displacement, leading to a larger particle acceleration and ultimately (for wave motion) a larger speed of sound. Solids are stiffer than liquids, and liquids are stiffer than gases, which explains why the speed of sound is largest in solids and lowest in gases.

The density factor in the denominator in Equation 13.3 means that the speed of sound increases as the medium becomes "lighter." The effect of density is especially noticeable for high-density solids; for example, lead has a very high density, and its speed of sound is much less than that found for aluminum, which has a low density (for a metal). A similar effect is found with gases; for example, the speed of sound in helium gas is much greater than the speed of sound in air at the same pressure and temperature. This result can again be understood from Newton's second law: for a given force, a smaller mass has a larger acceleration and hence (for wave motion) a larger velocity.

The speed of sound in a fluid can also be derived quantitatively from Newton's laws. The stiffness of a liquid is measured by its bulk modulus (Chapter 11), and we find

$$v_{sound} = \sqrt{\frac{B}{\rho}} \quad \text{(speed of sound in a liquid)} \tag{13.4}$$

The speed of sound in a gas depends on its composition and also on temperature. (We'll discuss the properties of gases in Chapter 15.) The most important case is air near room temperature and pressure, for which

$$v_{sound} \approx 343 + 0.6(T - 20°C) \text{ m/s} \quad \text{(speed of sound in air)} \tag{13.5}$$

where T is the temperature in Celsius.

The general relation $v_{sound} = f\lambda$ (Eq. 13.1) connects the speed of sound with the frequency and wavelength. The speed of sound is a property of the medium and is independent of both the frequency and the amplitude of the wave. The frequency and amplitude of a sound wave depend on how the sound is generated. For example, the frequency of the sound wave in Figure 13.1 equals the oscillation frequency of the loudspeaker diaphragm. This frequency can vary over an extremely wide range, so there is no absolute lowest or highest frequency for a sound wave. The range of normal human hearing extends from approximately 20 Hz to about 20,000 Hz. These limits are only approximate and vary from person to person. The upper limit also tends to decrease with age due to a gradual stiffening of the eardrum and the other flexible connections within the ear. Sound waves can have both higher and lower frequencies than this 20–20,000 Hz range. Frequencies below 20 Hz are called *infrasonic* and can be generated and detected by large animals such as elephants. People can sometimes sense such very low frequencies as vibrations rather than as sound detected by the ears. Frequencies above 20,000 Hz are called *ultrasonic* and can be heard by many species, such as dogs and bats.

Musical Tones and Pitch

A sound wave described by a single frequency is called a *pure tone*. Although it is possible to generate a pure tone, most sounds are a combination of many pure tones. For example, the sounds associated with speech or with the banging of a drum are complex combinations of pure tones.

In Section 12.8, we discussed standing waves on a guitar string and saw that they have frequencies given by $f_1, 2f_1, 3f_1, \ldots$. This pattern is called a *harmonic sequence*, and f_1 is called the *fundamental frequency*. Many sounds are a combination of frequencies that are harmonically related. For example, with normal playing, the standing wave vibrations of a guitar string are transferred to the body of the guitar, and the vibrating surfaces of the guitar act like the loudspeaker diaphragm in Figure 13.1. The guitar then produces a combination tone containing waves of frequency $f_1, 2f_1, 3f_1, \ldots$.

One way to characterize such a combination tone is by a property called *pitch*. Pitch is a subjective quality that depends on how the human brain processes sound. This processing is complicated and is still not completely understood. Under most conditions, the pitch of a pure tone is equal to its frequency, whereas the pitch of a combination of harmonically related tones (as would be produced by a plucked guitar string) is associated with the fundamental frequency. More complex sounds, as might be produced by an orchestra, are usually interpreted by the brain as a collection of tones with different pitches.

| EXAMPLE 13.1 | The Wavelength of Sound in Air |

The frequency range that can normally be detected by a human ear is 20–20,000 Hz. What is the corresponding range of wavelengths for sound in air?

RECOGNIZE THE PRINCIPLE

The wavelength is related to the speed of sound and the frequency through $v_{sound} = f\lambda$ (Eq. 13.1). Given v_{sound} and f, we can thus find λ.

SKETCH THE PROBLEM

No figure is necessary.

IDENTIFY THE RELATIONSHIPS

Solving for the wavelength in terms of the speed of sound and the frequency gives

$$\lambda = \frac{v_{sound}}{f}$$

Here v_{sound} is the speed of sound in air at room temperature; according to Table 13.1 and Equation 13.5, $v_{sound} = 343$ m/s.

SOLVE

For the lower end of the given frequency range, we have

$$\lambda = \frac{v_{sound}}{f} = \frac{343 \text{ m/s}}{20 \text{ Hz}}$$

$$\lambda = 17\,\frac{\text{m/s}}{\text{Hz}} = 17\,\frac{\text{m/s}}{1/\text{s}} = \boxed{17 \text{ m}}$$

For the high-frequency end of the range of human hearing,

$$\lambda = \frac{v_{sound}}{f} = \frac{343 \text{ m/s}}{20{,}000 \text{ Hz}} = 0.017 \text{ m} = \boxed{1.7 \text{ cm}}$$

What does it mean?

The ear canal is only a few millimeters in diameter, so it is smaller than the wavelengths of the sounds that people can hear.

13.2 | AMPLITUDE AND INTENSITY OF A SOUND WAVE

The oscillating pressure associated with a sound wave is sketched in Figure 13.2B, where the amplitude of the pressure oscillation is denoted p_{amp}. When a sound wave reaches your ear, the pressure on the outside of your eardrum is the sum of atmospheric pressure plus the oscillating sound pressure, $P_{atm} + p$, whereas the pressure inside your eardrum is P_{atm} (Fig. 13.3). The total force on your eardrum equals the difference in these two pressures multiplied by the area of the eardrum A. Hence, the total force equals pA, and your ear detects the sound pressure p while being insensitive[1] to P_{atm}.

The ***intensity*** I of the sound wave equals the power carried by the wave per unit area of wave front. We showed in Chapter 12 that the intensity of a wave is proportional to the square of the amplitude, so for sound we have

$$I \propto p_{amp}^2 \tag{13.6}$$

In many cases, sound waves propagate outward in all directions away from their source. In Section 12.4, we discussed the spherical wave fronts of such waves and found that the intensity at a distance r from the source varies as

$$I \propto \frac{1}{r^2} \tag{13.7}$$

Combining this expression with Equation 13.6, we see that the pressure amplitude of a spherical sound wave falls with distance as $p_{amp}^2 \propto 1/r^2$, which is equivalent to

$$p_{amp} \propto \frac{1}{r} \tag{13.8}$$

CONCEPT CHECK 13.1 | The Intensity of a Sound Wave Falls Off with Increasing Distance from the Source

A chirping bird sits at the top of a tree, and the chirping sound has an intensity of I_0 at a certain distance from the bird. If the distance to the bird is increased by a factor of three, is the new intensity (a) $I_0/3$, (b) $I_0/9$, or (c) I_0?

Intensity is proportional to the square of the amplitude.

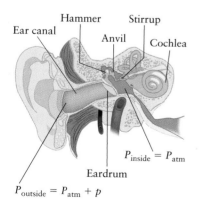

Figure 13.3 Your ear detects sound through the pressure on the eardrum. The oscillating pressure of the sound wave produces an oscillating force on the eardrum.

Ear canal · Hammer · Anvil · Stirrup · Cochlea · $P_{inside} = P_{atm}$ · Eardrum · $P_{outside} = P_{atm} + p$

---

[1]Sometimes—often in airplanes—P_{atm} outside your ear changes without a corresponding change in the inside pressure, which can lead to an uncomfortable force on your eardrum (a "pop").

Decibels

The sensitivity of the ear depends on frequency. In the most favorable range of frequencies, the ear can detect sound intensities as small as about 10^{-12} W/m². Because values this small are rather cumbersome, another unit, the *decibel* (abbreviated dB), is widely used to measure sound intensity. This unit is computed in the following way. Suppose the intensity of a sound wave as measured in watts per square meter is I. We now introduce a constant $I_0 \equiv 1.00 \times 10^{-12}$ W/m², which is (approximately) the lowest intensity detectable by the human ear. We then define a quantity called the *sound intensity level β* (the Greek letter beta) by

$$\beta = 10 \log\left(\frac{I}{I_0}\right) \quad \text{(sound intensity level in decibels)} \qquad (13.9)$$

Threshold for human hearing and the decibel scale

The intensity level β is given in decibels, but you can see from Equation 13.9 that since β is defined through a ratio of intensities, it is, in a strict mathematical sense, unitless! When working with β and Equation 13.9, it is helpful to recall the properties of common (base 10) logarithms as reviewed in Appendix B. In particular, if $a = 10^b$, then $b = \log(a)$. We also have $\log(1) = 0$, $\log(10) = 1$, $\log(100) = 2$, and so forth.

To see how the definition of sound intensity level in Equation 13.9 is applied, consider a sound that is just at the threshold of human hearing; that is, a sound with an intensity of $I = 1.0 \times 10^{-12}$ W/m². The intensity level in decibels is then

$$\beta = 10 \log\left(\frac{I}{I_0}\right) = 10 \log\left(\frac{1.0 \times 10^{-12} \text{ W/m}^2}{1.00 \times 10^{-12} \text{ W/m}^2}\right) = 10 \log(1)$$

Because $\log(1) = 0$, we find

$$\beta = 0 \text{ dB}$$

The intensity of this sound is thus 0 dB = "zero decibels." An intensity level of $\beta = 0$ dB does *not* mean that there is an absence of sound, however. It merely indicates that the intensity I is equal to the reference intensity I_0, which corresponds to a very faint sound, a sound at the limit of human hearing.

Now suppose we have another sound of intensity $I = 1.0 \times 10^{-11}$ W/m² = $10 \times I_0$, which is 10 times greater than the nominal threshold for human hearing. The intensity level of this sound as measured in decibels is

$$\beta = 10 \log\left(\frac{I}{I_0}\right) = 10 \log\left(\frac{10 \times I_0}{I_0}\right) = 10 \log(10)$$

Because $\log(10) = \log(10^1) = 1$, we get

$$\beta = 10 \text{ dB}$$

which is the result for a sound that has an intensity 10 times the threshold for hearing. If a sound has an intensity 100 times this threshold ($I = 1.0 \times 10^{-10}$ W/m²), we have

$$\beta = 10 \log\left(\frac{I}{I_0}\right) = 10 \log\left(\frac{100 \times I_0}{I_0}\right) = 10 \log(100)$$

Because $\log(100) = 2$, we get

$$\beta = 20 \text{ dB}$$

Thus, for every increase in intensity by a factor of 10, the intensity level as measured in decibels increases by 10 dB. The intensity levels of some familiar sounds are listed in Table 13.2.

CONCEPT CHECK 13.2 | Sound Intensity Level and Decibels

Suppose you measure a sound intensity level of 50 dB. What is the intensity of this sound?

(a) 1.0×10^{-12} W/m²

(b) 1.0×10^{-10} W/m²

(c) 1.0×10^{-7} W/m²

(d) 5.0×10^{-12} W/m²

Table 13.2 | Intensity Level of Some Common Sounds

Sound	Intensity Level (dB)	Sound	Intensity Level (dB)
Threshold of pain for humans	130	City traffic	70
Jet takeoff at a distance of 100 m	90–120	Normal conversation	60–70
Rock concert	110–120	Whisper	20
Thunder (nearby)	110	Pin dropping	10
Factory noise	80	Threshold of human hearing	0

CONCEPT CHECK 13.3 | How Loud Is It?

The sound intensity level at a rock concert can be as high as 120 dB, whereas normal speech corresponds to about 70 dB. What is the ratio of the intensities of these two sounds?

   (a) A factor of 190   (c) A factor of 5
   (b) A factor of 50   (d) A factor of 10^5

Human Perception of Sound ✖

Decibels are based on a ratio involving I_0 and logarithms, and you might now wonder why this way of measuring sound intensity was invented. The motivation comes from a feature of human hearing illustrated in Figure 13.4, which shows several contour lines obtained from extensive studies of human hearing. The lowest line plots the intensity of the minimum detectable sound as a function of the sound frequency. In the range of 500 Hz to 5000 Hz, this curve is roughly constant (horizontal) with a value close to $I_0 = 1.00 \times 10^{-12}$ W/m^2. (See the right-hand scale in Fig. 13.4.) When expressed in terms of intensity level in decibels (the left-hand scale), this curve corresponds to approximately 0 dB. The other contour curves show how the intensity varies for a fixed value of the *perceived* loudness, with each successive contour curve corresponding to an increase by a factor of two in the loudness. These results were obtained by asking many people to judge sounds of different loudness, and it was found that doubling the perceived loudness corresponds to moving from one contour to the next. Figure 13.4 shows averages obtained from many different individuals, and these contour curves represent normal human hearing. Loudness is a subjective quality that depends on how the ear functions at different frequencies, so it is fascinating that different people can agree on a loudness scale!

Figure 13.4 Contours of constant perceived loudness as a function of frequency for normal human hearing.

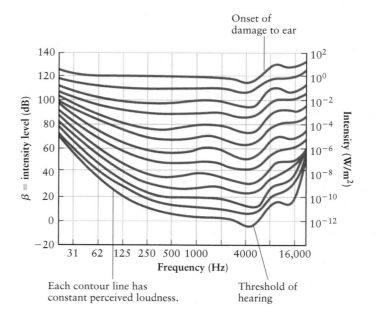

Over the range of 250 Hz to 8000 Hz, most of the contours in Figure 13.4 are approximately flat, so a constant perceived loudness corresponds to a roughly constant intensity I. Moreover, the contours have intensity levels β that differ by 10 dB. Hence, a factor of two increase in the perceived loudness corresponds to about a 10-dB increase in the intensity level, regardless of the initial loudness. That is why the sound intensity level measured in decibels with its logarithmic definition is a very useful way to describe the intensity of a sound.

The highest contour curve in Figure 13.4 has a loudness at which the ear is damaged. Such damage occurs at an intensity level $\beta \approx 120$ dB, which is approximately 120 dB larger than the threshold for hearing. Each change in the intensity level by 10 dB corresponds to a factor of 10 change in the intensity, so an increase of β by 120 dB corresponds to an increase in the intensity by a factor of 10^{12}! Hence, as a mechanical device, the ear is able to function over a very wide range of forces.

Another important feature of Figure 13.4 is that the loudness contours depend somewhat on frequency. That is, the ear's sensitivity to sound pressure varies with frequency; this effect is most noticeable at low frequencies, where we see that the sensitivity decreases. For example, to have the same perceived loudness (and thus fall on the same contour curve), a sound at 60 Hz must have a larger intensity than a sound at 1000 Hz.

The Ear as a Pressure Detector ⊗

Quantitative measurements of sound often involve values of the intensity, but it is also interesting to consider the corresponding values of the sound pressure. Intensity is proportional to the square of the pressure amplitude (Eq. 13.6), so there is a direct relation between the threshold intensity for human hearing I_0 and the smallest pressure oscillation that can be detected by the ear. A sound intensity of $I_0 = 1.00 \times 10^{-12}$ W/m^2 corresponds to a pressure amplitude of $p_0 = 2 \times 10^{-5}$ Pa. (This result can be obtained by applying Newton's laws to calculate the molecular motion of a sound wave, but the derivation is more than we can include here.) Hence, the smallest pressure amplitude that can be detected by the human ear is about $p_{amp} = p_0 = 2 \times 10^{-5}$ Pa. Since atmospheric pressure is $P_{atm} = 1.01 \times 10^5$ Pa, the ear is therefore able to detect variations in pressure that are $p_0/P_{atm} = (2 \times 10^{-5} \text{ Pa})/(1.01 \times 10^5 \text{ Pa}) = 2 \times 10^{-10}$ times smaller than atmospheric pressure! The ear is thus a very impressive mechanical device.

| EXAMPLE 13.2 | The Pressure Amplitude at Rock Concert |

One of the loudest sounds encountered in "normal" living is found at a rock concert. According to Table 13.2, the sound intensity level at a rock concert can reach values as high as 120 dB. What is the approximate value of the corresponding pressure amplitude? Use the fact that the pressure amplitude at $\beta = 0$ dB is $p_0 = 2 \times 10^{-5}$ Pa.

RECOGNIZE THE PRINCIPLE

We first use the definition of decibels (Eq. 13.9) to convert the given sound intensity level β to an intensity value I. We can then use the relationship between I and p_{amp} to find the corresponding pressure amplitude.

SKETCH THE PROBLEM

No figure is necessary.

IDENTIFY THE RELATIONSHIPS

We first convert the intensity level at a rock concert ($\beta = 120$ dB) into intensity I in units of W/m^2. Using Equation 13.9, we can express the given sound intensity level as

$$\beta = 120 = 10 \log\left(\frac{I_{rc}}{I_0}\right)$$

where I_{rc} is the sound intensity at the rock concert (in watts per square meter) and I_0 is known to be 1.00×10^{-12} W/m^2. We thus have

$$\log\left(\frac{I_{rc}}{I_0}\right) = 12 \qquad (1)$$

According to the properties of logarithms, if $\log(a) = b$, then $a = 10^b$. Applying this fact to Equation (1), we get

$$\frac{I_{rc}}{I_0} = 10^{12}$$

$$I_{rc} = 10^{12} I_0 \qquad (2)$$

Now use the relation between the intensity I and pressure amplitude p_{amp}. At an intensity of I_0 (corresponding to 0 dB), the pressure amplitude is $p_0 = 2 \times 10^{-5}$ Pa. We also know that intensity is proportional to the *square* of the pressure amplitude (Eq. 13.6). We can therefore replace the intensities in Equation (2) by the squares of the corresponding pressure amplitudes and write

$$p_{rc}^2 = 10^{12} p_0^2$$

where p_{rc} is the pressure amplitude at the rock concert.

SOLVE

Inserting the value of p_0 leads to

$$p_{rc} = \sqrt{10^{12}}\, p_0 = (10^6)(2 \times 10^{-5}\ \text{Pa}) = \boxed{20\ \text{Pa}}$$

What does it mean?
Although a rock concert produces quite a loud sound, the resulting pressure amplitude is still only a very tiny fraction of atmospheric pressure ($P_{atm} = 1.01 \times 10^5$ Pa).

13.3 | STANDING SOUND WAVES

In Figures 13.1 and 13.2, we considered sound waves that travel in one direction inside a tube. We next want to consider what happens when these waves reach the end of the tube. This is the key to understanding how many musical instruments work, including flutes, trombones, and organs, and it is very similar to the case of standing waves on a guitar string studied in Chapter 12.

Standing Waves in a Pipe Closed at Both Ends

A standing wave is created when two waves travel in opposite directions in the same medium. As a first example, consider sound waves propagating back-and-forth in a pipe that is closed at both ends. A standing wave is possible only for certain wavelengths; some number of sound waves must "fit into" or match the length of the pipe. To understand this matching process, we must consider how a sound wave behaves at the ends of the pipe. Because sound is a longitudinal wave, the motion of the molecules would normally be directed along the pipe (the x direction in Fig. 13.5). The wall prevents this motion, however; for example, the right-hand wall in Figure 13.5 prevents the adjacent molecules from moving to the right. These molecules will also not move to the left (away from the wall) because that would produce a low-pressure region that pulls them back, so the velocity of the air molecules ($\vec{v}_{molecule}$ in Fig. 13.1) is zero at a wall, that is, at the closed end of a pipe. Because $\vec{v}_{molecule} = 0$ at a wall, the displacement of the molecules $d_{molecule}$ is also zero at the ends of the pipe.

A standing sound wave must therefore have zero displacement at both ends of the pipe, leading to the results in Figure 13.5. Figure 13.5A shows the displacement

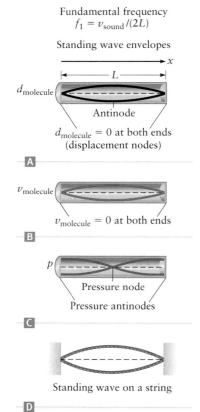

$\lambda = 2L$

Fundamental frequency
$f_1 = v_{sound}/(2L)$

Standing wave envelopes

$d_{molecule} = 0$ at both ends
(displacement nodes)

A

$v_{molecule} = 0$ at both ends

B

Pressure node

Pressure antinodes

C

Standing wave on a string

D

Figure 13.5 Standing sound waves in a pipe that is closed at both ends. **A** Molecular displacement, **B** velocity, and **C** sound pressure for the lowest-frequency standing wave. This is the fundamental frequency of the pipe. **D** Displacement for a standing wave on a string; it is identical to the molecular displacement in **A**.

pattern ($d_{molecule}$) for a possible standing wave. These curves are the standing wave envelopes; that is, they show how the amplitude of the oscillation varies with position along the pipe. Places where the displacement is zero are called displacement *nodes*, while the displacement is largest at the displacement *antinodes*. The wavelength of this standing wave is $2L$, where L is the length of the pipe. The standing wave pattern in Figure 13.5A is identical to the pattern found for a standing wave on a string (Fig. 13.5D; compare with Fig. 12.23B). Both have nodes at the ends of a pipe or string and an antinode in the center. The molecular velocity forms a standing wave described by the same pattern (Fig. 13.5B). The amplitude of the molecular velocity is zero at the ends of the pipe and large at the center.

A sketch of the sound pressure p is shown in Figure 13.5C. Notice that p also forms a standing wave. However, the behavior of the pressure at the ends is different from that of the molecular velocity and displacement. We have already mentioned that the air molecules cannot move past the end of the pipe (because of the walls). During a portion of the wave's cycle, air molecules that are initially in the interior of the pipe move toward a wall, which increases the pressure at the wall to a large value because the molecules are "trapped" there. During the opposite portion of the wave's cycle, these air molecules move away from the wall, producing a very low pressure at the wall. As a result, the pressure oscillations are large at a closed end, forming a *pressure antinode* in the standing wave (Fig. 13.5C). Hence, both the displacement and pressure form standing waves, with nodes and antinodes. The nodes of the displacement are located at the antinodes of p and vice versa.

Other possible standing wave patterns for the displacement and pressure are shown in Figure 13.6 for standing waves with $\lambda = L$ and $\lambda = 2L/3$. We notice again that these standing wave patterns are identical to those found for a wave on a string (Fig. 12.23).

Now that we understand how a standing sound wave can "fit" into a pipe, we can see that there is a mathematical pattern to the standing waves. Figures 13.5 and 13.6 show the three standing waves with the longest possible wavelengths: $\lambda = 2L$, L, and $2L/3$. There are many more possible values of the wavelength (an infinite number), and they follow the pattern

$$\lambda_n = 2L, \; L, \; \frac{2L}{3}, \; \frac{2L}{4}, \; \frac{2L}{5}, \ldots$$

or

$$\lambda_n = \frac{2L}{n} \quad \text{with } n = 1, 2, 3, \ldots \quad \text{(standing waves in a pipe closed at both ends)} \; (13.10)$$

Since frequency and wavelength are connected by $v_{sound} = f\lambda$, the frequencies of these standing waves are

$$f_n = n\frac{v_{sound}}{2L} \tag{13.11}$$

$$f_n = nf_1 \quad \text{with } n = 1, 2, 3, \ldots$$

$$\text{(standing waves in a pipe closed at both ends)} \quad (13.12)$$

where $f_1 = v_{sound}/(2L)$ is called the **fundamental frequency** (also known as the first harmonic). The standing wave with a frequency $f_2 = 2f_1$ is called the second harmonic, the standing wave with frequency f_3 is called the third harmonic, and so on.

Standing Waves in a Pipe Open at One End and Closed at the Other

Some musical instruments—trumpets, clarinets and many others—have an opening at one end. Let's therefore consider standing waves in a pipe that is *open* at one end and closed at the other. To analyze this situation, we must consider what happens to a sound wave at the open end of the pipe. We have already seen that there is a displacement node and a pressure antinode at the closed end of a pipe (Fig. 13.5). The behavior at an open end is quite different. At an open end, a pipe connects to a

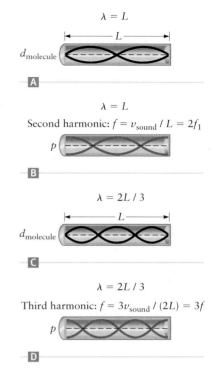

Figure 13.6 A and B: Displacement and pressure for a standing wave in a pipe closed at both ends, at the frequency of the second harmonic. C and D: Displacement and pressure at the frequency of the third harmonic.

Frequencies of standing sound waves in a closed pipe

Standing wave envelopes

$d_{molecule}$ Open end
Closed end

A

p — Pressure node
at open end

|←1/4 of a complete →|
wavelength

B

$\lambda = 4L$
Fundamental frequency: $f_1 = v_{sound} / (4L)$

Figure 13.7 Standing sound waves in a pipe that is closed at one end and open at the other. **A** Molecular displacement and **B** pressure at the fundamental frequency of the pipe. There is a pressure node at the open end, which makes the standing wave frequencies different from those found in a pipe that is closed at both ends (Fig. 13.5).

very large volume of air—for example, the rest of the room—where the pressure is equal to atmospheric pressure P_{atm}. This outside volume is much larger than the volume of air in the pipe, so the flow of air molecules out of or into the pipe due to a sound wave has a negligible effect on the outside pressure. The pressure at the open end of the pipe is therefore held fixed at P_{atm} by the large body of air in the room, and the pressure amplitude of a standing sound wave at an open end must be $p_{amp} = 0$. Hence, there is a pressure *node* at the open end of the pipe.

Figures 13.7 and 13.8 show some of the possible standing waves for a pipe open at one end and closed at the other. There is a pressure node at the open end and (as we have seen in Fig. 13.5C) a pressure antinode at the closed end. The behaviors of the sound pressure and displacement are linked, so there is a displacement antinode at the open end and a displacement node at the closed end. Standing sound waves must always satisfy these conditions at the ends of a pipe. The standing wave with the longest possible wavelength is shown in Figure 13.7; it has $\lambda = 4L$. The next standing waves (in order of decreasing wavelength) are shown in Figure 13.8; they have wavelengths of $\lambda = 4L/3$ and $\lambda = 4L/5$. These and the other possible standing waves for this tube follow the pattern

$$\lambda_n = 4L, \frac{4L}{3}, \frac{4L}{5}, \frac{4L}{7}, \ldots$$

$$\lambda_n = \frac{4L}{n} \quad \text{with } n = 1, 3, 5, \ldots \quad \text{(standing waves in a pipe open at one end)} \quad (13.13)$$

The frequencies of these standing waves are $f_n = v_{sound} / \lambda_n$, so

$$f_n = n \frac{v_{sound}}{4L} \qquad (13.14)$$

Frequencies of standing sound waves in a pipe open at one end

$$f_n = nf_1 \quad \text{with } n = 1, 3, 5, \ldots \quad \text{(standing waves in a pipe open at one end)} \quad (13.15)$$

and the fundamental frequency is $f_1 = v_{sound}/(4L)$. This pattern of standing wave frequencies is quite different from the result for a pipe closed at both ends (Eqs. 13.11 and 13.12). The fundamental frequency is lower for the pipe with one end open (if we compare two pipes of the same length L), and the pattern of harmonics is also different. For the pipe with both ends closed, the allowed frequencies form the sequence f_1, $2f_1$, $3f_1$, ..., whereas with one end open, the sequence is f_1, $3f_1$, $5f_1$, ... Hence, the even-numbered harmonics are missing for a pipe that is open at one end. In Figure 13.9, the pattern of standing wave frequencies in the two cases is shown.

Important properties of standing sound waves

The allowed frequencies of standing sound waves in pipes can be predicted using three main facts:

1. The *closed* end of a pipe is a pressure *antinode* for the standing wave.
2. The *open* end of a pipe is a pressure *node* for the standing wave.
3. A pressure node is always a displacement antinode, and a pressure antinode is always a displacement node.

Figure 13.8 Other possible standing sound waves in a pipe that is closed at one end and open at the other. Compare with Figure 13.7, which shows the behavior at the fundamental frequency.

$d_{molecule}$

A

p

B

$\lambda = 4L/3$
Third harmonic: $f = 3v_{sound} / (4L) = 3f_1$

$d_{molecule}$

C

p

D

$\lambda = 4L/5$
Fifth harmonic: $f = 5v_{sound} / (4L) = 5f_1$

Composition of a Real Musical Tone

Pipes like those shown schematically in Figures 13.5 through 13.8 are the basis of many musical instruments, including flutes, trumpets, and organs. The sound we hear originates as a standing sound wave inside the instrument. This standing wave does not have a single frequency, but rather is a combination (a *superposition*) of standing waves with different frequencies. The allowed standing wave frequencies follow a pattern such as Equation 13.11 or 13.14, depending on whether the ends of the instrument are open or closed. Hence, these instruments produce combination tones as described above in our discussion of pitch. These combination tones are often said to have a property called **timbre**, or **tone color**. The musical timbre of a tone depends on the mix of frequencies (as in Fig. 13.9) and on the relative amplitudes of these different frequency components. The timbre is different for different instruments, and it also depends on how the instrument is played. (A "wailing" saxophone will have a different tone color than one played less forcefully.) These variations in timbre help make musical tones interesting to a listener.

EXAMPLE 13.3 | **The Lowest Note of a Clarinet**

A clarinet is a musical instrument we can model as a pipe open at one end and closed at the other (Fig. 13.10A). A standard clarinet is approximately $L = 66$ cm long. What is the frequency of the lowest tone such a clarinet can produce?

RECOGNIZE THE PRINCIPLE

According to our work with pipes, there must be a pressure antinode at the closed end of the clarinet and a pressure node at the open end. Frequency and wavelength are related by $v_{sound} = f\lambda$, so the standing wave with the lowest frequency will be the one with the longest wavelength.

SKETCH THE PROBLEM

Figure 13.10B shows the clarinet modeled as a pipe open at one end. The allowed standing waves will be just like those found in Figures 13.7 and 13.8.

IDENTIFY THE RELATIONSHIPS

The standing wave with the longest wavelength is sketched in Figure 13.7. Its frequency is given by Equation 13.14, with $n = 1$:

$$f_1 = \frac{v_{sound}}{4L}$$

This is the fundamental frequency for the clarinet with all tone holes closed.

SOLVE

The speed of sound in air (at room temperature) is $v_{sound} = 343$ m/s, and the length of the tube is $L = 0.66$ m, so we have

$$f_1 = \frac{v_{sound}}{4L} = \frac{343 \text{ m/s}}{4(0.66 \text{ m})} = \boxed{130 \text{ Hz}}$$

What does it mean?
This result is the lowest frequency—that is, the lowest "note"—a clarinet can produce according to our simple model. (To get a more accurate value, we would have to allow for the length of the mouthpiece and other factors.) To play other notes, the musician uncovers one of the many different tone holes along the body of the clarinet. If an open hole is sufficiently large, the pressure node moves from the end of the clarinet to a spot very near the tone hole (Fig. 13.10C), thereby shortening the wavelength of the fundamental standing wave and increasing f_1. In this way, a musician can use the many tone holes to produce many different notes.

Pipe closed at both ends

$f = f_1, 2f_1, 3f_1, \ldots$
(All harmonics present)

Pipe open at one end

$f = f_1, 3f_1, 5f_1, \ldots$
(Only odd harmonics present)

Figure 13.9 The pattern of standing wave frequencies in a pipe depends on whether the ends of the pipe are open or closed.

Figure 13.10 Example 13.3. 🄰 A clarinet. 🄱 Physicist's model of a clarinet, as a pipe closed at one end and open at the other. The standing waves will be the same as shown in Figures 13.7 and 13.8. 🄲 If a tone hole is opened, it will act as the open end of the pipe as far as the standing wave is concerned, changing the frequency of the standing waves.

CONCEPT CHECK 13.4 | Designing a Clarinet

Suppose you design a new type of clarinet that is half the length of the ordinary clarinet in Example 13.3. How will this change affect the frequency of the lowest note on this new instrument?
  (a) The frequency of the lowest note will decrease by a factor of two.
  (b) The frequency of the lowest note will increase by a factor of two.
  (c) The frequency of the lowest note will not change.

EXAMPLE 13.4 | Standing Waves in a Pipe Open at Both Ends

Consider a hypothetical musical instrument that uses a pipe open at both ends. If the pipe is $L = 1.0$ m long, what is the longest wavelength that can produce a standing wave? Sketch how the pressure of this standing wave varies along the tube.

RECOGNIZE THE PRINCIPLE

There must be a pressure node at the open end of a pipe (see page 416 and also the important properties of standing sound waves listed there). Hence, a pipe open at both ends must have pressure nodes at both ends.

SKETCH THE PROBLEM

Figure 13.11 shows the problem.

IDENTIFY THE RELATIONSHIPS AND SOLVE

The standing wave with the longest wavelength is sketched in Figure 13.11. For a pipe of length L, the wavelength is $\lambda = 2L$. We thus have

$$\lambda = 2L = 2(1.0 \text{ m}) = \boxed{2.0 \text{ m}}$$

What does it mean?
Comparing Figure 13.11 with the standing wave pattern for a pipe with both ends closed (Fig. 13.5), we find that the fundamental frequency is the same for these two cases.

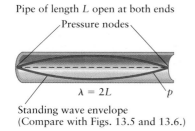

Pipe of length L open at both ends
Pressure nodes
$\lambda = 2L$ p
Standing wave envelope
(Compare with Figs. 13.5 and 13.6.)

Figure 13.11 Example 13.4.

CONCEPT CHECK 13.5 | Playing a Funky Organ

An organ uses standing sound waves in pipes of different lengths to produce different notes. Your friend is an organist, and you play a trick on him by filling his organ pipes with helium gas. The velocity of sound in helium is larger than the velocity of sound in air by a factor of $v_{\text{helium}}/v_{\text{air}} = 2.9$. (See Table 13.1.) How will your prank affect the tones produced by the organ?
  (a) The notes will all go up in frequency by this factor.
  (b) The notes will all go down in frequency by this factor.
  (c) There will be no change.

Note: An experiment similar to this one is often performed at carnivals, when a performer fills his vocal tract with helium gas and then demonstrates his "new" voice. The pitch of the human voice is determined by the frequencies of standing waves in the vocal tract.

13.4 | BEATS

Suppose you are asked to listen to two different pure tones and judge whether or not they have the same frequency. If the two tones are played one after the other, you will find that you are able to detect a minimum frequency difference Δf. The precise value of Δf varies somewhat with the overall intensity and also depends on the average frequency f of the tones, but for most people, Δf is approximately 1% of f. That is, $\Delta f \approx 0.01 \times f$. This is the smallest detectable frequency difference for tones that are played consecutively. If the two tones are played simultaneously, however, you can detect much smaller frequency differences.

When two tones are played at the same time, the principle of superposition tells us that the total sound pressure is the sum of the sound pressures of the individual tones. Figure 13.12A plots the variation of the pressure as a function of time for two hypothetical pure tones, one at 500 Hz and the other at 600 Hz. Figure 13.12B shows the sum of these two pressure waves, which would be the total sound pressure when the two tones are played simultaneously. This figure also shows how the amplitude of this combination pressure wave varies back-and-forth (oscillates) from a large value when the two waves are giving constructive interference to near zero when there is nearly destructive interference. These amplitude oscillations are called **beats**. The frequency of these oscillations in this case is 100 Hz, which is called the **beat frequency**. In general, whenever two tones of frequencies f_1 and f_2 are played simultaneously, they "beat against" each other with a beat frequency of

$$f_{\text{beat}} = |f_1 - f_2| \qquad (13.16)$$

The human ear is very sensitive to these beats, which are heard as variations in loudness with a frequency of f_{beat}. Most people are able to detect beat frequencies as low as about 0.2 Hz, even when the two simultaneous frequencies are near 1000 Hz or even higher. Hence, beats can be used to detect very small frequency differences (0.02% or even smaller), provided the tones are played simultaneously.

Beats are familiar to most musicians and can be used to tune an instrument. When the different instruments in an orchestra are "in tune," they each produce precisely the same fundamental frequency f_1 when the musicians are playing notes with the same pitch. Musicians adjust their instruments (prior to the performance!) by listening to the beat frequency between their own tone and a reference tone, and making that beat frequency as small as possible.

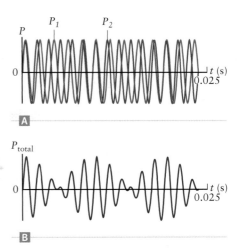

Figure 13.12 Beating of two sound waves. **A** Pressure as a function of time for two separate sound waves of frequency 500 Hz and 600 Hz. **B** The sum of these two pressure waves produces a pattern of "beats" in which the amplitude of the resulting wave oscillates with time.

EXAMPLE 13.5 | Beat Frequency of Two Guitar Tones

A standard guitar has six strings, so it is possible to play six tones at any particular moment. The fundamental frequency of a particular string is determined by the standing wave with the lowest possible frequency. This lowest standing wave frequency depends on the vibrating length of the string as discussed in Chapter 12 and illustrated in Figure 13.5D. By holding a string against a particular fret (Fig. 12.24), it is possible to vary the vibrating length of the string so that two different strings have the same fundamental frequency and thus play the same note. Two strings on your guitar that are *supposed* to produce the same note are actually found to produce 4 beats per second. If one note has a frequency of 247 Hz, what is the frequency of the other note?

RECOGNIZE THE PRINCIPLE

If the two fundamental frequencies are f_1 and f_2, the beat frequency is $f_{\text{beat}} = |f_1 - f_2|$ (Eq. 13.16). Given the value of f_{beat} and that one of the frequencies is $f_1 = 247$ Hz, we can then solve for f_2.

SKETCH THE PROBLEM

No figure is necessary.

IDENTIFY THE RELATIONSHIPS AND SOLVE

If there are 4 beats per second, the beat frequency is

$$f_{\text{beat}} = 4 \text{ Hz}$$

We know that $f_1 = 247$ Hz and $f_{\text{beat}} = |f_1 - f_2|$, so the other tone must have a frequency of either $247 + 4 = \boxed{251 \text{ Hz}}$ or $247 - 4 = \boxed{243 \text{ Hz}}$. If we are only allowed to listen to the beats, there is no way to tell which of these answers applies to a particular guitar.

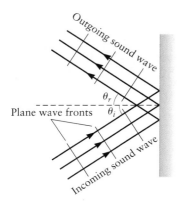

Plane wave fronts

Figure 13.13 Reflection of a sound wave with plane wave fronts from a flat wall. The outgoing (reflected) angle θ_r is equal to the incoming (incident) angle θ_i.

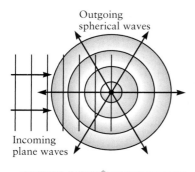

Outgoing spherical waves

Incoming plane waves

Scattering is important when λ_{sound} is comparable to or larger than the size of the object.

Figure 13.14 When a wave reflects from a small object, the reflected wave travels outward in all directions. When the object is very small compared to the wavelength, the outgoing wave fronts are approximately spherical.

What does it mean?
In practice, a guitarist will listen to the beats and adjust the tension of one of the strings to make the beat frequency as low as possible. (Recall from Chapter 12 that the fundamental frequency of waves on a string depends on the string tension.)

13.5 | REFLECTION AND SCATTERING OF SOUND

A standing wave can be viewed as a combination of two waves traveling in opposite directions. For the standing sound wave in a pipe, these waves travel back-and-forth as they are reflected from the ends of the pipe. Let's now examine the phenomenon of reflection for other geometries and also consider a related process called *scattering*.

Figure 13.13 shows a sound wave with a plane wave front reflecting from a flat surface. It is very similar to the reflecting wave fronts we sketched in Figure 12.19. Such a *plane wave* reflects in a mirrorlike manner from a flat surface. The incoming angle θ_i, called the *angle of incidence* (see Fig. 13.13), equals the outgoing angle θ_r, called the *angle of reflection*. We call this reflection mirrorlike because this situation is also encountered when light reflects from a mirror.

A mirrorlike reflection will occur whenever a plane wave encounters a flat surface as long as the size of the reflecting surface is large compared to the wavelength of the sound.[2] A typical sound might have a frequency of 200 Hz, in which case the wavelength is $\lambda = v_{sound}/f = (343 \text{ m/s})/(200 \text{ Hz}) = 1.7$ m. For sound in a typical room, most of the room surfaces are not large compared to this wavelength, so most of the reflections are not like the one shown in Figure 13.13. Instead, the reflected sound waves behave as in Figure 13.14, which shows an incoming plane wave reflecting from a small object. Here, there is at least a small amount of reflected sound in all directions. It is often called scattered sound because, loosely speaking, the original wave is "scattered" in many different directions. This result is found whenever a sound wave encounters an object whose size is approximately equal to, or smaller than, the wavelength of the sound. In such cases, reflected sound waves propagate out in virtually all directions from the scattering source.

13.6 | THE DOPPLER EFFECT

Consider the sound emitted by the siren of an ambulance, and for simplicity, assume the siren emits a pure tone of frequency f_{source}. If the ambulance is parked so that it is not moving relative to a listener standing nearby, the sound waves emitted by the siren will be as sketched in Figure 13.15A. Here we show two wave crests as they move from the source to the observer (listener). These wave crests travel at a speed v_{sound}. The source emits one wave crest during each period of the wave (one period $= 1/f_{source}$), so consecutive crests are separated by a distance equal to the wavelength $\lambda_{source} = v_{sound}/f_{source}$. The crests arrive at the observer spaced in time by $\lambda_{source}/v_{sound} = 1/f_{source}$, so one wave crest arrives at the observer every $1/f_{source}$ seconds. In other words, the frequency *as heard by the observer* is equal to f_{source}. The frequency the observer measures thus equals the frequency emitted by the siren, as you should have expected.

Let's now use the same approach to analyze the case of a siren moving toward the observer. The siren emits sound waves whose crests again travel toward the observer. These wave crests move at a speed v_{sound} *relative to the medium*, which in this case is the air. Once the sound wave leaves the source, the wave's motion is determined by

---

[2]This requirement does *not* apply to a sound wave traveling in a very narrow pipe (as in Fig. 13.5 or with the clarinet in Fig. 13.10). When the diameter of the pipe is small compared to wavelength, the sound wave can only be reflected backward because it literally has nowhere else to go.

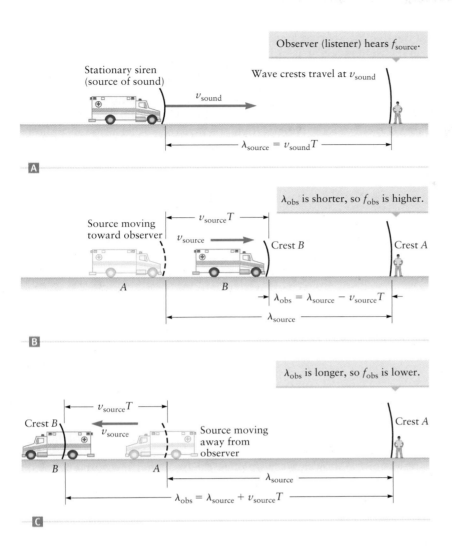

Observer (listener) hears f_{source}.

Stationary siren (source of sound)

Wave crests travel at v_{sound}

v_{sound}

$\lambda_{source} = v_{sound}T$

A

λ_{obs} is shorter, so f_{obs} is higher.

Source moving toward observer

$v_{source}T$

v_{source}

Crest B

Crest A

A B

$\lambda_{obs} = \lambda_{source} - v_{source}T$

λ_{source}

B

λ_{obs} is longer, so f_{obs} is lower.

$v_{source}T$

Crest B

v_{source}

Source moving away from observer

Crest A

B A

λ_{source}

$\lambda_{obs} = \lambda_{source} + v_{source}T$

C

Figure 13.15 The Doppler effect. **A** When a source of sound is at rest relative to a stationary observer, the frequency heard by the observer is equal to the frequency emitted by the source. **B** When the source is moving toward the observer, the observed frequency is higher than the emitted frequency. **C** When the source is moving away from the observer, the observed frequency is lower than the emitted frequency.

the medium; the observer in Figure 13.15B is stationary relative to the air, so he will find that the wave crests move at v_{sound} no matter what speed the source may have.

We must now account for the motion of the siren between the emission of each wave crest. Figure 13.15B shows two successive wave crests along with the position of the siren when it emitted these crests. This snapshot shows the siren at point B, having just emitted the wave crest also labeled B. The previous wave crest labeled A was emitted when the siren was at location A, one period earlier. The frequency of the siren, as would be measured by someone sitting in the ambulance, is still f_{source}, so the siren emits wave crests with the same period as when the ambulance is not moving, and this period equals $T = 1/f_{source}$. Wave crest A has traveled for this amount of time at a speed v_{sound}, so it has traveled a distance $v_{sound}T = v_{sound}/f_{source}$ from the point where it was emitted.

Let's call the speed of the ambulance (and attached siren) v_{source}. By the time the siren emits wave crest B, it has traveled a distance $v_{source}T$ since emitting crest A. The spacing between crest A and crest B is thus smaller than was the case with a stationary siren. From Figure 13.15B, we see that the new distance between crests is

$$\lambda_{obs} = v_{sound}T - v_{source}T$$

The period T is equal to $1/f_{source}$, so

$$\lambda_{obs} = \frac{v_{sound} - v_{source}}{f_{source}} \tag{13.17}$$

Because the distance between crests is smaller when the ambulance moves toward the observer than when it is stationary, the crests arrive more frequently at the observer and the frequency as *measured by the observer* is *higher*. This phenomenon, in which the frequency heard by an observer is different from the frequency emitted by a moving source, is called the **Doppler effect** in honor of Christian Doppler (1803–1853), the person who discovered it. (We'll see below that some animals were using the effect long before Doppler's discovery.) Let's now use the result in Equation 13.17 to derive the frequency found by the observer.

The wave crests move at the speed of sound v_{sound}. Since they are separated in space by a distance λ_{obs}, they are spaced in time by an amount T_{obs}, where

$$T_{\text{obs}} = \frac{\lambda_{\text{obs}}}{v_{\text{sound}}}$$

Substituting the observed wavelength λ_{obs} from Equation 13.17 leads to

$$T_{\text{obs}} = \frac{v_{\text{sound}} - v_{\text{source}}}{v_{\text{sound}} f_{\text{source}}} = \frac{1 - (v_{\text{source}}/v_{\text{sound}})}{f_{\text{source}}} \tag{13.18}$$

In words, wave crest B arrives at the observer a time T_{obs} after crest A, which is the period of the wave measured by the observer. The observer's frequency is related to Equation 13.18 by $f_{\text{obs}} = 1/T_{\text{obs}}$, so we have

$$f_{\text{obs}} = \frac{f_{\text{source}}}{1 - (v_{\text{source}}/v_{\text{sound}})} \quad \text{(source moving toward observer)} \tag{13.19}$$

The difference between the frequency heard by the observer and the source frequency is called the **Doppler shift**. Notice that Figure 13.15B assumes $v_{\text{sound}} > v_{\text{source}}$, so the wave crests move faster than the source. Hence, the denominator in Equation 13.19 can never be zero (or negative).

This result for f_{obs} shows that in this case the frequency measured by the observer is higher than the frequency emitted by the siren. You are likely familiar with this result; a siren traveling toward you appears to have a higher pitch—that is, a higher frequency—than when the siren is at rest. If the siren is moving away from the observer, the same type of analysis shows that the spacing between wave crests is increased relative to their spacing when the siren is stationary, as shown in Figure 13.15C. Here the source emits wave crest A and then moves away from the observer by a distance $v_{\text{source}}T$ before emitting the next wave crest B. The distance between the adjacent wave crests is now increased, causing the frequency heard by the observer to be lower. Taking the same approach that led to Equation 13.19, we find that the result for f_{obs} in this case is

$$f_{\text{obs}} = \frac{f_{\text{source}}}{1 + (v_{\text{source}}/v_{\text{sound}})} \quad \text{(source moving away from observer)} \tag{13.20}$$

Let's estimate the size of the Doppler shift for a typical case. We assume the source is a siren whose frequency when at rest is $f_{\text{source}} = 500$ Hz. Suppose the siren is on a fire truck racing to an emergency at a speed of 27 m/s (about 60 mi/h). For a listener standing on a nearby sidewalk, what frequency does the siren have? If the truck is moving toward the observer, the Doppler-shifted frequency is given by Equation 13.19, where the speed of the source is $v_{\text{source}} = 27$ m/s. Inserting these values along with the speed of sound $v_{\text{sound}} = 343$ m/s into Equation 13.19 gives

$$f_{\text{obs}} = \frac{f_{\text{source}}}{1 - (v_{\text{source}}/v_{\text{sound}})} = \frac{500 \text{ Hz}}{1 - [(27 \text{ m/s})/(343 \text{ m/s})]} \tag{13.21}$$

$$f_{\text{obs}} = 540 \text{ Hz}$$

Hence, the Doppler shift is $f_{\text{obs}} - f_{\text{source}} = 40$ Hz, which is nearly 10% of the source frequency. In terms of notes on a musical scale, this value is approximately the separation between two consecutive whole notes (i.e., "do" and "re").

So far, we have considered the Doppler effect for a stationary observer. Motion of the observer also gives a Doppler shift. In the most general case, in which the observer has a speed v_{obs} and the source has a speed v_{source}, the observed frequency is

$$f_{obs} = f_{source}\left[\frac{1 \pm (v_{obs}/v_{sound})}{1 \pm (v_{source}/v_{sound})}\right] \qquad (13.22)$$

The choices of plus and minus signs here depend on whether the source and observer are moving toward or away from each other. The signs can always be remembered by noting that the frequency measured by the observer f_{obs} is increased when the observer and source are moving toward each other, whereas f_{obs} is decreased when they move away from each other.

EXAMPLE 13.6 | Doppler Shift, Coming and Going

A fire truck approaches you from one direction, passes by, and then continues on away from you, with the siren emitting sound the entire time. You have a friend (a music student) with a very keen sense of pitch who tells you that the frequency of the siren drops by precisely 50 Hz when the truck passes your location on the street. If the truck has a speed of 20 m/s, what is the frequency of the siren at rest? Take the speed of sound to be $v_{sound} = 343$ m/s.

RECOGNIZE THE PRINCIPLE

This problem is similar to our analysis of the Doppler shift for a moving truck (Eq. 13.21), but now we have to take into account *two* Doppler shifts, one as the truck moves toward you and another as it moves away from you. The observed frequency will be shifted upward to a higher frequency when the truck is moving toward you and downward to a lower frequency after the truck passes. Your friend has used his sense of pitch to estimate the difference between the "coming" and "going" frequencies.

SKETCH THE PROBLEM

The problem is described by parts B and C of Figure 13.15.

IDENTIFY THE RELATIONSHIPS

When the truck is approaching, the observed frequency is (from Eq. 13.19)

$$f_{coming} = \frac{f_{source}}{1 - (v_{source}/v_{sound})}$$

and as it moves away the frequency is (from Eq. 13.20)

$$f_{going} = \frac{f_{source}}{1 + (v_{source}/v_{sound})}$$

Your friend senses the difference $\Delta f = f_{coming} - f_{going}$, which is given by

$$\Delta f = \frac{f_{source}}{1 - (v_{source}/v_{sound})} - \frac{f_{source}}{1 + (v_{source}/v_{sound})} \qquad (1)$$

SOLVE

We now want to solve for f_{source}. Rearranging Equation (1) leads to

$$\Delta f = f_{source}\left[\frac{1}{(1 - v_{source}/v_{sound})} - \frac{1}{1 + (v_{source}/v_{sound})}\right]$$

$$f_{source} = \frac{\Delta f}{\left[\frac{1}{(1 - v_{source}/v_{sound})} - \frac{1}{1 + (v_{source}/v_{sound})}\right]}$$

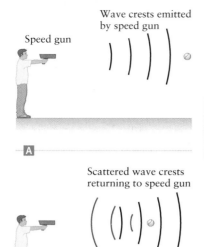

Speed gun

Wave crests emitted by speed gun

A

Scattered wave crests returning to speed gun

B

Speed gun detecting returning wave crests.

99

C

Figure 13.16 A "speed gun" (sometimes also called a "radar gun") uses the Doppler effect to measure the speed of a moving object. A Waves are emitted by the speed gun and strike the object. B These waves reflect from the object, and some of the reflected waves return to the speed gun. C The speed gun measures the difference between the frequencies of the emitted and reflected waves.

Figure 13.17 Some bats use reflected sound to judge the location and motion of nearby objects, including their prey.

Inserting the given values of Δf, v_{source}, and v_{sound}, we get

$$f_{\text{source}} = \frac{50\text{ Hz}}{\left[\dfrac{1}{1 - [(20\text{ m/s})/(343\text{ m/s})]} - \dfrac{1}{1 + [(20\text{ m/s})/(343\text{ m/s})]}\right]} = \boxed{430\text{ Hz}}$$

What does it mean?

The Doppler shift can also be used to measure the speed of the source, as we explore in the next subsection.

Speed Guns, Bats, and the Doppler Effect

So far, we have discussed the Doppler effect using the example of sound waves from a moving siren, but the Doppler effect applies to other types of waves as well. For example, a "speed gun" used in sporting events such as baseball (Fig. 13.16) uses the Doppler effect with reflected electromagnetic waves to measure the speed of a moving object. The electromagnetic waves from the speed gun are reflected by the ball, and some of the reflected (i.e., scattered) waves then return to the speed gun. (Compare with the scattered waves in Fig. 13.14.) As the waves strike it, the ball acts like a moving "observer," whereas for the reflected waves, it acts like a moving source. The motion of the ball thus causes the reflected wave crests to be more closely spaced than if the ball were at rest, just as we found with the moving siren in Figure 13.15B, and there is again a Doppler shift. The speed gun uses the Doppler-shifted frequency of the reflected wave to deduce v_{source}, the speed of the ball.

Bats employ an approach similar to that used by the speed gun to help them navigate as they fly (Fig. 13.17). Bats are most active at night, so they cannot use light (i.e., conventional vision) to judge their surroundings. Instead, bats generate pulses of sound waves at frequencies well above the range of human hearing (in the ultrasonic range). These ultrasonic sound pulses are reflected by objects in the bat's environment, and the bat uses the reflected sound to judge the location and motion of nearby objects. Bats have a very keen sense of timing and are able to determine accurately the time delay between an emitted pulse and its reflection. A long reflection time means the object is far away, and a small reflection time means the object is nearby (similar to the way radar works as discussed in Chapter 12). In addition to the reflection time, some bats also use the Doppler effect.[3] Suppose the bat's ultrasonic sound wave is reflected by a moth moving through the air.[4] The reflected sound is then Doppler-shifted, and the bat uses this frequency shift to sense the moth's speed and obtain a meal.

Moving Sources and Shock Waves

In all our examples involving moving sources of sound, the speed of the source has (so far) been much less than the speed of sound, but it is also possible for the speed of a wave's source to be greater than the speed of the wave. For example, a jet can fly faster than the speed of sound, which is sometimes referred to as the "sound barrier" (although it is not a barrier at all). Figure 13.18 shows the wave crests of sound emitted by a hypothetical jet in three cases. Figure 13.18A shows the behavior when the jet's speed is less than the speed of sound; this arrangement of wave crests is the one we considered in our analysis of the Doppler effect in Figure 13.15. Figure 13.18B shows what happens when the jet's speed is equal to the speed of sound. The

---

[3]You can learn more about bats and the Doppler effect in C. F. Moss and S. R. Sinha, "Neurobiology of echolocation in bats," *Current Opinion in Neurobiology* 13:751–758 (2003).

[4]Because the bat is also moving, there is also a Doppler effect due to the bat's motion. The bat is able to compensate for Doppler shifts caused by his motion and isolate the shift due solely to the moth's motion.

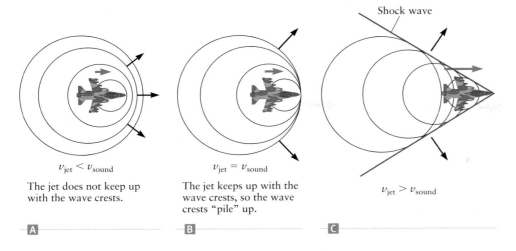

Figure 13.18 Wave fronts emitted by a moving source. **A** The speed of the source is smaller than the speed of sound, leading to the Doppler effect (Fig. 13.15). **B** When the speed of the source is equal to the speed of sound, the leading part of the wave front never "escapes" from the source. **C** When the speed of the source is greater than the speed of sound, shock waves are emitted. The direction of the shock wave depends on the ratio of v_{source}/v_{sound}.

$v_{jet} < v_{sound}$

The jet does not keep up with the wave crests.

A

$v_{jet} = v_{sound}$

The jet keeps up with the wave crests, so the wave crests "pile" up.

B

Shock wave

$v_{jet} > v_{sound}$

C

jet now travels at the same speed as the wave crests, so the crests in front are not able to move away from it. The wave crests at the front of the jet thus "pile up" at this point. According to the principle of superposition, the total sound pressure is the sum of the pressures due to the individual sound waves, so there is a very large sound intensity at the front of the jet. Figure 13.18C shows the situation when the jet moves faster than the speed of sound. Sound waves emitted at past times pile up along a conical envelope (which appears as two sides of a triangle in the projection in Fig. 13.18C). At any point on this envelope, many different sound waves, emitted at many different previous times, arrive simultaneously, producing a very large sound intensity and forming a **shock wave**. Such a shock wave is the source of the **sonic booms** that can be heard when a supersonic jet passes overhead.

13.7 | APPLICATIONS

Applications involving sound are found in many places, including the home (such as the earphones of your MP3 player), in scientific research, and in medicine.

Using Sound to Study Global Warming

Table 13.1 lists the speed of sound in some typical materials at room temperature. The speed of sound depends on the material's temperature. For example, sound travels through salt water with a typical speed of $v_{sound} \approx 1500$ m/s, and its value increases by approximately 4.0 m/s when the temperature is increased by 1°C. This effect is now being used in studies of the oceans. Many scientists believe that the Earth's average surface temperature may be increasing, an effect known as "global warming." Because changes in the Earth's average temperature as small as 1°C can greatly affect life on this planet, it is very important to confirm that global warming is actually occurring and, if so, to measure the rate of warming. Studies of global warming are very difficult due to natural fluctuations in the Earth's temperature from place to place and from year to year. For example, the fluctuations of the average temperature over the surface of the entire Earth are typically 0.2°C from one year to the next.

A clever approach called "acoustic thermometry" has been developed to study global warming. This approach uses the temperature dependence of the speed of sound to measure changes in the temperature of the oceans. Special sound generators and detectors have been developed to send sound waves through the oceans across very large distances. For example, sound generated in the Pacific Ocean near Hawaii can be detected when it reaches the coast of California (Fig. 13.19).

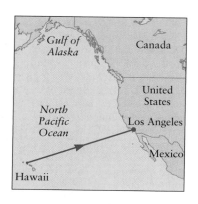

Figure 13.19 Sound can travel for very large distances through the oceans. Sound emitted in Hawaii can be detected in Los Angeles and is used to study the temperature of the Pacific Ocean.

The speed of sound in salt water is approximately 1500 m/s and the distance from Honolulu to Los Angeles is about 4100 km, so it takes about

$$t = \frac{L}{v_{\text{sound}}} = \frac{4.1 \times 10^6 \text{ m}}{1500 \text{ m/s}} = 2700 \text{ s} \approx 45 \text{ min}$$

for sound to travel this distance. The speed of sound depends on temperature, and because the sound travels over a very long distance, the travel time will depend on the average temperature of the seawater over this distance. Global warming would produce changes in this average temperature, so acoustic thermometry is ideal for studies of global warming. The travel time for sound can be measured very accurately; current precision approaches 20 ms (0.02 s), making it possible to measure very accurately how the average ocean temperature varies over the course of days and years. These experiments are in their initial stages and should yield important clues for understanding global warming.

EXAMPLE 13.7 Temperature Changes Measured by Acoustic Thermometry

What is the smallest temperature change that can be measured using acoustic thermometry? Assume the sound generator and detector are located near Honolulu and Los Angeles, and are thus separated by $L = 4100$ km, and the sound travel time can be measured with an accuracy of 0.02 s. The speed of sound in seawater varies by 4.0 m/s for each 1.0°C change in temperature.

RECOGNIZE THE PRINCIPLE

The travel time t is related to the travel distance and the speed of sound by $t = L/v_{\text{sound}}$. To change the travel time t by 0.02 s, there must be a corresponding change in v_{sound}. This change in v_{sound} can then be used to find the change in temperature.

SKETCH THE PROBLEM

No figure is needed.

IDENTIFY THE RELATIONSHIPS

On an "average" day, the travel time for this example is approximately

$$t = \frac{L}{v_{\text{sound}}} = \frac{4.1 \times 10^6 \text{ m}}{1500 \text{ m/s}} = 2700 \text{ s}$$

Now suppose the experiment is performed on two consecutive days and the travel times are measured to be $t_1 = 2700.00$ s and $t_2 = 2700.02$ s. We can rearrange the relation between t and v_{sound} to solve for the value of v_{sound} on each day. The results are

$$v_1 = \frac{L}{t_1} = \frac{4.1 \times 10^6 \text{ m}}{2700.00 \text{ s}} = 1518.519 \text{ m/s}$$

and

$$v_2 = \frac{L}{t_2} = \frac{4.1 \times 10^6 \text{ m}}{2700.02 \text{ s}} = 1518.507 \text{ m/s}$$

Here we have kept extra significant figures because we now need to compare v_1 and v_2. The difference in these velocities is

$$v_1 - v_2 = 0.012 \text{ m/s}$$

SOLVE

This velocity difference corresponds to a temperature change ΔT of

$$\Delta T = (0.012 \text{ m/s}) \times \frac{1.0°C}{4.0 \text{ m/s}} = \boxed{0.003°C}$$

What does it mean?
Acoustic thermometry can thus be used to measure *extremely* small changes in the average temperature of the ocean.

Imaging with Ultrasound ⊗

Another important application of sound is the technique of **ultrasonic imaging**. This technique uses sound waves to obtain images inside a material. The material could be virtually anything (such as an airplane wing), but the most familiar use of ultrasonic imaging produces views inside the human body. An instrument that generates sound is placed on the skin (Fig. 13.20A), and sound waves propagate into the body. A portion of the sound energy is reflected by bones and other tissue, and the reflected waves are used to construct an image (Fig. 13.20B). The depth of the bone (or whatever causes the reflection) is determined by the time it takes the sound to travel from the generator to the bone and then return to the detector.

Some medical ultrasonic imaging devices also use the Doppler effect. When ultrasound is used to construct an image of a blood vessel, sound waves are reflected by cells moving in the blood. The frequency of these reflected sound waves is Doppler-shifted as we found with the speed gun in Figure 13.16. By measuring this Doppler shift, physicians can measure the blood velocity (see Problem 73 at the end of this chapter). This information can be used to determine if blood is flowing properly and can reveal the presence of blockages or constrictions.

Figure 13.20 The method of ultrasonic imaging, also called ultrasound, uses reflected sound waves to reveal structures inside living tissue. **A** A transducer (a device that can both emit and detect sound) is placed on the abdomen of a pregnant woman. **B** Reflected waves are detected by the transducer and used to construct an image of the fetus.

CONCEPT CHECK 13.6 | Ultrasonic Imaging ⊗

A typical ultrasonic imaging device employs sound with a frequency of 10 MHz. If the speed of sound in tissue is approximately 1500 m/s, is the wavelength of these sound waves (a) 1.5 mm, (b) 0.15 mm, or (c) 1.5 cm?

SUMMARY | Chapter 13

KEY CONCEPTS AND PRINCIPLES

Sound is a longitudinal wave
The molecular motion associated with a sound wave is parallel to the wave velocity, so sound is a longitudinal wave. Sound can also be viewed as a **pressure wave**. Regions of high pressure are regions of condensation, whereas low-pressure regions are regions of rarefaction. The frequency and wavelength of a periodic sound wave are related by

$$v_{sound} = f\lambda \qquad \text{(13.1) (page 407)}$$

The speed of sound depends on the properties of the medium and is generally high in solids and low in gases. The speed of sound in air at atmospheric pressure and room temperature is approximately 343 m/s.

Human hearing
The normal range of human hearing is approximately 20–20,000 Hz. Sounds with lower frequencies are called *infrasonic*, and sounds with higher frequencies are *ultrasonic*.

(Continued)

Sound intensity

The *intensity* of a sound wave is proportional to the square of the amplitude. Typically, the human ear can hear sounds with intensities as small as 1×10^{-12} W/m^2, which corresponds to a pressure amplitude of about 2×10^{-5} Pa. Sound intensity is often measured in *decibels* (dB). The intensity level of a sound in decibels is given by

$$\beta = 10 \log\left(\frac{I}{I_0}\right) \qquad \textbf{(13.9)} \text{ (page 411)}$$

where $I_0 = 1.00 \times 10^{-12}$ W/m^2. The human ear can function (without damage) at intensity levels up to about 120 dB.

APPLICATIONS

Sound in pipes

Sound propagating back-and-forth within a pipe forms *standing waves* when the wavelength of the sound (or a multiple of the wavelength) matches the length of the pipe. Standing waves have pressure *nodes* where the pressure amplitude is zero and pressure *antinodes* where the pressure amplitude is largest. The open end of a pipe is always a pressure node, while the closed end is always a pressure antinode.

For a pipe closed at both ends, the possible standing waves have frequencies

$$f_n = n\,\frac{v_{\text{sound}}}{2L} = nf_1 \quad \text{with } n = 1, 2, 3, \ldots$$

$$\textbf{(13.11)} \text{ (page 415)}$$

where $f_1 = v_{\text{sound}}/(2L)$ is called the *fundamental frequency*, the lowest possible standing wave frequency.
For a pipe that is closed at one end and open at the other (as in many musical instruments, such as a clarinet), the standing wave frequencies are

$$f_n = n\,\frac{v_{\text{sound}}}{4L} = nf_1, \quad \text{with } n = 1, 3, 5, \ldots \qquad \textbf{(13.14} \text{ and } \textbf{13.15)} \text{ (page 416)}$$

with $f_1 = v_{\text{sound}}/(4L)$.

BEHAVIOR OF PRESSURE FOR STANDING WAVES

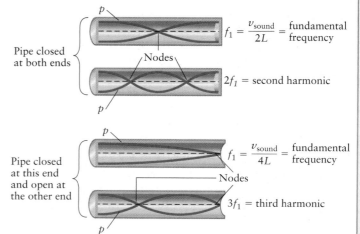

Pipe closed at both ends

$f_1 = \dfrac{v_{\text{sound}}}{2L} =$ fundamental frequency

Nodes

$2f_1 =$ second harmonic

Pipe closed at this end and open at the other end

$f_1 = \dfrac{v_{\text{sound}}}{4L} =$ fundamental frequency

Nodes

$3f_1 =$ third harmonic

Beats

When two sound waves that have nearly the same frequency are combined, a listener will hear *beats*. Beats are an oscillation of the overall intensity of the sound at a beat frequency equal to $|f_1 - f_2|$, where f_1 and f_2 are the frequencies of the two sound waves.

Doppler shift

When the source of a sound is moving, the frequency observed by a listener is *Doppler-shifted*. If the frequency of the source when stationary is f_{source}, then when the source moves at a speed v_{source} and the listener moves at a speed v_{obs}, the frequency heard by the listener is

$$f_{\text{obs}} = f_{\text{source}}\left[\frac{1 \pm (v_{\text{obs}}/v_{\text{sound}})}{1 \pm (v_{\text{source}}/v_{\text{sound}})}\right] \qquad \textbf{(13.22)} \text{ (page 423)}$$

The choices of plus and minus signs in Equation 13.22 depend on whether the source and observer are moving toward or away from each other and can always be remembered by noting that the frequency measured by the observer f_{obs} is increased when the observer and source are moving toward each other, whereas f_{obs} is decreased when they move away from each other.

SSM = answer in Student Companion & Problem-Solving Guide   ⊗ = life science application

1. Rank each substance according to how fast sound travels in that medium, from that with the largest velocity to that with the slowest.
   (a) Air
   (b) Water
   (c) Steel
   (d) Aluminum

2. ⊗ A typical dog can hear sounds with frequencies as high as 45 kHz. Assuming they can hear sounds as low as 40 Hz, make a qualitative sketch of the sensitivity curve for a dog (i.e., the canine version of the dashed curve in Fig. 13.4).

3. Musicians who play the guitar and other stringed instruments use beats to tune their instruments. The value of the beat frequency tells them the difference in the frequencies of two strings, but it does not tell them which one is higher (or lower). Describe an experiment musicians could do to resolve this question.

4. Two sound waves of equal frequency interfere constructively at a listener. Which of the following statements is true?
   (a) The frequency is increased by the interference.
   (b) The sound amplitude is increased by the interference.
   (c) The listener will hear beats.

5. How is it possible for a singer to shatter a wine glass by singing a note? Which is more important in this process, the frequency or intensity of the singing, or are they both important?

6. In this chapter, we discussed the Doppler effect that occurs when a source of sound moves directly toward or away from a listener. Do you think there will be a Doppler shift if the velocity of the source is perpendicular to a line that runs to the listener (Fig. Q13.6)? Explain why or why not.

Figure Q13.6

7. ⊗ Explain why the sounds emitted by large animals usually have a lower frequency than those from small animals. *Hint*: Consider the standing waves in the vocal tract.

8. ⊗ **Ocean in a shell.** What is it that you hear when you hold a seashell (or any closed chamber like a cup or jar) to your ear? We are almost always surrounded by sound that spans every frequency (white noise). What happens if you cup your hands around an ear and vary the size of the chamber? The pitch should change as you vary the chamber. What are you hearing and why?

9. Consider the method of acoustic thermometry using the ocean. In Example 13.7, we saw that very small changes in the travel time of sound can be detected and can be used to observe very small changes in the ocean temperature. There are, however, complications that we did not mention in that example. It turns out that the velocity of sound in water depends on the salinity of the water. Ocean water contains about 3.5% salt by volume, and the speed of sound changes by about 1.4 m/s for each 0.1% change in the salt concentration. Investigate (using the Internet or some other reference) how much the salinity of ocean water fluctuates

over time and discuss how much this fluctuation will affect the measurement of temperature changes with this technique.

10. Use the general relation for the speed of sound in Equation 13.3 together with the ideal gas law (see Chapter 15 if you need a refresher on the ideal gas law) to show that the speed of sound in a gas can be written as $v_{sound} \propto \sqrt{P/\rho}$ where P is the pressure. Can you give an intuitive explanation of why v_{sound} increases as the pressure is increased?

11. The first television remote controllers (invented in the 1950s) used ultrasonic waves. Some people claimed they could change channels by simply jiggling their keys. Is that possible? Explain why or why not.

12. SSM ⊗ A buzzer generates sound by vibrating with a frequency of 440 Hz in air. If this buzzer is placed underwater, what frequency would a fish hear? How does placing the buzzer underwater affect the wavelength of the sound?

13. One day, while hiking in a particular spot in your favorite cave, you notice that it takes a certain time t for an echo to return to you. The next day, you find that the echo time at the same spot is shorter. How might that happen?

14. When a police car is at rest, its siren has a frequency of 750 Hz. You are a thief, and just after finishing a night's work, you hear a siren with a frequency of 700 Hz. Your fellow thief tells you to relax and assures you that there is nothing to worry about. Is she right?

15. A radio produces sound waves in air, but they can also travel from air into water so that they can be heard by a swimmer. Is the frequency of the sound in air the same or different from the frequency in water? Is the wavelength in air the same or different? Explain.

16. ⊗ The manufacturer of a set of earplugs advertises that they will reduce the sound intensity level by 40 dB. By what factor do they reduce the intensity?

17. SSM When you blow across the top of a cola bottle, you can generate sound (Figure Q13.17). As you drink more and more of the cola and lower the liquid level, how does the frequency of the sound change? What is the lowest frequency (approximately) you can generate in this way? Explain.

Figure Q13.17

18. When you tune a flute or trombone, what causes the frequency to change?

19. Noise-canceling headphones make use of destructive interference of sound waves to eliminate "noise," while still playing music. Explain how they work.

PROBLEMS

SSM = solution in Student Companion & Problem-Solving Guide   ⊗ = life science application
★ = intermediate   ✪ = challenging   ® = reasoning and relationships problem

13.1 SOUND IS A LONGITUDINAL WAVE

1. A sound wave in air has a frequency of 220 Hz. What is its wavelength?

2. The distance along a sound wave between a point of condensation and the nearest rarefaction is 0.45 m. What is the wavelength of this sound wave?

3. Consider a sound wave in air with a frequency of 440 Hz. What is the distance between a region of condensation and an adjacent region of rarefaction?

4. Suppose a sound wave in air at room temperature has a wavelength of 400 m. What is its frequency?

5. Two ships are traveling on a very dark and foggy night and cannot see each other, even though they have their lights on. They therefore use their foghorns to broadcast their presence. If the ships are 50 m apart, how long does it take the sound from one ship to reach the other?

6. A battleship is using sonar (reflected underwater sound signals) to detect the presence of nearby submarines. It is found that a sonar reflection has a round-trip travel time (from the battleship to the submarine and back) of 17 s. How far away from the battleship is the submarine?

7. ★ A sound wave in an unknown gas has a frequency of 440 Hz and a wavelength of 2.2 m. What type of gas might it be? *Hint*: Use Table 13.1.

8. SSM ★ Consider two sounds of frequency 300 Hz, one traveling in carbon dioxide gas and the other in hydrogen gas. What is the ratio of the wavelengths in the two cases?

9. The lowest note on a piano has a fundamental frequency of about 27 Hz. What is the wavelength of this sound in air?

10. ✖ The ultrasonic sound waves used by owls for echolocation typically have a frequency of approximately 100 kHz. What is the wavelength?

11. ★ Consider a sound wave of frequency f generated in air by a loudspeaker. There is a nearby lake, and when this sound wave reaches the lake, the sound wave propagates into the water, where it has the same frequency f. What is the ratio of the wavelength of the wave in the water to the wavelength in air?

12. ✪ An automobile wheel of diameter 0.50 m is found to emit sound with a frequency of 10 Hz. What is the automobile's speed? *Hint*: Assume the frequency of the emitted sound is equal to the rotation frequency of the wheel.

13. ★ In western movies, the hero sometimes detects an approaching train by "listening" to the railroad tracks. If the tracks are made of steel, how many times faster does sound travel in the tracks than in air?

13.2 AMPLITUDE AND INTENSITY OF A SOUND WAVE

14. ✖ Use the data in Figure 13.4 to determine the approximate frequency at which the human ear is most sensitive to sound.

15. ✖ Use Figure 13.4 to find the sensitivity of the human ear at 50 Hz. That is, what is the lowest sound intensity the ear can detect at this frequency?

16. ★ A sound wave has an intensity level of 80 dB. What is the pressure amplitude of this wave?

17. SSM ✪ ✖ ℝ You are at a rock concert, and the sound intensity reaches levels as high as 130 dB. The sound pressure produces an oscillating force on your eardrum. Estimate the amplitude of this oscillating force.

18. ✪ ✖ The human auditory system is able to detect changes in the sound intensity level as small as 1 dB. What is the corresponding percentage change in the intensity I?

19. ★ To decrease the intensity of a wave by a factor of four, what must you do?
(a) Decrease the pressure amplitude by a factor of two.
(b) Decrease the pressure amplitude by a factor of four.
(c) Decrease the pressure amplitude by a factor of eight.
(d) Decrease the pressure amplitude by a factor of $\sqrt{2}$.

20. ★ Two sounds differ in intensity level by 6.0 dB. What is the ratio of the intensities of these two sounds?

21. ★ If a sound intensity level increases by 16 dB, by what factor does the intensity change?

22. ✪ ✖ ℝ Estimate the amplitude of the force on your eardrum from a sound that has an intensity level of 90 dB.

23. A sound wave has an intensity of 2.5×10^{-3} W/m². What is the intensity level in decibels?

24. ★ A source of sound has a total emitted power of 300 W. What is the intensity a distance of 10 m from the source? Assume the source emits spherical waves.

25. ✪ Suppose the sound intensity level 0.10 m from a loudspeaker is 110 dB. Assuming the speaker generates wave fronts that have a hemispherical shape (Fig. P13.25), what is the intensity 5.5 m from the speaker?

Figure P13.25

26. ★ The sound intensity level 20 m from a particular loudspeaker is 70 dB. Estimate the total power emitted by the speaker. Assume it emits spherical waves.

27. ★ What is the pressure amplitude of a sound wave if the intensity level is 40 dB?

28. ★ What is the intensity level of a sound wave if the pressure amplitude is 7.9 Pa?

29. ★ A point source emits sound with a power of 250 W. What is the intensity level in decibels at a distance of 35 m?

30. SSM ★ ✪ A person has severe hearing loss that reduces the intensity in his inner ear by 40 dB. To compensate, a hearing aid can amplify the sound pressure amplitude by a certain factor to bring the intensity level back to the value for a healthy ear. What amplification factor is needed for this person?

13.3 STANDING SOUND WAVES

31. For a standing sound wave, the closed end of an organ pipe forms which of the following? (More than one answer may be correct.)
(a) A displacement node
(b) A displacement antinode
(c) A pressure node
(d) A pressure antinode

32. For a standing sound wave, the open end of a flute forms which of the following? (More than one answer may be correct.)
(a) A displacement node
(b) A displacement antinode
(c) A pressure node
(d) A pressure antinode

33. Consider an organ pipe 1.2 m long that is closed at one end and open at the other. What is the fundamental frequency of the pipe?

34. The distance between a node and the nearest antinode of a standing sound wave in air is 0.24 m. What is the frequency of this wave?

35. If an organ pipe is filled with oxygen gas instead of air, will the frequency of the note go up or go down, and by what factor?

36. SSM ★ An organ pipe is open at one end and closed at the other. The pipe is designed to produce the note middle C (262 Hz). (a) How long is the pipe? (b) What is the frequency of the second harmonic? (c) How many pressure nodes of the second harmonic are found inside the pipe? Do not count nodes that may be at the ends of the pipe.

37. ★ A piccolo can be approximated as a tube that is open at both ends. If a piccolo is 20 cm long, what is the frequency of the lowest note it can play?

38. ✪ Consider an organ pipe that is closed at one end and open at the other. This pipe has resonant frequencies of 1200 Hz and 1500 Hz (among others). Find two possible values for the length of the pipe.

39. ✪ A pipe is open at both ends. The pipe has resonant frequencies of 528 Hz and 660 Hz (among others). Find two possible values for the length of the pipe.

40. ✪ ℝ An organ pipe that is closed at one end has a fundamental frequency of 175 Hz. There is a leak in the church roof, and some water gets into the bottom of the pipe as shown in Figure P13.40. The organist then

Figure P13.40

finds that this organ pipe has a frequency of 230 Hz. What is the depth of the water in the pipe?

41. Consider a steel rod that vibrates longitudinally. If the rod is 1.5 m long, what is the fundamental frequency? Assume the ends of the rod are displacement antinodes (i.e., it behaves as an organ pipe with both ends open).

42. ✪ A pipe has a pattern of standing wave frequencies given in Figure P13.42. Is this pipe (a) open at both ends, (b) closed at both ends, or (c) open at one end and closed at the other? What is the length of the pipe?

Figure P13.42

13.4 BEATS

43. Two sound waves of frequency 80 Hz and 83 Hz are played. How many beats per second are heard?

44. A guitarist plucks two strings simultaneously. One string is tuned to a (fundamental) frequency of 340.0 Hz. It is found that beats are audible, with a beat frequency of 0.5 Hz. What are possible values for the frequency of the other string?

45. SSM ☆ Two similar guitar strings have the same length and mass, but slightly different tensions. If the beat frequency is 1.5 Hz, what is the ratio of the tensions of the two strings? Assume they are both tuned to approximately 330 Hz. Give your answer to three significant figures.

46. ☆ A violin string tuned to 485.0 Hz is found to exhibit 2.0 beats per second with a guitar string. The guitar string is then tightened, and the beat frequency is found to increase. What was the original frequency of the guitar string?

47. ☆ A person can detect a beat rate as slow as 0.3 beat per second at a frequency of 1500 Hz. What is the percentage difference in frequency of the two notes?

48. ✪ Ⓡ One tuning fork vibrates at 440 Hz, and a second tuning fork vibrates at an unknown frequency. When the two forks are struck at the same time, beats with a frequency of 3 Hz are heard. (a) What is the frequency of the second tuning fork? Why are there two possible answers? (b) Describe an experiment you could perform with a small piece of wax to determine which of the possible answers in part (a) is correct.

13.5 REFLECTION AND SCATTERING OF SOUND

49. Ⓡ The sound from a church organ "reverberates" for many seconds after the organ stops producing sound due to reflections inside the church. That is, the sound from the organ bounces back-and-forth off the church walls many times before the sound intensity decays away completely. If the reverberation time is 10 s and the church has the (approximate) shape of a square box of length 25 m, approximately how many times does the sound reflect back-and-forth during the reverberation time?

50. SSM ✗ An owl is chasing a squirrel and is using echolocation (the reflection of sound) to aid in the hunt. If the squirrel is 25 m from the owl, how long does it take sound to travel from the owl to the squirrel and then back to the owl?

13.6 THE DOPPLER EFFECT

51. A siren has a frequency of 750 Hz when it is at rest, whereas it has an apparent frequency of 800 Hz when it is moving toward a stationary observer. What is the speed of the siren in the latter case?

52. SSM ☆ A siren has a frequency of 950 Hz when it and an observer are both at rest. The observer then starts to move and finds that the frequency he hears is 1000 Hz. (a) Is the observer moving toward or away from the siren? (b) What is the speed of the observer?

53. A guitar is tuned to play a note at 440 Hz. If the guitar is in a moving car and the note has a perceived frequency of 410 Hz, what is the speed of the car? Assume the car is moving directly away from a listener on the sidewalk.

54. A siren is approaching you at 35 m/s and is perceived to have a frequency of 1100 Hz. If the siren stops moving, what frequency do you now hear?

55. ✪ The siren in an ambulance has a frequency (according to you) of 1100 Hz as it moves toward you at a speed v. It then makes a U-turn and travels away from you with a speed v, and you hear a frequency of 950 Hz. What is the final speed v of the siren?

56. ☆ ✗ A bat is pursuing an insect and using echolocation of 60-kHz sound waves to track the insect. The insect is initially 30 cm from the bat, but then moves to a distance of 40 cm as it tries to escape. What is the difference in the two echo times the bat measures?

57. ✪ ✗ Consider again the insect in Problem 56. If the insect moves the 10-cm distance in 0.50 s, what is the magnitude of the Doppler shift in the bat's reflected sound wave? Assume the insect is moving directly away from the bat.

58. ✪ Two identical police cars are chasing a robber. When at rest, their sirens have a frequency of 500 Hz. A stationary observer watches as the two cars approach. The siren of one car (car 1) has a frequency of 600 Hz, and the other (car 2) has a frequency of 700 Hz. (a) Which car is moving faster? (b) What is the speed of each car? (c) If the robber hears a frequency of 650 Hz from car 2, is car 2 catching up to the robber? (d) What is the speed of the robber?

59. ✪ An ambulance with a siren emitting a whine at 1200 Hz overtakes and passes a cyclist pedaling a bike at 2.5 m/s. After being passed, the cyclist hears a frequency of 980 Hz. How fast is the ambulance moving?

13.7 APPLICATIONS

60. Ⓡ Some climate data suggest that during the period 1950 to 2000, the Earth's surface has warmed an average of about 1°C. If the oceans have warmed this amount, how much will this change the time it takes a sound wave to travel from San Francisco to Honolulu?

61. SSM ☆ A child drops a rock into a vertical mineshaft that is precisely 406 m deep. The sound of the rock hitting the bottom of the shaft is heard 10.3 s after the child drops the rock. What is the temperature of the air in the shaft? Assume the temperature is the same throughout the mine.

62. ☆ ✗ The human ear canal (Fig. 13.3) is typically about 2.4 cm long, is roughly cylindrical, and is closed at one end where it is capped by the eardrum. (a) What is the fundamental frequency of the ear canal? (b) Determine the range of frequencies where the ear is most sensitive (see Fig. 13.4). How does the ear canal's fundamental frequency compare?

63. ☆ ✗ While taking a walk around your neighborhood, you see a flash of lightning hit the Earth in the distance. You immediately look at your watch and then hear thunder approximately 9 s later. (a) How far away did the lightning strike the ground? (b) Find a conversion factor for a thunder delay (measured in seconds counted) to the distance away of a corresponding lightning strike (in kilometers).

64. SSM ☆ ✗ The lowest frequency an average human ear can perceive is about 15 Hz, and the highest about 18 kHz. How tall would an organ pipe have to be to produce the lowest note? How short a tube would be needed to play the highest note?

What harmonic (i.e., what value of *n* in Eq. 13.15) would the smaller tube excite in the larger one?

65. ✪ ✗ **Fundamental speech.** The sound of the talking or singing voice is produced by the vocal cords, which are located in an open-ended pipe called the vocal tract (Fig. P13.65A). Linguists have found that consonant and vowel sounds are comprised of a spectrum of certain emphasized frequencies called *formants*. The formants for three common vowel sounds are illustrated in Figure P13.65B. (a) Estimate the fundamental frequency of the human vocal tract, assuming that it is about 25 cm long. (b) Consider the harmonics involved in producing the sounds of the vowels of the "oo" as in *pool*, the "ah" as in *father*, and the "ee" as in *heed*. Rank the formants in terms of the fundamental frequency and its harmonic overtones in each vowel sound from the most dominant to the least. The vocal tract changes shape for each vowel, so your calculated frequencies are approximate.

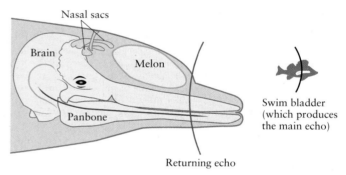

Figure P13.65 Ⓐ The human vocal tract. Ⓑ Loudness versus frequency for three common vowel sounds. Gray bands show location of formants.

66. ✪ **Drop in pitch.** A tuning fork that rings at 440 Hz is dropped down a deep well. How far down the well is the fork when the pitch is at 425 Hz? Don't forget to include the time it takes the sound to propagate to the top of the well.

67. ✪ Consider the falling tuning fork in Problem 66 with frequency 440 Hz. (a) You hear it splash into the water at the bottom of the well after 1.8 s. How deep is the well? (b) What frequency is heard just before the splash?

68. ✪ On a summer afternoon, a pipe that is open at both ends is held vertically with one end submerged in water as shown in Figure P13.68. If a 220-Hz tuning fork is held over the end of the tube, it induces resonance when the top of the tube is $h_1 = 40$ cm from the surface of the water and when it is $h_2 = 120$ cm from the surface. (a) What was the velocity of sound in the air at that time on that hot summer day? (b) What was the air temperature? Is this method a good way to measure temperature? Why or why not?

Figure P13.68

69. ✪ You buy a dog whistle that operates at 21 kHz, but when you blow into it, your dog does not show the slightest reaction. You decide to see if you can hear the whistle yourself. Two friends help you. One drives her car past you while the other blows the whistle from the car's window. (a) How fast

must your friend drive the car past you to allow you to hear the whistle, assuming you can hear a maximum frequency of 19 kHz? Give your answer in meters per second and miles per hour. (b) If the whistle instead operates at 24 kHz, what speed would the car then need? Would this experiment be feasible? How could you do the experiment with two cars?

70. ✪ ✗ Dolphins use sonar as a sixth sense, giving them a three-dimensional view of their surroundings via sound. A dolphin's hearing spans the range from 15 Hz to 150 kHz, making use of the 75–110 kHz range for its echolocation sonar. Sound is emitted in a narrow beam using an organ called the melon (Fig. P13.70), which has a dense core (high sound velocity) compared with the surrounding tissue. The melon acts as an acoustic lens that focuses the sound wave pulse, and the dolphin's jaw acts as a receiver that channels the appropriate frequencies of the reflected pulse to the ear. (a) Consider a dolphin chasing a fish as in Figure P13.70. If the fish is 50 cm away, how long does it take a wave pulse to travel from the dolphin to the fish and back? (b) What is the wavelength of the sonar pulse if the frequency of the emitted sound is 90 kHz? How does this wavelength compare to the size of the fish? Why is that important? (c) Suppose the dolphin is in pursuit of the fish, with the fish swimming at a speed of 10 m/s and the dolphin swimming at 25 m/s. What frequency for the return pulse does the dolphin hear? (d) The dolphin can sense the speed of the fish, not by the small shift in frequency away from 90 kHz, but by emitting a series of pulses that sound like clicks. Each click is a pulse of 90-kHz sound with a duration of about 50 μs. Such clicks are emitted at a rate of 900 per second. What click rate would a dolphin detect reflected from the fish in part (c)?

Figure P13.70

71. ★ Ⓡ A trumpet at room temperature (25°C) is tuned to play the note middle C, which has a fundamental frequency of 262 Hz. The trumpet is then taken outside, and after a few minutes, it is found that the same note has a frequency of 245 Hz. What is the outside temperature?

72. ★ Two adjacent notes in the musical scale used in Western music have fundamental frequencies whose ratio is approximately 1.059. For example, the fundamental frequencies of the notes C-sharp and C have this ratio. The frequency of a note produced by a flute, or any other woodwind instrument, depends on the temperature. How much must the temperature change to change the note C into C-sharp?

73. ✪ ✗ A blood flowmeter uses the Doppler effect to measure the speed of blood. Suppose a sound wave at 80,000 Hz is emitted into a vein where the speed of sound is about 1500 m/s. The wave is Doppler-shifted as it is absorbed by the red blood cells moving toward a stationary receiver. The wave is then Doppler-shifted again as it is reflected back toward the detector. Hence, there are two Doppler shifts, one as the red blood cells act as the "observer" and another as they act as a new source of sound. If the frequency of the returning sound is 80,020 Hz, what is the speed of the blood flow?

Temperature and Heat

These hot springs at Yellowstone National Park are a good example of heat flow and phase changes. Energy from the hot spring water is transferred to the rocks and to the air, and some of the water evaporates, turning into steam. (© Robert Harding Picture Library/ Alamy)

All our work on mechanics in Chapters 2 through 13 has been based on Newton's laws of motion. The central thread of that work was *force*. We discussed various types of forces (such as gravity) and how these forces cause a particle to move in various situations. A careful consideration of Newton's laws led us to other important concepts such as work and energy, but the main theme was the concept of force. In this chapter and Chapters 15 and 16, we move to a different area of physics known as *thermodynamics*, and our focus now shifts to *energy*. The key pillar of thermodynamics is the principle of conservation of energy. Although conservation of energy certainly has an important place in mechanics, it will now play an even more central role. In a nutshell, thermodynamics is about the transfer of energy between *systems* of particles and about the way changes in the energy of a system affect its properties.

For thermodynamics, we need several new quantities to describe the properties of systems and their interactions. Two of these quantities—temperature and heat—are the main subjects of this chapter.

Oxygen
molecule

A   B

Figure 14.1 These two hypothetical balloons have the same volume and gas density. Two such systems having equal density, pressure, and temperature are said to be in the same state, even though the precise positions and velocities of the gas molecules are different.

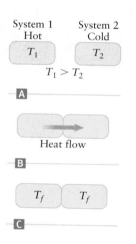

System 1   System 2
Hot   Cold
T_1 T_2

$T_1 > T_2$

A

Heat flow

B

T_f T_f

C

Figure 14.2 Heat (also referred to as heat energy) is energy transferred from one system to another due to a temperature difference. **A** Two systems, at different temperatures. **B** Energy flows from the hotter system at temperature T_1 to the cooler system at temperature T_2 when they are brought into contact. **C** If the systems are left in contact for a long time, they will eventually come into thermal equilibrium and have the same final temperature.

14.1 | THERMODYNAMICS: APPLYING PHYSICS TO A "SYSTEM"

In Chapter 2, we identified the key physical quantities associated with motion. Those quantities were position, velocity, and acceleration, which were all essential for our discussions of Newton's laws. We now consider the quantities to be used to describe the properties of a system composed of many particles. First, though, let's describe what we mean by a "system."

A thermodynamic *system* contains multiple particles, usually a very large number of particles. Examples are a balloon full of gas molecules, a drop of water, and a sugar cube, each of which contains many atoms and molecules. The particles in a system are able to exchange energy with one another (via collisions), and systems are able to exchange energy with other systems.

Figure 14.1A shows a balloon full of oxygen molecules. What quantities are useful for describing the physics of this system? We could describe the properties by specifying the positions and velocities of every oxygen molecule. We would then want to determine the forces exerted by each molecule in the balloon on all the other molecules in the balloon, along with all the external forces (e.g., the forces from outside the system such as gravity). With all this information, we could use Newton's laws to calculate the motion of each individual oxygen molecule. Although we could take this approach in principle, it is not feasible in practice because of the extremely large number of atoms or molecules in a typical sample of gas. Moreover, even if it were feasible, we are really not interested in this level of detail. We can describe all the important properties of this system in a much simpler way.

Figure 14.1B shows a second balloon that also contains oxygen gas. The two balloons in Figure 14.1 are "identical" in that they are the same size and shape and contain the same number of molecules. The positions and velocities of the molecules in the two balloons will certainly not be the same, however, so these two systems will not be identical on the molecular scale. Fortunately, we are usually not interested in knowing precisely where all the molecules are located. It is much more useful to know the values of various properties of the system *as a whole*.

To describe the physics of a system fully, only a small number of systemwide properties such as pressure and temperature are needed. Also called *macroscopic* properties, they describe the behavior on a scale that is much larger than the scale of an individual particle. Macroscopic properties contrast with *microscopic* variables such as the position and velocity of a particular molecule within the system. Newton's laws provide a description at the level of these microscopic variables. Our task in this chapter is to describe a system of many particles, such as a gas-filled balloon (Fig. 14.1) or a piece of steel, at the macroscopic level. We also explore the connection between microscopic and macroscopic descriptions and use Newton's laws to gain insight into the meanings of the macroscopic properties.

14.2 | TEMPERATURE AND HEAT

Temperature is not contained in or derivable from Newton's laws, which is one reason it is not easy to give a "simple" definition of temperature in the same way we gave definitions of velocity and acceleration in our work on mechanics. Temperature is a new type of quantity, requiring an entirely new unit of measure. We can approach the notion of temperature in several different ways, each yielding useful insights.

Temperature is connected with the "hotness" or "coldness" of a system, so let's first consider a macroscopic definition. Figure 14.2A shows two hypothetical systems, which might be balloons filled with gas. If each system is initially isolated from its surroundings for a long period of time, each will have its own temperature. In general, these two initial temperatures will not be the same, and we denote them by T_1 and T_2. Suppose $T_1 > T_2$; in ordinary language, we would say that system 1 is initially hotter than system 2. If we then bring the systems into contact (Fig. 14.2B), energy is transferred spontaneously from system 1 (the hotter system) to system 2 (the colder system), and their temperatures will change. If we wait for a long time (Fig. 14.2C), the two systems will reach **thermal equilibrium**. They then have the same final temperature T_f, which lies somewhere between T_1 and T_2. The energy that flows between the two systems is called **heat** or **heat energy**.

> **Heat** is energy that passes from one system to another by virtue of a temperature difference.

The terms *heat* and *heat energy* are often used interchangeably, especially in everyday conversation. When used in physics, these terms always refer to the transfer of energy between systems. Hence, you may also hear the term *heat flow* used to describe the process by which heat energy is transferred from one object to another. According to the principle of conservation of energy, the amount of heat energy that "leaves" system 1 in Figure 14.2B must equal the amount of heat energy that "enters" system 2. This transfer of energy can take place in different ways (as we'll describe in Sections 14.6 through 14.8), but the direction of transfer—the direction of heat flow—depends only on the temperature difference.

Units of Heat

Heat is measured with the SI unit of energy, the joule (J). A unit called the *calorie* is also widely used in measurements involving heat. The term *calorie* comes from the 1700s when some scientists thought that heat was a sort of fluid (called a "caloric fluid") that flowed between objects. Although it is not immediately obvious that the energy that flows into your finger when you touch a hot object is the same type of physical quantity as the kinetic energy of a falling brick, experiments in the mid-1800s using an apparatus like the one in Figure 14.3 showed that heat is, in fact, a form of energy.

While most physicists measure heat in units of joules, the calorie is widely used, especially in chemistry. These units are related by

$$1 \text{ cal} = 4.186 \text{ J} \tag{14.1}$$

A unit named the "Calorie" (uppercase "C") is used to measure the energy content of foods. This "food calorie" is actually equal to 1000 cal (lowercase "c"), so a careful scientist would call it a "kilocalorie."

Temperature: A Microscopic Picture

Our definition of temperature tells how to compare the temperatures of two different systems. We simply place the two systems in contact and then observe if energy is transferred from one to the other and in which direction (Fig. 14.2). This procedure, however, does not give us a way to measure the *value* of the temperature of a particular system. One way to measure the temperature is to place the system of interest in contact with a gas-filled container held at constant volume and density. The pressure in the gas is caused by collisions of the gas atoms and molecules with the walls of the container (Fig. 14.4). If the temperature is increased, the average speed of the gas atoms increases, causing the pressure to increase also. We'll see in Chapter 15 that for a dilute gas (one with a low density), the temperature is linearly proportional to the gas pressure, so measurement of the pressure gives a direct way to find the temperature (Fig. 14.5A). Such a device is called a **gas thermometer**.

The picture in Figure 14.4 gives insight into the microscopic meaning of temperature. The temperature of a system of particles (such as a gas) is related to the average particle speed. A high speed corresponds to a high temperature, and a low speed corresponds to a low temperature. Because high speeds also mean high kinetic energies, this relation suggests that the temperature of a system can be increased by adding energy to it. This is another example of the connection between temperature and energy. We'll derive the relation between temperature and energy for a gas in Chapter 15.

Figure 14.3 Apparatus used to measure the relation between heat energy and mechanical energy. See Insight 14.1.

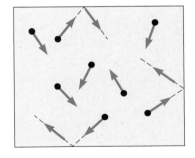

Figure 14.4 The pressure in a gas comes from collisions of the molecules with the surface of the container. Each collision produces a force on the walls of the container. If the temperature is increased, the molecular velocity increases, causing the collision forces and hence the pressure to increase.

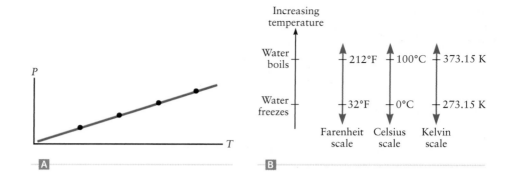

Figure 14.5 ▲ The pressure in a dilute gas (such as air) is proportional to the temperature. ▣ Temperature can be measured with the Fahrenheit, Celsius, or Kelvin scale.

Temperature Scales

The pressure in a gas thermometer is proportional to the temperature, but to make a practical thermometer, we must still decide how to calibrate the temperature axis in Figure 14.5A. Three different *temperature scales* are in common use: the *Fahrenheit* scale, the *Celsius* scale, and the *Kelvin* scale. The Kelvin scale is the one employed in the SI system of units, but all three scales are widely used, so you should be able to convert temperature values from one scale to any other. This conversion process can be understood from Figure 14.5B.

Two important reference points are the temperatures at which water freezes and boils. The Celsius scale is defined so that at atmospheric pressure, water freezes at exactly 0°C and boils at precisely 100°C. In words, these temperatures are "zero degrees Celsius" and "100 degrees Celsius." On the Fahrenheit scale, these processes occur at 32°F and 212°F, respectively, referred to as "32 degrees Fahrenheit" and so on. On the Kelvin scale, these temperatures are 273.15 K and 373.15 K, respectively. These values are referred to as "273.15 kelvin" and so on; by convention, the term *degrees* is not used in connection with the Kelvin scale.

The relations between the different temperature scales are illustrated in Figure 14.5B. Because all three scales are linear, the values of the temperatures at which water freezes and boils on each scale define how to convert a temperature value from one scale to another. For example, to convert from a value on the Celsius scale T_C to a value on the Kelvin scale T_K, we have (according to Fig. 14.5B)

Conversion between Kelvin and Celsius scales

$$T_K = T_C + 273.15 \quad \text{(conversion from Celsius to Kelvin)} \tag{14.2}$$

Notice that the sizes of a temperature unit are equal in the Celsius and Kelvin scales, so a change of 1 "degree" on the Celsius scale is equal to a change of 1 kelvin on the Kelvin scale.

Conversions involving the Fahrenheit scale are not quite as simple because a 1° change on the Fahrenheit scale is not equal to a 1° change on the Celsius scale. From Figure 14.5B, we see that a change of 100°C (going from freezing water to boiling water) corresponds to a change of $212 - 32 = 180°F$. A temperature change of 1°C is thus equivalent to a change of $180/100 = 9/5 = 1.8°F$. We also know that 0°C is equivalent to 32°F (the freezing point of water), which leads to the relation

Conversion between Fahrenheit and Celsius scales

$$T_F = \tfrac{9}{5}T_C + 32 \quad \text{(conversion from Celsius to Fahrenheit)} \tag{14.3}$$

| **EXAMPLE 14.1** | The Boiling Temperature of Liquid Nitrogen |

At atmospheric pressure, liquid nitrogen boils at a temperature of 77 K (Fig. 14.6). Convert this value to the Celsius and Fahrenheit scales.

RECOGNIZE THE PRINCIPLE

Given the temperature of liquid nitrogen, we can use Figure 14.5B and Equations 14.2 and 14.3 to find the corresponding temperatures on the Celsius and Fahrenheit scales.

SKETCH THE PROBLEM

Figure 14.5B shows the relationship between the Kelvin, Celsius, and Fahrenheit scales.

IDENTIFY THE RELATIONSHIPS

Rearranging Equation 14.2 to solve for the value on the Celsius scale (T_C) in terms of the Kelvin value, we have

$$T_C = T_K - 273$$

SOLVE

Inserting the value of T_K at the boiling point of nitrogen gives

$$T_C = T_K - 273 = 77 - 273 = \boxed{-196°C}$$

To get the value on the Fahrenheit scale, we use Equation 14.3:

$$T_F = \tfrac{9}{5}T_C + 32 = \tfrac{9}{5}(-196) + 32 = \boxed{-321°F}$$

What have we learned?
Equations 14.2 and 14.3 can be used to convert between any two temperature scales. If you forget the equations, you can always work out the conversions if you remember the values of the reference temperatures, the temperatures at which water freezes and boils.

Figure 14.6 Example 14.1. At atmospheric pressure, nitrogen boils at 77 K.

Very High and Very Low Temperatures: What Are the Limits?

The temperature axes in Figure 14.5B have arrows at both ends to indicate that the axes extend to both higher and lower temperatures. Many important temperature values lie well above the boiling point of water or well below the freezing point of water (Table 14.1). Several of these temperatures have negative values on either the Fahrenheit or Celsius scales (or both). Indeed, the boiling temperature of nitrogen, which was considered in Example 14.1, falls at such negative temperatures. However, there is a limit at the bottom ends of these scales; there are no negative values of temperature on the Kelvin scale listed in Table 14.1. The value $T_K = 0$ K is called *absolute zero*; in Chapter 16, we'll see that 0 K is a very special temperature.

Table 14.1 Values of Some Important Temperatures on the Kelvin, Celsius, and Fahrenheit Scales

	TEMPERATURE SCALE		
	K	°C	°F
Interior of the Sun	$\approx 1 \times 10^7$	$\approx 1 \times 10^7$	$\approx 2 \times 10^7$
Center of the Earth	≈ 7000	≈ 7000	$\approx 13{,}000$
Surface of the Sun	≈ 6000	≈ 6000	$\approx 10{,}000$
Lead melts	600	327	621
Water boils	373.15	100	212
Human body temperature	310	37	98.6
Water freezes	273.15	0	32
Nitrogen boils	77	−196	−321
Intergalactic space	3	−270	−454
Absolute zero	0	−273.15	−459.67

14.3 | THERMAL EQUILIBRIUM AND THE ZEROTH LAW OF THERMODYNAMICS

T_A T_B
A   B

T_C
C

ZEROTH LAW: If A and B are in thermal equilibrium and if B and C are in thermal equilibrium, then A and C are also in thermal equilibrium and $T_A = T_B = T_C$.

Figure 14.7 The zeroth law of thermodynamics involves the relationships between thermal equilibrium and temperature for three different systems.

In our discussion of temperature, we considered how heat flow between two systems depends on their temperatures (Fig. 14.2). Let's now extend that discussion to three different systems as sketched in Figure 14.7. We assume these three systems are initially isolated, with temperatures T_A, T_B, and T_C. Suppose we bring systems A and B into contact and find that there is no heat flow; that is, there is no transfer of energy between them. Using our definitions of temperature and heat, we would conclude that $T_A = T_B$. We then repeat this exercise with systems B and C, and again we find that there is no heat flow between these two systems; hence, $T_B = T_C$. What, then, would we say about the temperatures of A and C? The obvious conclusion is that $T_A = T_C$, which is called the *zeroth law of thermodynamics*.[1]

> *Zeroth law of thermodynamics:* If two systems A and B are in thermal equilibrium (so that $T_A = T_B$) and systems B and C are in thermal equilibrium ($T_B = T_C$), systems A and C are also in thermal equilibrium ($T_A = T_C$).

Although this statement seems almost trivial, it has some very important implications. In words, the zeroth law of thermodynamics tells us that the basic concept of temperature is meaningful and that temperature is a unique property of a system that is allowed to come into thermal equilibrium. Indeed, we took these facts for granted in our discussion of temperature and heat in Section 14.2. It is actually an amazing fact that *only one property* of a system, its temperature, determines when heat energy will flow between it and another system.

We can illustrate one implication of the zeroth law pictorially. Figure 14.8A shows an endless waterfall in which water circulates around a closed loop, flowing

Figure 14.8 🅰 In this famous lithograph *Waterfall* by M. C. Escher, water flows downhill endlessly around a closed loop. That is not possible in the real world because it is not consistent with the principle of conservation of energy. (It would violate the properties of gravitational potential energy.) 🅱 The zeroth law of thermodynamics states that this "circular" flow of heat energy between three different systems is not possible.

🅰

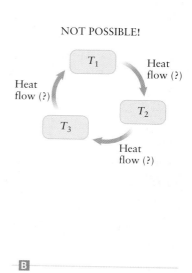

NOT POSSIBLE!

T_1
Heat flow (?)
Heat flow (?)
T_2
T_3
Heat flow (?)

🅱

---

[1]You might wonder why physicists would start counting such laws at the number zero. (After all, Newton did not have a zeroth law.) The logical need for the zeroth law of thermodynamics was realized only after the first, second, and third laws of thermodynamics were already discovered and named.

downhill during each part of the loop. Of course, such a waterfall is impossible because it would violate what we know about gravitational potential energy and the principle of conservation of energy. Figure 14.8B shows an "equivalent" (and still impossible) situation, this time involving heat flow. Here heat flows from system 1 to system 2, then flows from system 2 to system 3, and finally completes the loop by flowing from system 3 to system 1. The zeroth law and our understanding of heat flow tell us that such a circular flow of heat is impossible because heat must always flow from high temperature to low temperature. Figure 14.8 demonstrates again a connection between heat and other forms of energy.

14.4 | PHASES OF MATTER AND PHASE CHANGES

The three phases of matter—*solid*, *liquid*, and *gas*—are illustrated in Figure 14.9. The atoms in many solids are arranged in an orderly and repeating pattern called a crystalline lattice (Fig. 14.9A), in which each atom is held in place by the forces exerted by neighboring atoms. These forces are a result of the chemical bonds within the solid. For example, ionic bonds hold Na^+ and Cl^- within a lattice to form sodium chloride (NaCl), and covalent bonds hold silicon (Si) atoms in a crystalline lattice. The bonds between neighboring atoms in a solid behave as "springs," and the bond forces follow the mathematical form given in Hooke's law (Chapter 11). This force depends on the separation between atoms; if a particular bond is stretched or compressed, the atom experiences a restoring force proportional to its displacement. This force is directed back toward the "ideal" location of the atom within the lattice.

Figure 14.9A is a simplified picture, showing all the atoms in their ideal positions in the lattice. More realistically, the atoms vibrate about these positions as simple harmonic oscillators. The kinetic and potential energies associated with these oscillations can be understood using the ideas we encountered in Chapter 11.

Figure 14.9B shows another example of a solid phase called an amorphous solid. A familiar example is common window glass, which is composed mainly of silicon and oxygen. The atomic arrangement in an amorphous solid does not have the regular repeating structure found in a crystal. Even so, the atoms are still held in place by chemical bonds that behave as springs obeying Hooke's law.

Figure 14.9C shows the atomic arrangement found in a liquid. The difference between liquids and amorphous solids is evident when the atomic arrangement is followed for a period of time. The atoms in a liquid are not held in fixed locations by the forces from neighboring atoms. Instead, atoms in a liquid are able to move about, and the particular atoms that happen to be close neighbors at one moment are usually not neighbors a short time later. This motion at the atomic scale enables a liquid to flow. Although the bonds between different molecules in water (or in

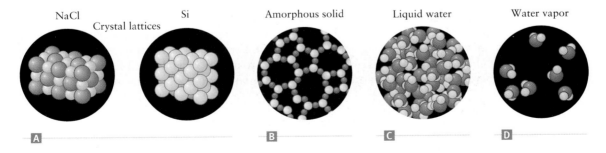

Figure 14.9 There are three phases of matter: solid, liquid, and gas. The atoms in a solid can be arranged in Ⓐ a crystal lattice or Ⓑ a somewhat random spatial arrangement called an amorphous solid. Ⓒ In a liquid and Ⓓ in a gas, the atoms do not have a regular arrangement and are able to move about.

A

B

Figure 14.10 A phase diagram shows the temperatures and pressures at which phase changes occur (solid lines). **A** Phase diagram for H_2O, and **B** phase diagram for CO_2. Note that 10^5 Pa is approximately atmospheric pressure.

other liquids) do not persist for long periods of time, some potential energy is still associated with the forces between different molecules and atoms in a liquid.

The arrangement of molecules in a gas is shown in Figure 14.9D. In some ways, a gas is very similar to a liquid. In both cases, the molecules are able to move easily over long distances, but the density of a gas is usually much lower than that of a liquid. The spacing between molecules in a gas is therefore larger, and the magnitude of the average intermolecular force (and potential energy) is much smaller. For this reason, most of the mechanical energy in a gas is found in the kinetic energies of the molecules.

Internal Energy

Consider a collection of H_2O molecules. What determines if these molecules form a solid, a liquid, or a gas? Figure 14.9 describes the difference between these phases in terms of the positions and motions of the molecules. Let's instead consider the energies of molecules when the system is in these different phases. We have already mentioned that the kinetic energy of a typical molecule is largest in the gas and smallest in the solid, so we can expect the mechanical energy of the molecules to be different in the different phases. This energy is called the ***internal energy*** of the system and is denoted by U. The internal energy of a system of molecules is the sum of the potential energies associated with all the intermolecular bonds plus the kinetic energies of all the molecules. The value of U increases as we go from the solid to the liquid to the gas phase. In general, the internal energy of all systems (not just water) increases as the temperature is increased.

Phase Changes

The transformation of a solid to a liquid, a liquid to a gas, and so on is called a ***phase change***. Phase changes can be produced by changing the temperature (by adding or removing heat energy) or by changing the system's pressure. It is therefore useful to consider the ***phase diagram*** of a substance. The phase diagram of H_2O (Fig. 14.10A) shows the phases found at different pressures and temperatures. For example, H_2O is a gas at low pressures and moderate to high temperatures, whereas it is a solid at low temperatures and moderate to high pressures. Phase changes occur at each of the solid lines in Figure 14.10A. For example, if we start with H_2O in the solid phase (ice) above a pressure of about 6×10^4 Pa and then increase the temperature (following, for example, path A in Fig. 14.10A), we eventually cross into the liquid phase. This phase change is called ***melting***. If instead we were to start in the liquid phase and then reduce the temperature, we would observe the phase change called ***freezing***. The phase diagram of H_2O includes other phase changes between liquid and gas (evaporation) and between solid and gas (sublimation). The line that separates liquid and gas ends at the ***critical point***. There is also a ***triple point***, where the solid, liquid, and gas phase regions all meet. The temperature of the triple point for water is 273.16 K; this value was used in the original definition of the Kelvin temperature scale.

Other substances have similar phase diagrams. For example, Figure 14.10B shows the phase diagram for CO_2. The overall arrangement of the phases is similar to the case of H_2O, although the values of temperature and pressure at which the phase changes occur vary from one substance to another. Table 14.2 lists the melting and evaporation temperatures of some common substances at atmospheric pressure.

Specific Heat and Heat Capacity

The phase diagram in Figure 14.10A shows a path labeled A along which a system (of water) is taken from a low to a high temperature. In practice, this change could be accomplished by adding heat energy to the system; we could place our system in contact with a second system at a higher temperature so that heat energy can flow between them. How much energy is required? Let's first consider what happens to a system when heat is added, but the system does not undergo a phase change. For

Table 14.2 Melting and Evaporation Temperatures for Some Common Substances at Atmospheric Pressure

SUBSTANCE	MELTING TEMPERATURE (K)	EVAPORATION TEMPERATURE (K)
Water	273	373
Aluminum	933	2720
Gold	1340	2930
Ethyl alcohol	159	351
Carbon dioxide	195	217
Nitrogen	63	77

example, our system might be a liquid (water) whose temperature increases but does not reach the evaporation (boiling) temperature. Let Q denote the heat energy added to the system and ΔT be the change in temperature. The ratio of these two quantities is called the **heat capacity**:

$$\text{heat capacity} = \frac{Q}{\Delta T} \qquad (14.4)$$

This ratio does not take account of the mass of the system; a large system has a large heat capacity because a large amount of energy Q is required to change the temperature by a given amount. (Likewise, a small system has a small heat capacity.) It is therefore useful to define a closely related quantity called the **specific heat** c by

$$\text{specific heat} = c = \frac{Q}{m\,\Delta T} \qquad (14.5)$$

The specific heat takes account of the size of the system, so c is the same for any sample of water (at a particular temperature and pressure), regardless of its mass.

We can rearrange Equation 14.5 as

$$Q = cm\,\Delta T \qquad (14.6)$$

Knowing the value of c thus allows us to calculate how much heat energy is needed to increase the temperature of a system of mass m by an amount ΔT. By convention, a positive value of Q corresponds to heat energy *added to* the system, and a negative value of Q indicates that energy flows *out of* the system. In our mechanical picture involving internal energy, adding or subtracting energy changes the internal energy U of the system (the "internal" kinetic and potential energies of the atoms and molecules). This change in U is related to changes in the temperature through the specific heat.

Values of the specific heat for a few common materials are given in Table 14.3. There, a single value of specific heat for each material is listed, but c can vary (usually by a small amount) with temperature. The values in Table 14.3 were (unless noted) determined at room temperature and pressure, with the volume of the system kept fixed; such values are called the specific heat at *constant volume*. Notice that c is always positive. Adding energy to a system (a positive value of Q in Eq. 14.6) always increases its temperature (a positive ΔT).

Values of the specific heat vary widely among different substances and tend to be larger for liquids than for solids or gases. In Table 14.3, the values of c are listed in SI units, which are J/(kg · K). You will often find values of specific heat quoted in terms of calories (instead of joules), especially in chemistry books, so you need to be able to convert between these two units. From Table 14.3, we have $c = 4186$ J/(kg · K) for water. Converting the joules to calories (according to Eq. 14.1), we have

$$c = 4186\,\frac{\text{J}}{\text{kg}\cdot\text{K}} \times \frac{1\text{ cal}}{4.186\text{ J}} \times \frac{1\text{ kg}}{1000\text{ g}} = 1.000\,\frac{\text{cal}}{\text{g}\cdot\text{K}} \quad (\text{specific heat of water})$$

Table 14.3 Values of the Specific Heat for Some Common Substances

SUBSTANCE	SPECIFIC HEAT J/(kg · K)
Water (liquid, 15°C)	4186
Ice (0°C)	2090
Steam (water vapor, 100°C)	1520
Aluminum	900
Ethyl alcohol	2450
Gold	129
Steel	450
Tungsten	130
Granite	840
Glass	840
Nitrogen gas	740
Human body (approximate)	3500

Note: All values for a constant volume, and at room temperature and atmospheric pressure, unless noted.

The value of the specific heat of water thus has a particularly "simple" (and easy to remember) value when expressed using calories. Indeed, the calorie was originally defined through the specific heat of water.

| CONCEPT CHECK 14.1 | Specific Heat of the Human Body

Even though the human body is not a pure substance, we can still define its specific heat through Equation 14.5. According to Table 14.3, the human body's specific heat is only a little smaller than the specific heat of water and is much larger than the specific heat of the other substances listed. Why?

Why Is Specific Heat Important?

With knowledge of the specific heat, we can calculate how the temperature of an object will change when a certain amount of heat energy is added or removed. One practical application is in understanding how oceans and lakes affect the weather. The value of c of water is much greater than the specific heat of most other substances. (Compare values in Table 14.3.) A large body of water thus has a very large heat capacity relative to the surrounding air and soil. As a result, the temperature of an ocean or lake changes relatively little even when relatively large amounts of heat energy are added in the summer or taken away in winter, and thus it moderates ("smoothes out") temperature fluctuations.

The value of the specific heat also gives insight into the internal energy U of a substance. According to the principle of conservation of energy, if an amount of heat Q is added to a system, then U must increase by that amount. The specific heat tells how the added heat changes the temperature; hence, c tells how changes in internal energy and temperature are related. This concept is important for understanding how the microscopic atomic motions in a substance change with temperature; we'll discuss this subject more in Chapter 15.

Calorimetry

Figure 14.11 In a typical calorimetry experiment, the system of interest is arranged so that the only heat flow into or out of the system is from a reference system.

We can find the specific heat of a system by adding a known amount of heat energy Q and measuring the resulting temperature change ΔT (Eq. 14.5). A measurement of this kind is called **calorimetry** and is shown schematically in Figure 14.11. The system of interest is placed in contact with a reference system; the reference system is often very large, in which case it is also called a **thermal reservoir**. For example, the system of interest might be a hot piece of metal, and the thermal reservoir could be a large bucket of water into which the metal is dropped (to cool it). When defining the specific heat, Q is the total heat energy added to the system (the metal); in this example, Q would come entirely from the thermal reservoir (the water). Most calorimetry experiments are designed to control the sources of heat flow into the system. In words, the system of interest is "thermally isolated" from all but the thermal reservoir.

Figure 14.11 and the general principles of calorimetry can also be used to analyze processes that do not involve a large thermal reservoir. For example, when a cold ice cube is dropped into a glass of warm water, heat flows between the water and the ice, changing the temperature of both. These temperature changes depend on the initial temperatures of both systems and their specific heats. In terms of conservation of energy, all the energy that leaves one system enters the other system, and that is the basis of a general way to attack all problems in calorimetry.

| PROBLEM SOLVING | Dealing with Specific Heat in Calorimetry

1. **RECOGNIZE THE PRINCIPLE.** All calorimetry problems are based on the principle of conservation of energy.

2. **SKETCH THE PROBLEM.** Your sketch should show the system or systems of interest and indicate how heat flows between them.

3. **IDENTIFY THE RELATIONSHIPS.** Determine (if possible) the initial and final temperatures of the system(s) and the heat energy Q added to each system. If energy flows into a system, Q is positive, whereas if energy flows out of a system, Q is negative.

4. **SOLVE.** Apply $Q = cm\,\Delta T$ (Eq. 14.6) to relate Q and ΔT for each system and solve for the quantities of interest.

Important note: Step 4 (and Eq. 14.6) assumes there are no phase changes; for example, the systems do not melt or evaporate. When we discuss the latent heat associated with phase changes later in this section, we'll see how to analyze such cases.

5. Always *consider what your answer means* and check that it makes sense.

EXAMPLE 14.2 Specific Heat in the Blacksmith's Shop: Part 1

A blacksmith uses the heat energy from an oven to heat steel and other metals. Increasing the temperature makes it easier to form a metal into a desired shape, such as a horseshoe. The blacksmith also uses cold water to cool a hot horseshoe to room temperature. A steel horseshoe of mass $m_{shoe} = 0.50$ kg at a temperature of 800 K ("red hot") is dropped into a bucket of water ($m_{water} = 1.0$ kg) at 300 K (about room temperature). What is the final temperature of the horseshoe and the water? For simplicity, assume none of the water evaporates, so none of the water is converted to steam.

RECOGNIZE THE PRINCIPLE

In this example, our two systems are the horseshoe and the water. When the horseshoe is dropped into the bucket, heat energy flows from the horseshoe to the water (Fig. 14.12). Because the horseshoe is completely immersed in the water, we can assume there is no other inflow or outflow of energy involving either system. We denote the energy that flows from the horseshoe into the water as Q. From conservation of energy principles, the energy "added" to the horseshoe is $-Q$; the negative sign here indicates that the final energy of the horseshoe is less than its initial energy. The initial temperatures of the horseshoe and the water are given, and the common final temperature T_f is the quantity we wish to find. We can find T_f by applying the specific heat relation between Q and ΔT for both systems.

SKETCH THE PROBLEM

Figure 14.12 shows the two systems along with their initial temperatures and the direction of heat flow.

IDENTIFY THE RELATIONSHIPS

We denote the specific heats of the horseshoe and the water by c_{steel} and c_{water}, respectively, and obtain their values from Table 14.3. Applying Equation 14.6 to the horseshoe, we get

$$Q_{shoe} = c_{steel}m_{shoe}\,\Delta T_{shoe} = c_{steel}m_{shoe}(T_f - T_{shoe,\,i}) \qquad (1)$$

Because the final temperature is less than the initial temperature of the steel horseshoe ($T_{shoe,\,i}$), Q_{shoe} will be negative, as anticipated above. Applying the same relation to the water, we get

$$Q_{water} = c_{water}m_{water}(T_f - T_{water,\,i}) \qquad (2)$$

Because the water temperature increases ($T_f > T_{water,\,i}$), Q_{water} will be positive. Energy is conserved, so

$$Q_{shoe} + Q_{water} = 0$$

and

$$Q_{shoe} = -Q_{water}$$

INITIAL

Horseshoe	Water
T_i = 800 K	T_i = 300 K

Q

| Horseshoe | Water |

When the horseshoe is placed in the water, energy will flow from the hot horseshoe to the cold water.

FINAL

| Horseshoe | Water |

T_f

Eventually, they reach the same final temperature.

Figure 14.12 Example 14.2.

SOLVE

Using our results from Equations (1) and (2) and solving for T_f leads to

$$c_{steel}m_{shoe}(T_f - T_{shoe,\,i}) = -c_{water}m_{water}(T_f - T_{water,\,i})$$

$$T_f(c_{steel}m_{shoe} + c_{water}m_{water}) = c_{steel}m_{shoe}T_{shoe,\,i} + c_{water}m_{water}T_{water,\,i}$$

$$T_f = \frac{c_{steel}m_{shoe}T_{shoe,\,i} + c_{water}m_{water}T_{water,\,i}}{c_{steel}m_{shoe} + c_{water}m_{water}}$$

Inserting the given values of the various quantities, including the specific heats from Table 14.3, gives

$$T_f = \frac{[450 \text{ J/(kg}\cdot\text{K})](0.50 \text{ kg})(800 \text{ K}) + [4186 \text{ J/(kg}\cdot\text{K})](1.0 \text{ kg})(300 \text{ K})}{[450 \text{ J/(kg}\cdot\text{K})](0.50 \text{ kg}) + [4186 \text{ J/(kg}\cdot\text{K})](1.0 \text{ kg})}$$

$$T_f = \boxed{330 \text{ K}}$$

What does it mean?
The final temperature is quite close to the initial temperature of the water and far from the initial temperature of the horseshoe. That is because the specific heat of water is much greater than the specific heat of steel, so the water acts as a kind of thermal reservoir in this case.

EXAMPLE 14.3 ⊗ Maintaining a Safe Body Temperature

A bicycle rider of mass $m = 60$ kg generates a power of approximately 400 W when riding at a moderate pace. Of this energy, about 25% is used to propel the bicycle, and the remaining 75% goes into heat generated within his body. If none of this heat were lost to the surroundings, how much would the cyclist's body temperature increase after 1 hour of riding?

RECOGNIZE THE PRINCIPLE

In this calorimetry problem, the system of interest is the cyclist, along with an "internal" source (the cyclist's metabolism) that adds an energy Q to this system. We must first find Q and then use it to compute the temperature rise of the system with Equation 14.6.

SKETCH THE PROBLEM

Figure 14.13 shows the system (the cyclist). An amount of heat Q is generated within, increasing his temperature.

IDENTIFY THE RELATIONSHIPS

Power (P) is energy per unit time, so the energy Q added to the cyclist in 1 h is $Q = P\,\Delta t$. Here P is the fraction of the power that goes into heat, which is $P = 0.75(400 \text{ W}) = 300 \text{ W}$, whereas $\Delta t = 1 \text{ h} = 3600 \text{ s}$. We thus get

$$Q = P\,\Delta t = (300 \text{ W})(3600 \text{ s}) = 1.1 \times 10^6 \text{ J} \qquad (1)$$

Equation (1) is related to the change in temperature of the cyclist by (using Eq. 14.6)

$$\Delta T = \frac{Q}{cm}$$

SOLVE

Inserting Q from Equation (1) along with the given value of m and the specific heat of the human body (Table 14.3) gives

$$\Delta T = \frac{Q}{cm} = \frac{1.2 \times 10^6 \text{ J}}{[3500 \text{ J/(kg}\cdot\text{K})](60 \text{ kg})} = 5.2 \text{ K} = \boxed{5.2°\text{C}}$$

Figure 14.13 Example 14.3. **A** When exercising (or even when just sitting on a couch), a person generates heat. **B** This heat will raise the body's temperature unless it escapes through the skin.

© Image Source/Jupiterimages

When a person's body temperature increases more than about 4°C (to give a body temperature above about 106°F), there is danger of heatstroke. It is therefore important that our cyclist, or anyone else engaged in strenuous exercise, have some way to pass this generated heat to his surroundings. We'll consider that problem in further examples below.

Latent Heat

The specific heat tells how much the temperature will increase if a certain amount of heat is added to a system, but the definition of specific heat assumes the system does not change from one phase to another as the heat is added. When the black-smith cooled the horseshoe in Example 14.2, we assumed none of the water evapo-rated (and the final answer confirmed this assumption). When a system undergoes a phase change such as melting or evaporation, we must also account for the *latent heat*. For example, for a block of ice at 0°C to undergo a phase change to liquid water at 0°C, an amount of heat equal to the *latent heat of fusion* must be added. Likewise, when water at a temperature of 0°C freezes to become ice at 0°C, this latent heat must be removed.

The latent heat associated with a phase change is connected with the energies of the atoms within the substance. For a solid, substantial energy is stored in the atomic bonds holding the atoms in place. These bonds store potential energy in a manner analogous to springs that obey Hooke's law (Chapter 6). For a solid to melt, energy is required so as to overcome this potential energy and break these bonds, enabling the atoms and molecules to move about freely as found in a liquid. That energy is provided by the latent heat.

By convention, the latent heat of fusion L_{fusion} is a positive quantity; for a system of mass m to melt, an amount of heat

$$Q_{melt} = +mL_{fusion} \qquad (14.7)$$

must be added. Likewise, when a system freezes (also called fusion), we have

$$Q_{freeze} = -mL_{fusion}$$

The negative sign indicates that energy must be removed from a system to make it freeze.

The *latent heat of vaporization* is associated with the phase change from a liquid to a gas (called vaporization or evaporation or boiling) and from a gas to a liquid (condensation). To vaporize an amount of liquid of mass m, we must add an amount of heat

$$Q_{vaporization} = mL_{vaporization} \qquad (14.8)$$

and for condensation, we have

$$Q_{condensation} = -mL_{vaporization}$$

Here the negative sign indicates that energy must be removed from a system to make it condense. On an atomic scale, the latent heat of vaporization can be understood as follows. When a liquid evaporates to become a gas, the weak atom–atom inter-actions in the liquid become even weaker in the gas as the atoms are moved even farther apart. This process again requires the addition of energy, now through the latent heat of vaporization.

Table 14.4 lists the latent heats of fusion and vaporization for some common substances. Notice that for a given substance, the latent heat of fusion is *not* equal to the latent heat of vaporization.

Calorimetry: Including the Latent Heat

Figure 14.14A shows the phase diagram for a hypothetical substance. Let's consider how much heat we must add to move along the path indicated by the horizontal dashed line. Starting from a point in the solid region of the phase diagram (point *A*),

Table 14.4 Latent Heats for Various Phase Changes

SUBSTANCE	LATENT HEAT OF FUSION (kJ/kg)	LATENT HEAT OF VAPORIZATION (kJ/kg)
Water	334	2,200
Aluminum	397	11,000
Gold	67	870
Ethyl alcohol	104	850
Copper	205	5,070
Lead	23	870

Figure 14.14 ◢ Phase diagram of a typical system. ◣ Results of a calorimetry experiment in which T is plotted as a function of the energy Q_{added} added to the system. The slopes in regions I, III, and V depend on the specific heat of the solid, liquid, and gas respectively. The slope is zero in regions II and IV.

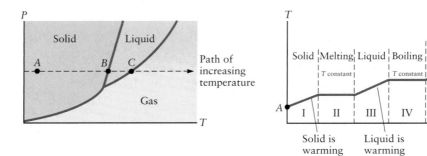

we add heat. Adding heat increases the temperature as shown in region I in Figure 14.14B, where we plot the temperature as a function of the amount of added heat Q_{added}. The temperature increase ΔT in the solid phase is described by the specific heat through $\Delta T = Q/(cm)$ (Eq. 14.6), with c equal to the specific heat of the solid.

Point B in Figure 14.14A is at the melting transition. If we continue to add heat, the system undergoes a phase change from solid to liquid and the amount of heat needed is given by the latent heat of fusion (Eq. 14.7). As this latent heat is added and the system is melting, the temperature remains constant (region II in Fig. 14.14B). If this substance is water, we would be converting a certain mass of ice at 0°C into the same amount of water at 0°C. When the system is completely melted, adding further heat increases the temperature according the specific heat of the liquid (region III). Eventually, the phase change to a gas is reached at point C. Now an amount of energy corresponding to the latent heat of vaporization must be added, and during this time, the temperature is again constant (region IV). If this substance is water, this latent heat converts water at 100°C to gas (water vapor) also at 100°C. Adding more heat would then increase the temperature of the gas.

Calorimetry calculations involving phase changes are similar to the problems we considered in Examples 14.2 and 14.3. The only difference is that we must now allow for the energy associated with the latent heat of any phase changes that occur.

PROBLEM SOLVING Dealing with Both Specific Heat and Latent Heat in Calorimetry

1. **RECOGNIZE THE PRINCIPLE.** All calorimetry problems are based on the principle of conservation of energy.

2. **SKETCH THE PROBLEM.** Your sketch should show the system or systems of interest and indicate how heat flows between them.

3. **IDENTIFY THE RELATIONSHIPS.** Determine (if possible) the initial and final temperatures of the system(s) and the heat energy Q added to each system. If energy flows into a system, Q is positive; if energy flows out of a system, Q is negative.

4. **SOLVE.**
   - Use $Q = cm \, \Delta T$ (Eq. 14.6) to relate Q and ΔT for temperature changes that do not involve a phase change.
   - If there is a phase change, the latent heat ($Q = \pm mL$, Eqs. 14.7 and 14.8) must be added to or subtracted from the system, depending on the phase change.

5. Always *consider what your answer means* and check that it makes sense.

EXAMPLE 14.4 Specific Heat in the Blacksmith's Shop: Part 2

Consider again the blacksmith in Example 14.2, who uses a bucket of water ($m_{water} = 1.0$ kg, $T_{water, i} = 300$ K) to cool a hot, steel horseshoe. This time, he has a much larger horseshoe ($m_{shoe} = 2.0$ kg, $T_{shoe, i} = 800$ K, and notices that when he drops the horseshoe in the water, some of the water evaporates as steam. Assuming the final temperature of the horseshoe plus water plus steam is 100°C (= 373 K), how much of the water is converted into steam?

RECOGNIZE THE PRINCIPLE

When the horseshoe is dropped into the water, an amount of heat energy Q flows from the horseshoe to the water (Fig. 14.15). From conservation of energy principles, the energy "added" to the horseshoe is $-Q$; the negative sign here indicates that the final energy of the horseshoe is less than its initial energy. The initial temperatures of the horseshoe and the water are given. In the end, an unknown amount of the water undergoes a phase change to steam, and the horseshoe, water, and steam have a common final temperature $T_f = 100°C = 373$ K. We can find how much water is converted to steam using the principle of conservation of energy.

SKETCH THE PROBLEM

Figure 14.15 shows the horseshoe and the water along with their initial temperatures and the direction of heat flow.

IDENTIFY THE RELATIONSHIPS

We are given the initial and final temperatures of the horseshoe, the water, and the steam; the only unknown is the mass of water converted to steam. The heat that flows out of the horseshoe can be found from the specific heat of the horseshoe and its temperature change. Applying Equation 14.6, we get (compare with Example 14.2)

$$Q_{shoe} = c_{steel} m_{shoe}(T_f - T_{shoe, i})$$

Inserting the given values, we find

$$Q_{shoe} = [450 \text{ J/(kg·K)}](2.0 \text{ kg})(373 \text{ K} - 800 \text{ K}) = -3.8 \times 10^5 \text{ J} \qquad (1)$$

The water begins at $T_{water, i}$, and it all reaches a final temperature of 373 K. The heat required to bring all the liquid to this temperature is

$$Q_{water} = c_{water} m_{water}(T_{water, f} - T_{water, i})$$

Inserting the given values, we find

$$Q_{water} = [4186 \text{ J/(kg·K)}](1.0 \text{ kg})(373 \text{ K} - 300 \text{ K}) = 3.1 \times 10^5 \text{ J} \qquad (2)$$

Next we write the conservation of energy condition. The sum of the energy Q_{shoe} plus the total energy that flows into the water must equal zero. The energy that flows into the water has two contributions, one from the energy needed to heat all the water to 373 K as done in Equation (2) and another from the latent heat of the water converted to steam. We thus have (compare again with Example 14.2)

$$Q_{shoe} + Q_{water} + Q_{steam} = 0 \qquad (3)$$

We can use the results from Equations (1) and (2) to find Q_{steam}:

$$Q_{steam} = -Q_{shoe} - Q_{water} = 3.8 \times 10^5 \text{ J} - 3.1 \times 10^5 \text{ J} = 7 \times 10^4 \text{ J} \qquad (4)$$

We want to find the mass of water m_{steam} converted to steam; it is related to Q_{steam} by (Eq. 14.8)

$$Q_{steam} = m_{steam} L_{vaporization}$$

SOLVE

Solving for m_{steam} and inserting the value of Q_{steam} from Equation (4) and the latent heat from Table 14.4, we find

$$m_{steam} = \frac{Q_{steam}}{L_{vaporization}} = \frac{7 \times 10^4 \text{ J}}{2.2 \times 10^6 \text{ J/kg}} = \boxed{0.03 \text{ kg}}$$

What have we learned?
The key to this problem, and to all calorimetry, is conservation of energy. In this example, it is expressed in Equation (3).

INITIAL

| Horseshoe $T_i = 800$ K | Water $T_i = 300$ K |

Q

| Horseshoe | Water |

FINAL

| Horseshoe | Water | Steam |

$T_f = 100°C = 373$ K

Figure 14.15 Example 14.4.

$L_0 + \Delta L$

Solid

L_0

A

ΔV

Fluid Volume = V_0

B

Figure 14.16 **A** If a solid object is heated, its length increases by an amount ΔL proportional to the coefficient of thermal expansion α. **B** Any substance (solid, liquid, or gas) may change volume when it is heated. This change in volume ΔV is proportional to the coefficient of volume expansion β.

Table 14.5 Coefficients of Linear and Volume Expansion

SUBSTANCE	α $(10^{-6}\,K^{-1})$	β $(10^{-6}\,K^{-1})$
Ethyl alcohol	—	1120
Mercury	—	182
Water	—	210
Ice (0°C)	51	153
Pyrex	3.3	9.8
Glass (ordinary)	9	27
Aluminum	22.5	69
Gold	14	42
Cement	12	36
Iron	12	36
Steel	12	36
Platinum	8.8	26
Invar[a]	0.9	2.7
Wood (typical)	5.4	16
Brass	18.7	56
Copper	17	51

Note: Values for room temperature and pressure unless noted.

[a]Invar is an alloy of iron and nickel that has very low values of the coefficients of thermal expansion, which makes it useful in applications.

CONCEPT CHECK 14.2 | Comparing Heat Energy with a Typical Kinetic Energy

The total heat energy required to cool the horseshoe in Example 14.4 is $|Q_{shoe}| = 3.8 \times 10^5$ J. Suppose this energy could be converted to the kinetic energy of a baseball ($m \approx 0.2$ kg). Would the speed of the baseball be (a) 50 m/s (about 100 mi/h), (b) 5 m/s, or (c) 2000 m/s?

14.5 | THERMAL EXPANSION

We have spent the past few pages discussing how the temperature of a system changes as heat energy is added. These temperature changes can affect many properties of a system. In particular, the size of a system will usually change with temperature, an effect called *thermal expansion*. Consider an object such as a piece of metal as in Figure 14.16A, with a length L_0. If the temperature of the object is increased by ΔT, the length changes by an amount ΔL. This change in length is proportional to the change in temperature ΔT and to the *coefficient of linear thermal expansion* α:

$$\frac{\Delta L}{L_0} = \alpha\,\Delta T \qquad (14.9)$$

As indicated in Figure 14.16A, the object's width and height also change. These changes are given by expressions similar to Equation 14.9, so both change by amounts proportional to α and ΔT, which leads to a *volume expansion* ΔV given by

$$\frac{\Delta V}{V_0} = \beta\,\Delta T \qquad (14.10)$$

where V_0 is the initial volume and β is the *coefficient of volume expansion*. Values of α and β for a variety of substances are given in Table 14.5. Values of α are only given for solids because liquids and gases cannot (on their own) maintain a fixed shape as in Figure 14.16A. We can still define a coefficient of volume expansion (see Fig. 14.16B) for liquids and gases, however, and they are listed in Table 14.5.

CONCEPT CHECK 14.3 | Linear Expansion (α) and Volume Expansion (β) Are Related

You can see from Table 14.5 that a substance with a large value of α also has a large value of β. Why?

EXAMPLE 14.5 Thermal Expansion of a Metal Bar

Consider a meterstick composed of platinum. It might be a copy of the standard meterstick as formerly employed to maintain the SI unit of length (Chapter 1). By what amount does the length of this meterstick change if the temperature increases by 1.0 K?

RECOGNIZE THE PRINCIPLE

This problem is a direct application of the definition of the thermal expansion coefficient. Given the change in temperature and the initial length, the change in length is $\Delta L = L_0 \alpha\,\Delta T$ (Eq. 14.9).

SKETCH THE PROBLEM

Figure 14.16A describes the problem.

IDENTIFY THE RELATIONSHIPS AND SOLVE

From Table 14.5, we have $\alpha = 8.8 \times 10^{-6}$ K^{-1} for platinum. Inserting into Equation 14.9 gives

$$\Delta L = L_0 \alpha\,\Delta T = (1.0\text{ m})(8.8 \times 10^{-6}\text{ K}^{-1})(1.0\text{ K}) = \boxed{8.8 \times 10^{-6}\text{ m}}$$

What does it mean?

This increase in the length of the bar is rather small (a bit smaller than the diameter of a typical human hair). For some applications, however, changes of this magnitude cannot be tolerated, especially when high precision is required. That is why standards of length that use the wavelength of light (instead of the length of a metal bar) have been developed. We'll explain in Chapter 25 how the wavelength of light can be measured very accurately.

EXAMPLE 14.6 | Area Expansion

Consider a square hole in an aluminum plate (Fig. 14.17). If the aluminum is heated, will this hole become smaller or larger?

RECOGNIZE THE PRINCIPLE

A metal bar becomes longer when it is heated; hence, the ends of the bar are farther apart. Therefore, any two points on the bar will also be farther apart after heating.

SKETCH THE PROBLEM

Figure 14.18 shows how two points on a metal bar move farther apart when the bar is heated.

IDENTIFY THE RELATIONSHIPS AND SOLVE

Consider two points A and B at the edges of the square hole in Figure 14.17. These points will be farther apart after the aluminum plate is heated, so the hole will get | larger |.

What does it mean?

The same reasoning applies to a circular hole and explains why heating the metal lid on a glass jar makes it easier to remove a "stuck" lid. The metal lid expands when it is heated. The glass jar also expands, but because the coefficient of linear thermal expansion of most metals is greater than the coefficient for glass (Table 14.5), the opening in the metal lid expands more than the glass jar, making it easier to remove the lid.

Figure 14.17 Example 14.6. What happens to the area of a hole when this plate is heated?

Figure 14.18 Example 14.6. When a bar expands, points C and D move farther apart.

Effects of Thermal Expansion

The thermal expansion of the metal bar in Example 14.5 is not large, but the effect of thermal expansion can be quite significant in certain situations. For example, consider the problem of designing a large bridge. One of the first bridges built to span the Mississippi River near St. Louis, the Eads Bridge, was built with steel in the 1870s. This bridge was one of the first applications of steel, which was then a new invention. The bridge used several spans, each with a length of about 500 ft (about 150 m). Let's calculate the change in length of one span from a cold winter night (−10°C) to a hot summer day (40°C). From Equation 14.9, we have

$$\Delta L = L_0 \alpha \, \Delta T = L_0 \alpha (T_f - T_i)$$

The temperature change is $40°C - (-10°C) = 50°C = 50$ K. Using the coefficient of linear thermal expansion for steel from Table 14.5 gives

$$\Delta L = (150 \text{ m})(12 \times 10^{-6} \text{ K}^{-1})(50 \text{ K}) = 0.090 \text{ m} \qquad (14.11)$$

This change in length is about 3.5 in. and is thus quite significant; it would certainly not be desirable for a bridge that carries trains or cars to have a gap of this size. For this reason, bridges, buildings, and many other structures are designed with special expansion joints that allow for expansion and contraction without forming unsafe gaps or breaking (Fig. 14.19).

© Richard Choy/Peter Arnold Inc.

Figure 14.19 Many bridges contain expansion joints so that they can expand and contract without breaking away from their supports or forming unsafe gaps.

Changes in T cause the mercury to expand or contract, and the meniscus moves a distance $\Delta L \propto \Delta T$.

Mercury

Figure 14.20 A mercury thermometer uses the thermal expansion of liquid mercury to measure temperature.

Another example of thermal expansion is sketched in Figure 14.20. A mercury thermometer uses the thermal expansion of mercury to measure changes in temperature. When T changes, the mercury (which is a liquid) expands and contracts, whereas the glass holding the mercury expands (or contracts) relatively little because the coefficients of thermal expansion of glass are small, producing a change in the level of the meniscus ΔL in the thermometer tube (Fig. 14.20). Changes in the meniscus position are proportional to changes in temperature, just as are changes in the length of the bar in Example 14.5. The magnitude of the changes, though, are much larger for the thermometer because there is a (relatively) large pool of mercury in the bulb of the thermometer, and all this mercury expands into a very narrow tube when the temperature increases.

EXAMPLE 14.7 Thermal Expansion and Forces on a Bridge

Figure 14.21 shows a steel bridge, similar to the one considered in connection with Equation 14.11. In that calculation, we found that a steel beam with an initial length of $L_0 = 150$ m will expand an amount $\Delta L = 0.090$ m when heated by 50 K. Suppose this steel beam has a cross-sectional area of $A = 1.0$ m $\times$ 0.10 m = 0.10 m^2 and is part of a bridge. The ends of the bridge will hold the ends of the steel beam fixed in place so that the ends cannot move. What force must the supports exert on the ends of the beam?

RECOGNIZE THE PRINCIPLE

If there are no forces on the ends of the beam, it will, due to thermal expansion, expand by an amount ΔL when it is heated. To prevent this expansion, the supports must exert a force that compresses the beam by an amount ΔL. This compression is described by the elastic properties of the beam and is proportional to Young's modulus (Chapter 11).

SKETCH THE PROBLEM

Figure 14.21 shows the problem along with the forces exerted by the supports on the ends of the beam.

Steel beam

F F

L_0

Figure 14.21 Example 14.7.

IDENTIFY THE RELATIONSHIPS

If a force F is applied to each end of the bar, it will compress an amount ΔL, with (Eq. 11.23)

$$\frac{F}{A} = Y\frac{|\Delta L|}{L_0} \qquad (1)$$

Here Y is Young's modulus and A is the cross-sectional area of the bar. We have used the absolute value of ΔL because the applied forces compress the bar, reducing its length (making ΔL negative). The bridge supports hold the ends fixed, so this change in length must compensate for the change in length due to thermal expansion in Equation 14.11. Hence, $|\Delta L| = 0.090$ m in Equation (1).

SOLVE

Young's modulus for steel is 2.0×10^{11} Pa (Table 11.1). Solving for the force, we get

$$F = AY\frac{|\Delta L|}{L_0} = (0.10 \text{ m}^2)(2.0 \times 10^{11} \text{ Pa})\frac{(0.090 \text{ m})}{(150 \text{ m})} = \boxed{1.2 \times 10^7 \text{ N}}$$

What does it mean?

This force is quite substantial and is why metal beams and bridges sometimes "buckle." Architects and mechanical engineers are paid to prevent such buckling.

Thermal Expansion of Water

Table 14.5 lists the coefficients of thermal expansion for a number of different substances near room temperature. Strictly speaking, α and β both vary with temperature, but this variation is often small and can usually be ignored. Therefore, the length and volume of an object both change linearly with temperature, and we can usually use Equations 14.9 and 14.10 to calculate changes in length and volume even when ΔT is large (i.e., many tens or even hundreds of kelvins). There is, however, one important substance whose expansion is not a simple linear function. Figure 14.22 shows how the density of H_2O varies with temperature; Figure 14.22A shows the behavior over a wide temperature range, and Figure 14.22B shows an expanded view near the freezing temperature ($T = 0°C$). At temperatures well above freezing, the density decreases as T increases. That is just the usual thermal expansion; an expanding sample has a lower density. At temperatures near and above room temperature (around 20°C), this expansion can be described by Equation 14.10 with $\beta \approx 2.1 \times 10^{-4}$ K^{-1}. Just above freezing, however, the density of water exhibits a maximum and cannot be described by Equation 14.10 with a constant value of β. What's more, the density of the solid (ice) near freezing is less than the density of the liquid, and H_2O thus contracts when it melts. That is why ice floats in water, as described by Archimedes's principle (Chapter 10). Most substances do not behave in this way; in nearly all cases, a solid does not float in its own liquid. This unusual behavior of water has many important consequences. For example, in winter a lake freezes first at the surface, which is good for ice skaters. It also enables fish to live comfortably in the water underneath and still have access to food at the lake bottom.

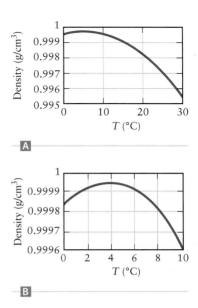

Figure 14.22 ◩ Density of water as a function of temperature. ◩ Expanded view of ◩ near the freezing temperature (0°C).

EXAMPLE 14.8 ✖ **Effect of Freezing on a Cell**

The fluid inside a cell is composed mainly of water, so when a cell freezes, the fluid inside expands. This expansion causes a large ("outward") pressure on the cell wall and usually causes the cell to rupture. Consider a spherical cell with an initial radius of $r_i = 5.0$ μm. Calculate the increase in pressure inside the cell when it freezes. If we assume the fluid inside the cell is mainly water, we can use the data for water from Figure 14.22; the density of water at 0°C is 999.8 kg/m³ (= 0.9998 g/cm³), and the density of ice is 917.0 kg/m³.

RECOGNIZE THE PRINCIPLE

When the water in the cell freezes, its volume expands by an amount ΔV_{ice}, which we can calculate from the given values of the densities of water and ice. We then need to calculate what pressure must be applied to the cell to compress it back to its original volume; this pressure is the extra pressure that would be exerted on the cell wall (assuming the cell does not rupture). We can find it using the bulk modulus of ice (Chapter 11).

SKETCH THE PROBLEM

Figure 14.23 describes the problem. The volume expansion that would occur due to freezing must be compensated for by an increase in the pressure.

IDENTIFY THE RELATIONSHIPS

If ΔV_{ice} is the amount the water would expand when it freezes, the pressure P from the cell wall must produce a volume change $\Delta V = -\Delta V_{ice}$. In Chapter 11, we learned that the pressure P required to compress a material to give a volume change ΔV is (Eq. 11.25)

$$P = -B \frac{\Delta V}{V} \tag{1}$$

where B is an elastic constant called the bulk modulus. In this example, we need to compress the ice, so ΔV is negative and P in Equation (1) will be positive. The bulk modulus for ice is approximately $B = 9 \times 10^9$ Pa.

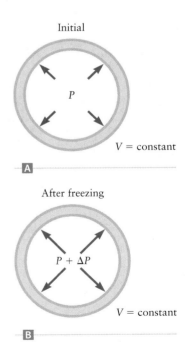

Figure 14.23 Example 14.8. If water in a cell freezes, it produces an increased pressure on the cell wall.

Mass is equal to the product of density and volume, so $m = \rho V$. The mass of the cell is the same at the start (i) and end (f), giving

$$\rho_i V_i = \rho_f V_f \qquad (2)$$

The cell is (for simplicity) spherical; hence,

$$V_i = \tfrac{4}{3}\pi r_i^3 = \tfrac{4}{3}\pi (5.0 \times 10^{-6}\,\text{m})^3 = 5.2 \times 10^{-16}\,\text{m}^3$$

Inserting this result into Equation (2) along with the given initial and final densities, we find

$$V_f = \frac{\rho_i}{\rho_f} V_i = \frac{999.8\,\text{kg/m}^3}{917.0\,\text{kg/m}^3}(5.2 \times 10^{-16}\,\text{m}^3) = 5.7 \times 10^{-16}\,\text{m}^3$$

The change of volume of the cell is thus

$$\Delta V_{\text{ice}} = V_f - V_i = 5 \times 10^{-17}\,\text{m}^3$$

Using the relation in Equation (1), we then find

$$P = -B\frac{(-\Delta V_{\text{ice}})}{V} = (9 \times 10^9\,\text{Pa})\frac{5 \times 10^{-17}\,\text{m}^3}{5.2 \times 10^{-16}\,\text{m}^3} = \boxed{9 \times 10^8\,\text{Pa}}$$

What does it mean?
Atmospheric pressure is only about 1×10^5 Pa, so this pressure is nearly 10,000 times atmospheric pressure! That is why cells usually rupture when they freeze.

14.6 | HEAT CONDUCTION

Heat is the energy that flows between systems at different temperatures. This energy transfer can take place in three ways: *conduction*, *convection*, and *radiation*. We now consider heat conduction; we'll discuss the other two methods of heat transfer in the following sections.

Figure 14.24A shows a solid bar connecting two separate systems. These systems might be balloons of gas, containers of liquid, or perhaps large pieces of metal. We assume these systems are at different temperatures T_1 and T_2, so heat energy flows through the bar from one end to the other. If $T_1 > T_2$, energy flows from the system on the left (the hotter one) to the one on the right, and we say that the bar *conducts* heat. The bar itself cannot be described by a single temperature value. Instead, the bar's temperature varies smoothly from T_1 on the left to T_2 on the right (Fig. 14.24B).

We can understand this heat flow in terms of the vibrations of the atoms in the bar. (If it is a metal bar, electrons outside the ion cores in the bar will also contribute to heat flow.) In any particular region within the bar, the atoms closer to the hot end (at the left) are at a slightly higher temperature than the neighboring atoms that are nearer the cold end (the right). The atoms at a higher temperature have a larger vibration amplitude, corresponding to larger potential and kinetic energies. As the atoms vibrate, some of the extra vibrational energy of the "hotter" atoms is transferred to the nearby "colder" atoms, similar to the way a wave transfers energy from place to place without transferring matter; energy thus "flows" from the hot end of the bar to the cold end.

The amount of energy that flows through the bar in Figure 14.24 depends on the area of the bar A, the length of the bar L, and the temperature difference $\Delta T = T_1 - T_2$. Heat flow also depends on a property of the bar called the *thermal conductivity*, κ. Heat flow is measured by the energy Q that arrives per unit time t at the cold end. This heat flow is the rate at which energy leaves the hot end of the bar and arrives at the cold end and is given by

Atoms at higher temperature are moving faster than the slightly cooler ones to the right.

Metal bar   Area = A

T_1 T_2

L

A

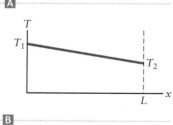

B

Figure 14.24 A When the ends of a solid bar are placed in contact with systems at different temperatures, the bar conducts energy from the hot end to the cold end. B Plot of the temperature as a function of position along the bar.

$$\frac{Q}{t} = \kappa A \frac{\Delta T}{L} \qquad (14.12)$$

Values of the thermal conductivity for a variety of materials are listed in Table 14.6. Values of κ vary widely, with small values for gases (such as air) and high values for metals.

CONCEPT CHECK 14.4 | Thermal Conductivity of Gases

The values in Table 14.6 show that the thermal conductivities of gases are generally much smaller than the thermal conductivities of solids. Use the microscopic picture of heat conduction involving atomic vibrations (Fig. 14.24) to explain why.

CONCEPT CHECK 14.5 | Designing a Better Building

The walls of buildings are made from many different materials, including concrete, wood, glass, and steel. If solid walls with the same area and thickness are made from these materials, which would allow the most heat conduction through the wall, (a) concrete, (b) wood, (c) glass, or (d) steel?

EXAMPLE 14.9 Conduction of Heat through a Windowpane

Consider a glass windowpane (Fig. 14.25) with an area $A = 0.50$ m^2, a thickness $L = 4.0$ mm, and a temperature difference $\Delta T = 30°$C between the two sides (i.e., between the inside of the house and the outside). What is the total amount of heat energy that flows through the windowpane in one day?

RECOGNIZE THE PRINCIPLE

We are given the dimensions of the windowpane, and we know the thermal conductivity of glass from Table 14.6. We can therefore use Equation 14.12 to calculate the rate at which heat flows through the windowpane, and from that we can calculate the total energy that flows through in one day.

SKETCH THE PROBLEM

Figure 14.25 describes the problem.

IDENTIFY THE RELATIONSHIPS

The value of a temperature difference on the Celsius and Kelvin scales is the same, so $\Delta T = 30$ K. Inserting these values into Equation 14.12 with the value of the thermal conductivity κ for glass from Table 14.6, we find

$$\frac{Q}{t} = \kappa A \frac{\Delta T}{L} = \left(0.80 \frac{\text{W}}{\text{m} \cdot \text{K}}\right)(0.50 \text{ m}^2)\frac{30 \text{ K}}{0.0040 \text{ m}} = 3000 \text{ W} \qquad (1)$$

The total energy that flows out through the window in one day can be calculated from Equation (1) using $t = 1$ day $= 86,400$ s. We get

$$Q = (3000 \text{ W})t = (3000 \text{ J/s})(86,400 \text{ s}) = \boxed{2.6 \times 10^8 \text{ J}} \qquad (2)$$

What does it mean?
This energy comes (ultimately) from the house's furnace. If it is an electric furnace, the cost of electrical energy is paid to the local utility company, with the typical charge being about $0.02 for 10^6 J. The total cost of the energy Q in Equation (2) is thus approximately

$$\text{cost} = (2.6 \times 10^8 \text{ J})(\$0.02/10^6 \text{ J}) = \$5$$

A cost of $5 per window per day is quite substantial because most houses have many windows; that is why energy-efficient windows (see Problem 38 at the end of this chapter) are popular.

Table 14.6 Thermal Conductivities of Some Common Substances

MATERIAL	κ W/(m · K)
Argon gas	0.016
Air	0.024
Wood	0.15
Asbestos	0.17
Cotton cloth (approximate)	0.2
Wool	2.2
Goose down	0.025
Styrofoam	0.010
Body fat	0.20
Snow	0.25
Glass	0.80
Water	0.56
Concrete	1.7
Aluminum	240
Iron	79
Steel	50
Copper	400
Silver	430
Ice (0°C)	2.2

Note: All substances are at atmospheric pressure and room temperature unless noted.

Figure 14.25 Conduction of heat through a windowpane.

Suppose the height and width of the windowpane in Figure 14.25 are both increased by a factor of two. By what factor does the heat flow through the window change?

(a) There is no change.

(b) It increases by a factor of two.

(c) It increases by a factor of four.

EXAMPLE 14.10 | Heat Flow and Cooling Your Coffee

The author likes to bring a thermos bottle of coffee to his office each morning. The walls and lid of a thermos bottle have low thermal conductivities, so the rate of heat flow out of the bottle is small (Fig. 14.26A). Suppose the coffee inside has a mass $m = 0.30$ kg and an initial temperature of 65°C. Coffee is mostly water, so its specific heat is about the same as that of water. If the average rate of heat flow through the walls of the thermos bottle is $Q/t = 3.0$ W, approximately how long will it take the coffee to cool to a final temperature of 55°C? Take room temperature to be 25°C.

RECOGNIZE THE PRINCIPLE

We use the general approach outlined in the problem-solving strategies on calorimetry. We identify two systems, the coffee and the room around it. These systems are initially at temperatures $T_{coffee, i} = 65$°C and $T_{room} = 25$°C, and the final temperature of the coffee is $T_{coffee, i} = 55$°C. Because the room is much larger than the coffee, the temperature of the room does not change significantly. (The room acts as a thermal reservoir.) The heat energy that must flow from the coffee into the room to produce this change in temperature can be found from the specific heat of the coffee and Equation 14.6. This energy equals the rate of heat flow (given to be $Q/t = 3.0$ W) multiplied by the total time.

SKETCH THE PROBLEM

Figure 14.26 describes the problem and shows the direction of heat flow.

IDENTIFY THE RELATIONSHIPS

Using the definition of specific heat (Eq. 14.6) gives

$$Q = cm\,\Delta T = cm(T_{coffee, f} - T_{coffee, i})$$

Inserting the given values for the temperatures, mass, and specific heat of water (Table 14.3), we find

$$Q = [4186 \text{ J/(kg} \cdot \text{K)}](0.30 \text{ kg})(55 - 65) \text{ K} = -1.3 \times 10^4 \text{ J} \qquad (1)$$

The negative sign indicates that energy flows out of the coffee. Here, Q equals the rate at which heat energy travels through the walls of the thermos bottle multiplied by the total time. The rate is $rate = |Q|/t = 3.0$ W, so we have

$$|Q| = (rate)(t_{total})$$

SOLVE

Using the result for Q from Equation (1) leads to

$$t_{total} = \frac{|Q|}{rate} = \frac{1.3 \times 10^4 \text{ J}}{3.0 \text{ W}} = \boxed{4300 \text{ s}}$$

What does it mean?

This result is more than 1 hour, which gives plenty of time to enjoy the coffee. A better thermos bottle would have a smaller heat flow rate and would thus keep the coffee warm longer.

Coffee
$T_i = 65$°C

Room
$T_i = 25$°C

Low thermal conductivity

A

Coffee → Room
Q

B

Figure 14.26 Example 14.10.
A The coffee in the author's thermos bottle is initially much warmer than room temperature.
B Energy flows via conduction from the coffee to the room, thus cooling the coffee.

Why Do Metals "Feel" Cold? ⊗

The concepts of thermal conductivity and thermal resistance can clarify a well-known aspect of heat conduction. Experience tells us that a metal object "feels" much colder than a piece of Styrofoam. For example, on a wintry day, an outdoor metal flagpole feels colder to the touch than Styrofoam. If the flagpole and the Styrofoam are at the same outdoor temperature, however, why do they feel different?

The temperature-sensitive nerves in your skin sense the difference between your inside body temperature and your skin temperature. We thus have a heat-flow problem involving a series of materials (Fig. 14.27). The thermal conductivity κ of a metal is high (Table 14.6), whereas κ for Styrofoam is low. So, according to the definition of thermal conductivity (Eq. 14.12), when your finger is in contact with a metal, the rate of heat flow Q/t from your finger to the metal is much larger than if you were touching Styrofoam instead. This increased heat flow into the metal causes your skin to have a lower temperature than with Styrofoam. That is why a metal feels colder to the touch than Styrofoam.

$Q_A > Q_B$ because metal is a better conductor of heat than Styrofoam. It makes T_{skin} lower in **A**.

Figure 14.27 Heat flows from inside your body into your surroundings. **A** If the outer layer of skin is in contact with a metal (a good thermal conductor), the rate of heat flow is greater than in **B**, where the skin is in contact with Stryofoam (a poor thermal conductor). The higher heat-flow rate in **A** makes the outer layer of skin colder than in **B**.

14.7 | CONVECTION

Convective heat flow is based on the phenomenon of thermal expansion. Consider a pot of water being heated on a stove (Fig. 14.28A). If the pot is heated gently so that no gas bubbles are formed, heat energy is carried through the water in two ways. One way is by conduction as discussed in Section 14.6, but much more energy can be transported via the motion of water sketched in Figure 14.28A. Water near the bottom becomes hotter due to the conduction of heat from the burner. Thermal expansion then causes the density of this warmer water to be less than the density of the colder water that is above it in the pot. This warm low-density water moves upward due to the buoyant force associated with Archimedes's principle. As the warm water moves upward, it cools through conduction of heat to the cooler part of the pot and the air above; eventually, this water returns to the bottom of the pot, following the convective (circular) flow pattern in Figure 14.28A. A similar convective heat-flow pattern involving the air in a house causes the upper floors to be warmer than the lower floors. Convective flow also plays an important role in transporting heat energy within the oceans and the atmosphere.

Wind Chill ⊗

A process similar to convection leads to the "wind chill" effect familiar on a cold winter day. Your skin is moist, and water from it evaporates slowly, placing a small amount of water vapor into the layer of air next to your skin. Evaporation is a phase change, requiring an amount of energy equal to the latent heat. This energy comes from your skin and flows to the nearby air, causing your skin to cool (Fig. 14.29A). This process is called *evaporative cooling*. In our previous discussion of the latent heat of vaporization of water, we assumed water was at the boiling temperature

Figure 14.28 Convection transports energy through the movement of matter. This movement is caused by thermal expansion.

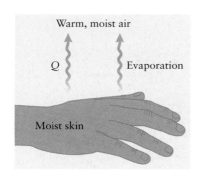

Warm, moist air

Q Evaporation

Moist skin

A

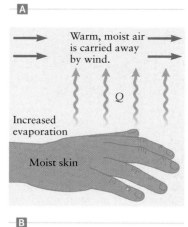

Warm, moist air is carried away by wind.

Q

Increased evaporation

Moist skin

B

Figure 14.29 **A** When moisture evaporates from the body, the latent heat of evaporation comes from your skin, cooling your body and forming a layer of moist and relatively warm air near your skin. **B** On a windy day, this moist layer of air is rapidly blown away, increasing the evaporation rate, which increases the amount of latent heat removed from your body and lowers the skin temperature relative to **A**.

Q Q

Figure 14.30 Example 14.11. Convection or just wind carries moist air away from this cyclist, increasing the rate at which sweat evaporates.

(100°C), but there is a latent heat of vaporization at all points along the liquid–gas phase change line in Figure 14.10. Evaporative cooling can thus occur over a wide temperature range.

On a windy day, the layer of water-vapor-filled air near your skin is quickly carried away, leaving a "dry" layer in its place. The rate of evaporation is faster into this dry layer than it would be into a moist layer of air. As a result, more latent heat is removed from your skin than if the air layer were filled with water vapor (Fig. 14.29B). Your skin thus loses energy faster on a windy day than on a calm day.

EXAMPLE 14.11 ⊗ Cooling Your Body by Sweating

The bicycle rider in Example 14.3 generated 300 J of heat each second. If all this energy is removed from his body by sweating, how much water would he lose each hour by sweating?

RECOGNIZE THE PRINCIPLE

When sweat evaporates from a cyclist's skin, it carries away an amount of energy equal to the latent heat of vaporization. The motion of the bicycle provides the "wind" that carries away the water vapor from a cyclist's skin.

SKETCH THE PROBLEM

Figure 14.30 describes the problem. Heat energy flows from the cyclist into the surrounding air.

IDENTIFY THE RELATIONSHIPS

Let m be the mass of water evaporated as sweat in 1 h. The corresponding latent heat is

$$Q = mL_{\text{vaporization}} \qquad (1)$$

which must equal the amount of heat generated by the cyclist. The cyclist generates 300 J each second, so the total heat generated in 1 h (= 3600 s) is

$$\text{total heat generated} = (300 \text{ J})(3600 \text{ s}) = 1.1 \times 10^6 \text{ J}$$

SOLVE

Setting the heat generated equal to the latent heat in Equation (1) gives

$$Q = mL_{\text{vaporization}} = 1.1 \times 10^6 \text{ J}$$

Solving for m and using the latent heat of vaporization of water from Table 14.4, we get

$$m = \frac{1.1 \times 10^6 \text{ J}}{2.2 \times 10^6 \text{ J/kg}} = \boxed{0.50 \text{ kg}}$$

What does it mean?
This mass corresponds to about half a liter of water, which is a typical amount of water a person should consume during 1 h of strenuous exercise.

14.8 | HEAT FLOW BY RADIATION

Both heat-flow processes discussed so far—conduction and convection—involve matter. Conduction involves the transfer of energy between vibrating atoms (and electrons) within the conducting substance, and convection involves the net motion of matter within a fluid. A third type of heat flow has an entirely different basis. *Radiative* heat flow involves energy carried by electromagnetic radiation.

Electromagnetic radiation is a type of wave and is thus characterized by frequency, wavelength, and wave speed. Like other types of waves, electromagnetic waves carry energy. The wavelength of a particular electromagnetic wave depends on how the wave is generated, and it can vary over an extremely wide range. Visible light is the most familiar type of electromagnetic radiation; it has a wavelength in the range of about 400 nm to 700 nm, which corresponds to frequencies of about 8×10^{14} Hz to 4×10^{14} Hz. Other types of electromagnetic radiation correspond to different wavelengths and frequencies, including ultraviolet and infrared radiation, which are also important in heat flow.

Radiative heat flow involves two objects, which might be the Sun and a person on the Earth as sketched in Figure 14.31A. If these two objects are at different temperatures, there is a net transfer of energy—heat energy—from the hotter object (in this case the Sun) to the cooler one (the person).

To understand radiative heat transfer, we must consider several questions. First, why does an object, such as the Sun in Figure 14.31, emit electromagnetic radiation, and how is this radiation absorbed by another object? Second, what are the frequencies of the radiated electromagnetic waves, and how do they depend on the temperature of the emitting object? Finally, how does the amount of radiation absorbed by the cooler object depend on the temperature of that object?

Electromagnetic radiation is generated whenever a particle with an electric charge, such as an electron, proton, or ion, vibrates or undergoes an acceleration in some way. Hence, the atomic vibrations sketched in Figure 14.24A generate electromagnetic radiation that carries away energy. The vibration amplitude of the atoms in an object depends on the temperature, so the radiated energy depends on temperature. This radiation—that is, the energy in these electromagnetic waves—is absorbed by another object when that radiation produces a force on the electric charges in the object.

Radiation and the Notion of a "Blackbody"

When electromagnetic radiation, including visible light, bombards an object, some of the radiation energy is absorbed and some is reflected. When you look at an object such as a wall or this book, your eye detects light reflected by the object, and this reflected light determines the color of the object. An object that completely absorbs all visible light appears black. Physicists have extended this notion and defined a **blackbody** as an object that absorbs all electromagnetic radiation at all frequencies. A perfect blackbody does not exist, but the concept of an ideal blackbody is very useful.

The total amount of energy an object radiates depends on its temperature. An object with a surface area A and a temperature T radiates energy at a rate Q/t given by the **Stefan–Boltzmann law**,

$$\frac{Q}{t} = \sigma e A T^4 \qquad (14.13)$$

Stefan–Boltzmann law

where $\sigma = 5.67 \times 10^{-8}$ W/(m² · K⁴) is called Stefan's constant. The factor e in Equation 14.13 is the **emissivity** of the object, a measure of how efficiently it radiates energy. The value of the emissivity must be less than or equal to 1. Objects for which $e = 1$ are perfect blackbodies. Hence, a blackbody is not only a perfect absorber of radiation, it is also the best possible radiator of energy.

The Stefan–Boltzmann law gives the rate at which energy is emitted as electromagnetic radiation—that is, the radiated power—at all wavelengths. This power is distributed as a function of wavelength as shown in Figure 14.32. For an object at temperature T, the radiated power is largest at a wavelength λ_{max}, which is given by **Wien's law**:

$$\lambda_{max} = (2.90 \times 10^{-3} \text{ m} \cdot \text{K})/T \qquad (14.14)$$

Wien's law

Figure 14.32 shows a plot of the radiated power as a function of wavelength for blackbody objects at different temperatures. According to Wien's law, the

Energy radiated by the Sun is absorbed by objects on Earth.

Earth

A

Sun $\xrightarrow{Q_{sun}}$ Person
$\xleftarrow{Q_{person}}$

$Q_{Sun} > Q_{person}$ because $T_{Sun} > T_{person}$.

B

Figure 14.31 **A** Electromagnetic radiation from the Sun is absorbed by a person on the Earth. **B** The person also emits radiation, so there is a flow of energy in both directions, from the Sun to the person and from the person to the Sun.

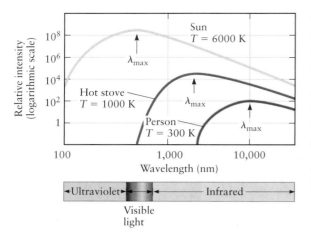

wavelength λ_{max} at which the radiated power is largest depends on the temperature of the object. The Sun has a surface temperature of approximately 6000 K, and λ_{max} falls in the range of visible light. Some stars are much hotter than our Sun, with temperatures as high as 20,000 K, and for these stars, λ_{max} is shorter than the visible range of wavelengths. These stars appear blue to the human eye because they emit more radiation at the blue end of the visible range than at the red end. For a somewhat cooler object, such as the burner of an electric stove at a typical temperature of 1000 K, λ_{max} lies in the infrared. The stove burner appears red because it emits some light at the red end of the visible spectrum in addition to its infrared radiation.

Figure 14.32 According to Wien's law, the wavelength at which an object emits radiation most strongly (λ_{max}) depends on its temperature, and that determines the "color" at which it glows. Notice how the total amount of radiation emitted increases very rapidly as T increases.

EXAMPLE 14.12 ⊗ Radiation Emitted by the Human Body

The human body emits electromagnetic radiation according to the body's temperature. Estimate λ_{max} for the human body.

RECOGNIZE THE PRINCIPLE

The peak intensity of the blackbody radiation from an object occurs at a wavelength that depends on its temperature according to Wien's law. Knowing the temperature of the body, we can find λ_{max}.

SKETCH THE PROBLEM

No sketch is needed.

IDENTIFY THE RELATIONSHIPS

The temperature of the human body is approximately 37°C, which must be converted to the Kelvin scale before insertion into Wien's law. We find $T = 37 + 273.15 = 310$ K.

SOLVE

Using this value of the temperature in Wien's law (Eq. 14.14), we find

$$\lambda_{max} = (2.90 \times 10^{-3} \text{ m} \cdot \text{K})/T = (2.90 \times 10^{-3} \text{ m} \cdot \text{K})/(310 \text{ K}) = \boxed{9.4 \times 10^{-6} \text{ m}}$$

What does it mean?

The wavelength λ_{max} lies far into the infrared region of the spectrum. It is well outside the visible range, so your eye cannot detect this emitted radiation. The radiation people emit, however, can be used to "see" them in an otherwise dark environment. Night-vision equipment is able to detect this radiation and convert it into wavelengths that can be detected by the human eye (Fig. 14.33A). Also, the "ear thermometers" used by medical professionals (Fig. 14.33B) measure your body temperature by detecting the infrared radiation emitted by your eardrum.

Figure 14.33 Example 14.12. A Night-vision goggles convert the infrared radiation emitted by an object to visible light, making it possible to see objects in the dark. B An "ear thermometer" uses the blackbody radiation from the eardrum to find the body's temperature.

The Stefan–Boltzmann Law and Heat Flow

The Stefan–Boltzmann law (Eq. 14.13), which gives the total power radiated by an object at temperature T, has several noteworthy features. First, the total power varies as the fourth power of the temperature, so the radiated energy increases very rapidly as T is increased. Second, the power is proportional to the emissivity e; this factor is equal to unity for an ideal blackbody, but is smaller than 1 for any real object. In general, the value of e depends on the detailed properties of the material and will even be a function of frequency. Many objects are not far from the blackbody limit, however, so the Stefan–Boltzmann law with an emissivity factor of $e = 1$ provides an approximate description of most radiating objects. Third, the Stefan–Boltzmann law applies to all objects, at all temperatures. The Sun radiates energy that reaches a person on the Earth, but the person also radiates energy that reaches the Sun (Fig. 14.31B). Because the hotter object has a larger emitted power than the cooler one, the net transfer of energy is from the hotter to the cooler object, as we expect for heat flow. Fourth, the Stefan–Boltzmann law also describes how radiation is absorbed by an object. When an object of surface area A is located in an environment that has a temperature T, the heat absorbed by the object is

$$\left(\frac{Q}{t}\right)_{\text{absorbed}} = \sigma e A T^4 \qquad (14.15)$$

Radiation from the Sun and the Temperature of the Earth

Let's apply the Stefan–Boltzmann law to the energy radiated by the Sun. The Sun has a surface temperature of approximately 6000 K, and its emissivity is close to unity $e \approx 1$. The radius of the Sun is $r = 7.0 \times 10^8$ m, and the surface area through which the radiation is emitted is just the area of a sphere with this radius, $A = 4\pi r^2 = 6.2 \times 10^{18}$ m^2. Inserting these values into the Stefan–Boltzmann law (Eq. 14.13), we have

$$\frac{Q}{t} = \sigma e A T^4 = [5.67 \times 10^{-8} \text{ W/(m}^2 \cdot \text{K}^4)](1)(6.2 \times 10^{18} \text{ m}^2)(6000 \text{ K})^4$$

$$\frac{Q}{t} = 4.6 \times 10^{26} \text{ W} \qquad (14.16)$$

which is an enormous amount of power. Equation 14.16 gives the rate at which energy leaves the Sun, so to find the total radiated energy, we must multiply it by the time interval of interest.

We can use our result for the heat radiated by the Sun to compute the Earth's temperature. This calculation is based on the principle of conservation of energy. According to Figure 14.34, energy arrives at the Earth via radiation from the Sun, and the Earth itself then radiates energy back into space.[2] The amount of energy radiated by the Earth depends on its temperature T_E, and the value of T_E must be such that the total energy *radiated by* the Earth is equal to the energy it *absorbs* from the Sun. We'll leave the details of this calculation to the next example.

EXAMPLE 14.13 ⓡ Temperature of the Earth

Estimate the temperature of the Earth according to the picture of heat flow in Figure 14.34.

RECOGNIZE THE PRINCIPLE

We now complete the calculation we began above. We need to compute the amount of energy radiated by the Sun that is absorbed by the Earth, Q_{Sun}, and the energy radiated by the Earth, Q_{Earth}. We have already calculated the total power radiated by the Sun in Equation 14.16, but only a fraction of this power reaches the Earth. That fraction is given by the area of the Earth's *disk* (not the spherical surface area of the Earth) divided by the surface area of a sphere that contains the Earth's orbit (Fig. 14.35).

SKETCH THE PROBLEM

Figure 14.35 shows the problem.

Insight 14.2

THE RELATION BETWEEN RADIATION AND ABSORPTION

The same relation gives the amount of radiation emitted by an object at a particular temperature (Eq. 14.13) and the amount of radiation absorbed by an object from an environment at that same temperature (Eq. 14.15). This follows from the notion of thermal equilibrium. If an object is in thermal equilibrium with its environment, both are at the same temperature. The total energy of the object is then constant, and the rate at which the object emits energy (through radiation) must equal the rate at which it absorbs energy (also through radiation) from its environment. Hence,

$$\left(\frac{Q}{t}\right)_{\text{absorbed}} = \left(\frac{Q}{t}\right)_{\text{emitted}}$$

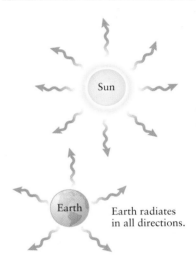

Figure 14.34 The Earth absorbs energy through radiation from the Sun and emits radiation according to its own temperature. The balance of these energies determines the temperature of the Earth.

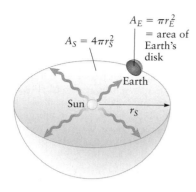

Figure 14.35 Example 14.13. The Sun emits radiation that propagates outward with circular wave fronts. The spherical wave front that intercepts the Earth has an area $A_S = 4\pi r_S^2$, where r_S is the Earth–Sun distance. The area of the Earth exposed to the Sun is $A_E = \pi r_E^2$, where r_E is the radius of the Earth. Notice that A_E is *not* the (spherical) surface area of the Earth.

IDENTIFY THE RELATIONSHIPS

Denoting the radius of the Earth by r_E, the area of the Earth's disk is $A_E = \pi r_E^2$. If r_S is the Earth–Sun distance, the surface area of the sphere that contains the Earth's orbit (Fig. 14.35) is $A_S = 4\pi r_S^2$, and the fraction of the Sun's radiated power that strikes the Earth is the ratio A_E/A_S. The radiated power from the Sun that actually intercepts the Earth is thus

$$\frac{Q_{\text{Sun}}}{t} = \frac{A_E}{A_S}\frac{Q_{\text{total}}}{t} = \frac{\pi r_E^2}{4\pi r_S^2}\frac{Q_{\text{total}}}{t}$$

where Q_{total}/t is given in Equation 14.16. Inserting values for r_E, r_S (from Appendix A, Table A.3), and Q_{total}, we obtain

$$\frac{Q_{\text{Sun}}}{t} = \frac{\pi r_E^2}{4\pi r_S^2}\frac{Q_{\text{total}}}{t} = \frac{\pi(6.4 \times 10^6\text{ m})^2}{4\pi(1.5 \times 10^{11}\text{ m})^2}(4.6 \times 10^{26}\text{ W}) = 2.1 \times 10^{17}\text{ W} \quad (1)$$

The energy radiated by the Earth is also given by the Stefan–Boltzmann law, Equation 14.13. The factor A in this application of the Stefan–Boltzmann law is the area of the Earth's spherical surface, which is $4\pi r_E^2$. (It is *not* the same as the area of the Earth's disk in Fig. 14.35.) The temperature of the Earth T_E is the quantity that we wish to find. Inserting all these values into the Stefan–Boltzmann law gives

$$\frac{Q_{\text{Earth}}}{t} = \sigma e A T_E^4$$

SOLVE

The rate at which the Earth absorbs energy must equal the rate at which it radiates; otherwise, the Earth would be heating up or cooling down over time. Hence,

$$\frac{Q_{\text{Sun}}}{t} = \frac{Q_{\text{Earth}}}{t} = \sigma e A T_E^4$$

Solving for T_E leads to

$$T_E^4 = \frac{Q_{\text{Sun}}/t}{\sigma e A}$$

$$T_E = \left(\frac{Q_{\text{Sun}}/t}{\sigma e A}\right)^{1/4}$$

Taking the Earth's emissivity as $e = 1$, inserting Q_{Sun}/t from Equation (1), and using $r_E = 6.4 \times 10^6$ m gives

$$T_E = \left[\frac{2.1 \times 10^{17}\text{ W}}{[5.67 \times 10^{-8}\text{ W/(m}^2 \cdot \text{K}^4)](1)(4\pi r_E^2)}\right]^{1/4} = \boxed{290\text{ K}}$$

What does it mean?

Our result for T_E is approximately 20°C. In obtaining this result, we assumed the Earth is a perfect blackbody, so $e = 1$. In fact, the Earth absorbs only about 70% of the radiation from the Sun and reflects the rest. In addition, our calculation actually gives the temperature of the atmosphere, which is somewhat cooler than the Earth's surface. Even with these approximations, our result is quite close to the actual value, confirming that the Sun is responsible for maintaining the Earth's temperature.

Medical Uses of Heat Radiation ⊗

An important application of heat radiation is medical thermography. This technique measures the heat radiation emitted by the body and forms an image based on the wavelength at which the radiated power is largest. According to Wien's law (Eq. 14.14), this wavelength depends on temperature, so these images show where a body is warmest and where it is coolest. Some examples are given in Figure 14.36; in these images, whiter regions are warmer than redder areas. Figure 14.36A shows that the nose is the coolest place on a person's face (not a surprising result!). There can be significant variations in temperature near joints; the image in Figure 14.36B shows arthritis associated with a knee joint. Other medical problems can be detected in this way; for example, cancerous tumors are often warmer than normal tissue. Other medical devices, such as ear thermometers (Example 14.12 and Fig. 14.33B), also use the body's blackbody radiation.

The Greenhouse Effect ⊗

As a first approximation, we can treat most real objects as perfect blackbodies. Such was the approach that we took in Example 14.13, when we estimated the temperature of the Earth. A more realistic model of the Earth's temperature must go beyond this approximation, which leads us to consider the *greenhouse effect*.

Although a complete discussion of the greenhouse effect is too complex to include here, we can understand the qualitative behavior using Figure 14.37, which shows how radiated energy is absorbed and reemitted by the Earth. This sketch is similar to Figure 14.34, but now we have included the effect of the atmosphere. We also indicate separately the infrared and visible radiation coming from the Sun; the atmosphere is not an ideal blackbody, and it absorbs these radiations differently. The Earth's atmosphere allows most of the radiation in the visible through to the surface, whereas it absorbs much of the infrared radiation, allowing only a little of it to reach the Earth. Because much of the Sun's radiation is in the visible range, most of the energy from the Sun still passes through to the Earth's surface.

When we estimated the temperature of the Earth in Example 14.13, we had to account for radiation emitted by the Earth. Because its temperature is much lower than that of the Sun, most of the power radiated by the Earth is in the infrared (Fig. 14.32). The atmosphere absorbs some of this radiation, effectively trapping it and not letting this energy escape back into space. This pattern of energy flow is similar to what takes place in a greenhouse, where the glass roof plays the role of the atmosphere. (The analogy with a true greenhouse is not perfect, but the effect is still widely referred to as "the" greenhouse effect.) The infrared radiation trapped by the atmosphere makes the Earth's surface warmer than would be found if the atmosphere were not present.

The absorption of infrared radiation by the atmosphere is caused by certain molecules such as CO_2, which are called *greenhouse gases*. If the concentration of these gases in the atmosphere increases, the atmosphere absorbs infrared radiation more strongly and traps even more of the Earth's emitted radiation. The result is an increase in the Earth's temperature. The concentration of greenhouse gases in the atmosphere can be altered by human processes. It appears that the burning of fossil fuels is now causing an increase in the concentration of greenhouse gases, and many scientists believe that it will cause a significant increase in the Earth's temperature in the coming decades.

The details of radiative heat flow can be quite involved; even Figure 14.37 is a highly simplified picture of the heat-flow processes that determine the Earth's temperature. Our simple blackbody approximation is a very useful first step, but more accurate models are much more complex.[3]

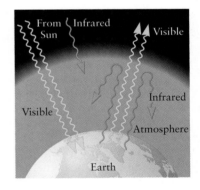

Figure 14.36 All objects, including people, emit radiation according to their temperature. With special detection equipment, this radiation can be used to deduce how the temperature of the skin varies from place to place on the body. **A** The skin temperature is lowest at the nose. **B** An arthritic knee (left) is warmer than a healthy one.

Figure 14.37 Greenhouse model of radiative heat flow between the Sun, the Earth, and the Earth's atmosphere.

---

[3] See T. R. Karl, N. Nicholls, and J. Gregory, "The Coming Climate," *Scientific American* May:79 (1997).

Zeroth law of thermodynamics

If two systems A and B are in thermal equilibrium ($T_A = T_B$) and systems B and C are in thermal equilibrium ($T_B = T_C$), systems A and C are also in thermal equilibrium ($T_A = T_C$).

In words, the zeroth law of thermodynamics asserts that temperature is a unique property of a system that determines when two systems are in thermal equilibrium. Three different *temperature scales* are in common use: *Fahrenheit, Celsius,* and *Kelvin.*

Heat

Heat flow is the transfer of energy from one object to another by virtue of a temperature difference. The energy transferred is also called heat energy or simply heat. Heat flow can occur in three ways: *conduction, convection,* and *radiation.*

- *Heat conduction* is determined by the *thermal conductivity* κ, with

$$\frac{Q}{t} = \kappa A \frac{\Delta T}{L} \qquad \text{(14.12) (page 453)}$$

where Q is the heat energy transferred in a time t, L is the length of the system, and A is the cross-sectional area of the system.

- *Convective heat flow* involves the movement of mass and is caused by thermal expansion and contraction.
- *Radiative heat transfer* involves electromagnetic radiation. The total amount of energy an object radiates depends on its temperature. An object with a surface area A and temperature T radiates energy at a rate Q/t given by the *Stefan–Boltzmann law,*

$$\frac{Q}{t} = \sigma e A T^4 \qquad \text{(14.13) (page 457)}$$

The wavelength at which this radiation is strongest depends on temperature through *Wien's law,*

$$\lambda_{\max} = (2.90 \times 10^{-3} \text{ m} \cdot \text{K})/T \qquad \text{(14.14) (page 457)}$$

That is why the color of a glowing object depends on its temperature.

Specific heat and latent heat

If an amount of heat energy Q is added to a substance of mass m, the temperature of the system increases by an amount ΔT, where

$$Q = cm \, \Delta T \qquad \text{(14.6) (page 441)}$$

and c is the *specific heat* of the substance. Equation 14.6 applies provided that there is not a phase change. If there is a phase change such as melting or evaporation, a certain amount of energy called the *latent heat* must be added to or taken from the system.

Thermal expansion

When the temperature of a substance changes, its physical dimensions usually change. Changes in length are proportional to the *coefficient of thermal expansion* α, with

$$\frac{\Delta L}{L_0} = \alpha \, \Delta T \qquad \text{(14.9) (page 448)}$$

and changes in volume are proportional to the *coefficient of volume expansion* β

$$\frac{\Delta V}{V_0} = \beta \, \Delta T \qquad \text{(14.10) (page 448)}$$

SSM = answer in Student Companion & Problem-Solving Guide

 = life science application

1. When the metal cap on a glass jar is stuck (i.e., when the cap is too tight to open), it sometimes helps to hold the jar and cap under hot water. Explain why.

2. Consider a hole in a block of steel such as the cylinder in a steel engine block. Will the diameter of this hole get larger or smaller when the steel is heated? By how much will it change?

3. Common bulb thermometers use liquid mercury (the familiar silver liquid) or liquid methanol (with red dye for coloring) as the working substance enclosed in glass. What are the limitations in using these devices? Could a mercury or methanol thermometer measure the temperature of dry ice? (CO_2 is a solid below $-78°C$ at atmospheric pressure.) Why or why not? Could either type of thermometer measure the temperature of molten lead?

4. A bar of steel is attached to a bar of brass side by side to form one thicker bar of the same length (Fig. Q14.4). You find that when you heat this bar of two metals, it bends. Why? Will it bend toward the brass or toward the steel? Which way would it bend if the bar were cooled instead of heated?

Figure Q14.4

5. The pressure–temperature phase diagram for a typical substance is shown in Figure Q14.5. The line that separates the solid and liquid phases has a positive slope; that is, the melting temperature increases as the pressure is increased. Give a qualitative explanation of this observation, assuming that a material expands when it is heated.

6. When you step out of a swimming pool on a windy day, you usually feel chilly. Explain why.

Figure Q14.5

7. On a hot summer day, an electric fan can make a difference in comfort. The fan does not cool the air, however. In fact, it slightly raises the temperature of the air that it circulates. How is it that a fan can make you cooler?

8. Most internal combustion engines in modern automobiles have a cooling system that uses a circulating fluid. In designing such a cooling system, should the liquid coolant have a small or a large heat capacity? Which would be best, a large or a small thermal conductivity?

9. Consider a winter mountain survival scenario. Why is it a bad idea to eat snow, even if you have no way of melting it for drinking?

10. Given the value of α (e.g., for steel), can you calculate the value of β? Derive a general relation between α and β.

11. Explain why getting burned with a certain mass of steam at $100°C$ is more painful and harmful than getting burned with the same mass of water (liquid) at $100°C$.

12. When you take a cookie sheet out of the oven, you use a pot holder because touching the hot cookie sheet would result in injury. The air in the oven, however, is at exactly the same temperature. Why doesn't the air burn your hand the way the cookie sheet would?

13. You pour a cup of coffee, intending to drink it a few minutes later. If you want your coffee to be as hot as possible at this later time, should you add the cream when you pour the coffee or just before you drink it? Why?

14. SSM Fruit trees bloom in the spring, and their blossoms eventually turn into oranges, apples, and so forth. Blossoms are susceptible to cold weather, so when there is the possibility of a frost, orchard owners often spray their trees with water. Explain how this practice can help prevent the blossoms from freezing.

15. Metal cookware made from steel or aluminum often has a copper bottom. Why?

16. A potato will bake faster if a nail is stuck through it. Why? Which would make the potato cook faster, an aluminum nail or a steel nail?

17. Explain why a layer of fur (Fig. Q14.17) is a very poor conductor of heat.

18. There are many different types of glass; all are composed mainly of SiO_2, but the addition of other elements can change the color and other properties substantially. One type of glass, called Pyrex, incorporates small amounts of the element boron, which gives it a very small coefficient of thermal expansion. Most oven-safe dishes are made from Pyrex because it tends not to crack when heated or cooled. Explain why.

Figure Q14.17

19. On a camping trip in the mountains, you find that it takes longer to cook food such as rice and noodles than at sea level. Explain why.

20. SSM A good-quality thermos bottle is double-walled and evacuated between these walls, and the internal surfaces are like mirrors with a silver coating. How does this configuration combat heat loss from all three transfer methods and keep the bottle's contents—your coffee—hot?

21. Table 14.6 lists the thermal conductivities of several substances. Give qualitative arguments to explain the following questions. (a) Why is the thermal conductivity of goose down about the same as the thermal conductivity of air? (b) Goose down is very effective for making winter coats. Explain why goose down must be kept dry to be effective. Can you predict the value of the thermal conductivity when it is soaked with water?

22. Heat energy is "lost" from the body through the skin, so Q_{lost} is proportional to the area of the skin. Heat energy is produced throughout the body as food is metabolized, muscles flex, and so on, so $Q_{produced}$ is proportional to the body volume. Explain why these two facts tend to make it easier for a large animal to stay warm compared with a small animal.

23. An ice cube can exist for many minutes while floating in water, even when the temperature of the water is greater than $0°C$. Explain why.

24. On a hot summer day, why is it easier to keep cool by wearing white clothing instead of black clothing?

25. When a wire is strung tightly between two supports, it is found that transverse waves on the wire have a speed v. If the wire is heated, does the wave speed increase or decrease? Explain.

26. Why does a lake freeze first at its surface?

27. Explain why it takes longer to cook a hard-boiled egg in Denver than in San Francisco. *Hint*: Consider the phase diagram of water and how the boiling temperature depends on pressure.

28. Climates near an ocean tend to be milder than climates far inland. Explain why.

14.2 TEMPERATURE AND HEAT

1. The melting point of tin is 232°C. What is this temperature on the Fahrenheit and Kelvin scales?

2. Liquid oxygen boils at 90 K at atmospheric pressure. What is this temperature in degrees Celsius?

3. What is the value of the "absolute zero" of temperature on the Fahrenheit scale?

4. What is the melting temperature of gold on the Celsius scale?

5. The temperature inside a typical house is 70°F. What is this temperature on the Celsius and Kelvin scales?

6. ★ You and a friend each have a thermometer; yours reads in Celsius, whereas your friend's is an old-fashioned Fahrenheit model. You go outside and notice that the two thermometers read the same value. Is it hot or cold outside? What is the outside temperature?

7. The lowest temperature ever recorded outdoors is approximately −130°F. What is this temperature on the Celsius scale?

8. SSM ★ Consider Joule's experiment described in Insight 14.1. Assume the mass $m = 2.0$ kg and the fluid is water with a volume of 1.0 L. What is the increase in the temperature of the water when the mass moves vertically a distance $z = 1.0$ m?

14.4 PHASES OF MATTER AND PHASE CHANGES

9. How much heat is required to raise the temperature of a 3.0-kg block of steel by 50 K?

10. ★ You have 5.0 g of an unknown substance. To identify the substance, you decide to measure its specific heat and find that it requires 16 J of heat to increase the temperature of your sample by 25 K. What might this substance be?

11. While working in a research lab, you find that it takes 400 J of heat to increase the temperature of 0.12 kg of a material by 30 K. What is the specific heat of the material?

12. How much heat is required to increase the temperature of a glass sphere of mass 400 g by 45 K?

13. ✪ An ice cube of volume 10 cm^3 is initially at a temperature of −15°C. How much heat is required to convert this ice cube into steam?

14. ✪ You drop 500 g of solid ice at −35°C into a bucket of water (1200 g) at 45°C. Eventually, the ice all melts, and the entire system comes into thermal equilibrium. Find the final temperature of the system.

15. SSM ★ If you combine 35 kg of ice at 0°C with 75 kg of steam at 100°C, what is the final temperature of the system?

16. ★ Ⓡ A cup of coffee is initially at a temperature of 70°C and then cools to room temperature. (a) Estimate how much heat flows out of the cup during this time. (b) If this energy were converted to the kinetic energy of a baseball, what would be the speed of the ball?

17. ✪ You are a blacksmith and have been working with 12 kg of steel. When you are finished shaping it, the steel is at a temperature of 400°C. To cool it off, you drop it into a bucket containing 5.0 kg of water at 50°C. How much of this water is converted to steam? Assume the steel, the water, and the steam all have the same final temperature.

18. ★ Ⓡ (a) How much energy is required to evaporate all the water in a swimming pool of area 100 m^2 and depth 2.0 m on a typical summer day? (b) The intensity of sunlight is about 1000 W/m^2. If all this energy is absorbed by the water in the pool, how long will it take the water to evaporate?

14.5 THERMAL EXPANSION

19. One section of a steel railroad track is 25 m long. If its temperature increases by 25 K during the day, how much does the track expand?

20. ★ Ⓡ A new cement bridge will be 50 m long. The bridge contains expansion joints, so it will not crack when the temperature changes. Suppose you are the designer of this bridge, and your job is to design the expansion joints. Estimate how much change in length the joints will have to accommodate.

21. Ⓡ The Golden Gate Bridge is a famous suspension bridge in San Francisco. The main segment is approximately 1400 m long and is composed of steel. Estimate how much the length of the bridge increases as time passes from morning to afternoon on a typical day.

22. ★ Your swimming pool is square and 5.0 m on a side. It is 3.0 m deep in the morning. If the temperature changes by 20°C during the afternoon, how much does the depth of the water increase?

23. A steel building is 120 m tall when the outside temperature is 0°C. How tall is the building on a hot summer day (30°C)? The change in height is small, so express your answer with five significant figures.

24. ★ You are given the job of designing a bridge that is to be made of concrete slabs that rest on a steel support frame. The total length of the bridge is 300 m, and you want to be sure that the expansion joints are sufficiently large. If the temperature increases by 25°C, what is the difference in the new lengths of the concrete and steel components of the bridge? Which of these components will have the largest increase in length?

25. A material of unknown composition with an initial length of 2.0 m is found to expand by 0.33 mm when it is heated by 15 K. What is the coefficient of thermal expansion of this material?

26. SSM ★ You are a high-precision carpenter and use a high-precision steel ruler. If you want the dimensions of your work to be accurate to 1.0 mm over distances of 15 m, you will have to keep the temperature of your ruler constant to within an uncertainly of ΔT. Find ΔT.

27. ★ A rectangular aluminum plate has an area of 0.40 m^2 at 15°C. If it is heated until its area has increased by 4.0×10^{-6} m^2, what is the final temperature of the plate?

28. ★ The author's gold wedding ring is stuck on his finger, and he must remove it. The ring has an inside diameter of 2.00 cm, but the knuckle of the finger in question has a diameter of 2.05 cm. The author decides to get the ring off by heating it. How much must the temperature of the ring be increased for it to slide past the knuckle? Assume the author's finger does not change size when it is heated.

29. A carpenter constructs a house of length 15 m out of wood. If the temperature changes by 5.0°C, how much does the length of the house change?

30. ★ Consider a steel rod of diameter 1.0 cm and length 1.0 m. If the temperature of this rod is increased by 25°C, its length will change by a certain amount ΔL. The length can also be changed by applying a tensile force (see Chapter 11). How much force must be applied to change the length of the rod by the same amount ΔL?

31. ✪ Ⓡ Design a mercury thermometer. Choose dimensions for the tube and the reservoir and then estimate how much the meniscus will move in response to changes in temperature; that is, calculate how far the meniscus moves when the temperature changes by 1°C.

32. ✪ Ⓧ A rock has a narrow crack (width 1.0 mm) into which water can penetrate. When this water freezes, it expands. (a) Estimate the pressure on the rock when the water freezes.

(b) What is the corresponding force on a rock surface that has an area of 1.0 cm² ? (c) Compare the force to your weight.

33. Ⓡ The Eiffel Tower in Paris (Fig. P14.33) is approximately 300 m tall and is composed of iron. How much does its height increase from midnight to noon on a typical summer day?

Figure P14.33

34. ✴ A pendulum clock uses a rock fastened to a very thin steel wire. This clock is designed so that one "tick" of the clock (one complete period of the pendulum) is equal to precisely 1 s. The clock is then heated by 10°C. Will it now run "fast" or "slow"? Find the new value of the period of the clock. Assume the thermal expansion coefficients of the rock are both zero.

35. ✪ An aluminum beam 80 m long and with a cross-sectional area of 0.10 m² is used as part of a bridge. The beam is clamped rigidly at both ends. (a) If the temperature of the beam is increased by 30°C, what is the extra force exerted by the beam on one of its supports? (b) What is the force if the temperature of the beam is decreased by 40°C?

14.6 HEAT CONDUCTION

36. ✖ A person has a core (inner) body temperature of 37°C, and the air temperature is 20°C. Estimate the rate at which heat flows via conduction through the person's skin. Assume a surface area of 1.5 m².

37. ✴ ✖ Ⓡ A cold-weather sleeping bag uses goose down to keep the inside warm. The rate at which a sleeping camper generates heat energy is about 70 W. Estimate the temperature of the air inside the bag and around the camper when the outside temperature of the bag is 0°C.

38. **Energy-efficient windows.** Energy-efficient windows are constructed with two panes of glass separated by an air gap (Fig. P14.38). In a typical window of this type, the air gap makes a huge difference in the rate of heat flow. The heat flow rate is much smaller for an energy-efficient window than for a single pane of glass, because the thermal conductivity of the air gap is much less than the thermal conductivity of a glass pane. Suppose the air gap is 1.0 cm thick. How thick would a pane of glass have to be to give the same rate of heat flow as through the air gap, assuming the same temperature difference?

Figure P14.38

39. SSM ✖ A small dog has thick fur, with a thermal conductivity of 0.040 W/(m · K). The dog's metabolism produces heat at a rate of 40 W and its internal (body) temperature is 38°C. If all of this heat flows out through the dog's fur, what is the outside temperature? Assume the dog has a surface area of 0.50 m², and that the length of the dog's hair is 1.0 cm.

40. ✴ Ⓡ Estimate the heat-flow rate for a solid cement wall of a typical-size house and compare it with the heat flow rate for the window in Example 14.9. Assume the wall is 5.0 m wide and 2.5 m tall.

41. ✖ A winter hiker's body generates heat at a rate up to a maximum value of 120 W. She is hiking on a brisk day in which the outdoor temperature is −10°C, while wearing a jacket with goose down insulation. How thick must the jacket be so that she can maintain a normal body temperature? Assume her surface area for heat flow (the jacket) is 1.5 m² and that all of her body heat flows out through the jacket.

14.7 CONVECTION

42. SSM ✖ When a dog "pants," it exhausts water vapor through its mouth (Fig. P14.42). This process converts liquid water (inside the dog) into water vapor, removing heat from the dog via the evaporation of water. If 10 g of water evaporates from the dog each minute, how much heat energy does it lose during this time?

Figure P14.42

43. ✖ A jogger generates heat energy at a rate of 800 W. If all this energy is removed by sweating, how much water must evaporate from the jogger's skin each hour?

14.8 HEAT FLOW BY RADIATION

44. ✴ In Example 14.13, we calculated the temperature of the Earth by assuming it is a perfect blackbody with an emissivity $e = 1$ (and we also ignored the atmosphere). Suppose the emissivity is $e = 0.70$. Find the temperature of the Earth (according to this simplified model) with this value of e. Explain why your answer is nearly the same as found in Example 14.13.

45. A typical incandescent lightbulb has a filament temperature of approximately 3000 K. At what wavelength is the intensity of the emitted light highest?

46. ✴ ✖ Ⓡ A snake is sunning itself on a rock. (a) If the snake is 0.3 m long, what is the approximate amount of radiation it absorbs from the Sun in 10 min? (b) If no heat flows out of the snake via conduction or by evaporative cooling, how much does its temperature increase during 10 min? Assume all the Sun's energy goes into heating the snake and the specific heat of the snake is equal to the specific heat of water.

47. ✪ The intensity of sunlight at the Earth's surface is about 1000 W/m². What is the total power emitted by the Sun?

48. ✴ Ⓡ Estimate the temperature of the surface of Mars. Assume Mars acts as a perfect blackbody and the only heat input comes from the Sun. How does your result compare with the Earth's temperature?

49. ✪ Ⓡ Repeat Problem 48, but now consider Pluto. Compare your answer with the measured temperature of Pluto. How do you think astronomers are able to measure the temperature of Pluto?

50. SSM ✴ An incandescent lightbulb emits 100 W of radiation. If the filament is at a temperature of 3000 K, what is the area of the filament?

51. ✪ Ⓡ Estimate the temperature of the Moon's surface in two cases: (a) when it is facing toward the Sun and (b) when it is facing away from the Sun. Explain qualitatively why your answer in part (b) is probably lower than the actual temperature. *Hint*: Consider the specific heat of the surface of the Moon.

52. ✴ ✖ Ⓡ What is the approximate energy emitted as blackbody radiation by a human being each second? What is the blackbody energy absorbed by a person in 1 s in a room? Why are the answers different?

53. ✪ Ⓡ A typical coal-burning power plant produces energy at the rate of 1000 megawatts. How large in area must a solar power plant be to produce the same amount of energy during daylight hours?

ADDITIONAL PROBLEMS

54. SSM ✴ Ⓡ An ice-cube tray containing 600 g of water at room temperature is placed into a freezer. It is found that the water becomes solid after 60 min. What is the average rate at which heat energy flows out of the water into the freezer?

55. ✪ Ⓡ An incandescent lightbulb produces light using a thin tungsten filament (wire) heated to about 5000 K by an electric current that passes through it. A typical filament has a diameter of 10 μm and a length of 2.0 cm. (a) What is the rate at which

heat energy is emitted by the filament? (b) When the current through the filament is turned off, the filament cools as heat is radiated away. Estimate how long it takes the filament to cool 100 K to a final temperature of 4900 K.

56. ⊗ A pot of very cold water (0°C) is placed on a stove with the burner adjusted for maximum heat. It is found that the water just begins to boil after 2.0 min. How much longer will it take the water to completely boil away?

57. ✪ An aluminum block slides along a horizontal surface. The block has an initial speed of 10 m/s and an initial temperature of 10°C. The block eventually slides to rest due to friction. Assuming all the initial kinetic energy is converted to heat energy and all this energy stays in the block, what is the final temperature of the block?

58. For the aluminum block in Problem 57, it is likely that at least a little bit of the heat energy will go toward heating the horizontal surface on which the block slides. Will this effect make the final temperature of the block larger or smaller than found in Problem 57?

59. [SSM] ✪ **The Rankine scale.** Named after William John Macquorn Rankine (a Scottish engineer and physicist who proposed it in 1859), the Rankine scale is similar to the Kelvin scale in that the zero point is placed at absolute zero, but the size of temperature differences are the same as that of the Fahrenheit scale (e.g., $1°F = 1°R$). (a) Determine the conversion formula to go from the Fahrenheit to the Rankine scale. (b) Find the formula to convert from Kelvin to Rankine. (c) What is the temperature of the freezing point of water on the Rankine scale? What is room temperature on this scale?

60. ☯ **Dry ice.** At standard atmospheric pressure, the solid form of carbon dioxide called "dry ice" undergoes a phase change not to a liquid, but straight to a gas. This process is called *sublimation*, and like other phase transitions, heat energy is required. In this case, it is the latent heat of sublimation, which for carbon dioxide is 573 kJ/kg at the sublimation temperature of −78.5°C. If 150 g of dry ice is dropped into 0.50 L of water at room temperature (20°C), how much of the water will turn to ice by the time all the dry ice has sublimated? *Note:* When dry ice becomes a gas, the gas bubbles out of the system.

61. ☯ Measuring the temperature of very cold substances can be challenging. Here is one way to measure the temperature of liquid nitrogen. A 50-g flake of aluminum is submerged in liquid nitrogen and left immersed until it is in thermal equilibrium. The flake is then removed and placed in a well-insulated container with 500 ml of water at 30°C. When water and flake attain thermal equilibrium, the temperature of the water is found to be 25°C. What was the temperature of the liquid nitrogen?

62. ✪ ⊗ **Winter survival.** If you are stranded in a cold and snowy environment, it is never a good idea to eat snow for hydration. Estimate the heat energy needed to convert 0.50 kg of snow to water and then heat the water to body temperature. Assume the snow is at 0°C in the survival snow cave you built as a shelter. State your answer in SI units (J) and in food calories (Cal).

63. ☯ ⊗ **Phase change hand warmer.** Sodium acetate in solution has a freezing point at 58°C and a latent heat of freezing of 210 kJ/kg, but a solution of this compound can remain in liquid form at much lower temperatures. This condition, known as *supercooling*, occurs in substances that form specific crystalline structures when in solid form. Such systems need a seed crystal to start the solidification process. Reusable hand warmers like that in Figure P14.63 are sold in sporting goods stores and contain about

Figure P14.63 Commercial sodium acetate reusable hand warmer.

60 g of sodium acetate solution. Clicking a piece of metal in the solution generates sound waves. The resulting increase in pressure causes a seed crystal to form, leading to full crystallization of the hand warmer within 4 s. A hand warmer is activated on a particularly cold morning when the solution was at a temperature of 10°C. Phase transitions occur at constant temperature, and the liquid is mostly water. (a) Calculate the heat generated by the phase transition. (b) What is the final temperature of the hand warmer?

64. ☯ **Freeze plugs.** The internal combustion gasoline engines used in automobiles produce much heat, which is mostly a nuisance and must be removed from the engine to prevent deformation and melting. The heat is carried away by a cooling system, typically using a fluid that flows through the walls of the engine block and out to a radiator, where flowing air cools the fluid. The fluid is mostly water with a mix of chemicals to prevent the fluid from freezing in cold climates when the engine is not in use. Freeze plugs, a standard part of any engine block, are brass alloy plugs that are pressed into holes in the engine block (Fig. P14.64). In principle, in the event of a hard freeze of the cooling system fluid, the freeze plugs will pop out and prevent the engine block from cracking. (a) Consider a carelessly maintained engine that has only water in its cooling system. On a cold winter night in Chicago, the water becomes frozen. How much pressure is exerted on a freeze plug? (b) If the freeze plug is 5.0 cm in diameter, how much force is exerted on the plug by the frozen water?

Figure P14.64 Problems 64 and 65.

65. ☯ One of the freeze plugs in Problem 64 is 3.0 cm thick and made of brass and just fits inside the hole in the steel of the engine block when the block is at room temperature and the brass was brought to the temperature of dry ice. (a) If the bulk modulus of brass is 6.1×10^{10} N/m², what pressure does the freeze plug exert on the block when both the freeze plug and the engine block are at room temperature? (b) If the coefficient of friction between brass and steel is 0.45, what force would it take to push the plug out of the block? If water were to freeze in the cooling system, would the plug pop out? See part (b) of Problem 64.

66. ☯ Figure P14.66 shows a modern electronic circuit board. It features standard miniature components such as the resistor (R516), which measures 1.6 mm long, 0.80 mm wide, and 0.20 mm high. The ceramic material of the resistor has a Young's modulus of 1.2×10^{10} Pa. When electricity flows through the circuit, the components, especially resistors, can heat up substantially. The resistor has a coefficient of thermal

Figure P14.66

expansion of $2.2 \times 10^{-6} \text{ K}^{-1}$, and that of the circuit board it is soldered to is $6.0 \times 10^{-6} \text{ K}^{-1}$. (a) If the temperature of the circuit rises 30°C, how much longer will the circuit's "footprint" be than the resistor that is soldered to it? (b) If the resistor is stretched by this amount by the circuit board, how much stress is the resistor subjected to? (c) The resistor has a maximum stress of 25 MPa before it will break. If a power glitch raises the temperature of the circuit by 120°C, will the resistor suffer damage from the stress?

67. ✪ A bimetal strip can be used as a thermometer and thermostat, a device that converts small changes in temperature to mechanical motion (Fig. P14.67A). A thermostat can be constructed by joining two pieces of metal, brass and steel, along their length as shown in Figure P14.67B. The brass and steel are both 50 cm long and 0.10 mm thick. If this bimetal strip is heated, the brass will expand more than the steel, causing the strip to bend. What change in temperature is needed to make the bimetal strip curl into a full circle as shown in Figure P14.67C? *Hint:* The difference in the radius of the two metals making the strip is just the thickness of one metal strip.

Figure P14.67

68. ✪ The owners of a cabin in the mountains forgot to turn off and purge the water out of the plumbing system before the winter season. One particularly cold night, a pipe near an outside wall freezes. Consider a length of "$^{1}/_{2}$-in." copper plumbing pipe filled with water. Although called $^{1}/_{2}$ in., the outside diameter of this standard pipe is $^{5}/_{8}$ in. and the inside diameter is 0.537 in. (a) Water freezes in a 50-cm length of this pipe capped at both ends by closed valves such that the expansion of the ice is constrained by the pipe. Calculate the pressure on the wall of the pipe. (b) Is the diameter of the pipe just before the water freezes significantly different from the diameter at room temperature? What is the percent change? (c) The pipe has a connection (Fig. P14.68). Assume the water freezes so that the separation between ice and water occurs right in the middle of the connector. Calculate the force that pulls on the connector when the water freezes.

connection

Figure P14.68

69. ✪ **Mass wasting.** Water filling a thin crack in a rock wall and then freezing can put large forces on the rock, making the cracks bigger (Fig. P14.69). (a) Calculate the force exerted by a 10-m² area of a thin (0.50 mm) sheet of water freezing between two slabs of rock. (b) Take the sheet of water to be oriented in a horizontal plane (unlike the case in Fig. P14.69), and assume a large slab of rock sits on top of the ice. By how much is the rock on top of the water displaced vertically? (c) If the column of rock is 15 m tall and has a density of 5.2 g/cm³, how much work is done by the expansion of the ice on this rock mass?

Figure P14.69

70. ✪ An extreme thermos bottle might use silica aerogel as an insulator. Figure P14.70 demonstrates the insulating properties of the aerogel compound. Aerogel can protect human skin from a blowtorch at nearly point-blank range, made possible by its very low thermal conductivity (only 0.003 W/(m · K)) and a melting point of 1200°C. Let's see if we can improve on the thermos bottle used by the author in Example 14.10, where we found that it took 1.2 h for his coffee to cool from 65°C to 55°C. (a) If the thermos bottle is a cylinder having an inner diameter of 7.0 cm, inner length of 12 cm, and wall thickness of 1.2 cm, what is the thermal conductance of the thermos material in the example? (b) If we replace the material with aerogel, how long will it take the coffee to cool the same amount? Assume the aerogel is coated with a thin protective layer of negligible thickness.

Figure P14.70 The insulating property of silica aerogel.

71. ✪ ℝ **Firewalking.** A man strides across a bed of glowing embers as seen in Figure P14.71, and due to some very specific physical circumstances, he emerges with no physical harm. Assume the glowing charcoal is covered with a thin layer of ash 0.50 mm thick and is at a temperature of 650°C. The thermal conductivity of ash and charcoal is approximately 1.5 W/(m · K). (a) Estimate the heat transferred to the feet of the firewalker as he traverses the bed of embers in six quick steps. Assume the area of the bottom of each foot comes into only 50% contact with the embers (a rough surface) and the contact lasts 0.10 s for each step. Is the quantity of heat transferred great or small? (b) Assume he steps onto the fire from slightly wet grass. If his feet are protected by the conversion of water from a liquid to a gas, how much water is needed to carry away the heat from part (a)? (c) How thick would this layer of water on the sole of each foot need to be? (d) Describe what would happen if instead he were to take the same six quick steps across a slab of iron at the same temperature.

Figure P14.71

Chapter 15

Gases and Kinetic Theory

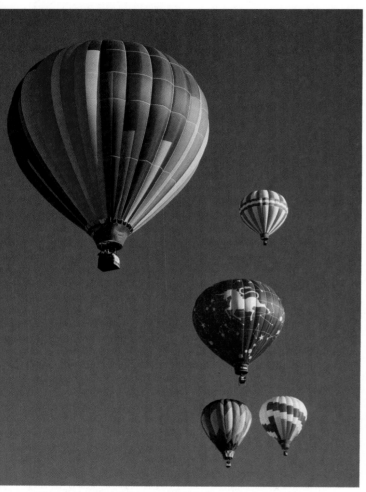

The properties of the air in this balloon and other gases are described by an area of physics called kinetic theory. (© Radius Images/Jupiterimages)

Newton's laws of motion are the centerpieces of mechanics. Newton and fellow physicists used these laws to explain the motion of many objects, including projectiles, planets and moons, and falling apples. Their attention also turned to the building blocks from which these objects are constructed: they wanted to apply their understanding of mechanics to deduce how matter is put together. We now know that matter is composed of atoms, so in modern terms, we would say that their goal was to discover the atomic theory of matter.

The simplest phase of matter is a gas, and in this chapter, we apply Newton's laws to analyze the behavior of gases. In Chapter 14, we argued that systems containing many particles are so complicated that an approach using Newton's laws is prohibitively difficult, and does not give a useful understanding of problems pertaining to heat flow and temperature. However, one special type of system—a dilute gas—is simple enough that Newton's laws *can* be applied. The atoms and molecules in a dilute gas are far apart, so they collide with one another only rarely. Hence, it is possible to apply Newton's laws in ways not possible

468

with liquids or solids. The analysis of the properties of a dilute gas leads to important connections between temperature and motion on an atomic scale. Many properties of dilute gases are independent of the composition, meaning that the same theory applies to a dilute gas of hydrogen, helium, oxygen, and so on. These ideas led to the notion of an *ideal gas* and an area of physics called *kinetic theory*. This theory has a very rich and interesting history and has seen contributions from many physicists, including Newton and Einstein.

15.1 | MOLECULAR PICTURE OF A GAS

Figure 15.1 is a molecular-scale picture of a gas. We consider a balloon containing a pure substance, so all the particles inside the balloon are identical. These particles might be atoms, such as a balloon full of helium atoms, or molecules, such as carbon dioxide. For simplicity, we often refer to the particles as "molecules," with the understanding that all our ideas apply just as well to gases composed of single atoms, such as helium or argon.

The molecules in Figure 15.1 are all in motion, colliding with other gas molecules and with the walls of the container. The snapshot in Figure 15.1 raises a number of questions. (1) How often does a gas molecule collide with other molecules? (2) Are these collisions elastic or inelastic? (3) How are *microscopic* properties of the gas such as the molecular velocities related to the *macroscopic* properties of the system such as temperature and pressure? (4) How many molecules are there in a typical sample of gas? Our job in this chapter is to answer all these questions.

A typical sample of gas contains a very large number of particles, usually on the order of 10^{23} molecules (or more). This very large number would seem to make the application of Newton's laws hopelessly complex. Actually, though, this large number of particles simplifies things by allowing us to apply statistical arguments, and because the number of particles is so large, the accuracy of a statistical analysis will be very good.

Let's begin by considering the origin of the number 10^{23} just mentioned. As you probably suspect, 10^{23} is connected with *Avogadro's number*, N_A. This pure number has the value

$$N_A = 6.02 \times 10^{23} \qquad (15.1)$$

where N_A is the number of particles in a *mole*.[1] Avogadro's number plays a central role in connecting the microscopic world of atoms and molecules to the macroscopic world. The *atomic mass* of an element is the mass (usually quoted in grams) of a natural sample of the element containing 1 mole of atoms. For example, the atomic mass of carbon is $M_C = 12.011$ g (sometimes the units are given as grams per mole). We denote atomic mass using a capital letter to distinguish it from the mass of a single carbon atom m_C; the units of m_C are grams per atom, or simply grams. From the definition of atomic mass, we have

$$m_C = \frac{M_C}{N_A} = \frac{12.011 \text{ g/mole}}{6.02 \times 10^{23} \text{ atoms/mole}} = 2.0 \times 10^{-23} \text{ g} = 2.0 \times 10^{-26} \text{ kg} \quad (15.2)$$

In Section 15.2, we'll show that a typical sample of gas particles contains on the order of 1 mole of particles. Thus, in discussions of kinetic theory and the properties of gases, N_A and factors like 10^{23} appear often.

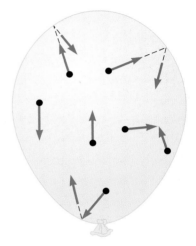

Figure 15.1 The molecules in a gas are in constant motion as they collide with one another and with the walls of their container.

Avogadro's number

---

[1]Avogadro's number, N_A, is defined as the number of atoms in a sample containing exactly 12 g of the carbon-12 isotope. We'll discuss isotopes in Chapter 30.

EXAMPLE 15.1 Ⓡ **The Number of Molecules in a Cup of Coffee**

How many molecules are in a typical cup of coffee? Coffee is composed mainly of water, so for simplicity assume the sample contains only water molecules.

RECOGNIZE THE PRINCIPLE

We can calculate the number of moles of water molecules in a cup of coffee if we know the total mass of water M_{total} and the mass of 1 mole of water molecules M_{water}. We can find M_{water} from the atomic masses of hydrogen and oxygen, and although M_{total} is not given, we can use our experience to estimate a value.

SKETCH THE PROBLEM

No figure is necessary.

IDENTIFY THE RELATIONSHIPS

A typical cup at your local coffee house contains about 300 g (300 milliliters) of water, so we take $M_{total} = 300$ g. Water has the molecular formula H_2O (two hydrogen atoms plus one oxygen atom). The atomic mass of hydrogen is $M_H = 1.01$ g, and the atomic mass of oxygen is $M_O = 16.00$ g. (See the periodic table at the back of this book.) The mass of 1 mole of water molecules is thus

$$M_{water} = 2M_H + M_O = 2(1.01 \text{ g}) + 16.00 \text{ g} = 18.02 \text{ g}$$

which can also be written as $M_{water} = 18.02$ g/mole.

SOLVE

To find the number of moles in our sample, we divide M_{total} by the mass of 1 mole. The number of moles of water molecules in our sample of 300 g is thus

$$n_{water} = \frac{M_{total}}{M_{water}} = \frac{300 \text{ g}}{18.02 \text{ g/mole}} = 17 \text{ moles}$$

and the number of molecules N_{water} is

$$N_{water} = n_{water} \times N_A = (28 \text{ moles})(6.02 \times 10^{23} \text{ molecules/mole})$$

$$\boxed{N_{water} = 1.0 \times 10^{25} \text{ molecules}}$$

What does it mean?

The number of water molecules in this sample is within two orders of magnitude of Avogadro's number. We'll find similar results for the numbers of atoms or molecules in typical macroscopic samples of gases, liquids, and solids.

15.2 | IDEAL GASES: AN EXPERIMENTAL PERSPECTIVE

Let's now step back from the microscopic (molecular scale) picture to consider some macroscopic properties of a simple gas. Experimental studies of these properties began around the time of Newton and led to the discovery of a number of different "gas laws." These laws provided the clues that eventually gave rise to kinetic theory and relate various macroscopic properties of a gas such as its temperature and pressure. Most of these gas laws apply accurately only for dilute samples, which means that the spacing between molecules is much larger than the size of an individual molecule. Fortunately, that is not a serious restriction because many important cases, including air at room temperature, can be closely approximated as a dilute gas.

Avogadro's law

For a sample of gas at constant pressure and temperature, the volume is proportional to the number of molecules in the sample.

Avogadro's law

In terms of our picture in Figure 15.1, Avogadro's law tells us that the average spacing between gas particles is constant, so the density is also constant, provided the pressure and temperature are held fixed.

Boyle's law

For a sample of gas at constant temperature, the product of the pressure and volume is constant.

In mathematical terms, Boyle's law is just

$$PV = \text{constant} \quad (\text{at constant } T) \qquad (15.3)$$

Boyle's law

According to Boyle's law, if we increase the pressure in a sample of gas held at constant temperature, the volume must decrease so that the product PV stays fixed. On the other hand, if we decrease the pressure, the volume must then increase, again keeping PV constant. In terms of Figure 15.1, Boyle's law says that changing the pressure changes the average spacing between particles.

CONCEPT CHECK 15.1 | Density of a Dilute Gas

A cylinder (Fig. 15.2) contains a dilute gas at pressure P_i and volume V_i. One end wall of the cylinder is movable (forming a piston); moving this wall changes the volume of the gas. Suppose the gas is compressed so that the final pressure is twice the initial pressure, $P_f = 2P_i$. If the temperature is constant (so that Boyle's law applies), what is the final density?
   (a) The final density is twice the initial density.
   (b) The final density is half the initial density.
   (c) The final density is equal to the initial density.

Figure 15.2 Concept Check 15.1.

An important part of Boyle's law is the constant on the right-hand side of Equation 15.3. This constant must be proportional to the number of particles in the gas to be consistent with Avogadro's law. Experiments show that this constant depends on temperature, as described by two more gas laws.

Charles's law

For a sample of gas at constant pressure, if the temperature is changed by a small amount ΔT, the volume changes by an amount ΔV, with

$$\Delta V \propto \Delta T \qquad (15.4)$$

Charles's law

Gay-Lussac's law

For a sample of gas held in a container with constant volume, changes in pressure are proportional to changes in temperature.

$$\Delta P \propto \Delta T \qquad (15.5)$$

Gay-Lussac's law

We have actually encountered Gay-Lussac's law in our discussion of gas thermometers in Chapter 14, where we found a linear relation between P and T in a dilute gas (Fig. 14.5A).

Absolute Temperature and the Kelvin Scale

Charles's law and Gay-Lussac's law both include temperature. We saw in Chapter 14 that temperature is a quantity that cannot be derived from Newton's laws, so it is interesting to ask how Charles and Gay-Lussac were able to define and measure temperature in their work. Charles and Gay-Lussac used *reference temperatures* determined by the properties of various substances. Two convenient temperatures are those at which ice melts and at which water boils. Charles and Gay-Lussac

As the temperature changes, the volume of the mercury changes due to thermal expansion. The position of the mercury meniscus can thus be used to find the temperature.

Ice water
$T = 0°C$

Boiling water
$T = 100°C$

Figure 15.3 A mercury thermometer uses thermal expansion of mercury to measure changes in temperature. This thermometer is calibrated using known reference temperatures such as the melting and boiling temperatures of water.

used these reference temperatures together with a material property such as thermal expansion (Chapter 14) to interpolate between and extrapolate beyond the reference range. That is how a mercury thermometer works; the basic principle is shown in Figure 15.3.

This approach to measuring temperature was very useful, but it did not provide a scientific definition of temperature. Scientists then had to grapple with some crucial questions. (1) What materials and properties can be used to make a thermometer (besides mercury), and can we be sure that all such thermometers will give the same results? (2) According to Equation 15.5, changes in P are proportional to changes in T. What happens to the temperature when $P = 0$, and does such a pressure (or temperature) even make sense?

Let's consider question 2 first. Careful measurements of pressure and temperature for a variety of gases give results like those sketched in Figure 15.4A; as temperature is reduced, the pressure becomes smaller in a linear fashion (in proportion to changes in T), according to Gay-Lussac's law (Eq. 15.5). Any particular sample of gas will eventually condense and form a liquid, at which point the pressure falls below the linear $P–T$ relation that holds at high temperature (Fig. 15.4B). This result does not conflict with Gay-Lussac's law because that relation applies only for dilute samples, and a gas is not dilute near its condensation temperature. Even though the linear behavior breaks down at low temperatures, we can still extrapolate the linear region from high temperature and pressure down to low pressures and find the temperature at which the pressure "would" be zero. If we do that with different samples of different gases, we find that they all extrapolate to zero at the same temperature, $T = -273.15°C$ (Fig. 15.4A). This result applies to all gases, regardless of composition, and also holds for samples with different densities. Pressure is defined as the magnitude of the force per unit area exerted by a fluid on a surface (see Eq. 10.1), so it is not possible for P to be negative. This fact helped lead to the creation of the Kelvin temperature scale. The Kelvin scale is defined so that $T = 0$ K when the pressure of a dilute gas becomes zero, which is also called the *absolute zero* of temperature.

Because the linear relation between P and T is a universal property of dilute gases, it can be used as a universal way to measure temperature. One can calibrate a *gas thermometer* by measuring P at two reference temperatures (such as the melting point of ice and the boiling point of water) and thereby obtain an absolute calibration of the $P–T$ relation for that particular thermometer. Most importantly, a gas thermometer can be made using any type of dilute gas. They all give the same results.

The laws of Charles and (especially) Gay-Lussac thus lead to the notion of absolute temperature and provide an experimental method for measuring it. When we deal with gases and kinetic theory, it is most convenient to work with the Kelvin temperature scale.

The Ideal Gas Law

Each of the gas laws discussed above concern a particular aspect of the behavior of a dilute gas. These laws are all contained in a single relation known as the *ideal gas law*.

Ideal gas law

For a dilute gas composed of any substance,

$$PV = nRT \tag{15.6}$$

This relation applies to a sample of a dilute gas containing n moles of gas particles, and R is the *universal gas constant* with the value

$$R = 8.31 \frac{J}{mole \cdot K} \tag{15.7}$$

We can also express the ideal gas relation in terms of the number of gas particles N instead of the number of moles n. Using the definition of the mole and Avogadro's number (Eq. 15.1), the total number of particles is $N = nN_A$, so $n = N/N_A$. Substituting for n in Equation 15.6, we get

$$PV = nRT = \frac{N}{N_A}RT$$

$$PV = N\left(\frac{R}{N_A}\right)T \tag{15.8}$$

The factor R/N_A is the ratio of the gas constant to Avogadro's number; it is called *Boltzmann's constant* k_B and has the value

$$k_B = \frac{R}{N_A} \tag{15.9}$$

$$k_B = 1.38 \times 10^{-23} \text{ J/K} \tag{15.10}$$

Using the number of gas particles N and Boltzmann's constant, the ideal gas law (Eq. 15.8) can be written as

$$PV = Nk_BT \tag{15.11}$$

The two versions of the ideal gas law, Equations 15.6 and 15.11, are equivalent; the only difference is that one involves the number of moles n, whereas the other involves the number of particles N.

The ideal gas law contains the laws of Avogadro, Boyle, Charles, and Gay-Lussac all as special cases. For example, when the temperature and number of particles are kept fixed, the ideal gas law predicts that the product PV is a constant, which is just Boyle's law (Eq. 15.3). Most importantly, the ideal gas law applies to all dilute gases, with the same value of the gas constant R and the same value of Boltzmann's constant k_B.

EXAMPLE 15.2 | Number of Gas Particles in a Balloon

You attend a carnival and win a balloon full of helium atoms. This balloon is spherical with a radius of 15 cm, and the gas pressure is twice atmospheric pressure. **(a)** How many helium atoms did you win? **(b)** What is the total mass of the molecules in the balloon? Assume the carnival was held on a nice day, when the outdoor temperature was 20°C.

RECOGNIZE THE PRINCIPLE

We can rearrange the ideal gas law to solve for N, the number of helium atoms:

$$N = \frac{PV}{k_BT} \tag{1}$$

We are given the pressure, volume, and temperature, so we can find N. The total mass can then be found using the atomic mass of helium.

SKETCH THE PROBLEM

No figure is needed.

IDENTIFY THE RELATIONSHIPS

The pressure is twice atmospheric pressure, and $P_{atm} = 1.01 \times 10^5$ Pa, so for the balloon, we get $P = 2P_{atm} = 2.0 \times 10^5$ Pa. The sample is a sphere of radius 0.15 m; hence, the volume is

$$V = \tfrac{4}{3}\pi r^3 = \tfrac{4}{3}\pi(0.15 \text{ m})^3 = 0.014 \text{ m}^3$$

Converting the temperature to the Kelvin scale, we have $T = 273 + 20 = 293$ K.

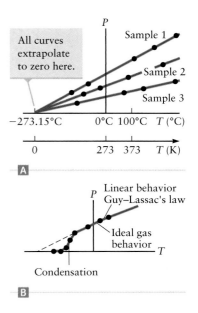

Figure 15.4 **A** Experiments show that the pressure of all dilute gases varies linearly with temperature when the temperature is well above the condensation temperature. *All* the data at high temperatures for *all* gases extrapolate to $P = 0$ at the *same* temperature, defined to be $T = 0$ on the Kelvin scale. **B** At sufficiently low temperatures, all gases condense and the pressure then drops below the linear P–T dependence found in **A**.

SOLVE

(a) Inserting these values for the pressure, volume, and temperature into our expression for N, Equation (1), gives

$$N = \frac{PV}{k_B T} = \frac{(2.0 \times 10^5 \text{ Pa})(0.014 \text{ m}^3)}{(1.38 \times 10^{-23} \text{ J/K})(293 \text{ K})} = \boxed{6.9 \times 10^{23} \text{ atoms}}$$

which corresponds to

$$n = \frac{N}{N_A} = \frac{6.9 \times 10^{23} \text{ atoms}}{6.02 \times 10^{23} \text{ atoms/mole}} = 1.1 \text{ moles}$$

(b) The total mass m_{total} of the helium atoms equals the number of moles n multiplied by the mass of 1 mole. According to the periodic table, the mass of 1 mole of helium atoms is 4.00 g. Combining this with the answer for n from part (a) gives

$$m_{total} = n \times (4.00 \text{ g/mole}) = (1.1 \text{ moles})(0.00400 \text{ kg/mole}) = \boxed{0.0044 \text{ kg}}$$

What does it mean?

The total mass is equal to the mass of about two (U.S.) pennies. When using the ideal gas law in either form (Eq. 15.6 or 15.11), T must be expressed in terms of the absolute temperature, the temperature on the Kelvin scale. The answer for part (a) illustrates again that the number of molecules in a typical macroscopic sample of gas is on the order of Avogadro's number.

CONCEPT CHECK 15.2 | The Ideal Gas Law and the Effect of Composition

Suppose the balloon in Example 15.2 contains nitrogen molecules instead of helium atoms. How does that affect the number of particles N in the balloon?
  (a) There is no change. The number of nitrogen molecules would be equal to the number of helium atoms found in Example 15.2.
  (b) The number of nitrogen molecules is twice the number of helium atoms found in Example 15.2.
  (c) The number of nitrogen molecules is half the number of helium atoms found in Example 15.2.

EXAMPLE 15.3 Particle Spacing in a Gas

Consider the balloon of helium gas in Example 15.2. What is the average distance between atoms?

RECOGNIZE THE PRINCIPLE

At any instant in time, the atoms in the balloon are distributed in an approximately random fashion throughout the balloon (Fig. 15.1). To find the average spacing, we can imagine that the atoms are arranged in a regular manner as sketched in Figure 15.5, with the distance between neighboring atoms given by L. This average arrangement has atoms at the corners of cubes with edges of length L. Hence, there is a volume L^3 associated with each atom. The total gas volume is then NL^3, where the number of atoms N was found in Example 15.2.

SKETCH THE PROBLEM

Figure 15.5 shows the atoms in a regular cubic arrangement. The atoms in a real gas will not be arranged in such a simple way, but if the average distance between atoms is L, the average volume associated with each atom will still be L^3.

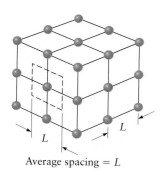

Average spacing = L

Figure 15.5 Example 15.3. To find the average molecular spacing, we can imagine the molecules are arranged in a regular array.

IDENTIFY THE RELATIONSHIPS AND SOLVE

The total gas volume from Example 15.2 is $V = 0.014 \text{ m}^3$. Setting this volume equal to NL^3 and using the result for the number of atoms N found in that example, we get $NL^3 = V$ and

$$L = \left(\frac{V}{N}\right)^{1/3} = \left(\frac{0.014 \text{ m}^3}{6.9 \times 10^{23}}\right)^{1/3} = \boxed{2.7 \times 10^{-9} \text{ m}}$$

What does it mean?
The radius of an atom is about 5×10^{-11} m, so the spacing between particles in a gas is *much* larger than the size of an atom. The spacing between atoms in a solid or liquid is typically 0.1 nm (= 0.1×10^{-9} m). The average distance between molecules in this balloon, and in a typical dilute gas at room temperature, is thus about 30 times larger than the spacing between atoms in a solid or liquid.

CONCEPT CHECK 15.3 | Changing the Pressure in a Gas

The helium-filled balloon in Example 15.2 contained $n = 1.1$ moles of helium atoms at room temperature and a pressure $P = 2 \times P_{\text{atm}}$ (twice atmospheric pressure). Atoms are added to the balloon so that the pressure doubles to $4 \times P_{\text{atm}}$ while the temperature and volume are kept fixed. Is number of moles in the balloon (a) 1.1 moles, (b) 2.2 moles, (c) 4.4 moles, or (d) 0.55 mole?

EXAMPLE 15.4 | Effect of Temperature on the Pressure of an Ideal Gas

A jar containing a gas of nitrogen molecules has a pressure $P_{\text{atm}} = 1.01 \times 10^5$ Pa at a temperature of 20°C. This jar is then put in a refrigerator where the temperature is 0°C. If the lid of the jar is circular with a radius of 5.0 cm, how much force is needed to open the lid?

RECOGNIZE THE PRINCIPLE

When the gas is cooled, the pressure inside the jar P_{in} will decrease according to the ideal gas law and will then be less than the pressure outside. The force required to open the jar will equal the difference in pressures multiplied by the area of the lid.

SKETCH THE PROBLEM

Figure 15.6 describes the problem.

IDENTIFY THE RELATIONSHIPS

The ideal gas law reads $PV = Nk_BT$, and in this example, the volume V and the number of molecules N are held fixed. Even though we do not know the values of N and V, we can still write

$$\frac{P}{T} = \frac{Nk_B}{V} = \text{constant}$$

(which is just another way of writing Gay-Lussac's law, Eq. 15.5). If we denote the initial pressure and temperature inside as $P_{\text{in},i}$ and T_i and denote the final values as $P_{\text{in},f}$ and T_f, we can now write

$$\frac{P_{\text{in},i}}{T_i} = \frac{P_{\text{in},f}}{T_f}$$

Rearranging to solve for the final pressure gives

$$P_{\text{in},f} = \left(\frac{P_{\text{in},i}}{T_i}\right)T_f$$

Figure 15.6 Example 15.4. The force on the lid of a jar is proportional to the difference in pressure between outside and inside the jar.

SOLVE

Inserting the given values of the initial pressure ($P_{in,i} = 1.01 \times 10^5$ Pa) and temperature ($T_i = 293$ K), along with $T_f = 0°C = 273$ K, we find

$$P_{in,f} = \left(\frac{P_{in,i}}{T_i}\right)T_f = \left(\frac{1.01 \times 10^5 \text{ Pa}}{293 \text{ K}}\right)(273 \text{ K}) = 9.4 \times 10^4 \text{ Pa}$$

With a pressure P_{atm} on the outside of the lid and a smaller pressure $P_{in,f}$ inside, the force required to open the lid is

$$F = (P_{atm} - P_{in,f})A$$

The area of the lid is $A = \pi r^2$, so we have

$$F = (P_{atm} - P_{in,f})A = (P_{atm} - P_{in,f})(\pi r^2)$$

$$F = (1.01 \times 10^5 \text{ Pa} - 0.93 \times 10^5 \text{ Pa})\pi(5.0 \times 10^{-2} \text{ m})^2 = \boxed{55 \text{ N}}$$

What does it mean?

The pressure $P_{in,f}$ is only slightly smaller than the initial pressure. The pressure change is relatively small as a percentage of the initial pressure because the change in temperature is only a small fraction of the absolute temperature. Even though the resulting pressure difference across the lid is small, the force is still substantial. You could do this experiment at home.

15.3 | IDEAL GASES AND NEWTON'S LAWS

We now return to some of the issues raised at the very beginning of the chapter. In particular, let's see how to account for the gas laws by applying Newton's laws to the microscopic molecular motions.

Pressure Comes from Collisions with Gas Molecules

The molecules in a gas move with different speeds and in all possible directions (Fig. 15.1). To simplify things, we will (for the moment) consider just one typical molecule of mass m and take its speed v to be the average molecular speed. Figure 15.7 shows this molecule as it is initially traveling along the $-x$ direction; it then collides with a container wall and travels back along the $+x$ direction. In a typical gas, such collisions are elastic, so the speed is the same before and after the collision. From Chapter 7, we know that the change in momentum of the molecule $\Delta\vec{p}$ is proportional to the force exerted by the wall on the molecule. According to the impulse theorem (Eq. 7.5), we have

$$\Delta\vec{p} = \vec{F}\,\Delta t \tag{15.12}$$

For the molecule in Figure 15.7, the initial and final momenta are along x and the force from the wall is parallel to the x direction. We can therefore write Equation 15.12 in terms of the components along x. From the geometry in the figure, the initial momentum is $-mv$ and the final momentum is $+mv$. Inserting these terms into Equation 15.12, we find

$$\Delta p = p_f - p_i = +mv - (-mv) = F\Delta t$$

$$2mv = F\Delta t$$

Solving for the force gives

$$F = \frac{2mv}{\Delta t} \tag{15.13}$$

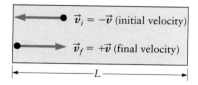

Figure 15.7 When a gas molecule collides with a wall, the force exerted by the wall leads to an impulse on the molecule, causing the molecule's momentum to change. The corresponding force exerted by the molecule on the wall contributes to the gas pressure.

Equation 15.13 gives the force exerted by the wall on the molecule. According to Newton's third law (the action–reaction principle), the force exerted by the molecule on the wall has a magnitude F. We also know that the force exerted by a molecule on the wall contributes to the pressure. To find the total pressure, we must add up the collision forces from all the molecules that strike the wall. Figure 15.7 shows one hypothetical molecule initially moving in the $-x$ direction; there will also be molecules moving in the $+x$, $+y$, $-y$, $+z$, and $-z$ directions. We thus have six different possible directions, so for a system with a total of N molecules, we can expect that one-sixth of them are initially moving along $-x$ and are thus candidates for striking the left-hand wall in Figure 15.7. Of these $N/6$ molecules, only those within a distance $v\,\Delta t$ strike the wall during a time interval Δt because only those molecules reach the wall during this time period. The container in Figure 15.7 has a length L, so the number of molecules that collide with the wall during this time interval is

$$\frac{N}{6} \times (\text{fraction that reach wall}) = \frac{N}{6}\frac{v\,\Delta t}{L} \qquad (15.14)$$

The total force exerted by all these molecules on the wall is the product of the force exerted by one molecule (Eq. 15.13) and the number of such molecules (Eq. 15.14). The result is

$$F_{\text{total}} = \left(\frac{2mv}{\Delta t}\right)\left(\frac{N}{6}\frac{v\,\Delta t}{L}\right) = \frac{Nmv^2}{3L}$$

Pressure is defined as the total force on the wall divided by the area A, so

$$P = \frac{F_{\text{total}}}{A} = \frac{Nmv^2}{3LA}$$

The volume of the container is $V = LA$, so the pressure can also be written as

$$P = \frac{Nmv^2}{3V}$$

or

$$PV = N\frac{mv^2}{3} \qquad (15.15)$$

The Microscopic Basis of Temperature

Equation 15.15 relates a *microscopic* property of the molecules, their speed v, to a *macroscopic* property, the pressure in the gas. We also notice that the mathematical form of this result is *very* similar to the ideal gas law. According to Equation 15.11,

$$PV = Nk_BT$$

Comparing this expression with Equation 15.15, we see that they agree provided that

$$\frac{mv^2}{3} = k_BT \qquad (15.16)$$

The kinetic energy of a molecule is $KE = \frac{1}{2}mv^2$, so Equation 15.16 can be written as $\frac{2}{3}KE = k_BT$ or, equivalently,

$$KE = \tfrac{3}{2}k_BT \qquad (15.17)$$

Average kinetic energy of a gas atom

While we derived this result for a dilute gas, it also holds for the average kinetic energy of atoms and molecules in a liquid or solid.

In our derivation of Equation 15.15, we assumed all the molecules move with the same speed v. An exact analysis allows for different molecules moving at different speeds and shows that v in Equations 15.15 and 15.16 is nearly equal to the average molecular speed (the difference is only a few percent). We can therefore still think of v as being the speed of a typical molecule.

The factor KE in Equation 15.17 is the average kinetic energy; that is, it is the kinetic energy of a typical molecule. In words, Equation 15.17 says that this kinetic energy is proportional to the temperature of the gas. This result is remarkable because it tells how temperature is related to the microscopic motion of the molecules in a gas. We have thus derived the ideal gas law from Newton's laws of mechanics, provided that temperature is related to the molecular kinetic energy by Equation 15.17. This result answers one of the big questions raised in Section 15.1 because we are now able to relate the macroscopic quantities in the ideal gas law to the microscopic quantities that appear in Newton's laws.

It is interesting to use Equation 15.16 to find the speed of a typical molecule in the atmosphere. Rearranging to solve for v, we find

$$v = \sqrt{\frac{3k_B T}{m}} \tag{15.18}$$

The Earth's atmosphere is composed mainly of nitrogen, so let's calculate v for N_2 in the atmosphere. The mass of a nitrogen molecule is $m = 2(14.01 \text{ g})/N_A = 4.7 \times 10^{-23}$ g $= 4.7 \times 10^{-26}$ kg. At room temperature, we have $T \approx 293$ K (corresponding to 20°C). Inserting these values into Equation 15.18 gives

$$v_{N_2} = \sqrt{\frac{3k_B T}{m}} = \sqrt{\frac{3(1.38 \times 10^{-23} \text{ J/K})(293 \text{ K})}{4.7 \times 10^{-26} \text{ kg}}} = 510 \text{ m/s} \tag{15.19}$$

which is approximately 1100 mi/h! Molecules having this speed are striking your body right now as you read this book.

15.4 | KINETIC THEORY

In Section 15.3, we saw how the ideal gas law could be understood in terms of Newton's laws. Those calculations take us to the heart of the mechanics of a dilute gas, a topic called **kinetic theory**. This theory is based on a small number of assumptions that were used either explicitly or implicitly in Section 15.3:

1. Gas atoms and molecules spend most of their time moving freely; that is, they move with a constant speed in a straight-line path. Collisions with other atoms and molecules in the gas are very infrequent.
2. Newton's laws can be used to describe the motion of individual gas particles.
3. The collisions between gas atoms and molecules, and with the walls of the container, are elastic. The average distance between collisions is called the **mean free path**, ℓ. The value of ℓ depends on the density of gas particles, their size, and temperature. The mean free path is not the same as the average spacing between particles (Fig. 15.8). In Example 15.3, we found that the average spacing between molecules in a typical dilute gas is approximately 3×10^{-9} m. The mean free path for a molecule in air is about 100×10^{-9} m, so in this case, ℓ is about 30 times longer than the average spacing.
4. Even a dilute gas contains a very large number of atoms or molecules, so we can use statistical analyses to calculate its properties. We have already discussed the large number of particles present in a typical gas and have seen that it is on the order of Avogadro's number.

These assumptions about the nature of an ideal gas are all satisfied very accurately by a dilute gas. In the last section, we saw that they lead to the ideal gas law. We next consider some other properties of ideal gases.

The mean free path depends on the molecular size.

A **B**

Figure 15.8 The average distance a gas molecule travels between collisions is called the mean free path. **A** If the molecules are very small, the mean free path is much longer than the average spacing between them. **B** As molecular size increases, the mean free path becomes shorter. See Insight 15.1.

Internal Energy of an Ideal Gas

The particles in an ideal gas spend most of their time moving freely, during which time the force on a particular molecule from all the other molecules is essentially zero. The mechanical energy then equals the kinetic energy, and (as we learned in Chapters 6 and 8) this kinetic energy has contributions from both the translational motion ($\frac{1}{2}mv^2$) and the rotational motion ($\frac{1}{2}I\omega^2$). For a monatomic gas such as helium, the rotational kinetic energy doesn't contribute to the gas properties and can be ignored. (For polyatomic gases such as N_2, one must include the rotational contribution; see below.) The translational kinetic energy KE_{trans} is $KE_{\text{trans}} = \frac{1}{2}mv^2$. We already calculated this kinetic energy in Section 15.3, where we found that (Eq. 15.17)

$$KE_{\text{trans}} = \tfrac{3}{2}k_B T \qquad (15.20)$$

The total energy of a system of N gas atoms is thus

$$KE_{\text{total}} = N(KE_{\text{trans}}) = \tfrac{3}{2}Nk_B T$$

which is the total energy of a monatomic ideal gas. In a sense, this energy is held "internally" by gas atoms and is not evident if we are simply given a container of the gas. For that reason, it is called the **internal energy**, denoted by U. Hence, for a monatomic ideal gas, we have

$$U = \tfrac{3}{2}Nk_B T \qquad (15.21)$$

Internal energy of an ideal gas

According to Equation 15.21, the internal energy of an ideal gas depends *only* on temperature and the number of gas atoms N. It is independent of other factors such as the molecular mass and pressure.

We can also express the internal energy in terms of the number of moles n in the system instead of the number of atoms N. The number of atoms is $N = n \times N_A$, so Equation 15.21 becomes

$$U = \tfrac{3}{2}(nN_A)k_B T = \tfrac{3}{2}n(N_A k_B)T$$

From the definition of Boltzmann's constant (Eq. 15.9), we have $N_A k_B = R$, so

$$U = \tfrac{3}{2}nRT \qquad (15.22)$$

The results in Equations 15.21 and 15.22 for the internal energy apply only for a monatomic ideal gas. Other systems such as liquids, solids, nondilute gases, and polyatomic ideal gases all have an internal energy, but in those cases, U is not the same as for a monatomic gas. For liquids and solids (and nondilute gases), the contribution from the potential energy is important and makes the internal energy a more complicated function than the simple result in Equation 15.21. We'll discuss a few such cases later in this chapter and also in Chapter 16, where the internal energy plays an important role in our work on thermodynamics.

CONCEPT CHECK 15.4 Internal Energy of Two Gases

One balloon contains exactly 3 moles of helium atoms, and a second balloon contains exactly 3 moles of argon atoms. The mass of an argon atom is about 9 times greater than the mass of a helium atom. If the internal energies of these two systems are equal, what can we conclude?
  (a) They must have the same volume.
  (b) They must have the same pressure.
  (c) They must have the same temperature.
  (d) They must have the same pressure, volume, and temperature.

EXAMPLE 15.5 | ® Internal Energy of a Gas: How Big Is It?

Helium-filled balloons are often sold at carnivals. What is the approximate internal energy of the gas in such a balloon?

RECOGNIZE THE PRINCIPLE

To calculate the internal energy of a dilute gas, we must know its temperature and the number of moles n. Neither value is given, but we can use our experience to find typical values. The temperature is about room temperature, so $T \approx 290$ K. A typical balloon has a volume of about $V = 3$ L $= 3 \times 10^{-3}$ m³ (Fig. 15.9), and the pressure is near atmospheric pressure ($P_{atm} = 1.01 \times 10^5$ Pa); using this information, we can use the ideal gas law to find n.

SKETCH THE PROBLEM

No sketch is needed; we have estimated the balloon's volume from the photo in Figure 15.9.

IDENTIFY THE RELATIONSHIPS

We first rearrange the ideal gas law (Eq. 15.6) to solve for n:

$$n = \frac{PV}{RT}$$

Inserting our estimates for P, V, and T leads to

$$n = \frac{P_{atm}V}{RT} = \frac{(1.01 \times 10^5 \text{ Pa})(3 \times 10^{-3} \text{ m}^3)}{[8.31 \text{ J/(mole} \cdot \text{K)}](290 \text{ K})} = 0.1 \text{ mole}$$

SOLVE

Using this value for n in our relation for the internal energy (Eq. 15.22) gives

$$U = \tfrac{3}{2}nRT = \tfrac{3}{2}(0.1 \text{ mole})[8.31 \text{ J/(mole} \cdot \text{K)}](290 \text{ K}) = \boxed{400 \text{ J}}$$

What does it mean?

This result is approximately equal to the kinetic energy of a baseball traveling at 100 mi/h, so it is a very significant amount of energy! It is quite surprising that a simple balloon of helium gas contains such a large amount of energy "internally."

Figure 15.9 Example 15.5.

Specific Heat of an Ideal Gas

In Chapter 14, we learned that when heat is added to a system, the temperature changes according to the value of the system's specific heat. Let's apply kinetic theory to find the specific heat of an ideal gas. Heat energy added to an ideal gas causes the molecular kinetic energy to increase, thus increasing the internal energy. So, if an amount of heat energy ΔQ is added to an ideal gas, the internal energy must increase by that amount and $\Delta Q = \Delta U$. Using Equation 15.21, we can find the corresponding change in temperature ΔT. We have

$$\Delta Q = \Delta U = \tfrac{3}{2} N k_B \Delta T$$

The heat capacity of the system (see Chapter 14) is equal to the ratio of ΔQ and ΔT, so

$$\text{heat capacity} = \frac{\Delta Q}{\Delta T} = \tfrac{3}{2} N k_B$$

This answer is the heat capacity at *constant volume* because our derivation depended on Figure 15.7, where we assumed the walls of the container were fixed in place so that the system's total volume is constant. For a system containing 1 mole of atoms, N is equal to Avogadro's number ($N = N_A$); using the relation $N_A k_B = R$ (Eq. 15.9), we thus get

$$\text{heat capacity of 1 mole} = \frac{\Delta Q}{\Delta T} = \tfrac{3}{2} N_A k_B = \tfrac{3}{2} R$$

which is also called the **specific heat** per mole at constant volume and is denoted C_V:

$$C_V = \tfrac{3}{2} R \qquad\qquad (15.23)$$

Specific heat per mole of a monatomic ideal gas

In our work on calorimetry in Chapter 14, we discussed the specific heat *per unit mass*, denoted by c. Here we denote the specific heat *per mole* by C_V (an uppercase C).

Polyatomic Gases

Our results for the internal energy (Eqs. 15.21 and 15.22), and the specific heat (Eq. 15.23), apply only for monatomic ideal gases. The derivations of those results assumed all the internal energy is in the form of translational kinetic energy. For gases composed of polyatomic molecules such as O_2 or N_2 or NH_3, we must also include the energy associated with the rotational and vibrational motion of the molecules. The internal energy of these gases can also be found using kinetic theory, but the analysis is more than we can include here. The results depend on the nature of the molecule—that is, if it diatomic or triatomic and so forth—and both the internal energy and the specific heat per mole of such gases are larger than the values for monatomic gases.

Distribution of Speeds in a Gas

So far, we have dealt with the speed of a typical (average) molecule. The speeds of individual molecules in a given sample are not all the same, however. Some molecules have speeds greater than the typical speed v in Equation 15.18, whereas others have smaller speeds. The problem of the molecular speed distribution in a gas is an important part of kinetic theory. We will not calculate this distribution here, but it is instructive to consider it qualitatively because it gives rise to some important effects.

Figure 15.10 shows the distribution of molecular speeds in a typical case. Called the **Maxwell–Boltzmann** distribution, it gives the probability that a molecule will have a particular speed. The typical speed v found in Equation 15.18 is near the average speed, but a significant number of molecules have speeds that are greater or smaller by a factor of two or even more. The distribution varies with temperature,

Figure 15.10 Distribution of speeds for a dilute gas of N_2. The vertical scale is the relative probability of finding a molecule moving with a particular speed v.

and if the temperature is increased, the entire distribution curve shifts to higher speeds. This temperature dependence makes intuitive sense (because "hotter" molecules have greater kinetic energy) and is responsible for the temperature dependence of the average molecular speed in Equation 15.18.

EXAMPLE 15.6 (Absence of) Helium in the Atmosphere

The Earth's atmosphere contains mostly nitrogen and oxygen, with smaller concentrations of other gases, but it does not contain any helium. Experiments show that a helium atom released into the atmosphere escapes to outer space, never to return. To understand how a helium atom is able to escape from the Earth's gravitational attraction, begin by calculating the typical speed v of a helium atom in a balloon held at atmospheric pressure and temperature. Compare your value of v with the escape velocity for an object near the Earth's surface, as calculated in Chapter 6. You will find that v is smaller than the escape velocity, so it would seem that a "typical" helium will not be able to escape. Given that fact, how can the distribution of molecular speeds in Figure 15.10 explain why all helium atoms do, eventually, escape from the atmosphere?

RECOGNIZE THE PRINCIPLE

We can find the speed v of a typical helium atom using the same approach we applied to a nitrogen molecule (which led to the result in Eq. 15.19). If v is greater than the escape velocity, a typical helium atom will be able to escape the Earth's gravitation attraction, just as would any projectile fired from the surface of the Earth. We must also recognize that, according to the Maxwell–Boltzmann speed distribution (Fig. 15.10), some helium atoms will have a speed much greater than v, and we also need to consider the behavior of these atoms.

SKETCH THE PROBLEM

The Maxwell–Boltzmann distribution (Fig. 15.10) is a key to this problem.

IDENTIFY THE RELATIONSHIPS

According to Equation 15.18, the speed of an atom or molecule in a gas depends on the particle's mass. The mass of a helium atom (see the periodic table) is $m_{He} = 4.0 \text{ g}/N_A = 6.6 \times 10^{-27}$ kg.

SOLVE

Using Equation 15.18 at room temperature ($T = 293$ K), we get

$$v_{He} = \sqrt{\frac{3k_B T}{m_{He}}} = \sqrt{\frac{3(1.38 \times 10^{-23} \text{ J/K})(293 \text{ K})}{6.6 \times 10^{-27} \text{ kg}}} = \boxed{1400 \text{ m/s}}$$

What does it mean?

This speed is greater than the typical speed of N_2 molecules in the atmosphere found in Equation 15.19 because helium has a much smaller mass. In Chapter 6, we found that a projectile at the Earth's surface with a speed greater than about 11,000 m/s has enough kinetic energy to "escape" from the Earth's gravitational attraction, so the speed v of a typical helium atom will still be *less* than the escape velocity. However, according to the distribution of molecular speeds in Figure 15.10, some helium atoms will have speeds much greater than the typical speed. A small fraction of the helium atoms will be moving faster than the escape velocity and hence will be able to escape from the atmosphere. All helium atoms released into the atmosphere have some probability of achieving such a high speed, and when they do, they will escape into space. That is why our atmosphere does not contain any helium gas. The helium used in carnival balloons and other applications was obtained from trapped underground deposits. What about atoms and molecules heavier than helium? A small percentage of such heavier molecules, including O_2 and N_2, will also escape from the Earth, but their rate of escape is extremely small; hence, these gases remain in the atmosphere indefinitely. (They are also replenished by natural processes.)

15.5 | DIFFUSION

Let's now focus on the motion of one particular molecule as it moves through the gas. The molecule follows a zigzag path as it collides successively with other molecules (Fig. 15.11). This type of motion is called *diffusion*. Similar motion is found for molecules moving through a liquid and through certain types of solids, including cell membranes.

Each particle in a gas follows a different random zigzag path, but these different trajectories can be described in an average way in terms of the typical molecular speed v between collisions and the average distance between collisions ℓ. For a molecule traveling through a gas, the speed of a typical molecule is given by Equation 15.18, so v depends on both the temperature and the mass of the molecule. The distance ℓ is the mean free path we introduced earlier in this chapter. The value of the mean free path depends on the density of the gas; if the density is high, the gas molecules are close together, the probability for a collision is high, and ℓ is short. If the density is low, the collision probability is also low and ℓ is long. An analysis of the collision mechanics shows that the mean free path varies inversely with the density. Because the density is proportional to the number of molecules per unit volume N/V, the average distance between collisions varies as

$$\ell \propto \frac{V}{N} \qquad (15.24)$$

for a gas in which N particles occupy a total volume V. For air at room temperature and pressure, the mean free path is approximately $\ell \approx 1 \times 10^{-7}$ m = 100 nm, which is about 30 times larger than the average spacing between gas molecules.

The mean free path is the average length of each segment of the zigzag molecular path in Figure 15.11. In most situations, we are interested in how the molecule travels over a relatively long distance such as when a gas molecule travels from one side of a room to the other. Figure 15.11 shows a typical diffusive path taken as a molecule travels from an initial point A to a final location B. The total distance traveled by a molecule is much longer than the length of the direct path that connects A and B. For this reason, diffusive motion between two points is much slower than motion without all the collisions.

Although we can never predict the precise path followed by a particular molecule, diffusion usually involves a large number of molecules making a large number of zigzag steps, so statistical arguments give an accurate description of an average random walk path. The magnitude of the average displacement Δr of a molecule after many steps taken over a time t is

$$\Delta r = \sqrt{Dt} \qquad (15.25)$$

The quantity D is called the *diffusion constant*. The result above can also be written as

$$t = \frac{(\Delta r)^2}{D} \qquad (15.26)$$

giving the time t required for an average molecule to diffuse a distance Δr.

The value of the diffusion constant depends on both v and ℓ, so it depends on temperature and on the mass of the diffusing particle as well as on the properties of the medium. The values of some typical diffusion constants are listed in Table 15.1. Diffusive motion is found in many situations, including molecules through a gas or liquid, and through a porous membrane such as a cell membrane.

Mean free path (average distance between collisions)

Start $(t = 0)$ A

ℓ

Δr

B

Position at time t

$\Delta r = \sqrt{Dt}$

Figure 15.11 The zigzag motion of a molecule in a gas or liquid is called diffusion. (Similar motion is found for an impurity molecule moving through a solid.) The average length of each step is the mean free path. Each change of direction of a molecule is caused by a collision. The steps are in random directions, so after many steps (i.e., many collisions), the magnitude of the net displacement (Δr) is much smaller than if there had been no collisions.

Distance traveled by a diffusing particle

Table 15.1 Values of Some Typical Diffusion Constants

DIFFUSING MOLECULE OR PARTICLE	MEDIUM	D (m²/s)
Oxygen (O_2)	Water	1×10^{-9}
Oxygen (O_2)	Air	2×10^{-5}
Nitrogen (N_2)	Air	2×10^{-5}
Hydrogen (H_2)	Air	6×10^{-5}
Oxygen (O_2)	Tissue	2×10^{-11}
Water (H_2O)	Water	3×10^{-9}
Sucrose	Water	5×10^{-10}
DNA	Water	1×10^{-12}
Hemoglobin	Water	7×10^{-11}
Pollen grain (5×10^{-7} m diameter)	Water	1×10^{-9}

Notes: All substances are at room temperature and pressure. Except for the pollen grain, the diffusing objects are all molecules.

A key feature of the result for Δr (Eq. 15.25) is that the net displacement is proportional to $\sqrt{t}$, the *square root* of the time. That is in contrast to motion with a constant velocity, for which the displacement increases *linearly* with time. This square root dependence means that a diffusing particle moves relatively slowly through the surrounding medium, as we explore in Example 15.7.

EXAMPLE 15.7 Ⓧ Ⓡ Diffusion in a Gas

You are at work in a chemical laboratory when a dangerous gas accidentally begins to leak from a container. Approximately how long do you have to evacuate the room safely, assuming the gas travels by diffusion alone?

RECOGNIZE THE PRINCIPLE

If the gas leaks from the center of the room, you will want to be gone long before a typical molecule has diffused to the door. This diffusion time is $t = (\Delta r)^2 / D$ (Eq. 15.26). The diffusion distance Δr is not given, so to get the approximate time, we must estimate the size of a typical room.

SKETCH THE PROBLEM

Figure 15.11 describes the problem, with the distance from A to B equal to the size of the room.

IDENTIFY THE RELATIONSHIPS AND SOLVE

The diffusion constant for a typical molecule in air is $D \approx 2 \times 10^{-5}$ m²/s (Table 15.1), and the size of a typical room is about $\Delta r = 3$ m. These values lead to

$$t = \frac{(\Delta r)^2}{D} = \frac{(3 \text{ m})^2}{2 \times 10^{-5} \text{ m}^2/\text{s}} = \boxed{5 \times 10^5 \text{ s}}$$

What does it mean?
This result is more than 100 h, so diffusion is a very slow process. It would thus seem that you have plenty of time to escape. In most cases, however, convection currents in the room will disperse the molecules much more quickly. In addition, some molecules will have speeds much greater than the typical speed (Fig. 15.10) and these will also arrive much sooner than the typical molecule considered here.

Using Diffusion in Medicine ⊗

The slow process of diffusion is employed in a medical application called ***transdermal drug delivery***. Sometimes a drug should be delivered into the body in a slow and steady manner, which can be accomplished by placing a "patch" containing the drug in direct contact with the skin (Fig. 15.12A). The drug diffuses through the patch membrane, through the skin, and into the body. By careful choice of the membrane thickness, the drug can be made to enter the body slowly, over the course of many days or weeks. Such a slow delivery is preferred in certain cases such as the use of nicotine to help a person stop smoking.

CONCEPT CHECK 15.5 Designing a Transdermal Patch

The membrane in a transdermal patch (Fig. 15.12B) is typically about 200 μm thick. Suppose the thickness is doubled. How does that affect the rate with which the drug diffuses through the membrane?

    (a) The rate is doubled.
    (b) The rate is cut in half.
    (c) The rate is reduced by a factor of four.

Isotope Separation

Another important application of diffusion is in the separation of different ***isotopes*** of a given element. As an example, consider the nucleus of a uranium atom. The two most common types of uranium nuclei are denoted $^{235}_{92}U$ and $^{238}_{92}U$; here the lower number is the number of protons in the nucleus, and the upper number is the number of neutrons plus protons. These two different nuclei, which contain different numbers of neutrons, are called isotopes of uranium. (We'll say much more about the properties of different nuclei in Chapter 30.) Naturally occurring uranium contains a mixture of $^{235}_{92}U$ and $^{238}_{92}U$, but only $^{235}_{92}U$ is able to undergo the nuclear reactions required in power plants (and nuclear bombs). To make nuclear fuel, the $^{235}_{92}U$ atoms must be removed from a uranium sample without having the other isotope come along with them. This process is called ***isotope separation*** and is used to make what is called "enriched" uranium. One way to accomplish isotope separation is with diffusion.

A diffusing molecule follows a zigzag path as sketched in Figure 15.11. On each straight portion, the molecule moves with a constant speed; for diffusion through a gas, the average speed on these different zigs and zags is just v in Equation 15.18. In this case and in diffusion through other substances, the average speed increases as the molecular mass decreases. Hence, a light atom or molecule diffuses faster than a heavy one.

Diffusion experiments with uranium usually employ molecules of $^{235}_{92}UF$ and $^{238}_{92}UF$. Since $^{235}_{92}UF$ has a slightly smaller mass, it diffuses slightly faster than a molecule of $^{238}_{92}UF$. This difference can be used separate isotopes as shown in Figure 15.13. The left side of this system contains an equal mixture of $^{235}_{92}UF$ and $^{238}_{92}UF$, and in the middle there is a medium (perhaps a thin membrane like the one used in a

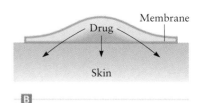

Figure 15.12 Ⓐ A transdermal patch delivers drugs into the body at a slow, steady rate. Ⓑ Structure of a transdermal patch.

$^{235}_{92}UF$ will diffuse to the right slightly faster than $^{238}_{92}UF$ because it has a smaller mass.

Figure 15.13 The process of diffusion can be used to separate different isotopes of uranium. For simplicity, we show roughly equal numbers of $^{235}_{92}UF$ and $^{238}_{92}UF$ on the left. Usually, the initial concentration of $^{235}_{92}UF$ is much smaller than the concentration of $^{238}_{92}UF$.

transdermal patch; Fig. 15.12) through which the molecules can diffuse. The lighter $^{235}_{92}$UF molecules (with the larger diffusion constant) move through the membrane slightly faster than the $^{238}_{92}$UF molecules, so there is a slightly higher concentration of $^{235}_{92}$UF on the right compared with the left. Hence, we have separated (at least partially) the different isotopes; additional diffusion steps can be used to make the separation more and more complete. A similar separation process takes place in plants when they process two isotopes of carbon, $^{12}_{6}$C and $^{13}_{6}$C, via photosynthesis.

CONCEPT CHECK 15.6 Relation of the Diffusion Distance to Time

Suppose we double the size of the room in Example 15.7. How does that affect the time required for an average molecule to diffuse across the room?
  (a) The time is doubled.
  (b) The time does not change.
  (c) The time increases by a factor of four.

Diffusion, Brownian Motion, and the Discovery of Atoms

Figure 15.11 shows the characteristic random-walk path followed by a diffusing particle. This figure is only schematic; there is no way to observe the motion of a molecule in a gas or liquid in such fine detail. It is possible, however, to observe the diffusive motion of larger objects. This experiment was actually carried out nearly 200 years ago by botanist Robert Brown in his studies of the motion of pollen grains in water. The grains were about 1 μm in size, much larger than a molecule but still small enough to make diffusion observable with a microscope. Brown found that pollen grains do indeed follow random zigzag paths like those sketched in Figure 15.11, a motion now called **Brownian motion** in his honor. Just as for the molecule in Figure 15.11, the zigzag motion of Brown's pollen grains was caused by collisions with molecules in the fluid. Brown made his observations before the discovery of atoms and molecules, and there was no real explanation of this motion during his lifetime. In hindsight, his experiments provided some of the very first evidence for the existence of atoms and molecules, and detailed studies of Brownian motion led to the first accurate measurements of Avogadro's number.

The Arrow of Time

Consider a video recording of an elastic collision of two particles; they might be two molecules in a dilute gas. We could play this video normally or in reverse (Fig. 15.14), and both cases yield motion consistent with our intuition. In fact, both collision trajectories are consistent with Newton's laws.

Such reversibility of the direction of time is not always true. For example, consider the mixing of two gases A and B in Figure 15.15. Here we begin at $t = 0$ with all the molecules of type A on the left side of a container and all the molecules of type B on the right side. After a period of time, the gases mix, and there is an equal chance of finding either type of molecule anywhere in the container. If we were to videotape this mixing process, the first few frames of the video would look like Figure 15.15A and the final frames would look like Figure 15.15B. If we were to play this video backward, it would show molecules in a fully mixed gas (Fig. 15.15B) spontaneously unmix so as to end up with the two separate regions in Figure 15.15A. Your intuition should tell you that such behavior is not possible, and indeed, such unmixing of two gases does not occur in real life. In this process, the direction of time does matter.

The hypothetical unmixing process involves many collisions between molecules, and we could use Newton's laws to analyze these collisions one at a time. Each collision is time reversible, just as with the collisions in Figure 15.14, so each individual unmixing collision would appear to be perfectly reasonable. Only when we string them all together do we run into problems.

This discussion reveals a deep aspect of physics. Newton's laws are time reversible, but many processes in nature are not! Processes that are not time reversible

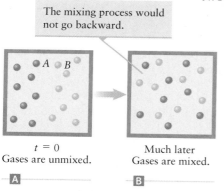

Both trajectories obey Newton's laws.

Forward collision   Backward collision

Figure 15.14 An elastic collision between two molecules in a gas is time reversible. Both forward and backward collision processes satisfy all the laws of physics (including conservation of energy and momentum).

The mixing process would not go backward.

$t = 0$
Gases are unmixed.
A

Much later
Gases are mixed.
B

Figure 15.15 The mixing of two gases is not time reversible.
A Two different types of molecules are initially separated (placed on different sides of a container).
B As time passes, the molecules mix, with both types of molecules spread evenly (with equal probability) throughout the container.

are possible in systems that contain a very large number of particles (such as a gas). The statistical improbability of different systemwide configurations such as the mixed and unmixed gas pictures in Figure 15.15 makes certain processes such as the unmixing of a gas time irreversible. This discovery was made possible through work in kinetic theory. We'll see in Chapter 16 that time-irreversible behavior of systems with many particles also plays a role in the area of physics called *thermodynamics*.

15.6 | DEEP PUZZLES IN KINETIC THEORY

In this chapter, we focused on the successes of classical kinetic theory. That theory is based on an application of classical mechanics—Newton's laws of motion—to the properties of gases. Classical kinetic theory is able to explain the ideal gas law, the meaning of temperature, and the universal values of the specific heats of dilute gases. However, certain properties of simple gases are *not* explained by classical kinetic theory. These failures were important in the history of physics and eventually led to the development of quantum theory. A kinetic theory based on quantum rather than classical mechanics successfully explains all the properties of dilute gases.

Another puzzle concerns the notion of an absolute temperature scale. What happens in a dilute gas, or any other phase of matter, when it is cooled to absolute zero? Do the atoms in such a gas or solid stop all motion? If so, what forces hold such a substance together? These important conceptual issues are not addressed by classical kinetic theory. We'll explore these issues later in this book, after we learn about thermodynamics and quantum theory.

SUMMARY | Chapter 15

KEY CONCEPTS AND PRINCIPLES

Kinetic theory

Kinetic theory is the application of Newton's laws to the mechanics of a gas of particles. This theory leads to the **ideal gas law**; for a dilute gas composed of any substance,

$$PV = nRT \qquad \text{(15.6) (page 472)}$$

where P is the pressure, V is the volume, n is the number of particles as measured in units of moles, and R is the **universal gas constant** with the value

$$R = 8.31 \frac{\text{J}}{\text{mole} \cdot \text{K}} \qquad \text{(15.7) (page 472)}$$

The ideal gas law can also be written in the form

$$PV = Nk_B T \qquad \text{(15.11) (page 473)}$$

where N equals the number of particles (i.e., molecules) and k_B is **Boltzmann's constant**,

$$k_B = 1.38 \times 10^{-23} \text{ J/K} \qquad \text{(15.10) (page 473)}$$

(Continued)

Internal energy

The temperature of a dilute gas is related to the average kinetic energy of a molecule by

$$KE = \tfrac{3}{2}k_B T \qquad\qquad \textbf{(15.17)} \text{ (page 477)}$$

The *internal energy* of a monatomic ideal gas is equal to the total kinetic energy of the molecules and is given by

$$U = \tfrac{3}{2}Nk_B T \qquad\qquad \textbf{(15.21)} \text{ (page 479)}$$

For an ideal gas, U is independent of all other factors such as the molecular mass and pressure.

APPLICATIONS

Gas laws

The ideal gas law is a general relation that contains several other gas laws as special cases.

Avogadro's law:
For a sample of gas at constant pressure and temperature, the volume is proportional to the number of molecules in the sample.

Boyle's law:
For a sample of gas at constant temperature, the product of the pressure and the volume is constant:

$$PV = \text{constant} \qquad\qquad \textbf{(15.3)} \text{ (page 471)}$$

Charles's law:
For a sample of gas at constant pressure, if the temperature is changed by a small amount ΔT, the volume changes by an amount ΔV, with

$$\Delta V \propto \Delta T \qquad\qquad \textbf{(15.4)} \text{ (page 471)}$$

Gay-Lussac's law:
For a sample of gas held in a container with constant volume, changes in pressure are proportional to changes in temperature:

$$\Delta P \propto \Delta T \qquad\qquad \textbf{(15.5)} \text{ (page 471)}$$

Specific heat of a gas

The specific heat per mole of a monatomic ideal gas at constant volume is given by

$$C_V = \tfrac{3}{2}R \qquad\qquad \textbf{(15.23)} \text{ (page 481)}$$

Diffusion

The random motion of individual gas molecules is described by *diffusion*. The average net distance that a molecule diffuses in a time t is given by

$$\Delta r = \sqrt{Dt} \qquad\qquad \textbf{(15.25)} \text{ (page 483)}$$

where D is the *diffusion constant*.

SSM = answer in Student Companion & Problem-Solving Guide   ⊗ = life science application

1. Show that Avogadro's law, Charles's law, and Gay-Lussac's law all follow from the ideal gas law (Eq. 15.6).

2. In Example 15.5, we claimed that the total kinetic energy of the gas is approximately equal to the kinetic energy of a baseball. Show that this claim is in fact true by working out the value of $\frac{1}{2}mv^2$ with realistic values of m and v for a baseball.

3. If you were to hammer on a piece of metal, the temperature of the metal would rise. Why? Describe from a microscopic point of view.

4. Is it possible to heat a can of soup by continuously lifting and dropping it on a hard surface? (Assume the can will take the abuse and not rupture.) Would it make a difference if the can were encased in an excellent thermal insulator?

5. Consider two identical containers, one full of helium and the other full of oxygen. If the molecules of both gases have the same average speed, what is the ratio of their temperatures? (Approximate both as ideal gases.)

6. SSM In Chapter 11, we learned about the bulk modulus B of a substance. The bulk modulus is related to changes in the volume by

$$\frac{\Delta V}{V} = -\frac{\Delta P}{B}$$

Use the ideal gas law to calculate the bulk modulus for an ideal gas. Assume the temperature is held constant.

7. An automobile cooling system typically operates at a pressure higher than atmospheric pressure. Why is that desirable?

8. A family drives from a low valley to a high-altitude mountain meadow for a picnic. When they arrive at the meadow, they find that their bag of potato chips has burst open and their bag of pretzels looks like an inflated balloon. Explain the behavior of the packaging.

9. SSM A container of gas under pressure has a pinhole leak. If the content of the container is oxygen, the pressure decreases at a certain rate as the gas escapes. If the content is hydrogen gas,

however, we find that the rate of pressure drop is greater under the same conditions. Why?

10. The diffusion constant D of a particle in a gas or liquid depends on temperature. If the temperature is increased, does D increase or decrease? Explain why.

11. In Example 15.7 we saw that diffusion is a very slow process. Calculate the *total distance* (not the net displacement Δr) traveled by a diffusing molecule in Example 15.7.

12. Compare the act of making tea with a tea bag in cold water versus steeping it in hot water. Which method will make the tea stronger faster? Why?

13. Consider an ideal gas composed of helium atoms. The collisions between atoms are similar to the collisions between billiard balls. That is, whenever the distance between two atoms is less than $2r$ (where r is the diameter of a helium atom), the two atoms undergo a collision. Show that the mean free path between collisions ℓ is proportional to $1/\rho$, where ρ is the density of the gas. *Hint*: Consider the motion of just one atom and assume the others are all "frozen" in place.

14. The mass of 1 mole of carbon atoms of isotope number 12 (so-called carbon 12) is defined to be exactly 12 g. The *atomic mass unit* is defined as one-twelfth the mass of a single carbon 12 atom. Use this fact to calculate the conversion factor that relates atomic mass units to kilograms.

15. ⊗ In Concept Check 15.5, we considered how to slow down the release of a drug in a transdermal patch by increasing the thickness of the membrane. Which of the following strategies will decrease the rate at which drug molecules enter the body?
(a) Lower the concentration of molecules in the patch.
(b) Increase the temperature.
(c) Use larger drug molecules.

16. The air in a room consists mainly of O_2 and N_2 molecules. Which has the larger average speed?

17. Why does the pressure in a car tire increase when the car is driven?

PROBLEMS

SSM = solution in Student Companion & Problem-Solving Guide   ⊗ = life science application
★ = intermediate   ✪ = challenging   ℝ = reasoning and relationships problem

15.1 MOLECULAR PICTURE OF A GAS

1. What is the mass of a single carbon atom?

2. How many oxygen atoms are in 1.0 g of water?

3. Find the mass of a molecule of carbon dioxide.

4. ★ ℝ Estimate the number of atoms in your body.

5. SSM ★ You have 6.0 moles of particles of an unknown (pure) substance with a mass of 240 g. What might the substance be?

15.2 IDEAL GASES: AN EXPERIMENTAL PERSPECTIVE

6. Consider a balloon containing nitrogen molecules at room temperature (20°C) and atmospheric pressure. If the balloon has a spherical shape with radius of 20 cm, what is the number of molecules in the balloon?

7. ★ A balloon having an initial temperature of 20°C is heated so that the volume doubles while the pressure is kept fixed. What is the new value of the temperature?

8. ★ A spherical balloon containing nitrogen gas has a radius of 20 cm. If the balloon is at atmospheric pressure and room temperature, what is the mass of nitrogen in the balloon?

9. ★ Find the average spacing between molecules in the balloon in Problem 8. How does the answer change when the balloon is heated, assuming the pressure does not change?

10. ★ ℝ Estimate the number of air molecules in a bicycle tire.

11. Imagine you are sitting in a room full of oxygen molecules at 300 K. If the room has a volume of 1.0×10^4 m^3, what is the number of oxygen molecules in the room?

12. You are given the job of calibrating a gas thermometer. You do some experiments and find that the gas pressure in the thermometer is 40,000 Pa at $T = 280$ K. If the pressure later reads 60,000 Pa, what is the new temperature?

13. For the gas thermometer in Problem 12, what is the pressure when $T = 230$ K?

14. SSM ★ The gas in a cylindrical container has a pressure of 1.0×10^5 Pa and is at room temperature (293 K). A piston (a movable wall) at one end of the container is then adjusted so

that the volume is reduced by a factor of three, and it is found that the pressure increases by a factor of five. What is the final temperature of the gas?

15. ⭐ A balloon contains 3.0 moles of helium gas at normal atmospheric pressure ($P_{atm} = 1.01 \times 10^5$ Pa) and room temperature (293 K). (a) What is the volume of the balloon? (b) The weather becomes very stormy (a hurricane is approaching), and the volume of the balloon increases by 15%. What is the new pressure?

16. How many molecules are there in 1.0 cm² of air? *Hint*: The average mass of a mole of molecules in the air is approximately 29 g.

17. ⊗ The lungs of a typical adult hold about 2 L of air (0.002 m³). How many moles of air molecules are in a typical adult's lungs?

18. ⭐ A weather balloon filled with helium gas has a volume of 300 m³ and pressure 1.4×10^5 Pa at the Earth's surface (where the temperature is 293 K). The balloon is then released and moves to a high altitude, where the pressure in the balloon is 25,000 Pa and the temperature is 250 K (−23°C). What is the new volume of the balloon?

19. ⭐ Ⓡ Approximately how many moles of gas molecules are in your classroom?

20. ⊗ A tank of compressed oxygen gas at a doctor's office has a pressure of 100 times atmospheric pressure. If the volume of the tank is 500 cm³, how many oxygen molecules does it contain?

21. An ultrahigh vacuum system can reach pressures in the laboratory that are less than 1.0×10^{-7} Pa. If that is the pressure in a system of volume 1000 cm³, how many molecules are in the system?

22. ✪ Ⓡ Estimate the number of helium molecules in the Goodyear blimp. *Hint*: Assume the pressure is a factor of two larger than atmospheric pressure (so that it will not collapse).

23. SSM ⭐ The pressure in a gas thermometer is 5000 Pa at the freezing point of water. (a) What is the pressure in this thermometer at the boiling point of water? (b) If the pressure is 6000 Pa, what is the temperature?

24. ⭐ ⊗ Some endurance athletes (such as cyclists and long-distance runners) use a hypobaric chamber to enhance their ability to use oxygen. A hypobaric chamber mimics the effect of being at a high altitude, where the amount of oxygen available to the lungs is smaller than at sea level. A typical hypobaric chamber simulates the environment at an altitude of 9000 ft above sea level, where the air pressure is approximately 7×10^4 Pa. What is the amount of oxygen available at this altitude compared with that at sea level? Express your answer as a fraction (without units).

15.3 IDEAL GASES AND NEWTON'S LAWS

15.4 KINETIC THEORY

25. What is the average speed of a molecule of H_2O in the atmosphere?

26. SSM ⭐ You place 80 moles of hydrogen gas in a balloon of volume 2.5 m³, and find the pressure to be 1.5 times atmospheric pressure. What is the typical speed of a hydrogen molecule?

27. ⭐ For the balloon in Problem 26, what is the total kinetic energy of all the hydrogen molecules in the balloon?

28. ⭐ You have a container of an ideal gas at room temperature (20°C) and pressure, and you want to use it to store energy in the form of the kinetic energy of the gas. If you want to double

the energy in the system, what must be the temperature's new value?

29. The surface of the Sun has a temperature of about 6000 K. What is the average speed of a hydrogen molecule in the Sun's atmosphere?

30. ⭐ What is the ratio of the speed of a hydrogen molecule at the Sun's surface to the speed of a hydrogen molecule in the Earth's atmosphere? The temperature at the surface of the Sun is approximately 6000 K.

31. What is the total internal energy of 15 moles of helium gas at room temperature? How does the answer change if the helium is replaced by argon?

32. SSM ⭐ A balloon of volume 1.5 m³ contains argon gas at a pressure of 1.5×10^5 Pa and room temperature (20°C). What is the total internal energy of the gas?

33. ✪ Calculate the average speed v of a hydrogen molecule in the Earth's atmosphere. Is v less than or greater than the escape velocity? Experiments show that a hydrogen molecule released into the atmosphere escapes into outer space. If v is less than the escape velocity, why is this so?

15.5 DIFFUSION

34. Approximately how long does it take a molecule of nitrogen to diffuse a distance of 1 cm through the atmosphere?

35. SSM ⭐ How long would it take a nitrogen molecule in the atmosphere to travel a distance of 1 cm if there were no collisions with other molecules?

36. Approximately how long does it take a water molecule to diffuse a distance of 3 cm through a glass of water?

37. About how far does a typical water molecule diffuse (in water) in 1 minute?

38. ⭐ ⊗ Approximately how long does it take an oxygen molecule to diffuse across the interior of a cell that is 100 μm in diameter?

39. SSM ✪ ⊗ Ⓡ What is the approximate time required for a hemoglobin molecule to diffuse through a cell membrane? *Hint*: The diffusion constant D is approximately proportional to $1/r$, where r is the diameter of the diffusing particle or atom. Estimate the diffusion constant of a hemoglobin molecule from the value of D for an oxygen molecule in Table 15.1.

40. ⭐ ⊗ One way for a protein to move from one side of a cell to another is through diffusion. If the diameter of a cell is increased by a factor of four, by what factor does the time required to diffuse across the cell increase?

41. ⊗ Consider an oxygen molecule as it diffuses in your blood. How far will a typical molecule diffuse in 1 second? Assume the value of D for oxygen in blood is the same as for oxygen in water.

42. Ⓡ Approximately how long does it take a hydrogen molecule to diffuse across a small room?

43. Ⓡ In our discussion of diffusion, we mentioned that between collisions, a molecule in air travels in a straight line with a constant speed. For a hydrogen molecule in air, this speed is approximately 3000 m/s. If the hydrogen molecule in Problem 42 was moving with this speed and did not collide with any air molecules, how long would it take to travel across the room?

44. ⊗ A cell membrane is typically 10 nm (1×10^{-8} m) thick. How long does it take an oxygen molecule to diffuse this distance through the membrane?

45. ⊗ Calculate the net distance a pollen grain of radius 5 × 10^{-7} m will diffuse in water in 1 minute.

ADDITIONAL PROBLEMS

46. ⭐ You place 3.5 g of water at room temperature in a pressure cooker (a sealed pot) of volume 2.5 L and heat the water to 200°C. What is the pressure in the cooker?

47. ✪ ⊗ Ⓡ **Caesar's last breath.** (a) How many molecules from Julius Caesar's dying breath are currently in your lungs? The total mass of the Earth's atmosphere is approximately

5×10^{18} kg, it is composed mainly of N_2 (78%) and O_2 (21%), and an average breath (called the tidal volume) is 1.5 L. (b) Is it possible that Caesar's last breath would be evenly diffused around the globe by now? Many processes would distribute the molecules of his last breath, including convection, absorption and release in water, and diffusion. Calculate how far a nitrogen molecule will travel by the process of diffusion alone in the approximately 2000 years since Caesar's demise. How much of a role does diffusion play in distributing and diluting emissions into the atmosphere?

48. ✪ Calculate the volume occupied by 1 mole of each of the following substances (in both cubic centimeters and liters): (a) water at room temperature, (b) ice at 0°C, (c) air at room temperature and atmospheric pressure, (d) the element aluminum at room temperature, and (e) helium at 0°C and atmospheric pressure. (f) Rank these substances from largest volume to smallest. (g) Rank these substances according to density from smallest to largest.

49. ✪ Ⓡ The continent of North America has an area of approximately 2.4×10^7 km². If the entire continent were evenly covered with Avogadro's number of BBs (spherical projectiles 4.5 mm in diameter), how deep would the layer of BBs be? *Hint*: If spheres are randomly poured into a jar, the fraction of the jar's volume occupied by spheres when the jar is "full" is about 65% if the diameter of the jar is large compared to the size of the spheres.

50. ☆ An air bubble ascends from the bottom of a lake 15 m deep. The temperature at the bottom of the lake is 4°C, and near the surface it is 20°C. How many times larger in volume is the bubble at the top of the lake compared with its volume at the bottom?

51. ✪ ⊗ Ⓡ A scuba diver descends to 8.2 m below the surface of the ocean. (a) What pressure will her regulator need to supply so that her lungs can fill as normal? (b) When the diver is at 8.2 m, how many times more molecules of air are in her lungs than at the surface? Treat air as an ideal gas and assume the temperature is the same in both instances, which will be somewhat less than body temperature.

52. ✪ ⊗ A diver uses an air cylinder with a volume of 11 L and filled to a pressure of 200 atm (Fig. P15.52). He dives at a leisurely pace, taking about 15 breaths per minute, and each breath is 1.2 L on average. (a) If he must leave 17% of the air for reserve, how long can he dive at 20 m deep? *Hint*: The pressure must be considered as in Problem 51. (b) How long will his air supply last if he dives at 30 m? Approximate air as an ideal gas.

Figure P15.52 A standard 80 diving tank: capacity 88 ft³ at 200 atm.

53. ✪ A steel cylinder is filled with an ideal gas at 20°C and pressure of 8.0 times atmospheric pressure. (a) The cylinder is submerged in a bath of boiling water and allowed to reach thermal equilibrium. Calculate the pressure inside the cylinder. (b) Some gas is then allowed to escape so that the pressure returns to its original value. Find what fraction of the original gas remains. (c) If the cylinder is then allowed to cool to its original temperature, what would the pressure inside be?

54. ✪ ⊗ Ⓡ The human nose is very sensitive to certain molecules. For example, it can sense the presence of the chemical

CH_3SH (methyl mercaptan) to 2 parts per billion. Another especially smelly substance is ammonia, NH_3. A student in a lab with very still air has two vials placed 20 cm from his nose; one contains methyl mercaptan, the other ammonia. He begins his experiment and removes the caps from both vials, and after 30 minutes, he notices the smell of ammonia. How soon after does his nose detect the methyl mercaptan? Is diffusion a good model for how fast odors travel?

55. ✪ ⊗ **Bug breath.** Insects do not have lungs or a blood circulatory system. Instead, a system of openings in the exoskeleton (spiracles) lead to branching tubes of decreasing diameter called trachea, the smallest of which are about 1 μm in diameter and connect to every cell in the insect's body (Fig. P15.55). For many insects, each cell's steady supply of oxygen is provided by diffusion of air along the trachea. The average distance to a cell along a trachea for a small beetle is 0.25 cm. (a) Starting from one of the insect's spiracle openings, how long does it take an oxygen molecule on average to reach a cell? (b) Repeat the calculation for a human being, assuming people obtain their needed oxygen by these means. Assume the average distance to a cell along a trachea is now 10 cm (since human beings are much larger than beetles). (c) These calculated diffusion times are proportional to the rate at which oxygen can be delivered to a cell. Why don't we ever see beetles the size of cows, or mosquitoes the size of birds?

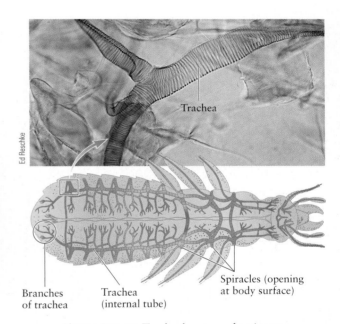

Figure P15.55 Tracheal system of an insect.

56. ✪ Find the average speed for Ne, Ar, and Kr atoms and for H_2, LiF, and Cl_2 molecules in the atmosphere at room temperature. Which ones do you expect to find in the atmosphere? Explain why or why not.

57. SSM ☆ ⊗ In our discussions of kinetic theory, we have focused on the motion of the molecules in a gas. The result for the typical speed in Equation 15.18, however, also applies to molecules in a liquid. Use this result to calculate the speed of the amino acid molecule glutamine ($C_5H_{10}N_2O_3$) in solution at room temperature.

Chapter 16

Thermodynamics

The operation of a steam engine is governed by the laws of thermodynamics. In fact, the need to understand the physics and engineering principles of the steam engine helped lead to the development of thermodynamics. (©Andy Moore/photolibrary/ Jupiterimages)

In Chapter 14, we discussed heat energy and how the flow of heat into or out of a system can affect its properties. Then, in Chapter 15, we developed the kinetic theory of a dilute gas and saw how the mechanics of molecular motion is connected to the temperature and pressure of a gas. The central thread of Chapters 14 and 15 was the principle of conservation of energy. In this chapter, we continue along that thread, exploring the area of physics called *thermodynamics*. Our goal now is more ambitious: we wish to understand the rules that govern the flow of heat and the exchange of other forms of energy between all types of systems, including gases, liquids, and solids. We are especially interested in *fundamental limits* on how heat can flow from one system to another. This understanding will lead us to some very surprising and powerful results, including the impossibility of building a perpetual motion machine, absolute limits on temperature, and the best possible efficiency of an engine.

16.1 | THERMODYNAMICS IS ABOUT THE WAY A SYSTEM EXCHANGES ENERGY WITH ITS ENVIRONMENT

Thermodynamics is concerned with the properties of systems composed of many particles such as a gas, liquid, or solid. Two such systems can interact with each other, so we need variables for describing the state of a system and for describing the interactions of one system with another. In mechanics, we describe the state of motion of a particle in terms of its position, velocity, and acceleration, and the interactions between particles are described by forces. However, a typical system such as 1 mole of a dilute gas contains far too many values of position, velocity, and acceleration to keep track of, even with any computers that are likely to be available in your lifetime. In addition, even if we had such a computer, it would be virtually impossible to make sense of all that information.

Fortunately, we can use a small number of macroscopic quantities such as temperature, pressure, and volume to describe systems and their interactions. In thermodynamics, the state of a system is fully specified by the values of such macroscopic quantities. For example, we might have two balloons containing helium gas as sketched in Figure 16.1. These balloons have the same shape and volume V and contain the same number of helium atoms N, with the same temperature T and pressure P. Although P, V, N, and T are each the same in the two balloons, the precise locations of the helium atoms and their velocities are certainly *not* the same, so these balloons are not microscopically identical. They are nevertheless macroscopically identical, and their properties as systems will be the same. For example, they will have the same values for the velocity of sound and will exhibit the same condensation temperature. Thermodynamics is based on a description involving such macroscopic quantities; the extra detail of the microscopic variables is not needed.

Let's now consider how to describe the interaction of a system with its surroundings. For example, our gas balloon may interact with the air around it as sketched in Figure 16.2. In such cases, when one of the systems (the air around the balloon) is very much larger than the other (the balloon), the larger system is often referred to as the ***environment***.

The interactions between two systems or between a system and its environment generally involve forces and the transfer of energy. These interactions can be described in terms of the work done by one system on another or in terms of the heat that flows between them. One goal of thermodynamics is to understand how energy is exchanged between systems.

As with Newton's laws of mechanics, thermodynamics is based on a small set of physical laws. The laws of thermodynamics each seem quite "innocent" or obvious. One of them (the first law) is a restatement of the principle of conservation of energy, and another (the second law) is a simple statement about the relation between heat flow and temperature. When the laws of thermodynamics are put together, however, they yield remarkable results, including fundamental limits on what is and is not possible.

Two systems are in the same state if their macroscopic properties are the same.

Figure 16.1 A system of many particles such as the gas in a balloon can be described by macroscopic quantities such as temperature and pressure.

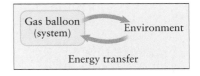

Figure 16.2 Thermodynamics is concerned with the interactions of a system with its environment. These interactions transfer energy between the two.

16.2 | THE ZEROTH LAW OF THERMODYNAMICS AND THE MEANING OF TEMPERATURE

There are four laws of thermodynamics, beginning with the "zeroth" law. The need for the zeroth law became apparent only after the others were developed (and named), and the law is a statement about the concept of temperature. We encountered the zeroth law in our study of heat in Chapter 14, but it is important to repeat it here.

Zeroth law of thermodynamics

Zeroth law of thermodynamics

If two systems A and B are both in thermal equilibrium with a third system, then A and B are in thermal equilibrium with each other.

This statement of the zeroth law is based on the notion of *thermal equilibrium* and hence involves the properties of heat. You should recall from Chapter 14 that heat is the energy that flows from one system to another as a result of a temperature difference. We can use this fact as the basis of an experiment to tell if two systems, which we can call A and B, are in thermal equilibrium with each other. The experiment begins with the two systems separated from each other so that no heat can flow between them. We then allow the systems to interact; for example, we could put them into direct contact so that heat can flow between them via conduction (Chapter 14). If our experiment finds that heat flows between A and B, then the two systems were not initially in thermal equilibrium; that is, they had different temperatures prior to the experiment. The notions of thermal equilibrium and temperature are thus intimately connected. Two systems in thermal equilibrium with each other have the same temperature. So, the zeroth law of thermodynamics can be restated as follows.

Zeroth law of thermodynamics (restatement)

Zeroth law of thermodynamics: alternative form

Suppose the temperature of system A is equal to the temperature of system C; that is, $T_A = T_C$ (so that A and C are in thermal equilibrium). In addition, the temperature of system B is equal to the temperature of system C; that is, $T_B = T_C$. Then $T_A = T_B$.

The zeroth law is illustrated schematically in Figure 16.3.

From the point of view of arithmetic, the statements of the zeroth law regarding the temperatures of three systems seem quite obvious. So why does it deserve to be called a law of physics? The answer is that the zeroth law declares that a property called *temperature* does indeed exist (!) and that this quantity determines the way heat flows between systems. This profound statement cannot be derived from Newton's laws or any other laws of physics.

Figure 16.3 The zeroth law of thermodynamics involves the notion of thermal equilibrium as applied to three (or more) systems. See also Figure 14.7.

16.3 | THE FIRST LAW OF THERMODYNAMICS AND THE CONSERVATION OF ENERGY

One property central to thermodynamics is *internal energy*, denoted by U. We first introduced internal energy in Chapter 14; it is the total energy associated with all the particles in a system. For an ideal monatomic gas, all this energy resides in the kinetic energy of the particles, whereas in other systems such liquids and solids, the chemical bonds or interactions between the particles also contribute to U. A physicist would say that these bonds and interactions give rise to potential energy because they can be modeled as springlike forces obeying Hooke's law. (See spring potential energy, Equation 6.26.) A chemist would say that the bonds store chemical energy because this energy can be released (or gained) in chemical reactions. Both views are correct. We'll adopt the physicist's point of view of the internal

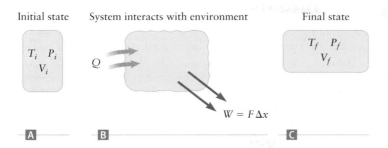

Figure 16.4 Schematic description of a thermodynamic process. **A** A system begins in some initial state. **B** The system then interacts with its environment. This interaction involves the flow of heat (Q) and mechanical work (W). **C** The system is then left in some final state.

energy of a system as the sum of the molecular kinetic energies plus the potential energies due to the interactions between particles within the system.[1]

The internal energy of a system is a function of just a few macroscopic variables such as temperature, pressure, and volume. For example, in Chapter 15 we showed that the internal energy of a monatomic ideal gas containing N atoms at temperature T is

$$U = \tfrac{3}{2}Nk_BT \tag{16.1}$$

where k_B is Boltzmann's constant. It is useful to write Equation 16.1 in terms of the number of moles n in the gas instead of the number of atoms. The result is

$$U = \tfrac{3}{2}nRT \tag{16.2}$$

where R is the gas constant. The internal energy of an ideal gas thus depends only on temperature and the number of particles and is independent of the pressure and volume. Because this internal energy function is so mathematically simple, we'll use the monatomic ideal gas in many of our thermodynamic examples; however, the same basic ideas apply to all systems, although with more complicated expressions for U.

Consider a system in some initial state described by temperature T_i, pressure P_i, and volume V_i as shown in Figure 16.4A. We now allow the system to interact with its environment as shown schematically in Figure 16.4B. These interactions with the environment might result in some heat flow into or out of the system. Forces might also be exerted between our system and the environment, causing the system to do work on the environment (Fig. 16.4B). As a result of the interactions, the system ends up in some final state (Fig. 16.4C), with a temperature T_f, pressure P_f, and volume V_f. This change in our system is called a **_thermodynamic process_**. The first law of thermodynamics is an application of the principle of conservation of energy to such processes.

Thermodynamic processes

First law of thermodynamics

If an amount of heat Q flows _into_ a system from its environment and an amount of work W is done _by_ the system on its environment, the internal energy of the system changes by an amount

First law of thermodynamics

$$\Delta U = U_f - U_i = Q - W \tag{16.3}$$

The First Law of Thermodynamics: The Meaning of _Q_ and _W_

From our previous studies of heat and work, we know that there are two ways to change the energy of a system. One way is to arrange for heat to flow into (or out of) the system. By convention, a positive value for Q means that heat flows into the system (and energy is added); hence, the internal energy increases (ΔU is positive) as indicated by Equation 16.3. Another way to change the internal energy is through

---

[1]Also, the internal energy does not include the kinetic energy of the system "as a whole," that is, the kinetic energy associated with either the motion of the center of mass or the rotational motion of the system.

the action of an external force, that is, a force exerted on the system by the environment. We saw in Chapter 6 that the work done *on* a system is equal to the change in the energy of the system. By convention, a positive value of W in Equation 16.3 means that a positive amount of work is done *by* the system on its environment, so if W is positive, the internal energy of a system must decrease. To apply Equation 16.3 correctly, you must remember that Q is the heat energy that flows *into the system*, whereas W is the work done *by the system*.

Nowadays, we take it for granted that heat is a form of energy, so it is natural to include heat energy along with kinetic and potential energy whenever we discuss the principle of conservation of energy. However, when the science of thermodynamics was being developed around 200 years ago, the nature of "heat" was a matter of much debate. For example, when you place your hand over a flame, you know that "something" flows from the flame to your hand. It is not obvious that the "stuff" that flows from the flame to your hand is similar to the kinetic energy of a moving baseball or the potential energy of a compressed spring. In fact, many scientists believed (incorrectly!) that heat was fundamentally different from kinetic and potential energy. Eventually, physicists discovered that heat is indeed a form of energy, so heat can be measured with either joules (the SI unit for energy) or calories (an alternative unit of energy defined in Chapter 14).

Definitions of Q and W in the first law of thermodynamics

EXAMPLE 16.1 | The First Law of Thermodynamics Applied to an Ideal Gas

Consider a flexible balloon containing $n \approx 3.5$ moles of an ideal monatomic gas such as helium. The balloon is placed over a flame so that $Q = 700$ J of heat flows into the gas, and its temperature increases by 10 K. How much work is done by the gas in the balloon on its environment during this process?

RECOGNIZE THE PRINCIPLE

A central thread of thermodynamics is energy and how it is exchanged between systems. The principle of conservation of energy is thus at the heart of all problem solving in thermodynamics. From the first law of thermodynamics,

$$\Delta U = Q - W \tag{1}$$

We also know that the internal energy of a monatomic ideal gas is given by $U = \frac{3}{2}nRT$ (Eq. 16.2), so changes in U are related to changes in temperature by

$$\Delta U = \tfrac{3}{2}nR\,\Delta T \tag{2}$$

In this example, the change in temperature is known, so we can use it to calculate the corresponding change in the internal energy ΔU. Because Q is given, we can then use the first law of thermodynamics, Equation (1), to find the work W done by the gas in the balloon.

SKETCH THE PROBLEM

Figure 16.5 shows the initial and final states of the balloon. This sketch shows the balloon as it is expanding. An expanding balloon does work on its environment because the pressure force and displacement are parallel, corresponding to a positive value of W in the first law of thermodynamics.

IDENTIFY THE RELATIONSHIPS

We first find the change in the internal energy using Equation (2):

$$\Delta U = \frac{3}{2}\,nR\,\Delta T = \frac{3}{2}(3.5 \text{ moles})\left(8.31\,\frac{\text{J}}{\text{mole}\cdot\text{K}}\right)(10 \text{ K}) = 440 \text{ J}$$

Notice that this value is positive. The internal energy of an ideal gas increases when its temperature is increased.

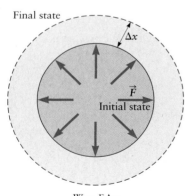

$W \approx F\,\Delta x$
If $W > 0$, balloon expands.

Figure 16.5 Example 16.1. If a balloon expands, the gas inside does work on the surface of the balloon. The thermodynamic quantity W is defined as the work done by the system (the gas) on its environment (the wall of the balloon), so for the balloon in this sketch, $W > 0$.

SOLVE

Inserting this result for ΔU with the given value of Q into Equation (1) leads to

$$W = Q - \Delta U = (700 - 440)\,\text{J} = \boxed{260\,\text{J}}$$

What does it mean?

Recall again that W is the work done *by* the gas in the balloon *on* its environment. For W to be positive, the force exerted by the gas in the balloon must be parallel to the associated displacement. Here that is the displacement Δx of the surface of the balloon, whereas the force F is produced by the pressure of the gas on this surface (see Fig. 16.5). A positive value for W thus means that the balloon *expands* as shown in the figure.

CONCEPT CHECK 16.1 | A Thermodynamic Process with $W = Q$

Consider a thermodynamic process involving a monatomic ideal gas in which, by some coincidence, $W = Q$. Which statement best describes the process?

(a) The temperature increases.
(b) The temperature decreases.
(c) The temperature does not change.
(d) The balloon expands.
(e) The balloon contracts.
(f) The volume of the balloon does not change.

The Signs of *Q* and *W* in the First Law of Thermodynamics

Most examples with the first law of thermodynamics, such as Example 16.1, are mathematically straightforward. However, a common pitfall involves the signs of Q and W in Equation 16.3. The signs of Q and W are such an important part of thermodynamics that it is worth repeating and emphasizing a few crucial points.

The first law of thermodynamics reads

$$\Delta U = Q - W \tag{16.4}$$

and is indicated pictorially in Figure 16.6A. That the signs in front of Q and W in Equation 16.4 are different can be a source of confusion. By convention, a *positive* value of Q indicates that heat flows *into* the system, which means that the energy of the system *increases*; hence, the change in the internal energy is positive. In any particular case, it is possible for Q to have a *negative* value, and if so, heat energy flows *out of* the system *into* the environment and the internal energy of the system decreases. These two cases are shown schematically in Figure 16.6B.

By convention, a *positive* value of W indicates that the system does a positive amount of work *on* its environment. The internal energy of the system must therefore decrease while the energy of the environment increases (Fig. 16.6C). It is also possible for W to have a *negative* value, which leads to an increase in the internal energy of the system.

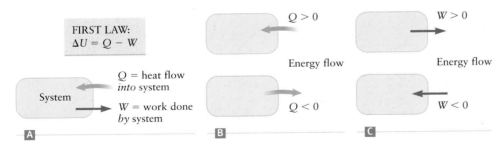

Figure 16.6 **A** In the first law of thermodynamics, Q is defined as the heat added to the system and W is the work done by the system. **B** When Q is positive, heat energy flows into the system from its environment. If $Q < 0$, heat energy flows out of the system into the environment. **C** W is defined as the work done by the system on its environment. If $W > 0$, the system does a positive amount of work on its environment, whereas if $W < 0$, the environment does a positive amount of work on the system.

Q is negative because heat flows from the cyclist to the environment.

$Q < 0$ $W > 0$

Work done by cyclist on pedals

Figure 16.7 Example 16.2.

These sign conventions for Q and W are a crucial part of the first law of thermodynamics. You can always remember (or re-derive) the correct signs in Equation 16.4 by using energy conservation principles as illustrated in Figure 16.6.

EXAMPLE 16.2 ✖ Thermodynamics and Your Diet

A human being is an example of a thermodynamic system. A person riding a bicycle at a moderate speed (Fig. 16.7A) does mechanical work on the pedals at a rate of about

$$P = \frac{W}{t} \approx 100 \text{ W} \qquad (1)$$

At the same time, she gives off some heat to her surroundings as she sweats and so forth. Because this heat is flowing out of the cyclist and into the environment, the term Q in the first law of thermodynamics is negative. Measurements on typical cyclists give

$$\frac{Q}{t} = -400 \text{ W} \qquad (2)$$

Suppose a cyclist rides for 1 h. What is the change in her internal energy?

RECOGNIZE THE PRINCIPLE

The principle of conservation of energy is at the heart of all problem solving in thermodynamics. From the first law of thermodynamics,

$$\Delta U = Q - W$$

To calculate the change in the internal energy, we need to find both Q and W. The value of Q is given, and W can be found from the power exerted by the cyclist.

SKETCH THE PROBLEM

From the point of view of thermodynamics (Fig. 16.7B), the cyclist does work on her environment (through the force exerted on the pedals), so W is positive. Heat is lost to the environment through sweating, so Q in the first law of thermodynamics is negative.

IDENTIFY THE RELATIONSHIPS

In 1 h ($t = 3600$ s), the total work done by the cyclist is, using Equation (1),

$$W = Pt = (100 \text{ W})(3600 \text{ s}) = +3.6 \times 10^5 \text{ J}$$

where we have included the positive sign explicitly to emphasize that the cyclist does a positive amount of work on the pedals. In 1 h, the total heat Q is

$$Q = (-400 \text{ W})(3600 \text{ s}) = -1.4 \times 10^6 \text{ J}$$

Because heat flows from the cyclist to her environment, Q is negative.

SOLVE

Inserting these values of W and Q into the first law gives

$$\Delta U = Q - W = (-1.4 \times 10^6 \text{ J} - 3.6 \times 10^5 \text{ J}) = \boxed{-1.8 \times 10^6 \text{ J}} \qquad (3)$$

What does it mean?

The negative value of ΔU means that the internal energy of the cyclist decreases as the result of her exercise, as expected. The cyclist can replenish this energy by eating. For example, according to dietary tables, the total energy available from a typical fast-food hamburger is about 2.0×10^6 J. So, our cyclist would have to eat about one hamburger to fuel her hour of exercise.

Thermal Reservoirs

Heat can flow via conduction from one system to another when the two systems are in contact. For example, suppose an ice cube is placed outside on a sidewalk on a warm day (Fig. 16.8). Because the ice cube is colder than its environment, heat flows from the air and the sidewalk into the ice cube and Q is positive. Eventually, the ice cube melts and the puddle of water reaches a final temperature equal to the temperature of its surroundings. In this example, the environment around the ice cube acts as a *thermal reservoir.* By definition, a reservoir is much larger than the system to which it is connected, so there is essentially no change in the temperature of the reservoir when heat enters or leaves it. The notion of a thermal reservoir comes up repeatedly in discussions of thermodynamic processes.

Calculating the Work Done in a Thermodynamic Process

To understand how to calculate the work associated with a thermodynamic process, consider a container filled with a gas (Fig. 16.9A). One wall of this container is a movable piston. Due to its pressure, the gas exerts a force on the piston, moving the piston to the right through a distance Δx (Fig. 16.9B). If F is the force exerted by the gas on the piston, the work done by the gas on the piston is

$$W = F \Delta x \tag{16.5}$$

The piston is part of the environment of the gas (the piston might be connected to other components of an engine), so W is also the work done by the gas on its environment.

If the pressure in the gas is P and the area of the piston is A, the force on the piston is $F = PA$ and the work done by the gas on the piston is (using Eq. 16.5)

$$W = F \Delta x = PA \Delta x$$

The product $A \Delta x$ equals the change in volume of the gas ΔV, so we have

$$W = P \Delta V \tag{16.6}$$

We have derived this result for a gas in a piston-type container, but the result is very general. The only restriction on Equation 16.6 is that the force and hence also the pressure are constant. Whenever the volume of a system changes by a small amount ΔV, the work done by the system is equal to $P \Delta V$.

We can apply this result to cases in which the volume changes by a large amount by viewing the process as plotted in Figure 16.10. Here the initial (i) and final (f) states of the system are indicated by points in the P–V plane. A particular thermodynamic process such as the expansion of the gas in Figure 16.9B corresponds to a line or curve in the P–V plane that connects the initial state in which the system has a pressure P_i and a volume V_i to the final state with pressure P_f and volume V_f. The entire process produces a large change in the volume, but we can also view it as a series of small changes and apply Equation 16.6 to each small change (Fig. 16.10A). The total work done during the entire process thus equals *the area under the corresponding curve in the P–V plane.*

In the hypothetical process in Figure 16.10A, the final volume is greater than the initial volume, so the gas expands. The force exerted by the system is thus parallel to Δx (Fig. 16.9A and B), so the work done by the gas on the environment is positive ($W > 0$). The work done by a system can also be negative. For W to be negative, the force exerted by the gas on the piston must be opposite to the piston's displacement. In Figure 16.9C, the piston moves to the left, which corresponds to compression of the gas, giving a negative value of ΔV in Equation 16.6. This situation can occur if there is an external force on the piston as might be exerted by a person pushing on

Figure 16.8 Example of a thermodynamic process involving a thermal reservoir. Here the system is the H_2O (the ice plus water). The air and the sidewalk act as a thermal reservoir; their temperature does not change during the thermodynamic process in which the ice cube melts.

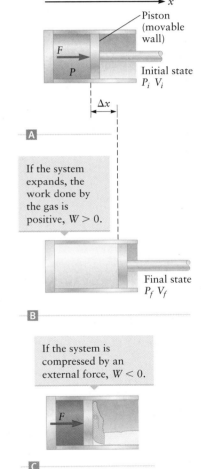

Figure 16.9 A gas expands as the piston moves to the right in going from **A** to **B**. The work done by the gas on the piston is thus positive, and $W > 0$ in Equation 16.5. **C** If an outside agent exerts a force on the piston so as to compress the gas, the work done by the gas is still $W = F \Delta x$ (where F is the force exerted by the gas on the piston) but now $W < 0$.

Figure 16.10 The work W is the area under the path in the P–V plane. **A** In this case, the gas expands ($V_f > V_i$) and does a positive amount of work on its environment ($W > 0$). **B** Here the gas is compressed ($V_f < V_i$), and $W < 0$.

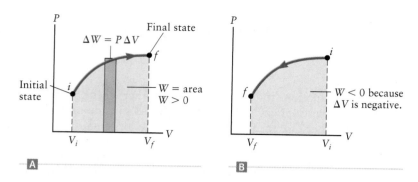

the piston. Such a process is shown in the P–V plane in Figure 16.10B. The magnitude of the work done by the system during this process is still equal to the area under the P–V curve, but we are traversing this curve from right to left, so ΔV is negative and hence W is negative.

CONCEPT CHECK 16.2 | Interpreting a P–V Diagram

Consider a hypothetical thermodynamic process involving an ideal gas that is described by the P–V diagram in Figure 16.11. Which of the following statements is true?

   (a) The gas contracts, and W (the work done by the gas) is negative.
   (b) The gas pressure drops, so W is negative.
   (c) The gas expands, and W is positive.

Figure 16.11 Concept Check 16.2.

Examples of Thermodynamic Processes

Thermodynamic processes are classified according to how quantities such as P, V, and T change or remain constant during the course of the process.

Changes at constant pressure (isobaric). Consider again a container with an ideal gas as sketched in Figure 16.9A and suppose the piston moves to the right as it is acted on by a force $F = PA$ from the gas. The movement of the piston produces a change ΔV in the volume of the gas. If the pressure is constant during this process, the process is called *isobaric*. When viewed in a P–V diagram, an isobaric process is a horizontal line as shown in Figure 16.12. In the hypothetical process shown in Figure 16.12A, the system expands and the work done by the gas is positive. The work W is equal to the area under the P–V curve, and because P is constant, we can write

$$W = P \, \Delta V = P(V_f - V_i) \quad \text{(isobaric process)} \tag{16.7}$$

For an expansion, $V_f > V_i$ so W is positive, whereas for a contraction, $\Delta V < 0$ and the work done is negative (Fig. 16.12B). An isobaric process may also involve some heat flow, as we illustrate in Example 16.3.

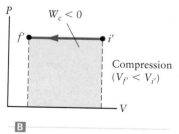

Figure 16.12 **A** P–V diagram for an isobaric (constant pressure) expansion. **B** An isobaric compression.

EXAMPLE 16.3 | Analyzing an Isobaric (Constant Pressure) Process

A cylinder contains $n = 5.0$ moles of a monatomic ideal gas at a pressure $P = 2.0 \times 10^5$ Pa (about twice atmospheric pressure). The gas is then compressed at constant pressure (isobarically) from an initial volume of $V_i = 0.090$ m^3 to $V_f = 0.040$ m^3 (Fig. 16.13). **(a)** What is the work W done by the gas? **(b)** What is the change in the internal energy of the gas? **(c)** How much heat energy Q flows into the gas during this process?

RECOGNIZE THE PRINCIPLE

We can find the work done by the gas using $W = P \, \Delta V$ (Eq. 16.7) along with the given initial and final volumes. The change in the internal energy can be calculated using

$U = \frac{3}{2}nRT$ for an ideal gas (Eq. 16.2) together with the initial and final temperatures, which can be found using the ideal gas law. When we know W and ΔU, we can use the first law of thermodynamics to get Q.

SKETCH THE PROBLEM

A P–V diagram describing the process is given in Figure 16.13. The work done by the gas is the area under the curve that connects the initial (i) and final (f) states.

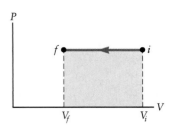

Figure 16.13 Example 16.3.

IDENTIFY THE RELATIONSHIPS AND SOLVE

(a) The magnitude of W is just the area under the P–V curve in Figure 16.13. We find

$$W = P\,\Delta V = P(V_f - V_i) = (2.0 \times 10^5 \text{ Pa})(0.040 \text{ m}^3 - 0.090 \text{ m}^3)$$

$$W = \boxed{-1.0 \times 10^4 \text{ J}}$$

Because $V_f < V_i$, ΔV is negative and W is negative.

(b) The internal energy of an ideal gas is

$$U = \frac{3}{2}nRT \qquad (1)$$

so to find the change in U, we must first calculate the initial and final temperatures. Rearranging the ideal gas law $PV = nRT$ to solve for T gives

$$T_i = \frac{PV_i}{nR} = \frac{(2.0 \times 10^5 \text{ Pa})(0.090 \text{ m}^3)}{(5.0 \text{ moles})[8.31 \text{ J/(mole} \cdot \text{K)}]} = 430 \text{ K}$$

For the final temperature we have

$$T_f = \frac{PV_f}{nR} = \frac{(2.0 \times 10^5 \text{ Pa})(0.040 \text{ m}^3)}{(5.0 \text{ mole})[8.31 \text{ J/(mole} \cdot \text{K)}]} = 190 \text{ K}$$

Using these values with Equation (1) leads to

$$\Delta U = \frac{3}{2}nR\,\Delta T = \frac{3}{2}nR(T_f - T_i)$$

$$\Delta U = \frac{3}{2}(5.0 \text{ moles})[8.31 \text{ J/(mole} \cdot \text{K)}](190 \text{ K} - 430 \text{ K}) = \boxed{-1.5 \times 10^4 \text{ J}}$$

(c) According to the first law of thermodynamics,

$$\Delta U = Q - W$$

Using the values of W and ΔU from parts (a) and (b), we find

$$Q = \Delta U + W = (-1.5 \times 10^4 \text{ J}) + (-1.0 \times 10^4 \text{ J}) = \boxed{-2.5 \times 10^4 \text{ J}}$$

What does it mean?
In an isobaric process, the pressure stays constant and the work can be found using $W = P\,\Delta V$. At the same time, the temperature may or may not change, depending on the value of W and on how much heat Q flows into or out of the system.

Changes at constant temperature (isothermal). Consider the expansion of an ideal gas in Figure 16.14A, in which the gas expands, moving the piston to the right. Let's also assume the walls of the container are in contact with a thermal reservoir. The container might be surrounded with air at room temperature, for example, or it might be immersed in water at a particular temperature. Either way, contact with the thermal reservoir keeps the temperature of the gas constant during the course of the expansion. A thermodynamic process in which the temperature is constant is called an *isothermal process*.

Even though the temperature is constant in an isothermal process, the pressure and volume can both change. Figure 16.14B shows an isothermal process for an

For an isothermal process, the temperature is constant.

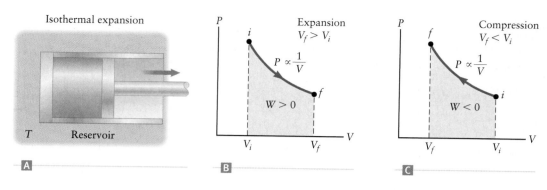

Figure 16.14 **A** In an isothermal process, the temperature of the system is held fixed, usually because it is in contact with a thermal reservoir. **B** Isothermal expansion of an ideal gas in the P–V plane. **C** Isothermal compression.

ideal gas, drawn as a path in a P–V diagram. The ideal gas law is $PV = nRT$, so if the temperature is constant, the pressure and volume are related by

$$P = \frac{nRT}{V} \propto \frac{1}{V} \quad \text{(isothermal process)} \tag{16.8}$$

which is the curve plotted in Figure 16.14B. The work done by the gas is again equal to the area under the P–V curve; for an expansion described by Equation 16.8 with T held constant, the result is[2]

$$W = nRT \ln\left(\frac{V_f}{V_i}\right) \quad \text{(isothermal process)} \tag{16.9}$$

Here, "ln" is the natural logarithm function, which is reviewed in Appendix B. In an expansion, the volume increases, so $V_f > V_i$ and the logarithmic term in Equation 16.9 is positive. Hence, the work done by the gas is positive in this case, as we would expect from Figure 16.14B, and from our previous discussions of the work done by an expanding gas. Conversely, for an isothermal compression (Fig. 16.14C), the factor $\ln(V_f/V_i)$ in Equation 16.9 is negative and the work done by the gas is also negative.

EXAMPLE 16.4 Analyzing an Isothermal (Constant-Temperature) Process

Consider the isothermal expansion of an ideal gas sketched in Figure 16.14B. The system contains $n = 4.0$ moles of argon gas at room temperature. If the system then expands to double its initial volume, what are (a) the work done by the system and (b) the amount of heat that flows into the system from its surroundings?

RECOGNIZE THE PRINCIPLE

Our discussion of isothermal processes involving an ideal gas has emphasized the work done (Eq. 16.9), but there may also be heat flow either into or out of the system during the process. In this example, the system expands to fill a greater volume. For the gas to maintain a constant temperature (because we have an isothermal process), some heat Q must flow into the gas from its environment. We can calculate Q using the first law.

SKETCH THE PROBLEM

Figure 16.14B describes an isothermal expansion. The work W is the area under the P–V curve, and because the gas expands, W is positive.

---

[2]This equation can be derived by using calculus to compute the area under the curve specified by Equation 16.8.

(a) IDENTIFY THE RELATIONSHIPS

The work done by an ideal gas in an isothermal process is given by Equation 16.9. We have

$$W = nRT \ln\left(\frac{V_f}{V_i}\right)$$

The system is at room temperature ($T \approx 293$ K), and the final volume is twice the initial volume ($V_f = 2V_i$), so

$$W = (4.0 \text{ moles})[8.31 \text{ J/(mole} \cdot \text{K})](293 \text{ K}) \ln\left(\frac{2V_i}{V_i}\right)$$

SOLVE

Evaluating this expression for W gives

$$W = \boxed{6800 \text{ J}}$$

(b) IDENTIFY THE RELATIONSHIPS

To find the heat that flows into the gas from its environment during this expansion, we use the first law of thermodynamics,

$$\Delta U = Q - W$$

The internal energy of an ideal gas is a function of just the temperature, with $U = \frac{3}{2}nRT$ for a monatomic ideal gas (such as argon). Because we are dealing with an isothermal process, the final temperature is equal to the initial temperature, and there is no change in the internal energy.

SOLVE

We thus have $\Delta U = 0$, and the first law of thermodynamics gives

$$\Delta U = 0 = Q - W$$

Using the result for W from part (a), we find

$$Q = W = \boxed{6800 \text{ J}}$$

What does it mean?
In this example, the internal energy of the system does not change. Hence, as expected from the principle of conservation of energy, the energy added to the system (Q) must equal the energy that leaves the system (W).

Changes with $Q = 0$ (adiabatic). Sometimes, a system is thermally isolated from its surroundings as shown schematically in Figure 16.15A. For example, the system might be the liquid in a thermos bottle. The wall of a thermos bottle contains a vacuum, making the heat flow through the wall extremely small. If there is no heat flowing into or out of a system during a thermodynamic process, then $Q = 0$ and the process is called *adiabatic.*

To analyze an adiabatic process for a particular system, we need to know the relation between temperature, pressure, and volume. Let's again consider an example with an ideal gas (because we know its internal energy function). The path for an adiabatic process in the P–V plane can be calculated using the ideal gas law and the expression for the internal energy (Eq. 16.2). We will omit the derivation, but the result is

$$PV^\gamma = \text{constant}$$

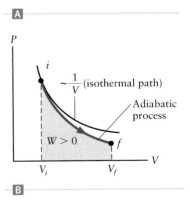

Figure 16.15 **A** In an adiabatic process, no heat flows into or out of the system ($Q = 0$). **B** Adiabatic expansion of an ideal gas in the P–V plane.

Isochoric (constant-volume) process: $V_i = V_f$

P

f ●

i ● Area = 0, so $W = 0$.

V

Figure 16.16 In an isochoric process, the volume does not change.

P

f ●

↑ Case (a)

i ●

V

P

i ●

↓ Case (b)

f ●

V

Figure 16.17 Concept Check 16.3.

where $\gamma = \frac{5}{3}$ for a dilute monatomic gas.[3] The pressure thus varies as

$$P \propto \frac{1}{V^\gamma} \quad \text{(adiabatic process)} \qquad (16.10)$$

In a P–V diagram, an adiabatic path (Fig. 16.15B) is steeper than the path for an isothermal process that begins from the same initial state. (Recall that an isothermal path follows $P \propto 1/V$; Eq. 16.8.)

Although $Q = 0$ for an adiabatic process, a nonzero work W can still be associated with the process. The value of W for an adiabatic process is again equal to the area under the P–V curve (Fig. 16.15B). According to the first law of thermodynamics, $\Delta U = Q - W$ and because $Q = 0$ for an adiabatic process, we have

$$\Delta U = -W \quad \text{(adiabatic process)} \qquad (16.11)$$

A hypothetical adiabatic process is the expansion of a gas against a piston. As the gas expands, it does a positive amount of work on the piston, and $W > 0$. According to Equation 16.11, the internal energy of the gas must decrease. For a monatomic ideal gas, $U = \frac{3}{2}Nk_BT$, meaning that the temperature decreases; that is, the gas cools. Although $Q = 0$ in an adiabatic process, other properties of the system such as the temperature can change during the process.

Changes at constant volume (isochoric). An *isochoric* process is one for which the volume does not change; it is also called an *isovolumetric* process. The corresponding path in a P–V plot is shown in Figure 16.16. A key feature of all isochoric processes is discussed in Example 16.5.

CONCEPT CHECK 16.3 | Heating a Gas at Constant Volume

Consider a gas held at constant volume. Figure 16.17 shows P–V plots for two different isochoric (constant volume) processes. In one case, the pressure increases, whereas in the other the pressure decreases. In which case does the temperature increase?

EXAMPLE 16.5 | A Constant Volume (Isochoric) Process

Consider an ideal gas that is heated under isochoric conditions. What is the work done by the gas?

RECOGNIZE THE PRINCIPLE

Because the volume is kept fixed, an isochoric process is a vertical line in the P–V plane.

SKETCH THE PROBLEM

The P–V diagram for this process is shown in Figure 16.16. In this example, the gas is heated, so the final pressure is greater than the initial pressure.

IDENTIFY THE RELATIONSHIPS AND SOLVE

The work done by the gas is the area under this curve, and that area is zero. So, for this process (and for all isochoric processes), $\boxed{W = 0}$.

What does it mean?
When analyzing a thermodynamic process, it is always useful to consider the P–V diagram. The work done by the system is always equal to the area under the P–V curve.

---

[3]The parameter γ is the ratio of the specific heats at constant pressure and constant volume. This ratio depends on the type of molecules in the gas. For a monatomic gas, $\gamma = \frac{5}{3}$, and for a diatomic gas, $\gamma = \frac{7}{5}$.

CONCEPT CHECK 16.4 | ⊗ What Kind of Process Is It?

A bear has a very thick fur coat, providing very good thermal insulation from its environment (Fig. 16.18). When the bear hibernates, is it an example of an isothermal or an adiabatic process, or both? Explain.

CONCEPT CHECK 16.5 | Work Done in a Thermodynamic Process

Figure 16.19 shows several thermodynamic processes. Without doing a calculation, determine if W is positive, negative, or zero for each process.

Figure 16.18 Concept Check 16.4.

Figure 16.19 Concept Check 16.5.

Some Properties of Q and W

A thermodynamic process can be described by a path in the P–V diagram that takes the system from some initial state i to a final state f. Many paths connect a particular initial state to a given final state; Figure 16.20 shows two paths that connect states i and f. One of these paths passes through point A, and we'll refer to it as "path A." The other passes through point B ("path B"). Let's consider how ΔU, W, and Q compare for these two paths, that is, for these two different thermodynamic processes.

We can write the change in internal energy as

$$\Delta U = U_f - U_i \qquad (16.12)$$

In words, the change in U depends only on the internal energies of the initial and final states, so ΔU is the same for *any* path that connects states i and f. This result is reminiscent of our work in Chapter 6, where we learned that changes in the potential energy are independent of the path taken.

The work done in going from state i to state f along path A is the sum of the work done in going from i to A and the work done in going from A to f; hence,

$$W_A = W_{iA} + W_{Af}$$

Here W_{iA} denotes the work done on the path from i to A in Figure 16.20 and so forth for W_{Af}. Work is the area under the P–V curve, so

$$W_{iA} = P\,\Delta V = P_i(V_A - V_i)$$

Figure 16.20 Two different thermodynamic processes (two paths in the P–V plane) can connect the same initial and final states.

and
$$W_{Af} = 0$$

The work W_{Af} along the portion of the path from A to f is zero because there is no change in the volume. The total work done along path A is thus

$$W_A = W_{iA} + W_{Af} = P_i(V_A - V_i)$$

The volume is the same at point A and f ($V_A = V_f$); hence, we get

$$W_A = P_i(V_f - V_i) \tag{16.13}$$

We can find the work done along path B in a similar way:

$$W_B = W_{iB} + W_{Bf}$$

with

$$W_{iB} = 0$$

because there is no volume change on this portion of the path, and

$$W_{Bf} = P_f(V_f - V_B)$$

The result for the work on path B is then

$$W_B = W_{iB} + W_{Bf} = P_f(V_f - V_B)$$

Using $V_B = V_i$, this expression becomes

$$W_B = P_f(V_f - V_i) \tag{16.14}$$

Comparing the results for W_A (Eq. 16.13) and W_B (Eq. 16.14), we see that the work done by the system along path A is *not* the same as the work done along path B. Unlike changes in the internal energy, the amount of work done *depends on the path*.

We can also consider the heat Q added to the system as it travels along paths A and B. Applying the first law of thermodynamics to path A, we have

$$\Delta U_A = Q_A - W_A \quad \text{(along path } A\text{)} \tag{16.15}$$

$$\Delta U_B = Q_B - W_B \quad \text{(along path } B\text{)} \tag{16.16}$$

We have already seen that changes in the internal energy are independent of the path, so $\Delta U_A = \Delta U_B$. We have also just shown that the work done is different along the two paths, so $W_A \neq W_B$. Using these facts with Equations 16.15 and 16.16, we see that

$$Q_A \neq Q_B$$

In words, the heat added to the system also *depends on the path* taken.

The principle of conservation of energy thus leads us to the following conclusions.

Properties of *U, W,* and *Q* for thermodynamic processes

1. The internal energy of a system depends only on the current state of the system. Changes in the internal energy are thus independent of the path taken in a thermodynamic process. The change in internal energy ΔU depends only on the internal energies of the initial and final states.
2. The work done during a thermodynamic process depends on the path taken. Even with the same initial and final states, two different thermodynamic paths can have different values of *W*.
3. The heat added to a system during a thermodynamic process depends on the path taken. Even with the same initial and final states, two different thermodynamic paths can have different values of *Q*.

EXAMPLE 16.6 | Analyzing a Cyclic Process

Consider the thermodynamic process described in Figure 16.21. The P–V diagram is similar to the one in Figure 16.20 except that now we begin and end at the same point (the same state). The system in Figure 16.21 starts in state 1, moves as indicated to state 2, state 3, and state 4, and then moves back to state 1. This set of changes is called a *cyclic process* because the system returns to its initial state. Find the work done during this process.

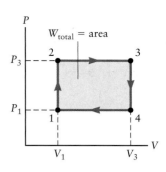

Figure 16.21 Example 16.6. A cyclic process is one that returns to its original state.

RECOGNIZE THE PRINCIPLE

We can use $W = P \, \Delta V$ (Eq. 16.6) to find the work done along each portion of the path. Adding those results gives the total work done for the entire process.

SKETCH THE PROBLEM

Figure 16.21 shows the P–V diagram.

IDENTIFY THE RELATIONSHIPS

The volume does not change along the path from state 1 to state 2 or along the path from state 3 to state 4, so $\Delta V = 0$ and the work done in these parts of the process is zero. Using $W = P \, \Delta V$ for the path from state 2 to state 3 gives

$$W_{23} = P \, \Delta V = P_3(V_3 - V_1)$$

Likewise, along the path from state 4 to state 1, we get

$$W_{41} = P_1(V_1 - V_3)$$

Notice that W_{41} is negative because $V_1 < V_3$.

SOLVE

Combining these results gives

$$W_{\text{total}} = W_{23} + W_{41} = P_3(V_3 - V_1) + P_1(V_1 - V_3) = \boxed{(P_3 - P_1)(V_3 - V_1)}$$

What does it mean?
From the geometry of Figure 16.21, W_{total} is just the area enclosed by the path in the P–V diagram.

Work Done by a Cyclic Process

Example 16.6 considered the work done in a cyclic process and showed that W is equal to the area enclosed by the path in the P–V diagram. The process analyzed in that example had an especially simple path, but the general result applies to all cyclic processes. The work done in any cyclic process equals the area enclosed by the path in the corresponding P–V diagram (Fig. 16.22). We'll use this result when we analyze the performance of heat engines and other devices in Section 16.6.

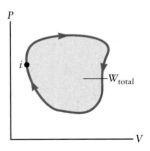

Figure 16.22 In any cyclic thermodynamic process, the work done by the system equals the area enclosed by the path in the P–V diagram.

| CONCEPT CHECK 16.6 | Thermodynamic Processes

Figure 16.23 shows several thermodynamic processes. Which of these processes could be (a) an isothermal expansion, (b) an adiabatic compression, (c) an isobaric expansion, and (d) an isobaric compression?

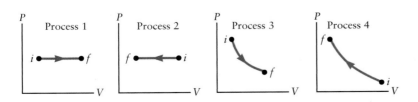

Figure 16.23 Concept Check 16.6.

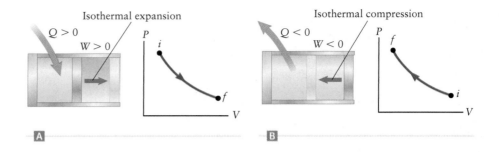

Figure 16.24 Schematic of a thermodynamically reversible process. ◣ Isothermal expansion from state *i* to state *f*. ◳ Isothermal compression from state *f* back to state *i*.

Isothermal expansion
$Q > 0$
$W > 0$

Isothermal compression
$Q < 0$
$W < 0$

A

B

16.5 | REVERSIBLE AND IRREVERSIBLE PROCESSES AND THE SECOND LAW OF THERMODYNAMICS

In our discussions of thermodynamic processes, we have focused so far on changes in the system. It is sometimes important to also consider what happens to the environment. Figure 16.24A shows an isothermal expansion of a dilute gas in which the system starts in an initial state *i* and moves to a final state *f*. The piston moves to the right as the gas does some amount of work $W > 0$, and during this time, it absorbs some heat $Q > 0$ from the environment. This example shows a thermodynamically *reversible* process. We could imagine a reversed process in which the gas begins in state *f* and is isothermally compressed back to state *i* (Fig. 16.24B). During this compression, the gas does a negative amount of work ($W < 0$) on the piston because an outside agent must push on the piston to compress the gas. At the same time, heat would flow out of the gas into the surroundings. The work done by the gas during the compression is equal in magnitude but opposite in sign compared with *W* during the expansion. Also, the amount of heat that flows out of the gas during compression is equal in magnitude to the heat that flows in during expansion. Hence, both the system and the environment are brought back to precisely their original states, which is the key feature of a thermodynamically reversible process.

An example of an *irreversible* process is sketched in Figure 16.25. Here a hockey puck is initially sliding to the right on a horizontal surface. There is some friction between the puck and the surface, causing the puck to come to a stop as all its initial kinetic energy is converted to heat energy. This heat energy flows from the region where the puck is in contact with the horizontal surface into its surroundings (Fig. 16.25A). Suppose we now try to reverse this process by arranging for heat to flow from the surroundings into the puck, perhaps with a torch as sketched in Figure 16.25B. Intuition tells us that this will not cause the puck to begin moving! Of course, we could start the puck in motion again by striking it with a hockey stick, but for a process to be thermodynamically reversible, it must be possible to return both the system (the puck) and its surroundings (the rest of the universe) back to their original states. There is nothing we could do to reverse the process in Figure 16.25; it is simply *not possible* to take the puck and the rest of the universe back to precisely their original states. This is the essential feature of an irreversible process. And, as in Figure 16.25, irreversible processes usually involve friction.

The distinction between reversible and irreversible processes is the subject of the second law of thermodynamics. There are several different ways of stating the second law of thermodynamics; all are equivalent, but because they emphasize different aspects of the second law, it is useful to discuss more than just one of them. Perhaps the simplest version is given as follows.

Second law of thermodynamics

Heat flows spontaneously from a warm body to a colder one. It is not possible for heat to flow spontaneously from a cold body to a warmer one.

Reversible versus irreversible processes

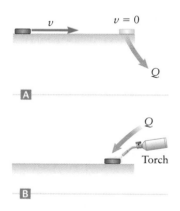

v
$v = 0$
Q
A

Q
Torch
B

Figure 16.25 ◣ A hockey puck sliding across a floor with friction is an example of a thermodynamically irreversible process. ◳ Heating a hockey puck will *not* cause it to move back to its original position!

Second law of thermodynamics

This statement certainly seems obvious. The part about heat flowing from a warm object to a colder one is implied by the zeroth law of thermodynamics. Two crucial aspects of the second law go beyond the zeroth law, however. The first is the notion of a *spontaneous* process. Such a process will essentially take place "on its own," without work done by any added or external forces. For example, the flow of heat from the surroundings into the ice cube in Figure 16.8 is a spontaneous process. The other crucial part of the second law of thermodynamics is the statement that heat cannot flow spontaneously from a cold object to a warmer one. This statement leads to many important results, including the impossibility of building a perpetual motion machine.

16.6 | HEAT ENGINES AND OTHER THERMODYNAMIC DEVICES

One of the discoverers of the second law of thermodynamics was an engineer named Sadi Carnot (1796–1832), who was interested in how the science of thermodynamics could be applied to practical devices such as steam engines and other marvels of the Industrial Revolution that were being developed around that time. Carnot's engineering approach leads to a rather different way of thinking about the second law of thermodynamics.

Imagine you are an engineer and you have been asked to invent the world's best heat engine. A heat engine takes heat energy, as might be generated by a coal-burning furnace, and converts this energy into work. For example, this work might be in the form of a rotating shaft as would be useful in a steamboat, locomotive, or automobile. With that in mind, we can draw a heat engine in the highly schematic form shown in Figure 16.26A. An amount of heat Q_H is extracted from the furnace and fed into some sort of mechanical device, which then does an amount of work W. It is your engineering challenge to design the mechanical device. The principle of conservation of energy and the first law of thermodynamics tell us that the work done cannot be larger than the energy input Q_H, but you might hope to design and patent a device that reaches this limit. The world's best heat engine would perform an amount of work $W = Q_H$, and from the principle of conservation of energy, you would be absolutely guaranteed that no one could design an engine better than that one.

That is the problem Carnot tackled. Although he was not able to construct the heat engine in Figure 16.26A, Carnot was able to show that such an engine is *impossible*! He showed that it is simply not possible to construct a device whose sole effect is to convert heat energy into work, that is, to convert a given amount of heat energy completely into mechanical energy. Instead, all heat engines must be diagrammed as in Figure 16.26B. Here we again have a mechanical device that takes an amount of heat energy Q_H from a hot reservoir and does an amount of

Figure 16.26 Schematic diagrams for a heat engine. **A** A hypothetical engine, which the laws of thermodynamics tell us is *not* possible. **B** All real heat engines must expel some heat to a cold reservoir.

work W. However, Carnot showed that all real heat engines must also expel some heat energy Q_C to a colder reservoir.

Applying conservation of energy to the heat engine in Figure 16.26B, we have

$$W = Q_H - Q_C \tag{16.17}$$

Your goal as an engine designer is get the most work (the largest W) possible out of the engine, so we define the *efficiency e* of the engine as

$$e = \frac{W}{Q_H} \tag{16.18}$$

Combining Equation 16.18 with Equation 16.17 gives

$$e = \frac{W}{Q_H} = \frac{Q_H - Q_C}{Q_H}$$

$$e = 1 - \frac{Q_C}{Q_H} \tag{16.19}$$

When comparing heat engines, a larger value of e means that the engine converts a larger fraction of the input energy Q_H into work.

Carnot's Engine

To maximize the efficiency e, one must make Q_C as small as possible, which makes W as large as possible. Carnot considered this problem for a heat engine that makes use of the reversible compression and expansion of an ideal gas. Because Carnot's engine uses only reversible processes, it is an example of a reversible heat engine. He was able to prove that all reversible heat engines have the same efficiency as his particular design. Carnot also showed that for a hot reservoir at temperature T_H (providing the heat input Q_H) and a cold reservoir at temperature T_C, the heat expelled to the cold reservoir satisfies

$$\frac{Q_C}{Q_H} = \frac{T_C}{T_H} \tag{16.20}$$

Here the temperatures T_H and T_C must be measured in Kelvin units (not degrees Celsius). Combining Equation 16.20 with the definition of efficiency (Eq. 16.19) leads to

Efficiency of a reversible heat engine

$$e = 1 - \frac{T_C}{T_H} \tag{16.21}$$

Equation 16.21 is an amazing result: it applies to *all* reversible engines, no matter how they are constructed. Carnot did not stop here, but discovered one more crucial point: no engine can have an efficiency that is better than the efficiency of a reversible engine. In practice, all heat engines will always be irreversible to some extent, so Carnot's result sets an *absolute limit* on the efficiency of all real heat engines. Carnot's results are the basis for an alternative statement of the second law of thermodynamics.

Second law of thermodynamics (alternative form)

The efficiency of a reversible heat engine is given by

Second law of thermodynamics: formulation involving heat engines

$$e = 1 - \frac{T_C}{T_H}$$

No heat engine can have a greater efficiency than this.

We have now seen two statements of the second law, this one involving the efficiency of heat engines and another (in Section 16.5) involving the direction of heat flow. Although it is not obvious, these two statements are *completely equivalent*. Given either one of them as a starting point, the other one can be derived!

The result in Equation 16.21 shows that the efficiency of an engine depends on the temperatures of the reservoirs. If the temperature of the cold reservoir is zero ($T_C = 0$ K), the efficiency $e = 1$. In this case, we also have $Q_C = 0$, and it would thus seem that we have created the hypothetical heat engine in Figure 16.26A—a heat engine that expels no heat—which we said was impossible. One way out of this apparent contradiction is to hypothesize that it is impossible to lower the temperature of a reservoir to $T = 0$ K, the temperature called absolute zero. We'll see this hypothesis again in Section 16.8 when we discuss the third law of thermodynamics.

EXAMPLE 16.7 | Efficiency of a Steam Engine

Steamboats were used extensively in the 1800s to carry people and goods (Fig. 16.27). These boats used heat engines: a hot reservoir contained steam ($T_H \approx 200°C$) that was heated by burning wood or coal, and the cold reservoir was provided by the air or river water ($T_C \approx 10°C$). The engine turned a shaft that drove the boat's paddle wheels. Suppose one of these steam engines does $W = 10,000$ J of work each second. If it is a reversible heat engine, how much heat Q_H does it absorb from the hot reservoir in 1 s?

Figure 16.27 Example 16.7. Steamboats in the 1800s used heat engines to drive their paddle wheels.

RECOGNIZE THE PRINCIPLE

The efficiency of any heat engine is defined as (Eq. 16.18)

$$e = \frac{W}{Q_H} \qquad (1)$$

For a reversible heat engine, Carnot showed that e is also equal to

$$e = 1 - \frac{T_C}{T_H} \qquad (2)$$

We can use these two relations to find the heat absorbed from the hot reservoir Q_H in terms of the given quantities W, T_H, and T_C.

SKETCH THE PROBLEM

Figure 16.27 shows a picture of a real steamboat, but the diagram of a heat engine in Figure 16.26B is more useful for analyzing the operation of the engine.

IDENTIFY THE RELATIONSHIPS

Combining Equations (1) and (2) gives

$$\frac{W}{Q_H} = 1 - \frac{T_C}{T_H}$$

Rearranging to solve for Q_H, we have

$$Q_H = \frac{W}{1 - (T_C/T_H)} \qquad (3)$$

SOLVE

When using Equation (3), we must be careful to use values of the temperature on the Kelvin scale; Carnot's result in Equation (2) assumes the temperatures are measured on this scale. Converting the given values of T_H and T_C gives $T_H = (200 + 273)$ K = 473 K and $T_C = (10 + 273)$ K = 283 K. Inserting these values of the reservoir temperatures and the given value of W into Equation (3), we find

$$Q_H = \frac{W}{1 - (T_C/T_H)} = \frac{10,000 \text{ J}}{1 - (283 \text{ K}/473 \text{ K})} = \boxed{25,000 \text{ J}}$$

Because 10,000 J of work were done by the engine, the energy expelled to the cold reservoir was $Q_C = Q_H - W = 15,000$ J. Most of the energy extracted from the hot reservoir was expelled to the cold reservoir and was thus "wasted." Also, when analyzing the operation of a heat engine, temperature values should always be expressed in Kelvin units.

Figure 16.28 Example 16.8. A real automobile engine is a type of heat engine following the general schematic form in Figure 16.26B. The laws of thermodynamics place rigorous limits on the efficiency of all real heat engines.

EXAMPLE 16.8 ® Efficiency of a Modern Automobile Engine

The internal combustion engine found in most cars is a type of heat engine (Fig. 16.28). A hot reservoir created by exploding gasoline provides heat energy, causing the gas in a piston to expand and do work as it rotates the driveshaft of the engine. Some heat is then expelled to a cold reservoir (the air) via the radiator and the exhaust. What is the maximum possible efficiency of such an engine?

RECOGNIZE THE PRINCIPLE

A real internal combustion engine is irreversible, so its efficiency will always be less than that of a reversible engine. This irreversibility is caused by friction between the moving parts inside the engine, the failure to "capture" and use all the heat energy created by combustion, and other factors. Nevertheless, an upper limit on the efficiency is set by the reversible case (Eq. 16.21)

$$e = 1 - \frac{T_C}{T_H} \tag{1}$$

To apply Equation (1) to an internal combustion engine, we need to know the temperatures of the two reservoirs T_C and T_H; these values are not given, so we need to use some common sense and knowledge about this type of engine to find approximate values.

SKETCH THE PROBLEM

The schematic heat engine in Figure 16.26B describes the problem. The corresponding hot and cold reservoirs for a real automobile engine are shown in Figure 16.28.

IDENTIFY THE RELATIONSHIPS

The cold reservoir (the radiator and exhaust) is at air temperature, so we estimate $T_C \approx 300$ K. For the hot reservoir, we can use reference data (from the Internet or elsewhere) to find that the temperature of exploding gas in a piston is about $T_H \approx 600$ K.

SOLVE

Inserting our estimates for the reservoir temperatures into Equation (1) gives

$$e = 1 - \frac{T_C}{T_H} = 1 - \frac{300 \text{ K}}{600 \text{ K}} = \boxed{0.5} \tag{2}$$

Insight 16.1
EFFICIENCY OF A DIESEL ENGINE

A diesel engine is similar to a gasoline internal combustion engine (Example 16.8). One difference is that a gasoline engine ignites the fuel mixture with a spark from a spark plug, whereas a diesel engine ignites the fuel mixture purely "by compression" (without a spark). The compression of the fuel mixture is therefore much greater in a diesel engine, which leads to a higher temperature in the hot reservoir. According to Equation 16.21 and Example 16.8, this higher temperature gives a higher theoretical limit on the efficiency of a diesel engine.

What have we learned?

Even in the best possible case, an automobile engine can only convert about half the chemical energy in the gasoline into useful work. This limit on the efficiency could be improved by increasing the temperature of the hot reservoir (the temperature in the piston), which can be accomplished by using a suitable blend of gasoline with a higher octane rating. Such high-performance engines and fuels are more expensive than typical engines and fuels, however.

In a real engine, a substantial amount of energy is lost to friction. In a typical car engine, only about 25% of the chemical energy in the gasoline is converted into work.

Perpetual Motion and the Second Law of Thermodynamics

The construction of a perpetual motion machine has been the quest of many inventors. Various laws of physics imply that such machines are impossible. One popular perpetual motion machine design sketched in Figure 16.29 makes use of the hypothetical (and impossible) heat engine in Figure 16.26A. The machine in Figure 16.29 takes some heat energy Q from a hot reservoir, which here is the atmosphere, and converts it all to an amount of work W. This work is then used to power a car. As the car moves, it loses energy to friction due to air drag, the wheels, and so on. This loss causes all the energy W to be converted to heat energy, which returns to the atmosphere. The net effect is that the car has moved, whereas the energy in the atmosphere is back to its original value; hence, the process can be repeated, and the car can move on forever.

Carnot's work and the second law of thermodynamics tell us that the perpetual motion machine in Figure 16.29 is *impossible* because a real heat engine must always expel some heat energy to a cold reservoir. Hence, all the heat energy Q extracted from the atmosphere cannot later be returned to the atmosphere. In this way, thermodynamics rules out many hypothetical perpetual motion designs. In fact, the second law along with conservation of energy principles rule out *all* perpetual motion machines.

Figure 16.29 Hypothetical (and *impossible*) design of a perpetual motion machine. This machine is not possible because it violates the second law of thermodynamics.

The Carnot Cycle

We have so far represented heat engines in a highly schematic form as in the block diagram in Figure 16.26B and have ignored how one might construct a real engine. Carnot showed how to use an ideal gas and piston to make a reversible heat engine. Figure 16.30A shows the operation of Carnot's engine as a path in the *P–V* plane. The corresponding changes of the gas and piston are shown in Figure 16.30B, where we show only the mechanical part of the heat engine; the thermal reservoirs could be air or water held at temperatures T_H and T_C.

The system begins in state 1 in the *P–V* diagram. It then moves progressively to states 2, 3, and 4 and finally back to state 1.

Change from state 1 to state 2: The system is placed in contact with the hot reservoir, during which time the gas absorbs an amount of heat Q_H, which causes the gas to expand to state 2. Because the system is in contact with the hot reservoir, this process is an *isothermal expansion* at temperature T_H.

Change from state 2 to state 3: The system is isolated from its surroundings (detached from the hot reservoir) and allowed to expand to state 3. Because no heat is absorbed from or expelled to the environment, this process is an *adiabatic expansion*.

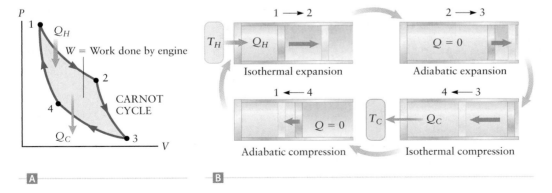

Figure 16.30 An example of a thermodynamically reversible heat engine. This engine uses a cyclic process called a Carnot cycle. **A** The work done during a Carnot cycle is equal to the area enclosed in the *P–V* diagram. **B** The four steps in the Carnot cycle.

Change from state 3 to state 4: We place the system in contact with the cold reservoir, and an amount of heat Q_C flows out of the gas and into that reservoir. This change causes the gas volume to decrease and is an *isothermal compression* because it occurs at the fixed temperature T_C of the cold reservoir.

Change from state 4 to state 1: The system is again isolated from its surroundings and *compressed adiabatically* back to its initial state 1.

This cyclic path in the *P–V* diagram is called a **Carnot cycle**. Each step in the cycle involves a reversible process, and we saw in Section 16.4 how each of these processes (isothermal expansions and compressions, and adiabatic expansions and compressions) can be analyzed. That is precisely the sort of analysis Carnot carried out when he discovered his results for reversible heat engines.

Because it is a cyclic thermodynamic process, we can find the work done during a Carnot cycle using the result we derived in connection with Figure 16.21. There and in Example 16.6, we showed that the total work done by the gas in one cycle is the area enclosed by the path in the *P–V* diagram. This area is the total work *W* done by the heat engine during one complete cycle and is indicated in Figure 16.30A for the Carnot cycle.

CONCEPT CHECK 16.7 | Work Done in a Carnot Cycle

Consider the steps in the Carnot cycle in Figure 16.30B. During which steps is the work *W* done by the system on the piston positive and during which steps is *W* negative?

Thermodynamic Devices: The Refrigerator

The Carnot heat engine in Figure 16.30 is reversible because each step in the cycle involves a reversible process. It is therefore possible to run this engine in reverse, that is, to run "backward" along the *P–V* path through states $1 \rightarrow 4 \rightarrow 3 \rightarrow 2 \rightarrow 1$. An amount of heat Q_C is then extracted from the cold reservoir, an amount of heat Q_H is expelled to the hot reservoir, and a certain amount of work *W* is done on the gas as indicated schematically in Figure 16.31A. This diagram is very similar to the heat engine in Figure 16.26B except that all the arrows showing the direction of energy flow are reversed.

Running a heat engine in reverse may seem like a crazy thing to do, but it is actually of great practical importance. Such a device is called a **refrigerator**. In a typical case (Fig. 16.31B), an electric motor does an amount of work *W* on a refrigeration "unit," which extracts an amount of heat Q_C from inside the refrigerator, keeping your food cold. The thermodynamic details of a refrigeration unit are more than we can discuss here, but they can be based on (you guessed it) cyclic processes involving the compression and expansion of gases. A refrigerator expels some heat Q_H to a hot reservoir, which in this case is the air around the refrigerator. Hence, refrigerators warm the air around them, usually underneath or behind the refrigerator. Applying conservation of energy to the refrigerator in Figure 16.31, we have

$$W + Q_C = Q_H \tag{16.22}$$

Carnot's results apply in this case, just as they apply to a heat engine, so we again have (from Eq. 16.20)

$$\frac{Q_C}{Q_H} = \frac{T_C}{T_H} \tag{16.23}$$

A good refrigerator extracts a great deal of heat Q_C from the cold reservoir while using as little work *W* as possible. The performance of a refrigerator is therefore measured in terms of the **refrigeration efficiency** e_{refrig} (also called the coefficient of performance), where

$$e_{\text{refrig}} = \frac{Q_C}{W} \tag{16.24}$$

A refrigerator is a heat engine run in reverse.

 A

© Cengage Learning/Charles D. Winters

B

Figure 16.31 **A** When a heat engine is operated in "reverse," it functions as a refrigerator. It extracts heat energy from a cold reservoir that could then be used to store food. **B** In a real refrigerator, the cold reservoir is the air inside the refrigerator, and the hot reservoir is the room air.

The goal of a refrigerator designer is to make e_{refrig} as large as possible. Thermodynamics, however, places a limit on this efficiency. As with heat engines, the highest efficiency is found with a reversible refrigerator, and we can use Equations 16.22 and 16.23 to calculate e_{refrig}. Combining those relations, we find

$$W + Q_C = Q_H = Q_C \frac{T_H}{T_C}$$

$$W = Q_C\left(\frac{T_H}{T_C} - 1\right) \qquad (16.25)$$

Inserting this result for W into the definition of the efficiency (Eq. 16.24) leads to

$$e_{refrig} = \frac{Q_C}{W} = \frac{Q_C}{Q_C[(T_H/T_C) - 1]} = \frac{1}{(T_H/T_C) - 1}$$

$$e_{refrig} = \frac{T_C/T_H}{1 - (T_C/T_H)} \qquad (16.26)$$

A refrigerator is used to keep things cold, so T_C is usually low, but a low value of T_C leads to a small value for e_{refrig}. We can see that by inserting $T_C \approx 0$ into Equation 16.26, which gives

$$e_{refrig} = \frac{T_C/T_H}{1 - (T_C/T_H)} \approx \frac{0/T_H}{1 - (0/T_H)} = 0$$

So, when the cold reservoir is at a very low temperature, the efficiency is also very low. That is an unavoidable consequence of the second law of thermodynamics.

The schematic diagram of a refrigerator in Figure 16.31 shows that heat is removed from a cold reservoir and expelled to a hot reservoir. Hence, a refrigerator causes heat energy to flow from a cold body to a warmer one. Doesn't that violate the second law of thermodynamics? The answer is no; it does not violate the second law because the second law applies to the *spontaneous* flow of heat energy. The heat flow in a refrigerator is possible only because of the work W done on the refrigerator, so the heat flow in Figure 16.31 is not spontaneous.

Heat Pumps

We have seen that a refrigerator must emit some heat; this heat is Q_H in Figure 16.31. A device called a **heat pump** uses this principle as part of a furnace for heating a house (Fig. 16.32). An amount of heat Q_C is extracted from a cold reservoir, which might be the outside air or perhaps water from underground. This device literally "pumps heat" from the cold reservoir into your house.

Are heat pumps really an efficient way to heat your house? For a heat pump to be effective, you want the amount of heat that goes into your house to be as large as possible while at the same time keeping W as small as possible. So, the efficiency of a heat pump is defined as

$$e_{pump} = \frac{Q_H}{W} \qquad (16.27)$$

and a good heat pump should have a large value of e_{pump}. Equations 16.22 and 16.23 apply to a heat pump as well as to a refrigerator (because a heat pump is thermodynamically identical to a refrigerator; compare Figs. 16.31A and 16.32), and we can use them to calculate e_{pump} for a reversible heat pump. They lead to

$$W = Q_H - Q_C = Q_H - Q_H \frac{T_C}{T_H}$$

$$W = Q_H[1 - (T_C/T_H)]$$

Figure 16.32 A heat pump is a device that can be used to heat homes and other buildings. It "pumps" heat from a cold reservoir into the building (the hot reservoir).

Inserting this result into the definition of e_{pump} (Eq. 16.27), we get

$$e_{pump} = \frac{Q_H}{W} = \frac{Q_H}{Q_H[1 - (T_C/T_H)]}$$

$$e_{pump} = \frac{1}{1 - (T_C/T_H)} \qquad (16.28)$$

Efficiency values much greater than 1 are mathematically possible and achievable in practice, as we explore in Example 16.9.

EXAMPLE 16.9 ⓡ **Heating the Author's House**

The author uses a heat pump to keep his house warm in the winter. The cold reservoir is water that comes from an underground well. What is the approximate efficiency of such a heat pump, assuming it is reversible?

RECOGNIZE THE PRINCIPLE

To find the efficiency of the heat pump (Eq. 16.28), we must know the reservoir temperatures. These temperatures are not given, so we need to estimate values from the description of the heat pump.

SKETCH THE PROBLEM

No sketch is needed.

IDENTIFY THE RELATIONSHIPS

The "hot" reservoir is the temperature inside the house, which is about 20°C = 293 K = T_H. The temperature of the cold reservoir is the temperature of the water in the well; for cold water from underground, it is about 10°C = 283 K = T_C.

SOLVE

Using these values for the reservoir temperatures, the efficiency is

$$e_{pump} = \frac{1}{1 - T_C/T_H} = \frac{1}{1 - (283 \text{ K})/(293 \text{ K})} \approx \boxed{30}$$

What does it mean?

The work W that acts as the "input" to the heat pump (Fig. 16.32) is provided by an electric motor. With the value of e_{pump} for this heat pump, each 1 J of energy used to run the electric motor leads to about 30 J of heat energy "pumped" into the house, which is quite a good deal! An alternative scheme, called "electric resistive heating," produces a maximum of 1 J of heat for each 1 J of electrical energy consumed. Hence, a heat pump can be a very efficient way to heat a house. Notice, however, that we have calculated the efficiency of an *reversible* heat pump. The efficiency of a real heat pump will be lower than this value. (See Problem 65.)

16.7 | ENTROPY

Before the work of Carnot and other scientists who developed the theory of thermodynamics, the nature of heat energy was not completely understood. Some physicists proposed (incorrectly) that heat was a sort of "fluid" that can flow from one object to another, much like water can flow from one container to another. This picture implied that an object "contains" only a certain amount of heat; if that amount of heat is somehow removed, it would then be impossible to extract any more heat from the object. We now know that this picture is not correct. However, there is a quantity called *entropy* connected with heat flow that does have some of these properties.

Entropy, denoted by S, is a macroscopic property of a system much like pressure and temperature. If a small amount of heat Q flows into a system, its entropy changes by

$$\Delta S = \frac{Q}{T} \qquad (16.29)$$

where T is the temperature of the system. The units of entropy are thus joules per kelvin. Let's apply Equation 16.29 to compute the total entropy change when heat flows between the two systems in Figure 16.33. The quantity Q in Equation 16.29 is the heat that flows *into* a system; in Figure 16.33, an amount of heat Q flows into system 2, so the entropy of this system changes by

$$\Delta S_2 = \frac{+Q}{T_2} \qquad (16.30)$$

Likewise, an amount of heat Q flows *out* of system 1, so we have

$$\Delta S_1 = \frac{-Q}{T_1} \qquad (16.31)$$

Hence, if heat flows from system 1 to system 2 as shown in Figure 16.33, the entropy of system 2 increases and the entropy of system 1 decreases.

A typical heat-flow experiment might involve a system at 373 K (the boiling temperature of water) in contact with a system at 273 K (the freezing temperature of water) as sketched in Figure 16.34. In this example, heat flows via conduction through a metal rod connecting the two systems. If a total amount of heat Q flows from the hot system (system 1) to the cold system (system 2), the total change in entropy is (using Eqs. 16.30 and 16.31)

$$\Delta S_{total} = \Delta S_1 + \Delta S_2 = \frac{-Q}{T_1} + \frac{Q}{T_2}$$

$$\Delta S_{total} = Q\left(-\frac{1}{T_1} + \frac{1}{T_2}\right) \qquad (16.32)$$

Inserting $T_1 = 373$ K and $T_2 = 273$ K gives

$$\Delta S_{total} = \frac{-Q}{T_1} + \frac{Q}{T_2} = Q\left(-\frac{1}{373\,\text{K}} + \frac{1}{273\,\text{K}}\right) \qquad (16.33)$$

The quantity in parentheses is positive, so ΔS_{total} is also positive.

Now suppose the temperatures of the two systems are nearly equal; in fact, let's assume we have an isothermal process so that $T_1 = T_2$. According to Equation 16.32, the total entropy change of both systems combined is zero. Such an isothermal process is reversible and gives an example of a process in which the entropy does not change. In fact, the total entropy change is zero for *all* reversible processes:

$$\Delta S_{total} = 0 \quad \text{(for a reversible process)} \qquad (16.34)$$

We could also say that the total entropy change of the universe is zero for any reversible process.

The heat-flow process in Figure 16.34 involves systems at different temperatures and is irreversible because heat flows spontaneously from the hot to the cold system but not in the reverse direction (according to the second law of thermodynamics). According to Equation 16.32, ΔS_{total} in this case is always positive, which leads to the result

$$\Delta S_{total} > 0 \quad \text{(for an irreversible process)} \qquad (16.35)$$

Equation 16.35 applies to *all* irreversible processes.

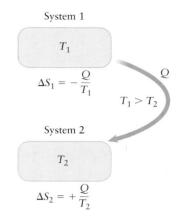

System 1

$$\Delta S_1 = -\frac{Q}{T_1}$$

$$T_1 > T_2$$

System 2

$$\Delta S_2 = +\frac{Q}{T_2}$$

Figure 16.33 Whenever heat flows into a system, the entropy of the system changes by an amount $\Delta S = Q/T$. The entropy change of a system can be positive or negative, depending on the direction of heat flow (the sign of Q).

$\Delta S_{total} > 0$

Boiling water $T = 373$ K

Ice $T = 273$ K

Metal rod

| Hot system (1) | Q | Cold system (2) |

Figure 16.34 Heat flow from a hot system (1) to a cold system (2). The entropy change for the entire process, including both systems, is ΔS_{total}.

The results in Equations 16.34 and 16.35 form the basis for another statement of the second law of thermodynamics.

Second law of thermodynamics (entropy version)
In any thermodynamic process, $\Delta S_{\text{universe}} \geq 0$.

We now have three separate versions of the second law of thermodynamics. All are equivalent; any one version can be used to prove the other two. Each version gives different insights into the meaning and consequences of the second law.

Entropy: A Microscopic View

Let's now describe entropy in terms of a microscopic picture. Consider an ice cube placed in a warm environment. The ice cube is initially at a lower temperature than its environment, so heat will flow from the surroundings into the ice cube, causing it to melt. Because heat flows into the ice cube, its entropy will (according to Eq. 16.29) increase. A molecular-scale view of this process is shown in Figure 16.35. How can we tell from this picture that the entropy has changed?

Before the ice cube melts, the water molecules are in regular positions characteristic of a solid, whereas after melting, the molecules are in the disordered arrangement characteristic of a liquid state. The system thus becomes much more random after the heat has been added and the entropy has increased. In microscopic terms, *entropy is a measure of the amount of disorder* or randomness in a system. Increasing the temperature of a gas or liquid increases the kinetic energy of the molecules and increases the amount of disorder. Hence, adding heat to a system always increases its entropy.

Entropy and Probability

Let's consider a system of gas molecules in a box and ask how the molecules are distributed. Are they all on one side of the box, are they distributed randomly throughout the box, or are they distributed in some other way? For simplicity, we can divide our box into two halves and ask how many molecules are in each half at any given "snapshot" in time. By symmetry, a particular molecule is equally likely to be on one side as on the other, so the probability of finding any given molecule on the left side is $p = \frac{1}{2}$. For a system containing only a single molecule, we thus have two possibilities—we call these two possible "states" of the system—as shown in Figure 16.36A.

We can do the same analysis with 2 molecules. Because each molecule can be on either the right or left side of the box, we now have $2 \times 2 = 4$ possible states for the system—four different possible arrangements of the molecules—shown in Figure 16.36B. Figure 16.36C shows all the different possible states with 4 molecules. Notice how the number of possible states grows very rapidly as we add molecules to the system. With 4 molecules, we already have 16 possible states, and by the time we reach 10 molecules, the number of states is more than 1000. A key assumption of thermodynamics is that each of these states is equally likely. With this assumption, we can calculate P_{left}, the probability of finding a molecule on the left side of the box; when $P_{\text{left}} = 0.5$, the molecules are equally distributed on the two sides. Hence, the expected random distribution of molecules corresponds to values of P_{left} near 0.5. Figure 16.37 shows results for 4 molecules, 10 molecules, and 100 molecules. As the number of molecules grows, the probability peak becomes narrower, meaning that the probability of finding the molecules equally distributed becomes very high.

What do the probability curves in Figure 16.37 have to do with entropy? Large values of the entropy correspond to a greater amount of randomness in a system. The results in Figures 16.36 and 16.37 show that there are many more ways of distributing the gas molecules so that they are randomly distributed throughout

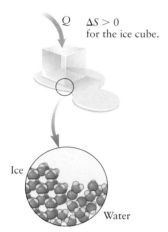

Q $\Delta S > 0$
for the ice cube.

Ice

Water

Figure 16.35 When the entropy of a system increases, the system becomes more disordered.

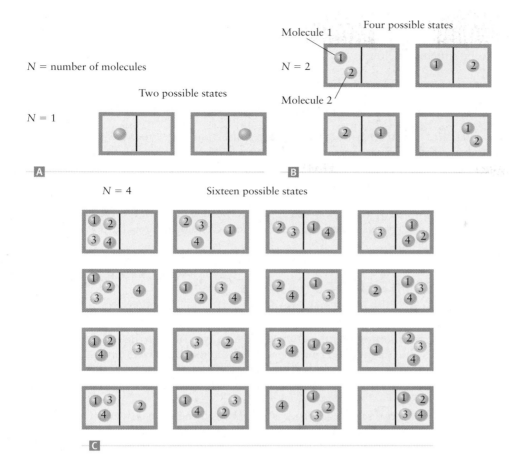

N = number of molecules

Two possible states

$N = 1$

Molecule 1

Four possible states

$N = 2$

Molecule 2

A

B

$N = 4$ Sixteen possible states

C

the box—that is, so that $P_{left} = 0.5$—than if we place all the molecules on one side or the other. Hence, the most likely state of the system (the state with the highest probability) is also the state with the greatest randomness and hence the state with the largest entropy.

16.8 | THE THIRD LAW OF THERMODYNAMICS AND ABSOLUTE ZERO

We defined the three common temperature scales (Fahrenheit, Celsius, and Kelvin) in Chapter 14. The Kelvin scale is most closely connected to kinetic theory and thermodynamics. For example, the ideal gas law has the mathematical form $PV = nRT$ when temperature is measured on the Kelvin scale, and the definition of entropy $\Delta S = Q/T$ also assumes T is given on the Kelvin scale. The implication is that there is something special about $T = 0$ on the Kelvin scale.

We have had other hints that something special must take place at $T = 0$ K, the temperature called absolute zero. One hint comes from kinetic theory (Chapter 15), where we found that the average translational kinetic energy of a gas molecule is related to the absolute temperature by

$$KE = \tfrac{3}{2}k_B T \qquad (16.36)$$

Thus, at $T = 0$, we expect the kinetic energy of a gas molecule to be zero! Because the kinetic energy cannot be less than zero, this finding implies that temperatures below absolute zero are not possible. Another hint comes from Carnot's work with heat engines and Equation 16.20. Carnot's statement that all heat engines must

Figure 16.36 ◨ A box containing a single gas molecule. We divide the box into two equal parts; the molecule might be found in either of these two possible "states." ◨ If the box contains 2 molecules, they can be arranged in the two parts of the box in four different ways. ◨ With 4 molecules, there are 16 different ways to arrange the molecules. There are thus 16 different possible states of the system, and all are equally likely to occur.

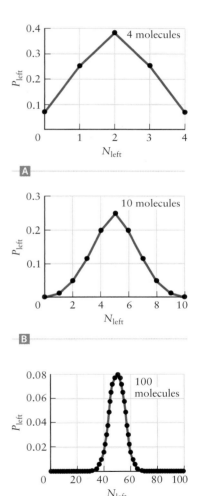

A

B

C

Figure 16.37 Probability for finding different numbers of molecules (N_{left}) on the left side of the box. ◨ For 4 molecules. When $N_{left} = 2$, there is an equal number of molecules on both sides. ◨ For 10 molecules. ◨ For 100 molecules.

expel some heat to a cold reservoir implies that a reservoir with $T = 0$ K is not possible. These notions all come together in the third law of thermodynamics, which can be stated as follows.

Third law of thermodynamics

Third law of thermodynamics
It is impossible for the temperature of a system to reach absolute zero.

The third law of thermodynamics has important consequences for our understanding of mechanics. According to Newton's mechanics, it should be possible to make the velocity of a particle zero. Indeed, if we were to do that for every particle in a gas, the temperature of the gas would then, according to Equation 16.36, be $T = 0$. The third law of thermodynamics, however, tells us that is not possible. Hence, the laws of thermodynamics appear to be at odds with Newton's laws! This conflict was not resolved until the development of quantum mechanics in the early twentieth century.

16.9 | ⊗ THERMODYNAMICS AND PHOTOSYNTHESIS

In our work on thermodynamics, we made extensive use of the ideal gas because its properties, such as the internal energy and the ideal gas law, have simple mathematical forms. Thermodynamics, however, can be applied to many different physical systems. In this section, we show how thermodynamic arguments can be applied to photosynthesis.

Figure 16.38 represents the process of photosynthesis in a way that shows its similarity to a heat engine. An amount of heat Q_H from the Sun enters a plant, and (in the spirit of a heat engine) the plant "does an amount of work" W. At the same time, the plant expels some heat Q_C to its surroundings (the cold reservoir). In physics and engineering, the heat engine is a mechanical device that does work (e.g., by turning a shaft). In photosynthesis, a plant stores chemical energy in certain molecules, so the product of the photosynthetic heat engine in Figure 16.38 is actually potential energy stored in chemical bonds. This chemical energy can be used later to produce mechanical work when, for instance, an animal eats the plant, so the analogy with a heat engine still holds.

Let's now analyze the efficiency of a photosynthetic heat engine. Assuming it is a reversible engine, this efficiency depends on the temperatures of the reservoirs and is given by $e = 1 - (T_C/T_H)$ (Eq. 16.21). The cold reservoir is the air or soil around the plant; hence, $T_C \approx 300$ K. The Sun serves as the hot reservoir, so our first estimate for the temperature of the hot reservoir is the temperature of the surface of the Sun, which is $T_H \approx 6000$ K as mentioned in Chapter 14. (The temperature of the hot reservoir is *not* the temperature of the plant. The leaves are *not* at 6000 K!) Inserting these values into Equation 16.21 gives

$$e_{\text{photosyn}} = 1 - \frac{T_C}{T_H} = 1 - \frac{300 \text{ K}}{6000 \text{ K}} = 0.95$$

A real photosynthetic engine does not approach this efficiency for several reasons, one of which can be seen from Figure 16.39. Radiation emitted by the Sun is distributed over a range of wavelengths throughout the infrared, visible, and ultraviolet regions (the blackbody distribution in Fig. 14.32). Plants absorb solar radiation mainly in the range of wavelengths $\lambda \approx 400$–700 nm and do not absorb much of the radiation outside this range. A mathematical analysis of the blackbody radiation curve in Figure 16.39 shows that plants absorb only about 50% of the Sun's total radiation, thus reducing the maximum efficiency of photosynthesis to approximately $0.95 \times 0.5 \approx 48\%$. There are other inefficiencies in photosynthesis, and the

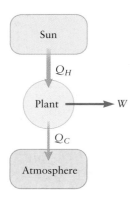

Figure 16.38 Photosynthesis is a thermodynamic process and can be viewed as a type of heat engine. Compare this diagram with Figure 16.26B.

Figure 16.39 Plants absorb light only in the visible portion of the spectrum. Visible radiation corresponds to only about half the total radiated energy from the Sun. Shown is the blackbody distribution from Figure 14.32, but here it is plotted on a linear scale.

actual efficiency of the entire process is somewhat lower than 48%.[4] The important point is that thermodynamics and the physics of blackbody radiation place absolute physical limits on the efficiency of a real process.

16.10 | CONVERTING HEAT ENERGY TO MECHANICAL ENERGY AND THE ORIGIN OF THE SECOND LAW OF THERMODYNAMICS

One key tenet of thermodynamics is that heat is a form of energy. The second law of thermodynamics, however, also tells us that heat energy differs from kinetic and potential energy in a subtle way. If you are given a certain amount of energy in the form of the kinetic energy of a baseball or the potential energy of a compressed spring, it is possible to convert all this energy to heat energy. Suppose, however, you are given the same amount of heat energy and asked to convert it to mechanical energy. The second law of thermodynamics as applied to a heat engine tells us that *only a fraction* of this energy can be converted to mechanical energy; this fraction is equal to the efficiency of a reversible heat engine (Eq. 16.21). A conversion from mechanical energy to heat can be done with 100% efficiency; why can't we convert heat energy to work with the same 100% efficiency?

To gain some insight into this difference, we first consider an example in which mechanical energy is converted to heat: the compression of an ideal gas. A molecular-scale view of this process is given in Figure 16.40, showing gas compressed by moving a piston. As the piston moves to the left (Figs. 16.40A and B), it does work on the gas molecules with which it collides. The work done by an external force acting on the piston is $F \Delta x$; this amount of energy is converted to additional kinetic energy of the molecules and shows up thermodynamically as an increase in the internal energy of the gas. If we now place this warm gas in contact with a colder thermal reservoir, heat flows from the gas into the reservoir (Fig. 16.40C). In this way, we could convert all the mechanical energy used to move the piston to heat energy.

Now consider the possibility (actually, the *impossibility*) of running the process in Figure 16.40 backward. To run in reverse, all the gas molecules in Figure 16.40C would have to collide with the piston in synchrony so as to exert a force on the piston to the right. The randomness of the gas makes such perfect synchrony impossible, however; some of the molecules may collide with the piston in this way, but the inherent disorder associated with heat energy prevents all the molecules from simultaneously "giving energy back" to the piston. For this reason, it is not possible to extract all the molecular kinetic energy—that is, all the heat energy—from the gas. The disordered motion of the gas molecules makes it impossible to convert heat energy completely to mechanical energy.

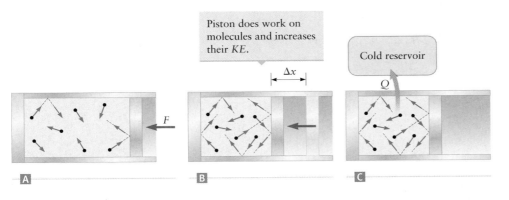

Piston does work on molecules and increases their *KE.*

Δx

Cold reservoir

Q

F

A

B

C

Figure 16.40 **A** Molecular-scale view of how the movement of a piston changes the energy of a gas. **B** When the gas is compressed, the piston does a net positive amount of work on all the gas molecules with which it collides. **C** This increased energy can later be transferred (via conduction) to a colder reservoir.

---

[4]For comparison, solar cells convert sunlight into electrical energy with an efficiency of typically 15% to 20%. Photosynthesis is thus more efficient than manufactured solar cells.

SUMMARY | Chapter 16

KEY CONCEPTS AND PRINCIPLES

Thermodynamics is about the properties of systems

Thermodynamics is concerned with the behavior of systems containing many particles such as a liquid, a gas, or a solid. Such systems interact with their environment, and the laws of thermodynamics describe these interactions.

Zeroth law of thermodynamics

If two systems A and B are both in thermal equilibrium with a third system, then A and B are in thermal equilibrium with each other.

First law of thermodynamics

If an amount of heat Q flows *into* a system from its environment and an amount of work W is done *by* the system on its environment, the internal energy of the system changes by an amount

$$\Delta U = U_f - U_i = Q - W \qquad \text{(16.3) (page 495)}$$

Second law of thermodynamics

The second law of thermodynamics can be stated in several different ways.

- *Second law of thermodynamics (formulation involving heat flow):* Heat flows spontaneously from a warm body to a colder one. It is not possible for heat to flow spontaneously from a cold body to a warmer one.
- *Second law of thermodynamics (formulation involving heat engines):* The efficiency of a reversible heat engine operating between reservoirs at temperatures T_H and T_C is

$$e = 1 - \frac{T_C}{T_H} \qquad \text{(16.21) (page 510)}$$

  No heat engine can have a greater efficiency.
- *Second law of thermodynamics (entropy version):* In any thermodynamic process, $\Delta S_{\text{universe}} \geq 0$.

Third law of thermodynamics

It is impossible for the temperature of a system to reach absolute zero.

APPLICATIONS

Thermodynamic processes

A *thermodynamic process* takes a system from some initial state with an initial temperature T_i, pressure P_i, and volume V_i to a final state described by T_f, P_f, and V_f. Such a process can be described as a path in a *P–V diagram*. The work done by the system on its environment is the area under this P–V curve. An *isothermal process* is one in which the temperature is constant ($T_i = T_f$). In an *adiabatic process*, there is no heat flow into or out of the system ($Q = 0$).

(Continued)

Heat engines

A *heat engine* takes heat energy from a hot reservoir, expels heat energy to a cold reservoir, and does an amount of work on its environment. The *efficiency of a heat engine* is defined as

$$e = \frac{W}{Q_H} = \frac{Q_H - Q_C}{Q_H}$$

The efficiency of a thermodynamically reversible heat engine is given by the laws of thermodynamics as

$$e = 1 - \frac{Q_C}{Q_H} = 1 - \frac{T_C}{T_H} \qquad \textbf{(16.19 and 16.21)} \text{ (pages 509 and 510)}$$

No heat engine can have a greater efficiency.

A heat engine absorbs heat from a hot reservoir and expels heat to a cold reservoir.

Refrigerators and heat pumps

In a *refrigerator*, work is done by the environment on the system, which then extracts heat from a cold reservoir and expels heat to a hot reservoir. The *efficiency of a refrigerator* is $e_{refrig} = Q_C/W$. A thermodynamically reversible refrigerator has an efficiency of

$$e_{refrig} = \frac{T_C/T_H}{1 - (T_C/T_H)} \qquad \textbf{(16.26)} \text{ (page 514)}$$

No refrigerator can have a greater efficiency.

Heat pumps use the same thermodynamic arrangement as in a refrigerator to cool your house. The cold reservoir in the case of a heat pump is the interior of your house.

A refrigerator is a heat engine run in reverse.

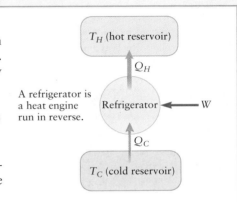

QUESTIONS

SSM = answer in Student Companion & Problem-Solving Guide

⊗ = life science application

1. ⊗ Two blocks, one of wood and the other aluminum, are at room temperature. When you touch them, the aluminum feels cold and the wood feels warm. Why? Is there a temperature at which both blocks would feel the same temperature to you? Explain.

2. It is a very hot day, and your friend decides to cool his apartment by opening the door of his refrigerator. Use thermodynamics to answer the following question: Will this method cool his apartment?

3. Can heat either flow into or out of a substance without the substance changing temperature? If not, why? If so, explain and give an example of such a process.

4. Consider an ideal gas that is expanding in a cylinder with a piston. As the gas expands, the piston moves such that a constant pressure is maintained. Does heat flow into or out of the gas in this process? Why or why not? Does the temperature rise or fall?

5. What quantity is represented by an area enclosed by a curve on a P–V diagram? What device is described by following such a curve in a clockwise direction? If the process follows the curve counterclockwise, what kind of device is being represented?

6. Consider a window air conditioner in terms of the thermodynamic block diagram in Figure 16.31A. Identify the hot reservoir, the cold reservoir, and the source of the mechanical work.

7. Explain how a heat pump can deliver more heat energy into a house than the pump consumes as work.

8. SSM Consider a thermodynamic process in which an ice cube melts to become a liquid (water). Is the change of the internal energy of the H_2O positive or negative? Is the change of the entropy of the H_2O positive or negative?

9. Heat engines need both a working fluid and a source of heat, often from a chemical process like combustion. Classify each substance as (i) a fuel or (ii) a working fluid. For which substances is this classification not exact? Why?
   (a) Air   (d) Coal   (f) Freon
   (b) Liquid water   (e) Gasoline   (g) Propane
   (c) Steam

10. Figure Q16.10 shows a thermodynamic process in which a system first expands and is then compressed. Indicate the area in the P–V diagram that corresponds to the work done by the system.

11. In Section 16.4, we asserted that the work done in any cyclic process is equal to the area enclosed by the path in a P–V plot. In Example 16.6, we proved this assertion for a cyclic process described by a rectangular path. Prove this result for a path of any shape. *Hint:* Use a graphical argument involving area.

Figure Q16.10

12. Give an example of a mechanical process that is reversible.

13. Give an example of a mechanical process that is irreversible. *Hint:* Include friction.

14. Suppose the second law of thermodynamics is *not* true and it is possible for heat to flow spontaneously from a cold body to a warmer one. Show that this process would then allow one to construct a perpetual motion machine. *Hint:* Consider a machine that functions in two or more separate steps, with one step being the flow of heat from a cold reservoir to a hot reservoir. Then, in a separate step, energy from the hot reservoir is used in some way.

15. Consider the entropy changes that occur during the Carnot cycle sketched in Figure 16.30. For each stage of the cycle, is the entropy change of the engine positive, negative, or zero? What is the total entropy change of the engine, and of the environment, during one complete cycle?

16. Under what conditions will the efficiency of a Carnot engine approach 100%?

17. If an ideal gas is compressed adiabatically, does the internal energy of the gas increase or decrease?

18. [SSM] Explain why heat pumps that use air as the cold reservoir are most efficient in mild climates. (They are called "air-to-air" heat pumps.) Why does a heat pump that uses water from deep underground as the cold reservoir have a thermodynamic advantage over air-to-air heat pumps when the weather is very cold?

19. In Example 16.9, we found that for typical temperatures of the hot and cold reservoirs, the theoretical efficiency of a home heat pump can be quite large. Investigate the efficiency obtained with real heat pumps (i.e., do some research on the Internet). How does the efficiency of a real heat pump compare with the theoretical limit in Equation 16.28? What factors make the efficiency of a real heat pump smaller than this theoretical limit?

20. Two ideal heat engines A and B operate using the same hot reservoir. It is found that engine B is more efficient that engine A. Explain how that can be.

21. Consider the common expressions (a) "There's no such thing as a free lunch" and (b) "You can't even break even." Which of the laws of thermodynamics might these expressions refer to? Explain.

22. A hydroelectric power plant converts potential energy of water stored behind a dam to electrical energy with an efficiency of close to 100%. A coal-burning (or nuclear) power plant only approaches 40% for conversion of the chemical (or fission) energy to electricity. Why?

PROBLEMS

16.3 THE FIRST LAW OF THERMODYNAMICS AND THE CONSERVATION OF ENERGY

1. What is the total internal energy of a monatomic ideal gas that contains 4.0 moles at a temperature of 400 K?

2. A system takes in 5000 J of heat and does 4000 J of work on its surroundings. What is the change in the internal energy of the system?

3. [SSM] ⬛ The temperature of an ideal monatomic gas increases by 100 K while 3000 J of heat are added to the gas. If the gas contains 4.0 moles, how much work does it do on its surroundings?

4. ⬛ The temperature of a sample of dilute argon gas with $n = 7.5$ moles decreases by 200 K. If 35,000 J of heat are extracted from the gas, what is the work done by the gas on its surroundings?

5. One gallon of gasoline releases approximately 1.2×10^8 J when it is "burned" (i.e., undergoes combustion) in a car engine. If the engine releases 4.0×10^7 J of heat during this time, what is the mechanical work done by the engine?

6. ⬛ Ⓧ A doughnut contains about 300 Calories. (*Note:* These are "food" Calories.) After eating this doughnut, you decide to compensate by taking the stairs instead of using the elevator. If each stair has a height of 20 cm, how many stairs must you climb to compensate for the doughnut? Assume all the energy in the doughnut can be converted to mechanical energy.

7. ⬢ Ⓧ Table 16.1 lists the energy that can be extracted from various foods, and Table 16.2 gives the rate at which work is done and heat is generated by a person engaged in various activities. Use these data to analyze the following cases.
(a) How much heat energy is generated during a normal night (8 h) of sleep? How many apples must a person consume to compensate for this energy loss? (Round to the next highest whole number.)
(b) What is the total energy expended when a person runs for 30 min at a moderate pace? How many soft drinks does this energy correspond to? (Round to the next highest whole number.)
(c) What is the total energy expended when an elite cyclist races for 6 h at a rapid pace (i.e., at the Tour de France bicycle race)?

How many hamburgers does this energy correspond to? (Round to the next highest whole number.)

Table 16.1 Energy Obtained by Eating Various Foods

FOOD	APPROXIMATE ENERGY OBTAINED (J)
Apple	2.2×10^5
Banana	4.6×10^5
Egg	2.7×10^5
Fast-food hamburger	2.0×10^6
Soft drink (12-oz size)	5.7×10^5

Table 16.2 Approximate Values of the Work W and Heat Q Associated with Various Human Activities

ACTIVITY	W/t (W)	Q/t (W)
Walking	50	200
Bicycle riding (moderate speed)	100	400
Bicycle racing	400	1600
Sleeping	0	80
Running (moderate pace)	200	800
Shoveling snow	15	500

Note: Work W and heat Q are both expressed in terms of the rate at which they are done.

8. ⬛ An ideal gas absorbs 400 J of heat from its environment and does 800 J of work on the environment. Find the change in the internal energy of the gas and the change in the internal energy of the environment.

16.4 THERMODYNAMIC PROCESSES

9. Match the following processes (at left) with the descriptions (at right).

    (a) Isobaric   (i)   The volume remains constant.

    (b) Isothermal   (ii)   The pressure remains constant.

    (c) Adiabatic   (iii)   No heat energy is added to or subtracted from the system.

    (d) Isochoric   (iv)   The temperature remains constant.

10. In which of these processes does the internal energy of an ideal gas not change?

    (a) An isothermal process

    (b) An adiabatic process

    (c) An isobaric process

    (d) A process at constant volume

11. A gas expands in an isobaric (constant pressure) process, with $P = 4.5 \times 10^5$ Pa. If the volume changes from 4.5 m³ to 7.8 m³, how much work is done by the gas?

12. Consider a system that undergoes an expansion from an initial state (i) to a final state (f) in Figure P16.12. Approximately how much work is done by the system in this process?

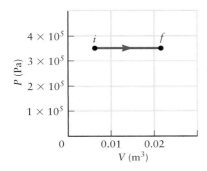

Figure P16.12 Problems 12, 14, and 15.

13. Approximately how much work is done by the system in the thermodynamic processes in the two parts of Figure P16.13? Are they expansions or contractions?

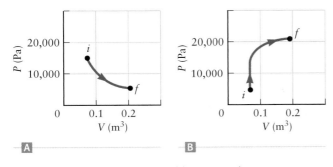

Figure P16.13 Problems 13 and 14.

14. ⭐ For the thermodynamic processes in Figures P16.12 and P16.13, what is the work done *on* the system?

15. ⭐ For the expansion in Figure P16.12, if the internal energy of the system decreases by 4500 J, what is the heat added to the system?

16. ⭐ A system is compressed adiabatically, during which time the environment does 3500 J of work on the system. What is the change in the internal energy of the system?

17. SSM ⭐ Consider a balloon of volume 2.0 m³ that contains 3.5 moles of helium gas at a temperature of 300 K. If the gas is compressed adiabatically so as to have a final temperature of 400 K, what is the work done by the gas?

18. ✪ Ⓡ A container holds 1.5 kg of water that is initially at 320 K and is converted to steam at 400 K. Assuming the volume

of the system does not change, what is the change in the internal energy of the system?

19. ⭐ Water vapor ($n = 1.0$ mole) is heated in a pressure cooker so that the volume is held constant. If the temperature increases from 500 K to 600 K, what is the change in the internal energy of the water vapor?

20. ✪ You have 3.2 moles of a monatomic ideal gas that are compressed isothermally. (a) If the pressure changes from 4.5×10^5 Pa to 8.5×10^5 Pa and the initial volume was 0.50 m³, what is the final volume of the system? (b) For the compression in this problem, what are the initial and final temperatures?

21. A system undergoes a two-step process as sketched in Figure P16.21. The system first expands at constant pressure from state i to state A and then it is compressed with the pressure again held fixed, ending with the system in state f. What is the approximate amount of work done by the system during each step?

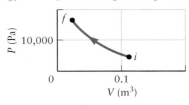

Figure P16.21 Problems 21 and 22.

22. ⭐ Consider the process described by Figure P16.21. If 500 J of heat are added to the system during the second step, what is the change in the internal energy of the system during this step?

23. SSM ⭐ Figure P16.23 shows a system that is compressed from an initial state i to a final state f. If the internal energy of the system decreases by 5000 J, approximately how much heat was added to the system?

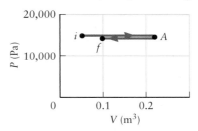

Figure P16.23

24. ✪ An ideal monatomic gas expands adiabatically from an initial temperature of 400 K and volume of 4.5 m³ to a final volume of 12 m³. If the initial pressure is $1.5 \times P_{atm}$, how much work is done on the gas?

25. ✪ You have 2.0 moles of helium gas initially at a temperature of 340 K and a pressure of $0.20 \times P_{atm}$. The gas is compressed isothermally to a pressure $0.50 \times P_{atm}$. (a) Find the final volume of the gas. Assume it behaves as an ideal gas. (b) Find the work done by the gas. (c) How much heat flows into the gas?

16.6 HEAT ENGINES AND OTHER THERMODYNAMIC DEVICES

26. A reversible heat engine operates between thermal reservoirs at 900 K and 200 K. If 9000 J are extracted from the hot reservoir, how much work is done by the heat engine?

27. An engine absorbs 3500 J of heat energy each cycle and expels 1200 J. (a) How much work does it do in one cycle? (b) If it is a reversible heat engine and the hot reservoir is at 750 K, what is the temperature of the cold reservoir?

28. ⭐ A gasoline engine absorbs 4000 J of heat and performs 3000 J of mechanical work for each turn of the engine shaft. How much heat is expelled from the engine after 500 turns of the shaft?

29. ✪ Consider a reversible heat engine that operates between a hot reservoir at 550 K and a cold reservoir at 150 K. If this engine has a power rating of 750 W, how much heat is expelled to the cold reservoir in 10 min?

30. A reversible heat engine operates between a hot reservoir at 580 K and a cold reservoir of unknown temperature. After 5 min, the engine has absorbed 1000 J of energy from the hot reservoir and expelled 500 J of energy to the cold reservoir. What is the temperature of the cold reservoir?

31. [SSM] [★] Assume the gasoline engine in your car is a heat engine operating between a hot reservoir at 800 K and a cold reservoir at 280 K and that your engine produces a peak power output of 250 hp (horsepower). If the temperature of the hot reservoir is increased to 900 K, what is the theoretical peak power output?

32. [★] Consider two heat engines: Engine 1 operates between thermal reservoirs at 450 K and 170 K, and engine 2 operates between reservoirs at 50 K and 10 K. Which engine has a greater maximum theoretical efficiency?

33. [★] A reversible heat engine consumes five times more heat Q_H from its hot reservoir than it expels to its cold reservoir, Q_C. What is the ratio T_H/T_C?

34. [★] Consider a heat engine that operates between thermal reservoirs at 500 K and 250 K. Suppose the engine does 5000 J of work while it expels 1000 J of energy to the cold reservoir. Is this engine allowed by the second law of thermodynamics?

35. [SSM] [★] [R] Suppose the refrigerator in your house is a reversible refrigerator. Estimate how much heat is extracted from the fresh-food compartment when 1000 J of work are done. *Hint:* You will first need to identify the hot and cold reservoirs and estimate their temperatures. Ignore the freezer compartment.

36. [✪] [R] It is a cold winter day in the author's neighborhood. One of the houses is heated with a reversible heat pump that uses the outside air as the cold reservoir. Suppose 40,000 J of energy are lost from the house due to heat conduction through the walls, windows, and one broken window. Estimate how much energy must be used by the heat pump (as input through mechanical work) to compensate for this loss of energy.

37. [★] A reversible heat pump operating between thermal reservoirs at 25°C and 10°C consumes 500 J of energy each second. How much heat energy is "pumped" into the house each second?

38. [✪] A heat pump operates between reservoirs at 293 K (inside the house) and 265 K (outside). It is not a reversible heat pump; it only pumps 75% of the maximum possible heat into the house. How much work must be done to "pump" 6000 J of heat energy into the house?

39. A reversible heat engine operates between thermal reservoirs at 600 K and 150 K. What is the efficiency of this engine?

40. Suppose the heat engine in Problem 39 is operated "in reverse" as a refrigerator. What is its refrigeration efficiency?

41. [✪] A Carnot engine extracts 2000 J of heat from a hot reservoir and does 1200 J of mechanical work. Suppose the cold reservoir is a very large block of ice at 0°C. What mass of ice would melt during this time?

42. A heat engine operates using the *P–V* cycle shown in Figure P16.42. Approximately how much work is done by the system during each cycle?

43. [★] For the heat engine in Figure P16.42, during which portion(s) of the cycle does the engine do positive work on its

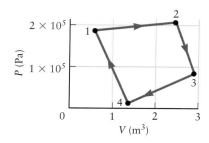

Figure P16.42 Problems 42 and 43.

environment and during which portion(s) does the environment do work on the system?

44. [SSM] [★] A heat engine absorbs 8000 J of heat from a hot reservoir and produces 5500 J of work. What is the efficiency of the engine?

45. If the heat engine in Problem 44 is reversible and the hot reservoir is at a temperature of 600 K, (a) what is the temperature of the cold reservoir? (b) How much heat is expelled to the cold reservoir?

16.7 ENTROPY

46. [★] Consider a reversible heat engine that employs a hot reservoir at a temperature of 750 K and a cold reservoir at 230 K. (a) What is the entropy change of the hot reservoir during a period in which 5000 J are extracted from the hot reservoir? (b) What is the change in the entropy of the cold reservoir? (c) Find the change in the entropy of the engine itself during this time.

47. [SSM] [★] Suppose 7000 J of heat flow out of a house that is at 293 K to the outside environment at 273 K. Find the change in entropy of the house, the change in the entropy of the environment, and the total change in entropy of the universe. Is this process reversible or irreversible?

48. [✪] Suppose 5000 J of heat flow from a reservoir at 400 K to a colder reservoir. If the total entropy change of the universe is +200 J/K, what is the temperature of the cold reservoir? *Note:* Because the reservoirs are assumed to be very large, their temperatures do not change during this process.

49. [★] A large piece of steel (initial temperature 400 K) is placed into a large bucket of ice water. If 4.0 J of heat flow from the steel into the ice water, what are the entropy changes of the steel, the ice water, and the universe? Assume this small amount of heat flow does not change the temperatures of the steel or the ice water by any significant amount.

50. [✪] A pot contains 250 g of water at 100°C. If this water is heated and all evaporates to form steam at 100°C, what is the change in the entropy of the H_2O?

51. [✪] A 1.0-kg block of ice is initially at 0°C. Heat is added to this system, and the entropy is increased by 300 J/K. How much of the ice melts?

52. [★] [R] The Sun emits heat via radiation. Estimate ΔS for the Sun during one day. The surface of the Sun is approximately 6000 K.

---

53. [★] Consider the following systems and how closely each approximates a reversible process. Rank them from approximately reversible to irreversible with greatest change in entropy. *Hint:* Consider a video of each in reverse.
(a) A child swinging on a swing
(b) Mars orbiting the Sun

(c) Making a milkshake in a blender
(d) A yo-yo climbing up its string

54. [SSM] [★] Figure P16.54 shows the *P–V* diagrams for a variety of different processes involving an ideal gas. (a) Which diagrams describe an expansion? (b) Which diagrams describe a compression? (c) Which diagrams might describe an adiabatic process?

(d) Which diagrams might describe an isothermal process?
(e) During which processes does the internal energy increase?

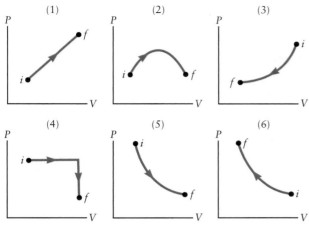

Figure P16.54

55. ⭐ Figure P16.55 shows several cyclic processes involving an ideal gas.
(a) Which processes might describe a heat engine?
(b) Which processes might describe a refrigerator?
(c) Which process might be a Carnot cycle?

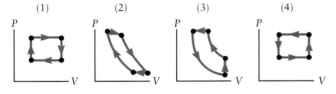

Figure P16.55

56. ⭐ (a) Calculate the efficiency of a Carnot engine running between a low temperature of 27.0°C and a high temperature of 387°C. Determine the change in efficiency if you were to (b) raise the high temperature by 10.0°C or (c) lower the low temperature by 10.0°C. Which improves the efficiency more? *Note:* Consider three significant figures in your calculations for this problem.

57. ✪ **One engine versus two.** A Carnot engine produces 1000 J of work when running between reservoirs of 470°C and 100°C. (a) How much heat flows from the hot reservoir? *Hint:* Consider the efficiency. (b) Find the heat flow to the cold reservoir. (c) Consider a second engine that uses the heat flow out of the first engine as the hot reservoir and has a cold reservoir at room temperature, 20°C. How much work can the second engine produce? (d) Finally, consider a third engine that runs between the hot reservoir of the first engine and the cold reservoir of the second engine. Calculate the work done by this engine and compare it with the sum of the work of the first two engines.

58. ✪ **The drinking bird.** A popular novelty bird toy consists of two semievacuated glass chambers (the head and the body), which are connected by a glass tube as shown in Figure P16.58 (left). The head is covered with an absorbent coating, and the bottom chamber is partially filled with a liquid that has a boiling point near room temperature. The bird goes through a cycle illustrated in Figure P16.58 (right). The bird starts with all the liquid in the bottom chamber and is suspended at an off-center pivot point so that it leans slightly forward. Next the fluid rises slowly up the tube, which is open at the bottom but submerged in the fluid. The fluid reaches the top chamber and the bird tips forward. Finally, at full tilt, the bottom of the tube breaks the surface and the liquid pours back into the bottom chamber. The

bird then returns to the upright position, and the cycle repeats. (a) Is work done by this cyclic process? Is the bird a heat engine? Describe. (b) The bird dips its head in the water during the cycle. Why is that important for the toy's operation?

Figure P16.58

59. ✪ Figure P16.59 shows the P–V diagrams for idealizations of three important real engines: the Brayton (gas turbine) engine, the Otto (four-stroke) automobile engine, and the diesel engine. For each of these cycles, there is no heat flow for the processes along the curves $2 \rightarrow 3$ and $4 \rightarrow 1$. (a) Using the diagram for each engine, describe the sequence and name each thermodynamic process that constitutes the cycle, starting at point 1. (b) Identify the points of highest and lowest temperature for each engine.

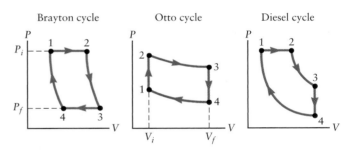

Figure P16.59 The Brayton cycle, Otto cycle, and diesel cycle for Problems 59 through 62.

60. ✪ The Brayton cycle shown in Figure P16.59 is the idealized cycle for a gas turbine or jet engine. Consider an engine using the Brayton cycle that operates on 0.070 mol of ideal gas between the pressures $P_i = P_1 = P_2 = 4.9$ atm and $P_f = P_3 = P_4 = 3.5$ atm with volumes $V_1 = 0.50$ L, $V_2 = 1.00$ L, $V_3 = 1.22$ L, and $V_4 = 0.61$ L. *Note:* Retain three significant figures in all calculations. (a) Determine the work done for the displacements $1 \rightarrow 2$ and $3 \rightarrow 4$. (b) Determine the temperature of the ideal gas at each point. (c) Calculate the work done for the adiabatic expansion $2 \rightarrow 3$ and contraction $4 \rightarrow 1$. *Hint:* Consider the change in internal energy. (d) What is the total work done by one cycle? (e) Determine the amount of heat flowing in, $Q_H = Q_{1 \rightarrow 2}$ during the isobaric expansion and the heat flowing out of the gas during the isobaric contraction, $Q_C = Q_{3 \rightarrow 4}$. *Hint:* The specific heat for an ideal gas under constant pressure is given by $C_P = C_V + nR$. (f) Determine the efficiency for this Brayton engine (Eq. 16.18). *Note:* The heat loss and friction in a real Brayton engine will reduce this ideal efficiency. (e) Calculate and compare to the efficiency of a Carnot engine running between the same high and low temperatures.

61. ✪ The Otto cycle shown in Figure P16.59 is the idealized cycle for an internal combustion automobile engine. Consider an engine using the Otto cycle that operates on 0.070 mol of ideal gas between the volumes $V_i = V_1 = V_2 = 0.2$ L and $V_f = V_3 = V_4 = 1.6$ L. (a) In which parts of the cycle ($1 \rightarrow 2$, $2 \rightarrow 3$, $3 \rightarrow 4$, $4 \rightarrow 1$) is work done on or by the system? In which parts does

heat flow into or out of the system? (b) The lowest temperature is 323 K at point 4, where the exhaust valve is open and the pressure is 1.0 atm. Find the temperature at point 1. *Hint:* Use Equation 16.10 to first find the pressure of the ideal gas. Assume that $\gamma = \frac{5}{3}$. (c) The combustion of fuel in stroke $1 \rightarrow 2$ supplies 1000 J of energy to the gas. Calculate the change in temperature of the gas and find the temperature at point 2. (d) What is the pressure at point 2? (e) Find the pressure and temperature at point 3, the end of the "power stroke." (f) Determine the heat loss during stroke $3 \rightarrow 4$. (g) Determine the efficiency for this Otto engine (Eq. 16.18). *Note:* The heat loss and friction in a real Otto engine will reduce this ideal efficiency. (h) Calculate and compare to the efficiency of a Carnot engine running between the same high and low temperatures.

62. ✪ **Compression ratio and engine efficiency.** The Otto cycle shown in Figure P16.59 is the idealized cycle for an internal combustion automobile engine. Corresponding to positions of the piston, the cycle is in four parts (two isochoric and two adiabatic). (a) Show that Equation 16.19 describing the efficiency can be expressed as

$$e = 1 - \left(\frac{T_3 - T_4}{T_2 - T_1} \right)$$

for the Otto cycle, where the temperatures are taken from the four points on the cycle. (b) If an ideal gas is used as the working fluid, it turns out that the efficiency for the Otto cycle is independent of the temperature and depends only on the ratio V_f / V_i, called the *compression ratio*. Show that the efficiency can also be written in terms of the initial and final volumes as

$$e = 1 - \left(\frac{V_i}{V_f} \right)^{\gamma - 1}$$

where $\gamma = \frac{5}{3}$ for an ideal gas. *Hint:* Use the ideal gas law and Equation 16.10 to write each temperature in terms of volume and pressure.

63. ☒ A cylinder and piston arrangement is made of an excellent conductor in contact with a thermal reservoir at 38°C and is filled with an ideal gas. The gas is slowly compressed in an isothermal process in which 80 J of work are done on the gas. (a) What is the change in entropy of the gas? (b) What is the change in entropy of the thermal reservoir? (c) What is the net change in entropy of the universe?

64. ✪ Ⓡ **Efficiency of heat pumps.** Consider two different heat pumps used to heat a house. One uses the outside air as the cold reservoir, and the other uses water from underground. (a) Compare their efficiencies when the outside air is at 10°C and at −10°C. Assume both are reversible heat pumps. (b) Now consider just the heat pump that uses the outside air and suppose it is not an ideal heat pump but instead loses 50% of its input energy (the work done on the heat pump) to friction. At what outdoor temperature does it no longer pay to use this heat pump compared to using electric resistive heating?

65. ✪ The efficiency of a reversible heat pump is given by Equation 16.28. Real heat pumps, however, may achieve an efficiency of only about 20% of this ideal limit. A heat pump with this reduced efficiency is used to heat a house for which $T_H = 300$ K. What is the lowest value of the temperature of the cold reservoir for which it makes sense to use this heat pump instead of just using electric resistive heating, which converts the input energy W directly to the same amount of heat energy Q_H?

Electric Forces and Fields

In this and the next two chapters, we focus on electric phenomena before moving on to describe magnetism in Chapter 20. Electric and magnetic forces both act on electric charges and currents. Electric charges and currents also act as *sources* of electric and magnetic fields. In fact, electricity and magnetism are really a unified phenomenon, often referred to as ***electromagnetism***, and this unified picture leads to a theory of electromagnetic radiation. Light is a type of electromagnetic radiation, so electromagnetism also forms the basis for the theory of ***optics***, which is discussed in Chapters 24 through 26. Our study of electricity and magnetism will thus take us quite a long way!

Benjamin Franklin's famous experiments with kites in thunderstorms showed that lightning involves the motion of electric charge. This motion and the associated electric and magnetic phenomena are the subject of Chapters 17 through 23. (© Gandee Vasan/Stone/Getty Images)

17.1 | EVIDENCE FOR ELECTRIC FORCES: THE OBSERVATIONAL FACTS

The discovery of electricity is generally credited to the Greeks and is thought to have occurred around 2500 years ago, which is approximately the era in which Aristotle lived. The Greeks observed electric charges and the forces between them in a variety of situations. Many of their observations made use of a material called amber, a plastic-like substance formed by allowing the sap from certain trees to dry and harden (Fig. 17.1A). The Greeks found that after amber is rubbed with a piece of animal fur, the amber can attract small pieces of dust. The same effect can be demonstrated with a piece of plastic and small bits of paper, although the Greeks did not have these materials to work with! This experiment (parts B and C of Fig. 17.1) is quite striking because neither the fur nor the plastic rod ordinarily attract paper. Somehow, the process of rubbing the two together "creates" an attractive force. Moreover, this force occurs even when the amber and the paper are *not* in contact, and there would be an attractive force even if the plastic rod and the paper were placed in a vacuum (although the Greeks did not try that).

The Greek word for amber is "elektron"; this is the origin of the terms *electricity* and *electron*. The Greeks did not discover the electron; even though the notion of a small charge-carrying particle was hypothesized by a number of early philosophers and scientists, there was no clear experimental demonstration of the existence of the electron until the late 1880s. The long history of scientific work on electricity is very interesting, but we will skip most of the details and simply state a few key results.

Basic properties of electric charge

- *There are two types of electric charge*, called **positive** and **negative**. The sub-atomic particle called a **proton** has a positive charge, and an **electron** has a negative charge.
- *Charge comes in quantized units. All* protons carry the *same* amount of charge $+e$, and *all* electrons carry a charge $-e$. We will discuss how charge is measured and the unit of electric charge below.
- *Like charges repel each other, unlike charges attract.* The electric force between two objects is repulsive if the objects carry "like" charge, that is, if both are positively charged or both are negatively charged (Fig. 17.2). The electric force is attractive if the two objects carry "unlike" charge. Here the terms *like* and *unlike* refer to the *signs* of the charges, not their magnitudes. So, the expression "like charges" means that the two charges are both positive or both negative. The expression "unlike charges" means that one charge is positive and the other is negative.
- *Charge is conserved.* The *total* charge on an object is the sum of all the individual charges (protons and electrons) carried by the object. The total charge can

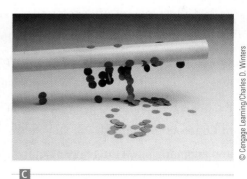

Figure 17.1 **A** Amber is the dried, hardened sap from certain trees. Small creatures or bits of plant material are often preserved inside. **B** When a plastic or amber rod is rubbed with fur, the rod acquires an electric charge. **C** A charged rod attracts small bits of paper and other objects.

be positive, negative, or zero. Charge can move from place to place, and from one object to another, but the total charge of the universe does not change.

What Is Electric Charge?

You are likely familiar with the term *electric charge* or simply *charge*. Charge is a fundamental property of matter: the amount of charge that is "on" or "carried by" a particle determines how the particle reacts to electric and magnetic fields. In many ways, it is similar to the quantity we call *mass*; the mass of a particle determines how it reacts when a force acts on it. The mass of a particle is a measure of the amount of matter it carries, whereas the charge of a particle is a measure of the amount of "electric-ness" it carries. Charge and mass are both primary entries in our dictionary of physics terms, but it is not possible to give definitions for these fundamental entries in our dictionary.

In the SI system of units, charge is measured in **coulombs** (C) in honor of French physicist Charles de Coulomb (1736–1806). The charge on a single electron is[1]

$$\text{electron charge} = -e = -1.60 \times 10^{-19}\ \text{C} \tag{17.1}$$

and the charge carried by a single proton is

$$\text{proton charge} = +e = +1.60 \times 10^{-19}\ \text{C} \tag{17.2}$$

The symbol e is usually used to denote the *magnitude* of the charge on an electron or a proton. We will always take e to be a *positive* quantity ($e = +1.60 \times 10^{-19}$ C). We use the symbols q and Q to denote charge in general, such as the total charge on a bit of paper. When discussing a charged particle, it is common to say that the particle "has" a charge q or that it "carries" a charge q. These two expressions simply mean that the total charge of the particle is q.

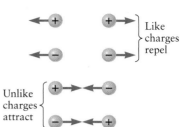

Figure 17.2 The electric force between two like charges (charges with the same sign) is repulsive, whereas the force between unlike charges (those of opposite sign) is attractive.

17.2 | ELECTRIC FORCES AND COULOMB'S LAW

Imagine that you travel to a distant star where the laws of physics seem different from the ones that describe physics on Earth. While on this star, you discover a new type of force acting between particles; the force can be attractive or repulsive, and it is *extremely* large. In fact, the force between two objects each having a mass of only 1 g is large enough to hold an entire planet in orbit around this alien star! This force sounds very different from anything you have encountered before, but it *already exists* on Earth. It is the *electric force* between two charged objects.

Consider two charged objects that are so tiny that they can be modeled as point particles. If the charges carried by the two objects are q_1 and q_2 and they are separated by a distance r (Fig. 17.3), the electric force between the objects has a magnitude

$$F = k \frac{q_1 q_2}{r^2} \tag{17.3}$$

Equation 17.3 is called **Coulomb's law**. The constant k has the value

$$k = 8.99 \times 10^9\ \text{N} \cdot \text{m}^2/\text{C}^2 \tag{17.4}$$

Do not confuse the constant k in Equation 17.4 with the spring constant in Hooke's law. (Most physicists use the same symbol for both.) The direction of the electric force on each of the charges is along the line that connects the two charges and is illustrated in Figure 17.3 for the case of two like charges. As already mentioned, this force is repulsive for like charges, corresponding to a positive value of F in

Like charges q_1 and q_2 repel.

COULOMB'S LAW:
$$|\vec{F}| = \frac{k q_1 q_2}{r^2} = \frac{q_1 q_2}{4 \pi \varepsilon_0 r^2}$$

Figure 17.3 The electric force between two point charges q_1 and q_2 is given by Coulomb's law.

---

[1]Here we give the value of e to just three significant figures, but it is known to much greater precision than that. The latest experiments give $e = 1.60217653 \times 10^{-19}$ C.

Equation 17.3, whereas the force is attractive for unlike charges and F in Equation 17.3 is negative in that case.

Strictly speaking, the value of F in Equation 17.3 applies only for two point charges, but it is a good approximation whenever the sizes of the particles are much smaller than their separation r. Later in this chapter, we'll see how to build on Equation 17.3 to find the electric force in other situations, including nonpoint particles.

The mathematical form of Equation 17.3 is very similar to Newton's law of gravitation, with the constant k playing a role analogous to the gravitational constant G. Another way to write Coulomb's law is

$$F = \frac{q_1 q_2}{4\pi\varepsilon_0 r^2} \tag{17.5}$$

where ε_0 is yet another physical constant called the **permittivity of free space**, having the value

$$\varepsilon_0 = 8.85 \times 10^{-12} \, \frac{C^2}{N \cdot m^2} \tag{17.6}$$

The values of ε_0 and k are related by

$$\frac{1}{4\pi\varepsilon_0} = k$$

so these two forms of Coulomb's law, Equations 17.3 and 17.5, are completely equivalent. Why do we bother to write Coulomb's law in the slightly more complicated form in Equation 17.5? We do so because other equations of electromagnetism have simpler forms when written in terms of ε_0 instead of k. For calculating the electric force between charges, either form of Coulomb's law (Eq. 17.3 or 17.5) can be used.

Features of Coulomb's Law

Coulomb's law has several important properties.

1. We have already seen that the electric force is repulsive for like charges and attractive for unlike charges (Fig. 17.2). Mathematically, this property results from the product $q_1 q_2$ in the numerator in Equations 17.3 and 17.5. The factor $q_1 q_2$ is positive for like charges, so F is positive and the force tends to push the charges farther apart. For unlike charges the product $q_1 q_2$ is negative and the value of F in Equations 17.3 and 17.5 is also negative, and the particles are attracted to each other.
2. We have already noted that the form of Coulomb's law is very similar to Newton's universal law of gravitation. Both laws exhibit a $1/r^2$ dependence on the separation of the two particles. Therefore, a negative charge can move in a circular orbit around a positive charge, just like a planet orbiting the Sun, and that was an early model for the hydrogen atom. There is one very important difference, however: gravity is always an attractive force, whereas the electric force in Coulomb's law can be either attractive or repulsive.
3. The magnitude of F in Equations 17.3 and 17.5 is the magnitude of the force exerted on *each* of the particles. That is, a force of magnitude F is exerted on charge q_1, and a force of equal magnitude and opposite direction is exerted on q_2. We should expect such a pair of forces, based on Newton's third law, the action–reaction principle.

Using Equation 17.3, let's now show that electric forces can be extremely large. Suppose you are given a box holding 1 g (0.001 kg) of pure electrons; that is, the box contains *only* electrons. The mass of a single electron is $m_e = 9.11 \times 10^{-31}$ kg, so the total number of electrons N_e in the box is

$$N_e = \frac{1 \times 10^{-3} \, \text{kg}}{9.11 \times 10^{-31} \, \text{kg/electron}} = 1.1 \times 10^{27}$$

Each of these electrons carries a charge $-e$, so the total charge is

$$Q_{total} = N_e(-e) = (1.1 \times 10^{27})(-1.60 \times 10^{-19} \text{ C}) = -1.8 \times 10^8 \text{ C}$$

Now suppose there are two of these boxes, each with charge Q_{total}, separated by a distance $r = 1$ m; for simplicity, we assume each box is small so that it can be modeled as a point particle. The magnitude of the total electric force is (Eq. 17.3)

$$F = \frac{kQ_{total}Q_{total}}{r^2} = \frac{(8.99 \times 10^9 \text{ N} \cdot \text{m}^2/\text{C}^2)(-1.8 \times 10^8 \text{ C})(-1.8 \times 10^8 \text{ C})}{(1 \text{ m})^2}$$

$$F = 3 \times 10^{26} \text{ N} \tag{17.7}$$

which is an *enormous* force. It is nearly a million times larger than the gravitational force exerted between the Sun and the Earth. All that from just two small containers of electrons!

The value found for F in Equation 17.7 is so large that you may be a little skeptical: if the electric force between two small pieces of matter is this large, there must be many staggering consequences. For example, the electric forces acting within ordinary matter, which contains both electrons and protons, must be huge. Why don't these forces dominate everyday life? The resolution of this apparent paradox is that it is essentially impossible to obtain a box containing only electrons. You probably know that a neutral atom contains equal numbers of electrons and protons. In fact, the term *neutral* means that the total charge is zero. If our two pointlike boxes had contained equal numbers of electrons and protons, their *total* charges would have been $Q_{total} = 0$ and the force in Equation 17.7 would be zero.

Ordinary matter always consists of equal, or nearly equal, numbers of electrons and protons. The total charge is then zero or very close to zero. At the atomic and molecular scale, however, it is common to have the positive and negative charges (nuclei and electrons) separated by a small distance. In this case, the electric force is not zero, and these electric forces are responsible for holding matter together. That is why solids can be quite strong.

EXAMPLE 17.1 | **Using Coulomb's Law to Clean the Air**

Coal-burning power plants produce large amounts of potential pollution in the form of small particles (soot). Modern smokestacks use devices called scrubbers to remove these particles from the smoke they emit. Scrubbers use a two-step process: electrons are first added to the soot particle, and an electric force then pulls the particle out of the smoke stream (Fig. 17.4). In this example, we analyze the force on a soot particle after electrons are added.

Consider a soot particle of mass $m_{soot} = 1.0$ nanogram ($= 1.0 \times 10^{-9}$ g $= 1.0 \times 10^{-12}$ kg, which corresponds to a diameter of a few microns) that has a charge q_{soot}. That is, some number of electrons have been added to give the particle this total charge. Suppose the collector has a total charge $q_{collector} = 1.0 \times 10^{-6}$ C and is small enough to be treated as a point charge at the rim of the smokestack. (a) If the separation between the collector and the soot particle is $r = 0.10$ m, what is the value of q_{soot} so that the electric force exerted on the particle is equal to its weight? (b) How many electrons must be added to the soot particle?

RECOGNIZE THE PRINCIPLE

The collector charge is assumed to be a point particle and the soot particle is also very small, so we can apply Coulomb's law for two point particles (Eq. 17.3). The electric force between the collector (charge $q_{collector}$) and the soot particle (charge q_{soot}) has a magnitude

$$F_{elec} = \frac{kq_{collector}q_{soot}}{r^2} \tag{1}$$

Figure 17.4 Example 17.1. A dust particle in a smokestack is attracted to the electric charge on the collector of a smokestack scrubber.

SKETCH THE PROBLEM

Figure 17.4 shows the soot particle and the collector charge. They are separated by a distance r, just like the charges in Figure 17.3.

IDENTIFY THE RELATIONSHIPS

We want the electric force on the particle in Equation (1) to be equal in magnitude to the particle's weight, so

$$F_{\text{elec}} = \frac{kq_{\text{collector}}q_{\text{soot}}}{r^2} = mg$$

SOLVE

(a) Solving for the charge q_{soot} leads to

$$q_{\text{soot}} = \frac{mgr^2}{kq_{\text{collector}}}$$

Inserting the given values of m and r, we find

$$q_{\text{soot}} = \frac{mgr^2}{kq_{\text{collector}}} = \frac{(1.0 \times 10^{-12}\ \text{kg})(9.8\ \text{m/s}^2)(0.10\ \text{m})^2}{(8.99 \times 10^9\ \text{N} \cdot \text{m}^2/\text{C}^2)(1.0 \times 10^{-6}\ \text{C})}$$

$$q_{\text{soot}} = \boxed{1.1 \times 10^{-17}\ \text{C}}$$

(b) The charge on a single electron has a magnitude $e = 1.60 \times 10^{-19}$ C (Eq. 17.1), so our value of q_{soot} corresponds to

$$N = \frac{q_{\text{soot}}}{e} = \frac{1.1 \times 10^{-17}\ \text{C}}{1.60 \times 10^{-19}\ \text{C/electron}} = \boxed{69\ \text{electrons}}$$

which is the number of extra electrons added to the soot particle.

What have we learned?
It is amazing that only this small number of electrons is needed for the scrubber to successfully remove a soot particle from air. The total number of atoms in the particle is about 4×10^{14} (see Question 13 at the end of this chapter), so the fraction of "unbalanced" electrons on the particle is only $69/(4 \times 10^{14}) \approx 2 \times 10^{-13}$!

Figure 17.5 The total force on q_2 is equal to the sum of the forces from charges q_1 and $2q_1$. This is the principle of superposition at work.

Electric forces and the principle of superposition

Superposition of Electric Forces

So far, we have discussed the electric force between two point particles; let's now consider how to use Coulomb's law to deal with more complicated charge distributions. Suppose two particles of charge q_1 and $2q_1$ are separated by a distance L as shown in Figure 17.5. What is the force on a third charge q_2 placed midway between these two charges?

We can deal with this problem by first using Coulomb's law to find the force exerted by charge q_1 on q_2 and then using Coulomb's law a second time to calculate the force exerted by the charge $2q_1$ on q_2. The total force on q_2 is the sum of these two separate contributions, which is an example of the *principle of superposition*. That is, the total force on q_2 is just the sum (the superposition) of the individual forces exerted on q_2 by all the other charges in the problem. Because force is a vector, we must be careful to add these forces as vectors.

From Figure 17.5, the separation between q_2 and the charge q_1 is $L/2$. Writing Coulomb's law for this pair of charges, we have

$$F_1 = \frac{kq_1q_2}{(L/2)^2} = \frac{4kq_1q_2}{L^2} \tag{17.8}$$

Here a positive value corresponds to a repulsive force (since like charges repel), and in the coordinate system in Figure 17.5 it corresponds to a force on q_2 in the $+x$ direction. Hence, F_1 in Equation 17.8 is the component of the force along the x axis, and the component along y is zero.

We can deal with the force from the charge $2q_1$ in a similar way. The separation of the charges is again $L/2$, so, applying Coulomb's law (Eq. 17.5), we find

$$F_2 = \frac{k(2q_1)q_2}{(L/2)^2} = \frac{8kq_1q_2}{L^2} \tag{17.9}$$

From the geometry of Figure 17.5 we can see that this result is the x component of the force on q_2 because the electric force acts along the line that connects the two charges and here that line is along the x direction. For this reason, the y component of the force on q_2 is again zero. Equation 17.9 gives the magnitude of the force F_2. If they are "like" charges (i.e., if q_1 and q_2 are both positive or both negative so that the product q_1q_2 is positive), the force F_2 is in the $-x$ direction.

The total force on q_2 is the sum of Equations 17.8 and 17.9, but we must account for the direction (the sign) of each. Assuming all charges are of "like sign," the result is

$$F_{\text{total}} = +\frac{4kq_1q_2}{L^2} - \frac{8kq_1q_2}{L^2} = -\frac{4kq_1q_2}{L^2}$$

The negative sign here tells us that if the product $q_1q_2 > 0$ (consistent with our "like sign" assumption), the force exerted on q_2 is along the $-x$ direction. This example leads to the following strategy.

PROBLEM SOLVING | Calculating Forces with Coulomb's Law

1. **RECOGNIZE THE PRINCIPLE.** The electric force on a charged particle can be found using Coulomb's law together with the principle of superposition.

2. **SKETCH THE PROBLEM.** Construct a drawing (including a coordinate system) and show the location and charge for each object in the problem. Your drawing should also show the directions of all the electric forces—$\vec{F}_1$, $\vec{F}_2$, and so forth—on the particle(s) of interest.

3. **IDENTIFY THE RELATIONSHIPS.** Use Coulomb's law (Eq. 17.3 or 17.5) to find the magnitudes of the forces $\vec{F}_1$, $\vec{F}_2$, ... acting on the particle(s) of interest.

4. **SOLVE.** The total force on a particle is the sum (the superposition) of all the individual forces from steps 2 and 3. Add these forces *as vectors* to get the total force. When adding these vectors, it is usually simplest to work in terms of the components of $\vec{F}_1$, $\vec{F}_2$, ... along the coordinate axes.

5. Always *consider what your answer means* and check that it makes sense.

CONCEPT CHECK 17.1 | Superposition and the Direction of the Electric Force

An electron and a proton are each placed on the x axis, a distance L from the origin as shown in Figure 17.6. Another electron (electron 2 in Fig. 17.6) is placed to the right of the proton as shown. Is the *direction* of the electric force on electron 2 (a) in the $+x$ direction or (b) in the $-x$ direction, or (c) is the force zero?

Figure 17.6 Concept Check 17.1.

EXAMPLE 17.2 | An Electric Dipole

Consider two point particles of charge $+q$ and $-q$ (with q positive) arranged as shown in Figure 17.7A. This arrangement of charge, called an *electric dipole*, is found in many molecules, so it is important in chemistry as well as physics. Suppose a third charge Q is placed on the x axis as shown in Figure 17.7A. Assuming Q is positive, what are the magnitude and the direction of the total electric force on Q?

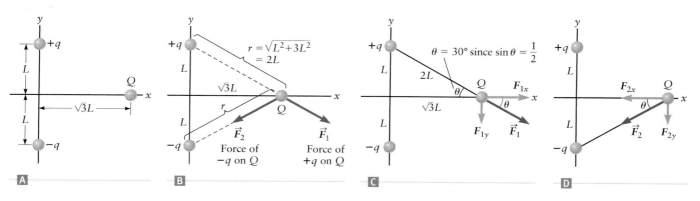

Figure 17.7 Example 17.2. Calculation of the force exerted by a dipole (charges $+q$ and $-q$) on a third charge Q.

RECOGNIZE THE PRINCIPLE

Our approach is based on Coulomb's law and the principle of superposition, following the general approach outlined in the "Calculating Forces with Coulomb's Law" problem-solving strategy.

SKETCH THE PROBLEM

We begin with the sketch in Figure 17.7A. This picture includes the coordinate axes and the positions of all the charges in the problem. We next draw in vectors $\vec{F}_1$ and $\vec{F}_2$ showing the forces on charge Q in Figure 17.7B. The force of charge $+q$ on charge Q is represented by $\vec{F}_1$. These charges are both positive, so the force is repulsive (away from $+q$) and $\vec{F}_1$ is directed to the lower right of the figure. The force exerted by $-q$ on Q is represented by $\vec{F}_2$. They are unlike charges, so this force is attractive (toward $-q$).

IDENTIFY THE RELATIONSHIPS

We apply Coulomb's law (Eq. 17.3) to find the magnitudes of $\vec{F}_1$ and $\vec{F}_2$. Using the Pythagorean theorem in Figure 17.7B, the distance from $+q$ to Q is $r = \sqrt{L^2 + 3L^2} = 2L$. Inserting into Equation 17.3 gives the magnitude of vector $\vec{F}_1$

$$F_1 = \frac{kqQ}{r^2} = \frac{kqQ}{(2L)^2} = \frac{kqQ}{4L^2}$$

The distance between $-q$ and Q is also r, and the magnitude of $\vec{F}_2$ is

$$F_2 = \frac{kqQ}{r^2} = \frac{kqQ}{4L^2} = F_1 \tag{1}$$

SOLVE

We now add $\vec{F}_1$ and $\vec{F}_2$ *as vectors* to get the total force on Q. The components of these vectors are shown in parts C and D of Figure 17.7. For the components of $\vec{F}_1$, we have

$$F_{1x} = F_1 \cos \theta$$

and

$$F_{1y} = -F_1 \sin \theta \tag{2}$$

where the angle θ is defined in Figure 17.7C. This angle is contained in a right triangle with $\sin \theta = \frac{1}{2}$, so $\theta = 30°$. Note that F_{1y} is negative; the negative sign in Equation (2) is due to the direction of $\vec{F}_1$ in parts B and C of Figure 17.7. We next write the components of $\vec{F}_2$. From Figure 17.7D, we get

$$F_{2x} = -F_2 \cos \theta$$

and

$$F_{2y} = -F_2 \sin \theta$$

We now sum the components of $\vec{F}_1$ and $\vec{F}_2$ to get the total force. For the x components, we find

$$F_{\text{total}, x} = F_{1x} + F_{2x} = F_1 \cos \theta - F_2 \cos \theta$$

Since $F_1 = F_2$ by Equation (1), we have

$$F_{\text{total}, x} = \boxed{0}$$

For the y components,

$$F_{\text{total}, y} = F_{1y} + F_{2y} = -F_1 \sin \theta - F_2 \sin \theta$$

$$F_{\text{total}, y} = -2F_1 \sin \theta = -\frac{kqQ}{2L^2} \sin \theta$$

We already found $\theta = 30°$, so

$$\sin \theta = \sin(30°) = \tfrac{1}{2}$$

Our final result is thus

$$F_{\text{total}, y} = -\frac{kqQ}{2L^2} \sin \theta = \boxed{-\frac{kqQ}{4L^2}}$$

What does it mean?
When Q is at this particular location, the force from the dipole on a positive charge Q is along the $-y$ direction. At other locations (e.g., away from the x and y axes), the force will have nonzero components along both x and y. Had the charge Q instead been $-Q$ (negative), the force would have the same magnitude as found here, but directed along $+y$.

CONCEPT CHECK 17.2 | Force on a Point Charge

Two particles of charge q and $-2q$ are located as shown in Figure 17.8. A third charge Q is placed on the x axis and it is found that the total electric force on Q is zero. In which region in Figure 17.8 is Q located?

(a) region A (b) region B (c) region C

Figure 17.8 Concept Check 17.2.

CONCEPT CHECK 17.3 | Holding Together Two Charges

Two particles, both with charge $q = +1$ C, are located 1 m apart as sketched in Figure 17.9. If q_1 is held fixed in place, what is the magnitude of the force that must be exerted on q_2 to keep it from moving?

(a) 9×10^9 N   (b) 1 N   (c) zero   (d) 9×10^{18} N

Figure 17.9 Concept Check 17.3.

17.3 | THE ELECTRIC FIELD

Coulomb's law gives the electric force between a pair of charges, but there is another way to describe electric forces. Imagine a single isolated point charge, that is, a point charge very far from any other charges as in Figure 17.10A. The presence of this charge produces an *electric field*, represented by the arrows in Figure 17.10A. This electric field is similar to the gravitational field near an isolated mass discussed in Chapter 5. A positive point charge produces field lines that emanate radially outward from the charge as sketched in Figure 17.10A. For a negative charge, the field lines are directed inward, toward the charge (Fig. 17.10B).

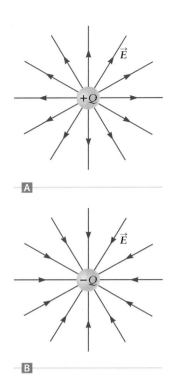

Figure 17.10 Electric field lines near a point charge placed at the origin. **A** If the charge is positive, the electric field is directed outward away from the charge, while **B** if the charge is negative, the field lines are directed inward toward the charge.

Electric field of a point charge

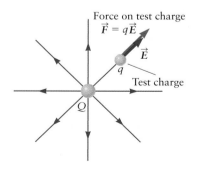

Figure 17.11 The electric field at a particular point in space is related to the electric force on a test charge q placed at that location by $\vec{F} = q\vec{E}$.

The electric field is a *vector*—it has a magnitude and a direction—and is denoted by $\vec{E}$. The associated field lines are parallel to the direction of $\vec{E}$, whereas the density of field lines is proportional to the magnitude of $\vec{E}$. In parts A and B of Figure 17.10, the field lines are most dense in the vicinity of the charges; in both cases, the magnitude of $\vec{E}$ increases as one gets closer to the charge.

Using a Test Charge to Measure $\vec{E}$

Consider a particular point in space where the electric field is $\vec{E}$ (Fig. 17.11). This field might be produced by a point charge or by some other arrangement of charges. If we now place a charge q at this location, it experiences an electric force given by

$$\vec{F} = q\vec{E} \tag{17.10}$$

The electric force is thus either parallel to $\vec{E}$ (if q is positive) or antiparallel (if q is negative). The charge q in Figure 17.11 (and Eq. 17.10) is called a ***test charge***. By measuring the force on a test charge, we can infer the magnitude and direction of the electric field at the location of the test charge. Because force is measured in units of newtons (N) and charge in coulombs (C), the electric field (according to Eq. 17.10) has units of newtons/coulombs.

The electric field $\vec{E}$ is related to electric force by Equation 17.10, so we can use a Coulomb's law calculation of the force to find the electric field in many situations. For example, let's calculate the electric field a distance r from the charge Q in Figure 17.11. Here q is the test charge, and we want to calculate the electric field produced by Q. According to Coulomb's law, the magnitude of the electric force exerted on the test charge q is

$$F = \frac{kQq}{r^2}$$

Inserting this expression into our relation for the electric field (Eq. 17.10) gives

$$F = \frac{kQq}{r^2} = qE$$

which leads to

$$E = \frac{kQ}{r^2} \tag{17.11}$$

This result is the magnitude of the electric field a distance r from a point charge Q. The direction of $\vec{E}$ is along the line that connects the charge producing the field to the point where the field is measured. The electric field is directed outward away from Q when Q is positive (as sketched in Fig. 17.11) and inward toward Q when Q is negative.

Electric Field Lines, the Inverse Square Law, and Action-at-a-Distance

According to Coulomb's law, the force between two point charges falls off as $1/r^2$, where r is the separation between the two charges. In a similar way, the electric field produced by a point charge (Eq. 17.11) also varies as $1/r^2$. The electric force thus obeys an ***inverse square law***, just as we found for the gravitational force in Chapter 5 (Fig. 5.30). Roughly speaking, a charged particle "sets up" field lines in its neighborhood, and the density of field lines is proportional to the amount of charge on the particle. These electric field lines produce a force on a nearby test charge and on any other nearby charges, and the magnitude of that force is proportional to the density of field lines at the test charge. For a point charge Q as in Figure 17.12, the density of field lines—that is, the number of field lines per unit area of space—falls as one goes away from the charge. Because the field lines emanating from Q spread out in a three-dimensional space, the number of field lines per unit area falls as $1/r^2$ as shown by the geometry in Figure 17.12.

Another interesting aspect of the electric force is the question of *action-at-a-distance*. How do two point charges that interact through Coulomb's law "know" about each other? In other words, how is the electric force transmitted from one charge to another? In terms of the field line picture, every charge generates (or carries with it) an electric field through which the electric force is transmitted.

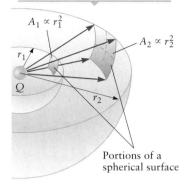

The field line area grows as r^2 because we live in a three-dimensional space.

Figure 17.12 As electric field lines emanate outward from a point charge, they intercept a larger and larger surface area. These surfaces are spherical, so their areas increase with radius as $A \propto r^2$.

EXAMPLE 17.3 | Electrons in a Television Tube

A conventional television picture tube called a cathode-ray tube (CRT) uses electrons to produce an image on its screen. A beam of electrons is produced at one end of the tube (Fig. 17.13) by a device called an "electron gun." The electrons have a very small velocity when they leave the vicinity of the gun and then move through a second region, where an electric field accelerates them to the final speed necessary for the CRT to operate properly.

The electric field in a typical TV tube has a value of 8.0×10^5 N/C. If the acceleration region has a length $L = 2.0$ cm, what is the speed of an electron when it exits this region?

RECOGNIZE THE PRINCIPLE

The electric field in the acceleration region produces a force on the electron according to Equation 17.10. If the force F and the electron's displacement L are parallel, the work done by the electric force on the electron is

$$W = FL$$

We can then use the work–energy theorem to calculate the final kinetic energy of the electron when it leaves the accelerator.

SKETCH THE PROBLEM

Figure 17.13 shows the directions of the electric field and the force. Since the charge on an electron is negative, the electric field must be directed to the left so as to give a force in the $+x$ direction (to the right).

IDENTIFY THE RELATIONSHIPS

The horizontal axis in Figure 17.13 is the $+x$ direction, and the electric field in the acceleration region is along $-x$ (to the left). We denote the x component of the electric field as E, so the value of E is negative. The charge on an electron is $-e$, so the x component of the force exerted on an electron is $F = -eE$. (Since the value of E is negative, F has a positive value; see Fig. 17.13.) This is the component of the force along x; the y component of the force in the acceleration region is zero. The work done on the electron by the electric force is

$$W = FL = -eEL$$

According to the work–energy theorem, this result equals the change in the electron's kinetic energy. Because the problem specified that the electron begins with a very small initial velocity, we assume the initial kinetic energy is approximately zero. If the electron's mass is m_e and its final speed (when it leaves the accelerator region) is v_f, we have

$$KE_f - KE_i = W = -eEL$$

$$KE_f = \tfrac{1}{2}m_e v_f^2 = -eEL \qquad (1)$$

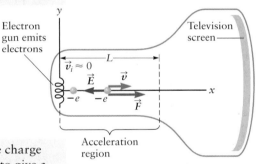

Figure 17.13 Example 17.3. A television picture tube. (Not to scale.)

Figure 17.14 Concept Check 17.4.

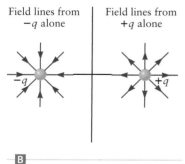

Figure 17.15 Example 17.4.

SOLVE

We can now solve for the final speed of the electron using Equation (1):

$$v_f^2 = -\frac{2eEL}{m_e}$$

$$v_f = \sqrt{-\frac{2eEL}{m_e}} \qquad (2)$$

The negative sign inside the square root might be a "sign" of trouble, but here it is correct. As we have already mentioned, the electric field is in the $-x$ direction and hence the value of E is negative. Inserting values of the various factors in Equation (2), including the electron mass from Appendix A, gives

$$v_f = \sqrt{-\frac{2eEL}{m_e}} = \sqrt{-\frac{2(1.60 \times 10^{-19} \text{ C})(-8.0 \times 10^5 \text{ N/C})(0.020 \text{ m})}{9.11 \times 10^{-31} \text{ kg}}}$$

$$v_f = \boxed{7.5 \times 10^7 \text{ m/s}}$$

What does it mean?

The electrons in a CRT have quite a high speed. In fact, our result for v_f is about 25% of the speed of light![2] In Section 17.5, we'll describe how the electric field in the acceleration region is generated.

CONCEPT CHECK 17.4 | Direction of an Electric Field

Three equal and positive charges q are located equidistant from the origin as shown in Figure 17.14. Is the direction of the electric field at the origin (a) along $+x$, (b) along $-x$, (c) along $+y$, or (d) along $-y$?

EXAMPLE 17.4 | Electric Field of a Dipole

Consider an electric dipole consisting of two charges $+q$ and $-q$ as sketched in Figure 17.15A. They might be an electron and a proton or negative and positive ions that make up part of a molecule. Make a qualitative sketch of the electric field lines in the vicinity of this dipole.

RECOGNIZE THE PRINCIPLE

The electric field due to a single point charge is sketched in Figure 17.10. To find the electric field of a dipole, we use the superposition principle: the field of the dipole is the *sum* of the fields produced by the individual charges.

SKETCH THE PROBLEM AND SOLVE

Figure 17.15B shows the field lines from the charges $+q$ and $-q$ separately, and Figure 17.16 shows their sum. The fields from the two point charges must be added as *vectors*. This addition is shown carefully at point A, where the separate fields from $+q$ and $-q$ are drawn along with their sum.

What does it mean?

The dipole field in Figure 17.16 is an important field pattern that arises in many applications. Many molecules, including H_2O, act as electric dipoles, and the forces between such molecules involve the dipole electric field. We'll see a similar field pattern when we study the magnetic field produced by a magnetic dipole.

---

[2]The speed of our TV electrons is so high that a high-accuracy calculation of their motion requires the use of special relativity (Chapter 27). Including relativity would change our result here by about 6%.

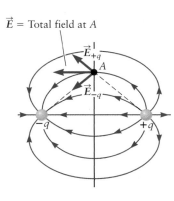

$\vec{E}$ = Total field at A

Figure 17.16 Example 17.4. The electric field of a dipole.

EXAMPLE 17.5 | Electric Field from Two Point Charges

Two point charges $Q_1 = 4.0\ \mu C$ and $Q_2 = -2.0\ \mu C$ are arranged as shown in Figure 17.17A. If $L = 3.0$ cm, what is the electric field at the origin?

SKETCH THE PROBLEM

Figure 17.17B shows the directions of the electric fields produced by the two charges along with the total field $\vec{E}_{total}$.

RECOGNIZE THE PRINCIPLE

The magnitude of the electric field from each charge can be calculated using Equation 17.11. The electric field at the origin is the vector sum of the field from each charge. We need to find both the magnitude and direction of $\vec{E}_{total}$.

IDENTIFY THE RELATIONSHIPS

The field from Q_1 has a magnitude (Eq. 17.11)

$$E_1 = \frac{kQ_1}{L^2} = \frac{(8.99 \times 10^9\ \text{N} \cdot \text{m}^2/\text{C}^2)(4.0 \times 10^{-6}\ \text{C})}{(0.030\ \text{m})^2} = 4.0 \times 10^7\ \text{N/C}$$

The field from Q_2 has a magnitude

$$E_2 = \frac{kQ_2}{L^2} = \frac{(8.99 \times 10^9\ \text{N} \cdot \text{m}^2/\text{C}^2)(2.0 \times 10^{-6}\ \text{C})}{(0.030\ \text{m})^2} = 2.0 \times 10^7\ \text{N/C}$$

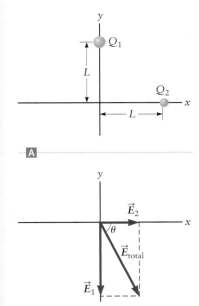

A

B

Figure 17.17 Example 17.5.

SOLVE

Adding the electric fields as vectors is done graphically in Figure 17.17B. The magnitude of the total field E_{total} is

$$E_{total} = \sqrt{E_1^2 + E_2^2} = \sqrt{(4.0 \times 10^7\ \text{N/C})^2 + (2.0 \times 10^7\ \text{N/C})^2}$$

$$E_{total} = \boxed{4.5 \times 10^7\ \text{N/C}}$$

From the trigonometry in Figure 17.17B, the angle that $\vec{E}_{total}$ makes with the x axis is

$$\theta = \tan^{-1}\left(\frac{E_1}{E_2}\right) = \tan^{-1}\left(\frac{4.0 \times 10^7\ \text{N/C}}{2.0 \times 10^7\ \text{N/C}}\right) = \boxed{63°}$$

What does it mean?
When adding the fields from different charges, the fields must always be added as vectors.

17.4 | CONDUCTORS, INSULATORS, AND THE MOTION OF ELECTRIC CHARGE

In Section 17.1, we described how the ancient Greeks used amber and fur to observe electric forces. To understand those experiments, we need to understand how charges can be transferred from one material to another.

The Structure of Matter from an "Electrical" Viewpoint

An atomic-scale picture of a *metal* such as copper is given in Figure 17.18A. Each Cu atom by itself is electrically *neutral*, with equal numbers of protons and electrons. When these neutral atoms come together to form a piece of copper metal, one

Insight 17.2
DRAWING ELECTRIC FIELD LINES
When you draw the electric field lines produced by a collection of point charges, the field lines must always begin and end on charges. As with the dipole in Figure 17.16, electric field lines begin at a positive charge and end at a negative charge.

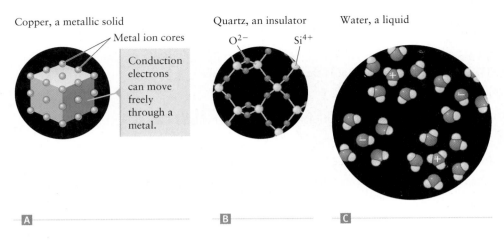

Copper, a metallic solid   Quartz, an insulator   Water, a liquid

Metal ion cores

Conduction electrons can move freely through a metal.

O²⁻   Si⁴⁺

A   B   C

Figure 17.18 **A** In a metal such as copper, some of the electrons (called conduction electrons) are able to move freely through the entire sample. **B** In an insulator such as quartz, all the electrons are bound to ions and there are no conduction electrons. **C** Liquids and gases contain relatively small numbers of mobile charges (mainly ions).

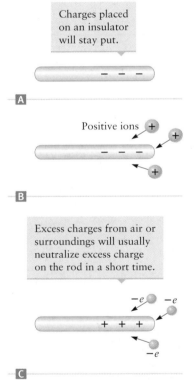

Charges placed on an insulator will stay put.

− − −

A

Positive ions   +   +

− − −

+

B

Excess charges from air or surroundings will usually neutralize excess charge on the rod in a short time.

−e   −e

+ + +

−e

C

Figure 17.19 **A** When excess electrons are placed on an insulator, this charge generally stays where it is placed for a period of time (unlike in a metal, where the charge will move immediately). **B** and **C** Excess electrons or stray ions in the surrounding air are attracted to the insulator's surface, so the insulator's excess charge will eventually be neutralized.

or more electrons from each atom are able to "escape" from its "parent" atom and move freely through the entire piece of metal. These electrons are called *conduction electrons*. They leave behind positively charged ion cores that are bound in place and not mobile. In addition, a piece of metal can accept some additional electrons or let some of the conduction electrons be taken away, so the entire piece of metal can acquire a net negative or positive charge.

Amber, plastic, and quartz are *insulators*; electrons in these materials are *not* able to move freely through the material. A typical example is quartz, which is composed of SiO_2 molecules, each of which is a combination of one Si^{+4} and two O^{-2} ions. (See Fig. 17.18B.) These ions are bonded in place within the solid and are not able to move about. Unlike in metals, electrons cannot "escape" from these ions, and there are no conduction electrons available to carry charge through the solid. Extra electrons can be placed onto a piece of quartz, but they are not able to move about freely. Instead, an added electron would tend to stay in one spot (wherever it was initially placed), although the small amount of moisture that is often present will allow excess charge to gradually move about on the surface.

An atomic-scale picture of another important type of matter is shown in Figure 17.18C. A drop or cup of water consists of a collection of H_2O molecules, each of which has a total charge of zero and is thus electrically neutral. Each water molecule consists of two H atoms and one O atom; these atoms and their electrons are usually tightly bound to each other within a particular molecule. If all the atoms and electrons were bound in this way, there would be no "free" charges available to carry charge throughout the liquid, and water would behave as a perfect insulator. However, a few water molecules in any sample are always dissociated into free H^+ and OH^- ions. These ions can carry charge from place to place in much the same way that conduction electrons carry charge in a metal. (H^+ ions may also combine with neutral water molecules to form hydronium ions, H_3O^+, which along with OH^- ions can carry charge from place to place in a sample of water.) Most samples of water also contain impurities that contribute additional ions such as Na^+ or Cl^-, which act as mobile charge carriers.

The picture in a gas is similar to that found in a liquid; the constituent atoms and molecules are mostly neutral, but a few are present as ions able to carry charge from place to place. A few free electrons are usually present in a gas as well.

Placing Charge on an Insulator

To understand the electrical behavior of an object, we must understand what happens when charge is placed on or taken off the object. Let's begin with the case of an insulator such as quartz, and, for simplicity, assume there is no moisture on it and the air around it is very dry. If a few electrons are placed onto such an insulator, they will stay where they are put (Fig. 17.19A) because there are no free ions in an insulator and the insulator does not allow the extra electrons to move about.

In real life, excess charge will not stay on an insulator indefinitely. If an insulator contains some excess electrons, they will attract stray positive ions from the surrounding air; these stray ions will move onto the insulator and cause it to become neutral, or the electrons may combine with the ions to form neutral atoms (Fig. 17.19B). The amount of time needed to neutralize the charge depends on the amount of moisture in the air. It can be many minutes on a dry (usually winter) day or only a few seconds on a humid day. Any small amount of moisture on an insulator's surface forms a thin layer of water that enables charge to move from one spot to another.

Excess Charge Goes to the Surface of a Metal

Unlike the case of an insulator, electrons can move easily through a metal. So, if we place electrons on one spot of a piece of metal, there is no guarantee they will stay there, and we might imagine that they could go anywhere within the metal. In fact, any excess electrons on a piece of metal will all be distributed *on the surface* of the metal as sketched in Figure 17.20A. To understand this result, consider a piece of metal that is electrically isolated from the rest of the universe so that no new charge can jump onto or off the metal. If we wait long enough, all mobile charges (electrons) in the metal will come to rest and be in static equilibrium. Let's now imagine that some of these excess electrons are not on the surface but are in the interior instead. These electrons would each produce an electric field (according to Fig. 17.10), and the resulting electric field would cause other electrons in the metal to move. This result, however, contradicts our assumption that the system is in static equilibrium and means that the excess electrons cannot be inside the metal. Hence, for a metal in equilibrium, any excess electrons must be *at the surface* and *the electric field is zero inside a metal in equilibrium* (Fig. 17.20C).

This argument applies only to the *excess* charges placed on the metal. The interior of a neutral, uncharged metal contains equal numbers of positive and negative charges. These charges are present inside the metal at all times, and their electric fields cancel, leaving zero electric field inside the metal. Only the excess charge resides at the metal's surface.

Parts A and B of Figure 17.20 show how excess charge is distributed on a metal. In Figure 17.20A, some extra electrons have been added to the metal, so the surface has a net negative charge. In Figure 17.20B, we imagine that some electrons have been removed from the metal, so its net charge is positive. It is customary to draw this situation as if there are positive charges at the surface of the metal, even though that does *not* correspond to placing positive ions or protons onto the metal!

Charging an Object by Rubbing It

The Greeks discovered electricity when they rubbed a piece of amber with fur. The act of rubbing caused some charge to be transferred between the amber and the fur, and we now know that electrons moved from the fur to the amber. The amber thus acquired a net negative charge (an excess of electrons), whereas the fur was left with a net positive charge (an excess number of positive ions) as illustrated in Figure 17.21.

There is nothing unique about amber; a similar result is found if a rubber or plastic rod is rubbed with fur. Likewise, there is nothing special about fur. For example, a glass rod can be charged by rubbing it with a silk cloth, although in this case the glass acquires a net positive charge because electrons leave the glass and move to the silk. The mechanical process of rubbing is thus a general way to move electrons from one material to another. This is the basis of the Van de Graaff generator (Fig. 17.22), which is found in many science museums.

All excess charge sits on the surface of a metal.

A

B

$$\vec{E} = 0$$
in equilibrium

C

Figure 17.20 In static equilibrium, the electric field inside a metal must be zero. All excess charges then reside on the metal's surface. **A** If the excess charge is negative—that is, if there are extra electrons on the metal—these electrons will rest on the surface. **B** If the excess charge is positive, there is a deficit of electrons. This deficit of electrons will be at the surface, so the surface will have a net positive charge. **C** In all cases, the electric field is zero inside a metal in static equilibrium.

Rubbing an amber rod
with fur transfers electrons
from the fur to the amber.

A

Rubbing glass with
silk transfers electrons
from the glass to the silk.

B

Figure 17.21 **A** When a piece of amber or a plastic rod is rubbed with fur, it becomes negatively charged. **B** A glass rod rubbed with silk becomes positively charged.

Figure 17.22 A Van de Graaff generator produces a large electric charge by rubbing an internal rubber belt. This belt transfers electric charge to the metal sphere at the top of the generator.

We mentioned in connection with Figure 17.1 that a charged object, such as a piece of amber, can be used to attract and move small pieces of paper and similar objects. Let's consider this attraction in a little more detail. Suppose we have a rubber rod and a small piece of paper, and that both are electrically neutral at the start of the experiment (Fig. 17.23). If the rubber rod is rubbed with fur, it will acquire a negative charge. If this charged rod is then brought near the paper, the paper is attracted to the rod. This effect is caused by electric forces between charges on the rod and the paper. The rod acquired its excess charge by rubbing, but the paper is electrically neutral (its net charge is zero), so why is an electric force exerted on the paper? Although the paper is electrically neutral, the presence of the rod nearby causes some movement of the charges in the paper. The rod is negatively charged, so electrons in the paper are repelled, and positive ions in the paper are attracted. Although these electrons and ions are not free to move very far, they can move a small amount while still being bound inside the paper. The electrons thus move a short distance away from the rod and the positive ions move toward the rod as sketched in Figure 17.23C, leaving a net positive charge on the portion of the paper nearest the rod and a net negative charge on the opposite side of the paper. The paper is then said to be *polarized*.

Because unlike charges attract, the positive side of the paper is attracted to the negatively charged rod. At the same time, the negative side of the paper is repelled from the rod. The positively charged side of the paper is closer to the rod, so because the electric force decreases as the distance between the charges increases, the magnitude of the attractive force is larger than the magnitude of the repulsive force. The net result is that the paper is *attracted* to the rod. In this way, there can be an electric force on an object even when the object is electrically neutral, provided the object is polarized.

The same principle is illustrated in Figure 17.24A, which shows water flowing from a container. The stream of water would ordinarily flow vertically downward due to the force of gravity, but a charged balloon placed near the stream exerts a force on the water and deflects the stream sideways. The electric field from the charged balloon polarizes the water molecules (Fig. 17.24B), leading to an attractive force as found with the paper in Figure 17.23.

CONCEPT CHECK 17.5 | Force on a Polarized Object

The negatively charged balloon in Figure 17.24A attracts the stream of water. Suppose a positively charged balloon was used instead. Which of the following statements would then be true?
  (a) The water would be repelled from the positively charged balloon.
  (b) The water would be attracted to the positively charged balloon.
  (c) There would be no force on the water.

Start:
Rod and paper are
both neutral.

Rubber rod

Charge the rod by rubbing
with fur, which places a
negative charge (electrons)
on the rod.

The paper is *polarized*
and attracted to the rod.

A　　　　　**B**　　　　　**C**

Figure 17.23 **A** A rubber rod that is initially neutral can **B** be charged negatively by rubbing it with fur. **C** This negatively charged rod can then polarize a nearby object such as a small piece of paper, resulting in electrical attraction between the rod and the paper.

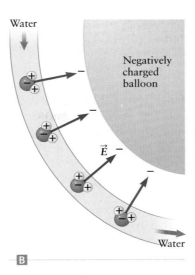

A   **B**

© Cengage Learning/Charles D. Winters

Water

Negatively
charged
balloon

$\vec{E}$

Water

Figure 17.24 **A** A stream of water is deflected by a nearby charged balloon. **B** The balloon polarizes the water, similar to the piece of paper in Figure 17.23C.

Insight 17.3

WHY DOES RUBBING TWO OBJECTS MAKE THEM CHARGED?

To understand why rubbing two objects together transfers electrons from one object to the other, we appeal to the concept of *electronegativity* that arises in chemistry. Electronegativity is a measure of the tendency of an atom to attract bonding electrons. When two objects are rubbed together, atoms from the two materials are brought into contact. Unless they are identical atoms in identical materials, their electronegativities will be different, causing the atoms in one material to "steal" electrons from the other material.

The Concept of Electrical "Ground"

Suppose a rubber rod is charged by rubbing so that it acquires some extra electrons and is then placed on a table. If we wait for some period of time (the time required depends on the amount of moisture in the room), we would find that the rod has lost its excess charge. If we watch very carefully with very sensitive electronic measuring equipment, we would find that the excess electrons originally on the rod move from the rod to the table and eventually into the ground below. The ground is generally quite moist, so it conducts charge well. Because the ground is literally everywhere, it provides a common path that excess charge can use to flow from one spot to another. In fact, the term *ground* is used generically to denote the path or destination of such excess charges, even if the true path does not involve any dirt. This notion of **electrical ground** plays an important role in many situations.

Charging by Induction

Suppose you are given a negatively charged rubber rod and are asked to transfer some of this charge to a piece of metal. You could accomplish this task by simply bringing the rubber rod into contact with the metal (Fig. 17.25). The negative charge on the rod is caused by an excess of electrons, and some of these electrons will move to the metal when it touches the rod; this process is just an example of charging by rubbing.

Now suppose you are given the same negatively charged rubber rod, but this time you are asked to give the metal a net *positive* charge. This task might seem to be impossible because the excess charges on the rod are negative, but an approach called **charging by induction** will accomplish it. This approach makes use of polarization and the properties of an electrical ground. The negatively charged rod is first brought near the metal (Fig. 17.26A), polarizing the metal. That is, some of the conduction electrons in the metal move to the opposite side—away from the rod—due to the repulsion of like charges, leaving the side of the metal near the rod with a net positive charge. A connection is then made from the piece of metal to electrical ground using a metal wire, so electrons are able to move even farther from the charged rod. In fact, some of the electrons will move off the original piece of metal and into the electrical ground region (Fig. 17.26B). The final step in the process is to remove the grounding wire, leaving the original piece of metal with a net positive charge (Fig. 17.26C). Notice that we have not put protons or any other

---

CHARGING AN OBJECT BY CONTACT

Before contact

Contact

After contact

Figure 17.25 If one charged object touches a second object, the second object will usually acquire some of the excess charge. Hence, the second object is charged "by contact."

Positive charge is
attracted to rod.

Charged rod

Negative charge
(electrons) will flow
to ground.

Ground −

B

Final stage: Excess
charge is positive.

C

Figure 17.26 An object can be
charged by the process of induction. **A** The object is first brought
near a charged rod, polarizing the
object. **B** The object is then connected to ground, and some electrons flow between it and ground.
C The object is left with an excess
charge when it is disconnected
from ground.

positive charges on the metal. Instead, we have *removed* electrons. The final result
is that the metal has a net positive charge.

17.5 | ELECTRIC FLUX AND GAUSS'S LAW

We can calculate the electric field due to a point charge using Equation 17.11. In
principle, we can use this result to deal with any conceivable distribution of charges
by treating the distribution as a collection of point charges and using superposition to find the total field. Fortunately, there is a simpler way to deal with complex
charge distributions, based on *Gauss's law*. To understand Gauss's law, we must
first define a quantity called the *electric flux*. In words, electric flux is equal to the
number of electric field lines that pass through a particular surface multiplied by the
area of the surface. Electric flux is denoted by the symbol Φ_E.

Some examples of electric flux calculations are given pictorially in Figure 17.27.
For simplicity, these examples assume the electric field is constant in both magnitude and direction. In Figure 17.27A, $\vec{E}$ is perpendicular to a flat surface having a
total area A. The electric flux through this surface is $\Phi_E = EA$. In this case, the flux
is just the magnitude of the electric field multiplied by the area of the surface. Flux
is a scalar, just an ordinary number; it is *not* a vector.

The value of Φ_E depends on how many field lines intersect the surface as illustrated in Figure 17.27B, where $\vec{E}$ is parallel to a surface. In this case, the field lines
do not cross the surface, and $\Phi_E = 0$. If the field makes an angle θ with the surface
(Fig. 17.27C), the flux is

$$\Phi_E = EA \cos \theta \qquad (17.12)$$

The three examples in Figure 17.27A–C all involve simple, flat surfaces. We'll
often be interested in the flux through a *closed* surface such as a box or a sphere.
The flux due to a constant field $\vec{E}$ through these closed surfaces is shown in Figure
17.27D and E. By convention, the flux through a surface is *positive* if the field is
directed *out* of the region contained by the surface, whereas the flux is *negative* if
$\vec{E}$ is directed *into* the region. For the cases with a constant field in parts D and E
of Figure 17.27, the total flux through the entire closed surface is *zero* in each case.
For the box in Figure 17.27D, there is a negative flux through the face of the box on
the left and a positive flux through the face on the right. The total flux is the sum
of these two contributions and is zero. Likewise, there is a negative flux through
the left side of the spherical surface in Figure 17.27E and a positive flux through the
right side, and the total flux is again zero.

Gauss's Law

Gauss's law asserts that the electric flux through any closed surface is proportional
to the total charge q inside the surface, with

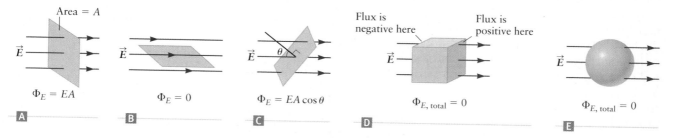

Figure 17.27 Finding the electric flux through a surface.

$$\Phi_E = \frac{q}{\varepsilon_0}$$ (17.13) Gauss's law

The constant of proportionality that relates flux and charge is ε_0, the same physical constant that enters Coulomb's law (Eq. 17.5).

Since electric flux depends on the magnitude and direction of the electric field, the left-hand side of Equation 17.13 depends on $\vec{E}$ while the right-hand side depends on the charge. We would like to use Gauss's law to calculate $\vec{E}$, but it is not immediately obvious how to do so. In particular, Φ_E is the total flux through a closed surface, so its value depends on the magnitude and direction of the electric field at *all* points on the surface. To see how to use Gauss's law, it is simplest to consider some examples.

Using Gauss's Law to Find $\vec{E}$ for a Point Charge

Let's first consider the familiar case of a single point charge. To apply Gauss's law, we must first choose the surface, called a ***Gaussian surface***, that will be used in the flux calculation. Equation 17.13 holds for *any* closed surface, so our strategy is to choose a surface that will make the calculation of Φ_E as simple as possible. To this end, we choose a surface that *matches the symmetry of the problem*. For a point charge, we know that the electric field lines have a spherical symmetry (Fig. 17.28). This spherical symmetry means that the magnitude of the electric field depends only on the distance from the charge r and that $\vec{E}$ must be directed radially, either outward or inward (and not sideways). A surface that matches this symmetry is a sphere centered on the charge as sketched in Figure 17.28. Because of the symmetry, the magnitude of $\vec{E}$ is the same at all points on the sphere, and $\vec{E}$ is perpendicular to the sphere at all points where it intersects the surface.

A calculation of the flux through the sphere in Figure 17.28 is similar to the flux computation in Figure 17.27A. When $\vec{E}$ is perpendicular to a surface, the flux is equal to the magnitude of $\vec{E}$ multiplied by the area of the surface. So, for the flux in Figure 17.28, we have

$$\Phi_E = EA_{sphere}$$

where A_{sphere} is the area of our spherical Gaussian surface. If the radius of this sphere is r, then $A_{sphere} = 4\pi r^2$, and the flux is

$$\Phi_E = 4\pi r^2 E$$

According to Gauss's law, this flux is proportional to the total charge contained within the surface. Using Equation 17.13, we have

$$\Phi_E = 4\pi r^2 E = \frac{q}{\varepsilon_0}$$

We can now solve for E and find

$$E = \frac{q}{4\pi\varepsilon_0 r^2}$$ (17.14)

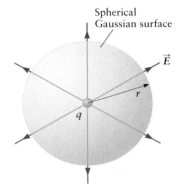

Figure 17.28 To calculate the electric field near a point charge using Gauss's law, we choose a spherical Gaussian surface centered on the point charge.

which agrees with our previous result obtained using Coulomb's law (Eq. 17.11). The key to this application of Gauss's law was our choice of the Gaussian surface. This choice made the calculation of Φ_E straightforward because E has the same value over the entire surface and the electric field's direction is always perpendicular to the surface.

The result in Equation 17.14 shows that Coulomb's law is actually a special case of Gauss's law. That is, Gauss's law applied to the case of a point charge together with the relationship between the electric field and the electric force (Eq. 17.10) can be used to derive Coulomb's law. Hence, all the results we have found using Coulomb's law also follow from Gauss's law. In addition, Gauss's law can readily deal with many other situations that are mathematically very difficult or awkward to attack with Coulomb's law. We will spend the rest of this section working through a few such cases, using the general approach summarized below.

PROBLEM SOLVING | Applying Gauss's Law

1. **RECOGNIZE THE PRINCIPLE.** To apply Gauss's law, you must first calculate the electric flux through a Gaussian surface. The choice of the Gaussian surface is key.

2. **SKETCH THE PROBLEM.** Your drawing should show the charge distribution. Use the symmetry of this distribution to also sketch the electric field. The Gaussian surface should match the symmetry of the electric field.

3. **IDENTIFY THE RELATIONSHIPS.** Your Gaussian surface should satisfy one or more of the following conditions:

   • $\vec{E}$ has constant magnitude over all or much of the surface and makes a constant angle with the surface.

The most convenient surface is one that is perpendicular to $\vec{E}$ at all (or most) points.
   • $\vec{E}$ may be zero over a portion of the surface. The flux through that part of the surface is zero.
   • $\vec{E}$ may be parallel to some part of the surface. The flux through that part of the surface is zero.

4. **SOLVE.** First calculate the *total* electric flux through the entire Gaussian surface. Then find the total electric charge inside the surface and apply Gauss's law (Eq. 17.13) to solve for $\vec{E}$.

5. Always *consider what your answer means* and check that it makes sense.

EXAMPLE 17.6 | Electric Field from a Spherical Charge

Consider the uniform spherical ball of charge in Figure 17.29A, with total charge Q and radius r_b. Find the electric field at points outside the ball.

RECOGNIZE THE PRINCIPLE

We follow the strategy described in the "Applying Gauss's Law" problem-solving strategy. The key is to find a Gaussian surface that matches the symmetry of the charge distribution (step 1 in the strategy). The charge is spherically distributed in Figure 17.29A, so the electric field must also have spherical symmetry.

SKETCH THE PROBLEM

Step 2 in the strategy: The spherical symmetry means that $\vec{E}$ must be directed radially pointing either away from or toward the center of the ball of charge. If Q is positive, the field will be as sketched in Figure 17.29B.

IDENTIFY THE RELATIONSHIPS

Step 3: Choose a Gaussian surface. Because of the spherical symmetry, we can also say that the magnitude of $\vec{E}$ will depend only on the distance from the center of the ball of charge. We therefore pick a spherical Gaussian surface as sketched in Figure 17.29A. The electric field will have a constant magnitude at all points on this surface, and $\vec{E}$ will be perpendicular to the surface. This situation is similar to what we had in Figure 17.28 when we applied Gauss's law to the case of a point charge.

SOLVE

We next calculate the electric flux through our chosen Gaussian surface. The Gaussian surface is a sphere of radius r, so its surface area is $A_{sphere} = 4\pi r^2$. Denoting the

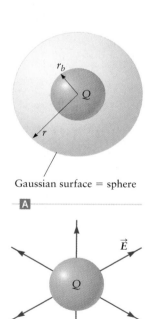

Gaussian surface = sphere

A

$\vec{E}$

Q

B

Figure 17.29 Example 17.6.

electric field on the sphere by E and again noting that $\vec{E}$ is perpendicular to all points on the surface, the electric flux is

$$\Phi_E = EA_{\text{sphere}} = 4\pi r^2 E$$

Gauss's law relates this flux to the total charge contained within the spherical surface. We are interested in locations outside the ball $(r > r_b)$, so the charge inside the Gaussian surface is the total charge of the ball Q. We thus have

$$\Phi_E = 4\pi r^2 E = \frac{Q}{\varepsilon_0}$$

which leads to

$$E = \boxed{\frac{Q}{4\pi\varepsilon_0 r^2}}$$

Since $1/(4\pi\varepsilon_0) = k$, we can also write this result as

$$E = \boxed{\frac{kQ}{r^2}}$$

What does it mean?

Our result for E is identical to the electric field of a point charge of magnitude Q located at the origin (Eq. 17.11). In fact, the electric field from *any* spherical distribution of charge is the *same* as the field from a point charge with the same total charge. Note that this result applies *only outside* the ball of charge.[3]

CONCEPT CHECK 17.6 | Finding the Electric Flux

Consider a point charge q located outside a closed surface as sketched in blue in Figure 17.30. The electric field lines from the charge penetrate the surface. What is the total electric flux through this surface?

(a) q/ε_0　　　(b) $-q/\varepsilon_0$　　　(c) 0

More Applications of Gauss's Law

Let's next discuss an application of Gauss's law that does not involve the spherical symmetry of a point charge or a ball of charge. Consider a very long, straight line of charge that lies along the x axis as sketched in Figure 17.31A. The line of charge has a total length L (where L is very large) and a total charge Q. How can we find the electric field produced by this charge distribution?

The first step in the "Applying Gauss's Law" problem-solving strategy is to recognize the symmetries of the charge distribution and of the electric field it produces so that you can choose a Gaussian surface. For step 2, the sketch in Figure 17.31B shows two observers who are studying the charged line and its electric field. These two observers are looking at the charged line from opposite directions. Because it is very long, the charged line will look the same to these observers, so the electric field must also look the same. Suppose (hypothetically) observer A finds that $\vec{E}$ points away from him and makes an angle θ with the x axis as sketched in Figure 17.31C. Since the observers are symmetrically arranged, observer B must find a similar result (Fig. 17.31D), with the electric field pointing away from him and also making an angle θ with the x axis. However, the two observers *must agree* on the magnitude and direction of $\vec{E}$, and they can only agree if $\theta = 90°$. The symmetry of

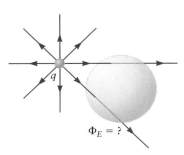

Figure 17.30 Concept Check 17.6. What is the total electric flux through this surface?

---

[3]We found a similar result in connection with gravitation in Chapter 5, where we noted that the gravitational force due to a spherically symmetric mass is the same as if all the mass were located at the center of the sphere.

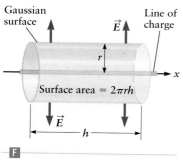

Figure 17.31 **A** Calculation of the electric field near a line of charge. **B**–**E** Symmetry requires that the electric field be perpendicular to the line. The magnitude of E depends only on the distance r from the line. **F** If we choose a cylindrical Gaussian surface, the electric field is constant in magnitude at the walls of the cylinder and $\vec{E}$ is parallel to the ends of the cylinder.

the line of charge means that $\vec{E}$ must be *perpendicular to the line*. The electric field could be directed away from the line or toward the line, but symmetry requires that it be perpendicular (Fig. 17.31E).

The next step is to choose a Gaussian surface that matches this symmetry. We wish (if possible) for $\vec{E}$ to be perpendicular to the surface, so we choose a cylinder centered on the line of charge (Fig. 17.31F). This cylinder can be divided into a curved section plus two flat, circular ends. The electric field is parallel to the circular ends, so the electric flux through the ends is zero. We want the magnitude of $\vec{E}$ to be the same at all points on the curved part of the cylinder. The magnitude of the field can only depend on the distance from the charged line, so E will be a constant on our Gaussian surface provided the charged line runs along the central axis of the cylinder.

Next (step 4), we calculate the flux through our chosen Gaussian surface. For a cylinder of radius r and length h, the area of the curved part of the cylinder is $A = 2\pi rh$. The flux through this curved surface is thus

$$\Phi_E = EA = 2\pi Erh$$

We now find the total charge inside the Gaussian surface and then apply Gauss's law. The total charge within the cylinder in Figure 17.31F is equal to the charge per unit length Q/L multiplied by the length of the cylinder h, so

$$q = \frac{Q}{L}h$$

Applying Gauss's law then gives

$$\Phi_E = 2\pi Erh = \frac{q}{\varepsilon_0} = \frac{Qh/L}{\varepsilon_0}$$

We can now solve for the magnitude of E and find

$$E = \frac{Q}{2\pi\varepsilon_0 Lr} \qquad (17.15)$$

CONCEPT CHECK 17.7 | Electric Field from a Line of Charge

If the charge Q on the line in Figure 17.31F is negative, is the direction of the electric field (a) perpendicular to the line and directed away from it, (b) perpendicular to the line and directed toward it, or (c) parallel to the line?

EXAMPLE 17.7 | Field from a Flat Sheet of Charge

Consider a large, flat sheet of charge as sketched in Figure 17.32A. If this sheet has a positive charge per unit area σ, find the electric field produced by the sheet.

RECOGNIZE THE PRINCIPLE

We again follow the steps outlined in the "Applying Gauss's Law" problem-solving strategy. We begin by considering the symmetry of the charge distribution and of the resulting electric field so that we can choose a Gaussian surface.

SKETCH THE PROBLEM

We again imagine how this charge distribution and field would appear to different observers who look at them from different directions (Fig. 17.32B). These observers must agree on the direction of $\vec{E}$, and as was the case with the charged line in Figure 17.31, the electric field must be perpendicular to the plane. If the charge on the plane is positive, the electric field points away from the plane.

IDENTIFY THE RELATIONSHIPS

Now that we have identified the direction of $\vec{E}$, we must choose a Gaussian surface. The cylinder's axis in Figure 17.32A is oriented perpendicular to the plane. The flux through the curved sidewall of the surface is zero because $\vec{E}$ is parallel to this part of the cylinder. The electric field $\vec{E}$ is perpendicular to the ends of the cylinder, so if the ends each have an area A, the flux is $\Phi_E = EA$ through each end of the cylinder. For a positively charged plane, the electric flux through each end of the cylinder is positive because the electric field lines pass outward through the cylinder. The total electric flux through the cylinder is thus

$$\Phi_E = 2EA$$

where the factor of two is needed to include the flux through both ends.

SOLVE

The total charge inside the Gaussian surface is equal to the charge per unit area on the sheet σ multiplied by the cross-sectional area of the cylinder A, so $q = \sigma A$. Inserting this into Gauss's law leads to

$$\Phi_E = \frac{q}{\varepsilon_0}$$

$$2EA = \frac{\sigma A}{\varepsilon_0}$$

Solving for the electric field, we get

$$E = \boxed{\frac{\sigma}{2\varepsilon_0}}$$

What does it mean?
The magnitude of the electric field is constant, independent of distance from the plane. For this reason, a large, flat sheet of charge is a very convenient way to produce a uniform electric field. You may think that a plane of charge is a rather artificial example, but such charged sheets are used often in practical devices such as the accelerator of the TV picture tube in Figure 17.13.

Electric Field between Two Metal Plates

Figure 17.33 shows two parallel metal plates that carry charges $+Q$ and $-Q$. This arrangement, called a *capacitor*, is used in many electric circuits. Let's calculate the electric field produced by this parallel-plate capacitor.

Recall from Section 17.4 that the excess charge on a metal in equilibrium is always located at the surface. For a single metal plate with a total charge of $+Q$ (Fig. 17.34), symmetry tells us that half the charge will reside on each side. (To be strictly correct, there would also be some charge on the edges, but if the plate is very thin, the amount of charge on the edges will be negligible.) When we consider two plates with opposite charges as in our capacitor (Fig. 17.33), the excess charges on the two plates will attract each other and draw all the excess charge to the inner sides of the two plates.

To find the electric field between the two capacitor plates in Figure 17.33, we use the result from Example 17.7 for the field from a single plane of charge. There we found

$$E = \frac{\sigma}{2\varepsilon_0} \quad \text{(magnitude of electric field from a single charged plane)} \quad (17.16)$$

where σ is the charge per unit area on the plane.

The capacitor in Figure 17.33 contains two plates; if each has an area A, they have a charge per unit area $\sigma_1 = +Q/A = +\sigma$ and $\sigma_2 = -Q/A = -\sigma$. According to the principle of superposition, the total field is the vector sum of the fields from each

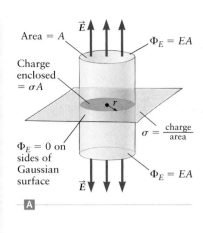

$\text{Area} = A$ $\vec{E}$ $\Phi_E = EA$

Charge enclosed $= \sigma A$

r

$\sigma = \dfrac{\text{charge}}{\text{area}}$

$\Phi_E = 0$ on sides of Gaussian surface

$\Phi_E = EA$

$\vec{E}$

A

Charged plane

All observers must agree on the direction of $\vec{E}$ so $\vec{E}$ must be perpendicular to the plane.

B

Figure 17.32 Example 17.7.
A Calculation of the electric field near a charged plane. **B** Symmetry requires that the electric field be perpendicular to the plane.

Field from upper plate $= E_1 = \dfrac{\sigma}{2\varepsilon_0}$

Total charge $= +Q$

$\sigma = \dfrac{+Q}{A}$

$-\sigma = \dfrac{-Q}{A}$

Total charge $= -Q$

Field from lower plate $= E_2 = \dfrac{\sigma}{2\varepsilon_0}$

Figure 17.33 Calculation of the electric field between two thin metal plates. This arrangement is called a parallel-plate capacitor.

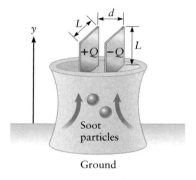

Figure 17.34 For a single thin metal plate, any excess charge is distributed evenly on the two sides.

plate. In the region between the plates, the fields from the two plates are parallel and add to give a total field

$$E = \frac{\sigma}{\varepsilon_0}$$ (magnitude of electric field between two charged plates)   (17.17)

The field between the two plates is thus twice the field from a single plate (Eq. 17.16). In terms of the charge Q and plate area A, the field is

$$E = \frac{Q}{\varepsilon_0 A}$$ (17.18)

We'll use this result in Chapters 18 and 19 as we analyze the properties of capacitors.

Figure 17.35 Example 17.8.

EXAMPLE 17.8 Electric Fields in a Smokestack Scrubber

In Example 17.1, we discussed a very simple model for a smokestack scrubber that removes undesirable particles from the air by first adding some excess electrons and then using electric forces to pull the particles out of the air. Consider again a soot particle of mass $m_{soot} = 1.0 \times 10^{-12}$ kg that travels upward in a smokestack. The scrubber in Example 17.1 added 69 electrons to the particle, and its charge was $q_{soot} = 1.1 \times 10^{-17}$ C. Assume the electric field in the scrubber is produced by two parallel, square plates of width $L = 1.0$ m and separation $d = 0.010$ m, with charges $\pm Q$. **(a)** What must be the value of the electric field between the plates so that force on the soot particle is equal to the weight of the particle? (A real scrubber would use a collection of many pairs of such plates in parallel.) **(b)** What charge Q on the scrubber's plates is required to produce the electric field in part (a)?

RECOGNIZE THE PRINCIPLE

We can find the magnitude of E by setting the electric force $q_{soot}E$ on a soot particle equal to its weight. We can then use our result for the electric field between two charged plates (Eq. 17.18) to relate the electric field to the charge $\pm Q$ on each plate.

SKETCH THE PROBLEM

The scrubber is shown in Figure 17.35. This arrangement is similar to the two metal plates in Figure 17.33, with $\vec{E}$ perpendicular to the plates.

(A) IDENTIFY THE RELATIONSHIPS

The electric force on a soot particle is

$$F_{elec} = q_{soot}E$$

where E is the magnitude of the field between the scrubber plates. This force is designed to equal the particle's weight, so we have

$$F_{elec} = q_{soot}E = m_{soot}g$$

SOLVE

Solving for E gives

$$E = \frac{m_{soot}g}{q_{soot}}$$

Inserting the mass and charge on the soot particle leads to

$$E = \frac{m_{soot}g}{q_{soot}} = \frac{(1.0 \times 10^{-12}\ \text{kg})(9.8\ \text{m/s}^2)}{1.1 \times 10^{-17}\ \text{C}} = \boxed{8.9 \times 10^5\ \text{N/C}}$$

(B) IDENTIFY THE RELATIONSHIPS

The electric field between two metal scrubber plates with charge $\pm Q$ is (Eq. 17.18)

$$E = \frac{Q}{\varepsilon_0 A}$$

For square plates with sides of length L, we have $A = L^2$ and

$$E = \frac{Q}{\varepsilon_0 L^2}$$

SOLVE

Solving for the charge Q, we find

$$Q = \varepsilon_0 E L^2$$

Inserting the given dimensions of our plates and the value of E from part (a) gives

$$Q = \varepsilon_0 E L^2 = [8.85 \times 10^{-12} \text{ C}^2/(\text{N} \cdot \text{m}^2)](8.9 \times 10^5 \text{ N/C})(1.0 \text{ m})^2$$

$$Q = \boxed{7.9 \times 10^{-6} \text{ C}}$$

What have we learned?

Our analysis of a scrubber has so far only considered the force and electric field that acts on the soot particle. We'll consider how to place the necessary charge Q on the plates when we discuss electric potential energy in Chapter 18.

17.6 | ⊗ APPLICATIONS: DNA FINGERPRINTING

Electric forces and fields are used in many applications, some of which are very old. The smokestack scrubber used for pollution control in Examples 17.1 and 17.8 was invented about 100 years ago. Let's now consider a relatively new application, the technique of DNA fingerprinting, that is widely used in biology. The DNA molecules to be studied are first broken into fragments using enzymes (Fig. 17.36A; a job for biochemists). The fragments have a range of sizes and form ions in solution with typical charges between $-100e$ and $+100e$. Each specific DNA molecule forms its own unique set of fragments. The DNA fragments are next placed at one end of a long tube containing a gel (Fig. 17.36B). An electric field is then turned on in the gel tube; this field can be produced using two metal plates as we analyzed in Section 17.5. Since the DNA fragments are ions, they experience an electric force that pushes them through the gel. In Figure 17.36B, we assume the ions are negatively charged, so the electric field is to the left and the force $\vec{F}_{\text{elec}} = q\vec{E}$ is to the right.

In addition to the electric force, there is also a drag force on the ions due to collisions with molecules in the gel. This drag force was described in Chapter 3; for a particle of radius r and velocity $\vec{v}$, the drag force is given by (Eq. 3.23)

$$\vec{F}_{\text{drag}} = -Cr\vec{v} \qquad (17.19)$$

where C is a constant that depends on the properties of the gel. The DNA fragment ions move through the gel in response to the forces $\vec{F}_{\text{elec}}$ and $\vec{F}_{\text{drag}}$. At first, their velocity is very small, so the drag force is also small and the fragment ions accelerate. After a short time, their velocity increases to the point at which $\vec{F}_{\text{drag}}$ is equal in magnitude and opposite in direction relative to $\vec{F}_{\text{elec}}$. The total force is then zero,

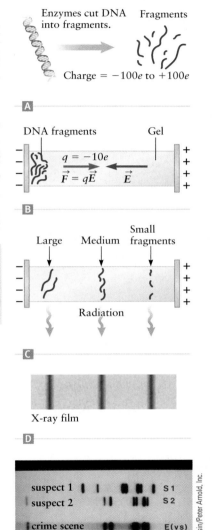

Figure 17.36 Ⓐ–Ⓓ DNA fingerprinting uses the electric force on fragments of DNA to separate the fragments according to their size and charge. Ⓔ In a forensic analysis, the DNA fingerprints of suspects are compared to a "known" sample (i.e., one collected at the crime scene). Here it appears that suspect 2 is guilty.

and the ions move at a constant velocity through the gel in a process very similar to a skydiver moving through the air at her terminal velocity (Chapter 3). When an ion in the gel travels at its terminal velocity, we thus have $\vec{F}_{\text{elec}} + \vec{F}_{\text{drag}} = 0$. Inserting our results for the components of these forces along the gel tube gives

$$qE - Crv = 0$$

Solving for the ion's speed, we find

$$Crv = qE$$

$$v = \frac{qE}{Cr} \tag{17.20}$$

The speed of the ion thus depends on its charge q and its size r, so different DNA fragments move through the gel at different speeds, with the smallest fragments moving fastest (assuming similar values of q). After some time, the fragments are distributed along the gel tube as sketched in Figure 17.36C, and the fragments are thus separated according to their size.

The final step in this process is to record the fragment locations, which is done by attaching radioactive isotopes (see Chapter 30) to the DNA before the fragments are made. These isotopes emit radiation that can be detected with photographic film or in other ways to record the position of the fragments within the gel tube (Fig. 17.36D).

The technique of separating ions according to their size by using electric forces to push the ions through a gel is called *electrophoresis*. This method has become an important tool in biology and forensics (Fig. 17.36E).

17.7 | "WHY IS CHARGE QUANTIZED?" AND OTHER DEEP QUESTIONS

In Section 17.1, we stated some "well-known" facts about electric charges. Let's now reexamine those facts in light of what we have learned about electric forces and fields.

How Are Protons Held Together?

We have mentioned many times that electric charge is carried by subatomic particles called protons and electrons. Indeed, we tend to take the existence of these particles for granted today, but what do they "look like" inside? Let's consider the proton first. We have seen that electric forces can be very large, and we know that like charges repel. If a proton is a small ball of positive charge as sketched in Figure 17.37A, the repulsive electric forces between different parts of a proton will tend to make these parts fly apart. How, then, is the proton held together?

Experiments have shown that a proton is actually an assembly of three particles called *quarks* as illustrated schematically in Figure 17.37B. The proton is composed of two particular types of quarks called "up" and "down." Each proton consists of one up quark of charge $-\frac{1}{3}e$, and two down quarks, each with charge $+\frac{2}{3}e$. The total charge of a proton is thus $-\frac{1}{3}e + \frac{2}{3}e + \frac{2}{3}e = +e$. That a proton is composed of three separate quarks does not really answer our initial question about how protons are held together because electric forces between the quarks will still tend to push them apart, especially since the down quarks are both positively charged. The large repulsive electric forces between quarks are overcome by an even more powerful attractive force called the *strong force*, which is a fundamental force of physics just as gravity and electromagnetism are fundamental forces. What's more, the strong force is so powerful that an individual quark is not able to escape from a proton under any circumstances. This is why we only see integral multiples of $\pm e$ in nature.

Asking simple questions about electric forces and Coulomb's law has thus led us to some deep questions about the fundamental structure of matter. The stability of

The Coulomb force tends to make a proton fly apart. There must be a strong attractive force to hold it together.

 A

Down quark's charge $= +\frac{2}{3}e$ each

Up quark's charge $= -\frac{1}{3}e$

B

Figure 17.37 **A** A "ball" of electric charge will tend to fly apart due to the electric forces between different parts of the ball. If this ball is to be in mechanical equilibrium, there must be another force that overcomes the electric force and holds the ball together. **B** A proton is such a charged ball, made up of three quarks held together by the strong force.

matter relies on a force we had so far not even known to exist! We'll say more about the strong force and quarks in Chapters 30 and 31.

Electrons Are Point Charges

We have just described the internal structure of the proton in terms of quarks. You might now expect that the electron has a similar internal structure, but nature has another surprise for us. As far as physicists can tell, the electron is truly a *point* particle; the mass and charge of an electron are located at an infinitesimal "point" in space. Your "everyday" intuition will probably have difficulty accepting that a particle can be a true "point" in space. There is no way to explain it except to say that this is how nature seems to have designed the electron!

Conservation of Charge

Electric charge is conserved. That is, the total charge of the universe is a constant. This does *not* mean that the total number of protons or electrons in the universe is constant. Indeed, in Section 7.8 we mentioned that in certain situations a neutron can decay to form a proton, an electron, and an antineutrino (see also Chapter 31). The neutron is neutral, so the charge before this decay is zero. The total charge of all the decay products is also zero since the antineutrino is neutral and the proton and electron carry charges of $+e$ and $-e$, respectively. Hence, the net charge is the same before and after the decay, even though the number of charged particles changes.

The principle of **conservation of charge** is believed to be an exact law of nature. No violations of this principle have ever been observed, but there is no fundamental understanding of why. Again, all we can say is that this is the way the laws of physics seem to operate.

SUMMARY | Chapter 17

KEY CONCEPTS AND PRINCIPLES

Electric charge and Coulomb's law
There are two types of electric charge, positive and negative. Protons have a positive charge $+e$, and electrons carry a negative charge $-e$. Charge is measured in **coulombs** (C), with

$$e = 1.60 \times 10^{-19} \text{ C} \qquad \text{(17.1) and (17.2) (page 531)}$$

The electric force between two point charges q_1 and q_2 is given by **Coulomb's law.** If the charges are separated by a distance r, this force has a magnitude

$$F = k \frac{q_1 q_2}{r^2} = \frac{q_1 q_2}{4\pi\varepsilon_0 r^2} \qquad \text{(17.3) and (17.5) (pages 531 and 532)}$$

and is directed along the line joining the two charges.

The electric force is repulsive for like charges (charges with the same sign) and attractive for unlike charges (those of opposite sign). Here, k is a constant of nature with the value

$$k = 8.99 \times 10^9 \text{ N} \cdot \text{m}^2/\text{C}^2 \qquad \text{(17.4) (page 531)}$$

and ε_0 is a universal constant called the permittivity of free space, with the value

$$\varepsilon_0 = 8.85 \times 10^{-12} \frac{\text{C}^2}{\text{N} \cdot \text{m}^2} \qquad \text{(17.6) (page 532)}$$

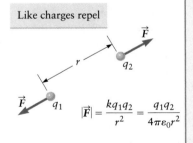

Like charges repel

$$|\vec{F}| = \frac{k q_1 q_2}{r^2} = \frac{q_1 q_2}{4\pi\varepsilon_0 r^2}$$

Unlike charges attract

(Continued)

Electric fields

The ***electric field*** $\vec{E}$ can be measured using a test charge q and is proportional to the electric force, with

$$\vec{F} = q\vec{E} \qquad \text{(17.10) (page 538)}$$

The electric field due to a point charge q can be determined from Coulomb's law. Its magnitude is

$$E = \frac{kq}{r^2} \qquad \text{(17.11) (page 538)}$$

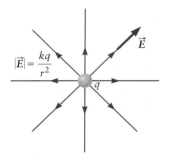

$|\vec{E}| = \dfrac{kq}{r^2}$

Electric flux and Gauss's law

When an electric field of magnitude E passes through a surface of area A while making an angle θ with the normal to the surface, the ***electric flux*** through the surface is

$$\Phi_E = EA \cos \theta \qquad \text{(17.12) (page 546)}$$

Gauss's law relates the electric flux Φ_E through a closed surface to the amount of electric charge inside the surface:

$$\Phi_E = \frac{q}{\varepsilon_0} \qquad \text{(17.13) (page 547)}$$

Gauss's law gives a powerful way to calculate the electric field in cases that are very symmetric. Coulomb's law can be derived from Gauss's law.

$\Phi_E = EA \cos \theta$

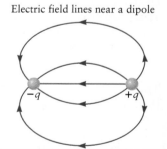

$\Phi_{E,\,\text{total}} = 0$

Conservation of charge

A fundamental conservation law of nature is that electric charge is conserved. Charge is also quantized; it comes in discrete amounts, such as electrons and protons, which have charges $-e$ (an electron) and $+e$ (a proton).

APPLICATIONS

Principle of superposition

Coulomb's law gives the electric field due to a single point charge (Eq. 17.11). For a collection of point charges, the principle of superposition states that the total electric field is the vector sum of the fields produced by the individual charges.

Electric field lines near a dipole

Conductors and insulators

In a metal, some of the electrons are free to move, whereas in an insulator, there are no mobile charges. For a metal in static equilibrium, the electric field inside is zero and all excess charges reside on the metal's surface. The electric field inside an insulator need not be zero.

DNA fingerprinting

The technique of ***electrophoresis*** uses the electric force to separate ions according to their size and charge. This is the basis of DNA fingerprinting.

SSM = answer in Student Companion & Problem-Solving Guide   ⊗ = life science application

1. When a glass rod is rubbed with silk, the rod becomes positively charged, but when a rubber rod is rubbed with fur, it becomes negatively charged. Suppose you have a charged object but don't know whether it carries a positive or a negative charge. Explain how you could use a glass rod and piece of silk to determine the sign of the charge on the unknown object.

2. Explain why two electric field lines cannot cross.

3. Determine the sign of each charge q_a, q_b, and q_c in Figure Q17.3.

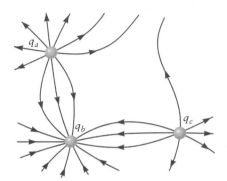

Figure Q17.3 Questions 3 and 4.

4. Consider the charges q_a, q_b, and q_c in Figure Q17.3. In terms of magnitude, which charge is the greatest? Which is the smallest? How do you know?

5. SSM Explain how two objects can be attracted due to an electric force, even when *one* object has *zero* net charge.

6. Explain how two objects can be attracted due to an electric force, even when *both* objects have *zero* net charge.

7. It is found that a charged point particle is repelled from a glass rod that has been charged by rubbing with silk. What is the sign of the charge on the particle?

8. The children in Figure Q17.8 are rubbing balloons on their hair and then placing the balloons on the wall and ceiling. If the rubbing process puts excess electrons on the balloon, how does the balloon then stay attached to the ceiling or wall?

Figure Q17.8

© Cengage Learning/Charles D. Winters

9. The end of a charged rubber rod will attract small pellets of Styrofoam that, having made contact with the rod, will move violently away from it. Describe why that happens.

10. Two point particles, each of charge Q, are located on the x axis at $x = \pm L$. Another charge q is now placed at the origin. (a) Show that the total force on q is zero. (b) Suppose q is now moved a very small distance away from the origin, along the x axis. What is the direction of the electric force on q? (c) Based on your result in part (b), explain why the origin is, or is not, a point of *stable* equilibrium for q.

11. Under normal atmospheric conditions, the surface of the Earth is negatively charged. What is the direction of the electric field near the Earth's surface?

12. When two objects (such as a glass rod and a silk cloth) are rubbed together, electrons can be transferred from one to the other. Can protons also be transferred? Explain why or why not.

13. In Example 17.1, we mentioned that there are approximately 4×10^{14} atoms in a dust particle of mass 1 ng. Estimate this number for yourself.

14. Explain why the child's hair in Figure 17.22 is "standing on end."

15. SSM Describe at least three ways in which Coulomb's law is similar to Newton's law of gravitation. Discuss at least two ways in which the two laws are different.

16. Explain why a test charge used to measure a field must be small in comparison with the field's source charge.

17. A charge is placed inside a partially inflated balloon. If the balloon is then further inflated to a larger volume and the associated surface area increases, does the electric flux though the balloon's surface change? Why or why not?

18. The electron was not discovered until well after the laws of electricity were determined. Protons were discovered even later, so scientists such as Coulomb and Gauss did not know that electric charge is quantized. How do you think that affected their way of thinking about electric phenomena?

19. An ion is released from rest in a region in which the electric field is nonzero. If the ion moves in a direction antiparallel (opposite) to the direction of the electric field, is the ion positively charged or negatively charged?

20. Figure Q17.20 shows the electric field lines outside several Gaussian surfaces, but does not show the electric charges inside. In which cases is the net charge inside positive, negative, or perhaps zero? How do you know?

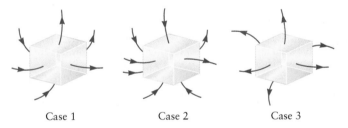

Case 1   Case 2   Case 3

Figure Q17.20

17.1 EVIDENCE FOR ELECTRIC FORCES: THE OBSERVATIONAL FACTS

1. What is the total charge of 1 mole of electrons?

2. You are given a container with 5.0 g of hydrogen atoms. What is the total charge on the electrons in the container? What is the total charge on the protons?

3. A piece of amber is charged by rubbing with a piece of fur. If the net excess charge on the fur is -8.5 nC (-8.5×10^{-9} C), how many electrons were added to the amber?

4. SSM ★ R Approximately how many electrons are in your body? What is the total charge of these electrons?

5. What is the net charge on an object that has an excess of 45 electrons?

6. ★ R Approximately how many electrons are in a penny? Assume the penny is made of pure copper.

7. The excess charge per unit area on a surface is $\sigma = -5.0 \times 10^{-3}$ C/m^2. How many excess electrons are there on a 1.0-mm^2 piece of the surface?

8. The charge per unit length on a glass rod is $\lambda = -7.5$ μC/m. If the rod is 2.5 m long, how many excess electrons are on the rod?

17.2 ELECTRIC FORCES AND COULOMB'S LAW

9. What is the magnitude of the electric force between two electrons separated by a distance of 0.10 nm (approximately the diameter of an atom)?

10. ★ Two point charges are separated by a distance r. If the separation is reduced by a factor of 1.5, by what factor does the electric force between them change?

11. What is the magnitude of the electric force between an electron and a proton in a hydrogen atom? Assume they are point charges separated by 0.050 nm.

12. Two particles with electric charges Q and $-3Q$ are separated by a distance of 1.2 m. (a) If $Q = 4.5$ C, what is the electric force between the two particles? (b) If $Q = -4.5$ C, how does the answer change?

13. ✪ Which graph in Figure P17.13, a, b, or c, best describes how the electric force between two point charges varies with their separation, r?

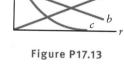

Figure P17.13

14. ★ A point charge with $q_1 = +2.5$ C is located at $x = -3.0$ m, $y = 0$, and a second point charge with $q_2 = +4.0$ C is at $x = +1.0$ m, $y = +2.0$ m. (a) What is the force exerted by q_1 on q_2? (b) What is the force exerted by q_2 on q_1?

Figure P17.14

Figure P17.15

15. ★ R Four point particles are located at the corners of a square (Fig. P17.15). The two particles on the left each carry a charge

$+Q$. If the electric field at the center of the square is directed to the right (along the $+x$ axis), what might the charges on the particles on the right be?

16. ✪ Two particles of charge Q and $5Q$ are located as shown in Figure P17.16. When a third charge is placed at the origin, it is found that the force on it is zero. Find α.

Figure P17.16

17. Two small, identical particles have charges $Q_1 = +3.0$ μC and $Q_2 = -5.0$ μC. If the electric force between the particles is 120 N, what is the distance between the particles?

18. ★ The particles in Problem 17 are conducting and are brought together so that they touch. Charge then moves between the two particles so as to make the excess charge on the two particles equal. If the particles are then separated by a distance of 5.3 mm, what is the magnitude of the electric force between them?

19. ✪ Three charges with $q = +7.5$ μC are located as shown in Figure P17.19, with $L = 25$ cm. (a) What are the magnitude and direction of the total electric force on the charge at the bottom? (b) What are the magnitude and direction of the total electric force on the charge at the right?

Figure P17.19 Problems 19, 20, and 47.

20. SSM ★ Three charges each with positive charge q are located as in Figure P17.19. What are the magnitude and direction of the force on an electron (charge $-e$) at the origin?

21. R You are given two boxes containing electrons, with 1.0 gram of electrons in each box. If these boxes are separated by 1.0 meter, estimate the gravitational force between them. Compare this force with the electric force found in Equation 17.7. Assume the boxes are both very small so that you can use the point-particle approximation.

22. ✪ Two electric dipoles with charges $\pm Q$, where $Q = 5.0$ C, are arranged as shown in Figure P17.22. The charges $\pm Q$ on the left are attached to each other by a rigid rod, and the two charges on the right are also rigidly attached to each other. Find the electric force between the two dipoles.

Figure P17.22

23. ✪ Particles of charge Q and $3Q$ are placed on the x axis at $x = -L$ and $x = +L$, respectively. A third particle of charge q is placed on the x axis, and it is found that the total electric force on this particle is zero. Where is the particle?

24. ✪ A point charge $q_1 = -1.5$ C is at the origin, and a second point charge $q_2 = +5.0$ C is at the point $x = 1.2$ m, $y = 2.5$ m. (a) Find the x and y coordinates of the position at which an electron would be in equilibrium. (b) Find the x and y coordinates of the position at which a proton would be in equilibrium. (c) Suppose the charges q_1 and q_2 were both doubled. How would your answers change? Why?

25. ✪ Consider an electron and a proton separated by a distance of 1.0 nm. (a) What is the magnitude of the gravitational force between them? (b) What is the magnitude of the electric force between them? (c) How would the *ratio* of these gravitational and electric forces change if the distance were increased to 1.0 m?

26. ✪ You are on vacation in an alternate universe where nearly all the laws of physics are the same as in your home universe, but you notice that the charge on an electron in the alternate universe is $-0.9999e$ and the charge on a proton is $+e$. Consider now two spheres of copper, each with mass 1.0 kg. (a) What is the total charge on one of the spheres? (b) What is the electric force between the two spheres when they are separated by a distance of 0.50 m? Is this force attractive or repulsive? (c) Compare the magnitude of the electric force with the magnitude of the gravitational force between the two spheres.

27. ✪ The mass of a typical car is $m = 1000$ kg, so its weight on the Earth's surface is $mg = 9800$ N. Suppose you have two containers, one with N electrons and another with N protons. These two containers are placed a distance 10 m apart, and the electric force between them is equal to the weight of the car. Find N.

28. ✪ A helium nucleus contains two protons. What is the approximate magnitude of the electric force between these two protons? Is this force attractive or repulsive? The radius of this nucleus is about 1.0×10^{-15} m. Assume the protons are point particles.

29. SSM ✪ An electric dipole is situated as shown in Figure P17.29, and the dipole charges are $\pm Q$, with $Q = 3.3$ μC. A point particle with charge q is now placed at location A as shown in the figure. If the force on the particle is 2.5 N and is directed toward the origin, find q.

Figure P17.29 Problems 29 and 30.

30. ✪ Consider again the dipole in Figure P17.29 with $Q = 3.3$ μC. A point particle of charge 6.9 μC is placed at location B, and the magnitude of the electric force on this particle is found to be 0.035 N. Find L.

31. ✪ Two point charges are a distance L apart, and the electric force exerted by one charge on the other is F_E. The charges are then moved farther apart, and the electric force between them decreases to $F_E/9$. What is the new separation between the charges?

17.3 THE ELECTRIC FIELD

32. A small piece of dust of mass $m = 1.0$ μg travels through an electric air cleaner in which the electric field is 500 N/C. The electric force on the dust particle is equal to the weight of the particle. (a) What is the charge on the dust particle? (b) If this charge is provided by an excess of electrons, how many electrons does that correspond to?

33. What is the magnitude of the electric field at a distance of 1.5 m from a point charge with $Q = 3.5$ C?

34. Find the electric field a distance of 1.0 nm from an electron. Is this field directed toward or away from the electron?

35. ✪ A point particle with charge $q = 4.5$ μC is placed on the x axis at $x = -10$ cm. A second particle of charge Q is now

placed on the x axis at $x = +25$ cm, and it is found that the electric field at the origin is zero. Find Q.

36. ✪ A point particle with charge $q = 4.5$ μC is placed on the x axis at $x = -10$ cm and a second particle of charge $Q = 6.7$ μC is placed on the x axis at $x = +25$ cm. (a) Determine the x and y components of the electric field due to this arrangement of charges at the point $(x, y) = (10, 10)$ (the units here are centimeters). (b) Determine the magnitude and direction of the electric field at this point.

37. ✪ A point charge with $q_1 = +2.5$ C is located at $x = -3.0$ m, $y = 0$, and a second point charge with $q_2 = +4.0$ C is at $x = +1.0$ m, $y = +2.0$ m. What are the components of the electric field at $x = +1.0$ m, $y = 0$?

38. A circular ring of radius 25 cm and total charge 500 μC is centered at the origin as shown in Figure P17.38. If the charge is distributed uniformly around the ring, what is the electric field at the origin?

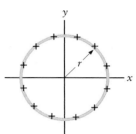

Figure P17.38

39. An electron is traveling through a region of space in which the electric field is along the $+x$ direction and has a magnitude of 2000 N/C. What is the acceleration of the electron? Give both the magnitude and direction of $\vec{a}$.

40. ✪ An interesting (but oversimplified) model of an atom pictures an electron "in orbit" around a proton. Suppose this electron is moving in a circular orbit of radius 0.10 nm (1.0×10^{-10}) and the force that makes this circular motion possible is the electric force exerted by the proton on the electron. Find the speed of the electron.

41. ✪ Sketch the electric field lines near the electric dipole in Figure P17.41. Assume Q is positive.

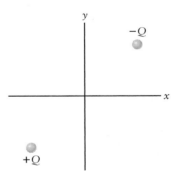

Figure P17.41

42. ✪ Sketch the electric field lines near the two charges in Figure P17.42. Assume Q is positive.

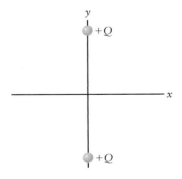

Figure P17.42

43. ✪ A small, plastic sphere of mass $m = 100$ g is attached to a string as shown in Figure P17.43. There is an electric field of 100 N/C directed along the $+x$ direction. If the string makes an angle $\theta = 30°$ with the y axis, what is the charge on the sphere?

Figure P17.43

44. ✪ An electron is traveling through a region between two metal plates in which there is a constant electric field of magnitude E directed along the y direction as sketched in Figure P17.44. This region has a total length of L, and the electron has an initial velocity of v_0 along the x direction. (a) How long does it take the electron to travel the length of the plates? (b) What are the magnitude and direction of the electric force on the electron while it is between the plates? (c) What is the acceleration of the electron? (d) By what distance is the electron deflected (Δy) when it leaves the plates? (e) What is the final velocity of the electron? (Give the magnitude and direction or give the components of the velocity along x and y.)

Figure P17.44

45. ☆ Two point particles with charges q_1 and q_2 are separated by a distance L as shown in Figure P17.45. The electric field is zero at point A, which is a distance $L/4$ from q_1. What is the ratio q_1/q_2?

Figure P17.45

46. ✪ Five point charges, all with $q = 7.5$ C, are spaced equally along a semicircle as shown in Figure P17.46. If the semicircle has a radius of 2.3 m, what are the magnitude and direction of the electric field at the origin?

47. SSM ☆ Three point charges all with $q = -8.2$ C are located as shown in Figure P17.19 with $L = 4.5$ m. What are the magnitude and direction of the electric field (a) at the origin and (b) at the point $y = 6.8$ m on the y axis?

Figure P17.46

17.4 CONDUCTORS, INSULATORS, AND THE MOTION OF ELECTRIC CHARGE

48. Consider a metal sphere onto which a large number (suppose 10^6 or more) of extra electrons is placed (Fig. P17.48). (a) Explain how these electrons will

Figure P17.48
Problems 48 and 50.

be distributed over the interior and the surface of the sphere. (b) Draw the electric field lines inside and outside the sphere.

49. SSM One million electrons are placed on a solid chunk of metal shaped as shown in Figure P17.49. Draw the electric field lines inside and outside the metal.

Figure P17.49

50. Assume the sphere in Figure P17.48 is a piece of plastic (an insulator). How will that affect (change) the answer to Problem 48?

17.5 ELECTRIC FLUX AND GAUSS'S LAW

51. Consider an electron and the surface that encloses it in Figure P17.51. What is the electric flux through this surface?

Figure P17.51
Problems 51 and 52.

52. ☆ (a) Explain how the answer to Problem 51 would change if the surface were increased in size (expanded) by a factor of two. (b) How would the answer to Problem 51 change if three more electrons were placed inside the surface?

53. An electric dipole is enclosed in a spherical surface. What is the electric flux through the surface?

54. ☆ A cylinder has a volume V and a charge density ρ (Fig. P17.54). What is the electric flux through a sphere of radius R that encloses the cylinder?

Figure P17.54

55. ☆ Three point charges are located near a spherical Gaussian surface of radius R (Fig. P17.55). One charge ($+3Q$) is inside the sphere, and the others are a distance $R/3$ outside the surface. What is the electric flux through this surface?

56. SSM ☆ Consider a very long, very thin plastic rod as sketched in Figure P17.56. Assume the rod has a length of $L = 1.0 \times 10^6$ m and a total charge of 1.0 C distributed evenly along the rod. Your job in this problem is to calculate the electric field a distance $r = 10$ cm from the rod. (a) What is the direction of the electric field? Give a qualitative answer. Is the field directed toward or away from the rod? (b) What Gaussian surface matches the symmetry of this problem? Can you pick a surface on which the electric field is constant and perpendicular to the surface or zero at all points on the surface? (c) What is the total charge inside your Gaussian surface? (d) Use Gauss's law to find the electric field.

Figure P17.55

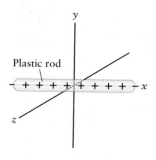

Figure P17.56

57. ✪ Use the step-by-step approach in Problem 56 to find the electric field produced by the hollow metal cylinder with charge per unit length $\lambda = 1.0 \times 10^{-6}$ C/m in Figure P17.57. Be sure to find the electric fields inside *and* outside the cylinder, assuming you are far from the ends and that the cylinder is extremely long.

Figure P17.57

58. ✿ Consider three very large, parallel planes of charge that are equally spaced as shown in Figure P17.58. These planes are insulators, and each is charged uniformly with $+10\ \mu C$ on every square centimeter of area on the top surface of each plane. Use Gauss's law to determine the field at points A, B, C, and D.

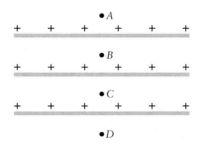

Figure P17.58 Problems 58 and 59.

59. ✿ Consider again the charged planes in Problem 58 (Fig. P17.58), but now assume the inner plane has a charge density of $-10\ \mu C/cm^2$ and the outer planes have charge densities $+10\ \mu C/cm^2$. Find the electric field at points A, B, C, and D.

60. ✿ A point charge q is located at the center of a cubical box with an edge length L. What is the electric flux through one face of the box?

61. ✪ Consider an arrangement of eight charges Q at the corners of a cube of size L. A spherical surface of radius $3L$ is arranged so as to completely contain these charges. What is the electric flux through this surface?

62. Consider the spherical Gaussian surface shown in Figure P17.62. It is situated near a point charge Q which is located a distance $2r$ from the center of the sphere and outside the Gaussian surface. Compute the net electric flux through this surface.

Figure P17.62

63. ✿ An insulating sphere has a total excess charge Q distributed uniformly throughout the sphere. Use Gauss's law to calculate the electric field inside the sphere at a distance r from the center of the sphere. Express your answer in terms of the radius of the sphere r_s and the total charge on the sphere Q.

64. ✪ Consider a solid metal sphere of radius R. An excess charge Q is placed on the sphere. (a) How will this excess charge be distributed on the sphere? (b) Make a qualitative sketch of the electric field lines outside the metal sphere. (c) Choose a Gaussian surface that matches the symmetry of the electric field. That is, choose a surface on which the electric field is a constant. (d) Apply Gauss's law to find the electric flux through your Gaussian surface. (e) Solve for the electric field a distance r from the center of the sphere, with $r > R$.

65. For the metal sphere in Problem 64, what is the electric field *inside* the sphere?

66. ✪ A spherical metal shell has a charge per unit area σ and a radius R. What is the magnitude of the electric field at a distance x from the surface of the sphere?

67. The electric flux through a large cardboard box is $500\ N \cdot m^2/C$. If the box is a cube whose edges are 1 m, what is the net charge inside the box?

68. ✿ Figure P17.68 shows a cylindrical capacitor; it consists of a solid metal rod of radius r_1 surrounded by a metal cylinder with inner radius r_2 and outer radius r_3. Suppose the capacitor has a length L (with L very large). Also suppose a charge $+Q$ is placed on the inner rod and a charge $-Q$ is placed on the outer cylinder. (a) Find the electric field inside the rod (i.e., for $r < r_1$). (b) Find the electric field in the space between the rod and the cylinder ($r_2 > r > r_1$). (c) Find the electric field outside the capacitor ($r > r_3$).

Figure P17.68

69. ✿ Consider a uniformly charged sphere, with radius R and charge per unit volume ρ. What is the magnitude of the electric field a distance $R/3$ from the center of the sphere?

17.6 APPLICATIONS: DNA FINGERPRINTING

70. ✖ An electrophoresis experiment is performed on two fragments of DNA with radius $r_1 = 20$ nm and charge $q_1 = -5e$, and $r_2 = 30$ nm. The speed of the smaller fragment is two times greater than the speed of the larger fragment. What is the charge on the larger fragment?

71. ✿ ✖ Consider two fragments of DNA of radius $r_1 = 30$ nm and $r_2 = 35$ nm when they are coiled up as they move through a gel. An electrophoresis analysis (Fig. 17.36) is carried out to separate these two fragments. Suppose the drag factor C in Equation 17.19 is $C = 5.0 \times 10^{-5}$ kg/(m · s), and assume each fragment has a charge $q = -10e$. (a) If the electric field has a magnitude of 1.0 N/C, what is the speed of each fragment in the gel? (b) How long will it take the smaller fragment to move a distance 1 cm (a typical distance used in DNA fingerprinting)? (c) After the time in part (b) has elapsed, what is the separation of the two different-size fragments in the electrophoresis tube?

72. SSM ✪ ✖ A sample of DNA is prepared for fingerprinting. It contains three fragments of different sizes, all having charge $-8e$. After a certain amount of time, the fingerprint pattern in Figure P17.72 is found. What are the relative sizes of the three fragments? That is, assuming the fragments are spheres with radii r_A, r_B, and r_C, what are the ratios r_B/r_A and r_C/r_A?

Figure P17.72

17.7 "WHY IS CHARGE QUANTIZED?" AND OTHER DEEP QUESTIONS

73. SSM A proton consists of three quarks of charge $+\frac{2}{3}e$, $+\frac{2}{3}e$, and $-\frac{1}{3}e$. The average spacing between the quarks is approximately 1.0×10^{-15} m. Assuming the quarks are arranged to form an equilateral triangle, find the magnitude of the total force on each quark due to the other two quarks. Do you think this arrangement of quarks is stable? Assume quarks are point charges.

74. Repeat Problem 73 for a neutron. A neutron consists of three quarks of charge $+\frac{2}{3}e$, $-\frac{1}{3}e$, and $-\frac{1}{3}e$.

75. ★ A balloon of N_2 at atmospheric pressure and room temperature has a volume of 1.5 m³. What is the total charge of the electrons in the balloon?

76. Two infinite, parallel planes have excess charge densities σ and -2σ. If $\sigma > 0$, what are the magnitude and direction electric field between the planes?

77. ✪ Consider a half-infinite line of charge with charge per unit length λ. This line begins at the origin and then goes to infinity to the left (Fig. P17.77). Find the component of the electric field in the direction shown (the y direction, perpendicular to the line) at a distance r from the end of the line.

Figure P17.77

78. ✪ Two infinite, parallel, charged planes have charge per unit area $\sigma_1 = +3\sigma$ and $\sigma_2 = -\sigma$, respectively, with $\sigma > 0$ (Fig. P17.78). What are the magnitude and direction of the electric field at points A, B, and C?

Figure P17.78   **Figure P17.79**

79. ★ Two infinite, parallel, charged planes have charge per unit area $\sigma_1 = +\sigma$ and $\sigma_2 = -2\sigma$, respectively (Fig. P17.79). A small and electrically neutral bit of paper is dropped so as to fall between the plates. Will the paper be deflected by the electric field? If so, in what direction?

80. ✪ Two very large parallel plates have charge per unit area $+2.0 \ \mu C/m^2$ and $-2.0 \ \mu C/m^2$, respectively. A small grain of pollen of mass $m = 200$ mg and charge q hangs by a slender thread as shown in Figure P17.80. If the thread makes an angle $\theta = 30°$ with the vertical direction, what is q?

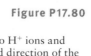

Figure P17.80

81. ✪ ✖ A water molecule (H_2O) is shaped as shown in Figure P17.81A. The hydrogen–oxygen bond length is about 9.6×10^{-11} m, and the bond angle is $\theta = 104°$. To a first approximation, the molecule is composed of three point ions (two H^+ ions and one O^{-2} ion). (a) What are the magnitude and direction of the electric field at the point midway between the O^{-2} ion and the line connecting the two H^+ ions? (b) We can model this system as the dipole in Figure P17.81B. If the electric field at the middle of this dipole is the same as your result in part (a), what is the

Figure P17.81

effective charge Q? (c) Use the effective dipole model to calculate the force between two water molecules that are separated by a distance of 0.30 nm (the typical separation in liquid water). Assume the dipoles of the two water molecules are oriented in opposite directions with their axes perpendicular to the line connecting the centers of the dipoles.

82. ★ **Charged planet.** Lightning is striking somewhere on the planet at all times of the day. This and other meteorological phenomena produce a sustained vertical electric field—pointing inward toward the center of the Earth—that averages about 140 N/C near the surface. What is the average net charge per unit area on the Earth's surface?

83. ★ A small drop of water, measuring 0.011 mm in diameter, hangs suspended above the ground due to the Earth's electric field (see Problem 82). How many extra electrons are on the drop of water?

84. ✪ Imagine that the Moon is locked into its orbit around the Earth by electrical forces and the gravitational force between the Earth and Moon is somehow "turned off." (a) What magnitude charge would need to be on each body? Approximate the Moon and the Earth as point particles and assume both have the same magnitude of charge. (b) How many moles of electrons would that be?

85. ✪ Repeat Problem 84. This time, assume the charges on the Moon and the Earth are not the same magnitude, but the ratio of the charges are the same as the ratio of their masses. Find (a) the charges on the Moon and the Earth and (b) the corresponding number of moles of electrons on each.

86. SSM ★ An electron gun in the picture tube of a television set accelerates electrons from rest to a speed of 4.0×10^7 m/s along a distance of 1.0 cm. What is the magnitude of the uniform electric field used by the gun?

87. ✪ An electron with an initial velocity of 3.1×10^6 m/s enters a parallel-plate capacitor at an angle of 40° through a small hole in the bottom plate as shown in Figure P17.87. (a) If the plates are separated by a distance $d = 2.4$ cm, what minimum electric field must be present to prevent the electron from hitting the top plate? (b) Which plate has to have the positive charge? Assume there is a vacuum between the plates.

Figure P17.87 Problems 87 and 88.

88. ✪ Reconsider part (a) of Problem 87, replacing the electron with a proton and reversing the direction of the electric field. Compare your answer with the results obtained with an electron.

89. ✪ Consider a beam of protons traveling though a vacuum, where each proton has a velocity of 6.0×10^6 m/s. You need to design a means to change the direction of the proton beam by 90°. Consider the charged parallel plates separated by a distance

$d = 60$ cm as depicted in Figure P17.89. Small holes have been drilled into the bottom plate at right angles to one another to allow the beams to follow the trajectory shown. Calculate the electric field between the plates.

Figure P17.89

90. ✪ **Spray it to say it.** A particular inkjet printer uses electrostatic forces to direct droplets of ink to create letters, words, and images at a resolution of 300 dots per inch. Figure P17.90 shows a schematic of the process. Using ultrasound, 100,000 droplets of ink are produced per second out of the droplet gun in the inkjet's head, which moves across the page horizontally. Each drop has a diameter of 70 μm and moves at 20 m/s. Vertical displacement of the drops is achieved by placing a charge on each drop with an electron gun and then deflecting the drop through a uniform electric field provided by charged parallel plates. (a) Assume ink has a density equal to that of water (1000 kg/m³). What is the mass of each drop? (b) Each ink drop is sprayed with electrons until it has a charge of -650 nC. How many excess electrons does that correspond to? (c) A particular drop experiences an electric field of 100 N/C as it passes through the 0.50-mm-long parallel plates. How much is it displaced vertically as it leaves the plates? (d) What is the vertical component of velocity as it leaves the plates? (e) If the plates are not charged, the drop would land in the center of the paper, which is an additional 1.5 mm away from the plates. How far below this point does the drop strike the paper?

Figure P17.90

91. ✪ A charge of 1.8 nC is be placed in the center of three solids of equal volume as shown in Figure P17.91. Calculate the electric flux out of *one face* of the (a) tetrahedron if the charge is placed at its center. Repeat for (b) the cube and (c) the dodecahedron.

Figure P17.91

92. ✪ Eight point particles, each of charge Q, are located on the corners of a cube. What is the direction of the electric force on one of the particles from the others?

93. ✪ Two particles with the same charge are attached to strings that are 80 cm long and that hang from a ceiling as shown in Figure P17.93. If the angle between the strings is $\theta = 60°$ and the particles each have a mass of 250 g, what is the charge on each particle?

Figure P17.93

94. ✪ Consider three particles of charge Q, $2Q$, and $4Q$, where $Q = 2.4$ C, arranged as shown in Figure P17.94. If a particle of charge $q = -1.5$ C is placed at the origin, (a) what are the components of the electric force on this particle along both the x and y directions? (b) What are the components of the electric force on the particle of charge $4Q$?

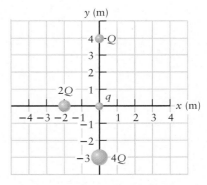

Figure P17.94

Chapter 18

Electric Potential

These sparks are one example of the motion of electric charges. This motion is determined by how the electric potential energy varies from place to place. (© Victor de Schwunberg/Photo Researchers, Inc.)

In Chapter 17, we discussed electric forces and electric fields, and learned how to calculate them using Coulomb's law and Gauss's law. From Chapter 6, we know that when a force acts on an object, the force may do mechanical work on the object. It should thus be no surprise that electric forces can do work on a charged object. Electrical work is related to *electric potential energy*, which is analogous in many ways to gravitational potential energy. Using electric potential energy, we will define a closely related quantity called the *electric potential*. These quantities and the underlying ideas are quite similar to mechanical potential energy and will lead us again to the principle of conservation of energy that we explored earlier in mechanical systems (Chapter 6). Extending these ideas to electric forces and systems will give us a more complete picture of electric phenomena and provide the basis for our work on electric circuits in Chapter 19.

18.1 | ELECTRIC POTENTIAL ENERGY

Let's begin by reviewing the relationship between force and work. Figure 18.1A shows a region of space in which the electric field is constant, so $\vec{E}$ has the same magnitude and direction at all points. A point charge q in this region experiences an electric force

$$\vec{F} = q\vec{E} \tag{18.1}$$

If the charge q is positive, this force is parallel to $\vec{E}$ as shown in Figure 18.1A.

Suppose this charge moves a distance Δx, starting at point A and ending up at point B, and for simplicity we assume this displacement is parallel to the electric force $\vec{F}$. According to the definition of work in Chapter 6, the work done by the electric force on the charge is

$$W = F\,\Delta x \tag{18.2}$$

The electric force is conservative, so the work done on the charge is independent of the path it takes to go from A to B. We can now define the *electric potential energy*, which we denote by PE_{elec}. From Chapter 6, we know that the change in the potential energy associated with a particular conservative force is equal to $-W$, where W is the work done by that force (see Eq. 6.12). So, if the electric force does an amount of work W on a charged particle, the change in the electric potential energy is

$$\Delta PE_{\text{elec}} = -W \tag{18.3}$$

Combining this equation with Equations 18.1 and 18.2, the change in electric potential energy when the charged particle moves from A to B in Figure 18.1A is

$$\Delta PE_{\text{elec}} = -W = -F\,\Delta x = -qE\,\Delta x \tag{18.4}$$

Change in electric potential energy

Equation 18.4 gives the *change* in the potential energy as the charge moves through a displacement Δx, in a region where the electric field is parallel to the displacement. In Equation 18.4, the change in the potential energy depends on the starting and ending locations, but does *not* depend on the path taken. In Figure 18.1A, the displacement is along a line, but the charge may move from A to B along many other paths without affecting ΔPE_{elec}. All conservative forces, including the electric force, have this property: changes in the potential energy and the work done are independent of the path taken between two points. In the simple example shown in Figure 18.1A, the electric field is a constant, but the same results hold when $\vec{E}$ varies with position. In this more general case, the potential energy function will have a different mathematical form than the one in Equation 18.4, but the electric force is still conservative.

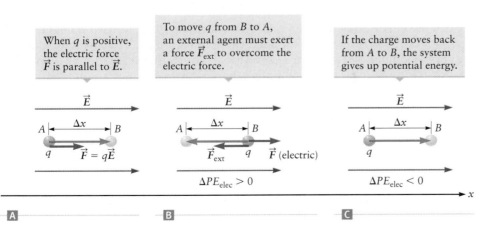

When q is positive, the electric force $\vec{F}$ is parallel to $\vec{E}$.

To move q from B to A, an external agent must exert a force $\vec{F}_{\text{ext}}$ to overcome the electric force.

If the charge moves back from A to B, the system gives up potential energy.

$\Delta PE_{\text{elec}} > 0$

$\Delta PE_{\text{elec}} < 0$

Figure 18.1 A The electric force on a charged particle can do work on the particle. B and C This work is connected with changes in the electric potential energy PE_{elec} of the particle. Here we assume a positively charged particle ($q > 0$).

A

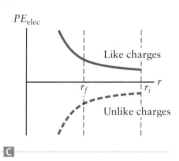

B

C

Figure 18.2 **A** Two point charges exert an electric force on each other. If the separation between the two charges changes, this force does work. **B** Plot of the force and **C** electric potential energy as functions of the separation between q_1 and q_2. In B a negative value of F indicates an attractive force.

Electric potential energy of two point charges

Potential Energy Is Stored Energy

Electric potential energy is stored through the "potential" effect of the electric field on an electric charge. This effect is illustrated in Figure 18.1B, where the charge q is moved from point B to point A by a force $\vec{F}_{ext}$ directed to the left; this force is exerted by an external "agent," which might be your hand. The electric force on q is directed to the right (we assume q is positive), so the displacement is opposite to the electric force. The work done *by the electric field* on the particle is thus *negative*, and according to Equations 18.3 and 18.4 the change in the electric potential energy is *positive*. In words, we have stored a positive amount of energy in the system that is composed of the charge q and the electric field. That energy came from the positive amount of work done by the external agent (and the external force $\vec{F}_{ext}$) that moved the charge from B to A.

The entire process can be reversed if we allow the charge to move from A back to B again (Fig. 18.1C). Now the electric field does a positive amount of work on the particle because the electric force and the particle's displacement are parallel and the change in the potential energy is negative (Eq. 18.4). Hence, stored energy is taken out of the system, and this energy might show up as an increase in the kinetic energy of the particle when it reaches B.

Potential Energy of Two Point Charges

Let's now consider the electric potential energy for the slightly more complicated case of two charges q_1 and q_2 in Figure 18.2A. From Coulomb's law, we know that for two point charges separated by a distance r, the magnitude of the electric force exerted by q_1 on q_2 is

$$F = \frac{kq_1q_2}{r^2} \qquad (18.5)$$

If we now assume they are like charges (i.e., both positive or both negative), the force is repulsive. The solid curve sketched in Figure 18.2B shows how F varies as a function of the separation r. If q_2 is brought closer to q_1, the electric force is directed opposite to the displacement (Fig. 18.2A), so the work done *by the electric field* on q_2 is negative. From our relation between work and potential energy (Eq. 18.3), we know that the electric potential energy *increases*, so $\Delta PE_{elec} > 0$. Mathematically, it can be shown that this electric potential energy is given by

$$PE_{elec} = \frac{kq_1q_2}{r} \qquad (18.6)$$

We can also write PE_{elec} in terms of the permittivity of free space ε_0. From Chapter 17, $k = 1/(4\pi\varepsilon_0)$; using this expression in Equation 18.6 leads to

$$PE_{elec} = \frac{q_1q_2}{4\pi\varepsilon_0 r} \qquad (18.7)$$

We can use either Equation 18.6 or 18.7 to calculate PE_{elec} for two point charges.

A plot of PE_{elec} is shown as the solid curve in Figure 18.2C. This is qualitatively similar to Coulomb's law for the electric force (Eq. 18.5), *except* that PE_{elec} varies as $1/r$, not $1/r^2$. If the charges are initially separated by a distance r_i and then brought together to a final separation r_f, the change in the potential energy is

$$\Delta PE_{elec} = PE_{elec,\,f} - PE_{elec,\,i} = \frac{kq_1q_2}{r_f} - \frac{kq_1q_2}{r_i} \qquad (18.8)$$

We know that only changes in potential energy are important, so we could add a constant value to PE_{elec} in Equation 18.6 without changing ΔPE_{elec}. Conventional practice, however, is to use Equation 18.6 without any added constant, so PE_{elec} approaches zero when the two charges are very far apart (i.e., when r becomes infinitely large). The electric force also approaches zero in this limit.

The general behavior of the electric potential energy for two like charges is shown as the solid curve in Figure 18.2C. In this case, PE_{elec} is *positive* and becomes larger and larger as the charges are brought together. On the other hand, if the charges have opposite sign, the electric force is attractive; that is, it is negative as plotted by the dashed curve in Figure 18.2B. The potential energy is then also *negative* (according to Eq. 18.6) and is shown by the dashed curve in Figure 18.2C.

Insight 18.1

SIMILARITY BETWEEN ELECTRIC AND GRAVITATIONAL POTENTIAL ENERGY

The electric potential energy of two point charges (Eq. 18.6) falls to zero as the separation between the charges is made very large (infinite). This behavior is similar to that of the gravitational potential energy between two masses (Eq. 6.22), which also varies as $1/r$ and falls to zero when the two masses are very far apart.

EXAMPLE 18.1	Potential Energy of a Hydrogen Atom

A very simple model of a hydrogen atom is sketched in Figure 18.3, which shows an electron traveling in a circular orbit around proton. (We'll discuss more realistic models of an atom in Chapter 29.) If the distance between the electron and the proton is $r = 5.0 \times 10^{-11}$ m, what is the electric potential energy of this atom?

RECOGNIZE THE PRINCIPLE

We can treat the electron and proton as two point charges. We know their charges and the separation r is given, so we can calculate the potential energy using Equation 18.6.

SKETCH THE PROBLEM

Figure 18.3 shows the problem.

IDENTIFY THE RELATIONSHIPS

The electric potential energy is given by Equation 18.6, with $q_1 = +e$ (the proton) and $q_2 = -e$ (the electron). We have

$$PE_{elec} = \frac{kq_1q_2}{r} = \frac{k(+e)(-e)}{r} = -\frac{ke^2}{r}$$

This potential energy is negative because the electron and proton have opposite charges. This energy holds the atom together and is related to the ionization energy of the atom, the energy required to separate an electron from the rest of the atom. In this case, it is the energy needed to move the electron an infinite distance from the proton.

SOLVE

Inserting the given value of r, we find

$$PE_{elec} = -\frac{ke^2}{r} = -\frac{(8.99 \times 10^9 \text{ N} \cdot \text{m}^2/\text{C}^2)(1.6 \times 10^{-19} \text{ C})^2}{5.0 \times 10^{-11} \text{ m}}$$

$$PE_{elec} = \boxed{-4.6 \times 10^{-18} \text{ J}}$$

What have we learned?
Our calculation has considered only the potential energy of the electron, but a better calculation of the ionization energy must also include the electron's kinetic energy (Section 29.1). Even so, let's compare our value of PE_{elec} with the measured ionization energy of hydrogen. It is conventional to quote positive values for the ionization energy, so our estimate is 4.6×10^{-18} J. The actual value for a hydrogen atom is approximately 2.2×10^{-18} J, so our simple calculation gives a value within about a factor of two of the correct answer.

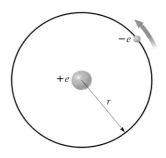

Figure 18.3 Example 18.1. A simplified model of an atom, with an electron in orbit around a proton.

Electric Potential Energy and Superposition

Our result for the electric potential energy of two point charges (Eq. 18.6) can be applied to other situations by using the principle of superposition. For example, if we have a collection of point charges, the total potential energy is the sum of the potential energies of each pair of charges. Note again that this electric potential energy arises from the interaction (the electric force) between two charges, so when we speak of potential energy we must keep in mind that it is the potential energy of a *system* of two particles.

The principle of superposition can be used to prove another important result. We know that the Coulomb's law force between two point charges is conservative, with a potential energy function given by Equation 18.6. More complicated charge distributions can always be treated as a collection of point charges arranged in some particular manner. We can therefore use superposition to conclude that electric forces between a collection of charges will always be conservative, no matter how complex the charge distribution or electric field may be.

Using the Principle of Conservation of Energy

In Chapter 6, we applied the principle of conservation of energy to solve various problems in mechanics, and we usually dealt with gravitational or elastic (spring) potential energy. The same approach can be applied to problems involving electric potential energy. The following problem-solving strategy is adapted from the one given in Chapter 6.

PROBLEM SOLVING Applying the Principle of Conservation of Energy to Problems with Electric Potential Energy

1. **RECOGNIZE THE PRINCIPLE.** First determine the system whose energy is conserved. It might be two point charges, a collection of point charges, or a single charge in an electric field.

2. **SKETCH THE PROBLEM.** Draw a figure showing the initial and final states of all the charges in the system.

3. **IDENTIFY THE RELATIONSHIPS.** Identify the change in the electric potential energy. For a charge in a constant electric field, you can find ΔPE_{elec} using $\Delta PE_{\text{elec}} = -qE\,\Delta x$ (Eq. 18.4), whereas for two point charges we have $PE_{\text{elec}} = kq_1q_2/r$ (Eq. 18.6).

4. **SOLVE.** The next step depends on the problem.

   - If there are no external forces on the system, the sum of the electric potential energy plus kinetic energy is conserved and $KE_i + PE_i = KE_f + PE_f$.
   - If there is a nonzero external force, the work done by that force will equal the change in potential plus kinetic energy.
   - The electric force is conservative, so the work done by the electric force is independent of the path and equals $-\Delta PE_{\text{elec}}$.

5. Always *consider what your answer means* and check that it makes sense.

We'll apply this general approach in the next two examples.

EXAMPLE 18.2 Moving Charges Around

Consider two point charges $+q$ and $-q$ separated by a distance $2r$ as shown in Figure 18.4. A third particle of charge Q is taken from rest very far away by an external force and brought to rest at the origin along the path shown in the figure. How much work is required to push or pull Q against the electric forces from the other charges?

RECOGNIZE THE PRINCIPLE

We start with step 1 in our strategy for applying the principle of conservation of energy to problems with electric potential energy. The charge Q moves in an electric field produced by the charges $\pm q$; these three charges are the system. The initial and final velocities of Q are zero, so its kinetic energy is zero at both the start and end. According to the principle of conservation of energy, the total work done on Q (the work done by the electric forces plus the work done by the external agent) must be zero. (The work done by one is the negative of the work done by the other.) So, the work done on Q by the external agent that moves it to the origin must equal ΔPE_{elec}, the change in its electric potential energy.

Figure 18.4 Example 18.2. A charge Q is taken from very far away and then moved to the origin. Like the gravitational force, the electric force is conservative; for both forces, the work done is independent of the path taken.

Step 2 in our strategy is the sketch in Figure 18.4, which shows one possible path the charge Q might take to get to the origin.

IDENTIFY THE RELATIONSHIPS

Step 3: We use the result for the change in the potential energy in Equation 18.8. We apply this relation twice, once for the pair of charges $+q$ and Q, and again for the pair $-q$ and Q. The charge Q begins very far from the others, so r_i is very large (infinite) in both cases, whereas the final location has $r_f = r$ for both pairs (since r is the final distance of charge Q from each of the other two charges).

SOLVE

Step 4: We solve for the change in the electric potential energy:

$$\Delta PE_{elec} = \frac{kqQ}{r} - \frac{kqQ}{r_i} + \frac{k(-q)Q}{r} - \frac{k(-q)Q}{r_i}$$

The terms involving r_i are each zero because r_i is infinite, whereas the other two terms (involving r) have opposite signs and cancel. Hence,

$$\Delta PE_{elec} = 0$$

The work done by the external agent that moves the charge to the origin is thus zero .

What does it mean?
The work required to move Q to a point midway between two particles of charge $+q$ and $-q$ is zero because the two particles have opposite charge, making their contributions to the total potential energy cancel. Note also that ΔPE_{elec} is independent of the path taken by Q.

EXAMPLE 18.3 Moving Charges (Again)

A point charge q with mass m is initially at rest, a distance L from a second charge Q (Fig. 18.5). Both charges are positive, so they repel each other. Charge Q is held fixed at the origin, whereas q is released and travels to the right. What is the speed of charge q when it reaches a distance $2L$ from the origin?

RECOGNIZE THE PRINCIPLE

We again follow our strategy for applying the principle of conservation of energy to problems with electric potential energy. Our system is just the two charges Q and q. Because there are no external forces on the system, its energy must be conserved. We can thus write

$$KE_i + PE_i = KE_f + PE_f \qquad (1)$$

Because charge Q is held fixed, KE is the kinetic energy of charge q, and PE is the electric potential energy of the two charges.

SKETCH THE PROBLEM

Step 2: Figure 18.5 shows the problem. This sketch shows both the initial and final states of the system.

IDENTIFY THE RELATIONSHIPS

Step 3: We can find the change in the potential energy using $PE_{elec} = kqQ/r$ (Eq. 18.6). We can then calculate the change in the kinetic energy and, from that, the final speed.

Initial state

Final state

Figure 18.5 Example 18.3.

Charge q is initially at rest, so the initial kinetic energy is $KE_i = 0$. The initial separation between the two charges is L, so the initial potential energy is (Eq. 18.6)

$$PE_i = PE_{\text{elec}, i} = \frac{kqQ}{L}$$

The final separation is $2L$, so

$$PE_f = PE_{\text{elec}, f} = \frac{kqQ}{2L}$$

Denoting the final speed by v_f, the final kinetic energy is

$$KE_f = \tfrac{1}{2}mv_f^2$$

SOLVE

Step 4: Inserting all this information in Equation (1) leads to

$$KE_i + PE_{\text{elec}, i} = 0 + \frac{kqQ}{L} = KE_f + PE_{\text{elec}, f} = \frac{1}{2}mv_f^2 + \frac{kqQ}{2L}$$

Solving for v_f gives

$$\frac{1}{2}mv_f^2 = \frac{kqQ}{L} - \frac{kqQ}{2L} = \frac{kqQ}{2L}$$

$$v_f^2 = \frac{2}{m}\left(\frac{kqQ}{2L}\right) = \frac{kqQ}{mL}$$

$$\boxed{v_f = \sqrt{\frac{kqQ}{mL}}}$$

What have we learned?

Electric forces are conservative, so we can use conservation of energy principles to analyze many situations involving electric forces and electric potential energy.

Figure 18.6 Concept Check 18.1.

CONCEPT CHECK 18.1 | Moving a Charge from Place to Place

A point charge Q is held fixed at the origin while an external force $\vec{F}_{\text{ext}}$ is used to move a second charge q from point A to point B along the three different paths shown in Figure 18.6. The charge q begins from rest and has zero velocity when it reaches B. Is the work done *by the external force* largest along (a) Path I, (b) Path II, or (c) Path III, or (d) is the work done by $\vec{F}_{\text{ext}}$ the same for all three paths?

18.2 | ELECTRIC POTENTIAL: VOLTAGE

Electric potential energy is a property of a *system* of charges or of a point charge in an electric field (where the field is created by other charges). In either case, this electric potential energy is not the property of a single charge alone. For example, in the case of two point charges, the potential energy is associated with the Coulomb's law force that acts between them. For this reason, the potential energy in Equation 18.6 depends on the values of both charges involved in the interaction. Now let's consider a slightly different way of viewing electric potential energy: as a property associated with just the electric field.

In Chapter 17, we introduced the notion of a *test charge* in our definition of the electric field. We can measure the electric field at a particular location in space by placing a test charge q at the location of interest (Fig. 18.7) and then measuring the force on the test charge. The electric force is related to the electric field by $\vec{F} = q\vec{E}$, so we can deduce the field from the measured force using (Eq. 17.10)

$$\vec{E} = \frac{\vec{F}}{q} \qquad (18.9)$$

If this field is produced by a second point charge Q, the magnitude of the electric field is

$$E = \frac{kQ}{r^2}$$

The value of this electric field does *not* depend on the test charge; the test charge merely gives us a convenient way to measure the field.

We now treat the electric potential energy in a similar manner. We place a test charge q at a particular location and measure the potential energy of this charge. We then define the *electric potential* V through the relation

$$V = \frac{PE_{elec}}{q} \qquad (18.10)$$

The electric potential is often referred to as simply "the potential." The SI unit of potential is the volt (V) in honor of Alessandro Volta, 1745–1827. According to Equation 18.10, the volt is related to other SI units by

$$\text{units of electric potential: } 1 \text{ V} = 1 \text{ J/C} = 1 \text{ N} \cdot \text{m/C} \qquad (18.11)$$

The volt is widely employed in the measurement of many electrical quantities. In particular, the electric field is usually measured in units of volts per meter. We thus have several different ways to express the units of the electric field. The ones most commonly used are

$$\text{units of electric field: } 1 \text{ V/m} = 1 \text{ N/C} \qquad (18.12)$$

Using an Electric Potential Difference to Accelerate Charged Particles

Many of the important concepts associated with electric potential are at work in a conventional (and old fashioned) television picture tube. A TV tube uses a beam of electrons to produce an image on a screen. As shown in Figure 18.8, this electron beam is produced by passing a stream of electrons through a region called the "accelerator," which consists of two parallel metal plates containing small holes that allow the electrons to enter and exit the central region. Some external electronic circuitry is attached to the metal plates, placing positive charge on one plate and negative charge on the other. We calculated the electric field between two parallel

Figure 18.7 The electric field at a particular point in space can be determined by measuring the force on a test charge q placed at that point. In a similar way, changes in the potential energy of a test charge as it is moved from place to place can be used to measure changes in the electric potential.

Electric potential V

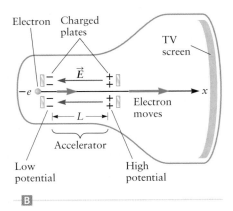

Figure 18.8 Electric field and potential inside a TV tube. **A** Hypothetical arrangement using the electric field to accelerate a positive test charge $+q$. **B** Realistic TV tube, with an electron moving through the accelerator from left to right.

charged plates in Chapter 17, where we found that for plates of a given charge and area this field is constant (Eq. 17.17).

Let's first analyze the case of a positive test charge moving through the accelerator shown in Figure 18.8A. Here the electric field is directed from left to right, which is the $+x$ direction in Figure 18.8A. Suppose the electric field has a magnitude E and the plates are separated by a distance L. We imagine that the test charge q is initially placed at the entry hole on the left and then moves toward the plate on the right. The magnitude of the electric force on the test charge is $F = qE$, and the charge moves a distance L from left to right. The work done on the test charge by the electric field is thus $W = FL = qEL$. From Equation 18.4, we then know that the corresponding change in the electric potential energy is

$$\Delta PE_{elec} = -qEL$$

From the definition of electric potential (Eq. 18.10), the change in the potential is

$$\Delta V = \frac{\Delta PE_{elec}}{q} = -EL \tag{18.13}$$

We say that there is a *potential difference* ΔV between the two plates. The electric field is directed to the right in Figure 18.8A, so the negative sign in Equation 18.13 means that the plate on the left is at a higher potential than the plate on the right. We say that the plate on the left is at "high" potential and the plate on the right is at "low" potential.

Now we replace the test charge by an electron and consider the motion of the electron as it travels through the accelerator. Since an electron is negatively charged, we must reverse the direction of the electric field; the electric field is now negative because it is along the $-x$ direction in Figure 18.8B. This change is needed because we want the electric force on the electron to push it from left to right through the accelerator region. Because the electron is negatively charged, it will tend to move from the region of low potential (the plate on the left) to the region of high potential (the plate on the right) in Figure 18.8B.

The potential difference ΔV between the plates is sometimes a more useful quantity to know than the electric field. Electrons enter the accelerator with a very small velocity, so for simplicity we'll assume their initial velocity is $v_i = 0$. We can use ΔV to find their final velocity v_f as they leave the accelerator by using the principle of conservation of energy. Conservation of energy tells us that

$$KE_i + PE_{elec,\, i} = KE_f + PE_{elec,\, f}$$

The initial kinetic energy is zero (since $v_i = 0$), so we have

$$KE_f = \tfrac{1}{2}mv_f^2 = PE_{elec,\, i} - PE_{elec,\, f} = -\Delta PE_{elec} \tag{18.14}$$

The change in potential energy is related to the change in the electric potential. Using Equation 18.10 gives

$$\Delta PE_{elec} = q\,\Delta V = -e\,\Delta V$$

Inserting this result into Equation 18.14 leads to

$$\tfrac{1}{2}mv_f^2 = -\Delta PE_{elec} = e\,\Delta V \tag{18.15}$$

$$v_f = \sqrt{\frac{2e\,\Delta V}{m}}$$

The potential difference in a TV tube is typically 15,000 V; inserting this value along with the values of the charge and mass of an electron, we find

$$v_f = \sqrt{\frac{2e\,\Delta V}{m}} = \sqrt{\frac{2(1.60 \times 10^{-19}\ \text{C})(1.5 \times 10^4\ \text{V})}{9.11 \times 10^{-31}\ \text{kg}}} = 7.3 \times 10^7\ \text{m/s}$$

So, these electrons are moving quite rapidly.[1] This high velocity and high kinetic energy enables the electrons to produce light when they strike the specially coated screen at the opposite face of the TV tube.

Relating the Electric Potential to the Electric Field

In Figure 18.8, we assumed the electric field is constant in the accelerator region, which is a good assumption in that case. In many situations, though, the electric field varies with position, so let's now ask how the electric field and the electric potential are related in these more general cases. We consider a very small region of space so that the electric field is approximately constant and apply the approach sketched in Figure 18.9, where a test charge q moves a distance Δx parallel to $\vec{E}$. The force on this charge is $F = qE$, and the work done by the field on the charge is $W = F \Delta x = qE \Delta x$. The change in the electric potential energy during this process is then (from Eq. 18.3)

$$\Delta PE_{elec} = -W = -qE \Delta x \qquad (18.16)$$

Using Equation 18.10 leads to an electric potential difference

$$\Delta V = \frac{\Delta PE_{elec}}{q}$$

which gives

$$\Delta V = -E \Delta x \qquad (18.17)$$

Equation 18.17 is a general relation between the electric field E and the variation in the electric potential V with position. If we know E, we can use Equation 18.17 to calculate how the potential changes as we move from place to place. Strictly speaking, this relation only holds for small steps Δx during which the field is constant. We can, however, combine the potential changes ΔV from many such small steps to find the change in the voltage over a large distance. We can also use Equation 18.17 to deduce the electric field from knowledge of how the potential changes with position. Rearranging that equation gives

$$E = -\frac{\Delta V}{\Delta x} \qquad (18.18)$$

In words, this says that the magnitude of the electric field is largest in regions where V is changing rapidly (and ΔV is large). Conversely, the electric field is zero in regions where V is constant. Notice that because of the negative sign in Equation 18.18 the electric field is directed from regions of high potential to regions of low potential.

The relation between the electric field and changes of the potential with position in Equation 18.18 involves the component of the electric field in the direction parallel to the displacement Δx. The electric field is a vector, so if we want to find E in a particular direction, we must consider how the potential V changes as we (or a test charge) move along that direction.

We now have a relation between the electric field and the variation of the potential V with position. This relation holds in simple cases when the electric field is constant as in the TV tube in Figure 18.8 and in more complicated cases in which the electric field varies with position.

| EXAMPLE 18.4 | Electric Field in a TV Tube

Consider the accelerator region of the TV tube in Figure 18.8. If the potential between the two plates is 15,000 V and the plate separation is $L = 2.0$ cm, what is the electric field in this region? Assume the electric field is constant in the accelerator.

---

[1] Our result for v_f is about 25% of the speed of light. We show in Chapter 27 that Newton's mechanics begins to break down at such high speeds. The corrections can be calculated using the theory of special relativity, and they amount to a few percent for these electrons.

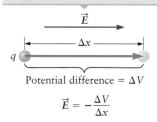

The electric field is proportional to the rate of change of V with position.

$$\vec{E} = -\frac{\Delta V}{\Delta x}$$

Figure 18.9 The magnitude and direction of the electric field are related to how the electric potential V changes with position.

Relation between E and changes in V with position

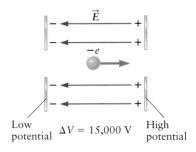

Figure 18.10 Example 18.4.

Low potential $\quad \Delta V = 15{,}000$ V $\quad$ High potential

We can find the electric field using the relation between the electric potential and E in Equation 18.18,

$$E = -\frac{\Delta V}{\Delta x} \qquad (1)$$

SKETCH THE PROBLEM

Figure 18.10 shows the problem. Because the electron has a negative charge, it moves from left to right, from low to high electric potential. The corresponding charge on the plates is also shown.

IDENTIFY THE RELATIONSHIPS

The electric field is constant in the accelerator, so we can take ΔV to be the total potential difference, and Δx is the distance between the plates; we thus have $\Delta V = 15{,}000$ V and $\Delta x = L = 2.0$ cm. Inserting these values in Equation (1), we find

$$E = -\frac{\Delta V}{\Delta x} = -\frac{1.5 \times 10^4 \text{ V}}{0.020 \text{ m}} = \boxed{-7.5 \times 10^5 \text{ V/m}}$$

What does it mean?
The negative sign indicates that the field is directed from high to low potential as shown in Figure 18.10. The electric field vector always points from regions of high V to low V because $\vec{E}$ is parallel to the direction a positive test charge would move if it were placed at that location.

The Electron-volt as a Unit of Energy

The volt is a unit of electric potential, and the product of charge times electric potential has units of energy (Eq. 18.10). We often deal with the motion of electrons or protons, both of which have a charge of magnitude e. Also, we are often concerned with the energy gained or lost as an electron or ion moves through a potential difference measured in volts as with the example in Figure 18.10. For these reasons, it is convenient to define a unit of energy called the ***electron-volt***, the energy gained or lost when an electron travels through a potential difference of 1 V. The electron-volt is abbreviated as "eV"; it is not part of the SI system of units, but we can relate the electron-volt to the joule. Using Equation 18.10, we have

$$V = \frac{PE_{\text{elec}}}{q}$$

which can be rearranged as

$$PE_{\text{elec}} = qV$$

In words, this equation reads

$$\text{energy} = (\text{charge})(\text{potential})$$

Applying this definition to that of the electron volt as a unit of energy, we find

$$1 \text{ eV} = e\,\Delta V = (1.60 \times 10^{-19} \text{ C})(1 \text{ V})$$

$$1 \text{ eV} = 1.60 \times 10^{-19} \text{ J} \qquad (18.19)$$

In Example 18.4 with the TV tube, the electrons gained a kinetic energy $e\,\Delta V = 15{,}000$ eV as they traveled through the accelerator.

CONCEPT CHECK 18.2 | Electron-volts and Potential Changes

In Figure 18.10, an electron moved from left to right through a region in which the electric potential changed by $\Delta V = 15{,}000$ V. The corresponding change in potential energy was $\Delta PE_{elec} = -e\,\Delta V = -15{,}000$ eV. Suppose the electron is replaced by an oxygen ion O^{2-}. What is the change in potential energy of the oxygen ion as measured in electron-volts, (a) -7500 eV, (b) $-15{,}000$ eV, or (c) $-30{,}000$ eV?

Electric Potential Due to a Point Charge

We have described the electric properties of a point charge thoroughly because this case is useful in many situations. In Equation 18.6, we found the electric potential energy of two point charges. We can combine that with the definition of electric potential (Eq. 18.10) to obtain the electric potential a distance r away from a single point charge q:

$$V = \frac{kq}{r} \tag{18.20}$$

Because the constants k and ε_0 are related by $k = 1/(4\pi\varepsilon_0)$, we can also write the potential from a point charge as

$$V = \frac{q}{4\pi\varepsilon_0 r} \tag{18.21}$$

We show this result as the solid curve in Figure 18.11 for a positive charge $(q > 0)$ and as the dashed curve for a negative charge $(q < 0)$.

These results reinforce what we learned about potential energy and the behavior of like and unlike charges. Suppose a charge $q > 0$ is placed at the origin; we'll call it a "source charge" because it produces a potential given by Equations 18.20 and 18.21. We then place a positive test charge nearby. Because both the source charge and the test charge are positive, they repel, and the test charge experiences a force that carries it "downhill" along the solid potential curve in Figure 18.11, away from the origin. On the other hand, if the source charge is negative, the potential is described by the dashed curve in Figure 18.11. A positive test charge would be attracted to this source charge (unlike charges) and would again be carried "downhill," this time along the dashed potential curve.

Only Changes in Electric Potential Matter

Only *changes* in potential energy are important, so we must always be clear about where the "reference point" for potential energy is chosen. Because electric potential is proportional to electric potential energy, we must also pay attention to the reference point for electric potential. For example, when dealing with a point charge and Equation 18.20, the standard convention is to choose $V = 0$ at an infinite distance from the source charge, that is, when $r = \infty$. In many other problems, the usual choice is to take the Earth as $V = 0$ because (from Chapter 17) the Earth is a good conductor of charge. That is the origin of the term **electrical ground** or, simply, **ground** and the convention that the ground is a point where $V = 0$.

Electric Potential and Field Near a Metal

Consider a solid metal sphere as in Figure 18.12A that carries an excess positive charge q, with the charge at rest. To understand the behavior of the electric potential and field, recall from Chapter 17 that the excess electric charge on a metal resides at the surface and the field inside the metal is zero. What about the electric field outside the sphere? Since the metal is spherical, the field must have spherical symmetry; hence, E outside the sphere must depend only on the distance r from the center of the sphere. In Chapter 17, we showed in Example 17.6 that the field outside any spherical ball of charge is given by

$$E = \frac{kq}{r^2} = \frac{q}{4\pi\varepsilon_0 r^2} \tag{18.22}$$

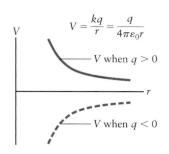

Figure 18.11 Electric potential near a point charge. The solid curve is for a positive point charge, and the dashed curve is for a negative point charge.

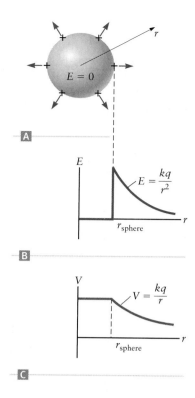

A

B

C

Figure 18.12 **A** Electric field lines near a charged metal sphere. **B** Magnitude of the electric field as a function of the distance r from the center of the sphere. **C** Electric potential as a function of r.

where q is the total charge and r is the distance from the center of the ball. This result is precisely the same as that for the electric field due to a point charge. Equation 18.22 holds only outside the metal sphere; the field inside is zero. The magnitude of the field, both inside and outside, is plotted in Figure 18.12B as a function of the distance r from the center of the sphere.

We can use this result for the electric field to derive the electric potential. Because the field outside the metal sphere is identical to the field of a point charge, the potential outside the sphere is the same as the potential due to a point charge. We thus have (from Eq. 18.20)

$$V = \frac{kq}{r} \quad \text{for } r > r_{\text{sphere}} \tag{18.23}$$

The field inside the sphere is zero, so a test charge placed inside the metal would experience no force. The work done in moving a test charge around inside the metal is thus also zero. From the relation between work and electric potential energy (Eq. 18.3), we know that both the electric potential energy and the electric potential are *constant inside a metal*. Neither the potential energy nor the electric potential can jump discontinuously (see Insight 18.2), so the potential at all locations in the metal sphere must equal the potential at the outer edge of the sphere, which we can obtain from Equation 18.23. Hence,

$$V = \frac{kq}{r_{\text{sphere}}} \quad \text{everywhere inside a metal sphere} \tag{18.24}$$

Figure 18.12C shows the electric potential as a function of distance r from the sphere's center.

This example reviews and illustrates several important points. (1) The excess charge on a metal always resides at the surface, as we established in Chapter 17. (2) Because the electric field inside a metal is zero, the potential is constant throughout a piece of metal. (3) The electric field lines approach perpendicular to the surface of the metal sphere, Figure 18.12A, as required by the spherical symmetry of this electric field. Moreover, electric field lines *always* approach perpendicular to a metal surface of any shape, spherical or not. Why? Suppose an electric field at the surface of a metal somehow had a component parallel to the surface. This parallel field component would produce a force on any charge at the surface, causing it to move (because excess charge moves very easily in a conductor). This statement, however, contradicts our assumption that all the excess charge is in static equilibrium. Thus, for a piece of metal with any shape, electric field lines near a metal in equilibrium must approach perpendicular to the metal's surface.

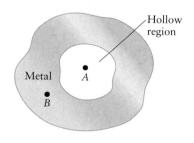

Figure 18.13 Concept Check 18.3.

CONCEPT CHECK 18.3 | Electric Field and Potential Inside a Metal

Consider a metal shell with a hollow region inside (Fig. 18.13). If the shell has a total charge Q, which of the following statements is true? (More than one statement may be correct.)

(a) The electric field is zero inside the hollow part of the shell (e.g., at point A).
(b) The electric potential is zero inside the hollow part of the shell (e.g., at point A).
(c) The electric field is zero in the metal part of the shell (e.g., at point B).
(d) The electric potential is zero in the metal part of the shell (e.g., at point B).

Electric Field Near a Lightning Rod

We can use our results for the electric field near the surface of a metal to understand the operation of a lightning rod. The stories about Benjamin Franklin's experiments with kites during thunderstorms are famous (even though we don't know for sure if these stories are true!). Franklin understood the connection between lightning and electricity and also showed the value of attaching a lightning rod (which he also invented) to the top of a building as in Figure 18.14A. When lightning strikes the rod, a large amount of charge (electrons) moves onto the rod; the rod is connected to the ground by a metal wire, allowing this charge to move safely to a region of low potential (the ground) through the metal wire rather than travel destructively through other portions of the building. Let's now consider why a lightning strike is "attracted" to a lightning rod.

Figure 18.14B gives a sketch of the electric field lines near a positively charged lightning rod. The electric field lines are perpendicular to the surface of the metal rod as we have just shown above in connection with Figure 18.12A. The field is largest near the sharp tip of the rod and smaller near the flat sides. That is another general property of the electric field near a metal: E is always largest near sharp metal corners and smaller near flat, metal surfaces. We can explain this by applying Equation 18.23 for the electric potential near a metal sphere to the simplified model of a lightning rod shown in Figure 18.14C. Here the rod is modeled as two metal spheres connected by a metal wire. The smaller sphere (radius r_1) represents the tip of the lightning rod, and the larger sphere (radius r_2) represents the flatter body of the rod. We also assume a net positive charge on the system. A portion Q_1 of this charge sits on the surface of the smaller sphere, and the rest of the charge Q_2 sits on the large sphere. The potential due to each sphere can be obtained from Equation 18.23. Strictly speaking, the result in Equation 18.23 applies only to a single sphere far from any other charges. However, if the two spherical parts of our model lightning rod are not too close together we can use Equation 18.23 to calculate the potential of each sphere. Because they are connected by a wire, the two spheres are part of a single piece of metal and must therefore have the same electric potential. Applying Equation 18.23 gives

$$V(\text{sphere 1}) = \frac{kQ_1}{r_1} = V(\text{sphere 2}) = \frac{kQ_2}{r_2}$$

Canceling common factors, we find that the charges on the two spheres are related by

$$\frac{Q_1}{r_1} = \frac{Q_2}{r_2} \qquad\qquad (18.25)$$

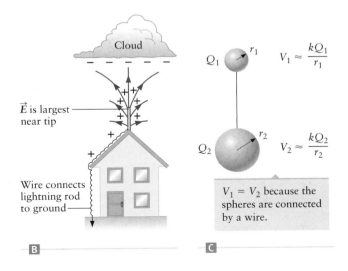

Figure 18.14 **A** Lightning rods on the roof of a building. **B** For the lightning rod to function properly, a wire must connect the rod to the ground. **C** A simplified model of a lightning rod. The small metal sphere represents the sharp tip of the rod.

We can use this equation to find the relative magnitudes of the electric fields produced by the two spheres. The electric field outside each sphere is given by $E = kQ/r^2$ (Eq. 18.22). For the field just outside sphere 1,

$$E_1 = \frac{kQ_1}{r_1^2}$$

and just outside sphere 2, we get

$$E_2 = \frac{kQ_2}{r_2^2}$$

The ratio of these fields is thus

$$\frac{E_1}{E_2} = \frac{kQ_1/r_1^2}{kQ_2/r_2^2} = \frac{Q_1 r_2^2}{Q_2 r_1^2}$$

Inserting the ratio of the charges from Equation 18.25, we find

$$\frac{E_1}{E_2} = \frac{Q_1}{Q_2}\frac{r_2^2}{r_1^2} = \frac{r_1}{r_2}\frac{r_2^2}{r_1^2}$$

$$\frac{E(\text{small sphere})}{E(\text{large sphere})} = \frac{E_1}{E_2} = \frac{r_2}{r_1} \tag{18.26}$$

Hence, the field is *larger* near the surface of the *smaller* sphere (because $r_2 > r_1$), and the field is largest near the sharp edges of a charged piece of metal.

This result explains how a lightning rod works. The large electric field near the tip causes nearby air molecules to be ionized in an effect known as **dielectric breakdown** (discussed in more detail in Section 18.5). The resulting mobile electrons and ions in the air near the tip are then able to carry charge from the atmosphere (the lightning) to the rod. The key point is that dielectric breakdown occurs preferentially near the sharp tip of the lighting rod, so the large current from a lightning bolt is carried only to the rod and not to the rest of the building.

Shielding Out the Electric Field

Our results for the electric field and potential near and inside the metal sphere in Figure 18.12 can be used to design a region in which the electric field is zero. This process is called "shielding out" the electric field. We already know how to make the electric field zero: we simply go inside a metal. This approach is sketched in Figure 18.15, where we show a piece of metal that contains a hollow cavity. A charge Q placed near the metal produces an electric field in its vicinity, but the metal prevents the field from penetrating into the cavity. The electric field in the cavity is zero no matter what charges or electric fields are present outside the metal.

A physical picture of this shielding is given in Figure 18.15. The field from a positive charge Q pulls electrons to the surface of the metal nearest Q, leaving a net positive charge on the surface farthest from Q. The fields of these induced charges resting on the outer surface of the metal precisely cancel the field from Q inside the metal and inside the cavity. This method of shielding has many applications, one of which you may have already experienced. During a thunderstorm, large and potentially deadly electric fields are associated with every "bolt" of lightning, but when you are in your car during a thunderstorm, you are protected from these large fields. The shell of your car acts as the piece of metal in Figure 18.15 while you sit in the cavity, and the shielding effect of the cavity protects you from the large electric fields of a lightning strike. The same principle is used to construct shielded cables used in high-end stereo systems. The shielding keeps stray electric fields produced by other appliances from penetrating inside the cable and causing "noise" in the signals carried to the stereo speakers.

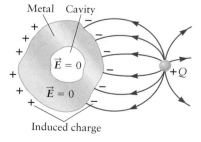

Figure 18.15 Shielding of the electric field in a cavity inside a metal.

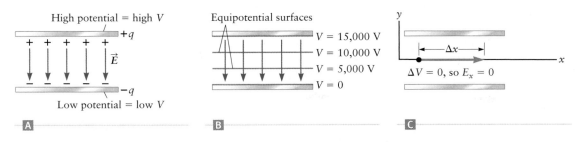

High potential = high V
+q
+ + + + +
$\vec{E}$
- - - - -
-q
Low potential = low V

Equipotential surfaces
$V = 15,000$ V
$V = 10,000$ V
$V = 5,000$ V
$V = 0$

y
Δx
x
$\Delta V = 0$, so $E_x = 0$

A   B   C

Figure 18.16 **A** Electric field between two parallel, charged metal plates. **B** In this case, the contours of equal V, called equipotential surfaces, are planes parallel to the metal plates. **C** The electric potential is constant along the x direction, so $E_x = 0$.

18.3 | EQUIPOTENTIAL LINES AND SURFACES

We have used drawings of electric field lines extensively to show how the electric field varies in space around various charge distributions. Another very useful way to visualize the electric field is through plots of **equipotential surfaces**. These are contours on which the electric potential is constant. Consider again the electric field in the region between two large, flat plates that carry charges $+q$ and $-q$ as sketched in Figure 18.16A. To be specific, assume they are the accelerator plates in the TV tube in Figure 18.8 and the potential difference between the two plates is 15,000 V. The electric field between the plates is constant and is directed from the plate with positive charge to the plate with negative charge. According to our relation between $\vec{E}$ and the electric potential (Eq. 18.17), the electric potential varies as we move from one plate to the other. We can now draw an imaginary surface between the plates such that $V = 10,000$ V at all points on this surface. Likewise, we can imagine a surface on which the potential is 5000 V (or any other value of interest). Several such surfaces are indicated in Figure 18.16B, labeled by their values of the potential. In this example, these surfaces are planes parallel to the charged plates.

Equipotential surfaces (also called equipotential *lines* in two-dimensional plots) are a very useful way to visualize the electric field and how it varies in space. Equipotential surfaces and lines are always perpendicular to the direction of the electric field. This property follows from the relationship between $\vec{E}$ and V. In Section 18.2, we showed that for a displacement along a particular direction, the change in V is related to the component of the electric field in that direction. If we consider a displacement along the x direction as in Figure 18.16C, the component of the field along x is

$$E_x = -\frac{\Delta V}{\Delta x} \qquad (18.27)$$

By definition, V is constant and $\Delta V = 0$ for motion parallel to an equipotential surface. From Equation 18.27, the electric field component parallel to the surface is thus zero. Hence, $\vec{E}$ is always perpendicular to an equipotential surface.

| CONCEPT CHECK 18.4 | Equipotential Surfaces and the Magnitude of E

Figure 18.17 shows equipotential surfaces near a hypothetical charge distribution. Is the magnitude of the electric field largest at (a) point A, (b) point B, or (c) point C? *Hint*: Because $E = -\Delta V/\Delta x$, the field is largest in regions and directions where the potential is changing most rapidly.

Some equipotential surfaces in the neighborhood of a point charge are shown in Figure 18.18. In this case, the electric field lines emanate radially outward from the charge. The equipotential surfaces are perpendicular to $\vec{E}$ and are thus a series of concentric spheres, with different spheres corresponding to different values of V.

Insight 18.3
EQUIPOTENTIAL SURFACES AND GRAVITY
The meaning of an equipotential surface can also be understood by analogy with the gravitational potential energy at different places on a mountain. As you climb up a mountain, your gravitational potential energy PE_{grav} increases, whereas PE_{grav} decreases when you go downhill. However, PE_{grav} stays constant if you follow a path that stays at a constant altitude. Such a path lies on a surface of constant gravitational potential energy and is thus an equipotential surface for gravitation.

V_1
B
C
V_2
V_3
A
Equipotential surfaces

Figure 18.17 Concept Check 18.4.

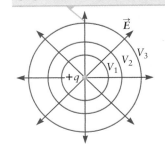

Equipotential surfaces are perpendicular to the field lines.

$\vec{E}$
V_3
V_2
V_1
$+q$

Figure 18.18 The equipotential surfaces around a point charge are spheres.

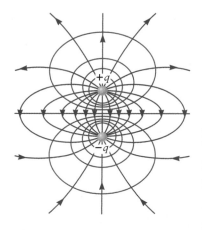

Figure 18.19 Equipotential surfaces near a dipole.

Another example of equipotential surfaces is shown in Figure 18.19. For this dipole consisting of charges $+q$ and $-q$, we plot both the electric field lines (from Example 17.4) and the equipotential surfaces. The result is a two-dimensional plot, so these surfaces appear as equipotential lines. Equipotential surfaces and lines are *always* perpendicular to the electric field.

| EXAMPLE 18.5 | Equipotential Surfaces Around a Line of Charge |

Consider a uniform line of positive charge as shown in Figure 18.20A. Make a plot showing some of the equipotential surfaces.

RECOGNIZE THE PRINCIPLE

Equipotential surfaces and lines are always perpendicular to the electric field. So, we first determine the direction of $\vec{E}$ and then find surfaces perpendicular to the field.

SKETCH THE PROBLEM

In Chapter 17, we showed that the electric field produced by a line of charge emanates radially outward from the line as sketched in Figure 18.20B.

IDENTIFY THE RELATIONSHIPS AND SOLVE

Equipotential surfaces must always be perpendicular to the electric field lines, so the equipotential surfaces are cylinders centered on the line of charge as shown in parts C and D of Figure 18.20.

What does it mean?
Because equipotential surfaces are, by their definition, always perpendicular to $\vec{E}$, they are often the same or very similar to the Gaussian surfaces we would choose in applying Gauss's law.

Figure 18.20 Example 18.5.
A A long line of positive charge.
B End view of electric field lines near the line of positive charge.
C The equipotential surfaces are cylinders centered on the line of charge.
D End view of the electric field lines and equipotential surfaces.

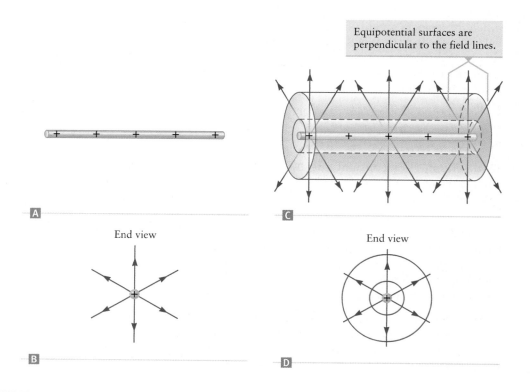

How would the sketches of the electric field and the equipotential surfaces in Figure 18.20 change if the charge on the line is negative instead of positive?

CONCEPT CHECK 18.6 | **Relating the Electric Field to the Equipotential Surfaces**

Figure 18.21 shows a family of equipotential surfaces. Sketch qualitatively the corresponding electric field lines. If $V_1 > V_2 > V_3 > V_4$, is the object in Figure 18.21 positively charged or negatively charged?

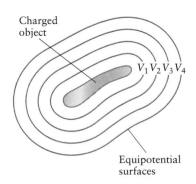

Figure 18.21 Concept Check 18.6.

18.4 | CAPACITORS

You have seen the example of two parallel, charged plates several times in our discussions of electric fields and potential. We now use this arrangement yet again to introduce the notion of *capacitance*. Figure 18.22 shows two flat, metal plates separated by a distance d, and for simplicity the region between the plates is a perfect vacuum. Recall from Chapter 17 that this is called a *parallel-plate capacitor*. It can be used to store electric charge and energy, as we will now demonstrate.

Let's connect the two metal plates of our capacitor to metal wires that can be used to carry charge onto or off the plates. Now suppose we place a charge $+Q$ onto the top plate and a charge $-Q$ on the bottom one. These charges attract each other and therefore sit on the inner surfaces of the two plates as indicated Figure 18.22.

In Chapter 17, we showed that a single plane with area A and charge Q produces an electric field of magnitude $E(\text{single plane}) = Q/(2\varepsilon_0 A)$. If the charge is positive, the field points away from the plane, whereas if the charge is negative, the field points toward the plane as sketched in Figure 18.22A. In the region between the plates (Fig. 18.22B), these fields add, giving a total field

$$E = \frac{Q}{\varepsilon_0 A} \tag{18.28}$$

which is the electric field inside a parallel-plate capacitor (Eq. 17.18). From our relation between the electric field and potential (Eq. 18.17), there must be a potential difference "across" the two plates. For a parallel-plate capacitor, the positively charged plate has a higher potential than the negatively charged plate, and the magnitude of the potential difference between the two plates is

$$\Delta V = Ed$$

where d is the distance between the plates. Inserting E from Equation 18.28, we find

$$\Delta V = Ed = \frac{Qd}{\varepsilon_0 A} \tag{18.29}$$

Hence, the potential difference across the capacitor plates is proportional to the charge Q. We now define a quantity C called the capacitance with

$$\Delta V = \frac{Q}{C} \tag{18.30}$$

Comparing Equations 18.29 and 18.30, C in this case is given by

$$C = \frac{\varepsilon_0 A}{d} \quad \text{(parallel-plate capacitor)} \tag{18.31}$$

where A is the area of a single plate and d is the plate separation.

The parallel-plate capacitor design in Figure 18.22 is a convenient one to analyze, but there are many other designs with different shapes and different spacing

Figure 18.22 ◭ A parallel-plate capacitor consists of two parallel metal plates. Wires connected to each plate allow charge to be placed on or taken off the plates. ◳ The electric field between the plates is related to the potential difference between the plates.

Definition of capacitance

Capacitance of a parallel-plate capacitor

geometries. All capacitors, though, employ two metal plates of some sort, and in *all* cases the charge on the capacitor plates is proportional to the potential difference across the plates as in Equation 18.30. The actual value of the capacitance depends on the design; Equation 18.31 applies to the parallel-plate design, whereas other plate geometries lead to other formulas for C (see Problem 46).

According to Equation 18.30, capacitance is the ratio of electric charge to the difference in electric potential across the plates, so the SI unit of capacitance is coulombs per volt, or C/V. This combination of units is called the farad (F) in honor of Michael Faraday, 1791–1867.

Storing Energy in a Capacitor

Capacitors find many applications in electronic circuits, including radios, computers, and MP3 players. All these uses depend on the ability of a capacitor to store electric charge and on the relation between charge and voltage (= potential difference), Equation 18.30.

When there is a nonzero potential difference between the two plates of a capacitor, energy is stored in the device. It is similar to the potential energy present in a system of two point charges as in Figure 18.2. To calculate the potential energy associated with the charges $+Q$ and $-Q$ on the two plates of a capacitor, we imagine that the capacitor is initially uncharged ($Q = 0$). We then transfer small amounts of charge ΔQ (perhaps by moving one electron at a time) from one plate to the other in Figure 18.23A. The capacitor plates are initially uncharged, so the potential difference is initially $\Delta V = 0$ and the electric field between the plates is zero. It therefore takes no energy to transfer the first packet of charge. As we continue to transfer charge between the plates, however, the total charge on the plates grows and, because Q and ΔV are proportional (Eq. 18.30), the potential difference across the plates grows too. This dependence is sketched in Figure 18.23B.

To move an amount of charge ΔQ through a potential difference ΔV requires an energy $(\Delta Q)(\Delta V)$, which corresponds to the area shown by the thin rectangle in Figure 18.23B. The total energy stored in the capacitor is equal to the energy required to move all the packets of charge ΔQ from one plate to the other, and it corresponds to the total area under the ΔV–Q line in Figure 18.23B. The area of this triangle is $\frac{1}{2} Q \, \Delta V$, where ΔV is now the final potential difference and Q is the final charge. Hence, the energy stored in a parallel-plate capacitor is

Energy stored in a capacitor

$$PE_{\text{cap}} = \tfrac{1}{2} Q \, \Delta V \qquad (18.32)$$

Using the definition of capacitance $\Delta V = Q/C$ (Eq. 18.30), we can write Equation 18.32 as

$$PE_{\text{cap}} = \frac{1}{2} C (\Delta V)^2 = \frac{1}{2} \frac{Q^2}{C} \qquad (18.33)$$

Figure 18.23 ◮ When charge is transferred from one capacitor plate to the other, the value of Q increases. ◮ The voltage across the capacitor plates increases with Q. The area under the graph of ΔV versus Q is equal to the potential energy stored in the capacitor PE_{elec}.

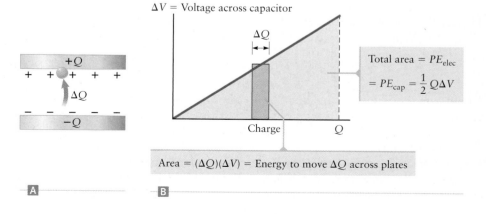

The relations for the stored energy in Equations 18.32 and 18.33 apply to all types of capacitors, not just parallel-plate capacitors. The energy stored in a capacitor is *potential energy*. It can be extracted from the capacitor and transformed into other forms of energy, or it can be used to do mechanical work.

EXAMPLE 18.6 Capacitance and Energy Storage in a Typical Capacitor

Consider a parallel-plate capacitor that is about the size of your fingernail (Fig. 18.24). The plates are squares with edges of length $L = 1.0$ cm, separated by $d = 10\ \mu m = 1.0 \times 10^{-5}$ m, which is about the diameter of a human hair. **(a)** Find the capacitance. **(b)** If the potential across the capacitor is $\Delta V = 12$ V, what is the energy stored?

RECOGNIZE THE PRINCIPLE

These calculations involve a direct application of our results for a parallel-plate capacitor and the energy stored in a capacitor. The values we find will be interesting.

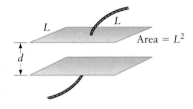

Figure 18.24 Example 18.6. A parallel-plate capacitor with two square plates.

SKETCH THE PROBLEM

The capacitor and its dimensions are shown in Figure 18.24.

IDENTIFY THE RELATIONSHIPS AND SOLVE

(a) The area of the capacitor plates is $A = L^2$. Inserting this information along with the given values of L and d into Equation 18.31 gives

$$C = \frac{\varepsilon_0 A}{d} = \frac{\varepsilon_0 L^2}{d} = \frac{[8.85 \times 10^{-12}\ C^2/(N \cdot m^2)](0.010\ m)^2}{(1.0 \times 10^{-5}\ m)}$$

$$C = \boxed{8.9 \times 10^{-11}\ F}$$

(b) The energy stored in a capacitor is (Eq. 18.33)

$$PE_{cap} = \tfrac{1}{2}C(\Delta V)^2$$

Inserting our result for C along with the given value of the potential difference ($\Delta V = 12$ V), we get

$$PE_{cap} = \tfrac{1}{2}C(\Delta V)^2 = \tfrac{1}{2}(8.9 \times 10^{-11}\ F)(12\ V)^2 = \boxed{6.4 \times 10^{-9}\ J}$$

What does it mean?
The value of C for this capacitor is only 89 picofarads (1 picofarad = 10^{-12} F). Typical capacitors used in electronic circuits are much less than 1 F, but the value of C for our fingernail-size capacitor is still quite small compared with most capacitors. We'll discuss ways to increase the value of C in the next section. The energy stored in this capacitor is also very small. As a comparison with familiar mechanical energies, PE_{cap} is much less than the change in the gravitational potential energy when a mosquito ($m \approx 10^{-6}$ kg) falls about 1 cm! One practical use of a capacitor is to store energy. To store useful amounts of energy, we must design capacitors with much larger values of C.

EXAMPLE 18.7 Using a Capacitor as a Computer Memory Element

Modern computer memories use parallel-plate capacitors to store information, and these capacitors are the basic elements of a random-access memory (RAM) chip. Assume one of these capacitors has plates with an area of $L \times L$, where $L = 1.0\ \mu m$ ($= 1.0 \times 10^{-6}$ m), and a plate separation of $d = 10$ nm ($= 1.0 \times 10^{-8}$ m). **(a)** Find the capacitance of such a capacitor. **(b)** Calculate the amount of charge that must

be placed onto the plates to obtain a potential difference of 5.0 V across the plates.
(c) How many electrons does this charge correspond to?

RECOGNIZE THE PRINCIPLE

Our RAM capacitor is just a parallel-plate capacitor, and we can find its capacitance using $C = \varepsilon_0 A / d$ (Eq. 18.31). Using the given value of ΔV with this value of C, we can then calculate the charge on the capacitor plates.

SKETCH THE PROBLEM

Our simplified model of a computer memory capacitor is described by Figure 18.24.

IDENTIFY THE RELATIONSHIPS AND SOLVE

(a) The capacitance of a parallel-plate capacitor is (Eq. 18.31)

$$C = \frac{\varepsilon_0 A}{d} = \frac{[8.85 \times 10^{-12} \text{ C}^2/(\text{N} \cdot \text{m}^2)](1.0 \times 10^{-6} \text{ m})^2}{1.0 \times 10^{-8} \text{ m}} = \boxed{8.9 \times 10^{-16} \text{ F}}$$

(b) Potential, charge, and capacitance are related by $Q = C \Delta V$ (Eq. 18.30), so obtaining a voltage of 5.0 V requires a charge

$$Q = C \Delta V = (8.9 \times 10^{-16} \text{ F})(5.0 \text{ V}) = \boxed{4.5 \times 10^{-15} \text{ C}}$$

(c) The charge on a single electron has a magnitude of e, so this value of Q corresponds to

$$N = \frac{Q}{e} = \frac{4.5 \times 10^{-15} \text{ C}}{1.60 \times 10^{-19} \text{ C/electron}} = \boxed{2.8 \times 10^4 \text{ electrons}}$$

What does it mean?
The number of electrons found in part (c) is not very large. In fact, the latest computer memory capacitors are somewhat smaller than we have assumed here, so they employ an even smaller number of electrons.

Capacitors in Series

Figure 18.25 shows a capacitor-type combination of three metal plates. This is just our usual parallel-plate capacitor (Fig. 18.22) with an added a metal plate in the middle, and for simplicity we assume this middle plate is separated from the top and bottom plates by a vacuum. Let's find the capacitance of this structure; our calculation will lead to some useful results for our work with electric circuits in Chapter 19.

A calculation of the capacitance relies on the definition of C as the charge on the capacitor plates divided by voltage (Eq. 18.30). We imagine that charges $+Q$ and $-Q$ are placed on the outermost plates (Fig. 18.25A), and we want to find the total potential difference ΔV_{total} between these two plates. The ratio $Q/\Delta V_{\text{total}}$ is then the

Figure 18.25 **A** Three parallel metal plates act as two capacitors sharing a common central plate. **B** When a charge $+Q$ is added to the top plate and $-Q$ to the bottom plate, these charges move to the "inner" surfaces as shown and induce charges $-Q$ and $+Q$ on the two surfaces of the inner plate.

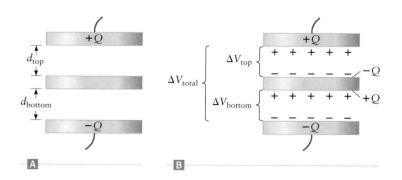

capacitance. To calculate ΔV_{total}, we must know the potential difference between each pair of plates. This potential difference will depend on the charges on the outer plates, which are just $+Q$ and $-Q$, and on the charges on the inner plate, which we now need to find.

The inner plate is neutral, with a net (excess) charge of zero. The charge $+Q$ on the top plate, however, will attract the mobile negative charges (electrons) in the inner plate. This attraction will give a net negative charge on the upper surface of the inner plate, leaving a net positive charge on the lower surface of the inner plate; the different surfaces of the three plates will then be charged as shown in Figure 18.25B. Comparing this figure with our sketch of a single capacitor (two plates) in Figure 18.22, we see that the three plates in Figure 18.25 act as two "consecutive" capacitors, each with charges $\pm Q$. To find the total capacitance of this structure, we need to know the potential difference ΔV_{total} between the top and bottom plates. This total potential difference equals the potential difference across the top two plates ΔV_{top} plus the potential difference across the bottom two plates ΔV_{bottom}. The top two plates form a capacitor, so we can write $\Delta V_{\text{top}} = Q/C_{\text{top}}$, where C_{top} is the capacitance of the upper capacitor, with a similar result for ΔV_{bottom}. We have

$$\Delta V_{\text{total}} = \Delta V_{\text{top}} + \Delta V_{\text{bottom}} = \frac{Q}{C_{\text{top}}} + \frac{Q}{C_{\text{bottom}}} \tag{18.34}$$

Capacitance is defined as the charge divided by the total potential difference, so

$$C_{\text{total}} = \frac{Q}{\Delta V_{\text{total}}} = \frac{Q}{\dfrac{Q}{C_{\text{top}}} + \dfrac{Q}{C_{\text{bottom}}}}$$

$$C_{\text{total}} = \frac{1}{\dfrac{1}{C_{\text{top}}} + \dfrac{1}{C_{\text{bottom}}}} \tag{18.35}$$

This result can be rewritten in an equivalent (and easier to remember) form:

$$\frac{1}{C_{\text{total}}} = \frac{1}{C_{\text{top}}} + \frac{1}{C_{\text{bottom}}} \tag{18.36}$$

An arrangement of two "consecutive" capacitors as in Figure 18.25 is referred to as *capacitors in series*. The result in Equation 18.36 tells us that when any two capacitors (which we can denote by C_1 and C_2) are placed in series, they are completely *equivalent* to a *single* capacitor whose capacitance C_{equiv} is given by

$$\frac{1}{C_{\text{equiv}}} = \frac{1}{C_1} + \frac{1}{C_2} \quad \text{(capacitors in series)} \tag{18.37}$$

We have derived this result for the case of two parallel-plate capacitors, but it applies to *any* type of capacitor. This result for capacitors in series will be very useful in our work with electric circuits in Chapter 19.

The result for capacitors in series from Equation 18.37 is shown schematically in Figure 18.26. Here we introduce the standard notation for drawing a capacitor in an electric circuit. The two parallel lines represent the two plates of a parallel-plate capacitor; this notation is used to represent all types of capacitors.

Capacitors in Parallel

Figure 18.26 shows one particular way of connecting two capacitors. Another way is illustrated in Figure 18.27, which shows two capacitors C_1 and C_2 connected in *parallel*. The capacitors C_1 and C_2 in Figure 18.27A are equivalent to a single capacitor C_{equiv}. To find the value of this equivalent capacitance, we imagine that charges Q_1 and Q_2 are placed on the two capacitors as shown in that figure. The

Figure 18.26 In a circuit diagram, capacitors are shown as two parallel lines, representing the plates of the capacitor. The top and middle plates in Figure 18.25 behave as a single capacitor, and the middle and bottom plates behave as a second capacitor. These two capacitors are equivalent to a single capacitance C_{equiv}.

Equivalent capacitance of capacitors in series

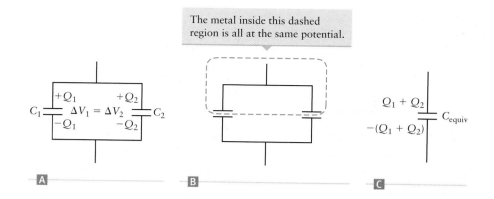

The metal inside this dashed region is all at the same potential.

voltage across each capacitor is related to its charge and capacitance, and, from Equation 18.30, we have

$$C_1 = \frac{Q_1}{\Delta V_1} \quad \text{and} \quad C_2 = \frac{Q_2}{\Delta V_2} \tag{18.38}$$

The top plates of these two capacitors are connected by a metal wire. A piece of metal in equilibrium is an equipotential object, so all the metal inside the upper dashed contour in Figure 18.27B is at the same potential. Because they are connected by a metal wire, the two upper plates of C_1 and C_2 are thus at the same potential. Likewise, the lower plate of C_1 is at the same potential as the lower plate of C_2. Hence, because of the way they are connected, the potential difference is the same across the two capacitors, and $\Delta V_1 = \Delta V_2$. For our equivalent capacitor (Fig. 18.27C), the total capacitance is equal to the ratio of the total charge and the potential difference across the plates with (from Eq. 18.30)

$$C_{\text{equiv}} = \frac{Q}{\Delta V}$$

The total charge on the equivalent capacitor is $Q = Q_1 + Q_2$, so we have[2]

$$C_{\text{equiv}} = \frac{Q_1 + Q_2}{\Delta V} = \frac{Q_1}{\Delta V} + \frac{Q_2}{\Delta V} \tag{18.39}$$

The voltage across the equivalent capacitor is also equal to the voltage across the individual capacitors $\Delta V = \Delta V_1 = \Delta V_2$. Using this information along with Equation 18.38 leads to

$$C_{\text{equiv}} = \frac{Q_1}{\Delta V_1} + \frac{Q_2}{\Delta V_2}$$

Equivalent capacitance of capacitors in parallel

$$C_{\text{equiv}} = C_1 + C_2 \quad \text{(capacitors in parallel)} \tag{18.40}$$

Equation 18.40 is the general rule for how capacitors combine in parallel to act as a single equivalent capacitor.

Combinations of Three or More Capacitors

Equations 18.37 and 18.40 describe how two capacitors combine in series or in parallel. What about combinations of three (or more) capacitors? To see how to deal with such cases, consider three capacitors in parallel as shown in Figure 18.28A. We can first combine capacitors C_1 and C_2 to get an equivalent capacitor C_{12} (Fig. 18.28B); using Equation 18.40, $C_{12} = C_1 + C_2$, which reduces the problem to two capacitors (C_{12} and C_3) in parallel. We can then apply Equation 18.40 again to obtain the effective capacitance of C_{12} and C_3 to get the equivalent capacitance of

---

[2]Notice that the charges Q_1 and Q_2 on the two capacitors need *not* be equal. As we saw with the lightning rod in Figure 18.14, two metal objects at the same potential need not have the same charge.

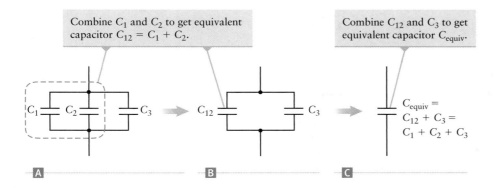

Combine C_1 and C_2 to get equivalent capacitor $C_{12} = C_1 + C_2$.

Combine C_{12} and C_3 to get equivalent capacitor C_{equiv}.

C_1 C_2 C_3

C_{12} C_3

$C_{equiv} = C_{12} + C_3 = C_1 + C_2 + C_3$

A   **B**   **C**

Figure 18.28 Three (or more) capacitors connected in parallel behave as a single equivalent capacitor.

all three capacitors $C_{equiv} = C_{12} + C_3 = C_1 + C_2 + C_3$. The general combination rule for many capacitors in parallel is

$$C_{equiv} = C_1 + C_2 + C_3 + \cdots \quad \text{(many capacitors in parallel)} \qquad (18.41)$$

For multiple capacitors in series, a similar approach gives

$$\frac{1}{C_{equiv}} = \frac{1}{C_1} + \frac{1}{C_2} + \frac{1}{C_3} \cdots \quad \text{(many capacitors in series)} \qquad (18.42)$$

EXAMPLE 18.8 | Combining Capacitors

Four capacitors are connected as shown in Figure 18.29A. If all have the same capacitance ($C_1 = C_2 = C_3 = C_4 = C$), what is the equivalent capacitance of this combination?

RECOGNIZE THE PRINCIPLE

We can apply the rules for combining capacitors in series and in parallel. We first combine C_1 and C_2 in parallel (using Eq. 18.42) and so forth for C_3 and C_4. We are left with two capacitors in series (Fig. 18.29B), which can be analyzed using the rule for a series combination.

SKETCH THE PROBLEM

Figure 18.29 shows the problem. We first combine C_1 and C_2 to get C_{12}, and C_3 and C_4 to get C_{34}. We then combine these results to get C_{equiv}.

IDENTIFY THE RELATIONSHIPS AND SOLVE

Figure 18.29B shows an equivalent arrangement. On the left, we have combined C_1 and C_2 in parallel to get an equivalent capacitance C_{12}. According to Equation 18.41, we have

$$C_{12} = C_1 + C_2 = 2C$$

Likewise, the parallel combination of C_3 and C_4 is

$$C_{34} = C_3 + C_4 = 2C$$

The final step is to combine C_{12} and C_{34} in series (Fig. 18.29C) to get the total capacitance. Applying the rule for combining capacitors in series (Eq. 18.42) gives

$$\frac{1}{C_{equiv}} = \frac{1}{C_{12}} + \frac{1}{C_{34}} = \frac{1}{2C} + \frac{1}{2C}$$

We thus find

$$C_{equiv} = \boxed{C}$$

C_1 C_3

C_2 C_4

A

$C_{12} = 2C$ $C_{34} = 2C$

B

$C_{equiv} = C$

C

Figure 18.29 Example 18.8.

What have we learned?

We can deal with most combinations of capacitors by applying the parallel and series combination rules to obtain a single equivalent capacitance.

Capacitor in a cylindrical "package"

Dielectric

Metal "plates"

Figure 18.30 **A** Capacitors can be constructed in many different ways and come in different shapes and sizes. **B** Many capacitors consist of two metal foils ("plates") wrapped up in a cylinder, with a dielectric between the foils.

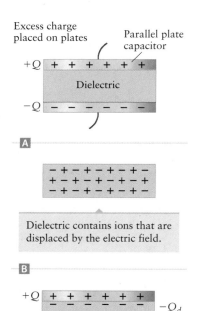

Excess charge placed on plates

Parallel plate capacitor

$+Q$

Dielectric

$-Q$

A

Dielectric contains ions that are displaced by the electric field.

B

$+Q$ $-Q_d$

$+Q_d$

$-Q$

C

Figure 18.31 **A** When a dielectric is placed between the plates of a capacitor, the value of the capacitance increases. **B** Most dielectrics are ionic. **C** The electric field inside the capacitor induces charge $\mp Q_d$ on the surfaces of the dielectric.

18.5 | DIELECTRICS

Most real capacitors are not as simple as the parallel-plate capacitor analyzed in Examples 18.6 and 18.7. Figure 18.30A shows a number of real capacitors. They all contain two metal "plates" (obscured by their outer casings in Fig. 18.30A) separated by a thin insulating region, but notice that many of the casings are cylindrical. These cylinders contain two thin, metal foils (the "plates") wrapped like a jellyroll. In the parallel-plate capacitors we have considered so far, the region between the plates is a vacuum. In most real capacitors, however, the space between the plates is filled with a solid material called a *dielectric* as illustrated in Figures 18.30B and 18.31A.

The Term *Dielectric* Is Used to Describe Any Insulating Material.

Inserting a dielectric material between the plates of a capacitor changes the value of the capacitance in proportion to a factor called the *dielectric constant* of the material, denoted by κ. If C_{vac} is the value of the capacitance with a vacuum in the gap between the plates, the capacitance with the dielectric present is

$$C_d = \kappa C_{\text{vac}} \qquad (18.43)$$

Note that κ is a dimensionless factor, so it has no units. Except for some very unusual materials, κ is greater than or equal to 1, so the presence of the dielectric *increases* the capacitance.

Why Does a Dielectric Change the Capacitance?

Charge on the plates of a capacitor produces an electric field between the plates. This electric field extends into a dielectric placed between the plates, causing electric forces on charges in the dielectric. "Good" dielectrics—that is, materials with large values of κ—are usually highly ionic, meaning that they contain positive and negative ions bound together as sketched in Figure 18.31B. The electric force on each ion causes the ion to shift slightly according to the direction of $\vec{E}$. In the dielectric in Figure 18.31, the negative ions move slightly toward the top capacitor plate (because they are attracted to the charge $+Q$) and the positive ions shift a small amount toward the bottom plate (because they are attracted to the charge $-Q$). This slight shift in the ions' positions leads to a small amount of negative charge on the top surface of the dielectric and a corresponding small amount of positive charge on the bottom surface of the dielectric.

An equivalent way to view a capacitor with dielectric inside is shown in Figure 18.31C. The metal plates have charges $\pm Q$ on their inner surfaces, while the surfaces of the dielectric carry charge $\mp Q_d$. If we lump these two together, we then have a total charge $+Q - Q_d$ at the top of the capacitor and $-Q + Q_d$ at the bottom. The dielectric thus reduces the magnitude of the charge on the capacitor.

The field between the plates of a parallel-plate capacitor containing a dielectric is given by

$$E_d = \frac{Q}{\kappa \varepsilon_0 A} \qquad (18.44)$$

where Q is the charge on just the metal plates. Since $\kappa \geq 1$, the electric field inside the capacitor is smaller with the dielectric present than when the gap is filled with a vacuum (compare with Eq. 18.28). Hence, the potential difference across the plates is also smaller, again by a factor of κ. Capacitance is defined as the ratio of the external charge placed on the capacitor plates divided by the potential difference. Because the potential difference ΔV is reduced by a factor of κ, the ratio $C = Q/\Delta V$ is larger by the same factor, which takes us to Equation 18.43.

These results apply to *any* type of capacitor. Adding a dielectric (1) increases the capacitance by a factor equal to the dielectric constant κ and (2) reduces the electric

Table 18.1 | Values of the Dielectric Constant and the Dielectric Breakdown Field for Some Common Materials

SUBSTANCE	κ	E(breakdown), V/m
Mica	5	100×10^6
Glass	6	14×10^6
Paper	4	16×10^6
Plexiglas	3.4	40×10^6
Quartz	3.8	40×10^6
Mylar	3	300×10^6
Teflon	2	60×10^6
Strontium titanate	230	8×10^6
Air (at room temperature and pressure)	1.0006	3×10^6
Helium gas (at room temperature and pressure)	1.000065	
Water	80.4	
Glycerine	42.5	
Benzene	2.28	

field inside the capacitor by the same factor. The actual value of the dielectric constant depends on the material. Materials that are very ionic tend to have the largest values of κ. Table 18.1 lists the values of κ for solids, liquids, and gases; all three phases of matter exhibit dielectric behavior.

Effects of Very Large Electric Fields

As more and more charge is added to a capacitor, the electric field between the plates increases in proportion to Q (Eq. 18.44). For a capacitor containing a dielectric, this field can be so large that it literally rips the ions in the dielectric apart, an effect called **dielectric breakdown**. The resulting free ions are able to move through the material much like electrons in a metal. Because there is a large electric field present (the field that caused the dielectric breakdown), these ions move rapidly toward the oppositely charged capacitor plate and destroy the capacitor.

The value of the electric field at which dielectric breakdown occurs depends on the material (see Table 18.1) and sets a practical limit on the maximum voltage that can be placed across a capacitor in applications. Dielectric breakdown can occur whenever the electric field is large (not just in capacitors), so it comes into play in other phenomena such as lightning, as we'll discuss in Section 18.6.

EXAMPLE 18.9 | Dielectric Breakdown of a Capacitor

Figure 18.30A shows capacitors used in practical devices such as a laptop computer or an MP3 player. A typical capacitor of this type has a capacitance of around 1.0 nF ($= 1.0 \times 10^{-9}$ F). Assume it is a parallel-plate capacitor with a plate area of $A = 1.0$ cm^2 that uses mica as the dielectric. Find (**a**) the spacing between the plates and (**b**) the voltage at which dielectric breakdown occurs (i.e., the voltage at which the capacitor will fail).

RECOGNIZE THE PRINCIPLE

For a parallel-plate capacitor containing a dielectric, the capacitance is given by

$$C_d = \kappa C_{\text{vac}} = \frac{\kappa \varepsilon_0 A}{d} \qquad (1)$$

(a combination of Eqs. 18.31 and 18.43). Given the value of C along with the plate area A and the dielectric constant, we can solve Equation (1) to find the spacing between the plates d. We can then calculate the electric field between the plates and compare it with the dielectric breakdown field for mica.

SKETCH THE PROBLEM

Figure 18.31A describes our capacitor; the dielectric in this case is mica.

IDENTIFY THE RELATIONSHIPS AND SOLVE

(a) Rearranging Equation (1) to solve for the spacing d between the plates, we find

$$d = \frac{\kappa \varepsilon_0 A}{C_d}$$

From Table 18.1, the dielectric constant of mica is $\kappa = 5$. Inserting this and the given values of A and C_d, we get

$$d = \frac{\kappa \varepsilon_0 A}{C_d} = \frac{(5)[8.85 \times 10^{-12} \text{ C}^2/(\text{N} \cdot \text{m}^2)](0.010 \text{ m})^2}{1.0 \times 10^{-9} \text{ F}}$$

$$d = 4 \times 10^{-6} \text{ m} = \boxed{4 \ \mu\text{m}}$$

(b) The dielectric breakdown field of mica is $E = 100 \times 10^6$ V/m (from Table 18.1). Using the relation between the electric field and potential difference between the plates (Eq. 18.17) leads to a breakdown voltage of magnitude

$$\Delta V = |-Ed| = (100 \times 10^6 \text{ V/m})(4 \times 10^{-6} \text{ m}) = \boxed{400 \text{ V}}$$

What does it mean?
Our result for d in part (a) is much smaller than the thickness of a human hair! Even so, pieces of mica this thin are easily obtained. The voltages used in consumer electronics (such as an MP3 player) are typically less than 10 or 20 V, so this capacitor will not be in danger of failure.

18.6 | ELECTRICITY IN THE ATMOSPHERE

One of nature's most impressive displays involves the motion of electric charge called lightning. During a lightning strike, large amounts of electric charge move between a cloud and the surface of the Earth or between two clouds (Fig. 18.32). This charge motion occurs when a large electric field is established between the cloud and the Earth or between two clouds, causing dielectric breakdown of the air. Molecules in the air are ripped apart to form free electrons, leaving behind positively charged ions. These charges are then accelerated by the electric field, and some travel to the Earth, where they arrive with very high velocities. This rapid motion of charge heats nearby air molecules, leading to the sound (thunder) associated with lightning.

Most of the charge motion in lightning involves electrons because they are much lighter than the positive ions that are also produced in dielectric breakdown of air. Which way do these electrons move? Do they travel from a cloud to the Earth, or vice versa? Experiments show that the electric field of a lightning strike is directed from the Earth to the cloud[3] as shown in Figure 18.33. Recall that $\vec{E}$ points in the direction a positive test charge would move. Thus, once dielectric breakdown occurs in the air, electrons travel from the cloud to the Earth's surface. A typical lightning bolt contains a charge of about $|Q| = 20$ C, corresponding to

Figure 18.32 Example of a lightning strike.

© Lionel Brown/Image Bank/Getty Images

[3]In fair weather, there is a small electric field in the opposite direction (downward, toward the Earth).

$$N = \frac{|Q|}{e} = \frac{20 \text{ C}}{1.60 \times 10^{-19} \text{ C/electron}} \approx 1 \times 10^{20} \text{ electrons/bolt} \qquad \text{(18.45)}$$

You might think that these electrons would take the shortest path and travel along a straight line to the Earth, but they actually follow a jagged path, giving a lightning bolt its familiar appearance (Fig. 18.32). The jaggedness is due to the complicated nature of dielectric breakdown in air.

Perhaps the most basic question concerning lightning is, Where does the electric field in Figure 18.33 come from in the first place? Interestingly, this problem is still not very well understood. So far, no one has a complete theory of how motion of air and clouds during a storm leads to the large electric field necessary to cause lightning, but the correct qualitative explanation is probably as follows. During a thunderstorm, the atmosphere is very unsettled, with large temperature and pressure differences between low and high altitudes. These differences cause air molecules, water droplets, dust, and small ice crystals to move vertically through the atmosphere. During the later phases of a thunderstorm, the water droplets and ice crystals move downward, acquiring charge as they "rub" against molecules and ions in the air (Fig. 18.33B). This process is similar to what we found in Chapter 17 when we saw that rubbing a piece of amber with fur causes the amber to acquire a negative charge. In a thunderstorm, rubbing gives the water droplets and ice crystals a negative charge that they carry to the bottom of the cloud, leaving the top charged positively. As negative charge accumulates at the bottom of the cloud, it repels electrons from the Earth's surface because the ground is a good conductor of charge and electrons can move freely from place to place in the soil. This process causes the Earth's surface to be positively charged (Fig. 18.33A), producing an electric field qualitatively similar to the electric field between two parallel plates of a capacitor. This electric field produces the dielectric breakdown, which is the lightning bolt.

Figure 18.33 **A** During an electrical storm, the top of a storm cloud is positively charged, while the bottom is negatively charged. The resulting electric field induces a positive electric charge at the Earth's surface. **B** The cloud's charge is produced by the motion of water droplets and ice crystals as they are pushed along by the wind.

| EXAMPLE 18.10 | ® Electric Field and Potential of a Lightning Strike |

Consider a thunderstorm involving the bottom surface of a cloud that is $h = 500$ m above the surface of the Earth. If a lightning bolt travels from this cloud to the Earth, estimate the electric potential of the bottom of the cloud just prior to the lightning bolt, relative to "ground" potential $V = 0$ at the Earth's surface.

RECOGNIZE THE PRINCIPLE

A lightning bolt occurs when there is dielectric breakdown of the air. So, just before the lightning bolt occurs, the electric field between the cloud and the Earth must equal the dielectric breakdown field of air.

SKETCH THE PROBLEM

Figure 18.34 shows the problem.

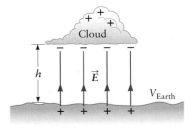

Figure 18.34 Example 18.10.

IDENTIFY THE RELATIONSHIPS

Using the relation between the electric field and the potential (Eq. 18.18) gives

$$E = -\frac{\Delta V}{\Delta x} = -\frac{V_{\text{cloud}} - V_{\text{Earth}}}{h}$$

The Earth is at "ground" potential ($V_{\text{Earth}} = 0$), so

$$V_{\text{cloud}} = -Eh$$

SOLVE

Inserting the dielectric breakdown field of air (Table 18.1) and the height of the cloud, we find

$$V_{\text{cloud}} = -E_{\text{breakdown}}h = -(3 \times 10^6 \text{ V/m})(500 \text{ m}) \approx \boxed{-2 \times 10^9 \text{ V}}$$

A

EKG voltage at different electrodes

Time

B

Figure 18.35 **A** A person having an EKG. **B** In an EKG, the electric potential as a function of time is recorded from electrodes at many different spots on the body. This recording shows when and how strongly different parts of the heart muscle are contracting. Typical EKG potentials are around 1 mV (10^{-3} V), about 10,000 times smaller than the potential produced by the battery of an MP3 player.

EEG Voltage

Alpha Wave

Delta Wave

~10^{-4} V

1 sec

Time

Figure 18.36 Typical EEG signals. The top signal was taken when the person was awake and alert and is called an alpha wave. The bottom EEG was recorded when the person was sleeping (a delta wave). Typical EEG potentials are about 10^{-4} V, somewhat smaller than in an EKG.

What does it mean?
The negative sign for V_{cloud} means that the cloud is negatively charged. Typical household voltages are between 1 V and 120 V, so the potential associated with lightning is *much* larger than the voltages found around the house!

18.7 | ⊗ BIOLOGICAL EXAMPLES AND APPLICATIONS

In the 1790s, an Italian scientist, Luigi Galvani, discovered that electricity could make muscles twitch. Galvani originally observed that sparks from static electricity can cause muscular contraction in a frog. Galvani's work was the first evidence that the nerves and muscles in frogs use an electrical potential and the movement of charge as part of normal physiological function. The same is true in other animals, including humans.

The heart is perhaps the most important muscle in the body. Like the muscles in a frog's leg, the heart can also be made to twitch using an externally applied potential. For example, a device called a *defibrillator* is used to shock the heart into a normal pattern of beating. A common diagnostic procedure is the *electrocardiogram*, called an ECG or EKG. This procedure uses a collection of electrodes to monitor the potential at various places on a person's chest (Fig. 18.35A). The electric potentials the body sends to various parts of the heart form a rhythmic pattern that causes different chambers of the heart to contract at different times, and the different electrode potentials in an EKG can reveal if a person's heart is (or is not) following a healthy rhythm (Fig. 18.35B). There is no way to use physics to calculate what a healthy rhythm should look like, but your doctor can tell from experience.

A related procedure is the *electroencephalogram* (EEG), in which electrodes on the scalp are used to detect potentials generated in the brain. These potentials oscillate with time, and the frequency of an EEG signal depends on the person's activity. The brain of an awake and alert person generates what are called alpha waves with frequencies of about 10–14 Hz, whereas during deep sleep one finds delta waves with a typical frequency of 2–4 Hz. (Fig. 18.36). Physiologists have also identified beta and gamma waves, which are generated in other brain states and which have other frequencies. The absence of any EEG signal is usually taken as evidence that a person is "brain dead."

18.8 | ELECTRIC POTENTIAL ENERGY REVISITED: WHERE IS THE ENERGY?

A certain amount of potential energy is associated with two point charges as a result of the electric interaction between them. We have described this potential energy as a property of the system of the two charges. Another way to view and think about this electric potential energy, however, is that potential energy is stored *in the electric field* itself. That is, whenever an electric field is present in a particular region of space, potential energy is located in that region. In almost all respects, this viewpoint is equivalent to the one we have taken in previous sections, so it does not affect any of our results to this point. Let's now calculate this field energy in a familiar geometry.

Consider two flat, metal plates separated by a vacuum; this structure is just a parallel-plate capacitor, and we have already worked out many of its properties. If the plates have an area A and are separated by a distance d, the electric field between the plates is (Eq. 18.28)

$$E = \frac{Q}{\varepsilon_0 A} \tag{18.46}$$

where the charge on the plates is $\pm Q$. The magnitude of the potential difference between the plates is $\Delta V = Ed$. We also found that the electric potential energy stored in the capacitor is (Eq. 18.32)

$$PE_{\text{elec}} = \tfrac{1}{2} Q \, \Delta V$$

Combining these last two results, we get

$$PE_{\text{elec}} = \tfrac{1}{2} QEd$$

We now eliminate Q using Equation 18.46, leading to

$$PE_{\text{elec}} = \tfrac{1}{2}(\varepsilon_0 EA)Ed = \tfrac{1}{2}\varepsilon_0 E^2(Ad)$$

The factor (Ad) is the volume between the plates, so we can also write

$$PE_{\text{elec}} = \tfrac{1}{2}\varepsilon_0 E^2(\text{volume}) \tag{18.47}$$

In words, this says that the electric field energy in the region between the capacitor plates is equal to $\tfrac{1}{2}\varepsilon_0 E^2$ multiplied by the volume of the region. Hence, the energy per unit volume, denoted by u_{elec}, is

$$u_{\text{elec}} = \tfrac{1}{2}\varepsilon_0 E^2 \tag{18.48}$$

We call u_{elec} the **energy density** in the electric field.

Although our derivation is for a parallel-plate capacitor, this result for u_{elec} applies for any arrangement of charges, including point charges. It tells us that potential energy is present wherever an electric field is present. Philosophically, this result is a major break from our previous view that potential energy is due to the interaction of charges. One might argue that electric fields are caused by charges, so the field E in Equation 18.48 is actually due to a charge. If this argument is correct, aren't we back with our original view that electric potential energy is always a property of a system of charges? In most cases, that is true, but in one extremely important situation it is not: when dealing with an electromagnetic wave. We know from our work on waves in Chapter 12 that waves transport energy. Electromagnetic waves transport energy from place to place, even through regions of space where *no charges are present* or even close by. The energy carried by such a wave is carried by electric and magnetic fields, and the electric part of this energy is given by Equation 18.48. (A portion of the energy is also contained in the magnetic field.) So, to explain the energy carried by an electromagnetic wave, it is essential that energy be carried by the field. Our original view that electric potential energy always involves pairs of charges is thus *not* the entire story. We'll come back to this problem in Chapter 23 when we discuss electromagnetic waves.

SUMMARY | Chapter 18

KEY CONCEPTS AND PRINCIPLES

Electric potential energy

The electric force is a conservative force, so there is an **electric potential energy** PE_{elec}. The potential energy of two point charges separated by a distance r is

$$PE_{\text{elec}} = \frac{kq_1 q_2}{r} \tag{18.6} \ (\text{page 566})$$

(Continued)

Electric potential

The *electric potential* V (also sometimes called just the *potential*) is proportional to the electric potential energy. If the electric potential energy of a charge q at a particular location is PE_{elec}, the electric potential at that point is

$$V = \frac{PE_{elec}}{q} \qquad \textbf{(18.10) (page 571)}$$

with V measured in units of *volts* (V).

Relation between the electric field and the electric potential

Suppose the potential changes by an amount ΔV over a distance Δx. The component of the electric field along this direction is then

$$E = -\frac{\Delta V}{\Delta x} \qquad \textbf{(18.18) (page 573)}$$

The electric field thus has units of volts per meter (V/m).

Capacitors

Two parallel metal plates form a *capacitor*. The *capacitance* C of this structure determines how easily charge can be stored on the plates. The charge on a capacitor is related to the magnitude of the *potential difference* between the plates by

$$\Delta V = \frac{Q}{C} \qquad \textbf{(18.30) (page 581)}$$

This relation holds for any type of capacitor. For a parallel-plate capacitor, the capacitance is

$$C = \frac{\varepsilon_0 A}{d} \quad \text{(parallel-plate capacitor)} \qquad \textbf{(18.31) (page 581)}$$

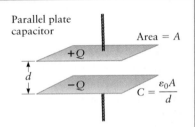

Energy stored in a capacitor and in an electric field

The electric potential energy stored in a capacitor is

$$PE_{cap} = \tfrac{1}{2}Q\,\Delta V = \tfrac{1}{2}C(\Delta V)^2 = \frac{1}{2}\frac{Q^2}{C} \qquad \textbf{(18.32), (18.33) (page 582)}$$

The energy per unit volume stored in an electric field E is

$$u_{elec} = \tfrac{1}{2}\varepsilon_0 E^2 \qquad \textbf{(18.48) (page 593)}$$

APPLICATIONS

Equipotential surfaces

An *equipotential surface* is a surface on which the electric potential is constant. The electric field $\vec{E}$ at a particular location is always perpendicular to the equipotential surface at that spot.

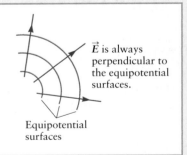

$\vec{E}$ is always perpendicular to the equipotential surfaces.

Equipotential surfaces

(*Continued*)

Electric potential in a metal

For a piece of metal in equilibrium, the electric potential is constant throughout the metal, while the electric field is zero inside the metal. There is an electric field just outside a charged metal, however, and this field is largest near places where the radius of curvature is smallest (i.e., near sharp corners). This result explains why a lightning rod works.

Capacitors in series and parallel

When two capacitors are connected in series, they act as a single equivalent capacitance with a value

$$\frac{1}{C_{equiv}} = \frac{1}{C_1} + \frac{1}{C_2} \quad \text{(capacitors in series)} \quad (18.37) \text{ (page 585)}$$

When two capacitors are connected in parallel, the equivalent capacitance is

$$C_{equiv} = C_1 + C_2 \quad \text{(capacitors in parallel)} \quad (18.40) \text{ (page 586)}$$

QUESTIONS

SSM = answer in Student Companion & Problem-Solving Guide   ⊗ = life science application

1. If the electric field is zero in a particular region of space, what does that tell you about the electric potential in that region? Is the potential zero, or constant, or something else? Explain.

2. Will the *electric field* always be zero at any point where the *electric potential* is zero? Why or why not?

3. Will the *electric potential* always be zero at any point where the *electric field* is zero? Why or why not?

4. **SSM** Make sketches of the equipotential surfaces around (a) a point charge, (b) an infinite line of charge, (c) an infinite plane of charge, (d) a finite line of charge, and (e) a charged plate of finite size.

5. An electron is released from rest at the origin and moves along the +x direction. Other experiments show that the electric field is uniform (i.e., constant). Make a sketch showing the direction of the electric field and the equipotential surfaces.

6. A particle of positive charge is released from rest and is found to move as a result of an electric force. Does the particle move to a region of higher or lower potential energy? Does the particle move to a region of higher or lower electric potential?

7. Repeat Question 6 for an electron.

8. A charged particle is released from rest in an electric field. Its motion does not follow the electric field lines. Why? Are there configurations of the electric field that would lead to motion along electric field lines? If so, describe some of them and explain what they have in common.

9. Two particles are at locations where the electric potential is the same. Do these particles necessarily have the same electric potential energy? Explain.

10. ⊗ Many species of fish generate electric fields and use these fields to sense their surroundings. If another entity with a dielectric constant different from that of water comes near the fish, the electric field lines generated by the fish are altered, and sensory organs on the fish can detect these slight changes. The electric field of most of these fishes is that of a dipole. Figure Q18.10 shows lines of equal potential around such a fish as seen looking down on the top of the fish. Assume the head of the fish is positively charged. (a) Sketch the electric field associated with these equipotential lines. (You may want to make a photocopy of the figure to proceed.) (b) If a sphere of metal (perhaps a lead fishing weight) were placed near the side of the fish, sketch how the field lines in the vicinity of the metal would change.

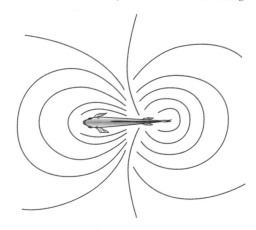

Figure Q18.10

11. How much work is required to move a charge from one spot on an equipotential surface to another location on the same equipotential surface?

12. **SSM** A charged particle obtains a speed v when accelerated through a potential difference of 5 V. How many times faster would it be going if accelerated though a potential difference of 20 V?

13. Consider the energy density in the electric field at a distance d away from a positive point charge. If the positive point charge is then replaced with a negative point charge of equal magnitude, how does the energy density change at that same point?

14. To increase the energy stored in a capacitor, what might you do? Explain your reasoning.
    (a) Increase the charge on the capacitor.
    (b) Insert a dielectric between the plates while holding the voltage fixed.
    (c) Move the plates closer together while keeping the charge fixed.
    (d) All the above.
    (e) None of the above.

15. Compared with the applied electric field, is the electric field inside a dielectric material (a) smaller by a factor of κ, (b) larger by a factor of κ, or (c) larger by a factor of κ^2? Or, does it (d) depend on the shape of the dielectric? Explain your reasoning.

16. Derive the result for the equivalent capacitance of many capacitors in series in Equation 18.42.

17. When one speaks of a fully charged capacitor, with what is it charged? How is the word *charged* being used here?

18. How does a charged capacitor differ from a capacitor that is not charged? Does the charged capacitor have a different net charge? Explain.

19. A parallel-plate capacitor is connected to a battery such that a constant electric potential difference is produced between the plates. If the plates are moved farther apart, which of the following quantities change? Do they get larger or smaller?
    (a) The magnitude of the electric field
    (b) The amount of charge on a plate
    (c) The capacitance
    (d) The energy stored in the capacitor

20. A certain amount of charge $\pm q$ is placed on the plates of a capacitor, and the plates are then disconnected from the outside world. If the capacitor plates are pulled apart, does the amount of electric potential energy stored in the capacitor increase or does it decrease? Explain your answer in terms of the principle of conservation of energy.

18.1 ELECTRIC POTENTIAL ENERGY

1. Two point particles of charge $Q_1 = 45\ \mu C$ and $Q_2 = 85\ \mu C$ are found to have a potential energy of 40 J. What is the distance between the charges?

2. Two particles with $Q_1 = 45\ \mu C$ and $Q_2 = 85\ \mu C$ are initially separated by a distance of 2.5 m and then brought closer together so that the final separation is 1.5 m. What is the change in the electric potential energy?

3. The nucleus of a helium atom contains two protons. In a simple model of this nucleus, the protons are viewed as point particles separated by a distance of 1.0 fm (1.0×10^{-15} m). What is the electric potential energy of the two protons?

4. Two point charges $Q_1 = 3.5\ \mu C$ and $Q_2 = 7.5\ \mu C$ are initially very far apart. They are then brought together, with a final separation of 2.5 m. How much work does it take to bring them together?

5. A simple model of a hydrogen atom pictures the electron and proton as point charges separated by a distance of 0.050 nm, and in Example 18.1 we calculated the associated electric potential energy. If the electron is initially at rest, how much work is required to "break apart" these two charges, that is, to separate them by a very large distance?

6. Consider an electric dipole consisting of charges $+Q$ and $-Q$ as sketched in Figure P18.6. How much work is required to place a charge q at the origin? Assume q starts from very far away.

Figure P18.6
Problems 6 and 16.

7. Two point charges are located as shown in Figure P18.7, with charge $q_1 = +2.5$ C at $x = -3.0$ m, $y = 0$,

Figure P18.7

and charge $q_2 = +4.0$ C at $x = +1.0$ m, $y = +2.0$ m. An electron is now taken from a point very far away and placed at the origin. How much work must be done on the electron to move it to the origin?

8. ★ The electric field in a particular region of space is found to be uniform, with a magnitude of 400 N/C and a direction parallel to the y axis. (a) What is the change in the electric potential energy of a charge $q = 3.5\ \mu C$ if it is moved from a location $(x, y) = (20$ cm, 45 cm$)$ to $(5$ cm, 30 cm$)$? (b) What is the change in the electric potential energy if the charge is moved the same distance along the x axis?

9. ✪ Three point charges $Q_1 = 2.5\ \mu C$, $Q_2 = 4.5\ \mu C$, and $Q_3 = -3.5\ \mu C$ are arranged as shown in Figure P18.9. What is the total electric potential energy of this system?

10. SSM ★ Consider again the three charges in Figure P18.9 with $Q_1 = 2.5\ \mu C$, $Q_2 = 4.5\ \mu C$, and $Q_3 = -3.5\ \mu C$. A fourth charge $q = -5.0\ \mu C$ is brought from very far away and placed at the origin. (a) How much work is required in this process? (b) How much work is required to move q from the origin to a distance that is very far away?

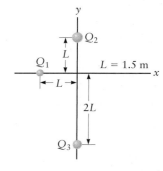

Figure P18.9 Problems 9 and 10.

11. A proton is placed at the origin. An electron that is initially very far away is then taken on the path shown in Figure P18.11 and eventually stops a distance of 3.0 nm ($= 3.0 \times 10^{-9}$ m) from the proton. How much work was done on the electron to move it along this path?

12. ★ An electron and a proton are a distance $r = 7.5 \times 10^{-9}$ m

Figure P18.11

apart. How much energy is required to increase their separation by a factor of two?

13. ⭐ Two point charges $Q_1 = +1.2 \ \mu C$ and $Q_2 = +3.2 \ \mu C$ are initially separated by 1.0 m and held fixed. The charges are then released, and both move in response to the electric force between them. What is the sum of the kinetic energies of the two particles when they are very far apart?

14. ⭐ The two electrons in a helium atom are separated by about 0.10 nm and the separation between either electron and the nucleus is about 0.050 nm. What is the total electric potential energy of the atom? Assume the nucleus is a point charge with $q = +2e$ and for simplicity do not include the potential energy associated with the Coulomb repulsion of the two protons in the nucleus. Compare your answer with the ionization energy of a helium atom and with the energy required to remove both electrons from a helium atom.

15. ⭐ An oxygen ion O^{2-} is a distance $r = 5.0 \times 10^{-10}$ m from a H^+ ion. How much energy is required to separate them completely? Treat both ions as point charges.

18.2 ELECTRIC POTENTIAL: VOLTAGE

16. Find the electric potential V at the center of the electric dipole in Figure P18.6. Express your answer in terms of Q and L.

17. An electron is moved from an initial location where the potential is $V_i = 30$ V to a final location where $V_f = 150$ V. What is the change in the electron's potential energy? Express your answer in joules and in electron-volts.

18. The electric potential difference between two infinite, parallel metal plates is V. If the plates are separated by a distance $L = 3.0$ mm and the electric field between the plates is $E = 250$ V/m, what is V?

19. ⭐ For the situation described in Problem 17, what is the average electric field along a line 10 cm long that connects the initial and final locations of the electron? Be sure to give both the magnitude and direction of $\vec{E}$.

20. ⭐ A proton moves from a location where $V = 75$ V to a spot where $V = -20$ V. (a) What is the change in the proton's kinetic energy? (b) If we replace the proton with an electron, what is the change in kinetic energy?

21. Consider a very small charged sphere that contains 35 electrons. At what distance from the center of the sphere is the potential equal to -60 V?

22. ⭐ How much work (as measured in joules) is required to push an electron through a region where the potential change is $+45$ V? Assume the electron moves from a region of low potential to a region of higher potential.

23. ⭐ The electrons in a TV picture tube are accelerated through a potential difference of 30 kV. (a) Do the electrons move from a region of high potential to a region of low potential, or vice versa? (b) What is the change in the kinetic energy of one of the electrons? (c) If the initial speed is very small, what is the final speed of an electron?

24. ⭐ The electric field in a particular region of space is constant with a magnitude of 300 V/m and is along the y direction. (a) Sketch how the electric potential varies along y, starting from the origin and ending at $y = 4$ m. Assume the potential is zero at the origin. (b) Sketch how V varies along the x direction, starting at the origin and ending at $x = 5$ m.

25. SSM ⭐ Four point charges, each with $Q = 4.5 \ \mu C$, are arranged at the corners of a square of edge length 1.5 m. What is the electric potential at the center of the square?

26. ⭐ The electric potential varies with x as sketched in Figure P18.26. Make a qualitative plot of the component of the electric field along the x direction as a function of x.

27. An infinite plane of charge (Fig. P18.27) has a charge per unit area of $\sigma = 2.0 \ \mu C/m^2$. What is the change in the electric potential between points that are 2.5 m (initial) and 4.5 m (final) from the plane?

Figure P18.26

σ = charge per unit area

Infinite sheet of charge

Figure P18.27 Problems 27, 28, and 36.

28. ⭐ Consider again the plane of charge in Figure P18.27 with $\sigma = 2.0 \ \mu C/m^2$. Calculate the change in potential experienced by an electron that starts a distance 2.5 m from the plane and moves a distance 5.5 m in a direction perpendicular to the y axis. Give an intuitive explanation of your answer.

29. Ten electrons are placed on a metal sphere. If the potential of the sphere is -35 V, what is the radius of the sphere?

30. ⭐ A point charge Q is located at the center of a very thin, spherical metal shell. The net excess charge of the shell is zero. If the radius the shell is r_{shell}, what is the electric potential of the shell?

31. ✪ A point charge Q is located at the center of a spherical metal shell that has an inner radius r_1 and an outer radius r_2. The net excess charge of the shell is zero. (a) What is the excess charge on the inner surface of the shell (at $r = r_1$)? (b) What is the electric potential at $r = r_1$?

32. SSM ⭐ Two infinite, parallel, uniformly charged plates with charge densities $+\sigma$ and -3σ are separated by a distance L (Fig. P18.32). If the plate on the right is at $V = 0$, what is the potential of the plate on the left? Express your answer in terms of σ and L.

$+\sigma$ $\quad$ -3σ

$\leftarrow L \rightarrow$

Figure P18.32

33. ⭐ A circular ring of charge has a radius R and a charge per unit length λ. What is the electric potential at the center of the ring? Take $V = 0$ at infinity.

34. ⭐ Consider a solid sphere of radius $R = 0.55$ m that is uniformly charged with $\rho = -2.5 \ \mu C/m^3$. What is the electric potential a distance 2.5 m from the center of the sphere?

35. ⭐ Two point particles with charges q_1 and q_2 are separated by a distance L as shown in Figure P18.35. The electric potential

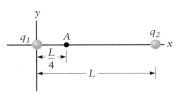

Figure P18.35

is zero at point A, which is a distance $L/4$ from q_1. What is the ratio q_1/q_2?

18.3 EQUIPOTENTIAL LINES AND SURFACES

36. ✪ Make a sketch of the equipotential surfaces near the plane of charge in Figure P18.27. *Hint*: How is the direction of $\vec{E}$ related to the orientation of an equipotential surface?

37. SSM ✪ The equipotential planes in a particular region of space are shown in Figure P18.37. What is the approximate magnitude and direction of the electric field at the origin? Consider only the components of the field along x and y.

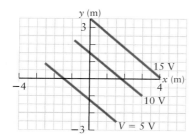

Figure P18.37 Problems 37 and 38.

38. ✪ If an electron placed at the point $(x, y) = (1.0\text{ m}, 2.0\text{ m})$ in Figure P18.37 is released from rest, in what direction will it move?

18.4 CAPACITORS

39. ✪ The dimensions of a parallel-plate capacitor are all increased by a factor of three. By what factor does the capacitance change?

40. Make a sketch of the equipotential surfaces between the plates of a parallel-plate capacitor.

41. A parallel-plate capacitor has square plates of edge length 1.0 cm and a plate spacing of 0.010 mm. If the gap between the plates is a vacuum, what is the capacitance?

42. A typical capacitor in an MP3 player has $C = 0.10\ \mu\text{F}$. If a charge $\pm 5.0\ \mu\text{C}$ is placed on the plates, what is the voltage across the capacitor?

43. A voltage of 12 V is placed on a capacitor with $C = 100$ pF (picofarads). (a) What is the charge on the capacitor? (b) How much energy is stored in the capacitor?

44. ✪ ⓡ Design a parallel-plate capacitor with $C = 5.0$ F. That is, find values of the plate area and the plate spacing that will give this capacitance. Assume there is a vacuum between the plates. Be sure to choose a practical value for the separation between the plates. Do you think that you could lift this capacitor?

45. ✪ A charge $\pm Q$ is placed on the plates of a capacitor. The separation between the plates is changed, and it is found that the voltage across the plates decreases. (a) Have the plates been moved closer together or farther apart? (b) If the new voltage is smaller than the original voltage by a factor of nine, what is the ratio of the new plate spacing to the original spacing?

46. ✪ Consider a hollow metal cylinder of radius 1.0 mm that has a thin wire of radius 0.50 mm running down the center as sketched in Figure P18.46. This structure is a capacitor, with

Figure P18.46

the wire acting as one "plate" and the cylinder acting as the other plate. Calculate the capacitance using the following steps. (a) Place charges $\pm Q$ on the wire and the cylinder and calculate the electric field between them. Assume the cylinder has a length L and assume L is very long. (b) Estimate qualitatively the electric potential difference between the wire and the cylinder, which can be done using a plot of the electric field as a function of radius. (c) Use the potential difference to find the capacitance C for $L = 10$ cm.

Figure P18.47

47. Three capacitors, all with capacitance C, are connected in parallel as shown in Figure P18.47. What is the equivalent capacitance of this combination?

48. Four capacitors, all with capacitance C, are connected in series as shown in Figure P18.48. What is the equivalent capacitance of this combination?

Figure P18.48

49. ✪ Three capacitors are connected as shown in Figure P18.49. What is the equivalent capacitance?

Figure P18.49

50. ✪ Design a combination of identical capacitors, each with capacitance C, for which the equivalent capacitance is $3C/4$.

51. SSM ✪ Four capacitors are connected as shown in Figure P18.51. What is the equivalent capacitance?

52. ✪ Two capacitors with $C_1 = 1.5\ \mu\text{F}$ and $C_2 = 2.5\ \mu\text{F}$ are connected in parallel. If the combined charge on both capacitors is $25\ \mu\text{C}$, what is the voltage across the capacitors?

Figure P18.51

53. ✪ An amount of charge $\pm Q$ is placed on the plates of a parallel-plate capacitor so that the potential across the plates is ΔV_{init}. The capacitor is then disconnected, and the separation between its plates is increased by a factor of three. If the charge on the plates is $\pm 35\ \mu\text{C}$ while the final capacitance is 2.0×10^{-9} F, what was ΔV_{init}?

54. ✪ Some charge is placed on a capacitor with $C_1 = 35\ \mu\text{F}$ so that $\Delta V = 12$ V. The capacitor is then attached in parallel to a second capacitor with $C_2 = 55\ \mu\text{F}$. What is the voltage across the two capacitors?

55. ✪ A charge $\pm Q$ is placed on the plates of a parallel-plate capacitor. The plate spacing is then increased from L to $2L$. (a) By what factor does the electric potential energy stored in this capacitor change? (b) Energy must be conserved, so where did this extra energy come from (or go to)?

18.5 DIELECTRICS

56. A parallel-plate capacitor has square plates of edge length 1.0 cm and a plate spacing of 0.10 mm. If the gap between the plates is filled with mica, what is the capacitance?

57. Consider a capacitor with the same dimensions as the capacitor in Example 18.9, but now suppose there is air between the plates. What is the maximum safe operating voltage of this capacitor? That is, what is the voltage at which there will be dielectric breakdown?

58. The space between the plates of a capacitor is filled with paper. By what factor does the paper change the capacitance relative to that found when the plates are filled with air?

59. In Example 18.7, we analyzed a parallel-plate capacitor as might be found in an integrated circuit (RAM chip), but we omitted that such a capacitor would usually contain a dielectric between the plates. This dielectric is composed of SiO_2 and has a dielectric constant close to that of glass. What is the capacitance of the RAM capacitor in Example 18.7 when it is filled with SiO_2?

60. ✪ ✖ Ⓡ A cell membrane is composed of lipid molecules and is approximately 10 nm thick. If the dielectric constant of a lipid is $\kappa \approx 5$, what is the approximate capacitance of a spherical cell that has a diameter of 10 μm?

61. Consider a parallel-plate capacitor with an area of 1.0 cm^2, with a plate spacing of 0.20 mm, and filled with mica. At what voltage will this capacitor exhibit dielectric breakdown?

62. ✪ ✖ Ⓡ When you walk across a carpeted floor while wearing socks on a dry day, your socks (and hence you) become charged by rubbing with the carpet. When your finger approaches a metal doorknob, you notice that a spark jumps across the air gap between your finger and the doorknob. What is the approximate magnitude of the electric field between your finger and the doorknob just before the spark jumps? What is the potential difference between your finger and the doorknob?

63. ✪ The dielectric in a capacitor is changed to a material with a dielectric constant that is larger by a factor of five. If the charge on the capacitor is held fixed, by what factor does the energy stored in the capacitor change? Explain why the energy is different in the two cases.

64. SSM ✪ A parallel-plate capacitor initially has Mylar between its plates and carries charge $\pm Q$. A different dielectric is then inserted between the plates without changing the charge. If the energy stored in the capacitor decreases to 30% of its initial value, what is the value of κ for the dielectric?

18.6 ELECTRICITY IN THE ATMOSPHERE

65. SSM ✪ Ⓡ Experiments have shown that in good weather the Earth has, on average, a negative charge of approximately 10^{-13} C on every square centimeter of surface area. What is the approximate total excess charge on the entire Earth? How many electrons does that correspond to? For simplicity, assume that charge is spread evenly over the entire surface of the Earth, including the oceans.

66. ✪ Ⓡ Use the data from Problem 65 to estimate the electric field at the Earth's surface.

67. Ⓡ Using the data from Problems 65 and 66, find the approximate electric potential 1.5 m above the Earth's surface (about eye level). Take the ground to be at $V = 0$.

68. Using the data from Problems 65 and 66, find the approximate separation between two equipotential surfaces with $\Delta V = 100$ V near the Earth's surface. Compare this distance to your height.

18.7 BIOLOGICAL EXAMPLES AND APPLICATIONS

69. ✪ ✖ A defibrillator containing a 20-μF capacitor is used to shock the heart of a patient in serious condition by attaching it to the patient's chest. Just prior to discharging, the capacitor has a potential difference of 10,000 V across its plates. (a) What is the energy released into the patient? (b) If the energy is discharged over 20 ms, what is the power output of the defibrillator?

18.8 ELECTRIC POTENTIAL ENERGY REVISITED: WHERE IS THE ENERGY?

70. Suppose the energy stored in the electric field in a particular region of space is 50 J in a region of volume 10 mm^3. What is the average electric field strength in this region?

71. A parallel-plate capacitor with $C = 10$ μF is charged so as to contain 1.2 J of energy. If the capacitor has a vacuum between plates that are spaced by 0.30 mm, what is the energy density (the energy per unit volume)?

72. SSM ✪ Figure P18.72 shows a point charge. At which point (A, B, or C) is the energy density (the energy per unit volume) largest?

73. ✪ If the charge Q in Figure P18.72 is doubled, by what factor does the energy density change at point A?

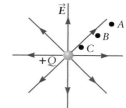

Figure P18.72
Problems 72 and 73.

74. ✪ Consider two isolated, charged conducting spheres: a large sphere and a second smaller sphere with a radius four times smaller than that of the large sphere, but with four times as much charge. (a) Compute the ratio of the electric potential at the surface of the large sphere to that of the small sphere. (b) Calculate the ratio (large to small) of the magnitudes of the electric fields at the surface of each sphere.

75. ✪ A proton is directed such that it comes within 1.8×10^{-15} m of a carbon nucleus. (a) How much kinetic energy must the proton have initially to get this close to (i.e., "collide" with) the carbon nucleus? (b) What velocity does that correspond to for the proton? (c) What potential would be needed to give the proton this much kinetic energy? Assume the target carbon nucleus is completely ionized (missing all electrons).

76. ✪ Two metal spheres, each of mass 10 g and initially at rest, are dropped from a height of 5.0 m in an evacuated chamber. One sphere has a charge of +100 μC, and the other has a charge −100 μC. Find the difference in final speeds of the two spheres (the speeds just before each one hits the ground). Recall that there is an electric field of about 150 V/m near the Earth's surface.

77. **Big cap.** The unit of capacitance is the farad. A capacitor with $C = 1$ F is a very large one. (a) To get a sense of how big a unit that is, find the area of a hypothetical 1.0-F parallel-plate air capacitor, where the plates are separated by a distance $d = 2.0$ mm. (b) Find the plate area if we fill the capacitor gap with the excellent dielectric strontium titanate. Give your answers in units of square meters, square kilometers, and square miles.

78. ✪ A homemade capacitor is made from a sandwich of a piece of standard printer paper (8.5 in. by 11 in., 20-lb bond, 0.0038 in. thick) between two equal area sheets of aluminum foil. (a) Calculate the capacitance of this device. (b) What maximum potential can be put across it? (c) How much energy can be stored in it at maximum voltage?

79. ✪ Ⓡ Some of the first capacitors constructed were called Leiden jars. These capacitors were actual jars with a layer of foil on the inside and outside of the glass as shown in Figure P18.79. A conducting sphere topped a metal rod that in turn connected to a dangling chain that made contact with the inner layer of foil, all held in place by a noncon- ducting lid made of wood or cork. The Leiden jar was developed in 1745 by Pieter van Musschenbroeck of the University of Leiden, the Netherlands, but Benjamin Frank- lin's experiments with these jars led to an understanding of how energy

Figure P18.79

is stored within the jar. The working assumption at the time was that electricity had a fluid nature, so the shape of a jar for stor- age fit well with this initial hypothesis. One of Franklin's jars measured 4 in. in diameter, with foil on the bottom and up the sides to a height of 9 in. (a) Find the approximate capacitance of his Leiden jar if the wall of the jar is $\frac{1}{8}$ in. thick. (Find the area of foil coverage and assume an equivalent parallel-plate capacitor.) (b) What is the maximum energy that could be stored in such a device?

80. ✪ Benjamin Franklin made use of a "bank" of Leiden jars (Fig. P18.80) for some of his experiments. (a) Are these 35 Leiden jar capacitors arranged in series or in parallel? *Hint:* Close inspec- tion of the photo shows that the central rods of the jars are all connected together. (b) Calculate the maximum capacitance obtained by such an arrangement using the capacitance for one such jar found in Problem 79. (c) What is the maximum amount of energy that can be stored in this bank of jars?

Figure P18.80 European-made bank of Leiden jars used by Benjamin Franklin in his 1747 experiments.

81. ✪ ✪ **The electric eel.** The knifefish species *Electrophorus electricus* shown in Figure P18.81 is not really an eel, but it can grow up to 2 m long and generate a potential difference of 600 V between a region just behind its head and its tail. The fish can stun or kill by sending large quantities of charge through its target. The charge comes from Hunter's organ, which in this instance is made up of 70 chains of 4000 stacked cells called electrocytes. Each disk-shaped electrocyte has a diameter of approximately 1.8 mm and a height of about 0.4 mm, and cells are stacked as shown in the detail of Figure P18.81. Assume the oily innards of the fish have a dielectric constant of $\kappa = 105$. (a) Model an electrocyte as a disk-shaped parallel-plate capaci- tor and find the capacitance of such a cell. (b) If the electrocytes in a chain are stacked so that the capacitance combines in paral-

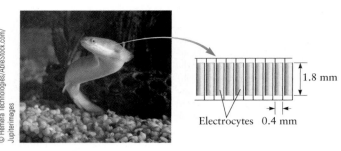

Figure P18.81 *Electrophorus electricus* and detail of the stack of electrocytes in Hunter's organ.

lel and the chains are in turn also combined in parallel, what is the total capacitance of Hunter's organ? (c) Use this model to find how much charge is on the head and tail of the fish when the maximum voltage is generated.

82. ✪ **Big shock.** The world's most advanced capacitor bank (Fig. P18.82), located at the Rossendorf Research Center in Dresden, Germany, can store 50 MJ of energy and release it in less than 5 ms. This high-energy pulse of electricity is used to produce the world's most powerful magnetic fields. (a) If the potential differ- ence at the terminals of the capacitor bank is 22 kV, how much charge is on the collective plates of the capacitor bank? (b) What is the total capacitance of the bank?

Figure P18.82

83. ✪ (a) A capacitor with C = 15 F is used to store energy. If this capacitor stores an amount of energy equal to the kinetic energy of a baseball (*m* = 0.22 kg) moving at 45 m/s (about 100 mi/h), what is the voltage across the capacitor? (b) If this structure is a parallel-plate capacitor with a plate area of 12 m², what is the electric field between the plates? Assume there is a vacuum between the plates.

84. SSM ✪ ✪ A useful model of a water molecule is sketched in Figure P18.84, with point charges of 5.2×10^{-20} C at the H sites and a point charge -10.4×10^{-20} C at the O site. The H–O bond length is 9.6×10^{-11} m, and the bond angle is $\theta = 104°$. (a) What is the total electric potential energy of this model of a water molecule? (b) How much work would be required to completely disassoci- ate the molecule?

Figure P18.84

Electric Currents and Circuits

Electric circuits are at the heart of MP3 music players and other devices and are thus an indispensable part of everyday life. (©Andreas Rentz/Getty Images News)

In Chapters 17 and 18, we considered electric forces, fields, and energy in many different situations involving static (stationary) arrangements of electric charge. We now turn our attention to the *motion* of charges and the concept of ***electric current***. When charge can flow in a closed path (a "loop"), we describe the path as an electric ***circuit***. Electric circuits are at the heart of all modern electronic devices such as televisions, CD players, and computers. In this chapter, we'll also examine some basic properties of circuit elements called resistors and capacitors. Once again, conservation principles will be very useful, and the principles of conservation of energy and charge will lead to general rules for analyzing electric circuits. In Chapter 20, we'll then study how the moving charges that form electric currents give rise to magnetic phenomena.

When charges move from one place to another, we say that there is an *electric current*. The term *current* probably comes from the analogy with the flow of water in a river. The strength of a river's current (Fig. 19.1A) depends on both the speed of the water molecules and the number of water molecules involved in the flow. In a similar way, we can consider the flow of electrons in a wire as sketched in Figure 19.1B. Electrons can move freely through a metal (much like the water in a river), and we will always use the term *wire* to refer to a metal wire.

The strength or magnitude of a current is measured by the amount of water that moves down the river (Fig. 19.1A) or the amount of electric charge that moves along the wire (Fig. 19.1B) in a particular amount of time. Electric current is denoted by the symbol I and defined as the amount of charge Δq that passes a given point on the wire per unit time Δt:

$$I = \frac{\Delta q}{\Delta t} \tag{19.1}$$

For historical reasons, current is defined in terms of net positive charge flow (Fig. 19.1C), even though electrons are responsible for the current in a metal. We'll discuss how to use Equation 19.1 to deal with a current carried by electrons shortly.

Current is measured in units of charge divided by time, or coulombs per second (C/s). This unit has been given the name *ampere* (A), in honor of André-Marie Ampère (1775–1836), who played a major role in discovering of the laws of magnetism. We have

$$\text{SI unit of electric current: } 1 \text{ A } = 1 \text{ C/s} \tag{19.2}$$

The ampere is a primary unit in the SI system. Other units involving electricity, such as the coulomb and the volt, are defined in terms of the ampere. For example, the coulomb is defined as the amount of charge carried by a current of 1 A in 1 s.

In Figure 19.1, we have drawn the electric current as confined to a wire (or river). The wire is not a necessary part of the picture, however, because an electric current is present whenever there are moving charges. For the case of a lightning bolt, in which charges move through the atmosphere, we can still discuss and measure the electric current carried by the bolt, even though the moving charges are not confined to a wire.

Which Way Does the Charge Move?

In our definition of current in Equation 19.1, Δq is the *net* electric charge that passes a particular point during a time interval Δt. This amount of charge could consist of a few particles each carrying a large electric charge or of many particles each having a small amount of charge. A current can also be produced by the motion of a combination of positive and negative charges such as a collection of positive and negative ions in solution. By convention, an electric current's direction is defined as shown in Figure 19.2. If the current is carried by positive charges moving with a velocity $\vec{v}$, the direction of the current is *parallel* to $\vec{v}$. On the other hand, if current is carried by negative charges, the direction of I is *opposite* to the charges' velocity. Thus, the direction of the current carried by electrons in a metal wire (an extremely common and important case) is *opposite* to the electron velocity (Fig. 19.2A).

A positive electric current can thus be caused by positive charges moving in one direction *or* by negative charges moving in the opposite direction. A measurement of the current alone cannot distinguish between these two alternatives.[1] Current in a metal is carried by electrons. In a liquid or gas, there will usually be both posi-

Definition of electric current

A

Electrons

B

$$I = \frac{\Delta q}{\Delta t} \text{ measured here}$$

C

Figure 19.1 ◢ When water flows, it forms a current as water molecules move along a river or stream. ◣ When electrons move through a wire, they form an electric current as charge moves along the wire. ◪ The electric current I is defined as the rate at which net positive charge passes by a point in the wire.

---

[1]A phenomenon called the *Hall effect* does distinguish between these two possibilities. The Hall effect relies on magnetic forces, as we describe in Chapter 20.

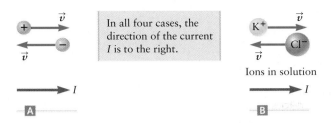

In all four cases, the direction of the current I is to the right.

Ions in solution

Figure 19.2 **A** When a positive charge moves to the right, the current is also directed to the right. When a negative charge moves to the left, the current is to the right. In both cases, the net positive charge on the right increases. **B** In a solution containing positive and negative ions, motion of these ions produces an electric current.

tive and negative mobile ions present; if so, the current is carried by a combination of positive ions moving in one direction and negative ions moving in the opposite direction (Fig. 19.2B).

EXAMPLE 19.1 | **How Many Electrons Does It Take to Light a Lightbulb?**

To emit a normal amount of light, a typical lightbulb requires a current of about $I = 0.50$ A. If this light is turned on for 1 hour, how many electrons pass through the lightbulb?

RECOGNIZE THE PRINCIPLE

According to the definition of electric current, $I = \Delta q/\Delta t$, so the total amount of charge that passes through the lightbulb in a time Δt is $\Delta q = I\,\Delta t$.

SKETCH THE PROBLEM

Figure 19.1B describes the problem.

IDENTIFY THE RELATIONSHIPS AND SOLVE

Our lightbulb is carrying current (i.e., is turned on) for 1 hour, so $\Delta t = 1$ h $= 3600$ s. We thus find

$$\Delta q = I\,\Delta t = (0.50 \text{ A})(3600 \text{ s}) = 1800 \text{ C}$$

which is the total charge in coulombs. To get the number of electrons, we divide by the magnitude of the electron's charge:

$$N = \frac{\Delta q}{e} = \frac{1800 \text{ C}}{1.6 \times 10^{-19} \text{ C/electron}} = \boxed{1.1 \times 10^{22} \text{ electrons}}$$

What does it mean?

Let's compare this number with the total number of atoms in the filament of a lightbulb (assuming an incandescent bulb). A typical filament is a piece of wire with a radius of about 50 μm and a length of about 1 cm. It is composed of tungsten, which contains approximately 6×10^{28} atoms/m^3. The number of atoms in the filament is thus

$$N_{atoms} = \pi r^2 L (6 \times 10^{28} \text{ atoms/m}^3)$$

$$= \pi (5 \times 10^{-5} \text{ m})^2 (0.01 \text{ m})(6 \times 10^{28} \text{ atoms/m}^3)$$

$$N_{atoms} \approx 5 \times 10^{18} \text{ atoms}$$

So, the number of electrons that pass through the filament during 1 hour is *much larger* that the number of atoms in the filament!

Figure 19.3 **A** Water flows downhill, from a region of high gravitational potential energy to a region of lower potential energy. **B** Electric charges in a wire move from a region of high electric potential energy to a region of lower potential energy. This electric potential difference can be produced by a battery.

19.2 | BATTERIES

Electric Current and Potential Energy

Current is the flow of charge, but how do we produce this charge motion? From our work in mechanics, we know that the current of water in a river (Fig. 19.3A) is connected with gravitational potential energy. The kinetic energy of the moving water is derived from the change in gravitational potential energy as water moves from the high end of the river to the low end.

Electric current is produced in a similar way. For charge to move along a wire, the electric potential energy at one end of the wire must be higher than the electric potential energy at the other end. Suppose the electric current in Figure 19.3B is carried by positive charges moving to the right in the copper cylinder representing the wire. The electric potential energy is related to the potential V by $V = PE_{elec}/q$ (Eq. 18.10). In dealing with current and electric circuits, electrical engineers and physicists often refer to this potential as simply the "voltage." For a positive charge, a region of high potential energy is also a region of high voltage. The current in Figure 19.3B is thus directed from a region of higher voltage to a region of lower voltage, that is, from high to low potential. On the other hand, the current in Figure 19.3B might be carried by electrons (negative charges) moving to the left in the copper wire. Since q is negative, the potential energy PE_{elec} for electrons will be higher at the right end than at the left, and the electrons move from a region of lower voltage to a region of higher voltage in going from right to left in Figure 19.3B. Either way, *the direction of the current I is always from high to low potential*, regardless of whether this current is carried by positive or negative charges.

Figure 19.3B shows the relative voltages at the two ends of the wire using the + and − symbols. The source of this potential difference is often a device called a *battery* (Fig. 19.4A). A typical battery possesses two terminals, labeled "positive" (+) and "negative" (−); these terms refer to the relative electric potential at the terminals. If the battery leads are attached to the ends of a wire as sketched in Figure 19.3B, there will be a potential difference "across" the wire, producing a current.

Construction of a Battery

Alessandro Volta (1745–1827) and his contemporaries discovered and developed the first batteries (see Insight 19.1). Before that time, all studies of electricity used static electricity produced by, for instance, rubbing amber with fur. Although some clever rubbing machines generating quite high electric potentials were constructed, the availability of batteries made possible many new experiments in electricity and magnetism. These experiments and the availability of batteries also led to the first practical applications of electricity such as the telegraph. Batteries illustrate conser-

Figure 19.4 **A** Some commonly available batteries. **B** A battery uses chemical potential energy to produce an electric potential difference across its terminals. **C** Circuit diagrams use this symbol to denote a battery. The longer line represents the positive terminal of the battery. **D** Design of the battery constructed by Volta.

vation of energy in an interesting way as they convert chemical energy to electrical energy using *electrochemical reactions*. These reactions are usually not discussed until the end of most introductory chemistry texts, so we will not describe them in detail here. For most of our work, we can take the "black box" view of batteries shown in Figure 19.4B. A typical battery contains two pieces of metal called electrodes, attached by metal wires to the battery terminals. The electrodes are in contact with specially chosen chemicals, and electrochemical reactions involving the electrodes and these chemicals produce an electric potential difference between the two electrodes. Different types of batteries employ different electrochemical reactions and are packaged in many different forms. In drawing an electric circuit, all types of batteries are represented by the symbol shown in Figure 19.4C, with terminals labeled + (positive potential) and − (negative potential).

The potential difference between a battery's terminals is called an *electromotive force*, or *emf* (pronounced "ee-em-eff"). This term was adopted before it was realized that batteries actually produce an *electric potential difference* (and not a force), but unfortunately we are stuck with the term *emf*. When dealing with a potential difference produced by a battery, we follow standard practice and denote this emf by $\mathcal{E}$, and we will often refer to it as simply a voltage. The value of $\mathcal{E}$ produced by a battery depends on the particular chemical reactions it employs and how the electrodes are arranged. Typical values are a few volts, but it is possible to connect batteries in various ways to produce much higher voltages.

Ideal Batteries and Real Batteries

A battery produces an emf, and if we connect the battery terminals to the two ends of a wire (Fig. 19.3B), this emf produces a current. When a battery is attached to a wire as sketched in Figure 19.5, electrons move out of the negative terminal of the battery through the wire and into the positive battery terminal. Hence, charge moves from one battery terminal to the other through this external *circuit*. At the same time, ions inside the battery move between the electrodes as part of the electrochemical reaction we mentioned in connection with Figure 19.4B; this reaction moves charge internally between the electrodes so that no net charge accumulates on the battery terminals while the current is present.

An *ideal battery* has two important properties. First, it *always* maintains a fixed potential difference (a constant emf) between its terminals. Second, this emf is maintained no matter how much current flows from the battery.

This picture of an ideal battery is useful, and we'll employ it in much of our work on electric circuits. *Real* batteries, however, have two practical limitations. One is that the emf decreases when the current is very high. The value of the current at which this decrease becomes significant depends on the battery size and design. The decrease in emf at high current occurs because the electrochemical reactions inside the battery do not happen instantaneously. A certain amount of time is needed for charge to move internally between the electrodes. Another limitation is that a real battery will "run down"; it will not work forever. Under normal operation, there is a certain total amount of charge that can be drawn from the battery, and when this total is reached, the battery ceases to function. In terms of Figure 19.4B, this limit is reached when the chemicals used in the electrochemical reaction are exhausted. At that point, all the available chemical energy has been extracted from the battery.

Insight 19.1
VOLTA'S BATTERIES
In Chapter 18, we mentioned Luigi Galvani's experiments with frogs and electricity. He found that frog muscle lying on an iron plate twitched when he touched it with a brass scalpel. This discovery was the basis for Volta's first batteries (Fig. 19.4D), which consisted of alternating sheets of zinc and copper separated by a piece of parchment soaked with salt water. (Salt water is similar to the fluid inside a frog.)

Figure 19.5 Because electrons are negatively charged, their velocity is directed opposite to the current.

| **EXAMPLE 19.2** | Using a Capacitor as a Source of Electric Energy |

There are some similarities between a battery and a capacitor. For instance, both can store energy. In a battery, the energy is stored as chemical energy, and in a capacitor, the energy is stored as electric potential energy. In both cases, this energy can be used to produce an electric current. In some applications in which very high currents are needed for a very short time (such as a camera flash), capacitors are preferred over batteries. A typical capacitor used in this way might have a capacitance of 1.0 F and an initial potential difference of 3.0 V between its plates. Compare the energy stored in this capacitor with the energy stored in an AA battery, which is about 7000 J.

RECOGNIZE THE PRINCIPLE

From Chapter 18, we know that the energy stored in a capacitor is given by

$$PE_{elec} = \frac{1}{2}C(\Delta V)^2$$

(Eq. 18.33). The values of C and ΔV are given, so we can calculate PE_{elec}.

SKETCH THE PROBLEM

No sketch is necessary.

RECOGNIZE THE PRINCIPLE AND SOLVE

Inserting the given values of capacitance and potential difference leads to

$$PE_{elec} = \frac{1}{2}C(\Delta V)^2 = \frac{1}{2}(1.0 \text{ F})(3.0 \text{ V})^2 = \boxed{4.5 \text{ J}}$$

What does it mean?

The energy stored in this capacitor is *much* less—by a factor of about 1500—than the amount of energy stored in an AA battery. A very large amount of chemical energy can be stored in a small volume, if one uses the proper chemicals! Although the capacitor stores less energy, its advantage in this application is that its energy can be extracted very quickly, which is essential for a camera flash.

19.3 | CURRENT AND VOLTAGE IN A RESISTOR CIRCUIT

Figures 19.3B and 19.5 show simple electric *circuits*, arrangements in which charge is able to flow around a closed path. In these examples, this path starts at one battery terminal, passes through a wire, continues to the other battery terminal, and then returns to the original terminal as it travels through the interior of the battery. In the previous section, we discussed the part of the path that is within the battery. Now let's focus on the current external to the battery and consider how this current is related to the electric potential difference provided by the battery.

Figure 19.6A shows an atomic-scale view of the charge motion within the wire. Assuming we are dealing with a metal wire, these charges are electrons and their motion is similar to the motion of gas molecules studied in our work on kinetic theory in Chapter 15. The electrons frequently collide with one another and with the stationary ions within the metal, resulting in the zigzag trajectories sketched in Figure 19.6A. Similarly, the molecules in a gas move in a zigzag manner as they collide with other gas molecules (Fig. 19.7A). If no electric field is present, the net result of these zigs and zags is that the average electron displacement after many collisions is zero; there is no net movement of charge and hence no current.

When a battery is connected to the ends of the wire, an electric potential difference is generated between those ends. To this point, we have denoted a potential difference as ΔV, using Δ (the Greek letter delta) to emphasize that we are dealing with a "difference" measured between two locations. When working with circuits, however, it is common practice to simply use V (instead of ΔV). This should cause no confusion as long as we choose a reference point in the circuit at which we take the potential to be zero (our reference potential). For the wire in Figure 19.6B, we can choose this reference point to be at one end of the wire. If the electric potential across the wire has a magnitude V, the electric field within the wire is

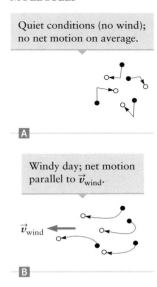

Figure 19.7 **A** Gas molecules in a room follow zigzag trajectories as they collide with other gas molecules, but the net (average) displacement is zero. **B** On a windy day, the average velocity of a gas molecule is equal to the wind velocity, even though the instantaneous velocity during the zigs and zags in part A is much greater.

Figure 19.6 The total current in a wire is due to the motion of electrons. **A** When $I = 0$, the *average* electron velocity is zero, but the *instantaneous* velocity of any particular electron is not zero. **B** An electric field accelerates *all* the electrons in the same direction. **C** This acceleration leads to a drift velocity that is the same for all electrons in the wire.

$$E = \frac{V}{L} \qquad (19.3)$$

where L is the length of the wire.

Equation 19.3 might seem to contradict Chapter 17, where we argued that the electric field inside a metal is zero. However, the electric field inside a metal is zero only when the charges in the metal are in static equilibrium, which is *not* the case here. For a metal to carry a current, there must be a nonzero electric field inside, and this electric field provides the force that maintains the current.

Drift Velocity and Current

The electric field in the wire in Figure 19.6B produces a force that pushes all the electrons to the left. Hence, the zigzag "swarm" of electrons now moves to the left, and the velocity of this swarm is called the **drift velocity**, $\vec{v}_d$. All electrons in the wire experience this extra "drifting" motion, giving a net movement of negative charge to the left in Figure 19.6B and thus an electric current to the right. The drift motion of the electrons is similar to the motion of gas molecules in a gentle wind. In that case, the molecular motion is a combination of the zigzag trajectories in Figure 19.7A with the wind's velocity, producing a "cloud" of molecules moving as shown in Figure 19.7B. The wind velocity of the molecules is thus similar to the drift velocity of the electrons.

To calculate the electric current in a wire, it is useful to focus on just the drift motion as shown in Figure 19.6C. The current is equal to the amount of charge that passes out the end of the wire per unit time. In a time Δt, all electrons move (on average) a distance $v_d \Delta t$, so all electrons that are within this distance from the left end will leave the wire in a time Δt. If the density of electrons per unit volume is n, the number of electrons that leave is equal to the number of electrons in the gray volume in Figure 19.6C. For a wire with cross-sectional area A, the number of electrons that exit the wire is thus

$$N = nAv_d \, \Delta t \qquad (19.4)$$

Each electron carries a charge $-e$, so the total charge that leaves the wire is

$$\Delta q = N(-e) = -nAev_d \, \Delta t \qquad (19.5)$$

The current is given by $I = \Delta q / \Delta t$, so we can divide both sides of Equation 19.5 by a factor of Δt to get

$$I = -neAv_d \qquad (19.6)$$

Ohm's Law

Equation 19.6 tells how the current is related to the drift velocity, but what we really want to know is how I is related to the potential difference across the ends, that is, to the battery's emf. To answer this question, we must consider the drift velocity v_d in a little more detail. Recall our discussion of the drag forces on a bacterium in a fluid (Chapter 3). An electron moving through a metal (Fig. 19.6A) is similar to a bacterium moving through a fluid. The bacterium undergoes many collisions with fluid molecules, just as electrons experience many collisions in the course of their zigzag trajectories. In both cases, there is a drag force[2] that is directed opposite to the velocity

$$\vec{F}_{\text{drag}} = -b\vec{v}$$

In Chapter 3, we showed that this drag force leads to a terminal velocity proportional to the force that pushes the *E. coli* through a fluid (Eq. 3.25). In the same way, the drag force on electrons leads to a drift velocity proportional to the force pushing the electrons through the metal. This electric force is proportional to E, so the drift velocity is proportional to the magnitude of the electric field

$$v_d \propto E \qquad (19.7)$$

According to Equation 19.3, E is itself proportional to the voltage across the wire, so the drift velocity is also proportional to V:

$$v_d \propto V \qquad (19.8)$$

Inserting Equation 19.8 into our expression for the current (Eq. 19.6) gives

$$I \propto V \qquad (19.9)$$

The constant of proportionality between I and V in Equation 19.9 involves a quantity called the electrical ***resistance*** of the wire R, which is defined by

Ohm's law

$$I = \frac{V}{R} \qquad (19.10)$$

This relation between resistance, current, and voltage is called ***Ohm's law***, and it applies to a very wide range of materials. According to Equation 19.10, R can be measured in units of voltage/current = volts/amperes = V/A. This combination of units has the name "ohm" and is denoted by the symbol Ω.

The value of a wire's resistance depends on its composition, size, and shape. It is useful to define a quantity ρ called the ***resistivity***, where the value of ρ depends on the material used to make the wire, but not on its size or shape. The resistance of a cylindrical wire of length L and cross-sectional area A (Fig. 19.8) is related to the resistivity by

$$R = \rho \frac{L}{A} \qquad (19.11)$$

A = Cross-sectional area

$$R = \rho \frac{L}{A}$$

Figure 19.8 The resistance R of a cylindrical wire is related to the resistivity ρ by $R = \rho L/A$.

[2]See Chapter 3 and Equation 3.23.

Table 19.1 Resistivities of Some Common Materials at Room Temperature

MATERIAL	$\rho \; (\Omega \cdot m)$
Metals:	
Copper	1.7×10^{-8}
Aluminum	2.7×10^{-8}
Gold	2.2×10^{-8}
Silver	1.6×10^{-8}
Lead	22×10^{-8}
Insulators:[a]	
Glass	$1 \text{ to } 1000 \times 10^{9}$
Rubber	$1 \text{ to } 100 \times 10^{12}$
Semiconductors[a]	
Silicon	$0.1 \text{ to } 100$
Germanium	$0.001 \text{ to } 1$

[a]The resistivities of insulators and semiconductors depend strongly on their purity.

The value of ρ is different for different metals; copper has a different resistivity than aluminum or gold, for example. Values of ρ can also depend on the metallurgical history (how the particular metal is heat treated, etc.). Typical resistivities of some common materials are listed in Table 19.1.

Ohm's law (Eq. 19.10) predicts a linear relationship between current and voltage. However, even though it is called a "law," it is not a fundamental law of physics in the sense of Newton's laws of mechanics. While the relation between electric current and voltage for many materials (including metals) is accurately described by Ohm's law, there are some important exceptions. For example, devices called diodes and transistors do not obey Ohm's law. These devices are essential for making amplifiers, electronic switches, and computer memory elements, and are thus extremely important in many applications. Even though Ohm's law does not apply to every type of circuit element, the lessons we'll learn from studying resistors and Ohm's law in this chapter are the foundation of all circuit theory.

EXAMPLE 19.3 | The Resistance of a Piece of Copper Wire

The wires commonly used in household wiring are made of copper and have a typical radius $r = 1.0$ mm. **(a)** Find the resistance of a piece of this wire that has a length $L = 5.0$ m. **(b)** Assume a battery with an emf of 12 V is attached to the ends of this wire. What is the current?

RECOGNIZE THE PRINCIPLE

We can use the relation between resistance and resistivity, $R = \rho L/A$, to find the resistance. Given the voltage across the wire, Ohm's law then gives the current.

SKETCH THE PROBLEM

Figure 19.8 describes the problem.

IDENTIFY THE RELATIONSHIPS AND SOLVE

(a) We use $R = \rho L/A$ (Eq. 19.11) to find the resistance of the wire. The wire's cross-sectional area is $A = \pi r^2 = 3.1 \times 10^{-6}$ m². Using this area and the length of the wire along with the resistivity of copper from Table 19.1, we find

$$R = \rho \frac{L}{A} = (1.7 \times 10^{-8} \ \Omega \cdot \text{m}) \frac{5.0 \ \text{m}}{3.1 \times 10^{-6} \ \text{m}^2} = \boxed{0.027 \ \Omega}$$

(b) The corresponding current through this wire can be found from Ohm's law (Eq. 19.10):

$$I = \frac{V}{R} = \frac{12 \ \text{V}}{0.027 \ \Omega} = \boxed{440 \ \text{A}}$$

What have we learned?
We'll compare these results for resistance and current to other common cases shortly. However, you should suspect that the result in part (b) is a very large current.

Figure 19.9 Resistors come in a wide variety of shapes and sizes. Larger ones are generally able to carry higher currents without overheating.

Wires and Resistors

Figure 19.9 shows a photo of some common types of electrical resistors, also referred to as simply *resistors*. The metal wires we considered above are types of resistors, but resistors that follow Ohm's law can be made in other shapes and from other materials. Each of these objects has a particular resistance R that determines how the current through and voltage "across" the resistor are related. If we connect the two ends of a resistor to the terminals of a battery, the voltage V across the resistor is equal to the battery's emf, hence, $V = \mathcal{E}$. Inserting into Ohm's law (Eq. 19.10), we find a current $I = \mathcal{E}/R$ through the resistor. This electric circuit is represented schematically in the circuit diagram in Figure 19.10A. The battery is represented using the symbol —|⊢— as shown in Figure 19.4C, and the resistor is represented in a circuit diagram by the symbol —W—. By convention, these symbols represent batteries and resistors in circuit diagrams. Table 19.2 lists the symbols for these and several other common circuit elements.

The straight lines in Figure 19.10A represent metal wires that attach the resistor to the battery's terminals. (The wires may not be straight in the actual circuit as shown in Fig. 19.10B, but bending a wire in this way does not significantly affect its resistance.) Metal wires are commonly referred to as simply "wires" or "leads." We saw in Example 19.3 that a metal wire has a nonzero resistance. Hence, strictly speaking, we should also refer to a wire as a resistor. However, the value of R for a typical metal wire is *much smaller* than for the resistors used in electronic devices, so when analyzing circuits it is a very good approximation to assume all wires are ideal having $R = 0$. We can account for the small but nonzero resistance of a real wire when necessary.

Dots in a circuit diagram represent connections between different circuit elements. In Figure 19.10A, they represent the points at which the wires attach to the battery, and where the wires attach to the resistor. It is sometimes useful to show the dots when one wants to be clear about where the different parts of a circuit begin and end. When that is not important or is obvious, it is usually simpler to omit the dots.

Figure 19.10 ◻A Circuit diagram for a battery connected to a resistor. ◻B Sketch of the actual physical circuit represented by the diagram in ◻A.

Speed of an Electric Current

If we attach a battery to a metal wire at time $t = 0$, when will electrons start to "flow" out of the end of the wire? From Figure 19.6, you might expect that this time is determined by the drift velocity, but that is *not* correct. To see why, let's

© Cengage Learning/Charles D. Winters

Table 19.2 Symbols for Some Commonly Used Circuit Elements

CIRCUIT ELEMENT	SYMBOL	ACTUAL APPEARANCE	CIRCUIT ELEMENT	SYMBOL	ACTUAL APPEARANCE
Battery		© Cengage Learning/ Charles D. Winters	Lightbulb		© David Graeme-Baker/Alamy
Resistor		©INSADCO Photography/Alamy	Wire		© Cengage Learning/ Charles D. Winters
Capacitor		© Hemera Technologies/Photo Objects/Jupiterimages	Connection between wires		© Cengage Learning/ Charles D. Winters
Switch		© Hemera Technologies/Photo Objects/Jupiterimages	Electrical ground		

calculate v_d in a typical case. The relation between the current and the drift velocity is (Eq. 19.6)

$$I = \frac{\Delta q}{\Delta t} = -neAv_d$$

Solving for v_d leads to

$$v_d = -\frac{I}{neA} \qquad\qquad (19.12)$$

For a household wire such as the copper wires that connect to a lightbulb, the current is typically $I = 1$ A. The cross-sectional area is about $A = 3 \times 10^{-6}$ m^2 (a cylindrical wire with a radius of 1 mm), and the density of electrons in copper is about[3] 2×10^{26} electrons/m^3. Inserting these values into Equation 19.12 gives

$$v_d = -\frac{I}{neA} = -\frac{1\ \text{A}}{(2 \times 10^{26}\ \text{electrons/m}^3)(1.60 \times 10^{-19}\ \text{C})(3 \times 10^{-6}\ \text{m}^2)}$$

$$v_d = -0.01\ \text{m/s}$$

where the negative sign means that the drift velocity is directed opposite to the current since the electron has a negative charge. The magnitude of this drift velocity is quite small; you can easily walk at a speed greater than v_d. A typical wire can

[3]This value assumes there is one current-carrying electron for each copper atom in the wire, which is a good approximation—accurate to within a factor of two or so—for copper and most metals.

be many meters long, so electrons moving at this drift velocity take many seconds to complete their path through a circuit. You might therefore expect to notice a considerable delay from the moment you "throw a switch" to the time a lightbulb begins to emit light.

You have no doubt done this experiment, so you know that there is in fact *no* perceptible time delay from the time you push a switch and a light turns on. How do we reconcile this observation with the very small value of the drift velocity? Consider an analogy with the flow of water through a garden hose when the hose is initially full of water. When water from a faucet enters one end of the hose, water leaves the other end almost immediately. Similarly, when the first electrons enter one end of a wire, other electrons leave the other end almost immediately. The first electrons travel through the entire length of the wire during a time determined by the drift velocity, but charges begin to leave the opposite end of the wire almost instantaneously once a potential difference is established across the ends of the wire. In fact, the speed of an electric current is equal to the speed of electromagnetic radiation in the wire, which is nearly the speed of light in a vacuum. (We'll return to this subject in Chapter 23.)

19.4 | DC CIRCUITS: BATTERIES, RESISTORS, AND KIRCHHOFF'S RULES

An electric circuit is a combination of connected elements such as batteries and resistors forming a complete path through which charge is able to move. Calculating the current in a circuit is a job electrical engineers call *circuit analysis*. In this chapter, we consider only the simplest "DC" circuits. The letters DC stand for "direct current," in contrast to circuits with an "AC" or "alternating current." Roughly speaking, an AC circuit is one in which the current oscillates at a particular frequency, while all other cases fall into the category of DC circuits. The current in a DC circuit is constant (independent of time) or approaches a constant if we wait long enough. Our main goal for the rest of this chapter is to learn how to analyze DC circuits.

What Is a "Circuit"?

In the circuit in Figure 19.10, two wires connect the battery to the resistor. Why do we need two wires? Why can't we carry charge from the battery to the resistor with a single wire? The reason two wires are required can be understood from analogy with the "water pipe circuit" in Figure 19.11, in which a water pump is used to push water through a pipe. This pipe carries water from the outlet (high-pressure) side of the pump and returns it to the inlet (low-pressure side), and the water current has the same magnitude throughout the circuit. This water circuit could be broken by inserting a wall into the pipe, which prevents water from flowing around the circuit and makes the current zero. Notice that the current is zero on *both* sides of the wall. The pressure produced by the water pump is now felt as a pressure difference between the two sides of the wall.

The same ideas apply to the electric circuit in Figure 19.11C. The electric current can be viewed as the motion of positive charge[4] out the high-potential (positive) terminal of the battery; this charge travels around the circuit passing through the resistor and returns at the low-potential (negative) terminal of the battery. As with the water circuit, the electric current is the same at all points in the circuit: *I* has the *same* value when this charge is "heading into" and "returning from" the resistor. If we now cut the wire (Fig. 19.11D), charge cannot return to the battery and we have an **open circuit**. The current is now zero *throughout the circuit*, and the potential

---

[4]For a metal wire, the charge would be carried by electrons (negative charges) going in the opposite direction, but that does not change our argument or conclusions.

difference generated by the battery appears across the open circuit (points *A* and *B* in Fig. 19.11D), just as the pressure difference from the water pump appears across the wall in Figure 19.11B.

The important message from Figure 19.11 is that *there must be a complete circuit for charge flow*. Simply connecting one wire from a battery to a resistor cannot produce a current. *There must always be a return path* for the current to return to the voltage source. This fact is illustrated in Figure 19.12A, which shows a bird sitting on an electric power line. A power line acts as a voltage source for your house and is at high potential. When a bird comes in contact with a power line, the bird is also at a high potential. If the bird does not simultaneously touch the ground, however, there is no return path to ground through the bird, and the current through the bird is negligible. We thus have essentially an open circuit with no current (Fig. 19.12B). As we'll discuss in Section 19.8, the health hazards of electricity are determined by the current; as long as very little current flows through the bird, its health is not endangered.

One-Loop Circuits: Kirchhoff's Loop Rule

The circuit in Figure 19.13 shows a single ideal battery with an emf $\mathcal{E}$ connected to a single resistor with resistance R. From Ohm's law (Eq. 19.10), the current in this circuit is

$$I = \frac{\mathcal{E}}{R} \qquad (19.13)$$

In words, we say that the battery produces an electric potential difference $\mathcal{E}$ "across" the resistor. We also refer to $\mathcal{E}$ as the "voltage across the resistor."

The electric current in Figure 19.13A consists of charges that travel along a complete loop, passing through the battery, through the attached wires and resistor, and back to the battery. Let's consider the electric potential energy of a test charge q as it moves through the circuit. Electric potential energy is related to the potential by $PE_{\text{elec}} = qV$, so we are really asking how the electric potential varies around the circuit. Suppose the test charge begins at the negative battery terminal. We can use this spot as the reference point for the potential and thus assign $V = 0$ at this point. We also recall the convention that $V = 0$ at an electrical "ground," so we call that our "ground point." In a circuit diagram, ground is indicated by the triangular symbol indicated in Figure 19.13B. Our charge q thus begins at the lower left in the figure

Figure 19.12 If a bird on a power line does not touch the ground, the circuit is not completed and the current is zero. The bird is spared.

Current through a resistor

Figure 19.13 ◳ This battery produces a clockwise current through the resistor. ◳ The electric potential changes as we move from one part of the circuit to another. The sum of all these voltage changes must be zero when we move completely around the circuit.

with a potential $V = 0$ and then travels through the battery, where the electrochemical reaction increases its potential by an amount $\mathcal{E}$, the battery's emf. When the charge reaches the wire on the top edge of the circuit, it thus has a potential $\mathcal{E}$. We assume the wires in this circuit are ideal, so their resistance is zero and the potential is the same everywhere within this top wire. The potential at the "top" end of the resistor is thus still $\mathcal{E}$. The charge q then passes through the resistor, moving from a point of high potential (the top end of the resistor) to a point of lower potential (the bottom of the resistor). According to Equation 19.10, the potential *decreases* by IR, so the voltage "drops" an amount IR as the charge moves between these two ends of the resistor. The final step takes the charge through the bottom wire back to the battery, and because it is another ideal resistanceless wire, there is no potential change along this wire.

At the completion of these four steps, the charge has returned to its starting point. The net change of the potential energy and hence also the net change in the electric potential must be *zero* when the charge returns to where it began, at the bottom of the battery. We thus have

$$\underset{\substack{\text{change in the} \\ \text{potential around} \\ \text{the circuit}}}{} = \overset{\substack{\text{potential increase} \\ \text{from the battery}}}{+\mathcal{E}} - \overset{\substack{\text{potential drop} \\ \text{across the resistor}}}{IR} = 0 \qquad (19.14)$$

This leads directly to

$$I = \frac{\mathcal{E}}{R}$$

which is just Equation 19.13.

Kirchhoff's loop rule

This simple example shows that conservation of energy is at the heart of circuit analysis. The steps in Figure 19.13B amount to the statement that *the change in potential energy of a charge as it travels around a complete circuit loop must be zero*. This statement is known as **Kirchhoff's loop rule**. Because the electric potential energy of a charge is proportional to the potential ($PE_{\text{elec}} = qV$), Kirchhoff's loop rule also means that the change in the electric potential around a closed-circuit path is zero. Kirchhoff's loop rule is one of our main tools for analyzing all types of circuits.

Dissipation of Energy in a Resistor

In Figure 19.13, the test charge gained energy (an increase in potential) when it passed through the battery and lost energy (a decrease in potential) as it passed through the resistor. If the potential energy of a charge decreases when it moves through a resistor, where does this energy go? The answer is that the energy of the charge is converted to heat energy inside the resistor. We say that this energy is *dissipated* as heat. In Figure 19.13B, the potential of the charge q decreases (i.e., drops) an amount $V = IR$ as it travels through the resistor. The energy of this charge thus decreases by an amount

$$\text{energy decrease} = qV$$

From our definition of current, $q = I\,\Delta t$, so

$$\text{energy decrease} = (I\,\Delta t)V$$

This energy shows up as heat energy in the resistor; hence,

$$\text{dissipated energy} = (I\,\Delta t)V$$

The amount of energy dissipated per unit time is the dissipated *power P*, so we get

$$P = \frac{\text{dissipated energy}}{\Delta t} = VI \qquad (19.15)$$

Power dissipated in a circuit element

Using our relation between voltage, current, and resistance ($I = V/R$, Eq. 19.10), Equation 19.15 can be written as

$$P = I^2R = \frac{V^2}{R} \qquad (19.16)$$

Power dissipated in a resistor

When there is current in a resistor, a certain amount of power given by Equations 19.15 and 19.16 is dissipated as heat. This energy comes from the chemical energy of the battery. So, the circuit in Figure 19.13 converts chemical energy (in the battery) to heat energy (in the resistor). We should also note that Equation 19.15 applies to *all* circuit elements, not just resistors. The power dissipated or transferred from a current I to a circuit element is always equal to the product VI, where V is the voltage across the circuit element.

Resistors in Series

Figure 19.14A shows a slightly more complicated circuit with a battery connected to two resistors. This arrangement, in which the current passes first through one resistor (R_1) and then another (R_2), is called *resistors in series*. All the charge that flows from the battery must pass through both resistors consecutively. Let's apply Kirchhoff's loop rule to this circuit, starting from a reference point (ground) at the negative terminal of the battery, where $V = 0$. If a charge q begins there and travels clockwise around the circuit, its potential first increases by $\mathcal{E}$ as it passes through the battery. The potential then decreases by an amount IR_1 as it passes through the first resistor and decreases by IR_2 when it passes through the second resistor. According to Kirchhoff's loop rule, the sum of these changes around the closed circuit loop must be zero:

$$\begin{array}{ccccccc} & \overbrace{}^{\substack{\text{potential increase}\\\text{in battery}}} & & \overbrace{}^{\substack{\text{potential drop}\\\text{across }R_1}} & & \overbrace{}^{\substack{\text{potential drop}\\\text{across }R_2}} & \\ \substack{\text{change in the}\\\text{potential around} =\\\text{the circuit}} & +\mathcal{E} & - & IR_1 & - & IR_2 & = 0 \end{array}$$

Figure 19.14 A To apply Kirchhoff's loop rule, we add up all the changes in electric potential as we move around the circuit. B When resistors are connected in series, they are equivalent to a resistor of value R_{equiv} as given in Equation 19.18.

Solving for the current, we find

$$I(R_1 + R_2) = \mathcal{E}$$

$$I = \frac{\mathcal{E}}{R_1 + R_2} \quad (19.17)$$

For the circuit with a single resistor (Fig. 19.13), we found $I = \mathcal{E}/R$ (Eq. 19.14). The combination of two resistors in Figure 19.14 is thus equivalent to a single resistor with a resistance of $R_{\text{equiv}} = R_1 + R_2$. In general, we can connect any number of resistors in series as in Figure 19.14B and they will be equivalent to a single resistor with

$$R_{\text{equiv}} = R_1 + R_2 + R_3 + \cdots \quad \text{(resistors in series)} \quad (19.18)$$

This result will be very useful when we analyze more complex circuits.

CONCEPT CHECK 19.1 | Potential in a One-Loop Circuit

Consider the one-loop circuit in Figure 19.15. What is the potential at point A?
    (a) $V_A = \mathcal{E}$ (b) $V_A = -\mathcal{E}$ (c) $V_A = 0$

CONCEPT CHECK 19.2 | Potential in a One-Loop Circuit (Revisited)

One of the wires in Figure 19.15 is cut, resulting in the open circuit shown in Figure 19.16. What is the potential difference between points B and C?
    (a) $V_B - V_C = \mathcal{E}$ (b) $V_B - V_C = -\mathcal{E}$ (c) $V_B - V_C = 0$

Combining resistors in series

Figure 19.15 Concept Check 19.1.

Figure 19.16 Concept Check 19.2.

Insight 19.3
WHAT IS AN "EQUIVALENT" CIRCUIT?
When we say that a resistor R_{equiv} is "equivalent" to a certain arrangement of several resistors, we mean that if that arrangement of resistors is replaced by R_{equiv}, the current through the rest of the circuit is unchanged. This notion of equivalence applies to many types of circuit elements, including capacitors.

EXAMPLE 19.4 | Batteries in Series

Consider the circuit in Figure 19.17A, in which two batteries with emfs $\mathcal{E}_1$ and $\mathcal{E}_2$ are connected in series to a single resistor. Use Kirchhoff's loop rule to find the current in this circuit.

RECOGNIZE THE PRINCIPLE

To apply Kirchhoff's loop rule, we start at a particular point in the circuit (the reference point at ground is a convenient choice) and move around the circuit, adding up the changes in the electric potential. According to Kirchhoff's loop rule, the total change in the potential around one complete loop is zero.

SKETCH THE PROBLEM

Figure 19.17B shows how the potential changes as we move around the circuit.

IDENTIFY THE RELATIONSHIPS

We start at the reference ground point, at the negative terminal of battery 1, and move clockwise in Figure 19.17B. There is an increase $\mathcal{E}_1$ in the potential when we pass through battery 1 and another increase $\mathcal{E}_2$ as we pass through battery 2, followed by a potential drop of magnitude IR as we move through the resistor.

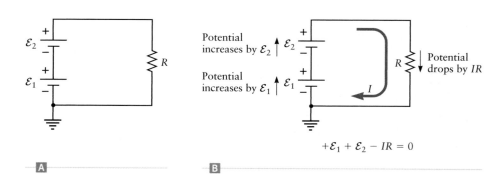

Figure 19.17 Example 19.4. Combining batteries in series.

According to Kirchhoff's loop rule, the sum of these changes in potential must be zero:

$$\frac{\text{change in potential}}{\text{around the circuit}} = \mathcal{E}_1 + \mathcal{E}_2 - IR = 0$$

Solving for the current, we find

$$I = \frac{\mathcal{E}_1 + \mathcal{E}_2}{R}$$

What does it mean?
This combination of two *batteries in series* is equivalent to a single battery with emf $\mathcal{E}_{\text{equiv}} = \mathcal{E}_1 + \mathcal{E}_2$.

Multibranch Circuits

Figure 19.18 shows a circuit in which the current from the positive battery terminal can return to the negative terminal along either of two paths: one path goes through resistor R_1, and the other path is through R_2. These different current paths are called *branches*, and this is an example of a multibranch circuit. The particular arrangement in Figure 19.18 is also referred to as *resistors in parallel*. The currents through these different branches need not be equal (unless $R_1 = R_2$), so now we have to solve for two unknown currents. In Figure 19.18A, we label the current through resistor 1 as I_1 and the current through resistor 2 as I_2. The current through the other parts of the circuit (the wires that connect to the battery terminals) is I_3. These three currents are not independent. Figure 19.18A shows expanded views of the points where these three currents are connected. Such points are called circuit "junctions" or "nodes." Electric charge cannot be created or destroyed at a junction (or anywhere else), so *the amount of current entering a junction must be equal to the amount of current leaving it.* This fact is called *Kirchhoff's junction rule*, and, for this particular circuit, it means that

Kirchhoff's junction rule

$$I_3 = I_1 + I_2 \tag{19.19}$$

Figure 19.18 **A** A circuit with two resistors connected in parallel. According to Kirchhoff's junction rule, the total current entering a junction must equal the total current leaving the junction. **B** and **C** Kirchhoff's loop rule still holds, so the sum of all changes in potential along any complete path (either path 1 or path 2) around a circuit must be zero.

To analyze this circuit and solve for all the currents, we must also apply Kirchhoff's loop rule. We start, as usual, from the negative terminal of the battery and sum the changes in electric potential as we move around a closed loop. Beginning from the negative battery terminal, two closed loops take us around this circuit and back to the battery as shown in parts B and C of Figure 19.18. Following the dashed loop in part B (path 1) in which we move through resistor 1, the changes in potential are

$$\underset{\text{around path 1}}{\text{change in potential}} = \overbrace{+\mathcal{E}}^{\substack{\text{battery} \\ \text{emf}}} - \overbrace{I_1 R_1}^{\substack{\text{potential drop} \\ \text{across } R_1}} = 0 \quad (19.20)$$

The dashed loop in Figure 19.18C (path 2) takes us through resistor 2, and the changes in potential in this case are

$$\underset{\text{around path 2}}{\text{change in potential}} = \overbrace{+\mathcal{E}}^{\substack{\text{battery} \\ \text{emf}}} - \overbrace{I_2 R_2}^{\substack{\text{potential drop} \\ \text{across } R_2}} = 0 \quad (19.21)$$

Solving for the two branch currents I_1 and I_2, we get

$$I_1 = \frac{\mathcal{E}}{R_1} \qquad I_2 = \frac{\mathcal{E}}{R_2} \quad (19.22)$$

The total current through the battery is then (according to Eq. 19.19)

$$I_3 = I_1 + I_2 = \frac{\mathcal{E}}{R_1} + \frac{\mathcal{E}}{R_2} \quad (19.23)$$

Kirchhoff's Rules

Kirchhoff's two rules are at the heart of *all* problems in circuit analysis. These two rules can be summarized as follows.

> *Kirchhoff's rules for circuit analysis*
>
> 1. *Kirchhoff's loop rule:* The total change in the electric potential around any closed circuit path must be zero.
> 2. *Kirchhoff's junction rule:* The current entering a circuit junction must equal the current leaving the junction.

Kirchhoff's rules are not new laws of physics, but are actually applications of more fundamental ideas. The loop rule follows from the principle of conservation of energy, while the junction rule is true because charge is conserved. These two general and fundamental conservation principles lead directly to Kirchhoff's very practical rules for circuit analysis.

So far, we have discussed Kirchhoff's rules using DC circuits containing batteries and resistors, but these rules apply to *all* types of circuits involving *all* types of circuit elements. The general approach to circuit analysis can be summarized as follows.

| PROBLEM SOLVING | General Rules for Circuit Analysis

1. **RECOGNIZE THE PRINCIPLE.** All circuit analysis is based on Kirchhoff's loop and junction rules.

2. **SKETCH THE PROBLEM.** Start with a circuit diagram and identify all the different circuit branches. The currents in each of these branches are usually the unknown variables in the problem.

3. **IDENTIFY** the different possible closed loops and junctions in the circuit.

  • Apply Kirchhoff's loop rule to each closed loop to get a set of equations involving the branch currents (such as I_1, I_2, and I_3 in Fig. 19.18) and quantities such as the battery emfs and the resistances in each loop.

• Apply Kirchhoff's junction rule to each junction to get equations relating the different branch currents.

4. **SOLVE** the resulting equations for the current in each branch.

5. Always *consider what your answer means* and check that it makes sense.

When we analyzed the circuit in Figure 19.18, we had three unknowns, the current in the three branches I_1, I_2, and I_3. To solve for these three unknowns, we needed three equations, which we obtained by applying the junction rule once and the loop rule twice. This circuit contains two junctions at the top and bottom of resistors R_1 and R_2, though. Could we apply the junction rule a second time to get another equation? Indeed, we could apply the junction rule to both of these junctions, obtaining the equations $I_3 = I_1 + I_2$ (Eq. 19.19) for the upper junction and $I_1 + I_2 = I_3$ for the lower junction. These two equations are identical, so the second application of the junction rule gave no new mathematical information. In general, if a circuit contains N junctions, the junction rule can be applied $N - 1$ times to obtain $N - 1$ independent equations.

Let's also consider how many times we can apply the loop rule to the circuit in Figure 19.18. Figure 19.19 shows three possible loops around the circuit, and next to each path is the resulting loop equation. We used the loop equations from paths 1 and 2 in our previous analysis of this circuit (Eqs. 19.20 and 19.21). The third loop equation—obtained using path 3 in Figure 19.19—is just the difference of the loop equations for paths 1 and 2. Hence, this third loop equation is not an independent equation. In general, a loop equation will be independent of other loop equations if it contains at least one circuit element that is not involved in the other loops.

PATH IN CIRCUIT	LOOP EQUATION
Path 1	$\mathcal{E} - I_1 R_1 = 0$
Path 2	$\mathcal{E} - I_2 R_2 = 0$
Path 3	$+I_1 R_1 - I_2 R_2 = 0$

Figure 19.19 Kirchhoff's loop rule can be applied for all three of these loops around the circuit. The loop equation in C is the difference of the loop equations in A and B, however, so there are only two independent loop equations.

Using Kirchhoff's Rules

Circuit elements can be connected in many different ways, so many different conceivable loops and branches are possible. While all such circuits can be attacked and solved using the above "General Rules for Circuit Analysis," it is useful to keep the following points in mind.

When applying Kirchhoff's loop rule, you must pay attention to the sign of the voltage drop across a circuit element. For example, when navigating clockwise around the loop in Figure 19.20A, the potential *drops* as we move through the resistor. The potential change across the resistor is thus $-IR$ because we are moving in a direction parallel to the direction of the current I. On the other hand, when applying Kirchhoff's loop rule, we are also free to move counterclockwise around the loop (Fig. 19.20B). The potential change in moving through the resistor is then $+IR$ because we are now moving from the low-potential end to the high-potential end. When writing Kirchhoff's loop equation, we can go around a loop in either direction; changing this direction changes the sign of every term in Kirchhoff's loop equation, so we are left mathematically with the same equation.

The signs of the potential changes in Figure 19.20 depend on the assumed direction of the current. Here and in our other examples with Kirchhoff's loop rule, we assume a direction for the current and stick with it as we write the loop rule equations. After deriving these equations, we can solve for the value of I. If I is found to be positive, our assumed direction for the current was correct, whereas if I has a negative value, the current is opposite to our assumed direction.

Figure 19.20 **A** The potential *drops* when passing through a resistor in the direction parallel to the current. **B** The potential *increases* when moving through a resistor in the direction opposite ("against") the current.

Resistors in parallel

Figure 19.21 Many resistors connected in parallel are equivalent to a resistance of value R_{equiv} as given by Equation 19.25.

When applying Kirchhoff's rules to find the currents through all the branches in a circuit, the number of equations must (according to the rules of algebra) equal the number of unknown currents. The repeated use of Kirchhoff's rules can sometimes lead to "repeated" equations as discussed earlier in connection with Figure 19.19. When solving for branch currents, we must use Kirchhoff's rules to get a set of independent equations, with the number of equations equal to the number of unknowns.

Resistors in Parallel

When we analyzed the circuit in Figure 19.14A, we found that two resistors R_1 and R_2 in series are equivalent to a single resistor of value $R_{equiv} = R_1 + R_2$. The resistors R_1 and R_2 can thus be replaced by this equivalent resistor without affecting the rest of the circuit; in particular, the current in the circuit will be the same. The resistors in Figure 19.21 are connected in parallel, and it is also possible to replace the parallel resistors R_1 and R_2 by a single equivalent resistor R_{equiv}. To find the value of this equivalent resistor, we recall our result for the total current through the battery in the parallel circuit (Eq. 19.23):

$$I_3 = \frac{\mathcal{E}}{R_1} + \frac{\mathcal{E}}{R_2} = \mathcal{E}\left(\frac{1}{R_1} + \frac{1}{R_2}\right)$$

For the equivalent circuit in Figure 19.21B, we have

$$I = \frac{\mathcal{E}}{R_{equiv}}$$

where I is again the current through the battery. For the circuits in parts A and B of Figure 19.21 to be truly equivalent, the current through the battery must be the same, so we have

$$\mathcal{E}\left(\frac{1}{R_1} + \frac{1}{R_2}\right) = \frac{\mathcal{E}}{R_{equiv}}$$

which gives

$$\frac{1}{R_{equiv}} = \frac{1}{R_1} + \frac{1}{R_2} \tag{19.24}$$

This result can be extended to any number of resistors in parallel (Fig. 19.21C). In the general case, the value of the equivalent resistance is

$$\frac{1}{R_{equiv}} = \frac{1}{R_1} + \frac{1}{R_2} + \frac{1}{R_3} + \cdots \quad \text{(resistors in parallel)} \tag{19.25}$$

We'll demonstrate this handy result in the next example.

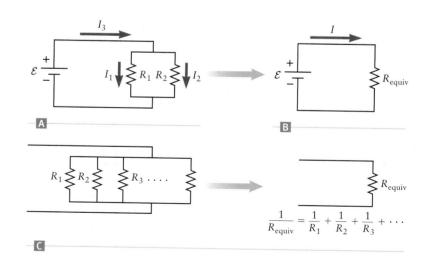

EXAMPLE 19.5 | Resistors in Series and in Parallel

Consider the circuit in Figure 19.22A. Find the current through resistor R_1.

RECOGNIZE THE PRINCIPLE

We could attack this problem using Kirchhoff's rules, applying the loop and junction rules to derive equations involving the currents through the different branches. In this case, however, we can instead use the rules for combining resistors in series and in parallel (Eqs. 19.18 and 19.25). These equations were derived from Kirchhoff's rules, so those rules are still the foundation of our calculation.

SKETCH THE PROBLEM

The three parts of Figure 19.22 show how the parallel and series addition rules can be applied to obtain a single equivalent resistor R_{equiv}.

IDENTIFY THE RELATIONSHIPS AND SOLVE

Resistors R_2 and R_3 in Figure 19.22A are connected in parallel, so we can replace them with a single resistor $R_{parallel}$ as shown in Figure 19.22B. That parallel combination of R_2 and R_3 is equivalent to a single resistor with (Eq. 19.25)

$$\frac{1}{R_{parallel}} = \frac{1}{R_2} + \frac{1}{R_3}$$

which can be rearranged to get

$$R_{parallel} = \frac{R_2 R_3}{R_2 + R_3}$$

We now have an equivalent circuit (Fig. 19.22B) containing two resistors R_1 and $R_{parallel}$ in series. We can replace these two resistors by a single new resistor R_{equiv} as shown in Figure 19.22C. To find the value of R_{equiv}, we use the rule for combining resistors in series (Eq. 19.18) to get

$$R_{equiv} = R_1 + R_{parallel}$$

Using our result for $R_{parallel}$ gives

$$R_{equiv} = R_1 + R_{parallel} = R_1 + \frac{R_2 R_3}{R_2 + R_3}$$

We have now arrived at a much simpler equivalent circuit that contains a battery with a single resistor R_{equiv}. The current in this circuit is given by our familiar result

$$I = \frac{\mathcal{E}}{R_{equiv}}$$

What have we learned?

Complicated combinations of resistors can often be analyzed by applying the rules for series and parallel resistors as we have done here. For some circuits, though, one has to start directly with Kirchhoff's rules instead as we'll see in Example 19.6.

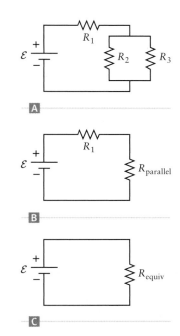

Figure 19.22 Example 19.5.
A Using the rule for combining resistors in parallel, we can replace R_2 and R_3 by an equivalent resistor $R_{parallel}$ in **B**. We can then combine R_1 and $R_{parallel}$ in series to get a resistance value R_{equiv} shown in **C** that is equivalent to the three original resistors.

CONCEPT CHECK 19.3 | Combining Resistors

Three resistors are connected as shown in Figure 19.23. What is the equivalent resistance of this combination?
    (a) R (b) $3R$ (c) $R/3$

Figure 19.23 Concept Check 19.3.

EXAMPLE 19.6 | A Multiloop Circuit

Figure 19.24 Example 19.6.
A A multiloop circuit with three identical resistors. **B** We label the currents through the three branches I_1, I_2, and I_3 and apply Kirchhoff's rules to solve for these currents.

Consider the circuit in Figure 19.24A. Apply Kirchhoff's rules to find the current I_2 though the central branch. For simplicity, assume all three resistors have the same resistance R.

RECOGNIZE THE PRINCIPLE

This circuit cannot be simplified using the rules for series and parallel combinations of resistors because two of the branches contain both a battery and a resistor. The rules for resistors in series or in parallel in Equations 19.18 and 19.25 were derived assuming the branches contain only resistors. We must therefore start with Kirchhoff's two rules, following Step 1 of the approach outlined in the "General Rules for Circuit Analysis" given earlier.

SKETCH THE PROBLEM

Step 2: We begin by labeling the currents through all branches of the circuit in Figure 19.24B and assigning positive directions for each current as indicated by arrows in the figure. We do not know the values of the emfs, so we cannot know at this point if the currents will be in the directions indicated by these arrows, but we can proceed with these assumed directions and solve for I_1, I_2, and I_3. A negative value for one of the currents means that current is directed opposite to the corresponding arrow in Figure 19.24B.

IDENTIFY THE RELATIONSHIPS

Step 3 of the strategy: We express Kirchhoff's loop rule for the two loops indicated by the dashed paths in the figure. Starting at the negative terminal of battery 1, we travel around path 1 in a clockwise direction. There is a potential increase of $\mathcal{E}_1$ at the battery, a drop of I_1R as we pass through resistor 1, and another potential drop I_2R when passing through resistor 2. These potential changes must add up to zero, so

$$\mathcal{E}_1 - I_1R - I_2R = 0 \tag{1}$$

We next travel around loop 2, starting at the negative terminal of battery 2 and moving around this loop in the counterclockwise direction, which takes us first through the battery with a potential increase $\mathcal{E}_2$. We then pass through a resistor "against" the assigned direction of I_3. We are moving from the low-potential end of the resistor to the high-potential end, so now we have a potential *increase* of I_3R. Continuing around loop 2, we have a potential drop I_2R on passing through the resistor in the central branch that carries current I_2. Summing all the potential changes along path 2, Kirchhoff's loop rule gives

$$\mathcal{E}_2 + I_3R - I_2R = 0 \tag{2}$$

We now have two equations, (1) and (2), and three unknowns (the currents through the three branches). We need one more equation (because the number of equations must equal the number of unknowns), and we can get it using Kirchhoff's junction rule. Applying this rule at the junction indicated at the top in Figure 19.24B, we find

$$I_1 = I_2 + I_3 \tag{3}$$

which gives us three equations. We can now solve for all three unknown currents.

SOLVE

Step 4: Adding Equations (1) and (2) gives

$$\mathcal{E}_1 - I_1R - I_2R + \mathcal{E}_2 + I_3R - I_2R = 0 \qquad (4)$$

Collecting the terms involving R leads to

$$\mathcal{E}_1 + \mathcal{E}_2 + R(-I_1 - I_2 + I_3 - I_2) = 0$$

$$\mathcal{E}_1 + \mathcal{E}_2 = (I_1 + 2I_2 - I_3)R$$

We can use Equation (3) to eliminate I_3 and find

$$\mathcal{E}_1 + \mathcal{E}_2 = (I_1 + 2I_2 - I_3)R = (I_1 + 2I_2 - I_1 + I_2)R$$

$$\mathcal{E}_1 + \mathcal{E}_2 = 3I_2R$$

Solving for the current I_2 gives our final result:

$$I_2 = \boxed{\dfrac{\mathcal{E}_1 + \mathcal{E}_2}{3R}}$$

What have we learned?
If we want to find the currents in the other branches, we could substitute this result for I_2 into Equation (1) to obtain I_1 and so forth, which would give us the complete solution for this circuit.

EXAMPLE 19.7 | ⓑ **Parallel and Series Combinations of Resistors (Revisited)**

You are given an unlimited number of identical resistors with $R = 10$ kΩ. **(a)** How could you use several of these resistors to construct an equivalent resistance with $R_{equiv} = 60$ kΩ? **(b)** How could you use several of these resistors to construct an equivalent resistance with $R_{equiv} = 2.5$ kΩ?

RECOGNIZE THE PRINCIPLE

We begin with some observations about the rules for combining resistors in series and in parallel. The general series combination rule is (Eq. 19.18)

$$R_{equiv} = R_1 + R_2 + R_3 + \cdots \qquad (1)$$

In words, this says that for a purely series combination of resistors, the equivalent resistance is always *larger* than *any* of the individual resistors. The parallel-resistor combination rule is (Eq. 19.25)

$$\frac{1}{R_{equiv}} = \frac{1}{R_1} + \frac{1}{R_2} + \frac{1}{R_3} + \cdots \qquad (2)$$

In words, this says that the equivalent resistance is always *smaller* than *any* of the individual resistances.

SKETCH THE PROBLEM

No figure is necessary.

IDENTIFY THE RELATIONSHIPS AND SOLVE

(a) The problem asked for a total resistance of 60 kΩ, but the components have $R = 10$ kΩ; according to Equation (1), we must connect six of the component resistors in series.

(b) If we connect n of these resistors in parallel, Equation (2) becomes

$$\frac{1}{R_{\text{equiv}}} = n\left(\frac{1}{R}\right)$$

Solving for n and inserting the desired value of the equivalent resistance, we get

$$n = \frac{R}{R_{\text{equiv}}} = \frac{10 \text{ k}\Omega}{2.5 \text{ k}\Omega} = 4$$

We should thus connect four resistors in parallel to get an equivalent resistance of 2.5 kΩ.

What does it mean?
Connecting resistors in series always gives a total (equivalent) resistance that is larger than the resistance of any of the component resistors. A parallel connection always gives an equivalent resistance smaller than any of the component resistances.

CONCEPT CHECK 19.4 | Comparing Series and Parallel Circuits

Incandescent lightbulbs, like the ones used in flashlights, act as resistors, and their light intensity depends on the current: a high current produces a high light intensity, while a small current gives a small light intensity. Figure 19.25 shows two circuits with identical batteries and identical lightbulbs (each with resistance R). In which circuit do the lightbulbs shine brightest?

Real Batteries

In Section 19.2, we mentioned that an ideal battery always maintains a constant voltage across its terminals; the value of this voltage is just the emf of the battery. An ideal battery maintains this constant voltage no matter how much current it provides to a circuit. Real batteries fall short of this ideal. The voltage provided by a real battery decreases at high currents, in part because of the time associated with the electrochemical reactions and the movement of charge inside the battery (Fig. 19.4B). This behavior can also be understood using the equivalent circuit picture in Figure 19.26A.

A real battery is equivalent to an ideal battery in series with a resistor R_{battery}, called the **internal resistance** of the battery. The current through the battery passes through this internal resistance, and the effect can be seen by analyzing the circuit in Figure 19.26B. The internal resistor R_{battery} and the external resistor R are connected in series; applying our relation for resistors in series (Eq. 19.18), the current through this circuit is

$$I = \frac{\mathcal{E}}{R + R_{\text{battery}}}$$

Usually, we are interested in the voltage (i.e., the potential difference) that appears across the battery terminals. The voltage across the real battery's terminals is also equal to the potential drop across the external resistor, which is IR. We thus have

$$\text{potential of real battery} = IR = \mathcal{E}\left(\frac{R}{R + R_{\text{battery}}}\right) \tag{19.26}$$

If the internal resistance of the battery is small compared to the external resistor ($R_{\text{battery}} \ll R$), the factor in parentheses in Equation 19.26 is approximately 1 and the voltage across the terminals is nearly equal to $\mathcal{E}$, the value for an ideal battery. However, if the internal resistance is not small compared with R, the voltage across the battery terminals can be much less than $\mathcal{E}$. Hence, in practical applications, the internal resistance of the battery should be small compared with the resistance of the circuit to which the battery is attached.

Circuit 1

Circuit 2

Figure 19.25 Concept Check 19.4.

A

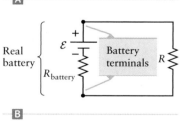

B

Figure 19.26 **A** All real batteries have a nonzero internal resistance R_{battery}. **B** We can think of a real battery as an ideal battery in series with a resistance R_{battery} and apply Kirchhoff's rules to this circuit model.

19.5 | DC CIRCUITS: ADDING CAPACITORS

In Chapter 18, we introduced capacitors and showed that when charges $\pm q$ are placed onto the plates of a capacitor, the potential difference V across the capacitor plates has a magnitude (Eq. 18.30)

$$V = \frac{q}{C} \qquad (19.27)$$

Here, as in all our work on circuits in this chapter, V is the potential *difference* across the circuit element (in this case, the capacitor). In Chapter 18, we denoted this potential difference by ΔV, but now we follow the convention widely used in circuit analysis and drop the "Δ."

Applying Kirchhoff's Rules to a Circuit with a Capacitor

Kirchhoff's rules for circuit analysis apply to all kinds of circuits, including those with capacitors. Figure 19.27 shows an RC circuit containing a resistor and a capacitor in series with a battery. A capacitor is represented in a circuit diagram as two parallel plates (Fig. 18.26); this symbol is designed to remind you of a parallel-plate capacitor, even though most real capacitors do not have this simple geometry. To apply Kirchhoff's loop rule, we start at the negative terminal of the battery and proceed clockwise around the circuit. The potential increases by $\mathcal{E}$ as we pass through the battery and then decreases by an amount IR (a potential drop) as we move across the resistor. The potential across a capacitor is given by Equation 19.27; because the top capacitor plate in Figure 19.27 is assumed to be positively charged ($+q$), the potential drops as we move from the top to the bottom capacitor plate (as we travel clockwise around the circuit). Collecting these terms and writing Kirchhoff's loop rule for this circuit, we get

$$
\begin{array}{c}
\text{change in} \\
\text{potential around} = \\
\text{the circuit}
\end{array}
\quad
\underbrace{+\mathcal{E}}_{\substack{\text{potential increase} \\ \text{in battery}}}
\quad - \quad
\underbrace{IR}_{\substack{\text{potential drop} \\ \text{across } R}}
\quad - \quad
\underbrace{\frac{q}{C}}_{\substack{\text{potential drop} \\ \text{across } C}}
\quad = 0
$$

As in our previous applications of Kirchhoff's rules, we want to solve this equation to find the current in the circuit. Now, though, we have a new complication: the value of q will change with time because current carries charge onto or off the capacitor plates. As a result, both I and q will be time dependent.

RC Circuits: A Qualitative Analysis

We can get a complete solution for I and q in an RC circuit using the relation between current and charge $I = \Delta q / \Delta t$ (Eq. 19.1) along with Kirchhoff's loop rule, but the mathematics is more complicated than we wish to tackle in this book. We will therefore resort to a qualitative analysis based on the general relationship between current and the charge on a capacitor.

Figure 19.28 shows a circuit that contains a single capacitor along with a resistor, a battery, and a *switch* denoted by S. A simple switch can be open, in which case it does not allow any current to pass, or closed, allowing current to pass with no resistance, like an ideal wire. Suppose the capacitor is initially uncharged and the switch is open (Fig. 19.28A). In this case, the current is zero and there is no voltage across the capacitor because $q = 0$. We now close the switch at $t = 0$ and consider what happens. Immediately after the switch is closed, there is a clockwise current as indicated in Figure 19.28B, carrying a positive charge to the top plate of the capacitor. The charge accumulating on the top plate repels charge from the bottom plate, causing charge to flow into the negative battery terminal and leaving the

Figure 19.27 The circuit symbol for a capacitor consists of two parallel lines representing the two plates of a parallel-plate capacitor. When charges $\pm q$ are placed on the capacitor plates, an electric potential difference $V = q/C$ is established between the two plates.

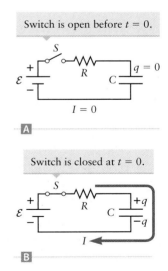

Figure 19.28 �**A** An RC circuit containing a battery, a resistor, a capacitor, and a switch. When the switch is open, no current can pass through it or through the rest of the circuit. **B** When the switch is closed, a current is established, taking charge to (or from) the plates of the capacitor.

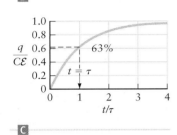

Figure 19.29 Variation of **A** the current, **B** the voltage across the capacitor, and **C** the charge on the capacitor as functions of time for the circuit in Figure 19.28. The switch was closed at $t = 0$.

bottom plate of the capacitor with a negative charge. At any particular instant, the capacitor plates are charged $\pm q$, and according to Equation 19.27 there is a nonzero voltage across the capacitor.

As time passes, the current carries more and more charge onto the capacitor, and we say that the capacitor is "charged" by the current. As q increases, so does the voltage across the capacitor. Figure 19.29 shows how the current through the circuit and the voltage across the capacitor vary with time. The voltage across the capacitor increases and asymptotically approaches $\mathcal{E}$, the battery emf. At the same time, the current through the circuit decreases to zero. Mathematically, the current in the circuit is described by

$$I = \frac{\mathcal{E}}{R} e^{-t/\tau} \tag{19.28}$$

and the voltage across the capacitor is

$$V_{cap} = \mathcal{E}(1 - e^{-t/\tau}) \tag{19.29}$$

From our relation between the voltage across and charge on a capacitor (Eq. 19.27), we get

$$q = CV_{cap} = C\mathcal{E}(1 - e^{-t/\tau}) \tag{19.30}$$

which is also plotted in Figure 19.29. The quantity τ is called the ***time constant*** and is given by

$$\tau = RC \tag{19.31}$$

The results in Equations 19.28 through 19.31 along with Figure 19.29 show how the charge on and voltage across a capacitor vary with time as the capacitor is "charged up." The amount of time required for this process depends on the time constant τ and hence depends on the product of R and C. After the switch is closed for an amount of time equal to τ—that is, for a period equal to "one time constant"—the current decreases to $e^{-t/\tau} = e^{-1} = 0.37 = 37\%$ of its value at $t = 0$. Likewise, V_{cap} and q increase to $1 - e^{-1} = 0.63 = 63\%$ of their asymptotic values after this time.

Although we will not derive the results for the current, voltage, and charge in Equations 19.28 through 19.31, we can still use Kirchhoff's loop rule to understand the circuit behavior just after the switch is closed (Fig. 19.30A). The initial charge on the capacitor is zero, and just after the switch is closed the charge will still be very small. If q is small, the voltage across the capacitor is also small, that is, $V_{cap} \approx 0$ just after closing the switch. In terms of a circuit analysis, the capacitor at this moment is nearly equivalent to a wire because there is almost no potential drop as we pass from one side of the capacitor to the other. The equivalent circuit thus looks as shown Figure 19.30B, containing a battery and a resistor, and the current is $I = \mathcal{E}/R$. This current is the initial ($t = 0$) value in Figure 19.29A (and Eq. 19.28).

We can analyze the circuit behavior when t is large—that is, a long time after the switch is closed—in a similar way. As t increases, the charge on the capacitor in Figure 19.30 becomes larger and larger. Eventually, it reaches a value for which the voltage across the capacitor is $V_{cap} \approx \mathcal{E}$, the battery emf. At this time, the equivalent circuit appears as shown in Figure 19.30D; the battery with emf $\mathcal{E}$ is now connected to the capacitor across which there is a voltage also equal to $\mathcal{E}$. However, the *polarity* of the capacitor voltage is such that it *opposes* the battery emf. We essentially have two voltage sources connected in opposition, and the net result is no current in the circuit. That is why the current approaches zero when t is large in an RC circuit (Fig. 19.29A).

Discharging a Capacitor

We have just analyzed how the current and charge in an RC circuit vary when a capacitor is charged. Let's now consider the behavior when a capacitor is discharged. An appropriate circuit is shown in Figure 19.31; it is identical to our original circuit

AT TIME $t \approx 0$

The capacitor initially acts like a short circuit.

AFTER A LONG TIME

The capacitor eventually acts as an emf $\mathcal{E}$.

Figure 19.30 **A** When the charge on the capacitor $q = 0$, the voltage across the capacitor is zero. The circuit is then equivalent to **B**, in which the only potential changes are across the resistor and the battery. **C** When $q > 0$, there is a voltage drop across the capacitor, which must be included when applying Kirchhoff's loop rule. **D** After a long time, the potential drop across the capacitor has a magnitude $\mathcal{E}$ and "opposes" the battery potential, so the current is zero.

(Fig. 19.28) except for an added second switch S_2. The capacitor can be charged by keeping switch S_1 closed for a very long time with switch S_2 open (Fig. 19.31A), and we have seen that the charge on the capacitor is then $q = C\mathcal{E}$ (Eq. 19.30). Now let's open switch S_1 and close S_2. This produces a counterclockwise current through the loop on the right in Figure 19.31B, carrying charge from the (initially) positive capacitor plate at the top to the (initially) negative plate at the bottom. After a long time, the charge on both plates is zero and the capacitor is "discharged." If switch S_2 is closed at $t = 0$ (i.e., we start our clock when the capacitor begins to discharge), the discharge process is described by

$$I = -\frac{\mathcal{E}}{R} e^{-t/\tau} \tag{19.32}$$

$$V_{cap} = \mathcal{E} e^{-t/\tau} \tag{19.33}$$

$$q = C\mathcal{E} e^{-t/\tau} \tag{19.34}$$

These results are plotted in Figure 19.32, where we see the same exponential behavior found during the charging process. Notice that the current is *negative* as the capacitor discharges. The current's direction is *opposite* to the direction found when the capacitor was charged (Fig. 19.28). The time constant for discharging is again given by $\tau = RC$ (Eq. 19.31). The same time constant describes both charging and discharging.

First, S_1 is closed and the capacitor is "charged."

A

Switch S_1 is opened and S_2 is closed. The capacitor is then discharged.

B

Figure 19.31 **A** Charging and **B** discharging a capacitor.

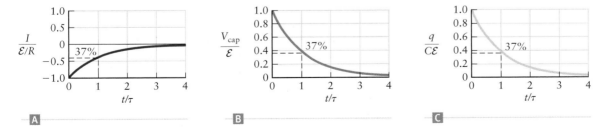

Figure 19.32 For the circuit in Figure 19.31, **A** the current through the capacitor, **B** the voltage across the capacitor, and **C** the charge on the capacitor as the capacitor discharges. Switch S_2 was closed at $t = 0$.

A

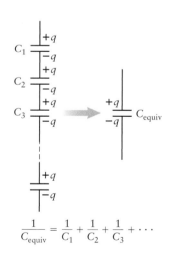

B

Figure 19.33 **A** A real capacitor such as the ones used in a DRAM circuit can be modeled **B** as an ideal capacitor in parallel with an ideal resistor. Typical DRAM integrated circuits are a few cm on a side.

EXAMPLE 19.8 Discharge Time for a Capacitor

A key quantity when charging and discharging a capacitor is the time constant τ. The value of τ can vary widely because values of R in practical circuits can range from less than 1 Ω to more than 10^{12} Ω, and the capacitance can vary from as much as 1 F to 10^{-12} F or even less. Capacitors are central elements of a computer's dynamic random-access memory (DRAM) integrated circuit chip (Fig. 19.33A). In normal operation, a DRAM capacitor is connected to a resistor (Fig. 19.33B), and this RC combination is attached to other circuitry on the same computer chip that produces a current to charge or discharge the capacitor. The charge on the capacitor is used to store information in binary form; the presence of a certain amount of charge denotes a "1" in the memory element, and the absence of charge denotes a "0." The RC circuit loop in Figure 19.33B is identical to the loop on the right in Figure 19.31B, so the resistance in parallel with the capacitor provides a discharge path. This discharging is actually a bad thing because immediately after charge is placed on the DRAM capacitor it starts to discharge, and the information stored in the memory element is eventually lost. Consider a DRAM chip for which $C = 10^{-15}$ F and $R = 10^{13}$ Ω. Calculate the discharge time for this RC circuit.

RECOGNIZE THE PRINCIPLE

The solution requires only that we evaluate the time constant τ.

SKETCH THE PROBLEM

The problem is described in Figure 19.33B.

IDENTIFY THE RELATIONSHIPS AND SOLVE

The time constant $\tau = RC$ (Eq. 19.31), so we have

$$\tau = RC = (10^{13} \ \Omega)(10^{-15} \ \text{F}) = 10^{-2} \ \text{s} = \boxed{0.01 \ \text{s}}$$

What does it mean?

The voltage across the capacitor in an RC circuit decays in a time approximately equal to τ, so a DRAM capacitor will "remember" its contents for only about 0.01 s. For this reason, a DRAM chip contains circuitry that recharges the capacitor (the technical term is "refresh") at regular intervals. Typical refresh rates are once every 0.001 s, which is much shorter than τ, and the capacitor is recharged before the memory information is lost. The refresh circuit requires a power supply to recharge the capacitors, so when your computer is turned off the DRAM information is lost. It is possible to make memory chips containing much larger resistances R so that the capacitors do not need to be refreshed. Called *static* RAM chips, these are more expensive to make than other chips, but are useful in devices that must be able to hold their memory even when no power is available. This type of memory chip is used in many MP3 players.

$$\frac{1}{C_{\text{equiv}}} = \frac{1}{C_1} + \frac{1}{C_2} + \frac{1}{C_3} + \cdots$$

Figure 19.34 When many capacitors are connected in series, they behave as a single equivalent capacitance given by Equation 19.35.

Capacitor Combinations

Two or more capacitors may be connected in either series or in parallel arrangements. We considered such problems in Chapter 18, and it is interesting to compare those results with the rules for combining resistors in Section 19.4.

Several capacitors connected in series are shown in Figure 19.34. Let's assume an external circuit has placed charge $+q$ on the upper plate of C_1 and charge $-q$ on the bottom plate of the lowest capacitor. In Chapter 18, we saw that charges $\mp q$ are then induced on the corresponding inner capacitor plates. We also showed (Eq. 18.42) that this arrangement is equivalent to a single capacitor with

Equivalent capacitance of capacitors in series

$$\frac{1}{C_{\text{equiv}}} = \frac{1}{C_1} + \frac{1}{C_2} + \frac{1}{C_3} + \cdots \quad \text{(capacitors in series)} \qquad \text{(19.35)}$$

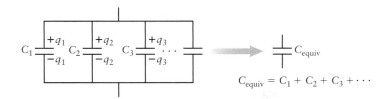

$$C_{equiv} = C_1 + C_2 + C_3 + \cdots$$

which has the same form as the rule for combining resistors in *parallel* (Eq. 19.25). This form also tells us that the equivalent capacitance C_{equiv} is *smaller* than *any* of the individual capacitances C_1, C_2,....

A combination of two or more capacitors connected in parallel is also equivalent to a single capacitor (Fig. 19.35). In Chapter 18, we derived the rule for combining capacitors in parallel (Eq. 18.41):

$$C_{equiv} = C_1 + C_2 + C_3 + \cdots \quad \text{(capacitors in parallel)} \quad (19.36)$$

Equivalent capacitance of capacitors in parallel

Hence, the equivalent capacitance is just the sum of the individual component capacitances and is therefore *greater* than *any* of the individual values. Notice that the rule for combining capacitors in parallel has the same form as the rule for combining resistors in *series* (Eq. 19.18).

CONCEPT CHECK 19.5 | **What Is the *RC* Time Constant?**

Suppose the *RC* time constant for the combination in Figure 19.36A is 2.4 ms. A second capacitor is then placed in parallel with the original capacitor as in Figure 19.36B. If these capacitors are identical, what is the new value of the time constant τ?

    (a) 2.4 ms   (b) 4.8 ms   (c) 1.2 ms

Figure 19.36 Concept Check 19.5.

19.6 | MAKING ELECTRICAL MEASUREMENTS: AMMETERS AND VOLTMETERS

We can use the circuit analysis rules described in Sections 19.4 and 19.5 to calculate the current in a particular circuit branch or the voltage across a particular circuit element, but how does one actually measure these currents and voltages? A device that measures current is called an ***ammeter***, while voltage is measured with a ***voltmeter***. Ammeters and voltmeters are both circuit elements; let's now consider how to incorporate them into a circuit analysis.

The purpose of an ammeter is to measure the current through a circuit branch, which can be accomplished by arranging for all the current to pass through the ammeter. Hence, an ammeter must be connected *in series* with the desired circuit branch as shown in Figure 19.37A. An ideal ammeter should have absolutely no effect on the circuit. That is, an ideal ammeter should measure the current without changing its value. From the properties of resistors in series, this means that an ammeter must have a very low resistance, ideally zero. Although an ammeter with a resistance of exactly zero is difficult to construct, in practice it is only necessary that the ammeter's resistance be much smaller than the resistance of the rest of the circuit branch. For the circuit in Figure 19.37A, the resistance of the ammeter should be much less than the resistance R_3.

The purpose of a voltmeter is to measure the voltage *across* a particular circuit element (or elements). The voltmeter must therefore be connected in a *parallel* arrangement as shown in Figure 19.37B. Here a voltmeter is connected in parallel with resistor R_3 so that it can measure the voltage across this resistor. The voltmeter should measure this voltage without affecting its value; a voltmeter should therefore

An ammeter must be inserted in series with the branch whose current is to be measured.

A voltmeter is inserted in parallel with the circuit element whose voltage drop (or increase) is to be measured.

Figure 19.37 Ⓐ An ammeter must be placed in series to measure the current in a particular branch of a circuit. This ammeter is placed in series with R_3 so that it can measure the current through R_3. Ⓑ A voltmeter must be placed in parallel with the circuit element of interest. This voltmeter is in parallel with R_3 so that it can measure the voltage across that resistor.

have a resistance much larger than the circuit element to which it is attached so that very little current passes through it. An ideal voltmeter has an infinite resistance so that the current through the voltmeter is exactly zero. Real voltmeters never have an infinite resistance, but in practice it is possible to make the internal resistance of a voltmeter very large; values of 10^{12} Ω or even more can be achieved.

19.7 | *RC* CIRCUITS AS FILTERS

It is often desirable to reduce or "filter out" time-dependent fluctuations in a voltage signal. Circuits that do so are called *filters* and can be constructed with a simple *RC* combination. Suppose the input voltage in Figure 19.38A is initially constant. From our work with *RC* circuits, we know that the capacitor will initially have a charge q such that the voltage across the capacitor is equal to the input voltage V_{in}. Now suppose at a certain time V_{in} increases very quickly as shown in Figure 19.38B. This increase causes the charge on the capacitor to change to a new value, and this charging is described by the time constant τ of the *RC* circuit. If the time constant is long, the voltage across the capacitor increases much more slowly than the abrupt change of the input voltage. In this way, the voltage across the capacitor, which serves as the output voltage for the circuit, removes rapid changes or fluctuations in the voltage and replaces them with more gradual variations.

A filter circuit is useful in many applications. For example, the noise in a radio signal ("static") can be removed using an *RC* filter, leaving the music signal largely unaffected. The amount of filtering depends on the value of $\tau = RC$, which can be varied over a wide range by the appropriate choice of R and C.

Electric Currents in Nerves ⊗

An interesting example of an *RC* filter is found in the nervous system of the human body and in many other animals. Many nerves are long and thin, much like a wire, carrying electric currents from one part of the body to another. Nerve fibers consist of a thin layer of lipid molecules enclosing a conducting solution containing a mix of ions (K^+, Na^+, Cl^-, and others) as sketched in Figure 19.39A. This figure describes the simplest nerve fibers as found in squids and other invertebrates. The conducting solution inside the fiber acts as a resistor and the lipid layer acts as a capacitor, so this structure and biochemical composition causes a nerve fiber to behave as an *RC* filter circuit, just as in Figure 19.38.

Sensory cells provide the input signal to some nerve fibers. For example, when a squid is touched by a predator, the squid's sensory cells produce an abrupt increase in the input voltage to a nerve fiber, similar to that sketched for V_{in} in Figure 19.38B. The output signal from the *RC* nerve filter increases more slowly than the input. It is in the squid's best interest for V_{out} to increase as rapidly as possible so that the sensory signal will reach the brain quickly and the squid can react rapidly. The *RC* time

Figure 19.38 Ⓐ An *RC* filter circuit takes an input voltage signal and produces an output voltage. Ⓑ Comparison of the input (solid) and output (dashed) voltages. Shown is a low-pass filter that "filters out" rapid changes in the input voltage and only passes slow changes to the output.

constant should therefore be as small as possible. In squids and in other animals, nerves that carry signals for which a rapid response is needed have values of R and C that make the time constant very small, guaranteeing a rapid response. The value of τ is estimated for a realistic nerve fiber in Example 19.9.

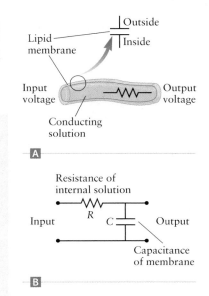

Figure 19.39 **A** A nerve fiber is electrically equivalent to the RC filter circuit in **B**. The membrane acts like a parallel-plate capacitor, and the resistance is provided by the ionic solution inside the nerve fiber.

| EXAMPLE 19.9 | **X** Filtering by a Squid Nerve Fiber |

Consider a squid nerve fiber with a radius of $r = 1$ mm and a length of 5 cm. What is the time constant τ for this RC filter? The resistivity of the internal solution is about $\rho = 2\ \Omega \cdot$ m, the thickness of the lipid layer is $d = 10$ nm, and the dielectric constant of the lipid is $\kappa = 10$.

RECOGNIZE THE PRINCIPLE

The time constant is $\tau = RC$ (Eq. 19.31), so to find τ we must first calculate the values of R and C.

SKETCH THE PROBLEM

Figure 19.39 shows how the conducting fluid inside a nerve fiber acts as a resistor and how the lipid layer can be modeled as a capacitor, thus giving an RC circuit.

IDENTIFY THE RELATIONSHIPS

The resistance of the nerve fiber can be found using $R = \rho L/A$ (Eq. 19.11). Inserting the resistivity along with the given values of the length and cross-sectional area ($A = \pi r^2$) of the fiber into Equation 19.11, we obtain

$$R = \rho \frac{L}{A} = (2\ \Omega \cdot \text{m}) \frac{0.05\ \text{m}}{\pi (0.001\ \text{m})^2} = 3 \times 10^4\ \Omega$$

To estimate the capacitance, we approximate the lipid layer as a parallel-plate capacitor. This capacitor's "plate" area is the cylindrical surface area of the fiber, $A_{\text{cap}} = 2\pi r L$. (This cylindrical surface area is *not* the same as the cross-sectional area A of the fiber, which we used to find the resistance.) Using the expression for the capacitance of a parallel-plate capacitor filled with dielectric (Eqs. 18.31 and 18.43), we find

$$C = \frac{\kappa \varepsilon_0 A}{d} = \frac{(10)[8.85 \times 10^{-12}\ \text{C}^2/(\text{N} \cdot \text{m}^2)][2\pi(0.001\ \text{m})(0.05\ \text{m})]}{1.0 \times 10^{-8}\ \text{m}} \tag{1}$$

$$C = 3 \times 10^{-6}\ \text{F}$$

SOLVE

The time constant is then

$$\tau = RC = (3 \times 10^4\ \Omega)(3 \times 10^{-6}\ \text{F}) = \boxed{0.09\ \text{s}}$$

What does it mean?
According to our results for an RC filter in Figure 19.38, this time constant is the approximate response time of a squid nerve fiber.

Designing a Better Nerve Fiber **X**

The time constant calculated in Example 19.9 is a rough measure of how fast the nerve transmits its information to the squid's brain. To keep this reaction time short, τ should be small. The squid's nerve fiber attains a fairly small time constant

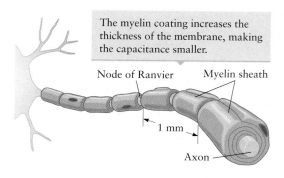

The myelin coating increases the thickness of the membrane, making the capacitance smaller.

Node of Ranvier   Myelin sheath

1 mm

Axon

Figure 19.40 Some vertebrate nerve fibers have very thick membranes due to a coating of myelin that increases the plate spacing of the membrane capacitor (Fig. 19.39) and hence reduces the capacitance. They are called myelinated fibers.

by making the resistance R small, and it has a small resistance because its radius is so large. Another way to keep the time constant short is to make the capacitance C small; that approach is used in many of the sensory and muscle nerve fibers in vertebrates, such as humans. Vertebrate nerve fibers are often coated with a relatively thick layer of a protein called myelin (Fig. 19.40). This coating increases the distance between the capacitor plates—the value of d in Equation (1) of Example 19.9—making C smaller. As a result, the value of the time constant is also reduced, leading to a fast response time for these nerve fibers. A faster response time is always better for the nervous system (and for survival).

19.8 | ⊗ ELECTRIC CURRENTS IN THE HUMAN BODY

Electric Current and Your Health

Your body is a moderately good conductor of electricity. Taken as a whole, your body's resistance measured for conduction from the fingers of one hand to the other is about 1500 Ω when the body is dry and typically 500 Ω when wet. The electric potential at a wall socket is 120 V, so if you inadvertently touch a high-potential house wire with one hand and ground with the other, the current through your body is approximately

$$I = \frac{V}{R} \approx \frac{120 \text{ V}}{1500 \text{ }\Omega} = 0.080 \text{ A} = 80 \text{ mA} \quad (\text{dry})$$

$$\approx \frac{120 \text{ V}}{500 \text{ }\Omega} = 240 \text{ mA} \quad (\text{wet}) \qquad (19.37)$$

This current is carried through different parts of the body; some flows in the skin, and some can flow through your heart or other organs.

The effect of current in your body is described in Table 19.3. A current of 1 mA is felt as a "tingling" sensation, but has no harmful effect.[5] Your body uses electric currents to control muscle contraction, and external currents also affect muscles. External currents of 10 mA to 20 mA can cause your muscles to contract involuntarily, which is what happens when a person "can't let go" of a high-potential wire. Your heart is a muscle, and currents above about 100 mA can cause your heart to beat irregularly or even stop. Comparing these values with the currents calculated in Equation 19.37, we see that the current produced by contact with a wall socket can be a definite health hazard.

Large currents can be beneficial in certain situations. For example, your heart can sometimes produce a series of rapid and irregular beats, a condition called fibrillation. A large and brief external current applied by a medical device called a defibrillator can "shock" the heart back into a normal rhythm.

Table 19.3 Effect of Electric Current on the Human Body

Current (mA)	Effect on Body
1	Threshold for sensing (can "feel it tingle")
5	Harmless
10–20	Involuntary muscle contraction
50	Pain
100–200	Disrupts the heart
300	Burns

19.9 | HOUSEHOLD CIRCUITS

Electric circuits are everywhere around us; for example, your house is filled with devices (computers, lights, televisions, etc.) connected to the electric wiring in the house, which in turn is connected to the electric power lines provided by your utility company. In this chapter, we have focused on DC electric circuits involving batteries. However, the voltage provided by your utility company is not constant; rather, it is an AC voltage oscillating at a frequency of 60 Hz. This is the frequency

---

[5]Your tongue is especially sensitive to small currents. Some people are known to test for a dead battery using their tongue (not recommended!).

© Comstock Images/ Jupiterimages

A

© Cengage Learning/ Charles D. Winters

B

© Phil Degginger/Alamy

C

© Comstock Images/Jupiterimages

D

Figure 19.41 Several safety devices are incorporated in household circuits, including **A** three-pronged plugs, **B** polarized plugs, **C** outlets, and **D** circuit breakers.

in the United States; the frequency may be different in other countries. We'll discuss AC circuits in Chapter 22. Here we will mention some electrical safety issues that apply to both AC and DC circuits.

Safety in Home Circuits

Figure 19.41 shows some components used in household circuits, including different types of electrical connectors ("plugs") that can be used to connect to an outlet. Most modern outlets and plugs (Fig. 19.41A) contain three connections: two are flat strips, and the third is round. The round one is connected to ground potential, whereas the other two connections carry current to and from devices in your house. The connector in Figure 19.41B is called a polarized plug; it has two flat connectors of different size so that it can be inserted into the wall outlet in only one way. The larger of the two flat strips connects to a low potential that is very close to ground potential.

In normal operation, an appliance such as a hair dryer is plugged into a wall outlet (Fig. 19.41C); current passes from the outlet, through the hair dryer, and back to the outlet, completing the electric circuit. The magnitude of this current can be large enough to pose a health hazard if it were to flow through your body. For example, a wire in the hair dryer might accidentally touch the handle while you are holding it, allowing a large amount of current to pass to the hair dryer handle and through your body. To prevent damage to the household circuitry and to a user, safety devices are inserted into the circuit connecting the wall outlet to the utility company's power line. Older houses employ *fuses*, in which current passes through a thin metal strip. This strip acts as a resistor with a small resistance, so under normal conditions only a small amount of power is dissipated in the fuse. If a failure causes the current to become very large, heat dissipated in the fuse is sufficient to melt and break the metal strip, stopping the current. Newer houses employ *circuit breakers* (Fig. 19.41D), which also stop the current when it exceeds a predetermined limit. A circuit breaker—unlike a fuse, which must be replaced after its metal strip melts—contains a switch that can be "reset" for continued use after the offending circuit has been repaired.

Electrical devices also contain safety features. The ground connection from the wall outlet is usually connected via the round piece of the plug to the outer case of an appliance, so even if there is an internal failure, a large and dangerous voltage does not appear on any surfaces that can be touched by a user. Older appliances sometimes do not use a three-pronged plug, so they cannot make use of this ground connection. In these appliances, the low-potential side of the polarized plug is often connected to the case, providing almost as much protection as a connection to ground. What do we mean by "almost"? When a device without a ground connection fails (perhaps a hair dryer), there may be a large amount of current I through the low-potential plug in the wall outlet (Fig. 19.42). The resistance R of the wires to your hair dryer is small but not zero, so this current causes the voltage at the low-potential side of the hair dryer to increase to a value IR relative to ground. If

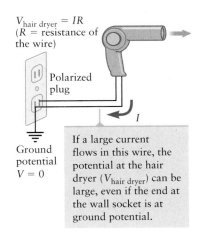

$V_{\text{hair dryer}} = IR$
(R = resistance of the wire)

Polarized plug

I

Ground potential $V = 0$

If a large current flows in this wire, the potential at the hair dryer ($V_{\text{hair dryer}}$) can be large, even if the end at the wall socket is at ground potential.

Figure 19.42 A hair dryer with a polarized plug is not as safe as one with a three-pronged plug. If the hair dryer with a polarized plug fails, it is possible for a large voltage to be present on the normally grounded part of the circuit.

I is very large, this voltage may be large enough to pose a health hazard, especially if it is connected to the hair dryer's handle so that a large current could then flow through a person holding the hair dryer. This arrangement is thus not as safe as using a three-pronged plug, which connects the handle of the hair dryer to ground potential through a separate wire. This separate "ground wire" in the hair dryer is designed so that it should never carry a large current, even if the dryer fails, and the handle should never be at a dangerous potential.

19.10 | TEMPERATURE DEPENDENCE OF RESISTANCE AND SUPERCONDUCTIVITY

In Section 19.3, we explained the atomic-scale origin of resistivity: when electrons move through a metal, they collide, especially with ions in the metal (Fig. 19.6). As temperature increases, these ions vibrate with larger and larger amplitudes, causing more frequent collisions and an increase in the resistance. For many metals near room temperature, this temperature dependence of the resistivity can be described by

$$\rho = \rho_0[1 + \alpha(T - T_0)] \qquad (19.38)$$

where α is called the ***temperature coefficient of the resistivity***. In Equation 19.38, ρ_0 is the resistivity at a reference temperature T_0. Usually, one picks T_0 to be room temperature, so ρ_0 is just the resistivity listed in Table 19.1. According to Equation 19.38, the resistivity varies linearly with temperature; because the resistance of a wire is proportional to ρ (Eq. 19.11), the resistance of a metal also varies linearly with temperature. This linear variation usually breaks down at very low temperatures, and the resistivity of metals such as copper approaches a nonzero value at very low temperatures (Fig. 19.43A).

For some metals, the resistivity does not quite follow the low-temperature behavior found for copper. In some cases, such as mercury and lead, the resistivity drops abruptly and is zero below a certain temperature as sketched in Figure 19.43B. The value of this "critical" temperature depends on the metal. Metals for which the resistivity is zero below a critical temperature are called ***superconductors***. Such behavior is most commonly found in metals cooled to within a few degrees of absolute zero, although there are a few examples of superconductivity at temperatures above 100 K. These materials have many important applications; for example, they can be used to construct very powerful magnets as we'll discuss in Chapter 20.

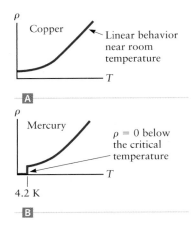

Figure 19.43 Qualitative behavior of the resistivity of **A** copper and **B** mercury as a function of temperature.

SUMMARY | Chapter 19

KEY CONCEPTS AND PRINCIPLES

Electric current
Electric current *I* is the amount of charge that passes a particular spot per unit time.

$$I = \frac{\Delta q}{\Delta t} \qquad \textbf{(19.1)} \text{ (page 602)}$$

where *I* is measured in units of ***amperes*** (A), with 1 A = 1 C/s.

(Continued)

Electrical resistance

The *resistance* R of a material of length L and cross-sectional area A is

$$R = \rho \frac{L}{A} \qquad\qquad \text{(19.11) (page 608)}$$

where ρ is the *resistivity* of the material (Table 19.1). For a circuit element called a *resistor*, the voltage across the circuit element and the current through the element are related by *Ohm's law*:

$$V = IR \qquad\qquad \text{(19.10) (page 608)}$$

Kirchhoff's rules

Kirchhoff's rules tell how to determine the current throughout an electrical circuit. These two rules are based on the principles of conservation of energy and charge:

1. *Kirchhoff's loop rule*: The total change in the electric potential around any closed circuit path must be zero.

2. *Kirchhoff's junction rule*: The current entering a circuit junction must equal the current leaving the junction.

Voltage drop across a capacitor

If a capacitor carries a charge $\pm q$, the voltage across the capacitor is

$$V = \frac{q}{C} \qquad\qquad \text{(19.27) (page 625)}$$

where C is the capacitance.

APPLICATIONS

Batteries

Batteries act as sources of energy in an electric circuit. The electric potential difference between the terminals of a battery is called the *electromotive force* or *emf* $\mathcal{E}$ produced by the battery.

Charge carriers

The current in a metal wire is carried by mobile electrons that move at the *drift velocity*.

(Continued)

Circuit analysis

Kirchhoff's rules tell how resistors and capacitors behave when connected in *series* and in *parallel* arrangements.

Series resistors: $R_{equiv} = R_1 + R_2 + \cdots$ Parallel resistors: $\dfrac{1}{R_{equiv}} = \dfrac{1}{R_1} + \dfrac{1}{R_2} + \cdots$

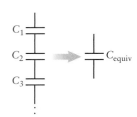

Series capacitors: $\dfrac{1}{C_{equiv}} = \dfrac{1}{C_1} + \dfrac{1}{C_2} + \cdots$ Parallel capacitors: $C_{equiv} = C_1 + C_2 + \cdots$

Charging and discharging a capacitor

When the switch is closed in an *RC* circuit, the current through the circuit and the voltage across the capacitor vary with time according to

$$I = \frac{\mathcal{E}}{R} e^{-t/\tau} \qquad \textbf{(19.28)} \text{ (page 626)}$$

$$V_{cap} = \mathcal{E}(1 - e^{-t/\tau}) \qquad \textbf{(19.29)} \text{ (page 626)}$$

where $\tau = RC$ is the **time constant** for the circuit.

QUESTIONS

SSM = answer in Student Companion & Problem-Solving Guide   ⊗ = life science application

1. Discuss how Kirchhoff's rules for circuit analysis are related to conservation principles.

2. Explain why a complete circuit is required to have a nonzero current in a circuit.

3. In a metal, is an electric current carried by (a) protons, (b) electrons, or (c) both protons and electrons?

4. In a solution of salt water (NaCl dissolved in water), is an electric current carried mainly by (a) electrons, (b) protons, (c) Na^+, (d) Cl^-, (e) Na^+ and Cl^-, or (f) electrons and protons?

5. Is the ratio of the voltage across a conductor to the current through the conductor called the (a) power, (b) capacitance, or (c) resistance?

6. Is the product of the voltage across a conductor and the current through the conductor called the (a) power, (b) capacitance, or (c) resistance?

7. In Chapter 17, we argued that the electric field inside a metal in equilibrium is always zero. In this chapter, we found that there is a nonzero electric field inside a current-carrying wire. How can both of these statements be correct?

8. Which of the plots in Figure Q19.8 shows how the current depends on voltage for an ordinary resistor?

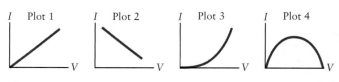

Figure Q19.8

9. SSM An incandescent lightbulb contains a filament that has a certain electrical resistance R. The brightness of the bulb depends on the current and increases as the current through the filament is increased. Consider the following situations in Figure Q19.9 on the next page.
   (a) Two identical lightbulbs are connected as shown to a single ideal battery as in circuit 2. Will the brightness of one of these bulbs be greater than, less than, or the same as the brightness of the bulb in circuit 1?
   (b) Repeat part (a) for the bulbs in circuits 3, 4, and 5, comparing their brightness to that of the bulb in circuit 1. Explain your answers.

10. If two circuit elements are connected in series, which of the following statements is true?

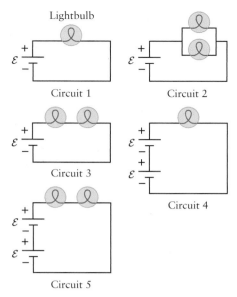

Circuit 1   Circuit 2

Circuit 3   Circuit 4

Circuit 5

Figure Q19.9

(a) The voltage across the circuit elements is the same.
(b) The current through the circuit elements is the same.
(c) The current is largest through the first circuit element.

11. If three circuit elements are connected in parallel, which of the following statements is always true?
(a) The voltage across the circuit elements is the same.
(b) The current through the circuit elements is the same.
(c) The power is the same in all three circuit elements.

12. You are given four resistors, each with a resistance R. Devise two ways to connect these resistors to get a total equivalent resistance that is greater than R.

13. You are given four resistors, each with a resistance R. Devise two ways to connect these resistors to get a total equivalent resistance less than R.

14. You are given four resistors, all with resistance R. Devise a way to connect these resistors to get an equivalent resistance of $2.5R$.

15. Show that the RC time constant in Equation 19.31 does indeed have units of time.

16. You are given four capacitors, each with a capacitance C. Devise two ways to connect these capacitors to get a total equivalent capacitance greater than C.

17. You are given four capacitors, each with capacitance C. Devise two ways to connect these capacitors to get a total capacitance less than C.

18. [SSM] In Example 19.5, we used the rules for combining resistors in series and in parallel to analyze the circuit in Figure 19.22. Take a different approach and use Kirchhoff's rules to derive equations for the currents through the different circuit branches. Show that these equations lead to the same result found in Example 19.5.

19. Figure Q19.19 shows two circuits which contain the same circuit elements, but the positions of the resistors are interchanged. Will that affect the current? Justify your answer using Kirchhoff's rules.

Figure Q19.19

20. In Example 19.2, we discussed how a capacitor can be used to store electrical energy. Investigate how capacitors are used in real camera flash circuits and discuss why they are better for this application than batteries.

21. In Example 19.1, we stated an estimate for the number of tungsten atoms in the filament of a lightbulb. Carry out a calculation of this number and compare your result with our estimate.

22. If one terminal of a battery is connected to an object, does any charge flow from the battery to the object? If there is charge flow, explain why it is very brief.

23. There will be a nonzero current in the circuit in Figure Q19.23. This current will carry electrons from one end of the resistor through the wire at the top to the positive terminal of the battery. We know that electrons always move from a region of low electric potential to a region of higher potential. We also know that an ideal wire has zero resistance, so there can be no potential drop across an ideal wire. Explain how to reconcile these two facts.

Figure Q19.23

24. When the author built his current house, he ran wires in the walls to be used for stereo speakers. He ran wires from several different rooms to a central location, but he forgot to mark the wires so he could tell which wires emerge in which room. Explain how he was able to solve this problem and identify which wire is which.

PROBLEMS

19.1 ELECTRIC CURRENT: THE FLOW OF CHARGE

1. A current of 3.5 A flows through a wire. How many electrons pass a particular point on the wire in 12 s?

2. (X) As a treatment for chronic back pain, a medical patient may be fitted with a device that passes a small electrical current (10 mA) through the muscles in the lower back as shown in Figure P19.2. If the current is supplied in short pulses 0.50 s in length, how many electrons pass through the muscles during one pulse?

3. A current of 0.75 A flows through a lightbulb for 1 h. How many electrons pass through the lightbulb in this time?

Figure P19.2

© Michael Krasowitz/Photographers Choice/Getty Images

4. Electrons move between points A and B in Figure P19.4 at a rate of 15 electrons per second. What is the current? Give the magnitude and direction of I.

Figure P19.4

5. [SSM] During a thunderstorm, a lightning "bolt" carries current between a cloud and the ground below. If a particular bolt carries a total charge of 20 C in 1.0 ms, what is the magnitude of the current? How many electrons are involved in this process?

19.2 BATTERIES

6. ☆ Car batteries are typically rated in ampere-hours, or A · h. (a) Show that ampere-hours are actually a unit of charge and determine the conversion from ampere-hours to coulombs. (b) If a 100 A · h battery can deliver a steady current of 5.0 A until it is completely depleted, what is the total time the battery can deliver current without being recharged? (c) How much total charge will the battery deliver in this time?

7. [SSM] ☆ ✖ Lithium-iodine batteries are particularly useful in situations in which small amounts of current are required over a long period of time. One such application is in a cardiac pacemaker (Fig. P19.7), where changing a battery requires a patient to undergo an operation. Typical pacemakers require 0.50 μA of continuous current. If a lithium-ion battery is capable of supplying 0.50 A · h of charge, what is the lifetime of the battery? Does this lifetime seem sufficiently long?

Figure P19.7 View of a cardiac pacemaker, both inside and outside a patient.

8. ☆ A 14.4-V battery for a cordless drill can supply 2.0 A of current. (a) If the drill is continuously operated for 5 min, what is the total charge passing through the battery? (b) What total amount of work does the battery do on the charge during these 5 min?

19.3 CURRENT AND VOLTAGE IN A RESISTOR CIRCUIT

9. Calculate the resistance of a piece of copper wire that is 1.0 m long and has a diameter of 1.0 mm.

10. ☆ If the length of a wire is increased by a factor of 4, by what factor does the resistance change?

11. ☆ If the diameter of a piece of wire is reduced by a factor of 2.5, by what factor does the resistance change?

12. A wire has a diameter of 1.0 mm and a length of 30 m, and is found to have a resistance of 1.2 Ω. What is the resistivity of the wire?

13. ☆ ✖ The electric eel (*Electrophorus electricus*) can generate a potential difference of 600 V between a region just behind its head and its tail. Suppose an unsuspecting fish swims into this area, resulting in a lethal current of 1.0 A passing through the length of its body. If the fish is approximately 25 cm in length and 5.0 cm in diameter, what is its average resistivity?

14. An aluminum wire has a diameter of 0.50 mm and a length of 7.5 m. What is its resistance?

15. It is possible to get copper wires with diameters as small as about 1 μm. If such a wire has a resistance of 5 Ω, how long is it?

16. If the current through a resistor is increased by a factor of 3, by what factor does the voltage change?

17. ☆ If the current through a resistor is increased by a factor of 3, by what factor does the power change?

18. A glass capillary tube with a diameter of 0.50 mm and length 10 cm is filled with a salt solution with a resistivity of 0.10 Ω · m. What is the resistance?

19. [SSM] ☆ A copper wire is made with the same diameter and length as the capillary tube in Problem 18. What is the ratio of the resistance of the capillary tube to the resistance of the copper wire?

20. There is a potential difference of $V = 20$ V across a resistor with $R = 50$ Ω. What is the current through the resistor?

21. If a current of 0.10 A flows through a resistor and the potential drop across the resistor is 1.5 V, what is the resistance?

19.4 DC CIRCUITS: BATTERIES, RESISTORS, AND KIRCHHOFF'S RULES

22. ☆ A resistor with $R = 50$ Ω is connected to a battery with emf $\mathcal{E} = 3.0$ V. (a) What is the current through the battery? (b) What is the current through the resistor? (c) What is the power dissipated in the resistor? (d) How much energy is dissipated in the resistor in 3 minutes?

23. ☆ The three resistors in Figure P19.23 all have the same resistance. When the switch in Figure P19.23 is closed, (a) by what factor does the current in resistor 1 change? (b) By what factor does the potential difference across resistor 2 change? (c) By what factor does the current in resistor 2 change?

Figure P19.23

24. ✖ Two resistors with $R_1 = 1500$ Ω and $R_2 = 3500$ Ω are connected in parallel as shown in Figure P19.24. If the battery emf is $\mathcal{E} = 12$ V, what is the current through each of the resistors?

Figure P19.24 Problems 24, 25, and 43.

25. What is the equivalent resistance of the two resistors in Problem 24?

26. Five resistors, all with resistance R, are connected in parallel. What is the equivalent resistance of this combination?

27. ☆ Two resistors with $R_1 = 1500$ Ω and $R_2 = 3500$ Ω are connected in series as shown in Figure P19.27. (a) If the battery emf is $\mathcal{E} = 12$ V, what is the current through each of the resistors? (b) What is the equivalent resistance of the two resistors?

Figure P19.27 Problems 27, 29, 32, and 45.

28. Seven resistors, all with resistance R, are connected in series. What is the equivalent resistance of this combination?

29. ☆ Consider the circuit in Figure P19.27 with $R_1 = 1500$ Ω and $R_2 = 3500$ Ω. If $\mathcal{E} = 12$ V, what is the power dissipated in R_2?

30. ✖ A typical lightbulb in your house is rated 100 W, which means that the bulb dissipates 100 W when connected to a DC voltage of 110 V. (a) What is the current? (b) If the voltage is reduced to 55 V, what is the dissipated power?

31. ✪ An electric heater consumes electrical energy at a rate of $P = 250$ W when connected to a battery with an emf of 120 V. (a) What is the current in the heater? (b) If you wish to increase the power consumption to 500 W, by what factor should the heater resistance be changed?

32. ✪ If the current in the circuit in Figure P19.27 is 0.15 A and the resistances are $R_1 = 1500$ Ω and $R_2 = 2500$ Ω, what is the emf of the battery?

33. ✪ Two appliances are connected in parallel to a 120-V battery and draw currents of $I_1 = 2.0$ A and $I_2 = 3.5$ A. If these appliances are instead connected in series to the same battery, what is the total current in the circuit?

34. ✪ The circuit in Figure P19.34 shows three identical lightbulbs attached to an ideal battery with $\mathcal{E} = 18$ V. If the resistance of each bulb is 200 Ω, what are (a) the current through bulb 1 and (b) the current through bulb 2? (c) If bulb 2 burns out, what happens to the brightness of bulbs 1 and 3?
  (i) The brightness of both bulbs stays the same.
  (ii) Bulbs 1 and 3 get brighter.
  (iii) Bulbs 1 and 3 both get dimmer.
  (iv) Bulb 1 gets dimmer while bulb 3 gets brighter.
  (v) Bulb 1 gets brighter while bulb 3 gets dimmer.
  Hint: The brightness of a bulb increases as its current increases.

Figure P19.34

35. ✪ You are given four resistors, each with resistance R. Devise a way to connect these resistors so that the total equivalent resistance is R. No cheating; you must use all the resistors so that there is current through each one if a battery is connected.

36. ✪ Consider the circuit in Figure P19.36. The resistors are identical with $R_1 = R_2 = R_3 = 2000$ Ω, and the battery voltage is $\mathcal{E}_1 = 9.0$ V. What is the current through resistor R_1?

Figure P19.36 Problems 36, 37, 38 and 39.

37. ✪ Consider the circuit in Figure P19.36 and assume all the resistors are identical. Only one of the following statements is true. Which one?
  (a) The power dissipated is the same in all three resistors.
  (b) The power dissipated in R_1 is greater than the power dissipated in R_2.
  (c) The power dissipated in R_1 is less than the power dissipated in R_2.
  (d) The power dissipated in R_1 is equal to the power dissipated in R_2.
  (e) The power dissipated in R_1 is equal to sum of the powers dissipated in R_2 and R_3.

In Problems 38 through 44, assume the resistance values are $R_1 = 2400$ Ω, $R_2 = 1400$ Ω, $R_3 = 4500$ Ω, and $R_4 = 6000$ Ω and the battery emfs are $\mathcal{E}_1 = 1.5$ V and $\mathcal{E}_2 = 3.0$ V.

38. SSM ✪ Analyze the circuit in Figure P19.36. (a) Express the resistors in parallel as a single equivalent resistor. What is the value of this resistor? (b) Combine this equivalent resistor with R_1 to get the total equivalent resistance of the circuit. What is the value of this resistor? (c) The circuit is now "reduced" to a

battery in series with a single equivalent resistance. Use your result from part (b) to find the current through the battery and resistor R_1.

39. ✪ Use your results from Problem 38 to find the current through R_2 in Figure P19.36. *Hint:* First find the voltage across the parallel combination of resistors.

40. ✪ Use the approach from Problem 38 to analyze the circuit in Figure P19.40. What is the current through the battery and resistor R_1? *Hint:* First redraw the circuit so that the parallel and series resistor combinations are more obvious.

Figure P19.40

41. ✪ Use Kirchhoff's rules to analyze the circuit in Figure P19.41. (a) Let I_1 be the branch current through R_1 and I_2 be the branch current through R_2. Write Kirchhoff's loop rule relation for a loop that travels through battery 1, resistor 1, and battery 2. (b) Write Kirchhoff's loop rule relation for a loop that travels through battery 2 and resistor 2. (c) You should now have two equations and two unknowns (I_1 and I_2). Solve for the two branch currents.

Figure P19.41
Problems 41 and 42.

42. ✪ Consider again the circuit in Figure P19.41. Write the Kirchhoff's loop rule relation for a loop that passes through battery 1, resistor 1, and resistor 2. Show that the resulting equation can be derived by combining the two loop equations from Problem 41.

43. ✪ Consider the resistors in parallel in Figure P19.24. If $R_1 = 2500$ Ω and $R_2 = 6500$ Ω, what is the *ratio* of the powers dissipated in the two resistors?

44. ✪ Use Kirchhoff's rules to analyze the circuit in Figure P19.44. (a) Let I_1 be the branch current though R_1, I_2 be the branch current through R_2, and I_3 be the branch current through R_3. Write Kirchhoff's loop rule relation for a loop that travels through battery 1, resistor 1, and resistor 3. (b) Write Kirchhoff's loop rule relation for a loop that travels through battery 2, resistor 2, and resistor 3. (c) Apply Kirchhoff's junction rule to the junction at A to get a relation between the three branch currents. (d) You should now have three equations and three unknowns (I_1, I_2, and I_3). Solve for the three branch currents.

Figure P19.44

45. ✪ Consider the resistors in series in Figure P19.27. If $R_1 = 2500$ Ω and $R_2 = 3500$ Ω, what is the *ratio* of the powers dissipated in the two resistors?

46. ✪ Consider the cube of resistors in Figure P19.46. If all the resistors have the same value, what is the equivalent resistance between points A and B? *Hint:* Start by considering a Kirchhoff's rule analysis and sketch the various branch currents. Can you find some relation(s) between these currents?

Figure P19.46

47. ★ The lightbulbs in Figure P19.47 are all identical. Which bulb will be brightest? Which one will be dimmest?

Figure P19.47

19.5 DC CIRCUITS: ADDING CAPACITORS

48. SSM ★ Consider the *RC* circuit in Figure P19.48, with $R_1 = 1000\ \Omega$, $R_2 = 3000\ \Omega$, and $C = 1.0\ \mu F$. When the switch is closed, the current will vary with time as sketched in Figure 19.29. What is the time constant for the circuit in Figure P19.48? *Hint:* Use equivalent circuit ideas to "reduce" this circuit to a form that looks like Figure 19.28.

Figure P19.48 Problems 48 and 49.

49. ★ For the circuit in the Problem 48, what is the current through the circuit (a) just after the switch is closed and (b) after the switch has been closed for a very long time?

50. What is the equivalent capacitance of the two capacitors in Figure P19.50? Take $C_1 = 7.5\ \mu F$ and $C_2 = 3.5\ \mu F$.

Figure P19.50

51. Five capacitors, all with the same capacitance *C*, are connected in parallel. What is the value of the equivalent capacitor?

52. ★ What is the equivalent capacitance of the three capacitors in Figure P19.52? Take $C_1 = 2.5\ \mu F$, $C_2 = 3.5\ \mu F$, and $C_3 = 1.5\ \mu F$.

Figure P19.52

53. Seven capacitors, all with the same capacitance *C*, are connected in series. What is the value of the equivalent capacitor?

54. ✪ What is the equivalent capacitance of the four capacitors in Figure P19.54? Take $C_1 = 2.5\ \mu F$, $C_2 = 4.5\ \mu F$, $C_3 = 3.3\ \mu F$, and $C_4 = 1.5\ \mu F$.

Figure P19.54

In the circuits in Figures P19.55, P19.58, and P19.60, assume the resistance values are $R_1 = 1500\ \Omega$ and $R_2 = 2400\ \Omega$, with capacitances $C_1 = 45\ \mu F$ and $C_2 = 25\ \mu F$, and with $\mathcal{E} = 5.5$ V. The switches are labeled *S*.

55. ★ What is the current in the circuit in Figure P19.55 the moment after the switch is closed? Assume the capacitor is initially uncharged.

56. ★ If the switch in Figure P19.55 is closed for a very long time, what is the charge on the capacitor?

Figure P19.55 Problems 55, 56, and 57.

57. ✪ What is the time constant for the circuit in Figure P19.55? Sketch how the current through the circuit and the voltage across the capacitor vary with time after the switch is closed.

58. ★ The capacitor in Figure P19.58 is initially uncharged, and both switches are open. What is the current through the battery the instant after switch S_1 is closed?

Figure P19.58 Problems 58 and 59.

59. ✪ Consider again the circuit in Figure P19.58. After switch S_1 is closed for a very long time, switch S_1 is opened and switch S_2 is simultaneously closed. (a) What is the current through switch S_2 the instant after it is closed? (b) Make a sketch of how this current varies with time. (c) What is the current through S_2 after it is closed for a very long time?

60. ★ The capacitors in Figure P19.60 are initially uncharged when the switch is closed. Make a sketch of how the current through the circuit varies with time after the switch is closed. What is the time constant for this circuit?

Figure P19.60 Problems 60 and 61.

61. ✪ Consider the circuit in Figure P19.60 and assume the switch has been closed for a very long time. What is the current through the circuit? What are the charges on the two capacitors?

19.6 MAKING ELECTRICAL MEASUREMENTS: AMMETERS AND VOLTMETERS

62. ✪ Although an ideal voltmeter has an infinite internal resistance, this theoretical ideal is usually not met in practice. The voltmeter in Figure P19.62 has an internal resistance of $10^9\ \Omega$ and is used to measure the voltage across a resistor with $R_2 = 150$ kΩ as shown. How much does this nonideal voltmeter affect the circuit? That is, attaching the voltmeter changes the voltage across R_2. What is the fractional change in the voltage across R_2 when the voltmeter is attached?

Figure P19.62

63. ✪ Although an ideal ammeter has an internal resistance of zero, this theoretical ideal is usually not met in practice. The ammeter in Figure P19.63 has an internal resistance of $0.010\ \Omega$ and is used to measure the current in a circuit containing a resistor with $R = 150\ \Omega$ as shown. How much does this non-ideal ammeter affect the circuit? That is, what is the fractional change in the current when the ammeter is attached?

Figure P19.63

64. Suppose an ammeter has an internal resistance of 1.0 mΩ. Find the current in the ammeter when it is properly connected to a 2.0-Ω resistor and a 12-V source.

65. SSM Suppose a voltmeter has an internal resistance of 50 kΩ. Determine the current through the meter when it is properly connected across a 50-Ω resistor that is connected to a 10-V source.

66. ✪ For the circuit in Figure P19.66, draw the proper placement of an ammeter or voltmeter

Figure P19.66

to measure (a) the current through and voltage across R_1, (b) the current through and voltage across ($R_1 + R_2$), (c) the current through and voltage across R_3, and (d) the total current through the battery.

67. ✪ Consider the circuit in Figure P19.67. Find the potential difference between points A and B. Express your answer in terms of $\mathcal{E}$ and R.

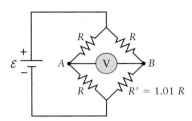

Figure P19.67

68. ✪ Suppose an ammeter and voltmeter are used to determine a resistor's resistance. If the ammeter is connected in series with the resistor and the voltmeter is connected across the resistor only, show that the correct resistance of the resistor is given by $R = V/[I - (V/R_V)]$, where V is the voltage measured by the voltmeter, I is the current measured by the ammeter, and R_V is the internal resistance of the voltmeter.

69. ✪ It is possible to devise clever circuit arrangements that behave as ideal meters. The circuit in Figure P19.69 is able to measure the voltage across resistor R_4 *without* affecting the original circuit. The resistor R_1 is an adjustable resistor whose value can be varied (e.g., by turning a knob). Analyze this circuit and show that for a certain value of the ratio $R_2/(R_1 + R_2)$, this voltmeter circuit has *no* effect on the current in the original circuit. Find the value of this ratio. Express your answer in terms of $\mathcal{E}_1$, $\mathcal{E}_2$, R_3, and R_4.

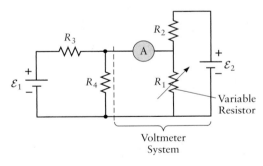

Figure P19.69

19.7 *RC* CIRCUITS AS FILTERS

70. ✪ ✪ ⓡ Consider the nerve fiber discussed in Example 19.9. Assume the fiber is coated with a layer of myelin that is 100 nm thick. If the dielectric constant of myelin is the same as for a lipid layer, what is the time constant of this fiber? Ignore the gaps in the myelin sheath in Figure 19.40. These gaps are called nodes of Ranvier.

19.8 ELECTRIC CURRENTS IN THE HUMAN BODY

71. ✪ Cell membranes contain openings called ion channels that allow ions to move from the inside of the cell to the outside. Consider a typical ion channel with a diameter of 1.0 nm and a length of 10 nm. If this channel has a resistance of 1.0×10^9 Ω, what is the resistivity of the solution in the channel?

72. SSM ✪ ✪ Consider a single ion channel (Problem 71) with $R = 10^9$ Ω. During an action potential, this channel is open for approximately 1.0 ms for the flow of Na$^+$ ions with a potential difference across the cell membrane of about 70 mV. How many Na$^+$ ions travel through the channel?

73. ✪ ✪ A defibrillator is used by emergency medical staff to shock the heart of an accident victim by applying a 10,000-V potential difference across the victim's chest. (a) If the defibrillator has an internal resistance of 10 Ω and the victim's heart has a resistance of 300 Ω, what is the current that passes through the victim's heart the moment the defibrillator is discharged? (b) If the defibrillator delivers this current over a 10-ms time interval, how many electrons pass through the heart during this process?

19.10 TEMPERATURE DEPENDENCE OF RESISTANCE AND SUPERCONDUCTIVITY

74. The temperature coefficient of resistivity for copper is $\alpha = 3.9 \times 10^{-3}$ K^{-1}. If a copper wire has a resistance of 350 Ω at 20 K, what is its resistance at 420 K?

75. SSM ✪ The filament of an incandescent lightbulb is a thin tungsten wire. Suppose a lightbulb has a resistance of 50 Ω at room temperature. During normal operation, the filament reaches a temperature of about 3000 K. (a) What is its resistance at that temperature? The temperature coefficient of resistivity for tungsten is $\alpha = 4.5 \times 10^{-3}$ K^{-1}. (b) If the bulb's brightness depends only on the current, and a larger current gives a greater brightness, will the bulb's brightness be greater or smaller after it has warmed up?

76. ✪ Platinum wires are often used as thermometers due to the change of the resistivity of platinum with temperature. The temperature coefficient of resistivity for platinum is $\alpha = 3.9 \times 10^{-3}$ K^{-1}. Suppose a platinum resistance thermometer has a resistance of 100 Ω at room temperature (293 K). If the temperature is changed so that the thermometer has a resistance of 55 Ω, what is the temperature?

ADDITIONAL PROBLEMS

77. Consider a lightbulb that dissipates 75 W when attached to a battery with an emf of 120 V. If this lightbulb is attached to a battery with a emf of 9.0 V, how many electrons pass through the bulb each second? Assume the resistance of the lightbulb is independent of the current through the bulb.

78. ✪ Suppose that electrical energy costs about \$0.060/(kW · hour) (that is what you pay your local electric company). In a recent month, the author's electric bill was \$150. (a) If the electric potential provided by the electric company was 120 V, what was the average current in all circuits in the author's house? (b) How many electrons did the author "purchase" from the electric company during this month?

79. ✪ The photo in Figure P19.79 shows an electric circuit containing resistors, batteries, and a capacitor. Draw a circuit diagram that corresponds to this circuit.

Figure P19.79

80. ⚀ Figure P19.80 shows a realistic sketch of a circuit. Draw a circuit diagram that corresponds to this circuit.

Figure P19.80

81. ✪ ✖ Consider a bird as it flies to and lands on a power line with a potential $V = 880$ V relative to ground potential. Just before the bird lands on the power line, the bird's potential is zero; after it lands, its potential is 880 V. Long after it lands, the current through the bird is zero, but there is a nonzero current just after the bird lands (called a "transient" current). (a) How much charge q must flow onto the bird (from the power line) to make the bird's potential equal 880 V? Estimate q by approximating the bird as one side (one "plate") of a capacitor, with the other plate at ground potential. Assume the value $C = 1.0 \ \mu$F. (b) If the bird's resistance is 1000 Ω, what is the time constant τ for the circuit? (c) The current I in this RC circuit flows for a time approximately equal to τ. Find the maximum value of I.

82. ✪ Design a parallel-plate capacitor that is able to store the same amount of energy as in the AA battery mentioned in Example 19.2. Give the area of the plates, the plate separation, and the composition of the dielectric and determine the voltage across the capacitor.

83. ⚀ A battery is connected to a single resistor (Fig. P19.83), and the current is measured with an ammeter. A second resistor having the same resistance is placed in parallel with the first resistor. Does the current through the ammeter increase or decrease?

Figure P19.83

84. ✪ A marine battery has an emf of approximately 12.5 V and an internal resistance of 0.075 Ω. The starter motor for the outboard engine requires 40 A while cranking the engine. (a) Determine the potential difference across the starter motor when it is cranking. (b) What is the resistance of the starter motor?

85. SSM ⚀ Several ammeters are placed in the circuit in Figure P19.85.
(a) Arrange the values of the potentials ($V_1, V_2, \ldots$) from highest to lowest.
(b) Arrange the values of the currents measured by the ammeters ($I_1, I_2, \ldots$) from highest to lowest.

Figure P19.85

86. ✪ Seven capacitors are connected to a 100-V battery as shown in Figure P19.86. (a) What is the total equivalent capacitance of the circuit? (b) Determine the total charge stored in the capacitors when they are fully charged. (c) Determine the total energy stored in the capacitors when they are fully charged. (d) Suppose the battery is then disconnected and replaced by a 100-Ω resistor. Determine the current through the resistor 10 s after the connection to the resistor is made.

Figure P19.86

87. ✪ Three resistors R_1, R_2, and R_3 are initially connected in parallel to a 12-V source. The total current supplied to the three resistors is 500 μA. Assume R_1 has a resistance of 40,000 Ω and R_2 has a current running through it of 100 μA. If the 12-V supply is kept in place but R_3 is moved from its current location and placed in series with R_1 and R_2, which are still kept in parallel, what will be the total current through the circuit?

88. ✪ A 70-Ω electrical appliance and a 200-Ω electrical appliance are plugged into the outlets of a house. By what factor will the power used change if an additional 400-Ω electrical device is plugged in?

89. ⚀ A 100-W lightbulb and a 60-W lightbulb are connected to a 120-V source. Determine the voltage across each lightbulb if they are connected (a) in series and (b) in parallel.

90. ✪ ✖ Ⓡ With the level of childhood obesity reaching epidemic proportions in the United States, one inventor decided to create a device to make children exercise to generate the electrical energy required for them to watch television. Attaching a small electric generator to a stationary exercise bike, the "Cyclevision" (Fig. P19.90) is able to generate enough electrical energy to supply a typical television. Assuming the generator is 80% efficient at converting the mechanical energy of the exercise bike to electrical energy, estimate the number of calories burned while watching 4 h of television (the average daily amount for a child in the United States).

Figure P19.90

91. ✪ Ⓡ **Lightning strike!** As discussed in Chapter 18, lightning is one of nature's most impressive displays. For a typical storm, a lightning bolt is generated when the average electric field is about 10% of the breakdown field for air. Consider a case in which the bottom of the storm cloud is 500 m above the Earth's surface and the cloud has an area of 200×200 m². At the moment of a lightning strike, determine (a) the potential difference between the cloud and the Earth's surface and (b) the total charge accumulated on the bottom of the cloud. (c) Estimate the current in the lightning strike. Assume the current flows for 0.1 ms.

92. ✪ Most recreational vehicles and boats have the capability of operating electrical appliances using a 12-V battery, which either requires the use of special appliances designed to run on 12-V DC or the use of an inverter that converts direct current to alternating current as used in our homes. Consider one such 12-V DC circuit in which the following appliances and devices are being operated simultaneously on the same circuit: a 40-W minirefrigerator, a 150-W hair dryer, a 200-W blender, and three 15-W lightbulbs. Determine (a) the current through each of these appliances/devices and (b) the minimum fuse required to handle this load. (Fuses are rated according to the maximum current they will allow.)

93. ✪ Ⓡ **Hydroelectric power.** Niagara Falls has one of the largest hydroelectric power-generating facilities in the world, providing one fourth of New York State's and all Ontario's electrical energy. The Canadian portion of the falls is called Horseshoe Falls based on its shape (Fig. P19.93). Upstream from the falls, water is diverted to the Canadian hydroelectric power-generating plants, which generate 2,000,000 kW of electrical power. The diverted water reenters the river downstream. Assuming the generators are 100% efficient at turning the mechanical energy of the flowing water into electrical energy, estimate the volume flow rate of the water diverted through the generating plants.

© Radius Images/Jupiterimages

Figure P19.93 Horseshoe Falls.

94. ✪ **Resistor puzzle 1.** Consider the configuration of resistors shown in Figure P19.94. Each connection can have a resistor that is disconnected or whose value is 5.0 Ω or 10.0 Ω. To determine the value of each resistor or disconnection point, you measure the following with an ohmmeter: $R_{AB} = 5.0$ Ω, $R_{AC} = 4.0$ Ω, $R_{CD} = 6.0$ Ω, $R_{AD} = 6.0$ Ω, $R_{BD} = 11.0$ Ω, and $R_{BC} = 9.0$ Ω. From these measurements, determine the value of each resistor or if it is disconnected.

95. ✪ **Resistor puzzle 2.** Consider again the configuration of resistors shown in Figure P19.94. As in Problem 94, each connection can have a resistor that is disconnected or whose value is 5.0 Ω or 10.0 Ω. This time, you measure a resistance of 6.25 Ω between points A and B, between A and C, between C and D, and between B and D. The resistance between points A and D is

Figure P19.94 Problems 94 and 95.

measured to be 5.0 Ω. From these measurements, determine the value of each resistor or if it is disconnected.

96. ✪ Ⓡ For reasons of safety, the interior light of most modern automobiles is designed to come on when you lift the exterior door handle and then stay on for a short period of time. Assuming the timing circuit for the light is controlled by an *RC* circuit, estimate the value of the capacitor used in this circuit.

97. ✪ ⓧ Ⓡ For muscle rehabilitation, patients can be fitted with a small electrical device designed to help strengthen muscle tissue. With this device, a small electrical current is passed through the muscle tissue, stimulating muscle contraction in that region. With involuntary muscle stimulation requiring a current of 10 mA, estimate the minimum required potential difference applied over the length of a bicep muscle needed to stimulate muscle contraction. Assume a bicep muscle has a diameter of 10 cm and a resistivity of 150 Ω · m.

98. ✪ Electrical wire comes in different sizes (different diameters), which are referred to according to their "gauge" number. For example, an 18-gauge wire has a diameter of approximately 1.02 mm and is often used for extension cords. Suppose an 8.0-m extension cord composed of 18-gauge copper wire is carrying 8.0 A of current to a compound miter saw. Determine the voltage drop along the length of the extension cord, remembering that the current must travel down the length of the cord and back.

99. ✪ Ⓡ **Immersion heater.** Tired of having his morning cup of coffee go cold, an enterprising young physics major decides to build an immersion heater out of batteries and a 1.0-m length of 32-gauge (0.20-mm diameter) copper wire. By coiling up the copper wire so that it can fit in a coffee cup and connecting the coil to several 9-V batteries connected in series, the coil of wire will heat up and in turn heat the coffee in the cup. Estimate the number of batteries this student will need to heat up a cup of coffee from room temperature to a reasonable drinking temperature (90°C) in a reasonable amount of time.

Chapter 20

Magnetic Fields and Forces

Placing a magnet next to a computer monitor or TV picture tube can distort the image and may do permanent damage to the display. In this chapter, we explain why. (© Cengage Learning/Charles D. Winters)

In Chapters 17 and 18, we studied electric fields and the forces they exert on charged particles. An electric field can be produced by a charge or a collection of charges, and electric fields lead to forces on an electric charge. We now consider the same kinds of questions for magnetic fields and forces. Magnetic fields are produced by *moving* electric charges and these fields exert forces on other *moving* charges. We have two main jobs in this chapter: (1) to understand the sources of magnetic fields and the fields they produce and (2) to see how to calculate the magnetic force on a charged particle. These two problems are intertwined. We begin with a qualitative discussion of how a magnetic field can be produced, and then we describe magnetic forces. We later return to a more quantitative treatment of how to calculate the magnetic field.

Magnetic forces act on charged particles, which suggests that magnetism and electricity are connected in some way, and that is indeed the case. In Chapters 21 and 22, we'll see that electric fields can in fact serve as a source of magnetic fields and vice versa. This close connection gives rise to electromagnetic waves and light, topics we'll explore in Chapters 23–26.

20.1 | SOURCES OF MAGNETIC FIELDS

The first observations of magnetic fields involved **permanent magnets**. The ancient Greeks, Romans, and Chinese all discovered that the mineral magnetite (also called lodestone) has the ability to attract or repel pieces of other magnetic materials, including iron and other pieces of magnetite. Permanent magnets are used in applications ranging from compass needles and "refrigerator magnets" to speakers, motors, and computer hard disks. Figure 20.1 shows a **bar magnet**, a permanent magnet made in the shape of a "bar," as well as small pieces of iron (called iron "filings") sprinkled around the bar magnet. These filings are themselves small, needle-shaped, permanent magnets. The needle directions indicate the direction of the magnetic field $\vec{B}$ near the bar magnet. Figure 20.1 shows some of the magnetic field lines deduced from the pattern of iron filings.

Figure 20.1 raises two questions. First, how can we understand this pattern of field lines? Second, why do the iron filings align along the direction of $\vec{B}$? Figure 20.2 compares the pattern of magnetic field lines near a bar magnet with the pattern of electric field lines near an electric dipole, and we see that the two patterns are quite similar. We have also indicated the **magnetic poles** at the ends of the bar magnet in Figure 20.2A. The magnetic poles are called north (N) and south (S), and are analogous to the positive and negative electric charges in Figure 20.2B. This analogy helps explain why the iron filings line up parallel to the magnetic field lines in Figure 20.1. Each iron filing is a small bar magnet, with its own north and south poles. The north pole of an iron filing is attracted to the south pole of the large bar magnet and is repelled from the north pole. At the same time, the south pole of an iron filing is attracted to the north pole of the large bar magnet and is repelled from the south pole. These magnetic forces cause the iron filing to align parallel to the magnetic field line and hence parallel to $\vec{B}$ as shown in Figure 20.2A. This is just like the situation with electric charges and electric fields studied in Chapter 17 and shown in Figure 20.2B.

The unit of magnetic field is a derived unit of measure in the SI system called the **tesla** (T). We'll show below how it is related to other SI units. The magnetic field close to one of the poles of a strong bar magnet has a magnitude of about 1 T. The Earth also acts approximately as a bar magnet, with a magnetic field of approximately 50 μT (= 5×10^{-5} T) near the Earth's surface. The Earth's magnetic field is responsible for the alignment of a compass needle, as we'll discuss in Section 20.9.

The magnetic field lines outside a permanent magnet "emanate" from the north pole and are directed toward the south pole, and the strength (magnitude) of the field decreases as one moves farther from the poles. We have already noted that the pattern of magnetic field lines of a bar magnet is similar to the pattern of electric field lines near two electric charges. However, the two field patterns in Figure 20.2 *differ* in one very important way: the magnetic field lines *inside* the bar magnet point from the south pole toward the north pole, whereas between the two charges of the electric dipole the electric field lines are in the *opposite* direction. As a result, the magnetic field lines form closed "loops." While electric field lines start on positive charges and end on negative changes, **magnetic field lines always close on themselves**, forming closed loops. This general property of magnetic fields is not limited to the magnetic field produced by a bar magnet.

Permanent magnets come in many shapes and sizes; the "horseshoe" magnet is another common example. A horseshoe magnet can be made by simply bending a

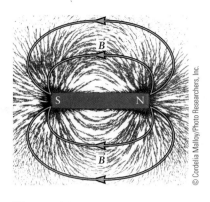

Figure 20.1 Iron filings are small slivers of iron. When placed near a bar magnet, they align along the magnetic field lines produced by the magnet. A few of these field lines are shown here. The letters "N" and "S" label the poles of the bar magnet.

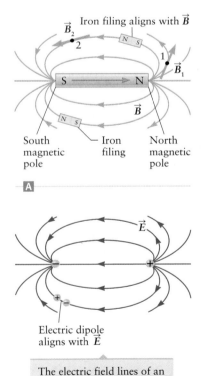

The electric field lines of an electric dipole are similar to the magnetic field lines of a bar magnet.

Figure 20.2 A Magnetic field lines near a bar magnet. B Electric field lines near an electric dipole.

Figure 20.3 A horseshoe magnet has a north pole and a south pole, and the lines of $\vec{B}$ form closed loops.

Figure 20.4 The magnetic field lines produced by a straight current-carrying wire form circles around the wire. The field lines can be visualized using iron filings. The magnitude of the field decreases as one moves away from the wire.

Right-hand rule number 1 relates the directions of the current and the magnetic field $\vec{B}$ produced by that current.

Figure 20.5 ◙ The direction of $\vec{B}$ is given by right-hand rule number 1. When the thumb of your *right* hand is aligned parallel to the current, your fingers will curl in the direction of $\vec{B}$. ◙ If the current is reversed, the lines of $\vec{B}$ circulate in the opposite direction.

bar magnet, and some of the resulting magnetic field lines are shown in Figure 20.3. There are north and south poles at the ends of the horseshoe, and the field lines again form closed loops as they circulate through the horseshoe. Permanent magnets with horseshoe shapes are used in many applications, including motors and generators. Key properties of these magnets are that the magnetic field is largest in the horseshoe gap and that it is directed across the gap from the north pole toward the south pole.

In the introduction to this chapter, we mentioned that magnetic fields are produced by moving electric charges. This statement also applies to the bar magnets and horseshoe magnets in Figures 20.1 through 20.3. The moving charges in these cases are electrons that move ("orbit") around the atoms in the magnetic material. We'll discuss the nature and size of the resulting currents, along with other properties of permanent magnet materials, in Section 20.8.

Magnetic Fields Produced by an Electric Current

We have already mentioned that moving charges produce a magnetic field. Since an electric current consists of moving charges, an electric current will also produce a magnetic field. Figure 20.4 shows long, straight wires carrying an electric current. Each wire passes through a hole in a piece of paper, and iron filings have been sprinkled on the paper. We have seen how iron filings line up parallel to the magnetic field lines produced by a bar magnet; their alignment in Figure 20.4 indicates the pattern of magnetic field lines near each wire. For the case of a straight wire, the magnetic field lines form circles as sketched in the figure and the direction of $\vec{B}$ is always tangent to these field line circles. In addition, the "strength" of the magnetic field (the magnitude of $\vec{B}$) decreases as one moves away from the wire.

So, the magnetic field lines circulate around a current-carrying wire, but how do we determine in which direction they circulate? The direction of the magnetic field produced by a current is given by the ***right-hand rule*** illustrated in Figure 20.5. To apply the right-hand rule, point the thumb of your *right* hand in the direction of the current, with your thumb parallel to the wire. Curling the fingers of your *right* hand around the wire then gives the direction of $\vec{B}$. *We call this method **right-hand rule number 1**.* (We'll encounter another right-hand-rule procedure below.)

If the direction of the current is reversed, the direction of $\vec{B}$ is also reversed. Figure 20.5B shows the same wire as in Figure 20.5A but with the current in the opposite direction. Applying right-hand rule number 1 shows that the lines of $\vec{B}$ now encircle the wire in the opposite sense.

Right-hand rule number 1 gives the direction of $\vec{B}$ but does not tell us its magnitude. We'll learn how to find the magnitude of $\vec{B}$ in Section 20.7.

Plotting Magnetic Fields and Field Lines

Since the magnetic field lines produced by a current-carrying wire encircle the wire, the relationship between I and $\vec{B}$ is inherently three dimensional, and three-

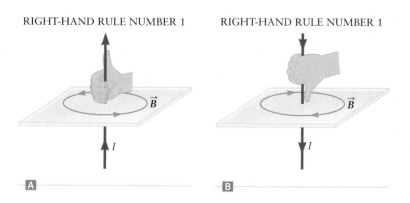

RIGHT-HAND RULE NUMBER 1   RIGHT-HAND RULE NUMBER 1

A   B

dimensional perspective plots likes those in Figures 20.4 and 20.5 will be extremely useful in our studies of magnetic fields. In many cases, however, we'll consider two-dimensional plots in which $\vec{B}$ points either into or out of the plane of the drawing. For example, Figure 20.6 shows the magnetic field of a long, straight wire. The three-dimensional representation of Figure 20.6A (similar to Fig. 20.5) shows how the field lines encircle the wire, whereas Figure 20.6B shows a two-dimensional plot of the same field. In the region above the wire, the magnetic field $\vec{B}$ points *out of* the x–y plane, whereas the field lines point *into* the x–y plane below the wire. In Figure 20.6, a large "dot" (•) denotes the tip of the $\vec{B}$ vector when it points out of the plane and a "cross" (×) denotes the tail of the $\vec{B}$ vector when it points into the plane.

Figure 20.6C shows a similar two-dimensional plot of the magnetic field produced by a current-carrying wire, but here the current is in the $-x$ direction. The magnetic field lines again encircle the wire, but are now directed out of the plane in the area below the wire and into the plane above the wire.

In Figures 20.5 and 20.6, we have drawn long, straight wires, but right-hand rule number 1 can also be used to find the direction of the magnetic field lines from a short piece of wire as sketched in Figure 20.7A. We again put the thumb of our right hand along the direction of I, and curling our fingers then gives the direction of $\vec{B}$. The magnetic field lines again form circles (as for a long wire), but now the magnitude of $\vec{B}$ falls off as we go farther from the short wire along the x axis in either direction.

The electric current in a short current-carrying wire can be modeled as a collection of positive electric charges moving with a velocity $\vec{v}$ parallel to the current direction (Fig. 20.7B). Each of these individual moving charges produces a magnetic field, and the direction of this magnetic field is given by right-hand rule number 1. Place your thumb of your right hand along the direction of $\vec{v}$; your fingers will then curl in the direction of the magnetic field (Fig. 20.7B). Figure 20.7C shows what happens to the magnetic field if the direction of the current is reversed. Here we have positive charges with $\vec{v}$ directed to the left, and applying right-hand rule number 1 shows that the field lines circulate opposite to those in Figure 20.7B. A positive charge moving to the left produces the same magnetic field as a negative charge moving to the right. Figure 20.7D shows such a moving negative charge; its magnetic field lines are the same as those produced by the positive charge in Figure 20.7C. These examples all illustrate how magnetic fields are produced by moving electric charges.

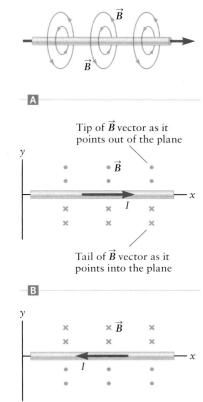

Figure 20.6 **A** Magnetic field near a long current-carrying wire. **B** and **C** In a two-dimensional plot, lines of $\vec{B}$ directed out of the plane are indicated by dots (the tip of the $\vec{B}$ vector), whereas lines of $\vec{B}$ pointing into the plane are denoted by crosses (the tail of the $\vec{B}$ vector).

CONCEPT CHECK 20.1 | Direction of the Magnetic Field

Part 1: A proton travels at a constant velocity, with $\vec{v}$ directed along the $+y$ direction as shown in Figure 20.8A. At a particular moment, the proton is at the origin. Is the direction of the magnetic field at point A (a) along the $+x$ direction, (b) along the $-x$ direction, (c) along the $+z$ direction, or (d) along the $-z$ direction?

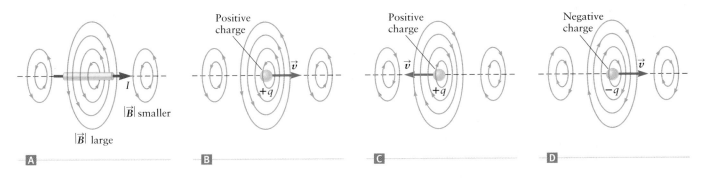

Figure 20.7 **A** Magnetic field lines produced by a short section of a current-carrying wire. **B**–**D** Magnetic fields produced by moving charges.

Figure 20.8 Concept Check 20.1.

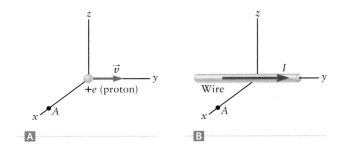

Part 2: The proton is replaced by a long, straight wire carrying a current in the $+y$ direction as shown in Figure 20.8B. Is the direction of the magnetic field at point A (a) along the $+x$ direction, (b) along the $-x$ direction, (c) along the $+z$ direction, or (d) along the $-z$ direction?

Magnetic Fields and Superposition

So far, we have considered the magnetic field lines produced by a bar magnet, by a current in a straight wire, and by a moving electric charge. The qualitative pattern of the magnetic field lines in most other situations can be understood using these results along with the *principle of superposition*.

The principle of superposition

> *Principle of superposition:* **The total magnetic field produced by two or more different sources is equal to the sum of the fields produced by each source individually.**

We can apply the principle of superposition to find the direction of the magnetic field produced by a "current loop." Consider a circular loop of wire carrying a current I as shown in Figure 20.9A. This case might seem very special and idealized, but current loops are actually found in many applications. We treat the loop as many small pieces and apply right-hand rule number 1 to find the field from each piece. The magnetic field lines encircle each segment of the wire, and for the current direction shown here, $\vec{B}$ is directed upward through the center of the loop and downward outside the loop. We then use the principle of superposition and add up the fields from all these pieces, giving the overall pattern of magnetic field lines in Figure 20.9B. The lines of $\vec{B}$ circulate through the current loop, all passing inside the loop and then circling around the outside.

The overall pattern of field lines from a current loop is very similar to the magnetic field produced by a bar magnet (Fig. 20.2A). The similarity between the magnetic fields of a current loop and a bar magnet is no accident. The field of a bar magnet and other permanent magnets is due to atomic-scale current loops, which we'll describe in Section 20.8.

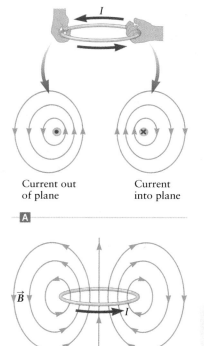

Current out of plane   Current into plane

A

$\vec{B}$ I

B

Figure 20.9 **A** Application of right hand rule number 1 to find direction of the field lines near the two sides of a current loop. **B** Magnetic field lines produced by a full current loop.

EXAMPLE 20.1 | Magnetic Field from Two Parallel Wires

Use the principle of superposition to sketch *qualitatively* the magnetic field lines near two long, straight wires, each parallel to the x axis and lying in the x–y plane. Assume each wire carries a current I, but with the currents in *opposite* directions.

RECOGNIZE THE PRINCIPLE

According to the principle of superposition, the total field is the sum of the fields from each wire. We use this principle along with our previous results for the magnetic field produced by a single wire (Figs. 20.5 and 20.6). Recall that for a single wire the field lines encircle the wire according to right-hand rule number 1 and the magnitude of $\vec{B}$ decreases as one moves away from the wire.

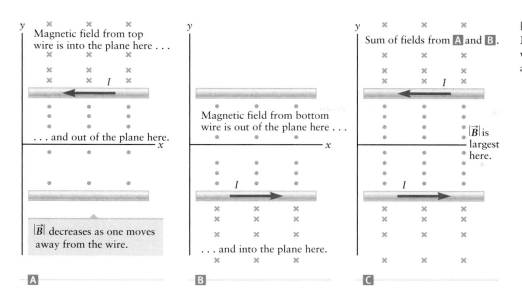

Figure 20.10 Example 20.1. Magnetic field due to **A** the top wire alone, **B** the bottom wire alone, and **C** both wires.

Figure A labels:
- Magnetic field from top wire is into the plane here . . .
- . . . and out of the plane here.
- $|\vec{B}|$ decreases as one moves away from the wire.

Figure B labels:
- Magnetic field from bottom wire is out of the plane here . . .
- . . . and into the plane here.

Figure C labels:
- Sum of fields from **A** and **B**.
- $|\vec{B}|$ is largest here.

SKETCH THE PROBLEM

Figure 20.10A shows the magnetic field produced by one of the wires, and Figure 20.10B shows the field produced by the other. The field lines encircle both wires and pass perpendicularly through the plane of the drawing (the x–y plane).

IDENTIFY THE RELATIONSHIPS

The sum of the fields from the two wires is shown qualitatively in Figure 20.10C. In the region between the wires, the fields from both wires pass upward out of the plane of the drawing, so the total field in this region is also directed out of the plane. In the region above the top wire, the field from the top wire is directed into the plane while the field from the bottom wire is out of the plane, so these fields are in opposite directions.

SOLVE

The top wire is closest to points in the upper region of Figure 20.10C, so the field of the top wire is larger (in magnitude) than the field from the bottom wire in that region. The sum of the two fields at the top of Figure 20.10C therefore points into the plane. The same reasoning shows that the field in the region below the bottom wire in Figure 20.10C is also directed into the plane.

What have we learned?
The principle of superposition can be used to find the pattern of magnetic field lines in virtually all situations. Most cases can be analyzed as a collection of bar magnets, straight wires, or current loops; the total magnetic field is then the sum of the fields produced by all the individual parts of the collection.

20.2 | MAGNETIC FORCES INVOLVING BAR MAGNETS

In Section 20.1, we described the magnetic fields produced by various sources. Let's now consider the magnetic *forces* that these fields produce, beginning with permanent magnets. Figure 20.11A shows a bar magnet situated in a constant magnetic field; when we say that $\vec{B}$ is constant, we mean that the magnetic field has a constant direction and a constant magnitude. To describe the total force on the bar magnet, we must include two forces, acting on the two poles of the bar. The force on the north pole is directed parallel to $\vec{B}$, while the force on the south pole is antiparallel

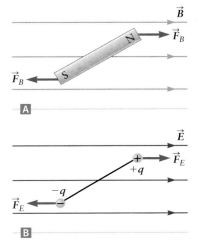

Figure 20.11 **A** Magnetic forces on a bar magnet placed in a uniform magnet field. **B** Electric forces on an electric dipole placed in a uniform electric field.

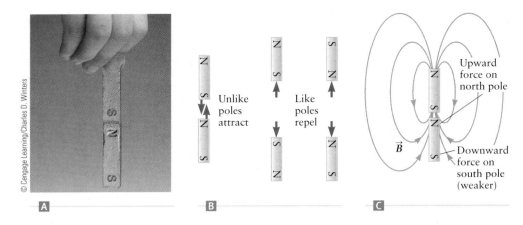

Figure 20.12 **A** and **B** Forces between two bar magnets. Unlike magnetic poles attract, and like poles repel. **C** The force on the north pole of the bottom magnet is upward, parallel to the field $\vec{B}$ of the upper magnet. The force on the south pole of the bottom magnet is downward. Since the field from the top magnet becomes smaller as one moves downward, the total (net) force on the bottom magnet is upward.

(opposite) to $\vec{B}$. For the case of a constant magnetic field, these two forces are equal in magnitude and in opposite directions. Although the total force is thus zero, these two forces produce a nonzero torque on the bar magnet, which acts to align the bar magnet along the field lines. That is why iron filings (Figs. 20.1 and 20.4) align with the magnetic field. (Recall that each iron filing is a small bar magnet.)

It is interesting to compare the magnetic forces and torque on the bar magnet in Figure 20.11A with the electric forces on an electric dipole as sketched in Figure 20.11B. The figure shows an electric dipole formed by positive and negative charges $+q$ and $-q$ attached to the ends of an uncharged rod. This electric dipole is placed in a constant electric field $\vec{E}$, which exerts a force directed to the right on $+q$ and a force directed to the left on $-q$. These forces thus produce a torque on the electric dipole, just like the torque on the bar magnet in Figure 20.11A. This similarity illustrates that a bar magnet possesses a ***magnetic moment***. This comparison also suggests that we can view the north and south poles of the bar magnet as a sort of "magnetic charge." The analogy with electric charge then implies that the north pole of one bar magnet should attract the south pole of another magnet (Fig. 20.12A), just as positive and negative electric charges attract each other. Likewise, we expect that "like" magnetic poles (two north poles or two south poles) will repel each other, just as found with two "like" electric charges (Fig. 20.12B). Such forces between the poles of permanent magnets do indeed occur and are one reason bar magnets make popular toys.

The forces between magnetic poles and between two bar magnets can also be analyzed and understood in terms of how the magnetic field produced by one magnet exerts a force on the other magnet. Figure 20.12C shows two bar magnets as well as the magnetic field produced by the top magnet. The central "axes" of the two magnets are aligned, so the magnetic field at the north pole of the lower magnet is directed upward, producing an upward force on this pole. The magnetic field at the bottom end of the lower magnet is also directed upward, which gives a downward force on the south pole. The field is weaker at the lower end, however, because it is farther from the upper magnet. As a result, the total force on the lower bar magnet is upward, and the two magnets attract.

The similar behavior of electric charges and magnetic poles, and electric and magnetic forces, in Figures 20.11 and 20.12 can help you visualize and understand magnetic fields and forces. However, there are some important differences between the electric and magnetic cases. In particular, north and south magnetic poles *always* occur in pairs. We'll see why it is not possible to obtain an isolated magnetic pole when we study what goes on inside a bar magnet in Section 20.8.

Magnetic forces act on electric charges, such as individual electrons, protons, or ions. The magnetic force depends on the velocity of the charge, and it is nonzero only if the charge is in motion. This is very different from gravity and the electric force, both of which are completely independent of velocity. The dependence of magnetic forces on velocity has some deep philosophical implications that we'll discuss in Section 20.11.

Figure 20.13A shows a particle with positive charge q moving with velocity $\vec{v}$. Here $\vec{v}$ is along the x direction, and the magnetic field $\vec{B}$ is parallel to y. The magnetic force on this charge is along the $+z$ direction, with a magnitude

$$F_B = qvB \sin \theta \qquad (20.1)$$

where θ is the angle between $\vec{v}$ and $\vec{B}$. In Figure 20.13, this angle is 90°, so $\sin \theta$ in Equation 20.1 is equal to 1. Notice that Equation 20.1 gives only the *magnitude* of the magnetic force. The *direction* of the force can be found using another right-hand rule illustrated in Figure 20.13B. Point the fingers of your right hand in the direction of $\vec{v}$ and curl your fingers in the direction of $\vec{B}$, taking the smallest angle between $\vec{v}$ and $\vec{B}$. Your thumb then points in the direction of the magnetic force $\vec{F}_B$. This procedure gives the direction of the force when q in Equation 20.1 is positive. For a positive charge ($q > 0$) with $\vec{v}$ and $\vec{B}$ directed as shown in Figure 20.13, the magnetic force is along the $+z$ direction. On the other hand, if q is negative, such as for an electron, the force direction is reversed and in this example is along $-z$. For a negative charge q, the value of F_B in Equation 20.1 is negative, so you can think of this as just reversing the direction of the magnetic force.

We'll call the procedure for finding the direction of the magnetic force on a moving charge in Figure 20.13B **right-hand rule number 2**. This procedure is similar to right-hand rule number 1 (Fig. 20.5), which gives the direction of the magnetic field from a current. You should practice using both rules.

Compared with other fundamental forces such as the gravitational and electric forces, the magnetic force in has several remarkable features. One is the dependence of $\vec{F}_B$ on velocity that we already noted. Also, the direction of $\vec{F}_B$ is always perpendicular to both the magnetic field and the particle's velocity. That is quite different from the gravitational and electric fields for which the force and the field are always parallel. Indeed, we originally introduced these fields as being simply proportional in magnitude and parallel in direction to the gravitational and electric forces. We now find that fields and forces (at least magnetic ones) need *not* be related in such a simple way.

CONCEPT CHECK 20.2 | Direction of the Magnetic Force

Part 1: Figure 20.14 shows a positive charge q moving with a nonzero speed v. If the magnetic field is directed to the right, is the direction of the magnetic force (a) into the plane of the drawing, (b) out of the plane of the drawing, (c) to the left, or (d) is the magnetic force zero?

Part 2: If the particle stops, is the direction of the magnetic force (a) into the plane of the drawing, (b) out of the plane of the drawing, (c) to the left, or (d) is the magnetic force zero?

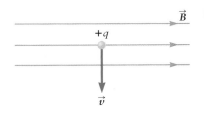

Figure 20.14 Concept Check 20.2.

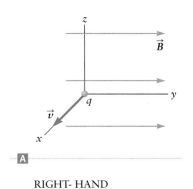

Magnetic force on a charged particle

A

RIGHT- HAND RULE NUMBER 2

If $q > 0$, the force will be along $+z$.

Right-hand rule number 2 gives the direction of magnetic force on a moving charged particle.

B

Figure 20.13 **A** A charged particle moving in a magnetic field. **B** The direction of the magnetic force exerted on the particle is given by right-hand rule number 2. Place the fingers of your right hand in the direction of $\vec{v}$ and wrap them in the direction of $\vec{B}$. If $q > 0$, your thumb then points in the direction of the magnetic force.

Magnetic Force on an Electron: How Big Is It?

To get a feeling for the size of magnetic forces, consider an electron moving near a bar magnet. The field very close to a bar magnet can be as large as 1 T or 2 T, so let's take $B = 1$ T for this example. The electron's speed depends on the situation; let's assume a typical terrestrial speed of 50 m/s (about 100 mi/h). The charge of an electron is $q = -e = -1.60 \times 10^{-19}$ C. For simplicity, we assume the angle θ between $\vec{v}$ and $\vec{B}$ is 90° and insert these values for q, v, and B into Equation 20.1. The magnitude of the force is

$$F_B = qvB \sin \theta = (1.60 \times 10^{-19}\ \text{C})(50\ \text{m/s})(1\ \text{T})\sin(90°)$$
$$= 8 \times 10^{-18}\ \text{C} \cdot \text{m} \cdot \text{T/s} \qquad (20.2)$$

Here, F_B must have units of force, so in the SI system the combination of units in Equation 20.2 must be equal to newtons (N). In terms of just the units, Equation 20.2 is thus

$$1\ \text{N} = 1\ \text{C} \cdot \text{m} \cdot \text{T/s}$$

We can now rearrange to solve for the tesla in terms of the primary SI units:

$$1\ \text{T} = 1\ \frac{\text{N} \cdot \text{s}}{\text{C} \cdot \text{m}} = 1\ \frac{(\text{kg} \cdot \text{m/s}^2) \cdot \text{s}}{\text{C} \cdot \text{m}} \qquad (20.3)$$

$$1\ \text{T} = 1\ \frac{\text{kg}}{\text{C} \cdot \text{s}} \qquad (20.4)$$

Returning to Equation 20.2, we find that the force on an electron moving at 50 m/s near a typical bar magnet is

$$F_B = 8 \times 10^{-18}\ \text{N} \qquad (20.5)$$

To judge if this force is large or small, let's compare F_B with the gravitational force on the electron. For an electron near the Earth's surface,

$$F_{\text{grav}} = mg = (9.11 \times 10^{-31}\ \text{kg})(9.8\ \text{m/s}^2) = 8.9 \times 10^{-30}\ \text{N}$$

Comparing with F_B in Equation 20.5, we see that the gravitational force on an electron is smaller than the magnetic force in this case by a factor of

$$\frac{F_{\text{grav}}}{F_B} = \frac{8.9 \times 10^{-30}}{8 \times 10^{-18}} \approx 1 \times 10^{-12}$$

So, this magnetic force is *much* larger that the gravitational force. Of course, this comparison depends on the electron's velocity and the magnetic field, but one typically finds that magnetic forces are very much larger than gravitational ones.

Motion of a Charged Particle in a Magnetic Field

Let's now consider the direction of $\vec{F}_B$ and how this force affects the motion of a charged particle in a little more detail. For simplicity, we assume the magnetic field is constant, so the magnitude and direction of $\vec{B}$ are the same everywhere. Figure 20.15A shows a charged particle moving parallel to the direction of $\vec{B}$. In this case, the angle between $\vec{v}$ and $\vec{B}$ is $\theta = 0$; the factor $\sin \theta$ in Equation 20.1 is then zero, so the magnetic force in this case is also zero. *If a charged particle has a velocity parallel to $\vec{B}$, the magnetic force on the particle is zero.*

Figure 20.15B shows a charged particle moving perpendicular to $\vec{B}$. Now we have $\theta = 90°$, so $\sin \theta = 1$. The magnitude of the magnetic force is thus $F_B = qvB$, and the force is perpendicular to $\vec{v}$. In Chapter 5, we learned that when a particle experiences a force of constant magnitude perpendicular to its velocity, the result is *circular motion* as sketched in Figure 20.15B. Hence, *if a charged particle is moving perpendicular to a constant magnetic field, the particle will move in a circle.* This circle lies in the plane perpendicular to the field lines. We can calculate the radius of

MOTION OF A POSITIVELY CHARGED PARTICLE

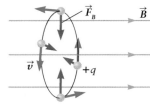

$\theta = 90°$, the charged particle moves in a circle that lies in a plane perpendicular to the plane of this page.

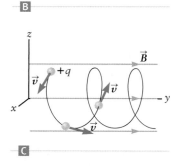

Figure 20.15 ◭ When the velocity of a charged particle is parallel to the magnetic field $\vec{B}$, the magnetic force on the particle is zero. ◨ When $\vec{v}$ makes a right angle with $\vec{B}$, the particle will move in a circle. ◧ When the velocity has nonzero components parallel and perpendicular to $\vec{B}$, the particle will move along a helical path.

the circle by recalling that for a particle to move in a circle of radius r, there must a force of magnitude mv^2/r directed toward the center of the circle (Eq. 5.6). Here the force producing circular motion is the magnetic force, so we have

$$F_B = \frac{mv^2}{r}$$

Inserting F_B from Equation 20.1 and using $\sin \theta = 1$ leads to

$$qvB = \frac{mv^2}{r}$$

Solving for r gives

$$r = \frac{mv^2}{qvB} = \frac{mv}{qB} \tag{20.6}$$

Let's calculate the value of r for the electron we considered in connection with Equation 20.5. In that example, we had $v = 50$ m/s and $B = 1$ T. Inserting those values into Equation 20.6, we find

$$r = \frac{mv}{qB} = \frac{(9.11 \times 10^{-31} \text{ kg})(50 \text{ m/s})}{(1.60 \times 10^{-19} \text{ C})(1 \text{ T})} \tag{20.7}$$

$$r = 3 \times 10^{-10} \text{ m} \tag{20.8}$$

which is *quite* a small radius, approximately the radius of an atom. A faster-moving electron (larger v) will move in a circle with a larger radius.

EXAMPLE 20.2 | Ⓡ Magnetic Fields and Television Picture Tubes

Some televisions and computer monitors produce images using an electron beam that strikes a fluorescent screen as shown in Figure 20.16. The electrons are steered to specific spots on the screen by magnetic fields, forming spots of different brightness and color according to the desired image. The magnetic field from an external bar magnet placed near the television can deflect the electron beam in an unwanted direction and thereby distort or ruin the image. (You can also produce "artistic" results; see the photo at the beginning of this chapter.) Assuming the electrons travel at a speed of approximately 5×10^7 m/s, find the approximate magnetic field necessary to disrupt a television image.

RECOGNIZE THE PRINCIPLE

We wish to find the field needed to just deflect the electrons a significant amount, so we do not need to calculate the precise electron trajectory. Instead, we know that an electron traveling perpendicular to a magnetic field is deflected into a circular path with a radius r given by Equation 20.6, so we'll calculate the value of B that gives a radius comparable to the size of a TV picture tube.

SKETCH THE PROBLEM

Figure 20.16 shows the problem. A beam of electrons passes through a region in the "neck" of the tube where there are current loops that produce a magnetic field directed perpendicular to the electron velocity.

IDENTIFY THE RELATIONSHIPS

A typical television picture tube has a size of about 50 cm, so we take that for our value of r. We have already discussed how the radius of a charged particle's circular motion is related to B. From Equation 20.6, we have

$$r = \frac{mv}{qB}$$

Insight 20.1

DEFLECTION COILS IN A TV

The television tube in Figure 20.16 uses magnetic fields in the "neck" of the tube to control the electron deflection. The electron velocity is to the right in this figure, and there are two deflecting fields; one field is vertical as shown in the figure, and another (not shown) is perpendicular to the plane of the drawing. The magnetic force $\vec{F}_B$ is perpendicular to both $\vec{v}$ and $\vec{B}$, so the vertical deflecting field in Figure 20.16 produces a *horizontal* deflection force that is perpendicular to the plane of the drawing. When the horizontal magnetic field is added, there is also a vertical deflection force. Controlling these two magnetic fields controls the overall deflection and determines the location at which the electron beam strikes the TV screen.

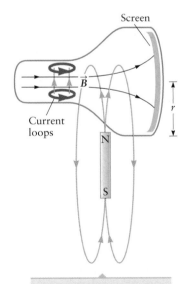

The external magnetic field from a bar magnet can deflect the electrons in a TV picture tube.

Figure 20.16 Example 20.2. Electron motion in a TV picture tube.

Rearranging to solve for B gives

$$B = \frac{mv}{qr}$$

We don't need to worry here about the charge on the electron being negative; it affects the direction in which the electron moves along its circular trajectory, but does not change the radius of the circle.

SOLVE

Inserting the given value of v, our estimate for r, and the known values of the charge and mass of an electron, we find

$$B = \frac{mv}{qr} = \frac{(9.11 \times 10^{-31}\ \text{kg})(5 \times 10^7\ \text{m/s})}{(1.60 \times 10^{-19}\ \text{C})(0.50\ \text{m})} = \boxed{6 \times 10^{-4}\ \text{T}}$$

What does it mean?

This answer is about 10 times larger than the Earth's field. Hence, it does not take a very strong magnet to disrupt the image on a television or a computer monitor; almost any bar magnet will do. We don't recommend that you try this experiment at home because it can do permanent damage to your TV or monitor!

© ScienceSource/Photo Researchers, Inc.

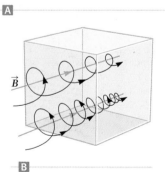

Figure 20.17 **A** Photo of the paths of charged particles as they move through a bubble chamber. **B** Schematic particle paths in three dimensions. In some cases, the particles lose energy as they move, causing the trajectories to spiral inward. As the particle speed decreases, the radius of the spiral trajectory also decreases.

In most cases, the velocity of a charged particle is not perfectly parallel or perpendicular to $\vec{B}$. Figure 20.15C shows the particle trajectory in such cases: the particle moves in a helical (spiral) path, with the axis of the helix parallel to $\vec{B}$. The helical motion can be understood as a combination of the results in parts A and B of Figure 20.15. For the component of $\vec{v}$ parallel to the magnetic field, there is no magnetic force and this component of the particle's velocity is constant. The perpendicular velocity component produces results as in Figure 20.15B: circular motion in a plane perpendicular to the field lines. The combination of circular motion in the x–z plane and a constant velocity along y gives a helical trajectory (Fig. 20.15C). Hence, *a charged particle will spiral around the magnetic field lines.*

A dramatic example of this spiral motion is shown in Figure 20.17A. This photograph is from a bubble chamber, a device used to study the trajectories of electrons, protons, and other particles. When charged particles travel through certain fluids at very high speed, they leave behind a trail of ionized fluid molecules, visible in the photograph as tracks of bubbles. The entire bubble chamber is inside a large magnet (Fig. 20.17B) whose magnetic field causes the tracks to be helical. Measuring the radius of these helical tracks gives information about a particle's charge, mass, and velocity through the relation in Equation 20.6.

CONCEPT CHECK 20.3 | Bubble Chamber Trajectories

Part 1: Suppose an electron and a proton with the same velocity are traveling in the same direction when they enter the bubble chamber in Figure 20.17. Each moves in a helical trajectory inside the chamber. How does the radius of the electron's helix compare to the radius of the proton's helix?

(a) The helix followed by the electron has a smaller radius.
(b) The helix followed by the electron has a larger radius.
(c) The electron and proton helices have the same radius.

Part 2: Sometimes the particle path in a bubble chamber is not a perfect helix, but instead the particle "spirals inward." (A few such paths are visible in Fig. 20.17.) What causes these paths to deviate from a simple helix?

(a) The particle loses energy and slows down as it travels through the bubble chamber.

(b) The particle gains energy and speeds up as it travels through the bubble chamber.

(c) Other forces (such as gravity) act on the particle.

| EXAMPLE 20.3 | Direction of the Magnetic Force on a Moving Charge |

Consider a positively charged particle traveling parallel to a current-carrying wire as sketched in Figure 20.18A. What is the *direction* of the magnetic force on this particle?

RECOGNIZE THE PRINCIPLE

Right-hand rule number 2 gives the direction of the magnetic force on a moving charge. To apply this rule, we must first establish the direction of the magnetic field produced by the wire, which we can find using right-hand rule number 1.

SKETCH THE PROBLEM

Figure 20.18 shows the direction of the current and the direction of $\vec{v}$, the particle's velocity.

IDENTIFY THE RELATIONSHIPS AND SOLVE

From our work with current-carrying wires and right-hand rule number 1 (Figs. 20.5 and 20.6), the magnetic field lines encircle the wire and are directed *into the plane* of the drawing at the particle's location (Fig. 20.18B). To find the magnetic force on this particle, we now apply right-hand rule number 2, placing our fingers along $\vec{v}$ and curling them in the direction of $\vec{B}$ as shown in Figure 20.18B. The thumb then points in the direction of the force, *toward* the wire as shown in the figure.

What have we learned?
To find the magnetic force on a moving charged particle, we must first find the direction of the magnetic field; here we used right-hand rule number 1 to find the direction of $\vec{B}$. We then applied right-hand rule number 2 to find the direction of the magnetic force. To deal with magnetic forces and fields, you *must* be able to apply the right-hand rules. Practice is strongly recommended.

Figure 20.18 Example 20.3. Finding the direction of $\vec{F}_B$.

CONCEPT CHECK 20.4 | Finding the Direction of $\vec{B}$

A positively charged particle is moving with a velocity $\vec{v}$ as shown in Figure 20.19. If the magnetic force on this particle is directed out of the plane of the drawing, is the possible direction of $\vec{B}$ (a) along direction 1 in the figure, (b) along direction 2, (c) along direction 3, or (d) directed into the plane of the drawing?

Applying the Right-Hand Rules

We have introduced two right-hand rules and applied them in several example calculations. Let's state them once more for future reference.

> *Right-hand rule number 1:* Finding the direction of $\vec{B}$ from an electric current.
>
> 1. Place the thumb of your right hand along the direction of the current.
> 2. Curl your fingers; they will then give the direction of $\vec{B}$ as the field lines encircle the current.
>
> *Right-hand rule number 2:* Finding the direction of the magnetic force on a moving charge q.
>
> 1. Point the fingers of your right hand along the direction of $\vec{v}$.

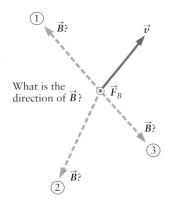

Figure 20.19 Concept Check 20.4.

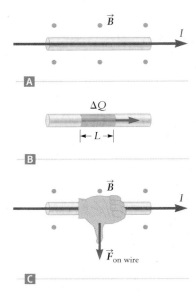

Figure 20.20 ◨ A current-carrying wire in an external magnetic field. ◨ The current in the wire is due to the motion of charge ΔQ along the wire. ◨ The magnetic force on this moving charge produces a force on the wire. The direction of the force is given by right-hand rule number 2. Place the fingers of your right hand in the direction of the current and wrap your fingers in the direction of $\vec{B}$. Your thumb then points in the direction of the magnetic force.

2. Curl your fingers in the direction of $\vec{B}$. Always curl through the smallest angle that connects $\vec{v}$ and $\vec{B}$.

3. If q is positive, the magnetic force on q is parallel to your thumb. If q is negative, the magnetic force is in the opposite direction.

20.4 | MAGNETIC FORCE ON AN ELECTRIC CURRENT

An electric current consists of a collection of moving charges, so our result for the magnetic force on a single moving charge (Eq. 20.1) can be used to find the magnetic force on a current-carrying wire. This force is important in many applications, including electric motors. Consider a current-carrying wire placed in an external magnetic field as sketched in Figure 20.20A, with $\vec{B}$ constant and perpendicular to the wire. This magnetic field is produced by some external source; it is *not* produced by the current in the wire. We can calculate the magnetic force on the wire due to this external field by adding up the magnetic forces on all the moving charges in the wire.

Let's focus on the segment of the wire of length L in Figure 20.20B. The current in this segment is

$$I = \frac{\Delta Q}{\Delta t} \qquad (20.9)$$

where ΔQ is the electric charge that passes by one end of the wire segment in a time Δt; this is just our usual relation between charge and current from Chapter 19. Applying Equation 20.1, the magnetic force on this moving charge is

$$F_B = qvB = (\Delta Q)vB \qquad (20.10)$$

The velocity of the charge is just

$$v = \frac{L}{\Delta t}$$

Inserting this into Equation 20.10 gives

$$F_B = (\Delta Q)vB = (\Delta Q)\frac{L}{\Delta t}B$$

Using the relation between current and charge in Equation 20.9, we get

$$F_B = \frac{\Delta Q}{\Delta t}LB = ILB$$

This force on the moving charge is really a force on the wire, so we arrive at

$$F_{\text{on wire}} = ILB$$

In Figure 20.20, we assumed the magnetic field is perpendicular to the wire. If $\vec{B}$ makes an angle θ with the wire, one finds

$$F_{\text{on wire}} = ILB \sin \theta \qquad (20.11)$$

Magnetic force on a current-carrying wire. The direction is given by right-hand rule number 2.

The direction of $F_{\text{on wire}}$ is given by (you guessed it) a right-hand rule as illustrated in Figure 20.20C. Begin with the fingers of your right hand in the direction of the current and curl them in the direction of the field. Your thumb then points in the direction of $\vec{F}_{\text{on wire}}$, downward in Figure 20.20C. The external magnetic field pulls this wire downward as long as the current is directed to the right. The magnetic force on a current is due to the force on a moving charge, so the right-hand rule used here is another example of right-hand rule number 2, with the direction of $\vec{v}$ for a moving charge replaced by the direction of the current.

CONCEPT CHECK 20.5 | Finding the Magnetic Force on a Wire

The wire in Figure 20.21 carries a current along the $+z$ direction. If there is an external magnetic field in the $+y$ direction, is the direction of the magnetic force on the wire along (a) $+x$, (b) $-x$, (c) $+y$, (d) $-y$, (e) $+z$, or (f) $-z$?

Figure 20.21 Concept Check 20.5.

EXAMPLE 20.4 | Magnetic Force between Two Wires

Consider two parallel wires, each carrying a current I as sketched in Figure 20.22A. Find the direction of the magnetic force that the top wire exerts on the bottom one. Then find the direction of the force that the bottom wire exerts on the top one.

RECOGNIZE THE PRINCIPLE

To find the force exerted on the bottom wire, we must first determine the direction of the magnetic field produced by the top wire, which we can do by using right-hand rule number 1. We can then apply right-hand rule number 2 to find the force produced by this field on the bottom wire.

SKETCH THE PROBLEM

Figure 20.22B shows the direction of the magnetic field produced by the top wire at the location of the bottom wire as well as the application of right-hand rule number 2 to find the force on the bottom wire. Figure 20.22C shows the application of right-hand rule number 2 to find the force on the top wire.

IDENTIFY THE RELATIONSHIPS AND SOLVE

From right-hand rule number 1, the magnetic field produced by the top wire is directed into the plane of the drawing near the bottom wire as shown in Figure 20.22B. (See also Fig. 20.6.) Applying right-hand rule number 2 then gives a magnetic force on the bottom wire directed upward, and the bottom wire is *attracted* to the top wire.

Now consider the force on the top wire. The magnetic field produced by the bottom wire in the vicinity of the top wire is directed out of the plane of the drawing (Fig. 20.22C), and applying right-hand rule number 2 shows that the force exerted on the top wire is directed downward, toward the bottom wire. Hence, the force on the top wire is downward and the force on the bottom wire is upward. Two wires carrying parallel currents are attracted to each other.

What does it mean?

The magnetic forces exerted by each wire on the other are an action–reaction pair of forces. Hence, in accord with Newton's third law (the action–reaction principle), these forces must be in opposite directions, which is precisely what we have found. You should expect that these forces must have equal magnitudes; we'll show this in Section 20.7.

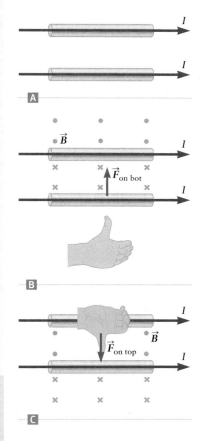

Figure 20.22 Example 20.4. A Two parallel current-carrying wires. B The magnetic force on the bottom wire. C The magnetic force on the top wire.

CONCEPT CHECK 20.6 | How Does the Force Depend on the Direction of I?

Consider again the two wires in Figure 20.22 where the currents are parallel and both directed to the right. Suppose the currents are instead in opposite directions, with I_{top} to the right and I_{bottom} to the left. Which of the following statements is correct?
  (a) The force on the bottom wire is still upward in Figure 20.22, but the force on the top wire is now also upward.
  (b) The force on the bottom wire is now downward and the force on the top wire is upward, so the overall effect is a repulsive force.
  (c) There is no change.

Figure 20.23 Torque on a current loop. **A** The magnetic forces on the sides of this loop tend to rotate the loop clockwise as viewed from the front, along the axis. **B** When the plane of the loop is perpendicular to $\vec{B}$, the torque is zero. **C** The torque on a current loop depends on the angle θ the field makes with the direction perpendicular to the loop.

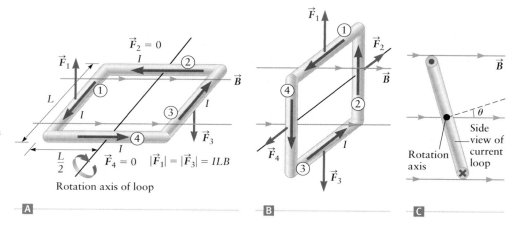

20.5 | TORQUE ON A CURRENT LOOP AND MAGNETIC MOMENTS

In Section 20.2, we saw that when a bar magnet is placed in a uniform magnetic field, there is a torque on the magnet (Fig. 20.11). A magnetic field can also produce a torque on a current loop. Figure 20.23 shows a square current loop with sides of length L carrying a current I in a constant magnetic field. In Figure 20.23A, the field is parallel to the plane of the loop and the forces on all four sides of the loop are shown. These forces can all be found using right-hand rule 2. In sides 2 and 4, the current is either parallel or antiparallel to $\vec{B}$, so the forces are zero. The forces on sides 1 and 3, denoted $\vec{F}_1$ and $\vec{F}_3$, respectively, are in opposite directions in Figure 20.23A and produce a torque around the loop's axis. The lever arms for these two forces are both $L/2$, and magnitude of this torque is

$$\tau = F_1\left(\frac{L}{2}\right) + F_3\left(\frac{L}{2}\right)$$

The force on each wire segment is $F_1 = F_3 = ILB$ (from Eq. 20.11), so

$$\tau = F_1\left(\frac{L}{2}\right) + F_3\left(\frac{L}{2}\right) = 2ILB\left(\frac{L}{2}\right)$$

$$\tau = IL^2B \tag{20.12}$$

Figure 20.23B shows the forces on the loop when its plane is perpendicular to the magnetic field. Now there is a nonzero force on all four sides of the loop; the four forces are of equal magnitude $F_1 = F_2 = F_3 = F_4 = ILB$ but are in different directions, and the total force on the loop is zero. The total torque is now also zero. Comparing parts A and B of Figure 20.23, we find that the torque on a current tends to align the plane of the loop perpendicular to $\vec{B}$.

So far, we have considered the torque when the plane of the loop is perpendicular or parallel to the field. When the angle between these two directions is θ (Fig. 20.23C), the torque is

$$\tau = IL^2B \sin\theta$$

The factor of L^2 is just the area of the square loop. For loops with different shapes, including circular, the torque on a loop of area A is

$$\tau = IAB \sin\theta \tag{20.13}$$

The torque on a current loop is very similar to the torque on a bar magnet; both the loop and the magnet tend to "line up" with the magnetic field. Both act as a ***magnetic moment***, with the direction of the magnetic moment either along the axis of the bar magnet or perpendicular to the current loop as shown in Figure 20.24.

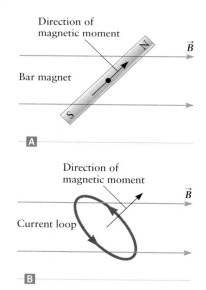

Figure 20.24 **A** The magnetic moment of a bar magnet is along the axis of the magnet, whereas **B** the magnetic moment of a current loop is perpendicular to the loop. In both cases, a magnetic field produces a torque that tends to align the magnetic moment with the field.

The magnetic torque tends to align the magnetic moment parallel to $\vec{B}$, and the strength of the torque depends on the magnitude of the magnetic moment. For a current loop, the magnitude of the magnetic moment equals IA, the product of the current and loop area.

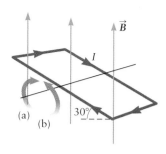

CONCEPT CHECK 20.7 | Torque on a Current Loop

Figure 20.25 shows a current loop in a magnetic field. Is there a torque on the loop and, if so, in what direction?
    (a) There is a torque that tends to rotate the loop in direction (a), clockwise.
    (b) There is a torque that tends to rotate the loop in direction (b), counterclockwise.
    (c) The torque is zero.

Figure 20.25 Concept Check 20.7.

20.6 | MOTION OF CHARGED PARTICLES IN THE PRESENCE OF ELECTRIC AND MAGNETIC FIELDS

We considered the magnetic force on a charged particle in a magnetic field in Section 20.3 and found that various types of motion, including circular and helical trajectories, are possible. Also of interest is the motion of an electric charge in the presence of a *combination* of magnetic and electric fields. We'll analyze a few such cases and describe some applications in this section.

The Mass Spectrometer

Scientists often want to separate ions according to their mass or charge. For example, archaeologists use a technique called carbon dating to determine an object's age. This technique is based on the observations that there are two isotopes of carbon atoms—denoted ^{12}C and ^{14}C (called "carbon 12" and "carbon 14," respectively)—and that the relative amounts of ^{12}C and ^{14}C in certain types of materials such as an animal bone can be used to determine the bone's age. (See Section 30.5.) One way to distinguish ^{12}C from ^{14}C is through the difference in their masses, which can be done using a mass spectrometer.

    Figure 20.26 shows the essential parts of a mass spectrometer. The charges are typically ions such as $^{12}C^+$ or $^{14}C^+$ that enter from the left, and for simplicity we'll assume they all have the same speed v. The ions then pass into a region in which the magnetic field is perpendicular to the velocity. In Figure 20.26, the magnetic field is out of the plane and perpendicular to the plane of the drawing, producing a magnetic force that causes the ions to move in a circle (as in Fig. 20.15B). The radius of this circle is (Eq. 20.6)

$$r = \frac{mv}{qB} \qquad (20.14)$$

where v is the velocity and q is the charge of the ion.

    Suppose the incoming ions are of two types, $^{12}C^+$ and $^{14}C^+$, with different masses m_{12} and m_{14}. According to Equation 20.14, the different values of m lead to different values of r, so the two types of ions follow different circular arcs through the mass spectrometer. A mass spectrometer has an ion detector positioned at a certain radius r as shown in Figure 20.26. The value of either v or B is then adjusted until the detector gives a maximum ion current, indicating that the ion's circular trajectory has the radius r of the detector. For our hypothetical example with $^{12}C^+$ and $^{14}C^+$, the mass spectrometer could be used to measure the ion currents for $^{12}C^+$ and $^{14}C^+$ separately and thereby determine the relative amounts of the two isotopes in an object, which would then be used to find the object's age.

    Another application of mass spectrometers is to "fingerprint" unknown molecules. Molecules of unknown composition are first ionized—by bombarding them

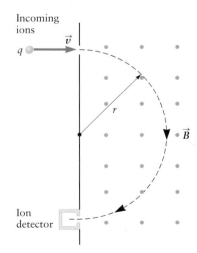

Figure 20.26 Design of a mass spectrometer.

with electrons, for instance—producing several different types of ions that are then fed to a mass spectrometer. The spectrometer then measures the values of v, B, and r corresponding to each ion. Equation 20.14 can be rearranged to give

$$\frac{q}{m} = \frac{v}{rB}$$

Because v, B, and r are all measured, the mass spectrometer gives q/m, the ratio of charge to mass for the ions. Chemists use this information to deduce the composition of each ion and hence the composition of the original molecule.

EXAMPLE 20.5 | ⊗ Using a Mass Spectrometer

A mass spectrometer (Fig. 20.26) is used in a carbon dating experiment. The incoming ions are a mixture of $^{12}C^+$ and $^{14}C^+$ having a velocity $v = 1.0 \times 10^5$ m/s, and the magnetic field is $B = 0.10$ T. The ion detector is first positioned to find the value of r for $^{12}C^+$ and then moved to find the value of r for $^{14}C^+$. How far must the detector move?

RECOGNIZE THE PRINCIPLE

The location of the detector depends on r, the radius of the ion's circular path. The velocity v and field B are given, and the ions $^{12}C^+$ and $^{14}C^+$ each have a charge $+e$. Equation 20.14 gives the value of r in terms of v, B, and m, the mass of the ion. If we can also find the mass m of each ion, we can therefore calculate r.

SKETCH THE PROBLEM

Figure 20.26 shows the problem. The ions follow a circular arc from where they enter the mass spectrometer to where they enter the detector.

IDENTIFY THE RELATIONSHIPS

The value of r in the mass spectrometer depends on the mass of the ion, so we need to find the masses of $^{12}C^+$ and $^{14}C^+$. The ion $^{12}C^+$ contains six protons, six neutrons, and five electrons. Consulting the table of fundamental constants on the inside front cover, we find a total mass of

$$m_{12} = 6m_p + 6m_n + 5m_e$$
$$m_{12} = 6(1.67 \times 10^{-27}\ \text{kg}) + 6(1.67 \times 10^{-27}\ \text{kg}) + 5(9.11 \times 10^{-31}\ \text{kg})$$
$$m_{12} = 2.0 \times 10^{-26}\ \text{kg}$$

A similar calculation gives the mass of $^{14}C^+$:

$$m_{14} = 2.3 \times 10^{-26}\ \text{kg}$$

SOLVE

From Equation 20.14, the radius in the mass spectrometer for $^{12}C^+$ is

$$r_{12} = \frac{m_{12}v}{qB}$$

and the charge is $q = +1.60 \times 10^{-19}$ C. Inserting the given values, we find

$$r_{12} = \frac{m_{12}v}{qB} = \frac{(2.0 \times 10^{-26}\ \text{kg})(1.0 \times 10^5\ \text{m/s})}{(1.60 \times 10^{-19}\ \text{C})(0.10\ \text{T})} = 0.13\ \text{m}$$

The same approach gives

$$r_{14} = 0.14\ \text{m}$$

The ion detector must move a distance equal to the difference in the diameters of the circular trajectories, so it must move a distance

$$\Delta r = 2(r_{14} - r_{12}) \approx \boxed{0.02\ \text{m}}$$

or about 2 cm.

Hall Effect: How Do We Know If the Charge Carriers Are Positive or Negative?

An electric current is produced by moving electric charges. In Chapter 19, we saw that the value of the current is proportional to the product of the charge q and velocity v of the charge carriers (Eq. 19.6), or

$$I \propto qv \qquad (20.15)$$

Suppose we reverse the direction of the velocity, changing from $+v$ to $-v$, and at the same time change the charge from $+q$ to $-q$. The new current is

$$I_{\text{new}} \propto (-q)(-v) = qv$$

which equals the old value. A particular value of the current can thus be produced by a positive charge moving to the right (Fig. 20.27A) or by a negative charge moving to the left (Fig. 20.27B). Because the current in a metal is carried by electrons (negative charges), the picture in Figure 20.27B should apply, but how do we really know? Is there some measurement we can perform that distinguishes between these two possibilities?

The answer to this question was given more than a century ago by Edwin Hall.[1] He suggested placing a current-carrying wire in a magnetic field directed perpendicular to the current. Suppose the current is produced by positive charges moving to the right (Fig. 20.28A). If $\vec{B}$ is directed into the plane of the drawing, the magnetic force on the moving charges is directed upward, causing the positively charged carriers to be deflected toward the wire's upper edge. The result is an excess of positive charge on the top edge of the wire and a deficit of positive charge on the bottom, giving an electric potential difference between the top and bottom edges. In Figure 20.28A, the upper edge has a positive electric potential relative to the bottom, which can be measured by attaching a voltmeter to the wire's top and bottom edges.

Now consider what happens if the current is carried instead by negative charges (electrons) moving to the left (Fig. 20.28B). Applying right-hand rule number 2 for negative charges, we find that the magnetic force is again directed toward the top edge of the wire. The magnetic force is thus in the *same* direction as with the positive charge carriers in Figure 20.28A, and the electrons in Figure 20.28B are

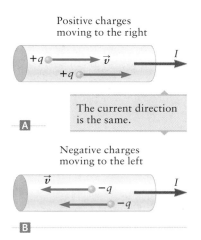

Figure 20.27 The Hall effect gives a way to measure if a current is caused by the motion of Ⓐ positive charges traveling to the right or Ⓑ negative charges traveling to the left. In both cases, the current is directed to the right.

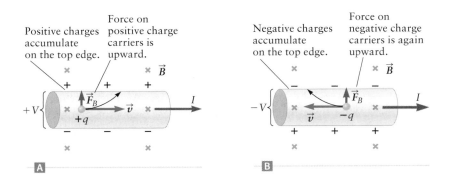

Figure 20.28 Hall's experiment. Motion of electric charges in a wire if the current is carried by Ⓐ positive charges traveling to the right, or Ⓑ negative charges traveling to the left.

---

[1] Hall had the idea for this experiment when he was a student! This question is still a good homework problem.

deflected toward the upper edge of the wire, leading to a *negative* potential on that edge relative to the bottom.

Hall's experiment thus gives a way to determine the sign of the charge carriers: we simply measure the potential difference between the two edges of the wire. Hence, we can indeed distinguish between a positive charge moving in one direction and a negative charge moving in the opposite direction. That is how we know that the current in metals such as copper and gold is carried by electrons.

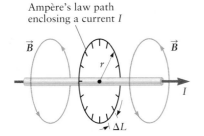

Figure 20.29 Ampère's law relates the magnetic field along a closed path to the current enclosed by that path. The path shown here is a circle of radius r.

Ampère's law

20.7 | CALCULATING THE MAGNETIC FIELD: AMPÈRE'S LAW

In Chapter 17, we discussed two approaches for calculating the electric field, Coulomb's law and Gauss's law. There are also several ways to calculate the magnetic field produced by a current. One approach, which treats each small piece of a wire as a separate source of $\vec{B}$, is similar in spirit to Coulomb's law but is mathematically too complicated for us to deal with here. Another approach, which uses *Ampère's law*, is very useful when the magnetic field lines have a simple symmetry. This approach is similar to Gauss's law for electric fields, which is most useful when the electric field is highly symmetric.

Consider a long, straight current-carrying wire encircled by magnetic field lines as shown in Figure 20.29, and imagine traveling around a closed path that also encircles the wire. Ampère's law relates the magnetic field *along* this path to the electric current *enclosed by* the path. Let's travel along the path in Figure 20.29 taking steps of length ΔL, and let $B_\parallel$ be the component of the magnetic field parallel to these steps. Ampère's law states that

$$\sum_{\substack{\text{closed} \\ \text{path}}} B_\parallel \Delta L = \mu_0 I_{\text{enclosed}} \tag{20.16}$$

where I_{enclosed} is the total current passing through the surface surrounded by the chosen closed path. The quantity μ_0 is a constant of nature called the *permeability of free space* whose value is

$$\mu_0 = 4\pi \times 10^{-7} \text{ T} \cdot \text{m/A} \tag{20.17}$$

Ampère's law thus relates the magnetic field *on the perimeter* of a region to the current that *passes through* the region. It is similar to Gauss's law for the electric field (Chapter 17), which relates the electric field *on a surface* to the electric charge *enclosed by* the surface.

Applying Ampère's Law: The Magnetic Field Produced by a Straight Wire

Although Ampère's law has a fairly simple mathematical form, it can be hard to apply in some cases because the sum $\sum B_\parallel \Delta L$ on the left-hand side of Equation 20.16 involves the field at all points along the Ampère's law path. If the value of $B_\parallel$ varies as one moves along the path, Ampère's law can be impossible to apply in practice. On the other hand, in some cases $B_\parallel$ is constant all along the chosen path; Ampère's law is then very handy.

Consider the magnetic field near a long, straight wire. We have already discussed how the magnetic field lines in this case form circles. From the symmetry in Figure 20.29, $B_\parallel$ must have the same value at all points along a circular path that is centered on the wire. We can thus write Ampère's law as

$$\sum_{\substack{\text{closed} \\ \text{path}}} B_\parallel \Delta L = B_\parallel \sum_{\substack{\text{closed} \\ \text{path}}} \Delta L = \mu_0 I_{\text{enclosed}}$$

The key here is that $B_\parallel$ is the same all along the path, so we can move it in front of the summation. If the circular path has a radius r, the total length of the path is $\sum_{\text{closed path}} \Delta L = 2\pi r$ and Ampère's law gives

$$\sum_{\substack{\text{closed} \\ \text{path}}} B_{\parallel} \Delta L = B_{\parallel} \sum_{\substack{\text{closed} \\ \text{path}}} \Delta L = B_{\parallel}(2\pi r) = \mu_0 I_{\text{enclosed}} \qquad (20.18)$$

The right-hand side of Ampère's law involves the total current passing through the chosen surface, which is just the current I in the wire. Inserting that into Equation 20.18 gives

$$B_{\parallel}(2\pi r) = \mu_0 I$$

Although $B_{\parallel}$ is the component of the field along the circular path, in this case it is also the total field, so we can denote it simply as B and find

$$B = \frac{\mu_0 I}{2\pi r} \qquad (20.19)$$

Magnetic field near a long, straight wire

With this result, we now have the full solution for the magnitude and direction of the magnetic field produced by a long, straight wire.

Let's use the result in Equation 20.19 to find the magnitude of the field produced by a current in a typical case. Suppose the current is $I = 1$ A (a common value in household appliances) and we are a distance $r = 1$ cm $= 0.01$ m from the wire. Inserting these values in Equation 20.19, we have

$$B = \frac{\mu_0 I}{2\pi r} = \frac{(4\pi \times 10^{-7} \text{ T} \cdot \text{m/A})(1 \text{ A})}{2\pi(0.01 \text{ m})} = 2 \times 10^{-5} \text{ T}$$

which is about half as large as the Earth's magnetic field. A substantial current of 1 A thus produces only a modest magnetic field, even at this small distance from the wire. This result for B also suggests that the direction of a compass needle can be noticeably affected by the magnetic field from the current in a nearby wire.

EXAMPLE 20.6 | Force between Two Current-Carrying Wires

Consider two long, straight, parallel wires each carrying a current $I = 2.0$ A, with the currents in the same direction. **(a)** Find the magnetic field at one wire produced by the other wire if the wires are separated by a distance $r = 1.0$ mm. **(b)** What is the magnitude of the magnetic force exerted by one of these wires on a 1.0-m-long section of the other wire?

RECOGNIZE THE PRINCIPLE

To find the force, we must first find the magnetic field produced by one wire at the location of the other. We can find the magnitude of B using Ampère's law applied to a wire with a current I (Fig. 20.29). Once we have B, we can use our previous result for the magnetic force on a current.

SKETCH THE PROBLEM

Figure 20.22B describes the problem; it shows the current directions and the magnetic field produced by the top wire.

IDENTIFY THE RELATIONSHIPS AND SOLVE

(a) From our Ampère's law result (Eq. 20.19), the magnetic field due to the top wire is

$$B = \frac{\mu_0 I_{\text{top}}}{2\pi r} \qquad (1)$$

where I_{top} is the current in the top wire. Inserting the given values of I_{top} and $r = 1.0$ mm, we get

$$B = \frac{\mu_0 I_{\text{top}}}{2\pi r} = \frac{(4\pi \times 10^{-7} \text{ T} \cdot \text{m/A})(2.0 \text{ A})}{2\pi(1.0 \times 10^{-3} \text{ m})} = \boxed{4.0 \times 10^{-4} \text{ T}}$$

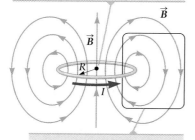

$B = \dfrac{\mu_0 I}{2R}$ at the center of the loop

$\vec{B}$

$\vec{B}$

R

I

$B_\parallel$ is not constant on this path (or on any other closed path around a current loop).

Figure 20.30 In some cases, such as this current loop, Ampère's law does not give a (mathematically) simple way to calculate $\vec{B}$. For a circular current loop, the magnitude of the magnetic field at the center of the loop is $B = \mu_0 I / 2R$.

Magnetic field at the center of a current loop

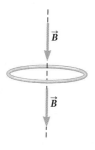

I

$\vec{B}$

I

$\vec{B}$

Figure 20.31 Concept Check 20.8.

(b) The field from the top wire is into the plane of the drawing at the location of the bottom wire (Fig. 20.22B), so the angle between $\vec{B}$ and the current in the bottom wire is $\theta = 90°$. The force on the bottom wire is (Eq. 20.11)

$$F_{\text{on bot}} = I_{\text{bot}} L B \sin \theta$$

where I_{bot} is the current in the bottom wire and $L = 1.0$ m is the section of the bottom wire on which we are asked to find the force. Substituting for B from part (a) and inserting $\sin \theta = 1$ (since $\theta = 90°$) gives

$$F_{\text{on bot}} = I_{\text{bot}} L B \sin \theta = (2.0 \text{ A})(1.0 \text{ m})(4.0 \times 10^{-4} \text{ T})(1) = \boxed{8.0 \times 10^{-4} \text{ N}}$$

What does it mean?
Although these two wires carry substantial currents and are quite close together, the magnetic force between them is still very modest. The wires in a typical household appliance or in the walls of your house carry similar currents and are sometimes spaced this closely. Our result for the force on the bottom wire shows that the magnetic force will not do any damage in such a case.

Field from a Current Loop

The mathematics of Ampère's law is manageable when one can find a path along which the magnetic field is constant. Unfortunately, that is not possible in many cases such as the current loop sketched in Figure 20.30, which shows the pattern of field lines described in Section 20.1. There is no closed path on which $B_\parallel$ (the component of the magnetic field along the path) is constant, so there is no simple way to apply Ampère's law to find $B_\parallel$ in this situation. Other methods for calculating $\vec{B}$ can be applied, and it is found that the field at the center of a circular current loop of radius R is

$$B = \frac{\mu_0 I}{2R} \tag{20.20}$$

The field at the center of the loop is perpendicular to the plane of the loop (Fig. 20.30) as given by right-hand rule number 1. The field lines encircle the loop as we already saw in Figure 20.9.

| **CONCEPT CHECK 20.8** | **Direction of the Field Near a Current Loop**

The current in the circular loop in Figure 20.31 produces a magnetic field on the axis that points downward as shown. As viewed from above, is the direction of the current in the loop (a) clockwise or (b) counterclockwise?

Field Inside a Solenoid

The pattern of magnetic field lines produced by a circular current loop is very similar to the field produced by a bar magnet. One handy feature of the current loop is that its field is adjustable simply by varying the current I in the loop. One disadvantage is that for reasonable values of I the field of a single current loop is much smaller than the field of a bar magnet, but this disadvantage can be overcome by stacking many current loops together. Figure 20.32A shows such a stack of current loops and the pattern of field lines they produce. This field can be understood in terms of the field of a single current loop (Fig. 20.30) along with the principle of superposition. By stacking the loops very close along the vertical axis, the field along this axis is much larger than the field of a single loop.

Usually, it is not practical to stack many separate current loops. Instead, a long piece of wire is wrapped in a very tight helix as shown in Figure 20.32B. Each loop of the helix forms a nearly circular loop, so each turn of the helix produces a field

Stacking many current loops produces a larger magnetic field than that of a single loop.

A

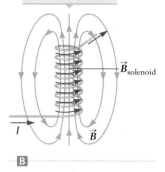

A tightly wound coil of wire is called a solenoid.

$\vec{B}_{solenoid}$

I

$\vec{B}$

B

Figure 20.32 **A** Stacking many current loops produces a magnetic field similar to that of a bar magnet. **B** Winding a wire in a "tight" helix produces a field similar to that in **A**.

Insight 20.3
AMPÈRE'S LAW IS ALWAYS TRUE (BUT MAY BE DIFFICULT TO APPLY).
In Figure 20.30, there is no path around the current loop for which the magnetic field component $B_{\parallel}$ is constant along the entire path. For this reason, Ampère's law does not lead to a simple way of finding the magnetic field in this case, but it *is still true* in such cases. Even though $B_{\parallel}$ may vary in a complicated fashion as one moves along a closed path, the left-hand side of Ampère's law (Eq. 20.16) is always equal to μ_0 multiplied by the total current through the surface enclosed by the path.

nearly identical to the circular loops in Figure 20.32A. Such a helical winding of wire is called a *solenoid*. The field lines circulate around the outside of the solenoid, leaving from one end and reentering at the other (Fig. 20.32B). Outside the solenoid, the field lines "expand" to fill a very large volume, so $\vec{B}$ is very much smaller outside than inside the solenoid. For a very long solenoid, it is a good approximation to assume the field is constant inside and zero outside. With this approximation, let's apply Ampère's law along the path shown in Figure 20.33. This path is a rectangle with one side (part 1) inside the solenoid and parts 2, 3, and 4 outside. We therefore split the sum on the left-hand side of Ampère's law into four parts:

$$\sum_{\substack{closed \\ path}} B_{\parallel} \Delta L = \sum_{part\,1} B_{\parallel}(1)\,\Delta L + \sum_{part\,2} B_{\parallel}(2)\,\Delta L + \sum_{part\,3} B_{\parallel}(3)\,\Delta L + \sum_{part\,4} B_{\parallel}(4)\,\Delta L$$

$$(20.21)$$

where $B_{\parallel}(1)$ is the field along part 1, and so on. Because the field outside the solenoid is approximately zero, $B_{\parallel}(2) \approx B_{\parallel}(3) \approx B_{\parallel}(4) \approx 0$ and we get

$$\sum_{\substack{closed \\ path}} B_{\parallel} \Delta L = \sum_{part\,1} B_{\parallel}(1)\,\Delta L \qquad (20.22)$$

Here, $B_{\parallel}(1)$ is just the field inside the solenoid, which is a constant along part 1 of the path; we can denote it as $B_{solenoid}$. If the solenoid has a length L, Equation 20.22 leads to

$$\sum_{\substack{closed \\ path}} B_{\parallel} \Delta L = \sum_{part\,1} B_{inside}\,\Delta L = B_{solenoid} \sum_{part\,1} \Delta L = B_{solenoid}L$$

Inserting this result into Ampère's law (Eq. 20.16) gives

$$B_{solenoid}L = \mu_0 I_{enclosed} \qquad (20.23)$$

where $I_{enclosed}$ is the total current passing through the surface enclosed by our Ampère's law path. A helical coil with N windings cuts through this surface N times, so if I is the current through the coil, then $I_{enclosed} = NI$. Inserting into Equation 20.23 leads to

$$B_{solenoid}L = \mu_0 NI$$

Solving for $B_{solenoid}$, we get

$$B_{solenoid} = \frac{\mu_0 NI}{L} \qquad (20.24)$$

This result for $B_{solenoid}$ is a good approximation for a solenoid whose length L is much greater than its diameter.

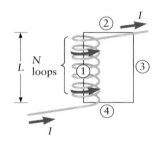

Figure 20.33 Application of Ampère's law to find the magnetic field inside a solenoid; we use the black rectangle as our Ampère's law path. The field outside a solenoid is very small (approximately zero), so only the inside field contributes to the sum of the field components in Ampère's law (Eq. 20.21).

Magnetic field inside a long solenoid

EXAMPLE 20.7 | Designing a Solenoid

You are given the task of designing a solenoid that must produce a field in its interior that is 20 times larger than the Earth's field ($B_{Earth} \approx 5 \times 10^{-5}$ T). This solenoid is to be 10 cm long and 1.0 cm in diameter, with $N = 500$ turns. What current must the wire be able to carry without failing?

RECOGNIZE THE PRINCIPLE

We have already used Ampère's law to find the field inside a solenoid (Eq. 20.24). We can use that result to find the current required to give the desired field in our solenoid.

SKETCH THE PROBLEM

Figure 20.33 describes our solenoid.

IDENTIFY THE RELATIONSHIPS

Equation 20.24 gives $B_{solenoid}$ in terms of the number of turns N, the length L, and the current I. Setting $B_{solenoid}$ equal to $20 \times B_{Earth}$, we find

$$B_{solenoid} = \frac{\mu_0 N I}{L} = 20 B_{Earth}$$

SOLVE

Solving for I and inserting the given values of the various quantities gives

$$I = \frac{L(20 B_{Earth})}{\mu_0 N} = \frac{(0.10 \text{ m})(20)(5 \times 10^{-5} \text{ T})}{(4\pi \times 10^{-7} \text{ T} \cdot \text{m/A})(500)} = \boxed{0.16 \text{ A}}$$

What does it mean?

This current can be safely carried by a wire with a diameter no smaller than about 0.2 mm. A tightly wrapped helix with 500 turns each 0.2 mm thick would be about 10 cm long, so this solenoid could indeed be built. If, however, we wanted a solenoid of this size to produce a field as large as that found near a common bar magnet (for which $B = 1$ T), a *much* larger current would be required. In fact, the current required in that case would probably melt the wire of the solenoid! For this reason, it is hard to use a solenoid like this one to obtain a field as large as can be attained near a typical bar magnet. A redesign of the solenoid is necessary, a problem we will consider further in the next section.

20.8 | MAGNETIC MATERIALS: WHAT GOES ON INSIDE?

We have mentioned several times that magnetic poles *always come in pairs. It is not possible to separate the poles of a magnet.* To understand why magnetic poles always come in pairs, we must consider the atomic origin of permanent magnetism. We have already noted that the pattern of magnetic field lines produced by a bar magnet is similar to the field produced by a current loop. The next example compares the magnitude of the field of a typical permanent magnet with the field near a current loop.

EXAMPLE 20.8 | Comparing the Magnetic Fields of a Permanent Magnet and a Current Loop

Let's calculate how much current is required to make the field at the center of a current loop comparable to the magnetic field near one of the poles of a permanent magnet. The field near a typical bar magnet is about $B_{\text{bar mag}} = 1$ T. Consider a single circular wire loop of radius $R = 0.50$ cm carrying a current I. What value of I is required to produce a field equal to $B_{\text{bar mag}}$ at the center of the loop?

RECOGNIZE THE PRINCIPLE

The magnetic field at the center of a current loop depends on the current and the radius of the loop and is given by Equation 20.20. We can apply this relation to a current loop with the given radius and then find the value of the current needed to give a field equal to $B_{\text{bar mag}}$.

SKETCH THE PROBLEM

Figures 20.2 and 20.9 show the fields of a bar magnet and a current loop.

IDENTIFY THE RELATIONSHIPS

The magnetic field at the center of the loop is (Eq. 20.20)

$$B = \frac{\mu_0 I}{2R}$$

Setting this result equal to $B_{\text{bar mag}}$ and rearranging to solve for I, we get

$$I = \frac{2R B_{\text{bar mag}}}{\mu_0}$$

SOLVE

Inserting the given values for R and $B_{\text{bar mag}}$ gives

$$I = \frac{2(0.0050 \text{ m})(1 \text{ T})}{4\pi \times 10^{-7} \text{ T} \cdot \text{m/A}} = \boxed{8000 \text{ A}} \tag{1}$$

What does it mean?
This current is *very* large. A current loop made from a copper wire of any reasonable size would not be able to safely carry this much current; the wire would melt instead. We have already suggested that a permanent magnet contains atomic-scale current loops. It is amazing that the atomic-scale currents can be this large.

Permanent Magnets: A Microscopic View

How can a permanent magnet generate such a strong magnetic field, especially when no obvious current is present? Atoms consist of a nucleus surrounded by one or more electrons, and in a classical picture the electrons are in constant motion (as in the Bohr model of the atom; Chapter 29). The motion of an electron around a nucleus can be pictured as a tiny current loop, with a radius of approximately the radius of the atom (Fig. 20.34A). The direction of the resulting magnetic field for a single atom is determined by the orientation of the current loop along with right-hand rule number 1 (Fig. 20.9).

An electron also produces a magnetic field due to an effect called *electron spin.* It is useful to imagine the electron as a spinning ball of negative charge. A spinning charge acts as a circulating electric current, producing a magnetic field as shown in Figure 20.34B. In analogy with the properties of a current loop, we also say that the electron has a *spin magnetic moment* just as the current loop in Figure 20.24 has a magnetic moment. And, like a bar magnet, when an electron is placed in a magnetic field it will tend to align its spin magnetic moment with the magnetic field.

This picture of an electron is very simplified; a correct explanation of electron spin requires quantum theory, which confirms that an atom can produce a magnetic field in the two ways sketched in Figure 20.34: (1) through the electron's orbital

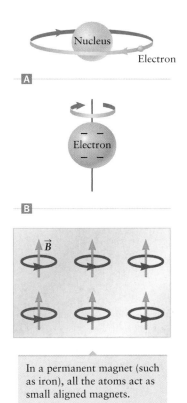

In a permanent magnet (such as iron), all the atoms act as small aligned magnets.

Figure 20.34 Ⓐ An electron in an atom acts as a small current loop, producing a magnetic field. Ⓑ A spinning charged particle also acts as a current loop, with the different regions of charge circulating around the rotation axis. Ⓒ The total magnetic field of a permanent magnet is the sum of the fields from parts A and B and from the alignment of the fields produced by many atoms.

The net effect is the same as that of one large current loop around the perimeter.

Figure 20.35 A collection of aligned atomic current loops in a "slice" of magnetic material is equivalent to one large current loop around the outside and hence one large magnet.

North and south magnetic poles cannot be separated.

current loop and (2) from the electron's spin. The total magnetic field produced by a single atom is the sum of these two fields.

Figure 20.34C shows a collection of atomic current loops, with each current loop representing a single atom. This collection of "small" current loops then acts as one "large" current loop (Fig. 20.35), producing the magnetic field of a magnetic material. In Example 20.8, we found that such a large loop must carry a very large current if it is to give the necessary field. Although the current in each atomic loop is very small, there are *many* atoms in a typical bar magnet, and the sum of all these atomic current loops does indeed give a net current similar to the value of I found in Example 20.8.

Lining Up the Atomic Magnets to Make a Permanent Magnet

Figure 20.34 shows how the electrons in an atom are capable of producing a magnetic field. Not all atoms will actually be magnetic, however, because the current loops of different electrons may point in different directions so that their magnetic fields cancel. The total magnetic field produced by a solid depends on how all the atomic magnetic fields are aligned. In some solids, the atomic magnets are not aligned; their individual fields point in different directions and the total field is zero. In materials that form permanent magnets, the atomic magnetic fields align to give a net magnetic field. A permanent magnet thus consists of many atomic-scale "bar magnets" all pointing in the same direction, giving rise to a large magnetic field.

This picture of a permanent magnet as a collection of aligned atomic-scale current loops explains why it is not possible to isolate the north and south magnetic poles of a magnet. One might think that the north and south poles of a permanent magnet could be separated by simply cutting the magnet in half. Figure 20.36 shows why this approach fails. Although one can certainly cut a bar magnet in half, each of the resulting pieces still produces the magnetic field of a "complete" bar magnet and hence still possesses a north pole and a south pole. It is thus not possible to produce an isolated magnetic pole.

Induced Magnetism and Magnetic Domains

Although permanent magnets can be formed from certain materials such as iron and magnetite, not all pieces of iron or magnetite behave as bar magnets. How can

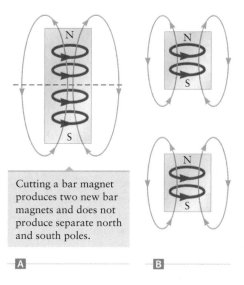

Cutting a bar magnet produces two new bar magnets and does not produce separate north and south poles.

Ⓐ   Ⓑ

Figure 20.36 Ⓐ Cutting a bar magnet in half Ⓑ produces two separate bar magnets.

Atomic magnets

The atomic magnets in different parts of a permanent magnet can point in different directions. Here the result is two side-by-side bar magnets.

Ⓐ   Ⓑ

Figure 20.37 Ⓐ If the atomic magnets within a permanent magnet are aligned in opposite directions, the magnetic field outside can be very small. Ⓑ It is equivalent to having two bar magnets side by side, but with opposite orientations. To an observer on the outside, this material would appear to be nonmagnetic.

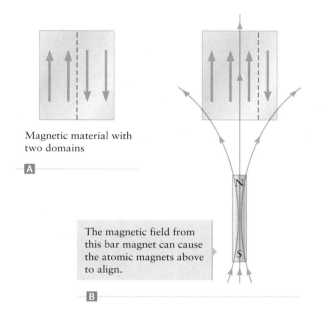

Magnetic material with
two domains

A

The magnetic field from
this bar magnet can cause
the atomic magnets above
to align.

B

Figure 20.38 **A** Oppositely aligned magnetic
domains. **B** A weak magnetic field can cause the
atomic magnets in a magnetic material to reorient.
Some domains then grow at the expense of the others.
The material at the top now behaves like a permanent
magnet.

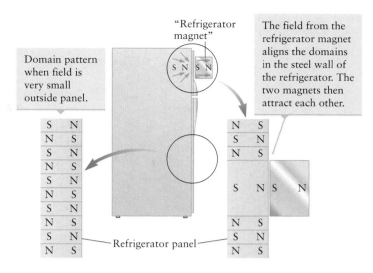

Domain pattern
when field is
very small
outside panel.

"Refrigerator
magnet"

The field from the
refrigerator magnet
aligns the domains
in the steel wall of
the refrigerator. The
two magnets then
attract each other.

Refrigerator panel

Figure 20.39 A "refrigerator magnet" is a permanent magnet
that "sticks" to a refrigerator panel through a magnetic force. The
refrigerator panel contains many domains oriented in opposite directions. The field of the refrigerator magnet aligns the domains so that
these two magnets then attract each other.

that be? Our atomic-scale view of a permanent magnet in Figure 20.34C shows the
atomic bar magnets all aligned in the same direction, but it is also possible for the
atomic magnets in different regions within a magnetic material to point in different directions. One possibility is shown in Figure 20.37A, which shows a material
in which the atomic magnets on one side all point up and those on the other side
all point down. This arrangement is equivalent to two bar magnets, side by side as
in Figure 20.37B. The magnetic field lines are also shown and are confined almost
totally within the material. An observer outside the material thus sees a very small
field, so this material would not appear to be a permanent magnet.

The two oppositely aligned regions in Figure 20.37 are called *magnetic domains*,
and they have some interesting properties. Figure 20.38A shows a material in which
the two domain regions are about equal in size. Application of a magnetic field from
a bar magnet can cause one of these regions to grow at the expense of the other. In
Figure 20.38B, the domain aligned with the magnetic field grows at the expense of
the other domain, and the material itself now acts like a bar magnet. This behavior
occurs whenever you stick a "refrigerator magnet" to your refrigerator. A refrigerator panel is made of steel and usually has magnetic domains that alternate in direction as shown in Figure 20.39. It is similar to the two oppositely directed domains
in Figure 20.37A, so your refrigerator appears to be nonmagnetic. When a magnet
is brought near, however, the domains in the refrigerator panel align (Fig. 20.39);
the panel now acts like a bar magnet and attracts the nearby magnet. That is why a
refrigerator magnet sticks to a refrigerator due to a magnetic force, even though the
refrigerator by itself does not appear to be magnetic.

20.9 | THE EARTH'S MAGNETIC FIELD

The largest magnet on the Earth is the Earth itself. About 400 years ago, William
Gilbert proposed that a compass needle orients along a north–south direction in

North geographic pole
(Earth's rotation axis)

South
magnetic pole

Compass needle aligns parallel to the Earth's magnetic field with its north pole pointing toward the north geographic pole.

North magnetic pole

South geographic pole

A

B

© Cengage Learning/Charles D. Winters

Figure 20.40 **A** Magnetic field of the Earth. **B** A compass needle is a small bar magnet and thus aligns parallel to the magnetic field. If there are no other sources of magnetic field nearby, the north end of a compass needle points toward the Earth's north geographic pole.

Figure 20.41 The Earth's magnetic poles move from year to year. The south magnetic pole (which is located near the north geographic pole) was near latitude 70° in northern Canada only 100 years ago.

response to the Earth's magnetic field (Fig. 20.40A). A compass needle (Fig. 20.40B) is a small bar magnet mounted so that it can swivel freely about its center. The Earth's magnetic field provides a torque on this magnet. From experiments with compass needles and some insightful reasoning, Gilbert suggested that the Earth acts as a very large magnet oriented as shown in Figure 20.40A.

A compass needle aligns with its north *magnetic* pole pointing approximately toward the Earth's north *geographic* pole, which is why the poles of a bar magnet were given the names "north" and "south." Because a north magnetic pole of one magnet (the compass needle) is attracted to the south magnetic pole of a second magnet, the Earth's *north geographic pole* is actually a *south magnetic pole*!

Gilbert was not able to explain what causes the Earth's magnetic field. That problem is still not completely solved, but several clues point to a qualitative explanation. First, we know that the location of the Earth's south magnetic pole does not quite coincide with the north geographic pole. In fact, the Earth's south magnetic pole actually moves slowly from day to day. Currently, it moves about 40 km/year, and within the last 100 years it has even been located in northern Canada (Fig. 20.41). Second, geological studies show that the Earth's field has completely reversed direction many times during the planet's history. The last such geomagnetic reversal occurred about 780,000 years ago. So, in the years immediately prior to that time, a compass needle would have pointed toward the Earth's *south geographic* pole.

This behavior of the Earth's magnetic field is probably caused by electric currents in the core. Recall (Chapter 12) that the Earth's core contains a liquid. This liquid conducts electricity, and the spin of the Earth about its axis causes the liquid to circulate much like the current in a conducting loop (Fig. 20.42). Circulation within the Earth's core is thought to have a complicated flow pattern that varies with time, and these variations are believed to cause the wandering of the Earth's magnetic poles. Scientists still do not have a complete understanding of these phenomena, however.

CONCEPT CHECK 20.9 | Direction of a Compass Needle

When we use a compass, we usually hold it "horizontally" so that the needle lies in a horizontal plane. However, the Earth's magnetic field has both horizontal and vertical components (Fig. 20.40). Your friend is an arctic explorer and pitches his tent directly over the Earth's south magnetic pole. Will his compass needle point (a) toward the Earth's north geographic pole, (b) downward, or (c) upward?

EXAMPLE 20.9 | ® Estimating the Electric Current in the Earth's Core

The strength of the Earth's magnetic field has an average value on the surface of about 5×10^{-5} T. Model this magnetic field by taking the Earth's core to be a current loop, with a radius equal to the radius of the core. What electric current must this current loop carry to generate the Earth's observed magnetic field? The Earth's core has a radius of approximately $R_{core} = 3 \times 10^6$ m (Fig. 20.42). *Hint*: Approximate the current in the core as a single current loop.

RECOGNIZE THE PRINCIPLE

To get an approximate value of the current, we treat the core as a simple current loop of radius of R_{core}; Equation 20.20 then gives the resulting magnetic field at the center of the loop. The field at the core's center will be larger than the field at the Earth's surface, but it will give an approximate value for the field seen by a compass at the surface.

SKETCH THE PROBLEM

Figure 20.42 shows the problem.

IDENTIFY THE RELATIONSHIPS

The field produced by our model current loop is

$$B = \frac{\mu_0 I}{2 R_{core}}$$

(Eq. 20.20). We can rearrange this equation to solve for the current:

$$I = \frac{2 R_{core} B}{\mu_0}$$

SOLVE

Inserting the radius of the core and setting B equal to the Earth's field (both given above), we find

$$I = \frac{2 R_{core} B}{\mu_0} = \frac{2(3 \times 10^6 \text{ m})(5 \times 10^{-5} \text{ T})}{4\pi \times 10^{-7} \text{ T} \cdot \text{m/A}} = \boxed{2 \times 10^8 \text{ A}}$$

What does it mean?

This current is enormous. A scientific explanation of the Earth's field must explain how such a large current is maintained. Physicists still do not have a full solution for this problem.

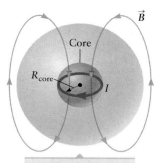

Figure 20.42 The Earth's field is believed to be caused by a current in the liquid core.

Cosmic Rays and the Earth's Magnetic Field

We usually think of "space"—the regions far above the Earth's atmosphere between the Earth, the Moon, and the Sun—as empty, but these regions actually contain many charged particles, including electrons and protons, that are emitted by the

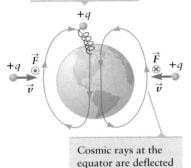

Cosmic ray particles at the poles spiral in along the field lines.

$+q$

$+q$ $\vec{F}$ $\vec{F}$ $+q$

$\vec{v}$ $\vec{v}$

Cosmic rays at the equator are deflected away from the Earth.

A

B

Figure 20.43 **A** The motion of a cosmic-ray particle is affected by the Earth's magnetic field. **B** The northern lights are caused by cosmic rays that ionize atoms and molecules as the cosmic rays travel through the atmosphere.

Sun or that reach us from outside our solar system. Such particles are called *cosmic rays*. Because they are charged, their motion is affected by the Earth's magnetic field. Figure 20.43A shows two positively charged cosmic-ray particles that approach the Earth's equator. Let's consider the magnetic force on the particle that approaches from the right. In this region, the magnetic field is directed parallel to the Earth's surface in a geographic south-to-north direction; applying right-hand rule number 2 shows that the magnetic force on this particle is directed perpendicular to the plane of this drawing and parallel to the Earth's surface in a west-to-east direction as shown in Figure 20.43A (into the plane of the drawing). Hence, the Earth's field deflects positively charged cosmic rays near the equator so that they tend to move in an easterly direction, away from the Earth's surface.

Cosmic-ray particles also bombard the Earth at the poles. In polar regions, the magnetic field is perpendicular to the surface, so these particles follow a helical path and spiral around the field lines (as in Fig. 20.15C). The Earth's magnetic field thus "steers" cosmic rays so that they tend to strike at the poles as opposed to locations at lower latitudes. As they travel through the atmosphere, these particles collide with and ionize air molecules, causing the molecules to emit light, much like the light in a fluorescent lightbulb. This phenomenon is observed as the *aurora borealis*, also called the "northern lights" (Fig. 20.43B). A similar phenomenon at the south geographic pole is called the *aurora australis* ("southern lights").

20.10 | APPLICATIONS OF MAGNETISM

So far, we have mentioned only a few applications of magnetism, including mass spectrometers and refrigerator magnets. In this section, we describe how magnetism is used by doctors, engineers, archaeologists, and even bacteria.

Blood-Flow Meters ⊗

An arrangement similar to the Hall effect experiment is used by surgeons to measure the blood velocity in arteries. Blood contains many ions that move at some speed v along the artery (Fig. 20.44A), just like the moving charges in the Hall effect conductor in Figure 20.28. When a magnetic field is directed perpendicular to the artery, the ions are deflected toward the sides of the artery, with positive ions deflected to one side and negative ions deflected to the opposite side (Fig. 20.44B). The result is an electric potential difference across the artery that can be measured with a voltmeter. Flowmeters based on this principle are used to measure the blood velocity. One disadvantage with these devices is that the voltmeter leads must be placed very close to the walls of the artery, so they are typically used in surgery when the artery is exposed. We'll consider the operation of a flowmeter and estimate the magnitude of the induced emf in Question 18 and Problems 92 and 93.

Figure 20.44 **A** Blood flowing through an artery contains many ions. **B** The magnetic force deflects the positive and negative ions to opposite sides of the artery. A blood-flow meter uses the electric potential difference across the artery to measure the speed of the blood.

A **B**

Figure 20.45 A relay uses the magnetic field from a solenoid to close a mechanical switch. This solenoid has been filled with a magnetic material whose domains are aligned by the field from the solenoid's current, producing a much larger field than is possible from the current alone.

In the figure: Spring, Steel bar (switch), Wire, $B = 0$, $I = 0$, "Off" (switch open), A — and — $\vec{F}$, $I \neq 0$, "On" (switch closed), B

Relays and Speakers

We have already seen how a solenoid (Fig. 20.32) produces a magnetic field when an electric current passes through it. In Example 20.7, we designed such an "electro-magnet" and found that even with a substantial current the field is rather modest. The field produced by a solenoid can be made much larger if the solenoid is filled with a magnetic material. The small field produced by the current can align the atomic magnets in the material (the magnetic domains described in Section 20.8), giving a large magnetic field due mainly to the magnetic material. In this way, the large field just outside the ends of a solenoid can be controlled with a relatively modest current.

The notion that a small magnetic field or current can control a large magnetic field is used to make an electromechanical switch called a **relay**, sketched in Figure 20.45. A relay is a type of "remote control" switch that uses a solenoid to exert a magnetic force on the movable part of a switch (a steel bar). A small current through the solenoid can thus control a much larger current through the switch. Although a switch is perhaps not a very "exciting" application, relays are widely used in a safety device called a ground fault interrupter (GFI), which is found in virtually all new homes; GFIs are discussed in more detail in Chapter 21.

A stereo speaker also employs a solenoidal coil wrapped around a magnetic material (Fig. 20.46). The current in the speaker's solenoid is the music "signal" played by the speaker. This current varies between positive and negative values according to the amplitude and frequency of the music. As the current varies, so does the magnetic alignment of the material inside the solenoid coil. Next to the solenoid is a stationary permanent magnet, so as the solenoid's magnetic poles vary back and forth (switching between north and south), the force on the solenoid varies and causes the solenoid to vibrate. The speaker "cone" is attached to the solenoid, and the vibrations of the cone produce sound.

Making an Electric Motor

In Figure 20.23, we described how a magnetic field can produce a torque on a current loop. If the loop is attached to a rotatable shaft, we have the "beginnings" of an electric motor. However, in a practical motor, the simple current loop is replaced by a solenoid containing a magnetic material (Fig. 20.47). This solenoid is similar to that used in a relay or a speaker (Figs. 20.45 and 20.46). When current is passed through the solenoid, the material inside forms a bar magnet, with the orientation of the poles determined by the direction of the current. The north and south poles of this magnet are attracted to the poles of a second stationary magnet (shaped like a horseshoe magnet), producing a torque on the solenoidal magnet and causing

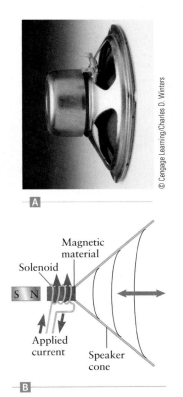

In figure B: Magnetic material, Solenoid, S N, Applied current, Speaker cone

Figure 20.46 A speaker contains a solenoid. The current through the solenoid varies with time according to the sound, producing a time-varying force on the solenoid. The speaker cone is attached to the solenoid, so both vibrate. The cone produces the sound that you hear.

Figure 20.47 A DC electric motor uses the torque on a solenoid magnet to cause a shaft to rotate.

Reverse current

A   B   C

the shaft to rotate (Fig. 20.47A). The torque on the solenoidal magnet is thus very similar to the torque on a current loop. The advantage of using the solenoid is that it gives a much larger torque, for the same reason that a bar magnet produces a much larger magnetic field than a current loop.

To make a working motor, we need to add one more "twist" to the design. When the solenoidal magnet rotates just past the horseshoe magnet, the direction of the current is reversed, switching the poles on the rotating magnet (Fig. 20.47B). These poles are now repelled from the poles of the horseshoe magnet, so the shaft continues to rotate. The current is reversed again every time the solenoid passes the poles of the horseshoe magnet, thus providing a continuing torque on the solenoidal magnet, and the motor's shaft continues to rotate.

Electric motors are a very important and widespread application of magnetism. A closely related application is the ***electric generator***. A generator produces an electric current by rotating a coil between the poles of a magnet; hence, a generator is a sort of "motor in reverse" as we'll discuss in Chapter 21. A combination of electric motor plus generator is at the heart of today's fuel-efficient "hybrid" automobiles such as the Toyota Prius and the Honda Insight.

Magnetic Bacteria ⊗

In addition to physicists and engineers, even bacteria seem to have an application for magnetism. ***Magnetotactic*** bacteria possess small grains of iron called magnetosomes, which are usually arranged as a "necklace" as shown in Figure 20.48. Magnetically, each grain acts like a bar magnet, and the north and south poles of the grains align just as found for bar magnets.

Bacteria use their magnetosomes to orient themselves with respect to the Earth's field. In the northern hemisphere, the Earth's field has a downward component as shown in Figure 20.40A; magnetic bacteria "prefer" to swim downward—toward the bottom of a lake, for instance—where the concentrations of oxygen and other nutrients are most favorable. It is believed that these bacteria use their magnetosomes to determine up from down.

Magnetic Dating

In Section 20.9, we mentioned that the direction of the Earth's field varies slightly from year to year. Occasionally, the direction even reverses completely. The most recent reversal occurred 780,000 years ago, and the dates of many previous reversals are also known (Fig. 20.49A). Archaeologists and geologists use these field reversals to "date" past events. When a volcano erupts, it produces molten rock, some of which solidifies to form magnetite and other magnetic materials. When molten, these materials are not magnetic because their atomic magnets point in random directions. Magnetite and other materials only become permanent magnets when they cool to normal surface temperatures. The orientation of these permanent magnets is set by the direction of the Earth's field at the time of cooling; hence, these materials record the direction of the Earth's field when they last cooled.

To understand how magnetic reversals are used for dating, consider the hypothetical deposits of material in Figure 20.49B. The material nearest the surface

Magnetosomes

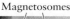

Figure 20.48 Some bacteria contain small grains of iron (shown here in yellow). It is believed that the bacteria use these iron magnets as a sort of compass to help determine the direction in which they swim.

© Dr. Dennis Kunkel/Phototake

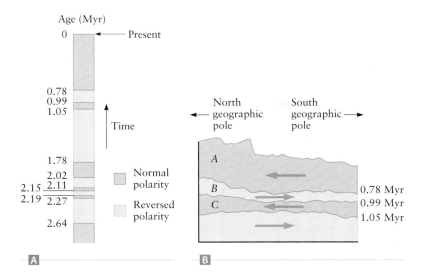

(layer A) is magnetized parallel to the current magnetic field lines, with $\vec{B}$ pointing toward the north geographic pole. As one digs deeper, one finds a layer (B) magnetized in the opposite direction and then a third layer (C) magnetized parallel to layer A. This layering can be understood in terms of the geomagnetic reversals in Figure 20.49A. The transition from layer A to layer B dates from the most recent reversal 780,000 years ago, whereas the transition from layer B to layer C occurred 990,000 years ago. Hence, archaeological artifacts found in layer B must have an age between 780,000 and 990,000 years.

The dates of more than 25 geomagnetic reversals going back more than five million years are known, so this approach can be used to date many artifacts of interest to archaeologists, anthropologists, and geologists.

20.11 | THE PUZZLE OF A VELOCITY-DEPENDENT FORCE

A remarkable feature of the magnetic force exerted on a moving charged particle is that it depends on the particle's velocity. Neither gravity (Newton's law of universal gravitation, Eq. 5.16) nor the electric force (Coulomb's law, Eq. 17.3 or 17.5) depend on velocity. The velocity dependence of the magnetic force has a very interesting consequence. Consider a particle of charge $+q$ moving with a velocity $\vec{v}$ perpendicular to a magnetic field as in Figure 20.50A. A stationary observer watching the particle's motion would see a force on the particle as indicated in the figure. (The observer would notice this force from the particle's acceleration.) This observer (observer 1) would explain the force using $F_B = qvB \sin\theta$ and right-hand rule number 2.

Now suppose this scene is viewed by a second observer (observer 2) who is moving at velocity $\vec{v}$. Observer 2 is moving along with the charged particle, so *relative to observer 2* the particle is not moving (Fig. 20.50B). This observer would also notice a force on the particle because she would also see that the particle is accelerated. But, according to this observer, the velocity of the particle is zero, so she would predict that the magnetic force on the particle is zero!

Our two observers would both agree that the particle is accelerated, but only observer 1 would say that there is a magnetic force acting on the particle. How does observer 2 interpret this scene? Such puzzles will always arise when the force depends on velocity. Einstein's theory of relativity provides the way out of this dilemma. In Chapter 27, we show that according to observer 2 there is an *electric field* present, so she will say that the force on the particle is an *electric force*. That is, observer 1 will claim that the force in Figure 20.50 is magnetic, whereas observer

Figure 20.50 When a charged particle moves through a magnetic field, a force is exerted on the particle. The interpretation of this force is different for different observers, however. A According to observer 1, the charge $+q$ is has a velocity $\vec{v}$ and therefore experiences a nonzero magnetic force given by Equation 20.1. B Observer 2 moves at the same velocity as charge $+q$, so relative to this observer this charge is at rest ($v = 0$). According to observer 2, the magnetic force on q is zero.

2 will measure a force of the same magnitude and direction but will claim that it is electric. There is thus a deep connection between electric and magnetic forces. This connection is a crucial part of the theory of *electromagnetism*, which we'll explore in the next chapter.

SUMMARY | Chapter 20

Sources of magnetic field
There are two ways to produce a *magnetic field*: a permanent magnet and an electric current. A bar magnet and a current loop produce similar patterns of magnetic field lines.

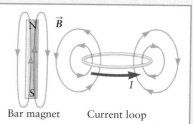

Bar magnet   Current loop

Permanent magnets and magnetic poles
The atoms in a permanent magnet material act as small bar magnets; their magnetic fields are produced by the circulating current of the atomic electrons and the spin magnetism of the electrons. A bar magnet contains a *north magnetic pole* and a *south magnetic pole*.

Ampère's law
Ampère's law can be used to calculate the magnetic field produced by a current. This law relates the field directed along a closed path on the edge of a surface with the current that passes through the surface:

$$\sum_{\substack{\text{closed} \\ \text{path}}} B_{\parallel}\, \Delta L = \mu_0 I_{\text{enclosed}} \qquad \textbf{(20.16)} \text{ (page 662)}$$

where B is measured in *teslas* (T). The constant μ_0 is the *permeability of free space* and has the value

$$\mu_0 = 4\pi \times 10^{-7}\ \text{T} \cdot \text{m/A} \qquad \textbf{(20.17)} \text{ (page 662)}$$

When applied to a long, straight wire carrying a current I, Ampère's law gives the field a distance r from the wire as

$$B = \frac{\mu_0 I}{2\pi r} \quad \text{magnetic field near a long wire} \quad \textbf{(20.19)} \text{ (page 663)}$$

Right-hand rule number 1
To find the direction of the field produced by a current,

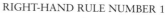
RIGHT-HAND RULE NUMBER 1

1. Place the thumb of your right hand along the direction of the current.
2. Curl your fingers; they will then give the direction of $\vec{B}$ as the field lines encircle the current.

(Continued)

Magnetic force on a moving charge

The magnitude of the magnetic force exerted on a charge moving with velocity $\vec{v}$ is given by

$$F_B = qvB \sin \theta \qquad \textbf{(20.1)} \text{ (page 651)}$$

where θ is the angle between $\vec{v}$ and $\vec{B}$.

Right-hand rule number 2

To find the direction of the magnetic force on a moving charge,

1. Point the fingers of your right hand along the direction of $\vec{v}$.
2. Curl your fingers in the direction of $\vec{B}$. Always curl through the smallest angle that connects $\vec{v}$ and $\vec{B}$.
3. If q is positive, the magnetic force on q is parallel to your thumb. If q is negative, the magnetic force is in the opposite direction.

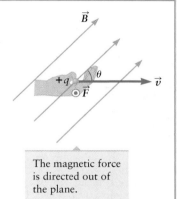

The magnetic force is directed out of the plane.

APPLICATIONS

Magnetic fields due to a current

The magnetic field at the center of a current loop is

$$B = \frac{\mu_0 I}{2R} \quad \text{magnetic field of a current loop} \quad \textbf{(20.20)} \text{ (page 664)}$$

A *solenoid* consists of N tightly wound turns of wire of finished length L. The field inside a solenoid is equal to

$$B_{\text{solenoid}} = \frac{\mu_0 NI}{L} \quad \text{magnetic field inside a solenoid} \qquad \textbf{(20.24)} \text{ (page 665)}$$

A permanent magnet material is sometimes inserted inside a solenoid; in this case, the total field will be the field due to the current (Eq. 20.24) plus the field from the permanent magnet.

Hall effect

The **Hall effect** involves a current-carrying wire that is placed in a magnetic field perpendicular to the current. The magnetic force on the current leads to an electric potential difference between the edges of the wire, which can be used to determine the sign (positive or negative) of the charges that carry the current.

Earth's magnetic field

The Earth is a large magnet, with a south magnetic pole located near the north geographic pole (and a north magnetic pole located near the south geographic pole). The Earth's field is produced by large, circulating electric currents in the core. The Earth's magnetic field aligns compass needles. The location of the Earth's magnetic poles varies with time, and throughout geological history (see Section 20.10) the Earth's field has reversed its direction many times.

The north geographic pole is near the south magnetic pole.

SSM = answer in Student Companion & Problem-Solving Guide

 = life science application

1. The photo in Figure Q20.1 shows a beam of electrons moving from right to left in a glass tube in which a vacuum has been established. This beam is deflected downward by a magnetic field. What is the direction of this magnetic field at the center of the tube?

Figure Q20.1

2. A long, straight, current-carrying wire is placed in a region where there is a magnetic field $\vec{B}$ that has a constant direction. If there is no magnetic force on the wire, what can you say about the direction of $\vec{B}$ relative to the wire?

3. Consider the moving cosmic-ray particle on the right in Figure 20.43A and suppose it has a negative charge. What is the direction of the magnetic force on this particle?

4. A positively charged particle moves with a velocity $\vec{v}$ in a region where there is a magnetic field $\vec{B}$ and a nonzero magnetic force $\vec{F}$ on the particle. Consider how these three vectors are oriented with respect to one another. (a) Which pairs of these vectors are always perpendicular? (b) Which pairs can have any angle between them?

5. Pairs of wires that allow current to flow into and out of a circuit are often twisted together. Figure Q20.5 shows a common type of cable with such twisted pairs. What is the advantage of pairing the wires that carry input currents with wires carrying output currents?

Figure Q20.5 An Ethernet cable.

6. SSM Make a qualitative sketch of the magnetic field near two bar magnets that are placed side by side as shown in Figure Q20.6.

Figure Q20.6

7. Figure Q20.7 shows the direction of $\vec{B}$ in a particular region of space. The density (i.e., spacing) of the crosses and dots indicate qualitatively the magnitude of $\vec{B}$. What is the possible source of this field?

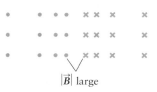

$|\vec{B}|$ large

Figure Q20.7

8. In our discussion of the solenoid in Figure 20.32B, we claimed that the field outside the solenoid is much smaller than the field inside and that it is a good approximation to assume the field outside a very long solenoid is zero. Explain how this approximation can be consistent with the property that magnetic field lines always form closed loops.

9. Two parallel wires are oriented perpendicular to the page as shown in Figure Q20.9. The wires carry equal currents, and the direction of the current is into the page for each. The vertices of an equilateral triangle are marked by the two wires and a point P as shown. (a) What is the direction of the magnetic field produced by the two wires at point P? (b) A third wire, parallel to the other two and carrying the same current, is placed at point P. Determine the direction of the force on the wire at this point.

I_1 ⊗

P •

I_2 ⊗

Figure Q20.9
Questions 9 and 10.

10. Repeat Question 9, but this time assume the current I_2 is directed out of the page and opposite to current I_1.

11. The medical technique called *magnetic resonance imaging* (MRI) uses the magnetic field produced by a large solenoid to obtain images from inside living tissue. These solenoids are large enough that a person can fit inside. Assuming the inside field is 1 T, design an MRI solenoid. Estimate the necessary length and radius of the solenoid assuming the wire can carry currents as large as 100 A. How many turns must this solenoid have to produce the desired field?

12. **The rail gun.** The simple configuration in Figure Q20.12 provides a means for launching projectiles, as long as the projectile can conduct electricity. Two copper rails are connected to a battery such that the current flows as indicated in the figure when a conducting rod is laid across the rails. What will happen to the conducting rod? If it begins to move, in what direction does it go? Consider (a) the direction of the magnetic field produced in the vicinity of the rod by the currents in the rails and rod and (b) the direction of the current in the rod. How could one make the rod move in the opposite direction?

Figure Q20.12

13. A bubble chamber is a device that makes the path of a charged particle visible. When a charged particle with sufficient kinetic energy passes through a substance such a liquid, it will knock electrons off the atoms in its path. If the liquid is at a tempera-

ture above its boiling point, these ionized atoms become the nucleation point for a bubble to form, thereby marking the path of the particle. For the images in Figure Q20.13, the magnetic field was directed out of the page. (a) Is the curve of the track in image 1 (on the left) consistent with that of a high-speed electron or of a high-speed proton? (b) The particle trajectory in image 1 spirals inward. Explain why. *Hint:* Each ion-creating encounter is a collision.

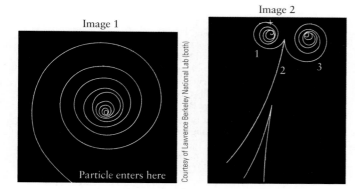

Image 1   Image 2

Particle enters here

Courtesy of Lawrence Berkeley National Lab (both)

Figure Q20.13 Questions 13 and 14, and Problem 98.

14. [SSM] The bubble chamber image on the right in Figure Q20.13 (image 2) shows three particles emanating from a point near the top of the image. The bubble chamber is in a magnetic field oriented so that the field lines point out of the page. (a) Identify the sign of the charge on each particle. (b) Rank the initial speeds of the particles from least to greatest. How do you know? Assume the particles all have the same mass.

15. The Earth's magnetic field is produced by the flow of electrons within its molten interior. If the flow is in the form of a ring of current in the plane of the equator, in which direction would the electrons have to flow to produce the observed polarity of the Earth's field?

16. Two 20-penny nails are wrapped with insulated wire in the shape of a solenoid. They are then laid down end to end with the heads almost touching as shown in Figure Q20.16. If a battery is connected to the solenoid of each nail as shown, will the nails be attracted or will they repel each other? Explain.

Figure Q20.16

17. Current is made to flow through a spring-shaped (helical) solenoid coil of thin wire. Will the loops of the coil tend to contract or tend to spread apart? Why?

18. ⊗ Blood plasma has a high content of Na^+ and Cl^- ions, and as the blood follows through a vessel, the ions travel along at the same speed. A magnetic field is applied such that the field is perpendicular to the vessel as shown in Figure Q20.18. (a) What effect will the magnetic field have on the distribution of positive ions? (b) What effect with the magnetic field have on how the negative ions are distributed? (c) Consider the overall distribution of charge in the presence of the magnetic field. Will an electric field be present? If so, in what direction is it oriented?

Figure Q20.18

19. Figure Q20.19 shows in part A an electric charge q moving past a stationary bar magnet and in part B a bar magnet moving past a stationary electric charge. In which case will there be a magnetic force on the charge? Will there be a force on the charge in the other case? Explain.

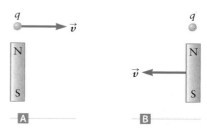

Figure Q20.19

20. A square current loop with current I is placed in a uniform magnetic field. Explain why the total force on the loop is zero, no matter how the loop is oriented relative to the field direction.

21. Figure Q20.21 shows a straight wire of length L carrying a current I in a magnetic field of magnitude B. In which case is the magnetic force on the wire largest? In which case is it smallest? Order the situations from largest to smallest according to the magnitude of the force.

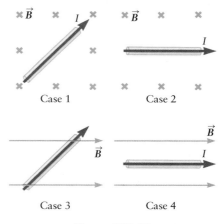

Figure Q20.21

22. Does a current-carrying wire placed in a magnetic field always experience a magnetic force? Explain.

23. A wire carries a current in the $-z$ direction. If there is a magnetic force in the $+x$ direction, what might be the direction of the magnetic field, (a) $+x$, (b) $-x$, (c) $+y$, (d) $-y$, (e) $+z$, or (f) $-z$?

SSM = solution in Student Companion & Problem-Solving Guide
★ = intermediate ✪ = challenging

✖ = life science application
Ⓡ = reasoning and relationships problem

20.1 SOURCES OF MAGNETIC FIELDS

1. The two long wires in Figure P20.1 carry parallel currents of equal magnitude I. Give the direction of the total magnetic field at points A, B, C, and D.

● A

● B

● C

● D

Figure P20.1 Problems 1 and 2.

2. Suppose the currents in Figure P20.1 are in opposite directions: the current in the top wire is from left to right and the current in the bottom wire is from right to left. Find the direction of the magnetic field at points A, B, C, and D.

3. Four long, straight wires each carrying a current I are oriented perpendicular to the plane of Figure P20.3, forming a square as shown. What is the direction of $\vec{B}$ at the center of the square (point A)?

Figure P20.3

4. An electron is traveling near a current-carrying wire as shown in Figure P20.4. If the magnetic force $\vec{F}_B$ is directed as shown, what is the direction of the current in the wire?

Figure P20.4

5. SSM ★ The magnetic field produced by the current loop in Figure P20.5 is along the $-z$ direction. What is the direction of the current in the loop? That is, is it clockwise or is it counterclockwise as viewed from above?

6. ✪ Use right-hand rule number 1 to find the direction of the magnetic field at point A near a wire that makes a right-angle bend as sketched in Figure P20.6.

Figure P20.5

Figure P20.6

20.2 MAGNETIC FORCES INVOLVING BAR MAGNETS

7. SSM ★ A bar magnet is placed in a uniform magnetic field (Fig. P20.7). (a) What is the direction of the total force on the bar magnet? (b) What is the direction (clockwise or counterclockwise) of the torque on the bar magnet about an axis perpendicular to the plane and passing through point A?

8. ★ In Figure P20.8, what is the direction of the force on bar magnet 3?

Figure P20.7

Figure P20.8

9. In Figure P20.9, bar magnet 1 is fixed in place and bar magnet 2 is free to move. (a) What is the direction of the total force on bar magnet 2? (b) What is the direction of the torque (clockwise or counterclockwise) on bar magnet 2 about an axis perpendicular to the plane of the drawing and through the bar magnet's center?

Figure P20.9

20.3 MAGNETIC FORCE ON A MOVING CHARGE

10. ★ The charged particle in Figure P20.10 has a velocity along $+y$ and experiences a force along the $+x$ direction. Does this particle have a positive charge or a negative charge?

11. A proton moves at a speed of 1000 m/s in a direction perpendicular to a magnetic field with a magnitude of 0.75 T. What is the magnitude of the magnetic force on the proton?

Figure P20.10

12. Repeat Problem 11, but now assume the particle is an electron.

13. A proton with $v = 300$ m/s is moving through a region in which the magnetic field is $B = 2.5$ T. If the magnitude of the force on the proton is 6.4×10^{-17} N, what angle does the proton's velocity make with $\vec{B}$?

14. ★ Ⓡ Consider a proton that approaches the Earth from outer space. If the proton is moving perpendicular to the Earth's surface near the equator, what is the direction of the magnetic force on the proton? The direction of the Earth's magnetic field is shown in Figure 20.40A.

15. ✪ An electron is moving in the x–y plane with its velocity along the $-y$ direction. If the magnetic force on the electron is in the $+z$ direction when $\vec{B}$ lies along one of the coordinate axes, what is the direction of the magnetic field?

16. ✪ A proton is at the origin and is moving in the $+z$ direction. What is the direction of the magnetic field due to the proton at a location on the x axis at $x = +1$ m?

17. [SSM] ✪ An electron is traveling near a current-carrying wire as shown in Figure P20.17. The electron's velocity is parallel to the $+x$ direction and the wire is parallel to the y axis. What is the direction of the force on the electron from the wire's magnetic field?

Figure P20.17

18. ✪ A particle of unknown charge has a velocity along the $+y$ direction. There is a constant magnetic field along the $-z$ axis, and it is found that the magnetic force on the particle is along $+x$. Does the particle have positive charge or a negative charge?

19. The particle in Figure P20.19 has a positive charge, and its initial velocity is to the right when it enters a region where $\vec{B}$ is directed into the plane of the drawing. Will the particle be deflected toward the top of the figure or toward the bottom?

Figure P20.19

20. ✪ The particle in Figure P20.20 has a negative charge, and its velocity vector lies in the x–y plane and makes an angle of 75° with the y axis. If the magnetic field is along the $+x$ direction, what is the direction of the magnetic force on the particle?

Figure P20.20

21. A particle of charge $q = 7.2\ \mu$C is moving at an angle of 25° with respect to the x axis with a speed of 250 m/s. There is a constant magnetic field of magnitude 0.55 T parallel to x. Find the magnitude of the magnetic force on the particle.

22. [SSM] ✪ A particle of charge $q = -8.3\ \mu$C has a velocity of 600 m/s that lies in the x–y plane and makes an angle of 65° with respect to the x axis as shown in Figure P20.22. If there is a constant magnetic field of magnitude 1.5 T parallel to y, what are the magnitude and direction of the magnetic force on the particle?

Figure P20.22

23. ✪ An alpha particle is a helium nucleus that consists of two protons bound with two neutrons. An alpha particle moving through a perpendicular magnetic field of 1.5 T experiences an acceleration of 2000 m/s². What is the alpha particle's speed?

24. ✪ Figure P20.24 shows an alpha particle (a helium atom, but without the two electrons) moving with a velocity in the $-x$ direction. If this particle experiences a force along the $+z$ direction, what can you conclude about the direction of $\vec{B}$?

Figure P20.24

25. ✪ An oxygen ion O^{-2} is moving with a speed of 300 m/s in a direction perpendicular to a magnetic field of magnitude B. If the acceleration of the O^{-2} ion is 1.5×10^9 m/s², what is B?

26. ✪ A particle of charge 23 μC and mass 1.3×10^{-16} kg moves at a speed of 2000 m/s, and its velocity vector lies in the x–y plane. There is a magnetic field of magnitude 0.88 T along the $+z$ direc-

tion. The magnetic force on the particle causes the particle to move in a circle that lies in the x–y plane. (a) Find the radius of this circle. (b) Find the period of the motion. (c) If the particle is moving parallel to $+y$ at $t = 0$, is the circular motion clockwise or is it counterclockwise as viewed from along the $+z$ axis?

27. ✪ Repeat part (c) of Problem 26, but now assume the particle's velocity is along the $-x$ direction at $t = 0$.

28. ✪ A proton with velocity $\vec{v}_A$ (which lies in the x–z plane) enters a region in which the magnetic field is along z as shown in Figure P20.28. Describe qualitatively the trajectory of the proton.

Figure P20.28 Problems 28–31.

29. ✪ Repeat Problem 28, but now assume the particle is an electron.

30. A proton has an initial velocity $\vec{v}_B$ when it enters a region in which $\vec{B}$ is parallel to the z direction (Fig. P20.28). (a) Show that the proton will move in a circle that lies in the x–y plane. (b) If $v_B = 500$ m/s and $B = 0.80$ T, what is the radius of this circle?

31. If the particle in Problem 30 is an electron instead of a proton, what is the radius of the trajectory?

32. ✪ Ⓡ A cosmic ray of charge $+5e$ approaches the Earth's equator with a speed of 2.2×10^7 m/s. If the acceleration of this particle is 8.0×10^{10} m/s², what is the approximate mass of the cosmic-ray particle?

33. Ⓡ A nitrogen ion N^+ is traveling through the atmosphere at a speed of 500 m/s. What is the approximate force on this ion from the Earth's magnetic field if the ion is moving perpendicular to the Earth's field?

34. ✪ A positively charged particle undergoes circular motion as a result of the magnetic force produced by a field that is perpendicular to the plane in Figure P20.34. Does this particle move clockwise or counterclockwise?

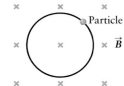

Figure P20.34

35. ✪ Ⓡ Cosmic rays undergo a helical (i.e., spiral) trajectory as they travel to the Earth. Consider a cosmic ray that is an iron ion Fe^+ traveling at 1×10^7 m/s. What is the approximate minimum radius of the helix when the ion is near the Earth's surface?

36. ✪ A bubble chamber is a device used to study the trajectories of elementary particles such as electrons and protons. Figure P20.36 shows the trajectories of some typical particles, one of which is indicated by a white arrow. Suppose this particle is moving in the direction indicated by the green arrow; the

Figure P20.36 Problems 36 and 37.

particle follows a curved path due to the presence of a magnetic field perpendicular to the plane of the photo. If $\vec{B}$ is directed into the plane, does this particle have a positive charge or a negative charge?

37. ✪ How would the trajectory in Figure P20.36 change if the charge on the particle were increased by a factor of 3? How would it change if the mass were decreased by a factor of 10?

38. ✪ A charged particle moves in the x–y plane, and a magnetic field B is directed into the plane as shown in Figure P20.38. The magnetic force on the particle causes the particle to move clockwise along a circle of radius R. (a) Does the particle have a positive charge or a negative charge? (b) The charge of the particle is increased by a factor of three, and the particle is found to move in a circle of radius R'. What is the ratio R'/R ? (c) The mass of the particle is decreased by a factor of four, and the particle is found to move in a circle of radius R'. What is the ratio R'/R ?

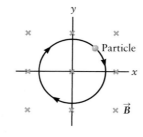

Figure P20.38

39. ✪ An electron and a proton are both moving in a circle in the x–y plane due to the force produced by a magnetic field perpendicular to the plane (Fig. P20.39). (a) If the electron moves clockwise and the proton moves counterclockwise, is the magnetic field directed into the plane or out of the plane of the drawing? (b) If the circular paths followed by the electron and proton have the same radii, what is the ratio of their speeds?

Figure P20.39

40. A positively charged particle moves as shown in Figure P20.40. Which statement best describes the direction of $\vec{B}$?
(a) It is into the plane of the figure.
(b) It is out of the plane of the figure.
(c) It is along the $+x$ direction.
(d) It is along the $+y$ direction.
(e) The direction of $\vec{B}$ varies as the particle moves.

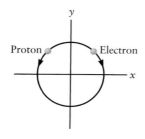

Figure P20.40

41. ✪ Figure P20.41 shows three identical particles, each with a positive charge and all moving at the same speed. In which case is $\vec{B}$ largest? Smallest? Assume $\vec{B}$ is perpendicular to the plane of the drawing.

Figure P20.41

42. ✪ A charged particle moves as shown in Figure P20.42. Does the particle have a positive charge or a negative charge?

Figure P20.42

43. ✪ Figure P20.43 shows two cases in which a positively charged particle is moving near a bar magnet. What is the direction of the magnetic force in each case?

Figure P20.43

44. ✪ Figure P20.44 shows an electron moving near a current loop. If the electron's velocity is parallel to the y axis, what is the direction of the magnetic force on the electron?

Figure P20.44

20.4 MAGNETIC FORCE ON AN ELECTRIC CURRENT

45. A long, straight wire of length 1.4 m carries a current of $I = 3.5$ A. If a magnetic field of magnitude $B = 1.5$ T is directed perpendicular to the wire, what is the magnitude of the force on the wire?

46. A long, straight wire carries a current of $I = 100$ A in a region where the magnetic field has a magnitude $B = 10$ T, but it is found that the force on the wire is zero. Explain how that can be.

47. ✪ A long, straight wire of length 0.75 m carries a current of $I = 1.5$ A in a region where $B = 2.3$ T. If the force on the wire is 1.4 N, what is the angle between the field and the wire?

48. ✪ Repeat Problem 47, but assume the force is 5.6 N. *Hint:* Is this possible?

49. ✪ Ⓡ The electric current carried in a typical household circuit is about 1 A, and the Earth's magnetic field is about 5×10^{-5} T. What is the approximate magnitude of the magnetic force on such a wire if its length is 1 m? Could you detect this force if you were holding the wire? Compare this force to the weight of some terrestrial object (such as a mosquito).

50. ✪ Ⓡ Consider a straight piece of copper wire of length 2 m and diameter 1 mm that carries a current $I = 3.5$ A. There is a magnetic field of magnitude B directed perpendicular to the wire, and the magnetic force on the wire is just strong enough to "levitate" the wire (i.e., the magnetic force on the wire is equal to its weight). Find B. *Hint:* The density of copper is 9000 kg/m^3.

51. ✪ A right-triangular current loop carries a current I, and there is a constant magnetic field B directed perpendicular to one edge of the loop (Fig. P20.51). If $\theta = 37°$, $L = 30$ cm, and $B = 2.3$ T, what is the total force on the loop?

Figure P20.51 Problems 51 and 52.

52. ✪ Consider again the force on the triangular current loop in Figure P20.51, but now work out the force as a function of L, θ, and B. Explain your answer.

53. A wire carries a current along the $+z$ direction and experiences a force along $-y$. If the magnetic field is perpendicular to the wire, what is the direction of $\vec{B}$?

54. ✪ A wire carries a current along the $+z$ direction and experiences a force that lies in the x–y plane as shown in Figure P20.54. What might be the direction of $\vec{B}$?

Figure P20.54

55. SSM ★ A long current-carrying wire is parallel to the y axis with the current $I = 4.5$ A along the $+y$ direction. A magnetic field of magnitude $B = 1.2$ T lies in the y–z plane, directed 30° away from the y axis as shown in Figure P20.55. (a) What is the direction of the magnetic force on the wire? (b) What is the magnitude of the magnetic force on a 1.0-m-long section of the wire?

Figure P20.55

56. A square current loop of edge length $L = 0.25$ m with $I = 4.5$ A is pulled from a region where the magnetic field is zero into a region where the magnetic field is perpendicular to the loop with a magnitude $B = 2.5$ T (Fig. P20.56). (a) What is the direction of the magnetic force on the loop? (b) What is the magnitude of the magnetic force on the loop? (c) If the loop's velocity is reversed, what is the direction of the magnetic force?

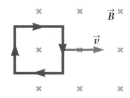

Figure P20.56

20.5 TORQUE ON A CURRENT LOOP AND MAGNETIC MOMENTS

57. ★ Figure P20.57 shows several different current loops in a magnetic field. In Case 1 the current is clockwise when viewed from the upper left. In Case 2 it is clockwise when viewed from the upper right. In Case 3 the current is clockwise when viewed from above. What is the direction of the torque on each current loop? Express your answer as either clockwise or counterclockwise as viewed along an axis perpendicular to the plane of the drawing.

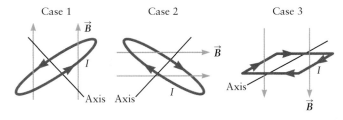

Figure P20.57 Problems 57 and 58.

58. SSM ★ Consider the square current loop, case 3 in Figure P20.57. It has an edge length $L = 0.33$ m and carries a current $I = 7.5$ A. The magnetic field $B = 0.22$ T is perpendicular to the plane of the loop. What is the magnitude of the torque on the loop?

59. ✪ Consider a square current loop ($L = 25$ cm) that carries a current $I = 5.6$ A (Fig. P20.59). A constant magnetic field $B = 0.25$ T makes an angle $\theta = 30°$ with the direction normal to the plane of the loop and perpendicular to the line connecting points A and B. (a) Find the total magnetic force on the loop. (b) Assume the loop is attached to a rotatable rod at points A and B. Find the torque on the loop.

Figure P20.59

60. The plane of a circular current loop is oriented at an angle θ relative to a magnetic field parallel to z (Fig. P20.60). If the current loop is free to rotate about an axis perpendicular to z, which of the following is true?
(a) The plane of the loop will rotate so as to be perpendicular to $\vec{B}$.
(b) The plane of the loop will rotate so that the field produced by the current loop is parallel to $\vec{B}$.
(c) Either statement (a) or statement (b) is possible, depending on the direction of the current.

Figure P20.60

61. ★ A ball of positive charge rotates about a vertical axis (Fig. P20.61). (a) If the rotation is clockwise as viewed from above, what is the direction of the magnetic moment of the ball? (b) If the charge on the ball is negative, what is the direction of the magnetic moment?

Figure P20.61

20.6 MOTION OF CHARGED PARTICLES IN THE PRESENCE OF ELECTRIC AND MAGNETIC FIELDS

62. ✪ **Velocity selector.** Consider a charged particle moving through a region in which the electric field is perpendicular to the magnetic field, with both fields perpendicular to the initial velocity of the particle (Fig. P20.62). Such a device is called a *velocity selector* because for one particular value of the particle speed v_{select} the particle will travel straight through the device without being deflected, whereas particles with speeds that are either less than or greater than v_{select} will be deflected. Consider a case with $E = 500$ V/m and $B = 0.25$ T. (a) When protons are used, what speed is selected? (b) What is the speed selected for Ca^{2+} ions? (c) If you were to use your selector with F^- ions, what change would be necessary?

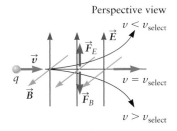

Figure P20.62 Problems 62 and 63.

63. ✪ You wish to design a velocity selector (see Problem 62 and Fig. P20.62) that will allow protons to pass through only if they have a speed of 500 m/s. (a) If the magnetic field is $B = 0.050$ T, what electric field do you need? (b) You decide to produce this electric field using a parallel-plate capacitor with a plate spacing of 1.0 mm. What is the electric potential across the plates? (c) Will your velocity selector also work for Li^+ ions? For Mg^{2+} ions? Explain why or why not.

64. ★ A mass spectrometer can separate ions according to their charges and masses. One simple design for such a device is shown in Figure 20.26. Ions of mass m, charge q, and speed v enter a region in which the magnetic field B is constant and perpendicular to the plane. The ions then travel in a circular arc and leave the spectrometer a distance $L = 2r$ from their entry point. Consider a hypothetical mass spectrometer used to study the isotopes of hydrogen. Find r for H^+, D^+ (deuterium), and T^+ (tritium).

65. SSM ★ Consider a mass spectrometer used to separate the two isotopes of uranium, $^{238}U^{3+}$ and $^{235}U^{3+}$. Assume the ions enter the magnetic field region with a speed of 6.0×10^5 m/s. What value of B is required to give a separation of 1.0 mm when the ions leave the spectrometer?

66. ✪ Some mass spectrometers use an electric field to give the incoming ions a certain known kinetic energy as they enter the magnetic field region. (a) If the ions have a charge q and a kinetic energy KE, what is the value of r in Figure 20.26 in terms of m, q, B, and KE? (b) How does the final kinetic energy of an ion (just before it reaches the ion detector) compare with the initial kinetic energy (just as the ion enters the region where the magnetic field is nonzero)?

67. ✪ Consider the Hall effect experiment in Figure P20.67 and suppose the current is carried by electrons. In Section 20.6, we saw that the magnetic field exerts a force on the moving electrons directed from the bottom edge of

Figure P20.67

the wire toward the top. This force causes a net negative charge to accumulate along the top edge, leaving a net positive charge along the bottom. The result is an electric field (called the Hall field) directed from bottom to top. The Hall field opposes the magnetic force on the electrons. The edge charge will grow until the electric force from the Hall field balances the magnetic force. Use this fact to calculate the magnitude of the Hall field. Express your answer in terms of the drift speed v of the electrons and the magnitude B of the magnetic field.

20.7 CALCULATING THE MAGNETIC FIELD: AMPÈRE'S LAW

68. A long, straight wire carries a current of 5.4 A. What is the magnitude of the magnetic field at a distance of 10 cm from the wire?

69. The Earth's magnetic field is approximately 5×10^{-5} T. If the field produced by the current in a long straight wire 1.0 mm from the wire is equal to the Earth's field, what is the current in the wire?

70. Two long, straight, parallel wires of length 4.0 m carry parallel currents of 3.5 A and 1.2 A. (a) If the wires are separated by a distance of 3.5 cm, what is the magnitude of the force between the two wires? (b) Is this force attractive or repulsive? (c) If the currents are in opposite directions (antiparallel), how do the answers to parts (a) and (b) change?

71. ✪ Consider a long wire that makes a right angle (90°) bend as shown in Figure P20.71. If the wire carries a current I, what is the magnetic field at point A? *Hint*: Use the result for the field from an infinitely long wire (Eq. 20.19) to find the field produced by a "semi"-infinite wire (i.e., half of an infinite wire).

Figure P20.71

Figure P20.72

72. SSM ✪ A wire is formed so as to make a circular loop connecting two long, straight sections as shown in Figure P20.72. If the radius of the loop is $R = 20$ cm and the wire carries a current $I = 2.4$ A, what are the magnitude and direction of the magnetic field at the center of the loop?

73. Consider a single circular loop of wire that carries a current $I = 1.5$ A. If the field due to the current in the loop at the loop's center is equal to the Earth's field, what is the radius of the loop?

74. ✪ Ⓡ A magnetic field $B = 1$ T is considered to be a strong field. If you want to produce this field using a single loop of wire wound around a pencil, what must the current be?

75. The magnetic field near a particular permanent magnet is 2.0 T. Consider a long wire carrying a current I. If the field produced by the current at a location 1.0 cm from the wire is equal to the field of this permanent magnet, what is I?

76. ✪ Design a solenoid that will produce a magnetic field of 0.025 T using a current $I = 30$ A. If the radius of the solenoid is 10 cm and the solenoid is 1.0 m long, how many turns of wire are required?

77. ✪ A long, straight wire (Fig. P20.77) carries a current $I_1 = 3.5$ A and is a distance $h = 15$ cm from a rectangular current loop that carries a current $I_2 = 1.5$ A. (a) If the loop has a length $L = 45$ cm and a width $w = 20$ cm, what are the magnitude and direction of the force exerted by the wire on the loop? (b) What is the force exerted by the current loop on the wire?

Figure P20.77   **Figure P20.78**

78. ✪ According to Ampère's law, the magnetic field produced by a long, straight wire depends only on the current enclosed by the Ampère's law path (Eq. 20.16). Use this fact to compute the magnetic field *inside* a wire (Fig. P20.78). Assume the wire has a radius $R = 1.0$ mm and carries a current $I = 2.5$ A. (a) Find the field a distance $R/2$ from the axis of the wire. (b) What is the ratio of the field in part (a) to the field at the surface of the wire?

79. ✪ Consider two long, straight, parallel wires (Fig. P20.79) carrying currents $I_1 = I_2$. At which point A, B, or C in the figure will the total magnetic field be largest?

Figure P20.79 Problems 79 and 80.

80. ✪ Consider again the two wires in Figure P20.79. For what value of the ratio I_1/I_2 is the magnetic field zero at (a) point A, (b) point B, and (c) point C. Do your answers to this problem depend on the value of d?

81. ✪ Two long, straight wires lie in the x–y plane and are parallel to the x direction as shown in Figure P20.81. The wires are both a distance of 2.0 m from the origin, and each wire carries a current of 25 A in the $+x$ direction. Find the magnitude of the magnetic field at the point $x = 0$, $y = 1.5$ m, $z = 0$.

Figure P20.81

82. ✪ Four very long wires are directed perpendicular to the plane of Figure P20.82. The points where these wires cross the plane of the drawing form a square of edge length 0.45 m, and each wire carries a current of magnitude 2.5 A. Find the magnetic field at the center of the square (i.e., at the origin).

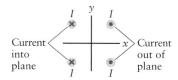

Figure P20.82

83. ✪ Three very long wires are directed perpendicular to the plane of Figure P20.83. Each passes through the x–y plane a distance L from the origin, and each carries a current I that is directed out of the plane of the drawing. An electron is at the origin and has a velocity directed along $+x$. What is the direction of the magnetic force on the electron?

Figure P20.83

84. ✪ Ⓡ A typical lightning bolt carries a peak current of 100 kA. (a) Estimate the peak magnetic field a distance of 500 m from where the lightning bolt strikes the ground. (b) If you have a device that can measure magnetic fields as small as $B = 100$ pT (10^{-10} T), how far can you be from the lightning bolt and still detect its magnetic field?

85. A high-voltage power line carries a current of 5000 A. If the power line is a height 20 m above the ground, what is the magnetic field strength at ground level? How does it compare with the Earth's magnetic field?

86. A solenoid has a length of 15 mm and a radius of 0.25 mm, and consists of 5000 circular turns. If a current of 0.35 A is passed through the solenoid, what is the magnetic field at the center (inside the solenoid)?

87. Consider the solenoid discussed in Example 20.7. What value of current is required to give a field of magnitude $B = 1.0$ T? Would this value be practical? Why or why not?

20.8 MAGNETIC MATERIALS: WHAT GOES ON INSIDE?

88. ✪ Ⓡ The U.S. Navy can detect a submarine by measuring the magnetic field produced by its metal hull (usually made of a magnetic metal). For the purpose of calculating its magnetic field, model a submarine as one large current loop located at the middle of the submarine carrying 8000 A, as calculated in Example 20.8 and with a diameter equal to the diameter of submarine. Assume that far from the submarine the magnitude of the magnetic field varies as $B_0(r/r_0)^{-3}$, where B_0 is the field at the center of the current loop, r_0 is the radius of the loop, and r is the distance from the center of the loop (and also the center of the submarine). What is the approximate value of the magnetic field a distance of 1000 m from the submarine?

89. ✪ Ⓡ Based on the results of Problem 88, the U.S. Navy has developed ways to make the magnetic field from its own submarines as small as possible. One way is by wrapping a solenoid around the hull and passing a current through the solenoid so that the field produced by the solenoid approximately cancels the field from the hull. Take the diameter of the submarine to be 20 m, and its length to be 50 m. Assume the field of the submarine alone to be 0.0010 T. Assuming a solenoid with 1000 turns, what is the approximate current required in the solenoid?

20.9 THE EARTH'S MAGNETIC FIELD

90. Ⓡ The Earth's magnetic field has a magnitude of approximately 5.0×10^{-5} T, and in New York City this field is directed toward the north geographic pole. If a proton is traveling in an easterly direction with a speed of 600 m/s, what are the magnitude and direction of the magnetic force on the proton?

91. SSM ✪ Ⓡ Suppose the Earth's magnetic field is produced by a single loop of wire that has a radius of R_E, where R_E is the radius of the Earth, and assume this loop lies in the equatorial plane. Estimate the current that must flow in this loop to produce a field equal to the Earth's field at the center of the loop.

20.10 APPLICATIONS OF MAGNETISM

92. ✪ ✖ **Blood-flow meter.** Blood plasma has a high content of Na^+ and Cl^- ions, and as the blood follows through a blood vessel, the ions travel along at the same speed as the fluid (Fig. P20.92, left). One method of measuring the blood's velocity is to place the vessel in a magnetic field and then measure the potential difference across the width of the vessel perpendicular to the magnetic field direction. Unlike the Hall effect, both positive and negative charges here are moving in the same direction, so the magnetic force pushes the positive charges to the top and the negative charges to the bottom as seen in Figure P20.92, right. (a) Consider the equilibrium condition where the magnetic force on the charges balances the electric force on the charges. Find an expression for the velocity of the blood v in terms of the width of the blood vessel, d, the magnitude of the supplied magnetic field, B, and the resulting measured electric potential, V. (b) In designing an actual measuring device for a typical blood-flow velocity of 30 cm/s, an engineer uses an available permanent magnet that supplies a uniform 0.20-T field. We then need to find how sensitive a voltmeter we would need to measure the flow in a small blood vessel. What potential difference would develop across a 1.2-cm-diameter blood vessel with this configuration?

Figure P20.92 Problems 92 and 93.

93. SSM ✪ ✖ Consider a blood-flow sensor placed around the artery in Figure P20.92. Assume the artery has a radius of 3.0 mm and the magnetic field is $B = 0.050$ T. (a) If the blood speed is 25 cm/s, what is the magnitude of the induced emf? (b) Which side of the blood vessel in Figure P20.92 is at a positive potential (relative to the other side)? (c) How does the induced emf change if the artery is smaller, with $r = 0.50$ mm? (d) Explain why this type of blood-flow meter is most useful for large arteries.

20.11 THE PUZZLE OF A VELOCITY-DEPENDENT FORCE

94. ✪ This problem leads you through a study of electric and magnetic fields as viewed by two different observers, that is, in two different reference frames.

Suppose a charge q_1 is moving with a constant velocity $\vec{v}_1$ (Fig. P20.94) and that its trajectory takes it to within a distance

Figure P20.94

L of stationary charge q_2. (a) What are the magnitude and the direction of the resulting electric current? (b) Equation 20.19 gives the magnetic field produced by a current-carrying wire. Assume this relation can also be applied here (it can!) and use it to find the magnetic field produced by the current in part (a). What is the resulting magnetic field a distance L from the path of q_1? (c) Now assume charge q_2 has a velocity $\vec{v}_2$ parallel to $\vec{v}_1$. Use the result from part (b) to show that the magnitude of the magnetic force on q_2 when the two charges are at their closest is $F = (\mu_0 q_1 q_2 v_1 v_2 / 2\pi L)$. (d) What is the direction of the force in part (c) if q_1 and q_2 are both positive charges? If one is positive and one is negative?

An observer watching q_1 and q_2 travel by would say that the force found in part (c) is a magnetic force. An observer in a moving reference frame would see things differently. If an observer is moving along with q_1 at a velocity $\vec{v}_1$, he or she would say that q_1 is not moving, so the current and hence the magnetic field and the magnetic force are zero. However, this observer would still find that there is a force between the two charges as given in part (c) and would attribute it to an extra electric field E_{new}. (e) Find the magnitude and direction of E_{new}. *Hint:* The Coulomb's law force produced by E_{new} must be equal to the force found in part (c). This electric field is a relativistic effect, and we'll discuss the ideas behind it in Chapter 27.

ADDITIONAL PROBLEMS

95. ✪ Electric power is transmitted from the generation plant to a municipality via high-voltage power lines such as those connected to the tower in Figure P20.95. Consider a case with towers typically placed at 0.50-km intervals, with three separate lines, spaced 20 m apart and operating at a high voltage (760 kV) with typical maximum currents of 500 A. (a) Find the force on a tower-to-tower length of one of the outside wires from the other two. (b) If the wire has a cross-sectional area of 200 mm² and an average density of 4000 kg/m³, compare the weight of a section of the wire to the force found in part (a).

Figure P20.95

© Comstock/Jupiterimages

96. ✪ Two aluminum rods of length 75 cm and mass 50 g are suspended by nonconducting cables of length 10 cm as shown in Figure P20.96. The rods are part of a circuit (not shown) that supplies a current I

Figure P20.96

through each rod as shown. If the angle between the two end cables is 4.0°, what is the magnitude of the current I? Assume the rods are very long.

97. ✪ A conducting rod of mass 20 g and length 45 cm has holes drilled near each end so that it is free to slide in a vertical direction along two conducting rails as shown in Figure P20.97. A current of 200 A is made to flow through an

Figure P20.97

identical rod at the bottom of the assembly and return through the sliding rod (notice the placement of an insulator). Determine the height of the sliding rod above the base rod when in equilibrium. Assume the horizontal rods are very long.

98. ✪ An electron enters a bubble chamber and produces a spiral track like that shown on the left in Figure Q20.13. The magnetic field is directed into the page, and the electron had an initial velocity of 2.7×10^8 m/s. (a) If the initial radius of the electron's trajectory is 5.0 cm, what is its velocity when it has spiraled into a radius of 2.0 cm? (b) What is the correspond-

ing change in kinetic energy? Give your answer in joules and electron-volts. (c) Calculate the magnitude of the magnetic field present in the bubble chamber. (d) What is the change in force on the electron in going from the initial radius to the final radius?

99. ✪ The *cyclotron* is a device that provides a beam of high-velocity charged particles. Consider the electron-accelerating cyclotron depicted in Figure P20.99. Two semicircular magnets provide a uniform magnetic field of 5.0 mT between the poles; in addition, an electric potential difference of 220 V is placed on each D-shaped section so that an accelerating electric field is produced in the gap between the semicircles. An electron will accelerate as it crosses the gap and enter the magnetic field, producing a circular orbit so that the electron is directed back into the gap. By the time the electron returns to the gap, circuitry will have reversed the direction of the electric field so that the electron is again accelerated. Ignore relativistic effects in the following. (a) How much does the energy of an electron increase each time it crosses the gap? If the radius of each D-shaped magnet is 5.0 cm and the electrons start in the center of the magnets at one edge of the gap, what are the (b) maximum speed and maximum kinetic energy that an electron can attain? (c) At what rate must the accelerating electric potential be reversed to make the device work?

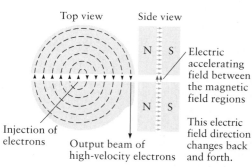

Figure P20.99

100. SSM ✪ You are standing directly under a high-voltage power line that is 4.5 m above, and a detailed map you have indicates that the power line runs exactly north to south. Your compass needle does not align with the power line, however, but instead makes an angle of 32° toward the east with respect to the power line's direction. (a) Knowing that the magnitude of the Earth's magnetic field is approximately 5.0×10^{-5} T, determine the magnitude and direction of the magnetic field produced by the power line at ground level. (b) How much current is being carried in the power line? (c) Is the current flowing toward the north or toward the south?

Magnetic Induction

An electric guitar uses magnetic induction to detect the motion of the guitar strings. This chapter explains how that works. (© Shepard Sherbell/Corbis)

Are electricity and magnetism separate phenomena, or are they related in some way? Perhaps the most obvious hint that they are connected is that electric and magnetic forces both act only on particles carrying an electric charge. Another clue is that moving electric charges (an electric current) create a magnetic field. But how do we know there is *really* an important connection between $\vec{E}$ and $\vec{B}$? After all, charged particles have mass, so they experience gravitational forces. Does that mean gravity is also connected to electricity and magnetism? (The answer is no, gravity does not play a role in electromagnetism.) In this chapter, we'll see that a changing magnetic field creates an electric field. This effect, called ***magnetic induction,*** links $\vec{E}$ and $\vec{B}$ in a truly fundamental way. Magnetic induction is also the key to many practical applications of magnetism, including motors, generators, loudspeakers, and cell phones.

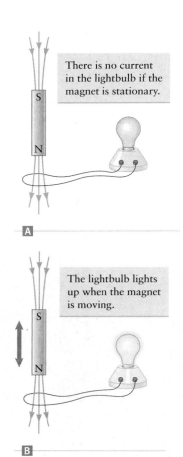

There is no current in the lightbulb if the magnet is stationary.

A

The lightbulb lights up when the magnet is moving.

B

Figure 21.1 **A** When a bar magnet is placed near a loop of wire, a magnetic field penetrates the loop. **B** When the bar magnet is in motion, an electric current is induced in the loop, lighting the lightbulb.

21.1 | WHY IS IT CALLED *ELECTROMAGNETISM*?

The discovery of electric charges and forces by the Greeks and other early civilizations involved static electricity, whereas the first observations of magnetic effects employed permanent magnets composed of the mineral magnetite as mentioned in Chapter 20. There was no obvious connection between the two effects until a chance observation by Danish physicist Hans Christian Ørsted in 1820. While giving a class lecture, he discovered that an electric current in a wire can exert a force on a compass needle. (Such magnetic forces were discussed in Chapter 20.) An electric current is produced by an electric field and a compass needle is a small bar magnet, so Ørsted's lecture experiment also showed that an electric field can lead to a force on a magnet. He therefore concluded that *an electric field can produce a magnetic field.*

When this connection between $\vec{E}$ and $\vec{B}$ was announced in 1820 it was a complete surprise and stimulated many further experiments to probe the connection between electricity and magnetism. One scientist engaged in this work was Michael Faraday, who reasoned that if an electric field can produce a magnetic field, perhaps a magnetic field can produce an electric field. Faraday attempted to observe such an "induced" electric field using an experiment like the one sketched in Figure 21.1A in which a bar magnet is positioned close to a loop of wire. This loop is connected to a lightbulb, so if there is current in the loop, the bulb will light up and thus act as a detector of the electric current. (Faraday didn't use a lightbulb—they weren't invented yet—instead, he used an ammeter to detect the current, with the same results.)

Faraday positioned a bar magnet near the loop of wire in Figure 21.1A so that the magnetic field would penetrate through and around the loop. If this magnetic field produces an electric field, the electric field would cause an electric current in the loop, lighting the bulb. Faraday was probably very disappointed that this experiment was not successful. No current was detected in the wire loop when the bar magnet was at rest near the loop. He did, however, discover a variation on this experiment in which the bulb does light up as shown in Figure 21.1B. If the bar magnet is *in motion* either toward or away from the loop, the bulb emits light, indicating an electric field within the wire. This electric field is produced *only when the magnet is moving.* If the magnet is stationary (as in Fig. 21.1A), the current and hence also the electric field are always zero.

Faraday devised another approach to this experiment in which a solenoid is positioned near a loop of wire containing a lightbulb (Fig. 21.2). Recall from Chapter 20 that when an electric current passes through a solenoid, a magnetic field is produced with field lines similar to those produced by a bar magnet. Faraday wanted to know if this magnetic field could also produce a current in his wire loop. He passed an electric current I through the solenoid by connecting it to a battery through a switch. Faraday found that when the current through the solenoid is constant (Fig. 21.2A), there is no current (the bulb did not light), but the bulb does light up the moment the switch is opened or closed (Fig. 21.2B). An electric current is produced

Figure 21.2 **A** When a current-carrying coil is placed near a loop of wire, the magnetic field produced by the coil penetrates the loop. When this magnetic field is constant, no current is induced in the loop. **B** When the magnetic field is changing with time (i.e., just as the switch is closed or opened), an electric current is induced in the loop.

Lightbulb is dark.

A

Lightbulb is lit just after switch is opened or closed.

B

during those instants in which the current through the solenoid is *changing* from zero to some nonzero value (when the switch was closed) or from a nonzero value to zero (as the switch was opened).

Faraday's experiments with the bar magnet (Fig. 21.1) and the solenoid (Fig. 21.2) show that an electric current is produced in the wire loop only when the magnetic field at the loop is changing. Hence, *a changing magnetic field produces an electric field.* An electric field produced in this way is called an *induced* electric field, and this phenomenon is called **magnetic induction.**

21.2 | MAGNETIC FLUX AND FARADAY'S LAW

Faraday's experiments in Figures 21.1 and 21.2 demonstrated qualitatively the effect of magnetic induction. Faraday also developed a quantitative theory of the effect, now called **Faraday's law,** that tells how to calculate the induced electric field in different situations. To use this law, we must first introduce a quantity called **magnetic flux.**

Suppose there is a magnetic field in some region of space and let A be the area of a particular surface as sketched in Figure 21.3A. If lines of $\vec{B}$ cross this surface, there is a magnetic flux through the surface. Magnetic flux is very similar to the concept of electric flux that we encountered in Chapter 17 in connection with Gauss's law.

Let's now calculate the magnetic flux in a few specific cases. Figure 21.3A shows a case in which the magnetic field has a constant magnitude B (the same everywhere in this region of space). The surface area A is flat, and $\vec{B}$ is perpendicular to the surface. The magnetic flux, denoted by Φ_B, is given by

$$\Phi_B = BA \quad \text{(when } \vec{B} \text{ is perpendicular to a surface)} \quad \text{(21.1)}$$

Qualitatively, you can think of the magnetic flux as a "count" of the number of field lines that pass through the surface.

Two other common situations are shown in parts B and C of Figure 21.3. In both cases, $\vec{B}$ has a constant magnitude and a constant direction. In Figure 21.3B, the magnetic field lines cut through the indicated surface, making an angle θ with the direction perpendicular (normal) to the surface. Here the magnetic flux is

$$\Phi_B = BA \cos \theta \quad \text{(when } \vec{B} \text{ makes an angle } \theta \text{ relative to the normal to a surface)} \quad \text{(21.2)}$$

Since θ is not zero, the flux in Figure 21.3B is smaller than when the field lines cut perpendicularly through the surface as in Figure 21.3A (for the same values of B and A). In Figure 21.3C, the magnetic field is parallel to the plane of the surface; no lines of $\vec{B}$ cut through the surface, so the magnetic flux is zero:

$$\Phi_B = 0 \quad \text{(when } \vec{B} \text{ is parallel to a surface)} \quad \text{(21.3)}$$

This case—with $\vec{B}$ parallel to the surface—is actually a special case of Equation 21.2, with $\theta = 90°$. Likewise, when $\vec{B}$ is perpendicular to the surface (Fig. 21.3A), $\theta = 0$ and the expression for the flux in Equation 21.2 reduces to Equation 21.1.

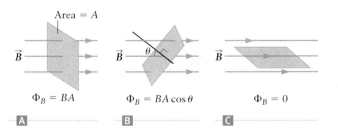

Figure 21.3 Ⓐ When a magnetic field passes through a surface with an area A, there is a magnetic flux through the surface. If $\vec{B}$ is perpendicular to the surface and the field's magnitude is the same at all points on the surface, the magnetic flux is $\Phi_B = BA$. Ⓑ If $\vec{B}$ makes an angle θ with a line drawn perpendicular to a surface of area A, the magnetic flux through this surface is $\Phi_B = BA \cos \theta$. Ⓒ If the magnetic field is parallel to the surface so that $\theta = 90°$, the magnetic flux through the surface is zero.

For simplicity, in the three examples in Figure 21.3 we have assumed the magnetic field is the same at all points on the surface and we are dealing with a flat surface. However, Φ_B can be defined in a similar way for any surface. Any complicated surface can be broken into small regions that are each approximately flat. The magnetic flux through each small region is equal to the magnitude of $\vec{B}$ multiplied by the surface area and the cosine of θ, the angle between $\vec{B}$ and the direction perpendicular to that surface region. Hence, Equation 21.2 can be applied to find the flux through any small region. The total flux is then the sum of the fluxes through all the individual pieces of the surface being considered.

When calculating electric flux in our applications of Gauss's law in Chapter 17, we were interested in closed surfaces, that is, surfaces that completely enclose a particular region of space. In magnetic induction, however, the surfaces of interest are *open* surfaces, like the ones in Figure 21.3. Such open surfaces always possess an edge or perimeter, which (as we will soon see) plays a big role in magnetic induction.

The units of magnetic flux can be deduced from Equation 21.2. The right-hand side of this equation has units of magnetic field multiplied by area, so magnetic flux has units of $T \cdot m^2$. This unit is named the *weber* (Wb):

$$\text{unit of magnetic flux: } 1 \text{ Wb} = 1 \text{ T} \cdot \text{m}^2 \qquad (21.4)$$

The weber is the official SI unit of magnetic flux and is used in some texts, but many physicists just use the combination of units $T \cdot m^2$.

Faraday's Law

Faraday's induction experiments in Figures 21.1 and 21.2 demonstrate that a changing magnetic field produces a current in a wire loop. *Faraday's law* tells us how to calculate the potential difference that produces this induced current. This law is written in terms of the magnetic flux and the electromotive force (emf) induced in the wire loop. Recall from Chapter 19 that an emf (denoted by $\mathcal{E}$) is not a force; rather, it is an electric potential difference, like the one that appears across the terminals of a battery. Faraday's law states that a changing magnetic flux produces an electric potential difference $\mathcal{E}$ given by

Faraday's law

$$\mathcal{E} = -\frac{\Delta\Phi_B}{\Delta t} \qquad (21.5)$$

This simple formula contains two important ideas. The first is that the *magnitude* of the induced emf equals the rate of change of the magnetic flux:

$$|\mathcal{E}| = \left|\frac{\Delta\Phi_B}{\Delta t}\right| \qquad (21.6)$$

The second important "idea" is the negative sign in Equation 21.5. In fact, this negative sign is so important that it has its own name: *Lenz's law.* We will devote Section 21.3 to a discussion of this negative sign and explain why it deserves to be called a law of physics. Before we do that, we'll consider some properties and situations that depend on just the magnitude of the induced emf (Eq. 21.6). After we understand that part of Faraday's law, we will be better prepared to understand the negative sign in Equation 21.5 and the importance of Lenz's law.

Let's consider the meaning of the two sides of Equation 21.6 in the context of the experiment in Figure 21.4. Here the magnetic field produced by a bar magnet passes through a loop of wire that encloses an area A. This arrangement is similar to the one in Figure 21.1 except for two small changes. First, suppose the wire loop is very small, so the magnetic field is constant across the area of the loop and the field is perpendicular to the plane of the loop. The magnetic flux is then simply $\Phi_B = BA$ (Eq. 21.1). However, the value of B changes as the bar magnet is moved closer or farther away from the loop, so the flux changes with time as the magnet moves. Second, we have replaced the lightbulb in Figure 21.1 with a voltmeter, and for simplicity we assume the voltmeter's input terminals are very close together and close to the wire loop so that the perimeter of the loop still defines our area A.

Let's now consider the terms in Faraday's law (Eq. 21.6) and what they mean.

1. The quantity $\mathcal{E}$ on the left-hand side is the induced emf that appears in the wire loop. The value of $\mathcal{E}$ would be indicated by the voltmeter in the wire loop circuit and has units of volts (V). This emf is an induced electric potential difference and is related to the electric field directed along and inside the wire loop. This induced potential difference produces the current in the lightbulb in Figures 21.1 and 21.2.

2. The right-hand side of Faraday's law contains the change in magnetic flux $\Delta\Phi_B$ during a short time interval Δt, divided by Δt. The ratio $\Delta\Phi_B/\Delta t$ is just the rate of change of the magnetic flux with time. In words, Faraday's law says that an emf (i.e., a voltage or electric potential difference) is produced in a circuit by *changes* in the magnetic flux through the circuit. A constant flux, no matter how large, does not produce an induced voltage, in accord with Faraday's unsuccessful experiments (Figs. 21.1A and 21.2A). Magnetic induction requires that the flux through the wire loop change with time. In Faraday's successful experiments (Figs. 21.1B and 21.2B), this requirement was accomplished by arranging for the magnitude of $\vec{B}$ to change with time. According to Equation 21.2, however, the magnetic flux also depends on the area A and the angle θ. Changes in A and θ can thus also produce a changing flux and hence an induced emf.

What about the *direction* of the induced current in a wire loop? The lightbulbs in Figures 21.1 and 21.2 were either lit or not, indicating current or no current, but that does not tell the current direction. This direction can be read from the *sign* of the voltmeter reading in Figure 21.4, which is at the heart of Lenz's law. The voltmeter reading (positive or negative) specifies the direction of the induced emf and hence the direction of the induced current and electric field, all of which are important in practical applications of Faraday's law.

| EXAMPLE 21.1 | Magnitude of $\mathcal{E}$ Produced by Moving a Bar Magnet |

Faraday's law tells us that a changing magnetic flux produces an induced emf, but how large is this emf in typical cases? Consider a wire loop of area $A = 1$ cm^2 initially far from any sources of magnetic field. A strong bar magnet with $B = 1$ T near its poles is suddenly inserted into the loop, filling the loop completely (Fig. 21.5). If the time required to insert the magnet is $\Delta t = 0.1$ s, what is the approximate value of the emf induced in the loop?

RECOGNIZE THE PRINCIPLE

The magnitude of an induced emf $\mathcal{E}$ is given by Faraday's law, with $\mathcal{E}$ equal to the change of the magnetic flux divided by the time over which this change of flux occurs.

SKETCH THE PROBLEM

Figure 21.5 shows the initial and final positions of the bar magnet, which correspond to the initial and final flux through the loop.

IDENTIFY THE RELATIONSHIPS

The change in magnetic flux is $\Delta\Phi_B = \Phi_f - \Phi_i$ (the final flux minus the initial flux). Because the wire loop is initially far from any magnetic fields, the initial flux through the loop is approximately zero and $\Phi_i = 0$. After inserting the bar magnet, the field through the loop is $B \approx 1$ T, so the final flux (according to Eq. 21.1) is $\Phi_f = BA$ and the change in flux is

$$\Delta\Phi_B = \Phi_f - \Phi_i = BA \qquad (1)$$

This flux change occurs over a time interval of $\Delta t = 0.1$ s (as given above).

Figure 21.4 If a voltmeter is attached to the wire loop in Figure 21.1 in place of the lightbulb, it can be used to measure the emf (i.e., the voltage) induced in the loop.

Figure 21.5 Example 21.1. If a bar magnet is inserted into this wire loop, the voltmeter will measure a nonzero induced electric potential.

Inserting the change in flux from Equation (1) into Faraday's law, along with the value of Δt, we find the magnitude of the emf to be

$$|\mathcal{E}| = \left|\frac{\Delta \Phi_B}{\Delta t}\right| = \frac{BA}{\Delta t} = \frac{(1 \text{ T})(1 \times 10^{-4} \text{ m}^2)}{(0.1 \text{ s})} = \boxed{0.001 \text{ V}}$$

What does it mean?
This voltage is rather modest. The emf of a typical flashlight battery is 1.5 V, for instance, and the voltage in a typical household circuit is about 120 V, so this induced emf would not be a practical replacement for the battery in a radio or MP3 player. An induced voltage of this size can be measured quite easily, however, and could be used in other applications. (See Example 21.9.)

CONCEPT CHECK 21.1 | Faraday's Law

Suppose the area A of the wire loop in Example 21.1 (Fig. 21.5) is increased and the bar magnet is also enlarged so that it still fills the loop. If the rest of the experiment (the values of B and Δt) stays the same, how is the magnitude of the induced emf affected?
(a) The induced emf is larger.
(b) The induced emf is smaller.
(c) The induced emf does not change because the magnetic field is the same.

Key Features of Faraday's Law

According to Faraday's law, the emf induced along a closed path such as a wire loop depends on changes in the magnetic flux through the area enclosed by that path. This fact leads to a number of important consequences.

1. Only *changes* in Φ_B matter. The magnetic flux might be very large, but if it changes only a small amount, the induced emf will still be small.
2. Rapid changes in magnetic flux produce larger values of $\mathcal{E}$ than do slow changes. This dependence on the *frequency* at which the flux changes means that the induced emf plays a key role in alternating-current (AC) circuits (Chapter 22).
3. The magnitude of $\mathcal{E}$ is proportional to the rate of change of the flux $\Delta \Phi_B/\Delta t$. If this rate of change is constant, then $\mathcal{E}$ is constant. However, achieving a constant rate of change $\Delta \Phi_B/\Delta t$ requires that Φ_B must eventually become very large, which is usually not possible in practice. So, in most applications, the rate of change of Φ_B is not constant and the induced emf varies with time, leading again to AC voltages (Chapter 22).
4. The induced emf is present even if there is no current in the path enclosing an area of changing magnetic flux. Hence, the voltmeter in Figure 21.4 will register an induced voltage whether or not a current is present in the loop.

Magnetic Flux through a Changing Area

Changes in the magnetic flux can be produced by changes in the magnitude of $\vec{B}$, by changes in area A of the surface, and by changes in the angle θ that the surface makes with respect to $\vec{B}$. An example in which this area is changing is sketched in Figure 21.6A, which shows a metal bar of length L that can slide on two frictionless, horizontal metal rails. The metal rails are connected through a resistor on the left, so the rails plus the bar form a closed electric circuit. A magnetic field is directed perpendicular to the plane of the rails and bar, with a constant magnitude B and direction throughout. Suppose the bar is being pushed to the right by a force from an external agent (perhaps a person is pulling on it), making the bar move with a constant speed v. Let's use Faraday's law to find the induced emf in this circuit.

We apply Faraday's law using the closed path indicated by the dashed loop in Figure 21.6A. The magnetic field is constant and perpendicular to the enclosed area, so the magnetic flux is (Eq. 21.1)

$$\Phi_B = BA = BwL \tag{21.7}$$

where w is the distance from the left side of the circuit to the bar. Because the bar is moving to the right, this distance and hence the magnitude of the flux increase with time. The speed v of the bar is constant and is $v = \Delta w/\Delta t$, so the flux in Equation 21.7 changes with time according to

$$\frac{\Delta \Phi_B}{\Delta t} = BL\frac{\Delta w}{\Delta t} = BLv \tag{21.8}$$

Inserting Equation 21.8 into Faraday's law (Eq. 21.6), we find the magnitude of the induced emf,

$$|\mathcal{E}| = \frac{\Delta \Phi_B}{\Delta t} = BLv \tag{21.9}$$

This induced emf acts much like a battery in the "equivalent" circuit shown in Figure 21.6B, with an emf given by Equation 21.9. The magnitude of the current I in this circuit is $I = \mathcal{E}/R$ (Eq. 19.10). Inserting our result for $\mathcal{E}$, we have

$$I = \frac{BLv}{R} \tag{21.10}$$

Recall from Chapter 19 that in circuits like the one in Figure 21.6B, power is dissipated in the resistor. The dissipated power is given by $P = I^2R$ (Eq. 19.16); hence,

$$P = I^2R = \left(\frac{BLv}{R}\right)^2 R = \frac{B^2L^2v^2}{R} \tag{21.11}$$

Conservation of Energy and the Induced emf

Let's now summarize in words what the result in Equation 21.11 means.

1. Since the metal bar in Figure 21.6A is moving, the area of the loop indicated by the dashed red line increases with time, producing a change in the magnetic flux through this loop.
2. According to Faraday's law of induction, an emf is induced in the circuit composed of the metal bar plus rails.
3. This induced emf acts just like a battery with an emf (or voltage) $\mathcal{E}$, producing an electric current in the circuit given by Equation 21.10.
4. This current leads to a power dissipation P in the circuit.

Where does this power come from? In mechanics, power equals the product of force and velocity, so to answer this question we must consider the forces acting on the metal bar. Actually, two forces act on the metal bar in Figure 21.6. One is a magnetic force due to the current in the bar. We have already calculated the magnitude of this current in Equation 21.10. The direction of this current is specified by Lenz's law; in Section 21.3, we'll show in detail how Lenz's law gives the current direction, from the lower end of the rod toward the upper end, sketched in Figure 21.6D. The magnetic field is perpendicular to the bar, so from Equation 20.11, the magnitude of the magnetic force on the bar is

$$F_B = ILB \tag{21.12}$$

We now apply right-hand rule number 2 from Chapter 20 to find the direction of this force. An upward current in Figure 21.6D corresponds to the motion of positive charges moving upward in the figure. If you place the fingers of your right hand upward along the bar and then curl them toward the direction of the magnetic field, your thumb must point to the left in the figure as indicated by the vector $\vec{F}_B$ in parts

A

B

Top view of bar

C

The direction of $\vec{F}_B$ comes from applying right-hand rule #2 to I and $\vec{B}$.

$\vec{F}_B$ opposes $\vec{v}$

D

Figure 21.6 Ⓐ The magnetic flux through a loop changes with time if the area of the loop changes. As the movable bar on the right slides along the rails, the enclosed area changes. Ⓑ The equivalent circuit is a battery connected to a resistor. The battery voltage represents the induced emf in the sliding bar. Ⓒ Forces acting on the sliding bar. Ⓓ The emf induced in the bar produces a current through the circuit. The resulting magnetic force $\vec{F}_B$ on the sliding bar opposes its motion.

C and D of Figure 21.6. Hence, the magnetic force on the bar *opposes* the motion of the bar.

Now we consider the second force acting on the bar. This force is from the external agent mentioned at the start of the problem, and we have assumed this agent acts so as to make the bar move to the right with a constant velocity. According to Newton's laws of mechanics, if the metal bar is moving with a constant velocity, the total force on the bar must be zero. Hence, the force of the external agent $\vec{F}_{\text{ext}}$ must be equal in magnitude and opposite in direction to the magnetic force in Equation 21.12. We thus have

$$F_{\text{ext}} = ILB \qquad (21.13)$$

and this external force must be directed to the right (Fig. 21.6C). This force must be provided by the external agent to keep the bar moving at a constant velocity. The power exerted by the external agent is then (Chapter 6) $P_{\text{ext}} = F_{\text{ext}}v$. Inserting F_{ext} from Equation 21.13 leads to

$$P_{\text{ext}} = F_{\text{ext}}v = ILBv$$

We know the current I from Equation 21.10. Inserting that gives

$$P_{\text{ext}} = \left(\frac{BLv}{R}\right)LBv = \frac{B^2 L^2 v^2}{R} \qquad (21.14)$$

which is identical to the electrical power dissipated in the resistor, Equation 21.11. Hence, the mechanical power "put into" the bar by the external agent is converted into electrical power that is "delivered to" the resistor. From a work–energy perspective, we should expect that these two must be equal, and our results in Equations 21.11 and 21.14 show that to be the case. While energy is converted from one form (mechanical) to another (electrical), we have shown that the total energy is conserved.

In Chapter 6, we saw how the principle of conservation of energy can be deduced for mechanical systems from Newton's laws of motion. Now we have found that conservation of energy is also obeyed by electromagnetic phenomena. Indeed, it is believed that the principle of energy conservation is obeyed by *all* physical processes and by all laws of physics. Whenever new laws of physics are proposed, one of the first tests is to check if the proposed law conserves energy. In this way, the principle of conservation of energy is at the very heart of physics.

EXAMPLE 21.2 | **Using Faraday's Law to Design a Generator**

A device that converts mechanical energy into electrical energy is called a **generator**. The system in Figure 21.6A is a very simple type of generator. A person pulling on the rod exerts a force on the sliding bar (doing mechanical work), converting mechanical energy into electrical energy as shown by the current in the circuit. Suppose we make a generator as in Figure 21.6A, with a bar of length $L = 1$ m and a magnetic field $B = 1$ T. To be useful for producing power for your house, the generator's output voltage (the induced emf) should be approximately 100 V. (The voltage in household wiring is typically 120 V.) How fast must the bar travel to produce this value for the induced emf? Is this way of making a generator practical?

RECOGNIZE THE PRINCIPLE

The combination of the bar and rails in Figure 21.6 acts as a generator since an emf is induced across the ends of the moving bar. According to Faraday's law, the magnitude of the emf induced across the bar equals the rate of change of the magnetic flux $\Delta\Phi_B/\Delta t$, which we calculated above in our discussion of Figure 21.6. We found that $\Delta\Phi_B/\Delta t$ is proportional to the speed of the bar v. So, given the desired value of $\mathcal{E}$, we can find the required value of v.

Figure 21.6 describes the problem.

IDENTIFY THE RELATIONSHIPS AND SOLVE

Faraday's law led to the magnitude of the induced emf in Equation 21.9:

$$|\mathcal{E}| = \frac{\Delta \Phi_B}{\Delta t} = BLv$$

Rearrange to solve for v gives

$$v = \frac{|\mathcal{E}|}{BL}$$

Inserting the given values of $\mathcal{E}$, B, and L, we have

$$v = \frac{100 \text{ V}}{(1 \text{ T})(1 \text{ m})} = \boxed{100 \text{ m/s}}$$

What does it mean?
Is this approach for generating electric power practical? This velocity is about 200 mi/h. If we turn the generator on for 10 s, the horizontal rails would have to be at least $vt = (100 \text{ m/s})(10 \text{ s}) = 1000$ m long! That would be an extremely impractical generator.

Designing a Practical Electrical Generator

The electrical generator in Example 21.2 could certainly be built, producing a constant emf as long as the bar slides at a constant speed, yet we concluded in Example 21.2 that such a DC generator is not a practical device. How can we make the rate of change of the flux $\Delta \Phi_B / \Delta t$ large enough to give a useful emf? The key is to use rotational motion instead of the linear motion of the bar in Figure 21.6. Generators based on this idea are found in all electric power plants, including nuclear power plants, hydroelectric plants, and those that burn coal or natural gas.

A simple generator based on rotational motion is sketched in Figure 21.7. A permanent magnet produces a constant magnetic field in the region between its two poles (north and south), and a wire loop is located in this region. This loop has a fixed area but is mounted on a rotating shaft, so the angle θ between the field and the plane of the loop changes as the loop rotates. According to the definition of magnetic flux, Φ_B is largest when the loop is perpendicular to the field (Fig. 21.7A) and zero when the plane of the loop is parallel to the field (Fig. 21.7B).

In practice, this generator would be driven mechanically by turning the shaft, which could be accomplished with a steam turbine (as in most power plants) or with a waterfall or windmill. If the shaft rotates with a constant angular velocity, the magnetic flux through the loop varies sinusoidally with time, leading to an induced emf that also varies sinusoidally with time (Fig. 21.7C).

The generator sketched in Figure 21.7 sounds reasonable, but is it really practical? Let's estimate the magnitude of the emf that could be produced by such an AC (alternating-current) generator. If the shaft rotates with a period T, the loop will turn from its perpendicular orientation in Figure 21.7A to the parallel orientation in Figure 21.7B in a time $T/4$ because this rotation is one-fourth of a complete turn of the rotating shaft. During this time, the magnetic flux changes from $\Phi_i = BA$ (Fig. 21.7A) to $\Phi_f = 0$ (Fig. 21.7B). The change in magnetic flux is

$$\Delta \Phi_B = \Phi_f - \Phi_i = -BA \tag{21.15}$$

during a time $\Delta t = T/4$. Inserting Equation 21.15 into Faraday's law (Eq. 21.6) gives

$$|\mathcal{E}| = \left| \frac{\Delta \Phi_B}{\Delta t} \right| = \frac{BA}{T/4} = \frac{4BA}{T} \tag{21.16}$$

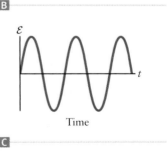

Horseshoe magnet

Area = A

$\vec{B}$

$\Phi_B = BA = $ maximum flux

A

$\vec{B}$

$\Phi_B = 0$

The flux is zero when the loop is in this orientation.

B

$\mathcal{E}$

t

Time

C

Figure 21.7 Design for an electrical generator based on Faraday's law of induction. **A** A wire loop is inserted between the poles of a magnet, creating a magnetic flux through the loop. If the loop is perpendicular to $\vec{B}$, the flux has a magnitude BA. **B** When the loop is parallel to $\vec{B}$, the flux is zero. If the loop rotates about its axis, the magnetic flux will vary with time, producing a time-varying induced voltage in the loop. **C** The emf induced in the loop as a function of time.

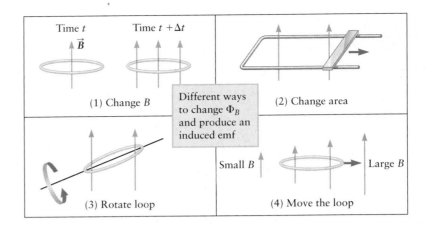

Figure 21.8 Different ways to produce an induced emf.

Time t Time $t + \Delta t$

$\vec{B}$

(1) Change B

Different ways to change Φ_B and produce an induced emf

(2) Change area

(3) Rotate loop

Small B Large B

(4) Move the loop

For a realistic generator, we might have $B = 1$ T (typical for a permanent magnet), and we take the area of the loop to be $A = 1$ m² as a fair comparison with the DC generator in Example 21.2. The period T depends on the angular speed of the shaft; a rotation rate of 1000 rpm is common (a car engine runs typically in the 2000–4000 rpm range). The corresponding period is

$$ T = \frac{1 \text{ min}}{1000 \text{ rev}} \times \frac{60 \text{ s}}{1 \text{ min}} = 0.060 \text{ s} $$

Inserting all these results into Equation 21.16 gives

$$ |\mathcal{E}| = \frac{4BA}{T} = \frac{4(1 \text{ T})(1 \text{ m}^2)}{0.060 \text{ s}} = 70 \text{ V} $$

This voltage is quite useful, so this design would indeed be practical. The voltage could easily be increased by increasing the rotation rate of the shaft or by making the area larger.

The basic design in Figure 21.7 is used in most real electrical generators, but it is still important to recall that because $\Phi_B = BA \cos \theta$, we can produce a change in magnetic flux and hence an induced emf in four different ways (Fig. 21.8): (1) if the magnitude of the magnetic field B changes with time, (2) if the area A changes with time, (3) if the loop rotates so that the angle θ between $\vec{B}$ and the loop changes with time (as in the generator in Fig. 21.7), or (4) if the loop moves from one region to another and the magnetic field is different in the two regions.

CONCEPT CHECK 21.2 | Designing a Better Generator

The generator in Figure 21.7 employs a loop that rotates with a period T. Suppose the rotation rate is increased so that the loop is made to rotate faster and the period is reduced by a factor of two. How will that affect the magnitude of the induced emf?

   (a) The magnitude of the induced emf will not change.
   (b) The magnitude of the induced emf will increase by a factor of two.
   (c) The magnitude of the induced emf will decrease by a factor of two.

21.3 | LENZ'S LAW AND WORK–ENERGY PRINCIPLES

In Section 21.2, we applied Faraday's law to calculate the magnitude of the induced emf in several different situations. Let's now consider the *sign* of this emf. Faraday's law reads (Eq. 21.5)

$$ \mathcal{E} = -\frac{\Delta \Phi_B}{\Delta t} \tag{21.17} $$

Algebraically, the sign of $\mathcal{E}$ is determined by the sign of $\Delta\Phi_B$ together with the negative sign on the right-hand side of Equation 21.17, so it would appear that determining the sign of the induced emf $\mathcal{E}$ is straightforward. However, there is a mathematical complication: the magnetic flux is produced by a field that is perpendicular to the surface of interest, whereas the induced current is along the perimeter of this surface (Fig. 21.9). The emf on the left-hand side of Faraday's law (Eq. 21.17) thus leads to a current *perpendicular* to the flux direction on the right-hand side. Hence, the two sides of Faraday's law involve quantities with different and perpendicular directions, which complicates the meaning of the negative sign. There are ways to deal with this complication, but it is simpler to make use of a principle called **Lenz's law**. This principle gives an easy way to determine the sign of the induced emf $\mathcal{E}$, and we'll see that it is closely connected with the principle of conservation of energy.

In many situations, the induced emf of Faraday's law produces a current leading to an induced magnetic field. Lenz's law is a statement about the direction of this induced field:

Lenz's law

The magnetic field produced by an induced current always opposes any changes in the magnetic flux.

Applying Lenz's Law

Let's apply Lenz's law to an arrangement in which the magnetic field passes through a metal loop with the field initially directed upward through the loop (Fig. 21.10A). We also suppose the magnitude of the field $|\vec{B}|$ increases with time (Fig. 21.10B) and ask two questions: (1) Is there a current induced in the loop? (2) What is the direction of this current?

Since we are concerned with the current in the loop, we let it be the perimeter of our Faraday's law surface. Here, $\vec{B}$ is perpendicular to the plane of the loop, so there is a magnetic flux directed upward through this surface. The magnitude $|\vec{B}|$ is increasing with time, so Φ_B is also increasing. This changing flux produces an induced emf, so there will indeed be an induced current in the loop. According to Lenz's law, the magnetic field produced by this induced current must *oppose the change* in flux, so the induced magnetic field must be *downward*. From our work on current loops (and right-hand rule number 1 in Chapter 20), the induced current must be *clockwise* in Figure 21.10C when looking at the loop from above to give a field in this direction.

The crucial feature of Lenz's law is that the induced field *opposes changes* in the flux as illustrated in Figure 21.11, which again shows a circular wire loop in a perpendicular magnetic field as in Figure 21.10. Now, however, we assume the magnitude of

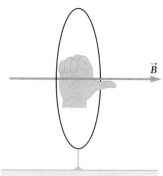

The induced emf is directed along the perimeter of the flux surface. The induced current is thus perpendicular to $\vec{B}$.

Figure 21.9 The magnetic field that produces the magnetic flux in Faraday's law is perpendicular to the plane in which the induced emf is sensed. Lenz's law gives the direction of the induced emf, that is, either clockwise or counterclockwise around the perimeter of the surface of interest.

Lenz's law

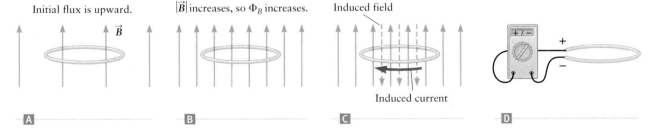

Figure 21.10 ◢ This magnetic field is directed through the plane of the loop, giving a flux "upward" through the loop. ◢ If the magnitude of B increases with time, the flux increases. ◢ According to Lenz's law, the induced emf in the loop leads to a magnetic field that opposes this change in flux. A downward magnetic field is created by a clockwise induced current. ◢ Even if the loop is broken, there is still an induced emf $\mathcal{E}$, which could be measured by attaching a voltmeter to the two open ends of the loop. The voltmeter reading indicates the polarity (+ or −) of $\mathcal{E}$.

Figure 21.11 ◨ $\vec{B}$ is directed upward through the loop, so the magnetic flux is also in the "upward" direction. ◨ If the magnitude of the field decreases with time, the flux through the loop decreases. ◨ The induced current opposes this change, so $I_{induced}$ produces a field that is upward. The required direction for the induced current is counterclockwise as viewed from above (according to right-hand rule number 1).

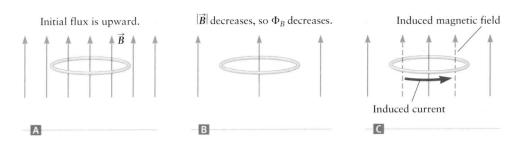

Initial flux is upward. $|\vec{B}|$ decreases, so Φ_B decreases. Induced magnetic field

Induced current

Ⓐ Ⓑ Ⓒ

the field is decreasing with time, so the "upward" flux is getting smaller. This change in flux again produces an induced emf, causing an induced current around the loop. The induced magnetic field produced by this current opposes the *change* in flux, so now the induced field is directed upward and the induced current is counterclockwise when viewed from above the loop.

We have stated Lenz's law in terms of the magnetic field produced by the induced current, but what if there is no induced current? For example, Figure 21.10D shows a loop that contains a small break, so current cannot flow all the way around the loop. Since there can be no current through this loop, it cannot produce a magnetic field to oppose the changing flux, yet there is still an induced emf. This emf could be measured by attaching a voltmeter across the open ends of the loop. In such cases, Lenz would say that the induced emf always attempts to oppose the change in flux.

PROBLEM SOLVING Applying Lenz's Law: Finding the Direction of the Induced emf

1. **RECOGNIZE THE PRINCIPLE.** The induced emf always opposes changes in flux through the Lenz's law loop or path.

2. **SKETCH THE PROBLEM,** showing a closed path that runs along the perimeter of a surface crossed by magnetic field lines.

3. **IDENTIFY** if the magnetic flux through the surface is increasing or decreasing with time.

4. **SOLVE.** Treat the perimeter of the surface as a wire loop; suppose there is a current in this loop and determine the direction of the resulting magnetic field. Find the current direction for which this induced magnetic field opposes the change in magnetic flux. This current direction gives the sign (i.e., the "direction") of the induced emf.

5. Always *consider what your answer means* and check that it makes sense.

EXAMPLE 21.3 Using Lenz's Law to Find the Direction of an Induced Current

Consider the sliding metal bar in Figure 21.12. The magnetic field is constant and is directed into the plane of the drawing. If the bar is sliding to the right, use Lenz's law to find the direction of the induced current.

RECOGNIZE THE PRINCIPLE

According to Lenz's law, the induced current produces a field that opposes the change in flux through the circuit loop.

SKETCH THE PROBLEM

Following our "Applying Lenz's Law" problem-solving strategy (step 2), the sketch in Figure 21.12 shows a dashed, rectangular path. We are interested in the current induced in this circuit, so we must consider the flux through this rectangle.

IDENTIFY THE RELATIONSHIPS

Step 3: The magnetic field is directed into the plane of this drawing, so the flux through the rectangular surface is directed into the plane. The area of this chosen sur-

Figure 21.12 Example 21.3.

face is wL, and the magnetic flux through the surface is $\Phi_B = BwL$. Because the bar is sliding to the right, Φ_B is increasing and is downward.

SOLVE

The induced emf produces an induced current that opposes the downward increase in the flux, so the induced magnetic field must be directed *upward*. Applying right-hand rule number 1 (Chapter 20), this field direction is produced by a *counterclockwise* induced current.

What does it mean?

The flux through a given area may be "upward" or "downward," and its magnitude may be increasing or decreasing with time. The induced emf always *opposes any changes* in the flux.

CONCEPT CHECK 21.3 | Applying Lenz's Law

Suppose the metal bar in Figure 21.12 is moving to the left instead of to the right as assumed in Example 21.3. How would this change affect the direction of the induced current?
  (a) There would be no change. The current through the loop will still be counterclockwise, directed from the bottom of the moving bar toward the top.
  (b) The direction of the current will be reversed, directed from top to bottom through the bar.
  (c) The current will be zero.

EXAMPLE 21.4 | A Child's Toy Based on Magnetic Induction

Consider the simple toy shown in Figure 21.13A, consisting of two wire loops. The lower loop is connected through a switch to a battery. Both loops are mounted on a vertical wooden rod, and the upper loop is able to slide freely along the rod. The current through the lower loop is initially zero. (The switch is open.) The switch is then closed, producing a counterclockwise current through the lower loop when viewed from above. (a) What is the direction of the current induced in the upper loop? (b) Will the upper loop be attracted to or repelled from the lower loop?

RECOGNIZE THE PRINCIPLE

When the current is switched on in the lower loop, there will be a change in flux through the upper loop. According to Lenz's law, the current induced in the upper loop will oppose this change in flux.

SKETCH THE PROBLEM

Following the "Applying Lenz's Law" problem-solving strategy (step 2), the sketch in Figure 21.13B shows the magnetic field produced by the counterclockwise current in the lower loop. This field is upward through the upper loop.

IDENTIFY THE RELATIONSHIPS

Step 3: The current in the lower loop produces a magnetic field that gives an upward flux through the upper loop. Hence, the flux increases upward when the switch is closed.

SOLVE

(a) Continuing with the "Applying Lenz's Law" problem-solving strategy, (step 4), the field induced in the upper loop must be downward to oppose the change in flux. This requires a *clockwise* induced current in the upper loop as viewed from above as sketched in Figure 21.13B.

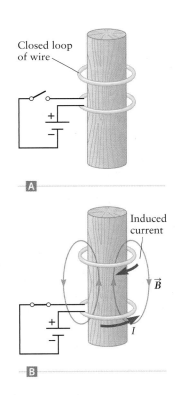

Closed loop of wire

Induced current

$\vec{B}$

I

A

B

Figure 21.13 Example 21.4.

(b) We now have two current-carrying loops. We can determine the direction of the force on the upper current loop in several different ways. For example, the currents through the two loops are in opposite directions, so the loops act as two oppositely directed bar magnets and thus *repel* each other. In addition, in Concept Check 20.6 we found that the force between two parallel current-carrying wires is repulsive if the currents are in opposite directions. The two current loops in Figure 21.13 are like parallel wires, so they repel for the same reason.

What does it mean?
The force on the upper loop will give it an upward acceleration and hence a nonzero velocity and a nonzero kinetic energy. Electrical energy from the lower loop is transferred to the mechanical energy of the upper loop, but how does this energy actually get to the upper loop? We'll discuss the amazing answer in Section 21.8.

Figure 21.14 Concept Check 21.4.

CONCEPT CHECK 21.4 | Lenz's Law and a Decreasing Magnetic Field

Consider a loop of wire in a horizontal plane with a magnetic field B directed downward through the loop (Fig. 21.14). If the magnitude of the field is decreasing, is the direction of the current induced in the loop (a) counterclockwise as viewed from above or (b) clockwise as viewed from above?

Lenz's Law and Conservation of Energy

Lenz's law involves the negative sign on the right-hand side of Faraday's law (Eqs. 21.5 and 21.17). This negative sign is *very* important. To see why, we return to the example of a metal bar sliding on two horizontal rails in Figure 21.15. A constant magnetic field is directed upward, and for simplicity there is no friction between the bar and the rails. The bar is initially at rest and is then given a push so that it starts to move with a very small velocity $\vec{v}$ to the right.

Let's use Lenz's law to analyze the resulting motion. The surface of interest is enclosed by the rails and the metal bar, and has an area A. The magnetic flux through this area is $\Phi_B = BA$. The bar is moving to the right, so this area and the flux through it are increasing. According to Lenz's law, the induced current opposes this change in flux; hence, the induced current is *clockwise* as viewed from above, producing a downward induced magnetic field. The resulting current through the bar will be as shown in Figure 21.15A. Because this current-carrying bar is in a magnetic field, there is a magnetic force $F_B = ILB$ exerted on the bar, in the direc-

Figure 21.15 When a conductor such as this metal bar moves through a magnetic field, the induced emf produces a current through the loop formed by the bar and rails. There is then a magnetic force on the sliding bar. **A** The negative sign in Faraday's law guarantees that the force on the bar opposes its motion; that is, $\vec{F}_B$ is directed opposite to the bar's velocity $\vec{v}$. **B** If Faraday's law did not contain a negative sign, the force would be parallel to $\vec{v}$, causing the bar to accelerate. This situation does *not* occur!

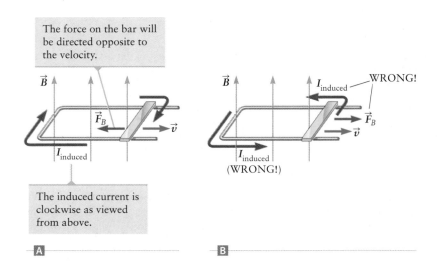

The force on the bar will be directed opposite to the velocity.

The induced current is clockwise as viewed from above.

tion given by right-hand rule number 2 (Chapter 20). Figure 21.15A shows that this force is directed to the left, so it *opposes the motion* of the bar. The force associated with magnetic induction will thus tend to bring the bar to a stop.

Now imagine what the behavior would be if the "polarity" of Lenz's law were reversed, that is, if we changed the negative sign in Faraday's law to a positive sign. This seemingly innocent change would have a very profound effect. The induced magnetic field would then add to the change in the magnetic flux, and the direction of the induced current would be as shown in Figure 21.15B. The induced current drawn here is counterclockwise (as viewed from above), and $\vec{F}_B$ would be in the same direction as the velocity of the bar. Hence, giving the bar a slight push and a small velocity would lead to a force $\vec{F}_B$ that *increases* this velocity. A larger $\vec{v}$ would then lead to a larger force, then to a larger velocity, and so forth, and the bar would accelerate to the right in Figure 21.15B, with no limit to the velocity! This acceleration would violate the principle of conservation of energy. We thus conclude that the correct version of Lenz's law, which states that the induced magnetic field always *opposes* changes in the flux, is actually *a consequence of the principle of conservation of energy*!

Mathematically, Lenz's law is just the negative sign in Faraday's law, so the principle of conservation of energy is contained in Faraday's law. This result is amazing because nowhere in Faraday's law or in the other laws of electricity and magnetism is there any explicit mention of energy or its conservation. When these laws are analyzed, however, we find that energy is indeed conserved in all electric and magnetic phenomena, suggesting that the principle of conservation of energy is in some sense a "deeper" principle of physics than Faraday's law or even Newton's laws of mechanics. In fact, as previously mentioned, physicists believe that *all* laws of physics must satisfy the principle of conservation of energy.

21.4 | INDUCTANCE

So far, when calculating the magnetic flux we have assumed that we can ignore the contribution of the induced field to the flux. This assumption is often correct, but in some situations we can't ignore the induced flux. Consider a coil of wire (a solenoid) connected to a battery through a switch as in Figure 21.16. Most real circuits and coils of wire have some nonzero electrical resistance R, shown as a separate resistor in the circuit. If there are no magnets or other sources of magnetic field nearby, the magnetic flux through the coil from all external sources is zero.

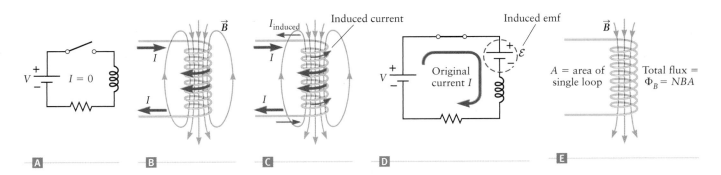

Figure 21.16 **A** A simple circuit containing a battery, switch, solenoid coil, and resistor. When current enters the coil just after the switch is closed, the rapidly increasing magnetic field lines thread through the coil as shown in **B**. This results in an increasing magnetic flux, inducing an opposing emf and current in the coil as shown in **C**. **D** The induced emf can be viewed as a battery ($\mathcal{E}$) with a polarity that opposes changes in current through the coil. A circuit element that acts in this way is called an inductor. **E** If each loop has an area A, the flux through each loop of a solenoid is BA. With N loops stacked up together, the total flux through all the loops is $\Phi_B = NBA$. The total flux determines the induced emf in Faraday's law.

Suppose the switch is initially open, so the current in the circuit is initially zero. If we close the switch, what happens?

When the switch is closed, there is a sudden change in the current in the coil. Figure 21.16B shows just the coil from Figure 21.16A and the current I produced by the battery. This current produces a magnetic field as sketched in the figure with a direction given by right-hand rule number 1 (Chapter 20). The magnetic field was zero before the switch was closed and is nonzero afterwards, so the magnetic flux through the coil increases with time. According to Faraday's law, this change in magnetic flux induces an emf in the coil. Lenz's law dictates that the induced emf must oppose the increase in flux, so the induced magnetic field is directed opposite to the magnetic field in Figure 21.16B. The induced current thus opposes the original applied current I produced by the battery (Fig. 21.16C). Figure 21.16D shows an equivalent way to understand what is happening. The induced emf can be viewed as a battery with a potential $\mathcal{E}$ and a polarity *opposite* to the polarity of the real battery in the circuit. This opposing emf $\mathcal{E}$ produces the opposing current $I_{induced}$ in Figure 21.16C.

The coil in Figure 21.16 is a type of circuit element called an **inductor**. Many inductors are constructed as small solenoids, but almost any coil or loop of wire will act as an inductor. Whenever the current through an inductor changes such as when the switch in Figure 21.16 is closed, a voltage is induced in the inductor that *opposes this change*. This phenomenon is called **self-inductance** because the changing current through a coil produces an induced current in the same coil. The induced current opposes the original applied current, as dictated by Lenz's law.

Inductance of a Solenoid

Let's now use Faraday's law to give a more quantitative description of inductance, based again on the solenoid in Figure 21.16. The induced emf across the solenoid is

$$\mathcal{E} = -\frac{\Delta \Phi_B}{\Delta t} \tag{21.18}$$

We assume the solenoid contains N loops (also called "turns"), and has a length ℓ and a cross-sectional area A. The magnetic field through the solenoid is (Eq. 20.24)

$$B = \frac{\mu_0 N I}{\ell} \tag{21.19}$$

This fields runs through the entire solenoid, whose loops are stacked as shown in Figure 21.16E. The area of a single loop is A, so the magnetic flux through one loop is $\Phi_1 = BA$. The N loops are stacked on top of one another, and there is flux Φ_1 through each loop. The total flux through the solenoid is thus

$$\Phi_B = NBA = \frac{\mu_0 N^2 I A}{\ell} \tag{21.20}$$

Here, N and A are constants, so changes in flux are produced by changes in the current I. Inserting this result for the flux into Faraday's law (Eq. 21.18), we get

$$\mathcal{E} = -\left(\frac{\mu_0 N^2 A}{\ell}\right)\frac{\Delta I}{\Delta t}$$

The factor in parentheses is called the **inductance L** of the solenoid:

$$L = \frac{\mu_0 N^2 A}{\ell} \quad \text{(inductance of a solenoid)} \tag{21.21}$$

The voltage across the solenoid is thus

Definition of inductance L

$$\mathcal{E} = -L\frac{\Delta I}{\Delta t} \tag{21.22}$$

Although the result for inductance in Equation 21.21 applies only for the particular case of a solenoid, the basic notion of inductance is contained in Equation 21.22. By its definition, the inductance L relates the induced emf to changes in the current. This result applies to all coils or loops of wire. Whenever a current produces a magnetic flux through a circuit element, the element will possess an inductance L. The value of L depends on the physical size and shape of the circuit element and on factors such as the number of loops and the area of each loop, just as we found for a solenoid. When the current in an inductor changes with time, the emf across the inductor is proportional to the inductance as given by Equation 21.22. The negative sign in Equation 21.22 indicates the polarity of the voltage. It is conventional practice to speak in terms of the potential "drop" across the inductor, which is given by

$$V_L = L \frac{\Delta I}{\Delta t} \qquad (21.23)$$

Potential drop across an inductor

Because V_L is defined as the voltage *drop*, there is no negative sign in Equation 21.23. This result for V_L will be needed when we construct a circuit theory that deals with inductors.

The unit of inductance is the **henry** (H). Equation 21.23 relates the unit of inductance to the units of voltage, current, and time. Rearranging Equation 21.23 as $L = V_L(\Delta t / \Delta I)$, the units are related by

$$1 \text{ H} = 1 \text{ V} \cdot \text{s/A} \qquad (21.24)$$

EXAMPLE 21.5 | Inductance of a Typical Coil

In a practical electric circuit, inductors are often coils of wire in the form of a solenoid. Consider a solenoid of length $\ell = 2.0$ cm consisting of 1000 turns of wire. Each turn is a circular loop with radius $r = 0.50$ cm. Calculate the value of L for this inductor.

RECOGNIZE THE PRINCIPLE

The inductance of a solenoid depends on its length, its cross-sectional area, and the number of turns. We can find the value of the inductance L by applying Equation 21.21. The point of this calculation is to get a feeling for the value of L in typical cases, which will be useful when we analyze circuits with inductors in Section 21.5.

SKETCH THE PROBLEM

Figure 21.16E describes the problem.

IDENTIFY THE RELATIONSHIPS AND SOLVE

Inserting the given values of the length and number of turns, with a loop area $A = \pi r^2$, we find

$$L = \frac{\mu_0 N^2 A}{\ell} = \frac{(4\pi \times 10^{-7} \text{ T} \cdot \text{m/A})(1000)^2 \pi (5.0 \times 10^{-3} \text{ m})^2}{0.020 \text{ m}}$$

$$L = 4.9 \times 10^{-3} \text{ H} = \boxed{4.9 \text{ mH}}$$

What does it mean?
This value of L is typical for common circuit inductors. An inductance with $L = 1$ H is very large, whereas a value of 1 microhenry (10^{-6} H) is rather small. An inductance of 1 mH (millihenry) is typical.

Mutual Inductance

In an inductor (i.e., a "coil"), the magnetic field produced by the coil produces an induced current in that coil. It is also possible for the magnetic field produced by

Current induced
in coil 2

I_1

I_2

I_2

I_1

Field produced
by coil 1

Figure 21.17 Two coils wound around each other exhibit a mutual inductance. Changes in the current in one coil (coil 1) induce an emf in the other (coil 2).

one coil to produce an induced current in a second coil. Figure 21.17 shows two solenoidal coils, one wrapped around the other. The inner coil (1, shown in black) is connected to a battery and switch. When the switch is closed and the current I_1 increases, this coil produces a magnetic field that gives an increasing flux in the outer coil (2, shown in gray). According to Faraday's law, the changing flux in coil 2 produces an induced emf in that coil. Hence, the current I_1 in coil 1 produces an emf in coil 2, even though the two coils are not "directly" connected by any wires. They are, however, "indirectly" connected through the magnetic flux. This effect is called ***mutual inductance*** because it involves the "mutual" behavior of two separate coils. Mutual inductance is used in devices called transformers, which we'll discuss in Chapter 22.

21.5 | *RL* CIRCUITS

We can apply Kirchhoff's circuit rules from Chapter 19 to circuits that contain inductors by including the voltage drop across an inductor given in Equation 21.23. The basic behavior of such circuits can be described using the following rules.

General behavior of *RL* circuits

Qualitative behavior of DC circuits with inductors

DC circuits may contain resistors, inductors, and capacitors. The voltage source in a DC circuit is a battery or some other source that provides a constant voltage across its output terminals.

1. **Immediately after any switch is closed or opened, the induced emfs keep the current through all inductors equal to the values they had the instant before the switch was thrown.**
2. **After a switch has been closed or opened for a very long time, the induced emfs are zero.**

The first rule above can be understood from the process shown in Figure 21.16. *Whenever there is an attempt to change* the current through an inductor (i.e., by closing or opening a switch), the induced emf across the inductor always opposes this change. Momentarily, this opposition prevents the current through the inductor from changing, and the current just after a switch is thrown is equal to the current just before. This result applies only for the *instant* after a switch is thrown, and is simply a statement of Lenz's law as applied to inductors.

After some time, the current through the inductor will change, eventually approaching a constant value. If I is constant, $\Delta I/\Delta t$ in Equation 21.23 is zero and all currents and magnetic fields are then constant. Hence, the voltage across the inductor is zero, leaving us with DC circuit rule 2.

These two rules apply only to DC circuits; in these circuits, the only time-dependent changes are initiated by the closing or opening of switches. Another important class of circuits, called AC circuits, contain power sources whose electric potentials vary sinusoidally with time. We'll discuss those circuits in Chapter 22, when we generalize our circuit rules to apply to both AC and DC circuits.

Let's use our two circuit analysis rules to study the behavior of the DC circuit sketched in Figure 21.18, which contains a battery, a switch, two resistors, and an inductor. The presence of the resistors and inductor make it an "*RL* circuit." The general circuit symbol for an inductor is a solenoid-like coil. In Figure 21.18A, the switch has been open for a very long time, so the current is zero through both branches of the circuit. At $t = 0$, the switch is closed, inducing a potential across the inductor that is equivalent to a battery with emf $\mathcal{E} = V_L$ (Fig. 21.18B). To find the values of I_1 (the current through resistor 1) and I_2 (the current through resistor 2) just after the switch is closed, we apply rule 1 from the boldfaced list above. At the instant after $t = 0$, the induced emf keeps the current through the inductor at the value it had before the switch was closed. Hence, $I_2 = 0$ just after $t = 0$; at that moment, the entire circuit is equivalent to the lower circuit in Figure 21.18B, and

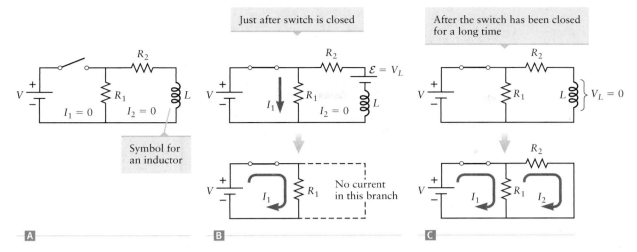

Just after switch is closed

After the switch has been closed for a long time

Symbol for an inductor

No current in this branch

Figure 21.18 **A** An *RL* circuit. **B** Just after the switch is closed, the emf induced in the inductor opposes any increase in current through it. At this instant, the current I_2 is zero. The equivalent circuit *just after* the switch is closed is shown below. **C** After the switch has been closed for a long time, the magnetic flux through the inductor is constant (not changing with time), so the induced emf is zero. The inductor then acts as a wire (a short circuit).

the circuit behaves as if L and R_2 have been removed. The result is a circuit with a battery connected to a single resistor carrying current $I_1 = V/R_1$.

Now let's find the currents after the switch in Figure 21.18 has been closed for a very long time. We apply rule 2 highlighted above and find that the voltage across the inductor is now zero, which is equivalent to replacing the inductor by a simple wire that has zero resistance (lower circuit in Fig. 21.18C). The voltage across each resistor is now the battery emf V, so the currents in the two circuit branches are

$$I_1 = \frac{V}{R_1} \quad \text{and} \quad I_2 = \frac{V}{R_2} \tag{21.25}$$

EXAMPLE 21.6 | Current in an *RL* Circuit

Consider again the circuit in Figure 21.18. Now assume the switch has been closed for a very long time, so the currents through R_1 and R_2 are as given in Equation 21.25 with the directions shown in Figure 21.18C. Find the directions of the currents through R_1 and R_2 the instant after the switch is reopened.

RECOGNIZE THE PRINCIPLE

Whenever a switch is opened or closed, the induced emf across an inductor always acts to keep the current the same as it was just before the switch was thrown. This principle is circuit rule 1 of "Qualitative behavior of DC circuits with inductors," page 704.

SKETCH THE PROBLEM

The current I_2 through the inductor remains at the value it had just prior to the change of the switch (Eq. 21.25), and it will also keep the same direction. Hence, at the instant the switch is opened the current through the branch containing R_2 and the inductor is directed clockwise as sketched in Figure 21.19B.

IDENTIFY THE RELATIONSHIPS AND SOLVE

Applying the equivalent circuit in Figure 21.19B, this current must flow "upward" through R_1, opposite to the direction found when the switch was closed in Figure 21.18B.

Switch is opened

This current flows clockwise around the circuit loop.

Figure 21.19 Example 21.6.

Figure 21.20 Concept Check 21.5.

Figure 21.21 Concept Check 21.6.

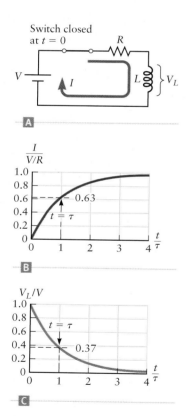

Figure 21.22 **A** An *RL* circuit with the switch closed. **B** Current through the circuit as a function of time. The switch was closed at *t* = 0. **C** Voltage across the inductor as a function of time. When this voltage is positive, it *opposes* the flow of current through the inductor.

What does it mean?

The induced emf in an inductor always opposes changes in the current. In this example, the initial current through *L* has a nonzero value; the induced emf maintains this nonzero current momentarily, even after the switch is opened. At that instant, the battery is disconnected leaving the circuit shown in Figure 21.19B with a nonzero current even though there is no battery! Where does the energy supplying this current come from? We'll explain that in Section 21.6 when we consider how an inductor can store energy.

CONCEPT CHECK 21.5 | Analyzing an *RL* Circuit (1)

Consider the circuit in Figure 21.20. The current is initially zero in both branches. The switch is then closed. Is the current through R_1 just after closing the switch (a) $I = V/R_1$, (b) $I = V/R_2$, (c) $I = V/(R_1 + R_2)$, or (d) $I = 0$?

CONCEPT CHECK 21.6 | Analyzing an *RL* Circuit (2)

Consider the circuit in Figure 21.21. The switch is in position 1 for a very long time and then changed to position 2. What is the direction of the current through the inductor just after the switch is changed?
(a) It is to the right.
(b) It is to the left.
(c) The current is zero just after the switch is thrown.
(d) The direction depends on the values of R_1, R_2, and R_3.

Quantitative Behavior of an *RL* Circuit

Our qualitative rules for dealing with inductors in DC circuits tell how to find the current the moment just after a switch is closed or opened and how to find the current after a switch has been closed or open for a very long time. The current between these two times can also be calculated using Kirchhoff's rules for circuit analysis. Figure 21.22A shows the simplest *RL* circuit, with a battery connected through a switch to a resistor *R* and an inductor *L*. The switch is open for a long time and then is closed at *t* = 0. The quantitative behavior of the current through and voltage across the inductor is shown in parts B and C of Figure 21.22. The current starts at *I* = 0 because the induced emf keeps the current through the inductor momentarily at the value it had before the switch was thrown (in accord with our circuit rule 1). After a very long time, the voltage across the inductor falls to zero (in accord with rule 2) and the current approaches *I* = *V/R*. Mathematically, the current is given by

$$I = \frac{V}{R}(1 - e^{-t/\tau})$$ (21.26)

where e^x is the exponential function and τ is called the ***time constant*** of the circuit. For an *RL* circuit in which a single resistor is in series with a single inductor as in Figure 21.22, the time constant is

$$\tau = \frac{L}{R}$$ (21.27)

The corresponding behavior of the voltage across the inductor is

$$V_L = Ve^{-t/\tau}$$ (21.28)

The time constant τ is the approximate time scale over which the current and voltage change in an *RL* circuit. From the properties of the exponential function (see Appendix B), the current in Equation 21.26 reaches approximately 63% of its final value when $t = \tau$ and the voltage across the inductor (Eq. 21.28) falls to approximately 37% of its initial value after this time. After three time constants

$(t = 3\tau)$, the current reaches 95% of its final value. The time constant τ in Equation 21.27 applies only to an RL circuit. It is *different* from the time constant we encountered for RC circuits in Chapter 19.

EXAMPLE 21.7 | Value of τ for a Typical RL Circuit

Consider an RL circuit with $R = 4000\ \Omega$ and $L = 3.0$ mH; these values are typical of those used in many electronic circuits such as radios and computers. Find the time constant for a circuit that uses these components.

RECOGNIZE THE PRINCIPLE

The time constant for an RL circuit is just the ratio L/R. The purpose of this example is to determine a "typical" value of the time constant τ. Is it hours, seconds, or microseconds? The value of τ is important in applications (see below).

SKETCH THE PROBLEM

No figure is necessary.

IDENTIFY THE RELATIONSHIPS AND SOLVE

We evaluate the time constant τ from Equation 21.27 using the given values of R and L:

$$\tau = \frac{L}{R} = \frac{3.0 \times 10^{-3}\ \text{H}}{4000\ \Omega} = 7.5 \times 10^{-7}\ \text{s} = \boxed{0.75\ \mu\text{s}}$$

What does it mean?

For this circuit, the time constant is approximately 1 microsecond (1 μs). In many applications of RL circuits (Chapter 22), this time constant is associated with the period of an oscillation, so the period of this circuit would correspond to a frequency of about

$$f = \frac{1}{\tau} \approx \frac{1}{1 \times 10^{-6}\ \text{s}} = 1 \times 10^6\ \text{Hz}$$

(1 megahertz). This frequency lies in the middle of the "AM band" on your radio, so such an RL circuit could be useful in an AM radio.

Insight 21.2

BUILDING A REAL INDUCTOR

We calculated L for a simple solenoid inductor in Equation 21.21, but most practical inductors are constructed by wrapping a wire coil around a magnetic material. From Chapter 20, we know that a small magnetic field can align the domains of a magnetic material, producing a much larger magnetic field. Filling a coil with magnetic material greatly increases the magnetic flux through the coil and therefore increases the induced emf. According to the definition of inductance Equation 21.22, the presence of magnetic material thus increases the inductance. For this reason, most inductors contain a magnetic material inside, which produces a larger value of L in a smaller package.

21.6 | ENERGY STORED IN A MAGNETIC FIELD

The RL circuit in Figure 21.23A is the same circuit we analyzed in Figures 21.18 and 21.19. We know that if the switch has been closed for a long time, currents I_1 and I_2 are established through the resistors as given in Equation 21.25. If the switch is then opened at $t = 0$, there is a current in the circuit loop formed by R_1, R_2, and L (Fig. 21.23B); the time dependence of this current is shown qualitatively in Figure 21.23C. After the switch is opened, the battery is completely disconnected from this circuit. Current thus exists even with the battery removed, and electrical

Figure 21.23 **A** If the switch in this RL circuit is closed for a long time, there will be current in the inductor. **B** When the switch is opened, the current through the inductor decays with time as plotted in **C**.

Figure 21.24 ◻A Voltage across and current through an inductor. ◻B If the voltage across the inductor is constant, the current increases linearly with time (Eq. 21.30). ◻C The power delivered to the inductor also increases linearly with time.

energy is dissipated as heat in the resistors during this time. Where does this energy come from if the battery has been disconnected? The answer is that this energy was originally (at $t = 0$) stored in the magnetic field of the inductor.

One way to analyze the energy stored in this magnetic field is to consider the energy required to establish the current in an inductor in the first place. Suppose the voltage across the inductor in Figure 21.24A has a constant value V. To keep V constant, the current must increase with time. We can see that from Equation 21.23; the voltage drop across the inductor is $V = L\,\Delta I/\Delta t$. Solving for the change in current, we have

$$\Delta I = \frac{V}{L}\Delta t \tag{21.29}$$

Hence, if V is constant and the current starts from zero at $t = 0$, Equation 21.29 means that the current through the inductor I must increase linearly with t (Fig. 21.24B), and

$$I = \frac{V}{L}t \tag{21.30}$$

The power P dissipated in a circuit element is equal to the product of the voltage across that element and the current (Eq. 19.15). Hence, the instantaneous power in the inductor—the rate at which energy is delivered to the inductor at time t—is

$$P = VI = V\left(\frac{V}{L}t\right) = \left(\frac{V^2}{L}\right)t \tag{21.31}$$

This linear relation between P and t is plotted in Figure 21.24C. We can use the result of Equation 21.31 to find the total energy that has been delivered to the inductor when the current reaches a particular value I_0 at time t_0. We denote this energy as PE_{ind} since we will soon see that it is a kind of potential energy. This total potential energy PE_{ind} is equal to the average of the power in Figure 21.24C multiplied by the total time t_0. The instantaneous power P varies linearly with time, so the average power is half the maximum power. Using Equation 21.31, we get

$$P_{ave} = \frac{1}{2}\frac{V^2 t_0}{L}$$

$$PE_{ind} = P_{ave}t_0 = \frac{V^2 t_0^2}{2L} \tag{21.32}$$

From Equation 21.30, we find $I_0 = Vt_0/L$, which we rearrange to give $t_0 = I_0 L/V$. Inserting into our expression for PE_{ind} leads to

$$PE_{ind} = \frac{V^2}{2L}\left(\frac{I_0 L}{V}\right)^2 = \frac{1}{2}LI_0^2$$

where I_0 is the final current through the inductor; we can just call it I and then write our final result as

Energy stored in an inductor

$$PE_{ind} = \tfrac{1}{2}LI^2 \tag{21.33}$$

Equation 21.33 is the general result for the magnetic energy PE_{ind} stored in an inductor. It has a form that is very similar to the electric energy stored in a capacitor, $PE_{cap} = \frac{1}{2}CV^2$ (Eq. 18.33). In the case of a capacitor, we saw that this energy is stored in the electric field that exists between the capacitor plates. In a similar way, the energy PE_{ind} is stored *in the magnetic field* of the inductor, and we can rewrite Equation 21.33 to make this more apparent. The inductance of a solenoid is $L = \mu_0 N^2 A/\ell$ (Eq. 21.21). Inserting into our expression for PE_{ind} (Eq. 21.33) gives

$$PE_{ind} = \frac{1}{2}LI^2 = \frac{1}{2}\left(\frac{\mu_0 N^2 A}{\ell}\right)I^2$$

which can be rearranged as

$$PE_{ind} = \frac{1}{2}\left(\frac{\mu_0 N^2 A}{\ell}\right)I^2 = \frac{1}{2\mu_0}\left(\frac{\mu_0 NI}{\ell}\right)^2 (A\ell)$$

We have rewritten the potential energy in this way because the factor involving I in parentheses is the solenoid's magnetic field ($B = \mu_0 NI/\ell$, Eq. 21.19), whereas the last factor $A\ell$ is the volume of the solenoid (the length times the cross-sectional area). So we get our final result,

$$PE_{mag} = \frac{1}{2\mu_0} B^2 (\text{volume}) \qquad (21.34)$$

Energy stored in a magnetic field

In Equation 21.34, we have now denoted the energy as PE_{mag}. It is the energy contained in the magnetic field of the inductor and applies not only to an inductor but to any region of space that contains a magnetic field. This energy is stored *in the magnetic field*. Because a magnetic field can exist in a vacuum, this potential energy can exist even in "empty" regions containing no matter! This fact, which also applies to electric fields and electric potential energy, will be important in understanding electromagnetic waves (Chapter 23).

Another way to express the potential energy in Equation 21.34 is to say that there is a certain *energy density* (the energy per unit volume) in the magnetic field. From Equation 21.34, we find

$$\text{energy density} = u_{mag} = \frac{PE_{mag}}{\text{volume}} = \frac{1}{2\mu_0} B^2 \qquad (21.35)$$

This expression for the magnetic energy density is very similar in form to the energy density contained in an electric field. In Chapter 18 we found that $u_{elec} = \frac{1}{2}\varepsilon_0 E^2$ (Equation 18.48), so there is an interesting parallel between our results for magnetic and electric energies. In both cases, the energy density is proportional to the square of the field.

EXAMPLE 21.8 | ⊗ ® Magnetic Field Energy: How Large Is It?

To get a feeling for the magnitude of the energy stored in a magnetic field, consider the energy stored in a magnet used for magnetic resonance imaging (MRI) (Fig. 21.25). A typical MRI magnet is a solenoid providing a field of about $B = 1.5$ T. **(a)** What is the approximate energy stored in this magnetic field? **(b)** If the current through the magnet during normal operation is $I = 100$ A, what is its inductance?

RECOGNIZE THE PRINCIPLE

The motivation for this example is to understand how the potential energy stored in a magnetic field compares with cases familiar from mechanics such as the gravitational potential energy or the kinetic energy of a moving object. To find the energy stored in an MRI magnet, we need to know its magnetic field and its inside volume. The volume is not given, but we can estimate its value because a person must be able to fit inside (Fig. 21.25).

SKETCH THE PROBLEM

Figure 21.25 shows a typical MRI magnet. We can estimate the volume from this photo.

IDENTIFY THE RELATIONSHIPS

We can find the stored magnetic energy using our result for the potential energy PE_{mag} (Eq. 21.34).

SOLVE

(a) From personal experience (and the photo in Fig. 21.25), we estimate that a typical MRI magnet is about 1.5 m long with an inside diameter of about 0.5 m so that

Figure 21.25 Example 21.8. The magnets used for magnetic resonance imaging (MRI) are very large, with a large amount of stored magnetic energy.

a person can fit inside. We have volume = $A\ell = (\pi r^2)\ell$, and inserting our estimates for r and ℓ gives

$$\text{volume} = (\pi r^2)\ell = \pi(0.25\ \text{m})^2(1.5\ \text{m}) = 0.3\ \text{m}^3$$

Using this volume in Equation 21.34 along with the given value of B, we find

$$PE_{\text{mag}} = \frac{1}{2\mu_0}B^2(\text{volume}) = \frac{1}{2(4\pi \times 10^{-7}\ \text{T} \cdot \text{m/A})}(1.5\ \text{T})^2(0.3\ \text{m}^3)$$

$$PE_{\text{mag}} = \boxed{3 \times 10^5\ \text{J}}$$

(b) The magnetic energy found in part (a) is stored in an inductor because the MRI magnet is just a solenoid. Therefore, $PE_{\text{mag}} = PE_{\text{ind}}$ is related to the inductance through Equation 21.33. Using our calculated value of PE_{mag} gives

$$\tfrac{1}{2}LI^2 = PE_{\text{mag}}$$

$$L = \frac{2PE_{\text{mag}}}{I^2} = \frac{2(3 \times 10^5\ \text{J})}{(100\ \text{A})^2} = \boxed{60\ \text{H}}$$

This inductance is large, much larger than we found for the small coil in Example 21.5.

What does it mean?
The amount of energy found in part (a) is quite large; it is approximately equal to the kinetic energy of a car of mass 1000 kg with a velocity of 50 mi/h (23 m/s)! An MRI magnet is carefully designed to be sure this energy is dealt with safely. In addition, patients undergoing an MRI exam as well as MRI technicians must take precautions such as removing all magnetic objects from their pockets.

21.7 | APPLICATIONS

In this section, we consider a few important applications of Faraday's law.

Bicycle Odometers

Most serious bicyclists use an odometer to monitor their speed and distance traveled. Figure 21.26A shows a photo of the odometer control unit, and a sketch of the working elements is shown in Figure 21.26B. A small, permanent magnet is

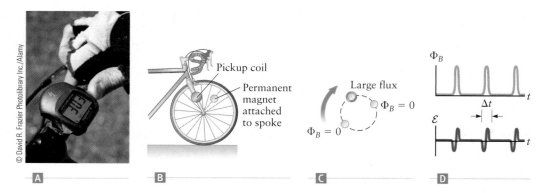

Figure 21.26 **A** This bicycle odometer uses Faraday's law to count the rotations of a bicycle wheel. **B** A small, permanent magnet is attached to a spoke, and a pickup coil is mounted on the bicycle frame. **C** When the magnet passes near the pickup coil, the time-varying magnetic flux induces an emf in the coil. **D** Qualitative sketch of the flux through the pickup coil as a function of time. This time-dependent flux produces an induced emf that varies with time as a series of pulses. The odometer counts these pulses.

attached to one of the spokes of a wheel, and a "pickup" coil is mounted on the axle support. On the author's bike, this pickup unit is contained in a piece of opaque plastic with a small coil of wire inside. As the wheel turns, the permanent magnet on the spoke travels in a circular path that takes it past the pickup coil once during each revolution of the wheel. When this magnet is close to the pickup coil (Fig. 21.26C), its magnetic field penetrates the coil, giving a large flux, whereas the flux through the coil is zero when the magnet is far from the coil. A qualitative sketch of the flux through the coil as a function of time is shown in the upper graph of Figure 21.26D. Because a pulse occurs each time the magnet passes over the pickup coil, there is one "pulse" in this plot for each revolution of the wheel.

From Faraday's law, we know that the time-varying magnetic flux through the pickup coil induces an emf in the coil. A corresponding plot of this induced emf is shown in the lower graph of Figure 21.26D. Because $\mathcal{E} = -\Delta\Phi_B/\Delta t$, the induced emf is negative while the flux is increasing (Lenz's law) and positive as Φ_B falls back to zero. The pickup coil is connected to an electronic circuit and a computer that keeps track of the number of pulses in the emf signal, and also measures the time between pulses. This information is then displayed as the bicycle's speed and distance traveled.

EXAMPLE 21.9 | Ⓡ Design of a Bicycle Odometer

Let's calculate the emf (voltage) induced in the odometer pickup coil in Figure 21.26 to show that this odometer design is actually practical. For a typical permanent magnet, the field very near one of the poles is about 1 T. In practice, the permanent magnet attached to the spoke must not be too close to the frame of the bicycle (so they don't hit each other), which reduces the field at the pickup coil. To be conservative, we assume a field $B = 0.01$ T at the coil. The odometer pickup on the author's bicycle has an area of about $A = 0.1$ cm^2 (about 3 mm on a side). We also assume the coil of wire inside consists of $N = 100$ loops.

RECOGNIZE THE PRINCIPLE

The odometer signal is due to Faraday's law: the voltage induced in the pickup coil is equal to the rate of change of the magnetic flux as the odometer magnet moves past. To find $\Delta\Phi_B/\Delta t$, we need to know B and the area of the coil, which are both given. We also need to know Δt, which is the approximate time the magnet "overlaps" with the coil. We can estimate Δt from the angular speed of the bicycle wheel.

SKETCH THE PROBLEM

Figure 21.26 describes the problem. Following our "Applying Lenz's Law" problem-solving strategy, Figure 21.26C shows the pickup coil. We will use this area when calculating the induced emf.

IDENTIFY THE RELATIONSHIPS

To find the change in flux, we need to calculate Φ_B when the odometer magnet is directly over the pickup coil and when it is far away. When the magnet far away, the flux is zero; when it is over the coil, the magnetic flux through one loop of the coil is BA. So, for N loops, the total flux is $\Phi_B = NBA$. Inserting our approximate values for N, B, and A, we get

$$\Phi_B = NBA = (100)(0.01 \text{ T})(1 \times 10^{-5} \text{ m}^2) = 1 \times 10^{-5} \text{ T} \cdot \text{m}^2 \qquad (1)$$

which is the flux at the peaks in Figure 21.26D and also the change in flux $\Delta\Phi_B$ when the magnet passes by the pickup coil.

To apply Faraday's law, we must now estimate the time Δt over which this flux change takes place. This time interval is indicated in Figure 21.26D; it depends on the speed of wheel, and it decreases as the bicycle travels faster. A slowly moving bicycle might travel at 5 mph, which is about 2 m/s. The wheels of the author's bike have a radius of about $R = 50$ cm, so the angular velocity at this speed is (from Eq. 8.43)

$$\omega = \frac{v}{R} = \frac{2 \text{ m/s}}{0.5 \text{ m}} = 4 \text{ rad/s}$$

The magnet on the spoke is approximately halfway from the edge of the wheel at $R_{magnet} = 0.25$ m, so its speed is

$$v_{magnet} = \omega R_{magnet} = (4 \text{ rad/s})(0.25 \text{ m}) = 1 \text{ m/s}$$

A reasonable diameter for the permanent magnet is about $d = 1$ cm (a value taken from the author's bike), so the time it takes for the magnet to move past the sensor is approximately

$$\Delta t = \frac{d}{v_{magnet}} = \frac{0.01 \text{ m}}{1 \text{ m/s}} = 0.01 \text{ s} \tag{2}$$

SOLVE

Inserting these results for $\Delta\Phi_B$ from Equation (1) and Δt into Faraday's law gives the approximate magnitude of the induced emf,

$$|\mathcal{E}| = \frac{\Delta\Phi_B}{\Delta t} = \frac{1 \times 10^{-5} \text{ T} \cdot \text{m}^2}{0.01 \text{ s}} = \boxed{0.001 \text{ V}}$$

What does it mean?
A voltage of this size can be easily measured (as your electrical engineering friends can verify), so this odometer design is indeed practical.

CONCEPT CHECK 21.7 | Changing Speeds with the Bicycle Odometer

If the speed of the bicycle in Example 21.9 is increased, how does the amplitude of the induced emf in the pickup coil change?
  (a) The induced emf gets smaller.
  (b) The induced emf gets larger.
  (c) The induced emf does not change.

Ground Fault Interrupters

A ground fault interrupter (GFI) is a safety device used in many household circuits (Fig. 21.27A). A GFI is often built into an electric power outlet in bathrooms, kitchens, and other places where there is an increased risk of accidental contact with a substantial electric potential. A GFI uses Faraday's law together with an electromechanical relay (Fig. 20.45). Suppose a hair dryer is plugged into a wall socket that contains a GFI as in Figure 21.27B. This connection contains two current-carrying wires, with current directed from the wall socket to the hair dryer through one wire and back to the wall socket through the other wire. It is just like the circuits we have drawn using batteries; all have two wires connecting the battery to the rest of the circuit. During normal (safe) operation of the hair dryer, the two connecting wires carry equal currents, but in the event of an accident, some current might pass through your body and then to ground. In that case, the current in the return wire in Figure 21.27B is much smaller than the current in the outgoing wire. A GFI senses this difference and immediately turns off power to the entire circuit so that no permanent harm is done.

A GFI accomplishes its task by using a relay to control the current through the circuit. In Chapter 20, we saw that relays use the current through a coil to exert a force on a magnetic metal bar. This bar is part of a switch, so the current through the coil controls whether the switch is open or closed. The relay in a GFI contains

Figure 21.27 **A** Ground fault interrupters (GFIs) use a specially designed inductor to detect circuit failure. **B** Most GFI units are built into a wall socket, so you plug an appliance, such as a hair dryer, directly into the GFI. **C** Two coils of wire, one carrying the outgoing current and another carrying the return current, are used to power a relay in the GFI. If an appliance fails, the current through the return coil drops to zero and the relay is activated, turning off current to the appliance.

two separate coils; the current through one coil is from the outgoing wire to the hair dryer, and the current through the other coil is from the return circuit. The two coils are wound in opposition so that their magnetic fields are in opposite directions. During normal (safe) operation, the currents in the two coils are equal and their fields cancel, giving zero magnetic field in the relay. The relay is designed so that its switch is closed in this case, allowing current in the hair dryer circuit. If there is an accident, however, the current in the return coil is smaller than the outgoing current, leading to a nonzero magnetic field from the coil. The relay switch then opens (Fig. 21.27C), turning off the current to the hair dryer.

The inductor in a GFI thus employs two separate coils of wire wound in opposite directions. This type of winding is used in other applications in which one wants to make the inductance of a device as small as possible.

Electric Guitars

An electric guitar (Fig. 21.28A) uses Faraday's law to sense the motion of the strings. A guitar string is made from steel, a magnetic material. The string passes near a "pickup" coil wound around a permanent magnet. The field from this magnet causes nearby regions of the string to be magnetized as shown in Figure 21.28B. This induced magnetism in the steel of the guitar string is just like the induced magnetism in the steel panel of a refrigerator or of any other magnetic material containing magnetic domains (see Fig. 20.39). The magnetized region of the string then produces its own magnetic field, which gives a magnetic flux through the pickup coil. As the string vibrates, this flux changes with time, leading to an induced emf (from Faraday's law) in the coil. This emf is sent to an amplifier, and the resulting signal can then be played through speakers.

Generators, Motors, and Hybrid Cars

In Chapter 20, we showed how a magnetic field can exert a torque on a current loop that can be used to make a motor and thus produce mechanical energy (Fig. 20.47). The "opposite" process takes place in a generator, which uses an externally applied mechanical torque on a current loop to produce an electric current and thus produce electrical energy. A motor and generator therefore provide examples of the conservation of energy and the conversion of energy from one type to another, mechanical to electrical and vice versa. So, it is no surprise that an ideal generator

Figure 21.28 **A** The strings of an electric guitar are made of steel, a magnetic material. **B** When the string vibrates, it produces a time-varying magnetic flux through the pickup coil (called a guitar "pickup"), producing a voltage in the coil that is converted into sound by the guitar's amplifier. In part A the pickups are the small circular regions, with three pickups under each string.

Figure 21.29 A hybrid car such as this Toyota Prius contains a gasoline powered motor along with an electric motor, a generator, and batteries.

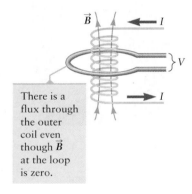

There is a flux through the outer coil even though $\vec{B}$ at the loop is zero.

Figure 21.30 This long, thin solenoid produces a magnetic flux through the larger-diameter loop that encircles it. If this flux varies with time, an emf is induced in the outer loop, even though there is essentially no magnetic field there.

can take a certain amount of mechanical energy and convert it to the same amount of electrical energy (and vice versa for the motor). This conversion back and forth is used in so-called hybrid cars to minimize the amount of energy "lost" to friction and hence improve fuel economy.

A hybrid car contains two motors (Fig. 21.29); one motor is a conventional gasoline powered motor (also called an "engine"; see Chapter 16), and the other is an electric motor powered by batteries. A hybrid car also contains a generator attached to the wheels, usually the front wheels, which are the ones connected to the motors. Starting from a standstill, either the gas motor or electric motor (or both) are used to accelerate the car. At low speeds, the electric motor is the more efficient motor, so the gas motor usually shuts off automatically for those periods. When the brakes are applied or the car travels downhill, the generator is switched on and the emf it produces is used to recharge the batteries. In an ordinary car, applying the brakes causes the car's kinetic energy to be converted to heat energy through friction. A hybrid car "recaptures" this energy, storing it in the batteries so that it can be used later. A hybrid car is thus a good practical example of the conversion between mechanical energy and electrical energy.

Of course, this conversion can never be perfect; no generator is perfect, so all the mechanical energy of braking can never be converted to electrical energy. Even so, the hybrid approach can cut a car's gasoline consumption by 30% or even more, which is a significant savings.

21.8 | THE PUZZLE OF INDUCTION FROM A DISTANCE

A solenoid carrying a current produces a large magnetic field in its interior and a very small field outside. By making the solenoid very long, the field outside can be made extremely small. Suppose a very long solenoid is inserted at the center of a single loop of wire as sketched in Figure 21.30. This example is of mutual inductance (Fig. 21.17) because the field of the inner solenoid produces a flux through the outer loop and changes in the current through the inner solenoid thus induce an emf in the loop. In Chapter 20, however, we saw that for a long solenoid, the field outside is very small. Hence, with the very long solenoid in Figure 21.30, the field from the solenoid *at the outer loop* is essentially zero. Even so, the field inside the solenoid at the center of the loop still produces a magnetic flux through the inner portion of the loop.

The solenoid field is proportional to the current I (Eq. 21.19), and let's assume I is changing and so this field is changing with time. The flux through the outer loop then changes with time, inducing an emf in the outer loop. This emf can be used to power an electric circuit (we might attach the loop to a lightbulb), so there is electrical energy associated with this induced emf. You should suspect that this energy was somehow transferred from the solenoid to the outside loop, but how is this energy transferred across the empty space between the two conductors? What's more, there is no magnetic field from the solenoid at the outside loop, so how does the solenoid have any influence at all on that loop?

The answer to this puzzle is that energy is carried from the solenoid to the outer loop by an ***electromagnetic wave.*** Just as with mechanical waves (Chapters 11 and 12), electromagnetic waves transport energy from one place to another. We'll explore electromagnetic waves and how they propagate in Chapter 23.

Magnetic flux and Faraday's law

A changing magnetic flux produces an electric field. This phenomenon is called *induction*. The induced electric potential is given by *Faraday's law,*

$$\mathcal{E} = -\frac{\Delta \Phi_B}{\Delta t} \qquad \textbf{(21.5)} \text{ (page 690)}$$

where $\mathcal{E}$ is the *induced emf* along a closed path and Φ_B is the *magnetic flux* through the area enclosed by that path.

The magnetic flux due to a field of magnitude B through a surface of area A is given by

$$\Phi_B = BA \cos \theta \qquad \textbf{(21.2)} \text{ (page 689)}$$

where θ is the angle that $\vec{B}$ makes with the direction normal to the surface.

A changing magnetic flux can be produced if B is changing, the size of the area is changing, or the angle between $\vec{B}$ and the area is changing. In all three cases, the change in flux produces an induced emf according to Faraday's law.

Magnetic flux
$\Phi_B = BA \cos \theta$

Lenz's law

Lenz's law says that the emf induced by a changing magnetic flux always tends to produce a magnetic field that opposes the change in flux. The key to Lenz's law is that the induced emf always opposes externally imposed changes.

Energy stored in a magnetic field

The energy stored in a magnetic field B in a particular volume of space is

$$PE_{\text{mag}} = \frac{1}{2\mu_0} B^2 (\text{volume}) \qquad \textbf{(21.34)} \text{ (page 709)}$$

Inductors

An *inductor* is a circuit element such as a coil or solenoid that generally consists of one or more loops of wire. When applied to an inductor with inductance L, Faraday's law gives the voltage drop across the inductor as

$$V_L = L \frac{\Delta I}{\Delta t} \qquad \textbf{(21.23)} \text{ (page 703)}$$

There is a nonzero voltage across an inductor only when the current is changing with time. The voltage induced across the inductor opposes changes in the applied current.

DC circuits with inductors

The qualitative behavior of a DC circuit containing inductors can be described using two general rules:

1. Immediately after any switch is closed or opened, the induced emfs keep the current through all inductors equal to the values they had the instant before the switch was thrown.
2. After a switch has been closed or opened for a very long time, the induced emfs are zero.

(Continued)

Time constant for an RL circuit

The time dependence of an RL circuit is characterized by the **time constant τ**.

$$\tau = \frac{L}{R} \qquad\qquad \textbf{(21.27)} \text{ (page 706)}$$

Quantitatively, when the switch in an LR circuit is closed at $t = 0$, the current varies as

$$I = \frac{V}{R}(1 - e^{-t/\tau}) \qquad\qquad \textbf{(21.26)} \text{ (page 706)}$$

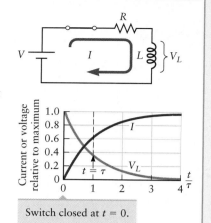

Switch closed at $t = 0$.

Energy stored in an inductor

The energy stored in an inductor (in its magnetic field) is

$$PE_{ind} = \tfrac{1}{2}LI^2 \qquad\qquad \textbf{(21.33)} \text{ (page 708)}$$

QUESTIONS

SSM = answer in Student Companion & Problem-Solving Guide   ⊗ = life science application

1. Describe an example in which the magnetic flux through a surface is zero and the magnetic field is not zero.

2. An aluminum disk is held between the poles of a powerful magnet and then pulled out. The person pulling on the disk feels an opposing force due to currents induced through Faraday's law. Explain why the magnitude of the force becomes larger as the disk is pulled faster.

3. SSM Modern coin vending machines use a configuration of magnets to help sort appropriate currency from fake or foreign coins. When a coin is dropped through the slot, it passes within a few millimeters of the magnets as it falls. Sensors (light sensitive) detect the passage of the coin as it breaks a beam at the top and bottom of the magnet assembly. How might the magnets help reject fake or foreign coins? How would the time of passage of a zinc coin compare with that of a coin made with copper in the presence of the magnetic field? (Modern pennies are copper-plated zinc, and modern dimes and quarters have a layer of copper sandwiched between the face and reverse.)

4. The inductance L of a solenoid is proportional to the square of the number of turns, whereas the energy stored in the solenoid PE_{ind} is proportional to L. (a) If the length of the solenoid is doubled, by what factors do L and PE_{ind} change? (b) Explain physically why the factors in part (a) are larger than 2.

5. The switch in Figure Q21.5 is initially closed for a very long time. Explain why Faraday's law can cause sparks when the

Figure Q21.5

switch is then opened. *Hint:* Assume the switch contacts involve two pieces of metal that are pulled apart.

6. Many trains use "magnetic brakes." Explain how Faraday's law could be used to give a "braking force." *Hint:* Consider the sliding bar in Figure 21.6 and imagine that it is part of the train.

7. A moving charged particle produces an electric current and hence an electric field. Suppose a positively charged particle passes over a circular current loop as sketched in Figure Q21.7. Make a qualitative sketch of the current through the loop as a function of time. Be sure to indicate the time at which the particle is directly over the center of the current loop.

Figure Q21.7

8. A conducting loop is pulled at constant velocity through a region in which the magnetic field is a constant, both inside and outside the loop (Fig. Q21.8). Explain why the induced emf is zero.

Figure Q21.8

9. Faraday and many other scientists expected that there should be a certain symmetry between electric and magnetic phenomena. Is there a magnetic analog of electric charge? That is, is there such a thing as "magnetic charge"?

10. You are given a length of wire L and told to form it into a current loop so as to produce the maximum possible emf when the loop is placed in a time-varying magnetic field. You investigate the three loop geometries in Figure Q21.10: a long, thin rectangle; a square; and a circle. Which one will give the largest emf?

Figure Q21.10

11. Consider a circular path as sketched in Figure Q21.11. This path forms the boundary for a disk and for a hemisphere as well as many other possible surfaces. If the magnetic flux through the disk is $\Phi_{B, \text{disk}}$ and the flux through the hemisphere is $\Phi_{B, \text{hemi}}$, how are these two fluxes related? Give a physical reason to explain your answer.

Figure Q21.11

12. A loop of copper passes through regions of magnetic field with a constant velocity as indicated in Figure Q21.12. (a) Determine the direction of the current (if any) induced in the copper loop at each point, 1 through 5, along its path. Assume the field is uniform and of constant magnitude inside the dashed boundary and zero outside. (b) Rank the currents according to their magnitude from greatest to least for points 1 through 5. Indicate ties, if any.

Figure Q21.12 Questions 12 and 13.

13. Consider again the copper loop in Figure Q21.12, but this time assume the loop has a constant acceleration. Again determine the direction of the current (if any) induced in the copper loop at each point 1 through 5 along its path. Again assume the field is uniform inside the dashed boundary and zero outside.

14. SSM A popular classroom demonstration of Lenz's law involves a vertical copper tube and a permanent magnet that fits very loosely inside the tube. The instructor first shows that a permanent magnet is not attracted to the copper of the tube. The magnet is then dropped down the tube and takes a surprisingly long time to come out the bottom end of the tube. When repeating the demonstration, a student looks down the tube and sees that the magnet actually moves at a constant and rather slow velocity and does not come in contact with the tube. Explain how this demonstration illustrates Lenz's law. Why doesn't the magnet come to a complete stop?

15. The demonstration described in Question 21.14 often has some additional components, sometimes including tubes of aluminum and lead with the exact dimensions of the copper tube. The magnet falls at a different rate through the different tubes. Why? Rank the tubes from least to greatest according to the time interval for the magnet to fall through the length of each. Table 19.1 may be useful.

16. You are hired by an amusement park and asked to design a ride similar to that sketched in Figure Q21.16. A horizontal metal rod slides along two sloped, frictionless rails with a constant

Figure Q21.16

magnetic field directed as shown. A rider sits on the bar and slides down along the rails. Magnetic induction leads to a force on the bar. Use Lenz's law to show that this force is directed upward along the rails and hence keeps the rider from reaching excessive speeds.

17. When a large, time-varying current flows through a coil, a piece of steel placed within the coil will quickly become red hot (Fig. Q21.17). Why? If a piece of copper were placed in the coil, would the effect be the same? Explain.

Figure Q21.17

18. Induction range. An alternative to gas or electric stove burners, which have an energy transfer efficiency of less than 50%, are magnetic induction cooktops (Fig. Q21.18), which can be 90% efficient. The "burner" surface does not get hot at all; only the pan gets hot. The main component of an induction range is a coil just under the surface. Describe how this device might work, taking into account that steel and iron cookware must be used. Although the surface never gets hot, is this method of heating safer than gas or conventional electric heating elements? (Consider the effect on a fork or spoon left misplaced while someone is cooking near the coil.)

Figure Q21.18 Chocolate in the metal cooking pan is melted while the chocolate in direct contact with the induction burner remains cool and solid.

19. In Chapter 19, we explained how an RC circuit could be used as a filter. The RL circuit in Figure Q21.19 can also be used as a filter. Give a qualitative argument to explain why this circuit will filter out rapid fluctuations in the input voltage.

Figure Q21.19

20. In an electric circuit, the largest induced voltage usually occurs just after a switch is opened or closed. Explain why.

21. The induced emf is an electric potential difference, so it should have units of volts (V). Show that the right-hand side of Faraday's law (Eq. 21.5) does indeed have the correct units.

22. Use the result for the current in an RL circuit in Equation 21.26 to derive the voltage across the inductor as given in Equation 21.28.

23. The roads near many traffic lights contain buried sensors that detect when a car is present (waiting at a stop light). These sensors are large loops of wire just beneath the surface of the road. (a) Explain how they can use Faraday's induced emf to detect the presence of a car. *Hint:* Cars contain a lot of steel, which is magnetic. (b) Explain why these sensors sometimes fail to detect a motorcycle and always fail to detect bicycles.

21.2 MAGNETIC FLUX AND FARADAY'S LAW

1. A rectangular loop of wire that is 35 cm wide and 15 cm long is placed in a region where the magnetic field is $B = 1.2$ T and directed perpendicular to the plane of the loop. What is the magnitude of the magnetic flux through the loop?

2. Suppose the wire loop in Problem 1 lies in the x–y plane and that the magnetic field now makes an angle of 60° with the z axis. Find the magnitude of the magnetic flux.

3. ⭐ The magnetic flux through a current loop is $\Phi_B = 5.5$ T · m². If the magnetic field has a magnitude $B = 0.85$ T, what is the smallest possible value for the area of the loop?

4. ⭐ The magnetic field in a particular region is independent of position, but varies in time as sketched in Figure P21.4. If a loop of area 0.24 m² is oriented perpendicular to this magnetic field, what is the approximate magnitude of the flux through the loop at $t = 0.25$ s?

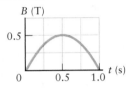

Figure P21.4
Problems 4, 5, and 10.

5. ✪ Consider again the current loop described in Problem 4. What is the approximate magnitude of the induced emf at (a) $t = 0$ and (b) $t = 0.7$ s? (c) How are the signs of these two emfs related?

6. ⭐ A bar magnet is thrust into a current loop as sketched in Figure P21.6. Before the magnet reaches the center of the loop, what is the direction of the induced current as seen by the observer on the right, clockwise or counterclockwise?

Observer

Figure P21.6 Problems 6, 7, and 8.

7. ⭐ Consider again the bar magnet and loop in Figure P21.6. Make a qualitative sketch of the current through the loop as a function of time. Assume the bar magnet starts on the far left and continues to the far right and take a clockwise current as viewed from the right as positive.

8. ⭐ Repeat Problem 7, but assume the leading end of the bar magnet is now the south magnetic pole.

9. SSM ⭐ ✖ ℝ Magnetic resonance imaging (MRI) devices use a very large magnet that produces a large magnetic field. Consider an MRI magnet that can accommodate a person. Typically, the person lies down and the field is parallel to her spine (Fig. 21.25). If the magnetic field has a magnitude $B = 1.2$ T, what is the approximate flux through the person?

10. ✪ ℝ A current loop with a resistance $R = 350\ \Omega$ and an area $A = 0.20$ m² is oriented perpendicular to a magnetic field that varies in time as sketched in Figure P21.4. What is the approximate current induced in the loop at (a) $t = 0.30$ s and (b) $t = 0.50$ s?

11. ✪ ℝ Consider a magnetic field that varies sinusoidally with time with an amplitude (peak value) of 0.30 T and a frequency of 400 Hz. If a current loop of area $A = 2.5$ m² is oriented perpendicular to this field, what is the approximate value of the maximum induced emf in the loop?

12. A single loop of wire is placed in a time-varying magnetic field, and an induced emf of 3.5 V is observed. Suppose 25 identical loops with the same orientation are now connected in series. What is the value of the total induced emf?

13. ✪ A square current loop (edge length 45 cm) is found to have an induced emf of 1.2 V at $t = 0$. (a) Assuming the magnetic field is perpendicular to the loop and that the magnitude of the field varies linearly with time, what is the change in the magnetic field from $t = 0$ to $t = 0.10$ s? (b) Explain why it is only possible to compute the change in B and not its absolute magnitude.

14. SSM ⭐ A metal bar of length $L = 1.5$ m slides along two horizontal metal rails as shown in Figure P21.14. There is a magnetic field of magnitude $B = 2.3$ T directed vertically. (a) If the bar is moving at a speed $v = 0.54$ m/s, what is the emf induced across the ends of the bar? (b) Which end of the bar is at the higher potential?

Figure P21.14 Problems 14 and 22.

15. ✪ The Earth's magnetic field is typically 5×10^{-5} T, but fluctuates with time. Suppose the fluctuations in the Earth's field are sinusoidal with a frequency of 10^4 Hz and an amplitude of 5% of the magnitude of the Earth's field. What is the approximate magnitude of the induced emf in a current loop of area 1 cm²?

21.3 LENZ'S LAW AND WORK–ENERGY PRINCIPLES

16. A metal loop is placed in a perpendicular magnetic field as sketched in Figure P21.16. (a) We wish to find the induced current through the loop. What closed path and which corresponding area should we choose for the application of Faraday's law? (b) What is the direction of the magnetic flux through the area chosen in part (a)? That is, is the flux upward (out of the plane of the figure) or downward (into the plane)? (c) If the magnitude of B is increasing with time, is this flux increasing or decreasing with time? (d) What is the direction of the magnetic field produced by the induced current? (e) Will the induced current in the loop be clockwise or counterclockwise?

Figure P21.16
Problems 16 and 17.

17. SSM ⭐ The loop in Figure P21.16 is rotated out of the plane while B is kept constant. (a) Does the flux through the loop increase or decrease with time? (Assume the loop has not had time to rotate by more than 90°.) (b) Will the induced current be clockwise or counterclockwise?

18. ⭐ A circular loop of wire lies in the x–y plane so that the axis of the loop lies along z (Fig. P21.18). A magnetic field of magnitude B is parallel to the z axis. This field is positive (i.e., along the +z direction) and is

Figure P21.18

the same everywhere within the area of the loop. If B is decreasing with time, what is the direction of the induced current in the loop as viewed from above?

19. A long, straight, current-carrying wire passes near a wire loop (Fig. P21.19). The current in the straight wire is directed to the right, and its magnitude is decreasing. (a) What is the direction of the magnetic flux through the loop? (b) Is the magnitude of the flux through the loop increasing or decreasing with time? (c) What is the direction of the magnetic field produced by the induced current in the loop? (d) What is the direction of the current induced in the loop? (e) This induced current leads to a magnetic force on the loop. What is the direction of this force? Is it directed toward the wire or away from it?

Wire loop

Figure P21.19 Problems 19, 20, and 21.

20. ☆ Consider again the situation in Figure P21.19, but suppose the current I is constant and the loop is pulled away from the wire. (a) What is the direction of the current induced in the loop? (b) This induced current leads to a magnetic force on the loop. What is the direction of this magnetic force? Is it directed towards the wire or away from it?

21. ✪ In Problem 19, there is a current I_{ind} induced in the loop. If there is a resistance R in the loop, there will be a certain amount of power $P = I_{ind}^2 R$ dissipated in the resistor. Does this energy come from (a) the force exerted on the loop by the magnetic field of the wire, (b) the voltage source that produces the current in the long straight wire, or (c) the gravitational potential energy of the loop?

22. Suppose the metal rails in Figure P21.14 are now connected on the left so that a complete conducting path (through the bar and the rails) is formed. (a) If we wish to find the induced current through the bar, what closed path should we choose for the application of Faraday's law? (b) What is the direction of the flux through the path chosen in part (a)? That is, is the flux upward or downward relative to the plane of the rails? (c) Is the magnitude of the flux increasing or decreasing with time? (d) What is the direction of the magnetic field produced by the induced current? (e) If the resistance of the loop is 1500 Ω, what are the magnitude and the direction of the current through the bar?

23. A metal bar is pulled through a region in which the magnetic field is perpendicular to the plane of motion (Figure P21.23). There is an induced emf across the bar, with the top of the bar positive relative to the bottom. Is $\vec{B}$ directed into or out of the plane of the drawing?

Figure P21.23

24. SSM ☆ A metal bar of length L is attached to two vertical metal rails as sketched in Figure P21.24. The connections are frictionless, but allow current to flow between the rails and the bar. There is a constant horizontal magnetic field directed perpendicular to the area enclosed by the bar and rails. (a) What closed path should we use for applying Faraday's law to find the current induced in the bar? (b) Suppose the bar is at a height z above the bottom. Find the magnitude and direction of the magnetic flux Φ_B. (c) If the bar is moving downward at a speed v, is the flux in part (b) increasing or decreasing with time? (d) What is the induced emf across the bar? (e) Use your answer to part (d) to get the current I in the circuit. Be sure to give the direction of I. (f) Now find the magnetic force on the bar. Notice that the magnitude of this force increases as v increases. (g) What is the direction of the force on the bar? (h) If the bar has a mass m, there will also be a gravitational force on the bar. Eventually, the bar will reach a

speed v at which the gravitational and magnetic forces cancel so that the total force on the bar is zero. Find v.

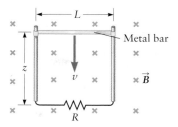

Metal bar

Figure P21.24

25. ✪ A metal bar of length $L = 0.60$ m slides freely on two frictionless, horizontal metal rails as shown in Figure P21.25. The entire circuit has a resistance $R = 200$ Ω. There is a magnetic field $B = 1.5$ T directed upward, perpendicular to the plane of the drawing and the rectangular area enclosed by the bars. An external force is applied to the bar such that it moves to the right with a speed $v = 25$ m/s. What are the magnitude and direction of the current induced in the bar?

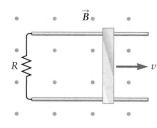

Figure P21.25

26. ✪ A circular metal hoop is dropped over a bar magnet as sketched in Figure P21.26. (a) Make a qualitative sketch of how the current induced in the hoop varies as a function of its height z. Start with the hoop well above the north pole of the magnet and let it fall all the way to large negative values of z ($z \ll L/2$). (b) Make a qualitative plot of the magnetic force on the hoop as a function of z. Interpret your result in terms of Lenz's law.

Figure P21.26
Problems 26 and 27.

27. ✪ Consider again the metal hoop in Figure P21.26. If a person grabs the hoop and pulls it upward, what is the direction of the magnetic force on the hoop? Give answers when the hoop is at (a) $z = +L/2$ and (b) $z = 0$.

28. ✪ Two circular current loops lie parallel to the x–y plane and are separated by a short distance along z (Fig. P21.28). The currents through these loops are positive if they are counterclockwise as viewed from the $+z$ direction and are negative if they are clockwise. (a) If the current I_1 through the bottom loop is increasing with time, what is the direction of the induced current in the top loop? (b) If the current I_2 through the top loop is decreasing with time, what is the direction of the induced current in the bottom loop?

Figure P21.28

29. ✪ A bar magnet is dropped through a current loop as shown in Figure P21.29. (a) When the bar magnet is above the plane of the loop, the current induced in the current loop is counterclockwise as viewed from above. Is the

Figure P21.29

orientation of the bar magnet north pole up or south pole up? (b) What is the direction of the resulting magnetic force on the bar magnet? (c) Suppose the orientation of the bar magnet is inverted. How does your answer to part (b) change?

21.4 INDUCTANCE

30. Calculate the inductance of a solenoid that is 1.5 cm long and 3.0 mm in diameter, with 300 turns of wire.

31. An inductor carries a current of 2.5 A, and the magnetic energy stored in the inductor is 0.0035 J. What is the inductance?

32. An inductor with $L = 1.5$ mH is connected to a circuit that produces a current increasing steadily from 1.5 A to 5.6 A over a time of 2.3 s. What is the voltage across the inductor?

33. SSM ✪ An inductor has the form of a coil with 2000 turns and a diameter of 1.5 mm. The inductor is placed in a magnetic field perpendicular to the plane of the coil and increasing at a rate of 0.35 T/s. The current in the inductor is zero at $t = 0$, and then increases to 6.5 mA at $t = 1.0$ s. What is the inductance?

34. ✪ If the number of turns in a solenoid is increased by a factor of three, (a) by what factor does the magnetic field inside the solenoid change? (b) By what factor does the inductance change?

35. ✪ An MRI magnet has an inductance of $L = 5.0$ H. What is the total flux through the magnet's coils when the current is $I = 100$ A?

36. The flux through an inductor with $L = 5.0$ mH changes at a rate of $\Delta\Phi_B/\Delta t = 0.0045$ T·m²/s. What is the rate of change $\Delta I/\Delta t$ of the current through the inductor?

37. ✪ Figure P21.37 shows a top view of a mutual inductor (compare with Fig. 21.17). The two coils are shown in different colors, and the top and bottom wire leads are also indicated. (a) If the current in coil 1 is directed as shown and I_1 is increasing, what is the sign of the induced emf V_2 in coil 2? ($V_2 = V_A - V_B$) (b) Both coils have a length $\ell = 12$ cm and diameter 1.5 mm. Coil 1 has 300 turns, and coil 2 has 3000 turns. If $\Delta I/\Delta t = 3.0$ A/s, what is the magnitude of V_2?

Top view

Figure P21.37

21.5 *RL* CIRCUITS

38. ✪ Consider an *RL* circuit (Fig. P21.38) with $V = 9.0$ V, $L = 5.5$ mH, and $R = 1500\ \Omega$. After the switch has been left open for a long time, the switch is closed at $t = 0$. (a) What is the current the instant after the switch closed? (b) What is the current a very long time after the switch is closed? (c) How long does it take the current in the circuit to reach 50% of its final (long time) value?

Figure P21.38 Problems 38–41.

39. ✪ Consider again the circuit in Figure P21.38. The switch is first closed for a very long time so that the current through the inductor is constant. The switch is then opened suddenly. What is the voltage across the inductor the instant after the switch is opened?

40. For the circuit in Figure P21.38, if $R = 4000\ \Omega$, what must the value of L be to have a time constant of 1.5 s?

41. SSM ✪ The switch in Figure P21.38 is open for a very long time and then closed at $t = 0$. What is the voltage across the resistor just after the switch is closed?

42. ✪ The switch in Figure P21.42 is open for a very long time and then closed at $t = 0$. (a) What is the current through R_1 and through R_2 immediately after the switch is closed? (b) What is the voltage across the inductor immediately after the switch is closed? (c) After the switch is closed for a very long time, what is the current through R_1, the current through R_2, and the voltage across L? (d) The switch is now opened at $t = t_{open}$. What are the magnitude and direction of the current through R_1 immediately after t_{open}? (e) What is the voltage across L just after t_{open}?

Figure P21.42 Problems 42 and 48.

43. Find the *RL* time constant for decay of the current in the circuit in Figure 21.23B. Express your answer in terms of the values of the resistances R_1 and R_2, and the inductance L.

44. ✪ Consider three inductors L_1, L_2, and L_3 connected in series as shown in Figure P21.44. These three inductors act as one equivalent inductance L_{equiv}. To find L_{equiv}, we first note that because the current flows through the inductors sequentially, the factor $\Delta I/\Delta t$ is the same for each. (a) What is the voltage across each inductor? Express your answers in terms of L_1, L_2, L_3, and $\Delta I/\Delta t$. (b) Use the result from part (a) to find the total voltage across all three inductors. (c) The voltage in part (b) is also equal to the voltage across the equivalent inductance L_{equiv}. Use this fact to find L_{equiv} in terms of L_1, L_2, and L_3. The general result for many inductors in series is $L_{equiv} = L_1 + L_2 + L_3 + \cdots$.

Figure P21.44

45. ✪ Consider two inductors L_1 and L_2 connected in parallel as shown in Figure P21.45. These two inductors act as one equivalent inductance L_{equiv}. To find L_{equiv}, we first notice that because they are connected in parallel, the voltage across L_1 and L_2 must be the same, but the rate at which the current changes with time $\Delta I/\Delta t$ is different for the two inductors. Use these facts to write the total voltage across the two inductors in the form of Equation 21.23, where L is now the equivalent inductance, and find L_{equiv}. *Note:* The general result for many inductors in parallel is $1/L_{equiv} = 1/L_1 + 1/L_2 + 1/L_3 + \cdots$.

Figure P21.45

21.6 ENERGY STORED IN A MAGNETIC FIELD

46. What is the energy stored in the inductor in Figure P21.46 after the switch has been closed for a very long time?

Figure P21.46

47. ✪ ✖ Consider an MRI (magnetic resonance imaging) magnet that produces a magnetic field $B = 1.5$ T at a current $I = 140$ A. Assume the magnet is a solenoid with a radius of 0.30 m and a length of 2.0 m. (a) What is the number of turns of the solenoid? (b) What is its inductance? (c) How much energy is stored in this magnet? (d) If all the energy in part (b) were converted to the kinetic energy of a car ($m = 1000$ kg), what would the speed of the car be?

48. [SSM] ☆ Consider again the circuit in Figure P21.42. (a) Suppose the switch is closed for a very long time. How much energy is stored in the inductor? (b) If the switch is then opened, where does this energy go? Explain.

49. ☆ Consider a region of space in which the energy density stored in the magnetic field (Eq. 21.35) is equal to the energy density in the electric field (Eq. 18.48). What is the ratio of B/E? Express your result in terms of the fundamental constants ε_0 and μ_0.

21.7 APPLICATIONS

50. ☆ Ⓡ A single circular loop of wire of radius $r = 2.0$ cm rotates at a frequency of 60 Hz in a constant magnetic field of magnitude $B = 1.5$ T. What is the approximate maximum emf induced in this generator?

51. ☆ Ⓡ Use the result from Problem 50 to design a generator that produces an induced emf of 500 V. *Hint:* You must choose values for the number of loops and the area of each loop.

52. [SSM] ☆ Ⓡ The pickup coil in an electric guitar contains 10,000 turns and has a diameter of 2.0 mm. When the pickup coil is very close to a guitar string, there is a field of magnitude $B = 0.10$ T in the coil. If this steel string is vibrating at a frequency of 200 Hz, what is the approximate maximum emf induced in the coil? For simplicity, assume the induced flux varies back and forth between zero and the value found when B in the coil is at its maximum value (found when the string is very close to touching the coil).

53. ☆ A guitar string can vibrate at different frequencies, depending on how it is held (Chapter 12). Consider a string that is being played at either 400 Hz or 800 Hz. If everything else is kept the same, how do the induced emfs at the two frequencies compare? That is, which one is larger, and by what factor? *Note to musicians:* Electric guitar amplifiers are designed to compensate for this frequency variation.

54. ✪ Ⓡ A metal detector uses a coil to find pieces of metal that are buried or otherwise obscured. Many metals are magnetic, and when a magnet is moved near a metal detector's coil, an emf is induced in the coil. Assume a coil of diameter 10 cm with 500 turns. This coil is scanned over a large piece of iron that is just below soil's surface. Estimate the magnitude of the emf induced in the coil. Assume the field near the iron is $B = 0.30$ T.

21.8 THE PUZZLE OF INDUCTION FROM A DISTANCE

55. In Section 21.8, we mentioned that an emf is induced in a distant coil or current loop via an electromagnetic wave. For the case in Figure 21.30, this electromagnetic wave travels from the inner solenoid to the outer loop. We will see in Chapter 23 that in a vacuum this wave travels at a speed $c = 3.00 \times 10^8$ m/s. Calculate the time Δt it takes this wave to travel a distance of 1.0 mm as might be typical in an electric circuit containing several inductors. Your result for Δt should show that in this case the induced emf appears nearly instantaneously.

ADDITIONAL PROBLEMS

56. ☆ A long, straight wire carrying a current I is next to a wire loop, with the straight wire lying in the plane of the loop (Fig. P21.56). The current I is increasing with time. What is the direction of the induced current in the loop, clockwise or counterclockwise as viewed in the figure?

57. ✪ A current loop (loop 1) lies in a horizontal plane. A second loop (loop 2) of the same size is positioned above loop 1 and is also oriented horizontally (Fig. P21.57). The current in loop 1 is constant and counterclockwise as viewed from above. If loop 2 falls toward loop 1, what is the direction of the induced current in loop 2?

Figure P21.56

Figure P21.57

58. [SSM] ☆ A current loop is placed in a magnetic field that varies with time according to the graph in Figure P21.58. Assume the loop is oriented perpendicular to the magnetic field. Sketch

the graph for the induced emf in the loop for the same time intervals.

Figure P21.58

59. ✪ Figure P21.59 shows four configurations of a wire loop placed near a long, straight wire with a time-varying current. Determine the direction of (a) the induced magnetic field in the wire loop (into or out of the page) and (b) the induced current (clockwise or counterclockwise) for each configuration (1) through (4).

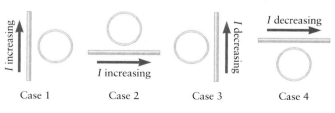

Figure P21.59

60. ✪ A 3.0-V flashlight bulb is connected to a square loop of wire that measures 60 cm on a side as shown in Figure P21.60. Assume the bulb will only light if the potential across its terminals is at least 3.0 V. The loop is oriented perpendicular to a uniform magnetic field of 2.5 T. The magnetic field is then made to decrease to zero over a period of time Δt such that the bulb is just barely lit. (a) Find how long the bulb will light at full output. (b) In which direction will the current flow around the loop?

Figure P21.60

61. ✪ The 3.0-V flashlight bulb assembly of Problem 60 is now attached to a wood dowel with a crank handle as shown in Figure P21.61, and the loop is rotated in a uniform magnetic field of 2.5 T. (a) Assume the crank arm rotates with a period T. What value of T is just small enough to make the bulb light? For simplicity, assume that if the loop makes 1/4 turn in a time $\Delta t_{1/4}$, the induced emf is equal to $\Delta\Phi/\Delta t_{1/4}$. (b) If the magnitude of the field is redued to $0.25T$, what is the smallest value of T that will work?

Figure P21.61

62. ✪ The switch in the circuit shown in Figure P21.62 has been open for a long time. The resistors in the circuit each have a value of 50 Ω, the inductors each have a value of 40 mH, and $V = 12$ V. (a) What is the total equivalent inductance of the circuit between points A and B? (b) What is the current through R_1 the instant after the switch is closed? (c) What is the current through R_1 long after the switch has been closed? *Hint:* See Problems 44 and 45.

Figure P21.62

63. ✪ Consider an RL circuit with resistance 45 Ω and inductance 250 mH placed in series. (a) If a 10-V battery is connected to the circuit, how long would it take the current to reach 10 mA? (b) What is the voltage drop across the inductor at this time?

64. ✪ Consider an RL circuit with resistance 100 Ω and inductance 200 mH. (a) If the battery is removed from the circuit and replaced by a connecting wire to complete the circuit, how long would it take for the potential difference across the resistor to drop to 37% of its initial value? (b) How long would it take to reach just 50%? (c) How long to get down to 10%?

65. ✪ The switch in Figure P21.65 is closed, and 5.5 μs later the current is measured to be 0.10 A. (a) What is the inductance L? (b) What is the maximum current in this circuit? (c) When is the current half of this maximum?

Figure P21.65

66. ✪ The circuit in Figure P21.66 has two identical resistors and three identical inductors. If $L = 30$ mH, what value must R have to give a time constant of 50 μs? *Hint:* See Problems 44 and 45.

Figure P21.66

67. ✪ An inductor ($L = 80$ mH) is subjected to a time-varying current according to the graph in Figure P21.67. Plot the corresponding emf as a function of time. What is the highest potential difference produced? What is the lowest? Assume the potential of the supply voltage is positive so that a decrease in current generates a positive drop across the inductor.

Figure P21.67

68. ✪ An inductor ($L = 50$ μH) is subjected to a time-varying current. The emf then produced by the inductor is graphed in Figure P21.68. The initial current is 40 mA. Plot the corresponding current as a function of time. Assume the potential of the supply voltage is positive so that a decrease in current generates a positive drop across the inductor.

Figure P21.68

69. ✪ Ⓡ The Earth's magnetic field is approximately 5×10^{-5} T. Suppose a square coil of edge length 0.10 m is rotated with a frequency $f = 400$ Hz about an axis perpendicular to the Earth's field (Fig. P21.69). The induced emf in this coil will vary with time according to $V = V_0 \sin(2\pi f t)$. What is the approximate value of V_0?

Figure P21.69

Alternating-Current Circuits and Machines

We discussed circuits containing resistors and capacitors in Chapter 19 and those with inductors in Chapter 21. The source of electrical energy in those circuits was a battery, which provides a constant voltage at its output terminals (Fig. 22.1A). If a battery-powered circuit contains only resistors, the current will be independent of time. In circuits containing capacitors (*RC* circuits) or inductors (*RL* circuits), the current can vary with time but always approaches a constant value a long time after closing or opening a switch. Such circuits are called **DC circuits,** where

Electrical energy is delivered to your house using AC voltages. The flow of this energy is controlled by transformers (shown here) and by other devices described in this chapter. (© Lester Lefkowitz/ Photographer's Choice/Getty)

"DC" stands for **direct current,** which is another way of referring to a constant current. In an **AC circuit,** the battery is replaced by a device that produces an electric potential (i.e., a voltage) that varies with time (Fig. 22.1B). Such a device is called an AC voltage source, where "AC" stands for **alternating current.**

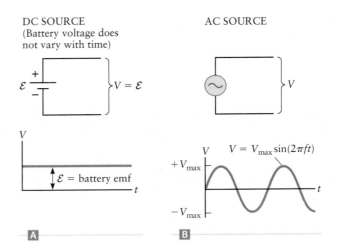

Figure 22.1 **A** A battery produces a constant electric potential (the emf) at its terminals. **B** The electric potential (voltage) at the terminals of an AC voltage source varies sinusoidally with time.

An AC voltage source is represented in a circuit diagram by the symbol ─(∼)─. The voltage at the output terminals of this device varies sinusoidally with time, with a frequency f and an amplitude V_{max}, also called the peak value, as shown in Figure 22.1B.

The electrical energy used in your house is provided by an AC source. If you were to measure the voltage at the terminals of an electrical socket in a house in the United States, you would find a time dependence like that shown in Figure 22.1B with a frequency of 60 Hz. The famous inventor Thomas Edison originally proposed the use of DC electrical systems for houses and cities, but around 1880 other engineers, including one named George Westinghouse, pushed for an AC system. In the end, Westinghouse's proposal won out, and a 60-Hz system was adopted in the United States in the 1890s.

In this chapter, we'll explore the advantages of AC power. We begin with a study of AC circuits containing the basic circuit elements: resistors, capacitors, and inductors. After describing some of the properties of AC circuits, we'll then be able to understand why we use alternating current instead of direct current in our homes.

22.1 | GENERATION OF AC VOLTAGES

One reason an AC system was adopted for the large-scale distribution of electrical energy is that generating an AC voltage is simpler than constructing and maintaining a large DC source. Most sources of AC electrical energy employ a generator based on magnetic induction. Figure 21.7 showed the general design of an AC generator. At a hydroelectric facility, water is used to turn a mechanical shaft; a steam turbine serves a similar function at a nuclear or coal-burning power plant (Fig. 22.2). This shaft is part of the generator and holds a coil with many loops of wire. The coil is positioned between the poles of a permanent magnet, so the magnetic flux through the coil varies with time as the shaft turns. According to Faraday's law of induction (Chapter 21), this changing flux induces a voltage in the coil, which is the generator's output. Faraday discovered his law of induction in 1831, and the first AC generators based on this principle were constructed in 1832! It did not take long to make practical use of Faraday's discovery.

The approach in Figure 21.7 produces an AC voltage with a frequency equal to the rotation frequency of the shaft, and electrical energy is produced as long as the shaft is rotating. In contrast, for a DC source such as a battery, one would have to replace the battery often (or replenish the chemicals in the battery), and for many practical applications this battery would need to be quite large. That is one reason an AC power system is simpler to implement than a DC one.

Devices that are commonly referred to as "generators" of electrical energy actually just convert the mechanical energy of the rotating shaft into electrical energy (an electric potential difference that can produce a current in a circuit). The principle of conservation of energy still applies, so for an ideal generator the amount of electrical energy dissipated in an attached circuit is equal to the mechanical energy used to rotate the generator shaft. In the same way, when we refer to the circuit element —(∼)— as a "source" of electrical energy, you should realize that this device simply enables the transfer of electrical energy from a generator to an attached circuit.

Figure 22.2 Large AC generators are used to produce AC voltages at hydroelectric dams and at other types of electric power plants.

| EXAMPLE 22.1 | Generating an AC Voltage |

Let's analyze the large AC generators that are used in facilities such as the Hoover Dam power station near Las Vegas. Estimate the amplitude of the voltage (V_{max} in Figure 22.1B) produced by rotating a coil containing one loop of wire in the generator. Assume the generator's magnetic field has a magnitude $B = 1.0$ T (typical of a large magnet) and the coil has an area $A = 5.0$ m^2. (Generators like the one in Fig. 22.2 are large.) Because this voltage is destined for household use, assume a frequency $f = 60$ Hz.

RECOGNIZE THE PRINCIPLE

Generators are based on Faraday's law. Because the coil rotates, the magnetic flux through it changes with time. This time-varying flux induces a voltage in the coil, which is the AC voltage produced by the generator.

SKETCH THE PROBLEM

Figure 22.3 shows how the magnetic flux changes as the loop in our coil rotates.

IDENTIFY THE RELATIONSHIPS AND SOLVE

The magnetic flux through a generator coil is largest when the face of the coil is perpendicular to the magnetic field (Fig. 22.3A). This maximum flux is $\Phi_B = BA$, where B is the magnitude of the magnetic field and A is the area of the coil (see Eq. 21.1). As the coil rotates one half turn, the flux alternates in sign from $+BA$ to $-BA$ (Fig. 22.3B) because the angle θ between the field and the coil's surface varies between 0° and 180°. The total flux change is

$$\Delta\Phi_B = -2BA$$

If the rotation period of the shaft is T, this flux change occurs in a time $\Delta t = T/2$. The rotation period is related to the frequency by $T = 1/f = 1/60$ s.

From Faraday's law (Eq. 21.6), the magnitude of the emf induced in the coil is

$$|\mathcal{E}| = \left|\frac{\Delta\Phi_B}{\Delta t}\right|$$

Strictly speaking, this equation applies only for small flux changes $\Delta\Phi_B$ over small time intervals Δt, but we can still use it to get an approximate value for the generator emf. Inserting our estimates for $\Delta\Phi_B$ and Δt, we get

$$|\mathcal{E}| = \left|\frac{\Delta\Phi_B}{\Delta t}\right| = \frac{2BA}{T/2} = \frac{2(1.0 \text{ T})(5.0 \text{ m}^2)}{(1/120) \text{ s}}$$

$$|\mathcal{E}| = \boxed{1200 \text{ V}}$$

$\theta = 0°$

$\Phi_B = BA \cos\theta = BA$

A

$\theta = 180°$

$\Phi_B = BA \cos\theta = -BA$

B

Figure 22.3 Example 22.1. As the wire loop makes a half turn around the rotation axis, the flux changes from $\Phi_B = +BA$ to $\Phi_B = -BA$.

CONCEPT CHECK 22.1 | Generating a Larger Voltage

Consider the Hoover Dam generator discussed in Example 22.1. Suppose you wish to produce an AC voltage with an amplitude of 50,000 V. Which of the following approaches would work? Explain your answer.

 (a) Connect many generators in series.
 (b) Connect many generators in parallel.
 (c) Add more turns to the coil.

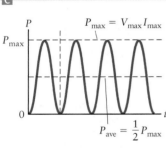

Figure 22.4 **A** Circuit with an AC voltage source connected to a resistor. **B** Voltage at the output terminals of the source as a function of time. **C** Current in the circuit as a function of time. **D** Instantaneous power dissipated in the resistor as a function of time. The average power (i.e., averaged over many cycles of the oscillation) is one-half the maximum power.

AC Circuits and Simple Harmonic Motion

An old-fashioned but very descriptive unit of frequency is the "cycle per second." In terms of the time dependence of an AC voltage in Figure 22.1B, a frequency of 1 Hz means that the voltage completes one full cycle each second. The voltage variation in Figure 22.1B should remind you of the simple harmonic oscillators in Chapter 11. Because of this similarity, the voltage and current in an AC circuit are said to "oscillate." We'll also find (Sections 22.5 and 22.6) a close connection between circuits with capacitors and inductors (called *LC* circuits) and simple harmonic oscillators.

22.2 | ANALYSIS OF AC RESISTOR CIRCUITS

Now that we've seen how AC voltages can be generated, let's consider how to calculate the current in an AC circuit. We begin with the simplest possible circuit consisting of an AC generator and a resistor *R* as sketched in Figure 22.4A.

The voltage across the output terminals of an AC source varies with time. If *f* is the frequency and V_{max} is the amplitude of the AC voltage, the output voltage has the form

$$V = V_{max} \sin(2\pi f t) \qquad (22.1)$$

In Equation 22.1, *V* is the *instantaneous* potential difference between the terminals of the AC source and is plotted as a function of time in Figure 22.4B. In this circuit, the voltage across the output terminals is also equal to the voltage across the resistor. To find the resulting current *I* through the resistor, we can use **Ohm's law** (Eq. 19.10):

$$I = \frac{V}{R} \qquad (22.2)$$

In Chapter 19, we applied Ohm's law for resistors in DC circuits, but it also holds for resistors in an AC circuit, where *V* and *I* are the instantaneous voltage and current. You can think of an AC voltage source as equivalent to a battery whose output voltage varies in time according to Equation 22.1. At each instant in time, the current through this AC circuit is given by *I* = *V/R*, just as for a DC circuit containing a battery and a resistor. As time passes, the output voltage *V* of the AC source varies sinusoidally, so the current through the resistor is similarly time dependent. Inserting *V* from Equation 22.1 into Ohm's law (Eq. 22.2), we find

$$I = \frac{V_{max}}{R} \sin(2\pi f t) \qquad (22.3)$$

which is shown in Figure 22.4C. The current thus oscillates with time, with the same frequency as the voltage. The voltage amplitude V_{max} is constant, so we can denote the current amplitude as I_{max} and write

$$I = I_{max} \sin(2\pi ft) \quad \text{with } I_{max} = \frac{V_{max}}{R} \tag{22.4}$$

Root-Mean-Square Values

The result in Equation 22.4 might be the current in your television or some other appliance. The appliance manufacturer usually specifies how much current and voltage the device is designed for. However, the voltage and current vary with time, so what numbers should the manufacturer provide? One possibility is to provide some kind of average value, but from Figure 22.4B and C we see that the average values of V and I taken over many oscillation cycles are both zero, so these values are not useful as specifications. That is one reason the notion of "root-mean-square" was adopted. Figure 22.5 shows a plot of the AC voltage $V = V_{max} \sin(2\pi ft)$ as well as how the square of the voltage varies with time. Since V^2 is always positive, its average is not zero. The average of V^2 is related to the voltage amplitude V_{max} by (see Fig. 22.5B)

$$(V^2)_{ave} = \tfrac{1}{2} V^2_{max} \tag{22.5}$$

We now define a quantity called the root-mean-square (rms) voltage, denoted as V_{rms}, through the relation $V_{rms} = \sqrt{(V^2)_{ave}}$. Here the word *mean* indicates an average, and "root" refers to a square root. We thus have

$$V_{rms} = \sqrt{(V^2)_{ave}} \tag{22.6}$$
square *root*   *square*   *mean* (average)

In words, V_{rms} is the square *root* of the *mean* (average) of the voltage *squared*. Inserting Equation 22.5 into the definition of the root-mean-square gives

$$V_{rms} = \sqrt{\frac{1}{2} V^2_{max}} = \frac{V_{max}}{\sqrt{2}} \tag{22.7}$$

Definition of rms voltage

So, the rms value of the voltage is equal to the voltage amplitude V_{max} divided by $\sqrt{2}$; hence, $V_{rms} \approx 0.71 \times V_{max}$. (See Fig. 22.6.)

The root-mean-square value can be defined for any quantity that varies with time, including the current in an AC circuit. So, for the oscillating current in Equation 22.4 we have

$$I_{rms} = \sqrt{\frac{1}{2} I^2_{max}} = \frac{I_{max}}{\sqrt{2}} \approx 0.71 \times I_{max} \tag{22.8}$$

Definition of rms current

The root-mean-square values of the voltage and current are widely used to specify the properties of an AC circuit. For example, in the United States the voltage in a household circuit has the value "120 V AC." This voltage is actually the rms value of the voltage across the terminals of a wall socket.

Power in an AC Circuit

The instantaneous power dissipated in a resistor is equal to the product of the instantaneous voltage across the resistor and the instantaneous current through the resistor, which is just an extension of Equation 19.15 to AC circuits. Hence, the instantaneous power dissipated in the resistor in Figure 22.4A is

$$P = VI \tag{22.9}$$

where V and I are the results in Equations 22.1 and 22.4. Because V and I in this circuit both vary with time, the power P is also a function of time. Inserting our results for V and I into Equation 22.9 gives

$$P = [V_{max} \sin(2\pi ft)][I_{max} \sin(2\pi ft)] = V_{max} I_{max} \sin^2(2\pi ft) \tag{22.10}$$

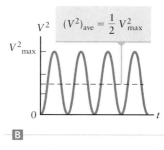

Figure 22.5 Variation of an AC voltage V and V^2 as functions of time.

which is plotted in Figure 22.4D. The instantaneous power oscillates between a maximum value of $V_{max}I_{max}$ and a minimum value of zero. These minima occur at moments when the instantaneous voltage and current are zero.

The instantaneous power in Figure 22.4D might be the power dissipated in your refrigerator or desktop computer, or anything else that is plugged into a wall socket. These and other household devices come with a power rating provided by the manufacturer—this rating is a single number that tells you about the power dissipated in the device. The power rating is useful because it gives you an idea of how much it will cost to run the device. How does the power rating relate to the instantaneous power curve in Figure 22.4D? Since the AC voltage and hence the instantaneous power oscillate rapidly (in the United States, the voltage has a frequency of 60 Hz), it makes sense to average the power over many cycles of its oscillation. The value of the *average power* P_{ave} can be seen from Figure 22.4D, which shows that the average power equals half the maximum power. According to Equation 22.10, the maximum power is $P_{max} = V_{max}I_{max}$ so the average power is

Average power dissipated in a resistor in terms of the voltage and current amplitudes

$$P_{ave} = \frac{V_{max}I_{max}}{2} \tag{22.11}$$

Using our results for the rms voltage V_{rms} and current I_{rms} (Eqs. 22.7 and 22.8), we can write the average power in Equation 22.11 as

$$P_{ave} = V_{rms}I_{rms} \tag{22.12}$$

One advantage of writing the average power in this way is that Equation 22.12 has precisely the same mathematical form as the power in a DC circuit (Eq. 19.15).

When dealing with the power dissipated in a resistor, the relation for the power (Eq. 22.12) can be simplified using Ohm's law, $I = V/R$. Since the Ohm's law relation between current, voltage, and resistance (Eq. 22.2) holds for instantaneous values, it also holds for rms values. We can thus write

$$I_{rms} = \frac{V_{rms}}{R}$$

Combining this relation with Equation 22.12 leads to

Average power dissipated in a resistor in terms of the rms voltage or current

$$P_{ave} = \frac{V_{rms}^2}{R} = I_{rms}^2 R \tag{22.13}$$

EXAMPLE 22.2 | Power and Voltage in a Household Circuit

The standard AC voltage found in homes in the United States has a frequency of 60 Hz and is often referred to as "120-V power," which is the value of the rms voltage; that is, $V_{rms} = 120$ V. What is the corresponding value of V_{max} for the voltage at an electrical outlet?

RECOGNIZE THE PRINCIPLE

The amplitude V_{max} of an AC voltage is also the "peak" value of the voltage (Fig. 22.6). V_{max} and V_{rms} are just two ways of describing the same AC voltage. They are related through the basic time dependence of an AC voltage $V = V_{max} \sin(2\pi ft)$ and Equation 22.7. The point of this example is to get an intuitive understanding of the relation between rms and peak values.

SKETCH THE PROBLEM

Figure 22.6 shows an AC voltage and indicates V_{max}.

IDENTIFY THE RELATIONSHIPS AND SOLVE

Applying Equation 22.7, we have

$$V_{rms} = \frac{V_{max}}{\sqrt{2}}$$

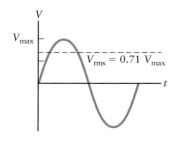

Figure 22.6 Example 22.2.

Solving for V_{max} and inserting the given value of V_{rms} gives

$$V_{max} = \sqrt{2}\,V_{rms} = \sqrt{2}(120\ V) = \boxed{170\ V}$$

What does it mean?

Figure 22.6 could represent a plot of an AC voltage with $V_{max} = 170$ V (the value found in this example). A DC voltage equal to the rms value of 120 V would give the same average dissipated power in the resistor.

EXAMPLE 22.3 | Voltage, Current, and Power in a Lightbulb

As a circuit element, we can assume that an incandescent lightbulb functions just like a resistor (Fig. 22.7), so the relations between the voltage, current, and power in a resistor apply also to a lightbulb. Household lightbulbs are specified by their power rating. For example, one might be rated as a "60-watt" bulb. This power rating refers to the average dissipated power P_{ave}, so $P_{ave} = 60$ W when a 60-W bulb is used in a normal household "120-V" circuit. Find the resistance of a 60-W lightbulb.

Figure 22.7 Example 22.3. In circuit analysis, an incandescent lightbulb can be modeled as a resistor.

RECOGNIZE THE PRINCIPLE

The AC voltage in a "120-V" household circuit has the rms value $V_{rms} = 120$ V. When connected to a lightbulb, it gives the same average power as if we had instead connected a battery with a constant emf of $V_{DC} = 120$ V. For a given value of the voltage, the average power dissipated in a resistor depends on the value of the resistance R. For a DC voltage, the relation is $P_{ave} = V_{DC}^2/R$, whereas for an AC voltage, it is $P_{ave} = V_{rms}^2/R$. The point of this example is to get a feeling for the value of the resistance of a common household "appliance."

SKETCH THE PROBLEM

Figure 22.7 describes the problem.

IDENTIFY THE RELATIONSHIPS

For an AC circuit, the average power dissipated in a resistor is (Eq. 22.13) $P_{ave} = V_{rms}^2/R$.

SOLVE

Solving for R, we get

$$R = \frac{V_{rms}^2}{P_{ave}}$$

For the AC voltage in a household circuit, $V_{rms} = 120$ V, which we can use along with the given value of P_{ave} for the lightbulb to get

$$R = \frac{V_{rms}^2}{P_{ave}} = \frac{(120\ V)^2}{60\ W} = \boxed{240\ \Omega}$$

What does it mean?

The average power is important because it determines the amount of energy used by the lightbulb over a long time, and that is how the power company computes your bill. The instantaneous intensity of a lightbulb depends on the instantaneous power dissipated in the filament. According to Figure 22.4D, the power oscillates, so the intensity must also oscillate. Hence, an ordinary lightbulb must "flicker." Why don't we notice this flickering? (*Hint*: See Concept Check 22.2.)

Table 22.1 | Useful Quantities and Notation for Describing AC Circuits

VOLTAGE	CURRENT	POWER
V = instantaneous voltage $= V_{max} \sin(2\pi ft)$	I = instantaneous current	P = instantaneous power = VI
V_{max} = voltage amplitude	I_{max} = current amplitude	P_{max} = power amplitude
V_{rms} = rms voltage = $\dfrac{V_{max}}{\sqrt{2}}$	I_{rms} = rms current = $\dfrac{I_{max}}{\sqrt{2}}$	—
$V_{ave} = 0$	$I_{ave} = 0$	P_{ave} = average power = $V_{rms}I_{rms}$

When dealing with AC circuits, it is important to distinguish between instantaneous and average values of the voltage, current, and power. The amplitudes of the AC variations and the rms values are also important. Table 22.1 collects these terms and summarizes the relations between them.

CONCEPT CHECK 22.2 | Flickering Lightbulbs

In Example 22.3, we noted that the intensity of an ordinary incandescent lightbulb flickers with time in the same way that the instantaneous power in a resistor varies with time (Fig. 22.4D). In the United States, the AC voltage across the lightbulb has a frequency of $f = 60$ Hz. Is the frequency of the flickering (a) 30 Hz, (b) 60 Hz, (c) 120 Hz, or (d) 240 Hz? How does the frequency of the flickering compare with the response time of your eyes?

CONCEPT CHECK 22.3 | Traveling in Europe

In most European countries, the AC voltage available at a wall outlet has an rms value of about 220 V and a frequency of 50 Hz. (Both of these values are different from those used in the United States.) Is the peak value (the amplitude) of this AC voltage (a) 220 V, (b) 170 V, or (c) 310 V?

Describing AC Voltages and Currents: Phasors

The sinusoidal variation of the voltage in an AC circuit,

$$V = V_{max} \sin(2\pi ft) \tag{22.14}$$

is very similar to what we encountered for circular motion (Chapter 5) and simple harmonic motion (Chapter 11). This similarity leads to a useful graphical way to analyze AC circuits. Imagine an arrow of length V_{max} as sketched in Figure 22.8. The arrow's tail is tied to the origin, while the tip moves along a circle of radius V_{max}. If this arrow makes an angle θ with the horizontal axis, the vertical component of the "voltage arrow" is

$$V = V_{max} \sin \theta \tag{22.15}$$

If the tip of the arrow undergoes uniform circular motion, the angle θ varies with time according to

$$\theta = 2\pi ft \tag{22.16}$$

where f is the rotation frequency. Inserting this result into Equation 22.15, we get $V = V_{max} \sin(2\pi ft)$, which is just the voltage in an AC circuit, Equation 22.14. The rotating arrow representing the voltage in Figure 22.8 is called a **phasor**. A phasor is *not* a vector; the voltage in an AC circuit is *not* a vector in the way that quantities such as position and velocity are vectors. Phasor diagrams simply provide a convenient way to illustrate and think about the time dependence in an AC circuit.

The current in an AC circuit can also be represented by a phasor. Figure 22.9A shows a resistor connected to an AC voltage source. A phasor diagram for this cir-

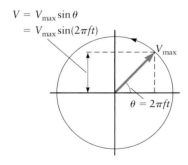

Figure 22.8 Phasor diagram of the voltage in an AC circuit. The voltage is represented by a "phasor," an arrow of length V_{max} that rotates in the coordinate plane. The instantaneous voltage in the circuit is equal to the vertical component of this arrow.

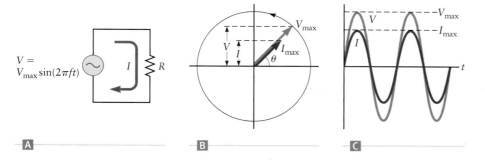

cuit is given in Figure 22.9B, which shows the voltage arrow as it rotates around the origin as well as the current as a rotating arrow of length I_{max}. In Equation 22.4, we found that $I = I_{max} \sin(2\pi ft) = I_{max} \sin\theta$, so the current arrow rotates in synchrony with the voltage arrow. These two arrows—that is, these two phasors—always make the same angle θ with the horizontal axis as time passes. Because the voltage and current phasors rotate together, we say that they are *in phase*. This phase relationship can also be seen from plots of V and I as functions of time in Figure 22.9C. When the instantaneous voltage V is at one of its maxima, the instantaneous current I is also a maximum. Likewise, the minima of V and I occur at the same times, and the voltage and current also pass through zero at the same time.

For the resistor circuit in Figure 22.9A, the phase relation is very simple: the voltage and the current always oscillate "in synchrony" with each other. The current and voltage relationships in circuits with capacitors and inductors are usually not this simple, and phasor diagrams are extremely useful for understanding those circuits.

22.3 | AC CIRCUITS WITH CAPACITORS

Figure 22.10A shows an AC circuit containing a single capacitor with capacitance C. We want to calculate the current when the capacitor is attached to an AC generator with $V = V_{max} \sin(2\pi ft)$. The plates of the capacitor carry charges $+q$ and $-q$, related to the potential difference V across the capacitor by $q = CV$ (which is just Eq. 18.30 with the potential difference or "voltage" across the capacitor denoted simply by V instead of ΔV). This relation between q and V holds at all times, so q is the *instantaneous* charge on the capacitor, just as V is the instantaneous voltage across the capacitor. Since V is also equal to the voltage produced by the AC source, we have

$$q = CV = CV_{max} \sin(2\pi ft) \tag{22.17}$$

Parts B and C of Figure 22.10 show the voltage and charge as functions of time. The capacitor's voltage and charge are *in phase* with each other; that is, their maxima occur at the same times.

To understand the time dependence of the current, we recall that the electric current is defined as (Chapter 19)

$$I = \frac{\Delta q}{\Delta t} \tag{22.18}$$

The instantaneous current I equals the rate at which charge flows onto the capacitor plates in a short time interval Δt. On a graph of q versus t, the current I is the *slope* of the q–t plot. The qualitative behavior of this slope is illustrated in Figure 22.10C, which shows one value of t when the slope $\Delta q/\Delta t$ is zero along with another value of t when the slope is a maximum. By estimating the slope in a graphical way, we can construct the qualitative plot of the current I as a function of time shown in Figure 22.10D. The current has its largest value when the charge on the plates is zero. On the other hand, the current is zero when the charge has its largest values

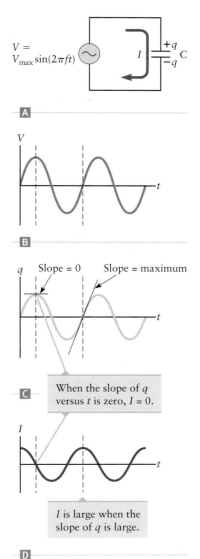

Figure 22.10 ◫ **A** An AC circuit with a capacitor. **B** Voltage, **C** charge on the capacitor, and **D** current as functions of time for the circuit in part A.

Figure 22.11 Phasor diagram showing the voltage and current through a capacitor. The current is represented by a rotating arrow of length I_{max}. The instantaneous current is the component of this arrow along the vertical axis. The current arrow always makes an angle of 90° ($\pi/2$) with the voltage, so the instantaneous current and voltage are always out of phase by 90°.

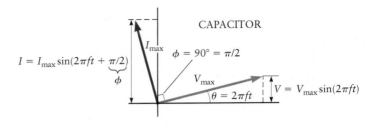

(in magnitude). Comparing the general plots of q and I, we see that although q is a sine function (Eq. 22.17), the current is proportional to a cosine function:

$$I = I_{max} \cos(2\pi ft) \tag{22.19}$$

The sine and cosine functions have the property[1] that for any angle α, $\cos(\alpha) = \sin(\alpha + \pi/2)$. We can therefore write the current in the equivalent forms

$$I = I_{max} \sin(2\pi ft + \pi/2)$$

$$I = I_{max} \sin(2\pi ft + \phi) \qquad \text{with } \phi = \pi/2 \tag{22.20}$$

A phasor diagram for voltage and current in the capacitor circuit in Figure 22.10A is given in Figure 22.11. We show the voltage as a rotating arrow (a phasor) making an angle $\theta = 2\pi ft$ with the x axis. The tip of this voltage phasor moves in a circle as in the resistor circuit (compare with Eq. 22.15 and Fig. 22.9B). The current in the capacitor circuit is represented by an additional rotating arrow of length I_{max} (which we have not yet calculated), and according to Equation 22.20 this phasor makes an angle $2\pi ft + \pi/2 = \theta + \pi/2$ with the horizontal axis. The current is therefore *out of phase with the voltage* (parts B and D of Fig. 22.10). Unlike the behavior we found in the resistor circuit, the maximum values of the voltage and current *do not occur at the same time*. The angle $\pi/2$ is called the **phase angle** ϕ between V and I. Since the angle $\pi/2$ corresponds to 90°, for this circuit the voltage and current are $\phi = 90°$ out of phase.

The graphical analysis in Figure 22.10 shows that the current and voltage are out of phase for a capacitor. This is quite different from a resistor, for which I and V are always in phase (Fig. 22.9). Why is a capacitor different from a resistor? They are different because the *charge* on a capacitor is proportional to the voltage (Eq. 22.17), whereas for a resistor the *current* is proportional to V (Eq. 22.2). This difference, together with the relation between current and charge $I = \Delta q/\Delta t$ (Eq. 22.18), makes the behavior of the AC capacitor circuit different from that of an AC resistor circuit.

Equation 22.20 and the phasor diagram in Figure 22.11 give the phase of the current through a capacitor. To find the peak value (the amplitude) I_{max}, we must calculate the slope of the q–t plot in Figure 22.10. This calculation requires more mathematics than we can include here. The result is

Amplitude of the current in a circuit with a capacitor

$$I_{max} = \frac{V_{max}}{X_C} \tag{22.21}$$

The factor X_C is called the **reactance** of the capacitor and is given by

Reactance of a capacitor

$$X_C = \frac{1}{2\pi fC} \tag{22.22}$$

The relation between the peak current and voltage for a capacitor in Equation 22.21 has the same mathematical form as the corresponding relation for a resistor $I = V/R$. Here, X_C (for a capacitor) and R (for a resistor) are both measures of how much the circuit "resists" a current. Both resistance and reactance are measured in ohms (Ω), but X_C differs from R in one very crucial way: the reactance of a capaci-

---

[1]See Appendix B.4 for a review of the properties of sine and cosine functions.

tor X_C depends on frequency. As a result, the peak value of the current through a capacitor, I_{max} in Equation 22.21, depends on frequency. The reason for this frequency dependence can be traced back to the relation between current and charge $I = \Delta q / \Delta t$. Current I is the rate of change of the charge q on the capacitor. If the frequency is increased, the charge oscillates more rapidly and Δt is smaller, giving a larger current. That is why the peak current in a capacitor is larger at high frequencies and why the reactance X_C becomes smaller at high frequencies.

EXAMPLE 22.4 | Properties of a Typical Capacitor

Figure 22.12 shows a photo of some typical capacitors, like ones commonly found in a radio or MP3 player, and some typical resistors, each with $R = 100\ \Omega$. Consider a capacitor with a capacitance $C = 1000$ pF (1.0×10^{-9} F). Find the frequency at which the reactance of the capacitor is equal to the resistance of one of these resistors.

RECOGNIZE THE PRINCIPLE

The reactance X_C of a capacitor depends on its capacitance C and the frequency. By adjusting the frequency, we can attain a reactance of 100 Ω. We'll see in Section 22.7 how this frequency dependence is used in applications of RC circuits; for example, it is used to tune a radio to different stations.

SKETCH THE PROBLEM

No figure is needed.

IDENTIFY THE RELATIONSHIPS

The reactance of a capacitor is (Eq. 22.22) $X_C = 1/(2\pi f C)$. We want to find the value of f at which this reactance is equal to R, so we set

$$X_C = \frac{1}{2\pi f C} = R$$

SOLVE

Solving for the frequency and substituting $R = 100\ \Omega$, we get

$$f = \frac{1}{2\pi RC} = \frac{1}{2\pi (100\ \Omega)(1.0 \times 10^{-9}\ \text{F})} = 1.6 \times 10^6\ \text{Hz} = \boxed{1.6\ \text{MHz}}$$

What does it mean?
At this frequency, the capacitor's reactance X_C and the resistor's resistance R are the same, but that is not true at any other frequency. The reactance of a capacitor, and hence the current in an AC circuit containing a capacitor, depends on frequency. For this combination of R and C, the frequency at which $X_C = R$ lies at the upper end of the AM radio band.

Figure 22.12 Example 22.4. Photo of typical capacitors (larger cylindrical objects) and resistors (smaller objects at lower right) from a small radio.

CONCEPT CHECK 22.4 | The Effect of Frequency on a Capacitor's Reactance

Consider an AC circuit containing a single capacitor C. If the frequency is increased by a factor of 4, how does the amplitude of the current change?
    (a) The current increases by a factor of 4.
    (b) The current increases by a factor of 16.
    (c) The current decreases by a factor of 4.

Power in a Circuit with Capacitors

The instantaneous power dissipated in a resistor is $P = VI$ (Eq. 22.9). The derivation of this relation in Chapter 19 (Eq. 19.15) calculated the power from the change in the potential energy of a charge as it moved through an electric potential difference

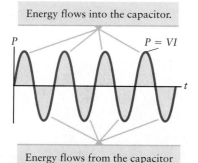

Energy flows into the capacitor.

$P = VI$

Energy flows from the capacitor back to the generator.

Figure 22.13 Instantaneous power P is the product of V (a sine wave) and I (a cosine); the result is another sine wave. At some portions of the cycle P is positive, whereas at other times it is negative. The average power over one complete cycle is zero.

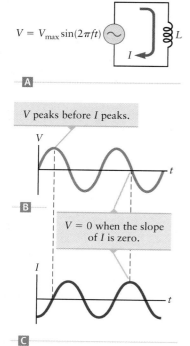

$V = V_{max} \sin(2\pi ft)$

A

V peaks before I peaks.

V

B

$V = 0$ when the slope of I is zero.

I

C

Figure 22.14 **A** An AC circuit with an inductor. **B** Voltage across the inductor as a function of time. **C** Current through the inductor as a function of time.

V across a circuit element. In Chapter 19, that circuit element was a resistor, but the general argument applies to *any* circuit element. Let's now use that result for P to find the instantaneous power in a circuit in which an AC voltage source is connected to a capacitor. Inserting the results for V (Eq. 22.14) and I (Eq. 22.19) into our expression for P, we find

$$P = VI = V_{max}I_{max} \sin(2\pi ft)\cos(2\pi ft) \tag{22.23}$$

This result is plotted in Figure 22.13, which shows how P oscillates between positive and negative values. These oscillations and the negative values of P can be traced to the oscillating behaviors of the sine and cosine functions in Equation 22.23. The *average* value of the power over many oscillation cycles is $P_{ave} = 0$. The voltage and current across the capacitor both oscillate with time, so there are portions of each cycle during which current is flowing into the generator and "against" the generator voltage. That is, at times during each cycle the voltage across the capacitor is positive while the current is negative. When the current and voltage have opposite signs, the power $P = VI$ is negative. At these moments, energy is being transferred from the capacitor back to the generator. In fact, during half of each cycle, energy flows from the generator to the capacitor (P is positive in Fig. 22.13), whereas during the other half of the cycle energy flows from the capacitor back into the generator (P is negative). The net result—the average power—is zero.

We say that electrical energy is *dissipated* in a resistor because the electrical energy supplied by the voltage source is converted to heat energy and hence is lost from the circuit. In a purely capacitive circuit, however, the energy is not converted to heat. Instead, energy is *stored* in the capacitor as electric potential energy, and this stored energy is later returned to the AC source. Hence, in a capacitor circuit we should really speak of the power P that is *delivered to* the capacitor. When P is positive, energy is being added to the capacitor, whereas when P is negative, energy is being removed from the capacitor. That is possible because a capacitor can store energy, but a resistor cannot.

22.4 | AC CIRCUITS WITH INDUCTORS

We next consider a circuit consisting of an AC generator and a single inductor as shown in Figure 22.14A. Our job again is to calculate the current in this circuit. In Chapter 21, we learned that the voltage across an inductor is proportional to the rate of change in the current. If L is the inductance and ΔI is the change in the current over a short time interval Δt, the voltage drop across the inductor is

$$V = L \frac{\Delta I}{\Delta t} \tag{22.24}$$

This voltage is also equal to output of the AC voltage source attached to the inductor, so $V = V_{max} \sin(2\pi ft)$. In words, Equation 22.24 states that V is proportional to the slope of the I–t relation. Because we are given V (Fig. 22.14B), we can infer how I varies with t; to do so, we must construct an I–t plot with a slope that satisfies Equation 22.24. We have sketched the answer qualitatively in Figure 22.14C. Here, $V = 0$ at times when the slope of I is zero, and V is largest and positive when I has its largest positive slope.

From Figure 22.14, we can see that V oscillates according to a sine function, whereas I oscillates with time according to a cosine function

$$I = -I_{max} \cos(2\pi ft) \tag{22.25}$$

We can use the trigonometric identity $\cos \alpha = -\sin(\alpha - \pi/2)$ to write Equation 22.25 as

$$I = I_{max} \sin(2\pi ft - \pi/2)$$

$$I = I_{max} \sin(2\pi ft + \phi) \quad \text{with } \phi = -\pi/2 \tag{22.26}$$

Figure 22.15 shows the current and voltage relationship for this inductor circuit in a phasor diagram. The current phasor makes an angle of $\phi = \pi/2 = -90°$ with the voltage, with the negative sign arising from Equation 22.26 and from our standard convention of measuring positive angles in the counterclockwise direction in the coordinate plane. In words, we say that the current is 90° out of phase from the voltage. This phasor diagram is very similar to what we found with the capacitor. For both capacitors and inductors, the current is 90° out of phase with the voltage. For a capacitor the phase angle is $\phi = +90°$, whereas for an inductor it is $\phi = -90°$.

The current and voltage are out of phase in an inductor because the voltage across it is proportional to the rate of change of the current, $V = L(\Delta I/\Delta t)$ (Eq. 22.24). This is in contrast to the case with a resistor, for which the current is proportional to the voltage (Eq. 22.2) so that I and V are in phase with each other.

Equation 22.26 and the phasor diagram in Figure 22.15 give the phase of the current through an inductor. An exact calculation of the peak current I_{max} shows that

$$I_{max} = \frac{V_{max}}{X_L} \tag{22.27}$$

where X_L is called the **reactance** of the inductor and is given by

$$X_L = 2\pi f L \tag{22.28}$$

The relation between the peak current and voltage for an inductor is very similar to the corresponding relations for a capacitor (Eq. 22.21) and a resistor (Eq. 22.2). Again, X_L is a measure of how strongly an inductor "resists" a current. Just as for a capacitor, the reactance of an inductor depends on frequency. Unlike in a capacitor, however, the reactance of an inductor X_L becomes larger as the frequency is increased. (Recall from Eq. 22.22 that X_C decreases at high frequencies.) This behavior of X_L can be traced back to the basic properties of an inductor. According to Lenz's law (Chapter 21), an inductor opposes changes in the flux through its coils. Increasing the frequency increases the rate of change of the flux and hence increases this opposing emf. For this reason, the current through an inductor decreases at high frequencies.

We have collected the results for the peak current in AC circuits with a resistor, a capacitor, or an inductor in Table 22.2.

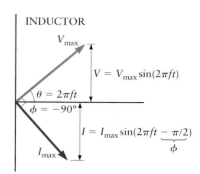

Figure 22.15 Phasor diagram of the voltage and current in an AC circuit with an inductor. The current through an inductor is 90° out of phase with the voltage.

CONCEPT CHECK 22.5 | Reactance of an Inductor

An inductor has a reactance $X_L = 2000\ \Omega$ when measured at a frequency of 1500 Hz. The frequency is then changed, and the reactance is found to be $6000\ \Omega$. Is the new frequency (a) 500 Hz, (b) 1000 Hz, (c) 3000 Hz, or (d) 4500 Hz?

Table 22.2 Properties of AC Circuits

CIRCUIT ELEMENT	RESISTANCE OR REACTANCE TO CURRENT FLOW	CURRENT THROUGH CIRCUIT ELEMENT	PHASE RELATION BETWEEN CURRENT AND VOLTAGE	AVERAGE POWER
Resistor ⎓W⎓	R	$I = V/R$	I and V are in phase	$P_{ave} = V_{rms}I_{rms} = \dfrac{V_{max}I_{max}}{2}$
Capacitor ⎓⊦	$X_C = \dfrac{1}{2\pi f C}$	$I = V/X_C$	I and V are out of phase by 90°	$P_{ave} = 0$
Inductor ⎓ⵕ⎓	$X_L = 2\pi f L$	$I = V/X_L$	I and V are out of phase by 90°	$P_{ave} = 0$

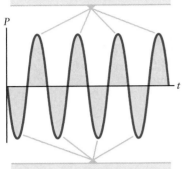

Energy flows into the inductor.

Energy flows from the inductor back to the generator.

Figure 22.16 The instantaneous power P delivered to the inductor is the product of V (a sine wave) and I (a cosine); the result is another sine wave, and the power delivered to an inductor in an AC circuit varies sinusoidally with time. At some portions of the cycle P is positive, whereas at other times it is negative. The average power over one complete cycle is zero.

Power in a Circuit Containing an Inductor

We can calculate the instantaneous power in a circuit containing an inductor using the result for the current in Equation 22.25 together with the general result for power in a circuit element $P = VI$. We find

$$P = VI = -V_{max}I_{max}\sin(2\pi ft)\cos(2\pi ft) \qquad (22.29)$$

This result is shown in Figure 22.16; it is very similar to our result for the power in a circuit with a capacitor (Fig. 22.13). The power delivered to the inductor oscillates between positive and negative values, and the average power over a complete cycle is zero. The generator delivers energy to the inductor during part of the AC cycle, and the inductor delivers an equal amount of energy back to the generator during other parts of the cycle. Energy is stored in the inductor as magnetic potential energy by virtue of the inductor's magnetic field.

22.5 | *LC* CIRCUITS

We have analyzed the simplest AC circuits containing only a single resistor, capacitor, or inductor. Most useful circuits contain multiple circuit elements, so now let's see what happens when resistors, capacitors, and inductors are connected together in different ways. In this section, we consider one particularly important example, the *LC* circuit.

Figure 22.17 shows the simplest possible *LC* circuit, consisting of a single inductor connected to a single capacitor. There is no AC generator in this circuit, but some excess charge is placed on the capacitor at $t = 0$ while the current is zero at that moment. That is, we start with a charge $+q$ on one plate and $-q$ on the other (Fig. 22.17A); a capacitor with a nonzero charge stores an energy $PE_{cap} = q^2/(2C)$, so we have "placed" this amount of energy in the circuit. The charge on the capacitor produces a voltage $V_C = q/C$ across the capacitor, and since the capacitor's leads are connected directly to the inductor, V_C is also the voltage across the inductor. In qualitative terms, the capacitor tries to "push" a current through the inductor as the charge on one capacitor plate attempts to flow through the inductor to the other plate. An inductor always resists changes in current, however; it does so through an induced emf and Lenz's law. Hence, at $t = 0$ there is an induced emf across the inductor opposing any current, and $I = 0$ at that instant.

This state of affairs with $I = 0$ is only maintained for an instant at $t = 0$; the current is nonzero after $t = 0$, just as for *RL* circuits (Chapter 21). So, at times $t > 0$ charge moves from one capacitor plate to the other and current passes through the inductor. Eventually, the charge on each capacitor plate falls to zero (Fig. 22.17B) and the corresponding voltage across the capacitor is $V_C = q/C = 0$. At this point you might expect that the current through the inductor would fall to zero, but the inductor again *opposes change* in the current, so the induced emf now acts to *maintain* the current at a nonzero value! This current continues to transport charge from one capacitor plate to the other even after reaching the point when $q = 0$, causing the capacitor's charge and voltage to reverse sign (Fig. 22.17C). The voltage across the capacitor then peaks at a negative value, leading to a current in the opposite direction through the inductor (Fig. 22.17D) as charge again moves from one plate to the other. Eventually, the charge on the capacitor plates returns to the original values found at $t = 0$ (Fig. 22.17E). The circuit has now returned to its original state (compare parts A and E of Fig. 22.17), and the process starts over again. The voltage and current in this circuit thus **oscillate** between positive and negative values. In fact, this circuit behaves as a **simple harmonic oscillator** with the current and charge given by

$$q = q_{max}\cos(2\pi ft) \quad \text{and} \quad I = I_{max}\sin(2\pi ft) \qquad (22.30)$$

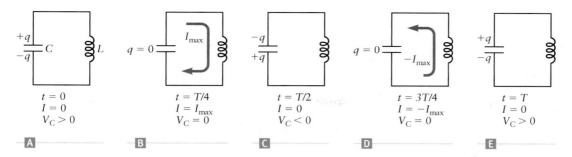

Figure 22.17 In an *LC* circuit with an initial charge, the current in the circuit and the voltage across the capacitor oscillate with a period *T*. This circuit behaves as a simple harmonic oscillator.

These results are identical to the time dependence of the position and velocity of a simple harmonic oscillator such as a mass on a spring as we studied in Chapter 11 (Eqs. 11.9 and 11.10).

Energy Conservation and the Oscillations in an *LC* Circuit

Capacitors and inductors can both store energy; the energy in a capacitor is stored in its electric field and depends on the charge, $PE_{cap} = \frac{1}{2}(q^2/C)$, while the energy in an inductor is stored in its magnetic field and depends on the current, $PE_{ind} = \frac{1}{2}LI^2$. Hence, as the charge and current oscillate in Figure 22.17, the energies stored in the capacitor and inductor also oscillate. We can compute these energies using the results for *q* and *I* from Equation 22.30 and find

$$PE_{cap} = \frac{1}{2}\frac{q^2}{C} = \frac{1}{2}\frac{q_{max}^2}{C}\cos^2(2\pi ft) \tag{22.31}$$

and

$$PE_{ind} = \frac{1}{2}LI^2 = \frac{1}{2}LI_{max}^2\sin^2(2\pi ft) \tag{22.32}$$

These results are shown in Figure 22.18. The energy oscillates back and forth between the capacitor and its electric field (PE_{cap}) and the inductor and its magnetic field (PE_{ind}). This behavior is very similar to that of a mechanical simple harmonic oscillator such as a pendulum or a mass on a spring. In a mechanical oscillator, the energy oscillates back and forth between kinetic and potential energy.

From the principle of conservation of energy, we expect that the *total* energy in the circuit $PE_{cap} + PE_{ind}$ must be constant. Hence, the maximum value of PE_{cap} must equal the maximum value of PE_{ind} as shown in Figure 22.18. Then, from Equations 22.31 and 22.32,

$$(PE_{cap})_{max} = (PE_{ind})_{max}$$

$$\frac{1}{2}\frac{q_{max}^2}{C} = \frac{1}{2}LI_{max}^2$$

$$I_{max} = \frac{1}{\sqrt{LC}}q_{max} \tag{22.33}$$

This equation tells us how the amplitudes of the current and charge oscillations in an *LC* circuit are related.

Frequency of Oscillations in an *LC* Circuit

Equation 22.33 relates the amplitudes of the current and charge oscillations in an *LC* circuit. We also want to know the *frequency* of these oscillations. In the *LC*

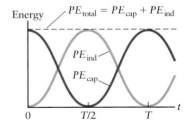

Figure 22.18 Plots of the energy in an *LC* circuit as a function of time, where PE_{cap} is the electric potential energy stored in the capacitor and PE_{ind} is the energy stored in the magnetic field of the inductor. The total energy $PE_{cap} + PE_{ind}$ is constant.

Figure 22.19 Metal detectors use changes in the resonant frequency of an LC circuit to sense when a piece of metal is nearby.

Frequency of an LC oscillator

circuit in Figure 22.17, the instantaneous voltages across the capacitor and inductor are always equal. Hence, the magnitude of the voltage across the capacitor must equal the magnitude of the voltage across the inductor. In terms of the reactances, these voltages are $V_C = IX_C$ (Eq. 22.21) and $V_L = IX_L$ (Eq. 22.27). Equating these magnitudes leads to

$$|V_C| = |IX_C| = |V_L| = |IX_L|$$

The reactances are always positive, so we can cancel the factors of I to get

$$X_C = X_L \qquad (22.34)$$

This condition was derived by assuming the current in the LC circuit is oscillating and hence applies *only* at the oscillation frequency. In words, Equation 22.34 says that at this oscillation frequency the reactances of the capacitor and inductor are equal.

Inserting the results for X_C and X_L from Equations 22.22 and 22.28 leads to

$$X_C = \frac{1}{2\pi f C} = X_L = 2\pi f L$$

The frequency here is the **resonant frequency** of the circuit. It is the oscillation frequency of the charge and current in Figure 22.17. Solving for this frequency, which we now denote as f_{res}, we get

$$f_{\text{res}} = \frac{1}{2\pi\sqrt{LC}} \qquad (22.35)$$

The oscillation frequency of an LC circuit thus depends on the values of L and C.

This frequency plays an important role in many applications of AC circuits. For example, a metal detector uses changes in the inductance of an LC circuit to sense the presence of a metal object. A metal detector used by beachcombers (Fig. 22.19) has a large coil of wire for its inductor. When the coil is placed near a metal object, the inductance of the coil changes, changing the resonant frequency of the circuit. The electronic unit of the metal detector is designed to register an audible signal when that happens. Metal detectors used for screening people at airports work in the same way.

22.6 | RESONANCE

Our analysis of an LC circuit in Section 22.5 explains how this circuit can act as a simple harmonic oscillator. Let's now consider a more realistic circuit containing an inductor, capacitor, and resistor along with an AC voltage source (Fig. 22.20). Our goal is to understand how the current in this LCR circuit depends on the amplitude of the AC voltage and on frequency. We will also show how phasors can be used to analyze the behavior of an AC circuit.

To analyze the LCR circuit in Figure 22.20, we appeal to Kirchhoff's loop rule (Chapter 19) and move around the circuit, adding up the voltage changes across the AC voltage source (V_{AC}), the resistor (V_R), the capacitor (V_C), and the inductor (V_L). According to the loop rule, the total change in the potential must be zero, so we have

$$V_{\text{AC}} = V_L + V_C + V_R \qquad (22.36)$$

The simple form of Equation 22.36 hides one crucial point—these voltages are not all in phase—so we must take account of both their magnitudes and the phase angles when we add them.

Figure 22.21 shows a series of phasor diagrams for the current in our LCR circuit and the voltage across each circuit element. The inductor, capacitor, and resistor are in series, so the current is the same through each one, and all the current phasors in Figure 22.21 have the same orientation. Figure 22.21A is for the

Figure 22.20 An LCR circuit.

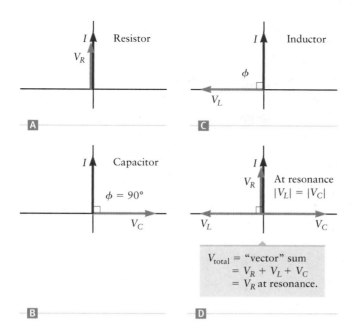

A — Resistor

C — Inductor

B — Capacitor

$\phi = 90°$

D — At resonance $|V_L| = |V_C|$

V_{total} = "vector" sum
= $V_R + V_L + V_C$
= V_R at resonance.

Figure 22.21 🄐–🄒 Phasor diagrams of the current and voltage across different components in an *LCR* circuit. 🄓 The total voltage across the circuit is the phasor sum (i.e., the "vector" sum) of the voltages across the resistor, the capacitor, and the inductor. Notice that V_C and V_L cancel at resonance.

resistor; in this case, the current and voltage are in phase with each other and the phasor arrows for I and V_R are parallel. This diagram is similar to Figure 22.9B except that we have placed these arrows along the vertical axis. Parts B and C of Figure 22.21 show the corresponding phasor diagrams for the voltages across the capacitor and inductor. As we saw in Sections 22.3 and 22.4, these voltages are 90° out of phase with the current. Figure 22.21D shows all three voltage phasors; the total voltage across all three circuit elements is the sum of these three phasors. The key point is that the phasors for V_L and V_C are in *opposite directions*, so they tend to cancel. In fact, we found in deriving Equation 22.34 that the two phasors V_C and V_L have the same magnitude (i.e., the same "length") when the frequency equals the resonant frequency f_{res}. Hence, at this frequency the voltages across the capacitor and inductor *cancel*. Only the resistor is left to "resist" the flow of current at this frequency. This perfect cancellation only occurs at f_{res}, so the current is highest when the frequency equals f_{res}.

Figure 22.22 shows how the current through an *LCR* circuit varies with frequency. Here, I is largest at f_{res} and decreases at both higher and lower frequencies. This is an example of ***resonance***, and it is very similar to the behavior of mechanical harmonic oscillators that we discussed in Chapter 11 (Fig. 11.28). By itself, an *LC* circuit acts as a simple harmonic oscillator with a resonant frequency given by Equation 22.35. Adding a resistor to the circuit (Fig. 22.20) adds damping and gives a damped harmonic oscillator, but the system still oscillates at the frequency f_{res}. When this oscillator is "driven" by attaching it to an AC voltage source, the resulting current depends on frequency. The current is largest when the frequency of the AC source matches the resonant frequency of the circuit, hence the peak in I in Figure 22.22 when $f = f_{res}$.

Applications of Resonance in Electronic Circuits

The resonant behavior of an *LCR* circuit is used in radios, televisions, and other similar applications. The input to a radio comes from an antenna that "picks up" signals from many different stations at many different frequencies. This composite signal is the AC voltage source for an *LCR* circuit. This circuit responds most strongly—that is, it gives the largest current—at frequencies that are near its resonant frequency. When you tune your radio, you are changing the value of the capacitance in an *LCR* circuit so that the resonant frequency matches the frequency of the station you want to listen to. The resulting current is then dominated by the

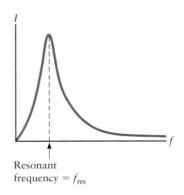

Resonant frequency = f_{res}

Figure 22.22 Current I in an *LCR* circuit as a function of frequency.

frequency of the desired station; this current is amplified and ultimately sent to your speakers or earphones. In this way, LCR circuits are used to construct devices that are frequency selective.

EXAMPLE 22.5 | Resonant Frequency of an LC Circuit

AM radio stations broadcast at frequencies in a range centered near 1.0 MHz (1.0×10^6 Hz). If the inductance in an LC circuit is $L = 50$ μH (5.0×10^{-5} H), what is the value of the capacitance that gives a resonant frequency at the middle of the AM range?

RECOGNIZE THE PRINCIPLE

We want to choose the value of the capacitance so that the resonant frequency equals the AM frequency $f = 1.0 \times 10^6$ Hz. The purpose of this problem is to see what particular values of L and C are needed to get the desired resonant frequency. The answer will tell us if an LC circuit is really practical for this application.

SKETCH THE PROBLEM

No figure is needed.

IDENTIFY THE RELATIONSHIPS

The resonant frequency of an LC circuit is given by (Eq. 22.35)

$$f_{\text{res}} = \frac{1}{2\pi\sqrt{LC}}$$

Rearranging to solve for C, we get

$$\sqrt{LC} = \frac{1}{2\pi f_{\text{res}}}$$

$$LC = \frac{1}{4\pi^2 f_{\text{res}}^2}$$

$$C = \frac{1}{4\pi^2 f_{\text{res}}^2 L}$$

SOLVE

Inserting the given values of f_{res} and L yields

$$C = \frac{1}{4\pi^2 (1.0 \times 10^6 \text{ Hz})^2 (5.0 \times 10^{-5} \text{ H})} = \boxed{5.1 \times 10^{-10} \text{ F}}$$

What does it mean?
This value for the capacitance is easily achieved with typical capacitors (see Example 22.4), so an LC circuit with a resonant frequency in the AM band is quite practical. Similar circuits are used in televisions, MP3 players, cell phones, and other applications in which one needs to distinguish between electrical signals according to their frequency.

22.7 | AC CIRCUITS AND IMPEDANCE

We introduced phasors as a convenient way to represent the phase relationship of current and voltage in a capacitor or inductor. We also used phasors in our discussion of resonance in LCR circuits. Now let's use them to find the current in some other AC circuits. Figure 22.23A shows an AC voltage source connected to an inductor. A typical inductor is constructed using a coil of wire, and real wires have

Figure 22.23 Real inductors have at least a small amount of resistance, so **A** an AC circuit with an inductor can be modeled as **B** a circuit with an ideal inductor in series with a resistor. **C** Phasor diagram for the current in the circuit and the voltages across the resistor and the inductor. **D** The total voltage across the inductor and the resistor is the phasor sum of the voltages in part C.

a nonzero resistance. A real inductor can therefore be modeled as an ideal inductor L in series with a resistor R as sketched in Figure 22.23B. We can calculate the current I in this circuit using phasors. The voltage across the inductor is $V_L = IX_L$ (Eq. 22.27), and the voltage across the resistor is $V_R = IR$ (Eq. 22.2). These two circuit elements are connected in series, so their currents are equal and the sum of their voltages equals the AC source voltage. It is tempting to just add our expressions for V_L and V_R, but that would *not* give the correct result. The voltages V_L and V_R must be added as phasors, so we must consider their phases when we take this sum.

Figure 22.23C is a phasor diagram showing the current I along with the voltages across L and R. For simplicity, we have placed the current phasor along the horizontal axis; this phasor describes the current through both the inductor and resistor because they are in series. The voltage across the resistor is then a phasor that also lies along the horizontal direction. (Compare with Fig. 22.9B.) The voltage across the inductor makes a phase angle $\phi = 90°$ with respect to the current in the "counterclockwise" sense (as in Figure 22.15), so the phasor representing V_L is along the upward vertical direction in this diagram. The total voltage V_{total} across the inductor and resistor combined is the sum of these two voltage phasors. This sum is sketched in Figure 22.23D, which shows that the total voltage has an amplitude

$$V_{\text{total}} = \sqrt{V_R^2 + V_L^2} \tag{22.37}$$

Inserting our results for V_R and V_L gives

$$V_{\text{total}} = \sqrt{V_R^2 + V_L^2} = \sqrt{(IR)^2 + (IX_L)^2} = I\sqrt{R^2 + X_L^2}$$

$$V_{\text{total}} = I\sqrt{R^2 + (2\pi fL)^2} \tag{22.38}$$

We now define a quantity called the **impedance** Z of the circuit through the relation

$$V_{\text{total}} = IZ \tag{22.39}$$ Impedance of an AC circuit

From Equation 22.38, we find

$$Z = \sqrt{R^2 + (2\pi fL)^2} \quad \text{(impedance for an } RL \text{ circuit)} \tag{22.40}$$

The impedance is a measure of how strongly a circuit "impedes" current. For a single resistor, the impedance is simply equal to the resistance. Likewise, for a single capacitor or inductor, the impedance is equal to the reactance of the capacitor or inductor. For combinations of these circuit elements, the impedance is given by a more complicated combination of these quantities. The result in Equation 22.40 applies only for a circuit consisting of a single resistor in series with a single inductor. The impedance of other circuits can be calculated by adding phasors, as we'll see in other examples below.

Figure 22.23D contains another important lesson. In all the AC circuits we have analyzed so far, the phase angle between the current and voltage was either zero or

±π/2. For an *LR* circuit, the phase angle has a value between zero and π/2, depending on the relative values of V_R and V_L, which themselves depend on R, L, and f. For most AC circuits, the phase angle depends on frequency and also on the specific values of the resistances, inductances, and capacitances in the circuit. We'll explore this topic further in Problems 52 and 53.

EXAMPLE 22.6 | Phasor Analysis of an *RC* Circuit

Consider the circuit in Figure 22.24, with a capacitor in series with a resistor. Given that the frequency of the AC source is f, find an expression for the impedance of this circuit. Express your answer in terms of C, R, and f.

RECOGNIZE THE PRINCIPLE

The impedance is proportional to the total voltage through Equation 22.39. To find the total voltage across the capacitor plus the resistor, we must add their voltages as phasors. The phasor diagrams in this case are similar to the case with an inductor (Fig. 22.23) except that the phasor representing V_C is downward along the vertical axis (Fig. 22.24B); compare this figure with the phasor diagram for a capacitor alone in Figure 22.11.

SKETCH THE PROBLEM

Figure 22.24 shows the circuit along with the corresponding phasor diagrams for the current and voltage across the capacitor and resistor.

IDENTIFY THE RELATIONSHIPS

We add the phasors representing V_R and V_C as shown in Figure 22.24C. The phasor arrows form a right triangle, so we have

$$V_{total} = \sqrt{V_R^2 + V_C^2}$$

Inserting $V_R = IR$ (Eq. 22.2) and $V_C = IX_C = I/(2\pi fC)$ (Eqs. 22.21 and 22.22) gives

$$V_{total} = \sqrt{V_R^2 + V_C^2} = \sqrt{(IR)^2 + \left(\frac{I}{2\pi fC}\right)^2} = I\sqrt{R^2 + \left(\frac{1}{2\pi fC}\right)^2}$$

SOLVE

Comparing this expression with the definition of impedance $V = IZ$ (Eq. 22.39), we find that for this circuit the impedance is

$$Z = \sqrt{R^2 + \left(\frac{1}{2\pi fC}\right)^2}$$

What does it mean?

The impedance of this circuit decreases at high frequencies, so for a given value of the AC voltage, the current is larger at high frequencies than at low frequencies. This is because the reactance X_C of the capacitor becomes smaller at high frequencies, reducing the impedance of the entire circuit.

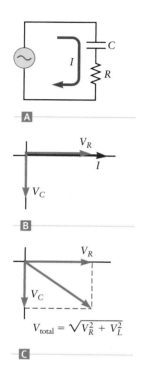

Figure 22.24 Example 22.6.

Impedance of an *LCR* Circuit

A phasor analysis for an *LCR* circuit is shown in Figure 22.21D at the resonant frequency only. We can find the impedance for the general case using a method like that in Figures 22.23 and 22.24. Beginning anew in Figure 22.25A, we again place the current phasor along the horizontal axis, so the phasor for the resistor's voltage is also in the horizontal direction. The phasor representing the voltage across the inductor is then upward along the vertical axis, and the phasor for the capacitor's voltage is downward. Collecting the horizontal and vertical components in Figure 22.25B, we add them to get the total voltage across the circuit:

$$V_{total} = \sqrt{(IR)^2 + I^2(X_L - X_C)^2} = \sqrt{(IR)^2 + I^2\left(2\pi fL - \frac{1}{2\pi fC}\right)^2}$$

$$V_{total} = I\sqrt{R^2 + \left(2\pi fL - \frac{1}{2\pi fC}\right)^2}$$

The impedance is defined through $V_{total} = IZ$, so for this circuit we get

$$Z = \sqrt{R^2 + \left(2\pi fL - \frac{1}{2\pi fC}\right)^2} \qquad (22.41)$$

If the voltage provided by the AC generator is $V_{total} = V_{max}\sin(2\pi ft)$, the amplitude of the current is $I_{max} = V_{max}/Z$.

Impedance and Resonance in an *LCR* Circuit

Let's examine more carefully our result for the impedance of a general *LCR* circuit (Eq. 22.41) and see how it and the current in the circuit behave near resonance.

From the general definition of impedance, the amplitude of the current is $I_{max} = V_{max}/Z$. Since the impedance varies with frequency, the current amplitude is also frequency dependent and the maximum value of the current occurs when the impedance Z is a minimum. This minimum in Z occurs when the factors involving L and C cancel. We have

$$Z = \sqrt{R^2 + \left(2\pi fL - \frac{1}{2\pi fC}\right)^2}$$

The minimum impedance thus occurs when

$$2\pi fL - \frac{1}{2\pi fC} = 0$$

Solving for the frequency gives

$$f = \frac{1}{2\pi\sqrt{LC}}$$

which is precisely the resonant frequency we found for an *LC* circuit in Section 22.5 (Eq. 22.35). The maximum current thus occurs at the resonant frequency, the frequency at which the *LCR* circuit responds most strongly to an applied AC voltage.

> **CONCEPT CHECK 22.6** | Phase Angle in an *LCR* Circuit at Resonance
>
> Is the phase angle between the current and voltage in an *LCR* circuit at resonance (a) 0, (b) 90°, or (c) 180°? (*Hint:* Consider the phasor diagrams in Figs. 22.21 and 22.25.)

22.8 | FREQUENCY-DEPENDENT BEHAVIOR OF AC CIRCUITS: A CONCEPTUAL RECAP

We can make better sense of the circuit theory in this chapter, with its many new concepts including reactance and impedance, if we have a clear qualitative picture of why the frequency is so important in AC circuits.

Circuit Behavior at Low and High Frequencies

Resistors in an AC circuit behave very much like resistors in a DC circuit. For both types of circuits, the current in a resistor is related to the voltage by $I = V/R$ independent of the frequency, and the current is always *in phase* with the voltage (Fig. 22.9). In contrast, the current through a capacitor or an inductor is *frequency dependent*, a key aspect of AC circuits that makes them useful in applications such

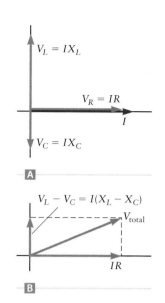

Figure 22.25 **A** Phasor diagram of the current in a series *LCR* circuit and the voltages across all three circuit elements for the general case in which $X_L \neq X_C$. Notice how the voltages across the capacitor and inductor are out of phase and hence in opposite directions. (Compare with Fig. 22.21.) **B** Phasor sum of the voltages.

Table 22.3 Qualitative Behavior of Capacitors and Inductors at Low and High Frequencies

CIRCUIT ELEMENT	RESISTANCE OR REACTANCE	BEHAVIOR AT VERY LOW FREQUENCIES	BEHAVIOR AT VERY HIGH FREQUENCIES
Resistor —W—	R	Independent of frequency	Independent of frequency
Capacitor —||—	$X_C = \dfrac{1}{2\pi f C}$	$X_C \approx \infty$; very little current passes	$X_C \approx 0$; current passes freely
Inductor —₀₀₀—	$X_L = 2\pi f L$	$X_L \approx 0$; current passes freely	$X_L \approx \infty$; very little current passes

Figure 22.26 Qualitative analysis of **A** an LR circuit. **B** At very low frequencies, the reactance of the inductor X_L is very small and the inductor acts as a short circuit (i.e., a simple wire). **C** At very high frequencies, X_L is very large, so the inductor acts as an open circuit. The original circuit thus removes low frequencies from the output and passes high frequencies.

as the radio tuner discussed in Section 22.6 and the filter circuits described later in this section.

The frequency-dependent behavior of AC circuits is due to the frequency dependence of the reactances X_C and X_L. For a capacitor,

$$I_{max} = \frac{V_{max}}{X_C} \quad \text{(for a capacitor)} \tag{22.42}$$

and X_C is a measure of how strongly the capacitor "resists" current. When the frequency is low, the oscillation period is long and more charge is "pushed" on and off a capacitor during each AC cycle. This larger charge produces a larger electric field in opposition; hence, the reactance X_C is largest at low frequencies. From Equation 22.42, the current through a capacitor is *smallest* at *low frequencies*. At very low frequencies, a capacitor thus approaches an "open circuit" and passes very little current, whereas at very high frequencies, a capacitor resembles a short circuit, passing current like a simple ideal wire with no resistance.

For an inductor, we have

$$I_{max} = \frac{V_{max}}{X_L} \quad \text{(for an inductor)} \tag{22.43}$$

and X_L is a measure of how strongly an inductor "resists" current. Inductance is a result of the laws of Faraday and Lenz because an inductor opposes changes in current. The rate of change of I is largest at high frequencies, making the reactance X_L largest at high frequencies. From Equation 22.43, this fact makes the current through an inductor *smallest* at *high frequencies*. At very low frequencies, an inductor thus passes current freely, like a simple wire with no resistance, whereas at very high frequencies, an inductor approaches an open circuit and passes little or no current.

These qualitative observations are summarized in Table 22.3. We can use these results to get a general sense of the behavior of a circuit that contains multiple circuit elements. Consider the circuit in Figure 22.26A, an RL circuit for which we indicate the "input" and "output" voltage signals. Although we show an AC generator providing the input voltage V_{in}, this voltage might instead be provided by an antenna (as in a radio or TV) or another AC circuit. The voltage across the inductor is the "output" voltage V_{out} and is typically connected to another portion of the circuit that is not shown here. Let's use our qualitative rules from Table 22.3 to estimate the output voltage at very low and at very high frequencies.

When the input frequency is very low, the reactance of the inductor is very small. The inductor thus acts as a simple wire, so we have redrawn the circuit with the inductor replaced by a wire in Figure 22.26B. From our work with resistors and wires in Chapter 19, we know that there is zero voltage drop across the ends of the wire, so V_{out} in Figure 22.26B is zero. Hence, even though the input voltage is generally not zero, the output voltage of this circuit is very small at low frequencies.

In the limit of very high frequencies, an inductor acts as an open circuit (Table 22.3) as sketched in Figure 22.26C. Because an open circuit passes no current, the current through the resistor and the voltage drop across it are approximately zero.

The output voltage is then equal to the input voltage, $V_{out} = V_{in}$. We thus find that the output voltage of the circuit in Figure 22.26A is zero at low frequencies but equal to the input voltage at high frequencies. This circuit thus acts as a *high-pass filter*, allowing high-frequency signals to pass through to the output while low frequencies are blocked. In the end-of-chapter problems, we'll explore how other combinations of inductors, resistors, and capacitors can be used to make other types of filters (e.g., low-pass filters).

| EXAMPLE 22.7 | ⓡ An *RC* Circuit at Low and High Frequencies |

Figure 22.27A shows an *RC* circuit. Use the qualitative frequency-dependent behavior of a capacitor as given in Table 22.3 to estimate the output voltage at (a) very low and (b) very high frequencies.

RECOGNIZE THE PRINCIPLE

The qualitative behavior at low and high frequencies can be found using the frequency dependence of the reactance of a capacitor (Table 22.3), which we can use to draw "equivalent circuits" for low and high frequencies. At low frequencies, the capacitor is replaced by an open circuit (Fig. 22.27B), whereas at high frequencies, the capacitor is replaced by a short circuit (Fig. 22.27C).

SKETCH THE PROBLEM

Figure 22.27 shows the circuit, along with equivalent circuits at low and high frequencies.

IDENTIFY THE RELATIONSHIPS AND SOLVE

(a) At very low frequencies, the capacitor acts approximately as an open circuit (Fig. 22.27B) and the current is very small. The output voltage then equals the input voltage, and we have

$$V_{out} = \boxed{V_{in}} \text{ at low frequencies}$$

(b) At very high frequencies, the capacitor acts as a short circuit, so we replace it by a wire in Figure 22.27C. The output voltage is equal to the voltage across this short circuit, which is zero. Hence,

$$V_{out} = \boxed{0} \text{ at high frequencies}$$

What does it mean?

This circuit prevents high frequencies from reaching the output (since $V_{out} = 0$ at high frequencies), so this circuit acts as a *low-pass filter*.

When Are We at Low or High Frequency?

The results from Table 22.3 are extremely useful in gaining a qualitative understanding of an AC circuit's behavior at low and high frequencies, but how do we know when a particular frequency is low enough or high enough to apply the limiting values of X_C and X_L from Table 22.3? For an *RL* circuit (Fig. 22.26), the input frequency must be compared with the corresponding *RL* time constant. The time constant of an *RL* circuit is given by $\tau_{RL} = L/R$ (Eq. 21.27). Qualitatively, you can think of this time constant as the period of an oscillation and define a corresponding frequency by

$$f_{RL} = \frac{1}{\tau_{RL}} = \frac{R}{L} \tag{22.44}$$

The high-frequency limit in Table 22.3 applies when the input or source frequency in an *RL* circuit is much greater than f_{RL}, whereas the low-frequency limit corresponds to frequencies much lower than f_{RL}. For qualitative estimates, any frequency

Insight 22.1

APPLICATIONS OF A LOW-PASS FILTER

A low-pass filter can be constructed by replacing the inductor in Figure 22.26A with a capacitor (see Example 22.7). A low-pass filter is used in radios and MP3 players. A music (audio) signal often contains "static," which comes from unwanted high-frequency components added to the desired music. These high frequencies can be filtered out using a low-pass filter.

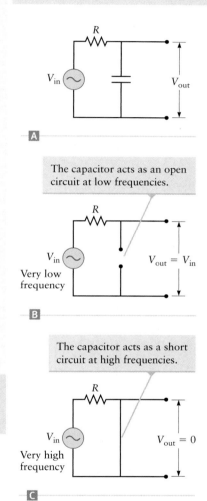

Figure 22.27 Example 22.7. Qualitative analysis of the frequency dependence of an *RC* circuit.

Figure 22.28 Concept Check 22.7.

greater than about $10 \times f_{RL}$ falls into the high-frequency limit, whereas the low-frequency limit is below about $f_{RL}/10$.

The approach is similar for an AC circuit containing a capacitor and a resistor. In this case, we have an RC time constant given by $\tau_{RC} = RC$ (Eq. 19.31), which leads to a corresponding frequency

$$f_{RC} = \frac{1}{\tau_{RC}} = \frac{1}{RC} \qquad (22.45)$$

The high- and low-frequency limits for a capacitor circuit then apply when the input or source frequency is much higher or much lower[2] than f_{RC}.

As a final case, we consider an AC circuit containing a capacitor and an inductor. In this case, the resonant LC frequency $f_{res} = 1/(2\pi\sqrt{LC})$ (Eq. 22.35) determines the boundary between the high- and low-frequency limits in Table 22.3.

CONCEPT CHECK 22.7 | Frequency Dependence of an LC Circuit

Consider the LC circuit in Figure 22.28. The voltage from the AC source is $V = V_{max} \sin(2\pi ft)$, with V_{max} held fixed while the frequency f is adjustable. Which of the following statements is correct?

(a) At very high frequencies, the voltage across the inductor is greater than the voltage across the capacitor.

(b) At very high frequencies, the voltage across the inductor is less than the voltage across the capacitor.

(c) The voltage across the inductor is equal to the voltage across the capacitor at all frequencies.

22.9 | TRANSFORMERS

Transformers are devices that can either increase or decrease the amplitude of an applied AC voltage. The ability to "transform" an AC voltage from small to large, or from large to small, is used by electric companies to transmit electrical energy efficiently over long distances. Transformers are also found in many household devices, including televisions and radios, and are used in the "power bricks" that connect your laptop computer to a wall socket (Fig. 22.29).

A simple transformer consists of two solenoidal coils with the loops arranged so that all or most of the magnetic field lines and flux generated by one coil pass through the other coil. This can be accomplished in several ways, two of which are illustrated in Figure 22.30. In Figure 22.30A, two coils are placed end to end, whereas Figure 22.30B shows one coil wrapped tightly around the other. The wires are covered with a nonconducting layer ("insulation") so that current cannot flow directly from one coil to the other. Even so, an AC current in one coil will induce an AC voltage across the other one. According to Faraday's law, the magnitude of the voltage drop across a coil is related to the magnetic flux through the coil by

$$V = \frac{\Delta\Phi}{\Delta t} \qquad (22.46)$$

In a transformer, an AC voltage source is typically attached to one of the coils—denoted as the "input" coil—and the other is called the "output" coil. Equation 22.46 applies for both coils, so we can write

$$V_{in} = \frac{\Delta\Phi_{in}}{\Delta t} \qquad V_{out} = \frac{\Delta\Phi_{out}}{\Delta t} \qquad (22.47)$$

The coils in Figure 22.30 are designed so all the field lines that pass through one coil also pass through the other. If the coils are identical (having the same diameters,

Figure 22.29 A "power brick" used with a laptop computer employs a transformer.

---

[2]As with an inductor, the high- and low-frequency limits for a capacitor are approximately $10 \times f_{RC}$ and $f_{RC}/10$, respectively.

Figure 22.30 Two simple transformers.

lengths, and numbers of turns), the flux is the same through each coil and $\Phi_{in} = \Phi_{out}$. The fluxes in Equation 22.47 are thus the same, so the voltages are also equal:

$$V_{in} = \frac{\Delta\Phi}{\Delta t} = V_{out} \qquad (22.48)$$

We have thus managed to "transfer" a voltage from the input coil to the output coil without any direct connection between the two. However, there is an indirect connection through the magnetic flux.

With the transformers in Figure 22.30, the output voltage is equal to the input voltage, but that could have been accomplished more simply by just connecting the input and output wires. How can we *change* the voltage? Consider again the two coils in either part of Figure 22.30, but this time let the output coil have twice as many turns as the input coil. If the flux through the input coil is Φ_{in}, the flux through the output coil will be $\Phi_{out} = 2\Phi_{in}$ because there are twice as many loops in the output coil. We can use this relation between the fluxes in Equation 22.47 to get a relation between the input and output voltages:

$$V_{out} = \frac{\Delta\Phi_{out}}{\Delta t} = 2\frac{\Delta\Phi_{in}}{\Delta t} = 2V_{in}$$

Hence, the output voltage is twice as large as the input voltage.

We can apply the same analysis to the general case of an input coil with N_{in} turns and an output coil with N_{out} turns. For a coil with N turns and a total flux Φ, the flux through a single turn is $\Phi_1 = \Phi/N$; this relation applies to both the input coil and the output coil. For the transformers in Figure 22.30 (and for most practical transformers), the flux through one turn of the output coil equals the flux through one turn of the input coil. The input and output fluxes are then related by

$$\Phi_{out} = \frac{N_{out}}{N_{in}}\Phi_{in}$$

Using this result with Equation 22.47, the output voltage is

$$V_{out} = \frac{N_{out}}{N_{in}}V_{in} \qquad (22.49)$$

Output voltage of a transformer

Since the ratio N_{out}/N_{in} can be larger or smaller than unity, we can thus "transform" the input voltage to a different value, either larger or smaller than the input voltage. That is why the device is called a transformer. Transformers are based on Faraday's law, so, to apply Equation 22.47, the flux must be changing with time. Hence, transformers cannot be used to "transform" a DC voltage.

The transformer designs in Figure 22.30 are simplest to understand, but the different arrangement shown in Figure 22.31 is used in most practical applications. The central regions of both coils are filled with a magnetic material such as iron. The field from the input coil aligns the atomic magnets in the magnetic material, producing a much larger flux than would be produced by the coil's field alone. This larger flux then yields larger voltages at both the input coil and the output coil. The ratio V_{out}/V_{in} is not affected by the presence of magnetic material in the transformer, however, and this ratio is still given by Equation 22.49.

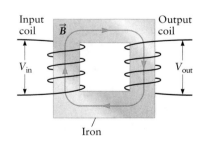

Figure 22.31 Many transformers use a magnetic material such as iron to "carry" the flux from the input coil to the output coil.

A transformer has an output coil with three times as many turns as the input coil. According to Equation 22.49, the output voltage is thus three times larger than the input voltage. If the input frequency is f, is the output frequency (a) f, (b) $3f$, or (c) $f/3$?

Applications of Transformers

One of the most important uses of transformers involves the "transmission" of electric power over large distances such as from a power plant to a city many miles away. Transformers are also used in many household appliances to convert the AC voltage at a wall socket (120-V rms) to the smaller voltages needed in radios, computers, and other devices that commonly require DC voltages around 9 V or 12 V. These DC voltages are produced in a two-step process by "power bricks" like the one in Figure 22.29 or by power supplies built into an appliance. The first step uses a transformer to convert the 120 V from a wall socket to a value near 9 V or 12 V. The second step is to convert this smaller AC voltage to direct current, using a circuit called a "rectifier." We don't have space here to describe how a rectifier works, but its output is a DC voltage, similar to the output of a battery. That is why a power brick can substitute for batteries for your laptop computer.

Transforming Voltage and Power

According to Equation 22.49, the output voltage of a transformer can be much larger than the input voltage by just arranging for the number of turns in the output coil to be much larger than the number of turns in the input coil. Electrical power delivered to a circuit is the product of the voltage and current (Eq. 22.9). So, if V_{out} is much larger than V_{in}, what happens to the output power? Is the output power also much larger than the input power? We can apply the principle of conservation of energy to answer these questions.

A transformer is like an inductor because it stores energy in the magnetic field produced by the two coils. This energy is *stored*; it is not "lost" or dissipated as heat as in a resistor. According to energy conservation principles, the energy delivered through the input coil must either be stored in the transformer's magnetic field or transferred to the output circuit. Over many AC cycles, the stored energy is constant, so the power delivered to the input coil must *equal* the output power. The input power is $P_{in} = V_{in}I_{in}$, whereas the output power is $P_{out} = V_{out}I_{out}$. Hence, if V_{out} is transformed to a value larger than V_{in}, the output current is simultaneously transformed to a smaller value than the input current.

This argument applies only to ideal transformers. The coils of a real transformer always have a small electrical resistance, causing some power dissipation. Hence, for a real transformer, the output power is always less than the input power, but usually by only a small amount.

22.10 | MOTORS

In Chapter 20, we described how the magnetic torque on a current loop can be used to make a DC motor (Section 20.10). Another way to make a motor is shown in Figure 22.32; this approach also uses the torque on a current loop, but now we use an AC voltage source to "power" the motor. The AC source is connected to a coil (called the input coil) wound around a horseshoe magnet. The input coil induces a magnetic field that circulates through the horseshoe magnet in the direction parallel to the field from the coil. As the current in the input coil oscillates, the direction of

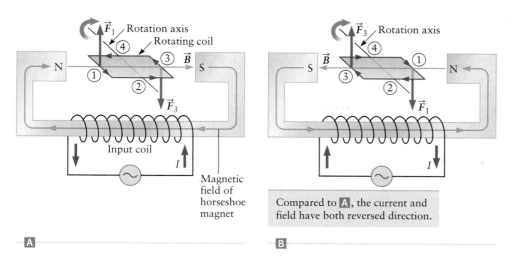

Figure 22.32 Design of an AC motor.

N   S $\vec{F}_1$ Rotation axis   Rotating coil $\vec{B}$ ① ② ③ ④ $\vec{F}_3$ Input coil I Magnetic field of horseshoe magnet

$\vec{F}_3$ Rotation axis $\vec{B}$ S   N   ① ② ③ ④ $\vec{F}_1$ I

Compared to **A**, the current and field have both reversed direction.

A   **B**

the magnetic field from the horseshoe magnet alternates back and forth as indicated in Figure 22.32.

A second coil of wire is mounted between the poles of the horseshoe magnet and attached to a rotating shaft. The rotating coil is connected to a source of DC voltage (i.e., a battery), so this coil carries a DC current. Figure 22.32A shows the rotating coil in a particular position and the north pole of the horseshoe magnet on the left. From Chapter 20, when a current-carrying wire—that is, sections of the rotating coil—is placed in a magnetic field, it experiences a force. Applying right-hand rule number 2 to calculate the direction of this force, we find the results shown in Figure 22.32A. The force $\vec{F}_1$ on section 1 of the coil is directed upward, whereas the force $\vec{F}_3$ on section 3 is down. There is no force on sections 2 and 4 because the current in those sections is parallel or antiparallel to the magnetic field. The forces $\vec{F}_1$ and $\vec{F}_3$ produce a torque on the coil, causing it and the shaft to rotate. Hence, the motor turns.

After a short time, the coil rotates to the position in Figure 22.32B and at the same time the AC current in the input coil changes direction so that the north pole of the horseshoe magnet is now on the right. Applying right-hand rule number 2 to calculate the forces on the current loop, we find that $\vec{F}_1$ is now downward and $\vec{F}_3$ is upward. Comparing the two parts of Figure 22.32, we see that the torques on the coil are in the same direction; both torques are clockwise as viewed from the front of the drawing. Hence, the motor shaft continues to turn in a clockwise manner.

As the coil continues to rotate, the field from the horseshoe magnet continues to change direction so that the torque exerted on the coil is always clockwise. In this way, the oscillations of an AC current and field make the shaft rotate. Practical AC motor designs are more complicated than shown here, with typically three separate magnets (mounted at different angles) and a slightly more complicated rotating coil, but the principle is the same as shown in Figure 22.32.

EXAMPLE 22.8 | Torque Produced by an AC Motor

Consider an AC motor like the one in Figure 22.32 and let the rotating coil be square with sides of length $L = 10$ cm, carrying a current $I = 1.0$ A. If the field of the horseshoe magnet is $B = 1.5$ T, what is the torque on the coil when it is oriented as shown in Figure 22.32A?

RECOGNIZE THE PRINCIPLE

The torque exerted on the coil is produced by the forces $\vec{F}_1$ and $\vec{F}_3$ as shown in more detail in Figure 22.33. In each case, we have a current-carrying wire in a

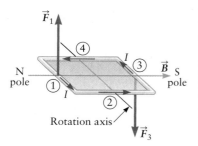

Figure 22.33 Example 22.8. Torque on the rotating coil of the AC motor in Figure 22.32A. The coil is composed of four straight sections.

perpendicular magnetic field, so from Chapter 20 each force has a magnitude $F = ILB$. The directions of these forces are given by right-hand rule number 2 and are shown in Figure 22.33; each gives a torque with a lever arm $L/2$ (the distance from the rotation axis to the point where the forces act). From Chapter 8 (Eq. 8.9), the torque associated with $\vec{F}_1$ is thus

$$\tau_1 = F_1 \times (\text{lever arm}) = \frac{F_1 L}{2}$$

In a similar way, the torque from $\vec{F}_3$ is

$$\tau_3 = F_3 \times (\text{lever arm}) = \frac{F_3 L}{2}$$

These torques combine to give the total torque on the shaft.

SKETCH THE PROBLEM

Figure 22.33 shows the problem.

IDENTIFY THE RELATIONSHIPS

The torques τ_1 and τ_3 both tend to make the shaft rotate clockwise, so the magnitude of the total torque is the sum

$$\tau_{\text{total}} = \tau_1 + \tau_3 = \frac{F_1 L}{2} + \frac{F_3 L}{2} = \frac{L}{2}(F_1 + F_3) \qquad (1)$$

To compute the total torque from Equation (1), we must find the forces F_1 and F_3. The force on a wire in a perpendicular magnetic field B is $F = ILB$ (from Chapter 20). Both sections of the wire have length L and carry the same current I, so we find

$$F_1 = ILB = F_3$$

Inserting this expression into Equation (1), we get

$$\tau_{\text{total}} = \frac{L}{2}(F_1 + F_3) = \frac{L}{2}(2ILB) = IL^2 B$$

SOLVE

Inserting the given values of the various quantities gives

$$\tau_{\text{total}} = IL^2 B = (1.0 \text{ A})(0.10 \text{ m})^2(1.5 \text{ T}) = \boxed{0.015 \text{ N} \cdot \text{m}}$$

What does it mean?
This torque is not very large, so this design would make a very weak motor. In more practical motors, the coil would consist of many turns (hundreds or thousands or more), giving a much larger torque.

22.11 | WHAT CAN AC CIRCUITS DO THAT DC CIRCUITS CANNOT?

At the beginning of this chapter, we mentioned that Thomas Edison proposed to wire cities and houses with DC circuits, whereas George Westinghouse and others pushed for an AC system. Westinghouse won the argument, but why? What advantages does alternating current have over direct current for such applications?

One advantage discussed in Chapter 21 is that large AC generators are easily built using a rotating shaft and permanent magnets. The biggest advantage of AC circuits over DC circuits, however, is in the systems that distribute electrical power across long distances. Most electrical power is generated at large facilities such as hydroelectric dams or coal-burning power plants. The electrical power produced at these facilities must then be distributed to distant cities as shown schematically in Figure 22.34A. The power plant acts as an AC generator, sending current through power lines (large, overhead wires) to the destination. These wires have a small but nonzero electrical resistance R_{line}, so if the power line carries an rms current I_{rms}, there is some power $P_{ave} = I_{rms}^2 R_{line}$ dissipated in the power line resistance. Because electrical energy costs money, the power company prefers to minimize the energy dissipated in the power line. The resistance of the power line usually can't be changed easily; hence, the only way to reduce the dissipated power is to make I_{rms} as small as possible. The power delivered to the destination city is $P_{city} = V_{city}I_{city}$, so if the power company decreases the current in the power line, it must increase the voltage so as to transmit the same amount of power to the city. This voltage increase is accomplished using a transformer.

The normal household voltage is "120-V" alternating current, an rms value. In Example 22.2, we showed that this value actually corresponds to a voltage amplitude of $V_{max} = 170$ V. A much larger AC voltage is used for power transmission. Typical power line voltages have $V_{max} = 500,000$ V or even higher. When the power line reaches your house, a transformer reduces the AC voltage amplitude to the value $V_{max} = 170$ V required by your household appliances (Fig. 22.34B). Even with a power line voltage of 500,000 V, typically 5% to 10% of the energy that leaves a power plant is dissipated in the resistance of the power line. The fraction lost would be much greater if the power line were operated at lower AC voltage amplitudes.

Transformers can readily change AC voltage amplitudes, but there is no simple way to accomplish the same changes with DC voltages. There is no efficient way to increase and decrease DC voltage levels without dissipating significant amounts of energy. That is the main reason alternating current was adopted for the electrical power system.

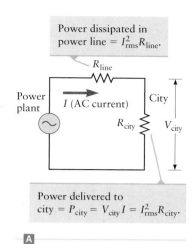

Power dissipated in power line = $I_{rms}^2 R_{line}$.

Power delivered to city = $P_{city} = V_{city}I = I_{rms}^2 R_{city}$.

A

Figure 22.34 **A** Circuit diagram showing a power line carrying current from an electric power plant to a city. **B** Photo of a transformer used to transform the high voltage at a power line to a lower voltage suitable for use in a house.

B

EXAMPLE 22.9 | Power Dissipated in an Electrical Transmission Line

An AC power line operates with a voltage $V_{rms} = 500,000$ V and carries an AC current with $I_{rms} = 1000$ A. **(a)** What is the average (rms) power carried by the power line? **(b)** If 10% of this power is dissipated in the power line itself, what is the resistance of the power line? **(c)** This power line is now operated with a voltage $V_{rms} = 250,000$ V and current $I_{rms} = 2000$ A. The product $V_{rms}I_{rms}$ is the same as in (a), so the power carried by the line is the same. What percentage of this power is now dissipated in the power line?

RECOGNIZE THE PRINCIPLE

The total power in any circuit element is $P = VI$, so the power carried by the power line is $P_{ave} = V_{rms}I_{rms}$ (Eq. 22.12). Some of this power is dissipated in the line, and the rest is delivered to the city. The power line can be modeled as a resistor R_{line}, and the power dissipated in this resistance is $P_{line} = I_{rms}^2 R_{line}$. We can use these relations for P_{ave} and P_{line} to (a) find the average power carried by the line, (b) find the line resistance, and (c) analyze the effect of changing the line voltage.

SKETCH THE PROBLEM

Figure 22.34A describes the problem.

IDENTIFY THE RELATIONSHIPS AND SOLVE

(a) The total power carried by the power line is

$$P_{ave} = V_{rms}I_{rms} = (500{,}000 \text{ V})(1000 \text{ A})$$

$$P_{ave} = 5.0 \times 10^8 \text{ W} = \boxed{500 \text{ MW}}$$

(b) Most of this power is delivered (and sold) to the city, whereas 10% is dissipated in the power line. As a resistor, the line dissipates an amount of power $P_{line} = I_{rms}^2 R_{line}$, which is equal to 10% of $P_{ave} = 0.10 \times P_{ave}$. We thus have

$$P_{line} = I_{rms}^2 R_{line} = (0.10)P_{ave}$$

Solving for R_{line},

$$R_{line} = \frac{(0.10)P_{ave}}{I_{rms}^2} = \frac{(0.10)(5.0 \times 10^8 \text{ W})}{(1000 \text{ A})^2} = \boxed{50 \text{ } \Omega}$$

(c) If the power line voltage is reduced by a factor of two (to 250,000 V) while the current is increased by the same factor (to 2000 A), the power delivered is unchanged and we still have $P_{ave} = 5.0 \times 10^8$ W; see part (a). The power dissipated in the line is now

$$P_{line} = I_{rms}^2 R_{line} = (2000 \text{ A})^2(50 \text{ } \Omega) = 2.0 \times 10^8 \text{ W}$$

The percentage of the total power now dissipated in the line is

$$\text{percent dissipated} = \frac{P_{line}}{P_{total}} \times 100 = \frac{2.0 \times 10^8 \text{ W}}{5.0 \times 10^8 \text{ W}} \times 100$$

$$\text{percent dissipated} = \boxed{40\%}$$

What does it mean?
A much larger fraction of the energy is "lost" in the power line in part (c), resulting in much smaller profits for the power company. The fraction of the energy dissipated in the line decreases when the system operates at higher voltages and hence lower currents. That is why power lines are operated with very high AC voltages.

SUMMARY | Chapter 22

KEY CONCEPTS AND PRINCIPLES

Root-mean-square (rms) voltage
The ***instantaneous voltage*** V in an AC circuit varies in time as

$$V = V_{max} \sin(2\pi ft) \qquad \textbf{(22.1)} \text{ (page 726)}$$

where V_{max} is the voltage ***amplitude*** and f is the frequency. The ***root-mean-square*** (rms) voltage is

$$V_{rms} = \frac{V_{max}}{\sqrt{2}} \qquad \textbf{(22.7)} \text{ (page 727)}$$

(Continued)

AC voltage and current in a resistor: phasors

In an AC circuit containing a resistor, the instantaneous current is

$$I = \frac{V}{R} = \frac{V_{max}}{R} \sin(2\pi ft) \qquad \textbf{(22.2)} \text{ and } \textbf{(22.3)} \text{ (page 726)}$$

This current is *in phase* with the voltage. The time dependences of V and I are represented in a *phasor diagram*.

Phasor diagram

The *average power* dissipated in the resistor is

$$P_{ave} = V_{rms}I_{rms} \qquad \textbf{(22.12)} \text{ (page 728)}$$

where I_{rms} is the rms current.

$$V = V_{max}\sin(2\pi ft)$$
$$I = I_{max}\sin(2\pi ft)$$

Reactance of a capacitor in an AC circuit

In an AC circuit containing a capacitor, the current and voltage are out of phase as shown in a phasor diagram. The amplitude of the current is

$$I_{max} = \frac{V_{max}}{X_C} \qquad \textbf{(22.21)} \text{ (page 732)}$$

where X_C is the *reactance* of the capacitor given by

$$X_C = \frac{1}{2\pi fC} \qquad \textbf{(22.22)} \text{ (page 732)}$$

The average power delivered to the capacitor is $P_{ave} = 0$.

Phasor diagram

$$V = V_{max}\sin(2\pi ft)$$
$$I = I_{max}\sin\left(2\pi ft + \frac{\pi}{2}\right)$$

Reactance of an inductor in an AC circuit

In an AC circuit containing an inductor, the current and voltage are again out of phase as shown in a phasor diagram The amplitude of the current is

$$I_{max} = \frac{V_{max}}{X_L} \qquad \textbf{(22.27)} \text{ (page 735)}$$

where X_L is the *reactance* of the inductor and is given by

$$X_L = 2\pi fL \qquad \textbf{(22.28)} \text{ (page 735)}$$

The average power delivered to the inductor is $P_{ave} = 0$.

Hence, for both capacitors and inductors, the circuit behavior *depends on the frequency*.

Phasor diagram

$$V = V_{max}\sin(2\pi ft)$$
$$I = I_{max}\sin(2\pi ft - \pi/2)$$

LC circuits

An LC circuit acts as a simple harmonic oscillator; both the charge on the capacitor and the current through the circuit oscillate with time:

$$q = q_{max} \cos(2\pi ft) \quad \text{and} \quad I = I_{max} \sin(2\pi ft) \qquad \textbf{(22.30)} \text{ (page 736)}$$

The frequency of the oscillations is

$$f_{res} = \frac{1}{2\pi\sqrt{LC}} \qquad \textbf{(22.35)} \text{ (page 738)}$$

These oscillations are just like the oscillations of the position and velocity of a mechanical simple harmonic oscillator.

Impedance

The amplitude of the current in a general AC circuit is related to the voltage amplitude by

$$V = IZ \qquad \textbf{(22.39)} \text{ (page 741)}$$

where Z is the **impedance** of the circuit. The impedance Z can be found from an AC circuit analysis using phasors.

APPLICATIONS

Frequency-dependent behavior of AC circuits

Reactances of capacitors and inductors are frequency dependent. At very low frequencies a capacitor acts as an open circuit and an inductor approximates a simple wire (a short circuit). In the opposite limit, at very high frequencies a capacitor acts as a short circuit and an inductor acts as an open circuit. These frequency dependences are due to the dependences of the reactances (Eqs. 22.22 and 22.28) on frequency and lead to applications such as high- and low-pass filters.

Transformers

A **transformer** changes the amplitude of an AC voltage. It is an essential element of power transmission circuits.

Generators and motors

Alternating-current machines can convert rotational kinetic energy into electrical energy (a **generator**) and electrical energy into rotational motion (a **motor**).

QUESTIONS

SSM = answer in Student Companion & Problem-Solving Guide ⊗ = life science application

1. Explain in words why (a) the reactance of a capacitor decreases as the frequency increases and (b) the reactance of an inductor increases as the frequency increases.

2. An AC circuit contains a capacitor in series with a resistor. If the frequency of the voltage source is increased while its amplitude is held fixed, does the current amplitude increase, decrease, or stay the same?

3. A resistor is connected in series with an inductor (Fig. Q22.3). If the frequency of the AC voltage source is increased while

its amplitude V_{max} is kept fixed, does the current amplitude increase, decrease, or stay the same?

Figure Q22.3

4. An AC source with a frequency of 60 Hz is connected to a light-bulb. The instantaneous brightness of the lightbulb depends on the magnitude of the instantaneous current. Explain why the brightness of the bulb flickers (i.e., oscillates with time) and find the frequency of the flicker. Why don't we notice the flicker?

5. The AC generator described in Section 22.1 uses a rotating coil to produce an AC voltage through Faraday's law of induction. (a) Explain how this approach can produce a voltage whose frequency is equal to the rotation frequency of the shaft. (b) Explain how you could modify the rotating coil so that the frequency of the AC voltage is twice the shaft rotation frequency. (c) Explain how you could modify the magnets in Figure 22.3 so that the frequency of the AC voltage is four times the rotation frequency of the shaft.

6. Distributed circuits. When applying Kirchhoff's loop rule to an AC circuit, one assumes the current in the loop is the same through all circuit elements at any particular instant. Changes in current travel at the speed of light (see Chapter 23), so a certain amount of time is actually required for the AC changes to propagate from one part of a circuit to another. (a) Estimate how much time Δt is needed for these changes to travel across a circuit 5 cm in size. Assume the propagation speed in a metal wire is approximately 3×10^8 m/s. (b) Use your value of Δt to find the lowest frequency at which the current can be a maximum in one part of the circuit and a minimum at another part. Cases like this one at high frequencies are called ***distributed circuits*** and must be analyzed using a generalized version of Kirchhoff's laws, which accounts for the propagation time.

7. Discuss why it is relatively easy to convert the amplitude of an AC voltage to a higher or lower value, but why it is difficult to change its frequency.

8. Most AC voltage sources produce a voltage that varies sinusoidally with time, but some produce a "square wave" as sketched in Figure Q22.8. Use the general definition of rms voltage to find the relation between the amplitude of the square wave and the rms voltage. That is, if the square wave amplitude is V_{sq}, find V_{rms}.

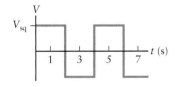

Figure Q22.8 Question 8 and Problem 6.

9. Consider two AC voltage sources connected in series with voltages $V_1 = V_{1,\,max} \sin(2\pi ft)$ and $V_2 = V_{2,\,max} \sin(2\pi ft + \pi/4)$, and $V_{1,\,max} = V_{2,\,max} = 30$ V. Show that these two sources in series are equivalent to a single source with $V_{series} = V_{equiv} \sin(2\pi ft + \phi)$ and find the values of V_{equiv} and ϕ.

10. If the voltage across a capacitor is $V = V_{max} \sin(2\pi ft)$, the charge on the capacitor is also proportional to $\sin(2\pi ft)$ (see Eq. 22.17). The current through the capacitor in this case is given by $I = I_{max} \sin(2\pi ft + \pi/2)$; hence, the phase angle is $\phi = \pi/2$ radians relative to the voltage. Suppose we begin instead by assuming the capacitor voltage is proportional to a cosine function $V = V_{max} \cos(2\pi ft)$. Use a graphical argument to show that the current is now given by a function of the form $I = I_{max} \cos(2\pi ft + \pi/2)$ so that the phase of the current relative to the voltage is again $\phi = \pi/2$ rad.

11. Show that the reactance of a capacitor X_C has units of ohms. Explain why it makes sense to compare the reactance of a capacitor with the resistance of a resistor.

12. The reactance of an inductor is given by $X_L = 2\pi fL$. Show that this reactance has units of ohms. Explain why it makes sense to compare the reactance of an inductor with the resistance of a resistor.

13. **SSM** Consider the circuit in Figure Q22.13. (a) Draw equivalent circuits that apply at high and low frequencies. (b) Estimate the ratio V_{out}/V_{in} at high and low frequencies. (c) Explain how this circuit acts as a filter that blocks high frequencies but lets low frequencies pass. It is one type of "low-pass" filter.

Figure Q22.13

14. A capacitor is connected in parallel with an inductor, and this circuit is connected to an AC voltage source of frequency f. It is found that the impedance of this circuit increases if the frequency is increased by a small amount. Is f above or below the resonant frequency of the circuit? Explain.

15. An LC circuit is called an oscillator. What, physically, is oscillating in such a circuit? What variables are periodic?

16. The equations describing an LC oscillator are very similar in form to those describing the mechanical harmonic oscillator of a mass on a spring. Identify the term in the equation for the mechanical oscillator that plays the role of C. What term plays the role of L? How are these corresponding pairs of variables similar?

17. An LCR circuit with a resonant frequency of 40 Hz is connected to a standard U.S. wall outlet of 120 V AC. Does the current in the circuit "lag" or "lead" the emf? That is, does the current have its maximum before or after the maximum in the voltage?

18. How is a phasor different from a vector? Is the emf a vector quantity?

19. In actual transformers (Fig. Q22.19), the iron cores are actually comprised of thin sheets of metal laminated together with insulating layers in between. Why would the cores be made this way?

Figure Q22.19 Transformer with a laminated core.

20. Although transformers are AC devices, is there some way to test a transformer to see if it works with a battery? Explain.

21. **SSM** **Boosting and bucking transformers.** Sometimes an electrical device may need an input voltage that is a few percent higher or lower than the supply voltage. A ***boosting transformer*** connects the output coil with the input voltage (here, a 120-V AC source) as shown on the left in Figure Q22.21. (a) Draw the sine waveforms of the input and output coils of the transformer and show how they add to produce the 150-V output. (b) If the leads from the output coil are connected as shown on the right in Figure Q22.21, a ***bucking transformer*** configuration results. (It has this name because two voltages are combined in "opposition" to make the output.) What output voltage would you expect from the circuit on the right in Figure Q22.21? Again draw the sine waveforms of the input and output coils, and the final output waveform. (Note that when the voltages from the input and output coils are combined, you must consider the phases of these voltages relative to each other.)

Figure Q22.21

22. Some resistors are made from a coil of wire. What is the best way to model the circuit behavior of such a resistor? Choose one answer and explain.
   (a) As an ideal resistor in series with an ideal inductor?
   (b) As an ideal resistor in parallel with an ideal inductor?
   (c) As an ideal resistor in parallel with an ideal capacitor?

23. For an ideal generator, the electrical energy generated equals the work W done in making the shaft rotate. For a real generator, is the electrical energy greater than, equal to, or less than W? Explain.

22.1 GENERATION OF AC VOLTAGES

1. Ⓡ Figure P22.1 shows the AC voltage produced by a particular source. Estimate (a) the frequency of this voltage, (b) the amplitude of this voltage, and (c) the rms value of this voltage. (d) If this AC voltage were attached to a lightbulb, would you notice the flicker of the bulb?

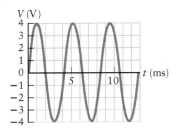

Figure P22.1

2. Some clocks that are powered by an AC voltage use the AC frequency to keep time. The power company makes this possible by arranging to have precisely the correct number of cycles occur every day, which keeps such clocks from gaining or losing time over long periods. (a) How many cycles of AC voltage are there each day in the United States? (b) Suppose a clock designed for the United States (where the AC frequency is 60 Hz) is taken to a different country where the frequency is 55 Hz. After one full day at its new location, this clock will be in error by an amount Δt. Find Δt.

3. SSM ★ Make a graph of an AC voltage that has a frequency of 300 Hz and an rms value of 45 V. Be sure to label the axes of your graph.

4. ★ Suppose the magnetic field B in the generator in Example 22.1 is reduced in magnitude by a factor of three. By what factor will the AC voltage change?
   (a) The voltage increases by a factor of three.
   (b) The voltage decreases by a factor of three.
   (c) The voltage increases by a factor of nine.
   (d) The voltage decreases by a factor of nine.
   (e) There is no change in the voltage.

5. ★ Suppose the loop in the generator in Example 22.1 is square, and that the length of each side of the loop is increased by a factor of two. By what factor will the AC voltage change?

6. Consider the square wave voltage in Figure Q22.8. What is its frequency?

7. Consider an AC voltage source that produces a voltage given by $V = V_{max} \sin(2\pi f t)$, with $V_{max} = 18$ V and $f = 6.2$ Hz. What is the instantaneous value of the voltage at $t = 1.4$ s?

22.2 ANALYSIS OF AC RESISTOR CIRCUITS

8. In an AC circuit, does the voltage across a resistor differ in phase from the current by a phase angle of (a) zero, (b) 30°, (c) 45°, (d) 90°, or (e) 180°?

9. SSM ★ A resistor with $R = 2.0$ kΩ is connected to an AC generator that operates at 60 Hz with an amplitude of 35 V. Find (a) the amplitude of the current through the resistor, (b) the maximum instantaneous power dissipated in the resistor, (c) the minimum instantaneous power dissipated in the resistor, and (d) the average power dissipated in the resistor. (e) How long will it take this resistor to dissipate 250 J?

10. ★ The value of R for the resistor in Problem 9 is increased by a factor of 2.5 while the current is kept fixed. By what factor does the average power change?

11. The average power dissipated in a 45-kΩ resistor in an AC circuit is 75 W. What is the amplitude of the AC voltage across the resistor?

12. What is the resistance of a lightbulb that has a rating of 300 W? Assume the lightbulb is designed for use in an AC circuit with $V_{rms} = 120$ V. *Hint*: In a circuit analysis, a lightbulb can be treated as a resistor.

13. ★ A heater runs on an AC voltage of 120 V and draws a power of 25 W. If this heater were run instead on a DC voltage of 75 V, how much current would flow through the heater? *Hint*: The heater can be modeled as a resistor.

14. Consider the AC circuit containing two resistors in Figure P22.14. If the amplitude of the AC voltage source is $V_{max} = 25$ V, $R_1 = 300$ Ω and $R_2 = 500$ Ω what is the amplitude of the current through R_1 and R_2?

Figure P22.14
Problems 14 and 15.

15. ★ Make a phasor diagram for the circuit considered in Problem 14, showing the voltage across each resistor and the current through the circuit. Do these three phasors all make the same angle with the horizontal axis? Explain why or why not.

16. ★ Make a phasor diagram for the circuit in Figure P22.16 showing the voltage across each resistor and the current through each resistor. Do these phasors make the same angle with the horizontal axis? Express the length of the phasors in terms of the amplitude V_{max} of the AC voltage source, and R_1 and R_2.

Figure P22.16 Problems 16 and 17.

17. ✪ Consider the circuit in Figure P22.16 and suppose $R_1 = 20\ \Omega$ and $R_2 = 40\ \Omega$. If the total average power dissipated in the two resistors is 55 W, what is the amplitude V_{max} of the AC voltage?

22.3 AC CIRCUITS WITH CAPACITORS

18. In an AC circuit, does the voltage across a capacitor differ in phase from the current by a phase angle of (a) zero, (b) 30°, (c) 45°, (d) 90°, or (e) 180°?

19. You are given two capacitors with capacitances of 20 pF and 20 μF. If the frequency is 5000 Hz, which capacitor has the larger reactance?

20. ✪ An AC voltage source with a frequency of 120 Hz and an amplitude $V_{max} = 25$ V is connected across a capacitor with $C = 1.5\ \mu$F (Fig. P22.20). Draw a phasor diagram showing the voltage across and the current through the capacitor. What is the length of the phasor that represents the current across the capacitor? Be sure that your answer has the correct units.

Figure P22.20
Problems 20, 26, and 27.

21. What is the reactance of a 1200-pF capacitor at 400 Hz?

22. ✪ If the frequency is increased from 35 Hz to 140 Hz, does the reactance of a capacitor increase or decrease? By what factor does it change?

23. A capacitor has a reactance of 450 Ω at 1200 Hz. What is its capacitance?

24. The AC voltage across a capacitor with $C = 1.5\ \mu$F has an amplitude of $V_{max} = 22$ V, and the current through the capacitor has an amplitude of 2.5 mA. What is the frequency?

25. ✪ Two capacitors are connected to an AC source as shown in Figure P22.25; assume $V_{AC} = V_{max} \sin(2\pi ft)$, with $V_{max} = 18$ V and a frequency $f = 2000$ Hz. (a) What is the amplitude (peak value) of the current through the capacitors? (b) What is the rms current?

Figure P22.25

26. ✪ For the circuit in Figure P22.20, find (a) the average power, (b) the maximum instantaneous power, and (c) the minimum instantaneous power delivered to the capacitor. Use $f = 400$ Hz, $V_{max} = 25$ V, and $C = 1.5\ \mu$F.

27. SSM ✪ Consider the AC circuit in Figure P22.20 and assume $V_{AC} = V_{max} \sin(2\pi ft)$ with $V_{max} = 12$ V, $f = 300$ Hz, and $C = 1.5\ \mu$F. The charge on the capacitor will also vary sinusoidally. Find (a) the amplitude of the charge oscillation and (b) the maximum energy stored in the capacitor. (c) At what time t is the energy stored a maximum? (Find just one value of t, but there will be an infinite number of values.)

28. Consider a circuit containing a resistor with $R = 1200\ \Omega$ and a capacitor with $C = 6.5\ \mu$F. At what frequency does the reactance of the capacitor equal the resistance of the resistor?

29. ✪ Two capacitors with $C_1 = 450$ pF and $C_2 = 820$ pF are connected in parallel. What is the reactance of the equivalent capacitance at 30 kHz?

30. ✪ Two capacitors with $C_1 = 4.5\ \mu$F and $C_2 = 1.9\ \mu$F are connected in series. What is the reactance of the equivalent capacitance at 350 Hz?

31. ✪ The phase angle between the voltage and current in an AC circuit is 90°. Is the average power delivered to the circuit from the AC source positive, negative, or zero?

22.4 AC CIRCUITS WITH INDUCTORS

32. In an AC circuit, does the voltage across an inductor differ in phase from the current by a phase angle of (a) zero, (b) 30°, (c) 45°, (d) 90°, or (e) 180°?

33. You are given two inductors with inductances of 40 μH and 80 mH. If the frequency is 5000 Hz, which inductor has the larger reactance?

34. An inductor has a reactance of 5500 Ω at a frequency of 2500 Hz. What is the value of L?

35. ✪ Find the reactance of an inductor with $L = 15$ mH at 22 kHz. If the frequency is increased to 66 kHz, by what factor does the reactance change?

36. Consider the circuit in Figure P22.36. (a) If the current through the inductor ($L = 25$ mH) at 400 Hz has an amplitude of 55 mA, what is the amplitude of the AC voltage? (b) What is the rms value of the AC voltage across the inductor?

Figure P22.36
Problems 36 and 37.

37. SSM ✪ Draw a phasor diagram that shows the voltage across and current through the inductor in Problem 36. See Figure P22.36.

38. The voltage source in Figure P22.38 has an amplitude of 5.0 V and a frequency of 75 Hz. What is the amplitude of the current through the inductors?

Figure P22.38

39. ✪ Figure P22.39 shows the voltage across an inductor with $L = 20$ mH. (a) At what time is the energy stored in the inductor a maximum? What is the energy at that moment? (b) At what time is the energy stored in the inductor a minimum? What is the energy at that moment?

Figure P22.39

40. Consider an inductor with $L = 50$ mH and a resistor with $R = 20$ kΩ. At what frequency is the reactance of the inductor equal to the resistance of this resistor?

22.5 LC CIRCUITS

22.6 RESONANCE

41. ✪ The resonant frequency of an LC circuit is f_0. (a) If the inductance is increased by a factor of eight, what is the new resonant frequency? (b) How should the capacitance be changed to change the resonant frequency back to f_0?

42. Consider an LC circuit with $L = 15$ mH and $C = 2500$ pF (Fig P22.42). (a) At what frequency is the reactance of the inductor equal to the reactance of the capacitor? (b) What is the resonant frequency of this circuit?

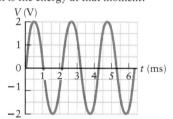

Figure P22.42 Problems 42, 43, and 46.

43. ✪ Consider the LC circuit in Problem 22.42 and Figure P22.42. At $t = 0$, the current is $I = 45$ mA and the charge q on the capacitor is zero. (a) What is the energy stored in the inductor at $t = 0$? (b) At some instant t_1, the current will be zero. What is the charge on the capacitor at t_1? (c) Find t_1. (There are many possible values of t_1; find the one closest to $t = 0$.) (d) Using the value of t_1 from part (c), find the magnitudes of I and q at $t = 3t_1$.

44. An LC circuit has a resonant frequency $f = 50$ kHz, and $L = 0.44$ mH. What is the capacitance of the capacitor?

45. ✪ An LC circuit has $L = 0.50$ mH and $C = 1.2\ \mu$F. If the total energy stored in this circuit is 2.5 μJ (2.5×10^{-6} J), what are

(a) the maximum charge q_{max} on the capacitor and (b) the maximum current? (c) When the charge on the capacitor is $q_{max}/2$, what is the magnitude of the current?

46. ✪ Consider again the LC circuit in Problem 42 and Figure P22.42. If the charge on the capacitor at a particular instant is 5.0×10^{-8} C, and the energy stored in the capacitor is equal to the energy stored in the inductor, what is the current at that moment?

47. SSM ✪ An LC circuit with $L = 55$ mH is used in a radio to determine the frequency of a selected radio station. The station's frequency is equal to the resonant frequency of the circuit. (a) The radio is initially set for a frequency $f = 1.05$ MHz. What is the value of C? (b) This capacitance is then increased by 10%. What is the new resonant frequency? (c) Can you calculate the answer to part (b) without first finding the value of C in part (a)?

48. ✪ The LCR circuit in Figure P22.48 is driven by a voltage source with $V_{AC} = V_{max} \sin(2\pi ft)$, where $V_{max} = 12$ V, $R = 150\ \Omega$, and $C = 25$ nF. (a) If the resonant frequency is $f_{res} = 95$ MHz, what is L? (b) What is the amplitude (the peak value) of the current when $f = f_{res}$?

Figure P22.48

49. ✪ You are given a collection of inductors and capacitors and are asked to construct LC circuits for several different applications. The inductances range from $L = 10\ \mu$H to 85 mH, and the capacitances are in the range of $C = 200$ pF to 1.5 mF. Is it possible to use these inductors and capacitors (one of each) to make LC circuits with the following frequencies? If so, give one set of possible values for L and C.
(a) $f_{res} = 1.5$ MHz for use in an AM radio
(b) $f_{res} = 95$ MHz for use in an FM radio
(c) $f_{res} = 60$ MHz for use in a television
(d) $f_{res} = 2.0$ GHz for use in a cell phone

22.7 AC CIRCUITS AND IMPEDANCE

50. SSM ✪ An inductor with $L = 35$ mH is connected in parallel with a capacitor with $C = 75\ \mu$F. (a) If they are connected to an AC voltage source with $f = 2.5$ kHz, where is the current larger, in the inductor or in the capacitor? (b) At what frequency is the current amplitude the same?

51. When an LC circuit is at resonance, is the impedance (a) zero, (b) $2\pi fL$, (c) $1/(2\pi fC)$, or (d) $1/\sqrt{LC}$?

52. ✪ Consider the RL circuit in Figure P22.52 with $R = 1500\ \Omega$, $L = 2.5$ mH, and an AC frequency of 70 kHz. What is the phase angle ϕ between the phasors that represent V_{AC} and the current?

Figure P22.52
Problems 52 and 54.

53. ✪ An LCR circuit contains a resistor with $R = 4500\ \Omega$, a capacitor with $C = 5500\ \mu$F, and an inductor with $L = 2.2$ mH, all connected in series. The AC source has a frequency of 75 kHz. (a) What is the impedance of the circuit? (b) What phase angle does the phasor describing the voltage across the voltage source make with the current phasor?

22.8 FREQUENCY-DEPENDENT BEHAVIOR OF AC CIRCUITS: A CONCEPTUAL RECAP

54. ✪ Consider the circuit in Figure P22.52. The voltage of the AC source is $V = V_{max} \sin(2\pi ft)$. If the frequency is extremely low (i.e., if f is very small), what is the amplitude of the current in the circuit? Express your answer in terms of V_{max}, R, and L.

55. ✪ Estimate the impedance of the LC circuit in Figure P22.55 at the very high frequency of 30 MHz. Will this impedance be dominated by the capacitor or the inductor? Will this impedance increase or decrease if the frequency is increased?

56. ✪ Estimate the impedance of the LC circuit in Figure P22.55 at the very low frequency of 30 Hz. Will this reactance be dominated by the capacitor or by the inductor? Will this reactance increase or decrease if the frequency is increased?

Figure P22.55
Problems 55 and 56.

57. An LCR circuit contains a resistor with $R = 450\ \Omega$ and a capacitor with $C = 5500\ \mu$F. If the resonant frequency is 33 kHz, what is the value of the inductance?

58. SSM ✪ An RC circuit containing a resistor with $R = 2500\ \Omega$ and a capacitor with $C = 1500\ \mu$F is attached to an AC generator with $V_{max} = 3.5$ V and $f = 25$ kHz. (a) What is the impedance of this circuit? (b) What is the amplitude of the current?

59. ✪ An RL circuit contains a resistor with $R = 6500\ \Omega$ and an inductor with $L = 1500\ \mu$H. If the impedance of this circuit is 150,000 Ω, what is the frequency?

60. ✪ The circuit in Figure P22.60 acts as a filter. (a) What is the ratio of the output voltage (amplitude) to the input voltage at very high frequencies? (b) What is the ratio of the output voltage to the input voltage at very low frequencies? (c) Is it a low-pass filter or a high-pass filter?

Figure P22.60

61. ✪ The circuit in Figure P22.61 acts as a filter. (a) What is the ratio of the output voltage (amplitude) to the input voltage at very high frequencies? (b) What is the ratio of the output voltage to the input voltage at very low frequencies? (c) Is it a low-pass filter or a high-pass filter?

Figure P22.61

62. ✪ ✪ A nerve fiber can be modeled electrically as an RC circuit as illustrated in Figure P22.62. The interior of the fiber acts as a resistor, and the resistance is determined by the solution inside the fiber, which has a resistivity of about $\rho = 0.1\ \Omega \cdot$m. The fiber's membrane acts as a capacitor; its capacitance per unit area of membrane is about 10^{-2} F/m². (a) Estimate the R and C of the equivalent circuit in Figure P22.62. (b) Is it a low-pass filter or a high-pass filter? (c) What frequency separates the high and low frequency limits for this circuit?

Figure P22.62

22.9 TRANSFORMERS

63. A transformer has an input coil with $N_{in} = 300$ turns and $N_{out} = 1000$ turns. If the voltage across the input coil has an amplitude of 2.0 V, what is the voltage across the output circuit?

64. Consider again the transformer in Problem 63. Suppose the two coils are reversed so that 2.0 V is put across the secondary coil. What is the voltage across the primary coil?

65. **European trip.** In France, the wall sockets provide an AC voltage with $V_{rms} = 230$ V. You want to use an appliance designed to operate in the United States ($V_{rms} = 120$ V) and decide to build a transformer to convert the power line voltage in France to the value required by your appliance. (a) Should you use a "step-down" transformer (to make V_{rms} smaller) or a "step-up" transformer (which makes V_{rms} larger)? (b) If the input coil of your transformer has 2000 turns, how many turns should the output coil have?

66. The transformer in a power brick (as used in a laptop computer) converts the AC voltage from a wall socket (V_{rms} = 120 V) to a value near 12 V. (a) If the output coil of the transformer has 4000 turns, how many turns does the input coil have? (b) If this transformer were accidentally connected backward (with the input and output coils reversed), what would the output voltage be?

67. ✪ The power brick for the author's laptop computer has a power rating of 45 W. It uses a step-down transformer to convert the AC voltage from a wall socket (V_{rms} = 120 V) to a value near 12 V. (a) What is the rms current in the input coil of the transformer? (b) What is the rms current in the output coil of the transformer?

22.10 MOTORS

68. ⭐ Consider the AC motor described in Example 22.8. You want to make this motor more powerful; that is, you want it to produce a larger torque. Evaluate the following proposed design changes. (a) If the number of turns in the coil were increased from one to five, how would the torque change? (b) If the size of the rotating coil were increased to L = 30 cm, how would the torque change? (c) If the magnetic field strength were increased

to B = 2.5 T, how would the torque change? (d) If all three of these changes were made, by what factor would the total torque increase?

69. SSM ⭐ A small electric motor such as the one in a home furnace is rated at "1/4 horsepower." Recall that "horsepower" (abbreviated hp) is a unit of power, with 1 hp ≈ 750 W. (a) How much work does this motor do in 1 s? In 1 h? (b) Assume it is an ideal motor and all the input electrical energy is converted to mechanical energy. If the motor operates with an AC voltage with V_{rms} = 120 V, what is the rms current in the motor?

22.11 WHAT CAN AC CIRCUITS DO THAT DC CIRCUITS CANNOT?

70. ✪ You are given the job of evaluating potential power line systems. One system is designed to operate at V = 3000 V, and another would operate at V = 12,000 V. Suppose the power line at the higher voltage would dissipate 5% of the power that it carries. What percentage of the power would be dissipated in the low-voltage power line?

71. SSM ⭐ Consider the power line system in Example 22.9. Suppose the operating voltage is increased to 750,000 V rms and a power of 500 MW is transmitted by the power line. What percentage of the power would now be dissipated in the power line?

ADDITIONAL PROBLEMS

72. ✪ When an inductor with L = 50 mH is attached to an AC voltage source, the inductive reactance is 8000 Ω. What is the capacitive reactance of a capacitor with C = 450 pF when attached to the same voltage source?

73. ✪ Figure P22.73 shows a model for a leaky capacitor, with a resistor in parallel with an ideal capacitor. Find the phase angle between the current through and voltage across the leaky capacitor in Figure P22.73. Express your answer in terms of R, C, and the frequency f.

Figure P22.73

74. ✪ (a) At what frequency will a 10-μF capacitor have the same reactance as a 10-mH inductor? (b) What is the reactance at this frequency? (c) If these components were assembled into a series LC circuit, what would the resonance frequency be?

75. ✪ (a) A generator supplies a current to a circuit containing an inductor. It is found that the current through the inductor reaches its maximum value 22 ms after the voltage across the inductor reaches its maximum value. What is the minimum possible operating frequency of the generator? (b) The generator is now connected to a lone capacitor. If the maximum current of the generator occurs at t = 0 s, how long after that will the voltage across the capacitor reach its maximum?

76. ✪ **The variac.** An indispensable device for any electronic workbench is a variable transformer or variac. Most variacs use a rotating contact to select a certain number of windings on the secondary coil, usually a solenoid in the shape of a toroid like that in Figure P22.76. Both the primary and secondary coils have 800 windings, but the rotor can connect to any of the individual loops of the secondary coil so that the output voltage is "tapped off" in as few as 100 turns on the secondary coil. (a) If the AC voltage across the primary is 120 V, what are the maximum and minimum voltage outputs from this device? (b) What is the smallest increment of change in voltage possible? (c) The dial on the variac needs calibration. What angle of rotation of the contact corresponds to 25 V? Assume the rotation angle varies from 0° (at the minimum output voltage) to 360° (at the maximum output voltage).

Figure P22.76 A variac with control dial and windings exposed.

77. ✪ Transformers are not limited to two sets of windings. The relationship for winding ratios holds true for multiple sets of coils as well. The AC voltage on the primary coil in Figure P22.77 is V_P = 120 V, and the two outputs supply V_{S1} = 220 V and V_{S2} = 12 V, respectively. If the 220-V secondary coil has 880 turns, what are the number of turns in the 12-V secondary coil and the primary coil?

Figure P22.77 A transformer with two secondary coils.

78. ✪ ✪ **Induction loop assistive listening system.** Magnetic induction is used to provide real-time amplification to hearing-impaired audience members. The system taps into the amplification system already in place in the room and routes the amplified electric waveforms from the speaker's microphone into two wire loops around the perimeter of the audience. (Figure P22.78 shows only one of these loops.) The bottom loop is at floor level, and the top loop is near the ceiling. The changing current in the loop induces changes in magnetic flux, which induces currents in smaller loops worn by audience members sitting within the main loop (photo), where the signal is then amplified and provided to headphones or earpieces. (a) What is the inductance of two loops, separated by 2.5 m, that fits the perimeter of a rectangular room 7.0 m wide by 15 m long? (Approximate it as a solenoid.) (b) The frequency of sound from speaking ranges from approximately 500 Hz to 4 kHz. Since the system should be sensitive to this range, what capacitance would give this system a resonance frequency of 2000 Hz? *Hint:* The expression for the inductance of a solenoid was derived in Chapter 20 (Eq. 20.21).

Figure P22.78 Diagram of an induction loop assistive listening system for a lecture hall and an audience member using an induction loop receiver.

79. ✪ **Crystal radio set.** The phenomenon of resonance makes it possible to listen to AM broadcasts with a very simple circuit and never need a battery. When a system is driven at its resonant frequency, even if by minute oscillations of current in the antenna, the resulting amplitude can be many times greater than the driving amplitude. The crystal radio, when *tuned* to a broadcast frequency, can produce a current large enough to operate an earphone speaker. A crystal radio is a simple *LCR* circuit with an antenna wire and a diode (the "crystal," involved in separating the signal from the carrier wave for the earphone). Figure P22.79 shows an illustration and circuit diagram of such a radio. The *LCR* circuit is made to match the broadcast frequency by varying the inductance, in this case by moving

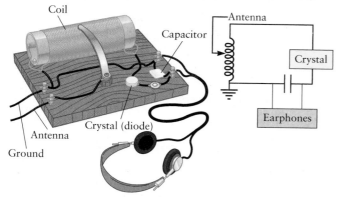

Figure P22.79 A crystal radio set and corresponding circuit diagram.

a wire to select how many turns of the solenoid are included in the circuit. (a) If the capacitance of the circuit is 950 pF, what inductance is needed to receive a broadcast at the highest AM frequency (1710 kHz)? At the lowest (520 kHz)? (b) The solenoid inductor's windings are made on a cylinder 6.0 cm in diameter and 20 cm long. How many turns are needed to provide the maximum and minimum inductances to cover the full AM band range? (c) Calculate the number of windings per length of solenoid (turns per millimeter). (d) AM radio specifications mandate that stations are spaced 10 kHz apart. Will this solenoid have sufficient resolution to separate two stations? (What change in the resonant frequency is produced by including one more winding?)

80. ✪ A series *LCR* circuit is comprised of a resistor, an inductor, and a capacitor of values 50 Ω, 87 mH, and 44 μF, respectively. (a) What is the generator frequency if the total impedance of the circuit measures 57 Ω? (b) What is the impedance of the circuit if it is plugged into a European outlet (50 Hz)? (c) At what frequency is the impedance of the circuit a minimum?

81. [SSM] ✪ **Simple inductive furnace.** An iron nut of mass 50 g is heated from room temperature to its melting point in 52 s through magnetic induction as shown in Figure P22.81. The circuit has a capacitance of 560 nF so that it will operate at its resonant frequency of 173 kHz with an input rms voltage of 1000 V. (a) If this **Figure P22.81** inductive heater is 90% efficient, what is the power supplied to the inductor of this circuit to heat the iron? *Hint:* The melting temperature of iron is 1810 K, and its specific heat is 440 J/(kg · K). (b) What is the inductance of the coil of thick copper wire? (c) What is the rms current through the coil? Is the thick copper wire needed? (d) The needed capacitance is provided by placing 12 identical capacitors in parallel. What is the capacitance of one of these capacitors?

82. ✪ **Fast Internet.** Copper telephone wires were originally designed to carry "commercial speech" using a band of frequencies from 300 Hz to 3400 Hz through a system called the PSTN (Public Switched Telephone Network). The ADSL (Asymmetric Digital Subscriber Line) uses frequencies very much higher than this speech band to carry fast data traffic using frequencies between 25 kHz and 1.1 MHz. To ensure that the higher-frequency transmissions do not interfere with normal phone electronics, filters like that shown in Figure P22.82 must be installed on all phone lines. (a) Calculate the frequencies that would be filtered out with the circuit configuration in the figure. Do they fall in the ADSL transmission range? (b) If the three inductors in series on each line were instead replaced by one equivalent inductor, would it alter the effectiveness of the filter? Why or why not?

Figure P22.82

Electromagnetic Waves

These radio telescopes detect electromagnetic waves that originate outside our galaxy. (Courtesy NRAO/AUI)

Theoretical understanding of electricity and magnetism developed rapidly during the early 1800s and seemed to be complete by about 1850. The laws of Coulomb and Gauss describing electric forces and fields and of Ampère and Faraday describing magnetic forces and fields were verified in many experiments and provided the basis of new technologies as key parts of the industrial revolution. Despite this success, some deep questions at the heart of electromagnetic theory remained unanswered. One concerned the nature of electric and magnetic field lines and the notion of action at a distance. (A similar question also applies to the gravitational field.) Consider two electric charges q_1 and q_2 located in distant space, far from all other matter. The electric field associated with q_1 is felt by (i.e., produces a force on) q_2. Now suppose q_1 moves at some instant in time to a new location. We expect the electric field lines to move along with q_1, but how fast will these field lines move? Is the motion of q_1 felt *instantaneously* at q_2, or is there some delay, and if so, what is the delay time? These questions are different aspects of the larger issue, "What are electric field lines?"

Physicist James Clerk Maxwell pondered some of these questions in the mid-1800s. His work led to what is probably the greatest physics discovery of the 19th century, the discovery of electromagnetic waves. Electromagnetic waves are used in many applications, including radios, televisions, radar, microwave ovens, cell phones, and the wireless Internet. Life today would certainly be very different without these applications of electromagnetic waves.

The general features of wave motion were introduced in Chapters 12 and 13. We'll rely heavily on that material in our work on electromagnetic waves in this chapter.

23.1 | THE DISCOVERY OF ELECTROMAGNETIC WAVES

In Chapters 20 and 21, we learned how Ampère's law and Faraday's law can be used to calculate magnetic fields and forces in different situations. An intriguing aspect of Faraday's law is that a time-varying magnetic field gives rise to an electric field. That is, a magnetic field *by itself* can produce an electric field. Many physicists then wondered, can an electric field by itself produce a magnetic field? Maxwell hypothesized that it should indeed be possible, which led him to propose a modification of Ampère's law. According to this modification, a time-varying electric field produces a magnetic field. Maxwell's new version of Ampère's law does not affect any of our work with Ampère's law in Chapter 20, and it also fixes a subtle problem with that law that had not previously been noticed. The most important result of Maxwell's hypothesis, however, is that it gives a new way to create a magnetic field; it also gives the equations of electromagnetism an interesting symmetry.

Symmetry of E and B

- **The correct form of Ampère's law (due to Maxwell) says that a changing electric flux produces a magnetic field. Since a changing electric flux can be caused by a changing E, we can also say that *a changing electric field produces a magnetic field*.**

- **Faraday's law says that a changing magnetic flux produces an induced emf, and an emf is always associated with an electric field. Since a changing magnetic flux can be caused by a changing B, we can also say that a *changing magnetic field produces an electric field*.**

There is thus a sort of "circularity" as illustrated in Figure 23.1: an electric field can (if it is changing) produce a magnetic field, and a magnetic field can (if it is changing) produce an electric field. Can this "loop" or oscillation involving E and B continue, with the new electric field producing a new magnetic field, producing a new electric field, and so on? The answer is yes; self-sustaining oscillations involving electric and magnetic fields are indeed possible. In fact, such oscillations are an *electromagnetic wave*. Electromagnetic waves are also referred to as *electromagnetic radiation*.

A key feature of the circularity in Figure 23.1 is that both the electric and magnetic fields must be changing with time. Faraday's law requires a time-varying magnetic field to produce an electric field, and Ampère's law (as modified by Maxwell) needs a time-varying electric field to produce a magnetic field. These requirements also affect how an electromagnetic wave can be created. Typically, a time-varying electric field is set up by an AC current

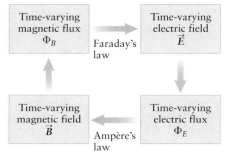

Figure 23.1 In schematic terms, a changing (time-varying) electric field produces a magnetic field, whereas a changing (time-varying) magnetic field produces an electric field.

in a wire or by moving electric charges, leading to the electric and magnetic fields in Figure 23.1. We'll come back to this idea in Sections 23.4 and 23.5.

Maxwell was the first to recognize that the equations of electromagnetism lead to the existence of electromagnetic waves, and he worked out the properties of these waves in great mathematical detail. Maxwell's prediction of the existence of electromagnetic waves was ahead of the experiments of the time. In fact, there was no experimental proof of the existence of electromagnetic waves until Heinrich Hertz demonstrated the generation and detection of radio waves in 1887, eight years after Maxwell's death. Hertz also showed that these waves travel at the speed of light (as predicted by Maxwell). Many other workers contributed to the work on radio, including Guglielmo Marconi, who developed the first practical radio system. (Marconi achieved trans-Atlantic radio communication in 1901.)

A full mathematical discussion of electromagnetic waves is more than we can include in this book, but we can use our understanding of Faraday's law to understand one important property of these waves. According to Faraday's law, a changing magnetic flux through a given area produces an induced emf directed along the perimeter of that area (Fig. 23.2A). This induced emf is associated with an induced electric field through our usual connection between electric potential and field (Chapter 18). Hence, the direction of the induced $\vec{E}$ is *perpendicular* to the magnetic field $\vec{B}$ that produced it. In a similar way, the magnetic field $\vec{B}$ produced by a changing electric field is perpendicular to the electric field that produced it (Fig. 23.2B). As a result, the electric and magnetic fields in an electromagnetic wave are perpendicular to each other. In addition, the direction of propagation of an electromagnetic wave is perpendicular to both $\vec{E}$ and $\vec{B}$. The directions of $\vec{E}$ and $\vec{B}$ and the propagation direction all depend on how the wave is generated, but these three directions are always perpendicular to one another.

Let's now summarize three fundamental properties of all electromagnetic waves.

Properties of electromagnetic waves
- **An electromagnetic wave involves *both* an electric field and a magnetic field.**
- **These fields are *perpendicular* to each other.**
- **The propagation direction of the wave is *perpendicular to both* $\vec{E}$ *and* $\vec{B}$.**

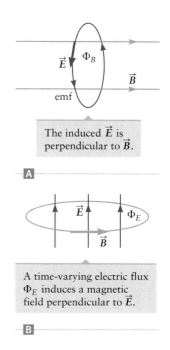

Figure 23.2 🅰 According to Faraday's law, a changing magnetic field produces an electric field. The induced electric field is perpendicular to the magnetic field. 🅱 A changing electric field produces a magnetic field, and these two fields are again perpendicular.

23.2 | PROPERTIES OF ELECTROMAGNETIC WAVES

In the previous section, we gave some qualitative arguments concerning the existence and properties of electromagnetic waves. The laws of electricity and magnetism can be used to derive *all* the properties of electromagnetic waves. Here are some of the key results.

Electromagnetic Waves Are Transverse Waves

We have already mentioned that the fields associated with an electromagnetic wave are perpendicular to each other and to the propagation direction. These relationships are illustrated in Figure 23.3A, which shows a snapshot of $\vec{E}$ and $\vec{B}$ for an electromagnetic wave at one instant in time. In this snapshot, the electric field is directed along x, the magnetic field is along y, and the wave propagates in the $+z$ direction.

To interpret Figure 23.3A, first focus on the electric field at the origin. Here, $\vec{E}$ is parallel to the x axis while the wave travels along z, and the tips of the electric field vectors form a pattern similar to a wave on a string. If we took another snapshot a short time later, we would find that the entire electric field pattern has moved a short distance along the z direction while maintaining the same wave profile. The behavior of the magnetic field in Figure 23.3A is similar, but $\vec{B}$ is along the

Figure 23.3 A Electric and magnetic fields of an electromagnetic wave. This plot shows a snapshot of $\vec{E}$ and $\vec{B}$ at a particular moment in time. B The fields $\vec{E}$ and $\vec{B}$ are perpendicular to each other. The propagation direction $\vec{v}$ of the wave is along the direction defined by right-hand rule number 2.

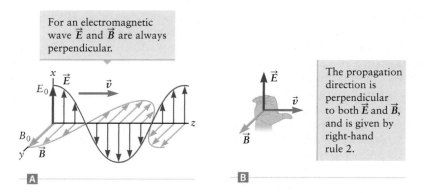

For an electromagnetic wave $\vec{E}$ and $\vec{B}$ are always perpendicular.

The propagation direction is perpendicular to both $\vec{E}$ and $\vec{B}$, and is given by right-hand rule 2.

A

B

y direction, so it is perpendicular to both $\vec{E}$ and the propagation direction. Because the fields are both perpendicular to the propagation direction, this wave is a ***transverse wave***.

The propagation direction can be determined with right-hand rule number 2 using the procedure shown in Figure 23.3B. Place the fingers of your right hand along the direction of $\vec{E}$ and curl your fingers in the direction of $\vec{B}$. Your thumb then points along the propagation direction.

> **CONCEPT CHECK 23.1** | Finding the Propagation Direction
>
> At a particular instant, the electric field associated with an electromagnetic wave is directed along $+y$ and the magnetic field is along $+z$. Is the propagation direction (a) $+x$, (b) $-x$, (c) $+y$, or (d) $+z$?

Light Is an Electromagnetic Wave

Maxwell found that the speed of an electromagnetic wave can be expressed in terms of two universal constants that appear in the theories of electricity and magnetism: the permittivity of free space ε_0 and the magnetic permeability of free space μ_0. The speed of an electromagnetic wave in a vacuum (free space) is commonly denoted by the symbol c:

Speed of electromagnetic waves in a vacuum

$$\text{speed of an electromagnetic wave} = c = \frac{1}{\sqrt{\varepsilon_0\mu_0}} \qquad (23.1)$$

When Maxwell inserted the known values of ε_0 and μ_0 into Equation 23.1, he found

$$c = \frac{1}{\sqrt{\varepsilon_0\mu_0}} = \frac{1}{\sqrt{[8.85 \times 10^{-12}\ \text{C}^2/(\text{N}\cdot\text{m}^2)](4\pi \times 10^{-7}\ \text{T}\cdot\text{m/A})}}$$

$$= 3.00 \times 10^8\ \text{m/s} \qquad (23.2)$$

This value for the speed c is precisely equal to the speed of light. When Maxwell derived this result in 1864, the nature of light was the subject of much debate. Most physicists suggested it was a wave while a few believed it was a particle, but no one could say what kind of wave or particle it might be. Maxwell had the answer: ***light is an electromagnetic wave***. He thus showed that the equations of electricity and magnetism also provide the theory of light.

It is difficult to overestimate the importance of Maxwell's discovery. Besides explaining the nature of light, his work led to the discovery of other types of electromagnetic waves, including radio waves. This work also led eventually to the work of Einstein and his theory of relativity (Chapter 27).

Electromagnetic Waves Can Travel through a Vacuum, All at the Same Speed

Although Maxwell worked out many of the basic properties of electromagnetic waves, including the value for their speed c, some results must have surprised even

Electromagnetic waves are transverse waves.

him. Recall from Chapter 12 that all mechanical waves travel in a medium of some kind. For waves on a string, the medium is the string, and sound waves travel in a material such as air. These waves "disturb" their medium, and the properties of the medium determine the wave speed. Many physicists searched for the corresponding medium for an electromagnetic wave. These waves can travel through material substances such as air or glass, but they can also travel through a vacuum, which is certainly *not* a material substance. Hence, electromagnetic waves are not like the mechanical waves we studied in Chapter 12. Roughly speaking, you can think of electric and magnetic fields as the "media" through which electromagnetic waves travel. These fields can exist in a vacuum, and electromagnetic waves can also travel through a vacuum.

Maxwell's theory does not specify the frequency or wavelength of an electromagnetic wave. Rather, frequency and wavelength are determined by the way the wave is produced. In Section 23.4, we'll describe examples of electromagnetic waves, including radio waves and X-rays. They *all* travel with speed c through a vacuum.

Electromagnetic Waves Travel Slower Than c in Material Substances

When an electromagnetic wave travels through a material substance such as glass, water, or the Earth's atmosphere, the speed of the wave depends on the properties of the substance. In such cases, the wave speed is always less than c, the speed in a vacuum (Eq. 23.2). We'll see in Chapter 24 that this difference in wave speed explains how a lens can focus light. What's more, the speed of an electromagnetic wave inside a substance depends on the wave's frequency. So, for example, blue light travels through glass at a higher speed than red light. This effect is called *dispersion* and gives rise to rainbows and other interesting effects, as discussed in Chapter 24.

> **CONCEPT CHECK 23.2 | Time Delay in a Cell Phone Conversation**
>
> A cell phone signal that travels from coast to coast in the United States must travel approximately 3000 miles (about 5×10^6 m). This signal is carried by different types of electromagnetic radiation (typically radio and microwaves) at a speed close to c. Is the approximate time required for your cell phone signal to propagate from coast to coast (a) 0.5 s, (b) 3 s, (c) 0.02 s, or (d) 0.0005 s? Can you notice this time delay?

23.3 | ELECTROMAGNETIC WAVES CARRY ENERGY AND MOMENTUM

When we first discussed waves in Chapter 12, we stated that waves transport energy without transporting matter. This general property of all waves is true for electromagnetic waves, too. For mechanical waves such as waves on a string or sound, this energy is contained in the medium through which the wave travels. For example, with a wave on a string, the energy of the wave is associated with the elastic energy and the tension in the string. With an electromagnetic wave, the energy is carried by the electric and magnetic fields associated with the wave. We can understand how these fields carry energy by considering what happens when an electromagnetic wave encounters a charged particle. Figure 23.4 shows the fields associated with an electromagnetic wave along with an electric charge $+q$ that is in the path of the wave. The charge will experience an electric force $\vec{F}_E = q\vec{E}$; as the wave passes by and the electric field oscillates along the x axis, the force on the charge will also oscillate. If this charge is able to move (which is true for electrons and ion cores in a solid), the electric force will do work on the charge, increasing its kinetic energy. In this way, energy is transferred from the wave to the charge; hence, the wave does indeed carry energy. You might also wonder about the magnetic force on the charge in Figure 23.4; in most cases, the

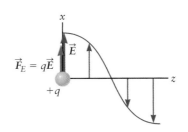

Figure 23.4 When an electromagnetic wave "strikes" an electric charge, the electric field associated with the wave produces a force on the charge. This force oscillates in direction as the wave passes by.

magnetic force is much smaller than the electric force, but we will see below that the magnetic field is crucial for understanding how an electromagnetic wave can carry momentum.

In Chapters 18 and 21, we discussed the potential energy contained in the electric and magnetic fields. For an electric field, the energy is proportional to E^2 (Eq. 18.48), and the potential energy per unit volume (the energy density) is

$$u_{\text{elec}} = \frac{1}{2}\varepsilon_0 E^2 \qquad (23.3)$$

For a magnetic field, the potential energy is proportional to B^2 (Eq. 21.35), and the magnetic potential energy per unit volume is

$$u_{\text{mag}} = \frac{1}{2\mu_0}B^2 \qquad (23.4)$$

These results apply also to electromagnetic waves, and the total energy per unit volume carried by an electromagnetic wave is the sum of its electric and magnetic energies, $u_{\text{elec}} + u_{\text{mag}}$. As the wave propagates, the electric and magnetic fields oscillate (Fig. 23.3), so u_{elec} and u_{mag} also oscillate. It also turns out that the electric and magnetic energies are equal, so $u_{\text{elec}} = u_{\text{mag}}$. Applying this relation when the fields have their peak values E_0 and B_0 (Fig. 23.3A) and using Equations 23.3 and 23.4, we get

$$\frac{1}{2}\varepsilon_0 E_0^2 = \frac{1}{2\mu_0}B_0^2 \qquad (23.5)$$

These terms are the peak energies carried by the electric and magnetic fields of the wave. In words, Equation 23.5 also says that the peak values of the fields E_0 and B_0 are proportional. Rearranging, we get

$$E_0 = \frac{1}{\sqrt{\varepsilon_0\mu_0}}B_0$$

The factor $1/\sqrt{\varepsilon_0\mu_0}$ is just the speed of light in a vacuum c (Eq. 23.1); hence,

Relation between E and B for an electromagnetic wave

$$E_0 = cB_0 \qquad (23.6)$$

| EXAMPLE 23.1 | Electric and Magnetic Fields of a TV Signal |

The signal from a radio station is an electromagnetic wave and is detected through its electric field. In a typical case, the electric field needed to give a "good" radio signal is about 1×10^{-2} V/m, the amplitude (peak value) E_0 associated with the wave. What is the corresponding magnetic field amplitude associated with this electromagnetic wave?

RECOGNIZE THE PRINCIPLE

For an electromagnetic wave, the amplitude of the electric and magnetic fields are proportional. Because the electric field amplitude E_0 is given, we can find the magnetic field amplitude B_0. The aim of this problem is to understand the magnitudes of these fields for a typical application.

SKETCH THE PROBLEM

No sketch is needed.

IDENTIFY THE RELATIONSHIPS

The relation between the peak electric and magnetic fields in an electromagnetic wave is $E_0 = cB_0$ (Eq. 23.6).

SOLVE

Inserting the given value of E_0, we find

$$B_0 = \frac{E_0}{c} = \frac{1 \times 10^{-2}\text{ V/m}}{3.00 \times 10^8\text{ m/s}} = \boxed{3 \times 10^{-11}\text{ T}}$$

What does it mean?

This value for B_0 certainly "seems" much smaller than the electric field E_0 (because the numerical values are very different in magnitude), but a careful comparison requires us to consider the relative magnetic and electric forces in a typical case. We'll do so after we discuss how an antenna works in Section 23.5; that analysis will show that the electric force is (in that case) much greater than the magnetic force. Thus, an electromagnetic wave is usually detected through its electric field and not by the effect of its magnetic field.

Intensity of an Electromagnetic Wave

The strength of an electromagnetic wave is usually measured in terms of its intensity, with units of watts per square meter. The intensity is the amount of energy the wave transports per unit time across a surface of unit area, and it equals the energy density (the amount of energy carried by the wave) multiplied by the wave speed (because a faster wave delivers energy at a faster rate). The wave speed is just c, the speed of light. Denoting the intensity by I, we have

$$I = \text{total energy density} \times c \tag{23.7}$$

The peak electric and magnetic energy densities are given in Equation 23.5. The total energy density carried by the wave is proportional to the sum of these two terms:

$$\text{total energy density} \propto \frac{1}{2}\varepsilon_0 E_0^2 + \frac{1}{2\mu_0}B_0^2 \tag{23.8}$$

We have already mentioned that the electric and magnetic contributions to the energy are equal, so we can rewrite Equation 23.8 as

$$\text{total energy density} \propto \frac{1}{2}\varepsilon_0 E_0^2 + \frac{1}{2\mu_0}B_0^2 = \varepsilon_0 E_0^2$$

Inserting into Equation 23.7, we get $I \propto \varepsilon_0 E_0^2 c$. In our derivation, we have dealt with E_0 and B_0, which are the peak values of the fields. To find the average intensity, we must average over a complete cycle of oscillation of the fields; this average intensity is half the peak intensity, so our final result for the intensity is

$$I = \frac{1}{2}\varepsilon_0 c E_0^2 \tag{23.9}$$

Intensity of an electromagnetic wave

This result is the *average* intensity of an electromagnetic wave in terms of the electric field amplitude (the maximum electric field) E_0. The intensity of an electromagnetic wave is thus proportional to the square of the electric field amplitude. Since $E_0 = cB_0$, the intensity is also proportional to the square of the magnetic field amplitude.

EXAMPLE 23.2 | The Electric Field of Sunlight

On a typical summer day, the intensity of sunlight at the Earth's surface is approximately $I = 1000$ W/m^2. What is the amplitude of the electric field associated with this electromagnetic wave?

RECOGNIZE THE PRINCIPLE

The intensity of an electromagnetic wave is proportional to E_0^2, where E_0 is the amplitude of the electric field. Given the intensity, we can thus find E_0. This calculation will give an understanding of the size of the electric field associated with an important source of electromagnetic radiation.

SKETCH THE PROBLEM

No sketch is necessary.

Insight 23.1

APPLICATIONS OF SOLAR ENERGY: SOLAR CELLS

Example 23.2 mentions that the intensity of sunlight on a typical sunny day is about 1000 W/m^2. Several approaches can be used to "capture" this energy. A *solar cell* converts the energy from sunlight into electrical energy that can be used to run the circuits in your house or to charge a battery. A typical house uses energy at an average rate of a few thousand watts. Because sunlight has an intensity of about 1000 W/m^2, you might think that a solar collector that captures sunlight over an area of only a few square meters would be sufficient. Current photovoltaic cells, however, capture only about 15% of the energy that strikes them, and one must also allow for nights and cloudy days during which the Sun's intensity is much lower than its peak value. The problem of making better and more practical solar cells is an important engineering challenge.

The electric field amplitude and the intensity are related by $I = \frac{1}{2}\varepsilon_0 c E_0^2$ (Eq. 23.9). Rearranging to solve for E_0 gives

$$E_0 = \sqrt{\frac{2I}{\varepsilon_0 c}}$$

SOLVE

Inserting the value of the intensity of sunlight, we find

$$E_0 = \sqrt{\frac{2(1000 \text{ W/m}^2)}{[8.85 \times 10^{-12} \text{ C}^2/(\text{N} \cdot \text{m}^2)](3.00 \times 10^8 \text{ m/s})}} = \boxed{870 \text{ V/m}}$$

What does it mean?
This electric field is large, especially compared with the field found in a typical electric circuit. This large value for E_0 explains how sunlight can produce large effects, such as sunburn!

Electromagnetic Waves Carry Momentum

In mechanics, a particle's momentum is equal to the mass of the particle times its velocity (Eq. 7.1). An electromagnetic wave does not have a mass, however, so how can it have momentum? The momentum of a particle is related to the force that the particle exerts in a collision with another object. Let's therefore consider the force that an electromagnetic wave exerts when it is absorbed by (i.e., collides with) an object as sketched in Figure 23.5. If the object is at rest before the collision, all the initial momentum is carried by the electromagnetic wave (Fig. 23.5A). Because the wave is absorbed by the object, all the final momentum is carried by the object. The absorption of the wave occurs through the electric and magnetic forces on charges in the object. Figure 23.5B and C shows the effects on a positive charge q (which might be an ion core) at the surface of the object. When the wave is absorbed, momentum is transferred to the charge q. Since total momentum is conserved in a collision, we can write

$$p_i(\text{wave}) = p_f(\text{charge } q) \qquad (23.10)$$

To analyze the effect of the electromagnetic wave on q, let's first consider the electric force. The electric field of the wave $\vec{E}$ is upward along $+x$ in Figure 23.5B, so the electric force $\vec{F}_E = q\vec{E}$ is also upward (assuming q is positive). This force accelerates the charge, giving it a velocity that is also upward. Now consider the magnetic force on this moving charge in Figure 23.5C. The magnetic field of the wave gives a magnetic force of magnitude $F_B = qvB$, and, according to right-hand rule number 2, this force is to the right along $+z$. As the electric field $\vec{E}$ oscillates up and down, the velocity in Figure 23.5B also oscillates up and down, but the magnetic force F_B in Figure 23.5C is always to the right because when $\vec{E}$ reverses direction, so does $\vec{B}$. (See Question 17 at the end of the chapter.)

We have thus shown that when a charge q absorbs an electromagnetic wave, there is a force on the charge in the direction of propagation of the original wave. According to the impulse theorem (Eq. 7.5), the force F_B on the charge q is related to the charge's change in momentum Δp along the z direction during a time Δt by

$$F_B = \frac{\Delta p}{\Delta t} \qquad (23.11)$$

Hence, the charge's final momentum along the z direction in Figure 23.5 is given through Equation 23.11. According to the principle of conservation of momentum, the final momentum of the charge must equal the initial momentum of the electromagnetic wave (Eq. 23.10), so the wave does indeed have momentum!

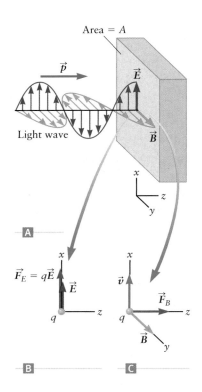

Figure 23.5 **A** An electromagnetic wave carries momentum. **B** and **C** This momentum can be understood in terms of the force exerted on charges in a material from the electric and magnetic fields of the wave. When electromagnetic radiation is absorbed by an object, the momentum of the object increases such that the total momentum of the system (radiation plus object) is conserved.

A full calculation of the momentum p of an electromagnetic wave shows that p is proportional to the total energy E_{total} carried by the wave, with the result

$$p = \frac{E_{total}}{c} \qquad\qquad (23.12)$$

Momentum carried by an electromagnetic wave

Radiation Pressure

The analysis in Figure 23.5 shows that when an electromagnetic wave is absorbed by an object, it exerts a force on the object. The total force on the object is proportional to its exposed area. Recalling that pressure is equal to the force per unit area, we can define a quantity called the **radiation pressure** $P_{radiation}$, which is equal to this electromagnetic force divided by the area. If I is the intensity of the wave, an analysis based on the ideas in Figure 23.5 gives

$$P_{radiation} = \frac{F}{A} = \frac{I}{c} \qquad\qquad (23.13)$$

Radiation pressure

In words, when an electromagnetic wave is absorbed by an object, the object experiences a radiation pressure $P_{radiation}$. As an example, the radiation pressure from sunlight could be used to propel a spacecraft, as we explore in the next example.

| EXAMPLE 23.3 | The Momentum Carried by Sunlight |

A spacecraft designer wants to use radiation pressure to propel a spacecraft as sketched in Figure 23.6. Near the Earth, the intensity of sunlight is about 1000 W/m^2 and the exposed area of the spacecraft is 500 m^2. If the sunlight is completely absorbed, what is the force on the spacecraft?

RECOGNIZE THE PRINCIPLE

The radiation pressure produced by an electromagnetic wave is proportional to its intensity. From the definition of pressure, the resulting force is then equal to the radiation pressure multiplied by the area of the spacecraft.

SKETCH THE PROBLEM

Figure 23.6 shows the problem.

IDENTIFY THE RELATIONSHIPS

The radiation pressure is related to the intensity by $P_{radiation} = I/c$, and the force on the spacecraft is $F = P_{radiation}A$. We thus have

$$F = P_{radiation}A = \frac{IA}{c}$$

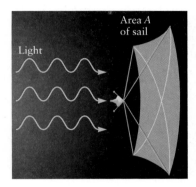

Figure 23.6 Example 23.3. A solar sail absorbs momentum from sunlight and uses the resulting force to accelerate a spacecraft.

SOLVE

Inserting the given values of the intensity and the area leads to

$$F = \frac{(1000 \text{ W/m}^2)(500 \text{ m}^2)}{3.00 \times 10^8 \text{ m/s}} = \boxed{1.7 \times 10^{-3} \text{ N}}$$

What does it mean?
This force is rather small; it is roughly equal to the weight of a penny at the Earth's surface. Achieving a larger force would require a larger exposed area (a larger "solar sail"). This approach does have one advantage, however: the fuel is free. Spacecraft that rely on solar sails for propulsion have not yet been built, but the effects of this force on the motions of several spacecraft (including *Mariner 10*) have been observed.

Figure 23.7 Concept Check 23.3.

CONCEPT CHECK 23.3 | Radiation Pressure and Reflected Light

When light is absorbed by an object, the radiation pressure is $P_{\text{radiation}} = I/c$ (Eq. 23.13). Suppose the exposed surface of an object has instead a reflective coating so that it reflects electromagnetic waves (Fig. 23.7). Which of the following statements is true? *Hint*: Compare absorption (Fig. 23.5) to a completely inelastic collision and reflection (Fig. 23.7) to an elastic collision.

(a) The force on the object due to radiation pressure is the same whether the light is absorbed or reflected.

(b) The force on the object due to radiation pressure is greater by a factor of two when the light is reflected.

(c) There is no force on the object due to radiation pressure when the light is reflected.

23.4 | TYPES OF ELECTROMAGNETIC RADIATION: THE ELECTROMAGNETIC SPECTRUM

All electromagnetic waves travel through a vacuum at the same speed c. The value of c has been measured with *great* precision and has the value

$$c = 2.99792458 \times 10^8 \text{ m/s} \qquad (23.14)$$

Of all the fundamental physical constants, the speed of electromagnetic waves in a vacuum is known the most accurately. In fact, c is actually *defined* to have the value in Equation 23.14, and the value of the SI standard meter is then derived through the value of c and the unit of time, the standard second. In most calculations, we will not need all the significant figures in Equation 23.14; the value $c = 3.00 \times 10^8$ m/s is accurate enough for the work in this book.

Electromagnetic waves are classified according to their frequency f and wavelength λ. The frequency, wavelength, and speed of an electromagnetic wave are related by

$$f\lambda = c \qquad (23.15)$$

which is similar to the general relation we found for mechanical waves in Chapter 12 (Eq. 12.2); as the frequency is increased, the wavelength decreases, and vice versa. Figure 23.8 lists a number of different types of electromagnetic waves, arranged according to their frequencies and wavelengths. In this listing, the frequency increases from bottom to top; frequency and wavelength are inversely related (Eq. 23.15), so the wavelength increases from top to bottom.

This range of all possible electromagnetic waves is called the ***electromagnetic spectrum***. There is no strict lower or upper limit for electromagnetic wave frequencies, which vary over an extremely large range, from below a few hertz (Hz) for ultralow-frequency radio waves to above 10^{22} Hz for gamma rays. The range of frequencies assigned to the different types of waves is somewhat arbitrary. For example, the ultraviolet and X-ray regions in Figure 23.8 overlap as do the regions assigned to microwaves and radio waves. The names of these different regions were chosen based on how the radiation in each frequency range interacts with matter and on how it is generated. We now describe some properties of these waves, starting at the low frequency end of the electromagnetic spectrum.

Radio Waves

Radio waves have frequencies from as low as a few hertz to about 10^9 Hz. The corresponding wavelengths vary from 10^8 m (around 10^5 miles!) to a few centimeters. Radio waves are usually produced by an AC circuit attached to an antenna. A simple wire can function as an antenna (as discussed in Section 23.5), but antennas containing multiple conducting elements (Fig. 23.9) are usually more efficient and are much more common than those consisting of a single wire. The frequency of a radio wave generated in this way is equal to the frequency of the current in the

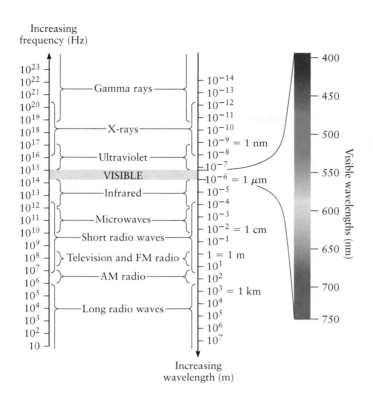

Increasing
frequency (Hz)

10^{23}		
10^{22}	10^{-14}	
10^{21}	10^{-13}	Gamma rays
10^{20}	10^{-12}	
10^{19}	10^{-11}	
10^{18}	10^{-10}	X-rays
10^{17}	$10^{-9} = 1$ nm	
10^{16}	10^{-8}	Ultraviolet
10^{15}	10^{-7}	
10^{14}	$10^{-6} = 1\ \mu m$	VISIBLE
10^{13}	10^{-5}	Infrared
10^{12}	10^{-4}	
10^{11}	10^{-3}	Microwaves
10^{10}	$10^{-2} = 1$ cm	Short radio waves
10^{9}	10^{-1}	
10^{8}	$1 = 1$ m	Television and FM radio
10^{7}	10^{1}	
10^{6}	10^{2}	AM radio
10^{5}	$10^{3} = 1$ km	
10^{4}	10^{4}	Long radio waves
10^{3}	10^{5}	
10^{2}	10^{6}	
10	10^{7}	

Visible wavelengths (nm): 400, 450, 500, 550, 600, 650, 700, 750

Increasing
wavelength (m)

Figure 23.8 The electromagnetic spectrum. Frequency increases from bottom to top, whereas wavelength increases from top to bottom. Near the middle of the spectrum is a narrow section corresponding to visible light; this section is expanded at the right, showing also how the color of visible light varies with frequency and wavelength.

attached AC circuit, which would usually contain inductors and capacitors as discussed in Chapter 22. Radio waves can be detected by an antenna similar to the one used for generation. These waves are used to carry conventional radio and television signals as well as signals for some cell phones and pagers.

In our discussion of the induced electric and magnetic fields in Figure 23.1, we emphasized that an electromagnetic wave is produced by *changing* electric and magnetic fields. For radio waves, the AC current in the antenna is produced by time-varying (oscillating) electric fields in the antenna. This time-varying electric field then produces a changing magnetic field and thus the electromagnetic wave. When the AC current in an antenna oscillates with time, the moving charges in the antenna are accelerated as they oscillate. Many other sources of electromagnetic radiation involve similar time-varying electric and magnetic fields produced by *accelerated* charges. In general, *when an electric charge is accelerated, it produces electromagnetic radiation.*

Figure 23.9 A set of parallel conductors (i.e., wires) arranged with the proper spacing acts as an antenna that can efficiently emit or absorb electromagnetic radiation. This antenna is designed for radio waves.

Microwaves

Microwaves have frequencies between about 10^9 Hz and 10^{12} Hz. The corresponding wavelengths vary from a few centimeters to a few tenths of a millimeter. This radiation interacts strongly with molecules such as H_2O and CO, making microwaves useful in the kitchen. Microwave ovens (Fig. 23.10) generate radiation with a frequency near 2.5×10^9 Hz. Waves of this frequency are absorbed strongly by H_2O molecules, causing the atoms in those molecules to vibrate just as masses connected by springs can vibrate. The term *absorption* means that some of or all the microwave energy is transferred to the water molecules, heating your food. Vibrating molecules can also emit electromagnetic radiation because they contain accelerated charges. For example, a vibrating H_2O molecule contains accelerated charges and thus emits electromagnetic radiation.

Infrared Radiation

Infrared radiation falls into the frequency range from around 10^{12} Hz to about 4×10^{14} Hz (wavelengths from a few tenths of a millimeter to a few microns). This

Figure 23.10 A microwave oven uses radiation in the microwave range to heat food.

Figure 23.11 Blackbody radiation (typically in the infrared range) indicates the temperature of an object. In this satellite photo, the temperature of the Earth's surface and atmosphere is measured by the wavelength at which the blackbody radiation has the highest intensity. (See Fig. 14.32.) In this satellite image, different temperatures are displayed as different colors.

radiation is what we sense as "heat," and much of the blackbody radiation from objects at temperatures near room temperature falls in the infrared range. Satellite photographs sensitive to infrared radiation are thus useful for monitoring the Earth's weather (Fig. 23.11). Infrared radiation interacts strongly with molecules and is absorbed by most substances. Infrared radiation can be produced by vibrating molecules or when electrons undergo transitions within an atom. We'll discuss these properties of electrons in atoms in Chapter 29.

Visible Light

The radiation we are able to detect with our eyes falls into the rather narrow frequency range from about 4×10^{14} Hz to near 8×10^{14} Hz, corresponding to wavelengths from about 750 nm to 400 nm. This radiation is called *visible light*. The color of visible light varies systematically with its frequency. Light at the low end of this frequency range (long wavelengths) is sensed as the color red; increasing frequencies then correspond to yellow, green, and blue, while the high-frequency (short-wavelength) end of the visible range is violet. A mixture of all these colors such as the light from a flashlight produces "white" light.

The speed of an electromagnetic wave inside a material substance has a value c' that is less than the speed c in a vacuum. For example, visible light travels through air at a speed about 0.03% smaller than c, whereas in water the speed of light is about 33% less than c. This variation of the speed of light from one material to another leads to many effects in optics, including refraction and the focusing of light by a lens (Chapter 24). The speed of light inside a material also depends on the frequency of the radiation. This effect is called *dispersion* because it causes a prism to "disperse" the different colors in white light (Fig. 23.12). Dispersion is also responsible for rainbows.

As with infrared radiation, visible light can be produced when electrons undergo atomic transitions (Chapter 29). Many sources of visible light such as the flashlight in Figure 23.12 produce radiation with a range of frequencies, but it is also possible to produce light consisting essentially of a single frequency. Such radiation is called *monochromatic*, meaning "single color." This term is usually used when describing visible light, but can also be applied to radiation in other parts of the spectrum. Lasers (Fig. 23.13) produce monochromatic radiation. We'll explain how lasers work in Chapter 29.

Ultraviolet Light

Proceeding to frequencies above the range of visible light takes us to *ultraviolet (UV) light*, with frequencies from about 8×10^{14} Hz to 10^{17} Hz; the associated wavelengths are from about 3 nm to 400 nm. Although we cannot detect this radiation with our eyes, it has some important health consequences. The UV portion of the spectrum is commonly subdivided into several regions: UV-A has wavelengths

Figure 23.12 White light from a flashlight is a combination of many different frequencies (i.e., colors). After passing through a prism, the different colors propagate in different directions. Newton used a prism to show that white light contains different colors.

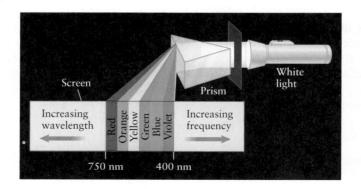

from 315 nm to 400 nm, UV-B falls in the wavelength range of 280 nm to 315 nm, and UV-C lies in the range of 200 nm to 280 nm. Even shorter wavelengths are referred to as the vacuum ultraviolet range because these wavelengths are so strongly absorbed by air and other substances that they can only propagate a significant distance in vacuum. As the wavelength becomes shorter, the damage done to exposed human skin increases. Thus, UV-C radiation has the greatest potential for damaging tissue. Fortunately, the atmosphere absorbs UV radiation most strongly at the shorter wavelengths, so most of the UV-C radiation from the Sun does not reach the Earth's surface.

Ultraviolet radiation has one major health benefit: it stimulates the production of vitamin D in the body. Nearly all the other effects of UV radiation are detrimental. Excessive exposure to UV light (Fig. 23.14) can cause sunburn, skin cancer, and cataracts. Many skin cancers (the type known as carcinoma) are treatable and often not fatal, but malignant melanoma is much more deadly. Persons who receive only two serious sunburns as children are found to have about twice the normal probability of developing melanoma as adults as those who did not receive such sunburns.

X-Rays

X-rays have frequencies from about 10^{17} Hz to around 10^{20} Hz. Wilhelm Röntgen discovered X-rays in 1895 and found that they can be used to form images from inside living tissue. Only one month after he announced this discovery, X-rays were being applied in medicine and other fields, and Röntgen became a worldwide celebrity. (He received the first Nobel Prize in Physics soon after, in 1901.) An early X-ray image of a human hand is reproduced in Figure 23.15A. The soft tissue, bones, and even a ring on one of the fingers are clearly visible. This image is possible because X-rays are weakly absorbed by skin and other soft tissue and strongly absorbed by dense materials such as bone, teeth, and metal. This difference in absorption provides the contrast seen in Figure 23.15B and makes it possible for X-rays to detect broken bones.

The X-ray images in Figure 23.15 are essentially "shadows" of the tissue that the X-rays pass through. All X-ray images were produced in this way until the 1970s, when a technique called computed axial tomography (abbreviated CT or CAT) was developed. Figure 23.16A shows how photographs taken at two different angles can be combined to reveal information about the thickness of an object. For example, if two bones overlap in an X-ray image, the bone behind is obscured. This effect is illustrated in Figure 23.16A, where a basketball is positioned in front of the person;

Figure 23.13 Ⓐ A "key chain" laser. Ⓑ A research laser. Both lasers produce monochromatic radiation (light of a single color).

Figure 23.14 Sunbathers use ultraviolet radiation from the Sun to get tan. Too much exposure to UV radiation can be hazardous to your health.

Figure 23.15 X-rays can pass through skin, but are absorbed strongly by dense material such as bone and metal.

Figure 23.16 ⓐ Shadow images taken at different angles can show the thickness of an object and can also reveal when one part of an object is in the shadow of another part. ⓑ Example of a CT scan apparatus as found in most hospitals.

here the basketball is obscured in the image on the screen behind. For any single image, there will always be parts of the person's body that are obscured in some way. This problem can be overcome by comparing two images taken from different angles; in Figure 23.16A, the basketball is visible when viewed with the X-ray source at the right. A CT scan (Fig. 23.16B) takes many X-ray images at many different angles and then employs a sophisticated computer analysis to combine these images into a three-dimensional representation of the object. The results can be used to measure accurately the shape and size of bones, tumors, and so forth within the body.

Although X-ray imaging has enormous health benefits, excessive exposure to X-ray radiation can have harmful effects, including burns. At high exposures, X-rays can cause cancer and other serious diseases.

Gamma Rays

Gamma rays lie at the high-frequency end of the electromagnetic spectrum, with frequencies above about 10^{20} Hz and wavelengths less than about 10^{-12} m. (As with other regions of the electromagnetic spectrum, there is no precise boundary between the X-ray and gamma ray parts of the spectrum.) Gamma rays are produced by processes occurring inside the nucleus of an atom, as we'll discuss in Chapter 30. Gamma rays are produced in nuclear power plants and the Sun. Gamma rays also reach the Earth from sources outside our solar system.

EXAMPLE 23.4 | ⊗ Wavelength of X-rays

The X-rays used in your dentist's office or local hospital have a frequency of approximately $f = 2.0 \times 10^{18}$ Hz. What is the wavelength of these X-rays, and how does that wavelength compare to the size of an atom?

RECOGNIZE THE PRINCIPLE

The frequency and wavelength of an electromagnetic wave are related through $f\lambda = c$. Since the wave speed c is known, we can use the given value of f to find the wavelength. These X-rays are created by processes within the atoms of metals such as copper, molybdenum, and tungsten.

SKETCH THE PROBLEM

No sketch is necessary.

IDENTIFY THE RELATIONSHIPS

Using the relation between speed, frequency, and wavelength (Eq. 23.15), we get $\lambda = c/f$.

SOLVE

Inserting the given value of the frequency and the known value of c yields

$$\lambda = \frac{c}{f} = \frac{3.00 \times 10^8 \text{ m/s}}{2.0 \times 10^{18} \text{ Hz}} = \boxed{1.5 \times 10^{-10} \text{ m}}$$

What does it mean?
The radius of an atom is about 5×10^{-11} m, so the wavelength of these X-rays is approximately the diameter of an atom. As mentioned above, these X-rays are produced by atomic processes (so-called atomic transitions; see Chapter 29). Hence, a process involving electrons in atoms produces X-rays whose wavelengths are comparable to the size of an atom.

| X-ray image | Visible light | Infrared | Radio image |

Figure 23.17 Photographs of electromagnetic radiation emitted by the Crab Nebula, taken using different types of radiation. The colors in these images indicate the intensity of the radiation and not its true color.

CONCEPT CHECK 23.4 | Types of Electromagnetic Radiation

Place the following types of electromagnetic radiation in order according to their wavelength. Start with the radiation with the longest wavelength.

(a) Infrared
(b) X-rays
(c) Blue light
(d) Radio
(e) Microwaves
(f) Red light

Astronomy and Electromagnetic Radiation

In the next few sections, we'll describe some of the applications of electromagnetic radiation. These applications use different types of radiation, depending on the needs of the application. For example, most cell phones use microwaves, the optical fibers used to carry signals between computers use infrared and visible light, and tanning salons use UV radiation. One application—astronomy—uses virtually all types of electromagnetic radiation. Many objects in the universe such as stars and supernovas emit radiation over a wide range of frequencies, from radio waves to visible light to X-rays and gamma rays. For example, Figure 23.17 shows images of an object called the Crab Nebula. It is the remnants of a supernova, a star that exploded in the year 1054. The images shown here were constructed using different types of radiation. Comparisons of the different images give astronomers important information about how mass is distributed in the nebula and how it generates radiation.

23.5 | GENERATION AND PROPAGATION OF ELECTROMAGNETIC WAVES

In Section 23.4, we described many different types of electromagnetic waves. Let's now consider in more detail how one particular type of electromagnetic waves—radio waves—can be generated. Figure 23.18 shows an AC voltage source connected to two wires. These wires act as an *antenna*; they are not connected to anything else, so this setup is different from the AC circuits discussed in Chapter 22. As the voltage of the AC source oscillates, the electric potentials of the two wires also oscillate, and we indicate these potentials with the positive and negative signs in Figure 23.18. These signs also indicate electric charges present on the two wires; these charges flow from the voltage source onto and off the wires as the AC voltage alternates from positive to negative, back to positive, back to negative, and so on.

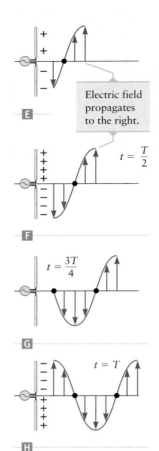

Figure 23.18 Electric field near a dipole antenna. **A** Variation of the AC voltage across the antenna as a function of time. **B** through **H** Snapshots of the electric field produced by the antenna as a function of time. As the electric field at the antenna oscillates, an electromagnetic wave propagates away from the antenna.

Figure 23.18A shows how the potential varies with time (just our usual AC voltage source), and the other parts of the figure show "snapshots" taken at different times as the potential difference between the two antenna wires oscillates. In part B of Figure 23.18, the potential difference is large; there are many negative charges on the upper wire and many positive charges on the bottom wire. These charges produce a large electric field directed upward as shown. As time passes (Fig. 23.18C), the potential difference becomes smaller in magnitude, so the amount of charge on the two wires decreases and the electric field is smaller. At the same time, the large electric field from part B has begun to propagate away from the wires. In Figure 23.18D, the potential difference is zero and the electric field at the wires is zero, and the fields generated at previous times (in parts B and C) continue to propagate away. Parts E through H of Figure 23.18 show the charge on the wires and the corresponding electric fields as time passes and the AC potential oscillates through one complete cycle.

The behavior of the electric field in Figure 23.18 is similar to the behavior of a wave on a string. When one end of a string is vibrated up and down, a wave is generated that propagates away along the string. Here the electric field acts in some ways like the string, and the AC voltage source generates a vibrating electric field at the antenna, similar to shaking the end of a string.

The pictures in Figure 23.18 also explain some other important features of electromagnetic waves. First, the charges move onto and off of the antenna wires in an oscillatory way. From our work on the motion of a simple harmonic oscillator (Chapter 11), we know that these charges are accelerated, confirming the claim we made in Section 23.2 that electromagnetic waves are produced whenever charges have a nonzero acceleration. Second, the charges in Figure 23.18 undergo harmonic motion at a frequency equal to the frequency of the AC voltage source, which is also the frequency of the electromagnetic wave that is generated. Whenever an electromagnetic wave with frequency f is observed, we can say that it was generated by an electromagnetic oscillator (such as an LC circuit) that oscillates at the same frequency f.

CONCEPT CHECK 23.5 | What Is the Direction of $\vec{B}$?

In Figure 23.18, we considered how charge on the wires produces electric fields that then generate an electromagnetic wave. At the same time, the motion of charges on and off the wires also gives rise to the magnetic field associated with the wave. The electromagnetic wave in Figure 23.18 propagates to the right. What is the direction of the magnetic field $\vec{B}$ near the antenna in the snapshot in Figure 23.18B? (a) Directed to the right, (b) directed to the left, (c) directed out of the plane (toward the reader), or (d) directed into the plane (away from the reader)?

Antennas

The basic approach in Figure 23.18, in which an AC voltage source is attached to a set of wires, is used to produce the waves that carry radio, television, and cell phone signals. The wires in Figure 23.18 are one type of antenna, but there are many other designs, some containing many wires (Fig. 23.9); the simple case with two wires is called a ***dipole antenna***. The reason for the name "dipole" can be seen from Figure 23.19, which shows a dipole antenna with the wires parallel to the vertical (z)

direction. At any particular moment, the two wires are oppositely charged, forming a sort of electric dipole. For the antenna in Figure 23.19, the electric field is along z and the electromagnetic waves propagate away in the x and y directions (and in other directions in the x–y plane).

In our discussion of antennas and the electric fields they produce (Figs. 23.18 and 23.19), we have emphasized the electric fields near the antennas. There is also an electromagnetic wave propagating inside the antenna wires, with an electric field that oscillates between positive and negative values as one moves away from the AC source. For a very long antenna, these oscillations tend to cancel, so a very long antenna does not necessarily produce a "stronger" electromagnetic wave than a short antenna. This cancellation becomes significant when one half of the antenna is longer than about $\lambda/4$ (one quarter of the wavelength). For this reason, most dipole antennas have a total length of $\lambda/2$.

Although other antennas may be more complicated than the dipole design, the cancellation effect mentioned above still occurs, so the size or length of antenna is usually comparable to the wavelength of the radiation. That is why the antenna for your cell phone is so small (Fig. 23.20A), whereas the antennas for an AM radio station are quite large (Fig. 23.20B).

Figures 23.18 and 23.19 describe how an antenna generates an electromagnetic wave, but the same antennas can be used to detect these waves. Figure 23.21 shows an electromagnetic wave interacting with a dipole antenna that might be in your cell phone. The electric field associated with the wave exerts a force on the electrons in the antenna, producing a current and an induced voltage across the antenna wires. This voltage is then the input to an AC circuit inside your cell phone. That AC circuit amplifies the voltage signal and ultimately sends it to the phone's speakers.

Propagation and Intensity

The sketch of charge motion in Figure 23.18 shows how accelerated charges give rise to an electromagnetic wave. This picture is useful for understanding many radiation sources, but the charge motion is sometimes more complicated. For example, an ordinary lightbulb produces light due to the vibrations of electrons and ion cores in the hot filament of the bulb. Unlike the case in Figure 23.18, these vibrations in the filament are not confined to one particular direction; rather, charges vibrate along all different directions. As a result, the bulb produces light that propagates outward in all directions, unlike the case in Figure 23.18, in which the waves propagate only in the horizontal direction.

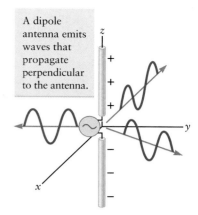

A dipole antenna emits waves that propagate perpendicular to the antenna.

Figure 23.19 Radiation from a dipole antenna propagates outward in directions perpendicular to the antenna's axis. There is essentially no radiation generated "off the ends" of the antenna. This antenna radiates strongly in all directions in the x–y plane, but does not emit radiation in the $+z$ or $-z$ directions.

A

Figure 23.20 In most applications, an antenna's length is comparable to $\lambda/2$. A For a cell phone, the wavelength is short, so the antenna (visible at the upper left) is also short. B The wavelength of the radiation used by an AM radio station is much longer, so its antenna is very large.

B

© Digital Vision/Getty Images

Figure 23.21 A dipole antenna can be used to detect an electromagnetic wave. The electric field of the wave induces a voltage across the wires of the antenna. This induced voltage is then amplified and processed by your cell phone or radio.

$\vec{E}$ $\vec{E}$ ΔV

Electromagnetic wave   To detector (e.g., cell phone)

In many other cases, a source produces waves that propagate outward in all directions. The "ideal" case of a very small source producing spherical wave fronts is called a "point source." These wave fronts describe how energy propagates outward from the source. In Chapter 12, we saw that the intensity of a spherical wave falls with distance as (Eq. 12.11)

$$I \propto \frac{1}{r^2} \tag{23.16}$$

The intensity falls as one moves away from the source because there is a constant amount of energy and power associated with each wave front, whereas the wave front area ($A = 4\pi r^2$) increases with distance. (See Fig. 12.12, page 389, or Fig. 17.12, page 539.) Intensity is the power per unit area, so the intensity decreases as this constant amount of energy is spread over an ever-increasing area. The decrease of intensity with distance is thus a direct consequence of the principle of conservation of energy.

The relation between intensity and distance in Equation 23.16 applies to many cases, including the falloff of intensity for a lightbulb, the falloff of intensity of sunlight with distance through the universe, and the "strength" (i.e., intensity) of a radio signal from a distant radio station or space satellite.

EXAMPLE 23.5 | Intensity of Sunlight

The intensity of sunlight near the Earth is about 1000 W/m². Since the Earth is very far from the Sun, the Sun acts approximately as a point source of radiation. What is the intensity of sunlight at the surface of Mars?

RECOGNIZE THE PRINCIPLE

Because the Sun acts as a point source of spherical waves, its intensity falls with distance as $1/r^2$. We can use the known distances between the Sun and the Earth and the Sun and Mars to get both the relative intensity and the absolute intensity at Mars.

SKETCH THE PROBLEM

Figure 23.22 describes the problem.

Figure 23.22 Example 23.5.

IDENTIFY THE RELATIONSHIPS

We denote the Sun–Earth distance as r_E and the Sun–Mars distance as r_M. Using Equation 23.16, we can write the intensity at Earth as $I_E = K/r_E^2$ and the intensity at Mars as $I_M = K/r_M^2$. The constant of proportionality K is the same in these two cases, so taking the ratio, we have

$$\frac{I_M}{I_E} = \frac{K/r_M^2}{K/r_E^2} = \frac{r_E^2}{r_M^2} \tag{1}$$

From Table A.3 of solar system data in Appendix A, we find $r_E = 1.5 \times 10^8$ km and $r_M = 2.3 \times 10^8$ km.

SOLVE

Rearranging Equation (1) to solve for the intensity at Mars gives

$$I_M = I_E\left(\frac{r_E}{r_M}\right)^2$$

Inserting our values of the intensity at the Earth and the distances r_E and r_M we get

$$I_M = (1000 \text{ W/m}^2)\left(\frac{1.5 \times 10^8 \text{ km}}{2.3 \times 10^8 \text{ km}}\right)^2 = \boxed{430 \text{ W/m}^2}$$

What does it mean?
Since the intensity at the Earth is about 1000 W/m², the intensity at Mars is only 43% of that value. As a result, the surface temperature of Mars is much less than that of the Earth.

23.6 | POLARIZATION

When discussing an electromagnetic wave, it is important to know its propagation direction since that tells the direction in which energy is traveling. We know that the electric field must be perpendicular to the propagation direction, but there are still many different possible directions for $\vec{E}$ that are perpendicular to the wave direction. The direction of $\vec{E}$ is important because it determines how the wave interacts with matter (including antennas).

Figure 23.19 shows how a dipole antenna can be used to produce an electromagnetic wave with an electric field $\vec{E}$ directed along $\pm z$. To indicate that the electric field is always parallel to z, we say that this wave is *linearly polarized* along the z direction. Electromagnetic radiation that is linearly polarized is in many ways the simplest case to describe. Most light is not polarized in this way, however. For example, sunlight and the light from a flashlight are both *unpolarized*. Light from these sources is a combination of many individual waves, all with different field amplitudes and polarization directions.

Polarizers

Polarized light can be created using a device called a *polarizer*. One type of polarizer is shown in Figure 23.23. It consists of a thin, plastic film that allows an electromagnetic wave to pass through only if the electric field of the wave is parallel to a particular direction called the axis of the polarizer. The polarizer absorbs radiation with electric fields that are not along the axis. Hence, when unpolarized light (perhaps from a flashlight) strikes a polarizer, the light that comes out is linearly polarized.

To evaluate the intensity of the light transmitted by a polarizer, let's first consider what happens when linearly polarized light strikes a polarizer. In Figure 23.24A, the incident light is polarized parallel to the axis of the polarizer and the outgoing electric field is equal in amplitude to the incoming field, so $E_{\text{out}} = E_{\text{in}}$; all the incident energy is thus transmitted through. In Figure 23.24B, the incident light is polarized perpendicular to the polarizer axis and no light is transmitted, so $E_{\text{out}} = 0$. In Figure 23.24C, the incident light is polarized at an angle θ relative to the axis of the polarizer. The outgoing light is now polarized along the direction defined by the polarizer with an electric field amplitude

$$E_{\text{out}} = E_{\text{in}} \cos \theta$$

In words, only the component of the electric field along the polarizer axis is transmitted. Since intensity is proportional to the square of the electric field amplitude (Eq. 23.9), we can also write

$$I_{\text{out}} = I_{\text{in}} \cos^2 \theta \qquad (23.17)$$

which is called *Malus's law*. When $\theta \neq 0$, the outgoing intensity is smaller than the incident intensity and the outgoing light is linearly polarized along the direction of the polarizer axis.

EXAMPLE 23.6 | Electric Field and Intensity Transmitted through a Polarizer

Linearly polarized light is incident on a polarizer, with an angle θ between the incoming electric field and the axis of the polarizer (Fig. 23.24C). Find the value of θ for which the transmitted intensity is half the incident intensity.

RECOGNIZE THE PRINCIPLE

When the incident light is linearly polarized, only the component of the electric field along the polarizer axis is transmitted, and the transmitted intensity is given by Malus's law (Eq. 23.17). We can use this relation to find the value of θ for which I_{out} is half of I_{in}.

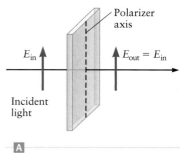

Figure 23.23 When unpolarized light is incident on a polarizer, only the component with the electric field parallel to the axis of the polarizer is transmitted.

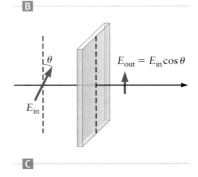

Figure 23.24 When linearly polarized light is incident on a polarizer, the transmitted intensity depends on the angle θ between the incident electric field E_{in} and the polarizer axis.

SKETCH THE PROBLEM

Figure 23.24C describes the problem. It shows the incident and transmitted electric fields.

IDENTIFY THE RELATIONSHIPS

From Malus's law, we have $I_{out} = I_{in} \cos^2 \theta$, which we can rearrange to read

$$\frac{I_{out}}{I_{in}} = \cos^2 \theta$$

SOLVE

We want the transmitted intensity to be half the incident intensity; hence, $I_{out}/I_{in} = \frac{1}{2}$, which leads to

$$\cos^2 \theta = \frac{I_{out}}{I_{in}} = \frac{1}{2}$$

$$\cos \theta = \frac{1}{\sqrt{2}}$$

$$\theta = \boxed{45°}$$

What does it mean?

When the angle θ between the polarization direction of the incident light and the axis of the polarizer is not zero, the transmitted intensity is smaller than the incident intensity. The transmitted light is still linearly polarized, but with a different polarization direction.

Malus's law applies when the incident light is linearly polarized, but it can also be used to understand what happens when unpolarized light strikes a polarizer. Unpolarized light can be thought of as a collection of many separate light waves, each linearly polarized but with different and random polarization directions. Each separate wave is transmitted through the polarizer according to Malus's law. The average outgoing intensity is then the average of I_{out} for all the incident waves. We can thus write

$$I_{out} = (I_{in} \cos^2 \theta)_{ave}$$

The average here is an average over all possible polarization angles for the incoming light waves. The factor $\cos^2 \theta$ varies between 0 and 1, and its average is $\frac{1}{2}$, so for unpolarized light the intensity transmitted through a polarizer is

$$I_{out} = \tfrac{1}{2} I_{in} \quad \text{(unpolarized light)} \tag{23.18}$$

An interesting application of Malus's law is sketched in Figure 23.25A, which shows light that is initially unpolarized before it passes through two polarizers in a row. After passing through polarizer 1, the light intensity is reduced by half (Eq. 23.18) and is now linearly polarized along z, the axis of polarizer 1. The axis of polarizer 2 is perpendicular to z, so the final transmitted intensity is zero.

Figure 23.25B shows a similar arrangement, but now we have inserted a third polarizer in the middle. The axis of polarizer 3 makes a 45° angle with the z axis, so some of the light from polarizer 1 is transmitted through polarizer 3. Moreover, after passing through polarizer 3 this light is polarized at a 45° angle with respect to polarizer 2. Hence, the final transmitted intensity after passing through all three polarizers is nonzero. Comparing this conclusion to the experiment in Figure 23.25A leads to the following "puzzle." In Figure 23.25A, the final transmitted intensity I_{final} is zero. In Figure 23.25B, we have added an extra polarizer, and now I_{final} is no longer zero! Although that may seem surprising, it can be understood by using two facts about polarizers. First, when analyzing light as it passes through

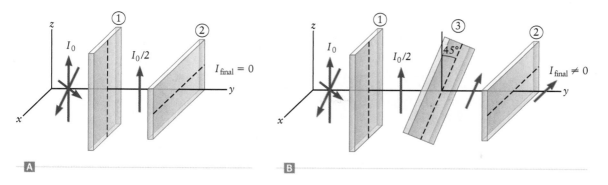

Figure 23.25 ⓐ Unpolarized light is incident on two polarizers whose axes are perpendicular. The final transmitted intensity is zero. ⓑ A third polarizer is placed between the polarizers in part A. Now the transmitted intensity is not zero!

several polarizers in succession, you must always analyze the effect of one polarizer at a time. Second, the light transmitted by a polarizer is always linearly polarized, with a polarization direction determined solely by the polarizer axis. The transmitted light wave has no "memory" of its original polarization.

| **CONCEPT CHECK 23.6** | Polarizers in "Series"

If the intensity of the initial unpolarized light beam in Figure 23.25B is I_0, is the final transmitted intensity (a) $I_0/2$, (b) $I_0/4$, (c) $I_0/8$, or (d) $I_0/16$?

How Does a Polarizer Work?

Polarizers have some interesting applications (including sunglasses and in the displays in watches and other devices) and can be made in several different ways. Most applications use a sandwich structure with certain types of long molecules (usually polyvinyl alcohol dyed with iodine) placed between two thin sheets of plastic.[1] When the molecules are aligned parallel to one another, the sheet acts as a polarizer, with the axis of the polarizer *perpendicular* to the direction of the molecules (Fig. 23.26).

Electrons in the polarizer molecules respond to electric fields as shown in Figure 23.26. When the electric field of an electromagnetic wave is parallel to the molecular chains, the electrons move back and forth along the molecule as the direction of $\vec{E}$ oscillates, producing an AC electric current along the molecular chains. This current leads to the absorption of energy, just like a resistor dissipates (absorbs) energy in an AC electric circuit. Light with a polarization direction parallel to the molecular direction is therefore absorbed. On the other hand, when $\vec{E}$ is perpendicular to the molecular direction, each electron is bound to a particular molecule and moves very little because it cannot "jump" from one molecule to another. The electrons then absorb very little energy, letting light with this polarization pass through. That is why the polarization axis is perpendicular to the molecular direction.

Polarization by Scattering and Reflection of Light

Most applications of polarized light use polarizers like those described in Figure 23.26, but polarized light is also found in nature. As shown in Figure 23.27A, light from the Sun is unpolarized, but sunlight can become polarized after it is scattered by molecules in the atmosphere through a series of events. The molecules act as antennas; the positive ion cores and electrons respond to sunlight by oscillating

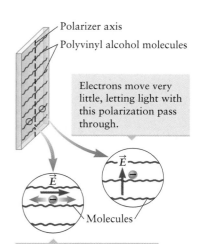

Figure 23.26 A film containing long molecules of polyvinyl alcohol aligned parallel to one another acts as a polarizer. The axis of the polarizer is perpendicular to the molecular direction. When the electric field is parallel to the molecules, electrons move easily along the molecules, producing an electric current that causes energy to be absorbed just like a resistor dissipates energy in an electric circuit. When the electric field is perpendicular to the molecular direction, the electrons move very little and most of the radiation passes through.

---

[1]This type of polarizer was invented by Edwin Land when he was an undergraduate student. He also invented "instant" film cameras (called "Polaroid" cameras).

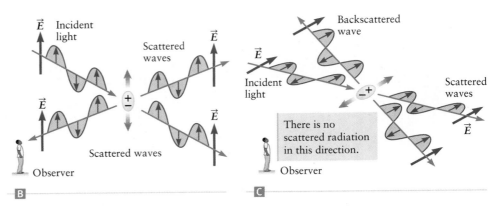

Figure 23.27 Polarization by scattering of light. **A** Sunlight is unpolarized. When light strikes molecules in the atmosphere, the electrons and ion cores oscillate in response to the electric force due to the light's electric field. **B** The vibrating charges act as a dipole antenna and produce scattered electromagnetic waves. The electric field of the scattered waves is parallel to the vibration direction. Here there is a scattered (and polarized) wave in the direction of the observer. **C** A dipole antenna does not emit radiation "off its ends," so radiation with this polarization does not propagate toward the observer in the position shown. Since the only radiation that reaches the observer has the electric field direction shown in part B, the scattered radiation is polarized even if the incident light is not.

back and forth in the direction of the electric field, like the charges in the antenna wires in Figure 23.18. Parts B and C of Figure 23.27 show this motion for light polarized in two different directions. In Figure 23.27B, the polarization direction and the direction of $\vec{E}$ are vertical; the ion cores and electrons thus vibrate up and down and produce new outgoing waves that are also polarized vertically. These outgoing waves are called *scattered waves*.

Figure 23.27C shows how the ion cores and electrons move when the incoming light is polarized horizontally. The ion cores and electrons now oscillate in a horizontal direction and again produce scattered light waves. Because dipole antennas do not radiate "off their ends" (compare with Fig. 23.19), no scattered light reaches the observer looking along the vibration direction.

Now let's combine the results in parts B and C of Figure 23.27 to understand how unpolarized sunlight is scattered. The component of this light polarized vertically (Fig. 23.27B) is scattered to the observer, but none of the horizontally polarized scattered light propagates in the observer's direction. Hence, the light that reaches the observer is said to be **polarized by reflection** (or by scattering). You can observe this polarization by looking at the sky through a polarizer and comparing to the case without a polarizer.

Similar effects can occur whenever unpolarized light is scattered. For example, sunlight scattered by dust particles on a windshield is polarized. This scattered light, called "glare," is usually a nuisance for a driver. It can be removed using the polarizers that are built into most sunglasses. We will return to the polarization of light in Chapter 24.

Optical Activity ⊗

When linearly polarized light passes through certain materials, the polarization direction is rotated as shown in Figure 23.28. This effect is called *optical activity*. Materials that are optically active usually contain molecules with a screwlike or helical structure. DNA is one of many examples; in fact, simple sugars such as fructose and dextrose are optically active molecules. In words, the electric field of linearly polarized light is "dragged along" the screw direction of the molecules. Dextrose and fructose molecules are both helical, but one forms a right-handed helix and the other is left-handed, so they rotate the polarization in different directions.

Figure 23.28 When linearly polarized light passes through an optically active material, the direction of polarization is rotated. The outgoing light is still linearly polarized, but with a different polarization direction. Optically active materials contain molecules with screwlike (helical) shapes. The screw direction determines the rotation direction.

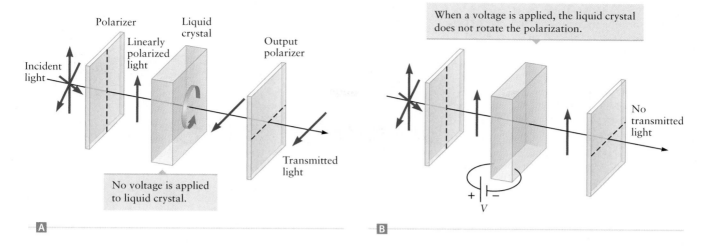

Figure 23.29 Light passing through a liquid crystal display. **A** When no voltage is applied to the liquid crystal layer, it rotates the plane of polarization by 90°, allowing light to pass through the output polarizer. **B** When a voltage is applied to the liquid crystal, it no longer rotates the plane of polarization and no light is transmitted through the output polarizer.

This effect is used in studies of sugar molecules in solution. The rotation direction gives information on the type of sugar present, while the amount of rotation is used to deduce the molecular concentration.

Applications of Polarized Light

The displays in wristwatches, cell phones, and many other devices make use of polarizers and optical activity. These displays are called LCDs ("liquid crystal displays") and work as follows. The incident light is linearly polarized with a polarizer sheet and then passes through a film containing an optically active material called a liquid crystal (Fig. 23.29A). The long, screw-shaped molecules in the liquid crystal rotate the plane of polarization by 90° so that the outgoing light can pass through a second "output" polarizer. To "turn off" the display, a voltage is applied to the liquid crystal, aligning the molecules in such a way that they no longer rotate the direction of polarization (Fig. 23.29B). The light that comes out of the liquid crystal is then polarized at 90° relative to the output polarizer, so no light is transmitted and the display is dark. By applying different voltages to different areas of the liquid crystal, a pattern of light and dark regions corresponding to different letters or numbers can be formed, resulting in the letters and numbers you see in the display of your wristwatch, cell phone, or MP3 player, or even the displays at gasoline station pumps. You can tell that the light from a display is polarized by placing another polarizer such as the polarizers in your sunglasses in front as shown in Figure 23.30. Try it.

Figure 23.30 A liquid crystal display on a wristwatch as viewed through a polarizer. The output light is polarized (Fig. 23.29A), so **A** when the plane of polarization is at a right angle with the axis of a polarizer placed in front of the display, no light is transmitted and the display is dark. **B** When the polarizer is rotated, light from the display passes through.

23.7 | DOPPLER EFFECT

In Section 13.6, we discussed the Doppler effect, a phenomenon in which the frequency of a wave is affected by the relative motion of the source and observer. The Doppler effect was used by the astronomer Edwin Hubble in his studies of the motion of stars and galaxies in the universe in the early 1900s. Light is emitted and absorbed by atoms of different elements at the surface of the star. From the work of physicists in the 1800s, Hubble knew that each atom emits and absorbs light of certain characteristic colors. Figure 23.31 (top) shows some of the colors absorbed by hydrogen, magnesium, and sodium on the surface of our Sun. Each vertical dark

Figure 23.31 Spectral lines showing absorption by hydrogen (H), magnesium (Mg), and sodium (Na) atoms. (Emission lines occur at the same wavelengths.) Top: Spectrum found when the atoms are at rest relative to the observer. Bottom: Spectral lines of the same element as we observe from a distant galaxy. Each line is shifted toward longer wavelengths (i.e., lower frequencies).

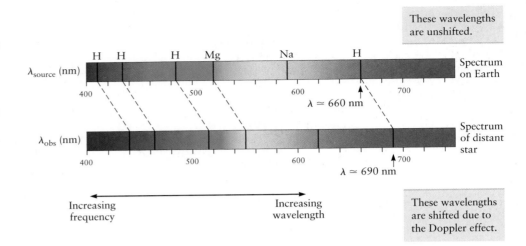

These wavelengths are unshifted.

These wavelengths are shifted due to the Doppler effect.

line in this plot corresponds to a color absorbed by these atoms, and the location of each line along the horizontal axis corresponds to a certain wavelength of light. This type of plot is called a *spectrum* of the starlight, and each vertical line (each color) is called a *spectral line*. A similar set of spectral lines is found if one measures the frequencies emitted by an atom; one finds that an atom absorbs and emits radiation at the same frequencies. The origin of these spectral lines was not understood until the discovery of quantum mechanics in the 1920s and will be explained in Chapter 29.

Hubble was using measurements of spectra to identify the elemental composition of distant galaxies when he observed cases in which the spectra did not seem to correspond to any known elements. One example is shown at the bottom of Figure 23.31. Hubble showed that spectra like this one are "shifted" versions of the expected spectra from known elements. In Figure 23.31, the spectral lines below can all be derived from those above by shifting each line to a longer wavelength (lower frequency), and the shift factor is the same for each line. In this example and in all others found by Hubble, the spectral lines of distant galaxies were always shifted to longer wavelengths relative to the wavelength of the same spectral lines emitted by atoms on Earth. Because red light is at the long-wavelength end of the visible part of the electromagnetic spectrum (Fig. 23.8), this effect is called a *red shift*.

To explain this result, Hubble proposed that the galaxy emitting the red-shifted light was moving away from the Earth. The Doppler effect then causes the frequency of the light as measured by an observer on Earth to be shifted to lower frequencies.[2] This shift is similar to the Doppler shift for the sound emitted by an ambulance siren when the ambulance is going away from you. The size of the frequency shift can be used to determine the velocity of the emitting galaxy. After many galaxies were studied in this way, it was found that most galaxies in the observable universe are moving away from the Earth.

At first glance, you might think that if most or all of the galaxies are moving away from us, the Earth must be at the "center" of the universe, but Hubble found that this picture is *not* correct. He showed that the farther a galaxy is from the Earth, the faster it is receding from the Earth, leading to the picture in Figure 23.32 of an expanding universe at two different times. The galaxy at point *B* might, for example, contain the Earth. As time passes, moving from early in the universe's history to later, the distance from *B* to galaxy *A* and from *B* to all other galaxies increases. Hence, *as viewed from the Earth*, all other galaxies are moving away from us. At the same time, observers in other locations find that other galaxies are also moving away from them. For example, the distances from galaxy *F* to galaxies *B*, *D*, and *E* also increase as time passes, so observers at *F* would also say that all other galaxies are moving away from them.

Early universe

Galaxies

Later universe

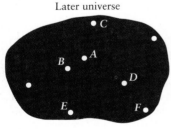

Figure 23.32 Schematic of an expanding universe. As the universe expands, the distance between any two points (i.e., between any two galaxies) increases with time. The relative velocity between two nearby points is less than the relative velocity between two widely separated points.

---

[2]We'll say more about Hubble's discovery in Chapter 31, where we'll explain how this red shift is similar to, but not quite the same as, an ordinary Doppler shift.

Measurements of the red shift are now used to determine the motion of many different objects in the universe and are the basis of much of modern astronomy and astrophysics. The analysis of red shift experiments is based on how the Doppler effect causes the frequency seen by an observer f_{obs} to be shifted relative to the frequency of the source f_{source}. In Chapter 13, we derived relations for the Doppler shift for sound waves. Due to the effect of relativity (Chapter 27), however, the Doppler shift relations for light are different. For Doppler shifts of light and other electromagnetic waves, the frequency measured by an observer is given by

$$f_{obs} = f_{source} \frac{\sqrt{1 - v_{rel}/c}}{\sqrt{1 + v_{rel}/c}} \qquad (23.19)$$

where v_{rel} is the velocity of the source relative to the observer. Here a positive value of v_{rel} corresponds to a source moving away from the observer.

| EXAMPLE 23.7 | ⓡ Doppler Shift of a Distant Galaxy |

Figure 23.31 shows how several spectral lines from a distant galaxy are Doppler-shifted due to the galaxy's motion relative to the Earth. Use the data in this figure to estimate how fast this galaxy is moving relative to the Earth.

RECOGNIZE THE PRINCIPLE

When a light source is moving away from an observer, the wavelength of the light as measured by the observer is longer than the wavelength emitted by the source. Hence, the frequency seen by the observer f_{obs} is lower than the source frequency f_{source}.

SKETCH THE PROBLEM

Figure 23.31 shows the wavelengths at the observer λ_{obs} and source λ_{source}. From the data given in the figure, we can measure the ratio $\lambda_{obs}/\lambda_{source}$.

IDENTIFY THE RELATIONSHIPS AND SOLVE

If the velocity of the source galaxy relative to an observer on the Earth is v_{rel}, the frequencies at the observer and source are related by (Eq. 23.19)

$$f_{obs} = f_{source} \frac{\sqrt{1 - v_{rel}/c}}{\sqrt{1 + v_{rel}/c}}$$

Rearranging leads to

$$\frac{\sqrt{1 - v_{rel}/c}}{\sqrt{1 + v_{rel}/c}} = \frac{f_{obs}}{f_{source}}$$

$$\frac{1 - v_{rel}/c}{1 + v_{rel}/c} = \left(\frac{f_{obs}}{f_{source}}\right)^2$$

Because $f = c/\lambda$, we also have

$$\frac{1 - v_{rel}/c}{1 + v_{rel}/c} = \left(\frac{\lambda_{source}}{\lambda_{obs}}\right)^2 \qquad (1)$$

Defining $y = (\lambda_{source}/\lambda_{obs})^2$, we get

$$1 - v_{rel}/c = y(1 + v_{rel}/c)$$

Next we solve for v_{rel}/c,

$$(v_{rel}/c)(1 + y) = 1 - y$$

$$v_{rel}/c = \frac{1 - y}{1 + y} = \frac{1 - 0.91}{1 + 0.91} = \boxed{0.05}$$

where we have inserted the value of $y = (\lambda_{source}/\lambda_{obs})^2 \approx 0.91$ derived from the red shift data in Figure 23.31. We thus get

$$v_{rel} \approx \boxed{2 \times 10^7 \text{ m/s}}$$

The positive value of v_{rel} indicates that the source is moving away from the observer.

What does it mean?
The value of v_{rel} is about 5% of the speed of light. Hubble discovered that the speed of a galaxy is related to its distance from the Earth by what is now known as Hubble's law. This law can be written as

$$d = \frac{v_{rel}}{H_0}$$

where H_0 is called Hubble's constant, which has the value $H_0 = 2.2 \times 10^{-18} \text{ s}^{-1}$. Inserting our value of v_{rel} into Hubble's law gives

$$d = \frac{v_{rel}}{H_0} = \frac{2 \times 10^7 \text{ m/s}}{2.2 \times 10^{-18} \text{ s}^{-1}} = 1 \times 10^{25} \text{ m}$$

This distance is incredibly large; it can be appreciated better if it is expressed in light-years, the distance that light travels in one year. We find

$$d = 1 \times 10^9 \text{ light-years}$$

The spectrum in Figure 23.31 thus involves light that spent about one billion years traveling to the Earth!

23.8 | DEEP CONCEPTS AND PUZZLES CONNECTED WITH ELECTROMAGNETIC WAVES

Maxwell's work and the subsequent experimental studies of electromagnetic waves by Hertz, Marconi, and others led to profound changes in the way we think about electromagnetism and the laws of physics.

Fields Are Real

We first introduced the notion of a *field* in our discussion of gravitation. When analyzing the force of gravity, lines of force led naturally to the concept of gravitational field lines. The gravitational field is certainly a convenient way to visualize and think about gravity, but, in the opinion of Newton and most of his contemporaries, forces were "real" and field lines were just a helpful picture.

We took a similar approach in our work with electromagnetism. There we started with electric forces given by Coulomb's law and then noticed that these forces could be conveniently described using the notion of electric field lines. Magnetic forces and magnetic fields could be understood in a similar way, but the question remains, "Are these fields real, or are they just a useful mathematical tool?"

The existence of electromagnetic waves means that electric and magnetic fields *are indeed real*. There is no other way to explain how an electromagnetic wave can propagate through a vacuum, carrying energy and momentum. An analogy with the gravitational field also suggests the existence of ***gravitational waves***. Physicists are now working to detect the gravitational waves believed to come to us from distant galaxies.

How Can Electromagnetic Waves Travel through a Vacuum?

Although Maxwell's theoretical work (based on the equations of electricity and magnetism) showed that electromagnetic waves can travel through a vacuum, physicists at the time had difficulty accepting this result. All mechanical waves travel through a material medium, and most physicists believed that there must be a similar medium called the *ether* to support electromagnetic waves. It was believed that the ether permeated all space, including material substances and the vacuum, too. The ether was assumed to have some amazing properties; for example, an object moving through a vacuum would experience no frictional force due to the ether. By postulating the existence of the ether, it was claimed that electromagnetic waves were mechanical waves through the ether. There were many experimental attempts to study the ether and its properties, but eventually (around 1900) the existence of the ether was disproved. Since there is no ether, electromagnetic waves need no medium for travel. The electric and magnetic fields associated with an electromagnetic wave can carry energy even through empty space (a vacuum). This finding opened the way to Einstein's work and his theory of relativity (Chapter 27), which completely changed the way physicists view the universe. Maxwell's work was thus a catalyst for much of 20th-century physics.

Electromagnetic Waves and Quantum Theory

What is light? This question was debated by scientists for many centuries. Newton and others proposed that light is made up of particles, whereas other physicists suggested that light is a wave. We'll discuss the experimental evidence for both points of view in the next several chapters. Maxwell's work seemed to demonstrate conclusively that light is indeed a wave, but that is not the end of the story. We mentioned earlier that Newton's laws break down when applied to very small objects such as electrons and protons and that such objects must be described using quantum theory (Chapters 28 and 29). Maxwell's equations must also be modified to describe light and other electromagnetic waves at the quantum level. Physicists now know that light has *both* wavelike *and* particle-like properties. The "particles" of light are called *photons*.

SUMMARY | Chapter 23

KEY CONCEPTS AND PRINCIPLES

A time-varying magnetic field produces an electric field
According to Faraday's law, a changing magnetic flux Φ_B produces an induced emf $\mathcal{E}$ with $\mathcal{E} = -\Delta\Phi_B/\Delta t$. An emf $\mathcal{E}$ is just an electric potential difference (voltage) ΔV, and there is always an electric field associated with a voltage. Hence, Faraday's law means that *a changing magnetic field produces an electric field*. Maxwell showed that *a changing electric field produces a magnetic field*. This interdependency of E and B leads to the existence of *electromagnetic waves*.

(Continued)

Properties of electromagnetic waves

Electromagnetic waves are **transverse waves**: their electric and magnetic fields are perpendicular to each other, and both $\vec{E}$ and $\vec{B}$ are perpendicular to the propagation direction. These waves can travel through a vacuum and through many material substances. In a vacuum, they travel at a speed c:

$$c = \frac{1}{\sqrt{\varepsilon_0 \mu_0}} = 3.00 \times 10^8 \text{ m/s} \qquad \textbf{(23.1)} \text{ and } \textbf{(23.2)} \text{ (page 764)}$$

The **intensity** of an electromagnetic wave (I) is related to the amplitude of the electric field E_0 by

$$I = \tfrac{1}{2}\varepsilon_0 c E_0^2 \qquad \textbf{(23.9)} \text{ (page 767)}$$

The electric and magnetic field amplitudes are related by

$$E_0 = cB_0 \qquad \textbf{(23.6)} \text{ (page 766)}$$

In addition to carrying energy, electromagnetic waves carry momentum, leading to a **radiation pressure**. When light (or any other electromagnetic wave) is absorbed by an object, the radiation pressure on the object is

$$P_{\text{radiation}} = \frac{I}{c} \qquad \textbf{(23.13)} \text{ (page 769)}$$

Electric field amplitude

Propagation direction

Magnetic field amplitude

Maxwell's equations

The equations of electromagnetism—the laws of Gauss, Coulomb, Ampère, and Faraday—are called **Maxwell's equations**.

APPLICATIONS

The electromagnetic spectrum

Electromagnetic waves are classified according to their frequency. This frequency ranges from a few Hz for **radio waves** to 10^{22} Hz and above for **gamma rays**. In between we have **microwaves**, **infrared radiation**, **visible light**, **ultraviolet (UV) light**, and **X-rays** (Fig. 23.8).

Polarization

Electromagnetic waves can be **polarized**. If the electric field vector points along a single direction, the wave is **linearly polarized**, whereas light from many sources (such as the Sun) is **unpolarized**. Light can be polarized by transmission through a polarizer film or by scattering and reflection.

Doppler effect

The frequency of an electromagnetic wave depends on the relative motion of the source and observer, the **Doppler effect** we studied in connection with sound in Chapter 13. This effect is used by astronomers to study the motion of distant galaxies.

1. A general rule of thumb when listening to and watching a thunderstorm is that the distance from a listener to the lightning strike can be estimated in the following way: First, count the number of seconds between the time you first see the lightning and the time you first hear the thunder. Then multiply this time (in seconds) by 1000 ft to get the distance to the lightning strike. Explain the origin of this rule of thumb.

2. Most vehicles have a radio antenna that is oriented along the vertical direction. Why? Is having an antenna mounted at a slanting angle an advantage or disadvantage? Explain.

3. Electromagnetic waves do not easily penetrate far into metal surfaces. Discuss why that is the case in terms of the known properties of metals, atoms, and electrons.

4. Microwave ovens heat food items by bathing them in electromagnetic waves, yet you can watch your food cook through a window in the door (Fig. Q23.4). How is it possible for the door, with only a thin layer of metal with a grid of holes, to contain the powerful electromagnetic radiation?

© Cengage Learning/Charles D. Winters

Figure Q23.4

5. A method for spacecraft propulsion called a "solar sail" has been proposed. (See Example 23.3 and Fig. 23.6.) It would use a sail exposed to the Sun in much the same way a sailboat's sail uses wind. The sail on such a spacecraft might reflect light, or it might absorb light. Which type of sail would yield the largest force on the sail? Explain your answer.

6. Consider the radiation pressure on an object in outer space. Explain why, all else being similar, this pressure has a larger effect on the motion of a small particle than on a large particle.

7. Explain how an AC circuit can generate electromagnetic radiation.

8. Consider the RC circuit in Figure Q23.8. The switch is closed at $t = 0$ and then kept closed for a very long time. Will this circuit emit electromagnetic radiation (a) just after the switch is closed? (b) After a very long time? Explain your answers. Can you think of ways you might have observed radiation when you opened or closed a switch?

$V_{battery}$

Figure Q23.8

9. In the United Kingdom (and many other European countries), residents are not allowed to use a TV without purchasing a yearly license. Fines of 1000 British pounds are levied against those who are caught using a TV without a license. The law is enforced through devices that can detect a TV in use from 20 m away or more. How might this device, such as the handheld unit shown in Figure Q23.9, work? *Hint*: All TVs have *LRC* circuits in them.

© Touhig Sion/Corbis Sygma

Figure Q23.9

10. SSM The glare off horizontal surfaces that you see while driving a car on a sunny day can be minimized if you wear sunglasses with polarizer films built into the lenses so that light polarized in a certain direction is not allowed to reach your eyes. Discuss how such sunglasses can reduce glare. Is the polarization axis of the glasses vertical or horizontal?

11. Suppose you are given a polarizer, but the axis of polarization is not marked. Devise an experiment you could do to determine the polarization axis.

12. Photographers often use a "haze filter" to remove (or at least minimize) the effects of haze on their photographs. Explain how these filters work.

13. It is a very dusty day, and you are standing by the side of the road looking directly across the road as a car is approaching (Fig. Q23.13). Due to the dust particles in the air, you can see light from the car's headlights as it scatters from the particles. Will this scattered light be (partially) polarized? If so, will the polarization axis be horizontal or vertical?

Dust

Figure Q23.13

14. Light from bulbs and other forms of indoor lighting is typically unpolarized, whereas radio waves picked up by radios or cell phones are mostly polarized. Explain why these two types of electromagnetic radiation are different in this way.

15. Can sound waves be polarized? What about ocean waves or waves propagating on a string? What fundamental characteristic allows waves to have a polarization?

16. Consider a bright sunny day at the beach. Why is it that polarizing sunglasses work well at reducing the glare off the sand and water except when you lie on your side (or tilt your head toward a shoulder)?

17. In Figure 23.5, we analyzed the force exerted by the electric and magnetic fields of an electromagnetic wave on a positive charge at the surface of a material when the electric field has the direction shown in Figure Q23.17A. (a) Assume the charge is instead negative (i.e., an electron). Show that the combination of electric and magnetic fields shown in Figure Q23.17A leads to a force

on the electron with a component to the right. (b) After a short time, the electric field of the wave at the location of the electron changes direction and is pointing down as shown in Figure Q23.17B. Consider again the combined electric and magnetic force on the electron and show that it again has a component to the right. The conclusion of this analysis is that although the electric and magnetic fields oscillate in direction, there is over time always a component of the force along the direction of propagation (except at instants when the force is zero). This force is connected with the wave's momentum.

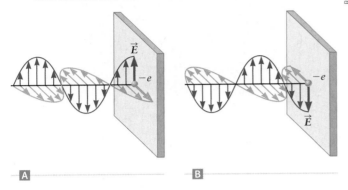

A   **B**

Figure Q23.17

18. Figure 23.18 shows the direction of the electric field produced by an antenna. Sketch the direction of the corresponding magnetic field near the antenna for parts B through H of the figure.

19. Fluorescent lights will glow faintly if brought under a high-tension power line as shown in Figure Q23.19. The lights glow even though they are not plugged into any fixture. Give an explanation of why they glow under the power lines.

Figure Q23.19

20. Figure Q23.20 shows linearly polarized light striking a polarizer, with different angles between the incoming electric field $\vec{E}_{in}$ and the polarizer axis. Is there a force on the polarizer due to the incoming light? If so, what is the direction of the force and in which case is the magnitude of the force largest? Explain.

Case 1   Case 2   Case 3

$\vec{E}_{in}$ $\vec{E}_{out}$ $\vec{E}_{in}$ $\vec{E}_{out}$ $\vec{E}_{in}$ $\vec{E}_{out} = 0$

Polarizer

Figure Q23.20

21. Air traffic controllers at an airport use radar to monitor the location of nearby planes. Their radar units deduce this distance from the time it takes a radio wave to travel from their radar antenna to the airplane and back. The radar signal also gives information about the size and type of the plane. How do they do it?

PROBLEMS

| SSM = solution in Student Companion & Problem-Solving Guide | ⊗ = life science application |
| ★ = intermediate ✪ = challenging | ℝ = reasoning and relationships problem |

23.2 PROPERTIES OF ELECTROMAGNETIC WAVES

23.3 ELECTROMAGNETIC WAVES CARRY ENERGY AND MOMENTUM

1. ℝ Approximately how long does it take a light wave to travel from your head to your toes? Assume you are standing straight up.

2. Figure P23.2 shows the directions of the electric and magnetic fields associated with several different electromagnetic waves. What are the directions of propagation of these waves?

Case 1   Case 2   Case 3

Figure P23.2

3. ✪ ⊗ The human eye can detect light intensities as small as about 1×10^{-23} W/m². (a) What is the corresponding electric field amplitude? (b) Suppose the eye senses this light by detect-

ing the force on electrons inside a photoreceptor cell within the eye. What is the electric force exerted on an electron by the electric field found in part (a)? Is it a large force or a small force? Explain your answer.

4. ★ The electric field of an electromagnetic wave lies along the x axis, whereas the magnetic field is parallel to the z axis. What is the direction of the wave? *Note:* There are two correct answers.

5. SSM ★ The light from a red laser pointer has a wavelength of about 600 nm. If this laser has a power of 0.10 mW, what is the momentum carried by a 3.0-s pulse of this radiation?

6. ★ ℝ How long does it take light to travel from (a) the Earth to the Moon? (b) The Sun to the Earth? (c) Jupiter to the Earth, assuming Jupiter is at its closest distance to the Earth? (d) Jupiter to the Earth, assuming Jupiter is at its farthest distance to the Earth? The difference between the times in parts (c) and (d) was used by Danish astronomer Ole Romer to give the first experimental measurement of the speed of light. (Romer did this work in the 1670s, about the time Newton was doing his work on mechanics.)

7. ★ ℝ The spacecraft that have landed on Mars send their information to the Earth via radio waves. How long do these waves take to reach the Earth when (a) Mars is at its closest to the Earth? (b) Mars is farthest from the Earth? This time delay is important for NASA when it sends a spacecraft to Mars.

8. ✪ ℝ Galileo devised an experiment to measure the speed of light. He and an assistant went to the top of nearby mountains and did the following. (1) Galileo switched on a lantern that was visible to his assistant. (2) Immediately after seeing the light from Galileo's lantern, the assistant switched on his lantern so that Galileo could see it. (3) Galileo then measured the interval between the time that he (Galileo) switched on his lantern and the time that he first saw the light from his assistant's lantern. Suppose the distance between Galileo and his assistant was 5 km. (a) What is the round-trip travel time for the light in this experiment? (b) Do you think that this experiment was successful? *Hint:* The typical human reaction time is about 0.2 s. (c) Now include the typical human reaction time of 0.2 s in your analysis and answer the following question. How far apart should Galileo and his assistant be in order to observe the contribution to the total time delay from the speed of light?

9. ✪ ⓡ In 2005, the spacecraft named *Huygens* landed on one of the moons of Saturn. The data gathered by *Huygens* were relayed by radio to the Earth by a second satellite in orbit about the moon. Estimate the time required for this radio signal to reach the Earth.

10. ✪ ⓡ A radio station called WWV broadcasts from Fort Collins, Colorado. This station broadcasts the time and has served as a "standard clock" for a many years. The radio waves transmitted by this station travel through the atmosphere at approximately the speed of light, so they reach different listeners at different times. Suppose the signal announcing 9:00 AM reaches a person in Chicago at time $t = 0$ (on her local stopwatch). At what time (on this stopwatch) will this signal reach (a) a listener in New York City? (b) A listener in Honolulu?

11. The intensity of the electromagnetic radiation from the Sun when it reaches the Earth's surface is approximately 1000 W/m^2. What is the amplitude of the associated magnetic field?

12. The intensity of moonlight when it reaches the Earth's surface is approximately 0.02 W/m^2 (for a full moon). What are the amplitudes of the associated electric and magnetic fields?

13. An electromagnetic wave has an electric field amplitude of 20 V/m. What is the magnetic field amplitude associated with this wave?

14. SSM ✪ The magnetic field associated with an electromagnetic wave has an amplitude of $1.5 \ \mu\text{T}$. (a) What is the amplitude of the electric field associated with this wave? (b) What is the intensity of this wave? (c) What is the radiation pressure associated with this wave? (d) How much momentum is carried by a 2.0-s pulse of this wave over an area of 20 m^2?

15. ⓡ What is the electric field amplitude associated with the light from a laser pointer? Assume a power of 0.10 mW with a wavelength of 600 nm.

16. ✪ What is the electric field amplitude at a distance of 0.75 m from a 100-W lightbulb? Assume all the power of the bulb goes into light of a single color with $\lambda = 500$ nm and assume the bulb produces a spherical wave.

17. ✪ ⓡ You are initially standing at rest on the surface of a frozen lake where the ice is perfectly frictionless. At $t = 0$, you turn on a laser pointer ($P = 1.0$ mW) and shine it in a horizontal direction. After a period of time, you notice that you are moving in a direction opposite to the light from the laser pointer. Explain why that is so and estimate your speed after 10 min.

18. SSM ✪ Compare the electric field amplitude for the laser pointer in Problem 15 with the electric field from a point charge. Consider a point charge Q and suppose the magnitude of the electric field at a distance of 1.0 cm from Q is equal to the electric field from the laser pointer. Find Q. Express your answer in coulombs and as an equivalent number of electrons.

19. Radar is used to scan for airplanes near an airport. If an airplane is 30 km from the airport, what is the round-trip travel time for the electromagnetic wave used by the radar system?

20. ✪ ⓡ The satellite TV system used by the author employs a satellite in geosynchronous orbit about the Earth. This satellite transmits an electromagnetic wave with a total power of approximately 40 W. Assume all this power is at a single frequency and is broadcast to an area of just the continental United States. Estimate the following quantities. (a) The author's satellite dish has a diameter of about $d = 50$ cm. What is the total power received by this antenna? (b) What is the amplitude of the electric field received by the antenna? (c) Estimate the magnitude of the voltage that is induced at the antenna. For simplicity, assume $|V| \approx Ed$.

21. ⓡ **Global Positioning System.** Global Positioning System (GPS) satellites transmit radio signals used by receivers on the Earth to determine position with very high precision. The operating principle is that a receiver gets signals simultaneously from several different satellites (typically at least four) and uses them to "triangulate" its position. This method relies on precise knowledge of the time and also the time it takes for a satellite signal to reach the receiving position. Until a few years ago, the U.S. government intentionally degraded the timing information so that it had an uncertainty of about 350 ns. Based on this timing error, estimate the precision in meters with which the position can be determined.

22. ✪ ⓡ The GPS satellites described in Problem 21 are placed in orbits for which the period is approximately 12 h. (They are thus not in geosynchronous orbits.) Estimate the time it takes for a GPS signal to travel to the Earth's surface from (a) a GPS satellite that is directly overhead and (b) a GPS satellite that is on the horizon.

23.4 TYPES OF ELECTROMAGNETIC RADIATION: THE ELECTROMAGNETIC SPECTRUM

23. The light used in a CD player has a frequency of about 5.0×10^{14} Hz. What is its wavelength?

24. ✪ ⓡ What is the approximate frequency at which the Sun's radiation has the highest intensity?

25. ⊗ The human eye is most sensitive to light with a wavelength of about 550 nm. What is this light's frequency?

26. The dial on an FM radio contains numbers ranging from about 88 to about 108. These numbers correspond to the frequency of the radio station as measured in megahertz. What is the corresponding range of wavelengths?

27. SSM X-rays are electromagnetic waves with very short wavelengths. Suppose an X-ray has a wavelength of 0.10 nm. What is its frequency?

28. ✪ The electromagnetic waves used in a microwave oven have a wavelength of 12.24 cm. (a) What is the frequency of this radiation? (b) Electromagnetic waves in a microwave can exhibit standing waves just as we discussed for other types of waves in Chapters 12 and 13. If there is a standing wave in your microwave oven, what is the distance between nodes? (c) Based on your answer in part (b), why do you think microwave ovens contain rotating carousels?

29. Some cordless telephones use radio waves with a frequency near 2.4 GHz to transmit to their "base station." What is the wavelength of these waves?

30. AM radio stations use frequencies near 1 MHz. What is the corresponding wavelength?

23.5 GENERATION AND PROPAGATION OF ELECTROMAGNETIC WAVES

31. The U.S. Navy uses extremely low frequency electromagnetic waves to communicate with submarines. Suppose the frequency of these waves is 50 Hz. What is the wavelength? Typically, the size of an antenna is equal to about half of the wavelength. Would this antenna fit in your backyard?

32. ⊠ ® Consider an antenna used to transmit your favorite FM radio station (f = 103 MHz). (a) How long would a dipole antenna be for this frequency? (b) Compare your result in part (a) to the height of a typical radio station tower. Does your answer to part (a) make sense?

33. ⊠ ® (a) What is the approximate size of the antenna in your cell phone? (b) Based on the length of this antenna, estimate the wavelength and frequency of the radiation used by the cell phone.

34. The "wireless" connection used by your laptop computer employs electromagnetic waves with a frequency near 2.4 GHz. (a) What is the wavelength of this radiation? (b) What is the approximate size of a dipole antenna for this frequency? (c) Will such an antenna fit easily into the case of your laptop computer?

35. A dipole antenna is oriented vertically as shown in Figure P23.35 and is used to generate an electromagnetic wave. A second dipole antenna is used to receive this radiation. How should the second antenna be oriented? (a) vertically along z, (b) horizontally along x, (c) horizontally along y, or (d) it doesn't matter. Explain your answer.

Figure P23.35

36. ⊠ The intensity of light from a lightbulb is 10 W/m² at a distance of 0.50 m from the bulb. What is the intensity across the room at a distance of 3.5 m?

37. SSM ⊠ The distance from the Sun to Jupiter is 5.2 times longer than the Earth–Sun distance. If the intensity of sunlight at the Earth is 1000 W/m², find (a) the intensity of sunlight at Jupiter and (b) the amplitude of the associated electric field at Jupiter.

38. ⊠ Light from the star Vega has an intensity of about 2 × 10⁻⁸ W/m². If Vega emits radiation with the same power as our Sun, how far is Vega from the Earth?

23.6 POLARIZATION

39. SSM Unpolarized light with an electric field amplitude of 0.25 V/m is incident on a polarizer. What is the electric field amplitude of the transmitted light?

40. Linearly polarized light propagating along the y direction is incident on a polarizer whose axis is parallel to the z direction. If the intensity of the transmitted light is equal to 35% of the incident intensity, what is the angle of polarization of the incident light?

41. ✪ Unpolarized light with intensity I_0 is incident on two polarizers as sketched in Figure P23.41. Suppose the angle between the axes of the two polarizers is θ = 60°. What is the intensity of the transmitted light? Express your answer as a fraction of I_0.

Figure P23.41 Problems 41 and 42.

42. ✪ For the polarizers in Figure P23.41, suppose the incident light is linearly polarized, the transmitted intensity (through both polarizers) is 0.15 × I_0, and the axis of second polarizer makes an angle θ = 60° with the axis of the first polarizer. What is the angle the initial direction of polarization makes with the first polarizer?

23.7 DOPPLER EFFECT

43. SSM ⊠ When viewed by an observer on the Earth, light from a distant galaxy is Doppler shifted to the red such that the ratio of the unshifted frequency to the observed frequency is 2.5. What is the galaxy's speed relative to the Earth?

44. We know from experience that the Doppler shift of sound from a moving car is easily detectable. Assuming light with a frequency of 6.0 × 10¹⁴ Hz, estimate the Doppler shift for *light* from a car moving toward you at a speed of 50 m/s (about 100 mi/h). Will this light be shifted to the blue (a higher frequency) or to the red (lower frequency)?

45. ⊠ A galactic accident causes a red giant star to be moving on a collision course with our solar system. This star is moving very fast, and its light is Doppler shifted. Which of the following statements might be true?
(a) Light from the star is observed as blue.
(b) Light from the star is observed as X-rays.
(c) Light from the star is observed in the infrared.
(d) Light from the star interferes with cell phone reception.

46. ⊠ Estimate the Doppler shift for a radio wave reflected from an airplane moving directly toward you at a speed of 250 m/s. Assume the radio wave has a frequency of 1.0 × 10⁷ Hz.

47. ✪ How fast would a star have to travel to make violet light (λ = 400 nm) appear to be red (λ = 600 nm)? Assume the star is moving along the line that connects the star to the Earth. Is this star moving toward you or away from you?

48. ⊠ Consider a Doppler radar system that monitors the wind speed of a tornado by measuring the shift in frequency of radio waves reflected from water droplets or other particles in the air. If this wind speed is 45 m/s (about 100 mi/h) and the transmitted radio wave has a frequency of 1.0 MHz (10⁶ Hz), what is the frequency of the reflected wave? Assume the droplets reflecting the wave signal are moving toward the antenna. *Hint:* Take all given numbers as "exact" and keep a large number of significant figures in your final calculation.

ADDITIONAL PROBLEMS

49. ⊠ A powerful AM radio station has a radiated power of 100,000 W. Suppose this energy is emitted into a region shaped as a half-sphere (Fig. P23.49). If you are a distance of 100 km from the station, what is the electric field amplitude there?

50. ✪ ® (a) Estimate the radiation pressure exerted by sunlight on the Earth. (b) What is the total force on the Earth due to this radiation pressure? (c) How does this force compare with the force of gravity exerted by the Sun on the Earth?

51. SSM ⊠ The intensity of sunlight near the Earth is approximately 1000 W/m². (a) If a spacecraft near the Earth uses a

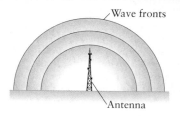

Figure P23.49

square solar sail 200 m on a side and absorbs all light that strikes it, what is the force on the spacecraft? (b) This spacecraft eventually reaches Jupiter, where its distance from the Sun is 5.2 times larger than the distance from the Sun to the Earth. What is the force on the sail now?

52. ⭐ The momentum carried by sunlight can be used by a solar sail to propel a spacecraft (Example 23.3). (a) Is the force on a sail directed toward or away from the Sun? (b) Suppose a solar sail is used for a spacecraft traveling from the Earth to Pluto. What is the ratio of the force on the sail when it is near Pluto to the force when it is near the Earth?

53. ✪ ✖ **LASIK eye surgery.** Lasers are used extensively in eye surgery to remove unwanted "material" from the eye or reshape the eye's lens. These lasers use pulses of very intense radiation to literally "blast" away material. These pulses are typically 10 ns long (10^{-8} s) and carry an energy of 2.0 mJ each. (a) What is the power during one of these pulses? (b) If this energy is delivered to a spot on the eye that is 1.0 mm in diameter, what is the intensity during a pulse? (c) If the radiation has a wavelength of 350 nm, what is the electric field amplitude during a pulse?

54. ✪ Consider the force of an electromagnetic wave on the electrons in an antenna. Assume the amplitude of the electric field associated with the wave is $E_0 = 200$ V/m. (a) What is the magnitude of the electric force on an electron? Express your answer in terms of e and E_0, and in newtons using the given value of E_0. (b) What is the magnitude of the magnetic force on an electron in the antenna? Express your answer in terms of e, B_0 (the magnetic field amplitude), and the speed v of the electron. (c) Suppose v is the drift velocity of an electron. In Chapter 19 (Section 19.3), we found that for typical values of the current the drift velocity is about 0.01 m/s. What is the magnetic force on the electron? What is the ratio of the magnetic force to the electric force found in part (a)? (d) At what speed v does the magnetic force on an electron equal the electric force?

55. ⭐ Ⓡ A college football stadium happens to be in your neighborhood, and you are watching the game at home on your TV. The team makes a touchdown, and the capacity crowd in the stadium cheers. Like an echo, 3.5 s after you see the touchdown and hear the cheering on your home TV, you then hear the cheer of the crowd through your window. Approximately how far away, in kilometers, is the stadium from your house? Assume there are no additional delays at the TV station.

56. ✪ **Lightning Detection and Ranging (LDAR).** Lightning is a rapid transfer of charge from a cloud to the ground (or other parts of the cloud) and constitutes a very large, varying current. When a lightning strike occurs, a large pulse of electromagnetic radiation propagates outward in a wide range of frequencies from light waves to radio waves. NASA has developed a system called LDAR that is implemented near Cape Canaveral to monitor the sometimes sudden and dangerous lightning storms that pass through Florida. The difference in time of arrival of these radio waves is used in trilateration ("triangulation") to measure the position of a lightning strike. Consider three antennas arranged in a right triangle formation as shown in Figure P23.56. Each antenna can pick up a radio pulse from as far away as 100 km. Placing antenna 1 at the origin, find the

Figure P23.56 An image of an LDAR antenna station and a geographic arrangement of three such stations.

x and y coordinates of a lightning strike given the following travel times to antennas 1, 2, and 3: 93 ns, 131 ns, and 51 ns, respectively.

57. ✪ Consider three polarizing filters oriented as shown in Figure P23.57. A beam of unpolarized light is directed through the three filters. The first filter has a polarization axis along the vertical, the second has its polarization axis rotated at an angle of 45° from the vertical, and the third has its axis along the horizontal. (a) What fraction of the light intensity comes out of the third filter? (b) What fraction makes it through the system if the middle filter is removed? (c) Explain how removing a filter can result in less light transmitted. What if the first or third filter were removed instead of the middle filter? Does the result change? Explain.

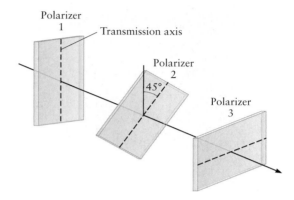

Figure P23.57

58. ✪ **Speed limit enforced by radar.** The use of Doppler radar in the enforcement of traffic law has become common in the United States and many parts of the world. A typical "radar gun" (Fig. P23.58) emits a 24-GHz beam. A highway patrol officer aims the beam at a bus on the highway. (a) If the reflected wave is 4.5 kHz higher in frequency than the emitted beam, how fast is the bus traveling in units of meters per second and miles per hour? (b) Is the bus approaching or moving away? *Hint:* Here there are two shifts to the emitted beam: one because the wave incident on the moving bus is Doppler shifted (as if the bus were stationary with the radar gun approaching) and a second shift from the beam reflected off the moving bus (equivalent to the bus sending out the frequency it sees).

Figure P23.58

59. SSM ⭐ Ⓡ How fast would a car need to travel to make a yellow light appear to be green as the car approached an intersection? Would a judge accept the excuse of such a Doppler shift for running a yellow light?

60. ✪ The radiation pressure from the Sun has a negligible effect on the orbit of a satellite, but it does pose a problem in terms of keeping a spacecraft pointing in the appropriate orientation. Consider the GOES-N geosynchronous weather satellite shown in Figure P23.60. The satellite body is roughly a cube 2.0 m on

a side with a 2.0-m by 4.0-m solar panel attached to the center of one face of the satellite's body. Assume the center of mass of the spacecraft is at the center of the cube. (a) Find the maximum force on the solar panel from the radiation pressure when the panel is perpendicular to the Sun's rays. (b) Calculate the torque on the spacecraft about its center of mass when the Sun's rays are perpendicular to the solar panel. (c) Estimate the moment of inertia about the center of mass of the satellite, assuming uniform density and assuming the solar panel and other projections are negligible compared with its overall mass of 900 kg. (d) What maximum rotational velocity, in degrees per hour, will the torque produce from a 5-min exposure to full sunlight? Assume the initial angular speed is zero.

Figure P23.60 GOES-N geosynchronous weather satellite.

61. **Shooting the Moon.** The *Apollo* astronauts placed special reflector arrays on the Moon (Fig. P23.61A). The array was comprised of "corner reflector" prisms that have the property of reflecting a beam of light exactly back along its incoming path. Using the Apache Point Observatory Lunar Laser-Ranging Operation (Fig. P23.61B), powerful laser pulses are directed at the Moon, and the time it takes them to reflect back is measured. (a) If the time of reflection can be measured to an accuracy of one hundredth of a nanosecond, what is the

uncertainty in the measured Earth–Moon distance? (b) A 2.3-W laser is used to get a very faint reflected pulse. When the laser's beam reaches the lunar surface, it is approximately 2.0 km in diameter. What is the intensity of the laser light at the Moon's surface, and how does it compare to sunlight there? Is shooting the Moon with this laser destructive to the lunar surface?

Figure P23.61 🅐 An *Apollo* reflector array on the lunar surface. 🅑 The Apache Point Observatory Lunar Laser-Ranging Operation. The bright object in the sky in this photo is the Moon. The photo is overexposed so as to show the laser beam used in the experiments (visible as it approaches the Moon from the right).

62. ✪ ⓡ Consider a cell phone that emits electromagnetic radiation with a frequency of 1.0 GHz (1.0×10^9 Hz) with a power of 1.0 W. (a) What is the approximate electric field amplitude a distance 1000 m from the phone? Assume it radiates as a point source. (b) What is the approximate electric field amplitude at your ear as you use the phone? Again assume it radiates as a point source. (c) If the phone can be turned on and transmitting for 3 h before discharging the battery, how much energy can be stored in the battery?

Geometrical Optics

The study of light is called **optics**. This area of physics dates to at least the third century BC. Eyeglasses were invented around AD 1300, and microscopes and telescopes were first developed around 1600. These applications are based on the ability of lenses and mirrors to focus light, and that is the main topic of this chapter. Light is an electromagnetic wave, and our discussions of optics in this chapter and in Chapters 25 and 26 will need to account for the wave nature of light using concepts such as wave fronts and rays. So, now would be a good time to review these and other basic properties of waves discussed in Chapter 12.

The optical images formed by mirrors and lenses play an important role in everyday life. Here two women play with their images in a funhouse mirror. (© Charles Gupton/Stone/Getty Images)

In this chapter, we consider **geometrical optics**, the regime in which light travels in straight-line paths and effects involving wave interference are not important. In general, geometrical optics describes cases in which the wavelength of light is much smaller than the size of objects in the light's path. The wavelength of visible light is less than 1 μm (10^{-6} m) and is about one hundred times smaller

Regime of GEOMETRICAL (RAY) OPTICS

Sharp edge between bright and dark regions

When the opening is large compared to λ, the light rays pass straight through.

Dark shadow

Screen

A

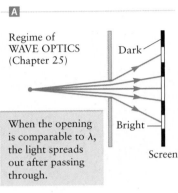

Regime of WAVE OPTICS (Chapter 25)

Dark

When the opening is comparable to λ, the light spreads out after passing through.

Bright

Screen

B

Figure 24.1 Rays of light passing through an opening and onto a screen. **A** A large opening casts an ordinary shadow on the screen. This is the regime of geometrical (also called ray) optics. **B** If the size of the opening is comparable to or smaller than the wavelength λ, the light "spreads out" as it passes through the opening. Instead of just a bright spot and a dark shadow, there is a pattern of bright and dark regions that can only be explained by taking into account the wave nature of light (Chapter 25).

than the diameter of a human hair. Most objects we work with in everyday life are much larger than that, so geometrical optics describes many everyday applications of optics, including the behavior of lenses and mirrors. In this chapter, we'll study how lenses and mirrors work, laying the foundation for our discussion of optical devices such as microscopes and telescopes in Chapter 26. The physics of **wave optics** is the subject of Chapter 25. When the wavelength of light is not small compared to the size of objects in its path, the wave nature of light leads to effects described by wave optics. Such effects give the colors of a thin soap film and are used to read information from CDs and DVDs.

24.1 | RAY (GEOMETRICAL) OPTICS

Figure 24.1A shows a light wave as it passes through an opening that is large compared to the wavelength of the light. This figure also shows **rays**, indicating the path and direction of propagation. We first introduced rays in Chapter 12 (see Fig. 12.12) when describing water waves and sound, but rays are useful for describing the propagation of any type of wave. Strictly speaking, rays indicate the path followed by a wave, but for convenience we will often say that a ray "travels" or "propagates" from one place to another, with the understanding that it is really a light wave that propagates. In Figure 24.1A, rays pass through a wide opening and make an ordinary shadow on the blocked regions of the screen; to a very good approximation, the rays follow straight lines that pass through the opening. If, however, the width of the opening is made very small (Fig. 24.1B), one would now observe that the light "spreads out" after passing through the opening. This spreading is due to the wave nature of light (including interference). We'll come back to this and related situations in Chapter 25 when we discuss wave optics.

The behavior seen in Figure 24.1B is found whenever a wave passes through an opening whose width is about the same size or smaller than the wavelength λ. The wavelength of visible light is less than 1 μm (10^{-6} m), so this behavior is found only with very narrow openings. In most familiar cases, such as light passing through a window, the rays are straight lines as in Figure 24.1A. This straight-line propagation is characteristic of geometrical optics, the regime we'll be discussing throughout the rest of this chapter.

Section 12.4 introduced the concept of a **wave front**. Two examples of wave fronts are shown in Figure 24.2. Wave front surfaces are determined by the crests and troughs of the wave, and are always perpendicular to the associated rays. The shape of a wave front thus depends on how the wave is generated and the distance from the source; wave fronts can be simple planes as in Figure 24.2A or spherical as in Figure 24.2B. When a point source is very far away, the wave fronts form very large spheres, which appear as nearly flat to an observer and can be used to form approximate plane waves (Fig. 24.2C).

Although the behavior of light rays and wave fronts can be derived mathematically from Maxwell's equations, two important properties of light were discovered through experiments several centuries before Maxwell. First, the motion of light along a light ray is *reversible*. That is, if light can travel in one direction along a ray that connects point A to point B, light can also propagate in the reverse direction from point B to point A. Second, the *perpendicular distance* between two wave fronts is proportional to the speed of light because of the way wave fronts are related to the crests and troughs of a wave (see parts A and B of Fig. 24.2). In Section 24.3, we'll build on the proportionality between wave front separation and wave speed to predict the shapes of wave fronts and the directions of rays in more general cases.

A collection of plane wave fronts

The wave fronts travel at the wave speed v.

A

The distance between wave fronts is proportional to the wave speed.

λ

Spherical wave fronts expanding outward at the wave speed v.

B

Spherical wave fronts

Light source

Opening

Wave fronts are approximately planar here ($\approx$ plane waves).

C

Figure 24.2 The shape of a wave front is determined by the crests and troughs of the wave. **A** A plane wave traveling from left to right. The blue arrows denote the corresponding rays; the wave fronts are planes perpendicular to the rays. **B** When a wave is emitted from a point source, the wave fronts are spherical. **C** When a point source is very far away, the wave fronts over any given (small) region are approximately planar and can be used to make an approximate plane wave.

Forming an Image: Ray Tracing

Light that emanates from an object is used by your eye to form an *image* of the object. Figure 24.3A shows how rays originating from two different points on an object reach your eye.[1] When your eye combines these rays to form an image, your brain extrapolates the rays back to their point of origin. This notion of extrapolating rays back to a common point is an essential part of all image formation. The method of following individual rays as they travel from an object to your eye or to some other point is called *ray tracing*. Ray tracing involves the extensive use of geometry; hence, ray optics is also called "geometrical optics." (Now might be a good time to study the quick review of geometry in Chapter 1 and Appendix B.)

Figure 24.3A shows rays from only two points on the object and only a few rays from each point. It would be more realistic to draw a very large ("infinite") number of rays originating from each point on an object. Figure 24.3B shows how rays from another point on the object reach your eye. The light waves associated with *all* these rays contribute to the image formed by your eye. In most of our ray diagrams, we draw only a few rays from the top or bottom of the object, but the complete image is formed by light that originates from all points on the object.

To understand how images are formed, we must consider two basic problems: (1) what happens to light rays when they reflect from a surface such as a mirror or a piece of glass, and (2) what happens to light rays when they pass across a surface from one material to another such as when they pass from air into a piece of glass. In each case, we must distinguish between a flat surface and a curved surface. We'll first consider flat surfaces in Sections 24.2 and 24.3, and then build on that work to deal with curved surfaces in later sections.

24.2 | REFLECTION FROM A PLANE MIRROR: THE LAW OF REFLECTION

Light rays travel in straight lines until they strike something. What happens next depends on what the rays strike. One possibility is that the rays are reflected as sketched in Figure 24.4A, showing the reflection of rays from a *plane mirror*, a flat surface that reflects all or nearly all the light that strikes it.

[1]We note again that rays themselves do not "travel" but rather show the direction of propagation of the corresponding light wave.

Eye

Object

A

Rays emanate from all points on object; unobstructed rays reach eye.

B

Figure 24.3 **A** An optical image is formed in your eye by rays of light that emanate from an object. **B** Many rays (an infinite number) emanate from each point on an object.

Figure 24.4 ◾ Rays reflecting from a flat surface. A surface can be considered flat if its roughness is smaller than λ. This mirror is called a plane mirror. ◾ The angle of reflection is equal to the angle of incidence.

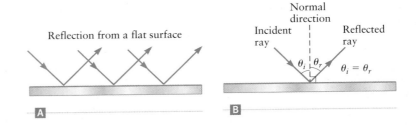

Reflection from a flat surface

A

Normal direction

Incident ray

Reflected ray

θ_i θ_r $\theta_i = \theta_r$

B

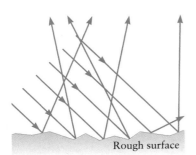

Rough surface

The reflected rays will not all be parallel to one another. Compare with Figure 24.4.

Figure 24.5 Rays are reflected in different directions by different parts of a rough surface, forming a diffuse reflection.

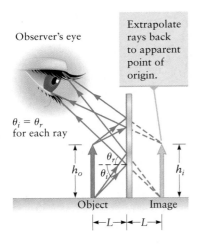

Observer's eye

Extrapolate rays back to apparent point of origin.

$\theta_i = \theta_r$ for each ray

h_o θ_r θ_i h_i

Object   Image

|←—L—→|←—L—→|

Figure 24.6 Forming an image with a plane mirror. Rays from the object (the arrow on the left) reflect from the mirror and reach your eye. These rays appear to have come from the image, on the right of the mirror. It is a virtual image because light does not actually pass through the image point. This sketch shows only rays from the top and bottom of the object, but rays from other points on the object also contribute to the image.

Figure 24.4A describes the case of an incoming plane wave, so the rays are parallel and strike the surface at many different points. We can characterize this reflection by considering only a single ray as shown in Figure 24.4B. (For a plane wave, the other rays are parallel as in Fig. 24.4A and reflect at the same angle.) The vertical dashed line in Figure 24.4B is drawn perpendicular (also called "normal") to the mirror, and we can measure the directions of the incoming and outgoing rays relative to this perpendicular direction. The incoming ray is called the ***incident*** ray, and the angle it makes with the surface normal is the angle of incidence θ_i. The outgoing ray is called the ***reflected*** ray, and its angle with the normal is the angle of reflection θ_r. For reflection from a flat surface, the angle of incidence is equal to the angle of reflection,

$$\theta_i = \theta_r \qquad (24.1)$$

This equation, called the ***law of reflection***, can be derived from our general principle concerning the reversibility of the propagation of light. Suppose the angle of reflection is smaller than θ_i. If we were to reverse the outgoing ray, its angle of reflection would be even smaller and the new reflected ray would not retrace the original incoming ray. The only way the reflection process can be reversible is for the angle of reflection to equal the angle of incidence (Eq. 24.1).

Reflections from a perfectly flat mirror (Fig. 24.4) are called ***specular*** reflections. If the reflecting surface is rough, we must consider reflections from all the individual pieces of the surface. In that case, an incident plane wave will give rise to many reflected rays propagating outward in many different directions (Fig. 24.5). This reflection is called a ***diffuse*** reflection.

Image Formation by a Plane Mirror

When you view an object through its reflection in a mirror, you are viewing an image of the object. An image formed by a plane mirror is shown in Figure 24.6. Here we show a particularly simple object (an arrow), but the object could be a tree or a person. We will usually use an arrow as our example object because we'll often be interested in the orientation of the image relative to the object. As in Figure 24.3, an infinite number light rays emanate from each point on the object, although we only show two rays emanating from the top and bottom here. Some of these rays strike the mirror and are reflected so as to reach your eye, which then uses the rays to form an image. To your eye, the location of this image—that is, the location of the arrow—is the point from which these rays appear to emanate. This point can be found by extrapolating the rays back in a straight-line fashion to their apparent point of origin as shown by the dashed lines in Figure 24.6. This is our first example of the ray-tracing procedure. Since each ray obeys the law of reflection (Eq. 24.1), we can use that together with some geometry to determine the image location. A few rays are shown in Figure 24.6 for an object located a distance L in front of the mirror. When the rays that reach the observer's eye are extrapolated back to their apparent common point of origin, the image is found to be a distance L behind the mirror.

In addition to locating the image, Figure 24.6 gives several other results. First, the size of an image formed by a plane mirror is equal to the size of the object; that is, the height of the image h_i is equal to the height of the object h_o. Second, because the image point is located behind the mirror, light does *not* actually pass through the image. For this reason, the image is called a **virtual image**.

EXAMPLE 24.1 | Reflection from a Corner Cube

Figure 24.7A shows a combination of two plane mirrors connected to form a right angle. In three dimensions, one can form a similar structure called a **corner cube** by combining mirrors to make three faces of a cube. The light ray in Figure 24.7A is incident at an angle θ_i relative to the direction normal (perpendicular) to mirror 1. Find the angle of the outgoing light ray θ_{out} relative to this normal direction.

RECOGNIZE THE PRINCIPLE

To analyze reflection from a single plane mirror, we must first determine the angle of incidence, the angle the incident ray makes with the direction normal to the mirror. According to the law of reflection, the angle of reflection is then equal to the angle of incidence. Here we must use this procedure first for the reflection from mirror 1 and then a second time for the reflection from mirror 2.

SKETCH THE PROBLEM

Figure 24.7A shows the incident ray along with the normal direction for mirror 1 (shown as the dashed line). Reflection from mirror 1 gives an outgoing ray making an angle θ_r with the normal to mirror 1 as sketched in Figure 24.7B. This ray is then the incident ray for reflection from the bottom mirror. We must apply the law of reflection (Eq. 24.1) for each reflection in Figure 24.7B.

IDENTIFY THE RELATIONSHIPS

The angle θ_i in Figure 24.7A is the angle the incoming ray makes with the normal to its reflecting surface, so it is also the angle of incidence in the law of reflection, Equation 24.1. The angle of reflection from this first surface is thus $\theta_r = \theta_i$. The two reflecting surfaces (the two mirrors) are at right angles, so the dashed lines showing the directions normal to each mirror in Figure 24.7B also form a right angle. Because the interior angles of a triangle add up to 180°, the angle of incidence for the second reflection is

$$\theta_{i2} = 90° - \theta_r = 90° - \theta_i$$

Applying the law of reflection to the ray as it reflects from mirror 2, we find

$$\theta_{r2} = \theta_{i2} = 90° - \theta_i \qquad (1)$$

SOLVE

From the geometry in Figure 24.7B, the angle the outgoing ray makes with the horizontal is

$$\theta_{out} = 90° - \theta_{r2}$$

Combining this with Equation (1) gives

$$\theta_{out} = 90° - \theta_{r2} = 90° - (90° - \theta_i)$$

$$\theta_{out} = \boxed{\theta_i}$$

The outgoing ray is thus parallel to the incident ray.

What does it mean?

An interesting feature of this result is that the outgoing ray is *always* parallel to the incident ray for *any* value of θ_i. A combination of mirrors called a corner cube, the three-dimensional version of Figure 24.7, has a similar property: its reflected rays are always parallel to the incident rays. Such devices, called **retroreflectors**,

A

B

Figure 24.7 Example 24.1. Reflection from two mirrors that form a right angle. **A** The incident ray. **B** Reflected rays and the corresponding angles of incidence and reflection.

Figure 24.8 Reflective clothing contains tiny corner cubes, each of which reflects light back parallel to the incident rays as in Figure 24.7B. This cyclist's vest seems to "glow" in the dark because it reflects light (such as from a car's headlights) back toward its source.

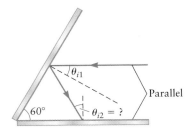

Figure 24.9 Concept Check 24.1.

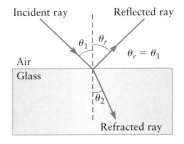

Figure 24.10 An incident light ray is both reflected and refracted when it strikes the surface of a transparent object such as a piece of glass. A portion of the light energy is reflected with $\theta_r = \theta_1$, whereas the rest is refracted as it passes into the glass.

are useful in many applications. For example, the Apollo astronauts who visited the Moon left a number of corner cube reflectors on its surface. When laser light from the Earth strikes one of these corner cubes, the reflected light travels in a path parallel to the incident ray and back to the Earth. The time delay between incident and reflected light pulses can be used to measure the Earth–Moon distance with high precision. (See Problem 23.61, page 794.) In principle, a single, flat mirror would reflect light in the same way, but the mirror would have to be aligned very precisely; otherwise, the reflected light would miss the Earth. With a retroreflector, the reflected light is always parallel to the incident ray, no matter what the alignment. A more down-to-Earth application of retroreflectors is found on many roads. Most roadside reflectors, and even some paints, contain tiny crystals that act as corner cubes to reflect the rays from your headlights back to you, thus giving a noticeable indication of the edges of the road. That is also how reflective clothing works (Fig. 24.8).

CONCEPT CHECK 24.1 | Reflection from Two Mirrors

Consider two plane mirrors that make an angle of 60° with each other (Fig. 24.9). An incident ray is parallel to the bottom mirror. This ray reflects from the mirror at the left and then strikes the bottom mirror. What angle of incidence θ_{i2} does the reflected ray make with the bottom mirror, (a) 30°, (b) 45°, or (c) 60°?

24.3 | REFRACTION

When a light ray strikes a transparent material like a windowpane, some of the light is reflected (according to the law of reflection described in Section 24.2), and the rest passes into the material. Figure 24.10 shows a light ray incident on a piece of glass with two flat, parallel sides. The ray that travels into the glass is called a *refracted* ray.

The Index of Refraction and Snell's Law

The direction of the refracted ray is measured using the angle θ_2 this ray makes with the surface normal. The value of θ_2 depends on the incident angle (which we now denote as θ_1) and also on the speed of light within the material. The speed of light in vacuum is $c = 3.00 \times 10^8$ m/s. When light travels through a material substance, however, the associated electric and magnetic fields interact with the atoms of the substance, and this interaction affects the speed of the wave. As a result, the speed of light inside a substance such as glass, water, or air is less than the speed of light in vacuum. Table 24.1 lists the speed of light in several substances.

Figure 24.11A shows a plane wave initially traveling in a vacuum and then striking the flat surface of a piece of glass; this sketch illustrates how the change in the speed of light in going from vacuum to glass affects the direction of the rays inside the glass. Here we show several rays along with some associated wave fronts in both the vacuum and the glass. The incoming rays make an angle θ_1 relative to the surface normal, and the incident wave fronts are perpendicular to these rays. The incident light travels at speed c, the speed of light in a vacuum. The light inside the glass, however, moves at speed v, which is less than c.

From Section 24.1 (Fig. 24.2), the perpendicular distance between wave fronts is proportional to the wave speed. The wave fronts in Figure 24.11A are drawn at successive, equally spaced moments in time, so the spacing between wave fronts is ct on the vacuum side and vt in the glass. Figure 24.11B shows an expanded view of two adjacent wave fronts, and two right triangles are indicated. On the vacuum side, the right triangle shaded in red has sides L and ct, with angle θ_1 adjacent to side L. From the geometry of this triangle, we have

Table 24.1 Speed of Light and Index of Refraction for Several Substances for Room Temperature and Yellow Light ($\lambda = 550$ nm)

SUBSTANCE	SPEED OF LIGHT (m/s)	INDEX OF REFRACTION
Vacuum	2.998×10^8	1.0000 (exactly)
Air	2.997×10^8	1.0003
Water (liquid)	2.26×10^8	1.33
Water (ice)	2.29×10^8	1.31
Benzene	2.00×10^8	1.50
Quartz crystal	2.05×10^8	1.46
Diamond	1.24×10^8	2.42
Pyrex glass	2.04×10^8	1.47
Plexiglas (plastic)	2.01×10^8	1.49
Regions in the eye:		
vitreous humor	2.23×10^8	1.34
aqueous humor	2.26×10^8	1.33
cornea	2.17×10^8	1.38
lens	2.13×10^8	1.41

$$\sin \theta_1 = \frac{ct}{L} \qquad (24.2)$$

There is a corresponding triangle shaded in yellow on the glass side with edge lengths L and vt and an angle θ_2 adjacent to side L, which leads to

$$\sin \theta_2 = \frac{vt}{L} \qquad (24.3)$$

We can rearrange each of these relations to solve for L, resulting in $L = ct/\sin \theta_1$ and $L = vt/\sin \theta_2$. Setting them equal, we find

$$\frac{ct}{\sin \theta_1} = \frac{vt}{\sin \theta_2}$$

The factor of t can be canceled, giving

$$\sin \theta_1 = \frac{c}{v} \sin \theta_2 \qquad (24.4)$$

A

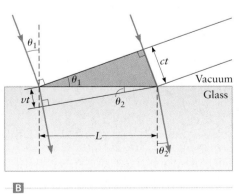

B

Figure 24.11 A Refraction occurs due to a difference in the wave speeds in two regions, which causes the rays to change direction at the interface. B Derivation of Snell's law. The speed of light in a vacuum is c, whereas the speed of light in glass is v.

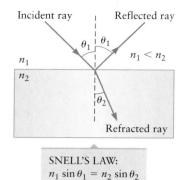

SNELL'S LAW:
$n_1 \sin \theta_1 = n_2 \sin \theta_2$

Figure 24.12 Snell's law gives the relationship between the angle of incidence and the angle of refraction.

The ratio c/v is called the *index of refraction* of the glass and is denoted by

$$n = \frac{c}{v} \qquad (24.5)$$

We can thus rewrite Equation 24.4 as

$$\sin \theta_1 = n \sin \theta_2 \qquad (24.6)$$

In Figure 24.11, we assumed there is vacuum on the incident side and glass on the refracted side. A common situation involves light passing between two substances such as air and glass or air and water. If the two substances have indices of refraction n_1 and n_2 (Fig. 24.12), the incident and refracted angles are related by

$$n_1 \sin \theta_1 = n_2 \sin \theta_2 \qquad (24.7)$$

This relation between the incident and refracted angles is called *Snell's law*. Table 24.1 lists the indices of refraction for some common substances. Since n is the ratio of two speeds (Eq. 24.5), the index of refraction is a dimensionless number.

EXAMPLE 24.2 | Refraction on a Pier

A lobster fisherman looks just over the edge of a pier and spots a lobster resting at the bottom. This fishing spot is $d = 3.5$ m deep. Judging from the angle at which he spots the lobster, the fisherman thinks the lobster is a horizontal distance $L_{app} = 6.0$ m from the shore, but when he drops his trap at that location, he does not catch the lobster. What is the true horizontal distance of the lobster from the shore?

RECOGNIZE THE PRINCIPLE

The fisherman determines the lobster's apparent position by extrapolating the ray that meets his eye back to its apparent point of origin at the bottom of the harbor. This extrapolation is shown by the dashed line in Figure 24.13. However, because the ray from the lobster is refracted when it passes from the water into the air, the actual position of the lobster is much closer to the shore. We can use Snell's law to find the angle θ_1 that the true ray from the lobster makes with the surface normal, and from that we can locate the true position of the lobster.

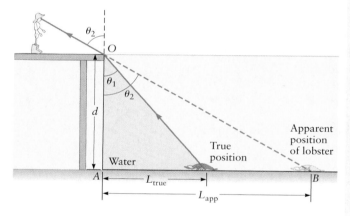

Figure 24.13 Example 24.2.

SKETCH THE PROBLEM

Figure 24.13 shows the problem, including the apparent position of the lobster.

IDENTIFY THE RELATIONSHIPS

To apply Snell's law, we denote the angle of incidence (on the water side of the water–air interface) as θ_1 and the angle of refraction (on the air side of the interface) as θ_2. The angle θ_2 is related to L_{app}, the apparent distance of the lobster from the shore, by

$$\tan \theta_2 = \frac{L_{app}}{d}$$

because L_{app} and d are two sides of the right triangle OAB in Figure 24.13. Inserting the given values $d = 3.5$ m and $L_{app} = 6.0$ m, we find

$$\theta_2 = \tan^{-1}\left(\frac{6.0 \text{ m}}{3.5 \text{ m}}\right) = 60°$$

which is also the angle of refraction for the ray that travels from the lobster to the fisherman. From Snell's law, we thus have $n_1 \sin \theta_1 = n_2 \sin \theta_2$, where n_1 is the index

Insight 24.2

REVERSIBILITY OF LIGHT RAYS AND REFRACTION

In Figure 24.12, light is incident at an angle θ_1 from side 1, leading to a refracted angle θ_2 given by Snell's law (Eq. 24.7). If the light had instead been incident from side 2 (from the bottom of the figure) at an incident angle θ_2, applying Snell's law would give a refracted angle θ_1. Hence, in accord with the general principle of reversibility of propagation direction stated in Section 24.1, the refraction of light is reversible.

of refraction of water and n_2 is the index of refraction for air. Using the values of n_1 and n_2 from Table 24.1 along with our value of the angle θ_2, we have

$$\sin \theta_1 = \frac{n_2}{n_1} \sin \theta_2 = \frac{1.00}{1.33} \sin(60°)$$

$$\theta_1 = 41°$$

SOLVE

Now that we have the value of θ_1, we can find the lobster's true distance from the shore, L_{true}. Using the right triangle shaded in green in Figure 24.13, we have

$$\tan \theta_1 = \frac{L_{\text{true}}}{d}$$

$$L_{\text{true}} = d \tan \theta_1 = (3.5 \text{ m}) \tan(41°) = \boxed{3.0 \text{ m}}$$

What does it mean?
The lobster is thus much closer to shore than it appears to the naive fisherman, which would affect where he drops his trap to have a chance of catching the lobster. This effect is illustrated in Figure 24.14, which shows a pencil inserted into a glass of water. The pencil appears to bend where it enters the water due to refraction of light rays as they leave the water.

Figure 24.14 This pencil is partially immersed in water. It appears to "bend" under the water's surface due to refraction of rays that pass from the water into the air.

Applying Snell's Law

When a light ray strikes a plane surface, the angle of the reflected ray is given by the law of reflection (Eq. 24.1) and the angle of the refracted ray is given by Snell's law (Eq. 24.7). Figure 24.12 shows the case in which the incident ray comes from the side with the smaller index of refraction, whereas in Figure 24.15 the light is incident from the side with the larger index. For instance, substance 1 (the lower substance in Fig. 24.15) might be water or glass and substance 2 could be air.

Snell's law reads the same whether light begins in the substance with the larger or smaller index of refraction,

$$n_1 \sin \theta_1 = n_2 \sin \theta_2 \qquad (24.8)$$

Possible angles of incidence always lie between zero and 90°, so $0° \leq \theta_1 \leq 90°$ and $0° \leq \theta_2 \leq 90°$. For these angles, the function $\sin \theta$ increases as θ is increased. Hence, according to Equation 24.8, the side with the *larger* index always has the *smaller* angle. In words, we say that light is refracted *toward* the normal direction when moving into the substance with the larger index of refraction (Fig. 24.12). Light is refracted *away* from the normal direction when moving into the substance with the smaller index of refraction (Fig. 24.15).

CONCEPT CHECK 24.2 | Snell's Law

Light travels from a vacuum ($n_1 = 1.00$) into a plate of glass with $n_2 = 1.55$, with an angle of incidence of 30°. Is the angle of refraction (a) larger than 30° or (b) smaller than 30°? Also, use Snell's law to calculate the angle of refraction.

Total Internal Reflection

Let's now consider light incident from the side with the larger index of refraction in Figure 24.15 (the bottom side) a little more carefully. Because the index n_1 is greater than n_2, the angle of refraction θ_2 is greater than θ_1 and light is refracted *away* from the normal direction. As the incident angle θ_1 is made larger and larger, the refracted angle θ_2 increases more and more. Eventually, θ_2 will reach 90° and the refracted ray will emerge *parallel* to the surface. The value of the incident angle at which that occurs is called the ***critical angle***. If the incident angle θ_1 in Figure 24.15

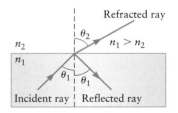

Figure 24.15 Reflection and refraction when light travels from a region of higher to a region of lower index of refraction. Here $n_1 > n_2$; region 1 at the bottom might be glass, and region 2 at the top might be air. Compare with Figure 24.12, in which $n_1 < n_2$.

Figure 24.16 An example of total internal reflection. Light from a laser at the lower left is reflected at the air–water interface. There is no refracted ray.

Courtesy of The Harvard Science Center. © The President & Fellows of Harvard College.

is increased beyond the critical angle, Snell's law has no solution for the refracted angle θ_2. Physically, there is no refracted ray, and all the energy from the incident light ray is reflected. This behavior, called ***total internal reflection***, is possible only when light is incident from the side with the larger index of refraction.

Total internal reflection

Figure 24.16 shows an example of total internal reflection. Here light from a laser travels into the side of a tank of water. The light is reflected at the interface between the water and the air, but the angle of incidence exceeds the critical angle, so there is no refracted ray. Hence, no light gets through this otherwise "transparent" interface!

We can find the value of the critical angle using Snell's law. When the angle of incidence equals the critical angle, the angle of refraction in Figure 24.15 is $\theta_2 = 90°$ (because the direction of the refracted ray would be parallel to the interface). Inserting this angle into Snell's law (Eq. 24.8) and using $\sin(90°) = 1$, we get

$$n_1 \sin \theta_1 = n_2 \sin \theta_2$$

$$n_1 \sin \theta_1 = n_2 \sin \theta_2 = n_2 \sin(90°) = n_2$$

$$\sin \theta_1 = \frac{n_2}{n_1}$$

Denoting the critical angle as θ_{crit} ($= \theta_1$) gives

$$\sin \theta_{crit} = \frac{n_2}{n_1}$$

or

$$\theta_{crit} = \sin^{-1}\left(\frac{n_2}{n_1}\right) \qquad (24.9)$$

The value of θ_{crit} thus depends on the ratio of the indices of refraction of the substances on the two sides of the interface. When the angle of incidence is equal to or greater than the critical angle, light is reflected completely at the interface. Amazingly, this phenomenon can occur even when the interface between the two substances would seem to be completely transparent (such as between water and air). Look for a reflection from such an interface the next time you go swimming or gaze through a window.

Total internal reflection is used in fiber optics. Optical fibers are composed of specially made glass and are used to carry telephone and computer communications signals. These signals are sent as light waves, using total internal reflection as sketched in Figure 24.17 to keep the light from "leaking" out the sides of the fiber. Optical fibers are designed to carry sig-

Total internal reflection

Optical fiber

Light is confined to an optical fiber by total internal reflection.

Ⓐ

© Tetra Images/Jupiterimages

Ⓑ

Figure 24.17 Ⓐ Total internal reflection of light traveling inside an optical fiber. Ⓑ Light escapes from an optical fiber only through the end.

nals over very long distances, so to minimize signal loss it is important that these reflections be as perfect as possible. (The model of an optical fiber shown here is highly simplified; we'll describe a more realistic picture in Chapter 26.) A mirage is another example of total internal reflection (Fig. 24.18). A mirage is formed by total reflection of light between layers of cool air and warm air.

CONCEPT CHECK 24.3 | When Is Total Internal Reflection Possible?

In which of the following cases is it possible to have total internal reflection?
  (a) Light traveling from water into air
  (b) Light traveling from air into water
  (c) Light traveling from glass into water
  (d) Light traveling from water into diamond
  (e) Light traveling from benzene into quartz (*Hint:* See Table 24.1.)

Figure 24.18 A mirage forms when there is a layer of warm air next to the ground. Light from this car undergoes total internal reflection from this layer.

© Stephen St. John/National Geographic/Getty Images

| EXAMPLE 24.3 | Total Internal Reflection between Water and Air |

Find the critical angle for total internal reflection between water and air.

RECOGNIZE THE PRINCIPLE

When we used Snell's law to derive Equation 24.9, we saw how to calculate the critical angle. The purpose of this problem is to get a feel for typical values of the critical angle.

SKETCH THE PROBLEM

Figure 24.19A shows light incident on the interface between water and air. The incident ray is at the critical angle, so the refracted ray (if it existed) would be parallel to the interface and $\theta_2 = 90°$.

IDENTIFY THE RELATIONSHIPS

We can find the critical angle using the result for θ_{crit} in Equation 24.9:

$$\theta_{crit} = \sin^{-1}\left(\frac{n_2}{n_1}\right) = \sin^{-1}\left(\frac{n_{air}}{n_{water}}\right)$$

We can get values for the indices of refraction for water (n_{water}) and air (n_{air}) from Table 24.1.

SOLVE

Inserting the values $n_{air} = 1.0003 \approx 1.00$, and $n_{water} = 1.33$, we get

$$\theta_{crit} = \sin^{-1}\left(\frac{n_{air}}{n_{water}}\right) = \sin^{-1}\left(\frac{1.00}{1.33}\right) = \sin^{-1}(0.75) = \boxed{49°}$$

What does it mean?
The incident ray in Figure 24.19A is drawn at $\theta_{crit} = 49°$, showing that the incident ray does not have to be "close" to the interface to achieve total internal reflection. You can test this result the next time you go swimming. If you are underwater and look in a direction for which the angle of incidence is greater than the critical angle (Fig. 24.19B), the water–air interface will appear as a perfect mirror. The swimmer in Figure 24.19B will see the fish reflected in this interface mirror. Try it.

Figure 24.19 Example 24.3. **A** Total internal reflection at an air–water interface. **B** When a swimmer (on the right) looks at light coming from the air–water interface, all light with an angle of incidence greater than the critical angle is reflected. There are no rays incident from the air that give rays at this angle; the interface thus acts as a perfect mirror when viewed at this angle.

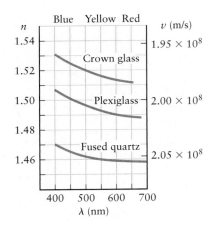

Blue Yellow Red

Figure 24.20 Variation of the speed of light (right axis) and the index of refraction (left axis) as a function of wavelength for several materials.

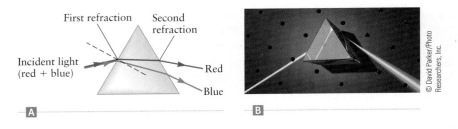

First refraction   Second refraction

Incident light (red + blue)

Red

Blue

© David Parker/Photo Researchers, Inc.

A   **B**

Figure 24.21 ◼ The angle of refraction depends on the wavelength (i.e., the color) of the light. Blue light and red light incident from the left will therefore refract at different angles and emerge from the prism propagating in different directions. ◼ This enables a prism to separate different colors in an incident beam of white light.

Dispersion

The direction of a refracted ray depends on the indices of refraction of the materials on the two sides of an interface, and these indices of refraction depend on the speed of light in each material (Eq. 24.5). The speed of light in a vacuum has the constant value c for all wavelengths, that is, for all colors. When light travels in a material, however, the speed depends on the color of the light. This dependence of wave speed on color is called *dispersion*, and the variation of the speed of light with wavelength for several materials is shown in Figure 24.20.

For example, the index of refraction for red light in quartz is $n_{red} = 1.46$ and that for blue light is $n_{blue} = 1.47$. The difference $n_{blue} - n_{red}$ is much smaller than the difference between the (average) index of refraction for quartz and water (Table 24.1). This difference, however, does mean that the angle of refraction is slightly different for different colors in quartz (and in other materials, too). This effect is used by a *prism* to separate a beam of light into its component colors.

Prisms are typically composed of glass, with a triangular cross section (Fig. 24.21). In Figure 24.21A, a beam of light is incident on one surface of a prism; we suppose this light is composed of two colors (red and blue), and we use two rays with different colors to denote the incident light. Each ray is refracted twice: once when it enters the prism and again when it exits. The first refraction occurs at the left-hand face of the prism. Because the refractive index of glass depends on the color of the light, incident beams of different color have slightly different angles of refraction. Blue light has a larger index of refraction (Fig. 24.20) than red light, so the blue ray inside the glass makes a smaller angle with the normal direction on the left than does the red ray. Hence, inside the glass the red and blue components travel in different directions. This difference in propagation direction is increased at the second refraction, when the light leaves the prism. The calculation of these outgoing angles involves two applications of Snell's law, one for each surface of the prism. (We leave this calculation for the end-of-chapter problems; see Problems 29–31.) The result is shown in Figure 24.21B. In words, we say that the prism has "dispersed" the light into different directions according to color. That is why the variation of the index of refraction with wavelength (Fig. 24.20) is called dispersion.

CONCEPT CHECK 24.4 | **Refraction of Rays with Different Colors**

A mixture of red light and blue light is incident from vacuum onto a thick plexiglass slab. If $n_{red} = 1.49$ and $n_{blue} = 1.51$, which color will have the larger angle of refraction, (a) the red light or (b) the blue light? Or, will they (c) have the same angle of refraction? *Hint:* You should be able to answer this question without working out values of the angles of refraction.

EXAMPLE 24.4 | Dispersing Light by Refraction

Figure 24.22 shows light incident from air onto a flat quartz slab. This light is a mixture of red and blue, denoted in the figure by the two incident rays with these colors. Both rays have an angle of incidence $\theta_1 = 45.0°$. Find the angles of refraction for the two rays in the quartz plate. Assume the indices of refraction for red and blue light in quartz are $n_{quartz, red} = n_{2, red} = 1.459$ and $n_{quartz, blue} = n_{2, blue} = 1.467$. Give your answers to three significant figures.

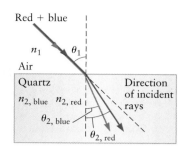

Figure 24.22 Example 24.4.

RECOGNIZE THE PRINCIPLE

We can find the angle of refraction for each color using Snell's law and the values of the index of refraction for each color. The point of this example is to get a feeling for how much the two rays are "dispersed." How large is the difference between the directions of the two rays?

SKETCH THE PROBLEM

Figure 24.22 shows the problem. The index of refraction in quartz is greater for blue light than for red light, so the blue ray is refracted closer to the normal as shown in the figure.

IDENTIFY THE RELATIONSHIPS

The indices of refraction for red and blue light in quartz are given. In air, the indices of refraction for both colors are the same to four significant figures (Table 24.1):
$n_{air, red} = n_{air, blue} = n_1 = 1.000$.

SOLVE

We can find the angles of refraction using Snell's law (Eq. 24.7). For the red light ray, we have

$$n_1 \sin \theta_1 = n_{2, red} \sin \theta_{2, red}$$

Solving for $\theta_{2, red}$, we get

$$\sin \theta_{2, red} = \frac{n_1}{n_{2, red}} \sin \theta_1 \qquad (1)$$

Inserting the values of the indices of refraction and θ_1 gives

$$\sin \theta_{2, red} = \frac{n_1}{n_{2, red}} \sin \theta_1 = \frac{1.000}{1.459} \sin(45.0°)$$

$$\theta_{2, red} = \boxed{29.0°} \qquad (2)$$

We have kept three significant figures because, as we'll see below, the difference between the refracted angles for the two rays will be small.

We compute the angle of refraction for the blue light ray in the same way. Equation (1) leads to

$$\sin \theta_{2, blue} = \frac{n_1}{n_{2, blue}} \sin \theta_1 = \frac{1.000}{1.467} \sin(45°)$$

$$\theta_{2, blue} = \boxed{28.8°} \qquad (3)$$

What does it mean?

The angles of refraction of the two rays differ by only a small amount, just 0.2°, but that is enough to give the dispersion of light by a prism in Figure 24.21B. (To find the angle of the outgoing ray for a prism, we must also include the refraction of the light rays at the second surface of the prism.) Also, the angle of refraction is *smaller* for blue light (see Concept Check 24.4), so the blue ray is *closer* to the normal direction. Thus, refraction causes blue light to be deflected *more* (with respect to the incoming direction) than red light.

In Section 24.2, we saw that a plane mirror can produce an image of an object (Fig. 24.6). That image and all others formed by plane mirrors are the same size as the original object. A curved mirror, however, can achieve a *magnified* image, an image that appears larger or smaller than the original object. Magnified images are used in many applications, ranging from telescopes to a car's rearview mirror, so let's now consider how to produce one.

Ray Tracing for Spherical Concave Mirrors

Consider the spherical mirror in Figure 24.23A in which the surface of the mirror forms a section of a spherical shell (like a beach ball). The radius R of this sphere is called the radius of curvature of the mirror. The mirror's *principal axis* (parts B and C of Fig. 24.23) is the line that extends from the center of curvature C to the center of the mirror. The mirror shown in Figure 24.23 is *concave*, curving toward objects placed in front of it.

Concave spherical mirrors have two important properties. First, consider Figure 24.23B, which shows incoming light rays directed parallel to the principal axis of the mirror. If these rays are close to the principal axis, then after reflecting at the surface of the mirror they all pass through the single point F indicated in Figure 24.23B. (We'll discuss the behavior of rays far from the axis in Section 24.8.) This point F, called the *focal point* of the mirror, is located a certain distance f from the mirror surface along the principal axis. The distance f is called the *focal length* of the mirror. Second, applying the general principle that the propagation of light is reversible, rays that originate at the focal point reflect from the mirror and propagate outward, parallel to the principal axis as in Figure 24.23C. The mirror thus "works" both ways: either focusing parallel light rays at the focal point F or generating a set of parallel rays from light that originates at point F and strikes the mirror.

When an object is placed in front of a spherical mirror, light from the object is reflected from the mirror and forms an *image*. This is illustrated in Figure 24.24, which shows photos of objects placed in front of a concave mirror. Accompanying each photo is a schematic showing the object, the mirror, and the "observer" (the camera that took the photo). In Figure 24.24A, the object—a candle—is placed fairly close to the mirror. The observer can see both the original candle and its image formed by the mirror. Here the image appears larger than the true candle. In Figure 24.24B, the object—a person's head—is somewhat farther from the mirror. The observer now sees the back of the person's head along with an image of the person's face. Note that the face is upside down; we call this an *inverted image*. Figure 24.24 shows two examples of images formed by curved mirrors. Let's now apply ray tracing to analyze some of the image properties.

Figure 24.25 is a ray diagram for an object placed in front of a curved mirror; here the object is the upward-pointing arrow. The figure shows rays that emanate from the tip of the arrow on the left. Although these rays reflect from different parts of the mirror, they all intersect at a single point, which is the tip of the arrow's image. A similar result is found for rays from other points on the object.

We can use the following ray-tracing procedure to find the location of the image formed by a spherical mirror. Figure 24.26 shows an object far from the mirror,

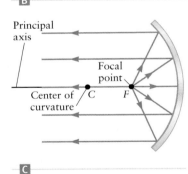

Figure 24.23 ◮ A spherical mirror matches the shape of a sphere. The radius R of this sphere is called the radius of curvature of the mirror. ◮ If the incoming rays are parallel to the principal axis of the mirror, all the reflected rays converge at the focal point of the mirror. ◮ If a light source is placed at the focal point, the outgoing rays reflected from the mirror are parallel to the axis.

A

B

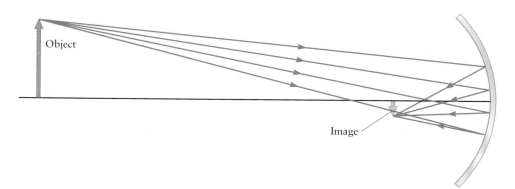

Figure 24.24 Images formed by a concave mirror. The sketches next to each image show the relative positions of the object, mirror, and observer (i.e., the camera that took the photo).

Figure 24.25 Ray diagram for a curved mirror. Rays from each point on the object are reflected by the mirror and intersect at a corresponding image point. Here we show only rays from the tip of the object.

meaning that the center of curvature and the focal point are between the object and the mirror (corresponding to the case in Fig. 24.24B.) The upward-pointing arrow on the left again represents the object, and for simplicity we place the bottom of the object arrow on the principal axis. Although many rays (an infinite number) emanate from each point on the object, three ray directions are particularly easy to draw. Figure 24.26 shows these three rays as they emanate from the tip of the object and intersect at the tip of the image.

One of these rays—called the *focal ray*—begins at the object's tip and passes through the focal point *F*. After reflection from the mirror, this ray travels parallel to the principal axis (compare with Fig. 24.23C). A second ray—called the *parallel ray*—begins at the object's tip and is initially parallel to the principal axis, so after reflection it passes through the focal point (compare with Fig. 24.23B). The third ray—called the *central ray*—begins at the object's tip and passes through the center of curvature C of the mirror. The central ray is directed along the radius of the spherical mirror and is therefore perpendicular to the mirror's surface. By the law of reflection, this ray reflects back on itself and passes back through C. These three rays all intersect (are "focused") at the tip of the arrow's image. Other rays that emanate from the tip of the object also intersect at the corresponding point on the image (see Fig. 24.25), but the focal, parallel, and central rays drawn from the object's tip provide the easiest way to locate the tip of the image and reveal its nature. These three rays are very useful in many applications of ray tracing, so we have drawn them with different colors here and in other figures. These ray colors are *not* the color of the light, but rather are intended to help you identify these rays.

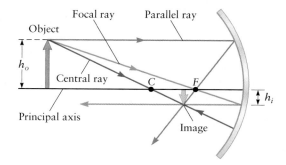

Figure 24.26 Using ray tracing to find the image formed by a concave spherical mirror. For an object outside the focal length, the image is real, inverted, and reduced. Rays emanating from the tip of the object intersect at the tip of the image. *Important note:* The colors of these rays do *not* indicate the color of the light.

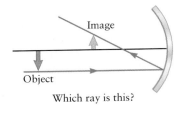

Figure 24.27 Concept Check 24.5.

CONCEPT CHECK 24.5 | Which Ray Is It?

Figure 24.27 shows an object in front of a spherical mirror along with the image and one ray that travels from the tip of the object to the tip of this image. Which ray is it, (a) the focal ray, (b) the parallel ray, or (c) the central ray?

Properties of an Image

The image in Figure 24.26 is smaller than the object and is inverted (upside down). The ratio of the height of the image h_i to the height of the object h_o is called the **magnification**, m:

Magnification

$$m = \frac{h_i}{h_o} \tag{24.10}$$

By convention, the image height h_i in Figure 24.26 is taken to be negative; the magnification in this example is thus also negative. Images smaller than the corresponding objects are said to be **reduced** ($|m| < 1$).

Real image

The rays that form the image in Figure 24.26 all pass through a point on the image. We call the image **real**. A real image differs from the virtual image in Figure 24.6 in two ways:

1. Light rays only *appear* to emanate from a virtual image; they do not actually pass through the image. For a real image, the rays *do* pass through the image.
2. An object and its real image are both on the same (front) side of the mirror. A virtual image is located behind the mirror, while the object is in front.

Forming a Virtual Image

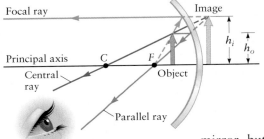

Figure 24.28 If an object is close to a concave mirror (inside the focal point), the outgoing light rays appear to emanate from behind the mirror, producing a virtual, upright image similar to the image in Figure 24.24A.

We can use ray tracing to find the image of an object placed close to a spherical mirror—that is, closer to the mirror than the focal point and the center of curvature—as shown in Figure 24.28. Here we draw just the focal ray, the parallel ray, and the central ray as they emanate from the object's tip (the arrow to the left of the mirror). You should remember, however, that there are actually many rays (an infinite number) that emanate from the tip of the arrow, reflect from the mirror, and then converge at the image. In Figure 24.28, the focal, parallel, and central rays do not intersect at any point on the left (the front) of the mirror, but if we extrapolate the rays back behind the mirror, we find that these extrapolations all intersect at a single "image point." If we were to observe the light reflected by the mirror as shown at the lower left in Figure 24.28, the rays would appear to emanate from this image point behind the mirror. This image is virtual because light does not actually pass through any point on the image. The object and its image are on different sides of the mirror; the image is upright (h_i is positive) and enlarged.

CONCEPT CHECK 24.6 | Magnification of an Image

Consider the virtual image formed by the mirror in Figure 24.28. Which of the following statements is correct?
(a) The magnification is positive and less than 1.
(b) The magnification is positive and greater than 1.
(c) The magnification is negative with an absolute value less than 1.
(d) The magnification is negative and with an absolute value greater than 1.

Rules for Ray Tracing with Mirrors

Figures 24.26 and 24.28 show two examples of how to use ray tracing to construct the image produced by a spherical concave mirror. The general approach can be summarized as follows.

Ray tracing applied to spherical mirrors

1. Construct a figure showing the mirror and its principal axis. The figure should also show the focal point and the center of curvature.

2. Draw the object at the appropriate point. One end of the object will often lie on the principal axis.

3. Draw three rays that emanate from the tip of the object:

   (a) The *focal ray* (or its extrapolation) passes through the focal point. After reflection, this ray will be parallel to the axis.

   (b) The *parallel ray* is initially parallel to the axis. After reflection, this ray (or its extrapolation) passes through the focal point.

   (c) The *central ray* (or its extrapolation) passes through the center of curvature of the mirror. After reflection, this ray passes back through the tip of the object.

4. The point where the focal, parallel, and central rays (or their extrapolations) intersect is the image point. This point may be in front of the mirror giving a *real image*, or it may be necessary to extrapolate the rays back behind the mirror to locate a *virtual image*.

5. This ray-tracing procedure can be repeated for any desired point on the object, thus locating other points on the image. It is usually sufficient to consider just the tip of the image, but additional points can be used if needed.

In many of our ray diagrams, we use different colors to clearly denote the focal, parallel, and central rays. This use of color is to help you distinguish the different rays; these ray colors are *not* the color of the light used to form the image!

Ray Tracing for Spherical Convex Mirrors

A mirror that curves away from the object as in Figure 24.29 is called a *convex* mirror. For a convex mirror, both C and the focal point lie *behind* the mirror. The location of the focal point is illustrated in Figure 24.29A, which shows a collection of parallel rays incident on the mirror. After striking the convex surface, the reflected rays diverge from the mirror axis. These rays, however, appear to come from a virtual image point behind the mirror, the focal point F. That is, all these rays extrapolate back to the focal point F, which lies behind the mirror.

Figure 24.29B shows an application of ray tracing to a convex mirror following the suggested ray tracing rules listed above. This sketch shows the principal axis of the mirror along with the center of curvature C and the focal point F. We also show the focal ray, the parallel ray, and the central ray. Notice that the focal ray is initially directed toward the focal point, but because F is behind the mirror, this ray is reflected before reaching the focal point. These three rays extrapolate to a point of intersection that lies *behind* the mirror; hence, we have a virtual image. This image is upright, so the image height h_i and magnification m are both positive.

Finding the Location of the Image: The Mirror Equation

While ray tracing is an instructive way to understand image formation, it is not a convenient way to find the image location quantitatively. Suppose we know the radius of curvature of a mirror and the distance from the object to the mirror. How can we calculate the location of the image? To answer this question, we must consider the geometry of the reflected rays. Figure 24.30A shows this geometry for a concave mirror, concentrating on two particular rays: the central ray that passes through C and a ray that reflects from the point where the principal axis meets the mirror. (This second ray is *not* a focal or parallel ray.) These two rays intersect at the image point, giving the height of the image h_i. The image here is inverted, so

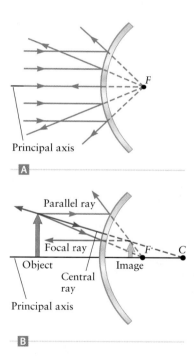

Figure 24.29 Ray diagram for a convex mirror. **A** If the incoming light is parallel to the axis of a convex mirror, the reflected rays appear to emanate from a focal point F behind the mirror. **B** Ray tracing with a convex mirror. The image is behind the mirror (a virtual image).

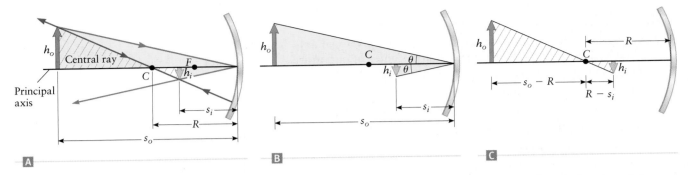

Figure 24.30 Geometry for deriving the mirror equation. **A** Finding the location of the image formed by a concave mirror. **B** Two similar triangles from part A. **C** Two other similar triangles from part A.

by our sign convention h_i is negative and h_o is positive. The figure also defines the distance s_o from the object to the mirror and the distance s_i between the image and the mirror.

The two triangles shaded in yellow in Figure 24.30A are redrawn in part B. They are both right triangles with the same values for the interior angle θ, so the ratios of their sides are equal, giving

$$\frac{|h_o|}{s_o} = \frac{|h_i|}{s_i} \tag{24.11}$$

We have used absolute values of the object and image heights because h_o and h_i can be negative if they lie below the principal axis, and Equation 24.11 holds for the magnitudes of the sides of the similar triangles shaded in yellow in Figure 24.30B. We next consider the pair of similar triangles crosshatched in red in Figure 24.30A. These triangles are redrawn in Figure 24.30C; the ratios of their sides are equal, giving

$$\frac{|h_o|}{s_o - R} = \frac{|h_i|}{R - s_i} \tag{24.12}$$

We now take the ratios of Equations 24.11 and 24.12:

$$\frac{|h_o|/s_o}{|h_o|/(s_o - R)} = \frac{|h_i|/s_i}{|h_i|/(R - s_i)}$$

Canceling the common factors of $|h_o|$ and $|h_i|$ and doing some rearranging leads to

$$\frac{R - s_o}{s_o} = \frac{s_i - R}{s_i}$$

$$s_i(R - s_o) = s_o(s_i - R)$$

$$s_i R - s_i s_o = s_o s_i - s_o R$$

$$s_o R + s_i R = 2 s_i s_o$$

Dividing both sides by the factor $s_i s_o R$, we finally get

$$\frac{1}{s_o} + \frac{1}{s_i} = \frac{2}{R} \tag{24.13}$$

which is called the ***mirror equation***. If we know a mirror's shape (its radius of curvature R) and are given the location of an object s_o, we can use the mirror equation to calculate s_i, the location of the image.

The Mirror Equation and Focal Length

Figure 24.31 shows rays for an object very far from a mirror so that the rays that strike the mirror are all approximately parallel to the axis. This object is a distance

Insight 24.3
WHAT IS THE "BEST" SHAPE FOR A CURVED MIRROR?
In our derivation of the mirror equation (Eq. 24.13), we assumed a spherical mirror, but other mirror shapes give better focusing properties for rays that are far from the principal axis. Mirrors with a parabolic shape are used in many high-performance telescopes because this shape gives smaller aberrations than a spherical mirror. (See Section 24.8.) It is more difficult and expensive to make a surface with a parabolic shape than a spherical one, however.

Mirror

Object is very far away ($s_o \approx \infty$).

Rays are approximately parallel at mirror.

Figure 24.31 When an object is very far away ($s_o \approx \infty$), the rays are approximately parallel to the principal axis. The object is said to be "at infinity."

$s_o \approx \infty$ from the mirror, so we call it an object "at infinity." Inserting an infinite object distance into Equation 24.13, we get

$$\frac{1}{\infty} + \frac{1}{s_i} = \frac{2}{R}$$

Because $1/\infty$ is zero, the image distance for an object at infinity is $s_i = R/2$. A collection of parallel rays would thus reflect from the mirror and intersect at the point a distance $R/2$ from the mirror. By definition, however, rays that are parallel to the principal axis all reflect and pass through the focal point F (compare with Fig. 24.23B), so we have just shown that the focal point is a distance $R/2$ from the mirror. This distance is called the *focal length, f,* of the mirror:

$$f = \frac{R}{2} \tag{24.14}$$

Focal length for a spherical mirror

Equation 24.14 enables us to rewrite the mirror equation in terms of the mirror's focal length:

$$\frac{1}{s_o} + \frac{1}{s_i} = \frac{1}{f} \tag{24.15}$$

Mirror equation in terms of focal length

Our analysis of the rays in Figure 24.30 has involved considerable geometry, so let's recap what we have found. Our main results are the mirror equation in its two forms, one using the radius of curvature of the mirror R (Eq. 24.13) and another written in terms of the focal length f (Eq. 24.15). These relations give us a way to calculate the location of an image produced by a spherical curved mirror. We have also shown that the focal length of such a mirror is related to its radius of curvature by $f = R/2$ (Eq. 24.14). So, if we know the curvature of a spherical mirror, we can now give a complete analysis of the images it produces.

Sign Conventions

We derived the mirror equation for a concave mirror, but that relation also applies to a convex mirror provided we use the following sign conventions for the object and image distances s_o and s_i, the focal length f, and the object and image heights h_o and h_i.

> *Sign conventions for the mirror equation*
> 1. **All diagrams with mirrors should be drawn with the light rays incident on the mirror from the left.**
> 2. **The object distance s_o is positive when the object is to the left of the mirror and negative if the object is to the right (behind the mirror[2]).**
> 3. **The image distance s_i is positive when the image is to the left of the mirror and negative if the image is to the right (behind the mirror). Hence, $s_i > 0$ for a real image and $s_i < 0$ for a virtual image.**
> 4. **The focal length is positive for a concave mirror and negative for a convex mirror. Hence, for a spherical mirror we have**
>
> *(Continued)*

---

[2]So far, all our objects have been in front of the mirror, but when two or more mirrors and lenses are combined, it is possible for an object to be behind a mirror. We'll see such examples in Chapter 26.

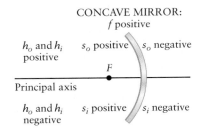

CONCAVE MIRROR:
f positive

h_o and h_i positive

s_o positive s_o negative

F

Principal axis

h_o and h_i negative

s_i positive s_i negative

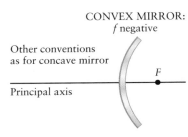

CONVEX MIRROR:
f negative

Other conventions as for concave mirror

Principal axis

F

Figure 24.32 Sign conventions for application of the mirror equation.

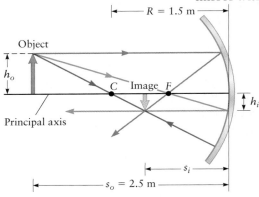

Object

h_o

C   Image *F*

Principal axis

h_i

s_i

s_o = 2.5 m

R = 1.5 m

Figure 24.33 Using ray tracing together with the mirror equation to find the location of an image.

$$\text{concave mirror: } f = \frac{R}{2}$$

$$\text{convex mirror: } f = -\frac{R}{2} \qquad (24.16)$$

5. **The object and image heights h_o and h_i are positive if the object/image is upright and negative if it is inverted.**

We'll follow these sign conventions (summarized in Fig. 24.32) throughout this book. Other books may follow different sign conventions, however, so you should always check the sign conventions before applying relations involving s_o, s_i, and so forth from other books.

CONCEPT CHECK 24.7 | Object and Image Distances

Figure 24.27 shows an object and its image as formed by a curved mirror. What are the signs (positive or negative) of (a) the object distance s_o, (b) the image distance s_i, and (c) the focal length f?

Applications of the Mirror Equation

To see the mirror equation in action, consider the problem sketched in Figure 24.33 in which an object is placed a distance s_o = 2.5 m in front (to the left) of a concave mirror with R = 1.5 m. Let's find the location of the image and also determine if the image is real or virtual. The distances s_o and s_i along with the three rays in our standard ray-tracing diagram are shown in Figure 24.33. (Compare with Fig. 24.26.) While this diagram gives information about the image, the mirror equation provides quantitative answers.

We first use the given value of R to find the focal length $f = R/2 = 0.75$ m and show the position of the focal point F in the ray diagram. (We could use the form of the mirror equation in Eq. 24.13 to find the image location s_i without first finding f, but it is usually a good idea to include the focal point F in all ray diagrams.) The mirror equation can then be used to find the image location. Solving Equation 24.15 for $1/s_i$, we get

$$\frac{1}{s_o} + \frac{1}{s_i} = \frac{1}{f}$$

$$\frac{1}{s_i} = \frac{1}{f} - \frac{1}{s_o}$$

Substituting values for the focal length and object distance gives

$$\frac{1}{s_i} = \frac{1}{0.75 \text{ m}} - \frac{1}{2.5 \text{ m}} = 0.933 \text{ m}^{-1}$$

$$s_i = 1.07 \text{ m}$$

Here we keep three significant figures to avoid roundoff problems later in this example. Since s_i is positive, we expect to find a real image on the left (in front) of the mirror, which is confirmed by the ray diagram. The ray diagram also shows that the image is inverted, so according to our sign conventions the image height h_i is negative. To find the value of h_i, we recall the triangles in Figure 24.30B, which led to Equation 24.11; now we must be careful about the signs of h_o and h_i. We see from Figure 24.33 that the object is above the principal axis, so according to our sign conventions h_o is positive. Since h_i is below the principal axis, h_i is negative. We thus have

$$\frac{-h_i}{s_i} = \frac{h_o}{s_o} \qquad (24.17)$$

Solving for the magnification $m = h_i/h_o$ (Eq. 24.10), we get

$$m = \frac{h_i}{h_o} = -\frac{s_i}{s_o} \qquad (24.18)$$

$$m = -\frac{1.07 \text{ m}}{2.5 \text{ m}} = -0.43$$

The negative value of m means that we have an inverted image (as already noted). Because $|m| < 1$, the image is reduced as shown in Figure 24.33.

EXAMPLE 24.5 | **Applying the Mirror Equation to a Convex Mirror**

Consider again the problem sketched in Figure 24.33, but now assume the mirror is convex. (a) Construct the ray diagram. (b) Is the image inverted or upright? (c) Find the image location s_i. (d) Find the size of the image. Assume the same values of R and s_o as in Figure 24.33 and assume the object height is $h_o = 1.0$ cm.

RECOGNIZE THE PRINCIPLE

A ray diagram should always be your first step in problems involving image formation. We first construct a ray diagram, similar to the ray diagram we used in dealing with a concave mirror. We next use our ray diagram to apply the sign conventions for the image and object distances and heights and for the focal length. The mirror equation can then be applied to find the image's location and properties.

SKETCH THE PROBLEM

(a) We follow our recommended procedure for ray tracing applied to spherical mirrors. Step 1: Figure 24.34A shows the mirror with its principal axis, center of curvature C, and focal point F. According to our sign conventions, the focal length for a convex mirror is negative, so F is behind the mirror. The focal length is (from Eq. 24.16) $f = -R/2 = -0.75$ m. Step 2: We add the object to our picture. Step 3: The focal ray, parallel ray, and central ray are added (Fig. 24.34B). The focal ray is initially directed toward the focal point; after reflecting from the mirror, it travels parallel to the axis. The parallel ray initially travels parallel to the principal axis and then reflects from the mirror so that it appears to emanate from the focal point. The central ray is initially directed toward the center of curvature of the mirror and reflects back toward the object. Step 4: Extrapolations of the rays in Figure 24.34B intersect at the image point, which is behind the mirror.

IDENTIFY THE RELATIONSHIPS AND SOLVE

(b) The ray diagram in Figure 24.34B shows that the image is behind the mirror (a virtual image), is upright, and is smaller than the object.

(c) We next use our sign conventions for image location. The object is in front of the mirror, so s_o is positive with $s_o = +2.5$ m. Using this value in the mirror equation (Eq. 24.15) along with the focal length determined above, we get

$$\frac{1}{s_i} = \frac{1}{f} - \frac{1}{s_o} = \frac{1}{-0.75 \text{ m}} - \frac{1}{2.5 \text{ m}}$$

$$s_i = \boxed{-0.58 \text{ m}}$$

The negative value of s_i indicates that the image is to the right of the mirror (behind), so it is a virtual image as we noted in connection with the ray diagram in Figure 24.34B.

(d) We can find the image height h_i using Equation 24.17; we get

$$-\frac{h_i}{s_i} = \frac{h_o}{s_o}$$

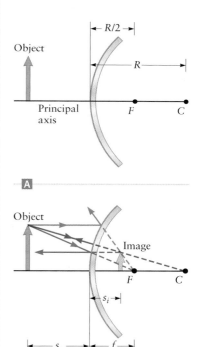

Figure 24.34 Example 24.5. Ray diagram for a convex mirror. **A** A sketch of just the object, mirror, focal point, and center of curvature, in preparation for ray tracing. **B** Same as part A, but now with the parallel, focal, and central rays included.

Note that in this example, h_i is positive while s_i is negative. Solving for h_i and using our result for s_i gives

$$h_i = -h_o \left(\frac{s_i}{s_o} \right) = -h_o \left(\frac{-0.58 \text{ m}}{2.5 \text{ m}} \right) = +0.23 h_o = \boxed{0.23 \text{ cm}} \qquad (1)$$

where we have used the given value of $h_o = 1.0$ cm.

According to our sign conventions, a positive value of h_o corresponds to an upright object as confirmed in Figure 24.34B. A positive value of h_o in Equation (1) gives a positive value for h_i, so we again find that the image is $\boxed{\text{upright}}$.

What does it mean?

When an object is placed in front of a convex mirror, the solution of the mirror equation for s_i is always negative, so the image will *always* be virtual. In addition, if the object is upright, the image will always be upright. This property is handy in many applications, one of which is discussed in Example 24.6.

EXAMPLE 24.6 | ℝ Designing a Car Rearview Side Mirror

The external rearview side mirrors on most cars carry the phrase "objects in the mirror are closer than they appear." **(a)** What does this phrase tell us about the magnification of the image? **(b)** Is this mirror convex or concave? **(c)** Estimate the radius of curvature of a typical rearview mirror.

RECOGNIZE THE PRINCIPLE

(a) The "apparent size" of an object gets smaller as the object is moved farther and farther away. When viewed by its reflection from a mirror, this apparent size is also just the image size. When judging the distance from your eye to an object, your brain compares the known size of the object (as stored in your memory) with the size of the image (the apparent size). So, when you judge an object to be farther away that it actually is, the image must be *smaller* than expected for that particular object distance. Hence, the image formed by the car's rearview side mirror must have a $\boxed{\text{magnification} |m| < 1}$.

SKETCH THE PROBLEM

(b) The next step is to draw a ray diagram, but should this diagram be for a concave mirror or a convex mirror? A car's external rearview mirror *always* produces an upright (noninverted) image, for any object distance. (You should check for yourself.) We have seen that the image produced by a concave mirror can be upright or inverted, depending on the object distance (see Figs. 24.26 and 24.28), so a rearview mirror cannot be concave. We have also seen that a convex mirror always produces a virtual image, as can be seen from the ray diagrams in Figure 24.29. Or, if we insert a negative value of f (as appropriate for a convex mirror) into the mirror equation together with a positive s_o, we find that s_i is always negative, which means that the image is behind the mirror and is thus a virtual image. (See also Example 24.5.) So, a rearview mirror must be $\boxed{\text{convex}}$, and the ray diagram will look like Figure 24.34B.

IDENTIFY THE RELATIONSHIPS

(c) Because $f = -R/2$ (Eq. 24.16 for a convex mirror), we can find the radius of curvature of a rearview mirror if we can estimate the focal length. If we know something about the object and image distances, the value of f can be found from the mirror equation. We have already argued that the magnification is less than 1, but what is its value? A value of $m = 0.1$ would be a very small image, whereas (based on the author's experience) the image is only a little smaller than the true object size. We therefore estimate a magnification of about $m \approx \frac{1}{3}$ for a car that is $s_o = 30$ m from the mirror. These estimates are only rough, but they should not be off by more than a factor of three. (You should think about how you could check these values for yourself.) The magnification is related to the object and image distances by Equation 24.18, giving $m = -s_i/s_o$ or $s_i = -ms_o$. Inserting this information into the mirror equation, we find

$$\frac{1}{s_o} + \frac{1}{s_i} = \frac{1}{f} = \frac{1}{s_o} - \frac{1}{ms_o}$$

and, solving for f, we get

$$f = s_o\left(\frac{m}{m-1}\right)$$

SOLVE

Using our estimates for m and s_o gives

$$f = s_o\left(\frac{m}{m-1}\right) = (30 \text{ m})\left(\frac{\frac{1}{3}}{\frac{1}{3}-1}\right) = -15 \text{ m}$$

The radius of curvature of the mirror is thus

$$R = -2f = \boxed{30 \text{ m}}$$

What does it mean?
This large value of R means that the mirror has just a very slight curvature, which is why it is easily mistaken for a plane mirror. But what is the purpose for using a curved mirror in the first place? Why not just use a flat mirror? The reason is that a curved mirror gives a larger "field of view," allowing the driver to see objects over a wider angle and thus spot cars and other things that might be missed with a plane mirror (Fig. 24.35).

24.5 | LENSES

We have seen how a mirror uses reflection to form an image. A *lens* uses refraction to accomplish the same thing. Typical lenses are composed of glass or plastic, and the refraction of light rays as they pass from the air into the lens and then back into the air causes the rays to be redirected as shown Figure 24.36. Refraction occurs at the two separate surfaces as shown in Figure 24.36A, but for simplicity our ray diagrams will usually be drawn as in parts B and C of the figure, showing a single deflection of each ray as it passes through the lens.

In the simplest case, the lens surfaces are spherical (Fig. 24.37). A careful treatment of the refraction of rays that are close to the principal axis shows that if the incident rays are parallel to the principal axis of the lens, then after passing through the lens the rays all pass through a point on the axis called the *focal point*, *F*. This is true for parallel rays incident from either side of the lens (parts B and C of Fig. 24.36). Hence, there are two focal points, and they are at identical distances from the lens. This symmetry is related to the reversibility principle of light propagation (Section 24.1).

Lenses can be classified according to the curvatures of their two surfaces (Fig. 24.38). Given the radius of curvature of each surface (Fig. 24.37), it is possible to use Snell's law to calculate the focal length. (We'll see how that works in Section 24.6.) For most applications, we only need to know two things about a lens: the value of the focal length f and whether the lens is *converging* or *diverging*. The lenses in Figure 24.36 are converging lenses: all the incoming rays intersect at the focal point on the opposite side. A ray diagram for a diverging lens is shown in Figure 24.39. In that case, parallel rays from the left diverge after passing through the lens, but they all extrapolate back to the focal point on the left (the same side as the incident rays).

Figure 24.35 Example 24.6. View through a car's rearview side mirror.

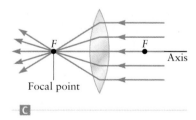

Figure 24.36 Ⓐ Light rays are refracted at *both* surfaces of a lens. When parallel light rays strike a converging lens, the refracted rays converge at the focal point *F* of the lens. Ⓑ For simplicity, most ray diagrams with lenses show only a single "deflection" of a ray as it passes through the lens. Ⓒ If light strikes the lens from the opposite side, the rays converge at a second focal point *F* on the left side of the lens.

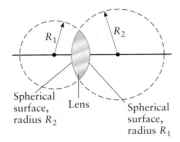

Figure 24.37 A spherical lens: both surfaces follow the shape of a sphere. The radii of curvature of the lens are the radii R_1 and R_2 of the corresponding spheres.

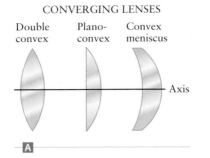

Figure 24.38 Some examples of **A** converging lenses and **B** diverging lenses.

Figure 24.39 With a diverging lens, parallel incident rays from the left are refracted away from the axis. The rays on the right appear to emanate from a point F on the left side of the lens. This point is one of the focal points.

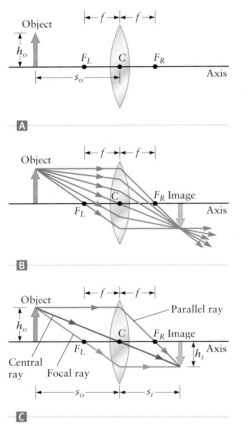

Figure 24.40 Ray tracing with a converging lens when the object is outside the focal point F_L. **A** Sketch showing the object, lens, and focal points. **B** Some typical rays from the tip of the object, intersecting at the tip of the image. **C** Ray diagram showing just the parallel ray, focal ray, and central ray, all emanating from the tip of the object. In this case, the image is real. (When the object is inside the focal point and $s_o < f$, the image is virtual instead.)

Forming an Image with a Lens: Ray Tracing

Our next job is to analyze the images produced by a lens. We'll start with a ray-tracing analysis and then consider how to calculate quantitatively the image properties. Let's begin with the case of a converging lens. Figure 24.40A shows the object along with the focal points on the left (F_L) and right (F_R) sides of the lens. In a typical case, we know the distance s_o from the object to the lens and the object's height h_o, and we want to find the location of the image and its size. Figure 24.40B shows many rays emanating from the object; these rays are refracted by the lens and then intersect to form the image. For simplicity, this diagram shows only rays that emanate from the tip of the object, but rays also emanate from other points on the object and intersect to form other points on the image. In Figure 24.40C, we show only three rays that emanate from the tip of the object; these three rays are particularly useful in a ray tracing analysis. The *parallel ray* is initially parallel to the principal axis of the lens. The lens refracts this ray so that it passes through the focal point F_R on the right. The *focal ray* passes through the focal point on the left. When the focal ray strikes the lens, the ray is refracted to be parallel to the axis on the right. The *central ray* passes straight through the center of the lens (denoted as C). If the lens is very thin (a good approximation for many lenses), this ray is not deflected by the lens. These three rays come together at the tip of the image on the right of the lens. In this case, the image is inverted, and because the rays pass through the image, the image is real.

This approach to ray tracing with lenses is described by the following rules.

Ray tracing applied to lenses

1. **Construct a figure showing the lens and the principal axis of the lens. The figure should also show the focal points on both sides of the lens.**

2. **Draw the object at the appropriate point. One end of the object will generally lie on the axis.**

3. **Draw three rays that emanate from the tip of the object:**

   (a) **The *parallel ray* is initially parallel to the axis. After passing through the lens, this ray or its extrapolation passes through one of the focal points.**

(b) The *focal ray* is directed at the other focal point. After passing through the lens, this ray is parallel to the axis.

(c) The *central ray* passes straight through the center of the lens and is not deflected.

4. These three rays or their extrapolations intersect at the image. If the rays actually pass through the image, the image is *real*; if they do not, the image is *virtual* (just as in the case of images formed by a mirror). When a lens forms a real image, the object and image are on opposite sides of the lens (Fig. 24.40C). In Example 24.7, we'll show that when a lens forms a virtual image, the object and image are on the same side of the lens.

This approach to ray tracing is very similar to the method we used for mirrors; the three rays in Figure 24.40C are similar to the rays used in ray diagrams for mirrors (Fig. 24.26). Our ray diagrams have emphasized the parallel, focal, and central rays because these rays are the simplest ones to draw. However, all other rays that pass through the lens will also pass through the image point as illustrated in Figure 24.40B.

| EXAMPLE 24.7 | Ray Tracing for a Diverging Lens |

Consider a diverging lens with an object to the left of the focal point on the left (incident) side of the lens. (a) Use ray tracing to construct the resulting image. (b) Is the image real or virtual?

RECOGNIZE THE PRINCIPLE

These questions can both be answered with a ray diagram, which we can construct using the rules for ray tracing applied to lenses.

SKETCH THE PROBLEM

Figure 24.41 shows the problem, including a full ray diagram.

IDENTIFY THE RELATIONSHIPS AND SOLVE

(a) We follow the suggested steps for making a ray diagram with lenses. Step 1: Figure 24.41A shows the lens, its axis, and the focal points. Step 2: We add the object to the drawing. Step 3: Figure 24.41B shows the parallel ray, the focal ray, and the central ray, drawn according to our ray-tracing rules. Step 4: These rays do not intersect at any point to the right of the lens. If the rays on the right are extrapolated back to the left side of the lens, however, the extrapolations do intersect. The point of intersection is the image point at the tip of the image.

(b) The image is | virtual | because the rays do not pass through the image point, and the image is on the same side of the lens as the object.

What does it mean?

For all problems involving the image formed by a lens, you should always begin with a ray diagram. Our ray tracing rules apply to both converging and diverging lenses.

Finding the Image Location for a Lens

Ray tracing enables us to describe qualitatively the image formed by a lens. We now want to derive a quantitative approach for calculating the image location. As with our work with the mirror equation, we must adopt sign conventions for dealing with the object and image distances, heights, and so forth. The conventions we'll follow are shown in Figure 24.42 and are listed below.

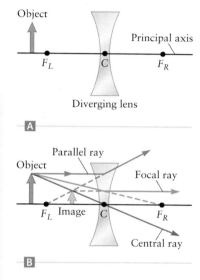

A

B

Figure 24.41 Example 24.7. Ray tracing with a diverging lens.

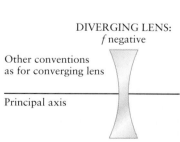

Figure 24.42 Sign conventions for application of the thin-lens equation.

Once again, remember that there are several different sign conventions for working with lenses, so be careful when comparing our equations with those from other books.

CONCEPT CHECK 24.8 | Object and Image Distances

Figure 24.41B shows an object and its image as formed by a diverging lens. What are the signs (positive or negative) of (a) the object distance s_o, (b) the image distance s_i, and (c) the focal length f?

The Thin-Lens Equation

Figure 24.43A shows a ray diagram for a converging lens. We can use the geometry of this sketch to find a mathematical relation for locating the image produced by a converging lens. Within this figure, we identify two pairs of similar right triangles. One pair is shaded in yellow in Figure 24.43A; one yellow triangle has side lengths h_o and s_o, and the other yellow triangle has sides of length h_i and s_i. The similarity of these triangles leads to

$$\frac{h_o}{s_o} = \frac{-h_i}{s_i} \qquad (24.19)$$

The negative sign here is due to our convention that h_i is negative because the image in this case points downward (extends below the axis). Another pair of similar triangles is shaded in red in Figure 24.43B and leads to

$$\frac{h_o}{f} = \frac{-h_i}{s_i - f} \qquad (24.20)$$

with the same sign convention for the image height as in Equation 24.19. Taking the ratio of Equations 24.19 and 24.20 gives

$$\frac{h_o/s_o}{h_o/f} = \frac{h_i/s_i}{h_i/(s_i - f)}$$

Canceling the factors of h_o and h_i, we have

$$\frac{f}{s_o} = \frac{s_i - f}{s_i}$$

and cross multiplying, we get

$$s_i f = s_o(s_i - f)$$

We can now rearrange to obtain

$$s_o f + s_i f = s_o s_i$$

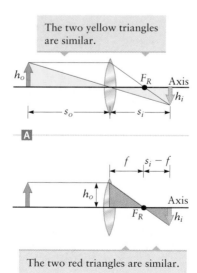

The two yellow triangles are similar.

h_o F_R Axis h_i s_o s_i

A

f $s_i - f$

h_o Axis F_R h_i

The two red triangles are similar.

B

Figure 24.43 Geometry for deriving the thin-lens equation.

---

[3]Notice, however, that when light passes through two or more lenses (as in a telescope or microscope), it is possible for the object distance s_o to be negative. We'll see examples of that in Chapter 26.

Dividing both sides by a factor of $fs_o s_i$ leads to our final result,

$$\frac{1}{s_o} + \frac{1}{s_i} = \frac{1}{f} \qquad\qquad (24.21) \qquad \text{Thin-lens equation}$$

Equation 24.21 is called the **thin-lens equation**. It can be used to calculate the image properties of both converging lenses and diverging lenses, provided we use the sign conventions described above.[4] Notice that the thin-lens equation is actually identical to the mirror equation (Eq. 24.15).

The thin-lens equation (Eq. 24.21) gives a relation between the object and image distances s_o and s_i, and can thus be used to find the image location. Equation 24.21 can also be used to obtain the magnification. As with mirrors, the magnification is the ratio of the image height to the object height,

$$m = \frac{h_i}{h_o} = -\frac{s_i}{s_o} \qquad\qquad (24.22)$$

Applying the Thin-Lens Equation

Let's now see how to use the thin-lens equation together with a ray diagram to locate and describe the image formed by a lens. Consider a converging lens with $f = +0.50$ m and an object at $s_o = +0.30$ m. We have included the positive signs here to emphasize that f is positive (a converging lens) and that s_o is positive because the object is on the left side of the lens. Figure 24.44, which also shows our standard ray diagram, is sketched to scale. We can find the image distance from the thin-lens equation,

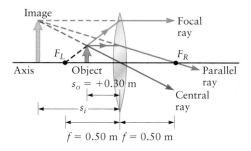

Figure 24.44 Applying the thin-lens equation to find the location of an image formed by a converging lens.

$$\frac{1}{s_o} + \frac{1}{s_i} = \frac{1}{f}$$

Solving for $1/s_i$ first, we get

$$\frac{1}{s_i} = \frac{1}{f} - \frac{1}{s_o} = \frac{s_o - f}{fs_o}$$

Rearranging to find s_i gives

$$s_i = \frac{fs_o}{s_o - f} \qquad\qquad (24.23)$$

Inserting the given values of f and s_o, we obtain

$$s_i = \frac{fs_o}{s_o - f} = \frac{(0.50 \text{ m})(0.30 \text{ m})}{0.30 \text{ m} - 0.50 \text{ m}} = -0.75 \text{ m}$$

According to our sign convention, this negative value for s_i means that the image is on the left side of the lens, so it is a virtual image. This result is confirmed by the ray diagram in Figure 24.44.

| CONCEPT CHECK 24.9 | Image Formed by a Converging Lens

Is the image in Figure 24.44 (a) real and upright, (b) virtual and upright, (c) real and inverted, or (d) virtual and inverted?

| EXAMPLE 24.8 | Image Formed by a Diverging Lens

Suppose the converging lens in Figure 24.44 is replaced by a diverging lens with a focal length $f = -0.50$ m. (a) If the object is again at $s_o = +0.30$ m (i.e., on the left side of the lens), what is the location of the image? (b) Is this image real or virtual?

---

[4]As the name "thin lens" implies, "thick" lenses also exist, and their image properties are described by a different relation. We will not consider thick lenses in this book.

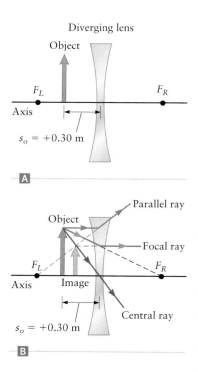

Diverging lens

Object

F_L

F_R

Axis

$s_o = +0.30$ m

A

Object

Parallel ray

Focal ray

F_L

F_R

Axis   Image

$s_o = +0.30$ m

Central ray

B

Figure 24.45   Example 24.8.

RECOGNIZE THE PRINCIPLE

We begin with a ray diagram, which will give us a qualitative understanding of the image. We then proceed to apply the thin-lens equation to find the actual value of s_i.

SKETCH THE PROBLEM

Figure 24.45 shows a ray diagram constructed by following our suggested procedures for ray tracing with lenses. Step 1: Figure 24.45A shows the lens and the focal points on the left and right. Step 2: We add the object to our drawing. Step 3: In Figure 24.45B, we add three rays. The parallel ray is initially parallel to the principal axis; after passing through the lens, this ray is refracted away from the principal axis since we have a diverging lens. Extrapolating this ray back to the left-hand side, we see that it passes through the focal point on the left, F_L. The focal ray is directed from the object toward the focal point to the right of the lens, F_R; this ray is refracted by the lens and like the parallel ray is deflected away from the principal axis. On the right side of the lens, the focal ray travels parallel to the principal axis. The central ray passes through the center of the lens and is not deflected by the lens. Step 4: These rays do not intersect at any point on the right side of the lens. Extrapolating them back, however, we find intersection at a common image point to the left of the lens. This ray diagram tells us much about the image and provides a check on results obtained with the thin-lens equation.

IDENTIFY THE RELATIONSHIPS AND SOLVE

(a) We apply the thin-lens equation (Eq. 24.21) and solve for s_i just as we did in deriving Equation 24.23:

$$s_i = \frac{fs_o}{s_o - f}$$

Substituting the given values of the focal length ($f = -0.50$ m) and object distance ($s_o = +0.30$ m), we get

$$s_i = \frac{fs_o}{s_o - f} = \frac{(-0.50 \text{ m})(0.30 \text{ m})}{0.30 \text{ m} - (-0.50 \text{ m})} = \boxed{-0.19 \text{ m}}$$

(b) According to the ray diagram, light does not pass through this image point, so the image is virtual. This conclusion is confirmed by our analysis; since the image distance s_i is negative, the image is indeed on the left side of the lens and is $\boxed{\text{virtual}}$.

What does it mean?

Because the focal length f for a diverging lens is negative, when s_o is positive (an object on the left side of the lens), the thin-lens equation will always give a negative value for s_i and thus a virtual image. This result is different from a converging lens, which can give either a real or a virtual image, depending on how far the object is from the lens.

Applications of the thin-lens equation are mathematically straightforward, provided you carefully follow the sign conventions in Figure 24.42. Constructing the image using ray tracing is *always* advisable, however, because it provides a check on results from the thin-lens equation and gives an intuitive understanding of the image properties.

24.6 | ✖ HOW THE EYE WORKS

Lenses and mirrors are used in many different kinds of optical instruments, including microscopes and telescopes, which we'll discuss in Chapter 26. In a sense, the oldest optical instrument is the eye. A sketch of a vertebrate eye is shown in Figure 24.46A; it is a complicated structure with several different regions, but its basic

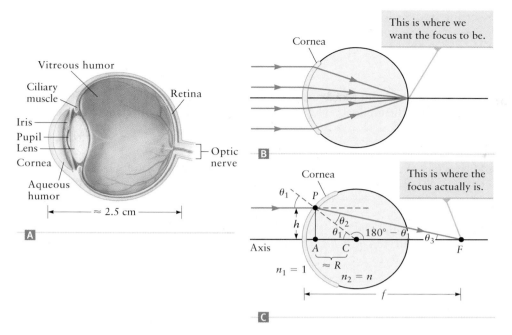

Figure 24.46 **Figure 24.46** **A** Structure of the eye. **B** Corneal refraction model of the eye. The curved surface of the cornea refracts and thus focuses light near the retina (the back surface of the eye). **C** The refraction at the cornea can be analyzed using Snell's law. Here we ignore refraction when the light leaves the right surface of the spherical eye.

function can be understood using the principles of refraction and the properties of lenses. Light enters through the cornea, passes through a liquid region called the aqueous humor, through a lens, into a gel-like region called the vitreous humor, and then to the retina at the back surface of the eye. The retina contains light-sensitive cells that convert light into electrical signals carried to the brain by the optic nerve.

A Simple Model of the Eye: Corneal Refraction

Different regions of the eye have different indices of refraction, so light rays are refracted and change direction as they pass from the air into the cornea and then from region to region within the eye. The indices of refraction of the different regions inside the eye are all in the range of approximately 1.33 to 1.40 (Table 24.1), so the biggest index change and largest refraction occur when light first enters the cornea. In the simplified model in Figure 24.46B, we approximate the eye as spherical, and we also treat all the different parts of the eye as a single region with a single index of refraction. In this model, the only refraction occurs at the surface of the cornea, and we want to calculate how this refraction focuses light. We are especially interested in the focal length of this corneal "lens" and how close (or how far away) the focal point is to the retina. Although this illustration is certainly a simplified version of the real eye, it will still help us understand how the eye works. We'll then consider improvements to the model that make it closer to the real thing.

Figure 24.46C shows a ray diagram for refraction by the cornea. The incident angle is θ_1, the angle the incoming ray makes with the normal to the surface. Here we consider a single incident ray that initially travels parallel to the principal axis; we want to find where its refracted ray intersects the axis since that will give the location of the focal point F. Because we have assumed the surface of the eye is spherical, the normal direction (the dashed line through points P and C) passes through the center of the eye at C. For simplicity, we assume the incoming ray is near the axis, so the height h and the angle θ_1 are small. The refracted angle is given by Snell's law (Eq. 24.7), $n_1 \sin \theta_1 = n_2 \sin \theta_2$. The outside region is air, so $n_1 = n_{air} \approx 1.00$, and we denote the index of refraction inside the eye as simply n. We also assume the angles are small so that $\sin \theta_1 \approx \theta_1$ (and so forth for θ_2). We now have

$$n_1 \sin \theta_1 = n_{air} \sin \theta_1 \approx (1.00)\theta_1 = \theta_1$$

$$n_2 \sin \theta_2 = n \sin \theta_2 \approx n\theta_2$$

Inserting into Snell's law gives

$$\theta_1 = n\theta_2$$

$$\theta_2 = \frac{\theta_1}{n} \qquad (24.24)$$

To locate the focal point and find the focal length f, we use the geometry in Figure 24.46C to find the angle θ_3. The sum of the interior angles of triangle PFC is 180°, so we have

$$180° - \theta_1 + \theta_2 + \theta_3 = 180°$$

and thus $\theta_3 = \theta_1 - \theta_2$. Inserting the result for θ_2 from Equation 24.24 gives

$$\theta_3 = \theta_1 - \theta_2 = \theta_1 - \frac{\theta_1}{n} = \theta_1\left(\frac{n-1}{n}\right) \qquad (24.25)$$

We next consider the two right triangles PFA and PCA. From the first of these triangles, we find[5] $\tan\theta_3 = h/f$. The angle θ_3 is small, so we can also use the approximation $\tan\theta_3 \approx \theta_3$ to get

$$\theta_3 \approx \tan\theta_3 = \frac{h}{f}$$

$$h \approx f\theta_3 \qquad (24.26)$$

Applying the same approach to triangle PCA leads to

$$\theta_1 \approx \tan\theta_1 = \frac{h}{R}$$

$$h \approx R\theta_1 \qquad (24.27)$$

Setting Equations 24.26 and 24.27 equal gives $f\theta_3 = R\theta_1$. The focal length of the corneal "lens" is then

$$f = R\frac{\theta_1}{\theta_3}$$

and inserting the result for θ_3 from Equation 24.25, we find

$$f = R\frac{\theta_1}{\theta_1(n-1)/n}$$

$$f = \frac{n}{n-1}R \qquad (24.28)$$

Properties of the Spherical Eye

Equation 24.28 gives the distance from the refracting surface (the cornea) to the focal point for our simplified model of the eye in Figure 24.46C. Where does this focal point lie? Is it inside the eye? We can estimate this focal length using the average index of refraction of the eye, which is about $n = 1.4$; we find

$$f = \frac{n}{n-1}R = \frac{1.4}{1.4-1}R = \frac{1.4}{0.4}R = 3.5R \qquad (24.29)$$

In our simple model of a spherical eye, the retina is a distance $2R$ from the front surface of the cornea, so the focal point found in Equation 24.29 is about $1.5R$ *behind* the retina.

We have worked through this simplified model of the eye for two reasons. First, it shows how to deal with refraction from a spherical surface. Second, the result for the focal length in Equation 24.29 shows that this simple spherical eye would not function very well because it does not form an image at the retina; indeed, that

---

[5]We assume point A is very close to the spherical refracting surface. This approximation is good when the rays are close to the axis (as we have assumed) and h is small.

explains why the eye actually deviates somewhat from a spherical shape. In the more realistic picture of Figure 24.46A, you will notice that the cornea protrudes slightly from the otherwise spherical profile of the eye. This protrusion makes the radius of curvature of the cornea smaller than the radius of the main spherical portion of the eye. The quantity R in Equation 24.29 is the radius of curvature of the *cornea*, so allowing for this more realistic shape will bring the focal point closer to the retina. Hence, our simple model has given some insight into why the cornea is shaped the way it is.

Refining the Model: Adding the Lens

Let's now add the eye's lens to our model. In a real eye, the lens is inside the eye (Fig. 24.46A), and light passes through this lens after being refracted by the cornea. We will instead consider a slightly simpler model in which the lens is in front of the cornea and hence just outside the eye. Our model will thus not apply strictly to a real eye, but it will yield some useful insights and can be a good way to understand the combination of an eye with a contact lens. So, in our simple model, the incoming light first passes through and is partially focused by the lens and then enters the cornea, which finishes the focusing job. By placing the lens in front of the cornea, we can more easily apply our results for lenses, including the thin-lens equation. We now ask two questions: "What kind of lens is needed for the eye to function properly (i.e., converging or diverging)?" and "What is the shape of the lens (i.e., its radius of curvature)?"

We begin with the ray diagrams for two different cases. In Figure 24.47A, the eye is focused on an object at infinity (i.e., on an object that is very far away). The incoming rays are parallel to the axis, and we assume the cornea has the correct radius of curvature so that these rays are focused on the retina. Hence, we assume the focusing for distant objects is done solely by the cornea using the refraction focusing of our first model in Figure 24.46B. The ray diagram in Figure 24.47B shows why we need the added lens. Here the object has been moved very close to the eye; the incoming rays are now diverging from the axis, and the refracted rays produced by the cornea alone intersect at a point behind the retina. We need the extra focusing of the added lens in Figure 24.47C to focus the image on the retina.

Now let's consider what properties this lens must have to make the ray diagram in Figure 24.47C work. We place the object a distance s_o in front of the lens and suppose the lens produces parallel, outgoing rays (i.e., an image at infinity) as shown in the figure. We know from Figure 24.47A that the cornea by itself can focus these parallel rays on the retina. Hence, the lens should be designed such that an object at s_o focuses at infinity (so that $s_i = \infty$), producing parallel rays for the cornea. Using the thin-lens equation (Eq. 24.21) with an assumed focal length of f_{lens}, we have

$$\frac{1}{f_{\text{lens}}} = \frac{1}{s_o} + \frac{1}{s_i} = \frac{1}{s_o} + \frac{1}{\infty} = \frac{1}{s_o} \tag{24.30}$$

and thus

$$f_{\text{lens}} = s_o \tag{24.31}$$

The eyes of a person with normal vision are able to focus on objects as close as $s_o = 25$ cm from each eye. The lens in Figure 24.47C must therefore have a focal length $f_{\text{lens}} = 25$ cm. Since f_{lens} is positive, the lens is a converging lens.

How must this lens be shaped to give this value for the focal length? The focal length is related to the radii of curvature of the two surfaces of a lens by a relationship[6] called the ***lens maker's formula***. For a converging lens with two convex surfaces with radii of curvature R_1 and R_2, this formula reads

$$\frac{1}{f} = (n-1)\left(\frac{1}{|R_1|} + \frac{1}{|R_2|}\right) \tag{24.32}$$

Lens maker's formula

---

[6]The lens maker's formula can be derived by considering the refraction from the two surfaces of the lens in succession, similar to the calculation we did for our model eye in Figure 24.46C.

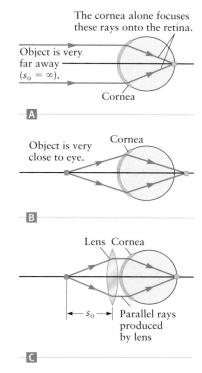

The cornea alone focuses these rays onto the retina.

Object is very far away ($s_o = \infty$).

Cornea

A

Object is very close to eye.

Cornea

B

Lens   Cornea

s_o Parallel rays produced by lens

C

Figure 24.47 Function of the eye's lens. **A** In this model, light from a very distant object is focused by the cornea alone on the retina. **B** When an object is very close to the eye, the refracted rays do not fall on the retina, but instead intersect behind it. **C** If we place a lens in front of the eye, light from a nearby object is focused by the combination of the eye's lens and the cornea so that an image is formed at the retina.

where n is the index of refraction of the lens. Equation 24.32 applies only to the double convex lens shape in Figure 24.37 and assumes the lens is in air. More general versions of the lens maker's formula apply to other lens shapes and to situations such as a lens under water. The absolute value terms in Equation 24.32 emphasize that both these terms are positive for the double convex lens considered here. For other lens shapes, one or both of these terms may be negative and hence describe a diverging lens with a negative value for f. A few such cases are discussed in the end-of-chapter problems. (See Problems 81 and 82.)

Applying the Lens Maker's Formula

Let's now apply the lens maker's formula to estimate the radius of curvature of the lens in our model of the eye in Figure 24.47C. The required focal length of this lens was estimated using Equation 24.31, where we found $f_{lens} \approx 25$ cm. Although the lens maker's formula involves the radii of curvature for both sides of the lens separately, in the spirit of our approximate treatment we assume these radii are equal and $R_1 = R_2 = R$. Using this assumption in Equation 24.32 gives

$$\frac{1}{f_{lens}} = (n-1)\left(\frac{1}{|R_1|} + \frac{1}{|R_2|}\right) = \frac{2(n-1)}{R}$$

or

$$f_{lens} = \frac{R}{2(n-1)}$$

where we have now dropped the absolute value signs and assume R is positive. Solving for R, we find

$$R = 2(n-1)f_{lens} \tag{24.33}$$

Inserting the value of $f_{lens} = 25$ cm given above and using the value of the index of refraction for the eye's lens $n = 1.41$ (see Table 24.1), we have

$$R = 2(n-1)f_{lens} = 2(1.41 - 1)(0.25 \text{ m}) = 0.21 \text{ m} \tag{24.34}$$

The radius of curvature of the lens is thus several times larger than the diameter of the eye. (Recall that the diameter of the eye is about 2.5 cm = 0.025 m.) This result for the radius of curvature is in accord with the realistic drawing of the eye in Figure 24.46A. The front surface of the lens within the eye, which is responsible for most of the focusing, has a rather flat surface, so its radius of curvature is much larger than the radius of the eye.

In addition to allowing the eye to focus on a nearby object, the eye's lens plays another crucial role. When the object is at a particular location, the focal length of the eye's lens has a value such that the final image from the cornea is properly focused on the retina. If the object is moved closer to or farther away from the eye, the focal length of the lens must change to keep the final image in focus. This change of shape is accomplished using muscles attached to the edges of the lens (Fig. 24.46A) to pull on the edges and change the lens' radius of curvature. The lens must be sufficiently flexible to allow its shape to change accordingly. As a person grows older, the lens becomes less flexible and the eye can no longer focus properly for objects that are both far away and very close. That is why the author wears glasses.

24.7 | OPTICS IN THE ATMOSPHERE

Rainbows

In Section 24.3, we showed that refraction of light by a prism causes rays to change direction and that the amount light is deflected depends on the color of the light.

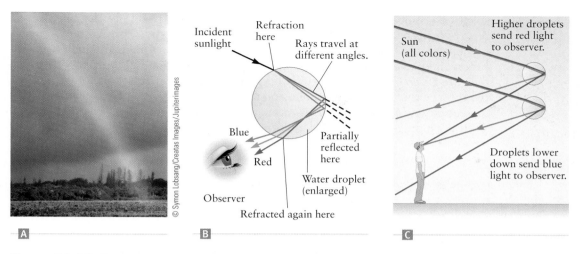

A **B** **C**

Figure 24.48 ◪ A typical rainbow. Rainbows are caused by refraction, dispersion, and reflection of light from water droplets in the atmosphere. ◪ When light enters a droplet, refraction and dispersion separate different colors just as in a prism. After reflecting from the back of the droplet (on the right), these rays are separated further by refraction and dispersion when they leave the droplet. They finally emerge from the droplet at different angles. ◪ Different colors of the rainbow come from different parts of the sky. The colors of these rays *do* represent the color of the light.

A similar effect is responsible for rainbows (Fig. 24.48A), but in this case the refracting object is a water droplet. Water droplets in the atmosphere are illuminated by light from the Sun, and these droplets disperse and redirect light to an observer according to the ray diagram in Figure 24.48B. An incident ray from the Sun is refracted when it enters a droplet; the refracted angles depend on the color of the light, so rays for different colors travel at different angles inside the droplet. When light reaches the back surface of the droplet, a portion is reflected, giving a set of reflected rays. (The rest is refracted and propagates out the back of the droplet.) The reflected rays are refracted when they leave the droplet; the refracted angles again depend on the color, so the final rays leave the droplet at different angles. The rays in Figure 24.48B enter droplets at one particular angle, but other rays will enter at different angles, so the outgoing rays will emerge over a range of angles. However, for each color there is one particular outgoing angle for which the outgoing rays have the greatest intensity, and this "best" outgoing angle depends on the color due to dispersion. As a result, the different colors of a rainbow appear at different (angular) positions in the sky (Fig. 24.48C).

Occasionally, a secondary rainbow is also visible, outside the primary brighter rainbow shown in Figure 24.48A. We'll consider the origin of the secondary rainbow in Problem 77.

Why Stars Twinkle

The index of refraction of a substance can depend on many factors. For example, the index of refraction of air depends on the pressure and temperature of the air. Wind and other effects cause the air temperature and pressure, and hence the index of refraction, to fluctuate from place to place in the atmosphere. These fluctuations cause each small region of air to act as a small lens or prism, deflecting light. As an example, Figure 24.49 shows starlight as it travels through the atmosphere and through one pocket of cool air on its way to an observer on the Earth. This pocket of air has a different index of refraction than the air around it, so rays from the star are refracted as they enter and leave this region, just as light is refracted by a lens. This "lens" of cool air thus deflects rays from the star and affects the apparent location of a star as seen by an observer on the Earth. These "lenses" change

Figure 24.49 Pockets of cool air or warm air refract starlight as it travels to an observer. These pockets shift with time, so the apparent position of the star changes with time and the stars appear to "twinkle." The amount a ray is deflected is much smaller than drawn here.

shape and come and go as air—both warm air and cool air—is blown from place to place. Thus, the deflection of the rays from a particular star changes from moment to moment, causing a star to "twinkle." A similar effect is found with light from the Moon and planets; these sources, however, appear much larger than stars (because they are much closer), so the twinkling effect is not as apparent as for stars. This twinkling effect is why astronomers prefer to have telescopes located on mountaintops or in space, with less of the atmosphere between the telescope and the star. We'll say more about this subject when we discuss telescopes in Chapter 26.

24.8 | ABERRATIONS

In our simple ray diagrams and our derivations of the mirror equation and the thin-lens equation, we found that rays emanating from a particular point on an object intersect (focus) at a single image point. That result relied on several assumptions, and although these assumptions are usually good approximations, they are still just approximations. The images formed by most real mirrors and lenses suffer from effects called ***aberrations***. Aberrations are deviations from the ideal images we have encountered so far; the most common is when all the light rays from a point on an object do not quite focus at a single image point.

Chromatic Aberration

Chromatic aberrations are related to the dispersion of light into different colors as in a prism. Because the index of refraction of a lens depends on the color of the light, different colors are refracted by different amounts. For a plane surface such as the prism in Figure 24.21, the way the rays of different colors disperse can be calculated using Snell's law (Example 24.4). For a lens, we can use the lens maker's formula to estimate the effect of chromatic aberration. According to this formula (Eq. 24.32), the focal length f depends on the index of refraction of the lens. Hence, f is different for different colors and the location of an image is different for each color as shown in Figure 24.50. Many optical instruments (as we'll describe in Chapter 26) employ multiple lenses, and it is sometimes possible to compensate for chromatic aberration by arranging the aberrations of separate lenses to cancel each other. Even with this approach, however, most optical systems have at least a small amount of chromatic aberration.

Spherical Aberration

Since it is relatively simple to make an accurate spherical shape, spherical surfaces for mirrors and lenses are convenient. In this chapter, we have assumed the surfaces of curved mirrors and lenses are spherical and have showed that the resulting focal lengths depend on the radii of curvature. Ray diagrams such as Figures 24.26 and 24.41 rely on the assumption that the rays drawn are very close to the axis of the mirror or lens. Mathematically, this assumption allows us to make the approximations $\sin \theta \approx \theta$ and $\tan \theta \approx \theta$, and if those approximations are accurate, the rays will focus at a single image point. (Those approximations were used in the derivations of the mirror equation and the thin-lens equation.) They are, however, only approximations; in any real situation, rays that are not exactly on the axis will not focus exactly at the ideal image points we found for mirrors or lenses. Such deviations are called ***spherical aberrations*** because they are due to the spherical shape of the mirror or lens.

Certain curved surfaces with nonspherical shapes give better focusing properties than those of spherical shapes. For this reason, high-precision optical components often employ parabolically shaped surfaces. Components of this shape cost more to manufacture, but they give better image properties for off-axis rays.

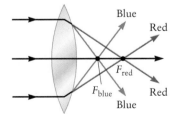

Figure 24.50 Chromatic aberration produced by a converging lens. Rays of different wavelengths (colors) focus at different points.

Ray optics (also called geometrical optics)

In the *ray optics* approximation, light travels along straight lines called *rays*. These rays change direction when light is reflected or refracted. Ray optics gives an accurate description of light propagation when light passes through openings or strikes objects that are large compared to its wavelength.

Reflection and refraction

For reflection and refraction, we measure the direction of a ray using the angle θ made by the ray with the direction normal (perpendicular) to the surface of interest.

According to the *law of reflection*, the angle of incidence is equal to the angle of reflection:

$$\theta_i = \theta_r \qquad \textbf{(24.1)} \text{ (page 798)}$$

Refraction is described by *Snell's law*,

$$n_1 \sin \theta_1 = n_2 \sin \theta_2 \qquad \textbf{(24.7)} \text{ (page 802)}$$

where the *index of refraction* $n = c/v$ is the ratio of the speed of light in a vacuum (c) to the speed of light in the refracting material (v). By applying the law of reflection and Snell's law, we can calculate how light interacts with plane mirrors, curved mirrors, lenses, and other refracting objects such as flat plates and water droplets.

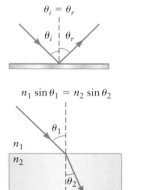

Ray tracing

Ray tracing is the process of using light rays that emanate from an object to derive the location and other properties of an image. Important properties of an image include the following:

- *Real image*: Light from the object passes through the image point.
- *Virtual image*: Light rays do not pass through the image point. The image point is found by extrapolating rays back to a common point behind a mirror or in front of a lens.
- *Upright image*: The image has the same orientation as the object.
- *Inverted image*: The image is upside down compared with the object.
- *Enlarged image*: The image is larger than the object (the magnification is $|m| > 1$).
- *Reduced image*: The image is smaller than the object ($|m| < 1$).

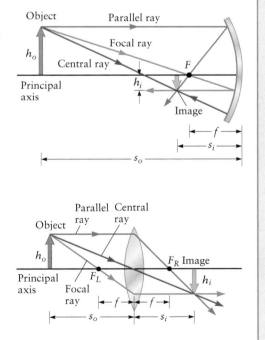

(Continued)

Magnification

Magnification is the ratio of the image height h_i to the object height h_o,

$$m = \frac{h_i}{h_o} \qquad \text{(24.10) (page 810)}$$

According to our **sign conventions** for h_i and h_o, the magnification can be either positive or negative. See pages 813–814 and 820, and Figures 24.32 and 24.42, for a description of these sign conventions.

Mirror equation

The **mirror equation** relates the object distance, the image distance, and the **focal length** for a curved mirror:

$$\frac{1}{s_o} + \frac{1}{s_i} = \frac{1}{f} \qquad \text{(24.15) (page 813)}$$

For a concave spherical mirror, f is given by

$$f = \frac{R}{2} \qquad \text{(24.14) (page 813)}$$

where R is the radius of curvature. For a convex spherical mirror, f has the same magnitude, but is negative. See Figure 24.32 for a description of the mirror sign conventions.

Thin-lens equation

The images formed by a lens are described by the **thin-lens equation**,

$$\frac{1}{s_o} + \frac{1}{s_i} = \frac{1}{f} \qquad \text{(24.21) (page 821)}$$

where f is the focal length of the lens. For a converging lens, f is positive; for a diverging lens, it is negative. See Figure 24.42 for a description of the sign conventions for lenses.

Aberrations

Real lenses do not form perfect images. **Chromatic aberrations** are deviations due to the dependence of the refractive index on color. **Spherical aberrations** result from rays that do not propagate close to the principal axis of a lens or mirror with spherical surfaces.

QUESTIONS

SSM = answer in Student Companion & Problem-Solving Guide

X = life science application

1. An angler observes a fish whose image is 10 m from the shore. Is this fish closer to or farther from the shore than its image? Explain. *Hint*: Draw a ray diagram.

2. Where is the image formed for an object in front of a plane mirror?

3. Does a plane mirror produce a real image or a virtual image? Explain.

4. What is the focal length of a plane mirror? Explain.

5. Explain why reflection from a window consisting of a single pane of glass produces two images.

6. The author's satellite TV dish receives signals from a satellite in geosynchronous orbit about the Earth. (For this orbit, the period is precisely one day, so the satellite is always "at" the same point in the sky.) The signal is picked up with a receiver located 40 cm from the dish's center as shown Figure Q24.6. Explain why this "dish" is concave and estimate its radius of curvature.

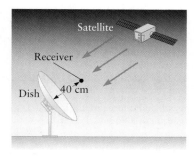

Figure Q24.6

7. When finding the image produced by a lens via ray tracing, it was suggested that you should always draw a ray that goes through the focal point (the focal ray). (In some cases, an extrapolation of the focal ray passes through the focal point.) Which of the following statements gives the most accurate description of this ray?
   (a) The focal ray passes through the center of the lens.
   (b) The focal ray always leaves the lens parallel to the axis of the lens.
   (c) The focal ray always leaves the object parallel to the axis of the lens.
   (d) The focal ray always passes through a virtual image.
   (e) There are two focal rays, one for each focal point.

8. In the ray diagram in Figure Q24.8, a combination of blue ($n = 1.60$) and red ($n = 1.50$) light strikes the surface of a flat glass plate. Because of dispersion, there are two different refracted rays inside the glass. Which of the following statements is correct?

   (a) Ray 1 is the blue ray, and ray 2 is the red ray.
   (b) Ray 1 is the red ray, and ray 2 is the blue ray.
   (c) Different rays reflect at different angles due to their color.
   (d) This diagram shows an example of total internal reflection.

   Figure Q24.8

   (e) The index of refraction in the glass is larger for red light.

9. Consider a diverging lens with an object in front of the lens. Prove that the image is always virtual.

10. Explain how a refraction experiment could be used to measure the index of refraction of a material as a function of the frequency of the light.

11. Imagine a light ray as it passes from one material to another (as in Fig. 24.11) and consider the electric and magnetic fields associated with these rays at the boundary between the materials. At (i.e., very near) the boundary, the component of the electric field directed parallel to the boundary must be the same on the two sides of the boundary. Use this fact to argue how the frequency of the light wave must change (or not change) when you cross the boundary.

12. Light travels from air into plastic. Which of the following statements is correct? The results from Question 11 will be useful here.
    (a) The frequency of the light is greater in air than in plastic.
    (b) The frequency of the light is smaller in air than in plastic.
    (c) The frequency of the light is the same in both air and plastic.
    (d) The wavelength of the light is greater in air than in plastic.
    (e) The wavelength of the light is smaller in air than in plastic.
    (f) The wavelength of the light is the same in both air and plastic.

13. **SSM** The image formed by a plane mirror appears to be a "reverse" of the object, with left and right interchanged, as illustrated in Figure Q24.13. It is actually more accurate to think of this reversal as front to back. (a) Use a ray diagram to explain why this front-to-back reversal occurs. (b) Use your ray diagram to explain why a top-to-bottom reversal does *not* also occur.

Figure Q24.13

14. When you look at yourself at the center of two plane mirrors forming a 90° corner, your image is not reversed from right to left. In other words, you see yourself like others see you. Using a ray diagram, explain this observation.

15. In Figure Q24.15, water is streaming out of a hole in a can, and it appears that light is also "pouring" out of the can. Explain what is actually happening.

Figure Q24.15

16. ⊗ Under water, the human eye cannot focus. Why?

17. In many convenience stores, a mirror is placed in a top corner of the store to allow the clerk to have a full view of the store. Is this mirror spherical concave or convex? Explain your answer.

18. Figure Q24.18 is a picture of a mirror one might find in a bathroom. The mirror is actually three mirrors with different curvatures so that you can view your image in different ways. The largest of the mirrors produces a normal image, whereas the smaller, two-sided mirror is used to magnify your image by two different amounts. Discuss the curvatures (plane, concave, convex) of the three mirrors required to produce these three images.

Figure Q24.18

19. When you observe a fish in a spherical fishbowl, the fish appears larger than it actually is. Why?

20. Funhouse mirrors distort your image so that you look shorter, taller, fatter, or thinner than you actually are. For the four images shown in Figure Q24.20, describe the curvature of the mirror.

Figure Q24.20

21. A magnifying glass produces an enlarged image that is upright and virtual as shown in Figure Q24.21. Does it use a converging or a diverging lens? How close must you hold the magnifying glass to the object to view the image in this way?

Figure Q24.21

22. [SSM] A plastic rod becomes invisible when inserted in a container of vegetable oil as shown in Figure Q24.22. What can you conclude about the plastic rod and the vegetable oil?

23. Explain how you could measure the index of refraction of a liquid.

24. Explain why dispersion is essential for producing a rainbow.

25. Explain how you could use Snell's law to measure the speed of light in a piece of plastic.

26. Explain how a person swimming under water could observe total internal reflection and how she would know when she saw it.

Figure Q24.22

PROBLEMS

24.1 RAY (GEOMETRICAL) OPTICS

1. [SSM] Light from a very tiny lightbulb passes through a hole of diameter 0.15 m in a dark screen that is 2.5 m away (Fig. P24.1). Assume the light travels as ideal rays as it passes from the bulb, through the hole, and onto a second screen on the far right in Figure P24.1. If the light rays illuminate a spot of diameter 0.65 m on the second screen, what is the spacing between the two screens?

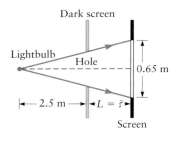

Figure P24.1

24.2 REFLECTION FROM A PLANE MIRROR: THE LAW OF REFLECTION

2. The light ray in Figure P24.2 emanates from a point on one wall of a room and is incident on a plane mirror that lies flat on the opposite wall. The reflected ray strikes the left-hand wall at point A, which is a distance $h = 1.4$ m below the light source. What is the angle of incidence θ_i?

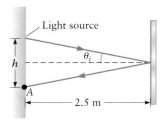

Figure P24.2

3. ★ Two plane mirrors meet at a right angle (Fig. P24.3). An incident ray strikes the bottom mirror at point B with an angle of incidence θ_i; this ray reflects from each mirror and strikes the screen on top at point A. (The screen is parallel to the bottom mirror.) Given the dimensions in Figure P24.3, what is θ_i?

Figure P24.3

4. A pencil is placed in front of a plane mirror and is oriented parallel to the plane of the mirror (Fig. P24.4). Construct a careful ray diagram and use it to find the image of the pencil. Be sure to draw several rays from at least two different points on the pencil (not just the tip).

5. What is the magnification of the pencil in Problem 4 and Figure P24.4? Is the image upright or inverted?

Figure P24.4
Problems 4 and 5.

6. ★ A light ray reflects from a mirror hanging vertically on a wall in your bathroom, and the ray strikes a particular point on the opposite wall (Fig. P24.6). The bottom end of the mirror is then tilted slightly away from the wall so that the mirror makes an angle of 5° with the wall. The spot where the reflected ray hits is found to move upward a distance of 0.25 m. What is the length L of the room?

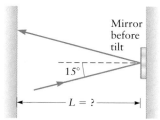

Figure P24.6

7. ✪ The two plane mirrors in Figure P24.7 come together as shown. Find two images of point *A* as viewed by an observer at *B*. Do your analysis graphically and assume the images are made by just a single reflection.

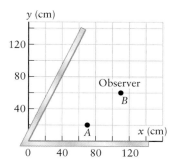

Figure P24.7

8. ✪ A marble is placed near a plane mirror as shown in Figure P24.8. (a) Where will the image be located? (b) How does the image location change as the observer moves closer to the mirror? (c) How does the image location change as the marble is moved away from the mirror?

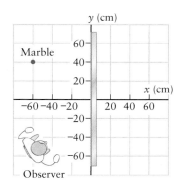

Figure P24.8

9. SSM ✪ Two plane mirrors are each *L* = 3.5 m long. They are parallel and are placed 0.25 m apart. If a light ray enters from one end as shown in Figure P24.9, how many reflections occur before it leaves the region between mirrors?

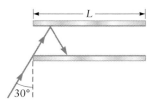

Figure P24.9

10. ✪ A plane mirror made of a very thin piece of glass lies flat on the ground. As shown in Figure P24.10, one end of the mirror is 2.0 m from you and the other end is 30 m from a nearby tree. You are 1.8 m tall, and the mirror has a length *L*. The mirror is arranged so that you can just see the image of the top of the tree at one edge of the mirror. (a) How tall is the tree? (b) If you can just see a point halfway up the tree at the other edge of the mirror, what is *L*?

Figure P24.10 Not to scale.

11. ✪ With an ideal corner cube, the reflected ray is precisely parallel to the incident ray. No corner cube will ever be perfect, however. The astronauts who visited the Moon left corner cubes so that scientists can do experiments in which light is reflected from the Moon and back to the Earth. Suppose one of these corner cubes has an angle of *θ* = 90.00001° (Fig. P24.11). By what distance will the reflected ray miss its target on the Earth?

Figure P24.11

12. ✪ In a typical changing room at a clothing store, you may find two mirrors on opposite walls, a setup that allows you to see how you look from the front and the back (Fig. P24.12). A consequence of this configuration is that you see a seemingly endless number of images of yourself getting farther and farther away. Assuming you're standing in the middle of the changing room and the mirrors are separated by a distance *d*, derive a general expression for the positions of the images you see.

Figure P24.12

24.3 REFRACTION

13. A light ray is incident from air onto the surface of a very quiet lake. If the angle of incidence is 35° and the angle of refraction is 25°, what is the index of refraction of the water?

14. The index of refraction of beer is about 1.35. What is the speed of light in beer?

15. Light is refracted through diamond from air. If the angle of incidence is 30° and the angle of refraction is 12°, what is the speed of light in diamond?

16. Suppose light with a wavelength of 560 nm in a vacuum is incident upon fused quartz, which has an index of refraction of 1.46. What will the wavelength of this light be inside the quartz?

17. A particular type of glass has an index of refraction of 1.45. What is the frequency of light with a wavelength of 450 nm inside this glass? Recall from Chapter 12 that the speed of a wave is equal to the product $f\lambda$.

18. Light from a helium–neon laser has a wavelength of 630 nm in a vacuum. What is its wavelength in water ($n = 1.33$)?

19. ⭐ Light is incident on a plane mirror with an angle of incidence of $\theta_i = 65°$. If this experiment takes place under water ($n = 1.33$), what is the angle of reflection?

20. What is the critical angle for light to stay inside a thin glass rod ($n = 1.40$) surrounded by air? What if the rod is surrounded by water?

21. A ray of light is incident upon a still water surface from above. What is the maximum possible angle of refraction?

22. A light beam containing several different colors is incident on a flat piece of glass with $\theta_i = 25°$. Which of the following statements is correct?

  (a) The angle of refraction will be greater for red light than for blue light.

  (b) The angle of refraction will be smaller for red light than for blue light.

  (c) The angle of refraction will be the same for red light as for blue light.

23. For Problem 22, if the angle of refraction for green light is 14°, what is the refractive index for this wavelength (i.e., color)?

24. ⭐ A thin film of water sits on top of a flat piece of glass, and light is incident from below (Fig. P24.24). If the angle of incidence is 30°, what is the angle of refraction of the final outgoing ray?

Figure P24.24

25. What is the critical angle for total internal reflection for light traveling between glass ($n = 1.65$) and water ($n = 1.33$)? Does the light start in the glass or the water?

26. ⭐ A combination of red light and blue light is incident on a flat piece of quartz (Fig. P24.26). The two colors refract at different angles. (a) Which refracted ray in Figure P24.26 is red, and which is blue? (b) What is the angle $\Delta\theta$ between the two rays? *Hint*: Use the data in Figure 24.20 and the discussion on page 806.

Figure P24.26

27. ✪ A light ray is incident on a plate of plastic ($n = 1.50$) that is $d = 2.5$ cm thick (Fig. P24.27). (a) Show that the outgoing ray is parallel to the incident ray. (b) If the angle of incidence is 40°, what is the displacement Δh of the outgoing ray relative to the incident ray?

28. ✪ Figure 24.14 shows a pencil that penetrates into a body of water. The pencil appears to be bent due to refraction. Explain this effect and calculate the angle that the pencil appears to be bent at the point it enters the water.

Figure P24.27

29. SSM ⭐ Monochromatic light (i.e., light of a single color) is incident on the symmetric triangular prism in Figure P24.29 as shown with $\theta = 55°$. If the index of refraction of the prism is $n = 1.60$, what angle does the outgoing ray make with the horizontal?

Figure P24.29
Problems 29 and 30.

30. ✪ Consider again the prism in Figure P24.29. White light is now incident on the prism. Calculate the angular separation of red light and blue light that is refracted by the prism. Assume the prism is composed of quartz and use the values for the indices of refraction in Figure 24.20.

31. ⭐ The prism shown in Figure P24.31 has an index of refraction $n = 1.75$ for a particular color of light. Four different incident rays are shown in the figure. Construct a ray diagram showing the outgoing rays in all four cases and also calculate the angles these outgoing rays make with a horizontal direction.

Figure P24.31

32. ⭐ The prism in Figure P24.32 is symmetric (i.e., forms an isosceles triangle). If the ray shown is perpendicular to the bottom surface of the prism and just barely undergoes total internal reflection at its first reflection (i.e., it strikes the interface at the critical angle) at the left face of the prism, what is the index of refraction of the prism?

Figure P24.32

33. ✪ A rectangular slab of glass has an index of refraction $n = 1.80$. Light is incident at 45°, striking the upper surface very close to the right edge as shown in Figure P24.33, and is refracted as it enters the glass forming ray 2. When ray 2 strikes the right side of the slab, will any light emerge from the glass?

Figure P24.33
Problems 33 and 34.

34. ✪ Consider again the rectangular slab of glass in Figure P24.33 with light incident on the top surface with an angle of incidence of 45°. The refracted ray (denoted ray 2) will undergo partial reflection at the right side of the slab and again at the bottom surface, then pass back through the top surface. What angle will the ray that exits the top surface make with the direction normal to that surface?

35. ⭐ When an observer looks into the empty swimming pool in Figure P24.35 and peers over the top edge, she finds that she can just see the far corner of the bottom of the pool. When the pool is filled with water, she can see a much bigger portion of the pool's bottom; that is, she can now see a point on the bottom a distance L from the far edge. Find L.

Figure P24.35

36. ⚙ Figure P24.36 shows light incident at an angle θ on a prism with index of refraction $n = 1.55$. It is found that light reflects from the upper right face of the prism and travels back along the path followed by the incident ray. Find θ.

Figure P24.36

24.4 REFLECTIONS AND IMAGES PRODUCED BY CURVED MIRRORS

37. An object 5.0 cm tall is 45 cm in front of a concave mirror with a radius of curvature of 15 cm. Construct a ray diagram to scale following the steps outlined in the procedure for "Ray Tracing Applied to Spherical Mirrors." (a) Indicate the parallel, focal, and central rays in your diagram. (b) Where is the image? (c) Is the image real or virtual? How big is it?

38. A cat is located 4.5 m in front of a concave mirror. The mirror has a radius of 1.5 m. Construct a ray diagram to scale following the steps outlined in the procedure for Ray Tracing Applied to Spherical Mirrors. (a) Where is the image of the cat? (b) Is this image real or virtual? (c) If the image of the cat is 5.0 cm tall, how tall is the cat?

39. If the cat in Problem 38 moves away from the mirror, does the cat's image get larger or smaller in magnitude?

40. Where should the cat in Problem 39 sit so as to appear as large as possible (i.e., to be as scary as possible) to a dog?

41. A concave mirror has a focal length of 10 cm. What is the magnification of an object placed 100 cm in front of this mirror?

42. A ball whose diameter is 0.25 cm is placed 50 cm in front of a convex spherical mirror. If the focal length of the mirror is −40 cm, what is the diameter of the ball as viewed in the mirror?

43. An object is placed 75 cm in front of a concave mirror. If the magnification is −0.75, what is the focal length of the mirror?

44. An object 22 mm tall is located 1.5 m in front of a convex mirror. If the image is 1.2 m behind the mirror, what is the height of the image?

45. An object 5.0 cm tall is 25 cm in front of a convex mirror with a radius of curvature of 25 cm. Construct a ray diagram to scale following the steps outlined in the procedure for "Ray Tracing Applied to Spherical Mirrors." (a) Indicate the parallel, focal, and central rays in your diagram. (b) Where is the image? (c) Is the image real or virtual? How big is it?

46. 🌟 A convex mirror has a focal length of −20 cm. Where is the image if the image is upright and one-fourth the size of the object?

47. A pencil is placed 45 cm in front of a convex mirror. Is the image in front of or behind the mirror?

48. 🌟 When an object is placed 12 cm in front of a mirror, an image is formed 25 cm behind the mirror. Is the mirror concave or convex? What is the mirror's focal length?

49. SSM 🌟 An object is 35 cm in front of a concave mirror and produces an upright image that is 2.5 times larger than the object. What is the focal length of the mirror?

50. A convex mirror has a radius of curvature of 85 cm. An image is produced 30 cm in front of the mirror. Where is the object?

51. 🌟 Suppose an object is somehow located a distance 35 cm *behind* a concave mirror with a radius of curvature of 1.50 m. Such a "virtual object" can be produced using the image of a second mirror. (a) Use the mirror equation to find the location of the image. (b) Is the image real or virtual? (c) What is magnification of the object?

52. A monster is 10 m in front of a convex mirror and appears with a magnification of 0.50. What is the radius of curvature of the mirror?

53. 🌟 Suppose the rearview mirror on a car is convex with a radius of curvature of 25 m. (a) If a motorcycle is 40 m behind the car, what is the magnification of the motorcycle? (b) How does your answer change when the motorcycle is 15 m behind the car?

54. An object 3.5 cm high is placed 20 cm in front of a convex mirror with a radius of 12 cm. (a) Make an accurate ray diagram showing the location and size of the image. Construct your ray diagram following the steps outlined in the procedures for "Ray Tracing Applied to Spherical Mirrors." (b) Check your results using the mirror equation.

55. An object is placed in front of a concave mirror with a focal length of 10 cm. What is the position of the resulting image if the image is upright and twice the size of the object?

56. ⚙ A meterstick lies along the optical axis of a convex mirror of focal length −40 cm, with its nearest end 60 cm from the mirror. How long is the image of the meterstick?

24.5 LENSES

57. An object 1.5 cm high is placed a distance of 10 cm in front of a converging lens with a focal length of 5.0 cm. (a) Follow the steps outlined in the procedures for "Ray Tracing Applied to Lenses" and make an accurate ray diagram to scale showing the location and size of the image. (b) Check your results using the thin-lens equation.

58. An object is a distance $2f$ from a converging lens. (a) Following the steps outlined in the procedures for "Ray Tracing Applied to Lenses," construct a ray diagram showing the location and orientation of the image. (b) Where is the image located? (c) Is it a real image or a virtual image?

59. Repeat Problem 58 for an object that is a distance $f/3$ in front of the lens.

60. An object is 45 cm to the left of a lens, and the image is 25 cm to the left of the lens. What is the magnification?

61. An object is placed in front of a lens. The image is upright, 35 cm to the left of the lens, and is half as tall as the object. What is the focal length of the lens?

62. An object is a distance $2|f|$ in front of a diverging lens. (a) Following the steps outlined in the procedures for "Ray Tracing Applied to Lenses," construct a ray diagram showing the location and orientation of the image. (b) Where is the image located? (c) Is it a real image or a virtual image?

63. An object 5.0 cm tall is 25 cm in front (on the left) of a converging lens with a focal length of 40 cm. Where is the image?

64. 🌟 A fly appears to be upside down and twice its actual size when viewed through a converging lens with a focal length of 5.5 cm. How far is the fly from the lens?

65. 🌟 If the fly in Problem 64 moves farther from the lens, will the image become larger or smaller?

66. ✪ Two converging lenses with $f_1 = 25$ cm and $f_2 = 35$ cm are located 1.50 m apart as shown in Figure P24.66. An object is placed 50 cm in front of the lens on the left. (a) Construct a ray diagram following the steps outlined in the procedures for "Ray Tracing Applied to Lenses." Find the location and size of the image produced by the first lens. (b) Construct a ray diagram for the second lens, using the image from part (a) as the object. Where is the image produced by the second lens? How big is it?

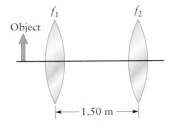

Figure P24.66

67. ✪ A converging glass lens ($n = 1.50$) is immersed in water. Does the focal length get longer or shorter and by what factor? Explain why. *Hint*: In the lens maker's formula in Equation 24.32, the term $n - 1$ is actually the difference between the index of refraction of the lens and the surrounding air.

68. A converging lens with a focal length of 25 cm is used to form an image on a screen that is 3.0 m from the lens. Where must the object be placed?

69. A converging lens has a focal length of 10 cm. How far from the lens should an object be placed to get (a) a real image with a magnification of 2.5 and (b) a virtual image with a magnification of 2.5?

70. SSM ⭐ For a converging lens of focal length f, determine the minimum object-to-image distance so that the image is (a) real and (b) virtual.

24.6 HOW THE EYE WORKS

71. Ⓧ A young child is just able to read a newspaper that is only 10 cm from her eyes. What is the focal length of the child's eye? Assume her retina is 1.5 cm in diameter.

72. ⭐ Ⓧ Your eye (diameter 2.5 cm) is able to focus the image of a tree on your retina. The tree is 25 m tall and 75 m away. What is the size of the image of the tree on your retina?

73. ⭐ Ⓧ A cell phone 7.0 cm tall is 40 cm in front of your eye (diameter 2.5 cm). (a) Find the size of the image of the cell phone on your retina. (b) The cell phone is now moved to a spot 60 cm from your eye. Is the image on the retina larger or smaller than the image in part (a)? By what factor?

74. ✪ Ⓧ Ⓡ **Horse's eye.** Have you ever noticed that as you approach a horse, it appears to raise its head in a way that seems aggressive? In fact, the horse is not usually showing aggression, but rather is raising its head to see you. The lens in the eye of a horse does not have the ability to change shape as in the human eye. Instead, the back of the eye (the retina) is curved in an elliptical shape so that the straight-line distance from the front of the eye to the retinal surface depends on where the light is hitting the retinal surface. The result is that when a horse is trying to view objects up close it must lift its head in such a way that it uses the upper portion of its retina. More distant objects are

viewed using the central portion of the retina in much the same way as a person who is presbyopic uses bifocal lenses to view both objects near and far. (a) Given the information provided in Figure P24.74, determine the near-point distance for the horse. The *near-point distance* is the shortest object distance for which the eye is able to focus. (b) For a typical horse, estimate the distance from its eye to the tip of its nose. Using this estimate, does your answer to part (a) seem reasonable? Explain.

24.7 OPTICS IN THE ATMOSPHERE

75. SSM A rainbow is formed by refraction of light as it passes through a water droplet. In the ray diagram in Figure 24.48B, which of the following statements best explains the origin of a rainbow? Explain your answers.
 (a) The water droplet must be exactly spherical for a rainbow to occur.
 (b) Different rays refract at different angles due to their color.
 (c) Different rays reflect at different angles due to their color.
 (d) The water droplet acts like a lens and focuses the light.
 (e) Light from the water droplet is polarized.

76. ✪ Figure 24.48B shows a ray diagram for light from the Sun as it passes through a water droplet and produces a rainbow. Construct a careful ray diagram of this type for light rays of different colors (e.g., red and blue) and use it to explain the order (sequence) of colors in a rainbow.

77. ✪ On rare occasions, it is possible to observe a "double" rainbow (Fig. P24.77A). (The secondary rainbow is always present, but is usually too faint to be observed.) The ray diagram in Figure P24.77B shows how the light rays that form the secondary rainbow are refracted and reflected by a water droplet. Use this ray diagram to explain why the order of colors in the secondary rainbow is opposite to the order of colors in the primary rainbow.

Figure P24.77

78. ✪ Ⓡ **Stellar illusion.** As we discussed in the chapter, the position of a star in the sky as viewed from the surface of the Earth is not where it actually is because the light from the star refracts as it enters the Earth's atmosphere. Approximating the Earth's atmosphere to be homogenous with a thickness of 100 km and having an index of refraction of 1.0003, determine the angular separation between the observed position of a star just barely above the horizon and the actual position of the star.

24.8 ABERRATIONS

79. ✪ A converging lens has a focal length of $f = 55$ cm for blue light ($n = 1.65$). The index of refraction of this lens for red light is $n = 1.55$. What is the focal length for red light?

80. ✪ Consider the lens in Problem 79. An object is placed 95 cm in front of the lens and is illuminated with white light. Chromatic aberration causes the image formed with blue light to be at a different position than the red image. How far apart are the two images?

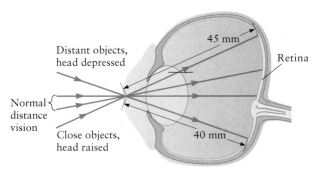

Figure P24.74

81. Lens maker's formula for a double concave lens. Equation 24.32 gives the focal length of a double convex lens (i.e., a lens for which both surfaces are convex) in terms of the radii of curvature of the two surfaces. For a double concave lens (Fig. P24.81), the lens maker's formula has a slightly different form:

$$\frac{1}{f} = (n-1)\left(-\frac{1}{|R_1|} - \frac{1}{|R_2|}\right)$$

This relation always gives a negative focal length f as expected for a double concave lens (because it is a diverging lens). Consider such a lens composed of quartz with $R_1 = R_2 = 50$ cm and $n = 1.56$ for blue light. What is its focal length for blue light?

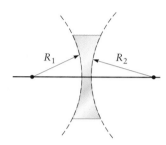

Figure P24.81

82. ✪ Compare the forms of the lens maker's formulas in Equation 24.32 and Problem 81 and devise a formula for a lens that has one concave surface and one convex surface.

83. ✪ In Chapter 23, we discussed the polarization of light by reflection. Figure P24.83 shows light that is both reflected and refracted as it passes from air into glass with $n = 1.60$. The incident light is unpolarized, containing components with the electric field perpendicular to the plane of the drawing and parallel to this plane. When the reflected and refracted rays are precisely perpendicular, the reflected ray will be purely linearly polarized, with $\vec{E}$ perpendicular to the plane of the drawing as indicated. Find the angle of the incident ray for which the refracted and reflected rays are perpendicular. It is called *Brewster's angle*.

Figure P24.83

84. SSM ✪ Ⓡ In Problem 83, we saw how reflection can lead to polarization of reflected light. Is this effect important for reflection of sunlight from a car's windshield? Investigate this question by working through the following steps. (a) Calculate Brewster's angle for glass. (b) Draw a ray diagram of the problem. Show the Sun, windshield, and driver along with the incident and refracted rays. The angle of incidence should equal Brewster's angle found in part (a).

85. Brewster's angle (see Problem 83) for a particular prism is found to be 65°. What is the index of refraction of the prism?

86. ⭐ A slide projector uses a converging lens to project images on a screen 4.0 m away. If the distance between the slide (image) and the lens is 5.0 cm, what are (a) the focal length of the lens and (b) the magnification of the image?

87. SSM ⭐ Prior to the development of the digital camera, the 35-mm film camera was the norm. The designation "35 mm" refers to the size of the film (screen) on which the image is formed. Suppose you wish to use one of these cameras to photograph a 10-m-tall tree that is 30 m away and have the tree's image fill the entire 35-mm film. What must the focal length of the camera lens be?

88. ✪ Suppose a ray of light is incident on three stacked layers of glass with different indices of refraction as shown in Figure P24.88. If the ray enters from air, what is the angle θ_3?

Figure P24.88

89. ✪ A ray of light is incident on the surface of a block of clear ice ($n = 1.309$) at an angle of 30° to the normal (Fig. P24.89). A portion of the light is reflected from the surface, and a portion is refracted into the ice. Determine the angle between the reflected and refracted light.

Figure P24.89

90. ✪ A laser beam strikes in the center of one end of a long, thin glass plate that has an index of refraction of 1.5 as shown in Figure P24.90. Determine the number of internal reflections the beam undergoes before it exits the plate at the other end.

Figure P24.90 Not to scale.

91. ✪ **Refracting surfaces.** A more general analysis of refracting surfaces would include the case of refraction of light at a spherical interface such as a fishbowl. Similar to the thin-lens equation, the expression relating the object distance, image distance, indices of refraction, and the radius of curvature of the surface is given by

$$\frac{n_1}{s_o} + \frac{n_2}{s_i} = \frac{n_2 - n_1}{R}$$

where n_1 and n_2 are indices of refraction where the object and image are located, respectively. For the configuration shown in Figure P24.91, the radius of curvature of the bowl is $R = -15$ cm and

Figure P24.91

the fish is 10 cm from the curved surface and 8.0 cm down from the top. We can neglect the glass bowl and only consider the refraction due to the water. Determine the image distance of the fish as viewed from (a) the side and (b) the top. *Hint:* You'll need to determine the radius of curvature of a flat surface to answer part (b).

92. ✴ A rock is dropped in a large tank of benzene and sinks to the bottom of the tank. If the apparent depth of the rock as viewed directly from above is 35 cm, what is the actual depth of the rock? *Hint:* See Problem 91.

93. ✴ A swimming pool constructed below ground level is 3.0 m square and 2.0 m deep (Fig. P24.93). A drain is at the bottom center of the pool. How far from the edge of the pool can a person whose eyes are 1.2 m above the ground stand and still see the drain?

Figure P24.93   **Figure P24.94**

94. ✴ Suppose a beam of white light is incident at a 45° angle upon a glass plate 10 cm thick. The index of refraction of the glass for blue light is 1.65, and for red light it is 1.60 (Fig. P24.94). Determine the distance between the blue light and red light as they exit the glass plate.

95. ✪ A thin, converging lens whose focal length is 80 cm is placed 5.0 m from a convex spherical mirror whose focal length is −50 cm (Fig. P24.95). For an object placed 1.0 m to the left of the lens, determine (a) the location of the image, (b) the magnification of the image, and (c) the nature of the image (real/virtual, upright/inverted, enlarged/reduced). *Hint:* Use the image formed by the lens as the object for the mirror.

Figure P24.95

96. ✪ A material with an index of refraction of 1.80 is machined into the shape of a quarter of a cylinder. The radius of the cylinder is 10 cm; it rests on a flat, horizontal surface; and it is surrounded by a vacuum. A beam of light is incident upon the cyl-inder as shown in Figure P24.96. Determine the distance *d* from the quarter-cylinder at which the exiting beam strikes the horizontal surface.

Figure P24.96

97. ✪ ✴ A birdwatcher observes a hummingbird at a distance of 10 m through a simple eyepiece (converging lens) of focal

length 25 cm (Fig. P24.97). As the hummingbird flies away at a speed of 2.5 m/s, how fast does the image move? Does the image move toward or away from the lens?

Figure P24.97

98. ✪ **Renaissance technology.** The Hockney–Falco theory claims that early Renaissance painters used concave mirrors to aid their depth perception (e.g., Jan van Eyck's *Portrait of Giovanni Arnolfini and his Wife*; Fig. P24.98A). Reportedly, van Eyck traced the reflected image from his subjects accord-ing to the schematic shown in Figure P24.98B. Calculate the diameter of the blown-glass sphere from which the reflec-tive mirror was cut and comment on the likelihood of such a theory.

Figure P24.98

99. ✪ An overhead projector produces an image of an object (transparency) on a distant screen by using a light source, a plane mirror, and a converging lens (*f* = 50 cm). The distance between the lens and the mirror is 15 cm. (a) If you wish to project a focused image on a screen 3.0 m away from the lens, what distance above the transparency must you set the center of the mirror–lens combination? (b) Describe how you must place the transparency on the projector so that the image is oriented properly.

100. ✪ Sometimes on hot summer days the road appears to be wet as shown in Figure P24.100. The black surface becomes

Figure P24.100

very hot, heating up the air just above the road. This increase in temperature decreases the density of the air and in turn decreases its index of refraction. The result is that light from a distant object can be totally internally reflected, producing a mirage. (a) If your eye level while driving your car is 2.0 m above the surface of the road and you observe a mirage 100 m ahead, what is the index of refraction of air just above the ground? (b) Describe how you might use a mirage to estimate the temperature of the air just above the ground.

101. ✪ **Camera obscura.** Prior to the development of the photographic camera, artists would use a camera obscura like the one shown in Figure P24.101 to draw the image of a subject. In this figure, light from a subject passes through a small hole or lens in the box, is reflected from a plane mirror, and is projected onto a translucent sheet of paper. The artist would then trace the image on the paper. Room-sized camera obscuras were also used. Suppose the distance from the pinhole to the

Figure P24.101

mirror is 1.0 m and the mirror-to-screen distance is 50 cm. (a) How far away should a 1.5-m-tall subject stand so that she fills a 25-cm-square screen? (b) Describe how adjusting the size of the pinhole will affect the clarity and brightness of projected image.

102. ✪ **Pinhole camera.** The earliest cameras were simply light-tight boxes with a very small hole in one side of the box. By uncovering the pinhole for a brief period of time, light from an object could pass through the hole, exposing a photographic plate (film) on the inside of the box. Later versions of this type of camera replaced the hole with a lens. For a pinhole camera, the image appears upside down on the photographic plate in relation to the object. (a) Draw a ray diagram to explain how a pinhole camera is able to form a clear image without the use of a lens. (b) If the distance between the photographic plate and the pinhole is 30 cm, what is the size of the image for an object 10 m away and 2.0 m tall?

103. ✪ Ⓡ **Solar eclipse.** Suppose you use the pinhole camera described in Problem 102 to photograph the Sun during an eclipse. Determine the size of the image of the Sun. Repeat the calculation for the Moon.

104. ✪ **Underwater stargazers.** Some creative amateur astronomers have been known to improve their view of the night sky by wearing scuba gear and observing the sky from the bottom of a swimming pool. Looking up from the bottom of a swimming pool, you are able to see a wider angle of the sky with a pool full of water than when the pool is empty. For a swimming pool 10 m wide and 5.0 m deep, determine the ratio of the angular field of view one sees when the pool is filled with water to that when the pool is empty.

Chapter 25

Wave Optics

This photograph shows two water waves undergoing **interference** with each other, producing regions of high and low wave intensity. Interference is the central subject of this chapter. (© Illustration-Nature/Alamy)

Is light a particle, or is it a wave? This question was debated for many centuries. Newton believed that light is a particle; he reasoned that since light travels in straight-line paths (as we found in our ray diagrams in Chapter 24), it must consist of particles moving with constant velocity. Newton (and others) developed a particle theory of light that was able to explain reflection, refraction, and all the other ray properties of geometrical optics discussed in Chapter 24. Over time, however, experiments revealed several properties of light that can only be explained with a wave theory. Maxwell's theory of electromagnetism then convinced physicists that light is an electromagnetic wave. In this chapter, we discuss properties of light that depend on its wave nature, the field known as ***wave optics***. We also show

how geometrical optics—ray tracing and the focusing of light by mirrors and lenses—fits together with the wave properties of light.

The wavelength of light λ plays a key role in determining when we can or cannot use geometrical optics. When discussing image formation and the properties of mirrors and lenses over distances much larger than λ, geometrical optics is extremely accurate, but for all discussions of interference effects and the propagation of light through small openings (openings whose sizes are comparable to or smaller than λ), wave optics is required. Hence, in their appropriate regimes, geometrical optics and wave optics both describe light.

However, there is still more to this story: new properties of light were discovered only a few decades after Maxwell. These properties could not be understood with Maxwell's theory and led to the development of the quantum theory of light in which light has properties of both particles *and* waves! We'll say a little more about this quantum theory at the end of this chapter and a lot more about it in Chapter 28. The quantum theory of light builds on Maxwell's theory, so everything we learn in this chapter about the wave properties of light is true also in the quantum theory.

25.1 | COHERENCE AND CONDITIONS FOR INTERFERENCE

How is a wave different from a particle? How can we tell one from the other? In some cases (such as water waves), the repeating oscillations of wave motion are obvious, but in other cases (such as light and sound), the wave nature of a phenomenon is not as apparent. How can we really prove that light and sound are wave phenomena? One property is unique to waves: only waves can exhibit *interference*. We described the interference of waves on a string and water waves in Chapter 12 and of sound in Chapter 13. In this chapter, we study the interference of light waves.

Let's first consider when light from two or more sources can and cannot exhibit interference. In particular, can any two light waves exhibit interference effects? If not, how can we tell in advance if light from two or more sources is capable of exhibiting interference? For example, can light from the Sun exhibit interference with light from a car's headlight?

We can answer these questions if we first consider the interference of sound waves emitted by two speakers as sketched in Figure 25.1A. The speakers emit sound waves of wavelength λ that come together and combine at a listener's location. If the sources emit identical waves and if the distances from the sources to the listener are equal ($L_1 = L_2$), the two waves undergo the same number of oscillations on their way to the listener. In this case, we say that the waves are *in phase* (Fig. 25.1B). When two waves are in phase, their maxima occur at the same time at a given point in space. The total wave displacement at the listener's location is the sum of the displacements of the two individual waves. If two waves are in phase, the sum of their displacements is large and the waves interfere constructively to produce a large amplitude and a large intensity.

Now suppose the listener moves relative to the speakers so that the distances L_1 and L_2 traveled by the two waves are no longer equal. If these distances differ by an amount $\lambda/2$, one of the waves travels an extra one-half wavelength on the way

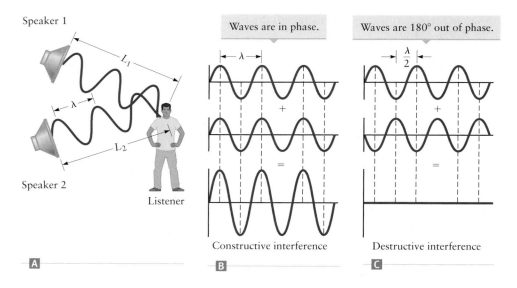

Figure 25.1 **A** Sound waves emitted by two separate sources (two speakers) combine when they reach a listener. These waves travel distances L_1 and L_2 on their way to the listener, where L_1 and L_2 may or may not be equal. **B** Here the two waves arrive in phase, and their sum is larger than either of the individual waves. **C** If the two waves are 180° out of phase when they reach the listener, their sum is zero, and no sound (and no energy) arrives at the listener.

Speaker 1

L_1

λ

L_2

Speaker 2

Listener

Waves are in phase.

λ

+

=

Constructive interference

Waves are 180° out of phase.

$\frac{\lambda}{2}$

+

=

Destructive interference

to the listener and the maximum of one wave will coincide with the minimum of the other (Fig. 25.1C). We say that these waves are[1] *out of phase*; in this example, the sum of the two wave displacements at the listener's location is zero and there is *destructive* interference.

Figure 25.1 illustrates several crucial aspects common to all interference experiments, including those involving light.

> *General conditions for interference.* Two waves can interfere if all the following conditions are met.
>
> 1. **Two or more interfering waves travel through different regions of space over at least part of their propagation from source to destination.**
> 2. **The waves are brought together at a common point.**
> 3. **The waves must have the same frequency and must also have a *fixed phase relationship*. This means that over a given distance or time interval the phase difference between the waves remains constant. Such waves are called *coherent*.**

Coherence

Interference conditions 1 and 2 are illustrated in Figure 25.1, but condition 3 and the notion of coherence warrants more discussion. When the ear detects a sound wave or the eye detects a light wave, it is actually measuring the wave intensity averaged over many cycles. For example, the frequency of blue light is about 7×10^{14} Hz. The eye cannot follow the variation of every cycle; instead, the eye averages the light intensity over time intervals of about 0.1 s, which corresponds to approximately 7×10^{13} cycles, and then sends this average signal to the brain. Suppose the waves in Figure 25.1 are two interfering light waves. In order for the waves to interfere constructively and give a large total intensity, they must stay in phase during the time the eye is averaging the intensity. Likewise, for the two waves in Figure 25.1C to interfere destructively, they must stay out of phase during this averaging time. Each scenario is possible only if the two waves have precisely the same frequency. If their frequencies differ, two waves that are initially in phase with each other (and interfere constructively) will be out of phase at a later time (and interfere destructively).

---

[1]Mathematically, we would say that the waves in Figure 25.1C are 180° out of phase, corresponding to a shift of one wave relative to another by $\lambda/2$. Figure 25.1 considers waves that are either perfectly in phase or completely (180°) out of phase, but other relationships (specified by other phase angles) are also possible.

Waves are in phase at start.

Two waves with slightly different frequencies

Waves are out of phase now.

Figure 25.2 Two waves with slightly different frequencies. Initially, they are in phase, but as time passes they become out of phase. The in-phase and out-of-phase portions average out over many cycles, so on average there is no interference. *Important note*: The different colors here do *not* indicate different colors of light, but rather denote the amplitudes of the different waves.

Figure 25.2 shows two waves with slightly different frequencies. The waves are initially in phase, as their maxima and minima occur at about the same times. After many cycles, however, one of the waves "gets ahead" of the other; eventually, the wave with the higher frequency has completed an extra one-half cycle, and the two waves are out of phase. The two waves thus go in and out of phase as time passes. When averaged over a very large number of cycles, the two waves display neither constructive nor destructive interference; on average, there is no interference at all. So, to exhibit interference, two light waves must have exactly the same frequency.

Most visible light is emitted by atoms; you might expect that light waves produced by two atoms of the same element would be "identical" and have the same frequency and could therefore exhibit interference. To exhibit interference, however, two waves must satisfy an additional condition. Consider atoms in a gas that are emitting light. These atoms undergo collisions with other atoms in the gas and with the walls of the container, and each collision causes the phase of the emitted light to jump abruptly to a new value. In other words, each collision "resets" the light wave, starting its oscillation over again. These collisions are different for each atom and spoil interference effects because they cause waves that may have originally been in phase to suddenly become out of phase and vice versa. Experiments show that phase jumps in typical light waves occur about every 10^{-8} s; hence, light from two different atoms of a given element are not coherent. (Certain special cases are an exception, such as when the atoms are all in the same laser. See Chapter 29.) One way to eliminate the effect of phase jumps is to derive both waves from the *same* source. In Section 25.2, we'll explain how to do that.

CONCEPT CHECK 25.1 | Conditions for Interference

Consider an (attempted) experiment to observe interference of light from two different helium–neon lasers. The light from these lasers is emitted by neon atoms. Is it possible to observe interference with these two light sources? Explain why or why not.

EXAMPLE 25.1 | Coherence Time for a Light Wave

The phase of light emitted by an atom changes (jumps) after a typical time interval $\Delta t \approx 10^{-8}$ s, called the *coherence time* of the light. How many cycles of a light wave occur during this coherence time? Assume the light is blue, with $f = 7.0 \times 10^{14}$ Hz.

RECOGNIZE THE PRINCIPLE

A wave's frequency is inversely related to its period T, with $f = 1/T$, and the period is the time it takes to undergo one cycle of oscillation. Given the frequency, we can find the period and from that calculate how many cycles are completed in the coherence time Δt.

SKETCH THE PROBLEM

No sketch is necessary.

Recall some basic facts about oscillations and waves from Chapters 11 and 12. The period of this light wave is (using Eq. 11.1)

$$T = \frac{1}{f} = \frac{1}{7.0 \times 10^{14} \text{ Hz}} = 1.4 \times 10^{-15} \text{ s}$$

which is the time required to complete one cycle of the wave. The number of cycles N contained in the coherence time Δt is thus

$$N = \frac{\Delta t}{T}$$

SOLVE

Inserting our values of Δt and T gives

$$N = \frac{\Delta t}{T} = \frac{1 \times 10^{-8} \text{ s}}{1.4 \times 10^{-15} \text{ s}} = \boxed{7 \times 10^6 \text{ cycles}}$$

What does it mean?

Although the coherence time Δt is very short, the frequency of visible light is very high, so there are more than one million oscillations—quite a large number—during this coherence time. The precise value of Δt depends on the type of atom and its environment; the value 10^{-8} s is only a typical estimate.

MICHELSON INTERFEROMETER

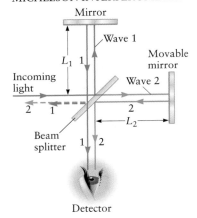

Figure 25.3 Schematic diagram of a Michelson interferometer. Light is incident from the left and split into two separate waves (also called beams) by partial reflection at the beam splitter in the center. These waves reflect from the mirrors at the top and right and arrive back at the beam splitter, where they recombine and travel together to the detector at the bottom. (Waves 1 and 2 also recombine and travel back to the left, but these beams are not used in the experiment.) The colors of the rays do *not* indicate the color of the light.

25.2 | THE MICHELSON INTERFEROMETER

An optical instrument called a *Michelson interferometer* is based on interference of reflected waves (Fig. 25.3). This clever device, invented by Albert Michelson in the late 1800s, played an important role in the discovery of relativity (Chapter 27). The device contains two reflecting mirrors mounted at right angles. At least one of these mirrors is movable; here the mirror on the far right can be moved along the horizontal axis. A third partially reflecting mirror called a "beam splitter" is mounted at a 45° angle relative to the other two. The beam splitter reflects half the light incident on it and lets the other half pass through.

Light incident from the left in Figure 25.3 strikes the beam splitter and is divided into two waves. One of these waves (denoted 1) travels to the mirror at the top where it is reflected and then returns to the beam splitter. When it reaches the beam splitter, it is again split into two waves, and one of these waves travels downward in Figure 25.3 to the detector as wave 1. The other wave from the beam splitter (wave 2) travels to the mirror on the right, where it is also reflected back to the beam splitter. When wave 2 returns to the beam splitter, a portion is reflected to the detector at the bottom of the figure. Waves 1 and 2 thus combine at the detector; they are the two waves that interfere, and they follow very similar paths as they travel through the interferometer. First, both waves are reflected once by the beam splitter. Second, both waves pass through the beam splitter once. Third, both waves are reflected once by a mirror (at the far right or top). The *only* difference between the two waves is that they travel different distances between their respective mirrors and the beam splitter. After the waves are created by the beam splitter, wave 1 travels a round-trip distance $2L_1$ and wave 2 travels a round-trip distance $2L_2$ before they recombine. The implications for interference are shown in Figure 25.4. The *path length difference* is

$$\Delta L = 2L_2 - 2L_1 \tag{25.1}$$

If we assume $L_2 > L_1$, wave 2 travels an extra distance ΔL and undergoes a number of extra oscillations before it reaches the detector. Each oscillation occurs over a distance equal to the wavelength λ, so the number of extra oscillations made by wave 2 is

$$N = \frac{\Delta L}{\lambda} \qquad (25.2)$$

If N is an integer, then N complete oscillations fit into the extra path length and the two waves are in phase when they recombine, producing constructive interference. On the other hand, if N is half integral (i.e., if $N = \frac{1}{2}, \frac{3}{2}, \ldots, m + \frac{1}{2}$, where m is an integer), wave 2 travels an "extra" one-half wavelength relative to wave 1 and the maxima of one wave will coincide with the minima of the other as in Figure 25.1C. That is the condition for destructive interference. The interference conditions for a Michelson interferometer are thus

$$\Delta L = m\lambda \qquad \text{(constructive interference)}$$

$$\Delta L = (m + \tfrac{1}{2})\lambda \qquad \text{(destructive interference)} \qquad (25.3)$$

where m is an integer. If the interference is constructive, the light intensity at the detector is large, whereas if the interference is destructive, the detector intensity is zero. Figure 25.5 shows how the light intensity at the detector varies as a function of ΔL. The places on the intensity curve where the interference intensity is greatest are called bright "fringes," and the places where the intensity is zero are called dark "fringes."[2]

The Michelson interferometer uses reflection to satisfy the general requirements for interference from Section 25.1. First, reflection at the central beam splitter produces two separate waves that travel through different regions of space. Second, these waves are brought back together by the beam splitter so that they recombine at the detector. Third, because they are produced from a common incident wave, the two interfering waves are coherent.

Using a Michelson Interferometer to Measure Length

The wavelengths of light emitted by various sources are known. For example, a helium–neon (He–Ne) laser emits light with a wavelength of approximately $\lambda_{He-Ne} = 633$ nm. Recall that 1 nm $= 1 \times 10^{-9}$ m and that the wavelength of visible light is in the range of about 400 nm to 750 nm. As a rough comparison, the wavelength of visible light is about one thousand times less than the thickness of a human fingernail.

Knowing the value of λ_{He-Ne}, an experimenter can use a Michelson interferometer to measure a displacement in the following way. Using light from the laser, the mirror on the far right in Figure 25.3 is adjusted to give constructive interference and its position is noted; this corresponds to one of the bright fringes in Figure 25.5. The mirror on the far right is then moved horizontally, changing the path length difference ΔL, and the intensity at the detector moves along the curve in Figure 25.5. The intensity changes from a high value to zero and back to a high value every time the path length difference ΔL changes by one wavelength of the light. According to Equation 25.3, moving the mirror from one bright fringe to the next bright fringe corresponds to starting at the condition for constructive interference with $\Delta L = m\lambda_{He-Ne}$ and increasing the path length difference to $\Delta L = (m + 1)\lambda_{He-Ne}$. If the interferometer mirror is moved so as to pass through N cycles from bright to dark and back to bright, the change in the path length difference is

$$(\Delta L)_{change} = N\lambda_{He-Ne} \qquad (25.4)$$

When the mirror moves a distance d, the distance traveled by the light changes by $2d$ because the light travels back and forth between the beam splitter and the mirror

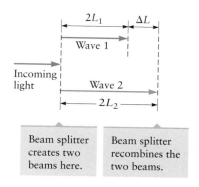

Figure 25.4 A Michelson interferometer with the waves redrawn. The two waves travel distances $2L_1$ and $2L_2$, respectively, as they travel between the beam splitter and the mirrors. The path length difference ΔL leads to interference; compare with Figure 25.1A.

Figure 25.5 The light intensity at the detector of a Michelson interferometer oscillates as a function of the path length difference $\Delta L = 2L_2 - 2L_1$, which is varied by moving one or both mirrors.

---

[2]We'll see why they are called fringes in Section 25.3, when we consider some examples of how interfering light waves appear when viewed by the eye.

(Fig. 25.3). Hence, if the mirror moves through N bright fringes, the distance d traveled by the mirror satisfies

$$(\Delta L)_{\text{change}} = 2d = N\lambda_{\text{He–Ne}}$$

so that

$$d = \frac{N}{2}\lambda_{\text{He–Ne}} \qquad (25.5)$$

The accuracy of this measurement of d depends on the accuracy with which the wavelength $\lambda_{\text{He–Ne}}$ is known. For this reason, physicists have devoted a lot of effort to measuring the wavelength of light from certain light sources. In fact, the wavelength produced by a specially constructed helium–neon laser is known to be

$$\lambda_{\text{He–Ne}} = 632.99139822 \text{ nm} \qquad (25.6)$$

Suppose this light source is used in a Michelson interferometer and one of the mirrors is moved a distance d such that exactly $N = 1{,}000{,}000$ bright fringes are counted. According to Equation 25.5, we have $d = N\lambda_{\text{He–Ne}}/2$. Since N can be counted directly (and is thus accurately known) and the wavelength in Equation 25.6 is also known with high accuracy, this approach provides a very precise way to measure length. Moreover, people in different laboratories can use helium and neon to construct lasers that produce the wavelength given in Equation 25.6, so they can compare their separate length measurements with high accuracy. The Michelson interferometer thus makes possible a very convenient standard for the measurement of length. A similar type of interference effect is used to read information from CDs and DVDs.

CONCEPT CHECK 25.2 | Analyzing the Michelson Interferometer

A Michelson interferometer using light from a helium–neon laser ($\lambda_{\text{He–Ne}} = 633$ nm) is adjusted to give an intensity maximum (a bright fringe) at the detector. One of the mirrors is then moved a very short distance so that the intensity changes to zero. How far was the mirror moved, (a) $\lambda_{\text{He–Ne}}/4$, (b) $\lambda_{\text{He–Ne}}/2$, (c) $\lambda_{\text{He–Ne}}$, or (d) $2\lambda_{\text{He–Ne}}$?

EXAMPLE 25.2 | Applications of a Michelson Interferometer: Detecting Gravitational Waves

An experiment called the LIGO (for *l*aser *i*nterferometer *g*ravitational wave *o*bservatory) is designed to detect very small vibrations associated with gravitational waves that arrive at the Earth from distant galaxies. The LIGO experiment involves several Michelson interferometers; in one of these interferometers, the mirrors are placed a distance $L_2 = 4$ km (kilometers) from the beam splitter. Suppose changes in the interference pattern corresponding to 1% ($\frac{1}{100}$) of a cycle in Figure 25.5 can be detected. What is the change in the mirror–beam splitter distance that can be detected by the LIGO experiment? Assume the light used in LIGO has a wavelength $\lambda = 500$ nm.

RECOGNIZE THE PRINCIPLE

For a general interference experiment, the path length difference ΔL must change by one wavelength λ to go from a bright fringe to a dark fringe and then back to a bright fringe. A 1% change of a fringe thus corresponds to a $\lambda/100$ change in ΔL. When a mirror of a Michelson interferometer moves a distance d, the path length difference changes by $2d$. We can first find the value of d that gives a 1% fringe change and from that get the change of the mirror–beam splitter distance.

SKETCH THE PROBLEM

Figures 25.3 through 25.5 show a Michelson interferometer and the interference conditions.

IDENTIFY THE RELATIONSHIPS

The change in the path length difference equals $2d$, so each complete intensity cycle from bright to dark to bright in Figure 25.5 corresponds to moving the mirror a distance $\lambda/2$. If we can detect $\frac{1}{100}$ of this change, the corresponding displacement of the mirror is

$$d = \frac{\lambda/2}{100} = \frac{\lambda}{200} \tag{1}$$

SOLVE

Inserting the given value of the wavelength into Equation (1), we get

$$d = \frac{\lambda}{200} = \frac{(500 \times 10^{-9} \text{ m})}{200} = \boxed{2.5 \times 10^{-9} \text{ m}}$$

What does it mean?

The value of d found here is 2.5 nm—approximately 10 times the diameter of an atom—which is quite impressive. This result also shows that when adjusting the mirrors of a Michelson interferometer to actually observe interference, an experimenter must be able to move the mirrors with a precision of a few nanometers. Such precision can be accomplished with carefully designed screw adjustments (a low-tech approach, but there are other ways, too).

25.3 | THIN-FILM INTERFERENCE

Soapy water is normally transparent and colorless, but the photo of a thin soap film in Figure 25.6 shows very bright colors resulting from interference of reflected waves from the film's two surfaces. The colorful "bands" in Figure 25.6 are called "fringes" and correspond to the locations of constructive and destructive interference for light waves with different wavelengths (colors) in the incident light.

To understand where these colors come from, we start with the case shown in Figure 25.7A, in which a thin soap film rests on a flat glass surface. For simplicity, we also assume the incident light is monochromatic, that is, that it has a single wavelength. (Later we'll consider the behavior with white light, which is light containing many different colors.) The upper surface of the soap film in Figure 25.7A is similar to the beam splitter in the center of the Michelson interferometer, reflecting part of the incoming light (ray 2) and allowing the rest of the incident light to be transmitted into the soap layer after refraction at the air–soap interface. This transmitted light is partially reflected at the bottom surface, producing the wave that travels back into the air as ray 3. The two outgoing rays denoted as 2 and 3 meet the conditions for interference: they travel through different regions (one travels the extra distance through the soap film), they recombine when they leave the film as parallel rays, and they are coherent because they originated from the same light source and initial wave, ray 1.

To determine whether these interfering waves are in phase or out of phase with each other, we must consider what happens in the extra distance traveled by ray 3. For simplicity, let's assume the incident and reflected rays are all approximately normal to the film (Fig. 25.7B), which corresponds to looking straight down at the soap film. For a film of thickness d, the extra distance traveled by ray 3 is just $2d$. We must next account for the index of refraction of the soap film. Recall that light propagates in a vacuum at a speed $c = 3.00 \times 10^8$ m/s. The frequency f and wavelength $\lambda = \lambda_{\text{vac}}$ are related by

$$\lambda_{\text{vac}} f = c\text{t} \tag{25.7}$$

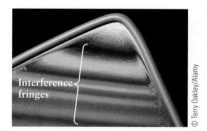

Figure 25.6 When a soap film is viewed with white light (as from the Sun), one observes colored interference fringes (colored bands). These colors are due to constructive interference; different colors (different wavelengths) give constructive interference for different values of the film thickness and for different viewing angles.

A   **B**   **C**

Figure 25.7 **A** Thin-film interference. Incident light (beam 1) is partially reflected at the air–soap film interface. The refracted light enters the film and is partially reflected at the bottom surface, in this case a soap film–glass interface. Beams 2 and 3 have traveled different distances and can interfere either constructively or destructively, depending on the thickness d of the soap film. Beams 2 and 3 both undergo a 180° phase change when they are reflected. (See ahead to Fig. 25.9.) Some light is also transmitted through the bottom surface of the glass. **B** Same as part A but with the incident and reflected rays normal to the film. **C** When the bottom layer is air, there is again interference of the two reflected beams, but now there is no phase change of beam 3 when it reflects at the bottom surface.

When light travels through a substance, its speed is reduced to $v = c/n$, where n is the index of refraction of the substance (Chapter 24). A soap film is mostly water and its index of refraction is about $n_{film} = 1.35$, so light slows down inside the soap film by 35% relative to c. The product of the wavelength inside the film λ_{film} and the frequency inside the film f_{film} equals the wave speed. (Compare with the case in vacuum, Eq. 25.7.) We then have

$$\lambda_{film} f_{film} = \frac{c}{n_{film}} \qquad (25.8)$$

Comparing with Equation 25.7, we see that either the wavelength or the frequency (or both) must change when light travels from a vacuum into the film. It turns out that the wavelength changes but the frequency does *not*. This can be seen from Figure 25.8, which sketches how the electric field associated with a light wave varies as the wave travels from air into a film. The fields just outside and just inside the film must stay in phase; otherwise, the wave fronts would "pile up" at the boundary. This is similar to the propagation of a wave on a string that travels between two strings of different diameters; the knot between the two strings then corresponds to the interface between air and the soap film in Figure 25.8. The strings on both sides of the knot must move with the same frequency; otherwise, the knot would come apart. Hence, the wave must have the same frequency in both strings. The same kind of argument shows that the frequency of light (or any other wave) does not change as it travels from one medium into another.

Although the frequency of light does not change when it enters the soap film in Figure 25.7, its wavelength does change. Inserting $f_{film} = f$ (the frequency in vacuum) into Equation 25.8, the wavelength in the film is

$$\lambda_{film} = \frac{v}{f} = \frac{c/n_{film}}{f}$$

Using Equation 25.7, this leads to

$$\lambda_{film} = \frac{\lambda_{vac}}{n_{film}} \qquad (25.9)$$

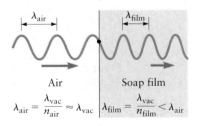

Figure 25.8 When a light wave passes from one medium to another (e.g., from air on the left into a soap film on the right), the waves on the two sides must stay in phase at the interface. That can only happen if the frequency is the same on both sides of the interface.

and the wavelength of light inside the film is thus shorter than its wavelength in a vacuum. The same is true for light traveling in air, with $\lambda_{air} = \lambda_{vac}/n_{air}$. Since n_{air} is very close to 1 ($n_{air} = 1.0003$), the wavelength of light in air is very close to its wavelength in a vacuum:

$$\lambda_{air} \approx \lambda_{vac} \qquad (25.10)$$

We can usually ignore the very small difference between the wavelengths of light in air and in a vacuum and will denote the wavelength in a vacuum as simply λ (without the subscript "vac").

Ray 3 in Figure 25.7B is traveling the distance $2d$ inside the soap film, and its wavelength is given by Equation 25.9. The extra number of wavelengths this wave travels is thus

$$N = \frac{2d}{\lambda_{\text{film}}} = \frac{2d}{\lambda/n_{\text{film}}} \tag{25.11}$$

If N is an integer, ray 3 will travel some number of complete, extra wavelengths compared with ray 2. As with the Michelson interferometer, the value of N will help determine how rays 2 and 3 interfere. Before we can say if the interference is constructive or destructive, however, we must consider how the reflections at the surfaces of the soap film affect the waves.

Effect of Reflection on the Phase of a Light Wave

The result in Equation 25.11 is analogous to what we found for the extra number of cycles in one of the interfering waves in a Michelson interferometer (Eq. 25.2). We might therefore expect similar conditions for interference as for a Michelson interferometer. However, we also must consider one additional piece of physics. In Chapter 12, we found that under certain conditions a reflected wave is inverted relative to the incident wave; this happens for waves on a string (Fig. 12.17) when the reflecting end of the string is tied to a rigid wall. Similarly, when a light wave reflects from a surface it *may* also be inverted, corresponding to a phase change of 180° as illustrated in Figure 25.9A. This phase change on reflection is found whenever the index of refraction on the incident side (n_1) is less than the index of refraction on the opposite side (n_2). On the other hand, if the index of refraction is larger on the incident side (Fig. 25.9B), the reflected light wave is not inverted and there is no phase change.

Phase change on reflection

In a thin-film interference experiment, we must take account of the *total* phase difference between the two interfering light waves; the total phase difference includes the phase difference due to the extra distance traveled by ray 3 plus phase changes that occur due to reflection. In Figure 25.7B, we have a soap film on top of a glass layer; to three significant figures, the indices of refraction are $n_{\text{air}} = 1.00$,

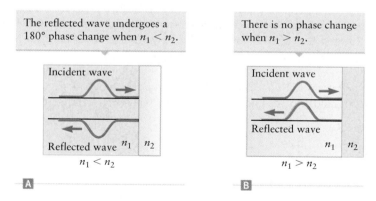

The reflected wave undergoes a 180° phase change when $n_1 < n_2$.

Incident wave

Reflected wave $\quad n_1 \quad n_2$

$n_1 < n_2$

There is no phase change when $n_1 > n_2$.

Incident wave

Reflected wave $\quad n_1 \quad n_2$

$n_1 > n_2$

A   **B**

Figure 25.9 Phase change upon reflection. When light strikes an interface at which the index of refraction changes, part of the wave is reflected even when both substances are transparent. **A** When the index of refraction n_1 of the first substance is less than the index of refraction n_2 of the second substance, the phase of the reflected wave changes by 180° and the reflected wave is inverted, similar to the behavior of a wave on a string when the string is tied to a wall (Fig. 12.17, page 392). **B** On the other hand, if $n_1 > n_2$, the phase of the reflected wave does not change and the wave is not inverted, similar to the behavior of a wave on a string when the end of the string is able to slide up and down (Fig. 12.20).

$n_{film} \approx 1.35$, and $n_{glass} \approx 1.50$. These values are typical and will vary slightly depending on the color of the light and the composition of the film and glass, but for a soap film on glass we'll always have $n_{air} < n_{film} < n_{glass}$. Hence, there are phase changes for both reflections at the soap film interfaces. The wave reflected at the upper surface of the film (wave 2) undergoes a 180° phase change because $n_{air} < n_{film}$, and the wave reflected at the bottom surface (wave 3) undergoes a 180° phase change because $n_{film} < n_{glass}$. Both interfering waves undergo the same phase change due to reflection, so their *relative* phase is not affected by the reflections and the nature of the interference is determined only by the extra path length $2d$ traveled by wave 3. Hence, we finally arrive at interference conditions similar to those found with the Michelson interferometer, with the only difference being the extra factor of n_{film} due to the change in wavelength inside the film.

Thin-film interference case 1: When both waves reflected by a thin film undergo a phase change. In this case the number of extra cycles traveled by the ray inside the film completely determines the nature of the interference. If the number of extra cycles N equals an integer, there is constructive interference; if N is a half integer, there is destructive interference. Using Equation 25.11, we have

$$2d = \frac{m\lambda}{n_{film}} \qquad \text{(constructive interference)}$$

$$2d = \frac{(m + \frac{1}{2})\lambda}{n_{film}} \qquad \text{(destructive interference)}$$

(25.12)

These interference conditions apply whenever the indices of refraction are related as in Figure 25.7B, with $n_{air} < n_{film} < n$ (substance below film).

Thin-film interference case 2: When only one wave reflected by a thin film undergoes a phase change. A different case, with different interference conditions, is shown in Figure 25.7C. Here a soap film is surrounded by air; that is, there is air above and below the film. That is also the situation in a soap bubble as in Figure 25.6. When ray 1 reflects from the top surface, there is a phase change because $n_{air} < n_{film}$ and the index of refraction is smaller on the incident side of the interface. For the reflection at the bottom surface of the film, however, there is no phase change because now the index is larger on the incident side. One of the interfering waves (2) thus undergoes a phase change due to reflection, and the other (3) does not. Figure 25.1C shows that a 180° phase change is equivalent to a shift of the wave by $\lambda/2$. This is equivalent to inserting an extra half cycle into the wave, which changes the conditions for achieving constructive interference and destructive interference.

The case of only one wave undergoing a phase change occurs in thin-film interference when the index of refraction of the film in the middle is greater than the indices of the regions on either side. Now only one of the interfering waves (wave 2 in Fig. 25.7C) undergoes a 180° phase change due to reflection, and wave 3 must travel an extra half wavelength to get back in phase with wave 2. The interference conditions are thus

$$2d = \frac{(m + \frac{1}{2})\lambda}{n_{film}} \qquad \text{(constructive interference)}$$

$$2d = \frac{m\lambda}{n_{film}} \qquad \text{(destructive interference)}$$

(25.13)

Thin-Film Interference with White Light

Thin-film interference is responsible for the colors often seen in thin layers of water or oil on flat pavement and in free-standing films (Fig. 25.6). This effect also gives the bright colors found on the wings of many insects and birds (Fig. 25.10). Those cases usually involve sunlight (so-called white light), whereas in the situation ana-

Figure 25.10 The colors of many insect wings are due to thin-film interference. For a wing of a particular thickness, constructive interference and hence a bright interference fringe occurs for only one or a few values of the wavelength, giving the wing its color when viewed in reflected light.

lyzed in Figure 25.7 we assumed light with a single wavelength. We can understand what happens with white light using Figure 25.11A, which shows light of just two different colors (red and blue) incident on a thin film. We also assume the condition for constructive interference is satisfied for blue light at normal incidence (on the left in the figure). The wavelength of red light is longer than the wavelength for blue light, so for the same angle of incidence the condition for constructive interference will usually *not* be satisfied for red light at the same time. However, if the angle of incidence θ_i for red light is increased (on the right in Fig. 25.11A), the path length difference between the two light waves increases. It is therefore possible to have constructive interference for red light, but *for a different angle of incidence* than for blue light.

When an observer looks at a soap film, white light from the sky illuminates the film over a range of angles (Fig. 25.11B). Each different color component of the sunlight can exhibit constructive interference for a different angle of incidence, giving a series of interference fringes with different colors and producing brightly colored fringes even with a soap film of constant thickness.

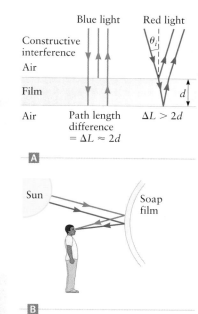

Figure 25.11 **A** On the right, light is incident on a thin film at an angle away from the normal direction. The path length difference between the reflected waves is greater than for light at normal incidence (on the left). **B** When you view a soap film (perhaps as part of a soap bubble) with reflected sunlight, different colors give constructive interference at different angles, producing the "colored" interference fringes in Figure 25.6.

| EXAMPLE 25.3 | Color and Thickness of a Soap Film |

Consider a bubble formed from an *extremely* thin, soapy film that looks blue when viewed at normal incidence. Estimate the thickness of the film. Assume the index of refraction of the soap film is $n_{\text{film}} = 1.35$ and blue light has a wavelength $\lambda_{\text{blue}} = 400$ nm. Also assume the film is so thin that thinner films are not able to give constructive interference.

RECOGNIZE THE PRINCIPLE

The blue color of the soapy film comes from interference of waves reflected from the two film surfaces as sketched in Figure 25.7C. The wave reflected from the top surface (denoted by ray 2) undergoes a phase change on reflection because the index of refraction on the incident side (air) is less than the index of the film. On the other hand, the wave reflected from the bottom surface (denoted by ray 3) does not undergo a phase change on reflection because the index on the incident side (the film) is greater than the index of the opposite side (air). (See Fig. 25.9.)

SKETCH THE PROBLEM

Figure 25.7C describes thin-film interference for a soap film with air on both sides.

IDENTIFY THE RELATIONSHIPS

The interference conditions that apply with a phase change at only one of the reflections are given by Equations 25.13. Since the film has a blue color, there must be constructive interference for blue light of the given wavelength $\lambda_{\text{blue}} = 400$ nm $= 4.0 \times 10^{-7}$ m.

Applying Equation 25.13 for constructive interference, we have

$$2d = \frac{(m + \tfrac{1}{2})\lambda_{\text{blue}}}{n_{\text{film}}} \tag{1}$$

Recall that m in Equation (1) is an integer that can have a value 0, 1, 2, ... Equation (1) thus has many solutions for the film thickness d, depending on the value of m. How can we determine the value of m? The problem states that the film is extremely thin and that thinner films do not give constructive interference. The smallest possible value of d corresponds to $m = 0$, so that case must apply for the film.

SOLVE

Using $m = 0$ in Equation (1), we find

$$2d = \frac{(m + \tfrac{1}{2})\lambda_{\text{blue}}}{n_{\text{film}}} = \frac{\lambda_{\text{blue}}}{2n_{\text{film}}}$$

Solving for d and inserting values for the wavelength and index of refraction gives

$$d = \frac{\lambda_{blue}}{4n_{film}} = \frac{4.0 \times 10^{-7} \text{ m}}{4(1.35)} = 7.4 \times 10^{-8} \text{ m} = \boxed{74 \text{ nm}}$$

What does it mean?
This value is about five hundred times smaller than the diameter of a human hair. It is amazing that a soap film this thin can be strong enough to form a bubble.

| **EXAMPLE 25.4** | Measuring the Diameter of a Thin Fiber |

Figure 25.12 shows how thin-film interference can be used to measure the diameter of a thin fiber such as a strand of hair. Two glass plates are in contact at one end but are held apart at the other end by the fiber. Light incident from above is reflected at the top and bottom of the "film" of air between the plates, producing interference. When viewed at normal incidence and with a small wedge angle θ, the path length difference between the interfering light waves is equal to twice the plate separation. This separation increases from left to right in Figure 25.12A, so the interference alternates between constructive and destructive. The resulting interference pattern is sketched as a function of position x along the plates in Figure 25.12B. The light intensity is zero at the left end (see Concept Check 25.3) and then passes through nine bright fringes (places where the intensity is a maximum), and the intensity at the right end next to the fiber is again zero. Find the diameter of the fiber. Assume the wavelength of the light is $\lambda = 600$ nm.

RECOGNIZE THE PRINCIPLE

The interference involves waves like those represented by ray 2 that reflect at the glass–air interface and waves represented by ray 3 that reflect at the air–glass interface. Ray 2 does not undergo a phase change at this reflection because $n_{air} < n_{glass}$, similar to the reflection from the bottom surface in Figure 25.7C. Ray 3 does undergo a 180° phase change on reflection, however, as found for the waves in Figure 25.7B. Hence, Equations 25.13 are the correct interference conditions for this case.

SKETCH THE PROBLEM

Figure 25.12 shows the problem.

IDENTIFY THE RELATIONSHIPS

The condition for destructive interference in this case is (Eq. 25.13)

$$2d = \frac{m\lambda}{n_{film}} \quad \text{(destructive interference)} \tag{1}$$

To pass from one dark fringe in the interference pattern to the next dark fringe, the path difference between the interfering waves must change by exactly one wavelength. In terms of the condition for destructive interference in Equation (1), this change corresponds to going from m to $m + 1$. We are told that there are nine complete fringes in Figure 25.12B, so if the separation between the plates on the right-hand side is equal to the diameter of the fiber d, the condition for destructive interference becomes

$$2d = \frac{m\lambda}{n_{film}} = \frac{9\lambda}{n_{air}}$$

In this case the "film" between the plates is filled with air.

SOLVE

Solving for d, we get

$$d = \frac{9\lambda}{2n_{air}} = \frac{9(6.0 \times 10^{-7} \text{ m})}{2(1.00)} = \boxed{2.7 \times 10^{-6} \text{ m}}$$

Figure 25.12 Example 25.4. **A** Two glass plates separated at one end by a thin fiber. **B** Reflected light intensity as a function of position along the plates.

What does it mean?
Figure 25.13 shows a photograph of the interference fringes produced by an air wedge. Each bright fringe is a "strip" on which the intensity is high, corresponding to the intensity peaks in Figure 25.12B. Similar interference methods are used to measure extremely small distances or heights in many applications, including the reading and writing of information on CDs and DVDs.

Figure 25.13 Photo of interference fringes (the horizontal bands) produced by two glass plates separated by a small amount at one end.

CONCEPT CHECK 25.3 | Reflection from an Air Wedge

Explain why the waves that reflect from the far left edge of the plates in Figure 25.12A exhibit destructive interference (zero intensity) at $x = 0$. The puzzle: At the far left edge, the plates are in contact, so the path length difference between beams 2 and 3 is zero. Why is there still destructive interference?

Antireflection Coatings

A partially reflecting mirror splits a light wave into two separate waves, which may later be recombined and interfere. Nearly any flat piece of glass will act in this way. You can readily demonstrate this fact by looking through the nearest window; even though the window glass is designed to let light pass through, a small amount (about 4%) of the light is reflected. That is not a problem for an ordinary window, but is bad for some applications. For example, light in a typical camera must pass through several lenses and filters; if 4% of the intensity is reflected at each interface, it would take only a few surfaces to reduce the light intensity by 25% or more, which would be bad for the camera's performance. To avoid this problem, most camera lenses have *antireflection coatings* that make a lens appear slightly dark in color when viewed in reflected light as in Figure 25.14A.

Antireflection coatings are often made from MgF_2, which has an index of refraction of about $n_{MgF_2} = 1.38$. Incident light reflects from the front surface of the MgF_2 film as in Figure 25.14B and again from the MgF_2–glass interface. There is a 180° phase change at both reflections (as with the soap film on glass in Fig. 25.7B), so the interference conditions in Equation 25.12 apply and destructive interference occurs when

$$2d = \frac{(m + \frac{1}{2})\lambda}{n_{MgF_2}}$$

Here, m can have the values $0, 1, 2 \ldots$, so the smallest possible value of d and hence the thinnest MgF_2 film that gives destructive interference corresponds to $m = 0$. We thus get

$$2d = \frac{\lambda}{2n_{MgF_2}}$$

If we want to design an antireflection coating of MgF_2 to operate for red light ($\lambda = 6.0 \times 10^{-7}$ m), we have

$$d = \frac{\lambda}{4n_{MgF_2}} = \frac{6.0 \times 10^{-7} \text{ m}}{4(1.38)} = 1.1 \times 10^{-7} \text{ m} = 110 \text{ nm}$$

MgF_2 is a popular material for such coatings in part because it can be made into very uniform films with this thickness. An antireflection coating will only work best at one particular wavelength in the visible range; ideally, it will give "perfect" destructive interference at only one wavelength, but it will also give "partially" destructive interference at nearby wavelengths. Antireflection coatings that function well over the entire range of visible wavelengths are made using multiple layers that give "perfect" destructive interference at different wavelengths.

A

B

Figure 25.14 **A** The lenses on most binoculars and cameras are coated with antireflection coatings. Although these coatings are transparent, they give the lens a dark tint. **B** Thin-film interference in an antireflection coating composed of MgF_2.

Figure 25.15 Reflection and interference of light from the edge of a "pit" in a CD or DVD.

Interference and the Design of CDs and DVDs

Thin-film interference is used in CDs and DVDs, which encode information as a series of pits in a layer of plastic. The depth of the pit is chosen to give destructive interference between light waves reflecting from the bottom and top surfaces at a pit edge (Fig. 25.15). Because of this destructive interference, the reflected intensity is zero when the incident light is over the edge of a pit. Light striking the plastic surface away from a pit edge, whether fully inside or fully outside the pit, is reflected to the detector with a large intensity. The presence or absence of a pit edge can thus be detected by the absence or presence of a reflected light wave. The design of CDs and DVDs is discussed in more detail in Chapter 26.

25.4 | LIGHT THROUGH A SINGLE SLIT: QUALITATIVE BEHAVIOR

The photo at the start of this chapter shows two interfering circular water waves. The analogous experiment with light can be performed by letting light shine onto a pair of slits in an opaque screen. To understand the resulting double-slit interference pattern, we must first consider carefully what happens when light passes through a single slit. In this section, we discuss the qualitative behavior; in the two following sections, we then give a more detailed analysis.

The basic experiment is sketched in Figure 25.16A. Light passes through a slit and then illuminates a screen. For simplicity, the incident light is assumed to be a plane wave, represented by the parallel beams on the left side of the slit. When the slit is very wide, the light that reaches the screen is a projection of the slit. If we plot the light intensity versus position along the x axis marked on the screen in Figure 25.16A, we find a "step" in intensity approximately equal to the width w of the slit as shown at the top of Figure 25.16B, but a very careful look reveals that the edges of this shadow are slightly "smeared out." As the slit is made narrow, the intensity pattern on the screen spreads out. If we keep reducing the slit width so that it is comparable to or even smaller than λ, the intensity pattern spreads more and more, and some additional intensity maxima become noticeable.

This behavior occurs for all types of waves, including light and sound; it is easy to observe with a water wave because the wavelength of a water wave is large enough for individual waves to be seen easily by eye. In Figure 25.17, a water wave is incident from the left and passes through an opening whose width is approximately equal to λ. The outgoing wave on the right spreads out as a broad beam, analogous to striking a screen over a broad region as in the lowermost pattern in Figure 25.16B. Figures 25.16 and 25.17 illustrate the phenomenon of **single-slit diffraction**. Dif-

Single-slit diffraction

Figure 25.16 A Light passing through an opening or slit and onto a screen. B If the slit is very wide compared with the wavelength λ, an ordinary shadow (projection of the slit) appears on the screen (top). If the slit width is comparable to or smaller than the wavelength, the wave intensity is spread over a larger region of the screen. As the slit width is made narrower, the amount of spreading increases.

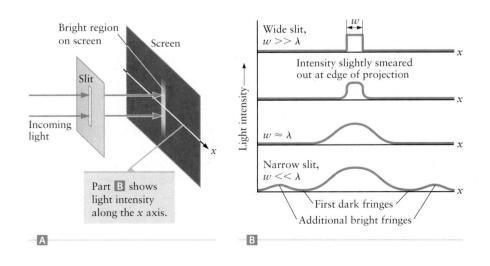

fraction is the bending or spreading of a wave when it passes through an opening. In single-slit diffraction, the opening is just a single slit.

To understand the qualitative behavior of how light passes through a very narrow slit, it is useful to draw the wave fronts and rays for the incident and diffracting waves. Figure 25.18 shows the results for an extremely narrow slit. In this case, the outgoing wave radiates from the slit over a very wide range of angles. This is an example of *Huygens's principle*, which can be stated as follows.

Huygens's principle

All points on a wave front can be thought of as new sources of spherical waves.

When the incoming wave encounters the narrow openings in Figures 25.17 and 25.18, each point inside the opening acts as a new source of spherical waves. In Figure 25.18, the opening is *very much smaller* than the wavelength; in this case, the opening acts as a single point source of waves, and the simple outgoing spherical waves sketched in Figure 25.18 are the result.

We will not attempt to derive or justify Huygens's principle. It was proposed (by Huygens) in about 1680, long before there was a detailed theory of light. This principle gives a qualitative way to understand how a very narrow slit can convert a plane wave into a spherical one. This conversion is key to understanding the effects of interference and diffraction when light passes through a small opening.

25.5 | DOUBLE-SLIT INTERFERENCE: YOUNG'S EXPERIMENT

We next consider the behavior of light passing through two very narrow slits, a problem called *double-slit interference*. We jump to the double-slit case before finishing the discussion of the single slit from Section 25.4 because a full analysis of light passing through a single slit is in some ways more complicated than that for two *very narrow* slits. When the two slits in the double-slit case are both very narrow, each slit acts as a simple point source of new waves, and the outgoing waves from each slit are like the simple spherical waves in Figure 25.18.

The double-slit experiment is famous in the history of physics because it demonstrated conclusively that light is a wave. This experiment, first carried out around 1800 by Thomas Young, is shown in Figure 25.19. Light is incident from the left onto two slits and after passing through them strikes a screen on the right. The incident light here is a plane wave; this important requirement is easy to achieve today using light from a laser. It can also be accomplished with a lens as described in Chapter 24.

The double-slit experiment in Figure 25.19 is very similar to the interference arrangement involving sound waves in Figure 25.1 and satisfies the general conditions for interference from Section 25.1. First, the interfering waves travel through different regions of space as they travel through different slits. Second, the waves come together at a common point on the screen where they interfere. Third, the waves are coherent because they come from the same source, the incoming plane wave on the left. We thus expect interference to determine how the intensity of light on the screen varies with position. For simplicity, we assume the two slits are very narrow, so according to Huygens's principle each slit acts as a simple source with circular wave fronts as viewed from above. If we cover one of the slits in Figure 25.19, a very wide area of the screen is illuminated with light as with the single slit in Figure 25.18, but when we allow light to pass through both slits, we get the very different result shown in Figure 25.19. The light intensity on the screen alternates between bright and dark as we move along the screen, signaling regions of constructive interference and destructive interference.

Figure 25.17 Diffraction of a water wave as it passes from left to right through a single narrow slit, producing an outgoing wave on the right. The outgoing wave front is approximately spherical.

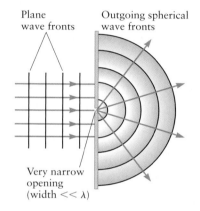

Figure 25.18 A plane wave incident on a single narrow slit. According to Huygens's principle, the slit acts as a new wave source, producing a spherical wave that propagates away from the slit.

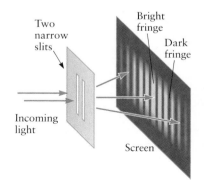

Figure 25.19 Interference of light produced by two slits. The pattern on the screen at right is a series of bright and dark interference fringes.

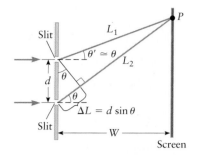

Figure 25.20 Analysis of double-slit interference. Waves that pass through the two slits travel different distances L_1 and L_2 on their way to point P on the screen. (Compare with Fig. 25.1.) Not to scale.

Double-slit interference conditions

Where Are the Fringes?

To analyze the interference pattern in Figure 25.19, we must determine the path length between each slit and the screen. Consider the path lengths for waves that arrive at point P on the screen in Figure 25.20. For simplicity, we assume the screen is very far away from the slits, so W is very large. The distance from the bottom slit to P is greater than the distance between P and the top slit. Also, because W is large, the angles θ and θ' that specify the directions from the slits to point P are approximately equal, and we denote both by θ.

If the slits are separated by a distance d, the extra path length traveled by the wave from the bottom slit is ΔL, which is given by

$$\Delta L = d \sin \theta \tag{25.14}$$

as shown in Figure 25.20. If this extra path length is equal to an integral number of complete wavelengths, the two waves will be in phase when they strike the screen; the interference is then constructive and the light intensity is large. On the other hand, if the extra path length is equal to $\lambda/2$, $3\lambda/2$, $5\lambda/2$, ..., the interference is destructive and the intensity at the screen is zero. The condition for constructive interference and a bright fringe in the interference pattern is thus

$$d \sin \theta = m\lambda \quad m = 0, \pm 1, \pm 2, \pm 3, \ldots \tag{25.15}$$

(constructive interference)

whereas the condition for destructive interference and a dark fringe is

$$d \sin \theta = (m + \tfrac{1}{2})\lambda \quad m = 0, \pm 1, \pm 2, \pm 3, \ldots \tag{25.16}$$

(destructive interference)

The double-slit intensity pattern at the screen is shown greatly enlarged in Figure 25.21. The angle θ varies as we move along the screen. At certain values of θ, the condition for constructive interference is satisfied, producing a maximum in the intensity (a bright interference fringe). Each bright fringe satisfies Equation 25.15 with a different value of the integer m. The value $m = 0$ gives $\theta = 0$ and corresponds to the center of the screen. Moving up or down from this point gives bright fringes with $m = +1, +2, +3, \ldots$ and $m = -1, -2, -3, \ldots$. Negative values of m indicate that the path to those points on the screen from the lower slit is shorter than the path to those points from the upper slit.

Now let's consider what the interference pattern will look like for particular values of the separation d between the slits and the perpendicular distance to the screen W. Suppose the slits are $d = 0.10$ mm apart and the screen is located $W = 50$ cm $= 0.50$ m from the slits. If we use light from a red laser with $\lambda = 630$ nm, what is the separation between the bright fringe at the center of the screen ($\theta = 0$) and the nearest bright fringe on either side in Figure 25.21? Two such bright fringes correspond to $m = 0$ (the central fringe) and $m = 1$ in Equation 25.15. The angle θ measured at the slits between the central fringe and the first fringe above it thus satisfies $d \sin \theta = \lambda$, and we have

$$\sin \theta = \frac{\lambda}{d} \tag{25.17}$$

Figure 25.21 Double-slit interference. Values of θ from Equation 25.15 give the locations of the bright fringes on the screen.

The wavelength is much smaller than the spacing d between the slits, so the ratio on the right-hand side of Equation 25.17 is very small. Hence, $\sin \theta$ is also very small, and we can use the approximation $\sin \theta \approx \theta$ (where the angle is measured in radians; see Appendix B) to get

$$\sin \theta \approx \theta \approx \frac{\lambda}{d} \tag{25.18}$$

To find the separation between the bright fringes on the screen, we use the right triangle in Figure 25.21 with sides of length W and h. The spacing h between the two fringes corresponding to $m = 0$ and $m = 1$ is

$$h = W \tan \theta$$

Using the approximation for small angles $\tan \theta \approx \theta$ (Appendix B), we get

$$h = W\theta$$

Our result for θ from Equation 25.18 then gives

$$h = W\theta = W\frac{\lambda}{d} \tag{25.19}$$

Inserting the values of W, d, and λ given above, we find

$$h = W\frac{\lambda}{d} = (0.50 \text{ m})\frac{6.3 \times 10^{-7} \text{ m}}{1.0 \times 10^{-4} \text{ m}} = 3.2 \times 10^{-3} \text{ m} = 3.2 \text{ mm} \tag{25.20}$$

This fringe spacing is large enough to be seen easily by the naked eye.

Our analysis of Young's double-slit experiment and the result in Equation 25.20 shows that the bright and dark fringes associated with interference are observable, provided the slits are fairly close together. The results for h in Equations 25.19 and 25.20 show that a slit separation of about 0.1 mm (or smaller) is a good choice. If the spacing is larger than a few millimeters or so, the fringes are quite close together and can be difficult to observe.

Insight 25.2

REMEMBER THE APPROXIMATIONS FOR SMALL ANGLES

Because the wavelength of light is small, the conditions for constructive interference and destructive interference often involve small angles. For small values of the angle θ, the sine and tangent functions can be approximated by

$$\sin \theta \approx \theta \quad \text{and} \quad \tan \theta \approx \theta$$

These relations are true only if θ is measured in radians. See Appendix B.

CONCEPT CHECK 25.4 | Double-Slit Interference

Suppose the distance between the slits in Figure 25.19 is reduced. How would this change affect the interference pattern?
(a) The bright fringes would be more closely spaced on the screen.
(b) The bright fringes would be more widely spaced on the screen.
(c) The fringe spacing on the screen would not change.

EXAMPLE 25.5 | Measuring the Wavelength of Light

Young's double-slit experiment shows that light is indeed a wave, and also gives a way to measure the wavelength. Suppose the double-slit experiment in Figure 25.21 is carried out with a slit spacing $d = 0.40$ mm and the screen at a distance $W = 1.5$ m. If the bright fringes nearest the center of the screen are separated by a distance $h = 1.5$ mm, what is the wavelength of the light?

RECOGNIZE THE PRINCIPLE

In a double-slit experiment, the bright fringes occur when light waves from the two slits interfere constructively. That happens when the path length difference ΔL (Fig. 25.20) is equal to an integral number of wavelengths. We used this condition to derive Equation 25.19, which gives a relation between h, W, d, and λ for the first bright fringe.

SKETCH THE PROBLEM

Figures 25.20 and 25.21 describe the problem.

IDENTIFY THE RELATIONSHIPS

Equation 25.19 relates the slit spacing d and the perpendicular distance W to the screen, along with h and λ. Rearranging to solve for the wavelength gives

$$\lambda = \frac{hd}{W}$$

SOLVE

Inserting the given values of h, d, and W, we find

$$\lambda = \frac{hd}{W} = \frac{(1.5 \times 10^{-3}\,\text{m})(4.0 \times 10^{-4}\,\text{m})}{1.5\,\text{m}} = 4.0 \times 10^{-7}\,\text{m} = \boxed{400\,\text{nm}}$$

What does it mean?

Young's double-slit experiment gives a way to measure even the very short wavelength of visible light. The values of h, W, and d can all be measured with an ordinary ruler, but together they enable us to measure the wavelength λ, which is much too small to be measured with a ruler. While technology has certainly advanced since Young's work two hundred years ago, methods involving interference are still widely used in precision measurements of the wavelength of light.

Interference with Monochromatic Light

The conditions for interference listed in Section 25.1 (page 842) state that the interfering waves must have the same frequency and hence the same value of the wavelength λ. Light with a single frequency is called **monochromatic** because it consists of a single color. Light from most sources, such as the Sun or an ordinary lightbulb, contains a distribution of wavelengths. Such light sources are generally not useful in double-slit interference experiments. Suppose a light source emits mainly at two colors, that is, at two wavelengths λ_1 and λ_2. The angles that give bright fringes in the double-slit experiment (Eq. 25.15) depend on the value of λ. Different values of the wavelength thus give bright fringes at different locations on the screen, and places on the screen where wavelength λ_1 gives a bright fringe will often coincide with places where wavelength λ_2 gives a dark fringe. If a light source emits a broad distribution of wavelengths, the total intensity pattern will be a "washed-out" sum of bright and dark regions, and no bright or dark fringes will be visible. For this reason, interference experiments with light usually work best with monochromatic waves.

25.6 | SINGLE-SLIT DIFFRACTION: INTERFERENCE OF LIGHT FROM A SINGLE SLIT

We began our treatment of how light passes through small openings by first discussing the qualitative behavior for a single slit (Section 25.4), including how the intensity pattern changes as we go from a very wide to a very narrow slit (Fig. 25.16). We next discussed the behavior with two slits (Section 25.5), with the simplifying assumption that the slits are very narrow so that each acts as a simple point source of new spherical waves (Huygens's principle). With the insight gained from the double-slit case, we are ready to tackle the behavior of a single slit in a more general way. We'll now see how to deal with slits that are narrow enough to exhibit diffraction but not so narrow that they can be treated as a single point source of waves.

Analysis of Single-Slit Diffraction

Figure 25.22A shows light incident on a single slit of width w. Light is diffracted as it passes through the slit and then propagates to the screen. The intensity pattern at the screen contains a central very bright fringe with dark fringes alongside as well as other fringes where the intensity peaks at values much smaller than the intensity of the central fringe. We qualitatively sketched this pattern at the bottom of Figure 25.16B. We want to explain the origin of these fringes and calculate where bright and dark fringes occur on the screen. The key to this calculation is Huygens's principle, which states that each point on a wave front acts as a new source of waves. All points across the slit in Figure 25.22 act as wave sources, and these different waves interfere at the screen.

Figure 25.22 A Light passing through a single slit and then on to a screen. B Each point within the slit $A_1, A_2, \ldots, B_1, B_2, \ldots$ acts as a source of spherical waves. C Analysis of the single-slit diffraction pattern. Notice that the slit widths in parts B and C are greatly expanded to show the geometry of the interfering waves. In a scale drawing, the slit widths would be much *smaller* than the width of the central diffraction peak in part C.

Where Are the Dark Fringes?

We divide the slit into two halves and denote points in one half by A and those in the other half by B (Fig. 25.22B). Each of these points is a source of Huygens (i.e., spherical) waves, including point A_1 at one edge of the slit and the corresponding source point B_1 in the other portion of the slit. Figure 25.22C shows the distance from these points to a point P on the screen and also shows the path length difference ΔL. We again assume the screen is very far away from the slit, so all the angles denoted by θ are approximately equal. Suppose the path length difference is $\Delta L = \lambda/2$ so that the waves from A_1 and B_1 interfere destructively when they reach a particular point P on the screen. If this condition for destructive interference is met for points A_1 and B_1, the waves from A_2 and B_2 will also interfere destructively and so on for points A_3 and B_3 and for *all* similar pairs of points within the slit. Destructive interference will thus produce a dark fringe on the screen. This condition can be written as $\Delta L = \lambda/2$. From Figure 25.22C, we see that $\Delta L = (w/2)\sin\theta$, so we have

$$\frac{w}{2}\sin\theta = \Delta L = \frac{\lambda}{2}$$

which gives the angle for the first dark fringe above the center of the screen. There is a first dark fringe below center, which corresponds to negative values of θ with

$$\frac{w}{2}\sin\theta = -\frac{\lambda}{2}$$

The first dark fringes are thus at the angles

$$w\sin\theta = \pm\lambda \quad \text{single-slit diffraction: first dark fringes} \qquad (25.21)$$

$$\text{(destructive interference)}$$

The position of the second dark fringe can be found using Figure 25.23. In Figure 25.23A, we divide the slit into four regions and denote the points in these four regions by A, B, C, and D. Figure 25.23B shows the path lengths from the top point in each region (A_1, B_1, C_1, and D_1) to a point P on the screen. Suppose the path length difference for the Huygens waves from A_1 and B_1 is $\Delta L = \lambda/2$, so these waves interfere destructively. The path length difference for C_1 and D_1 has the same value,

Figure 25.23 ◪ To locate the second dark fringe in the intensity pattern, we imagine the slit as split into four regions, *A*, *B*, *C*, and *D*. Points in these regions can be grouped in pairs, A_1 and B_1, A_2 and B_2, ..., C_1 and D_1, ..., similar to the pairing in Figure 25.22B. ◪ Analysis of the location of the second dark fringe in a single-slit diffraction pattern. Notice that the slit widths are greatly expanded to show the geometry of the interfering waves. In a scale drawing, the slit widths would be much *smaller* than the width of the central diffraction peak in part B.

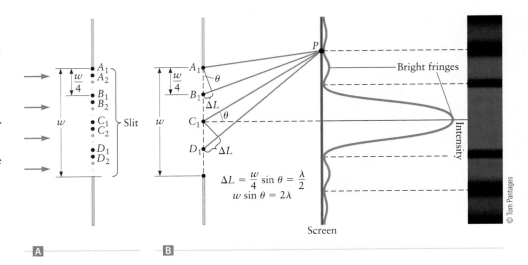

so these waves will also interfere destructively. Also note that because the figure is not to scale, the two path length differences ΔL for the pairs *AB* and *CD* appear to have very different lengths. However, for a small slit width *w*, they will be very close in length. The path length difference will be the same for waves from others pairs of points A_2 and B_2, C_2 and D_2, and so on (Fig. 25.23A), so these waves will also interfere destructively, and the total intensity on the screen is zero. For the case of the second dark fringe, we thus have $\Delta L = (w/4)\sin\theta$. Setting this expression equal to $\lambda/2$, the interference condition in this case is

$$w \sin\theta = \pm 2\lambda \quad \text{single-slit diffraction: second dark fringes}$$

$$(\text{destructive interference})$$

where the negative sign again corresponds to a fringe below the center of the screen. The same argument can be used to find the third, fourth, ... dark fringes by dividing the slit into six, eight, ... regions (following Figs. 25.22 and 25.23). The result for the *m*th dark fringe is

$$w \sin\theta = \pm m\lambda \quad \text{single-slit diffraction: } m\text{th dark fringes} \quad (25.22)$$

$$(\text{destructive interference})$$

where $m = 1, 2, 3, \ldots$. The values of θ from Equation 25.22 give the angles of all dark fringes in the diffraction pattern.

Analyzing the Complete Diffraction Pattern

The angles that give the intensity peaks in Figures 25.22 and 25.23 are not as easy to calculate as the angles of the dark fringes (Eqs. 25.21 and 25.22), and there is no simple formula for the angles at which the bright fringes occur. However, the intensity on the screen can be calculated by adding up all the Huygens waves that come from all the source points A_1, B_1, A_2, B_2, and so forth. The resulting intensity curve is sketched on the right in Figures 25.22C and 25.23B. There is a central bright diffraction fringe (the central maximum), with other "bright" fringes much lower in intensity. A full calculation of the intensity curve shows that the central bright fringe is about 20 times more intense than the bright fringes on either side. The width of the central bright fringe is given approximately by the angular separation of the first dark fringes on either side. The angles for these first dark fringes are given by Equation 25.21:

$$\frac{w}{2}\sin\theta = \pm\frac{\lambda}{2}$$

$$\sin\theta = \pm\frac{\lambda}{w}$$

If the angle θ is small (as it usually is), then $\sin\theta \approx \theta$, which gives

$$\theta = \pm\frac{\lambda}{w} \quad \text{(first dark fringes)} \tag{25.23}$$

These two dark fringes occur on either side of the central bright fringe in Figure 25.22A, so the *full* angular width of the central intensity is 2θ:

$$\text{full angular width of central bright fringe} = 2\frac{\lambda}{w} \tag{25.24}$$

As a rough rule of thumb, one can say that diffraction of a light wave as it passes through a slit produces a central bright fringe extending over the angular range given by Equation 25.24. If the slit is much wider than the wavelength, this angle is small. For example, if the slit width is equal to the thickness of a typical fingernail ($w \approx 0.5$ mm $= 5 \times 10^{-4}$ m) and we consider red light ($\lambda = 630$ nm), the angle is

$$\text{full angular width} = 2\frac{\lambda}{w} = \frac{2(6.3 \times 10^{-7}\text{ m})}{5 \times 10^{-4}\text{ m}} = 0.0025\text{ rad} = 0.14° \tag{25.25}$$

This angle is very small, so the light beam would essentially pass straight through the slit with almost no effect from diffraction. The central diffraction fringe is much broader when the slit width approaches λ. For example, suppose the slit width is only five times larger than the wavelength; for the red light considered in Equation 25.25, we get $w = 5\lambda = 3.2 \times 10^{-6}$ m $= 3.2$ μm (about 300 times smaller than 1 mm). The central bright diffraction fringe would then correspond to an angle

$$\text{full angular width} = \frac{2\lambda}{w} = \frac{2\lambda}{5\lambda} = \frac{2}{5} = 0.40\text{ rad} = 23°$$

so the spreading of the light beam due to diffraction is now quite substantial.

CONCEPT CHECK 25.5 | Single-Slit Diffraction

A single-slit diffraction experiment is performed with red light, and the width of the central diffraction fringe is measured. The experiment is then repeated with blue light. Which of the following statements describes the result?
   (a) The width of the central diffraction fringe is greater with blue light.
   (b) The width of the central diffraction fringe is smaller with blue light.
   (c) The width of the central diffraction fringe does not change.

Double-Slit Interference with Wide Slits

In our discussion of Young's double-slit experiment in Section 25.5, we assumed for simplicity that the slits are very narrow (widths much less than the wavelength of light); such slits produce an intensity pattern that varies approximately sinusoidally with position on the screen as shown in Figure 25.21. When the slits are not extremely narrow, the single-slit diffraction pattern produced by each slit is combined with the sinusoidal double-slit interference pattern. A full calculation of the intensity pattern is complicated, but the result can be understood by combining Figure 25.24A, which shows the double-slit pattern with two very narrow slits, and Figure 25.24B, which shows the single-slit pattern for a wide slit. For two double slits that are wide, the intensity pattern is essentially the product of these two patterns (Fig. 25.24C).

25.7 | DIFFRACTION GRATINGS

In the past few sections, we have discussed the interference patterns found when light passes through either a single slit or a double slit. As you should expect, similar interference effects also occur with three or more slits. In fact, physicists have

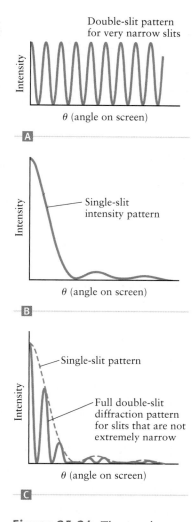

Figure 25.24 The complete interference pattern produced by two slits is a combination of ⒶA the double-slit pattern and ⒷB the single-slit pattern. The combination of these two gives the intensity pattern in ⒸC.

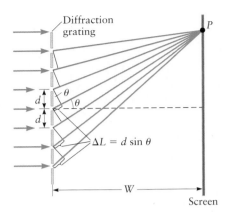

Figure 25.25 Interference experiment with a diffraction grating.

Diffraction from a grating: conditions for constructive interference

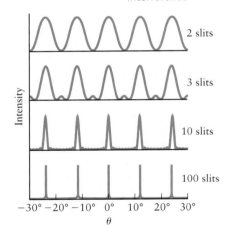

Figure 25.26 Interference patterns with 2, 3, 10, and 100 equally spaced slits. The main intensity peaks (the principal maxima) become narrower as the number of slits is increased. With 100 slits, these peaks are very narrow.

Figure 25.27 ◩ Sketch of a diffraction experiment with a grating using red light and green light. The colors of the rays here *do* indicate the color of the light. The bright fringes are at different angles for the different colors because their wavelengths are different. ◩ Diffraction experiment with both red laser light and green laser light shining on a grating.

found that systems with hundreds or even thousands of slits are quite useful; you probably own one without even knowing it as we'll explain below.

Figure 25.25 shows an arrangement of many slits called a ***diffraction grating***. When light passes through a diffraction grating, a pattern of light and dark fringes is produced on the screen. To analyze this pattern, we'll assume for simplicity that the slits are each very narrow (so that each one produces a single outgoing wave) and the screen is very far away. The rays connecting each slit with a point P on the screen are then approximately parallel, and all make an angle θ with the horizontal axis in Figure 25.25. If the slit-to-slit spacing is d, the path length difference for rays from two adjacent slits is

$$\Delta L = d \sin \theta$$

which is the same as in the case of a double slit (Fig. 25.20). If ΔL is equal to $0, \pm\lambda, \pm 2\lambda, \pm 3\lambda, \ldots$, two adjacent rays will interfere constructively at the screen. In addition, since the path length difference is the same for every pair of adjacent rays, *all* rays will interfere constructively. The condition for a bright fringe is thus

$$d \sin \theta = m\lambda \quad m = 0, \pm 1, \pm 2, \ldots \tag{25.26}$$

constructive interference for a diffraction grating

The condition for a bright fringe from a diffraction grating is identical to the condition for constructive interference from a double slit (Eq. 25.15). The angles that give constructive interference and the positions of the bright fringes on the screen are thus the same in the two cases. However, the overall intensity pattern depends on the number of slits in the grating. Figure 25.26 shows the intensity pattern found with two slits (just the double-slit pattern from Fig. 25.21), three slits, ten slits, and one hundred slits in the grating. The main intensity peaks are called the ***principal maxima*** of the pattern. These maxima occur at the same angles for 2, 3, or any number of slits. As the number of slits increases, however, smaller peaks occur between the main peaks. Most importantly, the main peaks become narrower as the number of slits increases, and when the number of slits is large, these peaks are *extremely* narrow. Compare the intensity pattern for a double slit (top of Fig. 25.26) with the result for a grating containing one hundred slits (bottom). The intensity peaks are *much* narrower with the grating.

Using a Grating to "Separate" Colors

The narrow intensity peaks for a grating with many slits make the grating a powerful tool for "separating" different light waves according to their wavelengths, that is, according to their colors. Figure 25.27A shows light composed of two different colors, red and green, incident on a grating. The grating produces an intensity pattern on the screen for each color, with the bright fringes given by the interference

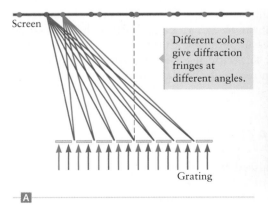

Different colors give diffraction fringes at different angles.

© Cengage Learning/Charles D. Winters

conditions in Equation 25.26. The two colors have different wavelengths, so the fringes for red and green light are at different angles and thus different places on the screen as illustrated in Figure 25.27B. The diffraction fringes appear here as spots on the screen because the laser illuminates only a small spot on the diffraction grating. Gratings are widely used to analyze the colors in a beam of light.

Diffraction and the "Color" of a CD

The diffraction gratings in Figures 25.25 and 25.27 contain straight slits arranged with a constant spacing. Similar diffraction effects occur in other situations in which many light waves interfere, such as with light reflected by a compact disc (CD). A CD contains "pits" in a reflecting layer, with the pits arranged in arcs ("tracks") as sketched in Figure 25.28A. Light reflected from these arcs acts as sources of Huygens waves, just like the slits in a grating. The reflected waves exhibit constructive interference (bright fringes) at certain angles, depending on the color (wavelength) of the light. Light reflected from a CD therefore has the colors "separated" as shown in Figure 25.28B.

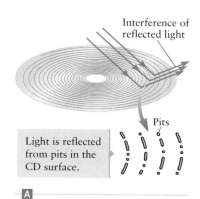

Light is reflected from pits in the CD surface.

A

© Digital Vision/Getty Images

B

| EXAMPLE 25.6 | Diffraction of Light by a Grating |

A diffraction experiment is carried out with a grating. Using light from a red laser ($\lambda = 630$ nm), the diffraction fringes are separated by $h = 0.15$ m on a screen that is $W = 2.0$ m from the grating. Find the spacing d between slits in the grating.

RECOGNIZE THE PRINCIPLE

The bright fringes in a grating's diffraction pattern occur when the waves from adjacent slits interfere constructively. The condition for constructive interference is given by Equation 25.26.

SKETCH THE PROBLEM

Figure 25.29 shows the problem and indicates the fringe spacing h and the distance W from the grating to the screen.

IDENTIFY THE RELATIONSHIPS

From the geometry of the diffraction pattern in Figure 25.29, we can calculate the angle θ corresponding to the first bright diffraction fringe. We can then use Equation 25.26 with $m = 1$ to find the slit spacing. From Figure 25.29, h and W form two sides of a right triangle containing θ, which leads to

$$\tan \theta = \frac{h}{W}$$

SOLVE

Inserting the given values of h and W and solving for θ gives

$$\tan \theta = \frac{0.15 \text{ m}}{2.0 \text{ m}} = 0.075$$

$$\theta = 4.3°$$

From Equation 25.26, the angle for the first diffraction fringe (corresponding to $m = 1$) is

$$d \sin \theta = m\lambda = \lambda$$

Rearranging to find d and using our value of θ, we get

$$d = \frac{\lambda}{\sin \theta} = \frac{630 \times 10^{-9} \text{ m}}{\sin(4.3°)} = 8.4 \times 10^{-6} \text{ m} = \boxed{8.4 \ \mu\text{m}}$$

Figure 25.28 **A** The pits in a CD have a regular spacing. Reflections from these pits act as separate but coherent sources of light, just like the slits of a grating, and produce diffraction fringes at different angles for different colors. **B** Hence, CDs appear brightly colored when viewed with reflected light.

Figure 25.29 Example 25.6.

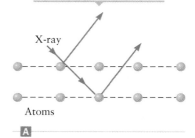

Waves are reflected by each atomic plane. These reflected waves interfere.

X-ray

Atoms

A

Omikron/Photo Researchers

B

Figure 25.30 **A** The atomic planes in a crystal reflect electromagnetic radiation. These reflected waves interfere in a manner similar to that found with visible light and a grating. Diffraction experiments with a crystal are usually done with X-rays. The angles at which X-rays interfere constructively can be used to deduce the distance between atomic planes and the arrangement of atoms in the crystal. **B** This X-ray diffraction pattern was obtained with a DNA crystal and led to the discovery of the double-helix structure of DNA. [See J. D. Watson and F. H. C. Crick, "Molecular structure of nucleic acid," *Nature* 4356:737 (1953).]

Diffraction of X-rays by a Crystal

In our discussion of gratings, we have so far considered the diffraction of light waves. Similar diffraction effects also occur with other types of electromagnetic waves (and with any other type of wave, such as sound). The atoms in a crystal are arranged in a periodic array, forming planes of atoms (Fig. 25.30A). These planes reflect electromagnetic radiation, leading to interference of the reflected waves, just as the light reflected by a CD can act as a grating (Fig. 25.28).

In a diffraction grating designed to work with light, the spacing between slits is typically 10^{-5} m to 10^{-4} m, which is 100 to 1000 times larger than the wavelength of light. For diffraction from the atomic planes in a crystal, the effective slit spacing is the distance between atomic planes, which is typically 3×10^{-10} m and hence much smaller than the spacing in a diffraction grating. To observe diffraction using the atomic planes, one therefore needs to use wavelengths much shorter than visible light, and X-rays with wavelengths on the order of 10^{-10} m are a convenient choice. Experiments of this type are called *X-ray diffraction*. By measuring the angles that give constructive interference of reflected X-rays, one can determine the spacing between atomic planes, just as the spacing between the slits in a diffraction grating can be determined from the diffraction pattern produced by a grating (Example 25.6). Because the atomic planes consist of atoms instead of simple slits, however, the intensity peaks appear as diffraction "spots" instead of fringes. An example is shown in Figure 25.30B, which shows the intensity pattern produced by diffraction of X-rays from a crystal of DNA. This diffraction pattern is very famous and led to the discovery of the structure of DNA. Because each DNA molecule has a helical structure, a crystal of DNA molecules gives a more complicated array of diffraction spots than most crystals. Through careful analysis of this pattern (along with other evidence), James Watson and Francis Crick deduced the double helix structure of DNA.

25.8 | OPTICAL RESOLUTION AND THE RAYLEIGH CRITERION

Diffraction causes light to spread out after passing through a narrow slit. A similar diffractive spreading is found whenever a beam of light passes through an opening of any shape, including the aperture of a telescope or microscope or the pupil of your eye. The result in Equation 25.23 gives the angle between the central bright fringe and the first dark fringe for the case of a slit. For a circular opening of diameter D, the result is almost the same, but the circular geometry leads to an additional numerical factor of 1.22:

$$\theta = \frac{1.22\lambda}{D} \qquad (25.27)$$

where we assume D is much larger than the wavelength. The angle in Equation 25.27 is the angle between the central bright maximum and the first minimum in the diffraction pattern. Figure 25.31A shows the diffraction pattern produced by a circular opening, and Figure 25.31B shows the light intensity as a function of angle on the screen. The first dark fringe occurs at $\theta = 1.22\lambda/D$ away from the central maximum.

This result is important in many situations. For example, suppose you are looking through a telescope at a distant star. Light from the star must pass through a circular opening (an aperture) in the telescope before it reaches a screen or your eye, as sketched in parts A and B of Figure 25.32. Diffraction at the opening produces a circular diffraction spot like the one in Figure 25.31. Now suppose there are actually two light sources (two stars) viewed again through the telescope. If these two stars are at slightly different angles with respect to the central axis of the opening, your eye will see them at slightly different viewing angles. Because they are two separate light sources, their waves are incoherent and do not interfere. Light from each source produces its own diffraction pattern, however, resulting in two diffraction spots that may overlap at your eye.

Figure 25.33 shows schematically how the two light sources in Figure 25.32C would actually appear through a telescope. Both sources produce a circular diffraction spot, and if the two sources are sufficiently far apart, two separate diffraction spots are observable as shown in Figure 25.33A. If the sources are too close together, however, their diffraction spots will overlap so much that they appear as a single spot as in Figure 25.33C. As a rough criterion, two sources will be resolved as two distinct sources of light if their angular separation is greater than the angular spread of a single diffraction spot. This result is called the **Rayleigh criterion**. For a circular opening, the Rayleigh criterion for the angular resolution is (using Eq. 25.27)

$$\theta_{min} = \frac{1.22\lambda}{D} \qquad (25.28)$$

In words, Equation 25.28 says that two objects will be resolved when viewed through an opening of diameter D (which might be a telescope or your eye) if the light rays from the two objects are separated by an angle at least as large as θ_{min}.

Figure 25.31 ◭ When a diffraction opening is circular, a circular diffraction spot is created. ◮ The intensity curve has a shape similar to that of a single slit, with the first intensity minima at an angle given by Equation 25.27.

Figure 25.32 ◭ The aperture of a telescope diffracts the incoming light. ◮ If light from a single point source passes through an opening, it produces a single diffraction spot like the one in Figure 25.31. ◰ If there are two nearby sources, their diffraction spots may overlap on the screen.

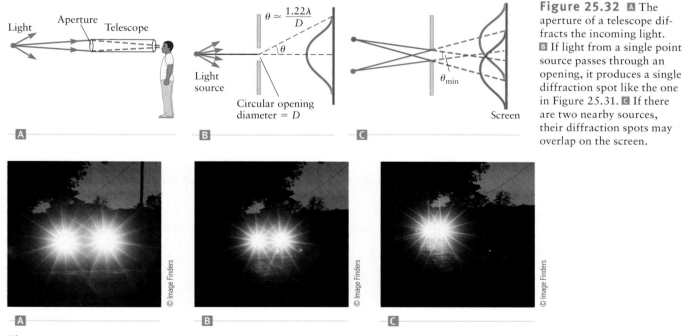

Figure 25.33 Light from two car headlights demonstrates the Rayleigh criterion. ◭ When the headlights are far apart, they can easily be resolved as two separate light sources. ◮ As the headlights are brought close together their images overlap. Here they can still just be resolved as separate sources. This corresponds to approximately the angular separation given by the Rayleigh criterion (Eq. 25.28). ◰ When the headlights are very close together, they cannot be resolved as separate sources.

EXAMPLE 25.7 ⊗ ℞ Limits of Resolution of the Eye

The Rayleigh criterion sets a limit on the resolution of the eye. Light enters the eye through a circular opening called the pupil, resulting in a diffraction spot when the light is focused on the retina. **(a)** Estimate the limiting angular resolution of the eye as measured at the retina. Assume green light is incident with $\lambda = 500$ nm. *Note:* The wavelength λ in the Rayleigh criterion is the wavelength after the waves have passed through the opening, inside the eye in this example. We must therefore allow for the index of refraction of the eye and use the wavelength found in the same way we did for light in a film in Equation 25.9. Use the (approximate) value $n_{\text{eye}} = 1.3$ for the index of refraction of the eye. **(b)** Using your result for the angular resolution in part (a), suppose two very tiny pieces of dust are separated by 50 μm (approximately the diameter of a human hair) and are located a distance $L = 25$ cm from your pupil. Can you resolve them as two separate dust particles?

(a) Let's consider part (a) of this example before moving on to part (b).

RECOGNIZE THE PRINCIPLE

The pupil is approximately circular and diffracts light, just like the aperture of a telescope (Fig. 25.31A). A true point source of light will be imaged as a "spot" on the retina; the size of this spot determines how well your eye can resolve two tiny objects that are close together. Using the Rayleigh criterion, we can determine how this effect limits the resolution of the eye.

SKETCH THE PROBLEM

Figure 25.34 shows a simplified sketch of the eye. Light is diffracted by the pupil, forming a diffraction spot on the retina.

IDENTIFY THE RELATIONSHIPS

The angular resolution limit from the Rayleigh criterion is

$$\theta_{\text{min}} = \frac{1.22\lambda_{\text{eye}}}{D}$$

(Eq. 25.28), where λ_{eye} is the wavelength of the light as it travels inside the eye. By analogy with Equation 25.9, we have

$$\lambda_{\text{eye}} = \frac{\lambda}{n_{\text{eye}}} \quad (1)$$

where λ is the wavelength outside the eye (given as 500 nm). We are given the wavelength λ and n_{eye}, but must estimate D to apply the Rayleigh criterion. Looking at your own eye in a mirror, you can estimate the pupil diameter as about $D = 5$ mm.

SOLVE

Substituting the λ_{eye} from Equation (1) and the estimated pupil diameter gives

$$\theta_{\text{min}} = \frac{1.22(\lambda/n_{\text{eye}})}{D} = \frac{1.22(5.0 \times 10^{-7} \text{ m})/1.3}{5 \times 10^{-3} \text{ m}} = \boxed{9 \times 10^{-5} \text{ rad}} \quad (2)$$

Notice that the units here are radians because the expression for θ_{min} is a dimensionless ratio.

What does it mean?
The calculated angle is only 9×10^{-5} rad = 0.005°! The angular resolution of the human eye is quite good; two beams of light separated by only 0.005° will be seen as separate spots on the retina.

Pupil of eye   Retina

D

Diffraction causes the image from a point object to spread over an angle θ_{min} given by Rayleigh's criterion.

Figure 25.34 Example 25.7. The angular resolution of the eye is determined by the Rayleigh criterion.

(b) Now let's apply the result from part (a) to determine how well the eye can resolve two small pieces of dust.

SKETCH THE PROBLEM

The geometry is shown in Figure 25.35. The dust particles are a distance L from the eye and are separated by a distance h. Given the values of L and h, we want to know if the particles can be resolved as separate pieces of dust.

IDENTIFY THE RELATIONSHIPS

When you view the dust particles, your retina is detecting light reflected by the particles. The dust particles thus act as separate sources of light, similar to the two sources of light in Figure 25.32C. From Figure 25.35, we see that light waves from the dust have an angular separation at the retina of $\Delta\theta_{source}$, where

$$\tan(\Delta\theta_{source}) \approx \frac{h}{L}$$

Because the dust particles are very close together, $\Delta\theta_{source}$ is very small and we can use the small-angle approximation $\tan(\Delta\theta) \approx \Delta\theta$.

SOLVE

We are left with

$$\tan(\Delta\theta_{source}) \approx \Delta\theta_{source} \approx \frac{h}{L}$$

Inserting the values of h and L, we find

$$\Delta\theta_{source} = \frac{h}{L} = \frac{50 \times 10^{-6} \text{ m}}{0.25 \text{ m}} = 2.0 \times 10^{-4} \text{ rad} \qquad (3)$$

The limiting value of θ_{min} in Equation (2) is smaller than the angular separation of the dust particles $\Delta\theta_{source}$ in Equation (3). Hence, $\boxed{\text{the eye can resolve}}$ the pieces of dust as two separate objects.

What does it mean?
Our estimation of the eye's resolution using the Rayleigh criterion is only an approximate treatment, but even though we have ignored factors such as the separation of the detector cells (called rods and cones) in the retina, our estimate matches the actual resolution of the eye to within about a factor of two. You should check for yourself. You can do so by carefully placing two strands of hair side by side and pushing them as close as possible, but not so close that they appear as one strand of hair. Then use a microscope or magnifying glass to measure the distance between the strands.

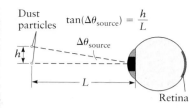

Figure 25.35 Example 25.7. Application of the Rayleigh criterion. Can your eye resolve the dust particles?

CONCEPT CHECK 25.6 | **The Rayleigh Criterion and the Diameter of a Telescope**

The Rayleigh criterion determines the ultimate resolution of a telescope. This resolution is commonly measured in terms of the smallest angular separation θ_{min} between two objects that the telescope can resolve (Fig. 25.32C). If the diameter of the telescope aperture is increased, does θ_{min} (a) get smaller, (b) get larger, or (c) stay the same?

Limits on Focusing

Suppose you are asked to focus a light beam to a very tiny spot using a very high quality (and expensive) lens. Is there any limit to how small you can make this spot? According to the ray optics approximation of Chapter 24, a perfect lens will

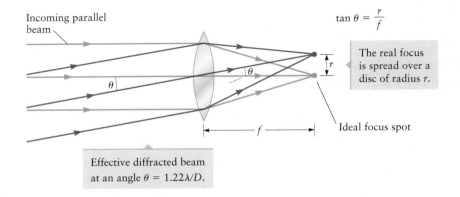

Figure 25.36 Focusing light from a distant object. The lens acts as an opening as in Figure 25.32, causing the image to be broadened ("smeared out") by diffraction. This sets limits on focusing and resolution of the lens. *Note:* The colors of these rays do *not* represent the colors of the light waves.

focus a very narrow parallel beam of light to a precise point at the focal point of the lens. However, the ray optics approximation ignores diffraction. If the lens has a diameter D, it acts just like the openings in Figures 25.31 and 25.32 and according to the Rayleigh criterion produces a diffracted beam spread over a range of angles up to $\theta \approx 1.22\lambda/D$ (Eq. 25.27). This spreading is modeled in Figure 25.36, which shows the original beam (drawn in green) traveling along the principal axis of a lens. The diffraction spreading is modeled by the second incident beam (drawn in red), which makes an angle $\theta \approx 1.22\lambda/D$ relative to the original beam. Without diffraction, the rays would all focus (intersect) at a point on the axis a distance f from the lens, where f is the focal length. Diffraction spreads this point over a disk of radius r as shown near the intersection of the rays drawn in red. From the geometry in Figure 25.36, we find $\tan\theta = r/f$. The angle θ is usually small, so we can use the small-angle approximation to get

$$\tan\theta \approx \theta \approx \frac{r}{f}$$

which can be rearranged to solve for r:

$$r = f\theta \tag{25.29}$$

Inserting θ from Equation 25.27 gives

$$r = f\theta = f\frac{1.22\lambda}{D} \tag{25.30}$$

The lens thus focuses parallel incoming light to a "smeared-out" circular spot of radius r. According to Equation 25.30, the value of r depends on the diameter D of the lens and its focal length f, but these two quantities are not independent of each other. The lens maker's formula from Chapter 24 tells how f is related to the radius of curvature R of the lens. It is not physically possible to make the diameter of the lens opening larger than $2R$. (See Fig. 24.37.) For a lens with an index of refraction n and two convex surfaces both with a radius of curvature R, the focal length is

$$\frac{1}{f} = \frac{2(n-1)}{R}$$

as obtained from the lens maker's formula (Eq. 24.32) with $R_1 = R_2 = R$. We have just argued that the upper limit on the opening of a lens is $D = 2R$, so the limit on focal length is

$$\frac{1}{f} = \frac{2(n-1)}{D/2}$$

which leads to

$$f = \frac{D}{4(n-1)}$$

For a typical lens material such as glass, $n \approx 1.5$ (see Table 25.1), and we get

$$f \approx \frac{D}{2} \qquad (25.31)$$

Inserting this into Equation 25.30 gives the radius of the focus spot (also called the focal disk) as

$$r = f \frac{1.22\lambda}{D} \approx \frac{D}{2} \frac{1.22\lambda}{D} = \frac{1.22\lambda}{2} \qquad (25.32)$$

Our result for r is only approximate because it depends on our estimate for how the lens opening D depends on its radius of curvature R. In the spirit of an estimation problem, we can thus write

$$r \approx \lambda \qquad (25.33)$$

In words, Equation 25.33 says that the wave nature of light limits the focusing qualities of even a perfect lens. *It is not possible to focus a beam of light to a spot smaller than approximately the wavelength.*

This fundamental limit on focusing determines the ultimate performance of microscopes and other optical instruments, as we'll explain in Chapter 26. Equation 25.33 defines yet again the boundary between the regimes of ray optics and wave optics. As a general rule, the ray approximation of geometrical optics can only be applied at size scales much greater than the wavelength. When a slit or a focused beam of light is made so small that its dimensions are comparable to λ, we must expect that diffraction effects will be important. These effects are the essence of wave optics.

25.9 | WHY IS THE SKY BLUE?

The wave nature of light leads to interesting effects when light reflects from very small objects. When light reflects from a plane surface, the angle of reflection is equal to the angle of incidence (the law of reflection, Eq. 24.1), but this result is true only when the size of the reflecting surface is much greater than the wavelength. Figure 25.37 shows the opposite situation, when λ is larger than the reflecting object. The reflected waves now travel away in all directions and are called *scattered* waves. This behavior is found for all types of waves, including sound (Fig. 13.14).

The amplitude of the scattered wave depends on the size of the scattering object relative to the wavelength. For particles much smaller than the wavelength of light, the amplitude of the scattered wave grows rapidly as the wavelength is decreased (Fig. 25.38). Blue light and ultraviolet light are thus scattered much more strongly than red light. This effect is called *Rayleigh scattering.*

Figure 25.38 helps answer the very simple everyday question, "Why is the sky blue?" The Earth's atmosphere is composed mainly of oxygen and nitrogen molecules, which are both colorless, so one might expect the atmosphere to be colorless. We know that the atmosphere is blue on a normal clear day, however, and Figure 25.39 shows why. The light that we see from the sky overhead is sunlight scattered by N_2, O_2, and other molecules in the atmosphere. These molecules are much smaller than the wavelength of visible light, so (according to Fig. 25.38) they scatter blue light more strongly than red light, giving the atmosphere its blue color.

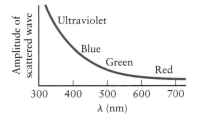

Figure 25.38 When light is scattered by small particles in the atmosphere, the amplitude of the scattered wave in Figure 25.37 increases as the wavelength decreases.

Table 25.1 Index of Refraction of Some Common Substances for Yellow Light

SUBSTANCE	INDEX OF REFRACTION
Air	1.0003
Water	1.33
Glass	1.5[a]
MgF_2	1.38

[a]Typical value. The precise value depends on composition.

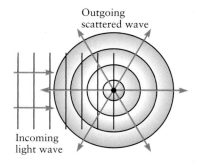

Figure 25.37 When a plane light wave strikes an object that is much smaller than the wavelength, a spherical wave propagates outward in all directions from the object. This outgoing wave is called scattered light. A similar phenomenon occurs for sound waves; see Section 13.5 and Figure 13.14.

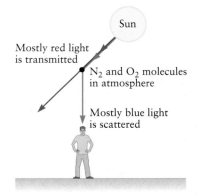

Figure 25.39 Scattering of sunlight by N_2, O_2, and other molecules in the air is responsible for the color of the atmosphere. Since these molecules are smaller than the wavelength of visible light, they scatter short-wavelength light most strongly, giving the atmosphere its blue color.

EXAMPLE 25.8 | Color of the Sun

In Chapter 14, we learned that the Sun emits light over a range of wavelengths (colors) as shown in Figure 14.32. The maximum intensity falls close to the middle of the visible range, and the Sun appears white when it is overhead (Fig. 25.40A), but when the Sun is close to the horizon (at sunrise or sunset), it has a distinctly reddish color (Fig. 25.40B). Explain this effect in terms of Rayleigh scattering.

Figure 25.40 Example 25.8.
A When the Sun is high in the sky, it has a white to pale yellow color due to its blackbody spectrum, which peaks near the middle of the visible range. **B** When the Sun is low on the horizon, it appears more reddish in color.

A   B

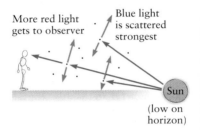

More red light gets to observer

Blue light is scattered strongest

Sun (low on horizon)

Figure 25.41 The red color is a result of scattering by molecules and small particles in the atmosphere, removing most of the blue light. Compare with Figure 25.39.

SKETCH THE PROBLEM

Figure 25.41 shows light traveling from the Sun to an observer on the Earth.

IDENTIFY THE RELATIONSHIPS AND SOLVE

When the Sun is near the horizon, the sunlight you observe travels a greater distance through the atmosphere than when the Sun is directly overhead. Because of Rayleigh scattering, molecules and other small particles in the air scatter short wavelengths (blue light) more strongly than long wavelengths (red light). By the time the light from a rising or setting Sun reaches an observer, a substantial fraction of the blue component has been scattered away. The remaining light (Fig. 25.41) contains a greater proportion of red light than when the Sun is high in the sky.

What does it mean?

The color of an object depends on both the colors of light the object emits and what happens to the different colors (wavelengths) as they travel to an observer. The scattering of light can have a big effect on "the" color of an object.

CONCEPT CHECK 25.7 | Color of the Atmosphere

Why does a forest fire tend to give the sky a pronounced reddish color?
  (a) The small particles of soot and molecules produced by combustion from a fire usually have a red color.
  (b) The small particulates from a fire scatter red light from the Sun less than blue light.
  (c) Heat from the fire removes water vapor from the air.

Insight 25.3

WHY IS THE SKY BLUE AND NOT VIOLET?

The Rayleigh scattering of light by molecules in the atmosphere gets stronger as the wavelength decreases. We have claimed that this scattered light gives the sky its blue color (Fig. 25.39), but violet light has a shorter wavelength than blue light, so why isn't the sky violet instead? Two factors combine to make the sky blue instead of violet. First, the Sun emits more strongly in the blue than in the violet range (Fig. 14.32). Second, our eyes are more sensitive to blue light than to violet light. Hence, even though violet light is scattered more strongly than blue light, we still perceive the sky as blue.

25.10 | THE NATURE OF LIGHT: WAVE OR PARTICLE?

The observations of double-slit interference and of diffraction by a single slit show convincingly that light has wave properties. However, certain properties of light and optical phenomena can *only* be explained with the particle theory of light. One very common effect that can only be correctly explained by the particle theory of light is color vision.

✖ Origin of Color Vision

Human and many other vertebrate eyes are sensitive to color; that is, they are able to distinguish light waves according to their wavelength. This is accomplished by cells in the retina called *cones*.[3] There are three types of cones, and Figure 25.42 shows that these different cone cells are sensitive to light of different colors. Some cones are most sensitive to red light—that is, they send a large signal to the brain when exposed to red light and a much smaller signal with either blue or green light—and other cones are sensitive to mainly blue or green light.

How can we explain the behavior of different cone cells? The energy carried by an ordinary wave (such as a sound wave) depends on its amplitude; a large amplitude corresponds to a high intensity and vice versa. So, we should always be able to arrange for two light waves with different colors (say, red and blue) to carry the same total energy just by adjusting their relative intensities, but even when two such waves carry the same total energy into the eye, the red cone cells still respond mainly to red light and the blue cone cells respond mainly to blue light. How can a cone cell be much more sensitive to light of one color (i.e., one wavelength) than another despite such differences in intensity? How can a red cone cell be almost completely insensitive to blue and violet light, or a blue cone cell insensitive to red light? Studies of these and related questions led to the quantum theory of light (Chapter 28). According to that theory, particles of light called *photons* carry an amount of energy that depends on the frequency of the light, with blue photons carrying more energy than red photons. A complete explanation of color vision involves the properties of photons along with the quantum properties of molecules inside cone cells, and we'll come back to this problem when we describe quantum theory in Chapter 28.

So, although Newton did not realize it, the evidence for his particle theory of light was literally right in front of him. No one else realized it either until about 1905, when the beginnings of a quantum theory of light were first developed by Einstein. If light is a particle, though, how do we explain interference and diffraction? After all, aren't they strictly wave phenomena? It would thus seem that we have strong evidence that light is both a particle *and* a wave! This concept is called "particle-wave duality." Reconciling these notions is a major part of quantum theory, as we'll discuss in Chapter 28.

The study of light has been a central thread in the development of physics as a science, connecting Newton and mechanics to the theory of waves, to Maxwell's work on electromagnetism, and all the way to quantum theory. In Chapter 28, we'll see how quantum theory extends the concept of particle-waves to electrons, protons, and other objects that a classical physicist (Newton!) would have classified as particles. The discovery that light has properties of both classical particles and waves thus opened the way to 20th-century physics.

Figure 25.42 Color vision is due to light detectors in the eye called cones. The three types of cones cells are sensitive to light from different regions of the visible spectrum.

---

[3]Cells in the retina called *rods* also detect light, but rods are not sensitive to color.

SUMMARY | Chapter 25

KEY CONCEPTS AND PRINCIPLES

Conditions for interference
Light is a wave and exhibits interference if certain conditions are met.
1. Two or more waves must travel through different regions of space.
2. The waves must be brought together at a common point.
3. The waves must be *coherent*.

(Continued)

Constructive and destructive interference
Interference of two waves is *constructive* if the waves are *in phase* and is *destructive* if the waves are 180° *out of phase*.

Phase change upon reflection
When light reflects from a surface, the phase of the reflected wave changes by 180° if the index of refraction on the incident side is smaller than the index on the opposite side. This effect, called a *phase change on reflection*, is important in thin-film interference.

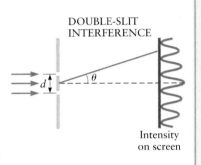

Phase change on reflection if $n_1 < n_2$

Huygens's principle
All points on a wave front act as sources of new spherical waves. Huygens's principle explains how light passing through a narrow opening is *diffracted*.

Double-slit interference
Young's double-slit experiment demonstrates that light passing through two narrow slits can interfere. The interference pattern on a distant screen shows constructive or destructive interference, depending on the outgoing angle. Constructive interference (and thus a bright fringe in the interference pattern) is found when

$$d \sin \theta = m\lambda \quad m = 0, \pm 1, \pm 2, \pm 3, \ldots \quad \textbf{(25.15)} \text{ (page 856)}$$

Destructive interference resulting in a dark fringe is found when

$$d \sin \theta = (m + \tfrac{1}{2})\lambda \quad m = 0, \pm 1, \pm 2, \pm 3, \ldots \quad \textbf{(25.16)} \text{ (page 856)}$$

DOUBLE-SLIT INTERFERENCE

Intensity on screen

Single-slit diffraction
Single-slit diffraction occurs when light passes though a narrow opening. Dark fringes (destructive interference) occur when

$$w \sin \theta = \pm m\lambda \quad m = \pm 1, \pm 2, \ldots \quad \textbf{(25.22)} \text{ (page 860)}$$

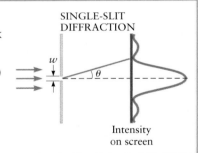

SINGLE-SLIT DIFFRACTION

Intensity on screen

Michelson interferometer
The *Michelson interferometer* uses the interference of light waves to measure small displacements. It was first used in fundamental studies of light and tests of relativity (Chapter 27) and is still widely used in studies of the optical properties of matter.

Thin-film interference
Light waves reflected from the two surfaces of a thin film can interfere to produce brightly colored interference fringes. The relative phases of the interfering waves are affected by the different path lengths traveled by the two waves and by any phase changes on reflection.

(Continued)

Rayleigh criterion

When light passes through a circular opening, diffraction causes the light beam to spread. The **Rayleigh criterion** states that two light beams can be resolved only if their directions differ by an amount as large as or greater than

$$\theta_{min} = \frac{1.22\lambda}{D} \qquad \textbf{(25.28)} \text{ (page 865)}$$

This result means that it is impossible to focus a light beam to a spot size much smaller than the wavelength.

Rayleigh scattering

Rayleigh scattering of light by small particles in the atmosphere produces the color of the atmosphere. This scattering is strongest for shorter wavelengths and gives the atmosphere its normal blue color.

QUESTIONS

1. Is it possible for light waves of different color to exhibit constructive or destructive interference? Explain.

2. Why don't you observe interference between light waves produced by two lightbulbs?

3. Figure Q25.3 shows the bands of color that are produced when a thin soap-bubble film is exposed to white light. What mechanism produces the colors seen here? Is the bubble of uniform thickness? If not, where is the bubble thinnest? Thickest? How do you know?

Figure Q25.3

4. If a soap film suspended in air is extremely thin, what color will it be when viewed in reflected light?

5. Explain why two light waves can interfere only if they are coherent.

6. Yellow light is used to make a double-slit interference pattern with two very narrow slits. If the same slits are used with blue light, will the separation between two adjacent bright fringes be larger or smaller?

7. **SSM** How does a double-slit interference pattern change when the following changes are made? Do the interference fringes become closer together or farther apart? Explain your answers.
   (a) The two slits are brought closer together.
   (b) The two slits are moved farther apart.

(c) The wavelength is increased.
(d) The wavelength is decreased.
(e) The entire experiment is immersed in water.

8. Green light is used to make a single-slit diffraction pattern with a particular slit. If the same slit is used with red light, will the width of the central bright fringe be wider or narrower?

9. Explain how a thin coating on a lens (called an antireflection coating) can reduce the amount of reflected light. Most expensive lenses have these coatings.

10. Some vivid rainbows, called supernumerary rainbows, have extra inner arcs of color (Fig. Q25.10) that do not follow the normal pattern in Figure 24.48. The many internal reflections that occur within a spherical raindrop can produce rays of different colors leaving the drop along the same path. Given this information, can you suggest a mechanism that could produce these extra bows?

Figure Q25.10

11. Design an experiment in which light from a helium–neon laser ($\lambda = 630$ nm) can be used to measure (a) the width of a single slit and (b) the distance between two nearby slits.

12. In Section 25.2, we assumed the Michelson interferometer was operated with monochromatic light (light of a single wavelength). Suppose a source of white light is used instead. Is it

possible to have constructive interference and hence a bright fringe with white light? If your answer is yes, what value(s) of the path length difference ΔL is required for constructive interference and thus a bright fringe?

13. Some small, bright lightbulbs on a strand (Fig. Q25.13) are grouped in pairs according to color and then spaced 5 cm apart. (a) From a long distance, it becomes difficult to tell if each color is coming from one bulb or from a pair. Why? (b) There is a distance from the lights where the pairs of blue bulbs can be resolved but the pairs of red bulbs would look like one point. What feature of the red and blue light would account for this difference?

Figure Q25.13

14. **SSM** The star in the bottom of the Northern Cross asterism is actually a binary pair. The photograph in Figure Q25.14 came from a telescope with a mirror 10 inches in diameter. Which feature of the telescope is more important in enabling it to resolve the pair when your eye sees them as one point of light, (a) magnification or (b) the size of mirror or primary lens? (You can assume fluctuations in the atmosphere as described in Section 24.7 are negligible in this case.)

Figure Q25.14

15. A guitar player is at a rock concert and watches the fingers of the lead guitarist from a few rows back (Fig. Q25.15). She notes that she can see the guitarist's chord positions clearly when a yellow or blue spotlight falls on him through the smoke, but when a bright purple light is used, the scene becomes blurry. Why can she see less detail when the purple light is used?

Figure Q25.15

16. ⊗ **Eyeshine.** The eyes of cats, foxes, raccoons, and many other mammals brightly reflect light that shines into them back toward the source and makes the animal's eyes seem to glow as seen in Figure Q25.16. The reflection occurs within a thin membrane behind the retina called the *tapetum lucidum*, which allows the animal's eye more sensitivity at night. There is no metallic material within the eye, and membrane does not look metallic under magnification. Describe a mechanism that might produce the bright reflection.

Figure Q25.16

17. You are designing a billboard that will be viewed at night. The letters used on this billboard are made of an array of bright lights like that shown in Figure Q25.17. If you want to make the sign legible from the greatest distance, which color should you choose for the lights? Why?

Figure Q25.17

18. Which single-slit diffraction experiment will give the widest central bright fringe, (a) a 900-nm slit with light of wavelength 500 nm or (b) a 600-nm slit with light of wavelength 450 nm?

19. The sky has a blue color due to the scattering of light by molecules in the air (Fig. 25.39). Explain why the sky's color can be a deeper blue when viewed through polarizing sunglasses.

20. Figure 25.21 shows the intensity pattern produced in a double-slit interference pattern with two very narrow slits. Suppose one of the slits is covered so that light can only pass through one slit. Sketch what the new intensity pattern would look like.

21. Constructive interference occurs when the crests of two coherent waves overlap (Fig. 25.1B). What happens when the troughs of two coherent waves overlap? Is the interference constructive or destructive? Explain.

SSM = solution in Student Companion & Problem-Solving Guide
★ = intermediate   ✪ = challenging

Ⓧ = life science application
Ⓡ = reasoning and relationships problem

25.1 COHERENCE AND CONDITIONS FOR INTERFERENCE

1. Light waves are emitted coherently from two sources and detected at a spot equidistant from the two sources. Do these waves interfere constructively or destructively?

2. Sound waves with a wavelength of 0.25 m are emitted coherently (i.e., in phase with each other) from two sources. These waves are found to interfere constructively at a point P that is 1.60 m from one of the sources. What is the distance from P to the other source? *Note:* There is more than one correct answer. Give three of them.

3. Coherent sound waves with $\lambda = 0.75$ m are emitted from points P_1 and P_2 in Figure P25.3. Do these waves interfere constructively or destructively at point P_3?

4. Two sources produce coherent light waves that come together at a detector located 1.50000 mm from one source and 1.50500 mm from the other. If the two waves interfere constructively, what is a possible value for the wavelength? Assume the light is visible light.

5. The detection point in Problem 4 is moved so that it is 1.50000 mm from one source and 1.50150 mm from the other. If the two waves interfere destructively, what is a possible value for the wavelength? Assume the light is visible light and keep three significant figures in your calculation.

6. SSM ★ Coherent sound waves are emitted from points P_1 and P_2 in Figure P25.6. The waves have a wavelength $\lambda = 1.5$ m. (a) Find the point on the x axis nearest the origin at which the waves interfere constructively. (b) Find the point on the x axis nearest the origin at which the waves interfere destructively.

Figure P25.6

7. Waves from a radio station have a wavelength of 250 m. These waves can travel directly from the antenna to a receiver or can reflect from a nearby mountain cliff and then reach the receiver (Fig. P25.7). If the distance from the receiver to the cliff is $L = 1000$ m, is there constructive or destructive interference at the receiver? Assume there is no phase change when the radio wave reflects from the cliff.

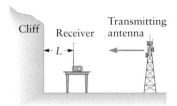

Figure P25.7 Problems 7 and 8.

8. ✪ Consider again the interfering radio waves in Problem 7, but now assume the amplitude of the reflected wave is exactly half the amplitude of the wave that arrives directly. If the amplitude

of the direct wave is A, what is the intensity at the receiver when there is (a) constructive interference or (b) destructive interference?

9. ✪ An omnidirectional speaker that emits sound with a wavelength λ is positioned between two vertical walls spaced 2.00 m apart as shown in Figure P25.9. A listener approaches as shown and finds that there is constructive interference of the waves that reflect from the two walls when he is a distance $L = 5.00$ m from the speaker. Find the longest possible wavelength.

Figure P25.9

25.2 THE MICHELSON INTERFEROMETER

10. One of the mirrors in a Michelson interferometer is moved a distance of 2.0 mm. Meanwhile, the interference intensity moves through 7000 dark fringes. What is the wavelength of the light?

11. Suppose the wavelength in Problem 10 is increased by a factor of 1.2. How many dark fringes would now be found when the mirror moves the same distance?

12. ✪ Both arms of a Michelson interferometer are enclosed in pipes so that the air can be removed from the regions between the mirrors and the beam splitter. There is initially a vacuum in both regions, and the mirrors are the same distance from the beam splitter. Air is then let into one of the pipes, and light that travels to and from one mirror now propagates in air while the other part of the interferometer is still in a vacuum. If the interference pattern shifts to the 4000th bright fringe when the air is added, what is the distance from a mirror to the beam splitter? Assume $\lambda = 600$ nm.

13. ✪ Consider a Michelson interferometer that uses a sodium lamp as its source. A sodium lamp gives a large intensity at two wavelengths, 589.00 nm and 589.59 nm, in the yellow range of the visible spectrum. The mirrors are adjusted so that there is constructive interference with the Nth bright fringe for one of these wavelengths and the $(N + 1)$st bright fringe for the other wavelength. Find N. Be careful with significant figures in your calculation, and give your answer for N to two significant figures.

14. SSM ★ Consider a Michelson interferometer operated with light from a sodium lamp with $\lambda_{Na} = 589$ nm. Suppose the interferometer is used to check the accuracy of a meterstick by moving one of the mirrors a distance of precisely 1.00 m (as determined by the meterstick) and counting bright fringes as they appear at the detector. How many bright fringes will be observed while the mirror is being moved?

15. ✪ A Michelson interferometer is used to check the length of a "standard" platinum meterstick. The interferometer is operated with light from a specially constructed helium–neon laser, which gives red light with λ = 632.99139822 nm. (a) The standard meterstick is placed next to one of the interferometer's arms, and the mirror is moved from one end of the meterstick to the other. How many interference fringes N are counted as the mirror is moved? (b) Suppose the wavelength λ is known with an accuracy of ±0.00000002 nm. Also suppose the number of interference fringes N is known to an accuracy of ±1. What determines the uncertainty in the measured length of the meterstick, the uncertainty in the wavelength or the uncertainty in N? (c) In a real experiment, you would not simply count the number of bright fringes but instead would measure "fractions of a fringe" by measuring the intensity and comparing it with the sinusoidal curve in Figure 25.5. If the number of fringes could be measured to ±0.0001, would the uncertainty in N still be a limiting factor?

25.3 THIN-FILM INTERFERENCE

16. An extremely thin film of soapy water ($n = 1.35$) sits on top of a flat glass plate with $n = 1.50$. The soap film has a red color when viewed at normal incidence. What is the thickness of the film? ($\lambda_{red} = 600$ nm.)

17. Repeat Problem 16 assuming the soap film rests on a material having an index of refraction $n = 1.10$.

Figure P25.18

18. ⭐ A plastic film of thickness 250 nm appears to be green ($\lambda_{green} = 500$ nm) when viewed in reflection at normal incidence (Fig. P25.18). What is the plastic's index of refraction?

19. ✪ A very thin sheet of glass ($n = 1.55$) floats on the surface of water ($n = 1.33$). When illuminated with white light at normal incidence, the reflected light consists predominantly of the wavelengths 560 nm and 400 nm. How thick is the glass?

20. A compact disk uses the interference of reflected waves from the CD to encode information. Light from a red laser ($\lambda = 630$ nm) is reflected from small pits in the CD, and waves from the bottom of a pit and the adjacent top edge interfere (Fig. P25.20). What is the minimum pit depth for which the two waves interfere destructively? Assume this pit is on the surface of the CD so that the region above the pit is air.

Figure P25.20 Problems 20 and 21.

21. ⭐ In a real CD, the pits in Figure P25.20 are covered with a layer of plastic so that surface scratches do not spoil the pits. If the index of refraction of the plastic coating is $n = 1.70$, what minimum depth should a pit have to give destructive interference using light with $\lambda = 630$ nm?

22. SSM ⭐ Two glass plates are arranged as in Figure 25.12 so that light reflected from the bottom surface of the top plate interferes with light reflected from top surface of the bottom plate. As one moves along the x axis, the total reflected intensity alternates between bright and dark fringes. If there are a total of 25 bright fringes across the entire width of the plates, what is the plate spacing d at the right edge? Assume the source of illumination is blue light with a wavelength of 420 nm.

23. ✪ A device called a Fabry–Perot interferometer contains two parallel mirrors as shown in Figure P25.23. Multiple reflec-

tions occur between the inner mirror surfaces, and the waves emitted at the top can interfere. Assume the light waves travel nearly perpendicular to the mirrors, the emitted waves interfere constructively, and the spacing between the mirrors is exactly 3.5 μm. What is a possible value for the wavelength in the visible range between 600 and 700 nm?

Figure P25.23

25.4 LIGHT THROUGH A SINGLE SLIT: QUALITATIVE BEHAVIOR

24. SSM Single-slit diffraction can be observed with any type of electromagnetic wave (not just light). Suppose you want to make a diffraction slit whose width is five times larger than the wavelength for the following cases. How wide would the slit be?
(a) A radio wave for your favorite FM station ($f = 100$ MHz)
(b) The waves that carry your cell phone signals ($f = 2.0$ GHz)
(c) Red light ($f = 5.0 \times 10^{14}$ Hz)
(d) The X-rays used in your dentist's office ($f \approx 4 \times 10^{20}$ Hz)

25.5 DOUBLE-SLIT INTERFERENCE: YOUNG'S EXPERIMENT

25. A double-slit interference experiment is performed with two very narrow slits separated by 0.15 mm. The experiment uses red light with a wavelength of 600 nm and projects the interference pattern onto a screen 7.5 m away from the slits. (a) What is the distance between two nearby bright fringes on the screen? (b) What is the distance between two nearby dark fringes on the screen? Assume these fringes are all near $\theta = 0$.

26. Two small stereo speakers are 2.5 m apart and act as a double-slit interference experiment with sound waves. If the opposite wall of the room is 5.5 m away and the sound frequency is 1.5 kHz, what is the distance between the two interference maxima nearest the center of the wall? Assume the speakers act as point sources of sound.

27. ⭐ In Problem 26, you were told to assume the speakers could be treated as point sources. Which of the following statements justifies this assumption?
(a) Small speakers could be much smaller than the wavelength of the sound waves.
(b) The speakers could be larger than the wavelength but smaller than the size of the room.
(c) The wavelength is smaller than the size of the room.
(d) The wavelength is smaller than the spacing between the speakers.

28. SSM ⭐ Monochromatic light ($\lambda = 500$ nm) passes through two slits and onto a screen 2.5 m away. (a) If two nearby bright fringes are separated by 12 mm, what is the slit spacing? (b) If the slit spacing is reduced by a factor of three, what is the new distance between bright fringes?

29. Two slits are separated by 0.35 mm and are used to perform a double-slit experiment with a screen 1.5 m away from the slits. If the distance between a dark region and the nearest bright spot on the screen is 1.0 mm, what is the wavelength?

30. ✪ Two narrow slits are used to produce a double-slit interference pattern with monochromatic light. The slits are separated by 0.90 mm, and the interference pattern is projected onto a screen 9.5 m away from the slits. The central bright fringe is at a certain spot on the screen. Using a ruler with one end placed at the central fringe, you move along the ruler passing by two more bright fringes and find that the next bright fringe is 20 mm away from the central fringe. What is the wavelength of the light?

31. Two narrow slits are used to produce a double-slit interference pattern with monochromatic light ($\lambda = 450$ nm). The slits are

separated by 0.90 mm, and the interference pattern is projected onto a screen 7.5 m away from the slits. (a) What is the distance between two nearby bright fringes on the screen? (b) What is the distance between two nearby dark fringes on the screen? Assume the fringes are near the center of the interference pattern.

32. ✪ A Young's double-slit interference experiment gives bright fringes separated by 2.5 cm on a screen when the experiment is performed in air. The entire apparatus then placed under water ($n = 1.33$). What is the new spacing between bright fringes? Assume the fringes are near the center of the interference pattern.

33. ✪ Two narrow slits are used to produce a double-slit interference pattern with monochromatic light. It is found that the third bright fringe from the center with blue light ($\lambda_1 = 420$ nm) occurs at the same place on the screen as the second bright fringe with light of an unknown wavelength λ_2. Find λ_2. Here the central bright fringe is fringe number "zero."

25.6 SINGLE-SLIT DIFFRACTION: INTERFERENCE OF LIGHT FROM A SINGLE SLIT

34. Red light ($\lambda = 600$ nm) is diffracted by a single slit, and the first intensity minimum (the first dark fringe) occurs at $\theta = 6.9°$. What is the width of the slit?

35. SSM ✪ ℝ A doorway acts as a single slit for light that passes through it, but a doorway is much larger than the wavelength of visible light, so the diffraction angles are very small. Estimate the angle of the 10th diffraction minimum for a typical doorway.

36. A slit with width 0.15 mm is located 10 m from a screen. If light with $\lambda = 450$ nm is passed through this slit, what is the distance between the central bright fringe and the third dark fringe on the screen?

37. Light with $\lambda = 550$ nm passes through a single slit and then illuminates a screen 2.0 m away. If the distance on the screen from the first dark fringe to the center of the interference pattern is 5.5 mm, what is the width of the slit?

38. A helium–neon laser ($\lambda = 630$ nm) is used in a single-slit experiment with a screen 3.0 m away from the slit. If the slit is 0.25 mm wide, what is the width of the central bright fringe on the screen? Measure this width using the locations where there is destructive interference.

25.7 DIFFRACTION GRATINGS

39. The slits in a diffraction grating are spaced 25 μm apart. What is the diffraction angle for the second bright fringe away from $\theta = 0$ with red light ($\lambda = 600$ nm)?

40. ✪ A diffraction grating is a square 1.0 cm on a side and contains 3000 slits. What is the angle of the first intensity maximum in the diffraction pattern with blue light ($\lambda = 450$ nm)?

41. ✪ A sodium lamp produces yellow light with wavelengths λ and $\lambda + \Delta\lambda$, where $\lambda = 589.00$ nm and $\Delta\lambda = 0.59$ nm. This light is incident on a diffraction grating with a slit spacing of 15 μm, and the diffracted light illuminates a screen 2.5 m away. What is the distance on the screen between the fifth bright fringes for these two wavelengths?

42. ✪ Light from a helium–neon laser is incident on a diffraction grating and then illuminates a screen that is 10 m away. If the bright diffraction fringes near $\theta = 0$ are spaced 0.50 m apart, what is the spacing between the slits in the grating?

43. SSM ✪ The atomic planes in a crystal are spaced $d = 0.50$ nm apart. These planes are used in a diffraction experiment with X-rays with $\lambda = 0.05$ nm. Assume the bright fringes in the diffraction pattern are at the same angles as for a grating with slits with the same value of d. What are the diffraction angles for the first three bright fringes?

44. ✪ Consider again diffraction from a crystal in Problem 43, but now assume light with $\lambda = 600$ nm is used. What is the diffraction angle for the first bright fringe? Your answer explains why visible light cannot be used in diffraction experiments with crystals.

45. ✪ You are designing a lecture demonstration on diffraction gratings. You plan to use light from a laser pointer ($\lambda = 630$ nm) and shine the light from the grating on the far wall of the lecture room ($W = 9.0$ m). You want the separation between the central diffraction spot and one of the spots closest to it on the wall to be 1.0 m apart (Fig. 25.27B). The grating store offers gratings with a specified number of slits per millimeter. How many slits per millimeter do you want for your demonstration?

25.8 OPTICAL RESOLUTION AND THE RAYLEIGH CRITERION

46. ✪ ℝ Is the resolution of an optical instrument greater with blue light or with red light? Explain your answer. By approximately what factor are the resolutions different?

47. ✪ ℝ The camera on a spy satellite has a lens with a diameter of 1.5 m. This satellite is in low-Earth orbit about 2.9×10^5 m above the surface of the Earth (see Chapter 5, Eq. 5.28). What is the approximate size of the smallest feature the camera can resolve when taking a picture of something on the Earth's surface? Assume blue light with a wavelength of 400 nm. Ignore the effect of the Earth's atmosphere on the resolution.

48. ✪ A terrestrial telescope (a "spyglass") with a diameter of 2.0 cm is attached to a rifle and used for target practice with targets that are 300 m away. What is the smallest feature on the target that can be resolved with this telescope? Assume $\lambda = 400$ nm. Ignore the effect of the Earth's atmosphere on the resolution.

49. SSM ✪ Radio telescopes use radio waves (instead of light) to produce images. If radio waves with a frequency of 500 MHz are used together with a radio-wave mirror of diameter 1000 m, what is the best possible angular resolution that can be achieved? Ignore the effect of the Earth's atmosphere on the resolution.

50. A large optical telescope has a mirror with a diameter of 15 m. What is the angular resolution of this telescope when used to form an image with blue light? *Hint:* The diameter of the mirror is the size of the telescope's "opening" (also called its aperture). Ignore the effect of the Earth's atmosphere on the resolution.

51. ✪ ℝ An amateur astronomer's telescope containing a lens with a diameter of 15 cm is used to study the Moon. What is the smallest object on the Moon that can be resolved with this telescope using blue light? Ignore the effect of the Earth's atmosphere on the resolution.

52. ✪ (a) A large optical telescope with an aperture of diameter 15 m is used to study the Moon. What is the size of the smallest feature on the Moon that this telescope can resolve using blue light? (b) Calculate the size of the smallest object on Mars that this telescope can resolve. Assume Mars is at its closest distance from the Earth. Ignore the effect of the Earth's atmosphere on the resolution.

25.9 WHY IS THE SKY BLUE?

53. ✪ ℝ The Sun appears red when it is on the horizon due to scattering of light by molecules in the atmosphere (Fig. 25.41). The Sun does not appear as reddish in color when it is overhead because the thickness of the atmosphere through which sunlight must travel is smaller when the Sun is overhead than when it is near the horizon. (a) Take the approximate thickness of the atmosphere to be 2×10^5 m. What distance through the atmosphere must light travel to reach us when the Sun is at the horizon? (b) Estimate the ratio of the distances that sunlight travels through the atmosphere in these two cases.

54. ✪ You want to design an antireflection coating for a lens. You want this coating to work well at two wavelengths, $\lambda_1 = 450$ nm and $\lambda_2 = 650$ nm, which you accomplish using two separate layers. One layer uses MgF_2 and is designed to remove the reflections at wavelength λ_1 ($n = 1.38$). The second coating is composed of a material with $n = 1.55$ and is designed to remove the reflections at λ_2. If the MgF_2 coating is deposited on a lens whose index of refraction is $n = 1.60$ and the other coating sits on top of the MgF_2, what is the thickness of each layer?

55. SSM ✪ A very thin soap bubble is floating in air. If the bubble strongly reflects green light ($\lambda = 520$ nm), what is the minimum possible thickness of the bubble? Assume the index of refraction of the bubble is $n = 1.35$.

56. ✪ A challenge astronomers face is that any telescope on the ground must look through many kilometers of air to peer into space. The air is not of uniform refractive index and is also turbulent, which limits the angular resolution of any telescope to approximately 1 arc second (1/3600 of a degree). (a) What is the maximum diameter of a ground-based telescope where the resolution limit by diffraction (Rayleigh limit) is equal to that of atmospheric turbulence? Assume a wavelength of 450 nm (blue light) for the calculation. (b) Find the angular separation of Pluto and its moon Charon, which are separated by approximately 2.3×10^4 km. Assume Pluto and the Earth are at their closest. Would this telescope be able to resolve Pluto from Charon?

57. ✪ If a laser is directed at a thin wire or hair, a diffraction pattern will occur similar to that produced by a slit of the same width. If the distance from the thin object to the screen on which the diffraction pattern is cast is known, a measurement of the spacing between peaks in the diffraction pattern can be used to determine the object's width. This method is used in the wire-making industry to monitor the production of wire with a uniform width. A hair is placed a distance $L = 1.2$ m away from a screen as shown in Figure P25.57A and illuminated with a green 532-nm laser pointer. The resulting diffraction pattern is shown in Figure P25.57B, where the distance to the first dark fringe is $d = 1.4$ cm. What is the diameter of the hair?

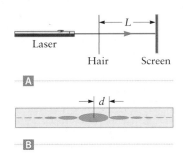

Figure P25.57

58. ✪ An oil tanker has collided with a smaller vessel, resulting in an oil spill in a large, calm-water bay of the ocean. You are investigating the environmental effects of the accident and need to know the area of the spill. The tanker captain informs you that 20,000 liters of oil have escaped and that the oil has an index of refraction of $n = 1.1$. From the deck of your ship, you note that in the sunlight the oil slick appears to be blue. A spectroscope confirms that the dominant wavelength from the surface of the spill is 420 nm. Assuming a uniform thickness, what is the total area of the oil slick?

59. ✪ An astronomical interferometer is an apparatus that combines a number of telescopes into a much larger equivalent light-gathering instrument. The very first extrasolar planet to be discovered was imaged in 2005 (Fig. P25.59) with the interferometer in the mountains of northern Chile known as the Very Large Telescope (VLT) project, which consists of four separate optical telescopes each 8.3 m in diameter. The interferometer operates with infrared wavelengths (1.4 μm), which are much less affected by atmospheric turbulence than other wavelengths. (a) The star system with the extrasolar planet was found to be at a distance of 200 light-years, and the planet orbits the star at a distant of 55 astronomical units (55 times the distance between the Earth and the Sun). What is the angular separation of the planet and star? Give your answer in units of arc seconds. (b) The interferometer has an angular resolution of 0.0020 arc second. What is its effective aperture? That is, a telescope of what diameter would have such resolution?

Figure P25.59 The first optical image of an extrasolar planet found in April 2005 with the ESO's Very Large Telescope Array in the mountains of northern Chile. This star is the brown dwarf 2M1207.

60. ⊗ ✪ **Visual acuity.** Optometrists quantify the resolution of a patient's vision using a concept called *visual acuity*, which is calculated as the inverse of the angular resolution of the eye measured in arc minutes (where 60 arc minutes = 1°) under standard illumination. (a) What is the visual acuity of a patient who can resolve two periods separated by 1.0 mm at a distance of 35 cm from the page? (b) Another patient has a visual acuity of 0.25. Calculate the farthest distance from which she could read 12-point lettering. Assume a needed resolution of one point to be able to discern letter shapes. (In 12-point type, a period has a diameter of about 0.35 mm.)

61. ⊗ ✪ **Resolution of the eye.** The angular resolution of the eye is determined by two factors, diffraction (the Rayleigh criterion) and the density of photoreceptor cells within the retina. Here we investigate which one is more important in limiting resolution. The center circular area of the retina, roughly 0.3 mm in diameter, is called the *fovea centralis* and is the most sensitive part. The fovea has photoreceptor cells spaced about 1 μm apart center to center, compared with a spacing of 3 μm outside this region. The diagrams in Figure P25.61 illustrate how the angular resolution is limited by receptor density. Notice that for two objects to be resolved, there must be an unstimulated cell between the image of the objects on the retina. In the upper diagram of Figure P25.61, the objects A and B are seen as one; in the lower diagram, there is an unstimulated receptor between, so the objects are seen as two distinct objects. The distance from the lens to the retina is about 16 mm. (a) What is the angular resolution of the eye due to the spacing of the photoreceptors in the fovea? (b) Under bright light conditions, the pupil is about 2.0 mm in diameter. What is the angular resolution of an eye for blue light (450 nm) for a pupil of this diameter? Assume n(eye) = 1.3. (c) What is the resolution for blue light under dim light conditions where the pupil's diameter is 6.0 mm? (d) Do bright light conditions lead to better resolution? Why or why not? (e) At what pupil size is the angular resolution due to diffraction less than that due to receptor spacing?

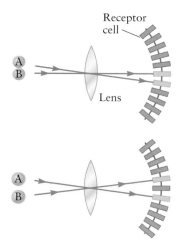

Figure P25.61 Resolution of two objects and receptor cell density.

62. ✪ **CD or DVD?** A simple experiment can tell the difference between a CD and a higher-data-density DVD. The tracks of pits on an optical data disc can act like a grating when laser light is reflected from its surface. Figure P25.62 shows a simple experimental configuration to produce an interference pattern from a data disc using a red laser pointer ($\lambda = 650$ nm). In this particular experiment, the point on the disc struck by the laser is positioned 10 cm above the tabletop. When the disc is oriented at a 45° angle with respect to the vertical, the reflected laser light shows an interference pattern where the first maximum is 4.5 cm beyond the reflected beam (directly below the disc). (a) What is the spacing of the tracks on the CD? (b) If the CD is replaced with a DVD, will the first maximum in the interference pattern move closer to or farther from the reflected beam? Explain. (c) The standard track spacing on a DVD is 0.74 μm. What is the distance from the reflected beam to the first interference maximum using this experimental configuration?

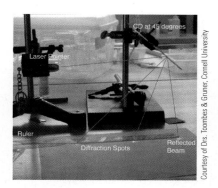

Figure P25.62

63. Ⓡ ☆ To make transistors and other circuit elements that are very small, modern integrated circuits (like those used in a computer) are produced using very high resolution optical patterning. The process uses a very expensive lens to produce the pattern. If the smallest circuit feature size is 200 nm, what does the Rayleigh criterion predict for the approximate wavelength of the light used to make the pattern? How does that compare with the wavelength of blue light?

Chapter 26

Applications of Optics

The Hubble Space Telescope is one of the most advanced optical instruments ever constructed. In this chapter, you'll find out how it works. (NASA)

It is difficult to imagine a world without cameras, microscopes, CDs and DVDs, and (especially) eyeglasses and contact lenses. These and many other devices are based on the principles of optics developed in the last two chapters. In this chapter, we discuss a few of these optics-based inventions and analyze how they work.

Eyeglasses were invented in the late 1200s and are perhaps the oldest optical instruments. Microscopes and telescopes were developed around 1600, CDs and DVDs were developed much more recently, and new optical devices are constantly being invented. In addition, there have been recent improvements to telescopes and microscopes. All these devices are discussed in this chapter.

26.1 | ⊗ APPLICATIONS OF A SINGLE LENS: CONTACT LENSES, EYEGLASSES, AND THE MAGNIFYING GLASS

In Section 24.6, we described how light from an object is refracted by the cornea and other regions in the eye to form an image on the retina. For simplicity, we can model the entire eye as a single lens with a focal length f_{eye}. Figure 26.1A shows a ray diagram for a "normal" eye; light emanating from a point on the object is focused to a corresponding point on the retina. The object is a distance s_o from the front of the eye, and the image is formed on the retina at an image distance $s_i = d_{eye}$, where d_{eye} is the diameter of the eye ($d_{eye} \approx 2.5$ cm in a typical case). To understand the eye, we need to know the value of the focal length f_{eye}. We can find it using the thin-lens equation (Eq. 24.21),

$$\frac{1}{s_o} + \frac{1}{s_i} = \frac{1}{f} \qquad (26.1)$$

Thin-lens equation

Substituting $s_i = d_{eye}$ and $f = f_{eye}$ gives

$$\frac{1}{s_o} + \frac{1}{d_{eye}} = \frac{1}{f_{eye}} \qquad (26.2)$$

and solving for the focal length of the eye leads to

$$f_{eye} = \frac{s_o d_{eye}}{s_o + d_{eye}} \qquad (26.3)$$

The object distance s_o in Equation 26.3 varies depending on the location of the object. For a young adult, the eye can typically focus on objects as near as about 25 cm. This distance, called the **near-point** distance, is denoted as s_N. Objects closer to the eye than s_N cannot be focused on the retina; rays from such an object extrapolate to an intersection point behind the retina as shown in Figure 26.1B. This ray diagram corresponds to placing an object, such as this book, very close to your eyes; the page then appears "out of focus" on your retinas.

Near point of the eye

A typical young adult can also focus on objects that are very far away, that is, with $s_o \approx \infty$ (at "infinity"). Hence, the value of s_o in Equation 26.3 can vary from s_N to infinity. To satisfy Equation 26.3 at these two values of the object distance and at all values in between, the eye must adjust its focal length, which it does by using muscles that deform and change the shape of the eye's lens (Fig. 24.46).

Let's estimate the focal length of the eye when an object is at the typical near-point distance $s_N = s_o = 25$ cm. Inserting this value along with $d_{eye} \approx 2.5$ cm (as mentioned above) into Equation 26.3 gives

$$f_{eye,\,near} = \frac{s_N d_{eye}}{s_N + d_{eye}} = \frac{(25\ \text{cm})(2.5\ \text{cm})}{25\ \text{cm} + 2.5\ \text{cm}} = 2.3\ \text{cm} \qquad (26.4)$$

On the other hand, for at object at infinity we have $s_o \approx \infty$, which when inserted into Equation 26.2[1] leads to

$$\frac{1}{f_{eye,\,distant}} = \frac{1}{s_o} + \frac{1}{d_{eye}} = \frac{1}{\infty} + \frac{1}{d_{eye}} = \frac{1}{d_{eye}}$$

$$f_{eye,\,distant} = d_{eye} = 2.5\ \text{cm} \qquad (26.5)$$

Hence, as an object moves from the near point to a point very far away, the eye must adjust its focal length from $f_{eye} = 2.3$ cm to 2.5 cm, about a 10% change.

Eyeglasses and Contact Lenses: Adding a Lens in Front of the Eye

Figure 26.1 and our applications of the thin-lens equation in Equations 26.2 through 26.5 describe the optics of a "normal" eye. In many cases, such as when a person is nearsighted or farsighted, the eye is aided by eyeglasses or contact lenses, which are simply lenses placed in front of the eye. To understand how eyeglasses and contact lenses work, we must deal with optical "systems" that contain two or more lenses; here one of the lenses is the lens of the eye and the other is the lens in the eyeglass or

A

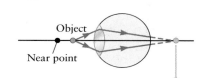

When an object is placed inside the near point, the focus point is behind the retina.

B

Figure 26.1 Ⓐ When properly focused, the eye forms an image on the retina. Ⓑ If an object is located closer than the near point, the eye is unable to produce an image at the retina. Instead, the refracted rays extrapolate to a point behind the retina.

---

[1]We could get the same result by inserting $s_o = \infty$ into Equation 26.3 and carefully canceling terms in the numerator and denominator.

Figure 26.2 Sign conventions for working with the thin-lens equation.

Figure 26.2 Sign conventions for working with the thin-lens equation.

RAY DIAGRAM FOR A
FAR-SIGHTED PERSON

Without glasses, the image is behind the retina.

The image produced by this lens is outside the near point.

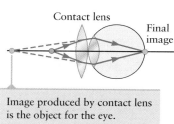

Image produced by contact lens is the object for the eye.

Figure 26.3 Ⓐ For a farsighted person, the near point is farther away than is convenient for reading or other close work. Rays from an object within the near point focus behind the retina. Ⓑ When a converging lens (such as a contact lens) is placed in front of the eye, a virtual image is produced by the lens. Ⓒ The image produced by the contact lens acts as the object for the eye. If this image is at or outside the near-point distance, the eye can form an image on the retina.

contact lens. Learning how to deal with systems containing multiple lenses is very useful because telescopes, microscopes, and many other optical devices contain two or more lenses.

In Chapter 24, we learned how to use ray tracing along with the mirror equation and the thin-lens equation to analyze the images produced by curved mirrors and lenses. In a typical case, an object is placed in front of a lens or mirror and the location of the image is then calculated. For a system containing two lenses, the general approach is as follows.

> *Analyzing an optical system with two (or more) lenses*
>
> 1. **Draw a picture showing the object of interest and the lenses in the problem.**
>
> 2. **Use the rules for ray tracing along with the thin-lens equation from Chapter 24 to find the location and magnification of the image produced by the first lens in the system.**
>
> 3. **The *image produced by the first lens* then acts *as the object for the second lens* in the system.**
>
> 4. **Use ray tracing and the thin-lens equation a second time to find the image produced by the second lens in the system.**

We'll illustrate this general approach in a number of examples throughout this chapter. Most of these examples involve lenses, so Figure 26.2 shows our sign conventions (from Fig. 24.42) for applying the thin-lens equation to a single lens. A positive image distance s_i corresponds to an image on the right of the lens, whereas a negative value of s_i describes an image to the left of the lens (a "virtual image"). Notice also that a positive object distance s_o corresponds to an object to the left of the lens. In all examples in Chapter 24, the object distance was positive. Our sign conventions also allow for a negative value of s_o, which means that the object is to the right of the lens (a "virtual object"). We'll encounter such negative values of s_o in systems with two lenses, when the image produced the first lens is "beyond" the second lens.

Rather than dwell on the rules for analyzing systems of multiple lenses, we'll illustrate the key concepts by working through a series of examples.

Designing a Pair of Contact Lenses

As you age, the lenses in your eyes usually become less flexible and are not able to change focal length over the range found in Equations 26.4 and 26.5. The focal length is instead confined to a narrower range of values, reducing your ability to focus. In some cases, the near-point distance increases and is greater than the near-point distance for a normal eye; objects located closer than this near-point distance cannot be focused on the retina (Fig. 26.3A), and you must move things (such as a newspaper or this book) to a greater distance to see them clearly. You are then *farsighted*. To compensate, you can place a lens (a contact lens) immediately in front each eye. Let's now see how the required focal length of this lens depends on a person's near-point distance.

Suppose the near-point distance for a certain individual is $s_N = 75$ cm, which is much greater than the "normal" near-point distance of 25 cm. We wish to design a contact lens that enables this person to focus on objects at the normal near-point distance of 25 cm. The contact lens is placed just in front of the eye as in Figure 26.3B; it forms the first lens of our "system," and the eye acts as the second lens. We want this system of two lenses to be able to take an object at $s_o = 25$ cm from the contact lens and produce a final image on the retina. We follow our general rules for analyzing optical systems with two lenses and analyze the problem one lens at a time. The contact lens produces an image (Fig. 26.3B) that then acts as the object for the second "lens," the eye, but where should the image formed by the contact lens be located? We are given that the near-point distance for the eye is 75 cm, which means that the eye alone can focus an object that is 75 cm away to produce an image on the retina. So, if the image produced by the contact lens is at this distance (or greater, Fig. 26.3C), the eye can produce a final image at the retina. We thus want the contact lens to take an object at $s_o = 25$ cm and produce an image that is 75 cm to the left of the eye. Since the contact lens is touching the eye, the image distance is $s_i = -75$ cm; this image distance is negative because it is to the left of the contact lens (and is thus a virtual image; see the sign conventions in Fig. 26.2).

We now apply the thin-lens equation (Eq. 26.1) to the contact lens; we substitute $s_o = 25$ cm and $s_i = -75$ cm to find

$$\frac{1}{f_{\text{lens}}} = \frac{1}{s_o} + \frac{1}{s_i} = \frac{1}{+25 \text{ cm}} + \frac{1}{-75 \text{ cm}} \quad (26.6)$$

and solve for f_{lens}:

$$f_{\text{lens}} = 38 \text{ cm} \quad (26.7)$$

which is the focal length required for the contact lens. Since f_{lens} is positive, this lens is a converging lens.

If the person's near-point distance is larger than 75 cm, Equation 26.6 shows that a lens with a shorter focal length is needed. In words, we say that a person with a large near-point distance needs a "stronger" lens than a person with a short s_N. The strength of a lens is sometimes measured in terms of its **refractive power**, defined as

$$\text{refractive power} = \frac{1}{f_{\text{lens}}} \quad (26.8)$$

The unit of refractive power is m^{-1}, which is called a **diopter**, and

$$1 \text{ diopter} = 1 \text{ m}^{-1}$$

The contact lens with the focal length in Equation 26.7 thus has a refractive power of $1/f_{\text{lens}} = 1/(0.37 \text{ m}) = 2.7$ diopters.

Let's now consider what type of contact lens is needed by a **nearsighted** person. Such a person is unable to focus light from distant objects on the retina as shown in the ray diagram in Figure 26.4A. The incoming rays from an object very far away are approximately parallel to the axis (an object at "infinity"); a nearsighted eye produces an image in front of the retina.

A nearsighted person is able to properly focus objects that are within a certain distance. Suppose a person can focus objects within 2.0 m on the retina. What kind of contact lens does this person need? We follow the approach we took for a farsighted person in Figure 26.3. We want an object that is very far away ($s_o = \infty$) to lead to a final image at the retina. It is also given that the eye alone can focus an object that is 2.0 m away on the retina. So, the contact lens must take an object at $s_o = \infty$ and give an image at $s_i = -2.0$ m (Fig. 26.4B).

Applying the thin-lens equation to the contact lens, we have $s_o = \infty$ and $s_i = -2.0$ m, and we get

$$\frac{1}{f_{\text{lens}}} = \frac{1}{s_o} + \frac{1}{s_i} = \frac{1}{\infty} + \frac{1}{-2.0 \text{ m}}$$

RAY DIAGRAM FOR A
NEAR-SIGHTED PERSON

Without glasses, the image is in front of the retina.

Distant object

$s_o = \infty$

Image

A

Virtual image produced by diverging lens

s_i

B

Image produced by contact lens

Final image

The object for the eye gives the image on the retina.

Diverging lens (contact lens for a near-sighted person)

C

Figure 26.4 **A** When a person is nearsighted, rays from an object very far away (at infinity) focus in front of the retina. **B** Placing a diverging lens in front of the eye produces a virtual image to the left of the lens. **C** The image formed by the lens acts as the object for the eye. The nearsighted eye is able to focus this image on the retina.

Solving for f_{lens}, we have

$$f_{lens} = -2.0 \text{ m}$$

which is the focal length required for the contact lens. Since f_{lens} is negative, this lens is a diverging lens.

CONCEPT CHECK 26.1 | What Kind of Image Is It?

Figures 26.3 and 26.4 show ray diagrams for different types of contact lenses. Which of the following statements correctly describes the images formed by these contact lenses?
   (a) A contact lens always forms a real image.
   (b) A contact lens always forms a virtual image.
   (c) A contact lens can form a real or virtual image, depending on the type of lens.

CONCEPT CHECK 26.2 | ⊗

A farsighted person has a very long near-point distance. What should the focal length of this person's contact lenses be so that he can read a newspaper at a normal distance of (a) 25 cm, (b) 400 cm, or (c) 4.0 diopters? *Hint*: Assume the near-point distance is *extremely* long so that $s_N \approx \infty$.

Designing a Pair of Eyeglasses

The problem of selecting the proper lens for a set of eyeglasses is very similar to the contact lens problems considered in Figures 26.3 and 26.4. The main difference is that the eyeglass lens is a short distance in front of the eye, rather than touching it (Fig. 26.5). This distance must be accounted for when relating the image distance from the eyeglass lens and the object distance for the eye. The rest of the analysis is the same as with the contact lens problem and is illustrated in Example 26.1.

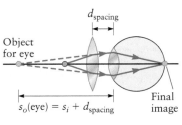

Figure 26.5 Image produced by the combination of an eyeglass lens plus the eye. **A** The eyeglass lens produces an image that then acts as the object for the eye. **B** When calculating the object distance for the eye, we must account for the distance $d_{spacing}$ between the eyeglass lens and the eye.

| **EXAMPLE 26.1** | ⊗ What Kind of Eyeglass Lenses Do You Need? |

Consider a farsighted person with a near-point distance of 75 cm. In Figure 26.3 (and Eqs. 26.6 and 26.7), we designed a contact lens that would enable this person to read a book located at the "normal" near-point distance of 25 cm. The same person now wants a set of eyeglasses that he can use to read this book at 25 cm. What is the required focal length of the eyeglass lenses? Assume the glasses are positioned about $d_{spacing} = 2$ cm in front of the person's eye.

RECOGNIZE THE PRINCIPLE

With an object 25 cm in front of the eye, we want the eyeglass lens to produce an image 75 cm in front of the eye. This image then forms the object for the eye of our farsighted person, and we know that for this distance he will then be able to focus the final image on his retina.

SKETCH THE PROBLEM

The problem is described by the ray diagrams in Figure 26.5.

IDENTIFY THE RELATIONSHIPS

Since the eyeglass lens is $d_{spacing} = 2$ cm in front of the eye and the object is 25 cm in front of the eye, the object and image distances *for the eyeglass lens* are $s_o = 25 - d_{spacing} = 23$ cm and $s_i = -73$ cm (Fig. 26.5A). This image is a virtual image (s_i is negative) because light does not actually pass through this point.

SOLVE

Using the thin-lens equation with $s_o = 23$ cm and $s_i = -73$ cm, we get

$$\frac{1}{f_{lens}} = \frac{1}{s_o} + \frac{1}{s_i} = \frac{1}{23\text{ cm}} + \frac{1}{-73\text{ cm}}$$

Solving for f_{lens} gives

$$f_{lens} = \boxed{34\text{ cm}}$$

Since f_{lens} is positive, this lens is a converging lens.

What does it mean?
Comparing this result for f_{lens} with the result in Equation 26.7 for a contact lens, we see that the eyeglass lenses have a focal length about 10% shorter. That is why your prescription for contact lenses is slightly different than for your eyeglasses.

The Magnifying Glass

The magnifying glass is one of the most basic of all optical devices and is a component in microscopes, telescopes, and other instruments. The simplest magnifying glass is a single lens, used to view things that are too small to examine with the naked eye (Fig. 26.6). To analyze this optical "instrument," we must again consider a system consisting of two lenses—the magnifying glass lens and the eye behind it—so the problem is similar to our analysis of contact lenses and eyeglasses. In those applications, the goal was to choose a lens that would allow the eye to produce an image at the retina. With a magnifying glass, the goal is to produce a greatly magnified image at the retina. We thus want the image on the retina to be as large as possible.

Consider first the case without a magnifying glass. To examine a small object, the natural thing to do is to place the object as close as possible to your eye. Figure 26.7A shows how the position of an object in front of the eye affects the size of its image on the retina. As the object is brought closer and closer to the eye from point *A* to *B*, the image on the retina grows larger and the apparent size of the object increases, but there is a limit to how close you can bring the object and still keep it in focus. The largest clearly focused image for the unaided eye results when the object is at the near point. The object's apparent size when it is located at the near point can be measured using the angle θ defined in Figure 26.7B; we'll call it the image angle.

To make the object even larger on the retina than in Figure 26.7B, we need to move it closer to the eye and hence inside the near point. The eye by itself cannot then produce an image on the retina, but it can do so with the help of the magnifying glass lens.

Image Properties with a Magnifying Glass

A magnifying glass is a single converging lens placed in front of the eye as shown in Figure 26.7C. The object is positioned inside the focal length of this lens, that is, so that the focal point *F* of the lens is to the left of the object in Figure 26.7C. We showed in Chapter 24 (Fig. 24.44) that this position leads to an upright virtual image at a point farther from the eye, on the far left in Figure 26.7C. As far as the eye is concerned, light from the object emanates from this virtual image, and if this image produced by the magnifying glass is at the near point, the eye is able to focus this light onto the retina. Comparing the ray diagrams in Figures 26.7B and C shows that the image angle with the magnifying glass θ_M (Fig. 26.7C) is greater than the image angle for the eye alone (θ in Fig. 26.7B), so the image *on the retina* is enlarged by the magnifying glass.

Figure 26.6 A magnifying glass produces an upright virtual image.

The size of an image on the retina determines the object's apparent size.

A

B

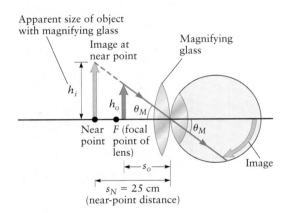

C

Figure 26.7 **A** The size of an object as perceived by the eye depends on the size of the image on the retina. If an object is brought closer to the eye (from *A* to *B*), the rays from its tip make a larger angle with the axis and the image on the retina becomes larger. **B** The closest object distance at which the eye can focus the image on the retina is the near point, which for a typical young person is approximately 25 cm from the eye. **C** A magnifying glass produces a magnified (enlarged) image at the near point of the eye. The eye focuses the light from this image on the retina.

The enlargement of the image on the retina is given by the *angular magnification* m_θ, defined by

$$m_\theta = \frac{\theta_M}{\theta} \qquad (26.9)$$

The angular magnification is related to (but slightly different from) the linear magnification m defined in Chapter 24. The linear magnification involves the height of the object and image, whereas the angular magnification involves the angle the image makes on the retina.

We can calculate the angular magnification m_θ by applying the thin-lens equation (Eq. 26.1) to the lens of the magnifying glass (Fig. 26.7C):

$$\frac{1}{s_o} + \frac{1}{s_i} = \frac{1}{f} \qquad (26.10)$$

As always, we must be careful to follow the proper sign conventions. Because the magnifying glass is a converging lens, its focal length f is positive. The object is on the left side of the lens, so the object distance s_o is positive; the image is also on the left side of the lens, so the image distance s_i is negative. For simplicity, let's assume the magnifying glass is very close to the eye; the image is at the near point and hence $s_i = -s_N$ in Equation 26.10. Inserting all this information leads to

$$\frac{1}{s_o} - \frac{1}{s_N} = \frac{1}{f}$$

$$\frac{1}{s_o} = \frac{1}{f} + \frac{1}{s_N} = \frac{s_N + f}{f s_N}$$

Solving for s_o, we get

$$s_o = \frac{s_N f}{s_N + f} \qquad (26.11)$$

We can now find the image angle with and without the magnifying glass in place. From the geometry in parts B and C of Figure 26.7, we have

$$\tan \theta = \frac{h_o}{s_N} \quad \text{and} \quad \tan \theta_M = \frac{h_i}{s_N}$$

These angles are usually small, so we can use the approximations $\tan \theta \approx \theta$ and $\tan \theta_M \approx \theta_M$, leading to

$$\theta = \frac{h_o}{s_N} \quad \text{and} \quad \theta_M = \frac{h_i}{s_N} \qquad (26.12)$$

Inserting these angles into the definition for the angular magnification (Eq. 26.9) gives

$$m_\theta = \frac{\theta_M}{\theta} = \frac{h_i}{h_o}$$

There are two right triangles in Figure 26.7C, one involving h_o and s_o and another involving h_i and s_N. These triangles are similar (having the same interior angles), so $h_i/s_N = h_o/s_o$. The angular magnification is thus

$$m_\theta = \frac{h_i}{h_o} = \frac{s_N}{s_o}$$

Inserting the result for s_o from Equation 26.11 gives our final result,

$$m_\theta = \frac{s_N}{s_o} = \frac{s_N}{(s_N f)/(s_N + f)}$$

$$m_\theta = \frac{s_N + f}{f} = \frac{s_N}{f} + 1 \qquad (26.13)$$

Practical magnifying glass lenses have focal lengths that are smaller than the near-point distance by a factor of 10 or more, so the ratio s_N/f in Equation 26.13 is much greater than 1. Hence, to a good approximation we can write

$$m_\theta = \frac{s_N}{f} \qquad (26.14)$$

Magnification of a magnifying glass

For a typical magnifying glass, this result is accurate to 10% or better (and we can always use the result for m_θ in Eq. 26.13 if greater accuracy is needed). This expression for m_θ is also found when a magnifying glass is used as the eyepiece of a microscope or telescope, which we'll explore in Example 26.2.

The near-point distance s_N varies from person to person, so the magnification actually obtained with a particular magnifying glass (having a particular value of f) will depend on the person. Because the "typical" near-point distance for a young adult is 25 cm, this value is used as a convention when specifying magnifying glasses, so

$$m_\theta = \frac{s_N}{f} = \frac{25 \text{ cm}}{f}$$

For a magnifying glass specified by the manufacturer as providing $m_\theta = 10$, the focal length is thus

$$f = \frac{25 \text{ cm}}{m_\theta} = \frac{25 \text{ cm}}{10} = 2.5 \text{ cm}$$

The angular magnification of a typical magnifying glass or eyepiece is usually 10 or 20.

EXAMPLE 26.2 | Angular Magnification for an Object at the Focal Point

Figure 26.8A shows an object at the focal point of a magnifying glass. This arrangement is slightly different from the one we considered in Figure 26.7C, where the object was inside the focal point. Now the image from the magnifying glass is at infinity instead of at the near point of the eye (Fig. 26.7C). Show that the angular magnification in this case is given by $m_\theta = s_N/f$ (Eq. 26.14).

RECOGNIZE THE PRINCIPLE

To find the angular magnification, we must compare the angle θ that the object makes with the principal axis of the eye when the object is at the near point (without the magnifying glass; Fig. 26.7B), with the angle θ_f when the object is at the focal point of the magnifying glass (Fig. 26.8B). The magnification is the ratio of θ_f to θ. (Compare with Eq. 26.9.)

$$m_\theta = \frac{\theta_f}{\theta} \qquad (1)$$

SKETCH THE PROBLEM

Figure 26.8B shows a ray diagram. The object is at the focal point of a converging lens. From Chapter 24, we know that the lens will form an image at infinity ($s_i = \infty$); hence, the rays between the magnifying glass and the eye are parallel as shown in Figure 26.8B.

IDENTIFY THE RELATIONSHIPS AND SOLVE

To find θ_f, we consider the right triangle shaded in red in Figure 26.8B. We assume the magnifying glass is very close to the eye, so the red triangle has sides f and h_o and contains the angle θ_f. From the geometry of this triangle, $\tan \theta_f = h_o/f$. Using the small-angle approximation ($\tan \theta \approx \theta$) gives $\theta_f = h_o/f$. From Figure 26.7B, an object at

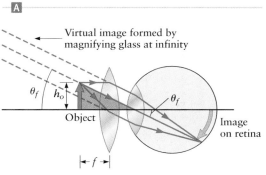

Figure 26.8 Example 26.2. **A** An object at the focal point of a magnifying glass. **B** The lens forms a virtual image an infinite distance (at "infinity") in front of the eye. The light from this virtual image forms a parallel bundle of rays that are then focused by the eye to form an image on the retina.

the near point viewed with the unaided eye has an angular size θ with $\tan \theta = h_o/s_N$. Using the small-angle approximation here leads to $\theta = h_o/s_N$. Inserting these results for θ_f and θ into the expression for m_θ in Equation (1), we get

$$m_\theta = \frac{\theta_f}{\theta} = \frac{h_o/f}{h_o/s_N} = \boxed{\frac{s_N}{f}}$$

which is Equation 26.14.

What does it mean?

As viewed by the eye, the object in Figure 26.8 appears to be at a location very far away (at "infinity") because the magnifying glass produces a collection of parallel rays that enter the eye. The magnifying glass is a component in many other optical instruments, including microscopes and telescopes, where it also produces a similar set of parallel rays that are viewed by an observer. This result for the magnification will help us analyze the role of the magnifying glass in those instruments.

| CONCEPT CHECK 26.3 |

Does a magnifying glass form a real image or a virtual image? Explain.

26.2 | ⊗ MICROSCOPES

A magnifying glass is very useful for viewing small objects, but there is a practical limit to the magnification that can be obtained with a single lens. Lenses with a focal length of less than a few millimeters are difficult to make, which leads to a limit on the angular magnification (Eq. 26.14) of around 100 or so for a typical near-point distance. Most practical magnifying glasses have angular magnifications of only 10 or 20. A more useful way to achieve higher magnification is by using two lenses arranged as a *compound microscope*. The image produced by one lens is used as the object for the second lens, and the image produced by the second lens is then viewed by the eye of an observer. The total magnification is the product of the magnifications of the two lenses; an angular magnification as high as $m_\theta = 1000$ is readily attained in this way.

Figure 26.9A shows a compound microscope and the usual vertical path followed by the light, and Figure 26.9B shows a ray diagram with the light traveling from left to right according to our usual convention. The two lenses in a compound microscope are called the *objective* and the *eyepiece*. The objective is a converging lens, so according to our sign conventions (Fig. 26.2) it has a positive focal length that we denote by f_{obj}. The eyepiece is simply a magnifying glass; it also has a positive focal length that we denote by $f_{eyepiece}$. To analyze the image produced by a compound microscope, we first apply ray tracing and the thin-lens equation to find the image produced by the objective lens. This image acts as the object for the eyepiece; the image produced by the eyepiece is then viewed by the eye.

The object to be studied is placed just outside (i.e., just to left) of the focal point of the objective lens (Fig. 26.9B). In Chapter 24, we showed that an object located outside the focal length of a converging lens gives an image that is real and inverted (Fig. 24.40C). The image produced by the objective in Figure 26.9B is therefore drawn a distance s_i to the right of the objective lens. The image produced by the objective lens forms the object for the eyepiece. The eyepiece then behaves in the same way as the magnifying glass discussed in Section 26.1. The distance between the objective lens and the original object is adjusted so that the image produced by the objective falls at the focal point of the eyepiece, which then gives a final virtual image for the observer. (Compare the rays produced by the eyepiece in Fig. 26.9B with the corresponding rays in Fig. 26.8B.)

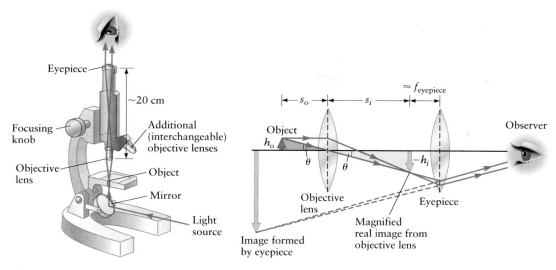

Figure 26.9 🅰 A compound microscope contains two lenses. The object of interest is placed close to the objective lens, and the observer's eye is close to the eyepiece. 🅱 Ray diagram for a compound microscope. The object is placed just outside the focal point of the objective lens, producing a magnified real image. The image from the objective lens is then the object for the eyepiece, which produces a final virtual image much larger than the original object.

To find the total magnification of the microscope, we must first calculate s_i. Figure 26.9B shows two similar triangles, one shaded in red with sides h_o and s_o and another shaded in yellow with sides $-h_i$ and s_i. (The negative sign in front of h_i is due to our sign convention for lenses, which says an inverted image has a negative value for its height.) Both of these right triangles have an interior angle θ, so we have

$$\frac{h_o}{s_o} = \tan \theta = \frac{-h_i}{s_i}$$

The object is placed very close to the focal point of the objective; hence, $s_o \approx f_{obj}$. In Figure 26.9B, the image produced by the objective is larger than the original object. The ratio of the image and object heights is called the **linear magnification**, which we discussed in Chapter 24. For the objective lens in Figure 26.9B, we have a linear magnification $m_{obj} = h_i/h_o$. Putting it all together, we find

$$m_{obj} = \frac{h_i}{h_o} = -\frac{s_i}{f_{obj}} \tag{26.15}$$

The final virtual image produced by the eyepiece is a greatly magnified version of the original object. The total magnification of the microscope is the product of the linear magnification of the objective and the angular magnification of the eyepiece. If the eyepiece has a focal length $f_{eyepiece}$, its angular magnification according to Equation 26.14 is $m_{\theta, eyepiece} = s_N/f_{eyepiece}$, where $s_N = 25$ cm is the value (by convention) of the near-point distance. Combining this expression with Equation 26.15, we have

$$m_{total} = m_{obj} m_{\theta, eyepiece}$$

$$m_{total} = -\frac{s_i}{f_{obj}} \frac{s_N}{f_{eyepiece}} \tag{26.16}$$

Magnification of a compound microscope

Notice that m_{total} is negative, which means that the image is inverted.

Using a compound microscope is a good way to attain a large magnification. For a typical student microscope like the one in Figure 26.9A, the focal lengths might be $f_{obj} = 4$ mm and $f_{eyepiece} = 2.5$ cm (a so-called 10× eyepiece), and the distance between the objective and the eyepiece is about 20 cm. Inserting these values into Equation 26.16 gives

$$m_{\text{total}} = -\frac{s_i}{f_{\text{obj}}}\frac{s_N}{f_{\text{eyepiece}}} = -\frac{(0.20\ \text{m})(25\ \text{cm})}{(0.004\ \text{m})(2.5\ \text{cm})} = -500$$

which is quite a useful value.

Advances in Microscope Design

The first compound microscopes, which were invented around the time of Galileo, followed the simple design in Figure 26.9 with two lenses, but this design has the following problem. The index of refraction of the glass used to make lenses is slightly different for light of different colors, which, according to the lens maker's formula (Eq. 24.32), makes the focal length f slightly different for different colors and affects the focusing properties of a microscope. When light of a particular color is in focus, light of a different color will not quite be in focus. This chromatic aberration can be minimized by making the objective lens out of several separate lenses, composed of different types of glass with different indices of refraction. Clever optical design approximately cancels the aberrations from these individual lenses. Such lenses are called *achromatic* and are used in most modern high-quality microscopes.

Although most microscopes are thus not quite as simple as the one sketched in Figure 26.9A, the basic result for the total magnification in Equation 26.16 still applies, but now f_{obj} is the focal length of the objective lens system and f_{eyepiece} is the focal length of the complete eyepiece. Hence, to obtain a high magnification, these two focal lengths should be as short as possible. This aspect of microscope design is explored in Example 26.3.

EXAMPLE 26.3 | ⊗ ⓡ Designing a Compound Microscope

You are given the task of designing a compound microscope like the one in Figure 26.9A. The eyepiece is to have an angular magnification $m_{\theta,\ \text{eyepiece}} = 10$, and you want the total magnification to be $m_{\text{total}} = -250$. To achieve this magnification, what should the focal length of the objective lens f_{obj} be?

RECOGNIZE THE PRINCIPLE

The total magnification of a compound microscope depends on several factors, including the focal lengths of the objective and eyepiece and the distance between the two lenses. We are not given this distance, but we can estimate a typical value from the information in Figure 26.9A. We can also find the focal length of the eyepiece from its magnification (which is given).

SKETCH THE PROBLEM

Figure 26.9 shows the geometry of a compound microscope.

IDENTIFY THE RELATIONSHIPS

The total magnification of a compound microscope is (Eq. 26.16)

$$m_{\text{total}} = -\frac{s_i}{f_{\text{obj}}}\frac{s_N}{f_{\text{eyepiece}}} \tag{1}$$

The desired magnification is given, and by convention the near-point distance is $s_N = 25$ cm. The angular magnification of the eyepiece is also given, and from Equation 26.14 we have $m_{\theta,\ \text{eyepiece}} = s_N/f_{\text{eyepiece}}$. Hence, to use Equation (1) to find f_{obj} we need to estimate s_i in Figure 26.9B. This distance is nearly equal to the distance between the objective and the eyepiece (because s_i is usually much larger than the focal length of the eyepiece); from Figure 26.9A, we see that this distance is also the approximate length of the body of the microscope. From our experience with student microscopes (and Fig. 26.9A), we estimate $s_i = 20$ cm. Rearranging to solve for f_{obj} leads to

$$f_{obj} = -\frac{s_i}{m_{total}}\frac{s_N}{f_{eyepiece}}$$

Using $m_{\theta,\,eyepiece} = s_N/f_{eyepiece}$ this equation can also be written as

$$f_{obj} = -\frac{s_i}{m_{total}}m_{\theta,\,eyepiece} \qquad (2)$$

SOLVE

Inserting the values of the various quantities gives

$$f_{obj} = -\frac{s_i}{m_{total}}m_{\theta,\,eyepiece} = -\left(\frac{20\ cm}{-250}\right)(10) = 0.80\ cm = \boxed{8.0\ mm}$$

What does it mean?

This value for f_{obj} and the general result in Equation (2) lead to some useful insights. If we want a higher total magnification, we need an objective with an even smaller focal length. Because the object must be placed a distance f_{obj} from the approximate *center* of the objective lens (see Fig. 26.9B), however, a practical limit arises. A real lens is at least several millimeters thick, so we cannot make f_{obj} smaller than that.

Insight 26.1
FOCUSING A COMPOUND MICROSCOPE
A typical compound microscope like the one in Figure 26.9A has several interchangeable objective lenses, allowing the user to obtain different magnifications. According to the result for the total magnification in Equation 26.16, if we wish to increase the magnification, we must make the focal length of the objective very short. The object is positioned just outside the focal point of the objective (Fig. 26.9B), and since f_{obj} is small, the object must be very close to the objective lens. For an objective lens with a magnification of -100, the distance between the object and the surface of the objective lens can be less than 1 mm. This distance is adjusted with the microscope's focusing knob.

Ultimate Resolution of a Microscope

Example 26.3 mentioned some practical limits on the total magnification of a compound microscope. Might a very clever physicist devise a different microscope design that can achieve much greater magnifications, or is there a fundamental limit to the magnification that any microscope can achieve? One can indeed use additional lenses, more powerful eyepieces, and so forth and make the magnitude of the total magnification much greater than the value of 500 to 1000 that is typical for compound microscopes. Even with such a high magnification, however, there is still a fundamental limit to the **resolution** that can be achieved with *any* microscope that relies on focusing.[2] This limit is due to the diffraction of light passing through the aperture of the microscope (i.e., through the lens openings). In Section 25.8, we described the Rayleigh criterion and showed that even when a perfect lens is used to focus light, diffraction prevents the size of the focused spot from being less than a value approximately equal to the wavelength of the light (Fig. 25.36 and Eq. 25.33).

The same diffraction effect limits the resolution achievable with a microscope. Suppose a microscope is used to observe two very closely spaced features on an object. Figure 26.10 shows light from these two features as it travels through the objective lens. The aperture of the objective leads to a slight spreading of the outgoing light rays from the two features, just the "inverse" of the spreading found when focusing light to a spot. An analysis similar to the one given in Section 25.8 shows that it is possible to resolve the outgoing light from the two features only if they are separated by a distance greater than approximately the wavelength λ. If they are closer, it is not possible to tell that there are two separate features. Hence, the ultimate optical resolution of a microscope is set by diffraction and is approximately equal to the wavelength of the light that is used. Applications requiring the best possible resolution use blue or even ultraviolet light because these colors have the shortest wavelengths compared with other colors of visible light.

---

[2]A microscope that does not rely on focusing is not restricted by this limit; one such microscope is described in Section 26.7.

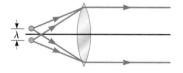

Objects are indistinguishable in the outgoing beam.

Figure 26.10 When two objects are close together, their diffraction spots overlap. If they are too close together (separated by about λ or less), they appear as a single object instead of two separate objects when viewed through a lens.

Figure 26.11 Ⓐ A confocal microscope is similar in design to the compound microscope in Figure 26.9, but with the addition of a small pinhole (aperture) in front of the observer. This pinhole selects light from a specific distance below the objective lens. By scanning the objective lens back and forth and also up and down, one can construct a three-dimensional image. Ⓑ Confocal light micrograph of the retina. The tissue has been fluorescently stained and the colors shown do not correlate with the colors of the rays in part A. Optic nerve fibers are shown in red and blood vessels are shown in blue. The region shown here is about 0.6 mm on a side.

The Confocal Microscope

Biologists examining samples of cells or tissue often want to resolve two nearby features according to their depth within a semitransparent object. Figure 26.11A shows a schematic diagram of a *confocal microscope*. This instrument is designed so that features at only one particular depth form the final image, which is accomplished by placing a very small opening, called a pinhole or aperture, in front of the observer. This pinhole allows mainly light from the selected depth to reach the observer and blocks most of the light from other depths. By combining images made at different depths, a confocal microscope yields a three-dimensional image of a sample. Confocal microscopes provide an extremely powerful way to examine features inside transparent objects such as a living cell (Fig. 26.11B). The depth resolution of a confocal microscope is again limited by diffraction effects. Separate features can be resolved only if their depths differ by more than approximately λ.

Although the optical principles of a confocal microscope have been understood since the time of Galileo, modern computers and lasers make confocal microscopes practical. The pinhole in front of the observer blocks much of the light from the object, so a very intense light source is needed. For this reason, most confocal microscopes use lasers, and lasers have been available for this kind of application only since the 1960s. Second, collecting many images at different depths requires that the objective lens be scanned side to side and also up and down in Figure 26.11A. Computers are essential for automating the scanning and for analyzing the large amount of image data needed for a three-dimensional image.

26.3 | TELESCOPES

When you observe two stars with your naked eyes or with a telescope, light rays extending from the stars to you are nearly parallel to one another. Figure 26.12 shows light rays emanating from two stars; here rays from one star (star 1) make a very small angle θ with the axis of a telescope, whereas light from star 2 travels parallel to the axis. The purpose of the telescope is to increase the angular separation of rays from these two stars so that your eye can distinguish one star from the other.

Refracting Telescopes

The simplest telescopes employ two lenses called an objective and an eyepiece as sketched in Figure 26.12. They are called *refracting* telescopes because they use light refracted by the objective lens. This type of telescope was invented around 1600 and then refined by several physicists, including Galileo who used it in his studies of the solar system. In a way, this arrangement of two lenses is similar to that used in a compound microscope (Fig. 26.9), but now the object to be studied is very far from the objective. The objective lens forms an image of a star or whatever is being studied. This image then acts as the object for a second lens (the eyepiece), which forms the final image viewed by an observer.

Let's consider light from star 1 as it moves through the telescope in Figure 26.12. This star is very far away (at "infinity") and the objective is a converging lens, so a real image is formed. For an object distance $s_o = \infty$, the thin-lens equation (Eq. 26.1) gives an image distance $s_i = f_{obj}$, where f_{obj} is the focal length of the objective lens. The rays from star 1 make a small angle θ with the axis, so the image of star 1 is a small distance h off the principal axis. The telescope's eyepiece is located such that the image formed by the objective is very close to the focal point of the eyepiece. The image from the objective is now the object for the eyepiece; since this new object is a distance $f_{eyepiece}$ from the eyepiece lens, the rays emanating from this object are refracted by the eyepiece to form a "bundle" of nearly parallel rays that are perceived by the observer.

The telescope's magnification is determined by the angles that rays from star 1 make with the axis; before reaching the telescope (on the far left in Fig. 26.12), this angle is θ; after passing through the eyepiece, it is θ_T. The total magnification of the telescope is the ratio of these two angles,

$$m_\theta = \frac{\theta_T}{\theta} \tag{26.17}$$

This is actually an angular magnification because it involves the two angles, but it is usually called just the "magnification." A large magnification m_θ means that rays from the two stars in Figure 26.12 are separated by a large amount and hence an observer can more easily resolve that there are actually two separate stars present.

We can calculate the value of m_θ from Figure 26.12. The red right triangle in Figure 26.12 has sides of length h and f_{obj} and contains the angle θ; it leads to $\tan\theta = h/f_{obj}$. Using the small-angle approximation ($\tan\theta \approx \theta$), we get

$$\theta = \frac{h}{f_{obj}} \tag{26.18}$$

The height h is also one side of the yellow right triangle in Figure 26.12 containing the angle θ_T and leads to

$$\tan\theta_T = \frac{h}{f_{eyepiece}}$$

The angle θ_T is greatly exaggerated in Figure 26.12. In practice, this angle is still small, so we can again use the small-angle approximation ($\tan\theta_T \approx \theta_T$), yielding

$$\tan\theta_T \approx \theta_T = \frac{h}{f_{eyepiece}} \tag{26.19}$$

Figure 26.12 A refracting telescope uses two lenses. Light from a distant object is focused very close to the focal point of the objective lens. The eyepiece then forms a final virtual image for the observer.

Inserting these results for the angles θ and θ_T into Equation 26.17, we have

$$m_\theta = \frac{\theta_T}{\theta} = \frac{h/f_{\text{eyepiece}}}{h/f_{\text{obj}}}$$

$$m_\theta = \frac{f_{\text{obj}}}{f_{\text{eyepiece}}} \qquad (26.20)$$

Magnification of a refracting telescope

Since one usually wants the magnification to be as large as possible, Equation 26.20 suggests that the focal length of the objective lens should be as large as possible. That is why long refracting telescopes are usually preferred, and such telescopes with lengths of 10 m or more were common in the 1700s and 1800s (Fig. 26.13).

© UC Regents/Lick Observatory

Figure 26.13 The magnification of a refracting telescope is proportional to the focal length of the objective lens, which motivated people to construct some very long telescopes. This one, at Lick Observatory in California, is more than 9 m long.

EXAMPLE 26.4 | **ℝ Finding the Magnification of a Refracting Telescope**

The refracting telescope in Figure 26.13 has a total length of approximately 9.0 m. Suppose this telescope has an eyepiece with a magnification of 20. Estimate the total angular magnification of this telescope. *Hint*: The eyepiece is just a magnifying glass.

RECOGNIZE THE PRINCIPLE

The total magnification of a refracting telescope equals the ratio of the focal length of the objective lens f_{obj} to the focal length of the eyepiece f_{eyepiece} (Eq. 26.20),

$$m_\theta = \frac{f_{\text{obj}}}{f_{\text{eyepiece}}} \qquad (1)$$

The eyepiece is a magnifying glass, and given its magnification we can find f_{eyepiece}. From our discussion of the refracting telescope, the sum of the two focal lengths is approximately equal to the telescope's total length, which gives us a way to find f_{obj}.

SKETCH THE PROBLEM

Figures 26.12 and 26.13 describe the problem.

IDENTIFY THE RELATIONSHIPS

The magnification of the eyepiece is given by (Eq. 26.14) $m_\theta = s_N/f_{\text{eyepiece}}$, where $s_N = 25$ cm is the normal value of the near-point distance. Rearranging to solve for f_{eyepiece} leads to

$$f_{\text{eyepiece}} = \frac{s_N}{m_\theta}$$

Inserting the given value $m_\theta = 20$ gives

$$f_{\text{eyepiece}} = \frac{s_N}{m_\theta} = \frac{25 \text{ cm}}{20} = 1.25 \text{ cm}$$

where we keep an extra significant figure to avoid problems with roundoff errors below. According to the ray diagram for a refracting telescope in Figure 26.12, the sum of f_{obj} and f_{eyepiece} is equal to distance between the lenses, and this distance is approximately the total length of the telescope, which is given as 9.0 m. We thus have $f_{\text{eyepiece}} + f_{\text{obj}} = 9.0$ m, which leads to

$$f_{\text{obj}} = (9.0 \text{ m}) - f_{\text{eyepiece}} = (9.0 \text{ m}) - (1.25 \text{ cm}) \approx 9.0 \text{ m}$$

SOLVE

Inserting our values of the focal lengths in Equation (1), the total magnification of the telescope is

$$m_\theta = \frac{f_{\text{obj}}}{f_{\text{eyepiece}}} = \frac{9.0 \text{ m}}{0.0125 \text{ m}} = \boxed{720}$$

Reflecting Telescopes

The refracting telescope was an important experimental tool in the history of physics and astronomy. Galileo and others used it to observe the motion of planets and moons in the solar system, and their findings refuted the Earth-centered theory of the universe. Newton was also a serious astronomer, and during his work he noticed the chromatic aberration produced by a single objective lens (Section 24.8). This aberration led him to devise another telescope design, shown in Figure 26.14A; it is called a ***reflecting telescope*** because it replaces the objective lens in Figure 26.12 with a mirror. The particular telescope design in Figure 26.14A is called a Newtonian reflector.

Newton realized that, unlike lenses, a mirror would not have any chromatic aberration, and that is a big advantage for reflecting telescopes. Other advantages are even more important today. For instance, it is easier to make a large, high-quality mirror than an equally large lens of the same quality because the lens requires two carefully made surfaces and the glass must be absolutely transparent. Also, for a given diameter, a mirror is lighter and easier to support than a comparable lens, which is important because modern telescopes have mirrors that are 10 m in diameter or even larger. A lens of this size would, if supported by its edges, sag considerably under its own weight and be useless for a refracting telescope.

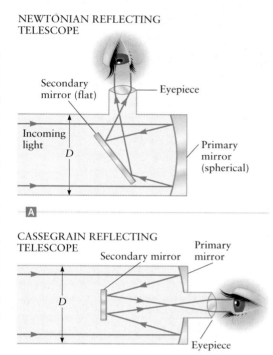

Magnification of a Reflecting Telescope

Figure 26.14A shows how light travels through a Newtonian reflecting telescope. By itself, the concave mirror at the far right would form a real image of a distant object at a point very close to the focal point of the mirror (see Fig. 24.23B). A second mirror is positioned in front of the focal point of the concave mirror, reflecting the light to an eyepiece and on to an observer. The Newtonian reflector's magnifying elements are similar to those in the refracting telescope, so the total angular magnification is given by a result similar to Equation 26.20. If the concave mirror has a focal length f_M, the total magnification is

$$m_\theta = \frac{f_M}{f_{eyepiece}} \qquad (26.21)$$

Resolution of a Telescope

In addition to magnification, another important issue in the performance of a telescope is its resolution. Resolution determines how close together in angle two stars can be and yet still be seen as two separate stars (see Fig. 26.12). The resolution of a telescope is limited by two main factors: diffraction at the telescope's aperture (the "opening") and atmospheric turbulence. As in a compound microscope (Section 26.2), diffraction spreads the light rays when they enter a telescope. The size of a telescope aperture is normally equal to the diameter D of the objective lens or

Figure 26.14 A reflecting telescope uses a concave mirror (called the primary mirror) in place of the objective lens. **A** The design shown here was invented by Newton and also uses a flat mirror (called the secondary mirror). **B** A Cassegrain reflecting telescope uses a slightly different design.

Insight 26.2
THE CASSEGRAIN REFLECTING TELESCOPE
Nowadays, many reflecting telescopes (especially those used by hobbyists) follow the Cassegrain design (Fig. 26.14B). Light from a large concave mirror (the primary mirror) strikes a second mirror and is reflected along the axis and through a small hole in the center of the primary mirror. The light then travels through an eyepiece and to the observer. The Hubble Space Telescope is a Cassegrain reflecting telescope.

Figure 26.15 This reflecting telescope on Mount Palomar has a primary mirror with a diameter of 5.0 m.

Insight 26.3

SIZE AND WEIGHT OF THE HALE REFLECTING TELESCOPE

The Hale Observatory on Mount Palomar contains a typical reflecting telescope used in astronomical research (Fig. 26.15). The mirror of the Hale telescope has a diameter of 5.0 m and a thickness of about 0.5 m, with a total mass of approximately $m = 25,000$ kg. The mirror's weight is thus $mg \approx 2.5 \times 10^5$ N, which is the weight of about 25 average-sized cars! The mirror's weight is so great that the supporting structure must be specially designed to prevent the mirror from sagging or bending as the telescope is rotated to view different parts of the sky.

primary mirror, and the Rayleigh criterion for the limiting angular resolution set by diffraction is (Eq. 25.28)

$$\theta_{min} = \frac{1.22\lambda}{D} \tag{26.22}$$

As an example, one of the largest reflecting telescopes in the world, with a mirror diameter $D = 10$ m, is at the Keck Observatory in Hawaii. Inserting this diameter in Equation 26.22 and assuming blue light ($\lambda = 400$ nm) gives

$$\theta_{min} = \frac{1.22\lambda}{D} = \frac{1.22(4 \times 10^{-7} \text{ m})}{10 \text{ m}} = 5 \times 10^{-8} \text{ rad} \tag{26.23}$$

which is a very small angle. To get an idea of how small this angle θ_{min} is, let's convert it to degrees. For historical reasons, astronomers work in degrees, and fractions of a degree called arc minutes and arc seconds, with

$$1° = 60 \text{ arc minutes} = 60'$$

$$1 \text{ arc minute} = 60 \text{ arc seconds} = 60''$$

Converting from radians in Equation 26.23 to arc seconds, we find

$$\theta_{min} = 5 \times 10^{-8} \text{ rad } \times \frac{360°}{2\pi \text{ rad}} \times \frac{60'}{1°} \times \frac{60''}{1'} = 0.01'' \tag{26.24}$$

Unfortunately, most telescopes do not attain the resolution limit θ_{min} set by diffraction. Starlight must pass through many kilometers of air (the atmosphere) before reaching an observer on the Earth, and turbulent motion of the air causes fluctuations in the refractive index from place to place. These fluctuations act like lenses and refract the incoming light from the star (Fig. 24.49). These "lenses" are constantly changing, so the direction of the starlight changes too, making the star "twinkle." Twinkling is a random process, causing light rays from a star to fluctuate through an angle that depends on the wind speed and other meteorological factors. For a location at the Earth's surface, the angular spread caused by these atmospheric fluctuations is typically

$$\Delta\theta \approx 1 \text{ arc second} = 1'' \tag{26.25}$$

This is usually much larger than the resolution limit due to diffraction (Eq. 26.24), so most ground-based telescopes are limited more by atmospheric fluctuations than by diffraction.

The value of $\Delta\theta$ due to the atmosphere becomes smaller at high altitudes because there is less air between the Earth's surface at high altitudes and outer space, which is why most observatories are placed on mountaintops. Even at a good observatory, however, the resolution limit due to the atmosphere is rarely smaller than $\Delta\theta = 0.5''$. This value is still much larger than the typical diffraction limit, so, for example, the resolution of the 10-m Keck telescope considered in Equation 26.24 is limited by the atmosphere and not by diffraction at the telescope aperture. Why, then, did astronomers bother making such a large telescope? The answer is that the amount of light a telescope collects for an image depends on the telescope's size. Although the resolution of a 10-m telescope is not any better than the resolution of a 5-m telescope, the larger telescope collects light from a larger area and can thus observe fainter stars.

It is interesting to calculate the telescope diameter at which the diffraction limit for $\Delta\theta$ is equal to the atmospheric limit. Converting the atmospheric limit of about $\Delta\theta = 1$ arc second to radians gives

$$\Delta\theta = 1'' \times \frac{1°}{3600''} \times \frac{2\pi \text{ rad}}{360°} = 5 \times 10^{-6} \text{ rad}$$

Setting $\Delta\theta$ equal to the diffraction limit (Eq. 26.22) and solving for the diameter of the telescope, we find

$$\frac{1.22\lambda}{D} = 5 \times 10^{-6} \text{ rad}$$

$$D = \frac{1.22\lambda}{5 \times 10^{-6} \text{ rad}}$$

Assuming $\lambda = 400$ nm (blue light) gives a diameter of

$$D = \frac{1.22\lambda}{5 \times 10^{-6} \text{ rad}} = \frac{1.22(4 \times 10^{-7} \text{ m})}{5 \times 10^{-6} \text{ rad}}$$

$$D = 0.1 \text{ m} = 10 \text{ cm} \qquad (26.26)$$

So, if you have a telescope with this diameter (or larger), your resolution is as good as *any* telescope located near sea level. Many amateur astronomers have telescopes that are at least this large.

This discussion shows that the resolution of a telescope located on the Earth's surface is severely limited by the atmosphere, which has prompted astronomers to move their telescopes to orbits around the Earth. Several such telescopes are now in operation and many more are in the planning stages, but the best known is the Hubble Space Telescope (see the photo at the beginning of this chapter). The Hubble telescope's resolution is limited by diffraction, as considered in Example 26.5.

| EXAMPLE 26.5 | **Resolution of the Hubble Space Telescope** |

The Hubble Space Telescope is a Cassegrain reflector (Fig. 26.14B), with a mirror diameter $D = 2.4$ m. How does its resolution compare with that of a ground-based telescope of the same size? Assume the observer is using blue light with $\lambda = 400$ nm.

RECOGNIZE THE PRINCIPLE

Because there is no atmosphere around the Hubble telescope, its resolution is limited only by diffraction and is given by Equation 26.22. Given the diameter of the Hubble's mirror, we can find its resolution limit. A ground-based telescope of the same size would be limited to about one arc second of resolution by the atmosphere (Eq. 26.25).

SKETCH THE PROBLEM

Figure 26.14B shows the design of a Cassegrain reflecting telescope. The resolution is determined by the diameter D of the primary mirror.

IDENTIFY THE RELATIONSHIPS AND SOLVE

The resolution limit of the Hubble telescope is given by $\theta_{min} = 1.22\lambda/D$ (Eq. 26.22). Inserting the given values $D = 2.4$ m and $\lambda = 400$ nm $= 4.0 \times 10^{-7}$ m, we get

$$\theta_{min} = \frac{1.22\lambda}{D} = \frac{1.22(4.0 \times 10^{-7} \text{ m})}{2.4 \text{ m}}$$

$$\theta_{min} = 2.0 \times 10^{-7} \text{rad} \times \frac{360°}{2\pi \text{ rad}} \times \frac{3600''}{1°} = 0.041''$$

This value is the Hubble telescope's resolution limit as set by diffraction. The atmospheric limit for a telescope on the Earth is about 1 arc second $= 1''$, so the Hubble telescope's resolution is better than ground-based resolution by about

$$\frac{1''}{0.041''} = \boxed{24 \text{ times}}$$

What does it mean?
The superior resolution of the Hubble Space Telescope means that it can resolve much more detail in photos of distant galaxies as illustrated in Figure 26.16,

Figure 26.16 Example 26.5. Because it is above the Earth's atmosphere, the Hubble Space Telescope has much better resolution than a ground-based telescope. These images are of the same regions of space; much more fine detail is visible in the Hubble telescope's images when compared to images obtained with the Subaru Telescope, an 8.2-m telescope in Hawaii.

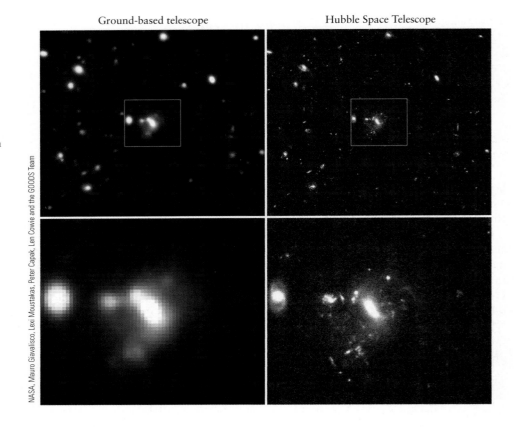

Ground-based telescope   Hubble Space Telescope

NASA, Mauro Giavalisco, Lexi Moustakas, Peter Capak, Len Cowie and the GOODS Team

which compares photos of a particular region of space taken using the Hubble telescope and a ground-based telescope. Telescopes in space thus have a natural advantage over conventional telescopes on the Earth's surface. We'll next see how an "unconventional" ground-based telescope can overcome the resolution limit caused by atmospheric fluctuations.

Adaptive Optics

The distortion of starlight by the atmosphere is due to the refraction of light by small pockets of the atmosphere that have a different refractive index than the neighboring air. This refraction is illustrated very schematically in Figure 26.17. Figure 26.17A shows a hypothetical case with no atmospheric distortion, and the light is focused perfectly by the telescope's mirror. Figure 26.17B shows a pocket of air acting as a lens, refracting some of the light rays in a different direction. Any focusing done by one lens can always be undone by a second lens, so a very clever astronomer could place a second lens in front of the telescope that would completely correct for distortion caused by the atmosphere (Fig. 26.17C). Atmospheric distortion changes as the winds and so forth change, however, so the corrective lens must change too. In practice, this corrective procedure is not done with a lens, but rather through very small changes in the shape of the mirror, made by pushing and pulling on the mirror's mounting platform (Fig. 26.17D). The mirror shape is adjusted many times each second to match fluctuations in the atmosphere by monitoring the image of a "reference star." The technology of building telescopes with adjustable mirrors to compensate for atmospheric distortion is called *adaptive optics*.

A "reference star" is an object known to appear as a point source at the telescope (so the object cannot be a planet or galaxy). If there were no distortion from the atmosphere, light from the reference star would focus to an image point limited only by diffraction. As the atmosphere causes the image of the reference star to be smeared out, the telescope mirror is adjusted to make this image as perfect as pos-

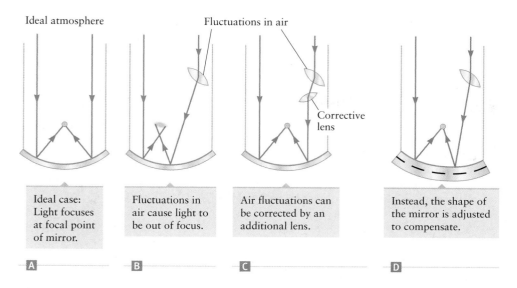

Figure 26.17 Adaptive optics. **A** Under ideal conditions, a telescope focuses light from a star onto a spot that is limited only by diffraction. **B** Under real conditions, fluctuations in the atmosphere refract some of the incoming light, producing a larger focus "spot." **C** These fluctuations can in theory be corrected by using a lens to "undo" the refraction of atmosphere. Since the atmosphere fluctuates rapidly with time, the focal length of such a (hypothetical) lens must be rapidly adjusted to match the atmospheric fluctuations. **D** A practical approach is to adjust the shape of the mirror instead of adding an extra lens.

Labels in figure 26.17:

- Ideal atmosphere
- Fluctuations in air
- Corrective lens
- **A** Ideal case: Light focuses at focal point of mirror.
- **B** Fluctuations in air cause light to be out of focus.
- **C** Air fluctuations can be corrected by an additional lens.
- **D** Instead, the shape of the mirror is adjusted to compensate.

sible. Rapid and accurate control of the mirror shape requires modern computers. The technology of adaptive optics is fairly new and is still being improved, but in some cases it is currently possible to reduce atmospheric distortion by a factor of three or more.

26.4 | CAMERAS

Cameras are extremely common optical devices. In its simplest version (Fig. 26.18), a camera consists of a single lens positioned in front of a light-sensitive material. In an old-fashioned camera, this material is called **film**. The chemistry of photographic film is surprisingly complicated, but roughly speaking, the regions of film exposed to light change their chemical properties so as to record how much light energy has been absorbed and the color of this light. Digital cameras replace film with an electronic detector that also records information on the light intensity and color, which can then be transferred to a computer. In both types of camera, the lens forms an image on the light-sensitive detector, and an aperture (the shutter) is opened for a short time to allow sufficient light energy to enter.

The simplest cameras use lenses consisting of a single piece of glass, but many camera lenses contain multiple lens elements. In either case, their overall behavior is like a single converging lens, and a camera functions in many ways like the eye. However, the eye adjusts its focus by changing the shape of its lens, while the focus of a camera is adjusted by moving the lens relative to the film (or detector) as indicated in Figure 26.18.

Design and Operation of a Film Camera

The distance between the camera's lens and the film determines which objects are in focus and which are not. For a typical film camera, the film has dimensions of approximately 24×35 mm^2 and is known as a "35-mm camera." The standard lens for such a camera has a focal length of about $f = 40$ mm, but other lenses can be purchased with both larger and smaller values of f. To understand how these different focal lengths affect the resulting photographs, consider the ray diagram in Figure 26.19, which shows the image formed by an object that is very far from the camera. The thin-lens equation (Eq. 26.1) gives a relation between the object distance s_o, the image distance s_i, and the focal length f. We have

$$\frac{1}{s_o} + \frac{1}{s_i} = \frac{1}{f} \tag{26.27}$$

Labels: Lens moves to adjust focus; Film or electronic detector; Object; Image

A

B

Figure 26.18 **A** A camera is focused by moving the lens relative to the film or an electronic (CCD) detector. **B** Typical camera used by an amateur photographer.

Camera

Object

h_o

θ

$s_i \approx f$

s_o

θ

h_i

Film

A

Film

f

A telephoto lens has a large focal length f. This lens must be farther from the film to focus the image properly.

B

Figure 26.19 **A** The size of the image on the film depends on the focal length of the lens. **B** Increasing the focal length by using a telephoto lens produces a larger image. These drawing are not to scale. The object (the tree) is typically far from the camera.

The object (the tree) in Figure 26.19A is very far from the camera, so s_o is very large, much larger than both f and s_i. The term $1/s_o$ in Equation 26.27 is thus much smaller than the other terms, and to a very good approximation we have

$$\frac{1}{s_i} \approx \frac{1}{f}$$

and

$$s_i \approx f \tag{26.28}$$

This result tells us that the lens should be positioned a distance of approximately f from the film. It also enables us to calculate the size of the image on the film. There are two right triangles in Figure 26.19A involving the angle θ. One side of the red triangle is the height of the object h_o and yields

$$\tan \theta = \frac{h_o}{s_o} \tag{26.29}$$

A similar analysis of the right triangle containing the image distance s_i and the image height h_i (shaded in yellow in Fig. 26.19A) leads to

$$\tan \theta = \frac{-h_i}{s_i} \tag{26.30}$$

where the negative sign comes from our sign convention for geometric optics (Fig. 26.2) that a downward-pointing image has a negative height. Inserting $s_i = f$ from Equation 26.28 and setting the results for $\tan \theta$ in Equations 26.29 and 26.30 equal gives

$$\frac{h_o}{s_o} = -\frac{h_i}{f}$$

The linear magnification of the image is the ratio of the image height to the object height, $m = h_i/h_o$, and we find

$$m = \frac{h_i}{h_o} = -\frac{f}{s_o} \tag{26.31}$$

The magnification is negative since the image is inverted. The object distance s_o is much larger than the focal length, so the magnification is much less than one and the image of a tree or other large object can thus fit on a piece of 35-mm film.

This result for the magnification tells us when a lens with a large value of the focal length (i.e., a telephoto lens) would be handy. Telephoto lenses usually contain several component lenses, and the overall focal length can be adjusted by changing the spacing between these lenses. Suppose a normal lens with $f = 40$ mm is used in Figure 26.19A, while Figure 26.19B shows the corresponding diagram using a telephoto lens with $f = 200$ mm. With a longer focal length, the lens must be positioned farther from the film (because $s_i \approx f$, Eq. 26.28), giving a larger magnification for a given value of the object distance s_o (Eq. 26.31). A telephoto lens is thus convenient for photographs of distant objects.

CONCEPT CHECK 26.4 | The Image Produced by a Camera

Is the image formed by the camera in Figure 26.19 a real image or a virtual image? Explain why a camera could not work if it formed a virtual image.

EXAMPLE 26.6 Designing a Camera

You are asked to design a film camera using a lens with $f = 40$ mm and want to determine how much the lens will have to move to change the focus. Your camera must focus on objects very far away and also on objects that are as close as 1.00 m. Calculate how much the distance between the lens and the film must change to adjust the focusing from one case to the other.

RECOGNIZE THE PRINCIPLE

The camera's lens must form an image on the film for two different object distances, $s_o = \infty$ and $s_o = 1.00$ m. The focal length f is given, so we must find the image distances that correspond to these two object distances. We can do so using the thin-lens equation.

SKETCH THE PROBLEM

Figure 26.19A shows the problem.

IDENTIFY THE RELATIONSHIPS

We already analyzed this problem for the case of a very distant object and in that case found $s_i = f$ (Eq. 26.28). We thus only need to consider the case with $s_o = 1.00$ m and find the corresponding image distance. The thin-lens equation reads

$$\frac{1}{s_o} + \frac{1}{s_i} = \frac{1}{f}$$

Rearranging to solve for the image distance yields

$$\frac{1}{s_i} = \frac{1}{f} - \frac{1}{s_o}$$

and doing some arithmetic gives

$$\frac{1}{s_i} = \frac{1}{f} - \frac{1}{s_o} = \frac{s_o}{s_o f} - \frac{f}{s_o f} = \frac{s_o - f}{s_o f}$$

SOLVE

We can now solve for s_i:

$$s_i = \frac{s_o f}{s_o - f}$$

Inserting the given value of f and $s_o = 1.00$ m for the smallest object distance, we find

$$s_i = \frac{s_o f}{s_o - f} = \frac{(1.00 \text{ m})(0.040 \text{ m})}{1.00 \text{ m} - 0.040 \text{ m}} = 0.042 \text{ m} = 42 \text{ mm}$$

This is the image distance when the object is very close; it is also the distance between the lens and the film. When the object is very far away, we had $s_i = f = 40$ mm. So, we only need to move the lens $(42 - 40)$ mm = $\boxed{2 \text{ mm}}$ farther from the film to focus properly.

What does it mean?
The focusing knob or the auto-focus mechanism of a camera moves the lens relative to the film. It only has to move the lens a few millimeters.

Digital Cameras

The digital camera has developed rapidly in recent years and is now the camera of choice for most amateurs and many professionals. Digital cameras are made possible by a light-sensitive detector called a charge-coupled device (CCD), which replaces the film in a film camera. A CCD uses a type of capacitor to detect light and record its intensity. The optical system (lens and aperture) of a digital camera is basically the same as that of a film camera, but the cameras differ in a few important ways.

Detecting Light with a CCD: What Is a Pixel?

A CCD is fabricated in an integrated circuit chip; current CCD chips are about the size of your thumbnail (Fig. 26.20A). A CCD chip contains many capacitors

"Window" (light enters here)

Silicon e^- } V_{CCD}

A   **B**   **C**

Figure 26.20 **A** A CCD integrated circuit chip. Light enters through the "window" on the chip. **B** The CCD chip contains many capacitors constructed on a piece of silicon. When light is absorbed, an electron is "released" from one of the covalent bonds in the silicon, leading to a charge (and hence a voltage) on the nearest capacitor. **C** The CCD chip contains an array of millions of these capacitors.

arranged in a grid over the surface of the chip as sketched in parts B and C of Figure 26.20. The structure of the capacitors in Figure 26.20B is somewhat simplified, showing only the two capacitor plates and a dielectric material in between. When light strikes the chip, it is absorbed in the dielectric layer and ejects some electrons from their normal chemical bonds. These electrons then move to the capacitor plates, leading to a voltage across the capacitor that is closest to where the light was absorbed. This voltage, denoted V_{CCD} in Figure 26.20B, is then detected by additional circuitry, and its value is stored in a computer memory in the camera. The magnitude of V_{CCD} depends on the light intensity because a greater intensity ejects more electrons and gives a larger voltage. The pattern of voltages on the capacitors in the grid (Fig. 26.20C) thus gives the light intensity as a function of position on the surface of the CCD, just as ordinary film records the intensity as a function of position on the surface of a piece of film.

The capacitors in Figure 26.20B do not distinguish the color of the light; a charge on the capacitors indicates that light was absorbed, but does not tell its color. One way to measure the color is to combine the information from four adjacent capacitors as sketched in Figure 26.21. A filter is placed in front of each of the capacitors; one of the filters allows red light to pass through to the CCD capacitor behind it, and other filters allow blue light and green light to pass. From the voltages on the four capacitors, the camera's computer can estimate the average color over this region. The CCD contains many such regions, called *pixels* (an abbreviation for "picture element"). The image produced by the CCD chip is thus stored by the camera as a set of intensity and color values for each pixel.

An important specification for a digital camera is the number of pixels in each photograph. A larger number of pixels indicates a finer level of detail in the photograph, so generally speaking, better cameras produce photos with larger numbers of pixels.

Optics of a Digital Camera

The optics and ray diagram for a digital camera are very similar to those for a film camera. Figure 26.19 thus still applies, but there is one important difference between the two types of cameras. The film for a 35-mm camera is (as noted above) approximately 24×35 mm². The CCD detector is much smaller, typically about 6×8 mm². Our derivation of the magnification for a film camera applies just as well to a digital camera, so we can again apply Equation 26.31:

$$m = \frac{h_i}{h_o} = -\frac{f}{s_o}$$

The CCD detector is smaller than a piece of 35-mm film; hence, the image height h_i must be smaller so that the image still fits on the detector. For a given value of the object distance, the focal length f of the lens must therefore be smaller for a digital

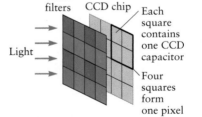

Color filters   CCD chip   Each square contains one CCD capacitor

Light   Four squares form one pixel

Figure 26.21 A CCD detector detects colors by using filters in front of the CCD capacitors. Each filter (red, green, or blue) lets only one color of light pass to the capacitor behind it, so each capacitor is sensitive to one particular color. The voltages from four adjacent capacitors are analyzed by the camera's computer to find the overall color of the light striking that region. These four adjacent capacitors form a pixel.

camera than for a film camera. The CCD detector is about four times smaller on each side, so the focal length is also about four times smaller.

In our analysis of the camera in Figure 26.19, we also showed that the distance s_i between the lens and the film is approximately equal to the focal length f (Eq. 26.28). Since it has a shorter focal length, the lens in a digital camera must be closer to the detector, which is why a digital camera can be much thinner and more compact than a conventional film camera (Fig. 26.22).

As with a film camera, the magnification of a digital camera can be changed by moving the lens relative to the CCD detector. (Compare again with Fig. 26.19.) For a digital camera, this magnification is called "optical zoom." It is also possible to use a "digital zoom" to enlarge an image. The digital zoom process constructs the entire photo using just the image data from near the center of the CCD grid. Hence, a photo taken with the digital zoom uses fewer pixels and has poorer resolution than a photo obtained with the digital zoom turned off.

Figure 26.22 Digital cameras can be very compact.

f-Number

Two important camera settings that apply to both film and digital cameras are the shutter speed and the *f-number*. The shutter "speed" is the amount of time the film is exposed to light from the object, and the *f*-number is associated with the camera's **aperture**, an opening that controls the open area of the lens (Fig. 26.23). A large aperture admits light from a large area of the lens, whereas a small aperture admits only a small amount of light (passing through the center of the lens) to the detector (the film or CCD). There are thus two ways to change the total amount of light energy that reaches the film: changing the shutter speed or changing the diameter of the aperture. The *f*-number is the ratio of the focal length to the aperture diameter D,

$$f\text{-number} = \frac{f}{D} \tag{26.32}$$

A larger aperture (a larger value of D) lets more light reach the film or CCD and gives a smaller *f*-number. A large *f*-number (small D) allows less light to get to the film or CCD. To have a properly exposed photograph (not too bright and not too dark), the total light energy must lie within a certain range. If you reduce the shutter speed and thus keep the shutter open longer, you must compensate by using a smaller aperture and hence a larger *f*-number.

| **CONCEPT CHECK 26.5** | Aperture and Shutter Speed for a Camera

A photographer is taking photos using an *f*-number of 4 and shutter speed setting that keeps the aperture open for $1/100 = 0.010$ s. The photographer then increases the *f*-number to 16. How long should the shutter now be open so as to have the same amount of light energy reach the detector, (a) the same amount of time, (b) 0.0025 s, (c) 0.04 s, or (d) 0.16 s?

Depth of Focus

Because of the trade-off between shutter speed and *f*-number, a photographer is often faced with the question of how to choose these two settings.[3] Using a short shutter speed reduces problems with vibrations when holding a camera and also makes it easier to photograph moving objects. To get enough light to the detector when using a short shutter speed, you need a small *f*-number (large aperture). The *f*-number, however, also affects a property called the **depth of focus** illustrated in Figure 26.24, which shows two photographs of the same scene using the same camera. Figure 26.24A was taken with a large *f*-number, and all the objects are fairly well focused. Figure 26.24B was taken with a smaller *f*-number, and only objects in the center of the scene are in focus, whereas those farthest from and closest to the

Large aperture/small *f*-number

Small aperture/large *f*-number

Figure 26.23 The adjustable opening in front of a camera lens is called the aperture.

---

[3]Some inexpensive cameras make this choice for you; others allow you to make choices such as "indoor" or "action" and then set the shutter speed and *f*-number accordingly.

Figure 26.24 Changing the aperture of a camera changes the *f*-number. **A** A larger *f*-number gives a larger depth of focus. Here objects from a range of distances in front and behind the best focus distance are all well focused on the film. **B** With a small *f*-number, only objects a particular distance from the camera are in focus. Objects farther away or closer than this distance are now far out of focus.

Charles D. Winters (both)

A   **B**

Figure 26.25 Ray diagrams for **A** a small aperture (large *f*-number) and **B** a large aperture (small *f*-number). With the small aperture the rays diverge more slowly from the central ray, making the depth of focus larger.

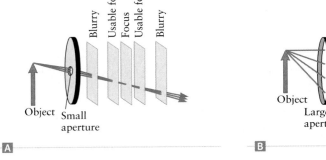

Large *f*-number—large depth of focus

Small *f*-number—small depth of focus

A   **B**

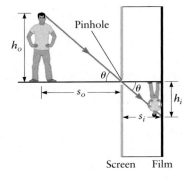

Figure 26.26 A pinhole camera forms an image without a lens. The lens of a conventional camera (Fig. 26.19) is replaced by a small pinhole. The magnification is determined by the ratio of the image distance s_i and the object distance s_o.

camera are badly out of focus. A photographer would say that the depth of focus is much larger in Figure 26.24A. Having a large depth of focus means that objects that are not at the best focusing point will produce images that are still close to ideal.

The *f*-number and depth of focus are related as shown by the ray diagrams in Figure 26.25. When the aperture is large (a small *f*-number; Fig. 26.25B), some rays make a large angle with the central ray, so they diverge more quickly as one moves away from the image point. With a small aperture (large *f*-number; Fig. 26.25A), the diverging angle is smaller, and the blurring of images away from the best focus is smaller than with a large aperture.

The Pinhole Camera

The ray diagrams in Figure 26.25 show that the "sharpness" of a photograph improves as the aperture is made smaller and smaller. In fact, if we make the aperture extremely tiny, we can make an extremely good image even without a lens! The *pinhole camera* uses this principle. This device, invented in the 1500s and sometimes called[4] the *camera obscura*, is sketched in Figure 26.26. Light from the object passes through the aperture (a tiny hole) in a screen and forms an image when it strikes a piece of film that lies behind the screen (most pinhole cameras use film rather than a CCD). Since the aperture is very small, the intensity at the film is very low and the film must be exposed to the image for a long time. Pinhole cameras are easy to make and are used mainly by hobbyists. They can also be used to view very intense light sources safely; for example, they are a safe way to view a solar eclipse and to search for sunspots (Fig. 26.27).

---

[4]Film and CCDs had not yet been invented at that time, so the first pinhole cameras were used to form images on a screen or wall behind the pinhole.

EXAMPLE 26.7 | Magnification of a Pinhole Camera

A pinhole camera can be made from a shoebox. Assume the distance from the pinhole to the film is $s_i = 20$ cm (about the size of a shoebox). If the object is $s_o = 20$ m from the camera, what is the magnification of the camera?

RECOGNIZE THE PRINCIPLE

There is no lens in a pinhole camera, so the object and image distances do not depend on any focal lengths and the thin-lens equation does not apply. We must rely on the basic definition of magnification as the ratio of the image height to the object height.

SKETCH THE PROBLEM

Figure 26.26 shows the problem and defines the image and object distances as well as the image and object heights.

IDENTIFY THE RELATIONSHIPS

Consider the two right triangles with interior angle θ in Figure 26.26, one on the object side of the pinhole with sides s_o and h_o, and the other on the image side of the pinhole with sides s_i and h_i. These triangles have the same interior angles, so the ratios of their corresponding sides are equal and we have

$$\frac{h_o}{s_o} = \frac{-h_i}{s_i}$$

The negative sign here comes from our sign convention that an inverted image is assigned a negative height (Fig. 26.2).

SOLVE

The magnification is $m = h_i / h_o$; hence,

$$m = \frac{h_i}{h_o} = -\frac{s_i}{s_o}$$

Inserting the given values for s_o and s_i gives

$$m = -\frac{s_i}{s_o} = -\frac{0.20 \text{ m}}{20 \text{ m}} = \boxed{-0.010}$$

What does it mean?
The negative value for the magnification shows that the image is inverted, as we knew from Figure 26.26. The image is also greatly reduced ($|m| \ll 1$).

Figure 26.27 Ⓐ Landscape photograph taken with a pinhole camera. Ⓑ Viewing the Sun during an eclipse through a pinhole.

26.5 | CDS AND DVDS

Most optical devices, including eyeglasses, microscopes, telescopes, and cameras, form images using lenses and mirrors and can therefore be analyzed using the principles of geometrical optics. Some of their performance limits, however, can only be understood by applying wave optics as in the Rayleigh criterion. We now consider several applications based on the wave nature of light that can only be understood using the ideas of wave optics from Chapter 25. The CD and DVD are applications of this type. CDs and DVDs operate through similar principles, so we'll focus mainly on the CD.

Figure 26.28 shows the structure of a CD; it is like a sandwich with the thickest layer composed of a plastic similar to Plexiglas. The bottom surface of the plastic in Figure 26.28A is smooth, but the top surface contains a pattern of pits used to

Figure 26.28 The information on a compact disc is encoded in a series of pits buried within the CD. **A** Light from a laser is incident from below and reflects from the aluminum layer. **B** The pits are arranged in a long spiral track. The presence or absence of a pit edge represents a binary 1 or 0.

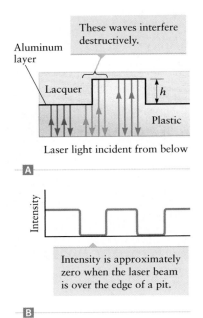

Figure 26.29 **A** Light from a laser comes in from below a CD; see Figure 26.28. *Note*: The colors of these rays do *not* represent the color of the light. Light waves reflected from the top and bottom of a pit edge interfere destructively, so there is no reflected intensity at the edge of a pit (where the interfering waves are denoted in green). Reflections from elsewhere on the CD (denoted by the red and blue rays) give a large intensity. **B** The reflected intensity thus encodes the presence or absence of a pit edge, representing a binary 1 or 0.

encode information in the CD. This top surface is coated with a thin layer of aluminum to make it reflecting and then covered with a protective layer of lacquer (a varnishlike coating), and the label of the CD is placed on top of the lacquer.

The pits are arranged in a long spiral "track" with sequential pits as close as 0.83 μm and neighboring paths on the spiral about 1.6 μm apart (Fig. 26.28B). Information encoded in the pits is read by reflecting a laser beam from the aluminum surface. Laser light passes in and out through the bottom surface of the plastic in Figure 26.28A, so that surface must be kept clean. The technology of focusing a laser beam onto a rapidly rotating CD is quite impressive. This beam must focus to a spot size smaller than a single pit and the CD player must control the spot's position on the spiral, all in the presence of vibrations and dirt. These challenging problems were solved by clever engineers. Here we consider just the optical problem of how the laser light actually detects the presence or absence of a pit.

The thin layer of aluminum in the CD acts as a mirror, reflecting the laser light. The pits influence this reflection through the thin-film interference effect illustrated in Figure 26.29. When the laser beam is positioned directly over the edge of a pit (the green rays in Fig. 26.29A), some of the light reflects from the bottom of the pit, and the rest reflects from the level outside the pit. These reflected beams are very similar to reflections from a soap film (Section 25.3). Destructive interference occurs if the difference in path lengths of the two reflected beams is $\lambda/2$, $3\lambda/2$, $5\lambda/2$, and so on.[5] The pit depth is designed to produce destructive interference, so there is *no* reflected light when the laser beam is over a pit edge, whereas the intensity is large when the laser beam is over the center of a pit or is outside a pit. As the laser beam travels along a track, the reflected light intensity varies between zero and a large value depending on the presence or absence of pit edges as sketched in Figure 26.29B. These high and low values of the intensity correspond to ones and zeros in a binary encoding of information on the CD.

EXAMPLE 26.8 | Designing a CD

The CD drive in your music player uses laser light with $\lambda_{air} = 780$ nm in air (in the infrared part of the spectrum). What is the smallest pit depth h in Figure 26.29 that will result in destructive interference for a laser beam positioned at the edge of a pit? The index of refraction of the plastic portion of the CD is $n_{plastic} = 1.55$.

RECOGNIZE THE PRINCIPLE

To have destructive interference with the smallest possible pit depth, the path lengths of the two reflected light beams must differ by

$$\Delta L = \frac{\lambda_{plastic}}{2}$$

---

[5]There is also a phase change of the light wave on reflection from the aluminum surface. This phase change occurs for both reflected light beams in Figure 26.29A and hence does not affect their *relative* phases.

where $\lambda_{plastic}$ is the wavelength of the light when it travels in the plastic. Recall that the wavelength of light in a substance (such as plastic) depends on the index of refraction of the substance (Eq. 25.9). The index of refraction of air is $n_{air} = 1.00$, so for light traveling within the CD we have

$$\lambda_{plastic} = \frac{\lambda_{air}}{n_{plastic}} \qquad (1)$$

SKETCH THE PROBLEM

Figure 26.29 shows the problem. The two interfering waves are represented by the green rays near the pit edge on the left.

IDENTIFY THE RELATIONSHIPS

Using the given values of λ_{air} and $n_{plastic}$ in Equation (1) gives

$$\lambda_{plastic} = \frac{\lambda_{air}}{n_{plastic}} = \frac{780 \text{ nm}}{1.55} = 503 \text{ nm}$$

The path length difference is equal to *twice* the depth of a pit h (Fig. 26.29A), so to get destructive interference we must have

$$h = \frac{\Delta L}{2} = \frac{\lambda_{plastic}}{4}$$

SOLVE

Inserting our value for $\lambda_{plastic}$, we find

$$h = \frac{\lambda_{plastic}}{4} = \frac{503 \text{ nm}}{4} = 126 \text{ nm} = \boxed{1.26 \times 10^{-7} \text{ m}}$$

What does it mean?
The pit depth in a CD is very small, about five hundred times smaller than the thickness of a typical human hair. Covering the reflecting side of the pits with a layer of plastic protects them from scratches that would otherwise ruin the CD.

To store as much information as possible on a CD, the area of a pit must be as small as possible. Here again we encounter a limit set by wave optics. The minimum size of a focused spot of light is approximately equal to the wavelength (Eq. 25.33). CDs use the light from a laser with $\lambda_{air} = 780$ nm and $\lambda_{plastic} = 503$ nm (see Example 26.8), so the CD pits cannot be smaller than about $\lambda_{plastic}$ in width and length, as you can confirm from Figure 26.28B. DVDs can store information more densely than CDs because of several design changes. Most importantly, DVD players today have a green laser with $\lambda_{plastic} = 420$ nm. The shorter wavelength allows the focus spot to be smaller, which in turn allows the pits in a DVD to be more closely spaced than on a CD. New DVD designs have been developed that can store even more information by using multiple aluminum layers in different layers of pits that are probed with laser beams from both sides instead of in a single layer as in current DVDs. These new-style DVDs can store at least 30 times more information than a CD.

CONCEPT CHECK 26.6 | Designing a DVD

The depth of a pit in a CD or DVD is based on the wavelength of the light used to read the information. If a CD uses light with $\lambda_{plastic} = 503$ nm and a DVD uses $\lambda_{plastic} = 420$ nm, how does the depth of a pit in a DVD compare to the depth in a CD?

   (a) The pits in a DVD are deeper by a factor of 503/420 = 1.20.
   (b) The pits are the same depth.
   (c) The pits in a DVD are shallower by a factor of 503/420 = 1.20.

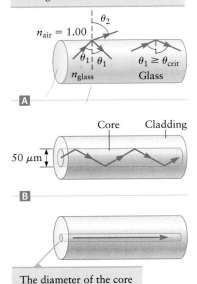

In an optical fiber, light always undergoes total internal reflection.

A

Core   Cladding

50 μm

B

The diameter of the core of a modern optical fiber is comparable to λ.

C

Figure 26.30 **A** Light traveling inside a glass fiber is reflected at the fiber's surface. The angle of incidence is always greater than the critical angle for total internal reflection, so the light does not leave the fiber. **B** Real optical fibers consist of two (or more) types of glass with different indices of refraction. The light is confined to the core of the fiber. **C** When the diameter of the core is comparable to the wavelength (as in current fibers), the light wave travels directly along the axis of the fiber.

26.6 | OPTICAL FIBERS

Optical fibers are flexible strands of glass that conduct and transmit light using total internal reflection. We described them briefly in Section 24.3, but provide more details here. Recall that light undergoes partial reflection and refraction when it encounters an interface between materials with different indices of refraction. Figure 26.30A shows a simplified model of an optical fiber. Although the fiber is cylindrical, we assume its diameter is large compared to the wavelength of light so that reflection of light at the glass–air interface is just like reflection from a flat interface. If a light ray traveling inside the fiber meets the sidewall of the fiber with an angle of incidence θ_1, there is a reflected ray with an angle of reflection equal to θ_1. If it exists, the refracted ray in the air travels at an angle θ_2. This angle of refraction is given by Snell's law (Eq. 24.7),

$$n_{\text{glass}} \sin \theta_1 = n_{\text{air}} \sin \theta_2 \qquad (26.33)$$

which can be arranged to get

$$\sin \theta_2 = \frac{n_{\text{glass}}}{n_{\text{air}}} \sin \theta_1 \qquad (26.34)$$

The value of n_{glass} depends on the composition of the glass, but is typically around $n_{\text{glass}} \approx 1.5$, while $n_{\text{air}} \approx 1.00$. Hence, the ratio $n_{\text{glass}}/n_{\text{air}}$ is greater than 1, so if $\sin \theta_1$ is large enough, the right-hand side of Equation 26.34 is greater than 1. Since the sine of an angle cannot be larger than 1, Snell's law does not have any solution for the angle of refraction θ_2 in this case. This is the phenomenon of total internal reflection described in Section 24.3, where we showed that there is no refracted ray when the angle of incidence is greater than a certain critical angle θ_{crit}. Whenever the angle of incidence is greater than the critical angle, all the light is reflected at the surface, which allows optical fibers to carry light over long distances without any leakage of light energy out of the fiber.

The optical fiber in Figure 26.30A is only a simplified design. All practical optical fibers consist of at least two different types of glass as sketched in Figure 26.30B. The central core is surrounded by an outer layer called the cladding. The core and the cladding are both made of glass, but with different compositions and different indices of refraction. The index of refraction of the cladding is smaller than the index of the core, $n_{\text{cladding}} < n_{\text{core}}$, enabling total internal reflection of light within the core. In an application, light from a laser enters the fiber at one end, and the entry angle is arranged so that the light always undergoes total internal reflection at the boundary between the core and the cladding. Signals (such as a telephone signal) are carried from one end of the fiber to the other by pulses of laser light within the core. These fibers can be very long (many kilometers), and a single fiber can carry many more simultaneous signals than is possible with a conventional metal wire.

Some early optical fibers were designed as in Figure 26.30B, and although this approach works, it has the following flaw. The laser light carrying the signal consists of pulses that encode information in a binary fashion (ones and zeros, corresponding to "on" and "off" values of the light intensity). There is a certain delay between the time a pulse enters the fiber and the time it exits at the other end. For the optical fiber in Figure 26.30B, a pulse is carried by a collection of light waves that reflect at different angles as they move along the fiber. Since they propagate at different angles, some waves travel a greater total path length than others; all travel at the same speed, so some reach the end of the fiber before others, causing the pulses to be smeared out in time, thus limiting the amount of information the fiber can carry.[6]

---

[6]Shorter pulses transmit information at a higher rate because more of them can be sent in a given amount of time, but the time between pulses must be longer than the smearing in time to avoid losing information; hence, pulse smearing limits the transmission rate.

This problem is overcome using the design sketched in Figure 26.30C. This fiber is similar to the one in Figure 26.30B except that the diameter of the core is much smaller. In fact, the core diameter is so small that ray optics can no longer be used to describe how light travels along the fiber. An analysis using wave optics shows that a light wave travels directly along the axis of the core, with no side-to-side reflections as for the larger core in Figure 26.30B. Fibers with very small diameter cores are called "single-mode" fibers because there is only one way for light to propagate in the fiber. These fibers greatly reduce the smearing of light pulses, allowing the fibers to carry more information than the "multimode" fiber in Figure 26.30B.

CONCEPT CHECK 26.7 | What Happens When You Bend an Optical Fiber?

The simplest optical fibers consist of a single strand of glass. If such a fiber is straight, the angle of incidence θ_i for a ray inside the glass is large enough that there is total internal reflection and no light leaves the fiber (Fig. 26.31A). However, if the fiber is bent through a large enough angle, the angle of incidence can become small enough that some light does "escape" from the fiber (Fig. 26.31B). If, starting from a straight fiber, the bending angle of the fiber is gradually increased, which color of light escapes first?
  (a) The longest wavelengths (red light) escape first.
  (b) The shortest wavelengths (blue and violet light) escape first.
  (c) All wavelengths escape at the same time.

Figure 26.31 Concept Check 26.7.

EXAMPLE 26.9 | Analysis of an Optical Fiber

Consider an optical fiber like the one in Figure 26.30B, with length $L = 10$ km and core diameter $d = 40$ μm. Suppose the index of refraction of core and cladding are $n_{core} = 1.80$ and $n_{cladding} = 1.40$. Assuming light travels along the fiber as rays as sketched in Figure 26.32, estimate (a) the minimum and (b) the maximum travel times of light rays along the fiber.

RECOGNIZE THE PRINCIPLE

The minimum travel time will be for light rays parallel to the axis of the fiber, which can be calculated from the length of the fiber and the speed of light. The maximum travel time occurs for rays at the critical angle for total internal reflection (Fig. 26.32), and we can find that angle using Snell's law. Using geometry, we can compare the minimum and maximum travel distances and therefore the minimum and maximum travel times.

SKETCH THE PROBLEM

Figure 26.32 shows the problem. One ray (shown in blue) lies along the axis of the fiber, and the other ray (shown in green) makes an angle θ_{crit} (the critical angle) with the direction normal to the wall of the fiber.

IDENTIFY THE RELATIONSHIPS AND SOLVE

(a) We first find the minimum travel time. The speed of light in a vacuum is $c = 3.00 \times 10^8$ m/s, but in the core this speed is reduced to $v_{core} = c/n_{core}$. The time for a light ray to travel the length of the fiber L is

$$t_{min} = \frac{L}{v_{core}} = \frac{L}{c/n_{core}}$$

Inserting the given values, we find

$$t_{min} = \frac{L}{c/n_{core}} = \frac{10 \times 10^3 \text{ m}}{(3.00 \times 10^8 \text{ m/s})/1.80} = 6.0 \times 10^{-5} \text{ s} = \boxed{60 \ \mu s} \qquad (1)$$

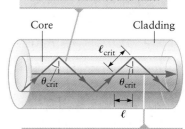

Light on this path travels the maximum distance and time.

Core   Cladding

ℓ_{crit}

θ_{crit} θ_{crit}

ℓ

Light traveling straight through the fiber travels the minimum distance and time.

Figure 26.32 Example 26.9.

(b) To find the maximum travel time, we first find the maximum travel distance, which depends on the critical angle θ_{crit} (Fig. 26.32). The critical angle for reflection at the interface between the core and the cladding can be found by applying Snell's law (Eq. 26.33) to the green ray in Figure 26.32. At the critical angle, we have an angle of refraction $\theta_2 = 90°$ in the cladding. We get

$$n_{cladding} \sin \theta_2 = n_{core} \sin \theta_{crit}$$

Inserting our value of θ_2 and the indices of refraction, we can solve for θ_{crit}:

$$n_{cladding} \sin \theta_2 = n_{cladding}(1) = n_{core} \sin \theta_{crit}$$

$$\sin \theta_{crit} = \frac{n_{cladding}}{n_{core}} = \frac{1.40}{1.80} = 0.78$$

$$\theta_{crit} = 0.89 \text{ rad} = 51°$$

The path length traveled along a ray that makes an angle θ_{crit} with the normal to the interface is longer than the path parallel to the axis of the fiber. The critical-angle path in Figure 26.32 has a length ℓ_{crit}, whereas the path along the axis has a length ℓ. From the geometry in Figure 26.32,

$$\ell_{crit} = \frac{\ell}{\sin \theta_{crit}}$$

The corresponding travel time for light along this path is thus longer than the minimum (direct) travel time by a factor

$$\frac{\ell_{crit}}{\ell} = \frac{1}{\sin \theta_{crit}}$$

Light traveling along the axis takes a time t_{min} as found in Equation (1), so light at the critical angle takes a longer time:

$$t_{max} = t_{min}\left(\frac{1}{\sin \theta_{crit}}\right)$$

Using our results above for t_{min} and $\sin \theta_{crit}$, we find

$$t_{max} = \frac{t_{min}}{\sin \theta_{crit}} = \frac{60 \ \mu s}{0.78} = 7.7 \times 10^{-5} \text{ s} = \boxed{77 \ \mu s}$$

The difference in travel times is thus

$$\Delta t = t_{max} - t_{min} = 77 \ \mu s - 60 \ \mu s = \boxed{17 \ \mu s}$$

What does it mean?
The value of Δt is nearly one-third of the minimum travel time, so the laser pulses will be quite smeared out. That is why single-mode fibers were invented.

26.7 | ⊗ MICROSCOPY WITH OPTICAL FIBERS

The wavelength of light sets a limit on the resolution of a compound microscope because light cannot be focused to a spot size smaller than approximately λ. However, light can still be used to perceive details smaller than λ using an approach called *near-field scanning optical microscopy* (NSOM). The NSOM technique uses an optical fiber to illuminate a very tiny region at the end of the fiber. The light is most intense at the opening of the tip, which can be much smaller than the wavelength. The tip is positioned very close to the object to be studied and is then is scanned over one of its surfaces (Fig. 26.33A). Because the tip is so close to the object, only a very small area near the tip is strongly illuminated; the resolution is then determined by the spacing from the surface and the tip diameter, both of which can be much smaller than the wavelength. By measuring the scattered or reflected

Core — Cladding

A

Courtesy of Nanonics Imaging Ltd.

400 nm

B

Figure 26.33 **A** The diameter of an optical fiber tip can be much smaller than the wavelength of light, producing a very tiny region just outside the tip where the light intensity is high. The tiny region at the fiber's tip is used in a near-field optical microscope by scanning it across the surface of the object of interest. **B** Image of a specially patterned layer of gold on the surface of silicon taken with a near-field scanning optical microscope. The scale bar at the lower left is 400 nm, the approximate wavelength of blue light. The small features resolved in this image are thus much smaller than the wavelength of visible light.

light, one can construct an image of the surface based on the scattered intensity as a function of the tip's position.

Current NSOM fibers have openings of 25 nm or even smaller, much less than the wavelength of visible light. (Blue light has a wavelength of about 400 nm.) The resolution of the NSOM microscope is constantly improving, and the method is now being used to create images of objects as small as individual molecules! Conventional microscopes are limited by the ability of a lens to focus, so their resolution is set by the Rayleigh criterion (Fig. 26.10) and is approximately equal to the wavelength of the light that is used. An NSOM microscope can thus resolve much smaller objects and features than is possible with a conventional microscope.

SUMMARY | Chapter 26

Optics and the eye

The eye is an essential part of many optical instruments, including magnifying glasses, microscopes, and telescopes. The purpose of these devices is to produce an image on the retina.

A *magnifying glass*, also called an *eyepiece*, is a converging lens that produces an upright virtual image when an object is placed near its focal point. This image appears at the *near point* of the eye, which is approximately $s_N = 25$ cm from the front of the eye.

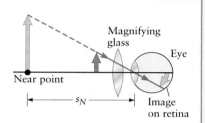

Diffraction and the resolution of microscopes and telescopes

The ultimate resolution of compound microscopes and telescopes is determined by diffraction. A compound microscope can resolve two nearby objects only when they are separated by a distance greater than approximately the wavelength of light.

Optical instruments

Microscopes and refracting *telescopes* employ (at a minimum) an *objective* lens together with an *eyepiece*. In a reflecting telescope, the objective lens is replaced by a mirror; this design has some practical advantages when the telescope is very large.

A *camera* produces an image on a light-sensitive detector, either film or a CCD. Focusing of a camera is done by moving the lens relative to the detector.

Although microscopes, telescopes, and cameras were all invented long ago, optical technology is still improving. CDs, adaptive optics, optical fibers, and near-field scanning optical microscopes have all been developed recently (during the author's lifetime!), and all can be understood using the principles of geometrical and wave optics.

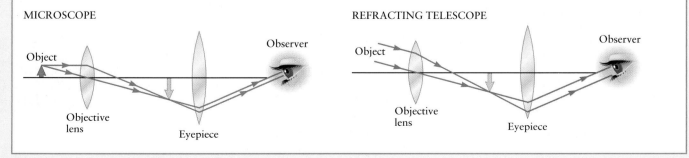

1. Explain why a magnifying glass uses a convex (converging) lens.

2. A magnifying glass uses a lens with a focal length of magnitude $|f| = 300$ cm. Is f positive or negative?

3. Will a particular magnifying glass produce the same magnification for any user? Why or why not? If a magnifying glass is marked 5×, what does that imply?

4. In Section 26.1, we showed that the angular magnification of a magnifying glass is equal to the ratio of the near-point distance to the focal length of the lens ($m_\theta = s_N/f$, Eq. 26.14). Hence, by making the focal length sufficiently short, a very large magnification can be achieved in principle. However, other practical considerations limit the maximum useable value of m_θ. Explain why it is hard to find magnifying glasses with magnifications greater than about 20.

5. X Your friend has eyeglasses that make his eyes appear larger than they really are. Is your friend nearsighted or farsighted? Explain.

6. X If you are nearsighted, should your eyeglasses contain converging lenses or diverging lenses? What if you are farsighted?

7. X Bifocals are eyeglasses that contain two different lenses for each eye (Fig. Q26.7). Each of these lenses is made from one piece of glass. One lens (lens 1) is used when looking at things far away, and the other (lens 2) is used for reading. The author uses bifocals because, without glasses, his eyes can only focus correctly on objects that are about 1 m away. (a) Is lens 1 a converging lens or a diverging lens? (b) Is lens 2 a converging lens or a diverging lens?

Figure Q26.7

8. X Surgery can be used to correct the vision of a person who is nearsighted. Should the surgeon increase or decrease the curvature of the cornea? What should the surgeon do if the person is farsighted?

9. SSM For typical cameras, the f-number can be selected from a set of fixed values such as 2, 2.8, 4, 5.6, 8, 11, and 16. Explain why this combination of f-numbers is used. *Hint*: The diameter of the aperture is inversely proportional to the f-number (Eq. 26.32), while the total intensity of the light that reaches the film is proportional to the area of the aperture.

10. X You want to increase the magnification of a particular compound microscope. Which of the following changes would achieve that goal?
    (a) Increase the focal length of the objective lens.
    (b) Decrease the focal length of the objective lens.
    (c) Increase the focal length of the eyepiece.
    (d) Decrease the focal length of the eyepiece.
    (e) Increase the spacing between the objective lens and the eyepiece.

11. Is the final image produced by a refracting telescope real or virtual? Explain.

12. Many of the refracting telescopes designed and built in the 18th and 19th centuries were extremely long (Fig. 26.13). Explain why they were made in this way; that is, what desirable property of the telescope becomes enhanced when it is made very long? What are the disadvantages of this design approach?

13. What kind of telescope did Galileo build and use in his studies? Do some investigating and try to find answers to the following questions. (See also Problem 30.)
    (a) How long was his telescope?
    (b) What was the total magnification?
    (c) What was the focal length of the objective lens?
    (d) What was the magnification of the eyepiece?

14. X Will the image produced by the contact lenses designed in Section 26.1 (and Eq. 26.7) be upright or inverted? What about the image produced by glasses for a nearsighted person? Explain.

15. X When an optometrist determines your eyeglass prescription, she always places a set of "test" lenses in front of your eyes as if they were your real glasses. The optometrist is interested in learning what lenses work best for you, but she is also measuring several other properties of your eyes. What are they?

16. X Why does squinting often allow you to read at a distance? For example, even if you forget to wear corrective lenses, forming small slits by squinting your eyes often allows you to read a clock across a room or a sign across a street. Why does squinting help?

17. Some recent water fountain landscape installations include arcs of water exhibiting laminar flow like those shown in Figure Q26.17. The arc of water is illuminated at the base, and the light appears to follow along with the flow of the stream into the pond. Why does the light stay mostly within the arcing stream of water?

Figure Q26.17 Question 17 and Problem 58.

18. The largest telescopes are all the reflector variety. Describe and explain two main design considerations in the construction of large-diameter telescopes that favor a reflector over a refractor.

19. Consider two telescopes, one a reflector and the other a refractor. Assume both telescopes are of equal diameter, are of equal magnification, and have the same focal length. In what ways is the reflector superior to the refractor and vice versa?

20. X The pupil of the eye (Fig. Q26.20) defines the aperture of the eye and thus determines how much light enters the eye and strikes the retina. If the pupil "opens up," does the f-number of the eye increase or decrease?

21. Does a camera use a diverging lens or a converging lens? Does a camera produce a real image or a virtual image?

Figure Q26.20

22. SSM A photographer intends to take a photograph of a country landscape with a large depth of field such that the flowers in the foreground of the picture and a barn in the distance are both in sharp focus. What conditions are more suitable to such a photo, a bright and sunny day or a dim and overcast day? Explain your choice.

23. DVD players use a laser with a higher frequency than that of CD players. Why can DVD players can read CDs, but CD players cannot read DVDs?

24. You are asked to design an optical fiber. Should the refractive index of the core material be larger or smaller than that of the surrounding cladding? For the fastest transmission times, do you want a high or a low refractive index for the core material?

25. You have two unlabeled laser discs, one a CD and the other a DVD. Describe how you might use a laser pointer to tell which is which.

Figure Q26.26

26. Small binoculars or opera glasses (Fig. Q26.26) are useful in seeing the stage in detail. No prisms or mirrors are used, and yet the magnified image in these glasses is not inverted. Are the eyepieces converging lenses or diverging lenses?

PROBLEMS

SSM = solution in Student Companion & Problem-Solving Guide   ☒ = life science application
★ = intermediate   ✪ = challenging   ℝ = reasoning and relationships problem

26.1 APPLICATIONS OF A SINGLE LENS: CONTACT LENSES, EYEGLASSES, AND THE MAGNIFYING GLASS

1. ☒ You are nearsighted and can only focus clearly on things that are closer than 2.5 m. What should the focal length be for contact lenses that correct this problem?

2. ☒ Your friend is farsighted with a near-point distance of 500 cm. What should the focal length be for the lenses in his reading glasses? Assume the glasses are placed 2.0 cm in front of his eyes.

3. SSM ★ ☒ A person is nearsighted and can clearly focus on objects that are no farther than 3.0 m away from her eyes. She borrows a friend's glasses and finds that she can now focus on objects as far away as 4.5 m. What is the focal length of the glasses? Do they use converging lenses or diverging lenses? Assume the glasses are placed 2.0 cm in front of her eyes.

4. ★ ☒ Consider again the nearsighted person in Problem 3, but now assume the borrowed glasses make things worse; that is, the person can now focus only on things that are within 1.5 m away. Do these glasses contain converging lenses or diverging lenses? What is the focal length of the glasses?

5. ☒ The distance from the front of the eye to the retina is about 2.5 cm. If this eye is focused on a newspaper that is 40 cm in front of the eye, what is the focal length of the eye?

6. ★ ☒ A person uses eyeglasses that have a focal length $f = -25$ cm and are mounted 1.5 cm in front of his eyes. (a) If he switches to contact lenses that are mounted in contact with the eye, what focal length should the contact lenses have? (b) Make a qualitative sketch of the shape of the lenses. Are both lens surfaces concave, convex, or . . .?

7. ✪ ☒ ℝ The author is nearsighted and (at present) is unable to focus clearly on objects that are farther than about 2.0 m from his eyes. (a) What type of lens (converging or diverging) should his eyeglasses contain so that he can focus on more distant objects? (b) What should be the focal length of his eyeglass lenses so that he can focus on a very distant object? Assume the glasses are placed 2.0 cm in front of the eyes. (c) What is the refractive power of the lenses in part (b) in units of diopters?

8. ★ ☒ ℝ The author's near-point distance is about 40 cm. What is the focal length of his reading glasses if using them allows him to bring his newspaper to within 10 cm of his eyes? Assume the glasses are placed 2.0 cm in front of his eyes.

9. ☒ ✪ **Helping your optometrist.** In Section 26.1, we designed a set of contact lenses. For a person with a particular value of the near-point distance, we found that lenses with focal length $f = 38$ cm are required (Eq. 26.7). An optometrist needs instructions on how to make these lenses; specifically, he needs to know the radii of curvature R_1 and R_2 of the two lens surfaces. In Chapter 24, we used the lens maker's formula to relate the focal length of a lens to R_1 and R_2. For a double convex lens, the lens maker's formula reads

$$\frac{1}{f} = (n - 1)\left(\frac{1}{|R_1|} + \frac{1}{|R_2|}\right)$$

where n is the index of refraction of the glass used to make the lens. To apply the lens maker's formula to our contact lens, assume one side of the lens is flat (which is not very realistic) so that its radius of curvature is $R_2 = \infty$. Find the radius of curvature R_1 of the other surface of the lens. Assume the refractive index is $n = 1.55$. Does your value for R_1 seem reasonable? How does it compare with your experience with real contact lenses?

10. ✪ ☒ **Designing an eyeglass lens.** In Problem 9, we calculated the shape (i.e., the radius of curvature) of a contact lens. In this problem, you will do the same for an eyeglass lens. One side of a contact lens must be flat (or nearly flat) so that it can rest against the surface of the eye, but with an eyeglass lens both surfaces can be convex or concave. Your job is to design a double convex lens with a focal length $f = 38$ cm (the same as in Problem 9). For simplicity, assume both surfaces have the same radius of curvature R and the lens is made from glass with $n = 1.55$. Find R. *Hint*: Use the lens maker's formula given in Problem 9.

11. What is the refractive power of the lens whose focal length is $f = 37$ cm? Give your answer in diopters.

12. ☒ The cornea of an average human eye has a refractive power of about 40 diopters. To what focal length does this value correspond? How does this focal length compare to the size of the eye?

13. ★ ☒ Design a set of eyeglasses for a nearsighted person. Assume the person is able to focus only on objects closer than 2.0 m. What type of lens (converging or diverging) with what focal length is required so that the person can view an object at infinity? Assume the glasses are placed 2.0 cm in front of the eyes.

14. ✪ ☒ A pair of glasses is designed for a person with a far-point distance of 3.0 m so that she can read street signs 20 m away. (The far-point distance is the distance from the eye at which you are just able to properly focus a distant object.) (a) If the glasses are to be worn 1.0 cm from her eyes, what is the needed focal length? (b) Compare this focal length (find the percent difference) to the focal length she would need if she chooses a style of glasses that fit on her face so that the lenses are instead 2.0 cm away from her eyes. For a nearsighted person, is the position of the glasses from the eyes as important as that for a farsighted person?

15. ⭐ ✖ **Accommodation in the eye.** A relaxed human eye has a focal length of about 2.5 cm. When looking at something very close, however, the refractive power of the eye can change by up to 16 diopters for a young adult. The shaping of the lens of the eye through muscles that stretch and compress is known as the process of *accommodation*. (a) Does the refractive power increase or does it decrease by this amount when viewing an object close up? (b) What is the focal length of the eye when focusing on something that is very close?

16. A magnifying glass gives an actual magnification of 10 for a person with a near-point distance of $s_N = 35$ cm. What is the focal length of the lens?

17. SSM ⭐ ✖ An entomologist with a near-point distance of 30 cm has two lenses that she uses to examine small beetles. One lens has a focal length $f_1 = 2.3$ cm, and the second has a focal length $f_2 = 7.5$ cm. (a) Which of the two lenses would allow her to see the smallest detail? (b) Calculate the maximum magnification available to her for each lens. (c) She examines a beetle that is 3.7 mm in length. What is the apparent length of the beetle under the highest magnification available to her?

18. ✖ A person with a near-point distance of 25 cm is able to get a usable magnification of 20 with a particular magnifying glass. Her friend is only able to get a usable magnification of 15. What is the friend's near-point distance?

19. ✖ Will you achieve a larger magnification with a particular magnifying glass if you have a near-point distance of 25 cm or of 15 cm? If the focal length of the magnifying glass is 7.5 cm, what is the ratio of the magnifications in the two cases?

20. ⭐ ✖ Ⓡ Suppose you have a near-point distance of 25 cm. The angular resolution of the eye is determined by diffraction and the Rayleigh criterion (Chapter 25) and is approximately 1 arcmin = 1′. What is the smallest object your eye can resolve without a magnifying glass?

21. ⭐ ✖ Ⓡ Consider again Problem 20. With a near-point distance of 25 cm, what is the smallest object a person can resolve using a magnifying glass with a magnification of 10?

26.2 MICROSCOPES

22. ✖ A compound microscope in which the objective and eyepiece are 20 cm apart uses an objective lens with $f_{obj} = 1.0$ cm and an eyepiece with $f_{eyepiece} = 5.0$ cm. What is the magnification of the microscope?

23. ✖ For the microscope in Problem 22, how would the magnification change if the focal length of the objective lens were decreased by a factor of three?

24. ⭐ ✖ The distance between the objective lens and the eyepiece of a microscope is 30 cm. If the eyepiece has a magnification of 20 and the total magnification of the microscope is $|m| = 250$, what is the focal length of the objective lens?

25. SSM ⭐ ✖ A compound microscope has an objective lens that by itself produces a magnification of −15. The total magnification of the microscope is designed to be −150. (a) What is the focal length of the eyepiece? (b) Why is the value of the magnification negative?

26. ✖ A compound microscope has a total magnification $|m| = 500$, using an eyepiece whose magnification is 20. What is the focal length of the objective lens?

27. ⭐ ✖ Figure P26.27 shows a typical student microscope with a rotating platform that allows the user to select from three objective lenses. A common model has objective focal lengths of 18 mm, 4.0 mm, and 1.6 mm. Each objective lens

Figure P26.27

forms an image 10 cm beyond its focal point. If the eyepiece has an angular magnification of 10×, calculate the highest and lowest overall angular magnifications available.

26.3 TELESCOPES

28. Consider a refracting telescope like that described in Figure 26.12. If it has an objective lens with $f_{obj} = 1.5$ m and an eyepiece with $f_{eyepiece} = 2.0$ cm, what is the angular magnification of this telescope?

29. ⭐ You wish to design a telescope using an objective lens with $f_{obj} = 0.75$ m. If you use an eyepiece with a magnification of 20, what are (a) the total magnification of the telescope and (b) the length of the telescope?

30. ✪ Galileo's telescope consisted of two lenses, an objective and an eyepiece. Unlike the telescope in Figure 26.12, however, Galileo used a diverging lens as his eyepiece (Fig. P26.30). This telescope was designed for the object (i.e., the Moon) to be very far away (at infinity) and produced an image that was also at infinity. In contrast to the case with a converging lens for an eyepiece, Galileo's telescope produced an *upright* image. (a) Construct a ray diagram using Figure P26.30, showing how an object very far to the left will produce an upright image that is also very far to the left. *Hint*: Consider the situation in which the image produced by the objective lens is to the right of the eyepiece. The image produced by the objective then acts as the object for the eyepiece lens. (b) Suppose $f_{obj} = 1.5$ m and $f_{eyepiece} = -1.0$ cm. What is the approximate total length of the telescope? (c) What is the total magnification?

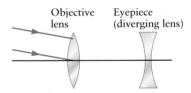

Objective lens   Eyepiece (diverging lens)

Figure P26.30

31. ⭐ Ⓡ Galileo was famous for his use of a telescope to study the solar system. Among his many discoveries, he showed that there are mountains on the surface of the Moon. Estimate the size of the smallest object that Galileo's telescope could resolve on the Moon. *Hint*: Galileo used a refracting telescope with an aperture of approximately 37 mm.

32. SSM ⭐ Ⓡ The Hubble Space Telescope is a Cassegrain reflecting telescope (Fig. 26.14B) with mirror of diameter of 2.4 m. What is the size of the smallest object that this telescope can resolve (a) on Mars, (b) on the Moon, and (c) on the surface of one of the moons of Saturn?

33. ⭐ One of the largest refracting telescopes made to date has an objective lens with a focal length of approximately 20 m. When this telescope is used to look at the Moon, what is the angle subtended by the edges of the Moon in the image formed by the objective? If the eyepiece has a focal length of 5.0 cm, what is the total magnification?

34. Consider a reflecting telescope described by the ray diagram in Figure 26.14A. If f_M is the focal length of the mirror and $f_{eyepiece}$ is the focal length of the eyepiece, the total magnification of the telescope is (Eq. 26.21)

$$m_\theta = \frac{f_M}{f_{eyepiece}}$$

This telescope has a main mirror with $f_M = 550$ mm. If an eyepiece with a magnification of 25 is used, what is the total magnification of the telescope?

26.4 CAMERAS

35. A film camera uses a single lens with a focal length $f = 60$ cm.
(a) Calculate the image distance for an object at infinity.
(b) Calculate the image distance for an object that is 2.5 m away.
(c) How far must the lens be moved with respect to the film when an object moves from infinity to a point 2.0 m from the camera?

36. The distance between a camera lens and the film in a particular camera is 5.0 cm. If this lens is focused on an object that is very far away, what is the focal length of the lens?

37. ✪ The *f*-number on a camera is changed from 4 to 8. If the photographer wants to have the same amount of light energy strike the film, should she increase or decrease the shutter speed, and by what factor?

38. ✪ The shutter speed on a camera is increased from 1/60 s to 1/500 s. If the *f*-number is originally 2.8, what new setting would you recommend?

39. Ⓡ For a camera with a shutter speed set to "250" as read on the camera, how long is the film exposed to light?

40. A photograph of a particular scene is nicely exposed with an *f*-number of 5.6 and an exposure time of 0.010 s. If the exposure time is changed to 0.020 s, what *f*-number should be used?

41. A camera has an aperture of 2.0 cm and a lens with $f = 4.2$ cm.
(a) What is the *f*-number? (b) If the *f*-number is reduced by a factor of three, by what factor should the exposure time be changed to have the same total light energy strike the film?

42. SSM ✪ A 35-mm camera uses film that is 24 mm tall and 35 mm wide. If it is used to take a photograph of a car that is 2.5 m long and 11 m away from the camera and the car just fills the photo, what is the focal length of the lens?

43. ✪ A digital camera uses a CCD detector that is 7.2 mm wide and 5.3 mm tall. (a) If the detector is rated for "4.0 megapixels" and the pixels are equally spaced, what is the distance between the centers of adjacent pixels on the CCD chip? (Use Figure 26.21 to model the arrangement of CCD pixels and ignore any space between the edges of adjacent pixels.) (b) When an object is at infinity, the lens is found to be 1.5 cm from the CCD detector. What is the focal length of the lens? (c) If an object is placed 30 cm from the lens, what spacing between the lens and the CCD detector is required for the image to be at the detector?

44. ✪ The author has an inexpensive digital camera that has the following specifications: "5× optical zoom, 7× digital zoom." (a) The "lens" is actually a combination of several lenses, and the value of the optical zoom is adjusted by changing the focal length *f* of the lens system. If the magnification is changed by a factor of five (the maximum optical zoom), by what factor does *f* change? (b) The digital zoom is adjusted by using the image from only a portion of the CCD detector. When the digital zoom is "off," all the CCD's pixels are used. What fraction of the pixels is used when the digital zoom is set for its maximum value (7×)?

26.5 CDS AND DVDS

45. SSM A hypothetical CD player uses a laser with $\lambda = 880$ nm in air. Inside the plastic coating of a hypothetical CD the wavelength is 550 nm. What is the index of refraction of the plastic?

46. ✪ Ⓡ A real CD can hold approximately 600 Mbytes of information. Each byte is an 8-bit binary number, so it takes eight digits of zeros and ones to make 1 byte. In order that each pit edge can be optically resolved by the laser beam that reflects from the CD (Fig. 26.28), the length of a pit must be no smaller than the wavelength. Assume each pit on a CD stores 2 bits of

information (two edges), the pits are 3λ long, and $\lambda = 503$ nm in the plastic of the CD. Approximately how many megabytes of information can be stored on this model of CD? Explain why your number agrees or does not agree with the storage capacity of a real CD.

47. Ⓡ The light used in a CD has a wavelength of 503 nm (inside the plastic coating), whereas a Blu-ray disc uses a wavelength of 260 nm (again, inside the plastic coating). Based solely on the effects of diffraction, a Blu-ray disc can contain more pits and thus hold more information than a CD. Estimate the ratio of the number of pits that can be on a Blu-ray disc to the number on a CD. How does this answer compare with what you know (or can look up) about the amount of information that can be stored on a CD and a Blu-ray disc?

26.6 OPTICAL FIBERS

48. SSM ✪ The optical fiber in Figure P26.48 has a core of diameter 50 μm and a total diameter of 60 μm. The indices of refraction are $n_{core} = 1.65$ and $n_{cladding} = 1.55$. What is the maximum value of θ in Figure P26.48 that gives total internal reflection? This is called the "acceptance" angle of the fiber.

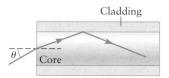

Figure P26.48 Problems 48 and 49.

49. ✪ Consider the simple optical fiber in Figure 26.30A that does not have a cladding and suppose the fiber has an index of refraction $n = 1.75$. The acceptance angle θ is the maximum angle that an incoming ray can have with the end of the fiber and still achieve total internal reflection inside the fiber. (See Fig. P26.48.) (a) What is the acceptance angle of this fiber? (b) If the fiber is immersed in water (perhaps as part of an undersea cable that leaks), what is the new acceptance angle of the fiber? Assume that the ends of the fiber are in air while the sides are immersed in water.

50. ✪ In Example 26.9, we discussed how light pulses in a multimode optical fiber are "smeared out" in time due to the different angles of travel that are possible within the fiber. This problem does not occur in single-mode fibers because all light travels along the axis of the fiber, but there is another smearing effect in single-mode fibers due to dispersion. The laser light traveling in the fiber has a small spread in wavelength; hence, a single pulse consists of components of different colors. Since the index of refraction depends on the wavelength, different components of a pulse travel at different speeds, and eventually the faster components of one pulse will "overtake" the slower components of the pulse ahead of it. Consider a light pulse with just two components with wavelengths (in the fiber) of $\lambda_1 = 503.0$ nm and $\lambda_2 = 504.0$ nm, with corresponding indices of refraction $n_1 = 1.5500$ and $n_2 = 1.5510$, traveling in a fiber of length 20 km. (a) Which wavelength travels fastest? (b) How long will it take each component to travel the entire length of the fiber? (c) Two very short pulses, one with wavelength λ_1 and another with λ_2, are started simultaneously at one end of the fiber. When the faster component reaches the end of the fiber, how far behind is the slower component?

51. ✪ ✖ Ⓡ An objective lens has a focal length of 0.50 cm. Use it to design a compound microscope with a total magnification of 500. First identify the quantities whose values you must choose for the design and then find a set of appropriate values.

52. ✪ Ⓡ The author has a Cassegrain reflecting telescope (Fig. 26.14B) with a mirror diameter of 10 cm that he is using to study the many rings of Saturn. Figure P26.52 shows a photo taken with the Hubble Space Telescope showing various features of the rings. The most prominent (brightest) are the A and B rings. Can the author's telescope resolve the gap between Saturn and the inner edge of the B ring? Estimate the data you need from the figure. The distance from the Earth to Saturn is approximately 1.3×10^9 km.

Figure P26.52

53. ✪ You have lenses with focal lengths $f_1 = 4.0$ cm and $f_2 = 12$ cm. Assume your near point distance is 25 cm and consider the following configurations of these lenses. (a) For magnifying glasses, what magnification could you achieve with each lens? (b) What is the maximum magnification you could achieve if you used these lenses in a compound microscope of length 25 cm? (c) What maximum magnification could you achieve if you made them into a telescope? What is the separation distance between the lenses for the telescope configuration?

54. ✪ Two lenses are mounted in a tube such that the distance between them can be varied. This device can be used as either a telescope or a compound microscope. One lens has a focal length of 27 cm. (a) Find the focal length of the second lens such that a telescope made from the two lenses would have the same magnification as a microscope made from the same two lenses. *Hint:* It will be a low-magnification microscope, so you will need to substitute Equation 26.13 for $m_{\theta, \text{eye}}$ in deriving a result to replace Equation 26.16. (b) What is the magnification achieved?

55. Completed in 1895, the Yerkes telescope operated by the University of Chicago remains the biggest (largest-diameter) refracting telescope ever built. Figure P26.55 shows a photograph of the telescope taken in 1921, when it was the most important astronomical instrument in the world. The refractor has an *f*-number of 19 and an objective lens that measures 1.02 m across. What are the focal length and approximate length of this telescope? Do your calculations agree with an estimate made from the photograph?

56. ✪ Ⓡ A digital camera has an 8.3-megapixel CCD array with the pixels evenly spaced over a rectangular region, 7.2 mm × 5.3 mm. The camera has a lens 2.0 cm in diameter with a focal length of 3.5 cm. (a) What is the resolution limit due to pixel size? (b) What is the approximate resolution limit due to aperture size? (c) If an object 1.5 m high is 200 m away, how many pixels high will the image be? (d) What is the maximum distance from the camera that this page could be photographed in order to read this paragraph from the digital photo? (e) In a movie theater, the screen is 4.0 m wide and 2.2 m tall. How far back from the screen would one have to sit to capture an image of the whole screen with this camera?

57. ✪ Ⓡ **Blu-ray disc system data storage.** As the name implies, the Blu-ray disc has the same diameter and refractive index as a DVD or CD, but the laser light used is in the blue–violet region with a wavelength in air of 405 nm. The index of refraction of the plastic coating on the disc is $n = 1.55$. (a) What is the wavelength of this light within the disc? (b) How deep are the pits in a Blu-ray disc? Are they deeper or shallower than the pits in a CD or DVD? (The laser light used by a CD has a wavelength of 780 nm while for a DVD it is 650 nm.) (c) Tracks (adjacent rows of pits) are separated by 1.6 μm in a CD, 0.74 μm in a DVD, and 0.32 μm in a Blu-ray disc. Calculate the approximate amount of memory in megabytes (1 byte = 8 bits, and 1 Mb = 10^6 bytes) per square centimeter of disc area for each format. Note that each pit edge, going from the bottom of a pit to the upper surface or vice versa, counts as 1 bit.

58. ✪ Consider the water fountain in Figure Q26.17, where the stream of water acts like an optical fiber. What is the minimum angle of total internal reflection in the stream of water for blue light with $\lambda = 450$ nm?

59. SSM ✪ The constellation Orion is approximately 800 light-years from Earth. (One light-year is the distance traveled by light in a vacuum in one year.) Assume that one of the stars in this constellation has a planet orbiting at the same distance as the Earth orbits the Sun. How large should the aperture of a telescope be so that it can resolve the star from its planet using light with $\lambda = 450$ nm?

Figure P26.55 Albert Einstein and observatory staff stand in front of the Yerkes 40-in. refracting telescope in 1921.

Relativity

The propagation of light through the universe has some surprising features that are explained by Einstein's theories of relativity. This is a photo of the Sombrero Galaxy—it is 28 million light-years from the Earth and is believed to have a black hole at its center. (NASA/JPL-Caltech/University of Arizona/STScI)

Newton discovered his laws of mechanics around 1700. Over the next two centuries, these laws were applied to a wide variety of problems and seemed to work perfectly. With the development of Maxwell's equations in the late 1800s, the physics of electromagnetism and light fell into place. By 1900, many physicists thought that physics was completely understood with nothing more to be done, but how wrong they were! Two completely new and unanticipated subfields of physics—relativity and quantum theory—were discovered in the first decades of the 1900s. For historical reasons, these topics are called "modern physics," although it seems peculiar to use the term *modern* to describe work nearly one hundred years old. Relativity was developed mainly by Albert Einstein (although other physicists had important roles), and his theory of relativistic mechanics has profoundly changed the way we think about space and time.

There are actually two types of relativity theory. **Special relativity** is concerned with objects and observers moving with constant velocity, the case we will consider through most of this chapter. *General relativity* applies to situations in which an object or observer is accelerated, and it has some surprising implications for the understanding of gravitation.

27.1 | NEWTON'S MECHANICS AND RELATIVITY

In physics, the term *relativity* arises when we describe a situation from two different points of view. Figure 27.1 shows a person (Ted) standing on an open railroad car. In Figure 27.1A, the car is at rest while Ted throws a ball straight up into the air and then catches it when it falls back down to him. Figure 27.1B shows a similar event, but now the railroad car is moving with constant speed v to the right and a person on the ground next to the car (Alice) is watching. According to Ted—that is, *relative to Ted*—the ball travels straight up and then back down, just as in Figure 27.1A. Relative to Alice, the ball has a horizontal component of velocity equal to that of the railroad car and follows a parabolic trajectory. The ball's motion can be analyzed by both Ted and Alice using the results for motion with constant acceleration (projectile motion) from Chapter 4.

These two observers are in different *reference frames*. You can think of a reference frame as a set of coordinate axes. In this example, Alice is at rest relative to the $x–y$ axes, whereas Ted would describe the motion of the ball using the coordinate axes $x'–y'$ that travel along with his railroad car.

Reference frames that move with a constant velocity, like the reference frames in Figure 27.1, are called *inertial reference frames*. The idea that the laws of motion should be the same in all inertial frames can be traced to Galileo and is called the *principle of Galilean relativity*. Newton certainly understood that his laws of mechanics obey Galilean relativity; that is, if Newton's laws of mechanics are obeyed in one inertial reference frame, they are obeyed in all inertial frames. For example, Newton's second law reads

$$\sum \vec{F} = m\vec{a} \qquad (27.1)$$

where $\vec{a}$ is the acceleration of a particle of mass m and $\sum \vec{F}$ is the total force on the particle. Adding or subtracting a constant velocity does not change the acceleration of a particle. In Figure 27.1, Ted would say that the horizontal component of the ball's velocity (the component along x') is zero, but Alice would disagree; in her reference frame, the horizontal (x) component of the ball's velocity is v. Both Ted and Alice would agree, however, that the ball's acceleration is directed downward (along either y' or y) with magnitude g, and both would agree that the ball's horizontal acceleration is zero. They would also agree that the only force on the ball is the force of gravity, so both would say that Newton's second law (Eq. 27.1) is obeyed.

(left margin) Inertial reference frame: a set of coordinate axes that move at constant velocity relative to other inertial reference frames

Figure 27.1 A When Ted throws a ball upward or drops a ball downward, he observes that the ball's motion is purely along the vertical (y') direction. B When Ted's railroad car has a speed v relative to a second observer (Alice), the ball undergoes projectile motion with a nonzero displacement along both x and y in Alice's reference frame. However, as viewed by Ted using his reference frame and coordinates x' and y', the ball's motion is still purely vertical as in part A.

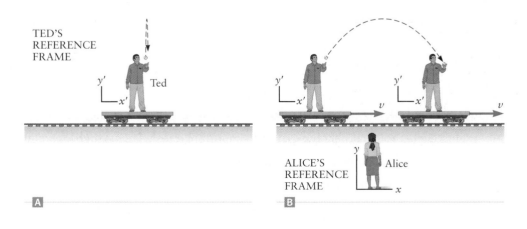

Newton's second law is thus obeyed in both of these inertial reference frames. This result is similar to what we found in our analysis of the airplane accelerometer in Chapter 4; there again, observers in different inertial reference frames both found that Newton's second law is obeyed.

The principle of Galilean relativity is certainly reasonable. It agrees with our everyday intuition and the experiment in Figure 27.1, and was believed to be an exact law of physics for more than two hundred years. The first hint of a problem came from Maxwell's work in electromagnetism. According to Maxwell's equations, the speed of light is (Eq. 23.1)

$$ c = \frac{1}{\sqrt{\varepsilon_0 \mu_0}} \tag{27.2} $$

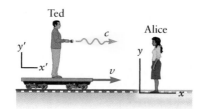

Figure 27.2 Ted moves at speed v relative to Alice when his flashlight emits a light pulse that moves at speed c relative to him. What is the speed of the light pulse relative to Alice?

Maxwell also showed that this result is independent of the motion of both the source of the light and the observer. To see why this finding is surprising, suppose Ted is again moving with a speed v relative to Alice when he turns on a flashlight that generates a light pulse traveling to the right in Figure 27.2. This is a one-dimensional situation, so we need consider only the component of velocity along the horizontal (x) direction. The light wave in Figure 27.2 has a speed c relative to Ted and his flashlight. Since Ted's speed relative to Alice is v, Newton's mechanics predicts that the speed of the light wave relative to Alice should be $c + v$. According to Maxwell's theory, however, Ted and Alice should *both* observe the light wave to move with speed c. Maxwell's theory of electromagnetism is thus not consistent with the predictions of Galilean relativity for observers in different inertial frames.

Which prediction is correct? Initially, Maxwell and other physicists thought that there must be a problem with Maxwell's theory, but an experimental test is difficult. For example, if Ted in Figure 27.2 is moving at 100 m/s (about 200 mi/h), Galilean relativity predicts that Alice will observe a speed of

$$ v(\text{Alice}) = c + v(\text{Ted}) = [(3.00 \times 10^8) + 100]\, \text{m/s} \tag{27.3} $$

This differs from c by less than one part in one million, and such a small difference would be very hard to measure using the technology available in Maxwell's time. Successful experiments were carried out only well after Maxwell's death and showed that Maxwell's theory has it right: the speed of light in a vacuum is always c, and the prediction of Galilean relativity for how the speed of light depends on the motion of the source is wrong. In 1905, Einstein worked out the correct theory, which had profound implications for all physics.

27.2 | THE POSTULATES OF SPECIAL RELATIVITY

The theory Einstein developed to analyze the predicament of Ted and Alice in Figure 27.2 is called *special relativity*. According to Einstein, his work on this theory was not motivated by any particular experiment; indeed, at the time he developed his theory (1905), there were not yet any clear experimental results to show that Galilean relativity was wrong. Those experiments came only after Einstein's theory. Instead, Einstein suspected that Maxwell's result—that the speed of light is the same in all reference frames—is correct, and he then worked out what that implies for the other laws of physics such as Newton's laws. In a sense, Einstein's work was similar to that of Newton: both proposed basic laws of physics and then worked out the consequences. For Newton, these basic laws were his three laws of motion. For Einstein, the basic laws are now known as two "postulates" about the laws of physics. We'll first state the postulates and then spend most of the rest of this chapter working out the consequences. Experiments performed after 1905 showed that Newton's theory breaks down for fast-moving objects (such as light) and that Einstein's theory gives a correct description of motion in such a regime.

Observer
$\vec{v}_{boat}$
Water waves
on a lake
$\vec{v}_{wave}$

Figure 27.3 According to our intuition, water waves on a lake obey Galilean relativity. The velocity of these waves relative to the water (the wave medium) is $\vec{v}_{wave}$, whereas the velocity of a boat relative to the water is $\vec{v}_{boat}$. In this example, the velocity of the waves relative to the boat is $\vec{v}_{wave} - \vec{v}_{boat}$.

Postulates of Special Relativity

1. **All the laws of physics are the same in all inertial reference frames.**
2. **The speed of light in a vacuum is a constant, independent of the motion of the light source and all observers.**

The first postulate can be traced to the ideas of Galileo and Newton on relativity, but this postulate goes further than Galileo because it applies to *all* physical laws, not just mechanics. The second postulate of special relativity is motivated by Maxwell's theory of light, which we have seen is *not* consistent with Newton's mechanics. Newton would have predicted that the speed of the light pulse relative to Alice in Figure 27.2 is $c + v$ (Eq. 27.3), whereas Ted (who is in a different inertial reference frame) would measure a speed c. According to the second postulate of special relativity, both observers will measure the *same* speed c for light.

The postulates of special relativity will lead us to a new theory of mechanics that corrects and extends Newton's laws. Postulate 2 also tells us something very special and unique about light. Light is a wave, and all the other waves we have studied travel in a mechanical medium. For example, Figure 27.3 shows an observer traveling in a boat with velocity $\vec{v}_{boat}$ relative to still water. If there are waves traveling at velocity $\vec{v}_{wave}$ relative to still water and if the boat is traveling along the wave propagation direction, Newton's mechanics (and Galilean relativity) predict that the velocity of the waves *relative to this observer* is $\vec{v}_{wave} - \vec{v}_{boat}$. An observer who is stationary relative to the water would thus measure a different wave speed than the observer in the boat. This is *not* what the two observers in Figure 27.2 find; they measure the *same* speed for the light wave.

The conclusion from Figures 27.2 and 27.3 is that our everyday experience with conventional waves *cannot* be applied to light. According to postulate 2 of special relativity, the speed of a light wave is independent of the velocity of the observer. What role does the medium have in this case? Light does not depend on having a conventional material medium in which to travel. For a light wave, the role of the medium is played by the electric and magnetic fields, so a light wave essentially carries its medium with it as it propagates (which is why light can travel through a vacuum). The lack of a conventional medium is surprising and difficult to reconcile with one's intuition, and Maxwell's results were therefore slow to be accepted. Experiments, however, show that nature does work this way; the speed of light *is* independent of the motion of the observer. This example is just one of many for which conventional intuition fails. We'll come to more such failures as we study special relativity.

Finding an Inertial Reference Frame

Inertial reference frames played a role in our work with Newton's mechanics, and they play a crucial role in special relativity. We have stated that an inertial frame is one that moves with constant velocity, but relative to what? Newton believed that the heavens and stars formed a "fixed" and absolute reference frame to which all other reference frames could be compared. We now know that stars are also in motion, so we need a better definition of what it means to be "inertial." Nowadays, we define an inertial reference frame as one in which Newton's first law holds. Recall that Newton's first law states that if the total force on a particle is zero, the particle will move with a constant velocity, in a straight line with constant speed. So, we can test for an inertial reference frame by observing the motion of a particle for which the total force is zero. If the particle moves with a constant velocity, the reference frame in question—the reference frame we use to make the observation—is an inertial frame.

Alternate definition of an inertial reference frame

This definition of how to find an inertial reference frame is thus tied to Newton's first law and the concept of inertia, so the definition may seem a bit circular. However, the notion of Galilean relativity also asserts that Newton's other laws of mechanics are valid in all inertial frames. Hence, Newton's second law ($\sum \vec{F} = m\vec{a}$) and third law (the action–reaction principle) should apply in all inertial frames, too.

The Earth as a Reference Frame

Inertial reference frames move with constant velocity; hence, their acceleration is zero. Since the Earth spins about its axis as it orbits the Sun, all points on the Earth's surface have a nonzero acceleration. Strictly speaking, then, a person standing on the surface of the Earth is not in an inertial reference frame. However, the Earth's acceleration is small enough that it can be ignored in most cases, so in most situations we can consider the Earth to be an inertial reference frame.

27.3 | TIME DILATION

Einstein's two postulates seem quite "innocent." The first postulate—that the laws of physics must be the same in all inertial reference frames—is in accord with Newton's laws, so it does not seem that this postulate can lead to anything new for mechanics. The second postulate concerns the speed of light, and it is not obvious what it will mean for objects other than light. Einstein, however, showed that these two postulates together lead to a surprising result concerning the very nature of time. He did so by considering in a very careful way how time can be measured.

Let's use the postulates of special relativity to analyze the operation of the simple clock in Figure 27.4. This clock keeps time using a pulse of light that travels back and forth between two mirrors. The mirrors are separated by a distance ℓ, and light travels between them at speed c. The time required for a light pulse to make one round trip through the clock is thus $2\ell/c$. That is the time required for the clock to "tick" once.

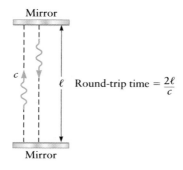

Figure 27.4 A light clock. Each round-trip motion of the light pulse between the two mirrors corresponds to one tick of the clock.

Analysis of a Moving Light Clock

We now place our light clock on Ted's railroad car in Figure 27.5A, so the clock is moving at constant velocity with speed v relative to the ground. How does that affect the operation of the clock? Let's first analyze the clock from Ted's viewpoint—that is, in Ted's reference frame—as we ride along on the railroad car in Figure 27.5B. In this reference frame, the operation of the clock is identical to that shown in Figure 27.4; the light pulse simply travels up and down between the two mirrors. The separation of the mirrors is still ℓ, so the round-trip time is still $2\ell/c$.

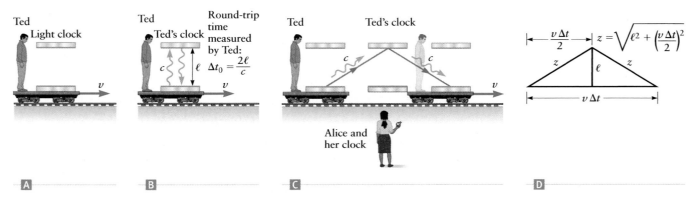

Figure 27.5 **A** A light clock traveling with Ted on his railroad car. **B** According to Ted, light pulses travel back and forth in the clock just as in Figure 27.4. Each tick of the clock takes a time $\Delta t_0 = 2\ell/c$. According to Ted, the operation of the clock is the same whether or not his railroad car is moving. **C** Motion of the light pulses in Ted's clock as viewed by Alice, who is at rest on the ground. **D** According to Alice, the round-trip travel distance for a light pulse is $2z$, where $z = \sqrt{\ell^2 + (v\,\Delta t/2)^2}$, which is longer than the round-trip distance 2ℓ seen by Ted in part B.

If Δt_0 is the time required for the clock to make one "tick" as measured by Ted, we have the result

$$\Delta t_0 = \frac{2\ell}{c} \tag{27.4}$$

A second observer, Alice, is standing on the ground watching Ted's railroad car travel by and sees things differently. In her reference frame, Ted's clock is moving horizontally, so from her point of view the light pulse does not simply travel up and down between the mirrors, but must travel a longer distance as shown in Figure 27.5C. According to postulate 2 of special relativity, the speed of light is the same for Alice as it is for Ted. Because light travels a longer distance in Figure 27.5C than in Figure 27.5B, according to Alice the light will take longer to travel between the mirrors. Let's use Δt to denote the round-trip time as observed by Alice; that is the time it takes for the clock to complete one "tick" in Alice's reference frame. We can find Δt by using a little geometry. In a time Δt (one "tick"), Alice sees the clock move a horizontal distance $v\,\Delta t$. The path of the light pulse forms each hypotenuse z of the back-to-back right triangles in Figure 27.5D. Using the Pythagorean theorem,

$$z^2 = \ell^2 + \left(\frac{v\,\Delta t}{2}\right)^2 \tag{27.5}$$

Since z is half the total round-trip distance, Alice finds

$$z = \frac{c\,\Delta t}{2}$$

or

$$z^2 = \frac{c^2(\Delta t)^2}{4} \tag{27.6}$$

Combining Equations 27.5 and 27.6 gives

$$\frac{c^2(\Delta t)^2}{4} = \ell^2 + \frac{v^2(\Delta t)^2}{4}$$

We next solve for Δt:

$$(\Delta t)^2 = \frac{4\ell^2}{c^2} + \frac{v^2}{c^2}(\Delta t)^2$$

$$(\Delta t)^2\left(1 - \frac{v^2}{c^2}\right) = \frac{4\ell^2}{c^2}$$

$$(\Delta t)^2 = \frac{4\ell^2/c^2}{1 - v^2/c^2}$$

Taking the square root of both sides and using Equation 27.4 finally leads to

$$\Delta t = \frac{\Delta t_0}{\sqrt{1 - v^2/c^2}} \tag{27.7}$$

Recall that Δt and Δt_0 are the times required for one tick of the light clock as observed by Alice and Ted, respectively. In words, Equation 27.7 thus says that these times are *different*! The operation of this clock *depends on the motion of the observer*. Let's now consider the implications of Equation 27.7 in more detail.

Moving Clocks Run Slow

The clock in Figure 27.5 is at rest relative to Ted, and he measures a time Δt_0 for each tick. The same clock is moving relative to Alice, and according to Equation 27.7 she measures a longer time Δt for each tick. This result is not limited to light clocks. Postulate 1 of special relativity states that *all* the laws of physics must be the same in *all* inertial reference frames. We could use a light clock to monitor or time any process in any reference frame. Since Equation 27.7 holds for light clocks, it must therefore apply to *any* type of clock or process, including biological ones.

According to Equation 27.7, the ratio of Δt (the time measured by Alice) to the time Δt_0 (measured by Ted) is

$$\frac{\Delta t}{\Delta t_0} = \frac{1}{\sqrt{1 - v^2/c^2}} \tag{27.8}$$

Assuming v is less than the speed of light c (discussed further below), the factor on the right-hand side is always greater than 1. Hence, the ratio $\Delta t/\Delta t_0$ is larger than 1, and Alice measures a longer time than Ted does. In words, a moving clock will, according to Alice, take longer for each tick. Hence, special relativity predicts that *moving clocks run slow*. This effect is called **time dilation**.

This result seems very strange; our everyday experience does not suggest that a clock (such as your wristwatch) traveling in a car gives different results than an identical clock at rest. If Equation 27.8 is true (and experiments definitely show that it *is* correct), why haven't you noticed it before now? Figure 27.6 shows a plot of the ratio $\Delta t/\Delta t_0 = 1/\sqrt{1 - v^2/c^2}$ as a function of the speed v of the clock. At ordinary terrestrial speeds, v is much smaller than the speed of light c and the ratio $\Delta t/\Delta t_0$ is very close to 1. For example, when $v = 50$ m/s (about 100 mi/h), the ratio is

$$\frac{\Delta t}{\Delta t_0} = \frac{1}{\sqrt{1 - v^2/c^2}} = \frac{1}{\sqrt{1 - (50 \text{ m/s})^2/(3.00 \times 10^8 \text{ m/s})^2}}$$

$$\frac{\Delta t}{\Delta t_0} = 1.000000000000014 \tag{27.9}$$

The result in Equation 27.9 is extremely close to 1, so for all practical purposes the times measured by Ted and by Alice are the same if Ted moves at 50 m/s relative to Alice. For typical terrestrial speeds, the difference between Δt and Δt_0 is completely negligible.

Nature's Speed Limit

A curious feature of the time-dilation factor in Equation 27.8 is that its value is imaginary when v is greater than the speed of light. For example, if we insert $v = 2c$ into Equation 27.8, we get

$$\frac{\Delta t}{\Delta t_0} = \frac{1}{\sqrt{1 - (2c)^2/c^2}} = \frac{1}{\sqrt{-3}}$$

This result is an imaginary number! Does it mean that special relativity predicts that some time intervals Δt can be imaginary numbers? No, it does not. Speeds greater than c have never been observed in nature. We'll come back to this issue in Section 27.9 when we discuss work and energy in special relativity, and we'll see why it is not possible for an object to travel faster than the speed of light.

| EXAMPLE 27.1 | Time Dilation for an Astronaut |

The astronauts who traveled to the Moon in the Apollo missions hold the record for the highest speed traveled by people, with $v = 11{,}000$ m/s. What is the ratio $\Delta t/\Delta t_0$ for the Apollo astronauts' clock?

RECOGNIZE THE PRINCIPLE

The Apollo astronauts have the role of Ted in Figure 27.5 because we are interested in a clock that travels with them, while an observer on the Earth has the role of Alice. The time measured by the astronaut's clock thus reads the time interval Δt_0, and an observer on the Earth measures a longer time interval Δt.

Time dilation: moving clocks run slow.

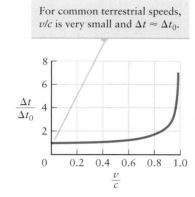

Figure 27.6 Time dilation factor $\Delta t/\Delta t_0$ as a function of v/c.

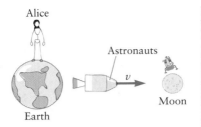

Alice

Astronauts

v

Earth

Moon

Figure 27.7 Example 27.1.

This problem is described by Figure 27.7. We assume the astronauts are carrying a light clock with them to the Moon, just as Ted carried a clock in his railroad car in Figure 27.5.

IDENTIFY THE RELATIONSHIPS

We can find $\Delta t/\Delta t_0$ using our analysis of Figure 27.5 (and Eq. 27.8), substituting $v = 1.1 \times 10^4$ m/s and the known speed of light.

SOLVE

Inserting the given values, we have

$$\frac{\Delta t}{\Delta t_0} = \frac{1}{\sqrt{1 - v^2/c^2}}$$

$$= \frac{1}{\sqrt{1 - (1.1 \times 10^4 \text{ m/s})^2/(3.00 \times 10^8 \text{ m/s})^2}} = \boxed{1.00000000067}$$

When v is such a small fraction of the speed of light, you may be limited by the number of significant figures your calculator can display. In such cases, you can use one of the approximations given in Insight 27.1. The second approximation gives

$$\frac{\Delta t}{\Delta t_0} \approx 1 + \frac{v^2}{2c^2} = 1 + \frac{(1.1 \times 10^4 \text{ m/s})^2}{2(3.00 \times 10^8 \text{ m/s})^2} \approx \boxed{1 + 6.7 \times 10^{-10}} \qquad (1)$$

What does it mean?
Time dilation is a very small effect, even at this (relatively) high speed, yet it is possible to make clocks that are accurate enough to observe the small amount of slowing down in Equation (1). Experiments have shown that the time dilation predicted by special relativity is indeed correct. This result for Δt applies to all clocks, including the biological clocks of the Apollo astronauts. Hence, these astronauts aged slightly less than other people who stayed behind!

Proper Time

We derived Equation 27.8 from an analysis of a light clock, but the result applies to all time intervals measured with any type of clock. The time interval Δt_0 for a particular clock or process is measured by an observer *at rest relative to the clock* (Ted in Fig. 27.5). The quantity Δt_0 is called the **proper time**. The proper time is always measured by an observer at rest relative to the clock or process that is being studied. So, while Ted is moving on his railroad car in Figure 27.5, the clock is moving along with Ted. Hence, Ted is at rest relative to this clock and he measures the proper time. On the other hand, Alice in Figure 27.5 is moving *relative to the clock*, so she does not measure the proper time. The time interval Δt measured by a moving observer (Alice) for the same process is always longer than the proper time.

When an observer is at rest relative to a clock or process, the start and end of the process occur at the same location for this observer. For the light clock in Figure 27.5B, Ted might be standing next to the bottom mirror, so from his viewpoint the light pulse starts and ends at the same location. By comparison, for Alice in Figure 27.5C, the light pulse begins at the bottom mirror when the clock is at the left; the pulse returns to this mirror when the clock is in a different location (relative to Alice), and Alice measures a longer time interval Δt. The proper time is always the shortest possible time that can be measured for a process, by any observer.

CONCEPT CHECK 27.1 | Measuring Proper Time

Ted is traveling in his railroad car with speed v relative to Alice, who is standing on the ground nearby (Fig. 27.8). Ted is playing with his yo-yo and uses a clock

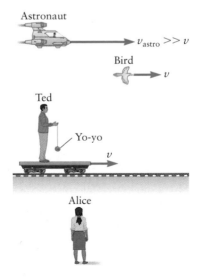

Astronaut

$v_{\text{astro}} \gg v$

Bird

v

Ted

Yo-yo

v

Alice

Figure 27.8 Concept Check 27.1.

on the railroad car to measure the time it takes for the yo-yo to complete one up-and-down oscillation. The yo-yo is also observed by Alice, by a bird flying nearby (also with speed v), and by an astronaut who is cruising by at a very high speed v_{astro}. Which observer measures the proper time for the yo-yo's period, (a) Ted, (b) Alice, (c) the bird, or (d) the astronaut?

EXAMPLE 27.2 | Time Dilation for a Muon

Subatomic particles called muons are created when cosmic rays collide with atoms in the Earth's atmosphere. Muons created in this way have a typical speed $v = 0.99c$, very close to that of light. Muons are unstable, with an average lifetime of about $\tau = 2.2 \times 10^{-6}$ s before they decay into other particles. That is, physicist 1 at rest relative to the muon measures this lifetime τ. Another (physicist 2) studies the decay of muons that are moving through the atmosphere with a speed of $0.99c$ relative to her laboratory (Fig. 27.9). What lifetime would physicist 2 measure?

RECOGNIZE THE PRINCIPLE

The muon acts as a sort of "clock" in which its lifetime corresponds to one "tick." Our results for a light clock, including Equation 27.8, apply to this muon "clock" since the results of special relativity apply to *all* physical processes. A clock moving along with the muon measures the proper time Δt_0, just as Ted in Figure 27.5B measures the proper time of a clock that travels along with him in his railroad car. The muon is moving relative to physicist 2, so that physicist is just like Alice in Figure 27.5C. Hence, that physicist measures a longer time Δt for the muon's lifetime.

SKETCH THE PROBLEM

Figure 27.9 shows the problem schematically.

IDENTIFY THE RELATIONSHIPS

Applying the time dilation result from Equation 27.7, we have

$$\Delta t = \frac{\Delta t_0}{\sqrt{1 - v^2/c^2}}$$

The lifetime for the muon at rest (i.e., measured by a clock at rest relative to the muon) is $\Delta t_0 = \tau$. We are given $v = 0.99c$.

SOLVE

The lifetime of the moving muon is

$$\Delta t = \frac{\tau}{\sqrt{1 - v^2/c^2}} = \frac{\tau}{\sqrt{1 - (0.99c)^2/c^2}}$$

$$\Delta t \approx \boxed{7.1 \times \tau}$$

What does it mean?

According to physicist 2, the moving muon exists for a much longer time than a muon at rest. Experiments with muons show that this result is correct: moving muons do indeed "live" longer before decaying than muons at rest in the laboratory. This is another surprising and counterintuitive result of special relativity.

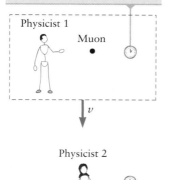

A clock moving with the muon measures the proper time for the muon's lifetime.

Physicist 1   Muon

v

Physicist 2

Clock at rest on the Earth

Figure 27.9 Example 27.2.

The Twin Paradox

Example 27.2 describes the effect of time dilation on the lifetime of a muon, but a similar result applies in other cases, including the lifetime of a person. Consider an astronaut (Ted) who is on a mission to travel to the nearby star named Sirius[1] and

---

[1]Sirius is actually a double star, but that does not affect the mission; Ted gets to visit both stars.

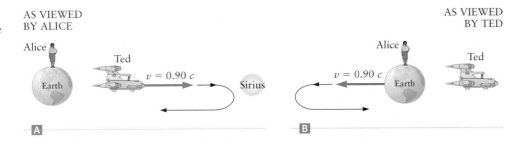

Figure 27.10 The twin paradox. **A** Astronaut Ted travels to the distant star Sirius and then returns while his trip is monitored by his twin, Alice. **B** As viewed by Ted (in his reference frame), Alice and the Earth take a journey in the direction opposite to that in part A. Not to scale!

then return to the Earth (Fig. 27.10A). Sirius is 8.6 light-years (ly) from the Earth, which means that light takes 8.6 years to travel from the Earth to Sirius. Ted must therefore travel in a very fast spaceship to complete his mission before he is too old to be an astronaut, so NASA has given him a ship that travels at a speed of $0.90c$. Ted's mission is being tracked by a second astronaut, Alice, who stays behind on Earth. Suppose Alice is Ted's twin, and one of her jobs is to monitor Ted's health and study how he ages during the trip. According to Alice, the round-trip distance is $2 \times 8.6 = 17.2$ ly, so the trip takes a time Δt with

$$\Delta t = \frac{17.2 \text{ ly}}{0.90c} = 19 \text{ years} \tag{27.10}$$

Alice also knows about time dilation and realizes that Ted's body acts as a clock (just like the muon in Example 27.2). Ted's body clock measures the proper time Δt_0 since it travels along with him. From our work on time dilation, we know that the proper time is shorter than Alice's time in Equation 27.10, with (from Eq. 27.7)

$$\Delta t_0 = (\Delta t)\sqrt{1 - v^2/c^2} \tag{27.11}$$

Inserting $v = 0.90c$ and our value of Δt gives

$$\Delta t_0 = (\Delta t)\sqrt{1 - v^2/c^2} = (19 \text{ years})\sqrt{1 - (0.90c)^2/c^2} = 8.3 \text{ years}$$

which is the time the trip takes according to Ted's body clock; in other words, Ted ages only 8.3 years, whereas his twin, Alice, ages by 19 years during the trip. When they compare notes at the end of the journey, Ted will be younger than his twin! This result can be understood in simple terms from the basic statement about time dilation: moving clocks (in this case, Ted himself) run slow.

It is interesting to ask now how Ted views the trip. According to Ted, Alice and the Earth both travel at a speed $0.90c$ *relative to his spaceship*, returning to him at the same speed at the end of his journey as sketched in Figure 27.10B. Ted then reasons that he can apply the results for time dilation described above to calculate that, according to his clock, Alice's trip will take 19 years while her body clock will age by the proper time $\Delta t_0 = 8.3$ years. Ted thus concludes that when they get back together, Alice will be younger than he is!

This problem is called the twin "paradox", as it appears that time dilation has led to contradictory results. Alice and Ted cannot both be right; only one can be the younger twin at the end of the journey. Alice's analysis is the correct one: Ted ages less than she does during the trip. The mistake in Ted's analysis is that special relativity applies only to inertial reference frames. Alice stays on the Earth, so she is always in an inertial reference frame and she can apply the results of special relativity. On the other hand, Ted spends some of his time in an accelerating reference frame when his spaceship turns around at Sirius to return to the Earth, and during this time he cannot use special relativity to analyze how Alice ages. That is why Ted's conclusion is wrong, and is the resolution of this apparent paradox.

How Do We Know That Time Dilation Really Happens?

Our applications of time dilation to analyze the decay of a muon in Example 27.2 and the aging of two twins in the twin paradox (Fig. 27.10) are good examples of special relativity, but they don't seem very relevant to everyday life. A similar application of time dilation, however, does have important practical applications. The Global Positioning System (GPS) consists of about 30 satellites that orbit the Earth twice each day (Fig. 27.11). These satellites send signals to receivers on the Earth, and the receivers use the signals to "triangulate" their position with an accuracy of about 10 m. Each satellite contains a very accurate clock, and the GPS receivers compare their clocks with the time signal from each satellite to do this triangulation. The GPS satellite clocks are moving in orbit, so they run slow by the factor $\sqrt{1 - v^2/c^2}$ according to Equation 27.11. The GPS satellites move at a speed of about 4000 m/s, which is much less than the speed of light. Even so, because of time dilation the GPS clocks run slow by about 7×10^{-6} s per day. The satellite signals travel at the speed of light, so the corresponding distance is more than 200 m, which is much larger than the theoretical accuracy of the GPS system. Only by accounting for the effect of time dilation on the satellite clocks can the GPS system successfully determine a position with an uncertainty of only 10 m. (We'll apply time dilation to analyze the GPS performance in Problem 70 at the end of the chapter.)

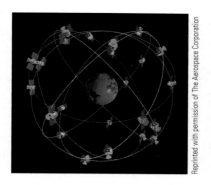

Figure 27.11 The Global Positioning System (GPS). Approximately 30 GPS satellites orbit the Earth, sending signals to the Earth. These signals are used by GPS receivers to determine their location with an accuracy of better than 10 m.

27.4 | SIMULTANEITY IS NOT ABSOLUTE

Two events are simultaneous if they occur at the same time. Our everyday experiences and intuition suggest that the notion of simultaneity is "absolute"; that is, two events are either simultaneous or they are not, for all observers. However, to determine if two events are simultaneous (or not), involves the measurement of time, and our studies of time dilation show that different observers do not always agree on measurements involving clocks and time intervals. So, let's examine what special relativity implies for the notion of simultaneity. If two events are judged simultaneous by one observer, will other observers also find them to be simultaneous? We can answer this question by analyzing the situation in Figure 27.12; Ted is standing at the middle of his railroad car, moving with a speed v relative to Alice, when two lightning bolts strike the ends of the car. The lightning bolts leave burn marks on the ground (points A and B), which indicate the locations of the two ends of the car when the bolts struck. We now ask, "Did the two lightning bolts strike simultaneously?"

We first ask this question of Alice, who notices that she is midway between the burn marks at A and B. Alice also observes that the light pulses from the lightning bolts reach her at the same time (Fig. 27.12C). Since she is midway between points A and B and the light pulses reach her at the same time, Alice concludes that the lightning bolts struck Ted's railroad car at the same time. *As viewed by Alice*, the bolts are simultaneous.

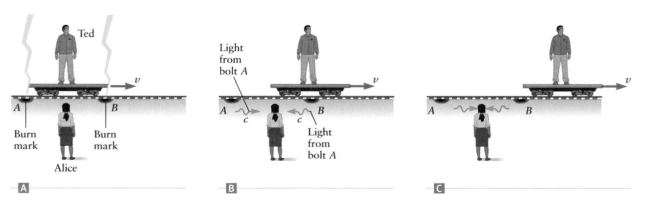

Figure 27.12 An experiment to study simultaneity. **A** Two lightning bolts strike at the ends of Ted's moving railroad car, leaving burn marks on the ground. **B** According to Alice, the lightning bolts are simultaneous. She comes to this conclusion because she is midway between the two burn marks and **C** the light pulses from the two bolts reach her at the same time.

What does Ted have to say? Ted stands at the middle of his railroad car, so (like Alice) he is also midway between the places where the bolts strike. Hence, if the two events are simultaneous *as viewed by Ted*, the light pulses should reach him at the same time. Do they? Alice can also answer this question! Although the light flashes are traveling to Alice, Ted's railroad car is moving to the right. Alice realizes that since Ted is moving, the flash emitted from *B* will reach him *before* the flash from point *A* in Figure 27.12B reaches him. Two observers must always agree on the order of two events that occur at the same point in space. (See Insight 27.2.) Hence, Ted will agree that the light pulse from *B* reaches him first. Ted therefore concludes that the lightning bolt at *B* struck *before* the bolt at *A*. In Ted's reference frame, the two lightning bolts are *not* simultaneous. The two lightning bolts in Figure 27.12 are therefore simultaneous for one observer (Alice) but not for another observer (Ted). The question of simultaneity is thus "relative" and can be different in different reference frames.

Time dilation and the relative nature of simultaneity mean that special relativity conflicts with many of our intuitive notions about time. Measurements of time intervals and judgments about simultaneity depend on the motion of the observer. That is very different from Newton's picture, in which "time" is an absolute, objective quantity, the same for all observers.

27.5 | LENGTH CONTRACTION

In the past few sections, we have seen that special relativity forces us to give up the notion of absolute time. Measurements of time and simultaneity are "relative"; that is, they can be different for different observers. Time is just one aspect of a reference frame; reference frames also involve measurements of position and length. What does relativity have to say about these quantities?

Let's consider how Ted and Alice might work together to measure a length or distance. Suppose Alice marks two spots *A* and *B* on the ground and measures these spots to be a distance L_0 apart on the *x* axis (Fig. 27.13). Ted travels along the *x* direction at constant speed *v*, and as he passes point *A* he reads his clock. Ted reads his clock again when he passes point *B* and calls the difference between the two readings Δt_0. This is a proper time interval because Ted measures the start and finish times at the same location (the center of his railroad car) with the same clock. From Section 27.3 (and Eq. 27.7), we know that when Alice measures the time it takes for Ted to travel from *A* to *B* with her clock, she will find a value

$$\text{time measured by Alice} = \Delta t = \frac{\Delta t_0}{\sqrt{1 - v^2/c^2}} \qquad (27.12)$$

Since Ted is traveling relative to Alice at speed *v* and points *A* and *B* are a distance L_0 apart, Alice concludes that

$$v = \frac{L_0}{\Delta t}$$

which can be rearranged as

$$\text{distance measured by Alice} = L_0 = v\,\Delta t \qquad (27.13)$$

Ted will calculate the distance from *A* to *B* in the same way, but he will use Δt_0, the time interval measured with his clock, so

$$\text{distance measured by Ted} = L = v\,\Delta t_0 \qquad (27.14)$$

Comparing Equations 27.13 and 27.14, we see that since Δt is different from Δt_0 due to time dilation, the lengths measured by Alice and Ted will also be different. Using Equation 27.11, we get

$$L = v\,\Delta t_0 = v(\Delta t\,\sqrt{1 - v^2/c^2})$$
$$L = L_0\sqrt{1 - v^2/c^2} \qquad (27.15)$$

Figure 27.13 Ted can measure the distance between points *A* and *B* by using a clock on his railroad car to measure the time Δt_0 it takes for him to travel from *A* to *B*, together with his known speed *v*.

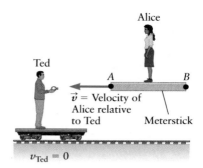

Alice

Ted

A B

$\vec{v}$ = Velocity of
Alice relative
to Ted

Meterstick

$v_{\text{Ted}} = 0$

Figure 27.14 A hypothetical experiment in which Ted is at rest and Alice travels on a meterstick with speed v relative to Ted. Ted finds that the moving meterstick is shorter than the length measured by Alice. This situation is very similar to Figure 27.13 because the meterstick (whose ends represent the points A and B) is at rest relative Alice, while it has speed v relative to Ted.

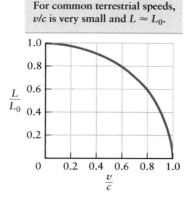

For common terrestrial speeds, v/c is very small and $L \approx L_0$.

$\frac{L}{L_0}$

Figure 27.15 Length contraction factor $L/L_0 = \sqrt{1 - v^2/c^2}$ as a function of v/c.

Hence, the length L measured by Ted is *shorter* than Alice's length L_0. This effect is called **length contraction**.

Proper Length

Length contraction also plays a role in the situation in Figure 27.14. Points A and B are now the two ends of a meterstick, and Ted is measuring the length of the meterstick as it moves past him. The meterstick is at rest relative to Alice, who might be standing on it, just as she is at rest relative to the marks A and B on the ground in Figure 27.13; she measures the meterstick in Figure 27.14 to have a length of exactly $L_0 = 1$ m. Ted measures a length L (Eq. 27.15), which is shorter than L_0. We can thus say that *moving metersticks are shortened.* The quantity L_0 is called the **proper length** because it is measured by an observer (Alice) *who is at rest relative to the meterstick.*

Length contraction is described by the factor

$$\frac{L}{L_0} = \sqrt{1 - v^2/c^2} \qquad (27.16)$$

Length contraction: moving metersticks are contracted.

This factor is plotted in Figure 27.15; it is very close to 1 when the speed is small and approaches zero as v approaches c.

Notice that we denote the proper time by Δt_0 and the proper length by L_0, in both cases using a subscript zero. In both cases, they are measurements made by an observer who is at rest relative to the "thing" being measured. Proper time is measured by an observer (Ted in Fig. 27.5B) who is at rest relative to the clock used for the measurement. Proper length is measured by an observer (Alice in Fig. 27.14) who is at rest relative to the object whose length is being measured.

Proper time and proper length

| EXAMPLE 27.3 | **Length Contraction of a Moving Car** |

Consider a race car measured by Ted to be 4.0 m long when moving past him at a speed of 100 m/s (about 200 mi/h). How long is the car when the race is finished and the car is stopped?

RECOGNIZE THE PRINCIPLE

This problem is an example of length contraction, with the race car playing the role of the meterstick in Figure 27.14. The length measured during the race (when $v = 100$ m/s) by an observer (Ted) watching from trackside is the contracted length $L = 4.0$ m (Fig. 27.16A). The length of the race car at the end of the race (when it is not moving) is the proper length L_0; that is the length that would be measured by an observer at rest relative to the car, such as Ted in Figure 27.16B.

SKETCH THE PROBLEM

Figure 27.16 describes the problem.

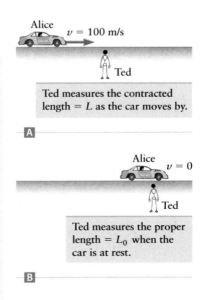

Alice $v = 100$ m/s

Ted

Ted measures the contracted length $= L$ as the car moves by.

A

Alice $v = 0$

Ted

Ted measures the proper length $= L_0$ when the car is at rest.

B

Figure 27.16 Example 27.3.

IDENTIFY THE RELATIONSHIPS

Length contraction is described by Equation 27.15; we have

$$L = L_0\sqrt{1 - v^2/c^2}$$

Rearranging to solve for the proper length gives

$$L_0 = \frac{L}{\sqrt{1 - v^2/c^2}}$$

SOLVE

Inserting the given values of L and v, we find

$$L_0 = \frac{L}{\sqrt{1 - v^2/c^2}} = \frac{4.0 \text{ m}}{\sqrt{1 - (100 \text{ m/s})^2/(3.00 \times 10^8 \text{ m/s})^2}}$$

$$L_0 = \boxed{4.000000000000022 \text{ m}} \tag{1}$$

Your calculator may not be able to show this many zeros to the right of the decimal point. If that is the case, we can again use the results in Insight 27.1 (as shown in Example 27.1) and write

$$L_0 = \frac{L}{\sqrt{1 - v^2/c^2}} \approx L\left(1 + \frac{v^2}{2c^2}\right)$$

Inserting the values of v and c gives

$$L_0 \approx L\left(1 + \frac{v^2}{2c^2}\right) = (4.0 \text{ m})\left[1 + \frac{(100 \text{ m/s})^2}{2(3.00 \times 10^8 \text{ m/s})^2}\right] \approx \boxed{4.0 + 2.2 \times 10^{-13} \text{ m}}$$

What does it mean?
We had to include many places to the right of the decimal point in Equation (1) to show the effect of length contraction in this example. Length contraction is quite negligible in this case and is always extremely small when v is a typical terrestrial speed. Length contraction is usually only important and measurable when v approaches the speed of light c. (See Fig. 27.15.)

EXAMPLE 27.4 | Length Contraction and a Moving Muon

Length contraction can be important for subatomic particles such as muons because these particles often move at very high speeds. Example 27.2 discussed muons created by cosmic rays in the Earth's atmosphere. Consider a muon created at point A in Figure 27.17 moving downward toward the Earth at speed $v = 0.99c$, decaying after a time $\tau = 2.2 \times 10^{-6}$ s. How far will the muon travel relative to the Earth before it decays? Give the answer from **(a)** the point of view of the muon and **(b)** from the point of view of an observer, Alice, standing on the ground.

RECOGNIZE THE PRINCIPLE

The muon travels a distance $= v \times$ its lifetime before it decays. Different observers measure different values for this lifetime and hence find different values for the distance traveled. For part (a), consider an observer, Ted, riding along with the muon so that the muon is at rest relative to Ted. The muon exists for a time $= \Delta t_0 = \tau$, which is a proper time because Ted records both the muon's creation and its decay on his clock at the same location (Ted's location). According to Ted, the Earth moves toward him and the muon through a distance $v \Delta t_0 = v\tau$ during the muon's lifetime. For part (b), let Alice be an observer on the ground. She measures a longer muon lifetime than τ due to time dilation (Example 27.2), so she will measure a longer travel distance than Ted. Notice that Alice measures the proper length between the muon's starting point at A and ground level at B because she is at rest relative to the points where the muon is created and decays.

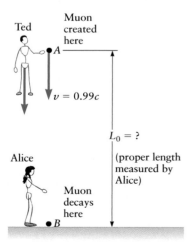

Figure 27.17 Example 27.4.

SKETCH THE PROBLEM

Figure 27.17 shows the problem.

IDENTIFY THE RELATIONSHIPS AND SOLVE

(a) Riding along with the muon, Ted measures the travel distance as

$$L = v \, \Delta t_0$$

Inserting the given values of v and $\Delta t_0 = \tau$, we get

$$L = v \, \Delta t_0 = 0.99c \, \Delta t_0 = 0.99(3.00 \times 10^8 \text{ m/s})(2.2 \times 10^{-6} \text{ s})$$

$$L = \boxed{650 \text{ m}} \tag{1}$$

(b) From Example 27.2, Alice observes the lifetime of the moving muon as

$$\Delta t = \frac{\Delta t_0}{\sqrt{1 - v^2/c^2}} = \frac{\tau}{\sqrt{1 - (0.99c)^2/c^2}} = 7.1\tau \tag{2}$$

During this time, the muon travels a distance

$$L_0 = v \, \Delta t$$

Inserting the values of v and Δt leads to

$$L_0 = v \, \Delta t = (0.99c)(7.1\tau) = 0.99(3.00 \times 10^8 \text{ m/s})(7.1)(2.2 \times 10^{-6} \text{ s})$$

$$L_0 = \boxed{4600 \text{ m}} \tag{3}$$

What does it mean?

Suppose Alice finds that the muon just reaches the ground when it decays (point *B* in Fig. 27.17). Ted will agree with her finding because he is traveling with the muon and will find himself on the ground when it decays. Comparing the answers for parts (a) and (b), we see that the distance traveled by the muon is different for Ted and Alice. How can they agree that the muon reaches the ground? Ted and the muon observe the *contracted* length between *A* and *B*, which is (Eq. 27.15)

$$L = L_0\sqrt{1 - v^2/c^2} \tag{4}$$

Inserting the values of L from Equation (1) and L_0 from Equation (3) along with the speed of the muon gives

$$\text{left side of Equation (4)} = 650 \text{ m}$$

$$\text{right side of Equation (4)} = (4600 \text{ m})\sqrt{1 - (0.99c)^2/c^2} = 650 \text{ m}$$

Equation (4) is thus satisfied by the results in parts (a) and (b) due to the effect of length contraction. Hence, Ted and Alice agree that the muon reaches the ground when it decays. Ted explains it in terms of length contraction and claims that the muon started only 650 m above the Earth's surface at point *A*. Alice explains the result in terms of time dilation and Equation (2).

CONCEPT CHECK 27.2 | Length Contraction and Muons

Suppose the speed of the muon in Example 27.4 is increased to $0.995c$. As viewed by Ted (and the muon), how far does the muon travel before it decays, (a) 6600 m, (b) 4600 m, or (c) 660 m?

What Is "Relative" and What Is Not?

The postulates of special relativity seem quite "innocent," but they have forced us to give up the notions of absolute time and space. These notions were dear to Newton and are ingrained in our intuition, so you should ask, "Are these predictions of special relativity really true?" They are indeed true; a large number of experiments have shown that time dilation and length contraction actually do occur. However, we have seen in Examples 27.1 and 27.3 that at ordinary terrestrial speeds these effects are negligibly small. The reason can be traced to the factor $\sqrt{1 - v^2/c^2}$ that appears in calculations of both time dilation and length contraction. At the typical terrestrial speed of 100 m/s (about 200 mph), this factor is approximately

$$\sqrt{1 - v^2/c^2} = \sqrt{1 - (100 \text{ m/s})^2/(3.00 \times 10^8 \text{ m/s})^2} \approx 1 - 6 \times 10^{-14}$$

The effects of relativity are proportional to how much this factor differs from 1, so relativity is completely negligible in nearly all everyday situations. We can thus continue to use our normal intuition at such "low" speeds, but must realize that these intuitive notions of time and length break down for objects moving at speeds approaching the speed of light.

27.6 | ADDITION OF VELOCITIES

Special relativity has forced us to give up some of our intuitive notions about space and time. Other quantities in mechanics, such as velocity and energy, involve displacement and time, so we must also revisit the behavior of these quantities.

The second postulate of special relativity—that the speed of light is the same for all observers—involves speed, so let's consider how the speed and velocity of an object appear to observers in different reference frames. The "object" could be a particle, a conventional wave (such as a water wave), or a light pulse. In Figure 27.18, Ted is traveling on his railroad car at constant speed v_{TA} relative to Alice when he throws an object, which might be a baseball. We now use the subscript *TA* to indicate that it is the velocity of Ted as measured by Alice, that is, the velocity of Ted relative to Alice. In this section, we will be concerned mainly with the component of the velocity along the x axis. Because this component can be positive or negative, we'll refer to it as the *velocity* rather than the speed. For simplicity, we ignore gravity and assume the object (the baseball in Fig. 27.18) travels horizontally, parallel to the velocity of the railroad car. If the velocity of the baseball *relative to Ted* is v_{OT}, what is the velocity v_{OA} of the ball relative to Alice who is at rest on the ground? (Notice how we again use subscripts to indicate which velocity we are referring to and who is measuring it.)

Newton's answer to this question was discussed in Section 27.2 and also in Chapter 4. For the baseball in Figure 27.18, Newton would predict that

$$v_{OA} = v_{OT} + v_{TA} \qquad (27.17)$$

or, in words,

velocity of the object (the ball) relative to Alice (v_{OA}) =

velocity of the object relative to Ted (v_{OT}) + velocity of Ted relative to Alice (v_{TA})

This result agrees with our everyday (classical) intuition, and it would be quite an accurate description for a person throwing a real baseball from a real railroad car. Equation 27.17, however, is inconsistent with the postulates of special relativity when the speeds are very high. To see the problem, suppose an object's speed relative to Ted is $v_{OT} = 0.9c$ and that Ted's railroad car is traveling very fast with $v_{TA} = 0.9c$. Inserting these values into Equation 27.17 gives a velocity $v_{OA} = v_{OT} + v_{TA} = 1.8c$. Newton's theory thus gives a velocity greater than the speed of light. Indeed, nothing in Newton's mechanics prevents Ted or a baseball or any other object from traveling at a speed greater than c. For an object or reference frame traveling at speed v, the factor that appears in time dilation (Eq. 27.7) and length contraction (Eq. 27.15) is $\sqrt{1 - v^2/c^2}$. If $v > c$, the result is an imaginary number, suggesting that Newton's formula for the addition of velocities (Eq. 27.17) cannot be correct for speeds close to the speed of light.

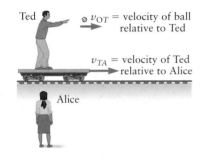

Figure 27.18 While he is moving with velocity v_{TA} relative to Alice, Ted throws a baseball with velocity v_{OT} relative to his railroad car. According to Galilean relativity and Newton's mechanics, the velocity of the ball relative to Alice is $v_{TA} + v_{OT}$. Special relativity gives a different result. Compare this situation with the light from Ted's flashlight in Figure 27.2.

Relativistic Addition of Velocities

It is possible to analyze the situation in Figure 27.18 using the postulates of special relativity. We will not give the derivation, but simply state the result of special relativity for the addition of velocities:

$$v_{OA} = \frac{v_{OT} + v_{TA}}{1 + \dfrac{v_{OT} v_{TA}}{c^2}}$$
(27.18)   Relativistic addition of velocities

This relation involves the component along a particular direction (x) of three different velocities, and it is easy to confuse them.

1. v_{OT} is the velocity of an object relative to observer an (e.g., Ted).
2. v_{TA} is the velocity of one observer (Ted) relative to a second observer (Alice).
3. v_{OA} is the velocity of the object relative to the second observer (Alice).

When the velocities v_{OT} and v_{TA} are much less than the speed of light, the factor $v_{OT}v_{TA}/c^2$ in the denominator in Equation 27.18 is much less than 1. In such cases, the relativistic addition of velocities formula (Eq. 27.18) gives nearly the same result as Newton's mechanics (Eq. 27.17). For example, suppose Ted is able to throw the baseball in Figure 27.18 with velocity $v_{OT} = 0.10c$, and his railroad car is moving at velocity $v_{TA} = 0.10c$ relative to Alice. Inserting these values into Equation 27.18 gives

$$v_{OA} = \frac{v_{OT} + v_{TA}}{1 + \dfrac{v_{OT} v_{TA}}{c^2}} = \frac{0.10c + 0.10c}{1 + \dfrac{(0.10c)(0.10c)}{c^2}} = 0.198c$$
(27.19)

We have kept an extra significant figure here to emphasize that the result is very close but not quite equal to the prediction of Newton's formula (Eq. 27.17), which gives

$$v_{OA} = v_{OT} + v_{TA} = 0.10c + 0.10c = 0.200c$$

This answer differs from the relativistic result (Eq. 27.19) by only a small amount (approximately 1%). Hence, for speeds less than about 10% of the speed of light, the Newtonian velocity addition formula works very well (although it is not exact!).

The results are quite different for objects traveling near the speed of light. Suppose $v_{OT} = 0.90c$ and $v_{TA} = 0.90c$. Inserting these values into Equation 27.18, we find

$$v_{OA} = \frac{v_{OT} + v_{TA}}{1 + \dfrac{v_{OT} v_{TA}}{c^2}} = \frac{0.90c + 0.90c}{1 + \dfrac{(0.90c)(0.90c)}{c^2}} = 0.994c$$
(27.20)

We have given the result to three significant figures to emphasize that v_{OA} is just slightly smaller than c, which is quite different from Newton's prediction for this case, which is $v_{OA} = v_{OT} + v_{TA} = 1.8c$. Experiments with particles moving at very high speeds show that the relativistic result is correct.

Relativistic Velocities and the Speed of Light as a "Speed Limit"

The relativistic formula for the addition of velocities (Eq. 27.18) applies when the velocity of the object is parallel (or antiparallel) to the relative velocity of the two observers $\vec{v}_{TA}$ as is the case for Ted, Alice, and the baseball in Figure 27.18. A slightly different result applies when the object is moving perpendicular to $\vec{v}_{TA}$. It is again found that if v_{OT} and v_{TA} are both less than c, then v_{OA} is also less than c. In general, *if an object has a speed less than c for one observer, its speed is less than c for all other observers.* Since no experiment has ever observed an object with a speed greater than the speed of light, c is again a sort of universal "speed limit."

As a final application of Equation 27.18, suppose the object leaving Ted's hand in Figure 27.18 is not a baseball but a pulse of light. Its velocity is then $v_{OT} = c$. What speed will Alice find? Inserting $v_{OT} = c$ into Equation 27.18 gives

$$v_{OA} = \frac{v_{OT} + v_{TA}}{1 + \dfrac{v_{OT}v_{TA}}{c^2}} = \frac{c + v}{1 + \dfrac{cv}{c^2}} = \frac{c + v}{1 + \dfrac{v}{c}}$$

Multiplying the numerator and denominator by c leads to

$$v_{OA} = \frac{c(c + v)}{c\left(1 + \dfrac{v}{c}\right)} = \frac{c(c + v)}{c + v} = c \qquad (27.21)$$

Alice thus finds a speed $v_{OA} = c$ regardless of Ted's speed v. In words, if an object moves at the speed of light for one observer, it moves at the speed of light for all observers. The only known "object" that can move at this speed is light; hence, this result is just a restatement of postulate 2 of special relativity.

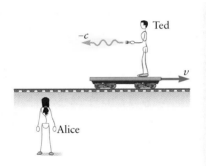

Figure 27.19 Example 27.5.

EXAMPLE 27.5 | Addition of Velocities

Ted is moving at a velocity $+v$ in his railroad car when he aims his flashlight backward as shown in Figure 27.19. What velocity will Alice find for the light from Ted's flashlight?

RECOGNIZE THE PRINCIPLE

According to the second postulate of special relativity, the speed of light must be the same for all observers, so Alice must measure a speed c for the light from Ted's flashlight. This result can also be obtained by applying the addition of velocities relation of special relativity.

SKETCH THE PROBLEM

The problem is described by Figure 27.19.

IDENTIFY THE RELATIONSHIPS

Light from Ted's flashlight moves at a velocity $v_{OT} = -c$ relative to Ted, whereas Ted moves at a velocity $v_{TA} = v$ relative to Alice. We wish to find the velocity of v_{OA} of the light relative to Alice.

SOLVE

Inserting these velocities into Equation 27.18 gives

$$v_{OA} = \frac{v_{OT} + v_{TA}}{1 + \dfrac{v_{OT}v_{TA}}{c^2}} = \frac{-c + v}{1 - \dfrac{cv}{c^2}} = \frac{-c + v}{1 - \dfrac{v}{c}}$$

Multiplying the numerator and denominator by c leads to

$$v_{OA} = \frac{c(-c + v)}{c\left(1 - \dfrac{v}{c}\right)} = \frac{-c(c - v)}{c - v} = \boxed{-c}$$

What does it mean?

The speed measured by Alice is again equal to the speed of light. The negative sign means that the light pulse is moving in the $-x$ direction in Alice's reference frame.

CONCEPT CHECK 27.3 | Riding on a Light Beam

Einstein said that some of the ideas for relativity theory occurred to him when he tried to imagine how the universe would look to a person "riding on a beam of light." Consider an experiment in which Ted rides on a light pulse with both Ted and the light traveling at speed c (Fig. 27.20). Ted is carrying his own flashlight and uses it to send a second light pulse to the right. According to postulate 2 of special relativity, the second light pulse travels at speed c relative to Ted. Is the speed of the second pulse as measured by Alice (a) $2c$, (b) c, or (c) 0?

Alice (At rest relative to flashlight 1)

27.7 | RELATIVISTIC MOMENTUM

According to Newton's mechanics, a particle of mass m_0 moving with a speed v has a momentum

$$\text{Newton's mechanics: } p = m_0 v \qquad (27.22)$$

Figure 27.20 Concept Check 27.3. What is the speed of the light beam from Ted's flashlight relative to Alice?

In Chapter 7, we showed that the total momentum of two colliding particles is conserved. We also found that when there are no external forces on a system of particles, the total momentum of the system is conserved. The principle of conservation of momentum is one of the fundamental conservation rules in physics and is believed to be satisfied by all the laws of physics, including the theory of special relativity. In Equation 27.22, however, there is a problem with the expression for momentum.

We can write Newton's result for the momentum of a single particle as

$$p = m_0 \frac{\Delta x}{\Delta t} \qquad (27.23)$$

From our analyses of time dilation and length contraction, we know that measurements of both time Δt and length Δx can be different for observers in different inertial reference frames. Should we use the proper time Δt_0 or the proper length Δx_0 to calculate the momentum? Einstein showed that we should use the proper time Δt_0 to calculate the momentum, which amounts to using a clock that travels along with the particle. At the same time, we should use the measurement of length Δx taken by an observer who watches the particle move by with speed v. Because $\Delta t_0 = \Delta t\sqrt{1 - v^2/c^2}$ (from Eq. 27.7), the result is

$$\text{special relativity: } p = m_0 \frac{\Delta x}{\Delta t_0} = m_0 \frac{\Delta x}{\Delta t \sqrt{1 - v^2/c^2}}$$

and since $v = \Delta x/\Delta t$, we get

$$\text{special relativity: } p = \frac{m_0 v}{\sqrt{1 - v^2/c^2}} \qquad (27.24)$$

Relativistic momentum

How do we know that this method is the correct way to calculate momentum? Einstein showed that when the momentum is calculated using Equation 27.24, the principle of conservation of momentum is obeyed exactly. Equation 27.24 is indeed the correct expression for momentum and applies even for particles moving at high speeds (i.e., close to the speed of light).

Comparing Equations 27.22 and 27.24, we see that the mass m_0 in Newton's expression is replaced by the factor $m_0/\sqrt{1 - v^2/c^2}$, which suggests that special relativity is trying to tell us something about the concept of mass. Before we discuss that, however, let's analyze the relativistic result for momentum in a little more detail.

When a particle's speed is small compared with the speed of light, the relativistic momentum (Eq. 27.24) becomes

$$\text{(relativistic momentum when } v \ll c) = \frac{m_0 v}{\sqrt{1 - v^2/c^2}} \approx \frac{m_0 v}{\sqrt{1 - 0}} = m_0 v$$

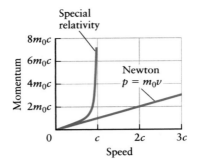

Figure 27.21 Momentum of a particle of mass m_0 as a function of the particle's speed v.

The relativistic momentum is thus equal to Newton's result for the momentum for "slowly" moving objects. As v approaches the speed of light, however, the relativistic result for momentum is very different from Newton's result as shown in Figure 27.21, which compares the two expressions for momentum (Eqs. 27.22 and 27.24).

Newton's result gives a linear relation between p and v. In contrast, the result from special relativity shows that p becomes extremely large as the particle's speed approaches c. A particle's momentum can be increased by the application of forces, which could occur in a collision or in some other way. By applying a very large force or arranging for a very violent collision, the particle's momentum can be made very large, and there is no limit to how large the momentum can be. However, even when the momentum is very large, the particle's speed never quite reaches the speed of light. Hence, when the speed is large, a particle responds to forces and impulses *as if its mass had increased* relative to m_0.

EXAMPLE 27.6 | Relativistic Momentum of an Electron

In experiments to study the properties of subatomic particles, physicists routinely accelerate electrons to speeds quite close to the speed of light. For an electron moving with speed $v = 0.99c$, compare its actual momentum to the momentum predicted by Newton's mechanics. Express your answer as the ratio p(special relativity)/p(Newton).

RECOGNIZE THE PRINCIPLE

Since the electron's speed is very close to the speed of light, special relativity will have a very large effect. According to Figure 27.21, the true momentum (as predicted by special relativity) is greater than the momentum predicted by Newton's mechanics.

SKETCH THE PROBLEM

Figure 27.21 shows the difference between the classical (Newton's) momentum and the true momentum (from special relativity).

IDENTIFY THE RELATIONSHIPS

This example is a direct application of the relativistic formula for momentum for an object (an electron) with speed $v = 0.99c$. We take the ratio of the relativistic momentum in Equation 27.24 to p in Equation 27.22. The mass of the electron cancels, leaving only terms that depend on v and c.

SOLVE

Taking the ratio of Equations 27.24 and 27.22 gives

$$\frac{p(\text{special relativity})}{p(\text{Newton})} = \frac{m_0 v / \sqrt{1 - v^2/c^2}}{m_0 v} = \frac{1}{\sqrt{1 - v^2/c^2}}$$

Inserting the given value of v we find

$$\frac{p(\text{special relativity})}{p(\text{Newton})} = \frac{1}{\sqrt{1 - v^2/c^2}} = \frac{1}{\sqrt{1 - (0.99c)^2/c^2}} = \boxed{7.1}$$

What does it mean?
When an electron moves at 99% of the speed of light relative to an observer (physicists in the laboratory), the true (relativistic) momentum is more than seven times larger than that predicted by Newton. When a particle's speed approaches the speed of light, the effects of special relativity become very significant.

CONCEPT CHECK 27.4 | Relativistic Momentum of a Proton

Consider again Example 27.6, but now the particle is a proton instead of an electron. The mass of a proton is about 1800 times greater than the mass of an electron. If the proton is moving at a speed $v = 0.99c$, is the ratio of p(special relativity)/p(Newton) for the proton

(a) the same as found for an electron with the same speed,

(b) greater than that found for an the electron with the same speed, or

(c) less than that found for an electron with the same speed?

27.8 | WHAT IS "MASS"?

The concept of "mass" is basic to all physics. When we first discussed mass in Chapter 1, we appealed to your intuition and suggested in very rough terms that mass is the "amount of matter" carried by a particle. This definition is admittedly imprecise, but it is surprisingly difficult to give a better one. Indeed, our intuitive notion of mass can be traced to Newton; in words, his laws of mechanics say that the mass of a particle determines how it will move (i.e., accelerate) in response to applied forces. According to Newton's second law, the acceleration of a particle of mass m_0 is given by

$$a = \frac{\sum F}{m_0} \tag{27.25}$$

so m_0 is just the constant of proportionality that relates acceleration and force. Newton's laws implicitly assume this constant of proportionality is really a constant,[2] but how do we know that this is true?

The only way to prove that Newton's second law is correct is through experiments, and experiments show that it does work with very high accuracy *provided the object's speed is small compared with the speed of light*. In the early part of the 1900s, however, experiments began to probe the regime in which a particle's speed approaches c, and these experiments showed that Newton's second law breaks down at such high speeds. When the postulates of special relativity are applied to this problem, we find that acceleration and force are related instead by

$$a = \frac{\sum F}{m_0/(1 - v^2/c^2)^{3/2}} \tag{27.26}$$

Comparing this relation with Newton's second law (Eq. 27.25), we have replaced m_0 by the factor $m_0/(1 - v^2/c^2)^{3/2}$. That is,

$$m_0 \quad \rightarrow \quad \frac{m_0}{(1 - v^2/c^2)^{3/2}} \tag{27.27}$$

At low speeds ($v \ll c$), the quantity on the right approaches m_0, so the two acceleration equations (Eqs. 27.25 and 27.26) are indistinguishable. However, as v approaches the speed of light, the denominator in Equation 27.27 becomes very small. When $v \approx c$, the acceleration in Equation 27.26 is thus very small, even when the force is very large. Hence, the particle responds to a force *as if it had a mass larger than m_0*. What's more, this "enhancement of the mass" depends on the particle's speed.

We encountered a similar effect when we considered relativistic momentum (Eq. 27.24). There we noticed that the relativistic result for p can be obtained by replacing the mass m_0 in Newton's expression by the term $m_0/\sqrt{1 - v^2/c^2}$. Then, in our interpretation of the relativistic formula, we noted that at high speeds v the particle responds to impulses and forces "as if" its mass were larger than m_0.

So, when dealing with both acceleration and momentum, a particle behaves "as if" it had a mass larger than expected from its behavior at low speeds. For this reason, physicists have introduced the term **rest mass**, denoted by m_0, for the mass measured by an observer who is moving very slowly relative to the particle (so that

---

[2]Assuming, of course, the particle does not break apart, for example.

$v \approx 0$ in Eqs. 27.24 and 27.26). Some authors call it the "proper mass" in analogy with the proper length of a meterstick at rest, but the term *rest mass* is much more widely used in physics, so we'll use it in this book.

When dealing with momentum and acceleration for particles moving at very high speeds, we can solve any problem we might encounter by simply using the results from special relativity in Equations 27.24 and 27.26, but we need to understand what these results mean: that Newton's simple notions about mass break down for particles moving at high speeds. The best way to describe the "mass" of a particle is through its rest mass m_0, which describes the motion of a particle for an observer who is moving very slowly relative to the particle.

27.9 | MASS AND ENERGY

As we saw in the last two sections, the concepts of mass and momentum are not as simple as Newton had assumed. You should therefore not be surprised that we must also reexamine our concepts of work and energy. That will lead us to what is probably the most famous equation in physics.

In Chapter 6, we derived a relation between mechanical work and kinetic energy and found that the kinetic energy of a particle of mass m_0 moving at speed v is

$$\text{Newton's kinetic energy: } KE = \tfrac{1}{2}m_0 v^2 \tag{27.28}$$

We can derive the corresponding relativistic expression for kinetic energy by applying work–energy ideas. According to the work–energy theorem (Chapter 6), the work done on a particle is equal to the change in the particle's kinetic energy. The work done by an applied force F is

$$W = F \, \Delta x$$

where Δx is the particle's displacement. Inserting the relation between F and acceleration according to special relativity (Eq. 27.26), we get

$$W = F \, \Delta x = \frac{m_0 a}{(1 - v^2/c^2)^{3/2}} \, (\Delta x) \tag{27.29}$$

which is similar to our derivation of Newton's kinetic energy in Chapter 6 (Eqs. 6.6 through 6.10). Simplifying the right-hand side of Equation 27.29 to find W involves more mathematics than we can include here, so we'll simply state the answer for a particle that starts at rest and comes to a final speed v. In this case, the work done equals the particle's relativistic kinetic energy KE, and the result is

Relativistic kinetic energy

$$KE = \frac{m_0 c^2}{\sqrt{1 - v^2/c^2}} - m_0 c^2 \tag{27.30}$$

Equation 27.30 looks quite different from Newton's result ($KE = \tfrac{1}{2}m_0 v^2$), but the two are actually closely related. To see the connection, we first rewrite Equation 27.30 as

$$KE = \frac{m_0 c^2}{\sqrt{1 - v^2/c^2}} - m_0 c^2 = m_0 c^2 \left(\frac{1}{\sqrt{1 - v^2/c^2}} - 1 \right)$$

According to Insight 27.1, for small v ($v \ll c$) we have

$$\frac{1}{\sqrt{1 - v^2/c^2}} \approx 1 + \frac{v^2}{2c^2}$$

Inserting this result into our expression for the kinetic energy gives

$$KE \approx m_0 c^2 \left(1 + \frac{v^2}{2c^2} - 1 \right)$$

$$KE \approx \tfrac{1}{2}m_0 v^2 \tag{27.31}$$

which is identical to Newton's kinetic energy. Hence, for small v (particles moving at speeds much less than the speed of light), we get our familiar result for KE as we should have expected. As v approaches c, however, the relativistic result for the kinetic energy has a different behavior than does Newton's expression. Figure 27.22 shows the kinetic energy as a function of the particle speed v (Eq. 27.30). At high speeds, v runs into the "speed limit" c. Although the kinetic energy can be made very large, the particle's speed never quite reaches the speed of light. As the speed increases, the particle responds as if it had a larger mass than at low speeds.

The expression for the kinetic energy in Equation 27.30 is the change in energy of a particle due to work done on the particle. We can also think of it as the difference between the final energy and the initial energy,

$$KE = \overbrace{\frac{m_0 c^2}{\sqrt{1 - v^2/c^2}}}^{\text{final energy}} - \overbrace{m_0 c^2}^{\text{initial energy}}$$

The last term, $m_0 c^2$, is a constant, however. What is the meaning of this initial energy term? Einstein proposed that this term is the energy of the particle, even when it is at rest. We now call it the **rest energy** of the particle:

$$\text{rest energy} = m_0 c^2 \qquad (27.32)$$

In words, Equation 27.32 asserts that a particle has an amount of energy $m_0 c^2$ even when its speed v is zero. When a particle is in motion, it also has kinetic energy; to get the total energy of the particle, we must then add the rest energy to the kinetic energy to get the total relativistic energy TE:

$$TE = \frac{m_0 c^2}{\sqrt{1 - v^2/c^2}} \qquad (27.33)$$

The result for the rest energy in Equation 27.32 forces us to rethink our notion of mass. The rest energy relation of special relativity implies that **mass is a form of energy**. This suggests that it is possible to convert an amount of energy $m_0 c^2$ into a particle of rest mass m_0 or to convert a particle of rest mass m_0 into an amount of energy $m_0 c^2$. So, the principle of conservation of energy must be extended to include this type of energy.

Since the speed of light is a very large number, the magnitude of the rest energy can be very large even when the rest mass m_0 is small. A rest mass of only 1 kg corresponds to a rest energy of

$$m_0 c^2 = (1 \text{ kg})(3.00 \times 10^8 \text{ m/s})^2 = 9 \times 10^{16} \text{ J} \qquad (27.34)$$

which is a tremendous amount of energy. Approximately 3×10^{17} J of energy is consumed each day in the United States. According to Equation 27.34, this energy could be obtained by converting a rest mass of just 3 kg into energy. Special relativity tells us that this much energy is available in principle, but it does not tell us how to actually convert mass into energy. Nuclear reactions (Chapter 30) are one way to do so.

| CONCEPT CHECK 27.5 | Relativistic Kinetic Energy

A particle has a kinetic energy that is twice its rest energy. What is the speed of the particle, (a) $c/2$, (b) $c/4$, or (c) $(2\sqrt{2}\,c)/3$?

| EXAMPLE 27.7 | Mass–Energy and Nuclear Weapons

The nuclear bomb dropped on Hiroshima, Japan, in 1945 released an amount of energy equivalent to that found in an explosion of approximately 20 kilotons of trinitrotoluene (TNT), a chemical explosive that releases about 4.2 MJ/kg when it explodes. How much rest mass was converted to energy at Hiroshima?

Figure 27.22 Kinetic energy of a particle as a function of its speed v. Even if the particle has a very large kinetic energy, its speed never quite reaches the speed of light.

Rest energy

Total relativistic energy

Insight 27.3

THE SPEED OF LIGHT IS A UNIVERSAL SPEED LIMIT. Several results of special relativity suggest that speeds greater than c are not possible.

1. The factor $\sqrt{1 - v^2/c^2}$ that appears in time dilation and length contraction is an imaginary number if $v > c$.
2. The relativistic momentum of a particle (Eq. 27.24) becomes infinite as $v \to c$, suggesting that an infinite force or impulse is required for a particle to reach the speed of light.
3. The total energy of a particle (Eq. 27.33) becomes infinite as $v \to c$, suggesting again that an infinite amount of mechanical work is required to accelerate a particle to the speed of light.

This notion that c is a "speed limit" is *not* one of the postulates of special relativity. Postulate 2 of special relativity states only that the speed of light is the *same* for all inertial reference frames. It is amazing that combining postulate 2 with postulate 1 (that the laws of physics are the same in all inertial frames) leads to the conclusion that it is not possible for a particle to travel faster than the speed of light.

RECOGNIZE THE PRINCIPLE

This problem involves the relativistic relation between mass and energy. Given the properties of TNT as an explosive, we can find how much energy was released at Hiroshima. We can then use the relation (rest energy) = $m_0 c^2$ to find the corresponding amount of rest mass that was converted to energy in the explosion.

SKETCH THE PROBLEM

No sketch is necessary.

IDENTIFY THE RELATIONSHIPS

We first compute the mass of 20 tons of TNT, arriving at a value in kilograms. Then we can find the amount of energy released by using the fact that 1 kg of TNT yields an energy of 4.2 MJ. The expression for the rest energy (Eq. 27.32) then gives the amount of mass required (i.e., consumed in the explosion).

SOLVE

A weight of 1 ton is equal to 2000 lb (U.S. customary units). We can obtain the mass of TNT in kilograms using the fact that a force of 1 lb equals 4.45 N. This corresponds to a mass m in kilograms given by 1 lb = 4.45 N = mg = m × (9.8 m/s^2). We can thus write:

$$\text{mass (in kilograms)} = \text{weight (in pounds)} \times \left(\frac{4.45}{9.8}\right) = \text{weight (in pounds)} \times 0.45$$

$$m_{\text{TNT}} = (20 \times 10^3 \text{ ton})\left(\frac{2000 \text{ lb}}{1 \text{ ton}}\right)(0.45) = 1.8 \times 10^7 \text{ kg}$$

An explosion with this amount of TNT releases an energy

$$E = m_{\text{TNT}}\left(\frac{4.2 \text{ MJ}}{1 \text{ kg}}\right) = (1.8 \times 10^7)(4.2 \times 10^6) \text{ J} = 7.6 \times 10^{13} \text{ J}$$

This result equals the amount of energy released by the uranium fission bomb that exploded at Hiroshima, so it is the amount of rest energy released. We thus have

$$E = \text{rest energy} = m_0 c^2 = 7.6 \times 10^{13} \text{ J}$$

Solving for m_0 gives

$$m_0 = \frac{7.6 \times 10^{13} \text{ J}}{c^2} = \frac{7.6 \times 10^{13} \text{ J}}{(3.00 \times 10^8 \text{ m/s})^2} = \boxed{8.4 \times 10^{-4} \text{ kg}}$$

What does it mean?

Less than 1 g of matter was converted to energy in this explosion! Even a small amount of rest mass contains a very large rest energy.

Mass–Energy Conversion and Chemical Reactions

The conversion of mass into energy is an important effect in nuclear reactions (Example 27.7), but it also occurs in other cases. Consider a chemical reaction in which a hydrogen atom is dissociated. The ionization energy of a hydrogen atom is 13.6 eV (electron-volts), meaning that when a proton and electron are bound together in a hydrogen atom their total energy is 13.6 eV lower than when they are separated. The principle of conservation of energy then implies that

total energy of hydrogen atom =

total energy of electron plus total energy of proton − 13.6 eV

If all these particles are observed at rest, the total energies of the hydrogen atom and of the electron and proton are just the rest energies. We thus have

$$m_0(\text{hydrogen atom})c^2 = m_0(\text{electron})c^2 + m_0(\text{proton})c^2 - 13.6 \text{ eV} \quad (27.35)$$

The mass of a hydrogen atom must therefore be less than the sum of the masses of an electron and proton. So, mass is *not conserved* when a hydrogen atom dissociates. In this case and in other chemical reactions, the change in the total rest mass Δm_0 in the reaction is small. For hydrogen dissociation (Eq. 27.35), we get

$$\Delta m_0 c^2 = 13.6 \text{ eV}$$

Expressing Δm_0 in SI units, we find

$$\Delta m_0 = \frac{13.6 \text{ eV}}{c^2} = \frac{(13.6 \text{ eV})(1.60 \times 10^{-19} \text{ J/eV})}{(3.00 \times 10^8 \text{ m/s})^2} = 2.4 \times 10^{-35} \text{ kg}$$

The mass of a proton is 1.7×10^{-27} kg, so Δm_0 is only about 0.000001% of the mass of the proton.

EXAMPLE 27.8 | **Energy Released in Electron–Positron Annihilation**

A positron is a subatomic particle with precisely the same mass as an electron, but with a positive charge $+e$; that is, its charge is opposite that of the electron. A positron is an example of **antimatter** (Chapter 31), and it is the antimatter "twin" of an electron. When an electron comes into close contact with a positron, they can annihilate each other, converting all their rest mass into energy. Typically, this annihilation reaction results in electromagnetic radiation. How much energy is released when an electron and a positron annihilate each other?

RECOGNIZE THE PRINCIPLE

In electron and positron annihilation, all the mass of both the electron and the positron are converted into energy. Both of these particles have a rest mass $m_0 = m_e$ (the mass of the electron), so we can use the known mass of the electron to find the amount of energy released from the relation (rest energy) $= m_0 c^2$.

SKETCH THE PROBLEM

Figure 27.23 describes the problem.

IDENTIFY THE RELATIONSHIPS

The (rest) mass for an electron is $m_e = 9.11 \times 10^{-31}$ kg (see Appendix A, Table A.1), and the positron's rest mass also equals m_e. We can use Equation 27.32 to get the total rest energy of the two particles:

$$E_{\text{total}} = \text{rest energy} = 2(m_e c^2)$$

SOLVE

This rest energy is the energy released in the annihilation process. Inserting the values of m_e and c gives

$$E_{\text{total}} = 2(m_e c^2) = 2(9.11 \times 10^{-31} \text{ kg})(3.00 \times 10^8 \text{ m/s})^2 = \boxed{1.6 \times 10^{-13} \text{ J}}$$

What does it mean?

When an electron and positron annihilate, their combined electric charge also disappears. Charge is conserved because the total charge of the two particles is $+e - e = 0$ before the annihilation. Although rest mass is not conserved, all reactions must conserve both electric charge and total energy.

CONCEPT CHECK 27.6 | **Relativistic Mass and Energy**

Ted takes a walk on a very dark night. It is so dark that he uses his flashlight to find his way home. Is the total mass of the flashlight (including its batteries) when Ted gets home (a) less than, (b) greater than, or (c) exactly the same as its mass just before he left?

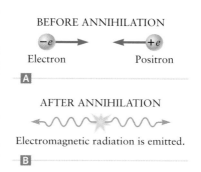

BEFORE ANNIHILATION

$-e \longrightarrow$ $\longleftarrow +e$
Electron   Positron

A

AFTER ANNIHILATION

Electromagnetic radiation is emitted.

B

Figure 27.23 Example 27.8. In electron–positron annihilation, all the rest mass of the two particles is converted to energy.

Insight 27.4

CONSERVATION PRINCIPLES

In Newton's mechanics, mass is a conserved quantity, so the total mass of a closed system cannot change. Special relativity tells us that mass is in fact not conserved. The principle of conservation of energy must instead be extended to include the rest energy (Eq. 27.32). What about our other conservation rules? Momentum is still conserved in collisions, but we must use the relativistic expression for momentum in Equation 27.24. Electric charge is also conserved, but it is possible to create or annihilate charges (as in Example 27.8) as long as the total charge does not change.

Special relativity is concerned with the laws of physics in inertial reference frames; these are reference frames that move at constant velocity. A *noninertial* reference frame is one that has a nonzero acceleration. Physics in noninertial frames is described by *general relativity*. Most of the results of special relativity, such as time dilation and length contraction, can be worked out using algebra, but general relativity involves much more complicated mathematics. We will therefore only describe the main ideas and results of the general theory.

General relativity is based on a postulate known as the *equivalence principle*:

Equivalence principle

The effects of a uniform gravitational field are identical to motion with constant acceleration.

An example of this equivalence is shown in Figure 27.24. In Figure 27.24A, Ted stands in an elevator compartment near the Earth's surface, with the elevator at rest. Ted feels the normal force exerted by the floor on his feet; he has mass m_0, so this force has magnitude $m_0 g$. From his study of physics, Ted deduces that g is a result of the Earth's gravity and hence that he is situated in a gravitational field.

Figure 27.24B shows the same elevator, but now it is located in distant space, far from any planets or stars, and the elevator compartment has an acceleration $a = g$. Since Ted's acceleration is a, Newton's second law tells us that a force $F = m_0 a = m_0 g$ is exerted by the elevator floor on Ted's feet.[3] The force in Figure 27.24B is thus not due to a gravitational force. According to the equivalence principle, however, there is no way for Ted to tell the difference between the effects of the gravitational field in Figure 27.24A and the accelerated motion in Figure 27.24B; they are completely equivalent.

This equivalence argument seems simple and innocent, but it has the following profound consequences.

1. *Equivalence of inertial and gravitational mass.* The concept of mass enters Newton's mechanics in two places. The mass in Newton's second law is called the *inertial mass* $m_{inertial}$, while the masses in the law of gravitation (Eq. 5.16) are called *gravitational mass* m_{grav}. Newton assumed these two types of mass are the same, but why should that be? In the elevator in a gravitational field (Fig. 27.24A), the force exerted on Ted is due to gravity, so $F = m_{grav} g$. In the accelerated elevator in Figure 27.24B, we have $F = m_{inertial} a = m_{inertial} g$. According to the equivalence principle, these two forces are precisely the same, so *the inertial mass must equal the gravitational mass.* We can also turn this argument around: the observation (by Newton and many others) that the inertial and gravitational masses are the same leads us to the equivalence principle.

2. *Deflection of light by gravity.* Figure 27.25 shows another experiment with Ted's elevator. A light pulse travels into the elevator car through a window on the left. In Figure 27.25A, the elevator is in distant space and has zero acceleration ($a = 0$). An observer (Ted) in the elevator then sees the light pulse travel in a straight-line path across the car. In Figure 27.25B, the elevator is accelerating upward; Ted now finds that *relative to him* the light beam travels on a downward arc. A nonaccelerated observer outside the elevator would say that the light pulse travels in a straight line, but *relative to the elevator* the light travels along a curved path. In Figure 27.25C, the experiment is repeated in a gravitational field and with no acceleration. According to the equivalence principle, the results must be the same as in the accelerated case in Figure 27.25B. So, Ted

Figure 27.24 A person in an elevator feels a normal force $m_0 g$ on the bottoms of his feet. He cannot tell if he is **A** at rest in a gravitational field so that the normal force is the result of a gravitational force or if **B** there is no gravitational field and the elevator has an acceleration $a = g$. According to the equivalence principle, there is no way to tell the difference between these two situations.

---

[3]We assume the elevator's speed is much smaller than the speed of light, so the difference between Newton's second law (Eq. 27.25) and the relativistic result in Equation 27.26 is negligible.

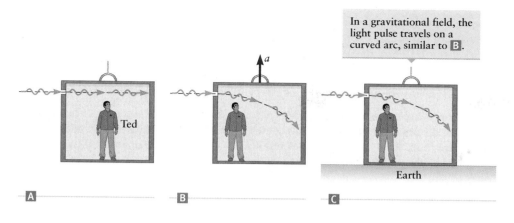

In a gravitational field, the light pulse travels on a curved arc, similar to **B**.

Ted

a

Earth

A   B   C

Figure 27.25 A light ray moving through the elevator compartment in Figure 27.24. **A** When the acceleration of the car is $a = 0$, the light travels in a straight line through the elevator. In **B**, the elevator is accelerating upward. According to an observer in the elevator, the light beam travels in a downward arc. **C** According to the equivalence principle, the situation in part B is identical to the effect of a gravitational field. Hence, a gravitational field must also deflect a light beam.

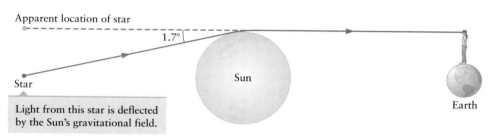

Apparent location of star

1.7°

Star

Sun

Earth

Light from this star is deflected by the Sun's gravitational field.

Figure 27.26 Deflection of light by the Sun's gravitational field. The deflection is largest for light beams that travel nearest the Sun. Light that just grazes the surface of the Sun is deflected through an angle of about 1.7°. This effect is easiest to measure when the Sun's light is blocked by the Moon during a solar eclipse.

must again observe that the light follows a curved path, but now this curvature is produced by the gravitational field. In words, we say that light is "bent" by gravity. This conclusion is a straightforward result of the equivalence principle, but when Einstein proposed that light is bent by a gravitational field his prediction was met with considerable skepticism. Einstein's ideas were confirmed in experiments in 1919, in which light passing near the Sun during an eclipse (Fig. 27.26) was found to be deflected by the predicted amount. This result made Einstein an international celebrity.

Black Holes

The most massive object in our solar system is the Sun, but even its large gravitational field only deflects light a small amount (Fig. 27.26). However, the universe contains other, much more massive objects. *Black holes* contain so much mass that light is not able to escape from their gravitational attraction. They are called "black" because any light that passes nearby is attracted to the black hole and never escapes. Hence, a region of space containing a black hole appears dark to an observer on the Earth. How, then, do we "see" a black hole? One way to find a black hole is through its effect on the motion of nearby objects. Stars near a black hole move along curved trajectories due to the gravitational attraction of the black hole. Astronomical observations of these curved trajectories give the mass and location of the black hole.

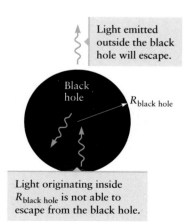

Light emitted outside the black hole will escape.

Black hole

$R_{\text{black hole}}$

Light originating inside $R_{\text{black hole}}$ is not able to escape from the black hole.

Figure 27.27 Example 27.9.

EXAMPLE 27.9 | Size of a Black Hole

One way to measure the size of a black hole is to study the motion of light emitted a distance R from the center of the black hole. If R is small, the gravitational force of the black hole is strong enough that all light is attracted to the black hole and cannot escape. On the other hand, if R is large, the light can escape. This problem is similar to finding the escape velocity for a rocket launched from the surface of a planet into outer space (Chapter 6), with light now playing the role of the rocket. We calculated the velocity required for a rocket to escape from a planet of radius R in Equation 6.24. Use that result to find the value of R at which light can just barely escape from a black hole of mass M.

RECOGNIZE THE PRINCIPLE

In Chapter 6, we considered the motion of a particle with speed v that is "fired" into space at a distance R from the Earth's center. Using the principle of conservation of energy, we found that such a particle could escape to an infinite distance if its speed exceeds the escape velocity, which is given by

$$v_{escape} = \sqrt{\frac{2GM}{R}} \tag{1}$$

where M is the mass of the Earth and R is the initial distance from the Earth's center. The same result for the escape velocity also applies to light near a black hole. Since light moves at speed c, it can escape from a black hole if the escape velocity is less than c.

SKETCH THE PROBLEM

Figure 27.27 describes the problem.

IDENTIFY THE RELATIONSHIPS AND SOLVE

Setting $v_{escape} = c$ in Equation (1) and solving for R leads to

$$v_{escape} = c = \sqrt{\frac{2GM}{R}}$$

$$c^2 = \frac{2GM}{R}$$

$$\boxed{R = \frac{2GM}{c^2}} \tag{2}$$

What does it mean?

In Chapter 30, we'll discuss how stars use nuclear reactions to generate their energy. It is believed that black holes are produced by stars that collapse to a very small size when they exhaust their nuclear fuel. Let's calculate the value of R for a black hole with the mass of the Sun ($M \approx 2.0 \times 10^{30}$ kg). Inserting this value into Equation (2) gives

$$R = \frac{2GM}{c^2} = \frac{2(6.67 \times 10^{-11} \text{ N} \cdot \text{m}^2/\text{kg}^2)(2.0 \times 10^{30} \text{ kg})}{(3.00 \times 10^8 \text{ m/s})^2} \approx 3000 \text{ m}$$

Considering the mass and (current) radius of the Sun, this size is incredibly small. For comparison, a typical airport runway is also about 3000 m in length. A black hole is thus an extremely dense object.

Detecting a Black Hole: Gravitational Lensing

Another way to detect a black hole is through *gravitational lensing*. This effect is illustrated in Figure 27.28A, which shows light from a star as it passes by a black hole on its way to an observer on the Earth. If the black hole is between the star and the Earth, light from the star can pass by either side of the black hole and still be bent by gravity so as to reach the Earth. This figure shows a view in only one plane; light from the star can also pass above and below the plane of the drawing. Hence, light from a single star can produce multiple images and even a set of images along a ring or arc. Astronomers have identified several such gravitational lenses; one example is shown in Figure 27.28B. The positions and shapes of the images can be

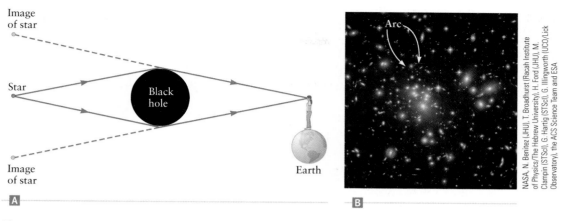

Figure 27.28 **A** A very massive object such as a black hole acts as a gravitational "lens." Here light rays from a star pass on both sides of a black hole and arrive at an observer, forming two separate images of the star. There will also be rays that come out of the plane of the drawing, producing additional images or a ring of images. **B** Example of gravitational lensing observed with the Hubble Space Telescope. Each circular arc of light (one is indicated by the arrows) comes from a single star. These arcs are due to light rays that travel out of the plane in part A.

NASA, N. Benitez (JHU), T. Broadhurst (Racah Institute of Physics/The Hebrew University), H. Ford (JHU), M. Clampin (STScI), G. Hartig (STScI), G. Illingworth (UCO/Lick Observatory), the ACS Science Team and ESA

used to deduce the mass of the black hole. From this and other types of astronomical observations, a number of black holes have been found. In fact, a black hole resides at the center of the Milky Way, our galaxy.

CONCEPT CHECK 27.7 | Size of the Black Hole at the Center of the Milky Way

Astronomers have found that the black hole at the center of the Milky Way has a mass about three million times larger than our Sun. Hence, this black hole has a mass $M = (3 \times 10^6) \, M_{Sun} = 6 \times 10^{36}$ kg. Is the radius of this black hole (a) 3000 m, (b) 9×10^9 m, (c) 3×10^9 m, or (d) 9.5×10^6 m? Is it larger or smaller than our Sun? *Hint:* Use the result in Example 27.9.

27.11 | RELATIVITY AND ELECTROMAGNETISM

Special relativity is all about how the physical world looks to inertial observers moving at different speeds. So far, our two observers, Ted and Alice, have focused on several problems in mechanics. Let's now consider a problem in electromagnetism and give them the job of calculating the force between an infinite line of charge with charge per unit length λ and a point charge $+q$ located a distance r away. Figure 27.29A shows Alice's analysis; she is at rest relative to both the charged line and the point charge. Alice applies results from Chapter 17 and finds that the line of charge produces an electric field $E = \lambda/(2\pi\varepsilon_0 r)$ directed away from the line, so the force on the point charge is

$$F_E(\text{Alice}) = \frac{q\lambda}{2\pi\varepsilon_0 r} \qquad (27.36)$$

Let's now consider how Ted would analyze this situation from his moving railroad car. Ted is moving to the left at speed v in Figure 27.29B, so from his point of view the charged line and the point charge are both moving to the right with speed v as shown in Figure 27.29C. In Ted's reference frame, the moving charged line acts as a current, so according to him this line produces both an electric field *and* a magnetic field. Since the test charge $+q$ is moving through this magnetic field, Ted concludes that there is also a magnetic force on this test charge. The current is to the right in Figure 27.29D, so the magnetic field at the test charge is directed into the page; applying right-hand rule 2 (Chapter 20) gives a magnetic force F_B (Ted) directed toward the line.

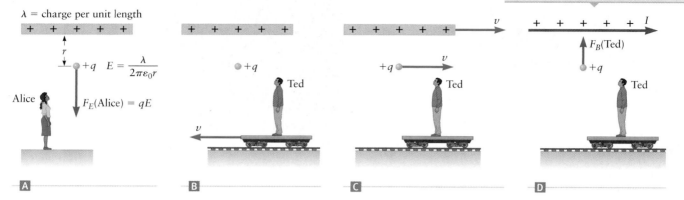

In Ted's reference frame, the moving line of charge is a current and produces a magnetic field.

λ = charge per unit length

$E = \dfrac{\lambda}{2\pi\varepsilon_0 r}$

$F_E(\text{Alice}) = qE$

Figure 27.29 **A** An infinite line of charge at rest relative to Alice. This line produces an electric field $E = \lambda/(2\pi\varepsilon_0 r)$ and hence an electric force on the test charge $+q$. **B** The same situation viewed by Ted as he moves past on his railroad car. **C** In Ted's frame of reference, both the charged line and the test charge move with speed v. **D** According to Ted, the moving charged line is equivalent to an electric current. This current produces a magnetic field and thus also a magnetic force on the test charge.

Ted will thus say that there is an electric force *and* a magnetic force on the particle. On the other hand, according to Alice there is *only* an electric force. Who is correct? The answer is that they are *both* correct!

How can Ted possibly agree with Alice? According to Ted, the moving charged line is contracted (length contraction) as discussed in Section 27.5. Squeezing this charge into a smaller length *increases* the charge per unit length and thus *increases* the electric force calculated by Ted when he uses Equation 27.36. Thus, Ted observes a *larger* electric field and a *larger* electric force directed away from the line than that found by Alice. We have also seen, however, that Ted observes a magnetic force directed *toward* the line. A careful analysis shows that this magnetic force precisely cancels the extra electric force, so Ted and Alice agree on the total force on the particle. Alice will attribute it completely to an electric force, whereas Ted will attribute it to a combination of electric and magnetic forces.

This example shows that the values of the electric and magnetic fields in a particular situation depend on the motion of the observer; here Ted finds a nonzero magnetic field, while for Alice the magnetic field is zero. It is interesting that Maxwell's equations already contain this effect; his equations of electromagnetism were already consistent with the theory of special relativity, but Maxwell (and others) didn't fully appreciate it until Einstein discovered special relativity.

27.12 WHY RELATIVITY IS IMPORTANT

Generally speaking, relativistic effects become important for objects or observers whose speed approaches the speed of light. Although there are a few everyday situations, such as the Global Positioning System (GPS) discussed in Section 27.3, where the effect of special relativity is important, relativity plays no significant role in most terrestrial situations. Even so, relativity is an extremely important part of physics for the following reasons.

1. The relation between mass and energy and the possibility that mass can be converted to energy (and vice versa) mean that mass is not conserved. Instead, we have a more general view of energy and its conservation. The three hallmark

conservation principles of physics are thus (a) conservation of energy, (b) conservation of momentum, and (c) conservation of charge. It is believed that *all* the laws of physics must obey these three principles.

2. The rest energy of a particle is huge, which has important consequences for the amount of energy available in processes such as nuclear reactions.

3. Relativity changes our notions of space and time. At the start of this book, we mentioned that time and position are two primary quantities in physics and that it is not possible to give precise definitions of such quantities. The best we can do is use our intuition to understand the meaning of space and time. Now we have found that our everyday intuition breaks down when applied to special relativity. While this may seem surprising, there is no reason why an intuition developed from terrestrial experience should be reliable when applied to stars and black holes.

4. Relativity plays a key role in understanding how the universe was formed and how it is evolving. Black holes can't be understood without relativity.

5. Relativity shows that Newton's mechanics is not an exact description of the physical world. Instead, Newton's laws are only an approximation that works very well in some cases, but not in others. So, we shouldn't discard Newton's mechanics, but we should understand its limits.

SUMMARY | Chapter 27

KEY CONCEPTS AND PRINCIPLES

The postulates of special relativity
Special relativity is based on two postulates:

1. **The laws of physics are the same in all inertial reference frames.**
2. **The speed of light in a vacuum is constant, independent of the motions of the source and observer.**

Time dilation and length contraction are consequences of these two postulates.

General relativity
General relativity is based on the **equivalence principle**: the effects of a uniform gravitational field are identical to motion with constant acceleration.

One consequence of the equivalence principle is that light rays are deflected by a gravitational field.

APPLICATIONS

Time dilation
The **proper time** Δt_0 is the time interval between two events as measured by an observer for which the two events occur at the same location (i.e., an observer at rest relative to the events). A second observer moving at a relative speed v measures a longer time,

$$\Delta t = \frac{\Delta t_0}{\sqrt{1 - v^2/c^2}}$$

(27.7) (page 922)

In words, we say that *moving clocks run slow.*

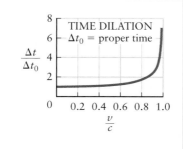

(Continued)

Length contraction

The **proper length** of an object is the length L_0 measured by a person at rest relative to the object. If a second observer is moving with a speed v relative to the object, that observer will measure a shorter length,

$$L = L_0\sqrt{1 - v^2/c^2}$$ **(27.15)** (page 928)

An equivalent situation occurs when the object is moving with speed v relative to an observer. In words, we say that *moving metersticks are contracted.*

Relativistic addition of velocities

Observer Alice is at rest while Ted moves with a velocity v_{TA} relative to Alice. Ted observes an object moving with a velocity v_{OT} relative to him. When $v_{TA} \ll c$ and $v_{OT} \ll c$, the velocity of the object relative to Alice is given by the Galilean addition of velocities relation (the result from Newton's mechanics),

v_{OT} = velocity of object relative to Ted

v_{TA} = velocity of Ted relative to Alice

v_{OA} = velocity of object relative to Alice

$$v_{OA} = v_{OT} + v_{TA}$$ **(27.17)** (page 932)

For speeds that are not small compared with c, we must use the relativistic addition of velocities relation,

$$v_{OA} = \frac{v_{OT} + v_{TA}}{1 + \dfrac{v_{OT}v_{TA}}{c^2}}$$ **(27.18)** (page 933)

If both $v_{TA} < c$ and $v_{OT} < c$, then $v_{OA} < c$. No object can move faster than c.

Relativistic momentum

In Newton's mechanics, the momentum of a particle of mass m_0 moving with a velocity v is

Newton's mechanics: $p = m_0 v$ **(27.22)** (page 935)

Special relativity gives a different result:

$$\text{special relativity: } p = \frac{m_0 v}{\sqrt{1 - v^2/c^2}}$$ **(27.24)** (page 935)

where m_0 is called the **rest mass** of the particle.

Relativistic energy

In Newton's mechanics, the kinetic energy of a particle of mass m_0 is

Newton's kinetic energy: $KE = \frac{1}{2}m_0 v^2$ **(27.28)** (page 938)

In special relativity, one considers instead the total energy of a particle:

$$\text{total relativistic energy: } TE = \frac{m_0 c^2}{\sqrt{1 - v^2/c^2}}$$ **(27.33)** (page 939)

When $v = 0$, the total energy is $TE = m_0 c^2$, which is called the **rest energy** of the particle. This result implies that rest mass is a form of energy. When $v > 0$, the total energy is larger than the rest energy; the energy in excess of the rest energy is the **relativistic kinetic energy,**

$$KE = \frac{m_0 c^2}{\sqrt{1 - v^2/c^2}} - m_0 c^2$$ **(27.30)** (page 938)

1. It is impossible for a particle of rest mass m_0 to travel at a speed greater than c. Is there an upper limit to the momentum or kinetic energy of the particle? Explain why or why not.

2. You are traveling in a windowless spacecraft, far from any planets or stars. Describe an experiment you could do to tell whether you are in an inertial or a noninertial frame.

3. SSM A constant force F is applied to a spacecraft of mass m. The spacecraft is initially at rest. Sketch the speed of the spacecraft as a function of time according to Newton's laws (a) without allowing for relativity and (b) allowing for relativity. Compare the two results and explain why they are different.

4. Give an example of an inertial frame of reference.

5. Give four examples of noninertial reference frames.

6. The speed of light inside a substance depends on the index of refraction n. For water, $n \approx 1.33$. (a) What is the speed of light in water? (b) Is it possible for a particle to travel faster than your result in part (a)? Explain why or why not.

7. Length contraction applies to the length of an object measured in the direction of motion. Consider the length of an object measured in a direction perpendicular to its velocity. Give an argument that explains why there is no contraction along the perpendicular direction.

8. A particle is "extremely" relativistic if its speed is very close to c. What is the ratio of the total energy to the momentum for such a particle?

9. Two observers move at different speeds. Which of the following quantities will they agree on? In each case, give a reason for your answer.
(a) The length of an object
(b) The time interval between two events
(c) The speed of light in a vacuum
(d) The speed of a moving object
(e) The relative speed between the observers

10. At a typical everyday speed such as 65 mi/h, does length contract, time dilate, and mass increase by amounts that could be easily measured? Explain.

11. A wind-up toy has a coil spring that stores spring potential energy. As the toy is wound, does its mass increase, decrease, or stay the same? Explain.

12. SSM Travelers onboard a spaceship are moving at $0.9c$ from one star to another. The stars are at rest relative to each other, and this is the speed of the spaceship relative to the stars. Which of the following statements are true?
(a) Observers at rest with respect to the stars measure a shorter distance between the stars than the travelers do.
(b) The observers see the travelers moving in slow motion relative to themselves.
(c) The observers see the travelers age more slowly than they themselves do.
(d) The travelers measure a shorter distance between the stars than the observers do.
(e) The travelers believe they are aging more slowly than the observers are.

13. Two spaceships, each traveling at $0.5c$ relative to an observer on the Earth, approach each other head on. One ship fires its laser beam weapon at the other ship. What does the other ship measure for the speed of the laser light?

14. Suppose an object moves away from a plane mirror at a speed of $0.9c$. Does the image recede from the source at a speed of $1.8c$? Explain.

15. You are standing on the Earth and observe a spherical spacecraft fly past at very high speed. Sketch the shape of the ship as observed by you.

16. Two identical clocks are synchronized. One stays on the Earth while the other goes in orbit around the Earth for 1 year as measured by the clock on the Earth. Which of the following statements are true after both clocks are again on the Earth?
(a) The clocks are still synchronized.
(b) The clock that made the trip runs slower after it returns.
(c) The clock that made the trip does not have the same time as the clock that stayed on the Earth.
(d) The clock that stayed on the Earth has the wrong time.
(e) The clock that orbited the Earth has the wrong time.

17. If you are traveling very close to the speed of light and hold a mirror in front of your face, what will you see?

18. If the Earth were compressed to the density of a black hole, would it be about the size of (a) an atom, (b) a grape, (c) an orange, or (d) a basketball?

27.1 NEWTON'S MECHANICS AND RELATIVITY

1. SSM Ted travels in a railroad car at constant velocity while his motion is watched by Alice, who is at rest on the ground (Fig. P27.1). Ted's speed v is much less than the speed of light. Ted releases a ball from his hand and observes that in his reference frame the ball falls directly downward. Hence, according to Ted, the component of the ball's velocity along the horizontal direction is zero. (a) According to Alice, what is the ball's velocity along x just after the ball is released? (b) According to Alice, what is the ball's velocity along x just before the ball lands at Ted's feet? (c) According to Alice, what is the acceleration of the ball along x? (d) According to Ted, what is the acceleration of the ball along x? (e) What is the force F_x on the ball along x? Do

Ted and Alice agree on the value of F_x? Explain how this answer is expected from the principle of Galilean relativity.

Figure P27.1

27.2 THE POSTULATES OF SPECIAL RELATIVITY

2. A person travels in a car with a speed of 75 m/s. To get the driver's attention, a stationary police officer behind the car shines his spotlight at the car. If the officer is 300 m behind the car when he turns on his spotlight, how long does it take for the light to reach the car?

3. The spotlight in Problem 2 emits light at a wavelength $\lambda = 450$ nm as viewed by the police officer. The wavelength seen by the driver is λ'. What is the percentage difference between these two wavelengths? *Hint*: Use the results for the Doppler effect in Chapter 23.

4. SSM Identify inertial and noninertial reference frames in the following list.
(a) A spinning ice skater
(b) A car turning a corner
(c) A spacecraft moving at constant speed in a straight line
(d) A spacecraft moving with constant velocity
(e) A spacecraft landing on the Moon
(f) A spacecraft traveling with its engines off, far from any stars or planets

27.3 TIME DILATION

5. Ted is traveling on his railroad car (Fig. P27.5) with speed $0.85c$ relative to Alice. Ted travels for 30 s as measured on his watch. (a) Who measures the proper time, Ted or Alice? (b) How much time elapses on Alice's watch during this motion?

Ted and his watch

$v = 0.85c$

Alice and her watch

Figure P27.5 Problems 5 and 6.

6. ⊗ Consider again the time intervals measured by Ted and Alice in Figure P27.5, but now suppose Alice measures the time required for her heart to beat 75 times and measures an elapsed time of 80 s with her watch. (a) Who measures the proper time, Ted or Alice? (b) How much time elapses on Ted's watch?

7. Ted is traveling in his railroad car at speed v relative to Alice. He is also carrying a light clock in his luggage. Alice compares notes with Ted and finds that each tick of the light clock takes 1.0 s in Ted's reference frame, whereas Alice measures 3.0 s per tick. Find v.

8. Consider the motion of a muon as it moves through the Earth's atmosphere (see Example 27.4). A particular muon is created 15,000 m above the Earth's surface and just reaches the ground before it decays. What is the speed of the muon?

9. ⊗ An astronaut travels at a speed of $0.98c$ from the Earth to a distant star that is 4.5×10^{17} m away. If the astronaut is 25 years old when she begins her trip, how old is she when she arrives at the star?

10. A subatomic particle has an unknown lifetime τ. The particle is created at $t = 0$ and travels around the Earth's equator at speed $0.999c$. If the particle decays at the moment it completes one trip around the Earth, what is τ?

11. SSM ☆ An astronaut is traveling to the Moon at a speed of $0.85c$. When his spacecraft is 2.0×10^8 m from the Moon, there is an explosion on the Moon. How long does it take for light from the explosion to reach the astronaut as measured on the astronaut's clock?

12. ☆ An astronaut travels away from the Earth at a speed of $0.95c$ and sends a light signal back to the Earth every 1.0 s as measured by his clock. An observer on the Earth finds that the arrival time between consecutive signals is Δt. Find Δt.

27.4 SIMULTANEITY IS NOT ABSOLUTE

13. SSM Ted is traveling on his railroad car (length 25 m as measured by Ted) at a speed of $0.95c$. Alice arranges for two small explosions to occur on the ground next to the ends of the railroad car (Fig. P27.13). According to Alice, the two explosions occur simultaneously, and she uses the burn marks on the ground to measure the length of Ted's railroad car. (a) According to Ted, do the two explosions occur simultaneously? (b) If not, then according to Ted which explosion occurs first?

Ted

v

Alice

Figure P27.13

14. Ted is traveling on his railroad car (length 25 m, as measured by Ted) at a speed of $0.95c$. Ted arranges for two small explosions to occur on the ends of the railroad car (Fig. P27.14). According to Ted, the two explosions occur simultaneously, and he asks Alice to use the burn marks on the ground to measure the length of the railroad car. (a) According to Alice, do the two explosions occur simultaneously? (b) If not, then according to Alice which explosion occurs first?

Ted

v

Alice

Figure P27.14

27.5 LENGTH CONTRACTION

15. A spaceship moves past you at speed v. You measure the ship to be 300 m long, whereas an astronaut on the ship measures a length of 400 m. Find v.

16. A muon is created at the top of Mount Everest (height 8900 m) and travels at constant speed to sea level, where it decays. According to a person traveling on the muon, what is the height of Mount Everest? Assume the muon travels vertically downward.

17. A meterstick moves toward you at speed $0.93c$. What length do you measure for the meterstick?

18. Ted is traveling on his railroad car (length 25 m) at speed $v = 0.95c$ (Fig. P27.18). As measured on Alice's clock, how long does it take for Ted's railroad car to pass Alice?

Ted

v

Alice

Figure P27.18

19. A car of rest length 2.5 m moves past you at speed v, and you measure its length to be 2.2 m. What is the car's speed?

20. An astronaut travels from a distant star to the Earth at speed $0.999c$. The astronaut ages 4.0 years during his journey. What is the distance from the star to the Earth as measured by an observer on the Earth?

21. Parking in a hurry. Your car is 5.0 m long, but you need to fit into a parking space that is only 4.5 m long. How fast must your car be moving so that you will not get a parking ticket from the police officer observing your predicament?

22. [SSM] ★ A military jet that is $L_0 = 35$ m long (as measured on its runway prior to taking off) travels at a speed of 600 m/s. What is the length L of the jet as measured by an observer on the ground? Keep extra significant figures so that you can obtain the *difference* in length $\Delta L = L_0 - L$ to two significant figures. Compare ΔL to the diameter of the hydrogen atom (approximately 5×10^{-11} m).

23. In studies of elementary particles such as electrons, physicists use machines called "accelerators" in which the particles move at very high speeds. Suppose an electron in one of these machines has a speed of $0.9999c$ and it travels along a straight line for a distance of 500 m as measured by physicists doing an experiment. How long is the accelerator as measured by a person riding along with the electron?

27.6 ADDITION OF VELOCITIES

24. Traveling on a light beam. When he was first thinking about relativity (and before he developed his theory), Einstein considered how the universe would look to a person who traveled on a light beam. Suppose you are traveling on a light beam when you encounter a friend who is traveling on a second light beam in the opposite direction. What is your friend's speed relative to your reference frame?

25. An electron (electron 1) moves to the left with speed $0.95c$ while a second electron (electron 2) moves to the left with speed $0.70c$. What is the speed of electron 2 as measured by an observer sitting on electron 1?

26. A spacecraft travels at a speed of $0.75c$ relative to the Earth. The spacecraft then launches a probe at speed v relative to the spacecraft (Fig. P27.26). If the probe has a speed of $0.90c$ relative to the Earth, what is v?

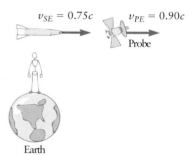

Figure P27.26

27. [SSM] Two electrons are moving in a collision course, each with speed $0.99c$ relative to a stationary observer. What is the speed of one electron as measured by an observer traveling on the other? Express your answer to five significant figures.

28. Spacecraft 1 is moving at speed $0.85c$ and is catching up with spacecraft 2, which is moving at speed $0.75c$ (both measured by an observer on the Earth). What is the speed of spacecraft 2 as measured by the pilot of spacecraft 1?

29. If the two spacecrafts in Problem 28 are initially 5.0×10^{10} m apart according to an observer on Earth, how long does it take for them to meet as measured by the Earthly observer?

30. Two asteroids are traveling along a common line and are on course for a head-on collision. Astronomers on the Earth observe that one asteroid has a speed of $0.75c$ and the other has a speed of $0.95c$. What is the relative speed of the two asteroids as measured by an observer on one of the asteroids?

31. An ice skater travels on a frozen lake at a speed of 30 m/s relative to the shore. The skater then throws a baseball in the direction parallel to her velocity with a speed of 20 m/s relative to the skater. What is the speed v of the baseball relative to the shore? Keep enough significant figures in your calculation so that you can calculate the difference $v - 50$ m/s to two significant figures.

32. A distant galaxy is moving at a speed of $0.93c$ away from the Earth. The galaxy ejects some material toward the Earth with speed $0.85c$ relative to the Earth. What is the speed of the material relative to the galaxy?

27.7 RELATIVISTIC MOMENTUM

33. An electron moves with speed $0.95c$. (a) What is the momentum of the electron according to Newton's mechanics? (b) What is the correct value of the electron's momentum according to special relativity? (c) Explain in words why the answer to part (b) is larger than the answer to part (a).

34. A baseball has a mass of about 0.14 kg; if thrown by a major league pitcher, it has a speed of about 50 m/s. Now suppose an electron is given a very large speed v so that it has the same momentum as the baseball. Find v. Express your answer as $c - v$.

35. [SSM] A proton has a relativistic momentum 10 times larger than its classical (Newton's law) momentum. What is the speed of the proton? Express your answer with three significant figures.

36. At what value of v is the relativistic momentum of a particle twice its classical (Newton's law) momentum?

37. The factor $m_0/\sqrt{1 - v^2/c^2}$ for an electron (see Eq. 27.24) is equal to five times its rest mass. What is the speed of the electron?

38. The factor $m_0/\sqrt{1 - v^2/c^2}$ for an electron (see Eq. 27.24) is equal to the rest mass of a neutron. Find the speed of the electron.

39. The relativistic momentum of a particle of rest mass m_0 and speed v is equal to $5m_0v$. What is the speed of the particle?

27.8 WHAT IS "MASS"?

27.9 MASS AND ENERGY

40. The kinetic energy of a particle is equal to the rest energy of the particle. What is the speed of the particle?

41. ★ An electron is accelerated through an electric potential difference ΔV such that its kinetic energy is equal to its rest energy. Find ΔV.

42. Annihilate this. The antiproton is a type of antimatter and is the antimatter "cousin" of the proton; the two particles have the same rest mass. It is possible for a proton and an antiproton to annihilate each other, producing pure energy in the form of electromagnetic radiation (Chapter 31). How much energy is released when a proton and an antiproton annihilate each other? Assume the two particles are rest just prior to the annihilation.

43. An asteroid of mass 2500 kg has a (relativistic) kinetic energy of 1.5×10^{20} J. What is the speed of the asteroid?

44. ★ An electron in a television picture tube has a classical kinetic energy of 30 keV, which is the kinetic energy that Newton would calculate using the measured speed and rest mass of the electron. What is the actual kinetic energy of the electron; that is, what is the value found using the relativistic result for the kinetic energy?

45. Cosmic ray protons have been observed with kinetic energies claimed to be as high as 10^{20} eV. What is speed of these protons? Express your answer as $c - v$ where v is the speed of the protons.

46. A proton has a kinetic energy of 1.5×10^7 eV. What is its speed?

47. [SSM] ✪ A chemical reaction that produces water molecules is

$$2H_2 + O_2 \rightarrow 2H_2O$$

This reaction releases 570 kJ for each mole of oxygen molecules that is consumed. The relativistic relation between mass and energy then implies that the mass of two hydrogen molecules plus the mass of one oxygen molecule is slightly larger than the mass of two water molecules. Find this mass difference.

48. ✪ The most common helium nucleus in nature contains two protons and two neutrons. The mass of this nucleus is 6.64466×10^{-27} kg, while the mass of a proton is 1.67262×10^{-27} kg and the mass of a neutron is 1.67493×10^{-27} kg. Find the binding energy of this helium nucleus.

49. ✪ One kg of the chemical explosive TNT releases approximately 4.2×10^6 J of energy when it explodes. How much of the initial mass of TNT is converted to energy in such an explosion?

50. In 2005, the total energy "consumed" by people on Earth was about 5×10^{20} J. About 85% of this was from the burning of fossil fuels. How much fossil fuel mass was converted to energy in 2005?

27.10 THE EQUIVALENCE PRINCIPLE AND GENERAL RELATIVITY

51. ✪ In Chapter 23, we learned that an accelerating electric charge produces electromagnetic waves. In the same way, the general theory of relativity predicts that an accelerated mass produces gravitational waves. As the Earth moves in orbit around the Sun, it undergoes accelerated motion (i.e., uniform circular motion), and it has been predicted that the power carried away by the resulting gravitational radiation is about 0.001 W. How long will it take for the Earth to fall into the Sun? That is, how long will it take for the Earth's orbit to have the same radius as the Sun's radius?

ADDITIONAL PROBLEMS

52. ✪ The muon in Example 27.4 has a relativistic momentum of 4.0×10^{-19} kg · m/s. The rest mass of a muon is $m_0 = 1.9 \times 10^{-28}$ kg. Find the kinetic energy of the muon.

53. What is the relativistic momentum of an electron whose relativistic kinetic energy is 4.0×10^{-22} J?

54. ✪ The Tevatron is an accelerator at Fermi National Laboratory near Chicago. It uses the collisions between protons and antiprotons to generate and study quarks and other elementary particles. If the kinetic energy of a proton in the Tevatron is 1.0 TeV (= 1.0×10^{12} eV), what are (a) the speed of the proton and (b) the momentum of the proton? (c) Compare your result to the momentum of a mosquito ($m = 1$ mg) flying at a speed of 1 m/s.

55. ✪ The kinetic energy of a particle of rest mass m_0 is equal to three times its rest energy. What is the momentum of the particle? Express your answer in terms of m_0.

56. ✪ The space shuttle has a length of 37 m. (a) In its normal orbital motion around the Earth, the shuttle has a speed of about 8000 m/s and an orbital period of about 90 minutes. Allowing for length contraction, what is the length of the shuttle as viewed by an observer at rest on the ground? (b) Suppose the shuttle increases its speed so that it has a length of 25 m when viewed by a stationary observer. How long would it take to complete one orbit?

57. An electron has a momentum equal to that of a baseball ($m = 0.22$ kg) moving at 45 m/s (about 100 mi/h). What is the kinetic energy of the electron?

58. ✪ ℝ Suppose two highly precise, identical clocks are synchronized and one clock is placed on the North Pole and the other on the equator. After 100 years, how much will the clocks differ in time? Assume the Earth is a perfect sphere.

59. [SSM] ✪ ℝ Like all stars, the Sun converts mass into energy that radiates out in all directions. The average rate at which this radiant energy reaches the Earth is approximately 1.4×10^3 W/m². (a) Calculate the rate at which the Sun is losing mass. (b) Assuming this rate remains constant, estimate the lifetime of the Sun.

60. Consider a proton moving at relativistic speed. (a) Determine the proton's rest energy in electron volts (1 eV = 1.60×10^{-19} J). (b) Suppose the total energy of the proton is three times its rest energy. With what speed is the proton moving? (c) At this speed, determine the kinetic energy of the proton in electron-volts. (d) What is the magnitude of the proton's momentum?

61. ✪ ℝ As we'll discuss in Chapter 30, radioactive decay involves the nucleus of an atom decaying by emitting either a particle or energy or both. The ^{216}Po nucleus decays to ^{212}Pb by emitting an alpha particle, which is a helium nucleus, ^{4}He. Using the mass data given in Appendix A, find (a) the mass change in this decay and (b) the energy that this mass represents. (c) Given that no other particles or energy are emitted in this process, where must the energy go?

62. ✪ **Falling light.** According to general relativity and the equivalence principle, light is bent by gravity. Suppose you stand two tall, perfectly reflecting mirrors exactly 1 m apart and facing each other. A beam of light is directed horizontally through a hole in one of the mirrors 10 m above the ground. (a) Determine the time it takes for the light to strike the ground. (b) The light will undergo N reflections (i.e., N/2 reflections from each mirror) before it strikes the ground. Find N.

63. [SSM] ✪ Suppose an electron is accelerated from rest through a potential difference of 100,000 volts. Determine the electron's final kinetic energy, speed, and momentum (a) ignoring relativistic effects and (b) including relativistic effects.

64. ✪ Suppose a rocket ship leaves the Earth in the year 2020. One of a set of twins born in 2000 remains on the Earth while the other rides in the rocket. The rocket ship travels at 0.90c in a straight line path for 10 years as measured by its own clock, turns around, and travels straight back at 0.90c for another 10 years as measured by its own clock before landing back on the Earth. (a) What year is it on the Earth? (b) How old is each twin? (c) How far away from the Earth did the rocket ship travel as measured by each twin?

65. ✪ One consequence of general relativity is that a clock in a gravitational field runs slower. This time dilation is distinct from the time dilation from relative motion (special relativity) and is given by $\Delta t = \Delta t_0 / \sqrt{1 - 2GM/(Rc^2)}$, where Δt is the dilated time, Δt_0 is the time when no gravitational field is present, G is the universal gravitational constant, M is the mass of the object producing the gravitational field, R is the distance from the center of the mass, and c is the speed of light. In 1976, the Smithsonian Astrophysical Observatory sent aloft a Scout rocket to a height of 10,000 km to confirm this effect. Show that at this height a clock should run faster than an identical clock on the Earth by a factor of 4.5×10^{-10}.

66. ✪ Thomas Jefferson National Accelerator Facility in Newport News, Virginia, is host to a continuous electron beam. Within the facility, electrons travel around a $\frac{7}{8}$-mi-long tunnel five times in 23.5 millionths of a second. (Actually, the facility is capable of producing even higher speeds.) (a) What is the speed of the electrons? (b) Determine the length around the tunnel as measured in the reference frame of the electrons. (c) Determine the time it takes for the electrons to complete one lap around the tunnel in the reference frame of the electrons. (d) What are the momentum and kinetic energy of the electrons?

67. ✪ **Gravitational red shift.** An important consequence of Einstein's general relativity is that gravity must affect a light wave's frequency and wavelength. As light moves upward from the Earth's surface, the wavelength of the light gets longer and the frequency gets lower as gravity "drains" the light of some energy. In a famous experiment in 1960, Robert Pound and Glen Rebka of Harvard University successfully tested this effect to within 10% of the predicted value. Later, in 1964, they improved the agreement to within 1%. The experimental procedure involved placing a detector 22.6 m above a radioactive source placed on the ground. Quantized "particles" of light called photons (see Chapter 28) are given off by the source with an energy of 14.4 keV. Using a highly sensitive technique called the Mössbauer effect, they were able to measure a small shift in the energies of the photons as they moved upward through the Earth's gravitational field. Although photons are massless, we can treat their energy as rest energy and find a corresponding "effective mass" for the photons. Using this approach, determine the shift in the photons' energy as they move the 22.6 m up to the detector. This shift is called a *gravitational red shift* because it results in a shift toward a lower frequency and longer wavelength.

68. ✴ Ⓡ By what amount is the diameter of the Moon shortened (as measured by a stationary observer on the Earth) due to its orbital motion around the Earth?

69. ✪ Suppose you take a trip to Mars, which is 80 million km from the Earth. You head directly toward Mars onboard your spaceship, traveling at $0.80c$. Unbeknownst to you, your mortal enemy placed a bomb on your ship that has a 4-minute timer that was initiated upon your departure. Will you live long enough to reach Mars? Answer the problem first without relativity and then using relativity from both your point of view and that of your Earth-bound enemy.

70. ✪ Ⓡ **GPS and relativity.** The Global Positioning System (GPS) consists of a network of about 30 satellites in orbit, each carrying atomic clocks on board. The orbital radius of the satellites is about four Earth radii (26,600 km). The orbits are nearly circular, with a typical eccentricity of less than 1%. The onboard atomic clocks keep highly accurate time, and oscillators produce signals at several different frequencies. The frequency used by nonmilitary GPS receivers is 1575.42 MHz. Two relativistic corrections must be made to this signal so that accurate position readings on the Earth are possible: first, special relativity predicts that moving clocks will appear to tick slower than nonmoving ones, and second, general relativity predicts that clocks in a stronger gravitational field will tick at a slower rate. Determine (a) the orbital period and (b) the orbital speed of these satellites. (c) Apply special relativity to determine the correction to the frequency of the signal received on the Earth. (d) Applying general relativity and following the method described in Problem 67, determine the correction to the frequency of the signal received on Earth. *Hint:* The energy of a photon is proportional to its frequency.

Chapter 28

Quantum Theory

A modern reconstruction of the apparatus used in experiments by J. J. Thomson in his discovery of the electron in 1897. The bright horizontal beam is produced by electrons as they travel from left to right inside the glass tube. Thomson's discovery and other work described in this chapter led to the development of quantum theory. (© Charles D. Winters/Photo Researchers, Inc.)

In this text, our work on mechanics, electricity, and magnetism has been based on two pillars of *classical* physics: Newton's laws and Maxwell's equations. Classical physics deals with the macroscopic world, including baseballs, automobiles, and planets. We have already mentioned that Newton's laws fail when applied to the atomic-scale world of electrons and atoms. Similarly, Maxwell's equations correctly describe electromagnetic phenomena in the macroworld, but fail when applied at the atomic scale. This atomic-scale world is called the *quantum* regime. In this chapter, we describe how quantum behavior is different from anything we have seen before. "Quantum" refers to a very small increment or parcel of energy. These parcels, called *quanta*, are a central aspect of the quantum world.

The discovery and development of quantum theory began in the late 1800s and continued during the first few decades of the twentieth century. As we introduce the key ideas of quantum theory, we'll also mention some of the very interesting history of this work.

28.1 | PARTICLES, WAVES, AND "PARTICLE-WAVES"

In the macroworld of Newton and Maxwell, energy can be carried from one point to another by only two types of "objects": particles (such as baseballs or bullets) and waves (such as sound or light). We have an intuitive understanding of particles and waves through our everyday experiences, and it is natural to use this macroworld intuition when we consider the behavior of things in the microworld such as electrons and atoms. We'll see, however, that this intuition is completely incapable of describing the quantum regime!

How Do Waves and Particles Differ?

In Chapter 25, we studied the interference and diffraction of light. Let's now consider the double-slit experiment in Figure 28.1A in which light is incident on an opaque barrier containing two very narrow openings. A double-slit interference pattern, consisting of a series of bright and dark fringes, is formed on the screen on the far right. The bright fringes are produced by constructive interference between waves that pass through the two slits, and the dark fringes are at locations where there is destructive interference. A similar result would be found with other types of waves, including sound and water waves.

Figure 28.1B shows another double-slit experiment, this time using bullets instead of light. Our macroworld intuition correctly predicts what will happen in this case: only bullets that pass through one or the other of the two slits will reach the screen on the far right; the other bullets (the ones that strike the barrier) are stopped by the barrier. The pattern of bullets (or holes!) on the screen is quite different from the pattern found with light waves; the bullet pattern corresponds only to the "shadows" of the slits. There is no constructive or destructive interference with bullets or other classical particles such as baseballs or rocks.

The behaviors sketched in Figure 28.1 illustrate some important classical differences between particles and waves.

Properties of waves and particles in the classical regime:

- Waves exhibit interference; particles do not.

- Particles often deliver their energy in discrete amounts. For example, when a bullet in Figure 28.1B strikes the screen, all its energy is deposited on the screen (assuming the bullet does not bounce off). Energy arrives at the screen in discrete parcels, with each parcel corresponding to the kinetic energy carried by a single bullet.

- The energy carried or delivered by a wave is not discrete, but varies in a continuous manner. Recall from Chapter 23 that the energy carried by a wave is described by its intensity, which equals the amount of energy the wave transports per unit time across a surface of unit area. Hence, for the light wave in Figure 28.1A, the amount of energy absorbed by the screen depends on the intensity of the wave and the absorption time. The amount of absorbed energy can thus take on any nonnegative value. Classically, wave energy is not delivered in discrete parcels.

An Interference Experiment with Electrons

In the world of classical physics, the experiment in Figure 28.1 can be used to distinguish between particles and waves. According to classical physics, only these two types of behavior are possible: waves exhibit interference; particles do not.

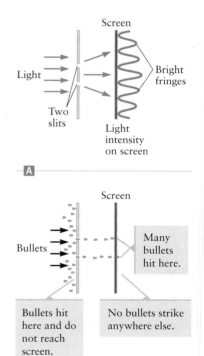

Figure 28.1 **A** When light passes through a double slit, constructive and destructive interference produce bright and dark fringes on the screen, evidence for the wave nature of light. **B** Classical particles such as bullets do not undergo interference when they pass through a double slit.

However, this tidy separation of particles versus waves is *not* found in the quantum world. Figure 28.2 shows a hypothetical double-slit experiment performed with electrons. A beam of electrons, all with the same speed, are incident from the left and pass through a pair of slits. The electrons then travel on to a screen on the right, which records where each electron strikes. Case 1 in Figure 28.2 shows the result after 20 electrons have passed through the slits. Each dot shows where a particular electron arrived at the screen, and the arrival points seem to be distributed randomly. If this experiment is repeated with another 20 electrons, the precise arrival points would be different, but the general appearance would be the same. If we proceed with the experiment and wait until 100 electrons have arrived, we find the result shown in case 2. The arrival points are still spread out, but we can now see that the electrons are more likely to strike at certain points than at others. By the time 300 electrons have reached the screen in case 3, it is clear that electrons are much more likely to hit certain points on the screen. Case 4 on the far right shows the results after a very large number of electrons have passed through the slits. This sketch shows the probability that electrons will arrive at different points. This probability curve has precisely the same form as the variation of light intensity in the double-slit interference experiment in Figure 28.1A. The experiment shows that electrons undergo constructive interference at certain locations on the screen, giving a large probability for electrons to arrive at those locations. At other places, the electrons undergo destructive interference, and the probability for an electron to reach those locations is very small or zero.

The results in Figure 28.2 show that electrons can exhibit interference, a property that classical theory says is possible only for waves. This experiment also shows aspects of particle-like behavior since the electrons arrive one at a time at the screen, with each dot in cases 1 through 3 of the figure corresponding to the arrival of a single electron as it deposits its parcel of energy on the screen.

The behavior in Figure 28.2 is characteristic of the quantum regime and shows that electrons behave in some ways as *both a classical particle and a classical wave*. In fact, *all* objects in the quantum world behave in this way. Electrons, protons, atoms, and molecules have all been found to give the results shown in Figure 28.2. Moreover, even light waves exhibit particle-like behavior. The clear-cut distinction between particles and waves thus breaks down in the quantum regime. It may be more helpful for your classical intuition to call these things "particle-waves," all of which exhibit the following properties.

Properties of waves and particles in the quantum regime:

- *All* objects, including light and electrons, can exhibit interference.
- *All* objects, including light and electrons, carry energy in discrete amounts. These discrete "parcels" are called quanta.

We explore the implications of these ideas for light and for electrons in the next two sections, beginning with an experiment showing that light energy is quantized.

Figure 28.2 An interference experiment with electrons. Each dot in cases 1 through 3 indicates where an electron strikes the screen. Case 1: pattern formed by the first 20 electrons. Case 2: pattern formed by the first 100 electrons. Case 3: After 300 electrons, it is clear that electrons tend to arrive at certain places on the screen and not at others. Case 4: When the experiment is carried out with a very large number of electrons, the probability for an electron to reach the screen forms a double-slit interference pattern, similar to the pattern for light in Figure 28.1A.

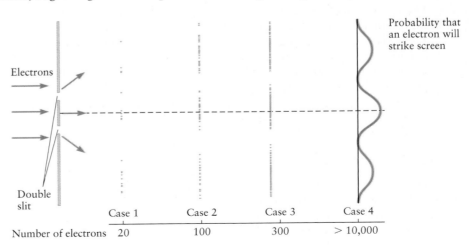

Around 1800, Thomas Young performed his double-slit interference experiment, which provided the first clear evidence that light is a wave. Maxwell's theory of electromagnetic waves was worked out about 60 years later, so physicists then had a detailed theory of light as an electromagnetic wave and believed that the nature of light was well understood. In the 1880s, however, studies of what happens when light is shined onto a metal gave some very puzzling results that could not be understood with the wave theory of light. To appreciate these experiments, we must first recall that a metal contains electrons that are free to move around within the metal. However, these electrons are still bound as a whole to the metal because of their attraction to the positive charges of the metal atom nuclei; that is why these electrons do not just "fall out of" the metal. The minimum energy required to remove a single electron from a piece of metal is called the **work function**, W_c. The value of the work function is different for different metals; it can be measured by applying an electric potential difference between the metal of interest and a second metal plate (all in vacuum) and observing the value of the potential at which electrons first leave the metal (Fig. 28.3). If V is the electric potential at which electrons begin to jump across the vacuum gap in Figure 28.3, the work function is $W_c = eV$, where e is the magnitude of the electron's charge ($e = 1.60 \times 10^{-19}$ C).

The work function has units of energy, so it can be measured in joules. Since the experiment in Figure 28.3 involves electrons, it is sometimes convenient to measure W_c in units of electron-volts (eV). The conversion factor 1.60×10^{-19} J/eV is listed on the page facing the inside front cover; see also the definition of the electron-volt in Eq. 18.19. Values of the work function in both units are listed in Table 28.1.

Figure 28.3 Measuring the work function W_c of a metal by applying an electric potential. When $eV > W_c$, electrons are ejected from the metal on top and jump to the metal plate below. Measuring the smallest V able to eject electrons gives the value of W_c.

The Photoelectric Effect

Another way to extract electrons from a metal is by shining light onto it. Light striking a metal is absorbed by the electrons, and if an electron absorbs an amount of light energy greater than W_c, it is ejected from the metal as sketched in Figure 28.4A. This phenomenon is called the **photoelectric effect**.

Experimental studies of the photoelectric effect carried out around 1900 found that no electrons are emitted unless the light's frequency is greater than a critical value f_c. When the frequency is above f_c, the kinetic energy of the emitted electrons varies linearly with frequency f as shown in Figure 28.4B. Physicists initially tried to explain these results using the classical wave theory of light, but there were two difficulties with the classical explanations. First, experiments show that the critical

Photoelectric effect: ejection of electrons from a metal after absorbing energy from light

Table 28.1 Work Function of Several Metals

METAL	WORK FUNCTION (J)	WORK FUNCTION (eV)
Na (sodium)	3.8×10^{-19}	2.4
Al (aluminum)	6.9×10^{-19}	4.3
Ca (calcium)	4.5×10^{-19}	2.8
Fe (iron)	6.9×10^{-19}	4.3
Cu (copper)	7.0×10^{-19}	4.4
Mo (molybdenum)	6.7×10^{-19}	4.2
Ag (silver)	6.9×10^{-19}	4.3
Pt (platinum)	1.0×10^{-18}	6.4
Pb (lead)	6.4×10^{-19}	4.0

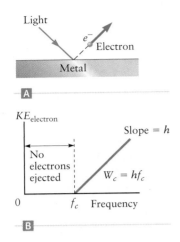

Figure 28.4 ◮ The photo-electric effect: electrons are ejected when light strikes a metal. ◪ Experiments show that the kinetic energy of the ejected electrons ($KE_{electron}$) depends on the frequency of the light. When the frequency is below a certain critical value f_c, no electrons are ejected. The work function is related to the critical frequency by $W_c = hf_c$, where h is Planck's constant defined in Equation 28.2.

frequency f_c is *independent* of the intensity of the light. According to classical wave theory, the energy carried by a light wave is proportional to the intensity, so it should always be possible to eject electrons by increasing the intensity to a sufficiently high value. Experiments found that when the frequency is below f_c, however, there are no ejected electrons no matter how great the light intensity. Second, the kinetic energy of an ejected electron is independent of the light intensity. Classical theory predicts that increasing the intensity will cause the ejected electrons to have a higher kinetic energy, but the experiments show no such result. In fact, the experiments show that the electron kinetic energy depends on the light's frequency (Fig. 28.4B) instead of the intensity.

The classical wave theory of light is thus not able to explain the photoelectric effect experiments. Albert Einstein surprised the physics world in 1905 when he offered the following explanation.[1] He proposed that light carries energy in discrete quanta, now called **photons**. According to Einstein, each photon carries a parcel of energy

$$E_{photon} = hf \qquad (28.1)$$

where h is a constant of nature called **Planck's constant**, which has the value

$$h = 6.626 \times 10^{-34} \, \text{J} \cdot \text{s} \qquad (28.2)$$

and f is the frequency of the light.

Planck's constant had been introduced a few years earlier by Max Planck to explain another unexpected property of electromagnetic radiation (the blackbody radiation spectrum, page 962). Let's now see how Einstein's photon theory explains the photoelectric effect; we'll also see why the slope of the kinetic energy curve in Figure 28.4B is equal to Planck's constant h.

Einstein suggested that a beam of light should be thought of as a collection of particles (photons), each of which has an energy that depends on frequency according to Equation 28.1. If the intensity of a monochromatic (single-frequency) light beam is increased, the number of photons is increased but the energy carried by each photon does not change. This theory explains the two puzzles associated with photoelectric experiments. First, the absorption of light by an electron is just like a collision between two particles, a photon and an electron. The photon (according to Eq. 28.1) carries an energy hf that is absorbed by the electron. If this energy is less than the work function, the electron is not able to escape from the metal. For monochromatic light, increasing the light intensity increases the number of photons that arrive each second, but if the photon energy hf is less than the work function, even a high intensity will not eject electrons. The energy of a single photon—and hence the energy gained by any particular electron—depends on frequency f but not on the light intensity. Second, Einstein's theory also explains why the kinetic energy of ejected electrons depends on light frequency but not intensity. The critical frequency in the photoelectric effect (Fig. 28.4B) corresponds to photons whose energy is equal to W_c:

$$hf_c = W_c$$

Such a photon has barely enough energy to eject an electron from the metal, but the ejected electron then has no kinetic energy. If a photon has a higher frequency and thus a greater energy, the extra energy above the work function goes into the kinetic energy of the electron. We have

$$KE_{electron} = hf - hf_c = hf - W_c \qquad (28.3)$$

which is the equation of a straight line; hence, the kinetic energy of an ejected electron should be linearly proportional to f. This linear behavior is precisely what is

---

[1]Einstein made a number of other monumental discoveries in the same year. He was awarded the Nobel Prize in Physics in 1921 for his theory of photons and the photoelectric effect.

found in experiments as shown in Figure 28.4B. The slope of this line is the factor multiplying f in Equation 28.3, which is just Planck's constant h. Thus, photoelectric experiments give a way to measure h, and the values found agreed with the value known prior to Einstein's theory.

Photons Carry Energy and Momentum

Einstein's photon theory asserts that light energy can only be absorbed or emitted in discrete parcels, that is, as single photons. Each photon carries an energy $E_{\text{photon}} = hf$. From Chapter 23, a light wave with an energy E also carries a certain amount of momentum $p = E/c$ (Eq. 23.12). Identifying the energy E with E_{photon}, Einstein's theory thus predicts that the momentum carried by a single photon is

$$p_{\text{photon}} = \frac{hf}{c} \tag{28.4}$$

Momentum of a photon

Wavelength is related to frequency by $f\lambda = c$, so Equation 28.4 can also be written as

$$p_{\text{photon}} = \frac{h}{\lambda} \tag{28.5}$$

The quantum theory of light thus predicts that "particles" of light called photons carry a discrete amount (a quantum) of both energy and momentum. Notice that we put the term *particles* in quotation marks. We did so because photons have two important properties that are quite different from classical particles: photons do not have any mass (!), and they exhibit interference effects (as in the double-slit interference experiment in Figure 28.1A). Photons are thus unlike anything we have encountered in classical physics.

Let's use Equation 28.1 to calculate the energy carried by a single photon. The result depends on the frequency of the light, so consider the light from a green laser pointer. This light has a wavelength of about $\lambda = 530$ nm, so its frequency is

$$f = \frac{c}{\lambda} = \frac{3.00 \times 10^8 \text{ m/s}}{530 \times 10^{-9} \text{ m}} = 5.7 \times 10^{14} \text{ Hz}$$

Inserting this frequency in Equation 28.1 gives

$$E_{\text{photon}} = hf = (6.63 \times 10^{-34} \text{ J} \cdot \text{s})(5.7 \times 10^{14} \text{ Hz}) = 3.8 \times 10^{-19} \text{ J} \tag{28.6}$$

which is a very small amount of energy, much smaller than is normally encountered in the macroworld (as shown in Example 28.1 and Problem 24). In most applications, one detects the presence or absence of light (the presence or absence of one or more photons) through the energy carried by the light. Hence, to detect a single photon of green light it is necessary to detect amounts of energy as small as 3.8×10^{-19} J. In Section 28.7, we describe how a cell in the human retina can respond to just a single photon! The eye is thus an exquisitely sensitive detector of light.

EXAMPLE 28.1	Number of Photons Emitted by a Lightbulb

The light emitted by an ordinary lightbulb consists of photons, each of which has an energy given by Equation 28.1. Consider a 60-W lightbulb and for simplicity assume it emits only green light with $\lambda = 530$ nm as considered in Equation 28.6. How many photons does the lightbulb emit in 1.0 s?

RECOGNIZE THE PRINCIPLE

This example is an application of the relation between photon energy and frequency. The power rating P of the lightbulb gives the rate at which it emits energy. Power is

Each photon carries an energy $E_{photon} = hf$

Figure 28.5 Example 28.1.

the energy emitted per unit time, so from the value of P we can get the energy emitted in 1.0 s. We know the energy of a single photon from Equation 28.6, so we can then find the total number of photons emitted in this time.

SKETCH THE PROBLEM

Figure 28.5 shows a lightbulb emitting photons. The energy emitted each second equals the energy of a single photon times the number of photons emitted in 1 second.

IDENTIFY THE RELATIONSHIPS

The lightbulb has a power rating of $P = 60$ W $= 60$ J/s, so the total energy emitted in $t = 1.0$ s is

$$E_{total} = Pt = (60 \text{ J/s})(1.0 \text{ s}) = 60 \text{ J}$$

This energy is carried by N photons, each of which has the energy E_{photon} found in Equation 28.6. We thus have

$$E_{total} = NE_{photon}$$

SOLVE

Solving for N, we get

$$N = \frac{E_{total}}{E_{photon}} = \frac{60 \text{ J}}{3.8 \times 10^{-19} \text{ J}} = \boxed{1.6 \times 10^{20} \text{ photons}}$$

What does it mean?

In this example as in most ordinary situations, the number of photons involved is extremely large, so the typical observer would not notice that the energy in a light beam actually does arrive as discrete quanta.

CONCEPT CHECK 28.1 | Energy of a Radio Frequency Photon

Your favorite AM radio station has a frequency near $f_{AM} \approx 1$ MHz, whereas the author's favorite FM station has $f_{FM} \approx 100$ MHz. Radio waves are electromagnetic radiation, just like visible light, so both AM and FM signals are carried by photons. Which of the following statements is true?

(a) The AM photons have a higher energy than the FM photons.
(b) The FM photons have a higher energy than the AM photons.
(c) They are both radio waves, so they have the same photon energy.

EXAMPLE 28.2 | The Photoelectric Effect with Aluminum

A photoelectric experiment is planned using aluminum (Al). (a) What is the minimum photon frequency that will eject electrons? (b) If the experiment is performed using blue light with $\lambda = 450$ nm, will there be any ejected electrons? If so, find the kinetic energy of an ejected electron.

RECOGNIZE THE PRINCIPLE

This example is a direct application of the photoelectric effect. (a) The minimum photon frequency f_c required to eject an electron is proportional to the work function. (b) We can determine if a photon will eject an electron by comparing its frequency with f_c.

SKETCH THE PROBLEM

No sketch is necessary.

IDENTIFY THE RELATIONSHIPS

The energy of a photon of frequency f is $E_{photon} = hf$ (Eq. 28.1), so we can use the value of the work function for Al from Table 28.1 to find the minimum photon frequency. We can then compare this frequency with the energy of a photon of blue light to decide if such a photon will eject an electron. (The variation of the frequency and wavelength across the electromagnetic spectrum is also shown in Fig. 23.8.) From Table 28.1, the work function for Al is $W_c = 4.3$ eV (electron-volts), or, expressed in units of joules, $W_c = (4.3 \text{ eV})(1.60 \times 10^{-19} \text{ J/eV}) = 6.9 \times 10^{-19}$ J. The work function W_c equals the energy of a photon that will just barely eject an electron, so

$$hf_c = W_c$$

SOLVE

(a) Solving for f_c gives

$$f_c = \frac{W_c}{h} = \frac{6.9 \times 10^{-19} \text{ J}}{6.63 \times 10^{-34} \text{ J} \cdot \text{s}} = \boxed{1.0 \times 10^{15} \text{ Hz}}$$

(b) The frequency of a photon with $\lambda = 450$ nm $= 4.5 \times 10^{-7}$ m is

$$f = \frac{c}{\lambda} = \frac{3.00 \times 10^8 \text{ m/s}}{4.5 \times 10^{-7} \text{ m}} = 6.7 \times 10^{14} \text{ Hz}$$

This frequency is smaller than f_c from part (a), so this photon will not have enough energy to eject an electron.

What does it mean?

To eject an electron requires a photon with a higher frequency (and hence shorter wavelength) than the blue photon considered here. So, only ultraviolet light can eject electrons from aluminum.

CONCEPT CHECK 28.2 | The Photoelectric Effect

A photoelectric effect experiment finds that the minimum photon frequency required to eject an electron from a metal is half that found for copper. What is the work function of the metal, (a) 4.4 eV, (b) 3.5 eV, (c) 2.2 eV, or (d) 8.8 eV?

EXAMPLE 28.3 | Photons and Chemical Reactions

Light plays a role in chemical processes called *photochemical reactions*. For example, the absorption of light can cause a hydrogen atom to dissociate, a process in which the atom absorbs a single photon and the electron is ejected (i.e., the hydrogen atom literally breaks apart). If the energy required for dissociation is 13.6 eV, what is the smallest photon frequency that can cause this reaction?

RECOGNIZE THE PRINCIPLE

This example is another application of the relation between photon frequency and energy. The minimum energy required for dissociation is called the ionization energy, which for hydrogen is $E_{ionization} = 13.6$ eV. To dissociate a hydrogen atom, the energy of an incoming photon must be equal to or greater than this value.

SKETCH THE PROBLEM

Figure 28.6 shows the problem.

Figure 28.6 Example 28.3. Ionization of a (highly schematic) hydrogen atom. See Chapter 29 for a more accurate description of atomic structure. **A** Incoming photon. **B** The atom's proton and electron have dissociated.

The photon frequency must satisfy the condition

$$E_{photon} = hf \geq E_{ionization} \qquad (1)$$

We know $E_{ionization}$ and Planck's constant h, so we can solve for the minimum photon frequency.

SOLVE

The lowest photon frequency that satisfies Equation (1) is

$$f = \frac{E_{ionization}}{h}$$

We are given the value $E_{ionization} = 13.6$ eV. Converting to units of joules, we use the conversion factor listed on the page facing the inside front cover and get $E_{ionization} = (13.6 \text{ eV})(1.60 \times 10^{-19} \text{ J/eV}) = 2.2 \times 10^{-18}$ J, which leads to

$$f = \frac{E_{ionization}}{h} = \frac{2.2 \times 10^{-18} \text{ J}}{6.63 \times 10^{-34} \text{ J} \cdot \text{s}} = \boxed{3.3 \times 10^{15} \text{ Hz}}$$

This frequency lies in the ultraviolet range and is not visible to the eye.

What does it mean?

Photochemical reactions are quite common in nature. In photosynthesis, for example, the absorption of a photon initiates a chain of reactions that leads to the storage of a portion of the photon's energy within a cell.

Insight 28.1

DETECTING SINGLE PHOTONS

Although the energy carried by a photon of visible light is small (see Eq. 28.6), it is nevertheless possible to detect single photons. Since the energy carried by a photon increases as the frequency increases, it is easiest to detect single photons of high-frequency radiation such as visible light, X-rays, and gamma rays. It is much more challenging to detect individual infrared, microwave, and radio frequency photons because they carry much less energy.

Blackbody Radiation

Planck's constant has an important role in Equation 28.1 and the photon theory of light. This constant was introduced by Max Planck in 1901 when he was studying the problem of *blackbody radiation*. Section 14.8 introduced the concept of a blackbody and described how the absorption and emission of light by a blackbody depends on its temperature. The specific problem that puzzled Planck is typified by the glowing oven in Figure 28.7A. This oven emits blackbody radiation over a range of wavelengths (λ) and frequencies as sketched in Figure 28.7B. To the eye, the color of the cavity is determined by the wavelength at which the radiation intensity is

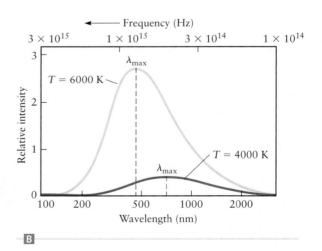

Figure 28.7 **A** This oven is an approximate blackbody. Light emitted from an ideal blackbody follows the blackbody spectrum in **B**. The wavelength λ_{max} at which the blackbody radiation intensity is largest depends on the temperature of the blackbody.

largest. This wavelength is denoted as λ_{max} in Figure 28.7B and is determined by the temperature T of the blackbody through Wien's law (Eq. 14.14),

$$\lambda_{max} = \frac{2.90 \times 10^{-3} \text{ m} \cdot \text{K}}{T} \qquad (28.7)$$

Here T is measured in kelvins and the numerical constant in Equation 28.7 is a combination of several fundamental constants. According to Wien's law, the value of λ_{max} and hence the color of the blackbody depend on the temperature. A hotter object (higher T) has a smaller value of λ_{max}, and the entire blackbody curve in Figure 28.7B shifts to shorter wavelengths when the temperature is increased. A flame that appears blue is therefore hotter than one that is red.

Experiments prior to Planck's work showed that the intensity curve in Figure 28.7B has the same shape for a wide variety of objects: the blackbody intensity falls to zero at both long and short wavelengths, corresponding to low and high frequencies, with a peak in the middle at the location given by Wien's law (Eq. 28.7). Planck tried to explain this behavior. At that time, physicists knew that electromagnetic waves form standing waves as they reflect back and forth inside an oven's cavity. Such standing waves are just like the standing waves we discussed in Chapter 12, where we found that standing waves on a string have frequencies following the pattern $f_n = nf_1$, where f_1 is the fundamental frequency and $n = 1, 2, 3, \ldots$. Standing electromagnetic waves in a cavity follow the same mathematical pattern. There is no limit to the value of n (as long as it is an integer), so the frequency f_n of a standing electromagnetic wave in a blackbody cavity can be infinitely large. According to classical physics, each of these standing waves carries energy, and as their frequency increases, so does their energy. As a result, the classical theory predicts that the blackbody intensity should become *infinite* as the frequency approaches infinite values!

Planck's Hypothesis and Quanta of Electromagnetic Radiation

It was obvious to Planck and other physicists of the time that any theory predicting that the intensity of blackbody radiation is infinite could not possibly be correct. Such a theory would also be in conflict with the experimental intensity curves in Figure 28.7B because the true blackbody intensity falls to zero at high frequencies. (It would also violate "common sense"; nearly all objects act as approximate blackbodies, and it is very hard to imagine how all objects could emit an infinite amount of energy and still be consistent with our ideas about conservation of energy.) Despite much effort, however, physicists were not able reconcile classical theory with the observed blackbody behavior. This disagreement was called the "ultraviolet catastrophe" since the predicted infinite intensity is found at high frequencies and high-frequency visible light falls at the ultraviolet end of the electromagnetic spectrum.

Planck proposed to resolve this catastrophe by assuming the energy in a blackbody cavity must come in discrete parcels (quanta), with each parcel having an energy $E = hf_n$, where f_n is one of the standing wave frequencies and h is the constant given in Equation 28.2. Planck showed that this assumption fixes the theory of blackbody radiation so that it correctly produces the blackbody spectrum in Figure 28.7B, but he could give no reason or justification for his assumption about standing wave quanta. His theory could fit the experiments perfectly, but no one (including Planck) knew *why* it worked. That question was answered in part by Einstein: the standing electromagnetic waves in a blackbody cavity consist of photons with quantized energies given by Equation 28.1. While Einstein's photon theory supported Planck's result, it would still be several decades before the photon concept was fully worked out and understood.

Planck's theory of blackbody radiation was developed in 1900 and is generally cited as the beginning of quantum theory, giving Planck an important place in the history of physics. Blackbody radiation also has several practical uses. Figure 14.33B (page 458) shows an "ear thermometer," which detects the blackbody radiation from inside your ear and uses this measurement to calculate your body temperature. According to Wien's law (Eq. 28.7), the wavelength at which the blackbody radiation intensity is largest depends on temperature. This thermometer measures λ_{max} and then uses Wien's law to find your body temperature.

| EXAMPLE 28.4 | Temperature of the Sun

The Sun emits radiation over a wide range of wavelengths, following an approximate blackbody curve. To the eye, the Sun appears yellow, and yellow light has a wavelength near $\lambda_{max} = 5.5 \times 10^{-7}$ m (550 nm). What is the approximate temperature of the Sun?

RECOGNIZE THE PRINCIPLE

The color of a blackbody, such as the Sun, depends on its temperature. The color is determined by the wavelength at which the intensity has its maximum, which is the wavelength λ_{max} given by Wien's law.

SKETCH THE PROBLEM

Figure 28.7B gives the blackbody spectrum at two temperatures, and shows how this spectrum shifts to shorter wavelengths as the temperature of the blackbody increases.

IDENTIFY THE RELATIONSHIPS

According to Wien's law (Eq. 28.7), the location of the intensity peak is related to the temperature of the blackbody by

$$\lambda_{max} = \frac{2.90 \times 10^{-3} \text{ m} \cdot \text{K}}{T}$$

We know λ_{max} for the Sun, so we can find its temperature T.

SOLVE

Rearranging Wien's law to solve for the temperature gives

$$T = \frac{2.90 \times 10^{-3} \text{ m} \cdot \text{K}}{\lambda_{max}}$$

Inserting the observed value for λ_{max}, we find

$$T = \frac{2.90 \times 10^{-3} \text{ m} \cdot \text{K}}{5.5 \times 10^{-7} \text{ m}} = 5.3 \times 10^3 \text{ K} = \boxed{5300 \text{ K}}$$

What does it mean?
The temperature of any star can be determined from its blackbody spectrum (and value of λ_{max}). Stars with a reddish color are thus cooler than the Sun, whereas stars that have a blue color are hotter.

CONCEPT CHECK 28.3 | ⊗ Blackbody Radiation from the Body

The human ear emits blackbody radiation. At what wavelength is the intensity of blackbody radiation from the ear largest, (a) 0.93 m, (b) 3.2×10^{-5} m, (c) 9.1×10^{-6} m, or (d) zero? *Hint*: The temperature of the human body is approximately 320 K.

Particle-Wave Nature of Light

We have discussed two phenomena, the photoelectric effect and blackbody radiation, that can only be understood in terms of the particle nature of light. While light thus has some properties like those of a classical particle, it also has wave properties at the same time. Electromagnetic radiation still exhibits some of the characteristic properties of classical waves, including interference. Hence, light has both wave-like and particle-like properties. In Figure 28.2, we described a double-slit experiment with electrons and described how the electrons arrive one at a time at the screen. Precisely the same behavior is found with light. Consider a double-slit experiment performed with light at very low intensity and suppose the screen responds to the arrival of individual photons by emitting light from the spot where the photon strikes. (The properties of this *fluorescent* screen are explained in Chapter 29.) The results would then appear just as in Figure 28.2, with the full interference pattern at the far right becoming visible only after many photons have reached the screen, thus demonstrating both the particle nature of light (the arrival of individual photons) and the wave nature (interference) in the same experiment.

28.3 | WAVELIKE PROPERTIES OF CLASSICAL PARTICLES

In Section 28.1, we described in a general way the wavelike properties of electrons and introduced the concept of "particle-waves." The notion that the properties of both classical waves and classical particles are present at the same time is also called *wave–particle duality* and is essential for understanding the microscale world of electrons, atoms, and molecules. Let's now discuss what wave–particle duality means in more quantitative terms. By the early 1920s, the photon theory of light was well established, but physicists were still struggling with how to describe particles such as electrons in the quantum regime. In particular, physicists did not yet realize that electrons are capable of the interference behavior sketched in Figure 28.2. This possibility was first proposed by Louis de Broglie in 1924, who suggested that *all* classical particles have wavelike properties. At the time, this idea must have seemed crazy because the experimental evidence for interference with electrons (similar to Fig. 28.2) had not yet been discovered. That did not stop de Broglie, however, who developed his theory in analogy with the behavior of photons. We have seen that a photon has a momentum given by (Eq. 28.5)

$$p = \frac{h}{\lambda}$$

De Broglie turned this result around and suggested that if a particle has a momentum p, its wavelength is

$$\lambda = \frac{h}{p} \qquad (28.8)$$

If a particle has a wavelength, it should be able to exhibit interference just as waves do, so the test of de Broglie's hypothesis was to look for interference involving classical particles. Such an experiment is easiest if the wavelength is long, and according to Equation 28.8, a long wavelength implies a small value of momentum p. For a classical particle, $p = mv$, so a particle with a very small mass is required, and the lightest known particle (at that time) was an electron. It is thus not surprising that the first observation of wavelike behavior with a classical particle came from an experiment done with electrons by Clinton Davisson and Lester Germer (Fig. 28.8). A beam of electrons was aimed at a crystal target whose atoms formed a regular array (as is typical of many solids), acting as a series of slits for the electrons. Hence, it is really a "multislit" interference experiment similar to the diffraction gratings for light discussed in Chapter 25. Just as with a diffraction grating, the Davisson–Germer experiment exhibits interference when the wavelength of the electrons is comparable to the spacing between the slits (i.e., the spacing between atoms). In Figure 28.8, we sketch the interference of electrons that pass through a crystal. Davisson and Germer actually looked at the electrons reflected by a crystal,

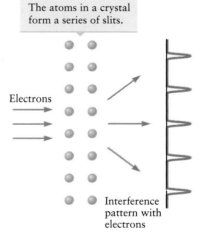

Figure 28.8 Schematic of an interference experiment with electrons, similar to the one first done by Davisson and Germer in 1927. When electrons pass through a thin crystal, the atoms in the crystal act as series of slits, producing interference and demonstrating the wave nature of electrons. Davisson and Germer's studies actually employed reflected electrons, which also form an interference pattern.

but the basic idea is the same. There is interference in both cases, but the reflection experiment is simpler to carry out.

Davisson and Germer's demonstration of interference with electrons showed conclusively that electrons have wavelike properties. Careful measurements of the interference pattern allowed them to deduce the electron wavelength, with a result in good agreement with de Broglie's theory (Eq. 28.8).

EXAMPLE 28.5 | **Wavelength of an Electron**

In their studies of interference with electrons, Davisson and Germer used electrons with a kinetic energy of about $KE = 50$ eV $= 8.0 \times 10^{-18}$ J. **(a)** What was the wavelength of these electrons? Express your answer in nanometers (1 nm $= 10^{-9}$ m). **(b)** The spacing between atoms in a typical crystal is about 0.3 nm. How does this spacing compare with the wavelength of the electrons used by Davisson and Germer?

RECOGNIZE THE PRINCIPLE

Given the kinetic energy, we can find the speed of the electrons, and from that, we can get their momentum ($p = mv$). We then use the de Broglie relation (Eq. 28.8) for the wavelength of a particle-wave to calculate the electron wavelength.

SKETCH THE PROBLEM

No sketch is necessary.

IDENTIFY THE RELATIONSHIPS

The kinetic energy of an electron of mass m and speed v is

$$KE = \tfrac{1}{2}mv^2$$

Rearranging to solve for v gives

$$v = \sqrt{\frac{2(KE)}{m}}$$

SOLVE

(a) Inserting the mass of an electron and the given value for the kinetic energy, we find

$$v = \sqrt{\frac{2(KE)}{m}} = \sqrt{\frac{2(8.0 \times 10^{-18} \text{ J})}{9.11 \times 10^{-31} \text{ kg}}} = 4.2 \times 10^6 \text{ m/s} \qquad (1)$$

The wavelength of an electron with this speed is

$$\lambda = \frac{h}{p} = \frac{h}{mv}$$

Inserting our result for v gives

$$\lambda = \frac{h}{mv} = \frac{6.63 \times 10^{-34} \text{ J} \cdot \text{s}}{(9.11 \times 10^{-31} \text{ kg})(4.2 \times 10^6 \text{ m/s})} = 1.7 \times 10^{-10} \text{ m} = \boxed{0.17 \text{ nm}}$$

(b) This wavelength is similar to the spacing between atoms in a solid (which is typically 0.3 nm). Hence, the conditions for observing interference are indeed satisfied, which is why the Davisson–Germer experiment was successful.

What does it mean?

The natural "slits" formed by atoms in a crystal (Fig. 28.8) are very useful for studying interference effects with particles such as electrons, protons, neutrons, and even atoms. Notice also that for electrons the speed found in Equation (1) is low enough that the classical (Newton's law) relations for the kinetic energy and momentum are accurate, and the relativistic results from Chapter 27 are not required.

Why Don't You See Wave Interference with Baseballs?

De Broglie's ideas were first confirmed with electrons, but he proposed that they apply to all classical particles. Equation 28.8 does indeed hold for all particles, including electrons, protons, and even baseballs, all of which exhibit wavelike properties under suitable conditions. Davisson and Germer demonstrated this with electrons, but the experiment is more difficult with other particles for the following reason. Consider a particle (such as an electron or proton) with mass m, speed v, and kinetic energy KE. Using the classical expression for the kinetic energy, we have

$$KE = \tfrac{1}{2}mv^2$$

Rearranging, we can solve for v:

$$v = \sqrt{\frac{2(KE)}{m}}$$

The momentum of the particle is then

$$p = mv = m\sqrt{\frac{2(KE)}{m}} = \sqrt{2m(KE)}$$

We can use this momentum in de Broglie's relation (Eq. 28.8) to get

$$\lambda = \frac{h}{p} = \frac{h}{\sqrt{2m(KE)}} \tag{28.9}$$

This result shows that for a fixed value of the kinetic energy, the wavelength becomes smaller as the mass of the particle is increased. The mass of a proton is about 1830 times greater than the mass of an electron, so assuming the same kinetic energy, the proton's wavelength will be shorter than that of an electron by a factor of approximately $\sqrt{1830} \approx 43$.

Will an interference experiment work with the much shorter wavelength of such protons? For a diffraction grating (Chapter 25), the condition for constructive interference is

$$d \sin \theta = n\lambda$$

where d is the slit spacing, θ is the diffraction angle, and $n = 0, 1, 2, \ldots$ is an integer. (Here we use n to denote an integer to avoid confusion with the particle mass m.) Had Davisson and Germer used protons, λ would have been much smaller (by a factor of 43); this small size could be compensated for by increasing n by the same factor, which corresponds to the 43rd maximum in an interference pattern such as Figure 28.2. In other words, with protons there would be 43 diffraction maxima squeezed in the same angle as one of the diffraction peaks found with electrons. For this reason, it would be much more difficult, although not completely impossible, to observe interference with protons.

In general, it becomes more and more difficult to observe interference as the mass of the particle is increased because if all else is kept fixed, a larger mass leads to a shorter wavelength (Eq. 28.9). In principle, one could observe interference with baseballs, but this experiment has not yet been done. (See Example 28.6.) This should remind you of the difference between geometrical optics and wave optics. Geometrical optics (Chapter 24) applies when the wavelength is very short, and in this regime light moves in straight lines just like classical particles. On the other hand, wave optics (Chapter 25) applies when the wavelength is relatively long and comparable to the size of any slits or openings, and in that case interference effects are important.

CONCEPT CHECK 28.4 | The de Broglie Wavelength
and the Mass of a Particle

An electron and a proton have the same de Broglie wavelength. Which of the following statements is true? *Note:* The mass of a proton is about 1800 times larger than the mass of an electron.

(a) The electron and proton have the same kinetic energy.
(b) The electron's kinetic energy must be larger than the proton's kinetic energy.
(c) The electron's kinetic energy must be smaller than the proton's kinetic energy.

EXAMPLE 28.6 De Broglie's Theory and the Wavelength of a Baseball

What is the wavelength for a baseball traveling at a speed of 100 mi/h (approximately 45 m/s)? This speed is attained by some major league pitchers. The mass of a baseball is approximately 0.14 kg.

RECOGNIZE THE PRINCIPLE

De Broglie's theory connects the momentum and wavelength of a particle-wave. The momentum of a baseball is $p = mv$, so given the mass and speed we can find its momentum. The wavelength of the baseball is then given by the de Broglie relation.

SKETCH THE PROBLEM

Figure 28.9 describes the problem. While a baseball is a particle-wave, its wavelength is *very* short.

IDENTIFY THE RELATIONSHIPS AND SOLVE

Combining de Broglie's relation for the wavelength of a particle-wave (Eq. 28.8) with $p = mv$ leads to

$$\lambda = \frac{h}{p} = \frac{h}{mv}$$

Using the given values of m and v for the baseball, we find

$$\lambda = \frac{h}{mv} = \frac{6.63 \times 10^{-34}\,\text{J}\cdot\text{s}}{(0.14\,\text{kg})(45\,\text{m/s})} = \boxed{1.0 \times 10^{-34}\,\text{m}}$$

What does it mean?

This wavelength is *extremely* short. For comparison, the diameter of a proton is about 10^{-15} m, so the wavelength of a baseball is smaller than the size of a proton by the enormous factor of 10^{19}. Extremely small values of the wavelength are found for other macroscopic objects such as hockey pucks, planets, and people. For this reason, the wave properties of macroscopic objects are almost always completely negligible and the particle description of Newton's mechanics works with very high accuracy. The wavelike behavior of a baseball and of most other macroscopic objects will probably never be observed.

28.4 | ELECTRON SPIN

We have seen that electrons behave in some ways like classical particles and in other ways like classical waves. Are there any "new" or "extra" properties that show up in the quantum regime? That is, do electrons have any properties not found in either classical particles or waves? The answer is yes: electrons have another quantum property that involves their magnetic behavior. In Section 20.8 (page 667), we mentioned that an electron has a magnetic moment, a property associated with the phenomenon of *electron spin*. In a classical model, the electron's magnetic moment can be understood by picturing the electron as a spinning ball of charge (Fig. 28.10A). This spinning ball of charge acts as a collection of current loops, producing a magnetic field and thus acting like a small bar magnet as shown in Figure 28.10B.

Figure 28.9 Example 28.6. A baseball is a particle-wave, but its wavelength is *very* short as given by the de Broglie relation.

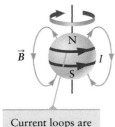

Figure 28.10 **A** Classical picture of the electron's spin. **B** A spinning ball of charge (the electron) acts like a collection of current loops, producing a magnetic field.

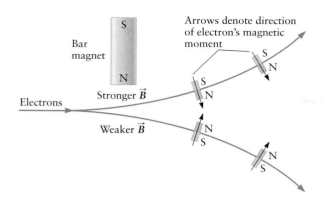

Bar magnet

Arrows denote direction of electron's magnetic moment

Stronger $\vec{B}$

Weaker $\vec{B}$

Electrons

Figure 28.11 Electrons have an intrinsic magnetism called the "spin magnetic moment." When a beam of electrons passes near one end of a bar magnet, it is found that the spin magnetic moment is either "up" or "down," so the electrons are either attracted to or repelled from the bar magnet. According to quantum theory, the electron's magnetic moment can only point in these two directions, up or down, but *not* "sideways" due to the quantization of the electron's spin.

Since an electron behaves as a small bar magnet, it is attracted to or repelled from the poles of other magnets. The magnetic properties of an electron were revealed in an experiment first performed by Otto Stern and Walther Gerlach. A simplified version of their experiment is sketched in Figure 28.11, showing a beam of electrons passing near one pole of a large (macrosize) magnet. (Stern and Gerlach used a slightly different arrangement of macrosize magnets, but the result is the same.) The electrons are attracted to or repelled from the large magnet according to the direction of the electron's north and south "poles." Two examples are shown in Figure 28.11; one electron is attracted to the large magnet's north pole and is thus deflected upward, and the other electron is repelled and deflected downward. The surprise in this experiment is that the outgoing electrons form just two beams. If an electron behaves like a bar magnet, a line connecting the north and south poles defines the axis of the magnet. These axes are indicated by arrows in Figure 28.11, with the arrowheads at the north ends of the magnets. These arrows also denote the direction of the magnetic moment of the electron, which is similar to the magnetic moment of a bar magnet or current loop as discussed in Chapter 20.

Classical theory predicts that the magnetic moment of an electron may point in any direction (up, down, sideways, or at any angle), and if that is true, the electrons in a Stern–Gerlach experiment should be deflected not just as two beams, but over a range of angles. However, only the two possibilities in Figure 28.11 actually occur; experiments always give just two outgoing beams of electrons. Hence two and only two orientations for the electron magnetic moment with respect to the direction of the applied magnetic field are possible. In words, the electron magnetic moment is *quantized*, with only two possible values.

The Stern–Gerlach experiment in Figure 28.11 shows that the quantization of the electron's magnetic moment applies to both the ***direction*** and ***magnitude*** of its magnetic moment. Hence, all electrons under all circumstances act as identical bar magnets.

Quantization of Electron Spin

To understand the implications of the Stern–Gerlach experiment, we consider again the picture in Figure 28.10 showing how an electron can act as a small bar magnet. Classically, we can picture an electron as a spinning ball of electric charge and acting as a collection of current loops. The electron in Figure 28.10A spins clockwise as viewed from the top of the figure, so the circulating charge acts as a counterclockwise current loop as viewed from above (because the electron is negatively charged). These current loops then produce a magnetic field in the same way that the current loops in Section 20.1 produce a magnetic field, and the result is called the ***spin magnetic moment*** of the electron. We also say that the electron has ***spin angular momentum***. This angular momentum is similar to the angular momentum of a macroscopic spinning ball.

This classical picture explains how the electron can have a magnetic moment, but it does *not* explain the Stern–Gerlach experiment. The rotation axis of a classical particle such as a spinning baseball can point in any direction, so the classical picture in Figure 28.10 suggests that an electron's magnetic moment can be oriented in any direction. This classical picture also predicts that the electron's magnetic moment can be large, small, or even zero, depending on the angular speed of the electron since a classical ball can spin fast, slow, or not at all. However, the Stern–Gerlach result indicates that there can be only two values for the electron magnetic moment! Physicists now understand that this property is a result of quantum theory: the electron's spin axis and the magnitude of its magnetic moment are both quantized, with only two possible values. These two states are called spin "up" and spin "down." In the Stern–Gerlach experiment, one of the outgoing beams of electrons in Figure 28.11 corresponds to electrons with spin up, whereas electrons in the other beam have spin down.

Other quantum particles, such as protons and neutrons, also have a spin angular momentum and a resulting magnetic moment. In all cases, they are quantized, although the details can be more complicated than for an electron.

The quantization of spin angular momentum and the spin magnetic moment is a purely quantum effect, and the simple classical model in Figure 28.10 gives only an intuitive picture. There is no classical explanation for why spin is quantized (i.e., why there are only two outgoing beams in the Stern–Gerlach experiment). All we can say is that is the way electrons work. Recognizing the existence of electron spin is essential for understanding the structure of atoms and the organization of the periodic table of the elements (Chapter 29).

28.5 | THE MEANING OF THE WAVE FUNCTION

Newton's laws of mechanics are based on the idea that the motion of a particle can be described completely in terms of the particle's position, velocity, and acceleration. After all our work with Newton's mechanics, it is easy to take this notion for granted, but it is only an assumption about how the laws of physics are structured. This assumption works well in the classical world as proven by the outstanding success of Newton's laws in describing the mechanics of macroscale objects, but it fails in the quantum world of particle-waves. In a quantum description, the motion of a particle-wave is described by its *wave function*. The full mathematics of the wave function is beyond the scope of this book, but we can describe a few of its properties.

The wave function can be calculated using an equation developed by Erwin Schrödinger. He was one of the inventors of quantum theory and in 1933 received the Nobel Prize in Physics for this work. Schrödinger's equation plays a role similar to Newton's laws of motion. In quantum theory, one uses the Schrödinger equation to find the wave function and how it varies with time, in much the same way that one uses Newton's laws of mechanics to find the position, velocity, and acceleration of a particle.

In many situations, the solutions of the Schrödinger equation are mathematically similar to standing waves. As an example, consider an electron confined to a particular region of space as sketched in Figure 28.12A. In classical terms, you can think of this region as a very deep canyon or box from which the electron cannot escape. A classical particle would travel back and forth inside the box, bouncing from the walls. The wave function for a particle-wave (such as an electron) in such a box is described by standing waves, similar to the standing waves on a string. Figure 28.12B shows two possible wave function solutions corresponding to electrons with different kinetic energies. The wavelengths of these standing waves are different, as predicted by the de Broglie picture, since the wavelength of an electron depends on its kinetic energy (Eq. 28.9). In a classical wave-on-a-string picture, these two solutions correspond to two standing waves with different wavelengths. (See Fig. 12.23.)

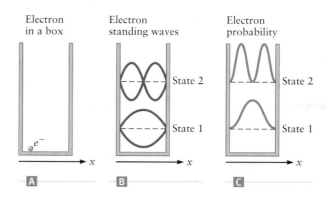

Electron in a box   Electron standing waves   Electron probability

Figure 28.12 ◰ An electron trapped in a box. ◳ The electron wave function forms a standing particle-wave similar to the standing waves on a string fastened to the walls of the box. The electron wavelength must therefore "fit" into the box as it would for a standing wave. ◱ Quantum mechanical probabilities of finding the electron at different locations in the box, corresponding to each of the two wave functions in part B.

After finding the wave function for a particular situation such as for the electron in Figure 28.12B, one can calculate the position and velocity of the electron. However, the results do not give a simple single value for x. Instead, the wave function allows the calculation of the *probability* of finding the electron at different locations in space. This probability for each of the wave functions in Figure 28.12B is shown in Figure 28.12C for our electron in a box. The probability of finding the electron at certain values of x is large in some regions and small in others, corresponding to the antinodes and nodes of the standing wave. The probability distribution (how the probability varies with position in the box) is different for each wave function.

CONCEPT CHECK 28.5 | Energy of an Electron in a Box

Figure 28.12B shows the wave functions for two different electron standing waves. Which of these electrons has the higher kinetic energy?

The Heisenberg Uncertainty Principle

In classical physics, we are used to calculating or measuring the position and velocity of a particle with great precision, but for a particle-wave, quantum effects place fundamental limits on this precision. For example, the electron in Figure 28.12 is described by a standing wave. The associated probability of finding the electron at different places in the box can be calculated from its wave function, yet what is the position of this electron? In a quantum sense, the standing waves in Figure 28.12B *are* the electron, so there is an inherent uncertainty in its position. There is some probability for finding the electron at virtually any spot in the box, and the uncertainty Δx in the electron's position is approximately the size of the box. This uncertainty is due to the wave nature of the electron.

We can gain insight into the uncertainty Δx from the experiment described in Figure 28.13 in which electrons are incident from the left on a narrow slit. This is very similar to the case of single-slit diffraction we studied with light waves in Section 25.6, and the electron wave is diffracted as it passes through the slit. If we carry out this experiment with many electrons, we can observe the associated interference pattern by measuring where the electrons strike the screen to the right of the slit. In quantum terms, the interference pattern gives a measure of how the wave function of the electron is distributed throughout space after it passes through the slit. The electrons coming in from the left in Figure 28.13 are all traveling parallel to the y axis, so their initial momentum is purely along the y direction. However, after being diffracted by the slit, an electron is likely to strike the screen some distance away from the y axis. Hence, a diffracted electron acquires a nonzero momentum along x.

Figure 28.13 also shows how the width Δx of the slit affects the interference pattern. The wide slit in Figure 28.13A produces a narrow electron distribution along the screen (the x direction in the figure). The narrow slit in Figure 28.13B produces a broader distribution as the wave function is more spread out along x.

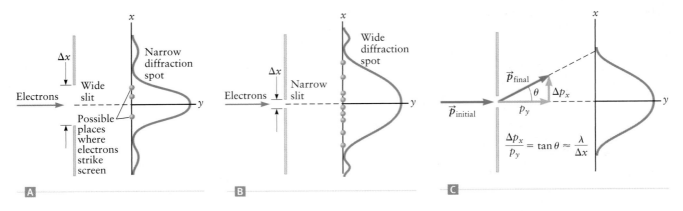

Figure 28.13 Single-slit diffraction with electrons. **A** If an electron passes through a wide slit, the diffraction spot on the screen is narrow. **B** If the slit is made narrower, the diffraction spot becomes wider. **C** The spreading of an electron wave as it passes through a slit gives a way to relate the uncertainty Δx in the electron's position to the uncertainty Δp in its momentum.

Relating the Uncertainties in Position (Δx) and Momentum (Δp)

The electrons pass through the slit, so the position of the electron at the moment it passes through the slit is known with an uncertainty Δx equal to the width of the slit. Since an electron may strike the screen over a range of angles defined approximately by the central diffraction maximum, the outgoing electrons have a spread in their momentum along x and we can say that there is some uncertainty Δp_x in the x component of the momentum. Most of the electrons arrive somewhere within the width of the central diffraction maximum, so the final momentum along the x direction is uncertain by an amount approximately equal to the x component of the momentum of an electron at the edge of the diffraction spot. The uncertainty in momentum Δp_x is related to the uncertainty in position Δx. A wide slit (Fig. 28.13A) has a large value of Δx and a smaller value of Δp_x (because the diffraction "spot" is relatively narrow). On the other hand, a narrow slit (Fig. 28.13B) has a small uncertainty in position Δx but a larger uncertainty in momentum Δp_x.

We can derive a quantitative relationship between Δx and Δp_x from Figure 28.13C. The diffraction angle θ is contained in a right triangle with sides p_y and Δp_x, where p_y is the momentum along y. We have

$$\frac{\Delta p_x}{p_y} = \tan \theta$$

If the angle is small, we can use the approximation $\tan \theta \approx \theta$, resulting in

$$\frac{\Delta p_x}{p_y} = \theta \tag{28.10}$$

The analysis of single-slit diffraction in Chapter 25 (Eq. 25.23) showed that for light of wavelength λ passing through a slit of width w, the first minimum in the diffracted intensity occurs at an angle

$$\theta = \frac{\lambda}{w}$$

This angle θ also describes the spread of the electron diffraction spot in Figure 28.13C, where the slit width w is equal to the uncertainty in the electron's position Δx. In the electron case, we have

$$\theta = \frac{\lambda}{\Delta x} \tag{28.11}$$

Setting the results in Equations 28.10 and 28.11 equal leads to

$$\frac{\Delta p_x}{p_y} = \frac{\lambda}{\Delta x}$$

or

$$\Delta x \, \Delta p_x = \lambda p_y \tag{28.12}$$

If we assume the uncertainty in the momentum Δp_x is small, then p_y is approximately equal to the total momentum, which we can denote by simply p. We will also follow standard notation and drop the subscript on Δp_x, referring to it as just the spread (i.e., uncertainty) in the electron's momentum. According to the de Broglie relation (Eq. 28.8), $p = h/\lambda$. Inserting that into Equation 28.12 gives

$$\Delta x \, \Delta p = \lambda p = \lambda \frac{h}{\lambda} = h$$

Thus, the uncertainty Δx in the electron's position is connected with the uncertainty Δp in its momentum in the x direction. Our analysis was only approximate; a more careful treatment gives nearly the same result,

$$\Delta x \, \Delta p = \frac{h}{4\pi} \tag{28.13}$$

The uncertainties Δx and Δp in Equation 28.13 are the absolute limits set by quantum theory. In many situations, there are additional contributions to these uncertainties resulting from experimental or measurement error. Hence, Equation 28.13 gives a lower limit on the product of Δx and Δp. We can allow for this lower limit by writing

$$\Delta x \, \Delta p \geq \frac{h}{4\pi} \tag{28.14}$$

Heisenberg uncertainty relation for position and momentum

This is called the **Heisenberg uncertainty principle**, in this case expressed as a relation between position and momentum. We derived this relation between Δx and Δp for the special case in Figure 28.13, but it holds for *any* quantum situation and for any particle-wave (not just electrons).

Position, Momentum, and Energy of a Particle: How Accurately Can We Know Them?

In classical physics, we often wish to calculate the position and momentum of a particle, as when finding the trajectory of a projectile using Newton's laws. Assuming there are no mistakes in our mathematics, we can find x and p to any desired *classical* accuracy. The Heisenberg uncertainty principle, however, dictates that in the quantum regime the uncertainties in x and p are connected. Under the very best of circumstances, the product of Δx and Δp is a constant, proportional to Planck's constant h. So, if you measure a particle-wave's position with great accuracy (small Δx), you must accept a large uncertainty Δp in its momentum. On the other hand, if you know the momentum very accurately (small Δp), you must accept a large position uncertainty Δx. You cannot make both uncertainties small at the same time. This key point is shown in Figure 28.13, where a wide slit (large Δx) leads to a small diffraction width (Δp) as in Figure 28.13A, whereas a narrow slit (small Δx) leads to a larger uncertainty Δp and a wider diffraction width as in Figure 28.13B.

The same ideas that lead to Equation 28.14 can also be used to derive a relation between the uncertainties in the energy ΔE of a particle and the time interval Δt over which this energy is measured or generated. The Heisenberg energy–time uncertainty principle is

$$\Delta E \, \Delta t \geq \frac{h}{4\pi} \tag{28.15}$$

Heisenberg uncertainty relation for energy and time

If the energy of a particle or system is measured over a time period Δt, Equation 28.15 leads to a minimum uncertainty in the measured energy. This minimum uncertainty is negligibly small for a macroscale object, but it can be important in atomic and nuclear reactions.

Philosophical and Practical Implications of the Uncertainty Principle

The Heisenberg uncertainty principle has profound implications for our physical description of the universe. According to Newton's mechanics, it is possible (if not always easy) to know the position and momentum of a particle with any desired precision. However, quantum theory and Heisenberg's uncertainty relation in Equation 28.14 mean that there is always a trade-off between the uncertainties (i.e., precision) in x and p. It is not possible, *even in principle*, to have perfect knowledge of both x and p, which suggests that there is always some inherent "uncertainty" in our knowledge of the physical universe. The best we can do is compute the probability for finding a particle (e.g., an electron) at a particular position or the probability that a particle will have a certain momentum. Quantum theory thus says that the world is inherently *unpredictable*, in sharp contrast to the perfect predictability of classical physics.

We should also ask what effect the uncertainty principle has on the use of Newton's mechanics. The value of Planck's constant is very small ($h = 6.63 \times 10^{-34}$ J · s). For any macroscale object such as a baseball, a planet, or even an amoeba, the uncertainties in the real measurement will always be much larger than the inherent uncertainties due to the Heisenberg uncertainty relation. (For a numerical example, see Example 28.7.) As a general rule, the Heisenberg uncertainty relations only become important on the scale of electrons, atoms, molecules, and other microscale objects.

| EXAMPLE 28.7 | Applying the Heisenberg Uncertainty Principle to a Brick |

Consider a typical macroscale object such as a brick with $m = 1.0$ kg sitting on a table. A very careful measurement of the brick's location has an uncertainty of $\Delta x = 0.1$ nm (approximately the diameter of an atom). What is the corresponding uncertainty in the speed of the brick?

RECOGNIZE THE PRINCIPLE

This example involves a direct application of the Heisenberg uncertainty relation between the uncertainty in position and momentum. A given uncertainty Δx in the brick's position leads to a minimum uncertainty Δp in the brick's momentum. Since $p = mv$, there is a corresponding uncertainty in the brick's speed.

SKETCH THE PROBLEM

Figure 28.14 shows the problem.

Figure 28.14 Example 28.7. Applying the Heisenberg uncertainty principle to a brick.

IDENTIFY THE RELATIONSHIPS

According to the Heisenberg uncertainty principle (Eq. 28.14),

$$\Delta x \, \Delta p \geq \frac{h}{4\pi}$$

Because we know Δx, we can solve for Δp. The momentum is $p = mv$ and we are given the brick's mass, so we can then find the uncertainty Δv in its speed.

SOLVE

Rearranging Equation 28.14 to solve for Δp gives

$$\Delta p \geq \frac{h}{4\pi(\Delta x)}$$

An uncertainty in the momentum corresponds to an uncertainty in the speed according to

$$\Delta p = m\, \Delta v \geq \frac{h}{4\pi(\Delta x)}$$

Solving for Δv and inserting the given values of m and Δx gives

$$\Delta v \geq \frac{h}{4\pi m\, \Delta x} = \frac{6.63 \times 10^{-34}\,\text{J} \cdot \text{s}}{4\pi(1.0\,\text{kg})(0.1 \times 10^{-9}\,\text{m})}$$

$$\Delta v \geq \boxed{5 \times 10^{-25}\,\text{m/s}}$$

What does it mean?

This uncertainty is *extremely* small. For all practical purposes, this quantum limit on the uncertainty is so small that it could never be measured. For bricks and other macroscale objects, the limits set by the Heisenberg uncertainty principle are unobservable, but they can be important in atomic and nuclear physics experiments, as we explore in Example 28.8.

EXAMPLE 28.8 | Applying the Uncertainty Principle to a Hydrogen Atom

A simple picture of the hydrogen atom has an electron moving around a proton as a sort of "electron cloud," with the density of the cloud proportional to the electron probability (Fig. 28.15). We'll see how this cloud picture is related to the electron's wave function in Chapter 29. Even without detailed knowledge of the wave function, however, we can still apply the Heisenberg uncertainty principle (Eq. 28.14), taking the uncertainty Δx in the electron's position equal to the diameter of the cloud. **(a)** If the diameter of a hydrogen atom is 1.0×10^{-10} m ($= 0.10$ nm), what is the minimum uncertainty Δp in the electron's momentum? **(b)** The momentum of the electron must be at least comparable to the uncertainty Δp, so we can use Δp as an estimate for the total momentum ($p \approx \Delta p$). With this approximation, use your result for Δp to find the electron's kinetic energy and compare it with the known ionization energy of a hydrogen atom, 13.6 eV.

RECOGNIZE THE PRINCIPLE

Part (a) is a direct application of the Heisenberg uncertainty relation between the uncertainty in position and momentum, with Δx equal to the given diameter of a hydrogen atom. For part (b), we take p to be equal to the uncertainty Δp in the momentum, and from that we can get the kinetic energy of the electron.

SKETCH THE PROBLEM

Figure 28.15 describes the problem.

IDENTIFY THE RELATIONSHIPS AND SOLVE

(a) Heisenberg's uncertainty relation (Eq. 28.14) gives

$$\Delta x\, \Delta p \geq \frac{h}{4\pi}$$

Solving for Δp, we find

$$\Delta p \geq \frac{h}{4\pi\, \Delta x} = \frac{6.63 \times 10^{-34}\,\text{J} \cdot \text{s}}{4\pi(1.0 \times 10^{-10}\,\text{m})} = \boxed{5.3 \times 10^{-25}\,\text{kg} \cdot \text{m/s}}$$

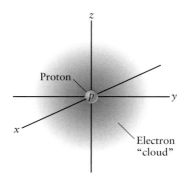

Figure 28.15 Example 28.8. Simple model of a hydrogen atom. The "cloud" represents the probability for finding the electron as a function of position near the proton.

(b) The momentum of an electron is $p = mv$, while the kinetic energy is $KE = \frac{1}{2}mv^2$. Combining these two relations gives

$$KE = \frac{p^2}{2m}$$

Using $p \approx \Delta p$ from part (a) and the known mass of an electron, we find

$$KE = \frac{p^2}{2m} = \frac{(5.3 \times 10^{-25} \text{ kg} \cdot \text{m/s})^2}{2(9.11 \times 10^{-31} \text{ kg})} = \boxed{2.9 \times 10^{-19} \text{ J}}$$

Converting this energy to units of electron-volts (eV) gives

$$KE = (2.9 \times 10^{-19} \text{ J})\left(\frac{1 \text{ eV}}{1.60 \times 10^{-19} \text{ J}}\right) = \boxed{1.8 \text{ eV}}$$

What does it mean?
This kinetic energy is about a factor of three smaller than the ionization energy of the hydrogen atom (13.6 eV). The fact that these energies are the same order of magnitude suggests that an accurate theory of the hydrogen atom must be based on quantum theory (the main topic of Chapter 29).

28.6 | TUNNELING

According to classical physics, the electron in Figure 28.12 is trapped in the box and cannot escape. However, a quantum effect called *tunneling* allows such an electron to escape under certain circumstances. Figure 28.16A shows the wave function for this electron, with the region near the wall of the box shown on an expanded scale. Quantum theory allows the electron's wave function to penetrate a short distance *into* the wall. We say that the wave function extends a short distance into the *classically forbidden* region because according to Newton's mechanics the electron must stay completely inside the box and cannot go into the wall. In most cases, the quantum penetration distance is short and of little consequence, but if two boxes are very close together so that the wall between them is very thin, the wave function can extend from the inside of one box and through the wall to the box on the other side (Fig. 28.16B). Since the wave function extends through the wall, the electron has some probability for passing through the wall. If bullets were quantum objects, it would be like a bullet passing through a wall without leaving a hole!

Using Quantum Mechanics to Make a New Kind of Microscope

Tunneling is important in certain nuclear decay processes (Chapter 30) and in the motion of certain molecules. It is also used in the operation of a *scanning tunnel-*

Figure 28.16 ◙ An electron wave function (standing wave) such as for the electron in a box in Figure 28.12B. The electron wave function actually penetrates a short distance into the walls of the box, although the wave function is very small in this region. ◙ If the walls of two boxes are brought close together, the wave functions from the two boxes can overlap, allowing an electron to tunnel from one box to the other.

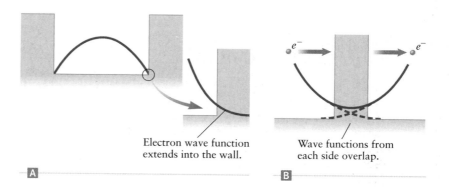

Electron wave function extends into the wall.

Wave functions from each side overlap.

A   **B**

ing microscope (STM). The basic design of an STM is shown in Figure 28.17A. A very sharp metal tip is positioned near a conducting surface, and an electric potential difference is applied between the tip and the surface. If the separation is large, the space (usually a vacuum) between the tip and the surface acts as a barrier for electron flow. This barrier is similar to the wall in Figure 28.16 since it prevents electrons from leaving the metal (i.e., the box). However, if the tip is brought very close to the surface, electrons may tunnel between them, producing a tunneling current. By measuring this current as the tip is scanned over the surface, it is possible to construct an image of how atoms are arranged on the surface. As the tip is scanned, it moves from regions where it is directly over an atom to other regions where the nearest atoms are much farther away as in Figure 28.17B. The tunneling current is largest when the tip is closest to an atom because the wave function overlap needed for tunneling is largest in this case. An STM image is based on variations in the size of the tunneling current as the tip is scanned across the surface, with high image intensity when the current is large and vice versa. A typical STM image is shown in Figure 28.18, with each "bump" corresponding to an individual iron atom on the surface of a piece of copper. The electric field between the tip and the surface atoms can also be used to push atoms around on the surface. In this experiment the STM tip was used to arrange iron atoms that were initially scattered around the surface (upper left) into a circle (lower right).

Tunneling plays a dual role in the operation of an STM. First, the detector current is produced by tunneling, so without tunneling there would be no image. Second, tunneling is essential for obtaining such high resolution. The STM tip is specially made to be very sharp, but even the best tips are rounded at the end. This rounding typically corresponds to a radius of curvature of 10 nm (10^{-8} m) or so, which is somewhat larger than the size of an atom. (An atom has a diameter of about 0.1 nm.) The effect of tip rounding on tunneling is shown in Figure 28.17C. Electrons can tunnel across the gap along many different paths, with the tunneling probability for each path depending on the size of the wave function in the tunneling gap. The wave function in the gap region falls off rapidly with distance (see Fig. 28.16B), so the shortest tunneling paths have the largest wave functions near the tip. Therefore, the vast majority of electrons that tunnel follow the shortest path, reaching a very small area on the surface, an area much smaller than the tip radius. The STM can then form images of individual atoms, even though the tip is larger than they are.

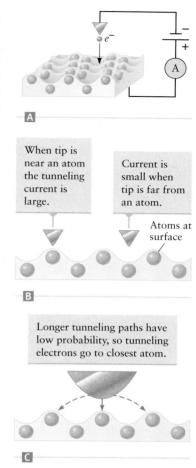

Figure 28.17 **A** In a scanning tunneling microscope (STM), a very sharp tip is brought very close to a surface, detecting the probability for an electron to tunnel between the tip and the surface by measuring the tunneling current. **B** The tunneling probability is largest for the shortest tunneling paths, so there is a large tunneling current when the tip is directly over an atom. **C** The STM is most sensitive to atoms that are directly below the tip.

Figure 28.18 STM images of iron atoms on the surface of copper. The electric field from an STM tip can be used to push atoms around on the surface of a metal. In the lower right photo, the iron atoms are arranged to form a circle.

28.7 | ⊗ DETECTION OF PHOTONS BY THE EYE

Human vision depends on the refractive properties of the eye as discussed in Chapters 24 and 26, treating light with Maxwell's wave theory, but a complete understanding of human vision also depends on the particle theory of light. In particular, the wave theory of light cannot explain color vision. Light is detected at the back surface of the eye, in a region called the retina. (Review Fig. 24.46.) The retina contains two kinds of light-sensitive cells, called *rods* and *cones*. When these cells absorb light, they generate an electrical signal that travels through the optic nerve to the brain. The rod cells are more sensitive to low light intensities and are used predominantly at night, while the cone cells are responsible for color vision.

Can the Eye Detect Single Photons?

Detection of low-intensity light is the job of the rod cells. When light enters your eye, only about 10% of it actually reaches the retina. The other 90% is reflected or absorbed by the cornea and other parts of the eye. Hence, for every ten photons that enter the eye, only one (on average) strikes the retina. Experiments show that the absorption of only a single photon by a rod cell causes the cell to generate a small electrical signal. The signal of an individual rod cell is not sent directly to the brain, however. Instead, the eye combines the signals from many rod cells before passing that combination signal along the optic nerve. Measurements show that for the combination signal to register at the brain, approximately five photons must be absorbed by rod cells within a period of about 0.1 s. So, although an individual rod cell is sensitive to single photons of visible light, your eye must receive approximately 50 photons within about 0.1 s for the brain to know that light has actually arrived at the eye.

Color Vision

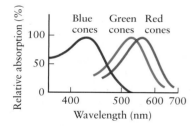

Figure 28.19 Pigment molecules inside the three types of cone cells absorb light preferentially at certain wavelengths (i.e., at certain frequencies).

The retina contains three types of cone cells, which respond to light of different colors (Fig. 28.19). The brain deduces the color of light by combining the signals from all three types of cones. For example, blue light triggers a large response from the blue cone cells and much smaller responses from the green and the red cone cells. A large signal from the blue cones together with small signals from the green and red cones tell the brain that the incident light lies in the blue part of the spectrum.[2]

The key property of the cones is that each type of cone cell is most sensitive to light with a particular color—that is, frequency—independent of the light intensity. For example, the green cone cells are most sensitive to green light and less sensitive to both red light (a lower frequency) and blue light (a higher frequency), even if the red or the blue light is very intense.

The correct explanation of color vision relies on two aspects of quantum theory. First, light arrives at the eye as photons whose energy depends on the frequency of the light. According to Equation 28.1, a photon of blue light carries more energy than a photon of green light, which in turn carries more energy than a photon of red light. When an individual photon is absorbed by a cone cell, the energy of the photon is taken up by a pigment molecule within the cell. Second, the energy of a pigment molecule is quantized, just as the energy of a photon is quantized. Figure 28.20A shows a simplified sketch of the quantized energies of blue, green, and red pigments. When a photon of the correct frequency is absorbed by a green pigment molecule, the molecule ends up in the upper energy level in Figure 28.20B, initiating a series of chemical reactions that eventually send an electrical signal to the brain. This photon absorption is possible because the difference in energy levels in the

---

[2]The precise way the brain combines these three signals is surprisingly complex. Some of this complexity was first revealed through clever experiments by Edwin Land, who also invented polarizing film (Chapter 23).

Figure 28.20 ◻A Pigment molecules have quantized energies. ◻B The molecules preferentially absorb photons whose energy matches the difference between two energy levels. For this reason, a green pigment molecule readily absorbs photons of green light, but not photons of red light or blue light. ◻C More realistic energy-level diagram. Each pigment molecule responds to a range of photon energies. These ranges overlap somewhat and give the broad absorption curves in Figure 28.19.

green pigment matches the energy of the photon. The quantized energy levels of a pigment molecule thus allow only photons with a certain energy to be absorbed.

According to the simplified energy-level diagram in Figure 28.20A, a pigment molecule can absorb a photon only if the photon energy precisely matches the pigment energy level. A more realistic energy diagram is shown in Figure 28.20C. The pigment molecule sits among other molecules in the retina, and interactions between the pigment and these molecules enable the absorption of photons over a range of energies, not just at the single value of the energy indicated in Figure 28.20A. The pigment molecule still responds most strongly to frequencies near the center of its preferred range, but it also gives a nonzero (although weaker) response to photons with both higher and lower energies, corresponding to nearby colors. The other pigments are sensitive to different energy ranges, so, by combining the response of all three types of pigments, the brain is still able to determine the color of the incident light. Hence, quantum theory and the existence of quantized energies for both photons and pigment molecules lead to color vision.

28.8 | THE NATURE OF QUANTA: A FEW PUZZLES

Quantum theory forces us to rethink our classical view of the world. Newton's laws cannot be applied at the atomic scale, and we must learn to deal with particle-waves. This rethinking leads to many new concepts, but a few of the central foundations of classical physics remain. The principles of conservation of energy, momentum, and charge are believed to hold true under *all* circumstances, but we must now allow for the existence of quanta. According to our understanding of the photon (Eq. 28.1) and the de Broglie theory (Eq. 28.8), its energy and momentum come in discrete quantized units. Electric charge also comes in quantized units, a result we have accepted throughout our study of electricity and magnetism. Indeed, the notion that charge is quantized in units of $\pm e$ is so familiar that it is easy to forget that the electron's existence was not known to Ampère, Faraday, Maxwell, and the other physicists who developed the theory of electricity and magnetism.

Does the discovery of quantum theory and the true nature of electrons and photons as particle-waves mean that we have now arrived at the "final" theory of nature? The answer is a definite *no*. Many puzzles remain, including (but not limited to) the following.

1. The relation between gravity and quantum theory is a major unsolved problem. No one really knows how Planck's constant enters the theory of gravitation or what a quantum theory of gravity looks like.

2. We know that electric charge is quantized in units of $\pm e$, but why are there two kinds of charge, positive and negative? In addition, why do the positive and negative charges come in the same quantized units? Physicists currently have no answer.

3. What new things happen in the regime where the micro- and macroworlds meet? According to quantum theory, all objects—even baseballs and bricks—behave as particle-waves. The quantum behavior of these macroscale objects is for all practical purposes unobservable, however, as we have seen in Examples 28.6 and 28.7. What about objects that are much smaller than baseballs and bricks? The quantum behavior of objects that are intermediate in size between the micro- and macroscales is now being intensively studied. There are also some interesting and unanswered questions about how quantum theory and the uncertainty principle apply to living things. For example, does the unpredictability associated with the Heisenberg uncertainty principle have any connection with consciousness and human free will? Some physicists think so, but the answer to this question is far from settled.

SUMMARY | Chapter 28

KEY CONCEPTS AND PRINCIPLES

Photons

Light energy is carried in quantized parcels called *photons*. For light with frequency f, the energy of a photon is

$$E_{\text{photon}} = hf \qquad \textbf{(28.1)} \text{ (page 958)}$$

The quantity h is *Planck's constant*, with the value

$$h = 6.626 \times 10^{-34} \text{ J} \cdot \text{s} \qquad \textbf{(28.2)} \text{ (page 958)}$$

Wave–particle duality

Electrons and other particles have a dual nature. They behave in some ways like classical particles and in other ways like classical waves. We call them *particle-waves*. The de Broglie wavelength of a particle-wave with momentum p is

$$\lambda = \frac{h}{p} \qquad \textbf{(28.8)} \text{ (page 965)}$$

This relation applies to electrons and all other particle-waves.

Electron spin

Electrons have *spin angular momentum* and behave as small bar magnets, with small magnetic moments. The direction of an electron magnetic moment is *quantized*; it can only point "up" or "down" (i.e., parallel or antiparallel to a particular direction). The magnitude of this magnetic moment is also quantized.

(Continued)

Wave functions

Schrödinger's equation is used to calculate the behavior of objects such as electrons in the quantum regime. The solution of Schrödinger's equation gives the **wave function** of an object. The probability of finding the object at different points in space can be calculated from the wave function.

Heisenberg uncertainty principle

The *Heisenberg uncertainty relations* give fundamental lower limits on the accuracy with which the position, momentum, and energy of an object can be determined, with

$$\Delta x \, \Delta p \geq \frac{h}{4\pi} \qquad \text{(28.14) (page 973)}$$

and

$$\Delta E \, \Delta t \geq \frac{h}{4\pi} \qquad \text{(28.15) (page 973)}$$

APPLICATIONS

Photoelectric effect

The *photoelectric effect* is a process in which light is absorbed by a metal. An electron is ejected from the metal when the photon energy is equal to or greater than the work function of the metal. This experiment can only be explained by the photon theory of light.

Tunneling

Tunneling is a process in which an electron (or any other type of particle-wave) passes through a barrier such as a wall that is classically impenetrable. Tunneling cannot be explained by Newton's mechanics, but is accounted for by quantum theory.

QUESTIONS

SSM = answer in Student Companion & Problem-Solving Guide   ⊗ = life science application

1. Estimate the de Broglie wavelength of a car that (a) has a speed of 100 mi/h, (b) has a speed of 10,000 mi/h, and (c) is at rest.

2. You have two lightbulbs: one gives off green light with a very low intensity, and the other emits red light with a very high intensity. Which one emits photons with a higher energy? If these two lightbulbs emit the same intensity, which one emits more photons each second?

3. ⊗ Consider a microscope that uses electrons instead of photons. Under what conditions will this microscope have better resolution than a microscope that uses visible light? *Hint:* Consider the electron's energy and wavelength.

4. What has greater energy, an ultraviolet photon or an X-ray photon?

5. Physicists often conduct experiments in which photons are counted one by one as they arrive at a detector. Explain why these experiments are easiest for photons of visible light and X-rays and are hardest for radio waves.

6. No electrons are ejected when a dim source of red light is directed onto a metallic surface. The intensity of the red light source is increased by a factor of 100. Is it now possible for electrons to be ejected via the photoelectric effect? Why or why not?

7. The (old-fashioned) film used for black-and-white photography is not affected by infrared light, is somewhat sensitive to red light, and exposes rapidly to blue light to the extent that blue filters are sometimes used to obtain appropriate contrast. Give a possible explanation for why the film has this color dependency.

8. ⊗ **Sunscreen.** Blocking ultraviolet radiation is effective in preventing sunburn. Using the concept of photons, describe why human skin is sensitive to ultraviolet frequencies, but much less so to those of visible light.

9. ⊗ Color vision can only be understood using a particle (photon) model of light. Compare color vision to the photoelectric effect, explaining how they are similar and how they differ.

10. Apply the Heisenberg uncertainty principle to a car. If you are asked to measure the position of the car, estimate the best accuracy you could expect to achieve with a ruler. Then calculate the minimum possible uncertainty in the car's momentum. Do you think it is feasible to measure the momentum with this accuracy?

11. [SSM] Consider the following types of radiation: visible light, infrared radiation, gamma rays, X-rays, radio waves, and ultraviolet radiation. From smallest to largest, order them according to (a) their frequency, (b) their wavelength, and (c) their photon energy.

12. The peak (maximum-intensity) wavelength emitted by a glowing piece of metal is found to be decreasing as time passes. Is the metal being heated, or is it being cooled?

13. A blacksmith heats a piece of metal in a furnace. The metal initially glows red and later yellow, and then it gets white hot while glowing brighter as more heat is applied. Is it possible for the metal to get "violet hot" and glow purple if it is hot enough? Why or why not?

14. ⊗ Astronomers classify the color of stars as red, yellow, white, and blue. Why are green stars never seen? *Hint*: Consider how the eye perceives a mixture of different colors of light.

15. Which of the following experiments or phenomena are evidence for the wave nature of light, and which are evidence for the existence of photons?
    (a) Single-slit diffraction
    (b) Photoelectric effect
    (c) Double-slit interference
    (d) Color vision

16. Explain why the wave nature of baseballs and cars is not apparent in everyday life.

17. Explain why the particle nature of light is not apparent in everyday life.

18. In Section 28.4, we gave a classical explanation of the electron's spin magnetic moment using the picture of an electron as a spinning ball of electric charge. The neutron has zero net charge, yet it also has a magnetic moment. Explain how a spinning ball with zero net charge can still have a magnetic moment due to its spin.

19. An electron and a proton have the same kinetic energy. Which one has the larger momentum?

20. [SSM] ⊗ Human color vision is based on the absorption of light by three different types of cone cells, which are most sensitive to red light, green light, and blue light. (a) Explain how the output of these cells can be used to determine the color of the incident light. (b) Could an eye with just two types of cone cells be used to give color vision? Explain why or why not. If your answer is yes, are there any examples in nature?

21. ⊗ Suppose the human eye was so sensitive that it could detect a single photon. With input of just a single photon, could our system of rods and cones detect the color of this light?

22. Explain why the existence of a cutoff frequency in the photoelectric effect cannot be explained with the wave theory of light.

23. Explain why a classical picture cannot account for the Stern–Gerlach experiment with electrons (Fig. 28.11). The Stern–Gerlach experiment can also be done with nuclei, and it is found that certain nuclei give rise to three outgoing beams. What does that imply for the number of possible directions for the magnetic moment of that particular nucleus?

24. Planck's constant was first used to explain what phenomenon?

PROBLEMS

28.1 PARTICLES, WAVES, AND "PARTICLE-WAVES"

28.2 PHOTONS

1. ℝ Find the approximate photon energy for (a) an FM radio signal, (b) the radiation in your microwave oven, (c) your cell phone signal, and (d) the light emitted by a burning match.

2. A helium–neon laser emits red light with $\lambda = 632.8$ nm. Find (a) the energy of a single photon and (b) the momentum carried by a single photon.

3. A helium–neon laser ($\lambda = 632.8$ nm) emits radiation with a power of 100 mW. How many photons does it emit in 1 s?

4. ⊗ X-rays used by your dentist have a wavelength near 0.070 nm. (a) What is the frequency of this radiation? (b) What is the energy of a single X-ray photon with this wavelength?

5. The highest-energy photons emitted by a hydrogen atom have an energy of 13.6 eV. Find the energy, in joules, of one of these photons.

6. ★ ℝ The color of a blackbody is related to its temperature through Wien's law (Eq. 28.7). Estimate the energy of a typical photon emitted by (a) an ice cube, (b) a human being, and (c) an oven baking a cake.

7. A photon has an energy of 6.0 eV. What is the frequency of the radiation?

8. Suppose a blackbody emits photons most strongly with a frequency near 2.0 GHz. (a) What is the photon energy? (b) What is the approximate temperature of the blackbody? (c) In what region of the electromagnetic spectrum does this radiation fall?

9. ★ ⊗ ℝ The intensity of the Sun at the Earth's surface is approximately 1000 W/m². Estimate the number of photons from the Sun that strike an area of 1 m² each second.

10. How many photons of red light ($\lambda = 600$ nm) does it take to have a total energy of 1 J?

11. Radiation from outer space that reaches the Earth follows a blackbody spectrum with a temperature of about 2.7 K. (a) What is the energy of a typical photon from outer space? (b) Where does this photon fall in the electromagnetic spectrum?

12. ★ An FM radio station transmits at a frequency of 95 MHz and a total power of 100 kW. How many photons does this station emit each second?

13. The energy carried by 300 photons is 2.5×10^{-16} J. What is the frequency of one of these photons? Where do these photons fall in the electromagnetic spectrum?

14. A hydrogen atom that is initially at rest absorbs a photon with an energy of 2.0 eV. What is the momentum of the atom after it absorbs the photon?

15. A sodium lamp emits light with a wavelength of 590 nm. If the light radiates with a power of 30 W, how many photons does it emit each second?

16. How many photons does a green laser ($\lambda = 510$ nm) with a power rating of 2.0 W emit in 1 microsecond?

17. The work function for a metal is 5.0 eV. What is the minimum photon frequency that can just eject an electron from the metal?

18. **SSM** ☆ The work function of gold is 4.6 eV. Photons of a certain energy are found to eject electrons with a kinetic energy of 2.3 eV. (a) What is the speed of the ejected electrons? (b) What is the photon frequency? (c) If the photon energy is increased by a factor of two, what is the kinetic energy of the ejected electrons?

19. ℝ Crystals can be used to do interference experiments with photons (as for electrons in Fig. 28.8). Consider a crystal with a 0.40-nm spacing between atoms. (a) If the photon wavelength must be approximately this size to observe interference, what is the energy of this photon? (b) Where is this radiation in the electromagnetic spectrum?

20. A device called a *photocell* detects light by letting it fall onto a metal and then measuring the current from the ejected electrons. You are considering a photocell to work with visible light and are considering the use of either aluminum (work function = 4.3 eV) or cesium (work function = 2.1 eV). (a) Which one is a better choice? (b) What is the lowest photon frequency that can be measured with your photocell? (c) Where does this frequency fall in the electromagnetic spectrum?

21. Molybdenum has a work function of 4.2 eV. If light of frequency 2.0×10^{15} Hz strikes the surface of molybdenum, what is the energy of the ejected electrons?

22. A nuclear fission bomb reaches a temperature of about 10^8 K when it explodes. What is the approximate photon energy at which the blackbody radiation has its highest intensity?

23. ℝ What is the approximate surface temperature of stars with the color (a) blue and (b) red?

24. ☆ A "mosquito push-up" is the work done by a mosquito when it lifts its body ($m \approx 1$ mg) through a distance of 1 mm (the approximate length of a mosquito's "arms"). This amount of mechanical energy is very small. (a) How many mosquito push-ups does it take to equal the energy in a baseball ($m = 0.14$ kg) moving at 45 m/s (about 100 mi/h)? (b) How many photons of yellow light ($\lambda = 580$ nm) does it take to equal one mosquito push-up?

25. ✪ The ionization energy of a hydrogen atom is 13.6 eV. When a hydrogen atom absorbs a photon with this energy, the electron is ejected from the atom. (a) What are the frequency and wavelength of a photon with this energy? Where does this photon lie in the electromagnetic spectrum? Could you see light with this frequency? (b) When a photon has an energy greater than 13.6 eV, the "extra" energy (the energy in excess of 13.6 eV) goes into kinetic energy of the ejected electron. If a hydrogen atom absorbs a photon with an energy of 15.0 eV, what are the kinetic energy and speed of the ejected electron? (c) When the light intensity is very high, it is possible for a hydrogen atom to absorb two photons simultaneously. If a hydrogen atom absorbs two photons of equal energy and the atom is just barely ionized, what are the frequency and wavelength of these photons? Could you see this light?

26. Two photons have different frequencies f_1 and f_2. If the ratio of their momenta is 3.0, what is the ratio of their frequencies?

27. **SSM** ☆ A hydrogen atom emits a photon of energy 3.4 eV. If the atom is at rest before emitting the photon, what is the speed of the atom after emission?

28. ☆ A red laser that operates at a wavelength of 650 nm is directed at a black surface, which completely absorbs the beam. If the laser exerts a force of 10 nN (1.0×10^{-8} N) on the surface, how many photons are striking the surface every second?

29. ✪ A red laser emits photons with a wavelength of 620 nm. (a) If this laser has a power of 30 W, how many photons does it emit in 1.0 s? (b) If the light emitted by the laser is absorbed by a black sail, what is the force on the sail? (c) If the light from this laser is reflected from a mirror, what is the force on the mirror? You may recall that in Chapter 23 (Ex. 23.3) we discussed the use of such a sail to propel a spacecraft.

28.3 WAVELIKE PROPERTIES OF CLASSICAL PARTICLES

30. The kinetic energy of an electron is increased by a factor of two. By what factor does the wavelength change?

31. Consider two particles with masses that differ by a factor of 210. If they have the same wavelengths, what is the ratio of their kinetic energies?

32. ☆ The kinetic energy of an electron in an atom such as hydrogen is typically around 10 eV. (a) What is the wavelength of an electron with this energy? (b) How does this wavelength compare with the size of an atom? (c) If the energy were decreased by a factor of 10, how much would the wavelength change?

33. ☆ An electron has the same energy as a photon of blue light ($\lambda = 400$ nm). (a) What is the momentum of the electron? (b) What is the ratio of the momentum of the electron to the momentum of the photon?

34. ✪ ✗ ℝ An electron microscope uses electrons instead of light to form images. There are two general types of electron microscopes. A *scanning electron microscope* (SEM) forms an image of the surface of an object by reflecting electrons from it. (a) If the electron energy is 100 keV, what is the wavelength? (b) Based on our discussion of light microscopes in Chapter 26, what resolution do you expect to achieve with this SEM? That is, what is the approximate size of the smallest object that you expect to resolve? *Note:* Due to other limitations, SEMs do not achieve this resolution.

35. ✪ ✗ In addition to the scanning electron microscope described briefly in Problem 34, a second type of electron microscope, called a *transmission electron microscope* (TEM), operates by firing electrons through an object. Suppose the electrons in a TEM have energies of 5 MeV. (a) Find the electron wavelength. (b) Do you expect this TEM to have better resolution than the SEM in Problem 34? Explain why or why not. (c) How does your result from part (a) compare with the spacing between atoms in a solid (about 0.3 nm)? With the size of an atom (about 0.05 nm)?

36. An electron and a neutron have the same wavelength. What is the ratio of (a) their kinetic energies and (b) their momenta? Assume the speeds are low enough that you can ignore relativity.

37. (a) What is the wavelength of a photon whose momentum is equal to that of an electron with a speed of 2000 m/s? (b) What is the ratio of their energies?

38. If the wavelength of an electron is 150 nm, what is its speed?

39. A "thermal neutron" is a neutron whose kinetic energy is equal to $k_B T$, with T = room temperature ≈ 300 K. (a) What is the speed of a thermal neutron? (b) What is the wavelength of a thermal neutron? (c) How does the wavelength compare with the spacing between atoms in a solid (about 0.3 nm)? *Note:* To study the structure of solids, thermal neutrons are used in experiments similar to those of Davisson and Germer.

40. ✪ ℝ Oxygen molecules (O_2) in the atmosphere are described by kinetic theory (Chapter 15). What is the average de Broglie wavelength of these molecules at room temperature?

41. **SSM** ☆ Photons with a frequency of 3.0×10^{15} Hz shine on a piece of aluminum (work function = 4.3 eV). What are the energy and the de Broglie wavelength of the ejected electrons?

42. ☆ X-rays, electrons, and neutrons can all be used in diffraction experiments with crystals. In all three cases, the wavelength must be smaller than the spacing between atoms. Assuming this spacing is $d = 0.30$ nm, calculate the energies of an X-ray photon, an electron, and a neutron with wavelengths equal to $d/5$.

43. The mass of an electron is approximately 1800 times smaller than the mass of a proton. If an electron and proton have the same wavelength, what is the ratio of their energies?

44. ☆ The de Broglie relation for the wavelength of a particle-wave (Eq. 28.8) is valid for relativistic particles (i.e., particles moving at speeds approaching the speed of light c) provided

the relativistic expression for momentum (Eq. 27.24) is used to compute p. What is the wavelength of an electron with a kinetic energy of 5.0 MeV?

45. ☆ Ⓡ In principle, a baseball could be used to do a diffraction experiment. If this experiment were to use a doorway as the diffraction slit, estimate the approximate speed that would be required for the baseball. *Hint:* First estimate the necessary wavelength for the baseball.

46. Electrons are used in a single-slit diffraction experiment. We saw in Chapter 25 that such an experiment works well if the slit has a width of 10 times the wavelength. If the slit in an experiment with electrons has a width of 10 nm, what is the electron energy?

47. SSM ☆ Consider a double-slit interference experiment employing electrons. If the separation between two very narrow slits is 150 nm and the electron energy is 10 eV, what is the angle at which the first bright interference fringe is found?

48. Crystals can be used to do interference experiments with electrons. Consider a crystal with a spacing between atoms of 0.30 nm. If the electron wavelength must be approximately this size to observe interference, what is the electron energy?

28.4 ELECTRON SPIN

49. SSM ✪ The spin s of an electron is a type of angular momentum. It is quantized and is equal to $s = \pm\frac{1}{2}\hbar$, where the symbol h with a slash through the top is the *reduced* Planck's constant $\hbar = h/(2\pi)$, a quantity often found in equations concerning spin. As discussed in Section 28.4, a positive value of s denotes "spin up" orientation and a negative value means "spin down." This intrinsic property determines the behavior of an electron under the influence of the magnetic field in a Stern–Gerlach experiment. Spin is related to the **magnetic moment**, which is given by the expression

$$\mu = -\frac{e}{m}s$$

where e and m are the charge and mass of the electron, respectively. (a) Find the value of the magnetic moment of the electron. It is a fundamental quantity called the **Bohr magneton**. (b) Show that the units of the Bohr magneton can be expressed as J/T (joules per tesla).

28.5 THE MEANING OF THE WAVE FUNCTION

50. If a measurement of an electron's energy takes 3.0 ns (3.0×10^{-9} s), what is the minimum possible uncertainty in the energy?

51. If an electron is confined to a region of size 1.0 nm, what is the minimum possible uncertainty in its momentum?

52. ✪ An atom has a size of approximately 0.1 nm. Take this size as Δx, the uncertainty in the position of an electron in the atom. (a) Use the Heisenberg uncertainty relation to compute the uncertainty Δp in the electron's momentum. (b) If the electron's total momentum is equal to Δp, what is its kinetic energy? (c) How does your answer to part (b) compare with the binding energy of a typical atom? (d) Suppose the atomic size is reduced to 0.05 nm. This size might correspond to the extent of the wave function of the most tightly bound electrons in an atom such as argon. What is the estimated binding energy in this case? Does your result imply that these electrons are more "tightly bound" than the electron in part (c)?

53. ☆ A proton in an atom is confined to the nucleus, which has a diameter of approximately 1 fm (1×10^{-15} m). This diameter is also the uncertainty Δx in the proton's position. Compute the uncertainty in the proton's momentum. What is the corresponding kinetic energy of the proton? We'll use this result in Chapter 30 when we study nuclear physics.

54. ☆ Use the Heisenberg uncertainty relation to find the approximate kinetic energy of an electron that is trapped in a region the size of an atomic nucleus (1 fm $= 1 \times 10^{-15}$ m). Is this electron moving relativistically?

55. ✪ Suppose an electron is confined to a region of width 10 nm. The electron wave function forms a standing wave with nodes at the ends of the region. (a) What is the lowest-energy electron that can form such a standing wave? (b) What is the energy of an electron that forms a standing wave with three nodes (one at each end of the region and one in the middle)?

56. SSM ☆ A beam of electrons is directed along the x axis and through a slit that is parallel to the y axis and 10 μm wide. The beam then hits a screen 1.5 m away. The electrons in the beam have a kinetic energy of 70 eV. (a) What is the de Broglie wavelength of the electrons? (b) What is the uncertainty in the y component of their momentum after passing through the slit? (c) How long does it take the electrons to reach the screen? (d) What is the uncertainty in the y position when they hit the screen?

28.6 TUNNELING

57. ✪ When a particle such as an electron in an STM tunnels through a barrier, the Heisenberg uncertainty relation can be applied to get an approximate upper limit on the tunneling distance. Suppose an electron with energy E_1 attempts to tunnel through the barrier sketched in Figure P28.57. For an STM, the width of the barrier is the distance between the STM tip and the object being studied, which is typically 0.5 nm. (a) Take this value for the barrier width as Δx and use the Heisenberg uncertainty relation to find the uncertainty Δp in the momentum of the electron. (b) Assuming Δp is the electron's total momentum, compute the speed v of the electron. (c) How long does it take the electron to travel through the barrier? (d) Use the time found in part (c) in the Heisenberg uncertainty relation (Eq. 28.15) to find the uncertainty ΔE in the energy of the electron. If ΔE is the total energy of the electron, compare it with the kinetic energy calculated with the value for the speed found in part (b). The tunneling probability will be significant if the energy uncertainty ΔE is comparable to or larger than the barrier "height" (measured in terms of the electron's potential energy).

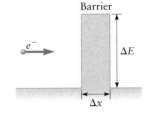

Figure P28.57

58. SSM ✪ Consider a hypothetical STM in which the tunneling particles are protons instead of electrons. Assuming an electron STM has a tunneling gap of 0.5 nm, estimate the size of the gap required for a proton STM. Assume all other parameters (including the kinetic energy) are the same.

28.7 ✪ DETECTION OF PHOTONS BY THE EYE

59. ✪ Estimate the photon energies at which the blue, green, and red cone cells in Figures 28.19 and 28.20 are most sensitive.

60. SSM ☆ ✪ Given that the threshold for a human eye to sense the presence of light is 50 photons within 0.1 s, what is the minimum power output of a red ($\lambda = 635$ nm) laser diode that would make it visible in a darkened room?

61. A photoelectric experiment uses light with a wavelength $\lambda = 550$ nm, and the ejected electrons are found to have a kinetic energy of 0.25 eV. What is the work function of the metal?

62. **SSM** ☒ ⓡ (a) What is the de Broglie wavelength of a helium atom with kinetic energy 2.0 eV? (b) Could you observe diffraction of these helium atoms using the atoms in a crystal to form the diffraction slits? (c) If your answer to part (b) is no, what is the approximate maximum speed for the helium atoms for which this diffraction experiment would work?

63. The mass of a mosquito is about 1 mg. If its speed is 2 m/s, what is its wavelength?

64. ✪ ⓧ ⓡ The threshold intensity for a human eye to sense the presence of light is 50 photons falling on the retina within a period of 0.1 s. The Sun radiates an output of 2.0×10^{26} W in the visible wavelengths, with a peak at $\lambda = 550$ nm. Estimate how far away a star like the Sun can be and still be visible to the unaided eye on a dark night.

65. Consider the following two quantum processes, each of which always emits a photon of a specific wavelength: (1) a 1.45-nm gamma ray from the nuclear process of cobalt-56 radioactive decay and (2) the 546-nm photon from the de-excitation of mercury in a street lamp (in the regime of chemical binding energies). (a) Determine the energy of the emitted photon (in units of joules and electron-volts) for each process and (b) calculate the ratio of the photon energies. How does the photon energy released from nuclear processes compare with that of chemical processes?

66. ✪ ⓡ In the mid 1960s, the Stanford Linear Accelerator was the first to attain the ability to accelerate electrons to a kinetic energy of 20 GeV (giga-electron-volts). A series of experiments involved directing the beam of electrons at protons. By examining the angular distribution of the electrons after the interaction, it was discovered that protons are not fundamental particles, but instead are made of smaller pointlike entities, later named quarks. What is the approximate diameter of the smallest object this beam of electrons could be used to detect, and how does it compare with the approximate size of a proton? *Hint*: Consider the de Broglie wavelength.

67. ☒ (a) Calculate the electric potential difference needed to accelerate a proton so that it has a wavelength of 1.0×10^{-12} m. (b) Repeat the calculation for an electron. *Hint*: Here you must use the relativistic expressions for energy and momentum. (c) Compare the energies of these two particles (ratio of proton energy to electron energy) when they have this same wavelength.

68. ⓧ ✪ ⓡ Microwave ovens operate at a wavelength of 12.24 cm and have a typical output of 700 W. (a) How many photons are emitted by the oven's microwave generator per second? (b) Approximately what is the minimum number of photons needed to heat 0.15 L of coffee (mostly water) from 25°C to 90°C?

69. ✪ **Heavy light.** In 1983, a hypothesized heavy force-carrying particle called the Z boson was discovered by studying the collisions of protons at the European Particle Physics laboratory, CERN. Just as a photon transmits the electromagnetic force, the Z boson carries the weak nuclear force (Section 31.5). Consider an experiment in which many Z bosons are produced, and the mass of each can be calculated by looking at the kinetic energies and trajectories of the particles into which it decays. The mass of each Z boson is then plotted in a histogram like that in Figure P28.69, which gives an average value of approximately 91 proton masses! We find that the actual mass of the particle can be quite different from one Z particle to another, and no matter how great the precision of the measurement, the distribution of the mass is spread out with an uncertainty of about 2.5 proton masses. What is the approximate lifetime of this particle? *Hint*: Consider the mass–energy relationship from Chapter 27. Can short-lived particles have a well-defined mass? What is the relationship?

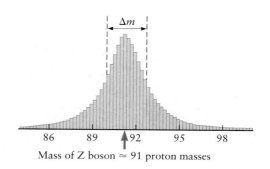

Figure P28.69

70. ✪ Using the Apache Point Observatory Lunar Laser-Ranging Operation (Fig. P23.61, page 794), the distance to the Moon is determined by how long it takes for a laser pulse to travel from the Earth to the Moon and then reflect back to the Earth from special reflector arrays placed there by the Apollo astronauts. A 2.3-W laser sends 20 pulses per second (each only 100 picoseconds long) at a wavelength of 532 nm. Due to atmospheric absorption and dispersion of the beam, only a very faint reflected pulse arrives back at the observatory with a power of 3.0×10^{-17} W. (a) How many photons does the observatory send in an outgoing pulse? (b) How many photons are received in a returning pulse? (c) How many photons are sent for every one that is detected upon return?

71. ✪ Electrons are accelerated through a potential difference of 20 kV in a vacuum tube to create a beam directed along the y axis. The beam passes through a 1.0-μm-wide slit and hits a fluorescent screen 25 cm away from the slit (as in Fig. 28.13C). (a) What is the de Broglie wavelength of the electrons in the beam as the beam enters the slit? (b) Consider single-slit diffraction and calculate the expected width along the x direction of the central bright spot on the screen. (Consider the distance to the first minimum in the diffraction pattern.) (c) What is the uncertainty in the x component of an electron's momentum in the beam after passing through the slit? (d) How long does it take for the electrons to reach the screen? (e) What is the uncertainty in the x position when the electrons hit the screen? (f) How does your answer in part (e) compare with that of part (b)? *Hint*: Consider the derivation in Section 28.5.

Chapter 29

Atomic Theory

The operation of a laser is based on the quantum properties of the atom. This laser employs helium and neon atoms, producing red light. (© Stockbyte/Alamy)

Matter is composed of atoms, and atoms are themselves assembled from electrons, protons, and neutrons. These facts are such a fundamental part of elementary science courses that it is hard to imagine a world without atoms. You should realize, however, that atoms were discovered *after* Galileo, Newton, Maxwell, and most of the other physicists mentioned so far had passed from the scene. Indeed, electrons were discovered barely one hundred years ago, and the discovery of protons and neutrons came even later. In this chapter, we review some of the history and development of atomic theory and emphasize how key aspects of quantum theory explain the way atoms are put together. The central goal of atomic theory is to understand why different elements have different properties. With this understanding, we can explain the organization of the periodic table and provide a basis for most of chemistry.

29.1 | STRUCTURE OF THE ATOM: WHAT'S INSIDE?

By about 1890 or so, it was apparent to most physicists and chemists that matter is composed of atoms, and it was also widely believed that atoms are indivisible. Perhaps the strongest evidence for this "atomic picture" of matter came from the gas laws (Chapter 15) and chemistry's "law of definite proportions" (Insight 29.1). At the end of the 1800s, about 90 elements were known, suggesting there were the same number of fundamental building blocks of matter. We now know that picture is not quite correct. At last count, there were more than 110 elements but each is composed of just three different types of particles: *electrons*, *protons*, and *neutrons*.

We encountered electrons and protons in our work on electricity (Chapter 17); they are subatomic particles that carry electric charges of $-e$ and $+e$, respectively. Neutrons are a third type of subatomic particle; they have no net electric charge and have a mass approximately the same as that of a proton. Two key questions are at the center of atomic theory: (1) What are the basic properties of these atomic building blocks; for example, what are the mass, charge, and size of an electron, a proton, and a neutron? (2) How do just these three building blocks combine to make so many different kinds of atoms?

The properties of these building blocks were determined by experiments between about 1890 and 1932, which also showed that the behavior of these particles cannot be understood in terms of Newton's mechanics. An understanding of how electrons, protons, and neutrons combine to form atoms required the invention of a new type of mechanics, called *quantum mechanics*. Some key aspects of quantum mechanics were described in Chapter 28, where the notion of a particle-wave was introduced. In this chapter, we apply the ideas of quantum theory to understand the structure of the atom.

Plum-Pudding Model of the Atom

Electrons were the first of the building-block particles to be discovered. In about 1897, experiments involving electricity showed that electrons carry a negative charge and that they are contained within atoms. These experiments also showed that electrons have a very small mass (compared with an atom). Since atoms are electrically neutral—that is, they have zero net charge—there must be some source of positive charge inside them. It was initially suggested that the positive charge is distributed as a sort of "pudding," with electrons suspended within positively charged pudding throughout the atom as sketched in Figure 29.1A. This analogy is known as the "plum-pudding" model of the atom, a name chosen by English physicists who proposed the model and named it after a popular dessert of the time.

The plum-pudding model of the atom raises an interesting conceptual question. A neutral atom has a total electric charge of zero, so the negative charge of the electrons must be exactly canceled by the positive charge of the pudding. Hence, an atom must contain a precisely measured amount of positive pudding, but how that could be accomplished was not clear. To learn more about the positive charge in the atom, physicists studied how atoms collide with other atomic-scale particles. The most famous of these experiments was carried out by Ernest Rutherford and his students Hans Geiger[1] and Ernst Marsden. They arranged for alpha particles to collide with a thin sheet of gold atoms. At that time, it was known that alpha particles are emitted spontaneously from certain radioactive elements, but the details of radioactivity were not understood. We now know that an alpha particle is actually the nucleus of a helium atom (Chapter 30).

Even though the true nature of an alpha particle was not understood at the time of Rutherford's work, prior experiments had shown that alpha particles behave as simple

[1]Inventor of the Geiger counter. (See Insight 30.3.)

Insight 29.1

CHEMISTRY, ATOMS, AND THE LAW OF DEFINITE PROPORTIONS

In the early 1800s, chemists discovered that when a compound is completely broken down into its constituent elements, the masses of the constituents always have the same *proportions*, no matter what the quantity of the original substance. This similarity can be explained by assuming the constituents are atoms that combine in definite proportions to make molecules. For example, when water is broken down into hydrogen and oxygen, the number of hydrogen atoms released is always twice the number of oxygen atoms, the mass of hydrogen released is always one-ninth of the mass of the original water, and the mass of oxygen gives the remaining eight-ninths of the mass. This is an example of the *law of definite proportions*.

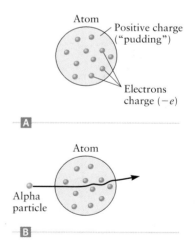

Figure 29.1 Ⓐ According to the plum-pudding model, the positive charge in an atom is distributed continuously, and the electrons sit in this "pudding." Ⓑ The pudding was thought to have a very low density, so the plum-pudding model predicts that an alpha particle should be deflected very little when it collides with (passes through) an atom.

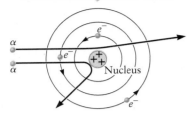

Nucleus *not* drawn to scale. The nucleus is actually about 1000 times smaller than the region occupied by electrons.

Figure 29.2 Planetary model of the atom, with all the positive charge located in the nucleus. When alpha particles are fired at an atom, they occasionally collide with the nucleus and are deflected through very large angles, but since the alpha particle and the nucleus are both very small, most alpha particles pass right through the atom.

Planetary model of the atom

Atomic number = number of protons in the nucleus

atomic-scale "bullets" when they collide with an atom. Rutherford's idea was to fire these "bullets" at an atom and then infer how the atom is put together by observing what comes out. Electrons have a very small mass, and Rutherford expected that the positively charged pudding would have a low density since (according to the plum-pudding model) this charge is spread throughout the entire atom. Using his knowledge of collisions and conservation of momentum (Chapter 7), Rutherford expected that the relatively massive alpha particles would pass freely through the plum-pudding atom as sketched in Figure 29.1B. For example, in a collision between a car (the alpha particle) and a Ping-Pong ball (an electron), the car is deflected only a very small amount because the momentum of a massive particle is much larger than the momentum of a low-mass particle with the same velocity. Likewise, an alpha particle should pass right through a wall of pudding because the pudding supposedly had a very small density compared with an alpha particle. When he carried out the experiment, Rutherford found that most of the alpha particles did indeed pass freely through the atom, but he also found that a small number of alpha particles were deflected a very *large* amount, some even bouncing directly "backward" toward the source of alpha particles. This surprising result could not be explained by the plum-pudding model. After much study, Rutherford realized that all the positive charge in an atom must be concentrated in a very small volume, with a mass and density about the same as an alpha particle. Most alpha particles in a collision experiment miss this dense region and pass right through the atom. Occasionally, though, an alpha particle collides with the dense region, giving a large deflection. Hence, Rutherford concluded that atoms contain a *nucleus* that is positively charged and has a large mass, much greater that the mass of an electron.

The Atomic Nucleus

After his discovery of the nucleus, Rutherford suggested that the atom is a sort of miniature solar system, with electrons orbiting the nucleus just as planets orbit the Sun (Fig. 29.2). This is called the *planetary model of the atom*. In contrast to the plum-pudding model, the electrons in the planetary model are not stationary. Indeed, the electrons must move in orbits to avoid "falling" into the nucleus as a result of the electric force.

We now know that the atomic nucleus contains protons and that the charge on a proton is precisely equal to $+e$. The charge carried by an electron is $-e$, so the total charge carried by a hydrogen atom (symbol H), which contains one electron along with a nucleus consisting of a single proton, is precisely zero. The total charge of neutral atoms of all other elements is also zero, so the number of protons in the nucleus is equal to the number of electrons in the neutral atom. This number, called the *atomic number* of the element, is denoted by Z.

Except for the H nucleus, which is a single proton, all other nuclei contain neutrons. As its name implies, the neutron is a neutral particle, carrying zero net electric charge. The neutron was not discovered until the 1930s, and we'll discuss some of its properties in Chapter 30. Even without a detailed understanding of the neutron, however, it is easy to understand why all nuclei (except H) contain both protons and neutrons. Protons are positively charged particles, repelling one another according to Coulomb's law, so a hypothetical nucleus containing just two or more protons would fly apart spontaneously. Protons are attracted to neutrons by an additional force that overcomes this Coulomb repulsion and thereby holds the nucleus together (described in Chapter 30).

Rutherford's discovery of the nucleus led to the planetary model of the atom. In the solar system, the gravitational force exerted by the Sun on the planets keeps the planets in their orbits; in the atom, the Coulomb force exerted by the nucleus on the electrons keeps the electrons in orbit. Both of these forces (gravity and Coulomb's law) follow the inverse square law, so physicists immediately tried to apply Newton's mechanics to the motion of electrons as they orbit within the atom. Their attempts were not successful as these "planetary" properties of electrons do not correctly predict atomic behavior. Even so, this approach provides some useful insights into the correct quantum theory that was developed later. With that in mind, let's estimate some atomic properties using this planetary model.

The Planetary Model: Energy of an Orbiting Electron

Figure 29.3 shows the planetary model as applied to hydrogen. It is the simplest atom, containing just[2] one proton and one electron. For simplicity, let's assume the electron's orbit is circular with a radius r. To estimate the properties of this orbit, we note that r is the approximate size of the hydrogen atom. Chemists have found that the typical distance between hydrogen atoms in various molecules suggests a radius of about $r \approx 1.0 \times 10^{-10}$ m. Since the electron in Figure 29.3 is moving in a circle, there must be a force directed toward the center of the circle. For an electron of mass m moving with speed v, this attractive force has a magnitude (Eq. 5.6)

$$F = \frac{mv^2}{r} \qquad (29.1)$$

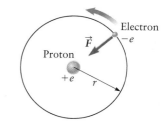

Figure 29.3 Planetary model of the hydrogen atom. A single electron is in orbit around a proton.

The force in Figure 29.3 is the electric force (from Coulomb's law) exerted by the proton on the electron, which is given by (Eq. 17.3)

$$F = \frac{kq_{proton}q_{electron}}{r^2} \qquad (29.2)$$

The charge on an electron is $q_{electron} = -e$, while the charge on a proton is $+e$. Inserting into Equation 29.2 gives $F = k(+e)(-e)/r^2 = -ke^2/r^2$. The negative sign indicates an attractive force, and we can drop it when considering just the magnitude of F. Inserting this result for F in Equation 29.1 leads to

$$\frac{ke^2}{r^2} = \frac{mv^2}{r}$$

The charge and mass of the electron are known and we have an estimate for the orbital radius r, so we can now solve for the electron's speed v. We get

$$v^2 = \frac{ke^2}{mr} \qquad (29.3)$$

$$v = \sqrt{\frac{ke^2}{mr}} \qquad (29.4)$$

Inserting values for the various quantities, including our estimate for r, we find

$$v = \sqrt{\frac{ke^2}{mr}} = \sqrt{\frac{(8.99 \times 10^9 \text{ N} \cdot \text{m}^2/\text{C}^2)(1.60 \times 10^{-19} \text{ C})^2}{(9.11 \times 10^{-31} \text{ kg})(1.0 \times 10^{-10} \text{ m})}}$$

$$v = 1.6 \times 10^6 \text{ m/s} \qquad (29.5)$$

This orbiting electron has quite a high speed (approximately 4×10^6 mi/h!). Given this value of the speed, the associated kinetic energy is

$$KE = \tfrac{1}{2}mv^2 = \tfrac{1}{2}(9.11 \times 10^{-31} \text{ kg})(1.6 \times 10^6 \text{ m/s})^2 = 1.2 \times 10^{-18} \text{ J}$$

We'd like to compare this roughly estimated value of the electron's orbital kinetic energy to some energy value associated with a hydrogen atom in a chemical reaction. Chemists measure the *ionization energy*, the energy required to remove an electron from an atom in the gas phase. In mechanical terms, the ionization energy is the energy needed to completely separate the two charges; that is, to take the electron in hydrogen from its original orbit to an infinite distance from the proton. For

---

[2]It is possible to form atoms called deuterium and tritium, which each contain one proton and one electron. A deuterium nucleus also contains one neutron, and a tritium nucleus contains two neutrons.

ease of comparison, let's convert the kinetic energy found above from joules to units of electron-volts (eV). From Equation 18.19, $1 \text{ eV} = 1.60 \times 10^{-19}$ J, which leads to

$$KE = (1.2 \times 10^{-18} \text{ J})\left(\frac{1 \text{ eV}}{1.60 \times 10^{-19} \text{ J}}\right) = 7.5 \text{ eV} \qquad (29.6)$$

In fact, the measured ionization energy of a hydrogen atom is 13.6 eV, so the electron kinetic energy found with the planetary model is indeed of the same order of magnitude as the ionization energy. For this reason, Rutherford's planetary model received strong initial support. In Example 29.1, we'll consider another aspect of the planetary model.

EXAMPLE 29.1 | Ionization Energy of a Hydrogen Atom

In our analysis of the planetary model of the hydrogen atom leading to Equation 29.6, we considered only the kinetic energy of the electron. To calculate the total energy required to remove the electron from the proton, we must also consider the potential energy. Assuming again the electron's orbit has a radius $r = 1.0 \times 10^{-10}$ m, find the electric potential energy in this model of the hydrogen atom. What is the change in this potential energy when the atom is ionized?

RECOGNIZE THE PRINCIPLE

The electric potential energy of two charged particles depends on the value of each charge and their separation. In the planetary model, the electron's orbit is circular, so the separation between the charges for the atom is just the radius of the orbit. When the atom is ionized, the electron is very (infinitely) far from the proton.

SKETCH THE PROBLEM

Figure 29.4 shows the problem. The electron is initially in orbit at a distance r_i from the proton (Fig. 29.4A), with a speed found in Equation 29.4. When the atom is ionized, the electron is at a distance $r_f = \infty$ from the proton (Fig. 29.4B).

IDENTIFY THE RELATIONSHIPS

The electron is initially in orbit at $r_i = 1.0 \times 10^{-10}$ m. The electric potential energy is given by (Eq. 18.6)

$$PE_{\text{elec, }i} = \frac{kq_{\text{proton}}q_{\text{electron}}}{r_i}$$

The charge on the proton is $q_{\text{proton}} = +e$ and the charge on the electron is $-e$, so

$$PE_{\text{elec, }i} = \frac{kq_{\text{proton}}q_{\text{electron}}}{r_i} = \frac{-ke^2}{r_i} \qquad (1)$$

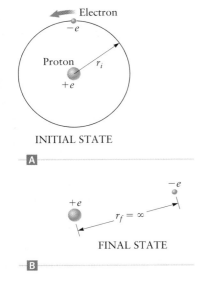

Figure 29.4 Example 29.1. Calculating the potential energy of an electron in the planetary model of the hydrogen atom.

SOLVE

Using our estimate for r_i, we find

$$PE_{\text{elec, }i} = \frac{-ke^2}{r_i} = \frac{-(8.99 \times 10^9 \text{ N} \cdot \text{m}^2/\text{C}^2)(1.60 \times 10^{-19} \text{ C})^2}{1.0 \times 10^{-10} \text{ m}}$$

$$PE_{\text{elec, }i} = -2.3 \times 10^{-18} \text{ J}$$

Converting to units of electron-volts as in Eq. 29.6 gives

$$PE_{\text{elec, }i} = (-2.3 \times 10^{-18} \text{ J})\left(\frac{1 \text{ eV}}{1.60 \times 10^{-19} \text{ J}}\right) = \boxed{-14 \text{ eV}} \qquad (3)$$

When the atom is ionized, the separation between the electron and proton is infinite. Using Equation (1), the corresponding potential energy is $PE_{\text{elec, }f} = 0$. Hence, the change in potential energy when the atom is ionized is

$$PE_{\text{elec, }f} - PE_{\text{elec, }i} = \boxed{+14 \text{ eV}}$$

Because the value of r_i was not known precisely, this result is only a rough estimate, but it does confirm that the kinetic and potential energies predicted by the planetary model are both close to the measured ionization energy of hydrogen, which is 13.6 eV. We'll return to this problem in Section 29.3.

Problems with the Planetary Model of the Atom

Calculations with Rutherford's planetary model (like the one in Example 29.1) initially suggested that Newton's mechanics could be used to describe the atom, but some fundamental problems with this model soon became apparent. The biggest problem concerns the stability of an electron orbit. The orbital motion of a planet such as the Earth's motion around the Sun is extremely stable. Astronomers believe that the Earth has been traveling in its present orbit around the Sun for billions of years, with relatively little change in the orbital radius and period during that time. Likewise, atoms such as hydrogen can also be extremely stable. (If chemical reactions are avoided, an individual atom can "last" indefinitely.) The electrons in Rutherford's planetary model, however, are undergoing accelerated motion, and in Chapter 23 we saw that accelerated charges emit electromagnetic radiation that carries away energy. If the electron in a hydrogen atom loses energy in this way, it will spiral inward to the nucleus (Fig. 29.5). Hence, according to classical physics (i.e., according to Newton and Maxwell), Rutherford's atom is inherently unstable!

A careful analysis in terms of Newton's laws shows that an electron in the planetary model must spiral into the nucleus in a very small fraction of a second. Hence, according to classical physics, such planetary model atoms cannot exist. Despite much trying, physicists found no way to fix the planetary model to make the atoms in this model stable. This problem was not resolved until the development of quantum theory.

Quantum theory avoids the problem of unstable electron orbits by replacing them with discrete energy levels, just like the discrete energy levels of an electron in a box (Fig. 28.12) and the discrete energies of the pigment molecules involved in color vision (Section 28.7). Quantum theory rejects the notion of the electron as a simple particle that obeys Newton's laws (and spirals into the nucleus in Fig. 29.5). Instead, an electron is a particle-wave described by a wave function with discrete energy levels. Electrons lose or gain energy only when they undergo a transition between energy levels.

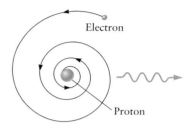

Figure 29.5 Maxwell's theory of electromagnetic waves predicts that an orbiting electron emits radiation, causing it to lose energy and spiral into the nucleus. If this picture is correct, all atoms would collapse, which is clearly *not* the case.

29.2 | ATOMIC SPECTRA

We have just claimed that an electron in an atom can exist only in discrete energy levels. The best evidence for this claim comes from the radiation an atom emits or absorbs when an electron undergoes a transition from one energy level to another. This radiation is key to another question that was studied intensely in the late 1800s: what gives an object its color? Physicists of that time knew about blackbody radiation (Chapters 14 and 28), including the relationship between the temperature of an object and its color. This relationship is described by Wien's law (Eqs. 14.14 and 28.7). The approximate blackbody spectrum of the Sun shown in Figure 29.6A is a continuous curve with a smooth distribution of intensity over a wide range of wavelengths and frequencies.

The spectrum in Figure 29.6A describes the radiation emitted by the Sun in a general way. Careful observations, however, show that the Sun's spectrum also contains sharp dips superimposed on the otherwise smooth blackbody curve as shown schematically in Figure 29.6B. These dips are called spectral "lines" because of their appearance when the spectrum is dispersed by a prism or diffraction grating (Chapters 24 and 25). Examination of the Sun's spectrum with a prism gives a band

Figure 29.6 **A** Ideal blackbody spectrum of the Sun. The emitted intensity is a smooth function of the wavelength. **B** The actual spectrum of the Sun contains sharp "dips" on top of the blackbody spectrum in part A. Another way of displaying these dips is to use a prism to disperse sunlight (Chapter 24); the dips then show up as dark "lines," having the same wavelengths as the dips in the spectrum.

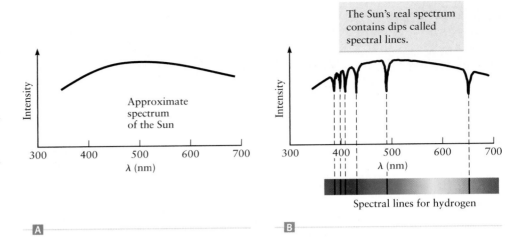

The Sun's real spectrum contains dips called spectral lines.

Spectral lines for hydrogen

A   **B**

of colors extending from red (long wavelength) to blue (short wavelength) as shown below the graph in Figure 29.6B. The dips in the graph of the figure show up as dark lines in the band of colors, indicating wavelengths at which the light intensity is much lower than the expected blackbody value.

The origin of the dark spectral lines is illustrated in Figure 29.7A; when light from a pure blackbody source passes through a gas, atoms in the gas absorb light at certain wavelengths, producing dips in the spectrum at those wavelengths. Some of the dark lines in the Sun's spectrum are produced when hydrogen atoms in the Sun's atmosphere absorb light. This can be confirmed by an experiment using a blackbody source and hydrogen atoms on the Earth, which gives the spectral lines shown in the top band in Figure 29.7B; these lines have precisely the same wavelengths as those found in the Sun's spectrum.

The dark spectral lines are called **absorption lines** because they result from the absorption of light, in this case by hydrogen atoms. These same atoms can also be made to *emit* light. When the spectrum of light emitted by atoms is analyzed, it is found that the emission occurs only at certain wavelengths. The **emission lines** for hydrogen are shown in the lower band in Figure 29.7B, demonstrating that these emission and absorption lines occur at the same wavelengths.

Another example of spectral lines is given in Figure 29.8, which shows the emission and absorption spectra for sodium (Na) atoms. Here again the emission and absorption lines occur at the same wavelengths. The pattern of spectral lines is different for each element. In fact, by analyzing the wavelengths at which these lines occur, physicists in the 1800s determined the composition of the Sun's atmosphere.

These observations of atomic spectra lead to several questions. Why do absorption and emission occur only at certain special wavelengths? Why do the absorption and emission lines for a particular element occur at the same wavelengths? What determines the pattern of the absorption and emission wavelengths, and why are they different for different elements?

Many physicists attempted to answer these questions by applying Newton's mechanics to Rutherford's planetary model of the atom, but all ran into the fol-

Atoms emit and absorb light at discrete frequencies.

Figure 29.7 **A** When the Sun's blackbody radiation passes through its atmosphere, atoms in the atmosphere absorb at certain discrete frequencies, producing dark spectral absorption lines and causing the dips in the spectrum in Figure 29.6B. **B** Absorption and emission spectra for hydrogen. The absorption and emission lines occur at the same wavelengths. This plot shows just the spectral lines for hydrogen that lie between 400 nm and 700 nm. Figure 29.6B shows these and some other lines with wavelengths just below 400 nm.

Figure 29.8 Sodium (Na) has two closely spaced emission lines in the yellow part of the visible spectrum. Sodium atoms also absorb strongly at the same wavelengths as the emission lines.

lowing problem. When a photon is emitted by an atom, the photon carries away a certain amount of energy; for a photon with frequency f, we have (Eq. 28.1)

$$E_{\text{photon}} = hf \qquad (29.7)$$

Total energy must be conserved, so the final energy of the atom is lower than the initial energy by an amount E_{photon}. Atomic emission occurs only at certain discrete wavelengths (i.e., discrete frequencies), which suggests that the energy of the orbiting electron can have only certain discrete values. The kinetic and potential energies of an orbiting electron depend on the radius r of the orbit as we already derived in Equation 29.6 and Example 29.1. According to Newton's mechanics, the orbital radius can be very large or very small, so the total energy of an orbiting electron can have a continuous range of values. Based on Newton's mechanics, there is no way for this planetary orbit picture to give discrete electron energies and no way to explain the existence of discrete spectral lines. This problem is resolved in quantum theory through the description of the electron's state in terms of a wave function instead of an orbit.

Atoms Have Quantized Energy Levels

Spectra like those in Figures 29.7 and 29.8 show that under certain conditions, the photons emitted or absorbed by an atom have only certain discrete wavelengths or frequencies. The implication is that the energy of the atom itself can have only certain discrete values. That is, the energy of an atom is **quantized**. In the language of quantum theory, we say that the energy of an absorbed or emitted photon is equal to the difference in energy between two discrete atomic energy levels (Fig. 29.9). Through Equation 29.7, the frequencies of the emission and absorption lines give the spacing between atomic energy levels. Hence, the study of atomic spectra gives a direct window into the atom's structure.

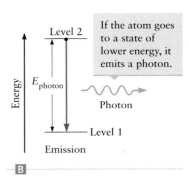

Figure 29.9 The discrete absorption and emission lines of an atom are due to discrete atomic energy levels. An atom **A** absorbs or **B** emits a photon when it makes a transition between energy levels.

| EXAMPLE 29.2 | Energy of Photons Emitted by the Hydrogen Atom |

Studies of the spectral lines of hydrogen show that the highest frequency of electromagnetic radiation emitted by a hydrogen atom is about 3.28×10^{15} Hz. **(a)** What is the energy of one of the corresponding photons? **(b)** Does this light fall in the infrared, visible, or ultraviolet part of the electromagnetic spectrum? *Hint*: The approximate short wavelength end of the visible range is at $\lambda \approx 400$ nm. **(c)** How does this photon energy compare with the ionization energy of the hydrogen atom, 13.6 eV?

RECOGNIZE THE PRINCIPLE

For part (a), this problem asks for a calculation of photon energy given the frequency, which can be done with Equation 29.7. For part (b), if we are given either the frequency or wavelength, we can use results from Chapter 23 and Figure 23.8 to find where this radiation falls within the electromagnetic spectrum. For part (c), the comparison will be apparent once we know the photon energy and convert it to electron-volts.

SKETCH THE PROBLEM

A spectral line is connected with photon emission or absorption between two discrete energy levels. Figure 29.10 shows this schematically for a hydrogen atom.

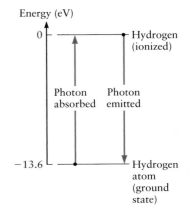

Figure 29.10 Example 29.2.

IDENTIFY THE RELATIONSHIPS AND SOLVE

(a) This problem begins with the relation $E_{photon} = hf$ (Eq. 29.7). Inserting the given value of f leads to

$$E_{photon} = hf = (6.63 \times 10^{-34} \text{ J} \cdot \text{s})(3.28 \times 10^{15} \text{ Hz}) = \boxed{2.17 \times 10^{-18} \text{ J}} \quad \textbf{(1)}$$

(Here we keep three significant figures because they will be useful below.)

(b) To determine where this light falls in the electromagnetic spectrum, let's calculate its wavelength. The speed of light is $c = 3.00 \times 10^8$ m/s, so we have

$$\lambda = \frac{c}{f} = \frac{3.00 \times 10^8 \text{ m/s}}{3.28 \times 10^{15} \text{ Hz}} = 91.5 \text{ nm}$$

According to the hint (and also Fig. 23.8), the visible range ends at about $\lambda \approx 400$ nm (corresponding to violet light), so this photon falls well into the $\boxed{\text{ultraviolet}}$.

(c) To compare the energy of the photon in Equation (1) with a typical atomic energy, it is useful to convert this energy to units of electron-volts. We get

$$E_{photon} = (2.17 \times 10^{-18} \text{ J})\left(\frac{1 \text{ eV}}{1.60 \times 10^{-19} \text{ J}}\right) = 13.6 \text{ eV}$$

Hence, this photon energy $\boxed{\text{is equal to}}$ the ionization energy.

What does it mean?
We have quoted the value of E_{photon} to three significant figures because it is a very special energy value in chemistry and physics. This result for E_{photon} is precisely equal to the ionization energy of hydrogen, the energy required to remove the electron from a hydrogen atom and thus "break" the atom apart. Our result for E_{photon} confirms the close connection between the way an electron is bound in an atom and the way the atom emits light. These processes involve the same two energy levels as sketched in Figure 29.10. The levels in hydrogen differ in energy by 13.6 eV, and by convention the upper level, corresponding to an ionized atom, has an energy $E = 0$.

29.3 | BOHR'S MODEL OF THE ATOM

The experimental facts described in the previous two sections showed that Rutherford's planetary model of the atom—and indeed *any* model based on Newton's mechanics—is a failure. The correct quantum theory of the atom did not "appear" instantaneously; there were first some intermediate proposals that eventually led to a complete quantum theory. One of these intermediate proposals is known as the ***Bohr model***, invented by Danish physicist Niels Bohr (1885–1962). (Bohr later engaged Albert Einstein in famous debates over the philosophical implications of quantum theory.) Although Bohr's model has some flaws, it also contains some of the most important and most revolutionary ideas of the correct quantum theory and is an important conceptual step along the way.

Quantum States and Energy-Level Diagrams

Bohr based his thinking on the planetary model. For simplicity, he first assumed the electron orbits are circular, which allowed him to use the results for the electron kinetic and potential energies that we derived in Section 29.1 for a hypothetical hydrogen atom (consisting of just one electron and one proton). However, to explain the discrete spectral lines, Bohr needed several changes to the standard results based on Newton's laws. First, Bohr postulated that only certain electron orbits are allowed. That is, only certain specific (and special) values of the orbital radius r are permitted. If only certain discrete values of r are allowed, the electron's

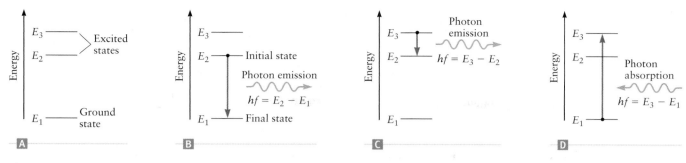

Figure 29.11 ◪ Hypothetical energy-level diagram for an atom with three quantum states. ◪ An atom that is initially in state 2 can undergo a transition to a state with lower energy (state 1) and emit a photon. ◪ Photon emission due to a transition between excited states. ◪ An atom can absorb a photon and undergo a transition to a state of higher energy.

kinetic and potential energies can have only certain values, so the total energy of the electron—the total energy of the atom—can have only certain discrete values. In the modern language of quantum theory, we say that the energy of the atom is *quantized*, and we refer to each of these allowed orbits as a *quantum state* of the atom. One way to view these energies and states is in terms of an *energy-level diagram*. Figure 29.11A shows a hypothetical diagram for an atom that can be in one of three different states. In the language of the Bohr model, there are three different allowed values of the orbital radius, and the corresponding total energy of an electron in these allowed orbits is E_1, E_2, or E_3. In this energy-level diagram, energy is plotted on the vertical axis and each state is represented by a horizontal line.

> Quantized atomic states are shown on an energy-level diagram.

With this postulate of discrete energy levels, Bohr could explain the discrete spectral emission and absorption lines described in Section 29.2. Let's assume an atom is initially in the quantum state with energy E_2 in Figure 29.11B, corresponding to a particular orbital radius r_2. This electron may "jump" to a different orbit with energy E_1 as shown on the energy-level diagram by a vertical arrow that connects the initial and final states. After doing so, the atom has decreased in energy by an amount $\Delta E = E_2 - E_1$. According to the principle of conservation of energy, this energy cannot just vanish; total energy is conserved as the atom emits a photon with precisely this energy ΔE. The discrete set of atomic energy levels produces the discrete frequencies found in the emission spectrum of an atom (Figs. 29.7 and 29.8). We say that the atom described in Figure 29.11B was initially in an *excited state* and made a transition to the *ground state*, the state of lowest possible energy for this atom. The photon emitted in the transition has a wavelength matching one of the lines in the atom's emission spectrum. It is also possible for an atom to emit a photon and undergo a transition between two excited states such as the states with energies E_3 and E_2 in Figure 29.11C.

Transitions between the discrete energy levels in Bohr's picture also explain the absorption spectrum of an atom. The energy-level diagram in Figure 29.11D shows an atom that starts in the ground state with energy E_1 and then absorbs a photon as it undergoes a transition to an excited state with energy E_3. To conserve energy, the absorbed photon's energy must be $\Delta E = hf = E_3 - E_1$, the energy gained by the atom. This picture explains why the frequencies of the emission lines are precisely the same as the frequencies of the absorption lines. Both are determined by the difference in energy between two atomic states.

CONCEPT CHECK 29.1 | Spectral Lines and Energy Levels

The spectra for Na in Figure 29.8 show two closely spaced lines in both absorption and emission. Consider now a different (and hypothetical) atom. Is the *minimum* number of atomic energy levels required to give spectral lines at two or more different frequencies (a) two, (b) three, or (c) four?

Quantization of Angular Momentum Leads to Quantized States with the Correct Energies

Central to the Bohr model is the postulate that electrons can orbit at only certain allowed values of the radius r, but what determines these values? Here Bohr made a complete break with Newton's mechanics. Bohr proposed that the orbital angular momentum L of the electron could only have certain values given by the relation

$$L = n\frac{h}{2\pi} \tag{29.8}$$

where $n = 1, 2, 3, \ldots$ is an integer and h is Planck's constant. According to Equation 29.8, the allowed values of angular momentum are quantized in units of $h/(2\pi)$. What motivated Bohr to make this guess? One motivation is that this guess about L leads to the correct results for the frequencies of the emission and absorption lines for hydrogen (Figs. 29.6 and 29.7). Let's now see how Bohr's guess about the angular momentum leads to quantized energy levels.

From Chapter 7, the angular momentum of a particle of mass m traveling with speed v in a circular orbit of radius r is

$$L = mvr \tag{29.9}$$

Inserting this into Bohr's angular momentum relation, Equation 29.8, gives

$$L = mvr = n\frac{h}{2\pi}$$

Rearranging to solve for the electron's orbital speed v leads to

$$v = \frac{nh}{2\pi mr} \tag{29.10}$$

Bohr kept enough of Rutherford's planetary model to use the relationship between speed and radius for an electron moving in a circular orbit that we found in Section 29.1 (Eq. 29.3),

$$v^2 = \frac{ke^2}{mr}$$

Inserting the result for v from Equation 29.10, we obtain

$$v^2 = \left(\frac{nh}{2\pi mr}\right)^2 = \frac{ke^2}{mr}$$

We can now solve for the orbital radius r:

$$\frac{n^2h^2}{4\pi^2m^2r^2} = \frac{ke^2}{mr}$$

$$r = n^2\left(\frac{h^2}{4\pi^2mke^2}\right) \tag{29.11}$$

The quantity in parentheses in Equation 29.11 is a combination of fundamental constants—Planck's constant, the mass and charge of an electron, and the constant k from Coulomb's law—so this term is also a constant. The variable n is an integer, so the factor n^2 on the right side of Equation 29.11 can have the values 1, $2^2 = 4$, $3^2 = 9$, and so on. In words, Equation 29.11 says that the orbital radius of an electron in a hydrogen atom can have only these particular quantized values. The smallest value of this radius is found when $n = 1$. This value is called the **Bohr radius** of the hydrogen atom and is the smallest orbit allowed in the Bohr model. We'll consider its value in Example 29.3.

Bohr radius of the hydrogen atom

Quantized Energies of the Bohr Atom

We can now continue with Bohr's approach and calculate the corresponding atomic energies using results from Section 29.1 (and Example 29.1). The kinetic energy of an orbiting electron is $KE = \frac{1}{2}mv^2$. Inserting the result for v^2 from Equation 29.3, we get

$$KE = \frac{1}{2}mv^2 = \frac{1}{2}m\frac{ke^2}{mr} = \frac{1}{2}\frac{ke^2}{r} \qquad (29.12)$$

The potential energy of this electron–proton pair is due to the electric force and is

$$PE_{elec} = -\frac{ke^2}{r} \qquad (29.13)$$

The total energy is thus

$$E_{tot} = KE + PE_{elec} = \frac{1}{2}\frac{ke^2}{r} - \frac{ke^2}{r} = -\frac{1}{2}\frac{ke^2}{r} \qquad (29.14)$$

We have already obtained the orbital radius r in Equation 29.11. Inserting that result, we find

$$E_{tot} = -\frac{1}{2}ke^2\left(\frac{1}{r}\right) = -\frac{1}{2}ke^2\left(\frac{4\pi^2mke^2}{n^2h^2}\right)$$

After some rearranging, we get

$$E_{tot} = -\left(\frac{2\pi^2k^2e^4m}{h^2}\right)\frac{1}{n^2} \qquad (29.15)$$

The factor in parentheses is a constant (because it is a combination of fundamental constants). The only variable is the integer n, which can have values $n = 1, 2, 3, \ldots$. Hence, the energy of the hydrogen atom in the Bohr model can have only certain *quantized* values corresponding to $n = 1, 2, 3, \ldots$ in Equation 29.15. In this way, Bohr's postulate about quantized angular momentum in Equation 29.8 leads to quantized energy levels for the atom.

The values of the energy levels predicted from Equation 29.15 can be used to derive the frequencies of the absorption and emission lines of hydrogen. To see how that works, we first find the values of E_{tot} predicted by Equation 29.15. Inserting the values of the various fundamental constants leads to

$$E_{tot} = -\left(\frac{2\pi^2k^2e^4m}{h^2}\right)\frac{1}{n^2}$$

$$E_{tot} = -\left[\frac{2\pi^2(8.99 \times 10^9 \text{ N} \cdot \text{m}^2/\text{C}^2)^2(1.60 \times 10^{-19} \text{ C})^4(9.11 \times 10^{-31} \text{ kg})}{(6.63 \times 10^{-34} \text{ J} \cdot \text{s})^2}\right]\frac{1}{n^2}$$

$$E_{tot} = -\frac{2.17 \times 10^{-18} \text{ J}}{n^2}$$

To compare with our previous calculations, we again convert to units of electron-volts and find

$$E_{tot} = -\frac{13.6 \text{ eV}}{n^2} \qquad (29.16)$$

Allowed energies of hydrogen in the Bohr model

We have given the value of E_{tot} to three significant figures so that we can more accurately compare it with the measured ionization energy of hydrogen.

The first few values in Equation 29.16 are plotted on the vertical axis in the energy-level diagram in Figure 29.12. The horizontal lines show levels corresponding to $n = 1, 2, 3, \ldots$. The level with $n = 1$ has the lowest energy and is the ground state of Bohr's hydrogen atom. Inserting $n = 1$ into Equation 29.16 gives $E_1 = -13.6$ eV. As n increases to $2, 3, \ldots$, the energies increase, with $E_2 = -3.4$ eV, $E_3 = -1.5$ eV, and so on, and at $n = \infty$, we have $E_\infty = 0$. All the levels except the highest one thus have negative energies. What do these negative values mean? How can an orbiting electron have a negative energy? The energy in Equation 29.16 is the total mechanical energy of Bohr's atom, the sum of the kinetic and potential energies. The kinetic energy is positive, but the potential energy is negative since the Coulomb force exerted between the proton and electron is attractive, and our convention for

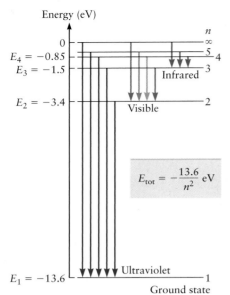

Energy (eV)

$E_4 = -0.85$
$E_3 = -1.5$

$E_2 = -3.4$

$$E_{tot} = -\frac{13.6}{n^2}\ eV$$

$E_1 = -13.6$

Infrared

Visible

Ultraviolet

Ground state

Figure 29.12 Energy-level diagram for the Bohr model of hydrogen.

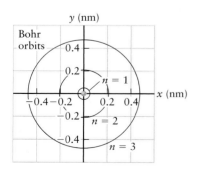

Figure 29.13 Example 29.3. Bohr orbits for the $n = 1$, 2, and 3 states of hydrogen. Note that the nucleus is not drawn to scale.

electric potential energy has $PE_{elec} = 0$ when the electron is infinitely far from the proton. The level with $E_{tot} = 0$ in Figure 29.12 and Equation 29.16 corresponds to completely removing the electron from the proton, or ionization of the atom. The energy required to take an electron from the ground state E_1 and remove it from the atom is the ionization energy. Bohr's theory thus predicts an ionization energy of 13.6 eV, in excellent agreement with the measured value.

The arrows in Figure 29.12 show some possible atomic transitions leading to emission of a photon. The energies of these transitions and the energies of the associated photons are equal to the difference in the energies of the levels at the start and end of each transition. Equation 29.16 predicts that there are an infinite number of energy levels (because $n = 1, 2, \ldots, \infty$), giving an infinite number of spectral emission lines. All these emission frequencies agree with the experimentally observed values, measured long before Bohr's work, adding convincing evidence that Bohr was indeed on the right track. More work was needed, but Bohr's ideas took physics much of the way toward the correct quantum theory of the atom.

EXAMPLE 29.3 | The Bohr Model and the Size of a Hydrogen Atom

In Section 29.1, we used an approximate orbital radius $r = 1.0 \times 10^{-10}$ m for the hydrogen atom as estimated from measurements of the atomic spacing in molecules and solids. Calculate r from the Bohr model for the lowest three energy levels, that is, the levels corresponding to $n = 1$, 2, and 3.

RECOGNIZE THE PRINCIPLE

The orbital radius r in the Bohr model depends on the value of n. According to the Bohr model (and Eq. 29.11), the radius is smallest for the quantum state with $n = 1$ and increases as n increases.

SKETCH THE PROBLEM

No sketch is needed to start the problem, but a plot of the results for the orbits is given in Figure 29.13.

IDENTIFY THE RELATIONSHIPS AND SOLVE

The result of Bohr's theory for the orbital radius is (Eq. 29.11)

$$r = n^2 \left(\frac{h^2}{4\pi^2 m k e^2} \right)$$

Inserting the values of the constants in parentheses, we find

$$r = n^2 \left[\frac{(6.63 \times 10^{-34}\ J \cdot s)^2}{4\pi^2 (9.11 \times 10^{-31}\ kg)(8.99 \times 10^9\ N \cdot m^2/C^2)(1.60 \times 10^{-19}\ C)^2} \right]$$

$$r = n^2 \times 5.3 \times 10^{-11}\ m$$

Evaluating for the given values of n leads to

$$n = 1 \text{ (ground state): } r = 5.3 \times 10^{-11}\ m = \boxed{0.053\ nm} \qquad (1)$$

$$n = 2: r = 2.1 \times 10^{-10}\ m = \boxed{0.21\ nm}$$

$$n = 3: r = 4.8 \times 10^{-10}\ m = \boxed{0.48\ nm}$$

What does it mean?

The value of r for the ground state given in Equation (1) differs by only a factor of two from the value we used in our earlier calculations with the planetary model (Example 29.1), so our earlier estimates were fairly accurate. The value of r in Equation (1) is the value of the Bohr radius of the hydrogen atom and is used extensively in many estimates in atomic theory (by both physicists and chemists). Our results also show that the orbital radius increases rapidly as n increases, so electrons in these different Bohr orbits are well separated in space (Fig. 29.13).

| EXAMPLE 29.4 | Bohr Theory and the Frequency of an Emission Line |

Consider the transition in the Bohr model of a hydrogen atom from the $n = 2$ level to the $n = 1$ state. What is the wavelength of the photon emitted when the atom makes this transition?

RECOGNIZE THE PRINCIPLE

The wavelength of an emission line is related to its frequency, which is proportional to the difference in energy of the two levels. We thus need to find the energy separation of the $n = 2$ and $n = 1$ levels, which we can get from the result for the Bohr model energy in Figure 29.12 or by using Equation 29.16.

SKETCH THE PROBLEM

Figure 29.12 shows the Bohr model energy levels.

IDENTIFY THE RELATIONSHIPS

The energy of the transition is $\Delta E = E_2 - E_1$, and that equals the photon energy hf. We thus have

$$hf = \Delta E = E_2 - E_1$$

$$f = \frac{\Delta E}{h}$$

The wavelength is then

$$\lambda = \frac{c}{f} = \frac{hc}{\Delta E}$$

SOLVE

We can read the values of E_1 and E_2 from Figure 29.12 and find

$$\Delta E = E_2 - E_1 = (-3.4 \text{ eV}) - (-13.6 \text{ eV}) = 10.2 \text{ eV}$$

Converting from electron-volts to joules gives

$$\Delta E = 10.2 \text{ eV} \times \frac{1.60 \times 10^{-19} \text{ J}}{1 \text{ eV}} = 1.63 \times 10^{-18} \text{ J}$$

The wavelength is thus

$$\lambda = \frac{(6.63 \times 10^{-34} \text{ J} \cdot \text{s})(3.00 \times 10^8 \text{ m/s})}{1.63 \times 10^{-18} \text{ J}} = 1.22 \times 10^{-7} \text{ m} = \boxed{122 \text{ nm}}$$

What does it mean?
This wavelength lies in the ultraviolet; recall that the visible range ends at a wavelength of about 400 nm. Hence, this radiation would not be visible to the human eye.

Generation of X-rays by Atoms

The ionization energy of the hydrogen atom is 13.6 eV. We have seen that this is the energy of the hydrogen emission line with the highest photon energy, corresponding

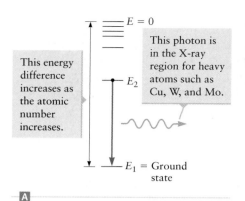

Figure 29.14 🅰 For heavy atoms, the emission lines can lie in the X-ray part of the electromagnetic spectrum. 🅱 The X-rays used by a dentist are photons generated by atomic transitions of atoms such as (typically) molybdenum and tungsten.

This energy difference increases as the atomic number increases.

$E = 0$

This photon is in the X-ray region for heavy atoms such as Cu, W, and Mo.

E_2

E_1 = Ground state

A   **B**

to a transition from the state with $E = 0$ to E_1 in Figure 29.12. The wavelength of this photon is $\lambda = 91.5$ nm (Example 29.2), which lies in the ultraviolet part of the electromagnetic spectrum. Although it is the highest photon energy that can be emitted by a hydrogen atom, other atoms can emit much more energetic photons. As the charge on the nucleus is increased (by going from hydrogen to helium and to heavier atoms), the magnitude of the electric potential energy of an electron also increases. As a result, for helium and heavier atoms the energy required to remove an electron from the $n = 1$ Bohr orbit (i.e., a tightly bound, "inner-shell" electron) is much larger than for hydrogen. The corresponding photon energies then increase into the X-ray region. Many applications employ X-ray photons generated when an electron undergoes a transition from the E_2 state to the E_1 state (Fig. 29.14A). These photons are denoted as "K X-rays," and their energies for a few elements are listed in Table 29.1. The X-ray photons used by your dentist or doctor are produced by these atomic transitions, usually involving atoms of tungsten or molybdenum (Fig. 29.14B).

CONCEPT CHECK 29.2 | K X-rays and the Energy Levels of an Atom

The energy of a K X-ray of copper (Cu) is 8.0 keV (Table 29.1). According to the Bohr model, is the energy required to remove an electron from the $n = 1$ Bohr orbit of a copper atom (and ionize the atom) (a) equal to, (b) less than, or (c) greater than 8.0 keV?

Continuous Spectra

The existence of discrete spectral lines is an essential part of quantum theory. In certain situations, however, the absorption may not involve discrete photon frequencies. Figure 29.15 shows a hydrogen atom that is initially in its ground state and then absorbs a photon. This photon has an energy greater than 13.6 eV, so it has more than enough energy to remove (eject) the electron from the atom. The extra energy goes into the kinetic energy of the ejected electron. Because the ejected electron's final kinetic energy can have a continuous range of values, the absorbed photon can also have a continuous range of energies. Hence, when an atom is ionized, the absorbed energy can have a range of values, producing what is called a continuous spectrum. Not everything is quantized in the Bohr model or in quantum mechanics.

Energy

e^- v

Kinetic energy of electron

$E = 0$

Photon

Figure 29.15 When an atom absorbs a photon and is ionized, the final kinetic energy of the electron is not quantized. Hence, the photon can have a range of energies, and the absorption energy in this case is *not* quantized.

Table 29.1	Energy of K X-rays Produced by a Few Elements	
ELEMENT	ATOMIC NUMBER (Z)	K X-RAY ENERGY (keV)
Ne (neon)	10	0.85
Al (aluminum)	13	1.6
Cu (copper)	29	8.0
Mo (molybdenum)	42	17
W (tungsten)	74	58

Why Is Angular Momentum Quantized?

Bohr's theory provided the first successful calculation of the frequencies of the spectral lines of any atom, but at first no one knew *why* it worked. What is the Bohr model telling us about the quantum world? Several of Bohr's assumptions can be traced to Einstein's theory of the photon and the notion that energy is conserved in atomic transitions, but Bohr's suggestion that the angular momentum of the electron is quantized (Eq. 29.8) was completely new. This assumption was needed to give quantized atomic levels, without which the Bohr model would have many of the same problems as Newton's mechanics and the planetary model. Bohr's assumption about angular momentum can be understood using the de Broglie theory of particle-waves. (But note that the de Broglie theory came about 10 years after Bohr's work.) De Broglie proposed (Chapter 28) that electrons and all other particles have a wave character, with a wavelength λ given by (Eq. 28.8)

$$\lambda = \frac{h}{p} \qquad (29.17)$$

where p is the momentum of the particle. Figure 29.16 shows such a particle-wave orbiting a nucleus, corresponding to an electron moving in a circular orbit. The allowed electron orbits in the Bohr model correspond to standing waves that fit precisely into the orbital circumference. To form a standing wave, the circumference of the orbit must be an integer multiple of the electron wavelength. Figure 29.16A shows a hypothetical case in which the electron wavelength does not match the orbit; for this value of r and λ, it is not possible to form a standing wave, and Bohr's condition for the angular momentum (Eq. 29.8) cannot be satisfied. This orbital radius is not allowed in the Bohr model. Figure 29.16B shows two cases in which the radius and wavelength match so that an integral number of wavelengths fit into one circumference and a standing wave is formed. These standing waves are two of the allowed quantized states of the Bohr model.

Figure 29.16 explains the qualitative origin of the discrete states in the Bohr model. We can also use this picture to derive Bohr's angular momentum relation in Equation 29.8 by setting the circumference of the circular orbit equal to an integer number of wavelengths. The circumference of an orbit is $2\pi r$, so if n is an integer, we require

$$2\pi r = n\lambda$$

The wavelength is given by de Broglie as $\lambda = h/p$ (Eq. 29.17), so

$$2\pi r = n\lambda = n\frac{h}{p} \qquad (29.18)$$

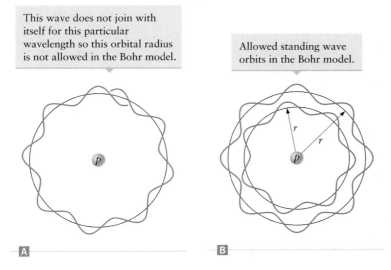

This wave does not join with itself for this particular wavelength so this orbital radius is not allowed in the Bohr model.

Allowed standing wave orbits in the Bohr model.

Figure 29.16 Bohr's quantization relation for angular momentum can be justified using the particle-wave nature of the electron, with the electron forming a standing wave in its orbit about the nucleus. The Bohr theory requires that an integer multiple of electron wavelengths equal the orbital circumference. In **A**, this condition is not met, so this orbital radius is not allowed in the Bohr theory for the wavelength shown. **B** Two allowed orbits. Note that the protons are not drawn to scale.

According to Newton's mechanics, the momentum of a particle is $p = mv$, which when inserted into Equation 29.18 gives

$$2\pi r = n\frac{h}{p} = n\frac{h}{mv}$$

Rearranging and noting that the angular momentum is $L = mvr$, we arrive at

$$2\pi r = n\frac{h}{mv}$$

$$L = mvr = n\frac{h}{2\pi} \qquad (29.19)$$

which is precisely Bohr's condition for the angular momentum, Equation 29.8.

Problems with the Bohr Model: Where Do We Go Next?

The success of the Bohr model of the hydrogen atom inspired Bohr and others to apply the same ideas to other atoms, and the Bohr theory was found to work well when only a single electron is present, as in the ions He^+ and Li^{2+}. This model does not, however, correctly explain the properties of atoms or ions that contain two or more electrons. Physicists eventually concluded that the Bohr model is not the correct quantum theory, but rather a "transition theory" that helped pave the way from Newton's mechanics to modern quantum mechanics. The Bohr model gives very useful insights into why an atom has quantized states, but the correct quantum theory contains even more radical ideas.

29.4 | WAVE MECHANICS AND THE HYDROGEN ATOM

Bohr based his work on Newton's mechanics, to which he added a few new assumptions about the nature of electron motion, but his approach still relied on classical ideas based on electrons moving in mechanical orbits. The quantum theory developed by Schrödinger, Heisenberg, and others in the 1920s provided a much more radical break with Newton. Modern quantum mechanics, also called *wave mechanics*, is based not on mechanical variables such as position and velocity, but on wave functions and probability densities as explained in Section 28.5. Although we will not go into the details here, one solves a problem in quantum mechanics using *Schrödinger's equation*. The solution of this equation gives the wave function, including its dependence on position and time. In this section, we describe the wave function and some features of the quantized states of the hydrogen atom. In Section 29.5, we'll apply these ideas to multielectron atoms to explain the structure of the periodic table of the elements.

In Bohr's theory, the electron levels are described by a single integer with allowed values $n = 1, 2, 3, \ldots$ corresponding to different quantized electron states. The integer n is called a *quantum number*. In Schrödinger's quantum theory, a full description of the quantum states of electrons in an atom requires *four* quantum numbers. These quantum numbers are collected in Table 29.2, which gives the standard notation and name for each. Each allowed electron energy level is specified by a set of values for all four quantum numbers and corresponds to one of the quantum states.

The Four Quantum Numbers for Electron States in Atoms

Quantum numbers for electron states in an atom

The four quantum numbers for electron states in atoms are n, ℓ, m, and s and are defined as follows.

n is the *principal quantum number*. It can have the values $n = 1, 2, 3, \ldots$. This quantum number is roughly similar to Bohr's quantum number. As n increases,

Table 29.2 Quantum Numbers for Allowed Electron States in an Atom

QUANTUM NUMBER	NAME	POSSIBLE VALUES
n	Principal quantum number	$n = 1, 2, 3, \ldots$
ℓ	Orbital quantum number	$\ell = 0, 1, 2, \ldots, n - 1$
m	Orbital magnetic quantum number	$m = -\ell, -\ell + 1, \ldots, 0, \ldots, \ell - 1, \ell$
s	Spin quantum number	$s = -\frac{1}{2}$ or $+\frac{1}{2}$

the average distance from the electron to the nucleus increases. (See also Example 29.3.) The states with a particular value of n are referred to as a "shell." For example, all the states with $n = 2$ make up the "$n = 2$ shell."

ℓ is the *orbital quantum number*, with allowed values $\ell = 0, 1, \ldots, n - 1$. The angular momentum of the electron is proportional to ℓ. An electron state with $\ell = 0$ is called an "s state," while states with $\ell = 1$ are called "p states." The shorthand letters for other states are (in order) "d" ($\ell = 2$) and "f" ($\ell = 3$) as listed in Table 29.3. States with $\ell = 0$ have zero angular momentum.

m is the *orbital magnetic quantum number*, with allowed values $m = -\ell$, $-\ell + 1, \ldots, -1, 0, +1, \ldots, +\ell$. Intuitively, you can think of m as giving the direction of the angular momentum of the electron in a particular state.

s is the *spin quantum number*. While the other quantum numbers all have integer values, the spin quantum number of an electron is $s = +\frac{1}{2}$ or $-\frac{1}{2}$, often referred to as "spin up" and "spin down." This quantum number gives the direction of the electron's spin angular momentum as discussed in Section 28.4.

Warning! Do not confuse the spin quantum number s with the notion of an "s state" ($\ell = 0$); they refer to completely different quantum numbers. This notation came about for historical reasons, and we are now (unfortunately) stuck with it.

Electron Shells and Probability Distributions

A particular quantized electron state is specified by all four of the quantum numbers n, ℓ, m, and s. Solution of the Schrödinger equation also gives the wave function of each quantum state, and from the wave function we can calculate the probability for finding the electron at different locations around the nucleus. The electron probability distributions for a few states of the hydrogen atom are shown in Figure 29.17; these probability plots are often called "electron clouds."

The ground state of hydrogen is specified by $n = 1$. According to the rules in Table 29.2, the only allowed values of the orbital and orbital magnetic quantum numbers for $n = 1$ are $\ell = 0$ (an "s state") and $m = 0$, but the electron can be in either the spin-up ($s = +\frac{1}{2}$) or spin-down ($s = -\frac{1}{2}$) state. The probability of finding the electron at a particular location does not depend on the value of s, so the spin-up and spin-down probabilities are the same. A plot of the electron probability for the ground state of hydrogen is shown at the far left of Figure 29.17A. The electron probability distribution forms a spherical "cloud" around the nucleus. For this state, the probability is largest at the nucleus (where the cloud is "darkest" in Fig. 29.17), so the electron has the highest probability of being found at or near the nucleus. In this ground state, the electron occupies the level with the lowest possible energy. There are two such states, with $n = 1$, $\ell = 0$, $m = 0$, and $s = +\frac{1}{2}$ or $-\frac{1}{2}$. Together they make up the $n = 1$ "shell" and are called the $1s$ states of the atom.

Figure 29.17 also shows the electron probability for the states with principal quantum number $n = 2$, called the $n = 2$ shell. The probability for the state with $n = 2$, $\ell = 0$, and $m = 0$ has a spherical component (a spherical "shell" where the

Table 29.3 Decoding the Configuration Shorthand for States with Various Orbital Quantum Numbers

ORBITAL QUANTUM NUMBER	CONFIGURATION LETTER
$\ell = 0$	s
$\ell = 1$	p
$\ell = 2$	d
$\ell = 3$	f
$\ell = 4$	g
$\ell = 5$	h

Figure 29.17 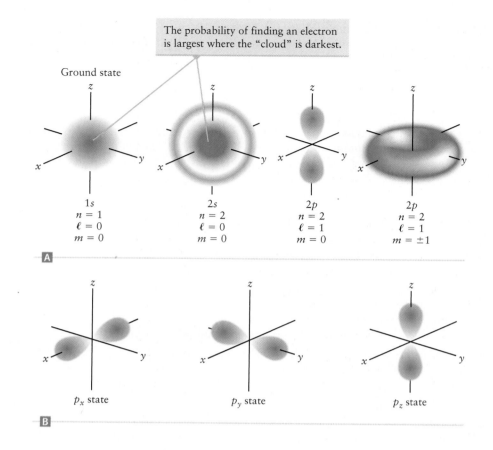 Electron probability distributions for the 1s, 2s, and 2p states of hydrogen, calculated using the Schrödinger equation. The 2p states can be pictured in two different ways. Here in part A, the 2p electron probability clouds for $m = \pm 1$ are shown with "doughnut" shapes. In **B**, the same 2p states are redrawn in an equivalent form popular in chemistry, called p_x, p_y, and p_z states.

The probability of finding an electron is largest where the "cloud" is darkest.

Ground state

1s
$n = 1$
$\ell = 0$
$m = 0$

2s
$n = 2$
$\ell = 0$
$m = 0$

2p
$n = 2$
$\ell = 1$
$m = 0$

2p
$n = 2$
$\ell = 1$
$m = \pm 1$

A

p_x state

p_y state

p_z state

B

probability is large) near the nucleus and a second "layer" (another spherical shell where the probability is high) farther away. The spin quantum number for electrons in this state can again have the values $s = +\frac{1}{2}$ and $-\frac{1}{2}$, forming two 2s states of the atom. The $n = 2$ shell also contains states with $\ell = 1$, which are called the 2p states. The value of the orbital quantum number ℓ specifies the "subshell," so the 2s and 2p states are said to be in the same shell (because they have the same value of n) but in different subshells. According to the quantum number rules in Table 29.2, there are 2p states with $m = +1$, 0, and -1. There are two ways to display the probability clouds for the 2p states with $m = \pm 1$. Many physicists prefer to show them as a doughnut-like electron cloud (on the far right in Fig. 29.17A). However, the two states with $m = \pm 1$ can also be combined to make what chemists call p_x and p_y states (Fig. 29.17B). While it is not obvious, both plots show the *same* quantum states. The chemists' way of displaying these electron probabilities is useful for understanding the directionality of chemical bonding.

The electron probability distributions for all these states are independent of the value of the spin quantum number. For the hydrogen atom, the electron energy depends only on the value of n and is independent of the values of ℓ, m, and s (except for some tiny effects we will not discuss in this book). That is not the case for atoms containing more than one electron, as we'll see in the next section.

29.5 | MULTIELECTRON ATOMS

The quantum states we have discussed for hydrogen can be used to describe the states of atoms containing more than one electron; they are called *multielectron atoms*. The electron energy levels of these atoms follow the same pattern as found for hydrogen, with the same quantum numbers listed in Tables 29.2 and 29.3. The electron probability distributions are also similar. There are two main quantitative differences between the electron states in a multielectron atom and in hydrogen. First, the values of the electron energies are different for different atoms. For example, the energy of the 1s state in helium is different from the energy of the 1s state in hydrogen. Second, the spatial extent of the electron probability clouds varies from element to element. For example, the 1s electron probability "cloud" for helium is closer on average to the nucleus than it is for hydrogen.

While the electron levels of all atoms are thus similar to those of hydrogen, one crucial feature is important for atoms with more than one electron: *each quantum state can be occupied by only one electron*. That is, each electron in an atom must occupy its own quantum state, different from the states of all other electrons. This is called the **Pauli exclusion principle**. Since each quantum state is characterized by a unique set of quantum numbers, each electron is described by a unique set of quantum numbers.

Pauli exclusion principle

Figure 29.18 shows how electrons are distributed among the possible energy levels in several different cases. The diagram in Figure 29.18A shows the levels of an "empty" atom, that is, an atom before we have added any electrons. For an "empty" atom and for hydrogen, all quantum states for a given value of Z and with a particular value of the principal quantum number n have almost exactly the same energy, so we only show a single horizontal line for each value of n.

In Figure 29.18B, we apply the energy-level diagram to hydrogen, in which there is only one electron to consider. We show this electron in the ground state (occupying the lowest energy level) as an arrow, with the direction of the arrow denoting the value of the electron's spin. (Here we have arbitrarily shown the electron as spin up, but it could just as well be spin down.) In words, the diagram in Figure 29.18B indicates that for a hydrogen atom in the ground state, the electron occupies the $n = 1$, $\ell = 0$, $m = 0$, $s = +\frac{1}{2}$ energy level. This is also called a "$1s^1$" *electron configuration*. This notation is a useful shorthand and is explained in Equation 29.20.

$$\text{electron configuration for H: } 1s^1 \qquad (29.20)$$

with labels: $n = 1$, one electron, $\ell = 0$

The first number in the configuration shorthand indicates the value of the principal quantum number n, while the letter indicates the value of ℓ using the code in Table 29.3. In Equation 29.20, the letter "s" indicates that $\ell = 0$. The superscript indicates that one electron is in the energy level with these values of n and ℓ. The value of the electron spin is not explicitly given in this notation.

Parts C through E of Figure 29.18 give energy-level diagrams for several other atoms. Helium (He) has two electrons, occupying levels with the quantum numbers $n = 1$, $\ell = 0$, $m = 0$, and $s = \pm\frac{1}{2}$, which satisfies the Pauli exclusion principle because

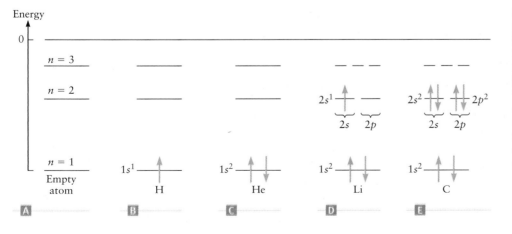

Figure 29.18 Filling the energy levels of an atom. **A** Empty levels. **B** Electron configuration of a hydrogen atom in its ground state. Configurations for **C** helium (He), **D** lithium (Li), and **E** carbon (C). These diagrams show the *order* of energy levels for each type of atom (each value of Z). The energies of these levels (e.g., the ground state and the excited states) are different for different atoms.

the electrons have different values of their spin quantum number. The electrons are indicated by the two arrows in Figure 29.18C, one pointing up (spin up, denoting $s = +\frac{1}{2}$) and the other pointing down (spin down, denoting $s = -\frac{1}{2}$). In the configuration shorthand, we have

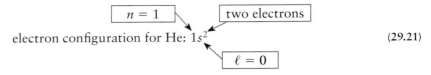

$$\text{electron configuration for He: } 1s^2 \tag{29.21}$$

This configuration differs from that of hydrogen (Eq. 29.20) only in the superscript; here the superscript 2 indicates that in helium there are two electrons in the energy levels with $n = 1$ and $\ell = 0$.

The energy-level diagram for the ground state of lithium (Li) is shown in Figure 29.18D; a lithium atom has three electrons, indicated by the three arrows. As with helium, two of these electrons occupy the levels with $n = 1$, $\ell = 0$, $m = 0$, and $s = \pm\frac{1}{2}$. According to the rules in Table 29.2, there are no other possible energy levels with $n = 1$. Since each electron must have its own unique set of quantum numbers (according to the Pauli exclusion principle), the third electron in lithium must occupy a higher energy level. This third electron is indicated by the arrow at the $n = 2$ level in Figure 29.18D. The lowest energy level in a shell has $\ell = 0$, so Figure 29.18D shows this third electron in the $2s$ state. The full set of quantum numbers for this electron is thus $n = 2$, $\ell = 0$, $m = 0$, and $s = +\frac{1}{2}$. The corresponding configuration for lithium is then

$$\text{electron configuration for Li: } 1s^2 2s^1$$

Notice that the configuration shorthand does not indicate the value of m for any of the electrons.

Carbon (C) contains six electrons. The third, fourth, fifth, and sixth electrons in a carbon atom occupy states with $n = 2$ as indicated in Figure 29.18E. The corresponding configuration is

$$\text{electron configuration for C: } 1s^2 2s^2 2p^2$$

In words, there are two electrons (spin up and spin down) in levels with $n = 1$ and $\ell = 0$; two electrons (spin up and spin down) in levels with $n = 2$, $\ell = 0$ (the $2s$ levels); and two electrons (spin up and spin down) in levels with $n = 2$ and $\ell = 1$ (the $2p$ levels).

The energies of levels in the diagrams in Figure 29.18 depend mainly on the value of n, with the energy increasing as n becomes larger. This is approximately true at small values of n, but for atoms containing many electrons, the order of energy levels is more complicated as n increases, and for shells higher than $n = 2$, the energies of subshells from different shells begin to overlap. In general, the energy levels fill with electrons in the following order:

$$1s \quad 2s \quad 2p \quad 3s \quad 3p \quad 4s \quad 3d \quad 4p \quad 5s \quad 4d \quad 5p \quad 6s \quad 4f \tag{29.22}$$

The $1s$ levels are filled first, followed by the $2s$ levels, then the $2p$ levels, and so on, with the overall order of the levels shown in Figure 29.19. This filling pattern is followed in Figure 29.18 and can also be used to get the results in Table 29.4 (page 1008), which lists the electron configurations for the ground states of all elements up to sodium. However, when one reaches the $3d$ levels the order of levels becomes even more complicated and can vary as electrons are added in going from one value of Z to the next.

CONCEPT CHECK 29.3 | What Element Is It?

The ground state of an atom has the configuration $1s^2 2s^2 2p^6 3s^2 3p^5$. What element is it? *Hint:* You may wish to consult the periodic table facing the last page of the index in the back of this book.

Energy

$6s$ $n = 6$ $\ell = 0$

$4f$ $n = 4$ $\ell = 3$

$5p$ $n = 5$ $\ell = 1$

$5s$ $n = 5$ $\ell = 0$

$4d$ $n = 4$ $\ell = 2$

$4p$ $n = 4$ $\ell = 1$

$4s$ $n = 4$ $\ell = 0$

$3d$ $n = 3$ $\ell = 2$

$3p$ $n = 3$ $\ell = 1$

$3s$ $n = 3$ $\ell = 0$

$2p$ $n = 2$ $\ell = 1$

$2s$ $n = 2$ $\ell = 0$

$1s$ $n = 1$ $\ell = 0$

s states p states d states f states

Figure 29.19 Order of energy levels in an atom.

EXAMPLE 29.5 | Electron Configuration and Energy-Level Diagram for Silicon

A silicon atom contains 14 electrons. Determine its electron configuration and construct the corresponding diagram of occupied levels.

RECOGNIZE THE PRINCIPLE

According to the Pauli exclusion principle, no two electrons can occupy the same quantum state, that is, have the same set of quantum numbers. To find the configuration of silicon, we assign the 14 electrons to the 14 quantum states that have the lowest energy.

SKETCH THE PROBLEM

We start with Figure 29.19 and add electrons to the lowest energy levels. Adding 14 electrons gives the diagram of filled energy levels in Figure 29.20.

IDENTIFY THE RELATIONSHIPS

We fill the lowest 14 energy levels in Figure 29.19 according to the rules in Table 29.2. The $1s$, $2s$, and $3s$ levels can each hold two electrons (one electron with spin up and another with spin down), and the $2p$ and $3p$ states can each hold six electrons.

SOLVE

The 14 electrons in silicon will fill completely the $1s$ (two electrons), $2s$ (two electrons), $2p$ (six electrons), and $3s$ (two electrons) energy levels, leaving two electrons for the $3p$ state. The resulting configuration is $\boxed{1s^2 2s^2 2p^6 3s^2 3p^2}$.

What does it mean?

The energy-level diagram in Figure 29.19 is the key to understanding the electron configurations of all the elements. The filled electron states of a particular atom are found by simply adding electrons to the lowest available energy levels.

Filled electron states for Si

$3p$ $n = 3$ $\ell = 1$

$3s$ $n = 3$ $\ell = 0$

$2p$ $n = 2$ $\ell = 1$

$2s$ $n = 2$ $\ell = 0$

$1s$ $n = 1$ $\ell = 0$

Energy

Figure 29.20 Example 29.5.

29.6 | CHEMICAL PROPERTIES OF THE ELEMENTS AND THE PERIODIC TABLE

It is probably fair to say that the periodic table of the elements is the key to all chemistry. One important triumph of quantum theory is that it explains why the periodic table has its structure. This table was first assembled by Dmitry Mendeleyev in the late 1860s. At that time, there were about 60 known elements, and Mendeleyev (along

Table 29.4 Electron Configurations for Several Elements

ELEMENT	NUMBER OF ELECTRONS	ELECTRON CONFIGURATION	QUANTUM NUMBERS OF OCCUPIED ELECTRON LEVELS
H (hydrogen)	1	$1s^1$	$n = 1, \ell = 0, m = 0, s = +\frac{1}{2}$
He (helium)	2	$1s^2$	$n = 1, \ell = 0, m = 0, s = +\frac{1}{2}$
			$n = 1, \ell = 0, m = 0, s = -\frac{1}{2}$
Li (lithium)	3	$1s^2 2s^1$	Same as He plus
			$n = 2, \ell = 0, m = 0, s = +\frac{1}{2}$
Be (beryllium)	4	$1s^2 2s^2$	Same as He plus
			$n = 2, \ell = 0, m = 0, s = +\frac{1}{2}$
			$n = 2, \ell = 0, m = 0, s = -\frac{1}{2}$
B (boron)	5	$1s^2 2s^2 2p^1$	Same as Be plus
			$n = 2, \ell = 1, m = 0, s = +\frac{1}{2}$
C (carbon)	6	$1s^2 2s^2 2p^2$	Same as Be plus
			$n = 2, \ell = 1, m = 0, s = +\frac{1}{2}$
			$n = 2, \ell = 1, m = 0, s = -\frac{1}{2}$
N (nitrogen)	7	$1s^2 2s^2 2p^3$	Same as C plus
			$n = 2, \ell = 1, m = 1, s = +\frac{1}{2}$
O (oxygen)	8	$1s^2 2s^2 2p^4$	Same as C plus
			$n = 2, \ell = 1, m = 1, s = +\frac{1}{2}$
			$n = 2, \ell = 1, m = 1, s = -\frac{1}{2}$
F (fluorine)	9	$1s^2 2s^2 2p^5$	Same as O plus
			$n = 2, \ell = 1, m = -1, s = +\frac{1}{2}$
Ne (neon)	10	$1s^2 2s^2 2p^6$	Same as O plus
			$n = 2, \ell = 1, m = -1, s = +\frac{1}{2}$
			$n = 2, \ell = 1, m = -1, s = -\frac{1}{2}$
Na (sodium)	11	$1s^2 2s^2 2p^6 3s^1$	Same as Ne plus
			$n = 3, \ell = 0, m = 0, s = +\frac{1}{2}$

with other chemists) had noticed that many elements could be grouped according to their chemical properties. For example, the elements lithium, sodium, and potassium form very similar chemical compounds and undergo similar reactions. Mendeleyev organized his table by grouping such related elements in the same column. He also placed the lightest (i.e., low atomic weight) elements at the top of each column and the heaviest ones at the bottom. His table was essentially identical to the modern version of the periodic table given in Figure 29.21 and facing the last page of the index in this book. The main difference was that Mendeleyev's table had a number of "holes" (i.e., openings) because many elements had not been discovered yet.

While Mendeleyev discovered an intriguing regularity in the elements and their properties, he could not explain why the periodic table has the form it does. That is, why does it contain just two elements in the first row, eight in the second and third rows, and even more elements in the lower rows? The answer is contained in the organization of electron energy levels given in Table 29.4 and Figure 29.19. To appreciate this answer, we must first understand how the electron energy levels and the electron configuration of an atom are responsible for its chemical properties.

When an atom participates in a chemical reaction, some of its electrons combine with electrons from other atoms to form chemical bonds. The bonding electrons are those that are most easily removed from an atom; hence, they are the electrons occupying the highest energy levels. For example, the ground state of lithium has

Figure 29.21 Periodic table of the elements.

the configuration $1s^2 2s^1$ (Fig. 29.18D and Table 29.4). The $2s$ electron occupies the state with the highest energy, so that electron is more weakly bound than the two electrons in the $1s$ state. This $2s$ electron forms bonds with other atoms and is called a *valence* electron. In contrast, the two electrons in the $1s$ state form a *closed shell*, a shell with all possible states filled. These $1s$ electrons have a much lower energy than the $2s$ electron. The $1s$ electrons are much more difficult to remove from the atom and do not participate in bonding (in chemical reactions).

The same ideas apply to other atoms. For example, the bonding of a sodium atom involves the $3s$ electron (Table 29.4). The remaining electrons form a closed $1s$ shell, a closed $2s$ subshell, and a closed $2p$ subshell. The configuration of the closed shells is thus $1s^2 2s^2 2p^6$. Likewise, potassium also has a single bonding electron, occupying the $4s$ state outside a $1s^2 2s^2 2p^6 3s^2 3p^6$ configuration. The elements lithium, sodium, and potassium thus have the same number of valence electrons (one), all in s states. As a result, these elements have similar chemical bonding properties, which is why Mendeleyev placed them in the same column in the periodic table (Fig. 29.21).

Closed Shells and the Periodic Table

The last column in the periodic table includes helium, neon, and argon (He, Ne, and Ar). Atoms of these elements all contain completely filled shells. Since they have no valence electrons, it is very difficult for them to form chemical bonds. Such elements are largely inert, almost never participating in chemical reactions. Atoms of these elements also interact very weakly with each other, resulting in low condensation and freezing temperatures. For this reason, they are called *inert gases* or *noble gases*. In Mendeleyev's time, elements were discovered through their chemical reactions and compounds, so very little was known about these elements when Mendeleyev did his work in the 1860s. In fact, most inert gases were discovered after Mendeleyev's time, so his version of the periodic table did not even contain that column.

Structure of the Periodic Table

In his periodic table, Mendeleyev grouped elements into *columns* according to their common bonding properties and chemical reactions. These properties rely on the valence electrons and can thus be traced to the electron configurations in Table 29.4. What, though, determines the number of elements in each *row* of the table, and why do different rows have different numbers of elements? Each row in the periodic table corresponds to a particular value of the principal quantum number n. The row containing hydrogen and helium corresponds to $n = 1$, the row beginning with lithium has $n = 2$, and so forth. The number of elements in a given row is equal to the number of electrons needed to fill completely a particular shell. Because the $n = 1$ shell can hold only two electrons, this row contains just two elements. The $n = 2$ shell can hold eight electrons (two in the $2s$ states and six more in the $2p$ states), so this row contains eight elements. The number of elements in all the other rows can be found using the rules for allowed quantum numbers in Table 29.2.

CONCEPT CHECK 29.4 | Valence Electrons for Barium

How many valence electrons does barium have, and what energy levels do they occupy?
   (a) One valence electron in the $6s$ state
   (b) Two valence electrons in $6s$ states
   (c) No valence electrons

CONCEPT CHECK 29.5 | Valence Electrons of an Atom

Which atoms from this group—(a) Sr, (b) F, (c) Mg, and (d) K—have the same number of valence electrons?

29.7 | APPLICATIONS

The existence of discrete spectral lines played a major role in the development of quantum theory and also makes possible a number of important applications.

Atomic Clocks and the Definition of the Second

Atomic clocks are used as global (and U.S.) time standards (Fig. 29.22). These clocks are based on the accurate measurement of certain spectral line frequencies.

Figure 29.22 ◻A Early atomic clock. This clock dates from the 1950s. ◻B Modern atomic clocks are much more compact and fit on the corner of a desk.

Since all atoms of a given element are identical, they all have precisely the same set of spectral lines at precisely the same frequencies.

Atomic clocks have been constructed using several different types of atoms, with cesium (Cs) being a popular choice. Two of the cesium levels differ by an energy of approximately 3.8×10^{-5} eV ($\approx 6.1 \times 10^{-24}$ J). These two levels are much closer in energy than the typical spacing of a few electron-volts seen for the levels in hydrogen. The spectral emission line produced when a cesium atom undergoes a transition between these two levels is very strong; that is, cesium atoms emit a relatively large intensity at this frequency. The frequency of this spectral line can also be measured with *extremely* high accuracy.

To make an atomic clock, a collection of cesium atoms in the gas phase is placed in a container. The atoms are excited by heating or other means so that many of them are in the higher of the two energy levels mentioned above. These atoms then undergo transitions to the lower state, all emitting photons with the same frequency. Separate calibration experiments have shown that this frequency is

$$f_{\text{Cs clock}} = 9{,}192{,}631{,}770 \text{ Hz} \tag{29.23}$$

This frequency is used as the "tick rate" of the atomic clock. A cesium clock thus ticks 9,192,631,770 times per second, forming a very accurate clock because the rate of ticking is extremely fast and is known very accurately. What's more, all cesium clocks have the same ticking rate, making them ideal for performing and comparing scientific measurements made in different laboratories at different times. In fact, in 1960 the cesium clock with Equation 29.23 was adopted as the SI definition of the second. One second is now *defined* to be the time it takes a cesium clock to complete precisely 9,192,631,770 ticks.

Atomic clocks are widely used in situations where very accurate time measurements are required. Applications such as GPS (Global Positioning System), which involves precisely timed radio signals from a set of satellites orbiting the Earth, are made possible by such clocks.

Fluorescent Lights

Not all applications of quantum and atomic theory are as exotic as atomic clocks. You probably see and use one of the applications every day: "neon" and fluorescent lighting.

There are two basic types of lightbulbs. The ***incandescent bulb***, developed by Thomas Edison, contains a very thin wire filament that carries a large electric current. The electrical energy dissipated in the filament heats it to a high temperature, and the filament then acts as a blackbody and emits radiation. The other type of lightbulb uses a gas of atoms in a glass container. In common terms, bulbs filled with a gas are often called either "neon" bulbs or "fluorescent" bulbs, but the spectra emitted by neon and fluorescent bulbs are somewhat different. A neon bulb contains a gas of Ne atoms. An electric current passes through the gas, producing ions and high-energy electrons as shown schematically in Figure 29.23A. The electrons, ions, and neutral atoms in the gas undergo many collisions, causing many of the Ne atoms to be in excited states. These atoms then decay back to their ground state, emitting light in the process as discussed in Section 29.2. Neon lamps thus emit light only at certain discrete wavelengths, that is, only with certain colors. By using different types of atoms, including neon and sodium, these lightbulbs can be used to make brightly colored signs (Fig. 29.23B).

Fluorescent lightbulbs (Fig. 29.23C) also contain a gas of atoms, often mercury (Hg), inside a glass bulb, but the inside of the bulb is coated with a fluorescent material. The Hg atoms are excited and emit photons in the same way as the Ne atoms in a neon bulb emit light. Mercury, however, emits strongly in the ultraviolet, and those photons are not detected by the eye, so Hg atoms alone would not make a useful lightbulb. This problem is overcome by the fluorescent material; the photons emitted by Hg are absorbed by the fluorescent coating, exciting those atoms and

Figure 29.23 A Neon bulbs contain a gas of Ne atoms in a glass tube. The atoms are excited by an electric current, and light is emitted when the Ne atoms undergo transitions from an excited state back to the ground state. B Neon bulbs emit red light and are used in many advertisements. The other colors seen in a "neon" sign come from atoms of other elements. C In a fluorescent lamp, the ultraviolet photons from an atom (here Hg) are absorbed by a coating on the inside of the tube and then reemitted as photons with a wide range of wavelengths in the visible range.

molecules to higher energy levels. When these atoms and molecules undergo transitions to lower energy states, they emit new photons. The fluorescent material is designed to emit photons throughout the visible spectrum, so a fluorescent lamp produces "white" light, that is, light of virtually all colors. Fluorescent bulbs were invented in the late nineteenth century, but their use has expanded greatly since about 2000 due to advances in the design of compact fluorescent bulbs (involving circuitry built into the base of the bulb).

EXAMPLE 29.6 | Photons Emitted by a Sodium Lamp

Sodium (Na) lamps emit light when the atoms undergo a transition from an excited $3p$ level to a $3s$ level. There are actually two levels in the $3p$ group with different values of the quantum numbers m and s and with slightly different energies, explaining why sodium emits two spectral lines that are very close in frequency (Fig. 29.8). The spectral line with a wavelength of $\lambda = 589.0$ nm is the stronger (more intense) of the two lines. What is the energy of these photons? Express your answer in units of electron-volts and compare it with the ionization energy of a hydrogen atom.

RECOGNIZE THE PRINCIPLE

This problem involves the relation between photon wavelength and energy. (See also Example 29.2.) We use the given photon wavelength λ to find the frequency f. The energy of the photon can then be found using $E_{\text{photon}} = hf$ (Eq. 29.7).

SKETCH THE PROBLEM

Figure 29.24 shows the transitions that produce the two yellow Na spectral lines.

IDENTIFY THE RELATIONSHIPS

The frequency of a photon produced by a Na lamp is

$$f = \frac{c}{\lambda}$$

where the wavelength is given above. Combining this equation with Equation 29.7, we find that the energy, frequency, and wavelength of a photon are related through

$$E = hf = \frac{hc}{\lambda}$$

SOLVE

Inserting the given value of the wavelength yields

$$E = \frac{hc}{\lambda} = \frac{(6.63 \times 10^{-34}\,\text{J} \cdot \text{s})(3.00 \times 10^{8}\,\text{m/s})}{5.890 \times 10^{-7}\,\text{m}} = 3.4 \times 10^{-19}\,\text{J}$$

Energy levels of Na

$\lambda = 589.0$ nm $\lambda = 589.6$ nm

3p

3s

Figure 29.24 Example 29.6. These three energy levels are involved in the yellow spectral lines emitted by a sodium atom.

which expressed in units of electron-volts is

$$E = (3.4 \times 10^{-19}\,\text{J})\left(\frac{1\,\text{eV}}{1.60 \times 10^{-19}\,\text{J}}\right) = \boxed{2.1\,\text{eV}} \tag{1}$$

The ionization of energy of H is 13.6 eV, so the energy of this spectral line in Na is much smaller.

What does it mean?

The energy of this photon emitted by Na lies in the yellow part of the visible spectrum, making it useful for lightbulbs. So-called high-pressure sodium bulbs are used in applications such as street lights, where very high intensities are required. The high pressures used in these bulbs cause them to emit light over a range of wavelengths (not just yellow).

Lasers

Lasers are another application of quantum and atomic theory and are commonplace in early-twenty-first-century life. Lasers depend on the coherent emission of light by many atoms, all at the same frequency.

In an ordinary fluorescent lightbulb, the transitions from the excited atomic states to lower energy levels occur in a random fashion. That is, once an atom is put into an excited state, it is impossible to predict when it will emit a photon, and the emitted photons are radiated randomly in all directions. In this *spontaneous emission* process, each atom emits photons independently of the other atoms. In a laser, an atom undergoes a transition and emits a photon in the presence of many other photons that have energies equal to the atom's transition energy (Fig. 29.25A). A process known as *stimulated emission* causes the light emitted by this atom to propagate in the same direction and with the same phase as surrounding light waves. Such a light source is called a laser, an acronym for *l*ight *a*mplification by *s*timulated *e*mission of *r*adiation. The light from a laser is thus a coherent source (Chapter 25) and is very useful for experiments involving the interference of light.

The design of many lasers is similar to that of a neon lightbulb, but there are mirrors at the ends of the bulb (called the laser "tube"). Light emitted by the gas atoms is reflected by the mirrors and travels back and forth along the tube (Fig. 29.25B). One of these mirrors is designed to let a small amount (typically a few percent) of the light pass through; that is the light you see from the laser. Just as in the neon bulb, atoms in the laser tube are in excited energy levels (produced typically by passing current through the gas). The excited atoms emit photons that stimulate the emission of more photons with the same energy, and all these photons are reflected repeatedly inside the tube. Photons resulting from stimulated emission are precisely in phase with the stimulating radiation. In this way, a laser can produce a very intense beam of light that is highly directional.

Lasers can be made with a variety of different atoms. A popular design uses a mixture of He and Ne gas and is called a helium–neon (He–Ne) laser. It was one of the first laser types invented (in 1960), making use of the energy levels shown in Figure 29.26. Helium atoms are excited by an electric current to an energy level

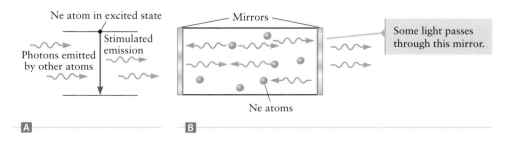

Figure 29.25 ⒶThe emission of photons by an atom can be stimulated by the presence of other photons with the same frequency. ⒷA laser uses mirrors to reflect light back and forth within the laser tube. A small amount of light is allowed to escape through one of the mirrors; that is the light you see coming from a laser.

Figure 29.26 Energy-level diagram for a helium–neon laser.

After emitting a photon, the Ne atom undergoes a transition back to its ground state.

about 23 eV above the ground state. There is an energy level of Ne at nearly the same energy, and when an excited He atom collides with a Ne atom, there is a high probability that the He atom's energy will be transferred to the Ne atom, leaving the Ne atom in an excited state, with an electron in its $3s$ level. The laser action makes use of photons emitted when electrons in these excited Ne atoms decay to a $2p$ level as sketched in Figure 29.26. After emitting a photon, the Ne atoms decay quickly to their ground state and are ready to begin the process again.

The photons emitted by a helium–neon laser have a wavelength of about 633 nm, which is in the red part of the visible spectrum. These lasers are used in many physics laboratories, often in lecture demonstrations. Another common type of laser is based on light produced by light-emitting diodes (called LEDs). These lasers typically have a wavelength near 650 nm (which is also in the red part of the spectrum) and are used in optical barcode scanners at many stores. Figure 29.27 shows some contemporary devices that use lasers.

CONCEPT CHECK 29.6 | Operation of a CO_2 Laser

The helium–neon laser described in Figure 29.26 emits visible light, but some lasers emit photons in the infrared. A popular type of infrared laser uses photons emitted by CO_2, which have a wavelength of 10.6 μm = 1.06×10^{-5} m. This wavelength is longer than that of the Na photon in Example 29.6 by a factor of 18. Which of the following statements is correct?
(a) The energy of a photon from a CO_2 laser is greater than that of a photon from a Na lamp by a factor of 18.
(b) The energy of a photon from a CO_2 laser is less than that of a photon from a Na lamp by a factor of 18.

The Force between Two Atoms ®

How much force does it take to pull two atoms apart? This question is important because physicists have recently developed ways to manipulate individual atoms, and one application of this work is to construct (or deconstruct) molecules one atom at a time. This question is also relevant for understanding the atomic force microscope (AFM) in which atoms from a sharp tip are scanned near atoms on the surface of another object. The AFM uses the force between atoms to produce an image of the surface (Chapter 11). An accurate quantum mechanical calculation of the force between two atoms would be very complex, but we can estimate it using what we know about the relation between potential energy and force, and using typical values of the ionization energy of an atom.

Consider the two hypothetical atoms in Figure 29.28A and assume they are bound together to form a molecule. The binding energy of a molecule is the energy required to break the chemical bond between the two atoms. This energy comes from the "sharing" of valence electrons, and we can estimate this energy from the spacing between energy levels in one of the atoms. Qualitatively, the sharing of valence electrons lowers the energy of one or more of the electrons by an amount that is approximately equal to the spacing between energy levels in an atom. For a hydrogen atom, this spacing is about 1 eV to 10 eV, and it will be greater for heavier atoms, so a typical bond energy is about 10 eV. That is the change in potential energy when comparing the molecule to two unbound atoms, so we have $\Delta PE \approx 10$ eV.

If this atom is now pulled apart by separating the atoms a distance Δx, there is an associated force F between the atoms. From the relation between potential energy and force (Sections 6.3 and 6.8), the magnitude of the force F is

$$F = \left| \frac{\Delta PE}{\Delta x} \right|$$

The radius of a hydrogen atom is about 0.05 nm (Example 29.3), so separating the atoms in Figure 29.28B a distance $\Delta x \approx 1$ nm should be large enough to break the chemical bond. Our estimate for the force between two atoms is thus

Figure 29.27 Some applications of lasers. A Some flatscreen televisions use laser light sources. B A barcode reader.

$$F = \left|\frac{\Delta PE}{\Delta x}\right| \approx \frac{10 \text{ eV}}{1 \text{ nm}}$$

An energy of 10 eV is equal to $(10 \text{ eV})(1.60 \times 10^{-19} \text{ J/eV}) = 1.6 \times 10^{-18}$ J. The force is thus

$$F \approx \frac{1.6 \times 10^{-18} \text{ J}}{1 \text{ nm}} = 1.6 \times 10^{-9} \text{ N} \tag{29.24}$$

This value is only an estimate for the force, but atomic force microscope experiments show that the force between two atoms is indeed close to that found in Equation 29.24. The keys to our calculation were the connection between force and potential energy, and our knowledge of the potential energy associated with a chemical bond.

Energy of a typical molecular bond is $|PE| \approx 10$ eV

A

$$F \approx \left|\frac{\Delta PE}{\Delta x}\right|$$

B

Figure 29.28 Estimating the force between two atoms.

29.8 | QUANTUM MECHANICS AND NEWTON'S MECHANICS: SOME PHILOSOPHICAL ISSUES

Newton's mechanics fails when applied to objects such as electrons and atoms. In that regime, quantum theory is required. Does that mean we should discard Newton's laws and redo all the physics in this book using quantum theory? To answer this question, we first note that Newton's laws work extremely well in the classical regime, that is, when applied to macroscopic objects such as baseballs and planets. Quantum theory can also be applied to such macroscopic objects, giving results that are virtually identical to Newton's mechanics. How can quantum theory, which is based on a particle-wave description of all objects, be consistent with Newton's particle theory for objects such as baseballs?

The answer depends on the wavelength of the particle-waves. The wavelength λ and momentum p of a particle-wave are connected by de Broglie's relation,

$$\lambda = \frac{h}{p} \tag{29.25}$$

An electron is a very light object (i.e., has a small mass), so its momentum is small. Equation 29.25 therefore tells us that λ for an electron is large. It is more accurate to say that the electron's wavelength is *relatively* large because we have seen that the electron wavelength in an atom is typically 0.1 nm (1×10^{-10} m). For an object such as a baseball, p is much larger than for an electron, so the wavelength is *much* smaller. For example, a baseball has a mass $m = 0.14$ kg and a speed of typically 50 m/s (about 100 mi/h). Inserting these values in Equation 29.25, the de Broglie wavelength of a baseball is

$$\lambda = \frac{h}{p} = \frac{h}{mv} = \frac{6.63 \times 10^{-34} \text{ J} \cdot \text{s}}{(0.14 \text{ kg})(50 \text{ m/s})} = 9.4 \times 10^{-35} \text{ m} \tag{29.26}$$

which is an *extremely* short wavelength. When a wave such as light has a very short wavelength, it moves in a straight line described by a simple ray (Chapter 24). The same is true of a baseball. It will move as a simple ray; that is, it will move as a simple particle. The wave properties of the baseball are only apparent on length scales comparable to the wavelength in Equation 29.26. Hence, for all practical purposes a baseball behaves as a classical particle, and its wave properties will never be observable. That is why Newton's mechanics works so well for classical objects. Such objects always have extremely short wavelengths, making the quantum theory description in terms of particle-waves unnecessary.

Where Quantum Theory Meets Newton

If Newton's mechanics works well for macroscopic objects while quantum theory is needed to describe behavior at the atomic scale, does anything interesting happen where these two regimes meet? Physicists are now actively studying this problem. One interesting question concerns the quantum behavior of a living entity. Objects such as viruses are both very small and "alive." Do the quantized energy levels of a living thing lead to any new behavior? What if the living thing is conscious? Can we understand the wave function of biological objects such as a brain? These intriguing questions are but a few of the ones physicists and biologists are now studying. The answers to some of these questions will likely emerge in the coming years. Stay tuned.

SUMMARY | Chapter 29

KEY CONCEPTS AND PRINCIPLES

The nuclear atom

Atoms contain *electrons*, *protons*, and *neutrons*. The positive charge in an atom is located in the *nucleus*, which contains protons and neutrons. The *Rutherford planetary model* pictures the atom as a miniature solar system in which electrons orbit the nucleus.

Atomic spectra

Atoms emit and absorb light (and other electromagnetic radiation) at certain discrete frequencies. These *spectral lines* result from transitions between discrete energy levels of an atom. When an electron undergoes a transition from a quantum state with energy E_2 to another with energy E_1, a photon of energy

$$hf = E_2 - E_1$$

is emitted. Here f is the frequency of the photon.

Bohr model

The *Bohr model* of the atom was an early attempt at incorporating quantum ideas into an atomic theory. A key assumption of the Bohr model is that the orbital angular momentum of an electron is quantized with

$$L = n\frac{h}{2\pi} \qquad \text{(29.8) (page 996)}$$

where n is an integer ($n = 1, 2, 3, \ldots$) and h is Planck's constant. This assumption implies that electrons form standing waves as they orbit the nucleus, with wavelengths given by the de Broglie theory of particle-waves ($\lambda = h/p$, where p is the momentum; Eq. 28.8). The Bohr model leads to discrete atomic energy levels with energies

$$E_{\text{tot}} = -\left(\frac{2\pi^2 k^2 e^4 m}{h^2}\right)\frac{1}{n^2} \qquad \text{(29.15) (page 997)}$$

The Bohr model predicts the correct values of the energy levels of hydrogen, but does not correctly describe atoms with two or more electrons.

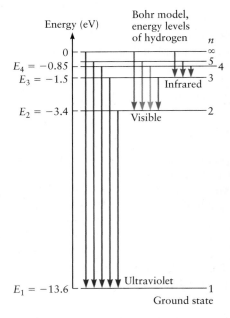

Allowed standing wave orbits in the Bohr model

Bohr model, energy levels of hydrogen

(Continued)

The four quantum numbers of the atom

The correct quantum theory is based on the Schrödinger equation. Each energy level in an atom is described by four *quantum numbers*.

1. n is the **principal quantum number**. Possible values are $n = 1, 2, 3, \ldots$.
2. ℓ is the **orbital quantum number**. Allowed values are $\ell = 0, 1, \ldots, n - 1$.
3. m is the **orbital magnetic quantum number**. Allowed values are $m = -\ell$, $-\ell + 1, \ldots, -1, 0, +1, \ldots, \ell$.
4. s is the **spin quantum number**. Allowed values are $s = +\frac{1}{2}$ ("spin up") and $s = -\frac{1}{2}$ ("spin down").

The Pauli exclusion principle

Each electron in an atom occupies a separate energy level and thus has a unique set of quantum numbers.

APPLICATIONS

Atomic clocks and lasers use transitions between discrete energy levels to produce photons with precisely defined frequencies. They are used in many applications, including the Global Positioning System, bar code readers, and CD and DVD players. The force between two atoms is also large enough to be measured in an atomic force microscope (AFM), which uses this force to form an image of atoms on the surface of an object.

QUESTIONS

SSM = answer in Student Companion & Problem-Solving Guide   ⊗ = life science application

1. Explain why He, Ne, and Ar have similar chemical properties.

2. Explain why some periodic tables list H twice, once above Li and again above the column containing F.

3. Consider the He$^+$ ion. Like a hydrogen atom, He$^+$ has one electron, but it has two protons, so the charge of the nucleus is $+2e$. Do you expect the ionization energy of He$^+$ to be greater than or less than that of H? Explain.

4. Is there an upper limit to the wavelength of light that a hydrogen atom can emit? Explain why or why not.

5. SSM Of the following configurations, which ones are *not* allowed for the outermost shell of an atom by quantum theory, (a) $4s^4$, (b) $3d^7$, (c) $4f^9$, or (d) $2d^3$? Explain why not.

6. Explain why the plum-pudding model of the atom is inconsistent with experimental results.

7. Explain why Rutherford's planetary model of the atom is inconsistent with classical physics and what we know about atoms.

8. Explain why the total energy of an electron in the Bohr model is negative.

9. Show that $4\ell + 2$ electrons can occupy a subshell with orbital quantum number ℓ.

10. Explain why F, Cl, and Br have similar chemical properties.

11. Explain why He, Ne, Ar, and Kr rarely form stable molecules.

12. Explain why the Pauli exclusion principle is not needed to understand the quantum states of a hydrogen atom.

13. Derive a general expression for the de Broglie wavelength of an electron for the Bohr model in state n.

14. Explain how different "neon" bulbs can emit light with different colors. Use the Internet to look up what kinds of atoms are used in bulbs that emit red, green, blue, and yellow light.

15. Explain how the elements Cr through Cu in the periodic table can have fairly similar chemical properties, even though they are not arranged in the same column of the periodic table.

16. Explain why it is difficult to chemically separate the elements Ce (cerium) through Lu (lutitium) from each other.

17. SSM What would the spectrum of hydrogen look like if the electrons could move between orbits of any radius in Bohr's model?

18. What experiment would you do to determine if there is oxygen in a distant star?

19. What experiment would you do to determine if there is oxygen in the atmosphere of a planet going around a distant star?

20. When illuminated with ultraviolet light, many minerals glow in the visible part of the electromagnetic spectrum. Why?

29.1 STRUCTURE OF THE ATOM: WHAT'S INSIDE?

1. The nucleus of a hydrogen atom is a single proton. The radius of a proton is approximately 1 fm (1×10^{-15} m). (a) What is the approximate volume of a proton? (b) A hydrogen atom has an approximate radius of 5×10^{-11} m (called the Bohr radius). What is the approximate volume of a hydrogen atom? (c) What fraction of a hydrogen atom is occupied by the nucleus?

2. SSM ★ ⊗ Suppose a hydrogen atom was somehow enlarged so that the nucleus is the size of a baseball (diameter approximately 8 cm). What is the approximate diameter of the volume occupied by the orbit of the electron in the planetary model? How does it compare with the diameter of the Earth?

3. ✪ Consider a Rutherford scattering experiment in which an alpha particle is fired at a sheet of atoms only one atomic layer thick (Fig. P29.3). The radius of an alpha particle is about 1 fm (1×10^{-15} m) and assume the radius of the nuclei in Figure P29.3 has the same value. Also assume the spacing between atoms in the sheet is 0.3 nm. (a) If alpha particles are fired at random at the sheet, what is the probability that one will strike a nucleus? (b) If one trillion (10^{12}) alpha particles are fired, about how many will hit a nucleus?

Alpha particles

3×10^{-10} m

Nucleus

Figure P29.3 Not to scale.

29.2 ATOMIC SPECTRA

4. An atom emits photons with an energy of 2.5 eV. (a) What is the spacing, in joules, between the energy levels involved in this atomic transition? (b) Where are these photons in the electromagnetic spectrum (i.e., infrared, visible, ultraviolet)?

5. A hypothetical atom contains energy levels as sketched in Figure P29.5. What are the energies of all possible emission lines involving these levels?

6. SSM ★ A hypothetical atom absorbs photons at the following energies (all given in electron-volts): 2.0, 4.0, 6.0, 8.0, and 10.0. Construct an energy-level diagram that explains this spectrum.

7. What is the minimum frequency a photon can have so as to ionize a hydrogen atom? Assume the atom begins in its ground state.

E (eV)

0

−2.5

−5.0

−7.5

Figure P29.5

29.3 BOHR'S MODEL OF THE ATOM

8. According to the Bohr model, what is the orbital radius of an electron in the $n = 10$ state of hydrogen?

9. Find the energy of an electron in the $n = 3$ state of hydrogen. If this electron undergoes a transition to the ground state, what is the energy, in electron-volts, of the photon that is emitted?

10. What is the smallest energy photon that can be absorbed by a hydrogen atom in its ground state?

11. How much energy is needed to ionize a hydrogen atom that is initially in its $n = 4$ state?

12. ★ In the Bohr model of a hydrogen atom, what is the electric force exerted between the proton and the electron when (a) the electron is in the $n = 1$ state and (b) the electron is in the $n = 5$ state?

13. ★ A hydrogen atom in the ground state absorbs a photon with an energy of approximately 12.8 eV. What is the value of n for the final state of the atom?

14. What is the energy needed to cause an electron to undergo a transition from the $n = 3$ state to the $n = 5$ state in hydrogen?

15. ★ A hydrogen atom emits a photon of energy about 2.55 eV. What are the values of n for the initial and final states involved in this transition?

16. ★ Find the de Broglie wavelength for an electron in the $n = 1$ state of hydrogen.

17. ★ Calculate the de Broglie wavelength of an electron in the $n = 3$ state of hydrogen. Is this wavelength larger or smaller than for the $n = 1$ state?

18. ✪ In the Bohr model, the energy of the different electron states depends on the value of n (Eq. 29.16). (a) Compute the energies of the states with $n = 1, 2, 3,$ and 4. (b) Plot your results from part (a) in an energy-level diagram like the one in Figure 29.12. (c) Compute the energies of all the possible emission lines involving these energy levels.

19. The radius of an electron orbit in the Bohr model depends on the value of n. (a) Find the value of n for which $r \approx 1\ \mu m$. These atoms are called Rydberg atoms. This orbit is relatively large; objects of this size are visible with an optical microscope. (b) Find the energy (according to the Bohr model) for the electron in part (a).

20. SSM ★ (a) What is the energy of the $n = 5$ state of hydrogen? (b) A hydrogen atom that is initially in some unknown state undergoes a transition and ends up in the $n = 5$ state. What is the highest energy photon that can possibly be *emitted*? (c) Does this photon fall in the visible part of the spectrum?

21. ★ A hydrogen atom in its ground state absorbs a photon with an energy of 21.0 eV, and the electron is ejected. What is the kinetic energy of the ejected electron?

29.4 WAVE MECHANICS AND THE HYDROGEN ATOM

22. How many electron states in the hydrogen atom have the quantum numbers (a) $n = 2$ and $\ell = 1$? (b) $n = 3$ and $\ell = 2$?

23. What is the total number of electron states in the $n = 3$ shell?

24. What are the allowed values of ℓ for electrons in the $n = 5$ shell?

25. SSM ★ An electron is in a subshell that contains 18 states. What is the value of ℓ?

29.5 MULTIELECTRON ATOMS

26. An atom has the electron configuration $1s^2 2s^2 2p^6 3s^2 3p^3$. What element is it?

27. Identify the following atoms from their electron configurations.
(a) $1s^2 2s^2 2p^6 3s^2 3p^6 3d^{10} 4s^1$
(b) $1s^2 2s^2 2p^6 3s^2 3p^6 3d^{10} 4s^2 4p^5$
(c) $1s^2 2s^2 2p^6 3s^2 3p^6 3d^{10} 4s^2 4p^6 5s^2$

28. How many electrons are in (a) the $4p$ subshell of Br, (b) the $n = 4$ shell of Rb, (c) the $4f$ subshell of U, and (d) the $3d$ subshell of Cs?

29. SSM Give the electron configurations of the following atoms in their ground states: (a) Ge, (b) Ar, (c) Mg, (d) Si, and (e) Br.

30. ✪ Table 29.4 lists the electron configurations of various atoms in their ground (lowest energy) states. The configuration notation can also describe atoms in their excited states. Which of the following configurations describe an excited state? Name the atom in each case.
(a) $1s^2 2s^2 2p^6 3s^1 3p^2$
(b) $1s^2 2s^2 2p^6 3s^2 3p^6 3d^3 4s^1$
(c) $1s^2 2s^2 2p^6 3s^2 3p^6 3d^5 4s^1$

31. How many electrons can occupy (a) the $3d$ subshell, (b) the $5s$ subshell, (c) the $n = 4$ shell, (d) the $4f$ subshell, (e) the $5g$ subshell, and (f) any h subshell?

29.6 CHEMICAL PROPERTIES OF THE ELEMENTS AND THE PERIODIC TABLE

32. SSM List all the known elements whose chemical properties are similar to those of fluorine (F).

33. Which elements have chemical properties similar to calcium (Ca)?

29.7 APPLICATIONS

34. Most helium–neon lasers emit visible light with a wavelength of 633 nm. These lasers can also emit photons in the infrared with a wavelength of 1.15 μm. What is the frequency of these photons?

35. SSM ✪ A ruby laser uses the photons emitted by Cr^{+3} ions that are embedded in a crystal of Al_2O_3 (instead of in a gas, as with a helium–neon laser). The light emitted by a ruby laser has a wavelength of approximately 694 nm. (a) What is the frequency of this light? (b) What is the spacing, in electron-volts, between the energy levels in Cr^{+3} that are responsible for these photons?

ADDITIONAL PROBLEMS

36. ✪ We applied the Bohr model to hydrogen, but it can be extended to treat ions with one electron such as He^+ and Li^{+2}. Follow the derivation in Equations 29.10 through 29.15, but now let the nucleus have a charge $+Ze$ (for a nucleus with atomic number Z containing Z protons) instead of just $+e$ (a single proton). Derive an expression for the energy of the ground state of the ion. Compare your result for the ionization energy of He^+ with the measured value of approximately 54 eV.

37. ✪ When an atom undergoes a transition from one level to another, the energy needed is usually provided by a photon, but it may also come from the kinetic energy of a collision. Consider the collision of two hydrogen atoms, one with an initial speed v and another initially at rest. What value of v is needed to give the moving hydrogen atom a kinetic energy of 13.6 eV? If all this energy can be transferred to one of the electrons, this collision could cause one of the atoms to ionize.

38. ✪ Consider an electron in the $n = 1$ state hydrogen. (a) Find the speed v of this electron and the radius r of its orbit in the Bohr model. (b) What is the time needed for the electron to complete one orbit? (c) This orbiting electron acts as a current loop. What is the value of the current? (d) What is the magnetic field produced by this current loop at the nucleus?

39. ✪ Scientists are able to study atoms much larger than a hydrogen atom in its ground state. Consider a hydrogen atom with $n = 200$. (a) Determine the diameter of this atom (i.e., the diameter of this Bohr orbit). (b) Compare the diameter of this atom to that of a dust particle (about 10^{-6} m).

40. Determine the dissociation energy for hydrogen's electron for the states with the following principal quantum number: (a) $n = 2$, (b) $n = 10$, (c) $n = 100$.

41. ✪ Consider a hydrogen atom in its $n = 5$ state. Suppose it decays to the $n = 2$ state before proceeding to its ground state. (a) Determine the wavelengths of the emitted photons. (b) To what region of the electromagnetic spectrum do these photons correspond?

42. ✪ Using the Bohr model of the hydrogen atom, show that the speed of an electron in an orbit is given by $v_n = (2.2 \times 10^6)/n$ m/s.

43. ✪ (a) How many different possible sets of quantum numbers are there for the $n = 3$ shell? (b) Write the explicit values of all the quantum numbers (n, ℓ, m, s) for electrons in this shell.

44. ✪ How many different possible sets of quantum numbers are there for the subshells with (a) $\ell = 0$, (b) $\ell = 2$, and (c) $\ell = 3$?

45. For each of the following ground-state electron configurations, identify the atom.
(a) $1s^2 2s^2 2p^6$
(b) $1s^2 2s^2 2p^6 3s^2 3p^4$
(c) $1s^2 2s^2$

46. ✪ Imagine that electron spin did not exist. (a) How many electrons would occupy the $1s$ state? Why? (b) If there were no electron spin, what would be the first two noble gases in the periodic table?

47. ✪ Suppose the electron spin had three possible values instead of two. For this situation, determine the electron configuration of the ground state of a lithium atom.

48. ✪ Suppose a hydrogen atom is in the $6h$ state. Determine (a) the principal quantum number, (b) the energy of the state, (c) the orbital angular momentum and its quantum number ℓ, and (d) the possible values for the magnetic quantum number.

49. ✪ The energy required to remove the "outermost" electron from a boron atom is 8.26 eV. As an approximation, you can model the quantum state of this electron as that of a single electron in a hydrogen atom, but with an "effective" charge on the nucleus that is different from the charge of a single proton ($+e$) as found for a real hydrogen atom. (a) Use the Bohr model to estimate the effective charge seen by this electron. (b) Estimate the average orbital radius for this electron.

50. ✪ Use the Bohr model to estimate the wavelength and frequency of the photon emitted during the $n = 3$ to $n = 1$ transition in molybdenum. The measured wavelength is 0.063 nm. Why do we not expect agreement?

51. ✪ The Bohr model can be accurately applied to a singly ionized helium atom. This atom has two protons in the nucleus ($Z = 2$) and a single electron orbiting the nucleus. When the electron makes a transition from a higher energy level to the $n = 4$ level, some of these transitions produce light in the visible part of the electromagnetic spectrum (about 380 nm to 750 nm). Determine which of these transitions correspond to these visible lines.

52. ✪ ✪ Consider laser eye surgery, in which laser light with a wavelength of 514 nm at a power output of 1.5 W is used to reattach a detached retina. Suppose the laser is pulsed for 0.050-s time intervals. (a) During this time, how many photons are emitted by the laser? (b) What is the difference in energy between the two energy levels involved in this laser transition?

53. ✪ In a muonic atom, the electron is replaced by a negatively charged particle called a muon. The muon has the same charge as an electron, but its mass is 207 times the mass of an electron.

Apply the Bohr model to a muonic hydrogen atom. (a) Calculate the radius of the orbit of the muon in the ground state, first excited state, and second exited state. (b) Calculate the energy of these states. (c) Determine the wavelength of the photon emitted when the muon undergoes a transition from the $n = 3$ to $n = 1$ state.

54. ✪ Ⓡ Suppose two hydrogen atoms, both initially in their ground state, are accelerated up to high speed and then collide head on. (a) Derive an expression for the speed necessary to raise both atoms to the same excited state, n. Express your answer as a function of n. (b) Evaluate this expression for $n = 2$. (c) To what temperature would you need to heat hydrogen gas to make this collision likely?

55. SSM ✪ LASIK, an acronym for *laser*-assisted in *situ* ker-atomileusis, is a form of laser eye surgery performed by ophthalmologists to correct myopia, hyperopia, and astigmatism. The procedure uses an excimer laser, which produces photons with a wavelength of 193 nm to remodel the corneal stroma. The photons vaporize thin layers of tissue less than 1 μm in thickness. Typically, the output power of the laser is around 5.0 mW, and it is pulsed for 10 nanoseconds. (a) During one of these pulses, how much energy is transferred to the cornea? (b) How many photons are contained in one pulse of the laser? (c) How many photons from a helium–neon laser would be required to deposit the same amount of energy? (d) What is the difference in energy between the two energy levels involved in this laser transition for the excimer laser? (e) Could you see the light from the LASIK laser?

56. ✪ The wavelength of the light from a sodium-vapor lamp is 589 nm. Assume a 100-W sodium-vapor lamp radiates its energy uniformly in all directions. (a) At what rate are photons emitted from the lamp? (b) At what distance from the lamp will the average flux of photons be 1 photon/(cm² · s)? (c) What is the difference in energy between the two energy levels involved in the atomic transition that produces these photons?

57. ✪ Ⓡ **Microprobe.** Energy dispersive spectroscopy (EDS) is a technique for identifying and quantifying the elemental composition of a sample. Volumes as small as a few cubic micrometers can be probed using EDS. The characteristic X-rays of the elements in a sample are produced when the sample is bombarded with electrons in an electron beam instrument such as a scanning electron microscope. For elements heavier than hydrogen, we can approximate the energy of an electron in the $n = 1$ state (called a K-shell electron) using

$$E_K = -\frac{(13.6 \text{ eV})(Z - 1)^2}{n^2}$$

with $n = 1$. Here Z is the atomic number of the element, and the $(Z - 1)$ term corrects for the partial cancellation of the charge

the electron sees as a result of the other K-shell electron. Suppose we want to detect the presence of aluminum, copper, and tungsten in a sample. (a) Estimate the minimum potential difference through which the bombarding electron beam must be accelerated to produce the characteristic X-rays of each of these elements. (b) From the data provided in Table 29.1, estimate the energy of an electron in the $n = 2$ state for each of these elements.

58. ✪ Suppose a certain ionized atom produces an emission line in accordance with the Bohr model, but the number of protons in the nucleus Z is not known. A group of lines in the spectrum forms a series in which the shortest wavelength is 22.8 nm and the longest is 41.0 nm. Using a modified version of Equation 29.15, find the second longest wavelength in this series of lines (See Problem 57 for a hint.).

59. ✪ A quantum system (a molecule) has the energy levels shown in Figure P29.59. A large number of these molecules in their ground states make up a gas. (a) Suppose we want to excite these molecules to the state labeled E_2 by shining light on the gas. What wavelength of light should we use? (b) If we expose the molecules to a continuous spectrum of light, photons of particular wavelengths might be absorbed. What wavelength photons would you expect to be absorbed? Do not assume that the molecules all start in their ground state. Explain why that assumption makes sense.

Figure P29.59

60. ✪ Ⓧ Ⓡ **Bioluminescence.** The process by which fireflies glow, called bioluminescence, is defined as the emission of light from living organisms. In fireflies, bioluminescence occurs in cells called photocytes that are located in the last two segments of the abdomen. Fireflies produce light through a two-step chemical process. During the first step, ATP (adenosine triphosphate), luciferin, and luciferase react to form luciferyl adenylate-luciferase and pyrophosphate. During the second step, luciferyl adenylate-luciferase decomposes in the presence of oxygen to form oxyluciferin and luciferase, releasing energy in the form of photons (visible light). These photons are in the yellow-green part of the spectrum with wavelengths around 550 nm. Typically, fireflies produce light in pulses up to a few seconds in length and sequence the pulses in a pattern designed to attract a mate. (a) Comparing the brightness of a firefly with other light sources with which you are familiar, estimate the power output of the firefly's light pulse. (b) About how much energy is released by the firefly in the form of light during a 1-s pulse? (c) Approximately how many photons are released during one pulse?

Nuclear Physics

Marie Curie was a pioneer in the study of radioactivity. She received two Nobel Prizes for her work: one in physics (along with her husband, Pierre, and French physicist Henri Becquerel) in 1903 and a second prize in chemistry in 1911. (© Photo Researchers/Alamy)

In Chapters 28 and 29, we described the development of quantum theory and discussed the quantum properties of electrons and photons. Rutherford's discovery of the atomic nucleus raised the obvious question, "How is the nucleus itself put together?" In Rutherford's time, approximately 85 different elements had been identified. Did every element have a completely different kind of nucleus? We now know that things are not that complicated: every nucleus is composed of just two different building blocks, particles called **protons** and **neutrons**. Different elements have different numbers of these two particles in their nuclei.

The protons and neutrons in most nuclei are bound together very tightly; that is, the potential energy of binding is large. Because a great deal of energy is needed to break apart a nucleus, the nucleus can usually be treated as a simple, unbreakable "particle" when dealing with phenomena such as chemical reactions and atomic spectra. In *nuclear reactions*, however, a nucleus breaks apart or is assembled when smaller nuclei and particles combine. Nuclear reactions can release very large amounts of energy that can be applied for great societal benefit, although they can also be used destructively. In this chapter, we describe the structure and properties of the nucleus and explain some applications of nuclear physics.

30.1 | STRUCTURE OF THE NUCLEUS: WHAT'S INSIDE?

The discovery of the nucleus is usually credited to Rutherford as a result of his scattering experiments described in Chapter 29. The experiments of Rutherford and others showed that all nuclei are composed of two particles, protons and neutrons. These two types of particles, called *nucleons*, are both much more massive than an electron. Their masses are

The nucleus is composed of nucleons (protons and neutrons).

$$m_p = 1.673 \times 10^{-27} \text{ kg} \quad \text{(mass of a proton)} \tag{30.1}$$

$$m_n = 1.675 \times 10^{-27} \text{ kg} \quad \text{(mass of a neutron)} \tag{30.2}$$

The values of m_p and m_n are thus very similar, and both are about 1800 times more massive than an electron. Perhaps the most important difference between protons and neutrons is that the neutron is electrically neutral, whereas the proton carries a positive charge $+e$. The charge on an electron is $-e$, so the total charge carried by one electron plus one proton is precisely zero.

Every nucleus can be specified by the numbers of protons and neutrons it contains. The number of protons in a nucleus is called the *atomic number*, Z. Because each element has a particular and unique number of protons in its nucleus, each has a characteristic value of Z. For example, the atomic number of He (helium) is $Z = 2$, while the atomic number of O (oxygen) is $Z = 8$. Since atoms are electrically neutral, Z is also equal to the number of electrons in an atom.

Definition of atomic number

The number of neutrons in a nucleus is usually denoted by the symbol N. The value of N for a particular element can vary. For example, He nuclei can have $N = 1$, $N = 2$, or $N = 4$, and other values are also possible. The number of neutrons in atomic nuclei generally increases as the atomic number increases.

The *mass number*, A, of a nucleus is the sum of the number of protons and neutrons, so

Definition of mass number

$$A = Z + N \tag{30.3}$$

All this information is described in the following shorthand notation:

$$\boxed{\text{mass number}} \longrightarrow {}^{A}_{Z}\text{X} \longleftarrow \boxed{\text{symbol for element}} \qquad \boxed{\text{atomic number}} \tag{30.4}$$

The letter "X" in Equation 30.4 denotes an element whose chemical symbol should be substituted from the periodic table. As an example, a He nucleus containing $N = 2$ neutrons is denoted

$$\boxed{A = 4 \text{ (2 protons + 2 neutrons)}} \longrightarrow {}^{4}_{2}\text{He} \longleftarrow \boxed{\substack{\text{symbol for element} \\ \text{(helium)}}} \qquad \boxed{Z = 2 \text{ (2 protons)}}$$

The various helium nuclei having $N = 1$, 2, or 4 are written as

$${}^{3}_{2}\text{He} \qquad {}^{4}_{2}\text{He} \qquad {}^{6}_{2}\text{He}$$

In Chapter 29, we found that a typical atomic-scale energy is around 10 eV. (The ionization energy of the hydrogen atom is 13.6 eV.) Equation 30.7 indicates that a typical nuclear-scale energy is larger than that by a factor of approximately 100,000! The potential energy associated with a nucleus is thus *much* larger than the energies associated with electrons, molecules, and chemical reactions.

Forces in the Nucleus

When Rutherford first did the calculation that leads to Equation 30.7, he must have been somewhat puzzled. The electric potential energy in Equation 30.7 is positive since all protons are positively charged. The protons in a nucleus thus repel one another with a very large force. To make a nucleus stable, there must be another even larger force—what physicists call the *strong force*—at work in the nucleus that is capable of producing an enormous attractive force between nucleons, strong enough to overcome the electrical repulsion. The strong force attracts any two nucleons to each other. Attraction due to the strong force occurs between a pair of protons, a pair of neutrons, and a proton and a neutron (Fig. 30.3). The strong force has a very large and approximately constant magnitude when the two nucleons are about 1 fm apart, but it is negligible when they are separated any farther (Fig. 30.4).

The strong force gives a large attractive force between nucleons and holds the nucleus together, but it does not act on electrons. So, when an electron approaches a nucleus, it experiences essentially just an electric force (Coulomb's law) due to the protons in the nucleus.[1] The strong force does not play any role in the binding of electrons in an atom.

Stability of the Nucleus

The simplest nucleus is ^1_1H (hydrogen), consisting of just a single proton. All other nuclei contain both protons and neutrons. For example, there are several nuclei of helium, including ^3_2He and ^4_2He, all containing neutrons. Why is there not a nucleus of helium that contains just two protons? The reason is that the strong force between two protons is not large enough to overcome the repulsive electric force in that case, so a nucleus containing just two protons is not stable (Fig. 30.3C). The addition of one or more neutrons provides the extra attractive force needed to overcome the Coulomb repulsion of the protons. If a neutron is positioned between two protons (Fig. 30.5), the strong force attracts the neutron to each of the protons. The presence of this neutron also increases the separation between the protons, reducing the repulsive Coulomb force. Neutrons are thus essential for the stability of the nucleus. The number of neutrons generally increases as the atomic number increases; as more and more protons are added to the nucleus, more and more neutrons are needed to keep all the protons sufficiently separated (at a "safe distance") so that the Coulomb force does not cause the nucleus to fly apart.

The picture in Figure 30.5 indicates that neutrons are an essential "glue" that holds the nucleus together. Based on this picture, you might also think that adding more glue (neutrons) should always result in a more stable nucleus, but adding more and more neutrons will eventually lead to an unstable nucleus. This result can be explained by the quantum theory of the nucleus called the *nuclear shell model*, which gives the energy levels of nucleons inside a nucleus, similar to the energy levels of electrons within an atom (Chapter 29).

Mass and Energy Scales in Nuclear Physics

We have seen that the electric potential energy of protons in a nucleus is extremely large, on the order of MeV (Eq. 30.7). The strong force in the nucleus must overcome this positive potential energy, so its associated potential energy is negative and is also on the order of MeV.

---

[1]An electron also experiences another, much weaker force, called the "weak force," which we'll discuss in Chapter 31.

The strong force is an attractive force between nucleons.

THE STRONG FORCE ACTS BETWEEN ANY TWO NUCLEONS

Two neutrons are attracted to each other by the strong force.

$\vec{F}_S \ \vec{F}_S$

n ⟶ ⟵ n

A

A neutron and a proton are attracted by the strong force.

$\vec{F}_S \ \vec{F}_S$

n ⟶ ⟵ p

B

The total force between two protons is the sum of the electric and strong forces.

$\vec{F}_E \qquad \vec{F}_S \ \vec{F}_S \qquad \vec{F}_E$

C

Figure 30.3 The strong nuclear force leads to an attraction between any pair of nucleons. The neutron has a net electric charge of zero, so the dominant force between **A** a pair of neutrons or **B** a neutron and a proton is the strong force. **C** Since protons are charged, there is a repulsive electric force between two protons along with the attractive strong force. If only two protons are involved, the electric force is larger in magnitude than the strong force, so a nucleus containing just two protons is not stable.

The strong force between nucleons is negligible when the separation is larger than about 1 fm.

F_S

$r \approx 1$ fm

r

The strong force is attractive for separations in this range.

Figure 30.4 The strong force is important only when the two nucleons are closer than about 1 fm. At larger separations, the strong force between two nucleons is negligible. At very small separations, the strong force is repulsive, so two nucleons can't get too close together.

The strong force attracts these protons to a neutron located between them.

$\vec{F}_S$ $\vec{F}_S$

$\vec{F}_E$ p   n   p $\vec{F}_E$

Figure 30.5 A neutron located between two protons acts as a sort of "glue" to bind them all together. Each proton is attracted to the neutron, and since the protons are now farther apart than they would be if the neutron were not present, the repulsive electric force is smaller than in Figure 30.3C. As a result, a nucleus with two protons and one neutron is stable. The nucleus shown here is $^3_2$He.

Another important energy scale in nuclear physics is the rest energy associated with the proton and neutron. According to the theory of relativity (Chapter 27), the rest energy of a particle of mass m is

$$E = mc^2 \qquad (30.8)$$

where c is the speed of light. For a proton, we find

$$E_p = m_p c^2 = (1.67 \times 10^{-27} \text{ kg})(3.00 \times 10^8 \text{ m/s})^2 = 1.50 \times 10^{-10} \text{ J}$$

$$E_p = 938 \text{ MeV} \qquad (30.9)$$

The mass and rest energy of any particle are related though Equation 30.8, which can be rearranged to read $m = E/c^2$. Nuclear physicists often specify the mass of a particle in units of (energy)/c^2 = MeV/c^2. In these units, the mass of a proton is (using Eq. 30.9)

$$m_p = 938 \text{ MeV}/c^2 \qquad (30.10)$$

The masses of the proton, neutron, and electron are collected in Table 30.1 in units of kilograms and MeV/c^2 and also in terms of the *atomic mass unit*, denoted by "u." This unit is defined as 1/12th of the mass of a carbon atom with the nucleus $^{12}_6$C; this nucleus contains 6 neutrons and 6 protons. An atomic mass unit is related to other units of mass by

$$1 \text{ u} = 1.66 \times 10^{-27} \text{ kg} = 931.5 \text{ MeV}/c^2 \qquad (30.11)$$

Atomic mass units are handy in nuclear physics because 1 u is approximately equal to the mass of a nucleon (either a proton or a neutron).

The values in Table 30.1 include more significant figures than we have commonly used in this book because (as we will soon see) small changes in mass in a nuclear reaction correspond to large amounts of released energy, which can be calculated using the relation between mass and energy in Equation 30.8. When working problems in nuclear physics, we need to keep track of small differences and changes in mass.

30.2 | NUCLEAR REACTIONS: SPONTANEOUS DECAY OF A NUCLEUS

There are two types of nuclear reactions: *spontaneous* and *induced*. In a nuclear reaction, one or more nuclei and nucleons "react" to form new nuclei, perhaps emitting nucleons and releasing energy. For example, a collision with an alpha particle can sometimes cause a nucleus to break into fragments. This reaction is called an induced nuclear reaction because the alpha particle caused the decay. We'll discuss several important examples of induced reactions when we describe the processes of nuclear fission and fusion in Section 30.3, but let's first discuss reactions that occur spontaneously.

Radioactivity

In addition to Rutherford, many famous scientists, including the husband-and-wife team of Marie Curie and Pierre Curie, played important roles in the development

Table 30.1 Properties of the Proton, Neutron, and Electron

PARTICLE	CHARGE	MASS (kg)	MASS (atomic mass units[a] = u)	MASS (MeV/c^2)
Proton	$+e$	1.6726×10^{-27}	1.0073	938.27
Neutron	0	1.6749×10^{-27}	1.0087	939.57
Electron	$-e$	9.109×10^{-31}	5.486×10^{-4}	0.5110

[a]1 u corresponds to 931.5 MeV/c^2.

of nuclear physics. The Curies and others found that some nuclei are unstable; that is, certain nuclei decay spontaneously into two or more particles. This process is called *radioactive decay*, and the term *radioactivity* refers to the process in which a nucleus spontaneously emits either particles or radiation.

When a nucleus decays, it can emit particles with mass such as electrons or neutrons, or it can emit photons (electromagnetic radiation). At first, physicists did not know the identities of the particles emitted in radioactive decay, so these unknown decay products were initially given the names *alpha*, *beta*, and *gamma*. Eventually, the nature of these particles was deduced from experiments, but these somewhat mysterious names are still widely used. Even though their identity was initially unknown, these three types of decay products could be distinguished according to their properties, including their electric charge and their ability to penetrate matter.

Loosely speaking, we sometimes refer to alpha, beta, and gamma radiation emitted in radioactive decay. The term *radiation* is reminiscent of electromagnetic radiation (Chapter 23). We now know that only gamma radiation, also called "gamma rays," is actually a form of electromagnetic radiation, while alpha and beta "radiation" consists of particles with mass.

Alpha particles. An alpha particle is composed of two protons and two neutrons. It is thus the nucleus of a He atom and is also denoted $^4_2$He. An alpha particle produced by nuclear decay usually does not have any bound electrons, so alpha particles almost always have a charge of $+2e$ (the charge of two protons). A typical radioactive decay involving an alpha particle is that of the radium nucleus $^{226}_{88}$Ra. This process is written in a form very similar to a chemical reaction:

$$\underset{\text{parent nucleus}}{^{226}_{88}\text{Ra}} \quad \rightarrow \quad \underset{\text{daughter nucleus}}{^{222}_{86}\text{Rn}} \quad + \quad \underset{\text{alpha particle}}{^4_2\text{He}} \tag{30.12}$$

Alpha decay: emission of an alpha particle from a nucleus

This decay reaction (Fig. 30.6A) begins with a *parent nucleus* of $^{226}_{88}$Ra containing 88 protons and $N = A - Z = 226 - 88 = 138$ neutrons. This nucleus spontaneously decays into two particles called *decay products*. One decay product is an isotope of radon, $^{222}_{86}$Rn, which contains $N = A - Z = 222 - 86 = 136$ neutrons. In this reaction, $^{222}_{86}$Rn is called the *daughter nucleus*. The other decay product in Equation 30.12 is an alpha particle, $^4_2$He. The total number of nucleons is *conserved* (does not change) in the reaction. Since the reaction products contain 222 nucleons ($^{222}_{86}$Rn) plus 4 nucleons ($^4_2$He), there are a total of 226 nucleons (protons plus neutrons) at the start of the reaction and the same number at the end. This particular reaction also conserves the numbers of protons and neutrons separately. That is, there are 88 protons at the start (in $^{226}_{88}$Ra) and 88 at the end (86 in the radon daughter nucleus and 2 in the alpha particle). You can view this reaction as simply ejecting two protons and two neutrons (the alpha particle) from the original radium nucleus. In his scattering experiments, Rutherford obtained his alpha particles from the alpha decay of nuclei such as $^{214}_{84}$Po.

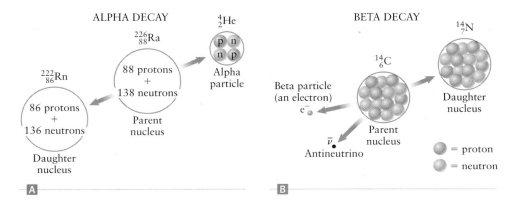

Figure 30.6 **A** Example of alpha decay: the original parent nucleus ($^{226}_{88}$Ra) splits to form $^{222}_{86}$Rn and an alpha particle ($^4_2$He). **B** Example of beta decay: the original $^{14}_6$C nucleus produces $^{14}_7$N plus an electron (a beta particle) and an antineutrino ($\bar{\nu}$).

The particular alpha decay reaction in Equation 30.12 has important medical consequences. It contributes to the (ill) health of miners and can also lead to a health hazard in homes. For more details, see Section 30.4.

Beta particles. There are two varieties of beta particles. One variety is negatively charged and is simply an electron. The other type is identical to an electron except it has a positive charge of precisely $+e$. This positive beta particle is called a *positron* (denoted e^+, with the superscript indicating its positive charge) and is an example of *antimatter* (Chapter 31).[2] Electrons and positrons have the same mass and are both point charges.

A typical radioactive decay producing a beta particle is

parent   daughter   beta particle   antineutrino
nucleus   nucleus   (electron)

$$\,^{14}_{6}\text{C} \;\rightarrow\; \,^{14}_{7}\text{N} \;+\; e^- \;+\; \bar{\nu} \tag{30.13}$$

In this reaction (shown schematically in Figure 30.6B), the emitted beta particle is an electron denoted by e^-; the superscript negative sign indicates that the particle is negatively charged. This reaction also produces a particle called an *antineutrino*, denoted by $\bar{\nu}$. The antineutrino is another example of antimatter and is thus a "cousin" of the positron. We'll discuss neutrinos and antimatter further in Chapter 31.

As with alpha decay, beta decay also conserves the total number of nucleons. It is useful to rewrite Equation 30.13 to indicate explicitly both the number of protons and neutrons and the electric charge at the start and end of the reaction:

number of protons: $6 \rightarrow 7$

number of neutrons: $8 \rightarrow 7$

$$\,^{14}_{6}\text{C} \rightarrow \,^{14}_{7}\text{N} + e^- + \bar{\nu}$$

charge: $+6e \rightarrow +7e - e + 0$ (30.14)

Hence, this reaction converts a neutron in the parent $^{14}_{6}\text{C}$ nucleus into a proton, conserving the total number of nucleons (neutrons plus protons). At the same time, electric charge is conserved since the total charge is $+6e$ at the beginning and at the end of the reaction. The positive charge of the new proton is balanced by the creation of an electron.

The beta decay reaction in Equation 30.13 is widely used in archeology in the technique called *carbon dating*. We'll discuss this application of radioactivity in Section 30.5.

Gamma decay. Gamma decay produces gamma "rays," which are photons, quanta of electromagnetic radiation denoted by the Greek letter γ. One example of a nuclear decay that produces gamma rays is

parent nucleus   daughter   gamma
(in excited state)   nucleus   ray

$$\,^{14}_{7}\text{N}^* \;\rightarrow\; \,^{14}_{7}\text{N} \;+\; \gamma \tag{30.15}$$

The asterisk on the left in this equation denotes that the nucleus is in an excited state. In Chapter 29, we saw that an atom in excited state can undergo a transition to a state of lower energy and emit a photon in the process. In the same way, a nucleus that is initially in an excited state can undergo a transition to a state of lower energy and emit a gamma ray photon. A particular nucleus is often placed in an excited state as a result of alpha or beta decay. For example, a $^{14}_{7}\text{N}^*$ nucleus in an excited state (as on the left in Eq. 30.15) can be produced by beta decay (although we did not indicate it in Eq. 30.13). A highly simplified energy-level diagram for the reaction in Equation 30.15 is shown in Figure 30.7. This diagram shows only the initial and final states in Equation 30.15. The decay from the excited to the ground

---

[2]Here and throughout this chapter we follow the standard notation of nuclear physics and denote particles such as an electron by a "non-italic" letter.

Beta decay: emission of a beta particle (an electron or positron) from a nucleus, along with an antineutrino or a neutrino

Gamma decay: emission of a gamma ray photon from an excited nucleus

state in Figure 30.7 produces a gamma ray with an energy equal to the difference in the energies of the two levels.

A more realistic energy-level diagram, this time for the excited states of $_{28}^{60}$Ni, is shown in Figure 30.8. A $_{28}^{60}$Ni* nucleus can emit gamma rays with many different energies, depending on which excited and final states are involved. For the states shown here, the emitted gamma rays have energies of a few MeV.

Gamma ray energies are much higher than those of visible light or X-ray photons. The energy of a gamma ray photon depends on the particular nuclear decay that produces it, but typical gamma rays have energies from 10 keV (10^4 eV) to 100 MeV (10^8 eV) or even higher. Recall that the spectral emission of atoms (Chapter 29) involved photons with energies of typically 10 eV. This comparison shows again that the overall energy scale of nuclear decay—and of nuclear processes in general—is much greater than a typical atomic scale energy.

CONCEPT CHECK 30.2 | Wavelength of a Gamma Ray Photon

According to the energy-level diagram for $_{28}^{60}$Ni in Figure 30.8, this nucleus can emit a gamma ray photon with an energy of 1.33 MeV. What is the wavelength of this photon?
  (a) 9.3×10^{-13} m $= 9.3 \times 10^{-4}$ nm
  (b) 2.7×10^{-31} m $= 2.7 \times 10^{-23}$ nm
  (c) 2.3×10^{2} m
How does the wavelength of the gamma ray compare with the wavelength of visible light?

Conservation Rules in Nuclear Reactions

The radioactive processes of alpha, beta, and gamma decay can be understood in terms of the following conservation principles.

Conservation of mass–energy. The principle of conservation of energy applies to *all* physical processes, including nuclear reactions. The total energy at the start of a reaction such as the gamma decay in Equation 30.15 must equal the total energy at the end. Indeed, this principle allows us to use an energy-level diagram (as in Figs. 30.7 and 30.8) to calculate the energy of the gamma ray produced in such a reaction. In addition, we must account for the results of special relativity and allow for the conversion of mass to energy and vice versa. We'll discuss some specific examples later in this section.

Conservation of momentum. All nuclear reactions must conserve momentum. The application of momentum conservation in nuclear physics is similar to our work on collisions in Chapter 7.

Conservation of electric charge. All nuclear reactions conserve electric charge, but the total number of charged particles may change in a reaction. For example, the beta decay in Equation 30.14 produces a proton and an electron. The total charge produced is thus zero (so the total charge does not change), but the number of charged particles increases.

Conservation of nucleon number. The number of nucleons—that is, the number of protons plus neutrons—does not change. For example, in the beta decay reaction in Equation 30.14, a neutron is converted into a proton, but the total number of neutrons plus protons is unchanged by the reaction.

Radioactive Decay Series

When a nucleus undergoes radioactive decay through the emission of an alpha particle or a beta particle, it is converted into another type of nucleus. An important example is the alpha decay of $_{92}^{238}$U:

$$_{92}^{238}\text{U} \rightarrow _{90}^{234}\text{Th} + _{2}^{4}\text{He} \qquad (30.16)$$

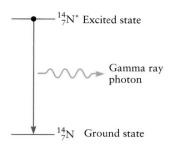

Figure 30.7 A nucleus in an excited state ($_{7}^{14}$N*) can decay to its ground state ($_{7}^{14}$N) by emitting a gamma ray photon.

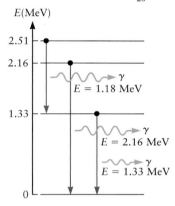

Figure 30.8 Energy-level diagram showing a few of the energy levels of the nucleus $_{28}^{60}$Ni. This nucleus can emit gamma ray photons with several different energies, depending on the initial and final states.

Figure 30.9 Sequence of nuclear decays starting with the parent nucleus $^{238}_{92}$U. The horizontal arrows denote beta decay, and the diagonal arrows show alpha decay.

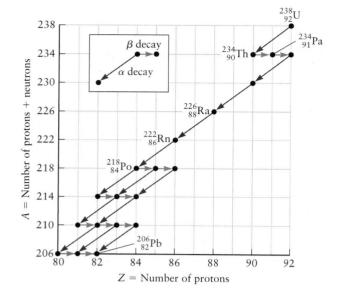

This decay converts uranium (U) into thorium (Th). The ability to convert one element into another was a goal of many chemists and alchemists throughout the Middle Ages. This conversion is precisely what happens in alpha and beta decay, but it is not an economical approach for producing any element on a commercial scale (and certainly not for producing the gold so eagerly sought by medieval alchemists).

Although Equation 30.16 describes a complete nuclear reaction, the decay reaction often continues. That is, after $^{238}_{92}$U decays into $^{234}_{90}$Th, this thorium nucleus decays further. Figure 30.9 shows a sequence of nuclear decays starting from $^{238}_{92}$U, with each decay indicated by an arrow. A horizontal arrow pointing to the right denotes a beta decay similar to the one in Equation 30.13, in which the number of protons, Z, increases by one. For example, the second decay in Figure 30.9 is the reaction

$$^{234}_{90}\text{Th} \rightarrow {}^{234}_{91}\text{Pa} + \text{e}^- + \bar{\nu} \tag{30.17}$$

in which thorium (Th) is converted into protactinium (Pa) plus an electron and an antineutrino. Each beta decay (horizontal arrow) in Figure 30.9 follows the same pattern. The diagonal arrows in the figure denote alpha decay; for example, the diagonal arrow at the upper right links $^{238}_{92}$U to $^{234}_{90}$Th in the reaction in Equation 30.16. Each alpha decay reduces the proton number by two and the neutron number by two. Some nuclei, such as $^{218}_{84}$Po, can undergo either beta decay or alpha decay, so both a horizontal arrow and a diagonal arrow originate at these nuclei.

In the radioactive decay series in Figure 30.9, $^{238}_{92}$U is the original parent nucleus while many unstable daughter nuclei are produced, and the decay proceeds until reaching the nucleus $^{206}_{82}$Pb. This nucleus is a stable final product of the decay of $^{238}_{92}$U. This particular decay series is especially important; among the daughter nuclei are $^{226}_{88}$Ra (radium), which is one of the radioactive nuclei discovered by Pierre and Marie Curie, and $^{222}_{86}$Rn (radon), which is a significant health hazard. We'll investigate other radioactive nuclei produced by other decay series in the end-of-chapter questions.

CONCEPT CHECK 30.3 Alpha Decay of Radium

Which of the following reaction equations correctly describes the alpha decay of $^{223}_{88}$Ra?

(a) $^{223}_{88}\text{Ra} \rightarrow {}^{227}_{90}\text{Th} + {}^4_2\text{He}$

(b) $^{227}_{90}\text{Th} \rightarrow {}^{223}_{88}\text{Ra} + {}^4_2\text{He}$

(c) $^{223}_{88}\text{Ra} \rightarrow {}^{219}_{86}\text{Rn} + {}^4_2\text{He}$

CONCEPT CHECK 30.4 Beta Decay

Consider a nucleus that undergoes beta decay and emits an electron. Which of the following statements is true?

    (a) The atomic number Z decreases by one.
    (b) The atomic number Z decreases by two.
    (c) The atomic number Z increases by one.
    (d) The atomic number Z increases by two.
    (e) Z and A are unchanged.

Calculating the Binding Energy of a Nucleus

We have mentioned several times that the mass–energy relation of special relativity is important for understanding nuclear binding energies. Recall from Chapter 27 that the rest energy of a particle (such as a nucleus) with rest mass m is

$$E = mc^2 \tag{30.18}$$

Let's now consider carefully how this relation applies to the nucleus ^4_2He, an alpha particle. With an understanding of the mass and binding energy in this case, we will be ready to discuss the energy released in nuclear reactions and how to calculate it.

An alpha particle consists of two protons and two neutrons, and from Table 30.1 we can add up the masses of two isolated protons plus two isolated neutrons to get

$$m(2\text{ protons} + 2\text{ neutrons}) = 2(1.0073 \text{ u}) + 2(1.0087 \text{ u}) = 4.0320 \text{ u} \tag{30.19}$$

Here we have included many more significant figures than usual because small mass differences will be extremely important in our final result. The result in Equation 30.19 is slightly larger than the mass of an alpha particle, which is 4.0015 u. (This is the mass of just a helium nucleus, a helium atom without the two electrons.) The difference is due to the binding energy of the alpha particle, E_{binding}. The mass difference is

$$\Delta m = 4.0015 - 4.0320 \text{ u} = -0.0305 \text{ u} \tag{30.20}$$

Applying the relation between mass and energy from special relativity (Eq. 30.18), the mass difference in Equation 30.20 corresponds to an energy

$$E_{\text{binding}} = (\Delta m)c^2$$

To evaluate E_{binding}, we note that 1 atomic mass unit (1 u) corresponds to 931.5 MeV/c^2 (Eq. 30.11). Using this conversion factor with the value of Δm in Equation 30.20 gives

$$E_{\text{binding}} = (-0.0305 \text{ u})\left(\frac{931.5 \text{ MeV}/c^2}{\text{u}}\right)c^2 = -28.4 \text{ MeV} \tag{30.21}$$

This binding energy is negative, indicating that an alpha particle has a *lower* energy than a collection of separated protons and neutrons. An alpha particle is therefore more stable than a collection of separate protons and neutrons and so will not decay spontaneously into protons and neutrons.

Finding the Energy Released in a Nuclear Reaction

Let's now apply our understanding of nuclear binding energy as illustrated for an alpha particle in Equations 30.19 through 30.21 to calculate the energy released in a nuclear decay process. The alpha decay reaction of radium was studied by Marie and Pierre Curie and plays an important role in many decay processes like the one diagramed in Figure 30.9:

$$^{226}_{88}\text{Ra} \quad \rightarrow \quad ^{222}_{86}\text{Rn} \quad + \quad ^4_2\text{He} \tag{30.22}$$

$$m = 226.0254 \text{ u} \rightarrow 222.0176 \text{ u} + 4.0026 \text{ u}$$

$$\text{total mass} = 226.0254 \text{ u} \rightarrow \qquad 226.0202 \text{ u}$$

(Note that the masses here and in other examples below include the electrons for each atom, unlike Eq. 30.20 where the mass value for the alpha particle was for the

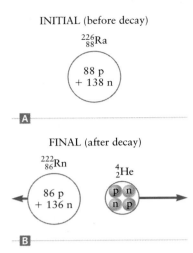

INITIAL (before decay)

$^{226}_{88}$Ra

88 p
+ 138 n

A

FINAL (after decay)

$^{222}_{86}$Rn

86 p
+ 136 n

$^{4}_{2}$He

p n
n p

B

Figure 30.10 When a $^{226}_{88}$Ra nucleus undergoes alpha decay, it emits an alpha particle ($^{4}_{2}$He) and a $^{222}_{86}$Rn nucleus. This decay leaves these two particles with some kinetic energy. To conserve momentum, the reaction products are emitted in opposite directions.

nucleus only.) For simplicity, Figure 30.10A shows the parent nucleus $^{226}_{88}$Ra as being at rest. Will the reaction products $^{222}_{86}$Rn and $^{4}_{2}$He also be at rest after the reaction? The answer is no; a nucleus such as $^{226}_{88}$Ra will decay spontaneously only if the total rest energy of the final products is less than the rest energy of the original parent nucleus. Total energy must be conserved, so the reaction in Equation 30.22 will release energy, usually in the form of kinetic energy of the decay products. (A portion of the energy is sometimes also used to put a daughter nucleus into an excited state, similar to the excited $^{14}_{7}$N* nucleus produced by beta decay; Eqs. 30.13 and 30.15.) That is why Figure 30.10B shows the $^{222}_{86}$Rn and $^{4}_{2}$He nuclei traveling away after the reaction.

To calculate the energy released in the decay of $^{226}_{88}$Ra, we must consider the masses of the nuclei in Equation 30.22. These masses are listed in Table A.4 and are given in Equation 30.22. The mass at the start of the reaction is $m_{\text{start}} = m(^{226}_{88}\text{Ra})$ = 226.0254 u, whereas the total mass at the end is $m_{\text{end}} = m(^{222}_{86}\text{Rn}) + m(^{4}_{2}\text{He})$ = 226.0202 u. The change in mass in the reaction is thus

$$\Delta m = m_{\text{end}} - m_{\text{start}} = 226.0202 - 226.0254 \text{ u} = -0.0052 \text{ u} \quad (30.23)$$

which corresponds (from Eq. 30.18) to an energy $E_{\text{reaction}} = \Delta mc^2$. Using the definition of the mass unit u, we get

$$E_{\text{reaction}} = \Delta mc^2 = -(0.0052 \text{ u})\left(\frac{931.5 \text{ MeV}/c^2}{1 \text{ u}}\right)c^2 = -4.8 \text{ MeV} \quad (30.24)$$

This energy is negative, indicating that, as expected, the nuclei of the reaction products have a lower rest energy than the parent nucleus. This reaction releases 4.8 MeV, typically as the kinetic energy of the products. We'll explore this issue in the next example.

EXAMPLE 30.2 | Speed of an Alpha Particle

The alpha decay of $^{226}_{88}$Ra releases 4.8 MeV (Eq. 30.24). If one-tenth of this energy goes into kinetic energy of the emitted alpha particle, what is the speed of the alpha particle? Use the classical relation for the kinetic energy ($KE = \frac{1}{2}mv^2$).

RECOGNIZE THE PRINCIPLE

The kinetic energy of the alpha particle is related to its speed by $KE = \frac{1}{2}mv^2$, so given the kinetic energy we can find v.

SKETCH THE PROBLEM

Figure 30.10B describes the problem.

IDENTIFY THE RELATIONSHIPS

We are given that the alpha particle has a kinetic energy of

$$KE = \frac{1}{2}m_\alpha v^2 = \frac{1}{10}(4.8 \text{ MeV}) = 4.8 \times 10^5 \text{ eV} \quad (1)$$

We must convert this energy to SI units (joules) so that we can obtain v in meters per second. The mass of the alpha particle is given in atomic mass units, and we must also convert this to SI units (kilograms) using Equation 30.11.

SOLVE

Rearranging Equation (1) to solve for v gives

$$v = \sqrt{\frac{2KE}{m_\alpha}} \quad (2)$$

Converting KE to joules gives

$$KE = (4.8 \times 10^5 \text{ eV})\left(\frac{1.60 \times 10^{-19} \text{ J}}{1 \text{ eV}}\right) = 7.7 \times 10^{-14} \text{ J}$$

Converting m_α to kilograms, we have

$$m_\alpha = (4.00 \text{ u})\left(\frac{1.66 \times 10^{-27} \text{ kg}}{1 \text{ u}}\right) = 6.6 \times 10^{-27} \text{ kg}$$

Substituting these values in Equation (2), we find

$$v = \sqrt{\frac{2KE}{m_\alpha}} = \sqrt{\frac{2(7.7 \times 10^{-14} \text{ J})}{6.6 \times 10^{-27} \text{ kg}}} = \boxed{4.8 \times 10^6 \text{ m/s}}$$

What does it mean?
While this speed is only about 2% of the speed of light, the alpha particle still carries a lot of energy compared to the binding energy of electrons in an atom or molecule. For this reason, nuclear decay products can do significant damage when they travel through human tissue (Section 30.4).

Half-life

An important aspect of spontaneous nuclear decay is the decay time; that is, if you are given a $^{226}_{88}\text{Ra}$ nucleus, how long will you have to wait for it to decay by the process in Equation 30.22? Experiments show that individual nuclei decay one at a time, at *random* times. It is not possible to predict when a particular $^{226}_{88}\text{Ra}$ nucleus will decay. This randomness is a feature of quantum theory; there is no classical analogy.

Although one cannot predict when a particular nucleus will decay, one can give the *probability* for decay. This probability is specified in terms of the **half-life** of the nucleus, $T_{1/2}$. Suppose a total of N_0 nuclei are present at time $t = 0$. Half of these nuclei will decay during a time equal to one half-life $t = T_{1/2}$. Hence, half of the original nuclei will remain at $t = T_{1/2}$ as shown graphically in Figure 30.11. If we then wait until another half-life has passed—that is, until $t = 2T_{1/2}$—the number of original, undecayed, nuclei will fall to $N_0/4$. Half of the nuclei decay during each time interval of length $\Delta t = T_{1/2}$.

The decay curve in Figure 30.11 is described by the exponential function. We define the **decay constant** λ in such a way that

$$N = N_0 e^{-\lambda t} \qquad (30.25)$$

The value of the decay constant for a particular isotope can be determined through experimental measurements of N as a function of time and comparing the results to the exponential function in Equation 30.25. The decay constant λ is closely related to the half-life. From the definition of $T_{1/2}$, the number of nuclei is $N = N_0/2$ at $t = T_{1/2}$. Inserting these values into Equation 30.25 leads to

$$\frac{N_0}{2} = N_0 e^{-\lambda T_{1/2}}$$

$$e^{-\lambda T_{1/2}} = \tfrac{1}{2} \qquad (30.26)$$

Taking the natural logarithm of both sides and doing some rearranging leads to

$$\ln(e^{-\lambda T_{1/2}}) = -\lambda T_{1/2} = \ln(1/2) = -\ln 2$$

$$\lambda T_{1/2} = \ln 2$$

$$T_{1/2} = \frac{\ln 2}{\lambda} \approx \frac{0.693}{\lambda} \qquad (30.27)$$

using $\ln 2 \approx 0.693$. Equation 30.27 thus relates the value of the decay constant to the half-life. This relation is an inverse one: a large decay constant means a short half-life and vice versa.

Values of $T_{1/2}$ vary widely; a half-life can be as short as a tiny fraction of a second or longer than the age of the Earth. Values for a few important nuclei are given in Table

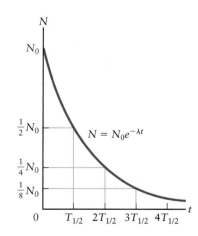

Figure 30.11 Radioactive decay is described by an exponential decay curve. The curve shows the number N of nuclei remaining after a time t. If there are N_0 nuclei at $t = 0$, only half are left at time $t = T_{1/2}$.

Table 30.2 Half-lives of Nucleons and Some Important Nuclei

NUCLEUS	HALF-LIFE
n (a free neutron)	10.4 min
^1_1H (a proton)	Stable
^2_1H (deuterium)	Stable
^3_1H (tritium)	12.33 yr
$^{14}_6C$	5730 yr
$^{60}_{27}Co$	5.27 yr
$^{123}_{53}I$	13.2 h
$^{222}_{86}Rn$	3.82 days
$^{226}_{88}Ra$	1600 yr
$^{235}_{92}U$	7.04×10^8 yr
$^{238}_{92}U$	4.47×10^9 yr

30.2, and a more complete listing is found in Table A.4. The value of $T_{1/2}$ is important in many effects and applications of radioactivity (Sections 30.4 and 30.5).

CONCEPT CHECK 30.5 | ⊗ PET Scans and the Half-Life of $^{20}_9F$

The isotope $^{20}_9F$ is used in a medical procedure called positron emission tomography (PET). The half-life of $^{20}_9F$ is approximately 110 minutes. If your doctor has a sample with 16 g of pure $^{20}_9F$ at $t = 0$, how much $^{20}_9F$ will he have 330 minutes later, (a) 12 g, (b) 8.0 g, (c) 4.0 g, (d) 2.0 g, or (e) 1.0 g?

EXAMPLE 30.3 | Half-life and the Radioactive Decay of Tritium

Tritium is an isotope of hydrogen containing two neutrons and is denoted by ^3_1H. It is a component in some fusion bombs (i.e., "hydrogen bombs") as we'll discuss in the next section. Tritium is radioactive and decays into an isotope of helium (^3_2He) with a half-life of $T_{1/2} = 12.33$ yr. Suppose a fusion bomb contains 10 kg of tritium when it is first assembled. How much tritium (in kilograms) will it contain after 30 years on the shelf?

RECOGNIZE THE PRINCIPLE

If there are N_0 tritium nuclei at $t = 0$, the number remaining after a time t is

$$N = N_0 e^{-\lambda t} \tag{1}$$

We weren't given the value of λ for tritium, but we can find it from the given half-life using the relation between λ and $T_{1/2}$ in Equation 30.27.

SKETCH THE PROBLEM

Figure 30.11 describes the problem. After a time $T_{1/2}$ passes, half of the radioactive nuclei have decayed.

IDENTIFY THE PRINCIPLES

We first rearrange Equation 30.27 to find the decay constant for tritium in terms of the half-life:

$$\lambda = \frac{\ln 2}{T_{1/2}} = \frac{0.693}{12.33 \text{ yr}} = 0.056 \text{ yr}^{-1}$$

SOLVE

Using this value of λ in Equation (1), we find that the fraction N/N_0 of tritium nuclei that remain after $t = 30$ yr is

$$\frac{N}{N_0} = e^{-\lambda t} = e^{-(0.056 \text{ yr}^{-1})(30 \text{ yr})} = 0.19$$

Hence, only 19% of the original tritium nuclei will remain. The initial mass was 10 kg, so the final mass of tritium is 0.19×10 kg = $\boxed{1.9 \text{ kg}}$.

What does it mean?
Because ^3_2He is not useful for making bombs, the decay of tritium will cause a fusion bomb to stop working after a certain period of time. As a result, fusion bombs containing tritium must be periodically "reloaded."

EXAMPLE 30.4 | Isotope Abundance and $^{235}_{92}U$

The isotope $^{235}_{92}U$ has a half-life for spontaneous fission of 7.0×10^8 yr. In deposits of naturally occurring uranium, this isotope is only about 0.72% of the total fraction of uranium nuclei. Hence, if you had a sample containing 100 g of pure uranium, only 0.72 g would be $^{235}_{92}U$. The Earth is believed to be about 4.5 billion $= 4.5 \times 10^9$ years old, or about $(4.5 \times 10^9)/(7.0 \times 10^8) \approx 6$ half-lives of $^{235}_{92}U$. How many grams of $^{235}_{92}U$ were in your sample when the Earth was formed 4.5×10^9 yr ago?

RECOGNIZE THE PRINCIPLE

As we go back in time, the amount of $^{235}_{92}U$ increases by a factor of two for every half-life. Hence, if we go back in time by one half-life, the amount of $^{235}_{92}U$ is doubled; if we go back in time another half-life, the amount of $^{235}_{92}U$ doubles again, and so on.

SKETCH THE PROBLEM

Figure 30.12 shows how the amount of $^{235}_{92}U$ in our sample grows as we go back in time by six half-lives.

Figure 30.12 Example 30.4. As we go back in time, the amount of $^{235}_{92}U$ doubles every $T_{1/2} = 7.0 \times 10^8$ yr.

IDENTIFY THE RELATIONSHIPS AND SOLVE

According to Figure 30.12, the total mass of $^{235}_{92}U$ after going back in time six half-lives is approximately $\boxed{46 \text{ g}}$.

We can be a bit more mathematical by noting that after one half-life, the mass increases by a factor of two, after two half-lives it increases by $2^2 = 4$, and after six half-lives it increases by $2^6 = 64$, which again gives

$$m_{start} = 2^6 m_{end} = 2^6(0.72 \text{ g}) = \boxed{46 \text{ g}}$$

What does it mean?
Geophysicists continue to investigate the composition of the Earth when it first formed. At that time, the uranium was more equally apportioned between $^{238}_{92}U$ and $^{235}_{92}U$. At present (about six half-lives later), most of the original $^{235}_{92}U$ has decayed. Notice that the half-life of $^{238}_{92}U$ is about 4.5×10^9 yr, so only about half of the original $^{238}_{92}U$ has decayed since the formation of the Earth.

How Do We "Measure" Radioactivity?

The half-life of a nucleus tells how quickly (on average) the nucleus will undergo radioactive decay. Each decay can produce potentially harmful products such as the alpha particle produced in the decay of $^{226}_{88}Ra$ (Eq. 30.22). A nucleus with a short half-life is more likely to decay, so it is potentially more dangerous than a nucleus with a long half-life. Since a sample with more radioactive nuclei will be more dangerous than one with a small number of such nuclei, to fully assess such dangers we must also know how many nuclei are present in a sample. For this reason, the "strength" of a radioactive sample is measured using a property called the *activity*, which can be measured with a Geiger counter (Fig. 30.13). The activity of a sample is proportional to the number of nuclei that decay in 1 second. If all other factors are similar, a sample with a high activity is more dangerous than a sample with a low activity. We often say that a sample with a high activity level is "hot."

Figure 30.13 A Activity can be measured with a Geiger counter. B Charged particles passing through the detector chamber ionize gas molecules, causing a short pulse of current to flow. Some Geiger counters give an audible "click" to indicate each current pulse.

One common unit of activity is the curie (Ci), which is defined as

$$1 \text{ Ci} = 3.7 \times 10^{10} \text{ decays/s} \tag{30.28}$$

Hence, a sample of radioactive material has an activity of 1 Ci if it exhibits 3.7×10^{10} decays each second; 1 g of naturally occurring radium has an activity of approximately 1 Ci. In practice, a sample with an activity of 1 Ci is extremely dangerous due to the large number of alpha and beta particles and gamma rays that it emits. Most studies or medical procedures with radioactive substances involve samples with activities of millicuries ($\text{mCi} = 10^{-3}$ Ci) or microcuries ($\mu\text{Ci} = 10^{-6}$ Ci).

The official SI unit of activity is the becquerel (Bq), named in honor of Henri Becquerel (1852–1908), who shared the Nobel Prize in Physics with Marie and Pierre Curie for his discovery of spontaneous radioactivity. The becquerel is defined by

$$1 \text{ Bq} = 1 \text{ decay/s} \tag{30.29}$$

As time passes and nuclei in a sample decay, the number of remaining unstable nuclei drops. Hence, the activity of a sample decreases with time. We thus say (loosely speaking) that a "hot" radioactive sample (such as the nuclear reactor at Chernobyl in Ukraine) becomes "cooler" with time and therefore safer to deal with or clean up.

While a sample with a high activity level is generally more dangerous than one with low activity, there are complicating factors due to the variable amount of energy carried by decay products as we'll discuss in Section 30.4.

CONCEPT CHECK 30.6 | How Does the Activity Level of a Radioactive Sample Change with Time?

Consider a radioactive material that decays with a half-life of $T_{1/2}$. How long does it take for the activity of the sample to decrease by a factor of four, (a) $T_{1/2}$, (b) $2T_{1/2}$, (c) $3T_{1/2}$, or (d) $4T_{1/2}$?

EXAMPLE 30.5 Activity of Tritium

You are given a sample containing exactly 1 mole of tritium (^3_1H). The tritium decays with a half-life of 12.33 yr. What is the activity of the sample? Express your answer in units of decays per second. *Hint*: Inserting $t = 1$ s in the exponential decay law $N = N_0 e^{-\lambda t}$ gives the number of nuclei remaining after 1 s. The activity is the number of nuclei that *decay* in the same 1-s time period.

RECOGNIZE THE PRINCIPLE

If a sample initially contains N_0 nuclei, the number remaining after $t = 1$ s is

$$N_{\text{remaining}} = N_0 e^{-\lambda t}$$

Hence, the number that decay during this time is

$$N_{\text{decays}} = N_0 - N_0 e^{-\lambda t} = N_0(1 - e^{-\lambda t}) \tag{1}$$

The activity equals the number of nuclei that decay per second.

SKETCH THE PROBLEM

Figure 30.14 shows how the number of nuclei remaining decreases with time as well as the number of nuclei that decay in $t = 1$ s.

IDENTIFY THE RELATIONSHIPS

To evaluate Equation (1) for our given tritium sample, we must find both N_0 and the decay constant λ. The sample originally contains exactly 1 mole of tritium, so $N_0 = 6.02 \times 10^{23}$ (Avogadro's number). From Equation 30.27, the decay constant is related to the half-life through $\lambda = 0.693/T_{1/2}$.

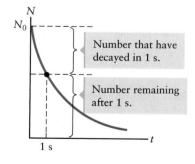

Figure 30.14 Example 30.5. Not to scale.

Inserting the known half-life of tritium (12.33 yr) gives

$$\lambda = \frac{0.693}{T_{1/2}} = \frac{0.693}{12.33 \text{ yr}} = \frac{0.693}{3.9 \times 10^8 \text{ s}} = 1.8 \times 10^{-9} \text{ s}^{-1}$$

SOLVE

Inserting our values of N_0 and λ as well as $t = 1$ s into Equation (1) leads to

$$N_{\text{decays}} = N_0(1 - e^{-\lambda t}) = (6.02 \times 10^{23})[1 - e^{-(1.8 \times 10^{-9} \text{ s}^{-1})(1 \text{ s})}]$$

The exponential factor is very close to one, so we have to keep extra significant figures in performing this calculation. We find

$$\text{activity} = N_{\text{decays}} = 1.1 \times 10^{15} \text{ per second} = \boxed{1.1 \times 10^{15} \text{ Bq}}$$

What does it mean?
Nuclei with a shorter half-life decay more rapidly (and have a larger value of λ) than those with a long half-life. Hence, when comparing two samples, the one with a short half-life will have a higher activity (and hence be more dangerous) than the one with a long half-life, assuming the number of nuclei in the two samples is the same.[3]

30.3 | STABILITY OF THE NUCLEUS: FISSION AND FUSION

To be stable, a nucleus containing two or more protons must contain neutrons. Neutrons are thus necessary for the stability of essentially all matter. Figure 30.15 shows a plot of the neutron number N as a function of the number of protons Z (i.e., the atomic number) for all known stable nuclei. The dashed line shows the function $N = Z$; nuclei that lie on this line have equal numbers of neutrons and protons, while nuclei above this line contain an excess of neutrons. Low-mass nuclei such as He and C tend to have equal numbers of protons and neutrons and are thus on or near the dashed line corresponding to $N = Z$ in Figure 30.15. Heavy nuclei, however, have many more neutrons than protons; for example, Pb (lead) nuclei contain around 40 more neutrons than protons.

The reason these additional neutrons are needed can be seen by considering the forces between protons and neutrons. Figure 30.16A shows a highly schematic nucleus with two protons and two neutrons, representing ^4_2He. This nucleus is extremely stable, so this arrangement of protons and neutrons is particularly favored. We now add another proton to make lithium (Li). To bind this additional proton and achieve a stable arrangement of protons and neutrons similar to ^4_2He, we need two more neutrons as shown in Figure 30.16B. This highly simplified picture leaves out the possibility that the protons in Li might rearrange to better "share" the attraction of the neutrons, just as two atoms share electrons when they form a chemical bond. In reality, ^6_3Li and ^7_3Li are both stable isotopes. However, this simple picture does explain why nuclei with large numbers of protons (high Z) require greater numbers of neutrons to be stable, accounting for the trend that $N > Z$ for heavy nuclei in Figure 30.15. This trend is essential for understanding nuclear fission.

Nuclear Binding Energy

We have discussed how the mass of a nucleus is related to its binding energy, and as an example we calculated the binding energy of an alpha particle (Eq. 30.21). We

---

[3]Notice also that our calculation gives the *average* activity during a 1-s interval, but because the half-life here is much greater than 1 s, the *instantaneous* activity will have almost exactly the same value.

Figure 30.15 Stable nuclei usually have a greater number of neutrons than protons, so $N > Z$ for all but the lightest nuclei. This trend is especially noticeable at high Z.

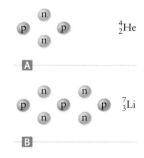

Figure 30.16 ◢ The isotope ^4_2He contains two protons and two neutrons. This nucleus is stable because the neutrons can (roughly speaking) sit between the protons, holding the nucleus together. (Compare with Fig. 30.5.) ◣ If a third proton is added, more neutrons are needed to keep the protons apart and provide the needed binding due to the strong force. As the number of protons grows, extra neutrons are needed. For this reason, ^7_3Li (pictured here) is more stable than ^6_3Li.

found that the mass of the nucleus $_2^4$He, which contains two protons and two neutrons, is less than the total mass of two isolated protons plus two isolated neutrons. The same is true for other stable nuclei; the mass of a nucleus containing N neutrons and Z protons is *less* than the total mass of N isolated neutrons plus Z isolated protons. This mass difference is connected to the binding energy of the nucleus through the relativistic relation between mass and energy ($E = mc^2$). By measuring the mass of a nucleus and then subtracting the mass of the same numbers of isolated protons and neutrons, we can find the binding energy of the nucleus. This approach has been used to find the binding energy of every nucleus in Figure 30.15, and the results for the binding energy per nucleon are shown in Figure 30.17. In the figure, the horizontal axis is mass number A. Different isotopes (nuclei having the same Z but different A) can have different binding energies; here we plot an approximate average value for each element, that is, for each value of A. Values of the binding energy are plotted as negative because each of these nuclei has a lower energy than a collection of separated protons and neutrons. The magnitude of the binding energy is largest for nuclei close to $A = 60$, which is near Fe and Ni in the periodic table, meaning that these nuclei are the *most stable* ones. This fact and the relationship between binding energy and Z are the basis for two important nuclear phenomena: fission and fusion.

In our other discussions of binding energies in previous chapters, including the binding energy of an atom, the energy of a stable bound system has also been negative. For example, the gravitational potential energy of two masses is negative (Eq. 6.22), and the energy of a bound atom such as hydrogen is also negative (Fig. 29.12). When an attractive force exists between two or more particles, you must add energy to separate them. Hence, binding lowers the potential energy relative to the state in which the particles are at infinite separation.

Nuclear Fission

Figure 30.9 shows how a parent nucleus (in that case, $_{92}^{238}$U) can undergo spontaneous radioactive decay to form a series of daughter nuclei. Radioactive decay can also be *induced* by a collision between two nuclei or by the collision of a neutron with a nucleus. A particularly important example occurs when $_{92}^{235}$U collides with a neutron. A typical reaction of this kind is

one neutron induces
fission of $_{92}^{235}$U   three neutrons
are produced

$$_0^1\text{n} + {}_{92}^{235}\text{U} \rightarrow {}_{56}^{139}\text{Ba} + {}_{36}^{94}\text{Kr} + 3_0^1\text{n} \tag{30.30}$$

Here a neutron is denoted by $_0^1$n since it "contains" no protons ($Z = 0$) and has a total of one nucleon ($A = 1$). The process in Equation 30.30 is called **nuclear fission**, a reaction whose main products are two nuclei both roughly half the size of

Nuclear fission

Figure 30.17 Binding energy per nucleon for nuclei with different mass numbers. A binding energy of high magnitude corresponds to a tightly bound nucleon, so nuclei falling in the shaded region of the graph are the most stable. The binding energy is negative because a nucleus is more stable and has a lower energy than a collection of separated protons plus neutrons.

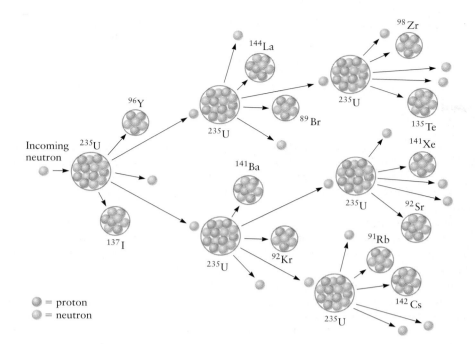

Figure 30.18 The fission of $^{235}_{92}U$ can be initiated by a collision with a neutron. Here the original $^{235}_{92}U$ nucleus splits into two fission fragments, $^{96}_{39}Y$ and $^{137}_{53}I$, along with three neutrons. These neutrons then collide with other $^{235}_{92}U$ nuclei and cause them to fission. This is an example of a chain reaction.

the original parent nucleus. (By contrast, in alpha and beta decay one of the decay products is very close in mass to the original parent nucleus.)

Fission reactions have several important features. First, there can be many different possible reaction products. For example, when $^{235}_{92}U$ collides with a neutron, it may produce $^{139}_{56}Ba$ and $^{94}_{36}Kr$ as in Equation 30.30, but it may also produce a different pair of nuclei. Another possible fission reaction involving $^{235}_{92}U$ is

$$^{1}_{0}n + {}^{235}_{92}U \rightarrow {}^{95}_{38}Sr + {}^{138}_{54}Xe + 3{}^{1}_{0}n$$

As in the reaction in Equation 30.30, the collision with a neutron causes $^{235}_{92}U$ to split into two smaller nuclei and also emit several neutrons. There are approximately 90 different fission reactions involving $^{235}_{92}U$.

A second key feature of fission is that it produces more free neutrons than it uses. For example, the reaction in Equation 30.30 begins with one neutron, but there are three neutrons at the end. Some fission reactions with $^{235}_{92}U$ emit three neutrons, whereas others emit two or four. On average, the fission of a single $^{235}_{92}U$ nucleus caused by the collision with a neutron results in about 2.5 neutrons at the end of the reaction. These neutrons can then induce the fission of other nearby $^{235}_{92}U$ nuclei, leading to a ***chain reaction*** as sketched in Figure 30.18. In this way, the fission of a single $^{235}_{92}U$ nucleus can induce the fission of an enormous number of additional $^{235}_{92}U$ nuclei. Each fission event produces additional free neutrons because the ratio of neutrons to protons in a heavy nucleus such as $^{235}_{92}U$ is greater than in either of the two nuclei that are produced (e.g., $^{139}_{56}Ba$ and $^{94}_{36}Kr$ in Eq. 30.30). This fact can be traced to the shape of the stability curve in Figure 30.15.

A third important feature of fission reactions is that they release energy as can be seen from the binding energy curve in Figure 30.17. The magnitude of the binding energy for $^{235}_{92}U$ is about 7.6 MeV/nucleon, while the magnitudes of the binding energies of $^{139}_{56}Ba$ and $^{94}_{36}Kr$ (the fission products in Eq. 30.30) are above 8.0 MeV/nucleon. This fission reaction releases the extra binding energy in the form of gamma rays and through the kinetic energies of the particles produced (Fig. 30.19). The shape of the binding energy curve guarantees that the magnitude of the binding energy per nucleon of $^{235}_{92}U$ is always smaller in magnitude than the binding energies of its fission products. The same is true for other nuclei that undergo fission such as $^{239}_{94}Pu$ (plutonium) . That is why fission produces large amounts of energy.

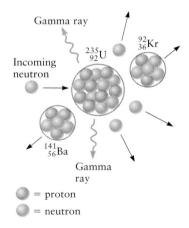

Figure 30.19 When a nucleus undergoes fission, some gamma ray photons are usually emitted along with the two fission fragments and neutrons. Typically, these particles all have large kinetic energies.

Consider the fission reaction

$$^{235}_{92}\text{U} + ^{1}_{0}\text{n} \rightarrow ^{144}_{56}\text{Ba} + ^{90}_{36}\text{Kr} + ?^{1}_{0}\text{n}$$

How many neutrons does it produce?

Energy from Fission

Nuclear fission is the basis for two modern technologies: fission weapons (bombs) and nuclear reactors for producing electrical power. Both are based on nuclear chain reactions like the one in Figure 30.18. In a nuclear fission bomb, the chain of fission reactions is allowed to run out of control, rapidly releasing extremely large amounts of energy.

Figure 30.18 implies that it's easy to set off an uncontrolled chain reaction: just fire one neutron at a sample of $^{235}_{92}$U and the 2 to 4 neutrons (2.5 on average) produced in the initial fission event go on to split other $^{235}_{92}$U nuclei, releasing enough energy to cause a huge explosion. Fortunately, it is not that easy. First of all, natural uranium is more than 99% $^{238}_{92}$U, an isotope that does not readily fission. The uranium used in weapons is "enriched" to increase the concentration of scarce $^{235}_{92}$U, and enrichment is a very difficult engineering process. Second, if the piece of uranium is small, many of the nuclei are close enough to the surface that many of the released neutrons escape from the sample before inducing another fission event. In such a case, the reaction may stop before a significant number of $^{235}_{92}$U nuclei undergo fission. Increasing the size (and hence also the mass) of a uranium sample lowers the surface-to-volume ratio and lowers the probability that a free neutron will reach the surface; it is then more likely that the neutron will encounter a $^{235}_{92}$U nucleus and induce fission. The minimum amount of nuclear fuel material needed to sustain a chain reaction is called the *critical mass*. For uranium, the critical mass is typically about 50 kg (corresponding to a uranium sphere with a radius of about 9 cm), but depending on how the sample is shaped and how the surface is prepared it can be as small as 15 kg.

When $^{235}_{92}$U undergoes fission, it produces two nuclei of roughly equal mass and several neutrons as described in Equation 30.30. We can estimate the amount of energy released using the binding energy curve in Figure 30.17. The binding energy per nucleon of $^{235}_{92}$U is about -7.6 MeV, whereas for many of the nuclei produced by fission it is about -8.5 MeV per nucleon. The difference is $(-8.5 + 7.6)$ MeV $= -0.9$ MeV per nucleon. Since there are 235 nucleons in one $^{235}_{92}$U nucleus, the total amount of energy released by the fission of one such nucleus is $(0.9 \text{ MeV}) \times 235 \approx 210$ MeV. To get a feeling for the scale of this value, let's calculate how much energy we can expect from the fission of 1 g of pure $^{235}_{92}$U (about the mass of a paper clip).

The atomic mass of pure $^{235}_{92}$U is 235 g (from the definition of the atomic mass unit), so 235 g of $^{235}_{92}$U contains 1 mole of nuclei. Our 1-g sample thus contains

$$N_{\text{U-235}} = (1 \text{ g}) \times \frac{6.02 \times 10^{23}}{235 \text{ g}} = 2.6 \times 10^{21} \text{ nuclei}$$

Each nucleus releases about 210 MeV when it undergoes fission, so the total energy released by the fission of 1 g of $^{235}_{92}$U is

$$E_{\text{fission}} = (2.6 \times 10^{21})(210 \text{ MeV})\left(\frac{10^6 \text{ eV}}{1 \text{ MeV}}\right)\left(\frac{1.60 \times 10^{-19} \text{ J}}{1 \text{ eV}}\right)$$

$$E_{\text{fission}} \approx 9 \times 10^{10} \text{ J per gram of } ^{235}_{92}\text{U} \qquad (30.31)$$

This is an *enormous* amount of energy. By comparison, 1 g of the chemical explosive TNT releases about 4200 J. The energy released by 1 g of $^{235}_{92}$U is more than 10 million times larger!

The amount of energy released in a fission bomb is usually measured in terms of the mass of TNT that would release the same amount of energy. The fission bombs that were exploded in World War II each released an amount of energy equivalent to more than 10,000 *tons* of TNT.

EXAMPLE 30.6 | Energy Released in a Fission Bomb

The fission bomb that was dropped on the Japanese city of Hiroshima in 1945 contained approximately 64 kg of $^{235}_{92}U$ and released an amount of energy equivalent to the explosion of approximately 15,000 tons (1.5×10^{10} g) of TNT (called a "15-kiloton bomb"). Did all this $^{235}_{92}U$ undergo fission? If not, what is the mass of $^{235}_{92}U$ that did?

RECOGNIZE THE PRINCIPLE

In our analysis of fission, we found that the energy released by the fission of 1 g of $^{235}_{92}U$ is about 9×10^{10} J (Eq. 30.31). We also noted that 1 g of TNT releases about 4200 J. We can use these values to compute what mass of $^{235}_{92}U$ is needed to give the energy released by 15,000 tons of TNT.

SKETCH THE PROBLEM

Fission releases energy in the form of the kinetic energy of the reaction products plus the energy of gamma rays that are emitted (Fig. 30.19).

IDENTIFY THE RELATIONSHIPS

Let's first find the energy released by 15,000 tons = 1.5×10^{10} g of TNT. From the given information, we have

$$E = (1.5 \times 10^{10} \text{ g})\left(\frac{4200 \text{ J}}{\text{g}}\right) = 6.3 \times 10^{13} \text{ J}$$

SOLVE

Using the value of $E_{\text{fission}} = 9 \times 10^{10}$ J/g of $^{235}_{92}U$ from Equation 30.31, this energy corresponds to

$$\text{mass of } ^{235}_{92}U = (6.3 \times 10^{13} \text{ J})\left(\frac{1 \text{ g of } ^{235}_{92}U}{9 \times 10^{10} \text{ J}}\right) = \boxed{700 \text{ g}}$$

What does it mean?

Only about 1% of the $^{235}_{92}U$ in the Hiroshima weapon actually underwent fission. Even so, an enormous amount of energy was released. So-called modern nuclear weapons yield much greater amounts of energy.

Nuclear Power Plants

Nuclear fission is used in a productive and controlled way in nuclear power plants. In a power reactor, fission takes place in a region called the reactor core, shown schematically in Figure 30.20A. The important components of the core are fuel rods, control rods, and moderators.

Fuel rods contain $^{235}_{92}U$ that can be added or removed from the core while the reactor is operating. The fuel rods are removed when the $^{235}_{92}U$ is exhausted or when operators need to reduce the fission rate. *Control rods* contain materials (e.g., $^{113}_{48}Cd$) that absorb neutrons. They can be inserted or removed from the reactor core, allowing the reactor operators to control the number of free neutrons and thus adjust the fission rate. The *moderator* (often water) circulates through the reactor core. One of the moderator's functions is to slow the neutrons down, which increases their probability of inducing a fission event. (Roughly speaking, a slow neutron passing near a uranium nucleus has more time to induce fission.) The water also carries away heat

Insight 30.4

SAFETY AT NUCLEAR FISSION REACTORS

Safety is always a major concern at a nuclear fission reactor, particularly those used as power reactors to generate electricity. All reactors are designed with extensive safety features to prevent a "meltdown" of the reactor core or any similar hazard. The latest reactor designs contain many "passive" safety features that do not require operator action or even electronic feedback to deal with emergency situations. For example, these safety features work even in the event of a complete electrical failure, making these reactors extremely safe.

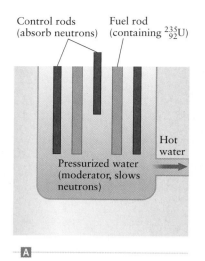

Control rods (absorb neutrons) Fuel rod (containing $^{235}_{92}$U)

Hot water

Pressurized water (moderator, slows neutrons)

A

Pressurized water

Hot steam in

Turbine

Electric generator

Cool steam out

Heat exchanger

Pump

Reactor Water

Water out

Pump Condenser

B

Figure 30.20 **A** Schematic of the core of a nuclear reactor. The fuel rods contain $^{235}_{92}$U (or some other fissionable nuclei). Neutrons emitted during fission collide with nuclei in the moderator (here, water), slowing the neutrons. Control rods containing nuclei that readily absorb neutrons are used to control the rate of fission. **B** Heat generated in the moderator (water) is used to power a steam turbine and thereby generate electric power.

Figure 30.21 The steam engines in a nuclear power reactor must cool the steam as part of the thermodynamic cycle (Chapter 16). This cooling is accomplished with large cooling towers or with a source of cool water from an ocean or river. This photo shows the Diablo Canyon Nuclear power plant in California. The two large domes each house a reactor, and the cooling water is taken from the ocean.

to a separate steam engine (Fig. 30.20B), where the energy generated by the reactor is converted to electrical energy using a generator (Figs. 21.7 and 22.3). This heat engine operates according to the principles of thermodynamics, so it requires a cold reservoir. Nuclear power plants are often located near bodies of water such as lakes that provide this cold reservoir (Fig. 30.21).

As mentioned previously, naturally occurring uranium contains a mixture of the isotopes $^{235}_{92}$U (0.7%) and $^{238}_{92}$U (99.3%). The fission reactions in a reactor involve only $^{235}_{92}$U , so the uranium in fuel rods is usually enriched to increase the concentration of $^{235}_{92}$U. However, this concentration is always kept low enough that the fuel rod does not exceed the critical mass, so it is not possible for the core of a nuclear reactor to undergo an uncontrollable chain reaction and explode like a nuclear fission bomb. It is possible, though, to have an accident called a "meltdown" in which the fuel rods become so hot that they and the walls of the reactor melt and the water in the core rapidly turns to steam. The result can be a conventional explosion (due to the high-pressure steam), as occurred at Chernobyl in Ukraine in 1986. This accident allowed the escape of significant amounts of radioactive material into the environment and still affects the safety and health of nearby residents.

Nuclear Fusion

In **nuclear fusion**, two nuclei join together to produce one new particle. Consider two $^{6}_{3}$Li nuclei; they have $Z = 3$ and are on the left side of the binding energy curve in Figure 30.17. If they fuse, the new nucleus has $Z = 6$ and contains a total of $A = 12$ nucleons, which is $^{12}_{6}$C. According to Figure 30.17, $^{12}_{6}$C has a greater binding energy per nucleon (in magnitude) than $^{6}_{3}$Li and is thus more stable. Hence, this fusion reaction releases energy:

Nuclear fusion

$$^{6}_{3}\text{Li} + ^{6}_{3}\text{Li} \rightarrow ^{12}_{6}\text{C} + \text{energy} \qquad (30.32)$$

This energy is in the form of gamma rays plus the kinetic energy of the $^{12}_{6}$C nucleus. We can use Figure 30.17 to estimate how much energy this reaction will release. The binding energy per nucleon is about -5.4 MeV for $^{6}_{3}$Li and about -7.9 MeV for $^{12}_{6}$C. Hence, the energy *released* in this fusion reaction is approximately $7.9 - 5.4 = 2.5$ MeV per nucleon. The fission of $^{235}_{92}$U releases approximately 0.9 MeV per nucleon, so this fusion reaction releases even more energy (per nucleon) that fission, which is one reason (unfortunately) bombs employing fusion are capable of even more explosive power than fission weapons. (We'll explore this topic more in Question 10 at the end of this chapter.)

The Sun's Power Comes from Fusion

Nuclear fusion is the process that powers stars such as our Sun. When first formed, a typical star consists mainly of hydrogen—that is, protons and electrons—with a small amount of $^4_2$He and much smaller amounts of heavier elements. Through a series of reactions, these protons fuse to produce $^4_2$He (alpha particles), which in turn fuse to form heavier elements. Our own Sun is hot enough to produce elements up to about carbon in this way. All the heavier elements in the universe were created in either hotter stars or in supernova explosions, which can trigger even more energetic fusion reactions.

Fusion reactions can take place in a star due to the very high temperatures and pressures in its core. These conditions cause nuclei to come into very close contact when they collide, which is necessary for fusion to occur. To create a fusion reaction on Earth, we must approach or duplicate stellar conditions, giving the initial nuclei very high speeds and keeping them very close together. In a fusion bomb, this is accomplished using a fission bomb as a trigger. That is, a bomb that uses the fission of $^{235}_{92}$U or some other nucleus is exploded in such a way that the fusion ingredients are compressed very rapidly, initiating a fusion reaction.

Fusion has one potential useful application on the Earth: as the basis for a power plant. Current (but so far hypothetical) designs for a fusion power reactor use $^2_1$H (deuterium) or $^3_1$H (tritium) as the starting nuclei in a fusion reaction that releases very large amounts of energy to produce electrical power. The fuel for this reactor could be obtained from water since naturally occurring water has a small but significant concentration of deuterium. The fuel for a fusion reactor would thus be very inexpensive and virtually limitless. Moreover, current designs for these reactors would produce virtually no harmful radiation or by-products. Although a practical fusion power plant has not yet been built, several exploratory engineering projects have been. The latest is the ITER (International Tokomak Engineering Reactor), which uses what is called "tokomak" geometry for confining the nuclei while they undergo fusion. The ITER will become operational in 2016 and should produce 500 MW (500,000,000 W) of power during 10-minute fusion periods, during which it consumes about 0.5 g of a deuterium–tritium mixture. The ITER will still be an experimental project, but the long-term hope is that it will lead to the construction of practical fusion-based power plants that would provide abundant amounts of cheap and clean electrical energy.

| EXAMPLE 30.7 | Energy of a Fusion Reaction

A fusion reaction that might be used in a power reactor involves deuterium, $^2_1$H:

$$^2_1\text{H} + ^2_1\text{H} \rightarrow ^3_1\text{H} + ^1_1\text{H} \tag{1}$$

In this reaction, two deuterium nuclei fuse to form $^3_1$H plus a proton, $^1_1$H. Use the masses of these nuclei from Table A.4 to calculate the energy released in this reaction. Express the answer in MeV.

RECOGNIZE THE PRINCIPLE

This reaction converts some of the mass of the original nuclei into energy. The amount of energy can be found from the mass–energy relation of special relativity $E = \Delta mc^2$, where Δm is the difference in the mass of the masses on the left- and right-hand sides of Equation (1).

SKETCH THE PROBLEM

No sketch is necessary, but it is useful to rewrite Equation (1) to also show the rest masses of the nuclei; see Equation (2) below.

IDENTIFY THE RELATIONSHIPS

According to Table A.4, the masses in Equation (1) are

$$m(^1_1H) = 1.007825 \text{ u}$$

$$m(^2_1H) = 2.014102 \text{ u}$$

$$m(^3_1H) = 3.016049 \text{ u}$$

We include extra significant figures because the change in mass in Equation (1) will be small, but will still lead to a substantial energy.

SOLVE

Let's rewrite Equation (1) to show the masses before and after the reaction:

$$^2_1H \quad + \quad ^2_1H \quad \rightarrow \quad ^3_1H \quad + \quad ^1_1H$$

$$2.014102 \text{ u} + 2.014102 \text{ u} \rightarrow 3.016049 \text{ u} + 1.007825 \text{ u}$$

$$4.028204 \text{ u} \rightarrow 4.023874 \text{ u} \tag{2}$$

The mass difference is thus

$$\Delta m = m(\text{right side}) - m(\text{left side}) = -0.004330 \text{ u} \tag{3}$$

which corresponds to an energy of

$$E = \Delta m c^2 \tag{4}$$

From Equation 30.11, atomic mass units can be converted to units of MeV/c^2 using the relation 1 u = 931.5 MeV/c^2. From Equations (2) and (3), we get

$$E = (-0.004330 \text{ u})\left(\frac{931.5 \text{ MeV}/c^2}{\text{u}}\right)c^2 = \boxed{-4.033 \text{ MeV}}$$

The negative value means that this is the energy *released* by the reaction.

What does it mean?
Since deuterium occurs naturally in ordinary water, this source of energy is potentially cheap. We will explore this topic in the end-of-chapter problems.

30.4 | ⊗ BIOLOGICAL EFFECTS OF RADIOACTIVITY

The biological effects of radioactivity result from the way decay products or reaction products interact with atoms and molecules. From Chapter 29, the typical binding energy of an electron in an atom is on the order of 10 eV. The energy released in a nuclear reaction is much larger (typically several or even hundreds of MeV), often appearing as the kinetic energy of an emitted alpha or beta particle. If one of these particles collides with an atomic electron or with a molecule, there is ample energy to eject the electron from the atom or to break a chemical bond in molecules such as DNA, making these molecules biologically inactive or even harmful. That is how radiation damages living tissue.

The amount of damage that a particular particle is capable of doing is a complicated issue because alpha, beta, and gamma radiation have different masses and charges and therefore interact with tissue in different ways. We defined the curie (Ci) and becquerel (Bq) as units of activity that measure the number of radioactive decays per unit time in Equations 30.28 and 30.29. Some particles produced in radioactive decay carry a great deal of kinetic energy, while others carry little. To account for this difference, we introduce a unit called the *rad*. By definition, 1 rad is the amount of radiation that deposits 10^{-2} J of energy into 1 kg of absorbing material. Hence, this unit accounts for both the amount of energy carried by the particle and the efficiency with which this energy is absorbed.

The name "rad" denotes *radiation absorbed dose*. The definition of the rad and the term *dose* both indicate that the total amount of radiation received is important when assessing biological effects, but even the rad unit does not capture the full complexity of radiation damage. Different types of particles can do different amounts of biological damage even if they deposit the same amount of energy and hence have the same strength as measured in rads. For this reason, another unit called the *relative biological effectiveness* factor is also very useful. This factor, denoted *RBE*, measures how efficiently a particular type of particle damages tissue. Some typical *RBE* factors for various kinds of radiation from nuclear reactions are given in Table 30.3. Notice that the *RBE* value tends to increase as the mass of the particle increases. For example, the *RBE* of an alpha particle (a helium nucleus) is greater than the *RBE* of a beta particle (an electron or positron).

Another widely used measure of radiation experienced by biological organisms is the unit called the *rem*, which is defined as[4]

$$\text{dose in rem} = (\text{dose in rads}) \times (RBE) \qquad (30.33)$$

Since the rem includes the total amount of energy delivered by the radiation (measured in rads) and the efficiency with which this radiation damages real tissue, it combines the usefulness of both the rad and the *RBE* factor. The rem is the most widely used unit in discussions of radiation exposure and damage.

Table 30.3 Relative Biological Effectiveness (*RBE*) for Different Types of Radiation

RADIATION	RBE
Alpha particles	10–20
Beta particles	1.0–1.7
Gamma rays and X-rays	1.0
Slow neutrons	4–5
Protons	5
Fast neutrons	10
Heavy ions	20

Definition of the rem

| EXAMPLE 30.8 | ✖ Comparing the Dose from Different Types of Radiation |

A person is exposed to a radiation dose of 0.25 mrad (millirad) in a chest X-ray. What dose of fast neutrons would produce the same amount of tissue damage?

RECOGNIZE THE PRINCIPLE

The amount of damage done by radiation is determined by the product of the dose (which can be measured in rads) and the *RBE* (relative biological effectiveness):

$$\text{radiation damage} = \text{dose} \times RBE \qquad (1)$$

Hence, to compare the damage done by X-rays and fast neutrons, we must take their *RBE* values into account.

SKETCH THE PROBLEM

No sketch is necessary.

IDENTIFY THE RELATIONSHIPS

From Table 30.3, the *RBE* for X-rays is 1.0, while for fast neutrons we find $RBE = 10$. Hence, to achieve the same damage with these two types of radiation, the fast neutron dose must be 10 times smaller than the X-ray dose.

SOLVE

If a dose of fast neutrons produces the same amount of damage as a 0.25-mrad chest X-ray, we can use Equation (1) to write

$$\text{dose(X-rays)} \times RBE(\text{X-rays}) = \text{dose(fast neutrons)} \times RBE(\text{fast neutrons})$$

---

[4]The name "rem" stands for *r*öntgen *e*quivalent in *m*an. Wilhelm Röntgen was the discoverer of X-rays.

which leads to

$$\text{dose(fast neutrons)} = \frac{RBE(\text{X-rays})}{RBE(\text{fast neutrons})} \times \text{dose(X-rays)}$$

Inserting RBE values along with the given value of the X-ray dose, we find

$$\text{dose(fast neutrons)} = \frac{1.0}{10} \times (0.25 \text{ mrad}) = \boxed{0.025 \text{ mrad}}$$

What does it mean?
When comparing the radiation damage from different sources, one must always take the RBE of the radiation into account.

Some Facts about Radiation Damage

The physiology and cell biology of radiation damage is a complex and interesting subject. Here we can only mention some of the important facts.

1. When the radiation dose is low, cells are sometimes able to repair the damage, especially if the dose is absorbed over a long period of time. Hence, small amounts of radiation generally do not cause significant harm to living cells.
2. If the radiation dose is very large, cells can be completely destroyed (cell death).
3. At intermediate radiation doses, cells survive but often malfunction as a result of the damage or in their attempt to repair the damage. A typical result is that the affected cells reproduce in an uncontrolled fashion, leading to cancer.
4. Radiation damage is usually most severe for quickly dividing cells. Many types of blood and bone marrow cells fall into this category, along with cells in an embryo or fetus. Cancerous cells are also quickly dividing, so radiation can be used as a tool to selectively destroy cancer cells (Section 30.5).
5. The amount of damage can depend strongly on where the radiation source is located. For example, alpha particles have a very short range and can be stopped by a sheet of paper. So, if a source of alpha particles is outside the body, the particles will be stopped in the outer layer of skin and do relatively little damage. On the other hand, if a person somehow ingests an alpha source, the source may become lodged in his or her lungs or some other place, and we say that the person is *contaminated*. In this case, the alpha particles can do a great deal of damage to nearby cells, damaging their DNA and so forth.

Sadly, many of the early scientists who studied radioactivity suffered from their exposure to hazardous amounts of radiation. Röntgen (the discoverer of X-rays) died from bone cancer, and Marie Curie died from leukemia that was probably caused by her exposure to many radioactive substances. Her husband, Pierre, also suffered from radiation-induced illnesses (although he died after being run over by a horse-drawn carriage; go figure). In fact, the Curies' scientific papers are reportedly still quite radioactive.

People who work with radiation are normally required to use radiation monitors to record their exposure. These monitors typically contain a piece of photographic film (Fig. 30.22). Exposure to radiation exposes the film, and the degree of film exposure measures the radiation dose. In the event of an accidental overdose, the worker can then quickly get the appropriate medical treatment.

Sources of Radioactivity in Everyday Life

Table 30.4 lists the radiation doses of several common procedures and shows that we are all exposed to radiation during the course of everyday life. Even so, the

Figure 30.22 The radiation and particles emitted by radioactive substances cause photographic film to be exposed. To monitor their exposure to harmful radiation, people who work with radioactive materials wear small "film badges."

Table 30.4 Radiation Doses in Perspective (Typical Values)

ACTIVITY	DOSE (mrem)
Airline flight from New York City to Los Angeles	3
Dental X-ray	10
Chest X-ray	5–10
Mammogram	50–100
Approximate annual dose from natural background sources	300
Recommended maximum annual dose in addition to background dose	500
Apollo XVI astronauts	500
1-in-30 risk of cancer	10,000
Radiation sickness possible	60,000

benefits from, for instance, a chest X-ray, dental X-ray, or mammogram usually far outweigh possible risks.

There are several natural sources of radiation. Cosmic rays—a collection of many different types of particles that come to the Earth from outer space—are a significant source, especially at high altitudes. They are responsible for the increased radiation dose absorbed by astronauts and by persons at altitude (e.g., during an airplane flight or in a city such as Denver).

One of the largest natural sources of radiation exposure involves radon gas produced by the radioactive decay of $^{238}_{92}$U. There are very small concentrations of $^{238}_{92}$U in the soil and rocks at virtually all places on this planet. These nuclei decay with a half-life of 4.5×10^9 yr, initiating the series of decays in Figure 30.9. One of these daughter nuclei is $^{222}_{86}$Rn (radon), and being a gas, it diffuses through pores in the surrounding soil or rock. Radon gas thus seeps into mines and into the lungs of miners. These $^{222}_{86}$Rn nuclei then decay further, producing nuclei that are absorbed in the lining of the lungs of anyone who breathes the gas, leading to alpha and beta radiation that can ultimately cause lung cancer. Radon gas is a major health hazard for miners, and in some homes. It can seep into basements and crawl spaces, where it can collect to unsafe levels, just as in mines. Fortunately, there are simple ways to remediate this hazard, typically by improving the ventilation in the mine or home (Fig. 30.23).

Figure 30.23 Simple radon detector for home use. This detector contains a small canister of activated charcoal, which absorbs radon gas (along with other gases). After being exposed to the atmosphere for a specified period of time, the canister is sent to a laboratory for analysis.

EXAMPLE 30.9 ⊗ Radiation Exposure from a CT Scan

In Chapter 23, we described how a CT (or CAT) scan uses X-rays to form an image of bones and soft tissue within the body. The X-ray photons in a CT scan produce radiation damage similar to that produced by gamma rays. The radiation dose for a CT scan depends on many factors, but is typically about 150 mrem for a scan of the head. How many CT scans can your doctor prescribe before the radiation damage of the scan itself becomes a potential health hazard?

RECOGNIZE THE PRINCIPLE

The "normal" dose from radiation sources in the environment (cosmic rays, the small amount of radioactive material in the soil, etc.) is about 300 mrem per year (Table 30.4). To be safe, it is recommended that the dose from other sources such as a CT scan should be less than about 500 mrem.

SKETCH THE PROBLEM

No sketch is necessary, but the data in Table 30.4 are very useful.

IDENTIFY THE RELATIONSHIPS AND SOLVE

We want the total dose from all the CT scans in a year to be no larger than the recommended maximum dose of 500 mrem. The safe number of CT scans is therefore

$$\text{number of scans} = \frac{500 \text{ mrem}}{\text{CT dose}} = \frac{500 \text{ mrem}}{150 \text{ mrem}} = \boxed{3}$$

What does it mean?

You might now conclude that it is unsafe to have more than three CT scans per year. Our analysis is a bit simplistic, however, since a doctor must carefully weigh the potential benefits (e.g., from locating a suspected tumor) in assessing whether or not to prescribe a CT scan.

30.5 | ⊗ APPLICATIONS OF NUCLEAR PHYSICS IN MEDICINE AND OTHER FIELDS

Although radiation can damage living tissue, it has many positive uses in medicine and other fields. We'll now describe a few of these applications.

Radioactive Tracers

A *radioactive tracer* is a chemical that contains radioactive nuclei. The movement of tracer nuclei through the body can be monitored by observing the radiation they emit through radioactive decay. Tracers are widely used in medicine and many other fields.

For example, in a method known as nuclear stress testing, a small amount of a radioactive liquid containing (typically) $^{201}_{81}\text{Th}$ or $^{99}_{43}\text{Tc}$ is injected into a vein near the heart. These isotopes emit gamma rays, which are detected by special cameras placed over the heart. Gamma rays from blood containing the tracer molecules cause the blood to "light up" in the images. These images thus show how blood flows (or does not flow) in and around the heart and are used to diagnose heart disease. In this and other applications of radioactive tracers, the dose of radioactivity is kept low so that it does no significant harm. The tracer isotopes usually have short half-lives, so their activity decays to a very low level after a few hours or days.

The method of positron emission tomography (called PET scanning) works in a similar way. A glucose solution containing typically $^{18}_{9}\text{F}$ (the isotopes $^{11}_{6}\text{C}$, $^{15}_{8}\text{O}$, and $^{13}_{7}\text{N}$ are also used) is injected into a patient's blood. These nuclei decay by the emission of a positron (a type of beta decay), and these positrons in turn produce gamma rays when they encounter an electron. (We'll discuss how and why in Chapter 31.) These gamma rays are observed by a collection of detectors positioned around the patient (Fig. 30.24). By combining the images from many different detectors (the "tomography" part of the process), medical professionals construct a three-dimensional image of the positions of the radioactive nuclei. These images are used to monitor blood flow near and inside the heart and brain. Some types of cancer cells absorb large amounts of glucose, so PET scans can also be used to find these cancers.

Another use of tracers is to fight thyroid cancer. Iodine tends to accumulate in the thyroid gland, so by using a tracer containing the radioactive isotopes $^{123}_{53}\text{I}$ or $^{131}_{53}\text{I}$ and detecting the radiation they emit, one can measure how effectively the thyroid absorbs iodine. Cancerous thyroid cells can be killed using the electrons emitted in beta decay as well as by the gamma radiation emitted by these iodine nuclei.

Radioactive tracers thus have many applications. The best choice of tracer nucleus depends on the application, but all tracers should have a short half-life so that their activity decreases to negligible levels soon after use. Also, tracer nuclei cannot be stored for future use, but must be made just before they are used. Tracer nuclei are usually made by nuclear reactions induced in a nuclear reactor core.

Gamma ray detectors

Figure 30.24 Positron emission tomography (PET scanning) is a widely used medical procedure. After ingesting a radioactive tracer, the patient enters an apparatus like the one shown here. This device measures the gamma rays emitted when positrons annihilate with electrons and determines the sites from which they are emitted.

© Jean-Claude Revy/ISM/Phototake USA

Using Radiation to Fight Cancer

Because quickly dividing cells are generally the ones most damaged by radiation, radiation treatments are effective in treating many types of cancers. For example, radiation is widely used to kill breast cancer cells. This therapy uses X-rays, so strictly speaking it does not involve radioactivity. Breast cancer is sometimes also treated by implanting radioactive nuclei, a treatment called *brachytherapy*. These nuclei (usually $^{90}_{38}\text{Sr}$) emit high-energy electrons (via beta decay), gamma rays, or both, killing nearby cancer cells. A similar treatment is used in dealing with prostate cancer in men. Radioactive "seeds" containing typically $^{125}_{53}\text{I}$ or $^{103}_{46}\text{Pd}$ are implanted in the prostate gland, and the gamma rays they emit kill nearby cancerous cells. These isotopes have relatively short half-lives, so their activity is very small after a few months.

Carbon Dating

One of the most important uses of radioactivity and tracers outside of medicine is the technique of *carbon dating* developed in 1947 by Willard Libby, who received the Nobel Prize in Chemistry for this work. Of the carbon nuclei occurring naturally on the Earth, approximately 99% are the isotope $^{12}_{6}\text{C}$ and about 1% are $^{13}_{6}\text{C}$. In addition to these two stable isotopes, the radioactive isotope $^{14}_{6}\text{C}$ is produced in the Earth's atmosphere when cosmic rays collide with $^{14}_{7}\text{N}$. The half-life of $^{14}_{6}\text{C}$ is $T_{1/2} = 5730$ yr as it undergoes beta decay, producing $^{14}_{7}\text{N}$ (Eq. 30.13). This decay is balanced by the cosmic-ray–induced production of $^{14}_{6}\text{C}$, resulting in a carbon isotopic ratio in the atmosphere[5] equal to $^{14}_{6}\text{C}/^{12}_{6}\text{C} \approx 1.3 \times 10^{-12}$.

Carbon-containing molecules including proteins and DNA are the building blocks of life. When an animal is living, it accumulates carbon from the environment as it eats and breathes, while plants take in atmospheric CO_2 during photosynthesis and absorb carbon-containing nutrients from the water and soil. Since there is a small amount of $^{14}_{6}\text{C}$ in the atmosphere and in their nutrients, organisms absorb $^{14}_{6}\text{C}$ until the percentage of $^{14}_{6}\text{C}$ in the organism equals the $^{14}_{6}\text{C}$ percentage in the atmosphere. When the organism dies, it stops accumulating new $^{14}_{6}\text{C}$, and the $^{14}_{6}\text{C}$ that it contained at death undergoes radioactive decay. Hence, after one half-life (5730 yr), the amount of $^{14}_{6}\text{C}$ in the organism is equal to half of what it was at the time the organism died. Likewise, after a time equal to $2T_{1/2}$, the amount of $^{14}_{6}\text{C}$ is a fourth of the original amount. The behavior of the $^{14}_{6}\text{C}$ concentration as a function of time is given by the exponential relation in Equation 30.25 and is plotted in Figure 30.25. Since the ratio $^{14}_{6}\text{C}/^{12}_{6}\text{C}$ at the time of death is known, the ratio measured today in a sample can be used to determine how much time has passed since death. In this way, the age of an animal or plant specimen can be accurately determined. This technique is now an essential part of experimental archaeology.

Carbon dating is done by first measuring the isotopic ratio of $^{14}_{6}\text{C}/^{12}_{6}\text{C}$ in a specimen, often using a mass spectrometer (Fig. 20.26) to measure the separate amounts of $^{14}_{6}\text{C}$ and $^{12}_{6}\text{C}$ that are present. Since $^{12}_{6}\text{C}$ is a stable isotope, the $^{14}_{6}\text{C}/^{12}_{6}\text{C}$ ratio decreases with time according to the decay of $^{14}_{6}\text{C}$ alone, and the isotopic ratio follows the decay curve in Figure 30.25. Given the measured isotopic ratio, the age of the sample can then be read off the curve in Figure 30.25.

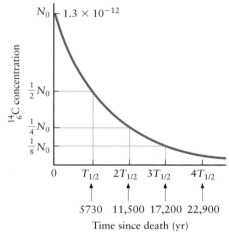

Figure 30.25 Carbon dating uses the concentration of $^{14}_{6}\text{C}$ (the ratio $^{14}_{6}\text{C}/^{12}_{6}\text{C}$) in a sample from a formerly living organism to deduce the time at which the organism died. After death, the concentration of $^{14}_{6}\text{C}$ falls with time due to its radioactive decay, with a half-life of 5730 yr.

EXAMPLE 30.10 ⊗ Carbon Dating

Some ancient plant material is found to have a fraction of $^{14}_{6}\text{C}$ given by $^{14}_{6}\text{C}/^{12}_{6}\text{C} = 6.0 \times 10^{-14}$. How old is it? The $^{14}_{6}\text{C}/^{12}_{6}\text{C}$ ratio at the time of death was equal to 1.3×10^{-12} (as explained above).

---

[5]Nuclear weapons testing in the 1950s and 1960s nearly doubled the $^{14}_{6}\text{C}/^{12}_{6}\text{C}$ ratio in the atmosphere. This ratio is now returning to its pre-1950s level.

RECOGNIZE THE PRINCIPLE

The radioactive decay of $^{14}_{6}C$ is described by

$$N = N_0 e^{-\lambda t} \qquad (1)$$

with a half-life of 5730 yr, whereas $^{12}_{6}C$ is stable and does not decay. The $^{14}_{6}C/^{12}_{6}C$ thus decreases with time according to

$$\frac{^{14}_{6}C}{^{12}_{6}C} \text{ at time } t = \left(\frac{^{14}_{6}C}{^{12}_{6}C} \text{ at time of death} \right) \times e^{-\lambda t} \qquad (2)$$

SKETCH THE PROBLEM

The decay behavior in Equations (1) and (2) is described graphically in Figure 30.25.

IDENTIFY THE RELATIONSHIPS

The given $^{14}_{6}C/^{12}_{6}C$ ratio is smaller than the ratio at the time of death by

$$\frac{^{14}_{6}C/^{12}_{6}C \text{ (at present)}}{^{14}_{6}C/^{12}_{6}C \text{ (at death)}} = \frac{6.0 \times 10^{-14}}{1.3 \times 10^{-12}} = 0.046 = 4.6\% \qquad (3)$$

We can now approach this problem in two ways. One would be to use the graph in Figure 30.25 and read off the time at which the $^{14}_{6}C/^{12}_{6}C$ ratio has dropped to 4.6% of its original value, but this small value is hard to locate accurately on the vertical scale in Figure 30.25. Instead, let's use the exponential relation in Equation (1). At an unknown time t, only 4.6% of the original $^{14}_{6}C$ remains, so if N is the number of $^{14}_{6}C$ we have

$$N(\text{at present}) = N(\text{at death})e^{-\lambda t} \qquad (4)$$

where λ is the decay constant for $^{14}_{6}C$. The decay constant is related to the half-life by (Eq. 30.27)

$$T_{1/2} = \frac{0.693}{\lambda}$$

SOLVE

Solving for λ, we get

$$\lambda = \frac{0.693}{T_{1/2}} = \frac{0.693}{5730 \text{ yr}} = 1.2 \times 10^{-4} \text{ yr}^{-1}$$

Rearranging Equation (4) and using the $^{14}_{6}C/^{12}_{6}C$ ratio in Equation (3) leads to

$$\frac{N(\text{at present})}{N(\text{at death})} = 0.046 = e^{-\lambda t}$$

We now take the logarithm of both sides and use the definition of logarithms $\ln(e^x) = x$. We find

$$\ln(0.046) = -\lambda t$$

$$t = -\frac{\ln(0.046)}{\lambda} = -\frac{\ln(0.046)}{1.2 \times 10^{-4} \text{ yr}^{-1}} = \boxed{26{,}000 \text{ yr}}$$

What does it mean?

Carbon dating is a precise way to date many kinds of material. However, one limitation on this technique is set by the half-life of $^{14}_{6}C$. If the material is very much older than the half-life, there will be very little $^{14}_{6}C$ left to measure. In practice, carbon dating is useful for material that is no older than about 50,000 years.

Magnetic Resonance Imaging

Magnetic resonance imaging (MRI) makes use of the magnetic properties of nuclei. Hydrogen has the simplest nucleus—a single proton—and a proton has an intrinsic spin angular momentum similar to the spin angular momentum of the electron discussed in Chapter 28. As with electrons, a proton has just two spin states, "spin up" and "spin down," and in a magnetic field these two states have different energies. The presence of a proton can be detected by observing the absorption of a photon that induces a transition from the lower- to the higher-energy spin state (Fig. 30.26A). This is just like the absorption of a photon by an atom as it undergoes a transition between energy levels (Chapter 29).

The absorption of a photon only occur if the separation between energy levels matches the photon energy, and this energy separation depends on the magnetic field. An MRI magnet is designed so that these energies match only at a particular spot within the body, so the MRI signal gives the density of protons (and other properties) at just that spot. By then scanning this spot around the body, a three-dimensional image can be constructed, giving a unique picture of the body's internal structure.

Most applications of MRI employ a magnet with coils in a unit large enough for a person to fit inside (Fig. 30.26B). After only a few minutes, an image with remarkable resolution can be obtained (Fig. 30.26C). MRI images are now widely used to diagnose many different types of medical conditions (including the author's knee injuries). This method does not employ radioactive tracers, but uses the magnetic properties of naturally occurring nuclei (usually hydrogen nuclei) in the patient and radio frequency photons (which have very low energies). As a result, there do not appear to be any health risks associated with MRI, making this technique a very safe and effective medical tool.

30.6 | QUESTIONS ABOUT THE NUCLEUS

In this chapter, we have described nuclear properties and nuclear reactions. We now ask *why* the nucleus has these properties. As usual, this "why" question is much more difficult than discovering the actual properties, but it also leads us to fundamental issues.

Conservation rules (Section 30.2) are fundamental principles of physics that apply to a wide range of situations. Nuclear physics reveals a new conservation rule, the conservation of nucleon number. There are just two types of nucleons—protons and neutrons—and *all* known nuclear reactions conserve the number of nucleons. The *total* number of nucleons at the start of a reaction is equal to the total number of nucleons at the end, which you can confirm for yourself by reviewing the examples of alpha, beta, and gamma decay described in Section 30.2. Physicists have spent much time and effort studying this problem, and as far as we can tell, nucleon number is always conserved. Why? We don't know. All we can say is that nucleon number conservation is a property of all known processes.

Such number conservation rules do not apply to all particles. For example, the number of electrons is *not* conserved in all reactions. Beta decay (e.g., Eq. 30.13) creates an electron where one did not previously exist, so electrons are different from nucleons in some fundamental way. We'll explore this topic further in Chapter 31.

We can also ask about the structure of a nucleon. Is there anything inside a proton or a neutron? Physicists have spent years working on this question. Scattering experiments similar (in principle) to those of Rutherford have shown that nucleons are composed of particles called *quarks*. There are several different types of quarks, all which carry an electric charge of $\pm e/3$ or $\pm 2e/3$, where $-e$ is the charge carried by an electron. Protons and neutrons are each composed of three quarks. In a proton, the quark charges add up to $+e$; in a neutron, they add up to zero. So, although a neutron is electrically neutral, it contains charged particles! In Chapter 31, we'll

Figure 30.26 **A** When a proton (a hydrogen nucleus) is placed in a magnetic field, the spin-up and spin-down states have different energies. A proton can absorb a radio frequency photon and undergo a transition from one level to the other. **B** Magnetic resonance imaging (MRI) uses the spin magnetism of hydrogen nuclei to measure the density and other properties of tissue within the body. This procedure requires that the patient be placed inside a large magnet. **C** MRI image of a person's abdomen.

discuss the behavior of quarks and how they combine to make protons and neutrons, but we still don't know *why* nature is put together this way.

SUMMARY | Chapter 30

Nuclear structure

Nuclei contain **protons** and **neutrons**, also called **nucleons**. The number of protons in a nucleus is the **atomic number** Z, and the total number of protons plus neutrons is the **mass number** A. The number of neutrons N is thus related to Z and A by

$$A = Z + N \qquad \text{(30.3) (page 1022)}$$

Different **isotopes** of a particular element contain the same number of protons, but different numbers of neutrons.

All this information is described in the following shorthand notation:

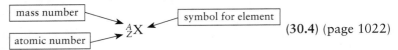

$$\qquad \text{(30.4) (page 1022)}$$

Nuclear size

Nuclei have a radius r of a few femtometers (1 fm = 10^{-15} m). The radius depends on the number of nucleons A in the nucleus through the relation

$$r = r_0 A^{1/3} \qquad \text{(30.5) (page 1024)}$$

with $r_0 \approx 1.2$ fm = 1.2×10^{-15} m.

The strong force

Nuclei are held together by the **strong force**, an attractive force that acts between pairs of nucleons. The strong force does not act on electrons, and it is responsible for the stability of the nucleus.

Radioactive decay

Some nuclei are unstable and decay spontaneously, a process called **radioactive decay**. In **alpha decay**, an alpha particle (a helium nucleus, $^{4}_{2}$He) is emitted. In **beta decay**, an electron or a positron is emitted. In **gamma decay**, a gamma ray (a high-energy photon) is produced. The **half-life** is the time it takes for half of the nuclei in a sample to decay. Half-lives can be less than 1 second or longer than the age of the Earth.

(Continued)

Nuclear fission

In nuclear *fission*, a nucleus splits into two nuclei that are each about half the size of the original nucleus and several neutrons. This reaction releases energy; it is the basis for nuclear weapons and nuclear power plants.

NUCLEAR FISSION (a heavy nucleus splits)

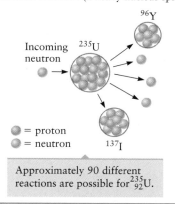

Approximately 90 different reactions are possible for $^{235}_{92}$U.

Nuclear fusion

In nuclear *fusion*, two low-mass nuclei combine to form one heavier nucleus. Fusion reactions also release large amounts of energy and are responsible for the energy output of the Sun.

NUCLEAR FUSION

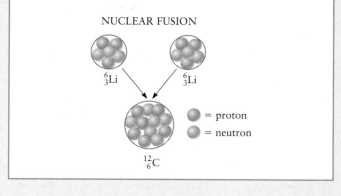

APPLICATIONS

Energy in nuclear reactions

The energy released in a typical nuclear reaction is typically on the order of MeV. It is much larger than a typical atomic or molecular energy (about 10 eV), and the particles emitted by a nuclear reaction can damage living tissue.

Units of radioactivity

Several units are used to measure the amount of radioactivity in a sample.

The *activity* of a sample is proportional to the number of nuclei in the sample that undergo radioactive decay in 1 second. One unit of activity, the curie (Ci), is defined by

$$1 \text{ Ci} = 3.7 \times 10^{10} \text{ decays/s} \qquad \textbf{(30.28)} \text{ (page 1036)}$$

Activity is also measured with the SI unit becquerel (Bq), with

$$1 \text{ Bq} = 1 \text{ decay/s} \qquad \textbf{(30.29)} \text{ (page 1036)}$$

Biological effects

The amount of damage caused by radiation is determined in part by the energy deposited by the radiation in tissue. A radiation dose of 1 *rad* deposits 10^{-2} J in 1 kg of material. Different types of radiation can do more or less damage. The *RBE* (relative biological effectiveness) factor measures this difference. The unit called the *rem* takes all these factors into account:

$$\text{dose in rem} = (\text{dose in rads}) \times (RBE) \qquad \textbf{(30.33)} \text{ (page 1045)}$$

Benefits of nuclear radiation

Radiation presents many health hazards, but it also has useful applications, including carbon dating and the treatment of many types of cancer.

1. Why do different isotopes of a given element have, for most practical purposes, the same chemical properties? That is, why do different isotopes form the same molecules and compounds even though they have different numbers of neutrons and different masses?

2. Different isotopes of an element essentially share chemical properties. How, then, is it possible to separate different isotopes?

3. What is heavy water? Would an equal number of moles of heavy water actually weigh more than plain water?

4. Was evidence for the nucleus was first obtained through (a) carbon dating, (b) the spectral lines of hydrogen, (c) alpha particle scattering, (d) cosmic rays, or (e) electromagnetic radiation?

5. What is (a) an alpha particle? (b) a beta particle? (c) a gamma ray? (d) an X-ray? (e) a daughter nucleus?

6. Is the mass of a nucleus larger or smaller than the masses of the nucleons of which it is made? Explain.

7. SSM In the mid 1800s, the elements of the periodic table were being identified and discovered, yet for many decades element 43, between molybdenum and ruthenium (Fig. Q30.7), remained unfound. Technetium (Tc) was finally discovered almost 100 years later, in 1936. (a) What is special about this element, and why was it so hard to find? (b) Where can technetium be found today?

Cr 24	Mn 25	Fe 26
51.996	54.938	55.845
$3d^54s^1$	$3d^54s^2$	$3d^64s^2$
Mo 42	43	Ru 44
95.94	?	101.07
$4d^55s^1$		$4d^75s^1$
W 74	Re 75	Os 76
183.84	186.21	190.23
$5d^46s^2$	$5d^56s^2$	$5d^66s^2$

Figure Q30.7 Element 43.

8. Why is the element plutonium not found in significant amounts in nature (on the Earth)?

9. Compare how a fission bomb works with how a fusion bomb works. Be sure to discuss what each would use for fuel and the waste products produced by each.

10. There is a critical mass for the creation of a spontaneous chain reaction involving $^{235}_{92}$U. Why would it be difficult to make a bomb that is much larger than the critical mass? Why does this limit not exist for a fusion bomb?

11. X Explain why carbon dating can in general only be used to measure the age of material that was once alive.

12. Do you expect that $^{56}_{26}$Fe would undergo fusion? Explain why or why not. *Hint*: Use the binding energy curve in Figure 30.17.

13. Consider the radioactive decay of $^{235}_{92}$U. Construct a series of alpha and beta decays (similar to Fig. 30.9) that include the daughter nuclei $^{223}_{88}$Ra, $^{219}_{86}$Rn, and $^{211}_{83}$Bi, and terminates with $^{207}_{82}$Pb. Why can this decay series not terminate with $^{206}_{82}$Pb or $^{208}_{82}$Pb?

14. X In the treatment of cancer using radioactive "seeds" (brachytherapy), the radioactive material must be placed very close to the cancer cells. (a) Explain why cells farther away from the radioactive seeds are not significantly harmed by the emitted radiation. (b) Explain why placing the radioactive seeds outside the body is not as effective as placing them inside or next to a cancerous tumor.

15. A nucleus of $^{245}_{96}$Cm that is initially at rest undergoes alpha decay. What is the daughter nucleus? Which of the reaction products has the greater kinetic energy, the alpha particle or the daughter nucleus?

16. Explain why the neutrons emitted in a nuclear reaction can penetrate very thick sheets of lead, but the alpha and beta particles cannot.

17. Which decays change the parent nucleus into a nucleus of a different element: alpha decay, beta decay, or gamma decay?

18. X An archaeologist friend proposes to determine the age of a dinosaur bone using carbon dating. Why does your friend need to take a physics course? That is, what is wrong with this proposal?

19. SSM X Scientists in areas such as geology, paleontology, and archaeology use "radiodating" techniques (such as carbon dating) in the study of the history of the Earth and its inhabitants. There are several radiodating techniques, each using different nuclei: (i) carbon 14, (ii) potassium–argon 40, (iii) rubidium–strontium, and (iv) uranium–lead. (a) Approximately how old an object can be dated with each of these techniques? (b) Conversely, what are the youngest objects that can be reliably dated with each technique? (c) Which techniques are best for archaeology? For paleontology? *Note*: You may have to look up the half-lives of the relevant parent isotopes for the elements used in each method.

20. In a volume of soil that measures 1 mi^2 and 1 ft deep there are roughly 2 g of radium, predominantly ^{236}Ra. But ^{236}Ra has a half-life of just 1600 yr, while the Earth is billions of years old. Where did this radium come from?

21. The nuclei $^{235}_{92}$U, $^{253}_{100}$Fm, and $^{14}_{6}$C have very different half-lives. Explain how samples of these materials can still have the same activity.

22. How might the $^{239}_{94}$Pu used in nuclear weapons be created? *Hint*: Consider reactions in which a nucleus captures a neutron.

23. The sketch in Figure 30.10 shows a nuclear decay in which two product nuclei are produced. Why do they travel away in opposite directions?

24. X Which of the following objects could be dated with carbon dating?
   (a) The logs used to make George Washington's house
   (b) The bricks used to make George Washington's house
   (c) Animal bones found in George Washington's barn
   (d) The blade of George Washington's pocket knife

30.1 STRUCTURE OF THE NUCLEUS: WHAT'S INSIDE?

1. Is the atomic number (a) the number of neutrons in the nucleus, (b) the number of protons in the nucleus, (c) the number of elec- trons in the nucleus, or (d) the sum of the number of neutrons and protons in the nucleus?

2. Is the mass number (a) the number of neutrons in the nucleus, (b) the number of protons in the nucleus, (c) the number of elec-

trons in the nucleus, or (d) the sum of the number of neutrons and protons in the nucleus?

3. Do different isotopes of the same element have the same number of (a) protons, (b) neutrons, (c) protons plus neutrons, or (d) electrons (assuming a neutral atom)? (More than one answer may be correct.)

4. Name three different nuclei that contain 120 neutrons.

5. What are Z, N, and A for the nuclei (a) $^{12}_{6}C$, (b) $^{34}_{16}S$, (c) $^{208}_{82}Pb$, and (d) $^{56}_{26}Fe$?

6. (a) A nucleus has $Z = 17$ and $N = 18$. What is the symbol for this nucleus? (b) If four neutrons are added to $^{208}_{82}Pb$, what element is created? Give the symbol for this nucleus.

7. If two protons are removed from $^{206}_{82}Pb$, what element is created? Give the symbol for this nucleus.

8. **SSM** ✪ **R** Estimate the electric potential energy of all the protons in ^6_3Li. Assume the distance between each pair of protons is 1 fm.

9. Use Equation 30.5 to estimate the radius of (a) an alpha particle, (b) $^{34}_{16}S$, and (c) $^{56}_{26}Fe$.

10. What is the approximate density of (a) an alpha particle and (b) $^{56}_{26}Fe$?

11. What is the ratio of the radius of $^{206}_{82}Pb$ to the radius of an alpha particle?

12. A nucleus has a radius of 4.5×10^{-15} m. What nucleus might it be?

13. An alpha particle has a kinetic energy of 1.2 MeV. What is its speed?

14. A beta particle emitted in nuclear decay has a kinetic energy of 50,000 eV. What is its speed?

15. What is the speed of a gamma ray?

16. Which two of the following forces are most important in a nucleus?
(a) electric force
(b) magnetic force
(c) the strong force
(d) friction
(e) gravity

17. Which of the following forces is (most) crucial for holding a nucleus together?
(a) electric force
(b) magnetic force
(c) the strong force
(d) friction
(e) gravity

18. Name the dominant force for (a) two protons far apart, (b) a neutron and a proton in a nucleus, and (c) a proton and an electron in a nucleus.

19. An electric potential difference ΔV is used to give an alpha particle a kinetic energy of 1.5 MeV. What is the value of ΔV?

20. **SSM** ✪ A proton is accelerated through a potential difference ΔV_p, and an alpha particle is accelerated through a potential difference ΔV_α. If the two particles have the same kinetic energy, is the ratio $\Delta V_p / \Delta V_\alpha$ (a) 4, (b) 2, (c) 1, (d) 1/2, or (e) 1/4?

30.2 NUCLEAR REACTIONS: SPONTANEOUS DECAY OF A NUCLEUS

21. After beta decay, the daughter nucleus differs from the parent nucleus in which of the following ways?
(a) The daughter nucleus has one more proton.
(b) The daughter nucleus has one less proton.
(c) The daughter nucleus has one more neutron.

22. After alpha decay, what is the difference in mass number between the daughter and the parent nuclei?
(a) The mass number of the daughter nucleus is smaller by four.

(b) The mass number of the daughter nucleus is larger by four.
(c) The mass number of the daughter nucleus is smaller by two.

23. If $^{242}_{94}Pu$ decays via beta decay, what nucleus is produced? If it decays by alpha decay, what nucleus is produced?

24. If $^{210}_{82}Pu$ is produced by alpha decay, what was the parent nucleus?

25. **SSM** ✪ Complete (i.e., balance) the following reactions. Also label each reaction as alpha decay, beta decay, or gamma decay.
(a) $^{234}_{90}Th \rightarrow ^{230}_{88}Ra + ?$
(b) $^{234}_{90}Th \rightarrow ^{234}_{91}Pa + ?$
(c) $^{234}_{90}Th^* \rightarrow ^{234}_{90}Th + ?$

26. Write the reaction for the alpha decay of $^{214}_{84}Po$.

27. Write the reaction for the beta decay of $^{60}_{27}Co$.

28. Write the reaction for the gamma decay of $^{210}_{82}Pb^*$.

29. If 16 g of radon gas with a half-life of 4 days is present in a balloon at $t = 0$, what is the mass of radon gas present after (a) 4 days, (b) 16 days, and (c) 25 days?

30. The half-life of $^{224}_{88}Ra$ is 3.7 days. What is its decay constant? Express your answer is s^{-1}.

31. The decay constant of a nucleus is 25 s^{-1}. What is its half-life?

32. ✪ After 25 years, 75% of a radioactive material decays. What is the half-life?

33. ✪ The isotope $^{19}_{10}Ne$ has a half-life of 17.2 s. If you have 45 g of it at $t = 0$, how much is left after 45 s?

34. **SSM** ✪ A particular nucleus has a half-life of 36 minutes. How long does it take for 95% of this material to decay?

35. ✪ At $t = 0$, 950 g of a radioactive material is present. Twenty-five years later it is found that only 23 g of the material remains. What is its half-life?

36. ✪ At present, the concentration of $^{235}_{92}U$ in naturally occurring uranium deposits is approximately 0.72%. What will the concentration be one billion years from now?

37. ✪ Balance the following nuclear reactions.
(a) $^1_1H + ^6_3Li \rightarrow ^4_2He + ?$
(b) $^1_0n + ^{235}_{92}U \rightarrow ^{133}_{51}Sb + ? + 4^1_0n$
(c) $^{14}_6C \rightarrow ? + e^- + \bar{\nu}$

38. Use the mass data in Table A.4 to calculate the binding energy of (a) $^{12}_6C$, (b) $^{56}_{26}Fe$, (c) $^{235}_{92}U$, (d) $^{238}_{92}U$, and (e) ^3_1H.

39. What is the magnitude of the binding energy per nucleon of $^{208}_{82}Pb$? What is the magnitude of the *total* binding energy of this nucleus? Give your answers in MeV.

40. What is the approximate energy needed to remove one neutron from $^{55}_{26}Fe$?

41. The isotope $^{89}_{36}Kr$ decays by beta emission. What nucleus is produced?

42. A piece of radium has an activity of 4.5 mCi. (a) What is the mass of the radium? (b) What is its activity in Bq?

43. A sample has an activity of 30 mCi (millicuries). If the sample is divided into 12 equal pieces, what is the activity of each piece?

30.3 STABILITY OF THE NUCLEUS: FISSION AND FUSION

44. How many neutrons are released when $^{235}_{92}U$ undergoes fission to produce $^{132}_{50}Sn$ and $^{100}_{42}Mo$? Do not count the neutron that initiated the fission.

45. ✪ In Example 30.7, we considered the energy produced by the fusion of two deuterium nuclei (^2_1H). (a) In the example, we found the energy in MeV. Convert this energy to joules. (b) In seawater, deuterium has a concentration of 0.0003% relative to that of normal hydrogen (^1_1H). If all the deuterium in 1.0 g of seawater undergoes fusion (as in Example 30.7), how much energy would be released?

46. Balance the fission reaction $^{235}_{92}U + ^1_0n \rightarrow ^{142}_{55}Cs + ^{90}_{37}Rb + ?^1_0n$.

47. ✪ The fission bomb that exploded at Hiroshima, Japan, in 1945 released an energy of about 6×10^{13} J. (The value is known only approximately.) Consider a fusion bomb that releases energy by the fusion of two deuterium nuclei (Example 30.7). What mass of deuterium nuclei is required to release the energy that was released at Hiroshima?

48. SSM ✯ The largest fusion bomb ever detonated (called the Tsar Bomba) was tested by the Soviet Union in 1961. The explosion from the device shown in Figure P30.48 released an energy of approximately 2.4×10^{17} J. (a) How many grams of matter were converted to energy? (b) If this bomb used the deuterium–deuterium fusion reaction (Example 30.7), how many grams of deuterium were required?

Figure P30.48 The Tsar Bomba weighed 2.7×10^4 kg and had a design yield equivalent to 50 megatons of TNT.

49. In the fission reaction $^{235}_{92}\text{U} + ^{1}_{0}\text{n} \rightarrow ^{138}_{54}\text{Xe} + ^{95}_{38}\text{Sr} + 3^{1}_{0}\text{n}$, the isotope $^{235}_{92}\text{U}$ has a mass of 235.04392 u, the mass of $^{138}_{54}\text{Xe}$ is 137.91395 u, and the mass of $^{95}_{38}\text{Sr}$ is 94.91936 u. How much energy does this reaction release?

50. ✪ Ⓡ **Energy production in the Sun.** The Sun produces energy through a series of fusion reactions, starting with protons ($^{1}_{1}\text{H}$) and ending with alpha particles (helium nuclei, $^{4}_{2}\text{He}$). The first reaction produces deuterium ($^{2}_{1}\text{H}$) plus a positron and a neutrino,

$$^{1}_{1}\text{H} + ^{1}_{1}\text{H} \rightarrow ^{2}_{1}\text{H} + e^{+} + \nu \qquad (1)$$

and the second reaction fuses a proton with deuterium to produce $^{3}_{2}\text{He}$ (a stable isotope of helium),

$$^{1}_{1}\text{H} + ^{2}_{1}\text{H} \rightarrow ^{3}_{2}\text{He} + \gamma \qquad (2)$$

These helium nuclei are then converted to alpha particles through either of the following reactions:

$$^{1}_{1}\text{H} + ^{3}_{2}\text{He} \rightarrow ^{4}_{2}\text{He} + e^{+} + \nu \qquad (3)$$

$$^{3}_{2}\text{He} + ^{3}_{2}\text{He} \rightarrow ^{4}_{2}\text{He} + ^{1}_{1}\text{H} + ^{1}_{1}\text{H} \qquad (4)$$

(a) Use the masses given in Table A.4 to find the energy released in each of the reactions (1) through (4). (Do not include any subsequent reactions involving the positron.) (b) What is the total energy released when one alpha particle is produced using the process in reactions (1), (2), and (3)? (c) What is the total energy released when one alpha particle is produced using the process in reactions (1), (2), and (4)? (d) The Sun is currently converting approximately 600,000,000 tons of hydrogen into alpha particles each second. How many years will it take to consume all the Sun's fuel? Assume all the Sun's current mass ($m_{\text{Sun}} = 2.0 \times 10^{30}$ kg) is made up of protons.

30.4 BIOLOGICAL EFFECTS OF RADIOACTIVITY

51. ⊗ A person absorbs a dose of 15 rads of protons. What dose of alpha particles causes the same amount of damage? Assume that the RBE for these protons is 15.

52. ⊗ Use the RBE factors in Table 30.3 to compare the damage produced by alpha particles, gamma rays, and fast neutrons.

For the same dose as measured in rads, rank these three types of radiation in order of least damage to most damage.

53. ⊗ Alpha particles with an RBE of 15 deliver 25 mrem of radiation to a patient. What dose does the patient receive? Express your answer in rad.

54. What is the activity of 5.0 g of $^{235}_{92}\text{U}$? Express your answer in decays/s and Ci.

55. ✯ ⊗ A person with a mass of 85 kg is exposed to a dose of 30 rad. How many joules of energy are deposited in the person's body?

56. SSM ✯ ⊗ A 125-rad dose of radiation is administered to a cancer patient (mass = 80 kg). If all this radiation is absorbed by the tumor, how much energy is absorbed?

57. ✯ ⊗ Ⓡ A typical exposure from a dental X-ray is 10 mrem. (See Table 30.4.) (a) How much energy is deposited to your head when you get a dental X-ray? (b) A weight of 1 N falls on your head. From how high above your head would the weight need to be dropped to impart the same energy as the X-ray?

58. A radioactive material has an activity of 2.5 Bq. What is its activity in Ci?

59. A radioactive sample has an initial activity of 4.5 mCi (millicuries). Its activity 15 hours later is 1.8 mCi. What is its half-life?

60. SSM ✯ A radioactive sample has an initial activity of 800 decays/s. Its activity 24 hours later is 200 decays/s. What is its half-life?

61. ⊗ An average annual dose from natural radiation sources is about 500 mrem, of which about 40 mrem comes from ^{40}K, an isotope found all around us and in our food. What is the maximum energy ($RBE = 1.7$) deposited in a 65-kg individual from the emitted beta rays of potassium each year?

62. ✪ ⊗ Ⓡ Approximate the activity (in units of Bq and Ci) in one banana from its ^{40}K content (Fig. P30.62). An average banana has about 600 mg of potassium. *Hint:* You will need to find the natural abundance and half-life of the ^{40}K isotope.

Figure P30.62 Bananas contain some ^{40}K, making them slightly radioactive.

30.5 APPLICATIONS OF NUCLEAR PHYSICS IN MEDICINE AND OTHER FIELDS

63. ⊗ A patient with thyroid problems has radioactive iodine ($^{123}_{53}\text{I}$) deposited in his thyroid gland. If the $^{123}_{53}\text{I}$ has an initial activity of 2.5 MBq, what is its activity after 20 days?

64. ✯ ⊗ A sample of material is found to be 25,000 years old using carbon dating. What fraction of the original $^{14}_{6}\text{C}$ remains?

65. ⊗ An archaeologist uses carbon dating to find the age of some ancient plant material. If the concentration of $^{14}_{6}\text{C}$ is 2.8×10^{-14} smaller than the concentration of $^{12}_{6}\text{C}$, how old is the material?

66. ⊗ A piece of very old wood is found to have a $^{14}_{6}\text{C}/^{12}_{6}\text{C}$ concentration ratio that is 0.10 times smaller than the concentration found in a tree that died in 1900. How old is the older piece of wood?

67. SSM ✯ ⊗ In 1991, hikers in the Alps discovered the remains of a man who had been trapped in a glacier (Fig. P30.67). Carbon dating revealed the $^{14}_{6}C/^{12}_{6}C$ ratio in the remains to be 6.8×10^{-13}. How old are the remains?

Figure P30.67

68. ⊗ The tracer $^{123}_{53}I$ is used to treat thyroid problems. This tracer nucleus is produced by bombarding xenon with protons in the reaction $^{1}_{1}H + ^{124}_{54}Xe = ^{123}_{53}I + ?$ What else does this reaction produce?

69. ✦ Some smoke detectors (Fig. P30.69) use the radiation produced by the decay of $^{241}_{95}Am$ (half-life 432 yr) to ionize air molecules, which in turn produces a steady current accross two electrodes. If smoke enters the region of ionization, the alpha particles are absorbed by the smoke particulates, reducing the current accross the electrodes. The electronic circuitry then sounds the alarm when it senses any current reduction. How long does it take for the activity of $^{241}_{95}Am$ to drop to 1% of its initial value? Would it pose a problem if the smoke detector is discarded in a landfill?

Figure P30.69

70. ✦ ⊗ In positron emission tomography (PET), a positron is created by the decay of a nucleus attached to a glucose molecule. This positron then annihilates an electron (because the positron is antimatter; that is, it is an anti-electron). The annihilation reaction produces two gamma rays, traveling off in opposite directions. When the gamma rays are detected, their origin in the body reveals how and where glucose is absorbed. (a) Find the energy, in MeV, of one of these gamma rays. (b) Find the frequency of the gamma ray in part (a). (c) Explain why there are two gamma rays emitted in opposite directions.

71. ⊗ A common nucleus used in PET scans is $^{18}_{9}F$, which has a half-life of 110 min. What fraction of the original activity remains in the body after 3 h?

ADDITIONAL PROBLEMS

72. ✦ **Heavy water.** The compound D_2O is called heavy water. It is essentially a water molecule in which each of the hydrogen atoms has a deuterium nucleus. (a) How much heavier, by percentage, is heavy water from regular water? (b) Estimate the density of heavy water.

73. ✦ A nuclear power plant with a 1000-MW output is 40% efficient in converting the energy released from fission into electricity. How much $^{235}_{92}U$ is consumed by the power plant in one day? Express your answer in kg and in atoms.

74. ✦ Ⓡ An MX "Peacekeeper" missile carries 10 warheads, each rated at 20 megatons. (a) How many times greater is the energy release from one of these warheads than that of the Hiroshima explosion? (b) What quantity of mass is used to produce this energy?

75. SSM ✦ The half-life of $^{223}_{88}Ra$ is 11.4 days. (a) Calculate its decay constant. (b) If the activity of a piece of $^{223}_{88}Ra$ is 0.75 Ci today, how long will it take for its activity to fall to 0.10 Ci?

76. ✦ Ⓡ Ordinary soil contains typically 1 part per million (ppm) of uranium by mass. (a) How many uranium nuclei are in the top 10 m of soil under a typical house (20 m × 20 m)? (b) Only 0.72% of the nuclei in part (a) are $^{235}_{92}U$; the rest are $^{238}_{92}U$, which has a much longer half-life. How many $^{235}_{92}U$ nuclei are in the soil under this house? How many $^{238}_{92}U$ nuclei are in the soil under this house? (c) How many of the $^{235}_{92}U$ nuclei in part (b) undergo radioactive decay over a period of one day? (d) How many of the $^{238}_{92}U$ nuclei in part (b) undergo radioactive decay over a period of one day? (e) The decay of $^{238}_{92}U$ leads to radon gas ($^{222}_{86}Rn$; see Fig. 30.9). If all the $^{238}_{92}U$ decays to make $^{222}_{86}Rn$, how many $^{222}_{86}Rn$ nuclei are produced under the house in one month? (f) If all the $^{222}_{86}Rn$ nuclei in part (e) seep into the house above, what is the activity in the house?

77. ✦ The binding energy of a typical nucleus is comparable to the electric potential energy $PE_{elec} \approx 2 \times 10^{-13}$ J found in Equation 30.7. If that is the magnitude of the binding energy of a single helium nucleus, what is the approximate magnitude of the total binding energy of 1 g of helium?

78. ✦ ⊗ One way to date ancient wood is with carbon dating. An instrument dealer comes to you with what he says is a violin made by the master violin maker Antonio Stradivari (1644–1736). You have a very small piece of wood from the violin analyzed, and the ratio of $^{14}_{6}C/^{12}_{6}C$ is found to be 96.4% of the ratio found in trees that were cut down in 1950. Could this violin have been made by Stradivari?

79. ✦ Tritium is used in nuclear weapons and has a half-life of 12.3 yr. Suppose a nuclear weapon will cease to function if the amount of tritium is less than 20% of the amount present when the weapon is first built. What is the maximum age such a weapon can have and still be dangerous?

80. ✦ Ⓡ A neutron star is a star that has collapsed so that it is essentially one big nucleus, and its radius is related to its "mass number" by Equation 30.5. If our Sun collapsed to make a neutron star, what would its radius be?

81. ✦ Two radioactive samples are composed of two different types of nuclei. The samples have the same number of radioactive nuclei, but the activity of sample 1 is four times greater than the activity of sample 2. What is the ratio of the half-lives of the two different types of nuclei?

82. ✦ Ⓡ A Rutherford scattering experiment is performed in which alpha particles are directed at a flat sheet of carbon. This sheet contains a single layer of carbon atoms arranged on a square grid (Fig. P30.82), with an atomic spacing of 0.25 nm. Estimate the probability that an alpha particle will collide with a carbon nucleus. Assume the carbon nuclei have a diameter of 2.0 fm and that the center of an alpha particle must overlap with a carbon nucleus in order for a collision to occur (since the strong force has a very short range).

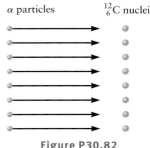

Figure P30.82

83. ✪ Ⓡ Consider again the Rutherford scattering experiment in Figure P30.82, but now assume there are many sheets (i.e., many layers) of carbon atoms. Assume the distance between sheets is 0.25 nm, the total thickness of the sheets is 1.0 cm, and make the same assumptions about collisions between an alpha particle and a carbon nucleus as described in Problem 82. Estimate the probability that an alpha particle will collide with a carbon nucleus as it travels through all the layers.

84. ✪ Ⓡ According to the Heisenberg uncertainty principle (Chapter 28), the uncertainties in the position Δx and momentum Δp of a particle are related by

$$\Delta x \Delta p \geq \frac{h}{4\pi}$$

This relation can be used to estimate an upper bound on the kinetic energy and velocity of a proton or neutron inside a nucleus by taking Δx to be the diameter of the nucleus and Δp the neutron's total momentum. (a) Use the uncertainty principle to estimate the kinetic energy of a proton inside an alpha particle. (b) What is the speed of the proton?

85. ✪ The critical mass of pure $^{235}_{92}U$ can be as small as about 15 kg, but a much smaller amount has a significant activity. (a) What is the activity of a sample of $^{235}_{92}U$ with a mass of 1.0 mg? (b) A Geiger counter with a detector area of 5.0 cm^2 is placed 20 cm from the $^{235}_{92}U$ sample in part (a). How many decays will the Geiger counter measure?

86. ✪ Chicago Pile One. In 1942, the world's first nuclear fission reactor was built and operated by team of physicists headed by Enrico Fermi under the bleachers of an abandoned stadium (Fig. P30.86) at the University of Chicago. The reactor was simply a pile of graphite bricks and uranium metal, measuring 20 ft high by 25 ft wide, with cadmium rods as a moderator. The reaction lasted 33 min and produced 0.5 W of power, confirming that a nuclear chain reaction had been achieved. (a) How much energy was released during this first fission reaction? (b) How many neutrons were produced and left the pile during the reaction?

Courtesy of Department of Energy (both)

Figure P30.86 (Left) The Stagg Field bleachers. (Right) The Chicago Pile One fission reactor stack of graphite bricks and uranium within the bleacher enclosure.

87. ✪ Ⓡ Supernova 1987A. At a relatively close 51.4 kiloparsecs away from the Earth, the supernova of 1987 was the closest nova to occur since telescopes were invented, and it allowed a unique opportunity to confirm that radioactive decay rates are constant over time. Theoretical considerations of the fusion processes in the extreme environment of a supernova blast predict that many isotopes will be created, including a large amount of the short-lived isotope ^{56}Co with a half-life of 77.1 days. Figure P30.87 shows the intensity of the light measured from SN1987A as a function of time. A few weeks after the peak intensity, the light from the explosion dims logarithmically as a function of time, as one would expect if the light were being generated due to radioactive decay of the created ^{56}Co. (a) How long ago did the explosion of the supernova take place? (b) Using the light

Figure P30.87 The light curve for Supernova 1987A.

curve in Figure P30.87, estimate the half-life of this ^{56}Co and compare it with the known half-life. What does this estimate tell you about the decay rate of ^{56}Co?

88. ✪ Geochronology and potassium-argon dating. One method of dating rocks uses the decay of ^{40}K to ^{40}Ar with a half-life of 1.26×10^9 yr. Most of the time, ^{40}K decays to ^{40}Ca, and 11% of the time it decays to ^{40}Ar. Initially, when the minerals in the rock form, there is very little argon included because argon is an inert noble gas. Potassium is abundant in nature, and most rocks have trace amounts of ^{40}K. To find the age of a rock, a sample is crushed into a fine powder and the released argon is carefully measured. The amount of ^{40}K is then extracted from the same sample and precisely measured. (a) A rock sample taken from atop the canyon wall is found to have one argon atom for every 52 atoms of ^{40}K. Approximately how old is the rock from the top strata? (b) Another sample of rock is taken from the bottom of a canyon at the base of this cliff wall. This sample is found to have one argon atom to every ten ^{40}K atoms. How much older is the rock at the canyon bottom than the rock at the top?

89. ✪ Consider conservation of energy and show that a ^{12}C nucleus cannot spontaneously decay into three alpha particles. *Hint:* Compare the masses of the ^{12}C nucleus and the three alpha particles.

90. ✪ Space probe power supply. The sunlight that reaches the outer solar system is too feeble to power spacecraft with solar panels. Space probes like *Cassini*, currently exploring Saturn and its moons, use a radioisotope thermoelectric generator (RTG), which converts the heat generated from radioactive decay to electricity. The RTG on *Cassini* (Fig. P30.90) generates 13 kW of heat using 23.4 kg of ^{238}Pu, which emits an alpha particle with 5.6 MeV of kinetic energy upon each decay. (a) Estimate the activity in units of Bq and Ci contained in the RTG. (b) *Cassini* was designed for an 11-yr mission. If the half-life of ^{238}Pu is 87.7 yr, how much power will be available from the RTG at the very end of the mission?

© NASA/JPL

Figure P30.90 The *Cassini* probe currently in orbit around Saturn. The RTG power supply is mounted on a long boom extending from the spacecraft.

Physics in the 21st Century

Toward the end of the 19th century, some physicists believed that all the laws of physics had been discovered and that the main job left for physicists was to measure quantities like the fundamental constants and the orbital parameters of the Moon more accurately. These physicists were proven wrong, with the discoveries of relativity and quantum theory leading to a century of fantastic progress in physics. Advances in physics during the 20th century led to many important applications, including computers, space travel,

The LIGO experiment (laser interferometer gravitational wave observatory) employs a Michelson interferometer (Example 25.2) to measure extremely small changes in the distance between mirrors placed several kilometers apart. Physicists hope to use several observatories like this one, placed around the world, to detect gravitational waves generated in distant galaxies. (Courtesy of LIGO Laboratory)

cell phones, nuclear power, lasers, CD and DVD players, and magnetic resonance imaging. What's more, physics continues to be an active research area, with no end in sight. In this chapter, we describe several topics of current research in physics. We start with elementary particle physics and consider how protons, neutrons, and other subatomic particles are put together. This field began with the study of cosmic rays in the early part of the 1900s, so we'll briefly describe that history. We then discuss some topics in the field of astrophysics, which is concerned with the origin and fate of the universe. In the

Figure 31.1 Some experiments use balloons to carry particle detectors to high altitudes and detect cosmic rays as they enter the atmosphere from outer space. Here NASA launches a balloon that will carry a package of instruments to the upper atmosphere.

last section of this chapter, we briefly describe how physicists are teaming up with biologists, chemists, and engineers to address interdisciplinary questions, including the new field of nanoscience.

31.1 | COSMIC RAYS

Electrons, protons, and neutrons were discovered and identified as atomic building blocks in the early 1900s. It was also found that these and other particles bombard the Earth from outer space. Particles that arrive from space are called *cosmic rays*, a name chosen when physicists mistakenly believed that cosmic rays were a type of electromagnetic radiation. Cosmic rays are actually a mix of different particles, including protons (about 89%) and alpha particles (10%); the rest are the nuclei of elements more massive than He and other types of particles. Cosmic rays are often studied in experiments in which balloons carry particle detectors to high altitudes (Fig. 31.1). Other experiments examine the particles created when cosmic rays collide with atoms in the Earth's atmosphere. In fact, several types of particles, including positrons and muons, were first discovered in cosmic-ray experiments.

A key question for physicists is, "How are cosmic rays generated?" It is believed that some cosmic rays are created when stars collide. These violent collisions can produce cosmic-ray particles with extremely high kinetic energy. Cosmic rays are also created when a massive star uses up its fuel. The star then collapses (due to gravity) and triggers what is called a supernova explosion (Fig. 31.2). These explosions are also thought to be the main source of elements heavier than oxygen in the universe.

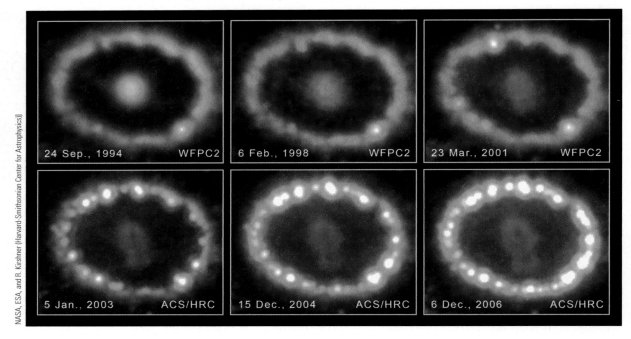

24 Sep., 1994 WFPC2	6 Feb., 1998 WFPC2	23 Mar., 2001 WFPC2
5 Jan., 2003 ACS/HRC	15 Dec., 2004 ACS/HRC	6 Dec., 2006 ACS/HRC

Figure 31.2 Physicists believe that cosmic-ray particles are generated in supernova explosions. These photographs of Supernova 1987A were obtained with the Hubble Space Telescope shortly after the supernova exploded in 1987. The bright ring around the star is due to a ring of gas that formed long before the explosion. When the supernova exploded, particles were ejected from the star; light from these particles is visible as violet in these photos. In the first photo (upper left), these particles were near the star (at the center of the ring). As time passed, the particles moved outward, and when they reached the gas ring, they produced the bright white spots. The bright spots at the upper-left and lower-right parts of the ring (seen most clearly in the image at the upper right) are due to stars between 1987A and the Earth.

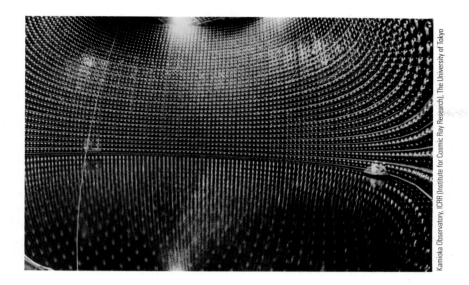

Figure 31.3 This detector is called Super-Kamiokande. Located in Japan, it is designed to detect neutrinos that arrive at the Earth from space. This detector is a large container of water with walls covered with photodetectors that detect light generated when incoming particles interact with the water. This photo of the detector was taken during maintenance, when some of the water was drained out. Notice the people in a small boat (doing repair work).

Typical cosmic-ray energies are 100 MeV (1×10^8 eV), but some particles with energies above 10^{20} eV have been detected. This is an incredibly high energy—approximately the kinetic energy of a baseball moving at 100 mi/h—and is carried by a particle with about the mass of a proton! It is still not completely understood how particles are accelerated to such high energies.

New, more sensitive detectors are continually being built, with the goal of better understanding cosmic rays and other particles that arrive from outer space. Figure 31.3 shows the Super-Kamiokande detector located in Japan, built as part of a joint experiment with physicists from the United States and other countries. Super-Kamiokande consists of a large volume of water (approximately 64,000 m^3) in a container whose walls are lined with photodetectors. The photodetectors detect light produced when cosmic rays and other particles from space collide with water molecules. The Super-Kamiokande experiment is designed to study neutrinos, which are generated as a result of fusion reactions in our Sun (Chapter 30) and in supernova explosions. These neutrinos give physicists a "glimpse" of the kinds of particle reactions that take place in these systems.

31.2 | MATTER AND ANTIMATTER

One of the developers of quantum physics was P. A. M. Dirac; around 1930, he formulated a theory of quantum mechanics that combined the quantum theory of Schrödinger and Heisenberg with the postulates of special relativity. Dirac's theory gave two important results: (1) it predicted the phenomenon of electron spin, and in Chapter 29 we saw how electron spin is essential for explaining the periodic table of the elements, and (2) it predicted the existence of a completely new particle having the same mass as the electron and an electric charge of the same magnitude as an electron but with opposite sign. Several years later, this particle, named the *positron*, was discovered in cosmic rays. The positron is denoted by the symbol e^+ to distinguish it from the ordinary electron, which is denoted by e^-. The superscripts indicate that the electron carries a negative charge, while the charge on the positron is positive. The positron is an example of **antimatter**, a particle of matter having the same mass and opposite charge as the corresponding particle of ordinary matter. Other examples of antimatter particles (**antiparticles**) are the antiproton and the antineutron.

Particles such as electrons, positrons, and protons undergo reactions, much like the reactions involving nuclei. When describing these reactions, we denote each of these particles by a symbol, similar to the symbols used in Chapter 30 to denote

Table 31.1 Some Properties of Electrons, Protons, Neutrons, and Their Antiparticles

PARTICLE	SYMBOL	MASS (kg)	MASS (MeV/c^2)	CHARGE
Electron	e^-	9.109×10^{-31}	0.511	$-e$
Positron	e^+	9.109×10^{-31}	0.511	$+e$
Proton	p	1.673×10^{-27}	938	$+e$
Antiproton	$\bar{p}$	1.673×10^{-27}	938	$-e$
Neutron	n	1.675×10^{-27}	940	0
Antineutron	$\bar{n}$	1.675×10^{-27}	940	0

different nuclei. Protons are denoted p and neutrons as n; the corresponding anti-particles are denoted with an overbar, with $\bar{p}$ for the antiproton (pronounced "p bar") and $\bar{n}$ for the antineutron. Most other particles and antiparticles are denoted in a similar way, except for the positron, which is denoted by e^+ without a bar on top. For convenience, Table 31.1 lists the mass and electric charge of the electron, proton, and neutron along with the values for the corresponding antiparticles.

The annihilation and creation of antiparticles give clear examples of special relativity at work. For example, when an electron encounters its antiparticle (a positron), the two can undergo a reaction in which both particles are annihilated. This reaction is written as

$$e^- + e^+ \rightarrow \gamma + \gamma \tag{31.1}$$

In words, this reaction states that an electron plus a positron react to form two gamma rays, that is, two photons. This reaction must satisfy the conservation of energy, so the total energy before the reaction (the energies of the original electron and positron) is equal to the final energy (of the two gamma ray photons). The total initial energy includes the kinetic energies of the electron and positron plus their rest energies. According to special relativity (Chapter 27), the rest energy of an electron (rest mass m_e) is $m_e c^2$. The positron also has a rest mass m_e, so it has the same rest energy. By measuring the energies of the two gamma rays emitted in this reaction, one can determine the rest energies of the electron and positron and thus check the predictions of special relativity. The analysis is carried out in Example 31.1.

EXAMPLE 31.1 | Electron–Positron Annihilation

An electron and positron initially at rest annihilate each other, producing two gamma rays. Assume the gamma rays have the same energies and are emitted in opposite directions as sketched in Figure 31.4. Find the energy of one of the gamma ray photons.

RECOGNIZE THE PRINCIPLE

The annihilation reaction must conserve energy, so

$$\text{initial energy } (e^- + e^+) = \text{final energy } (\gamma \text{ rays})$$

The energy of the electron–positron pair consists of their kinetic energies (which are zero) and their rest energies. A gamma ray photon's energy is related to its frequency, with $E_{\text{photon}} = hf$.

SKETCH THE PROBLEM

See Figure 31.4. The electron and positron are at rest, so the initial momentum is zero. Since the final momentum must also be zero, the outgoing gamma ray photons are

Initial electron and positron

Gamma ray photons produced by annihilation

Figure 31.4 Example 31.1.

traveling in opposite directions. The two photons must also have the same frequency. This is the only way two photons can have a total momentum of zero.

FIND THE RELATIONSHIPS

We have

$$\text{initial energy } (e^- + e^+) = KE + \text{rest energy}$$

and the initial kinetic energy is zero because the electron and positron are initially at rest. The initial rest energy is $m_e c^2 + m_e c^2 = 2m_e c^2$ since the electron and positron have the same mass m_e. We are left with

$$\text{initial energy } (e^- + e^+) = 2m_e c^2$$

SOLVE

The energy of each of the gamma ray photons must therefore be

$$E_{\text{photon}} = m_e c^2$$

From Table 31.1, we have

$$E_{\text{photon}} = \boxed{0.511 \text{ MeV}}$$

What does it mean?
This energy is *much* greater than that produced by atomic transitions. For example, in Chapter 29 we saw that the highest energy photon a hydrogen atom can emit is 13.6 eV.

Conservation Rules for Particle Reactions

The electron–positron reaction in Equation 31.1 obeys the principles of conservation of energy and conservation of momentum (Example 31.1). It also obeys another conservation law, the conservation of electric charge, as the total electric charge is the same (zero) before and after the reaction. All other particle reactions also conserve energy, momentum, and charge. These conservation laws are nothing new—we learned about them in our work on classical physics—but we'll soon encounter several other important conservation laws that go far beyond our classical laws of physics.

> **CONCEPT CHECK 31.1 | Electron–Positron Creation?**
>
> Your friend says that he has observed an electron–positron reaction in which a single photon forms an electron and positron as shown in Figure 31.5. He also claims that the two particles are at rest after the reaction and that the energy of the photon is equal to the sum of the rest energies of the two particles. Is this reaction (a) possible or (b) not possible? Explain why.

Figure 31.5 Concept Check 31.1.

The Stability of Antimatter

A positron is completely stable if it is kept away from electrons. Similarly, an antiproton is stable if it is kept away from protons. It is even possible to bring together a positron and an antiproton to form an atom of antihydrogen. Such atoms of antimatter have been studied in physics experiments, but it is very difficult to create large amounts of antimatter because it is difficult to contain it sufficiently far from regular matter to prevent annihilation.

The amount of energy released by the annihilation of an electron is very large (Example 31.1), and since a proton has a much larger mass than an electron, the

Figure 31.6 Richard Feynman (1918–1988) was one of the developers of quantum electrodynamics (QED). He also made important contributions to the Manhattan project in World War II, to our understanding of liquid helium, and to many other areas of physics. Feynman was a very entertaining and enthusiastic teacher of science to the general public.

energy released in the annihilation of a hydrogen atom with antihydrogen is much greater than calculated in Example 31.1. For this reason, antimatter is a popular fuel in science fiction stories. Unfortunately, it is not a practical energy source. All known methods for creating antiparticles require an amount of energy much greater than the rest energies of the particles that are created. The problem of making a suitable container hasn't been solved yet either.

The Earth and our galaxy are composed almost entirely of matter. One might expect to find an equal amount of antimatter in the universe, perhaps in the form of antimatter galaxies, but as far as astrophysicists can tell, the amount of matter in the universe is much greater than the amount of antimatter. Physicists are now trying to understand this puzzle.

31.3 | QUANTUM ELECTRODYNAMICS

The theory of the photon as a quantum of electromagnetic energy (i.e., a quantum of electromagnetic radiation) was first discussed by Einstein in 1905 and was an important step in the development of quantum theory (Chapter 28). Einstein's discovery of the photon was only part of a complete quantum theory of light, however. Electromagnetic waves (radiation) have an associated electric and magnetic field. Since the photon is a quantum of electromagnetic radiation, photons also "carry" electric and magnetic fields as described by Einstein's theory. Einstein's ideas were developed further by other physicists, including Richard Feynman (Fig. 31.6), Julian Schwinger, and Shinichiro Tomonaga, who developed a theory called *quantum electrodynamics* (QED), which combines electromagnetism and Maxwell's equations with quantum mechanics. According to QED, photons are the elementary quanta of electric and magnetic fields, and they play a role not just in electromagnetic radiation, but in all electric and magnetic forces.

Consider the interaction of two charged particles through Coulomb's law. A classical picture of this interaction is sketched in Figure 31.7A, showing the electric field produced by the charge on the left (q_1). The electric force on the right-hand charge (q_2) is due to this electric field. The QED picture of this force involves photons as sketched in Figure 31.7B; here the interaction between two charged particles is due to the exchange of photons, the elementary quanta of the electromagnetic field. We say that the electric force is "mediated" by photons. In this way, the photon is responsible for all electromagnetic forces.

Figure 31.7C gives a classical picture of how the exchange of particles can lead to a force between two objects. Here the "objects" are two children playing catch with a ball. The person throwing the ball experiences a recoil force during the throw

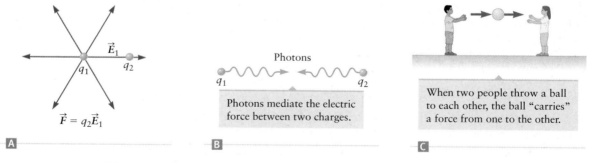

Figure 31.7 Force between two charged particles. **A** In the classical picture in Chapter 17, the force on one charge q_2 is due to the electric field produced by the charge q_1. **B** In the quantum mechanical picture of QED, the electric force is mediated (or "carried") from one charge to another by photons. **C** The force carried by photons is similar to the case of two people playing catch with a ball. The ball "carries" a force between the two people.

while the person catching the ball experiences a recoil force when she catches it. This ball plays the role of a photon in the electromagnetic force as both the ball and the photon mediate (or "carry") a force between two objects.[1] It is believed that all the fundamental forces in nature are mediated by elementary particles.

The photon is an example of an elementary particle, but it differs in two ways from the other particles we have discussed so far. First, the photon has zero rest mass. Second, the photon is its own antiparticle, so there is no "antiphoton."

31.4 | ELEMENTARY PARTICLE PHYSICS: THE STANDARD MODEL

Studies of nuclear and particle physics began in the early 1900s, when Rutherford and other physicists conducted many experiments in which various particles were fired at atoms and nuclei (Chapters 29 and 30), revealing that nuclei are composed of protons and neutrons. For a brief period, it was thought that electrons, protons, and neutrons are the fundamental particles from which all matter is composed, but that simple picture did not last long. Many other particles were discovered in cosmic-ray studies, in collision experiments, and in nuclear decay processes. It is now clear that protons and neutrons are themselves composed of particles called *quarks*, truly fundamental point particles whose charges are multiples of $\pm e/3$, one-third of the electron charge. Particles that are composed of quarks—such as protons and neutrons and their antiparticles—form a family of particles called *hadrons*. Not all particles are members of the hadron family, however. There is a second class of particles called *leptons*, which include electrons and positrons. The behavior of hadrons and leptons is described by what is called the *standard model* of elementary particles. Let's now consider the properties of hadrons and leptons.

The standard model describes the properties of fundamental particles and the interactions between them.

Quarks Bind Together to Form Hadrons

There are six different kinds of quarks, named "up" (denoted by the symbol u), "down" (d), "charm" (c), "strange" (s), "top" (t), and "bottom" (b). These names do not refer to any physical properties of the quarks; rather, they are just whimsical names coined by physicists. Each of these quarks has a corresponding antiquark; denoted $\bar{u}$, $\bar{d}$, and so on. Some properties of quarks are listed in Table 31.2.

Quarks were first discovered in collision experiments involving protons. When a high-energy electron collides with a proton, the way the electron scatters (i.e., its outgoing direction and energy) gives information about how mass and charge are distributed inside the proton. This is similar to Rutherford's experiment (Chapter 29), in which the way alpha particles are scattered from an atom indicate that a massive nucleus is located at the atom's center. Collision experiments with protons show that there are three pointlike particles inside. Table 31.3 gives the quark composition of the proton: two up quarks and one down quark. The total charge on the proton is the sum of the charges of the constituent quarks, so the proton's charge is

$$\text{proton charge} = 2 \times (\text{charge of u quark}) + (\text{charge of d quark})$$

$$= 2\left(\frac{+2e}{3}\right) + \left(\frac{-e}{3}\right)$$

$$= +e$$

as expected.

---

[1]This analogy is a very qualitative way to think about how photon exchange leads to forces and has its limitations. In particular, it is hard to give a classical explanation of how the exchange of particles can lead to an *attractive* force. The full theory of QED correctly describes both repulsive and attractive electromagnetic forces.

Table 31.2 Quarks and Their Properties

QUARK	SYMBOL	CHARGE	MASS (MeV/c^2)	ANTIPARTICLE
Up	u	$+2e/3$	4	$\bar{u}$
Down	d	$-e/3$	8	$\bar{d}$
Charm	c	$+2e/3$	1,500	$\bar{c}$
Strange	s	$-e/3$	150	$\bar{s}$
Top	t	$+2e/3$	176,000	$\bar{t}$
Bottom	b	$-e/3$	4,700	$\bar{b}$

Quarks can combine to form a hadron in two ways. Hadrons composed of three quarks are called **baryons** (Fig. 31.8A). Protons and neutrons are baryons; Table 31.3 lists a few other baryons, and dozens of other baryons have been observed and their component quarks identified. It is also possible for a quark and an antiquark to combine to form a particle. Hadronic particles composed of just two quarks are called **mesons** (Fig. 31.8B), and a few are listed in Table 31.4.

All hadrons are composed of quarks, so the interactions between quarks determine the properties of hadrons and how they interact with one another. The two most important hadrons are the proton and neutron, so the behavior of quarks also determines the properties of nuclei. Quarks are charged, so they interact via the electric (Coulomb) force. They also interact via the strong force mentioned in Chapter 30. The strong force holds quarks together to form protons and neutrons (nucleons) and is also responsible for holding protons and neutrons together to make nuclei.

Table 31.3 Properties of Some Baryons

PARTICLE	SYMBOL	CONSTITUENT QUARKS	LIFETIME (s)	MASS (MeV/c^2)
Proton	p	uud	Stable	938
Neutron	n	udd	890	940
Lambda zero	Λ^0	sud	2.6×10^{-10}	1116
Sigma plus	Σ^+	uus	0.8×10^{-10}	1189
Sigma zero	Σ^0	sud	6.0×10^{-20}	1193
Sigma minus	Σ^-	dds	1.5×10^{-10}	1197
Hyperion	Ξ^-	ssd	1.6×10^{-10}	1321

Note: There are many other baryons, which are composed of other combinations of three quarks and antiquarks.

Table 31.4 Properties of Some Mesons

PARTICLE	S SYMBOL	CONSTITUENT QUARKS	LIFETIME (s)	MASS (MeV/c^2)
Pion (pi plus)	π^+	$u\bar{d}$	2.6×10^{-8}	140
Pi zero*	π^0	$d\bar{d}/u\bar{u}$	8.4×10^{-17}	135
Kaon (K plus)	K^+	$u\bar{s}$	1.2×10^{-8}	494
Kaon (K minus)	K^-	$\bar{u}s$	1.2×10^{-8}	494
Phi	φ	$s\bar{s}$	1.6×10^{-22}	1020

Note: There are many other mesons, which are composed of other combinations of a quark and an antiquark.

*The π^0 is a quantum mechanical combination of the $d\bar{d}$ and $u\bar{u}$ quark states.

Insight 31.1

PARTICLES AND ANTIPARTICLES

Mesons all consist of a quark and an antiquark (Table 31.4), leading to some interesting consequences. For example, the phi meson (φ) is a combination of a strange quark s and the antiquark $\bar{s}$. That is,

$$\varphi = s\bar{s}$$

The antiphi meson is then

$$\bar{\varphi} = \bar{s}s$$

According to the rules of antiparticles, the antiparticle of the $\bar{s}$ is just an s quark, so $\bar{\bar{s}} = s$, and because the order in which the quarks are listed does not matter, we have

$$\bar{\varphi} = s\bar{s} = \bar{s}s = \varphi$$

The antiparticle of the phi meson is thus itself!

BARYONS

p $\bar{p}$ n

u u / d $\bar{u}$ $\bar{u}$ / $\bar{d}$ u d / d

A

MESONS

π^+ K^+

$\bar{d}$ u $\bar{s}$ u

B

Figure 31.8 **A** Baryons and antibaryons are composed of three quarks. The three particles represented here are the proton (p), the antiproton ($\bar{p}$), and the neutron (n). **B** Mesons are composed of one quark and one antiquark. These mesons are called pions and kaons.

Rules for "Making" Baryons and Mesons, and Their Reactions

The standard model places restrictions on the ways quarks can combine to form baryons and mesons and on the kinds of particle reactions that are possible. One of these restrictions has to do with the allowed electric charges of baryons and mesons. The charge of any particular baryon or meson can be found by simply adding up the charges of its constituent quarks. The quark charges are listed in Table 31.2, and the corresponding antiquarks have the opposite charges (so the $\bar{d}$ has a charge of $+e/3$, which is opposite to that of the d quark, etc.). Working out the charges of the baryons and mesons in Tables 31.3 and 31.4, you'll find that they all have electric charges of $+e$, $-e$, or zero; some other baryons (not listed here) have charges $+2e$ and $-2e$. The standard model guarantees that an isolated particle cannot have a fractional charge, that is, a charge of $\pm e/3$ or $\pm 2e/3$. Since quarks themselves have fractional charge, the implication is that individual quarks cannot be isolated. We'll explain why when we consider the strong force in more detail below.

In addition to this restriction on the charges of baryons and mesons, the standard model also predicts that **baryon number** must be conserved in all particle reactions. Each quark has a baryon number of $+\frac{1}{3}$, so the baryon number of a proton (which is composed of three quarks) is $+1$. Antiquarks have a baryon number of $-\frac{1}{3}$, so all mesons have a baryon number of zero. We'll discuss some examples of baryon number conservation below, after we describe leptons.

CONCEPT CHECK 31.2 | Which of these hadrons are allowed?

Which quark combination is *not* an allowed meson or baryon, (a) $c\bar{c}$, (b) bbs, (c) uud, (d) uu$\bar{d}$, or (e) $\overline{uu}\bar{d}$?

| EXAMPLE 31.2 | Rest Mass of the Proton |

The proton is composed of two up quarks and a down quark (p = uud). Compare the total rest mass of these three quarks with the rest mass of the proton. Are they equal?

RECOGNIZE THE PRINCIPLE

The rest masses of the proton and the u and d quarks are listed in Tables 31.2 and 31.3. We can simply add the appropriate masses and compare the results.

SKETCH THE PROBLEM

No sketch is needed.

FIND THE RELATIONSHIPS

The u quark has a mass of about 4 MeV/c^2 and the d quark mass is around 8 MeV/c^2, so the total mass of the quarks that make up a proton is

$$\text{mass of u} + \text{u} + \text{d} = (4 + 4 + 8) \text{ MeV/}c^2 = 16 \text{ MeV/}c^2 \tag{1}$$

From Table 31.3, the rest mass of the proton is

$$\text{rest mass of proton} = 938 \text{ MeV/}c^2 \tag{2}$$

SOLVE

Comparing Equations (1) and (2), the rest mass of the proton is *much* greater than the combined masses of its constituent quarks, so the answer is no .

What does it mean?

The difference between the masses in Equations (1) and (2) is due to the quark kinetic energies within the proton and from potential energy resulting from the

---

Insight 31.2

MEASURING THE MASS OF A QUARK

One way to measure the mass of a quark is through Newton's second law, $\Sigma F = ma$ (or the corresponding result of quantum mechanics). However, quarks are always confined in baryons or mesons, so it is impossible to study how a free quark would respond to an applied force. The best we can do is probe the motion or properties of quarks inside a baryon or meson and then deduce the quark mass using the theory of the strong force. Such an analysis must allow for the contributions of quark kinetic energy and the potential energy of the strong force, and these contributions are difficult to calculate with high accuracy. For this reason, the masses of the quarks are not known with the same level of precision as the masses of particles such as the proton, neutron, and electron.

Leptons

We have just discussed the hadron family of particles, which are all assembled from quarks. Let's now consider a second family of particles called leptons. There are six fundamental leptons (Table 31.5) plus their corresponding antiparticles. These particles group naturally into three pairs: the electron (e^-) and the electron neutrino (ν_e); the muon (μ^-) and the muon neutrino (ν_μ); and the tau (τ^-) and the tau neutrino (ν_τ). The muon (lifetime = 2.2×10^{-6} s) and the tau (lifetime 2.9×10^{-13} s) are not stable. Electrons are stable, while the behavior of neutrinos is more complicated. Recent experiments indicate that as they travel through space, neutrinos change from one to another of the three types of neutrinos listed in Table 31.5, an effect called "neutrino oscillations." Another interesting property of neutrinos is that they have very small masses; the best experiments to date give only an approximate mass for the electron neutrino between 0.05 and 2 eV/c^2. That is more than 100,000 times less than the mass of an electron!

Leptons play important roles in certain reactions involving hadrons. For example, an "isolated" neutron (a neutron outside a nucleus) decays to form a proton, an electron, and an anti-electron neutrino:

$$n \rightarrow p + e^- + \bar{\nu}_e \tag{31.2}$$

Leptons are also produced in many nuclear decay reactions. In fact, neutrinos were first observed in the 1950s in studies of the particles emitted by a nuclear power reactor. A very large source of the neutrinos observed on the Earth are the nuclear fusion processes that power the Sun. Studying these solar neutrinos provides one of the best ways to understand the nuclear reactions that take place inside the Sun.

Reactions involving leptons, such as neutron decay (Eq. 31.2), must satisfy conservation of *lepton number*. Each lepton has a lepton number of +1, whereas each antilepton has a lepton number of −1. Since neutron decay involves both baryons and leptons, this process must conserve both baryon number and lepton number. Notice that all baryons have a lepton number of zero and all leptons have a baryon number of zero. For Equation 31.2, we have

$$
\begin{aligned}
\text{n} \quad &\rightarrow \quad \text{p} \quad + \quad e^- \quad + \quad \bar{\nu}_e \\
\text{baryon number:} \quad +1 &\rightarrow +1 + 0 \quad + 0 \quad = +1 \\
\text{lepton number:} \quad 0 \ &\rightarrow 0 \ + 1 \ - 1 = 0
\end{aligned}
\tag{31.3}
$$

Hence, although this reaction produces a lepton and an antilepton, it still conserves both baryon number and lepton number.

Table 31.5 Leptons and Their Properties

PARTICLE	SYMBOL	CHARGE	MASS/c^2	ANTIPARTICLE
Electron	e^-	$-e$	0.511 MeV	e^+
Electron neutrino	ν_e	0	0.05 eV $< m <$ 2 eV	$\bar{\nu}_e$
Muon	μ^-	$-e$	106 MeV	μ^+
Muon neutrino	ν_μ	0	$<$ 0.19 MeV	$\bar{\nu}_\mu$
Tau	τ^-	$-e$	1780 MeV	τ^+
Tau neutrino	ν_τ	0	$<$ 18 MeV	$\bar{\nu}_\tau$

EXAMPLE 31.3

EXAMPLE 31.3 | Review Problem: Studying the Decay of a Tau Lepton

Elementary particles are produced and studied in collision experiments at laboratories like the one shown in Figure 31.9. Among the particles studied in this way is the tau lepton. Tau leptons have a lifetime of approximately $\Delta t_0 = 2.9 \times 10^{-13}$ s measured in the tau's reference frame (i.e., measured by an observer at rest relative to the tau). Suppose a certain collision experiment creates a tau particle with a speed $v = 0.99c$ relative to a physics laboratory. How far will this particle travel relative to the laboratory before it decays? Give the distance as measured in (a) the reference frame of the tau and (b) the reference frame of a physicist in the laboratory.

Figure 31.9 This large accelerator is at Fermi National Accelerator Laboratory, located near Chicago. These two large circular rings, each with a diameter of about 2 km, are part of a chain of accelerators in which protons and antiprotons travel with energies near 1 TeV (10^{12} eV). Collisions between protons and antiprotons are used to create and study quarks and other particles.

RECOGNIZE THE PRINCIPLE

The tau's speed is sufficiently high that we must consider the effects of special relativity. This example is a time-dilation problem very similar to Examples 27.2 and 27.4, in which a high-speed muon decayed in the Earth's atmosphere. According to an observer traveling with the tau (i.e., in the tau's reference frame), the lab moves at a speed of $0.99c$ for a time Δt_0. The physicist sees the tau moving by at $0.99c$, but observes a dilated lifetime. According to Equation 27.7,

$$\Delta t = \frac{\Delta t_0}{\sqrt{1 - v^2/c^2}} \tag{1}$$

SKETCH THE PROBLEM

Figure 31.10 shows the tau lepton along with two observers. One observer travels with the tau and measures a lifetime Δt_0. The other observer (the physicist at rest in the lab) measures a dilated lifetime Δt. (Compare with Fig. 27.9.)

FIND THE RELATIONSHIPS

(a) The value Δt_0 of the tau lifetime given above is the proper time because it is measured in the tau's reference frame. The distance d_0 traveled by the tau as measured in its own reference frame is

$$d_0 = v \,\Delta t_0 \tag{2}$$

where $v = 0.99c$.

(b) A physicist studying the moving tau measures a dilated lifetime Δt given by Equation (1). The distance traveled by the tau in the physicist's frame of reference is thus

$$d = v\Delta t = \frac{v\,\Delta t_0}{\sqrt{1 - v^2/c^2}} \tag{3}$$

SOLVE

(a) Inserting the given values of v and Δt_0 from Equation (2) leads to

$$d_0 = (0.99c)\Delta t_0 = (0.99)(3.00 \times 10^8 \text{ m/s})(2.9 \times 10^{-13} \text{ s})$$

$$d_0 = 8.6 \times 10^{-5} \text{ m} = \boxed{0.086 \text{ mm}}$$

(b) Inserting the values for v and Δt_0 again from Equation (3), the distance traveled in the laboratory reference frame is

$$d = \frac{(0.99c)\Delta t_0}{\sqrt{1 - (0.99c)^2/c^2}} = \frac{(0.99)(3.00 \times 10^8 \text{ m/s})(2.9 \times 10^{-13} \text{ s})}{\sqrt{1 - (0.99)^2}}$$

$$d = 6.1 \times 10^{-4} \text{ m} = \boxed{0.61 \text{ mm}}$$

Observer traveling with τ^- measures proper time Δt_0.

$v = 0.99c$

τ^-
Tau lepton

Physicist in lab measures dilated time

$\Delta t = \dfrac{\Delta t_0}{\sqrt{(1 - v^2/c^2)}}$.

Figure 31.10 Example 31.3.

31.5 | THE FUNDAMENTAL FORCES OF NATURE

In Section 31.4, we described a few of the properties of hadrons and leptons such as their charge and mass and some of the reactions they can undergo. To get a complete understanding, we must also consider the forces that act between them. There are four fundamental forces, as listed and compared in Table 31.6.

The Strong Nuclear Force Only Acts on Quarks

Quarks carry electric charge, so they experience electric and magnetic forces, but the largest force between two quarks is the strong force. The strong force actually has two main "roles": it binds quarks together to form particles such as protons and neutrons, and it holds protons and neutrons together to form nuclei. After the quark model was proposed, physicists searched unsuccessfully for free particles with charge $\pm e/3$ or $\pm 2e/3$. Despite much effort, a single isolated quark has never been observed, which can be explained by the theory of the strong force and the standard model. One striking prediction of these theories is that the energy required to separate two quarks grows larger as the separation increases; eventually, at large separations, this energy is large enough to produce new quark–antiquark pairs that then bind to form baryons and mesons. This phenomenon is called *confinement* and is why an individual quark is never observed in isolation.

A proton is composed of two up quarks and a down quark, a combination denoted by uud. Quarks are quantum particles obeying the Pauli exclusion principle (Chapter 29), so these two quarks cannot occupy the same quantum state. Quarks have been found to possess a new quantum property called *color* that helps satisfy the Pauli exclusion principle in the case of a proton. All quarks carry a *color quantum number* with the possible values red, green, and blue, like the spin quantum number of an electron. From Section 29.5, two electrons can occupy the 1*s* state of a helium atom provided that they have different spin values. In the same way, the two up quarks in a proton must have different color values.

Table 31.6   The Fundamental Forces in Nature

Force	Force-Carrying Particle(s)	Acts on	Responsible for	Comments
Electromagnetism (QED)	Photon	All particles that carry electric charge	Electric and magnetic forces	Unified with the weak force as the electroweak force
Weak force	W^+, W^-, Z	Leptons and quarks	Beta decay and other nuclear decay processes	
Strong force (QCD)	Gluons	Quarks	Binding of quarks to form baryons and mesons	Also responsible for binding of protons and neutrons in the nucleus
Gravitation	Graviton (not yet observed)	All particles[a]	Gravity	

[a]From general relativity, gravity acts on photons.

The Strong Force Is Mediated by Gluons

In the language of quantum electrodynamics introduced in Section 31.3, photons are said to "mediate" electromagnetic interactions because they "carry" the electromagnetic force acting on charged particles. In the same way, the strong force is mediated or carried by particles called **gluons**, and there are eight (!) different types of gluons. The force exerted between two quarks depends on the type of each quark.

The theory of the strong force is called **quantum chromodynamics**, or QCD. One prediction of QCD is that baryon number is always conserved in a reaction as we mentioned in connection with the decay of a neutron (Eq. 31.3). All known reactions obey the conservation of baryon number, now believed to be a fundamental conservation principle.

Forces Acting on Leptons: Electromagnetism and the Weak Force

Three of the fundamental leptons—the electron, muon, and tau—are charged, so they experience electric and magnetic forces. These forces are described by the theory of QED. The strong force that is so important for quarks and hadrons does not act on leptons. All leptons, however, do experience what is known as the **weak force**, which is carried by three different particles called the W^+, the W^-, and the Z. The weak force acts on both leptons and quarks (and hence on all hadrons). This force has an extremely short range; two particles must be within about 10^{-18} m = 0.001 fm to experience this force. This distance is about one one-thousandth the diameter of a proton! Two particles rarely come this close together; hence, the name "weak" for this force. Even so, the weak force is responsible for some important processes such as the decay of the neutron (Eq. 31.2):

$$n \rightarrow p + e^- + \bar{\nu}_e$$

The neutron and proton are composed of quarks, so we can also write this reaction as

$$udd \rightarrow uud + e^- + \bar{\nu}_e \tag{31.4}$$

Hence, the decay of the neutron converts one of the down quarks (d) in the neutron into an up quark (u). Processes in which one type of quark is converted into another type always involve the weak force.

EXAMPLE 31.4 | Particle Decay and Conservation Rules

The negatively charged pion π^- is a meson composed of two quarks $\bar{u}d$. The π^- meson can decay into a muon and an antimuon neutrino:

$$\pi^- \rightarrow \mu^- + \bar{\nu}_\mu \tag{1}$$

Show that this decay conserves both charge and lepton number.

RECOGNIZE THE PRINCIPLE

To determine if this decay conserves charge, we need to compare the total charge before and after the decay. We analyze the conservation of lepton number in the same way.

SKETCH THE PROBLEM

No sketch is necessary.

IDENTIFY THE RELATIONSHIPS

The π^- meson has a charge of $-e$. Mesons are hadrons, so the lepton number of the π^- is zero. The μ^- muon has a charge of $-e$ and a lepton number of $+1$. The antimuon neutrino is neutral, and since it is an antiparticle, its lepton number is -1.

SOLVE

We can rewrite the decay in Equation (1) as

$$\pi^- \rightarrow \mu^- + \bar{\nu}_\mu$$

		π^-	μ^-	$\bar{\nu}_\mu$
charge	=	$-e$	$-e$	0
lepton count	=	0	+1	−1

Hence, both charge and lepton number are indeed conserved .

What does it mean?

We have also seen that particle decays must conserve baryon number. In Equation (1), the final baryon number is zero because the muon and antimuon neutrino are both leptons. The π^- meson contains a $\bar{u}$ and a d quark; since the $\bar{u}$ is an antiparticle, the total baryon number of the π^- meson is zero, so baryon number is also conserved in this decay.

Insight 31.3
WHAT DOES IT MEAN WHEN TWO FORCES ARE "UNIFIED"?
Two forces are "unified" when it is shown that they are two aspects of the same "underlying" force. For example, in Section 27.11 we showed that the electromagnetic force on a moving charge can be viewed as either an electric force or a magnetic force, depending on the relative motion of the charge and an observer. Hence, we say that these are two aspects of the *electromagnetic* force. Maxwell's theory gives a unified theory of electromagnetism.

Unification of the Weak Force and Electromagnetism

In the 1970s, it was discovered that the weak force and electromagnetism are actually two different aspects of the same force, similar to the relation between electricity and magnetism. Prior to Maxwell, physicists thought that electricity and magnetism were two completely different types of forces, but Maxwell showed that they are really two aspects of the same phenomenon (hence the name "electromagnetism"). Maxwell thus provided a unified theory of electricity and magnetism. In the same way, Sheldon Glashow, Abdus Salam, and Steven Weinberg developed a theory that unifies electromagnetism and the weak force. This combined theory of the *electroweak force* explains phenomena that cannot be accounted for by either separate theory. This theory predicted the existence of new particles that carry the electroweak force; these particles—the W^+, the W^-, and the Z (Table 31.6)—were then found in subsequent experiments.

Gravitation

Newton's theory of gravitation (Chapter 5) was extended by Einstein in his work on general relativity (Chapter 27). Gravitation is another fundamental force in nature. It plays an important role in our everyday terrestrial lives and is also responsible for the motion of the planets and stars in the universe. Even so, in some ways gravity is much weaker than the other fundamental forces of nature. For example, the electric force between two electrons is much larger in magnitude than the gravitational force between them.

  The particles that mediate the electromagnetic, strong, and weak forces have all been observed and their properties measured (Table 31.6). It is believed that gravity is also carried by a particle; although this particle has not yet been directly observed, physicists believe it must exist and have already given it a name: the *graviton*.

EXAMPLE 31.5 | Comparing the Electromagnetic and Gravitational Forces

The two up quarks (u) in a proton are separated by a distance $r \approx 1$ fm ($= 1 \times 10^{-15}$ m). These quarks each have an electric charge $+2e/3$, so they repel each other through an electric force obeying Coulomb's law. They also attract each other through a gravitational force. What is the ratio of the magnitudes of the electric and gravitational forces between these two quarks?

RECOGNIZE THE PRINCIPLE

The gravitational force between two point particles of mass m separated by a distance r is

$$F_{grav} = \frac{Gmm}{r^2}$$

while the electric force between two point charges q is

$$F_{elec} = \frac{kqq}{r^2}$$

We wish to calculate F_{grav}/F_{elec} for two quarks in a proton.

SKETCH THE PROBLEM

Figure 31.11 shows a classical picture of two u quarks separated by a distance r. The electric force and the gravitational force between them are both proportional to $1/r^2$.

IDENTIFY THE RELATIONSHIPS

Let m_u be the mass of an up quark. The force of gravity between two up quarks is then

$$F_{grav} = \frac{Gm_u m_u}{r^2}$$

The charge on an up quark is $q = +2e/3$ (Table 31.2), so the electric force between two up quarks is

$$F_{elec} = \frac{kqq}{r^2} = \frac{k(2e/3)^2}{r^2} = \frac{4ke^2}{9r^2}$$

The ratio of these two forces is

$$\frac{F_{grav}}{F_{elec}} = \frac{Gm_u^2/r^2}{4ke^2/(9r^2)} = \frac{9Gm_u^2}{4ke^2} \tag{1}$$

To evaluate Equation (1), we need the mass of the u quark, m_u. From Table 31.2, the mass of the up quark is $m_u = 4$ MeV/c^2. Converting this value to kilograms, we get

$$m_u = \frac{4 \text{ MeV}}{c^2} = \left[\frac{4 \times 10^6 \text{ eV}}{(3.00 \times 10^8 \text{ m/s})^2} \right] \left(\frac{1.60 \times 10^{-19} \text{ J}}{1 \text{ eV}} \right) = 7 \times 10^{-30} \text{ kg}$$

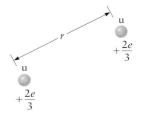

Figure 31.11 Example 31.5.

SOLVE

Inserting our value of m_u into Equation (1) along with values of the other factors leads to

$$\frac{F_{grav}}{F_{elec}} = \frac{9Gm_u^2}{4ke^2} = \frac{9(6.67 \times 10^{-11} \text{ N} \cdot \text{m}^2/\text{kg}^2)(7 \times 10^{-30} \text{ kg})^2}{4(8.99 \times 10^9 \text{ N} \cdot \text{m}^2/\text{C}^2)(1.60 \times 10^{-19} \text{ C})^2}$$

$$\frac{F_{grav}}{F_{elec}} = \boxed{3 \times 10^{-41}}$$

What does it mean?

The value of r canceled when we computed the ratio F_{grav}/F_{elec}, so the ratio has this value for all values of r. The gravitational force between two up quarks is thus always *much* smaller than the electric force. That is why gravity is considered the weakest of the fundamental forces.

The Four Forces in Nature and Their Unification

We have referred to the four fundamental forces in nature, but as indicated in Table 31.6 the electromagnetic and weak forces are two aspects of the same electroweak force. So, one could also say that there are only three fundamental forces in nature.

In addition, many physicists are attempting to unify the electroweak and strong forces; such a **grand unified theory** does not yet exist and is the subject of much current research. Going one step further, some physicists suspect that all forces in nature can be unified into one "theory of everything." A candidate for such a theory is string theory in which all particles are stringlike objects in a high-dimensional space. The details of such unified theories are still very uncertain. Einstein was working on this problem up until his death in 1955 and many others are working on it now, so progress has been slow.

31.6 | ELEMENTARY PARTICLE PHYSICS: IS THIS THE FINAL ANSWER?

Throughout this book, we have emphasized the simplicity of the laws of physics. In Newton's mechanics, all motion could be understood using just three laws (Newton's laws). Prior to quantum mechanics, all electromagnetic phenomena could be explained using Maxwell's four equations. At an intuitive level, scientists believe that nature is inherently "simple" and that the fundamental forces and particles should reflect this simplicity. However, adding up the number of quarks and leptons in Tables 31.2 and 31.5 along with the photon and other force-carrying particles in Table 31.6 brings the total number of fundamental particles to more than two dozen. Some physicists (including the author) suspect that a simple universe should not contain this many "fundamental" particles. For this reason, many physicists believe that our current understanding of fundamental particles and forces is not complete. The search for the ultimate laws of physics is thus still in progress. While debate over the correct laws of physics continues, it should not overshadow the theories we *do* have. These theories very accurately describe the quantum properties of atoms and molecules, the behavior of nuclei, the production of energy in the Sun, the behavior of the laser in your CD player, the motion of your car when you apply the brakes, and the propagation of the electromagnetic waves used by your cell phone. Our current theories may turn out to be only approximate, just as Newton's mechanics is an approximate theory, but our present theories do work *extremely* well.

31.7 | ASTROPHYSICS AND THE UNIVERSE

The Big Bang and the Expansion of the Universe

How did the universe begin, and how will it end? Did it even have a beginning? Will it ever end? Questions about the origin and fate of the universe have deep implications for philosophy and many other fields. Such questions came within reach of physics in the early 1900s when astronomers observed that the spectrum of light from distant galaxies can be used to determine the motion of those galaxies relative to the Earth. In Chapter 13, we discussed the Doppler effect, which causes the frequency of a wave to shift to either higher or lower values when the source and observer are in relative motion. As a result, the frequency of light or other electromagnetic radiation is shifted upward when the source and observer are moving toward each other and is shifted to a lower frequency when they are moving apart. (See also Chapter 23 and Example 23.7.)

In the early 1900s, astronomers discovered that light from all galaxies visible from the Earth is shifted to lower frequencies compared with what would be observed if their sources were at rest. This effect is called a **red shift** because red light is at the low frequency end of the visible spectrum. Using this measured frequency shift together with the Doppler effect formula, one can calculate the *apparent* velocity of a particular galaxy relative to the Earth. (We'll explain why we use the term *apparent* shortly.) Astronomer Edwin Hubble collected such data for

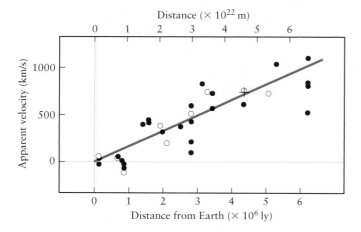

many galaxies and plotted their apparent velocities as a function of their distances from the Earth, obtaining the result in Figure 31.12. His data showed that galaxies farthest away have the largest apparent velocities, and all galaxies are moving away from the Earth.

Hubble's results in Figure 31.12 (and much data gathered more recently) follow a linear behavior, with the apparent velocity v of a galaxy proportional to its distance d from the Earth:

$$v = H_0 d \tag{31.5}$$

Hubble's law

This is called **Hubble's law**, and H_0 is called the Hubble constant. The best current estimate for the Hubble constant is $H_0 = 21$ (km/s)/(10^6 ly), where 1 ly = 1 light-year is the distance light travels in 1 year. So, a galaxy that is 10^6 ly (one million light-years) from the Earth is moving away at an apparent velocity of 21 km/s.

Hubble's law is evidence for the **Big Bang** model of the universe, which states that the universe originated with an explosion. At the instant of this explosion all the universe was an infinitesimal point, and since this "bang," the universe has expanded as sketched in Figure 31.13. All galaxies appear to be moving away from us, which would seem to suggest that the Earth is at the center of it all. Figure 31.13, however, shows that an observer at any location in an expanding universe will find that all galaxies are moving away from him. Galaxies with the largest apparent velocities relative to the Earth are farthest away from the Earth. Since light takes a certain amount of time to travel from these galaxies to the Earth, this light was emitted when the universe was younger than at present. In this way, studying the most distant galaxies gives a picture of the universe shortly after it was formed.

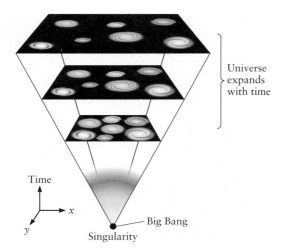

Figure 31.13 Schematic of the Big Bang. As time passes, the universe is expanding, carrying galaxies farther apart.

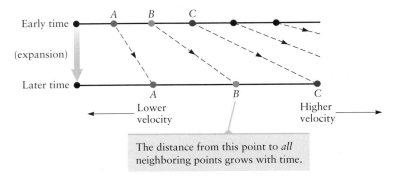

Figure 31.14 Schematic of how the relative distance between any two points in an expanding universe increases with time. As time increases (going from the top line to the bottom), the distance between any pair of points increases. For an observer at any point, all other points are moving farther away, and each observer perceives that he or she is at the "center" of the expansion.

Early time

(expansion)

Later time

Lower velocity

Higher velocity

The distance from this point to *all* neighboring points grows with time.

Figure 31.15 Arno Penzias and Robert Wilson were using this antenna to study how radio and microwaves propagate through the atmosphere. They discovered blackbody radiation arriving at the Earth from outer space, with an apparent blackbody temperature of approximately 3 K.

Now let's explain why we refer to the quantity v in Hubble's law as an *apparent* velocity. In our normal work in mechanics, we define and measure the velocity of an object in terms of its position relative to a fixed coordinate system, that is, relative to "markers" in space such as a coordinate axis. With an expanding universe (Fig. 31.14), these markers are themselves expanding along with the universe. That is, the "fabric" of space itself expands with time, carrying us and all other galaxies along with it. Although in some ways we can think of the expanding universe as involving "motion," it is a very different type of motion than in normal mechanics, which is why astronomers usually refer to the astronomical red shift as giving an "apparent" velocity. This picture also gives a more accurate way to understand the astronomical red shift. Light emitted by a distant galaxy has a certain wavelength. As that light travels to the Earth, the universe expands, "stretching" space and thereby increasing the wavelength of the light. This increase is similar to a Doppler red shift, but now the shift to a longer wavelength (lower frequency) is due to the expansion of space rather than the simple motion of the source.

One of the most important pieces of evidence in support of the Big Bang theory was discovered by two scientists, Arno Penzias and Robert Wilson, who were studying satellite communications systems at Bell Laboratories in the 1960s. They found that their antenna (Fig. 31.15) was detecting a small amount of radiation even when aimed at "empty" space. The intensity of this radiation was the same as that produced by a blackbody at a temperature of approximately 3 K, and this radiation from space is now known as the cosmic microwave background radiation. The existence of this radiation and its blackbody temperature are key features supporting the Big Bang model. According to the model, this radiation was emitted when the universe was only a few hundred thousand years old and then red-shifted due to subsequent expansion so that it now has an effective blackbody temperature

Figure 31.16 Recent experiments have studied the cosmic background radiation in great detail. This plot shows how the radiation intensity varies slightly from point to point in the sky, with the red regions being hottest and the blue regions coolest. The temperature variations are quite small, with the hottest regions only about 10^{-5} K higher in temperature than the average. These variations were produced by slight variations of the density of matter throughout space shortly after the Big Bang. These density variations eventually became stars and galaxies.

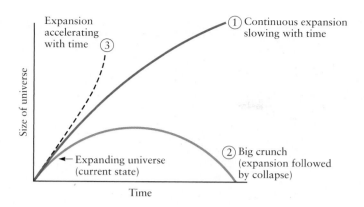

Figure 31.17 Different possible scenarios for the fate of the universe. Curve 1 shows a universe that expands forever, while curve 2 shows a universe that collapses back to a point (the big crunch). Our universe is now on the far left side of this plot and is expanding. Curve 3 shows a universe that is expanding at an ever-increasing rate ("accelerating"). Recent observations suggest that our universe is following a curve like curve 3.

of 3 K. In more recent experiments, physicists have measured tiny variations in the intensity of the cosmic microwave background radiation across the sky (Fig. 31.16). These variations give information on the state of the early universe when stars and galaxies were first forming.

Fate of the Universe

Hubble's law together with the cosmic microwave background radiation and other observations are strong evidence that the Big Bang occurred approximately 14 billion years ago. Physicists are also interested in the ultimate fate of the universe. For many years, it was believed that there were just two possibilities, labeled curves 1 and 2 in Figure 31.17. Curve 1 shows a universe that continues to expand forever, but with the rate of expansion slowing due to the gravitation attraction between stars. On the other hand, the gravitational attraction may be large enough so that the entire universe collapses back to an infinitesimal point; this model, called the "big crunch," is illustrated by curve 2 in Figure 31.17.

These "options" for the ultimate fate of the universe depend how much matter it contains; if the amount of mass is large enough, the mutual gravitational attraction will lead to the big crunch. This issue has prompted astronomers to do careful measurements of the total mass contained in the universe, and the results show that the universe contains a considerable amount of *dark matter*, matter that does not emit or absorb enough radiation to be observed directly. While dark matter cannot be "seen" with an ordinary telescope, its presence can still be measured by studying how it exerts a gravitational force on nearby galaxies. In fact, the best current measurements suggest that as much as 80% to 90% of the mass in the universe is dark matter! The nature of dark matter is not yet known. Speculations are that it could be composed of hitherto undiscovered fundamental particles, but that is not yet proven.

Another big surprise in recent years has been the discovery that the universe is actually described by curve 3 in Figure 31.17, which shows that the expansion of the universe is *accelerating*. According to the standard Big Bang model, all matter was given an initial velocity by the Big Bang. If there were no forces on a particular galaxy, it would continue forever with a constant velocity and the universe would expand forever at a constant rate. The effects of gravity should eventually cause the relative speed of a galaxy to decrease with time (as in curves 1 and 2 in Fig. 31.17), but the most recent observations indicate that galaxies are instead accelerating.

The explanation of this puzzle is not yet known. Some physicists have suggested that the acceleration is due to a phenomenon called *dark energy*. Currently, there is no agreed-upon theory of dark energy or even a precise definition of what it is. The rough idea is that an entirely new form of energy is present in "empty" space and that this energy is imperceptible to most observations ("dark"). The density of this energy is very low and is normally not noticed. In the very large volume of the

Figure 31.18 Image of the surface of Fe atoms on the surface of Cu, obtained using a scanning tunneling microscope (STM), one of the new tools of nanoscience. Each pyramid-like "bump" is an individual Fe atom. The STM is able to form images of atomic scale structure and to move atoms from place to place, producing the atomic arrangement shown here.

universe, however, dark energy produces significant effects and is responsible for repelling matter and for the accelerating expansion of the universe. The existence of dark energy and its properties are now being debated, so the picture of fundamental forces and elementary particles given in this chapter may soon need to be revised.

31.8 | PHYSICS AND INTERDISCIPLINARY SCIENCE

Physicists are actively studying many areas besides elementary particle physics and astrophysics. In fact, there are so many interesting research topics in physics that there is no way we can even list them all. We'll mention just a few of these topics and give only very brief descriptions.

Nanoscience is the study of objects with sizes in the range of roughly 1 nm to 1000 nm. It spans the regime of atoms and molecules up to objects such as cells, which can be seen with an optical microscope. There are several things that make this "nanoworld" interesting. First, quantum effects are important, so some of the "mysteries" of quantum theory play a key role in the properties of nanoscale objects. Second, nanoscientists have developed ways to manipulate matter atom by atom (Fig. 31.18), thus making things not found in nature, including materials that are both stronger and lighter than steel. Third, the new properties of nanoscale objects are beginning to lead to entirely new kinds of applications. For example, computers that store information using the quantum properties of individual electrons are now being developed; such computers, if successful, could be much faster and more powerful that those we have today.

Another promising area is *biophysics*, in which physics principles and methods are applied to problems in biology. This work is giving important new insights into a wide range of problems, including photosynthesis, molecular motors, how the nose detects odors, and how the brain works. Some of these advances involve new techniques from nanoscience. For example, nanoscientists may soon be able to probe and assemble DNA and other biomolecules atom by atom (Fig. 31.19), which could lead to the design of custom drugs, breakthrough medical therapies, new species of plants with higher yields than previously possible, and much more.

Physics in the 21st century thus has much to offer humanity and promises to be a very exciting journey.

Figure 31.19 Image of strand of DNA obtained using an atomic force microscope. The "bumps" on the DNA strand are individual base pairs.

Fundamental particles

There are two families of *fundamental particles*:

There are six types of *leptons*. electron (e^-) electron neutrino (ν_e) muon (μ^-) muon neutrino (ν_μ) tau (τ^-) tau neutrino (ν_τ) All leptons are point particles.	All *hadrons* are composed of *quarks*. There are six types of quarks: up (u) strange (s) down (d) top (t) charm (c) bottom (b) Quarks are point particles. Hadrons composed of three quarks are called *baryons*. They include protons (denoted p and composed of two up quarks and one down quark, uud), and neutrons (n, quark composition udd). Hadrons composed of two quarks are called *mesons*.

Matter and antimatter

For each fundamental particle, there is a corresponding antiparticle that has the same mass but opposite electric charge. For example, the antiparticle of the electron is the positron (denoted e^+). The electron has charge $-e$, whereas the positron has charge $+e$.

Fundamental forces

There are four fundamental forces in nature, and it is believed that each is carried or "mediated" by one or more particles.

1. *Electromagnetism* is described by Maxwell's equations. The quantum theory of this force is called *quantum electrodynamics* (QED). The electromagnetic force is carried by the photon.
2. The *strong force* is an attractive force between quarks and holds quarks together to form baryons (including protons and neutrons) and mesons. The strong force also attracts protons and neutrons and holds them together in the nucleus. The strong force is carried by particles called *gluons* and is described by *quantum chromodynamics* (QCD).
3. The *weak force* acts on all leptons and hadrons. The weak force is much smaller in magnitude than the strong force, but is responsible for processes such as beta decay of a nucleus. The weak force is carried by the W^+, W^-, and Z particles.
4. *Gravity* is the weakest of the fundamental forces. There is no quantum theory of gravity yet, but it is believed that the gravitational force is transmitted by a particle called the *graviton*. Gravitons have not yet been observed.

The *standard model* is a quantum mechanical theory for the behavior of the electromagnetic, strong, and weak forces and the interactions between leptons and hadrons.

(Continued)

QUESTIONS

SSM = answer in Student Companion & Problem-Solving Guide   ⊗ = life science application

1. In Figure 31.4, an electron and a positron annihilate and produce electromagnetic radiation in the form of two gamma ray photons, each of energy 0.511 MeV. Explain why it is not possible for this reaction to give a single gamma ray photon with energy $2 \times (0.511 \text{ MeV}) = 1.022 \text{ MeV}$.

2. Explain why there are no doubly charged mesons, that is, mesons with an electric charge of $+2e$ or $-2e$.

3. How many different hadrons are stable outside the nucleus? *Hint*: They are all listed in Table 31.3.

4. Consider the hypothetical reaction in Figure Q31.4 in which an electron (e^-) annihilates a positron (e^+), with the positron initially at rest. Can this reaction result in a single emitted photon? Explain why or why not.

Figure Q31.4

5. The Σ^+ is composed of two up quarks (u) and one strange quark (s). The mass of an up quark is about 4 MeV/c^2, while the mass of a strange quark is about 150 MeV/c^2. Does the mass of the Σ^+ come mainly from the masses of these quarks or from the potential energy associated with binding? Explain.

6. Design an experiment that could distinguish a neutron from an antineutron.

7. Are mesons composed of (a) a total of three quarks and antiquarks, (b) three leptons, or (c) a quark and an antiquark?

8. Are baryons composed of (a) three quarks, (b) three leptons, or (c) two quarks?

9. How many quarks and antiquarks are present in an antimeson?

10. Why is it that we never see a single electron decay to a pair of photons? Could an electron decay to a pair of neutrinos? Why or why not? Name the explicit conservation rule each of these supposed decay schemes would violate.

11. SSM A particle with nonzero velocity enters a region of space that has a uniform magnetic field directed into the page as depicted in Figure Q31.11. The resulting path (trajectory) fol-lowed by the particle will depend on the particle's charge and mass. Sketch qualitatively the trajectory of an electron, a positron, a proton, and a neutron. Assume all have the same initial velocity.

Figure Q31.11

12. At sea level, a significant number of cosmic-ray particles are muons with relativistic velocities. If the muons have a mean lifetime of only 2.2 microseconds, where do they come from and how are they produced?

13. Describe some naturally occurring common processes that produce antimatter. How common are anti-electrons (positrons) on the Earth? (Consider radioactivity.) How common are antimuons on the Earth?

14. It has been suggested that Newton's law of gravity brought about the first unification of physics principles. In what way does his law of gravity unify the motion of heavenly bodies with the motion of earthly objects?

15. Unification of physical laws often leads to new concepts and novel insights. Describe some of the consequences from the unification of the physics of electricity with that of magnetism. How was the concept of light altered?

16. Does the universe have a center? Does the universe have edges? If not, explain how that could be possible.

17. Consider the science of some hypothetical intelligent life-forms inhabiting a planet in a galaxy many millions of light-years away from the Earth. Would they measure the same Hubble constant as we do?

18. Given the current age of the universe and a telescope powerful enough, how far away is the most distant object we could observe? What kind of object is it?

19. SSM Due to the cosmological red shift, the photons that make up the cosmic microwave background are, as the name implies, in the microwave region of the electromagnetic spectrum. In what region of the electromagnetic spectrum were these photons when they were first released?

SSM = solution in Student Companion & Problem-Solving Guide   ✖ = life science application
★ = intermediate   ✪ = challenging   ℝ = reasoning and relationships problem

31.1 COSMIC RAYS

1. A cosmic-ray proton has a kinetic energy of 10^{19} eV. What is its speed? *Hint*: You will have to use the relativistic expression for the kinetic energy from Chapter 27.

2. The collisions of cosmic-ray particles with the Earth's atmosphere often produce muons that subsequently decay into an electron and two neutrinos according to the reaction

$$\mu^- \rightarrow e^- + \nu_\mu + \bar{\nu}_e$$

What is the maximum possible kinetic energy of the electron? Assume the muon is at rest just before it decays and the masses of the neutrinos are negligible.

3. A cosmic-ray proton has a de Broglie wavelength of 0.020 fm. What is its kinetic energy? *Hint*: Use the relativistic relations for momentum and energy along with de Broglie's relation for the wavelength of a particle-wave in Equation 28.9.

4. SSM ★ At sea level, some hundreds of muons go though your body each second. A muon has a mean lifetime of 2.2 microseconds, so why do we find so many near the Earth's surface? Consider a muon with a kinetic energy of 150 MeV that is created in the upper atmosphere when a cosmic ray collides with a nitrogen atom. (a) Determine the velocity of the muon. (b) Neglect relativity and calculate how far the muon can move from its creation assuming it decays within 2.2 microseconds. (c) Now take into consideration time dilation and calculate the distance the muon will travel.

31.2 MATTER AND ANTIMATTER

5. A proton and antiproton, both initially at rest, annihilate each other, producing two gamma ray photons. What is the frequency of one of the photons?

6. An electron–positron pair is produced by a photon with an energy of 11 MeV. (a) Name two quantities this process must conserve. (b) What is the total kinetic energy of the electron and positron combined? (c) What is the total momentum of the electron and positron combined?

7. An electron with speed $0.80c$ collides head-on with a positron with the same speed, producing two photons. What is the energy of each photon?

8. An unknown particle collides with its antiparticle and the two are annihilated, emitting two photons, each with energy 3.0 MeV. What is the rest mass of the particle, in kilograms? Assume the unknown particles are moving very slowly before annihilating.

9. SSM (a) Calculate the energy released when a proton annihilates an antiproton. Express your answer here and in the rest of this problem in MeV. (b) What is the approximate energy released in the fission of one $^{235}_{92}$U nucleus? (c) From your answers to parts (a) and (b), calculate the energy per unit mass of (1) a collection of equal numbers of protons and antiprotons and (2) a sample of $^{235}_{92}$U.

31.4 ELEMENTARY PARTICLE PHYSICS: THE STANDARD MODEL

10. At present, the most powerful accelerator employs protons and antiprotons at energies of about 1 TeV. (a) What is the kinetic energy of such a proton, in joules? (b) What is its de Broglie wavelength?

11. A pi meson (π^0) decays into two photons. What is the energy of each photon?

12. SSM ★ Describe the quark content of a particle that has a charge of $-e$ and contains one strange quark. Is this particle a baryon or a meson? Note: There is more than just one answer (but you only need to give one).

13. ★ Determine the quark content of a particle that is composed of 3 or fewer quarks and has a charge of $-2e$. Is this particle matter or antimatter? A baryon or a meson? Note: There is more than just one answer (but you only need to give one).

14. The following reactions are forbidden because they violate one or more conservation laws. Which conservation law(s) do they violate? The π^+ particle is a meson.
(a) $p \rightarrow e^+ + \nu_e$
(b) $p \rightarrow e^+ + \gamma$
(c) $n \rightarrow p + e^- + \nu_\mu$
(d) $p \rightarrow e^- + \pi^+$
(e) $\pi^+ \rightarrow \pi^+ + n$

15. The pi mesons π^+ and π^- are each composed of two quarks. The quark composition of the π^+ is $u\bar{d}$, while the composition of the π^- is $\bar{u}d$. (a) What is the antiparticle of the π^+? (b) The π^- can decay into a muon and an antimuon neutrino according to

$$\pi^- \rightarrow \mu^- + \bar{\nu}_\mu$$

What are the final quarks involved in this decay? (c) Does this decay conserve lepton number? If so, what is the final lepton number?

16. Find the charge on each of these baryons or antibaryons and identify them: (a) uud, (b) $\bar{u}\bar{u}\bar{d}$, and (c) $\bar{u}\bar{d}\bar{d}$.

31.5 THE FUNDAMENTAL FORCES OF NATURE

17. ℝ The strong force holds the quarks in a proton together, overcoming the repulsion between the positively charged u quarks. For simplicity, ignore the effect of the d quark and estimate the magnitude of the (attractive) strong force between two u quarks needed to keep them bound in a proton.

18. SSM ★ Complete the following reactions.
(a) $K^0 \rightarrow \pi^+ + ?$
(b) $p + \bar{p} \rightarrow \gamma + ?$
(c) $? + n \rightarrow p + e^-$

31.7 ASTROPHYSICS AND THE UNIVERSE

19. ★ A red shift measurement shows that the speed of a distant galaxy relative to an astronomer on the Earth is $v = 0.25c$, where c is the speed of light. (a) How far away is the galaxy? (b) If this galaxy emits blue light with $\lambda = 450$ nm (as viewed by an observer at rest in the galaxy), what is the wavelength observed on the Earth? *Hint*: Consider the Doppler shift formulas for light in Chapter 23.

20. (a) At what frequency f_{max} does the cosmic microwave background radiation from the Big Bang have its greatest intensity? (b) Our galaxy moves at a speed of about 300 km/s relative to the source of the cosmic microwave background radiation, shifting f_{max} relative to the value found for an observer at rest relative to the source. What are the magnitude and sign of this frequency shift?

21. SSM Based on his data in Figure 31.12, Hubble proposed that the velocity v of a galaxy relative to the Earth is related to its distance d from the Earth by what is now known as Hubble's law (Eq. 31.5). The red shift of light from the most distant galaxies known gives a relative velocity of about $v = 0.96c$. What is their distance from the Earth?

22. A galaxy is found to be 2.0×10^6 ly away from the Earth. Assume this galaxy does not have any motion relative to us other than that due to the expansion of space. (a) What is the galaxy's recession speed due to the expansion of space? (b) How much farther away from us will this galaxy be by next year?

ADDITIONAL PROBLEMS

23. **SSM** **R** How much energy would be released if a baseball were annihilated with an antibaseball? State your answer in joules and in megatons of TNT. (See Example 30.6.)

24. A Σ^+ particle moving with a kinetic energy of 500 MeV decays to a pion and neutron, $\Sigma^+ \rightarrow n + \pi^+$. What is the total kinetic energy of the decay products?

25. ✪ The probability for an individual particle to have *not yet* decayed after a time t since its creation is given by $\text{prob}(t) = e^{-t/\tau}$, where τ is the mean lifetime of the particle. (a) What is the probability that a π^+ at rest will still exist after 0.10 μs? (b) What is the probability that a π^+ of momentum 500 MeV/c can travel a distance of 20 m before decaying? *Hint*: Consider time dilation. (c) What is the probability that a φ particle with the same momentum could travel the same distance? (See Table 31.4.)

26. Consider the decay of an excited hyperion described in Figure P31.26. This figure shows a "cascade" of decay reactions, producing many particles along the way. (a) How many particles, including photons, are there in the final outcome? (b) Is charge conserved in the decay? That is, do the sum of the charges in the final particles add up to the initial charge? (c) Is baryon number conserved overall?

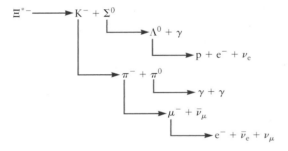

Figure P31.26 Decay scheme of an excited hyperion.

27. ✪ Shortly after the discovery of the expansion of space, physicists made many attempts to measure Hubble's constant. These measurements are difficult, and values of H_0 ranging from 12 to 28 (km/s)/Mly were found. (a) Find the age of the universe for this minimum and maximum value of H_0. (b) The best current value of H_0 comes from the WMAP probe of the cosmic background radiation, giving a value of 21 (km/s)/(Mly) with an uncertainty of about 4%. What age of the universe does this measured value suggest?

28. ✪ Consider a hypothetical cluster of galaxies distributed in space as shown in the top part of Figure P31.28, where each galaxy is positioned on a grid with a spacing of two million light-years. Four billion years later, the universe has expanded such that the grid spacing is now three million light-years as seen in the bottom part of Figure P31.28. (a) Assume your planet is in galaxy A. Calculate the *change in distance* to galaxies B, C, D, and E as seen from A over the 4.0-billion-year period. (b) Which galaxy moves the most with respect to galaxy A? (c) Now assume your planet is in galaxy D. Find the change in distance to galaxies C, B, A, and F. (d) Which galaxy moves the most with respect to galaxy D over this period of time? (e) Find the velocities of galaxies B, C, D, and E that would be measured if your planet were in galaxy A. (f) Find the velocities of galaxies C, B, A, and F as seen from a planet in galaxy D. (g) Finally, calculate the Hubble constant as measured from a planet in galaxy A and compare it to what an observer in galaxy D would measure.

Figure P31.28

29. ✪ **The Planck time.** It is believed that there is a period of time so short that the theory of gravity (or any of the known interactions) cannot make meaningful predictions. Interestingly, there exists an algebraic arrangement of three fundamental constants that gives units of time. The constants are G (Newton's gravitational constant), h (Planck's constant), and c (the speed of light). (a) Use dimensional analysis to find this period of time (called the Planck time). Our current understanding suggests that for the first instant of the Big Bang, from $t = 0$ to the Planck time, the known laws of physics did not apply, and this epoch remains a mystery. (b) The shortest time interval that can be currently measured is about 10^{-18} s. How many Planck time intervals would occur during this minimum measurement time?

Reference Tables

Table A.1 Some Useful Physical Constants and Data

QUANTITY	SYMBOL	APPROXIMATE VALUE
Universal gravitational constant	G	6.674×10^{-11} N $\cdot$ m^2/kg^2
Acceleration due to gravity near the Earth's surface	g	9.8 m/s^2
Speed of light in vacuum	c	$2.997\ 924\ 58 \times 10^8$ m/s
Mass of the electron	m_e	$9.109\ 382 \times 10^{-31}$ kg
Mass of the proton	m_p	$1.672\ 621\ 6 \times 10^{-27}$ kg
Mass of the neutron	m_n	$1.674\ 927\ 2 \times 10^{-27}$ kg
Magnitude of the electron charge	e	$1.602\ 176\ 5 \times 10^{-19}$ C
Permittivity of free space	ε_0	$8.854\ 187\ 817 \times 10^{-12}$ C^2/(N $\cdot$ m^2)
Coulomb constant	$k = \dfrac{1}{4\pi\varepsilon_0}$	$8.987\ 551\ 788 \times 10^9$ N $\cdot$ m^2/C^2
Permeability of free space	μ_0	$4\pi \times 10^{-7}$ T $\cdot$ m/A
Boltzmann's constant	k_B	$1.380\ 650 \times 10^{-23}$ J/K
Planck's constant	h	$6.626\ 069 \times 10^{-34}$ J$\cdot$s
Avogadro's number	N_A	$6.022\ 142 \times 10^{23}$ mol^{-1}
Universal gas constant	R	$8.314\ 472$ J/(mol $\cdot$ K)
Stefan−Boltzmann constant	σ	$5.670\ 40 \times 10^{-8}$ W/(m^2 $\cdot$ K^4)
Atomic mass unit	u	$1.660\ 538\ 8 \times 10^{-27}$ kg

Note: 2006 CODATA recommended values.

SI Base Units

Quantity	Unit	Symbol	Quantity	Unit	Symbol
Length	meter	m	Electric current	ampere	A
Mass	kilogram	kg	Temperature	kelvin	K
Time	second	s	Luminous intensity	candela	cd

Table A.3 Solar System Data

Body	Radius at equator (km)	Mass (kg)	Average density (kg/m³)	Orbital semimajor axis (km)	Orbital period (years)	Orbital eccentricity
Mercury	2,440	3.30×10^{23}	5430	5.79×10^{7}	0.241	0.205
Venus	6,050	4.87×10^{24}	5240	1.08×10^{8}	0.615	0.007
Earth	6,370	5.97×10^{24}	5510	1.50×10^{8}	1.000	0.017
Mars	3,390	6.42×10^{23}	3930	2.28×10^{8}	1.881	0.094
Jupiter	71,500	1.90×10^{27}	1330	7.78×10^{8}	11.86	0.049
Saturn	60,300	5.68×10^{26}	690	1.43×10^{9}	29.46	0.057
Uranus	25,600	8.68×10^{25}	1270	2.87×10^{9}	84.01	0.046
Neptune	24,800	1.02×10^{26}	1640	4.50×10^{9}	164.8	0.011
Pluto	1,200	1.3×10^{22}	1750	5.91×10^{9}	247.7	0.244
Sun	6.95×10^{5}	1.99×10^{30}	1410			
Moon	1,740	7.35×10^{22}	3340	3.84×10^{5}	27.3 days	0.055

Note: From NASA Planetary Fact Sheet (2006).

Atomic Number Z	Element	Chemical Symbol	Atomic Mass (u)	Mass Number A	Atomic Mass (u)	Percent Abundance	Half-Life (if radio-active) $T_{1/2}$	Primary Decay Modes*
0	(Neutron)	n		1	1.008 665		10.4 min	β^-
1	Hydrogen	H	1.007 94	1	1.007 825	99.988 5		
	Deuterium	D		2	2.014 102	0.011 5		
	Tritium	T		3	3.016 049		12.33 years	β^-
2	Helium	He	4.002 602	3	3.016 029	0.000 137		
				4	4.002 603	99.999 863		
3	Lithium	Li	6.941	6	6.015 122	7.5		
				7	7.016 004	92.5		
4	Beryllium	Be	9.012 182	7	7.016 929		53.1 days	EC
				9	9.012 182	100		
5	Boron	B	10.811	10	10.012 937	19.9		
				11	11.009 306	80.1		
6	Carbon	C	12.010 7	10	10.016 853		19.3 s	EC, β^+
				11	11.011 434		20.4 min	EC, β^+
				12	12.000 000	98.93		
				13	13.003 355	1.07		
				14	14.003 242		5 730 years	β^-
7	Nitrogen	N	14.006 7	13	13.005 739		9.96 min	EC, β^+, β^-
				14	14.003 074	99.632		
				15	15.000 109	0.368		
8	Oxygen	O	15.999 4	15	15.003 065		122 s	EC, β^+
				16	15.994 915	99.757		
				18	17.999 160	0.205		
9	Fluorine	F	18.998 403 2	19	18.998 403	100		
10	Neon	Ne	20.179 7	20	19.992 440	90.48		
				22	21.991 385	9.25		
11	Sodium	Na	22.989 77	22	21.994 437		2.60 years	EC, β^+
				23	22.989 770	100		
				24	23.990 963		14.96 h	β^-
12	Magnesium	Mg	24.305 0	24	23.985 042	78.99		
				25	24.985 837	10.00		
				26	25.982 593	11.01		
13	Aluminum	Al	26.981 538	27	26.981 539	100		
14	Silicon	Si	28.085 5	28	27.976 926	92.229 7		
15	Phosphorus	P	30.973 761	31	30.973 762	100		
				32	31.973 907		14.26 days	β^-

*EC denotes decay by electron capture, SF denotes spontaneous fission.

Atomic Number Z	Element	Chemical Symbol	Atomic Mass (u)	Mass Number A	Atomic Mass (u)	Percent Abundance	Half-Life (if radioactive) $T_{1/2}$	Primary Decay Modes*
16	Sulfur	S	32.066	32	31.972 071	94.93		
				35	34.969 032		87.3 days	β^-
17	Chlorine	Cl	35.452 7	35	34.968 853	75.78		
				37	36.965 903	24.22		
18	Argon	Ar	39.948	40	39.962 383	99.600 3		
19	Potassium	K	39.098 3	39	38.963 707	93.258 1		
				40	39.963 999	0.011 7	1.28×10^9 years	EC, β^+, β^-
20	Calcium	Ca	40.078	40	39.962 591	96.941		
21	Scandium	Sc	44.955 910	45	44.955 910	100		
22	Titanium	Ti	47.867	48	47.947 947	73.72		
23	Vanadium	V	50.941 5	51	50.943 964	99.750		
24	Chromium	Cr	51.996 1	52	51.940 512	83.789		
25	Manganese	Mn	54.938 049	55	54.938 050	100		
26	Iron	Fe	55.845	56	55.934 942	91.754		
27	Cobalt	Co	58.933 200	59	58.933 200	100		
				60	59.933 822		5.27 years	β^-
28	Nickel	Ni	58.693 4	58	57.935 348	68.076 9		
				60	59.930 790	26.223 1		
29	Copper	Cu	63.546	63	62.929 601	69.17		
				65	64.927 794	30.83		
30	Zinc	Zn	65.39	64	63.929 147	48.63		
				66	65.926 037	27.90		
				68	67.924 848	18.75		
31	Gallium	Ga	69.723	69	68.925 581	60.108		
				71	70.924 705	39.892		
32	Germanium	Ge	72.61	70	69.924 250	20.84		
				72	71.922 076	27.54		
				74	73.921 178	36.28		
33	Arsenic	As	74.921 60	75	74.921 596	100		
34	Selenium	Se	78.96	78	77.917 310	23.77		
				80	79.916 522	49.61		
35	Bromine	Br	79.904	79	78.918 338	50.69		
				81	80.916 291	49.31		
36	Krypton	Kr	83.80	82	81.913 485	11.58		
				83	82.914 136	11.49		
				84	83.911 507	57.00		
				86	85.910 610	17.30		

*EC denotes decay by electron capture, SF denotes spontaneous fission.

Atomic Number Z	Element	Chemical Symbol	Atomic Mass (u)	Mass Number A	Atomic Mass (u)	Percent Abundance	Half-Life (if radio-active) $T_{1/2}$	Primary Decay Modes*
37	Rubidium	Rb	85.467 8	85	84.911 789	72.17		
				87	86.909 184	27.83	4.75×10^{10} years	β^-
38	Strontium	Sr	87.62	86	85.909 262	9.86		
				88	87.905 614	82.58		
				90	89.907 738		158 s	β^-
39	Yttrium	Y	88.905 85	89	88.905 848	100		
40	Zirconium	Zr	91.224	90	89.904 704	51.45		
				91	90.905 645	11.22		
				92	91.905 040	17.15		
				94	93.906 316	17.38		
41	Niobium	Nb	92.906 38	93	92.906 378	100		
42	Molybdenum	Mo	95.94	92	91.906 810	14.84		
				95	94.905 842	15.92		
				96	95.904 679	16.68		
				98	97.905 408	24.13		
43	Technetium	Tc		98	97.907 216		4.2×10^6 years	β^-
				99	98.906 255		2.1×10^5 years	β^-
44	Ruthenium	Ru	101.07	99	98.905 939	12.76		
				100	99.904 220	12.60		
				101	100.905 582	17.06		
				102	101.904 350	31.55		
				104	103.905 430	18.62		
45	Rhodium	Rh	102.905 50	103	102.905 504	100		
46	Palladium	Pd	106.42	104	103.904 035	11.14		
				105	104.905 084	22.33		
				106	105.903 483	27.33		
				108	107.903 894	26.46		
				110	109.905 152	11.72		
47	Silver	Ag	107.868 2	107	106.905 093	51.839		
				109	108.904 756	48.161		
48	Cadmium	Cd	112.411	110	109.903 006	12.49		
				111	110.904 182	12.80		
				112	111.902 757	24.13		
				113	112.904 401	12.22	7.7×10^{15} years	β^-
				114	113.903 358	28.73		
49	Indium	In	114.818	115	114.903 878	95.71	4.4×10^{14} years	β^-

*EC denotes decay by electron capture, SF denotes spontaneous fission.

Atomic Number Z	Element	Chemical Symbol	Atomic Mass (u)	Mass Number A	Atomic Mass (u)	Percent Abundance	Half-Life (if radioactive) $T_{1/2}$	Primary Decay Modes*
50	Tin	Sn	118.710	116	115.901 744	14.54		
				118	117.901 606	24.22		
				120	119.902 197	32.58		
51	Antimony	Sb	121.760	121	120.903 818	57.21		
				123	122.904 216	42.79		
52	Tellurium	Te	127.60	126	125.903 306	18.84		
				128	127.904 461	31.74	2.2×10^{24} years	β^-
				130	129.906 223	34.08	7.9×10^{20} years	β^-
53	Iodine	I	126.904 47	127	126.904 468	100		
				129	128.904 988		1.6×10^7 years	β^-
54	Xenon	Xe	131.29	129	128.904 780	26.44		
				131	130.905 082	21.18		
				132	131.904 145	26.89		
				134	133.905 394	10.44		
				136	135.907 220	8.87	$\geq 2.36 \times 10^{21}$ years	β^-
55	Cesium	Cs	132.905 45	133	132.905 447	100		
56	Barium	Ba	137.327	137	136.905 821	11.232		
				138	137.905 241	71.698		
57	Lanthanum	La	138.905 5	139	138.906 349	99.910		
58	Cerium	Ce	140.116	140	139.905 434	88.450		
				142	141.909 240	11.114	$> 2.6 \times 10^{17}$ years	β^-
59	Praseodymium	Pr	140.907 65	141	140.907 648	100		
60	Neodymium	Nd	144.24	142	141.907 719	27.2		
				144	143.910 083	23.8	2.3×10^{15} years	α
				146	145.913 112	17.2		
61	Promethium	Pm		145	144.912 744		17.7 years	EC, α
62	Samarium	Sm	150.36	147	146.914 893	14.99	1.06×10^{11} years	α
				149	148.917 180	13.82		
				152	151.919 728	26.75		
				154	153.922 205	22.75		
63	Europium	Eu	151.964	151	150.919 846	47.81		
				153	152.921 226	52.19		
64	Gadolinium	Gd	157.25	156	155.922 120	20.47		
				158	157.924 100	24.84		
				160	159.927 051	21.86		
65	Terbium	Tb	158.925 34	159	158.925 343	100		

*EC denotes decay by electron capture, SF denotes spontaneous fission.

ATOMIC NUMBER Z	ELEMENT	CHEMICAL SYMBOL	ATOMIC MASS (u)	MASS NUMBER A	ATOMIC MASS (u)	PERCENT ABUNDANCE	HALF-LIFE (IF RADIO-ACTIVE) $T_{1/2}$	PRIMARY DECAY MODES*
66	Dysprosium	Dy	162.50	162	161.926 796	25.51		
				163	162.928 728	24.90		
				164	163.929 171	28.18		
67	Holmium	Ho	164.930 32	165	164.930 320	100		
68	Erbium	Er	167.6	166	165.930 290	33.61		
				167	166.932 045	22.93		
				168	167.932 368	26.78		
69	Thulium	Tm	168.934 21	169	168.934 211	100		
70	Ytterbium	Yb	173.04	172	171.936 378	21.83		
				173	172.938 207	16.13		
				174	173.938 858	31.83		
71	Lutecium	Lu	174.967	175	174.940 768	97.41		
72	Hafnium	Hf	178.49	177	176.943 220	18.60		
				178	177.943 698	27.28		
				179	178.945 815	13.62		
				180	179.946 549	35.08		
73	Tantalum	Ta	180.947 9	181	180.947 996	99.988		
74	Tungsten	W	183.84	182	181.948 206	26.50		
				183	182.950 224	14.31		
				184	183.950 933	30.64	$> 3 \times 10^{19}$ years	α
				186	185.954 362	28.43		
75	Rhenium	Re	186.207	185	184.952 956	37.40		
				187	186.955 751	62.60	4.1×10^{10} years	β^-
76	Osmium	Os	190.23	188	187.955 836	13.24		
				189	188.958 145	16.15		
				190	189.958 445	26.26		
				192	191.961 479	40.78		
77	Iridium	Ir	192.217	191	190.960 591	37.3		
				193	192.962 924	62.7		
78	Platinum	Pt	195.078	194	193.962 664	32.967		
				195	194.964 774	33.832		
				196	195.964 935	25.242		
79	Gold	Au	196.966 55	197	196.966 552	100		
80	Mercury	Hg	200.59	199	198.968 262	16.87		
				200	199.968 309	23.10		
				201	200.970 285	13.18		
				202	201.970 626	29.86		

*EC denotes decay by electron capture, SF denotes spontaneous fission.

ATOMIC NUMBER Z	ELEMENT	CHEMICAL SYMBOL	ATOMIC MASS (u)	MASS NUMBER A	ATOMIC MASS (u)	PERCENT ABUNDANCE	HALF-LIFE (IF RADIO-ACTIVE) $T_{1/2}$	PRIMARY DECAY MODES*
81	Thallium	Tl	204.383 3	203	202.972 329	29.524		
				205	204.974 412	70.476		
				208	207.982 005		3.053 min	β^-
				210	209.990 066		1.30 min	β^-
82	Lead	Pb	207.2	204	203.973 029	1.4	$\geq 1.4 \times 10^{17}$ years	α
				206	205.974 449	24.1		
				207	206.975 881	22.1		
				208	207.976 636	52.4		
				210	209.984 173		22.2 years	α, β^-
				211	210.988 732		36.1 min	β^-
				212	211.991 888		10.64 h	β^-
				214	213.999 798		26.8 min	β^-
83	Bismuth	Bi	208.980 38	209	208.980 383	100		
				211	210.987 258		2.14 min	α, β^-
84	Polonium	Po		210	209.982 857		138.38 days	α
				214	213.995 186		164 μs	α
85	Astatine	At		218	218.008 682		1.5 s	α, β^-
86	Radon	Rn		222	222.017 570		3.823 days	α, β^-
87	Francium	Fr		223	223.019 731		22 min	α, β^-
88	Radium	Ra		226	226.025 403		1 600 years	α
				228	228.031 064		5.75 years	β^-
89	Actinium	Ac		227	227.027 747		21.77 years	α, β^-
90	Thorium	Th	232.038 1					
				228	228.028 731		1.912 years	α
				232	232.038 050	100	1.40×10^{10} years	α
91	Protactinium	Pa	231.035 88	231	231.035 879		3.28×10^4 years	α
92	Uranium	U	238.028 9	232	232.037 146		69 years	α
				233	233.039 628		1.59×10^5 years	α
				235	235.043 923	0.720 0	7.04×10^8 years	α, SF
				236	236.045 562		2.34×10^7 years	α, SF
				238	238.050 783	99.274 5	4.47×10^9 years	α
93	Neptunium	Np		237	237.048 167		2.14×10^6 years	α
94	Plutonium	Pu		239	239.052 156		2.412×10^4 years	α
				242	242.058 737		3.75×10^5 years	α
				244	244.064 198		8.1×10^7 years	α, SF

*EC denotes decay by electron capture, SF denotes spontaneous fission.

Sources: Chemical atomic masses are from T. B. Coplen, "Atomic weights of the elements 1999," a technical report to the International Union of Pure and Applied Chemistry and published in *Pure and Applied Chemistry* 73(4):667−683 (2001). Atomic masses of the isotopes are from G. Audi and A. H. Wapstra, "The 1995 update to the atomic mass evaluation," *Nuclear Physics* A595(4):409−480 (December 25, 1995). Percent abundance values are from K. J. R. Rosman and P. D. P. Taylor, "Isotopic compositions of the elements 1999," a technical report to the International Union of Pure and Applied Chemistry and published in *Pure and Applied Chemistry* 70(1):217−236 (1998). Data on decay modes and half-lives are from Lawrence Berkeley Laboratory Isotopes Project Listing and the National Nuclear Data Center, Brookhaven National Laboratory.

Mathematical Review

B.1 | MATHEMATICAL SYMBOLS

The following symbols are used to express mathematical relationships.

SYMBOL	MEANING	EXAMPLE	EXPLANATION
$+$	Addition	$a + b = c$	a plus b equals c
$-$	Subtraction	$a - b = c$	a minus b equals c
$\times$ or $\cdot$	Multiplication	$a \times b = c$	a times b equals c
		$a \cdot b = c$	
/ or $\div$	Division	$a/b = c$	a divided by b equals c
		$a \div b = c$	
$>$	Greater than	$a > b$	a is greater than b
$<$	Less than	$a < b$	a is less than b
$\geq$	Greater than or equal to	$a \geq b$	a is greater than or equal to b
$\leq$	Less than or equal to	$a \leq b$	a is less than or equal to b
$\approx$	Approximately equal to	$a \approx b$	a is approximately equal to b
$\lim\limits_{x \to 0}$	Limit as x approaches 0	$y = \lim\limits_{x \to 0} f$	y equals the value of f as the variable x approaches zero

The basic rules of algebra are reviewed in Section 1.6. When solving an equation to find the value of a single variable, you should rearrange the equation to put the *unknown* variable (the variable you wish to solve for) on one side of the equation and the *known* quantities on the other side. For example, suppose the variable a in the equation

$$4a + 7 = T - 9$$

is the unknown. The goal is to isolate a on one side of the equal sign (usually the left). In this example, we can subtract 7 from both sides to get

$$4a = T - 16$$

We next divide both sides by the constant factor 4 to get the solution

$$a = \frac{T}{4} - 4$$

When dealing with two equations containing two unknowns, the usual approach is to use one equation to eliminate one of the unknowns from the other equation. For example, suppose a and T are unknown in the equations

$$3a = T - 9 \tag{B2.1}$$

$$2a = -T + 24 \tag{B2.2}$$

We can rearrange Equation B2.2 to get

$$T = -2a + 24$$

We then substitute this expression for T in Equation B2.1 to find

$$3a = T - 9 = (-2a + 24) - 9 = -2a + 15 \tag{B2.3}$$

We can now solve for a:

$$5a = 15$$

$$a = 3$$

This value can then be used in Equation B2.1 to find the value of T; we get $T = 18$.

Another approach when working with two equations and two unknowns is to simply add the two equations. For example, if we add Equations B2.1 and B2.2, we will eliminate T:

$$
\begin{aligned}
3a &= T - 9 \\
2a &= -T + 24 \\
\hline
5a &= -9 + 24
\end{aligned}
$$

which is another way to get Equation B2.3.

The equations considered in Equations B2.1 through B2.3 all involve the unknowns (e.g., a and T) raised to the first power. In a *quadratic equation,* the unknown quantity is raised to the second power. If x is an unknown, an equation of this type is

$$ax^2 + bx + c = 0$$

The solution is given by the *quadratic formula,*

$$x = \frac{-b \pm \sqrt{b^2 - 4ac}}{2a} \tag{B2.4}$$

Note that there are two solutions as indicated by the appearance of the $\pm$ sign.

B.3 | SCIENTIFIC NOTATION

Scientific notation is a convenient way to express extremely large or extremely small numbers. For example, the number 640,000 is written as 6.4×10^5 in scientific notation. To express a number in scientific notation, move the decimal point in the original number to obtain a new number between 1 and 10. Count the number of places the decimal point has been moved; this number will become the exponent of 10 in scientific notation. If you started with a number greater than 10 (such as 150,000,000,000), the exponent of 10 is positive (1.5×10^{11}). If you started with a number less than 1 (such as 0.000055), the exponent is negative (5.5×10^{-5}).

B.4 | GEOMETRY AND TRIGONOMETRY

One way to define and measure angles is shown in Figure B.1. The angle θ equals the ratio of the arc length s measured along the circle divided by the radius r of the circle:

$$\theta = \frac{s}{r} \tag{B4.1}$$

This relation gives the value of θ in **radians.** To convert to degrees, we use the fact that one complete "trip" around the circle in Figure B.1 corresponds to both 2π radians and to 360°. We thus have

$$\theta \text{ (in radians)} \times \frac{360°}{2\pi \text{ radians}} = \theta \text{ (in degrees)} \tag{B4.2}$$

Figure B.2 shows a right triangle with sides x, y, and r, where r is the hypotenuse. According to the **Pythagorean theorem,**

$$x^2 + y^2 = r^2 \tag{B4.3}$$

The trigonometric functions **sine, cosine,** and **tangent** are defined as

$$\sin \theta = y/r \tag{B4.4}$$

$$\cos \theta = x/r \tag{B4.5}$$

$$\tan \theta = y/x \tag{B4.6}$$

The Pythagorean theorem (Eq. B4.3) implies that

$$\sin^2 \theta + \cos^2 \theta = 1 \tag{B4.7}$$

for any value of the angle θ. Equations B4.3 through B4.6 also lead to a number of other trigonometric relations (called **identities**), some of which are the following

$$\sin(-\theta) = -\sin \theta \quad \cos(-\theta) = \cos \theta \quad \tan(-\theta) = -\tan \theta$$

$$\sin(\theta + 90°) = \sin(\theta + \pi/2) = \cos \theta$$

$$\cos(\theta + 90°) = \cos(\theta + \pi/2) = -\sin \theta$$

$$\sin(\alpha \pm \beta) = \sin \alpha \cos \beta \pm \cos \alpha \sin \beta$$

$$\cos(\alpha \pm \beta) = \cos \alpha \cos \beta \mp \sin \alpha \sin \beta \tag{B4.8}$$

The relations in Equation B4.8 hold for any values of the angles θ, α, and β. If $\alpha = \beta$, the relation for $\sin(\alpha + \beta)$ becomes

$$\sin(\alpha + \alpha) = \sin(2\alpha) = \sin \alpha \cos \alpha + \cos \alpha \sin \alpha$$

We can also replace the angle α with θ; rearranging then leads to

$$\sin \theta \cos \theta = \tfrac{1}{2}\sin(2\theta) \tag{B4.9}$$

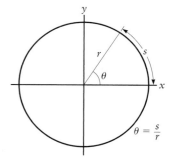

Figure B.1 The angle θ measured in radians equals the ratio of the length s measured along the circular arc divided by the radius r of the circle.

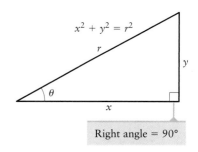

Right angle = 90°

Figure B.2 The Pythagorean theorem is a relation between the lengths of the sides of a right triangle.

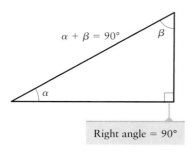

Right angle = 90°

Figure B.3 The sum of the three interior angles of a right triangle is 180°. Hence, $\alpha + \beta = 90°$.

Equations B4.4 through B4.6 tell how to calculate the sine, cosine, and tangent of an angle from the lengths of the sides of the associated right triangle. We can also use this approach to find the value of the angle. For example, if we know the values of y and r, we can solve for the angle as

$$\sin \theta = y/r$$

$$\theta = \sin^{-1}(y/r) \tag{B4.10}$$

The function $\sin^{-1}$ is the **inverse** sine function, also known as the **arcsine** (= arcsin = $\sin^{-1}$). In words, Equation B4.10 says that θ equals the angle whose sine is y/r. Scientific calculators all contain this function (as a single "button"). If the values of y and r are known, you must only take the ratio y/r and then compute the arcsine of this value to find θ in Equation B4.10. Likewise, there are also functions for the inverse cosine ($\cos^{-1}$ = arccos) and inverse tangent ($\tan^{-1}$ = arctan):

$$\theta = \cos^{-1}(x/r) \quad \text{and} \quad \theta = \tan^{-1}(y/x) \tag{B4.11}$$

We often need to deal with triangles like the one in Figure B.3, which shows a right triangle with interior angles α and β. The sum of the three interior angles of a triangle is always 180°. For the triangle in Figure B.3, one of the angles is 90° (it is a right triangle); hence,

$$\alpha + \beta = 90° \tag{B4.12}$$

Two angles whose sum is 90° are called complementary angles. From the definitions of the sine and cosine functions, for any pair of complementary angles α and β we have

$$\sin \alpha = \cos \beta$$

$$\cos \alpha = \sin \beta$$

Two approximate relations involving trigonometric functions are also useful. When the angle θ is small,

$$\sin \theta \approx \theta \quad \text{and} \quad \tan \theta \approx \theta \tag{B4.13}$$

which are good approximations (accurate to a few percent or better) when θ is smaller than about 30°. Notice that the relations in Equation B4.13 apply only when the angle is measured in radians.

B.5 | VECTORS

A vector quantity has both a magnitude and a direction. There are two ways to do calculations with vectors; one is a graphical approach, which is useful for approximate or qualitative calculations, while the other involves the components of the vectors.

The graphical approach is illustrated in Figure B.4, which shows two vectors $\vec{A}$ and $\vec{B}$ along with their sum ($\vec{A} + \vec{B}$) and difference ($\vec{A} - \vec{B}$). Multiplying a vector

Figure B.4 Examples showing the effect of multiplying a vector by a scalar, adding two vectors, and subtracting one vector from another.

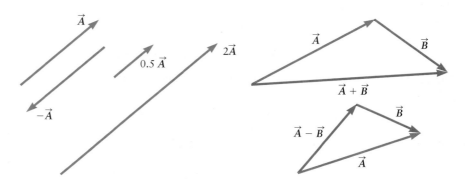

by a scalar (e.g., 0.5 or 2) changes the length but does not change the direction. Multiplying by a negative number (e.g., −1) reverses the direction. The vectors $-\vec{A}$, $0.5\vec{A}$, and $2\vec{A}$ are also shown in Figure B.4.

The components of a vector can be obtained using trigonometry. The length of a vector $\vec{A}$, also called its **magnitude,** is denoted by either A or $|\vec{A}|$. If this vector makes an angle θ with the x axis, the **components** are

$$A_x = A \cos \theta = |\vec{A}| \cos \theta \text{ and } A_y = A \sin \theta = |\vec{A}| \sin \theta \qquad \text{(B5.1)}$$

The sum of two vectors can be found by adding components, as illustrated in Figure B.5. If

$$\vec{C} = \vec{A} + \vec{B} \qquad \text{(B5.2)}$$

in terms of the components of these vectors we then have

$$C_x = A_x + B_x \text{ and } C_y = A_y + B_y \qquad \text{(B5.3)}$$

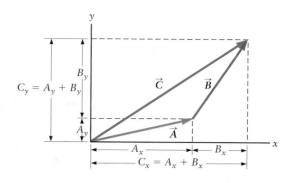

Figure B.5 To compute the sum of two vectors, we add their components. Here we add the x components of $\vec{A}$ and $\vec{B}$ to get the x component of $\vec{C}$. We follow the same procedure to find the y component of $\vec{C}$.

B.6 | EXPONENTIAL FUNCTIONS AND LOGARITHMS

Exponential functions and logarithms play a role in many areas of physics, including the behavior of fluids (Chapter 10), sound (Chapter 13), electric circuits (Chapters 19 and 22), and radioactive decay (Chapter 30).

The number $e = 2.71828\ldots$ plays an important role. Useful numerical values involving the number e are $e^{-1} \approx 0.37$ and $1 - e^{-1} \approx 0.73$.

There are two versions of the logarithm function. The **natural logarithm** function is denoted by the symbol "ln" and is related to the exponential function by

$$\ln(e^a) = a \qquad \text{(B6.1)}$$

where a is a number. Other useful properties of these functions are

$$e^a \times e^b = e^{a+b}$$

$$\frac{e^a}{e^b} = e^{a-b}$$

$$\ln(a) + \ln(b) = \ln(ab) \qquad \text{(B6.2)}$$

$$\ln(a) - \ln(b) = \ln(a/b)$$

$$\ln(a^b) = b \ln a$$

The **common logarithm** function (also called the base 10 logarithm) is denoted by the symbol "log"; the log function has properties similar to the natural logarithm, but it is based on the number 10 instead of e. These properties include

$$\log(10^a) = a \qquad \text{(B6.3)}$$

$$10^a \times 10^b = 10^{a+b}$$

$$\frac{10^a}{10^b} = 10^{a-b}$$

$$\log(a) + \log(b) = \log(ab) \qquad \text{(B6.4)}$$

$$\log(a) - \log(b) = \log(a/b)$$

$$\log(a^b) = b \log a$$

Exponential functions may also involve base numbers besides e and 10. For example, for any number y, one has

$$y^a \times y^b = y^{a+b} \quad \text{and} \quad \frac{y^a}{y^b} = y^{a-b} \tag{B6.5}$$

B.7 | SOME RELATIONS USEFUL IN SPECIAL RELATIVITY

In special relativity, factors such as v/c and v^2/c^2, where c is the speed of light and v is the speed of an object, appear often. The speed of light is very large, so the ratio v/c is usually very small. When that is the case, certain algebraic expressions can be simplified. For example, when A is a very small number, the approximation

$$\frac{1}{1+A} \approx 1 - A \tag{B7.1}$$

is very accurate. If we replace A by v/c, we find

$$\frac{1}{1+v/c} \approx 1 - v/c \tag{B7.2}$$

and in a similar way, we get

$$\frac{1}{1+(v/c)^2} \approx 1 - (v/c)^2 \tag{B7.3}$$

Another expression that arises in special relativity, especially in problems involving time dilation and length contraction, is $\sqrt{1-(v/c)^2}$. When v/c is very small, to a good approximation we have

$$\sqrt{1-v^2/c^2} \approx 1 - \frac{v^2}{2c^2} \tag{B7.4}$$

and

$$\frac{1}{\sqrt{1-v^2/c^2}} \approx 1 + \frac{v^2}{2c^2} \tag{B7.5}$$

The accuracy of these approximations depends on the value of v/c. If $v/c = 0.1$, the approximation in Equation B7.2 is good to about 1%, whereas Equations B7.4 and B7.5 are accurate to about 0.005%. The accuracy improves for smaller values of v/c.

CHAPTER 1

Concept Checks

1.1 | Significant Figures

Quantity	Length or Distance in Meters (m)	Number of Significant Figures
Diameter of a proton	1×10^{-15}	1
Diameter of a red blood cell	8×10^{-6}	1
Diameter of a human hair	5.5×10^{-5}	2
Thickness of a piece of paper	6.4×10^{-5}	2
Diameter of a compact disc	0.12	2
Height of the author	1.80	3
Height of the Empire State Building	443.2	4
Distance from New York City to Chicago	1,268,000	4
Circumference of the Earth	4.00×10^7	3
Distance from Earth to the Sun	1.5×10^{11}	2

1.2 | Using Prefixes and Powers of 10

Quantity	Length or Distance in Meters (m)	Same Length or Distance Using Prefix
Diameter of a proton	1×10^{-15}	1 femtometers = 1 fm
Diameter of a red blood cell	8×10^{-6}	8 micrometers = 8 μm
Diameter of a human hair	5.5×10^{-5}	55 micrometers = 55 μm
Thickness of a sheet of paper	6.4×10^{-5}	64 micrometers = 64 μm
Diameter of a compact disc	0.12	1.2 decimeters = 1.2 dm = 12 centimeters = 12 cm
Height of the author	1.80	1.80 meters = 1.80 m
Height of the Empire State Building	443.2	443.2 meters = 443.2 m
Distance from New York City to Chicago	1,268,000	1.268 megameters = 1.268 Mm
Circumference of the Earth	4.00×10^7	40.0 megameters = 40.0 Mm
Distance from Earth to the Sun	1.5×10^{11}	150 gigameters = 150 Gm

Problems

1. ® 7×10^6 pieces of paper
3. ® Approximately 2×10^6 steps
5. Table 1.2:

Quantity	Number of Significant Figures for Each Time Listed
Time for light to travel 1 meter	3
Time between heartbeats (approximate)	1
Time for light to travel from the Sun to Earth	1
1 day	exact
1 month (30 days)	exact
Human lifespan (approximate)	1
Age of the universe	1

Table 1.3:

Object	Number of Significant Figures for Each Mass Listed
Electron	2
Proton	2
Red blood cell	1
Mosquito	1
Typical person	1
Automobile	2
Earth	2
Sun	2

7. 8.9 km
9. 5×10^3 kg/m³
11. (a) 2.99792458×10^8 m/s (b) 2.998×10^8 m/s
13. ® 5×10^5 km (more than 3 times the diameter of Jupiter)
15. (a) 6.37×10^6 m (b) 3.84×10^5 km (c) 5.97×10^{24} kg
17. ® Assuming the measurements are made to a precision of two significant figures, the approximate volume of the book is 0.003 m³ or 3000 cm³.
19. ® An official NBA basketball has a volume of 7.10×10^{-3} m³.
21. 20 billion years
23. (a) 23 m (b) 2.3×10^4 mm
25. (a) 10^4 cm² (b) 10^6 mm²
27. 2.0×10^3 mm
29. (a) 10^6 cm³ (b) 10^9 mm³
31. The Corvette C4 engine has 7.44 L of combustion volume, larger than that of the Corvette Z06 with just 7.0 L. Yes, the newer car has more power despite having a smaller engine.
33. (a) ML/T (b) L²/T² (c) L²/T
35. 380 m/s

37. 5.0 boxes of cupcakes and 2.0 bags of donuts
39. (a) $3:00 \rightarrow 90°$ or $\pi/2$ rad
    (b) $6:00 \rightarrow 180°$ or π rad
    (c) $6:30 \rightarrow 195°$ or $13\pi/12$ rad
    (d) $9:00 \rightarrow 270°$ or $3\pi/2$ rad
    (e) $11:10 \rightarrow 335°$ or 5.85 rad
41. 65° and 90°
43. 1.3 rad
45. 47°, 43°
47. (a)

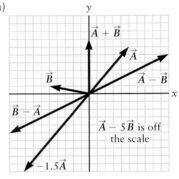

(b) $|\vec{A} + \vec{B}| = 2.1$ m, $|\vec{A} - \vec{B}| = 3.3$ m, $|\vec{B} - \vec{A}| = 3.3$ m, $|\vec{A} - 5\vec{B}| = 8.8$ m, $|-1.5\vec{A}| = 3.5$ m
49. 1.5 m; 2.0 m
51. The vector pairs (a)(c) and (b)(h) have approximately the same direction.
53. −37° or −143°
55. 47 km, 108°
57. (a) 37, 53° (b) 18, −74° (c) 110, 65° (d) 190, 230°
59. 51°, 120 m
61. 4900 m
63. ® (a) 40 cm (b) 44 m (c) 1.7 km ≈ 1 mi
65. 97
67. 27° south of east
69. ® 1×10^{-8} m/s
71. 49 m
73. 1300

CHAPTER 2

Concept Checks

2.1 (a) is inconsistent, (b) is inconsistent, (c) may be consistent.
2.2 Yes, at $t \approx 1.6$ s in Figure 2.9.
2.3 (a) graph 3; (b) graph 2; (c) graph 1.
2.4 (b) and (d)
2.5 (a) and (c) are not an action–reaction pairs.

Problems

1. (a), (d), (e), and (f)
3. (b), (c), (d), and (f)
5. 10 m/s
7. (a) 2.3 m/s (b) $v_{max} = 3.5$ m/s; $v_{min} = 0$. The average is closer to the maximum velocity.

9.

11.

13.

15.

17.

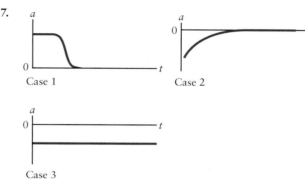

19. Case 1: A car starting from a traffic light (at rest), accelerating to the speed limit, and traveling at that speed; Case 2: A car slowing to a stop at a traffic light; Case 3: A ball thrown upward from the top of a building. The ball starts with some initial speed that decreases linearly due to gravity.

21. Case 1: A driver keeps his foot steady on the gas pedal to keep the car moving at constant acceleration. The driver then shifts gears and the car is still accelerating, but the value of a is then smaller than at the start. Case 2: A motorist applies his brakes so as to slow down to match the speed of a car in front of him. Case 3: A car is moving initially at constant velocity in a lane on the freeway; the driver then decides to momentarily accelerate and change to a faster-moving lane.

23. (a) Matches Case 2 in Figure P2.23
    (b) Matches Case 4 in Figure P2.23
    (c) Matches Case 3 in Figure P2.23
    (d) Matches Case 1 in Figure P2.23

25. (a) About 0.1 m/s (b) About 0.4 m/s

27. -10 m/s^2

29. 0.3 m/s^2, 0.2 m/s^2

31. (a)   (b)

(c)

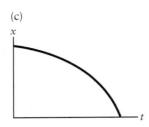

33. −5.0 m/s, −15 m/s, −10 m/s²

35. (a) 1.5 m/s² (b) About 1.0 m/s² (c) About 5.0 m/s² (d) 0

37. (a) 340 s (b) 1200 s

39. (a) Grass field:

(b) Asphalt street:

(c) Hardwood floor:

(d) Ice rink:

(e) Ideal, frictionless surface:

41. (a), directly below the drop point. From the point of view of the man on the railcar, the path of the ball is identical to what it would look like were he to drop the ball while standing on a sidewalk.

43. 2.0 m/s²

45. (c), (d), and (e)

47. 670 N

49. 11,000 kg

51. (a) Highway 99; 4 min (b) 5.5 min

53. (a) 3.0 m/s (b) 4.5 m/s (c) 1.5 m/s (d) The average of the answers found for parts (b) and (c) must be equal to the answer to part (a).

55. (a) 2.4 s
(b) (c)

57. 5.8 × 10⁹ m

59. 16 m

CHAPTER 3

Concept Checks

3.1 In Figure 3.6, $\vec{F}_{grav}$ and $\vec{N}$ are *not* an action–reaction pair.

3.2 The driver can drive with an acceleration smaller than $a_{crate, max}$ for a very long time and thereby reach a very large speed, and the crate will not slip.

3.3 (a) Opposite; the velocity is to the right, while the acceleration is directed to the left.
(b) Parallel; $\vec{v}$ and $\vec{a}$ are both directed to the right.
(c) Parallel; both $\vec{v}$ and $\vec{a}$ are directed downward.

3.4 Only statement (a) is true.

3.5 The bumper experiences a force equal to three times the weight of the car!

3.6 (d)

Problems

1. 0.015 m/s²

3. 140 m

5. 6.4 m/s²

7. +12,000 N

9. 260 N

11. (a) 1.3 m/s² (b) 710 m

13. 1.4 m

15. (a) 1.3 × 10⁻² mi/s (b) 5.8 × 10⁻² mi/s or 210 mi/h

17.

19. 16 m

21. Yes

23. (a) 35 m/s (b) 39 m/s

25. 2300 m

27. 51 kg

29. 1600 N

31. -4.0 m/s^2

33. (a) 50 N (b) 770 s

35. (a) 500 N (b) 530 N

37. 0.51

39. 75 N

41. 0.49 m/s^2

43. (a) 0.18 (b) 0.14

45. (a) 0.62 (b) Static friction

47. 0.66

49. $+35$ m/s or -35 m/s. There are two solutions because the baseball passes this height twice.

51. 9.2 m

53. 3.3 s

55. 7.3 m

57. 260 m

59. 6.2 m/s

61. 8.3 m/s^2 downward

63. 26 N, 90 N

65. 250 N

67. Ⓡ Approximately 900 N

69. About 0.04 m/s; about 4 min

71. Ⓡ Approximately 2000 N

73. Ⓡ Approximately 0.6 s

75. Ⓡ Rank (b), (d), (e), (f), (a), (c)

77. 22 m/s or 49 mi/h. You would not likely survive impact with the ground, but would have a good chance in the case of a water impact.

79. Ⓡ Approximately 14 m/s

81. Ⓡ Approximately 20 μm/s

83. (a) 0.30 s

(b)

Ct (s)	Cx (cm)
0.14	10
0.16	13
0.18	16
0.20	20
0.22	24

85. 0.76

87. Ⓡ About 0.25 s

89. (a) (b) 0.20 N

91. (a) 11 m/s^2 (b) 4.6 s (c) Changing the mass of the block does not alter the needed acceleration.

93. (a) 9.9 N (b) 17 N

95. (a) 5.6×10^5 m/s^2 (b) 5.4×10^{-4} s (c) 5.6×10^3 N

97. (a) 22 m/s (b) 61 m/s^2

99. Ⓡ About 6 N

CHAPTER 4

Concept Checks

4.1   (b)

4.2   (a) The acceleration is *never* zero. (b) $v_y = 0$ at the instant the ball is at the highest point on its trajectory. (c) The force is *never* zero. (d) The speed is *never* zero.

4.3   The acceleration of m_1 is *not* the same in the two cases.

4.4   (b)

4.5   Suppose a spacecraft is sitting at rest in deep space. Now consider the spacecraft as viewed from a noninertial reference frame; that is, it is viewed by an observer who has an acceleration $+a$ along a particular direction. As viewed by this observer, the spacecraft will appear to have an acceleration $-a$. Since there are no forces acting on the spacecraft, Newton's first law is violated in this noninertial reference frame.

Problems

1. 54 N, $-17°$

3. 46 N, $-22°$

5.

$$\vec{N}_{\text{floor}}$$
$$\vec{N}_{\text{wall}} \quad \vec{F}_{\text{friction}}$$
$$\vec{F}_{\text{grav}}$$

7. $T_1 = 260$ N, $T_2 = 440$ N, $T_3 = 510$ N

9. 27 N, 64°

11. 17°

13. 20 kg

15. (a) 2.3 s (b) 45 m (c) 30 m/s, $-48°$

17. 19 m/s

19. (a) 8.2 m (b) 27 m/s (c) 70 m

21. Yes

23. (a) 320 m/s (b) 320 m/s

25. 0.83 m, 2.7 ft

27. 25 m/s, 72°

29. Ⓡ 20 m/s, 2 s

31. 48°

33. (a) 5.1 m (b) It is about double the record. The high jumper's velocity must include both horizontal and vertical components if he is to clear the bar.

35. 95 m/s

37. 90 m

39. 2100 s

41. (a) 240 m/s (b) 240 m/s (c) 240 m/s

43. (a) 1.3 m/s^2, $-0.47°$ (b) 7.7 s

45. 25 m

47. 0.12

49. 0.27

51. 34°

53. 82 m

55. Ⓡ 6×10^5 N

57. $\theta = \tan^{-1}\left(\dfrac{a_x}{g}\right)$

59. Ⓡ 20 m/s

61. Ⓡ 20 m/s

63. Ⓡ 1×10^7 N

65. 1.2×10^3 N, 32°

67. (a) 670 N (b) 15 kg

69. (a) 250 ft (b) 86 mi/h; goggles are recommended.

71. (a) Perpendicular to the shore (b) 2.0 m/s (c) 81° from a line perpendicular to the shore (d) 50 s (e) 100 m

73. It is better to have Joe pull and Paul push (a_x = 5.0 m/s^2). If Joe pushes and Paul pulls, then a_x = 3.3 m/s^2.

75. 3.0 m/s^2

77. (a) 37 m/s (b) −8.6° (c) 2.6 s

79. (a) 0 (b) 0 (c) 4.5 N

81. Ⓡ 20 m/s

83. (a) 9.4 m/s (b) 11 m/s

85. (a) 58 m (b) It increases by 39 m, to a new distance of 97 m. (c) 43 m

87. 15°

89.

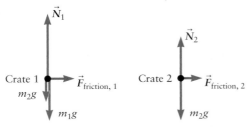

Magnitudes of forces are m_1g = 340 N, m_2g = 150 N, $F_{\text{friction, 1}}$ = 60 N, N_1 = 490 N, $F_{\text{friction, 2}}$ = 26 N, and N_2 = 150 N.

91. Ⓡ (a) −x direction (b) 150 mi/h (c) 1.7 (d) A 70% difference would certainly be significant.

CHAPTER 5

Concept Checks

5.1   Statements (a), (b), and (c) are all correct.

5.2   (c)

5.3   The observers in (a) and (b) are moving at constant velocities, so their accelerations are zero. Hence, they are both inertial frames. In (c) and (d), the frames are accelerating, so they are noninertial frames.

5.4   (d)

5.5   The answer is shown in the accompanying figure.

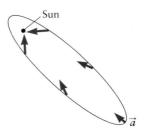

5.6   (b)

Problems

1. (a) 12 m/s^2 (b) 1.0 s

3. Ⓡ (a) 470 m/s (b) 350 m/s

5. 0.17 m/s^2

7. 130 s

9. (a) 3.4×10^{-2} m/s^2 (b) 2.9×10^{-2} m/s^2

11. (a) 0.75 (b) Static friction

13. Ⓡ 30 m/s^2, about $3 \times g$

15. 7.2 m/s

17. 2.2 m/s

19. 8.1 m/s

21. (a) 4200 N (b) 20 m/s

23. 20 m

25. (a)

Top   Side

(b)

The forces on the rock are tension (from the string) and gravity. (c) $T \cos \theta$ (d) 8.4°

27. 8.6 m/s

29. 7.0×10^3 N

31. 29 N

33. Ⓡ 4×10^4 s

35. 1.8×10^{-8} N

37. 4.0×10^{-9} N, 34° below the negative x axis

39. 3.5×10^{22} N

41. Ⓡ 8×10^{-8} N

43. 3.5×10^8 m

45. 0.9998

47. 470 N

49. 1.4×10^{12} m

51. 4.4×10^9 s or about 140 years

53. 2.7×10^4 s or 7.6 h

55. Satellite 3100 m/s, Earth 3.0×10^4 m/s

57. Ⓡ 0.94

59. 6.7×10^{-4} percent

61. Less, 0.997

63. Ⓡ Doctor's head 1.6×10^{-8} N, Mars 2.4×10^{-8} N

65. (a) 0.073 (b) 1.2 loops/s (c) Top 2.0×10^4 N, bottom 2.3×10^4 N

67. Ⓡ 1200 N

69. (a) 20 s (b) 31 m/s (c) $0.75mg$ (d) $1.3mg$ (e) With the rotation

71. (a) 8.8 m/s^2, 7700 m/s, 5500 s (b) 7600 m/s, 5800 s

73. (a) $v_{max} = \sqrt{\dfrac{T_{max}r^2}{mL}}$ (b) $N = 9$

75. (a) 26 m (b) 230 m

77. Ⓡ (a) 1.1 N (b) 1.3×10^{-10} m/s^2 (c) 6.6×10^4 m (d) The asteroid will hit the Earth.

CHAPTER 6

Concept Checks

6.1 (1) corresponds to case (a); (2) corresponds to case (c); (3) corresponds to case (b).

6.2 (c)

6.3 As the roller coaster in Figure 6.13 moves higher on the track, approaching point B, its velocity will decrease. If it is just barely able to make it to point B, it will do so with an extremely small velocity. In such a case, the kinetic energy at point B will be zero. Applying the conservation of energy condition then gives

$$PE_A + KE_A = PE_B + KE_B = PE_B$$
$$KE_A = \tfrac{1}{2}mv_A^2 = PE_B - PE_A = mgh_B - mgh_A$$

This can then be solved for h_B.

6.4 All three answers are correct.

6.5 (3)

Problems

1. 250 J

3. −510 J

5. 2400 J

7. −2400 J

9. −140,000 J

11. (a) 23,000 J (b) −23,000 J (c) 45,000 J

13. (a) 5500 J (b) 0 J

15. 16 J

17. 390,000 J

19. 9.9 m/s

21. 94 m/s

23. 40 m/s

25. 64 m/s

27. −8200 J

29. 11 kg

31. (a) 220,000 J (b) 63 m/s

33. (a) Yes

(b)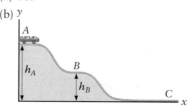

(c) mgh_A (d) $\tfrac{1}{2}mv_B^2 + mgh_B$ (e) $v_B = 17$ m/s; $v_C = 24$ m/s

35. -7.3×10^6 J

37. $h = 5.7$ m

39. (a) 8.9 m/s (b) 1600 N (c) 0 (d) 2.2 m above the top edge of the half-pipe

41. (a) Yes. The ramp is frictionless, so the only force acting on the puck is the force of gravity, which is conservative. (b) In principle, the origin of the y axis can be placed anywhere. For this problem, it is convenient to place $y = 0$ at ground level.

(b)

8.0 m

3.0 m

(c) 20 J (d) 20 J (e) 9.9 m/s (f) 7.7 m

43. 8.6 m/s

45. (a)

(b) $KE_{1i} = 0$, $KE_{2i} = 0$; taking $y = 0$ to be ground level, we have $PE_{1i} = (390 \text{ N})h$, $PE_{2i} = 0$
(c) $\Delta PE_1 = -(390 \text{ N})h$; $\Delta PE_2 = (150 \text{ N})h$ (d) 4.7 m/s

47.

Total mechanical energy

Energy

PE

KE

t

49. (a) 1.5 m/s (b) 39 N

51. (a) $v_f = \sqrt{v_0^2 + 2gh}$ (b) Since energy is a scalar quantity, the final magnitude of the bullet's velocity will depend only on the magnitude of the initial velocity.

53. (a) 1.1×10^4 m/s (b) Fuel savings would be less than 0.01%.

55.

57. $v_i = 9700$ m/s

59. 3.3×10^{-2} kg

61. 22 J

63. Ⓡ $k \approx 10,000$ N/m

65. Ⓡ $k \approx 2000$ N/m

67. (a) 0.14 J (b) 0.3 m

69. 0.2 J. The lump of clay dissipates all its KE (2.0 J) as heat and in deformation. The Super Ball dissipates only one-tenth its initial KE by similar processes.

71. 37 m/s

73. 450 W

75. 1.8×10^7 m, more than three orders of magnitude greater than the height of Mount Everest (about 8.85×10^3 m).

77. 0.29 m/s

79. 60 m/s

81. 20 pN

83. 5×10^{-18} W

85. Ⓡ The weight of a typical amino acid is about 3×10^{-24} N. The myosin molecular motor can apply a force about 12 orders of magnitude greater than that.

87. Ⓡ About half a kilometer

89. ≈ 1

91. (a) 1.1 kN (b) Yes, 21°

93. (a) 3500 J (b) 4300 J (c) PE_{max} is almost one-third greater than the maximum kinetic energy from running. One reason is that some of the energy stored in the pole comes from the bending of the pole just before takeoff.

95. (a) 7.0 cm (b) 80 cm/s (c) 16 cm

97. 8.7×10^4 W = 87 kW = 1.2×10^2 hp

99. ≈ 630 mi/h; it is very close to the reported value, suggesting that air drag at these altitudes is indeed quite small.

101. (a) 20 kJ (b) 66 W (c) 275 W (d) 0.24 (e) Her energy was dissipated by controlled frictional resistance in the device.

CHAPTER 7

Concept Checks

7.1   (a)

7.2   (b)

7.3   (b)

7.4   (f)

7.5   (d)

7.6   (a)

7.7   (c)

7.8   (c)

Problems

1. 6.3 kg·m/s

3. 18,000 kg·m/s

5. (a) 5.0 kg·m/s (b) 42 m/s

7. (a) 2.4 kg·m/s, directed upward (b) 0.13 s

9. (a) −0.85 kg·m/s (b) −0.85 kg·m/s

11. (a) 12 kg·m/s, $\theta = 26°$ (b) $F = 1500$ N, more than 100 times the ball's weight (1.4 N) (c) 12 kg·m/s, $\theta = 206°$

13. (a) The impulses are equal. (b) The average force on the egg must be greater than the force on the tomato because the collision time for the egg is smaller.

15. Ⓡ ≈ 360 N

17. Ⓡ ≈ 7 N·s

19. Ⓡ ≈ 3 N·s

21. (a) $v_f = -1.7$ m/s (b) 2.0×10^2 J

23. $v_{1f} = -15.3$ m/s; $v_{2f} = 4.2$ m/s

25. 8.9 m/s, 34° above the x axis

27. (a) 8.7 m/s (b) −28°

29. 140 kg

31. 0.50 m/s southward

33. −0.088 m/s

35. 680 kg

37. 6.9 m/s

39. 180 kg

41. $x = -0.75$ m

43. (a) Object 1 (b) $\dfrac{m_1}{m_2} = \dfrac{7}{5}$

45. Ⓡ At about (1, −1)

47. 4.7×10^6 m; the radius of the Earth is about 6.4×10^6 m, so the center of mass of the Earth–Moon system lies just inside the Earth.

49. Ⓡ $\approx 5 \times 10^{-25}$ m

51. 53 m/s; 101° clockwise from the x axis

53. Ⓡ $\approx 20,000$ N

55. Ⓡ 100 N

57. 1.3 kg

59. (a) 0.013 m (b) 0.22 m

61. (a) $v_{1f} = 0$, $v_{2f} = v_0$ (b) $v_{1f} = -0.96\, v_0$, $v_{2f} = 0.038\, v_0$ (c) $v_{1f} = 0.96\, v_0$, $v_{2f} = 2.0\, v_0$ (d) $v_{1f} = -v_0$, $v_{2f} = 0$ (e) $v_{1f} = v_0$, $v_{2f} = 2v_0$

63. Ⓡ (a) 2.0×10^4 N (b) 2.0×10^3 N (c) 11 ms

65. 0.97 km

67. (a) 5.6 m (b) 1.4 m

69. 3.4×10^2 N

71. 400 m/s

CHAPTER 8

Concept Checks

8.1 (h)

8.2 (c)

8.3 (b)

8.4 The torques are (a) negative, (b) positive, (c) positive, and (d) positive.

8.5 (b)

8.6 (d)

8.7 (a)

Problems

1. 1.5 rad

3. (a) 3.5 rad/s (b) 0.62 m/s

5. 750 m

7. −67 rad/s²

9. (a) 7.3×10^{-5} rad/s (b) 7.3×10^{-5} rad/s

11. 8.3×10^{-7} rad/s

13. 1.0×10^{-1} rad/s, 60 s; 5.0 cm

15. 2.2×10^{-2} rad/s

17. −5300 N·m

19. (a) $\phi = 28°$ (b) Positive

21. By a factor of six

23. Ⓡ $\approx -50{,}000$ N·m

25. 130 N; 60 N

27. 231 N, 25° above x axis

29. 40°

31. Positive

33. 290 N

35. Ⓡ ≈ 81 kg

37. Ⓡ 45°

39. 5.3 m from the left end of the board

41. (a) 14 kg·m² (b) 36 kg·m² (c) 22 kg·m² (d) 14 kg·m²

43. 2800 kg·m²

45. Ⓡ 0.07 kg·m²

47. 5.1/1

49. $\frac{1}{3}mL^2$

51. 13 rad/s

53. 15 km

55. 3.3×10^{-2} rad/s²; 60 rad

57. (a)

(b) Left crate: $\Sigma F = -m_1 g + T_1 = m_1 a$; right crate: $\Sigma F = -T_2 + m_2 g = m_2 a$; pulley: $\Sigma \tau = T_2 R_{pulley} - T_1 R_{pulley} = I_{pulley}\alpha$
(c) $a = \alpha R_{pulley}$ (d) $a = 2.5$ m/s²
(e) $T_1 = 62$ N; $T_2 = 66$ N

59. 4.5 m/s²

61. 5.4 km

63. -1.2×10^{-3} rad/s²; 3.0 rev

65. (a) 0.38 m/s² (b) 1.5 rad/s² (c) 1200 rad (d) 300 m

67. 300 rad/s

69. 10 m/s²

71. 32 cm

73. (a) 1.1 m/s² (b) $R_b = 45$ cm, $R_f = 15$ cm

75. (a) $\frac{11}{12}L$ (b) $\frac{25}{24}L$ (c) 31

77. (a) 1.2 kN (b) 26 N·m

79. 890 N

81. 96 N

83. $\alpha = \dfrac{6g}{[(m/m_p) + 3]L}$

CHAPTER 9

Concept Checks

9.1 (b)

9.2 (d)

9.3 (c)

9.4 (c)

9.5 In both cases, the system is able to rotate even though the angular momentum is zero and there are no external torques.

Problems

1. 0.16 J

3. 1/3

5. 29 N

7. Ⓡ 5.3 J

9. Ⓡ $\approx 91{,}000$ J

11. Ⓡ 0.13

13. 8.1×10^5

15. 1.7 m/s

17. 2.3 m/s²

19. $v_B = 0.71$ m/s, $v_C = 1.4$ m/s

21. $h = 14$ m

23. 13 m/s

25. 12 m/s

27. 2

29. (a) 7.9 J (b) 3.9 J

31. 5.3 m/s

33. 9.0

35. 4.8 kg·m²/s

37. Ⓡ ≈ 31 kg·m²/s

39. (a) 2.4 m/s (b) 0.88 J

41. 6.0×10^{-4} kg·m²/s

43. Ⓡ (a) 11 rad/s (b) $KE_{\text{arms out}}$ = 50 J, $KE_{\text{arms in}}$ = 110 J

45. Ⓡ 870 kg·m²/s

47. (a) 1200 kg·m²/s (b) 1200 kg·m²/s (c) The mass has the same tangential speed at a distance of 2.0 m from any point located along the x axis.

49. 0.30

51. Ⓡ $\approx 3 \times 10^{34}$ kg·m²/s

53. Into the page

55. $\approx 7.1 \times 10^{33}$ kg·m²/s along the rotation axis and toward the North Pole

57.

59. 11.8 m

61. Ⓡ $\approx$ 1.9 rad/s

63. (a) 4.6 J (b) 9.1×10^{-2} kg·m²/s

65. (a) $\approx$ 1 day (b) 22 days

67. (a) I = 6.9 kg·m², $M = 3.4 \times 10^2$ kg (b) 8.0 J (c) 0.19 m/s (d) 0.70 m

69. (a) 1.4×10^{-11} rad/s (b) 0.02 s

71. (a) 22 J (b) 0.42 kg·m²/s (c) This angular momentum is too small to notice.

73. 90 rad/s

75. 2.5

77. (a) 4/5 (b) 14/15

79. Ⓡ There will be no measurable change in the rotation rate of the Earth.

CHAPTER 10

Concept Checks

10.1   (a)

10.2   (c)

10.3   (a)

10.4   (b)

10.5   (a)

10.6   (a)

Problems

1. 5500 kg/m³

3. 2.1 m²

5. Ⓡ 200 kg

7. Ⓡ 7×10^{-2} m³

9. 1.1×10^5 N

11. 1.1×10^2 Pa

13. 300 cm²

15. Ⓡ 1000 Pa

17. 720 lb/in.²

19. Ⓡ 1000 kg

21. 1.3×10^5 Pa

23. 4.3×10^6 N

25. Ⓡ 5×10^{10} N

27. 10 cm

29. 2.1×10^6 Pa

31. Ⓡ 2×10^4 Pa, equal to the maximum systolic blood pressure

33. Ⓡ 4×10^5 Pa

35. $\geq$ 41 m

37. 1:10

39. 200 N

41. 650 kg/m³

43. 6.3 cm

45. 1.3×10^3 kg/m³; the object will sink.

47. 5.9×10^3 N

49. 1.4×10^3 m³

51. 3.1×10^4 N

53. 2.2×10^5 N

55. 85%

57. 40 m/s

59. (a) 50 s (b) 32 cm/s

61. 47 cm

63. 45 cm³

65. Ⓡ 7 m/s

67. 13 m/s

69. 4.6 Pa

71. Ⓡ 2×10^{-5} m/s

73. 1.5×10^7 Pa, 150:1

75. 1.5×10^{-1} cm³/s

77. The flow rate is reduced by a factor of 16.

79. The correct answer is (b). The steel ball bearing now displaces two liquids. $x = 0.46$

81. 10 m

83. The water undergoes free fall after it leaves the faucet, so we can use the relations for free fall to relate the speeds at points 1 and 2. The speeds and areas at these two points are also related by the equation of continuity. This gives enough information to find the flow rate. 48 cm³/s

85. (a) 92.8 m (b) No; the maximum depth is 90.5 m.

87. Ⓡ 10 N

89. 50 Pa. Yes; the measurement is good to at least three significant figures.

91. (a) 7.9×10^4 Pa (b) 1.8×10^7 N (c) 2.6×10^7 N

93. (a) It tends to push the window out of the house. (b) 52 N

95. (a) The pressure in the tank exceeds atmospheric pressure by 550 Pa. (b) 23 Pa

CHAPTER 11

Concept Checks

11.1   The velocity of the bungee jumper in Figure 11.1 is definitely *not* constant. There are forces on the jumper due to the bungee cord and gravity.

11.2 (c)

11.3 (a) Point *B* (b) Point *B*

11.4 Kinetic energy

11.5 Air drag and friction between the mass and the table

11.6 (c)

Problems

1. (a) 0.60 s (b) 1.7 Hz

3. (a) Period ≈ 4.0 s, frequency ≈ 0.25 Hz,
   amplitude ≈ 10 cm (b) At about $t ≈ 3.0$ s and $t ≈ 7.0$ s
   (c) At about $t ≈ 1.0$ s and $t ≈ 5.0$ s

5. (a) $v(t = 4.0$ s$) ≈ 0.15$ m/s, $v(t = 7.0$ s$) ≈ 0.025$ m/s
   (b) At about $t = 0$ s and $t = 4$ s

7. 140 m/s

9. It is not a simple harmonic oscillator because it does not
   move in a sinusoidal path; $f = 250$ Hz.

11. (a) ≈ ±0.18 m (b) 3.3 Hz

13. 1.8 s

15. (a) 4.9 N/m (b) $v(t = 4.0$ s$) ≈ 15$ cm/s, $v(t = 5.0$ s$) ≈$
    0 cm/s (c) −17 cm/s² (d) ≈ 24 cm/s²

17. 2.9 Hz

19. (a) 1.0 m/s² (b) 0.50 m/s

21. Ⓡ 16 m/s²; larger

23. 3.0 m

25. 6.2 s

27. $f_{new} = 0.58 f_{old}$

29. 0.029 Hz

31. 0.015 Hz

33. $v_{max} = 3.1 × 10^4$ mm/s, $a_{max} = 4.9 × 10^8$ mm/s²

35. (a) 13 J (b) 13 J

37. (a) 0.89 kg (b) 0.84 s (c) 43 m/s (d) 810 J (e) 810 J

39. $2.1 × 10^{-4}$ s

41. 0.010 J

43. (a) 0.136 J (b) 21 cm (c) 0.29 m/s (d) The answers would
    remain the same.

45. $3.0 × 10^6$ N/m

47. (a)

49. Strain = $1.0 × 10^{-4}$; diameter = 7.0 cm

51. 0.2497 m

53. $ΔL = 7.1 × 10^{-7}$ m

55. (a) $6.7 × 10^{-4}$ m (b) 140,000 N

57. $9.1 × 10^9$ Pa

59. 12 kg

61. Ⓡ About 75% of the energy is lost with each cycle.

63. $1.8 × 10^{-9}$ N; it is less than 1% of the weight of a
    mosquito.

65. (a) 2.0 m (b) 4.1 m/s (c) Period = 0.20 s,
    amplitude = 0.20 m (d) $x(t) ≈ (0.20$ m$)\sin[(31$ rad/s$)t]$,
    $v(t) ≈ (6.3$ m/s$)\cos[(31$ rad/s$)t]$ (e) 6.3 m/s (f) 2.0 m

(g)

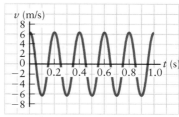

67. (a) 0.70 s (b) 25 cm

(c)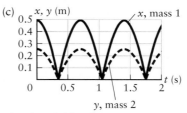

69. 10 m/s = 36 km/h

71. (a) 150 N/m (c) $1.0 × 10^5$ Pa

CHAPTER 12

Concept Checks

12.1 (c)

12.2 (b)

12.3 The frequency and amplitude depend on the way
     the wave is generated. The speed is a property of the
     medium. The wavelength depends on both the medium
     and the way the wave is generated because it depends
     on both the frequency and the speed.

12.4 Five nodes and four antinodes.

Problems

1. 13 m/s

3. Ⓡ ≈ 0.40 m/s; perpendicular to the wave velocity

5. $1.4 × 10^{11}$ m

7. 26 cm/s

9. 3.4 m

11. 11 m/s

13. $y = 0.50 \sin(250t + 0.75x)$ m

15. 0.010 s

17. $f = 480$ Hz, $λ = 14$ mm

19. 300 m/s

21. 0.0026 kg

23. $2.0 × 10^{-5}$ kg

25. Ⓡ ≈ 100 N

27. 5.0 m

29. By a factor of $\sqrt{3}$

31. 260 s (= 4.3 min)

33. 200 MHz

35. 0.39 W/m^2

37. 17 m; the answer for hemispherical wave fronts would be the same.

39. 440 Hz

41. 18,000 m

43. $v_1 < v_2$

45. (a)

47. 260 Hz; middle C

49. 2.8×10^{-4} kg

51. 46 N

53. The frequency will increase by a factor of 2.5.

55. 0.038 m

57. 140 m/s

59. 4/5

61. Surface waves = 1.8 h, P waves = 0.54 h

63. (a) 8.8 m/s (b) 1.8 s

65. (a) The tension needs to increase by 10%. (b) The tension must be decreased by 21%.

67. 28 s

69. (a) 0.58 Hz (b) 128th harmonic (c) 7.9 m/s

71. $\mu_G/\mu_E = 11$, $\mu_D/\mu_E = 5.0$, $\mu_A/\mu_E = 2.2$

CHAPTER 13

Concept Checks

13.1 (b)

13.2 (c)

13.3 (d)

13.4 (b)

13.5 (a)

13.6 (b)

Problems

1. 1.6 m

3. 0.39 m

5. 0.15 s

7. Helium

9. 13 m

11. $\lambda_{water}/\lambda_{air} = 4.3$

13. 17

15. About 1×10^{-8} W/m^2

17. Ⓡ About 0.006 N

19. (a)

21. It increases by a factor of 40.

23. 94 dB

25. 3.3×10^{-5} W/m^2

27. 0.002 Pa

29. 100 dB

31. (a) and (d)

33. 71 Hz

35. The frequency will go down to 96% of the frequency in air.

37. 860 Hz

39. 1.3 m and any multiple thereof, such as 2.6 m

41. 1.9×10^3 Hz

43. 3

45. 0.998 or 1.002

47. 0.02%

49. Ⓡ The sound makes about 140 trips across the church.

51. 21 m/s

53. 25 m/s

55. 25 m/s

57. 70 Hz

59. 37 m/s

61. 13.4°C

63. (a) 3 km (b) 0.34 km/s

65. (a) 480 Hz (b) "oo" in *pool*: 480 Hz, 960 Hz, and a bit of 2800 Hz; "ah" as in *father*: 960 Hz, 1440 Hz, and 2800 Hz; "ee" as in *heed*: 480 Hz and equal parts 2400 Hz and 3360 Hz

67. (a) 15 m (b) 420 Hz

69. (a) 35 m/s = 81 mi/h (b) 90 m/s = 200 mi/h; few automobiles can go this speed. Using two cars driven in opposite directions, each with half of this speed (100 mi/h) on a two-lane road, would still be very dangerous.

71. Ⓡ −12°C

73. 0.19 m/s

CHAPTER 14

Concept Checks

14.1 The body is composed mainly of water, so its specific heat is close to that of water.

14.2 (c)

14.3 The coefficient of thermal expansion α determines how the length of an object changes when it is heated. It applies to the length, width, and height, and if all three change a large amount, the corresponding changes in the volume and hence the value of β are also large.

14.4 Figure 14.24A shows how heat conduction is due to the transfer of energy between vibrating atoms. Solids have a higher density of atoms than gases, so (if all else is similar) solids can conduct a greater amount of heat.

14.5 (d)

14.6 (c)

Problems

1. $T_F = 450°F$ and $T_K = 505$ K

3. −460°F

5. $T_C = 21°C$ and $T_K = 294$ K

7. −90°C

9. 6.8×10^4 J

11. 110 J/(kg·K)

13. 2.7×10^4 J

15. 100°C

17. 0.26 kg

19. 7.5×10^{-3} m

21. Ⓡ 0.30 m

23. 120.04 m

25. 1.1×10^{-5} K^{-1}

27. 155°C

29. 4.1×10^{-4} m

31. Ⓡ 3×10^{-4} m

33. Ⓡ 4×10^{-2} m

35. (a) -4.7×10^6 N (compression)
    (b) 6.3×10^6 N (tension)

37. Ⓡ 20°C

39. 18°C

41. Ⓡ 1.5 cm

43. 1.3 kg/h

45. 9.6×10^{-7} m

47. 2.8×10^{26} W

49. Ⓡ 46 K; this estimate is very much in line with our best estimates of temperature from Wien's law, which range from 40 K to 60 K.

51. Ⓡ (a) Approximately 330 K (b) Approximately 3 K; the calculation for the dark side of the Moon is low because we assumed that all the heat input is coming from the 3-K background in space. The surface of the Moon facing the Sun gets very warm by Earth standards during the lunar "day." (Our 330-K estimate is actually low in equatorial regions, which have an equilibrium temperature more like 107°C!) The rock will retain some of that heat at "night," especially because there is no atmosphere and cooling takes place by radiation only.

53. Ⓡ 1×10^6 m^2

55. Ⓡ (a) 20 W (b) 2×10^{-5} s

57. 10.06°C

59. (a) $T_R = T_F + 459$ (b) $T_R = \frac{9}{5}T_K$ (c) Freezing point of water = 491°R; room temperature ≈ 530°R

61. -208°C

63. (a) 1.3×10^4 J (b) 60°C

65. (a) 2.2×10^8 Pa, with the engine block pushing on the plug (b) 4.7×10^5 N; yes

67. 190 K

69. **(a) 8.1×10^9 N (b) 0.00055 m (c) 4.2×10^3 J**

71. Ⓡ (a) 4 kJ, which is a small amount of energy that will raise the temperature of 1 kg of water by only about 1°C. (b) 2 g (c) 0.1 mm (d) The walker would almost certainly receive severe burns from a slab of iron.

CHAPTER 15

Concept Checks

15.1 (a)

15.2 (a)

15.3 (b)

15.4 (c)

15.5 (c)

15.6 (c)

Problems

1. 2.0×10^{-26} kg

3. 7.3×10^{-26} kg

5. Two possibilities are elemental calcium (Ca) and magnesium oxide (MgO).

7. 313°C

9. 3.4×10^{-9} m; the average spacing will increase with heating.

11. 2.44×10^{29} molecules

13. 33,000 Pa

15. (a) 0.072 m^3 (b) 8.8×10^4 Pa

17. 0.083 mole

19. Ⓡ 12,000 moles

21. 2.5×10^{10} molecules

23. (a) 6830 Pa (b) 55°C

25. 640 m/s

27. 5.7×10^5 J

29. **8.6×10^3 m/s**

31. 55 kJ; changing the gas to argon does not change the internal energy.

33. 1900 m/s, well below the escape velocity of the Earth. Hydrogen escapes because it has a distribution of speeds as explained in Example 15.6.

35. 2.0×10^{-5} s

37. 4.2×10^{-4} m

39. Ⓡ 5×10^{-7} s

41. 3.2×10^{-5} m

43. Ⓡ 8×10^{-4} s

45. 2.4×10^{-4} m

47. Ⓡ (a) ≈ 10 molecules for each breath (b) Distributed but not diffused; N$_2$ diffuses ≈ 1 km in 2000 years. Gaseous emissions disperse much more quickly than typical diffusion times; convection and other meteorological phenomena play a much larger role than does diffusion in mixing atmospheric constituents.

49. Ⓡ ≈ 2000 m

51. (a) 1.8 atm (b) 1.8

53. (a) 10 atm (b) 0.80 (c) 6.3 atm

55. (a) 0.31 s (b) 8.3 min (c) Such animals would be too large to allow sufficient oxygen to diffuse throughout their bodies at a rate fast enough to sustain metabolism.

57. 220 m/s

CHAPTER 16

Concept Checks

16.1 (c)

16.2 (c)

16.3 (a)

16.4 Both. Because bear fur is very thick, very little heat flows between the bear and its environment, so this situation is approximately adiabatic. During hibernation, the bear produces very little heat because its metabolic rate is very low. The little heat produced must flow out to the bear's environment (through its fur) to keep the bear at an approximately constant temperature. This balance between the heat it creates and the heat flow out through its fur keeps the bear approximately isothermal.

16.5 (a) W is positive. (b) W is negative. (c) $W = 0$. (d) W is positive. (e) W is negative.

16.6 (a) Process 3 could be an isothermal expansion. (b) Process 4 could be an adiabatic compression. (c) Process 1 is an isobaric expansion. (d) Process 2 is an isobaric compression.

16.7 W is positive during steps $1 \to 2$ and $2 \to 3$. W is negative during steps $3 \to 4$ and $4 \to 1$.

Problems

1. 20 kJ
3. -2000 J
5. 8.0×10^7 J
7. (a) 2.3×10^6 J; 11 apples (b) 1.8×10^6 J; 3 soft drinks (c) 4.3×10^7 J; 22 hamburgers
9. (a) Description ii (b) Description iv (c) Description iii (d) Description i
11. 1.5×10^6 J
13. A: 1300 J, expansion; B: 2300 J, expansion
15. $+1100$ J
17. -4400 J
19. 1200 J
21. $W_1 \approx 2600$ J; $W_2 \approx -1800$ J
23. -6000 J
25. (a) 0.11 m^3 (b) -5200 J (c) -5200 J
27. (a) 2300 J (b) 257 K
29. 1.7×10^5 J
31. 270 hp
33. 5:1
35. ℝ 20,000 J
37. 9900 J
39. 0.75
41. 2.4 g
43. During transitions $1 \to 2$ and $2 \to 3$, the system does positive work on the environment. During transitions $3 \to 4$ and $4 \to 1$, the environment does positive work on the system.
45. (a) 190 K (b) 2500 J
47. $\Delta S_{house} = -24$ J/K; $\Delta S_{environment} = +26$ J/K; $\Delta S_{universe} = +1.8$ J/K; irreversible process
49. $\Delta S_{steel} = -0.010$ J/K; $\Delta S_{ice\ water} = +0.015$ J/K; $\Delta S_{universe} = +0.005$ J/K
51. 0.25 kg
53. (b) < (a) < (d) < (c)

55. (a) Processes 1 and 2 (b) Processes 3 and 4 (c) Process 2
57. (a) 2000 J (b) 1000 J (c) 220 J (d) 1200 J. The work done by the first two engines combined is 1000 J + 220 J = 1220 J ≈ 1200 J, which is the same (to within rounding errors) as the work done by the engine operating between the highest and lowest reservoir temperatures.
59. (a) **Brayton cycle:** From states $1 \to 2$: isobaric expansion; from states $2 \to 3$: adiabatic expansion; from states $3 \to 4$: isobaric compression; from states $4 \to 1$: adiabatic compression. **Otto cycle:** From states $1 \to 2$: isochoric heating; from states $2 \to 3$: adiabatic expansion; from states $3 \to 4$: isochoric cooling; from state $4 \to 1$: adiabatic compression. **Diesel cycle:** From states $1 \to 2$: isobaric expansion; from states $2 \to 3$: adiabatic expansion; from states $3 \to 4$: isochoric cooling; from states $4 \to 1$: adiabatic compression. (b) **Brayton cycle:** Gas reaches its highest temperature at state 2 and its lowest temperature at state 4. **Otto cycle:** Gas reaches its highest temperature at point 2 and its lowest temperature at point 4. **Diesel cycle:** Diesel fuel reaches its highest temperature at state 2 and its lowest temperature at state 4.
61. (a) Work done on system: $4 \to 1$; work done by system: $2 \to 3$; heat flow into system: $1 \to 2$; heat flow out of system: $3 \to 4$ (b) 1100 K (c) $\Delta T = 1100$ K; $T_2 = 2300$ K (d) 6.6×10^6 Pa (e) 2.1×10^5 Pa; 560 K (f) -250 J (g) 0.75 (h) 0.88. The Carnot engine is an idealization that exceeds what can be done in a practical, running engine.
63. (a) $\Delta S_{gas} = -0.26$ J/K (b) $\Delta S_{reservoir} = 0.26$ J/K (c) $\Delta S_{universe} = 0$ J
65. ℝ 240 K ≈ $-27°$F

CHAPTER 17

Concept Checks

17.1 (b)
17.2 (a)
17.3 (a)
17.4 (d)
17.5 (b)
17.6 (c)
17.7 (b)

Problems

1. -9.6×10^4 C
3. 5.3×10^{10}
5. -7.2×10^{-18} C
7. 3.1×10^{10}
9. 2.3×10^{-8} N
11. 9.2×10^{-8} N
13. Graph c
15. ℝ Equal and negative, or equal and positive, but smaller than $+Q$.
17. 0.034 m

19. (a) 5.7 N downward (b) 5.7 N at 30.4° above horizontal

21. $F_{grav} = 6.7 \times 10^{-17}$ N $= 2.4 \times 10^{-43} F_{elec}$; the gravitational force is extremely small (or weak) compared with the electric force.

23. At $x = -0.27L$

25. (a) 1.0×10^{-49} N (b) 2.3×10^{-10} N (c) The ratio of the forces is the same at 1.0 m as it is at 1.0 nm.

27. 6.5×10^{16}

29. -5.1×10^{-5} C

31. $3L$

33. 1.4×10^{10} N/C

35. 2.8×10^{-5} C

37. $E_x = 1.4 \times 10^9$ N/C, $E_y = -9.0 \times 10^9$ N/C

39. 3.5×10^{14} m/s² in the $-x$ direction

41.
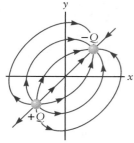

43. -5.7×10^{-3} C

45. 1/9

47. (a) 3.6×10^9 N/C downward (b) 2.4×10^9 N/C downward

49. Outside the conductor, the field is as shown. Within the conductor, the field is zero.

51. -1.8×10^{-8} N · m²/C

53. 0

55. $+3Q/\varepsilon_0$

57. (a) The electric field is directed radially outward from the cylinder. (b) A cylindrical Gaussian surface of length L (c) $q_{enclosed} = \lambda L$ if the Gaussian surface has a radius larger than that of the metal cylinder; $q_{enclosed} = 0$ if the Gaussian surface has a radius smaller than that of the cylinder. (d) Outside the cylinder, $E = \lambda/2\pi R\varepsilon_0 = (1.8 \times 10^4)/R$ N/C (with R in units of meters). The electric field is zero inside the cylinder and then falls off with the inverse of the radius outside the cylinder.

59. $E_A = 5.6 \times 10^9$ N/C, $E_B = -5.6 \times 10^9$ N/C, $E_C = 5.6 \times 10^9$ N/C, $E_D = -5.6 \times 10^9$ N/C

61. $8Q/\varepsilon_0$

63. $\dfrac{kQ}{r_s^3} r$

65. 0

67. 4.4×10^{-9} C

69. $\dfrac{\rho R}{9\varepsilon_0}$

71. (a) $v_1 = 11 \times 10^{-7}$ m/s, $v_2 = 9.1 \times 10^{-7}$ m/s (b) 9.3×10^3 s (c) 1.5×10^{-3} m

73. The magnitude of the force on the quark with charge $-e/3$ is 51 N. The magnitude of the force on one of the quarks with charge $2e/3$ is 88 N. If the only force acting between the quarks is the electric force, this arrangement of quarks is not stable.

75. -8.4×10^7 C

77. $\dfrac{\lambda}{\varepsilon_0(2\pi R)}$

79. The paper will move toward the right plate.

81. (a) 3.4×10^{12} N/C downward (b) 1.7×10^{-19} C (c) 3.1×10^{-10} N

83. 3.0×10^5

85. (a) $q_{Moon} = 6.3 \times 10^{12}$ C, $q_{Earth} = 5.2 \times 10^{14}$ C (b) $n_{Moon} = 6.6 \times 10^7$ mol, $n_{Earth} = 5.4 \times 10^9$ mol

87. (a) 470 N/C (b) The bottom plate

89. 3.1×10^5 N/C

91. (a) 17 N · m²/C (b) 34 N · m²/C (c) 17 N · m²/C

93. 1.0×10^{-5} C

CHAPTER 18

Concept Checks

1. (d)

2. (c)

3. (a) and (c)

4. (a)

5. If the line carries a negative charge, the electric field lines are directed radially and point toward the line. The equipotential surfaces are unchanged (cylinders centered on the line of charge).

6. A qualitative sketch of the field lines is as shown. Electric field lines are always perpendicular to equipotential surfaces. Because the electric field is directed from regions of high potential to regions of low potential, the object at the center must have a positive charge.

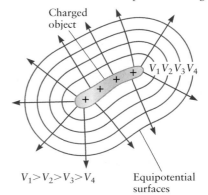

Problems

1. 0.86 m
3. 2.3×10^{-13} J
5. $+4.6 \times 10^{-18}$ J
7. -3.8×10^{-9} J
9. -7.2×10^{-3} J
11. -7.7×10^{-20} J
13. 3.5×10^{-2} J
15. 9.2×10^{-19} J
17. -1.9×10^{-17} J, -120 eV
19. 1200 V/m directed from the final position to the initial position of the electron
21. 8.4×10^{-10} m
23. (a) The electron moves from a region of lower potential to a region of higher potential. (b) 4.8×10^{-15} J (c) 1.0×10^{8} m/s
25. 1.5×10^{5} V
27. -2.3×10^{5} V
29. 4.1×10^{-10} m
31. (a) $-Q$ (b) $k \dfrac{Q}{r_2}$
33. $2\pi k\lambda$
35. $-1/3$
37. 2.5 V/m pointing from the 10 V equipotential surface toward the 5 V equipotential surface, perpendicular to both
39. A factor of three
41. 8.9×10^{-11} F
43. (a) 1.2×10^{-9} C (b) 7.2×10^{-9} J
45. (a) Closer together (b) 1/9
47. 3C
49. 8.3 μF
51. 15 μF
53. 5.8×10^{3} V
55. (a) PE_{elec} increases by a factor of two. (b) Work must be done to pull the plates farther apart.
57. 13 V
59. 5.3×10^{-15} F
61. 2.0×10^{4} V
63. The stored energy with the new dielectric is smaller by a factor of five. The capacitor does work to pull in the new dielectric, decreasing the energy stored in the capacitor.
65. Ⓡ -5.1×10^{5} C; 3.2×10^{24} electrons
67. Ⓡ 170 V
69. (a) 1.0×10^{3} J (b) 5.0×10^{4} W
71. 12 J/m^3
73. A factor of four
75. (a) 7.7×10^{-13} J (b) 3.0×10^{7} m/s (c) 4.8×10^{6} V
77. (a) 2.3×10^{8} m^2 = 230 km^2 = 87 mi^2
    (b) 1.0×10^{6} m^2 = 1.0 km^2 = 0.39 mi^2
79. Ⓡ (a) 1.4 nF (b) 1.4 J
81. (a) 5.9 pF (b) 1.7 μF (c) 990 μC
83. (a) 5.4 V (b) 7.7×10^{11} V/m

CHAPTER 19

Concept Checks

1. (c)
2. (a)
3. (c)
4. Circuit 2
5. (b)

Problems

1. 2.6×10^{20}
3. 1.7×10^{22}
5. 2.0×10^{4} A; 1.3×10^{20}
7. 3.6×10^{9} s; this time is just over 114 years, so it is quite sufficient!
9. 2.2×10^{-2} Ω
11. Increases by a factor of 6.3
13. 4.7 $\Omega \cdot$ m
15. 2×10^{-4} m
17. Increases by a factor of nine
19. 5.9×10^{6}
21. 15 Ω
23. (a) A factor of $\frac{4}{3}$ (b) A factor of $\frac{2}{3}$ (c) A factor of $\frac{2}{3}$
25. 1.1×10^{3} Ω
27. (a) 2.4×10^{-3} A (b) 5000 Ω
29. 2.0×10^{-2} W
31. (a) 2.1 A (b) A factor of $\frac{1}{2}$
33. 1.3 A
35. Two sets of two series resistors in parallel or, equivalently, two sets in series of two parallel resistors
37. (b)
39. 3.4×10^{-4} A
41. (a) $\mathcal{E}_1 - I_1 R_1 - \mathcal{E}_2 = 0$ (b) $\mathcal{E}_2 - I_2 R_2 = 0$
    (c) $I_1 = -6.3 \times 10^{-4}$ A, $I_2 = 2.1 \times 10^{-3}$ A
43. 2.6
45. 0.71
47. Bulb 1 is brightest; bulbs 3 and 4 are equally the dimmest.
49. (a) $\dfrac{\mathcal{E}}{R_1 + R_2}$ (b) 0
51. 5C
53. C/7
55. 3.7×10^{-3} A

57. 6.8×10^{-2} s

59. (a) -2.3×10^{-3} A; a negative value indicates counterclockwise current in Figure P19.58.

(b)

(c) 0

61. 0; 8.8×10^{-5} C

63. -6.7×10^{-5}

65. 0.0002 A

67. $0.002\mathcal{E}$

69. $\dfrac{\mathcal{E}_1}{\mathcal{E}_2}\left(\dfrac{R_3}{R_3 + R_4}\right)$

71. $7.9 \times 10^{-2}\ \Omega \cdot \text{m}$

73. (a) 32 A (b) 2.0×10^{18}

75. (a) 660 Ω (b) The bulb's brightness will decrease as the bulb warms up. (A warm bulb draws slightly less current and is slightly less bright than a cool bulb.)

77. 2.9×10^{17}

79.

81. (a) 8.8×10^{-4} C (b) 10^{-3} s (c) 0.88 A

83. Increase

85. (a) $V_1 > V_2 > V_3 > 0$ (b) $I_1 = I_2 = I_3$

87. 80 μA

89. (a) $V_{60} = 75$ V, $V_{100} = 45$ V (b) $V_{100} = V_{60} = 120$ V

91. Ⓡ (a) 1.5×10^8 V (b) 0.1 C (c) 1000 A

93. Ⓡ 4000 m³/s

95. $R_{AB} = R_{AC} = R_{CD} = R_{BD} = R_{AD} = 10\ \Omega$

97. Ⓡ 40 V

99. Ⓡ Two or three batteries

CHAPTER 20

Concept Checks

20.1   Part 1: (d), Part 2: (d)

20.2   Part 1: (b), Part 2: (d)

20.3   Part 1: (a), Part 2: (a)

20.4   (a)

20.5   (b)

20.6   (b)

20.7   (a)

20.8   (a)

20.9   (b)

Problems

1. A = out of the page, B = into the page, C = out of the page, D = into the page

3. To the left

5. Clockwise

7. (a) The force is zero. (b) Counterclockwise

9. (a) To the left (b) Counterclockwise

11. 1.2×10^{-16} N

13. 32°

15. Along the $-x$ direction

17. The force is zero.

19. Toward the top

21. 4.2×10^{-4} N

23. 2.8×10^{-5} m/s

25. 0.42 T

27. Clockwise

29. Counterclockwise helical when viewed from the $+z$ direction

31. 3.6×10^{-9} m

33. Ⓡ 4×10^{-21} N

35. Ⓡ 100 km

37. The radius would be decreased by a factor of 3; the radius would be increased by a factor of 10.

39. (a) Into the plane (b) The ratio of electron speed to proton speed is 1800:1.

41. Largest = case 1; smallest = case 3

43. Case 1 = force is into the page, case 2 = force is upward (toward the top of the page).

45. 7.4 N

47. 33°

49. Ⓡ 5×10^{-5} N; it is unlikely that you could detect the force holding the wire. The force is equal to the weight of about 5 mosquitoes.

51. 0

53. Along the $-x$ direction

55. (a) Along the $+x$ direction (b) 2.7 N

57. Case 1: The axis will rotate clockwise about the direction perpendicular to the plane of the drawing. Case 2: The axis will rotate counterclockwise about the direction perpendicular to the plane of the drawing. Case 3: There is no torque.

59. (a) 0 (b) 4.4×10^{-2} N · m

61. (a) Downward along the axis (b) Upward along the axis

63. (a) 25 V/m (b) 2.5×10^{-2} V (c) The velocity selector will work for both Li⁺ and Mg²⁺ ions because the velocity selector equation has no dependence on mass. It would even work with negative charges.

65. 12 T

67. $E = vB$

69. 0.25 A

71. $\dfrac{\mu_0 I}{4\pi r}$

73. 1.9×10^{-2} m

75. 2.5 A

77. (a) 1.8×10^{-6} N toward the long, straight wire
   (b) 1.8×10^{-6} N toward the current loop

79. The field is zero at point B. It has the same (nonzero) magnitude at points A and C.

81. 8.6×10^{-6} T

83. Out of the page

85. 5.0×10^{-5} T; it is about equal to the Earth's magnetic field.

87. 159 A; this very high current would likely melt the wire, making this setup impractical.

89. ⓡ 40 A

91. ⓡ 5×10^8 A

93. (a) 7.5×10^{-5} V (b) The top of the blood vessel (c) It would now be 1.3×10^{-5} V. (d) The narrower the artery, the smaller the induced emf that must be measured.

95. (a) 1.9 N (b) The weight of the wire is 3290 N, much heavier than the magnetic force on it.

97. 1.8 cm

99. (a) 3.5×10^{-17} J (b) 4.4×10^7 m/s, 8.8×10^{-16} J
   (c) 280 MHz

CHAPTER 21

Concept Checks

21.1   (a)

21.2   (b)

21.3   (b)

21.4   (b)

21.5   (c)

21.6   (a)

21.7   (b)

Problems

1. 0.063 Wb $= 0.063$ T $\cdot$ m^2

3. 6.5 m^2

5. (a) 0.5 V (b) 0.3 V (c) Opposite each other

7.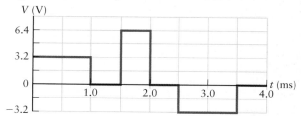

9. ⓡ 0.05 Wb $= 0.05$ T $\cdot$ m^2

11. ⓡ 1200 V

13. (a) 0.59 T (b) The induced emf depends only on *changes* in flux (Faraday's law).

15. 2×10^{-4} V

17. (a) The magnetic flux decreases with time. (b) Clockwise when viewed looking into the plane of Figure P21.16

19. (a) Into the page (b) Decreasing (c) Into the page (d) Clockwise (e) Toward the wire

21. (b)

23. Into the plane of the drawing

25. 0.11 A; clockwise when viewed looking into the plane of Figure P21.25

27. (a) Downward (b) The force is zero.

29. (a) South pole up (b) Upward (toward $+z$)
   (c) The answer to part (b) does not change.

31. 1.1 mH

33. 190 mH

35. 500 Wb $= 500$ T $\cdot$ m^2

37. (a) + (positive) (b) 5.0×10^{-5} V

39. At $t = 0$, the voltage is $V_L = 0$.

41. $V_R = 0$

43. $\tau = \dfrac{L}{R_1 + R_2}$

45. $V_{\text{equiv}} = \left(\dfrac{L_1 L_2}{L_1 + L_2}\right)\dfrac{\Delta I}{\Delta t}$, $L_{\text{equiv}} = \dfrac{L_1 L_2}{L_1 + L_2}$

47. (a) 17,000 (b) 51 H (c) 5.0×10^3 J (d) 32 m/s

49. $\sqrt{\varepsilon_0 \mu_0}$

51. ⓡ 1100 loops with a radius of 2.0 cm

53. The induced emf at 800 Hz is double the induced emf at 400 Hz.

55. 3.3×10^{-12} s

57. Clockwise as viewed from above

59. (a) (1) Out of the page, (2) into the page, (3) out of the page, (4) into the page (b) (1) Counterclockwise, (2) clockwise, (3) counterclockwise, (4) clockwise

61. (a) 1.2 s (b) 0.12 s

63. (a) 0.26 ms (b) 9.6 V

65. (a) 6.0×10^{-4} H (b) 0.60 A (c) 2.1×10^{-5} s

67. Highest emf = 6.4 V; lowest emf = -3.2 V

69. ⓡ 8×10^{-4} V

CHAPTER 22

Concept Checks

22.1   (a) and (c)

22.2   (c)

22.3   (c)

22.4   (a)

22.5 (d)

22.6 (a)

22.7 (a)

22.8 (a)

Problems

1. ⓡ (a) 240 Hz (b) 4.0 V (c) 2.8 V (d) No

3.

5. It will increase by a factor of four.

7. −16 V

9. (a) 0.018 A (b) 0.61 W (c) 0 (d) 0.30 W (e) 820 s

11. 2600 V

13. 0.13 A

15.

All four phasors point along the same direction because all four circuit quantities are in phase with one another. The relative lengths of these phasors depend on the values of V_{max}, R_1, and R_2.

17. 38 V

19. The 20-pF capacitor

21. $3.3 \times 10^5 \ \Omega$

23. 2.9×10^{-7} F

25. (a) 0.15 A (b) 0.11 A

27. (a) 18 μC (b) 1.1×10^{-4} J (c) 0.0033 s

29. 4200 Ω

31. Zero

33. The 80-mH inductor

35. 2100 Ω; it increases by a factor of three.

37.

39. (a) The maximum energy occurs at about $t = 1.0$ ms, 2.2 ms, 3.2 ms, and so on when the voltage is zero; the maximum energy is 1.1×10^{-5} J.

(b) The minimum energy occurs at about $t = 0.5$ ms, 1.6 ms, 2.7 ms, and so on when current is zero; the minimum energy is zero.

41. (a) $\dfrac{1}{2\sqrt{2}} f_0$

(b) $C_{new} = \dfrac{C}{8}$ (reduce capacitance by factor of eight)

43. (a) 1.5×10^{-5} J (b) 2.8×10^{-7} C (c) 9.6×10^{-6} s
(d) $I = 0$; $|q| = 2.8 \times 10^{-7}$ C

45. (a) 2.4×10^{-6} C (b) 0.10 A (c) 0.086 A

47. (a) 4.2×10^{-13} F (b) 1.0 MHz (c) Yes

49. Only (a) is possible; possible values are $L = 85$ mH and $C = 1.5$ mF.

51. (a)

53. (a) 4600 Ω (b) 13°, voltage leading

55. $4.7 \times 10^6 \ \Omega$; by the inductor; increase

57. 4.2×10^{-9} H

59. 1.6×10^7 Hz

61. (a) 0 (b) 1 (c) Low-pass filter

63. 6.7 V

65. (a) Step-down transformer (b) 1040 turns

67. (a) 0.38 A (b) 3.8 A

69. (a) 190 J; 6.8×10^5 J (b) 1.6 A

71. 4.4%

73. $\phi = \tan^{-1}(2\pi f C R)$

75. (a) 11 Hz (b) 22 ms later

77. The primary has 480 turns, and the secondary has 48 turns.

79. (a) 9.1 μH; 99 μH (b) A maximum of 75 and a minimum of 7 turns (c) 0.38 turns/mm (d) No. Changing the number of turns by one can select between stations for the low frequencies, but not for the high frequencies. One more turn = +4 kHz for low frequencies and −120 kHz for high frequencies.

81. (a) 0.71 kW (b) 1.5 μH (c) 0.71 A; yes, the thick copper wire is needed. (d) 47 nF

CHAPTER 23

Concept Checks

23.1 (a)

23.2 (c) This is much smaller than the maximum recommended delay for phone service, which is about 0.2 s. This is why you can have a pleasant conversation with a friend on the opposite coast.

23.3 (b)

23.4 Using Figure 23.8, we get the following order: (d) radio, (e) microwaves, (a) infrared, (f) red light, (c) blue light, (b) X-rays.

23.5 (c)

23.6 (c)

Problems

1. ⓡ 6×10^{-9} s

3. (a) 9×10^{-11} N/C (b) 1×10^{-29} N. It is a large force on an electron because it causes the electron to accelerate at about 15 m/s^2; it is also about equal to the weight of an electron.

5. 1.0×10^{-12} kg · m/s

7. ⓡ (a) 270 s, or about $4\frac{1}{2}$ min (b) 1300 s, or about 22 min

9. ⓡ 5300 s, or about 88 min

11. 2.9×10^{-6} T

13. 6.7×10^{-8} T

15. ⓡ 270 N/C

17. ⓡ Light from the laser pointer has momentum, and conservation of momentum requires that you have momentum of equal magnitude and opposite direction; $v_{10\,min} \approx 3 \times 10^{-11}$ m/s

19. 2.0×10^{-4} s

21. ⓡ 110 m

23. 600 nm

25. 5.5×10^{14} Hz

27. 3.0×10^{18} Hz

29. 0.12 m

31. 6.0×10^{6} m; no

33. ⓡ (a) 8 to 10 cm (b) $\lambda \approx 20$ cm or 0.2 m; $f \approx 1.5 \times 10^{9}$ Hz

35. (a). For the best reception when using a dipole antenna, orient the receiving antenna along the same axis as the transmitting antenna.

37. (a) 37 W/m² (b) 170 N/C

39. 0.18 V/m

41. $I_0/8$

43. $0.72c$

45. Statement (a) is most likely, although statement (b) is possible.

47. $0.39c$ away from you

49. 0.035 V/m

51. (a) 0.13 N (b) 4.9×10^{-3} N

53. (a) 2.0×10^{5} W (b) 2.5×10^{11} W/m² (c) 1.4×10^{7} V/m

55. ⓡ 1.2 km

57. (a) 1/8 (b) Zero (c) Removing the middle filter reduces the light intensity to zero because the first and third filters have transmission axes at right angles. In each situation, you must consider the effect of one polarizer at a time. If the first or third filter were removed instead, the final intensity would be 1/4 of the incident intensity.

59. ⓡ 2.1×10^{7} m/s or about 7% of c; no man-made object has ever achieved this speed, and a judge would not be impressed with the argument.

61. (a) 3.0 mm (b) $I_{laser} = 7.3 \times 10^{-7}$ W/m², which is about one billion times less intense than sunlight ($I_{sunlight} = 1000$ W/m², about the same as on the Earth) and is therefore not destructive.

CHAPTER 24

Concept Checks

24.1 (a)

24.2 (b); 19°

24.3 Total internal reflection is possible in cases (a), (c), and (e) because in these cases light is traveling into the material with the larger index of refraction.

24.4 (a)

24.5 (b)

24.6 (b)

24.7 (a) positive; (b) positive; (c) positive

24.8 (a) positive; (b) negative; (c) negative

24.9 (b)

Problems

1. 8.3 m

3. 70°

5. 1; upright

7. The two images are at (70 cm, −20 cm) and (−20 cm, 80 cm).

9. 24

11. 130 m

13. 1.36

15. 1.2×10^{8} m/s

17. 4.6×10^{14} Hz

19. 65°

21. 49°

23. 1.7

25. 54°; the light starts in the glass.

27. (b) 0.70 cm

29. 21° below the horizontal

31. Ray 1: $\theta_{1,\,out} = 24°$ below the horizontal

Ray 2: $\theta_{2,\,out} = 31°$ below the horizontal

Ray 3: $\theta_{3,\,out} = 61°$ below the horizontal

Ray 4: $\theta_{4,\,out} = 4°$ below the horizontal

33. No light emerges.

35. 0.49 m

37. (a)

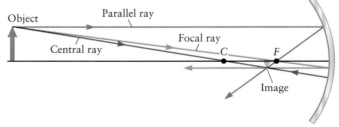

(b) $s_i = 0.090$ m = 9.0 cm; on same side of mirror as object (c) real (but inverted); 1.0 cm tall

39. Smaller

41. −0.11

43. 32 cm

45. (a)

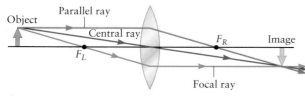

(b) $s_i = -0.083$ m $= -8.3$ cm; behind the mirror
(c) Virtual; 1.7 cm tall

47. Behind the mirror

49. 0.58 m = 58 cm

51. (a) +0.24 m (in front of the mirror) (b) Real (c) 0.69

53. (a) 0.24 (b) Now m is 0.45.

55. $s_i = -10$ cm (image is behind mirror)

57. (a)

(b) $s_i = +10$ cm; $h_i = -1.5$ cm

59. (a)

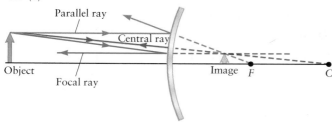

(b) $s_i = -f/2$ (a distance $f/2$ in front of the lens)
(c) Virtual

61. −70 cm

63. $s_i = -67$ cm (in front of the lens)

65. Smaller

67. The focal length increases by a factor of 2.9. The closer the index of refraction of the lens and the surrounding medium, the smaller is the refraction, so the rays travel a longer distance before crossing at the focal point.

69. (a) 14 cm from the lens (b) 6.0 cm from the lens

71. 1.3 cm

73. (a) $h_i = -0.44$ cm (inverted) (b) Smaller; it is only two-thirds as large as before.

75. (b); refraction and reflection take place in the droplets, but only refraction depends on wavelength. Different colors refract at different angles due to dispersion.

77.

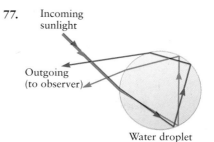

Consider incoming red and blue rays as shown. Both are reflected twice and are refracted twice, but the angles of refraction for the blue rays are greater. That gives the outgoing blue ray a steeper angle than the outgoing red ray, which is opposite to the order of colors in the primary rainbow (Fig. 24.48B and C).

79. 65 cm

81. −45 cm

83. 58°

85. 2.1

87. 10 cm

89. 128°

91. (a) −9.0 cm (inside the bowl)
(b) −6.0 cm (inside the bowl)

93. 1.6 m

95. (a) $s_{i2} = -33$ cm (behind the mirror) (b) −1.3
(c) Virtual, inverted, enlarged

97. 1.6 mm/s toward the lens

99. (a) 45 cm (b) The transparency should be upright on the projector if the screen is at your back.

101. (a) 9.0 m (b) As the pinhole becomes smaller, the image becomes sharper but the intensity of light decreases.

103. Ⓡ $h_{Sun} = 28$ mm; $h_{Moon} = 27$ mm

CHAPTER 25

Concept Checks

25.1 It is *not* possible to observe interference with light from two *different* helium–neon lasers. Light from the two lasers is emitted by different atoms, and light emitted by independent sources is not coherent.

25.2 (a)

25.3 Wave 2 in Figure 25.12A is reflected when it reaches the air interface at the bottom of the uppermost glass plate. The index of refraction is larger for glass than for air, similar to the reflection in Figure 25.9B, so there is no phase change at this reflection. On the other hand, wave 3 reflects back into air from the air–glass interface at the top of the lower plate, with a phase change of 180°. At the far left edge of the wedge in Figure 25.12 the path length difference between the two reflected waves is zero, so the only phase difference comes from the 180° phase change of wave 3. The interference is thus destructive, and the leftmost fringe is dark.

25.4 (b)
25.5 (b)
25.6 (a)
25.7 (b)

Problems

1. Constructively
3. Constructively
5. Two possible visible wavelengths are $\lambda = 600$ nm and $\lambda = 429$ nm.
7. Constructive interference
9. 0.73 m
11. 5800
13. 1000
15. (a) 3.2×10^6 (b) The uncertainty in N (c) Yes
17. If film appears red at normal incidence, $d_{\text{film}} = 110$ nm.
19. 450 nm
21. 93 nm
23. 640 nm or 700 nm
25. (a) 3.0 cm (b) 3.0 cm
27. (a)
29. 470 nm
31. (a) 3.8 mm (b) 3.8 mm
33. 630 nm
35. Ⓡ $(3 \times 10^{-4})°$
37. 200 μm
39. 2.8°
41. 5.2×10^{-4} m
43. 0, 5.7°, and 12°
45. 180 slits/mm
47. Ⓡ 9.4 cm
49. 7.3×10^{-4} rad = 0.042°
51. Ⓡ 1300 m
53. Ⓡ (a) 1.6×10^6 m (b) 8:1
55. 96 nm
57. 46 μm
59. (a) 0.89 arc second (b) 180 m
61. (a) 0.007° (b) 0.012° (c) 0.0040° (d) No (e) 3.4 mm
63. Ⓡ $\lambda \approx 200$ nm, which is a factor of two smaller than the wavelength of blue light

CHAPTER 26

Concept Checks

26.1 (c)
26.2 (a)
26.3 A magnifying glass produces a virtual upright image as can be seen from Figure 26.6. We can predict the nature of the image from the thin-lens equation,

$$\frac{1}{s_o} + \frac{1}{s_i} = \frac{1}{f}$$

The converging lens of the magnifying glass has a positive focal length f. The object distance is positive for an object to the left of the lens in Figure 26.7C (according to our sign conventions). The object is very close to the lens, so $s_o < f$. The solution for s_i from the thin-lens equation then always gives a negative image distance ($s_i < 0$), corresponding to a virtual image. Applying the result for the magnification from Chapter 24 (Eq. 24.18) shows that this virtual image is always upright.

26.4 The image is real because light passes through the image point. Light must strike the film or CCD detector to be recorded, so all cameras must produce a real image.

26.5 (d)
26.6 (c)
26.7 (b)

Problems

1. -2.5 m
3. -8.9 m; diverging
5. 2.4 cm
7. Ⓡ (a) Diverging (b) -2.0 m (c) -0.51 diopter
9. 20 cm; it is large for a contact lens, but the assumption that $R_2 = \infty$ is not realistic.
11. 2.7 diopters
13. Diverging lens with $f = -2.0$ m
15. (a) Increase (b) 1.8 cm
17. (a) The lens with $f = 2.3$ cm (b) $m_{\theta, 1} = 14$; $m_{\theta, 2} = 5$ (c) 5.2 cm
19. 25 cm; 1.4
21. Ⓡ 7.3 μm
23. It would increase by a factor of three to a new value of 300.
25. (a) 2.5 cm (b) The objective lens produces an inverted image that serves as the object for the eyepiece.
27. Highest is 640×; lowest is 66×
29. (a) 60× (b) 76 cm
31. Ⓡ About 6 km
33. 0.72 rad = 41°; 400×
35. (a) 60 cm (b) 79 cm (c) 26 cm
37. Decrease shutter speed (increase exposure time) by factor of four.
39. Ⓡ 0.0040 s
41. (a) 2.1 (b) Reduced by factor of nine
43. (a) 3.1 μm (b) 1.5 cm (c) 1.6 cm
45. 1.6
47. Ⓡ There are 3.7× as many pits in a Blu-Ray disc. A single-layer Blu-Ray disc can store 36 times as much information as a CD, so diffraction effects do not completely explain the difference in storage capacity.
49. (a) 90° (b) 59°
51. Ⓡ Focal length of eyepiece, $f_{\text{eyepiece}} = 2.5$ cm; magnification of objective, $m_{\text{obj}} = 50×$; magnification of eyepiece, $m_{\text{eyepiece}} = 10×$

53. (a) $m_{\theta,1} = 7.3$; $m_{\theta,2} = 3.1$ (b) $11\times$ (c) $3.0\times$; 16 cm

55. The focal length and approximate telescope length are both 19 m; yes.

57. Ⓡ (a) 261 nm (b) 65 nm; shallower (c) For a CD, 8 Mb/cm^2; for a DVD, 26 Mb/cm^2; for a Blu-Ray disc, 120 Mb/cm^2

59. 28 m

CHAPTER 27

Concept Checks

27.1 (a) and (c)

27.2 (c)

27.3 (b)

27.4 (a)

27.5 (c)

27.6 (a)

27.7 (b). R is about 10 times the radius of the Sun.

Problems

1. (a) v in $+x$ direction (b) v in $+x$ direction (c) 0 (d) 0 (e) 0; yes. The laws of motion are the same in all inertial reference frames. Since Newton's second law is obeyed in Ted's frame, it will be obeyed in Alice's frame.

3. 2.5×10^{-5} percent

5. (a) Ted (b) 57 s

7. $0.94c$

9. 35 yr

11. 0.19 s

13. (a) No (b) The explosion at the front of the railroad car

15. $0.66c$

17. 0.37 m

19. $0.47c$

21. $0.44c$

23. 7.1 m

25. $0.75c$

27. $0.99995c$

29. 1700 s

31. 49.99999999999967 m/s

33. (a) 2.6×10^{-22} kg · m/s (b) 8.3×10^{-22} kg · m/s (c) At high speeds, the electron responds to forces and collisions as if its mass were greater than the rest mass.

35. $0.995c$

37. $0.98c$

39. $0.98c$

41. 5.1×10^5 V

43. $0.80c$

45. $(4.4 \times 10^{-23})c$

47. 1.1×10^{-35} kg

49. 5.1×10^{-11} kg

51. 5.7×10^{38} s or 1.8×10^{31} yr

53. 2.7×10^{-26} kg · m/s

55. $\sqrt{15}mc$

57. 3.0×10^9 J

59. Ⓡ (a) 4.4×10^9 kg/s (b) 1.4×10^{13} yr

61. Ⓡ (a) 1.2×10^{-29} kg (b) 1.1×10^{-12} J (c) The energy goes into kinetic energies of the emitted ^{212}Pb nucleus and the alpha particle.

63. (a) 1.6×10^{-14} J; 1.9×10^8 m/s; 1.7×10^{-22} kg · m/s (b) 1.6×10^{-14} J; 1.6×10^8 m/s; 1.8×10^{-22} kg · m/s

67. 3.54×10^{-11} eV

69. Without relativity, you die before reaching Mars in either reference frame. If relativity is included, you live.

CHAPTER 28

Concept Checks

28.1 (b)

28.2 (c)

28.3 (c)

28.4 (b)

28.5 The electron in state 2 has the higher kinetic energy.

Problems

1. Ⓡ (a) 7×10^{-26} J (b) 1.6×10^{-24} J (c) 5.7×10^{-25} J (d) 4.0×10^{-19} J

3. 3.2×10^{17}

5. 2.2×10^{-18} J

7. 1.4×10^{15} Hz

9. Ⓡ About 3×10^{21}

11. (a) 1.9×10^{-22} J (b) Microwave

13. 1.3×10^{15} Hz; ultraviolet

15. 8.9×10^{19}

17. 1.2×10^{15} Hz

19. Ⓡ (a) 5.0×10^{-16} J (b) X-ray

21. 4.1 eV

23. Ⓡ (a) 7300 K (b) 4500 K

25. (a) $f = 3.3 \times 10^{15}$ Hz, $\lambda = 91$ nm; ultraviolet; no (b) 1.4 eV ($= 2.2 \times 10^{-19}$ J), 7.0×10^5 m/s (c) $f = 1.6 \times 10^{15}$ Hz, $\lambda = 180$ nm; no (still ultraviolet)

27. 1.1 m/s

29. (a) 9.4×10^{19} (b) 1.0×10^{-7} N (c) 2.0×10^{-7} N

31. 210:1

33. (a) 9.5×10^{-25} kg · m/s (b) 570

35. (a) 5.5×10^{-13} m (b) Yes; the wavelength of the electrons is smaller for the TEM. (c) The electron wavelength is much smaller than either.

37. (a) 360 nm (b) $E_{photon}/E_{electron} = 3.0 \times 10^5$

39. (a) 2200 m/s (b) 0.18 nm (c) The wavelength is smaller than the spacing between atoms by a factor of almost two.

41. 1.3×10^{-18} J ($= 8.1$ eV), 0.43 nm

43. 1800:1

45. Ⓡ 5×10^{-33} m/s

47. $0.15°$ ($= 0.0026$ rad)

49. (a) 9.2×10^{-24} C · J · s/kg

51. 1.1×10^{-25} kg · m/s

53. 5×10^{-20} kg · m/s; 8×10^{-13} eV

55. (a) 6.0×10^{-22} J (b) 2.4×10^{-21} J

57. (a) 1.1×10^{-25} kg · m/s (b) 1.2×10^5 m/s (c) 4.3×10^{-15} s (d) $\Delta E = 1.2 \times 10^{-20}$ J. The kinetic energy calculated from the speed in part (b) is 0.61×10^{-20} J. It is within a factor of two.

59. Blue, 4.6×10^{-19} J; green, 3.8×10^{-19} J; red, 3.5×10^{-19} J

61. 2.0 eV

63. 3.3×10^{-28} m

65. (a) Gamma-ray photon, 1.4×10^{-16} J $= 860$ eV; visible photon, 3.6×10^{-19} J $= 2.3$ eV (b) $E_\gamma/E_{\text{visible}} = 380$; nuclear processes have higher energies.

67. (a) 820 V (b) 0.83 MV (c) $KE_{\text{proton}}/KE_{\text{electron}} = 9.9 \times 10^{-4}$

69. 2.8×10^{-25} s; the shorter the lifetime, the less well defined the mass.

71. (a) 0.0087 nm (b) 2.2 μm (c) 5.3×10^{-29} kg · m/s (d) 3.0 ns (e) 0.17 μm (f) The spread of the beam as it hits the screen as calculated by diffraction is about a factor of 4π greater than that calculated using the uncertainty principle, in agreement with the discussion in Section 28.5.

CHAPTER 29

Concept Checks

29.1 (b)

29.2 (c)

29.3 chlorine (Cl)

29.4 (b)

29.5 (a) and (c)

29.6 (b)

Problems

1. (a) 4×10^{-45} m^3 (b) 5×10^{-31} m^3 (c) 8×10^{-15}

3. (a) 3×10^{-10} (b) 300

5. 2.5 eV, 5.0 eV, and 7.5 eV

7. 3.28×10^{15} Hz

9. $E_3 = -1.5$ eV; $E_{\text{photon}} = 12$ eV

11. 0.85 eV

13. 4

15. $n_{\text{initial}} = 4$, $n_{\text{final}} = 2$

17. 1.0×10^{-9} m; larger

19. (a) 137 (b) -7.2×10^{-4} eV

21. 7.4 eV

23. 18

25. 4

27. (a) Cu (copper) (b) Br (bromine) (c) Sr (strontium)

29. (a) $1s^2 2s^2 2p^6 3s^2 3p^6 3d^{10} 4s^2 4p^2$ (b) $1s^2 2s^2 2p^6 3s^2 3p^6$ (c) $1s^2 2s^2 2p^6 3s^2$ (d) $1s^2 2s^2 2p^6 3s^2 3p^2$ (e) $1s^2 2s^2 2p^6 3s^2 3p^6 3d^{10} 4s^2 4p^5$

31. (a) 10 (b) 2 (c) 32 (d) 14 (e) 18 (f) 22

33. Be (beryllium), Mg (magnesium), Sr (strontium), Ba (barium), and Ra (radium)

35. (a) 4.3×10^{14} Hz (b) 1.8 eV

37. 5.1×10^4 m/s

39. (a) 4.2×10^{-6} m (b) It is about four times larger than the diameter of a dust particle.

41. (a) $\lambda_{5 \to 2} = 430$ nm; $\lambda_{2 \to 1} = 120$ nm (b) The $5 \to 2$ transition is in the visible; the $2 \to 1$ transition is in the ultraviolet.

43. (a) 18

(b)

n	ℓ	m	s
3	0	0	$\frac{1}{2}$
3	0	0	$-\frac{1}{2}$
3	1	-1	$\frac{1}{2}$
3	1	-1	$-\frac{1}{2}$
3	1	0	$\frac{1}{2}$
3	1	0	$-\frac{1}{2}$
3	1	1	$\frac{1}{2}$
3	1	1	$-\frac{1}{2}$
3	2	-2	$\frac{1}{2}$
3	2	-2	$-\frac{1}{2}$
3	2	-1	$\frac{1}{2}$
3	2	-1	$-\frac{1}{2}$
3	2	0	$\frac{1}{2}$
3	2	0	$-\frac{1}{2}$
3	2	1	$\frac{1}{2}$
3	2	1	$-\frac{1}{2}$
3	2	2	$\frac{1}{2}$
3	2	2	$-\frac{1}{2}$

45. (a) Ne (neon) (b) S (sulfur) (c) Be (beryllium)

47. $1s^3$

49. (a) 2.5×10^{-19} C $=$ about 1.6× the charge on a proton (b) 1.4×10^{-10} m

51. Any transition down to $n = 4$ starting in states $n = 6$ through $n = 21$ will result in a visible wavelength.

53. (a) Ground state $r_1 = 2.6 \times 10^{-13}$ m; first excited state $r_2 = 1.0 \times 10^{-12}$ m; second excited state $r_3 = 2.3 \times 10^{-12}$ m (b) $E_1 = -4.5 \times 10^{-16}$ J $= 2800$ eV; $E_2 = -1.1 \times 10^{-16}$ J $= 700$ eV; $E_3 = -5.0 \times 10^{-17}$ J $= 310$ eV (c) 0.50 nm

55. (a) 5.0×10^{-11} J (b) 4.9×10^7 (c) 1.6×10^8 (d) 6.4 eV (e) No

57. Ⓡ (a) For aluminum, 2.0 kV; for copper, 11 kV; for tungsten, 72 kV (b) E_2 (aluminum) $= -3.6 \times 10^2$ eV; E_2 (copper) $= -2.7 \times 10^3$ eV; E_2 (tungsten) $= -1.4 \times 10^4$ eV

59. (a) 500 nm and 2500 nm (b) $\lambda_{2 \to 1} = 2500$ nm; $\lambda_{1 \to 0} = 620$ nm; $\lambda_{2 \to 0} = 500$ nm

CHAPTER 30

Concept Checks

30.1 $^{14}_{6}$C

30.2 (a) The wavelength of blue light is approximately 400 nm, so gamma rays have a *much* shorter wavelength.

30.3 (c)

30.4 (c)

30.5 (d)

30.6 (b)

30.7 2

Problems

1. (b)

3. (a) and (d)

5. (a) $Z = 6$, $N = 6$, $A = 12$ (b) $Z = 16$, $N = 18$, $A = 34$ (c) $Z = 82$, $N = 126$, $A = 208$ (d) $Z = 26$, $N = 30$, $A = 56$

7. Mercury, $^{204}_{80}\text{Hg}$

9. (a) 1.9 fm (b) 3.9 fm (c) 4.6 fm

11. 3.7

13. 7.6×10^6 m/s

15. 3.00×10^8 m/s

17. (c)

19. 7.5×10^5 V

21. (a)

23. Beta decay produces $^{242}_{95}\text{Am}$; alpha decay produces $^{238}_{92}\text{U}$.

25. (a) $^{234}_{90}\text{Th} \rightarrow ^{230}_{88}\text{Ra} + ^4_2\text{He}$; alpha decay

(b) $^{234}_{90}\text{Th} \rightarrow ^{234}_{91}\text{Pa} + e^- + \bar{\nu}$; beta decay

(c) $^{234}_{90}\text{Th}^* \rightarrow ^{234}_{90}\text{Th} + \gamma$; gamma decay

27. $^{60}_{27}\text{Co} \rightarrow ^{60}_{28}\text{Ni} + e^- + \bar{\nu}$

29. (a) 8.0 g (b) 1.0 g (c) 0.21 g

31. 0.028 s

33. 7.3 g

35. 4.7 yr

37. (a) $^1_1\text{H} + ^6_3\text{Li} \rightarrow ^4_2\text{He} + ^3_2\text{He}$

(b) $^1_0\text{n} + ^{235}_{92}\text{U} \rightarrow ^{133}_{51}\text{Sb} + ^{99}_{41}\text{Nb} + 4^1_0\text{n}$

(c) $^{14}_6\text{C} \rightarrow ^{14}_7\text{N} + e^- + \bar{\nu}$

39. 7.9 MeV/nucleon for a total binding energy of 1600 MeV

41. $^{89}_{37}\text{Rb}$

43. 2.5 mCi

45. (a) 6.5×10^{-13} J (b) 6.5×10^4 J

47. 0.62 kg

49. 180 MeV

51. 5.0 rads

53. 1.7 mrad

55. 26 J

57. Ⓡ (a) 5×10^{-4} J (b) 0.5 mm

59. 11 h

61. 0.015 J

63. 28 μBq

65. 32,000 yr

67. 5400 yr

69. It would take 2900 yr, which would be a problem because discarded smoke detectors will keep emitting radiation for a long time.

71. 0.32

73. 2 kg or 6×10^{24} atoms

75. (a) 7.0×10^{-7} s^{-1} (b) 33 days

77. 3×10^{10} J

79. 29 yr

81. 1:4

83. Ⓡ 0.002

85. (a) 80 decays/s = 80 Bq (b) 0.080 count/s

87. Ⓡ (a) 168,000 yr ago (b) About 80 days, the same as the currently measured half-life

89. The mass of the three alpha particles is greater than the mass of ^{12}C, so the reaction cannot occur spontaneously.

CHAPTER 31

Concept Checks

31.1 (b)

31.2 (d)

Problems

1. $\approx 3.0 \times 10^8$ m/s; very nearly equal to the speed of light

3. 6.1×10^{10} eV

5. 2.3×10^{23} Hz

7. 0.85 MeV

9. (a) 1880 MeV (b) 180.0 MeV (c) For protons and antiprotons, 9×10^{16} J/kg; for $^{235}_{92}\text{U}$, 9×10^{13} J/kg

11. 67.5 MeV

13. Three up, charm, or top antiquarks; antimatter; baryon

15. (a) π^- (b) There are no quarks in the final state. (c) Yes; zero

17. Ⓡ 100 N

19. (a) 3.6×10^9 ly (b) 581 nm

21. 1.4×10^{10} ly

23. Ⓡ 2.6×10^{16} J = 6.2 megatons TNT

25. (a) 2.1% (b) 48% (c) 0

27. (a) 25 billion years, 11 billion years (b) 14 billion years

29. (a) $t_{\text{Planck}} = \sqrt{\dfrac{Gh}{c^5}} = 1.3 \times 10^{-43}$ s (b) 7.7×10^{24}

Ice (*Continued*)
 refraction index in, 801*t*
 specific heat of, 441*t*
 thermal conductiveness of, 453*t*
Ideal batteries, 605, 605*f*
Ideal fluids, 328
Ideal gas law, 472–473, 473*f*
Ideal gases, 470–476, 471*f*–475*f*
Image formation
 by plane mirror, 798–799, 798*f*
 with lens, 818–819, 818*f*
Image location, for lens, 819–820, 819*f*
Image properties, 810
Imaging, ultrasound, 427, 427*f*
Impedance, AC circuits and, 740–743,
 741*f*, 742*f*
Impulse
 average force and, 210–211, 211*f*
 in baseball, 209–210, 210*f*
 theorem, 208–209
 variable force and, 209–210, 209*f*–210*f*
In phase, 841
Incidence, angle of, 420
Incident ray, 798, 798*f*
Incompressible fluids, 313
Incus bone, 255*f*
Independence, of vertical and horizontal
 motion, 101–102
Index of refraction, 800–803, 801*f*, 801*t*,
 802*f*
Induced magnetism, 668–669
Inductance, 701–704, 701*f*, 704*f*
Induction, charging by, 545–546, 545*f*,
 546*f*
Inductor, 702
Inductor circuits, AC, 734–736,
 734*f*–736*f*
Inductor reactance, 735–736, 735*f*–736*f*
Industrial Revolution, 508
Inelastic collisions, 213, 217–219,
 217*f*–219*f*, 223–226, 223*f*–226*f*
Inertia
 Aristotle and, 38
 Galileo and, 38–39
 mass and, 40–41
 moment of, 249, 260–263, 261*f*, 262*t*,
 263*f*
 Newton's first law and, 40
 principle of, 39
 rotational motion and, 249, 260–263,
 261*f*, 262*t*, 263*f*
Inertial mass, 63, 157
Inertial reference frames, 118–119, 119*f*,
 143–144, 144*f*, 918, 918*f*, 920
Infrared radiation, 771–772, 772*f*
Infrasonic frequencies, 409
Instantaneous acceleration, 34
Instantaneous velocity, 31–34
Insulators, 541–546, 588
Intensity
 of electromagnetic wave, 767
 of sound, 410–414
 of sunlight, 778

propagation and, 777–778
 wave, 389, 390
Interactions, forces and, 61–62
Interdisciplinary science, 1078
Interference
 CDs and, 855, 855*f*
 conditions for, 841–844, 842*f*–844*f*
 constructive, 391, 391*f*
 destructive, 391, 391*f*
 double-slit, 855–858, 855*f*, 856*f*, 861,
 861*f*
 DVDs and, 855, 855*f*
 thin-film, 847–854
 wave, 390–392, 390*f*–392*f*
 with monochromatic light, 858
Interferometer, 844–847, 844*f*, 845*f*
Internal energy, 439*f*, 440, 494–495
Internal reflection, total, 803–805, 803*f*,
 804*f*
Internal resistance, of battery, 624, 624*f*
International Tokomak Engineering
 Reactor, 1043
Inverse sine function, 16, 16*f*
Inverse square law, 156–157, 156*f*–157*f*,
 538–539, 539*f*
Inverse tangent function, 20
Inverted image, 808
Ionization energy, 989–990
Iris, 823*f*
Iron, 453*t*
Irreversible processes, 507–508,
 507*f*–508*f*
Isobaric processes, 500, 500*f*
Isochoric processes, 504, 504*f*
Isothermal processes, 501–503,
 501*f*–502*f*
Isotope separation, 485–486, 485*f*
Isotopes, 1023
ITER, 1043

J

Jets, supersonic, 424–425, 425*f*
Joule (J), 167, 574–575
Jumping from ladder, 75–77
Jumping from merry-go-round, 290–291,
 291*f*
Jupiter, 147*t*, 152

K

Kaons, 1066*t*
Kelvin scale, 436–437, 436*f*, 471–472
Kepler's first law, 151
Kepler's second law, 152, 152*f*, 293–294,
 293*f*–294*f*
Kepler's third law, 152–153
Kilogram, 11
Kinetic energy
 angular momentum and, 292–293
 as sum of rotational and translational
 energies, 280–281, 281*f*
 collisions and, 218–219
 distribution of, 282
 electric charge and, 568–569

harmonic motion and, 360–362,
 361*f*–362*f*
 mass and, 172
 momentum and, 208
 of falling object, 171, 171*f*
 of rotation, 280–284, 280*f*–284*f*
 rolling motion and, 282
 rotational, 280–281, 281*f*
 speed and, 172
 torque and, 282–283, 282*f*
 translational, 280–281, 281*f*
 work–energy theorem and, 169–170
Kinetic friction, coefficient of, 64
Kinetic theory, 478–483, 479*f*–481*f*, 998
Kinocilium, 118*f*
Kirchoff's junction rule, 617–618, 617*f*
Kirchoff's loop rule, 613–614, 614*f*, 618,
 619

L

Ladder
 jumping off, 75–77
 leaning on wall, 257–258, 257*f*–258*f*
Lambda zero particle, 1066*t*
Landing, 105
Laser interferometer gravitational wave
 observatory, 846–847
Lasers, 1013–1014, 1013*f*
Latent heat, 445, 445*t*
Lattices, 439, 439*f*
Law of definite proportions, 987
Law(s)
 Avogadro's, 471
 Boyle's, 471
 Charles's, 471
 Gay-Lussac's, 471
 Hooke's, 184, 184*f*
 ideal gas, 472–473, 473*f*
 in overview of physics, 2
 of motion
 Aristotle's, 27–28
 Kepler's, 150–155, 293–294,
 293*f*–294*f*
 Newton's, 40–45
 of thermodynamics, 493
 first, 494–498, 494*f*–498*f*
 second, 507–508, 507*f*–508*f*
 third, 519
 Poiseuille's, 334–335, 334*f*
 Stefan–Boltzmann, 457, 458–459
 Stokes's, 336–337
 Wien's, 457–458
LC circuits, 736–738, 737*f*, 738*f*, 740
LCDs (liquid crystal displays), 783, 783*f*
LCR circuit, 738–739, 738*f*, 739*f*,
 742–743, 743*f*
Lead, 314*t*, 408*t*, 437*t*, 445*t*, 609*t*
Leaning, on wall, 257–258, 257*f*–258*f*
Length contraction, 928–932, 928*f*–930*f*
Lens, of eye, 801*t*, 823*f*, 825–826, 825*f*
Lens applications, single, 880–888
Lens maker's formula, 825–826
Lenses, 817–822

Multielectron atoms, 1004–1007, 1005f, 1007f
Multiloop circuits, 622–623, 622f
Multiplication
  significant figures and, 6–7
  vectors and, 18
Muon neutrinos, 1068t
Muons, 925, 925f, 930–931, 930f, 1068t
Muscles, molecular motors in, 195–196, 195f
Musical tones, 395–396, 395f, 409–410, 417, 417f
Mutual inductance, 703–704
Myelin, 632, 632f
Mylar, 589t
Myosin, 195–196, 195f

N

Nanoscience, 1078
National Institute of Standards and Technology, 10–11, 11f, 11t
Natural abundance, of isotopes, 1023
Natural frequency, 367–368
Near-field scanning optical microscopy, 910–911, 910f
Near-point distance, 881
Negative charge, 530
Negative torque, 251
Neon, 1000t, 1008t
Neptune, 147t
Nerves, currents in, 630–632, 631f
Neutrino
  anti-, 232
  discovery of, 231, 231f
Neutron, discovery of, 231, 231f
Newton's apple, 150, 150f
Newton's first law of motion, 40
Newton's law of gravitation, 145–150
Newton's laws of motion, 40–45
Newton's mechanics, relativity and, 918–919, 918f, 919f
Newton's second law of motion, 41–43
Newton's third law of motion, 43–44
Newton, Isaac, 2–3, 2f, 26
Nitrogen, 1008t
  boiling point of liquid, 436–437, 437f
  specific heat of, 441t
Node
  of wave, 394
  pressure, 415
Node of Ranvier, 632f
Nonconservative forces, 176, 189–193, 192–193, 192f
Noninertial reference frames, 143–144, 144f
Nonzero acceleration, constant, 55–57
Normal forces, 59–63
Northern lights, 671–672, 672f
Notation, scientific, 4–5, 5t
Nuclear fission, 1037–1044
Nuclear fusion, 1037–1044
Nuclear power plants, 1041–1042, 1042f
Nuclear reactions, 1026–1037

Nuclear weapons, 939–940
Nucleon number conservation, 1029
Nucleus, atomic
  binding energy in, 1031, 1037–1038
  decay of, 1026–1037
  electric potential energy of, 1024–1025, 1024f
  forces in, 1025
  in atomic theory, 988–989
  size of, 1023–1024, 1024f
  stability of, 1025, 1037–1044
  structure of, 1022–1026, 1024f–1026f
Number(s)
  algebra and, 14–15
  atomic, 988, 1022
  color quantum, 1070
  in physics, 4–9
  mass, 1022
  scientific notation of, 4–5, 5t
  significant figures, 6–8

O

Octave, 395
Odometers, 710–712, 710f
Ohm's law, 608–610, 608f, 609t
One-dimensional motion, 29
One-loop circuits, 613–614, 614f
Optic nerve, 823f
Optical activity, 782–783, 782f
Optical fibers, 908–911, 908f–910f
Optical resolution, 864–869
Optics
  adaptive, 898–899, 899f
  cameras in, 899–905
  CDs and, 905–907, 906f
  contact lenses in, 881–885
  DVDs and, 905–907, 906f
  eyeglasses in, 881–885
  geometrical
    aberrations in, 828, 828f
    curved mirrors in, 808–817
    dispersion in, 806, 806f
    focal length in, 812–813, 813f
    image properties in, 810
    in atmosphere, 826–828, 827f
    in eye, 822–826, 823f, 825f
    lens maker's formula in, 825–826
    lenses in, 817–822
    mirror equation in, 811–812, 812f, 814–817, 815f
    overview of, 796–797, 796f, 797f
    rainbows in, 826–827, 827f
    ray tracing in, 797, 797f
    reflection in, 797–800, 798f–800f
    refraction in, 800–803, 800–808, 801f, 801t, 802f
    sign conventions in, 813–814
    Snell's law in, 800–803, 801f, 801t, 802f
    spherical concave mirrors in, 808–809, 808f, 809f
    spherical convex mirrors in, 811, 811f
    stars in, 827–828, 828f

    thin-lens equation in, 820–821, 820f, 821f
    total internal reflection in, 803–805, 803f, 804f
    virtual image in, 810, 810f
    wave front in, 796, 797f
  magnifying glass in, 885–888, 885f–887f
  microscopes in, 888–892
  of digital camera, 902–903
  single lens applications in, 880–888
  telescopes in, 892–898
  wave
    antireflection coatings in, 853, 853f
    coherence in, 842–844, 843f
    diffraction gratings and, 861–864, 862f–864f
    double-slit interference and, 855–858, 855f, 856f, 861, 861f
    Huygen's principle and, 855
    interference conditions in, 841–844, 842f–844f
    light through single slit and, 854–855, 854f
    Michelson interferometer in, 844–847, 844f, 845f
    path length difference in, 844
    principal maxima in, 862
    Rayleigh criterion and, 864–869
    resolution and, 864–869
    single-slit diffraction in, 858–861, 859f–861f
    thin-film interference in, 847–854
Orbits. See Planetary motion
Oscillation, 736
Oscillator
  damping in, 366–368, 366f–368f
  harmonic, 354–355, 354f, 359–360, 360f
  torsional, 359, 359f
Oscillatory motion, 349, 349f, 355–356, 355f
Otoliths, 118f
Out of phase, 842, 842f
Oval window, in ear, 255f
Oxygen, 1008t
  diffusion constant, 484t
  speed of sound in, 408t

P

P waves, 397–398, 398f
Paper, 589t
Parachutes, 79–82
Paradox, twin, 925–926, 926f
Parallel
  capacitors in, 585–586, 586f
  resistors in, 617–618, 617f, 620, 620f, 621, 621f, 623–624
Parallel ray, 809
Parallel-plate capacitor, 551–552, 551f, 581–582, 581f
Parent nucleus, 1027
Particle physics, 1065–1070, 1074

rotational motion and, 294–297,
   294f–297f
velocity as, 27
Velocity
   acceleration and, 35–36
   addition of, 932–935, 932f, 934f
   angular, 241–244, 242f–244f
   as function of time, 351–352, 352f
   as vector, 27
   average, 30, 31–34
   background, 406
   calculation of, 103–104, 103f
   constant nonzero acceleration and,
      55–57
   direction and, 30
   escape, 182–183
   graphical method for finding, 33–34
   harmonic motion and, 351–352, 352f
   in Aristotelian mechanics, 27–28
   instantaneous, 31–34
   momentum and, 206
   motion and, 29–30
   of medium, 407, 407f
   of wave, 380–381, 380f
   position and, 30–31
   power and, 195
   relative
      overview of, 108–109, 108f
      reference frames and, 107–108
   terminal, 79–82
   units of, 30
   vs. speed, 30
   work and, 166
Velocity-dependent force, 675–676,
   675f
Venus, 147t
Vertical vs. horizontal motion, 101–102
Vestibule (ear), 255f
Viewpoints, reference frames as, 143
Violin string, 386
Virtual image, 810, 810f
Viscosity, 334–339, 334f–338f
Visible light, 772
Vision, color, 871, 871f, 978–979, 979f
Vitreous humor, 801t, 823f
Volt (V), 571
Voltage
   electric potential and, 570–578
   filtration of, 630–632, 630f, 631f
   generation of AC, 724–726, 725f
   in lightbulb, 729, 729f
   in resistor circuit, 606–612
   measuring, 629–630, 630f
   notations, 730t
   Ohm's law and, 608–610, 608f, 609t
   phasors and, 730–731, 731f
   transformers and, 746–748, 746f, 747f
Voltmeters, 629–630, 630f
Volume
   in isochoric processes, 504, 504f
   units of, 12
Von Guericke, Otto, 311
Vortices, 339, 339f

W

Walking, friction and, 68
Water
   as dielectric, 589t
   boiling point of, 437t
   density of, 314t
   diffusion constant, 484t
   electric charge and, 542
   floating in, 326–327, 326f
   from faucet, 333–334, 333f–334f
   latent heat of, 445t
   modular values of, 363t
   reflection with air, 805, 805f
   refraction index in, 801t
   specific heat of, 441t
   speed of sound in, 408t
   surface tension of, 338t
   thermal conductivity of, 453t
   thermal expansion of, 451, 451f
   waves, 388, 388f
Wave fronts, 389–390, 389f, 796, 797f
Wave function, 970–976
Wave mechanics, 1002–1004, 1003t,
   1004f
Wave optics
   antireflection coatings in, 853, 853f
   coherence in, 842–844, 843f
   diffraction gratings and, 861–864,
      862f–864f
   double-slit interference and, 855–858,
      855f, 856f, 861, 861f
   Huygen's principle and, 855
   interference conditions in, 841–844,
      842f–844f
   light through single slit and, 854–855,
      854f
   Michelson interferometer in, 844–847,
      844f, 845f
   path length difference in, 844
   principal maxima in, 862
   Rayleigh criterion and, 864–869
   resolution and, 864–869
   single-slit diffraction in, 858–861,
      859f–861f
   thin-film interference in, 847–854
Wave-particle duality, 965–968
Wavelength, 381–382, 382f
   de Broglie, 967–968
   measurement of, 857–858
   of electron, 966
Waves and wave motion
   amplitude of, 382, 390
   analyzing equation of, 384–385
   antinodes of, 394
   coherence in, 842–844, 843f
   crests, 381, 381f
   definition of, 378, 379, 379f
   describing, 381–385, 381f–384f
   direction of, 382
   electromagnetic, 388
      as energy carriers, 765–770
      as transverse waves, 763–764, 764f
      astronomy and, 775

discovery of, 762–763, 762f
Doppler effect and, 783–786, 784f
electromagnetic spectrum and,
   770–775
gamma rays as, 774
generation of, 775–778, 776f, 777f
in vacuum, 787
intensity of, 767, 777–778
light as, 764
Malus's law and, 779–780, 779f
microwaves as, 771, 771f
momentum in, 765–770
optical activity and, 782–783, 782f
polarization and, 779–783, 779f,
   781f–783f
propagation of, 775–778, 776f,
   777f
properties of, 763–765, 764f
quantum theory and, 787
radiation pressure and, 769
radio waves as, 770–771
speed of, 764–765
ultraviolet light as, 772–773, 773f
visible light as, 772
X-rays as, 773–774, 773f, 774f
electrons as, 955–956, 955f, 956f
energy transport in, 379
equation of, 381–382, 382f
examples of, 385–388, 386f–388f
finding properties of, 383–384, 384f
frequency of, 381–382, 382f
geometry of, 389–390, 389f
gravitational, 786, 846–847
in phase, 841
in radar, 393–394, 393f
in solids, 387–388, 387f
intensity of, 389, 390
interference, 390–392, 390f–392f
light as, 870–871, 964–965
longitudinal, 380, 406–410
mechanical, 380
medium of, 380
musical tones and, 395–396, 395f
nodes of, 394
on string, 385–386, 386f
out of phase, 842, 842f
P, 397–398, 398f
particles as, 965–968
periodic, 379
plane, 389f, 390
power of, 389
propagation of, 379
properties of, 955
pulse, 379, 379f
radio, 770–771
reflection of, 392–394, 392f–393f
refraction of, 394, 394f
repeat distance of, 382
repeat time of, 382
S, 397–398, 398f
scattered, 782
seismic, 397–398, 398f
shock, 424–425, 425f

Periodic Table of the Elements

Transition elements

Key:

Symbol	Ca	20	— Atomic number
Atomic mass†	40.078		
	$4s^2$		— Configuration of outermost electrons

Main Group Elements

Group I	Group II		Group III	Group IV	Group V	Group VI	Group VII	Group 0
H 1 — 1.0079 — $1s^1$								He 2 — 4.00260 — $1s^2$
Li 3 — 6.941 — $2s^1$	Be 4 — 9.0122 — $2s^2$		B 5 — 10.811 — $2p^1$	C 6 — 12.011 — $2p^2$	N 7 — 14.007 — $2p^3$	O 8 — 15.999 — $2p^4$	F 9 — 18.998 — $2p^5$	Ne 10 — 20.180 — $2p^6$
Na 11 — 22.990 — $3s^1$	Mg 12 — 24.305 — $3s^2$		Al 13 — 26.982 — $3p^1$	Si 14 — 28.086 — $3p^2$	P 15 — 30.974 — $3p^3$	S 16 — 32.066 — $3p^4$	Cl 17 — 35.453 — $3p^5$	Ar 18 — 39.948 — $3p^6$

Transition Elements and Period 4–7 Main Groups

Group I	Group II	III											Group III	Group IV	Group V	Group VI	Group VII	Group 0
K 19 — 39.098 — $4s^1$	Ca 20 — 40.078 — $4s^2$	Sc 21 — 44.956 — $3d^14s^2$	Ti 22 — 47.867 — $3d^24s^2$	V 23 — 50.942 — $3d^34s^2$	Cr 24 — 51.996 — $3d^54s^1$	Mn 25 — 54.938 — $3d^54s^2$	Fe 26 — 55.845 — $3d^64s^2$	Co 27 — 58.933 — $3d^74s^2$	Ni 28 — 58.693 — $3d^84s^2$	Cu 29 — 63.546 — $3d^{10}4s^1$	Zn 30 — 65.41 — $3d^{10}4s^2$	Ga 31 — 69.723 — $4p^1$	Ge 32 — 72.64 — $4p^2$	As 33 — 74.922 — $4p^3$	Se 34 — 78.96 — $4p^4$	Br 35 — 79.904 — $4p^5$	Kr 36 — 83.80 — $4p^6$
Rb 37 — 85.468 — $5s^1$	Sr 38 — 87.62 — $5s^2$	Y 39 — 88.906 — $4d^15s^2$	Zr 40 — 91.224 — $4d^25s^2$	Nb 41 — 92.906 — $4d^45s^1$	Mo 42 — 95.94 — $4d^55s^1$	Tc 43 — (98) — $4d^55s^2$	Ru 44 — 101.07 — $4d^75s^1$	Rh 45 — 102.91 — $4d^85s^1$	Pd 46 — 106.42 — $4d^{10}$	Ag 47 — 107.87 — $4d^{10}5s^1$	Cd 48 — 112.41 — $4d^{10}5s^2$	In 49 — 114.82 — $5p^1$	Sn 50 — 118.71 — $5p^2$	Sb 51 — 121.76 — $5p^3$	Te 52 — 127.60 — $5p^4$	I 53 — 126.90 — $5p^5$	Xe 54 — 131.29 — $5p^6$
Cs 55 — 132.91 — $6s^1$	Ba 56 — 137.33 — $6s^2$	57–71*	Hf 72 — 178.49 — $5d^26s^2$	Ta 73 — 180.95 — $5d^36s^2$	W 74 — 183.84 — $5d^46s^2$	Re 75 — 186.21 — $5d^56s^2$	Os 76 — 190.23 — $5d^66s^2$	Ir 77 — 192.2 — $5d^76s^2$	Pt 78 — 195.08 — $5d^96s^1$	Au 79 — 196.97 — $5d^{10}6s^1$	Hg 80 — 200.59 — $5d^{10}6s^2$	Tl 81 — 204.38 — $6p^1$	Pb 82 — 207.2 — $6p^2$	Bi 83 — 208.98 — $6p^3$	Po 84 — (209) — $6p^4$	At 85 — (210) — $6p^5$	Rn 86 — (222) — $6p^6$
Fr 87 — (223) — $7s^1$	Ra 88 — (226) — $7s^2$	89–103**	Rf 104 — (261) — $6d^27s^2$	Db 105 — (262) — $6d^37s^2$	Sg 106 — (266)	Bh 107 — (264)	Hs 108 — (277)	Mt 109 — (268)	Ds 110 — (271)	Rg 111 — (272)							

*Lanthanide series

La 57	Ce 58	Pr 59	Nd 60	Pm 61	Sm 62	Eu 63	Gd 64	Tb 65	Dy 66	Ho 67	Er 68	Tm 69	Yb 70	Lu 71
138.91	140.12	140.91	144.24	(145)	150.36	151.96	157.25	158.93	162.50	164.93	167.26	168.93	173.04	174.97
$5d^16s^2$	$5d^14f^16s^2$	$4f^36s^2$	$4f^46s^2$	$4f^56s^2$	$4f^66s^2$	$4f^76s^2$	$4f^75d^16s^2$	$4f^85d^16s^2$	$4f^{10}6s^2$	$4f^{11}6s^2$	$4f^{12}6s^2$	$4f^{13}6s^2$	$4f^{14}6s^2$	$4f^{14}5d^16s^2$

**Actinide series

Ac 89	Th 90	Pa 91	U 92	Np 93	Pu 94	Am 95	Cm 96	Bk 97	Cf 98	Es 99	Fm 100	Md 101	No 102	Lr 103
(227)	232.04	231.04	238.03	(237)	(244)	(243)	(247)	(247)	(251)	(252)	(257)	(258)	(259)	(262)
$6d^17s^2$	$6d^27s^2$	$5f^26d^17s^2$	$5f^36d^17s^2$	$5f^46d^17s^2$	$5f^67s^2$	$5f^77s^2$	$5f^76d^17s^2$	$5f^86d^17s^2$	$5f^{10}7s^2$	$5f^{11}7s^2$	$5f^{12}7s^2$	$5f^{13}7s^2$	$5f^{14}7s^2$	$6d^15f^{14}7s^2$

Note: Atomic mass values given are averaged over isotopes in the percentages in which they exist in nature.

$\dagger$ For an unstable element, mass number of the most stable known isotope is given in parentheses.

$\dagger\dagger$ From the International Union of Pure and Applied Chemistry, June 2007.
See *old.iupac.org/reports/periodic_table/index.html.*

Powers of 10

Prefix	Factor
atto (a)	10^{-18}
femto (f)	10^{-15}
pico (p)	10^{-12}
nano (n)	10^{-9}
micro (μ)	10^{-6}
milli (m)	10^{-3}
centi (c)	10^{-2}
deci (d)	10^{-1}
kilo (k)	10^{3}
mega (M)	10^{6}
giga (G)	10^{9}
tera (T)	10^{12}
peta (P)	10^{15}
exa (E)	10^{18}

Useful Mathematical Formulas

Circumference of a circle $= 2\pi r$

Area of a circle $= \pi r^2$

Area of a triangle $= \frac{1}{2}(\text{base} \times \text{height})$

Surface area of a sphere $= 4\pi r^2$

Volume of a sphere $= \frac{4}{3}\pi r^3$

Volume of a cylinder $= \pi r^2 L$

$$ax^2 + bx + c = 0 \Rightarrow x = \frac{-b \pm \sqrt{b^2 - 4ac}}{2a}$$

Degrees and Radians

2π rad $= 360°$

1 revolution $= 1$ rev $= 360° = 2\pi$ rad

$1° = 60$ arcmin $= 60'$

1 arcmin $= 60$ arcsec $= 60''$

$45° = \pi/4$ rad

$90° = \pi/2$ rad

Exponential Functions and Logarithms

$x = e^y \Leftrightarrow \ln x = y$

$e^{xy} = e^x e^y$

$\ln(xy) = \ln x + \ln y$

$\ln\left(\dfrac{x}{y}\right) = \ln x - \ln y$

$x = 10^y \Leftrightarrow \log x = y$

$e^0 = 1$

$\ln(1) = 0$

$10^0 = 1$

$\log(1) = 0$

$\log(10) = 1$